LDA	lithium diisopropylamide
LDMAN	lithium 1-(dimethylamino)naphthalenide
LHMDS	= LiHMDS
LICA	lithium isopropylcyclohexylamide
LiHMDS	lithium hexamethyldisilazide
LiTMP	lithium 2,2,6,6-tetramethylpiperidide
LTMP	= LiTMP
LTA	lead tetraacetate
lut	lutidine
m-CPBA	*m*-chloroperbenzoic acid
MA	maleic anhydride
MAD	methylaluminum bis(2,6-di-*t*-butyl-4-methylphenoxide)
MAT	methylaluminum bis(2,4,6-tri-*t*-butylphenoxide)
Me	methyl
MEK	methyl ethyl ketone
MEM	(2-methoxyethoxy)methyl
MIC	methyl isocyanate
MMPP	magnesium monoperoxyphthalate
MOM	methoxymethyl
MoOPH	oxodiperoxomolybdenum(pyridine)-(hexamethylphosphoric triamide)
mp	melting point
MPM	= PMB
Ms	mesyl (methanesulfonyl)
MS	mass spectrometry; molecular sieves
MTEE	methyl *t*-butyl ether
MTM	methylthiomethyl
MVK	methyl vinyl ketone
n	refractive index
NaHDMS	sodium hexamethyldisilazide
Naph	naphthyl
NBA	*N*-bromoacetamide
nbd	norbornadiene (bicyclo[2.2.1]hepta-2,5-diene)
NBS	*N*-bromosuccinimide
NCS	*N*-chlorosuccinimide
NIS	*N*-iodosuccinimide
NMO	*N*-methylmorpholine *N*-oxide
NMP	*N*-methyl-2-pyrrolidinone
NMR	nuclear magnetic resonance
NORPHOS	bis(diphenylphosphino)bicyclo[2.2.1]-hept-5-ene
Np	= Naph
PCC	pyridinium chlorochromate
PDC	pyridinium dichromate
Pent	*n*-pentyl
Ph	phenyl
phen	1,10-phenanthroline
Phth	phthaloyl
Piv	pivaloyl
PMB	*p*-methoxybenzyl
PMDTA	*N,N,N′,N″,N″*-pentamethyldiethylene-triamine
PPA	polyphosphoric acid
PPE	polyphosphate ester
PPTS	pyridinium *p*-toluenesulfonate
Pr	*n*-propyl
PTC	phase transfer catalyst/catalysis
PTSA	*p*-toluenesulfonic acid
py	pyridine
RAMP	(*R*)-1-amino-2-(methoxymethyl)pyrrolidine
rt	room temperature
salen	bis(salicylidene)ethylenediamine
SAMP	(*S*)-1-amino-2-(methoxymethyl)pyrrolidine
SET	single electron transfer
Sia	siamyl (3-methyl-2-butyl)
TASF	tris(diethylamino)sulfonium difluorotrimethylsilicate
TBAB	tetrabutylammonium bromide
TBAF	tetrabutylammonium fluoride
TBAD	= DBAD
TBAI	tetrabutylammonium iodide
TBAP	tetrabutylammonium perruthenate
TBDMS	*t*-butyldimethylsilyl
TBDPS	*t*-butyldiphenylsilyl
TBHP	*t*-butyl hydroperoxide
TBS	= TBDMS
TCNE	tetracyanoethylene
TCNQ	7,7,8,8-tetracyanoquinodimethane
TEA	triethylamine
TEBA	triethylbenzylammonium chloride
TEBAC	= TEBA
TEMPO	2,2,6,6-tetramethylpiperidinoxyl
TES	triethylsilyl
Tf	triflyl (trifluoromethanesulfonyl)
TFA	trifluoroacetic acid
TFAA	trifluoroacetic anhydride
THF	tetrahydrofuran
THP	tetrahydropyran; tetrahydropyranyl
Thx	thexyl (2,3-dimethyl-2-butyl)
TIPS	triisopropylsilyl
TMANO	trimethylamine *N*-oxide
TMEDA	*N,N,N′,N′*-tetramethylethylenediamine
TMG	1,1,3,3-tetramethylguanidine
TMS	trimethylsilyl
Tol	*p*-tolyl
TPAP	tetrapropylammonium perruthenate
TBHP	*t*-butyl hydroperoxide
TPP	tetraphenylporphyrin
Tr	trityl (triphenylmethyl)
Ts	tosyl (*p*-toluenesulfonyl)
TTN	thallium(III) nitrate
UHP	urea–hydrogen peroxide complex
Z	= Cbz

UNIVERSITY OF THE PACIFIC
1851
STOCKTON, CALIFORNIA

Handbook of Reagents for Organic Synthesis

Oxidizing and Reducing Agents

OTHER TITLES IN THIS COLLECTION

Reagents, Auxiliaries and Catalysts for C–C Bond Formation
Edited by Robert M. Coates and Scott E. Denmark
ISBN 0 471 97924 4

Acidic and Basic Reagents
Edited by Hans J. Reich and James H. Rigby
ISBN 0 471 97925 2

Activating Agents and Protecting Groups
Edited by Anthony J. Pearson and William R. Roush
ISBN 0 471 97927 9

Handbook of Reagents for Organic Synthesis

Oxidizing and Reducing Agents

Edited by

Steven D. Burke
University of Wisconsin at Madison

and

Rick L. Danheiser
Massachussetts Institute of Technology, Cambridge, MA

JOHN WILEY & SONS

Chichester • New York • Weinheim • Brisbane • Toronto • Singapore

Baffins Lane, Chichester
West Sussex PO19 1UD, UK

National 01243 779777
International (+44) 1243 779777
e-mail (for orders and customer service enquiries): cs-books@wiley.co.uk
Visit our Home Page on
http://www.wiley.co.uk
or http:/www.wiley.com

Reprinted February 2000, May 2001

Other Wiley Editorial Offices

John Wiley & Sons Inc., 605 Third Avenue,
New York, NY 10158-0012, USA

Wiley-VCH Verlag GmbH, Pappelallee 3,
D-69469 Weinheim, Germany

Jacaranda Wiley Ltd, 33 Park Road, Milton,
Queensland 4064, Australia

John Wiley & Sons (Asia) Pte Ltd, 2 Clementi Loop #02-01,
Jin Xing Distripark, Singapore 129809

John Wiley & Sons (Canada) Ltd, 22 Worcester Road,
Rexdale, Ontario M9W 1L1, Canada

Library of Congress Cataloguing-in-Publication Data

Handbook of reagents for organic synthesis.
p. cm.
Includes bibliographical references.
Contents: [1] Reagents, auxiliaries, and catalysts for C–C bond formation / edited by Robert M. Coates and Scott E. Denmark [2] Oxidising and reducing agents / edited by Steven D. Burke and Riek L. Danheiser [3] Acidic and basic reagents / edited by Hans J. Reich and James H. Rigby [4] Activating agents and protecting groups / edited by Anthony J. Pearson and William R. Roush
ISBN 0-471-97924-4 (v. 1). ISBN 0-471-97926-0 (v. 2)
ISBN 0-471-97925-2 (v. 3) ISBN 0-471-97927-9 (v. 4)
1. Chemical tests and reagents. 2. Organic compounds–Synthesis.
QD77.H37 1999
547′.2 dc 21 98-53088
CIP

British Library Cataloguing in publication Data

A catalogue record for this book is available from the British Library

ISBN 0 471 97926 0

Typset by Thomson Press (India) Ltd., New Delhi
Printed and bound in Great Britian by Antony Rowe, Chippenham, Wilts
This book is printed on acid-free paper responsibly manufactured from sustainable forestry, in which at least two trees are planted for each one used in paper production.

Editorial Board

Contents

Preface

As stated in its Preface, the major motivation for our undertaking publication of the *Encyclopedia of Reagents for Organic Synthesis* was "to incorporate into a single work a genuinely authoritative and systematic description of the utility of all reagents used in organic chemistry." By all accounts, this reference compendium has succeeded admirably in attaining this objective. Experts from around the globe contributed many relevant facts that define the various uses characteristic of each reagent. The choice of a masthead format for providing relevant information about each entry, the highlighting of key transformations with illustrative equations, and the incorporation of detailed indexes serve in tandem to facilitate the retrieval of desired information.

Notwithstanding these accomplishments, the editors have since recognized that the large size of this eight-volume work and its cost of purchase have often served to deter the placement of copies of the *Encyclopedia* in or near laboratories where the need for this type of insight is most critically needed. In an effort to meet this demand in a cost-effective manner, the decision was made to cull from the major work that information having the highest probability for repeated consultation and to incorporate same into a set of handbooks. The latter would also be purchasable on a single unit basis.

The ultimate result of these deliberations is the publication of the *Handbook of Reagents for Organic Synthesis* consisting of the following four volumes:

Reagents, Auxiliaries and Catalysts for C–C Bond Formation
Edited by Robert M. Coates and Scott E. Denmark

Oxidizing and Reducing Agents
Edited by Steven D. Burke and Rick L. Danheiser

Acidic and Basic Reagents
Edited by Hans J. Reich and James H. Rigby

Activating Agents and Protecting Groups
Edited by Anthony J. Pearson and William R. Roush

Each of the volumes contains a complete compilation of those entries from the original *Encyclopedia* that bear on the specific topic. Ample listings can be found to functionally related reagents contained in the original work. For the sake of current awareness, references to recent reviews and monographs have been included, as have relevant new procedures from *Organic Syntheses*.

The end product of this effort by eight of the original editors of the *Encyclopedia* is an affordable, enlightening set of books that should find their way into the laboratories of all practicing synthetic chemists. Every attempt has been made to be of the broadest synthetic relevance and our expectation is that our colleagues will share this opinion.

Leo A. Paquette
Columbus, Ohio USA

Introduction

The aim of this volume of the *Handbook of Reagents for Organic Synthesis* ("*HROS*") is to provide the practicing synthetic chemist with a convenient compendium of information concerning the most important and frequently employed reagents for the oxidation and reduction of organic compounds. Modification of the oxidation state of organic compounds constitutes one of the most common transformations encountered in synthetic schemes, and literally hundreds of reagents are available for this purpose. For this volume, we have selected the 145 most important oxidizing and reducing agents that were previously included in the *Encyclopedia of Reagents for Organic Synthesis* ("*EROS*"). Reproduced in this Handbook is the full *EROS* article for each reagent, including the discussion of synthetic transformations effected by that reagent that do not involve oxidation or reduction.

The usefulness of each article in *HROS* has been enhanced by the incorporation of a new **Related Reagents** section which focuses on alternative reagents that have been employed and found to be effective for similar oxidative and reductive transformations. In these Related Reagents sections, reference is made not to individual reagents, but rather to classes of reagents which are defined in the listings found at the end of this Introduction. Each of these classes represents a general type of oxidation or reduction of importance in organic synthesis. For each class are listed (alphabetically) all of the reagents included in *EROS* that have utility for that particular transformation. These focused classifications will directly guide the user to alternative reagents in both this Volume and *EROS*. Those reagents that are included among the 145 reagents in this Handbook are highlighted by having their names printed in boldface type.

Also included in this Handbook is a Bibliography of Reviews and Monographs covering the period 1993–1997. All important review articles and monographs on the subject of oxidizing and reducing agents are listed here in order of their date of publication. Most of these articles appeared subsequent to the publication of *EROS* and hence are not referenced in the individual reagent articles found in the Handbook. Following this bibliography is a compilation of *Organic Syntheses* procedures that illustrate, with tested experimental details, oxidations and reductions by reagents included in this volume. The *Organic Syntheses* procedures are presented in alphabetical order, with separate sections for oxidizing and reducing agents. All references to oxidizing and reducing agents in volumes 69 through 75 of *Organic Syntheses* are included in this section.

We hope you will find this to be a useful and convenient Handbook, describing the most widely employed reagents for oxidation and reduction of organic compounds. The selected original *EROS* articles, augmented by the Related Reagents Classifications, the Recent Review and Monograph bibliographies, and the *Organic Syntheses* illustrations, should provide a unique resource for the practicing organic chemist.

Steven D. Burke
University of Wisconsin-Madison

Rick L. Danheiser
Massachusetts Institute of Technology

Classification of Oxidizing Agents

Class O-1: Reagents for the Oxidation of Alcohols to Aldehydes and Ketones.

Class O-2: Reagents for the Oxidation of Alcohols and Aldehydes to Carboxylic Acids.

Class O-3: Reagents for the Oxidation of Aldehydes to Carboxylic Esters.

Class O-4: Reagents for the Oxidation of Organohalogen Compounds to Carbonyl Compounds.

Class O-5: Reagents for the Oxidation of Phenols to Quinones.

Class O-6: Reagents for the Oxidation of Phenol Ethers to Quinones.

Class O-7: Reagents for the Oxidation of Thiols to Disulfides.

Class O-8: Reagents for the Oxidation of Sulfides to Sulfoxides and Sulfones.

Class O-9: Reagents for the Oxidation of Amines to Nitro Compounds.

Class O-10: Reagents for the Oxidation of Amines to Azo Compounds.

Class O-11: Reagents for the Oxidation of Amines to Amine Oxides.

Class O-12: Reagents for Dehydrogenation to Form Aromatic Compounds.

Class O-13: Reagents for Dehydrogenation to Form Unsaturated Carbonyl Compounds.

Class O-14: Reagents for Epoxidation.

Class O-15: Reagents for Baeyer-Villiger Oxidation of Ketones and Aldehydes.

Class O-16: Reagents for the Oxidation of Alkenes to 1,2-Diols.

Class O-17: Reagents for the Oxidative Cleavage of Alkenes.

Class O-18: Reagents for the Oxidative Cleavage of 1,2-Diols.

Class O-19: Reagents for the Oxidation of Alkenes to Ketones.

Class O-20: Reagents for the Oxidation of Enol Derivatives to α-Hydroxy Ketones and Derivatives.

Class O-21: Reagents for Oxidation at Allylic and Benzylic C–H Bonds.

Class O-22: Reagents for the Oxidative Cyclization of Alcohols to Cyclic Ethers.

Class O-23: Reagents for the Oxidation of Unactivated C–H Bonds.

Class O-24: Reagents for the Oxidation of Carbanions and Organometallic Compounds to Alcohols.

Class O-1: Reagents for the Oxidation of Alcohols to Aldehydes and Ketones.

4-Acetamido-2,2,6,6-tetramethyl-1-piperidinyloxyl; Ammonium Peroxydisulfate; 1,1′-(Azodicarbonyl)dipiperidine; Barium Manganate; Benzeneseleninic Acid; **1,4-Benzoquinone**; Benzyltriethylammonium Chlorochromate; Benzyltriethylammonium Permanganate; 2,2′-Bipyridinium Chlorochromate; Bismuth(III) Oxide; Bis(pyridine)silver(I) Permanganate; Bis (tetrabutylammonium) Dichromate; **Bis(tri-*n*-butyltin) Oxide**; Bis(trichloromethyl) Carbonate; **Bis(trimethylsilyl) Peroxide**; **Bromine**; Bromine-1,4-Diazabicyclo[2.2.2]octane; Bromine Trifluoride; *N*-Bromoacetamide; *N*-Bromosuccinimide; *N*-Bromosuccinimide–Dimethyl Sulfide; ***t*-Butyl Hydroperoxide**; ***t*-Butyl Hypochlorite**; Calcium Hypochlorite; **Cerium(IV) Ammonium Nitrate**; Cerium(IV) Ammonium Nitrate–Sodium Bromate; Cerium(IV)–Nafion 511; Cerium(IV) Trifluoromethanesulfonate; Cerium(IV) Trihydroxide Hydroperoxide; Cerium(IV) Trifluoroacetate; Chloramine; **Chlorine**; Chlorine-Pyridine; 1-Chlorobenzotriazole; *m*-Chloroperbenzoic Acid-2,2,6,6-Tetramethylpiperidine Hydrochloride; ***N*-Chlorosuccinimide–Dimethyl Sulfide**; **Chromic Acid**; **Chromium(VI) Oxide**; **Chromium(VI) Oxide–3,5-Dimethylpyrazole**; Chromium(VI) Oxide–Quinoline; Chromium(VI) Oxide–Silica Gel; **Copper(II) Bromide**; Copper(I) Chloride–Oxygen; Copper(II) Permanganate; Copper(II) Sulfate–Pyridine; Di-*t*-butyl Chromate; Di-*t*-butyl Chromate–Pyridine; **2,3-Dichloro-5,6-dicyano-1,4-benzoquinone**; 4-(Dimethylamino)pyridinium Chlorochromate; Dimethyldioxirane; Dimethyl Sulfide–Chlorine; **Dimethyl Sulfoxide–Acetic Anhydride**; Dimethyl Sulfoxide–Dicyclohexylcarbodiimide; Dimethyl Sulfoxide–Oxalyl Chloride; Dimethyl Sulfoxide–Methanesulfonic Anhydride; **Dimethyl Sulfoxide–Oxalyl Chloride**; Dimethyl Sulfoxide–Phosgene; Dimethyl Sulfoxide–Phosphorus Pentoxide; Dimethyl Sulfoxide–Sulfur Trioxide–Pyridine; Dimethyl Sulfoxide–Trifluoroacetic Anhydride; Dimethyl Sulfoxide–Triphosgene; Dinitratocerium(IV) Chromate Dihydrate; **Dipyridine Chromium(VI) Oxide**; Hydrogen Peroxide–Ammonium Heptamolybdate; Imidazolium Dichromate; Iron(III) Nitrate–K10 Montmorillonite Clay; **Manganese Dioxide**; ***N*-Methylmorpholine *N*-Oxide**; Methyl-(trifluoromethyl)dioxirane; Nickel(II) Peroxide; μ–Oxobis(chlorotriphenylbismuth); **Oxygen–Platinum Catalyst**; **Ozone**; Phenyliodine(III) Dichloride; Poly(4-vinylpyridinium dichromate); Potassium *t*-Butoxide–Benzophenone; Potassium Dichromate; Potassium Ferrate; **Potassium Permanganate**; Potassium Ruthenate; **Pyridinium Chlorochromate**; Pyridinium Chlorochromate–Alumina; **Pyridinium Dichromate**; Pyridinium Fluorochromate; **Ruthenium(VIII) Oxide**; **Silver(I) Carbonate on Celite**; Sodium Bromate; Sodium Bromite; Sodium Dichromate; **Sodium Hypochlorite**; Sodium Permanganate; Tetra-*n*-butylammonium Chlorochromate; **2,2,6,6-Tetramethylpiperidin-1-oxyl**; **Tetra-*n*-propylammonium Perruthenate**; **1,1,1-Triacetoxy-1,1-dihydro-1,2-benziodoxol-3(1*H*)-one**; Triphenylbismuth Dichloride; Tris[trinitratocerium(IV)]Paraperiodate.

Class O-2: Reagents for the Oxidation of Alcohols and Aldehydes to Carboxylic Acids.

Benzyltriethylammonium Permanganate; **Chromic Acid**; **Chromium(VI) Oxide**; 2-Hydroperoxyhexafluoro-2-propanol; **Oxygen–Platinum Catalyst**; Peroxyacetyl Nitrate; **Potassium Permanganate**; Potassium Ruthenate; **Pyridinium Dichromate**; **Ruthenium(VIII) Oxide**; **Silver(I) Oxide**; **Silver(II) Oxide**; **Sodium Chlorite**; **Sodium Hypochlorite**; **2,2,6,6-Tetramethylpiperidin-1-oxyl**.

Class O-3: Reagents for the Oxidation of Aldehydes to Carboxylic Esters.

3-Benzylthiazolium Bromide; **Bromine**; ***t*-Butyl Hydroperoxide**; **Chromic Acid**; **Manganese Dioxide**; Monoperoxysulfuric Acid; **Ozone**; **Silver(I) Carbonate on Celite**; **Silver(II) Oxide**; Sodium Cyanide.

Class O-4: Reagents for the Oxidation of Organohalogen Compounds to Carbonyl Compounds.

Bis(tetrabutylammonium)Dichromate; Dimethyl Selenoxide; Dimethyl Sulfoxide–Silver Tetrafluoroborate; Hexamethylenetetramine; ***N*-Methylmorpholine *N*-Oxide**; Potassium Chromate; Potassium Ruthenate; Pyridine N-Oxide; **Ruthenium(VIII) Oxide**; Sodium 4,6-Diphenyl-1-oxido-2-pyridone; **Trimethylamine *N*-Oxide**.

Class O-5: Reagents for the Oxidation of Phenols to Quinones.

Ammonium Peroxydisulfate; Benzeneseleninic Acid; Benzenetellurinic Anhydride; Bis(*N*-propylsalicylideneaminato)copper(II); Bis(*N*-propylsalicylideneaminato)cobalt(II); Bromine Trifluoride; *N*-Bromosuccinimide–Dimethylformamide; Cerium(IV) Ammonium Sulfate; Cerium(IV) Trihydroxide Hydroperoxide; ***N*-Chlorosuccinimide–Dimethyl Sulfide**; Chromyl Chloride; **Copper(II) Chloride**; Copper(I) Chloride–Oxygen; (Diacetoxyiodo)benzene; Dimethylsuccinimidosulfonium Tetrafluoroborate; Dinitratocerium(IV) Chromate Dihydrate; Diphenylselenium Bis(trifluoroacetate); Iodosylbenzene–Dichlorotris(triphenylphosphine)ruthenium; Lead(IV) Oxide; **Mercury(II) Oxide**; Methyl(trifluoromethyl)dioxirane; **Phenyliodine(III) Bis(trifluoroacetate)**; **Potassium Nitrosodisulfonate**; **Potassium Superoxide**; Salcomine; **Silver(I) Oxide**; **Silver(II) Oxide**; Sodium Bromate; **Sodium Hypochlorite**; **Sodium Periodate**; Thallium(III) Perchlorate Hexahydrate; **Thallium(III) Trifluoroacetate**.

Class O-6: Reagents for the Oxidation of Phenol Ethers to Quinones.

Cerium(IV) Ammonium Nitrate; Nitric Acid; **Silver(II) Oxide**; **Thallium(III) Trifluoroacetate**.

Class O-7: Reagents for the Oxidation of Thiols to Disulfides.

Ammonium Peroxy disulfate; Barium Manganate; Benzenetellurinic Anhydride; Benzoyl Nitrate; Bis(2,2′-bipyridyl)copper(II) Permanganate; Copper(II) Nitrate–K10 Bentonite Clay; Dimethyl Dithiobis(thioformate); Dinitratocerium(IV) Chromate Dihydrate; Iron(III) Nitrate–K10 Montmorillonite Clay; **Oxygen**; **Potassium Superoxide**; **Pyridinium Chlorochromate**; 2,4,4,6-Tetrabromo-2,5-cyclohexadienone; Tetra-*n*-butylammonium Chlorochromate; Thallium(III) Acetate.

Class O-8: Reagents for the Oxidation of Sulfides to Sulfoxides and Sulfones.

Acetyl Nitrate; Ammonium Peroxydisulfate; Benzyltriethylammonium Permanganate; 2,2-Bipyridinium Chlorochromate; **Bis(tri-*n*-butyltin) Oxide**; Bis(trimethylsilyl) Monoperoxysulfate; **Bis(trimethylsilyl) Peroxide**; Bromine-1,4-Diazabicyclo[2.2.2]octane; ***t*-Butyl Hydroperoxide**; ***t*-Butyl Hypochlorite**; (Camphorylsulfonyl)oxaziridine; **Cerium(IV) Ammonium Nitrate**; 1-Chlorobenzotriazole; ***m*-Chloroperbenzoic Acid**; Cumyl Hydroperoxide; Dimethyldioxirane; **Hydrogen Peroxide**; Hydrogen Peroxide-Tellurium Dioxide; Hydrogen Peroxide–Tungstic Acid; 2-Hydroperoxyhexafluoro-2-propanol; Iodosylbenzene–Dichlorotris(triphenylphosphine)ruthenium; **Monoperoxyphthalic Acid**; Nitronium Tetrafluoroborate; **Ozone**; **Perbenzoic Acid**; Phenyliodine(III) Dichloride; *N*-(Phenylsulfonyl)(3,3-dichlorocamphoryl)oxaziridine; (±)-*trans*-2-(Phenylsulfonyl)-3-phenyloxaziridine; **Potassium Permangate**, **Potassium Monoperoxysulfate**; **Potassium Superoxide**; **Pyridinium Chlorochromate**; **Ruthenium(VIII) Oxide**; **Singlet Oxygen**; Sodium Bromite; **Sodium Periodate**; Tetra-*n*-butylammonium Periodate; **Tetra-*n*-propylammonium Perruthenate**; **Thallium(III) Nitrate Trihydrate**; **Trifluoroperacetic acid**; Triphenylmethyl Hydroperoxide; **Vanadyl Bis(acetylacetonate)**.

Class O-9: Reagents for the Oxidation of Amines to Nitro Compounds.

Dimethyldioxirane; **Fluorine**; Hypofluorous Acid; Ozone–silica gel; Peroxymaleic Acid; **Potassium Permanganate**; **Potassium Monoperoxysulfate**; Sodium Perborate; **Trifluoroperacetic Acid**; **Vanadyl (Bis(acetylacetonate)**.

Class O-10: Reagents for the Oxidation of Amines to Azo Compounds.

Bis(pyridine)silver(I) Permanganate; **(Diacetoxyiodo)benzene**; **Lead(IV) Acetate**; **Manganese Dioxide**; **Potassium Superoxide**; **Silver(I) Carbonate on Celite**; **Silver(II) Oxide**.

Class O-11: Reagents for the Oxidation of Amines to Amine Oxides.

Bis(trimethylsilyl) Monoperoxysulfate; ***t*-Butyl Hydroperoxide-m–Chloroperbenzoic Acid**; ***m*-Chloroperbenzoic Acid**; **Hydrogen Peroxide**; **Hydrogen Peroxide–Urea**; 2-Hydroperoxyhexafluoro-2-propanol; **Monoperoxyphthalic Acid**; **Peracetic Acid**; (±)-*trans*-2-(Phenylsulfonyl)-3-phenyloxazirdine; **Potassium Monoperoxysulfate**; **Trifluoroperacetic Acid**; **Vanadyl Bis(acetylacetonate)**.

Class O-12: Reagents for Dehydrogenation to Form Aromatic Compounds.

Barium Manganate; Cadmium Chloride; Cerium(IV) Ammonium Sulfate; **2,3-Dichloro-5,6-dicyano-1,4-benzoquinone**; **Manganese Dioxide**; Nitrosylsulfuric Acid; Palladium on Carbon; **Sulfur**; **Triphenylcarbeneium Tetrafluoroborate**.

Class O-13: Reagents for Dehydrogenation to Form Unsaturated Carbonyl Compounds.

Benzeneseleninic Acid; Chloranil; **Copper(II) Bromide**; **2,3-Dichloro-5,6-dicyano-1,4-benzoquinone**; Iodylbenzene; Thallium(III) Acetate.

Class O-14: Reagents for Epoxidation.

Benzeneperoxyseleninic Acid; Bis(acetonitrile)chloronitropalladium(II); *N,N*-Bis(salicylidene)ethylenediaminenickel(II); ***t*-Butyl Hydroperoxide**; ***m*-Chloroperbenzoic Acid**; *m*-Chloroperbenzoic Acid-2,2,6,6-Tetramethylpiperidine Hydrochloride; Chromyl Acetate; Cumyl Hydroperoxide; **1,1-Di-*t*-butyl Peroxide**; Dimethyldioxirane; 2,4-Dinitrobenzeneseleninic Acid; *O*-Ethylperoxycarbonic Acid; **Fluorine**; **Hydrogen Peroxide**; Hydrogen Peroxide-Ammonium Heptamolybdate; Hydrogen Peroxide–Tungstic Acid; **Hydrogen Peroxide–Urea**; Hypofluorous Acid; **Iodosylbenzene**; *p*-Methoxycarbonylperbenzoic Acid; Methyl(trifluoromethyl)dioxirane; Monoperoxyphosphoric Acid; **Monoperoxyphthalic Acid**; *o*-Nitrobenzeneseleninic Acid; *p*-Nitroperbenzoic Acid; **Peracetic Acid**; **Perbenzoic Acid**; Peroxyacetimidic Acid; Peroxyacetyl Nitrate; Potassium *o*-Nitrobenzeneperoxysulfonate; **Potassium Permanganate**; **Potassium Monoperoxysulfate**; **Sodium Hypochlorite**; Sodium Hypochlorite-*N,N′*-Bis(3,5-di-*t*-butylsalicylidene)-1,2-cyclohexane–diaminomanganese(III) Chloride; Sodium Perborate; **Trifluoroperacetic Acid**; Triphenylmethyl Hydroperoxide; **Vanadyl Bis(acetylacetonate)**.

Class O-15: Reagents for Baeyer-Villiger Oxidation of Ketones and Aldehydes.

Benzeneperoxyseleninic Acid; Bis(trimethylsilyl) Monoperoxysulfate; **Bis(trimethylsilyl) Peroxide**; **Cerium(IV) Ammonium Nitrate**; Cerium(IV) Ammonium Sulfate; ***m*-Chloroperbenzoic Acid**; **Hydrogen Peroxide**; Hydrogen Peroxide-Boron Trifluoride; **Hydrogen Peroxide–Urea**; 2-Hydroperoxyhexafluoro-2-propanol; Hypofluorous Acid; *p*-Methoxycarbonylperbenzoic Acid; Monoperoxyphosphoric Acid; **Monoperoxyphthalic Acid**; Monoperoxysulfuric Acid; *p*-Nitroperbenzoic acid; **Peracetic Acid**; **Perbenzoic Acid**; Peroxymaleic Acid; Potassium Dichromate; **Potassium Monoperoxysulfate**; Sodium Perborate; Trifluoroacetic Acid; **Trifluoroperacetic Acid**.

Class O-16: Reagents for the Oxidation of Alkenes to 1,2-Diols.

***t*-Butyl Hydroperoxide**; **Hydrogen Peroxide**; Iodine–Copper(II) Acetate; Iodine–Silver Benzoate; ***N*-Methylmorpholine *N*-Oxide**; **Osmium Tetroxide**; **Osmium Tetroxide–*t*-Butyl Hydroperoxide**; **Osmium Tetroxide-*N*-Methylmorpholine *N*-Oxide**; Osmium Tetroxide–Potassium Ferricyanide; **Potassium Permanganate**; **Selenium(IV) Oxide**; *o*-Sulfoperbenzoic Acid; **Trifluoroperacetic Acid**.

Class O-17: Reagents for the Oxidative Cleavage of Alkenes.

Bis(2,2′-bipyridyl)copper(II) Permanganate; **Chromium(VI) Oxide**; Hydrogen Peroxide–Tungstic Acid; **Ozone**; **Osmium Tetroxide–*N*-Methylmorpholine *N*-Oxide**; **Singlet Oxygen**; **Sodium Periodate–Osmium Tetroxide**; Sodium Periodate–Potassium Permanganate.

Class O-18: Reagents for the Oxidative Cleavage of 1,2-Diols.

Calcium Hypochlorite; **Chromic Acid**; **Lead(IV) Acetate**; **Manganese Dioxide**; **Pyridinium Chlorochromate**; **Ruthenium(VIII) Oxide**; **Silver(I) Carbonate on Celite**; Sodium Bismuthate; **Sodium Periodate**; **Tetra-*n*-propylammonium Perruthenate**; Triphenylbismuth Carbonate.

Class O-19: Reagents for the Oxidation of Alkenes to Ketones.

Bis(acetonitrile)chloronitropalladium(II); Bis(acetonitrile)dinitropalladium(II); Bis(trimethylsilyl) Monoperoxysulfate; Palladium *t*-Butyl Peroxide Trifluoroacetate; Palladium(II) Chloride; Palladium(II) Chloride–Copper(I) Chloride; Palladium(II)–Trifluoroacetate; **Rhodium(III) Chloride**; **Thallium(III) Nitrate Trihydrate**.

Class O-20: Reagents for the Oxidation of Enol Derivatives to α-Hydroxy Ketones and Derivatives.

(Camphorylsulfonyl)oxaziridine; ***m*-Chloroperbenzoic Acid**; Chromyl Chloride; **(Diacetoxyiodo)benzene**; Dimethyldioxirane; **Iodosylbenzene**; Iodosylbenzene–Boron Trifluoride; **Lead(IV) Acetate**; μ-Oxobis(chlorotriphenylbismuth); **Oxodiperoxymolybdenum(Pyridine)–(hexamethylphosphoric triamide)**; **Oxygen**; **(Phenyliodine(III) Bis(trifluoroacetate)**.

Class O-21: Reagents for Oxidation at Allylic and Benzylic C–H Bonds.

2,2-Bipyridinium Chlorochromate; ***t*-Butyl Hydroperoxide**; Cerium(III) Methanesulfonate; Cerium(IV) Pyridinium Chloride; **Chromium(VI) Oxide**; **Chromium(VI) Oxide–3,5-Dimethylpyrazole**; **Copper(II) Acetate**; Copper(I) Chloride–Oxygen; Di-*t*-Butyl Chromate; **2,3-Dichloro-5,6-dicyano-1,4-benzoquinone**; **Dipyridine Chromium(VI) Oxide**; Iodylbenzene; Iron(II) Phthalocyanine; **Lead(IV) Acetate**; Mercury(II) Acetate; Nitric Acid; Potassium *o*-Nitrobenzeneperoxysulfonate; **Potassium Permanganate**; **Potassium Superoxide**; **Pyridinium Chlorochromate**; **Pyridinium Dichromate**; **Selenium(IV) Oxide**; **Selenium(IV) Oxide–Butyl Hydroperoxide**; **Singlet Oxygen**; Sodium Dichromate.

Class O-22: Reagents for the Oxidative Cyclization of Alcohols to Cyclic Ethers.

Bromine; Bromine–Silver(I) Oxide; **Cerium(IV) Ammonium Nitrate**; Di-*t*-butyl Chromate; **Lead(IV) Acetate**; Lead(IV) Acetate–Iodine; **Mercury(II) Oxide**; Mercury(II) Oxide–Iodine.

Class O-23: Reagents for the Oxidation of Unactivated C-H Bonds.

Ammonium Peroxydisulfate; Benzyltriethylammonium Permanganate; **Chromium(VI) Oxide**; Chromyl Acetate; Dichlorotris(triphenylphosphine)-ruthenium(II); Dimethyldioxirane; Disodium Tetrachloroplatinate(II); Fluorodimethoxyborane Diethyl Etherate; Hypofluorous Acid; Iodine Tris(trifluoroacetate); Iron(II) Chloride; Iron(II) Sulfate-Oxygen; Lead(IV) Trifluoroacetate; Methyl(trifluoromethyl)dioxirane; *p*-Nitroperbenzoic acid; Ozone-Silica Gel; Potassium Monoperoxysulfate; **Ruthenium(VIII) Oxide**.

Class O-24: Reagents for the Oxidation of Carbanions and Organometallic Compounds to Alcohols.

Bis(trimethylsilyl) Peroxide; ***t*-Butyl Hydroperoxide**; (Camphorylsulfonyl)oxaziridine; **Cerium(IV) Ammonium Nitrate**; **Copper(II) Acetate**; **Oxygen**; (±)-*trans*-2-(Phenylsulfonyl)-3-phenyloxazirdine.

Classification of Reducing Agents

Class R-1: Reagents for Reduction of Acetals.

Class R-2: Reagents for Reduction of Aldehydes or Ketones to Alcohols.

Class R-3: Reagents for Reduction of Alkenes.

Class R-4: Reagents for Reduction of Alkynes.

Class R-5: Reagents for Reduction of Amides, Imines, or Iminium Ions to Amines.

Class R-6: Reagents for Reduction of Anhydrides or Imides.

Class R-7: Reagents for Reduction of Aromatic Carbocycles.

Class R-8: Reagents for Reduction of Aromatic Heterocycles.

Class R-9: Reagents for Reduction of Azides, Azo Compounds, Hydrazones, or Oximes to Amines.

Class R-10: Reagents for Chemoselective Reduction of Carbonyl Compounds.

Class R-11: Reagents for Enantioselective Reduction of Carbonyl Compounds.

Class R-12: Reagents for Stereoselective Reduction of Carbonyl Compounds.

Class R-13: Reagents for Reduction of Carboxylic Acids, Esters or Derivatives to Alcohols.

Class R-14: Reagents for Reduction of Carboxylic Acids, Esters or Derivatives to Aldehydes or Hemiacetals.

Class R-15: Reagents for Conjugate Reduction of α,β-Unsaturated Carbonyl Compounds.

Class R-16: Reagents for Reductive Deoxygenation of Epoxides to Alkenes.

Class R-17: Reagents for Reduction via Hydroboration.

Class R-18: Reagents for Enantioselective Hydrogenation.

Class R-19: Reagents for Catalysis of Hydrogenation or Hydrogenolysis.

Class R-20: Reagents for Reduction of Nitriles to Imines or Amines.

Class R-21: Reagents for Reduction of Nitro Compounds to Amines or Oximes.

Class R-22: Reagents for Reduction of Quinones.

Class R-23: Reagents for Reductive Cleavage of Allylic, Benzylic or α-Carbonyl Functionality.

Class R-24: Reagents for Reductive Cleavage of N-O, N-N, O-O, O-S or S-S Bonds.

Class R-25: Reagents for Reductive Coupling or Cyclization.

Class R-26: Reagents for Reductive Decarbonylation, Decarboxylation or Decyanation.

Class R-27: Reagents for Reductive Dehalogenation.

Class R-28: Reagents for Reductive Deoxygenation of Alcohols or Derivatives.

Class R-29: Reagents for Reductive Deoxygenation of Aldehydes or Ketones.

Class R-30: Reagents for Reductive Desulfurization.

Class R-31: Reagents for Reductive Elimination or Fragmentation.

Class R-32: Reagents for Reductive Cleavage of Epoxides or Ethers to Alcohols.

Class R-33: Reagents for Reductive Metallation.

Class R-34: Reagents for Reduction of Sulfoxides or Sulfones to Sulfides.

Class R-35: Reagents for Reductive Amination of Carbonyls.

Class R-1: Reagents for Reduction of Acetals.

Aluminum Amalgam; **Aluminum Hydride**; **Aluminum Isopropoxide**; **Borane–Tetrahydrofuran**; Dibromoalane; Dichloroalane; Dichloroborane Diethyl Etherate; Diiodosilane; **Diisobutylaluminum Hydride**; **Lithium Aluminum Hydride**; Lithium Aluminum Hydride–Boron Trifluoride Etherate; **Lithium Triethylborohydride**; Monochloroalane; Monochloroborane–Dimethyl Sulfide; Potassium Naphthalenide; **Sodium Cyanoborohydride**; Tin(II) Bromide; **Triethylsilane**; **Zinc Borohydride**.

Class R-2: Reagents for Reduction of Aldehydes or Ketones to Alcohols.

Aluminum Amalgam; Aluminum Ethoxide; **Aluminum Hydride**; Ammonium Formate; Ammonium Sulfide; (*S*)-2-

(Anilinomethyl)pyrrolidine; Bis(bicyclo[2.2.1]hepta-2,5-diene)-rhodium Perchlorate; Bis(η^5-cyclopentadienyl)dihydridozirconium; 2,6-Bis[(*S*)-4′-isopropyloxazolin-2′-yl](pyridine)rhodium Trichloride; 9-Borabicyclo[3.3.1]nonane Dimer; Bis(trifluoroacetoxy)borane; Bis(triphenylphosphine)copper(I) Borohydride; Bis(triphenylphosphine)copper(I) Cyanoborohydride; **Borane–Dimethyl Sulfide**; **Borane–Tetrahydrofuran**; **Catecholborane**; (+)-*B*-Chlorodiisopinocampheylborane; Chloro(thexyl)-borane-Dimethyl Sulfide; Copper Chromite; Diborane; Dichloroalane; Dichlorotris(triphenylphosphine)ruthenium(II); Dicyclohexylborane; **Diisobutylaluminum Hydride**; **Diisopinocampheylborane**; 9-*O*-(1,2;5,6-Di-*O*-isopropylidene-α;-*D*-glucofuranosyl)-9-boratabicyclo[3.3.1]nonane, Potassium Salt; Dimethyl(phenyl)silane; Diphenylsilane–Cesium Fluoride; Diphenylstannane; **Disiamylborane**; Hexadecacarbonylhexarhodium; Hydrogen Selenide; Iron(III) Chloride–Sodium Hydride; **(2,3-*O*-Isopropylidene)-2,3-dihydroxy-1,4-bis(diphenylphosphino)butane**; **Lithium**; **Lithium Aluminum Hydride**; **Lithium Aluminum Hydride–2,2′-Dihydroxy-1,1′-binaphthyl**; Lithium 9-boratabicyclo[3.3.1]nonane; Lithium 9-boratabicyclo[3.3.1]nonane; **Lithium Borohydride**; Lithium *t*-Butyl(diisobutyl)aluminum Hydride; Lithium *n*-Butyl(hydrido)cuprate; **Lithium 4,4′-Di-*t*-butylbiphenylide**; Lithium 9,9-Dibutyl-9-borabicyclo[3.3.1]nonanate; **Lithium–Ethylamine**; Lithium Pyrrolidide; **Lithium Tri-*t*-butoxyaluminum Hydride**; **Lithium Triethylborohydride**; Lithium Trisiamylborohydride; Monochloroalane; Monoisopinocampheylborane; **Nickel Boride**; **Nickel(II) Chloride**; **Palladium on Carbon**; *B*-3-Pinanyl-9-borabicyclo[3.3.1]nonane; **Platinum on Carbon**; **Potassium**; **Potassium–Graphite Laminate**; **Potassium Tri-*s*-butylborohydride**; Potassium Triisopropoxyborohydride; Raney Nickel; **Rhodium on Alumina**; **Ruthenium Catalysts**; **Samarium(II) Iodide**; **Sodium**; **Sodium–Alcohol**; **Sodium–Ammonia**; **Sodium Bis(2-methoxyethoxy)aluminum Hydride**; **Sodium Borohydride**; **Sodium Cyanoborohydride**; **Sodium Dithionite**; Sodium Hydride–Nickel(II) Acetate–Sodium *t*-Pentoxide; Sodium Hydroxymethanesulfinate; Sodium Hypophosphite; **Sodium Triacetoxyborohydride**; Sodium Trimethoxyborohydride; Tetra-*n*-butylammonium Borohydride; Tetra-*n*-butylammonium Cyanoborohydride; **Tetrahydro-1-methyl-3,3-diphenyl-1*H*,3*H*-pyrrolo[1,2-*c*][1,3,2]oxazaborole**; **Tetramethylammonium Triacetoxyborohydride**; **Thexylborane**; Thiourea Dioxide; Trichlorosilane; Triethoxysilane; **Triethylsilane**; Triisobutylaluminum; Triphenylstannane; **Tris(trimethylsilyl)silane**; Urushibara Nickel; **Zinc**; **Zinc Borohydride**; Zinc–Copper(II) Acetate–Silver Nitrate; Zinc Chloride Etherate in Dichloromethane; Zinc–Dimethylformamide.

Class R-3: Reagents for Reduction of Alkenes.

Aluminum Amalgam; **Baker's Yeast**; (Bicyclo[2.2.1]hepta-2,5-diene)[1,4-bis(diphenylphosphino)butane]rhodium(I) Tetrafluoroborate; Bis(η^5-cyclopentadienyl)dihydridozirconium; **1,2-Bis(2,5-diethylphospholano)benzene**; (+)-*trans*-(2*S*,3*S*)-Bis(diphenylphosphino)bicyclo[2.2.1]hept-5-ene; **(*R*)- & (*S*)-2,2′-Bis(diphenylphosphino)-1,1′-binaphthyl**; **Borane–Tetrahydrofuran**; Calcium-Ammonia; **Catecholborane**; Chlorotris(triphenylphosphine)cobalt; **Chlorotris(triphenylphosphine)rhodium(I)**; Chromium(II) Sulfate; Cobalt Boride; (*R*)-(+)-Cyclohexyl(2-anisyl)methylphosphine; (1,5-Cyclooctadiene)(tricyclohexylphosphine)(pyridine)iridium(I) Hexafluorophosphate; Diborane; **Diimide**; **Diisopinocampheylborane**; **Disiamylborane**; (*R,R*)-[Ethylene-1,2-bis(η^5-4,5,6,7-tetrahydro-1-indenyl)]titanium (*R*)-1,1′-Bi-2,2′-naphtholate; (-)-[Ethylene-1,2-bis(η^5-4,5,6,7-tetrahydro-1-indenyl)]zirconium (*R*)-1,1′-Bi-2,2′-naphtholate; (±)-1,1′-Ethylenebis(4,5,6,7-tetrahydro-1-indenyl)zirconium Dichloride; Hexadecacarbonylhexarhodium; **Hydrazine**; Iron(III) Chloride–Sodium Hydride; **(2,3-*O*-Isopropylidene)-2,3-dihydroxy-1,4-bis(diphenylphosphino)butane**; **Lithium Aluminum Hydride**; Lithium Aluminum Hydride–Nickel(II) Chloride; Lithium Aluminum Hydride–Titanium(IV) Chloride; Lithium Aluminum Hydride–Cobalt(II) Chloride; **Lithium–Ethylamine**; **Lithium Naphthalenide**; **Lithium Triethylborohydride**; **Nickel Boride**; Nickel–Graphite; **Nickel(II) Chloride**; **Palladium on Barium Sulfate**; **Palladium on Carbon**; Palladium–Graphite; Palladium–Triethylamine–Formic Acid; **Platinum on Carbon**; **Platinum(IV) Oxide**; **Rhodium on Alumina**; **Ruthenium Catalysts**; **Sodium**; **Sodium–Alcohol**; Sodium Hypophosphite; Sodium Hydride–Nickel(II) Acetate–Sodium *t*-Pentoxide; **Sodium Triacetoxyborohydride**; **Titanium**; **Triethylsilane**; 2,4,6-Triisopropylbenzenesulfonylhydrazide; Urushibara Nickel.

Class R-4: Reagents for Reduction of Alkynes.

Aluminum Hydride; Bis(η^5-cyclopentadienyl)dihydridozirconium; Bis(diisopropylamino)aluminum Hydride; **Borane–Dimethyl Sulfide**; Calcium–Ammonia; **Catecholborane**; Chloro(thexyl)borane–Dimethyl Sulfide; **Chlorotris(triphenylphosphine)rhodium(I)**; **Chromium(II) Chloride**; Chromium(II) Sulfate; Cobalt Boride; Diborane; **Dibromoborane–Dimethyl Sulfide**; Dicarbonylbis(cyclopentadienyl)titanium; Dichloroborane Diethyl Etherate; Dicyclohexylborane; **Diimide**; **Diisobutylaluminum Hydride**; **Disiamylborane**; Ethylmagnesium Bromide–Copper(I) Iodide; **Hexa-μ-hydrohexakis(triphenylphosphine)hexacopper**; **Hydrazine**; Iron(III) Chloride–Sodium Hydride; **Lithium**; **Lithium Aluminum Hydride**; Lithium Aluminum Hydride–Cobalt(II) Chloride; Lithium Aluminum Hydride–Nickel(II) Chloride; Lithium Diisobutyl(methyl)aluminum Hydride; **Lithium–Ethylamine**; **Lithium Tri-*t*-butoxyaluminum Hydride**; Magnesium Hydride–Copper(I) Iodide; Nickel(II) Acetate; **Nickel Boride**; Nickel Catalysts (Heterogeneous); **Nickel(II) Chloride**; Nickel–Graphite; Niobium(V) Chloride–Zinc; Palladium(II) Acetate; **Palladium on Barium Sulfate**; **Palladium on Calcium Carbonate (Lead Poisoned)**; **Palladium on Carbon**; Palladium–Graphite; Palladium on Poly(ethylenimine); Palladium–Triethylamine–Formic Acid; **Platinum on Carbon**; **Potassium Tri-*s*-butylborohydride**; **Raney Nickel**; **Rhodium on Alumina**; **Sodium**; **Sodium–Ammonia**; **Sodium Bis(2-methoxyethoxy)aluminum Hydride**; Sodium Hydride–Nickel(II) Acetate–Sodium *t*-Pentoxide; Sodium Hydride–Palladium(II) Acetate–Sodium *t*-Pentoxide; Sodium Hypophosphite; **Sodium–Potassium Alloy**; **Thexylborane**; Tricarbonyl-(naphthalene)chromium; Urushibara Nickel; Ytterbium(0); **Zinc**; Zinc–Copper(II) Acetate-Silver Nitrate; **Zinc/Copper Couple**; Zinc–1,2-Dibromoethane.

Class R-5: Reagents for Reduction of Amides, Imines, or Iminium Ions to Amines.

Aluminum Amalgam; **Aluminum Hydride**; Bis(trifluoroacetoxy)borane; Borane–Ammonia; **Borane–Dimethyl Sulfide**; Dichlorotris(triphenylphosphine)ruthenium(II); **Diisobutylaluminum Hydride**; Dodecacarbonyltriiron; (*R,R*)-[Ethylene-1,2-bis(η^5-4,5,6,7-tetrahydro-1-indenyl)]titanium (*R*)-1,1′-Bi-2,2′-naphtholate; **(2,3-*O*-Isopropylidene)-2,3-dihydroxy-1,4-bis-(diphenylphosphino)butane**; **Lithium Aluminum Hydride**; **Lithium Aluminum Hydride–2,2′-Dihydroxy-1,1′-binaphthyl**; Lithium 9-boratabicyclo[3.3.1]nonane; **Lithium Borohydride**; **Lithium Tri-*t*-butoxyaluminum Hydride**; **Lithium Tri-*s*-butylborohydride**; **Lithium Triethylborohydride**; Monochloroalane; **Palladium on Calcium Carbonate (Lead Poisoned)**; **Platinum on Carbon**; **Sodium**; **Sodium Borohydride**; **Sodium Cyanoborohydride**; **Sodium Dithionite**; Sodium Telluride; **Sodium Triacetoxyborohydride**; Sodium Trifluoroacetoxyborohydride; Sodium Tris(trifluoroacetoxy)borohydride; **Tetrahydro-1-methyl-3,3-diphenyl-1*H*,3*H*-pyrrolo[1,2-*c*][1,3,2]oxazaborole**; Tetramethylammonium Triacetoxyborohydride; Tin(IV) Chloride; Titanium(IV) Chloride; Trichlorosilane; Triethoxysilane; Ytterbium(0); **Zinc**.

Class R-6: Reagents for Reduction of Anhydrides or Imides.

Aluminum Amalgam; **Lithium Aluminum Hydride**; **Lithium Tri-*s*-butylborohydride**; **Lithium Triethylborohydride**; **Lithium Trisiamylborohydride**; **Potassium Tri-*s*-butylborohydride**; **Sodium Bis(2-methoxyethoxy)aluminum Hydride**; **Sodium Borohydride**; **Sodium Triacetoxyborohydride**; **Tetramethylammonium Triacetoxyborohydride**; Zinc–Acetic Acid.

Class R-7: Reagents for Reduction of Aromatic Carbocycles.

Calcium–Ammonia; **Lithium**; **Lithium–Ethylamine**; **Palladium on Carbon**; **Platinum(IV) Oxide**; **Potassium**; **Potassium–Graphite Laminate**; **Rhodium on Alumina**; **Ruthenium Catalysts**; **Sodium–Alcohol**; **Sodium–Ammonia**; **Triethylsilane**; Ytterbium(0).

Class R-8: Reagents for Reduction of Aromatic Heterocycles.

Bis(trifluoroacetoxy)borane; Copper Chromite; **Lithium Borohydride**; **Lithium Triethylborohydride**; **Nickel Boride**; **Palladium on Carbon**; **Platinum on Carbon**; **Platinum(IV) Oxide**; **Raney Nickel**; **Rhodium on Alumina**; **Ruthenium Catalysts**; **Sodium–Alcohol**; **Sodium Amalgam**; **Sodium Dithionite**; **Sodium Triacetoxyborohydride**; Sodium Tris(trifluoroacetoxy)borohydride; Triethylsilane–Trifluoroacetic Acid.

Class R-9: Reagents for Reduction of Azides, Azo Compounds, Hydrazones, or Oximes to Amines.

Aluminum Amalgam; **1,2-Bis(2,5-diethylphospholano)benzene**; **Borane–Dimethyl Sulfide**; **Borane–Tetrahydrofuran**; **Diisobutylaluminum Hydride**; **Hydrogen Sulfide**; Magnesium–Methanol; Monochloroalane; **Nickel Boride**; **Nickel(II) Chloride**; Norephedrine–Borane; **Palladium on Calcium Carbonate (Lead Poisoned)**; **Palladium on Carbon**; **Palladium(II) Hydroxide on Carbon**; **Platinum(IV) Oxide**; 1,3-Propanedithiol; **Rhodium on Alumina**; **Sodium**; **Sodium–Alcohol**; **Sodium Amalgam**; **Sodium Bis(2-methoxyethoxy)aluminum Hydride**; **Sodium Borohydride**; **Sodium Cyanoborohydride**; **Sodium Dithionite**; Sodium Hydride; Sodium Hypophosphite; Sodium Telluride; **Sodium Thiosulfate**; Sodium Tris(trifluoroacetoxy)borohydride; **Tetrahydro-1-methyl-3,3-diphenyl-1*H*,3*H*-pyrrolo[1,2-*c*][1,3,2]oxazaborole**; **Tetramethylammonium Triacetoxyborohydride**; Tin(II) Chloride; Tri-*n*-butylhexadecylphosphonium Bromide; **Triphenylphosphine**; Vanadium(II) Chloride; **Zinc**; Zinc–Acetic Acid; **Zinc/Copper Couple**; Zinc–Dimethylformamide.

Class R-10: Reagents for Chemoselective Reduction of Carbonyl Compounds.

Aluminum Amalgam; **Aluminum Isopropoxide**; **1,2-Bis(2,5-diethylphospholano)benzene**; **(*R*)- & (*S*)-2,2′-Bis(diphenylphosphino)-1,1′-binaphthyl**; 9-Borabicyclo[3.3.1]nonane Dimer; **Borane–Dimethyl Sulfide**; **Borane–Tetrahydrofuran**; **Catecholborane**; Cerium(III) Chloride; **Chlorotris(triphenylphosphine)rhodium(I)**; Cobalt Boride; Diisobutylaluminum 2,6-Di-*t*-butyl-4-methylphenoxide; **Diisobutylaluminum Hydride**; **Hexa-μ-hydrohexakis(triphenylphosphine)hexacopper**; Lanthanum(III) Chloride; **Lithium Borohydride**; Lithium *t*-Butyl(diisobutyl)aluminum Hydride; Lithium *n*-Butyl(diisobutyl)aluminum Hydride; Lithium 9,9-Dibutyl-9-borabicyclo[3.3.1]nonanate; **Lithium Tri-*t*-butoxyaluminum Hydride**; **Lithium Tri-*s*-butylborohydride**; **Lithium Triethylborohydride**; **Lithium Trisiamylborohydride**; **Nickel Boride**; **Potassium Tri-*s*-butylborohydride**; **Sodium Bis(2-methoxyethoxy)aluminum Hydride**; **Sodium Borohydride**; **Sodium Dithionite**; **Sodium Triacetoxyborohydride**; Sodium Trimethoxyborohydride; Tetra-*n*-butylammonium Borohydride; **Tetrahydro-1-methyl-3,3-diphenyl-1*H*,3*H*-pyrrolo[1,2-*c*][1,3,2]oxazaborole**; **Tetramethylammonium Triacetoxyborohydride**; Triethoxysilane; **Zinc Borohydride**.

Class R-11: Reagents for Enantioselective Reduction of Carbonyl Compounds.

2-Amino-3-methyl-1,1-diphenyl-1-butanol; (*S*)-4-Anilino-3-methylamino-1-butanol; (*S*)-2-(Anilinomethyl)pyrrolidine; **Baker's Yeast**; 2-[2-[(Benzyloxy)ethyl]-6,6-dimethylbicyclo-[3.3.1]-3-nonyl]-9-borabicyclo[3.3.1]nonane; **1,2-Bis(2,5-diethylphospholano)benzene**; (+)-*trans*-(2*S*,3*S*)-Bis(diphenylphosphino)bicyclo[2.2.1]hept-5-ene; **(*R*)- & (*S*)-2,2′-Bis(diphenylphosphino)-1,1′-binaphthyl**; 2,6-Bis[(*S*)-4′-isopropyloxazolin-2′-yl](pyridine)rhodium Trichloride; **Borane–Dimethyl Sulfide**; **Borane–Tetrahydrofuran**; **Catecholborane**; (+)-*B*-Chlorodiisopinocampheylborane; **Diborane**; **Diisobutylaluminum Hydride**; **Diisopinocampheylborane**; 9-*O*-(1,2;5,6-Di-*O*-isopropylidene-α-*D*-glucofuranosyl)-9-boratabicyclo[3.3.1]nonane, Potassium Salt; (*R,R*)-2,5-Dimethylborolane; Ephedrine-borane; **Lithium Aluminum Hydride**; **Lithium Aluminum Hydride–2,2′-Dihydroxy-1,1′-binaphthyl**;

Lithium Tri-*t*-butoxyaluminum Hydride; Monoisopinocampheylborane; Norephedrine–Borane; Potassium Triisopropoxyborohydride; **Raney Nickel**; **Sodium Borohydride**; **Tetrahydro-1-methyl-3,3-diphenyl-1*H*,3*H*-pyrrolo[1,2-*c*][1,3,2]oxazaborole**; Tin(II) Chloride.

Class R-12: Reagents for Stereoselective Reduction of Carbonyl Compounds.

Aluminum Hydride; **Aluminum Isopropoxide**; **Baker's Yeast**; *trans*-2,5-Bis(methoxymethyl)pyrrolidine; **Borane–Tetrahydrofuran**; **Catecholborane**; Chlorodiisopropylsilane; **Chlorotris(triphenylphosphine)rhodium(I)**; Dicyclohexylborane; Diisobutylaluminum 2,6-Di-*t*-butyl-4-methylphenoxide; **Diisobutylaluminum Hydride**; (*R,R*)-2,5-Dimethylborolane; Dimethyl(phenyl)silane; **Disiamylborane**; Erbium(III) Chloride; **Hexa-μ-hydrohexakis(triphenylphosphine)hexacopper**; **Lithium**; **Lithium Aluminum Hydride**; **Lithium Aluminum Hydride–2,2′-Dihydroxy-1,1′-binaphthyl**; Lithium Aluminum Hydride–Titanium(IV) Chloride; Lithium 9-boratabicyclo[3.3.1]nonane; **Lithium Borohydride**; Lithium *n*-Butyl(diisobutyl)aluminum Hydride; Lithium 9,9-Dibutyl-9-borabicyclo[3.3.1]nonanate; **Lithium–Ethylamine**; **Lithium Tri-*t*-butoxyaluminum Hydride**; **Lithium Tri-*s*-butylborohydride**; **Lithium Triethylborohydride**; Lithium Trisiamylborohydride; **Palladium on Carbon**; **Platinum on Carbon**; Potassium 9-Siamyl-9-boratabicyclo[3.3.1]nonane; **Potassium Tri-*s*-butylborohydride**; **Raney Nickel**; **Rhodium on Alumina**; **Ruthenium Catalysts**; **Sodium–Alcohol**; **Sodium–Ammonia**; **Sodium Bis(2-methoxyethoxy)aluminum Hydride**; **Sodium Borohydride**; **Sodium Dithionite**; **Sodium Triacetoxyborohydride**; **Tetrahydro-1-methyl-3,3-diphenyl-1*H*,3*H*-pyrrolo[1,2-*c*][1,3,2]oxazaborole**; **Tetramethylammonium Triacetoxyborohydride**; **Thexylborane**; Triisobutylaluminum; **Tris(trimethylsilyl)silane**; Urushibara Nickel; **Zinc Borohydride**; Zinc Complex Reducing Agents.

Class R-13: Reagents for Reduction of Carboxylic Acids, Esters or Derivatives to Alcohols.

Aluminum Hydride; **Borane–Dimethyl Sulfide**; **Borane–Tetrahydrofuran**; Copper Chromite; **Diborane**; **Diisobutylaluminum Hydride**; Diphenylsilane–Cesium Fluoride; Ethyl Chloroformate; *N*-Ethyl-5-phenylisoxazolium-3′-sulfonate; **Lithium Aluminum Hydride**; **Lithium Aluminum Hydride–2,2′-Dihydroxy-1,1′-binaphthyl; Lithium Tri-*t*-butoxyaluminum Hydride**; **Lithium Tri-*s*-butylborohydride**; **Lithium Triethylborohydride**; Lithium 9-boratabicyclo[3.3.1]nonane; Lithium *t*-Butyl(diisobutyl)aluminum Hydride; Monochloroalane; **Potassium Tri-*s*-butylborohydride**; **Sodium**; **Sodium–Ammonia**; **Sodium Bis(2-methoxyethoxy)aluminum Hydride; Sodium Borohydride**; **Sodium Triacetoxyborohydride**; Titanium(IV) Chloride; Titanium Tetraisopropoxide; Triethoxysilane; Trimethyl Borate.

Class R-14: Reagents for Reduction of Carboxylic Acids, Esters or Derivatives to Aldehydes or Hemiacetals.

Borane–Dimethyl Sulfide; Chlorodiisopropylsilane; Chloro(thexyl)borane–Dimethyl Sulfide; Dichlorobis(cyclopentadienyl)titanium; **Diisobutylaluminum Hydride**; **(2-Dimethylaminomethylphenyl)phenylsilane**; **Disiamylborane**; **Lithium**; **Lithium Aluminum Hydride**; Lithium 9-boratabicyclo[3.3.1]nonane; **Lithium Borohydride**; Lithium *t*-Butyl(diisobutyl)aluminum Hydride; Lithium *n*-Butyl(diisobutyl)aluminum Hydride; **Lithium–Ethylamine**; **Lithium Tri-*s*-butylborohydride**; Lithium Trisiamylborohydride; **Nickel Boride**; **Palladium on Barium Sulfate**; **Sodium Amalgam**; **Sodium Bis(2-methoxyethoxy)aluminum Hydride**; Tetrakis(triphenylphosphine)palladium(0); **Thexylborane**; **Tri-*n*-butylstannane**; **Triethylsilane**; **Zinc Borohydride**.

Class R-15: Reagents for Conjugate Reduction of α,β-Unsaturated Carbonyl Compounds.

Aluminum Amalgam; **Baker's Yeast**; Benzeneselenol; (1*S*, 9*S*)-1,9-Bis{[(*t*-butyl)dimethylsiloxy]methyl}-5-cyanosemicorrin; **1,2-Bis(2,5-diethylphospholano)benzene**; Bis(diisopropylamino)aluminum Hydride; **(*R*)- & (*S*)-2,2′-Bis(diphenylphosphino)-1,1′-binaphthyl**; Calcium–Ammonia; **Catecholborane**; **Chlorotris(triphenylphosphine)rhodium(I)**; Copper(I) Bromide–Lithium Trimeth-oxyaluminum Hydride; Copper(I) Bromide–Sodium Bis(2-methoxyethoxy)aluminum Hydride; (1,5-Cyclooctadiene)(tricyclohexylphosphine)(pyridine)iridium(I) Hexafluorophosphate; **Diimide**; **Diisobutylaluminum Hydride**; 1,3-Dimethyl-2-phenylbenzimidazoline; Dimethyl(phenyl)silane; Diphenylsilane-Tetrakis(triphenylphosphine)palladium(0)-Zinc Chloride; **Disiamylborane**; Disodium Tetracarbonylferrate(-II); **Hexa-μ-hydrohexakis(triphenylphosphine)hexacopper**; **Hydrogen Sulfide**; **Lithium**; Lithium Aluminum Hydride-Copper(I) Iodide; Lithium *n*-Butyl(diisobutyl)aluminum Hydride; Lithium *n*-Butyl(hydrido)cuprate; Lithium Diisobutyl(methyl)aluminum Hydride; **Lithium Tri-*t*-butoxyaluminum Hydride**; **Lithium Tri-*s*-butylborohydride**; **Lithium Triethylborohydride**; Magnesium–Methanol; Mesitylcopper(I); **Nickel Boride**; **Nickel(II) Chloride**; Nickel–Graphite; **Potassium**; **Potassium–Graphite Laminate**; **Potassium Tri-*s*-butylborohydride**; **Sodium**; **Sodium Amalgam**; **Sodium–Ammonia**; **Sodium Bis(2-methoxyethoxy)aluminum Hydride**; **Sodium Borohydride**; **Sodium Cyanoborohydride**; **Sodium Dithionite**; Sodium Tetracarbonylcobaltate; Sodium Tetracarbonylhydridoferrate; **Sodium Thiosulfate**; Tetra-*n*-butylammonium Borohydride; **Tin**; **Titanium**; **Titanium(III) Chloride**; **Tri-*n*-butylstannane**; Tri-*n*-butyltin Trifluoromethanesulfonate; Triethylborane; **Triethylsilane**; Triethylsilane-Trifluoroacetic Acid; Triisobutylaluminum; Ytterbium(0); Ytterbium(II) Iodide; **Zinc**; Zinc–Acetic Acid; **Zinc/Copper Couple**; Zinc/Nickel Couple.

Class R-16: Reagents for Reductive Deoxygenation of Epoxides to Alkenes.

Aluminum Amalgam; Aluminum Iodide; Diethyl[dimethyl(phenyl)silyl]aluminum; Diethyl Phosphonite; Dimethyl Diazomalonate; Dimethylphenylsilyllithium; Diphosphorus Tetraiodide; Iodotrimethylsilane; Lithium Aluminum Hydride–Titanium(IV) Chloride; Methyltriphenoxyphosphonium Iodide; Phosphorus(III) Iodide; Potassium Selenocyanate; Sodium

O,O-Diethyl Phosphorotelluroate; Trifluoroacetic Anhydride-Sodium Iodide; Trimethylsilylpotassium; **Triphenylphosphine**; Triphenylphosphine–Iodine; Triphenylphosphine–Iodoform–Imidazole; Triphenylphosphine Selenide; Triphenylphosphine-2,4,5-Triiodoimidazole; Tungsten(VI) Chloride–*n*-Butyllithium; Zinc–Acetic Acid.

Class R-17: Reagents for Reduction via Hydroboration.

Borane-Dimethyl Sulfide; **Borane–Tetrahydrofuran**; **Catecholborane**; Chloro(thexyl)borane–Dimethyl Sulfide; **Chlorotris(triphenylphosphine)rhodium(I)**; **Diborane**; **Dibromoborane–Dimethyl Sulfide**; Dichloroborane Diethyl Etherate; Dicyclohexylborane; **Lithium Borohydride**; **Lithium Tri-*s*-butylborohydride**; **Lithium Triethylborohydride**; **Sodium Triacetoxyborohydride**; **Tetrahydro-1-methyl-3,3-diphenyl-1*H*,3*H*-pyrrolo[1,2-*c*][1,3,2]oxazaborole**; **Thexylborane**.

Class R-18: Reagents for Enantioselective Hydrogenation.

(Bicyclo[2.2.1]hepta-2,5-diene)[1,4-bis(diphenylphosphino)butane]rhodium(I) Tetrafluoroborate; Bis(bicyclo[2.2.1]hepta-2,5-diene)dichlorodirhodium; Bis(1,5-cyclooctadiene)rhodium Tetrafluoroborate-(*R*)-2,2′-Bis(diphenylphosphino)-1,1’-binaphthyl; **1,2-Bis(2,5-diethylphospholano)benzene**; (+)-*trans*-(2*S*,3*S*)-Bis(diphenylphosphino)bicyclo[2.2.1]hept-5-ene; ***(R)*- & *(S)*-2,2′-Bis(diphenylphosphino)-1,1′-binaphthyl**; 10-Camphorsulfonic Acid; (*R*)-(+)-Cyclohexyl(2-anisyl)methylphosphine; (*R*)-*N*-[2-(*N*,*N*-Dimethylamino)ethyl]-*N*-methyl-1-[(*S*)-1′,2-bis-(diphenylphosphino)ferrocenyl]ethylamine; (*R*,*R*)-[Ethylene-1,2-bis(η^5-4,5,6,7-tetrahydro-1-indenyl)]titanium (*R*)-1,1′-Bi-2,2′-naphtholate; (-)-[Ethylene-1,2-bis(η^5-4,5,6,7-tetrahydro-1-indenyl)]zirconium (*R*)-1,1′-Bi-2,2′-naphtholate; (±)-1,1′-Ethylenebis(4,5,6,7-tetrahydro-1-indenyl)zirconium Dichloride; **(2,3-*O*-Isopropylidene)-2,3-dihydroxy-1,4-bis(diphenylphosphino)-butane**; 5-Phenyl-5*H*-benzophosphindole; Pivalic Acid.

Class R-19: Reagents for Catalysis of Hydrogenation or Hydrogenolysis.

(Bicyclo[2.2.1]hepta-2,5-diene)[1,4-bis(diphenylphosphino)butane]rhodium(I) Tetrafluoroborate; (Bicyclo[2.2.1]hepta-2,5-diene)[1,4-bis(diphenylphosphino)butane]rhodium(I)Tetrafluoroborate; Bis(benzonitrile)dichloropalladium(II); Bis(bicyclo[2.2.1]hepta-2,5-diene)rhodium Perchlorate; Bis(η^5-cyclopentadienyl)dihydridozirconium; Chlorotriethylsilane; Chlorotris(triphenylphosphine)cobalt; **Chlorotris(triphenylphosphine)rhodium(I)**; Cobalt Boride; Copper Chromite; (1,5-Cyclooctadiene)bis(methyldiphenylphosphine)iridium(I) Hexafluorophosphate; (1,5-Cyclooctadiene)(tricyclohexylphosphine)(pyridine)iridium(I) Hexafluorophosphate; Dicarbonylbis(cyclopentadienyl)titanium; Dichlorobis(triphenylphosphine)platinum(II)–Tin(II) Chloride; Dichlorotris(triphenylphosphine)ruthenium(II); (*R*,*R*)-[Ethylene-1,2-bis(η^5-4,5,6,7-tetrahydro-1-indenyl)]titanium (*R*)-1,1′-Bi-2,2′-naphtholate; (-)-[Ethylene-1,2-bis(η^5-4,5,6,7-tetrahydro-1-indenyl)]zirconium (*R*)-1,1′-Bi-2,2′-naphtholate; (±)-1,1′-Ethylenebis(4,5,6,7-tetrahydro-1-indenyl)zirconium Dichloride; Hexadecacarbonylhexarhodium; Lithium Aluminum Hydride-Bis(cyclopentadienyl)nickel; Nafion-H; Nickel(II) Acetate; Nickel(II) Acetylacetonate; **Nickel Boride**; Nickel Catalysts (Heterogeneous); Nickel–Graphite; Octacarbonyldicobalt; Palladium(II) Acetate; Palladium(II) Acetylacetonate; **Palladium on Barium Sulfate; Palladium on Calcium Carbonate (Lead Poisoned)**; **Palladium on Carbon**; Palladium–Graphite; **Palladium(II) Hydroxide on Carbon**; Palladium on Poly(ethylenimine); Palladium–Triethylamine-Formic Acid; **Platinum on Carbon; Platinum(IV) Oxide**; Platinum (Sulfided) on Carbon; **Potassium–Graphite Laminate**; **Raney Nickel; Rhodium on Alumina**; **Ruthenium Catalysts**; Sodium Hydride–Nickel(II) Acetate–Sodium *t*-Pentoxide; Tetrachlorotris[bis(1,4-diphenylphosphino)butane]diruthenium; Tricarbonyl(naphthalene)chromium; Urushibara Nickel.

Class R-20: Reagents for Reduction of Nitriles to Imines or Amines.

Aluminum Hydride; **Borane–Dimethyl Sulfide**; **Borane–Tetrahydrofuran**; **Chlorotris(triphenylphosphine)rhodium(I)**; Cobalt Boride; Copper Chromite; **Diborane**; **Diisobutylaluminum Hydride**; **Lithium Aluminum Hydride**; **Lithium Borohydride**; **Lithium Triethylborohydride**; Monochloroalane; Nickel Catalysts (Heterogeneous); **Palladium on Carbon**; **Platinum on Carbon**; **Platinum(IV) Oxide**; **Potassium–Graphite Laminate**; Potassium 9-Siamyl-9-boratabicyclo[3.3.1]-nonane; **Rhodium on Alumina**; **Ruthenium Catalysts**; **Sodium–Alcohol**; **Sodium Bis(2-methoxyethoxy)aluminum Hydride**; **Sodium Borohydride**; Sodium Hypophosphite; **Sodium Triacetoxyborohydride**; Sodium Trifluoroacetoxyborohydride; Sodium Trimethoxyborohydride; **Tin**; Tin(II) Chloride; Titanium; **Zinc**; Zinc/Nickel Couple.

Class R-21: Reagents for Reduction of Nitro Compounds to Amines or Oximes.

Aluminum Amalgam; Ammonium Formate; Ammonium Sulfide; Bis(benzonitrile)dichloropalladium(II); Bis(trimethylsilyl) Sulfide; **Chromium(II) Chloride**; Copper Chromite; Dichlorotris(triphenylphosphine)ruthenium(II); **Diisobutylaluminum Hydride**; Dodecacarbonyltriiron; Hexadecacarbonylhexarhodium; Hydrogen Selenide; **Hydrogen Sulfide**; Hypophosphorous Acid; **Lithium Aluminum Hydride**; Magnesium Amalgam; **Nickel Boride**; **Nickel(II) Chloride**; Palladium(II) Acetate; **Palladium on Calcium Carbonate (Lead Poisoned)**; **Palladium on Carbon**; Palladium–Graphite; **Palladium(II) Hydroxide on Carbon**; Palladium–Triethylamine-Formic Acid; **Platinum on Carbon**; **Platinum(IV) Oxide**; Platinum (Sulfided) on Carbon; **Ruthenium Catalysts**; **Sodium Bis(2-methoxyethoxy)aluminum Hydride**; **Sodium Borohydride**; Sodium Disulfide; **Sodium Dithionite**; Sodium Hydrogen Sulfide; Sodium Hypophosphite; Sodium Sulfide; Sodium Telluride; Sodium Tetracarbonylhydridoferrate; **Titanium(III) Chloride**; Zinc–Acetic Acid; Zinc Amalgam; Zinc Complex Reducing Agents; Zinc/Nickel Couple.

Class R-22: Reagents for Reduction of Quinones.

Aluminum Amalgam; **Chlorotris(triphenylphosphine)rhodium(I)**; Hydrogen Iodide; **Sodium Dithionite**; **Sodium Thiosulfate**; **Tin**; Vanadium(II) Chloride; Zinc–Acetic Acid; Zinc–Zinc Chloride.

Class R-23: Reagents for Reductive Cleavage of Allylic, Benzylic or α-Carbonyl Functionality.

Aluminum Amalgam; **Aluminum Hydride**; Aluminum Iodide; Ammonium Formate; **Borane–Tetrahydrofuran**; **Chromium(II) Chloride**; Diiodosilane; **Diisobutylaluminum Hydride**; 1,3-Dimethyl-2-phenylbenzimidazoline; Diphenylsilane–Tetrakis(triphenylphosphine)palladium(0)–Zinc Chloride; **Hydrogen Sulfide**; Iodotrimethylsilane; Iron–Graphite; **Lithium**; **Lithium Aluminum Hydride**; **Lithium–Ethylamine**; **Lithium Tri-*s*-butylborohydride Lithium Triethylborohydride**; Methyltrichlorosilane; **Nickel Boride**; **Nickel(II) Chloride**; Nonacarbonyldiiron; Palladium(II) Acetate; Palladium(II) Acetylacetonate; **Palladium on Barium Sulfate**; **Palladium on Carbon**; **Palladium(II) Hydroxide on Carbon**; Palladium-Triethylamine-Formic Acid; Pentacarbonyliron; **Platinum(IV) Oxide**; **Raney Nickel**; **Samarium(II) Iodide**; **Sodium Amalgam**; **Sodium Cyanoborohydride**; Sodium *O,O*-Diethyl Phosphorotelluroate; **Sodium Dithionite**; Sodium Hypophosphite; Sodium Iodide; **Sodium Naphthalenide**; Sodium Telluride; Sodium Tris(trifluoroacetoxy)borohydride; Tetrakis(triphenylphosphine)palladium(0); **Tin**; Tin(II) Chloride; **Titanium(III) Chloride**; **Triethylsilane**; **Triphenylphosphine**; **Zinc**; Zinc–Acetic Acid; **Zinc Borohydride**; **Zinc/Copper Couple**.

Class R-24: Reagents for Reductive Cleavage of N-O, N-N, O-O, O-S or S-S Bonds.

Aluminum Amalgam; Aluminum Iodide; Aminoiminomethanesulfonic Acid; **Chromium(II) Chloride**; Dimethyl Sulfide; **Hydrogen Sulfide**; Iron(II) Sulfate; **Lithium Aluminum Hydride**; Lithium Aluminum Hydride–Nickel(II) Chloride; **Lithium Tri-*t*-butoxyaluminum Hydride**; **Nickel(II) Chloride; Palladium on Carbon**; **Platinum(IV) Oxide**; Platinum (Sulfided) on Carbon; **Potassium Tri-*s*-butylborohydride**; Potassium Triisopropoxyborohydride; **Raney Nickel**; **Sodium**; **Sodium Amalgam**; **Sodium–Ammonia**; **Sodium Thiosulfate**; **Tin**; **Titanium(III) Chloride**; **Triphenylphosphine**; **Zinc**; Zinc–Acetic Acid.

Class R-25: Reagents for Reductive Coupling or Cyclization.

Aluminum Amalgam; **Chromium(II) Chloride**; Dichlorobis(cyclopentadienyl)titanium; **Diisobutylaluminum Hydride**; **Disiamylborane**; **Hydrogen Sulfide**; **Lithium**; Lithium Aluminum Hydride–Titanium(IV) Chloride; Magnesium; Magnesium Amalgam; **Nickel(II) Chloride**; **Potassium**; **Potassium–Graphite Laminate; Samarium(II) Iodide**; **Sodium**; Sodium Hydride–Nickel(II) Acetate–Sodium *t*-Pentoxide; **Sodium Naphthalenide**; **Sodium–Potassium Alloy**; **Tin**; **Titanium**; **Titanium(III) Chloride**; **Titanium(III) Chloride–Potassium**; **Titanium(III) Chloride-Zinc/Copper Couple**; **Tri-*n*-butylstannane**; Ytterbium(0); Ytterbium(II) Iodide; **Zinc**; **Zinc/Copper Couple**; Zinc–Zinc Chloride.

Class R-26: Reagents for Reductive Decarbonylation, Decarboxylation or Decyanation.

Bis[1,3-bis(diphenylphosphino)propane]rhodium Tetrafluoroborate; *trans*-Carbonyl(chloro)bis(triphenylphosphine)iridium(I); Carbonyl(chloro)bis(triphenylphosphine)rhodium(I); **Chlorotris(triphenylphosphine)rhodium(I)**; Copper Chromite; Diphenyl Phosphorazidate; 1-Methyl-2-pyrrolidinone; **Palladium on Carbon**; **Potassium**; Potassium on Alumina; **Raney Nickel**; **Sodium**; **Sodium Borohydride**; **Titanium**; **Tri-*n*-butylstannane**; Tris(acetylacetonato)iron(III); **Tris(trimethylsilyl)silane**.

Class R-27: Reagents for Reductive Dehalogenation.

Aluminum Amalgam; **Aluminum Hydride**; Bis(η^5-cyclopentadienyl)dihydridozirconium; **Chromium(II) Chloride**; Dichlorobis(cyclopentadienyl)titanium; **Diisobutylaluminum Hydride**; 1,3-Dimethyl-2-phenylbenzimidazoline; Hexacarbonyltungsten; Iron–Graphite; **Lithium**; **Lithium Aluminum Hydride**; Lithium Aluminum Hydride–Cobalt(II) Chloride; Lithium Aluminum Hydride–Nickel(II) Chloride; Lithium *n*-Butyl(hydrido)cuprate; **Lithium 4,4′-Di-*t*-butylbiphenylide**; Lithium 9,9-Dibutyl-9-borabicyclo[3.3.1]nonanate; **Lithium Naphthalenide**; **Lithium Tri-*t*-butoxyaluminum Hydride; Lithium Tri-*s*-butylborohydride**; **Lithium Triethylborohydride**; Magnesium; Magnesium Amalgam; Magnesium–Methanol; **Nickel Boride**; **Nickel(II) Chloride**; Nonacarbonyldiiron; **Palladium on Barium Sulfate**; **Palladium on Carbon**; Pentacarbonyliron; **Potassium**; **Potassium–Graphite Laminate**; **Potassium Naphthalenide**; **Potassium Tri-*s*-butylborohydride**; Pyridine; **Samarium(II) Iodide**; **Sodium**; **Sodium–Alcohol**; **Sodium Amalgam**; **Sodium–Ammonia**; **Sodium Bis(2-methoxyethoxy)aluminum Hydride**; **Sodium Borohydride**; **Sodium Cyanoborohydride**; Sodium *O,O*-Diethyl Phosphorotelluroate; **Sodium Dithionite**; Sodium Hydride–Palladium(II) Acetate–Sodium *t*-Pentoxide; Sodium Iodide; Sodium Phenanthrenide; **Sodium-Potassium Alloy**; Sodium Sulfide; Sodium Tetracarbonylcobaltate; **Sodium Thiosulfate**; Sodium Trimethoxyborohydride; Tetrakis(triphenylphosphine)palladium(0); **Tin**; **Titanium**; **Titanium(III) Chloride**; **Tri-*n*-butylstannane**; **Triethylsilane**; Trimethyl Phosphite; **Triphenylphosphine**; **Tris(trimethylsilyl)silane**; **Zinc**; Zinc–Acetic Acid; **Zinc Borohydride**; Zinc Complex Reducing Agents; **Zinc/Copper Couple**; Zinc/Silver Couple.

Class R-28: Reagents for Reductive Deoxygenation of Alcohols or Derivatives.

Bis(dimethylamino) Phosphorochloridate; 1,4-Bis(diphenylhydrosilyl)benzene; Bis(trifluoroacetoxy)borane; **Diisobutylaluminum Hydride**; *N*-Hydroxypyridine-2-thione; **Lithium**;

Lithium Aluminum Hydride; Lithium Aluminum Hydride–Cobalt(II) Chloride; Lithium Aluminum Hydride-Titanium(IV) Chloride; **Lithium–Ethylamine**; **Lithium Triethylborohydride**; Methyl Chlorooxalate; **Nickel Boride**; **Potassium**; **Raney Nickel**; **Samarium(II) Iodide**; **Sodium**; **Sodium Borohydride**; **Sodium Cyanoborohydride**; 1,1′-Thiocarbonyldiimidazole; Thionocarbonates; **Titanium**; **Titanium(III) Chloride–Potassium**; *O*-*p*-Tolyl Chlorothioformate; **Tri-*n*-butylstannane**; **Triethylsilane**; Triethylsilane–Trifluoroacetic Acid; Triphenylsilane; **Tris(trimethylsilyl)silane**.

Class R-29: Reagents for Reductive Deoxygenation of Aldehydes or Ketones.

Ammonium Formate; Bis(benzoyloxy)borane; Bis(dimethylamino) Phosphorochloridate; Bis(trifluoroacetoxy)borane; Bis(triphenylphosphine)copper(I) Borohydride; **Catecholborane**; **Diborane**; Diethyl Phosphorochloridate; 2,4-Dinitrophenylhydrazine; 1,2-Ethanedithiol; **Hydrazine**; **Hydrogen Sulfide**; **Lithium**; Lithium Aluminum Hydride–Diphosphorus Tetraiodide; **Palladium on Carbon**; **Platinum on Carbon**; **Platinum(IV) Oxide**; 1,3-Propanedithiol; **Sodium–Alcohol**; **Sodium Cyanoborohydride**; **Sodium Triacetoxyborohydride**; Sodium Tris(trifluoroacetoxy)borohydride; **Titanium(III) Chloride–Potassium**; *p*-Toluenesulfonylhydrazide; **Triethylsilane**; Triethylsilane–Trifluoroacetic Acid; 2,4,6-Triisopropylbenzenesulfonylhydrazide; Triisopropyl Phosphite; **Zinc**; Zinc Amalgam.

Class R-30: Reagents for Reductive Desulfurization.

Aluminum Amalgam; **Aluminum Hydride**; *N*,*N*-Dimethyldithiocarbamoylacetonitrile; 1,3-Dimethyl-2-phenylbenzimidazoline; Dodecacarbonyltriiron; Hexabutyldistannane; Hexacarbonyltungsten; Hexachlorodisilane; **Hydrazine**; Lead(II) Acetate; **Lithium**; Lithium Aluminum Hydride-(2,2′-Bipyridyl)(1,5-cyclooctadiene)nickel; **Lithium 4,4′-Di-*t*-butylbiphenylide**; Lithium 1-(Dimethylamino)naphthalenide; **Lithium–Ethylamine**; **Lithium Naphthalenide**; **Nickel Boride**; **Nickel(II) Chloride**; Nickel Complex Reducing Agents; Nonacarbonyldiiron; **Potassium–Graphite Laminate**; **Raney Nickel**; **Sodium**; **Sodium Amalgam**; Sodium Cyanide; **Sodium Dithionite**; Sodium Hydride–Nickel(II) Acetate-Sodium *t*-Pentoxide; *p*-Toluenesulfonyl Isocyanate; **Tri-*n*-butylstannane**; Triethyl Phosphite; Trimethyl Phosphite; Triphenylstannane; Urushibara Nickel.

Class R-31: Reagents for Reductive Elimination or Fragmentation.

Aluminum Amalgam; **Borane–Tetrahydrofuran**; Dichlorobis(cyclopentadienyl)titanium; **Diisobutylaluminum Hydride**; **Lithium 4,4′-Di-*t*-butylbiphenylide**; **Lithium–Ethylamine**; Magnesium–Methanol; **Nickel Boride**; **Potassium**; **Potassium–Graphite Laminate**; **Potassium Naphthalenide**; **Samarium(II) Iodide**; **Sodium**; **Sodium Amalgam**; **Sodium–Ammonia**; **Sodium Dithionite**; **Sodium Iodide**; **Sodium Naphthalenide**; Sodium Phenanthrenide; **Sodium–Potassium Alloy**; Sodium Sulfide; Sodium Telluride; Sodium Tetracarbonylcobaltate; Sodium Trimethoxyborohydride; **Tin**; **Titanium**; **Titanium(III) Chloride-Potassium**; **Tri-*n*-butylstannane**; **Triethylsilane**; **Triphenylphosphine**; **Tris(trimethylsilyl)silane**; **Zinc**; Zinc–Acetic Acid; Zinc–Zinc Chloride.

Class R-32: Reagents for Reductive Cleavage of Epoxides or Ethers to Alcohols.

Aluminum Amalgam; **Aluminum Hydride**; **Borane–Tetrahydrofuran**; Calcium-Ammonia; Chromium(II) Acetate; Diisobutylaluminum 2,6-Di-*t*-butyl-4-methylphenoxide; **Diisobutylaluminum Hydride**; **Hydrazine**; Iron–Graphite; **Lithium**; **Lithium Aluminum Hydride**; Lithium Aluminum Hydride–Nickel(II) Chloride; Lithium 9-boratabicyclo[3.3.1]-nonane; **Lithium Borohydride**; Lithium *t*-Butyl(diisobutyl)-aluminum Hydride; **Lithium 4,4′-Di-*t*-butylbiphenylide**; Lithium 9,9-Dibutyl-9-borabicyclo[3.3.1]nonanate; **Lithium-Ethylamine**; **Lithium Tri-*t*-butoxyaluminum Hydride**; **Lithium Triethylborohydride**; Monochloroalane; **Palladium on Barium Sulfate**; **Palladium(II) Hydroxide on Carbon**; **Potassium Tri-*s*-butylborohydride**; **Raney Nickel**; **Samarium(II) Iodide**; **Sodium**; **Sodium Bis(2-methoxyethoxy)aluminum Hydride**; **Sodium Borohydride**; **Sodium Cyano-borohydride**; Sodium Hypophosphite; **Sodium Naphthalenide**; Sodium Telluride; Titanium Tetraisopropoxide; **Tri-*n*-butylstannane**; **Triethylsilane**; **Zinc**; **Zinc Borohydride**; Zinc Complex Reducing Agents; **Zinc/Copper Couple**.

Class R-33: Reagents for Reductive Metallation.

Lithium 4,4′-Di-*t*-butylbiphenylide; **Lithium Naphthalenide**; Magnesium; Magnesium Amalgam; **Nickel(II) Chloride**.

Class R-34: Reagents for Reduction of Sulfoxides or Sulfones to Sulfides.

Acetyl Chloride; 9-Borabicyclo[3.3.1]nonane Dimer; Bromodimethylborane; Cobalt Boride; Dicarbonylbis(cyclopentadienyl)titanium; Dichloroborane Diethyl Etherate; **Diisobutylaluminum Hydride**; Dimethyl Sulfide; Ethylmagnesium Bromide–Copper(I) Iodide; Hexamethylphosphorous Triamide; **Hydrogen Sulfide**; Lithium Aluminum Hydride–Titanium(IV) Chloride; **Lithium Borohydride**; **Samarium(II) Iodide** Thiourea Dioxide; Titanium(IV) Chloride; **Titanium(III) Chloride**; Trichlorosilane; Tris(phenylseleno)borane.

Class R-35: Reagents for Reductive Amination of Carbonyls.

Borane–Pyridine; Hexadecacarbonylhexarhodium; Platinum (Sulfided) on Carbon; **Raney Nickel**; **Sodium Cyanoborohydride**; Sodium Telluride; **Sodium Triacetoxyborohydride**; Tetra-*n*-butylammonium Cyanoborohydride; Titanium Tetraisopropoxide.

Reviews and Monographs on Oxidizing Agents, 1993–97

(1) Krow, G. R. The Bayer-Villiger Oxidation of Ketones and Aldehydes. *Org. React.* **1993**, *43*, 251–798.

(2) Simandi, L. I.; Barna, T.; Szeverenyi, Z.; Nemeth, S. Homogeneous Catalytic Oxidation of *o*-Substituted Anilines with Dioxygen in the Presence of Cobalt and Manganese Complexes. *Pure Appl. Chem.* **1992**, *64*, 1511–18.

(3) Lohray, B. B. Recent Advances in the Asymmetric Dihydroxylation of Alkenes. *Tetrahedron Asymmetry* **1992**, *3*, 1317–49.

(4) Heaney, H. Novel Organic Peroxygen Reagents for Use in Organic Synthesis. *Topics Curr. Chem.* **1993**, *164*, 1–19.

(5) Sheldon, R. A. Homogeneous and Heterogeneous Catalytic Oxidations with Peroxide Reagents. *Topics Curr. Chem.* **1993**, *164*, 21–43.

(6) Adam, W.; Hadjiarapoglou, L. Dioxiranes: Oxidation Chemistry Made Easy. *Topics Curr. Chem.* **1993**, *164*, 45–62.

(7) Höft, E. Enantioselective Epoxidation with Peroxidic Oxygen. *Topics Curr. Chem.* **1993**, *164*, 63–77.

(8) Warwel, S.; Sojka, M.; Klaas, M. Synthesis of Dicarboxylic Acids by Transition-Metal Catalyzed Oxidative Cleavage of Terminal-Unsaturated Fatty Acids. *Topics Curr. Chem.* **1993**, *164*, 79–98.

(9) Fossey, J.; LeFort, D.; Sorba, J. Peracids and Free Radicals: A Theoretical and Experimental Approach. *Topics Curr. Chem.* **1993**, *164*, 99–113.

(10) Ando, W. *Organic Peroxides*; Wiley: Chichester, U.K., 1992.

(11) Strukul, G. *Catalytic Oxidations with Hydrogen Peroxide as Oxidant*; Kluwer Academic Publ.: Dordrecht, The Netherlands, 1992.

(12) Heaney, H. Oxidation Reactions Using Magnesium Monoperphthalate and Urea Hydrogen Peroxide. *Aldrichimica Acta* **1993**, *26*, 35–45.

(13) Davis, F. A.; Reddy, R. T.; Han, W.; Reddy, R. E. Asymmetric Synthesis Using *N*-Sulfonyloxaziridines. *Pure Appl. Chem.* **1993**, *65*, 633–40.

(14) Adam, W.; Richter, M. J. Metal-Catalyzed Direct Hydroxy–Epoxidation of Olefins. *Acc. Chem. Res.* **1994**, *27*, 43–50.

(15) Reiser, O. Oxidation of Weakly Activated C–H Bonds. *Angew. Chem., Int. Ed. Engl.* **1994**, *33*, 69–72.

(16) Palou, J. Oxidation of Some Organic Compounds by Aqueous Bromine Solutions. *Chem. Soc. Rev.* **1994**, *23*, 357–62.

(17) Ley, S. V.; Norman, J.; Griffith, W. P.; Marsden, S. P. Tetrapropylammonium Perruthenate, $Pr_4N^+RuO^4_-$, TPAP: A Catalytic Oxidant for Organic Synthesis. *Synthesis* **1994**, *7*, 639–666.

(18) Besse, P.; Veschambre, H. Chemical and Biological Synthesis of Chiral Epoxides. *Tetrahedron*, **1994**, *50*, 8885–927.

(19) Kolb, H. C.; VanNieuwenhze, M.S.; Sharpless, K. B. Catalytic Asymmetric Dihydroxylation. *Chem. Rev.* **1994**, *94*, 2483–547.

(20) Kleindienst, T. E. Recent Developments in the Chemistry and Biology of Peroxyacetyl Nitrate. *Res. Chem. Intermed.* **1994**, *20*, 335–84.

(21) Berrisford, D. J.; Bolm, C.; Sharpless, K. B. Ligand-Accelerated Catalysis. *Angew. Chem., Int. Ed. Engl.* **1995**, *34*, 1059–70.

(22) McKillop, A.; Sanderson, W. R. Sodium Perborate and Sodium Percarbonate: Cheap, Safe and Versatile Oxidising Agents for Organic Synthesis. *Tetrahedron* **1995**, *51*, 6145–66.

(23) Mori, T.; Suzuki, H. Ozone-mediated Nitration of Aromatic Compounds with Lower Oxides of Nitrogen (The Kyodai Nitration). *Synlett* **1995**, *5*, 383–92.

(24) Muzart, J. Sodium Perborate and Sodium Percarbonate in Organic Synthesis. *Synthesis* **1995**, *11*, 1325–46.

(25) Prakash, O. Organo Iodine(III) and Thallium(III) Reagents in Organic Synthesis: Useful Methodologies Based on Oxidative Rearrangements. *Aldrichimica Acta* **1995**, *28*, 63–71.

(26) Ganeshpure, P. A.; Adam, W. α-Hydroxy Hydroperoxides (Perhydrates) as Oxygen Transfer Agents in Organic Synthesis. *Synthesis* **1996**, *2*, 179–88.

(27) Jones, G. R.; Landais, Y. The Oxidation of the Carbon-Silicon Bond. *Tetrahedron* **1996**, *52*, 7599–662.

(28) Katsuki, T.; Martin, V. S. Asymmetric Epoxidation of Allylic Alcohols: The Katsuki-Sharpless Epoxidation Reaction. *Org. React.* **1996**, *48*, 1–299.

(29) Art, S. J. H. F.; Mombarg, E. J. M.; Van Bekkum, H.; Sheldon, R. A. Hydrogen Peroxide and Oxygen in Catalytic Oxidation of Carbohydrates and Related Compounds. *Synthesis* **1997**, *6*, 597–613.

(30) van Deurzen, M. P. J.; van Rantwijk, F.; Sheldon, R. A. Selective Oxidation Catalyzed by Peroxidases. *Tetrahedron* **1997**, *53*, 13183–221.

Reviews and Monographs on Reducing Agents, 1993–97

(1) Fürstner, A. Chemistry of and with Highly Reactive Metals. *AG(E)*, **1993**, *32*, 164–89.

(2) Bolm, C. Enantioselective Transition Metal-catalyzed Hydrogenation for the Asymmetric Synthesis of Amines, *AG(E)*, **1993**, *32*, 232–3.

(3) Brown, J. M. Selectivity and Mechanism in Catalytical Asymmetric Synthesis. *CSR*, **1992**, *22*, 25–41.

(4) Rabideau, P. W.; Marcinow, Z. The Birch Reduction of Aromatic Compounds. *OR*, **1993**, *42*, 1–334.

(5) Heinekey, D. M.; Oldham, W. J., Jr. Coordination Chemistry of Dihydrogen. *CRV*, **1993**, *93*, 913–26.

(6) Imamoto, T. Synthesis and Reactions of New Phosphine-boranes. *PAC*, **1993**, *65*, 655–60.

(7) Nugent, W. A.; RajanBabu, T. V.; Burk, M. J. Beyond Nature's Chiral Pool: Enantioselective Catalysis in Industry. *Science*, **1993**, *259*, 479–83.

(8) Cintas, P. Activated Metals in Organic Synthesis. Monograph, CRC Press: Boca Raton, FL, **1993**.

(9) Reetz, M. T. Structural, Mechanistic, and Theoretical Aspects of Chelation-controlled Carbonyl Addition Reactions. *ACR*, **1993**, *26*, 462–8.

(10) Noyori, R. Asymmetric Catalysis by Chiral Metal Complexes. *CHEMTECH*, **1992**, *22*, 360–7.

(11) Ojima, I., Ed. Catalytic Asymmetric Synthesis. Monograph, VCH: New York, **1993**.

(12) Zhu, Q. C.; Hutchins, R. O.; Hutchins, M-G. K. Asymmetric Reductions of Carbon-nitrogen Double Bonds. *OPP*, **1994**, *26*, 193–236.

(13) Ranu, B. C. Zinc Borohydride – A Reducing Agent with High Potential. *SL*, **1993**, 855–92.

(14) Barton, D. H. R. Half a Century of Free Radical Chemistry. Monograph, Cambridge University Press: Cambridge, UK, **1993**.

(15) Dawson, G. J.; Williams, J. M. J. Catalytic Applications of Transition Metals in Organic Synthesis. *Contemp. Org. Synth.*, **1994**, *1*, *77*.

(16) Brown, H. C.; Ramachandran, P. V. Recent Advances in the Boron Route to Asymmetric Synthesis. *PAC*, **1994**, *66*, 201–12.

(17) Brandukova, N. E.; Vygodskii, Y. S.; Vinogradova, S. V. The Use of Samarium Diiodide in Organic and Polymer Synthesis. *RCR*, **1994**, *63*, 345.

(18) Noyori, R. Organometallic Ways for the Multiplication of Chirality. *T*, **1994**, *50*, 4259–92.

(19) Brunner, H.; Zettlmeier, W. Handbook of Enantioselective Catalysis. Vol. I: Products and Catalysts; Vol. II: Ligands, References. Monograph, VCH: Weinheim, Germany, **1993**.

(20) Chaloner, P. A. Homogeneous Hydrogenation. (Catalysis by Metal Complexes, Vol. 15). Monograph, Kluwer, Dordrecht, The Netherlands, **1994**.

(21) Noyori, R. Asymmetric Catalysis in Organic Synthesis. Monograph, Wiley: New York, **1994**.

(22) Wills, M.; Studley, J. R. The Asymmetric Reduction of Ketones. *CI(L)*, **1994**, 552–5.

(23) Turner, N. J. Biocatalytic Reductions. *CI(L)*, **1994**, 592–5.

(24) Faller, J. W.; Mazzieri, M. R.; Nguyen, J. T.; Pan, J.; Tokunaga, M. Controlling Stereochemistry in C– and C–H Bond Formation with Electronically Asymmetric Organometallics and Chiral Poisons. *PAC*, **1994**, *66*, 1463–9.

(25) Lubineau, A.; Augé, J.; Queneau, Y. Water Promoted Organic Reactions. *S*, **1994**, 741–60.

(26) Bateson, J. H.; Mitchell, M. B., Eds. Organometallic Reagents in Organic Synthesis. Monograph, Academic: London, UK, **1994**.

(27) Dhar, R. K. Diisopinocamphenylchloroborane, (DIP-Chloride), an Excellent Chiral Reducing Reagent for the Synthesis of Secondary Alcohols of High Enantiomeric Purity. *Aldrichim. Acta*, **1994**, *27*, 43–51.

(28) Azerad, R. Application of Biocatalysts in Organic Synthesis. *BSF(2)*, **1995**, *132*, 17–51.

(29) Glese, B.; Damm, W.; Batra, R. Allylic Strain Effects in Stereoselective Radical Reactions. *Chemtracts: Organic Chemistry*, **1994**, *7*, 355–70.

(30) Hegedus, L. S. Transition Metals in the Synthesis of Complex Organic Molecules. Monograph, University Science Books: Mill Valley, CA **1994**.

(31) Kabalka, G. W. Current Topics in the Chemistry of Boron. Monograph, Royal Society of Chemistry: Cambridge, UK, **1994**.

(32) Casson, S.; Kocienski, P. The Hydrometallation, Carbometallation, and Metallometallation of Heteroalkynes. *Contemp. Org. Synth.* **1995**, *2*, 19–34.

(33) Ungvary, F. Transition-metals in Organic Synthesis - Hydroformylation, Reduction and Oxidation - Annual Survey Covering the Year 1993. *Coord. Chem. Rev.* **1995**, *141*, 371–493.

(34) Abel, E. W.; Stone, F. G. A.; Wilkinson, G., Eds. Comprehensive Organometallic Chemistry II: A Review of the Literature 1982–1994. Vol. 1–14. Monograph, Elsevier Science Ltd.: New York, **1995**.

(35) Kosak, J. R.; Johnson, T. A. Catalysis of Organic Reactions. Monograph, Marcel Dekker: New York, **1994**.

(36) Chatgilialoglu, C. Structural and Chemical Properties of Silyl Radicals. *CRV*, **1995**, *95*, 1229–51.

(37) Otake, M. Emerging Japanese Catalytic Technologies. *CHEMTECH*, **1995**, *25*, 36–41.

(38) Cha, J. S.; Kim, E. J.; Kwon, O. O.; Kwon, S. Y.; Seo, W. W.; Chang, S. W. β-Hydroxydiisopinocamphenylborane as a Mild, Chemoselective Reducing Agent for Aldehydes. *OPP*, **1995**, *27*, 541–5.

(39) van Andel-Scheffer, P. J. M.; Barendrecht, E. Review on the Electrochemistry of Solvated Electrons. Its Use in Hydrogenation of Monobenzenoids. *RTC*, **1995**, *114*, 259–65.

(40) Drauz, K.; Waldmann, H., Eds. Enzyme Catalysis in Organic Synthesis: A Comprehensive Handbook. Monograph, VCH: Weinheim, Germany, **1994**.

(41) Thayer, J. S. Not for Synthesis Only: The Reactions of Organic Halides with Metal Surfaces. *Adv. Organomet. Chem.*, **1995**, *38*, 59–78.

(42) Molander, G. A.; Harris, C. R. Sequencing Reactions with Samarium(II) Iodide. *CRV*, **1996**, *96*, 307–38.

(43) Molander, G. A. Reductions with Samarium(II) Iodide. *OR*, **1995**, *46*, 211–368.

(44) Sul'man, E. M. Selective Hydrogenation of Unsaturated Ketones and Acetylene Alcohols. *RCR*, **1994**, *63*, 923–36.

(45) Savel'ev, S. R.; Noskova, N. F. Metal-Complex Catalysts in the Hydrogenation of Unsaturated Glycerides of Natural Oils. *RCR*, **1994**, *63*, 937–44.

(46) Burk, M. J.; Gross, M. F.; Harper, T. G. P.; Kalbert, C. S.; Lee, J. R.; Martinez, J. P. Asymmetric Catalytic Routes to Chiral Building Blocks of Medicinal Interest. *PAC*, **1996**, *68*, 37–44.

(47) Schmid, R.; Broger, E. A.; Cereghetti, M.; Crameri, Y.; Foricher, J.; Lalonde, M.; Müller, R. K.; Scalone, M.; Schoettel, G.; Zutter, U. New Developments in Enantioselective Hydrogenation. *PAC*, **1996**, *68*, 131–8.

(48) Cerveny, L.; Belohlav, Z.; Hamed, M. N. H. Catalytic-Hydrogenation of Aromatic-Aldehydes and Ketones Over Ruthenium Catalysts. *Res. on Chem. Intermed.*, **1996**, *22*, 15–22.

(49) Curran, D. P.; Porter, N. A.; Giese, B., Eds. Stereochemistry of Radical Reactions: Concepts, Guidelines, and Synthetic Applications. Monograph, VCH: Weinheim, Germany, **1995**.

(50) Trost, B. M., Ed. Stereodirected Synthesis with Organoboranes. Monograph, Springer: Berlin, Germany, **1995**.

(51) Davies, I. W.; Reider, P. J. Practical Asymmetric Synthesis. *CI(L)*, **1996**, 412–5.

(52) Birch, A. J. The Birch Reduction in Organic Synthesis. *PAC*, **1996**, *68*, 553–6.

(53) Kumobayashi, H. Industrial Application of Asymmetric Reactions Catalyzed by BINAP-Metal Complexes. RTC, **1996**, *115*, 201–10.

(54) Tsuji, J.; Mandai, T. Palladium-catalyzed Hydrogenolysis of Allylic and Propargylic Compounds with Various Hydrides. *S*, **1996**, 1–24.

(55) Doyle, M. P., Ed. Advances in Catalytic Processes: Asymmetric Chemical Transformations. Monograph, JAI: Greenwich, CT, **1995**.

(56) Hudlicky, M. Reductions in Organic Chemistry. Monograph, The Royal Society of Chemistry: Cambridge, UK, **1996**.

(57) Dawson, G. J.; Bower, J. F.; Williams, J. M. J. Catalytic Applications of Transition Metals in Organic Synthesis. *Contemp. Org. Synth.*, **1996**, *3*, 277–93.

(58) Yoon, N. M. Selective Reduction of Organic Compounds with Aluminum and Boron Hydrides. *PAC*, **1996**, *68*, 843–8.

(59) Klabunovskii, E. I. Catalytic Asymmetric Synthesis of B-Hydroxyacids and Their Esters. *RCR*, **1996**, *65*, 329–44.

(60) Procter, G. Asymmetric Synthesis. Monograph, Oxford University Press: Oxford, UK, **1995**.

(61) Roberts, S. M.; Turner, N. J.; Willetts, A. J.; Turner, M. K. Introduction to Biocatalysis Using Enzymes and Micro-organisms. Monograph, Cambridge University Press: Cambridge, UK, **1995**.

(62) Stephenson, G. R., Ed. Advanced Asymmetric Synthesis. Monograph, Chapman and Hall: London, UK, **1996**.

(63) Fürstner, A.; Bogdanovíc, B. New Developments in the Chemistry of Low-Valent Titanium. *AG(E)*, **1996**, *35*, 2442–69.

(64) Gribble, G. W. A Reduction Powerhouse [Using Sodium Borohydride and Carboxylic Acids]. *CHEMTECH*, **1996**, *26*, 26–31.

(65) Molander, G. Reductions with Samarium (II) Iodide. *OR*, **1994**, *46*, 211–368.

(66) Ponec, V. Selective De-oxygenation of Organic Compounds. *RTC*, **1996**, *115*, 451–5.

(67) Abdel-Magid, A. F., Ed. Reductions in Organic Synthesis: Recent Advances and Practical Applications. ACS Symposium Series. Monograph, The Royal Society of Chemistry: Cambridge, UK, **1996**.

(68) Ager, D. J.; East, M. B., Eds. Asymmetric Synthetic Methodology. Monograph, CSC: Boca Raton, FL, **1996**.

(69) Fürstner, A., Ed. Active Metals. Monograph, VCH: Weinheim, Germany, **1995**.

(70) Gawley, R. E.; Aube, J., Eds. Principles of Asymmetric Synthesis. Monograph, Elsevier: Amsterdam, The Netherlands, **1996**.

(71) Noyori, R.; Hashiguchi, S.; Iwasawa, Y. Asymmetric Transfer Hydrogenation Catalyzed by Chiral Ruthenium Complexes. *ACR*, **1997**, *30*, 97–102.

(72) Zard, S. Z. On the Trail of Xanthates: Some New Chemistry from an Old Functional Group. *AG(E)*, **1997**, *36*, 672–85.

(73) Quiclet-Sire, B.; Zard, S. Z. Riding the Tiger: Using Degeneracy to Tame Wild Radical Processes. *PAC*, **1997**, *69*, 645–50.

(74) Faber, K., Ed. Biotransformation in Organic Chemistry, 3rd Completely Revised Edition. Monograph, Springer: Berlin, Germany, **1997**.

(75) Beletskaya, I.; Pelter, A. Hydroborations Catalyzed by Transition Metal Complexes. *T*, **1997**, *53*, 4957–5026.

(76) Brown, H. C.; Ramachandra, P. V. Asymmetric Syntheses via Chiral Organoboranes Based on α-Pinene. Adv. Asymm. Synth., Vol. 1, Hassner, A., Ed. JAI: Greenwich, CT, **1995**.

(77) Kieslich, K.; Crout, D.; Dalton, H.; Schneider, M., Eds. Biotransformation. Monograph, Chapman and Hall: London, UK, **1996**.

(78) Ruchardt, C.; Gerst, M.; Ebenhoch, J. Uncatalyzed Transfer Hydrogenation and Transfer Hydrogenolysis: Two Novel Types of Hydrogen-Transfer Reactions. *AG(E)*, **1997**, *36*, 1406–30.

(79) Patz, M.; Fukuzumi, S. Electron Transfer in Organic Reactions. *J. of Phys. Org. Chem.*, **1997**, *10*, 129–37.

(80) Giese, B.; Kopping, B.; Gobel, T.; Dickhaut, J.; Thoma, G.; Kulicke, K. J.; Trach, F. Radical Cyclization Reactions. *OR*, **1996**, *48*, 301–856.

(81) Ager, D. J.; Laneman, S. A. Reductions of 1,3-Dicarbonyl Systems with Ruthenium-Biarylbisphosphine Catalysts. *TA*, **1997**, *8*, 3327–55.

(82) Turner, N. J.; Roberts, S. M. The Application of Microbial Methods to the Synthesis of Chiral Fine Chemicals. *Adv. Asymm. Synth.*, Vol. 2. Hassner, A., Ed. JAI: Greenwich, CT, **1997**.

(83) Jedlinski, Z. Novel Electron-Transfer Reactions Mediated by Alkali Metals Complexed by Macrocyclic Ligand. *ACR*, **1998**, *31*, 55–61.

(84) Wadepohl, H. Boryl Metal Complexes, Boron Complexes, and Catalytic (Hydro)boration. *AG(E)*, **1997**, *36*, 2441–4.

(85) Chatani, N.; Murai, S. $HSiR_3$/CO as the Potent Reactant Combination in Developing New Transition-Metal-Catalyzed Reactions. *SL*, **1996**, 414–24.

(86) Guindon, Y.; Jung, G.; Guerin, B.; Ogilvie, W. W. Hydrogen and Allylation Transfer Reactions in Acyclic Free Radicals. *SL*, **1998**, 213–20.

(87) Brandsma, L.; Vasilevsky, S. F.; Verkruijsse, H. D. Application of Transition Metal Catalysts in Organic Synthesis. Monograph, Springer: Berlin, Germany, **1997**.

Organic Syntheses Procedures Involving Oxidizing Agents, Volumes 69–75

m-Chloroperbenzoic Acid

Horiguchi, Y.; Nakamura, E.; Kuwajima, I. *Org. Synth.* **1996**, *73*, 123–133.

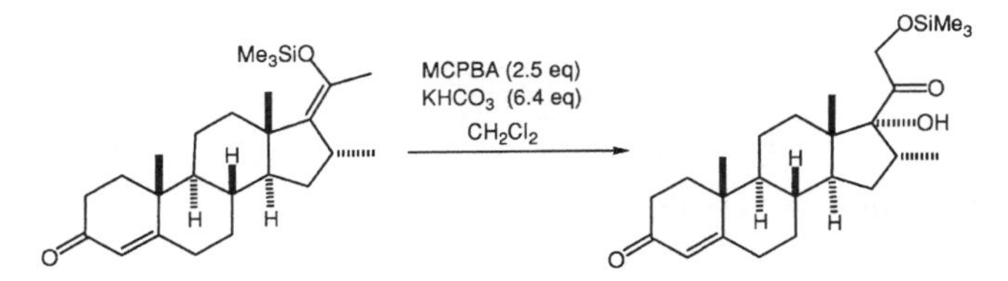

m-Chloroperbenzoic Acid

Begue, J.-P.; Bonnet-Delpon; Kornilov, A. *Org. Synth.* **1997**, *75*, 153–160.

Chromic Acid

Zibuck, R.; Streiber, J. *Org. Synth.* **1993**, *71*, 236–242.

Chromic Acid

Tschantz, M. A.; Burgess, L. E.; Meyers, A. I. *Org. Synth.* **1996**, *73*, 215–220.

Dimethyldioxirane

Murray, R. W.; Singh, M. *Org. Synth.* **1997**, *74*, 91–100.

Hydrogen Peroxide

Tamao, K.; Ishida, N.; Ito, Y.; Kumada, M. *Org. Synth.* **1990**, *69*, 96–105.

Hydrogen Peroxide

Murahashi, S.-I.; Shiota, T.; Imada, Y. *Org. Synth.* **1991**, *70*, 265–271.

Hydrogen Peroxide

Tamao, K.; Nakagawa, Y.; Ito, Y. *Org. Synth.* **1996**, *73*, 94–109.

Hydrogen Peroxide

Padwa, A.; Watterson, S. H.; Ni, Z. *Org. Synth.* **1997**, *74*, 147–157.

Iron(III) Chloride

Poupart, M.-A.; Lassalle, G.; Paquette, L. A. *Org. Synth.* **1990**, *69*, 173–179.

Lead(IV) Acetate

Webb, K. S.; Asirvatham, E.; Posner, G. H. *Org. Synth.* **1990**, *69*, 188–198.

Lead(IV) Acetate

Lenz, G. R.; Lessor, R. A. *Org. Synth.* **1991**, *70*, 139–147.

Lead(IV) Acetate

Sivik, M. R.; Stanton, K. J.; Paquette, L. A. *Org. Synth.* **1993**, *72*, 57–61.

Manganese Dioxide

Paquette, L. A.; Heidelbaugh, T. M. *Org. Synth.* **1996**, *73*, 44–49.

Osmium Tetroxide-*N*-Methylmorpholine-*N*-Oxide

McKee, B. H.; Gilheany, D. G.; Sharpless, K. B. *Org. Synth.* **1991**, *70*, 47–53.

Ozone

Tietze, L. F.; Bratz, M. *Org. Synth.* **1993**, *71*, 214–219.

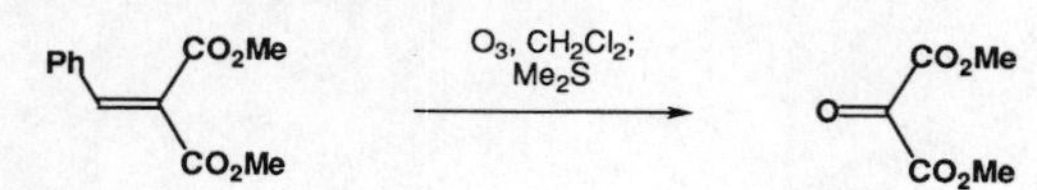

Peracetic Acid

Chen, B.-C.; Murphy, C. K.; Kumar, A.; Reddy, R. T.; Clark, C.; Zhou, P.; Lewis, B. M. Gala, D.; Mergelsberg, I.; Scherer, D.; Buckley, J.; DiBenedetto, D.; Davis, F. A. *Org. Synth.* **199**, *73*, 160–173.

Potassium Ferricyanide

Oi, R.; Sharpless, K. B. *Org. Synth.* **1996**, *73*, 1–12.

Potassium Monoperoxysulfate (Oxone)

Towson, T. C.; Weismiller, M. C.; Lal, G. S.; Sheppard, A. C.; Davis, F. A. *Org. Synth.* **1990**, *69*, 158–168.

Potassium Monoperoxysulfate (Oxone)

Bell, T. W.; Cho, Y.-M.; Firestone, A.; Healy, K.; Liu, J.; Ludwig, R.; Rothenberger, S. D.; *Org. Synth.* **1990**, *69*, 226–236.

Potassium Monoperoxysulfate (Oxone)

McCarthy, J. R.; Mathews, D. P.; Paolini, J. P. *Org. Synth.* **1993**, *72*, 209–215.

Potassium Permanganate–Copper(II) Sulfate

Jefford, C. W.; Li, Y.; Wang, Y. *Org. Synth.* **1993**, *71*, 207–213.

Pyridinium Chlorochromate

Paquette, L. A.; Earle, M. J.; Smith, G. F. *Org. Synth.* **1996**, *73*, 36–43.

Pyridinium Dichromate

Lee, T. V.; Porter, J. R. *Org. Synth.* **1993**, *72*, 189–198.

Selenium(IV) Oxide

Dahnke, K. R.; Paquette, L. A. *Org. Synth.* **1993**, *71*, 181–188.

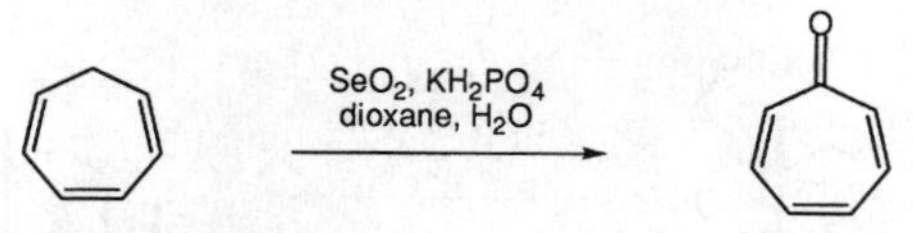

Selenium(IV) Oxide

Chen, B.-C.; Murphy, C. K.; Kumar, A.; Reddy, R. T.; Clark, C.; Zhou, P.; Lewis, B. M.; Gal, D.; Mergelsberg, I.; Scherer, D.; Buckley, J.; DiBenedetto, D.; Davis, F. A. *Org. Synth.* **199**, *73*, 160–173.

Sodium Dichromate

Sivik, M. R.; Stanton, K. J.; Paquette, L. A. *Org. Synth.* **1993**, *72*, 57–61.

Sodium Perborate

Kabalka, G. W.; Maddox, J. T.; Shoup, T. *Org. Synth.* **1996**, *73*, 116–122.

Sodium Periodate

Carrasco, M.; Jones, R. J.; Rapoport, H.; Truong, T. *Org. Synth.* **1991**, *70*, 29–34.

Sodium Periodate

Hubschwerien, C.; Specklin, J.-L.; Higelin, J. *Org. Synth.* **1993**, *72*, 1–5.

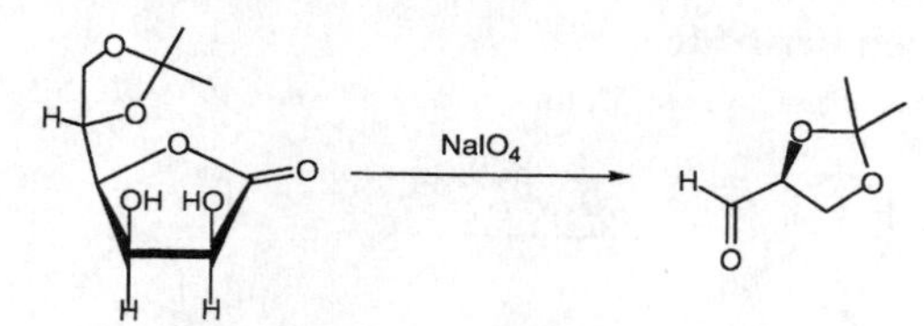

Sodium Periodate

Schmid, C. R.; Bryant, J. D. *Org. Synth.* **1993**, *72*, 6–13.

Sodium Periodate

Steuer, B.; Wehner, V.; Lieberknecht, A.; Jäger, V. *Org. Synth.* **1997**, *74*, 1–12.

BnO, OH, OH, OH → ($NaIO_4$) → BnO, O, OH

Sodium Periodate–Osmium Tetroxide

Behrens, C.; Paquette, L. A. *Org. Synth.* **1997**, *75*, 106–115.

O, N, $CH_2C_6H_5$ → (OsO_4–$NaIO_4$) → O, O, N, $CH_2C_6H_5$

2,2,6,6-Tetramethylpiperidin-1-oxyl

Anelli, P. L.; Montanari, F.; Quici, S. *Org. Synth.* **1990**, *69*, 212–219.

OH → (cat. N, O•; aq NaOCl, pH 9.5, cat KBr) → CHO

Organic Syntheses Procedures Involving Reducing Agents, Volumes 69–75

2,2′-Bis(diphenylphosphino)-1,1′-binaphthyl (BINAP)

Kitamura, M.; Tokunaga, M.; Ohkuma, T.; Noyori, R. *Org. Synth.* **1992**, *71*, 1–13.

A. 1/2 $[RuCl_2(benzene)]_2$ + (*R*)-BINAP $\xrightarrow{DMF}$ (*R*)-BINAP—Ru(II)

B. $\xrightarrow[CH_3OH]{H_2,\ (R)\text{-BINAP—Ru(II)}}$

2,2′-Bis(diphenylphosphino)-1,1′-binaphthyl (BINAP)

Takaya, H.; Ohta, T.; Inouse, S.-i.; Tokunaga, M.; Kitamura, M.; Noyori, R. *Org. Synth.* **1993**, *72*, 74–85.

A. 1/2 $[RuCl_2(benzene)]_2$ $\xrightarrow[DMF]{(R)\text{-BINAP}}$ $\xrightarrow[DMF-CH_3OH]{NaOCOCH_3}$ $Ru(OCOCH_3)_2[(R)\text{-BINAP}]$

B.

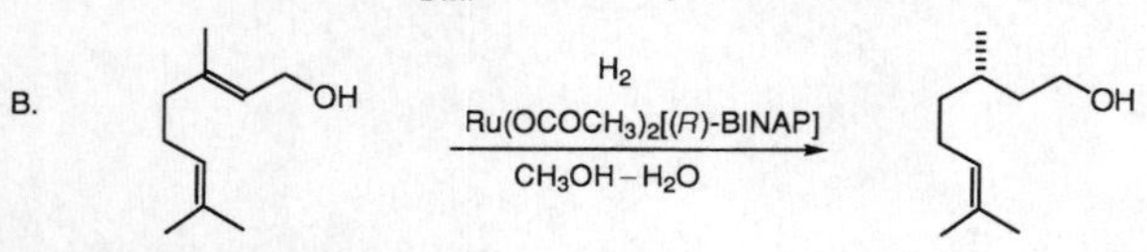

Borane–Dimethyl Sulfide

Soderquist, J. A.; Negron, A. *Org. Synth.* **1991**, *70*, 169–176.

$\xrightarrow[(CH_2OMe)_2]{BH_3 \cdot SMe_2}$

Borane–Dimethyl Sulfide

Denmark, S. E.; Marcin, L. R.; Schnute, M. E.; Thorarensen, A. *Org. Synth.* **1996**, *74*, 33–49.

$\xrightarrow[THF,\ 40\,°C]{BH_3 \cdot SMe_2}$

Borane–Dimethylsulfide

Xavier, L. C.; Mohan, J. J.; Mathre, D. J.; Thompson, A. S.; Carroll, J. D.; Corley, E. G.; Desmond, R. *Org. Synth.* **1996**, *74*, 50–71.

$\xrightarrow{2/3\ (CH_3BO)_3}$ $\xrightarrow[\Delta,\ -MeB(OH)_2]{xylene}$

$\xrightarrow[crystallize,\ -Me_2S]{Me_2S-BH_3,\ xylene/hexane}$

$\xrightarrow{CH_2Cl_2,\ Me_2S-BH_3,\ -20°C}$

Borane–Tetrahydrofuran

Krakowiak, K. E.; Bradshaw, J. S. *Org. Synth.* **1991**, *70*, 129–138.

+ $C_6H_5CH_2NH_2$ → 1) CH_3CN, NaI, Na_2CO_3; 2) $BH_3 \cdot THF$; 3) HCl, then NH_4OH

Borane–Tetrahydrofuran

Jefford, C. W.; Li, Y.; Wang, Y. *Org. Synth.* **1992**, *71*, 207–213.

1) $BH_3 \cdot THF$; 2) H_2O_2, NaOH; $KMnO_4/CuSO_4$

Borane–Tetrahydrofuran

Furuta, K.; Gao, Q.-z.; Yamamoto, H. *Org. Synth.* **1993**, *72*, 86–94.

+ $BH_3 \cdot THF$ → BLn* Chiral (Acyloxy) Borane Catalyst

Borane–Tetrahydrofuran

Light, J.; Breslow, R. *Org. Synth.* **1993**, *72*, 199–208.

$CH_3OCH_2CH_2OCH_2CH{=}CH_2$ $\xrightarrow[2.\ Br_2]{1.\ BH_3 \cdot THF}$ $CH_3OCH_2CH_2OCH_2CH_2CH_2Br$

Borane–Tetrahydrofuran

Kabalka, G. W.; Maddox, J. T.; Shoup, T. *Org. Synth.* **1995**, *73*, 116–122.

$\xrightarrow[THF]{BH_3 \cdot THF}$

Chromium (II) Chloride

Takai, K.; Sakogawa, K.; Kataoka, Y.; Oshima, K.; Utimoto, K. *Org. Synth.* **1993**, *72*, 180–188.

$\xrightarrow[DMF,\ 25°C]{CrCl_2,\ cat.\ NiCl_2}$

Diisobutylaluminum Hydride

Garner, P.; Park, J. M. *Org. Synth.* **1991**, *70*, 18–28.

$\xrightarrow[-78°C]{DIBAL}$

Diisobutylaluminum Hydride

Dahnke, K. R.; Paquette, L. A. *Org. Synth.* **1992**, *71*, 181–188.

CH_3Li, CuI, $(i\text{-}Bu)_2AlH$, THF–HMPA, −50°C to 0°C

Diisobutylaluminum Hydride

Marek, I.; Meyer, C.; Normant, J.-F. *Org. Synth.* **1996**, *74*, 194–204.

1) DIBAL-H / –78°C; 2) BuMgX

1) DIBAL-H / –78°C; 2) 0°C, 15 min; 3) BuMgBr

1) DIBAL-H / –78°C, CH_2Cl_2 / 30 min; 2) MeOH / –80°C; 3) $NaO_2C–(CHOH)_2–CO_2K$

(+)-Diisopinocampheylborane

Kabalka, G. W.; Maddox, J. T.; Shoup, T.; Bowers, K. R. *Org. Synth.* **1995**, *73*, 116–122.

$BH_3 \cdot THF$, THF

$NaBO_3 \cdot 4\ H_2O$, H_2O

Hydrazine

Barton, D. H. R.; Chen, M.; Jaszberenyi, J. Cs.; Taylor, D. K. *Org. Synth.* **1996**, *74*, 101–107.

H_2NNH_2, EtOH, Et_3N, 100°C, 2–3 d

Hydrazine

Reid, J. R.; Dufresne, R. F.; Chapman, J. J. *Org. Synth.* **1996**, *74*, 217–226.

N_2H_4

Lithium

Dickhaut, J.; Giese, B. *Org. Synth.* **1991**, *70*, 164–168.

$Me_3SiCl + SiCl_4 \xrightarrow[THF]{Li} (Me_3Si)_4Si$

Lithium

Wender, P. A.; White, A. W.; McDonald, F. E. *Org. Synth.* **1991**, *70*, 204–214.

Li°, Et_2O, 0°C

Lithium

Crabtree, S. R.; Mander. L. N.; Sethi, S. P. *Org. Synth.* **1991**, *70*, 256–264.

Li, NH_3 (liq.), $(CH_3)_3COH$

$NCCO_2Me$, Et_2O

Lithium

Pikul, S.; Corey, E. J. *Org. Synth.* **1992**, *71*, 22–29.

1. Li, THF–NH_3, EtOH
2. HCl
3. NaOH

(±)

Lithium

Mudryk, B.; Cohen, T. *Org. Synth.* **1993**, *72*, 173–179.

THF, 0°C, 4 h

LDBB

Lithium

Barrett, A. G. M.; Flygare, J. A.; Hill, J. M.; Wallace, E. M. *Org. Synth.* **1995**, *73*, 50–60.

$PhMe_2SiCl$: 1) Li, THF; hexanal; 2) CCl_4, PPh_3, THF

Lithium

Tschantz, M. A.; Burgess, L. E.; Meyers, A. I. *Org. Synth.* **1995**, *73*, 221–230.

Li, NH_3, EtOH / THF

Lithium Aluminum Hydride

Weismiller, M. C.; Towson, J. C.; Davis, F. A. *Org. Synth.* **1990**, *69*, 154–157.

$LiAlH_4$, THF, reflux

Lithium Aluminum Hydride

Buchwald, S. L.; LaMaire, S. J.; Nielsen, R. B.; Watson, B. T.; King, S. M. *Org. Synth.* **1992**, *71*, 77–82.

$Cp_2ZrCl_2 \xrightarrow[THF]{LiAlH_4} Cp_2Zr(H)Cl + Cp_2ZrH_2$; CH_2Cl_2 wash

Lithium Aluminum Hydride

Oi, R.; Sharpless, K. B. *Org. Synth.* **1995**, *73*, 1–12.

$LiAlH_4$, Ether–THF

Lithium Aluminum Hydride

Steuer, B.; Wehner, V.; Lieberknecht, A.; Jager, V. *Org. Synth.* **1996**, *74*, 1–12.

$LiAlH_4$, $AlCl_3$; Et_2O, CH_2Cl_2

Lithium Aluminum Hydride

Kann, N.; Bernardes, V.; Greene, A. E. *Org. Synth.* **1996**, *74*, 13–22.

$Pd/BaSO_4$, Pyridine, H_2; $LiAlH_4$, THF

Lithium Aluminum Hydride

Beresis, R. T.; Solomon, J. S.; Yang, M. G.; Jain, N. F.; Panek, J. S. *Org. Synth.* **1997**, *75*, 78–88.

$LiAlH_4$, 0°C, Et_2O

Lithium 4,4′-Di-*t*-butylbiphenylide

Mudryk, B.; Cohen, T. *Org. Synth.* **1993**, *72*, 173–179.

THF, 0°C, 4 h; LDBB; 2 LDBB, THF, −78°C; 1. *i*-PrCHO, −78°C; 2. H_2O

Palladium on Barium Sulfate

Kann, N.; Bernardes, V.; Greene, A. E. *Org. Synth.* **1996**, *74*, 13–22.

$Pd/BaSO_4$, Pyridine, H_2; $LiAlH_4$, THF

Palladium on Carbon

Furuta, K.; Gao, Q.-z.; Yamamoto, H. *Org. Synth.* **1993**, *72*, 86–94.

H_2, Pd/C, EtOAc

Palladium on Carbon

Modi, S. P.; Oglesby, R. C.; Archer, S. *Org. Synth.* **1993**, *72*, 125–134.

10% Pd–C, $HCOONH_4$

Palladium on Carbon

Saito, S.; Komada, K.; Moriwake, T. *Org. Synth.* **1995**, *73*, 184–200.

H_2, Pd–C/Boc_2O, EtOAc

Palladium on Carbon

Hutchison, D. R.; Khau, V. V.; Martinelli, M. J.; Nayyar, N. K.; Peterson, B. C.; Sullivan, K. A. *Org. Synth.* **1997**, *75*, 223–234.

H_2 / Pd–C, EtOH

Palladium (II) Hydroxide on Carbon

Smith, A. B., III; Yager, K. M.; Phillips, B. W.; Taylor, C. M. *Org. Synth.* **1997**, *75*, 19–30.

H_2, $Pd(OH)_2$, EtOH, 25°C (82%); (≥99% ee)

Potassium

Schultz, A. G.; Alva, C. W. *Org. Synth.* **1995**, *73*, 174–183.

1) K/NH_3, *t*-BuOH; 2) NH_4Cl

Raney Nickel

Sun, R. C.; Okabe, M. *Org. Synth.* **1993**, *72*, 48–56.

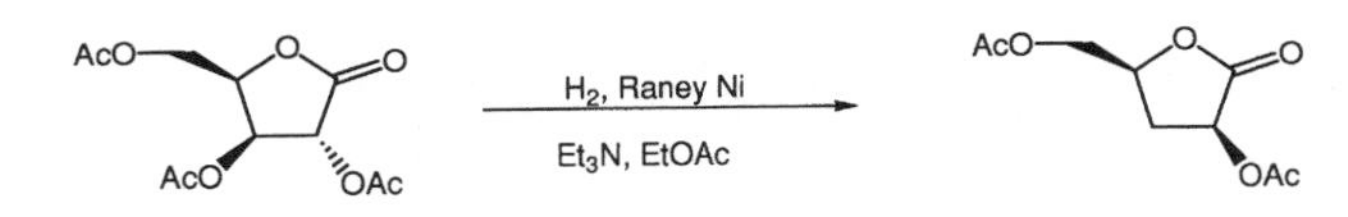

Ruthenium Catalysts

Kitamura, M.; Tokunaga, M.; Ohkuma, T.; Noyori, R. *Org. Synth.* **1992**, *71*, 1–13.

Ruthenium Catalysts

Takaya, H.; Ohta, T.; Inoue, S.-i.; Tokunaga, M.; Kitamura, M.; Noyori, R. *Org. Synth.* **1993**, *72*, 74–85.

Sodium

Hansen, T. K.; Becher, J.; Jorgensen, T.; Varma, K. S.; Khedekar, R.; Cava, M. P. *Org. Synth.* **1995**, *73*, 270–277.

Sodium Bis(2-methoxyethoxy)aluminum Hydride (RED-Al)

Meyers, A. I.; Berney, D. *Org. Synth.* **1990**, *69*, 55–65.

Sodium Borohydride

Meyers, A. I.; Flanagen, M. E. *Org. Synth.* **1992**, *71*, 107–117.

Sodium Borohydride

Dondoni, A.; Merino, P. *Org. Synth.* **1993**, *72*, 21–31.

Sodium Borohydride

Wang, X.; de Silva, S. O.; Reed, J. N.; Billadeau, R.; Griffen, E. J.; Chan, A.; Snieckus, V. *Org. Synth.* **1993**, *72*, 163–172.

Sodium Borohydride

Light, J.; Breslow, R. *Org. Synth.* **1993**, *72*, 199–208.

Sodium Borohydride

Gonzalez, J.; Foti, M. J.; Elsheimer, S. *Org. Synth.* **1993**, *72*, 225–231.

Tetrahydro-1-methyl-3,3,-diphenyl-1*H*,3*H*-pyrrolo[1,2-c][1,3,2]oxazaborole

Xavier, L. C.; Mohan, J. J. Mathre, D. J.; Thompson, A. S.; Carroll, J. D.; Corley, E. G.; Desmond, R. *Org. Synth.* **1996**, *74*, 50–71.

Tetrahydro-1-methyl-3,3-diphenyl-1*H*,3*H*-pyrrolo[1,2-c][1,3,2]oxaza-borole

Denmark, S. E.; Marcin, L. R.; Schnute, M. E.; Thorarensen, A. *Org. Synth.* **1996**, *74*, 33–49.

Tri-*n*-butylstannane

Giese, B.; Groninger, K. S. *Org. Synth.* **1990**, *69*, 66–71.

Tri-*n*-butylstannane

McCarthy, J. R.; Matthews, D. P.; Paolini, J. P. *Org. Synth.* **1993**, *72*, 216–224.

Triethylsilane

Tschantz, M. A.; Burgess, L. E.; Meyers, A. I. *Org. Synth.* **1995**, *73*, 221–230.

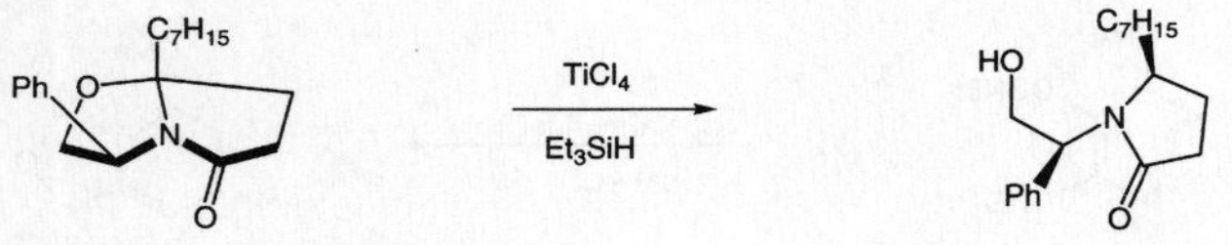

Triphenylphosphine

Pansare, S. V.; Huyer, G.; Arnold, L. D.; Vederas, J. C. *Org. Synth.* **1991**, *70*, 1–9.

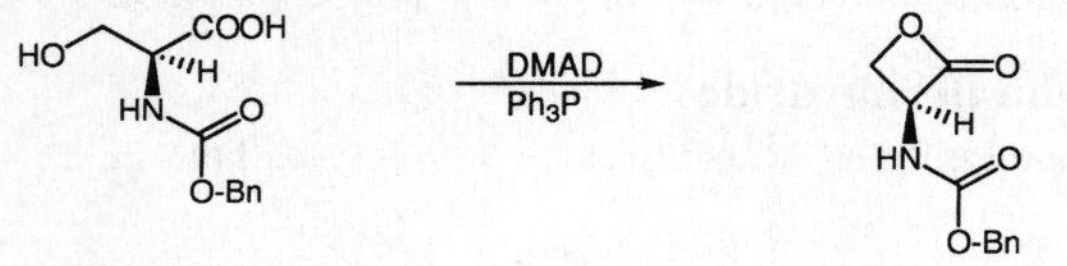

Triphenylphosphine

Pansare, S. V.; Huyer, G.; Arnold, L. D.; Vederas, J. C. *Org. Synth.* **1991**, *70*, 10–17.

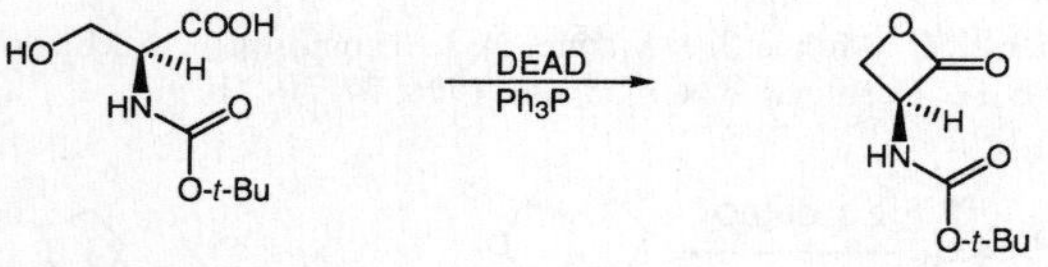

Triphenylphosphine

Dodge, J. A.; Nissen, J. S.; Presnell, M. *Org. Synth.* **1995**, *73*, 110–115.

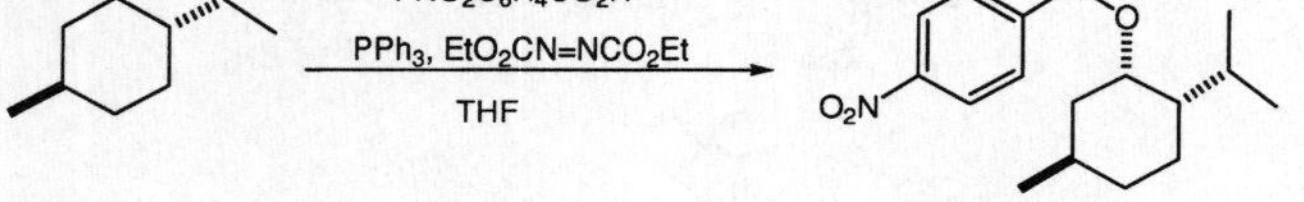

Zinc

Boger, D. L.; Panek, J. S.; Patel, M. *Org. Synth.* **1991**, *70*, 79–92.

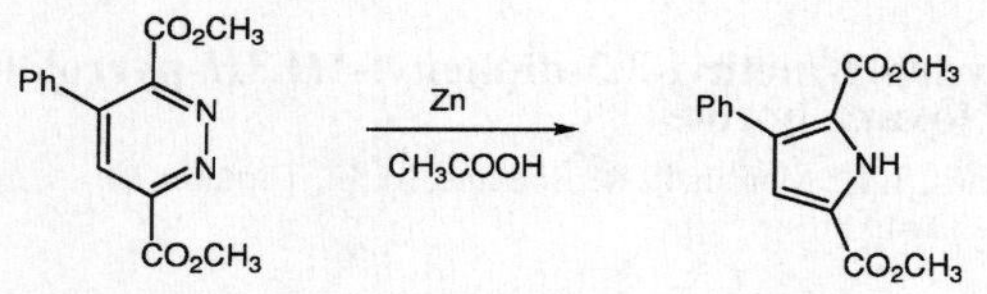

Zinc

Yeh, M. C. P.; Chen, H. G.; Knochel, P. *Org. Synth.* **1991**, *70*, 195–203.

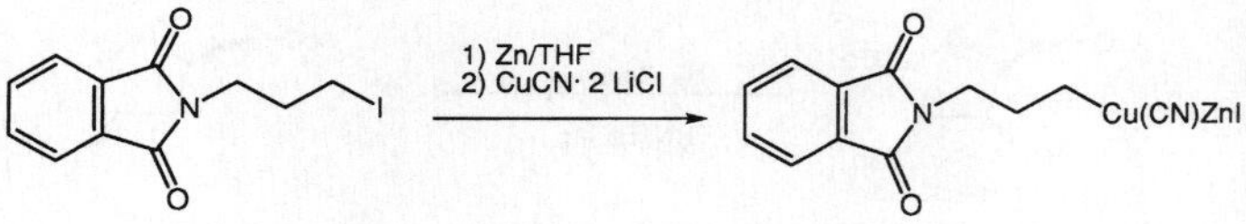

Zinc

Takai, K.; Kataoka, Y.; Miyai, J.; Okazoe, T.; Oshima, K.; Utimoto, K. *Org. Synth.* **1995**, *73*, 73–84.

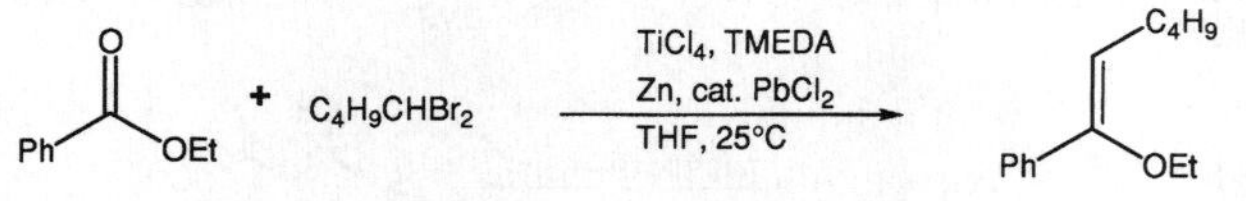

Zinc

Oda, M.; Kawase, T.; Okada, T.; Enomoto, T. *Org. Synth.* **1995**, *73*, 253–261.

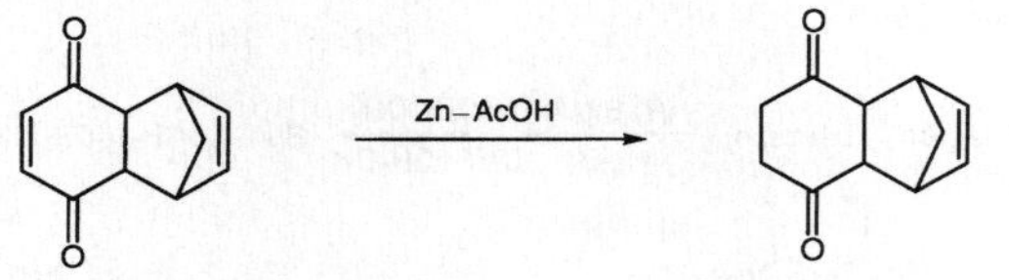

Zinc

Rosini, G.; Confalonieri, G.; Marotta, E.; Rama, F.; Righi, P. *Org. Synth.* **1996**, *74*, 158–168.

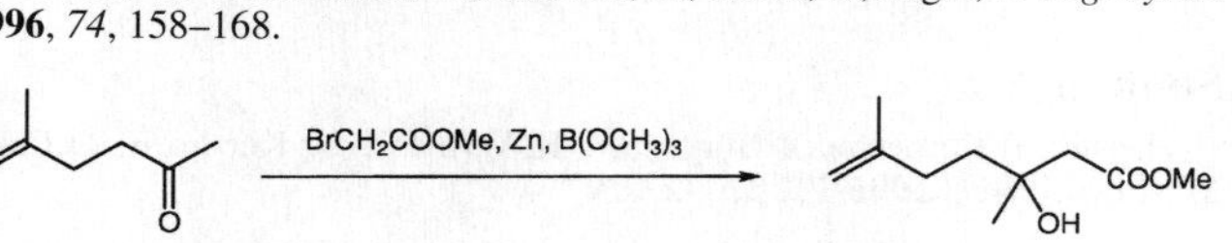

Zinc

Boudreault, N.; Leblanc, Y. *Org. Synth.* **1996**, *74*, 241–247.

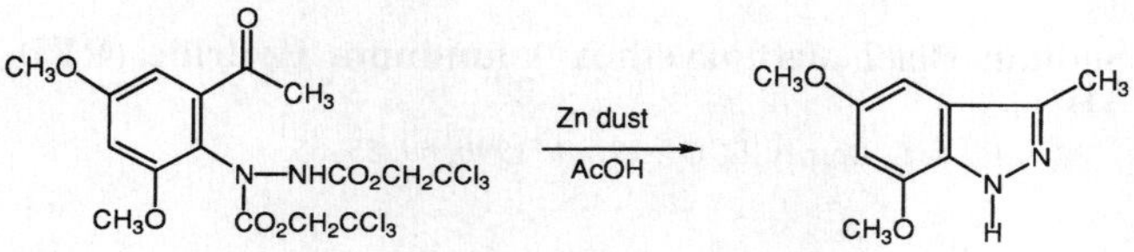

Aluminum Amalgam[1]

Al–Hg

[11146-30-8] Al (MW 26.98)

(reducing agent for many functional groups,[1] effects reductive dimerization of unsaturated compounds, can cleave carbon–element and element–element bonds.)

Physical Data: shiny solid.

Preparative Methods: ***Aluminum*** turnings (oil-free) are etched with dilute ***Sodium Hydroxide*** to a point of strong hydrogen evolution and the solution is decanted. The metal is washed once with water so that it retains some alkali, then treated with 0.5% ***Mercury(II) Chloride*** solution for 1–2 min, and the entire procedure is repeated. The shiny amalgamated metal is washed rapidly in sequence with water, ethanol, and ether and used at once.[2] Aluminum amalgam (**1**) reacts vigorously with water with liberation of hydrogen in the amount equivalent to the amount of aluminum present and can be used to dry organic solvents (ether, ethanol).[2a] Aluminum amalgam can be also prepared from aluminum foil, which is cut into strips ~10 cm × 1 cm and immersed, all at once, into a 2% aqueous solution of $HgCl_2$ for 15 s. The strips are rinsed with absolute alcohol and then with ether and cut immediately with scissors into pieces ~1 cm square, directly into the reaction vessel.[2b] Before immersion, each strip may be rolled into a cylinder ~1 cm in diameter. Each cylinder is amalgamated, rinsed successively with ethanol and ether and then placed in the reaction vessel.[2c]

Handling, Storage, and Precautions: moisture-sensitive. Precautions should be taken as it readily reacts with water with hydrogen evolution. Can be stored under dry ether. Toxic.

C=C Bond Reduction. Alkenic substrates are transformed into saturated compounds upon reaction with Al–Hg. Substrates with carbon–carbon double bonds activated by electron-withdrawing substituents are most easily reduced.[3] The reaction may proceed with asymmetric induction and high chemoselectivity, leaving other functionalities unchanged (eq 1).[4] Reduction of 1,3-dienes[5a] and α-nitroalkenes[5b] results in 1,4-addition of hydrogen to the π-system. Aluminum amalgam can also promote reductive dimerization of α,β-unsaturated acid esters.[6a,b]

C=O Bond Reduction. Cycloalkanones and aldehydes are reduced to the corresponding alcohols (eq 2). Acyclic ketones remain almost inert.[7]

In spite of the ability of carboxylic acid esters to be reduced to alcohols,[8] in oxosuccinic acid esters only the ketone carbonyl group is reduced.[9]

Ultrasound-promoted reduction of the C=O group of *N*-substituted phthalimides leads to hydroxylactams. The ultrasonic irradiation provides rapid fragmentation of the amalgam, giving a reactive dispersion and accelerates the reaction owing to the increase of mass transport between the solution and the Al/Hg surface where reduction occurs. The reaction is highly sensitive to substrate structure and *N*-benzylglutarimide and *N*-benzylsuccinimide are not reduced.[10]

Al–Hg, DME–H_2O (7.5:1), 0 °C, 6 h, 61%; HCl, reflux, 6 h, 74%; 17% ee (1)

10 equiv Al–Hg, THF–H_2O (9:1), −10 °C → rt, 70% (2)

Reductive dimerization of carbonyl compounds to pinacols does not always effectively compete with reduction to alcohols, but in certain cases it becomes the main process (eq 3).[7,11] Factors which determine reduction–dimerization ratios include steric inhibition, torsion strain, and angle strain.[7]

Al–Hg (3)

CH_2Cl_2, reflux, 1–4 h, 21–38%
C_6H_6–EtOH (1:1), reflux, 4 h, 95%
THF–H_2O (9:1), −10 °C → rt, 94%

A special case of reduction involves removal of the carbonyl oxygens from anthraquinones[12a,b] and related compounds[13] with rearomatization (eq 4).[12b]

Al–Hg, EtOH–H_2O (4:9), sat aq NH_4OH, 60–65 °C, 2 h, 78% (4)

C=N Bond Reduction. Aluminum amalgam reduces Schiff bases to the corresponding amines.[14] Among these reactions of great importance is the reduction of Δ^2-thiazolines to thiazolidines,[15–17] widely used in the synthesis of aldehydes (eq 5),[15] β-hydroxy aldehydes, and homoallylic alcohols.[16]

Aluminum amalgam also induces reductive dimerization of Schiff bases[14a,18] to produce 1,2-diamines. This process has been used in macrocyclic ring closure.[18a] In respect to reductive dimerization of Δ^1-pyrrolines, aluminum amalgam is much more effective than ***Zinc*** in aqueous NH_4Cl.[18b] Similar to Schiff bases, oximes are also readily reduced to corresponding amines,[19] while reduction of hydrazides provides hydrazines.[20]

1. BuLi, THF, –78 °C; 2. RI, THF, –78 °C; 74%. 10–12 equiv Al–Hg, Et_2O–H_2O, reflux, 1 h or rt, 4 h; 90%. R = $Ph(CH_2)_2CH_2$. $HgCl_2$, $MeCN$–H_2O, rt, 2 h; 71% (5)

NO_2 and N_3 Group Reduction. Aluminum amalgam reduction is an excellent method for deoxygenation of aliphatic and aromatic nitro groups to produce amines[21] (eq 6).[21d] Nitro alkenes can be chemoselectively reduced, retaining the C=C bond.[21a] In moist ether the reduction can be stopped at the stage of hydroxylamine formation.[22]

Al–Hg, ether–MeOH, 100% (6)

Organic azides are also readily reduced to the corresponding amines. The procedure has been used in a general synthesis of α,β-unsaturated α-amino acids.[23]

Reactions of Carbon–Halogen Bonds. Organic halides exhibit diverse behavior in reactions with Al–Hg. Thus a trichloromethyl group has been reduced to a dichloromethyl group.[24] At the same time, Al inserts into the C–Br bond of silylated propargyl bromide affording the allenylaluminum reagent, whereas only direct metalation without rearrangement has been observed in metalation with ***Zinc Amalgam*** (eq 7).[25]

1. Al–Hg, dry THF, reflux, 3 h; 2. cyclohexanone, 0 °C, 1 h; 70%. 1. Zn–$HgCl_2$, dry THF, 0–25 °C, 5–8 h; 2. cyclohexanone, 0 °C, 1 h; 75% (7)

Mild reductive deacetoxybromination of glycosyl bromides offers an approach to glucals bearing acid-sensitive substituents (eq 8).[26]

Al–Hg, 2% aq THF, rt, 6 h, 85% (8)

C–O Bond Cleavage. Ether linkages of various substrates, for example glycosides,[27] 1,3-dioxolanes,[28] tetrahydrofurans,[29,30] and oxiranes,[31] undergo reductive cleavage with formation of alcohols. Among the reactions listed, the reductive cleavage of epoxides (eq 9)[32a] is presumably of the most importance. It has been widely used in the synthesis of prostaglandins,[31a,b,32] steroids,[31c,d] erythronolide B,[33] and vitamin precursors.[34]

Al–Hg, EtOH–H_2O–THF–sat aq $NaHCO_3$ (87:48:30:3), 0 °C, 3 h, 99% (9)

Deoxygenation of certain terpene oxides with Al foil activated by $HgCl_2$ has been reported instead of reduction to the desired alcohols.[35]

N–O and N–N Bond Cleavage. On exposure to excess Al–Hg in aqueous THF at 0 °C for several hours, N–O bonds in bicyclic Diels–Alder adducts are readily cleaved.[36] Mild conditions provide a highly chemoselective process and carbon–carbon double bonds and acid labile functional groups survive, in contrast to the alternative methods employed such as catalytic hydrogenolysis or reduction with ***Zinc–Acetic Acid***.[36a] The reaction occurs with high stereoselectivity and the product is formed with a *cis* disposition of N- and O-containing substituents (eq 10).[36]

1. Al–Hg (excess), THF–H_2O (10:1), 0 °C; 2. Ac_2O, py–DMAP; 41% (10)

Aluminum analgam is also highly effective in reductive cleavage of N–N bonds to produce amines.[37]

Reductive Desulfurization. The ability of Al–Hg to reduce C–S bonds is widely used in organic synthesis in conjunction with methodology involving reactions of highly reactive α-sulfinyl- and α-sulfonylalkyl carbanions[2b,38] as well as (α-sulfoximidoyl)alkyl carbanions.[39] Removal of the activating sulfur substituent is frequently accomplished using Al–Hg. The methodology offers facile synthetic approaches to ketones,[2b,38,40] enones,[41] di- and triketones,[40,42] hydroxy ketones (eq 11),[43] unsaturated acids,[44] and γ-oxo-α-amino acids.[45]

$PhSO_2C_8H_{17}$: 1. 2 BuLi, dry THF, –90 °C; 2. β-propiolactone, 0 °C; 3. aq NH_4Cl, 0 °C; 61%. Al–Hg, THF–H_2O (9:1), reflux, 3–6 h; 99% (11)

Readily removed on treatment with Al–Hg asymmetric sulfinyl and sulfonimidoyl groups may serve as chiral auxiliaries in enantiocontrolled synthesis of 3-substituted cycloalkanones[46] and 3-hydroxy[47] and 3-arylcarboxylic acid esters.[48]

Aluminum amalgam mediated cleavage of C–S bonds plays an important role in sulfoximine-based alkenation of carbonyl compounds via β-hydroxysulfoximines[49] (eq 12).[49b]

Ph–CH$_2$CH$_2$C(O)CH$_3$ → (FCH$_2$S(O)(NMe)Ph, LDA; THF, –85 °C, under N_2; 91%) → Ph–CH$_2$CH$_2$C(OH)(CH$_3$)CH(F)S(O)(NMe)Ph → (Al–Hg; AcOH–THF–H_2O (15:15:1) 4 h; 76%) → Ph–CH$_2$CH$_2$C(CH$_3$)=CHF, E:Z = 1:1 (12)

Aluminum amalgam can also be employed in reductive elimination of phenylthio groups from 2-phenylthioalkanones,[50] stereospecific reduction of sulfoximines,[51] selective cleavage of the sulfinyl sulfur–methylene carbon bond in the presence of a disulfide moiety,[52] and reductive scission of S–S and S–N bonds.[53,54]

For some reactions mediated by aluminum activated with mercury(II) chloride, also see ***Aluminum***.

Related Reagents. See Classes R-1, R-2, R-3, R-5, R-6, R-9, R-10, R-15, R-16, R-21, R-22, R-23, R-24, R-25, R-27, R-30, R-31, and R-32, pages 1–10.

1. Smith, M. *Reduction Techniques and Applications in Organic Synthesis*; Augustine, R. L., Ed.; Dekker: New York, 1968; pp 95–170.
2. (a) Wislecenus, H.; Kaufmann, L. *CB* **1895**, *28*, 1323. (b) Corey, E. J.; Chaykovsky, M. *JACS* **1965**, *87*, 1345. (c) Calder, A.; Forrester, A. R.; Hepburn, S. P. *OS* **1972**, *52*, 77.
3. (a) Ghatak, U.; Saha, N. N.; Dutta, P. C. *JACS* **1957**, *79*, 4487. (b) Tankard, M. H.; Whitehurst, J. S. *JCS(P1)* **1973**, 615. (c) Stahly, G. P.; Jackson, A. *JOC* **1991**, *56*, 5472.
4. Tamura, M.; Harada, K. *BCJ* **1980**, *53*, 561.
5. (a) Miller, R. E.; Nord F. F. *JOC* **1951**, *16*, 1380. (b) Mladenov, I.; Boeva, R.; Aleksiev, D.; Lyubcheva, M. *God. Vissh. Khim.-Tekhnol. Inst., Burgas, Bulg.* **1977/1978**, *2*, 25; *CA* **1979**, *90*, 137 409t.
6. (a) Crombie, L.; Hancock, J. E. H.; Linstead, R. P. *JCS* **1953**, 3496. (b) Leraux, Y. *CR(C)* **1971**, *273*, 178.
7. Hulce, M.; LaVaute, T. *TL* **1988**, *29*, 525.
8. Ray, J. N.; Mukherji, A.; Gupta, N. D. *JIC* **1961**, *38*, 705.
9. Nguyen, D. A.; Cerutti, E. *BSF(2)* **1976**, 596
10. Luzzio, F. A.; O'Hara, L. C. *SC* **1990**, *20*, 3223.
11. (a) Schreibmann, A. A. P. *TL* **1970**, 4271. (b) Stocker, J. H.; Walsh, D. J. *JOC* **1979**, *44*, 3589.
12. (a) Bilger, C.; Demerseman, P.; Royer, R. *JHC* **1985**, *22*, 735. (b) Petti, M. A.; Shepadd, T. J.; Barrans, Jr., R. E.; Dougherty, D. A. *JACS* **1988**, *110*, 6825.
13. Atwell, G. J.; Rewcastle, G. W., Baguley, B. C.; Denny, W. A. *JMC* **1987**, *30*, 664.
14. (a) Thies, H.; Schönenberger, H.; Bauer, K. H. *AP* **1960**, *293*, 67. (b) Takeshima, T.; Muraoka, M.; Asaba, H.; Yokoyama, M. *BCJ* **1968**, *41*, 506.
15. Meyers, A. I.; Durandetta, J. L. *JOC* **1975**, *40*, 2021.
16. Meyers, A. I.; Durandetta, J. L.; Munavu, R. *JOC* **1975**, *40*, 2025.
17. Cooper, R. D. G.; Jose, F. L. *JACS* **1972**, *94*, 1021.
18. (a) Bastian, J.-M.; Jaunin, R. *HCA* **1963**, *46*, 1248. (b) Bapat, J. B.; Black, D. St. C. *AJC* **1968**, *21*, 2497.
19. (a) Berlin, K. D.; Claunch, R. T.; Gaudy, E. T. *JOC* **1968**, *33*, 3090. (b) Hosmane, R. S.; Lim, B. B. *TL* **1985**, *26*, 1915. (c) Muratake, H.; Okabe, K.; Natsume, M. *T* **1991**, *40*, 8545.
20. Gilchrist, T. L.; Hughes, D.; Wasson, R. *TL* **1987**, *28*, 1537.
21. (a) Boyer, J. H.; Alul, H. *JACS* **1959**, *81*, 2136. (b) Corey, E. J.; Andersen, N. H.; Carlson R. M.; Paust, J.; Vedejs, E.; Vlattas, I.; Winter, R. E. K. *JACS* **1968**, *90*, 3247. (c) Kraus, G. A.; Fraizer, K. *TL* **1978**, 3195. (d) Trost, B. M.; King, S. A.; Schmidt, T. *JACS* **1989**, *111*, 5902.
22. Healey, K.; Calder, I. C. *AJC* **1979**, *32*, 1307.
23. (a) Shin, C.; Yonezawa, Y.; Yoshimura, J. *CL* **1976**, 1095. (b) Smodiš, J.; Zupet, R.; Petric, A.; Stanovnik, B.; Tišler, M. *H* **1990**, *30*, 393.
24. Inoi, T.; Gericke, P.; Horton, W. J. *JOC* **1962**, *27*, 4597.
25. Daniels, R. G.; Paquette, L. *TL* **1981**, *22*, 1579.
26. Jain, S.; Suryawanshi, S. N.; Bhakuni, D. S. *IJC(B)* **1987**, *26*, 866.
27. Kennedy, R. M.; Abiko, A.; Masamune, S. *TL* **1988**, *29*, 447.
28. Johnson, C. R.; Penning, T. D. *JACS* **1986**, *108*, 5655.
29. Vandewalle, M.; Van der Eycken, J.; Oppolzer, W.; Vullioud, C. *T* **1986**, *42*, 4035.
30. Arai, Y.; Kawanami, S.; Koizumi, T. *CL* **1990**, 1585.
31. (a) Schneider, W. P.; Bundy, G. L., Lincoln, F. H. *CC* **1973**, 254. (b) Corey, E. J.; Ensley, H. E. *JOC* **1973**, *38*, 3187. (c) Narwid, T. A.; Blount, J. F.; Iacobelli, J. A.; Uskokovic, M. R. *HCA* **1974**, *57*, 781. (d) Hossain, A. M. M., Kirk, D. N.; Mitra, G. *Steroids* **1976**, *27*, 603. (e) Brough, P. A.; Gallagher, T.; Thomas, P.; Wonnacott, S.; Baker, R.; Abdul Malik, K. M.; Hursthouse, M. B. *CC* **1992**, 1087.
32. (a) Greene, A. E.; Teixeira, M. A.; Barreiro, E.; Cruz, A.; Crabbe, P. *JOC* **1982**, *47*, 2553. (b) Schneider, W. P.; Bundy, G. L., Lincoln, F. H.; Daniels, E. G.; Pike, J. E. *JACS* **1977**, *99*, 1222. (c) Danieli, R.; Martelli, G.; Spunta, G.; Rossini, S.; Cainelli, G.; Panunzio, M. *JOC* **1983**, *48*, 123.
33. Corey, E. J.; Trybulski, E. J.; Melvin, L. S.; Nicolaou, K. C.; Secrist, J. A.; Lett, R.; Sheldrake, P. W.; Falck, J. R.; Brunelle, D. J.; Haslanger, M. F.; Kim, S.; Yoo, S. *JACS* **1978**, *100*, 4618.
34. Solladie, G.; Hutt, J. *JOC* **1987**, *52*, 3560.
35. Mitchell, P. W. D. *OPP* **1990**, *22*, 534.
36. (a) Keck, G. E.; Fleming, S.; Nickell, D.; Weider, P. *SC* **1979**, *9*, 281. (b) King, S. B.; Ganem, B. *JACS* **1991**, *113*, 5089.
37. (a) Mellor, J. M.; Smith, N. M. *JCS(P1)* **1984**, *2927*. (b) Atkinson, R. S.; Edwards, P. J.; Thomson, G. A. *CC* **1992**, 1256.
38. Corey, E. J.; Chaykovsky, M. *JACS* **1964**, *86*, 1639.
39. Johnson, C. R. *ACR* **1973**, *6*, 341.
40. (a) Stetter, H.; Hesse, R. *M* **1967**, *98*, 755. (b) Fulmer, T. D.; Bryson, T. A. *JOC* **1989**, *54*, 3496. (c) He, X. S.; Eliel, E. L.; *T* **1987**, *43*, 4979.
41. Ohtsuka, Y.; Sasahara, T.; Oishi, T. *CPB* **1982**, *30*, 1106.
42. Cannon, J. R.; Chow, P. W.; Fuller, M. W.; Hamilton, B. H.; Metcalf, B. W.; Power, A. J. *AJC* **1973**, *26*, 2257.
43. Cavicchioli, S.; Savoia, D.; Trombini, C.; Umani-Ronchi, A. *JOC* **1984**, *49*, 1246.
44. Ohnuma, T.; Hata, N.; Fujiwara, H.; Ban, Y. *JOC* **1982**, *47*, 4713.
45. Baldwin, J. E.; Adlington, R. M.; Codfrey, C. R. A.; Gollins, D. W.; Smith, M. L.; Russel, A. T. *SL* **1993**, 51.
46. (a) Posner, G. H.; Mallamo, J. P.; Hulce, M.; Frye, L. L. *JACS* **1982**, *104*, 4180. (b) Posner, G. H.; Hulce, M. *TL* **1984**, *25*, 379. (c) Posner, G. H.; Weitzberg, M.; Hamill, T. G.; Asirvatham, E.; Cun-heng, H.; Clardy, J. *T* **1986**, *42*, 2919.
47. (a) Mioskowski, C.; Solladie, G. *T* **1980**, *36*, 227. (b) Fujisawa, T.; Fujimura, A.; Sato, T. *BCJ* **1988**, *61*, 1273.
48. Pyne, S. G. *JOC* **1986**, *51*, 81.
49. (a) Johnson, C. R.; Shanklin, J. R.; Kirchhoff, R. A. *JACS* **1973**, *95*, 6462. (b) Boys, M. L.; Collington, E. W.; Finch, H.; Swanson, S.; Whitehead, J. F. *TL* **1988**, *29*, 3365.
50. Monteiro, H. J. *JOC* **1977**, *42*, 2324.
51. Johnson, C. R.; Jonsson, E. U.; Wambsgans, A. *JOC* **1979**, *44*, 2061.

52. Block, E.; O'Connor, J. *JACS* **1974**, *96*, 3929.
53. Kornblum, N.; Widmer, J. *JACS* **1978**, *100*, 7086.
54. Arzeno, H. B.; Kemp, D. S. *S* **1988**, 32.

Emmanuil I. Troyansky
Institute of Organic Chemistry, Russian Academy of Sciences, Moscow, Russia

Aluminum Hydride[1]

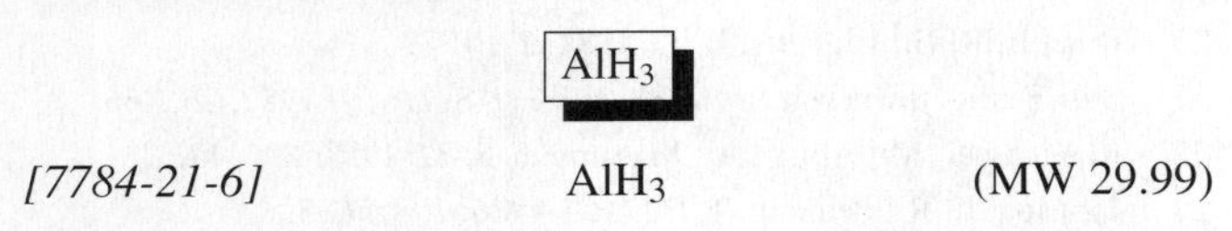

[7784-21-6] AlH_3 (MW 29.99)

(reducing agent for many functional groups; used in hydroalumination of alkynes; allylic rearrangements)

Alternate Name: alane.

Physical Data: colorless, nonvolatile solid in a highly polymerized state; mp 110 °C (dec). X-ray data have also been obtained.[2]

Solubility: sol THF and ether; precipitates from ether after standing for approximately 30 min depending upon method of preparation.

Analysis of Reagent Purity: hydride concentration can be determined by hydrolyzing aliquots and measuring the hydrogen evolved.[3]

Preparative Methods: can be prepared by treating an ether solution of ***Lithium Aluminum Hydride*** with ***Aluminum Chloride*** (eq 1).[4] This affords an ether solution of AlH_3 after precipitation of LiCl. Solutions have to be used immediately otherwise AlH_3 precipitates as a white solid which consists of a polymeric material with ether. The solvent can be removed and the solid redissolved in THF.[5] Alternatively, THF solutions can be prepared directly according to one of the reactions shown in eqs 2–4.[5,6]

$$3\,LiAlH_4 + AlCl_3 \longrightarrow 4\,AlH_3 + 3\,LiCl \quad (1)$$

$$2\,LiAlH_4 + BeCl_2 \longrightarrow 2\,AlH_3 + LiBeH_2Cl_2 \quad (2)$$

$$2\,LiAlH_4 + H_2SO_4 \longrightarrow 2\,AlH_3 + Li_2SO_4 + 2\,H_2 \quad (3)$$

$$2\,LiAlH_4 + ZnCl_2 \longrightarrow 2\,AlH_3 + 2\,LiCl + ZnH_2 \quad (4)$$

Handling, Storage, and Precautions: solutions of AlH_3 are not spontaneously inflammable.[3] However, since AlH_3 has reactivity comparable to $LiAlH_4$, one should follow similar handling and precautions as those exercised for $LiAlH_4$. Solutions of AlH_3 are prepared in situ but are known to degrade after 3 days, and long term storage of solutions is not possible. Use in a fume hood.

Functional Group Reductions. Reductions by alane take place primarily by a two-electron mechanism.[7,8] However, evidence for a SET pathway exists.[9] AlH_3 will reduce a wide variety of functional groups.[4] These include aldehydes, ketones, quinones, carboxylic acids, anhydrides, acid chlorides, esters, and lactones, from which the corresponding alcohol is the isolated product. Amides, nitriles, oximes, and isocyanates are reduced to amines. Nitro compounds are inert to AlH_3. Sulfides and sulfones are unreactive, but disulfides and sulfoxides can be reduced. Tosylates are also not reduced by AlH_3.

The reduction of ketones with AlH_3 has selectivity different from other hydride reagents (eqs 5 and 6).[10] The hydroxymethylation of ketones via a two-step procedure has also been accomplished (eq 7).[11] The conversion of α,β-unsaturated ketones to allylic alcohols can be carried out with very good selectivity using AlH_3 (eq 8);[12] however, DIBAL is the reagent of choice for this transformation (see ***Diisobutylaluminum Hydride***).[12c]

(5)
$LiAlH_4$ *trans*:*cis* = 1.9:1
AlH_3 7.3:1

(6)
$LiAlH(O\text{-}t\text{-}Bu)_3$ 0:100
AlH_3 91:9

(7)
1. NaH, EtO_2CH
2. NaH, AlH_3

(8)
AlH_3, 76%

Carboxylic acids and esters are reduced more rapidly by AlH_3 than by $LiAlH_4$, whereas the converse is true for alkyl halides. As a result, acids and esters can be reduced in the presence of halides (eq 9). In addition, esters can be reduced in the presence of nitro groups (eq 10). This stands in contrast to $LiAlH_4$ in which nitro groups are reduced. Acetals can also be reduced to the half protected diol as illustrated in eq 11.[13]

(9)
$Cl\text{–}CH_2CH_2\text{–}CO_2H \xrightarrow[89\%]{AlH_3} Cl\text{–}CH_2CH_2CH_2\text{–}OH$
AlH_3, 93%

(10)
AlH_3, 80%

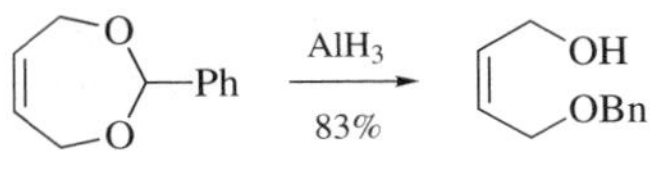

(11)

In reductions of amides to amines there is a competition between C–O and C–N bond cleavage which depends upon the reaction conditions. This complication does not occur with AlH_3. A quantitative yield of amine is obtained with short reaction times. Conjugated amides can be cleanly reduced to the allylic amine (eq 12).[6]

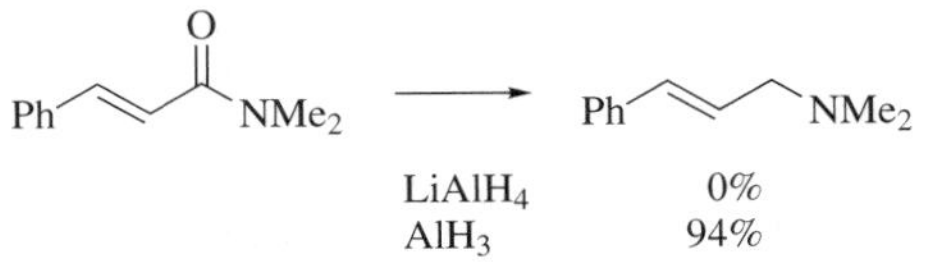

(12)

$LiAlH_4$	0%
AlH_3	94%

The less basic AlH_3 appears to be better than $LiAlH_4$ for reducing nitriles with relatively acidic α-hydrogens to amines, and enolizable keto esters to diols (eq 13).[6]

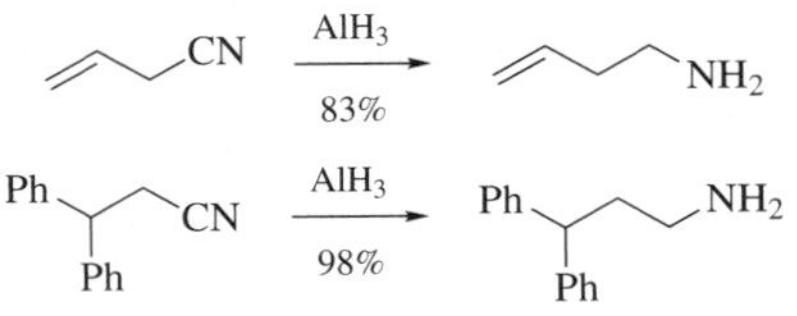

(13)

The reduction of β-lactams to azetidines can be accomplished with AlH_3 (eq 14),[14] while ring opening was observed with $LiAlH_4$. AlH_3 can also convert enamines to the corresponding alkenes (eq 15).[15]

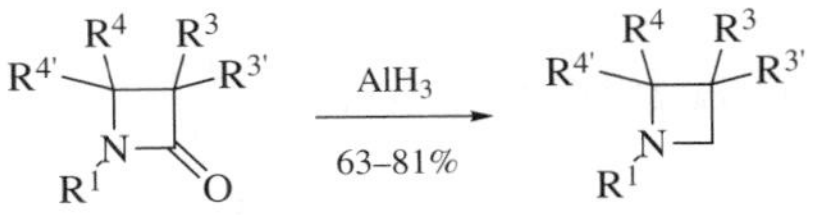

(14)

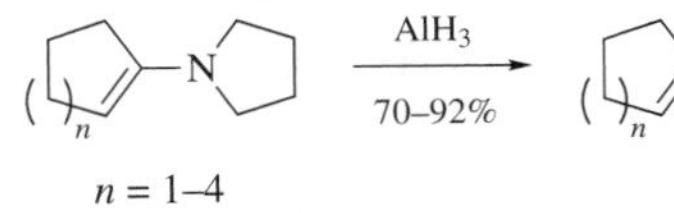

(15)

$n = 1–4$

While alkyl halides are usually inert to AlH_3, the reduction of cyclopropyl halides to cyclopropanes (eq 16)[16] and glycosyl fluorides to tetrahydropyrans (eq 17)[17] are known.

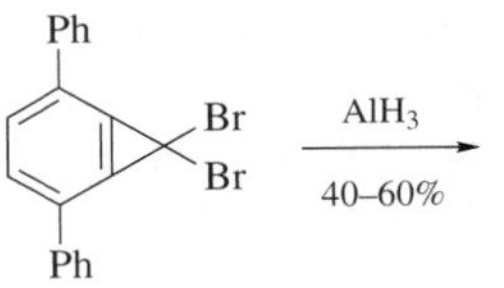

(16)

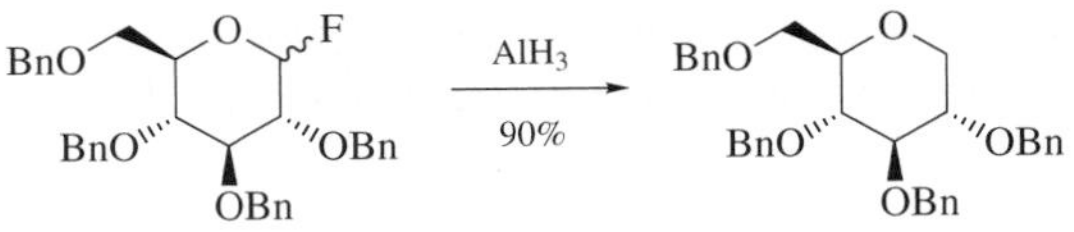

(17)

Desulfurization of sultones is rapid and proceeds in good yield with AlH_3 while $LiAlH_4$ affords poor yields with long reaction times (eq 18).[18]

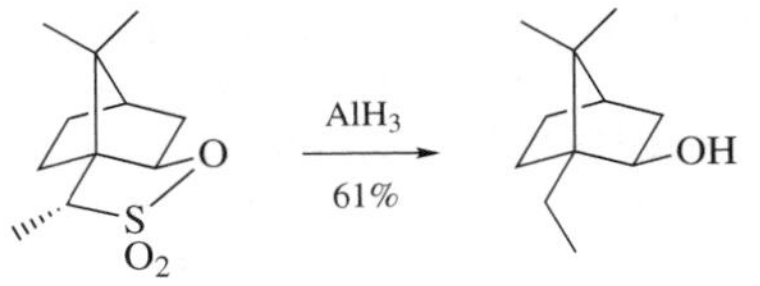

(18)

Epoxide Ring Opening. With most epoxides, hydride attack occurs at the least sterically hindered site to give the corresponding alcohol (eq 19).[19] However, due to the electrophilic nature of AlH_3 compared to $LiAlH_4$, it is possible for ring opening to occur at the more hindered site. With phenyl substituted epoxides, mechanistic studies have shown that attack at a benzylic carbenium ion or a 1,2-hydride shift followed by hydride attack gives products with the same regiochemistry but with different stereochemistry (eq 20).[6,20] The stereoselectivity of AlH_3 mediated epoxide openings has been studied in depth.[21]

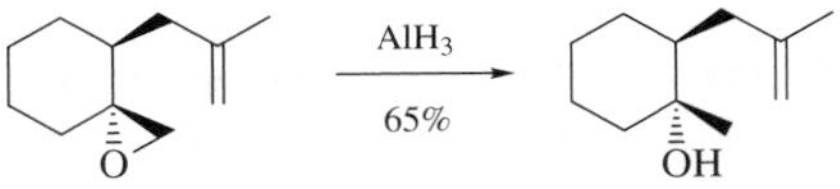

(19)

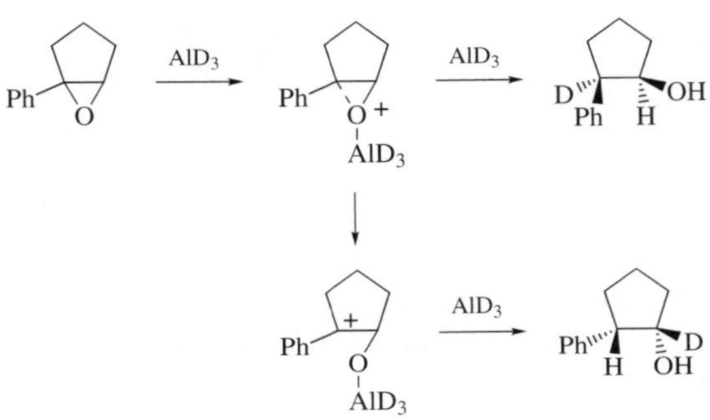

(20)

Hydroalumination. The addition of AlH_3 across a triple bond has been shown to occur in propargylic systems.[22] When the reactions are quenched with ***Iodine***, AlH_3 gives the 2-iodo-(*E*)-alkene while $LiAlH_4$ gives the 3-iodo-(*E*)-alkene (eq 21). AlH_3 can also be used in conjunction with ***Titanium(IV) Chloride*** to carry out a reaction similar to a hydroboration.[23] Thus 1-hexene is converted to hexane upon aqueous work-up or to the corresponding alcohol upon exposure of the reaction intermediate to oxygen (eq 22). Similar results were obtained with nonconjugated dienes.

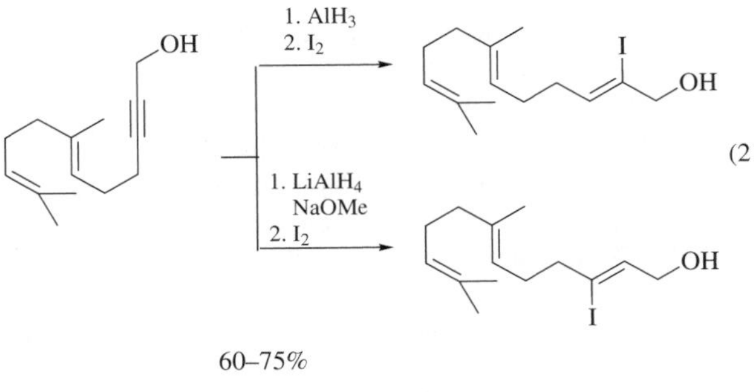

(21)

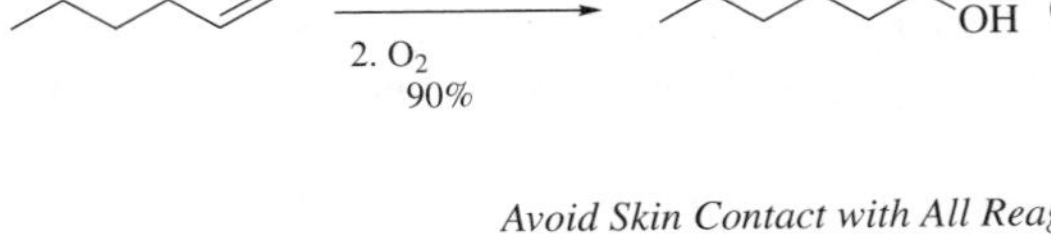

60–75%

(22)

Allylic Rearrangements (S_N2'). The S_N2' displacement of a good leaving group to give the rearranged allylic system can be carried out with AlH_3.[24] This reaction appears not to be sterically demanding as a variety of displacements are possible (eq 23).

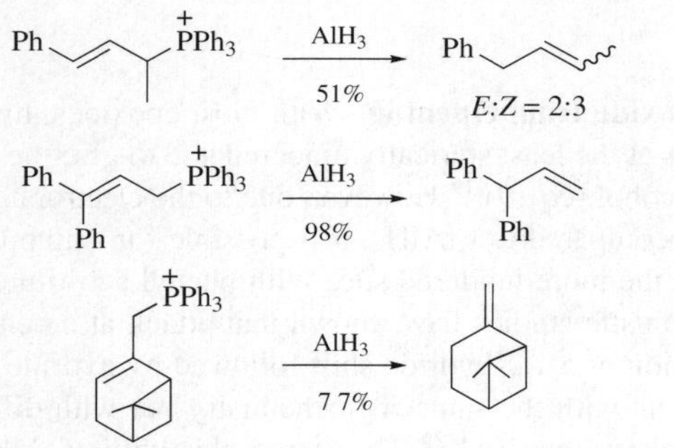

(23)

Preparation of allenes from propargylic systems can also be accomplished.[24] Most systems show a preference for *syn* elimination; however, mesylates prefer an *anti* mode of elimination (eq 24). This same procedure has been used to prepare fluoroallenes (eq 25).[25]

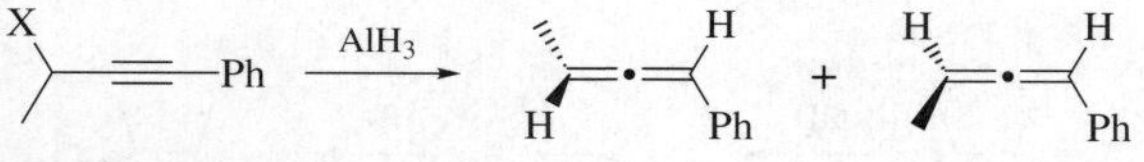

(24)

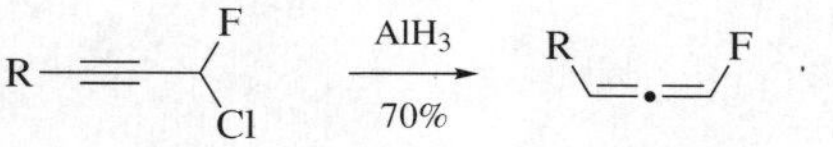

(25)

Dialkylaluminum hydrides also behave as hydroalumination reagents (see ***Diisobutylaluminum Hydride***) and are more commonly used than AlH_3.

Related Reagents. See Classes R-1, R-4, R-5, R-12, R-13, R-20, R-23, R-27, R-30, and R-32, pages 1–10.

---

1. (a) Gaylord, N. G. *Reduction With Complex Metal Hydrides*; Interscience: New York, 1956. (b) Semenenko, K. N.; Bulychev, B. M.; Shevlyagina, E. A, *RCR* **1966**, *35*, 649. (c) Rerick, M. N. *Reduction Techniques and Applications in Organic Synthesis*, Augustine, R. L., Ed.; Dekker: New York, 1968. (d) Cucinella, S.; Mazzei, A.; Marconi, W. *ICA Rev.* **1970**, 51. (e) Walker, E. R. H. *CSR* **1976**, *5*, 23. (f) Brown, H. C.; Krishnamurthy, S. *T* **1979**, *35*, 567. (g) Hajos, A. *Complex Hydrides and Related Reducing Agents in Organic Synthesis*; Elsevier: Amsterdam, 1979. (h) Seyden-Penne, J. *Reduction by the Alumino- and Borohydrides in Organic Synthesis*; VCH: New York, 1991.
2. Turley, J. W.; Rinn, H. W. *IC* **1969**, *8*, 18.
3. Yoon, N. M.; Brown, H. C. *JACS* **1968**, *90*, 2927.
4. Finholt, A. E.; Bond, A. C.; Schlesinger, H. I. *JACS* **1947**, *69*, 1199.
5. (a) Brown, H. C.; Yoon, N. M. *JACS* **1966**, *88*, 1464. (b) Browner, F. M.; Matzek, N. E.; Reigler, P. F.; Rinn, H. W.; Roberts, C. B.; Schmidt, D. L.; Snover, J. A.; Terada, K. *JACS* **1976**, *98*, 2450.
6. Ashby, E. C.; Sanders, J. R.; Claudy, P.; Schwarts, P. *JACS* **1973**, *95*, 6485.
7. Laszlo, P.; Teston, M. *JACS* **1990**, *112*, 8751.
8. (a) Park, S.-U.; Chung, S.-K.; Newcomb, M. *JOC* **1987**, *52*, 3275. (b) Yamataka, H.; Hanafusa, T. *JOC* **1988**, *53*, 773.
9. (a) Ashby, E. C.; Goel, A. B. *TL* **1981**, *22*, 4783. (b) Ashby, E. C.; DePriest, R. N.; Pham, T. N. *TL* **1983**, *24*, 2825. (c) Ashby, E. C.; DePriest, R. N.; Goel, A. B.; Wenderoth, B.; Pham, T. N. *JOC* **1984**, *49*, 3545. (d) Ashby, E. C.; Pham, T. N. *JOC* **1986**, *51*, 3598. (e) Ashby, E. C.; Pham, T. N. *TL* **1987**, *28*, 3197.
10. (a) Ayres, D. C.; Sawdaye, R. *JCS(P2)* **1967**, 581. (b) Ayres, D. C.; Kirk, D. N.; Sawdaye, R. *JCS(P2)* **1970**, 505. (c) Guyon, R.; Villa, P. *BSF* **1977**, 145. (d) Guyon, R.; Villa, P. *BSF* **1977**, 152. (e) Martinez, E.; Muchowski, J. M.; Velarde, E. *JOC* **1977**, *42*, 1087.
11. Corey, E. J.; Cane, D. *JOC* **1971**, *36*, 3070.
12. (a) Jorgenson, M. J. *TL* **1962**, 559. (b) Brown, H. C.; Hess, H. M. *JOC* **1969**, *34*, 2206. (c) Wilson, K. E.; Seidner, R. T.; Masamune, S. *CC* **1970**, 213. (d) Dilling, W. L.; Plepys, R. A. *JOC* **1970**, *35*, 2971. (e) Ashby, E. C.; Lin, J. J. *TL* **1976**, 3865.
13. (a) Danishefsky, S.; Regan, J. *TL* **1981**, *22*, 3919. (b) Takano, S.; Akiyama, M.; Sato, S.; Ogasawara, K. *CL* **1983**, 1593. (c) Richter, W. J. *JOC* **1981**, *46*, 5119.
14. Jackson, M. B.; Mander, L. N.; Spotswood, T. M. *AJC* **1983**, *36*, 779.
15. Coulter, J. M.; Lewis, J. W.; Lynch, P. P. *T* **1968**, *24*, 4489.
16. Muller, P. *HCA* **1974**, *57*, 704.
17. Nicolaou, K. C.; Dolle, R. E.; Chucholowski, A.; Randal, J. L. *CC* **1984**, 1153.
18. (a) Wolinsky, J.; Marhenke, R. L.; Eustace, E. J. *JOC* **1973**, *38*, 1428. (b) Smith, M. B.; Wolinsky, J. *JOC* **1981**, *46*, 101.
19. Maruoka, K.; Saito, S.; Ooi, T.; Yamamoto, H. *SL* **1991**, 255.
20. Lansbury, P. T.; Scharf, D. J.; Pattison, V. A. *JOC* **1967**, *32*, 1748.
21. Elsenbaumer, R. L.; Mosher, H. S.; Morrison, J. D.; Tomaszewski, J. E. *JOC* **1981**, *46*, 4034.
22. Corey, E. J.; Katzenellenbogen, J. A.; Posner, G. H. *JACS* **1967**, *89*, 4245.
23. Sato, F.; Sato, S.; Kodama, H.; Sato, M. *JOM* **1977**, *142*, 71.
24. Claesson, A.; Olsson, L.-I. *JACS* **1979**, *101*, 7302.
25. Castelhano, A.; Krantz, A. *JACS* **1987**, *109*, 3491.

Paul Galatsis
University of Guelph, Ontario, Canada

Aluminum Isopropoxide[1]

$Al(O\text{-}i\text{-}Pr)_3$

[555-31-7] $C_9H_{21}AlO_3$ (MW 204.25)

(mild reagent for Meerwein–Ponndorf–Verley reduction;[1] Oppenauer oxidation;[13] hydrolysis of oximes;[16] rearrangement of epoxides to allylic alcohols;[17] regio- and chemoselective ring opening of epoxides;[20] preparation of ethers[21])

Alternate Name: triisopropoxyaluminum.
Physical Data: mp 138–142 °C (99.99+%), 118 °C (98+%); bp 140.5 °C; d 1.035 g cm^{-3}.
Solubility: sol benzene; less sol alcohols.
Form Supplied in: white solid (99.99+% or 98+% purity based on metals analysis).
Preparative Methods: see example below.
Handling, Storage, and Precautions: the dry solid is corrosive, moisture sensitive, flammable, and an irritant. Use in a fume hood.

NMR Analysis of Aluminum Isopropoxide. Evidence from molecular weight determinations indicating that aluminum isopropoxide aged in benzene solution consists largely of the tetramer

(**1**), whereas freshly distilled molten material is trimeric (**2**),[2] is fully confirmed by NMR spectroscopy.[3]

(**1**) (**2**)

Meerwein–Ponndorf–Verley Reduction. One use of the reagent is for the reduction of carbonyl compounds, particularly of unsaturated aldehydes and ketones, for the reagent attacks only carbonyl compounds. An example is the reduction of crotonaldehyde to crotyl alcohol (eq 1).[1] A mixture of 27 g of cleaned ***Aluminum*** foil, 300 mL of isopropanol, and 0.5 g of ***Mercury(II) Chloride*** is heated to boiling, 2 mL of carbon tetrachloride is added as catalyst, and heating is continued. The mixture turns gray, and vigorous evolution of hydrogen begins. Refluxing is continued until gas evolution has largely subsided (6–12 h). The solution, which is black from the presence of suspended solid, can be concentrated and the aluminum isopropoxide distilled in vacuum (colorless liquid) or used as such. Thus the undistilled solution prepared as described from 1.74 mol of aluminum and 500 mL of isopropanol is treated with 3 mol of crotonaldehyde and 1 L of isopropanol. On reflux at a bath temperature of 110 °C, acetone slowly distills at 60–70 °C. After 8–9 h, when the distillate no longer gives a test for acetone, most of the remaining isopropanol is distilled at reduced pressure and the residue is cooled and hydrolyzed with 6 N sulfuric acid to liberate crotyl alcohol from its aluminum derivative.

CHO + OH —Al(O-*i*-Pr)$_3$, 60%→ OH + O (1)

The Meerwein–Ponndorf–Verley reduction of the ketone (**3**) involves formation of a cyclic coordination complex (**4**) which, by hydrogen transfer, affords the mixed alkoxide (**5**), hydrolyzed to the alcohol (**6**) (eq 2).[4] Further reflection suggests that under forcing conditions it might be possible to effect repetition of the hydrogen transfer and produce the hydrocarbon (**7**). Trial indeed shows that reduction of diaryl ketones can be effected efficiently by heating with excess reagent at 250 °C (eq 3).[5]

A study[6] of this reduction of mono- and bicyclic ketones shows that, contrary to commonly held views, the reduction proceeds at a relatively high rate. The reduction of cyclohexanone and of 2-methylcyclohexanone is immeasurably rapid. Even menthone is reduced almost completely in 2 h. The stereochemistry of the reduction of 3-isothujone (**8**) and of 3-thujone (**11**) has been examined (eqs 4 and 5). The ketone (**8**) produces a preponderance of the *cis*-alcohol (**9**). The stereoselectivity is less pronounced in the case of 3-thujone (**11**), although again the *cis*-alcohol (**12**) predominates. The preponderance of the *cis*-alcohols can be increased by decreasing the concentration of ketone and alkoxide.

(**3**) —Al(O-*i*-Pr)$_3$→ (**4**) → (**5**) —H$^+$→ (**6**); —250 °C→ (**7**) (2)

—Al(O-*i*-Pr)$_3$, 250 °C, 95%→ (3)

anthraquinone —75%→ anthracene

anthrone —92%→ anthracene

(**8**) → (**9**) + (**10**) 7:1 (4)

(**11**) → (**12**) + (**13**) 3.2:1 (5)

This reducing agent is the reagent of choice for reduction of enones of type (**14**) to the α,β-unsaturated alcohols (**15**) (eq 6). Usual reducing agents favor 1,4-reduction to the saturated alcohol.[7]

(**14**) —Al(O-*i*-Pr)$_3$→ (**15**) (6)

BPC = biphenylcarbonyl; (**15**) 20%, each isomer

The Meerwein–Ponndorf–Verley reduction of pyrimidin-2(1*H*)-ones using ***Zirconium Tetraisopropoxide*** or aluminum isopropoxide leads to exclusive formation of the 3,4-dihydro isomer (eq 7).[8] The former reducing agent is found to be more effective.

—Zr(O-*i*-Pr)$_4$ or Al(O-*i*-Pr)$_3$, *i*-PrOH, 90 °C, 2 days, 20–91%→ (7)

R = H, halide

Reductions with Chiral Aluminum Alkoxides. The reduction of cyclohexyl methyl ketone with catalytic amounts of aluminum alkoxide and excess chiral alcohol gives (*S*)-1-cyclohexylethanol in 22% ee (eq 8).[9]

Al(OR)$_3$ (8) 22% ee

Isobornyloxyaluminum dichloride is a good reagent for reducing ketones to alcohols. The reduction is irreversible and subject to marked steric approach control (eq 9).[10]

(9) 70% ee

Diastereoselective Reductions of Chiral Acetals. Recently, it has been reported that ***Pentafluorophenol*** is an effective accelerator for Meerwein–Ponndorf–Verley reduction.[11] Reduction of 4-*t*-butylcyclohexanone with aluminum isopropoxide (3 equiv) in dichloromethane, for example, is very slow at 0 °C (<5% yield for 5 h), but in the presence of pentafluorophenol (1 equiv), the reduction is cleanly completed within 4 h at 0 °C (eq 10). The question of why this reagent retains sufficient nucleophilicity is still open. It is possible that the *o*-halo substituents of the phenoxide ligand may coordinate with the aluminum atom, thus increasing the nucleophilicity of the reagent.

Al(O-*i*-Pr)$_3$ (3 equiv), C_6F_5OH (1 equiv), CH_2Cl_2, 0 °C (10)

Chiral acetals derived from (−)-***(2R,4R)-2,4-Pentanediol*** and ketone are reductively cleaved with high diastereoselectivity by a 1:2 mixture of diethylaluminum fluoride and pentafluorophenol.[11] Furthermore, aluminum pentafluorophenoxide is a very powerful Lewis acid catalyst for the present reaction.[12] The reductive cleavage in the presence of 5 mol % of Al(OC_6F_5)$_3$ affords stereoselectively retentive reduced β-alkoxy ketones. The reaction is an intramolecular Meerwein–Ponndorf–Verley reductive and Oppenauer oxidative reaction on an acetal template (eq 11).

Et_2AlF–2C_6F_5OH (1.2 equiv) or Al(OC_6F_5)$_3$ (5 mol %), CH_2Cl_2 or toluene; K_2CO_3; $R^1 > R^2$ (11)

The direct formation of α,β-alkoxy ketones is quite useful. Removal of the chiral auxiliary, followed by base-catalyzed β-elimination of the resulting β-alkoxy ketone, easily gives an optically pure alcohol in good yield. Several examples of the reaction are summarized in Table 1.

Table 1 Reductive Cleavages of Acetals Using Al(OC_6F_5)$_3$ Catalyst

Al(OC_6F_5)$_3$ (5 mol %), CH_2Cl_2

R^1	R^2	Yield (%)	Ratio (*S*:*R*)
C_5H_{11}	Me	83	82:18
i-Bu	Me	61	73:27
i-Pr	Me	90	94:6
Ph	Me	71	>99:1
Ph	Et	78	92:8
c-Hex	Me	89	95:5
CH_2 CH_2 (4-*t*-Bu-cyclohexylidene)		67	81:19 (*trans*:*cis*)

Although the detailed mechanism is not yet clear, it is assumed that an energetically stable tight ion-paired intermediate is generated by stereoselective coordination of Al(OC_6F_5)$_3$ to one of the oxygens of the acetal; the hydrogen atom of the alkoxide is then transferred as a hydride from the retentive direction to this departing oxygen, which leads to the (*S*) configuration at the resulting ether carbon, as described (eq 12).

Al(OC_6F_5)$_3$; rotation; $R^1 > R^2$; L = OC_6F_5 (12)

Oppenauer Oxidation[13]. Cholestenone is prepared by oxidation of cholesterol in toluene solution with aluminum isopropoxide as catalyst and cyclohexanone as hydrogen acceptor (eq 13).[14]

Al(O-*i*-Pr)$_3$, cyclohexanone, toluene, 72–74% (13)

A formate, unlike an acetate, is easily oxidized and gives the same product as the free alcohol.[15] For oxidation of (**16**) to (**17**) the combination of cyclohexanone and aluminum isopropoxide and a hydrocarbon solvent is used: xylene (bp 140 °C at 760 mmHg) or toluene (bp 111 °C at 760 mmHg) (eq 14).

(16) → (17); $Al(O\text{-}i\text{-}Pr)_3$, cyclohexanone, xylene or toluene, 86% (14)

Hydrolysis of Oximes.[16] Oximes can be converted into parent carbonyl compounds by aluminum isopropoxide followed by acid hydrolysis (2N HCl) (eq 15). Yields are generally high in the case of ketones, but are lower for regeneration of aldehydes.

$$R^1C(=NOH)R^2 \xrightarrow[2.\ HCl,\ H_2O]{1.\ Al(O\text{-}i\text{-}Pr)_3} R^1C(=O)R^2 \quad (15)$$

Rearrangement of Epoxides to Allylic Alcohols. The key step in the synthesis of the sesquiterpene lactone saussurea lactone (**21**) involved fragmentation of the epoxymesylate (**18**), obtained from α-santonin by several steps (eq 16).[17] When treated with aluminum isopropoxide in boiling toluene (N_2, 72 h), (**18**) is converted mainly into (**20**). The minor product (**19**) is the only product when the fragmentation is quenched after 12 h. Other bases such as potassium *t*-butoxide, LDA, and lithium diethylamide cannot be used. Aluminum isopropoxide is effective probably because aluminum has a marked affinity for oxygen and effects cleavage of the epoxide ring. Meerwein–Ponndorf–Verley reduction is probably involved in one step.

(18) → $Al(O\text{-}i\text{-}Pr)_3$, 68% → (19) 9% + (20) 68%; (20) → 5 steps → (21) (16)

α-Pinene oxide (**22**) rearranges to pinocarvenol (**23**) in the presence of 1 mol % of aluminum isopropoxide at 100–120 °C for 1 h.[18] The oxide (**22**) rearranges to pinanone (**24**) in the presence of 5 mol % of the alkoxide at 140–170 °C for 2 h. Aluminum isopropoxide has been used to rearrange (**23**) to (**24**) (200 °C, 3 h, 80% yield) (eq 17).[19]

(22) → $Al(O\text{-}i\text{-}Pr)_3$, 100 °C, ~85% → (23); (22) → $Al(O\text{-}i\text{-}Pr)_3$, 150 °C, 69% → (24) (17)

Regio- and Chemoselective Ring Opening of Epoxides. Functionalized epoxides are regioselectively opened using trimethylsilyl azide/aluminum isopropoxide, giving 2-trimethylsiloxy azides by attack on the less substituted carbon (eq 18).[20]

Epoxide → Me_3SiN_3 (1.5 equiv), $Al(O\text{-}i\text{-}Pr)_3$ (0.1 equiv), CH_2Cl_2, 59–93% → TMSO/N_3 product (18)

Preparation of Ethers. Ethers ROR′ are prepared from aluminum alkoxides, $Al(OR)_3$, and alkyl halides, R′X. Thus EtCHMeOH is treated with Al, $HgBr_2$, and MeI in DMF to give EtCHMeOMe (eq 19).[21]

$$Al(OR)_3 + R^1X \xrightarrow[20\text{–}80\%]{DMF,\ reflux,\ 2\ days} ROR^1 \quad (19)$$

R, R^1 = alkyl; X = halide

Related Reagents. See Classes R-1, R-10, and R-12, pages 1–10.

1. Wilds, A. L. *OR* **1944**, *2*, 178.
2. Shiner, V. J.; Whittaker, D.; Fernandez, V. P. *JACS* **1963**, *85*, 2318.
3. Worrall, I. J. *J. Chem. Educ.* **1969**, *46*, 510.
4. Woodward, R. B.; Wendler, N. L.; Brutschy, F. J. *JACS* **1945**, *67*, 1425.
5. Hoffsommer, R. D.; Taub, D.; Wendler, N. L. *CI(L)* **1964**, 482.
6. Hach, V. *JOC* **1973**, *38*, 293.
7. Picker, D. H.; Andersen, N. H.; Leovey, E. M. K. *SC* **1975**, *5*, 451.
8. Høseggen, T.; Rise, F.; Undheim, K. *JCS(P1)* **1986**, 849.
9. Doering, W. von E.; Young, R. W. *JACS* **1950**, *72*, 631.
10. Nasipuri, D.; Sarker, G. *JIC* **1967**, *44*, 165.
11. Ishihara, K.; Hanaki, N.; Yamamoto, H. *JACS* **1991**, *113*, 7074.
12. Ishihara, K.; Hanaki, N.; Yamamoto, H. *SL* **1993**, 127; *JACS*, **1993**, *115*, 10 695.

13. Djerassi, C. *OR* **1951**, *6*, 207.
14. Eastham, J. F.; Teranishi, R. *OSC* **1963**, *4*, 192.
15. Ringold, H. J.; Löken, B.; Rosenkranz, G.; Sondheimer, F. *JACS* **1956**, *78*, 816.
16. Sugden, J. K. *CI(L)* **1972**, 680.
17. Ando, M.; Tajima, K.; Takase, K. *CL* **1978**, 617.
18. Scheidl, F. *S* **1982**, 728.
19. Schmidt, H. *CB* **1929**, *62*, 104.
20. Emziane, M.; Lhoste, P.; Sinou, D. *S* **1988**, 541.
21. Lompa-Krzymien, L.; Leitch, L. C. *Pol. J. Chem.* **1983**, *57*, 629.

Kazuaki Ishihara & Hisashi Yamamoto
Nagoya University, Japan

B

Baker's Yeast

(microorganism used as biocatalyst for the reduction of carbonyl groups and double bonds,[1] either under fermenting conditions, immobilized, or ultrasonically stimulated)

Solubility: insol cold and warm H_2O; used as a slurry.
Form Supplied in: yellowish pressed cakes, commercially available as cubes from bakeries or supermarkets, usually produced by brewery companies.
Handling, Storage, and Precautions: the wet cake must be stored in the refrigerator (0–4 °C) and used within the date indicated by the manufacturer.

Baker's Yeast-Mediated Biotransformations. Baker's yeast (BY, *Saccharomyces cerevisiae*) is readily available and inexpensive, and its use does not require any special training in microbiology. For these reasons, this biocatalyst has enjoyed a wide popularity among organic chemists, so that it can be considered as a microbial reagent for organic synthesis.[2] BY is generally used as whole cells, in spite of the problems connected with rates of penetration and diffusion of the substrates into, and the product from, the cells. However, the crude system is an inexpensive reservoir of cofactor-dependent enzymes such as oxidoreductases.[3] These benefits overcome the complication caused by undesired enzymatic reactions which lead to the formation of byproducts. Well-defined experimental procedures for BY-mediated bioreductions can be found in *Organic Syntheses*.[4] In the usual applications the biotransformations lead to optically active compounds with variable, but generally high, enantioselectivity.[5] The reaction is easily carried out in a heterogeneous medium containing a slurry of the yeast in tap water, in aerobic and fermenting conditions. Typically, the experimental conditions require a variable yeast:substrate ratio (1–40 g $mmol^{-1}$). The yeast is suspended in an aqueous solution of glucose or sucrose (0.1–0.3 M) to start the fermentation and to the fermenting yeast the substrate is added neat or in a suitable solvent, and therefore dispersed into the heterogeneous medium. The reaction is kept at 25–30 °C and, if necessary, additional fermenting BY can be added. At the end, the yeast is filtered off through Celite and the product extracted with organic solvents.

Carbonyl Group Reductions. Early applications of BY date back to the end of the 19th century and the first examples are reductions of carbonyl compounds.[1c] The widespread applications of this biotransformation are based on some systematic investigations on various ketones[6] and the stereochemical outcome of the reaction is generally described by the so-called Prelog's rule[7] which successfully applies to a great number of structures (eq 1).

The structural variety of carbonyl compounds appears to be almost unlimited since aliphatic, aromatic, and cyclic ketones are good substrates for the bioreduction.[1,5] Also, organometallic carbonyl compounds such as $Cr(CO)_3$-complexed aromatic aldehydes (eq 2)[8] or ketones (eq 3)[9] are enantioselectively reduced by BY.

(1)

BY

46%, 81% ee

48%, 66% ee

(2)

BY 96%

>99% ee

(3)

In general, the enantiomeric excess and the configuration of the optically active alcohols are strongly dependent on the structure of the starting carbonyl compound; many examples of diastereoselective reduction have also been reported.[10] The reduction of an epoxy ketone is accompanied by a stereocontrolled epoxide hydrolytic opening to afford a racemic triol, diastereomerically pure (eq 4).[11]

BY 51–74% + enantiomer (4)

R = *n*-Bu, *n*-Pr, Et

>99% de

Many experimental procedures have been developed in order to influence the enantioselectivity and the stereochemistry of the products: use of organic media,[12] the addition of various compounds to the incubation mixture,[13] or enclosure in a dialysis tube[14] can be helpful. Immobilized BY can be used in water or in organic solvents for the same purpose.[15] Slight modifications of the substrate can obtain the same result and many examples are available.[16] Several other groups can be present in the carbonyl-containing substrate.[5] For instance, the asymmetric reduction of keto groups in compounds containing a cyclopropyl moiety has been achieved (eq 5).[17]

BY 71% (5)

>96% ee

β-Keto esters are reduced to the corresponding hydroxy esters but, since more oxidoreductases are present in the yeast,[18] occasionally different stereochemistry or lowered enantioselectivity are observed. This is well illustrated by the stereochemical outcome of the reduction of a β-keto ester such as ethyl

4-chloroacetoacetate,[16b] when compared to ethyl acetoacetate (eq 6).[4a]

Cl–C(O)–CO_2Et → BY → Cl–CH(OH)–CO_2Et (S) 55% ee (6)

$CH_3C(O)CH_2CO_2Et$ → BY, 57–67% → $CH_3CH(OH)CH_2CO_2Et$ (S) 84–87% ee

Both γ- and δ-keto acids are reduced to hydroxy acids, which directly cyclize to the corresponding lactones in the incubation media.[19] The pheromone (*R*)-(+)-hexadecanolide has been prepared in this way by reduction of the corresponding δ-keto acid (eq 7).[20]

HO_2C…C(O)$C_{11}H_{23}$ → BY → HO_2C…CH(OH)$C_{11}H_{23}$ → H^+ → lactone ($C_{11}H_{23}$) (7)

overall 40%
>98% ee

α-Hydroxy ketones are good substrates for the bioreduction and several optically active 1,2-diols have been prepared.[21] The monobenzoate of dihydroxyacetone is reduced to the corresponding optically pure glycerol derivative (eq 8).[22] In many instances, simple protection of the α-hydroxy group may afford the opposite enantiomer.[23]

PhOCO–CH_2C(O)CH_2–OH → BY, 99% → PhOCO–CH_2CH(OH)CH_2–OH (8)

99% ee

Activated Double-Bond Hydrogenation. Fermenting BY is able to carry out the hydrogenation of double bonds which bear certain functional groups. A compound containing an unsaturated acetal and an ester function is directly transformed enantioselectively in a hydroxy acid, later chemically cyclized to the corresponding lactone (eq 9).[24] Other α,β-unsaturated alcohols and aldehydes are efficiently and enantioselectively converted to the corresponding saturated alcohols.[25] 2-Chloro-2-alkenoates (eq 10)[26] or nitroalkenes (eq 11)[27] are enantioselectively hydrogenated, the stereochemistry of the reaction depending on the double bond configuration.

MeO–CH(OMe)–C(Me)=CH–CO_2Et → BY → HO–…–CO_2Et → H^+ → lactone (9)

overall 34%
97% ee

Cl–CH(Cl)–CH=C(Cl)–CO_2Me → 1. BY; 2. CH_2N_2, 58-71% → Cl_2CH–CH_2–CH(Cl)–CO_2Me (10)

98% ee

Ph(R)C=CH–NO_2 → BY → Ph–CH(R)–CH_2–NO_2 (11)

R = Me, 50%, 98% ee
R = Et, 64%, 97% ee

Acyloin Condensations. The condensation between furfural or benzaldehyde and a two-carbon unit to afford a hydroxy ketone has long been known.[1b] The extension of the reaction to α,β-unsaturated aldehydes has provided access to optically active functionalized diols (eq 12),[28] which are used as chiral intermediates for the synthesis of natural products.[1b]

PhCH=C(R)–CHO → BY → PhCH=C(R)–CH(OH)–C(O)–CH_3 → PhCH=C(R)–CH(OH)–CH(OH)–CH_3 (12)

R = H, Me

overall 25%
85–95% ee

Cyclization of Squalene-like Substrates. Ultrasonically stimulated BY is a source of sterol cyclase, which catalyzes the cyclization of squalene oxide and squalenoid compounds to lanosterol derivatives (eq 13).[29]

epoxide–…–CO_2Me → BY (ultrasonically stimulated) 31% → HO–lanosterol–CO_2Me (13)

96% ee

Hydrolyses. The presence of hydrolytic enzymes in BY is well documented.[30] However, the use of the yeast for biocatalytic ester hydrolysis suffers because of the availability of commercially available purified hydrolases. Nonetheless, the hydrolytic ability of fermenting BY has been proposed for the resolution of various amino acid esters (eq 14).[31] The BY-mediated enantioselective hydrolysis has also been applied to the resolution of acetates of hydroxyalkynes[32] and a hydroxybutanolide.[33]

NHCOMe R–CH–CO_2Et (R,S), R = Me, Et, Bn —BY→ (R)-R–CH(NHCOMe)–CO_2Et <40%, 98–100% ee + (S)-R–CH(NHCOMe)–CO_2H >60% (14)

Oxidations. Reductions are by far the most exploited reactions carried out in the presence of BY. However, a few interesting examples of oxidations are available,[34] such as the dehydrogenation of thiastearates.[35] The regeneration of protected functional groups is possible with BY, which can effect the deprotection of hydrazones[36] and, if ultrasonically stimulated, may release from the oximes the corresponding carbonyl compounds, without further reduction to the alcohols (eq 15).[37]

R^1R^2C=NOH —BY→ R^1R^2C=O (15)

R^1 = Ph, R^2 = H, 96%
R^1 = C_5H_{11}, R^2 = H, 95%
R^1 = Et, R^2 = Me, 93%

An attractive reaction has been reported for BY, which is able in ethanol to catalyze the oxidative coupling of various thiols to disulfides (eq 16).[38]

PhSH —BY, 97%→ PhSSPh (16)

Miscellaneous Reactions. Many other reactions can be realized in the presence of BY. An interesting hydrolysis–reduction process transformed a derivative of secologanin into two different cyclic compounds, depending on the pH of the incubation.[39] Here, a glucosidase activity afforded the intermediate aldehyde, which could be reduced or rearranged to different products (eq 17).

(17) Secologanin derivative (dioxolane, MeO_2C, OGlucose) —BY→ intermediate dialdehyde (CHO, CHO); pH 7.5 → cyclic product (CHO, MeO_2C, OH); pH 6.4 → cyclic product (MeO_2C, OH)

With BY, the reduction of α,β-unsaturated aldehydes can act together with a hydration process, affording optically active diols in acceptable yields (eq 18).[40]

RO–CH₂–CH=CH–CHO —BY, 25%→ RO–CH₂–CH(OH)–CH₂–CH₂–OH 90% (18)

R = $PhCH_2$, PhCO

Some cycloaddition reactions have also been carried out in the presence of BY.[41] The asymmetric 1,3-dipolar cycloaddition of benzonitrile *N*-oxides to various dipolarophiles led to optically active 2-oxazolines (eq 19).[42]

Ar–C≡N→O + CH_2=CH–R —BY→ 3-Ar-5-R-isoxazoline (N–O) (19)

R = pyridine, 85%, 64% ee
R = carbazole, 78%, 51% ee

The regio- and enantioselectivity of the reactions depend on the structure of dipolarophiles and the addition of β-cyclodextrin. BY is also able to carry out a Diels–Alder condensation,[41] and a few Michael-type additions are enantioselectively performed in the presence of BY.[43] For example, the addition of amines to α,β-unsaturated esters affords optically active β-amino acid esters (eq 20).[44]

Ph–CH=CH–CO_2Et + $PhCH_2NH_2$ —BY, 70%→ Ph–CH($NHCH_2Ph$)–CH_2–CO_2Et 72.5% ee (20)

Related Reagents. See Classes R-3, R,-11, R-12, and R-15, pages 1–10.

1. (a) Sih, C. J.; Chen, C.-S. *AG(E)* **1984**, *23*, 570. (b) Servi, S. *S* **1990**, 1. (c) Csuk, R.; Glänzer, B. I. *CRV* **1991**, *91*, 49.
2. *Microbial Reagents in Organic Synthesis*; Servi, S., Ed.; Kluwer: Dordrecht, 1992.
3. Ward, O. P.; Young, C. S. *Enz. Microb. Technol.* **1990**, *12*, 482.
4. (a) Seebach, D.; Sutter, M. A.; Weber, R. H.; Züger, M. F. *OS* **1985**, *63*, 1. (b) Mori, K.; Mori, H. *OS* **1990**, *68*, 56.
5. Santaniello, E.; Ferraboschi, P.; Grisenti, P.; Manzocchi, A. *CRV* **1992**, *92*, 1071.
6. (a) MacLeod, R.; Prosser, H.; Fikentscher, L.; Lanyi, J.; Mosher, H. S. *B* **1964**, *3*, 838. (b) Červinka, O.; Hub, L. *CCC* **1966**, *31*, 2615.
7. Prelog, V. *PAC* **1964**, *9*, 119.
8. (a) Top, S.; Jaouen, G.; Gillois, J.; Baldoli, C.; Maiorana, S. *CC* **1988**, 1284. (b) Top, S.; Jaouen, G.; Baldoli, C.; Del Buttero, P.; Maiorana, S. *JOM* **1991**, *413*, 125.
9. Gillois, J.; Jaouen, G.; Buisson, D.; Azerad, R. *JOM* **1989**, *367*, 85.
10. (a) Ticozzi, C.; Zanarotti, A. *LA* **1989**, 1257. (b) Itoh, T.; Fukuda, T.; Fujisawa, T. *BCJ* **1989**, *62*, 3851. (c) Fujisawa, T.; Yamanaka, K.; Mobele, B. I.; Shimizu, M. *TL* **1991**, *32*, 399.
11. Fouché, G.; Horak, R. M.; Meth-Cohn, O. *Molecular Mechanisms in Bioorganic Processes*; Bleasdale, C.; Golding, B. T., Eds.; Royal Society of Chemistry: London, 1990; p 350.
12. (a) Haag, T.; Arslan, T.; Seebach, D. *C* **1989**, *43*, 351. (b) Nakamura, K.; Kondo, S.; Kawai, Y.; Ohno, A. *TL* **1991**, *32*, 7075.
13. (a) Nakamura, K.; Kawai, Y.; Miyai, T.; Ohno, A. *TL* **1990**, *31*, 3631. (b) Nakamura, K.; Kawai, Y.; Ohno, A. *TL* **1990**, *31*, 267. (c) Nakamura, K.; Kawai, Y.; Oka, S.; Ohno, A. *BCJ* **1989**, *62*, 875. (d) Ushio, K.; Ebara, K.; Yamashita, T. *Enz. Microb. Technol.* **1991**, *13*, 834.
14. Spiliotis, V.; Papahatjis, D.; Ragoussis, N. *TL* **1990**, *31*, 1615.
15. (a) Nakamura, K.; Inoue, K.; Ushio, K.; Oka, S.; Ohno, A. *JOC* **1988**, *53*, 2589. (b) Nakamura, K.; Kawai, Y.; Oka, S.; Ohno, A. *TL* **1989**, *30*, 2245. (c) Naoshima, Y.; Maeda, J.; Munakata, Y. *JCS(P1)* **1992**, 659.
16. (a) Nakamura, K.; Ushio, K.; Oka, S.; Ohno, A.; Yasui, S. *TL* **1984**, *25*, 3979. (b) Zhou, B.-N.; Gopalan, A. S.; VanMiddlesworth, F.; Shieh, W.-R.; Sih, C. J. *JACS* **1983**, *105*, 5925.

17. Tkachev, A. V.; Rukavishnikov, A. V.; Gatilov, Y. V.; Bagrjanskaja, I. Yu. *TA* **1992**, *3*, 1165.
18. Shieh, W.-R.; Gopalan, A. S.; Sih, C. J. *JACS* **1985**, *107*, 2993.
19. (a) Muys, G. T.; Van der Ven, B.; de Jonge, A. P. *Nature* **1962**, *194*, 995. (b) Gessner, M.; Günther, C.; Mosandl, A. *ZN(C)* **1987**, *42c*, 1159. (c) Utaka, M.; Watabu, H.; Takeda, A. *JOC* **1987**, *52*, 4363. (d) Aquino, M.; Cardani, S.; Fronza, G.; Fuganti, C.; Pulido-Fernandez, R.; Tagliani, A. *T* **1991**, *47*, 7887.
20. Utaka, M.; Watabu, H.; Takeda, A. *CL* **1985**, 1475.
21. (a) Levene, P. A.; Walti, A. *OSC* **1943**, *2*, 545. (b) Guetté, J.-P.; Spassky, N. *BSF(2)* **1972**, 4217. (c) Barry, J.; Kagan, H. B. *S* **1981**, 453. (d) Kodama, M.; Minami, H.; Mima, Y.; Fukuyama, Y. *TL* **1990**, *31*, 4025. (e) Ramaswamy, S.; Oehlschlager, A. C. *T* **1991**, *47*, 1145.
22. Aragozzini, F.; Maconi, E.; Potenza, D.; Scolastico, C. *S* **1989**, 225.
23. (a) Manzocchi, A.; Fiecchi, A.; Santaniello, E. *JOC* **1988**, *53*, 4405. (b) Ferraboschi, P.; Grisenti, P.; Manzocchi, A.; Santaniello, E. *JCS(P1)* **1990**, 2469.
24. Leuenberger, H. G. W.; Boguth, W.; Barner, R.; Schmid, M; Zell, R. *HCA* **1979**, *62*, 455.
25. (a) Gramatica, P.; Manitto, P.; Poli, L. *JOC* **1985**, *50*, 4625. (b) Gramatica, P.; Manitto, P.; Monti, D.; Speranza, G. *T* **1986**, *42*, 6687. (c) Gramatica, P.; Manitto, P.; Monti, D.; Speranza, G. *T* **1988**, *44*, 1299. (d) Fuganti, C.; Grasselli, P.; Servi, S.; Högberg, H.-E. *JCS(P1)* **1988**, 3061. (e) Högberg, H.-E.; Hedenström, E.; Fägerhag, J.; Servi, S. *JOC* **1992**, *57*, 2052.
26. Utaka, M.; Konishi, S.; Mizuoka, A.; Ohkubo, T.; Sakai, T.; Tsuboi, S.; Takeda, A. *JOC* **1989**, *54*, 4989.
27. Ohta, H.; Kobayashi, N.; Ozaki, K. *JOC* **1989**, *54*, 1802.
28. Fuganti, C.; Grasselli, P. *CI(L)* **1977**, 983.
29. (a) Bujons, J.; Guajardo, R.; Kyler, K. S. *JACS* **1988**, *110*, 604. (b) Medina, J. C.; Kyler, K. S. *JACS* **1988**, *110*, 4818. (c) Medina, J. C.; Guajardo, R.; Kyler, K. S. *JACS* **1989**, *111*, 2310. (d) Xiao, X.-Y.; Prestwich, G. D. *TL* **1991**, *32*, 6843.
30. (a) Rose, A. H. *The Yeast*; Harrison, J. S., Ed.; Academic Press: London, 1969, Vol. I; 1971, Vol III. (b) Glänzer, B. I.; Faber, K.; Griengl. H.; Roehr, M.; Wöhrer, W. *Enz. Microb. Technol.* **1988**, *10*, 744.
31. (a) Glänzer, B. I.; Faber, K.; Griengl, H. *TL* **1986**, *27*, 4293. (b) Glänzer, B. I.; Faber, K.; Griengl, H. *T* **1987**, *43*, 771.
32. Glänzer, B. I.; Faber, K.; Griengl, H. *T* **1987**, *43*, 5791.
33. Glänzer, B. I.; Faber, K.; Griengl, H. *Enz. Microb. Technol.* **1988**, *10*, 689.
34. Sato, T.; Hanayama, K.; Fujisawa, T. *TL* **1988**, *29*, 2197.
35. (a) Buist, P. H.; Dallmann, H. G.; Rymerson, R. T.; Seigel, P. M. *TL* **1987**, *28*, 857. (b) Buist, P. H.; Dallmann, H. G. *TL* **1988**, *29*, 285. (c) Buist, P. H.; Dallmann, H. G.; Rymerson, R. T., Seigel, P. M.; Skala, P. *TL* **1988**, *29*, 435.
36. Kamal, A.; Rao, M. V.; Meshram, H. M. *TL* **1991**, *32*, 2657.
37. Kamal, A.; Rao, M. V.; Meshram, H. M. *JCS(P1)* **1991**, 2056.
38. Rao, K. R.; Sampath Kumar, H. M. *BML* **1991**, *1*, 507.
39. Brown, R. T.; Dauda, B. E. N.; Santos, C. A. M. *CC* **1991**, 825.
40. Fronza, G.; Fuganti, C.; Grasselli, P.; Poli, G.; Servi, S. *JOC* **1988**, *53*, 6153.
41. Rao, K. R.; Srinivasan, T. N.; Bhanumathi, N. *TL* **1990**, *31*, 5959.
42. (a) Rao, K. R.; Bhanumathi, N.; Sattur, P. B. *TL* **1990**, *31*, 3201. (b) Rao, K. R.; Bhanumathi, N.; Srinivasan, T. N.; Sattur, P. B. *TL* **1990**, *31*, 899. (c) Rao, K. R.; Nageswar, Y. V. D.; Sampathkumar, H. M. *JCS(P1)* **1990**, 3199.
43. Kitazume, T.; Ishikawa, N. *CL* **1984**, 1815.
44. Rao, K. R.; Nageswar, Y. V. D.; Sampath Kumar, H. M. *TL* **1991**, *32*, 6611.

Enzo Santaniello, Patrizia Ferraboschi, & Paride Grisenti
Università di Milano, Italy

1,4-Benzoquinone

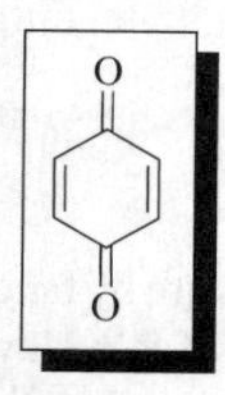

[106-51-4] $C_6H_4O_2$ (MW 108.10)

(useful as an oxidizing[1] or dehydrogenation agent;[2] can function as a dienophile[3] in the Diels–Alder reaction or as a dipolarophile to prepare 5-hydroxyindole derivatives[4])

Alternate Name: *p*-benzoquinone.
Physical Data: mp 115.7 °C; *d* 1.318 g cm^{-3}.
Solubility: slightly sol water; sol alcohol, ether, hot petroleum ether, and aqueous base.
Form Supplied in: yellowish powder; widely available.
Handling, Storage, and Precautions: the solid has an irritating odor and can cause conjunctivitis, corneal ulceration, and dermatitis. In severe cases, there can be necrotic changes in the skin. Use in a fume hood.

As Oxidizing or Dehydrogenation Agent. The ease of reduction of 1,4-benzoquinone to hydroquinone by various compounds renders it useful as an oxidizing or dehydrogenation agent.[2,5] The literature shows that 1,4-benzoquinone prefers to oxidize conjugated primary allylic alcohols over other alcohols (eq 1).[6] Kulkarni has demonstrated the selective oxidation of cinnamyl alcohols to cinnamaldehydes in the presence of secondary or benzylic alcohols. Primary alcohols have been oxidized to the corresponding aldehydes by using 1,4-benzoquinone as the hydrogen acceptor and hydrous zirconium(IV) oxide[7] as a catalyst (eqs 2 and 3).[1]

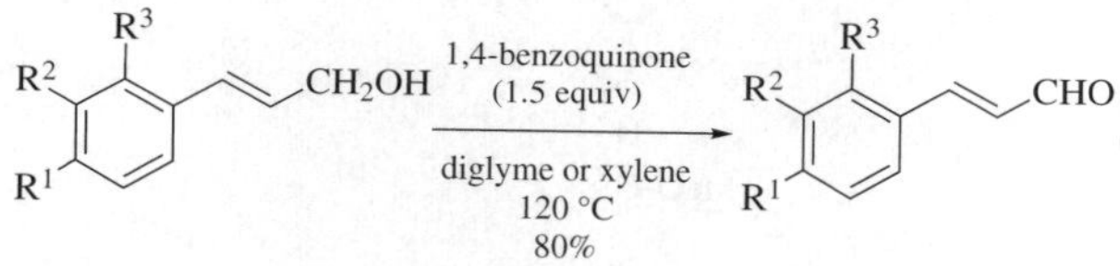

R³, R², R¹-substituted cinnamyl alcohol (CH_2OH) → [1,4-benzoquinone (1.5 equiv); diglyme or xylene, 120 °C; 80%] → R³, R², R¹-substituted cinnamaldehyde (CHO) (1)

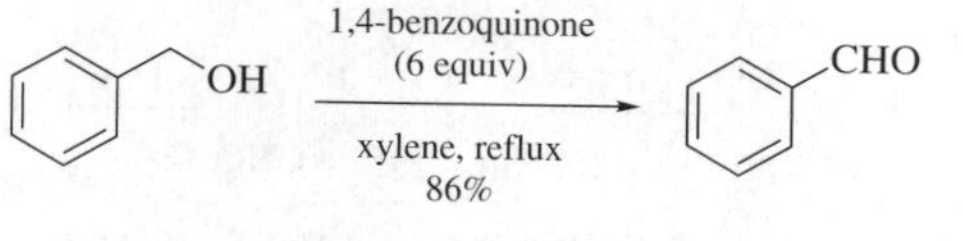

Benzyl alcohol (OH) → [1,4-benzoquinone (6 equiv); xylene, reflux; 86%] → benzaldehyde (CHO) (2)

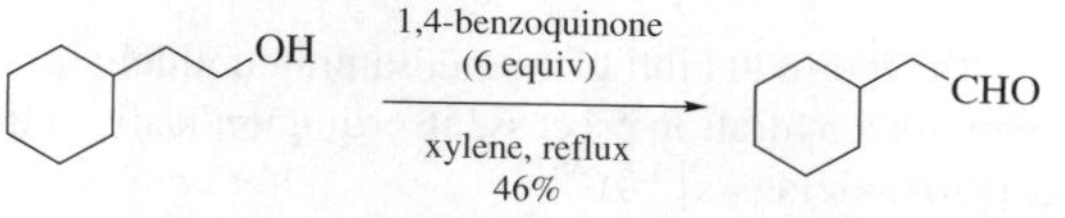

2-Cyclohexylethanol (OH) → [1,4-benzoquinone (6 equiv); xylene, reflux; 46%] → cyclohexylacetaldehyde (CHO) (3)

Nitrogenous compounds can be oxidized by 1,4-benzoquinone in refluxing benzene.[8,9] Aurich reported the conversion of hydroxylamine (**1**) to nitrone (**2**) through the reaction with 1,4-

benzoquinone (eq 4).[9] Wiberg used 1,4-benzoquinone to transform tetrazene (**3**) into molecular nitrogen (eq 5).[10] Rossazza reported the oxidation of leurosine at the carbon α to nitrogen (eq 6).[11]

(1) → 1,4-benzoquinone, benzene, rt, 47% → (2) (4)

(3) → 1,4-benzoquinone (2 equiv), benzene, 80 °C, ~100% → 2 TMSO–C6H4–OTMS + 2 N_2 (5)

1,4-benzoquinone, 0.2 M potassium phosphate buffer (pH = 7.0), 50% (6)

In the oxidation of alkenes with palladium(II) acetate, 1,4-benzoquinone serves as a cooxidant to reoxidize palladium(0) to palladium(II).[12,13] Davidson reported the oxidation of alkene (**4**) to a vinyl acetate with 0.1 equiv of palladium acetate and 1 equiv of 1,4-benzoquinone (eq 7).[14]

(4) → $Pd(OAc)_2$, 1,4-benzoquinone, AcOH, 25 °C, 50% (7)

Backvall reported that the reaction of 1,3-cyclohexadiene with ***Palladium(II) Acetate*** and 1,4-benzoquinone gave 1,4-diacetoxy-2-cyclohexene in high yield. The stereochemistry of the products was influenced by additives. In the presence of lithium acetate, the major *trans*-diacetate was obtained in 90% yield.[15] Without the addition of lithium acetate, a 1:1 mixture of *trans* and *cis* isomers was produced. However, the *cis*-diacetate was the major (>93%) product when both lithium acetate and lithium chloride were added (eq 8).[15] When the reaction was carried out in acetic acid containing trifluoroacetic acid and lithium trifluoroacetate, the *trans* isomer of 1-acetoxy-4-trifluoroacetoxy-2-cyclohexene was the major product in 67% yield (eq 9).[16] Similar oxidative 1,4-additions of other dienes, such as 1,3-butadiene, also can be accomplished in a regio- and stereoselective fashion (eq 10).[17] However, 1,4-benzoquinone has a high tendency to undergo Diels–Alder reactions when used as a cooxidant for $Pd(OAc)_2$ in such reactions. Many reactions of this type use only catalytic amounts of 1,4-benzoquinone and stoichiometric amounts of external oxidants, such as $Ce(SO_4)_2$, $Tl(OAc)_3$, and MnO_2.[18]

cat. $Pd(OAc)_2$, 1,4-benzoquinone, AcOH, LiOAc, 68% → >90% *trans*; cat. $Pd(OAc)_2$, 1,4-benzoquinone, 74% → >93% *cis* (8)

$Pd(OAc)_2$ (5 mol%), 1,4-benzoquinone, CF_3CO_2H, LiO_2CF_3, AcOH, 67% → CF_3CO_2–C6H8–OAc (9)

$Pd(OAc)_2$, 1,4-benzoquinone, AcOH, LiOAc, LiCl, 78% → Cl–CH2CH=CHCH2–OAc (10)

The combination of 1,4-benzoquinone and a catalytic amount of palladium acetate can transform silyl enol ethers into conjugated enones.[19] This reaction is not only regiospecific (eq 11) but also stereospecific to give the more stable *trans* acyclic enone (eq 12).

$Pd(OAc)_2$ (0.5 equiv), 1,4-benzoquinone (0.5 equiv); 94%; 85% (11)

E/Z mixture → $Pd(OAc)_2$, 1,4-benzoquinone, 97% → only *E* isomer (12)

Antonsson used ***Manganese Dioxide*** as the oxidant and 0.05 equiv of palladium acetate and 0.20 equiv of 1,4-benzoquinone as catalyst for oxidative ring closure of 1,5-hexadienes to give cyclopentane derivatives in good yield (eq 13).[20] With 1,4-benzoquinone and 10 mol % of ***Palladium(II) Chloride***, a homoallylic alcohol underwent oxidative ring closure to give a γ-butyrolactol (eq 14).[21]

$Pd(OAc)_2$ (0.05 equiv), 1,4-benzoquinone (0.2 equiv), MnO_2 (1 equiv), AcOH, 70% (13)

$PdCl_2$ (1 mol%), 1,4-benzoquinone (3 equiv), wet DMF, 64% (14)

The combination of palladium acetate and 1,4-benzoquinone was used in the ring opening of an α,β-epoxysilane to give 19%

of oct-2-enal in the presence of oxygen gas (eq 15).[22] Similar reagents are effective in the oxidative coupling of carbon monoxide in methanol to give methyl oxalate (eq 16).[12]

TMS; $Pd(OAc)_2$; 1,4-benzoquinone; O_2, DMF; 19%; CHO (15)

CO + MeOH; $Pd(OAc)_2$ (1 mol%); 1,4-benzoquinone; PPh_3 (3 mol%); 83%; MeO; OMe (16)

As Dienophile. 1,4-Benzoquinone is an excellent dienophile in the Diels–Alder reaction toward electron rich dienes.[23,24] The bicyclic cycloadducts of these reactions are frequently used as the starting materials for the synthesis of natural products. A typical example is demonstrated by Mehta in a total synthesis of capnellene, in which 1,4-benzoquinone is allowed to react with 1-methyl-1,3-cyclopentadiene to give compound (**5**) (eq 17).[25] The cycloaddition of 1,4-benzoquinone and another activated diene is shown in eq 18.[26]

THF, rt; 70%; (**5**); H (17)

H_3O^+; 75~80%; OTMS; H; OR (18)

Asymmetric Diels–Alder reactions of 1,4-benzoquinone have been reported. Several examples are shown in eqs 19–21.[3,27,28]

OTMS; OR; benzene; rt; H; OTMS; OR (19); 89:11

R = OAc; OAc; OAc OAc

MeOCH$_2$O; toluene; 80 °C; 58%; X

X = OTBDMS; MeOCH$_2$O; 75:25 (20)

toluene; 25 °C; OR; 15.7:1 (21)

R = HO

Under high pressure, even electron deficient dienes react with 1,4-benzoquinone. Dauben reported the asymmetric cycloaddition of a chiral dienic ester and 1,4-benzoquinone to produce chiral adducts with moderate enantioselectivity (eq 22).[29]

R; 15 kbar; 20 °C; 62 %; COR; H; 50% ee (22)

R = Ph; O

Preparation of 5-Hydroxyindoles. The reaction of 1,4-benzoquinone with certain enamines has been used to prepare 5-hydroxyindole derivatives.[4,30] The first example was reported by Nenitzescu, in which he successfully prepared ethyl 5-hydroxy-2-methylindole-3-carboxylate from 1,4-benzoquinone and ethyl 3-aminocrotonate (eq 23).[4] The ease of the reaction and the availability of the reactants have made it a popular method for indole synthesis. However, the yields of this type of reaction vary over a wide range (5–90%).

H; CO_2Et; H_2N; HO; N; + H_2O (23)

As Reaction Promoter. In the hydrogenation of nitrobenzene with platinum catalyst in DMSO, addition of 1,4-benzoquinone accelerates the reaction rate.[31]

Related Reagents. See Class O-1, pages 1–10.

1. Kuno, H.; Shibagaki, M.; Takahashi, K.; Matsushita, H. *BCJ* **1991**, *64*, 312.
2. Walker, D.; Hiebert, J. D. *CRV* **1967**, *67*, 153.
3. Tripathy, R.; Carroll, P. J.; Thornton, E. R. *JACS* **1990**, *112*, 6743.
4. Nenitzescu, C. D. *Bull. Soc. Chim. Rom.* **1929**, *11*, 37 (*CA* **1930**, *24*, 110).
5. Turner, A. B.; Ringold, H. J. *JCS(C)* **1967**, 1720.
6. Kulkarni, M. G.; Mathew, T. S. *TL* **1990**, *31*, 4497.
7. Shibagaki, M.; Takahashi, K.; Matsushita, H. *BCJ* **1988**, *61*, 3283.
8. Fujita, S.; Sano, K. *JOC* **1979**, *44*, 2647.
9. Aurich, H. G.; Mobus, K. D. *T* **1989**, *45*, 5815.
10. Wiberg, N.; Bayer, H.; Vasisht, S. K.; Meyers, R. *CB* **1980**, *113*, 2916.
11. Goswami, A.; Schaumberg, J. P.; Duffel, M. W.; Rosazza, J. P. *JOC* **1987**, *52*, 1500.
12. Current, S. P. *JOC* **1983**, *48*, 1779.
13. Backvall, J.-E. *ACR* **1983**, *16*, 335.
14. Brown, R. G.; Chaudhari, R. V.; Davidson, J. M. *JCS(D)* **1977**, 183.
15. Backvall, J.-E.; Nordberg, R. E. *JACS* **1981**, *103*, 4959.
16. Backvall, J.-E.; Vagberg, J.; Nordberg, R. E. *TL* **1984**, *25*, 2717.
17. (a) Bäckvall, J.-E.; Nordberg, R. E.; Nyström, J. E. *TL* **1982**, *23*, 1617. (b) Backvall, J. E.; Nyström, J.-E.; Nordberg, R. E. *JACS* **1985**, *107*, 3676.
18. Backvall, J.-E.; Bystrom, S. E.; Nordberg, R. E. *JOC* **1984**, *49*, 4619.
19. Ito, Y.; Hirao, T.; Saegusa, T. *JOC* **1978**, *43*, 1011.
20. Antonsson, T.; Heumann, A.; Moberg, C. *CC* **1986**, 518.
21. Nokami, J.; Ogawa, H.; Miyamoto, S.; Mandai, T.; Wakabayashi, S.; Tsuji, J. *TL* **1988**, *29*, 5181.
22. Hirao, T.; Murakami, T.; Ohno, M.; Ohshiro, Y. *CL* **1991**, 299.
23. (a) Marchand, A. P.; Allen, R. W. *JOC* **1974**, *39*, 1596. (b) Hill, R. K.; Newton, M. G.; Pantaleo, N. S.; Collins, K. M. *JOC* **1980**, *45*, 1593. (c) Jurczak, J.; Kozluk, T.; Filipek, S.; Eugster, C. H. *HCA* **1983**, *66*, 222. (d) Kozikowski, A. P.; Hiraga, K.; Springer, J. P.; Wang, B. C.; Xu, Z.-B. *JACS* **1984**, *106*, 1845. (e) Burnell, D. J.; Valenta, Z. *CC* **1985**, 1247. (f) Pandey, B.; Zope, U. R.; Ayyangar, N. R. *SC* **1989**, *19*, 585.
24. Danishefsky, S.; Craig, T. A. *T* **1981**, *37*, 4081.
25. Mehta, G.; Reddy, D. S.; Murty, A. N. *CC* **1983**, 824.
26. Krohn, K. *TL* **1980**, *21*, 3557.
27. Gupta, R. C.; Raynor, C. M.; Stoodley, R. J.; Slawin, A. M. Z.; Williams, D. J. *JCS(P1)* **1988**, 1773.
28. McDougal, P. G.; Jump, J. M.; Rojas, C.; Rico, J. G. *TL* **1989**, *30*, 3897.
29. Dauben, W. G.; Bunce, R. A. *TL* **1982**, *23*, 4875.
30. Monti, S. A. *JOC* **1966**, *31*, 2669.
31. Kushch, S. D.; Izakovich, E. N.; Khidekel, M. L.; Strelets, V. V. *IZV* **1981**, *7*, 1500 (*CA* **1981**, *95*, 149 595s).

Teng-Kuei Yang & Chi-Yung Shen
National Chung-Hsing University, Taichung, Republic of China

1,2-Bis(2,5-diethylphospholano)benzene

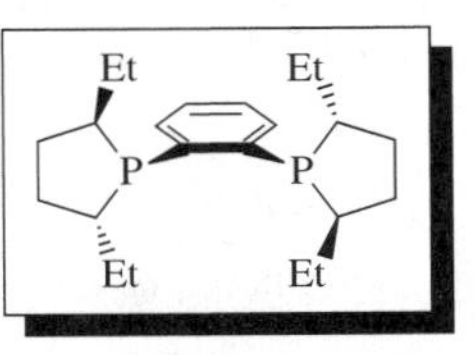

[136705-64-1] $C_{22}H_{36}P_2$ (MW 362.47)

(ligand for asymmetric catalysis;[1] rhodium complexes are efficient catalyst precursors for highly enantioselective hydrogenation of α-(*N*-acylamino)acrylates,[1,2] enol acetates,[1] and *N*-acylhydrazones[3,4])

Alternate Name: (*R,R*)- and (*S,S*)-Ethyl-DuPHOS.
Physical Data: bp 138–143 °C/0.045 mmHg; (*R,R*)-Ethyl-DuPHOS, $[\alpha]_D^{25} = -265°$ (*c* 1, hexane).
Form Supplied in: colorless viscous liquid; both enantiomers available commercially.
Analysis of Reagent Purity: optical rotation, ^{1}H NMR, ^{31}P NMR, ^{13}C NMR.
Preparative Methods: preparation of (*R,R*)-Ethyl-DuPHOS requires the use of (3*S*,6*S*)-3,6-octanediol.[1,2] Enantiomerically pure (3*S*,6*S*)-octanediol is obtained in 35–45% overall yields via a simple three-step procedure[5,6] involving the Ru–(*S*)-BINAP-catalyzed asymmetric hydrogenation[7] of methyl propionylacetate to methyl (3*S*)-3-hydroxypentanoate (95% yield, 99% ee), followed by quantitative hydrolysis to the corresponding β-hydroxy acid, and subsequent electrochemical Kolbe coupling (eq 1). This sequence has been used for the synthesis of multigram quantities of (3*S*,6*S*)-octanediol, as well as a series of related chiral 1,4-diols.[2,6] Antipodal (3*R*,6*R*)-octanediol was prepared in a similar fashion by employing the (*R*)-BINAP–Ru catalyst in the first step.[6]

Et–CO–CH₂–CO_2Me → (1. Ru–(*S*)-BINAP, H_2; 2. KOH)

Et–CH(OH)–CH₂–CO_2H → (Kolbe coupling; 35–45% overall yield) (3*S*,6*S*)-octanediol (1)

The crystalline (3*S*,6*S*)-octanediol (mp 51–52 °C; $[\alpha]_D^{25} = +22.8°$ (*c* 1, $CHCl_3$)) is next converted to the corresponding (3*S*,6*S*)-octanediol cyclic sulfate (mp 80–81 °C; $[\alpha]_D^{25} = +28.6°$ (*c* 1, $CHCl_3$)) through reaction with ***Thionyl Chloride***, followed by oxidation with ***Sodium Periodate*** and a catalytic amount (0.1 mol%) of ***Ruthenium(III) Chloride*** (eq 2).[2] The final step involves successively treating 1,2-bis(phosphino)benzene[8] with ***n-Butyllithium*** (2 equiv 1.6 M in hexane), followed by (3*S*,6*S*)-octanediol cyclic sulfate (2 equiv), and then *n*-BuLi (2.2 equiv) to provide the product (*R,R*)-Ethyl-DuPHOS in 78% yield after purification by distillation (eq 2). The use of (3*R*,6*R*)-octanediol cyclic sulfate in eq 2 allows the analogous preparation of (*S,S*)-Ethyl-DuPHOS. In addition to Ethyl-DuPHOS (R = Et), a series of other DuPHOS derivatives (R = Me, Pr, *i*-Pr, Cy, Bn) have been prepared in this manner.[1,2]

1. $SOCl_2$; 2. $RuCl_3$, $NaIO_4$

1. 2 BuLi; 2. 2 x cyclic sulfate; 3. 2 BuLi

(*R,R*)-Et-DuPHOS (2)

Handling, Storage, and Precautions: somewhat air sensitive and should be handled and stored in a nitrogen or argon atmosphere. Metal complexes generally are sensitive to oxygen in solution. Use in a fume hood.

Catalyst Precursors: Rhodium Complexes. The cationic rhodium complexes $[(cod)Rh(Ethyl\text{-}DuPHOS)]^+X^-$ (X = OTf, PF_6, BF_4, SbF_6) serve as efficient catalyst precursors for both enantioselective hydrogenation[1–6] and intramolecular hydrosilylation[9] reactions. These complexes are most conveniently prepared by reacting the ligand, either (*R,R*)-Ethyl-DuPHOS or (*S,S*)-Ethyl-DuPHOS, with the complexes $[(cod)_2Rh]^+X^-$ in THF.[2,10] Since the solid rhodium catalysts are less air-senstive than the ligands, they may be weighed quickly in air, although storage under nitrogen or argon is recommended.

Enantioselective Hydrogenations.

α-(N-Acylamino)acrylates. The cationic Ethyl-DuPHOS–Rh catalysts are particularly well-suited for highly enantioselective hydrogenation of α-(*N*-acylamino)acrylates to α-amino acid derivatives (eq 3).[1,2]

$R^1CH{=}C(CO_2R^3)N(H)COR^2 \xrightarrow[H_2\ (1\text{–}2\ atm)]{(R,R)\text{-Et-DuPHOS–Rh cat.}} R^1CH_2CH(CO_2R^3)N(H)COR^2$ (3)

The reactions proceed under mild conditions (1 atm H_2, 25 °C, MeOH) and are extremely efficient (substrate-to-catalyst ratios S/C up to 50 000 have been demonstrated). The breadth of the Ethyl-DuPHOS–Rh catalyst is noteworthy; extremely high enantioselectivities (≥99% ee) are achieved over a broad range of substrates (Table 1). Accordingly, the Ethyl-DuPHOS–Rh catalysts can provide practical access to a wide variety of natural, unnatural, nonproteinaceous, and labeled α-amino acids. The absolute configurations of the products are very predictable; (*R,R*)-Ethyl-DuPHOS–Rh complexes consistently afford products of (*R*) absolute configuration, while (*S,S*)-Ethyl-DuPHOS–Rh complexes provide (*S*)-α-amino acid derivatives. Similarly high ees are obtained with the corresponding carboxylic acid substrates (R^3 = H), as well as with *N*-benzoyl (R^2 = Ph) and *N*-Cbz (R^2 = OBn) α-(*N*-acylamino)acrylates. Significantly, the Ethyl-DuPHOS–Rh catalysts allow hydrogenation of both (*Z*) and (*E*) isomeric α-(*N*-acetylamino)acrylates in high enantiomeric excess (>99% ee) to afford products with the same absolute configuration. Many desirable α-(*N*-acylamino)acrylates are unavoidably synthesized as a mixture of (*E*) and (*Z*) isomers, and a separation step generally is required prior to hydrogenation. When using the Ethyl-DuPHOS–Rh catalysts, however, the need to separate isomeric substrates is often eliminated, thus providing a practical route to many amino acid derivatives.

Table 1 Asymmetric Hydrogenation of α-(*N*-acetamido)acrylates (R^2,R^3 = Me) with Ethyl-DuPHOS–Rh Catalyst

R^1	% ee
H	99.8
Me	99.6
Et	99.7
Pr	99.6
i-Pr	99.4
t-Bu	96.2
Ph	>99
1-Naphthyl	>99
2-Naphthyl	>99
2-Thienyl	>99
Ferrocenyl	>99

Enol Acetates. Several enol acetates are hydrogenated with high enantioselectivities using the Ethyl-DuPHOS–Rh catalysts (eqs 4 and 5).[1] The selectivities are significantly higher than any previously reported for these substrates.[11] The scope of these reactions and substrates bearing β-substituents have not yet been examined.

$EtO_2C\text{–}C(=CH_2)OAc \xrightarrow[H_2]{(R,R)\text{-Et-DuPHOS–Rh cat.}} EtO_2C\text{–}CH(CH_3)OAc$ (4)
>99% ee

$3\text{-}ClC_6H_4\text{–}C(=CH_2)OAc \xrightarrow[H_2]{(R,R)\text{-Et-DuPHOS–Rh cat.}} 3\text{-}ClC_6H_4\text{–}CH(CH_3)OAc$ (5)
91% ee

N-Benzoylhydrazones. The C=N double bond of *N*-benzoylhydrazones may be hydrogenated with high enantioselectivities using the Ethyl-DuPHOS–Rh catalyst systems (eq 6).[3,4]

$R^1R^2C{=}N\text{–}NHC(O)Ph \xrightarrow[H_2]{(R,R)\text{-Et-DuPHOS–Rh cat.}} R^1R^2CH\text{–}NH\text{–}NHC(O)Ph$ (6)

Under optimized conditions (0 °C, 60 psi H_2, *i*-PrOH, S/C 500), the (*R,R*)-Et-DuPHOS–Rh catalyst predictably provides a variety of *N*-benzoylhydrazine products in high enantiomeric excess and with (*S*) absolute configuration (Table 2).

An interesting and potentially useful property of the Et-DuPHOS–Rh catalyst system is the high level of chemoselectivity exhibited in the hydrogenation of *N*-benzoylhydrazones. Little or no reduction of various functional groups including alkenes, alkynes, ketones, aldehydes, nitriles, imines, carbon–halogen, and nitro groups is observed under conditions required for complete reduction of the hydrazones.

Table 2 Asymmetric Hydrogenation of *N*-Benzoylhydrazones with Ethyl-DuPHOS–Rh Catalyst

R^1	R^2	% ee
Ph	Me	95
Ph	CH_2Ph	84
p-$NO_2C_6H_4$	Me	97
p-BrC_6H_4	Me	96
2-Naphthyl	Me	95
CO_2Et	Me	89
CO_2Me	Et	91
CO_2Me	Ph	90
$P(O)(OEt)_2$	Ph	90
Cy	Me	72
Et	Me	43

Enantioselective hydrogenation of *N*-benzoylhydrazones represents a key transformation in a three-step asymmetric catalytic reductive amination process that involves: (1) treating a ketone with benzoic acid hydrazide; (2) Ethyl-DuPHOS–Rh-catalyzed hydrogenation of the *N*-benzoylhydrazone; and (3) hydrolysis of the product *N*-benzoyl group (3 M HCl) to provide the corresponding hydrazine, or reductive cleavage of the product N–N bond with ***Samarium(II) Iodide*** to directly afford the corresponding amine (eq 7).[4]

NH_2 R^1 R^2 ← SmI_2 — H–N(H)–N–C(O)Ph (R^1, R^2) — 3M HCl → N(H)NH_2 R^1 R^2 (7)

Related Reagents. See Classes R-3, R-9, R-10, R-11, R-15, and R-18, pages 1–10.

1. Burk, M. J. *JACS* **1991**, *113*, 8518.
2. Burk, M. J.; Feaster, J. E.; Nugent, W. A.; Harlow, R. L. *JACS* **1993**, *115*, 10 125.
3. Burk, M. J.; Feaster, J. E.; *JACS* **1992**, *114*, 6266.
4. Burk, M. J.; Feaster, J. E.; Cosford, N.; Martinez, J. P. *T* **1994**, *50*, 4399.
5. Burk, M. J.; Feaster, J. E.; Harlow, R. L. *OM* **1990**, *9*, 2653.
6. Burk, M. J.; Feaster, J. E.; Harlow, R. L. *TA* **1991**, *2*, 569.
7. (a) Noyori, R.; Ohkuma, T.; Kitamura, M.; Takaya, H.; Sayo, N.; Kumobayashi, H.; Akutagawa, S. *JACS* **1987**, *109*, 5856. (b) Kitamura, M.; Ohkuma, T.; Inoue, S.; Sayo, N.; Kumobayashi, H.; Akutagawa, S.; Ohta, T.; Takaya, H.; Noyori, R. *JACS* **1988**, *110*, 629. (c) Kawano, H.; Ikariya, T.; Ishii, Y.; Saburi, M.; Yoshikawa, S.; Uchida, Y.; Kumobayashi, H. *JCS(P1)* **1989**, 1571.
8. Kyba, E. P.; Liu, S.-T.; Harris, R. L. *OM* **1983**, *2*, 1877.
9. Burk, M. J.; Feaster, J. E. *TL* **1992**, *33*, 2099.
10. Schrock, R. R.; Osborn, J. A. *JACS* **1971**, *93*, 2397.
11. Koenig, K. E.; Bachman, G. L.; Vineyard, B. D. *JOC* **1980**, *45*, 2362.

Mark J. Burk
Duke University, Durham, NC, USA

(*R*)- & (*S*)-2,2′-Bis(diphenylphosphino)-1,1′-binaphthyl[1]

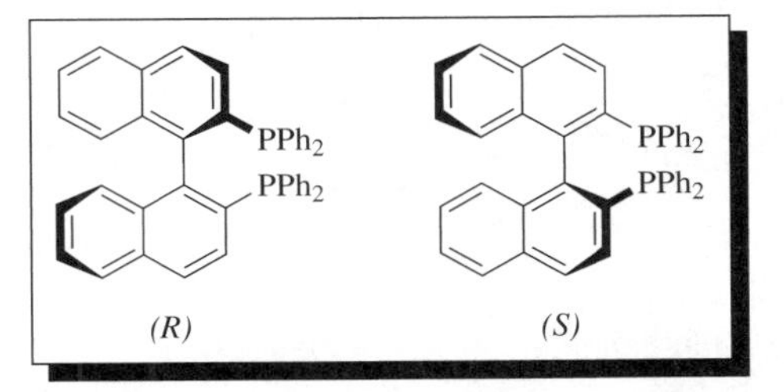

[76189-55-4] $C_{44}H_{32}P_2$ (MW 622.70)

(chiral diphosphine ligand for transition metals;[2] the complexes show high enantioselectivity and reactivity in a variety of organic reactions)

Alternate Name: BINAP.
Physical Data: mp 241–242 °C; $[\alpha]_D^{25}$ −229° (sc = 0.312, benzene) for (*S*)-BINAP.[3]
Solubility: sol THF, benzene, dichloromethane; modestly sol ether, methanol, ethanol; insol water.
Form Supplied in: colorless solid.
Analysis of Reagent Purity: GLC analysis (OV-101, capillary column, 5 m, 200–280 °C) and TLC analysis (E. Merck Kieselgel 60 PF_{254}, 1:19 methanol–chloroform); R_f 0.42 (BINAPO, dioxide of BINAP), 0.67 (monoxide of BINAP), and 0.83 (BINAP). The optical purity of BINAP is analyzed after oxidizing to BINAPO by HPLC using a Pirkle column (Baker bond II) and a hexane/ethanol mixture as eluent.[3]
Preparative Methods: enantiomerically pure BINAP is obtained by resolution of the racemic dioxide, BINAPO, with camphorsulfonic acid or 2,3-di-*O*-benzoyltartaric acid followed by deoxygenation with ***Trichlorosilane*** in the presence of ***Triethylamine***.[3]
Handling, Storage, and Precautions: solid BINAP is substantially stable to air, but bottles of BINAP should be flushed with N_2 or Ar and kept tightly closed for prolonged storage. BINAP is slowly air oxidized to the monoxide in solution.

BINAP–Ru^{II} Catalyzed Asymmetric Reactions. Halogen-containing BINAP–Ru complexes are most simply prepared by reaction of $[RuCl_2(cod)]_n$ or $[RuX_2(arene)]_2$ (X = Cl, Br, or I) with BINAP.[4] Sequential treatment of $[RuCl_2(benzene)]_2$ with BINAP and sodium carboxylates affords $Ru(carboxylate)_2(BINAP)$ complexes. The dicarboxylate complexes, upon treatment with strong acid HX,[5] can be converted to a series of Ru complexes empirically formulated as $RuX_2(BINAP)$. These Ru^{II} complexes act as catalysts for asymmetric hydrogenation of various achiral and chiral unsaturated compounds.

α,β-Unsaturated carboxylic acids are hydrogenated in the presence of a small amount of $Ru(OAc)_2(BINAP)$ to give the corresponding optically active saturated products in quantitative yields.[6] The reaction is carried out in methanol at ambient temperature with a substrate:catalyst (S:C) ratio of 100–600:1. The sense and degree of the enantioface differentiation are profoundly affected by hydrogen pressure and the substitution pattern of the substrates. Tiglic acid is hydrogenated quantitatively with a high enantioselectivity under a low hydrogen pressure (eq 1), whereas naproxen, a commercial anti-inflammatory agent, is obtained in 97% ee under high pressure (eq 2).[6a]

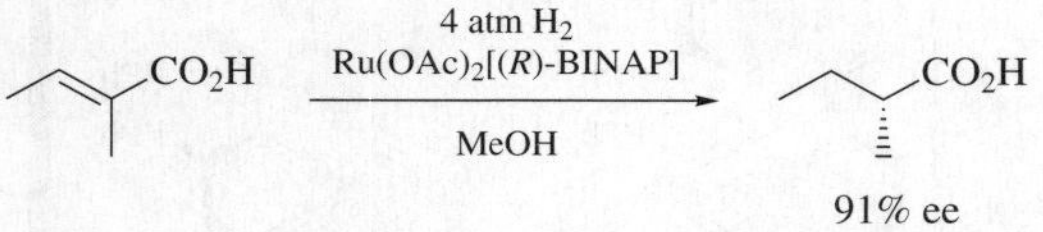

4 atm H_2, $Ru(OAc)_2[(R)\text{-BINAP}]$, MeOH (1)

91% ee

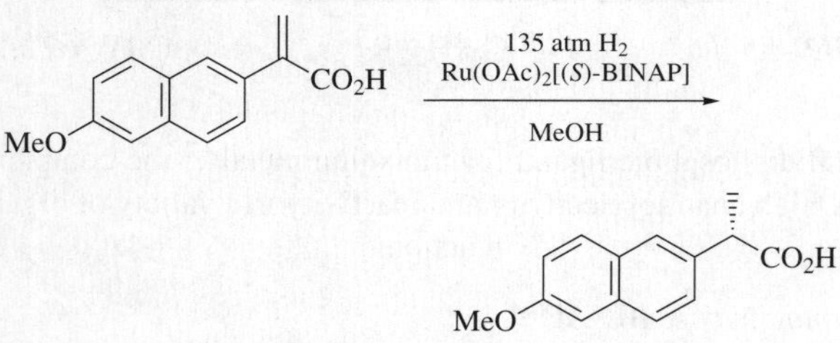

135 atm H_2, $Ru(OAc)_2[(S)\text{-BINAP}]$, MeOH (2)

97% ee

Enantioselective hydrogenation of certain α- and β-(acylamino)acrylic acids or esters in alcohols under 1–4 atm H_2 affords the protected α- and β-amino acids, respectively (eqs 3 and 4).[2a,7] Reaction of *N*-acylated 1-alkylidene-1,2,3,4-tetrahydroisoquinolines provides the 1*R*- or 1*S*-alkylated products. This method allows a general asymmetric synthesis of isoquinoline alkaloids (eq 5).[8]

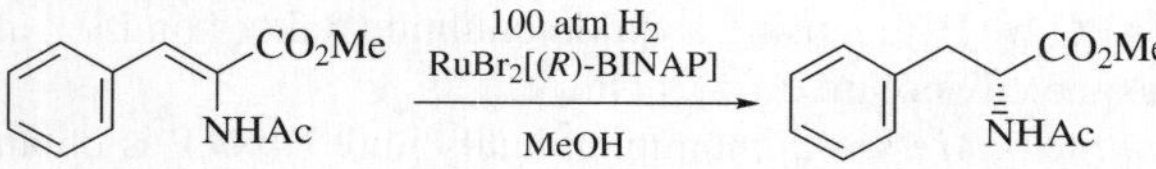

100 atm H_2, $RuBr_2[(R)\text{-BINAP}]$, MeOH (3)

85% ee

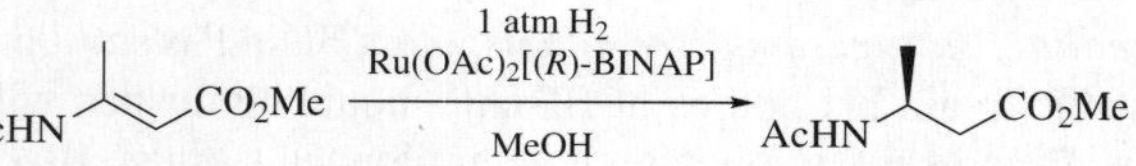

1 atm H_2, $Ru(OAc)_2[(R)\text{-BINAP}]$, MeOH (4)

96% ee

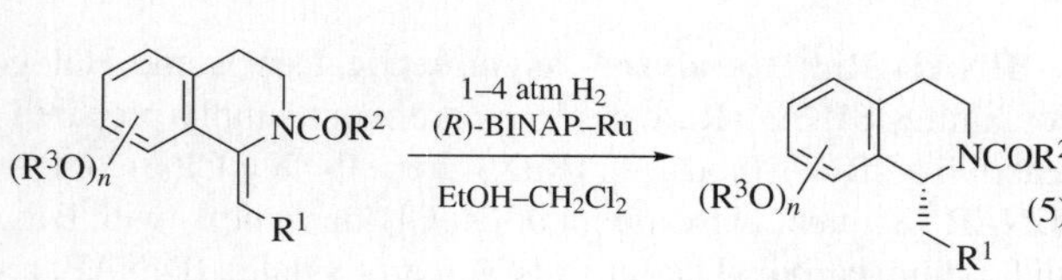

1–4 atm H_2, (R)-BINAP–Ru, $EtOH–CH_2Cl_2$ (5)

92–100% ee

Geraniol or nerol can be converted to citronellol in 96–99% ee in quantitative yield without saturation of the C(6)–C(7) double bond (eq 6).[9] The S:C ratio approaches 50 000. The use of alcoholic solvents such as methanol or ethanol and initial H_2 pressure greater than 30 atm is required to obtain high enantioselectivity. Diastereoselective hydrogenation of the enantiomerically pure allylic alcohol with an azetidinone skeleton proceeds at atmospheric pressure in the presence of an (*R*)-BINAP–Ru complex to afford the β-methyl product, a precursor of 1β-methylcarbapenem antibiotics (eq 7).[10] Racemic allylic alcohols such as 3-methyl-2-cyclohexenol and 4-hydroxy-2-cyclopentenone can be effectively resolved by the BINAP–Ru-catalyzed hydrogenation (eq 8).[11]

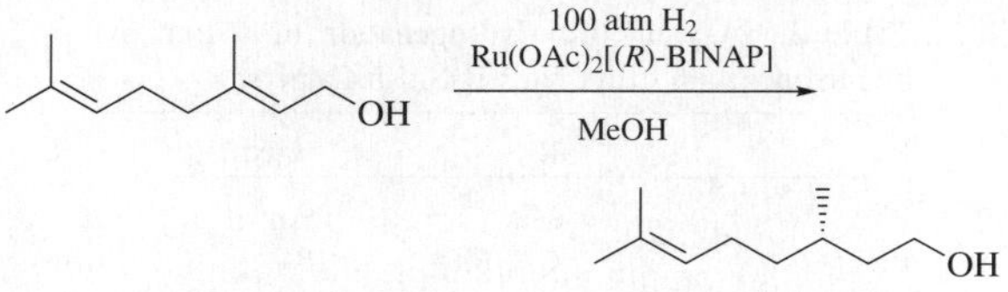

100 atm H_2, $Ru(OAc)_2[(R)\text{-BINAP}]$, MeOH (6)

99% ee

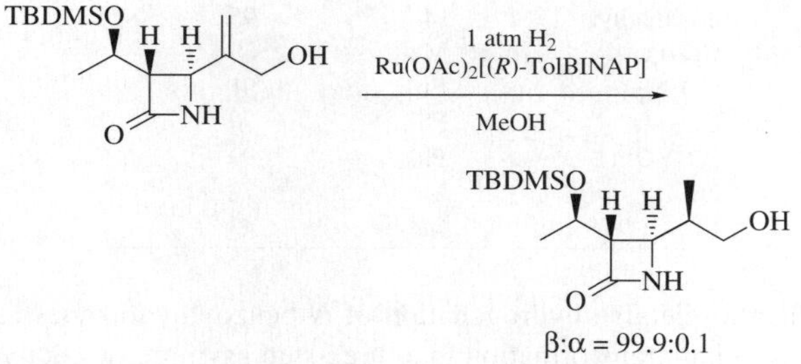

1 atm H_2, $Ru(OAc)_2[(R)\text{-TolBINAP}]$, MeOH (7)

β:α = 99.9:0.1

(±)- 4 atm H_2, $Ru(OAc)_2[(R)\text{-BINAP}]$, MeOH (8)

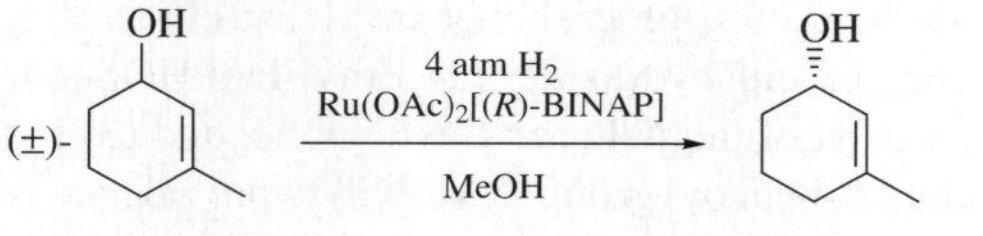

46% recovery
>99% ee

Diketene is quantitatively hydrogenated to 3-methyl-3-propanolide in 92% ee (eq 9). Certain 4-methylene- and 2-alkylidene-4-butanolides as well as 2-alkylidenecyclopentanone are also hydrogenated with high enantioselectivity.[12]

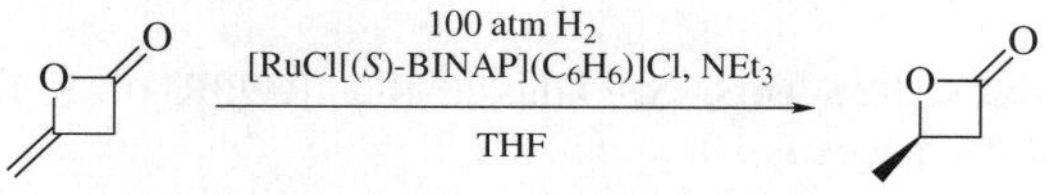

100 atm H_2, $[RuCl[(S)\text{-BINAP}](C_6H_6)]Cl$, NEt_3, THF (9)

92% ee

Hydrogenation with halogen-containing BINAP–Ru complexes can convert a wide range of functionalized prochiral ketones to stereo-defined secondary alcohols with high enantiomeric purity (eq 10).[13] 3-Oxocarboxylates are among the most appropriate substrates.[13a,4d] For example, the enantioselective hydrogenation of methyl 3-oxobutanoate proceeds quantitatively in methanol with an S:C ratio of 1000–10 000 to give the hydroxy ester product in nearly 100% ee (eq 11). Halogen-containing complexes $RuX_2(BINAP)$ (X = Cl, Br, or I; polymeric form) or $[RuCl_2(BINAP)]_2NEt_3$ are used as the catalysts. Alcohols are the solvents of choice, but aprotic solvents such as dichloromethane can also be used. At room temperature the reaction requires an initial H_2 pressure of 20–100 atm, but at 80–100 °C the reaction proceeds smoothly at 4 atm H_2.[4c,4d]

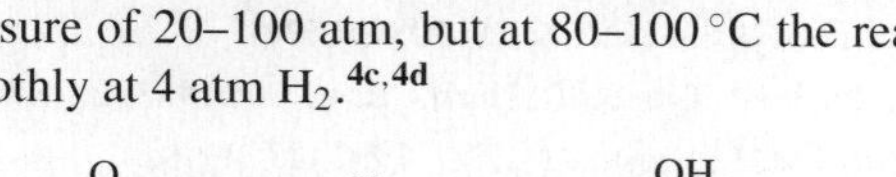

$$R^1C(O)R^2 \xrightarrow[(R)\text{-BINAP–Ru}]{H_2} R^1CH(OH)R^2 \quad (10)$$

R^1 = alkyl, aryl; R^2 = CH_2OH, CH_2NMe_2, CH_2CH_2OH, CH_2Ac, CH_2CO_2R, CH_2COSR, CH_2CONR_2, $CH_2CH_2CO_2R$, etc.

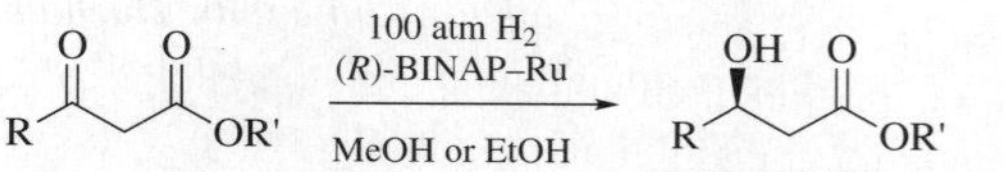

$$RC(O)CH_2CO_2R' \xrightarrow[\text{MeOH or EtOH}]{100\text{ atm } H_2,\ (R)\text{-BINAP–Ru}} RCH(OH)CH_2CO_2R' \quad (11)$$

98–100% ee

R = Me, Et, Bu, *i*-Pr; R' = Me, Et, *i*-Pr, *t*-Bu

3-Oxocarboxylates possessing an additional functional group can also be hydrogenated with high enantioselectivity by choosing appropriate reaction conditions or by suitable functional group modification (eq 12).[13b,13c]

100 atm H_2, $RuCl_2$[(*S*)-BINAP], EtOH, 100 °C, 5 min (12)
97% ee

The pre-existing stereogenic center in the chiral substrates profoundly affects the stereoselectivity. The (*R*)-BINAP–Ru-catalyzed reaction of (*S*)-4-(alkoxycarbonylamino)-3-oxocarboxylates give the statine series with (3*S*,4*S*) configuration almost exclusively (eq 13).[14]

100 atm H_2, $RuBr_2$[(*R*)-BINAP], MeOH or EtOH (13)
syn:*anti* = >99:1

Hydrogenation of certain racemic 2-substituted 3-oxocarboxylates occurs with high diastereo- and enantioselectivity via dynamic kinetic resolution involving in situ racemization of the substrates.[15] The (*R*)-BINAP–Ru-catalyzed reaction of 2-acylamino-3-oxocarboxylates in dichloromethane allows preparation of threonine and DOPS (anti-Parkinsonian agent) (eq 14).[16] In addition, a common intermediate for the synthesis of carbapenem antibiotics is prepared stereoselectively on an industrial scale from a 3-oxobutyric ester (**1**) with an acylaminomethyl substituent at the C(2) position.[16a] The second-order stereoselective hydrogenation of 2-ethoxycarbonylcycloalkanones gives predominantly the *trans* hydroxy esters (**2**) in high ee, whereas 2-acetyl-4-butanolide is hydrogenated to give the *syn* diastereomer (**3**).[17]

100 atm H_2, $RuBr_2$[(*R*)-BINAP], CH_2Cl_2 (14)
syn:*anti* = 99:1
92–98% ee

(**1**) *syn*:*anti* = 94:6, 98% ee
(**2**) R = CH_2, $(CH_2)_2$, $(CH_2)_3$; *trans*:*cis* = 93:7–99:1; 90–93% ee
(**3**) *syn*:*anti* = 98:2, 94% ee

Certain 1,2- and 1,3-diketones are doubly hydrogenated to give stereoisomeric diols. 2,4-Pentanedione, for instance, affords (*R*,*R*)- or (*S*,*S*)-2,4-pentanediol in nearly 100% ee accompanied by 1% of the *meso* diol.[13b]

A BINAP–Ru complex can hydrogenate a C=N double bond in a special cyclic sulfonimide to the sultam with >99% ee.[18]

The asymmetric transfer hydrogenation of the unsaturated carboxylic acids using formic acid or alcohols as the hydrogen source is catalyzed by Ru(acac-F_6)(η^3-C_3H_5)(BINAP) or $[RuH(BINAP)_2]PF_6$ to produce the saturated acids in up to 97% ee (eq 15).[19]

HCO_2H, NEt_3 or *i*-PrOH, (*S*)-BINAP–Ru, THF (15)
93–97% ee

BINAP–Ru complexes promote addition of arenesulfonyl chlorides to alkenes in 25–40% optical yield.[20]

BINAP–Rh[I] Catalyzed Asymmetric Reactions. The rhodium(I) complexes $[Rh(BINAP)(cod)]ClO_4$, $[Rh(BINAP)(nbd)]ClO_4$, and $[Rh(BINAP)_2]ClO_4$, are prepared from $[RhCl(cod)]_2$ or ***Bis(bicyclo[2.2.1]hepta-2,5-diene)dichlorodirhodium*** and BINAP in the presence of $AgClO_4$.[21] $[Rh(BINAP)S_2]ClO_4$ is prepared by reaction of [Rh(BINAP)-(cod or nbd)]ClO_4 with atmospheric pressure of hydrogen in an appropriate solvent, S.[21a] BINAP–Rh complexes catalyze a variety of asymmetric reactions.[2]

Prochiral α-(acylamino)acrylic acids or esters are hydrogenated under an initial hydrogen pressure of 3–4 atm to give the protected amino acids in up to 100% ee (eq 16).[21a] The BINAP–Rh catalyst was used for highly diastereoselective hydrogenation of a chiral homoallylic alcohol to give a fragment of the ionophore ionomycin.[22]

3–4 atm H_2, [Rh((*R*)-BINAP)]ClO_4, EtOH (16)
100% ee

The cationic BINAP–Rh complexes catalyze asymmetric 1,3-hydrogen shifts of certain alkenes. Diethylgeranylamine can be quantitatively isomerized in THF or acetone to citronellal diethylenamine in 96–99% ee (eq 17).[23] This process is the key step in the industrial production of (−)-menthol. In the presence of a cationic (*R*)-BINAP–Rh complex, (*S*)-4-hydroxy-2-cyclopentenone is isomerized five times faster than the (*R*) enantiomer, giving a chiral intermediate of prostaglandin synthesis.[24]

[Rh((*S*)-BINAP)(cod)]ClO_4, THF (17)
99% ee

Enantioselective cyclization of 4-substituted 4-pentenals to 3-substituted cyclopentanones in >99% ee is achieved with a cationic BINAP–Rh complex (eq 18).[25]

Rh[(*S*)-BINAP]ClO_4, CH_2Cl_2 (18)
>99% ee

Reaction of styrene and catecholborane in the presence of a BINAP–Rh complex at low temperature forms, after oxidative workup, 1-phenylethyl alcohol in 96% ee (eq 19).[26]

1. RhL*, DME; 2. H_2O_2 (19)

RhL* = $[Rh(cod)_2]BF_4$ + (*R*)-BINAP

96% ee

Neutral BINAP–Rh complexes catalyze intramolecular hydrosilylation of alkenes. Subsequent ***Hydrogen Peroxide*** oxidation produces the optically active 1,3-diol in up to 97% ee (eq 20).[27]

1. (*S*)-BINAP–Rh, acetone; 2. H_2O_2 (20)

97% ee

BINAP–Pd Catalyzed Asymmetric Reactions. BINAP–Pd^0 complexes are prepared in situ from ***Bis(dibenzylideneacetone)palladium(0)*** or $Pd_2(dba)_3{\cdot}CHCl_3$ and BINAP.[28] BINAP–Pd^{II} complexes are formed from ***Bis(allyl)di-μ-chlorodipalladium***, ***Palladium(II) Acetate***, or $PdCl_2(MeCN)_2$ and BINAP.[29–31]

A BINAP–Pd complex brings about enantioselective 1,4-disilylation of α,β-unsaturated ketones with chlorinated disilanes, giving enol silyl ethers in 74–92% ee (eq 21).[29]

+ $Me_3Si–SiPhCl_2$; $PdCl_2[(R)\text{-BINAP}]$, benzene (21)

92% ee

A BINAP–Pd^{II} complex catalyzes a highly enantioselective C–C bond formation between an aryl triflate and 2,3-dihydrofuran (eq 22).[30] The intramolecular version of the reaction using an alkenyl iodide in the presence of $PdCl_2[(R)\text{-BINAP}]$ and ***Silver(I) Phosphate*** allows enantioselective formation of a bicyclic ring system (eq 23).[31]

$Pd(OAc)_2$, (*R*)-BINAP, *i*-Pr_2NEt, benzene (22)

93% ee

$PdCl_2[(R)\text{-BINAP}]$, Ag_3PO_4, $CaCO_3$, NMP (23)

80% ee

Enantioselective electrophilic allylation of 2-acetamidomalonate esters is effected by a BINAP–Pd^0 complex (eq 24).[32]

+ $NaC(CO_2Me)_2NHAc$; $[PdCl(\eta^3\text{-}C_3H_5)]_2$, (*S*)-BINAP, THF (24)

94% ee

A BINAP–Pd^0 complex catalyzes hydrocyanation of norbornene to the *exo* nitrile with up to 40% ee.[28]

BINAP–Ir^I Catalyzed Asymmetric Reactions. $[Ir(BINAP)(cod)]BF_4$ is prepared from $[Ir(cod)\text{-}(MeCN)_2]BF_4$ and BINAP in THF.[33]

A combined system of the BINAP–Ir complex and bis(*o*-dimethylaminophenyl)phenylphosphine or (*o*-dimethylaminophenyl)diphenylphosphine catalyzes hydrogenation of benzylideneacetone[33a] and cyclic aromatic ketones[33b] with modest to high enantioselectivities (eq 25).

50–57 atm H_2, $[Ir((R)\text{-BINAP})(cod)]BF_4$, $PPh(C_6H_4\text{-}o\text{-}NMe_2)_2$, MeOH–dioxane (25)

95% ee

Related Reagents. See Classes R-3, R-10, R-11, R-15, and R-18, pages 1–10.

1. (a) Miyashita, A.; Yasuda, A.; Takaya, H.; Toriumi, K.; Ito, T.; Souchi, T.; Noyori, R. *JACS* **1980**, *102*, 7932. (b) Noyori, R.; Takaya, H. *CS* **1985**, *25*, 83.
2. (a) Noyori, R.; Kitamura, M. In *Modern Synthetic Methods*; Scheffold, R., Ed.; Springer: Berlin, 1989; p 115. (b) Noyori, R. *Science* **1990**, *248*, 1194. (c) Noyori, R.; Takaya, H. *ACR* **1990**, *23*, 345. (d) Noyori, R. *Chemtech* **1992**, *22*, 360.
3. Takaya, H.; Akutagawa, S.; Noyori, R. *OS* **1988**, *67*, 20.
4. (a) Ikariya, T.; Ishii, Y.; Kawano, H.; Arai, T.; Saburi, M.; Yoshikawa, S.; Akutagawa, S. *CC* **1985**, 922. (b) Ohta, T.; Takaya, H.; Noyori, R. *IC* **1988**, *27*, 566. (c) Kitamura, M.; Tokunaga, M.; Ohkuma, T.; Noyori, R. *TL* **1991**, *32*, 4163. (d) Kitamura, M.; Tokunaga, M.; Ohkuma, T.; Noyori, R. *OS* **1992**, *71*, 1.
5. Kitamura, M.; Tokunaga, M.; Noyori, R. *JOC* **1992**, *57*, 4053.
6. (a) Ohta, T.; Takaya, H.; Kitamura, M.; Nagai, K.; Noyori, R. *JOC* **1987**, *52*, 3174. (b) Saburi, M.; Takeuchi, H.; Ogasawara, M.; Tsukahara, T.; Ishii, Y.; Ikariya, T.; Takahashi, T.; Uchida, Y. *JOM* **1992**, *428*, 155.
7. Lubell, W. D.; Kitamura, M.; Noyori, R. *TA* **1991**, *2*, 543.
8. (a) Noyori, R.; Ohta, M.; Hsiao, Y.; Kitamura, M.; Ohta, T.; Takaya, H. *JACS* **1986**, *108*, 7117. (b) Kitamura, M.; Hsiao, Y.; Noyori, R.; Takaya, H. *TL* **1987**, *28*, 4829.
9. (a) Takaya, H.; Ohta, T.; Sayo, N.; Kumobayashi, H.; Akutagawa, S.; Inoue, S.; Kasahara, I.; Noyori, R. *JACS* **1987**, *109*, 1596, 4129. (b) Takaya, H.; Ohta, T.; Inoue, S.; Tokunaga, M.; Kitamura, M.; Noyori, R. *OS* **1994**, *72*, 74.
10. Kitamura, M.; Nagai, K.; Hsiao, Y.; Noyori, R. *TL* **1990**, *31*, 549.
11. Kitamura, M.; Kasahara, I.; Manabe, K.; Noyori, R.; Takaya, H. *JOC* **1988**, *53*, 708.

12. Ohta, T.; Miyake, T.; Seido, N.; Kumobayashi, H.; Akutagawa, S.; Takaya, H. *TL* **1992**, *33*, 635.
13. (a) Noyori, R.; Ohkuma, T.; Kitamura, M.; Takaya, H.; Sayo, N.; Kumobayashi, H.; Akutagawa, S. *JACS* **1987**, *109*, 5856. (b) Kitamura, M.; Ohkuma, T.; Inoue, S.; Sayo, N.; Kumobayashi, H.; Akutagawa, S.; Ohta, T.; Takaya, H.; Noyori, R. *JACS* **1988**, *110*, 629. (c) Kitamura, M.; Ohkuma, T.; Takaya, H.; Noyori, R. *TL* **1988**, *29*, 1555. (d) Kawano, H.; Ishii, Y.; Saburi, M.; Uchida, Y. *CC* **1988**, 87. (e) Ohkuma, T.; Kitamura, M.; Noyori, R. *TL* **1990**, *31*, 5509.
14. Nishi, T.; Kitamura, M.; Ohkuma, T.; Noyori, R. *TL* **1988**, *29*, 6327.
15. (a) Kitamura, M.; Tokunaga, M.; Noyori, R. *JACS* **1993**, *115*, 144. (b) Kitamura, M.; Tokunaga, M.; Noyori, R. *T* **1993**, *49*, 1853.
16. (a) Noyori, R.; Ikeda, T.; Ohkuma, T.; Widhalm, M.; Kitamura, M.; Takaya, H.; Akutagawa, S.; Sayo, N.; Saito, T.; Taketomi, T.; Kumobayashi, H. *JACS* **1989**, *111*, 9134. (b) Genet, J. P.; Pinel, C.; Mallart, S.; Juge, S.; Thorimbert, S.; Laffitte, J. A. *TA* **1991**, *2*, 555. (c) Mashima, K.; Matsumura, Y.; Kusano, K.; Kumobayashi, H.; Sayo, N.; Hori, Y.; Ishizaki, T.; Akutagawa, S.; Takaya, H. *CC* **1991**, 609.
17. Kitamura, M.; Ohkuma, T.; Tokunaga, M.; Noyori, R. *TA* **1990**, *1*, 1.
18. Oppolzer, W.; Wills, M.; Starkemann, C.; Bernardinelli, G. *TL* **1990**, *31*, 4117.
19. (a) Brown, J. M.; Brunner, H.; Leitner, W.; Rose, M. *TA* **1991**, *2*, 331. (b) Saburi, M.; Ohnuki, M.; Ogasawara, M.; Takahashi, T.; Uchida, Y. *TL* **1992**, *33*, 5783.
20. Kameyama, M.; Kamigata, N.; Kobayashi, M. *JOC* **1987**, *52*, 3312.
21. (a) Miyashita, A.; Takaya, H.; Souchi, T.; Noyori, R. *T* **1984**, *40*, 1245. (b) Toriumi, K.; Ito, T.; Takaya, H.; Souchi, T.; Noyori, R. *Acta Crystallogr.* **1982**, *B38*, 807.
22. Evans, D. A.; Morrissey, M. M. *TL* **1984**, *25*, 4637.
23. (a) Tani, K.; Yamagata, T.; Otsuka, S.; Akutagawa, S.; Kumobayashi, H.; Taketomi, T.; Takaya, H.; Miyashita, A.; Noyori, R. *CC* **1982**, 600. (b) Inoue, S.; Takaya, H.; Tani, K.; Otsuka, S.; Sato, T.; Noyori, R. *JACS* **1990**, *112*, 4897. (c) Yamakawa, M.; Noyori, R. *OM* **1992**, *11*, 3167. (d) Tani, K.; Yamagata, T.; Tatsuno, Y.; Yamagata, Y.; Tomita, K.; Akutagawa, S.; Kumobayashi, H.; Otsuka, S. *AG(E)* **1985**, *24*, 217. (e) Otsuka, S.; Tani, K. *S* **1991**, 665.
24. Kitamura, M.; Manabe, K.; Noyori, R.; Takaya, H. *TL* **1987**, *28*, 4719.
25. Wu, X.-M.; Funakoshi, K.; Sakai, K. *TL* **1992**, *33*, 6331.
26. (a) Hayashi, T.; Matsumoto, Y.; Ito, Y. *JACS* **1989**, *111*, 3426. (b) Sato, M.; Miyaura, N.; Suzuki, A. *TL* **1990**, *31*, 231. (c) Zhang, J.; Lou, B.; Guo, G.; Dai, L. *JOC* **1991**, *56*, 1670.
27. Tamao, K.; Tohma, T.; Inui, N.; Nakayama, O.; Ito, Y. *TL* **1990**, *31*, 7333.
28. Hodgson, M.; Parker, D. *JOM* **1987**, *325*, C27.
29. Hayashi, T.; Matsumoto, Y.; Ito, Y. *JACS* **1988**, *110*, 5579.
30. Ozawa, F.; Hayashi, T. *JOM* **1992**, *428*, 267.
31. (a) Sato, Y.; Sodeoka, M.; Shibasaki, M. *CL* **1990**, 1953; Sato, Y.; Sodeoka, M.; Shibasaki, M. *JOC* **1989**, *54*, 4738. (b) Ashimori, A.; Overman, L. E. *JOC* **1992**, *57*, 4571.
32. Yamaguchi, M.; Shima, T.; Yamagishi, T.; Hida, M. *TL* **1990**, *31*, 5049.
33. (a) Mashima, K.; Akutagawa, T.; Zhang, X.; Takaya, H.; Taketomi, T.; Kumobayashi, H.; Akutagawa, S. *JOM* **1992**, *428*, 213. (b) Zhang, X.; Taketomi, T.; Yoshizumi, T.; Kumobayashi, H.; Akutagawa, S.; Mashima, K.; Takaya, H. *JACS* **1993**, *115*, 3318.

Masato Kitamura & Ryoji Noyori
Nagoya University, Japan

Bis(tri-*n*-butyltin) Oxide

$(n\text{-}Bu_3Sn)_2O$

[56-35-9] $C_{24}H_{54}OSn_2$ (MW 596.20)

(promotes the oxidation of secondary alcohols and sulfides with Br_2; O- and N-activations; dehydrosulfurizations; hydrolysis catalyst)

Physical Data: bp 180 °C/2 mmHg; *d* 1.170 g cm^{-3}.
Solubility: sol ether and hexane.
Form Supplied in: colorless oil.
Handling, Storage, and Precautions: $(Bu_3Sn)_2O$ should be stored in the absence of moisture. Owing to the toxicity of organostannanes, this reagent should be handled in a well-ventilated fume hood. Contact with the eyes and skin should be avoided.

Oxidations. Benzylic, allylic, and secondary alcohols are oxidized to the corresponding carbonyl compounds by using $(Bu_3Sn)_2O$–***Bromine***.[1] This procedure is quite useful for selective oxidation of secondary alcohols in the presence of primary alcohols, which are inert under these conditions (eqs 1–3).[2] $(Bu_3Sn)_2O$–***N-Bromosuccinimide*** can also be applied to the selective oxidation of secondary alcohols (eq 4).[3]

(1) $(Bu_3Sn)_2O$, Br_2, CH_2Cl_2, rt; 76%

(2) $(Bu_3Sn)_2O$, Br_2, CH_2Cl_2, rt; 86%

(3) $(Bu_3Sn)_2O$, Br_2, CH_2Cl_2, rt; 68%

(4) 1. $(Bu_3Sn)_2O$ 2. NBS; 80%

$(Bu_3Sn)_2O$–Br_2 oxidizes sulfides to sulfoxides in CH_2Cl_2 without further oxidation to sulfones, even in the presence of excess reagent (eq 5).[4] This procedure is especially useful for sulfides having long, hydrophobic alkyl chains, for which solubility problems are often encountered in the ***Sodium Periodate*** oxidation in aqueous organic solvents. Oxidation of sulfenamides to sulfinamides can be achieved without formation of sulfonamides using the reagent (eq 6).[4]

$$PhCH_2SCH_2Ph \xrightarrow[92\%]{(Bu_3Sn)_2O,\ Br_2,\ CH_2Cl_2,\ rt} PhCH_2SOCH_2Ph \quad (5)$$

$$\text{PhS-N(morpholine)} \xrightarrow[92\%]{(Bu_3Sn)_2O,\ Br_2,\ CH_2Cl_2,\ rt} \text{PhSO-N(morpholine)} \quad (6)$$

$(Bu_3Sn)_2O$–Br_2–***Diphenyl Diselenide*** in refluxing $CHCl_3$ transforms alkenes into α-seleno ketones (eq 7).[5]

$$\text{PhCH=CH}_2 \xrightarrow[74\%]{(Bu_3Sn)_2O,\ Br_2,\ (PhSe)_2,\ CHCl_3,\ reflux} \text{PhCOCH}_2\text{SePh} \quad (7)$$

O- and N-activations. $(Bu_3Sn)_2O$ has been used in the activation of hydroxy groups toward sulfamoylations, acylations, carbamoylations, and alkylations because conversion of alcohols to stannyl ethers enhances the oxygen nucleophilicity. Tributylstannyl ethers are easily prepared by heating the alcohol and $(Bu_3Sn)_2O$, with azeotropic removal of water. Sulfamoylation of alcohols can be achieved via tributyltin derivatives in high yields, whereas direct sulfamoylation gives low yields (eq 8).[6] This activation can be used for selective acylation of vicinal diols (eq 9).[7] In carbohydrate chemistry this approach is extremely useful for the regioselective acylation without the use of a blocking–deblocking technique (eq 10).[8] The order of the activation of hydroxy groups on carbohydrates has been investigated, and is shown in partial structures (**1**), (**2**), and (**3**).[9] Regioselective carbamoylation can also be accomplished by changing experimental conditions (eq 11).[10] On the other hand, alkylations of the tin derivatives are sluggish and less selective than acylations under similar conditions. Regioselective alkylation of sugar compounds, however, can be carried out in high yield by conversion to a tributyltin ether followed by addition of alkylating agent and quaternary ammonium halide catalysts (eq 12).[11]

(8): HO–CH₂ sugar (F, acetonide, B) → $(Bu_3Sn)_2O$ → Bu_3SnO derivative → H_2NSO_2Cl, 87% → H_2NSO_3 derivative

(9): HO–CH₂CH(OH)(CH₂)₄OH → $(Bu_3Sn)_2O$ → Bu_3Sn---OH chelate → BzCl, 67% → BzO–CH₂CH(OH)(CH₂)₄OH

(10): methyl glycoside (OH, HO, HO, OH, OMe) → $(Bu_3Sn)_2O$ → O–SnBu₃ / Bu_3Sn stannylene → BzCl → OBz product (HO, HO, OH, OMe)

(1) most reactive; (2) next most reactive; (3) least reactive

(11): HO–CH₂ furanoside (B, HO, OH) → 1. $(Bu_3Sn)_2O$ (1 equiv), 2. PhNCO (2 equiv) → $PhNHCO_2$–CH₂ product (HO, OH); or $(Bu_3Sn)_2O$ (0.5 equiv), PhNCO (2 equiv) → HO–CH₂ product ($PhNHCO_2$, OH)

(12): TrO, OH, HO, HO, OBn galactoside → 1. $(Bu_3Sn)_2O$; 2. allyl bromide, Bu_4NBr, 77% → 3-O-allyl + 2-O-allyl products, 80:20

This O-activation is also effective for intramolecular alkylations such as oxetane synthesis (eq 13).[12] Similar N-activation has been used in the synthesis of pyrimidine nucleosides (eq 14).[13]

$$AcO(CH_2)_3Br + (Bu_3Sn)_2O \xrightarrow{80\ °C} Bu_3SnO(CH_2)_3Br \xrightarrow{240\ °C} \text{oxetane (33\%)} + Bu_3SnBr \quad (13)$$

(14): 5-bromouracil → 1. $(Bu_3Sn)_2O$; 2. BzO-protected ribofuranosyl acetate (OAc, BzO, OBz) → 5-bromouridine tribenzoate (BzO, BzO, OBz)

Dehydrosulfurizations. The thiophilicity of tin compounds is often utilized in functional group transformations. Thus conversion of aromatic and aliphatic thioamides to the corresponding nitriles can be accomplished by using $(Bu_3Sn)_2O$ in boiling benzene under azeotropic conditions (eq 15).[14]

(15): TrO-protected pyrrole thioamide (NH_2, S) → $(Bu_3Sn)_2O$, 95% → nitrile (CN, NH_2)

Hydrolysis. Esters are efficiently hydrolyzed with $(Bu_3Sn)_2O$ under mild conditions (eq 16).[15]

$$\text{Br}_2\text{-penam-CO}_2\text{CH}_2\text{OCO-}t\text{-Bu} \xrightarrow[61\%]{(\text{Bu}_3\text{Sn})_2\text{O},\ 25\ ^\circ\text{C}} \text{Br}_2\text{-penam-CO}_2\text{H} \quad (16)$$

Transformation of primary alkyl bromides or iodides to the corresponding primary alcohols is achieved in good yield by using $(Bu_3Sn)_2O$–***Silver(I) Nitrate*** (or ***Silver(I) p-Toluenesulfonate***) (eq 17),[16] whereas this method is not applicable to secondary halides due to elimination.

$$\text{MeCO}_2(\text{CH}_2)_4\text{I} \xrightarrow[96\%]{(\text{Bu}_3\text{Sn})_2\text{O, AgTos, DMF, 20 °C}} \text{MeCO}_2(\text{CH}_2)_4\text{OH} \quad (17)$$

$(Bu_3Sn)_2O$ is a useful starting material for the preparation of tributyltin hydride, which is a convenient radical reducing reagent in organic synthesis. Thus ***Tri-n-butyltin Hydride*** is easily prepared by using exchange reactions of $(Bu_3Sn)_2O$ with polysiloxanes (eq 18).[17]

$$(\text{Bu}_3\text{Sn})_2\text{O} + \left(\text{Si(Me)(H)–O}\right)_n \longrightarrow \text{Bu}_3\text{SnH} + \left[\text{Si(Me)–O–Si(Me)–O}\right]_n \quad (18)$$

Related Reagents. See Classes O-1 and O-8, pages 1–10.

1. Saigo, K.; Morikawa, A.; Mukaiyama, T. *CL* **1975**, 145.
2. Ueno, Y.; Okawara, M. *TL* **1976**, 4597.
3. Hanessian, S.; Roy, R. *CJC* **1985**, *63*, 163.
4. Ueno, Y.; Inoue, T.; Okawara, M. *TL* **1977**, 2413.
5. Kuwajima, I.; Shimizu, M. *TL* **1978**, 1277.
6. Jenkins, I. D.; Verheyden, J. P. H.; Moffatt, J. G. *JACS* **1971**, *93*, 4323.
7. (a) Ogawa, T.; Matsui, M. *T* **1981**, *37*, 2363. (b) David, S.; Hanessian, S. *T* **1985**, *41*, 643.
8. (a) Crowe, A. J.; Smith, P. J. *JOM* **1976**, *110*, C57. (b) Blunden, S. J.; Smith, P. J.; Beynon, P. J.; Gillies, D. G. *Carbohydr. Res.* **1981**, *88*, 9. (c) Ogawa, T.; Matsui, M. *Carbohydr. Res.* **1977**, *56*, C1. (d) Hanessian, S.; Roy, R. *JACS* **1979**, *101*, 5839. (e) Arnarp, J.; Loenngren, J. *CC* **1980**, 1000. (f) Ogawa, T.; Nakabayashi, S.; Sasajima, K. *Carbohydr. Res.* **1981**, *96*, 29.
9. Tsuda, Y.; Haque, M. E.; Yoshimoto, K. *CPB* **1983**, *31*, 1612.
10. (a) Ishido, Y.; Hirao, I.; Sakairi, N.; Araki, Y. *H* **1979**, *13*, 181. (b) Hirao, I.; Itoh, K.; Sakairi, N.; Araki, Y.; Ishido, Y. *Carbohydr. Res.* **1982**, *109*, 181.
11. (a) Alais, J.; Veyrières, A. *JCS(P1)* **1981**, 377. (b) Veyrières, A. *JCS(P1)* **1981**, 1626.
12. Biggs, J. *TL* **1975**, 4285.
13. Ogawa, T.; Matsui, M. *JOM* **1978**, *145*, C37.
14. Lim, M.-I.; Ren, W.-Y.; Klein, R. S. *JOC* **1982**, *47*, 4594.
15. Mata, E. G.; Mascaretti, O. A. *TL* **1988**, *29*, 6893.
16. Gingras, M.; Chan, T. H. *TL* **1989**, *30*, 279.
17. Hayashi, K.; Iyoda, J.; Shiihara, I. *JOM* **1967**, *10*, 81.

Hiroshi Sano

Gunma University, Kiryu, Japan

Bis(trimethylsilyl) Peroxide[1]

$$\text{Me}_3\text{Si–O–O–SiMe}_3$$

[5796-98-5] $C_6H_{18}O_2Si_2$ (MW 178.38)

(a masked form of 100% hydrogen peroxide;[2] synthon of HO^+ for electrophilic hydroxylation;[3,4] source of Me_3SiO^+ for electrophilic oxidations;[4–6] a versatile oxidant for alcohols, ketones, phosphines, phosphites, and sulfides[1,2,7])

Physical Data: bp 41 °C/30 mmHg;[7] n_D^{20} 1.3970.[1b]
Solubility: highly sol aprotic organic solvents.
Form Supplied in: colorless oil; used as an anhydrous and protected form of hydrogen peroxide; 10% solution in hexane or in CH_2Cl_2.
Analysis of Reagent Purity: ^{1}H NMR (s) δ 0.18 ppm;[3] ^{29}Si NMR δ 27.2 ppm.[8]
Preparative Methods: obtained in 80–96% yields by reaction of ***Chlorotrimethylsilane*** with ***1,4-Diazabicyclo[2.2.2]-octane***·$(H_2O_2)_2$,[3] ***Hexamethylenetetramine***·H_2O_2,[8] or ***Hydrogen Peroxide–Urea*** in CH_2Cl_2.[9]
Handling, Storage, and Precautions: thermally stable; can be handled in the pure state and distilled;[8] rearranges at 150–180 °C.[1c]

Electrophilic Oxidations. Bis(trimethylsilyl) peroxide (**1**) functions as an electrophilic hydroxylating agent for aliphatic, aromatic, and heteroaromatic anions.[3] Reaction of their lithium or Grignard compounds with (**1**) often affords trimethylsiloxy intermediates, which undergo desilylation with HCl in methanol to give the corresponding alcohols in good yields. In the presence of a catalytic amount of ***Trifluoromethanesulfonic Acid***, (**1**) reacts with aromatic compounds to produce the corresponding phenols after acidic workup.[10] In these reactions the Me_3SiO^+ moiety is considered as a synthon of the hydroxyl cation (i.e. OH^+).

Lithium or Grignard enolates of vinyl compounds react with (**1**) to produce the corresponding α-hydroxy ketones upon acidic workup.[4] Treatment of enolate anions derived from carboxylic acids and amides with (**1**) in THF at rt gives the corresponding α-hydroxy derivatives in 31–58% yields.[11] Furthermore, stereoselective synthesis of silyl enol ethers with retention of configuration is accomplished by oxidation of (*E*)- and (*Z*)-vinyllithiums, prepared from the corresponding bromides and *s*-BuLi, with (**1**) in THF at −78 °C.[6] Heterocyclic silyl enol ethers, such as 3-(trimethylsiloxy)furan and 3-(trimethylsiloxy)thiophene, can be obtained by reaction of (**1**) with 3-lithiofuran and 3-lithiothiophene, respectively.[5] In some reactions involving nucleophiles and (**1**), silylation may compete with siloxylation;[3] the outcome depends upon the counterion.[4]

Oxidative Desulfonylation and Selective Baeyer–Villiger Oxidation. Aliphatic, alicyclic, and benzylic phenyl sulfones react with *n*-BuLi and then with (**1**) in situ to give aldehydes or ketones in 66–91% yields (eq 1).[12] On the other hand, reaction of ketones with (**1**) in the presence of a catalyst, such as ***Trimethylsilyl Trifluoromethanesulfonate***,[2] ***Tin(IV) Chloride***, or ***Boron***

Trifluoride Etherate,[7] in CH_2Cl_2 gives esters in good to excellent yields (eq 2). This Baeyer–Villiger oxidation proceeds in a regio- and chemoselective manner: the competing epoxidation of a C–C double bond does not occur.

$$RR'CH-SO_2Ph \xrightarrow[2.\ Me_3SiOOSiMe_3]{1.\ n\text{-BuLi, THF}} RR'C{=}O + Me_3SiOSiMe_3 + PhSO_2Li \quad (1)$$

$$\xrightarrow[\text{catalyst, } CH_2Cl_2,\ 75\text{–}80\%]{Me_3SiOOSiMe_3} \quad (2)$$

catalyst = $Me_3SiSO_3CF_3$, $SnCl_4$, $BF_3{\bullet}OEt_2$

Oxidation of Alcohols, Sulfur, and Phosphorus Compounds as well as Si–Si and Si–H Bonds. Bis(trimethylsilyl) peroxide acts as an effective oxidant for alcohols in the presence of ***Pyridinium Dichromate*** or $RuCl_2(PPh_3)_3$ complex as the catalyst in CH_2Cl_2.[13] By this method, primary allylic and benzylic alcohols can be selectively oxidized to α-enals in the presence of a secondary alcohol.

Sulfur Trioxide reacts with (**1**) in CH_2Cl_2 at −30 °C, affording ***Bis(trimethylsilyl) Monoperoxysulfate***, an oxidant useful for the Baeyer–Villiger oxidation.[14] In addition, sulfides can be converted to sulfoxides or sulfones by use of (**1**) in benzene at reflux.[1b,15,16] This peroxide can also oxidize phosphines and phosphites to the corresponding oxyphosphoryl derivatives with retention of configuration at the phosphorus center in high yields.[1b,17] Furthermore, it converts the P=S and P=Se functionalities to the P=O group with inversion of configuration at the phosphorus center.[17] The $CF_3SO_3SiMe_3$ or Nafion–$SiMe_3$ catalyzed oxidation of nucleoside phosphites to phosphates under nonaqueous conditions is applied to the solid-phase synthesis of oligonucleotides.[18,19]

Reactions between (**1**) and disilanes containing fluorine atoms or possessing ring strain proceed readily at ambient temperature to generate the corresponding disiloxanes.[20] Oxidation of hydrosilanes (R_3SiH) with (**1**) gives a mixture of R_3SiOH and $R_3SiOSiMe_3$. The ratio of R_3SiOH to $R_3SiOSiMe_3$ tends to decrease in the order $Et_3SiH > PhMe_2SiH > Me_3SiSiHMe_2$.

Isomerization of Allylic Alcohols and Formation of 1-Halo-1-alkynes. Isomerization of primary and secondary allylic alcohols to tertiary isomers proceeds in CH_2Cl_2 at 25 °C in the presence of a catalyst that is prepared in situ by activation of ***Vanadyl Bis(acetylacetonate)*** or $MoO_2(acac)_2$ with (**1**) (eq 3).[21] On the other hand, terminal alkynes react with (**1**) in the presence of copper or zinc halides in THF at −15 °C to afford terminal 1-halo-1-alkynes in 40–85% yields.[22] By the same method, 1-cyano-1-alkynes are obtained in 65% yield by use of copper cyanide.

$$\xrightarrow[CH_2Cl_2,\ 25\ ^\circ C,\ 76\text{–}92\%]{Me_3SiOOSiMe_3,\ VO(acac)_2 \text{ or } MoO_2(acac)_2} \quad (3)$$

Related Reagents. See Classes O-1, O-8, O-15, and O-24, pages 1–10.

1. (a) Brandes, D.; Blaschette, A. *JOM* **1973**, *49*, C6. (b) Brandes, D.; Blaschette, A. *JOM* **1974**, *73*, 217. (c) Alexandrov, Y. A. *JOM* **1982**, *238*, 1. (d) Huang, L.; Hiyama, T. *Yuki Gosei Kagaku Kyokaishi* **1990**, *48*, 1004 (*CA* **1991**, *114*, 102 079x).
2. Suzuki, M.; Takada, H.; Noyori, R. *JOC* **1982**, *47*, 902.
3. Taddei, M.; Ricci, A. *S* **1986**, 633.
4. Camici, L.; Dembech, P.; Ricci, A.; Seconi, G.; Taddei, M. *T* **1988**, *44*, 4197.
5. Camici, L.; Ricci, A.; Taddei, M. *TL* **1986**, *27*, 5155.
6. Davis, F. A.; Lal, G. S.; Wei, J. *TL* **1988**, *29*, 4269.
7. Matsubara, S.; Takai, K.; Nozaki, H. *BCJ* **1983**, *56*, 2029.
8. Babin, P.; Bennetau, B.; Dunogues, J. *SC* **1992**, *22*, 2849.
9. Jackson, W. P. *SL* **1990**, *9*, 536.
10. Olah, G. A.; Ernst, T. D. *JOC* **1989**, *54*, 1204.
11. Pohmakotr, M.; Winotai, C. *SC* **1988**, *18*, 2141.
12. Hwu, J. R. *JOC* **1983**, *48*, 4432.
13. Kanemoto, S.; Matsubara, S.; Takai, K.; Oshima, K.; Utimoto, K.; Nozaki, H. *BCJ* **1988**, *61*, 3607.
14. Adam, W.; Rodriguez, A. *JOC* **1979**, *44*, 4969.
15. Kocienski, P.; Todd. M. *CC* **1982**, 1078.
16. Curci, R.; Mello, R.; Troisi, L. *T* **1986**, *42*, 877.
17. Kowalski, J.; Wozniak, L.; Chojnowski, J. *PS* **1987**, *30*, 125.
18. Hayakawa, Y.; Uchiyama, M.; Noyori, R. *TL* **1986**, *27*, 4191.
19. Hayakawa, Y.; Uchiyama, M.; Noyori, R. *TL* **1986**, *27*, 4195.
20. Tamao, K.; Kumada, M.; Takahashi, T. *JOM* **1975**, *94*, 367.
21. Matsubara, S.; Okazoe, T.; Oshima, K.; Takai, K.; Nozaki, H. *BCJ* **1985**, *58*, 844.
22. Casarini, A.; Dembech, P.; Reginato, G.; Ricci, A.; Seconi, G. *TL* **1991**, *32*, 2169.

Jih Ru Hwu & Buh-Luen Chen
Academia Sinica & National Tsing Hua University, Taiwan, Republic of China

Borane–Dimethyl Sulfide[1]

$BH_3{\cdot}SMe_2$

[13292-87-0] C_2H_9BS (MW 75.96)

(hydroborating and reducing agent)

Alternate Name: BMS.
Physical Data: $d_4^{20} = 0.801\ g\ cm^{-3}$.
Solubility: sol dichloromethane, benzene, toluene, xylene, hexane, diethyl ether, diglyme, DME, and ethyl acetate; insol but reacts slowly with water; reacts with alcohols, acetone.
Form Supplied in: neat complex, colorless liquid, ca. 10 M in BH_3; contains slight excess of dimethyl sulfide.
Analysis of Reagent Purity: active hydride is determined by hydrolysis of an aliquot in glycerol–water–methanol mixture and measuring the hydrogen evolved according to a standard procedure.[3a] ^{11}B NMR (CH_2Cl_2) δ −20.1 ppm (q, $J_{B-H} = 104$ Hz).[4]

Preparative Method: **Diborane** generated by the reaction of **Sodium Borohydride** with **Boron Trifluoride Etherate** in diglyme[3] is absorbed in **Dimethyl Sulfide**.

Purification: commercial reagent can be purified and freed of excess dimethyl sulfide by vacuum transfer.[2]

Handling, Storage, and Precautions: use in a fume hood; flammable liquid with stench; reacts with atmospheric moisture forming a crust of boric acid. Store and handle under nitrogen or argon. Stable indefinitely when kept at 0 °C. Stable for prolonged periods at rt.

Introduction. Borane–dimethyl sulfide (BMS) parallels ***Borane–Tetrahydrofuran*** in hydroboration and reduction reactions. Its advantages are stability, solubility in various solvents, and higher concentration.

Hydroboration. Directive effects and stereoselectivity in the hydroboration of representative alkenes with BMS are shown in Figure 1.[5,6]

Oxidation of organoborane intermediates with a standard alkaline hydrogen peroxide in the presence of dimethyl sulfide requires a higher concentration of alkali to suppress slow concurrent oxidation of dimethyl sulfide. When its presence is not desirable, it can be removed by selective oxidation with sodium hypochlorite.[7] The oxidation of primary trialkylboranes with ***Pyridinium Chlorochromate*** in methylene chloride and with ***Chromium(VI) Oxide*** in acetic acid provides the corresponding aldehydes and acids, respectively. The method works best for α-alkyl-substituted alkenes undergoing hydroboration with high regioselectivity (eq 1).[8] BMS finds application for the synthesis of various hydroborating and reducing agents derived from hindered alkenes and other precursors.[9]

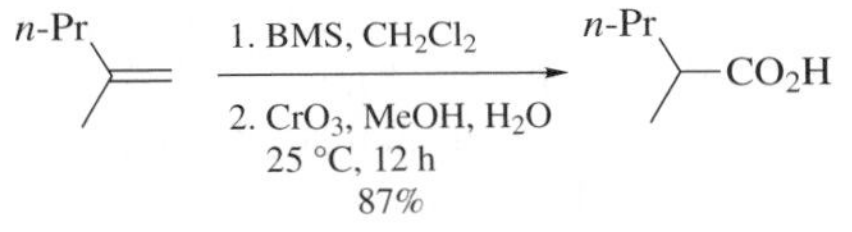

1. BMS, CH_2Cl_2; 2. CrO_3, MeOH, H_2O, 25 °C, 12 h; 87% — n-Pr, CO_2H (1)

The reaction of internal alkynes with BMS (also with ***Borane–Tetrahydrofuran***) can be controlled to give the corresponding (*Z*)-alkenylborane. Photochemical isomerization of such alkenylboranes has been reported (eq 2).[10]

The hydroboration of cyclic dienes with BMS gives mixtures of products. Dihydroboration predominates for seven- and eight-membered ring dienes, whereas cyclopentadiene and 1,3-cyclohexadiene give mainly the homoallylic and allylic organoborane, respectively.[11] Acyclic α,ω-dienes are transformed into bora heterocycles by cyclic hydroboration with ***9-Borabicyclo[3.3.1]nonane*** and BMS (eq 3).[9c,12]

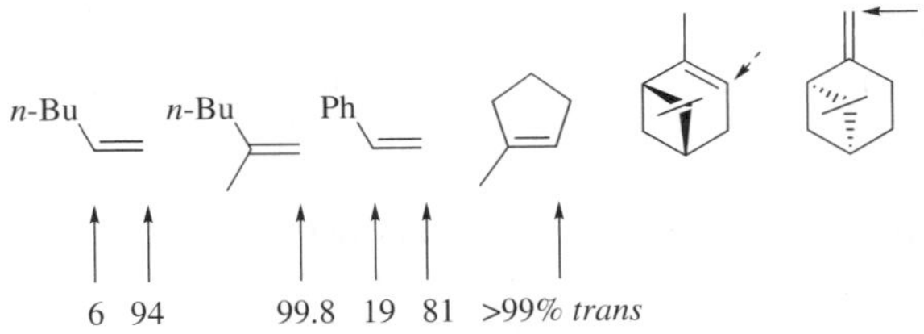

n-Bu, n-Bu, Ph — 6 94; 99.8; 19 81; >99% *trans*

Figure 1

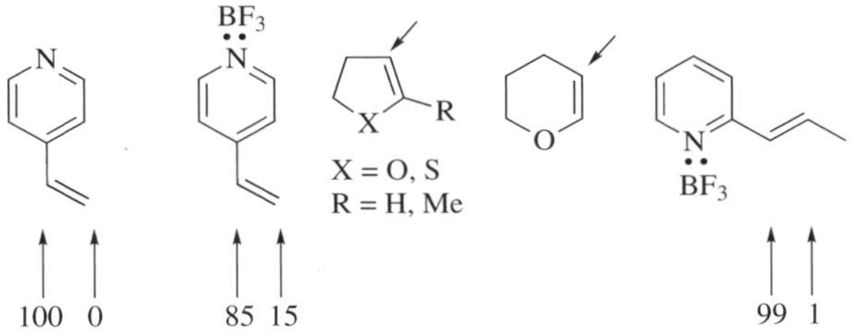

BF_3; X = O, S; R = H, Me; BF_3 — 100 0; 85 15; 99 1

Figure 2

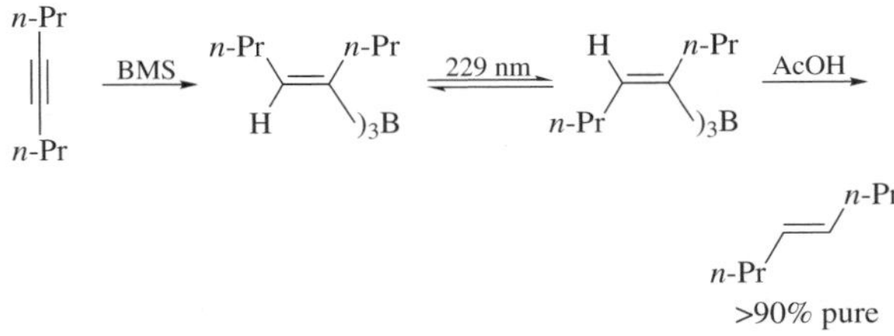

n-Pr, BMS, $)_3B$, 229 nm, AcOH, >90% pure (2)

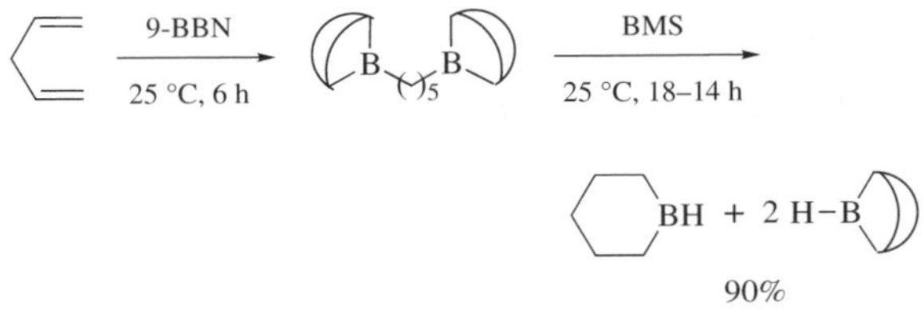

9-BBN, 25 °C, 6 h; BMS, 25 °C, 18–14 h; BH + 2 H–B; 90% (3)

The hydroboration reaction tolerates many functional groups. A high level of acyclic diastereoselection is achieved in certain highly functionalized alkenes with BMS (eq 4)[13a] and also with borane–THF.[13b,c] Vinylic and allylic derivatives containing oxygen, sulfur, or nitrogen substituents react with high regioselectivity, placing boron at the β-position (Figure 2).[6,14] Although such organoboranes are prone to elimination reactions, pure products can often be obtained by a careful control of the reaction conditions.

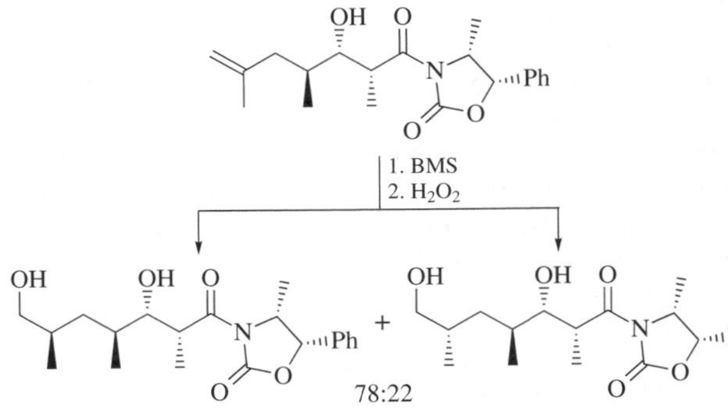

1. BMS; 2. H_2O_2 — OH, O, N, Ph, 78:22 (4)

Generally, the hydroboration of allylic amines and sulfides proceeds only to the monoalkylborane stage due to intramolecular complexation.[15,16] The amino group can be protected by Boc, benzyloxycarbonyl, trimethylsilyl, or phosphoramido groups[14,17,18] (eq 5).[17] Allylic *N*-phosphoroamidates are readily transformed into γ-haloamines via hydroboration–halogenolysis.[18]

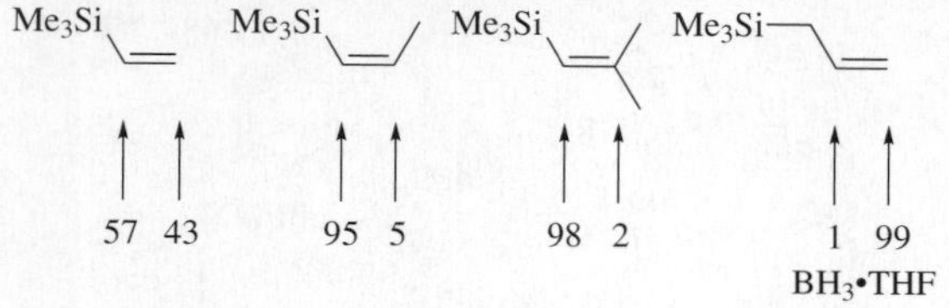

Me_3Si (57, 43); Me_3Si (95, 5); Me_3Si (98, 2); Me_3Si (1, 99) BH_3•THF

Figure 3

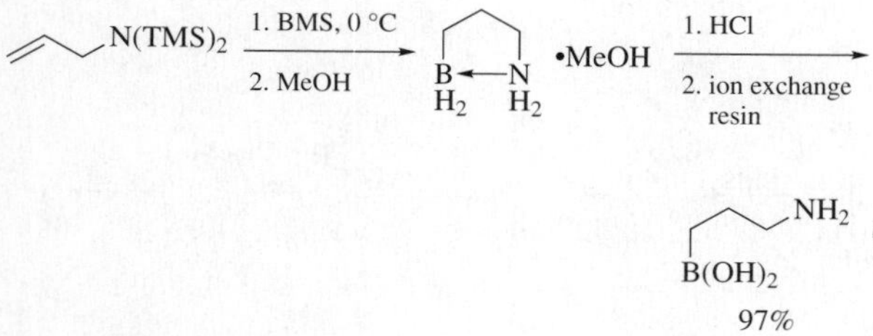

$$\text{CH}_2\text{=CHCH}_2\text{N(TMS)}_2 \xrightarrow[\text{2. MeOH}]{\text{1. BMS, 0 °C}} \text{cyclic B} \leftarrow \text{N} \ \cdot\text{MeOH} \xrightarrow[\text{2. ion exchange resin}]{\text{1. HCl}} \text{H}_2\text{N(CH}_2)_3\text{B(OH)}_2 \ \ 97\% \quad (5)$$

The hydroboration of β-monosubstituted enamines with BMS in THF affords [(2-dialkylamino)alkyl]boranes as the major product, which can be transformed into the corresponding boronates, boronic acids, amino alcohols, or alkenes, depending on the reaction conditions (eq 6).[16,19] Aldehydes and ketones can be stereoselectively transformed into (*E*)- or (*Z*)-alkenes by this method (eq 6).[19]

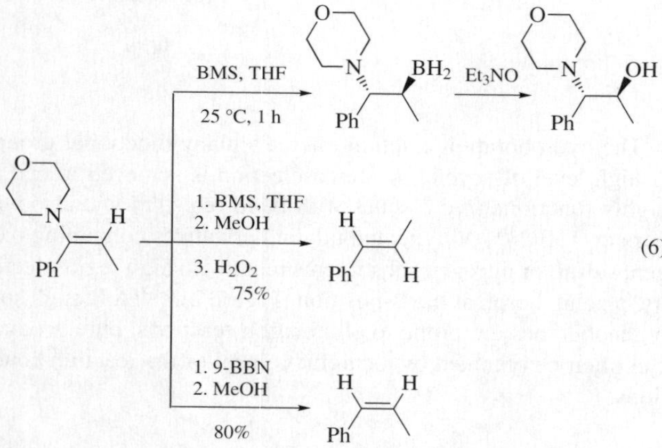

(6): Morpholino enamine (Ph) — BMS, THF, 25 °C, 1 h → BH_2 adduct — Et_3NO → amino alcohol (OH); 1. BMS, THF 2. MeOH 3. H_2O_2, 75% → (*E*)-alkene; 1. 9-BBN 2. MeOH, 80% → (*Z*)-alkene

Directive effects in the addition of BMS to vinylic and allylic silicon derivatives are opposite to those observed for the oxygen and nitrogen derivatives (Figure 3).[20]

Several stereodefined organoboranes containing two different metals, boron and silicon, in a 1:3 relationship have been synthesized and transformed into 1,3-diols.[21] The hydroboration of alkynylsilanes provides a convenient access to acylsilanes (eq 7).[20b,22]

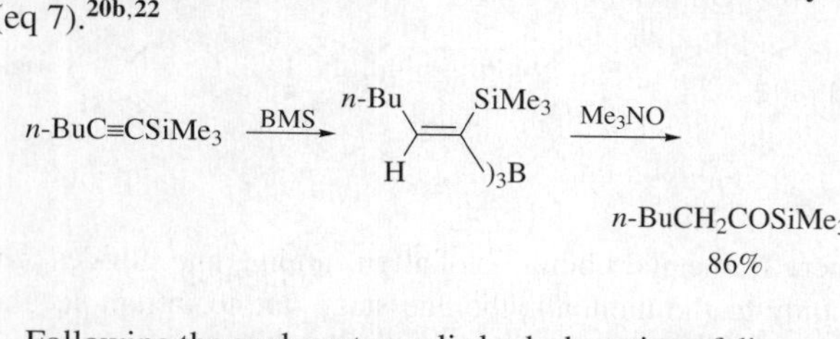

$$n\text{-BuC}\equiv\text{CSiMe}_3 \xrightarrow{\text{BMS}} (n\text{-Bu(H)C=C(SiMe}_3))_3\text{B} \xrightarrow{\text{Me}_3\text{NO}} n\text{-BuCH}_2\text{COSiMe}_3 \ \ 86\% \quad (7)$$

Following the analogy to cyclic hydroboration of dienes with 9-BBN/BMS, pharmacologically important 1-silacyclohexan-4-one has been synthesized via cyclic hydroboration of the divinylsilane derivative.[23,24]

Reduction of Functional Groups:[1,25,26] Carboxylic Acids and Derivatives. Carboxylic acids,[16,27] thioacids,[28] and amino acids[29] are transformed into the corresponding alcohols, thiols, and amino alcohols. Aromatic acids are reduced in the presence of trimethoxyborane.[30] They may also undergo reduction to hydrocarbons.[31] Oxidation of the intermediate alkoxyboranes yields aldehydes.[27a] Since alkoxyboranes are readily prepared from alcohols,[32] the PCC oxidation of alkoxyboranes may be advantageous if water-sensitive groups are present (eq 8).[27a,32b]

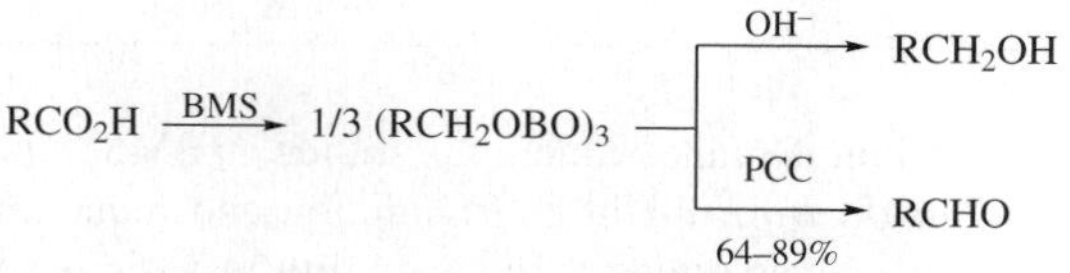

$$\text{RCO}_2\text{H} \xrightarrow{\text{BMS}} 1/3\ (\text{RCH}_2\text{OBO})_3 \begin{cases} \xrightarrow{\text{OH}^-} \text{RCH}_2\text{OH} \\ \xrightarrow{\text{PCC}} \text{RCHO} \ \ 64\text{–}89\% \end{cases} \quad (8)$$

Carboxylic groups can be protected from BMS reduction as trialkylsilyl esters.[33] Aliphatic esters and lactones are rapidly reduced in refluxing THF. Aromatic esters react at a slower rate.[26] Site selective reductions of α-hydroxy esters in the presence of other ester groups have been achieved in both aliphatic and aromatic systems[34] (eq 9).[34a,b]

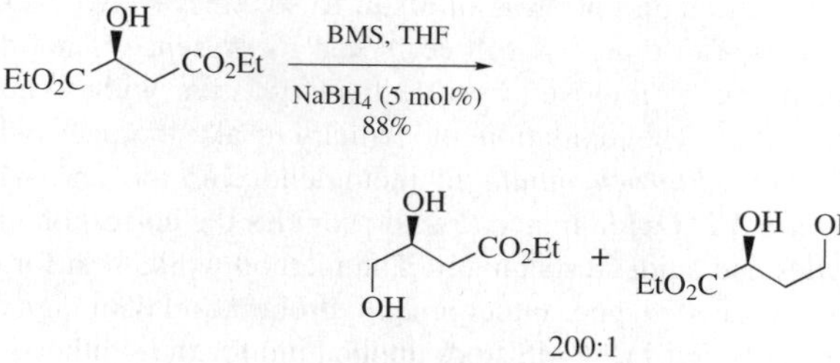

$$\text{EtO}_2\text{C–CH(OH)–CH}_2\text{–CO}_2\text{Et} \xrightarrow[\text{NaBH}_4\ (5\ \text{mol\%})\ 88\%]{\text{BMS, THF}} \text{HOCH}_2\text{–CH(OH)–CH}_2\text{CO}_2\text{Et} + \text{EtO}_2\text{C–CH(OH)–CH}_2\text{CH}_2\text{OH} \quad 200:1 \quad (9)$$

Amides and lactams are reduced to amines[35] (eq 10).[35b] α,ω-Amido esters undergo reductive cyclization.[36] Formylation–reduction of primary amines provides convenient access to monomethylated amines.[37] For the isolation of amines, a decomplexing agent such as BF_3 is added, or the amine is isolated as a hydrochloride salt.[26] Alternatively, if a theoretical amount of BMS is used in refluxing toluene, neither distillation of dimethyl sulfide nor a decomplexing agent is necessary.[38] Both aliphatic and aromatic nitriles undergo fast reduction.[26]

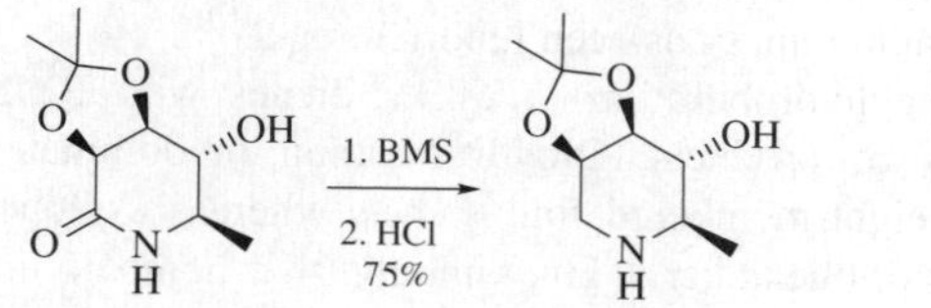

(10): acetonide-protected hydroxy lactam (OH, N–H) → 1. BMS 2. HCl, 75% → corresponding piperidine

Ketones and Derivatives. BMS is used as a source of borane in asymmetric reductions of prochiral ketones and derivatives catalyzed by oxazaborolidines (see ***Borane–Tetrahydrofuran***). It is also used for the synthesis of the catalysts.[39]

In stoichiometric ratios, BMS is a highly selective reducing agent for α-halo ketones and α-halo imines.[40] In the reduction of aldoximes and ketoximes, the reagent offers the advantage of a simpler isolation procedure as compared to reductions with borane–THF.[1b] Selective reduction of an aldehyde in the presence of a ketone and α,β-unsaturated aldehydes and ketones to the corresponding allylic alcohols has been demonstrated.[1a] Acetals are tolerated by the reagent and serve as protective groups for ketones in selective hydroborations and reductions. Efficient acetal opening is achieved upon activation with ***Trimethylsilyl Trifluoromethanesulfonate*** (eq 11).[41]

TMSOTf, BMS; CH_2Cl_2, –78 °C; 77%
1. BMS; 2. H_2O_2, NaOH; 80%
(11)

Cleavage of Single Bonds. Simple ethers appear to be stable to BMS, for example THF is not cleaved when refluxed with BMS for long periods. However, methoxy groups attached to aromatic rings may undergo cleavage to phenols.[35c]

BMS is an efficient reagent for the direct reduction of ozonides to alcohols in methylene chloride solution.[42]

Other Applications. BMS is used for the synthesis of various borane complexes, e.g. with amines,[43] phosphines,[44] azo compounds,[45] and oligonucleotides.[46] A novel staining method for transmission electron microscopy is based on the reduction (BMS)–oxidation of ester groups in polymeric materials.[47]

Related Reagents. See Classes R-2, R-4, R-5, R-9, R-10, R-11, R-13, R-14, R-17, and R-20, pages 1–10.

---

1. (a) Hutchins, R. O.; Cistone, F. *OPP* **1981**, *13*, 225. (b) Lane, C. F. *Aldrichim. Acta* **1975**, *8*, 20. (c) Lane, C. F. In *Synthetic Reagents*; Pitzey, J. S., Ed.; Ellis Horwood: Chichester, 1977; Vol. 3. (d) Lane, C. F. *CR* **1976**, *76*, 773. (e) Pelter, A.; Smith, K.; Brown, H. C., *Borane Reagents*, Academic: London, 1988.
2. Shiner, C. S.; Garner, C. M.; Haltiwanger, R. C. *JACS* **1985**, *107*, 7167.
3. (a) Brown, H. C. *Organic Syntheses via Boranes*, Wiley: New York, **1975**, p 18. (b) Zweifel, G.; Brown, H. C. *OR* **1963**, *13*, 1.
4. Young, D. E.; McAhran, G. E.; Shore, S. G. *JACS* **1966**, *88*, 4390.
5. (a) Lane, C. F. *JOC* **1974**, *39*, 1437. (b) Lane, C. F.; Daniels, J. J. *OSC* **1988**, *6*, 719.
6. Brown, H. C.; Vara Prasad, J. V. N.; Zee, S-H. *JOC* **1986**, *51*, 439.
7. Brown, H. C.; Mandal, A. K. *JOC* **1980**, *45*, 916.
8. (a) Brown, H. C.; Kulkarni, W. U.; Rao,. C. G.; Patil, V. D. *T* **1986**, *42*, 5515. (b) Brown, H. C.; Kulkarni, S. U.; Khanna, V. V.; Patil, V. D.; Racherla, U. S. *JOC* **1992**, *57*, 6173.
9. (a) Brown, H. C.; Mandal, A. K.; Kulkarni, S. U. *JOC* **1977**, *42*, 1392. (b) Schwier, J. R.; Brown, H. C. *JOC* **1993**, *58*, 1546. (c) Brown, H. C.; Pai, G. G. *JOM* **1983**, *250*, 13. (d) Brown, H. C.; Joshi, N, N, *JOM* **1988**, *53*, 4059. (e) Brown, H. C.; Vara Prasad, J. V. N.; Zaidlewicz, M. *JOM* **1988**, 2911. (f) Simpson, P.; Tschaen, D.; Verhoeven, T. R. *SC* **1991**, *21*, 1705. (g) Soderquist, J. A.; Negron, A. *OS* **1992**, *70*, 169.
10. Gano, J. E.; Srebnik, M. *TL* **1993**, *34*, 4889.
11. Brown, H. C.; Bhat, K. S. *JOC* **1986**, *51*, 445.
12. Brown, H. C.; Pai, G. G.; Naik, R. G. *JOC* **1984**, *49*, 1072.
13. (a) Evans, D. A.; Bartoli, J.; Godel, T. *TL* **1982**, *23*, 4577. (b) Nicolaou, K. C.; Pavia, M. R.; Seitz, S. P. *JACS* **1982**, *104*, 2027. (c) Schmid, G.; Fukuyama, T.; Akasaka, K.; Kishi, Y. *JACS* **1979**, *101*, 259.
14. (a) Brown, H. C.; Vara Prasad, J. V. N.; Zee, S-H. *JOC* **1985**, *50*, 1582. (b) Brown, H. C.; Vara Prasad, J. V. N. *H* **1987**, *25*, 641.
15. (a) Braun, R. A.; Brown, D. C.; Adams, R. M. *JACS* **1971**, *93*, 2823. (b) Polivka, Z.; Kubelka, V.; Holubova, N.; Ferles, M. *CCC* **1970**, *35*, 1131.
16. Goralski, C. T.; Singaram, B.; Brown, H. C. *JOC* **1987**, *52*, 4014.
17. Dicko, A.; Mountry, M.; Baoboulene, M. *SC* **1988**, *18*, 459.
18. Bonmaarouf-Khallaayoun, Z.; Babulene, M.; Speziale, V.; Lattes, A. *PS* **1988**, *36*, 181.
19. Singaram, B.; Rangaishenvi, M. V.; Brown, H. C.; Goralski, C. T.; Hasha, D. L. *JOC* **1991**, *56*, 1543.
20. (a) Brown, H. C.; Rangaishenvi, M. V. In *Chemistry and Technology of Silicon and Tin*; Das, K., Ed.; Oxford University Press: Oxford, 1992, p 3. (b) Soderquist, J. A.; Lee, S. J. H. *T* **1988**, *44*, 4033.
21. Fleming, J.; Lawrence, N. J. *JCS(P1)* **1992**, 3309.
22. Miller, J. A.; Zweifel, G. *S* **1981**, 288.
23. Soderquist, J. A.; Negron, A. *JOC* **1987**, *52*, 3441.
24. Soderquist, J. A.; Negron, A. *JOC* **1989**, *54*, 2462.
25. Braun, L. M.; Braun, R. A.; Crissman, H. R.; Opperman, M.; Adams, R. M. *JOC* **1971**, *36*, 2388.
26. Brown, H. C.; Choi, Y. M.; Narasimhan, S. *JOC* **1982**, *47*, 3153.
27. (a) Brown, H. C.; Rao, C. G.; Kulkarni, S. U. *S* **1979**, 704. (b) Kraus, J-L.; Attardo, G. *S* **1991**, 1046. (c) Jordis, U.; Sauter, R.; Siddiqu, S. M.; Künnenburg, B.; Bhattacharya, K. *S* **1990**, 925.
28. Jabre, I.; Saquet, M.; Thuillier, A. *JCR(S)* **1990**, 106.
29. (a) Smith, G. A.; Gawley, R. E. *OS* **1984**, *63*, 136. (b) Becker, Y.; Eisenstadt, A.; Stille, J. K. *JOC* **1980**, *45*, 2145. (c) Gage, J. R.; Evans, D. A. *OSC* **1993**, *8*, 528.
30. Lane, C. F.; Myatt, H. C.; Daniels, J.; Hopps, H. B. *JOC* **1974**, *39*, 3052.
31. Le Deit, H.; Cron, S.; Le Corre, M. *TL* **1991**, *32*, 2759.
32. (a) Masuda, Y.; Nunokawa, Y.; Hoshi, M.; Arase, A. *CL* **1992**, 349. (b) Brown, H. C.; Kulkarni, S. U.; Rao, C. G. *S* **1979**, 702. (c) Haken, J. K.; Abraham, F. *J. Chromatogr.* **1991**, *550*, 155.
33. (a) Larson, G. L.; Ortiz, M.; Rodriquez de Roca, M. *SC* **1981**, *11*, 583. (b) Kabalka, G. W.; Bierer, D. *OM* **1989**, *8*, 655.
34. (a) Daito, S.; Hasegawa, T.; Inaba, M.; Nishida, R.; Fujii, T.; Nomizu, S.; Moriwake, T. *CL* **1984**, 1389. (b) Saito, S.; Ishikawa, T.; Kuroda, A.; Koga, K.; Moriwake, T. *T* **1992**, *48*, 4067. (c) Plaumann, H. P.; Smith, J. G.; Rodrigo, R. *CC* **1980**, 354.
35. (a) Hendry, D.; Hough, L.; Richardson, A. C. *TL* **1987**, *28*, 4601. (b) Fairbanks, A. J.; Carpenter, N. C.; Fleet, G. W.; Ramsden, N. G.; de Bello, I. C.; Winchester, B. G.; Al-Daher, S. S.; Nagahashi, G. *T* **1992**, *48*, 3365. (c) Hisadea, Y.; Ihara, T.; Ohno, T.; Murakami, Y. *TL* **1990**, *31*, 1027.
36. Venuti, M. C.; Ort, O. *S* **1988**, 985.
37. Krishnamurthy, S. *TL* **1982**, *23*, 3315.
38. Bonnat, M.; Hercouet, A.; Le Corre, M. *SC* **1991**, *21*, 1579.
39. Deloux, L.; Srebnik, M. *CRV* **1993**, *93*, 763.
40. (a) Jensen, B. L.; Jewett-Bronson, J.; Hadley, S. B.; French, L. G. *S* **1982**, 732. (b) De Kimpe, N.; Stevens, C. *T* **1991**, *47*, 3407.
41. Hunter, R.; Bartels, B.; Michael, J. P. *TL* **1991**, *32*, 1095.
42. Flippin, L. A.; Gallagher, D. W.; Jalali-Araghi, K. *JOC* **1989**, *54*, 1430.
43. (a) Farfan, N.; Contreras, R. *JCS(P2)* **1988**, 1787. (b) Kaushal, P.; Mok, P. L. H.; Roberts, B. P. *JCS(P2)* **1990**, 1663.
44. Bedel, C.; Foucaud, A. *TL* **1993**, *34*, 311.
45. Hünig, S.; Kraft, P. *CB* **1990**, *123*, 895.
46. Spielvogel, B. F. *PAC* **1991**, *63*, 415.
47. Huong, D. M.; Drechsler, M.; Cantow, H. J.; Moeller, M. *Macromolecules* **1993**, *23*, 864.

Marek Zaidlewicz
Nicolaus Copernicus University, Torun, Poland

Borane–Tetrahydrofuran[1]

$\cdot BH_3$

[14044-65-6] $C_4H_{11}BO$ (MW 85.94)

(hydroborating and reducing agent)

Physical Data: $d_4^{20} = 0.898$ g cm^{-3} for 1 M solution in THF.
Solubility: sol THF; reacts violently with water.
Form Supplied in: solution in THF 1 M in BH_3, stabilized with 0.005 M sodium borohydride.
Analysis of Reagent Purity: ^{11}B NMR: δ −0.7 ppm ($J_{B-H} = 103$ Hz).[5] Active hydride is determined by hydrolysis of an aliquot in a mixture of glycerol and water (1:1) and measuring the hydrogen evolved according to a standard procedure.[6a] Boric acid formed by hydrolysis is determined by titration with standard sodium hydroxide in the presence of mannitol.
Preparative Methods: ***Diborane*** is conveniently generated by the reaction of ***Sodium Borohydride*** with ***Boron Trifluoride Etherate*** in diglyme and absorbed in THF.[2] The solution, ca. 2 M in BH_3, may contain traces of boron trifluoride, which poses problems only on rare occasions.[3] Diborane synthesis has been reviewed.[4]
Handling, Storage, and Precautions: use in a fume hood; air- and moisture-sensitive, flammable liquid; handle and store under nitrogen or argon. Commercial solutions are stable over prolonged periods when stored at 0 °C under nitrogen. Unstabilized solutions can be kept for several weeks under those conditions. However, in such solutions the BH_3 concentration may slowly decrease due to the cleavage of THF by borane.[6b]

Introduction. The stoichiometry, directive effects, stereochemical and mechanistic aspects of the hydroboration of alkenes, dienes, alkynes, and functional derivatives are discussed in detail in monographs and reviews[1,7,8] (see also ***Borane–Dimethyl Sulfide***). Here, transformations of major synthetic importance involving organoboranes produced by hydroboration with borane–THF are outlined.

Hydroboration–Oxidation: Synthesis of Alcohols. Oxidation of the intermediate organoborane with alkaline ***Hydrogen Peroxide***, the standard procedure, proceeds with complete retention of configuration to give alcohols in excellent yields (eq 1).[9]

3 + $BH_3\cdot THF$ —(THF, 10 °C)→ $)_3B$ —(H_2O_2, NaOH)→ 3 OH (1)

93%
>99% *exo*

Many functional groups are tolerated.[10] Other oxidation procedures are known.[11] Generally, functional groups in isolated positions do not affect the hydroboration of double or triple bonds. However, proximate substituents may dramatically influence the regioselectivity of addition and stability of the organoborane product.[12–15]

Hydroboration–Elimination: Synthesis of Alkenes. Enamines,[16] certain homoallylic alcohols,[17] and heterocyclic compounds[18] have been transformed into alkenes or dienes by elimination or fragmentation of the organoborane intermediate (eq 2).[17]

OSO_2Me —(1. $BH_3\cdot THF$; 2. OH^-; 90%)→ + (2)

85:15

Synthesis of Hydroperoxides. The low-temperature autoxidation of organoboranes in THF leads to the formation of diperoxyboranes, which provide the corresponding alkylhydroperoxides in excellent yields upon treatment with hydrogen peroxide. Two of the three alkyl groups on boron are utilized (eq 3).[19] The reaction involves a radical chain and the configurational integrity of the alkyl group is not retained.

$$R_3B + O_2 \xrightarrow[-78\ ^\circ C]{THF} (ROO)_2BR \xrightarrow{H_2O_2} 2\ ROOH + ROH \quad (3)$$

81–95%

Hydroboration–Protonolysis: Noncatalytic Stereospecific Hydrogenation. Trialkylboranes undergo protonolysis by treatment with carboxylic acids.[20] The first alkyl group is readily cleaved. Complete removal of alkyl groups requires heating with acid. The protonolysis proceeds with retention of configuration (eq 4). Internal alkynes can be transformed into (*Z*)-alkenes (see also ***Borane–Dimethyl Sulfide***).

—(1. $BH_3\cdot THF$)→ $)_3B$ —(2. $EtCO_2D$, reflux)→ D (4)

38%

Contrathermodynamic Isomerization of Alkenes. Thermal isomerization of alkylboranes at 150–170 °C results in migration of the boron atom to the least hindered position.[21] In aliphatic systems, boron migrates along the carbon chain to the terminal position; however, it cannot pass quarternary centers. Treatment of the isomerized organoborane with a high boiling alkene brings about displacement,[22] and oxidation yields the primary alcohol (eq 5).[23]

$$Me(CH_2)_{13}CH{=}CH(CH_2)_{13}Me \xrightarrow[\text{diglyme, 160 °C, 12 h; 3. } H_2O_2\text{, NaOH}]{1.\ BH_3\cdot THF;\ 2.\ 2\ C_{14}H_{29}CH{=}CH_2}$$

$$Me(CH_2)_{28}CH_2OH + s\text{-}C_{30}H_{61}H + C_{30}H_{60} \quad (5)$$

66% 13% 21%

Functional groups unreactive to borane, e.g. ethers, do not interfere; however, the direction of migration may be influenced by reactive groups, e.g. hydroxy.[24]

Amination: Synthesis of Primary and Secondary Amines. Trialkylboranes are converted by hydroxylamine-*O*-sulfonic acid to primary amines. The reaction proceeds with complete retention of configuration (eq 6).[25a]

1. BH_3•THF; 2. $H_2NNHOSO_3H$; 45% (6)

Other aminating reagents are also known.[26] Secondary amines can be prepared by the reaction of trialkylboranes with organic azides.[27] For better utilization of R groups in these reactions, $RBMe_2$ and $RBCl_2$ can be used, respectively.[25b,26]

Halogenolysis. Alkylboranes are readily converted into the corresponding alkyl chlorides by radical reaction with ***Trichloramine***.[28a] The reaction is not stereospecific; however, with other reagents, e.g. dichloramine T, the stereochemical integrity of the alkyl group is retained.[28b] Brominolysis and iodinolysis of trialkylboranes proceed under mild conditions in the presence of a base.[29] The yields of primary alkyl bromides and iodides are excellent. Three alkyl groups are utilized in the brominolysis and two in the iodinolysis. Secondary alkyl groups react with predominant inversion of configuration (eq 7). In contrast, the dark bromination proceeds with complete retention of configuration (eq 7).[30]

1. BH_3•THF → $)_3B$; Br_2, dark, CH_2Cl_2 → >99% *exo*; X_2, rt → (7)

X = Br, base = NaOMe	75%	25%
X = I, base = MeOH	80%	20%

Mercuration. Mercury(II) salts react with primary trialkylboranes to yield organomercurials (eq 8).[31] Mercuration of secondary alkylboranes is sluggish.

$$RCH{=}CH_2 \xrightarrow[\text{3. } H_2O,\ NaCl;\ 91\text{–}97\%]{\text{1. } BH_3\cdot THF;\ \text{2. } Hg(OAc)_2} 3\ RCH_2CH_2HgCl \quad (8)$$

Sulfuridation: Synthesis of Thioethers, Alkyl Thio- and Selenocyanates. Trialkylboranes are cleaved by dialkyl and diaryl sulfides, producing mixed thioethers (eq 9).[32] The reaction is catalyzed by air. Alkyl thio- and selenocyanates are obtained by treatment of trialkylboranes with iron(III) thio- and selenocyanate, respectively.[33]

$$R_3B + MeSSMe \xrightarrow[74\text{–}94\%]{} RSMe + R_2BSMe \quad (9)$$

Synthesis of Mono- and Dialkylboranes.[34] Several valuable hydroborating and reducing agents,[34] e.g. ***Thexylborane***, ***Disiamylborane***, ***Dicyclohexylborane***, ***9-Borabicyclo[3.3.1]nonane*** (9-BBN, eq 10),[35] and borinane,[36] are synthesized by hydroboration of the corresponding alkenes or dienes with the reagent.

1. BH_3•THF; 2. reflux; 70–80% → 9-BBN (10)

Single Carbon Homologation: Carbonylation, Cyanidation, Dichloromethyl Methyl Ether (DCME) Reaction; Synthesis of Alcohols, Aldehydes, and Ketones. Trialkylboranes react with carbon monoxide at 100–125 °C in diglyme. The reaction proceeds stepwise and the migration of alkyl groups can be controlled to give products of one, two, or three group transfers (eq 11).[37] Cyanidation[38] and the DCME reaction[39a] lead to the same products under milder conditions. Highly hindered tertiary alcohols can be prepared. Annulation and spiroannulation of 1-allylcyclohexenes have been achieved.[39b] All reactions proceed with retention of configuration of alkyl groups. Functional groups, such as ether, ester, nitro, and chloro are tolerated. Unsymmetrically substituted ketones can be obtained by the reaction of trialkylboranes with acyl carbanion equivalents.[40]

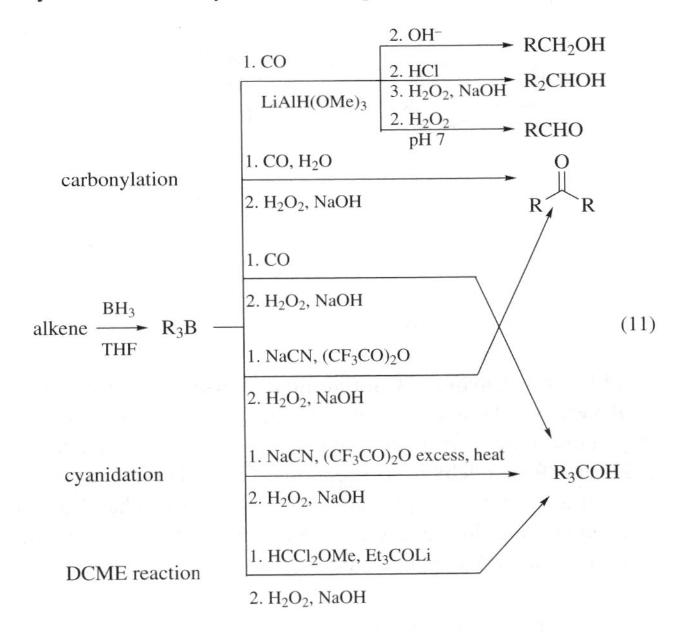

Coupling; Synthesis of Alkanes, Cycloalkanes and (*E*)-Alkenes. Treatment of trialkylboranes with alkaline ***Silver(I) Nitrate*** solution results in coupling of the alkyl groups.[41] Yields for coupling of primary and secondary groups are in the range of 60–80% and 35–50%, respectively.

α-Alkylation of Esters, Ketones and Nitriles. Trialkylboranes react with carbanions generated from α-halo esters, ketones, and nitriles, transferring the alkyl group from boron to carbon.[42] Two different groups can be introduced consecutively, starting with α,α-dihalonitriles.[42d] The anions are generated by hindered bases, e.g. *t*-butoxide, or better, 2,6-di-*t*-butylphenoxide for sensitive compounds (eq 12).

$$R_3B + XCH^-C(O)Y \rightarrow XCH(\bar{B}R_3)C(O)Y \rightarrow RCH(BR_2)C(O)Y \xrightarrow{H_2O} RCH_2C(O)Y \quad (12)$$

X = halogen; Y = H or alkyl

α-Diazocarbonyl compounds react with trialkylboranes directly in the absence of bases at −25 °C.[43] Mild conditions are advantageous when functionalities labile to bases are present (eq 13).[43a]

$$(\text{c-C}_5H_9)_3B + N_2CHCHO \xrightarrow[\text{THF, 25 °C; 98\%}]{H_2O} \text{c-C}_5H_9CH_2CHO \quad (13)$$

α-Bromination–Alkyl Group Transfer: Synthesis of Hindered Tertiary Alcohols. The photochemical α-bromination of trialkylboranes with bromine proceeds readily. Weak bases such as water or THF are sufficient to induce the migration of alkyl groups from boron to the α-brominated carbon. All three alkyl groups may be utilized. It is possible to halt the reaction after the first group migration. Highly hindered tertiary alcohols can be prepared (eq 14).[44]

$$(s\text{-Bu})_3B \xrightarrow[h\nu]{Br_2} \text{(α-bromo)}(s\text{-Bu})_2B \xrightarrow[86\%]{H_2O} R_2B\text{–OH} \xrightarrow[46\%]{1.\ Br_2,\ H_2O,\ h\nu;\ 2.\ [O]} R_3C\text{–OH} \quad (14)$$

Addition to Carbonyl Compounds: Synthesis of Unsymmetrical Ketones. Unlike Grignard and alkyllithium compounds, trialkylboranes are inert to carbonyl compounds. The air-catalyzed addition to formaldehyde is exceptional.[45] Alkyl borates are more reactive and can transfer primary alkyl groups to acyl halides. The reaction provides highly chemoselective boron-mediated synthesis of unsymmetrical ketones (eq 15).[46]

$$s\text{-}R^1{}_3B + R^2Li \rightarrow s\text{-}R^1{}_3\bar{B}R^2\ Li^+ \xrightarrow{R^3COX} R^3COR^2 + s\text{-}R^1{}_3B + LiX \quad (15)$$

57–88%

β-Alkylation of Conjugated Carbonyl Compounds and Oxiranes. Trialkylboranes undergo radical conjugate addition to various α,β-unsaturated carbonyl compounds, such as enals, enones, quinones, unsaturated nitriles, crotonaldimines, and ynones, and also to alkenyl- and alkynyloxiranes[47] (eq 16).[47b]

$$R_3B + CH_2{=}C(Br)CHO \xrightarrow[25\ °C]{THF} RCH_2C(Br){=}CHOBR_2 \xrightarrow{H_2O} RCH_2CH(Br)CHO \quad (16)$$

65–85%

Functional Group Reductions.[48] The reactivity order of representative functional groups toward borane–THF is[49a] carboxylic acids > aldehydes > ketones > alkenes ≫ nitriles > epoxides > esters > acid chlorides. Acid anhydrides, amides, lactones, acetals, oximes, oxime ethers, imines, and hydrazones are also reduced. Nitro compounds, organic halides, sulfones, sulfonic acids, disulfides, thiols, alcohols, phenols (hydrogen evolution), and amines are stable to borane–THF.[49]

Carboxylic Acids and Derivatives. Borane–THF is the reagent of choice for the reduction of carboxylic acids to alcohols.[50–56] Selective reduction in the presence of esters, halogen derivatives, nitrileLane, C.s, amides, lactones, nitro compounds, amino, phenolic, and other groups is possible (eq 17).[51] Carboxylic acid salts are reduced with 2 equiv of $BH_3 \cdot THF$.[52]

$$EtO_2C\text{–CH=CH–}CO_2H \xrightarrow[\text{8–10 h; 67\%}]{BH_3\cdot THF,\ -10\ °C\ \text{to rt}} EtO_2C\text{–CH=CH–CH}_2OH \quad (17)$$

Aromatic acids containing electron-donating groups may undergo overreduction to hydrocarbons.[57] Other carboxylic groups can also be converted into the methyl group via the reduction of *N*-acyl-*N'*-tosylhydrazines.[58] Isotopically labeled chiral methyl groups have been obtained by a sequence of reactions involving $BD_3 \cdot THF$.[59]

Acid anhydrides are reduced to alcohols at a slower rate than carboxylic acids.[60] Aliphatic esters and lactones react slowly (6–12 h at 0 °C) to give alcohols or ethers.[49,61–65]

The reduction of amides to amines is another major synthetic application of the reagent. The reactivity order is tertiary > secondary ≫ primary. All types of amides and lactams are reduced rapidly and quantitatively by excess of borane in refluxing THF.[66–69]

A large-scale preparation of an intermediate in the synthesis of nitrobenzyl-DOTA, an efficient chelating agent for metal ions, used in biochemical studies, is shown in eq 18.[70]

$$O_2N\text{–C}_6H_4\text{–CH}_2CH(NH_2\cdot TFA)C(O)NHCH_2C(O)NHCH_2C(O)NHCH_2C(O)OH \xrightarrow[\text{reflux; 93\%}]{BH_3\cdot THF} O_2N\text{–C}_6H_4\text{–CH}_2CH(NH_2)CH_2NHCH_2CH_2NHCH_2CH_2NHCH_2CH_2OH \quad (18)$$

Nitriles are readily reduced with an excess of borane in refluxing THF to give the corresponding amines upon acid hydrolysis of the intermediate borazines.[66,71] Less reactive functional groups are tolerated. The initial intermediate *N*-borylimine can be alkylated to give carbinamines (eq 19).[72] Cyanohydrins are reduced to amino alcohols; however, α-aminonitriles undergo decyanation.[73]

$$\text{RC≡N} \xrightarrow[20\,°C,\ 1\,h]{BH_3•THF} \text{R(H)C=N–BH}_2 \xrightarrow[2.\ H_2O]{1.\ R'Li,\ -80\,°C} \text{RR'CHNH}_2\ \ 53–95\% \quad (19)$$

Ketones and Derivatives. Although aldehydes and ketones are readily reduced by borane,[49] other reagents of higher selectivity are widely used for that purpose. In some cases, however, the use of borane may be advantageous, as demonstrated in the synthesis of ciramadol involving a stereoselective reduction of 2-(α-dimethylamino-*m*-hydroxybenzyl)cyclohexanone[74a] and in highly diastereoselective keto boronate reduction (eq 20).[74b]

(20) 1. BH_3•THF, –78 °C; 2. NaOH, H_2O_2; 3. Ac_2O, pyridine; 71%; OAc/OAc diastereomers 34:1

The asymmetric reductions of prochiral ketones are now dominated by catalytic reactions[75] and highly selective stoichiometric reagents.[76] Borane–THF reacts with chiral amino alcohols to give asymmetric reducing agents and serves as a source of BH_3 in enantioselective reductions of prochiral ketones catalyzed by oxazaboralidines and oxazaphospholidines (eq 21).[77] Catalysts of type 2 are best studied and new modifications are prepared.[75,78]

$$\text{RC(O)R'} + BH_3•THF \xrightarrow[THF]{catalyst} \text{RCH(OH)R'} \quad (21)$$

R,R' = Ph, Me cat. 1, 94% ee; cat. 2, 97% ee
R,R' = *i*-Pr, Me cat. 1, 60% ee; cat. 3, 92% ee

cat. 1 cat. 2 cat. 3

Oximes react with borane in refluxing THF to give the corresponding hydroxylamine derivatives (see also ***Borane–Dimethyl Sulfide***).[79] Oximes of aryltrifluoromethyl ketones are reduced at room temperature (not with BMS).[80] Oxime ethers and acetates react under mild conditions.[81] ***Aluminum Hydride*** (alane) is also a convenient reagent for these reductions.[82] Reagents for asymmetric reduction of prochiral ketoxime *O*-alkyl ethers and imines are prepared from borane–THF and chiral amino alcohols.[75,77a,83]

Cleavage of Single Bonds. Epoxides react slowly with borane–THF to yield mixtures of products.[84] In the presence of small amounts of sodium borohydride or boron trifluoride, the reaction is greatly accelerated and the epoxide ring is cleanly opened. Although anti-Markovnikov products predominate, the regioselectivity is high only in certain cases, e.g. styrene and indene epoxides give 2-phenylethanol and indan-2-ol, respectively.[84,85] In contrast, the reduction of *s*-cisoid α,β-unsaturated epoxides with borane–THF proceeds stereoselectively with the double bond shift to give allylic alcohols of (*Z*) configuration (eq 22).[86] Similarly, α,β-unsaturated aziridines are reduced to (*Z*) allylic amines.[87] Acetals and tetrahydropyranyl ethers undergo reductive cleavage[88] (eq 23).[88c]

(22) 1. BH_3•THF; 2. NaOH; 64%; → CH_2OH

(23) BH_3•THF; 97%

Imidazolidines and oxazolidines are also cleaved,[89] whereas thiazolidines appear to be stable to the reagent at rt.[90]

Nitro Compounds. The nitro group itself is stable to the reagent; however, aci nitro salts and α,β-unsaturated nitro compounds are reduced to the corresponding hydroxylamines.[91,92]

Other Applications. Borane–THF reacts with Grignard reagents, arylmercury, arylthalium, and allyl- and propargyllithium compounds to give organoboranes which can be oxidized to the corresponding alcohols, phenols, and 1,3-diols.[93] The reagent is used for blocking tertiary amine groups in Friedel–Crafts cyclizations and oxidative phenol coupling reactions in alkaloid syntheses.[94] It activates α,β-unsaturated acids in the reaction with 1,3-dienes[95] and is used in the synthesis of catalysts for asymmetric Diels–Alder and aldol reactions.[75] Trialkylboranes are useful intermediates in the synthesis of compounds labeled with various isotopes for medical applications.[96]

Related Reagents. See Classes R-1, R-2, R-3, R-9, R-10, R-11, R-12, R-13, R-17, R-20, R-23, R-31, and R-32, pages 1–10. Borane–Ammonia; Borane–Dimethyl Sulfide; Borane–Pyridine; Diborane.

1. (a) Brown, H. C. *Hydroboration*; Benjamin: New York, 1962. (b) Brown, H. C. *Boranes in Organic Chemistry*; Cornell University Press: Ithaca, NY, 1972. (c) Brown, H. C. *Organic Syntheses via Boranes*; Wiley: New York, 1975. (d) Pelter, A.; Smith, K.; Brown, H. C. *Borane Reagents*; Academic: London, 1988. (e) Lane, C. In *Synthetic Reagents*, Pitzey, J. S., Ed.; Ellis Horwood: Chichester, 1977; Vol. 3, p 1.
2. (a) Ref. 1c. (b) Zweifel, G.; Brown, H. C. *OR* **1963**, *13*, 1.
3. Biswas, K. M.; Jackson, A. H. *JCS(C)* **1970**, 1667.
4. Long, A. H. *Progr. Inorg. Chem.* **1972**, *15*, 1.
5. Phillips, W. D.; Miller, H. C.; Muetterties, E. L. *JACS* **1959**, *81*, 4496.
6. (a) Ref. 1c, p 241. (b) Kollonitsch, J. *JACS* **1961**, *83*, 1515.
7. Brown, H. C.; Chandrasekharan, J. *G* **1987**, *117*, 517.
8. (a) Ref. 1. (b) Mikhailov, B. M.; Bubnov, Yu. N. *Organoboron Compounds in Organic Synthesis*; Horwood: London, 1984. (c) Smith, K.; Pelter, A. *COS* **1991**, *8*, 703. (d) Zaidlewicz, M. In *Comprehensive Organometallic Chemistry*; Wilkinson, G.; Stone, F. G. A.; Abel, E. W., Ed.; Pergamon: Oxford, 1981; Vol. 7, p 143.
9. (a) Ref. 1c, p 23. (b) Brown, H. C.; Zweifel, G. *JACS* **1961**, *83*, 2544.

10. Ref. 8d, p 229.
11. Pelter, A.; Smith, K. Ref. 8c, Vol. 7, p 593.
12. (a) Brown, H. C.; Sharp, R. L. *JACS* **1968**, *90*, 2915. (b) Brown, H. C.; Gallivan, Jr., R. M. *JACS* **1968**, *90*, 2906. (c) Ref. 8d, p 239.
13. Larson, G. L.; Hernandez, D.; Hernandez, A. *JOM* **1974**, *76*, 9.
14. Borowitz, I. J.; Williams, G. J. *JOC* **1967**, *32*, 4157.
15. Dunkelblum, E.; Levene, R.; Klein, J. *T* **1972**, *28*, 1009.
16. (a) Lewis, J. W.; Pearce, A. A. *JCS(B)* **1969**, 863. (b) Froborg, J.; Magnuson, G.; Thoren, S. *TL* **1975**, 1621.
17. Marshall, J. A. *S* **1971**, 229.
18. Zweifel, G.; Plamondon, J. *JOC* **1970**, *35*, 898.
19. Brown, H. C.; Midland, M. M. *T* **1987**, *43*, 4059.
20. (a) Ref. 8c, p 724. (b) Brown, H. C.; Murray, K. J. *T* **1986**, *42*, 5497.
21. (a) Brown, H. C.; Zweifel, G. *JACS* **1966**, *88*, 1433. (b) *JACS* **1967**, *89*, 561.
22. Brown, H. C.; Bhatt, M. V.; Munekata, T.; Zweifel, G. *JACS* **1967**, *89*, 567.
23. Maruyama, K.; Terada, K.; Yamamoto, Y. *JOC* **1980**, *45*, 737.
24. Sisido, K.; Naruse, M.; Saito, A.; Utimoto, K. *JOC* **1972**, *37*, 733.
25. (a) Rathke, M. W.; Inoue, N.; Varma, K. R.; Brown, H. C. *JACS* **1966**, *88*, 2870. (b) Brown, H. C.; Kim, K-W.; Srebnik, M.; Singaram, B. *T* **1987**, *43*, 4071.
26. Kabalka, G. W.; Goudgaon, N. M.; Liang, Y. *SC* **1988**, *18*, 1363.
27. Brown, H, C.; Midland, M. M.; Levy, A. B.; Suzuki, A.; Sono, S.; Itoh, M. *T* **1987**, *43*, 4079.
28. (a) Brown, H. C.; De Lue, N. R. *T* **1988**, *44*, 2785. (b) Nelson, D. J.; Soundararajan, R. *JOC* **1987**, *53*, 5664.
29. (a) Brown, H. C.; Lane, C. F. *T* **1988**, *44*, 2763. (b) Brown, H. C.; Rathke, M. W.; Rogic′, M. M.; De Lue, N. R. *T* **1988**, *44*, 2751.
30. Brown, H. C.; Lane, C. F.; De Lue, N. R. *T* **1988**, *44*, 2773.
31. Larock, R. C. *T* **1982**, *38*, 1713.
32. Brown, H. C.; Midland, M. M. *JACS* **1971**, *93*, 3291.
33. Arase, A.; Masuda, Y. *CL* **1976**, 1115.
34. (a) Ref. 1c, p 37. (b) Ref. 8d, p 161. (c) Brown, H. C.; Negishi, E. *T* **1977**, *33*, 2331.
35. (a) Brown, H. C.; Knights, E. F.; Scouten, C. G. *JACS* **1974**, *96*, 7765. (b) Köster, R.; Yalpani, M. *PAC* **1991**, *63*, 387.
36. Negishi, E.; Burke, P. L.; Brown, H. C. *JACS* **1972**, *94*, 7431.
37. (a) Brown, H. C.; *ACR* **1969**, *2*, 65. (b) Ref. 1c, p 126. (c) Ref. 1d, p 272.
38. (a) Pelter, A. *CSR* **1982**, *11*, 191. (b) Ref. 1c, p 132. (c) Ref. 1d, p 280.
39. (a) Brown, H. C.; Carlson, B. A. *JOC* **1973**, *38*, 2422. (b) Akers, J. A.; Bryson, T. A. *TL* **1989**, *30*, 2187.
40. Ncube, S.; Pelter, A.; Smith, K. *TL* **1979**, 1895.
41. (a) Brown, H. C.; Verbrugge, C.; Snyder, C. H. *JACS*. **1962**, *83*, 1002. (b) Avasthi, K.; Ghosh, S. S.; Devaprabhakara, D. *TL* **1976**, 4871. (c) Murphy, R.; Prager, R. H. *TL* **1976**, 463. (d) Ref. 1c, p 125.
42. Ref. 1c, p 135. Ref. 1d, p 261.
43. (a) Hooz, J.; Morrison, G. F. *CJC* **1970**, *48*, 868. (b) Kono, H.; Hooz, J. *OSC* **1988**, *6*, 919.
44. Lane, C. F.; Brown, H. C. *JACS* **1971**, *93*, 1025.
45. Miyaura, N.; Itoh, M.; Suzuki, A.; Brown, H. C.; Midland, M. M.; Jacob, III, P. *JACS* **1972**, *94*, 6549.
46. Negishi, E.; Chiu, K. W.; Yosida, T. *JOC* **1975**, *40*, 1676.
47. (a) Ref. 1d, p 301. (b) Brown, H. C.; Kabalka, G. W.; Rathke, M. W.; Rogic′, M. M. *JACS* **1968**, *90*, 4165.
48. (a) Lane, C. F. *CR* **1976**, *76*, 773. (b) Pelter, A.; Smith, K. In *Comprehensive Organic Chemistry*; Barton, D. H. R.; Ollis, W. D., Eds.; Pergamon: Oxford, 1979; Vol. 3, p 695.
49. (a) Brown, H. C.; Heim, P.; Yoon, N. M. *JACS* **1970**, *92*, 1637. (b) Brown, H. C.; Korytnyk, N. *JACS* **1960**, *82*, 3866.
50. (a) Yoon, N. M.; Pak, C. S.; Brown, H. C.; Krishnamurthy, S.; Stocky, T. P. *JOC* **1973**, *38*, 2786. (b) Brown, H. C.; Stocky, T. P. *JACS* **1977**, *99*, 8218.
51. Kende, A. S.; Fludzinski, P. *OS* **1986**, *64*, 104.
52. Yoon, N. M.; Cho, B. E. *TL* **1982**, *23*, 2475.
53. Artz, S. P.; Cram, D. J. *JACS* **1984**, *106*, 2160.
54. Doxsee, K. M.; Feigel, M.; Stewart, K. D.; Canary, J. W.; Knobler, C. B.; Cram, D. J. *JACS* **1987**, *109*, 3098.
55. Choi, Y. M.; Emblidge, R. W.; Kucharczyk, N.; Sofia, R. D. *JOC* **1989**, *54*, 1194.
56. Huang, F-C.; Hsu Lee, L. F.; Mittal, R. S. D.; Ravikumar, P. R.; Chan, J. A.; Sih, C. J.; Caspi, E.; Eck, C. R. *JACS* **1975**, *97*, 4144.
57. Littel, R.; Allen, Jr.; G. R. *JOC* **1973**, *38*, 1504.
58. Attanasi, O.; Caglioti, L.; Gasparrini, F.; Misiti, D. *T* **1975**, *31*, 341.
59. O’Connor, E. J.; Kobayashi, M.; Floss, H. G.; Gladysz, J. A. *JACS* **1987**, *109*, 4837.
60. Yoon, N. M.; Lee, W. S. *Bull. Korean Chem. Soc.* **1986**, *7*, 296.
61. Pettit, G. R.; Piatak, D. M. *JOC* **1962**, *27*, 2127.
62. White, R. E.; Gardlund, Z. G. *JPS(A1)* **1972**, *8*, 1419.
63. Jackson, A. H.; Kenner, G. W.; Sach, G. S. *JCS(C)* **1967**, 2045.
64. Jackson, A. H.; Naidoo, B. *JCS(P2)* **1973**, 548.
65. (a) Pettit, G. R.; Dias, J. R. *JOC* **1971**, *36*, 3485. (b) Sitzmann, M. E.; Gilligan, W. H. *JHC* **1986**, *23*, 81.
66. Brown, H. C.; Heim, P. *JOC* **1973**, *38*, 912.
67. (a) Danishefsky, S. J.; Harrison, P. J.; Webb, R. R.; O’Neill, B. T. *JACS* **1985**, *107*, 1421. (b) Sammes, P. G.; Smith, S. *CC* **1982**, 1143.
68. Northrop, Jr., R. C.; Russ, P. L. *JOC* **1977**, *42*, 4148.
69. (a) Treadgill, M. D.; Webb, P. *SC* **1990**, *20*, 2319. (b) Denmark, S. E.; Marlin, J. E. *JOC* **1987**, *52*, 5742. (c) Nagarajan, S.; Ganem, B. *JOC* **1986**, *51*, 4856.
70. Renn, O.; Meatres, C. F. *Bioconjugate Chem.* **1992**, *3*, 563.
71. Hutchins, R. O.; Maryanoff, B. E. *OSC* **1988**, *6*, 223.
72. Itsuno, S.; Hachisuka, C.; Ito, K. *JCS(P1)* **1991**, 1767.
73. Ogura, K.; Shimamura, Y.; Fujita, M. *JOC* **1991**, *56*, 2920.
74. (a) Yardley, J. P.; Fletcher, III, H.; Russell, P. B. *Experientia* **1978**, *34*, 1124. (b) Molander, G. A.; Bobbitt, K. L.; Murray, C. K. *JACS* **1992**, *114*, 2759.
75. Deloux, L.; Srebnik, M. *CRV* **1993**, *93*, 763.
76. Brown, H. C.; Ramachandran, P. V. *ACR* **1992**, *25*, 16.
77. (a) Itsuno, S.; Nakano, M.; Miyazaki, K.; Msauda, H.; Ito, K.; Hirao, A.; Nakahama, S. *JCS(P1)* **1985**, 2039. (b) Corey, E. J.; Bakshi, R. K.; Shibata, S. *JACS* **1987**, *109*, 5551. (c) Brunel, J-M.; Pardigon, O.; Faure, B.; Buono, G. *CC* **1992**, 287.
78. Corey, E. J.; Link, J. O. *TL* **1992**, *33*, 4141.
79. Feuer, H.; Vincent, Jr., B. F.; Bartlett, R. S. *JOC* **1965**, *30*, 2877.
80. Kerdesky, F. A. J.; Horrom, B. W. *SC* **1991**, *21*, 2203.
81. (a) Itsuno, S.; Tanaka, K.; Ito, K. *CL* **1986**, 1133. (b) Ganem, B. *TL* **1976**, 1951.
82. (a) Yoon, N. M.; Brown, H. C. *JACS* **1968**, *90*, 2927. (b) Zaidlewicz, M.; Uzarewicz, I. G. *HC* **1993**, *4*, 73.
83. (a) Sakito, Y.; Yoneyoshi, Y.; Suzukamo, G. *TL* **1988**, *29*, 223. (b) Cho, B. T.; Chun, Y. S. *JCS(P1)* **1990**, 3200.
84. (a) Pasto, D. J.; Cumbo, C. C.; Hickman, J. *JACS* **1966**, *88*, 2201. (b) Marshall, P. A.; Prager, R. H. *AJC* **1977**, *30*, 141.
85. (a) Brown, H. C.; Yoon, N. M. *JACS* **1968**, *90*, 2686. (b) Brown, H. C.; Yoon, N. M. *CC* **1968**, 1549.
86. Zaidlewicz, M.; Uzarewicz, A.; Sarnowski, R. *S* **1979**, 62.
87. Chaabouni, R.; Laurent, A.; Marquet, B. *T* **1980**, *36*, 877.
88. (a) Bryson, T. A.; Akers, J. A.; Ergle, J. D. *SL* **1991**, 499. (b) Cossy, J.; Bellosta, V.; Müller, M. C. *TL* **1992**, *33*, 5045. (c) Castro, P. P.; Tihomirov, S.; Gutierrez, C. G. *JOC* **1988**, *53*, 5179.

89. Northrop, Jr., R. C.; Russ, P. L. *JOC* **1975**, *40*, 558.
90. Theobald, D. W. *JOC* **1965**, *30*, 3929.
91. Feuer, H.; Bartlett, R. S.; Vincent, Jr., B. F.; Anderson, R. S. *JOC* **1965**, *30*, 2880.
92. Mourad, M. S.; Varma, R. S.; Kabalka, G. W. *JOC* **1985**, *50*, 133.
93. (a) Breuer, S. W.; Broster, F. A. *JOM* **1972**, *35*, C5. (b) Breuer, S. W.; Leatham, M. J.; Thorpe, F. G. *CC* **1971**, 1475. (c) Santaniello, E.; Fiecchi, A.; Ferraboschi, P. *CC* **1982**, 1157. (d) Breuer, S. W.; Pickles, G. M.; Podesta, J. C.; Thorpe, F. G. *CC* **1975**, 36. (e) Medlik-Balan, A.; Klein. J. *T* **1980**, *36*, 299.
94. (a) Monkovic′, I.; Bachand, C.; Wong, H. *JACS* **1978**, *100*, 4609. (b) Schwartz, M. A.; Rose, B. F.; Vishnuvajjala *JACS* **1973**, *95*, 612. (c) Kupchan, S. M.; Kim, C-K. *JOC* **1976**, *41*, 3210. (d) Pesaro, M.; Bachmann, J-P. *CC* **1978**, 203.
95. Furuta, K.; Miwa, Y.; Iwanaga, K.; Yamamoto, H. *JACS* **1988**, *110*, 6254.
96. (a) Kabalka, G. W. *ACR* **1984**, *17*, 215. (b) *PAC* **1991**, *63*, 379.

Marek Zaidlewicz
Nicolaus Copernicus University, Torun, Poland

Herbert C. Brown
Purdue University, West Lafayette, IN, USA

Bromine[1]

Br_2

[7726-95-6] Br_2 (MW 159.81)

(powerful brominating and oxidizing agent; can initiate/participate in ring cleavage and rearrangement)

Physical Data: mp −7 °C; bp 59 °C; *d* 3.12 g cm^{-3}.

Solubility: sol H_2O, acetic acid, alcohol, ether, chloroform, carbon tetrachloride, carbon disulfide, hydrocarbon solvents (pentane, petroleum ether).

Form Supplied in: dark, red-brown, volatile liquid; also available as a 1 M solution in carbon tetrachloride.

Analysis of Reagent Purity: by iodometric titration.[79]

Purification: several methods have been described.[80]

Handling, Storage, and Precautions: bromine is an extremely corrosive and toxic reagent in both liquid and vapor form. As a liquid, it produces painful burns and blisters when spilled on the skin. Such burns should be flushed with water and neutralized with a 10% solution of sodium thiosulfate in water. Medical attention should be sought immediately. Protective clothing is therefore a must, including laboratory coat and apron, protective gloves, and a full-face respirator equipped with a NIOSH-approved organic vapor–acid gas canister. Bromine should be stored in a cool, dry area. It is incompatible with combustibles, liquid ammonia, alkali hydroxides, metals (including aluminum, mercury, magnesium, and titanium), and some types of rubber and plastic.[2] Use in a fume hood.

Halogenations. Bromine is a very powerful brominating agent that has found utility in a variety of systems. While the bromination of alkanes is usually not a viable synthetic method,[3] alkylbenzenes can be brominated at the benzylic position under radical conditions (eq 1).[4] ***N-Bromosuccinimide*** (NBS) can also be used for this transformation.

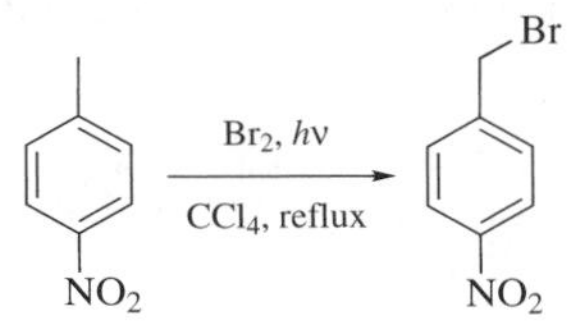

(1)

Electrophilic aromatic substitution occurs in the presence of Lewis acids to provide brominated aromatics.[5] Monobromination usually occurs due to the deactivating nature of the bromine. However, highly reactive aromatics, such as phenols, anilines, and polyalkylbenzenes, are frequently polybrominated, even in the absence of catalysts.

Bromine has been used in the bromination of heterocycles. However, the unique reactivity patterns of each heterocyclic ring system create a discussion beyond the scope of this review. The reader is directed to reviews on the subject.[1a,6]

The addition of bromine to alkenes[7] proceeds with formation of the cyclic bromonium ion (**1**), which can then be intercepted by an anionic species (Y^-), to give the product derived from *anti* addition (eq 2). In the case of bromine itself, this gives rise to *trans* vicinal dibromides (eq 2; Y = Br). Variations from ideality are not uncommon due to weakened, unsymmetrical bridging in the bromonium ion (eq 3),[8] transannular interactions (eq 4),[9] and substrates susceptible to rearrangements. Brominations in the presence of crown ethers[10] and zeolites[11] have been investigated to improve selectivity. Conjugated dienes give predominantly 1,4-addition, while alkynes are less susceptible to electrophilic attack.[12]

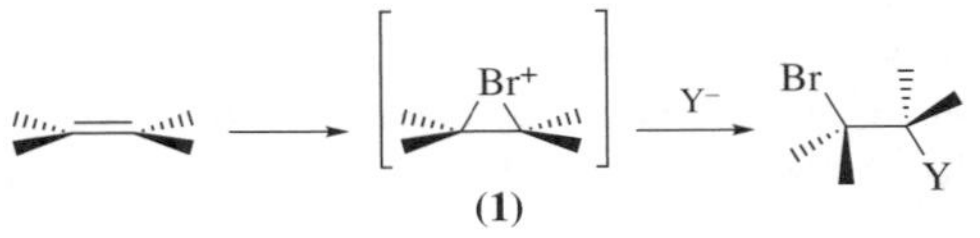

(2)

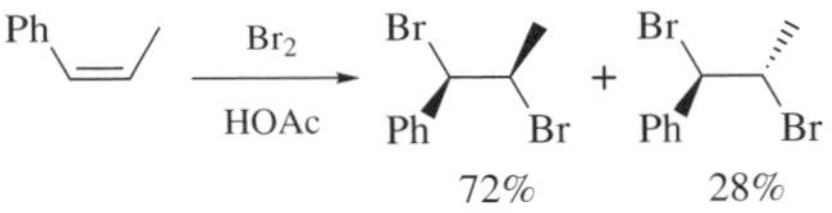

(3)

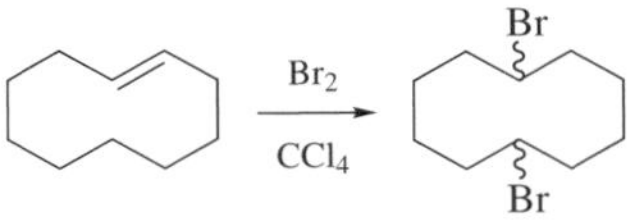

(4)

The addition of bromine to alkenes has been used as the first step in the oxidation of alkenes to 1,3-butadienes (eqs 5 and 6).[13,14] Alkenes can also be protected[15] or purified[16] by bromination and subsequent regeneration of the double bond.

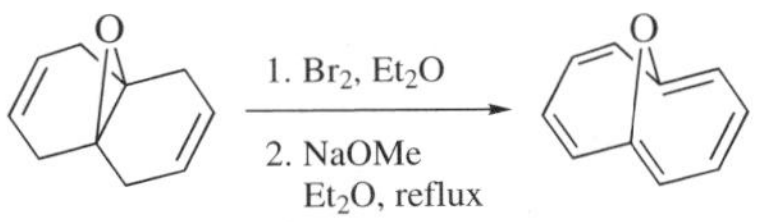

(5)

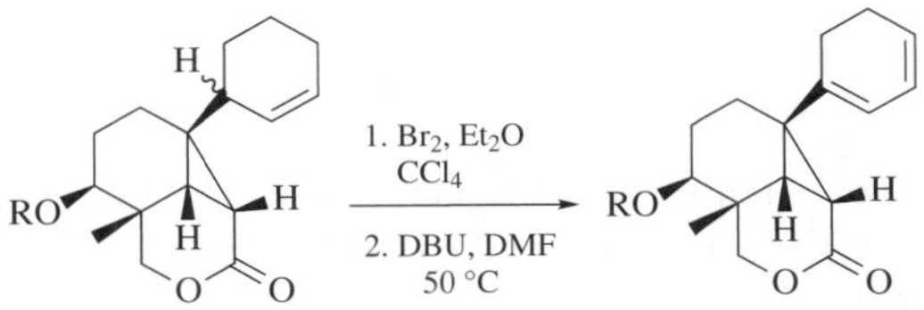

(6)

Alkenes bearing an electron-withdrawing group at one terminus are frequently converted to the α-bromo analogs via a bromination/dehydrobromination sequence (eqs 7 and 8).[17,18]

Br_2, Et_3N / CCl_4 (7)

1. Br_2, CCl_4 2. Et_3N, benzene (8)

Cyclic bromonium ions (**1**) can be opened by a variety of other nucleophiles. Thus bromination of alkenes in aqueous systems can lead to bromohydrins. However, NBS has been shown to be superior to bromine for this transformation, presumably due to the minimization of competing bromide ion in the reaction mixture.[19] Alcohols react to give vicinal bromo ethers, in both inter- and intramolecular fashion (eq 9).[20]

Br_2, CH_2Cl_2 (9)

Bromolactonization of alkenic acids has been the subject of extensive investigation.[21] Cyclizations can be performed on the carboxylic acid salts as well as the free acids. Thallium(I) salts have proven to be especially efficacious.[21b,22] Treatment of the mercury(II) salts with bromine proceeds via a radical mechanism and provides the expected products in substrates where normal bromolactonization conditions lead to rearrangement (eq 10).[23] Other sources of electrophilic bromine can also give different products (eq 11).[21b]

HgO, Br_2 / CCl_4; Br_2, Na_2CO_3 / H_2O (10)

Tl_2CO_3, Br_2 / CH_2Cl_2; NBS / DMF (11)

The enol lactonization of alkynic acids can be performed to give either the (*E*)- or (*Z*)-bromo enol isomers depending on reaction conditions (eq 12).[24]

NaOH, $AgNO_3$ / Br_2, MeCN → (*Z*); K_2CO_3, CH_2Cl_2 / NBS, H_2O → (*E*) (12)

Alkenic amides cyclize under standard conditions to form lactones rather than lactams.[25] Bromolactamization can be achieved, however, by introduction of substituents on the amide nitrogen that serve to lower its pK_a (eqs 13 and 14).[26,27]

Br_2, $NaHCO_3$ / CH_2Cl_2, H_2O (13)

Br_2, K_2CO_3 / MeCN, H_2O (14)

R = Me, *cis*:*trans* = 23:77
R = CbzNH, *cis*:*trans* = 70:30

Electron-rich alkenes, such as enol ethers[28] and enamines,[29] can be brominated to furnish the β-bromo compounds (eq 15).

1. Br_2, Et_2O 2. Et_3N, Et_2O (15)

Bromine has been used for brominations α to carbonyl groups.[30] Carboxylic acids are brominated in the presence of phosphorus or phosphorus trihalides in the classical Hell–Volhard–Zelinski reaction (eq 16).[31] Variations on this include brominations in thionyl chloride[32] and in polyphosphoric acid.[33] The less reactive carboxylic esters are frequently converted to the acid halide, α-brominated, and subsequently re-esterified in one pot.[34,35]

Br_2, PCl_3 (cat) / benzene, reflux (16)

The bromination of ketones is believed to occur via acid-catalyzed enolization, followed by electrophilic attack on the enol form.[30] Unsymmetrical ketones can give rise to mixtures of bromo ketones due to mixtures of enols, and several approaches to overcome this shortcoming have been reported. Radical bromination in the presence of epoxides (as acid scavengers) allows for substitution at the more highly substituted position (eq 17).[36] Silyl enol ethers of aldehydes and ketones react with bromine (or NBS) to give the α-brominated carbonyl compounds (eq 18).[37] This, combined with the ability to regiospecifically prepare silyl enol ethers (kinetic vs. thermodynamic), makes for an extremely useful technique for the preparation of α-bromo carbonyl compounds.

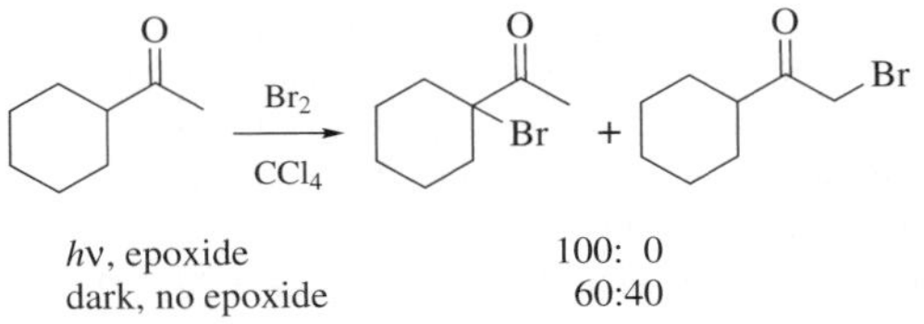

(17)

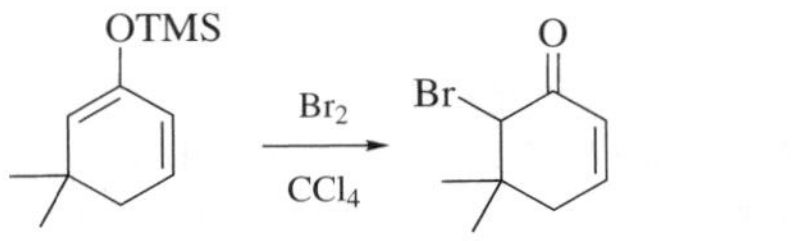

$h\nu$, epoxide	100: 0
dark, no epoxide	60:40

(18)

Sulfoxides are best α-brominated with a combination of bromine and NBS in pyridine.[38]

Bromine has been used in the halogenation of organometallic reagents. Organomagnesium,[39] organolithium,[40] and organoaluminum[41] reagents react to give the compounds in which the metal has been replaced by bromine. Organoboranes can react with bromine in several ways. Bromination in the presence of sodium methoxide gives the corresponding alkyl bromides.[42] Photobromination (in the absence of strong base) gives an initial α-bromo organoborane that can either give the corresponding alkyl bromide[43] or rearrange to a new organoborane.[44] Organoboranes can also be converted to alkyl bromides in aqueous media.[45] Alkenic bromides have been prepared from alkenylboronic esters with inversion of configuration (eq 19).[46]

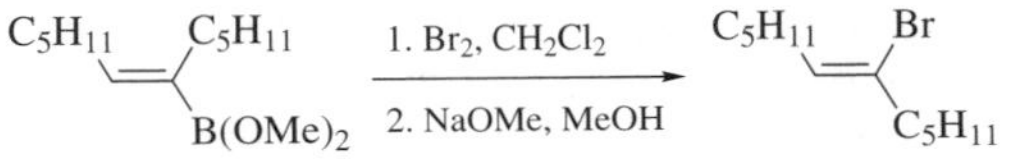

(19)

Enolates of ketones[47] and esters[48] can be brominated by treatment with bromine (eq 20), as can the anions of terminal alkynes.[49] The high reactivity of bromine, however, is sometimes a problem; milder sources of electrophilic bromine (such as ***1,2-Dibromoethane***) are occasionally used in its place.

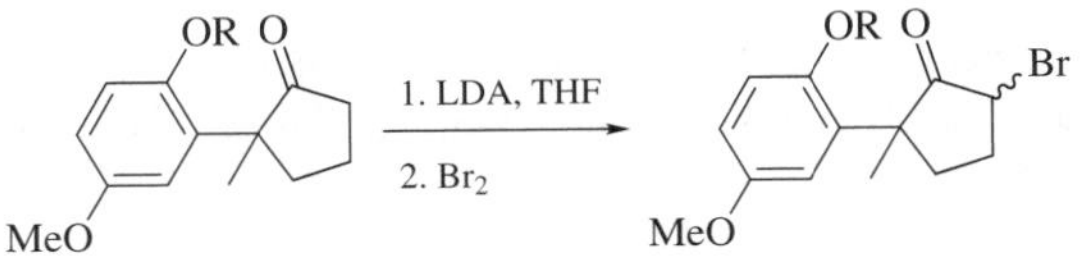

(20)

Oxidations. Bromine reacts with secondary alcohols to give ketones. Since ketones are subject to bromination themselves (see above), the α-bromo ketones can sometimes be undesirable byproducts. However, in cases where there are no α-protons, this can provide an excellent method of oxidation (eq 21).[50]

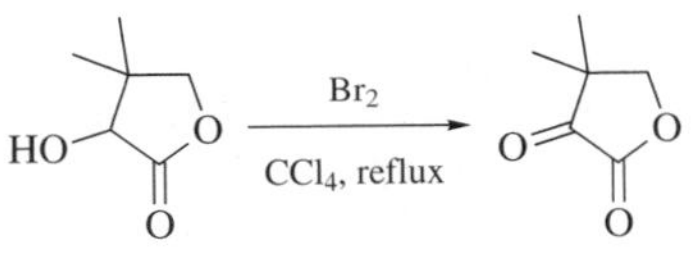

(21)

Primary alcohols are oxidized to either aldehydes or, more commonly, esters. An especially attractive corollary to this involves the oxidation of acetals to esters (eq 22).[51]

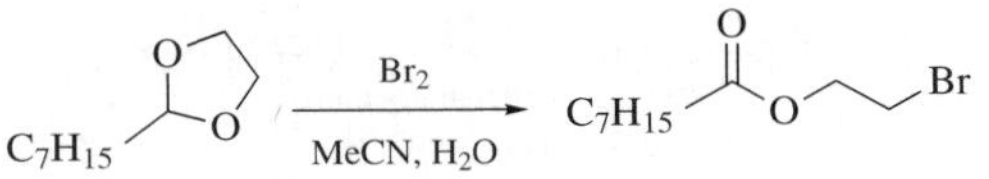

(22)

The addition of coreactants has provided a number of selective bromine-based oxidants. Both bromine/***Hexamethylphosphoric Triamide*** (HMPA)[52] and bromine/***Bis(tri-n-butyltin) Oxide*** (HBD)[53] have shown a preference for the oxidation of secondary vs. primary alcohols (eq 23), while bromine/nickel carboxylates[54] convert 1,4-diols to γ-butyrolactones by selective oxidation of the primary alcohols (eq 24).

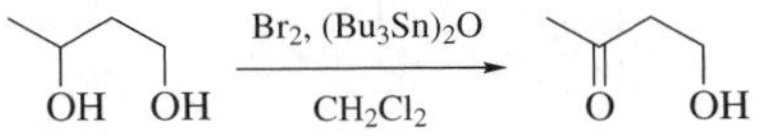

(23)

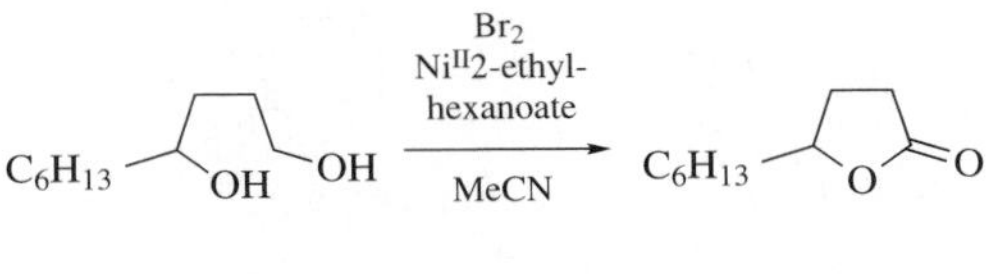

(24)

Tetrahydrofurans have been prepared from alcohols or diols by bromine/silver(I) salts (eq 25)[55] or bromine/DMSO,[56] respectively.

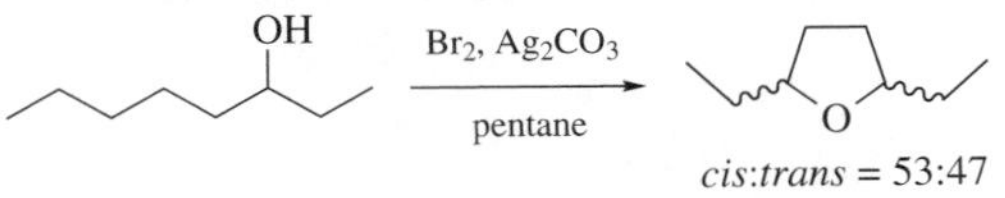

(25)

cis:*trans* = 53:47

Bromine has demonstrated its superiority in the oxidation of enediol bis-trimethylsilyl ethers to α-diketones (eq 26).[57]

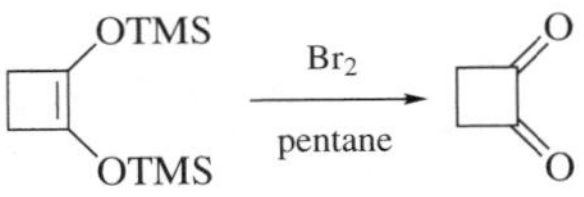

(26)

A number of other functional groups are oxidized by bromine; however, they do not appear to have gained widespread use. These include cyclohexenones,[58] ethers,[59] hydrazines,[60] oximes,[61] tertiary amines,[62] thiols,[63] sulfides,[64] and organoselenium reagents.[65]

Rearrangements. Bromine reacts with a number of functional groups to effect bond cleavage or other skeletal rearrangements. In the classical Hofmann rearrangement,[66] treatment of primary amides with bromine in the presence of base gives isocyanates, carbamates, or amines, depending on the reaction conditions (eq 27).[67]

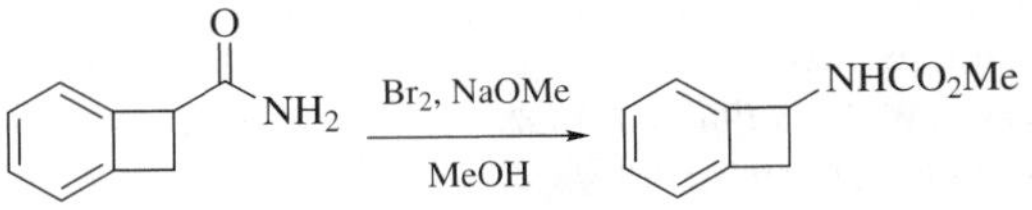

(27)

In the Hunsdiecker reaction,[68] treatment of silver salts of carboxylic acids with bromine furnishes the alkyl(aryl) bromides with one less carbon atom. Improvements that do not require the preparation of the dry silver salts include the use of mercury(II) salts (Cristol–Firth modification) (eq 28),[69] thallium(I) salts,[70] and photostimulation.[71]

$$\text{cyclopropyl-}CO_2H \xrightarrow[\text{tetrachloroethane}]{Br_2,\ HgO} \text{cyclopropyl-}Br \quad (28)$$

Three-membered rings are especially susceptible to reaction with bromine. Cyclopropanes are opened to give 1,3-dibromopropanes,[72] while cyclopropenylethanol derivatives rearrange in the presence of bromine to give 3-methylenetetrahydrofurans (eq 29).[73] Trimethylsilyl cyclopropyl ethers are opened to give β-bromo ketones (eq 30).[74] Epoxides can give either bromohydrins[75] or α-bromo ketones,[76] depending on the reaction conditions.

OH; Br_2; CH_2Cl_2; O; Br (29)

TMSO; Br_2; CH_2Cl_2; O; Br (30)

Mercury(II)-mediated cyclization of dienes allows access to bromine-containing natural products.[77] Thus, in the synthesis of aplysistatin, the key step was the cyclization of an acyclic precursor to the required bromoperhydro[1]benzoxepine ring system (eq 31).[78]

HO; OR; SPh; CO_2Me; 1. $Hg(TFA)_2$, $MeNO_2$; 2. KBr, H_2O; 3. Br_2, LiBr, O_2 pyridine; O; OR; SPh; CO_2Me; Br; H (31)

Related Reagents. See Classes O-1, O-3, and O-22, pages 1–10. Bromine–*t*-Butylamine; Bromine Chloride; Bromine–1,4-Diazabicyclo[2.2.2]octane; Bromine–1,4-Dioxane; Bromine–Silver(I) Oxide; Bromine–Triphenyl Phosphite; *N*-Bromosuccinimide; *N*-Bromosuccinimide–Dimethylformamide; *N*-Bromosuccinimide–Dimethyl Sulfide; *N*-Bromosuccinimide–Sodium Azide; Copper(II) Bromide; Hydrobromic Acid; Mercury(II) Oxide–Bromine; Phosphorus(III) Bromide; Pyridinium Hydrobromide Perbromide; Sodium Bromide; Thallium(III) Acetate–Bromine.

---

1. (a) Roedig, A. *MOC* **1960**, *V/4*, 1. (b) Buehler, C. A.; Pearson, D. E. *Survey of Organic Syntheses*; Wiley: New York, 1970; pp 329–410.
2. *The Sigma-Aldrich Library of Chemical Safety Data*, 2nd ed.; Lenga, R. E., Ed.; Sigma-Aldrich: Milwaukee, 1988; Vol. 2, p 2027.
3. Thaler, W. *JACS* **1963**, *85*, 2607.
4. Brewster, J. F. *JACS* **1918**, *40*, 406.
5. (a) Braendlin, H. P.; McBee, E. T. In *Friedel-Crafts and Related Reactions*; Olah, G. A., Ed.; Wiley: New York, 1964; Vol. 3, pp 1517–1593. (b) Carey, F. A.; Sundberg, R. J. *Advanced Organic Chemistry*, 2nd ed.; Plenum: New York, 1984; Part A, pp 505–511, and Part B, pp 377–381. (c) *Preparative Organic Chemistry*; Hilgetag, G.; Martini, A., Eds.; Wiley: New York, 1972; pp 118, 150–155. (d) March, J. *Advanced Organic Chemistry: Reactions, Mechanisms, and Structure*, 4th ed.; Wiley: New York, 1992; pp 531–533.
6. (a) de la Mare, P. B. D.; Ridd, J. H. *Aromatic Substitution, Nitration and Halogenation*; Butterworth: London, 1959; Chapter 15. (b) Eisch, J. J. *Adv. Heterocycl. Chem.* **1966**, *7*, 1.
7. (a) See Ref. 5(b), Part B, pp 147–154. (b) See Ref. 5(c), pp 105–117. (c) See Ref. 5(d), pp 734–755, 812–816. (d) House, H. O. *Modern Synthetic Reactions*, 2nd ed.; Benjamin: Menlo Park, CA, 1972; pp 422–446.
8. (a) Rolston, J. H.; Yates, K. *JACS* **1969**, *91*, 1469. (b) Rolston, J. H.; Yates, K. *JACS* **1969**, *91*, 1477.
9. Sicher, J.; Závada, J.; Svoboda, M. *CCC* **1962**, *27*, 1927.
10. Pannell, K. H.; Mayr, A. *CC* **1979**, 132.
11. Smith, K.; Fry, K. B. *CC* **1992**, 187.
12. Petrov, A. A. *RCR* **1960**, *29*, 489.
13. Vogel, E.; Klug, W.; Breuer, A. *OSC* **1988**, *6*, 862.
14. Corey, E. J.; Myers, A. G. *JACS* **1985**, *107*, 5574.
15. Barton, D. H. R.; Kumari, D.; Welzel, P.; Danks, L. J.; McGhie, J. F. *JCS(C)* **1969**, 332.
16. Fieser, L. F. *Experiments in Organic Chemistry*, 3rd ed.; Heath: Boston, 1957; pp 67–72.
17. (a) Guaciaro, M. A.; Wovkulich, P. M.; Smith, A. B. III, *TL* **1978**, 4661. (b) Smith, A. B. III; Branca, S. J.; Guaciaro, M. A.; Wovkulich, P. M.; Korn, A. *OSC* **1990**, *7*, 271.
18. (a) Distler, H. *AG(E)* **1965**, *4*, 300. (b) Aumaitre, G.; Chanet-Ray, J.; Durand, J.; Vessière, R.; Lonchambon, G. *S* **1983**, 816.
19. (a) Guss, C. O.; Rosenthal, R. *JACS* **1955**, *77*, 2549. (b) Sisti, A. J.; Meyers, M. *JOC* **1973**, *38*, 4431.
20. Woodward, R. B.; Bader, F. E.; Bickel, H.; Frey, A. J.; Kierstead, R. W. *T* **1958**, *2*, 1.
21. (a) Dowle, M. D.; Davies, D. I. *CSR* **1979**, *8*, 171. (b) Cambie, R. C.; Rutledge, P. S.; Somerville, R. F.; Woodgate, P. D. *S* **1988**, 1009.
22. Corey, E. J.; Hase, T. *TL* **1979**, 335.
23. Davies, D. I.; Dowle, M. D.; Kenyon, R. F. *S* **1979**, 990.
24. Dai, W.; Katzenellenbogen, J. A. *JOC* **1991**, *56*, 6893.
25. (a) Corey, E. J.; Fleet, G. W. J.; Kato, M. *TL* **1973**, 3963. (b) Clive, D. L. J.; Wong, C. K.; Kiel, W. A.; Menchen, S. M. *CC* **1978**, 379.
26. Biloski, A. J.; Wood, R. D.; Ganem, B. *JACS* **1982**, *104*, 3233.
27. (a) Rajendra, G.; Miller, M. J. *TL* **1985**, *26*, 5385. (b) Rajendra, G.; Miller, M. J. *JOC* **1987**, *52*, 4471.
28. Lau, K. S. Y.; Schlosser, M. *JOC* **1978**, *43*, 1595.
29. Duhamel, L.; Duhamel, P.; Enders, D.; Karl, W.; Leger, F.; Poirier, J. M.; Raabe, G. *S* **1991**, 649.
30. See Ref. 7(d), pp 459–478.
31. Carpino, L. A.; McAdams, L. V. III, *OSC* **1988**, *6*, 403.
32. (a) Schwenk, E.; Papa, D. *JACS* **1948**, *70*, 3626. (b) Reinheckel, H. *CB* **1960**, *93*, 2222.
33. Smissman, E. E. *JACS* **1954**, *76*, 5805.
34. Price, C. C.; Judge, J. M. *OSC* **1973**, *5*, 255.
35. Ziegler, H. J.; Walgraeve, L.; Binon, F. *S* **1969**, 39.
36. Calò, V.; Lopez, L.; Pesce, G. *JCS(P1)* **1977**, 501.
37. (a) Reuss, R. H.; Hassner, A. *JOC* **1974**, *39*, 1785. (b) Blanco, L.; Amice, P.; Conia, J. M. *S* **1976**, 194.
38. Iriuchijima, S.; Tsuchihashi, G. *S* **1970**, 588.
39. Kharasch, M. S.; Reinmuth, O. *Grignard Reactions of Nonmetallic Substances*; Prentice-Hall: New York, 1954; pp 1332–1335.
40. (a) Wakefield, B. J. *The Chemistry of Organolithium Compounds*; Pergamon: Oxford, 1974; pp 62–65. (b) Wakefield, B. J. *Organolithium Methods*; Academic: 1988; pp 143–148.
41. (a) Zweifel, G.; Whitney, C. C. *JACS* **1967**, *89*, 2753. (b) Mole, T.; Jeffery, E. A. *Organoaluminium Compounds*; Elsevier: New York, 1972; pp 16–17.
42. (a) Brown, H. C.; Lane, C. F. *JACS* **1970**, *92*, 6660. (b) Brown, H. C.; Lane, C. F. *T* **1988**, *44*, 2763.

43. Lane, C. F.; Brown, H. C. *JOM* **1971**, *26*, C51.
44. (a) Lane, C. F. *Aldrichim. Acta*, **1973**, *6(2)*, 21. (b) Lane, C. F. *Intra-Sci. Chem. Rep.* **1973**, *7*, 133 (*CA* **1974**, *80*, 96 052u).
45. Kabalka, G. W.; Sastry, K. A. R.; Hsu, H. C.; Hylarides, M. D. *JOC* **1981**, *46*, 3113.
46. (a) Brown, H. C.; Bhat, N. G.; Rajagopalan, S. *S* **1986**, 480. (b) Brown, H. C.; Bhat, N. G. *TL* **1988**, *29*, 21.
47. (a) Stotter, P. L.; Hill, K. A. *JOC* **1973**, *38*, 2576. (b) Anderson, W. K.; LaVoie, E. J.; Lee, G. E. *JOC* **1977**, *42*, 1045.
48. Rathke, M. W.; Lindert, A. *TL* **1971**, 3995.
49. Miller, S. I.; Ziegler, G. R.; Wieleseck, R. *OSC* **1973**, *5*, 921.
50. Ojima, I.; Kogure, T.; Yoda, Y. *OSC* **1990**, *7*, 417.
51. (a) Williams, D. R.; Klinger, F. D.; Allen, E. E.; Lichtenthaler, F. W. *TL* **1988**, *29*, 5087, and references therein. (b) Mingotaud, A-F.; Florentin, D.; Marquet, A. *SC* **1992**, *22*, 2401.
52. Al Neirabeyeh, M.; Ziegler, J-C.; Gross, B. *S* **1976**, 811.
53. Ueno, Y.; Okawara, M. *TL* **1976**, 4597.
54. Doyle, M. P.; Bagheri, V. *JOC* **1981**, *46*, 4806.
55. Mihailovic, M. L.; Gojkovic, S.; Konstantinovic, S. *T* **1973**, *29*, 3675.
56. Vlad, P. F.; Ungur, N. D. *S* **1983**, 216.
57. Denis, J. M.; Champion, J.; Conia, J. M. *OSC* **1990**, *7*, 112.
58. Shepherd, R. G.; White, A. C. *JCS(P1)* **1987**, 2153.
59. Deno, N. C.; Potter, N. H. *JACS* **1967**, *89*, 3550.
60. Wender, P. A.; Eissenstat, M. A.; Sapuppo, N.; Ziegler, F. E. *OSC* **1988**, *6*, 334.
61. (a) Olah, G. A.; Vankar, Y. D.; Prakash, G. K. S. *S* **1979**, 113. (b) Marchand, A. P.; Reddy, D. S. *JOC* **1984**, *49*, 4078.
62. Picot, A.; Lusinchi, X. *S* **1975**, 109.
63. Drabowicz, J.; Mikolajczyk, M. *S* **1980**, 32.
64. Drabowicz, J.; Midura, W.; Mikolajczyk, M. *S* **1979**, 39.
65. Reich, H. J.; Cohen, M. L.; Clark, P. S. *OSC* **1988**, *6*, 533.
66. Wallis, E. S.; Lane, J. F. *OR* **1946**, *3*, 267.
67. Radlick, P.; Brown, L. R. *S* **1974**, 290.
68. Wilson, C. V. *OR* **1957**, *9*, 332.
69. (a) Cristol, S. J.; Firth, W. C., Jr. *JOC* **1961**, *26*, 280. (b) Meek, J. S.; Osuga, D. T. *OSC* **1973**, *5*, 126.
70. Cambie, R. C.; Hayward, R. C.; Jurlina, J. L.; Rutledge, P. S.; Woodgate, P. D. *JCS(P1)* **1981**, 2608.
71. Meyers, A. I.; Fleming, M. P. *JOC* **1979**, *44*, 3405.
72. Ogg, R. A. Jr.; Priest, W. J. *JACS* **1938**, *60*, 217.
73. Al-Dulayymi, J. R.; Baird, M. S. *T* **1990**, *46*, 5703.
74. Murai, S.; Seki, Y.; Sonoda, N. *CC* **1974**, 1032.
75. (a) Alvarez, E.; Nuñez, M. T.; Martin, V. S. *JOC* **1990**, *55*, 3429. (b) Konaklieva, M. I.; Dahl, M. L.; Turos, E. *TL* **1992**, *33*, 7093.
76. Calò, V.; Lopez, L.; Valentino, D. S. *S* **1978**, 139.
77. Hoye, T. R.; Kurth, M. J. *JOC* **1979**, *44*, 3461, and references therein.
78. Hoye, T. R.; Kurth, M. J. *JACS* **1979**, *101*, 5065.
79. *Reagent Chemicals: American Chemical Society Specifications*, 8th ed.; American Chemical Society: Washington, 1993; pp 193–196.
80. Perrin, D. D.; Armarego, W. L. F. *Purification of Laboratory Chemicals*, 3rd ed.; Pergamon: New York, 1988; p 317.

R. Richard Goehring
Scios Nova, Baltimore, MD, USA

t-Butyl Hydroperoxide[1–3]

t-Bu–O–OH

[75-91-2] $C_4H_{10}O_2$ (MW 90.12)

(oxidizing agent used for the oxidation of alcohols and alkenes to allylic oxygenated compounds and epoxides[1,2])

Alternate Name: TBHP.

Physical Data: the following data are for a 90% aqueous solution: flash point 35 °C; *d* 0.901 g cm^{-3}; n_D^{20} 1.3960. The density of a 70% aqueous solution is 0.937 g cm^{-3}.

Solubility: sol alcohol, ether, chloroform; slightly sol H_2O, DMSO.

Form Supplied in: clear colorless liquid; widely available as 70–90% aqueous solutions, and anhydrous in hydrocarbon solvents. Aqueous solutions may be dried by a phase separation procedure, followed by azeotropic distillation to remove the last vestiges of water if necessary.[1]

Handling, Storage, and Precautions: eye protection and rubber gloves should be worn when handling this material; avoid skin contact; this reagent should be handled only in a fume hood. Eye and skin irritant; immediately flush with water if contact is made with the eyes. Flammable liquid; oxidizer; sensitive to shocks and sparks. May react explosively with reducing agents. Store in an explosion-proof container, and keep away from reducing materials and strong acids and bases. Avoid using high strength solutions; do not distill. The use of molecular sieves for drying is not recommended.

General Considerations. The title reagent is used in oxidations of various substrates to give epoxides, ketones, aldehydes, carboxylic acid esters, and nitro or azoxy compounds. The reagent and its metal complexes have been extensively reviewed.[1–3] This article describes representative applications to problems in organic synthesis.

Oxidations of Alkenes.

Hydroxylation. Under basic conditions (tetraethylammonium hydroxide, Et_4NOH), in the presence of catalytic amounts of ***Osmium Tetroxide***, TBHP vicinally hydroxylates alkenes (eq 1).[4] This method is preferable over the use of osmium tetroxide stoichiometrically due to the expense and toxicity of the latter (see also ***Osmium Tetroxide–t-Butyl Hydroperoxide***). This method is also preferable to the use of osmium tetroxide catalytically with ***Hydrogen Peroxide***[5] or metal chlorates,[6] both of which give lower yields for tri- and tetrasubstituted alkenes, and can lead to overoxidation. Et_4NOH can be replaced with Et_4NOAc, which allows this reaction to be carried out on alkenes containing base-sensitive functional groups, and often gives better yields than the use of Et_4NOH (eq 2).[7]

Chlorohydroxylation of nonfunctionalized alkenes can be accomplished (eq 3) through the reaction of TBHP with ***Titanium(IV) Chloride***.[8] Chlorohydroxylation can also be done asymmetrically to the alkenes of allylic alcohols using TBHP with

Dichlorotitanium Diisopropoxide and an asymmetric tartrate catalyst, and the stereochemistry can be controlled by the ratio of titanium to tartrate (eq 4) varying (see below).

t-BuOOH, Et_4NOH, 51% (1)

TBHP, Et_4NOAc, OsO_4, acetone, 83% (2)

TBHP, $TiCl_4$, CH_2Cl_2, 92% (3)

TBHP, 0 °C, $TiCl_2(O\text{-}i\text{-}Pr)_2$–(+)-diethyl tartrate (2:1), 76%, 73% ee (4)

Oxidation of Allylic, Benzylic, and Propargylic Carbons.[1,2] Alkenes with an allylic hydrogen can be selectively oxidized to the allylic alcohol by TBHP in the presence of ***Selenium(IV) Oxide*** (eq 5).[9] This is preferable to the oxidation using stoichiometric SeO_2 by itself, which leads to reduced forms of selenium and can make isolation and purification of the product difficult. Less substituted alkenes require 0.5 equivalents of SeO_2 while for more substituted alkenes it may be present in catalytic amounts. The regioselectivity of this reaction favors the more substituted site being oxidized. The addition of small amounts of carboxylic acids also aids this reaction with certain alkenes (eq 5). See also ***Selenium(IV) Oxide–t-Butyl Hydroperoxide***.

TBHP, salicylic acid, SeO_2 (cat.), 55% (5)

Alkenes can also be oxidized to give rearranged allylic alcohols using TBHP with phenylselenenic acid and ***Diphenyl Diselenide*** (eq 6).[10] The reaction proceeds through a β-hydroxyl phenylselenide adduct of the alkene, which then eliminates the selenide to give the allylic alcohol and a phenylselenenic acid byproduct. This method is preferable to the use of phenylselenenic acid with hydrogen peroxide, since the latter can lead to epoxidation of the alkene of the product to give the epoxy alcohol. This method also does not oxidatively remove the selenium from the phenylselenenic acid byproduct, as the H_2O_2 method does, allowing the phenylselenenic acid to be recovered and easily reduced back to diphenyl diselenide.

1. H_2O_2, PhSeSePh, CH_2Cl_2; 2. TBHP, 20 °C, 24 h; 87% (6)

Oxidations of allylic carbons to carbonyls to give enones may be effected by the reaction of the alkene with TBHP catalyzed by ***Hexacarbonylchromium*** (eq 7).[11] These reaction conditions are milder than other chromium reagents used for the same purpose, and are selective towards allylic carbons. This system has also been used for the oxidation of benzylic carbons to carbonyls, with much better yields than ***Chromic Acid*** oxidations.[12] Other chromium(VI) catalysts can also be used along with TBHP to oxidize allylic and benzylic carbons to the corresponding carbonyl.[13]

TBHP, $Cr(CO)_6$, MeCN, Δ, 60% (7)

Propargylic carbons can also be oxidized, using TBHP and SeO_2 (eq 8).[14] Unlike allylic systems, propargylic systems show a great tendency towards oxygenation on both sides of the triple bond, and are generally more reactive towards α oxygenation. A mixture of the propargylic alcohol, ketone, diol, and ketol will generally result from this reaction. If there are two sites possible for oxygenation, methine and methylene groups have about the same reactivity towards these conditions, while methyl groups show a lesser preference for oxidation. In symmetrical alkynes the diol is prevalent, and in some cases where the alkyne is in conjugation with other π systems the ketone is an important product, whereas in most other cases the ketone and ketol are minor products. If the alkyne has one methine and one methylene substituent, the enynone can be an important product.

TBHP, SeO_2, CH_2Cl_2, 30 h, total yield 70%: 60% + 30% + 7% + 3% (8)

If a chromium(VI) catalyst is used in the presence of TBHP, propargylic carbons will be oxidized to the alkynic ketone (eq 9).[15] The more highly substituted alkyl substituent on the alkyne is preferentially oxidized, and symmetrical alkynes give the monoketone accompanied by the diketone.

TBHP, CrO_3 (0.05 equiv), CH_2Cl_2 (9)

Oxidation of π-allylpalladium complexes can also occur with TBHP using a molybdenum(IV) catalyst to give the allylic alcohol (eq 10).[16] Hydroxyl attack will occur axially, *syn* to the complexed palladium. This conversion can also be carried out with peroxy acids or singlet oxygen, but these methods are not as selective.

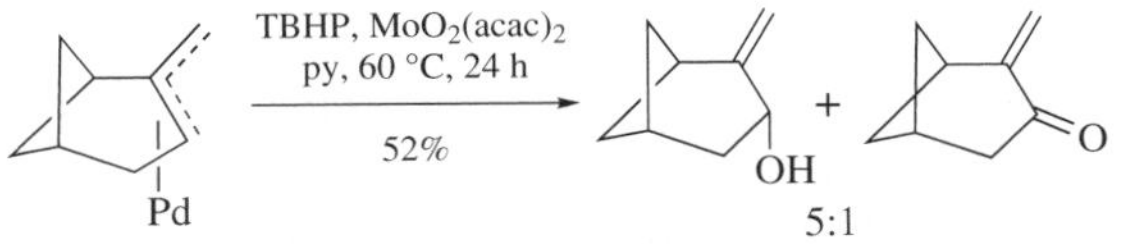

TBHP, $MoO_2(acac)_2$, py, 60 °C, 24 h; 52%; 5:1 (10)

Epoxidation.[1,2,17] TBHP is widely used as an epoxidizing agent, both synthetically and industrially.[18] TBHP has been used to effect regiospecific, stereospecific, and asymmetric epoxidations. In general, the rates of epoxidations using TBHP are slowed by polar solvents, and increased with higher alkyl substitution of the alkene. TBHP is considered superior to hydrogen peroxide for epoxidations, because it is soluble in hydrocarbon solvents, while hydrogen peroxide can readily transform the epoxide to the *vic*-glycol.

Epoxidations of simple alkenes can be carried out using TBHP with a vanadium or molybdenum catalyst (eq 11).[19]

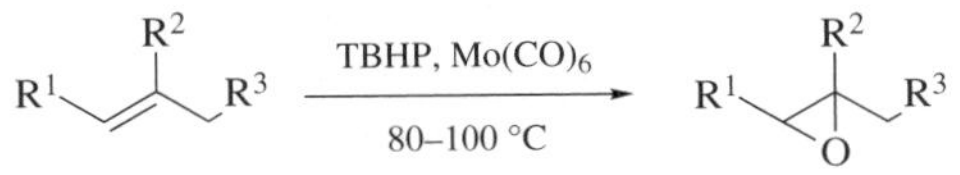

TBHP, $Mo(CO)_6$, 80–100 °C (11)

Epoxidations of alkenes in compounds containing other functional groups can also be accomplished using TBHP with a molybdenum catalyst (eq 12).[19,20] For nonconjugated dienes, more highly substituted alkenes can be selectively epoxidized over less substituted alkenes. Conjugated dienes are less susceptible to epoxidation than isolated alkenes, but the regioselectivity for the different double bonds of a conjugated system follows the same pattern as that for isolated alkenes (eq 13).

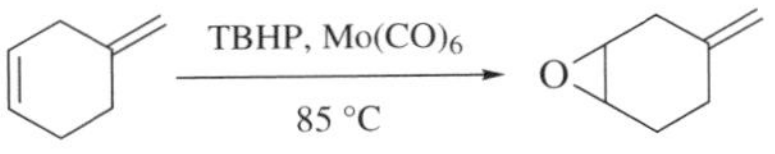

TBHP, $Mo(CO)_6$, 85 °C (12)

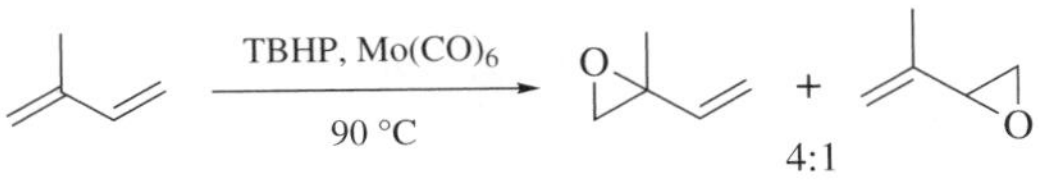

TBHP, $Mo(CO)_6$, 90 °C; 4:1 (13)

Epoxidations of compounds with functional groups in the allylic position can also be effected using TBHP and a molybdenum or vanadium catalyst, but the yields are not as high as those for isolated double bonds, and longer reaction times are required.[21] TBHP epoxidizes the alkenes of allylic and homoallylic alcohols stereoselectively with either molybdenum or vanadium catalysts (eqs 14 and 15).[21] With acyclic systems, vanadium-catalyzed epoxidations give predominantly the *erythro* product, and molybdenum-catalyzed epoxidations give predominantly the *threo* product.[22]

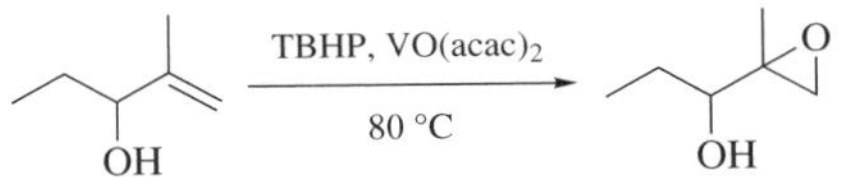

TBHP, $VO(acac)_2$, 80 °C (14)

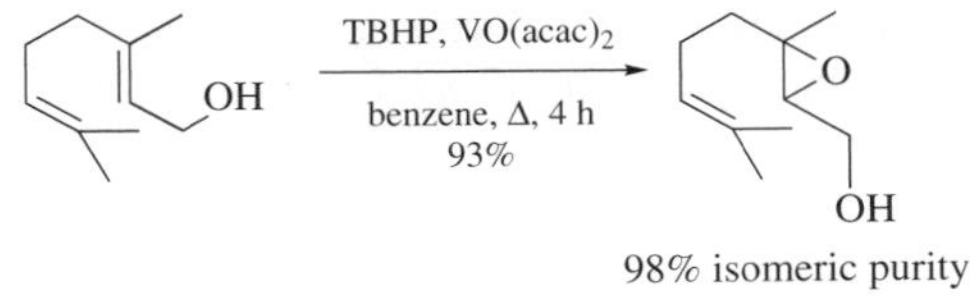

TBHP, $VO(acac)_2$, benzene, Δ, 4 h; 93%; 98% isomeric purity (15)

For cyclic systems, vanadium- and molybdenum-catalyzed reactions give predominantly the *cis* product (eq 16).[23] There are several factors that can affect the selectivity of this reaction for cyclic allylic alcohols. With increasing ring size, the selectivity decreases slightly for vanadium-catalyzed reactions, and more dramatically for molybdenum-catalyzed reactions. The selectivity was also observed to be better for cyclic allylic alcohols where the hydroxyl is in a quasi-axial position.

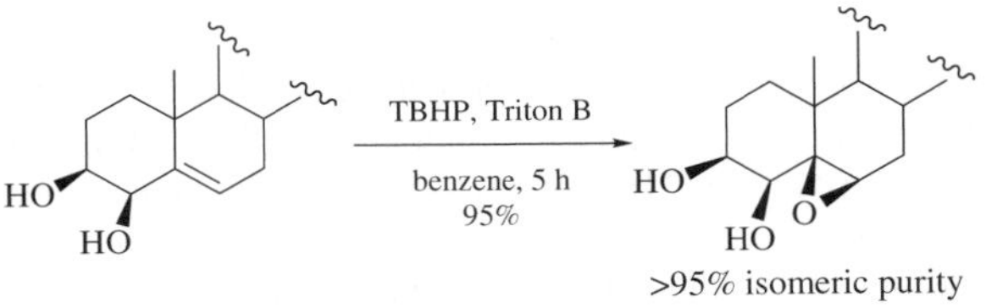

TBHP, Triton B, benzene, 5 h; 95%; >95% isomeric purity (16)

The epoxidation of α,β-unsaturated ketone and aldehyde compounds is accomplished by TBHP in the presence of catalytic amounts of Triton-B (***Benzyltrimethylammonium Hydroxide***) (eq 17).[24] This method has also been used for the synthesis of mono- and diepoxy-1,4-benzoquinones (eq 18).[25]

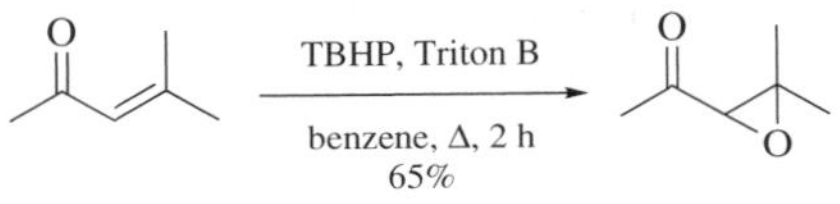

TBHP, Triton B, benzene, Δ, 2 h; 65% (17)

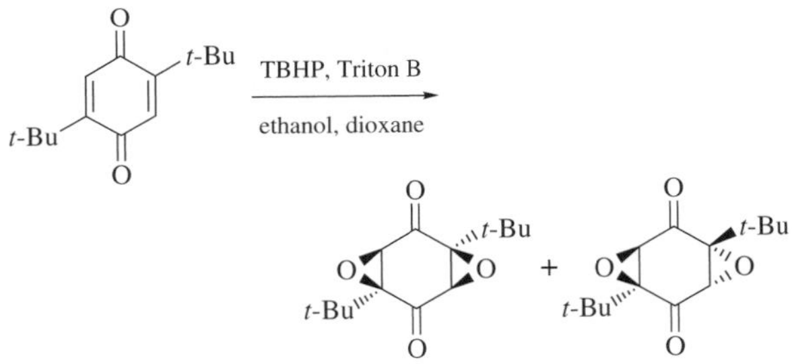

TBHP, Triton B, ethanol, dioxane (18)

Cyclic allylic alcohols conjugated to a second alkene react with TBHP catalyzed by vanadium to give not the epoxy but a bicyclic ether and new allylic alcohols, with the oxygen bridging the original allylic alcohol to the terminus of the conjugated diene (eq 19).[26]

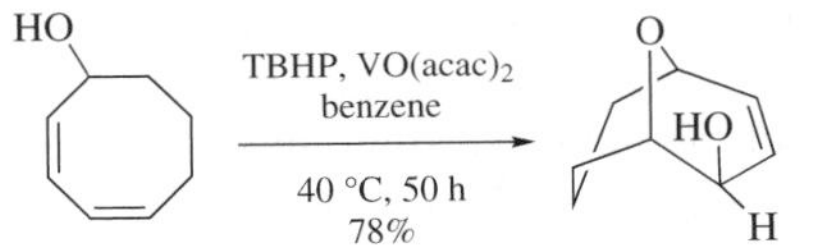

TBHP, $VO(acac)_2$, benzene, 40 °C, 50 h; 78% (19)

Stereoselective epoxidation of unactivated alkenes can be effected by remote chiral auxiliaries in the presence of molybdenum or vanadium catalysts. The configuration of the remote alcohol determines which face of the alkene is epoxidized (eq 20).[27]

TBHP, $Mo(CO)_6$, benzene, Δ, 8 h, 60% (20)

Allylic alcohols can be asymmetrically epoxidized with TBHP and stoichiometric quantities of a titanium–diethyl tartrate complex, generated in situ.[28] Either enantiomer can be formed by using either (+)- or (−)-diethyl tartrate, or by varying the ratio of titanium catalyst to (+)-diethyl tartrate (eq 21), because the nature of the titanium catalyst changes as the molar ratio of titanium to tartrate changes. This latter method is generally preferred due to the ready availability and relatively low cost of (+)-diethyl tartrate. In the presence of 3Å or 4Å molecular sieves, the titanium/tartrate complex may be used in catalytic quantities, but with somewhat lower product enantiomeric purities.[29] These methods can also be used for the stereospecific epoxidation of homoallylic alcohols, but in lower degrees of enantiomeric purity, because the hydroxyl group is further away from the reacting center, lessening its directing effects.[30]

TBHP, −20 °C; N,N-dibenzyl-(+)-tartramide–$Ti(O\text{-}i\text{-}Pr)_4$ (2:2.4) → 96% ee (21)
TBHP, −20 °C; N,N-dibenzyl-(+)-tartramide–$Ti(O\text{-}i\text{-}Pr)_4$ (2:1.0) → 82% ee

The use of TBHP and a titanium/tartrate complex in either stoichiometric or catalytic quantities is known as the Sharpless asymmetric epoxidation.[1] This method gives better stereo- and enantioselectivity than epoxidations using peroxy acids. Asymmetric epoxidations can be carried out using other transition metal catalysts and chiral ligands, but the enantioselectivities are not as high.[31] The Sharpless asymmetric epoxidation can also be used for the kinetic resolution of allylic alcohols.[32]

Asymmetric epoxidations of simple alkenes can be accomplished by Sharpless epoxidation of alkenylsilanols.[33] The alkenylsilanols are prepared from lithium alkenes, and after the Sharpless epoxidation the silyl group is removed by fluoride ion to give the simple epoxide (eq 22). Asymmetric epoxidations can also be done on alkenes without other functional groups using optically active diols as the solvent, but the ee values are generally very low.[34]

TBHP, (+)-diethyl tartrate, CH_2Cl_2, −20 °C, 50%; Et_4NF, MeCN, 90%; 85–95% ee (22)

Epoxidation of vinyl allenes by TBHP and ***Vanadyl Bis(acetylacetonate)*** catalyst leads to the formation of cyclopentenones (eq 23).[35] The intermediate in this reaction is an epoxide of the allene. The stereochemistry of the double bond can be retained. The stereoselectivity is kinetic in nature, and can be lost due to epimerization of the kinetic product if the reaction is continued for long periods of time.

TBHP, $VO(acac)_2$, CH_2Cl_2, rt, 40–70% (23)

Other Reactions with Alkenes. Double bonds of silyl enol ethers can be oxidatively cleaved to the corresponding ketones or carboxylic acids using TBHP with $MoO_2(acac)_2$ as a catalyst (eq 24).[36] The ease with which enols can be generated regiospecifically makes this a very powerful method in organic synthesis. This reaction selectively cleaves the double bond of silyl enol ethers in the presence of other double bonds within the molecule.

TBHP, $MoO_2(acac)_2$, benzene, CF_3CO_2H; THF, H_2O, 60 °C, 72 h, 85% (24)

The alkenes of allylic alcohols can also be cleaved under these conditions (eq 25).[37] In addition to cleaving the double bond of the allylic alcohol, the single bond between the alkene carbon and the allylic carbon bearing the hydroxyl group is also cleaved under these conditions. Vicinal diols are also cleaved to give the corresponding ketones or carboxylic acids (eq 26).[38]

TBHP, benzene, 70 °C, 48 h, 96% (25)

TBHP, MoO_2, chlorobenzene, 60 °C, 24 h, 62% (26)

Oxidation of α,β-unsaturated esters and ketones with palladium-catalyzed TBHP gives β-keto esters or 1,3-diketones (eq 27).[39] Hydrogen peroxide can also be used as the oxidant for this reaction.

TBHP, Na_2PdCl_4, acetic acid, H_2O, 68% (27)

Under basic conditions, TBHP can add in Michael fashion to a double bond that has an electron withdrawing group attached (eq 28).[40]

$$\text{CH}_2{=}\text{CH–C}{\equiv}\text{N} \xrightarrow{\text{TBHP, KOH}} t\text{-BuOO–CH}_2\text{CH}_2\text{–C}{\equiv}\text{N} \quad (28)$$

Reactions with Other Functional Groups.

Oxidation of Alcohols.[1,2] In the presence of catalytic amounts of diphenyl diselenide, TBHP oxidizes benzylic and allylic alcohols to the corresponding ketones (eq 29).[41] Saturated alcohols can be oxidized to the corresponding carbonyl compounds as well if bis(2,4,6-trimethylphenyl) diselenide is used as the catalyst and a small amount of a secondary or tertiary amine is present. These conditions do not affect other double bonds present in the substrate. This system can also be used for the oxidation of α-hydroxy selenides and thiols, selectively oxidizing the hydroxy function to the carbonyl.

bis(2,4,6-trimethylphenyl) diselenide, TBHP; benzene, Δ; 100% (29)

Saturated alcohol oxidation can also be achieved without oxidizing alkenes by the reaction of the alcohol with TBHP and benzyltrimethylammonium tetrabromooxomolybdate (BTMA-Mo) as a catalyst, and secondary alcohols will be oxidized preferentially over primary ones under these conditions (eq 30).[42] If the reaction time is lengthened, primary alcohols will be converted to the appropriate acid derivative depending upon the conditions used. Using $VO(acac)_2$ as the catalyst will oxidize secondary alcohols over primary ones selectively as well.[43]

TBHP, BTMA-Mo; THF, 60 °C; 60% (30)

Oxidation of alcohols can also be done using TBHP and a chromium(VI) catalyst (eq 31).[44] This system works best for allylic, benzylic, and propargylic alcohols, and will selectively oxidize these in the presence of other alcohols.

TBHP, CrO_3; CH_2Cl_2; 100% (31)

Oxidation of Sulfur-Containing Compounds.[2] Sulfides can be oxidized with TBHP to give sulfoxides.[45] If vanadium, molybdenum, or titanium catalysts are used, the addition of one equiv of TBHP will furnish the sulfoxide, while the use of excess TBHP will oxidize the sulfide to the sulfone (eqs 32 and 33). In the absence of metal catalysts the oxidation cannot be carried beyond the sulfoxide.[46] If only one equivalent of TBHP is used, sulfides will be preferentially oxidized over any alkenes present; however, excess TBHP will also oxidize alkenes. The oxidation of thiols with TBHP and Mo^{VI} or V^{V} catalysts produces sulfonic acids (eq 34).[47]

TBHP, $VO(acac)_2$; EtOH; 55% (32)

$$\text{PhSMe} \xrightarrow[\text{benzene, 98\%}]{\text{TBHP (excess), MoO}_2\text{(acac)}_2} \text{PhSO}_2\text{Me} \quad (33)$$

$$\text{RSH} \xrightarrow{\text{TBHP, V}^{\text{V}}} \text{RSO}_3\text{H} \quad (34)$$

Using Sharpless asymmetric epoxidation conditions (see above), modified by the addition of one mol equiv of water, unsymmetrical sulfides could be asymmetrically oxidized to sulfoxides (eq 35).[48] Asymmetric oxidations of sulfides to sulfoxides have also been carried out using optically active diols with TBHP and a molybdenum catalyst, but the enantioselectivities were very low (~10%).[49]

TBHP, Ti(O-*i*-Pr)$_4$, (+)-diethyl tartrate; H_2O, CH_2Cl_2; 75%; *R*, 90% ee (35)

Oxidation of Phosphines. Alkyl phosphines can be oxidized to the appropriate phosphine oxides by TBHP (eq 36).[50]

TBHP, 0 °C (36)

Oxidation of Selenides and Selenoxides. Alkenes can be produced oxidatively from selenides, through the selenoxides and elimination. This is done by stirring TBHP with basic alumina and the appropriate selenide (eq 37).[51] This transformation can also be accomplished by treatment of the selenide with ***Hydrogen Peroxide***, ***Ozone*** followed by ***Triethylamine***, periodate, or peroxy acids.

$Me(CH_2)_9$–SePh; TBHP, basic alumina; THF, 4.5 h; 86%; $Me(CH_2)_9$ alkenes, 65:35 (37)

Oxidation of Nitrogen-Containing Compounds.[2] Reactions of TBHP with compounds containing nitrogen have been used to effect a variety of oxidations, both of the nitrogen atom itself and of adjacent carbon atoms. Tertiary amines react with TBHP in the presence of vanadium and molybdenum catalysts to give amine oxides (eq 38).[52] This transformation can also be done with cumene and pentene hydroperoxides.

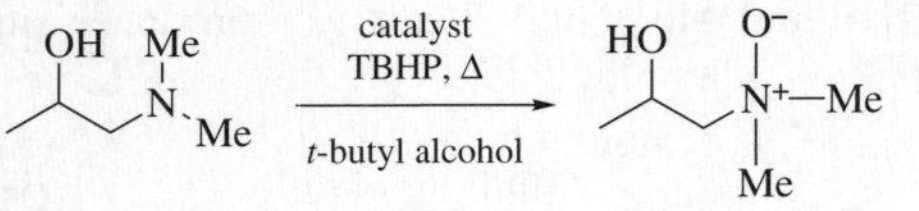

(38)

Secondary amines are oxidized to imines by TBHP in the presence of ruthenium(II) catalysts (eq 39).[53] Tertiary amines are oxidized by TBHP in the presence of ruthenium catalyst to give α-(*t*-butyldioxy)alkylamines, which decompose to iminium ion intermediates when treated with acid (eq 40).[54] *N*-Methyl groups are selectively oxidized when other *N*-alkyl or -alkenyl groups are present.

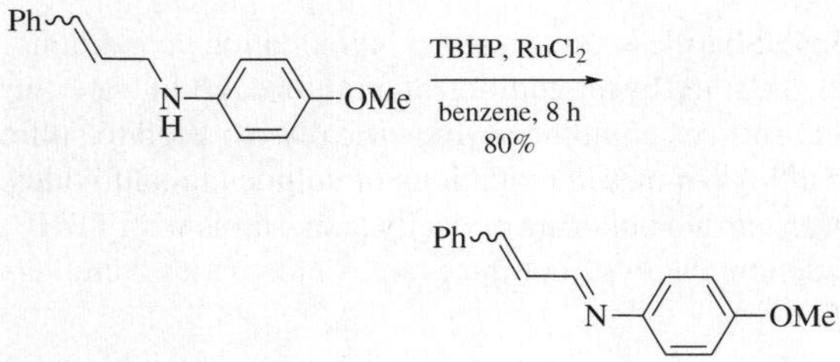

(39)

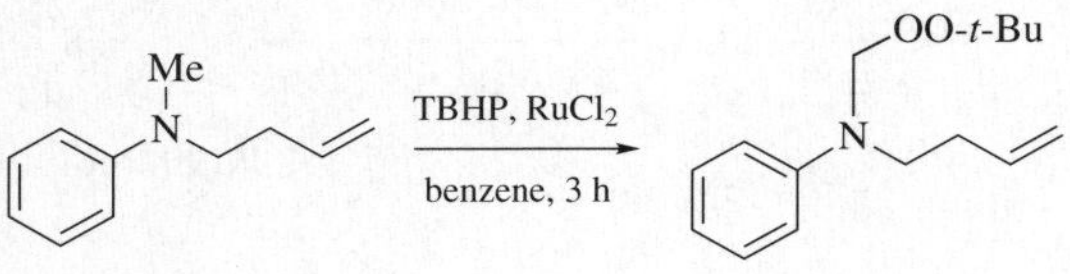

(40)

Amides are selectively oxidized to imides by TBHP, and other hydroperoxides, in the presence of cobalt or manganese salts as catalysts (eq 41).[55] The selectivity of this transformation is demonstrated by the oxidation of 3-ethoxycarbonyl-2-piperidone to the appropriate imide, with no other oxidation products. ***Peracetic Acid*** also effects this transformation, in many cases giving better yields and shorter reaction times, but the conditions for oxidation with TBHP are milder. Oxidation of amides with TBHP catalyzed by ruthenium gives the corresponding *t*-butylperoxy amide (eq 42).[56]

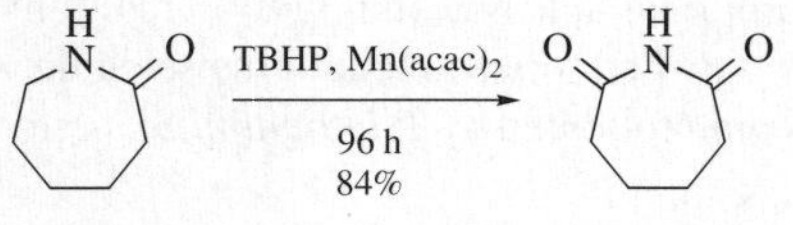

(41)

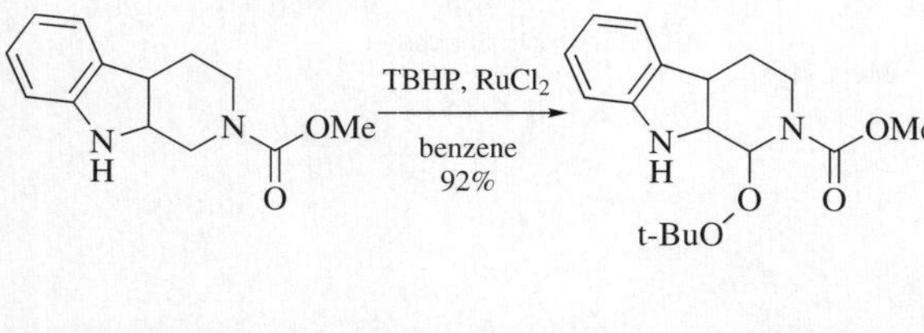

(42)

Nitronate anions, formed by deprotonation of nitro compounds, react with TBHP catalyzed by $VO(acac)_2$ or ***Hexacarbonylmolybdenum*** to give 1-hydroxy nitro compounds (eq 43). Analogous to α-hydroxyl carbonyl compounds, these collapse to give the carbonyl derivative and nitrous acid.[57]

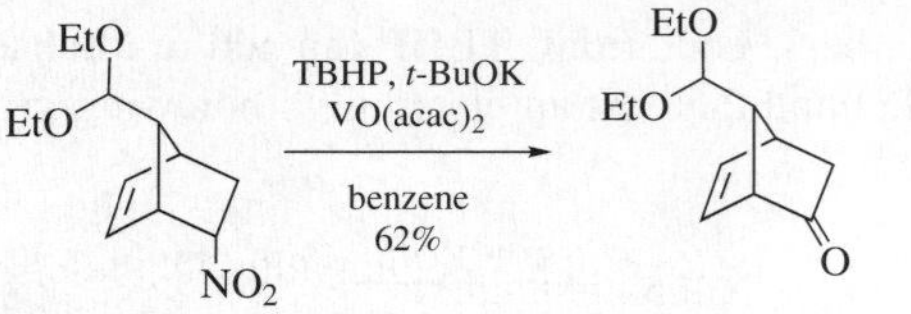

(43)

Introduction of Peroxy Groups into Organic Molecules.[58] Using catalytic amounts of copper, cobalt, or manganese salts, TBHP reacts with molecules that contain a slightly activated carbon–hydrogen bond, replacing the activated hydrogen with a peroxy group. This transformation can also be accomplished with other hydroperoxides. Carbon–hydrogen bonds α to an alkene (eqs 44 and 45),[59,60,61] phenyl groups,[60] carbonyls,[61] nitriles,[62] oxygen,[60,61] or nitrogen (eq 46)[63] atoms are activated towards this reaction. The primary function of the metal salts in these reactions is to initiate decomposition of the hydroperoxide.

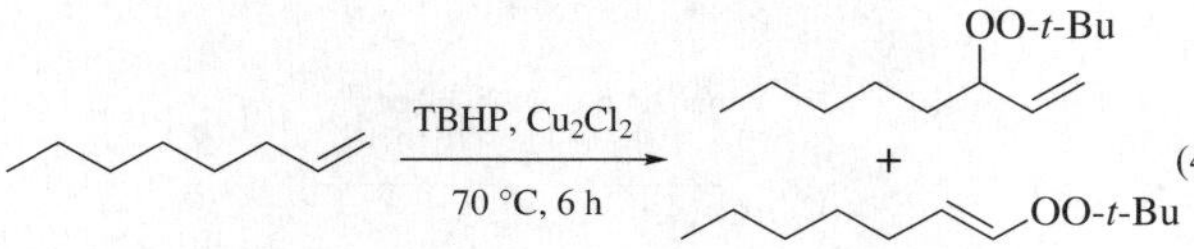

(44)

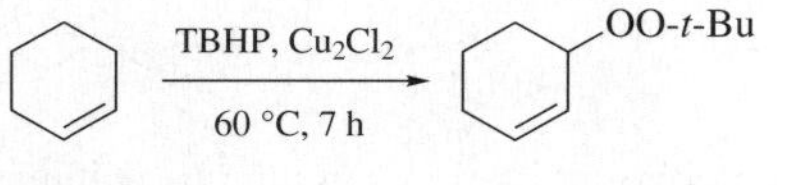

(45)

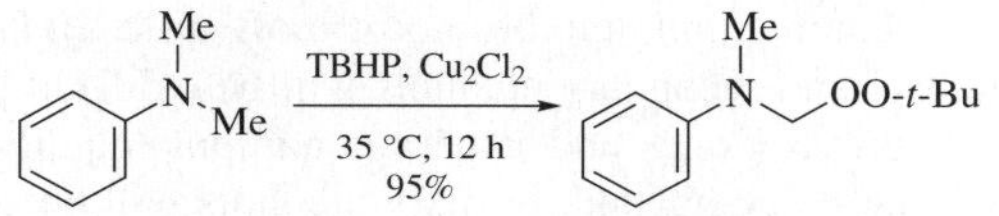

(46)

Peroxy groups may also replace alcohols (eq 47),[64] ethers,[65] or sulfates[66] directly, or be added to an alkene (with Markovnikov regioselectivity),[9] by reacting the functionalized organic compound with TBHP and concentrated sulfuric acid in acetic acid. Epoxides are transformed into β-hydroxy dialkyl peroxides using TBHP in the presence of base (eq 48).[67]

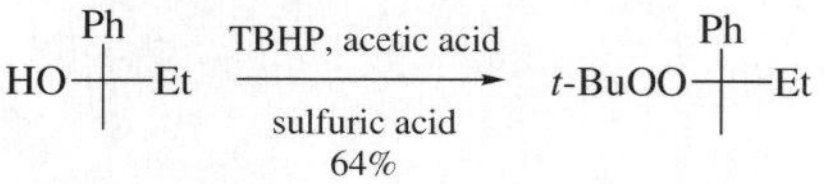

(47)

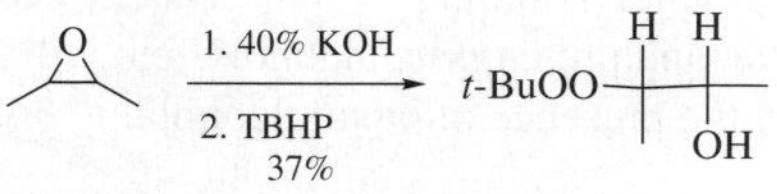

(48)

Peroxy *t*-butyl organosilanes can be prepared by reacting TBHP with the appropriate silyl chloride and pyridine, ammonia, or triethylamine (eq 49).[68] Peroxides of a number of other heteroatoms in organic compounds, such as germanium,[69] boron,[70] cadmium,[71] tin,[72] aluminum,[73] and mercury,[74] can also be synthesized using TBHP.

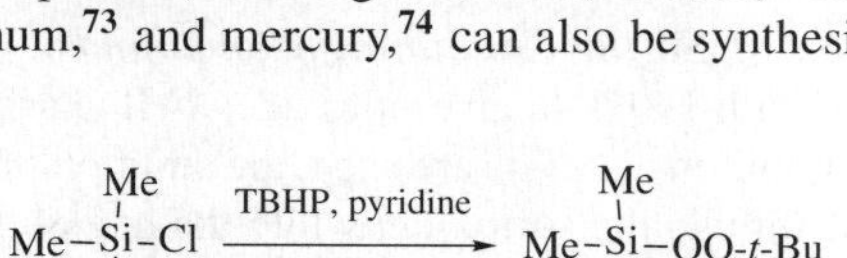

(49)

2,4,6-Substituted phenols react with TBHP to give 2- or 4-(t-butylperoxy)-2,4,6-trisubstituted quinones if the 4-substituent is not a methyl group, and 3,5,3′,5′-tetrasubstituted stilbene-4,4′-quinones if the 4-substituent is a methyl group (eq 50).[75]

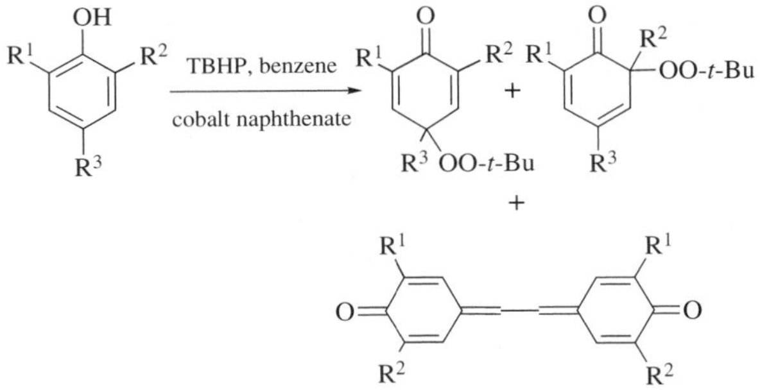

TBHP, benzene; cobalt naphthenate (50)

Reactions with Carbonyl Compounds. Aldehydes react with TBHP in the presence of catalytic amounts of copper, cobalt, or manganese salts to give the t-butyl ester (eq 51).[76] In the absence of a metal catalyst, benzaldehyde will react with TBHP to give a mixture of the *meso* and racemic forms of benzopinacol dibenzoate.

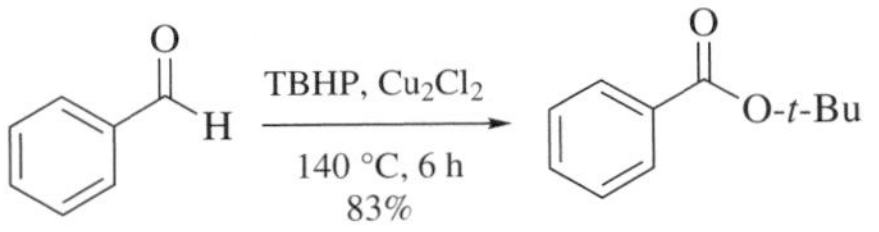

TBHP, Cu_2Cl_2; 140 °C, 6 h; 83% (51)

TBHP reacts with acid chlorides under basic conditions to give the appropriate t-butyl peroxy ester.[77] For small acids, 30% KOH is used, but for longer-chain acids, pyridine is substituted as the base (eq 52).[78] The use of pyridine as the base in this reaction allows the synthesis of carbamate peroxy esters from isocyanates and carbamic acid chlorides (eq 53).[79]

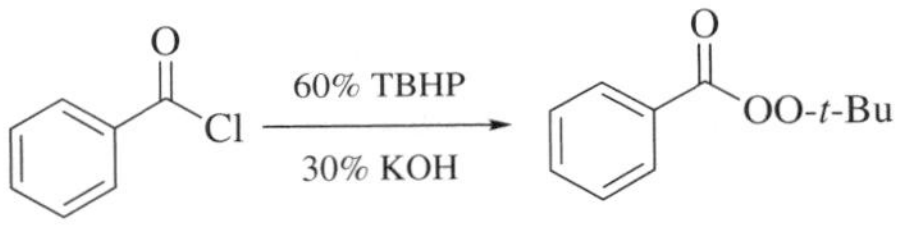

60% TBHP; 30% KOH (52)

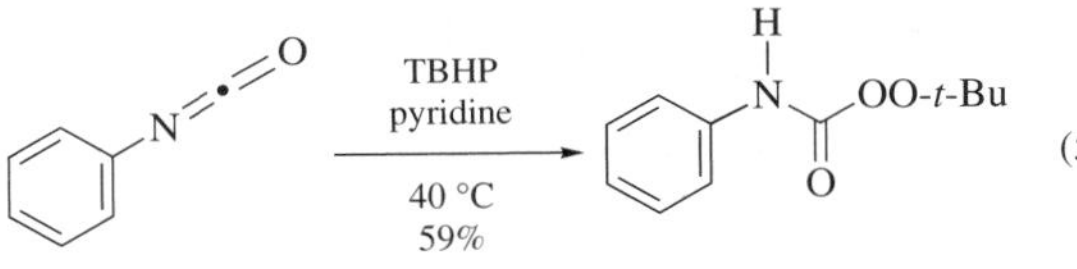

TBHP; pyridine; 40 °C; 59% (53)

TBHP reacts with ketones or aldehydes in the presence of a strong acid catalyst to give products with diperoxy groups in place of the carbonyl (eq 54), or in the absence of the acid catalyst to give an α-hydroxyl t-butyl peroxide (eq 55).[80]

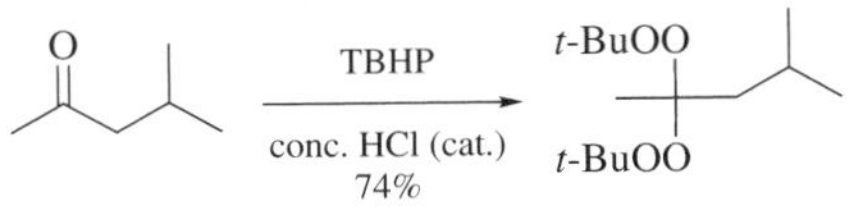

TBHP; conc. HCl (cat.); 74% (54)

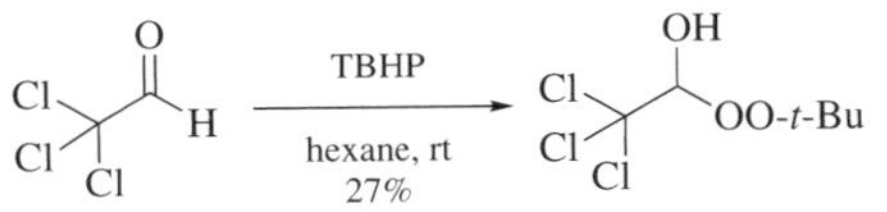

TBHP; hexane, rt; 27% (55)

Conversion of Halides to Alcohols. Grignard reagents react with TBHP to give the appropriate alcohol or phenol (eq 56).[81] This provides an alternative method for the conversion of halides to alcohols or phenols. Because the hydrogen of the peroxide is activated, either two equiv of Grignard reagent must be used or the magnesium salt of the hydroperoxide, prepared from the hydroperoxide and ethylmagnesium bromide.

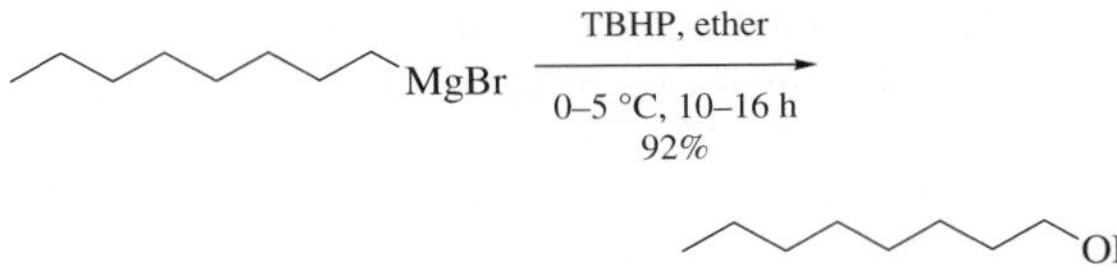

TBHP, ether; 0–5 °C, 10–16 h; 92% (56)

Conversion of Alcohols to Halides. In cases where traditional methods fail, alcohols can be converted to halides by a radical chain reaction.[82] This is accomplished by transforming the alcohol into a chloroglyoxylate, reacting it with TBHP, and warming this in the presence of a halogen donor such as CCl_4 or $BrCCl_3$, to initiate a radical reaction where first a t-butoxyl radical is eliminated, then CO_2 is eliminated twice in succession, leaving an alkyl radical which then reacts with the halogen donor to give the halide (eq 57).

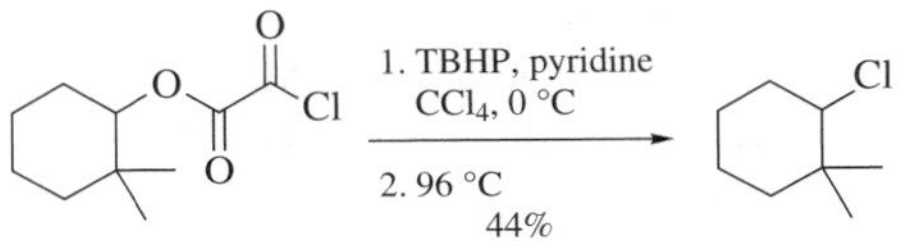

1. TBHP, pyridine CCl_4, 0 °C; 2. 96 °C; 44% (57)

Related Reagents. See Classes O-1, O-3, O-8, O-11, O-14, O-16, O-21, and O-24, pages 1–10.

1. Sharpless, K. B.; Verhoeven, T. R. *Aldrichim. Acta* **1979**, *12*, 63.
2. (a) Sheldon, R. A. In *Aspects of Homogeneous Catalysis*; Ugo, R., Ed.; Reidel: Boston, 1981; Vol. 4, p 3. (b) Sheldon, R. A.; Kochi, J. K. *Metal-Catalyzed Oxidations of Organic Compounds*; Academic: New York, 1981.
3. (a) Hawkins, E. G. E. *Organic Peroxides*; Van Nostrand: New York, 1961. (b) Davies, A. G. *Organic Peroxides*; Butterworths: London, 1961. (c) Richardson, W. H. In *The Chemistry of Peroxides*; Patai, S., Ed.; Wiley: New York, 1983; Chapter 5. (d) Sheldon, R. A. In *The Chemistry of Peroxides*; Patai, S., Ed.; Wiley: New York, 1983; Chapter 6.
4. Sharpless, K. B.; Akashi, K. *JACS* **1976**, *98*, 1986.
5. (a) Milas, N. A.; Sussman, S. *JACS* **1936**, *58*, 1302. (b) Milas, N. A.; Sussman, S. *JACS* **1937**, *59*, 2345.
6. Hoffman, K. A. *CB* **1912**, *45*, 3329.
7. Akashi, K.; Palermo, R. E.; Sharpless, K. B. *JOC* **1978**, *43*, 2063.
8. Klunder, J. M.; Caron, M.; Uchiyama, M.; Sharpless, K. B. *JOC* **1985**, *50*, 912.
9. Umbreit, M. A.; Sharpless, K. B. *JACS* **1977**, *99*, 5526.
10. Hori, T.; Sharpless, K. B. *JOC* **1978**, *43*, 1689.
11. (a) Pearson, A. J.; Chen, Y.-S.; Hsu, S.-Y.; Ray, T. *TL* **1984**, *25*, 1235. (b) Pearson, A. J.; Chen, Y.-S.; Han, G. R.; Hsu, S.-Y.; Ray, T. *JCS(P1)* **1985**, 267.
12. Pearson, A. J.; Han, G. R. *JOC* **1985**, *50*, 2791.
13. (a) Muzart, J. *TL* **1986**, *27*, 3139. (b) Muzart, J. *TL* **1987**, *28*, 2131. (c) Muzart, J. *TL* **1987**, *28*, 4665. (d) Chidambaram, N.; Chandrasekaran, S. *JOC* **1987**, *52*, 5048.
14. Chabaud, B.; Sharpless, K. B. *JOC* **1979**, *44*, 4202.
15. Muzart, J.; Piva, O. *TL* **1988**, *29*, 2321.
16. Jitsukawa, K.; Kaneda, K.; Teranishi, S. *JOC* **1983**, *48*, 389.
17. Sheldon, R. A. *J. Mol. Catal.* **1980**, *7*, 107.

18. Kollar, J. U. S. Patent 3 351 635 (*CA* **1967**, *68*, 21 821n)
19. (a) Kollar, J. U. S. Patent 3 350 422 (*CA* **1967**, *68*, 2922e). (b) Brill, W. F.; Indictor, N. *JOC* **1964**, *29*, 710. (c) Indictor, N.; Brill, W. F. *JOC* **1965**, *30*, 2074. (d) Sheng, M. N.; Zajacek, J. G. *Adv. Chem. Ser.* **1968**, *76*, 418.
20. Sheng, M. N.; Zajacek, J. G. *JOC* **1970**, *35*, 1839.
21. (a) Sharpless, K. B.; Michaelson, R. C. *JACS* **1973**, *95*, 6136. (b) Tanaka, S.; Yamamoto, H.; Nozaki, H.; Sharpless, K. B.; Michaelson, R. C.; Cutting, J. D. *JACS* **1974**, *96*, 5254.
22. (a) Mihelich, E. D. *TL* **1979**, 4729. (b) Rossiter, B. E.; Verhoeven, T. R.; Sharpless, K. B. *TL* **1979**, 4733.
23. (a) Dehnel, R. B.; Whitman, G. H. *JCS(P1)* **1979**, 953. (b) Itoh, T.; Jitsukawa, K.; Kaneda, K.; Teranishi, S. *JACS* **1979**, *101*, 159.
24. (a) Yang, N. C.; Finnegan, R. A. *JACS* **1958**, *80*, 5845. (b) Payne, G. C. *JOC* **1960**, *25*, 275.
25. Moore, H. W. *JOC* **1967**, *32*, 1996.
26. Itoh, T.; Jitsukawa, K.; Kaneda, K.; Teranishi, S. *TL* **1976**, 3157.
27. (a) Breslow, R.; Maresca, L. M. *TL* **1977**, 623. (b) Breslow, R.; Maresca, L. M. *TL* **1978**, 887.
28. (a) Katsuki, T.; Sharpless, K. B. *JACS* **1980**, *102*, 5974. (b) Rossiter, B. E.; Katsuki, T.; Sharpless, K. B. *JACS* **1981**, *103*, 464. (c) Klunder, J. M.; Ko, S. Y.; Sharpless, K. B. *JOC* **1986**, *51*, 3710. (d) Lu, L. D.-L.; Johnson, R. A.; Finn, M. G.; Sharpless, K. B. *JOC* **1984**, *49*, 728.
29. Hanson, R. M.; Sharpless, K. B. *JOC* **1986**, *51*, 1922.
30. Rossiter, B. E.; Sharpless, K. B. *JOC* **1984**, *49*, 3707.
31. (a) Helder, R.; Hummelen, J. C.; Laane, R. W. P. M.; Wiering, J. S.; Wynberg, H. *TL* **1976**, 1831. (b) Yamada, S. I.; Mashiko, T.; Terashima, S. *JACS* **1977**, *99*, 1988. (c) Michaelson, R. C.; Palermo, R. E.; Sharpless, K. B. *JACS* **1977**, *99*, 1990.
32. (a) Martin, V. S.; Woodard, S. S.; Katsuki, T.; Yamada, Y.; Ikeda, M.; Sharpless, K. B. *JACS* **1981**, *103*, 6237. (b) Gao, Y.; Hanson, R. H.; Klunder, J. M.; Ko, S. Y.; Masamune, H.; Sharpless, K. B. *JACS* **1987**, *109*, 5765.
33. Chan, T. H.; Chen, L. M.; Wang, D. *CC* **1988**, 1280.
34. Tani, K.; Hanafusa, M.; Otsuka, S. *TL* **1979**, 3017.
35. Kim, S. J.; Cha, J. K. *TL* **1988**, *29*, 5613.
36. Kaneda, K.; Kii, N.; Jitsukawa, K.; Teranishi, S. *TL* **1981**, *22*, 2595.
37. Jitsukawa, K.; Kaneda, K.; Teranishi, S. *JOC* **1984**, *49*, 199.
38. Kaneda, K.; Morimoto, K.; Imanaka, T. *CL* **1988**, 1295.
39. Tsuji, J.; Nagashima, H.; Hori, K. *CL* **1980**, 257.
40. Harman, D. U. S. Patent 2 508 256, 1950 (*CA* **1950**, *44*, 7341h).
41. (a) Shimizu, M.; Kuwajima, I. *TL* **1979**, 2801. (b) Kuwajima, I.; Shimizu, M.; Urabe, H. *JOC* **1982**, *47*, 837.
42. Masuyama, Y.; Takahashi, M.; Kurusu, Y. *TL* **1984**, *25*, 4417.
43. Kaneda, K.; Kawanishi, Y.; Jitsukawa, K.; Teranishi, S. *TL* **1983**, *24*, 5009.
44. Muzart, J. *TL* **1987**, *28*, 2133.
45. (a) Bateman, L.; Hargrave, K. R. *Proc. R. Soc. London A* **1954**, *224*, 389 and 399. (b) Hargrave, K. R. *Proc. R. Soc. London A* **1956**, *235*, 55. (c) Barnard, D. *JCS* **1956**, 489. (d) Johnson, C. R.; McCants, D., Jr. *JACS* **1965**, *87*, 1109.
46. (a) Kuhnen, L. *AG(E)* **1966**, *5*, 893. (b) List, F.; Kuhnen, L. *Erdoel Kohle, Erdgas, Petrochem.* **1967**, *20*, 192 (*CA* **1967**, *67*, 43 126w).
47. Sheng, M. N.; Zajacek, J. G. U. S. Patent 3 670 002, 1972 (*CA* **1972**, *77*, 88 089j).
48. (a) Pitchen, P.; Kagan, H. B. *TL* **1984**, *25*, 1049. (b) Pitchen, P.; Dunach, E.; Deshmukh, M. N.; Kagan, H. B. *JACS* **1984**, *106*, 8188.
49. di Furia, F.; Modena, G. *TL* **1976**, 4637.
50. (a) Horner, L.; Jurgerleit, W. *LA* **1955**, *591*, 138. (b) Horner, L.; Hoffman, H. *AG* **1956**, *68*, 473. (c) Walling, C.; Rabinowitz, R. *JACS* **1959**, *81*, 1243.
51. Labar, D.; Hevesi, L.; Dumont, W.; Krief, A. *TL* **1978**, 1141.
52. (a) Kuhnen, L. *CB* **1966**, *99*, 3384. (b) Sheng, M. N.; Zajacek, J. G. *JOC* **1968**, *33*, 588.
53. Murahashi, S. I.; Naota, T.; Taki, H. *CC* **1985**, 613.
54. Murahashi, S. I.; Naota, T.; Yonemura, K. *JACS* **1988**, *110*, 8256.
55. (a) Doumaux, A. R., Jr.; McKeon, J. E.; Trecker, D. J. *JACS* **1969**, *91*, 3992. (b) Doumaux, A. R., Jr.; Trecker, D. J. *JOC* **1970**, *35*, 2121.
56. Murahashi, S. I.; Naota, T.; Kuwabara, T.; Saito, T.; Kumobayashi, H.; Akutagawa, S. *JACS* **1990**, *112*, 7820.
57. Bartlett, P. A.; Green, F. R.; Webb, T. R. *TL* **1977**, 331.
58. Rawlinson, D. J.; Sosnovsky, G. *S* **1972**, 1.
59. Kharasch, M. S.; Simon, E.; Nudenberg, W. *JOC* **1953**, *18*, 322.
60. Kharasch, M. S.; Fono, A. *JOC* **1959**, *24*, 72.
61. Kharasch, M. S.; Fono, A. *JOC* **1958**, *23*, 324.
62. Kharasch, M. S.; Sosnovsky, G. *T* **1958**, *3*, 105.
63. (a) Rieche, A.; Schmitz, E.; Beyer, E. *CB* **1959**, *92*, 1206 (*CA* **1960**, *54*, 17 116b). (b) Rieche, A.; Schmitz, E.; Dietrich, P. *CB* **1959**, *92*, 2239 (*CA* **1960**, *54*, 2325d).
64. Davies, A. G.; Foster, R. V.; White, A. M. *JCS* **1954**, 2200.
65. Davies, A. G.; Feld, R. *JCS* **1956**, 4669.
66. Milas, N. A.; Perry, L. H. *JACS* **1946**, *68*, 1938.
67. Barusch, M. R.; Payne, J. Q. *JACS* **1953**, *75*, 1987.
68. Buncel, E.; Davies, A. G. *JCS* **1958**, 1550.
69. Davies, A. G.; Hall, C. D. *JCS* **1959**, 3835.
70. (a) Davies, A. G.; Moodie, R. B. *JCS* **1958**, 2372. (b) Maslennikov, V. P.; Gerbert, G. P.; Khodalev, G. F. *JGU* **1969**, *39*, 1854. (c) Gerbert, G. P.; Maslennikov, V. P. *JGU* **1970**, *40*, 1094.
71. (a) Davies, A. G.; Packer, J. E. *JCS* **1959**, 3164. (b) Razuvaev, G. A.; Pankratova, V. N.; Muraev, V. A.; Bykova, I. V. *JGU* **1969**, *39*, 2431.
72. Alleston, D. L.; Davies, A. G. *JCS* **1962**, 2465.
73. (a) Davies, A. G.; Hall, C. D. *JCS* **1963**, 1192. (b) Razuvaev, G. A.; Stepovik, L. P.; Dodonov, V. A. *JGU* **1969**, *39*, 1563.
74. Razuvaev, G. A.; Zhil'tsov, S. F.; Druzhkov, O. N.; Petukhov, G. G. *JGU* **1966**, *36*, 267.
75. Bickel, A. F.; Kooyman, E. C. *JCS* **1953**, 3211.
76. Kharasch, M. S.; Fono, A. *JOC* **1959**, *24*, 606.
77. Milas, N. A.; Surgenor, D. M. *JACS* **1946**, *68*, 642.
78. Silbert, L. S.; Swern, D. *JACS* **1959**, *81*, 2364.
79. (a) Davies, A. G.; Hunter, K. J. *JCS* **1953**, 1808. (b) Pederson, C. J. *JOC* **1958**, *23*, 252.
80. (a) Dickey, F. H.; Rust, F. F.; Vaughan, W. E. *JACS* **1949**, *71*, 1432. (b) Rieche, A. *AG* **1958**, *70*, 251.
81. Lawesson, S. O.; Yang, N. C. *JACS* **1959**, *81*, 4230.
82. Jensen, F. R.; Moder, T. I. *JACS* **1975**, *97*, 2281.

Andrew K. Jones & Timothy E. Wilson
Emory University, Atlanta, GA, USA

Sham S. Nikam
Warner-Lambert Company, Ann Arbor, MI, USA

t-Butyl Hypochlorite[1]

t-BuOCl

[507-40-4] C_4H_9ClO (MW 108.57)

(reagent for ionic[2] or radical[3] chlorination of hydrocarbons; N-chlorination;[4] oxidation of alcohols,[5] sulfides,[6] and selenides[7])

Physical Data: bp 79.6 °C/750 mmHg; *d* 0.9583 g cm^{-3}.
Solubility: sparingly sol H_2O; sol alcohol, CCl_4, and $CHCl_3$.
Form Supplied in: commercially available as neat liquid.
Preparative Method: usually prepared by the method of Mintz and Walling.[8]
Purification: distillation is possible but can be hazardous, and is reported not to improve purity.
Handling, Storage, and Precautions: avoid exposure of the reagent to light; store protected from light in a refrigerator or freezer.

Radical Chlorination. Under typical radical conditions, t-BuOCl can effect chlorination of a range of substrates, including alkanes, alkenes, ethers, epoxides, and aldehydes.[3] Reaction occurs by hydrogen abstraction from the substrate by the t-BuO· radical, this process showing the expected selectivity pattern of primary < secondary < tertiary (eq 1).[3a,c] The reaction can be initiated by irradiation or by the use of chemical initiators such as ***Azobisisobutyronitrile***[3a] or trialkylboranes.[3c] Reaction with alkenes results in allylic chlorination,[3d,e] whereas ethers[3b] and epoxides[3f] give products arising from attack at the position α to oxygen (eqs 2 and 3) and aldehydes give the corresponding acid chlorides (eq 4).[3b]

t-BuOCl, AIBN or hν or R_3B, 56% (1)

t-BuOCl, CCl_4, hν, 0 °C, 80% (2)

t-BuOCl, 90–95% (3)

t-BuOCl, 55% (4)

Ionic Reactions with Alkenes and Arenes. Allylic chlorination with double bond rearrangement can also be effected by t-BuOCl under nonradical conditions in nonprotic solvents.[2] This process appears facile with particularly electron-rich alkenes, or if a catalyst such as ***Boron Trifluoride Etherate*** is employed (eq 5). If an alcohol is used as the solvent then β-chloro ethers are obtained via electrophilic addition (eq 6).[9]

t-BuOCl, CH_2Cl_2, rt (5)

t-BuOCl, t-BuOH, $BF_3 \cdot OEt_2$, 49% (6)

Chlorination of aromatics can also be effected by use of the t-BuOCl–$BF_3 \cdot OEt_2$ reagent combination,[9] or by the use of t-BuOCl in the presence of silica[10] or certain zeolites (eq 7).[11] Activated aromatics react with t-BuOCl alone, whereas deactivated systems do not react, even in the presence of catalysts under forcing conditions. The modified *o/p* selectivity in the presence of zeolite is notable. Chlorination of other types of systems, such as aminoquinones[12] and indoles,[13] has also been reported (eqs 8 and 9).

t-BuOCl (7)

silica, CCl_4	65:35	100%
H^+, Na^+ Faujasite X, CH_2Cl_2	9:91	88%

t-BuOCl, Et_2O, 80–96% (8)

t-BuOCl, CCl_4, CH_2Cl_2, Et_3N (9)

Chlorination at Nitrogen. Chlorination of a wide range of nitrogen-containing systems to give chloroamines and related compounds, or products derived thereof, is an important use for t-BuOCl.[4] Examples include the hydroxylation of penicillins (eq 10)[4a–c] and a method for the asymmetric synthesis of amines via amino ester intermediates (eq 11).[4d] Both methods involve N-chlorination followed by dehydrochlorination and nucleophilic addition by the solvent.

PhLi, THF, t-BuOCl (10)

Oxidation Reactions. The oxidation of secondary alcohols, using *t*-BuOCl in the presence of ***Pyridine***, gives high yields of the corresponding ketones (eq 12),[5] whereas in the absence of pyridine both the ketone and α-chloro ketone are usually formed (eq 13),[3b] the latter products probably arising from the initially formed ketones via chlorination by chlorine generated in situ.

The oxidation of sulfides using *t*-BuOCl is very well established and can be used to prepare either *cis*- or *trans*-substituted sulfoxide products (eqs 14 and 15). Oxidation of selenides and tellurides is also possible in an analogous fashion, to give selenoxides and telluroxides (or their hydrates), respectively.[7]

Other Applications. Other types of reaction involving *t*-BuOCl which have been reported include cyclization of unsaturated peroxides to give chloroalkyl 1,2-dioxolanes (eq 16),[14] a ring expansion reaction of isopropenylcycloalkenols (eq 17),[15] and a dehydrogenation protocol for the synthesis of tetraethynylethylenes (eq 18).[16]

Related Reagents. See Classes O-1 and O-8, pages 1–10.

1. (a) Anbar, M.; Ginsburg, D. *CRV* **1954**, *54*, 925. (b) Hausweiler, A. *MOC* **1963**, *6/2*, 487.
2. (a) Meijer, E. W.; Kellogg, R. M.; Wynberg, H. *JOC* **1982**, *47*, 2005. (b) Ravindranath, B.; Srinivas, P. *IJC(B)* **1985**, *24*, 163.
3. (a) Walling, C.; Jacknow, B. B. *JACS* **1960**, *82*, 6108. (b) Walling, C.; Mintz, M. J. *JACS* **1967**, *89*, 1515. (c) Hoshi, M.; Masuda, Y.; Arase, A. *CL* **1984**, 195. (d) Walling, C.; Thaler, W. *JACS* **1961**, *83*, 3877. (e) Beckwith, A. L. J.; Westwood, S. W. *AJC* **1983**, *36*, 2123. (f) Walling, C.; Fredericks, P. S. *JACS* **1962**, *84*, 3326.
4. (a) Firestone, R. A.; Christensen, B. G. *JOC* **1973**, *38*, 1436. (b) Baldwin, J. E.; Urban, F. J.; Cooper, R. D. G.; Jose, F. L. *JACS* **1973**, *95*, 2401. (c) Koppel, G. A.; Koehler, R. E. *JACS* **1973**, *95*, 2403. (d) Yamada, S.; Ikota, N.; Achiwa, K. *TL* **1976**, 1001. (e) Awad, R.; Hussain, A.; Crooks, P. A. *JCS(P2)* **1990**, 1233. (f) Zey, R. L. *JHC* **1988**, *25*, 847.
5. Milovanovic, J. N.; Vasojevic, M.; Gojkovic, S. *JCS(P2)* **1988**, 533.
6. (a) Jalsovszky, I.; Ruff, F.; Kajtar-Peredy, M.; Kucsman, A. *S* **1990**, 1037. (b) Johnson, C. R.; McCants Jr., D. *JACS* **1965**, *87*, 1109. (c) Rigau, J. J.; Bacon, C. C.; Johnson, C. R.; *JOC* **1970**, *35*, 3655.
7. (a) Kobayashi, M.; Ohkubo, H.; Shimizu, T. *BCJ* **1986**, *59*, 503. (b) Detty, M. R. *JOC* **1980**, *45*, 274.
8. Mintz, M. J.; Walling, C. *OSC* **1973**, *5*, 183.
9. Walling, C.; Clark, R. T. *JOC* **1974**, *39*, 1962.
10. Smith, K.; Butters, M.; Paget, W. E.; Nay, B. *S* **1985**, 1155.
11. Smith, K.; Butters, M.; Nay, B. *S* **1985**, 1157.
12. Moore, H. W.; Cajipe, G. *S* **1973**, 49.
13. Büchi, G.; Manning, R. E. *JACS* **1966**, *88*, 2532.
14. Bloodworth, A. J.; Tallant, N. A. *TL* **1990**, *31*, 7077.
15. Johnson, C. R.; Herr, R. W. *JOC* **1973**, *38*, 3153.
16. Hauptmann, H. *AG(E)* **1975**, *14*, 498.

Nigel S. Simpkins
University of Nottingham, UK

(Camphorylsulfonyl)oxaziridine[1]

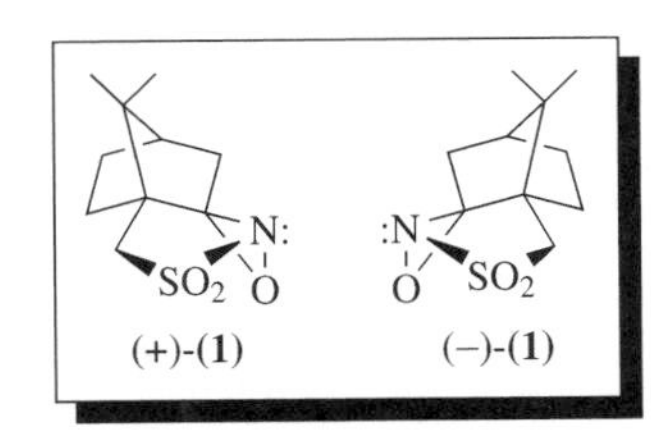

(+)-(**1**)
[104322-63-6] $C_{10}H_{15}NO_3S$ (MW 229.30)
(−)-(**1**)
[104372-31-8]

(neutral, aprotic, electrophilic, and asymmetric oxidizing agents for the chemoselective oxidation of many nucleophilic substrates such as sulfides, enamines, enol esters, carbanions, and enolates[1])

Physical Data: (+)-(**1**): mp 165–167 °C, $[\alpha]_D$ +44.6° ($CHCl_3$, *c* 2.2); (−)-(**1**): mp 166–167 °C, $[\alpha]_D$ −43.6° ($CHCl_3$, *c* 2.2).
Solubility: sol THF, CH_2Cl_2, $CHCl_3$; slightly sol isopropanol, ethanol; insol hexane, pentane, water.
Form Supplied in: commercially available as a white solid.
Analysis of Reagent Purity: by mp and specific rotation determination.
Preparative Methods: the enantiopure (+)- and (−)-(camphorylsulfonyl)oxaziridines (**1**) and [(8,8-dichlorocamphor)sulfonyl]oxaziridines (**2**) are commercially available. They can also be prepared on a large scale via the oxidation of corresponding camphorsulfonimines with buffered ***Potassium Monoperoxysulfate*** (Oxone)[2] or buffered peracetic acid.[3] Since oxidation takes place from the *endo* face of the C=N double bond, only a single oxaziridine isomer is obtained. The precursor camphorsulfonimines can be prepared in 3 steps (>80% yield) from inexpensive (+)- and (−)-***10-Camphorsulfonic Acids***. A variety of (camphorylsulfonyl)oxaziridine derivatives such as (**2**)–(**4**) are also readily available via the functionalization of the camphorsulfonimines followed by oxidation.[1,2–6]

(+)-(**2**) (+)-(**3**) (+)-(**4**)

Purification: by recrystallization.
Handling, Storage, and Precautions: indefinitely stable to storage at room temperature and to exposure to air.

Asymmetric Oxidation of Sulfides. Prochiral sulfides are oxidized by (camphorylsulfonyl)oxaziridine (**1**) to optically active sulfoxides. Over-oxidation to sulfones is not observed (eq 1).[7] However, the best chiral *N*-sulfonyloxaziridines for the asymmetric oxidation of sulfides to sulfoxides are the (+)- and (−)-***N-(Phenylsulfonyl)(3,3-dichlorocamphoryl)oxaziridines***.[8]

(+)-(1), CCl_4 — (*S*) 73% ee (1)

Oxidation of Enamines. Enamines are rapidly oxidized by (+)-(camphorylsulfonyl)oxaziridine (**1**). Disubstituted enamines give rise to racemic α-amino ketones, while trisubstituted enamines afford, after hydrolysis, α-hydroxy ketones (eq 2).[9] A mechanism involving initial oxidation of the enamine to an α-amino epoxide is suggested to account for these products.

(+)-(1); R^2 = H; R^2 H (2)

Oxidation of Oxaphospholenes. Reaction of oxaphospholene (**5**) with (+)-[(8,8-dichlorocamphoryl)sulfonyl]oxaziridine (**2**) affords β-hydroxy-γ-keto-phosphonate in 49% ee with undetermined absolute configuration (eq 3).[10] Higher temperatures accelerate the reaction but lower the stereoselectivity.

(+)-(2), THF, 5 °C, 88%, 49% ee (3)

Oxidation of Organolithium and Organomagnesium Compounds. Oxidation of phenylmagnesium bromide and phenyllithium with ***(±)-trans-2-(Phenylsulfonyl)-3-phenyloxaziridine*** or (camphorylsulfonyl)oxaziridine (**1**) gives phenol (eq 4).[11] Products are cleaner with the latter reagent because addition of the organometallic reagent to the C=N double bond of the imine is not observed. Oxidation of (*E*)- and (*Z*)-vinyllithium reagents with (+)-(**1**) affords enolates. The reaction is fast and represents useful methodology for the stereo- and regioselective formation of enolates.[12] While the enolates can be trapped with ***Chlorotrimethylsilane*** to give silyl enol ethers, better yields and higher stereoselectivity are obtained with ***Bis(trimethylsilyl) Peroxide*** (eq 5).[12]

$$\text{PhMgBr (or PhLi)} + (+)\text{-}(\mathbf{1}) \xrightarrow[-78\,^\circ\text{C}]{\text{THF}} \text{PhOH} \quad (4)$$

(+)-(1), THF, −78 °C; TMSCl (5)

Oxidation of Phosphoranes. Monosubstituted phosphoranes (ylides) are rapidly oxidized to *trans*-alkenes by (+)-(**1**), while disubstituted phosphoranes give ketones (eq 6). A mechanism involving initial attack of the carbanion of phosphorane to the electrophilic oxaziridine oxygen atom of (+)-(**1**) is proposed.[13]

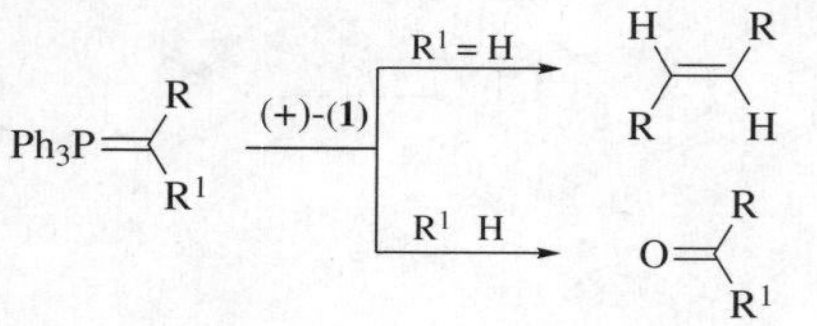

Asymmetric α-Hydroxylation of Enolates. α-Hydroxylation of enolates represents one of the simplest and most direct methods for the synthesis of α-hydroxy carbonyl compounds, a key structural unit found in many natural products.[1b] Enolate oxidations using (+)- and (−)-(**1**) and their derivatives generally effect this transformation in good to excellent yields with a minimum of side reactions (e.g. over-oxidation). Furthermore, these reagents are the only aprotic oxidants developed to date for the direct asymmetric hydroxylation of prochiral enolates to optically active α-hydroxy carbonyl compounds.

By choice of the appropriate reaction conditions and (camphorylsulfonyl)oxaziridine derivative, acyclic α-hydroxy ketones of high enantiomeric purity have been prepared.[1] An example is the oxidation of the sodium enolate of deoxybenzoin with (+)-(**1**). The reaction proceeds very fast at −78 °C, affording (+)-(*S*)-benzoin in 95% ee. Both benzoin enantiomers are readily available by choice of (+)- or (−)-(**1**), because the configuration of the oxaziridine controls the absolute stereochemistry of the product (eq 7).[14] Detailed studies have indicated that the generation of a single enolate regioisomer is a pre-condition for high enantioselectivity, although this does not necessarily always translate into high ee's. Hydroxylation of tertiary substituted acyclic ketone enolates usually gives lower stereoselectivities due to the formation of (*E*/*Z*) enolate mixtures (eq 8).[14] In addition to enolate geometry, the molecular recognition depends on the structure of the oxidant, the type of enolate, and the reaction conditions.[1b] Generally the stereoselectivity can be predicted by assuming that the oxaziridine approaches the enolate from the least sterically hindered direction.

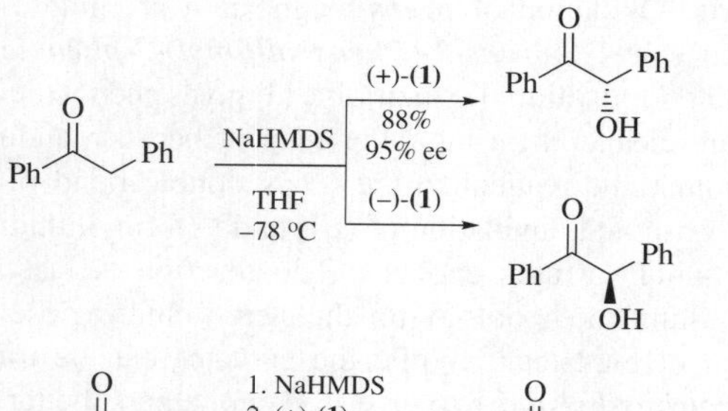

The asymmetric hydroxylation of cyclic ketone enolates, particularly the tetralone and 4-chromanone systems, has been studied in detail because the corresponding α-hydroxy carbonyl compounds are found in many natural products.[1b] Some general trends have been observed. 2-Substituted 1-tetralones having a variety of groups at C-2 (Me, Et, Bn) are best oxidized by chlorooxaziridine (**2**) in >90% ee (eq 9).[2c,15,16] However, substitution of a methoxy group into the 8-position lowers the stereoselectivity. For the 8-methoxytetralones, (8,8-dimethoxycamphorylsulfonyl)oxaziridine (**3**) is the reagent of choice. Similar trends have also been observed in 4-chromanones.[1b] Oxidation of the lithium enolate of (**6**) with (8,8-methoxycamphorsulfonyl)oxaziridine (**3**) affords 5,7-dimethyleucomol (**7**) in ≥96% ee (eq 10).[17] Hydroxylation of the enolate of 1-methyl-2-tetralone (**8**) to (**9**) gives poor to moderate stereoselectivities. The optimum result, 76% ee, is obtained using the sodium enolate and oxaziridine (+)-(**1**) (eq 11).[15]

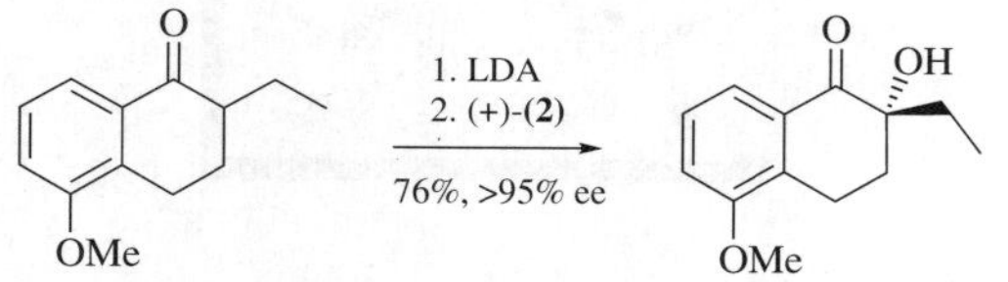

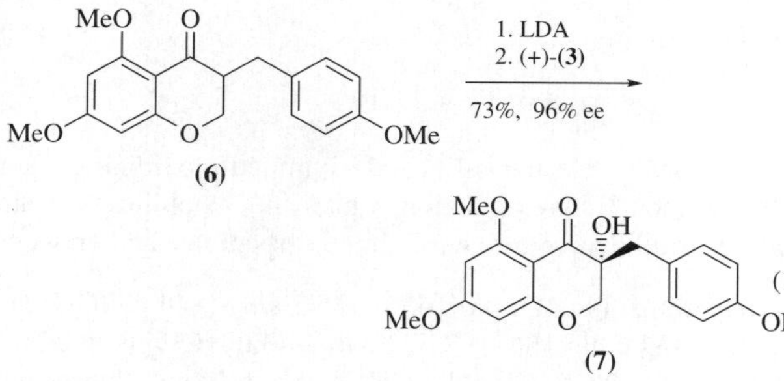

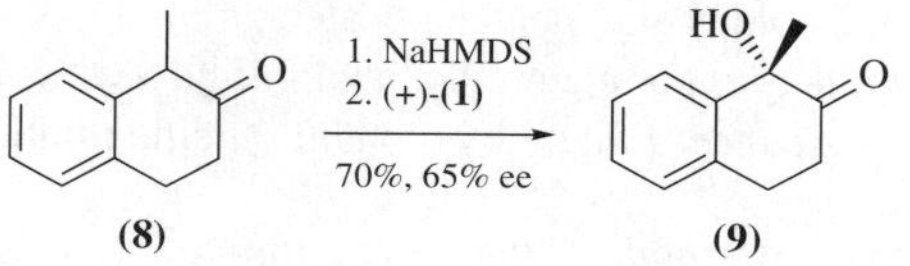

It should be pointed out that enolates are oxidized by the (camphorylsulfonyl)oxaziridine at a much faster rate than sulfides. An example is the preparation of α-hydroxy ketone sulfide (**9**), an intermediate for the total synthesis of (±)-breynolide (eq 12).[18]

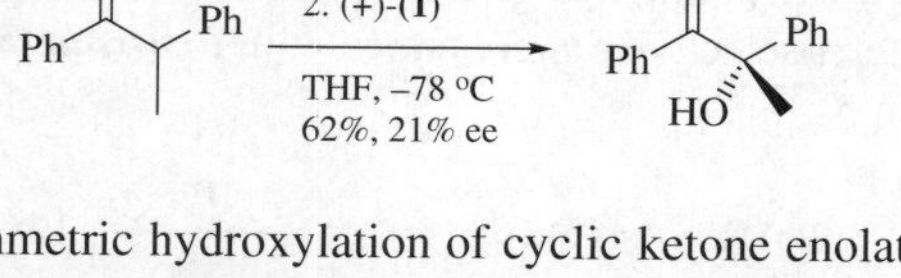

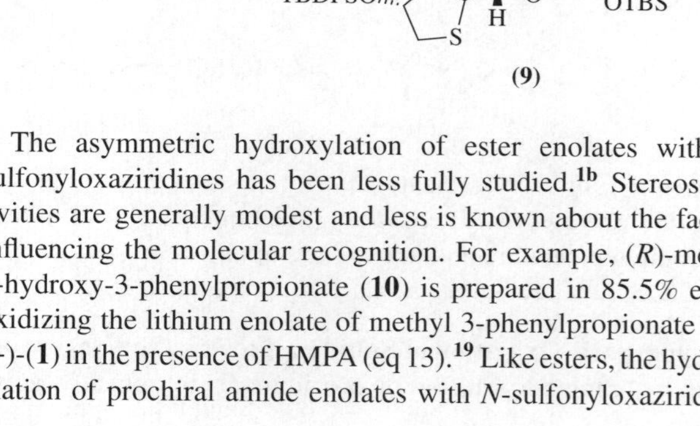

The asymmetric hydroxylation of ester enolates with *N*-sulfonyloxaziridines has been less fully studied.[1b] Stereoselectivities are generally modest and less is known about the factors influencing the molecular recognition. For example, (*R*)-methyl 2-hydroxy-3-phenylpropionate (**10**) is prepared in 85.5% ee by oxidizing the lithium enolate of methyl 3-phenylpropionate with (+)-(**1**) in the presence of HMPA (eq 13).[19] Like esters, the hydroxylation of prochiral amide enolates with *N*-sulfonyloxaziridines affords the corresponding enantiomerically enriched α-hydroxy amides. Thus treatment of amide (**11**) with LDA followed by addition of (+)-(**1**) produces α-hydroxy amide (**12**) in 60% ee (eq 14).[19] Improved stereoselectivities were achieved using double stereodifferentiation, e.g. the asymmetric oxidation of a chiral enolate. For example, oxidation of the lithium enolate of (**13**) with (−)-(**1**) (the matched pair) affords the α-hydroxy amide in 88–91% de (eq 15).[20] (+)-(Camphorsulfonyl)oxaziridine (**1**) mediated hydroxylation of the enolate dianion of (*R*)-(**14**) at −100 to −78 °C in the presence of 1.6 equiv of LiCl gave an 86:14 mixture of *syn*/*anti*-(**15**) (eq 16).[21] The *syn* product is an intermediate for the C-13 side chain of taxol.

Hydroxylation of the sodium enolate of lactone (**16**) with (+)-(**1**) gives α-hydroxy lactone in 77% ee (eq 17).[15] Kinetic resolution and asymmetric hydroxylation with (camphorsulfonyl)oxaziridines has been applied to the synthesis of enantiomerically enriched α-hydroxy carbonyl compounds having multiple stereocenters, which may otherwise be difficult to prepare.[22] Thus hydroxylation of the enolate of racemic 3-methylvalerolactone with substoichiometric amounts of (−)-(**1**) affords (2*S*,3*R*)-verrucarinolactone in 60% ee (eq 18) which on recrystallization is obtained enantiomerically pure.[22]

Oxidation of the dienolate of (**17**) with (+)-(**1**) affords α-hydroxy ester (**18**), a key intermediate in the enantioselective synthesis of the antibiotic echinosporin (eq 19),[23] whereas oxidation of enolates derived from 1,3-dioxin vinylogous ester (**19**) gives rise to both α′- and γ-hydroxylation depending on the reaction conditions (eq 20).[24] With (+)-(**1**) the lithium enolate of (**19**) gives primarily the α′-hydroxylation product (**20**), while the sodium enolate gives γ-hydroxylation product (**21**). Only low levels of asymmetric induction (ca. 16% ee) are found in these oxidations. Birch reduction products are also asymmetrically hydroxylated in situ by (+)-(**1**) (eq 21).[25]

	(20)	(21)
LDA	68%	4%
NaHMDS	3%	28%

Few reagents are available for the hydroxylation of stabilized enolates such as β-keto esters, e.g. Vedejs' MoOPH reagent (***Oxodiperoxymolybdenum(pyridine)(hexamethylphosphoric triamide)***) fails.[26] On the other hand, oxaziridines hydroxylate such enolates in good yield with good to excellent stereoselectivities.[1b] For example, enantioselective hydroxylation of the potassium enolate of the β-keto ester (**22**) with methoxyoxaziridine (−)-(**3**) affords (*R*)-(+)-2-acetyl-5,8-dimethoxy-1,2,3,4-tetrahydro-2-naphthol (**23**), a key intermediate in the asymmetric synthesis of the anthracycline antitumor agents demethoxyadriamycin and 4-demethoxydaunomycin (eq 22).[27] Hydroxylation of the sodium enolate of enone ester (**24**) furnishes kjellmanianone (**25**), an antibacterial agent isolated from marine algae (eq 23).[5] With (+)-(**1**) the ee's are modest (ca 40%), but improved to 69% ee with benzyloxaziridine (**4**).

(24) 1. NaHMDS 2. (+)-(4), 69% ee → (25) (23)

Related Reagents. See Classes O-8, O-20, O-24, pages 1–10.

1. (a) Davis, F. A.; Sheppard, A. C. *T* **1989**, *45*, 5703. (b) Davis, F. A.; Chen, B.-C. *CRV* **1992**, *92*, 919.
2. (a) Towson, J. C.; Weismiller, M. C.; Lal, G. S.; Sheppard, A. C.; Davis, F. A. *OS* **1990**, *69*, 158. (b) Towson, J. C.; Weismiller, M. C.; Lal, G. S.; Sheppard, A. C.; Kumar, A.; Davis, F. A. *OS* **1993**, *72*, 104. (c) Davis, F. A. Weismiller, M. C.; Murphy, C. K.; Thimma Reddy, R.; Chen, B.-C. *JOC* **1992**, *57*, 7274.
3. Mergelsberg, I.; Gala, D.; Scherer, D.; DiBenedetto, D.; Tanner *TL* **1992**, *33*, 161.
4. Davis, F. A.; Kumar, A.; Chen, B.-C. *JOC* **1991**, *56*, 1143.
5. Chen, B.-C.; Weismiller, M. C.; Davis, F. A.; Boschelli, D.; Empfield, J. R.; Smith, III, A. B. *T* **1991**, *47*, 173.
6. (a) Glahsl, G.; Herrmann, R. *JCS(P1)* **1988**, 1753. (b) Meladinis, V.; Herrmann, R.; Steigelmann, O.; Muller, G. *ZN(B)*. **1989**, *44b*, 1453.
7. Davis, F. A.; Towson, J. C.; Weismiller, M. C.; Lal, G. S.; Carroll, P. J. *JACS*. **1988**, *110*, 8477.
8. (a) Davis, F. A.; Thimma Reddy, R.; Weismiller, M. C. *JACS*, **1989**, *111*, 5964. (b) Davis, F. A.; Thimma Reddy, R.; Han, W.; Carroll, P. J. *JACS* **1992**, *113*, 1428.
9. Davis, F.; Sheppard, A. C. *TL* **1988**, *29*, 4365.
10. McClure, C. K.; Grote, C. W. *TL* **1991**, *32*, 5313.
11. Davis, F. A.; Wei, J.; Sheppard, A. C.; Gubernick, S. *TL* **1987**, *28*, 5115.
12. Davis, F. A.; Lal, G. S.; Wei, J. *TL* **1988**, *29*, 4269.
13. Davis, F. A.; Chen, B.-C. *JOC* **1990**, *55*, 360.
14. Davis, F. A.; Sheppard, A. C.; Chen., B.-C.; Haque, M. S. *JACS* **1990**, *112*, 6679.
15. Davis, F. A.; Weismiller, M. C. *JOC* **1990**, *55*, 3715.
16. Davis, F. A.; Kumar, A. *TL* **1991**, *32*, 7671.
17. Davis, F. A.; Chen, B.-C. *TL* **1990**, *31*, 6823.
18. Smith, III, A. B.; Empfield, J. R.; Rivero, R. A.; Vaccaro, H. A. *JACS* **1991**, *113*, 4037.
19. Davis, F. A; Haque, M. S.; Ulatowski, T. G.; Towson, J. C. *JOC* **1986**, *51*, 2402.
20. Davis, F. A.; Ulatowski, T. G.; Haque, M. S. *JOC* **1987**, *52*, 5288.
21. Davis, F. A.; Thimma Reddy, R.; Reddy, R. E. *JOC* **1992**, *57*, 6387.
22. Davis, F. A.; Kumar, A. *JOC*, **1992**, *57*, 3337.
23. Smith, III, A. B.; Sulikowski, G. A.; Fujimoto, K. *JACS* **1989**, *111*, 8039.
24. Smith III, A. B.; Dorsey, B. D.; Ohba, M.; Lupo Jr., A. T.; Malamas, M. S. *JOC* **1988**, *53*, 4314.
25. Schultz, A. G.; Harrington, R. E.; Holoboski, M. A. *JOC* **1992**, *57*, 2973.
26. (a) Vedejs, E.; Larsen, S. *OS* **1985**, *64*, 127. (b) Vedejs, E.; Engler, D. A.; Telschow, J. E. *JOC* **1978**, *43*, 188. Vedejs, E. *JACS* **1974**, *96*, 5944.
27. Davis, F. A.; Kumar, A.; Chen, B.-C. *TL* **1991**, *32*, 867. Davis, F. A.; Clark, C.; Kumar, A., Chen, B.-C. *JOC* **1994**, *59*, 1184.

Bang-Chi Chen
Bristol-Myers Squibb Company, Syracuse, NY, USA

Franklin A. Davis
Drexel University, Philadelphia, PA, USA

Catecholborane[1]

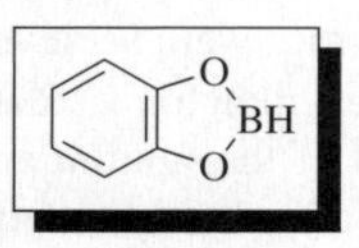

[274-07-7] $C_6H_5BO_2$ (MW 119.92)

(reducing agent for several functional groups;[2] used to prepare alkyl- and alkenylboronic acids and esters via hydroboration of alkenes and alkynes;[3] can be used for synthesis of amides and macrocyclic lactams from carboxylic acids[4])

Physical Data: mp 12 °C; bp 50 °C/50 mmHg; *d* 1.125 g cm^{-3}.
Solubility: sol diethyl ether, THF, CH_2Cl_2, $CHCl_3$, CCl_4, toluene, and benzene; reacts readily with water and other protic solvents.
Form Supplied in: available as a colorless liquid; 1.0 M solution in THF.
Analysis of Reagent Purity: the technical bulletin *Quantitative Analysis of Active Boron Hydrides*, available upon request from the Aldrich Chemical Company, Milwaukee, WI, USA, describes methodology to determine reagent purity by hydrogen gas evolution.
Purification: the neat reagent may be purified by distillation at reduced pressure.
Handling, Storage, and Precautions: should be stored in a cold room or refrigerator, without exposure to atmospheric moisture; cold storage has been found to minimize loss of hydrogen activity as well as pressure build-up; the neat reagent may be stored as a solid at 0 °C; a sample stored over one year at 0–5 °C showed no detectable loss in hydride activity;[1a] syringe and double-tipped needle techniques are recommended for reagent transfer. Use in a fume hood.

Functional Group Reductions. Catecholborane (CB) is one of the most versatile boron hydride reducing agents. This reagent possesses enhanced thermal stability and solubility characteristics as compared with other boron hydride reagents. Reductions can be carried out in several organic solvents including CCl_4, $CHCl_3$, benzene, toluene, diethyl ether, and THF, as well as in the absence of solvent.[1a]

Several functional groups do not react with catecholborane. Alkyl and aryl halides, nitro groups, sulfones, disulfides, thiols, primary amides, ethers, sulfides and alcohols are all inert toward the reagent. Nitriles, esters, and acid chlorides react slowly, while aldehydes, ketones, imines, and sulfoxides are readily reduced in a few hours at room temperature.[1a]

Catecholborane readily reduces tosylhydrazones[2] to the corresponding methylene derivatives (eqs 1 and 2). This mild method may be used to advantage with substrates possessing sensitive functional groups which preclude the use of the more conventional Wolff–Kishner and Clemmensen reductions.[5]

NNHTs 1. catecholborane 2. NaOAc, H_2O, Δ, 91% (1)

NNHTs — 1. catecholborane 2. NaOAc, H_2O, Δ — 92% (2)

Evans[6a] has demonstrated that catecholborane will undergo conjugate reduction of α,β-unsaturated ketones at room temperature (eqs 3 and 4).[6,7] This reaction is limited to α,β-unsaturated ketones that can adopt an *s-cis* conformation, while α,β-unsaturated imides, esters, and amides are unreactive under these conditions. It was also determined that catalytic quantities of ***Chlorotris(triphenylphosphine)rhodium(I)*** greatly accelerate the 1,4-addition process such that reduction occurs readily at −20 °C.

catecholborane, THF, $Rh(PPh_3)_3Cl$, −20 °C — 70% (3)

$PhSO_2$ … TESO, OPMB, OTIPS — catecholborane, THF — 94% (4)

Cyclic enones bearing an endocyclic alkene, e.g. (**1**) and (**2**), do not undergo 1,4-addition but are instead reduced exclusively at the carbonyl group. It is also possible to selectively reduce *trans*-1,2-disubstituted enones in the presence of 1,1-disubstituted systems. The trapping of the intermediate boron enolate by aldehyde electrophiles has also been demonstrated.[6]

CO_2Et

(**1**) (**2**)

Several diastereoselective and enantioselective reductions of ketones mediated by catecholborane have also been reported. Acyclic β-hydroxy ketones are stereoselectively reduced to *syn* 1,3-diols by catecholborane (eq 5).[8] In several instances it was found that stereoselectivity could be enhanced by performing the reaction with Rh^I catalysis. The high levels of diastereoselectivity are believed to be due to the ability of catecholborane to preorganize the substrate prior to intermolecular delivery of hydride by a second molecule of catecholborane. The incorporation of a methyl group between the hydroxyl and carbonyl groups can affect the *syn* diastereoselectivity, the extent depending upon the stereochemical relationship between the two substituents (eqs 6 and 7).

Chiral oxazaborolidines, initially reported by Itsuno[9] and later developed by Corey,[10] have been used for the enantioselective reduction of ketones. Corey has demonstrated that oxazaborolidine (**3**), prepared from ***α,α-Diphenyl-2-pyrrolidinem ethanol*** and butylboronic acid, is a highly effective catalyst for the asymmetric reduction of ketones using catecholborane as a stoichiometric reductant. This system was also shown to be superior for the asymmetric 1,2-reduction of enones.

OH O — catecholborane, THF — OH OH (5)

Conditions	Yield (*syn*:*anti*)
−10 °C	82% (10:1)
−10 °C, Rh^I	76% (12:1)
−35 °C	87% (6:1)
−35 °C, Rh^I	86% (20:1)

OH O — catecholborane, THF, −10 °C — 82% — OH OH, 35:1 (6)

OH O — catecholborane, THF, −10 °C — 77% — OH OH, 6:1 (7)

(**3**)

R_S, R_L + catecholborane (BH), 1.5–2.0 equiv — (**3**) (0.1 equiv), −78 °C, toluene — >95% — OH, R_S, R_L

Substrate	Absolute configuration	Enantiomeric excess
F_3C (9-anthryl trifluoromethyl ketone)	*R*	94%
(acetophenone)	*R*	94%
(2-methylcyclohexenone)	*R*	92%
Ph (benzylideneacetone)	*R*	93%

Hydroboration of Alkenes. Hydroborations of alkenes by catecholborane[3] are generally much slower than those employing dialkylboranes.[11] Elevated temperatures are usually required; however, the reaction rates may be enhanced by the use of Wilkinson's catalyst,[12] *N,N*-dimethylaniline–borane,[13] and ***Lithium Borohydride***.[14] The alkylboronic esters obtained are easily hydrolyzed to the corresponding alkylboronic acids or converted to aldehydes,[15] ketones,[16] carboxylic acids,[17] and alcohols.[3e,8] The esters may also be homologated, thus making possible the synthesis of substituted boronic esters, which have become increasingly useful as reagents for stereodirected synthesis.[18]

Catecholborane has been widely used for the transition metal catalyzed hydroborations of alkenes.[12,19] Both Rh^{I} and Ir^{I} have been used as the metal catalyst. Catalysis not only has a beneficial effect on rate, but has also been found to alter the chemo, regio- and stereochemical course of the hydroboration when compared to the uncatalyzed reaction.[19]

Evans has established that the rate of the transition metal catalyzed hydroboration is very sensitive to the alkene substitution pattern.[20,21] Terminal alkenes undergo complete hydroboration within minutes at room temperature, while 1,1- and 1,2-disubstituted alkenes require several hours. Trisubstituted alkenes are unreactive. The sensitivity of the catalyzed reaction to steric effects affords the possibility for selective hydroboration of the less hindered of two alkenyl groups in a given substrate (eq 8).[20]

catecholborane (2.5 equiv), 3% $Rh(PPh_3)_3Cl$, 20 °C, 90% (8)

Good to excellent levels of 1,2-asymmetric induction are obtained in the catalyzed hydroboration of chiral 1,1-disubstituted alkenes with catecholborane (eq 9). The extent of asymmetric induction has been found to depend on the size of X and, to a lesser extent, the size of R′. The catalyzed and uncatalyzed reactions are stereocomplementary, with the former favoring the formation of products with *syn* stereochemistry while the latter favors products with *anti* stereochemistry.[20,22] Allylic and homoallylic diphenyl phosphinite and amide functional groups have also been used to direct the regiochemistry of the catalyzed hydroboration reaction. Examples of this methodology with both cyclic and acyclic substrates have been reported.[20,23]

R_2BH or catecholborane → *anti* diastereomer

Rh^{I}, catecholborane → *syn* diastereomer (9)

X = OH, OTBDMS, OTBDPS, OAc, $OCONMe_2$, $OCOCF_3$, NHTs, NTsBn
R, R′ = alkyl, aryl

Several groups have reported catalyst systems for the asymmetric hydroboration of prochiral alkenes using the catalyzed hydroboration reaction with a chirally modified metal catalyst.[24] Most of the reactions employ a Rh^{I} catalyst modified with chirally modified bidentate phosphine ligands. Enantiomeric excesses obtained with several alkene substrates are in the good to excellent range.[24b,f]

Hydroboration of Alkynes. The hydroboration of alkynes with catecholborane is an efficient route to alkenylboronic esters and acids. The regioselectivity of the addition to unsymmetrical alkenes is very similar to that displayed by ***Disiamylborane***.[3c–e] The rate of reaction can be enhanced through use of *N,N*-dimethylaniline–borane complex.[13]

The alkenylboronic esters and acids obtained have proven to be valuable synthetic intermediates. Perhaps the most widespread use of these intermediates has been in the Pd^0 catalyzed cross-coupling reaction with alkenyl, alkynyl, and aryl halides to give the corresponding alkenes (eq 10).[25]

R—≡ → catecholborane → R′—I, Pd^0, base (10)

R′ = alkenyl, alkynyl or aryl

Protonolysis of alkenylboronic esters deriving from internal alkynes provides a route for the stereospecific synthesis of *cis*-disubstituted alkenes, while an oxidative workup will afford the corresponding ketone.[3d,e] Iodo- and bromoalkenes can also be prepared stereospecifically from alkenylboronic ester intermediates.[26] These compounds are useful in alkene cross-coupling reactions as well as in the preparation of alkenylmagnesium and alkenyllithium compounds. Haloalkenes may also be prepared from alkenylboronic acids via organomercurial intermediates (eq 11).[27]

$MeCO_2H$ ← $B(OR)_2$ → NaOH, H_2O_2

1. H_2O 2. NaOH, I_2 — 1. Br_2, −40 °C 2. NaOMe (11)

Reagent for Amide and Lactam Synthesis. Carboxylic acids react rapidly with catecholborane to produce 2-acyloxy-1,3,2-benzodioxaborolanes, e.g. (**4**). This reaction has been used as the carboxyl activation step for the synthesis of amides and macrocyclic lactams (eq 12).[4]

(**4**) → $PhCH_2NH_2$, THF, −78 °C, 92% (12)

Related Reagents. See Classes R-2, R-3, R-4, R-10, R-11, R-12, R-15, R-17, and R-29, pages 1–10. Bis(benzoyloxy)borane; Disiamylborane.

1. (a) Lane, C. F.; Kabalka, G. W. *T* **1976**, *32*, 981. (b) Wietelmann, U. *Janssen Chim. Acta* **1992**, *10*, 16. (c) Pelter, A.; Smith, K.; Brown, H. C. *Borane Reagents*; Academic Press: New York, 1988. (d) Brown, H. C. *Organic Synthesis via Boranes*; Wiley: New York, 1975. (e) Brown, H. C.; Chandrasekharan, J. *JOC* **1983**, *48*, 5080. (f) Kabalka, G. W. *OPP* **1977**, *9*, 131.
2. Kabalka, G. W.; Baker, J. D., Jr. *JOC* **1975**, *40*, 1834.
3. (a) *COS* **1991**, *8*, Chapter 3.10, p 703. (b) Suzuki, A.; Dhillon, R. S. *Top. Curr. Chem.* **1986**, *130*, 23. (c) Brown, H. C.; Gupta, S. K. *JACS* **1971**, *93*, 1816. (d) Brown, H. C.; Gupta, S. K. *JACS* **1972**, *94*, 4370. (e) Brown, H. C.; Gupta, S. K. *JACS* **1975**, *97*, 5249.
4. Collum, D. B.; Shen, S.-C.; Ganem, B. *JOC* **1978**, *43*, 4393.
5. *COS* **1991**, *8*, Chapter 1.13–1.14, pp 307 and 327.
6. (a) Evans, D. A.; Fu, G. C. *JOC* **1990**, *55*, 5678. (b) Matsumoto, Y.; Hayashi, T. *SL* **1991**, 349.
7. Other methods: (a) *COS* **1991**, *8*, Chapter 3.5, p 503 (b) Larock, R. C. *Comprehensive Organic Transformations*; VCH: New York, 1989, pp 8–17.
8. Evans, D. A.; Hoveyda, A. H. *JOC* **1990**, *55*, 5190.
9. (a) Hirao, A.; Itsuno, S.; Nakahama, S.; Yamazaki, N. *CC* **1981**, 315. (b) Itsuno, S.; Hirao, A.; Nakahama, S.; Yamazaki, N. *JCS(P1)* **1983**, 1673. (c) Itsuno, S.; Ito, K.; Hirao, A, Nakahama, S. *CC* **1983**, 469. (d) Itsuno, S.; Ito, K.; Hirao, A, Nakahama, S. *JOC* **1984**, *49*, 555. (e) Itsuno, S.; Nakano, M.; Miyazaki, K.; Masuda, H.; Ito, K.; Hirao, A.; Nakahama, S. *JCS(P1)* **1985**, 2039. (f) Itsuno, S.; Ito, K.; Maruyama, T.; Kanda, N.; Hirao, A.; Nakahama, S. *BCJ* **1986**, *59*, 3329.
10. (a) Corey, E. J.; Bakshi, R. K.; Shibata, S. *JACS* **1987**, *109*, 5551. (b) Corey, E. J.; Bakshi, R. K.; Shibata, S.; Chen, C.-P.; Singh, V. K. *JACS* **1987**, *109*, 7925. (c) Corey, E. J.; Shibata, S.; Bakshi, R. K. *JOC* **1988**, *53*, 2861. (d) Corey, E. J.; Jardine, P. D. S.; Rohloff, J. C. *JACS* **1988**, *110*, 3672. (e) Corey, E. J.; Gavai, A. V. *TL* **1988**, *29*, 3201. (f) Corey, E. J.; Bakshi, R. K. *TL* **1990**, *31*, 611.
11. (a) Fish, R. H. *JOC* **1973**, *38*, 158. (b) See also ref. 1e.
12. Männig, D.; Nöth, H. *AG(E)* **1985**, *24*, 878.
13. Suseela, Y.; Prasad, A. S. B.; Periasamy, M. *CC* **1990**, 446.
14. Arase, A.; Nunokawa, Y.; Masuda, Y.; Hoshi, M. *CC* **1991**, 205.
15. Brown, H. C.; Imai, T. *JACS* **1983**, *105*, 6285.
16. Brown, H. C.; Srebnik, M.; Bakshi, R. K.; Cole, T. E. *JACS* **1987**, *109*, 5420.
17. Brown, H. C.; Imai, T.; Desai, M. C.; Singaram, B. *JACS* **1985**, *107*, 4980.
18. Matteson, D. S. *T* **1989**, *45*, 1859.
19. Burgess, K.; Ohlmeyer, M. J. *CRV* **1991**, *91*, 1179.
20. Evans, D. A.; Fu, G. C.; Hoveyda, A. H. *JACS*, **1992**, *114*, 6671.
21. Mechanistic studies: (a) Evans, D. A.; Fu, G. C.; Anderson, B. A. *JACS* **1992**, *114*, 6679. (b) Burgess, K.; van der Donk, W. A.; Westcott, S. A.; Marder, T. B.; Baker, R. T.; Calabrese, J. C. *JACS* **1992**, *114*, 9350.
22. (a) Burgess, K.; Ohlmeyer, M. J. *TL* **1989**, *30*, 395. (b) Burgess, K.; Ohlmeyer, M. J. *JOC* **1991**, *56*, 1027.
23. Evans, D. A.; Fu, G. C.; Hoveyda, A. H. *JACS* **1988**, *110*, 6917.
24. (a) Burgess, K.; Ohlmeyer, M. J. *JOC* **1988**, *53*, 5178. (b) Hayashi, T.; Matsumoto, Y.; Ito, Y. *JACS* **1989**, *111*, 3426. (c) Sato, M.; Miyaura, N.; Suzuki, A. *TL* **1990**, *31*, 231. (d) Burgess, K.; van der Donk, W.; Ohlmeyer, M. J. *TA* **1991**, *2*, 613. (e) Matsumoto, Y.; Hayashi, T. *TL* **1991**, *32*, 3387. (f) Hayashi, T.; Matsumoto, Y.; Ito, Y. *TA* **1991**, *2*, 601. (g) Burgess, K.; Ohlmeyer, M. J.; Whitmire, K. H. *OM* **1992**, *11*, 3588.
25. (a) Miyaura, N.; Yamada, K.; Suzuki, A. *TL* **1979**, *20*, 3437. (b) Miyaura, N.; Suginome, H.; Suzuki, A. *TL* **1981**, *22*, 127. (c) Miyaura, N.; Yamada, K.; Suginome, H.; Suzuki, A. *JACS* **1985**, *107*, 972. (d) Suzuki, A. *PAC* **1985**, *57*, 1749. (e) Suzuki, A. *PAC* **1991**, *63*, 419.
26. Brown, H. C.; Hamaoka, T.; Ravindran, N. *JACS* **1973**, *95*, 6456.
27. Brown, H. C.; Larock, R. C.; Gupta, S. K.; Rajagopalan, S.; Bhat, N. G. *JOC* **1989**, *54*, 6079.

Michael S. VanNieuwenhze
The Scripps Research Institute, La Jolla, CA, USA

Cerium(IV) Ammonium Nitrate[1]

$(NH_4)_2Ce(NO_3)_6$

[16774-21-3] $H_8CeN_8O_{18}$ (MW 548.26)

(volumetric standard oxidant;[2] oxidant for many functional groups;[1] can promote oxidative halogenation[3])

Alternate Name: ammonium cerium(IV) nitrate; ceric ammonium nitrate; CAN.
Solubility: sol water (1.41 g mL^{-1} at 25 °C, 2.27 g mL^{-1} at 80 °C); sol nitric acid.
Form Supplied in: orange crystals; widely available.
Handling, Storage, and Precautions: solid used as supplied. No toxicity data available, but cerium is reputed to be of low toxicity.

Functional Group Oxidation. Ce^{IV} in acidic media is a stronger oxidant than elemental chlorine and is exceeded in oxidizing power only by a few reagents (F_2, XeO_3, Ag^{2+}, O_3, HN_3). The thermodynamically unstable solutions can be kept for days because of kinetic stability. CAN is a one-electron oxidant soluble in water and to a smaller extent in polar solvents such as acetic acid. Its consumption can be judged by the fading of an orange color to pale yellow, if the substrate or product is not strongly colored. Because of its extremely limited solubility in common organic solvents, oxidations are often carried out in mixed solvents such as aqueous acetonitrile. There are advantages in using dual oxidant systems in which Ce^{IV} is present in catalytic amounts. Cooxidants such as ***Sodium Bromate***,[4] ***t-Butyl Hydroperoxide***,[5] and ***Oxygen***[6] have been employed. Electrolytic recycling[7] of Ce^{IV} species is also possible.

Cerium(IV) sulfate and a few other ligand-modified CAN reagents have been used for the oxidation. The differences in their oxidation patterns are small, and consequently it is quite safe to replace one particular oxidizing system with another. More rarely employed is cerium(IV) perchlorate.

Oxidation of Alkenes and Arenes. The outcome of the oxidation of alkenes is solvent dependent, but dinitroxylation (eq 1)[8] has been achieved. Certain arylcyclopropanes are converted into the 1,3-diol dinitrates.[9]

CAN; MeCN, $h\nu$; 91% → O_2NO / ONO_2 (1)

CAN promotes benzylic oxidation of arenes,[10] e.g. methyl groups are converted into formyl groups but less efficiently when an electron-withdrawing group is present in the aromatic ring. A very interesting molecule, hexaoxo[1_6]orthocyclophane in an internal acetal form (eq 2),[11] has been generated via CAN oxidation. Regioselective oxidation is observed with certain substrates, e.g. 2,4-dimethylanisole gives 3-methyl-*p*-anisaldehyde. Oxidation may be diverted into formation of non-aldehyde products by using different media: benzylic acetates[12] are formed in glacial acetic acid, ethers[13] in alcohol solvents, and nitrates[14] in acetonitrile under photolytic conditions.

CAN, HOAc, Δ, 7 days (2)

hexaoxo[1_6]orthocyclophane

Polynuclear aromatic systems can be oxidized to quinones,[15] but unsymmetrical substrates will often give a mixture of products. It has been reported that mononitro derivatives were formed by the oxidation of polynuclear arenes with CAN adsorbed in silica,[16] whereas dinitro compounds and quinones were obtained from oxidation in solution.

Oxidation of Alcohols, Phenols, and Ethers. A primary alcohol can be retained when a secondary alcohol is oxidized to ketone.[17] Tetrahydrofuran formation (eq 3)[18] predominates in molecules with rigid frameworks, which are favorable to δ-hydrogen abstraction by an alkoxyl radical.

CAN, aq MeCN, 60 °C, 85% (3)

Tertiary alcohols are prone to fragmentation;[19] this process is facilitated by a β-trimethylsilyl group (eq 4)[20]. Other alcohols prone to fragmentation are cyclobutanols,[21] strained bicyclo[*x.y.z*]alkan-2-ols,[22] and homoallylic alcohols.[18d]

CAN, aq MeCN, 22 °C, 66% (4)

CAN converts benzylic alcohols into carbonyl compounds.[23] Even *p*-nitrobenzyl alcohol gives *p*-nitrobenzaldehyde in the catalytic oxidation system.[23c] Oxygen can be used[6] as the stoichiometric oxidant.

Catechols, hydroquinones, and their methyl ethers readily afford quinones on Ce^{IV} oxidation.[4,24] Partial demethylative oxidation is feasible, as shown in the preparation of several intramolecular quinhydrones (eq 5)[25] and a precursor of daunomycinone.[26] Sometimes the dual oxidant system of CAN–$NaBrO_3$ is useful. In a synthesis of methoxatin (eq 6)[27] the *o*-quinone moiety was generated from an aryl methyl ether.

CAN, aq MeCN, 22 °C, 42–97% (5)

n = 1–4

CAN, aq MeCN, 0 °C, 60% (6)

Oxidative regeneration of the carboxylic acid from its 2,6-di-*t*-butyl-4-methoxyphenyl ester[28] is the basis for the use of this auxiliary in a stereoselective α-hydroxyalkylation of carboxylic acids. The smooth removal of *p*-anisyl (eq 7)[29] and *p*-anisylmethyl[30] groups from an amidic nitrogen atom by Ce^{IV} oxidation makes these protective groups valuable in synthesis.

CAN, aq MeCN, 0 °C, 67% (7)

Simple ethers are oxidized[31] to carbonyl products and the intermediate from tetrahydrofuran oxidation can be trapped by alcohols.[32]

Vicinal diols undergo oxidative cleavage.[33] There is no apparent steric limitation as both *cis*- and *trans*-cycloalkane-1,2-diols are susceptible to cleavage. However, under certain conditions α-hydroxy ketones may be oxidized without breaking the C–C bond.[6]

Oxidation of Carbonyl Compounds. The Ce^{IV} oxidation of aldehydes and ketones is of much less synthetic significance than methods using other reagents. However, cage ketones often provide lactones (eq 8)[34,35] in good yield. Tetracyclones furnish α-pyrones.[36]

CAN, aq MeCN, 60 °C (8)

Concerning carboxylic acids and their derivatives, transformations of practical value are restricted to oxidative hydrolysis such as the conversion of hydrazides[37] back to carboxylic acids, transamidation of *N*-acyl-5,6-dihydrophenanthridines,[38] and decarboxylative processes, especially the degradation of α-hydroxymalonic acids (eq 9).[39] In some cases the Ce^{IV} oxidation is much superior to periodate cleavage. A related reaction is involved in a route to lactones.[40]

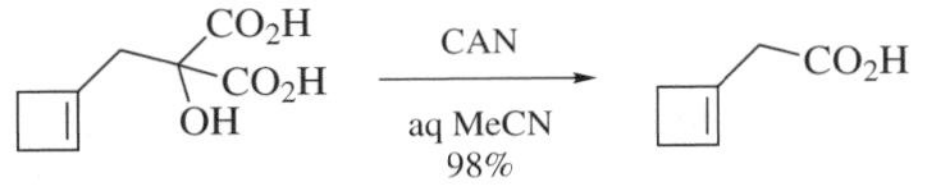

(9)

Nitrogenous derivatives of carbonyl compounds such as oximes and semicarbazones are oxidatively cleaved by Ce^{IV},[41] but only a few synthetic applications have been reported.[42]

Oxidation of Nitroalkanes. Ce^{IV} oxidation provides an alternative to the Nef reaction.[43] At least in the case of a ketomacrolide synthesis (eq 10),[44] complications arising from side reactions caused by other reagents are avoided.

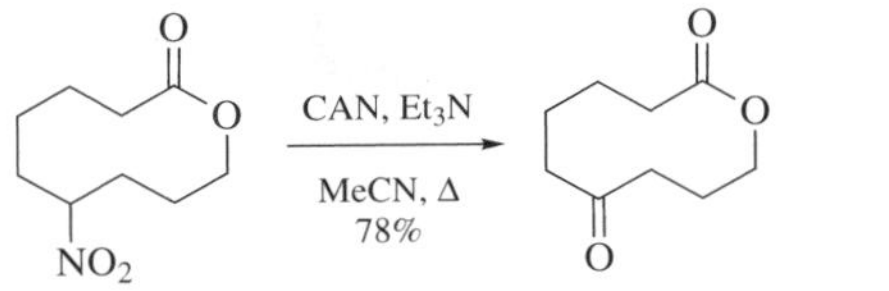

(10)

Oxidation of Organosulfur Compounds. Thiols are converted into disulfides using reagents such as ***Bis[trinitratocerium(IV)] Chromate***.[45] Chemoselective oxidation of sulfides by Ce^{IV} reagents to sulfoxides[4,46] is easily accomplished. Stoichiometric oxidation under phase transfer conditions[46b] and the dual oxidant[4] protocols permit oxidation of a variety of sulfides.

The reaction of dithioacetals including 1,3-dithiolanes and 1,3-dithianes with CAN provides a convenient procedure for the generation of the corresponding carbonyl group.[47] The rapid reaction is serviceable in many systems and superior to other methods, e.g. in the synthesis of acylsilanes.[48] In a series of compounds in which the dithiolane group is sterically hindered, the reaction led to enones, i.e. dehydrogenation accompanied the deprotection (eq 11).[49]

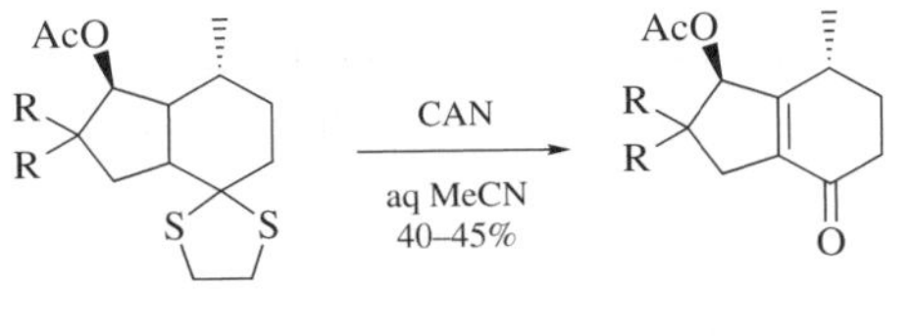

(11)

R = H, Me

Oxidative Cleavage of Organometallic Compounds. Oxidative deligation of both σ- and π-complexes by treatment with CAN is common practice. Ligands including cyclobutadiene and derivatives (eq 12)[50] and α-methylene-γ-butyrolactone (eq 13)[51] have been liberated successfully and applied to achieving the intended research goals. In the recovery of organic products from a Dötz reaction, CAN is often employed to cleave off the metallic species.[52]

Generation of α-Acyl Radicals. As a one-electron oxidant, Ce^{IV} can promote the formation of radicals from carbonyl compounds. In the presence of interceptors such as butadiene and alkenyl acetates, the α-acyl radicals undergo addition.[53] The carbonyl compounds may be introduced as enol silyl ethers, and the oxidative coupling of two such ethers may be accomplished.[54] Some differences in the efficiency for oxidative cyclization of δ,ε-, and ε,ζ-unsaturated enol silyl ethers using CAN and other oxidants have been noted (eq 14).[55]

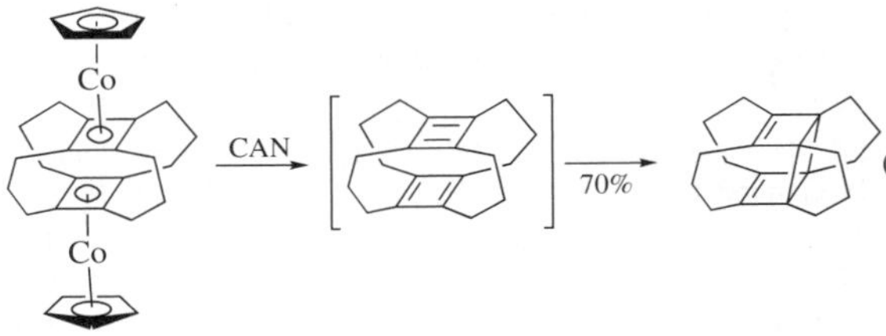

(12)

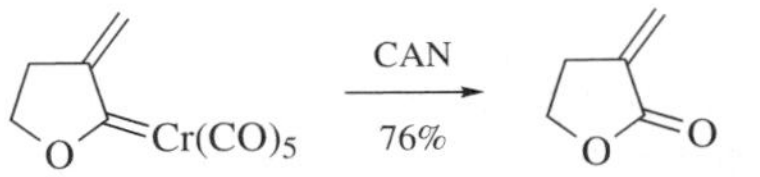

(13)

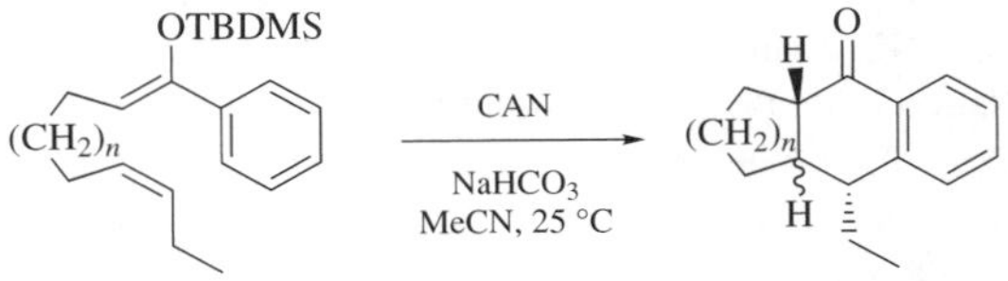

(14)

n = 1, 73% *cis*:*trans* = 20:1
n = 2, 42% *cis*:*trans* = 4.3:1

Oxidative Halogenation. Benzylic bromination[56] and α-iodination of ketones[3a] and uracil derivatives[3b] can be achieved with CAN as in situ oxidant.

Related Reagents. See Classes O-1, O-6, O-8, O-15, and O-24, pages 1–10. Cerium(IV) Ammonium Nitrate–Sodium Bromate; Iodine–Cerium(IV) Ammonium Nitrate.

1. (a) Richardson, W. H. In *Oxidation in Organic Chemistry*, Wiberg, K. B., Ed.; Academic: New York, 1965; Part A, Chapter IV. (b) Ho, T.-L. *S* **1973**, 347. (c) Ho, T.-L. In *Organic Syntheses by Oxidation with Metal Compounds*, Mijs, W. J.; de Jonge, C. R. H. I., Eds., Plenum: New York, 1986, Chapter 11.
2. Smith, G. F. *Cerate Oxidimetry*, G. Frederick Smith Chemical Co.: Columbus, OH, 1942.
3. (a) Horiuchi, C. A.; Kiji, S. *CL* **1988**, 31. (b) Asakura, J.; Robins, M. J. *JOC* **1990**, *55*, 4928.
4. Ho, T.-L. *SC* **1979**, *9*, 237.
5. Kanemoto, S.; Saimoto, H.; Oshima, K.; Nozaki, H. *TL* **1984**, *25*, 3317.
6. Hatanaka, Y.; Imamoto, T.; Yokoyama, M. *TL* **1983**, *24*, 2399.
7. Kreh, R. P.; Spotnitz, R. M.; Lundquist, J. T. *JOC* **1989**, *54*, 1526.
8. Baciocchi, E.; Giacco, T. D.; Murgia, S. M.; Sebastiani, G. V. *T* **1988**, *44*, 6651.
9. Young, L. B. *TL* **1968**, 5105.
10. (a) Syper, L. *TL* **1966**, 4493. (b) Laing, S. B. *JCS(C)* **1968**, 2915.
11. Lee, W. Y.; Park, C. H.; Kim, S. *JACS* **1993**, *115*, 1184.
12. Baciocchi, E.; Della Cort, A.; Eberson, L.; Mandolini, L.; Rol, C. *JOC* **1986**, *51*, 4544.
13. Della Cort, A.; Barbera, A. L.; Mandolini, L. *JCR(S)* **1983**, 44.
14. Baciocchi, E.; Rol, C.; Sebastiani, G. V.; Serena, B. *TL* **1984**, *25*, 1945.
15. (a) Ho, T.-L.; Hall, T.-W.; Wong, C. M. *S* **1973**, 206. (b) Periasamy, M.; Bhatt, M. V. *TL* **1977**, 2357.

16. Chawla, H. M.; Mittal, R. S. *S* **1985**, 70.
17. Kanemoto, S.; Tomioka, H.; Oshima, K.; Nozaki, H. *BCSJ* **1986**, *58*, 105.
18. (a) Trahanovsky, W. S.; Young, M. G.; Nave, P. M. *TL* **1969**, 2501. (b) Doyle, M. P.; Zuidema, L. J.; Bade, T. R. *JOC* **1975**, *40*, 1454. (c) Fujise, Y.; Kobayashi, E.; Tsuchida, H.; Ito, S. *H* **1978**, *11*, 351. (d) Balasubramanian, V.; Robinson, C. H. *TL* **1981**, 501.
19. Trahanovsky, W. S.; Macaulay, D. B. *JOC* **1973**, *38*, 1497.
20. Wilson, S. R.; Zucker, P. A.; Kim, C.; Villa, C. A. *TL* **1985**, *26*, 1969.
21. (a) Meyer, K.; Rocek, J. *JACS* **1972**, *94*, 1209. (b) Hunter, N. R.; MacAlpine, G. A.; Liu, H.-J.; Valenta, Z. *CJC* **1970**, *48*, 1436.
22. Trahanovsky, W. S.; Flash, P. J.; Smith, L. M. *JACS* **1969**, *91*, 5068.
23. (a) Trahanovsky, W. S.; Cramer, J. *JOC* **1971**, *36*, 1890. (b) Trahanovsky, W. S.; Fox, N. S. *JACS* **1974**, *96*, 7968. (c) Ho, T.-L. *S* **1978**, 936.
24. (a) Ho, T.-L.; Hall, T.-W.; Wong, C. M. *CI(L)* **1972**, 729. (b) Jacob, P., III; Callery, P. S.; Shulgin, A. T.; Castagnoli, N., Jr. *JOC* **1976**, *41*, 3627. (c) Syper, L.; Kloc, K.; Mlochowski, J. *S* **1979**, 521.
25. Bauer, H.; Briaire, J.; Staab, H. A. *AG(E)* **1983**, *22*, 334.
26. Hauser, F. M.; Prasanna, S. *JACS* **1981**, *103*, 6378.
27. Corey, E. J.; Tramontano, A. *JACS* **1981**, *103*, 5599.
28. Heathcock, C. H.; Pirrung, M. C.; Montgomery, S. H.; Lampe, J. *T* **1981**, *37*, 4087.
29. (a) Fukuyama, T.; Frank, R. K.; Jewell, C. F. *JACS* **1980**, *102*, 2122. (b) Kronenthal, D. R.; Han, CY.; Taylor, M. K. *JOC* **1982**, *47*, 2765.
30. Yamaura, M.; Suzuki, T.; Hashimoto, H.; Yoshimura, J.; Okamoto, T.; Shin, C. *BCJ* **1985**, *58*, 1413.
31. Olah, G. A.; Gupta, B. G. B.; Fung, A. P. *S* **1980**, 897.
32. Maione, A. M.; Romeo, A. *S* **1987**, 250.
33. (a) Hintz, H. L.; Johnson, D. C. *JOC* **1967**, *32*, 556. (b) Trahanovsky, W. S.; Young, L. H.; Bierman, M. H. *JOC* **1969**, *34*, 869.
34. Soucy, P.; Ho, T.-L; Deslongchamps, P. *CJC* **1972**, *50*, 2047.
35. Mehta, G.; Pandey, P. N.; Ho, T.-L. *JOC* **1976**, *41*, 953.
36. Ho, T.-L.; Hall, T.-W.; Wong, C. M. *SC* **1973**, *3*, 79.
37. Ho, T.-L.; Ho, H. C.; Wong, C. M. *S* **1972**, 562.
38. (a) Uchimaru, T.; Narasaka, K.; Mukaiyama, T. *CL* **1981**, 1551. (b) Narasaka, K.; Hirose, T.; Uchimaru, T.; Mukaiyama, T. *CL* **1982**, 991.
39. Salomon, M. F.; Pardo, S. N.; Salomon, R. G. *JACS* **1980**, *102*, 2473.
40. Salomon, R. G.; Roy, S.; Salomon, M. F. *TL* **1988**, *29*, 769.
41. Bird, J. W.; Diaper, D. G. M. *CJC* **1969**, *47*, 145.
42. (a) Oppolzer, W.; Petrzilka, M.; Bättig, K. *HCA* **1977**, *60*, 2964. (b) Oppolzer, W.; Bättig, K.; Hudlicky, T. *T* **1981**, *37*, 4359.
43. Olah, G. A.; Gupta, B. G. B. *S* **1980**, 44.
44. Cookson, R. C.; Ray, P. S. *TL* **1982**, *23*, 3521.
45. (a) Firouzabadi, H.; Iranpoor, N.; Parham, H.; Sardarian, A.; Toofan, J. *SC* **1984**, *14*, 717. (b) Firouzabadi, H.; Iranpoor, N.; Parham, H.; Toofan, J. *SC* **1984**, *14*, 631.
46. (a) Ho, T.-L.; Wong, C. M. *S* **1972**, 561. (b) Baciocchi, E.; Piermattei, A.; Ruzziconi, R. *SC* **1988**, *18*, 2167.
47. Ho, T.-L.; Ho, H. C.; Wong, C. M. *CC* **1972**, 791.
48. Tsai, Y.-M.; Nieh, H.-C.; Cherng, C.-D. *JOC* **1992**, *57*, 7010.
49. Lansbury, P. T.; Zhi, B. *TL* **1988**, *29*, 179.
50. (a) Watts, L.; Fitzpatrick, J. D.; Pettit, R. *JACS* **1965**, *87*, 3253. (b) Gleiter, R.; Karcher, M. *AG(E)* **1988**, *27*, 840.
51. Casey, C. P.; Brunsvold, W. R. *JOM* **1975**, *102*, 175.
52. Wulff, W. D.; Tang, P. C.; McCallum, J. S. *JACS* **1981**, *103*, 7677.
53. (a) Baciocchi, E.; Ruzziconi, R. *JOC* **1986**, *51*, 1645. (b) Baciocchi, E.; Civitarese, G.; Ruzziconi, R. *TL* **1987**, *28*, 5357. (c) Baciocchi, E.; Ruzziconi, R. *SC* **1988**, *28*, 1841.
54. Baciocchi, E.; Casu, A.; Ruzziconi, R. *TL* **1989**, *30*, 3707.
55. Snider, B. B.; Kwon, T. *JOC* **1990**, *55*, 4786.
56. Maknon'kov, D. I.; Cheprakov, A. V.; Rodkin, M. A.; Mil'chenko, A. Y.; Beletskaya, I. P. *Zh. Org. Chem.* **1986**, *22*, 30.

Tse-Lok Ho
National Chiao-Tung University, Hsinchu, Taiwan, Republic of China

Chlorine[1]

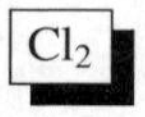

[7782-50-5] Cl_2 (MW 70.91)

(powerful oxidizing and chlorinating agent)

Physical Data: yellowish-green gas, mp −101 °C, bp −34 °C; *d* 3.21 g cm^{-3} (gas, rt), 1.56 g cm^{-3} (liq, −35 °C).
Solubility: sl sol water (0.7 g in 100 mL at 20 °C); sol acetic acid, benzene, aliphatic hydrocarbons, chlorinated solvents, DMF.
Form Supplied in: packaged in cylinders with stainless steel or monel regulators.
Preparative Methods: chlorine is commercially available, but small quantities can be generated in the laboratory. The most common procedure involves the treatment of solid $KMnO_4$ with conc HCl (0.89 g of $KMnO_4$ and 5.6 mL of conc HCl per g of chlorine required).[2] It is recommended that chlorine so generated be dried by passing in succession through gas-washing bottles containing H_2O (to remove HCl), concd H_2SO_4 (to remove H_2O), and glass wool (to remove spray).[3] Preparations using conc HCl/MnO_2[4] or by heating $CuCl_2$ (anhyd)[5] have also been reported.
Analysis of Reagent Purity: iodometric titration; the chlorine can be volatilized and the moisture and residue determined gravimetrically.[1a]
Purification: commercial chlorine should be purified with H_2SO_4, CaO, and P_2O_5 and subsequently condensed in a dry ice–acetone bath and vaporized, repeatedly, while the noncondensable gases are removed with a pump.[4]
Handling, Storage, and Precautions:[5] highly toxic, nonflammable gas. Forms explosive mixtures with hydrogen, acetylene, or anhydrous ammonia. It is corrosive when moist. Chlorine is a strong oxidant and reacts violently with combustible substances, reducing agents, organic compounds, phosphorus, and metal powders. Avoid skin contact. Use protective clothing and a full-face respirator equipped with a NIOSH-approved organic vapor-acid gas canister. Cylinders should be stored away from sources of heat. All reactions should be conducted in a well-ventilated fume hood. The amount of chlorine added to a reaction mixture can be determined by weighing the cylinder before and after addition, or by condensing the required volume into a calibrated vessel and subsequently allowing it to volatilize

while connected to the reaction vessel, or by generation from a known quantity of $KMnO_4$.

Substitution Chlorination. Chlorine atoms, obtained from the dissociation of chlorine molecules by thermal or photochemical energy, react with saturated hydrocarbons by a radical chain mechanism. Chlorine reacts with methane to form methyl chloride, methylene chloride, chloroform, and carbon tetrachloride.[1a] Trialkylboranes have also been used to induce the radical chlorination of alkanes, e.g. chloro-2,3-dimethylbutanes are produced from 2,3-dimethylbutane.[6] Ethers are also chlorinated by photoinduced radical substitution reactions (eq 1).[7a]

$$Et_2O + Cl_2 \xrightarrow{h\nu} \text{EtO–CH(Cl)–CH}_3 \qquad (1)$$

The chlorination of carboxylic acids with molecular chlorine is catalyzed by phosphorus and its trihalides.[7b] Chlorine has been used to chlorinate methyl esters of carboxylic acids to form monochloro esters.[8] Chlorinations with chlorine favor the ($\omega-1$)-position rather than the ($\omega-2$)-position obtained with ***Sulfuryl Chloride***.[8] Thus the chlorination of methyl heptanoate with chlorine results mainly in methyl 6-chloroheptanoate, whereas chlorination with sulfuryl chloride results mainly in methyl 5-chloroheptanoate.[8] Low temperatures are required for the chlorination of long-chain carboxylic acid methyl chlorides, as unsaturated compounds are formed at higher temperatures due to the elimination of HCl.

The main products of the chlorination of carbamates are the *N*-dichloro derivatives (eq 2).[7c]

$$H_2NCO_2R \xrightarrow{Cl_2\,(aq)} Cl_2NCO_2R \qquad (2)$$

Aromatic amines react readily with chlorine, e.g. aniline is chlorinated to give 2,4,6-trichloroaniline in high yield (eq 3).[7d]

$$\text{aniline} \xrightarrow[HCl,\ Cl_2,\ 160\,^\circ C]{CCl_4,\ EtOH} \text{2,4,6-trichloroaniline} \qquad (3)$$

Primary and secondary amides react with chlorine to give *N*-chloroamides and HCl.[7e] The reaction is reversible, and the products are favored by highly polar solvents.

α-Chlorination of aliphatic acids has been achieved with chlorine using enolizing agents like chlorosulfonic acid, H_2SO_4, HCl, or $FeCl_3$ with a radical trapper like *m*-dinitrobenzene, oxygen, or chloranil.[9] The imidyl hydrogen atoms of aldazines can be substituted with chlorine (eq 4).[7f]

$$\text{PhCH=N–N=CHPh} \xrightarrow{Cl_2,\ CCl_4} \text{PhC(Cl)=N–N=C(Cl)Ph} \qquad (4)$$

Addition Chlorination. Saturated chlorides are formed when chlorine reacts with alkenes, e.g. chlorination of ethylene results in ethylene dichloride and chlorination of vinyl chloride gives 1,1,2-trichloroethane.[1a] These alkyl chlorides are important synthetic intermediates, e.g. 1,1,2-trichloroethane can be converted into vinylidene chloride in an alkaline medium.[1a] The addition of chlorine to 1-trimethylsilyl-1-alkenes in CH_2Cl_2 at low temperatures, followed by the elimination of Me_3SiX with methanolic sodium methoxide at 25 °C, produces vinyl chlorides in good yields (eq 5).[10]

$$\text{Bu–CH=CH–SiMe}_3 \xrightarrow{Cl_2} \text{Bu–CH(Cl)–CH(Cl)–SiMe}_3 \xrightarrow[75\%]{MeONa} \text{Bu–CH=CH–Cl} \qquad (5)$$

The radical chain addition reactions of chlorine are initiated by light or the walls of the reaction vessel and inhibited by oxygen. Some ionic addition reactions are accelerated by ***Iron(III) Chloride***, ***Aluminum Chloride***, ***Antimony(V) Chloride***, or ***Copper(II) Chloride***.[11]

Chlorination of Aromatics. Aromatic compounds may be chlorinated in the presence of Lewis acids like iron and iron(III) chloride. Low temperatures favor monochlorination, while high temperatures (150–190 °C) favor dichlorinated products.[11] The chlorination of alkylbenzenes in alcoholic media can result in higher yields of monochloro derivatives than when $FeCl_3$ is used.[12] Chlorine diluted in water converts tyrosine into 3-chloro-4-hydroxybenzyl cyanide (eq 6);[13] larger amounts of chlorine give 3,5-dichloro-4-hydroxybenzyl cyanide and 1,3-dichloro-2,5-dihydroxybenzene. The latter product can be converted into 2,6-dichloro-*p*-benzoquinone with additional aqueous chlorine.

$$\text{tyrosine} \xrightarrow[H_2O]{Cl_2} \text{3-chloro-4-hydroxybenzyl cyanide} + \text{3,5-dichloro-4-hydroxybenzyl cyanide} + \text{1,3-dichloro-2,5-dihydroxybenzene} \qquad (6)$$

Toluene is chlorinated by the radical mechanism to give benzyl chloride, benzal chloride, and benzotrichloride.[11] Sulfuryl chloride, ***t-Butyl Hypochlorite***, ***Hydrogen Chloride*** (in the presence of a copper-salt catalyst), and ***N-Chlorosuccinimide*** have also been used to chlorinate aromatics.[11]

Chlorination of phenol yields 2-chlorophenol and 4-chlorophenol in a ratio of 0.45:0.49, which is higher than the *ortho*/*para* ratio obtained with *t*-butyl hypochlorite.[14] 4-Alkylphenols react with chlorine in various solvents to form mainly 4-alkyl-4-chlorocyclohexa-2,5-dienones (in yields of 19–100%) and substitution products.[15] Other chlorinating agents (alkyl hypochlorites, sulfuryl chloride, hypochlorous acid, and antimony pentachloride) have also been used, but they result in polychlorinated cyclohexadienones and cyclohexenones.[15] The chlorination of dimethylphenols with chlorine in acetic acid containing HCl results in polychlorinated cyclohexenones (eq 7).[16]

(7)

The α-monochlorination of alkyl aryl ketones by chlorine gas occurs readily in a variety of solvents (e.g. CH_2Cl_2, $CHCl_3$, CCl_4, HOAc).[17] α,α-Dichlorination of alkyl aryl ketones occurs with sodium acetate in refluxing acetic acid (5 h, 80–90% yield). Alternatively, DMF can be used as the catalyst (80–100 °C, 35–45 min) (eq 8).[18]

(8)

R = Me, Et, Pr, *t*-Bu, Ph; R^1 = H, Cl, Br, Me

Chlorination of phenylenedibenzenesulfonamides by chlorine in nitrobenzene results in the formation of a mixture of dichloro derivatives.[19] The more useful tetrachloro derivative can be prepared by successive oxidations and additions of HCl, or in one step using Cl_2 in DMF (eq 9).[19] The temperature must be kept below 60 °C when Cl_2/DMF is used, or a runaway thermal reaction can result.[20]

(9)

Chlorine has been used for the chlorination of heterocycles,[7i–k] but the varied reactivity of substrates makes discussion of these reactions too lengthy for this review.

Oxidation of Alcohols. Alcohols have been oxidized with complexes of chlorine with dimethyl sulfide,[21] DMSO,[22] iodobenzene,[21] pyridine,[23] and HMPA.[21] Secondary hydroxyl groups are more readily oxidized than primary hydroxyl groups when the ***Chlorine–Pyridine*** (eq 10)[23] and chlorine–HMPA complexes[21] are used. The same results are obtained with 3-iodopyridine dichloride.[23]

Hydroxythiols undergo chlorination reactions with Cl_2 in dichloromethane to form sultines and sulfinic esters after hydrolysis (eq 11).[24]

The oxidation of glycols usually results in C–C bond cleavage, but oxidation of the *s*-carbinol can be effected with a complex of a methyl sulfide (RSMe) and chlorine or NCS, or of DMSO and chlorine.[25] The tricyclic α-ketol (2) (eq 12) can be prepared from the glycol (1) using these reagents.[25]

(10)

$$HO(CH_2)_nSH \xrightarrow[2.\ H_2O\ (1\ equiv)]{1.\ Cl_2\ (2\ equiv)} \text{sultine } (CH_2)_n \text{ and (or) } [O(CH_2)_nSO]_x \quad (11)$$

(12)

Chlorinolysis. The C–C bond of short-chain hydrocarbons (<C_3 and any partially chlorinated derivatives) can be cleaved by chlorine at high temperatures to give chlorinated products. 1,2-Dichloroethane and 1,2-dichloropropane are cleaved by chlorine to give carbon tetrachloride and tetrachloroethylene with HCl as a byproduct.[11]

C–S Bond Cleavage. The benzylic group of alkyl benzyl sulfides can be selectively cleaved by chlorine in aqueous acetic acid to give alkanesulfonyl chlorides (eq 13).[26]

$$PhCH_2SMe \xrightarrow[75\%]{Cl_2,\ AcOH} MeSO_2Cl + PhCH_2Cl \quad (13)$$

Excess chlorine has been used to cleave the secondary C–S bond in lactam (3) (eq 14) to give the azetidin-2-one (4) in nearly quantitative yield.[27] When *N*-acyl groups are present, as in eq 15, the nitrogen lone pair electrons are inhibited, so cleavage of the tertiary C–S bond is favored over the azetidine C–S bond.[27]

(14)

Phth = phthalimido

(15)

Other Reactions. Chlorine and ***Sodium Bromide*** are used to produce ***Bromine Chloride***, which can be used in bromination reactions which are faster than those with elemental ***Bromine***, and which take place in aqueous solution rather than acidic solvents; the bromination of 4-nitrophenol to 2,6-dibromo-4-nitrophenol is one example.[28]

Thiocyanogen, prepared in anhydrous conditions from silver or lead thiocyanate and bromine, cannot be used for addition reactions with halogenated alkenes and is too expensive for most commercial processes.[29] Thiocyanogen can be prepared in a two-solvent system from ***Sodium Thiocyanate*** and chlorine.[29] The thiocyanogen is extracted into the toluene layer and can be used for addition reactions to vinyl halides in the synthesis of haloalkylene bisthiocyanates.

Several methods for the preparation of ***Phosgene*** are known, e.g. the gas-phase reaction of chlorine with carbon monoxide on activated carbon, the decomposition of trichloromethyl chloroformate, and the reaction of carbon tetrachloride with oleum.[30] Phosgene can be synthesized conveniently just before use by the reaction of chlorine with carbon monoxide in the presence of catalytic amounts of *t*-phosphine oxides, using carbon tetrachloride as the solvent.[30]

Chlorodimethylsulfonium chloride, generated in situ from ***Dimethyl Sulfide–Chlorine***, is a useful reagent for the conversion of epoxides to α-chloro ketones (eq 16)[31] and of aldoximes to nitriles,[32] in the presence of tertiary amines. Bromodimethylsulfonium bromide, generated in the same way, can be used to form α-bromo ketones, but the yields are lower than those obtained for α-chloro ketones.[31]

$$\text{R}_2\text{C(O)} \text{ epoxide} \xrightarrow[\text{2. Et}_3\text{N}]{\text{1. Cl(SMe)}_2^+\text{Cl}^-} \text{RC(=O)CHRCl} \quad (16)$$

Chlorine and ***Triphenylphosphine*** are used in the synthesis of lactams from cycloalkanone oximes in high yields (eq 17).[33]

$$\text{cycloalkanone}=\text{NOH} \xrightarrow[\text{2. H}_2\text{O}]{\text{1. PPh}_3\text{, Cl}_2} \text{lactam (NH)} \quad (17)$$

$n = 1, 2, 3, 4, 8$

Alkanesulfonyl chlorides, which are useful reagents and intermediates in organic synthesis, can be conveniently produced from dithiocarbonic acid esters (eq 18).[34]

$$\text{RS–C(=O)–SR} \xrightarrow[\text{5–10 °C; 93–100\%}]{\text{Cl}_2/\text{H}_2\text{O}} 2\ \text{RSO}_2\text{Cl} + \text{CO}_2 + \text{HCl} \quad (18)$$

The indirect oxidation of trialkyl phosphites to trialkyl phosphates can be achieved in high yield and purity with chlorine in the corresponding alcohol (eq 19).[35] The indirect oxidation can also be effected by carbon tetrachloride, bromotrichloromethane, ***Carbon Tetrabromide***, ***Chloroform***, and hexachlorocyclopentadiene in alcohol.[35] Cyclic phosphites may undergo ring opening during reactions with chlorine, depending on the size of the ring and its degree of substitution.[7g]

$$(\text{RO})_3\text{P} + \text{Cl}_2 + \text{ROH} \xrightarrow[\text{90–100\%}]{} (\text{RO})_3\text{PO} + \text{RCl} + \text{HCl} \quad (19)$$

Cyclooctanone oxime can be synthesized from cyclooctane by a photochemical reaction with chlorine and nitrous oxide (eq 20).[36] The oxime hydrochloride intermediate is converted into the oxime by aqueous sodium hydroxide.

$$\text{cyclooctane} \xrightarrow[\text{2. NaOH (aq)}]{\text{1. 3:1 NO:Cl}_2\text{, UV}} \text{cyclooctanone}=\text{NOH} \quad (20)$$

Chlorine can be used for the synthesis of hydrazines from ureas, via diaziridinone intermediates (eq 21).[7h]

$$\text{RNH–C(=O)–NHR}^1 \xrightarrow[\text{2. NaOH}]{\text{1. Cl}_2} \left[\text{RN–NR}^1 \text{ (C=O) diaziridinone}\right] \longrightarrow \text{RNHNHR}^1 \quad (21)$$

Related Reagents. See Class O-1, pages 1–10. Bromine; Chlorine–Chlorosulfuric Acid; Chlorine–Pyridine; Dimethyl Sulfide–Chlorine; Iodine; Sulfuryl Chloride.

---

1. (a) *Kirk-Othmer Encyclopedia of Chemical Technology*, 3rd ed.; Grayson, M., Ed.; Wiley: New York, 1978; Vol. 1, p 833. (b) Stroh, R.; Hahn, W. *MOC* **1962**, *5/3*, 503. (c) Hudlicky, M.; Hudlicky, T. In *The Chemistry of Halides, Pseudo-Halides and Azides*; Patai, S., Rappoport, Z., Eds.; Wiley: New York, 1983; Part 2, Chapter 22, pp 1066–1101.
2. Fieser, L. F. *Experiments in Organic Chemistry*, 3rd ed.; Heath: Boston, 1957; p 296.
3. Furniss, B. S.; Hannaford, A. J.; Smith, P. W. G.; Tatchell, A. R. *Vogel's Textbook of Practical Organic Chemistry*, 5th ed.; Wiley: New York, 1989; p 424.
4. Schmeisser, M. In *Handbook of Preparative Inorganic Chemistry*, 2nd ed.; Brauer, G., Ed.; Academic: New York, 1963; Vol. 1, p 272.
5. *Chemical Safety Sheets*; Zawierko, J., Ed.; Kluwer: Dordrecht, 1991; p 201.
6. Hoshi, M.; Masuda, Y.; Arase, A. *CL* **1984**, 195.
7. In *Comprehensive Organic Chemistry*; Pergamon: Oxford, 1979; (a) Vol. 1, p 840. (b) Vol. 2, p 642. (c) Vol. 2, p 1088. (d) Vol. 2, p 171. (e) Vol. 2, p 1021. (f) Vol. 2, p 460. (g) Vol. 2, p 1217. (h) Vol. 2, p 223. (i) Vol. 4, for example (cf. Vol. 6, pp 1110–1112). See also: (j) Ref. 1b, pp 1070–1076; (k) Ref. 1c, pp 1086–1087.
8. Korhonen, I. O. O.; Korvola, J. N. J. *ACS(B)* **1981**, *35*, 461.
9. Ogata, Y.; Harada, T.; Matsuyama, K.; Ikejiri, T. *JOC* **1975**, *40*, 2960.
10. Miller, R. B.; Reichenbach, T. *TL* **1974**, 543 (*FF* **1981**, *5*, 556).
11. See Ref. 1a, Vol. 5, p 668.
12. Bermejo, J.; Cabeza, C.; Blanco, C. G.; Moinelo, S. R.; Martínez, A. *J. Chem. Technol. Biotechnol.* **1986**, *36*, 129.
13. Shimizu, Y.; Hsu, R. Y. *CPB* **1975**, *23*, 2179.
14. Watson, W. D. *JOC* **1974**, *39*, 1160.
15. Fischer, A.; Henderson, G. N. *CJC* **1979**, *57*, 552.
16. Hartshorn, M. P.; Martyn, R. J.; Robinson, W. T.; Vaughan, J. *AJC* **1986**, *39*, 1609.
17. *FF* **1981**, *9*, 182.
18. De Kimpe, N.; De Buyck, L.; Verhé, R.; Wychuyse, F.; Schamp, N. *SC* **1979**, *9*, 575.
19. Adams, R.; Braun, B. H. *JACS* **1952**, *74*, 3171.
20. Woltornist, A. *Chem. Eng. News* **1983**, *61(6)*, 4.
21. Al Neirabeyeh, M.; Ziegler, J.-C.; Gross, B. *S* **1976**, 811.
22. Corey, E. J.; Kim, C. U. *JACS* **1972**, *94*, 7586.
23. Wicha, J.; Zarecki, A. *TL* **1974**, 3059.
24. King, J. F.; Rathore, R. *TL* **1989**, *30*, 2763.

25. Corey, E. J.; Kim, C. U. *TL* **1974**, 287.
26. Langler, R. F. *CJC* **1976**, *54*, 498.
27. Sheehan, J. C.; Ben-Ishai, D.; Piper, J. U. *JACS* **1973**, *95*, 3064.
28. Obenland, C. O. *J. Chem. Educ.* **1964**, *41*, 566.
29. Welcher, R. P.; Cutrufello, P. F. *JOC* **1972**, *37*, 4478.
30. Masaki, M.; Kakeya, N.; Fujimura, S. *JOC* **1979**, *44*, 3573.
31. Olah, G. A.; Vankar, Y. D.; Arvanaghi, M. *TL* **1979**, *38*, 3653.
32. (a) Ohno, M.; Sakai, I. *TL* **1965**, 4541. (b) Sakai, I.; Kawabe, N.; Ohno, M. *BCJ* **1979**, *52*, 3381.
33. Ho, T.-L.; Wong, C. M. *SC* **1975**, *5*, 423.
34. Barbero, M.; Cadamuro, S.; Degani, I.; Fochi, R.; Regondi, V. *S* **1989**, 957.
35. Frank, A. W.; Baranauckas, C. F. *JOC* **1966**, *31*, 872.
36. Müller, E.; Fries, D.; Metzger, H. *CB* **1957**, *90*, 1188.

Veronica Cornel
Emory University, Atlanta, GA, USA

m-Chloroperbenzoic Acid[1]

[937-14-4] $C_7H_5ClO_3$ (MW 172.57)

(electrophilic reagent capable of reacting with many functional groups; delivers oxygen to alkenes, sulfides, selenides, and amines)

Alternate Name: m-CPBA; MCPBA.
Physical Data: mp 92–94 °C.
Solubility: sol CH_2Cl_2, $CHCl_3$, 1,2-dichloroethane, ethyl acetate, benzene, and ether; slightly sol hexane; insol H_2O.
Form Supplied in: white powder, available with purity of 50%, 85%, and 98% (the rest is 3-chlorobenzoic acid and water).
Analysis of Reagent Purity: iodometry.[2]
Purification: commercial material (purity 85%) is washed with a phosphate buffer of pH 7.5 and dried under reduced pressure to furnish reagent with purity >99%.[3]
Handling, Storage, and Precautions: pure m-CPBA is shock sensitive and can deflagrate;[4] potentially explosive, and care is required while carrying out the reactions and during workup.[5] Store in polyethylene containers under refrigeration.

Functional Group Oxidations. The weak O–O bond of *m*-CPBA undergoes attack by electron-rich substrates such as simple alkenes, alkenes carrying a variety of functional groups (such as ethers, alcohols, esters, ketones, and amides which are inert to this reagent), some aromatic compounds,[6] sulfides, selenides, amines, and N-heterocycles; the result is that an oxygen atom is transferred to the substrate. Ketones and aldehydes undergo oxygen insertion reactions (Baeyer–Villiger oxidation).

Organic peroxy acids (**1**) readily epoxidize alkenes (eq 1).[1b] This reaction is *syn* stereospecific;[7] the groups (R^1 and R^3) which are *cis* related in the alkene (**2**) are *cis* in the epoxidation product (**3**). The reaction is believed to take place via the transition state (**4**).[8] The reaction rate is high if the group R in (**1**) is electron withdrawing, and the groups R^1, R^2, R^3, and R^4 in (**2**) are electron releasing.

(**1**) + (**2**) → (**3**) (1)

(**4**)

Epoxidations of alkenes with *m*-CPBA are usually carried out by mixing the reactants in CH_2Cl_2 or $CHCl_3$ at 0–25 °C.[9] After the reaction is complete the reaction mixture is cooled in an ice bath and the precipitated *m*-chlorobenzoic acid is removed by filtration. The organic layer is washed with sodium bisulfite solution, $NaHCO_3$ solution, and brine.[10] The organic layer is dried and concentrated under reduced pressure. Many epoxides have been purified chromatographically; however, some epoxides decompose during chromatography.[11] If distillation (*caution*: check for peroxides[12]) is employed to isolate volatile epoxides, a trace of alkali should be added to avoid acid-catalyzed rearrangement.

Alkenes having low reactivity (due to steric or electronic factors) can be epoxidized at high temperatures and by increasing the reaction time.[13] The weakly nucleophilic α,β-unsaturated ester (**5**) thus furnishes the epoxide (**6**) (eq 2).[13b] When alkenes are epoxidized at 90 °C, best results are obtained if radical inhibitor is added.[13a] For preparing acid-sensitive epoxides (benzyloxiranes, allyloxiranes) the pH of the reaction medium has to be controlled using $NaHCO_3$ (as solid or as aqueous solution),[14] Na_2HPO_4, or by using the *m*-CPBA–KF[9a] reagent.

(**5**) —[m-CPBA, CH_2Cl_2, reflux, 4.5 d; 92%]→ (**6**) (2)

Regioselective Epoxidations. In the epoxidation of simple alkenes (**2**) (eq 1), due to the electron-releasing effect of alkyl groups the reactivity rates are tetra- and trisubstituted alkenes > disubstituted alkenes > monosubstituted alkenes.[1a] High regioselectivity is observed in the epoxidation of diene hydrocarbons (e.g. **7**) having double bonds differing in degree of substitution (eq 3).[15] Epoxidation takes place selectively at the more electron-rich C-3–C-4 double bond in the dienes (**8**)[16] and (**9**).[17]

Diastereoselective Epoxidation of Cyclic Alkenes. π-Facial stereoselectivity (75% *anti*) is observed in the epoxidation of the allyl ether (**10a**) since reagent approach from the α-face is blocked by the allylic substituent; a higher diastereoselectivity (90% *anti* epoxidation) is observed when the bulkier O-*t*-Bu is located on the allylic carbon (eq 4).[18] Due to steric and other factors, the norbornene (**11**) undergoes selective (99%) epoxidation from the *exo* face.[19] In 7,7-dimethylnorbornene (**12**), approach to the *exo* face is effectively blocked by the methyl substituent at C-7, and (**12**) is epoxidized from the unfavored *endo* face, although much more slowly (1% of the rate of epoxidation of **11**).[19] The geminal methyl group at C-7 is able to block the approach of the peroxy acid even when the double bond is exocyclic to the norbornane ring system (for example, epoxidation of (**13**) proceeds with 86% *exo* attack, while (**14**) is oxidized with 84% *endo* attack). Folded molecules are epoxidized selectively from the less hindered convex side; *m*-CPBA epoxidation of the triene lactone (**15**) takes place from the α-face with 97% stereoselectivity.[20] The triepoxide (**16**) has been obtained in 74% yield by epoxidizing the corresponding triene;[21] in the epoxidation step, six new chiral centers are introduced stereoselectively as a result of steric effects.

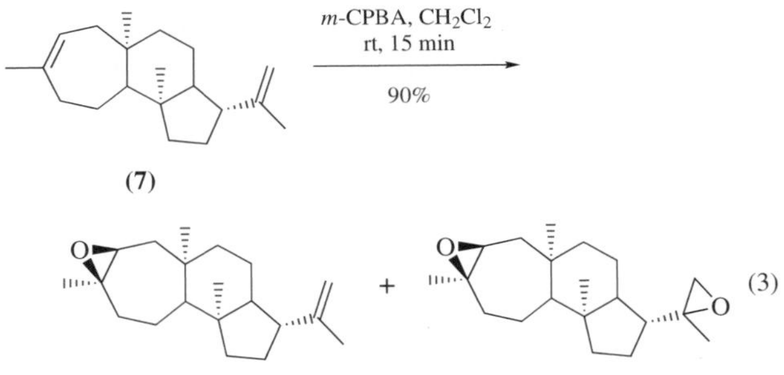

(7) —*m*-CPBA, CH_2Cl_2, rt, 15 min, 90%→ 86% + 4% (3)

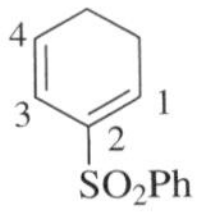

(8) 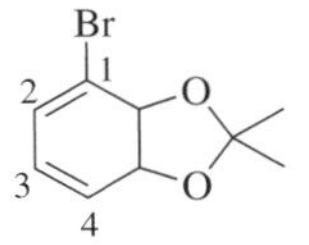 (9)

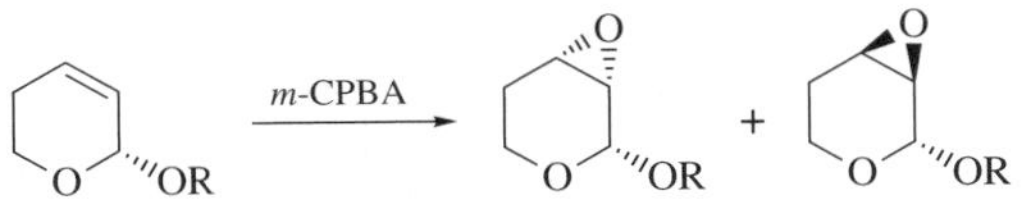

(10) —*m*-CPBA→ + (4)

(**a**) R = Me 1:3
(**b**) R = *t*-Bu 1:9

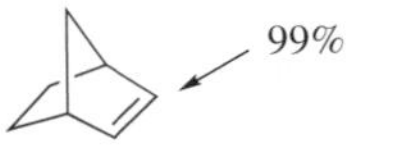

(11) 99% (12) 88%

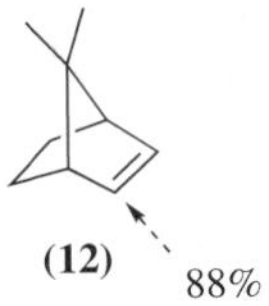

(13) 86% (14) 84%

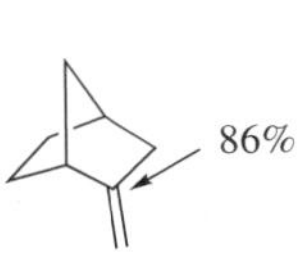

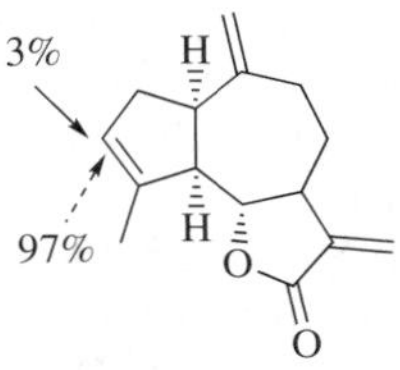

(15) 3%, 97% (16)

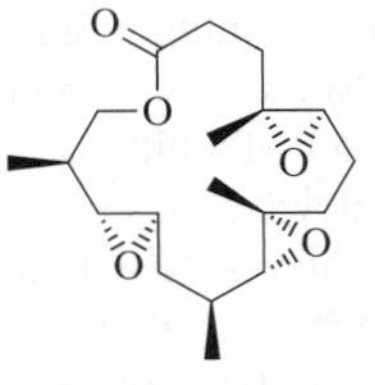

Unhindered methylenecyclohexanes and related compounds show a moderate preference for axial epoxidation. In the epoxidation of (**17**) the ratio of axial:equatorial attack is 86:14;[22] for the alkene (**18**) the ratio is 75:25.[23]

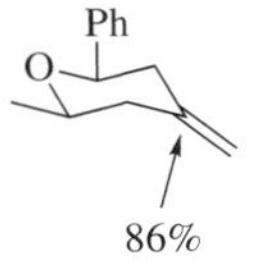

(17) 86% (18) 75%

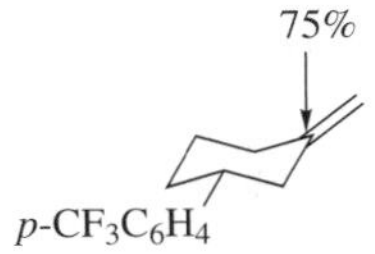

Epoxidation of Cyclic Alkenes having Directing Groups. Henbest showed that in the absence of severe steric interference, allylic cyclohexenols are epoxidized stereoselectively by organic peroxy acids to furnish *cis*-epoxy alcohols;[24a] a large number of *cis*-epoxy alcohols have been prepared by epoxidizing allylic cyclohexenols.[7] A mixture (5:1) of labile bisallylic alcohols (**19**) and (**20**) was reacted with *m*-CPBA (eq 5); from the reaction mixture diepoxide (**21**) was isolated as a single isomer.[25] Epoxidation of (*Z*)-cyclooct-2-en-1-ol (**22**) furnishes exclusively (99.8%) the *trans*-epoxide (**23**) (eq 6).[24b] Similar observations have been made subsequently.[26] This result, as well as the stereoselectivity observed during the epoxidation of other allylic alcohols, both cyclic and acylic, has been rationalized on the basis of transition state models.[24,27]

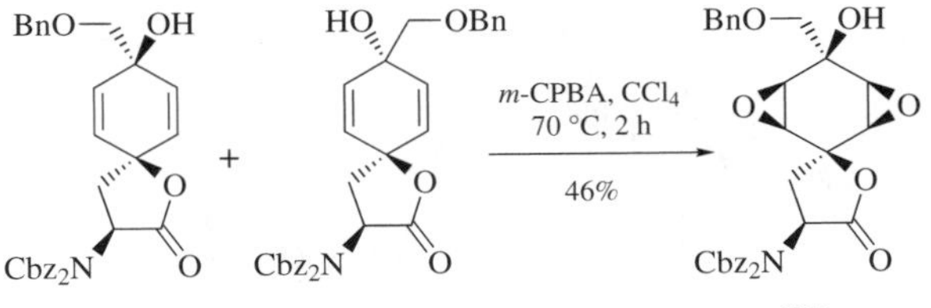

(19) + (20) —*m*-CPBA, CCl_4, 70 °C, 2 h, 46%→ (21) (5)

(22) —*m*-CPBA, CH_2Cl_2, 0 °C, 24 h, 81%→ (23) (6)

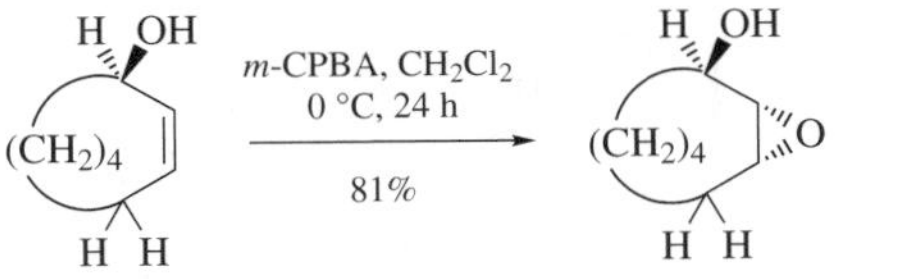

Stereoselectivity has been observed during the peroxy acid epoxidation of some homoallylic and bishomoallylic alcohols,[28] and the epoxidation of the allylic carbamate (**24**) is *syn* stereoselective (eq 7).[28]

Epoxidations of Acyclic Alkenes. Since acyclic systems normally are not rigid, high stereoselectivity has been observed only when special structural features are present. The presence of functional groups (OH, NH, CO, and ether) which form hydrogen bonds with the peroxy acid can facilitate stereoselective epoxidations by imparting rigidity to the system. High *anti* selectivity (>95%) has been observed in the epoxidation of both (**25**) and (**26**) each of which has a branched substituent adjacent to the carbon carrying the silicon group.[29] High *anti* selectivities have been noted during the epoxidation of (**27**) (95%),[30] (**28**) (96%),[31] (**29**) (95%),[32] and (**30**) (96%).[33] High *syn* selectivity has been observed in the reactions of (**31**) (98%)[33] and (**32**) (93%).[34] When the allyl alcohol (**28**) reacts with *m*-CPBA, in the transition state the reagent is hydrogen-bonded to the ether oxygen as well as to allylic hydroxyl. The high selectivity is due to the cooperative effect of the hydroxyl group and the ether oxygen.[31]

(**24**) — *m*-CPBA, CH_2Cl_2, 0 °C — *cis*:*trans* = 10:1 (7)

(**25**) (**26**) (**27**) (**28**) (**29**) (**30**) (**31**) (**32**) $R^1 = (CH_2)_6CO_2Me$, $R^2 = (CH_2)_5Me$

High stereoselectivity has also been observed in the epoxidation of some acyclic homoallylic alcohols.[35]

Oxidation of Enol Silyl Ethers and Furans. Epoxides of enol silyl ethers undergo facile ring opening and only in rare cases have stable epoxides been isolated.[36] α-Hydroxy enones have been prepared in two steps from α,β-unsaturated ketones; the enol silyl ether (**33**) prepared from the corresponding enone is treated with *m*-CPBA and the resulting product reacts with triethylammonium fluoride to furnish an α-hydroxy enone (**34**) (eq 8).[37] This method has also been used for the preparation of α-hydroxy ketones,[38] α-hydroxy acids,[39] and α-hydroxy esters. As illustrated in (eq 9), aldehydes have been converted to protected α-hydroxy aldehydes in a similar fashion.[40] Epoxidation of enol silyl ethers according to eq 10 has been used in synthesizing α,α′-dihydroxy ketones from methyl secondary alkyl ketones; the silyl ether (**36**) furnishes the corresponding dihydroxy ketone quantitatively upon brief acidic treatment.[41] Peroxy acid oxidation of furfuryl alcohols yields pyranones according to eq 11.[42,43] Furfurylamides also react similarly.[44]

(**33**) — 1. *m*-CPBA, hexane, −15 °C; 2. Et_3NHF, CH_2Cl_2; 70–73% — (**34**) (8)

C_8H_{17}–CH=CH–OTMS — *m*-CPBA, 84% — C_8H_{17}CH(OH)CH(OTMS)OC(O)Ar; Ar = 3-ClC_6H_4 (9)

(**35**) — 2.5 equiv *m*-CPBA, CH_2Cl_2, 10 $KHCO_3$, 0 °C, 1.5 h; 72% — (**36**) (10)

m-CPBA, CH_2Cl_2, rt, 3 h; 89% (11)

Baeyer–Villiger Rearrangement. Reaction of a ketone (**37**) with peroxy acid results in oxygen insertion to furnish the esters (**38**) and (**39**). This reaction, known as the Baeyer–Villiger rearrangement, has been reviewed recently.[45] Cyclobutanones undergo very facile rearrangement with peroxy acids, as well as with ***Hydrogen Peroxide*** in presence of base. The cyclobutanone (**40**) reacted readily with *m*-CPBA to furnish regio-, stereo-, and chemoselectively the lactone (**41**) (eq 12),[38b] which was elaborated to gingkolide. Baeyer–Villiger reaction of (**40**) with H_2O_2/base furnished a γ-lactone which was the regioisomer of (**41**). When 1,2,3,8,9,9a-hexahydro-1-methyl-3a,8-methano-3a*H*-cyclopentacycloocten-10-one, which has double bonds as well as a keto group, was treated with *m*-CPBA, exclusive alkene epoxidation was observed.[46] Ketones having stannyl groups on the β-carbon undergo a tin-directed Baeyer–Villiger reaction.[47]

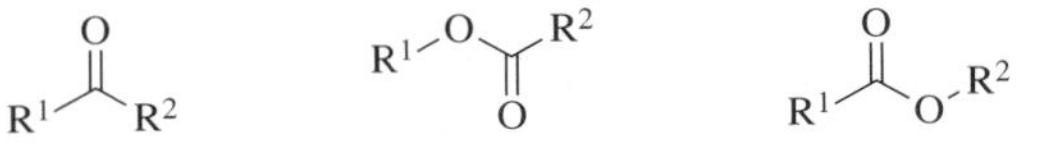

(37) (38) (39)

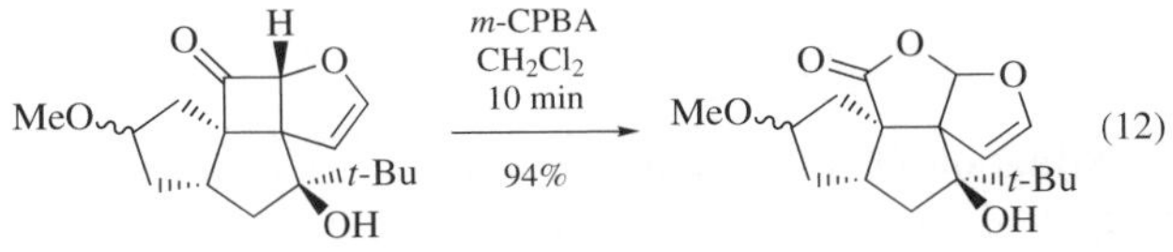

(40) → (41) (12) *m*-CPBA, CH_2Cl_2, 10 min, 94%

Oxidation of Nitrogen-Containing Compounds. Primary amines are oxidized by *m*-CPBA to the corresponding nitro compounds. One of the intermediates formed in this reaction is the corresponding nitroso compound, which reacts sluggishly with the reagent. High yields are obtained by carrying out the reaction at a high temperature (≈83 °C) and increasing the reaction time (3 hours). For example, *n*-hexylamine is oxidized to 1-nitrohexane in 66% yield.[48] When a substrate having the amino group at a chiral center was oxidized, the nitro compound was formed with substantial (≈95%) retention of configuration.[49]

m-CPBA oxidation of the sulfilimine (**42**) prepared from 2-aminopyridine, furnished 2-nitrosopyridine (**43**) (eq 13).[50]

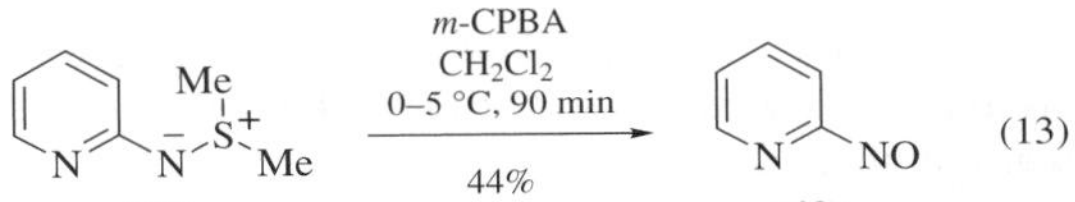

(42) → (43) (13) *m*-CPBA, CH_2Cl_2, 0–5 °C, 90 min, 44%

Secondary amines have been oxidized to hydroxylamines with *m*-CPBA.[26b] In this reaction, substantial amounts of nitrone as byproduct are expected. (The best method for the preparation of hydroxylamines is to oxidize the secondary amine with 2-(phenylsulfonyl)-3-aryloxaziridine (see e.g. ***(±)-trans-2-(Phenylsulfonyl)-3-phenyloxaziridine***) to the nitrone, and then to reduce the nitrone with ***Sodium Cyanoborohydride***).[51]

m-CPBA oxidation of N-heterocycles furnishes in high yields the corresponding *N*-oxides.[52] Several tertiary *N*-oxides have been prepared by the reaction of tertiary amines with *m*-CPBA in $CHCl_3$ at 0–25 °C and employing chromatography on alkaline alumina; for example, trimethylamine *N*-oxide was obtained in 96% yield.[53] When the optically pure tertiary amine (**44**) is oxidized with *m*-CPBA, the initially formed amine oxide rearranges to the hydroxylamine (**45**) with complete 1,3-transfer of chirality (eq 14).[54]

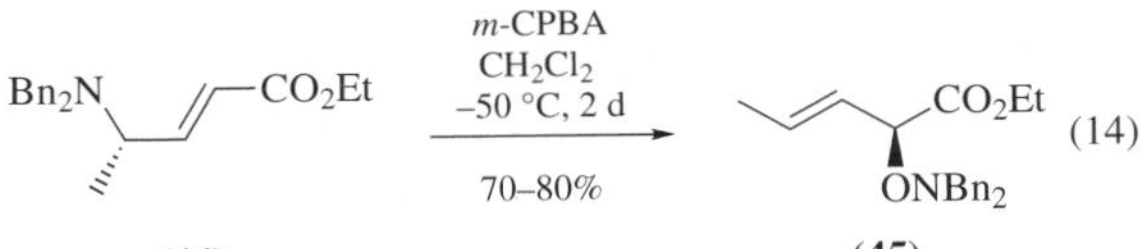

(44) → (45) (14) *m*-CPBA, CH_2Cl_2, –50 °C, 2 d, 70–80%

Reaction of *m*-CPBA with the isoxazole (**46**) furnishes the nitrone (**47**) (eq 15).[55] *m*-CPBA oxidation of (−)-isoxazole (**48**) and subsequent workup results in the formation of the (−)-cyclopentanone (**49**) (eq 16);[56] the initially formed nitrone is hydrolyzed during workup. The oxaziridine (**51**) has been prepared by epoxidizing the sulfonimine (**50**) (eq 17).[57]

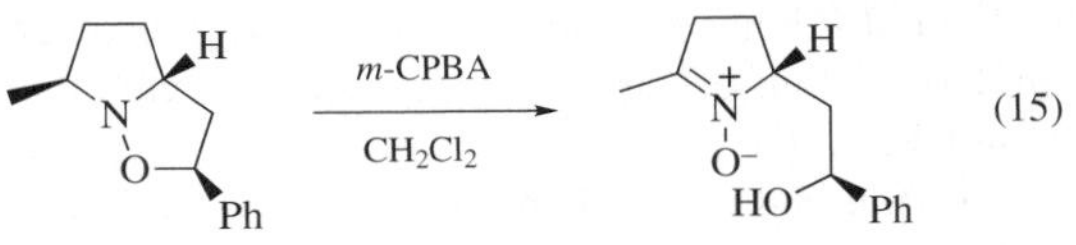

(46) → (47) (15) *m*-CPBA, CH_2Cl_2

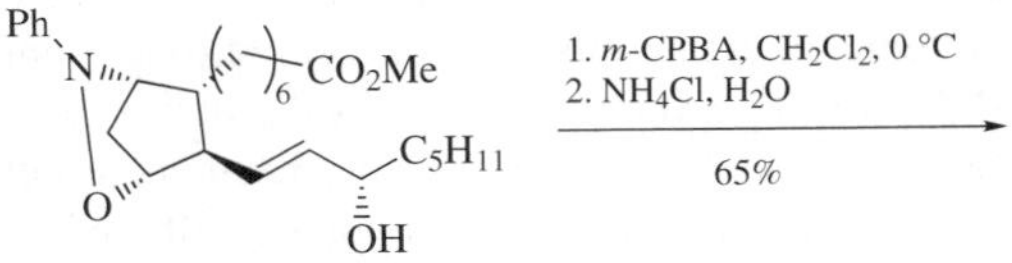

(−)-(48) 86% ee 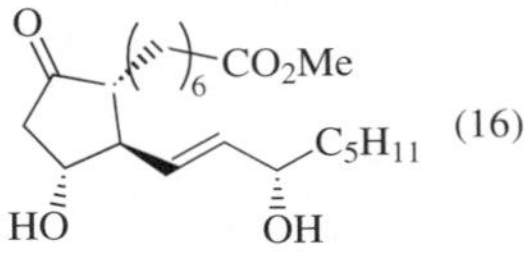 → (−)-(49) 86% ee (16) 1. *m*-CPBA, CH_2Cl_2, 0 °C 2. NH_4Cl, H_2O, 65%

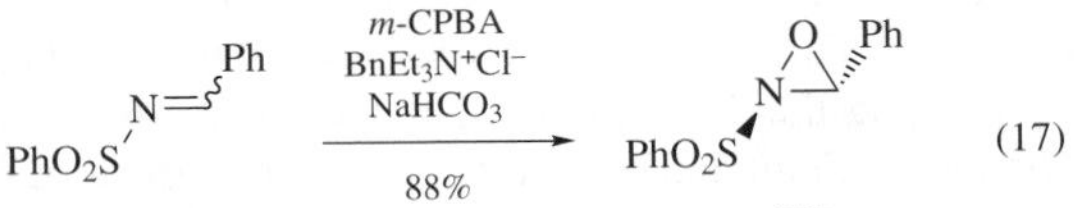

(50) → (51) (17) *m*-CPBA, $BnEt_3N^+Cl^-$, $NaHCO_3$, 88%

The cleavage of the *N,N*-dimethylhydrazone (**52**) proceeds rapidly in the presence of *m*-CPBA, even at low temperatures, to furnish the ketone (**53**), without isomerization to the more stable *cis* isomer (eq 18).[58]

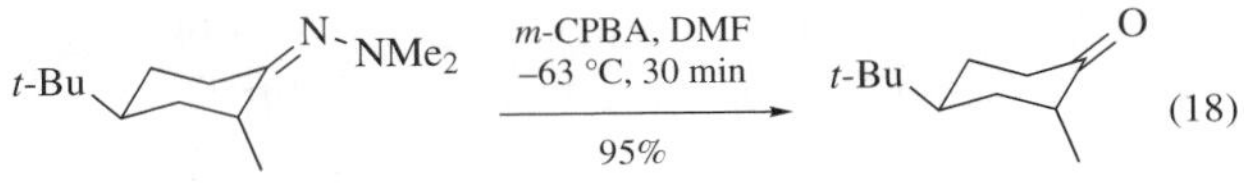

(52) → (53) (18) *m*-CPBA, DMF, –63 °C, 30 min, 95%

Oxidation of Phosphorus-Containing Compounds. *m*-CPBA oxidation of the phosphite (**54**) is stereospecific; it furnishes the phosphate (**55**) (eq 19).[59a] However, aqueous ***Iodine*** is the reagent of choice for the oxidation of nucleotidic phosphite triesters.[59b] *m*-CPBA oxidation of thiophosphate triesters furnishes the corresponding phosphate esters with retention of configuration.[60]

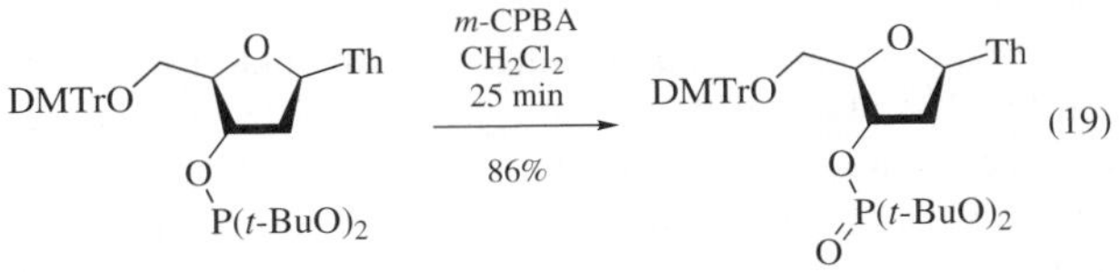

(54) → (55) (19) *m*-CPBA, CH_2Cl_2, 25 min, 86%

Oxidation of Sulfur-Containing Compounds. *n*-Butanethiol is oxidized by *m*-CPBA in CH_2Cl_2 at −30 °C to furnish in

82% yield *n*-butanesulfinic acid (*n*-$BuSO_2H$); other thiols react similarly.[61] Sulfides are oxidized chemoselectively to sulfoxides by *m*-CPBA; the reaction is fast even at −70 °C, and the product is free from sulfone.[62] Three reagents (*m*-CPBA, ***Sodium Periodate***, and ***Iodosylbenzene***) are regarded as ideal for the oxidation of sulfides to sulfoxides.[63] Good diastereoselectivity has been observed in the oxidation of the sulfide (**56**) (eq 20).[64] Sulfides carrying suitably located hydroxyl groups are oxidized diastereoselectively, due to the directing influence of the hydroxyl group.[65] A phenyl sulfide carrying a variety of functional groups (epoxide, hydroxyl, ether, carbamate, and enediyne) has been chemoselectively oxidized in 99% yield to the corresponding sulfone.[52]

(**56**) R = Ph —[*m*-CPBA, −20 °C, 36 h; 96%]→ 68% + 28% (20)

Allenyl chloromethyl sulfoxide (**57**) reacts with *m*-CPBA to furnish allenyl chloromethyl sulfone (**58**) (eq 21).[66] The enethiolizable thioketone (**59**) has been oxidized to the (*E*)-sulfine (**60**) (eq 22).[67] The 2′-deoxy-4-pyrimidinone (**62**) has been prepared by reacting the 2-thiopyrimidine nucleoside (**61**) with *m*-CPBA (eq 23).[68] Thioamides have been transformed to the amides in high yields.[69]

(**57**) —[*m*-CPBA]→ (**58**) (21)

(**59**) —[*m*-CPBA, CH_2Cl_2, 0 °C, 1–5 min; 100%]→ (**60**) (22)

(**61**) R = DMTrO —[*m*-CPBA, 10 min, CH_2Cl_2, py (9:1); 98%]→ (**62**) (23)

Oxidation of Selenides. Phenyl selenides react rapidly with *m*-CPBA at −10 °C to form phenyl selenoxides,[70a] which on warming to 0 °C or at rt undergo facile *cis* elimination. This procedure for introducing unsaturation under mild conditions has been used in the synthesis of thermally sensitive compounds; for an example see eq 24.[70b] The selenonyl moiety is a good leaving group, and its generation in the substrate can lead to the formation of cyclic compounds. The oxazoline (**64**) has been synthesized through oxidation of the selenide (**63**) and treatment of the oxidized material with base (eq 25).[71]

—[*m*-CPBA, CH_2Cl_2, 0 °C, 40 min; 93%]→ (24)

(**63**) —[1. *m*-CPBA; 2. KOH; 88%]→ (**64**) (25)

Oxidation of Allylic Iodides. *m*-CPBA oxidation of the primary allylic iodide (**65**) furnishes the secondary allylic alcohol (**66**) (eq 26);[72] this involves rearrangement of the iodoxy compound formed initially.

(**65**) —[*m*-CPBA, EtOAc, CH_2Cl_2, Na_2CO_3 (aq), 5–20 °C, 1 h; 65–67%]→ (**66**) (26)

Comparison with Other Reagents. To effect epoxidation, the most commonly used reagents are *m*-CPBA, ***Peracetic Acid*** (PAA), and ***Trifluoroperacetic Acid*** (TFPAA). TFPAA is not commercially available. *m*-CPBA is more reactive than PAA and is the reagent of choice for laboratory-scale reactions. For large-scale epoxidations the cheaper PAA is preferred. The highly reactive TFPAA is used for unreactive and heat-sensitive substrates; its reactivity permits the use of low reaction temperatures. The recently introduced reagent magnesium monoperphthalate (MMPP) (see ***Monoperoxyphthalic Acid***) is more stable than *m*-CPBA and has many applications.[4]

Epoxidations of hydroxyalkenes have been carried out with ***t-Butyl Hydroperoxide***/vanadium (TBHP/V). *m*-CPBA epoxidation of (*Z*)-cyclooct-2-en-1-ol is *anti* selective; with TBHP/V it is *cis* selective.[24b] Similar differences have been noticed in some acyclic systems.[27c] Since the directing effect of the hydroxyl group is larger in the TBHP/V system it is a better reagent for hydroxyl-directed regioselective epoxidations of polyunsaturated alcohols;[73] the TBHP/V system also exhibits higher hydroxyl-directed selectivity in highly hindered allylic alcohols.[74]

m-CPBA epoxidation of hindered alkenes takes place selectively from the less hindered side; the epoxide of opposite stereochemistry can be prepared by a two-step procedure involving initial preparation of bromohydrin, followed by base treatment.[28]

For the epoxidation of extremely unreactive alkenes[38b] and for the preparation of epoxides which are highly susceptible to nucleophilic attack, ***Dimethyldioxirane*** is the reagent of choice.[75] Electron-deficient alkenes such as α,β-unsaturated ketones are usually oxidized with ***Hydrogen Peroxide***/base.

Related Reagents. See Classes O-8, O-11, O-14, O-15 and O-20, pages 1–10. *m*-Chloroperbenzoic Acid–2,2,6,6-Tetramethylpiperidine Hydrochloride.

---

1. (a) Swern, D. *Organic Peroxides*; Wiley: New York, 1971; Vol 2, pp 355–533. (b) Plesnicar, B. *Organic Chemistry*; Academic: New York, 1978; Vol. 5 C, pp 211–294. (c) Rao, A. S. *COS* **1991**, *7*, Chapter 3.1.
2. McDonald, R. N.; Steppel, R. N.; Dorsey, J. E. *OS* **1970**, *50*, 15.
3. (a) Schwartz, N. N.; Blumbergs, J. H. *JOC* **1964**, *29*, 1976. (b) Nakayama, J.; Kamiyama, H. *TL* **1992**, *33*, 7539.

4. Brougham, P.; Cooper, M. S.; Cummerson, D. A.; Heany, H.; Thompson, N. *S* **1987**, 1015.
5. *Hazards in the Chemical Laboratory*; Luxon, S. G., Ed.; Royal Society of Chemistry: Cambridge, 1992.
6. (a) Ishikawa, K.; Charles, H. C.; Griffin, G. W. *TL* **1977**, 427. (b) Srebnik, M.; Mechoulam, R.; *S* **1983**, 1046.
7. Berti, G. *Top. Stereochem.*, **1973**, *7*, 93.
8. (a) Woods, K. W.; Beak, P. *JACS* **1991**, *113*, 6281. (b) Bartlett, P. D. *Rec. Chem. Prog.* **1950**, *11*, 47 (cited in Ref. 1a).
9. (a) Camps, F.; Coll, J.; Messeguer, A.; Pujol, F. *JOC* **1982**, *47*, 5402. (b) Chai, K.-B.; Sampson, P. *TL* **1992**, *33*, 585.
10. Paquette, L. A.; Barrett, J. H. *OS* **1969**, *49*, 62.
11. (a) Wender, P. A.; Zercher, C. K. *JACS* **1991**, *113*, 2311. (b) Philippo, C. M. G.; Vo, N. H.; Paquette, L. A. *JACS* **1991**, *113*, 2762.
12. Bach, R. D.; Knight, J. W. *OS* **1981**, *60*, 63.
13. (a) Kishi, Y.; Aratani, M.; Tanino, H.; Fukuyama, T.; Goto, T. *CC* **1972**, 64. (b) Valente, V. R.; Wolfhagen, J. L. *JOC* **1966**, *31*, 2509.
14. Anderson, W. K.; Veysoglu, T. *JOC* **1973**, *38*, 2267. (b) Imuta, M.; Ziffer, H. *JOC* **1979**, *44*, 1351.
15. Urones, J. G.; Marcos, I. S.; Basabe, P.; Alonso, C.; Oliva, I. M.; Garrido, N. M.; Martin, D. D.; Lithgow, A. M. *T* **1993**, *49*, 4051.
16. Bäckvall, J.-E.; Juntunen, S. K. *JOC* **1988**, *53*, 2398.
17. Hudlicky, T.; Price, J. D.; Rulin, F.; Tsunoda, T. *JACS* **1990**, *112*, 9439.
18. Marsh, E. A. *SL* **1991**, 529.
19. Brown, H. C.; Kawakami, J. H.; Ikegami, S. *JACS* **1970**, *92*, 6914.
20. Devreese, A. A.; Demuynck, M.; De Clercq, P. J.; Vandewalle, M. *T* **1983**, *39*, 3049.
21. Still, W. C.; Romero, A. G. *JACS* **1986**, *108*, 2105.
22. Schneider, A.; Séquin, U. *T* **1985**, *41*, 949.
23. Johnson, C. R.; Tait, B. D.; Cieplak, A. S. *JACS* **1987**, *109*, 5875.
24. (a) Henbest, H. B.; Wilson, R. A. L. *JCS* **1957**, 1958. (b) Itoh, T.; Jitsukawa, K.; Kaneda, K.; Teranishi, S. *JACS* **1979**, *101*, 159.
25. Wipf, P.; Kim, Y. *JOC* **1993**, *58*, 1649.
26. (a) Kim, G.; Chu-Moyer, M. Y.; Danishefsky, S. J.; Schulte, G. K. *JACS* **1993**, *115*, 30. (b) Fukuyama, T.; Xu, L.; Goto, S. *JACS* **1992**, *114*, 383.
27. (a) Sharpless, K. B.; Verhoeven, T. R. *Aldrichim. Acta* **1979**, *12*, 63. (b) Rossiter, B. E.; Verhoeven, T. R.; Sharpless, K. B. *TL* **1979**, 4733. (c) Adam, W.; Nestler, B. *JACS* **1993**, *115*, 5041.
28. Kočovský, P.; Starý, I. *JOC* **1990**, *55*, 3236.
29. Murphy, P. J.; Russel, A. T.; Procter, G. *TL* **1990**, *31*, 1055.
30. Fleming, I.; Sarkar, A. K.; Thomas, A. P. *CC* **1987**, 157.
31. Johnson, M. R.; Kishi, Y. *TL* **1979**, 4347.
32. Roush, W. R.; Straub, J. A.; Brown, R. J. *JOC* **1987**, *52*, 5127.
33. Sakai, N.; Ohfune, Y. *JACS* **1992**, *114*, 998.
34. Lewis, M. D.; Menes, R. *TL* **1987**, *28*, 5129.
35. Fukuyama, T.; Wang, C.-L. J.; Kishi, Y. *JACS* **1979**, *101*, 260.
36. Paquette, L. A.; Lin, H.-S.; Gallucci, J. C. *TL* **1987**, *28*, 1363.
37. (a) Rubottom, G. M.; Gruber, J. M. *JOC* **1978**, *43*, 1599. (b) Rubottom, G. M.; Gruber, J. M.; Juve, H. D.; Charleson, D. A. *OS* **1986**, *64*, 118.
38. (a) Herlem, D.; Kervagoret, J.; Yu, D.; Khuong-Huu, F.; Kende, A. S. *T* **1993**, *49*, 607. (b) Crimmins, M. T.; Jung, D. K.; Gray, J. L. *JACS* **1993**, *115*, 3146.
39. Rubottom, G. M.; Marrero, R. *JOC* **1975**, *40*, 3783.
40. Hassner, A.; Reuss, R. H.; Pinnick, H. W. *JOC* **1975**, *40*, 3427.
41. Horiguchi, Y.; Nakamura, E.; Kuwajima, I. *TL* **1989**, *30*, 3323.
42. Shimshock, S. J.; Waltermire, R. E.; DeShong, P. *JACS* **1991**, *113*, 8791.
43. Honda, T.; Kobayashi, Y.; Tsubuki, M. *T* **1993**, *49*, 1211.
44. Zhou, W.-S.; Lu, Z.-H.; Wang, Z.-M. *T* **1993**, *49*, 2641.
45. Krow, G. R. *COS* **1991**, *7*, Chapter 5, 1.
46. Feldman, K. S.; Wu, M.-J.; Rotella, D. P. *JACS* **1990**, *112*, 8490.
47. Bakale, R. P.; Scialdone, M. A.; Johnson, C. R. *JACS* **1990**, *112*, 6729.
48. Gilbert, K. E.; Borden, W. T. *JOC* **1979**, *44*, 659.
49. Robinson, C. H.; Milewich, L.; Hofer, P. *JOC* **1966**, *31*, 524.
50. Taylor, E. C.; Tseng, C.-P.; Rampal, J. B. *JOC* **1982**, *47*, 552.
51. Jasys, V. J.; Kelbaugh, P. R.; Nason, D. M.; Phillips, D.; Rosnack, K. J.; Saccomano, N. A.; Stroh, J. G.; Volkmann, R. A. *JACS* **1990**, *112*, 6696.
52. Nicolaou, K. C.; Maligres, P.; Suzuki, T.; Wendeborn, S. V.; Dai, W.-M.; Chadha, R. K. *JACS* **1992**, *114*, 8890.
53. Craig, J. C.; Purushothaman, K. K. *JOC* **1970**, *35*, 1721.
54. Reetz, M. T.; Lauterbach, E. H. *TL* **1991**, *32*, 4481.
55. Ali, Sk. A.; Wazeer, M. I. M. *T* **1993**, *49*, 4339.
56. Hwu, J. R.; Robl, J. A.; Gilbert, B. A. *JACS* **1992**, *114*, 3125.
57. Vishwakarma, L. C.; Stringer, O. D.; Davis, F. A. *OS* **1988**, *66*, 203.
58. Duraisamy, M.; Walborsky, H. M. *JOC* **1984**, *49*, 3410.
59. (a) Sekine, M.; Iimura, S.; Nakanishi, T. *TL* **1991**, *32*, 395. (b) Beaucage, S. L.; Iyer, R. P. *T* **1992**, *48*, 2223.
60. Cullis, P. M. *CC* **1984**, 1510.
61. Filby, W. G.; Günther, K.; Penzhorn, R. D. *JOC* **1973**, *38*, 4070.
62. Trost, B. M.; Salzmann, T. N.; Hiroi, K.; *JACS* **1976**, *98*, 4887.
63. Madesclaire, M. *T* **1986**, *42*, 5459.
64. Evans, D. A.; Faul, M. M.; Colombo, L.; Bisaha, J. J.; Clardy, J.; Cherry, D. *JACS* **1992**, *114*, 5977.
65. (a) Wang, X.; Ni, Z.; Lu, X.; Smith, T. Y.; Rodriguez, A.; Padwa, A. *TL* **1992**, *33*, 5917. (b) DeLucchi, O.; Fabri, D. *SL* **1990**, 287.
66. Block, E.; Putman, D. *JACS* **1990**, *112*, 4072.
67. Le Nocher, A. M.; Metzner, P. *TL* **1991**, *32*, 747.
68. Kuimelis, R. G.; Nambiar, K. P. *TL* **1993**, *34*, 3813.
69. Kochhar, K. S.; Cottrell, D. A.; Pinnick, H. W. *TL* **1983**, *24*, 1323.
70. (a) Reich, H. J.; Shah, S. K. *JACS* **1975**, *97*, 3250. (b) Danheiser, R. L.; Choi, Y. M.; Menichincheri, M.; Stoner, E. J. *JOC* **1993**, *58*, 322.
71. Toshimitsu, A.; Hirosawa, C.; Tanimoto, S.; Uemura, S. *TL* **1992**, *33*, 4017.
72. Yamamoto, S.; Itani, H.; Tsuji, T.; Nagata, W. *JACS* **1983**, *105*, 2908.
73. Boeckman, Jr. R. K.; Thomas, E. W. *JACS* **1979**, *101*, 987.
74. Sanghvi, Y. S.; Rao, A. S. *JHC* **1984**, *21*, 317.
75. (a) Murray, R. W. *CR* **1989**, *89*, 1187. (b) Halcomb, R. L.; Danishefsky, S. J. *JACS* **1989**, *111*, 6661. (c) Adam, W.; Hadjiarapoglou, L.; Wang, X. *TL* **1989**, *30*, 6497. (d) Adam, W.; Curci, R.; Edwards, J. O. *ACR* **1989**, *22*, 205.

A. Somasekar Rao & H. Rama Mohan
Indian Institute of Chemical Technology, Hyderabad, India

N-Chlorosuccinimide–Dimethyl Sulfide[1]

($R^1 = R^2 = Me$)
[39095-38-0] $C_6H_{10}ClNO_2S$ (MW 195.66)
($R^1 = Me$, $R^2 = Et$)
[54959-52-3] $C_7H_{12}ClNO_2S$ (MW 209.69)
($R^1 = Me$, $R^2 = n\text{-}Pr$)
[54959-53-4] $C_8H_{14}ClNO_2S$ (MW 223.72)
($R^1 = Et$, $R^2 = CH_2CH{=}CH_2$)
[59321-40-3] $C_9H_{14}ClNO_2S$ (MW 235.73)
($R^1 = Me$, $R^2 = CH_2CHMe_2$)
[54959-54-5] $C_9H_{16}ClNO_2S$ (MW 237.74)
($R^1 = R^2 = n\text{-}Pr$)
[59741-19-4] $C_{10}H_{18}ClNO_2S$ (MW 251.77)
($R^1 = Me$, $R^2 = Ph$)
[82661-92-5] $C_{11}H_{12}ClNO_2S$ (MW 257.73)
($R^1 = Et$, $R^2 = CH_2CH{=}CMe_2$)
[59321-43-6] $C_{11}H_{18}ClNO_2S$ (MW 263.78)
($R^1 = Me$, $R^2 = CH_2Ph$)
[65824-49-9] $C_{12}H_{14}ClNO_2S$ (MW 271.76)
($R^1 = Et$, $R^2 = CH_2Ph$)
[65824-51-3] $C_{13}H_{16}ClNO_2S$ (MW 285.79)
($R^1 = Et$, $R^2 = CH_2C_6H_4\text{-}4\text{-}Me$)
[65824-52-4] $C_{14}H_{18}ClNO_2S$ (MW 299.82)
($R^1 = Et$, $R^2 = CH_2(1\text{-}C_{10}H_7)$)
[65824-54-6] $C_{17}H_{18}ClNO_2S$ (MW 335.85)
($R^1 = R^2 = n\text{-}C_7H_{15}$)
[59741-21-8] $C_{20}H_{38}ClNO_2S$ (MW 392.04)

(oxidizing agent for alcohols to aldehydes and ketones,[1] catechols and hydroquinones to quinones,[2] aromatic amines to sulfilimines,[3] hydroxamic acids to acylnitroso compounds;[4] chlorination of allylic and benzylic alcohols,[5] *ortho* alkylation and formylation of phenols,[6] preparation of chloromethyl thioethers,[7] thioalkylation of pyrroles, indoles, and enamines;[8] preparation of sulfur ylides of active methylene compounds,[9] preparation of sulfonium salts from enamines and cyclopentadienyl anions,[10] dehydration to form keto enamines from β-diketones and nitriles from aldoximes[11])

Alternate Names: Corey–Kim reagent; dimethyl(succinimido)sulfonuim chloride.
Physical Data: $R^1 = R^2 = Me$: mp 70–72 °C.
Solubility: the reagents are slightly more sol in CH_2Cl_2, toluene, and THF than the NCS from which they are derived.
Form Supplied in: prepared in situ from NCS and dialkyl sulfides.
Handling, Storage, and Precautions: see ***N-Chlorosuccinimide***, ***Dimethyl Sulfide***, and other articles on alkyl sulfides.

***N*-Chlorosuccinimide–Dialkyl Sulfides.** These reagents are generally prepared in situ in solvents such as toluene, CH_2Cl_2, or THF. The reagents themselves are generally more soluble than the NCS from which they are derived and are often prepared at a temperature of about 0 °C to facilitate reaction with the limitedly soluble NCS, but because of their thermal instability are not prepared at higher temperature. They are often used at temperatures as low as −78 °C.

Oxidations. *N*-Chlorosuccinimide–dimethyl sulfide (NCS–DMS) is one of several reagents for converting alcohols to alkoxydimethylsulfonium salts, which in the presence of base convert to the corresponding carbonyl compounds via intramolecular proton transfer and loss of dimethyl sulfoxide.[1,12] These oxidations are among the mildest and most selective for conversions of alcohols to aldehydes and ketones. This reaction does not suffer the overoxidation to acids or the carbon–carbon bond cleavages which are often encountered in chromium(VI) or manganese(VII) oxidations (eq 1).[1,13]

t-Bu–C6H10–OH → (1. NCS–DMS, toluene; 2. Et_3N, toluene, −20 °C; 90–93%) → *t*-Bu–C6H10=O (1)

Several higher members of this class, including the NCS–diisopropyl sulfide reagent, are reported to show unusual selectivity for primary and secondary alcohols. Thus at 0 °C the reagent selectively oxidizes primary alcohols rather than secondary, whereas at −78 °C it selectively oxidizes secondary alcohols rather than primary (eq 2).[14]

NCS + $S(i\text{-}Pr)_2$, Et_3N, CH_2Cl_2: −78 °C, 65%; 0 °C, 70% (2)

These reagents are also effective in the oxidation of *s,t*-1,2-diols to α-ketols (eq 3).[13] Metal-based reagents give carbon–carbon bond cleavage with such diols.

1. NCS–DMS, toluene, 0 °C; 2. Et_3N, toluene, −20 °C; 86% (3)

Oxidation of 1,3-keto alcohols or 1,3-diketones gives 1,3-diketo sulfonium ylides, which may be reduced to the 1,3-diketones (eq 4).[9]

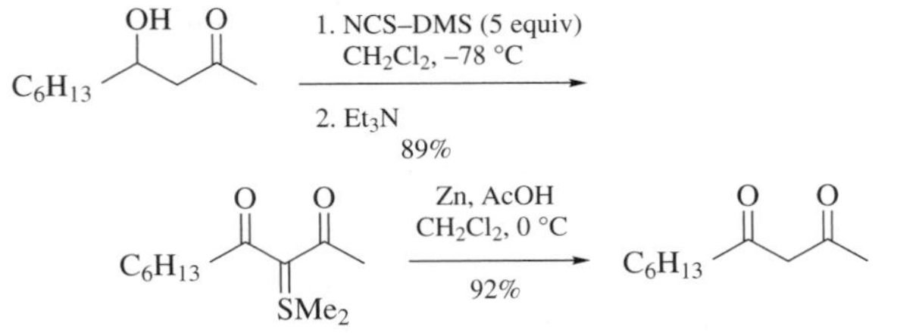

(4)

Catechols and hydroquinones are oxidized to quinones (eq 5),[3] and N-hydroxyureas are oxidized to nitroso compounds (eq 6).[4]

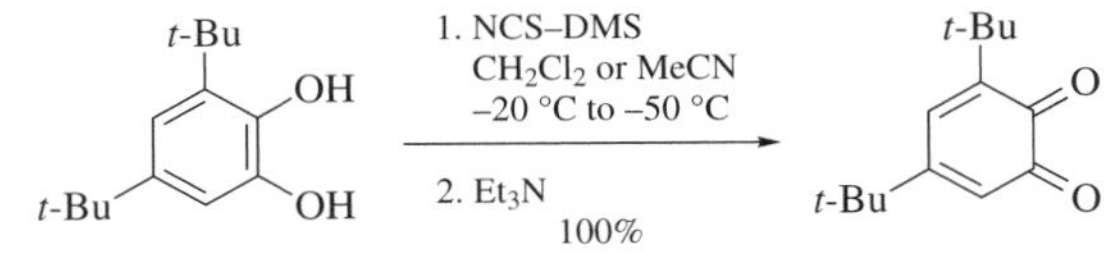

(5)

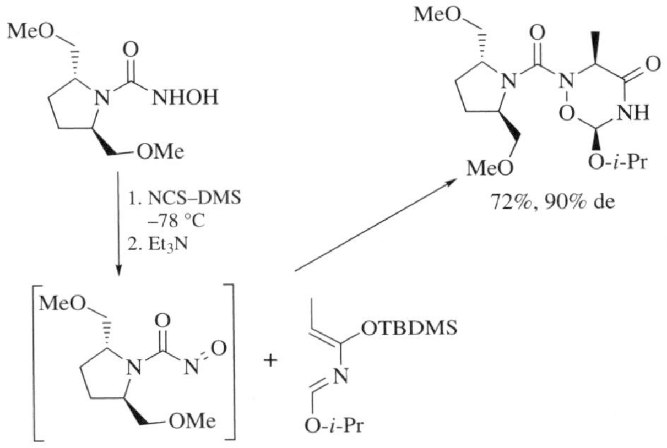

(6)

Halogenations. Allylic, benzylic, and cyclopropyl carbinyl alcohols are reported to undergo chlorination with or without allyl rearrangement, depending upon structure.[15] Thus primary and secondary allylic carbinols generally give the chlorides without rearrangement (eqs 7 and 8), whereas 2-formyl secondary allylic alcohols give the rearranged product (eq 9).[5]

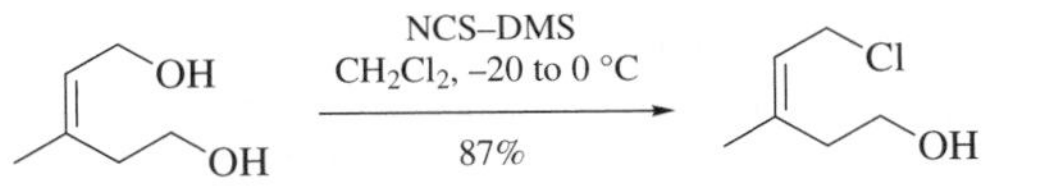

(7)

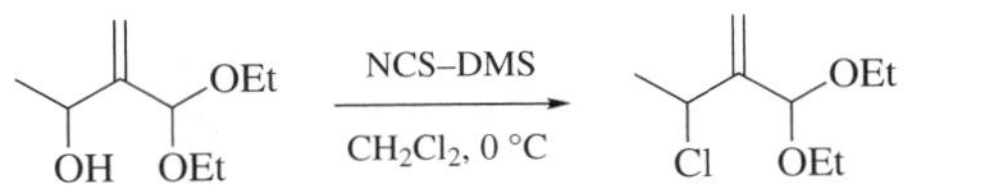

(8)

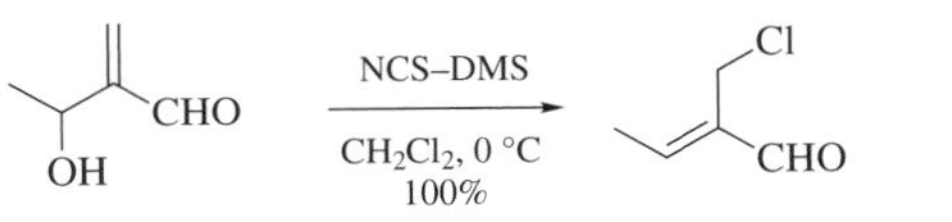

(9)

The NCS–dialkyl sulfide reagents are thermally labile, and depending on structure undergo a variety of transformations leading ultimately to α-chlorination of the sulfide (eq 10).[7]

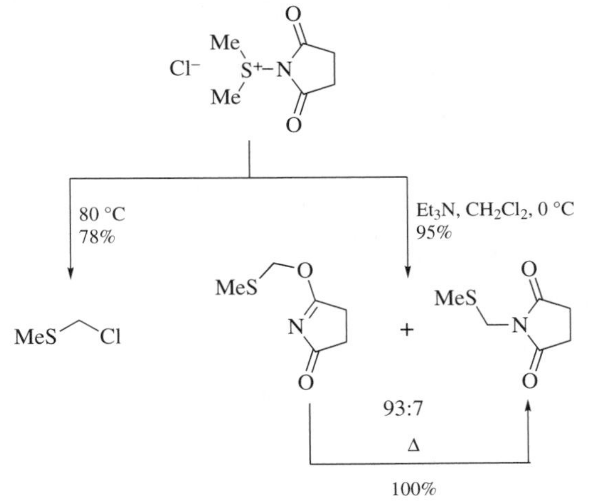

(10)

***O*-Alkylation of Anilines and Phenols.** Treatment of anilines with NCS–DMS leads to sulfilimines, which may be converted to 2-alkylanilines (eq 11).[3] Similarly, phenols give directly the 2-substituted product (eq 12).[6]

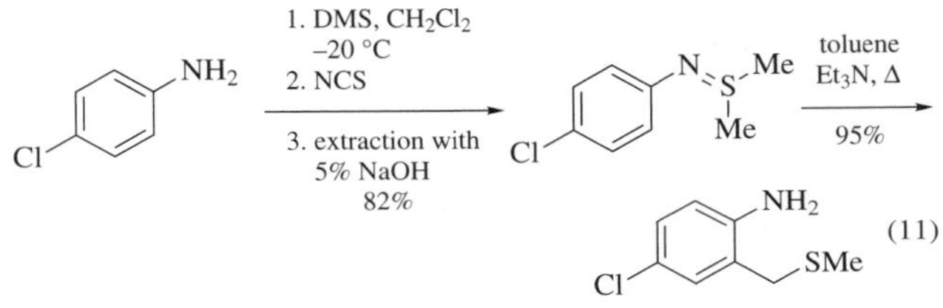

(11)

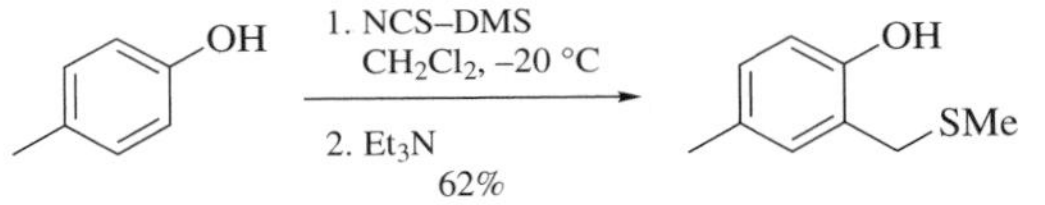

(12)

Other Alkylations. NCS and allylic sulfides react with 3-unsubstituted indoles at −20 °C to give initially 3-sulfonium salts which on warming to 20 °C rearrange to 2-allyl-3-thiomethylindoles (eq 13). These are readily desulfurized, either with or without concomitant reduction of the allylic double bond, to give 3-allyl or 3-alkyl substituted indoles.[8b,8c]

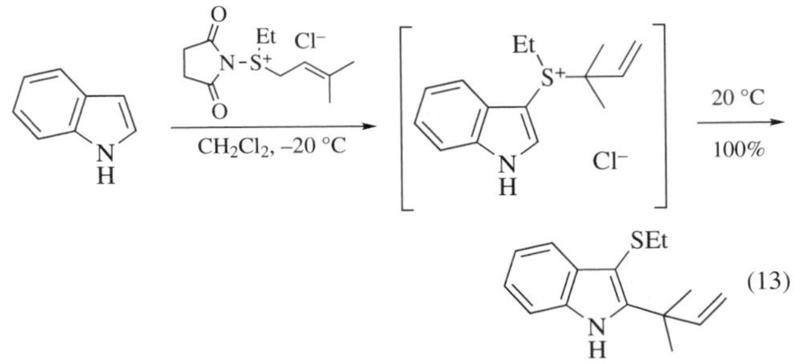

(13)

On the other hand, 3-alkyl-1*H*-indoles react with NCS–dialkyl sulfides in the presence of base to give 3-alkyl-3-alkylthioalkyl-3*H*-indoles (indolinenes, eq 14).[16]

NCS–DMS reacts with cyanoacetate anions nearly exclusively at nitrogen to give *N*-alkylthiomethylketenimines (eq 15).[17]

NCS–DMS, i-Pr_2NEt, CH_2Cl_2, –78 °C, 82% (14)

NCS–DMS, i-Pr_2NEt, CH_2Cl_2, –78 °C, 70% (15)

97:3

Thioalkylation. Treatment of pyrroles with NCS–DMS leads to 2-thiomethylpyrroles (eq 16). Indoles give 3-thioalkylindoles (eq 17).[8a–c]

NCS–DMS, CH_2Cl_2, –30 to –40 °C, 83%; $ClCH_2CH_2Cl$, 76 °C, 70% (16)

NCS–SEt_2, CH_2Cl_2, –20 to 20 °C, 100%; xylene, Δ, 80% (17)

NCS–DMS reacts with enamines to give sulfonium salts (eq 18),[10a,10b] and with cyclopentadienyl anions to give bis- and tris-substituted sulfonium salts (eq 19).[10c]

NCS–DMS, CH_2Cl_2, 73% (18)

1. NCS–DMS, CH_2Cl_2, –20 to –70 °C; 2. $LiClO_4$, 20–55% (19)

Dehydrations. While certain 1,3-diketones produce the sulfonium ylide, the treatment of N-acetylpiperidine-3,5-dione first with NCS–DMS and then with an amine in the presence of TFA resulted in the formation of the enaminone. The yields for this conversion were report to be significantly better than in the traditional method of azeotropic water removal (eq 20).[11a]

1. NCS–DMS, CH_2Cl_2, –25 °C; 2. TFA; 3. $PhNH_2$, –20 °C; 56% (20)

It was found during an attempt to produce N-methylthiomethyl nitrones that treatment of aldoximes with NCS–DMS resulted in high yields of nitriles (eq 21).[11b]

1. NCS–DMS, MeCN, 0 to 25 °C; 2. Et_3N; 80–90% (21)

Related Reagents. See Classes O-1 and O-5, pages 1–10. Dimethyl Sulfoxide–Oxalyl Chloride; Dimethyl Sulfoxide–Phosgene; Dimethyl Sulfoxide–Triphosgene.

1. (a) Tidwell, T. T. *OR* **1990**, *39*, 297. (b) Tidwell, T. T. *S* **1990**, 857. (c) Fisher, L. E.; Muchowski, J. M. *OPP* **1990**, *22*, 399.
2. Marino, J. P.; Schwartz, A. *CC* **1974**, 812.
3. (a) Claus, P. K.; Rieder, W.; Hofbauer, P.; Vilsmaier, E. *T* **1975**, *31*, 505. (b) Claus, P. K.; Vycudilik, W.; Rieder, W. *M* **1971**, *102*, 1571. (c) Gassman, P. G.; Gruetzmacher, G. D. *JACS* **1974**, *96*, 5487. (d) Gassman, P. G.; Parton, R. L. *TL* **1977**, 2055.
4. Gouverneur, V.; Ghosez, L. *TL* **1991**, *32*, 5349.
5. (a) Corey, E. J.; Kim, C. U.; Takeda, M. *TL* **1972**, 4339. (b) Depezay, J. C.; Le Merrer, Y. *TL* **1974**, 2751. (c) Depezay, J. C.; Le Merrer, Y. *TL* **1974**, 2755.
6. (a) Gassman, P. G.; Amick, D. R. *TL* **1974**, 889. (b) Gassman, P. G.; Amick, D. R. *TL* **1974**, 3463. (c) Gassman, P. G.; Amick, D. R. *JACS* **1978**, *100*, 7611.
7. (a) Vilsmaier, E.; Sprugel, W. *LA* **1971**, *747*, 151. (b) Vilsmaier, E.; Dittrich, K. H.; Sprugel, W. *TL* **1974**, 3601.
8. (a) Franco, F.; Greenhouse, R.; Muchowski, J. M. *JOC* **1982**, *47*, 1682. (b) Tomita, K.; Terada, A.; Tachikawa, R. *H* **1976**, *4*, 729. (c) Tomita, K.; Terada, A.; Tachikawa, R. *H* **1976**, *4*, 733. (c) Vilsmaier, E.; Sprugel, W.; Gagel, K. *TL* **1974**, 2475. (d) Vilsmaier, E.; Troger, W.; Sprugel, W.; Gagel, K. *CB* **1979**, *112*, 2997.
9. (a) Katayama, S.; Fukuda, K.; Watanabe, T.; Yamauchi, M. *S* **1988**, 178. (b) Katayama, S.; Watanabe, T.; Yamauchi, M. *CL* **1989**, 973. (c) Katayama, S.; Watanabe, T.; Yamauchi, M. *CPB* **1990**, *38*, 3314.
10. (a) Vilsmaier, E.; Sprugel, W.; Gagel, K. *TL* **1974**, 2475. (b) Vilsmaier, E.; Troger, W.; Sprugel, W.; Gagel, K. *CB* **1979**, *112*, 2997. (c) Schlingensief, K. H.; Hartke, K. *TL* **1977**, 1269.
11. (a) Tamura, Y.; Chen, L. C.; Fujita, M.; Kiyokawa, H.; Kita, Y. *CI(L)* **1979**, 668. (b) Dalgard, N. K. A.; Larsen, K. E.; Torssell, K. B. G. *ACS,B* **1984**, *38*, 423.
12. McCormick, J. P. *TL* **1974**, 1701.
13. Corey, E. J.; Kim, C. U.; Misco, P. F. *OSC* **1988**, *6*, 220.
14. Kim, K. S.; Cho, I. H.; Yoo, B. K.; Song, Y. H.; Hahn, C. S. *CC* **1984**, 762.
15. Murray, R. K.; Jr.; Babiak, K. A. *TL* **1974**, 311.
16. Katayama, S.; Watanabe, T.; Yamauchi, M. *CPB* **1992**, *40*, 2836.
17. Morel, G.; LeMoing-Orliac, M. A.; Khamsitthideth, S.; Foucaud, A. *T* **1992**, *38*, 527.

Robert C. Kelly
The Upjohn Company, Kalamazoo, MI, USA

Chlorotris(triphenylphosphine)-rhodium(I)[1,2]

$RhCl(PPh_3)_3$

[14694-95-2] $C_{54}H_{45}ClRh$ (MW 925.24)

(catalyst precursor for many reactions involving alkenes, alkynes, halogenated organics, and organometallic reagents; notably hydrogenations, hydrosilylations, hydroformylations, hydroborations, isomerizations, oxidations, and cross-coupling processes)

Alternate Name: Wilkinson's catalyst.

Physical Data: mp 157 °C. It exists in burgundy-red and orange polymeric forms, which have identical chemical properties (as far as is known).

Solubility: about 20 g L^{-1} in $CHCl_3$ or CH_2Cl_2, about 2 g L^{-1} in benzene or toluene; much less in acetic acid, acetone, methanol, and other aliphatic alcohols. Virtually insol in alkanes and cyclohexane. Reacts with donor solvents like DMSO, pyridine, and acetonitrile.

Form Supplied in: burgundy-red powder, possibly containing excess triphenylphosphine, triphenylphosphine oxide, and traces of rhodium(II) and -(III) complexes.

Analysis of Reagent Purity: ^{31}P NMR displays resonances for the complex in equilibrium with dissociated triphenylphosphine (CH_2Cl_2, approximate δ ppm: 31.5 and 48.0 {J values: Rh–P^1 −142 Hz; Rh–P^2 −189 Hz; P^1–P^2 −38 Hz} shifted in the presence of excess PPh_3).[3] Triphenylphosphine oxide contaminant can also be observed (CH_2Cl_2, δ ppm: 29.2) but paramagnetic impurities are generally not evident. In rhodium NMR a signal is observed at −1291 ppm.

Preparative Method: good quality material can be obtained using the latest *Inorganic Syntheses* procedure,[4] with careful exclusion of air. Recrystallization is *not* recommended.

Handling, Storage, and Precautions: the complex should be stored at reduced temperature under dinitrogen or argon. It oxidizes slowly when exposed to air in the solid state, and faster in solution. Such partial oxidation can influence the catalytic efficacy. Consequently, the necessary precautions are governed by the reaction in question. For mechanistic and kinetic studies, reproducible results may only be obtained if the catalyst is freshly prepared and manipulated in an inert atmosphere; even the substrate should be treated to remove peroxides. For hydrogenations of alkenes on a preparative scale, complex that has been handled in the air for very brief periods should be active, but competing isomerization processes may be enhanced as a result of partial oxidation of the catalyst. At the other extreme, exposure to air just before use is clearly acceptable for oxidations in the presence of O_2 and t-BuOOH.

Background.[1] In solution, Wilkinson's catalyst is in equilibrium with the 14e species $RhCl(PPh_3)_2$ (**1**) and triphenylphosphine. The 14e complex is far more reactive than the parent material; consequently it is the reactive entity most likely to coordinate with the substrate and/or the reagents. Generally, the catalytic cycles involving this material then proceed via a cascade of oxidative addition, migratory insertion, and reductive elimination reactions. The postulated mechanism for the hydrogenation of alkenes illustrates these features (Scheme 1), and is typical of the rationales frequently applied to comprehend the reactivity of $RhCl(PPh_3)_3$. Other types of transformations may be important (e.g. transmetalations), and the actual mechanisms are certainly more complicated in many cases; nevertheless, the underlying concepts are similar.

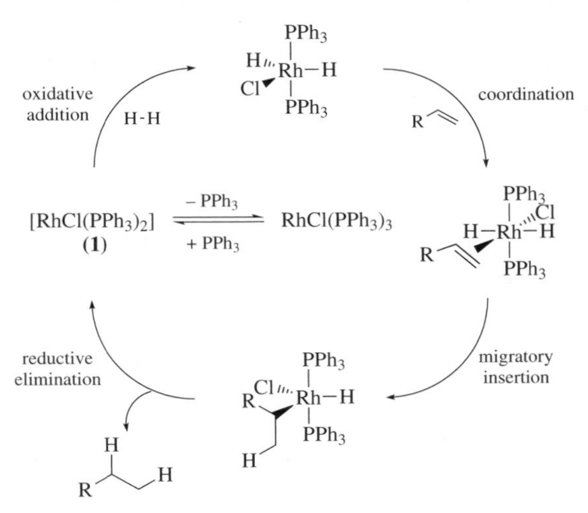

Scheme 1 Simplified mechanism for alkene hydrogenations mediated by $RhCl(PPh_3)_3$

Two important conclusions emerge from these mechanistic considerations. First, $RhCl(PPh_3)_3$ is not a catalyst in the most rigorous sense, but a catalyst precursor. This distinction is critical to the experimentalist because it implies that there are other ways to generate catalytically active rhodium(I) phosphine complexes in solution. Wilkinson's 'catalyst' is a convenient source of homogeneous rhodium(I); it has been extensively investigated because it is easily obtained, and because it was discovered early in the development of homogeneous transition metal catalysts. However, for any transformation there always may be better catalyst precursors than $RhCl(PPh_3)_3$. Secondly, reactions involving a catalytic cycle such as the one shown in Scheme 1 are inherently more complicated than most in organic chemistry. Equilibria and rates for each of the steps involved can be influenced by solvent, temperature, additives, and functional groups on the substrate. Competing reactions are likely to be involved and, if they are, the performance of the catalytic systems therefore is likely to be sensitive to these parameters. Consequently, the purity of the Wilkinson's catalyst used is an important factor. Indeed, less pure catalyst occasionally gives superior results because removal of a fraction of the triphenylphosphine in solution by oxidation to triphenylphosphine oxide gives more of the dissociation product (**1**).

In summary, practitioners of organometallic catalysis should consider the possible mechanistic pathways for the desired transformation, then screen likely catalyst systems and conditions until satisfactory results are obtained. Wilkinson's catalyst is one of the many possible sources of homogeneous rhodium(I).

Hydrogenations. Wilkinson's catalyst is highly active for hydrogenations of unconjugated alkenes at ambient temperatures and pressures. Steric effects are important insofar as less hindered alkenes react relatively quickly, whereas highly encumbered ones are not reduced (eq 1).[5] Hydrogen in the presence of $RhCl(PPh_3)_3$ under mild conditions does not reduce aromatic compounds, ketones, carboxylic acids, amides, or esters, nitriles, or nitro (eq 2) functionalities. Moreover, hydrogenations mediated by Wilkinson's catalyst are stereospecifically *cis* (eq 1). These characteristics have been successfully exploited to effect chemo-, regio-, and stereoselective alkene reductions in many organic syntheses (eqs 1–3). For instance, steric effects force delivery of dihydrogen to the least hindered face of the alkene in (eq 3).[6] Eq 4 illustrates that 1,4-cyclohexadienes can be reduced with little competing isomerization/aromatization,[7] unlike many other common hydrogenation catalysts.[5]

Et, *i*-Pr, AcO, H — D_2, cat. $RhCl(PPh_3)_3$ → Et, *i*-Pr, AcO, H, D, D (1)

Ph–CH=CH–NO_2 — H_2, cat. $RhCl(PPh_3)_3$ → Ph–CH_2CH_2–NO_2 (2)

OTr, O, OMe — H_2, benzene, cat. $RhCl(PPh_3)_3$ → OTr, O, OMe (3)

CO_2Me — H_2, cat. $RhCl(PPh_3)_3$ → CO_2Me + CO_2Me (4) 96:4; H_2, cat. PtO_2 → 49:26

Strongly coordinating ligands can suppress or completely inhibit hydrogenations mediated by Wilkinson's catalyst; examples include 1,3-butadiene, many phosphorus(III) compounds, sulfides, pyridine, and acetonitrile. Similarly, strongly coordinating substrates are not hydrogenated in the presence of Wilkinson's catalyst, presumably because they bind too well. Compounds in this category include maleic anhydride, ethylene, some 1,3-dienes, and some alkynes. Conversely, transient coordination of functional groups on the substrate can be useful with respect to directing $RhCl(PPh_3)_3$ to particular regions of the molecule for stereoselective reactions. However, in directed hydrogenations Wilkinson's catalyst is generally inferior to more Lewis acidic cationic rhodium(I) and iridium(I) complexes.[8] The activity of Wilkinson's catalyst towards hydrogenation of alkenes has been reported to be enhanced by trace quantities of oxygen.[9]

Hydrogenations of alkynes mediated by Wilkinson's catalyst generally give alkanes. *Cis*-alkene intermediates formed in such reactions tend to be more reactive than the alkyne substrate, so this is usually not a viable route to alkenes. Some alkynes suppress the catalytic reactions of $RhCl(PPh_3)_3$ by coordination. Nevertheless, hydrogenation of alkynes mediated by $RhCl(PPh_3)_3$ can be useful in some cases, as in eq 5 in which the catalyst tolerates sulfoxide functionalities and gives significantly higher yields than the corresponding reduction catalyzed by ***Palladium on Barium Sulfate***.[10]

C_6H_{13}–C≡C–S(O)–*p*-Tol — H_2, C_6H_6, cat. $RhCl(PPh_3)_3$ → C_6H_{13}–CH=CH–S(O)–*p*-Tol (5)

Wilkinson's catalyst can mediate the hydrogenation of allenes to isolated alkenes via reduction of the least hindered bond.[11] Di-*t*-butyl hydroperoxide is 'hydrogenated' to *t*-BuOH in the presence of $RhCl(PPh_3)_3$, though this transformation could occur via a radical process.[12]

Hydrogen Transfer Reactions. Wilkinson's catalyst should lower the energy barrier for dehydrogenations of alkanes to alkenes since it catalyzes the reverse process, but no useful transformation of this kind have been discovered. Presumably, the activation energy for this reaction is too great since alkanes have no coordinating groups. Alcohols and amines, however, do have ligating centers, and can dehydrogenate in the presence of Wilkinson's catalyst. These reactions have been used quite often, mostly from the perspective of hydrogen transfer from an alcohol or amine to an alkene substrate, although occasionally to dehydrogenate alcohols or amines.

2-Propanol solvent under basic conditions has been extensively used to transfer hydrogen to alkenes and other substrates. Elevated temperatures are usually required and under these conditions $RhCl(PPh_3)_3$ may be extensively modified prior to the catalysis. Ketones, alkenes (eq 6), aldimines (eq 7),[13] nitrobenzene, and some quinones are reduced in this way.

O — *i*-PrOH, KOH, cat. $RhCl(PPh_3)_3$ → OH — *i*-PrOH, KOH, cat. $RhCl(PPh_3)_3$ → OH (6)

NPh, CN — K_2CO_3, *i*-PrOH, reflux, cat. $RhCl(PPh_3)_3$ → NPh, CN (7)

Wilkinson's catalyst mediates a Cannizzaro-like process with benzaldehyde in ethanol; the aldehyde serves as a dihydrogen source to reduce itself, and the benzoic acid formed is esterified by the solvent (eq 8).[14] Pyrrolidine is *N*-methylated by methanol in the presence of $RhCl(PPh_3)_3$, a reaction that presumably occurs via hydrogen transfer from methanol, condensation of the formaldehyde formed with pyrrolidine, then hydrogen transfer to the iminium intermediate (eq 9).[15]

PhCHO — K_2CO_3, EtOH, reflux, cat. $RhCl(PPh_3)_3$ → PhC(O)OEt + $PhCH_2OH$ (8)

NH — MeOH, reflux, cat. $RhCl(PPh_3)_3$ → NMe (9)

Hydrosilylations.[16] Wilkinson's catalyst is one of several complexes which promote hydrosilylation reactions, and it often seems to be among the best identified.[17] However, hydrosilylations with $RhCl(PPh_3)_3$ tend to be slower than those mediated by H_2PtCl_6. Good turnover numbers are observed, the catalyst eventually being inactivated by P–C bond cleavage reactions at the phosphine,[18] and other unidentified processes. Catalysts without phosphine ligands may be even more robust than $RhCl(PPh_3)_3$ because they are unable to decompose via P–C bond cleavage.[19] Wilkinson's catalyst is relatively efficient with respect to converting silanes to disilanes.[20] The latter reaction could be useful in its own right but in the context of hydrosilylation processes it means that the product yields based on the silane are less than quantitative.

For hydrosilylation of alkenes, the reaction rate increases with temperature and hence many of these reactions have been performed at 100 °C. Higher reaction rates are obtained for silanes with very electronegative substituents and low steric requirements (e.g. $HSi(OEt)_3 > HSi(i\text{-}Pr)_3$). Terminal alkenes usually are hydrosilylated in an anti-Markovnikov sense to give terminal silanes. Internal alkenes tend not to react (e.g. cyclohexene), or isomerize to the terminal alkene which is then hydrosilylated (eq 10). Conversely, terminal alkenes may be partially isomerized to unreactive internal alkenes before the addition of silane can occur. 1,4-Additions to dienes are frequently observed, and the product distributions are extremely sensitive to the silane used (eq 11).

Et⌒ or Et(Z) —$HSiMe_2Ph$, cat. $RhCl(PPh_3)_3$→ Et⌒$SiMe_2Ph$ (10)

diene —$HSiR_3$, cat. $RhCl(PPh_3)_3$→ R_3Si⌒ + ⌒SiR_3 (11)

α,β-Unsaturated nitriles are hydrosilylated, even γ-substituted ones, to give 2-silyl nitriles with good regioselectivity (eq 12).[21] Secondary alkyl silanes are also formed in the hydrosilylation of phenylethylene. In fact, the latter reaction has been studied in some detail, and primary alkyl silanes, hydrogenation product (i.e. ethylbenzene), and *E*-2-silylphenylethylenes are also formed (eq 13).[22] Equimolar amounts of ethylbenzene (**2**) and *E*-2-silylphenylethylene (**4**) are produced, implying these products arise from the same reaction pathway. It has been suggested that this involves dimeric rhodium species because the relative amounts of these products increase with the rhodium:silane ratio; however, competing radical pathways cannot be ruled out. Certainly, product distributions are governed by the proportions of all the components in the reaction (i.e. catalyst, silane, and alkene), and the reaction temperature. Side products in the hydrosilylation of 1-octene include vinylsilanes and allylsilanes (eq 14).[23,24]

⌒CN —$HSiR_3$, cat. $RhCl(PPh_3)_3$→ CN / SiR_3 (12)

Hydrosilylation of alkynes gives both *trans* products (i.e. formally from *cis* addition), and *cis* products (from either isomerization or *trans* addition); H_2PtCl_6, however, gives almost completely *cis* addition to *trans* products.[25] Moreover, CC–H to CC–SiR_3 exchange processes can occur for terminal alkynes giving (**6**) (eq 15).[25–27] The product distribution in these reactions is temperature dependent, and other factors may be equally important. Nonstereospecific transition metal catalyzed hydrosilylations of alkynes are not confined to Wilkinson's catalyst, and the origin of the *trans* addition product has been investigated in detail for other homogeneous rhodium and iridium complexes.[19]

Ph⌒ —$HSiEt_3$, 50 °C, cat. $RhCl(PPh_3)_3$→

Ph⌒ (**2**) + Ph⌒$SiEt_3$ (**3**) + Ph⌒$SiEt_3$ (**4**) + Ph⌒($SiEt_3$) (**5**) (13)

mol % catalyst	(**2**):(**3**):(**4**):(**5**)
3.1	39: 34: 43: 3
0.12	3: 25: 2: 47

MeR_2SiH + ⌒C_6H_{11} —2.3 mol % $RhCl(PPh_3)_3$, 80–85 °C→

MeR_2Si⌒C_6H_{11} (60%) + ⌒C_6H_{11} + MeR_2Si⌒C_6H_{11} (32%) + MeR_2Si⌒C_5H_9 (5%) + MeR_2Si⌒C_5H_9 (3%) (14)

Ph≡ —$HSiEt_3$, cat. $RhCl(PPh_3)_3$→

Ph≡$SiEt_3$ (**6**) + Ph⌒$SiEt_3$ (**7**) + Ph⌒$SiEt_3$ (**8**) + Ph($SiEt_3$)= (**9**) (15)

temperature	(**6**):(**7**):(**8**):(**9**)
65 °C	4: 31: 57: 8
80 °C	3: 55: 35: 7

Hydrosilylation of terminal alkenes has been used in a polymerization process to form new polymeric organic materials.[28]

Hydrosilylation of α,β-unsaturated aldehydes and ketones gives silylenol ethers via 1,4-addition, even when the 4-position is relatively hindered.[29] Hydrolysis of the silyl enol ethers so formed gives saturated aldehydes. Combination of these reduction and hydrolysis steps gives overall reduction of alkenes conjugated to aldehydes, in selectivities which are generally superior to those obtained using hydridic reducing agents (eqs 16 and 17). Dihydrosilanes tend to reduce α,β-unsaturated carbonyl compounds to the corresponding alcohols, also with good regioselectivity (eq 18).

enone —$HSiEt_3$, cat. $RhCl(PPh_3)_3$→ $OSiEt_3$ enol ether —hydrolysis→ ketone (net conjugate reduction) (16)

enal —$DSiEt_3$, cat. $RhCl(PPh_3)_3$→ D, ⌒$OSiEt_3$ (17)

H_2SiEt_2, cat. $RhCl(PPh_3)_3$ (18)

Similar hydrosilylations of α,β-unsaturated esters are useful for obtaining silyl ketene acetals with over 98:2 (*Z*) selectivity (eq 19);[30] this transformation is complementary to the reaction of α-bromo esters with zinc and chlorotrialkylsilanes, which favors the formation of the corresponding (*E*) products.[30] In cases where (*E*):(*Z*) stereoselectivity is not an issue, ***Rhodium(III) Chloride*** ($RhCl_3 \cdot 6H_2O$) may be superior to Wilkinson's catalyst.[31] Unconjugated aldehydes and ketones are reduced by silanes in the presence of $RhCl(PPh_3)_3$; trihydrosilanes react quicker than di- than monohydrosilanes.[32,33]

$HSiPh_3$, cat. $RhCl(PPh_3)_3$ (19)

Alcohols (eq 20)[34] and amines (eq 21)[35] react with silanes in the presence of Wilkinson's catalyst to give the silylated compounds and, presumably, hydrogen. These reactions are useful in protecting group strategies.

$$Ph_2SiH_2 + MeOH \xrightarrow{\text{cat. RhCl(PPh}_3)_3} Ph_2SiH(OMe) + H_2 \quad (20)$$

$+ H_2N$–C_6H_4–Br, cat. $RhCl(PPh_3)_3$ (21)

N,N-Dimethylacrylamide and triethylsilane combine in the presence of Wilkinson's catalyst (50 °C) to give a *O,N*-silylketene acetal as the pure (*Z*) isomer after distillation; this reaction can be conveniently performed on a gram scale (eq 22). The products have been used in new aldol methodology.[36]

$HSiEt_3$, cat. $RhCl(PPh_3)_3$ (22)

Hydrostannylations. Hydrostannanes add to alkynes in uncatalyzed reactions at 60 °C. Phenylacetylene, for instance gives a mixture of (*E*)- and (*Z*)-vinylstannanes, wherein the tin atom has added to the terminal carbons. In the presence of Wilkinson's catalyst, however, the hydrostannylation proceeds at 0 °C to give mostly the regioisomeric vinylstannanes (eq 23).[37] Terminal stannanes in the latter process seem to result from competing free radical additions. This may not be a complication with some other catalysts; the complexes $PdCl_2(PPh_3)_2$ and $Mo(\eta^3\text{-allyl})(CO)_2(NCMe)_2$ also mediate hydrostannylations of alkynes, and they are reported to be 100% *cis* selective.[38] Hydrostannanes and thiols react in a similar way to silanes and alcohols (eq 24).[39]

$HSnBu_3$, cat. $RhCl(PPh_3)_3$; 88% + 12% (23)

$HSnBu_3$, cat. $RhCl(PPh_3)_3$ (24)

Hydroacylations. Alkenes with aldehyde functionality in the same molecule, but displaced by two carbon atoms, can cyclize via intramolecular hydroacylation reactions. Substituent effects can have a profound influence on these transformations. For instance, 3,4-disubstituted 4-pentenals cyclize to cyclopentanones without serious complications,[40] but 2,3-disubstituted 4-pentenals give a cyclopropane as a competing product (eqs 25 and 26).[41] Formation of the latter material illustrates two features which restrict the applicability of this type of reaction. First decarbonylation of the aldehyde can occur, in this case presumably giving a rhodium alkyl complex which then inserts the pendant alkene functionality. Secondly, decarbonylation reactions convert the catalyst into $RhCl(CO)(PPh_3)_2$, which tends to be inactive. Moreover, the reaction is only generally applicable to the formation of five-membered rings, and it is apparently necessary to use quite large amounts of Wilkinson's catalyst to ensure good yields (eq 27).[42] Rhodium(I) complexes other than $RhCl(PPh_3)_3$ can give better results in some cases.[43]

cat. $RhCl(PPh_3)_3$, CH_2Cl_2, 20 °C, 2.5 h (25)

cat. $RhCl(PPh_3)_3$; 1:1; $R = (CH_2)_7Me$ (26)

0.4 equiv $RhCl(PPh_3)_3$, CH_2Cl_2, 40 °C, 20 h (27)

Lactols can be cyclized under the typical hydroacylation conditions (eq 28), presumably via equilibrium amounts of the corresponding aldehyde.[40] Finally, intermolecular hydroacylation has been formally achieved in the reaction of a pyridyl aldimine with ethylene under pressure at 160 °C; here the pyridine functionality anchors the aldimine to the rhodium, and decarbonylation is impossible (eq 29).

cat. $RhCl(PPh_3)_3$ (28)

1. cat. $RhCl(PPh_3)_3$ 150 psi C_2H_4, 160 °C 2. hydrolysis (29)

Decarbonylations. Wilkinson's catalyst has been known for some time to decarbonylate aldehydes, even heavily functionalized ones, to the corresponding hydrocarbons.[44] Some examples are shown in eqs 30–33, illustrating high stereochemical retention in the decarbonylation of chiral, cyclopropyl, and unsaturated aldehydes.[45,46] Acid chlorides are also decarbonylated by $RhCl(PPh_3)_3$.

+ $RhCl(PPh_3)_3$ 130 °C − PPh_3 → + $RhCl(CO)(PPh_3)_2$ (30)

1.0 equiv $RhCl(PPh_3)_3$ (31) 93% retention

1.0 equiv $RhCl(PPh_3)_3$ (32) 94% retention

1.0 equiv $RhCl(PPh_3)_3$ (33) 100% retention

The problem with all these reactions is that stoichiometric amounts of the catalyst are required, and the process is inordinately expensive. Consequently, it has only been used by those wishing to illustrate a decarbonylation occurs for some special reason, or in the closing stages of small scale syntheses of complex organic molecules. Very recently, however, it has been shown that the reaction can be made catalytic by adding ***Diphenyl Phosphorazidate***.[47] The role of the latter is to decarbonylate the catalytically inactive $RhCl(CO)(PPh_3)_2$, regenerating rhodium(I) without carbonyl ligands. Examples of this catalytic process are shown in eqs 34 and 35. The path is now clear for extensive use of $RhCl(PPh_3)_3$ for catalytic decarbonylation reactions in organic synthesis.

$(PhO)_2P(O)N_3$, 25 °C < 5 mol % $RhCl(PPh_3)_3$ (34)

$(PhO)_2P(O)N_3$, 25 °C < 5 mol % $RhCl(PPh_3)_3$ (35)

Catalytic decarbonylations of a few substrates other than aldehydes have been known for some time, e.g. conversion of benzoic anhydrides to fluorenones at high temperatures (ca. 225 °C).[48]

Hydroformylations.[49] Carbon monoxide reacts rapidly with $RhCl(PPh_3)_3$ to give $RhCl(CO)(PPh_3)_2$. With hydrogen, in the presence of triphenylphosphine, the latter carbonyl complex affords some ***Carbonylhydridotris(triphenylphosphine)rhodium(I)***, and this very actively mediates hydroformylations.[50] Reactions wherein $RhCl(PPh_3)_3$ is used as a hydroformylation catalyst probably proceed via this route. A more direct means of hydroformylation is to use $RhH(CO)(PPh_3)_3$. Nevertheless, Wilkinson's catalyst (an unfortunate term here because Wilkinson also pioneered hydroformylations using $RhH(CO)(PPh_3)_3$) has been used to effect hydroformylations of some substrates. Eq 36 is one example and illustrates that transient coordination of the acyl group with rhodium apparently leads to predominant formation of a 'branched chain' aldehyde, whereas straight chain aldehydes are usually formed in these reactions.[51] Other hydroformylation catalysts that have been studied include cobalt and iridium based systems.[49]

H_2, CO cat. $RhCl(PPh_3)_3$ → + other minor products (36)

Hydroborations.[52] Addition of ***Catecholborane*** to alkenes is accelerated by Wilkinson's catalyst, and other sources of rhodium(I) complexes.[53] Unfortunately, the reaction of Wilkinson's catalyst with catecholborane is complex; hence if the conditions for these reactions are not carefully controlled, competing processes result. In the hydroboration of styrene, for instance, the secondary alcohol is formed almost exclusively (after oxidation of the intermediate boronate ester, eq 37); however, the primary alcohol also is formed if the catalyst is partially oxidized and this can be the major product in extreme cases.[54,55] Conversely, hydroboration of the allylic ether (**12**) catalyzed by pure Wilkinson's catalyst gives the expected alcohol (**13**), hydrogenation product (**14**), and aldehyde (**15**), but alcohol (**13**) is the exclusive (>95%) product if the $RhCl(PPh_3)_3$ is briefly exposed to air before use.[54] The *syn*-alcohol is generally the favored diastereomer in these and related reactions (eq 38), and the catalyzed reaction is therefore stereocomplementary to uncatalyzed hydroborations of allylic ether derivatives.[56–58]

1. catecholborane, cat. $RhCl(PPh_3)_3$; 2. H_2O_2, OH^- — primary alcohol if the catalyst is partially oxidized (37)

(12) → (13) + (14) + (15) (38)

syn product favored

Other sources of rhodium(I) are equally viable catalysts for hydroborations, notably $Rh(\eta^3\text{-}CH_2CMeCH_2)(i\text{-}Pr_2PCH_2CH_2P\text{-}i\text{-}Pr_2)$ which gives a much cleaner reaction with catecholborane than Wilkinson's catalyst.[59] Other catalysts for hydroborations are also emerging.[60–62]

Catecholborane hydroborations of carbonyl and related functionalities are also accelerated by $RhCl(PPh_3)_3$ (eqs 39–41); however, several related reactions proceed with similar selectivities in the absence of rhodium.[63–65]

1. catecholborane, cat. $RhCl(PPh_3)_3$; 2. hydrolysis (39)

syn:*anti* = 12:1

$C_{13}H_{27}$ ... OMe — 1. catecholborane, cat. $RhCl(PPh_3)_3$; 2. hydrolysis (40)

1. catecholborane, cat. $RhCl(PPh_3)_3$; 2. hydrolysis (41)

de 10:1

Cyclization, Isomerization, and Coupling Reactions. Inter- (eq 42)[66] and intramolecular (eq 43)[67] cyclotrimerizations of alkynes are mediated by Wilkinson's catalyst. This is an extremely efficient route to ring fused systems. Similarly, Diels–Alder-like [4+2] cyclization processes are promoted by $RhCl(PPh_3)_3$;[68] 'dienophile' components in these reactions need not be electron deficient, and they can be an alkene or alkyne (eqs 44 and 45). Allenes oligomerize in pathways determined by their substituents. For instance, four molecules of allene combine to give a spirocyclic system (eq 46), but tetraphenylallene isomerizes to give an indene (eq 47).[69]

cat. $RhCl(PPh_3)_3$ (42)

cat. $RhCl(PPh_3)_3$ (43)

cat. $RhCl(PPh_3)_3$ (44)

cat. $RhCl(PPh_3)_3$ (45)

4 =•= cat. $RhCl(PPh_3)_3$, 59% (46)

cat. $RhCl(PPh_3)_3$ (47)

Wilkinson's catalyst is also capable of mediating the formation of C–C bonds in reactions which apparently proceed via oxidative addition of an unsaturated organohalide across the metal (eq 48),[70] or via transmetalation from an organometallic (eq 49).[71] These two transformation types are very similar to couplings developed by Heck so, predictably, some palladium complexes also mediate these reactions (see ***Tetrakis(triphenylphosphine)palladium(0)*** and ***Palladium(II) Acetate***).

cat. $RhCl(PPh_3)_3$ (48)

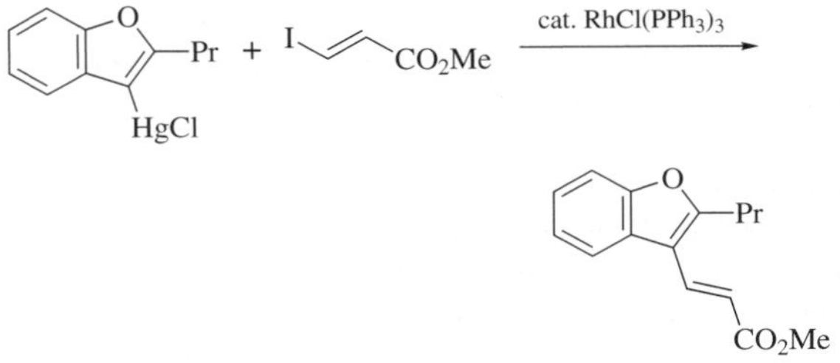

(49)

Intermolecular reactions of dienes, allenes, and methylenecyclopropanes with alkenes are mediated by $RhCl(PPh_3)_3$, although mixtures of products are usually formed (eqs 50, 51).[72–75]

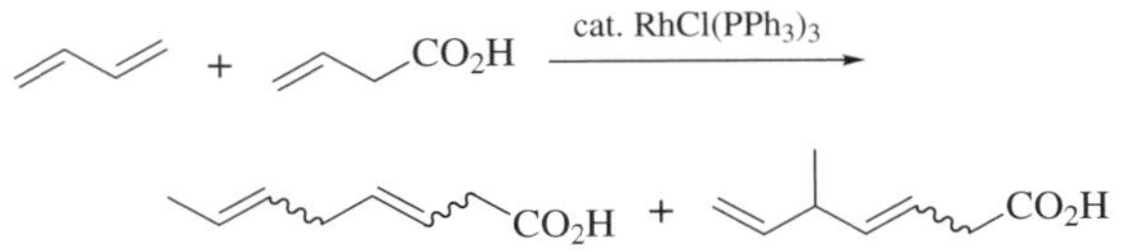

(50)

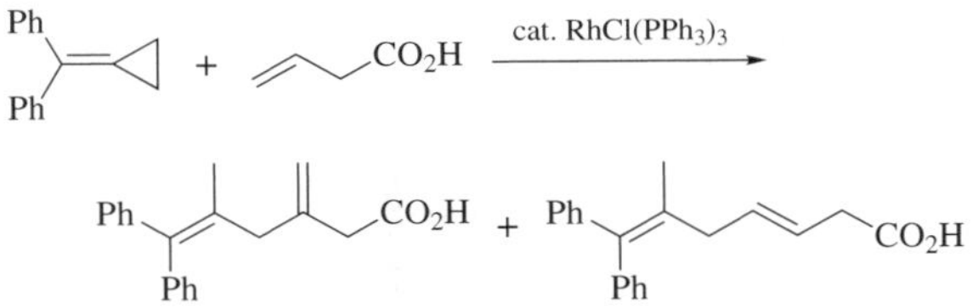

(51)

Wilkinson's catalyst mediates hydrogenation of 1,4-cyclohexadienes without double bond isomerization (see above), but at elevated temperatures in the absence of hydrogen it promotes isomerization to conjugated dienes (eq 52).[76] Isomerization of allylamines to imines followed by hydrolysis has also been performed using $RhCl(PPh_3)_3$ (eq 53),[77] although $RhH(PPh_3)_4$ and other catalysts are more frequently used for this reaction type.[78]

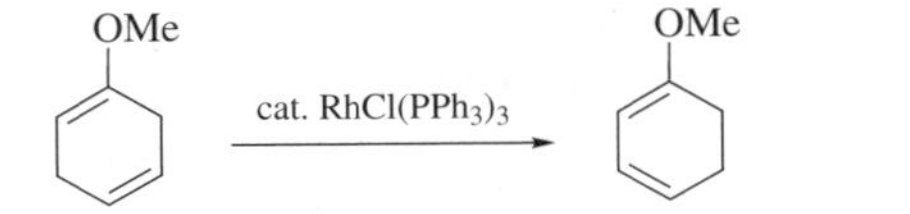

(52)

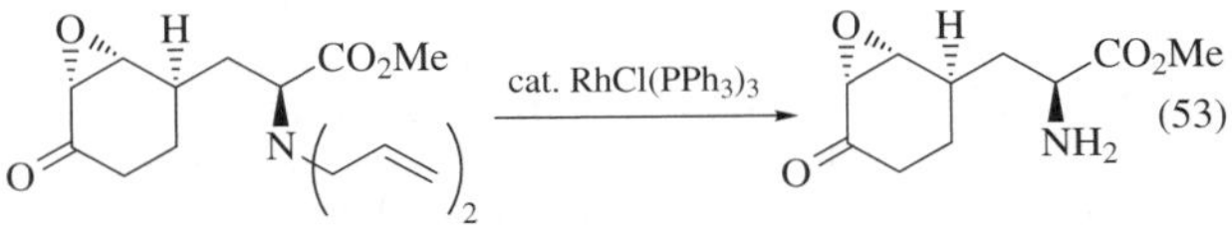

(53)

Oxidations. Cleavage of alkenes to aldehydes and ketones is promoted by Wilkinson's catalyst under pressures of air or oxygen,[79] but these reactions are inferior to ozonolysis because they tend to form a mixture of products. More useful are the oxidations of anthracene derivatives to anthraquinones in the presence of oxygen/***t*-Butyl Hydroperoxide** and catalytic $RhCl(PPh_3)_3$ (eq 54).[80,81] Wilkinson's catalyst reacts with oxygen to form an adduct so $RhCl(PPh_3)_3$ is clearly quite different from the true catalyst in all the reactions mentioned in this section.

Other Transformations. At high temperatures (>200 °C) aromatic sulfonyl chlorides are desulfonated to the corresponding aryl halides in the presence of Wilkinson's catalyst (eq 55).[82] Benzamides and malonamide also decompose under similar conditions, giving benzonitrile and acetamide, respectively.[83]

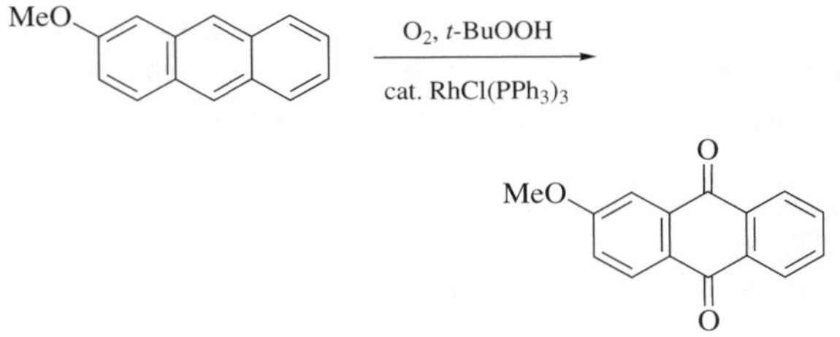

(54)

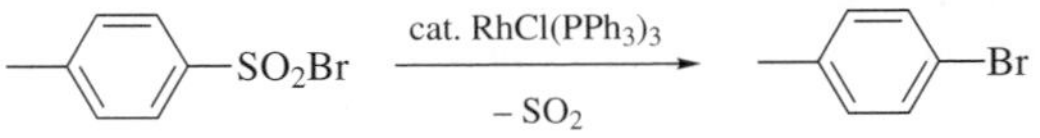

(55)

Diazonium fluoroborates are reduced to the corresponding unsubstituted aryl compounds by Wilkinson's catalyst in DMF; the solvent is apparently the hydride source in this reaction (eq 56).[84]

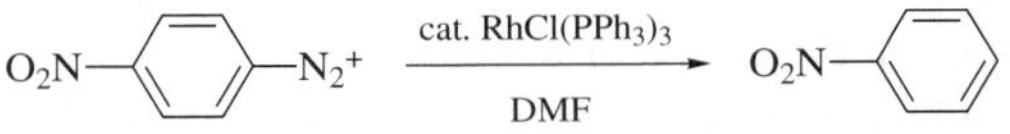

(56)

Finally, aryl group interchange between triarylphosphines is mediated by Wilkinson's catalyst at 120 °C, but a near statistical mixture of the exchanged materials is formed along with some byproducts.[85]

Related Reagents. See Classes R-3, R-4, R-10, R-12, R-15, R-17, R-19, R-20, R-22, and R-26, pages 1–10. Bis(bicyclo[2.2.1]hepta-2,5-diene)rhodium Perchlorate; [1,4-bis-(diphenylphosphino)-butane](norboradiene)rhodium tetrafluroborate Catecholborane; (1,5-Cyclooctadiene)[1,4-bis(diphenylphosphino)butane]iridium(I) Tetrafluoroborate; (1,5-Cyclooctadiene)(tricyclohexylphosphine)(pyridine)iridium(I) Hexafluorophosphate; Octacarbonyldicobalt; Palladium(II) Chloride; Tetrakis(triphenylphosphine)palladium(0).

1. Jardine, F. H. *Prog. Inorg. Chem.* **1981**, *28*, 63.
2. Osborn, J. A.; Jardine, F. H.; Young, J. F.; Wilkinson, G. *JCS(A)* **1966**, 1711.
3. Brown, T. H.; Green, P. J. *JACS* **1970**, *92*, 2359.
4. Osborn, J. A.; Wilkinson, G. *Inorg. Synth.* **1990**, *28*, 77.
5. Birch, A. J.; Williamson, D. H. *OR* **1976**, *24*, 1.
6. Sum, P.-E.; Weiler, L. *CJC* **1978**, *56*, 2700.
7. Birch, A. J.; Walker, K. A. M. *JCS(C)* **1966**, 1894.
8. Brown, J. M. *AG(E)* **1987**, *26*, 190.
9. van Bekkum, H.; van Rantwijk, F. van de Putte, T. *TL* **1969**, 1.
10. Kosugi,H.; Kitaoka, M.; Tagami, K.; Takahashi, A.; Uda, H. *JOC* **1987**, *52*, 1078.
11. Bhagwat, M. M.; Devaprabhakara, D. *TL* **1972**, 1391.
12. Kim, L.; Dewhirst, K. C. *JOC* **1973**, *38*, 2722.
13. Grigg, R.; Mitchell, T. R. B.; Tongpenyai, N. *S* **1981**, 442.
14. Grigg, R.; Mitchell, T. R. B.; Sutthivaiyakit, S. *T* **1981**, *37*, 4313.
15. Grigg, R.; Mitchell, T. R. B.; Sutthivaivakit, S.; Tongpenyai, N. *CC* **1981**, 611.
16. Speier, J. L. *Adv. Organomet. Chem.* **1979**, *17*, 407.
17. Ojima, I. In *The Chemistry of Organic Silicon Compounds*; Patai, S., Rappoport, Z., Eds.; Wiley: New York, 1989; Vol. 2, p 1479.
18. Garrou, P. E. *CRV* **1985**, *85*, 171.

19. Tanke, R. S.; Crabtree, R. H. *JACS* **1990**, *112*, 7984.
20. Brown-Wensley, K. A. *OM* **1987**, *6*, 1590.
21. Ojima, I.; Kumagai, M.; Nagai, Y. *JOM* **1976**, *111*, 43.
22. Onopchenko, A.; Sabourin, E. T.; Beach, D. L. *JOC* **1983**, *48*, 5101.
23. Onopchenko, A.; Sabourin, E. T.; Beach, D. L. *JOC* **1984**, *49*, 3389.
24. Millan, A.; Towns, E.; Maitlis, P. M. *CC* **1981**, 673.
25. Ojima, I.; Kumagai, M.; Nagai, Y. *JOM* **1974**, *66*, C14.
26. Dickers, H. M.; Haszeldine, R. N.; Mather, A. P.; Parish, R. V. *JOM* **1978**, *161*, 91.
27. Brady, K. A.; Nile, T. A. *JOM* **1981**, *206*, 299.
28. Crivello, J. V.; Fan, M. *J. Polym. Sci., Part A: Polym. Chem.* **1992**, *30*, 1.
29. Ojima, I.; Kogure, T. *OM* **1982**, *1*, 1390.
30. Slougui, N.; Rousseau, G. *SC* **1987**, *17*, 1.
31. Revis, A.; Hilty, T. K. *JOC* **1990**, *55*, 2972.
32. Ojima, I.; Kogure, T.; Nihonyanagi, M.; Nagai, Y. *BCJ* **1972**, *45*, 3506.
33. Ojima, I.; Nihonyanagi, M.; Kogure, T.; Kumagai, M.; Horiuchi, S.; Nakatsugawa, K.; Nagai, Y. *JOM* **1975**, *94*, 449.
34. Corriu, R. J. P.; Moreau, J. J. E. *CC* **1973**, 38.
35. Bonar-Law, R. P.; Davis, A. P.; Dorgan, B. J. *TL* **1990**, *31*, 6721.
36. Myers, A. G.; Widdowson, K. L. *JACS* **1990**, *112*, 9672.
37. Kikukawa, K.; Umekawa, H.; Wada, F.; Matsuda, T. *CL* **1988**, 881.
38. Zhang, H. X.; Guibé, F.; Balavoine, G. *JOC* **1990**, *55*, 1857.
39. Talley, J. J.; Colley, A. M. *JOM* **1981**, *215*, C38.
40. Sakai, K.; Ishiguro, Y.; Funakoshi, K.; Ueno, K.; Suemune, H. *TL* **1984**, *25*, 961.
41. Sakai, K.; Ide, J.; Oda, O.; Nakamura, N. *TL* **1972**, 1287.
42. Ueno, K.; Suemune, H.; Sakai, K. *CPB* **1984**, *32*, 3768.
43. Larock, R. C.; Oertle, K.; Potter, G. F. *JACS* **1980**, *102*, 190.
44. Andrews, M. A.; Gould, G. L.; Klaeren, S. A. *JOC* **1989**, *54*, 5257.
45. Walborsky, H. M.; Allen, L. A. *TL* **1970**, 823.
46. Walborsky, H. M.; Allen, L. E. *JACS* **1971**, *93*, 5465.
47. O'Connor, J. M.; Ma, J. *JOC* **1992**, *57*, 5075.
48. Blum, J.; Lipshes, Z. *JOC* **1969**, *34*, 3076.
49. Pruett, R. L. *Adv. Organomet. Chem.* **1979**, *17*, 1.
50. Jardine, F. H. *Polyhedron* **1982**, *1*, 569.
51. Ojima, I.; Zhang, Z. *JOM* **1991**, *417*, 253.
52. Burgess, K.; Ohlmeyer, M. J. *CRV* **1991**, *91*, 1179.
53. Männig, D.; Nöth, H. *AG(E)* **1985**, *24*, 878.
54. Burgess, K.; vander Donk, W. A.; Westcott, S. A.; Marder, T. B.; Baker, R. T.; Calabrese, J. C. *JACS* **1992**, *114*, 9350.
55. Evans, D. A.; Fu, G. C.; Anderson, B. A. *JACS* **1992**, *114*, 6679.
56. Evans, D. A.; Fu, G. C.; Hoveyda, A. H. *JACS* **1988**, *110*, 6917.
57. Burgess, K.; Cassidy, J.; Ohlmeyer, M. J. *JOC* **1991**, *56*, 1020.
58. Burgess, K.; Ohlmeyer, M. J. *JOC* **1991**, *56*, 1027.
59. Westcott, S. A.; Blom, H. P.; Marder, T. B.; Baker, R. T. *JACS* **1992**, *114*, 8863.
60. Evans, D. A.; Fu, G. C. *JACS* **1991**, *113*, 4042.
61. Harrison, K. N.; Marks, T. J. *JACS* **1992**, *114*, 9220.
62. Burgess, K.; Jaspars, M. *OM* **1993**, *12*, 4197.
63. Evans, D. A.; Hoveyda, A. H. *JOC* **1990**, *55*, 5190.
64. Evans, D. A.; Fu, G. C. *JOC* **1990**, *55*, 5678.
65. Kocieński, P.; Jarowicki, K.; Marczak, S. *S* **1991**, 1191.
66. Grigg, R.; Scott, R.; Stevenson, P. *TL* **1982**, *23*, 2691.
67. Neeson, S. J.; Stevenson, P. J. *T* **1989**, *45*, 6239.
68. Jolly, R. S.; Luedtke, G.; Sheehan, D.; Livinghouse, T. *JACS* **1990**, *112*, 4965.
69. Jones, F. N.; Lindsey, R. V.; Jr. *JOC* **1968**, *33*, 3838.
70. Grigg, R.; Stevenson, P.; Worakun, T. *CC* **1984**, 1073.
71. Larock, R. C.; Narayanan, K.; Hershberger, S. S. *JOC* **1983**, *48*, 4377.
72. Salerno, G.; Gigliotti, F.; Chiusoli, G. P. *JOM* **1986**, *314*, 231.
73. Salerno, G.; Gallo, C.; Chiusoli, G. P.; Costa, M. *JOM* **1986**, *317*, 373.
74. Chiusoli, G. P.; Costa, M.; Schianchi, P.; Salerno, G. *JOM* **1986**, *315*, C45.
75. Chiusoli, G. P.; Costa, M.; Pivetti, F. *JOM* **1989**, *373*, 377.
76. Harland, P. A.; Hodge, P. *S* **1983**, 419.
77. Laguzza, B. C.; Ganem, B. *TL* **1981**, *22*, 1483.
78. Stille, J. K.; Becker, Y. *JOC* **1980**, *45*, 2139.
79. Bönnemann, H.; Nunez, W.; Rohe, D. M. M. *HCA* **1983**, *66*, 177.
80. Müller, P.; Bobillier, C. *TL* **1981**, *22*, 5157.
81. Müller, P.; Bobillier, C. *TL* **1983**, *24*, 5499.
82. Blum, J.; Scharf, G. *JOC* **1970**, *35*, 1895.
83. Blum, J.; Fisher, A.; Greener, E. *T* **1973**, *29*, 1073.
84. Marx, G. S. *JOC* **1971**, *36*, 1725.
85. Abatjoglou, A. G.; Bryant, D. R. *OM* **1984**, *3*, 932.

Kevin Burgess & Wilfred A. van der Donk
Texas A & M University, College Station, TX, USA

Chromic Acid[1]

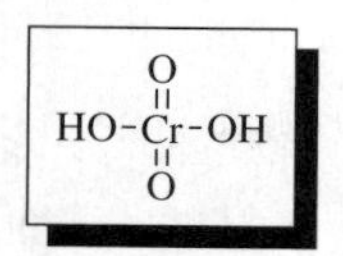

[7738-94-5] CrH_2O_4 (MW 118)

(reagent for oxidizing secondary alcohols to ketones, primary alcohols to carboxylic acids, and α-hydroxy ketones to α-diketones; for converting tertiary cyclobutanols to 1,4-ketols and 1,4-diketones; for cleaving 1,2-glycols to keto acids)

Alternate Name: Jones reagent.
Solubility: sol acetone and water.
Preparative Methods: the Jones reagent is a solution of chromium(VI) oxide and sulfuric acid in water,[2–4] in a typical procedure,[4a] 67g of CrO_3 is dissolved in 125 mL of H_2O and 58 mL of conc H_2SO_4 is then carefully added; the precipitated salts are dissolved by adding an additional (minimal) quantity of water (the total volume of the resultant solution should not exceed 225 mL); alternatively, 23.5 g of CrO_3 is dissolved in 21 mL of conc H_2SO_4 with cooling and then diluted with distilled water to give a total volume of 175 mL.[2,4b]
Handling, Storage, and Precautions: the mutagenicity of chromium(VI) compounds is well documented;[1f] handle with great care; use in a fume hood.

Oxidation of Secondary Alcohols to Ketones. The Jones reagent oxidizes cyclooctanol to cyclooctanone (92–96%) on a fairly large scale.[4] Nortricyclanol (**1**) is oxidized to the highly strained and reactive ketone nortricyclanone (**2**) (eq 1).[5] An epimeric mixture of 3,3-dimethyl-*cis*-bicyclo[3.2.0]heptan-2-ols is oxidized by Jones reagent to 3,3-dimethyl-*cis*-bicyclo[3.2.0]heptan-2-one (83–93%).[6] Jones reagent oxidizes alcohols (**3**) and (**5**) to the acid labile ketones (**4**) and (**6**), respectively (eqs 2 and 3).[7] A 90% yield of Δ^5-pregnene-3,20-dione is obtained from the Jones oxidation of pregnenolone.[8]

The Jones reagent, in the presence of oxalic acid, oxidizes 1,2-diphenylethanol and *t*-butylphenylmethanol to benzyl phenyl ketone and *t*-butyl phenyl ketone, respectively, in quantitative yield.[9] This procedure is also effective for the oxidation of 7-norbornanol to 7-norbornone. Chromic acid in aqueous acetic acid oxidizes *t*-butylphenylmethanol to *t*-butyl phenyl ketone, with formation of significant amounts of the cleavage products benzaldehyde and *t*-butanol.

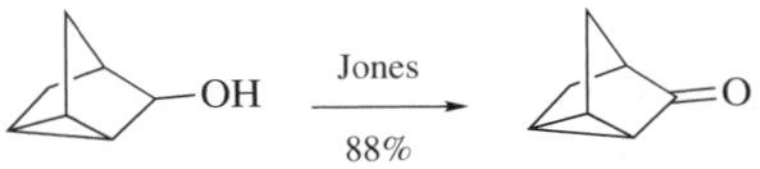

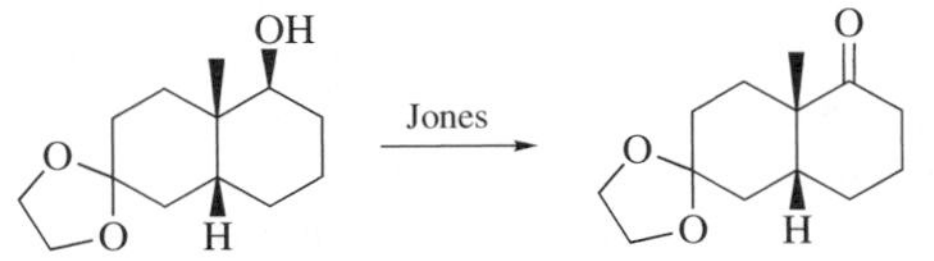

1,4-Ketols and 1,4-Diketones from tertiary Cyclobutanols. Jones reagent cleaves tertiary cyclobutanols to 1,4-ketols (eq 4) and 1,4-diketones (eqs 5 and 6).[10]

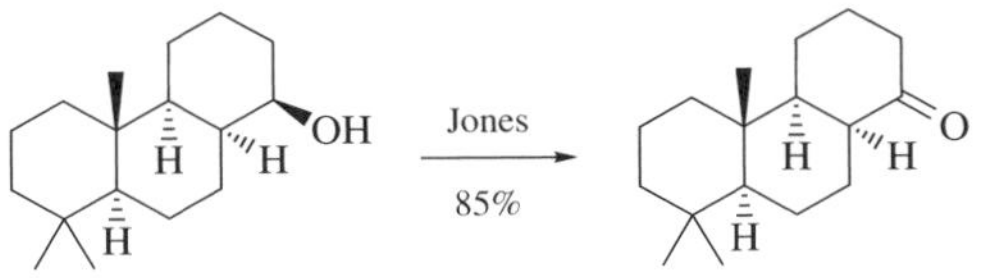

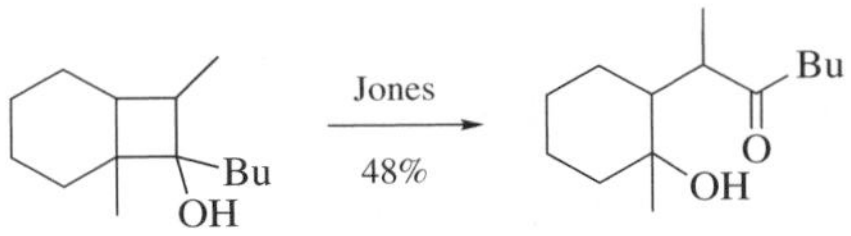

Oxidation of α,β-Unsaturated Alcohols. In general, allylic alcohols are easily converted to the corresponding α,β-unsaturated ketones. Ethyl 3-hydroxy-4-pentenoate (**7**) is oxidized to the volatile ketone ethyl 3-oxo-4-pentenoate (**8**) (eq 7),[11] which is an intermediate in the preparation of the Nazarov annulation reagent.[12] The Jones reagent was superior to ***Pyridinium Chlorochromate***, ***Pyridinium Dichromate***, and ***Dimethyl Sulfoxide–Oxalyl Chloride*** (Swern reagent) for the oxidation of this alcohol.

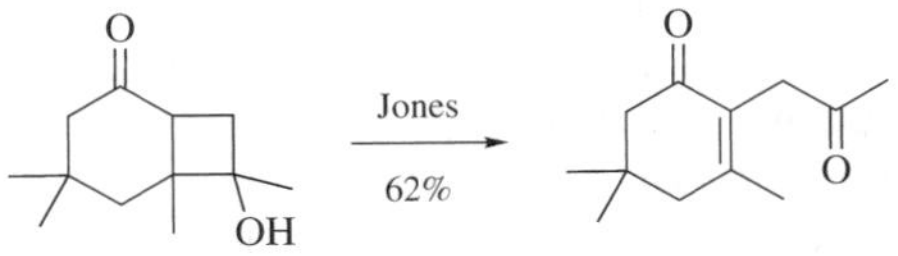

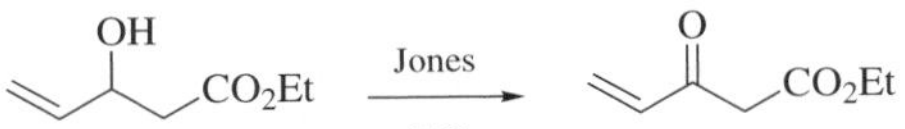

The steroidal lactone (**9**) and its isomeric allylic alcohol (**10**) are stereospecifically oxidized to the epoxy ketone (**11**) by the Jones reagent (eq 8).[13] Epoxidation occurs only when the hydroxyl group is axial and epoxidation is faster than oxidation of the hydroxyl group.

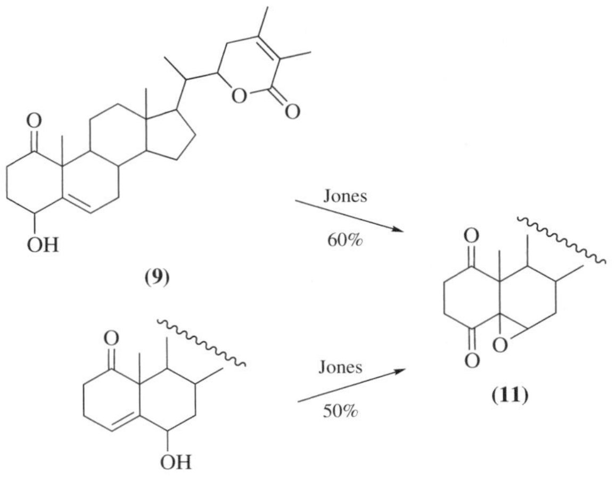

Oxidation of 6-hydroxy-2-(4-methoxyphenyl)-2-methyl-2*H*-pyran-3(6*H*)-ones (**12**) (eq 9) with Jones reagent affords diones (**13**), which are intermediates in the synthesis of 5-substituted 2-pyrrolidinones[14] and 1-oxaspiro[5.5]undecane derivatives.[15] The pyrrolidinone (γ-lactam) skeleton is found in molecules with great value in medicinal chemistry, and spiroacetals are subunits in numerous important natural products.

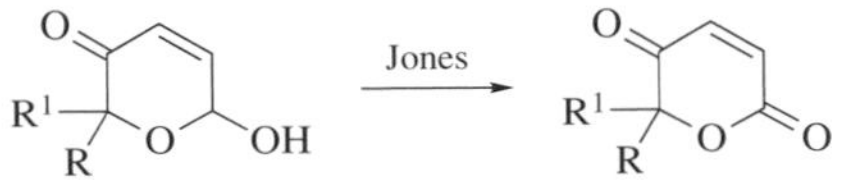

Jones oxidation of phosphorus-containing tertiary allylic alcohols such as (**14**) led to formation of the 1,3-carbonyl transposed β-substituted α,β-unsaturated ketones (**15**) (eq 10), which are useful intermediates for a variety of phosphorus-substituted and nonphosphorus-bearing heterocyclic systems.[16]

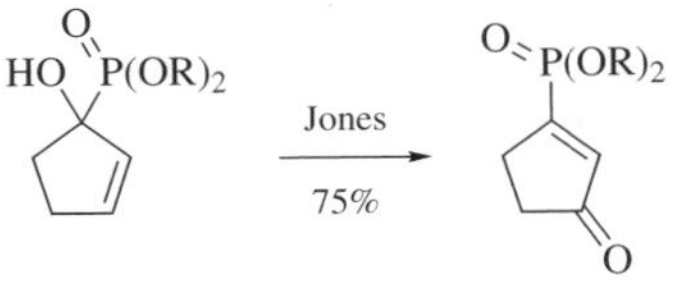

Alkynic ketones are prepared in 40–80% yield via the Jones oxidation of the corresponding secondary alcohols.[2] Alkynic glycols are oxidized to the diketones with Jones reagent.[2]

Cleavage of *s*,*t*-1,2-Glycols to Keto Acids. Jones reagent oxidatively cleaves *s*,*t*-1,2-glycol (**16**) to keto acid (**17**) (eq 11), and (**18**) or (**19**) to (**21**) (eq 12).[17]

Oxidation of Benzoins to Benzils. Benzoins are oxidized to benzils (90–95%) with Jones reagent in acetone.[18]

Oxidation of Primary Alcohols. Primary alcohols are rapidly oxidized by the Jones reagent to carboxylic acids. Double bonds and triple bonds are not oxidized.[19,20] Jones oxidation of alcohol (**22**) affords carboxylic acid (**23**), which is used in the synthesis of the antitumor agent (±)-sarkomycin (eq 13).[21]

(16) → Jones → (17) (11)

(18) → Jones 97% → (20) → Jones → (12)

(19) → Jones 85% → (21)

(22) → Jones → (23) (13)

Other Applications. Although ethers are generally inert toward chromic acid oxidation, the Jones reagent oxidizes benzyl ethers to the benzoic acid, esters, and ketones (eq 14).[22] The bicyclic acetoxy ether (**24**) is oxidized to the acetoxylactone (**25**) with Jones reagent at room temperature (eq 15).[23]

Jones → 79% + $PhCO_2H$ 61% + (14)

(24) → Jones 89% → (25) (15)

Oxidation of the alcohol (**26**) with Jones reagent gives the δ-lactone (**27**) instead of the ketone (eq 16).[24] The *p*-methoxy group is necessary for this oxidation of a methyl group.

(26) → Jones 33% → (27) (16)

Phenols substituted by at least one alkyl group in the *ortho* position are oxidized to *p*-quinones by a two-phase (ether/aqueous CrO_3) Jones reagent.[25] Chromic acid in 50% aqueous acetic acid, Collins reagent, and Jones reagent oxidatively deoximate ketoximes to the corresponding carbonyl compounds.[26] Jones reagent oxidizes 3-trimethylsilyl-3-buten-2-ol to the Michael acceptor 3-methylsilyl-3-buten-2-one.[27] A combination of a catalytic amount of ***Osmium Tetroxide*** and stoichiometric Jones reagent in acetone at room temperature oxidizes various types of alkenes into acids and/or ketones.[28]

Related Reagents. See Classes O-1, O-2, O-3 and O-18, pages 1–10.

---

1. (a) Wiberg, K. B. In *Oxidation in Organic Chemistry*; Wiberg, K. B., Ed; Academic: New York, 1965; Part A, Chapter 2. (b) Freeman, F. In *Organic Synthesis by Oxidation with Metal Compounds*; Miijs, W. J.; de Jonge, C. R. H. I., Ed; Plenum: New York, 1986; Chapter 2. (c) Lee, D. G. *The Oxidation of Organic Compounds by Permanganate Ion and Hexavalent Chromium*; Open Court: La Salle, IL, 1980. (d) Stewart, R. *Oxidation Mechanisms: Applications to Organic Chemistry*; Benjamin: New York, 1964. (e) Cainelli, G.; Cardillo, G. *Chromium Oxidations in Organic Chemistry*; Springer: Berlin, 1984. (f) Cupo, D. Y.; Wetterhahn, K. E. *Cancer Res.* **1985**, *45*, 1146 and references cited therein.
2. Bowden, K.; Heilbron, I. M.; Jones, E. R. H.; Weedon, B. C. L. *JCS* **1946**, 39.
3. Bowers, A.; Halsall, T. G.; Jones, E. R. H.; Lemin, A. J. *JCS* **1953**, 2548.
4. (a) Eisenbraun, E. J. *OSC* **1973**, *5*, 310. (b) Zibuck, R.; Streiber, J. *OS* **1993**, *71*, 236.
5. Meinwald, J.; Crandall, J.; Hymans, W. E. *OSC* **1973**, *5*, 866.
6. Salomon, R. G.; Ghosh, S. *OSC* **1990**, *7*, 177.
7. Church, R. F.; Ireland, R. E. *TL* **1961**, 493.
8. Djerassi, C.; Engle, R. R.; Bowers, A. *JOC* **1956**, *21*, 1547.
9. Müller, P.; Blanc, J. *HCA* **1979**, *62*, 1980.
10. Liu, H.-J. *CJC* **1976**, *54*, 3113.
11. Zibuck, R.; Streiber, J. M. *JOC* **1989**, *54*, 4717.
12. Nazarov, I. N.; Zavyalov, S. I. *Zh. Obshch. Khim.* **1953**, *23*, 1703; *Engl. Transl.* **1953**, *23*, 1793 (*CA* **1954**, *48*, 13 667).
13. Glotter, E.; Greenfield, S.; Lavie, D. *CC* **1968**, 1646.
14. Georgiadis, M. P.; Haroutounian, S. A.; Apostolopoulos, C. D. *S* **1991**, 379.
15. Georgiadis, M. P.; Tsekouras, A.; Kotretsou, S. I.; Haroutounian, S. A.; Polissiou, M. G. *S* **1991**, 929.
16. Öhler, E.; Zbiral, E. *S* **1991**, 357.
17. de A. Epifanio, R.; Camargo, W.; Pinto, A. C. *TL* **1988**, *29*, 6403.
18. Ho, T.-L. *CI(L)* **1972**, 807.
19. Rodin, J. O.; Leaffer, M. A.; Silverstein, R. M. *JOC* **1970**, *35*, 3152.
20. Cardillo, G.; Contento, M.; Sandri, S.; Panunzio, M. *JCS(P1)* **1979**, 1729.
21. Liu, Z.-Y., Shi, W.; Zhang, L. *S* **1990**, 235.

22. Bal, B. S.; Kochhar, K. S.; Pinnick, H. W. *JOC* **1981**, *46*, 1492.
23. Henbest, H. B.; Nicholls, B. *JCS* **1959**, 221.
24. Jones, R. A.; Saville, J. F.; Turner, S. *CC* **1976**, 231.
25. Liotta, D.; Arbiser, J.; Short, J. W.; Saindane, M. *JOC* **1983**, *48*, 2932.
26. Araújo, H. C.; Ferreira, G. A. L.; Mahajan, J. R. *JCS(P1)* **1974**, 2257.
27. Boeckman, R. K. Jr.; Blum, D. M.; Ganem, B.; Halvey, N. *OSC* **1988**, *6*, 1033.
28. Henry, J. R.; Weinreb, S. M. *JOC* **1993**, *58*, 4745.

Fillmore Freeman
University of California, Irvine, CA, USA

Chromium(II) Chloride

[10049-05-5] Cl_2Cr (MW 122.92)

(reducing agent for dehalogenation of organic halides, especially allylic and benzylic halides, and for transformation of carbon–carbon triple bonds leading to (*E*)-alkenes; conversion of dibromocyclopropanes to allenes; preparation and reaction of allylic chromium reagents; reduction of sulfur- or nitrogen-substituted alkyl halides to give hetero-substituted alkylchromium reagents)

Physical Data: mp 824 °C; d_4^{14} 2.751 g cm^{-3}.
Form Supplied in: off-white solid; commercially available.
Solubility: sol water, giving a blue solution; insol alcohol or ether.
Handling, Storage, and Precautions: very hygroscopic; oxidizes rapidly, especially under moist conditions; should be handled in a fume hood under an inert atmosphere (argon or nitrogen).

Reduction of Alkyl Halides.[1,2] Typically, the chromium(II) ion is prepared by reduction of chromium(III) salts with zinc and hydrochloric acid. Organochromium compounds produced in this way can subsequently be hydrolyzed to yield dehalogenated compounds (eq 1[3] and eq 2[4]). Anhydrous chromium(II) chloride is commercially available and can be used without further purification. The relative reactivities of various types of halide toward chromium(II) salts are shown in Scheme 1.

$$ClCH_2CO_2H \xrightarrow[HCl,\ H_2O]{CrCl_3,\ Zn} MeCO_2H \quad (1)$$

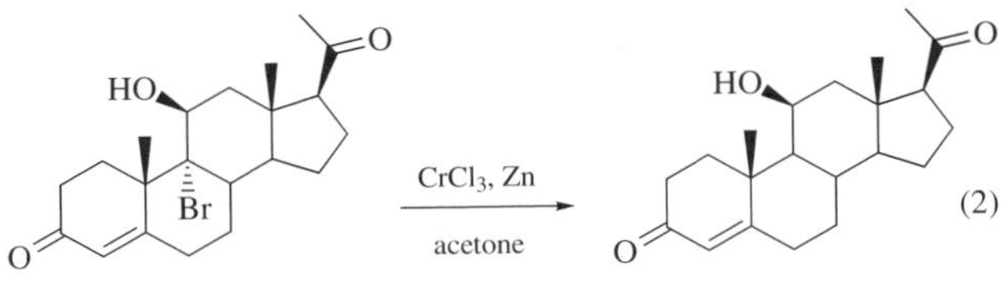

CrCl₃, Zn / acetone (2)

Conversion of Dihalocyclopropanes to Allenes. Reduction of geminal dihalides proceeds smoothly to give chromium carbenoids.[5] In the case of 1,1-dibromocyclopropanes, the intermediate carbenoids decompose instantaneously to give allenes (eq 3).[6,7]

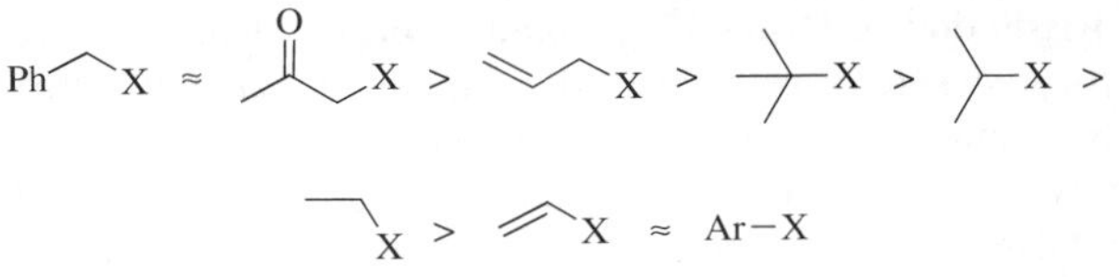

PhCH₂X ≈ MeCOCH₂X > allyl–X > t-Bu–X > i-Pr–X > n-alkyl–X > vinyl–X ≈ Ar–X

X = I > Br > Cl

Scheme 1

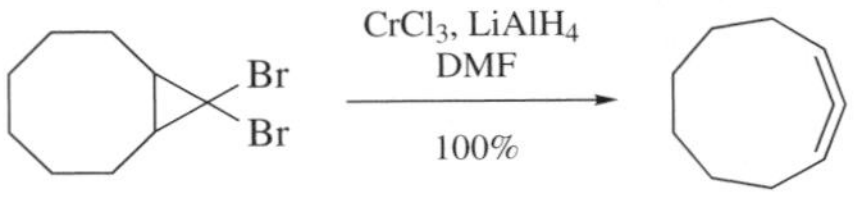

$CrCl_3$, $LiAlH_4$, DMF, 100% (3)

Reductive Coupling of Allylic and Benzylic Halides. Active halides, such as allyl and benzyl halides, are reduced with $CrCl_2$ smoothly to furnish homocoupling products. Allylic halides undergo coupling, forming mainly the head-to-head dimer (eq 4).[6,8]

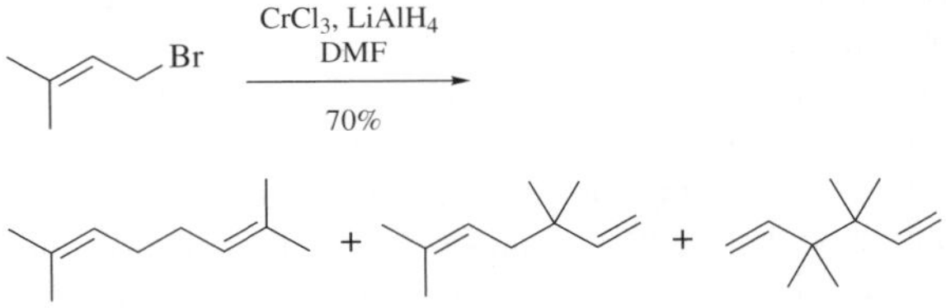

$CrCl_3$, $LiAlH_4$, DMF, 70% (4)

72:22:6

Formation of *o*-Quinodimethanes. α,α′-Dibrom o-*o*-xylenes are reduced with $CrCl_2$ in a mixed solvent of THF and HMPA to an *o*-quinodimethane, which can be trapped by a dienophile. The method has been applied to some anthracycline precursors (eq 5).[9]

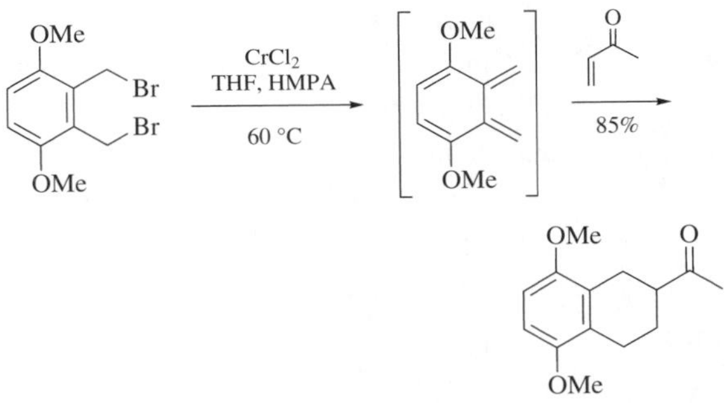

$CrCl_2$, THF, HMPA, 60 °C; 85% (5)

Reduction of Carbon–Carbon Unsaturated Bonds. The reduction of alkynes with chromium(II) salts in DMF leads to (*E*)-substituted alkenes.[10] The ease of reduction depends on the presence of an accessible coordination site in the molecule (eq 6). Chromium(II) chloride in THF/H_2O (2:1) (or ***Chromium(II) Sulfate*** in DMF/H_2O) is effective at reducing α-alkynic ketones to (*E*)-enones. Less than 2% of (*Z*)-enones are produced except in the case of highly substituted substrates, which also require longer reaction times.[11]

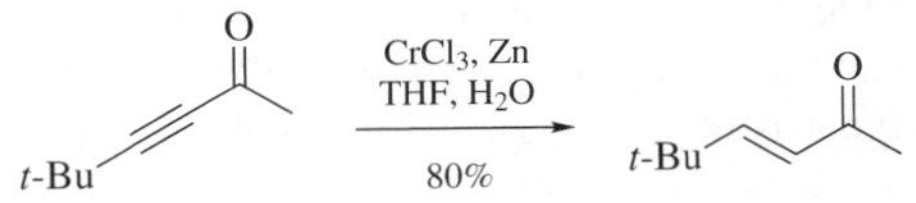

$CrCl_3$, Zn, THF, H_2O, 80% (6)

Reduction of Other Functional Groups. Deoxygenation of α,β-epoxy ketones proceeds with acidic solutions of $CrCl_2$ to form α,β-unsaturated ketones.[12] Chromium(II) chloride has been regularly used in the deoxygenation of the limonoid group of triterpenes, in which ring D bears an α,β-epoxy-δ-lactone.[13] Treatment of nitrobenzene derivatives with $CrCl_2$ in methanol under reflux gives anilines (eq 7), while aliphatic nitro compounds afford aldehydes under the same reaction conditions.[14] Reduction of a nitroalkene with acidic solutions of $CrCl_2$ resulted in the formation of an α-hydroxy oxime.[15]

$CrCl_3$, Zn, MeOH, H_2O, reflux; 85% (7)

Preparation of Allylic Chromium Reagents.[16] Allylic halides are reduced with low-valent chromium ($CrCl_3$–$LiAlH_4$) or $CrCl_2$ to give the corresponding allylic chromium reagents, which add to aldehydes and ketones in good to excellent yields.[17] The electronegativity of chromium is 1.6, almost the same as that of titanium (1.5). Therefore the nucleophilicity of organochromium reagents is not strong compared to the corresponding organolithium or -magnesium compounds. Chemoselective addition of allylic chromium reagents can be accomplished without affecting coexisting ketone and cyano groups (eq 8).[17] The reaction between crotylchromium reagents and aldehydes in THF proceeds with high diastereoselectivity (eq 9).[18,19] The *anti* (or *threo*) selectivity in the addition of acyclic allylic chromium reagents with aldehydes is explained by a chair-form six-membered transition state in which both R^1 and R^2 possess equatorial positions (**1** > **2**).

$CrCl_3$, $LiAlH_4$ or $CrCl_2$; THF; 66% (8)

PhCHO + crotyl bromide; $CrCl_2$ or $CrCl_3$, $LiAlH_4$; THF, 20 °C; 96–100% (9)

(**1**) (**2**)

Addition of crotylchromium reagents to aldehydes bearing a stereogenic center α to the carbonyl provides three of the four diastereomers (eq 10).[19] Excellent *anti* selectivity is observed with respect to the 1,2-positions, but the stereoselectivity with respect to the 2,3-positions (Cram/anti-Cram ratio) is poor.[20–22] High 1,2- and 2,3-diastereoselectivity is obtained with aldehydes having large substituents, especially a cyclic acetal group, on the α-carbon of the aldehyde (eq 11).[20,23] Reaction between chirally substituted acyclic allylic bromides and aldehydes proceeds with high stereocontrol (eq 12).[24]

THF (10)

$CrCl_2$ 90%, 62:31:7
$CrCl_3$, $LiAlH_4$ 78%, 64:29:7

$CrCl_3$, $LiAlH_4$; THF, rt; 86%; + isomers (11)

>20:1

$CrCl_3$, $LiAlH_4$; THF; 63% (12)

As with allylic halides, allylic diethylphosphates[25] and mesylates[26,27] are reduced with chromium(II) salts to give allylic chromium reagents which add to aldehydes regio- and stereoselectively. This transformation reveals conversion of the electronic nature of allylic phosphates (or mesylates) from electrophilic to nucleophilic by reduction with low-valent chromium.

The reaction of γ-disubstituted allylic phosphates with aldehydes mediated by $CrCl_2$ and a catalytic amount of LiI in DMPU is not stereoconvergent and proceeds with high stereoselectivity (eqs 13 and 14).[27] The presence of the two substituents at the γ-position slows down the process of equilibration between the intermediate allylic chromium reagents.

$C_5H_{11}CHO$, $CrCl_2$, LiI, DMPU; 93% (13)

94:6

$C_5H_{11}CHO$, $CrCl_2$, LiI, DMPU; 94% (14)

99:1

Because the coupling reaction between allylic halides and aldehydes proceeds under mild conditions, the reaction has been employed, in particular, in intramolecular cyclizations.[28–32] The intramolecular version also proceeds with high *anti* selectivity (eqs 15 and 16).

$CrCl_2$, THF, 25 °C
64%

4:1 (15)

$CrCl_2$, THF
67–76%

+ *trans* isomer 10–12% (16)

Functionalized and Hetero-Substituted Allylic Chromium Reagents. When functionalized allylic halides are employed as precursors of allylic chromium reagents, an acyclic skeleton bearing a foothold for further construction is produced. Reaction of α-bromomethyl-α,β-unsaturated esters with aldehydes mediated by $CrCl_2$ (or $CrCl_3$–$LiAlH_4$) affords homoallylic alcohols, which cyclize to yield α-methylene-γ-lactones in a stereoselective manner (eq 17).[33] Reaction between α-bromomethyl-α,β-un saturated sulfonates and aldehydes also proceeds with high stereocontrol.[34]

PhCHO + Bu(Br)C=CCO_2Et — $CrCl_3$, $LiAlH_4$, THF, 25 °C, 86% (17)

The reaction of 3-alkyl-1,1-dichloro-2-propene with $CrCl_2$ results in α-chloroalkylchromium reagents, which react with aldehydes to produce a 2-substituted *anti*-(*Z*)-4-chloro-3-buten-1-ol in a regio- and stereoselective manner (eq 18).[35] Vinyl-substituted β-hydroxy carbanion synthons are produced by reduction of 1,3-diene monoepoxides with $CrCl_2$ in the presence of LiI, which react stereoselectively with aldehydes to give (R^*,R^*)-1,3-diols having a quaternary center at C-2.[36] Reduction in situ of acrolein dialkyl acetals with $CrCl_2$ in THF provides γ-alkoxy-substituted allylic chromium reagents which add to aldehydes at the same position of the alkoxy group to afford 3,4-butene-1,2-diol derivatives. The reaction rate and stereoselectivity are increased by addition of ***Iodotrimethylsilane*** (eq 19).[37]

mixture
$CrCl_2$, THF, DMF
87% (18)

$CrCl_2$, TMSI
THF, –30 °C
99%

88:12 (19)

Preparation of Propargylic Chromium Reagents. Propargyl halides react with aldehydes or ketones in the presence of $CrCl_2$ with HMPA as cosolvent to give allenes stereoselectively (eq 20).[38] The reaction was modified to include polyfunctional propargylic halides by using $CrCl_2$ and ***Lithium Iodide*** in DMA, and allenic alcohols accompanied only by small amounts of homopropargylic alcohols are produced.[39]

$CrCl_3$, $LiAlH_4$
THF, 25 °C
80% (20)

Sulfur- and Nitrogen-Stabilized Alkylchromium Reagents. In combination with LiI, $CrCl_2$ reduces α-halo sulfides to (α-alkylthio)chromium compounds, which undergo selective 1,2-addition to aldehydes. Acetophenone is recovered unchanged under the reaction conditions. The (1-phenylthio)ethenylchromium reagents prepared in this way add to aldehydes under high stereocontrol in the presence of suitable ligands like 1,2-diphenylphosphinoethane (dppe) (eq 21).[40] The reaction of *N*-(chloromethyl)succinimide and -phthalimide with $CrCl_2$ provides the corresponding α-nitrogen-substituted organochromium reagents in the presence of LiI. These organochromium reagents react in situ with aldehydes, affording protected amino alcohols (eq 22).[41]

C_8H_{17}—CHO + $CrCl_2$, dppe, LiI, THF, 25 °C, 53%

>98:<2 (21)

$$\text{N-(chloromethyl)phthalimide} + \text{OHC-C}_6\text{H}_{11} \xrightarrow[\text{95\%}]{\text{CrCl}_2\text{, LiI, THF, 50 °C}} \text{PhthN-CH}_2\text{-CH(OH)-C}_6\text{H}_{11} \quad (22)$$

Preparation of Alkylchromium Reagents.[42] With the assistance of catalytic amounts of ***Vitamin B$_{12}$*** or cobalt phthalocyanine (CoPc), $CrCl_2$ reduces alkyl halides, especially 1-iodoalkanes, to form alkylchromium reagents which add to aldehydes without affecting ketone or ester groups. The chemoselective preparation of organochromium reagents can be done by changing either the catalyst or the solvent. Alkenyl and alkyl halides remain unchanged under the conditions of the preparation of allylchromium reagents; on the other hand, alkenyl- and alkylchromium reagents are produced selectively under nickel and cobalt catalysis, respectively (eqs 23 and 24).

$$\text{CH}_2\text{=CHCH}_2\text{Br} + \text{C}_{12}\text{H}_{25}\text{-Cl} + \text{C}_9\text{H}_{19}\text{C(=CH}_2\text{)Br} \xrightarrow[\text{83\%}]{\text{PhCHO, CrCl}_2\text{, THF, 25 °C}} \text{CH}_2\text{=CHCH}_2\text{CH(OH)Ph} + \underset{\text{(recov 97\%)}}{\text{C}_{12}\text{H}_{25}\text{-Cl}} + \underset{\text{(recov 97\%)}}{\text{C}_9\text{H}_{19}\text{C(=CH}_2\text{)Br}} \quad (23)$$

$$\text{C}_{12}\text{H}_{25}\text{-Cl} + \text{C}_9\text{H}_{19}\text{C(=CH}_2\text{)Br} \xrightarrow[\text{68\%}]{\text{PhCHO, CrCl}_2\text{, cat B}_{12}\text{, DMF, 30 °C}} \text{C}_{12}\text{H}_{25}\text{CH(OH)Ph} + \underset{\text{(recov 100\%)}}{\text{C}_9\text{H}_{19}\text{C(=CH}_2\text{)Br}}$$

$$\text{C}_{12}\text{H}_{25}\text{-Cl} + \text{C}_9\text{H}_{19}\text{C(=CH}_2\text{)Br} \xrightarrow[\text{82\%}]{\text{PhCHO, CrCl}_2\text{, cat NiCl}_2\text{, DMF, 25 °C}} \underset{\text{(recov 96\%)}}{\text{C}_{12}\text{H}_{25}\text{-Cl}} + \text{C}_9\text{H}_{19}\text{C(=CH}_2\text{)CH(OH)Ph} \quad (24)$$

Related Reagents. See Classes R-4, R-21, R-23, R-24, R-25, and R-27, pages 1–10. Chromium(II) Chloride–Haloform; Chromium(II) Chloride–Nickel(II) Chloride.

1. (a) Hanson, J. R.; Premuzic, E. *AG(E)* **1968**, *7*, 247; (b) Hanson, J. R. *S* **1974**, 1.
2. (a) Castro, C. E.; Kray, W. C., Jr. *JACS* **1963**, *85*, 2768. (b) Kray, W. C., Jr.; Castro, C. E. *JACS* **1964**, *86*, 4603. (c) Kochi, J. K.; Singleton, D. M.; Andrews, L. J. *T* **1968**, *24*, 3503.
3. Traube, W.; Lange, W. *CB* **1925**, *58*, 2773.
4. Beereboom, J. J.; Djerassi, C.; Ginsburg, D.; Fieser, L. F. *JACS* **1953**, *75*, 3500.
5. Castro, C. E.; Kray, W. C., Jr. *JACS* **1966**, *88*, 4447.
6. Okude, Y.; Hiyama, T.; Nozaki, H. *TL* **1977**, 3829.
7. Wolf, R.; Steckhan, E. *J. Electroanal. Chem.* **1981**, *130*, 367.
8. Sustmann, R.; Altevogt, R. *TL* **1981**, *22*, 5167.
9. Stephan, D.; Gorgues, A.; Le Coq, A. *TL* **1984**, *25*, 5649.
10. Castro, C. E.; Stephens, R. D. *JACS* **1964**, *86*, 4358.
11. Smith, A. B., III; Levenberg, P. A.; Suits, J. Z. *S* **1986**, 184.
12. Cole, W.; Julian, P. L. *JOC* **1954**, *19*, 131.
13. (a) Arigoni, D.; Barton, D. H. R.; Corey, E. J.; Jeger, O.; Caglioti, L.; Dev, S.; Ferrini, P. G.; Glazier, E. R.; Melera, A.; Pradhan, S. K.; Slhaffner, K.; Sternhell, S.; Templeton, J. F.; Tobinaga, S. *Experientia* **1960**, *16*, 41. (b) Akisanya, A.; Bevan, C. W. L.; Halsall, T. G.; Powell, J. W.; Taylor, D. A. H. *JCS* **1961**, 3705. (c) Ekong, D. E. U.; Olagbemi, O. E. *JCS(C)* **1966**, 944.
14. Akita, Y.; Inaba, M.; Uchida, H.; Ohta, A. *S* **1977**, 792.
15. (a) Hanson, J. R.; Premuzic, E. *TL* **1966**, 5441; (b) Rao, T. S.; Mathur, H. H.; Trivedi, G. K. *TL* **1984**, *25*, 5561.
16. Cintas, P. *S* **1992**, 248.
17. (a) Okude, Y.; Hirano, S.; Hiyama, T.; Nozaki, H. *JACS* **1977**, *99*, 3179. (b) Hiyama, T.; Okude, Y.; Kimura, K.; Nozaki, H. *BCJ* **1982**, *55*, 561.
18. Buse, C. T.; Heathcock, C. H. *TL* **1978**, 1685.
19. Hiyama, T.; Kimura, K.; Nozaki, H. *TL* **1981**, *22*, 1037.
20. (a) Nagaoka, H.; Kishi, Y. *T* **1981**, *37*, 3873. (b) Lewis, M. D.; Kishi, Y. *TL* **1982**, *23*, 2343.
21. Fronza, G.; Fganti, C.; Grasselli, P.; Pedrocchi-Fantoni, G.; Zirotti, C. *CL* **1984**, 335.
22. Evans, D. A.; Dow, R. L.; Shih, T. L.; Takacs, J. M.; Zahler, R. *JACS* **1990**, *112*, 5290.
23. Roush, W. R.; Palkowitz, A. D. *JOC* **1989**, *54*, 3009.
24. (a) Mulzer, J.; de Lasalle, P.; Freiler, A. *LA* **1986**, 1152. (b) Mulzer, J.; Kattner, L. *AG(E)* **1990**, *29*, 679. (c) Mulzer, J.; Kattner, L.; Strecker, A. R.; Schröder, C.; Buschmann, J.; Lehmann, C.; Luger, P. *JACS* **1991**, *113*, 4218.
25. Takai, K.; Utimoto, K. *J. Synth. Org. Chem. Jpn.* **1988**, *46*, 66.
26. Kato, N.; Tanaka, S.; Takeshita, H. *BCJ* **1988**, *61*, 3231.
27. Jubert, C.; Nowotny, S.; Kornemann, D.; Antes, I.; Tucker, C. E.; Knochel, P. *JOC* **1992**, *57*, 6384.
28. Still, W. C.; Mobilio, D. *JOC* **1983**, *48*, 4785.
29. Shibuya, H.; Ohashi, K.; Kawashima, K.; Hori, K.; Murakami, N.; Kitagawa, I. *CL* **1986**, 85.
30. Kato, N.; Tanaka, S.; Takeshita, H. *CL* **1986**, 1989.
31. Wender, P. A.; McKinney, J. A.; Mukai, C. *JACS* **1990**, *112*, 5369.
32. (a) Paquette, L. A.; Doherty, A. M.; Rayner, C. M. *JACS* **1992**, *114*, 3910. (b) Rayner, C. M.; Astles, P. C.; Paquette, L. A. *JACS* **1992**, *114*, 3926. (c) Paquette, L. A.; Astles, P. C. *JOC* **1993**, *58*, 165.
33. (a) Okuda, Y.; Nakatsukasa, S.; Oshima, Y.; Nozaki, H. *CL* **1985**, 481. (b) Drewes, S. E.; Hoole, R. F. A. *SC* **1985**, *15*, 1067.
34. (a) Auvray, P.; Knochel, P.; Normant, J. F. *TL* **1986**, *27*, 5091. (b) Auvray, P.; Knochel, P.; Vaissermann, J.; Normant, J. F. *BSF* **1990**, *127*, 813.
35. (a) Takai, K.; Kataoka, Y.; Utimoto, K. *TL* **1989**, *30*, 4389. (b) Wender, P. A.; Grissom, J. W.; Hoffmann, U.; Mah, R. *TL* **1990**, *31*, 6605. (c) Augé, J. *TL* **1988**, *29*, 6107.
36. Fujimura, O.; Takai, K.; Utimoto, K. *JOC* **1990**, *55*, 1705.
37. (a) Takai, K.; Nitta, K.; Utimoto, K. *TL* **1988**, *29*, 5263. (b) Roush, W. R.; Bannister, T. D. *TL* **1992**, *33*, 3587.
38. (a) Place, P.; Delbecq, F.; Gore, J. *TL* **1978**, 3801. (b) Place, P.; Venière, C.; Gore, J. *T* **1981**, *37*, 1359.
39. Belyk, K.; Rozema, M. J.; Knochel, P. *JOC* **1992**, *57*, 4070.
40. Nakatsukasa, S.; Takai, K.; Utimoto, K. *JOC* **1986**, *51*, 5045.
41. Knochel, P.; Chou, T.-S.; Jubert, C.; Rajagopal, D. *JOC* **1993**, *58*, 588.
42. Takai, K.; Nitta, K.; Fujimura, O.; Utimoto, K. *JOC* **1989**, *54*, 4732.

Kazuhiko Takai
Okayama University, Japan

Chromium(VI) Oxide[1]

CrO_3

[1333-82-0] CrO_3 (MW 99.99)

(reagent for oxidizing carbon–hydrogen bonds to alcohols, oxidizing alkylaromatics to ketones and carboxylic acids, converting alkenes to α,β-unsaturated ketones, oxidizing carbon–carbon double bonds, oxidizing arenes to quinones, oxidizing alcohols to aldehydes, ketones, acids, and keto acids)

Alternate Names: chromic anhydride; ***Chromic Acid*** in aqueous media.
Physical Data: mp 196 °C; d 2.70 g cm^{-3}.
Solubility: sol ether, H_2O, HNO_3, H_2SO_4, DMF, HMPA.
Form Supplied in: red crystals.
Handling, Storage, and Precautions: **caution**: chromium(VI) oxide is a highly toxic cancer suspect agent. All chromium(VI) reagents must be handled with care. The mutagenicity of Cr^{VI} compounds is well documented.[7] HMPA is also a highly toxic cancer suspect agent. Special care must always be exercised in adding CrO_3 to organic solvents. Add CrO_3 in small portions to HMPA in order to avoid a violent decomposition. This reagent must be handled in a fume hood.

Each mol of chromium(VI) oxide has 1.5 equivalents of oxygen. The oxidizing power of the reagent increases with decreasing water content of the solvent medium. The oxidizing medium may be aqueous acetic acid,[2,3] anhydrous acetic acid (Fieser reagent),[4] or concentrated[5] or aqueous[6] sulfuric acid.

Oxidation of Carbon–Hydrogen Bonds to Alcohols. Chromium(VI) oxide in 91% acetic acid oxidizes the methine hydrogen of (+)-3-methylheptane to (+)-3-methyl-3-heptanol with 70–85% retention of configuration.[8] 3β-Acetoxy-14α-hydroxyandrost-5-en-17-one is obtained by direct introduction of an α-hydroxyl group at C-14 in the dibromide of 3β-acetoxyandrost-5-en-17-one (eq 1).[9]

1. Br_2; 2. CrO_3, AcOH; 3. Zn; 44% (1)

Oxidation of Alkylaromatics to Ketones and Carboxylic Acids. Chromium(VI) oxide in concentrated sulfuric acid oxidizes 3,4-dinitrotoluene to 3,4-dinitrobenzoic acid (89%).[5] Under milder conditions, with longer alkyl chains, the benzylic position is converted to carbonyl. Chromium(VI) oxide in acetic acid oxidizes ethylbenzene to acetophenone and benzoic acid. More rigorous oxidizing experimental conditions convert longer chain alkyl groups to carboxyl, thus yielding benzoic acid or its derivatives. Methylene groups between two benzene rings are oxidized to carbonyl derivatives in preference to reaction at alkyl side chains.[10]

Indans are oxidized to 1-indanones by use of a dilute (10%) solution of chromium(VI) oxide in acetic acid at room temperature (eq 2).[11]

CrO_3, AcOH; 87–92% (2)

Allylic Oxidations. Allylic oxidations may be complicated by carbonyl formation at either one or both allylic positions. Although chromium(VI) oxide appears to be useful for allylic oxidation in steroid chemistry, better results may be obtained in other systems with ***Di-t-butyl Chromate*** or ***Dipyridine Chromium(VI) Oxide*** (Collins reagent). However, the ***Chromium(VI) Oxide–3,5-Dimethylpyrazole*** complex (CrO_3·DMP) is useful for allylic oxidations. The complex oxidized the allylic methylene group in (**1**) to the α,β-unsaturated ketone (**2**) which was used in the synthesis of the antibacterial helenanolide (+)-carpesiolin (eq 3).[12] Chromium(VI) oxide in glacial acetic acid oxidizes 3,21-diacetoxy-4,4,14-trimethyl-Δ^8-5-pregnene to the enetrione (eq 4).[13] Complex product mixtures are formed when epoxidation competes with the allylic oxidation.[14]

CrO_3•DMP, CH_2Cl_2; 60% (3)

(**1**) (**2**)

CrO_3, AcOH; 75% (4)

Oxidation of Carbon–Carbon Double Bonds. Chromium(VI) oxide in aqueous sulfuric acid generally cleaves carbon–carbon double bonds. Rearrangements may further complicate the oxidation. In anhydrous acetic acid, chromium(VI) oxide oxidizes tetraphenylethylene to the oxirane (70% yield) and benzophenone (11%).[15] The yield is lower and more double bond cleavage occurs in aqueous acetic acid. However, use of acetic anhydride as solvent (see ***Chromyl Acetate***) affords the oxiranes from tri- and tetrasubstituted alkenes in 50–88% yields, along with benzopinacols.[16,17] Many steroidal and terpenic cyclic alkenes react with chromium(VI) oxide in acetic acid to give oxiranes, and saturated, α,β-unsaturated, α-hydroxy, and α,β-epoxy ketones which arise from the initially formed oxirane.[18,19] A synthetically useful cleavage of double bonds involving chromium(VI) oxide is the Meystre–Miescher–Wettstein degradation[20] which shortens the side chain of a carboxylic acid by three atoms at one time. This procedure is a modification of the Barbier–Wieland degradation.[21,22]

Oxidation of Arenes to Quinones. In contrast to alkylaromatics, which undergo oxidation at the side chain with some chromium(VI) oxidants, polynuclear aromatic arenes undergo ring oxidation to quinones with chromium(VI) oxide. This chemoselectivity is shown in the chromium(VI) oxide in anhydrous acetic acid (Fieser reagent) oxidation of 2,3-dimethylnaphthalene to 2,3-dimethylnaphthoquinone in quantitative yield (eq 5).[23] In some cases, depending on experimental conditions, both benzylic and ring oxidations occur[24] or the alkyl groups may be eliminated (eq 6).[25] The oxidation of anthracene derivatives is important in the total synthesis of anthracycline antibiotics.[26,27]

CrO_3, AcOH, 100% (5)

Et; CrO_3, AcOH (6)

Oxidation of Alcohols to Aldehydes, Ketones, Acids, and Keto Acids. Chromium(VI) oxide in acetic acid oxidizes primary alcohols to aldehydes and acids, and secondary alcohols to ketones and keto acids (Fieser reagent) (eq 7).[28] Chromium(VI) oxide in water or aqueous acetic acid oxidizes primary alcohols to carboxylic acids.[29,30] Chromium(VI) oxide–***Hexamethylphosphoric Triamide*** (CrO_3·HMPA) selectively oxidized the primary hydroxyl group of strophanthidol (**3**) to an aldehyde group in the final step in the synthesis of strophanthidin (**4**) (eq 8).[31] The CrO_3·HMPA complex oxidizes saturated primary alcohols to aldehydes in about 80% yield.[32,33] The yields are lower with secondary alcohols and highest with α,β-unsaturated primary and secondary alcohols. It is possible to selectively oxidize certain allylic and benzylic hydroxyl groups in the presence of other unprotected saturated groups (eq 9; cf eq 8). Chromium(VI) oxide in DMF in the presence of catalytic amounts of sulfuric acid oxidizes steroidal alcohols to ketones.[34] Chromium(VI) oxide on graphite selectively oxidizes primary alcohols in the presence of secondary and tertiary alcohols.[35]

Ph OH; CrO_3, AcOH, 81%; Ph, O, CO_2H (7)

HO, OH, HO (**3**); CrO_3•HMPA, 35%; O, OH, HO (**4**) (8)

OH, HO; CrO_3•HMPA, 80%; O, HO (9)

Other Applications. Chromium(VI) oxide in aqueous acetic acid converts α-chlorohydrindene to α-hydrindanone (50–60%).[36] Suitably protected methylene or benzylidene acetals of alditols are cleaved by chromium(VI) oxide in glacial acetic acid to derivatives of ketoses.[37] Chromium(VI) oxide in anhydrous acetic acid converts methyl ethers into the corresponding formates, which can be hydrolyzed by base to alcohols (demethylation).[38]

Related Reagents. See Classes O-1, O-2, O-17, O-21, and O-23, pages 1–10. Chromium(VI) Oxide–3,5-Dimethylpyrazole; Chromium(VI) Oxide–Quinoline; Chromium(VI) Oxide–Silica Gel.

---

1. (a) Wiberg, K. B. *Oxidation in Organic Chemistry*; Wiberg, K. B., Ed.; Academic: New York, 1965; Part A, pp 131–135. (b) Freeman, F. *Organic Synthesis By Oxidation With Metal Compounds*; Miijs, W. J.; de Jonge, C. R. H. I., Eds.; Plenum: New York, 1986; Chapter 2. (c) Lee, D. G. *The Oxidation of Organic Compounds by Permanganate Ion and Hexavalent Chromium*; Open Court: La Salle, IL, 1980. (d) Stewart, R. *Oxidation Mechanisms: Applications to Organic Chemistry*; Benjamin: New York, 1964. (e) Cainelli, G.; Cardillo, G. *Chromium Oxidations in Organic Chemistry*; Springer: Berlin, 1984.
2. Schreiber, J.; Eschenmoser, A. *HCA* **1955**, *38*, 1529.
3. Braude, E. A.; Fawcett, J. S. *OSC* **1963**, *4*, 698.
4. Nakanishi, K.; Fieser, L. F. *JACS* **1952**, *74*, 3910.
5. Borel, E.; Deuel, H. *HCA* **1953**, *36*, 801.
6. Kuhn, R.; Roth, H. *CB* **1933**, *66*, 1274.
7. Cupo, D. Y.; Wetterhahn, K. E. *Cancer Res.* **1985**, *45*, 1146.
8. Wiberg, K. B.; Foster, G. *JACS* **1961**, *83*, 423.
9. St. André, A. F.; MacPhillamy, H. B.; Nelson, J. A.; Shabica, A. C.; Scholz, C. R. *JACS* **1952**, *74*, 5506.
10. Stephen, H.; Short, W. F.; Gladding, G. *JCS* **1920**, *117*, 510.
11. Harms, W. M.; Eisenbraun, E. J. *OPP* **1972**, *4*, 67.
12. Rosenthal, D.; Grabowich, P.; Sabo, E. F.; Fried, J. *JACS* **1963**, *85*, 3971.
13. Flatt, S. J.; Fleet, G. W. J.; Taylor, B. J. *S* **1979**, 815.
14. Barton, D. H. R.; Kulkarni, Y. D.; Sammes, P. G. *JCS(C)* **1971**, 1149.
15. Mosher, W. A.; Steffgen, F. W.; Lansbury, P. T. *JOC* **1961**, *26*, 670.
16. Hickinbottom, W. J.; Moussa, G. E. M. *JCS* **1957**, 4195.
17. Moussa, G. E. M.; Abdalla, S. O. *J. Appl. Chem.* **1970**, *20*, 256.
18. Birchenough, M. J.; McGhie, J. F. *JCS* **1950**, 1249.
19. Wintersteiner, O.; Moore, M. *JACS* **1950**, *72*, 1923.

20. Meystre, C.; Frey, H.; Wettstein, A.; Miescher, K. *HCA*, **1944**, *27*, 1815.
21. Barbier, P.; Loquin, R. *CR(C)* **1913**, *156*, 1443.
22. Wieland, H.; Schlichting, O.; Jacobi, R. *Z. Physiol. Chem.* **1926**, *161*, 80.
23. Smith, L. I.; Webster, I. M. *JACS* **1937**, *59*, 662.
24. Il'inskii, M. A.; Kazakova, V. A. *JGU* **1941**, *11*, 16 (*CA* **1941**, *35*, 5487).
25. Pschorr, R. *CB* **1906**, *39*, 3128.
26. Kende, A. S.; Curran, D. P.; Tsay, Y.; Mills, J. E. *TL* **1977**, 3537.
27. Broadhurst, M. J.; Hassall, C. H.; Thomas, G. J. *CC* **1982**, 158.
28. Fieser, L. F.; Szmuszkovicz, J. *JACS* **1948**, *70*, 3352.
29. Pattison, F. L. M.; Stothers, J. B.; Woolford, R. G. *JACS* **1956**, *78*, 2255.
30. Newman, M. S.; Arkell, A.; Fukunaga, T. *JACS* **1960**, *82*, 2498.
31. Crandall, J. K.; Heitmann, W. R. *JOC* **1979**, *44*, 3471.
32. Beugelmans, R.; Le Goff, M.-T. *BSF* **1969** 335.
33. Cardillo, G.; Orena, M.; Sandri, S. *S* **1976**, 394.
34. Snatzke, G. *CB* **1961**, *94*, 729.
35. Lalancette, J. M.; Rollin, G.; Dumas, P. *CJC* **1972**, *50*, 3058.
36. Pacaud, R. A.; Allen, C. F. H. *OSC* **1947**, *2*, 336.
37. Angyal, S. J.; Evans, M. E. *AJC* **1972**, *25*, 1513.
38. Harrison, I. T.; Harrison, S. *CC* **1966**, 752.

Fillmore Freeman
University of California, Irvine, CA, USA

Chromium(VI) Oxide–3,5-Dimethylpyrazole

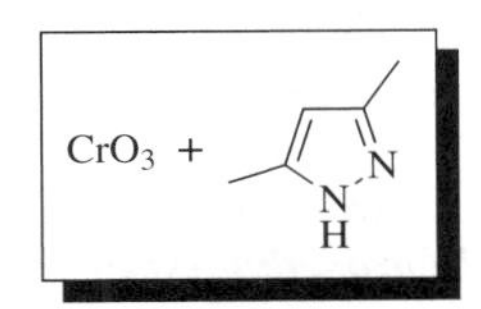

(CrO_3)		
[1333-82-0]	CrO_3	(MW 99.99)
($C_5H_8N_2$)		
[67-51-6]	$C_5H_8N_2$	(MW 96.13)

(oxidizing agent for alcohols;[1] oxidant for saturated carbons α to unsaturation[2–5])

Physical Data: see ***Chromium(VI) Oxide***.
Solubility: sol dichloromethane; slightly sol ether and pentane.
Form Supplied in: formed in situ from widely available reagents.
Preparative Method: drying of the CrO_3 over P_2O_5 is recommended; rapid addition of DMP (1 equiv) to a suspension of chromium(VI) oxide (1 equiv) in dry CH_2Cl_2 (−20 °C) results in a dark red homogeneous solution after 10 min; the solution is then treated with the organic substrate (0.05–0.5 equiv).[1,2]
Handling, Storage, and Precautions: chromium salts are carcinogenic; this reagent should be used in a fume hood.

Oxidation of Alcohols. The oxidation of primary and secondary alcohols is typically effected by treating the alcohol (1 equiv) with CrO_3·DMP (1–10 equiv) at rt (eqs 1–3).[1] Notably similar to ***Pyridinium Chlorochromate*** (PCC), the title reagent system has been shown to oxidize primary alcohols to aldehydes efficiently (eq 2). The ease of oxidation of both equatorial and axial alcohols using this complex has been demonstrated (eq 3); the reactions of both are complete within 40 min when stirred at room temperature. The authors suggest the intermediacy of a cyclic chromate ester species through which rapid intramolecular oxidation may occur.[1]

CrO3•DMP, 93% (1)

CrO3•DMP, 84% (2)

CrO3•DMP, 98% (3)

Oxidation of Unsaturated Alcohols.[1] Allylic, benzylic, and propargyl alcohols readily interact with CrO_3·DMP to provide the expected carbonyl compounds in good to excellent yield. Allylic alcohols can be oxidized with little or no competing oxidation of allylic methylene positions within the same molecule (eqs 4 and 5).

CrO3•DMP, 83% (4)

CrO3•DMP, 96% (5)

Benzylic alcohols are oxidized with comparable efficiency and the reaction is not sensitive to the electronic nature of the aromatic ring (eqs 6 and 7). Propargyl alcohols with both internal and terminal triple bonds are oxidized in good yield using CrO_3·DMP (eq 8).

CrO3•DMP, 100% (6)

CrO3•DMP, 98% (7)

CrO3•DMP, 75% (8)

Oxidation of Saturated Allylic and Benzylic Carbons. The CrO_3·DMP complex appears to offer two unique advantages over PCC: an empty Lewis acidic coordination site on chromium, and an internal basic nitrogen that may aid in cleavage of a carbon–hydrogen bond.[2] As a result, for allylic and benzylic oxidations, $CrO_3 \cdot DMP$ is superior to more conventional chromium oxidants which often suffer from low yields, practical complications, and/or inconvenient and extended reaction times. In the total synthesis of (±)-carpesiolin, $CrO_3 \cdot DMP$ was used to install an enone from an alkene precursor (eq 9).[3] It was used once more for a similar transformation in the total synthesis of (−)-retigeranic A and several of its derivatives, where it was found to be the most efficient reagent for the purpose (eq 10).[4]

CrO_3•DMP, CH_2Cl_2, −15 °C, 7 h, 60% (9)

Observations made in the course of studies toward the synthesis of Δ^5-7-keto steroids provide some insight into the mechanism of $CrO_3 \cdot DMP$-mediated allylic oxidation.[2] Here, axial hydrogens were found to be predisposed to more facile cleavage by this reagent, a common observation with chromium oxidants. The allylic oxidation of cholesteryl benzoate was complete within 30 min when $CrO_3 \cdot DMP$ was used in 20-fold excess. Alternatively, use of only a 12-fold excess of $CrO_3 \cdot DMP$ provided 74% conversion following stirring for 4 h at 0 °C. Benzylic methylenes are also susceptible to oxidation by this reagent.[5]

i-Pr, TBDMSO — CrO_3•DMP, CH_2Cl_2, −10 °C, 5 h, 49% → *i*-Pr, TBDMSO (10)

Related Reagents. See Classes O-1 and O-21, pages 1–10.

1. Corey, E. J.; Fleet, G. W. J. *TL* **1973**, 4459.
2. Salmond, W. G.; Barta, M. A.; Havens, J. L. *JOC* **1978**, *43*, 2057.
3. Kok, P.; DeClercq, P. J.; Vandewalle, M. E. *JOC* **1979**, *44*, 4553.
4. (a) Paquette, L. A.; Wright, J.; Drtina, G. J.; Roberts, R. A. *JOC* **1987**, *52*, 2960. (b) Wright, J.; Drtina, G. J.; Roberts, R. A.; Paquette, L. A. *JACS* **1988**, *110*, 5806.
5. McDonald, E.; Suksamrarn, A. *TL* **1975**, 4425.

Jeffrey N. Johnston
The Ohio State University, Columbus, OH, USA

Copper(II) Acetate[1]

$Cu(OAc)_2$

[142-71-2] $C_4H_6CuO_4$ (MW 181.64)

(oxidizes carbanions,[2] radicals[3] and hydrocarbons;[4] for oxidative coupling and solvolytic cleavage of Si–C,[5] Bi–C, Pb–C, and Sb–C bonds; rapid radical scavenger; catalyst for cyclopropanation of alkenes with diazo esters;[6] Lewis acid catalyst)

Alternate Name: cupric acetate.

Physical Data: blue crystals, mp 130–140 °C (dec); *d* 1.92–1.94 g cm^{-3}.[7]

Solubility: sol H_2O (6.79 g/100 mL, 25 °C); sol AcOH, pyridine; insol ether.

Form Supplied in: widely available; the anhydrous salt can be prepared from the usually available monohydrate $Cu(OAc)_2 \cdot H_2O$ *[6046-93-1]* by heating to 90 °C until constant weight[7,8] or by refluxing $Cu(OAc)_2 \cdot H_2O$ in acetic anhydride and washing the insoluble product with Et_2O.[9]

Analysis of Reagent Purity: iodometric titration;[10] atomic absorption spectroscopy.[11]

Purification: recrystallize (as monohydrate) from warm dil HOAc.[57]

Handling, Storage, and Precautions: must be stored in the absence of moisture; is decomposed on heating to hydrogen and Cu^IOAc.[7] Irritating to skin, eyes, and respiratory system. May be dissolved in a combustible solvent for incineration.

Oxidation of Carbanions. Oxidative coupling of terminal alkynes to diynes (eq 1) with $Cu(OAc)_2$ and ***Pyridine*** can be carried out in MeOH or in benzene/ether.[2] The reaction requires the presence of copper(I) salt; the rate-determining step corresponds to the formation of the Cu^I acetylide.[12]

$$R{-}C{\equiv}CH + py \rightleftharpoons py{-}H^+ + R{-}C{\equiv}C^-$$
$$R{-}C{\equiv}C^- + CuOAc \rightleftharpoons R{-}C{\equiv}C{-}Cu + AcO^-$$
$$R{-}C{\equiv}C{-}Cu + Cu(OAc)_2 \longrightarrow R{-}C{\equiv}C\bullet + 2\,CuOAc$$
$$2\,R{-}C{\equiv}C\bullet \longrightarrow R{-}C{\equiv}C{-}C{\equiv}C{-}R \quad (1)$$

While α-sulfonyl lithiated carbanions are oxidatively coupled with ***Copper(II) Trifluoromethanesulfonate*** (eq 2), $Cu(OAc)_2$ oxidizes them to the corresponding (*E*)-α,β-unsaturated sulfones (eq 3).[13]

PhO_2S, R, –Li + 2 $Cu(OTf)_2$ → SO_2Ph, R, R, SO_2Ph + 2 CuOTf + 2 LiOTf (2)

PhO_2S, R, –Li + 2 $Cu(OAc)_2$ → R, SO_2Ph + 2 CuOAc + LiOAc + AcOH (3)

Other carbanions can be coupled oxidatively by $Cu(OAc)_2$, as shown in the synthesis of β-lactams (eq 4).[14]

R, CO_2-*t*-Bu, O, N, R^1 — BuLi, −78 °C → R, OLi, O-*t*-Bu, LiO, N, R^1 — $Cu(OAc)_2$ → R, CO_2-*t*-Bu, O, N, R^1 (4)

In the presence of ***1,4-Diazabicyclo[2.2.2]octane*** in DMF, the complex of $Cu(OAc)_2$ and 2,2′-bipyridyl catalyzes the oxygenation of α-branched aldehydes with O_2 to ketones.[15]

Carbon–Hydrogen Bond Oxidations. *Ortho* hydroxylation of phenols with O_2 is catalyzed by a complex of $Cu(OAc)_2$ and ***Morpholine*** (soluble in EtOH).[16] In the absence of O_2, *ortho* acetoxylation of phenols can be induced with equimolar amounts of $Cu(OAc)_2$ in AcOH (eq 5).[17]

Cu(OAc)2 (1 equiv), AcOH, 115 °C, 24 h, 80% (5)

Allylic hydrogens are replaced by acyloxy groups by reaction of peroxy esters in the presence of catalytic amounts of copper salts, including $Cu(OAc)_2$.[18] The reaction probably proceeds via the formation of an allylic radical, which reacts quickly with Cu^{II} to form a Cu^{III} intermediate that generates the most substituted alkene, probably via a pericyclic transition state (eq 6).[19] Allylic oxidation can be enantioselective when performed in AcOH and ***Pivalic Acid*** in the presence of $Cu(OAc)_2$ and an L-amino acid.[20]

t-BuC(O)OOC(O)CH3 + CuOAc → t-BuCO• + Cu(OAc)2

t-BuCO• → [allyl radical] + t-BuCOH; Cu(OAc)2 → [CuOAc intermediate] → allylic acetate (OAc) + CuOAc (6)

Allylic oxidation of cyclohexene and related alkenes can be achieved with catalytic amounts of ***Palladium(II) Acetate***, $Cu(OAc)_2$, hydroquinone, and O_2 as oxidant in AcOH, leading to allylic acetates.[21] Methyl glyoxylate adducts of *N*-Boc-protected allylic amines cyclize, in the presence of catalytic $Pd(OAc)_2$ and an excess of $Cu(OAc)_2$ in DMSO at 70 °C, to 5-(1-alkenyl)-2-(methoxycarbonyl)oxazolidines (eq 7).[22]

MeO2C–CHO; Pd(OAc)2, Cu(OAc)2 (3 equiv), DMSO, 70 °C (7)

Methyl substituted benzene derivatives are oxidized in boiling AcOH to the corresponding benzyl acetates (eq 8) with sodium, potassium, or ***Ammonium Peroxydisulfate***, $Cu(OAc)_2 \cdot H_2O$, and NaOAc.[4] The peroxydisulfate radical is responsible for the primary oxidation, whereas $Cu(OAc)_2$ prevents dimerization of the intermediate benzylic radical by oxidizing it to benzyl acetate. The benzylic acetoxylation of alkyl aromatics can also be carried out with O_2 using $Pd(OAc)_2$ and $Cu(OAc)_2$ as catalysts.[23]

Cycloalkanes are transformed into the corresponding cycloalkenes by treatment with ***t-Butyl Hydroperoxide*** in pyridine/AcOH solution containing $Cu(OAc)_2 \cdot H_2O$. When Fe^{III} salts are used instead of $Cu(OAc)_2 \cdot H_2O$, the major product is the corresponding cycloalkanone.[24] Cyclohexanone is the main product of cyclohexane oxidation with H_2O_2, $Cu(OAc)_2 \cdot H_2O$ in pyridine, and AcOH (GoCHAgg system).[25] $Cu(OAc)_2$ also catalyzes the oxidation of secondary alcohols by ***Lead(IV) Acetate***.[26]

ArCH3 + $SO_4^{-}\bullet$ → [radical cation] + SO_4^{2-} → (– H+) benzylic radical → ($Cu(OAc)_2 \cdot H_2O$) ArCH2OAc + CuOAc (8)

Carbon–Metal Bond Oxidations. In MeOH and under O_2 atmosphere, a catalytic amount of $Cu(OAc)_2$ promotes the cleavage of the Si–C bond of (*E*)-alkenylpentafluorosilicates to give alkenyl ethers (eq 9). The reaction is highly stereoselective and leads to the (*E*)-enol ethers. In the presence of H_2O the corresponding aldehydes are obtained.[5]

$[RCH{=}CHSiF_5]^{2-}$ 2 K+ → (O_2, MeOH, $Cu(OAc)_2$) RCH=CHOMe (9)

$R = C_6H_{13}$, 56%
R = Ph, 51%
$R = MeO_2C(CH_2)_8$, 67%

In the presence of $Cu(OAc)_2$, 1,4-additions of alkylpentafluorosilicates to α,β-unsaturated ketones take place on heating (eq 10).[5] This reaction proceeds probably by initial one-electron oxidation with formation of an alkyl radical (eq 11), which then adds to the enone.

$$[C_8H_{17}SiF_5]K_2 + CH_2{=}CHCOCH_3 \xrightarrow[135\ °C]{Cu(OAc)_2,\ \text{sealed tube}} C_8H_{17}CH_2CH_2COCH_3 \quad (10)$$

$$RSiF_5^{2-} + CuX_2 \longrightarrow R\bullet + [XSiF_5]^{2-} + CuX \quad (11)$$

The monophenylation of 1,*n*-diols with ***Triphenylbismuth Diacetate***[27] is greatly accelerated by catalytic amounts of $Cu(OAc)_2$.[28] This reaction can be enantioselective in the presence of optically active pyridinyloxazoline ligands as cocatalysts (eq 12).[29] Reaction of alcohols (ROH) with ***Triphenylbismuthine*** and $Cu(OAc)_2$ gives the corresponding phenyl ethers (PhOR) and benzene.[30] The treatment of Ph_5Sb with a catalytic amount of $Cu(OAc)_2$ in toluene at 20 °C gave 100% yields of Ph_3Sb, Ph–Ph, and PhH.[31] $Cu(OAc)_2$ catalyzes the arylation of amines by diaryliodonium salts,[32] aryl halides,[33] $Ph_3Bi(OCOCF_3)_2$,[34] and aryllead triacetates.[35]

$$\text{diol} + Ph_3Bi(OAc)_2 \xrightarrow[\text{22 °C; 42\%, ee 30.4\%}]{\text{ligand (9 mol \%), } Cu(OAc)_2 \text{ (3 mol \%), } CH_2Cl_2} \text{OPh/OH product} \quad (12)$$

Fast Radical Scavenging and Oxidation. Rates of oxidative decarboxylation by $Pb(OAc)_4$ of primary and secondary carboxylic acids to alkenes[36] are enhanced in the presence of catalytic amounts of $Cu(OAc)_2$ or $Cu(OAc)_2 \cdot H_2O$. This effect is attributed to the fact that the rate of one-electron-transfer oxidation of alkyl radicals by Cu^{II} salts (eq 13) approaches a diffusion-controlled rate.[3] Oxidative decarboxylation of carboxylic acids can also be carried out with ***(Diacetoxyiodo)benzene*** in the presence of a catalytic amount of anhydrous $Cu(OAc)_2$.[37]

$$RCO_2Pb(OAc)_3 \longrightarrow R\bullet + CO_2 + Pb(OAc)_3$$
$$R\bullet + Cu(OAc)_2 \longrightarrow \text{alkene} + CuOAc + AcOH$$
$$CuOAc + RCO_2Pb(OAc)_3 \longrightarrow Cu(OAc)_2 + RCO_2Pb(OAc)_2$$
$$RCO_2Pb(OAc)_2 \longrightarrow R\bullet + CO_2 + Pb(OAc)_2 \quad (13)$$

The case of radical oxidation with $Cu(OAc)_2$ has been exploited by Schreiber[38] in the fragmentation of α-alkoxyhydroperoxides, as in eq 14.[38b]

$$\xrightarrow[\text{83\%}]{Fe^{II}SO_4,\ Cu(OAc)_2,\ MeOH} [\ \text{radical}\ ] + Fe^{III}(OH)SO_4 \longrightarrow \text{macrolide} \quad (14)$$

In an electrochemical system containing ***Manganese(III) Acetate***, acetic acid is added to butadiene to generate an allylic radical intermediate that is oxidized with $Cu(OAc)_2 \cdot H_2O$ to the corresponding allylic cation, leading to γ-vinyl-γ-butyrolactone (eq 15),[39] a precursor in the industrial synthesis of sorbic acid.

$$\text{butadiene} + MeCO_2H \xrightarrow{Mn^{3+}} [\ \text{radical}\ CO_2H \xrightarrow{Cu(OAc)_2 \cdot H_2O} \text{cation}\ CO_2H\ ] \longrightarrow \text{lactone} \quad (15)$$

β-Oxoesters are oxidized with $Mn(OAc)_3$ to the corresponding radicals that can add intermolecularly[40] or intramolecularly (eq 16)[41] to generate alkyl radicals. In the presence of $Cu(OAc)_2$ the latter are rapidly quenched and oxidized to give alkenes. Radical arylation with alkyl iodides can be induced with ***Dibenzoyl Peroxide***; the yield of the reaction can be improved using a catalytic amount of $Cu(OAc)_2 \cdot H_2O$,[42] which minimizes hydrogen abstraction by the intermediate radical but introduces a competitive electron-transfer oxidation of the intermediate radical. The oxidative addition of disulfides to alkenes (Trost hydroxysulfenylation[43]) can be promoted by catalytic amounts of $Cu(OAc)_2$.[44]

$$\xrightarrow{Mn(OAc)_3} [\ \text{radicals}\ ] \xrightarrow{Cu(OAc)_2} \text{bicyclic alkene} \quad (16)$$

Reoxidant in Palladium-Catalyzed Reactions. $Cu(OAc)_2$ has been used as a reoxidant in the Wacker oxidation ($CH_2{=}CH_2 + O_2 \rightarrow CH_3CHO$)[45] and in the $Pd(OAc)_2$-catalyzed alkenylation of aromatic compounds with alkenes[46] (eq 17).[47] $Pd(OAc)_2$ and $Cu(OAc)_2$ are effective catalysts for the reactions of nitrosobenzenes with carbon monoxide, dioxygen, and alcohols that give the corresponding *N*-alkylcarbamates.[48]

$$\text{uridine} + \text{maleimide (NR)} \xrightarrow[Pd(OAc)_2,\ Cu(OAc)_2]{AcOH,\ O_2} \text{5-substituted uridine} \quad (17)$$

Enantioselective Cyclopropanation. $Cu(OAc)_2$ has been used as procatalyst in the asymmetric cyclopropanation[49] of alkenes with alkyl diazoacetates with optically pure imines as cocatalyst (eq 18).[6]

$$R\text{-CH=CH}_2 + EtO_2C\text{-CH=}N_2 \xrightarrow[R^1CH{=}NR^*]{Cu(OAc)_2} \text{cis-}CO_2Et\ \text{cyclopropane} + \text{trans-}CO_2Et\ \text{cyclopropane} \quad (18)$$

$Cu(OAc)_2$ as Lewis Acid. Decarboxylation of L-tryptophan into L-tryptamine proved most effective in HMPA in the presence of $Cu(OAc)_2$.[50] In boiling MeCN and under $Cu(OAc)_2 \cdot H_2O$ catalysis, aldoximes are converted smoothly into nitriles.[51] In the presence of various Lewis acids including $Cu(OAc)_2$, cyclodeca-1,2,5,8-tetraene is rearranged to *cis,syn*-tricyclo[4.4.0.$0^{2,4}$]deca-5,8-diene (eq 19).[52]

$$\text{cyclodeca-1,2,5,8-tetraene} \xrightarrow[\text{Cu(OAc)}_2]{\text{AcOH}} \text{tricyclodecadiene} \quad (19)$$

The Michael reaction of $O_2NCH_2CO_2R$ ($R = Me$, Bn) with $R^1COCH{=}CHR^2$ (R^1 = Me, Et, $R^2 = H$; $R^1 = R^2 = Me$) is catalyzed by $Cu(OAc)_2$ and gives $R^1COCH_2CHR^2CH(NO_2)CO_2R$ in dioxane at 100 °C.[53] Knoevenagel condensation of *t*-butyl malonate with ***Paraformaldehyde*** to give di-*t*-butyl methylidenemalonate can be achieved in the presence of KOH and $Cu(OAc)_2$.[54] Lithium imine anions of α-amino esters undergo $Cu(OAc)_2$-catalyzed reactions with α,ω-dihalogenoalkanes to give the corresponding ω-halogenoalkylimines.[55] $Cu(OAc)_2$ catalyzes the coupling of PhYbI with *n*-BuI, giving *n*-BuPh and Ph–Ph.[56]

Acyl hydrazides are converted to the corresponding carboxylic acids by bubbling oxygen through a THF or MeOH solution containing the hydrazide and a catalytic amount of $Cu(OAc)_2$ (eq 20).[58]

$$\text{RC(O)NHNH}_2 \xrightarrow[\text{O}_2,\ \text{THF}]{\text{Cu(OAc)}_2} \text{RC(O)OH} \quad (20)$$

R = alkyl, aryl

Synthesis of Ynamines. Phenylacetylene reacts with dimethylamine under $Cu(OAc)_2$ catalysis to produce *N,N*-dimethyl-2-phenylethynylamine (eq 21).[59] The reaction is effected by bubbling oxygen through a benzene solution of the reagents and $Cu(OAc)_2$; in the absence of oxygen, 1,4-diphenylbutadiyne is the sole product. This may be suppressed by adding a reducing agent, such as hydrazine, to the reaction mixture.

$$\text{Ph}{-}{\equiv} + \text{HNMe}_2 \xrightarrow[\text{O}_2,\ \text{THF}]{\text{Cu(OAc)}_2} \text{Ph}{-}{\equiv}{-}\text{NMe}_2 \quad (21)$$

Related Reagents. See Classes O-21 and O-24, pages 1–10. Copper(I) Acetate; Copper(II) Acetate–Iron(II) Sulfate; Iodine–Copper(II) Acetate; Lead(IV) Acetate–Copper(II) Acetate; Manganese(III) Acetate–Copper(II) Acetate; Sodium Hydride–Copper(II) Acetate–Sodium *t*-Pentoxide; Zinc–Copper(II) Acetate–Silver Nitrate.

---

1. *FF*, **1967**, *1*, 157, 159; **1969**, *2*, 18, 84; **1972**, *3*, 65; **1974**, *4*, 105; **1975**, *5*, 156; **1977**, *6*, 138; **1979**, *7*, 126; **1982**, *10*, 103; **1986**, *12*, 140; **1990**, *15*, 99.
2. (a) Eglinton, G.; McCrae, W. *Adv. Org. Chem.* **1963**, *4*, 225. (b) Cresp, T. M.; Sondheimer, F. *JACS* **1975**, *97*, 4412. (c) Kashitani, T.; Akiyama, S.; Iyoda, M.; Nakagawa, M. *JACS* **1975**, *97*, 4424. (d) Boldi, A. M.; Anthony, J.; Knobler, C. B.; Diederich, F. *AG(E)* **1992**, *31*, 1240.
3. (a) Sheldon, R. A.; Kochi, J. K. *OR* **1972**, *19*, 279. (b) Jenkins, C. L.; Kochi, J. K. *JACS* **1972**, *94*, 843.
4. (a) Belli, A.; Giordano, C.; Citterio, A. *S* **1980**, 477. (b) Deardurff, L. A.; Alnajjar, M. S.; Camaioni, D. M. *JOC* **1986**, *51*, 3686. (c) Walling, C.; El-Taliawi, G. M.; Amarnath, K. *JACS* **1984**, *106*, 7573.
5. Yoshida, J.; Tamao, K.; Kakui, T.; Kurita, A.; Murata, M.; Yamada, K.; Kumada, M. *OM* **1982**, *1*, 369.
6. (a) Aratani, T. *PAC* **1985**, *57*, 1839. (b) Brunner, H.; Wutz, K. *NJC* **1992**, *16*, 57.
7. *Gmelins Handbuch der Anorganischen Chemie*; Verlag: Weinheim, 1961; Copper, Part B, p 679.
8. Davidson, A. W.; Griswold, E. *JACS* **1931**, *53*, 1341.
9. Späth, E. *Sitzungsber. Akad. Wiss. Wien, Math.-Naturwiss. Kl., Abt. 2B* **1911**, *120*, 117.
10. (a) Waser, J. *Quantitative Chemistry*; Benjamin: New York, 1964; p 343. (b) *Reagent Chemicals: American Chemical Society Specifications*, 8th ed.; American Chemical Society: Washington, 1993; p 277.
11. *Official Methods of Analysis of the Association of Official Analytical Chemists*, 15th ed.; Helrich, K., Ed.; AOAC: Arlington, VA, 1990; p 156.
12. Clifford, A. A.; Waters, W. A. *JCS* **1963**, 3056.
13. Baudin, J.-B.; Julia, M.; Rolando, C.; Verpeaux, J.-N. *TL* **1984**, *25*, 3203.
14. Kawabata, T.; Minami, T.; Hiyama, T. *JOC* **1992**, *57*, 1864.
15. (a) Van Rheenen, V. *TL* **1969**, 985. (b) Briggs, L. H.; Bartley, J. P.; Rutledge, P. S. *JCS(P1)* **1973**, 806.
16. Brackman, W.; Havinga, E. *RTC* **1955**, *74*, 937.
17. Takizawa, Y.; Tateishi, A.; Sugiyama, J.; Yoshida, H.; Yoshihara, N. *CC* **1991**, 104.
18. (a) Kharasch, M. S.; Fono, A. *JOC* **1958**, *23*, 324. (b) Kochi, J. K. *JACS* **1961**, *83*, 3162. (c) Kochi, J. K. *JACS* **1962**, *84*, 774.
19. Beckwith, A. L. J.; Zavitsas, A. A. *JACS* **1986**, *108*, 8230.
20. Muzart, J. *J. Mol. Catal.* **1991**, *64*, 381.
21. Byström, S. E.; Larsson, E. M.; Åkermark, B. *JOC* **1990**, *55*, 5674.
22. Van Benthem, R. A. T. M.; Hiemstra, H.; Speckamp, W. N. *JOC* **1992**, *57*, 6083.
23. Goel, A. B. *ICA* **1986**, *121*, L11.
24. (a) Barton, D. H. R.; Bévière, S. D.; Chavasiri, W.; Doller, D.; Hu, B. *TL* **1993**, *34*, 567. (b) Shul'pin, G. B.; Druzhinina, A. N. *React. Kinet. Catal. Lett.* **1992**, *47*, 207.
25. Barton, D. H. R.; Bévière, S. D.; Chavasiri, W.; Csuhai, E.; Doller, D. *T* **1992**, *48*, 2895.
26. Kapustina, N. I.; Popkov, A. Yu.; Gasanov, R. G.; Nikishin, G. I. *IZV* **1988**, *10*, 2327.
27. (a) David, S.; Thieffry, A. *TL* **1981**, *22*, 2885 and 5063. (b) David, S.; Thieffry, A. *JOC* **1983**, *48*, 441.
28. Barton, D. H. R.; Finet, J.-P.; Pichon, C. *CC* **1986**, 65.
29. Brunner, H.; Obermann, U.; Wimmer, P. *OM* **1989**, *8*, 821.
30. Dodonov, V. A.; Gushchin, A. V.; Brilkina, T. G.; Muratova, L. V. *ZOB* **1986**, *56*, 2714 (*CA* **1987**, *107*, 197 657b).
31. Dodonov, V. A.; Bolotova, O. P.; Gushchin, A. V. *ZOB* **1988**, *58*, 711 (*CA* **1988**, *109*, 231 186a).
32. Varvoglis, A. *S* **1984**, 709.
33. Lindley, J. *T* **1984**, *40*, 1433.
34. (a) Dodonov, V. A.; Gushchin, A. V.; Brilkina, T. G. *ZOB* **1985**, *55*, 466 (*CA* **1985**, *103*, 22 218z). (b) Barton, D. H. R.; Finet, J.-P.; Khamsi, J. *TL* **1988**, *29*, 1115.

35. Barton, D. H. R.; Donnelly, D. M. X.; Finet, J.-P.; Guiry, P. J. *TL* **1989**, *30*, 1377.
36. (a) Ogibin, Yu. N.; Katzin, M. I.; Nikishin, G. I. *S* **1974**, 889. (b) Nishiyama, H.; Matsumoto, M.; Arai, H.; Sakaguchi, H.; Itoh, K. *TL* **1986**, *27*, 1599. (c) Patel, D. V.; VanMiddlesworth, F.; Donaubauer, J.; Gannett, P.; Sih, C. J. *JACS* **1986**, *108*, 4603.
37. Concepción, J. I.; Francisco, C. G.; Freire, R.; Hernández, R.; Salazar, J. A.; Suárez, E. *JOC* **1986**, *51*, 402.
38. (a) Schreiber, S. L. *JACS* **1980**, *102*, 6163; (b) Schreiber, S. L.; Liew, W.-F. *JACS* **1985**, *107*, 2980.
39. (a) Coleman, J. P.; Hallcher, R. C.; McMackins, D. E.; Rogers, T. E.; Wagenknecht, J. H. *T* **1991**, *47*, 809; (b) Vinogradov, M. G.; Pogosyan, M. S.; Shteinschneider, A. Yu.; Nikishin, G. I. *IZV* **1981**, *9*, 2077.
40. Melikyan, G. G.; Vostrowsky, O.; Bauer, W.; Bestmann, H. J. *JOM* **1992**, *423*, C24.
41. (a) Snider, B. B.; Zhang, Q.; Dombroski, M. A. *JOC* **1992**, *57*, 4195. (b) Dombroski, M. A.; Snider, B. B. *T* **1992**, *48*, 1417. (c) Bertrand, M. P.; Sursur, J.-M.; Oumar-Mahamet, H.; Moustrou, C. *JOC* **1991**, *56*, 3089. (d) Breuilles, P.; Uguen, D. *TL* **1990**, *31*, 357.
42. Vismara, E.; Donna, A.; Minisci, F.; Naggi, A.; Pastori, N.; Torri, G. *JOC* **1993**, *58*, 959.
43. Trost, B. M.; Ochiai, M.; McDougal, P. G. *JACS* **1978**, *100*, 7103.
44. Bewick, A.; Mellor, J. M.; Milano, D.; Owton, W. M. *JCS(P1)* **1985**, 1045.
45. (a) Tsuji, J. *COS* **1991**, *7*, 449. (b) Bäckvall, J. E.; Awasthi, A. K.; Renko, Z. D. *JACS* **1987**, *109*, 4750.
46. (a) Moritani, I.; Fujiwara, Y. *S* **1973**, 524. (b) Fujiwara, Y.; Maruyawa, O.; Yoshidomi, M.; Taniguchi, H. *JOC* **1981**, *46*, 851.
47. Itahara, T. *CL* **1986**, 239.
48. Alper, H.; Vasapollo, G. *TL* **1987**, *28*, 6411.
49. Nozaki, H.; Moriuti, S.; Takaya, H.; Noyori, R. *TL* **1966**, 5239.
50. Kametani, T.; Suzuki, T.; Takahashi, K.; Fukumoto, K. *S* **1974**, 131.
51. Attanasi, O.; Palma, P.; Serra-Zanetti, F. *S* **1983**, 741.
52. Thies, R. W.; Boop, J. L.; Schiedler, M.; Zimmerman, D. C.; La Page, T. H. *JOC* **1983**, *48*, 2021.
53. (a) Coda, A. C.; Desimoni, G.; Invernizzi, A. G.; Righetti, P. P.; Seneci, P. F.; Taconi, G. *G* **1985**, *115*, 111. (b) Watanabe, K.; Miyazu, K.; Irie, K. *BCJ* **1982**, *55*, 3212.
54. (a) Ballesteros, P.; Roberts, B. W.; Wong, J. *JOC* **1983**, *48*, 3603. (b) De Keyser, J.-L.; De Cock, C. J. C.; Poupaert, J. H.; Dumont, P. *JOC* **1988**, *53*, 4859.
55. Joucla, M.; El Goumzili, M. *TL* **1986**, *27*, 1681.
56. Yokoo, K.; Fukagawa, T.; Yamanaka, Y.; Taniguchi, H.; Fujiwara, Y. *JOC* **1984**, *49*, 3237.
57. Perrin, D. D.; Armarego, W. L. F. *Purification of Laboratory Chemicals*, 3rd ed.; Pergamon: New York, 1988; p 321.
58. Tsuji, J.; Nagashima, T.; Nguyen, T. Q.; Takayanagi, H. *T* **1980**, *36*, 1311.
59. Peterson, L. I. *TL* **1968**, *51*, 5357.

Pierre Vogel
Université de Lausanne, Switzerland

Copper(II) Bromide

$CuBr_2$

[7789-45-9] Br_2Cu (MW 223.36)

(brominating agent; oxidizing agent; Lewis acid)

Alternate Name: cupric bromide.
Physical Data: mp 498 °C; *d* 4.770 g cm^{-3}.
Solubility: very sol water; sol acetone, ammonia, alcohol; practically insol benzene, Et_2O, conc H_2SO_4.
Form Supplied in: almost black solid crystals or crystalline powder; also supplied as reagent adsorbed on alumina (approx. 30 wt % $CuBr_2$ on alumina).
Purification: recryst from H_2O and dried in vacuo.[35]
Handling, Storage, and Precautions: anhydrous reagent is hygroscopic and should therefore be stored in the absence of moisture.

α-Bromination of Carbonyls. Copper(II) bromide is an efficient reagent for the selective bromination of methylenes adjacent to carbonyl functional groups.[1] Thus 2′-hydroxyacetophenone treated with a heterogeneous mixture of $CuBr_2$ in $CHCl_3$–EtOAc gives complete conversion to 2-bromo-2′-hydroxyacetophenone with no aromatic ring bromination (eq 1).[2]

OH O —($CuBr_2$, Δ; EtOAc–$CHCl_3$; 100%)→ OH O ... Br (1)

Similar selectivity is obtained with a homogeneous solution of the reagent in dioxane.[3] A limitation of the reaction is observed with 2′-hydroxy-4′,6′-dimethoxyacetophenone, which undergoes aromatic nuclear bromination with $CuBr_2$.[4] Steroidal ketones have been selectively α-brominated with $CuBr_2$ in the presence of a double bond without bromination of the alkene (eq 2),[5] while γ-bromination occurs in other steroidal enones.[1]

AcO ... O —($CuBr_2$; THF)→ AcO ... O ... Br (2)

Copper(II) bromide has been used to α-brominate diketotetraquinanes[6] and to introduce a double bond into a prostanoid nucleus in a one-pot bromination–elimination procedure (eq 3).[7] 3,7-Dibromo-2*H*,6*H*-benzodithiophene-2,6-diones (eq 4)[8] and 5-bromo-4-oxo-4,5,6,7-tetrahydroindoles (eq 5)[9] are prepared by the selective α-bromination of their respective ketone starting materials without bromination of the aromatic or heterocyclic rings. 4-Carboxyoxazolines are converted to the corresponding oxazoles using a mixture of $CuBr_2$ and ***1,8-Diazabicyclo[5.4.0]undec-7-ene*** (eq 6).[10]

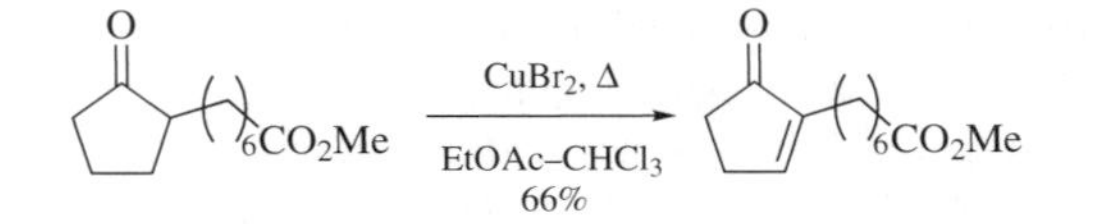

(3)

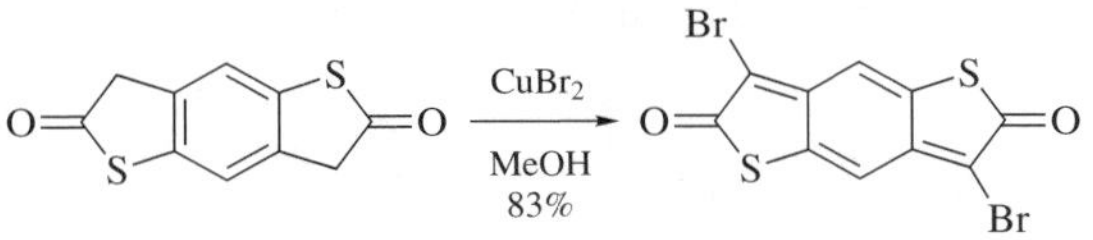

(4)

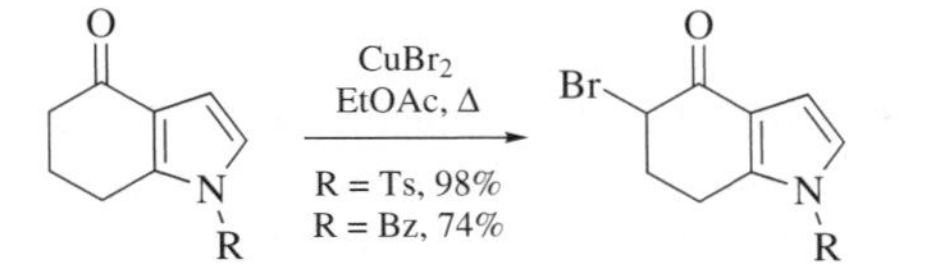

(5)

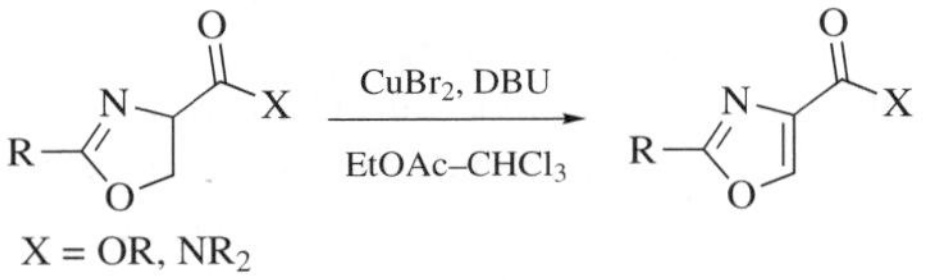

(6)

Bromination of Alkenes and Alkynes. Heating copper(II) bromide in methanol with compounds containing nonaromatic carbon–carbon multiple bonds leads to di- or tribromination.[11] For example, under these conditions allyl alcohol is converted to 1,2-dibromo-3-hydroxypropane in 99% yield (eq 7), while propargyl alcohol produces a mixture of *trans* di- and tribromoallyl alcohols (eq 8). 2′-Hydroxy-5′-methyl-4-methoxychalcone undergoes a bromination–ring-closure reaction, affording 3-bromo-6-methyl-5′-methoxyflavanone when heated with $CuBr_2$ in refluxing dioxane (eq 9).[12] The mechanism of the bromination of cyclohexene to 1,2-dibromocyclohexane with $CuBr_2$ has been studied.[13]

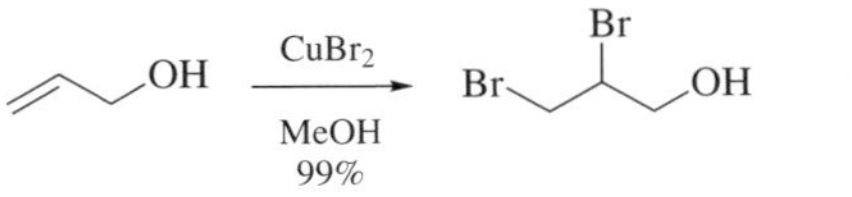

(7)

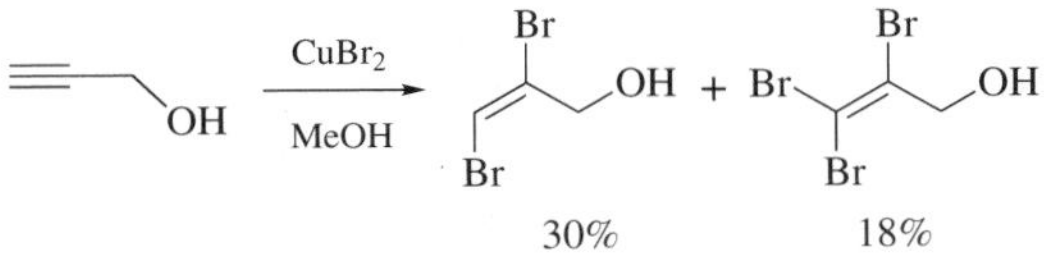

(8)

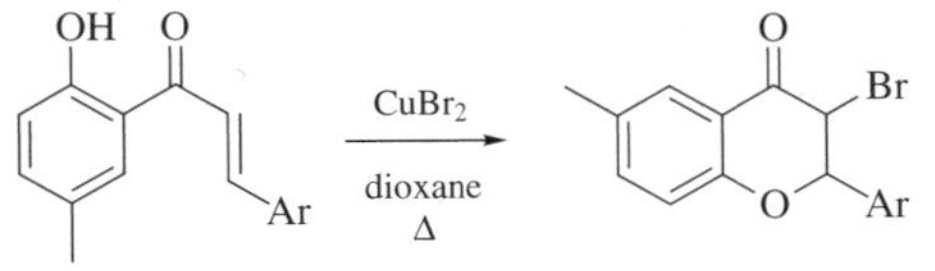

(9)

Bromination of Aromatics. Aromatic systems are brominated by copper(II) bromide. For example, 9-bromoanthracene is prepared in high yield by heating anthracene and the reagent in carbon tetrachloride (eq 10).[14] When the 9-position is blocked by a halogen, alkyl, or aryl group, the corresponding 10-bromoanthracene is formed.[15] Under similar conditions, 9-acylanthracenes give 9-acyl-10-bromoanthracenes as the predominant products.[16] The aromatic nuclear bromination of monoalkylbenzenes has been shown to proceed cleanly under strictly anhydrous conditions (eq 11).[17a] Polymethylbenzenes are efficiently and selectively converted to the nuclear brominated derivatives by $CuBr_2$/***Alumina***.[17b] In the absence of alumina, a mixture of products resulting from benzylic halogenation is isolated. 3-Acetylpyrroles are nuclear monobrominated at the 4-position in high yield by $CuBr_2$ in acetonitrile at ambient temperature (eq 12).[18] The reaction also proceeds with ethyl 3-pyrrolecarboxylates to give 4-bromopyrrole derivatives,[19] while an excess of brominating agent at 60 °C affords 4,5-dibromopyrroles.[20]

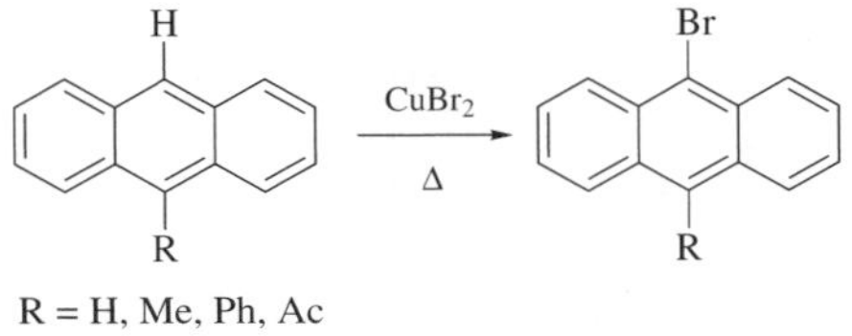

(10)

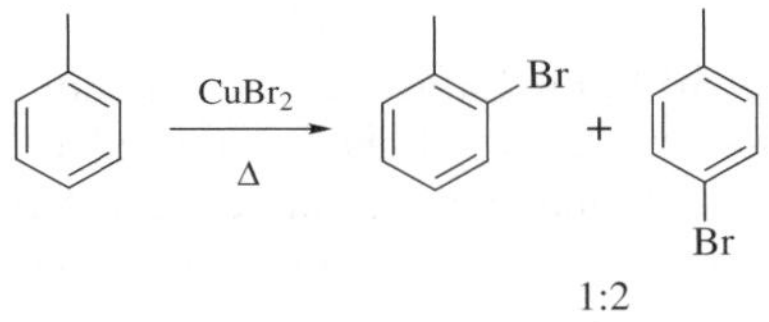

(11)

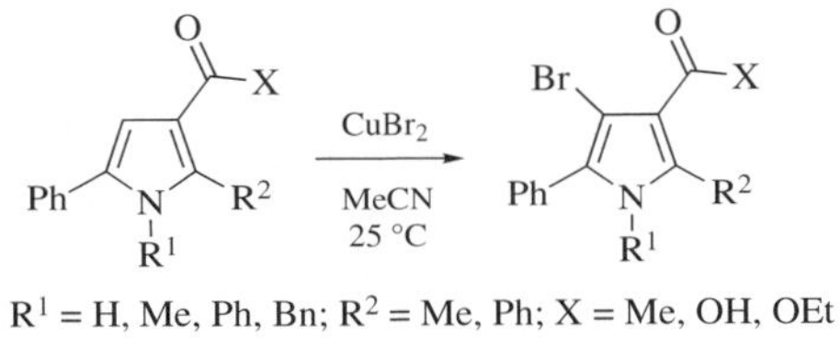

(12)

Bromination of Allylic Alcohols. Silica gel-supported copper(II) bromide has been used for the regioselective bromination of methyl 3-hydroxy-2-methylenepropanoates and 3-hydroxy-2-methylenepropanenitriles (eq 13).[21a] In the absence of silica gel, no reaction occurs between $CuBr_2$ and these substrates, while adsorption onto Al_2O_3, MgO, or TiO_2 leads to side reactions rather than the clean allylic bromination observed with $CuBr_2/SiO_2$. The reaction is stereoselective with respect to formation of the (*Z*) isomer.

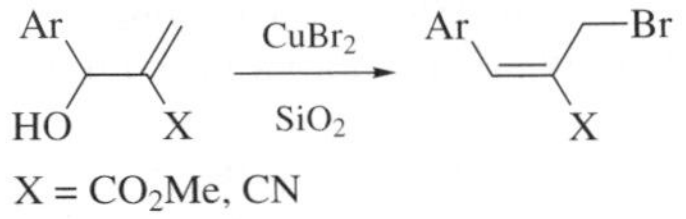

(13)

Benzylic Bromination. Toluene and substituted methylbenzenes undergo benzylic bromination using $CuBr_2$ and ***t-Butyl Hydroperoxide*** in acetic acid or anhydride (eq 14).[21b] While the yields (43–95%) are not quite as high as those obtained using ***N-Bromosuccinimide***, the copper(II) bromide procedure allows the benzylic bromination of compounds which are insoluble in nonpolar solvents.

Esterification Catalyst. Highly sterically hindered esters are prepared by the reaction of *S*-2-pyridyl thioates and alcohols in acetonitrile with copper(II) bromide as the catalyst.[22] The reaction proceeds at ambient temperature under mild conditions and affords high yields of a range of sterically crowded esters such as *t*-butyl 1-adamantanecarboxylate (eq 15).

$$\text{ArCH}_3 \xrightarrow[\text{AcOH, }\Delta;\ 43\text{–}95\%]{\text{CuBr}_2,\ t\text{-BuOOH}} \text{ArCH}_2\text{Br} \quad (14)$$

X = H, hal, CO_2H

(15) 2-pyridyl thioester of 1-adamantanecarboxylic acid + *t*-BuOH, $CuBr_2$, MeCN, 25 °C, 89% → *t*-butyl 1-adamantanecarboxylate

Conjugate Addition Catalyst. The 1,4-addition of Grignard reagents to α,β-unsaturated esters is promoted by catalytic $CuBr_2$ (1–5 mol%) with ***Chlorotrimethylsilane***/HMPA (eq 16).[23] Under these conditions the copper(II) species is not reduced by the Grignard reagent, resulting in high yields of the conjugate addition products.

(16) methyl cyclohex-1-enecarboxylate + $CuBr_2$, TMSCl, HMPA, *n*-BuMgBr, 99% → methyl 2-*n*-butylcyclohexanecarboxylate

Oxidation of Stannanes and Alcohols. Allylstannanes have been oxidized with copper(II) bromide in the presence of various nucleophilic reagents (H_2O, ROH, AcONa, RNH_2) to afford the corresponding allylic alcohols, ethers, acetates, and amines.[24] This chemistry has been extended to trimethylsilyl enol ethers, which undergo a $CuBr_2$-induced carbon–carbon bond forming process with allylstannanes (eq 17).[25] Alkoxytributylstannanes may be converted to the corresponding aldehyde or ketone with two equivalents of $CuBr_2$/***Lithium Bromide*** in THF at ambient temperature (path a, eq 18).[26] A combination of copper(II) bromide/***Lithium t-Butoxide*** oxidizes alcohols to carbonyl compounds quite rapidly and in high yield (path b, eq 18).[27]

(17) $RCH=CHCH_2SnBu_3$ + 1-(trimethylsilyloxy)cyclohexene, $CuBr_2$, $CHCl_3$, 25 °C → 2-($RCH=CHCH_2$)cyclohexanone

R = $Ph(CH_2)_2$, 57%

$$R^1R^2CH\text{–}OX \xrightarrow{\text{(a) or (b)}} R^1R^2C\text{=}O \quad (18)$$

(a) X = $SnBu_3$ **(a)** = $CuBr_2$, LiBr, Bu_3SnO-*t*-Bu
(b) X = H **(b)** = $CuBr_2$, LiO-*t*-Bu

Desilylbromination. β-Silyl ketones are desilylbrominated to α,β-unsaturated ketones with $CuBr_2$ in DMF.[28] This occurs spontaneously in cyclic ketones, while with open-chain ketones sodium bicarbonate is required to eliminate HBr from the β-bromo ketone thus formed. The carbon–silicon bond in organopentafluorosilicates prepared from alkenes and alkynes is cleaved with copper(II) bromide to give the corresponding alkyl and alkenyl bromides (eq 19).[29] The reaction is stereoselective; thus (*E*)-alkenyl bromides are obtained from (*E*)-alkenylsilicates.

$$R\text{–C≡C–}R^1 \xrightarrow[\text{2. KF}]{\text{1. HSiCl}_3,\ \text{H}_2\text{PtCl}_6} K_2[RCH\text{=}CR^1SiF_5] \xrightarrow{\text{CuBr}_2} RCH\text{=}CR^1Br \quad (19)$$

Reagent in the Sandmeyer and Meerwein Reactions. Diazonium salts of arylamines are converted to aryl halides (Sandmeyer reaction)[30] in the presence of copper(II) halides. Recent procedures have utilized *t*-butyl nitrite/$CuBr_2$[31] or *t*-butyl thionitrite/$CuBr_2$[32] combinations to afford aryl bromides from the corresponding arylamines in high yields (eq 20). The copper salt-catalyzed haloarylation of alkenes with arenediazonium salts (Meerwein reaction) also proceeds with copper(II) halides. For example, treatment of *p*-aminoacetophenone with *t*-butyl nitrite/$CuBr_2$ in the presence of excess acrylic acid gives *p*-acetyl-α-bromohydrocinnamic acid (59% yield, eq 21).[33] The intramolecular version of this reaction, which affords halogenated dihydrobenzofurans, has been accomplished by reacting arenediazonium tetrafluoroborates with $CuBr_2$ in DMSO (eq 22).[34]

$$\text{ArNH}_2 \xrightarrow[\text{MeCN, }\Delta]{\text{CuBr}_2,\ t\text{-BuONO}} \text{ArBr} \quad (20)$$

X = H, hal, CO_2R, NO_2, OMe, etc.

(21) *p*-aminoacetophenone + $CH_2=CHCO_2H$, $CuBr_2$, *t*-BuONO, MeCN, 25 °C, 59% → *p*-$AcC_6H_4CH_2CHBrCO_2H$

(22) 2-(allyloxy)benzenediazonium BF_4^-, $CuBr_2$, DMSO, 25 °C, 82% → 3-(bromomethyl)-2,3-dihydrobenzofuran

Related Reagents. See Classes O-1 and O-13, pages 1–10. Bromine; *N*-Bromosuccinimide; Copper(I) Bromide.

1. (a) *FF* **1967**, *1*, 161. (b) Bauer, D. P.; Macomber, R. S. *JOC* **1975**, *40*, 1990.
2. King, L. C.; Ostrum, G. K. *JOC* **1964**, *29*, 3459.
3. Doifode, K. B.; Marathey, M. G. *JOC* **1964**, *29*, 2025.
4. Jemison, R. W. *AJC* **1968**, *21*, 217.
5. Glazier, E. R. *JOC* **1962**, *27*, 4397.
6. Paquette, L. A.; Branan, B. M.; Rogers, R. D. *T* **1992**, *48*, 297.
7. Miller, D. D.; Moorthy, K. B.; Hamada, A. *TL* **1983**, *24*, 555.
8. Nakatsuka, M.; Nakasuji, K.; Murata, I.; Watanabe, I.; Saito, G.; Enoki, T.; Inokuchi, H. *CL* **1983**, 905.
9. Matsumoto, M.; Ishida, Y.; Watanabe, N. *H* **1985**, *23*, 165.

10. Barrish, J. C.; Singh, J.; Spergel, S. H.; Han, W.-C.; Kissick, T. P.; Kronenthal, D. R.; Mueller, R. H. *JOC* **1993**, *58*, 4494.
11. Castro, C. E.; Gaughan, E. J.; Owsley, D. C. *JOC* **1965**, *30*, 587.
12. Doifode, K. B. *JOC* **1962**, *27*, 2665.
13. Koyano, T. *BCJ* **1971**, *44*, 1158.
14. (a) See Ref. 1a, p 162. (b) Mosnaim, D.; Nonhebel, D. C. *T* **1969**, *25*, 1591.
15. Mosnaim, D.; Nonhebel, D. C.; Russell, J. A. *T* **1969**, *25*, 3485.
16. Nonhebel, D. C.; Russell, J. A. *T* **1970**, *26*, 2781.
17. (a) Kovacic, P.; Davis, K. E. *JACS* **1964**, *86*, 427. (b) Kodomari, M.; Satoh, H.; Yoshitomi, S. *BCJ* **1988**, *61*, 4149.
18. Petruso, S.; Caronna, S.; Sprio, V. *JHC* **1990**, *27*, 1209.
19. Petruso, S.; Caronna, S.; Sferlazzo, M.; Sprio, V. *JHC* **1990**, *27*, 1277.
20. Petruso, S.; Caronna, S.; *JHC* **1992**, *29*, 355.
21. (a) Gruiec, A.; Foucaud, A.; Moinet, C. *NJC* **1991**, *15*, 943. (b) Chaintreau, A.; Adrian, G.; Couturier, D. *SC* **1981**, *11*, 669.
22. Kim, S.; Lee, J. I. *JOC* **1984**, *49*, 1712.
23. Sakata, H.; Aoki, Y.; Kuwajima, I. *TL* **1990**, *31*, 1161.
24. Takeda, T.; Inoue, T.; Fujiwara, T. *CL* **1988**, 985.
25. Takeda, T.; Ogawa, S.; Koyama, M.; Kato, T.; Fujiwara, T. *CL* **1989**, 1257.
26. Yamaguchi, J.; Takeda, T. *CL* **1992**, 423.
27. Yamaguchi, J.; Yamamoto, S.; Takeda, T. *CL* **1992**, 1185.
28. (a) *FF* **1989**, *14*, 100. (b) *FF* **1980**, *8*, 196.
29. Yoshida, J.; Tamao, K.; Kakui, T.; Kurita, A.; Murata, M.; Yamada, K.; Kumada, M. *OM* **1982**, *1*, 369.
30. Dickerman, S. C.; DeSouza, D. J.; Jacobson, N. *JOC* **1969**, *34*, 710.
31. Doyle, M. P.; Siegfried, B.; Dellaria, J. F. *JOC* **1977**, *42*, 2426.
32. Oae, S.; Shinhama, K.; Kim, Y. H. *BCJ* **1980**, *53*, 1065.
33. Doyle, M. P.; Siegfried, B.; Elliott, R. C.; Dellaria, J. F. *JOC* **1977**, *42*, 2431.
34. Meijs, G. F.; Beckwith, A. L. J. *JACS* **1986**, *108*, 5890.
35. Perrin, D. D.; Armarego, W. L. F. *Purification of Laboratory Chemicals*, 3rd ed.; Pergamon: New York, 1988; p 321.

Nicholas D. P. Cosford
SIBIA, La Jolla, CA, USA

Copper(II) Chloride

$CuCl_2$

[7447-39-4] Cl_2Cu (MW 134.45)
($\cdot 2H_2O$)
[10125-13-0] $Cl_2CuH_4O_2$ (MW 170.48)

(chlorinating agent; oxidizing agent; Lewis acid)

Physical Data: anhydrous: *d* 3.386 g cm^{-3}; mp 620 °C (reported mp of 498 °C actually describes a mixture of $CuCl_2$ and CuCl); partially decomposes above 300 °C to CuCl and Cl_2; dihydrate *d* 2.51 g cm^{-3}; mp 100 °C.

Solubility: anhydrous: sol water, alcohol, and acetone; dihydrate: sol water, methanol, ethanol; mod sol acetone, ethyl acetate; sl sol Et_2O.

Form Supplied in: anhydrous: hygroscopic yellow to brown microcrystalline powder; dihydrate: green to blue powder or crystals; also supplied as reagent adsorbed on alumina (approx. 30 wt % $CuCl_2$ on alumina).

Analysis of Reagent Purity: by iodometric titration.[70]

Purification: cryst from hot dil aq HCl (0.6 mL g^{-1}) by cooling in a $CaCl_2$–ice bath.[71]

Handling, Storage, and Precautions: the anhydrous solid should be stored in the absence of moisture, since the dihydrate is formed in moist air. Irritating to skin and mucous membranes.

Chlorination of Carbonyls. Copper(II) chloride effects the α-chlorination of various carbonyl functional groups.[1] The reaction is usually performed in hot, polar solvents containing ***Lithium Chloride***, which enhances the reaction rate. For example, butyraldehyde is α-chlorinated in DMF (97% conversion, eq 1) while the same reaction in methanol leads to an 80% yield of the corresponding α-chloro dimethyl acetal (eq 2).[2]

$CuCl_2$; DMF, Δ; 97% (1)

$CuCl_2$; MeOH, Δ; 80% (2)

The process has been extended to carboxylic acids, anhydrides, and acid chlorides by using an inert solvent such as sulfolane.[3] 4-Oxo-4,5,6,7-tetrahydroindoles are selectively α-chlorinated, allowing facile transformation to 4-hydroxyindoles (eq 3).[4] The ability of the reaction to form α-chloro ketones selectively has been further improved by the use of trimethylsilyl enol ethers as substrates.[5] Recently, phase-transfer conditions have been employed in a particularly difficult synthesis of RCH(Cl)C(O)Me selectively from the parent ketones (eq 4).[6]

$CuCl_2$; aq AcOH, Δ; R = Ts, 86%; R = Bn, 81% (3)

$CuCl_2$, H_2O, $EtNH_3NO_3$; $C_5H_5N(CH_2)_{15}MeCl$; n = 4, 70% (4)

R = Me$(CH_2)_n$, n = 2–5, 8

Chlorination of Aromatics. Aromatic systems may be chlorinated by the reagent. For example, 9-chloroanthracene is prepared in high yield by heating anthracene and $CuCl_2$ in carbon tetrachloride (eq 5).[7] When the 9-position is blocked by a halogen, alkyl, or aryl group, the corresponding 10-chloroanthracenes are formed by heating the reactants in chlorobenzene.[8,9] Under similar conditions, 9-acylanthracenes give 9-acyl-10-chloroanthracenes as the predominant products.[10] Polymethylbenzenes are efficiently and selectively converted to the nuclear chlorinated derivatives by $CuCl_2$/***Alumina*** (eq 6).[11]

Reactions with Alkoxy and Hydroxy Aromatics. Hydroxy aromatics such as phenols and flavanones undergo aromatic nuclear chlorination with copper(II) chloride.[12] Thus heating

3,5-xylenol with a slight excess of the reagent in toluene at 90 °C gave a 93% yield of 4-chloro-3,5-xylenol (eq 7).[13] 2-Alkoxynaphthalenes are similarly halogenated at the 1-position.[14] Attempted reaction of $CuCl_2$ with anisole at 100 °C for 5 h gave no products; in contrast, it was found that alkoxybenzenes were almost exclusively *para*-chlorinated (92–95% *para*:0.5–3% *ortho*) using $CuCl_2/Al_2O_3$ (eq 8).[15] Anisole reacts with benzyl sulfides in the presence of equimolar $CuCl_2$ and ***Zinc Chloride*** to give anisyl(phenyl)methanes (*para*:*ortho* = 2:1, eq 9).[16,17]

(5) R = H, Me, Ph, Ac; $CuCl_2$, Δ

(6) $CuCl_2$, Al_2O_3; PhCl, Δ; 82%

(7) $CuCl_2$; PhMe, Δ; 93%

(8) $CuCl_2$, Al_2O_3; PhCl, Δ; 94%; <0.5%

(9) $CuCl_2$, $ZnCl_2$; BnSMe, Δ; 53%; 2:1

Reactions with Active Methylene-Containing Compounds. 9-Alkoxy(or acyloxy)-10-methylanthracenes react with $CuCl_2$ to give coupled products (eq 10), while the analogous 9-alkoxy(or acyloxy)-10-benzyl(or ethyl)anthracenes react at the alkoxy or acyloxy group to afford 10-benzylidene(or ethylidene)anthrones (eq 11).[18] The reactions are believed to proceed via a radical mechanism.

(10) R = Me, Ac; $CuCl_2$

(11) R = Me, Ph; $CuCl_2$

Under similar conditions, 9-alkyl(and aryl)-10-halogenoanthracenes give products resulting from replacement of the halogen, alkyl, or aryl groups with halogen from the $CuCl_2$.[19] Boiling toluene reacts with $CuCl_2$ to yield a mixture of phenyltolylmethanes.[20]

Lithium enolates of ketones[21] and esters[22] undergo a coupling reaction with copper(II) halides to afford the corresponding 1,4-dicarbonyl compounds. Thus treating a 3:1 mixture of *t*-butyl methyl ketone and acetophenone with ***Lithium Diisopropylamide*** and $CuCl_2$ gives a 60% yield of the cross-coupled product (eq 12).

(12) 3:1; 1. LDA; 2. $CuCl_2$; 60%

The intramolecular variant of this reaction producing carbocyclic derivatives has been reported.[23] Copper(II) chloride catalyzes the Knoevenagel condensation of 2,4-pentanedione with aldehydes and tosylhydrazones (eq 13).[24] The reagent also catalyzes the reaction of various 1,3-dicarbonyls with dithianes such as benzaldehyde diethyl dithioacetal to give the corresponding condensation products (eq 14).[25]

(13) R = alkyl, aryl; $CuCl_2$, THF, 25 °C; 48–95%

(14) $CuCl_2$, THF, 25 °C; R = Me, 65%; R = OEt, 46%

Catalyst for Conjugate Additions. The catalytic effect of copper(II) chloride on the 1,4-addition of β-dicarbonyl compounds to (arylazo)alkenes[26,27] and aminocarbonylazoalkenes[28,29] has been studied in some detail. The reactions proceed at ambient temperature in THF and afford the corresponding pyrrole derivatives (eq 15). This mild method requires no other catalyst and succeeds with β-diketones, β-ketoesters, and β-ketoamides. Copper(II) chloride also catalyzes the addition of water, alcohols, phenol, and aromatic amines to arylazoalkenes (eq 16).[30]

(15) $CuCl_2$; THF, 25 °C; X = alkyl, aryl, OR, NHR; R^1 = Ar, ArNHCO; R^2, R^3 = alkyl, aryl, CO_2R

$$\mathrm{ArNH_2 + R^1N{=}N\text{-}C(R^2){=}CHR^3 \xrightarrow[\text{THF, 25 °C}]{CuCl_2} R^1NH\text{-}N{=}C(R^2)\text{-}CH(R^3)NHAr} \quad (16)$$

$R^1, R^2, R^3 = Ar$

Oxidation and Coupling of Phenolic Derivatives. In the presence of oxygen, copper(II) chloride converts phenol derivatives to various oxidation products. Depending on the reaction conditions, quinones and/or coupled compounds are formed.[31] Several groups have examined different sets of conditions employing $CuCl_2$ to favor either of these products. Thus 2,3,6-trimethylphenol was selectively oxidized to trimethyl-*p*-benzoquinone with $CuCl_2$/amine/O_2 as the catalyst (eq 17),[32] while 2,4,6-trimethylphenol was converted to 3,5-dimethyl-4-hydroxybenzaldehyde using a catalytic system employing either acetone oxime or amine (eq 18).[33,34]

$CuCl_2$, O_2, Et_2NH; ROH, 25 °C (17)

76.7% + 0.9% coupled product

$CuCl_2$, O_2, Me_2CNOH; ROH, 25 °C (18)

CHO 85.6% + 6.1%

The oxidation of alkoxyphenols to the corresponding quinones has been studied,[35] and even benzoxazole derivatives are oxidized by a mixture of copper(II) chloride and ***Iron(III) Chloride*** (eq 19).[36] A $CuCl_2$/O_2/alcohol catalytic system has been used for the oxidative coupling of monophenols.[37]

$CuCl_2$, $FeCl_3$, HCl, H_2O, EtOH; Δ; 94% (19)

Copper(II) amine complexes are very effective catalysts for the oxidative coupling of 2-naphthols to give symmetrical 1,1′-binaphthalene-2,2′-diols.[38] Recent work has extended this methodology to the cross-coupling of various substituted 2-naphthols.[39,40] For example, 2-naphthol and 3-methoxycarbonyl-2-naphthol are coupled under strictly anaerobic conditions using $CuCl_2$/***t-Butylamine*** in methanol to give the unsymmetrical binaphthol in 86% yield (eq 20).

$CuCl_2$, *t*-$BuNH_2$, MeOH, Δ; 86% (20)

Other ligands such as methoxide are also effective; a mechanistic study indicates that the selectivity for cross- rather than homo-coupling is dependent upon the copper:ligand ratio.[41] A 1:1 mixture of 2-naphthol and 2-naphthylamine is cross-coupled with $CuCl_2$/benzylamine to give 2-amino-2′-hydroxy-1,1′-binaphthyl (68% yield, eq 21).[42] The cross-coupled products from these reactions are important in view of their use as chiral ligands for asymmetric synthesis.

$CuCl_2$, *t*-$BuNH_2$, MeOH, Δ; 68% (21)

Dioxygenation of 1,2-Diones. 1,2-Cyclohexanedione derivatives have been converted to the corresponding 1,5-dicarbonyl compounds by oxidation with O_2 employing copper(II) chloride as the catalyst.[43] More recently, $CuCl_2$–***Hydrogen Peroxide*** has been used to prepare terminal dicarboxylic acids in high yield.[44] While 1,2-cyclohexanedione afforded α-chloroadipic acid in 85% yield, 1,2-cyclododecanedione was converted to 1,12-dodecanedioic acid in 47% yield under identical conditions (eq 22).

1. $CuCl_2$, H_2O_2, MeOH, H_2O, 20 °C; 2. H_2SO_4; 47% → $HO_2C(CH_2)_{10}CO_2H$ (22)

Addition of Sulfonyl Chlorides to Unsaturated Bonds. The addition of alkyl and aryl sulfonyl chlorides across double and triple bonds is catalyzed by copper(II) chloride.[45–51] The reaction appears to be quite general and proceeds via a radical chain mechanism. The 2-chloroethyl sulfones produced in the reaction with alkenes undergo base-induced elimination to give vinyl sulfones (eq 23).[45–48] 1,3-Dienes similarly react, yielding 1,4-addition products (eq 24) which may be dehydrohalogenated to 1,3-unsaturated sulfones.[45,49]

Ph–CH=CH–CH$_3$ + $PhSO_2Cl$ → (1. $CuCl_2$, MeCN, Δ; 2. NEt_3; 87%) → SO_2Ph product (23)

cyclooctadiene + $PhSO_2Cl$ → ($CuCl_2$, 100 °C, 62%) → Cl, SO_2Ph product (24)

The stereoselectivity of the addition to alkynes can be controlled by varying the solvent or additive, and thus favoring either the *cis* or *trans* β-chlorovinyl sulfone.[50,51] For example (eq 25), when benzenesulfonyl chloride is reacted with phenylacetylene in acetonitrile with added triethylamine hydrochloride, the *trans*:*cis* ratio is 92:8, while the same reaction performed in CS_2 without additive favors the *cis* isomer (16:84).

$$\text{Ph}{-}{\equiv} + \text{PhSO}_2\text{Cl} \xrightarrow[100\,^\circ\text{C}]{\text{(a) or (b)}} \text{Ph(Cl)C}{=}\text{CH(SO}_2\text{Ph)}\ (\textit{trans}) + \text{Ph(Cl)C}{=}\text{CH(SO}_2\text{Ph)}\ (\textit{cis}) \quad (25)$$

(a) = $CuCl_2$, NEt_3HCl, MeCN (a) *trans:cis* = 92:8
(b) = $CuCl_2$, CS_2 (b) *trans:cis* = 16:84

Acylation Catalyst. *N*-Trimethylsilyl derivatives of (+)-bornane-2,10-sultam (Oppolzer's chiral sultam) and chiral 2-oxazolidinones (the Evans chiral auxiliaries) are *N*-acylated with a number of acyl chlorides including acryloyl chloride in refluxing benzene in the presence of $CuCl_2$.[52] The *N*-acylated products were prepared in high yields; the method does not require an aqueous workup, making it advantageous for large-scale preparations.

Racemization Suppression in Peptide Couplings. A mixture of copper(II) chloride and ***Triethylamine*** catalyzes the formation of peptide bonds.[53] Furthermore, when used as an additive, $CuCl_2$ suppresses racemization in both the carbodiimide[54] and mixed anhydride[55] peptide coupling methods. Recently it was shown that a combination of ***1-Hydroxybenzotriazole*** and $CuCl_2$ gives improved yields of peptides while eliminating racemization.[56,57]

Reaction with Palladium Complexes. π-Allylpalladium complexes undergo oxidative cleavage with copper(II) chloride to form allyl chlorides with the concomitant release of $PdCl_2$ (eq 26).[58]

$$[\pi\text{-allyl–Pd–Cl}]_2 \xrightarrow[\text{EtOH, 85\%}]{\text{CuCl}_2} \text{allyl chloride (Cl)} \quad (26)$$

This methodology has been used in the dimerization of allenes to 2,3-bis(chloromethyl)butadienes.[59] 1,5-Bismethylenecyclooctane was transformed into the bridgehead-substituted bicyclo[3.3.1]nonane system using $CuCl_2$/HOAc/NaOAc, while the same substrate produced bicyclo[4.3.1]decane derivatives (eq 27) with a ***Palladium(II) Chloride***/$CuCl_2$ catalytic system.[60]

$$\xrightarrow[\text{HOAc, NaOAc}]{\text{CuCl}_2} \text{bicyclo[3.3.1]nonane–X}; \quad \xrightarrow[\text{LiCl, O}_2\text{, HOAc}]{\text{PdCl}_2\text{, CuCl}_2} \text{X–bicyclo[4.3.1]decane–X} \quad (27)$$

X = Cl, OAc

While reaction of a steroidal π-allylpalladium complex with AcOK yields the allyl acetate arising from *trans* attack, treatment of a steroidal alkene with $PdCl_2/CuCl_2$/AcOK/AcOH gave the allyl acetate arising from *cis* attack.[61]

Reoxidant in Catalytic Palladium Reactions. Copper(II) chloride has been used extensively in catalytic palladium chemistry for the regeneration of Pd^{II} in the catalytic cycle. In particular, the reagent has found widespread use in the carbonylation of alkenes,[62–64] alkynes,[65] and allenes[66,67] to give carboxylic acids and esters using $PdCl_2/CuCl_2$/CO/HCl/ROH, and in the oxidation of alkenes to ketones with a catalytic $PdCl_2/CuCl_2/O_2$ system (the Wacker reaction).[68] The $PdCl_2/CuCl_2$/CO/NaOAc catalytic system has been used in a mild method for the carbonylation of β-aminoethanols, diols, and diol amines (eq 28).[69]

$$\text{R}^1\text{HN–CH}_2\text{–CH(OH)R}^2 \xrightarrow[\text{CO, NaOAc}]{\text{PdCl}_2\text{, CuCl}_2} \text{3-R}^1\text{-5-R}^2\text{-oxazolidin-2-one} \quad (28)$$

Cyclopropanation with $CuCl_2$–$Cu(OAc)_2$ Catalyst. ***Ethyl Cyanoacetate*** reacts with alkenes under $CuCl_2$–***Copper(II) Acetate*** catalysis to give cyclopropanes.[72] Thus heating cyclohexene in DMF (110 °C, 5 h) with this reagent combination gives a 53% yield of the isomeric cyclopropanes. The reaction also proceeds with styrene, 1-decene, and isobutene. Byproducts formed from the addition to the alkene are removed with ***Potassium Permanganate***.

Related Reagents. See Class O-5, pages 1–10. Chlorine; *N*-Chlorosuccinimide; Copper(I) Chloride; Copper(II) Chloride–Copper(II) Oxide; Copper(I) Chloride–Oxygen; Copper(I) Chloride–tetrabutylammonium Chloride; Copper(I) Chloride–Sulfur Dioxide; Iodine–Aluminum(III) Chloride–Copper(II) Chloride; Iodine–Copper(I) Chloride–Copper(II) Chloride; Iodine–Copper(II) Chloride; Methylmagnesium Iodide–Copper(I) Chloride; Palladium(II) Chloride–Copper(I) Chloride; Palladium(II) Chloride–Copper(II) Chloride; Phenyl Selenocyanate–Copper(II) Chloride; Vinylmagnesium Chloride–Copper(I) Chloride; Zinc–Copper(I) Chloride.

1. *FF* **1969**, *2*, 84.
2. Castro, C. E.; Gaughan, E. J.; Owsley, D. C. *JOC* **1965**, *30*, 587.
3. Louw, R. *CC* **1966**, 544.
4. Matsumoto, M.; Ishida, Y.; Watanabe, N. *H* **1985**, *23*, 165.
5. *FF* **1982**, *10*, 106.
6. Atlamsani, A.; Brégeault, J.-M. *NJC* **1991**, *15*, 671.
7. (a) *FF* **1967**, *1*, 163. (b) Nonhebel, D. C. *OSC* **1973**, *5*, 206.
8. Mosnaim, D.; Nonhebel, D. C. *T* **1969**, *25*, 1591.
9. Mosnaim, D.; Nonhebel, D. C.; Russell, J. A. *T* **1969**, *25*, 3485.
10. Nonhebel, D. C.; Russell, J. A. *T* **1970**, *26*, 2781.
11. Kodomari, M.; Satoh, H.; Yoshitomi, S. *BCJ* **1988**, *61*, 4149.
12. *FF* **1980**, *8*, 120.
13. Crocker, H. P.; Walser, R. *JCS(C)* **1970**, 1982.
14. *FF* **1975**, *5*, 158.
15. Kodomari, M.; Takahashi, S.; Yoshitomi, S. *CL* **1987**, 1901.
16. Mukaiyama, T.; Narasaka, K.; Hokonoki, H. *JACS* **1969**, *91*, 4315.
17. Mukaiyama, T.; Maekawa, K.; Narasaka, K. *TL* **1970**, 4669.
18. (a) *FF* **1969**, *2*, 86. (b) Mosnaim, A. D.; Nonhebel, D. C.; Russell, J. A. *T* **1970**, *26*, 1123.
19. Mosnaim, D. A.; Nonhebel, D. C. *JCS(C)* **1970**, 942.
20. Cummings, C. A.; Milner, D. J. *JCS(C)* **1971**, 1571.
21. (a) *FF* **1977**, *6*, 139. (b) Ito, Y.; Konoike, T.; Harada, T.; Saegusa, T. *JACS* **1977**, *99*, 1487.
22. Rathke, M. W.; Lindert, A. *JACS* **1971**, *93*, 4605.
23. (a) *FF* **1981**, *9*, 123. (b) Babler, J. H.; Sarussi, S. J. *JOC* **1987**, *52*, 3462.

24. Attanasi, O.; Filippone, P.; Mei, A. *SC* **1983**, *13*, 1203.
25. Mukaiyama, T.; Narasaka, K.; Maekawa, K.; Hokonoki, H. *BCJ* **1970**, *43*, 2549.
26. Attanasi, O.; Santeusanio, S. *S* **1983**, 742.
27. Attanasi, O.; Bonifazi, P.; Foresti, E.; Pradella, G. *JOC* **1982**, *47*, 684.
28. Attanasi, O.; Filippone, P.; Mei, A.; Santeusanio, S.; Serra-Zanetti, F. *S* **1985**, 157.
29. Attanasi, O.; Filippone, P.; Mei, A.; Santeusanio, S. *S* **1984**, 671.
30. Attanasi, O.; Filippone, P. *S* **1984**, 422.
31. Hewitt, D. G. *JCS(C)* **1971**, 2967.
32. Shimizu, M.; Watanabe, Y.; Orita, H.; Hayakawa, T.; Takehira, K. *BCJ* **1992**, *65*, 1522.
33. Shimizu, M.; Watanabe, Y.; Orita, H.; Hayakawa, T.; Takehira, K. *BCJ* **1993**, *66*, 251.
34. Takehira, K.; Shimizu, M.; Watanabe, Y.; Orita, H.; Hayakawa, T. *TL* **1990**, *31*, 2607.
35. Matsumoto, M.; Kobayashi, H. *SC* **1985**, *15*, 515.
36. Hegedus, L. S.; Odle, R. R.; Winton, P. M.; Weider, P. R. *JOC* **1982**, *47*, 2607.
37. Takizawa, Y.; Munakata, T.; Iwasa, Y.; Suzuki, T.; Mitsuhashi, T. *JOC* **1985**, *50*, 4383.
38. Brussee, J.; Groenendijk, J. L. G.; Koppele, J. M.; Jansen, A. C. A. *T* **1985**, *41*, 3313.
39. Hovorka, M.; Günterová, J.; Závada, J. *TL* **1990**, *31*, 413.
40. Hovorka, M.; Ščigel, R.; Gunterová, J.; Tichý, M.; Závada, J. *T* **1992**, *48*, 9503.
41. Hovorka, M.; Závada, J. *T* **1992**, *48*, 9517.
42. Smrčina, M.; Lorenc, M.; Hanuš, V.; Kočovský, P. *SL* **1991**, 231.
43. Utaka, M.; Hojo, M.; Fujii, Y.; Takeda, A. *CL* **1984**, 635.
44. Starostin, E. K.; Mazurchik, A. A.; Ignatenko, A. V.; Nikishin, G. I. *S* **1992**, 917.
45. Asscher, M.; Vofsi, D. *JCS* **1964**, 4962.
46. *FF* **1975**, *5*, 158.
47. Truce, W. E.; Goralski, C. T. *JOC* **1971**, *36*, 2536.
48. Truce, W. E.; Goralski, C. T.; Christensen, L. W.; Bavry, R. H. *JOC* **1970**, *35*, 4217.
49. Truce, W. E.; Goralski, C. T. *JOC* **1970**, *35*, 4220.
50. Amiel, Y. *TL* **1971**, 661.
51. *FF* **1974**, *4*, 107.
52. Thom, C.; Kocieński, P. *S* **1992**, 582.
53. *FF* **1975**, *5*, 158.
54. Miyazawa, T.; Otomatsu, T.; Yamada, T.; Kuwata, S. *TL* **1984**, *25*, 771.
55. Miyazawa, T.; Donkai, T.; Yamada, T.; Kuwata, S. *CL* **1989**, 2125.
56. Miyazawa, T.; Otomatsu, T.; Fukui, Y.; Yamada, T.; Kuwata, S. *CC* **1988**, 419.
57. Miyazawa, T.; Otomatsu, T.; Fukui, Y.; Yamada, T.; Kuwata, S. *Int. J. Pept. Prot. Res.* **1992**, *39*, 308.
58. Castanet, Y.; Petit, F. *TL* **1979**, *34*, 3221.
59. Hegedus, L. S.; Kambe, N.; Ishii, Y.; Mori, A. *JOC* **1985**, *50*, 2240.
60. Heumann, A.; Réglier, M.; Waegell, B. *TL* **1983**, *24*, 1971.
61. Horiuchi, C. A.; Satoh, J. Y. *JCS(P1)* **1982**, 2595.
62. Alper, H.; Woell, J. B.; Despeyroux, B.; Smith, D. J. H. *CC* **1983**, 1270.
63. Inomata, K.; Toda, S.; Kinoshita, H. *CL* **1990**, 1567.
64. Toda, S.; Miyamoto, M.; Kinoshita, H.; Inomata, K. *BCJ* **1991**, *64*, 3600.
65. Alper, H.; Despeyroux, B.; Woell, J. B. *TL* **1983**, *24*, 5691.
66. Alper, H.; Hartstock, F. W.; Despeyroux, B. *CC* **1984**, 905.
67. Gallagher, T.; Davies, I. W.; Jones, S. W.; Lathbury, D.; Mahon, M. F.; Molloy, K. C.; Shaw, R. W.; Vernon, P. *JCS(P1)* **1992**, 433.
68. Januszkiewicz, K.; Alper, H. *TL* **1983**, *24*, 5159.
69. Tam, W. *JOC* **1986**, *51*, 2977.
70. *Reagent Chemicals: American Chemical Society Specifications*, 8th ed.; American Chemical Society: Washington, 1993; p 279.
71. Perrin, D. D.; Armarego, W. L. F. *Purification of Laboratory Chemicals*, 3rd ed.; Pergamon: New York, 1988; p 322.
72. Barreau, M.; Bost, M.; Julia, M.; Lallemand, J.-Y. *TL* **1975**, 3465.

Nicholas D. P. Cosford
SIBIA, La Jolla, CA, USA

(Diacetoxyiodo)benzene[1-3]

$PhI(OAc)_2$

[3240-34-4] $C_{10}H_{11}IO_4$ (MW 322.10)

(transannular carbocyclization,[6] *vic*-diazide formation,[7] α-hydroxy dimethyl acetal formation,[8,10,11,13] oxetane formation,[9] chromone, flavone, chalcone oxidation,[11,12] arene–$Cr(CO)_3$ functionalization,[14] phenolic oxidation[16] and coupling,[17,18] lactol fragmentation,[19] iodonium ylides and intramolecular cyclopropanation,[20] oxidation of amines[24–28] and indoles,[30,31] hydrazine derivatives (diimide[32] and azodicarbonyls[33]) and radical type intramolecular oxide formation[44–46])

Alternate Names: phenyliodine(III) diacetate; DIB; iodobenzene diacetate; IBD.
Physical Data: mp 163–165 °C.
Solubility: sol AcOH, MeCN, CH_2Cl_2; in KOH or $NaHCO_3$/MeOH it is equivalent to $PhI(OH)_2$.
Form Supplied in: commercially available as a white solid.
Preparative Method: by reaction of iodobenzene with ***Peracetic Acid***.[4,5]
Purification: recrystallization from 5 M acetic acid.[4]
Handling, Storage, and Precautions: a stable compound which can be stored indefinitely.

Reactions with Alkenes. Reactions of simple alkenes with $PhI(OAc)_2$ are not synthetically useful because of formation of multiple products.

Transannular carbocyclization in the reaction of *cis,cis*-1,5-cyclooctadiene yields a mixture of three diastereomers of 2,6-diacetoxy-*cis*-bicyclo[3.3.0]octane, a useful precursor of *cis*-bicyclo[3.3.0]octane-2,6-dione (eq 1).[6]

(1) [cyclooctadiene → (PhI(OAc)2, AcOH) → diacetoxy bicyclooctane (OAc, AcO)]

$PhI(OAc)_2/NaN_3$/AcOH yields vicinal diazides (eq 2).[7]

(2) [benzofuran → (PhI(OAc)2, NaN3, AcOH) → 2,3-diazido-2,3-dihydrobenzofuran (N3, N3)]

Oxidation of Ketones to α-Hydroxyl Dimethyl Acetals. Ketones are converted to the α-hydroxy dimethyl acetal upon reaction with $PhI(OAc)_2$ in methanolic potassium hydroxide (eqs 3–5).[8]

$$\text{ArCOMe} \xrightarrow[\text{MeOH, KOH}]{\text{PhI(OAc)}_2} \text{ArC(OMe)}_2\text{CH}_2\text{OH} \quad (3)$$

(4) [2,6-diacetylpyridine → (PhI(OAc)2, MeOH, KOH) → 2,6-bis(hydroxyacetyl)pyridine (HO, OH)]

(5) [1,1'-diacetylferrocene (Fe) → (PhI(OAc)2, MeOH, KOH) → bis(hydroxyacetyl)ferrocene (OH)]

Several potentially oxidizable groups are unaffected in this reaction (eq 6).[13]

$$\text{R}^1\text{COCH}_2\text{R}^2 \xrightarrow[\text{MeOH, KOH}]{\text{PhI(OAc)}_2} \text{R}^1\text{C(OMe)}_2\text{CH(OH)R}^2 \quad (6)$$

R¹ = 2-pyridyl, R^2 = H, 61% R^1 = Ph, R^2 = CH_2–N(pyrrolidinyl) 25%
R¹ = 3-pyridyl, R^2 = H, 45% R^1 = Ph, R^2 = CH_2–N(piperidinyl) 50%
R¹ = 4-pyridyl, R^2 = H, 58% R^1 = Ph, R^2 = CH_2–N(thiomorpholinyl) 65%
R¹ = pyrazinyl, R^2 = H, 62% R^1 = Ph, R^2 = CH_2–N(morpholinyl) 60%
R¹ = 2-benzimidazolyl, R^2 = H, 65% R^1 = Ph, 4-$MeOC_6H_4$ R^2 = CH_2–N(morpholinyl) 44%

In the case of a 17α-hydroxy steroid the hydroxy group acts as an intramolecular nucleophile to yield the 17-spirooxetan-20-one. It is noteworthy that the 3β-hydroxy-Δ^5-system is unaffected (eq 7).[9]

(7) [17α-hydroxy-20-keto steroid (HO, OH, R) → (PhI(OAc)2, MeOH, KOH) → 17-spirooxetan-20-one (O, R, HO)]

R = H, NHAc, Me

cis-3-Hydroxyflavonone is obtained via acid-catalyzed hydrolysis of *cis*-3-hydroxyflavone dimethyl acetal, which is formed upon treatment of flavanone with $PhI(OAc)_2$ (eq 8).[10,11]

(8) [flavanone (O, Ph) → 3-hydroxy dimethyl acetal (OH, MeO, OMe) → 3-substituted flavanone (OR)]

R = H, Ac

α,β-Unsaturated ketones, such as chromone, flavone, chalcone, and flavanone, yield α-hydroxy-β-methoxy dimethyl acetal products (eqs 9–11).[12a]

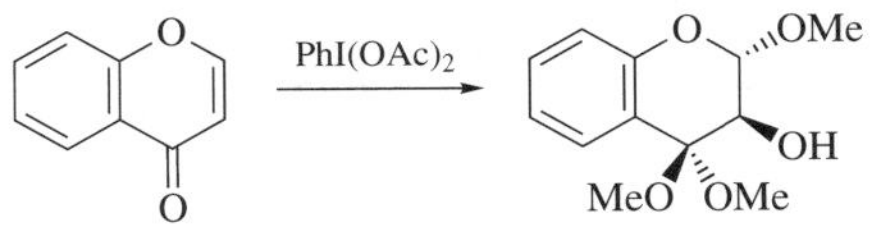

(9)

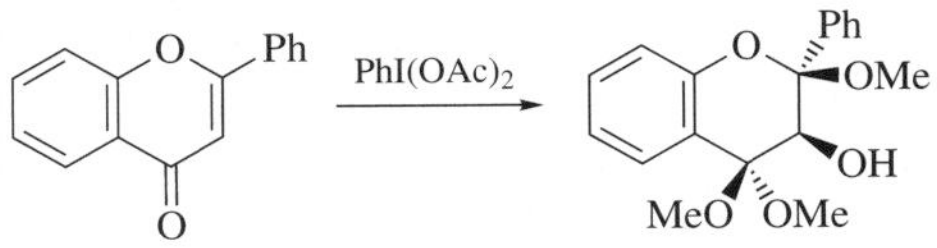

(10)

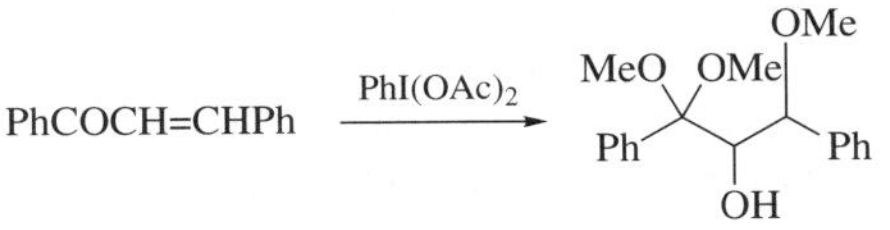

(11)

Intramolecular participation by the *ortho* hydroxy group occurs in the reaction of substituted *o*-hydroxyacetophenones, yielding the corresponding coumaran-3-ones (eq 12).[12b]

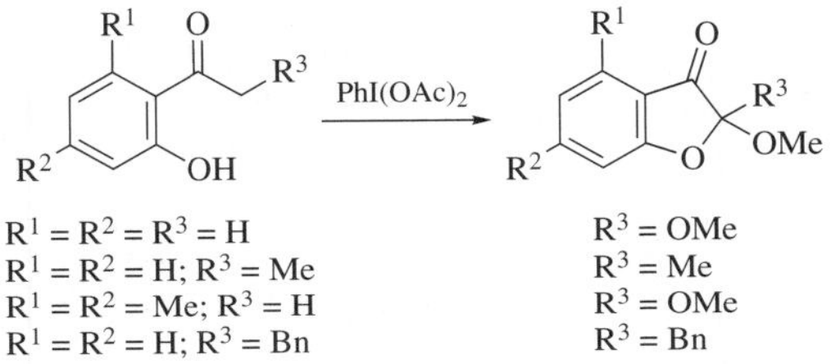

(12)

$R^1 = R^2 = R^3 = H$ → $R^3 = OMe$
$R^1 = R^2 = H; R^3 = Me$ → $R^3 = Me$
$R^1 = R^2 = Me; R^3 = H$ → $R^3 = OMe$
$R^1 = R^2 = H; R^3 = Bn$ → $R^3 = Bn$

Formation of the α-hydroxy dimethyl acetal occurs without reaction of the $Cr(CO)_3$ complex of η^6-benzo-cycloalkanones (eqs 13–15).[14]

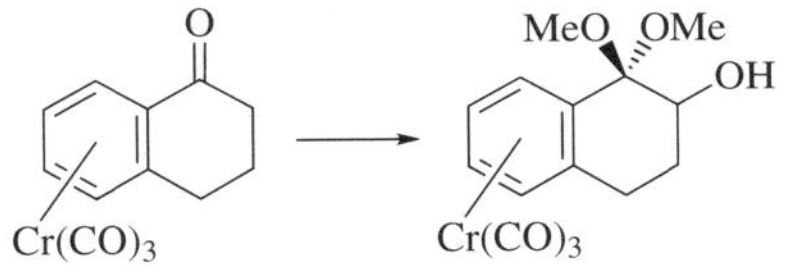

(13)

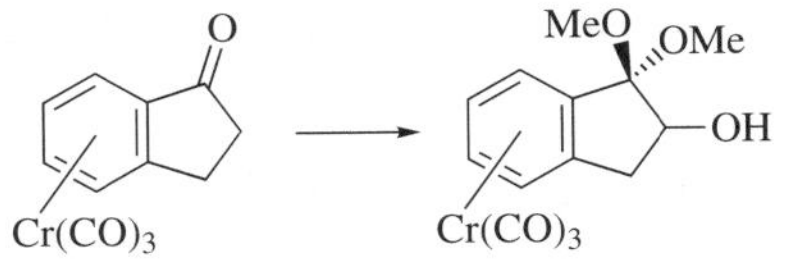

(14)

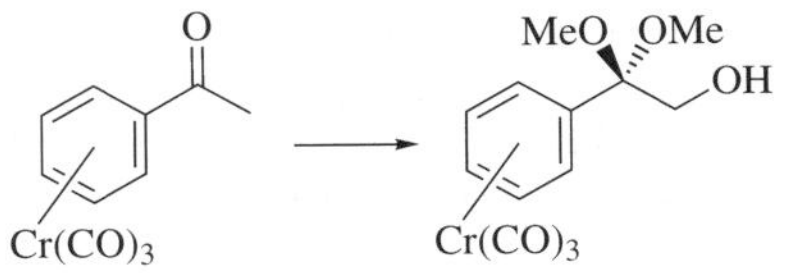

(15)

Carbon–Carbon Bond Cleavage with $PhI(OAc)_2$/$TMSN_3$. $PhI(OAc)_2$/***Azidotrimethylsilane*** reacts with unsaturated compounds even at −53 °C to yield keto nitriles (eq 16).[15]

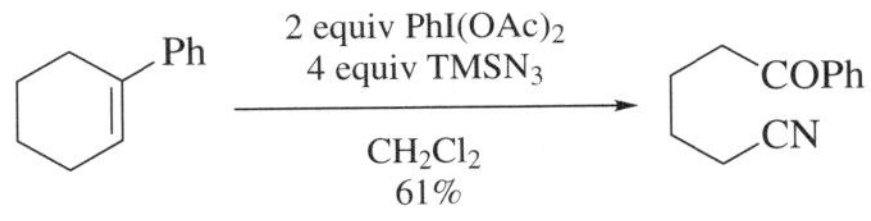

(16)

Oxidation of Phenols. Phenols are oxidized using $PhI(OAc)_2$ with nucleophilic attack by solvent (eqs 17 and 18),[16] or with intramolecular nucleophilic addition amounting to an overall oxidative coupling as with the bisnaphthol (eq 19)[17] and also with the conversion of reticuline to salutaridine (eq 20).[18]

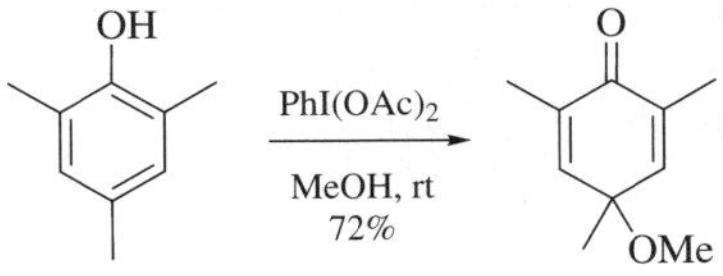

(17)

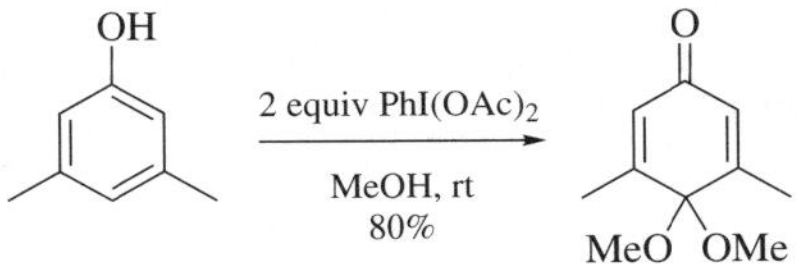

(18)

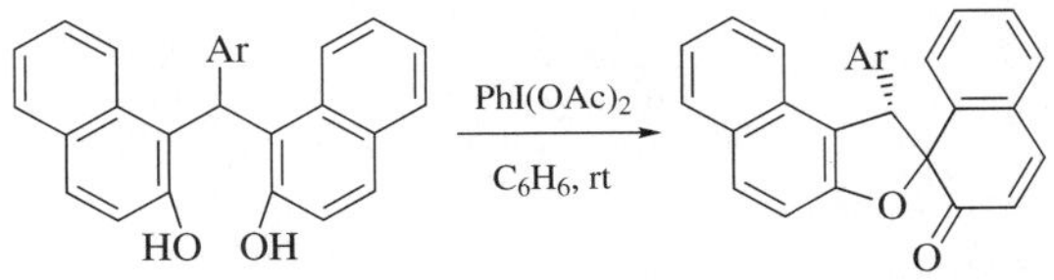

(19)

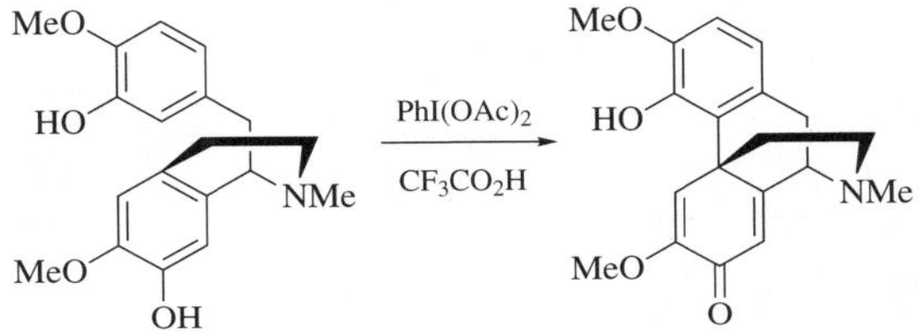

(20)

Fragmentation of Lactols to Unsaturated Medium-Ring Lactones. Ring cleavage to form a medium-sized ring lactone with a transannular double bond has been observed (eq 21).[19]

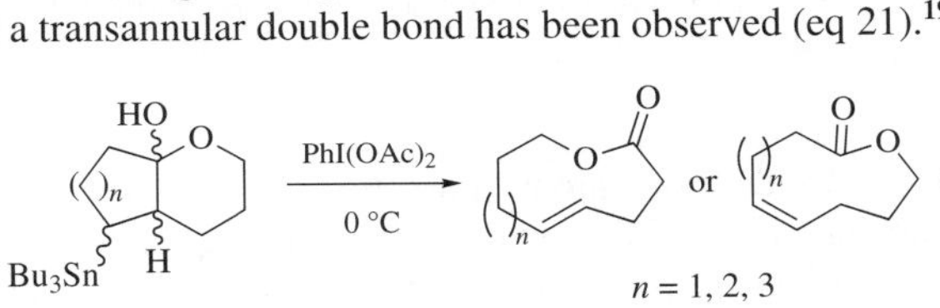

(21)

Reactions with β-Dicarbonyl Systems; Formation of Iodonium Ylides; Intramolecular Cyclopropanation. β-Dicarbonyl compounds upon reaction with $PhI(OAc)_2$ and KOH/MeOH at 0 °C yield isolable iodonium ylides (eq 22).[20] This is a general reaction which requires two stabilizing groups flanking the carbon of the C=I group, such as NO_2 and SO_2Ph.[21] Decomposition of unsaturated analogs in the presence of ***Copper(I) Chloride*** proceeds with intramolecular cyclopropanation (Table 1).[20]

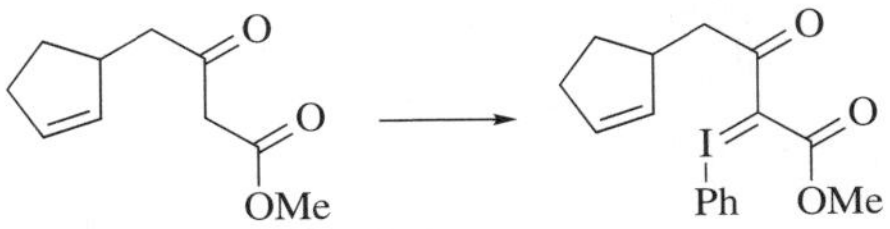

(22)

An asymmetric synthesis of a vitamin D ring A synthon employed this intramolecular cyclopropanation reaction (eq 23).[22]

Oxidation of Amines. Aromatic amines are oxidized with $PhI(OAc)_2$ to azo compounds in variable yield. $PhI(OAc)_2$ in benzene oxidizes aniline in excellent yield (eq 24);[23] however, substituted anilines give substantially lower yields.

Table 1 Intramolecular Cyclopropanation of Iodonium Ylides

Reactants	Iodonium ylide	Product	Yield (%)
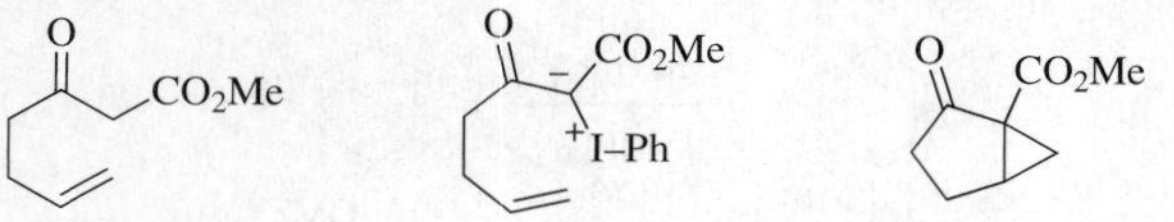			76
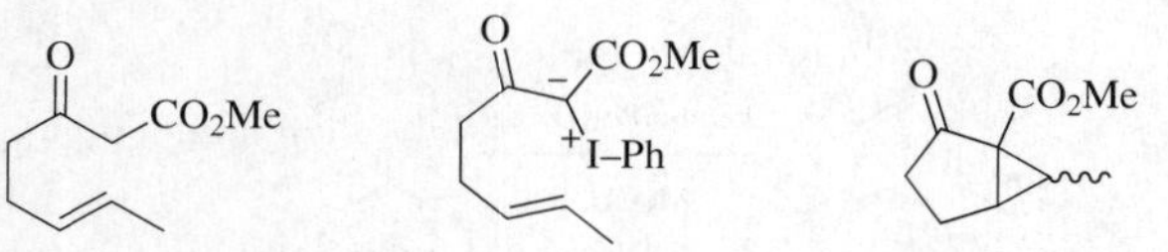			81
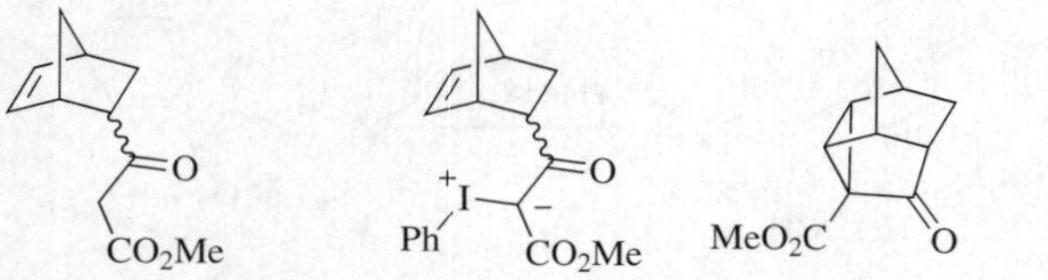			85
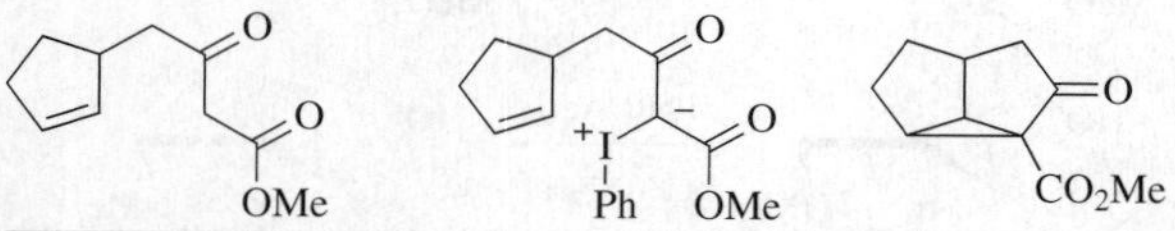			82

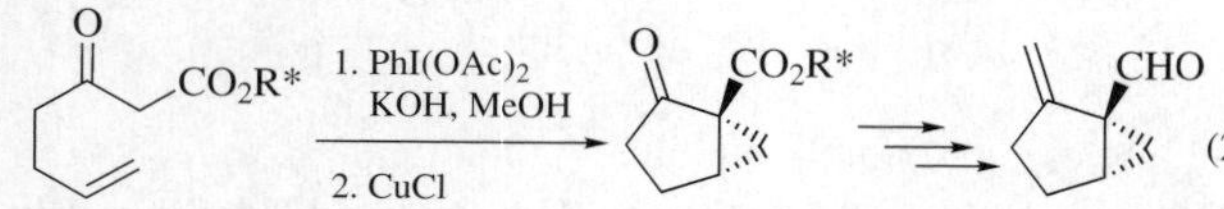

1. $PhI(OAc)_2$, KOH, MeOH; 2. CuCl (23)

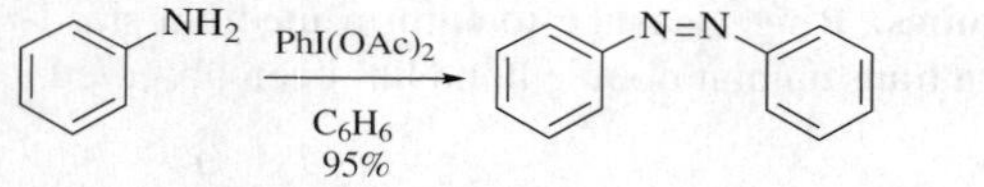

$PhI(OAc)_2$, C_6H_6, 95% (24)

Intramolecular azo group formation is a useful reaction for the formation of dibenzo[*c,f*]diazepine (eq 25).[24,25] Other *ortho* groups may react intramolecularly to yield the benzotriazole (eq 26), benzofuroxan (eq 27), or anthranil (eq 28) derivatives.[24–28]

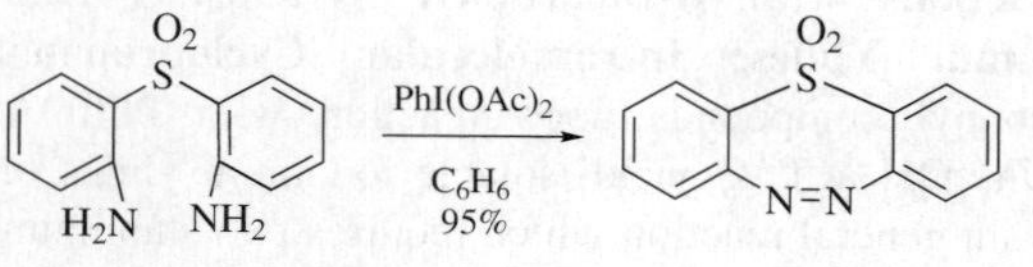

$PhI(OAc)_2$, C_6H_6, 95% (25)

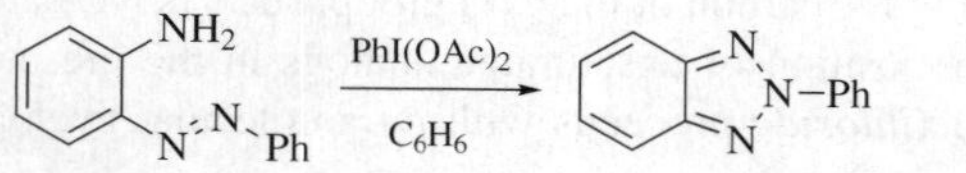

$PhI(OAc)_2$, C_6H_6 (26)

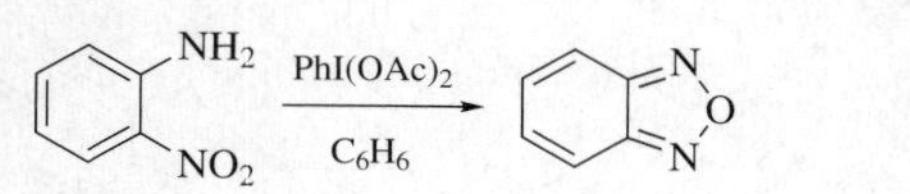

$PhI(OAc)_2$, C_6H_6 (27)

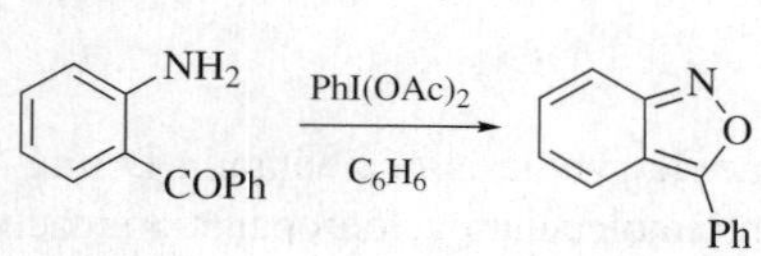

$PhI(OAc)_2$, C_6H_6 (28)

A number of examples of oxidative cyclization of 2-(2′-pyridylamino)imidazole[1,2-*a*]pyridines to dipyrido[1,2-*a*:2′,1′-*f*]-1,3,4,6-tetraazapentalenes with $PhI(OAc)_2/CF_3CH_2OH$ have been reported (eq 29).[29]

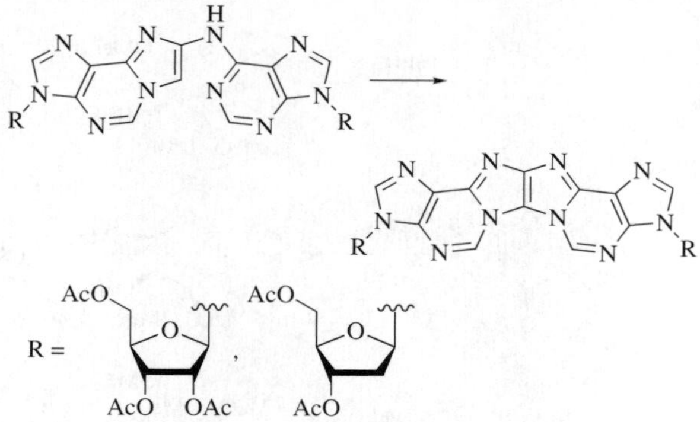

(29)

In the case of the oxidation of indole derivatives, nucleophilic attack by solvent may occur (eq 30).[30] Reserpine undergoes an analogous alkoxylation.[30] In the absence of a nucleophilic solvent, intramolecular cyclization occurs, an example of which is illustrated in the total synthesis of sporidesmin A (eq 31).[31]

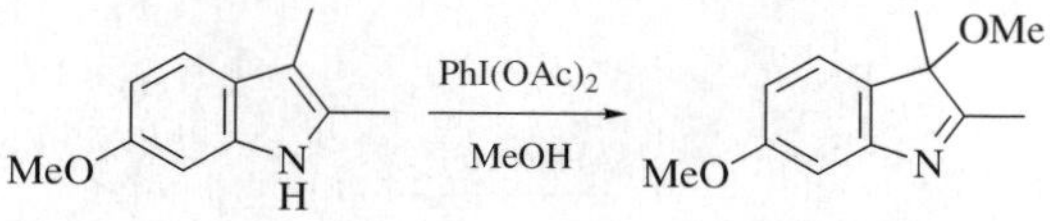

$PhI(OAc)_2$, MeOH (30)

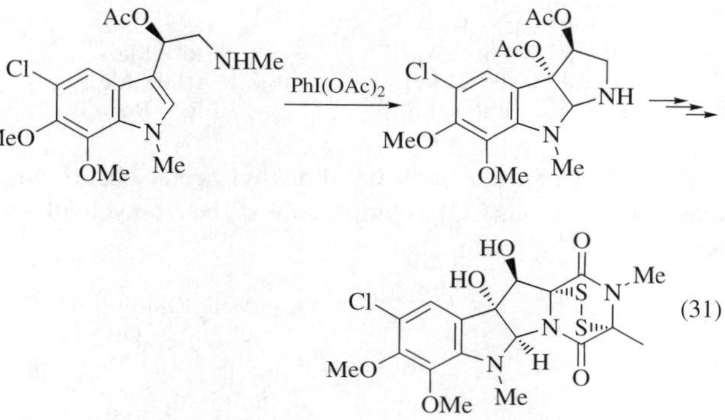

$PhI(OAc)_2$ (31)

Hydrazine is oxidized by $PhI(OAc)_2$ to diimide, which may be used to reduce alkenes and alkynes under mild conditions (Table 2).[32]

Table 2 Diimide Reduction of Various Compounds

Compound	Product	Yield (%)
$PhSCH{=}CH_2$	$PhSCH_2Me$	85
cis-$EtO_2CCH{=}CHCO_2Et$	$EtO_2CCH_2CH_2CO_2Et$	94
$EtO_2C{-}N{=}N{-}CO_2Et$	$EtO_2CNH{-}NHCO_2Et$	90
Maleic anhydride	$(MeCO)_2O$	83
PhC″CPh	*cis*-PhCH=CHPh	80
$PhCH{=}CHCO_2Et$	$PhCH_2CH_2CO_2Et$	96
$CH_2{=}CHCN$	$MeCH_2CN$	97

The hydrazodicarbonyl group is smoothly oxidized by $PhI(OAc)_2$ to the azodicarbonyl group (eqs 32 and 33).[33]

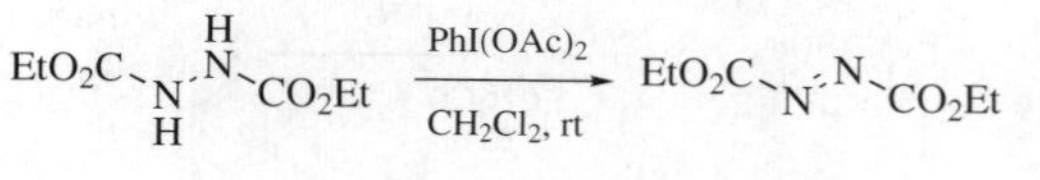

$EtO_2C{-}NH{-}NH{-}CO_2Et \xrightarrow[CH_2Cl_2,\ rt]{PhI(OAc)_2} EtO_2C{-}N{=}N{-}CO_2Et$ (32)

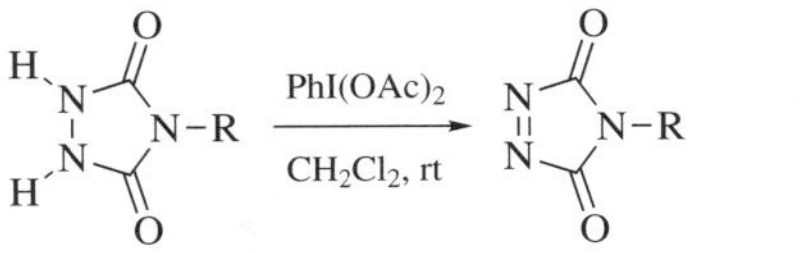

(33)

An intramolecular application of this reaction was used in a tandem sequence with $PhI(OAc)_2$ oxidation and a Diels–Alder reaction in the synthesis of nonpeptide β-turn mimetics (eq 34).[34,35]

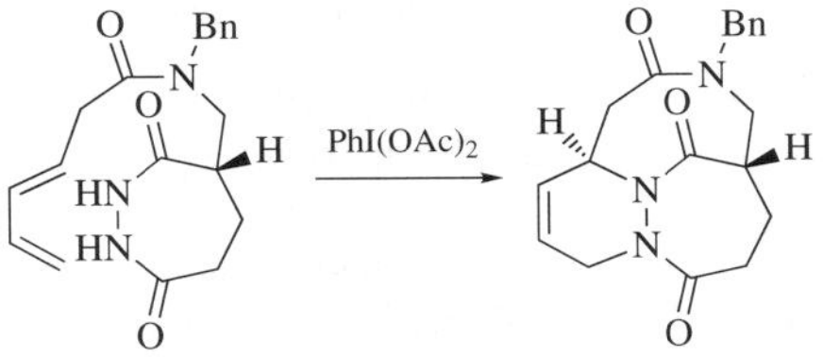

(34)

Oxidation of 5-Substituted Pyrazol-3(2*H*)-ones; Formation of Alkynyl Esters. Oxidation of various 5-substituted pyrazol-3(2*H*)-ones proceeded with fragmentative loss of molecular nitrogen to yield methyl-2-alkynoates (eq 35).[36] An analogous fragmentation process with pyrazol-3(2*H*)-ones occurs with ***Thallium(III) Nitrate***[37,38] and ***Lead(IV) Acetate***.[39]

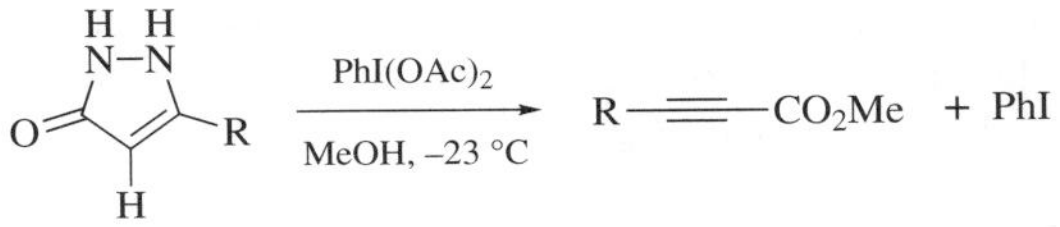

(35)

R = Me, 60% R = Ph, 59%
R = Et, 63% R = *p*-ClC_6H_4, 61%
R = CO_2Me, 59% R = *p*-MeC_6H_4, 59%
R = CH_2CO_2Me, 62% R = *p*-$MeOC_6H_4$, 64%

Oxidation of Hydrazones, Alkylhydrazones, *N*-Amino Heterocycles, *N*-Aminophthalimidates, and Aldazines. The oxidation of hydrazones to diazo compounds is not a generally useful reaction but it was uniquely effective in the oxidation of a triazole derivative (eq 36).[40]

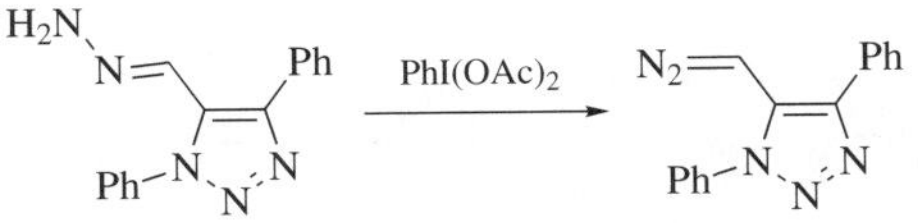

(36)

Oxidation of arylhydrazones proceeds with intramolecular cyclizations (eqs 37 and 38)[41] and aziridines may be formed via nitrene additions (eq 39).[42]

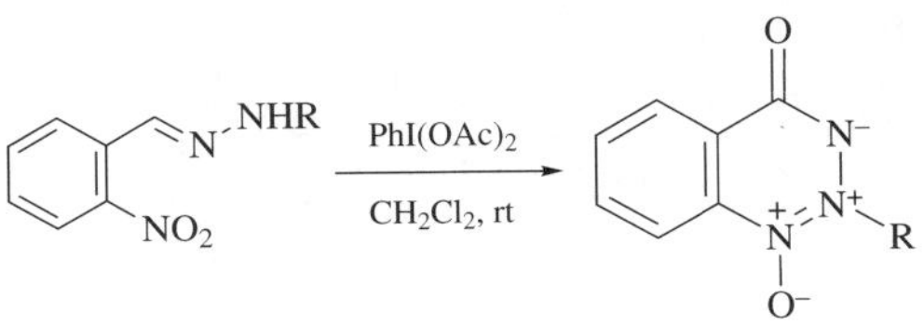

(37)

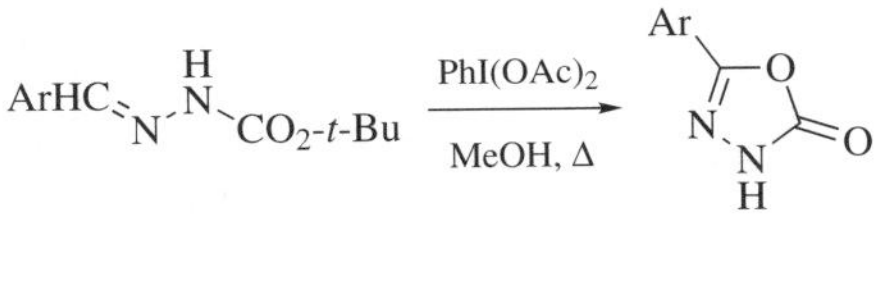

(38)

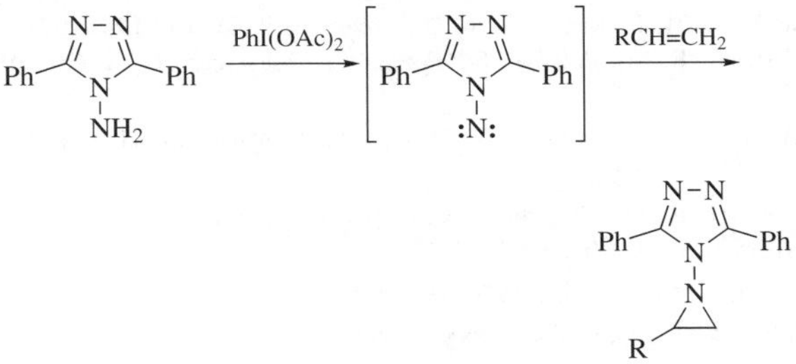

(39)

A linear tetrazane is formed in the oxidation of *N*-aminophthalimide (eq 40).[43]

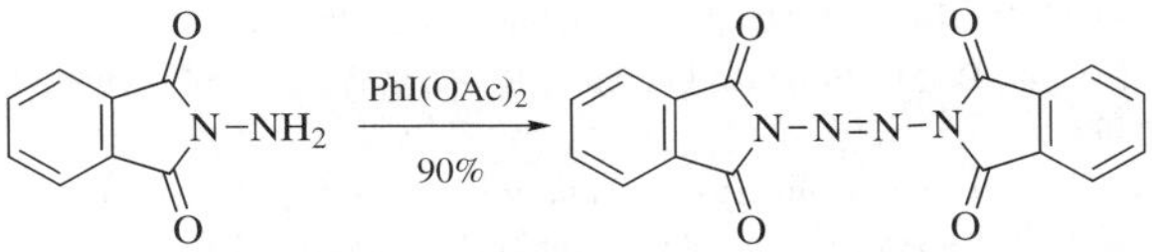

(40)

(Diacetoxyiodo)benzene/***Iodine*** is reported to be a more efficient and convenient reagent for the generation of alkoxyl radicals than Pb^{IV}, Hg^{II}, or Ag^{I}, and this system is useful for intramolecular oxide formation (eqs 41 and 42).[44]

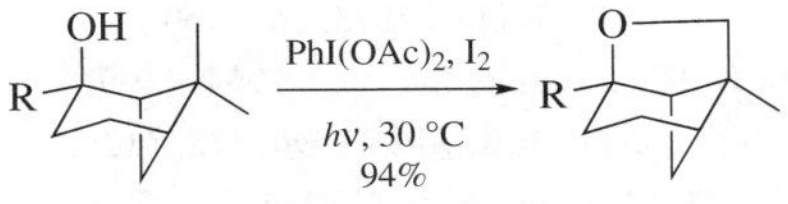

(41)

R = H, CH_2CO_2Et

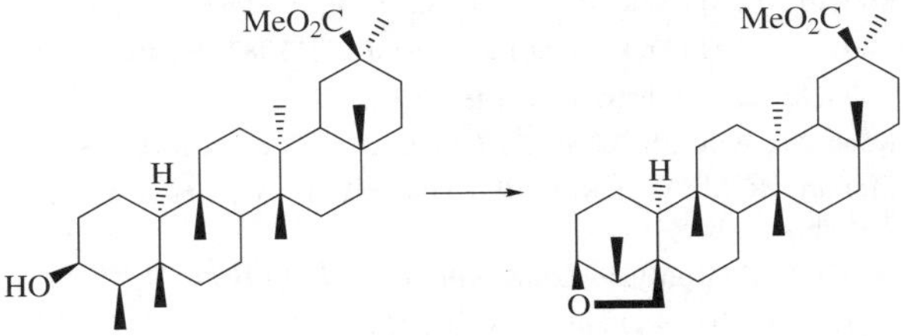

(42)

Fragmentation processes of carbohydrate anomeric alkoxyl radicals[45] and steroidal lactols[46] using $PhI(OAc)_2/I_2$ have been reported.

Related Reagents. See Classes O-5, O-10, O-20 and O-22, pages 1–10.

---

1. Moriarty, R. M.; Prakash, O. *ACR* **1986**, *19*, 244.
2. Moriarty, R. M.; Vaid, R. K. *S* **1990**, 431.
3. Varvoglis, A. *The Organic Chemistry of Polycoordinated Iodine*; VCH: New York, 1992; p 131.
4. Sharefkin, J. G.; Saltzman, H. *OSC* **1973**, *5*, 660.
5. Lucas, H. J.; Kennedy, E. R.; Formo, M. W. *OSC* **1955**, *3*, 483.
6. Moriarty, R. M.; Duncan, M. P.; Vaid, R. K.; Prakash, O. *OSC* **1992**, *8*, 43.
7. Moriarty, R. M.; Kamernitskii, J. S. *TL* **1986**, *27*, 2809.
8. Moriarty, R. M.; Hu, H.; Gupta, S. C. *TL* **1981**, *22*, 1283.
9. Turuta, A. M.; Kamernitzky, A. V.; Fadeeva, T. M.; Zhulin, A. V. *S* **1985**, 1129.
10. Moriarty, R. M.; Prakash, O. *JOC* **1985**, *50*, 151.
11. Moriarty, R. M.; Prakash, O.; Musallam, H. A. *JHC* **1985**, *22*, 583.

12. (a) Moriarty, R. M.; Prakash, O.; Freeman, W. A. *CC* **1984**, 927. (b) Moriarty, R. M.; Prakash, O.; Prakash, I.; Musallam, H. A. *CC* **1984**, 1342.
13. Moriarty, R. M.; Prakash, O.; Thachet, C. T.; Musallam, H. A. *H* **1985**, *23*, 633.
14. Moriarty, R. M.; Engerer, S. C.; Prakash, O.; Prakash, I.; Gill, U.S.; Freeman, W. A. *JOC* **1987**, *52*, 153.
15. Zbiral, E.; Nestler, G. *T* **1970**, *26*, 2945.
16. Pelter, A.; Elgendy, S. *TL* **1988**, *29*, 677.
17. Bennett, D.; Dean, F. M.; Herbin, G. A.; Matkin, D. A.; Price, A. W. *JCS(P2)* **1978**, 1980.
18. Szántay, C.; Blaskó, G.; Bárczai-Beke, M.; Pechy, P.; Dörnyei, G. *TL* **1980**, *21*, 3509.
19. Ochiai, M.; Iwaki, S.; Ukita, T.; Nagao, Y. *CL* **1987**, 133.
20. Moriarty, R. M.; Prakash, O.; Vaid, R. K.; Zhao, L. *JACS* **1989**, *111*, 6443.
21. Koser, G. F. *The Chemistry of Functional Groups, Supplement D*; S. Patai and Z. Rappoport, Eds.; Wiley: New York, 1983; p 721.
22. Moriarty, R. M.; Kim, J.; Guo, L. *TL* **1993**, *34*, 4129.
23. Pausacker, K. H. *JCS* **1953**, 1989.
24. Szmant, H. H.; Lapinski, R. L. *JACS* **1956**, *78*, 458.
25. Szmant, H. H.; Infante, R. *JOC* **1961**, *26*, 4173.
26. Dyall, L. K. *AJC* **1973**, *26*, 2665.
27. Dyall, L. K.; Kemp, J. E. *AJC* **1973**, *26*, 1969.
28. Pausacker, K. H.; Scroggie, J. G. *JCS* **1954**, 4499.
29. Devadas, B.; Leonard, N. J. *JACS* **1990**, *112*, 3125.
30. Awang, D. V. C.; Vincent, A. *CJC* **1980**, *58*, 1589.
31. Kishi, V.; Nakatsura, S.; Fukuyama, T.; Havel, M. *JACS* **1973**, *95*, 6493.
32. Moriarty, R. M.; Vald, R. K.; Duncan, M. P. *SC* **1987**, *17*, 703.
33. Moriarty, R. M.; Prakash, I.; Penmasta, R. *SC* **1987**, *17*, 409.
34. Kahn, M.; Bertenshaw, S. *TL* **1989**, *30*, 2317.
35. Kahn, M.; Wilke, S.; Chen, B.; Fujita, K. *JACS* **1988**, *110*, 1638.
36. Moriarty, R. M.; Vaid, R. K.; Ravikumar, V. T.; Hopkins, T. E.; Farid, P. *T* **1989**, *45*, 1605.
37. Taylor, E. C.; Robey, R. L.; McKillop, A. *AG(E)* **1972**, *11*, 48.
38. Myrboh, B.; Ile, H.; Junjappa, H. *S* **1992**, 1101.
39. Smith, P. A. S.; Bruckmann, E. M. *JOC* **1974**, *39*, 1047.
40. Boulton, A. J.; Devi, P.; Henderson, N.; Jarrar, A. A.; Kiss, M. *JCS(P1)* **1989**, *1*, 543.
41. Baumgarten, H. E.; Hwang, D.-R.; Rao, T. N. *JHC* **1986**, *23*, 945.
42. Schröppel, F.; Sauer, J. *TL* **1974**, 2945.
43. Anderson, D. J.; Gilchrist, T. L.; Rees, C. W. *CC* **1971**, 800.
44. Dorta, R. L.; Francisco, C. G.; Hernández, R.; Salazar, J. A.; Suárez, E. *JCR(S)* **1990**, 240.
45. deArmas, P.; Francisco, C. G.; Suárez, E. *AG(E)* **1992**, *31*, 772.
46. Freire, R.; Marrero, J. J.; Rodríguez, M. S.; Suárez, E. *TL* **1986**, *27*, 383.

Robert M. Moriarty, Calvin J. Chany II, &
Jerome W. Kosmeder II
University of Illinois at Chicago, IL, USA

Diborane

B_2H_6

[19287-45-7] B_2H_6 (MW 27.67)

(strong reducing agent for many functional groups; extremely efficient hydroborating reagent)

Physical Data: bp −92.5 °C; mp −165.5 °C; *d* 0.437 g cm^{-3} (liquid at −92.5 °C); heat of vaporization 3.41 kcal mol^{-1} (at −92.5 °C).
Solubility: slightly sol pentane and hexane; forms BH_3 adduct with DMS, THF, and other ethers; reacts with H_2O and protic solvents, releasing flammable hydrogen gas.
Form Supplied in: commercially available as a compressed gas.
Analysis of Reagent Purity: supplied as either 99.99% or >99.0% B_2H_6.
Handling, Storage, and Precautions: diborane is a toxic, pyrophoric gas and must be stored and handled accordingly. Feed lines and reactors should be flushed with N_2 and kept free of moisture. Cylinders should be hard-piped directly to reactor. When not in use, cylinder valve should be securely closed and capped. Cold (−20 to 0 °C) storage is recommended to ensure product purity. Decomposition products include higher boron hydrides and hydrogen gas.

Diborane and Borane Reagents. Borane complexes have been used in industrial-scale applications for years. In addition to their unique ability to add across carbon–carbon multiple bonds, boron hydrides reduce ketones, carboxylic acids, amides, and nitriles (eqs 1–3).[1] By the appropriate choice of borane reagent and reaction conditions, a significant degree of chemical selectivity can be achieved. This high level of selectivity and reactivity has found utility not only in classical organic synthesis,[2,3] but also in the area of asymmetric synthesis with the Corey–Itsuno catalyst serving as a prime example of a highly reactive and enantioselective reagent.[4]

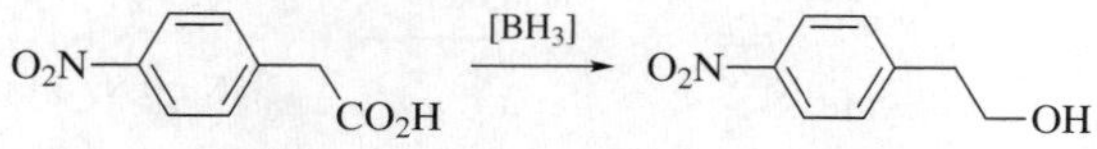

(1)

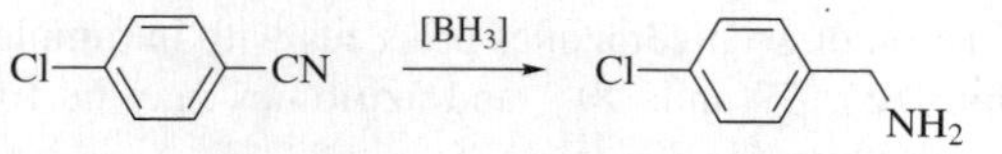

(2)

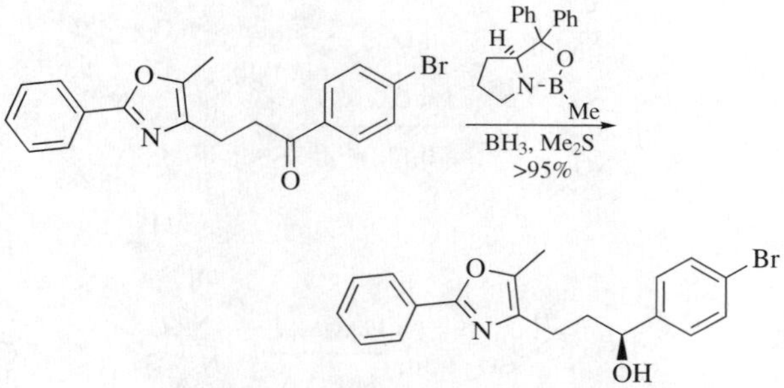

(3)

94% ee

Chemically, diborane is the parent of all the borane complexes. Diborane will react with a Lewis base to form an adduct in which the electron pair from the Lewis base is donated to the borane. This complexation alters the reactivity of the borane, subsequently changing its physical form. Diborane is a gas with a boiling point of −92.5 °C, whereas the borane complexes are either liquids or solids. As illustrated in the molecular model (**1**), diborane has some unusual bonding characteristics. The boron is so electron deficient that it resorts to sharing the electrons in the B–H bond, giving rise to bridging hydrogens. This electron deficiency allows formation of borane complexes with Lewis bases such as tetrahydrofuran (THF), dimethyl sulfide (DMS), and amines (Scheme 1).

○ = H
● = B

(**1**)

$R_2O{:}BH_3$ ← (R_2O) $0.5\ B_2H_6$ → (R_2S) $R_2S{:}BH_3$; → (R_3N) $R_3N{:}BH_3$

Scheme 1

In the search for better borane reagents, a variety of complexes have been synthesized; almost all fit under one of the three following classifications: ether–borane complexes, sulfide–borane complexes, or amine–borane complexes.

A general trend of borane complexes is that the stronger the Lewis base, the weaker the reducing power of the resulting borane.[5] ***Borane–Tetrahydrofuran*** (THFB) is more reactive than ***Borane–Dimethyl Sulfide*** (DMSB), which in turn is more reactive than most amine–boranes. When reducing a carboxylic acid or hydroborating a carbon–carbon double bond, THFB and DMSB are both effective reagents, while amine–boranes generally are not suitable choices.

Comparison of Borane Reagents. THFB and DMSB, however, do have drawbacks which can be of some concern in an industrial setting. The maximum workable concentration for THFB is 1 molar. Therefore a large reactor volume is required to achieve the desired reduction. The low loading in a given size reactor, and ultimately solvent waste, increase the cost-per-pound of product.

DMSB is commercially available as a neat solution (∼10M); thus solvent volume is determined by reactor loading of other reactants (substrate). However, anyone who has had the opportunity to use dimethyl sulfide can attest to the fact that the human nose can detect DMS in the ppm range. Various schemes have been used in an attempt to scrub the DMS from reaction streams, but none are truly effective.

When diborane is added to an ether solvent, ether–borane complexes are formed. If the solvent is THF, then THFB is formed in situ, and reacts in the same manner without the high dilution effect of using the 1M THFB solution. By using diborane rather than THFB, reactor loading can be increased, thereby reducing costs. Use of diborane will also reduce the amount of waste solvent remaining at the end of the process. The advantage of diborane over DMSB is the avoidance of odor control problems inherent in dimethyl sulfide, again reducing costs.

Although much of the literature describing borane reductions involves coordinating solvents such as ethers, diborane can be employed in nonpolar hydrocarbon solvents.[6–9] Table 1 shows an appreciable solubility of diborane in hexanes. It should be noted that the reactivity of diborane in hydrocarbons may be distinctly different from the reactivity of borane complexes. For example, because it is uncomplexed, diborane is generally much more reactive than sulfide boranes such as DMSB.

Table 1 Solubility of Diborane in Various Solvents

Solvent	Solubility (g B_2H_6/100 g solvent)	Temperature (°C)	B_2H_6 pressure (atm)
n-Pentane	0.996	19.6	0.958
n-Hexane	4.22	room	2.55
neo-Hexane	3.18	room	4
cis-Decalin	1.86	room	2.44
Mesitylene	0.323	25	0.653
Ethyl ether	1.1	20	1
1,4-Dioxane	3.37	room	2.38
Tetrahydrofuran	8.1	20	1

However, diborane should not be used in every reaction where THFB or DMSB are currently in use.

1. DMSB is a good choice if the reduction requires heat to push the reaction to completion, because DMSB is stable at higher temperatures than ether–borane complexes.
2. THFB is a good choice when addition of substrate to an excess of borane is required, and reactor loading is not a major concern. THFB solutions should not be heated above 50 °C, since diborane will be lost from the complex and reductive ring opening of the THF will occur.
3. Diborane is a good choice when it is desirable to add borane to the substrate in the reactor. With diborane, higher reactor loadings can be achieved than with either of the other two reagents.

Generally, if the reaction is fast at or below ambient temperature, diborane is an effective replacement for either THFB or DMSB.

The classic synthesis of ***9-Borabicyclo[3.3.1]nonane*** (9-BBN) from *cis,cis*-1,5-cyclooctadiene (cod) serves as a good model for hydroboration reactions (eq 4). The route described in the literature[10] advises the use of DMSB in dimethoxyethane (DME). Following the hydroboration and removal of DMS, crystalline 9-BBN precipitates out of the DME solvent. Unfortunately, the resulting product is contaminated with the odor of dimethyl sulfide which is not readily removed from the 9-BBN. The use of THFB as the hydroborating reagent does alleviate the DMS odor, but the product remains solubilized as a dilute solution in THF and DME. When diborane is added as a gas to a DME solution of cod, the resulting 9-BBN is obtained as a clean crystalline product with a typical purity of >96%. No further processing is required from the diborane-produced 9-BBN.[11]

$$\text{cod} \xrightarrow[\text{DMSB}]{0.5\ B_2H_6 \ /\ \text{THFB}} \text{9-BBN (B–H)} \quad (4)$$

Reduction of a substituted xanthone to a xanthene, shown in eq 5, was accomplished using three borane reagents.[12] THFB was an effective reagent for this transformation, and was the most desirable for small-scale preparations because of its ease of use. The reduction utilizing DMSB could not be driven to completion. Diborane proved to be the most effective reagent and was highly desirable for commercial-scale synthesis of the xanthene. With the higher loading achievable using diborane, the THFB route was considered to be economically unfeasible.

$$\text{xanthone (R}_4\text{)} \xrightarrow{[BH_3]} \text{xanthene (R}_4\text{)} \quad (5)$$

Practical Considerations. Introduction of diborane into a reactor requires some careful consideration of the reactor system. Diborane, like almost all hydrides, is very reactive. It is critical that the reactor be under an inert gas, such as nitrogen, since diborane is a pyrophoric material. The addition of diborane to a solution is best accomplished by feeding gaseous diborane under the surface of the solvent. If the product formed by reaction of diborane is a solid, diborane can be added to the reactor head space provided that a back pressure on the reactor keeps it contained. Given a reactor which is free of oxygen and moisture, diborane is used as any other compressed gas.

Diborane cylinders are packaged in dry ice for shipment. The vapor pressure of diborane at −78 °C is 15 psig (776 Torr). This allows for a convenient method to control the pressure of diborane gas being fed to a reactor. The rate at which diborane is fed into the reactor is dictated by the thermodynamics and kinetics of the reaction, and the temperature constraints of the system.

Conclusion. In many cases, diborane can be used as a cost effective reagent for reduction or hydroboration. Use of diborane in place of THFB can significantly increase the loading in a given reaction. Used in place of DMSB, it can eliminate the handling of dimethyl sulfide as a waste. All three reagents (diborane, THFB, and DMSB) have application in organic synthesis, and each should be evaluated for potential use.

Related Reagents. See Classes R-2, R-3, R-4, R-11, R-13, R-17, R-20, and R-29, pages 1–10. Borane–Ammonia; Borane–Dimethyl Sulfide; Borane–Pyridine; Borane–Tetrahydrofuran; Ephedrine–borane; Norephedrine–Borane.

1. Brown, H. C. *Boranes in Organic Chemistry*; Cornell University Press: Ithaca, NY, 1972.
2. Brown, H. C.; Krishnamurthy, S. *Aldrichim. Acta* **1979**, *12(1)*, 3.
3. Brown, H. C.; Choi, Y. M.; Narasimhan, S. *S* **1981**, *8*, 605.
4. Quallich, G. J.; Woodall, T. M. *TL* **1993**, *34*, 785.
5. Zaidlewicz, M. In *Comprehensive Organometallic Chemistry*; Wilkinson, G.; Stone, F. G. A.; Abel, E. W., Eds.; Pergamon: Oxford, 1982; Vol. 7, pp 162–163.
6. Schechter, W. H., Adams, R. M. and Jackson, C. B. *Boron Hydrides and Related Compounds*, Callery Chemical Company, 1951.
7. Schlesinger, H. I., University of Chicago, *Final Report Navy Contract N173s–9058 and 9820*, 1944–5.
8. Boldebuck, E. M.; Elliot, J. R.; Roedel, G. F.; Roth, W. L. *Solubility of Diborane in Ethyl Ether and in Tetrahydrofuran*; General Electric Company, Project Hermes Report No. 55288; Nov 19, 1948.
9. Reed, J. W.; Masi, J. F. *Solubility of Diborane in n-Pentane and in n-Butane*; Callery Chemical Company, Report No. CCC-454-TR-300; Sept 30, 1958.
10. Soderquist, J.; Negron, A. *OS* **1991**, *70*, 169.
11. Corella, J. A., Callery Chemical Company; unpublished results.
12. Burkhardt, E. R., Callery Chemical Company; unpublished results.

Joseph M. Barendt & Beth W. Dryden
Callery Chemical Company, Pittsburgh, PA, USA

Dibromoborane–Dimethyl Sulfide[1]

$BHBr_2{\cdot}SMe_2$

[55671-55-1] $C_2H_7BBr_2S$ (MW 233.77)

(hydroborating agent providing access to alkyl-[2] and alkenyl-dibromoboranes[3] and boronic acids[2,4])

Alternate Name: DBBS.
Physical Data: mp 30–35 °C; bp 75 °C/0.1 mmHg.
Solubility: sol dichloromethane, carbon disulfide, carbon tetrachloride.
Form Supplied in: white solid or liquid, 7.8 M in $BHBr_2$.
Preparative Methods: by redistribution of $BH_3{\cdot}SMe_2$ and $BBr_3{\cdot}SMe_2$ or $BH_3{\cdot}SMe_2$ and BBr_3;[5] by the reaction of bromine with $BH_3{\cdot}SMe_2$ in CS_2.[6]
Analysis of Reagent Purity: ^{1}H NMR (CCl_4) δ 2.48, (CS_2) 2.69 ppm; ^{11}B NMR (CCl_4) δ −7.3 (d, $J_{B\text{–}H}$ = 160 Hz),[5] (CS_2) −8.2 ppm.[6] Hydrolysis of an aliquot and measuring the hydrogen evolved according to the standard procedure.[7]
Handling, Storage, and Precautions: corrosive liquid; air and moisture sensitive; flammable; stench. Handle and store under nitrogen or argon. Stable indefinitely when stored under nitrogen at 25 °C. Reacts violently with water. This reagent should be handled in a fume hood.

Hydroboration of Alkenes. Dibromoborane–dimethyl sulfide hydroborates alkenes directly without need for a decomplexing agent (see ***Dichloroborane Diethyl Etherate***) (eq 1).[2]

$$RCH{=}CH_2 + BHBr_2{\bullet}SMe_2 \xrightarrow[71\text{–}93\%]{CH_2Cl_2} RCH_2CH_2BBr_2{\bullet}SMe_2 \quad (1)$$

Regioselectivity of DBBS in the hydroboration of alkenes and derivatives is high, approaching ***9-Borabicyclo[3.3.1]-nonane***, e.g. 1-hexene, styrene, 2-methyl-1-pentene, 2-methyl-2-butene and 4-(dimethylphenylsilyl)-2-pentene, react by placing the boron atom at the less hindered position with ≥99% selectivity.[2,8a] Lower regioselectivity of the hydroboration–oxidation is observed if an excess of alkene is used, due to hydrobromination, which is a side reaction in the hydrolysis–oxidation step.[8b] The reactivity of DBBS toward structurally different alkenes and alkynes is different from that of other hydroborating agents (Table 1).[3,9]

Table 1 Relative Reactivity of Representative Alkynes and Alkenes with $BHBr_2 \cdot SMe_2$, 9-BBN, and Sia_2BH

Compound	$BHBr_2 \cdot SMe_2$	9-BBN	Sia_2BH
1-Octene	100	100	100
(*Z*)-3-Hexene	20	0.55	1.25
1-Hexyne	290	15	345
3-Hexyne	5900	0.74	208

Reactions of Alkyldibromoboranes. Alkyldibromoboranes are versatile synthetic intermediates. They are resistant to thermal isomerization, a feature of considerable importance for the regio- and stereoselective synthesis of organoborane intermediates from highly labile alkenic structures.[10]

Standard oxidation of alkyldibromoboranes with alkaline ***Hydrogen Peroxide*** affords alcohols.[2,8b] Conversion of terminal alkenes to carboxylic acids using alkyldibromoboranes works well, although hydrolysis prior to oxidation is needed.[11] Chiral alkyldibromoboranes have been used as catalysts for the asymmetric Diels–Alder reaction.[12,13]

The hydridation–stepwise hydroboration procedure provides a convenient general approach to monoalkylbromoboranes, mixed dialkylbromoboranes, dialkylboranes, totally mixed trialkylboranes, ketones, alcohols (eq 2),[14] and stereodefined alkenes, dienes, and haloalkenes (see below). Mixed alkylalkenylalkynylboranes are also available by this methodology.[15]

$$\text{alkene A} \xrightarrow[\text{CH}_2\text{Cl}_2]{\text{BHBr}_2\bullet\text{SMe}_2} \text{R}^A\text{BBr}_2\bullet\text{SMe}_2 \xrightarrow[0.25\ \text{LiAlH}_4]{\text{alkene B}}$$

$$\text{R}^A\text{R}^B\text{BBr}\bullet\text{SMe}_2 \xrightarrow{\text{MeOH}} \text{R}^A\text{R}^B\text{BOMe} \xrightarrow[0.25\ \text{LiAlH}_4]{\text{alkene C}} \text{R}^A\text{R}^B\text{R}^C\text{B} \quad (2)$$

$$\text{R}^A\text{R}^B\text{BOMe} \xrightarrow{1.\ \text{Cl}_2\text{CHOMe};\ 2.\ \text{H}_2\text{O}_2,\ \text{NaOH}} \text{R}^A\text{R}^B\text{CO}\ (\sim70\%) \qquad \text{R}^A\text{R}^B\text{R}^C\text{B} \xrightarrow{1.\ \text{Cl}_2\text{CHOMe};\ 2.\ [\text{O}]} \text{R}^A\text{R}^B\text{R}^C\text{COH}\ (\sim70\%)$$

Hydrolysis and alcoholysis of alkyldibromoboranes provide simple access to alkylboronic acids and esters respectively,[4] which are important synthetic intermediates and reagents for protection of hydroxy groups of diols[16a–f] and derivatizing agents for GC and GC–MS analysis.[16g,h]

Selective Hydroboration of Dienes and Enynes. The opposite reactivity trends of DBBS and other hydroborating agents makes possible the selective hydroboration of dienes (eq 3)[9] and enynes.[9] In conjugated systems, however, bromoboration of the triple bond is observed.[17]

$$\text{diene} \xrightarrow[2.\ [\text{O}]]{1.\ (\text{Sia})_2\text{BH}} \text{HO-alcohol (terminal alkene hydroborated)}; \qquad \text{diene} \xrightarrow[2.\ [\text{O}]\ \ 70\%]{1.\ \text{BHBr}_2\bullet\text{SMe}_2} \text{alcohol-OH (1,1-disubstituted alkene hydroborated)} \quad (3)$$

Reactions of Alkenyldibromoboranes and Alkenylalkylbromoboranes.

Synthesis of Alkenylboronic Acids, Alkenes, Aldehydes, and Ketones. Alkenyldibromoboranes undergo many of the characteristic reactions of alkenylboranes. The presence of dimethyl sulfide does not interfere in their transformations. Protonolysis with acetic acid in refluxing dichloromethane gives the corresponding alkene. Oxidation leads to aldehydes or ketones (eq 4).[3] Hydrolysis and alcoholysis yields alkenylboronic acids and esters, respectively.

$$\text{R}^1\text{C}{\equiv}\text{CR}^2 \xrightarrow[77\text{–}84\%]{\text{BHBr}_2\bullet\text{SMe}_2} (\text{R}^1)(\text{H})\text{C}{=}\text{C}(\text{R}^2)(\text{BBr}_2\bullet\text{SMe}_2)$$

$$\xrightarrow[\sim30\%]{\text{AcOH}} (\text{R}^1)(\text{H})\text{C}{=}\text{C}(\text{R}^2)(\text{H}); \quad \xrightarrow[\sim95\%]{\text{H}_2\text{O, NaOH}} (\text{R}^1)(\text{H})\text{C}{=}\text{C}(\text{R}^2)(\text{B(OH)}_2); \quad \xrightarrow[3.\ \text{H}_2\text{O}_2\ \ \sim85\%]{1.\ \text{NaOH};\ 2.\ \text{phosphate buffer}} \text{R}^1\text{CH}_2\text{CHO}\ (\text{R}^2 = \text{H}) \quad (4)$$

$\text{R}^1 = \text{R}^2 = \text{alkyl}$; $\text{R}^1 = \text{alkyl}$, $\text{R}^2 = \text{H}$

The (*E*)-alkenylboronic acids are directly available from 1-alkynes by hydroboration–hydrolysis (eq 4).[4] The (*Z*)-isomers are prepared from 1-bromo-1-alkynes by hydroboration–hydride reduction (eq 5).[18]

$$\text{R}^1\text{C}{\equiv}\text{CBr} \xrightarrow[2.\ \text{R}^2\text{OH}\ \ 84\text{–}95\%]{1.\ \text{BHBr}_2\bullet\text{SMe}_2} (\text{R}^1)(\text{H})\text{C}{=}\text{C}(\text{Br})(\text{B(OR}^2)) \xrightarrow[\text{Et}_2\text{O}\ \ 74\text{–}94\%]{\text{KIPBH}} (\text{R}^1)(\text{H})\text{C}{=}\text{C}(\text{B(OR}^2))(\text{H}) \quad (5)$$

Synthesis of (E)- and (Z)-Alkenes, Trisubstituted Alkenes, 1,2-Disubstituted Alkenyl Bromides, Ketones, and Enolborates. The synthesis of (*Z*)-alkenes, according to eq 4, is a simple, convenient method, provided the alkynic precursor is readily available. A general Zweifel (*E*)- and (*Z*)-alkene synthesis starts with 1-alkynes and 1-bromo-1-alkynes, respectively. The precursors are hydroborated with monoalkylbromoborane to give the corresponding alkylalkenylbromoboranes. Migration of the alkyl group completes the formation of the carbon skeleton (eq 6).[19–21] The procedure allows full utilization of the alkyl group.

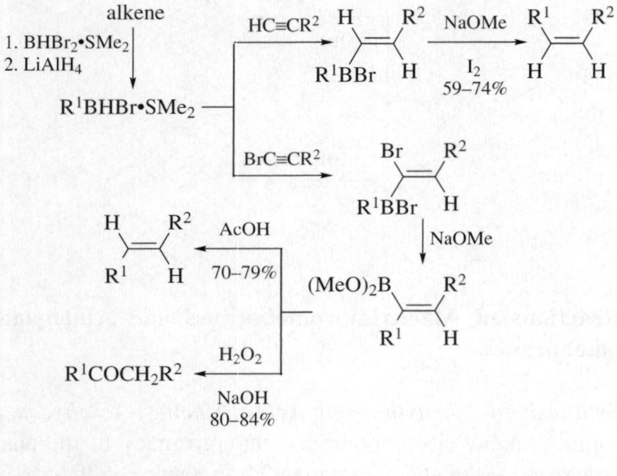

(6)

(Z)-9-Tricosene (muscalure), the sex pheromone of the housefly (*Musca domestica*), has been prepared by this method in 69% yield and >99% purity.[20,22] Extension of the methodology to trisubstituted alkenes is based on the iodine-induced migration of the second alkyl group R^3 (eq 7).[23]

The key alkenylboronate intermediate used for the introduction of the R^3 group (eq 7) is of the same structure as the one shown in eq 6. The procedure works well both for alkyl and aryl R^3 groups. However, a methyl group shows poor migratory aptitude in these reactions.[20,23] If the two *trans*-alkyl groups in the product alkene are the same or differ significantly in steric bulk, an internal alkyne may serve as a starting material (eq 8).[24]

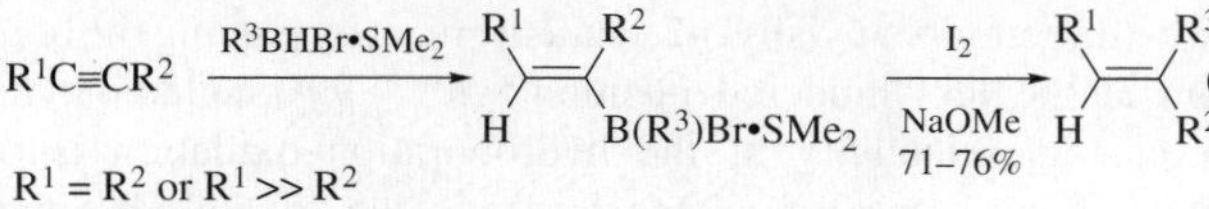

(8)

Other approaches to trisubstituted alkenes via organoboranes involve alkynyltrialkyl borates,[25] alkenyltrialkyl borates[26] or the cross-coupling reaction of alkenylboronic acids with alkyl halides.[27]

Both (*E*)- and (*Z*)-1,2-disubstituted alkenyl bromides can also be prepared by the methodology shown in eq 7.[28] The ***Boron Trifluoride Etherate***-mediated 1,4-addition of 1,2-disubstituted alkenylboronates affords γ,δ-unsaturated ketones (eq 9).[29] The boronates can also be converted into chiral enolborates for the enantioselective addition to aldehydes.[30]

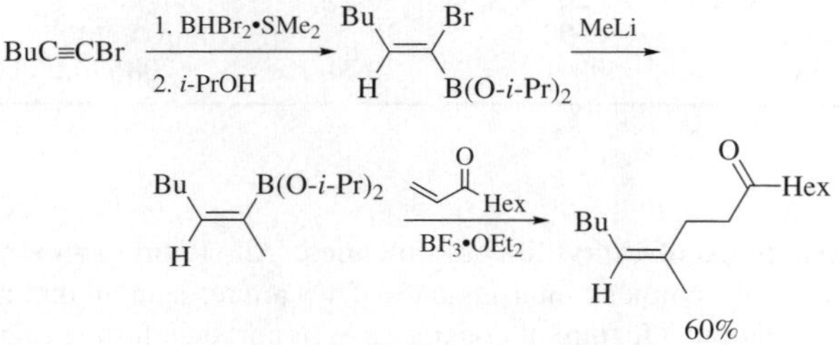

(9)

Synthesis of (E)- and (Z)-1-Halo-1-alkenes and α-Bromoacetals. Stereodefined alkenyl halides are important starting materials for the synthesis of alkenyl Grignard[31] and lithium[32] derivatives, pheromones,[33] and (*E,E*)-, (*E,Z*)-, (*Z,E*)- and (*Z,Z*)-dienes by the cross-coupling reaction.[34] An efficient, general (*E*)- and (*Z*)-1-halo-1-alkene synthesis starts with 1-alkynes and 1-halo-1-alkynes respectively, which are

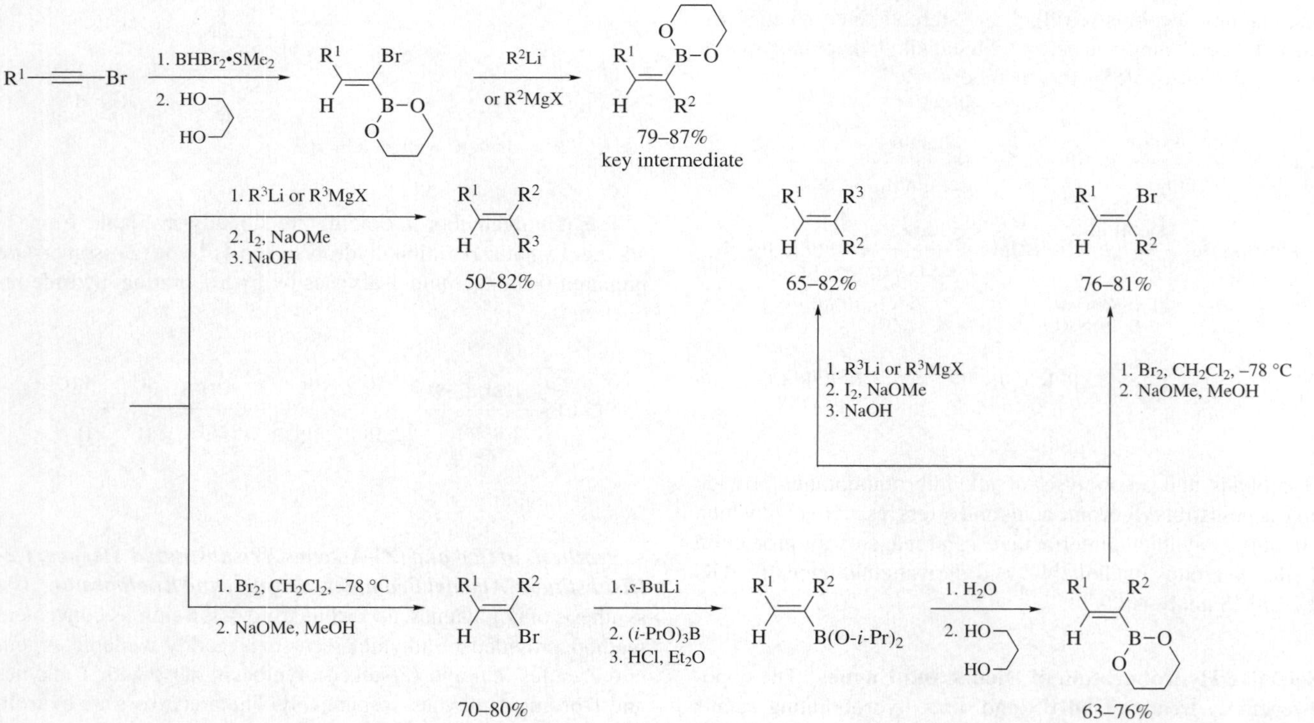

(7)

$$\text{(E)-}R^1CH{=}CHB(OR^2)_2 \xrightarrow{Br_2,\ NaOH \text{ or } NBS,\ NaBr} \text{(Z)-}R^1CH{=}CHBr\ ;\quad \xrightarrow{1.\ Hg(OAc)_2,\ THF,\ -30\ °C;\ 2.\ Br_2,\ pyridine,\ -30\ °C} \text{(E)-}R^1CH{=}CHBr$$
$$\xrightarrow{Cl_2,\ NaOMe} \text{(Z)-}R^1CH{=}CHCl\ ;\quad \xrightarrow{I_2,\ NaOH} \text{(E)-}R^1CH{=}CHI \text{ (also via 1. Hg(OAc), THF; 2. } I_2\text{, pyridine)}$$
$$\text{(Z)-}R^1CH{=}CHB(OR^2)_2 \xrightarrow{Br_2,\ NaOMe} \text{(E)-}R^1CH{=}CHBr\ ;\quad \xrightarrow{1.\ Cl_2;\ 2.\ Na_2SO_3} \text{(E)-}R^1CH{=}CHCl\ ;\quad \xrightarrow{I_2,\ NaOH} \text{(Z)-}R^1CH{=}CHI \qquad (10)$$

converted into the corresponding (*E*)- and (*Z*)-alkenylboronic acids or esters via alkenyldibromoboranes, according to eqs 4 and 5. The alkenylboronic acids and esters react with halogens directly[35–38] or via alkenylmercurials[39] to give the haloalkenes in high stereochemical purity in 70–100% yield (eq 10). In some of these halogenation reactions, alkenyldibromoboranes can be used directly.[35,36] Alternatively, (*Z*)-1-halo-1-alkenes are simply obtained by hydroboration–protonolysis of 1-halo-1-alkynes with 9-BBN or ***Disiamylborane***.[40]

Synthesis of Conjugated Dienes. The cross-coupling reaction of alkenylboronates with alkenyl halides is a general method for the synthesis of stereodefined 1,3-dienes (eq 11).[41]

$$BuC{\equiv}CBr \xrightarrow{1.\ BHBr_2{\cdot}SMe_2;\ 2.\ ROH;\ 3.\ KIPBH} \text{(E)-}BuCH{=}CHB(OR)_2 \xrightarrow[Pd(PPh_3)_4,\ NaOEt]{\text{(Z)-}BrCH{=}CHHex} \text{Bu}\diagdown\!{=}\!\diagup\!{=}\!\diagdown\text{Hex} \qquad (11)$$

87% yield
>99% pure

Other Applications. DBBS is an excellent precursor for the formation of bulk powders and ceramic fiber coatings of boron nitride.[42] It has been used for the synthesis of the small carborane *closo*-2,3-$Et_2C_2B_5H_5$,[43] 3-*O*-carboranylcarbene,[44] silver and sodium isocyanoborohydrides,[45] and dications based on the hydrotris(phosphonio)borate skeleton.[46]

Related Reagents. See Classes R-4 and R-17, pages 1–10. Dichloroborane–Dimethyl Sulfide.

1. (a) Brown, H. C.; Zaidlewicz, M. *Pol. J. Appl. Chem.* **1982**, *26*, 155. (b) Brown, H. C.; Kulkarni, S. U. *JOM* **1982**, *239*, 23. (c) Pelter, A.; Smith, K. *COS* **1991**, *8*, 703.
2. Brown, H. C.; Ravindran, N.; Kulkarni, S. U. *JOC* **1980**, *45*, 384.
3. Brown, H. C.; Campbell, Jr., J. B. *JOC* **1980**, *45*, 389.
4. Brown, H. C.; Bhat, N. G.; Somayaji, V. *OM* **1983**, *2*, 1311.
5. Brown, H. C.; Ravindran, N. *IC* **1977**, *16*, 2938.
6. Kingberger, K.; Siebert, W. *ZN(B)* **1975**, *30*, 55.
7. Brown, H. C. *Organic Syntheses via Boranes*; Wiley: New York, 1975, p 239.
8. (a) Fleming, I.; Lawrance, N. J. *JCS(P1)* **1992**, 3309. (b) Brown, H. C.; Racherla, U. S. *JOC* **1986**, *51*, 895.
9. Brown, H. C.; Chandrasekharan, J. *JOC* **1983**, *48*, 644.
10. Brown, H. C.; Racherla, U. S. *JOC* **1983**, *48*, 1389.
11. Brown, H. C.; Kulkarni, S. V.; Khanna, V. V.; Patil, V. D.; Racherla, U. S. *JOC* **1992**, *57*, 6173.
12. (a) Bir, G.; Kaufmann, D. *TL* **1987**, *28*, 777. (b) *JOM* **1990**, *390*, 1.
13. Kaufmann, D.; Boese, R. *AG(E)* **1990**, *29*, 545.
14. Kulkarni, S. U.; Basavaiah, D.; Zaidlewicz, M.; Brown, H. C. *OM* **1982**, *1*, 212.
15. Brown, H. C.; Basavaiah, D.; Bhat, N. G. *OM* **1983**, *2*, 1468.
16. (a) Reese, C. B. In *Protective Groups in Organic Chemistry*; McOmie, J. F. W., Ed., Plenum: New York, 1973, p 135. (b) Garlaschelli, L.; Mellerio, G.; Vidari, G. *TL* **1989**, *30*, 597. (c) Dahlhoff, W. V.; Köster, R. *H* **1982**, *18*, 421. (d) *JOC* **1977**, *42*, 3151. (e) Wiecko, J.; Sherman, W. R. *JACS* **1976**, *98*, 7631. (f) Fréchet, J. M. J.; Nuyens, L. J.; Seymour, E. *JACS* **1979**, *101*, 432. (g) Knapp, D. R. *Handbook of Analytical Derivatization Reactions*; Wiley: New York, 1979. (h) Schurig, V.; Wistuba, D. *TL* **1984**, *25*, 5633.
17. Rowley, E. G.; Schore, N. E. *JOC* **1992**, *57*, 6853.
18. Brown, H. C.; Imai, T. *OM* **1984**, *3*, 1392.
19. Brown, H. C.; Basavaiah, D.; Kulkarni, S. U.; Lee, H. D.; Negishi, E.; Katz, J. J. *JOC* **1986**, *51*, 5270.
20. Brown, H. C.; Basavaiah, D.; Kulkarni, S. U.; Bhat, N. G.; Vara Prasad, J. V. N. *JOC* **1988**, *53*, 239.
21. Brown, H. C.; Imai, T.; Bhat, N. G. *JOC* **1986**, *51*, 5277.
22. Brown, H. C.; Basavaiah, D. *JOC* **1982**, *47*, 3806.
23. Brown, H. C.; Bhat, N. G. *JOC* **1988**, *53*, 6009.
24. Brown, H. C.; Basavaiah, D. *JOC* **1982**, *47*, 5407.
25. (a) Pelter, A.; Bentley, T. W.; Harrison, C. R.; Subrahmanyam, C.; Laub, R. J. *JCS(P1)* **1976**, 2419. (b) Pelter, A.; Gould, K. J.; Harrison, C. R. *JCS(P1)* **1976**, 2428. (c) Pelter, A.; Harrison, C. R.; Subrahmanyam, C.; Kirkpatrick, D. *JCS(P1)* **1976**, 2435. (d) Pelter, A.; Subrahmanyam, C.; Laub, R. J.; Gould, K. J.; Harrison, C. R. *TL* **1975**, 1633.
26. (a) LaLima, Jr., N. J.; Levy, A. B. *JOC* **1978**, *43*, 1279. (b) Levy, A. B.; Angelastro, R.; Marinelli, E. R. *S* **1980**, 945.
27. (a) Satoh, M.; Miyaura, N.; Suzuki, A. *CL* **1986**, 1329. (b) Satoh, Y.; Serizawa, H.; Miyaura, N.; Hara, S.; Suzuki, A. *TL* **1988**, *29*, 1811.
28. (a) Brown, H. C.; Bhat, N. G. *TL* **1988**, *29*, 21. (b) Brown, H. C.; Bhat, N. G.; Rajagopalan, S. *S* **1986**, 480.
29. Hara, S.; Hyuga, S.; Aoyama, M.; Sato, M.; Suzuki, A. *TL* **1990**, *31*, 247.
30. Basile, T.; Biondi, S.; Boldrini, G. P.; Tagliavini, E.; Trombini, C.; Umani-Ronchi, A. *JCS(P1)* **1989**, 1025.
31. Normant, H. *Adv. Org. Chem.* **1960**, *2*, 1.
32. Dreiding, A. S.; Pratt, R. J. *JACS* **1954**, *76*, 1902.

33. (a) Rossi, R.; Carpita, A.; Quirici, M. G. *T* **1981**, *37*, 2617. (b) Rossi, R.; Carpita, A.; Quirici, M. G.; Gaudenzi, L. *T* **1982**, *38*, 631.
34. (a) Miyaura, N.; Suginome, H.; Suzuki, A. *TL* **1983**, *24*, 1527. (b) Suzuki, A. *PAC* **1991**, *63*, 419. (c) Cassani, G.; Massardo, P.; Piccardi, P. *TL* **1982**, *24*, 2513.
35. Brown, H. C.; Subrahmanyam, C.; Hamaoka, T.; Ravindran, N.; Bowman, D. H.; Misumi, S.; Unni, M. K.; Somayaji, V.; Bhat, N. G. *JOC* **1989**, *54*, 6068.
36. Brown, H. C.; Hamaoka, T.; Ravindran, N.; Subrahmanyam, C.; Somayaji, V.; Bhat, N. G. *JOC* **1989**, *54*, 6075.
37. (a) Brown, H. C.; Somayaji, V. *S* **1984**, 919. (b) Brown, H. C.; Hamaoka, T.; Ravindran, N. *JACS* **1973**, *95*, 5786.
38. (a) Kabalka, G. W.; Sastry, K. A. R.; Knapp, F. F.; Srivastava, P. C. *SC* **1983**, *13*, 1027. (b) Kabalka, G. W.; Sastry, K. A. R.; Somayaji, V. *H* **1982**, *18*, 157. (c) Kunda, S. A.; Smith, T. L.; Hylarides, M. D.; Kabalka, G. W. *TL* **1985**, *26*, 279.
39. Brown, H. C.; Larock, R. C.; Gupta, S. K.; Rajagopalan, S.; Bhat, N. G. *JOC* **1989**, *54*, 6079.
40. Brown, H. C.; Blue, C. D.; Nelson, D. J.; Bhat, N. G. *JOC* **1989**, *54*, 6064.
41. Miyaura, N.; Satoh, M.; Suzuki, A. *TL* **1986**, *27*, 3745.
42. Beck, J. S.; Albani, C. R.; McGhie, A. R.; Rothman, J. C.; Sneddon, L. G. *Chem. Mater.* **1989**, *1*, 433.
43. Beck, J. S.; Sneddon, L. G. *IC* **1990**, *29*, 295.
44. Li, J.; Caparrelli, D. J.; Jones, Jr., M. *JACS* **1993**, *115*, 408.
45. Györi, B.; Emri, J.; Fehér, I. *JOM* **1983**, *255*, 17.
46. (a) Schmidbaur, H.; Wimmer, T.; Reber, G.; Müller, G. *AG(E)* **1988**, *27*, 1071. (b) Schmidbaur, H.; Wimmer, T.; Grohmann, A.; Steigelman, O.; Müller, G. *CB* **1989**, *122*, 1607.

Marek Zaidlewicz
Nicolaus Copernicus University, Torun, Poland

Herbert C. Brown
Purdue University, West Lafayette, IN, USA

1,1-Di-*t*-butyl Peroxide[1–3]

t-Bu–O–O–*t*-Bu

[110-05-4] $C_8H_{18}O_2$ (MW 146.23)

(radical initiator, initiates anti-Markovnikov addition of HX to alkenes (X = halogen, S, Si, P);[4] initiates, by H-abstraction, the following radical reactions of compounds with an activated C–H bond: (a) dehydrodimerization;[5,6] (b) intra-[7] and intermolecular additions[3,8,9] to alkenes, alkynes, and carbonyl compounds; (c) addition to protonated heterocycles;[10,11] (d) fragmentation;[12,13] mediates alcohol deoxygenation via silane reduction of esters[14] and chlorohydrin formation by reaction with $TiCl_4$ and alkene[15])

Alternate Names: *t*-butyl peroxide; DTBP.
Physical Data: bp 109 °C/760 mmHg, 63 °C/119 mmHg; *d* 0.796 g cm^{-3}.
Solubility: freely soluble in organic solvents.
Form Supplied in: clear colorless liquid.
Analysis of Reagent Purity: Kropf describes various procedures (chemical and chromatographic) for the analysis of peroxides.[1b]
Handling, Storage, and Precautions: explosive; harmful if exposed by inhalation or skin contact; strong oxidizer; flammable; keep away from heat. Caution: All experiments involving peroxy compounds should be carried out behind a safety shield. Use in a fume hood.

General Discussion. Di-*t*-butyl peroxide is a commonly used initator for radical reactions. It undergoes facile unimolecular thermal decomposition to the *t*-butoxy radical, which in turn fragments to a methyl radical and acetone. The half-lives of DTBP are approximately 3 h at 140 °C and 24 h at 120 °C.[2] Accordingly, radical reactions that proceed at 110–150 °C can be initiated by DTBP, since a steady concentration of the initiating radical would be available for the reactions. Alkoxy radicals are electrophilic and they initiate the reaction by abstraction of a C–H bond α to a heteroatom. The resultant radicals undergo a variety of carbon–carbon bond forming reactions, including polymerization.[16] Scheme 1 shows the primary steps involved in useful C–C bond forming reactions. Use of precursors having an activated C–H bond is advantageous for this purpose, since the chain-transfer step also produces the primary adducts along with the propagating radical, which is also one of the reactants. However, because of the high bond energy associated with the C–H bond, these types of compound are poor H-atom donors. Chain lengths of these reactions are generally short; the use of a large excess of the C–H precursor and a steady supply of the initiator are important for success in many instances.

Initiation

t-Bu–O–O–*t*-Bu ⟶ 2 *t*-Bu–O• ⟶ Me• + Me_2CO

Initiator (In•) + XYCH–H ⟶ XYCH• + In–H

Propagation

XYCH• + CH₂=CHZ —addition→ XYCH–CH₂–CH•Z

XYCH–CH₂–CH•Z + CH₂=CHZ —telomerization→ XYCH–CH₂–CHZ–CH₂–CH•Z

XYCH• ⟶ Rearrangements or fragmentation followed by other radical reactions

Chain transfer

XYCH–CH₂–CH•Z + XYCH–H ⟶ XYCH–CH₂–CH₂Z + XYCH•

Termination

XYCH• —dimerization→ XYCH–CHXY

Other radical–radical reactions

Scheme 1 Primary radical processes

Abstractions of H adjacent to ethers and amines are important examples, the heteroatom adjacent to the site of reaction providing a favorable polar contribution to the transition state for abstraction by the electrophilic oxy radical, and providing stabilization for the product radical. Examples of the use of DTBP in this way include the study of anomeric radicals by Ingold and co-workers (eq 1),[17] and a synthesis of substituted tetrahydrofurans (eq 2).[18]

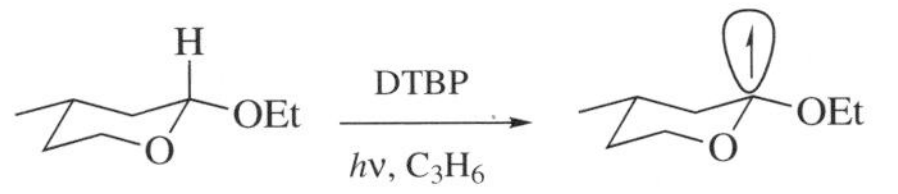

DTBP, $h\nu$, C_3H_6 (1)

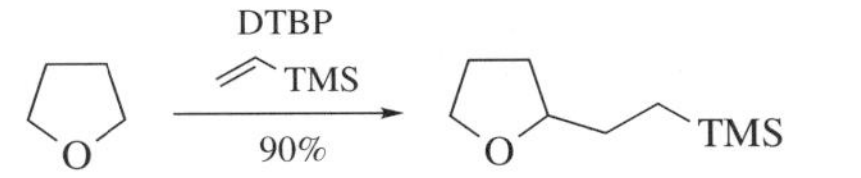

DTBP, TMS-ethylene, 90% (2)

In the first case the radicals were generated by UV photolysis of the reaction mixture in the cavity of the NMR instrument used for studying the configuration at the anomeric center. By contrast, the substitution of THF was carried out under more typical thermal conditions by heating in a sealed autoclave. In this study it was found that similar substitution of tetrahydro-2-furanone could also be carried out, although generally in modest yield.

Hydrogen atom abstraction from amino acid derivatives is especially facile, since the resulting captodative radicals are highly stabilized. Treatment of an alanine derivative with DTBP led to a mixture of an α-methylated product and a diastereomeric mixture of dimers (eq 3).[19] The first product is formed as a result of radical combination between the captodative amino acid radical and a methyl radical formed by scission of a *t*-butoxy radical. Substitution of protected dipeptides can be carried out using this approach, by including an alkylating agent such as toluene in the reaction medium (eq 4).[20] The reaction relies on the action of biacetyl (***2,3-Butanedione***) as a photoinitiator, hydrogen atom abstraction occurring preferentially at glycine residues.

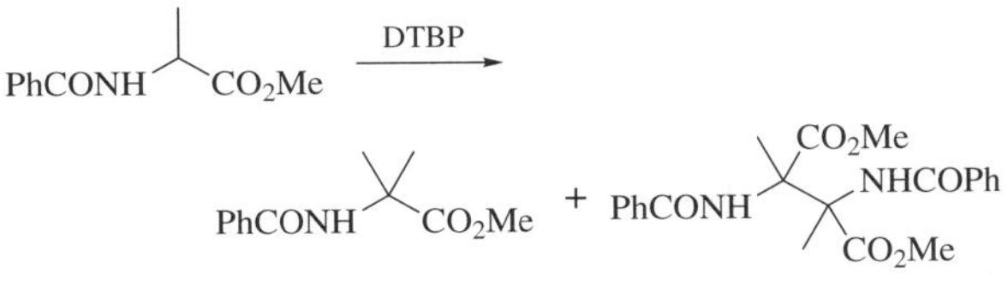

PhCONH–CH(CH3)–CO_2Me —DTBP→ PhCONH–C(CH3)2–CO_2Me (10%) + PhCONH–C(CH3)(CO_2Me)–C(CH3)(CO_2Me)–NHCOPh (20%) (3)

Tfa-Gly-Gly-OMe —DTBP, biacetyl, toluene→ Tfa-Phe-Gly-OMe (30%) + Tfa-Gly-Phe-OMe (29%) + Tfa-Phe-Phe-OMe (10%) (4)

Reactions involving phosphorus centered radicals derived from diethyl hydrogen phosphite are also initiated by DTBP under thermal conditions (eq 5).[21] Under modified reaction conditions, diphosphonate products, resulting from further addition to the initially formed unsaturated phosphonate, were also observed.

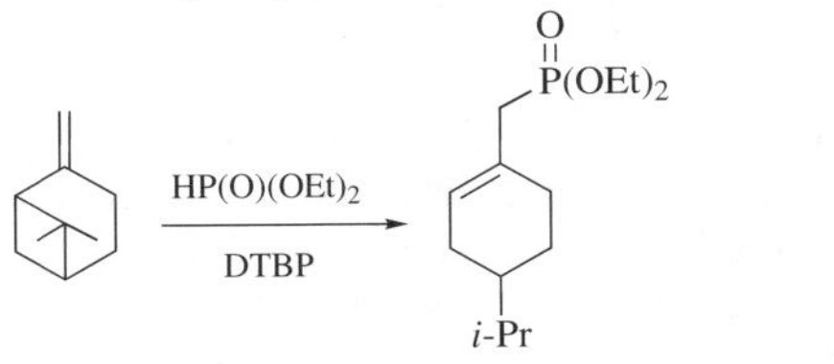

HP(O)$(OEt)_2$, DTBP → P(O)$(OEt)_2$ product, *i*-Pr (5)

Typical reactions of radicals with alkene acceptors, initiated by DTBP and used in synthesis, are listed below.

Intermolecular Additions. The radical chain nature and the anti-Markovnikov regiochemistry of radical addition reactions were originally discovered by Kharasch in the 1930s. Since then, these reactions have been used extensively for the formation of carbon–carbon[3,8] and carbon–heteroatom[4] bonds. Substrates that are suitable for the former include polyhalomethanes, alcohols, ethers, esters, amides, and amines. The prototypical examples compiled in Table 1 are from reviews by Walling[8] and Ghosez et al.[3]

Among the other notable applications are addition of the radical from diethyl malonate to alkynes and alkenes (eq 6),[9] addition of a 1,3-dioxalane-derived radical to formaldehyde (eq 7),[22] and addition of *s*-alcohols to alkenes (eq 8).[23] Novel radical mediated alkylations of a dipeptide makes use of DTBP as an initator (eqs 9 and 10).[20]

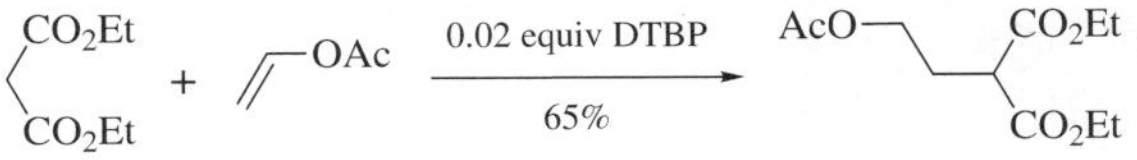

$CH_2(CO_2Et)_2$ + allyl OAc —0.02 equiv DTBP, 65%→ AcO–$(CH_2)_3$–$CH(CO_2Et)_2$ (6)

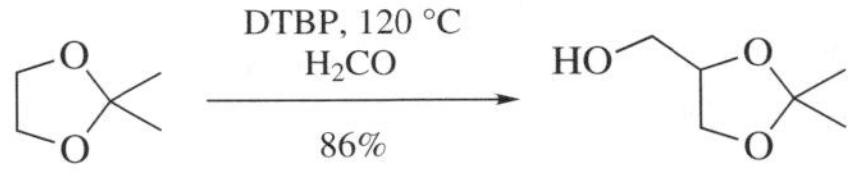

2,2-dimethyl-1,3-dioxolane —DTBP, 120 °C, H_2CO, 86%→ HO–CH_2–dioxolane (7)

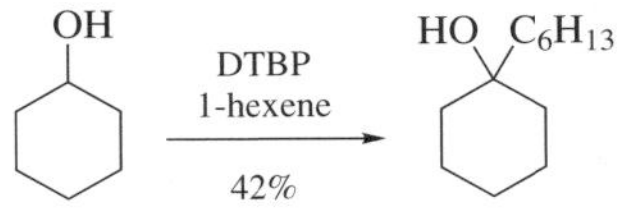

cyclohexanol —DTBP, 1-hexene, 42%→ 1-hexylcyclohexanol (HO, C_6H_{13}) (8)

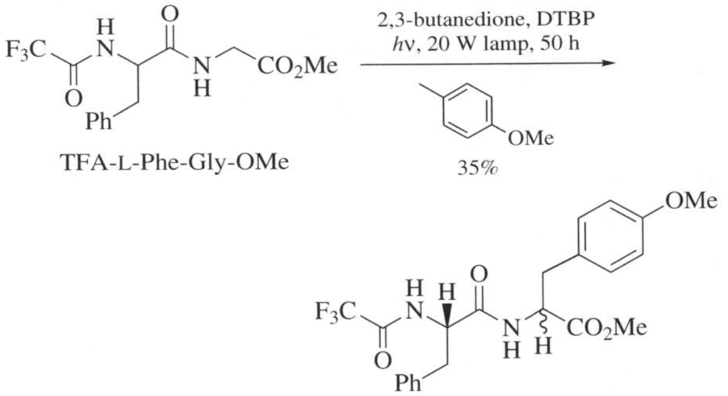

TFA-L-Phe-Gly-OMe —2,3-butanedione, DTBP, $h\nu$, 20 W lamp, 50 h; 4-methylanisole (OMe), 35%→ (9)

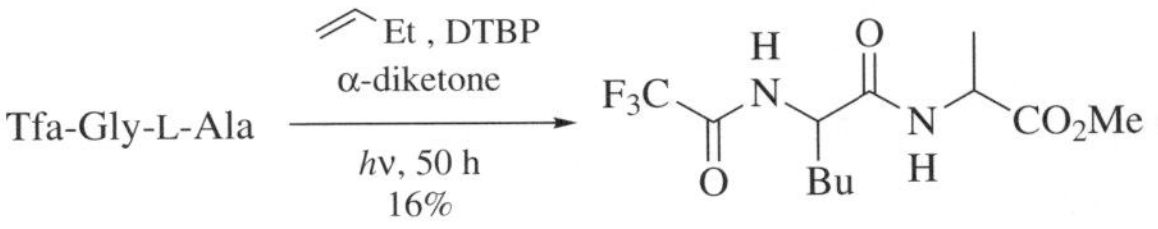

Tfa-Gly-L-Ala —Et-ethylene, DTBP, α-diketone, $h\nu$, 50 h, 16%→ F_3C–CO–NH–CH(Bu)–CO–NH–CH(CH3)–CO_2Me (10)

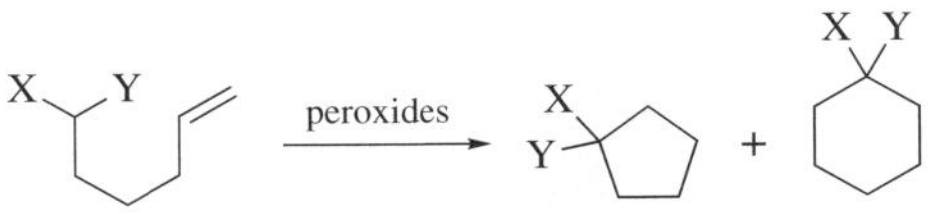

X, Y = CN, CO_2Et —peroxides→ cyclopentane + cyclohexane (X, Y) (11)

—2 equiv DTBP, cyclohexane, 150 °C→ 45% + 15% (CN, CO_2Et) (12)

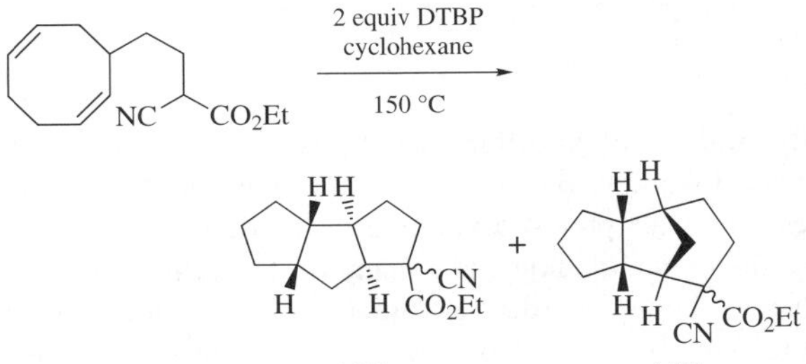

Table 1 Intermolecular C–C Bond Forming Reactions via Radicals

Precursor	Alkene	Reaction conditions	Product(s)
EtOH	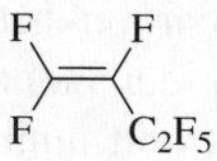 F F / F C_2F_5	DTBP, 120 °C, 48 h	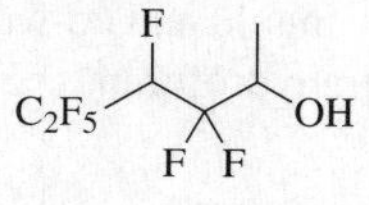 F, C_2F_5, OH, F F — 60%
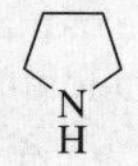 N H	 OH	DTBP, 120 °C, 48 h	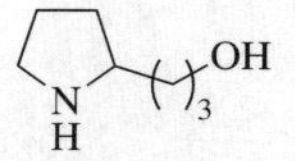 N H, OH, 3 — 54%
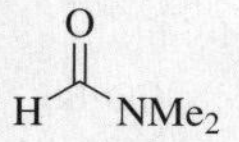 O, H NMe_2	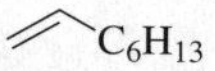 C_6H_{13}	DTBP, 132 °C, 18 h	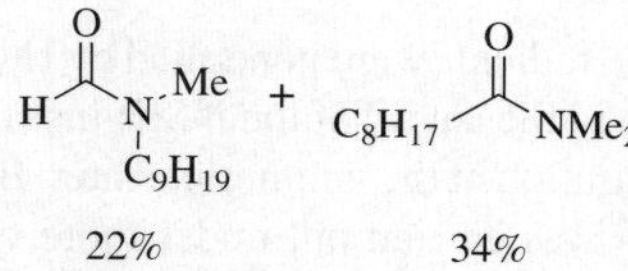 O, H, N, Me, C_9H_{19} + O, C_8H_{17}, NMe_2 — 22% 34%
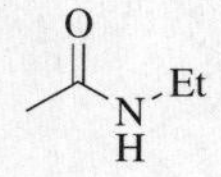 O, N H, Et	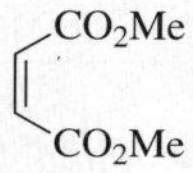 CO_2Me / CO_2Me	DTBP, 0.01 equiv, 6 h	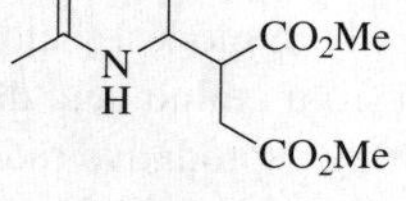 O, N H, CO_2Me, CO_2Me — 87%
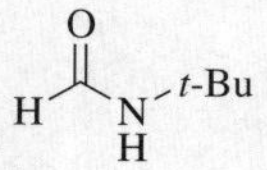 O, H, N H, *t*-Bu	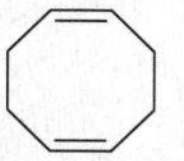	DTBP, 137 °C, 24 h	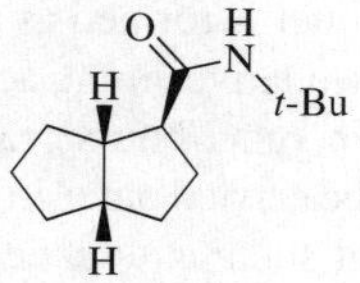 O, H, N, *t*-Bu, H, H
 C_5H_{11} CO_2Me	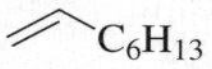 C_6H_{13}	DTBP, 150 °C	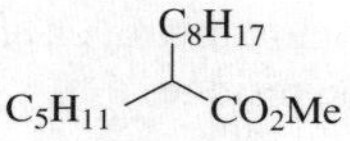 C_8H_{17}, C_5H_{11}, CO_2Me — 56%
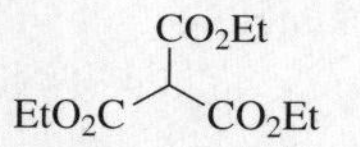 CO_2Et, EtO_2C CO_2Et	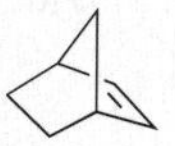	DTBP, 150 °C	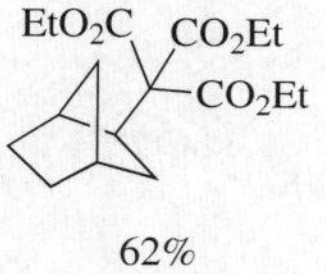 EtO_2C CO_2Et CO_2Et — 62%
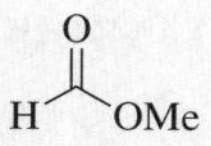 O, H OMe	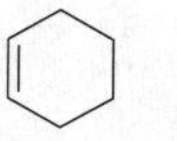	DTBP	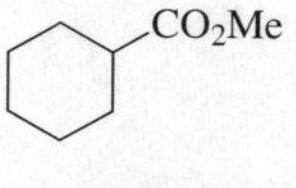 CO_2Me — 36%
EtCN	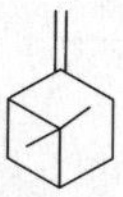	DTBP, 6 h, 140–150 °C	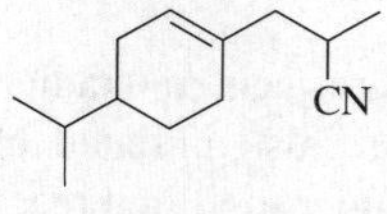 CN — 77%

Intramolecular Addition Reactions. Early studies by Julia and his co-workers on the cyclization of hexenyl (eq 11) and heptenyl radicals played a key role in the development of radical synthetic methodology, and many of the earlier studies were conducted with peroxides as initiators.[7] Cyclization of stabilized radicals such an malonates and cyanoesters are reversible, and the course of ring closure can be controlled by the appropriate choice of precursors and reaction conditions. Thus the cyanoacetate in eq 12[24] with 2 equiv of DTBP gives the products shown, whereas under kinetic conditions, using the tin hydride method, a different product distribution is obtained (eq 13).

Table 2 Dimerization of Radicals

Precursor	Reaction conditions	Products
$PhCH_2OH$	DTBP, 140 °C, 10 h	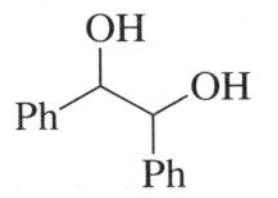 OH, Ph, OH, Ph 69%
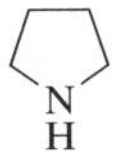 N H	DTBP, 135 °C	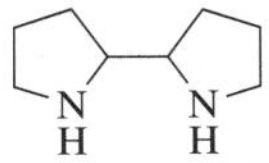 N H, N H 73%
$MeCONMe_2$	DTBP (5 mole %), 140 °C	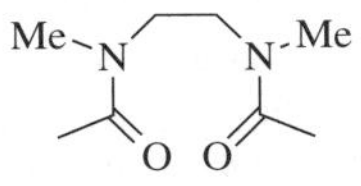 Me, N, N, Me, O, O 100%
 O	DTBP (5 mole %), 140 °C	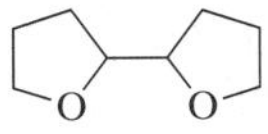 O, O 79%
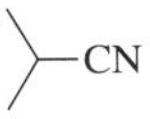 CN	DTBP (5 mole %), 140 °C	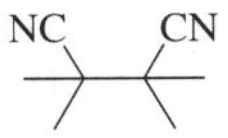 NC, CN 84%
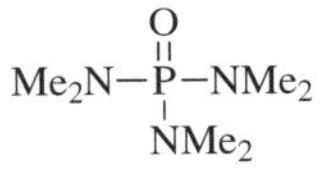 O, $Me_2N{-}P{-}NMe_2$, NMe_2	DTBP (5 mole %), 140 °C	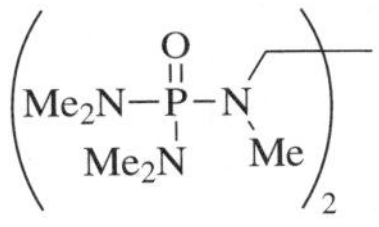 $(Me_2N{-}P(O)(NMe_2){-}N(Me)CH_2{-})_2$
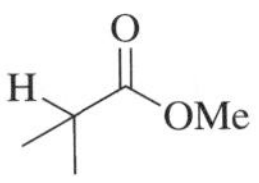 H, O, OMe	DBTP, 160 °C, 8 h	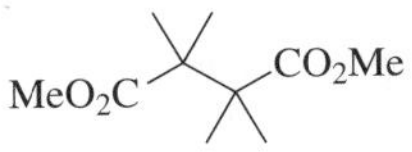 MeO_2C, CO_2Me
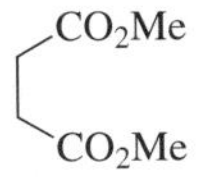 CO_2Me, CO_2Me	DBTP, 160 °C, 8 h	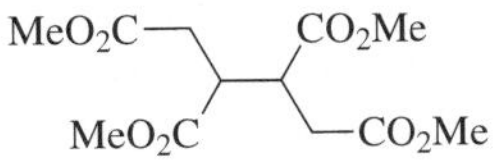 MeO_2C, CO_2Me, MeO_2C, CO_2Me

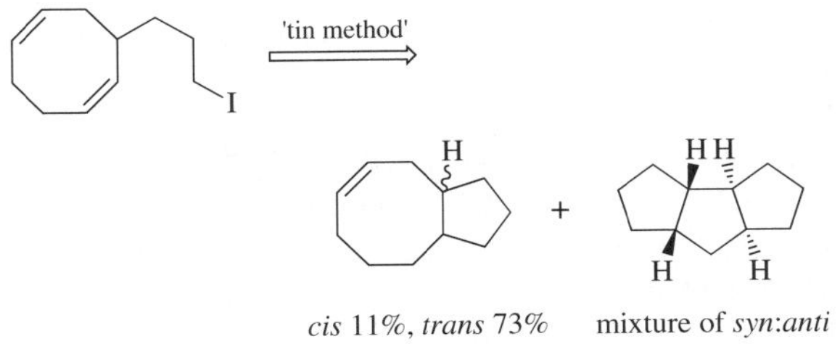

'tin method'

H + H H, H H (13)

cis 11%, *trans* 73% mixture of *syn*:*anti*

Dehydrodimerization. Radicals that are stabilized by an α-heteroatom, when produced in sufficiently high concentrations, will undergo dimerization. Use of DTBP is particularly effective for dehydrodimerizations of polyhaloalkanes,[25] alcohols, ethers,[5,25] amides, and esters (Table 2).[5,6] Viehe, who pioneered this work, has used α-*t*-butylmercaptoacrylonitrile as a trapping agent for the above mentioned C-centered radicals.[26] The adduct radical is stabilized by 'captodative effects'[27] and do not participate in further chain transfer chemistry. These radicals undergo ready dimerization, thereby providing a facile route to compounds with a four-carbon bridge between the original radicals (eq 14).

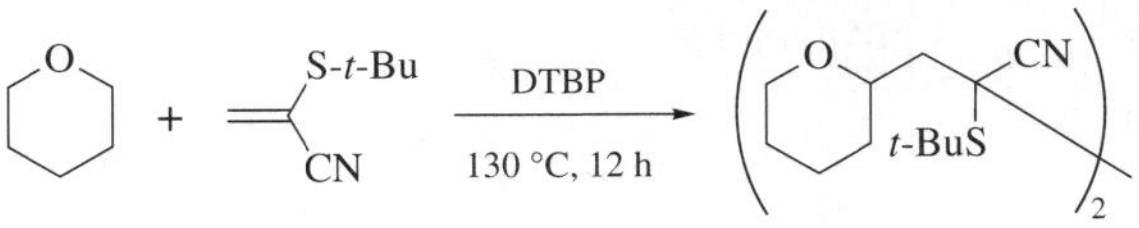

O + S-*t*-Bu, CN —DTBP, 130 °C, 12 h→ (O, CN, *t*-BuS)$_2$ (14)

Fragmentation Reactions. The tetrahydrofuranyl radical undergoes fragmentation at 140 °C to give an open-chain acyl radical. The THF radical as well as the rearranged radical are trapped by excess of alkene (eq 15).[12] Benzylidene acetals undergo similar fragmentation to give a benzoate ester (eq 16).[13]

Homolytic Substitution Reactions. Alkylation of electron-deficient heteroaromatic compounds developed by Minisci and co-workers is a powerful method for their functionalization.[10,28] Three examples are illustrated in eqs 17, 18 and 19. The product distribution often depends on the oxidant used. For example, as shown in eq 19, DTBP gives a 1:2 mixture of two products (A and B) upon alkylation of 4-methylquinoline. ***t*-Butyl Hydroperoxide** and Fe^{II} salts give almost exclusively the dimethylaminocarbonyl radical adduct A (eq 19).[11]

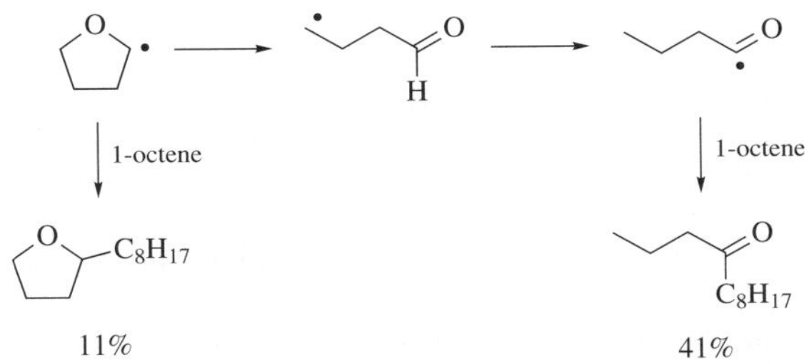

1-octene 1-octene (15)

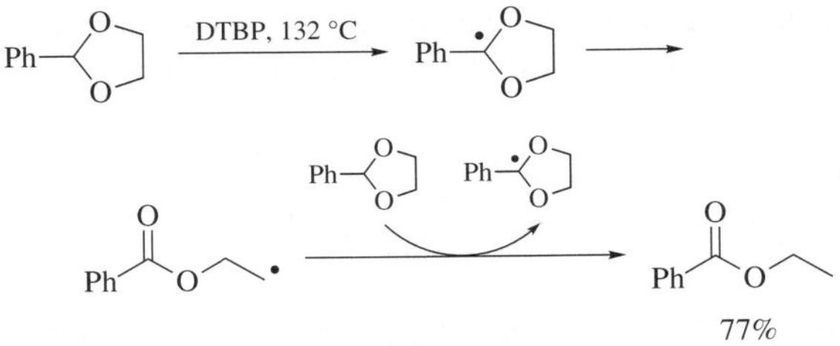

O, C_8H_{17} 11% O, C_8H_{17} 41%

Ph —DTBP, 132 °C→ Ph →

Ph, O, O, Ph → Ph, O, O (16) 77%

N, N + O —DTBP, H^+, Fe^{2+}, 0–20 °C, 45 min, 52%→ N, N, O (17)

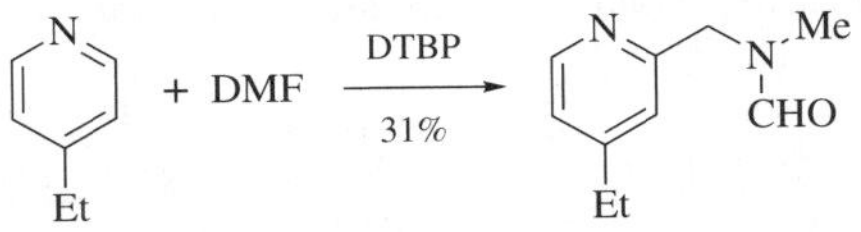

N, Et + DMF —DTBP, 31%→ N, N, Me, CHO, Et (18)

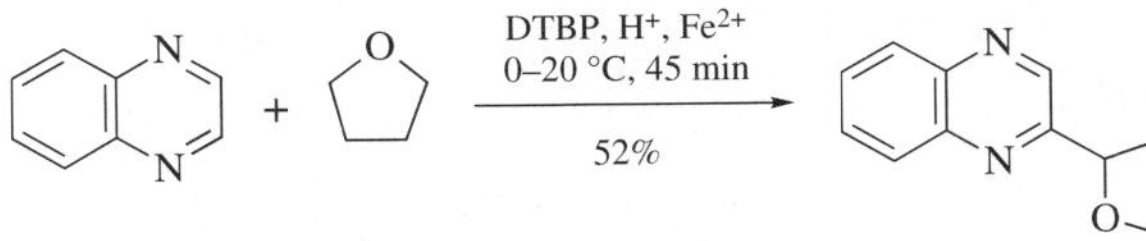

(19a) $H(C=O)NMe_2$ —[DTBP, 130 °C, DMF, lepidine, 57%]→ $\cdot C(=O)NMe_2$ → (A) 33%

(19b) → $HC(=O)N(Me)CH_2\cdot$ → (B) 67%

Miscellaneous Reactions. DTBP has been used as a hydrosilylation catalyst,[4] even though catalysis[29] by transition metal complexes have largely replaced the radical methods. DTBP has also been used as an oxidant for silanes.[30,31] Other applications of DTBP include its use as an initiator for radical mediated deoxygenation of alcohols via the corresponding chloroformate[32] or acetate ester[14] (eq 20). It has also been used as an initiator for the reduction of lactones and esters to ethers using ***Trichlorosilane***.[33] In a rare example of a nonradical reaction, DTBP has been used in conjunction with ***Titanium(IV) Chloride*** for the formation of chlorohydrin from alkenes (eq 21).[15]

(20) OAc —[$Ph_2(H)Si-C_6H_4-Si(H)Ph_2$, DTBP, 140 °C]→

(21) OMe —[1. DTBP, $TiCl_4$, −78 °C; 2. K_2CO_3, MeOH; 70%]→ OMe

cis:*trans* = 90:10
(with *m*-CPBA *cis*:*trans* = 1:1)

The carbonylation reaction of disulfides, catalyzed by ***Octacarbonyldicobalt***, normally leads to the production of thioesters. However, in the presence of DTBP and in the absence of CO the reaction takes an alternative course, with benzyl disulfides undergoing clean desulfurization to give the corresponding sulfides (eq 22).[34]

$$\text{BnSSBn} \xrightarrow[\text{DTBP}]{Co_2(CO)_8} \text{BnSBn} \qquad (22)$$

DTBP has also been employed in palladium catalyzed carbonylation reactions. Depending on the type of catalyst used, or on the reaction conditions, the carbonylation reaction of primary amines can be used to prepare either ureas (eq 23)[35] or carbamate esters (eq 24).[36]

$$R-NH_2 \xrightarrow[\text{MeOH, CO (1 atm)}]{\text{catalyst, } CuCl_2\text{, DTBP}} RNHC(O)NHR \qquad (23)$$

catalyst = montmorillonite(bipyridyl)palladium(II) acetate

$$R-NH_2 \xrightarrow[\text{MeOH, CO (1 atm)}]{PdCl_2, CuCl_2\text{, DTBP}} RNHCO_2Me \qquad (24)$$

Using the ***Palladium(II) Chloride*** system, the reaction involving secondary amines was found to give mixtures of carbamate ester and an oxamate ester resulting from double carbonylation.[36] Analogous carbonylations of alcohols can lead to a range of products, including dialkyl carbonates, oxalates, and succinates.[37]

Related Reagents. See Class O-14, pages 1–10.

1. (a) Sheldon, R. A. In *The Chemistry of Functional Groups, Peroxides*; Patai, S., Ed.; Wiley: New York, 1983; p 161. (b) Kropf, H. *MOC* **1988**; *E13*.
2. Walling, C. *T* **1985**, *41*, 3887.
3. Ghosez, A.; Giese, B.; Zipse, H. *MOC* **1989**, *EXIXa*, 533.
4. Stacey, F. W.; Harris, J. F., Jr. *OR* **1963**, *13*, 150.
5. Naarmann, H.; Beaujean, M.; Merényi, R.; Viehe, H. G. *Polym. Bull.* **1980**, *2*, 363.
6. Naarmann, H.; Beaujean, M.; Merényi, R.; Viehe, H. G. *Polym. Bull.* **1980**, *2*, 417.
7. Julia, M. *ACR* **1971**, *4*, 386. See also: Beckwith, A. L. J. *T* **1981**, *37*, 3073.
8. Walling, C.; Huyser, E. S. *OR* **1963**, *13*, 91.
9. Vogel, H. *S* **1970**, 99. For an attractive organometallic variation of several of the reactions described in this article, see Heiba, E. I.; Dessau, R. M.; Rodewald, P. G. *JACS* **1974**, *96*, 7977. See also: Fristad, W. E.; Peterson, J. R.; Ernst, A. B.; Urbi, G. B. *T* **1986**, *42*, 3429.
10. Minisci, F. *S* **1973**, 1.
11. Arnone, A.; Cecere, M.; Galli, R.; Minisci, F.; Perchinunno, M.; Porta, O.; Gardini, G. *G* **1973**, *103*, 13.
12. Wallace, T. J.; Gritter, R. J. *JOC* **1962**, *27*, 3067.
13. Huyser, E. S.; Garcia, Z. *JOC* **1962**, *27*, 2716.
14. Sano, H.; Takeda, T.; Migata, T. *CL* **1988**, 119. See also: Sano, H.; Ogata, M.; Migita, T. *CL* **1986**, 77.
15. Klunder, J. M.; Caron, M.; Uchiyama, M.; Sharpless, K. B. *JOC* **1985**, *50*, 912.
16. Polymerization is favored under low concentrations of chain transfer agents. The polymer forming reactions are beyond the scope of this article and more appropriate reviews and monographs should be consulted for further information. See for example: Hodge, P. In *Comprehensive Organic Chemistry*; Barton, D. H. R., Ed.; Pergamon: Oxford, 1991; Vol. 5, p 833 and references cited therein.
17. (a) Malatesta, V.; McKelvey, R. D.; Babcock, B. W.; Ingold, K. U. *JOC* **1979**, *44*, 1872. (b) Malatesta, V.; Ingold, K. U. *JACS* **1981**, *103*, 609.
18. Gevorgyan, V.; Priede, E.; Liepins, E.; Gavars, M.; Lukevics, E. *JOM* **1990**, *393*, 333.
19. Burgess, V. A.; Easton, C. J.; Hay, M. P. *JACS* **1989**, *111*, 1047.
20. Schwarzberg, M.; Sperling, J.; Elad, D. *JACS* **1973**, *95*, 6418.
21. Battiste, D. R.; Haseldine, D. L. *SC* **1984**, *14*, 993.
22. Sanderson, J. R.; Lin, J. J.; Duranleau, R. G.; Yeakey, E. L.; Marquis, E. T. *JOC* **1988**, *53*, 2859.
23. Urry, W. H.; Stacey, F. W.; Huyser, E. S.; Juveland, O. O. *JACS* **1954**, *76*, 450.
24. Winkler, J.; Sridar, V. *JACS* **1986**, *108*, 1708.
25. Schwetlick, K.; Jentzsch, J.; Karl, R.; Wolter, D. *JPR* **1964**, *25*, 95.
26. Mignani, S.; Beaujean, M.; Janousek, Z.; Merényi, R.; Viehe, H. G. *T (Suppl.)* **1981**, *37*, 111.
27. Viehe, H. G.; Janousek, Z.; Merényi, R.; Stella, R. *ACR* **1985**, *18*, 148.
28. Minisci, F.; Citterio, E.; Vismara, E.; Giordano, C. *T* **1985**, *41*, 4157.
29. Fleming, I. In *Comprehensive Organic Chemistry*; Barton, D. H. R., Ed.; Pergamon: Oxford, 1991; Vol. 3, p. 562 and references cited therein.
30. Curtice, J.; Gilman, H.; Hammond, G. S. *JACS* **1957**, *79*, 4754.

31. Sakurai, H.; Hosomi, A.; Kumada, M. *BCJ* **1967**, *40*, 1551.
32. Billingham, N. C.; Jackson, R. A.; Malek, F. *CC* **1977**, 344.
33. Nagata, Y.; Dohmaru, T.; Tsurugi, J. *JOC* **1973**, *38*, 795. See also: Nakao, R.; Fukumoto, T.; Tsurugi, J. *JOC* **1972**, *37*, 76 and Nakao, R.; Fukumoto, T.; Tsurugi, J. *JOC* **1972**, *37*, 4349.
34. Antebi, S.; Alper, H. *TL* **1985**, *26*, 2609.
35. Choudary, B. M.; Koteswara Rao, K.; Pirozhkov, S. D.; Lapidus, A. L. *SC* **1991**, 1923.
36. Alper, H.; Vasapollo, G.; Hartstock, F. W.; Mlekuz, M.; Smith, D. J. H.; Morris, G. E. *OM* **1987**, *6*, 2391.
37. Morris, G. E.; Oakley, D.; Pippard, D. A.; Smith, D. J. H. *CC* **1987**, 410.

T. V. (Babu) RajanBabu
The Ohio State University, Columbus, OH, USA

Nigel S. Simpkins
University of Nottingham, UK

2,3-Dichloro-5,6-dicyano-1,4-benzoquinone[1]

[84-58-2] $C_8Cl_2N_2O_2$ (MW 227.01)

(powerful oxidant, particularly useful for dehydrogenation to form aromatic[1a–d] and α,β-unsaturated carbonyl compounds;[1] oxidizes activated methylene[1a–c] and hydroxy groups[1b] to carbonyl compounds; phenols are particularly sensitive[1c])

Alternate Name: DDQ.
Physical Data: mp 213–216 °C; $E_0 \approx 1000$ mV.
Solubility: very sol ethyl acetate and THF; moderately sol dichloromethane, benzene, dioxane, and acetic acid; insol H_2O.
Form Supplied in: bright yellow solid; widely available.
Analysis of Reagent Purity: UV (λ_{max} [dioxane] 390 nm) and mp.
Purification: recrystallization from a large volume of dichloromethane.
Handling, Storage, and Precautions: indefinitely stable in a dry atmosphere, but decomposes in the presence of water with the evolution of HCN. Store under nitrogen in a sealed container.

Introduction. Quinones of high oxidation potential are powerful oxidants which perform a large number of useful reactions under relatively mild conditions. Within this class, DDQ represents one of the more versatile reagents since it combines high oxidant ability with relative stability[1] (see also ***Chloranil***). Reactions with DDQ may be carried out in inert solvents such as benzene, toluene, dioxane, THF, or AcOH, but dioxane and hydrocarbon solvents are often preferred because of the low solubility of the hydroquinone byproduct. Since DDQ decomposes with the formation of hydrogen cyanide in the presence of water, most reactions with this reagent should be carried out under anhydrous conditions.[1a]

Dehydrogenation of Hydrocarbons. The mechanism by which quinones effect dehydrogenation is believed to involve an initial rate-determining transfer of hydride ion from the hydrocarbon followed by a rapid proton transfer leading to hydroquinone formation.[1d] Dehydrogenation is therefore dependent upon the degree of stabilization of the incipient carbocation and is enhanced by the presence of functionality capable of stabilizing the transition state. As a consequence, unactivated hydrocarbons are stable to the actions of DDQ while the presence of alkenes or aromatic moieties is sufficient to initiate hydrogen transfer.[1d,2] The formation of stilbenes from suitably substituted 1,2-diarylethanes[3] and the synthesis of chromenes by dehydrogenation of the corresponding chromans (eq 1)[4] are particularly facile transformations. Similar reactions have also found considerable utility for the introduction of additional unsaturation into partially aromatized terpenes and steroids, where the ability to control the degree of unsaturation in the product is a particular feature of quinone dehydrogenations.[5] Moreover, the ability to effect exclusive dehydrogenation in the presence of sensitive substituents such as alcohols and phenols (eq 2)[5b] illustrates the mildness of the method and represents a further advantage.

DDQ, PhH, 80 °C, 20 h, 80% (1)

DDQ, 20 °C, 90% (2)

DDQ is a particularly effective aromatization reagent and is frequently the reagent of choice to effect facile dehydrogenation of both simple (eq 3)[6] and complex hydroaromatic carbocyclic compounds.[1d,7] Skeletal rearrangements are relatively uncommon features of quinone-mediated dehydrogenation reactions, but 1,1-dimethyltetralin readily undergoes aromatization with a 1,2-methyl shift when subjected to the usual reaction conditions (eq 4).[8] Wagner–Meerwein rearrangements have also been observed in the aromatization of steroids (eq 5), although in this instance considerably longer reaction times are required.[9] Such reactions provide a unique method for the aromatization of cyclic systems containing quaternary carbon atoms without the loss of carbon.

DDQ, PhH, 80 °C, 2 h, 100% (3)

DDQ, PhH, 80 °C, 5 h, 78% (4)

DDQ, TsOH dioxane 101 °C, 72 h 48% (5)

DDQ is also an effective reagent for the dehydrogenation of hydroaromatic heterocycles, and pyrroles,[10] pyrazoles,[11] triazoles,[12] pyrimidines,[13] pyrazines,[14] indoles,[15] quinolines,[16] furans,[17] thiophenes,[18] and isothiazoles[19] are among the many aromatic compounds prepared in this manner. Rearomatization of nitrogen heterocycles following nucleophilic addition across a C=N bond (eq 6)[13] is a particularly useful application of DDQ,[13,16] and similar addition and reoxidation reactions in acyclic systems have also been reported.[20]

1. MeLi, ether −30 °C 2. AcOH; DDQ, THF 0–20 °C 10 min 84% (6)

One particularly important use of DDQ has been in the dehydrogenation of reduced porphyrins, where the degree of aromatization of the product is highly dependent on the relative reagent:substrate stoichiometry.[1b] Under optimal conditions, excellent yields of partially or fully conjugated products may be isolated.[1b,21] The formation of porphyrins from tetrahydro precursors on reaction with 3 equiv of DDQ under very mild conditions (eq 7) typifies one of the more commonly described transformations.[21] More recently, DDQ has been used as part of a one-pot sequence for the formation of porphyrins from simple intermediates, although the overall yields in such reactions are generally comparatively low.[22]

In addition to the formation of neutral aromatic compounds, DDQ is also an effective agent for the preparation of the salts of stable aromatic cations. High yields of tropylium (eq 8) and triphenylcyclopropenyl (eq 9) cations have been isolated in the presence of acids such as perchloric, phosphoric, and picric acid,[23] and oxonium,[24] thioxonium,[23,25] and pyridinium[23,26] salts may be prepared in reasonable yields from appropriate starting materials under essentially similar conditions. The formation of the perinaphthyl radical has been reported on oxidation of perinaphthalene with DDQ under neutral conditions,[23] although such products are not usually expected.

Dehydrogenation of Carbonyl Compounds. DDQ and other high oxidation potential quinones are versatile reagents for the synthesis of α,β-unsaturated carbonyl compounds,[1e] a reaction that has found extensive application in the chemistry of 3-keto steroids.[1b] The regiochemical course of this dehydrogenation is highly dependent on the initial steroidal geometry; thus the 5α- and the 5β-series usually furnish Δ^1- and Δ^4-3-keto steroids, respectively (eq 10).[1b] The selection of one isomer over the other is likely to reflect the relative steric crowding of the C-4 hydrogen atom in the two series, but other factors may play a role in those instances where the anticipated product is not formed.[1b]

DDQ, MeCN–ether (degassed) 22 °C, 30 min 100% (7)

DDQ, $HClO_4$ AcOH 92% ClO_4^- (8)

DDQ, $HClO_4$ AcOH 95% ClO_4^- (9)

DDQ; 5α series; DDQ; 5β series (10)

A rather more unusual situation exists during the dehydrogenation of Δ^4-3-keto steroids where the product formed is dependent on the oxidizing quinone. Thus whereas DDQ gives the $\Delta^{4,6}$ ketone, chloranil and a number of other quinones yield only the $\Delta^{1,4}$ isomer (eq 11), a result that has been rationalized on the basis of DDQ proceeding via the kinetic enolate while less reactive quinones proceed via the thermodynamic enolate.[27]

DDQ; chloranil (11)

While DDQ is an effective reagent for the formation of α,β-unsaturated steroidal ketones, the dehydrogenation of cyclohexanones to the corresponding enone only proceeds well when the further reaction is blocked by *gem*-dialkyl substitution.[1b,28] Tropone, by contrast, has been prepared from 2,4-cycloheptadienone (eq 12), although the yield was somewhat low.[29] Heterocyclic enones such as flavones[30] and chromones[31] may be efficiently prepared from flavanones and chromanones, respectively, under similar conditions to those used for the dehydrogenation of steroids, and the dehydrogenation of larger ring heterocyclic ketones has been described.[32] Ketone enol ethers have also been shown to undergo facile dehydrogenation to α,β-unsaturated ketones with DDQ, although the nature of the product formed may be dependent on the presence or absence of moisture.[33] Prior formation of the silyl enol ether is a potentially more versatile procedure that has been shown to overcome the problems generally associated with the dehydrogenation of unblocked cyclohexanones (eq 13),[34] particularly when the acidic hydroquinone formed during the reaction is neutralized by the addition of ***N,O-Bis(trimethylsilyl)acetamide*** (BSA)[34] or a hindered base.[35] Preparation of the enone derived from either the kinetic or the thermodynamic enolate is possible in this manner.[34,35a]

DDQ, PhH
80 °C, 2 h
10% (12)

OTMS
DDQ, BSA, PhH
20 °C, 1 h
50% (13)

Quinone dehydrogenation reactions of carbonyl compounds are mostly limited to the more readily enolized ketones, and analogous reactions on esters[36] and amides[37] require stronger conditions and are far less common unless stabilization of the incipient carbonium ion is possible. Oxidation in the presence of the silylating agent bis(trimethylsilyl)trifluoroacetamide (BSTFA) considerably improves the dehydrogenation of steroidal lactams (eq 14) by facilitating the breakdown of the intermediate quinone–lactam complex.[38] Similar dehydrogenations of carboxylic acids are rare, but reaction of the α-anion of carboxylate salts generated in the presence of HMPA has given modest yields of a number of α,β-unsaturated fatty acids.[39]

DDQ, BSTFA
dioxane
20–110 °C, 22 h
85–90% (14)

Oxidation of Alcohols. Saturated alcohols are relatively stable to the action of DDQ in the absence of light, although some hindered secondary alcohols have been oxidized in reasonable yield on heating under reflux in toluene for extended periods of time (eq 15).[40] It has been suggested that oxidation proceeds in this instance as a result of relief of steric strain.[40] Allylic and benzylic alcohols, on the other hand, are readily oxidized to the corresponding carbonyl compounds,[1b,41] and procedures have been developed which utilize catalytic amounts of the reagent in the presence of a stoichiometric amount of a second oxidant.[42] Since the rate of oxidation of allylic alcohols is greater than that for many other reactions,[43] the use of DDQ provides a selective method for the synthesis of allylic and benzylic carbonyl compounds in the presence of other oxidizable groups.

DDQ, toluene
116 °C, 8 h
96% (15)

Benzylic Oxidation. The oxidation of benzylic alkyl groups proceeds rapidly in those instances in which stabilization of the incipient carbonium ion is possible[44,45] and a number of polycyclic aromatic compounds have been oxidized in good yield to the corresponding benzylic ketones on brief treatment with DDQ in aqueous acetic acid at rt.[44] The reaction is postulated to proceed via an intermediate benzylic acetate which is hydrolyzed and further oxidized under the reaction conditions.[44] It is interesting that 1-alkylazulenes, which are cleaved by many of the more common oxidants, are cleanly oxidized following a short exposure to DDQ in aqueous acetone (eq 16), while under the same conditions no oxidation of C-2 alkyl substituents takes place.[46] As expected, the oxidation was shown to be disfavored by the presence of strongly electron-withdrawing substituents.[46]

DDQ, acetone (aq)
20 °C, 5 min
93% (16)

The stabilization of benzylic carbonium ions is also a feature of arenes containing electron-donating substituents, especially those having 4-alkoxy or 4-hydroxy groups, and such compounds are particularly effective substrates for oxidation by DDQ. Thus 6-methoxytetralone has been prepared in 70% yield from 6-methoxytetralin on treatment with DDQ in methanol,[47] although it is possible to isolate intermediate benzylic acetates if the oxidation is carried out in acetic acid (eq 17).[48] An interesting variant of the oxidation in inert solvents in the presence of either ***Cyanotrimethylsilane***[49] or ***Azidotrimethylsilane***[50] results in the isolation of good to excellent yields of benzyl cyanides and azides, respectively.

DDQ, AcOH, argon
22 °C, 12 h
73% (17)

Benzylic oxidation of alkoxybenzyl ethers is particularly facile, and since some of the more activated derivatives are cleaved under conditions which leave benzyl, various ester, and formyl groups unaffected, they have found application in the protection of primary and secondary alcohols.[51] Deprotection with DDQ in dichloromethane/water follows the order: 3,4-dimethoxy > 4-methoxy > 3,5-dimethoxy > benzyl and secondary > primary, thus allowing the selective removal of one function in the presence of another.[51] 2,6-Dimethoxybenzyl esters are readily cleaved to the corresponding acids on treatment with

DDQ in wet dichloromethane at rt, whereas 4-methoxybenzyl esters are stable under these conditions.[52] Oxidative cleavage of N-linked 3,4-dimethoxybenzyl derivatives with DDQ has also been demonstrated.[53]

DDQ is a powerful oxidizing agent for phenols, and carbonium ion stabilization via the quinone methide makes benzylic oxidation of 4-alkylphenols a highly favored process.[47,54] With methanol as the solvent it is possible to isolate α-methoxybenzyl derivatives in reasonable yield.[55]

Phenolic Cyclization and Coupling Reactions. The oxidation of phenolic compounds which either do not possess benzylic hydrogen atoms, or which have an alternative reaction pathway, can result in a variety of interesting products. Cyclodehydrogenation reactions leading to oxygen heterocycles represent a particular application of phenolic oxidation by DDQ, and is common when intramolecular quenching of the intermediate phenoxyl radical is possible (eqs 18–20).[56–59] These reactions necessarily take place in nonpolar solvents and have given such products as coumarins,[56] chromenes (eq 18),[57] benzofurans (eq 19),[58] and spiro derivatives (eq 20).[59]

DDQ, ether, 2 h, 90% (18)

DDQ, PhH, 80 °C, 1 h, 63% (19)

DDQ, PhH, 16 h, 84% (20)

Phenols and enolizable ketones that cannot undergo α,β-dehydrogenation may afford intermolecular products arising from either C–C or C–O coupling on treatment with DDQ in methanol.[60] 2,6-Dimethoxyphenol, for example, results predominantly in oxidative dimerization (eq 21), while the hindered 2,4,6-tri-*t*-butylphenol generates the product of quinone coupling (eq 22).[60] Various other unusual products have been observed on DDQ oxidation of phenols and enolic compounds, their structure being dependent on that of the parent compound.[41,60,61]

DDQ, MeOH, 20 °C, 62% (21)

DDQ, MeOH, 20 °C, 62–68% (22)

Miscellaneous Reactions. In addition to the key reactions above, DDQ has been used for the oxidative removal of chromium,[62] iron,[63] and manganese[64] from their complexes with arenes and for the oxidative formation of imidazoles and thiadiazoles from acyclic precursors.[65] Catalytic amounts of DDQ also offer a mild method for the oxidative regeneration of carbonyl compounds from acetals,[66] which contrasts with their formation from diazo compounds on treatment with DDQ and methanol in nonpolar solvents.[67] DDQ also provides effective catalysis for the tetrahydropyranylation of alcohols.[68] Furthermore, the oxidation of chiral esters or amides of arylacetic acid by DDQ in acetic acid provides a mild procedure for the synthesis of chiral α-acetoxy derivatives, although the diastereoselectivity achieved so far is only 65–67%.[69]

While quinones in general are well known dienophiles in Diels–Alder reactions, DDQ itself only rarely forms such adducts.[70] It has, however, been shown to form 1:1 adducts with electron-rich heterocycles such as benzofurans and indoles where it forms C–O and C–C adducts, respectively.[71]

Related Reagents. See Classes O-1, O-12, O-13 and O-21, pages 1–10.

1. (a) Jackman, L. M. *Adv. Org. Chem.* **1960**, *2*, 329. (b) Walker, D.; Hiebert, J. D. *CRV* **1967**, *67*, 153. (c) Becker, H.-D. In *The Chemistry of the Quinonoid Compounds*; Patai, S., Ed.; Wiley: Chichester, 1974; Part 2, Chapter 7. (d) Fu, P. P.; Harvey, R. G. *CRV* **1978**, *78*, 317. (e) Buckle, D. R.; Pinto, I. L. *COS* **1991**, *7*, 119.
2. Asato, A. E.; Kiefer, E. F. *CC* **1968**, 1684.
3. Findlay, J. W. A.; Turner, A. B. *OS* **1969**, *49*, 53.
4. (a) Starratt, A. N.; Stoesl, A. *CJC* **1977**, *55*, 2360. (b) Ahluwalia, V. K.; Arora, K. K. *T* **1981**, *37*, 1437. (c) Ahluwalia, V. K.; Ghazanfari, F. A.; Arora, K. K. *S* **1981**, 526. (d) Ahluwalia, V. K.; Jolly, R. S. *S* **1982**, 74.
5. (a) Brown, W.; Turner, A. B. *JCS(C)* **1971**, 2057. (b) Turner, A. B. *CI(L)* **1976**, 1030. (c) Fu, P. P.; Harvey, R. G. *TL* **1977**, 2059. (d) Abad, A; Agulló, C.; Arnó, M.; Domingo, L. R.; Zaragozá, R. J. *JOC* **1988**, *53*, 3761.
6. Braude, E. A.; Brook, A. G.; Linstead, R. P. *JCS* **1954**, 3569.
7. (a) Muller, J. F.; Cagniant, D.; Cagniant, P. *BSF* **1972**, 4364. (b) Diederick, F.; Staab, H. A. *AG(E)* **1978**, *17*, 372. (c) Stowasser, B.; Hafner, K. *AG(E)* **1986**, *25*, 466. (d) Funhoff, D. J. H.; Staab, H. A. *AG(E)* **1986**, *25*, 742. (e) Di Raddo, P.; Harvey, R. G. *TL* **1988**, *29*, 3885.
8. Braude, E. A.; Jackman, L. M.; Linstead, R. P.; Lowe, G. *JCS* **1960**, 3123.
9. Brown, W.; Turner, A. B. *JCS(C)* **1971**, 2566.
10. Padwa, A.; Haffmanns, G.; Tomas, M. *JOC* **1984**, *49*, 3314.
11. (a) Bousquet, E. W.; Moran, M. D.; Harmon, J.; Johnson, A. L.; Summers, J. C. *JOC* **1975**, *40*, 2208. (b) Padwa, A.; Nahm, S.; Sato, E. *JOC* **1978**, *43*, 1664.
12. Gilgen, P.; Heimgartner, H.; Schmid, H. *HCA* **1974**, *57*, 1382.
13. Harden, D. B.; Mokrosz, M. J.; Strekowski, L. *JOC* **1988**, *53*, 4137.
14. Blake, K. W.; Porter, A. E. A.; Sammes, P. G. *JCS(P1)* **1972**, 2494.
15. Hayakawa, K.; Yasukouchi, T.; Kanematsu, K. *TL* **1986**, *27*, 1837.
16. Meyers, A. I.; Wettlaufer, D. G. *JACS* **1984**, *106*, 1135.
17. (a) Piozzi, F.; Venturella, P.; Bellino, A. *OPP* **1971**, *3*, 223. (b) Stanetty, P.; Purstinger, G. *JCR(M)* **1991**, 581.
18. (a) Schultz, A. G.; Fu, W. Y.; Lucci, R. D.; Kurr, B. G.; Lo, K. M.; Boxer, M. *JACS* **1978**, *100*, 2140. (b) Moursounidis, J.; Wege, D. *TL* **1986**, *27*, 3045. (c) Mazerolles, P.; Laurent, C. *JOM* **1991**, *35*, 402.
19. Howe, R. K.; Franz, J. E. *JOC* **1978**, *43*, 3742.
20. (a) Strekowski, L.; Cegla, M. T.; Kong, S.-B; Harden, D. B. *JHC* **1989**, *26*, 923. (b) Strekowski, L.; Cegla, M. T.; Harden, D. B.; Kong, S.-B. *JOC* **1989**, *54*, 2464.

21. (a) Kämpfen, U.; Eschenmoser, A. *TL* **1985**, *26*, 5899. (b) Barnett, G. H.; Hudson, M. F.; Smith, K. M. *JCS(P1)* **1975**, 1401.
22. (a) Hevesi, L.; Renard, M.; Proess, G. *CC* **1986**, 1725. (b) Proess, G.; Pankert, D.; Hevesi, L. *TL* **1992**, *33*, 269.
23. Reid, D. H.; Fraser, M.; Molloy, B. B.; Payne, H. A. S.; Sutherland, R. G. *TL* **1961**, 530.
24. Carretto, J.; Simalty, M. *TL* **1973**, 3445.
25. Nakazumi, H.; Ueyama, T.; Endo, T.; Kitao, T. *BCJ* **1983**, *56*, 1251.
26. Ishii, H.; Chen, I-S.; Ishikawa, T. *JCS(P1)* **1987**, 671.
27. Turner, A. B.; Ringold, H. J. *JCS(C)* **1967**, 1720.
28. (a) Kane, V. V.; Jones Jr, M. *OS* **1982**, *61*, 129. (b) Hagenbruch, B.; Hünig, S. *CB* **1983**, *116*, 3884. (c) Jeffs, P. W.; Redfearn, R.; Wolfram, J. *JOC* **1983**, *48*, 3861.
29. van Tamelen, E. E.; Hildahl, G. T. *JACS* **1956**, *78*, 4405.
30. (a) Amemiya, T.; Yasunami, M.; Takase, K. *CL* **1977**, 587. (b) Matsuura, S.; Iinuma, M.; Ishikawa, K.; Kagei, K. *CPB* **1978**, *26*, 305.
31. Shanka, C. G.; Mallaiah, B. V.; Srimannarayana, G. *S* **1983**, 310.
32. Cliff, G. R.; Jones, G. *JCS(C)* **1971**, 3418.
33. (a) Pradhan, S. K.; Ringold, H. J. *JOC* **1964**, *29*, 601. (b) Heathcock, C. H.; Mahaim, C.; Schlecht, M. F.; Utawanit, T. *JOC* **1984**, *49*, 3264.
34. Ryu, I.; Murai, S.; Hatayama, Y.; Sonoda, N. *TL* **1978**, 3455.
35. (a) Flemming, I.; Paterson, I. *S* **1979**, 736. (b) Fevig, T. L.; Elliott, R. L.; Curran, D. P. *JACS* **1988**, *110*, 5064.
36. (a) Cross, A. D. (Syntex Corp.), Neth. Patent 6 503 543, 1965 (*CA* **1966**, *64*, 5177). (b) Das Gupta, A. K.; Chatterje, R. M.; Paul, M. *JCS(C)* **1971**, 3367.
37. Tanaka, T.; Mashimo, K.; Wagatsuma, M. *TL* **1971**, 2803.
38. Bhattacharya, A.; DiMichele, L. M.; Dolling, U-H.; Douglas, A. W.; Grabowski, E. J. J. *JACS* **1988**, *110*, 3318.
39. (a) Cainelli, G.; Cardillo, G.; Ronchi, A. U. *CC* **1973**, 94. (b) Latif, N.; Mishriki, N.; Girgis, N. S. *CI(L)* **1976**, 28.
40. Iwamura, J.; Hirao, N. *TL* **1973**, 2447.
41. Becker, H.-D.; Bjork, A.; Alder, E. *JOC* **1980**, *45*, 1596.
42. Cacchi, S.; La Torre, F.; Paolucci, G. *S* **1978**, 848.
43. Burstein, S. H.; Ringold, H. J. *JACS* **1964**, *86*, 4952.
44. Lee, H.; Harvey, R. G. *JOC* **1988**, *53*, 4587.
45. (a) Creighton, A. M.; Jackman, L. M. *JCS* **1960**, 3138. (b) Oikawa, Y.; Yonemitsu, O. *H* **1976**, *4*, 1859.
46. Amemiya, T.; Yasunami, M.; Takase, K. *CL* **1977**, 587.
47. Findlay, J. W. A.; Turner, A. B. *CI(L)* **1970**, 158.
48. (a) Bouquet, M.; Guy, A.; Lemaire, M.; Guetté, J. P. *SC* **1985**, *15*, 1153. (b) Corey, E. J.; Xiang, Y. B. *TL* **1987**, *28*, 5403.
49. Lemaire, M.; Doussot, J.; Guy, A. *CL* **1988**, 1581.
50. Guy, A.; Lemor, A.; Doussot, J.; Lemaire, M. *S* **1988**, 900.
51. (a) Oikawa, Y.; Yoshioka, T.; Yonemitsu, O. *TL* **1982** *23*, 885, 889. (b) Oikawa, Y.; Tanaka, T.; Horita, K.; Yoshioka, T.; Yonemitsu, O. *TL* **1984**, *25*, 5393. (c) Nakajima, N.; Abe, R.; Yonemitsu, O. *CPB* **1988**, *36*, 4244. (d) Kozikowski, A. P.; Wu, J.-P. *TL* **1987**, *28*, 5125.
52. Kim, C. U.; Misco, P. F. *TL* **1985**, *26*, 2027.
53. Grunder-Klotz, E.; Ehrhardt, J.-D. *TL* **1991**, *32*, 751.
54. (a) Becker, H.-D. *JOC* **1965**, *30*, 982. (b) Findlay, J. W. A.; Turner, A. B. *JCS(C)* **1971**, 547.
55. (a) Buchan, G. M.; Findlay, J. W. A.; Turner, A. B. *CC* **1975**, 126. (b) Bouquet, M. *CR(II)* **1984**, *229*, 1389.
56. (a) Subba Raju, K. V.; Srimannarayana, G.; Subba Rao, N. V. *TL* **1977**, 473. (b) Prashant, A.; Krupadanam, G. L. D.; Srimannarayana, G. *BCJ* **1992**, *65*, 1191.
57. (a) Cardillo, C.; Cricchio, R.; Merlin, L. *T* **1971**, *27*, 1875. (b) Cardillo, G.; Orena, M.; Porzi, G.; Sandri, S. *CC* **1979**, 836. (c) Jain, A. C.; Khazanchi, R.; Kumar, A. *BCJ* **1979**, *52*, 1203.
58. Imafuku, K.; Fujita, R. *Chem. Express* **1991**, *6*, 323.
59. (a) Coutts, I. G. C.; Humphreys, D. J.; Schofield, K. *JCS(C)* **1969**, 1982. (b) Lewis, J. R.; Paul, J. G. *JCS(P1)* **1981**, 770.
60. Becker, H.-D. *JOC* **1965**, *30*, 982, 989.
61. (a) Schmand, H. L. K.; Boldt, P. *JACS* **1975**, *97*, 447. (b) Barton, D. H. R.; Bergé-Lurion, R.-M.; Lusinchi, X.; Pinto, B. M. *JCS(P1)* **1984**, 2077.
62. Semmelhack, M. F.; Bozell, J. J.; Sato, T.; Wulff, W.; Spiess, E.; Zask, A. *JACS* **1982**, *104*, 5850.
63. Sutherland, R. G.; Chowdhury, R. L.; Piórko, A.; Lee, C. C. *JOC* **1987**, *52*, 4618.
64. Miles, W. H.; Smiley, P. M.; Brinkman, H. R. *CC* **1989**, 1897.
65. (a) Begland, R. W.; Hartter, D. R.; Jones, F. N.; Sam, D. J.; Sheppard, W. A.; Webster, O. W.; Weigert, F. J. *JOC* **1974**, *39*, 2341. (b) Sugawara, T.; Masuya, H.; Matsuo, T.; Miki, T. *CPB* **1979**, *27*, 2544.
66. Tanemura, K.; Suzuki, T.; Horaguchi, T. *CC* **1992**, 979.
67. Oshima, T.; Nishioka, R.; Nagai, T. *TL* **1980**, *21*, 3919.
68. Tanemura, K.; Horaguchi, T.; Suzuki, T. *BCJ* **1992**, *65*, 304.
69. (a) Lemaire, M.; Guy, A.; Imbert, D.; Guetté, J.-P. *CC* **1986**, 741. (b) Guy, A.; Lemor, A.; Imbert, D.; Lemaire, M. *TL* **1989**, *30*, 327.
70. (a) Noyori, R.; Hayashi, N.; Kato, M. *TL* **1973**, 2983. (b) Kuroda, S.; Funamizu, M.; Kitahara, Y. *TL* **1975**, 1973.
71. Tanemura. K.; Suzuki, T.; Haraguchi, T. *BCJ* **1993**, *66*, 1235.

Derek R. Buckle
SmithKline Beecham Pharmaceuticals, Epsom, UK

Diimide[1]

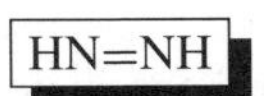

(*cis*)
[15626-42-3] H_2N_2 (MW 30.03)
(*trans*)
[15626-43-4]

(mild, noncatalytic reducing agent for the *syn* reduction of C=C and C≡C bonds; does not react with O–O, N–O, or other easily reduced single bonds, or with C=O, C=N, or aromatic π-bonds)

Alternate Name: diazene.
Form Supplied in: generated in situ.
Preparative Methods: the most widely used methods involve the Cu^I-catalyzed ***Oxygen*** or ***Hydrogen Peroxide*** oxidation of hydrazine, and the acid-catalyzed decarboxylation of ***Potassium Azodicarboxylate***.

Introduction. The reduction of C–C π-bonds in the presence of various N–N-containing compounds ultimately capable of producing diimide dates back to 1905;[2] however, it was not until the early 1960s that the role of *cis*-diimide (diazene) was recognized.[3–6] The results of stereochemical studies[7] and theoretical calculations[8] indicate that *cis*-diimide is the active reducing agent, and that the reduction occurs via a concerted, symmetry allowed transfer of the hydrogen atoms of *cis*-diimide to the C=C or C≡C bond, as illustrated in eq 1. In competition with the reduction of C–C π-systems, diimide undergoes a more facile disproportionation reaction to produce nitrogen and hydrazine[9] which requires the use of considerable excesses of the diimide precursor.

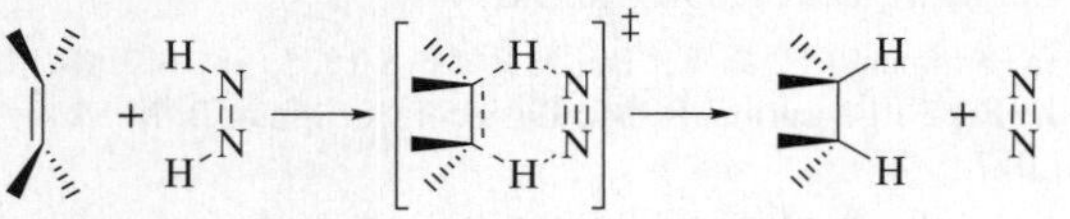

(1)

An important feature of the reduction of C–C π-systems is that many highly reactive functional groups can be present which would not survive under other chemical reducing or catalytic hydrogenation conditions. Such functional groups include allylic and benzylic derivatives, halides, sulfur-containing systems, O–O and N–O containing systems, and complex bioorganic molecules. Examples of the latter systems are shown in eqs 2 and 3.[10,11]

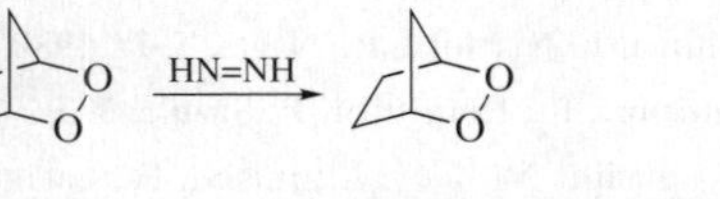

HN=NH (2)

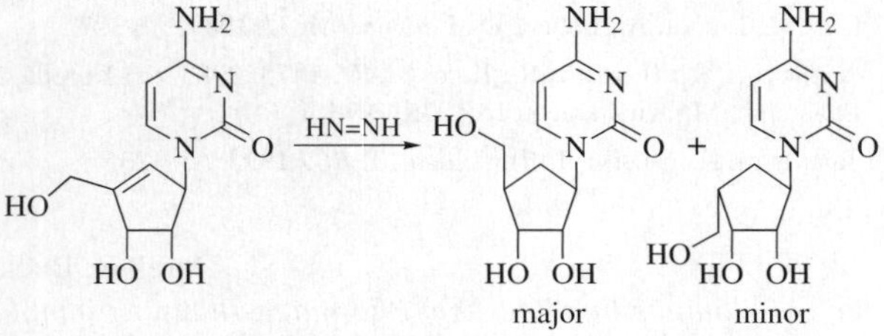

HN=NH (3)

major minor

Relative Reactivity Toward Reduction by Diimide. Carbon–carbon triple bonds, in general, are more reactive than double bonds toward reduction by diimide.[7,12] The relative reactivity of double bonds toward reduction decreases as the degree of alkyl substitution on the double bond increases,[12] and increases with increasing strain (see Table 1).[12] 1,3-Dienes are more reactive than monoenes, with the degree of substitution affecting the relative reactivity as observed with alkenes.[13] Examples of intramolecular selectivity are illustrated in eqs 4–6.[14–16]

Table 1 Relative Reactivities of Substituted Alkenes and 1,3-Dienes Toward Reduction by Diimide

Substrate	k_{rel}[12]
Cyclohexene	1.00
1-Pentene	20.2
trans-2-Pentene	2.59
cis-2-Pentene	2.65
2-Methyl-1-pentene	2.04
2-Methyl-2-butene	0.28
2,3-Dimethyl-2-butene	0.50
Bicyclo[2.2.1]heptene	450.0
Bicyclo[2.2.2]octene	29.0
1,3-Cyclohexadiene	47.0

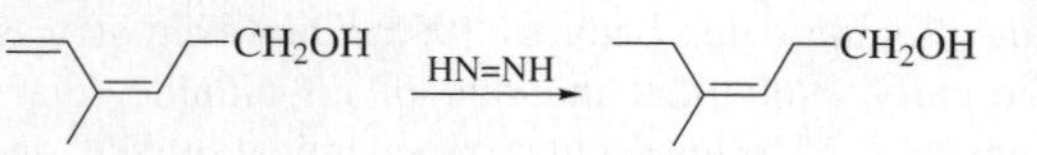

CH_2OH HN=NH CH_2OH (4)

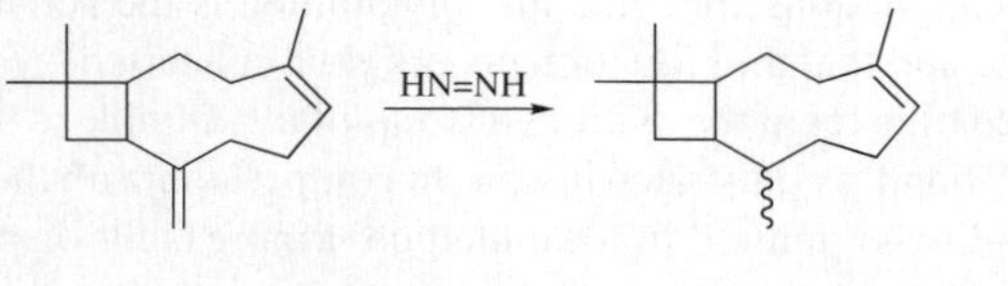

HN=NH (5)

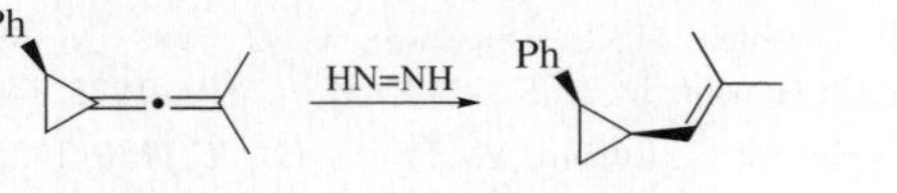

HN=NH (6)

Electronegatively substituted double bonds (e.g. α,β-unsaturated carboxylic acids) are more reactive toward reduction by diimide than are alkyl- or electron donating-substituted π-systems. (The attempted reduction of α,β-unsaturated aldehydes, ketones, and esters generally results in the formation of products deriving from the reaction of the carbonyl compound with hydrazine, which is formed by the disproportionation of the diimide.)

Facial Selectivity of Reduction. The results of experimental and theoretical studies suggest that the transition state for diimide reduction lies very early along the reaction coordinate with the *cis*-diimide approaching the less sterically hindered face of the π-system (eq 6).

Electronic factors also play a role in determining the chemoselectivity of reduction by diimide; this is illustrated in eq 7, in which *anti-exo* reduction occurs preferentially to *syn-exo* reduction.[17]

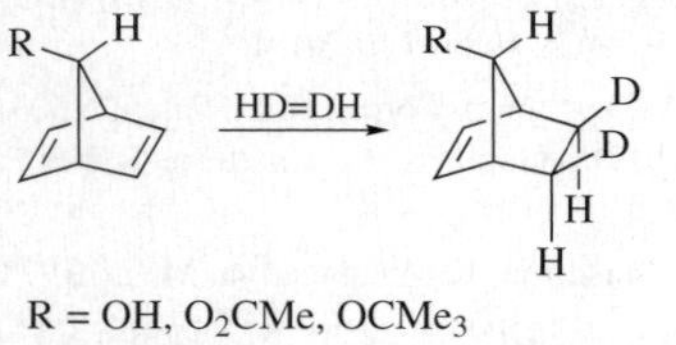

HD=DH (7)

R = OH, O_2CMe, $OCMe_3$

Methods of Generation of Diimide. Numerous methods have been discovered for the generation of diimide.[1a] Of these, only two processes find wide use: firstly, the air (oxygen) or ***Hydrogen Peroxide*** oxidation of hydrazine in the presence of copper(I) and a catalytic amount of a carboxylic acid (eq 8) (the carboxylic acid apparently being required to catalyze the isomerization of *trans*- to *cis*-diimide); secondly, the reaction of ***Potassium Azodicarboxylate*** with a carboxylic acid in aprotic solvents (eq 9).[18] The azodicarboxylate salt is formed by the reaction of ***Potassium Hydroxide*** with azodicarboxamide (commercially available). The latter procedure is especially useful for the position-specific and stereochemically controlled incorporation of deuterium or tritium into organic systems. An example is shown in eq 10, illustrating both the facial and stereoselectivity of the reaction.[19]

$$N_2H_4 \xrightarrow[\text{Cu}^I,\ RCO_2H]{O_2 \text{ or } H_2O_2} HN{=}NH \quad (8)$$

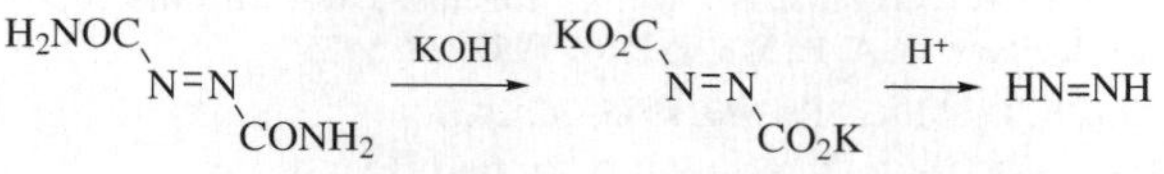

$$H_2NOC{-}N{=}N{-}CONH_2 \xrightarrow{KOH} KO_2C{-}N{=}N{-}CO_2K \xrightarrow{H^+} HN{=}NH \quad (9)$$

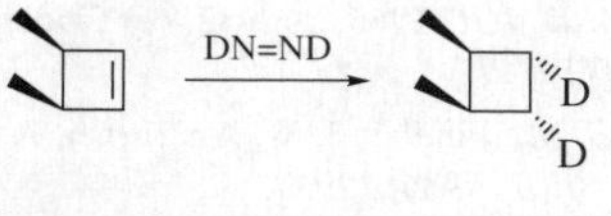

DN=ND (10)

Related Reagents. See Classes R-3, R-4, and R-15, pages 1–10.

1. (a) Pasto, D. J.; Taylor, R. T. *OR* **1991**, *40*, 91. (b) Pasto, D. J. *COS* **1991**, *8*, 471. (c) Hunig, S.; Müller, H. R.; Thier, W. *AG(E)* **1965**, *4*, 271. (d) Miller, C. E. *J. Chem. Educ.* **1965**, *42*, 254.

2. Hanus, J.; Vorisek, J. *CCC* **1929**, *1*, 223.
3. Corey, E. J.; Mock, W. L.; Pasto, D. J. *TL* **1961**, 347.
4. Hunig, S.; Muller, H.; Thier, W. *TL* **1961**, 353.
5. van Tamelen, E. E.; Dewey, R. S.; Timmons, R. J. *JACS* **1961**, *83*, 3725.
6. Aylward, F.; Sawistowska, M. *CI(L)* **1962**, 484.
7. Corey, E. J.; Pasto, D. J.; Mock, W. L. *JACS* **1961**, *83*, 2957.
8. Pasto, D. J.; Chipman, D. M. *JACS* **1979**, *101*, 2290.
9. Pasto, D. J.; *JACS* **1979**, *101*, 6852.
10. Adam, W.; Eggelte, H. J. *JOC* **1977**, *42*, 3987.
11. Russ, P. L.; Hegedus, L.; Kelley, J. A.; Barchi, J. J., Jr; Marquez, V. E. *Nucleosides Nucleotides* **1992**, *11*, 351.
12. Garbisch Jr., E. W.; Schildcrout, S. M.; Patterson, D. B.; Sprecher, C. M. *JACS* **1965**, *87*, 2932.
13. Siegel, S.; Foreman, M.; Fisher, R. P.; Johnson, S. E. *JOC* **1975**, *40*, 3599.
14. Mori, K.; Ohki, M.; Sato, A.; Matsui, M. *T* **1972**, *28*, 3739.
15. Rao, V. V. R.; Devaprabhakara, D. *T* **1978**, *34*, 2223.
16. Pasto, D. J.; Borchardt, J. K. *TL* **1973**, 2517.
17. Baird, W. C., Jr.; Franzus, B.; Surridge, J. H. *JACS* **1967**, *89*, 410.
18. For specific experimental procedures see Ref. 1 and references contained therein.
19. Srinivasan, R.; Hsu, J. N. C. *CC* **1972**, 1213.

Daniel J. Pasto
University of Notre Dame, IN, USA

Diisobutylaluminum Hydride[1]

i-Bu_2AlH

[1191-15-7] $C_8H_{19}Al$ (MW 142.22)

(reducing agent for many functional groups; opens epoxides; hydroaluminates alkynes and alkenes)

Alternate Names: DIBAL; DIBAL-H; DIBAH.

Physical Data: mp −80 to −70 °C; bp 116–118 °C/1 mmHg; *d* 0.798 g cm^{-3}; fp −18 °C.

Solubility: sol pentane, hexane, heptane, cyclohexane, benzene, toluene, xylenes, ether, dichloromethane, THF.

Form Supplied in: can be purchased as a neat liquid or as 1.0 and 1.5 M solutions in cyclohexane, CH_2Cl_2, heptane, hexanes, THF, and toluene.

Handling, Storage, and Precautions: neat liquid is pyrophoric; solutions react very vigorously with air and with H_2O, and related compounds, giving rise to fire hazards; use in a fume hood, in the absence of oxygen and moisture (see ***Lithium Aluminum Hydride*** for additional precautions); THF solutions should only be used below 70 °C, as above that temperature ether cleavage is problematic.

Reduction of Functional Groups.[2] Diisobutylaluminum hydride has several acronyms: DIBAH, DIBAL-H, and DIBAL. For the purpose of this article, DIBAL will be used.

In general, aldehydes, ketones, acids, esters, and acid chlorides are all reduced to the corresponding alcohols by this reagent. Alkyl halides are unreactive towards DIBAL. Amides are reduced to amines, while nitriles afford aldehydes upon hydrolysis of an intermediate imine. Isocyanates are also reduced to the corresponding imines. Nitro compounds are reduced to hydroxylamines. Disulfides are reduced to thiols, while sulfides, sulfones, and sulfonic acids are unreactive in toluene at 0 °C. Tosylates are converted quantitatively to the corresponding alkanes. Cyclic imides can be reduced to carbinol lactams.

Comparable stereochemistry[2] is observed in the reduction of ketones with DIBAL, ***Aluminum Hydride***, and ***Lithium Aluminum Hydride***. Excellent 1,3-asymmetric induction is possible with DIBAL (eq 1); however, this is strongly solvent dependent.[3a]

OH O ... Ph —DIBAL, 87%→ OH OH ... Ph + OH OH ... Ph (1)
93:7

In addition to the hydroxy directing group, amines and amides also showed excellent 1,3-asymmetric induction.[3b] In conjunction with chiral additives, DIBAL can reduce ketones with moderate to good enantioselectivity (eq 2).[4] The use of the Lewis acid ***Methylaluminum Bis(2,6-di-t-butyl-4-methylphenoxide)*** (MAD) along with DIBAL allows discrimination between carbonyl groups (eq 3).[5,6]

$R^1C(O)R^2$ —DIBAL, $SnCl_2$, (chiral diamine), 67–91%→ $R^1CH(OH)R^2$ 56–78% ee (2)

PhC(O)Me + PhC(O)*t*-Bu → PhCH(OH)Me (**1**) + PhCH(OH)*t*-Bu (**2**) (3)

DIBAL (99% yield) (**1**):(**2**) = 2.6:1
DIBAL + MAD (85% yield) (**1**):(**2**) = 1:16

Reduction of α-halo ketones to the carbonyl compounds can be accomplished with DIBAL in the presence of ***Tin(II) Chloride*** and ***N,N,N′,N′-Tetramethylethylenediamine*** (eqs 4 and 5).[7] Under these conditions, vicinal dibromides are converted to the corresponding alkenes.

$R^1C(O)CH(Br)R^2$ —DIBAL, $SnCl_4$, TMEDA, 81–90%→ $R^1C(O)CH_2R^2$ (4)

$R^1CH(Br)CH(Br)R^2$ —DIBAL, $SnCl_4$, TMEDA, 26–91%→ $R^1CH=CHR^2$ (5)

DIBAL is the reagent of choice (see also ***Aluminum Hydride***) for the reduction of α,β-unsaturated ketones to the corresponding allylic alcohols (eq 6).[8a] A reagent derived from DIBAL and

Methylcopper in HMPA alters the regiochemistry such that 1,4-reduction results (eq 7).[8b] Reductions of chiral β-keto sulfoxides occur with high diastereoselectivity.[9] The choice of reduction conditions makes it possible to obtain both epimers at the carbinol carbon (eqs 8 and 9).

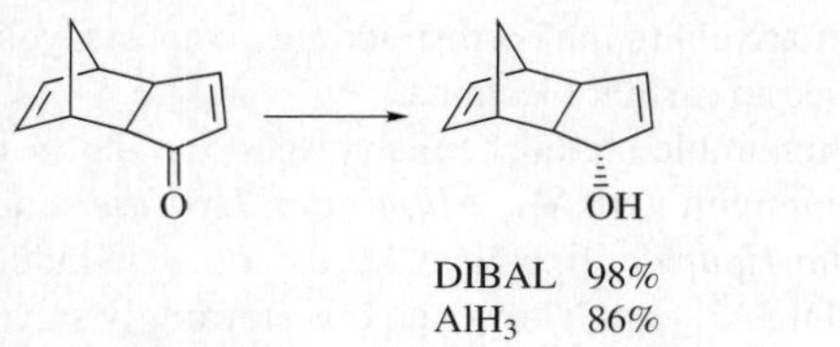

(6)

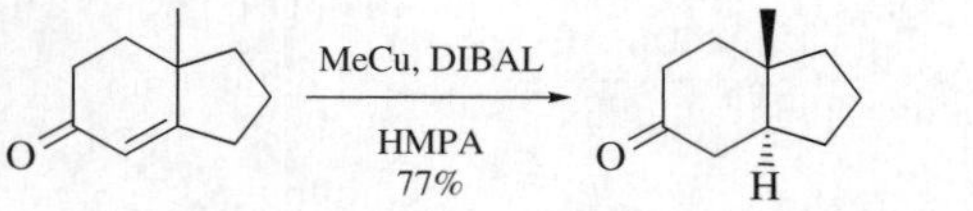

DIBAL 98%
AlH_3 86%

MeCu, DIBAL
HMPA
77%
(7)

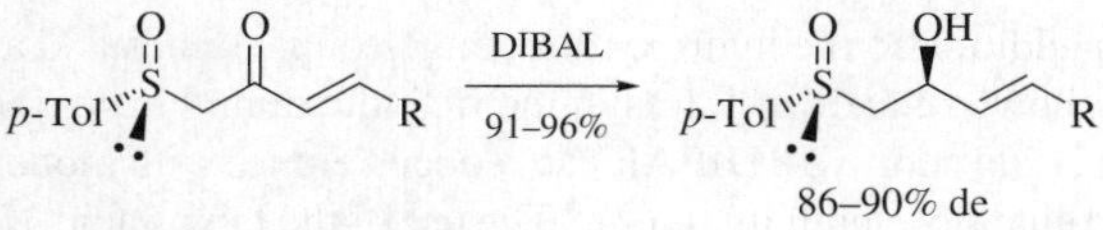

DIBAL
91–96%
(8) 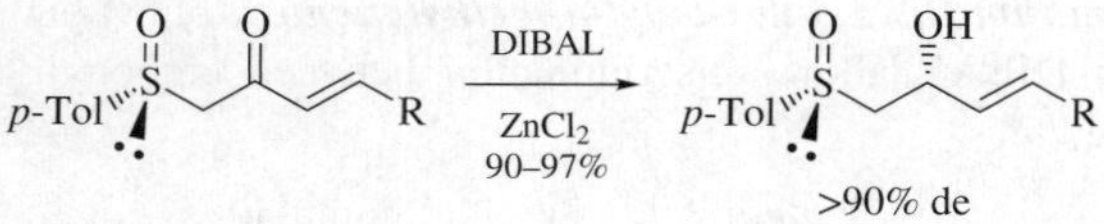
86–90% de

DIBAL
$ZnCl_2$
90–97%
(9)
>90% de

In conjuction with ***Triethylaluminum***, DIBAL has been used to mediate a reductive pinacol rearrangement[10] with enantiocontrol, as shown in eq 10.[22]

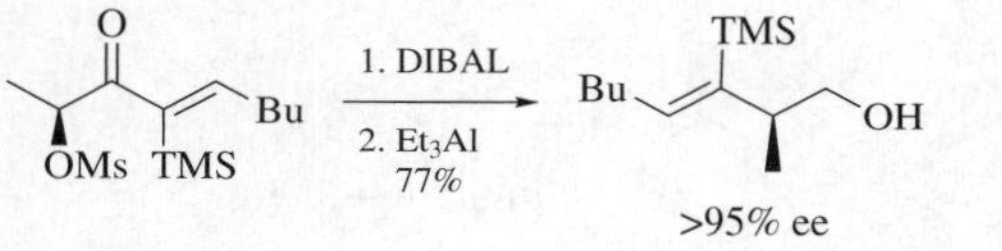

1. DIBAL
2. Et_3Al
77%
(10)
>95% ee

DIBAL is an excellent reagent for the reduction of α,β-unsaturated esters to allylic alcohols without complications from 1,4-addition (eq 11).[11]

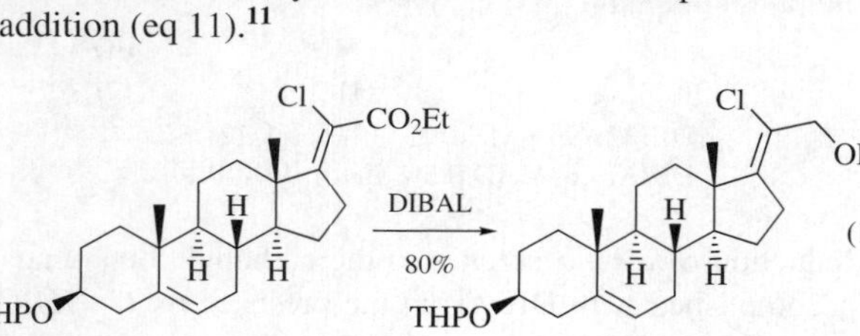

DIBAL
80%
(11)

Due to its Lewis acidity, DIBAL can be used in the reductive cleavage of acetals (see also ***Aluminum Hydride***). Chiral acetates are reduced with enantioselectivity (eq 12).[12] Oxidation of the intermediate alcohol followed by β-elimination gives an optically active alcohol, resulting in a net enantioselective reduction of the corresponding ketone.

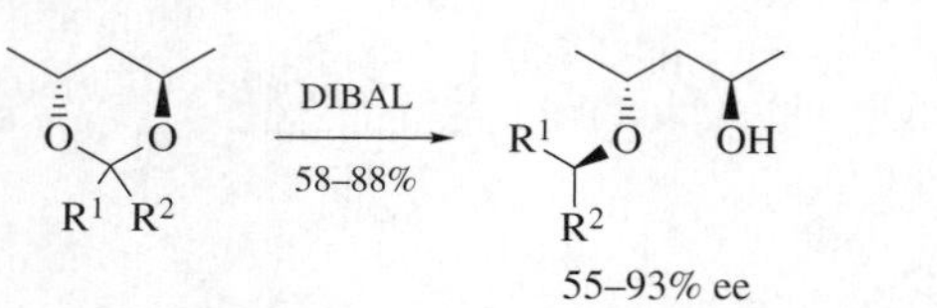

DIBAL
58–88%
(12)
55–93% ee

At low temperatures DIBAL converts esters to the corresponding aldehydes (eq 13)[13] and lactones to lactols (eq 14).[14] DIBAL reduction of α,β-unsaturated γ-lactones followed by an acidic work-up transforms the intermediate lactol to the furan (eq 15).[15] The reduction of nitriles can lead to aldehydes after hydrolysis of an intermediate imine (eq 16).[16] However, cyclic imines can be produced if a ***Sodium Fluoride*** work-up follows the reduction of halonitriles (eq 17).[17] Furthermore, imines can be reduced with excellent stereocontrol (eq 18).[18]

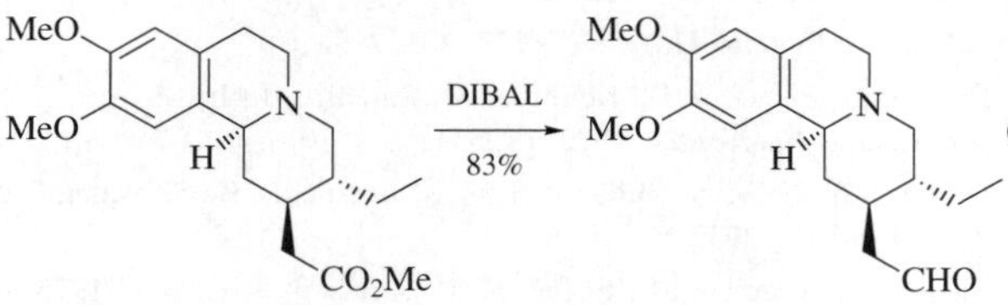

DIBAL
83%
(13)

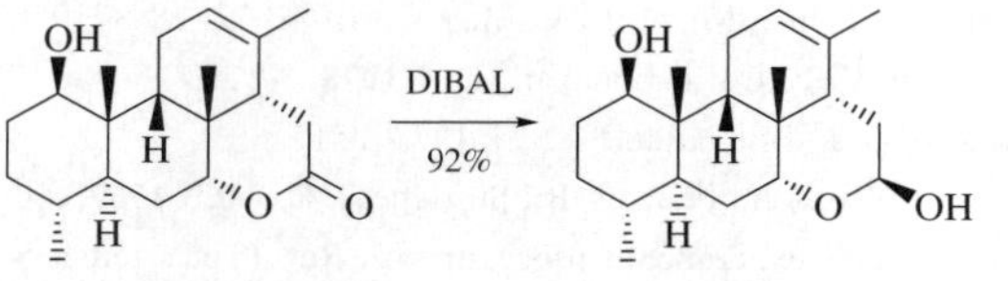

DIBAL
92%
(14)

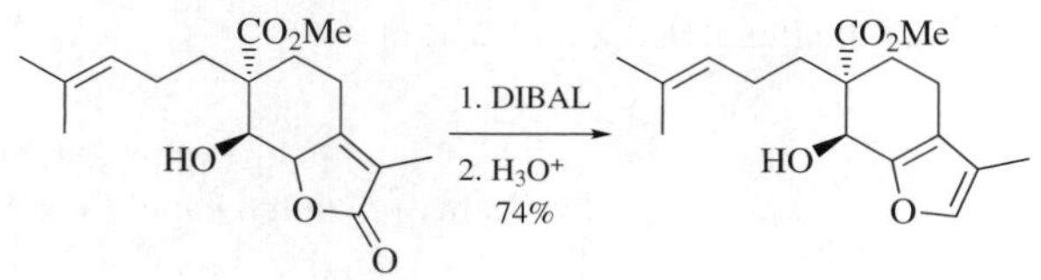

1. DIBAL
2. H_3O^+
74%
(15)

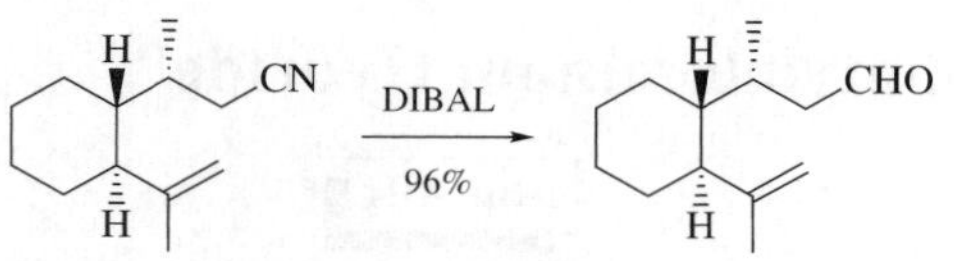

DIBAL
96%
(16)

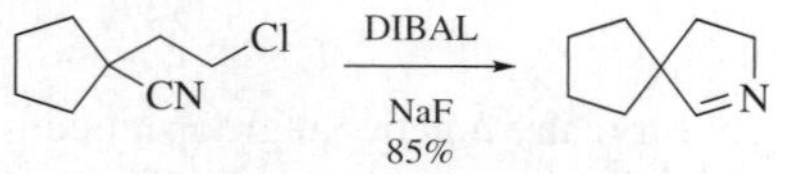

DIBAL
NaF
85%
(17)

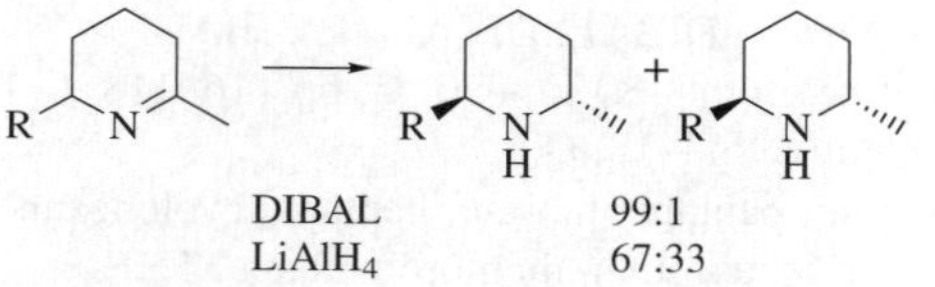

(18)
DIBAL 99:1
$LiAlH_4$ 67:33

DIBAL can also be used for the reductive cleavage of cyclic aminals and amidines (eq 19).[19] Oximes can be reduced to amines. Due to the Lewis acidity of DIBAL, however, rearranged products are obtained (eq 20).[20] This chemistry was used to prepare the alkaloid pumiliotoxin C via the Beckmann rearrangement/alkylation sequence shown in eq 21.[21]

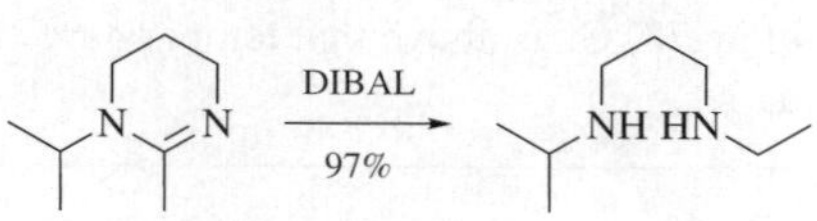

DIBAL
97%
(19)

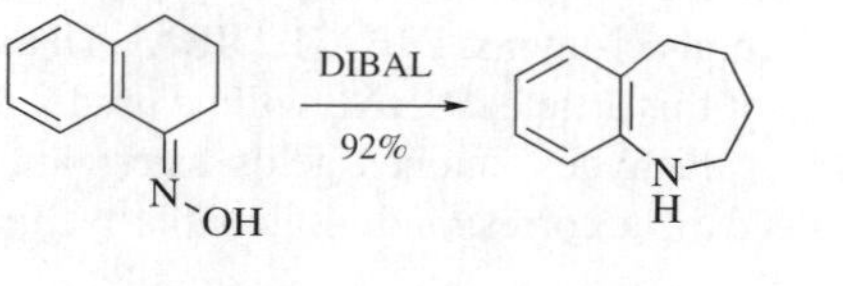

DIBAL
92%
(20)

1. n-Pr_3Al 2. DIBAL 60% (21)

While sulfones are unreactive with DIBAL at 0 °C in toluene,[2] reduction to the corresponding sulfide has been accomplished at higher temperatures.[22] This reaction can be accomplished with ***Lithium Aluminum Hydride***, but fewer equivalents are required and yields are better using DIBAL (eq 22).

$$R^1SO_2R^2 \xrightarrow[12\text{–}77\%]{\text{DIBAL}} R^1SR^2 \quad (22)$$

The reagent combination of DIBAL and ***n-Butyllithium***, which is most likely lithium diisobutylbutylaluminum hydride, has also been used as a reducing agent.[23]

Epoxide Ring Opening. As a result of its Lewis acidity, several reaction pathways are followed in the reductive ring opening of epoxides by DIBAL. Attack at the more hindered carbon via carbenium ion-like intermediates (see also ***Aluminum Hydride***) or S_N2' type reactions, are both known with vinyl epoxides. These modes stand in contrast to results with $LiAlH_4$ (eqs 23–25).[24]

$LiAlH_4$, Et_2O (23)

DIBAL, THF, 59%; OH + OH; 84:16 (24)

DIBAL, hexane, 68%; OH + OH; 95:5 (25)

Hydroalumination Reactions. DIBAL reacts with alkynes and alkenes to give hydroalumination products. *Syn* additions are usually observed; however, under appropriate conditions, equilibration to give a net *anti* addition is possible.[25] If the substrate has both an alkene and an alkyne group, then chemoselectivity for the alkyne is observed (eq 26).[26]

DIBAL; $Al(i\text{-}Bu)_2$ (26)

The intermediate alkenylalane can be used in several ways. If a protiolytic work-up is used, then formation of the corresponding (*Z*)-alkene is observed (eq 27).[27] Treatment of the alkenylalane with ***Methyllithium*** affords an ate complex which is nucleophilic and reacts with a variety of electrophiles, e.g. alkyl halides, CO_2, MeI, epoxides, tosylates, aldehydes, and ketones (eq 28).[1c,1h,26]

$Me(CH_2)_nC{\equiv}CCH_2C{\equiv}C(CH_2)_mCl$ —DIBAL→ $Me(CH_2)_n$ … $(CH_2)_mCl$ (27)

n	m	
5	6	82%
3	8	94%
2	9	87%

1. DIBAL 2. MeLi 3. ethylene oxide 81%; OH (28)

Hydroalumination of a terminal alkene followed by treatment of the intermediate alane with oxygen gives a primary alcohol (eq 29).[28]

1. DIBAL 2. O_2; H, OH (29)

Alkenylalanes can dimerize to afford 1,3-butadienes (eq 30),[29] and can be cyclopropanated under Simmons–Smith conditions to give cyclopropylalanes (eq 31),[30] which can be used for further chemistry.

Ph—≡— —DIBAL→ Ph, Ph (30)

Bu, $Al(i\text{-}Bu)_2$ —CH_2Br_2, Zn(Cu)→ Bu, $Al(i\text{-}Bu)_2$ (31)

Alkenylalanes can also be transmetalated (eq 32),[31] or coupled with vinyl halides via palladium catalysis (eq 33).[32]

R, H, H, $Al(i\text{-}Bu)_2$ —MX_n→ R, H, H, M (32)

M = B, Zr, Hg

Bu, $Al(i\text{-}Bu)_2$ + MeO_2C, Br —Pd, 75%→ Bu, CO_2Me (33)

Addition of DIBAL to allenes has been observed to occur at the more highly substituted double bond (eq 34).[33]

R^1, R^2 —DIBAL→ R^1, R^2 (34)

$R^1 = C_7H_{15}$; R^2 = H (88%), TMS (74%)

The conversion of unconjugated enynes to cyclic compounds has also been accomplished using DIBAL (eq 35).[34]

DIBAL, 79% (35)

Related hydroalumination reactions have been reported using ***Lithium Aluminum Hydride***.

Related Reagents. See Classes R-1, R-2, R-4, R-5, R-9, R-10, R-11, R-12, R-13, R-14, R-15, R-20, R-21, R-23, R-25, R-27, R-28, R-31, R-32, and R-34, pages 1–10.

1. (a) Mole, T.; Jeffery, E. A. *Organoaluminum Compounds*; Elsevier: Amsterdam, 1972. (b) Winterfeldt, W. *S* **1975**, 617. (c) Zweifel, G.; Miller, J. A. *OR* **1984**, *32*, 375. (d) Maruoka, K; Yamamoto, H. *AG(E)* **1985**, *24*, 668. (e) Maruoka, K.; Yamamoto, H. *T* **1988**, *44*, 5001. (f) Dzhemilev, V. M.; Vostrikova, O. S.; Tolstikov, G. A. *RCR* **1990**, *59*, 1157. (g) Seyden-Penne, J. *Reductions by the Alumino- and Borohydrides in Organic Synthesis*; VCH: New York, 1991. (h) Eisch, J. J. *COS* **1991**, *8*, 733. (i) Eisch, J. J. In *Comprehensive Organometallic Chemistry*; Wilkinson, G., Ed.; Pergamon: Oxford, 1982; vol. 1, p 555. (j) Zietz, J. R.; Robinson, G. C.; Lindsay, K. L. In *Comprehensive Organometallic Chemistry*; Wilkinson, G., Ed.; Pergamon: Oxford, 1982; vol. 7, p 365.
2. Yoon, N. M.; Gyoung, Y. S. *JOC* **1985**, *50*, 2443.
3. (a) Kiyooka, S.; Kuroda, H.; Shimasaki, Y. *TL* **1986**, *27*, 3009. (b) Barluenga, J.; Aguilar, E.; Fustero, S.; Olano, B.; Viado, A. L. *JOC* **1992**, *57*, 1219.
4. Oriyama, T.; Mukaiyama, T. *CL* **1984**, 2071.
5. (a) Maruoka, K.; Itoh, T.; Yamamoto, H. *JACS* **1985**, *107*, 4573. (b) Maruoka, K.; Sakurai, M.; Yamamoto, H. *TL* **1985**, *26*, 3853.
6. Maruoka, K.; Araki, Y.; Yamamoto, H. *JACS* **1988**, *110*, 2650.
7. Oriyama, T.; Mukaiyama, T. *CL* **1984**, 2069.
8. (a) Wilson, K. E.; Seidner, R. T.; Masamune, S. *CC* **1970**, 213. (b) Tsuda, T.; Kawamoto, T.; Kumamoto, Y.; Saegusa, T. *SC* **1986**, *16*, 639.
9. Solladie, G.; Frechou, G.; Demailly, G. *TL* **1986**, *27*, 2867.
10. Suzuki, K.; Katayama, E.; Matsumoto, T.; Tsuchihashi, G. *TL* **1984**, *25*, 3715.
11. Daniewski, A. R.; Wojceichowska, W. *JOC* **1982**, *47*, 2993.
12. Mori, A.; Fujiwara, J.; Maruoka, K.; Yamamoto, H. *TL* **1983**, *24*, 4581.
13. Szantay, C.; Toke, L.; Kolonits, P. *JOC* **1966**, *31*, 1447.
14. Vidari, G.; Ferrino, S.; Grieco, P. A. *JACS* **1984**, *106*, 3539.
15. Kido, F.; Noda, Y.; Maruyama, T.; Kabuto, C.; Yoshikoshi, A. *JOC* **1981**, *46*, 4264.
16. Marshall, J. A.; Andersen, N. H.; Schlicher, J. W. *JOC* **1970**, *35*, 858.
17. Overman, L. E.; Burk, R. M. *TL* **1984**, *25*, 5737.
18. Matsumura, Y.; Maruoka, K.; Yamamoto, H. *TL* **1982**, *23*, 1929.
19. Yamamoto, H.; Maruoka, K. *JACS* **1981**, *103*, 4186.
20. Sasatani, S.; Miyazaki, T.; Maruoka, K.; Yamamoto, H. *TL* **1983**, *24*, 4711.
21. Hattori, K.; Matsumura, Y.; Miyazaki, T.; Maruoka, K.; Yamamoto, H. *JACS* **1981**, *103*, 7368.
22. Gardner, J. N.; Kaiser, S.; Krubiner, A. Lucas, H. *CJC* **1973**, *51*, 1419.
23. Kim., S.; Ahn, K. H. *JOC* **1984**, *49*, 1717.
24. Lenox, R. S.; Katzenellenbogen, J. A. *JACS* **1973**, *95*, 957.
25. Eisch, J. J.; Foxton, M. W. *JOC* **1971**, *36*, 3520.
26. Utimoto, K.; Uchida, K.; Yamaya, M.; Nozaki, H. *TL* **1977**, 3641.
27. Gensler, W. J.; Bruno, J. J. *JOC* **1963**, *28*, 1254.
28. Ziegler, K.; Kropp, F.; Zosel, K. *LA* **1960**, *629*, 241.
29. Eisch, J. J.; Kaska, W. C. *JACS* **1966**, *88*, 2213.
30. Zweifel, G.; Clark, G. M.; Whitney, C. C. *JACS* **1971**, *93*, 1305.
31. (a) Negishi, E.; Boardman, L. D. *TL* **1982**, *23*, 3327. (b) Negishi, E.; Jadhav, K. P.; Daotien, N. *TL* **1982**, *23*, 2085.
32. Babas, S.; Negishi, E. *JACS* **1976**, *98*, 6729.
33. Monturi, M.; Gore, J. *TL* **1980**, *21*, 51.
34. Zweifel, G.; Clark, G. M.; Lynd, R. *CC* **1971**, 1593.

Paul Galatsis
University of Guelph, Ontario, Canada

Diisopinocampheylborane

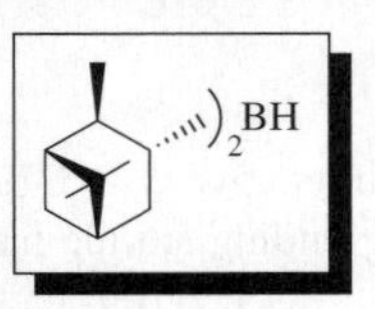

(+)
[21947-87-5] $C_{20}H_{35}B$ (MW 286.31)
(−)
[21932-54-7]

(chiral hydroborating reagent for asymmetric hydroboration of *cis*-alkenes to provide access to optically active secondary alcohols;[1] precursor for the preparation of a large number of chiral reagents for asymmetric synthesis.[1])

Alternate Name: Ipc_2BH.
Physical Data: white crystalline dimer.
Solubility: sparingly sol THF.
Analysis of Reagent Purity: active hydride is determined by hydrolysis of an aliquot and measuring the hydrogen evolved according to the standard procedure;[2] enantiomeric purity is determined by measuring the rotation of the α-pinene liberated in its reaction with 0.5 equiv of ***N,N,N′,N′-Tetramethylethylenediamine*** (TMEDA)[3] or reaction with aldehydes.[4]
Preparative Method: (+)-diisopinocampheylborane is prepared in high enantiomeric purity and good yield (Table 1) by hydroboration of commercially available (−)-α-pinene (of low enantiomeric purity) with ***Borane–Dimethyl Sulfide*** (BMS) complex, carried out by mixing the two reagents to make a solution of known molarity in THF at 0 °C or rt (eq 1); the mixture is left without stirring at 0 °C for ≈12 h for the development of crystals (the slow crystallization facilitates the incorporation of the major diastereomer in the crystalline product, leaving the undesired isomer in solution); the supernatant solution is decanted using a double-ended needle; the crystalline lumps are broken and washed with diethyl ether and dried under vacuum (≈12 mmHg) at rt.[3,4]

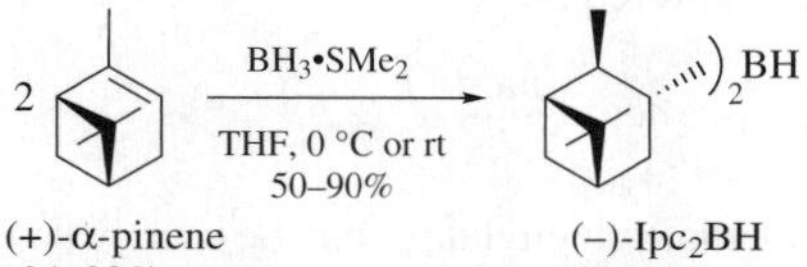

2 (+)-α-pinene, 84–92% ee —[$BH_3\cdot SMe_2$; THF, 0 °C or rt; 50–90%]→ (−)-Ipc_2BH, 98–99% ee (1)

Handling, Storage, and Precautions: air sensitive, reacting instantaneously with protic solvents to liberate hydrogen; must be handled under an inert atmosphere (N_2 or Ar); can be stored at 0 °C under inert atmosphere for several months without loss of hydride activity.[4]

Asymmetric Hydroboration. Brown and Zweifel originally carried out the hydroboration of α-pinene to study the sensitivity of the α-pinene structure towards rearrangement. Surprisingly, the hydroboration reaction proceeded without rearrangement and stopped at the dialkylborane (R_2BH) stage.[5] This important reaction (reported in 1961) thus gave birth to a unique reagent, diisopinocampheylborane (Ipc_2BH). The failure of this reagent to hydroborate a third molecule of α-pinene suggested the possibility of its application in asymmetric hydroboration of less sterically hindered alkenes.

Table 1 Synthesis of Diisopinocampheylborane (Ipc_2BH) of High Optical Purity via Selective Single Crystallization in THF (Optimized Conditions)

(+)-α-Pinene % ee	Molar ratio[a]	Molarity M (in borane)	Temp (°C)	Isolated (% yield)	(−)-Ipc_2BH % ee[b]
92.0	2.3:1	1	0	70–75	>99
91	2.5:1	1.25	20–25[c]	>90	>99
84	2:1	1	0	50–60	98.3

[a] Molar ratio of α-pinene to BMS. [b] Based on measuring the rotation of the (+)-α-pinene obtained from (−)-Ipc_2BH. [c] At times, Ipc_2BH starts precipitating immediately; in such cases the reaction mixture should be redissolved at 50–55 °C, followed by slow recrystallization.[4]

The first substrate which was asymmetrically hydroborated using Ipc_2BH was *cis*-2-butene, and the enantiomeric purity of the product 2-butanol (87% ee) obtained in this preliminary experiment was spectacular (eq 2), since Ipc_2BH was made from α-pinene of low optical purity.[5] This reaction represents the first nonenzymatic asymmetric synthesis for achieving high enantioselectivity. Its discovery marked the beginning of a new era of practical asymmetric synthesis obtained via reagent control.[1,5]

2 (~93% ee) + BH_3 —(DG, 0 °C)→ $(Ipc)_2BH$ → $(Ipc)_2B$-*sec*-Bu —[O]→ 2-butanol, (*R*)-(−) 87% ee (2)

Later, Brown and co-workers developed the method described above for the preparation of enantiomerically pure Ipc_2BH (>99% ee)[3,4] and applied the reagent in the asymmetric hydroboration of prochiral alkenes. Oxidation of the trialkylboranes provided optically active alcohols. In the case of *cis*-alkenes, secondary alcohols were obtained in excellent enantiomeric purity (Figure 1). The reaction is general for most types of *cis*-alkene, e.g. *cis*-2-butene forms (*R*)-2-butanol in 98.4% ee, and *cis*-3-hexene is converted to (*R*)-3-hexanol in 93% ee. However, the reagent is somewhat limited in reactions with unsymmetrical alkenes; e.g. *cis*-4-methyl-2-pentene yields 4-methyl-2-pentanol with 96% regioselectivity but only 76% ee (Figure 1).[6]

Asymmetric hydroborations of heterocyclic alkenes are highly regio- and enantioselective. For example, hydroboration of 2,3-dihydrofuran with Ipc_2BH followed by oxidation provides 3-hydroxyfuran in 83% ee, which can be upgraded to essentially the enantiomerically pure form (>99% ee) (Figure 2).[7]

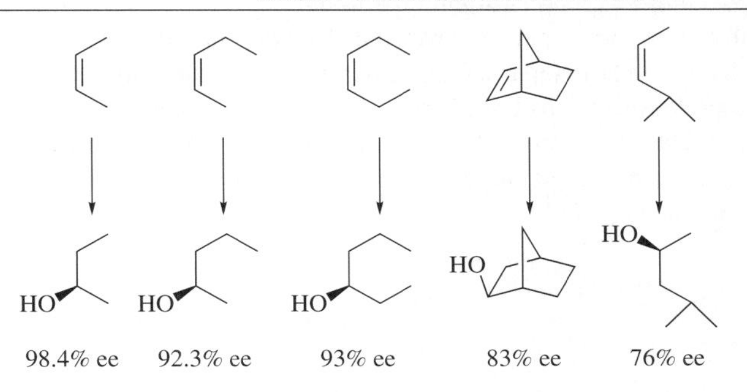

Figure 1 Asymmetric hydroboration–oxidation of *cis*-alkenes with Ipc_2BH

99% ee 99% ee 99% ee 99% ee 83 to >99% ee

Figure 2 Asymmetric hydroboration–oxidation of some heterocyclic alkenes with Ipc_2BH

Applications. The ability of Ipc_2BH to hydroborate *cis*-alkenes has been elegantly applied to the preparation of key intermediates which have been utilized in syntheses of valuable target molecules.[1a] For example, asymmetric hydroboration–oxidation of 5-methylcyclopentadiene to the corresponding optically active alcohol has been applied in the synthesis of loganin (eq 3).[8a] In another example, a prostaglandin precursor was obtained by the asymmetric hydroboration–oxidation reaction of methyl cyclopentadiene-5-acetate (eq 4).[8b] Ipc_2BH has also been used in the preparation of $PGF_{2\alpha}$.[9]

(+)-Ipc_2BH, [O] → 96% ee → → loganin (O-β-D-glucose, CO_2Me) (3)

(+)-Ipc_2BH, [O] → 92% ee → → (4)

Both the enantiomers of Ipc_2BH have been elegantly applied in the asymmetric hydroboration of safranol isoprenyl methyl ether for the synthesis of carotenoids (3*R*,3′*R*)-, (3*S*,3′*S*)-, and (3R,3′S;*meso*)-zeaxanthins (eq 5).[10] (3*S*,5*R*,3′*S*,5′*R*)-Capsorubin, a carotenoid found in the red paprika *Capsicum annuum*, was synthesized via a key step involving asymmetric hydroboration of the unsaturated acetal followed by an aldol condensation (eq 6).[11] Asymmetric hydroboration using Ipc_2BH was also applied in the stereocontrolled synthesis of a linearly fused triquinane, (+)-hirsutic acid (eq 7).[12]

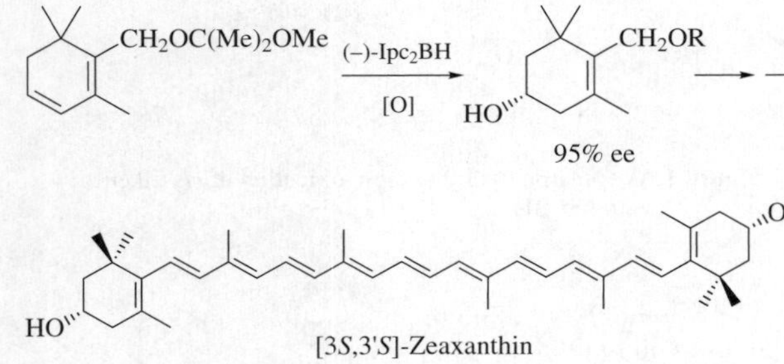

Diisopinocampheylborane is not an effective asymmetric hydroborating agent for 2-substituted 1-alkenes. High selectivities have, however, been achieved where one of the substituents is very bulky. This aspect has been elegantly demonstrated by the synthesis of both enantiomers of a precursor of tylonolide, the aglycone of tylosin, which is one of the members of the polyoxomacrolide antibiotics. In both cases the isomeric ratio was at least 50:1 (eqs 8 and 9).[13]

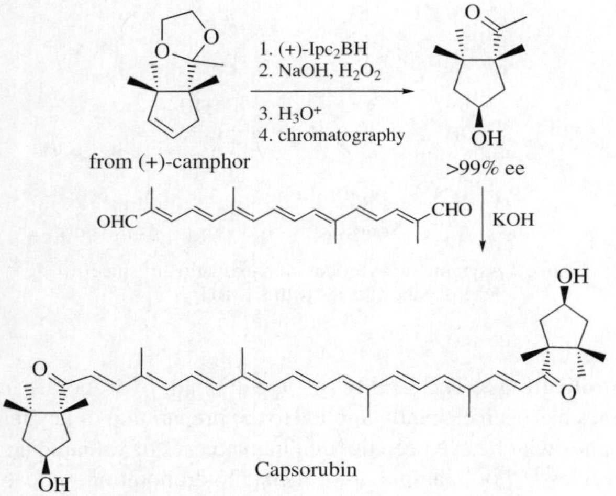

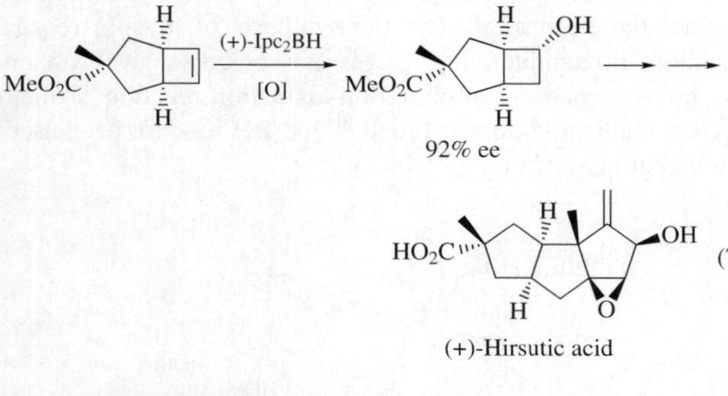

Application of Various Chiral Reagents Derived from Ipc_2BH. Diisopinocampheylborane does not normally yield satisfactory ee's in hydroboration reactions of 1,1-disubstituted alkenes, *trans*-alkenes, or trisubstituted alkenes. This problem has been partially solved by the introduction of ***Monoisopinocampheylborane***, $IpcBH_2$, which is derived from Ipc_2BH. $IpcBH_2$ handles *trans*-alkenes and trisubstituted alkenes effectively, since it is of lower steric requirement than Ipc_2BH (Table 2). Moreover, $IpcBH_2$ and Ipc_2BH provide an entry into the synthesis of a large variety of optically active borinate and boronate esters. These esters have been successfully converted into α-chiral aldehydes, acids, amines, α-chiral *cis*- and *trans*-alkenes, α-chiral alkynes, β-chiral esters, ketones,[1] etc.

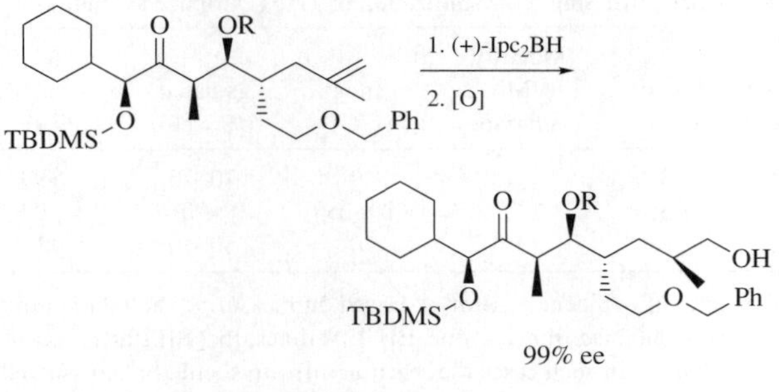

Table 2 Asymmetric Hydroboration of Alkenes with Ipc_2BH and $IpcBH_2$

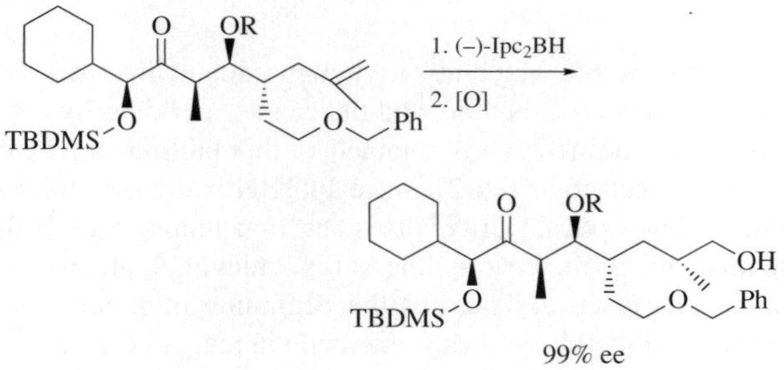

		% ee of alcohol	
Class[a]	Alkene	Ipc_2BH	$IpcBH_2$
I	2-Methyl-1-alkenes	~20	~1
II	*cis*-Alkenes	≥99	~25
III	*trans*-Alkenes	~20	70–90[b]
IV	Trisubstituted alkenes	~20	60≥99[b]

[a] Steric requirement increases from class I to class IV. [b] The ee of initial product can be upgraded to 99% ee via crystallization.

Other reagents which have been derived from Ipc_2BH include ***Diisopinocampheylboron Trifluoromethanesulfonate*** (Ipc_2BOTf),[14] ***B-Methoxydiisopinocampheylborane*** (Ipc_2BOMe), and ***(+)-B-Chlorodiisopinocampheylborane*** and its bromo- and iodo analogs (Scheme 1).[1] Ipc_2BOTf and Ipc_2BOMe reagents are used in stereoselective C–C bond forming reactions (aldol condensation and allylboration); Ipc_2BCl (DIP-chloride™) is used for asymmetric reduction of prochiral ketones, and Ipc_2BX(X = Br or I) for enantioselective opening of *meso*-epoxides to nonracemic halohydrins. Numerous applications of all these reagents have been reviewed in detail.[1]

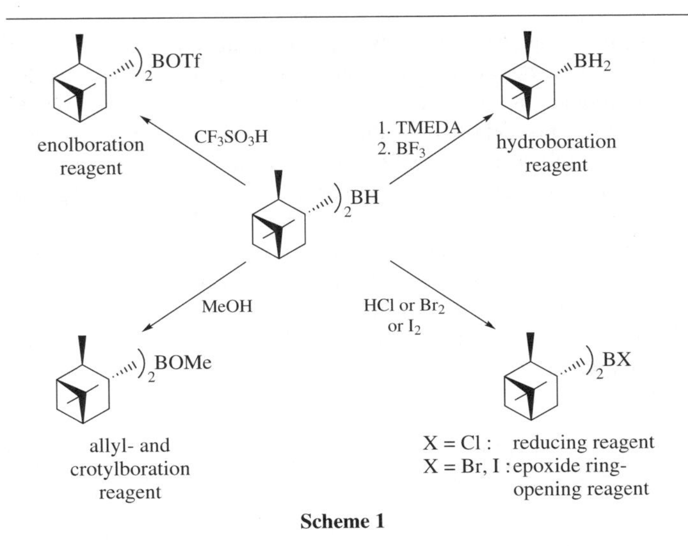

Scheme 1

Related Reagents. See Classes R-2, R-3, and R-11, pages 1–10. (+)-*B*-Chlorodiisopinocampheylborane; (*R*,*R*)-2,5-Dimethylborolane; Diisopinocampheylboron Trifluoromethanesulfonate; Dilongifolylborane; *B*-Methoxydiisopinocampheylborane; Monoisopinocampheylborane.

---

1. For some excellent reviews on synthetic applications of diisopinocampheylborane and related reagents, see: (a) Brown, H. C.; Ramachandran, P. V. *Advances in Asymmetric Synthesis*; Hassner, A., Ed.; JAI Press: Greenwich, CT, 1994; Vol. 2, in press. (b) Brown, H. C.; Ramachandran, P. V. *PAC* **1991**, *63*, 307. (c) Brown, H. C.; Singaram, B. *ACR* **1988**, *21*, 287. (d) Srebnik, M.; Ramachandran, P. V. *Aldrichim. Acta* **1987**, *20*, 9. (e) Matteson, D. S. *S* **1986**, 973.
2. Brown, H. C.; Kramer, G. W.; Levy, A. B.; Midland, M. M. *Organic Synthesis via Boranes*; Wiley: New York, 1975; p 239.
3. Brown, H. C.; Singaram, B. *JOC* **1984**, *49*, 945.
4. Brown, H. C.; Joshi, N. N. *JOC* **1988**, *53*, 4059.
5. Brown, H. C.; Zweifel, G. *JACS* **1961**, *83*, 486.
6. (a) Brown, H. C.; Desai, M. C.; Jadhav, P. K. *JOC* **1982**, *47*, 5065. (b) Brown, H. C.; Ayyangar, N. R.; Zweifel, G. *JACS* **1964**, *86*, 397.
7. Brown, H. C.; Prasad, J. V. N. V. *JACS* **1986**, *108*, 2049.
8. (a) Partridge, J. J.; Chadha, N. K.; Uskokovic, M. R. *JACS* **1973**, *95*, 532. (b) *JACS* **1973**, *95*, 7171.
9. Corey, E. J.; Noyori, R. *TL* **1970**, 311.
10. Ruttimann, A.; Mayer, H. *HCA* **1980**, *63*, 1456.
11. Ruttimann, A.; Englert, G.; Mayer, H.; Moss, G. P.; Weedon, B. C. L. *HCA* **1983**, *66*, 1939.
12. Greene, A. E.; Luche, M.-J.; Serra, A. A. *JOC* **1985**, *50*, 3957.
13. Masamune, S.; Lu, L. D.-L.; Jackson, W. P.; Kaiho, T.; Toyoda, T. *JACS* **1982**, *104*, 5523.
14. Paterson, I.; Goodman, J. M.; Lister, M. A.; Schumann, R. C.; McClure, C. K.; Norcross, R. D. *T* **1990**, *46*, 4663.

Raj K. Dhar
Aldrich Chemical Company, Sheboygan Falls, WI, USA

Dimethyldioxirane[1]

[74087-85-7] $C_3H_6O_2$ (MW 74.09)

(selective, reactive oxidizing agent capable of epoxidation of alkenes and arenes,[11] oxyfunctionalization of alkanes,[19] and oxidation of alcohols,[23] ethers,[21] amines, imines,[32] and sulfides[35])

Alternate Name: DDO.
Physical Data: known only in the form of a dilute solution.
Solubility: sol acetone and CH_2Cl_2; sol most other organic solvents, but reacts slowly with many of them.
Form Supplied in: dilute solutions of the reagent in acetone are prepared from Oxone and acetone, as described below.
Analysis of Reagent Purity: the concentrations of the reagent can be determined by classical iodometric titration or by reaction with an excess of an organosulfide and determination of the amount of sulfoxide formed by NMR or gas chromatography.
Preparative Methods: the discovery of a convenient method for the preparation of dimethyldioxirane has stimulated important advances in oxidation technology.[1] The observation[2] that ketones enhance the decomposition of the monoperoxysulfate anion prompted mechanistic studies that implicated dioxiranes as intermediates.[3] Ultimately, these investigations led to the isolation of dilute solutions of several dioxiranes.[4] DDO is by far the most convenient of the dioxiranes to prepare and use (eq 1). Several experimental set-ups for the preparation of DDO have been described,[4–6] but reproducible generation of high concentration solutions of DDO (ca 0.1M) is aided by a well-formulated protocol.[6] The procedure involves the portionwise addition of solid Oxone (***Potassium Monoperoxysulfate***) to a vigorously stirred solution of $NaHCO_3$ in a mixture of reagent grade ***Acetone*** and distilled water at 5–10 °C. The appearance of a yellow color signals the formation of DDO, at which point the cooling bath is removed and the DDO–acetone solution is distilled into a cooled (−78 °C) receiving flask under reduced pressure (80–100 Torr). After preliminary drying over reagent grade anhydrous $MgSO_4$ in the cold, solutions of DDO are stored over molecular sieves in the freezer of a refrigerator at −10 to −20 °C. In instances where the concentration of DDO is crucial, analysis is typically based on reaction with an excess of an organosulfide monitored by NMR.[4,7,8]

Oxone, H_2O, $NaHCO_3$, 5–10 °C (1)

Handling, Storage, and Precautions: solutions of the reagent can be kept in the freezer of a refrigerator (−10 to −20 °C) for as long as a week. The concentration of the reagent decreases relatively slowly, provided solutions are kept from light and traces of heavy metals. These dilute solutions are not known to decompose violently, but the usual precautions for handling peroxides should be applied, including the use of a shield. All reactions should be performed in a fume hood to avoid exposure to the volatile oxidant.

Introduction. Reactions with DDO are typically performed by adding the cold reagent solution to a cold solution of a reactant in acetone or some other solvent. CH_2Cl_2 is a convenient solvent which facilitates reaction in a number of cases. After the reactant has been consumed, as monitored by TLC, etc., the solvent and excess reagent are simply removed to provide a nearly pure product. An excess of DDO is often used to facilitate conversion, provided further oxidation is not a problem. Where the product is especially sensitive to acid, the reaction can be run in the presence of solid ***Potassium Carbonate*** as an acid scavenger and drying agent. When it is important to minimize water content, the use of powdered molecular sieves in the reaction mixture is recommended. Reactions can be run from ambient temperatures down to −78 °C.

Dimethyldioxirane is a powerful oxidant, but shows substantial selectivity in its reactions. It has been particularly valuable for the preparation of highly reactive products, since DDO can be employed under neutral, nonnucleophilic conditions which facilitate the isolation of such species. Whereas DDO performs the general conversions of more classic reagents like ***m-Chloroperbenzoic Acid***, it generates only an innocuous molecule of acetone as a byproduct. This is to be contrasted with peracids whose acidic side-products can induce rearrangements and nucleophilic attack on products. Although several other dioxiranes have been prepared, these usually offer no advantage over DDO. An important exception is ***Methyl(trifluoromethyl)dioxirane***, whose greater reactivity is advantageous in situations where DDO reacts sluggishly, as in the oxyfunctionalization of alkanes.

The need to prepare DDO solutions beforehand, the low yield of the reagent based on ***Potassium Monoperoxysulfate*** (Oxone) (ca. 5%),[6] and the inconvenience of making DDO for large-scale reactions are drawbacks that can be avoided when the product has good stability. In these instances, an in situ method for DDO oxidations is recommended.

Oxidation of Alkenes and Other Unsaturated Hydrocarbons. The epoxidation of double bonds has been the major area for the application of DDO methodology and a wide range of alkenes are effectively converted to epoxides by solutions of DDO.[4,7] Epoxidation is stereospecific with retention of alkene stereochemistry, as shown by the reactions of geometrical isomers; for example, (*Z*)-1-phenylpropene gives the *cis*-epoxide cleanly (eq 2), whereas the (*E*) isomer yields the corresponding *trans*-epoxide. Rate studies indicate that this reagent is electrophilic in nature and that alkyl substitution on the double bond enhances reactivity.[7] Interestingly, *cis*-disubstituted alkenes react 7–9 times faster than the *trans* isomers, an observation that has been interpreted in terms of a 'spiro' transition state.[9]

Ph (Z)-propenyl —[1.1 equiv DDO, in acetone, >95%]→ Ph epoxide (2)

From a preparative viewpoint, the use of DDO solutions, while efficient and easy to perform, are generally not needed for simple alkenes that give stable epoxides. Rather, in situ methodology is suggested. However, the extraordinary value of isolated DDO has been amply demonstrated for the generation of unstable epoxides that would not survive most epoxidation conditions.[1] A good example of this sort of application is the epoxidation of precocenes, as exemplified in eq 3.[10] A number of impressive epoxidations have been reported for oxygen-substituted alkenes, including enol ethers, silyl enol ethers, enol carboxylates, etc.[1] Examples include a number of 1,2-anhydro derivatives of monosaccharides.[11] Steric features often result in significant stereoselection in the epoxidation, as illustrated in eq 4.[11] Conversions of alkenes with two alkoxy substituents have also been achieved (eq 5), even when the epoxides are not stable at rt.[12]

(3): 1.2 equiv DDO, in acetone; MeCN, −40 °C; ca 100% (OMe)

(4): TBSO (×3) glycal; 1.2 equiv DDO, in acetone; CH_2Cl_2, 0 °C, 1 h; 96%

(5): excess DDO, in acetone; CH_2Cl_2, −70 °C, 1 h; ca 100%

Although reactions are much slower with conjugated carbonyl compounds, DDO is still effective for the epoxidation of these electron-deficient double bonds (eq 6).[13] Alkoxy-substitution on such conjugated alkenes can also be tolerated (eq 7).[14]

(6): CO_2H; excess DDO, in acetone; CH_2Cl_2, 23 h; 93%

(7): OEt; 1.5 equiv DDO, in acetone; CH_2Cl_2, −20 °C, 26 h; 98%

Allenes react with DDO by sequential epoxidation of the two double bonds to give the previously inaccessible, highly reactive allene diepoxides.[15] In the case of the *t*-butyl-substituted allene shown in eq 8, a single diastereomer of the diepoxide is generated, owing to steric control of the *t*-butyl group on reagent attack.

(8): *t*-Bu, H; 4 equiv DDO, in acetone; 10 min; 84%

Certain polycyclic aromatic hydrocarbons can be converted to their epoxides, as typified by the reaction of phenanthrene with DDO (eq 9).[4] Aromatic heterocycles like furans and benzofurans also give epoxides, although these products are quite susceptible to rearrangement, even at subambient temperatures (eq 10).[16] The oxidation of heavily substituted phenols by DDO leads to quinones, as shown in eq 11, which illustrates the formation of an orthoquinone.[17] The corresponding hydroquinones are intermediates in these reactions, but undergo ready oxidation to the quinones.

4.3 equiv DDO in acetone; MeCN, 25 °C, 45 min; 83% (9)

1.7 equiv DDO in acetone; CH_2Cl_2, –40 °C, 12 h; ca 100% (10)

1.7 equiv DDO in acetone; 53% (11)

Finally, preformed lithium enolates are converted to α-hydroxy ketones by addition to a cold solution of DDO (eq 12).[18]

addn to 1.7 equiv DDO in acetone; THF, pentane, –78 °C, 30 min; 77% (12)

Oxidation of Saturated Hydrocarbons, Ethers, and Alcohols. Surely the most striking reaction of dioxiranes is their ability to functionalize unactivated C–H bonds by the insertion of an oxygen atom into this σ-bond. This has opened up an important new area of oxidation chemistry.[1] While DDO has been used in a number of useful transformations outlined below, the more reactive ***Methyl(trifluoromethyl)dioxirane*** is often a better reagent for this type of conversion, despite its greater cost and difficulty of preparation.

The discrimination of DDO for tertiary > secondary > primary C–H bonds of alkanes is more pronounced than that of the *t*-butoxide radical.[19] Good yields of tertiary alcohols can be secured in favorable cases, as in the DDO oxidation of adamantane to 1-adamantanol, which occurs with only minor reaction at C-2 (eq 13). Of major significance is the observation that these reactions are stereospecific with high retention of configuration, as illustrated by the oxidation of *cis*-dimethylcyclohexane shown in eq 14; the *trans* isomer gives exclusively the diastereomeric alcohol. This and other data have been interpreted in terms of an 'oxenoid' mechanism for the insertion into the C–H bond. Several interesting applications in the steroid field involve significant site selectivity as well.[20] The slower reactions of DDO with hydrocarbons without tertiary hydrogens are less useful and lead to ketones owing to a rapid further oxidation of the initially formed secondary alcohol. For example, cyclododecane is converted to cyclododecanone.

0.6 equiv DDO in acetone; 22 °C, 18 h; 87% (13)

DDO in acetone; 22 °C, 18 h; ca 100% (14)

Ethers and acetals are slowly converted by DDO to carbonyl compounds. This serves as a nontraditional method for deprotection of these derivatives, an example of which is shown in eq 15.[21,22] Hemiacetals are presumed intermediates in these transformations.

2 equiv DDO in acetone; rt, 36 h; 87% (15)

While DDO has been little used for the oxidation of simple alcohols, it has found application in useful conversions of vicinal diols. The oxidation of tertiary–secondary diols to α-hydroxy ketones occurs without the usual problem of oxidative cleavage between the two functions (eq 16).[23] DDO has also been used to convert appropriate optically active diols selectively into α-hydroxy ketones of high optical purity; for example, see eq 17.[24]

1.2 equiv DDO in acetone; 0 °C, 4 h; 98% (16)

>98% ee; 1.2 equiv DDO in acetone; CH_2Cl_2, 0 °C, 4 h; 96%; >98% ee (17)

Finally, the Si–H bond of silanes suffers analogous oxidation to silanols upon reaction with DDO. This reaction takes place with retention of configuration and is, as expected, more facile than C–H oxidations.[25]

Oxidation of Nitrogen Functional Groups. Selective oxidations of nitrogen compounds are often difficult to achieve, but DDO methodology has been shown to be very useful in a number of instances. For example, one of the first applications of this reagent was in the conversion of primary amines to the corresponding nitro compounds (eq 18).[26] This process probably proceeds by successive oxidation steps via hydroxylamine and nitroso intermediates. Complications arise with unhindered primary aliphatic amines, owing to dimerization of the intermediate nitrosoalkanes and their tautomerization to oximes.[27] In oxidations of amino sugar and amino acid derivatives, it is possible to isolate the initially formed hydroxylamines (eq 19).[28]

5 equiv DDO in acetone; 22 °C, 30 min; 95% (18)

0.8 equiv DDO in acetone; –45 °C, 15 min; 75% (19)

The oxidation of secondary amines to hydroxylamines is readily achieved with 1 equiv of DDO (eq 20).[29] The use of 2 equiv of DDO results in further oxidation, the nature of which depends on the structure of the amine. Thus cyclic secondary amines which do not possess α-hydrogens are converted to nitroxides,[30] as illustrated in eq 21. Secondary benzylamines give nitrones (eq 22).[31]

$$(PhCH_2)_2NH \xrightarrow[0\ °C,\ 15\ min,\ 98\%]{1\ equiv\ DDO\ in\ acetone} (PhCH_2)_2NOH \quad (20)$$

(21): 2,2,5,5-tetramethyl-3-pyrroline-3-carboxamide (CONH₂, N–H) → 2 equiv DDO, in acetone, 0 °C, 30 min, 98% → corresponding nitroxide (N–O•)

$$PhCH_2NH\text{-}t\text{-}Bu \xrightarrow[0\ °C,\ 10\ min,\ 96\%]{2\ equiv\ DDO\ in\ acetone} PhCH{=}N(O)\text{-}t\text{-}Bu \quad (22)$$

A related transformation is the oxidation of imines to nitrones by DDO (eq 23).[32] It is interesting that the isomeric oxaziridines are not produced here, given that peracids favor these heterocycles.

$$C_6Me_5CH{=}NMe \xrightarrow[CH_2Cl_2,\ 0\ °C,\ 2\ h,\ 71\%]{1.1\ equiv\ DDO\ in\ acetone} C_6Me_5CH{=}N(O)Me \quad (23)$$

Reaction of α-diazo ketones with DDO leads to α-keto aldehyde hydrates (eq 24).[33] Oximes are converted to the free ketones by DDO.[34]

(24): 3-pyridyl-$COCHN_2$ → 1 equiv DDO, in acetone, 100% → 3-pyridyl-$COCH(OH)_2$

Oxidation of Sulfur Functional Groups. Dimethyldioxirane rapidly oxidizes sulfides to sulfoxides and converts sulfoxides to sulfones (eq 25).[4,35] The partial oxidation of sulfides to sulfoxides can be controlled by limiting the quantity of DDO. Since Oxone is one of the many reagents that can perform these reactions, the extra effort involved in preparing DDO solutions is often not warranted. An exception involves the transformation of thiophenes to the corresponding sulfones (eq 26).[36] A similar procedure gives α-oxo sulfones by DDO oxidation of thiol esters (eq 27).[37]

$$PhSMe \xrightarrow{DDO} PhSOMe \xrightarrow{DDO} PhSO_2Me \quad (25)$$

(26): 2,5-dimethylthiophene → >2 equiv DDO, in acetone, CH_2Cl_2, 93% → 2,5-dimethylthiophene 1,1-dioxide

(27): ArS–C(O)–Ar → >2 equiv DDO, in acetone, CH_2Cl_2, –30 °C, 2–4 h → Ar–SO_2–C(O)–Ar

Alkanethiols are selectively oxidized to alkanesulfinic acids by DDO (eq 28).[38] Air oxidation of an intermediate species appears to be important in this transformation.

$$Me(CH_2)_4SH \xrightarrow{DDO\ in\ acetone} \xrightarrow{O_2} Me(CH_2)_4SO_2H \quad (28)$$

Related Reagents. See Classes O-1, O-8, O-9, O-14, O-20, and O-23, pages 1–10.

---

1. (a) Adam, W.; Hadjiarapoglou, L. P.; Curci, R.; Mello, R. In *Organic Peroxides*; Ando W., Ed.; Wiley: New York, 1992; Chapter 4, pp 195–219. (b) Murray, R. W. *CRV* **1989**, *89*, 1187. (c) Curci, R. In *Advances in Oxygenated Processes*; Baumstark, A.; Ed; JAI: Greenwich, CT, 1990; Vol. 2, Chapter 1, pp 1–59. (d) Adam, W.; Edwards, J. O.; Curci, R. *ACR* **1989**, *22*, 205. (e) Adam, W.; Hadjiarapoglou, L. *Top. Curr. Chem.* **1993**, *164*, 45.
2. Montgomery, R. E. *JACS* **1974**, *96*, 7820.
3. Edwards, J. O.; Pater, R. H.; Curci, R. P.; Di Furia, F. *Photochem. Photobiol.* **1979**, *30*, 63.
4. Murray, R. W.; Jeyaraman, R. *JOC* **1985**, *50*, 2847.
5. Eaton, P. E.; Wicks, G. E. *JOC* **1988**, *53*, 5353.
6. Adam, W.; Bialas, J.; Hadjiarapoglou, L. *CB* **1991**, *124*, 2377.
7. Baumstark, A. L.; Vasquez, P. C. *JOC* **1988**, *53*, 3437.
8. Murray, R. W.; Shiang, D. L. *JCS(P2)* **1990**, *2*, 349.
9. Baumstark, A. L.; McCloskey, C. J. *TL* **1987**, *28*, 3311.
10. Bujons, J.; Camps, F.; Messeguer, A. *TL* **1990**, *31*, 5235.
11. Halcomb, R. L.; Danishefsky, S. J. *JACS* **1989**, *111*, 6661.
12. Adam, W.; Hadjiarapoglou, L.; Wang, X. *TL* **1991**, *32*, 1295.
13. Adam, W.; Hadjiarapoglou, L.; Nestler, B. *TL* **1990**, *31*, 331.
14. Adam, W.; Hadjiarapoglou, L. *CB* **1990**, *123*, 2077.
15. (a) Crandall, J. K.; Batal, D. J.; Sebesta, D. P.; Lin, F. *JOC* **1991**, *56*, 1153. (b) Crandall, J. K.; Batal, D. J.; Lin, F.; Reix, T.; Nadol, G. S.; Ng, R. A. *T* **1992**, *48*, 1427.
16. (a) Adger, B. M.; Barrett, C.; Brennan, J.; McGuigan, P.; McKervey, M. A.; Tarbit, B. *CC* **1993**, 1220. (b) Adger, B. M.; Barrett, C.; Brennan, J.; McKervey, M. A.; Murray, R. W. *CC* **1991**, 1553. (c) Adam, W.; Bialas, J.; Hadjiarapoglou, L.; Sauter, M. *CB* **1992**, *125*, 231.
17. (a) Crandall, J. K.; Zucco, M.; Kirsch, R. S.; Coppert, D. M. *TL* **1991**, *32*, 5441. (b) Altamura, A.; Fusco, C.; D'Accolti, L.; Mello, R.; Prencipe, T.; Curci, R. *TL* **1991**, *32*, 5445. (c) Adam, W.; Schönberger, A. *TL* **1992**, *33*, 53.
18. Guertin, K. R.; Chan, T. H. *TL* **1991**, *32*, 715.
19. Murray, R. W.; Jeyaraman, R.; Mohan, L. *JACS* **1986**, *108*, 2470.
20. (a) Bovicelli, P.; Lupattelli, P.; Mincione, E.; Prencipe, T.; Curci, R. *JOC* **1992**, *57*, 2182. (b) Bovicelli, P.; Lupattelli, P.; Mincione, E.; Prencipe, T.; Curci, R. *JOC* **1992**, *57*, 5052.
21. Curci, R.; D'Accolti, L.; Fiorentino, M.; Fusco, C.; Adam, W.; González-Nuñez M. E.; Mello, R. *TL* **1992**, *33*, 4225.
22. van Heerden, F. R.; Dixon, J. T.; Holzapfel, C. W. *TL* **1992**, *33*, 7399.
23. Curci, R.; D'Accolti, L.; Detomaso, A.; Fusco, C.; Takeuchi, K.; Ohga, Y.; Eaton, P. E.; Yip, Y. C. *TL* **1993**, *34*, 4559.
24. D'Accolti, L.; Detomaso, A.; Fusco, C.; Rosa, A.; Curci, R. *JOC* **1993**, *58*, 3600.
25. Adam, W.; Mello, R.; Curci, R. *AG(E)* **1990**, *102*, 890.
26. Murray, R. W.; Rajadhyaksha, S. N.; Mohan, L. *JOC* **1989**, *54*, 5783.
27. Crandall, J. K.; Reix, T. *JOC* **1992**, *57*, 6759.
28. Wittman, M. D.; Halcomb, R. L.; Danishefsky, S. J. *JOC* **1990**, *55*, 1981.
29. Murray, R. W.; Singh, M. *SC* **1989**, *19*, 3509.
30. Murray, R. W.; Singh, M. *TL* **1988**, *29*, 4677.
31. Murray, R. W.; Singh, M. *JOC* **1990**, *55*, 2954.
32. Boyd, D. R.; Coulter, P. B.; McGuckin, M. R.; Sharma, N. D. *JCS(P1)* **1990**, 301.

33. (a) Ihmels, H.; Maggini, M.; Prato, M.; Scorrano, G. *TL* **1991**, *32*, 6215. (b) Darkins, P.; McCarthy, N.; McKervey, M. A.; Ye, T. *CC* **1993**, 1222.
34. Olah, G. A.; Liao, Q.; Lee, C.-S.; Prakash, G. K. S. *SL* **1993**, 427.
35. Murray, R. W.; Jeyaraman, R.; Pillay, M. K. *JOC* **1987**, *52*, 746.
36. Miyahara, Y.; Inazu, T. *TL* **1990**, *31*, 5955.
37. Adam, W.; Hadjiarapoglou, L. *TL* **1992**, *33*, 469.
38. Gu, D.; Harpp, D. N. *TL* **1993**, *34*, 67.

Jack K. Crandall
Indiana University, Bloomington, IN, USA

Dimethyl Sulfoxide–Acetic Anhydride[1]

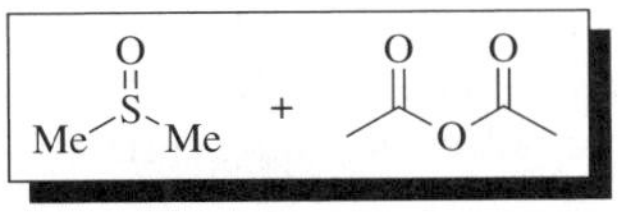

(DMSO)
[67-68-5] C_2H_6OS (MW 78.13)
(Ac_2O)
[108-24-7] $C_4H_6O_3$ (MW 102.09)

(oxidant for the conversion of primary and secondary alcohols to aldehydes and ketones, respectively; avoids overoxidation to carboxylic acids; suitable for large-scale oxidation; gives good yields with variable amounts of byproduct methylthiomethyl ethers)

Alternate Name: Albright–Goldman reagent.

Physical Data: DMSO mp 18.4 °C; bp 189 °C; *d* 1.101 g cm^{-3}. Ac_2O: mp −78 °C; bp 138–140 °C; *d* 1.082 g cm^{-3}.

Solubility: DMSO: sol H_2O, alcohol, acetone, THF, CH_2Cl_2. Ac_2O: sol ether, acetone, CH_2Cl_2.

Form Supplied in: colorless liquids; widely available, including 'anhydrous' grades of DMSO packed under N_2.

Preparative Method: the active reagent, presumably $Me_2\overset{+}{S}OAc$, forms slowly from a mixture of the coreactants over a period of hours at rt and reacts with the alcohol in situ.

Purification: DMSO: distillation from calcium hydride at 56–57 °C/5 mmHg[2a] or 83–85 °C/17 mmHg;[2b] storage over 3Å molecular sieves. Ac_2O: distillation from aluminum chloride or calcium carbide.

Handling, Storage, and Precautions: ***Dimethyl Sulfoxide*** is readily absorbed through the skin and should always be handled with gloves in a fume hood; its reactions form foul-smelling byproducts and should be carried out with good ventilation, and the waste byproducts and liquids used for washing should be treated with $KMnO_4$ solution to oxidize volatile sulfur compounds; DMSO undergoes appreciable disproportionation to dimethyl sulfide (stench!) and dimethyl sulfone above 90 °C;[2c] ***Acetic Anhydride*** is a corrosive lachrymator.

The title reagent is useful for the oxidation of primary or secondary alcohols to aldehydes and ketones, respectively, at rt without added base. The reagents are inexpensive and the procedure is adaptable to large-scale reactions, such as the oxidation of yohimbine on a 2.5 mol scale.[3] The mechanism of this reaction appears to involve formation of 'activated' DMSO by the rather slow reaction of DMSO with Ac_2O at rt to form the acyloxysulfonium ion (**1**); this then reacts with the alcohol to give the alkoxysulfonium ion (**2**) (eq 1). By analogy to other DMSO-based oxidations, (**2**) undergoes deprotonation to (**3**), which forms the oxidation product intramolecularly (eq 2).[1a,b] A common side reaction of alcohols ROH with activated DMSO is formation of methylthiomethyl ethers $ROCH_2SMe$ (**4**), which probably arise from dissociation of (**3**) or some other activated form of DMSO to form (**5**) due to the operational temperature of 25 °C and the long reaction times (12–24 h) (eq 3). Alternatively, (**5**) could be formed by an intramolecular proton abstraction in (**1**) via a six-membered transition state. The effect of pressure on the reaction has been examined, and the large negative entropy of activation is consistent with an associative mechanism.[4] The reaction is also greatly accelerated by pressure when carried out on a preparative scale.[4] Some examples of the procedure are shown in eqs 4–6.[5a,b,c]

$$Me_2SO + Ac_2O \longrightarrow \underset{(\mathbf{1})}{Me_2\overset{+}{S}O_2CMe} \xrightarrow{R^1R^2CHOH} \underset{(\mathbf{2})}{R^1R^2CHO\overset{+}{S}Me_2} \quad (1)$$

$$(\mathbf{2}) \xrightarrow{\text{base}} \underset{(\mathbf{3})}{R^1R^2CHO\overset{+}{S}MeCH_2^-} \longrightarrow R^1R^2CO + Me_2S \quad (2)$$

$$(\mathbf{3}) \longrightarrow \underset{(\mathbf{5})}{MeSCH_2^+} \xrightarrow{ROH} \underset{(\mathbf{4})}{ROCH_2SMe} \quad (3)$$

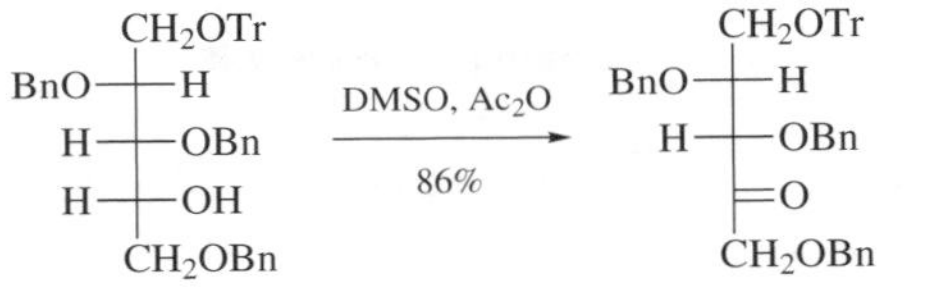

DMSO, Ac_2O, 86% (4)

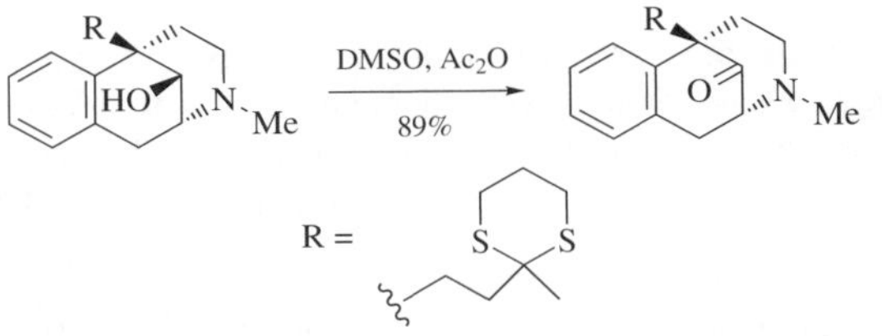

DMSO, Ac_2O, 89% (5)

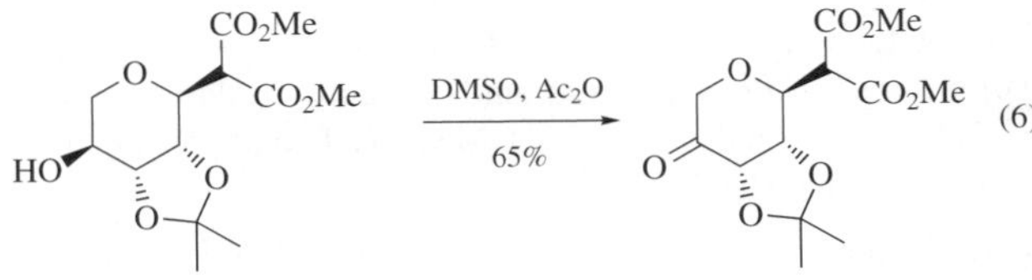

DMSO, Ac_2O, 65% (6)

Related Reagents. See Class O-1, pages 1–10. *N*-Chlorosuccinimide–Dimethyl Sulfide; Chromic Acid; Dimethyl Sulfide–Chlorine; Dimethyl Sulfoxide–Dicyclohexylcarbodiimide; Dimethyl Sulfoxide–Methanesulfonic Anhydride; Dimethyl Sulfoxide–Oxalyl Chloride; Dimethyl Sulfoxide–Phosphorus Pentoxide; Dimethyl Sulfoxide–Sulfur Trioxide/Pyridine; Dimethyl Sulfoxide–Trifluoroacetic Anhydride;

Dimethyl Sulfoxide–Triphosgene; Manganese Dioxide; Pyridinium Chlorochromate; Pyridinium Dichromate Ruthenium(IV) Oxide; Silver(I) Carbonate; 1,1,1-Triacetoxy-1,1-dihydro-1,2-benziodoxol-3(1*H*)-one.

1. (a) Tidwell, T. T. *OR* **1990**, *39*, 297. (b) Tidwell, T. T. *S* **1990**, 857. (c) Lee, T. V. *COS* **1991**, *7*, 291. (d) Haines, A. H. *Methods for the Oxidation of Organic Compounds*; Academic: London, 1988. (e) Hudlicky, M. *Oxidations in Organic Chemistry*; ACS: Washington, 1990. (f) Mancuso, A. J.; Swern, D. *S* **1981**, 165. (g) Moffatt, J. G. In *Oxidation*; Augustine, R. L.; Trecker, D. J., Eds.; Dekker: New York, 1971; Vol. 2, Chapter 1.
2. (a) Iwai, I.; Ide, J. *OSC* **1988**, *6*, 531. (b) Insalaco, M. A.; Tarbell, D. S. *OSC* **1988**, *6*, 207. (c) Corey, E. J.; Chaykovsky, M. *OSC* **1973**, *5*, 755.
3. (a) Albright, J. D.; Goldman, L. *JACS* **1965**, *87*, 4214. (b) Albright, J. D.; Goldman, L. *JACS* **1967**, *89*, 2416.
4. Isaacs, N. S.; Laila, A. H. *JPOC* **1991**, *4*, 639.
5. (a) Rabinsohn, Y.; Fletcher, H. G., Jr. In *Methods in Carbohydrate Chemistry*, Whistler, R. L.; BeMiller, J. N., Eds.; Academic: New York, 1972; Vol. 6; p 326. (b) Broka, C. A.; Gerlits, J. F. *JOC* **1988**, *53*, 2144. (c) Katagiri, N.; Akatsuka, H.; Haneda, T.; Kaneko, C.; Sera, A. *JOC* **1988**, *53*, 5464.

Thomas T. Tidwell
University of Toronto, Ontario, Canada

Dimethyl Sulfoxide–Oxalyl Chloride[1]

$$Me-S(=O)-Me \; + \; COCl-COCl$$

(DMSO)
[67-68-5] C_2H_6OS (MW 78.13)
($(COCl)_2$)
[79-37-8] $C_2Cl_2O_2$ (MW 126.93)

(oxidant for the conversion of primary and secondary alcohols to aldehydes and ketones, respectively avoids overoxidation to carboxylic acids; suitable for large-scale oxidation; gives good yields with minimal amounts of byproduct methylthiomethyl ethers)

Alternate Name: Swern reagent.
Physical Data: DMSO: mp 18.4 °C; bp 189 °C; *d* 1.101 g cm^{-3}. $(COCl)_2$: bp 63–64 °C/763 mmHg; *d* 1.455 g cm^{-3}.
Solubility: DMSO: sol H_2O, alcohol, acetone, THF, CH_2Cl_2. $(COCl)_2$: reacts H_2O; sol CH_2Cl_2, THF.
Form Supplied in: colorless liquids; widely available, including 'anhydrous' grades of 99%+ DMSO packed under N_2, and 2M $(COCl)_2$ in CH_2Cl_2 under N_2.
Preparative Method: the active reagent, $Me_2\overset{+}{S}Cl$, is formed rapidly from DMSO–$(COCl)_2$ at −78 °C in CH_2Cl_2.
Purification: DMSO: distillation from calcium hydride at 56–57 °C/5 mmHg[2a] or 83–85 °C/17 mmHg;[2b] storage over 3Å molecular sieves. $(COCl)_2$: distillation under N_2.
Handling, Storage, and Precautions: ***Dimethyl Sulfoxide*** is readily absorbed through the skin and should always be handled with gloves in a fume hood; its reactions form foul-smelling byproducts and should be carried out with good ventilation, and the waste byproducts and liquids used for washing should be treated with $KMnO_4$ solution to oxidize volatile sulfur compounds; DMSO undergoes appreciable disproportionation to dimethyl sulfide (stench!) and dimethyl sulfone above 90 °C;[2c] ***Oxalyl Chloride*** is corrosive and moisture-sensitive, and is said to react explosively with DMSO at rt.

DMSO–oxalyl chloride is the most widely used of the DMSO-based reagents for the oxidation of primary and secondary alcohols to aldehydes and ketones, respectively, and usually gives excellent yields with short reaction times and minimal formation of byproducts.[3] The active reagent (**1**) is generated in situ at low temperature by the addition of DMSO to $(COCl)_2$ in a solvent such as CH_2Cl_2, ether, or THF (eq 1). Addition of the alcohol to (**1**) gives the alkoxysulfonium ion (**2**) (eq 2), which on addition of an amine base is deprotonated to (**3**); the latter forms the carbonyl product (**4**) by intramolecular proton abstraction (eq 3).

$$Me_2S{=}O + (COCl)_2 \longrightarrow \underset{\mathbf{(1)}}{Me_2\overset{+}{S}Cl\ Cl^-} + CO + CO_2 \quad (1)$$

$$\mathbf{(1)} + RR^1CHOH \longrightarrow \underset{\mathbf{(2)}}{RR^1CHO\overset{+}{S}Me_2} \quad (2)$$

$$\mathbf{(2)} + \text{base} \longrightarrow \underset{\mathbf{(3)}}{RR^1CHO\overset{+}{S}(Me)CH_2^-} \longrightarrow \underset{\mathbf{(4)}}{RR^1C{=}O} + Me_2S \quad (3)$$

In a typical procedure using ***Triethylamine*** as base, DMSO (2.4 equiv) is added to $(COCl)_2$ (1.2 equiv) in CH_2Cl_2 cooled to −50 to −60 °C, and then geraniol (**5**) (1.0 equiv) is added, followed by Et_3N (5 equiv). Washing and distillation give geranial (**6**) in 94% yield (eq 4).[4a] The sensitive alcohol (**7**) is oxidized to the aldehyde (**8**) in 83% yield, with less than 8% racemization (eq 5).[4b] The use of ***Diisopropylethylamine*** (DIPEA) as the base helps prevent epimerization (eq 6).[4c] This procedure has been applied on a 1 mol scale in a conversion that gave only a 40% yield with ***Chromium(VI) Oxide*** and H_2SO_4 (eq 7).[4d]

(5) → (6): 1. DMSO, $(COCl)_2$; 2. Et_3N; 94% (4)

(7) → (8): 1. DMSO, $(COCl)_2$, −60 °C; 2. Et_3N; 83% (5)

TBDPSO…OH (OPMBOM) → TBDPSO…CHO (OPMBOM): 1. DMSO, $(COCl)_2$, −78 °C; 2. DIPEA (6)

PMBOM = 4-$MeOC_6H_4CH_2OCH_2$

Me-bicyclic alcohol (OH) → ketone (O): 1. DMSO, $(COCl)_2$; 2. Et_3N; 90–97% (7)

Diols are efficiently oxidized by this procedure (eqs 8 and 9),[4e,f] and selectivity for oxidation of less crowded hydroxy groups can be achieved (eq 10).[4g,h] Isolation of the ketoaldehyde (**10**) from oxidation of (**9**) proved difficult, so oxidation using the stronger base ***1,8-Diazabicyclo[5.4.0]undec-7-ene*** was employed to give (**11**) directly from (**9**) (eq 11).[4i] Aryl dimethanols are oxidized to *ortho*-phthalaldehydes.[4j]

1. DMSO, $(COCl)_2$, –78 °C; 2. Et_3N; 88–92% (8)

1. DMSO, $(COCl)_2$; 2. Et_3N; 88% (9)

DMSO, $(COCl)_2$; 84% (10)

1. DMSO, $(COCl)_2$; 2. DBU (**9**) → (**10**) → (**11**) 58% (11)

DMSO activated by $(COCl)_2$ without the addition of base has also been used for the conversion of β-triketides such as (**12**) to γ-pyrones (**13**) (eq 12).[5] A mechanism was proposed in which DMSO was regenerated (eq 13). In the presence of Et_3N the formation of (**14**) (30%) along with (**13**) (33%) was attributed to the intermediate (**15**) (eq 14).[5]

(**12**) → DMSO, $(COCl)_2$ (2 equiv), –30 to –15 °C, 69% → (**13**) (12)

(**12**) → –HCl → –DMSO, $–H^+$ → (**13**) (13)

(**12**) → (**15**) → $–Me_2S$ → (**14**) 30% (14)

Examples of oxidation of alcohols to carbonyl compounds by DMSO–$(COCl)_2$ followed by in situ reaction of the product with another reagent include the use of alkynyllithiums (eq 15),[6a] Wittig reagents (eq 16),[6b] and Mannich reagents (eq 17).[6c] Extraction of the crude product followed by reaction with ***Hydroxylamine*** gives oximes (eq 18),[6d] and oxidation in the presence of MeOH gave an ester (eq 19).[6e]

1. DMSO, $(COCl)_2$; 2. Et_3N; 3. $(i\text{-PrO})_3SiC{\equiv}CLi$; 88% → $Si(O\text{-}i\text{-}Pr)_3$ (15)

TMS⌒OH → 1. DMSO, $(COCl)_2$; 2. Et_3N; 3. $Ph_3P{=}CMeCO_2Et$; 54% → (*E*)-$TMSCH{=}CMeCO_2Et$ (16)

1. DMSO, $(COCl)_2$; 2. Et_3N; 3. $CH_2{=}\overset{+}{N}Me_2\ Cl^-$; 62% (17)

1. DMSO, $(COCl)_2$, CH_2Cl_2; 2. Et_3N; 3. extract, add EtOH H_2NOH, HCl; 90% (18)

The sequence of eqs 20 and 21 illustrates the use of consecutive Swern oxidations and further conversions.[6f] Reaction of trimethylsilylallyl alcohol on a 0.3 mol scale gave the ketone in 78% yield (eq 22).[6g]

Trimethylsilyl and triethylsilyl ethers are oxidized to carbonyl products by DMSO–$(COCl)_2$, but *t*-butyldimethylsilyl, *t*-butyldiphenylsilyl, and *t*-butoxydiphenylsilyl ethers are not (eq 23).[7a–d] Chloride attack on silicon is thought to assist the reaction.[7b] Primary triethylsilyl ethers are selectively oxidized by DMSO–$(COCl)_2$ in the presence of tertiary triethylsilyl ethers (eq 24).[7e] This subject has been reviewed.[7f]

Elimination can occur from the carbonyl products resulting from oxidation of sensitive substrates (eq 25).[8]

Allylic alcohols can be converted to chlorides by treatment with DMSO–$(COCl)_2$ without the use of added base, presumably by chloride displacement of DMSO from the alkoxysulfonium ion (eq 26).[9a,b] DMSO–$(COCl)_2$ also acts as an electrophilic chlorinating reagent toward ketones (eq 27).[9c] Oxidation to the nonchlorinated ketone was achieved with ***Dimethyl Sulfoxide–Acetic Anhydride***.[9c]

Oxidation of diaziridines with DMSO–$(COCl)_2$ gives diazirines (eq 28),[10a] and dibenzylamine gives partial conversion to an imine (eq 29).[10b]

A related reagent for the activation of DMSO for oxidation of alcohols is ***Phenyl Dichlorophosphate*** ($PhOP(O)Cl_2$),[11a–c] which is more effective than DMSO–$(COCl)_2$ in the oxidation of benzyl alcohol (eq 30).[11a] Cyanuric chloride also activates DMSO for oxidation of alcohols.[11d]

Related Reagents. See Class O-1, pages 1–10. *N*-Chlorosuccinimide–Dimethyl Sulfide; Chromic Acid; Dimethyl Sulfide–Chlorine; Dimethyl Sulfoxide–Acetic Anhydride; Dimethyl Sulfoxide–Dicyclohexylcarbodiimide; Dimethyl Sulfoxide–Iodine; Dimethyl Sulfoxide–Methanesulfonic Anhydride; Dimethyl Sulfoxide–Phosphorus Pentoxide; Dimethyl Sulfoxide–Sulfur Trioxide/Pyridine; Dimethyl Sulfoxide–Trifluoroacetic Anhydride; Dimethyl Sulfoxide–Triphosgene; Manganese Dioxide; Pyridinium Chlorochromate; Pyridinium Dichromate; Ruthenium(IV) Oxide; Silver(I) Carbonate; 1,1,1-Triacetoxy-1,1-dihydro-1,2-benziodoxol-3(1H)-one.

1. (a) Tidwell, T. T. *OR* **1990**, *39*, 297. (b) Tidwell, T. T. *S* **1990**, 857. (c) Lee, T. V. *COS* **1991**, *7*, 291. (d) Haines, A. H. *Methods for the Oxidation of Organic Compounds*; Academic: New York, 1988. (e) Hudlicky, M. *Oxidations in Organic Chemistry*; ACS: Washington, 1990. (f) Mancuso, A. J.; Swern, D. *S* **1981**, 165. (g) Moffatt, J. G. In *Oxidation*; Augustine, R. L.; Trecker, D. J., Eds.; Dekker: New York, 1971; Vol. 2, Chapter 1, p 1.
2. (a) Iwai, I.; Ide, J. *OSC* **1988**, *6*, 531. (b) Insalaco, M. A.; Tarbell, D. S. *OSC* **1988**, *6*, 207. (c) Corey, E. J.; Chaykovsky, M. *OSC* **1973**, *5*, 755.
3. (a) Omura, K.; Swern, D. *T* **1978**, *34*, 1651. (b) Mancuso, A. J.; Huang, S.-L.; Swern, D. *JOC* **1978**, *43*, 2480. (c) Mancuso, A. J.; Brownfain, D. S.; Swern, D. *JOC* **1979**, *44*, 4148.
4. (a) Leopold, E. J. *OSC* **1990**, *7*, 258. (b) Takai, K.; Heathcock, C. H. *JOC* **1985**, *50*, 3247. (c) Guanti, G.; Banfi, L.; Riva, R.; Zannetti, M. T. *TL* **1993**, *34*, 5483. (d) Salomon, R. G.; Sachinvala, N. D.; Roy, S.; Basu, B.; Raychaudhuri, S. R.; Miller, D. B.; Sharma, R. B. *JACS* **1991**, *113*, 3085. (e) Bishop, R. *OS* **1991**, *70*, 120. (f) Govindan, S. V.; Fuchs, P. L. *JOC* **1988**, *53*, 2593. (g) Vander Roest, J. M.; Grieco, P. A. *JACS* **1993**, *115*, 5841. (h) Sasaki, M.; Murae, T.; Takahashi, T. *JOC* **1990**, *55*, 528. (i) Boger, D. L.; Jacobson, I. C. *JOC* **1991**, *56*, 2115. (j) Farooq, O. *S* **1994**, 1035.
5. (a) Arimoto, H.; Nishiyama, S.; Yamamura, S. *TL* **1990**, *31*, 5619. (b) Arimoto, H.; Cheng, J.-F.; Nishiyama, S.; Yamamura, S. *TL* **1993**, *34*, 5781. (c) Arimoto, H.; Ohba, S.; Nishiyama, S.; Yamamura, S. *TL* **1994**, *35*, 4581.
6. (a) Berger, D.; Overman, L. E.; Renhowe, P. A. *JACS* **1993**, *115*, 9305. (b) Ireland, R. E.; Norbeck, D. W. *JOC* **1985**, *50*, 2198. (c) Takano, S.; Iwabuchi, Y.; Ogasawara, K. *CC* **1988**, 1204. (d) Smith, A. L.; Pitsinos, E. N.; Hwang, C.-K.; Mizuno, Y.; Saimoto, H.; Scarlato, G. R.; Suzuki, T.; Nicolaou, K. C. *JACS* **1993**, *115*, 7612. (e) Lichtenthaler, F. W.; Jarglis, P.; Lorenz, K. *S* **1988**, 790. (f) Marshall, J. A.; Andersen, M. W. *JOC* **1993**, *58*, 3912. (g) Danheiser, R. L.; Fink, D. M.; Okano, K.; Tsai, Y.-M.; Szczepanski, S. W. *OS* **1988**, *66*, 14.
7. (a) Afonso, C. M.; Barros, M. T.; Maycock, C. D. *JCS(P1)*, **1987**, 1221. (b) Tolstikov, G. A.; Miftakhov, M. S.; Adler, M. E.; Komissarova, N. G.; Kuznetsov, O. M.; Vostrikov, N. S. *S* **1989**, 940. (c) Hoffmann, R. W.; Dahmann, G. *CB* **1994**, *127*, 1317. (d) Kigoshi, H.; Imamura, Y.; Mizuta, K.; Niwa, H.; Yamada, K. *JACS* **1993**, *115*, 3056. (e) Hirst, G. C.; Johnson, T. O., Jr.; Overman, L. E. *JACS* **1993**, *115*, 2992. (f) Muzart, J. *S* **1993**, 11.
8. Groneberg, R. D.; Miyazaki, T.; Stylianides, N. A.; Schulze, T. J.; Stahl, W.; Schreiner, E. P.; Suzuki, T.; Iwabuchi, Y.; Smith, A. L.; Nicolaou, K. C. *JACS* **1993**, *115*, 7593.
9. (a) Kato, N.; Nakanishi, K.; Takeshita, H. *BCJ* **1986**, *59*, 1109. (b) Kende, A. S.; Johnson, S.; Sanfilippo, P.; Hodges, J. C.; Jungheim, L. N. *JACS* **1986**, *108*, 3513. (c) Smith, A. B. III; Leenay, T. L.; Liu, H.-J.; Nelson, L. A. K.; Ball, R. G. *TL* **1988**, *29*, 49.
10. (a) Richardson, S. K.; Ife, R. J. *JCS(P1)* **1989**, 1172. (b) Keirs, D.; Overton, K. *CC* **1987**, 1660.
11. (a) Liu, H.-J.; Nyangulu, J. M. *TL* **1988**, *29*, 3167. (b) Liu, H.-J.; Nyangulu, J. M. *TL* **1988**, *29*, 5467. (c) Liu, H.-J.; Nyangulu, J. M. *TL* **1989**, *30*, 5097. (d) Albright, J. D. *JOC* **1974**, *39*, 1977.

Thomas T. Tidwell
University of Toronto, Ontario, Canada

Dipyridine Chromium(VI) Oxide[1]

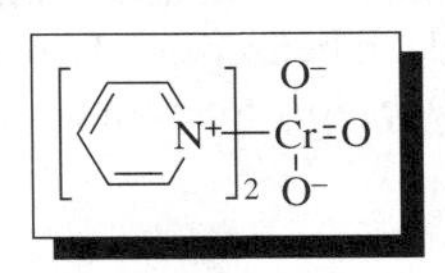

[26412-88-4] $C_{10}H_{10}CrN_2O_3$ (MW 258.22)

(reagent for oxidizing alcohols to carbonyl compounds)

Alternate Names: Collins reagent; chromium(VI) oxide–pyridine.
Solubility: sol CH_2Cl_2; (*Z*)-1,2-dichloroethylene; pyridine; $CHCl_3$.
Form Supplied in: red crystals; not commercially available.
Preparative Methods: prepared in 85–91% yield from ***Chromium(VI) Oxide*** and ***Pyridine***.[2,3] Caution: the reaction is extremely exothermic. The chromium(VI) oxide should be added to dry pyridine at such a rate that the temperature does not exceed 20 °C and in such a way that the oxide mixes rapidly with pyridine. Other chromium(VI) oxide–pyridine complexes are known. These include the Ratcliffe reagent (dipyridine chromium(VI) oxide prepared in situ in CH_2Cl_2),[4,5] and the Sarett reagent ($CrO_3 \cdot (C_5H_5N)_2$ in pyridine).[6] The preparation and workup of the Sarett reagent is sometimes tedious. The hygroscopic nature of the Collins reagent and its propensity to inflame may be avoided by the in situ preparation of the complex according to the Ratcliffe procedure.
Handling, Storage, and Precautions: Caution: the Collins reagent is extremely hygroscopic; exposure to moisture rapidly converts it to the yellow dipyridinium dichromate. The reagent should be stored at 0 °C under nitrogen or argon in a sealed container, protected from light. All chromium(VI) reagents must be handled with care; their mutagenicity is well documented.[7] This reagent should be prepared and handled in a fume hood.

Allylic Oxidation to Form α,β-Unsaturated Ketones. Although a solution of chromium(VI) oxide in pyridine is not very useful for allylic oxidation, the isolated dry chromium(VI) oxide–dipyridine complex in dichloromethane oxidizes allylic methylene groups to enones at rt in good to excellent yields. Δ^5-Androsten-7-one-3β,17β-diol diacetate is thus obtained (82%) from the oxidation of Δ^5-androstene-3β,17β-diacetate.[8] Attack at an allylic methine position yields the isomeric enone when possible; for example, 3-(4-fluorophenyl)cyclohexenol is oxidized to the isomeric enone (eq 1).[8] Similar rearrangements occur in methylene systems with steric hindrance.[8] If more than one allylic methylene group is present in a conformationally flexible molecule, isomeric enones resulting from attack at both positions are formed (eq 2).[8] Selectivity is observed in conformationally rigid molecules.[8] Methyl groups are not easily oxidized. In general, the Collins reagent gives higher yields and less overoxidation than ***Di-t-butyl Chromate*** or ***Chromium(VI) Oxide*** in acetic acid.

Alkynes are oxidized to conjugated alkynic ketones (ynones) by the Collins reagent.[9] The Collins reagent oxidizes 4-octyne to

4-octyn-3-one (eq 3). Oxidation of alkynes by ***t-Butyl Hydroperoxide*** and catalytic amounts of ***Selenium(IV) Oxide***[10] effects oxidation at both centers adjacent to a triple bond. A catalytic amount of chromium(VI) oxide in benzene and *t*-butyl hydroperoxide selectively oxidizes alkynes to ynones in about 50% yield.[11]

$CrO_3{\cdot}2py$, CH_2Cl_2, 84% (1)

$CrO_3{\cdot}2py$, CH_2Cl_2; 31%, 36% (2)

$CrO_3{\cdot}2py$, CH_2Cl_2, 42% (3)

Oxidation of Primary Alcohols to Aldehydes. The oxidation of alcohols is generally performed in dichloromethane with a sixfold excess of the Collins reagent. The Collins reagent[3] gives higher yields than the Sarett reagent[6] and comparable yields to the Ratcliffe reagent.[5] The Collins reagent oxidizes 1-heptanol to 1-heptanal in 70–84% yield[3] and the Ratcliffe reagent oxidizes 1-decanol to 1-decanal in 83% yield.[4,5] The Collins reagent oxidizes the primary allylic alcohols geraniol (eq 4) and nerol (eq 5) to geranial and neral, respectively, without isomerization.[12] The Ratcliffe reagent oxidizes cinnamyl alcohol to cinnamaldehyde in 96% yield.[5] The Collins reagent has been used to oxidize primary hydroxy groups of sugars to aldehydes in 50–75% yield.[13] Although the Sarett reagent is useful for the conversion of primary allylic and benzylic alcohols to their corresponding aldehydes, its use for primary saturated alcohols (with the exception of some steroidal ones) is less effective than the Collins or Ratcliffe reagent.

OH → CHO; $CrO_3{\cdot}2py$, 66% (4)

CH_2OH → CHO; $CrO_3{\cdot}2py$, 83% (5)

The oxidation of allylic alcohols to aldehydes is facilitated by use of Celite-supported Collins reagent (eq 6).[14,15] This method has been used to prepare intermediates in the synthesis of bulnesol[14] and guaiol.[15] A modified Ratcliffe reagent, $CrO_3{\cdot}2py$ in acetonitrile on Celite, was used to oxidize primary alcohols to aldehydes.[16]

CH_2OH → CHO; $CrO_3{\cdot}2py$, Celite 545, CH_2Cl_2, 60% (6)

The (*S*)-alcohol (**1**) is oxidized by the Collins reagent to the aldehyde (**2**) with no more than 5% racemization (eq 7).[17]

CH_2OH (**1**) → CHO (**2**); $CrO_3{\cdot}2py$ (7)

Primary alcohols are converted to the corresponding *t*-butyl esters by Collins reagent in CH_2Cl_2/DMF/Ac_2O and a large excess of *t*-butyl alcohol (eq 8).[18] This conversion is probably general except for aromatic aldehydes.

RCH_2OH → $RC(O)O$-*t*-Bu; $CrO_3{\cdot}2py$, CH_2Cl_2, DMF, AcOH, *t*-BuOH (8)

Oxidation of Secondary Alcohols to Ketones. The Collins reagent (eq 9),[19] the Ratcliffe reagent,[5] and the Sarett reagent[20,21] are effective in oxidizing secondary alcohols to ketones. Generally, acid sensitive functional groups such as acetals, double bonds, oxiranes, and thioethers are not affected, although there are exceptions. The Collins reagent oxidizes *exo*-7-hydroxybicyclo[4.3.1]deca-2,4,8-triene to bicyclo[4.4.1]deca-2,4,8-triene-7-one (64%).[22] Collins reagent oxidizes the secondary alcohol functional groups in β-hydroxy-(*Z*)-*O*-alkyloximes to the corresponding β-keto-*O*-alkyloximes.[23]

OBz, H, H, OH → OBz, H, H, O; $CrO_3{\cdot}2py$, 94% (9)

A number of alternative reagents are available for the oxidation of primary and secondary alcohols to aldehydes and ketones. Related chromium-based reagents include ***Chromium(VI) Oxide–3,5-Dimethylpyrazole***, ***Chromium(VI) Oxide-Quinoline***, ***Pyridinium Chlorochromate***, and ***Pyridinium Dichromate***.

Oxidation of Tertiary Allylic Alcohols to Epoxy Aldehydes. The Collins reagent oxidizes tertiary allylic alcohols to epoxy aldehydes (eq 10).[24]

OH → CHO (81%) + CHO (15%); $CrO_3{\cdot}2py$, CH_2Cl_2 (10)

Oxidation of Carbohydrates. Addition of acetic anhydride to the Collins reagent increases the yields (>90%) for the oxidation of secondary hydroxy groups to carbonyl groups in carbohydrates.[13,25,26]

Oxidation of β-Hydroxy Ketones to 1,3-Diketones. Collins reagent oxidizes β-hydroxy ketones to 1,3-diketones (eq 11).[26,27] Higher yields are generally obtained with ***Dimethyl Sulfoxide–Oxalyl Chloride*** (Swern reagent).

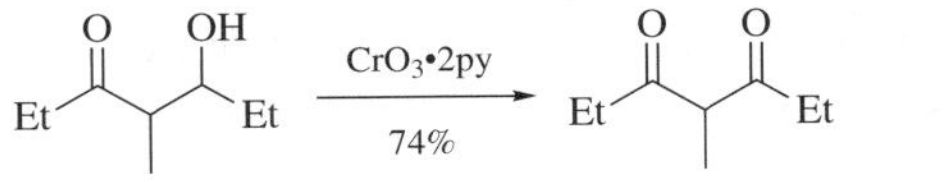

CrO_3•2py, 74% (11)

Oxidation of β-Hydroxy Esters to β-Keto Esters. Collins reagent oxidizes β-hydroxy esters to β-keto esters (eq 12).[27]

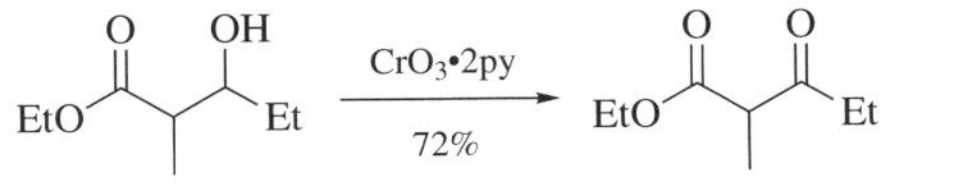

CrO_3•2py, 72% (12)

Oxidative Cyclization of 5,6-Dihydroxyalkenes. Collins reagent oxidizes the unsaturated diols (**3**) to the corresponding *cis*-tetrahydrofurandiols (**4**) (eq 13).[28]

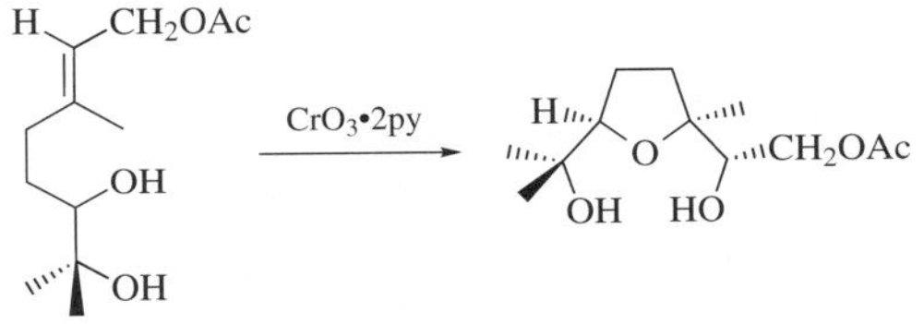

CrO_3•2py (13)

(**3**) (**4**)

Oxidation of 1,4-Dienes. Oxidation of the 1,4-diene (**5**) with Collins reagent or ***Di-t-butyl Chromate*** yields dienones (**6**) and (**7**) in the ratio of 1:3 (~65% total yield) (eq 14).[29] Complementary regioselectivity (**6**)/(**7**) = 9:1 (~70% total yield) is obtained with ***Pyridinium Chlorochromate*** (PCC).

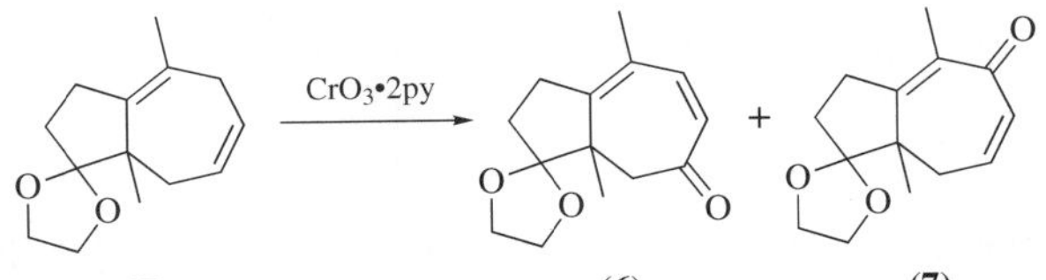

CrO_3•2py (14)

(**5**) (**6**) (**7**)

Other Applications. Collins reagent, Jones' reagent, and chromic acid in 50% acetic acid oxidatively deoximate ketoximes to the corresponding carbonyl compounds.[30]

Secondary alkylstannanes are converted into the corresponding carbonyl compounds by oxidation with Collins reagent (eq 15).[31,32] Mixtures of alcohols and dehydration products are obtained from tertiary alkylstannanes.

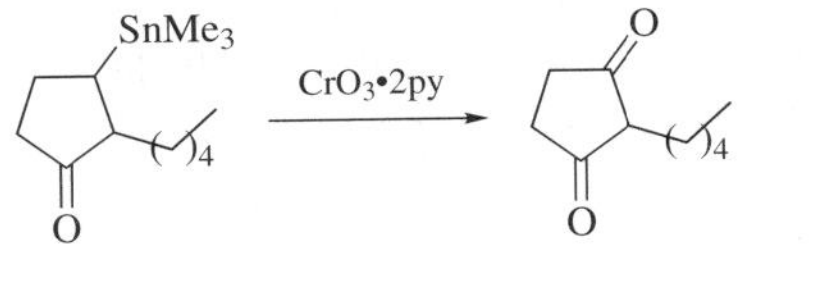

CrO_3•2py (15)

Collins reagent oxidizes trimethylsiloxy-substituted 1,4-cyclohexadienes to phenols (eq 16).[32]

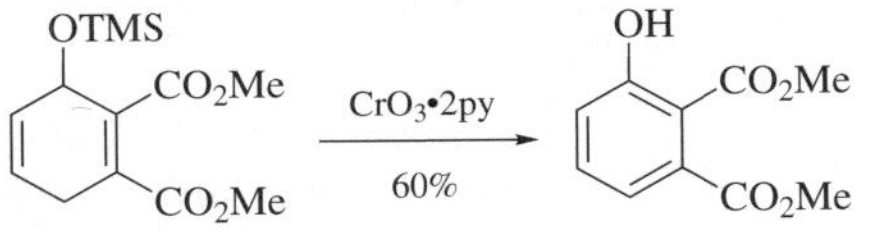

CrO_3•2py, 60% (16)

Collins reagent oxidizes steroidal tertiary amines to *N*-formyl derivatives (eq 17).[3]

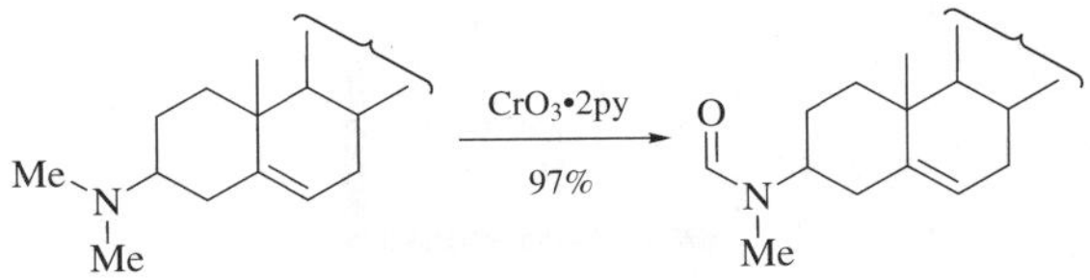

CrO_3•2py, 97% (17)

Related Reagents. See Classes O-1 and O-21, pages 1–10.

1. (a) Wiberg, K. B. *Oxidation in Organic Chemistry*; Wiberg, K. B., Ed.; Academic: New York, 1965; Part A, pp 131–135. (b) Freeman, F. *Organic Synthesis By Oxidation With Metal Compounds*; Miijs, W. J.; de Jonge, C. R. H. I., Eds.; Plenum: New York, 1986; Chapter 2. (c) Lee, D. G. *The Oxidation of Organic Compounds by Permanganate Ion and Hexavalent Chromium*; Open Court: La Salle, IL, 1980. (d) Stewart, R. *Oxidation Mechanisms: Applications to Organic Chemistry*; Benjamin: New York, 1964. (e) Cainelli, G.; Cardillo, G. *Chromium Oxidations in Organic Chemistry*; Springer: Berlin, 1984.
2. Collins, J. C.; Hess, W. W. *OSC* **1988**, *6*, 644.
3. Collins, J. C.; Hess, W. W.; Frank, F. J. *TL* **1968**, 3363.
4. Ratcliffe, R. W. *OSC* **1988**, *6*, 373.
5. Ratcliffe, R. W.; Rodehorst, R. *JOC* **1970**, *35*, 4000.
6. Poos, G. I.; Arth, G. E.; Beyler, R. E.; Sarett, L. H. *JACS* **1953**, *75*, 422.
7. Cupo, D. Y.; Wetterhahn, K. E. *Cancer Res.* **1985**, *45*, 1146 and references cited therein.
8. Dauben, W. G.; Lorber, M.; Fullerton, D. S. *JOC* **1969**, *34*, 3587.
9. Shaw, J. E.; Sherry, J. J. *TL* **1971**, 4379.
10. Chabaud, B.; Sharpless, K. B. *JOC* **1979**, *44*, 4202.
11. Muzart, J.; Piva, O. *TL* **1988**, *29*, 2321.
12. Holum, J. R. *JOC* **1961**, *26*, 4814.
13. Butterworth, R. F.; Hanessian, S. *S* **1971**, 70.
14. Andersen, N. H.; Uh, H. *SC* **1973**, *3*, 115.
15. Andersen, N. H.; Uh, H. *TL* **1973**, 2079.
16. Schmitt, S. M.; Johnston, D. B. R.; Christensen, B. G. *JOC* **1980**, *45*, 1135, 1142.
17. Evans, D. A.; Bartroli, J. *TL* **1982**, *23*, 807.
18. Corey, E. J.; Samuelsson, B. *JOC* **1984**, *49*, 4735.
19. Gilbert, J. C.; Smith, K. R. *JOC* **1976**, *41*, 3883.
20. Urech, J.; Vischer, E.; Wettstein, A. *HCA* **1960**, *43*, 1077.
21. Ellis, B.; Petrow, V. *JCS* **1956**, 4417.
22. Schröder, G.; Prange, U.; Putze, B.; Thio, J.; Oth, J. F. M. *CB* **1971**, *104*, 3406.
23. Shatzmiller, S.; Bahar, E.; Bercovici, S.; Cohen, A.; Verdoorn, G. *S* **1990**, 502.
24. Sundararaman, P.; Herz, W. *JOC* **1977**, *42*, 813.
25. Garegg, P. J.; Samuelsson, B. *Carbohydr. Res.* **1978**, *67*, 267.
26. Samano, V.; Robins, M. J. *S* **1991**, 283.
27. Smith, A. B. III.; Levenberg, P. A. *S* **1981**, 567.
28. Walba, D. M.; Stoudt, G. S. *TL* **1982**, *23*, 727.
29. Wender, P. A.; Eissenstat, M. A.; Filosa, M. P. *JACS* **1979**, *101*, 2196.

30. Araújo, H. C.; Ferreira, G. A. L.; Mahajan, J. R. *JCS(P1)* **1974**, 2257.
31. Still, W. C. *JACS* **1978**, *100*, 1481.
32. Still, W. C. *JACS* **1977**, *99*, 4836.

Fillmore Freeman
University of California, Irvine, CA, USA

Disiamylborane[1]

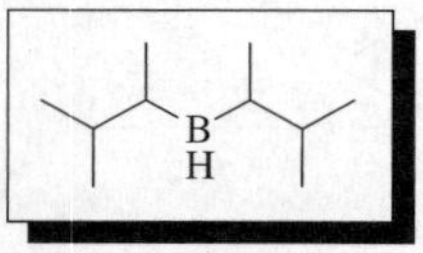

[1069-54-1] $C_{10}H_{23}B$ (MW 154.14)

(hindered organoborane for chemo-[2] and regioselective[3] hydroboration and chemo-[4] and stereoselective[5] reduction; can be used to mediate couplings of alkenes[6] and alkynes[7])

Alternate Name: bis(1,2-dimethylpropyl)borane.
Physical Data: mp 35–40 °C; fp −17°C.
Solubility: sol THF, ether, diglyme.
Form Supplied in: prepared in situ; kits are available.
Preparative Methods: generally prepared by reaction of ***Diborane*** with 2-methyl-2-butene in ethereal solvents at 0 °C;[8] alternatively by substituting borane-1,4-thioxane.[9]
Handling, Storage, and Precautions: flammable; very air- and moisture-sensitive. Generally prepared immediately prior to use. Handle in a fume hood.

Hydroboration of Alkenes. Compounds containing double bonds react with disiamylborane (Sia_2BH) to give tertiary boranes of the form Sia_2BR (eq 1). These trialkylboranes are generally not isolated, but are submitted to further reactions to provide a variety of products. The reactions of disiamylborane adducts are typical of organoboranes. If the intermediate is quenched with acid, the borane is replaced with a proton (eq 2); the net result is a *syn* hydrogenation.[10]

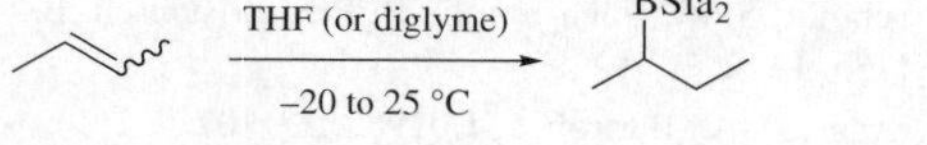

(1)

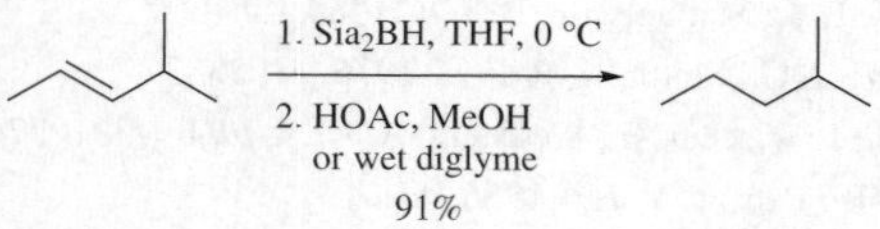

(2)

The organoborane may also be oxidized with alkaline peroxide to give an alcohol which retains the position and configuration of the borane adduct. Thus, for example, Sia_2BH is the reagent of choice for anti-Markovnikov hydration of ω-unsaturated esters due to its high selectivity for terminal hydroboration (eq 3).[11] Alternatively, direct oxidation of the intermediate resulting from hydroboration of a terminal alkene with ***Pyridinium Chlorochromate*** provides the aldehyde (eq 4).[12] Once again, Sia_2BH is the reagent of choice for this reaction because of its high selectivity for the alkenic terminus. Reaction of Sia_2BR with ***Iodine*** under basic conditions provides the corresponding iodide.[13] When R is primary, the formation of RI proceeds in high yield, because transfer of the siamyl groups occurs only slowly (eqs 5 and 6). A similar transformation may be accomplished using borane, but yields are generally lower because R_3B gives only partial conversion to RI.[14] This Sia_2BH-based approach was found to be clearly superior to a hydrozirconation route for the iodination of the (η^6-hexabutenylbenzene)(η^5-cyclopentadienyl)iron cationic complex illustrated in eq 7.[15]

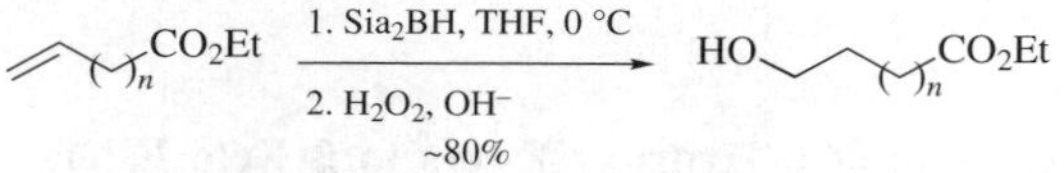

(3)

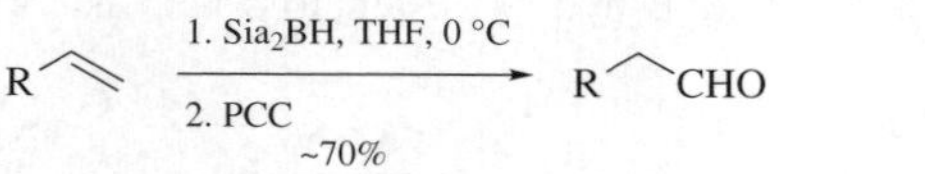

(4)

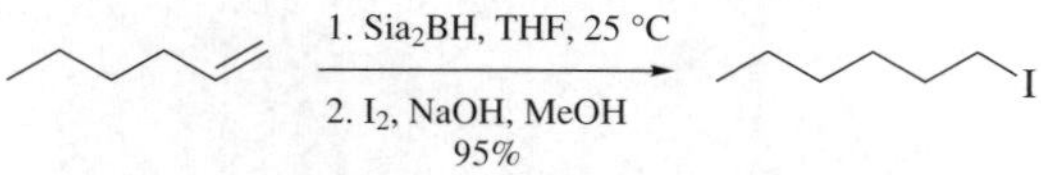

(5)

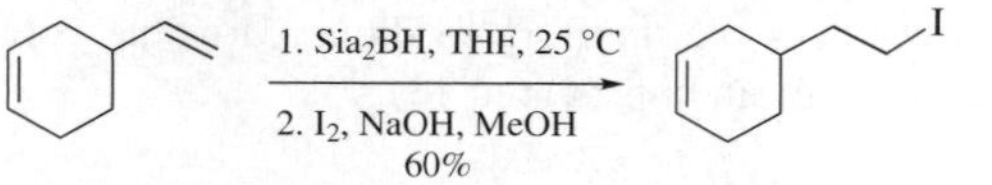

(6)

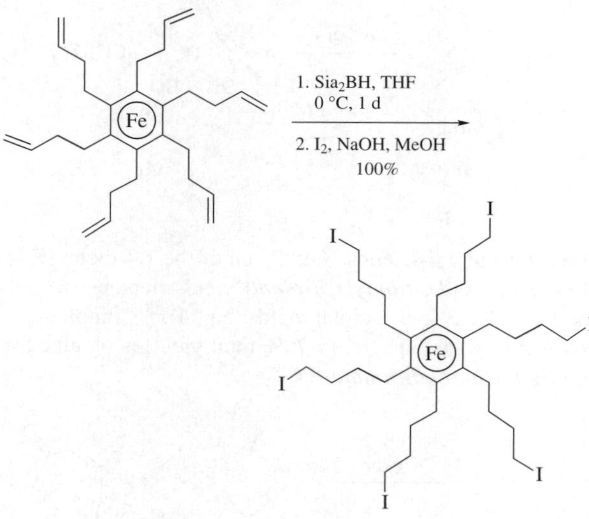

(7)

Unlike diborane, disiamylborane exists as a dimer even in solution in THF.[16] This failure to dissociate in a coordinating solvent, presumably due to the steric bulk of the siamyl groups which disfavor the borane/ether complex, causes Sia_2BH to act as a highly hindered hydroborating agent. As a general rule, Sia_2BH reacts preferentially with less sterically hindered alkenes. An extensive comparative study has determined the relative rates of reaction of double bonds with a variety of substitution patterns.[17]

Reaction occurs to place the borane at the least encumbered position. These differences in reaction rate are often large enough to be synthetically useful; thus 1-hexene reacts at the terminal position with >99:1 selectivity (with diborane the selectivity for the 1-position is ~15:1).[18] Sia_2BH even responds to the bulk of the substituents on the double bond; 4-methyl-*trans*-2-pentene reacts slowly with the reagent (12 h at 0 °C) to give mainly the 2-alcohol (Scheme 1).[18]

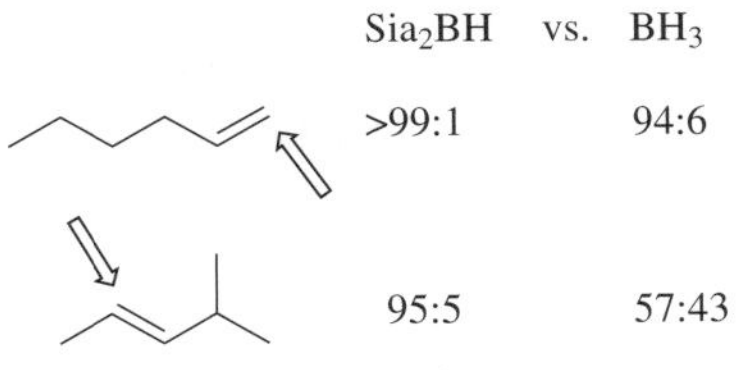

	Sia_2BH	vs. BH_3
(terminal alkene)	>99:1	94:6
(trisubstituted alkene)	95:5	57:43

Scheme 1

This high level of steric discrimination makes Sia_2BH a useful reagent for selective reaction with one double bond of a polyunsaturated compound; for example, hydroboration of limonene (eq 8) gives (after oxidative workup) a good yield of α-terpineol,[19] while the diene of eq 9 reacts selectively at one of the four possible alkenic positions.[20]

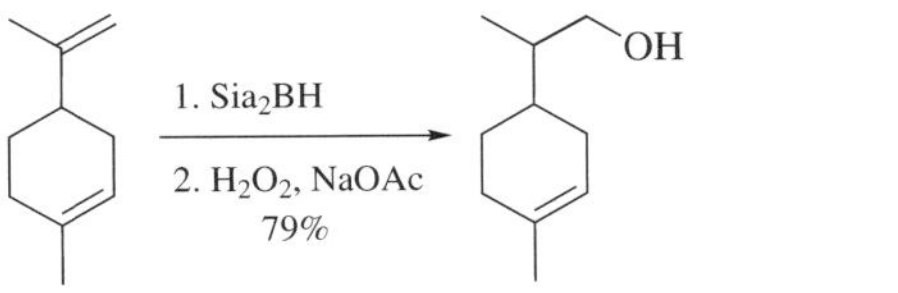

1. Sia_2BH; 2. H_2O_2, NaOAc; 79% (8)

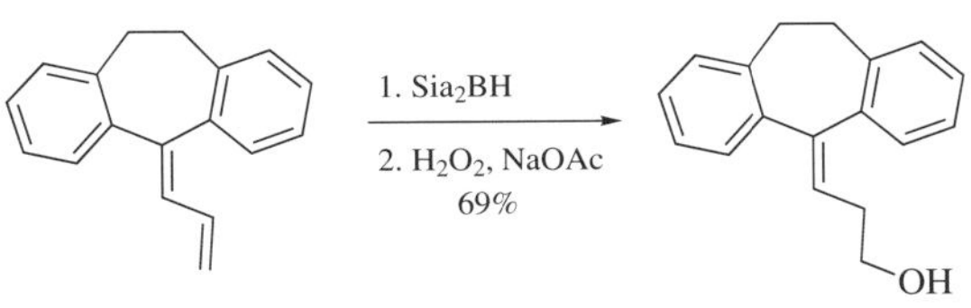

1. Sia_2BH; 2. H_2O_2, NaOAc; 69% (9)

Sia_2BH-mediated hydroboration is a key step in a unique carbocyclization reaction (eq 10).[6] Thus dienyl iodide is selectively hydroborated at the less-substituted alkene; oxidative cyclization of this intermediate provides a cycloalkene. The borane route provides a useful alternative to methods involving palladium catalysis.

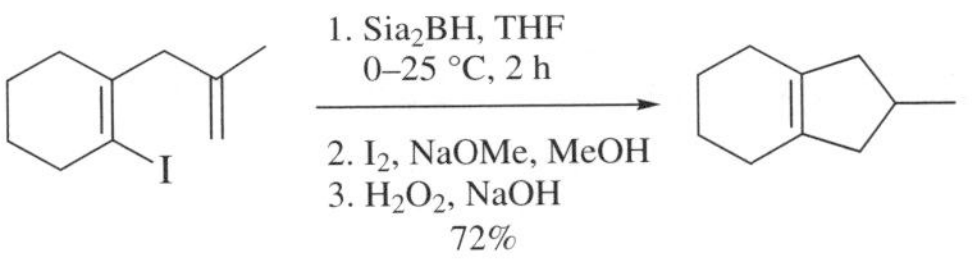

1. Sia_2BH, THF, 0–25 °C, 2 h; 2. I_2, NaOMe, MeOH; 3. H_2O_2, NaOH; 72% (10)

Hydroboration of Alkynes. Sia_2BH is a valuable reagent for the chemo- and regioselective hydroboration of alkynes. The reagent is more selective for terminal hydroboration than diborane; it is also more selective for monoaddition.[8] Reaction of 1-hexyne with Sia_2BH gives a vinylborane as an intermediate. Hydrolysis gives 1-hexene as the product (eq 11). In the case of internal alkynes, this process results in a net *syn* hydrogenation to give a (*Z*)-alkene (eq 14).[21] Oxidative workup of the vinylborane provides an aldehyde (eq 12).[8] The intermediate can also be oxidized with copper salts in the presence of cyanide ion to give an α,β-unsaturated nitrile (eq 13).[22] Sia_2BH is superior to ***Dicyclohexylborane*** for this transformation, which provides an alternative to the hydroalumination/cyanogen sequence.[23]

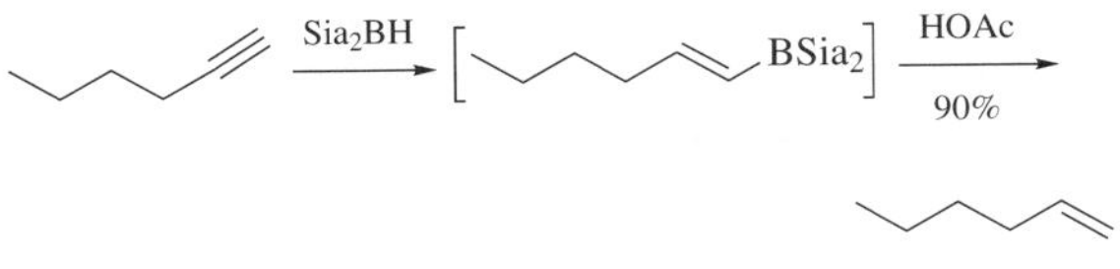

Sia_2BH → [$BSia_2$] → HOAc, 90% (11)

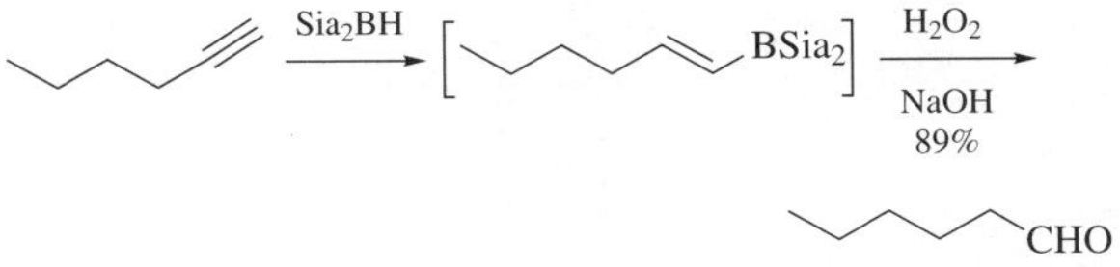

Sia_2BH → [$BSia_2$] → H_2O_2, NaOH, 89% → CHO (12)

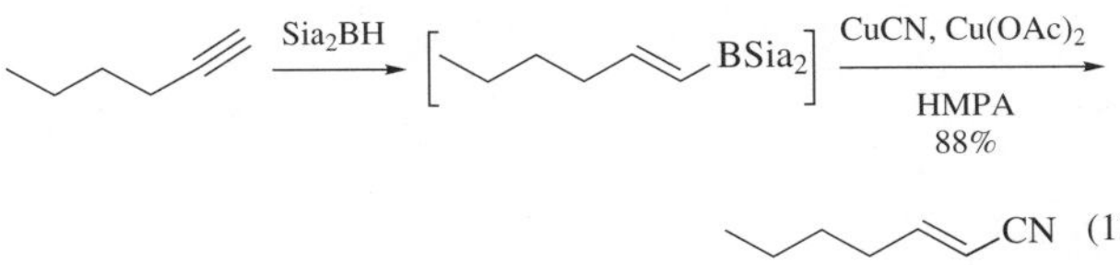

Sia_2BH → [$BSia_2$] → CuCN, $Cu(OAc)_2$, HMPA, 88% → CN (13)

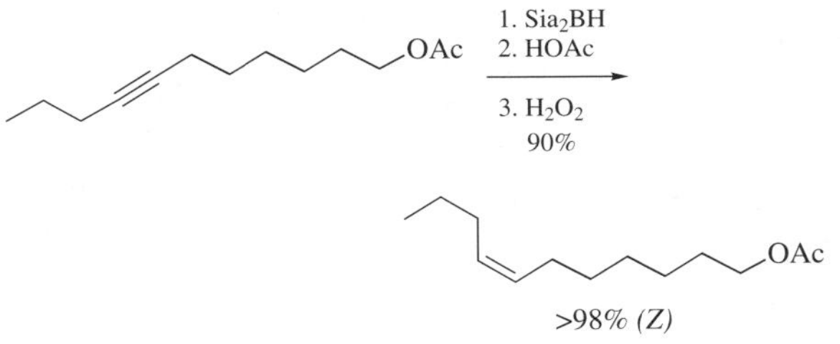

OAc → 1. Sia_2BH; 2. HOAc; 3. H_2O_2; 90% → OAc (14), >98% (*Z*)

The reaction of Sia_2BH with 1,4-dichloro-2-butyne, followed by treatment of the intermediate with a thiolate anion, provides a one-pot synthesis of 2-thioalkylbutadienes (eq 15).[24] A similar strategy, replacing the thiolate with an organolithium reagent, is unsuccessful for the synthesis of 2-alkylbutadienes, as the siamyl group migrates preferentially to primary alkyl.[25]

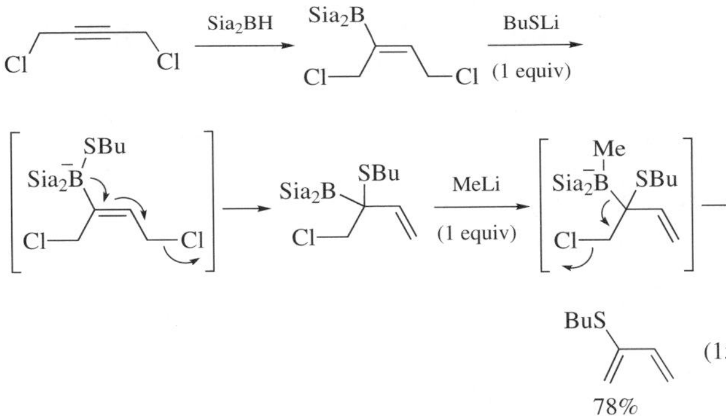

Cl, Cl → Sia_2BH → Sia_2B, Cl, Cl → BuSLi (1 equiv) → [SBu, Sia_2B, Cl, Cl] → Sia_2B, SBu, Cl → MeLi (1 equiv) → [Me, Sia_2B, SBu, Cl] → BuS (15), 78%

Sia_2BH adds regioselectively to a variety of asymmetrically substituted alkynes. Reaction with alkynyl sulfides occurs primarily α to sulfur; processing of the vinylborane intermediate by hydrolysis (of the corresponding ate complex) or oxidation leads to a vinyl sulfide (eq 17) or thiolester (eq 16), respectively.[26] Cyclohexylborane is more selective (83:17) for α-boration than Sia_2BH (72:28) in this reaction. The reagent also reacts selectively at the 2-position of 1-chloro-2-heptyne; acid hydrolysis leads to a (*Z*)-allyl chloride (eq 18), while base treatment gives the terminal allene (eq 19).[27] Similarly, alkynyl acetals react with Sia_2BH to place the borane proximal to the heteroatom.[28]

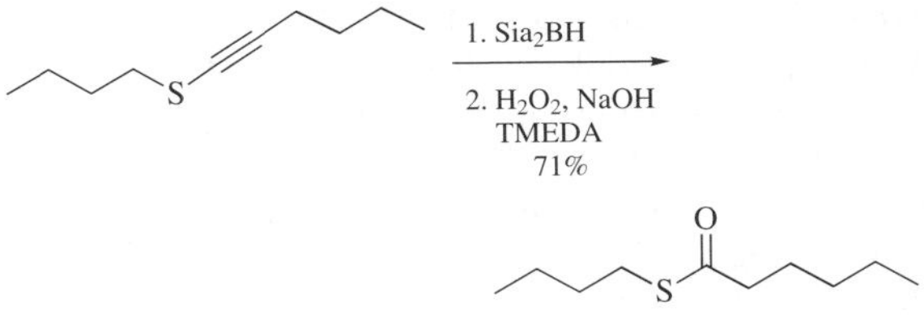

S → 1. Sia_2BH; 2. H_2O_2, NaOH, TMEDA; 71% → O, S (16)

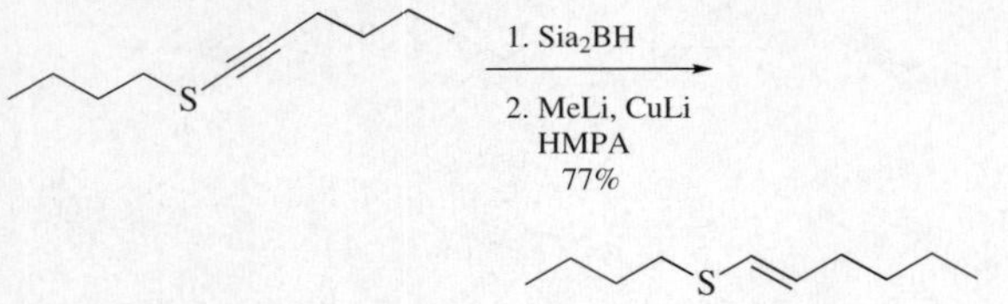

Hydroboration of alkynes with Sia_2BH is more rapid than the corresponding reaction of alkenes, allowing for the selective conversion of enynes into dienes (eq 21)[2] or enones (eq 20).[29]

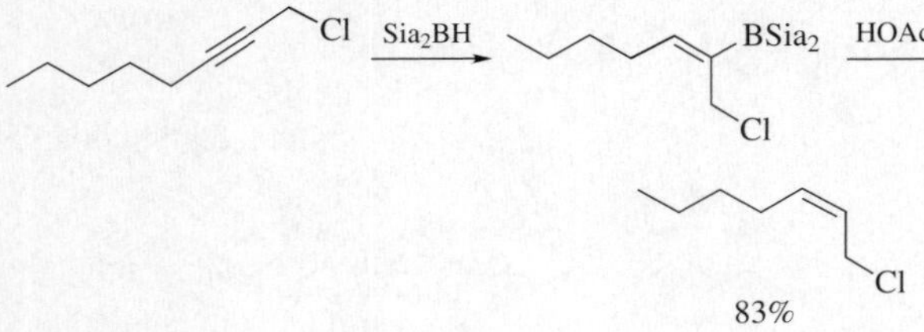

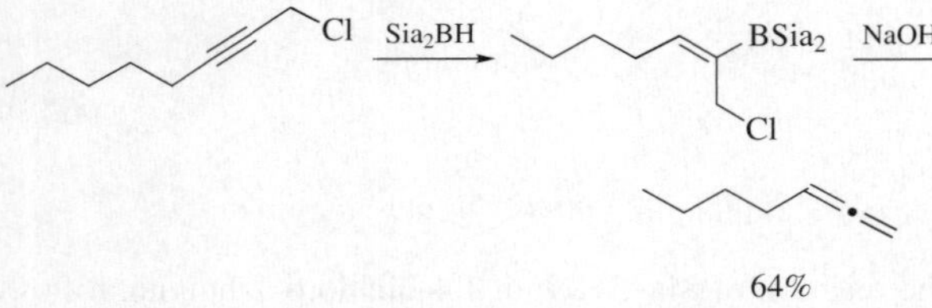

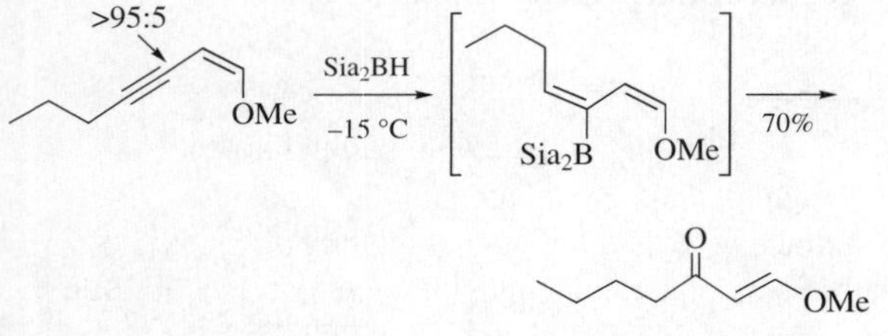

The enynes themselves can be prepared in Sia_2BH-mediated reactions. Hydroboration of diynes protected by bulky silyl groups occurs in a highly regioselective fashion; hydrolysis of the resultant organoboranes gives silylated enynes (eq 22).[30] Alternatively the vinylborane resulting from reaction of a terminal alkyne with Sia_2BH undergoes further reaction with an alkynyllithium to give an ate complex, oxidation of which with iodine leads to carbon–carbon bond formation (eq 23).[7]

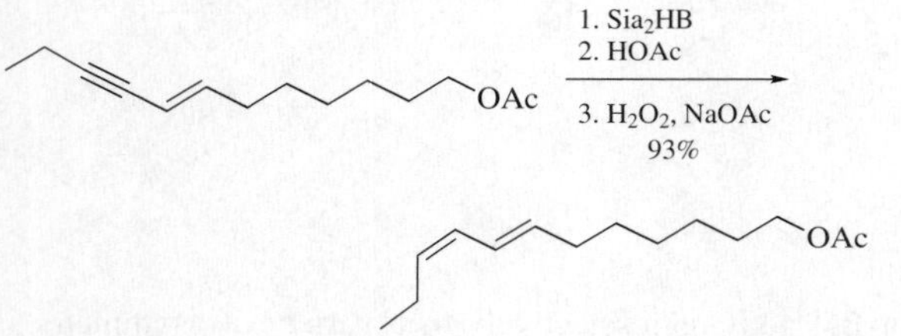

Functional Group Reductions. An unusual mix of chemo- and stereoselectivity makes Sia_2BH a useful reducing agent for carbonyl compounds. While aldehydes and ketones are reduced rapidly, a variety of other functionalities including carboxylic and sulfonic acids, acid chlorides, and sulfones are inert.[4] Ketone reduction is often highly stereoselective, with the bulky reducing reagent delivering hydride from the less hindered face of the carbonyl (eqs 24 and 25).[5]

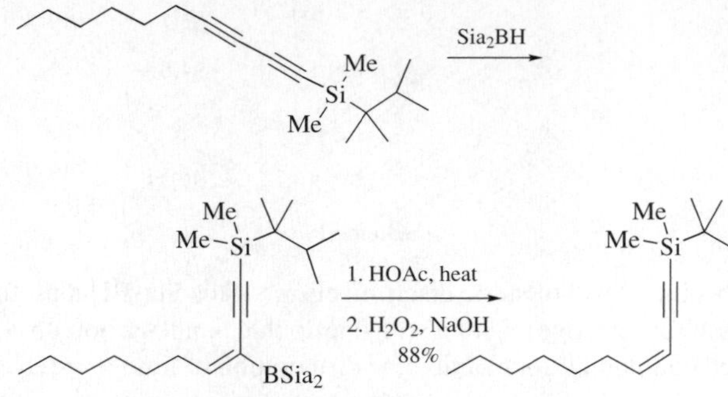

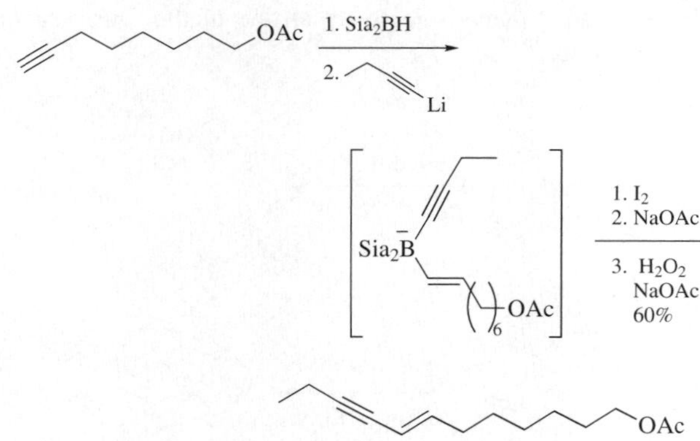

Sia_2BH is superior to ***Lithium Aluminum Hydride*** or borane for this transformation, but is generally less stereoselective than ***Diisopinocampheylborane***. While esters are unreactive, γ-lactones may be reduced to the corresponding lactols, a selectivity which has been used to advantage in carbohydrate synthesis (eq 26).[31] Amides also show a unique pattern of reactivity; while Sia_2BH fails to reduce primary amides, tertiary amides are converted to the corresponding aldehydes (eq 27).[31]

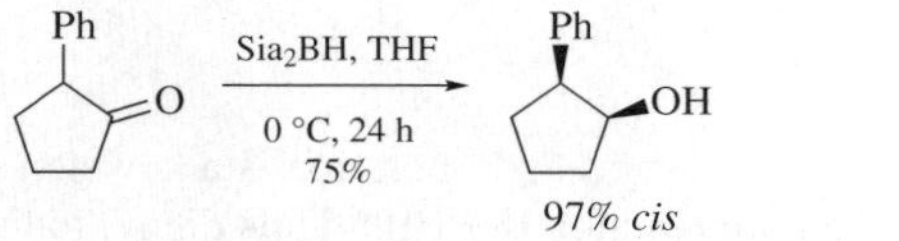

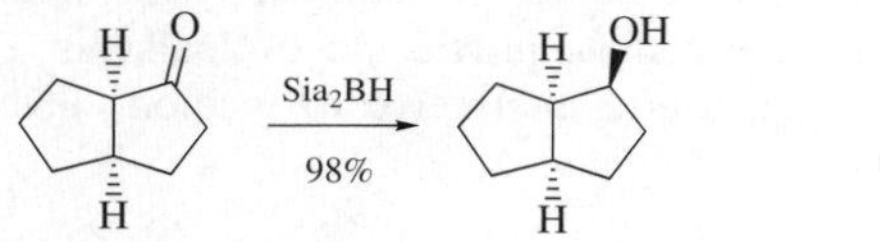

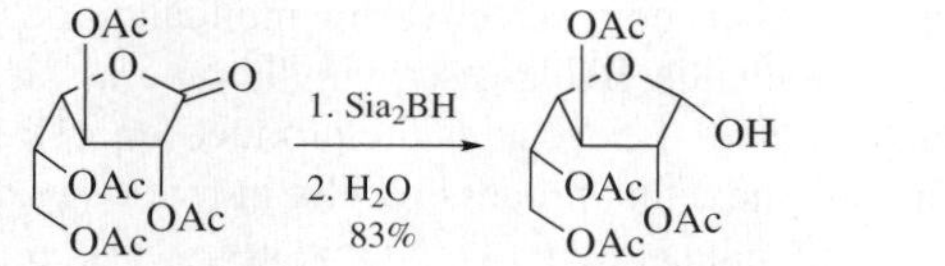

Sia_2BH also reacts with α,β-unsaturated ketones, with reduction occurring in a 1,4-sense.[32] The resultant boron enolate may be hydrolyzed to provide the saturated ketone (eq 28), or alternatively may be treated with an aldehyde to give *syn*-aldol products (eq 29).

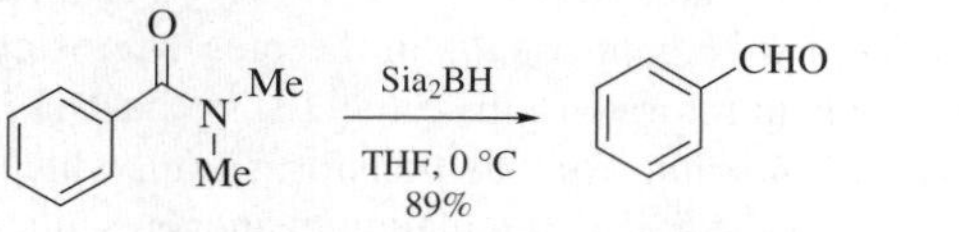

Sia_2BH, 0–20 °C; $OBSia_2$; H_2O, 92% (28)

<1% 1,2-reduction

Sia_2BH, 0–20 °C; $OBSia_2$; PhCHO, 0 °C, 18 h, 62%; OH O, Ph (29)

Dicyclohexylborane, diisopinocampheylborane, and diisocaranylborane all accomplish a similar transformation; 9-BBN is inferior because it leads to significantly more 1,2-reduction.

Related Reagents. See Classes R-2, R-3, R-4, R-12, R-14, R-15, and R-25, pages 1–10. Borane–Tetrahydrofuran; Diborane; Dicyclohexylborane.

---

1. Brown, H. C. *Organic Syntheses via Boranes*; Wiley: New York, 1975.
2. Negishi, E.; Yoshida, T; Abramovitch, A.; Lew, G.; Williams, R. M. *T* **1991**, *47*, 343.
3. Brown, H. C.; Zweifel, G. *JACS* **1961**, *83*, 1241.
4. Brown, H. C.; Bigley, D. B.; Arora, S. K.; Moon, N. M. *JACS* **1970**, *92*, 7161.
5. Brown, H. C.; Varma, V. *JOC* **1974**, *39*, 1631.
6. Negishi, E.; Sawada, H.; Tour, J. M.; Wei, Y. *JOC* **1988**, *53*, 913.
7. Negishi, E.; Abramovitch, A. *TL* **1977**, 411.
8. Brown, H. C.; Zweifel, G. *JACS* **1959**, *81*, 1512.
9. Brown, H. C.; Mandal, A. K. *JOC* **1992**, *57*, 4970.
10. Brown, H. C.; Murray, K. J. *JOC* **1961**, *26*, 631.
11. Brown, H. C.; Keblys, K. A. *JACS* **1964**, *86*, 1795.
12. Brown, H. C.; Kulkarni, S. U.; Rao, C. G. *S* **1980**, 151.
13. Brown, H. C.; Rathke, M. W.; Rogic, M. M.; deLue, N. R. *T* **1988**, *44*, 2751.
14. Brown, H. C.; Rathke, M. W.; Rogic, M. M. *JACS* **1968**, *90*, 5038.
15. Moulines, F.; Djakovitch, L.; Fillaut, J.-L.; Astruc, D. *SL* **1992**, 57.
16. Brown, H. C.; Klender, G. J. *IC* **1962**, *1*, 204.
17. Brown, H. C.; Zweifel, G. *JACS* **1961**, *83*, 1241.
18. Brown, H. C.; Zweifel, G. *JACS* **1960**, *82*, 3222.
19. Brown, H. C.; Zweifel, G. *JACS* **1960**, *82*, 3223.
20. Hoffsommer, R. D.; Taub, D.; Wendler, N. L. *JOC* **1963**, *28*, 1751.
21. Holan, G.; O'Keefe, D. F. *TL* **1973**, 673.
22. Masuda, Y; Hoshi, M.; Arase, A. *CC* **1991**, 748.
23. Zweifel, G. *OR* **1984**, *32*, 415.
24. Hoshi, M.; Masuda, W.; Arase, A. *CC* **1987**, 1629.
25. Arase, A.; Hoshi, M. *CC* **1987**, 531.
26. Hoshi, M.; Masuda, Y.; Arase, A. *BCJ* **1990**, *63*, 447.
27. Zweifel, G.; Horng, A.; Snow, J. T. *JACS* **1970**, *92*, 1427.
28. Zweifel, G.; Horng, A.; Plamondon, J. E. *JACS* **1974**, *96*, 316.
29. Zweifel, G.; Najafi, M. R.; Rajagopalan, S. *TL* **1988**, *29*, 1895.
30. Stracker, E. C.; Leong, W.; Miller, J. A.; Shoup, T. M.; Zweifel, G. *TL* **1989**, *30*, 6487.
31. Kohn, P.; Samaritano, R. H.; Lerner, L. M. *JACS* **1965**, *87*, 5475.
32. Boldrini, G. P.; Bortolotti, M.; Mancini, F.; Tagliavini, E.; Trombini, C.; Umani-Ronchi, A. *JOC* **1991**, *56*, 5820.

Thomas W. von Geldern
Abbott Laboratories, Abbott Park, IL, USA

Fluorine[1]

[7782-41-4] F_2 (MW 38.00)

(strong fluorinating agent and oxidizer; substitutes F for H in either radical[2] or electrophilic[2b,3] reactions; electrophilic substitution of F for other halides;[1c] adds to multiple bonds;[2b,4] oxidatively adds to centers of coordinative unsaturation;[5] used in synthesis of other (mostly electrophilic) fluorinating agents;[6] with aq MeCN, oxidizes tertiary H to alcohols,[7] alcohols to ketones,[8] ketones to esters,[8] aromatics to phenols or quinones,[9] amines to nitro derivatives,[10] and alkenes to oxiranes;[7,11] cyclotron-produced $^{18}F_2$ provides ^{18}F-labeled compounds for positron emission tomography[12])

Physical Data: mp −219.6 °C; bp −188.2 °C; vapor pressure at 77 K = 280 Torr.

Solubility: slightly sol $CFCl_3$, $CF_2ClCFCl_2$, $CHCl_3$, CH_2Cl_2, MeCN, perfluoroethers, perfluorocarbons; dec H_2O giving HF, O_2 and trace O_3, OF_2.[13]

Form Supplied in: faintly yellow compressed gas in steel cylinders at pressures of 160–400 psi, purity 97–99%; impurities HF, N_2, O_2, CF_4, SF_6, SiF_4; also in cylinders prediluted with inert gases.

Analysis of Reagent Purity: titration of Cl_2 liberated from NaCl, followed by GC for inert components; IR for HF. Rarely performed by user.

Preparative Methods: electrolysis of KF–HF mixtures.[14]

Purification: HF removed by passage through column of NaF pellets; other impurities removed only by cryogenic distillation.[15]

Handling, Storage, and Precautions: toxic; strong oxidizer. Pure fluorine should only be used by trained personnel! Prediluted F_2 is handled much more easily. Compatible materials: copper, brass, steel, stainless steel, nickel alloys (Monel, Inconel, Hastelloy), fluoropolymers (PTFE, FEP, PFA, Kel-F), *dry* glass. Vendors and handbooks[4a,14a,16] *must* be consulted for more detailed recommendations! Proper equipment cleaning and passivation are imperative to avoid ignition. Use fume hood, barricade, faceshield, leather gloves. Odor threshold: 20 ppb. TLV: 1 ppm. PEL: 0.1 ppm. IDLH: 25 ppm. Leaks easily detected with paper moistened with aq KI; monitors/detectors available. Use only fluorinated greases and oils for joints, bubblers, valve lubricants. Lubricate pipe threads with PTFE tape; permanent installations should be welded. Scrub effluent with soda lime, activated alumina, or 5–15% aq KOH. (More dilute base gives toxic OF_2; KF produced is more soluble than is NaF.) Users of solid scrubbers with vacuum pumps should consider an O_2-compatible pump fluid. NaF, aq base, and $CaCl_2$ are useful for workup of reactions producing HF. *Caution:* Many things will burn in pure fluorine given enough (sometimes insignificant) activation energy; even many materials (PTFE, metals, concrete) not usually considered to be 'fuels' can ignite in F_2.

Introduction of Fluorine.

Substitution of Fluorine for Hydrogen.

Perfluorination. The chief processes for perfluorination of organic compounds are the low-temperature gradient (LaMar) and aerosol fluorination methods. The goal of these is generally complete fluorination and saturation while minimizing fragmentation (eqs 1 and 2)[2a,17] although certain functional groups can be preserved (eq 3).[18] Perfluorination can also be achieved using inert solvents (eq 4).[19]

$$\text{Diamantane} \xrightarrow[8\%]{F_2,\ NaF} \text{Perfluorodiamantane} \quad (1)$$

$$\text{Naphthalene} \xrightarrow{F_2} \text{Perfluorodecalin} \quad (2)$$

$$Me_2C{=}O \xrightarrow[\text{Cu, −100 to −40 °C}\ 38\%]{F_2/He,\ NaF} (CF_3)_2C{=}O \quad (3)$$

$$MeOC(CF_3)_2OCH_2(CF_2)_4H \xrightarrow[\text{FC-72 (3M)}\ 71\%]{F_2,\ UV,\ -10\ °C} CF_3OC(CF_3)_2OCF_2(CF_2)_4F \quad (4)$$

NaF is sometimes added to trap the HF produced as $NaHF_2$. The above reactions are generally radical in character; complete fluorination sometimes requires photochemical 'polishing' to provide larger concentrations of F· since the organic material becomes increasingly unreactive as fluorine substitution proceeds (see ***Cobalt(III) Fluoride***).

Polymer Surface Modification. Direct fluorination with F_2 is used to convert polymers (e.g. coal-tar pitch)[20] to highly fluorinated materials. Fluorination also increases the activity of sulfonic acid catalyst resins in alkylation reactions[21] and, when used sparingly, increases the surface energy of polymers, producing improved adhesion.[22]

Electrophilic Fluorinations. Despite fluorine's extreme reactivity, its use in selective and electrophilic replacements is not only possible but synthetically useful when mild conditions and solvents such as $CFCl_3/CHCl_3$ mixtures are employed. Fluorine excels at replacement of unactivated tertiary hydrogens and complete retention of configuration is observed. High selectivity for tertiary H is due to the higher degree of p character in tertiary C–H bonds over that in primary and secondary bonds,[23] and this also influences the relative reactivity of competing tertiary positions (eq 5).[2b] Nearby electronegative centers, for example, deactivate tertiary H in this reaction. In contrast, LaMar fluorination of alkylcyclohexanes leaves some tertiary H as the only H remaining in the molecule,[24] perhaps for steric reasons. Direct electrophilic aromatic substitution is rare although recent improvements have involved the directing effects of Lewis acids.[25] Heteroaromatics can also be fluorinated (eq 6).[26] Both aliphatic and aromatic substitutions are often facilitated by the use of organometallic derivatives[27] in place of the parent molecules (eq 7), and 1,3-diketones can be fluorinated in the form of silyl enol ethers (eq 8).[28] Fluorination of fullerenes has been a topic of recent interest.[29]

F_2, 60% (5)

F_2, NaF; $CFCl_3$, −78 °C; 64% (6)

1. $[^{18}F]F_2$; 2. H^+ — 23% radiochem. yield (7)

5% F_2, N_2 — 74% from silyl enol ether (8)

Electrophilic Substitution of F for Other Halogens. While fluoride ion is useful for nucleophilic displacement of other halides, molecular fluorine uses an electrophilic mechanism to achieve this (eq 9).[30] Substitution proceeds most effectively when the intermediate carbocation is stabilized. The reactivity of the starting organic halide increases with the atomic weight of the departing species: Cl < Br < I.

1.5% F_2/N_2; $CHCl_3$, −75 °C; 99% (9)

Addition to Multiple Bonds.

Alkenes. Fluorine often adds smoothly in an electrophilic manner to alkenic double bonds, predominantly in *syn* orientation.[2b] Fluorine addition has even been applied to conversion of alkene impurities in HFC-134a (CF_3CH_2F) to more easily removed saturated products.[31] Many reactions which are formally replacement of vinylic H by F are addition reactions followed by dehydrofluorination (eq 10).[32] Under radical conditions, coupling of intermediate fluorocarbon radicals is often observed. Alkynes react to give tetrafluoro or difluoro products, depending on the conditions.[1b]

F_2, MeCN, −20 °C; Et_3N, CH_2Cl_2, rt — 84% overall (10)

Imines. Carbon–nitrogen double bonds react to form *N*-fluoro derivatives; for already highly fluorinated compounds, this is most often done with the aid of a metal fluoride catalyst (eq 11).[33] Subsequent dehydrofluorination is often observed if the vicinal proton is relatively acidic.[1b]

$$CF_3(CF_2)_5N{=}CF(CF_2)_4CF_3 \xrightarrow[22\ ^\circ C,\ 95\%]{F_2,\ CsF} [CF_3(CF_2)_5]_2NF \quad (11)$$

Carbonyls. Such catalytic pathways are also used to add fluorine across carbonyl groups, forming fluoroxy compounds (hypofluorites) (eq 12).[34] Some hypofluorites have found use as electrophilic fluorinating agents in their own right (see, for example, ***Trifluoromethyl Hypofluorite*** and ***Acetyl Hypofluorite***).

$$CF_3C(O)CF_2Cl \xrightarrow[-196\ \text{to}\ -10\ ^\circ C,\ 80\%]{F_2,\ CsF} CF_3CF(CF_2Cl)OF \quad (12)$$

Oxidative Addition. Fluorine is capable of oxidizing most elements to their highest valence state. In organic compounds, centers of coordinative unsaturation undergo oxidative addition with little fragmentation if conditions are mild. The sulfur(II) in thiols[5a] and sulfides, for instance, is transformed to sulfur(VI) in the form of $-SF_5$ or $-SF_4-$ groups (eq 13).[5b]

F_2/He, 80% (13)

Preparation of Other Fluorinating Agents. Most alternative electrophilic fluorinating agents are themselves prepared from elemental fluorine using variations of the reactions above, with extensions to substitution of H on N or O producing N–F or O–F bonds[6] (see for example, ***N-Fluoro-N-t-butyl-p-toluenesulfonamide***). Some hypofluorites such as $MeOF$[35] and *t*-BuOF[36] can be formed in situ and used to add RO–F across alkenic double bonds. Fluorination of diselenides can be used similarly[37] when conditions are mild enough to avoid oxidation. Oxidative addition is represented by the synthesis of reagents such as ***Xenon(II) Fluoride***, ***N-Fluoropyridinium Triflate***, *N*-fluoroquinuclidinium salts, and *N*-fluoro-1,4-diazabicyclo[2.2.2]octane salts.

Oxidations. Fluorine reacts with aq MeCN to form HOF stabilized by solvent complexation.[38] The complex can act as an efficient oxidizing agent and can be used for ^{18}O labeling.

Tertiary Hydrogen to Hydroxyl. Fluorine can be used selectively to hydroxylate tertiary hydrogens due to the interaction of the highly electrophilic oxygen atom in the HOF/MeCN complex with the relatively electron-rich tertiary H–C bond (eq 14).[7] Experiments with *cis*- and *trans*-decalin show that the hydroxylation occurs with full retention of configuration.

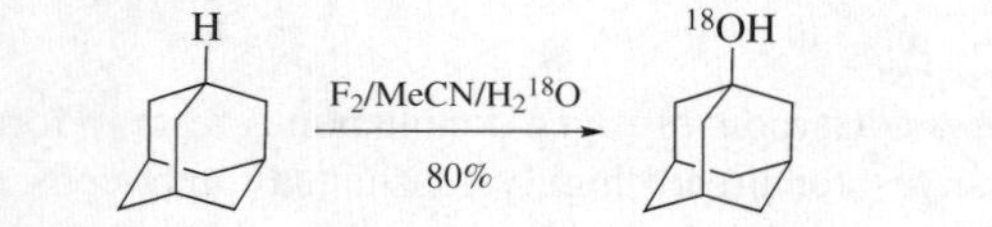
H
F_2/MeCN/$H_2{}^{18}O$
80%
^{18}OH
(14)

Alcohols to Ketones and Ketones to Esters.[8] Oxidation of secondary alcohols with stabilized HOF yields ketones (eq 15); primary alcohols are less reactive. Ketones are converted to esters more slowly and with a larger excess of reagent (eq 16).

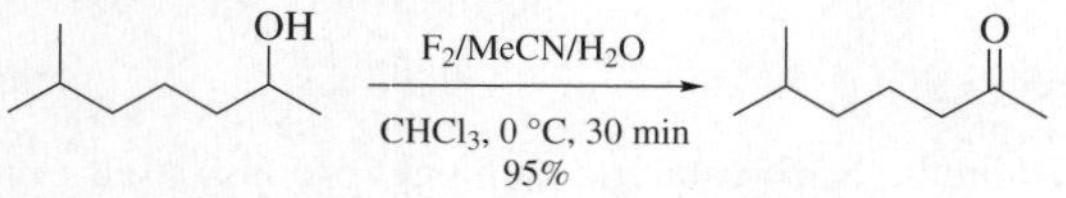
OH
F_2/MeCN/H_2O
$CHCl_3$, 0 °C, 30 min
95%
O
(15)

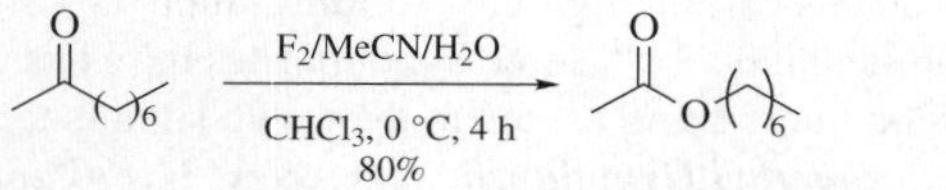
O
F_2/MeCN/H_2O
$CHCl_3$, 0 °C, 4 h
80%
O
(16)

Aromatics to Phenols and Quinones.[9] Aromatics and polynuclear aromatics are oxidized to the respective oxygenated derivatives. The reaction of mesitylene (eq 17) shows that phenols are likely intermediates to quinones. Quinones are quickly prepared in moderate yields (eq 18).

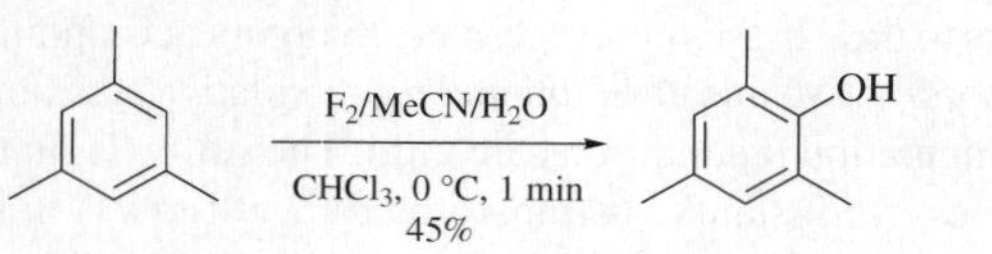
F_2/MeCN/H_2O
$CHCl_3$, 0 °C, 1 min
45%
OH
(17)

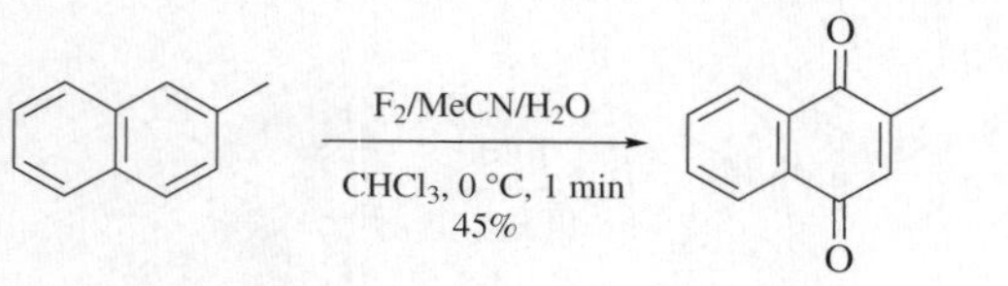
F_2/MeCN/H_2O
$CHCl_3$, 0 °C, 1 min
45%
O
O
(18)

Amines to Nitro Compounds. Primary aromatic[10a] and aliphatic[10b] amines are cleanly converted to nitro compounds without complications from other easily oxidized groups (eq 19).[10a] Amine salts are not reactive.[10b]

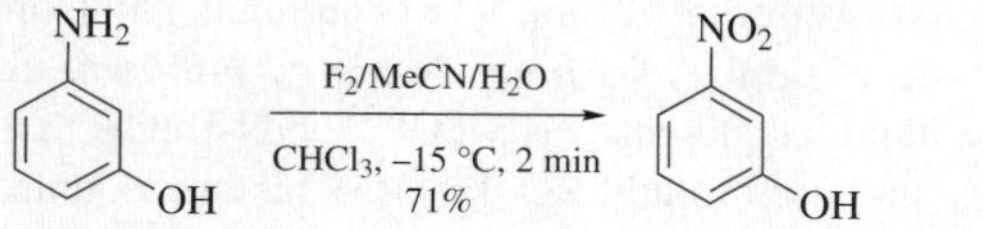
NH_2
OH
F_2/MeCN/H_2O
$CHCl_3$, –15 °C, 2 min
71%
NO_2
OH
(19)

Alkenes to Oxiranes (Epoxides)[7,8,11]. Fluorine/aq MeCN readily epoxidizes electron-rich alkenes (eq 20).[7] More electron-deficient alkenes require a large excess of reagent and extended reaction time (eq 21).[7] Full retention of configuration is characteristic, since the oxides from *cis*- and *trans*-stilbene are exclusively *cis* and *trans*, respectively.[7] The epoxidation occurs preferentially over hydroxyl oxidation, as shown for dihydrocarveol (eq 22).[8] Dienes can form bis-epoxides (eq 23),[11b] or one double bond can react preferentially, given sufficient differences in electron density (eq 24).[11b] Other popular reagents such as ***m-Chloroperbenzoic Acid*** and ***Trifluoroperacetic Acid*** failed to react with n-$C_4F_9CH{=}CH_2$, which was epoxidized by the F_2/MeCN/H_2O system in 63% yield.[11b]

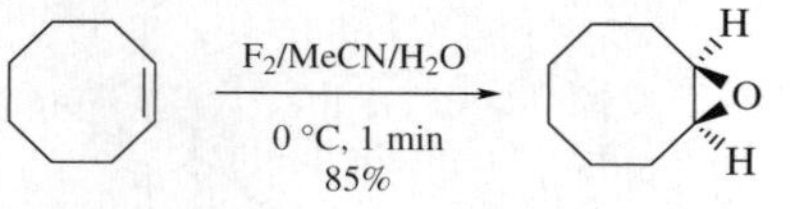
F_2/MeCN/H_2O
0 °C, 1 min
85%
H
O
H
(20)

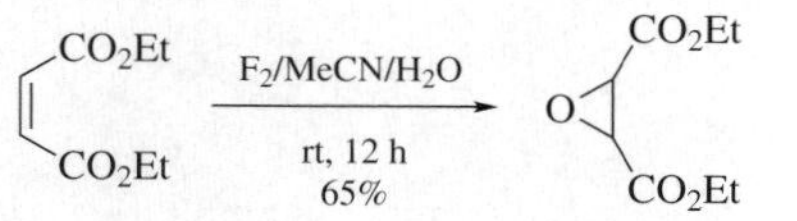
CO_2Et
CO_2Et
F_2/MeCN/H_2O
rt, 12 h
65%
CO_2Et
O
CO_2Et
(21)

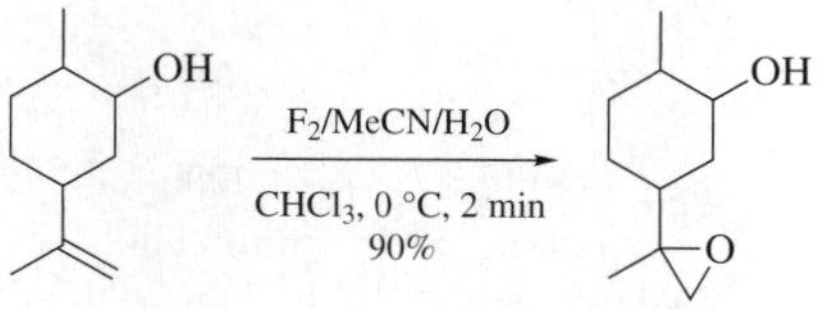
OH
F_2/MeCN/H_2O
$CHCl_3$, 0 °C, 2 min
90%
OH
O
(22)

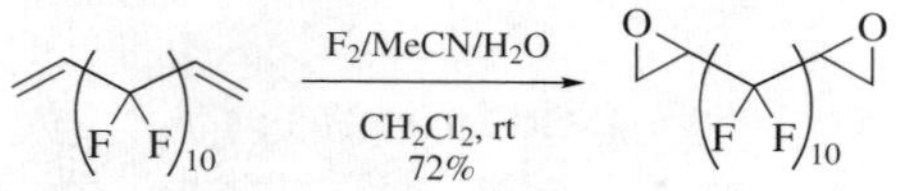
F F
10
F_2/MeCN/H_2O
CH_2Cl_2, rt
72%
O
O
F F
10
(23)

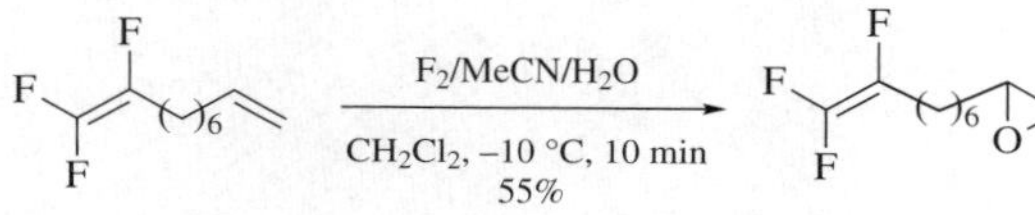
F
F
F
6
F_2/MeCN/H_2O
CH_2Cl_2, –10 °C, 10 min
55%
F
F
F
6
O
(24)

Related Reagents. See Classes O-9, O-14, and O-23, pages 1–10.

---

1. (a) Wilkinson, J. A. *CRV* **1992**, *92*, 505. (b) Purrington, S. T.; Kagen, B. S.; Patrick, T. B. *CRV* **1986**, *86*, 997. (c) Rozen, S. *ACR* **1988**, *21*, 307. (d) *Fluorine: The First Hundred Years (1886–1986)*; Banks, R. E.; Sharp, D. W. A.; Tatlow, J. C., Eds.; Elsevier: New York, 1986.
2. (a) Lagow, R. L.; Margrave, J. L. *Prog. Inorg. Chem.* **1979**, *26*, 161. (b) *Synthetic Fluorine Chemistry*; Olah, G. A.; Chambers, R. D.; Prakash, G. K. S., Eds.; Wiley: New York, 1992.
3. *Fluroine in Bioorganic Chemistry*; Welch, J. T.; Eswarakrishman, S., Eds.; Wiley: New York, 1991.
4. (a) Hudlický, M. *Chemistry of Organic Fluorine Compounds: A Laboratory Manual with Comprehensive Literature Coverage*, 2nd (rev) ed.; Ellis Horwood: New York, 1992. (b) Chambers, R. D. *Fluorine in Organic Chemistry*; Wiley: New York, 1973.
5. (a) Huang, H.-N.; Roesky, H.; Lagow, R. J. *IC* **1991**, *30*, 789. (b) Lin, W. H.; Lagow, R. J. *JFC* **1990**, *50*, 15.
6. *New Fluorinating Agents in Organic Synthesis*; German, L.; Zemskov, S., Eds.; Springer: New York, 1989.
7. *Selective Fluorination in Organic and Bioorganic Chemistry*; Welch, J. T., Ed.; Amercian Chemical Society: Washington, 1991.
8. Rozen, S.; Bareket, Y.; Kol, M. *T* **1993**, *49*, 8169.
9. Kol, M.; Rozen, S. *JOC* **1993**, *58*, 1593.
10. (a) Kol, M.; Rozen, S. *CC* **1991**, 567. (b) Rozen, S.; Kol, M. *JOC* **1992**, *57*, 7342.

11. (a) Hung, M.-H.; Rozen, S.; Feiring, A. E.; Resnick, P. R. *JOC* **1993**, *58*, 972. (b) Hung, M.-H.; Smart, B. E.; Feiring, A. E.; Rozen, S. *JOC* **1991**, *56*, 3187.
12. (a) *Organofluorine Compounds in Medicinal Chemistry and Biomedical Applications*; Filler, F.; Kobayashi, Y.; Yagupolskii, L. M., Eds.; Elsevier: New York, 1993. (b) Fowler, J. S.; Wolf, A. P. *The Synthesis of Carbon-11, Fluorine-18, and Nitrogen-13 Labeled Radiotracers for Biomedical Applications*; Nuclear Science Series, NAS-NS-3201, National Technical Information Center, U.S. Department of Energy: Washington, 1982. (c) Kilbourn, M. R. *Fluorine-18 Labeling of Radiopharmaceuticals*; Nuclear Science Series, NAS-NS-3203, National Academic Press: Washington, 1980.
13. Gambaretto, G. P.; Conte, L.; Napoli, M.; Legnaro, E.; Carlini, F. M. *JFC* **1993**, *60*, 19.
14. (a) Woytek, A. J. In *Kirk-Othmer Encyclopedia of Chemical Technology*, 3rd ed.; Wiley: New York, 1980; Vol. 10, pp 630–654. (b) Christe, K. O. *IC* **1986**, *25*, 3721.
15. Perrin, D. D.; Armarego, W. L. F. *Purification of Laboratory Chemicals*, 3rd ed.; Pergamon: New York, 1988; p 325.
16. (a) Braker, W.; Mossman, A. L. *Matheson Gas Data Book*, 6th ed.; Matheson: Lyndhurst, NJ, 1980; pp 330–335. (b) Compressed Gas Association. *Handbook of Compressed Gases*, 3rd ed.; van Nostrand Reinhold: New York, 1990; pp 352–359. (c) Jaccaud, M.; Faron, R.; Devilliers, D.; Romano, R. In *Ullmann's Encyclopedia of Industrial Chemistry*, 5th ed.; VCH: New York, 1988; Vol. A11, pp 293–305. (d) *Effects of Exposure to Toxic Gases – First Aid and Medical Treatment*, 3rd ed.; Scornavacca, F.; Mossman, A., Eds.; Matheson: Secaucus, NJ, 1988; pp 53–55. (e) L'Air Liquide. *Encyclopedie des Gaz*; Elsevier: New York, 1976; pp 815–822.
17. Adcock, J. L.; Luo, H. *JOC* **1992**, *57*, 2162.
18. Clark, W. D.; Lagow, R. J. *JFC* **1991**, *52*, 37.
19. Scherer, K. V., Jr.; Yamanouchi, K.; Ono, T. *JFC* **1990**, *50*, 47.
20. Maeda, T.; Fujimoto, H.; Yohikawa, M.; Saito, M. *Aromatikkusu* **1992**, *44*, 273 (*CA* **1993**, *119*, 141 596m.
21. Berenbaum, M. B.; Izod, T. P. J.; Taylor, D. R.; Hewes, J. D. (Allied-Signal Inc.) U.S. Patent 5 220 087, 1993 (*CA* **1993**, *119*, 184 566u).
22. (a) Krueger, G. *Adhaes.–Kleben Dichten* **1993**, *37*, 30 (*CA* **1993**, *119*, 226 984m). (b) Tarancon, G. (Liquid Carbonic Inc.) U.S. Patent 5 149 744, 1992 (*CA* **1992**, *117*, 235 125t).
23. Rozen, S.; Gal, C. *JOC* **1987**, *52*, 2769.
24. Lin, W.-H.; Lagow, R. J. *JFC* **1990**, *50*, 345.
25. Purrington, S. T.; Woodard, D. L. *JOC* **1991**, *56*, 142.
26. Bielefeldt, D.; Braden, R. Eur. Patent 499 930, 1992 (*CA* **1992**, *117*, 234 035b).
27. Namavari, M.; Satyamurthy, N.; Phelps, M. E.; Barrio, J. R. *Appl. Radiat. Isot.* **1993**, *44*, 527.
28. Bumgardner, C. L.; Sloop, J. C. *JFC* **1992**, *56*, 141.
29. (a) Tuinman, A. A.; Gakh, A. A.; Adcock, J. L.; Compton, R. N. *JACS* **1993**, *115*, 5885. (b) Selig, H.; Lifshitz, C.; Peres, T.; Fischer, J. E.; McGhie, A. R.; Romanow, W. J.; McCauley, J. P., Jr.; Smith, A. B., III *JACS* **1991**, *113*, 5475.
30. Rozen, S.; Brand, M. *JOC* **1981**, *46*, 733.
31. Guglielmo, G.; Gambaretto, G. (Ausimont S.p.A.) Eur. Patent 548 744, 1993 (*CA* **1993**, *119*, 180 360z).
32. Sato, M.; Kaneko, C.; Iwaoka, T.; Kobayashi, Y.; Iida, T. *CC* **1991**, 699.
33. Petrov, V. A.; DesMarteau, D. D. *IC* **1992**, *31*, 3776.
34. Randolph, B. B.; DesMarteau, D. D. *JFC* **1993**, *64*, 129.
35. Kol, M.; Rozen, S.; Appelman, E. *JACS* **1991**, *113*, 2648.
36. Appelman, E. H.; French, D.; Mishani, E.; Rozen, S. *JACS* **1993**, *115*, 1379.
37. Uneyama, K. (Nippon Zeon K. K.) Jpn. Patent 04 89 472, 1992 (*CA* **1992**, *117*, 69 591v).
38. Appelman, E. H.; Dunkelberg, O.; Kol, M. *JFC* **1992**, *56*, 199.

Stefan P. Kotun
Ohmeda, Murray Hill, NJ, USA

Hexa-μ-hydrohexakis(triphenylphosphine)hexacopper[1]

$[(Ph_3P)CuH]_6$

[33636-93-0] $C_{108}H_{96}Cu_6P_6$ (MW 1961.16)

(chemo- and stereoselective conjugate reduction;[2,3] catalytic, hydride-mediated, reduction of enones and ketones;[4] reduction of alkynes and propargyl alcohols to *cis*-alkenes[5])

Alternate Name: triphenylphosphine copper hydride hexamer.
Physical Data: mp 111 °C (dec).[1b]
Solubility: sol benzene, toluene, THF (~1 g/10 mL); reacts with CH_2Cl_2 and $CHCl_3$.
Form Supplied in: bright red crystals to dark red powders; commercially available.
Analysis of Reagent Purity: assay by means of hydrogen evolution.[1b] ^1H NMR spectroscopy (deaerated benzene-d_6) shows ligated and, if present, free Ph_3P.
Preparative Method: prepared by hydrogenolysis of Ph_3P-stabilized ***Copper(I) t-Butoxide***,[1c,d] which can be generated in situ from *t*-BuONa and ***Copper(I) Chloride***.[1e]
Purification: Ph_3P is removed by recrystallization from benzene layered with hexanes or acetonitrile under anaerobic conditions, or by trituration with deaerated hexanes or acetonitrile.[1e]
Handling, Storage, and Precautions: has an indefinite shelf life if stored under an inert atmosphere; can be handled briefly in the air without harm and is stable toward water. The reagent is highly air sensitive in solution; thus only rigorously deoxygenated solvents should be used. For bench top manipulation, transfer of the bulk reagent to a Schlenk tube that can be easily opened under a purge of inert gas and evacuated and backfilled after each use is recommended.

Stoichiometric Reductions. $[(Ph_3P)CuH]_6$ provides stoichiometric conjugate reduction of α,β-unsaturated ketones,[2,3] esters,[2a,3a] aldehydes,[2c] nitriles,[6a] sulfones, and sulfonates.[3b] The reagent is highly chemoselective. Isolated alkenes, carbonyl groups, halogens, and typical oxygenated functionality are not reduced under the reaction conditions (eq 1).[2,3]

When reducing substrates that are prone to aldol and Michael type reactions, it is advisable to include 5–20 equiv deaerated water in the reaction medium to hydrolyze the intermediate copper enolate. However, reductions run in 'wet' solvents or in the presence of a chlorotrialkylsilane require a slight excess of the reagent to completely consume starting material (0.18–0.5 equiv $[(Ph_3P)CuH]_6$).[2,3]

Initial 1,2-reduction is not competitive and further reaction of the carbonyl product resulting from conjugate reduction is generally only observed with enals. Over-reduction is prevented by running the reactions in the presence of a chlorotrialkylsilane (eq 2).[2c] Both ketone and aldehyde enolates are thus trapped as their corresponding silyl enol ethers. Alternatively, in the presence of water and excess hydride, some enals can be completely reduced to the saturated alcohol.[2c]

MeO, OCOPh, OTBDMS; 0.24 equiv $[(Ph_3P)CuH]_6$, C_6H_6, rt, 2 h, 83% (1)

(E):(Z) = 6:1; 0.32 equiv $[(Ph_3P)CuH]_6$, 2.5 equiv TBDMSCl, C_6H_6, rt, 1 h, 93%; OTBDMS, (E):(Z) = 5:1 (2)

Stoichiometric reductions are performed at rt in wet deaerated benzene or THF at or below the solubility limit of the reagent. Starting material may remain during reduction of sterically demanding substrates, especially at higher dilution (eq 3).[6b] Performing such reductions as concentrated suspensions and/or by adding the reagent dropwise to a heated solution of the substrate can help alleviate this difficulty as well as shorten the reaction time.

OAc; 0.5 equiv $[(Ph_3P)CuH]_6$, 10 equiv H_2O, C_6H_6, reflux, 1 h; 85%, *cis*:*trans* = 11:1; + starting material 7% (3)

Aside from offering broad chemoselectivity and complete regioselectivity, the reagent is highly stereoselective, delivering hydride to the less hindered face of the substrate (eqs 3 and 4).[2a,b,3] The selectivity exhibited towards 3,5-dimethylcyclohexanone (eq 4) is superior to that observed using either Boeckmans' hydridocuprate (9:1)[7a] or catalytic hydrogenation (15:1).[2a] The *cis* selectivity obtained when reducing tetra- and hexahydronaphthalen-6(7*H*)-ones (eq 3) parallels that observed with catalytic hydrogenation[8] and exceeds the selectivity observed when similar substrates are reduced with hydridocuprate reagents.[7] High selectivity for the *cis* A/B ring juncture is also observed when reducing 3-keto-$\Delta^{4,5}$ steroidal enones.[6b,9]

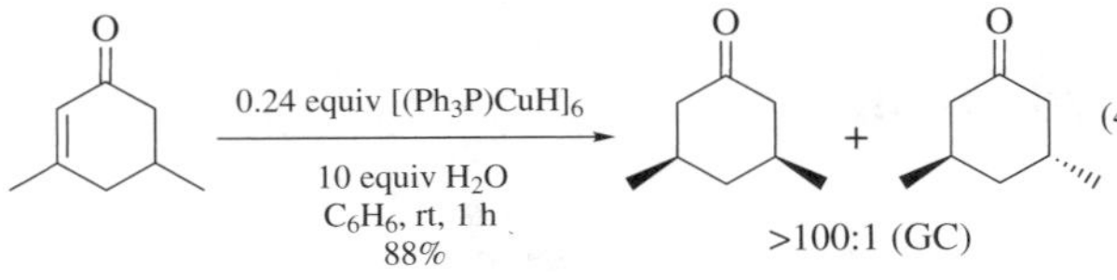

(4)

$[(Ph_3PCuH]_6$ is also compatible with γ-heteroatom-substituted enones (eqs 5 and 6).[2b] Conjugate reduction of such compounds without elimination of the γ-substituent is significant as they can suffer hydrogenolysis during catalytic hydrogenation[10] or elimination during reductions where electron transfer is mechanistically relevant.[11]

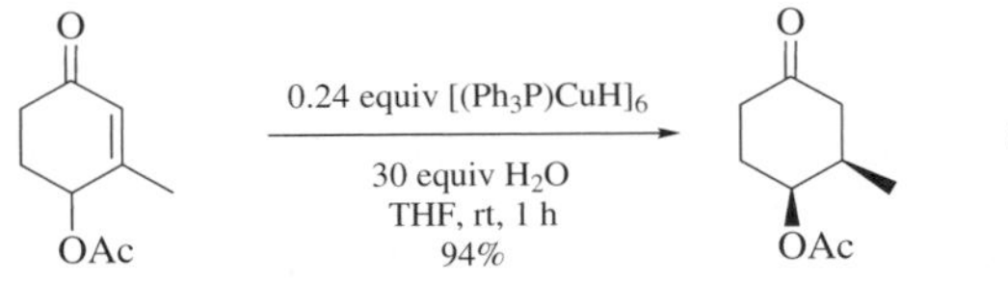

(5)

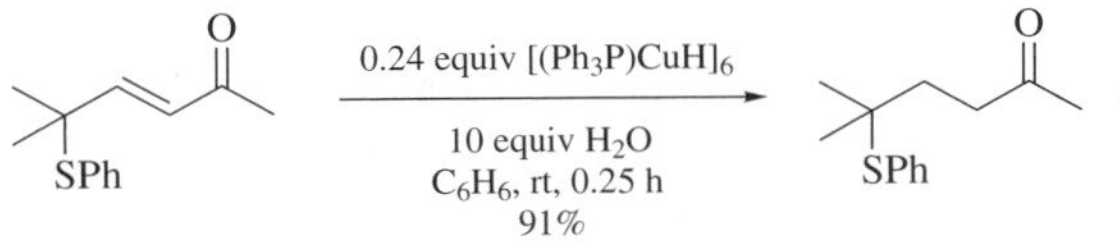

(6)

Conjugate reduction of β-heteroatom substituted enones is not straightforward.[12] However, endocyclic vinylogous esters[6b] and amides[3c] react under chlorotrialkylsilane-mediated conditions (eq 7) suggesting that they are reduced through the oxonium and iminium ion, respectively. All efforts to reduce exocyclic vinylogous esters and amides such as 3-substituted cyclohex-2-en-1-ones were unsuccessful.[6b]

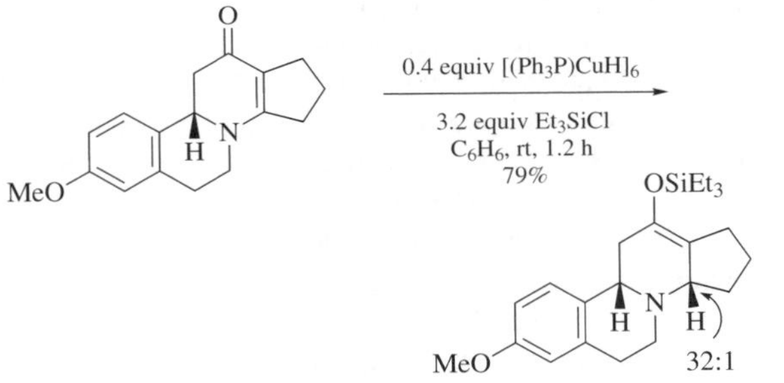

(7)

The workup requires exposure of the reaction to air and dilution with ether and/or hexanes. Addition of a small amount of silica gel facilitates decomposition of the copper containing byproducts. After being stirred under air for at least 1 h, the mixture is filtered through silica gel and purified by column chromatography. Alternatively, free Ph_3P can be removed from products of similar polarity by oxidation with 5% NaOCl. The resulting phosphine oxide is then removed by simple silica gel filtration. Isolation of the silyl enol ethers requires exposure of the reaction to air and dilution as outlined above, with the usual precautions taken when chromatographing these compounds.

Hydrometalation of Alkynes. $[(Ph_3P)CuH]_6$ reduces alkynes (eq 8) and propargyl alcohols (eq 9) to the *cis*-alkene.[5] Complications include overreduction (≤10%) with some substrates and partial fragmentation of unprotected secondary and tertiary propargyl alcohols. As with enone reductions mediated by $[(Ph_3P)CuH]_6$, alkyne reduction is concentration dependent, can be promoted by using higher temperatures, and benefits from the presence of water in the reaction medium.

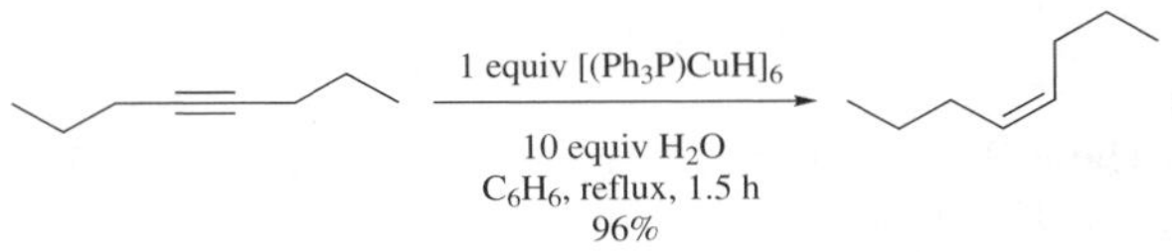

(8)

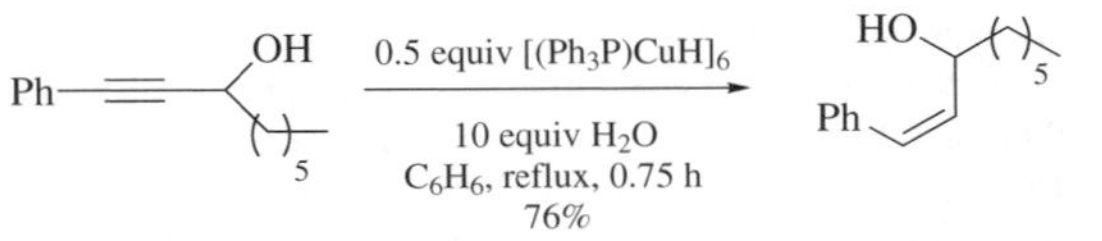

(9)

Catalytic Reductions. The presumed copper(I) enolate formed during conjugate reduction heterolytically activates H_2, allowing the reductions to proceed catalytically in copper.[4] In the presence of excess Ph_3P (4–6 equiv per Cu) and H_2 pressure (200–1000 psi) the catalytic enone reduction can be reasonably controlled, providing either conjugate reduction or complete reduction to saturated alcohol, depending on reaction conditions.[4] While work to provide a synthetically useful catalytic enone reduction continues, a practical catalytic reduction of ketones under one atmosphere of H_2 has been developed (eq 10).[6b] The catalytic reducing system is prepared simply by adding Me_2PPh (6–10 equiv per Cu) to $[(Ph_3P)CuH]_6$ followed by *t*-BuOH (10–20 equiv per Cu) and the reducible substrate (10–100 equiv per Cu) in dry, deaerated benzene (0.4–0.8 M in substrate) under an inert atmosphere. The solution is then subjected to one freeze–pump–thaw–degas cycle and an atmosphere of H_2 is admitted as the solution thaws. Alternatively, the catalyst can be prepared by substituting ***Copper(I) Chloride*** and NaO-*t*-Bu for $[(Ph_3P)CuH]_6$. Me_2PPh is superior to numerous other tertiary phosphines screened. This catalyst is quite robust, allowing the reactions to be monitored by TLC and displaying sustained turnover, as indicated by reduction of fresh substrate added to a completed reaction. The chemoselectivity is comparable to that of $[(Ph_3P)CuH]_6$ with the exception that this catalyst does reduce ketones effectively but reacts with enones to give mixtures of saturated and allylic alcohols. The reduction of cyclic ketones proceeds mainly by axial delivery of hydride, with stereoselectivity exceeding ***Sodium Borohydride*** and ***Lithium Aluminum Hydride*** in many cases (eq 11).[6b,13]

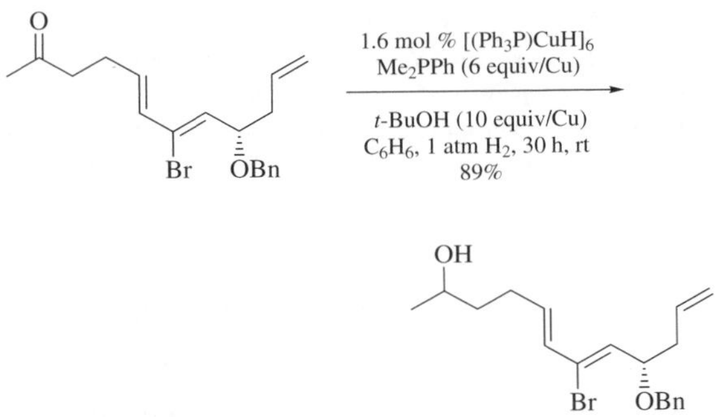

(10)

OAc → OAc (11)

0.8 mol % $[(Ph_3P)CuH]_6$

Me_2PPh (6 equiv/Cu)
t-BuOH (10 equiv/Cu)
C_6H_6, 1 atm H_2, 27 h, rt
93%

HO 16:1

Related Reagents. See Classes R-4, R-10, R-12, and R-15, pages 1–10. Lithium *n*-Butyl(hydrido)cuprate.

1. (a) Bezman, S. A.; Churchill, M. R.; Osborn, J. A.; Wormald, J. *JACS* **1971**, *93*, 2063. (b) Churchill, M. R.; Bezman, S. A.; Osborn, J. A.; Wormald, J. *IC* **1972**, *11*, 1818. (c) Goeden, G. V.; Caulton, K. G. *JACS* **1981**, *103*, 7354. (d) Lemmen, T. H.; Folting, K.; Huffman, J. C.; Caulton, K. G. *JACS* **1985**, *107*, 7774. (e) Brestensky, D. M.; Huseland, D. E.; McGettigan, C.; Stryker, J. M. *TL* **1988**, *29*, 3749.
2. (a) Mahoney, W. S.; Brestensky, D. M.; Stryker, J. M. *JACS* **1988**, *110*, 291, and references therein. (b) Koenig, T. M.; Daeuble, J. F.; Brestensky, D. M.; Stryker, J. M. *TL* **1990**, *31*, 3237. (c) Brestensky, D. M.; Stryker, J. M. *TL* **1989**, *30*, 5677.
3. (a) Ziegler, F. E.; Tung, J. S. *JOC* **1991**, *56*, 6530. (b) Musicki, B.; Widlanski, T. S. *TL* **1991**, *32*, 1267. (c) Meyers, A. I.; Elworthy, T. R. *JOC* **1992**, *57*, 4732. (d) Aicher, T. D.; Buszek, K. R.; Fang, F. G.; Forsyth, C. J.; Jung, S. H.; Kishi, Y.; Matelich, M. C.; Scola, P. M.; Spero, D. M.; Yoon, S. K. *JACS* **1992**, *114*, 3162. (e) Cai, S.; Stroud, M. R.; Hakomori, S.; Toyokuni, T. *JOC* **1992**, *57*, 6693.
4. (a) Mahoney, W. S.; Stryker, J. M. *JACS* **1989**, *111*, 8818. (b) Stryker, J. M.; Mahoney, W. S.; Brestensky, D. M.; Daeuble J. F. In *Catalysis of Organic Reactions*; Pascoe, W. E., Ed.; Dekker: New York, 1992; Vol. 47, pp 29–44. (c) Mahoney, W. S., Ph.D. Dissertation, Indiana University, 1989.
5. (a) Daeuble, J. F.; McGettigan, C.; Stryker, J. M. *TL* **1990**, *31*, 2397, and references therein.
6. (a) Brestensky, D. M., Ph.D. Dissertation, Indiana University, 1992. (b) Daeuble, J. F., Ph.D. Dissertation, Indiana University, 1993.
7. (a) Boeckman, R. K., Jr.; Michalak, R. *JACS* **1974**, *96*, 1623. (b) Review detailing reactivity of other hydridocuprates: Lipshutz, B. H.; Sengupta, S. *OR* **1992**, *41*, 135.
8. House, H. O. *Modern Synthetic Reactions*; Benjamin: Menlo Park, CA, 1972.
9. Hydrogenation of 3-keto-$\Delta^{4,5}$ steroidal enones over Cu^0/Al_2O_3 also gives the *cis* A/B ring juncture, but with lower selectivity: Ravasio, N.; Rossi, M. *JOC* **1991**, *56*, 4329.
10. (a) Rylander, P. N. *Catalytic Hydrogenation in Organic Synthesis*; Academic: New York, 1979; pp 235–250. (b) Rylander, P. N. *Hydrogenation Methods*; Academic: London, 1985; pp 148–183.
11. (a) Ruden, R. A.; Litterer, W. E.; *TL* **1975**, 2043. (b) Smith, R. A. J.; Hannah, D. J. *T* **1979**, *35*, 1183. (c) Nilsson, A.; Ronlá, A. *TL* **1975**, *16*, 1107. (d) Stork, G.; Rosen, P.; Goldman, N.; Coombs, R. V.; Tsuji, J. *JACS* **1965**, *87*, 275. (e) A recently reported hydridocuprate is also compatible with γ-hetero substituted enones: Lipshutz, B. H.; Ung, C. S.; Sengupta, S. *SL* **1989**, 64.
12. (a) Posner, G. H.; Brunelle, D. J. *CC* **1973**, *907*. (b) Nakamura, E.; Matsuzawa, S.; Horiguchi, Y.; Kuwajima, I. *TL* **1986**, *27*, 4029. (c) House, H. O.; Umen, M. J. *JOC* **1973**, *38*, 3893.
13. The octahydronaphthalen-6(7*H*)-one (eq 9) is reduced by $NaBH_4$ with the same facial selectivity for the equatorial alcohol (9:1): Lin, Y. Y.; Jones, J. B. *JOC* **1973**, *38*, 3575.

John F. Daeuble
Indiana University, Bloomington, IN, USA

Jeffrey M. Stryker
University of Alberta, Edmonton, Alberta, Canada

Hydrazine[1]

N_2H_4

(N_2H_4)		
[302-01-2]	H_4N_2	(MW 32.06)
(hydrate)		
[10217-52-4]		
(monohydrate)		
[7803-57-8]	H_5N_2O	(MW 49.07)
(monohydrochloride)		
[2644-70-4]	ClH_5N_2	(MW 68.52)
(dihydrochloride)		
[5341-61-7]	$Cl_2H_6N_2$	(MW 104.98)
(sulfate)		
[10034-93-2]	$H_6N_2O_4S$	(MW 130.15)

(reducing agent used in the conversion of carbonyls to methylene compounds;[1] reduces alkenes,[9] alkynes,[9] and nitro groups;[14] converts α,β-epoxy ketones to allylic alcohols;[32] synthesis of hydrazides;[35] synthesis of dinitrogen containing heterocycles[42–46])

Physical Data: mp 1.4 °C; bp 113.5 °C; *d* 1.021 g cm^{-3}.

Solubility: sol water, ethanol, methanol, propyl and isobutyl alcohols.

Form Supplied in: anhydrate, colorless oil that fumes in air; hydrate and monohydrate, colorless oils; monohydrochloride, dihydrochloride, sulfate, white solids; all widely available.

Analysis of Reagent Purity: titration.[1]

Purification: anhydrous hydrazine can be prepared by treating hydrazine hydrate with BaO, $Ba(OH)_2$, CaO, NaOH, or Na. Treatment with sodamide has been attempted but this yields diimide, NaOH, and ammonia. An excess of sodamide led to an explosion at 70 °C. The hydrate can be treated with boric acid to give the hydrazinium borate, which is dehydrated by heating. Further heating gives diimide.[1]

Handling, Storage, and Precautions: caution must be taken to avoid prolonged exposure to vapors as this can cause serious damage to the eyes and lungs. In cases of skin contact, wash the affected area immediately as burns similar to alkali contact can occur. Standard protective clothing including an ammonia gas mask are recommended. The vapors of hydrazine are flammable (ignition temperature 270 °C in presence of air). There have been reports of hydrazine, in contact with organic material such as wool or rags, burning spontaneously. Metal oxides can also initiate combustion of hydrazine. Hydrazine and its solutions should be stored in glass containers under nitrogen for extended periods. There are no significant precautions for reaction vessel type with hydrazine; however, there have been reports that stainless steel vessels must be checked for significant oxide formation prior to use. Use in a fume hood.

Reductions. The use of hydrazine in the reduction of carbonyl compounds to their corresponding methylene groups via the Wolff–Kishner reduction has been covered extensively in the literature.[1] The procedure involves the reaction of a carbonyl-containing compound with hydrazine at high temperatures in the presence of a base (usually ***Sodium Hydroxide*** or ***Potassium Hydroxide***). The intermediate hydrazone is converted directly to the fully reduced species. A modification of the original conditions was used by Paquette in the synthesis of (±)-isocomene (eq 1).[2]

$H_2NNH_2{\cdot}H_2O$, K_2CO_3, 200 °C; 80% (1)

Unfortunately, the original procedure suffers from the drawback of high temperatures, which makes large-scale runs impractical. The Huang–Minlon modification[3] of this procedure revolutionized the reaction, making it usable on large scales. This procedure involves direct reduction of the carbonyl compound with hydrazine hydrate in the presence of sodium or potassium hydroxide in diethylene glycol. The procedure is widely applicable to a variety of acid-labile substrates but caution must be taken where base-sensitive functionalities are present. This reaction has seen widespread use in the preparation of a variety of compounds. Other modifications[4] have allowed widespread application of this useful transformation. Barton and co-workers further elaborated the Huang–Minlon modifications by using anhydrous hydrazine and ***Sodium*** metal to ensure totally anhydrous conditions. This protocol allowed the reduction of sterically hindered ketones, such as in the deoxygenation of 11-keto steroids (eq 2).[4a] Cram utilized dry DMSO and ***Potassium t-Butoxide*** in the reduction of hydrazones. This procedure is limited in that the hydrazones must be prepared and isolated prior to reduction.[4b] The Henbest modification[4c] involves the utilization of dry toluene and potassium *t*-butoxide. The advantage of this procedure is the low temperatures needed (110 °C) but it suffers from the drawback that, again, preformed hydrazones must be used. Utilizing modified Wolff–Kishner conditions, 2,4-dehydroadamantanone is converted to 8,9-dehydroadamantane (eq 3).[5]

AcO; H_2NNH_2, RONa, Ac_2O; 69%; AcO (2)

(3)

Hindered aldehydes have been reduced using this procedure.[6] This example is particularly noteworthy in that the aldehyde is sterically hindered and resistant to other methods for conversion to the methyl group.[6a] Note also that the acetal survives the manipulation (eq 4). The reaction is equally useful in the reduction of semicarbazones or azines.

OHC; H_2NNH_2, KOH, $(HOCH_2CH_2)_2O$ (4)

In a similar reaction, hydrazine has been shown to desulfurize thioacetals, cyclic and acyclic, to methylene groups (eq 5). The reaction is run in diethylene glycol in the presence of potassium hydroxide, conditions similar to the Huang–Minlon protocol. Yields are generally good (60–95%). In situations where base sensitivity is a concern, the potassium hydroxide may be omitted. Higher temperatures are then required.[7]

R^2, R^1; H_2NNH_2, KOH, diethylene glycol; R^2, R^1 (5)

Hydrazine, via in situ copper(II)-catalyzed conversion to ***Diimide***, is a useful reagent in the reduction of carbon and nitrogen multiple bonds. The reagent is more reactive to symmetrical rather than polar multiple bonds (C=N, C=O, N=O, S=O, etc.)[8] and reviews of diimide reductions are available.[9] The generation of diimide from hydrazine has been well documented and a wide variety of oxidizing agents can be employed: oxygen (air),[10] ***Hydrogen Peroxide***,[10] ***Potassium Ferricyanide***,[11] ***Mercury(II) Oxide***,[11] ***Sodium Periodate***,[12] and hypervalent ***Iodine***[13] have all been reported. The reductions are stereospecific, with addition occurring *cis* on the less sterically hindered face of the substrate.

Other functional groups have been reduced using hydrazine. Nitroarenes are converted to anilines[14] in the presence of a variety of catalysts such as ***Raney Nickel***,[14a,15] platinum,[14a] ruthenium,[14a] ***Palladium on Carbon***,[16] β-iron(III) oxide,[17] and iron(III) chloride with activated carbon.[18] Graphite/hydrazine reduces aliphatic and aromatic nitro compounds in excellent yields.[19] Halonitrobenzenes generally give excellent yields of haloanilines. In experiments where palladium catalysts are used, significant dehalogenation occurs to an extent that this can be considered a general dehalogenation method.[20] Oximes have also been reduced.[21]

Hydrazones. Reaction of hydrazine with aldehydes and ketones is not generally useful due to competing azine formation or competing Wolff–Kishner reduction. Exceptions have been documented. Recommended conditions for hydrazone preparation are to reflux equimolar amounts of the carbonyl component and hydrazine in *n*-butanol.[22,23] A more useful method for simple hydrazone synthesis involves reaction of the carbonyl compound with dimethylhydrazine followed by an exchange reaction with hydrazine.[24] For substrates where an azine is formed, the hydrazone can be prepared by refluxing the azine with anhydrous hydrazine.[25] *gem*-Dibromo compounds have been converted to hydrazones by reaction with hydrazine (eq 6).[26]

Br Br, Br, Br, Br, Br; H_2NNH_2, Et_2O, MeOH; NH_2, N, Br, Br, Br, Br (6)

Hydrazones are useful synthetic intermediates and have been converted to vinyl iodides[27] and vinyl selenides (eq 7) (see also *p-Toluenesulfonylhydrazide*).[28]

RX, base (7)

R = I, PhSe; RX = I_2, PhSeBr
base = pentaalkylguanidine, triethylamine

Diazomalonates have been prepared from dialkyl mesoxylates via the ***Silver(I) Oxide***-catalyzed decomposition of the intermediate hydrazones.[29] Monohydrazones of 1,2-diketones yield ketenes after mercury(II) oxide oxidation followed by heating.[30] Dihydrazones of the same compounds give alkynes under similar conditions.[31]

Wharton Reaction. α,β-Epoxy ketones and aldehydes rearrange in the presence of hydrazine, via the epoxy hydrazone, to give the corresponding allylic alcohols. This reaction has been successful in the steroid field but, due to low yields, has seen limited use as a general synthetic tool. Some general reaction conditions have been set. If the intermediate epoxy hydrazone is isolable, treatment with a strong base (potassium *t*-butoxide or ***Potassium Diisopropylamide***) gives good yields, whereas ***Triethylamine*** can be used with nonisolable epoxy hydrazones (eq 8).[32]

(8)

Some deviations from expected Wharton reaction products have been reported in the literature. Investigators found that in some specific cases, treatment of α,β-epoxy ketones under Wharton conditions gives cyclized allylic alcohols (eq 9). No mechanistic interpretation of these observations has been offered. Related compounds have given the expected products, and it therefore appears this phenomenon is case-specific.[33]

H_2NNH_2 (9)

Cyclic α,β-epoxy ketones have been fragmented upon treatment with hydrazine to give alkynic aldehydes.[34]

Hydrazides. Acyl halides,[35] esters, and amides react with hydrazine to form hydrazides which are themselves useful synthetic intermediates. Treatment of the hydrazide with nitrous acid yields the acyl azide which, upon heating, gives isocyanates (Curtius rearrangement).[36] Di- or trichlorides are obtained upon reaction with ***Phosphorus(V) Chloride***.[37] Crotonate and other esters have been cleaved with hydrazine to liberate the free alcohol (eq 10).[38]

H_2NNH_2 (10)

Hydrazine deacylates amides (Gabriel amine synthesis) via the Ing–Manske protocol.[39] This procedure has its limitations, as shown in the synthesis of penicillins and cephalosporins where it was observed that hydrazine reacts with the azetidinone ring. In this case, ***Sodium Sulfide*** was used.[40]

Heterocycle Synthesis. The reaction of hydrazine with α,β-unsaturated ketones yields pyrazoles.[41,42] Although the products can be isolated as such, they are useful intermediates in the synthesis of cyclopropanes upon pyrolysis of cyclopropyl acetates after treatment with ***Lead(IV) Acetate*** (eq 11).

H_2NNH_2; $Pb(OAc)_4$, heat; 60% (11)

3,5-Diaminopyrazoles were prepared by the addition of hydrazine (eq 12), in refluxing ethanol, to benzylmalononitriles (42–73%).[43] Likewise, hydrazine reacted with 1,1-diacetylcyclopropyl ketones to give β-ethyl-1,2-azole derivatives. The reaction mixture must have a nucleophilic component (usually the solvent, i.e. methanol) to facilitate the opening of the cyclopropane ring. Without this, no identifiable products are obtained (eq 13).[44]

H_2NNH_2, ethanol, reflux, 42–73% (12)

Ar = Ph, 4-MeC_6H_4, 3-$NO_2C_6H_4$

H_2NNH_2; X^- (13)

R^1, R^2 = Me, Ph
X = Cl, Br, OMe, OEt, OPh, OAc, CN, etc.

In an attempt to reduce the nitro group of nitroimidazoles, an unexpected triazole product was obtained in 66% yield. The suggested mechanism involves addition of the hydrazine to the ring, followed by fragmentation and recombination to give the observed product (eq 14).[45]

Finally, hydrazine dihydrochloride reacted with 2-alkoxynaphthaldehydes to give a product which resulted from an intramolecular $[3^+ + 2]$ criss-cross cycloaddition (42–87%) (eq 15).[46]

Peptide Synthesis. Treatment of acyl hydrazides with nitrous acid leads to the formation of acid azides which react with amines to form amides in good yield. This procedure has been used in peptide synthesis, but is largely superseded by coupling reagents such as ***1,3-Dicyclohexylcarbodiimide***.[47]

Related Reagents. See Classes R-3, R-4, R-29, R-30, and R-32, pages 1–10.

1. (a) Todd, D. *OR* **1948**, *4*, 378. (b) Szmant, H. H. *AG(E)* **1968**, *7*, 120. (c) Reusch, W. *Reduction*; Dekker: New York, 1968, pp 171–185. (d) Clark, C. *Hydrazine*; Mathieson Chemical Corp.: Baltimore, MD, 1953.
2. Paquette, L. A.; Han, Y. K. *JOC* **1979**, *44*, 4014.
3. (a) Huang-Minlon *JACS* **1946**, *68*, 2487; **1949**, *71*, 3301. (b) Durham, L. J.; McLeod, D. J.; Cason, J. *OSC* **1963**, *4*, 510. (c) Hunig, S.; Lucke, E.; Brenninger, W. *OS* **1963**, *43*, 34.
4. (a) Barton, D. H. R.; Ives, D. A. J.; Thomas, B. R. *JCS* **1955**, 2056. (b) Cram, D. J.; Sahyun, M. R. V.; Knox, G. R. *JACS* **1962**, *90*, 7287. (c) Grundon, M. F.; Henbest, H. B.; Scott, M. D. *JCS* **1963**, 1855. (d) Moffett, R. B.; Hunter, J. H. *JACS* **1951**, *73*, 1973. (e) Nagata, W.; Itazaki, H. *CI(L)* **1964**, 1194.
5. Murray, R. K., Jr.; Babiak, K. A. *JOC* **1973**, *38*, 2556.
6. (a) Zalkow, L. H.; Girotra, N. N. *JOC* **1964**, *29*, 1299. (b) Aquila, H. *Ann. Chim.* **1968**, *721*, 117.
7. van Tamelen, E. E.; Dewey, R. S.; Lease, M. F.; Pirkle, W. H. *JACS* **1961**, *83*, 4302.
8. Georgian, V.; Harrisson, R.; Gubisch, N. *JACS* **1959**, *81*, 5834.
9. (a) Miller, C. E. *J. Chem. Educ.* **1965**, *42*, 254. (b) Hunig, S.; Muller, H. R.; Thier, W. *AG(E)* **1965**, *4*, 271. (c) Hammersma, J. W.; Snyder, E. I. *JOC* **1965**, *30*, 3985.
10. Buyle, R.; Van Overstraeten, A. *CI(L)* **1964**, 839.
11. Ohno, M.; Okamoto, M. *TL* **1964**, 2423.
12. Hoffman, J. M., Jr.; Schlessinger, R. H. *CC* **1971**, 1245.
13. Moriarty, R. M.; Vaid, R. K.; Duncan, M. P. *SC* **1987**, *17*, 703.
14. (a) Furst, A.; Berlo, R. C.; Hooton, S. *CRV* **1965**, *65*, 51. (b) Miyata, T.; Ishino, Y.; Hirashima, T. *S* **1978**, 834.
15. Ayynger, N. R.; Lugada, A. C.; Nikrad, P. V.; Sharma, V. K. *S* **1981**, 640.
16. (a) Pietra, S. *AC(R)* **1955**, *45*, 850. (b) Rondestvedt, C. S., Jr.; Johnson, T. A. *Chem. Eng. News* **1977**, 38. (c) Bavin, P. M. G. *OS* **1960**, *40*, 5.
17. Weiser, H. B.; Milligan, W. O.; Cook, E. L. *Inorg. Synth.* **1946**, 215.
18. Hirashima, T.; Manabe, O. *CL* **1975**, 259.
19. Han, B. H.; Shin, D. H.; Cho, S. Y. *TL* **1985**, *26*, 6233.
20. Mosby, W. L. *CI(L)* **1959**, 1348.
21. Lloyd, D.; McDougall, R. H.; Wasson, F. I. *JCS* **1965**, 822.
22. Schonberg, A.; Fateen, A. E. K.; Sammour, A. E. M. A. *JACS* **1957**, *79*, 6020.
23. Baltzly, R.; Mehta, N. B.; Russell, P. B.; Brooks, R. E.; Grivsky, E. M.; Steinberg, A. M. *JOC* **1961**, *26*, 3669.
24. Newkome, G. R.; Fishel, D. L. *JOC* **1966**, *31*, 677.
25. Day, A. C.; Whiting, M. C. *OS* **1970**, *50*, 3.
26. McBee, E. T.; Sienkowski, K. J. *JOC* **1973**, *38*, 1340.
27. Barton, D. H. R.; Basiardes, G.; Fourrey, J.-L. *TL* **1983**, *24*, 1605.
28. Barton, D. H. R.; Basiardes, G.; Fourrey, J.-L. *TL* **1984**, *25*, 1287.
29. Ciganek, E. *JOC* **1965**, *30*, 4366.
30. (a) Nenitzescu, C. D.; Solomonica, E. *OSC* **1943**, *2*, 496. (b) Smith, L. I.; Hoehn, H. H. *OSC* **1955**, *3*, 356.
31. Cope, A. C.; Smith, D. S.; Cotter, R. J. *OSC* **1963**, *4*, 377.
32. Dupuy, C.; Luche, J. L. *T* **1989**, *45*, 3437.
33. (a) Ohloff, G.; Unde, G. *HCA* **1970**, *53*, 531. (b) Schulte-Elte, K. N.; Rautenstrauch, V.; Ohloff, G. *HCA* **1971**, *54*, 1805. (c) Stork, G.; Williard, P. G. *JACS* **1977**, *99*, 7067.
34. (a) Felix, D.; Wintner, C.; Eschenmoser, A. *OS* **1976**, *55*, 52. (b) Felix, D.; Muller, R. K.; Joos, R.; Schreiber, J.; Eschenmoser, A. *HCA* **1972**, *55*, 1276.
35. ans Stoye, P. In *The Chemistry of Amides (The Chemistry of Functional Groups)*; Zabicky, J., Ed.; Interscience: New York, 1970; pp 515–600.
36. (a) *The Chemistry of the Azido Group*; Interscience: New York, 1971. (b) Pfister, J. R.; Wymann, W. E. *S* **1983**, 38.
37. (a) Mikhailov, Matyushecheva, Derkach, Yagupol'skii *ZOR* **1970**, *6*, 147. (b) Mikhailov, Matyushecheva, Yagupol'skii *ZOR* **1973**, *9*, 1847.
38. Arentzen, R.; Reese, C. B. *CC* **1977**, 270.
39. Ing, H. R.; Manske, R. H. F. *JCS* **1926**, 2348.
40. Kukolja, S.; Lammert, S. R. *JACS* **1975**, *97*, 5582 and 5583.
41. Freeman, J. P. *JOC* **1964**, *29*, 1379.
42. Reimlinger, H.; Vandewalle, J. J. M. *ANY* **1968**, *720*, 117.
43. Vequero, J. J.; Fuentes, L.; Del Castillo, J. C.; Pérez, M. I.; Garcia, J. L.; Soto, J. L. *S* **1987**, 33.
44. Kefirov, N. S.; Kozhushkov, S. I.; Kuzetsova, T. S. *T* **1986**, *42*, 709.
45. Goldman, P.; Ramos, S. M.; Wuest, J. D. *JOC* **1984**, *49*, 932.
46. Shimizu, T.; Hayashi, Y.; Miki, M.; Teramura, K. *JOC* **1987**, *52*, 2277.
47. Bodanszky, M. *The Principles of Peptide Synthesis*; Springer: New York, 1984; p 16.

Brian A. Roden
Abbott Laboratories, North Chicago, IL, USA

Hydrogen Peroxide[1]

HOOH

[7722-84-1] H_2O_2 (MW 34.02)

(nucleophilic reagent capable of effecting substitution reactions[2] and epoxidation of electron-deficient alkenes;[3] weak electrophile whose activity is enhanced in combination with transition metal oxides[4] and Lewis acids;[5] strong nonpolluting oxidant which can oxidize hydrogen halides[6])

Physical Data: 95% H_2O_2: mp −0.41 °C; bp 150.2 °C; *d* 1.4425 g cm^{-3} (at 25 °C). 90% H_2O_2: mp −11.5 °C; bp 141.3 °C; *d* 1.3867 g cm^{-3}. 30% H_2O_2: mp −25.7 °C; bp 106.2 °C; *d* 1.108 g cm^{-3}.

Solubility: sol ethanol, methanol, 1,4-dioxane, acetonitrile, THF, acetic acid.

Form Supplied in: clear colorless liquid widely available as a 30% aqueous solution and 50% aqueous solution; 70% and 90% H_2O_2 are not widely available.

Analysis of Reagent Purity: titration with $KMnO_4$ or cerium(IV) sulfate.[7]

Purification: 95% H_2O_2 (caution!) can be prepared from 50% solution by distilling off water in a vacuum at rt.[8]

Handling, Storage, and Precautions: H_2O_2 having a concentration of 50% or more is very hazardous and can explode violently, particularly in the presence of certain inorganic salts and easily oxidizable organic material. A safety shield should be used when handling this reagent.[9] After the reaction is complete, excess H_2O_2 should be destroyed by treatment with MnO_2 or Na_2SO_3 soln. Before solvent evaporation, ensure absence of peroxides. The use of acetone as solvent should be avoided.[10] The reagent should be stored in aluminum drums in a cool place away from oxidizable substances.

Synthesis of Peroxides via Perhydrolysis. H_2O_2 and the hydroperoxy anion are excellent nucleophiles which react with alkyl halides and other substrates having good leaving groups to furnish hydroperoxides. The hydroperoxide (**2**) has been prepared employing 98% H_2O_2 (eq 1).[11] To a stirred mixture of THF (50 mL), ***Silver(I) Trifluoromethanesulfonate*** (0.04 mol), and pyridine (0.02 mol) kept at 6 °C under argon and protected from light is gradually added 98% H_2O_2 (0.32 mol). The chloride (**1**) (0.02 mol) dissolved in THF (10 mL) is next added dropwise with cooling (6 °C). The reaction mixture is kept at rt for 24 h; the organic layer is separated by gravity filtration, diluted with ether, and washed with saturated aq $NaHCO_3$ at 0 °C. The organic layer is dried. The solvent as well as traces of pyridine and starting material are distilled out at rt under vacuum. The residual material is the hydroperoxide (**2**) which has been distilled in high vacuum using a bath maintained at 40 °C. The hydroperoxide (**3**) has been prepared in a similar fashion employing 30% H_2O_2 (eq 2).[12]

Tertiary alcohols $R^1R^2R^3COH$ and other alcohols which can readily furnish carbenium ion intermediates are solvolyzed by 90% H_2O_2 in the presence of acid catalysts to yield hydroperoxides $R^1R^2R^3COOH$.[13] Trimeric hydroperoxides having a nine-membered oxa heterocyclic ring have been prepared from ketones and hydrogen peroxide in the presence of acid catalysts.[14]

N-Alkyl-*N*′-tosyl hydrazides are oxidized by H_2O_2 and Na_2O_2 in THF at rt to the corresponding hydroperoxides; by employing this procedure, cyclohexyl hydroperoxide has been obtained in 92% yield.[15]

Several *gem* hydroperoxides have been prepared from acetals (eq 3).[16]

(1) → (2): H_2O_2, $AgOSO_2CF_3$, py, THF, rt, 24 h; 83% (1)

t-Bu–CH=CH–CH$_2$Br → t-Bu–CH=CH–CH$_2$OOH (3): 30% H_2O_2, MeOH, KOH, rt, 48 h; 42% (2)

30% H_2O_2, THF, H_2WO_4, 0 °C, 48 h; 86% (3)

The prostaglandin PGG_2 (**5**) has been synthesized from the dibromide (**4**) (eq 4).[17]

(4) → (5): 90% H_2O_2, Ag^+; 15–20% (4)

Perhydrolysis of acid anhydrides furnishes the corresponding peroxy acids (for an example, see ***Trifluoroperacetic Acid***). Perhydrolysis of acid chlorides also furnishes peroxy acids.[18] When an organic acid is mixed with H_2O_2 an equilibrium reaction is established, as shown in eq 5.[18] ***Methanesulfonic Acid*** has been used to accelerate the reaction and also to function as solvent (see preparation of ***Perbenzoic Acid***).

$$RCO_2H + H_2O_2 \overset{H^+}{\rightleftharpoons} RCO_3H + H_2O \quad (5)$$

A number of diacyl peroxides have been prepared in 90–95% yield by reacting the acid chloride (for example, phenylacetyl chloride) (1 equiv) with 30% H_2O_2 (0.55 equiv) in ether in the presence of pyridine (2 equiv) at 0 °C for 2 h.[19]

Reactions with Amides, Aldehydes, and Ketones. The oxazolidinone (**6**) is deacylated regioselectively on treatment with ***Lithium Hydroperoxide*** (eq 6).[20] For another example, see Evans.[21]

(6) $\xrightarrow[\text{100\%}]{\text{30\% } H_2O_2\text{, LiOH, THF, } H_2O\text{, rt, 1 h}}$ (6)

Aromatic aldehydes can be transformed to phenols by oxidizing with H_2O_2 in acidic methanol (eq 7).[22] Dilute alkaline H_2O_2 can convert only aldehydes having an hydroxyl in the *ortho* or *para* position to the corresponding phenols (Dakin reaction).[1b] *m*-CPBA is not useful for the preparation of phenol (**8**) from (**7**).[22]

(7) $\xrightarrow[\text{80\%}]{\text{31\% } H_2O_2\text{, MeOH, } KHSO_4\text{, rt, 4 h}}$ (8) (7)

Alkyl and aryl aldehydes are oxidized to the corresponding carboxylic acids in high yields via oxidation with H_2O_2 in the presence of ***Benzeneseleninic Acid*** as catalyst.[23] Cyclobutanones and other strained ketones undergo Baeyer–Villiger oxidation with H_2O_2. The cyclobutanone (**9**) has thus been oxidized to the γ-lactone (**10**) (eq 8).[24] Baeyer–Villiger oxidation of some cyclobutanones proceeds under very mild conditions (−78 °C).[25] Baeyer–Villiger reaction of ketones having isolated double bonds can be carried out with H_2O_2 without reaction at the double bond; however, when organic peroxy acids are used, the alkene often is oxidized.[26]

(9) $\xrightarrow[>90\%]{\text{30\% } H_2O_2\text{, glacial AcOH, 5–10 °C, 16 h}}$ (10) (8)

Epoxidation of α,β-Unsaturated Ketones and Acids. α,β-Unsaturated ketones furnish the corresponding α,β-epoxy ketones in high yields on treatment with H_2O_2 in the presence of a base.[3] In the cyclopentenone (**11**), approach to the β-face is sterically hindered. Epoxidation of (**11**) at −40 °C furnishes quantitatively a 94:6 mixture of α- and β-epoxides; the selectivity is less when the reaction is carried out at higher temperatures (eq 9).[27] Optically active epoxy ketones (about 99%) have been prepared with high ee by carrying out the epoxidation in the presence of a chiral catalyst such as polymer-supported poly(L-leucine).[28]

(11) $\xrightarrow[\text{−45 °C, 12 h}]{\text{30\% } H_2O_2\text{, MeOH, NaOH}}$ α:β = 94:6 (9)

α,β-Unsaturated acids have been epoxidized with 35% H_2O_2 using a catalyst prepared from 12-tungstophosphoric acid (WPA) and cetylpyridinium chloride (CPC) (pH 6–7, 60–65 °C); by this method, crotonic acid furnishes the α,β-epoxy acid in 90% yield.[29]

Synthesis of Epoxides, Vicinal Diols, Dichlorides, and Ketones from Alkenes. Terminal alkenes, as well as di- and trisubstituted alkenes, have been epoxidized at 25 °C using a molybdenum blue–***Bis(tri-n-butyltin) Oxide*** catalyst system (eq 10).[30] Epoxides have been prepared with 16% H_2O_2 using a (diperoxotungsto)phosphate catalyst (**12**) in a biphasic system.[31]

$\xrightarrow[\text{88\%}]{\text{60\% } H_2O_2\text{, Mo cat., } CHCl_3\text{, 3 h}}$ (10)

$[(C_8H_{17})_3NMe]_3^+[PO_4\{W(O)(O_2)_2\}_4]^{3-}$

(12)

Asymmetric epoxidation of 1,2-dihydronaphthalene has been achieved employing a chiral manganese(III) salen complex with an axial N-donor; even 1% H_2O_2 can be used as oxidant and the highest ee observed was 64%.[32]

Vicinal diols have been prepared from alkenes by oxidizing with H_2O_2 in the presence of Re_2O_7 catalyst, in dioxane at 90 °C for 16 h; the mole ratio of Re_2O_7:alkene:H_2O_2 is 1:100:120. The reaction proceeds via epoxidation followed by acid-catalyzed ring opening. Cyclohexene furnishes *trans*-cyclohexane-1,2-diol in 74% yield.[33]

Oxidative cleavage of ene–lactams takes place during oxidation with H_2O_2 in the presence of a selenium catalyst (eq 11).[34] The reaction proceeds under neutral and mild conditions. For the preparation of macrocyclic ketoimides, ***Palladium(II) Acetate*** is used as the catalyst.[34]

$\xrightarrow[\text{rt, 2 h, 82\%}]{\text{30\% aq } H_2O_2\text{, } SeO_2\text{, } CH_2Cl_2}$ (11)

Alkenes have been chlorinated with concentrated HCl/30% H_2O_2/CCl_4 in the presence of the phase-transfer catalyst ***Benzyltriethylammonium Chloride***. Side reactions take place when gaseous chlorine and sulfuryl chloride react with alkenes; under ionic conditions these side reactions are not favored. The method has also been applied for the bromination of alkenes.[6] 1-Octene furnishes 1,2-dichlorooctane in 56% yield.

Oxidation of Alcohols and Phenols. The system $H_2O_2/RuCl_3\cdot3H_2O$/phase-transfer catalyst (didecyldimethylammonium bromide) oxidizes a variety of alcohols selectively; the requirement of ruthenium is very low; ratio of substrate:$RuCl_3$ = 625:1.[35] By this method, *p*-methylbenzyl alcohol was oxidized to *p*-methylbenzaldehyde in 100% yield.

Vicinal diols are oxidized to α-hydroxy ketones by 35% H_2O_2 in the presence of peroxotungstophosphate (PCWP; 1.6 mol %) in a biphasic system using $CHCl_3$ as solvent. 1,2-Hexanediol has been oxidized in 93% yield to 1-hydroxy-2-hexanone.[36]

When 1,4-dihydroxybenzenes are reacted with stoichiometric quantities of iodine, the corresponding *p*-benzoquinones are formed in poor yields; however, they are oxidized in very good yields to *p*-quinones by reaction with 60% H_2O_2 in methanol or aq solution at rt in the presence of catalytic quantities of I_2 or HI. 2-Methyl-1,4-dihydroxynaphthalene has been oxidized to 2-methyl-1,4-naphthoquinone in 98% yield.[37]

Radical Reactions. Homolytic substitutions of pyrrole, indole, and some pyrrole derivatives have been carried out using electrophilic carbon centered radicals generated in DMSO by Fe^{2+}/H_2O_2 and ethyl iodoacetate or related iodo compounds; the substrate is taken in large excess (eq 12).[38]

35% H_2O_2, ICH_2CO_2Et, $FeSO_4{\cdot}7H_2O$, DMSO; 90% (12)

N-Acylpyrrolidines and -piperidines are oxidized by Fe^{II}/hydrogen peroxide in aqueous 95% acetonitrile to the corresponding pyrrolidin-2-ones and piperidin-2-ones;[39] *N*-phenylcarbamoyl-2-phenylpiperidine was oxidized to the corresponding lactam in 61% yield.

Oxidation of Organoboranes. Oxidative cleavage of the C–B bond with alkaline H_2O_2 to convert organoboranes to alcohols is a standard step in hydroboration reactions. In some procedures, organoboranes are formed in the presence of 1,4-oxathiane. When a mixture of tri-*n*-octylborane and 1,4-oxathiane in THF was treated initially with NaOH and subsequently with 30% H_2O_2, the organoborane was selectively oxidized to furnish in 98% yield a mixture (93:7) of octan-1-ol and octan-2-ol.[40]

Oxidation of Organosilicon Compounds. Organosilicon compounds having at least one heteroatom on silicon undergo oxidative cleavage of the Si–C bond when treated with H_2O_2 (eq 13).[41] For additional examples, see Roush[42a] and Andrey.[42b]

30% H_2O_2, $KHCO_3$, KF, MeOH, THF; rt, 2 h; 77% (13)

Oxidation of Amines. H_2O_2 in the presence of Na_2WO_4 has been used to oxidize (a) 2,4,4-trimethyl-2-pentanamine to the corresponding nitroso compound in 52% yield,[43] and (b) a primary amine (containing β-lactam and phenolic OH) to the corresponding oxime in 72% yield.[44]

The secondary amine 2-methylpiperidine (**13**) has been oxidized to the nitrone (**14**) with H_2O_2/Na_2WO_4 (eq 14);[45] the oxidation product also contains about 6–15% of the isomeric 2-methyl-2,3,4,5-tetrahydropyridine *N*-oxide (***Selenium(IV) Oxide*** is also an effective catalyst for this oxidation).[46] 1,2,3,4-Tetrahydroquinoline is oxidized to the 1-hydroxy-3,4-dihydroquinolin-2(1*H*)-one in 84% yield by H_2O_2/Na_2WO_4.[47] The flavin, $FlEt^+ClO_4^-$ (**15**) is a good catalyst for the H_2O_2 oxidation of secondary amines to nitrones.[48]

30% aq H_2O_2, $Na_2WO_4{\cdot}2H_2O$, H_2O, add. 30 min; rt, 3 h; 62–70% (14)

(**13**) (**14**)

(**15**) ClO_4^-

(**16**) *trans*:*cis* = 95:5

The tertiary amine *N*-methylmorpholine has been oxidized to the *N*-oxide in 84–89% yield; the reaction is carried out at 75 °C with 30% H_2O_2 and the reaction time (0.3 mol scale) is about 24 h.[49] The *trans*-*N*-oxide (**16**) has been obtained stereoselectively (*trans*:*cis* = 95:5) by reacting the corresponding *N*-methylpiperidine with 30% H_2O_2 in acetone at 25 °C.[50]

Oxidation of Sulfur-Containing Compounds. Oxidation of di-*n*-butyl sulfide with H_2O_2 in the presence of the catalyst $FlEt^+ClO_4^-$ (**15**) furnished the corresponding sulfoxide in 99% yield.[48] Sulfides have been oxidized to the corresponding sulfoxides with H_2O_2 in CH_2Cl_2 solution in the presence of the heterocycle (**17**); di-*n*-octyl sulfide yields *n*-octyl sulfoxide in 96% yield, and benzylpenicillin methyl ester is oxidized to the (*S*)-*S*-oxide in 90% yield.[51]

(**17**)

The oxidation of sulfides to sulfones proceeds in good yields when the reaction is catalyzed by tungstic acid; the cyclic sulfide thietane is oxidized to the sulfone (thietane 1,1-dioxide) in 89–94% yield.[52]

Oxidation of Selenium-Containing Compounds. Oxidation of the phenyl selenide (**18**) with H_2O_2 in THF furnishes the alkene (**19**) (eq 15);[53] the selenoxide initially formed through oxidation of (**18**) undergoes facile *syn* elimination (see also Grieco[54]).

30% H_2O_2, THF, 25 °C, 12 h; 90% (15)

(**18**) (**19**)

Hydrogen peroxide has a high (47%) active oxygen content and low molecular weight. It is cheap and is widely available. After delivering oxygen, the byproduct formed in H_2O_2 oxidations is the nonpolluting water. Hence the use of H_2O_2 in industry is highly favored. This reagent is able to oxidize SeO_2, WO_3, MoO_3, and several other inorganic oxides efficiently to the corresponding inorganic peroxy acids which are the actual oxidizing agents in many reactions described above.[4] Use of these oxides in catalytic amounts along with H_2O_2 as the primary oxidant reduces the cost of production, simplifies workup and minimizes the effluent disposal problem. Phase-transfer-catalyzed (PTC) reactions in a two-phase system are well suited for H_2O_2 oxidations and are widely used; epoxides are susceptible to ring opening by water and the PTC procedure allows the preparation of epoxides even with 16% aq H_2O_2 since the epoxide and water are in different phases.[31] Handling chlorine and bromine poses many problems, but HCl/H_2O_2 and HBr/H_2O_2 systems may be used as substitutes for chlorine and bromine, respectively.[6] The solids ***Sodium Perborate***, sodium percarbonate, and ***Hydrogen Peroxide–Urea***, which are prepared from H_2O_2, have wide applications since they release H_2O_2 readily.

Reactions with Nitriles. Treatment of nitriles (**20**) with $NaOH/H_2O_2$ in aqueous ethanol is a standard synthetic procedure for the preparation of amides (**21**); aromatic nitriles furnish amides in high yields but aliphatic nitriles give amides in moderate yields (50–60%).[55] It has been suggested[56] that addition of the hydroperoxy anion to the nitrile (**20**) furnishes the peroxycarboximidic acid (**22**) which reacts with H_2O_2 to give the amide (**21**) and molecular oxygen.

R—≡N (**20**) R–C(=O)NH₂ (**21**) R–C(=NH)OOH (**22**)

It has been observed[57] that in the reaction of nitriles with 30% H_2O_2 in the presence of 20% NaOH there is a significant increase in the reaction rate when *n*-tetrabutylammonium hydrogen sulfate (20 mol %) is used as phase-transfer catalyst. The reaction is carried out at 25 °C for 1–2 h employing CH_2Cl_2; aromatic as well as aliphatic amides are obtained in high yields (e.g. eq 16). This method cannot be used if the nitrile has an electron-withdrawing substituent on the carbon atom α to the cyano group.[57]

2-methylbenzonitrile (CN) —1.6 h, 97%→ 2-methylbenzamide (CONH₂) (16)

Treating a DMSO solution of a nitrile with an excess of 30% H_2O_2 in the presence of a catalytic amount of K_2CO_3 for 1–30 min at 25 °C furnishes the corresponding amide in high yields[58] (e.g. eq 17). Under these conditions, esters, amides, and urethanes do not react. α,β-Unsaturated nitriles furnish α,β-epoxy amides.[58] For other routes for the synthesis of amides from nitriles, see Cacchi[57] and Katritzky.[58]

Cl–C₆H₄–CN —30% H_2O_2, DMSO, K_2CO_3, 20 °C, 5 min; 85%→ Cl–C₆H₄–CONH₂ (17)

Related Reagents. See Classes O-11, O-14, O-15, and O-16, pages 1–10. Hydrogen Peroxide–Ammonium Heptamolybdate; Hydrogen Peroxide–Boron Trifluoride; Hydrogen Peroxide–Iron(II) Sulfate; Hydrogen Peroxide–Tellurium Dioxide; Hydrogen Peroxide–Tungstic Acid; Hydrogen Peroxide–Urea; Iron(III) Acetylacetonate–Hydrogen Peroxide; Perbenzoic Acid; Peroxyacetimidic Acid; Trifluoroperacetic Acid.

---

1. (a) *Kirk-Othmer Encyclopedia of Chemical Technology*; Wiley: New York, 1978; Vol. 3, p 944; Vol. 13, p 12; Vol. 2, p 264. (b) Fieser, L. F.; Fieser, M. *FF* **1967**, *1*, 456.
2. Swern, D. In *Comprehensive Organic Chemistry*; Barton, D. H. R., Ed.; Pergamon: Oxford, 1979; Vol. 1. pp 909–939.
3. Weitz, E.; Scheffer, A. *CB* **1921**, *54*, 2327.
4. Mimoun, H. In *Comprehensive Coordination Chemistry*; Wilkinson, G., Ed., Pergamon: Oxford, 1987, Vol. 6, p 317.
5. Olah, G. A.; Fung, A. P.; Keumi, T. *JOC* **1981**, *46*, 4305.
6. Ho, T.-L.; Gupta, B. G. B.; Olah, G. A. *S* **1977**, 676.
7. Swern, D. *Organic Peroxides*; Wiley: New York, 1970; Vol. 1, pp 475–516.
8. Cofre, P.; Sawyer, D. T. *IC* **1986**, *25*, 2089.
9. (a) Pagano, A. S.; Emmons, W. D. *OS* **1969**, *49*, 47. (b) *Hazards in the Chemical Laboratory*; Luxon, S. G., Ed.; Royal Society of Chemistry: Cambridge, 1992.
10. *Organic Peroxides*; Swern, D., Ed.; Wiley: New York, 1970; Vol. 1, pp 1–104.
11. Frimer, A. A. *JOC* **1977**, *42*, 3194.
12. Bloodworth, A. J.; Curtis, R. J.; Spencer, M. D.; Tallant, N. A. *T* **1993**, *49*, 2729.
13. Davies, A. G.; Foster, R. V.; White, A. M. *JCS* **1953**, 1541.
14. Story, P. R.; Lee, B.; Bishop, C. E.; Denson, D. D.; Busch, P. *JOC* **1970**, *35*, 3059.
15. Caglioti, L.; Gasparrini, F.; Palmieri, G. *TL* **1976**, 3987.
16. Jefford, C. W.; Li, Y.; Jaber, A.; Boukouvalas, J. *SC* **1990**, *20*, 2589.
17. Porter, N. A.; Byers, J. D.; Ali, A. E.; Eling, T. E. *JACS* **1980**, *102*, 1183.
18. Ogata, Y.; Sawaki, Y. *T* **1967**, *23*, 3327.
19. Kochi, J. K.; Macadlo, P. E. *JOC* **1965**, *30*, 1134.
20. (a) Gage, J. R.; Evans, D. A. *OS* **1990**, *68*, 83. (b) Evans, D. A.; Britton, T. C.; Ellman, J. A. *TL* **1987**, *28*, 6141.
21. Evans, D. A.; Britton, T. C.; Ellman, J. A.; Dorow, R. L. *JACS* **1990**, *112*, 4011.
22. Matsumoto, M.; Kobayashi, H.; Hotta, Y. *JOC* **1984**, *49*, 4740.
23. Choi, J.-K.; Chang, Y.-K.; Hong, S. Y. *TL* **1988**, *29*, 1967.
24. Corey, E. J.; Arnold, Z.; Hutton, J. *TL* **1970**, 307.
25. Crimmins, M. T.; Jung, D. K.; Gray, J. L. *JACS* **1993**, *115*, 3146.
26. Feldman, K. S.; Wu, M.-J.; Rotella, D. P. *JACS* **1990**, *112*, 8490.
27. Corey, E. J.; Ensley, H. E. *JOC* **1973**, *38*, 3187.
28. Itsuno, S.; Sakakura, M.; Ito, K. *JOC* **1990**, *55*, 6047.

29. Oguchi, T.; Sakata, Y.; Takeuchi, N.; Kaneda, K.; Ishii, Y.; Ogawa, M. *CL* **1989**, 2053.
30. Kamiyama, T.; Inoue, M.; Kashiwagi, H.; Enomoto, S. *BCJ* **1990**, *63*, 1559.
31. Venturello, C.; D'Aloisio, R. *JOC* **1988**, *53*, 1553.
32. Schwenkreis, T.; Berkessel, A. *TL* **1993**, *34*, 4785.
33. Warwel, S.; Rusch gen; Klaas, M.; Sojka, M. *CC* **1991**, 1578.
34. Naota, T.; Sasao, S.; Tanaka, K.; Yamamoto, H.; Murahashi, S.-I. *TL* **1993**, *34*, 4843.
35. Barak, G.; Dakka, J.; Sasson, Y. *JOC* **1988**, *53*, 3553.
36. Sakata, Y.; Ishii, Y. *JOC* **1991**, *56*, 6233.
37. Minisci, F.; Citterio, A.; Vismara, E.; Fontana, F.; Bernardinis, S. D. *JOC* **1989**, *54*, 728.
38. Baciocchi, E.; Muraglia, E.; Sleiter, G. *JOC* **1992**, *57*, 6817.
39. Murata, S.; Miura, M.; Nomura, M. *JCS(P1)* **1987**, 1259.
40. Brown, H. C.; Mandal, A. K. *JOC* **1980**, *45*, 916.
41. Tamao, K.; Ishida, N.; Ito, Y.; Kumada, M. *OS* **1990**, *69*, 96 and references cited therein.
42. (a) Roush, W. R.; Grover, P. T. *T* **1992**, *48*, 1981. (b) Andrey, O.; Landais, Y.; Planchenault, D. *TL* **1993**, *34*, 2927.
43. Corey, E. J.; Gross, A. W. *OS* **1987**, *65*, 166.
44. Salituro, G. M.; Townsend, C. A. *JACS* **1990**, *112*, 760.
45. Murahashi, S.-I.; Shiota, T.; Imada, Y. *OS* **1992**, *70*, 265.
46. Murahashi, S.-I.; Shiota, T. *TL* **1987**, *28*, 2383.
47. Murahashi, S.-I.; Oda, T.; Sugahara, T.; Masui, Y. *JOC* **1990**, *55*, 1744.
48. Murahashi, S. I.; Oda, T.; Masui, Y. *JACS* **1989**, *111*, 5002.
49. VanRheenen, V.; Cha, D. Y.; Hartley, W. M. *OS* **1978**, *58*, 44.
50. Shvo, Y.; Kaufman, E. D. *JOC* **1981**, *46*, 2148.
51. Torrini, I.; Paradisi, M. P.; Zecchini, G. P.; Agrosi, F. *SC* **1987**, *17*, 515.
52. Sedergran, T. C.; Dittmer, D. C. *OS* **1984**, *62*, 210.
53. Jones, G. B.; Huber, R. S.; Chau, S. *T* **1993**, *49*, 369.
54. Grieco, P. A.; Yokoyama, Y.; Gilman, S.; Nishizawa, M. *JOC* **1977**, *42*, 2034.
55. Noller, C. R. *OSC* **1943**, *2*, 586.
56. Wiberg, K. *JACS* **1953**, *75*, 3961.
57. Cacchi, S.; Misiti, D.; La Torre, F. *S* **1980**, 243.
58. Katritzky, A. R.; Pilarski, B.; Urogdi, L. *S* **1989**, 949.

A. Somasekar Rao & H. Rama Mohan
Indian Institute of Chemical Technology, Hyderabad, India

Hydrogen Peroxide–Urea[1]

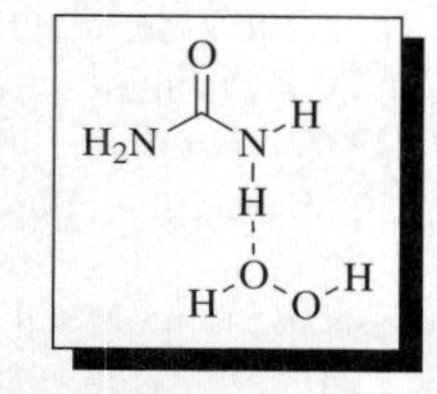

[124-43-6] $CH_6N_2O_3$ (MW 94.09)

(alternative to 90% hydrogen peroxide as a source of anhydrous hydrogen peroxide for use in oxidation reactions)

Alternate Name: UHP.
Physical Data: mp 84–86 °C (dec).[2]
Solubility: sol water and alcohols; low solubility in organic solvents such as dichloromethane.
Form Supplied in: white crystalline powder with urea as impurity; commercially available.
Drying: over calcium chloride in a desiccator.
Preparative Method: made by the recrystallization of ***Urea*** from aqueous ***Hydrogen Peroxide***.
Handling, Storage, and Precautions: the pure material should be stored at low temperature but the commericial material, which has a purity of ca. 90%, may be stored at rt. In a sufficiently forcing test, it can be made to explode; its decomposition is acceleratory above 82 °C. Conduct work with this reagent in an efficient fume hood behind a polycarbonate safety screen.

Epoxidation Reactions. The hydrogen peroxide–urea complex, which is normally known as urea–hydrogen peroxide (UHP), has been used in anhydrous organic solvents in combination with a number of carboxylic anhydrides together with disodium hydrogen phosphate for the epoxidation of a wide range of alkenes. The carboxylic anhydride of choice depends on the electron density in the double bond.[3] With electron-rich alkenes such as α-methylstyrene and α-pinene (eq 1), good yields of the expected products can be obtained by using ***Acetic Anhydride***. Other anhydrides have also been used; for example, the epoxide from *trans*-stilbene is obtained in 80% yield using ***Maleic Anhydride***.[4] Monoperoxymaleic acid, prepared by the reaction of maleic anhydride with 90% aqueous hydrogen peroxide, is more reactive than other common peroxy acids with the exception of ***Trifluoroperacetic Acid***.[5] With relatively nonnucleophilic and nonvolatile terminal alkenes such as 1-octene, a good yield of the epoxide can be obtained by substituting ***Trifluoroacetic Anhydride*** for the acetic anhydride (eq 2). The addition of ***Imidazole*** increases the rate at which epoxidation reactions proceed when using UHP and acetic anhydride.[6] Presumably *N*-acetylimidazole is formed and rapidly perhydrolyzed. A comparison of the diastereoselective epoxidations of a steroidal allylic alcohol has been carried out using a range of peroxy acids, including peracetic acid generated by the interaction of acetic anhydride with UHP.[7]

UHP, Na_2HPO_4, Ac_2O; CH_2Cl_2, rt, 15 h; 79% (1)

UHP, Na_2HPO_4, $(CF_3CO)_2O$; CH_2Cl_2, Δ, 0.5 h; 88% (2)

The epoxidation of electron-deficient alkenes like methyl methacrylate can also be achieved using the trifluoroacetic anhydride method. In the case of α,β-unsaturated ketones such as isophorone (eq 3) and nitro alkenes such as β-methyl-β-nitrostyrene (eq 4), alkaline hydrogen peroxide has been generated from UHP.

UHP, NaOH; MeOH; 68% (3)

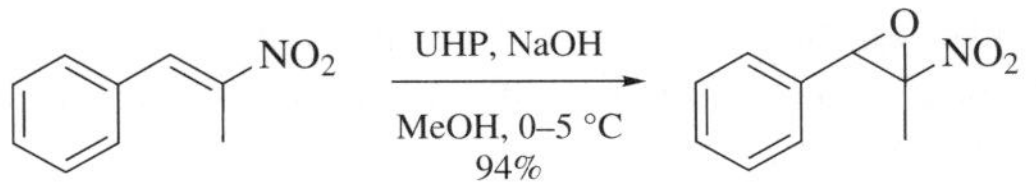

(4)

The selective epoxidation of compounds such as α-ionone can also be achieved. The result of an epoxidation reaction frequently depends on the reagent and precise reaction conditions. This is exemplified in the reactions shown in eq 5. When using peroxytrifluoroacetic acid generated conventionally, water that is always present diverts the reaction exclusively to the hemiacetal (**1**). However, when using the UHP method, the spiroacetal (**2**) predominates over the other product.[8]

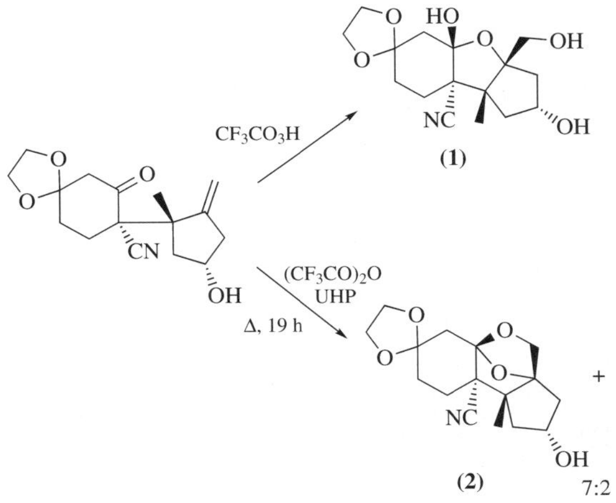

(5)

Baeyer–Villiger and Related Reactions. The oxidation of aldehydes and ketones to afford esters and lactones can be achieved using a variety of peroxycarboxylic acids. The ease with which the reaction occurs is related to the strength of the conjugate acid of the leaving group and so the stronger the carboxylic acid the more powerful is the peroxy acid in its oxidation reactions. Reactions involving ketones are frequently slow when using weakly acidic peroxy acids and so the majority of the reactions using UHP have been carried out with trifluoroacetic anhydride as the coreactant (eqs 6–8).

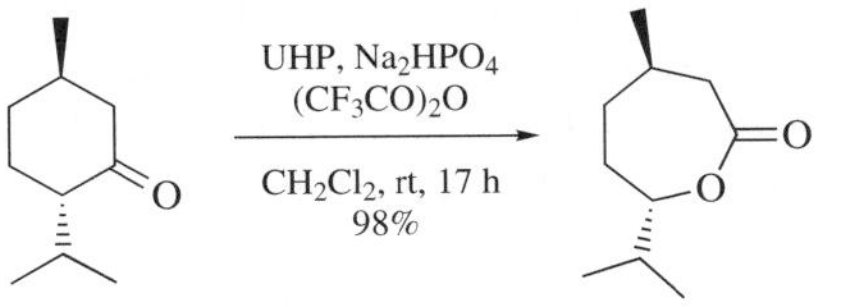

(6)

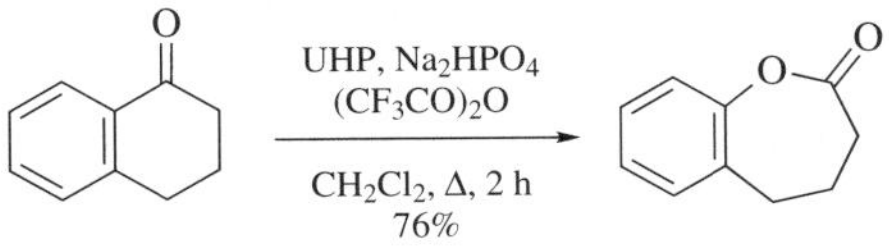

(7)

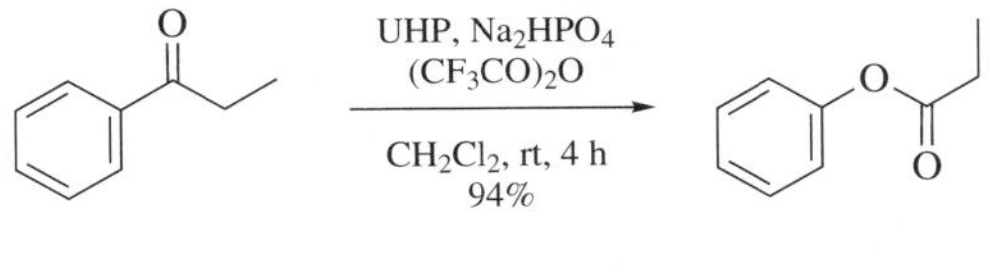

(8)

Dakin reactions, where an aromatic aldehyde has an electron releasing substituent either *ortho* or *para* to the formyl group, can be carried out using UHP–acetic anhydride as shown in eq 9.

In the absence of suitable activating substituents, hydrogen migration occurs in place of aryl migration and the product is then a carboxylic acid (eq 10). This is a potentially valuable way of converting a formyl group into a carboxyl group.

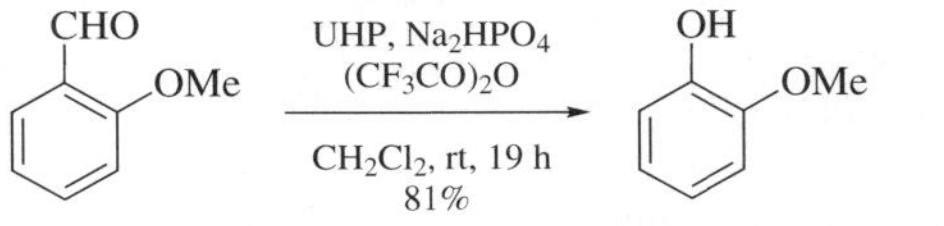

(9)

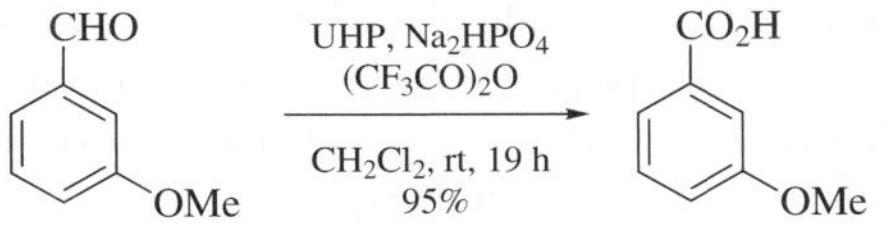

(10)

Heteroatom Oxidation. The first example of the use of UHP in organic chemistry involved the formation of the *N*-oxide shown in eq 11,[9] and, in a modification of the UHP method, phthalic anhydride has been used to oxidize 4-*t*-butylpyridine to the *N*-oxide in 93% yield.[10] The oxidation of aliphatic aldoximes using UHP–trifluoroacetic anhydride has been achieved in good yields with retention of configuration at neighboring chiral centers (eq 12).[11]

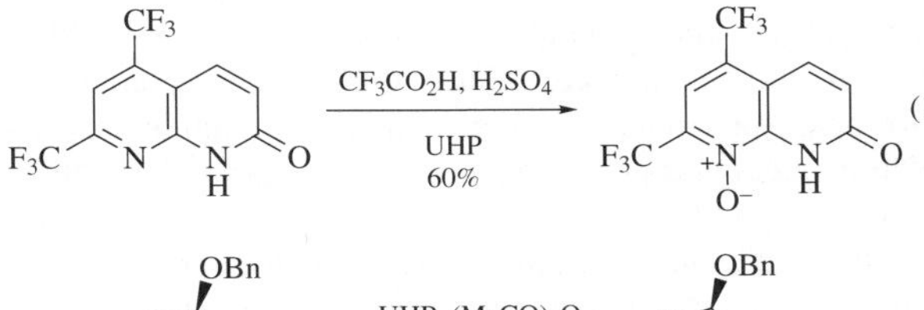

(11)

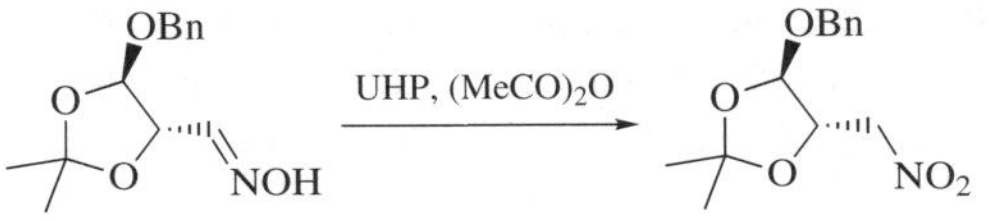

(12)

Related Reagents. See Classes O-11, O-14, and O-15, pages 1–10.

---

1. (a) Heaney, H. *Top. Curr. Chem.* **1993**, *164*, 1. (b) Heaney, H. *Aldrichim. Acta* **1993**, *26*, 35.
2. Lu, C.-S.; Hughes, E. W.; Giguère, P. A. *JACS* **1941**, *63*, 1507.
3. Cooper, M. S.; Heaney, H.; Newbold, A. J.; Sanderson, W. R. *SL* **1990**, 533.
4. Astudillo, L.; Galindo, A.; González, A. G.; Mansilla, H. *H* **1993**, *36*, 1075.
5. White, R. H.; Emmons, W. D. *T* **1962**, *17*, 31.
6. Rocha Gonsalves, A. M. d'A.; Johnstone, R. A. W.; Pereira, M. M.; Shaw, J. *JCR(S)* **1991**, 208.
7. Back, T. G.; Blazecka, P. G.; Krishna, M. V. *CJC* **1993**, *71*, 156.
8. (a) Ziegler, F. E.; Metcalf, C. A., III; Schulte, G. *TL* **1992**, *33*, 3117. (b) Ziegler, F. E.; Metcalf, C. A., III; Nangia, A.; Schulte, G. *JACS* **1993**, *115*, 2581.
9. Eichler, E.; Rooney, C. S.; Williams, H. W. R. *JHC* **1976**, *13*, 41.
10. Kaczmarek, L.; Balicki, R.; Nantka-Namirski, P. *CB* **1992**, *125*, 1965.
11. Ballini, R.; Marcantoni, E.; Petrini, M. *TL* **1992**, *33*, 4835.

Harry Heaney
Loughborough University of Technology, UK

Hydrogen Sulfide[1]

H_2S

[7783-06-4] H_2S (MW 34.09)

(nucleophile and precursor of Na_2S, NaSH, K_2S, and KSH for the synthesis of thiols,[2] thioethers,[3] thiocarbonyl compounds,[4] and heterocyclic compounds; reducing agent, particularly for the conversion (via S^{2-}, HS^-, $S_2{}^{2-}$) of aromatic nitro compounds to amines[5])

Physical Data: colorless gas; bp −60.75 °C; mp −85.60 °C; d (0 °C) 1.539 g L^{-1} (more dense than air); dipole moment (C_6H_6) 0.85–0.97 D. pK_{a1} (aq) 6.88, pK_{a2} (aq) 14.15; $E_0(H_2S \rightarrow S + 2H^+ + 2e^-)$ −0.14 V.

Solubility: sol H_2O (0.12 mol L^{-1} at 20.1 °C); sol polar and nonpolar organic solvents.

Form Supplied in: liquefied gas in steel cylinders.

Handling, Storage, and Precautions: extremely toxic, malodorous gas; use of a well-ventilated hood is a necessity; a scrubber flask containing 20% aqueous NaOH prevents loss of H_2S to the atmosphere. The sense of smell is paralyzed at concentrations of 150–250 ppm and death may ensue at concentrations above 300 ppm. An irritant to the eyes and mucous membranes; highly flammable; ignites spontaneously in air at around 250 °C; mixtures of H_2S (4.5–45%) and air are explosive; corrosive (especially in aqueous solutions) to many metals; anhydrous H_2S is unreactive to stainless steel at ambient temperatures.

Formation of Na_2S, NaSH, K_2S, and KSH. Passage of H_2S into excess aq NaOH or KOH yields ***Sodium Sulfide***, and K_2S, which can be used in situ.[6] ***Sodium Hydrogen Sulfide*** and ***Potassium Hydrogen Sulfide*** are produced in ethanolic base saturated with H_2S;[7] addition of, for example, excess KOEt yields K_2S.[7a] In the formation of KSH, H_2S is passed into an ethanolic KOH solution until the mixture does not test alkaline with phenolphthalein.[7c] Solutions of hydrogen sulfide salts are in equilibrium with the sulfide and H_2S: $2NaSH \rightleftharpoons Na_2S + H_2S$.

Thiols. Thiols or their salts are obtained by displacement reactions[2] by SH^- or S^{2-} (derived from H_2S) on primary or secondary halides or sulfonate esters, on oxiranes and aziridines, on quaternary ammonium salts, and on activated aromatic halides (eq 1).[8] The displacement of chloride ion shown in eq 1 takes precedence over the addition reaction of H_2S to the nitrile group (see below). Halonitrobenzenes may undergo both substitution of halogen and reduction of the nitro group.[2a,6,9] High temperatures are required for unactivated aryl halides (eq 2).[10] Elimination and thioether formation are side reactions with aliphatic substrates. To minimize thioether formation, an excess of H_2S is employed (see the above equilibrium).

Benzyl thiols can be obtained directly from the alcohol by the action of H_2S in the presence of ***Octacarbonyldicobalt*** (eq 3).[11a] Direct reaction between H_2S and alcohols normally requires high temperatures, high pressures, and special catalysts; this method is not generally suited for a laboratory preparation.[2b] The reaction of α-aryl-α-amino nitriles with H_2S under mild conditions gives high yields of thiols uncontaminated with α-amino thioamides (eq 4).[11b]

Pentachlorobenzonitrile → (H_2S, pyridine, NEt_3, 50 °C to rt, 5 h; 71%) 4-mercapto-2,3,5,6-tetrachlorobenzonitrile (1)

1-Chloronaphthalene + H_2S → (600 °C, quartz tube; 75.5%) 1-naphthalenethiol (2)

4-Et-C_6H_4-CH_2OH + H_2S (13 atm) → ($Co_2(CO)_8$, CO (47 atm), H_2O–hexane (2:1), 150 °C, 10 h (screw-cap autoclave)) 4-Et-C_6H_4-CH_2SH (60%) + 4-Et-C_6H_4-Me (30%) (3)

3,4-$(MeO)_2C_6H_3$-CH(CN)-NH-*i*-Pr → (1. H_2S, pyridine–Et_3N, 2 h; 2. rt, 16 h, sealed flask; 92%) 3,4-$(MeO)_2C_6H_3$-CH_2SH (4)

Thiolacids and their derivatives are obtained from the acid chloride or anhydride (eqs 5 and 6).[7c,12] Potassium thiotosylate, useful in the synthesis of thiotosylates, is obtained in a similar way from tosyl chloride (eq 7).[13] The solution must be saturated with H_2S. *gem*-Dithiols are obtained by treatment of ketones with H_2S and an organic base (or the preformed enamine may be used) (eq 8),[14] and best yields are obtained with cyclic ketones. Use of the ketimine or enamine gives satisfactory yields (53–83%) of *gem*-dithiols of acyclic ketones.[14b] Addition of H_2S to propionaldehyde in the presence of ***Chlorotrimethylsilane*** results in the silyloxythiol (eq 9).[15] A Michael-type of addition of H_2S to α,β-unsaturated systems provides β-mercapto derivatives that are readily desulfurized, thus providing a method for the selective reduction of the carbon–carbon double bond (eq 10).[16] Attempted reduction of this double bond (eq 10) by borohydride, hydride, and silane reagents was unsuccessful.

H_2S + KOH → (EtOH) KSH → (PhCOCl, 10–15 °C) PhC(O)SK → (I_2) $(PhC(O)S)_2$ 68–73% (5)

Ac_2O + H_2S → (NaOH (cat.), 50–55 °C, 6 h; 72–76%) AcSH (6)

H_2S + KOH → (H_2O) KSH → (TsCl, H_2O, 55–60 °C) TsSK 42–49% (7)

$$\text{cyclohexanone} + H_2S + \text{morpholine} \xrightarrow[\text{ice cold}]{\text{MeOH}} \text{cyclohexane-1,1-dithiol (80–83\%)} \quad (8)$$

$$\text{PrCHO} + \text{TMSCl} \xrightarrow[73\%]{H_2S,\ \text{pyridine},\ CH_2Cl_2,\ <15\ °C \text{ then rt, 2 h}} \text{PrCH(SH)OTMS} \quad (9)$$

$$\text{substrate (OCOPh; } MeO_2C\text{)} \xrightarrow[23\ °C,\ 1\ h]{H_2S,\ K_2CO_3,\ \text{DMSO}} \text{R–CH(SH)–CH}_2\text{–CO–CH}_2\text{CH}_2\text{CO}_2\text{Me} \xrightarrow{Bu_3P,\ h\nu,\ C_6H_6,\ 23\ °C,\ 7\ h} \text{R–CH}_2\text{–CH}_2\text{–CO–CH}_2\text{CH}_2\text{CO}_2\text{Me (65\% overall)} \quad (10)$$

Thioethers and Disulfides. Conversion of H_2S to sulfide ion, followed by treatment with a primary or secondary halide, sulfonate ester, quaternary ammonium salt, or an oxirane, aziridine, or activated aryl halide gives a good yield of symmetrical thioether via a thiolate intermediate.[3] Base-catalyzed addition of H_2S to electrophilic alkenes yields first the thiol which couples with excess alkene to give the thioether (eq 11).[17a] Disulfides are obtained via dithiol intermediates from addition of H_2S to aldehydes, ketones, and imines.[17b]

$$H_2C{=}CHCO_2Me \xrightarrow[71–81\%]{H_2S,\ NaOAc\cdot 3H_2O,\ 95\%\ EtOH,\ \text{reflux, 25 h}} (MeO_2CCH_2CH_2)_2S \quad (11)$$

Thiocarbonyl Compounds. Thioaldehydes, thioketones, thioesters, thioamides, and related compounds may be prepared by addition of H_2S (or SH^-, S^{2-}) to a C=X or a C=C–X bond where X is a group that is replaceable by sulfur.[4] In the total synthesis of chlorophyll, an important thioaldehyde intermediate was obtained by addition of H_2S to an imine salt (eq 12).[18] Related syntheses of thioaldehydes involve additions to Vilsmeier intermediates (iminium salts, etc).[19] The more common thioketones are prepared by addition of H_2S to ketones typically in the presence of HCl;[7b,14f,20] *gem*-dithiols[14] may be intermediates and they can be thermolyzed to thioketones. While aromatic thioketones are readily available (eq 13),[20a,b] their enethiolizable aliphatic analogs are often contaminated by the enethiol. Trimerization to 1,3,5-trithianes is a common side reaction. Methods for preparation of aliphatic thioketones (stable for 1 month at −15 °C) free from enethiols, and their conversion to pure enethiols free of the thioketone tautomer, have been reported (eq 14).[21] Aliphatic thioketones also are obtained in good yields (52–75%) by treatment of ketone anils with H_2S followed by removal of aniline from the initial adduct with ***Benzoic Anhydride***.[22] Thioketones are obtained by addition of H_2S to imines and iminium salts.[23]

$$\text{dipyrromethane iminium salt } (MeO_2C,\ MeO_2C,\ Et;\ \text{CH=N}^+\text{H–Et}\ Br^-) \xrightarrow[NaOMe,\ MeOH–C_6H_6]{H_2S} \text{RC(S)H} \quad (12)$$

$$Ph_2CO \xrightarrow[2.\ H_2S,\ 20\ h;\ 66–77\%]{1.\ H_2S,\ HCl,\ 95\%\ EtOH,\ \text{ice–salt cooling, 3 h}} Ph_2CS \quad (13)$$

$$R^1CH(R^2)C(OMe)_2R^3 \xrightarrow[36–67\%]{H_2S,\ ZnCl_2\ (0.05–0.10\ \text{equiv}),\ EtOH \text{ or } THF,\ 0\ °C,\ 30\ min} R^1CH(R^2)C(S)R^3 \quad (14)$$

Derivatives of thiocarboxylic acids[4c,24] are obtained from H_2S and orthoesters (eq 15),[25] carboxylic acid ester enolates (eq 16),[26] and imino esters.[27] A method for reduction of a tertiary alcohol to an alkane involves its conversion to a thioformate ester followed by treatment with ***Tri-n-butyltin Hydride*** (eq 17).[27b] Addition of H_2S to thioimino esters provides derivatives of dithiocarboxylic acids (eq 18).[4d,28] The thioamide function[4e,29] is obtained by addition of gaseous or liquid H_2S to nitriles (eq 19),[17b,30] ynamines,[31] various imines (eq 20),[32] and to orthoesters or chloroform in the presence of an amine (eq 21).[32f,33] Thioureas, thiobiurets, and thio heterocycles,[32d,34] thioimides (eq 22),[32b,35] and thioacylhydrazines[36] are obtained by similar reactions.

$$HC(OEt)_3 \xrightarrow[30–38\%]{H_2S,\ H_2SO_4,\ \text{ice bath}} HC(S)OEt \quad (15)$$

$$C_8H_{17}CH{=}CH(CH_2)_7CO_2Me \xrightarrow[89\%]{1.\ LDA,\ THF,\ -78\ °C,\ Ar;\ 2.\ TMSCl,\ 0\ °C;\ 3.\ \text{dry } H_2S,\ 25\ °C} C_8H_{17}CH{=}CH(CH_2)_7C(S)OMe \quad (16)$$

$$Me(CH_2)_{16}C(Me)_2OH + :C{=}N–C_6H_4–NMe_2 \xrightarrow[90\%]{CuO,\ 85\ °C} Me(CH_2)_{16}C(Me)_2–OCH{=}N–C_6H_4–NMe_2 \xrightarrow[90\%]{H_2S,\ \text{pyridine},\ 100\%\ H_2SO_4,\ -10 \text{ to } 0\ °C} Me(CH_2)_{16}C(Me)_2–OC(S)H \xrightarrow[83\%]{Bu_3SnH,\ C_6H_6,\ \text{reflux}} Me(CH_2)_{16}CH(Me)_2 \quad (17)$$

$$\text{Boc–NH–CH}_2\text{–C(S)NH}_2 \xrightarrow[MeOSO_2F]{CH_2Cl_2} \text{Boc–NH–CH}_2\text{–C(SMe)=}\overset{+}{N}H_2\ FSO_3^- \xrightarrow[0\ °C,\ 2\ h]{H_2S,\ 5{:}1\ THF–\text{pyridine}} \text{Boc–NH–CH}_2\text{–C(S)SMe (78\%)} \quad (18)$$

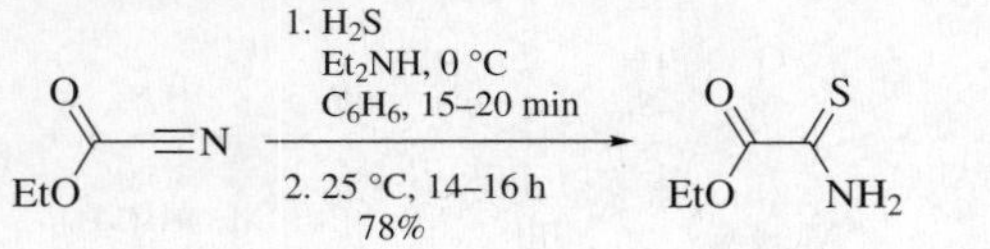

(19)

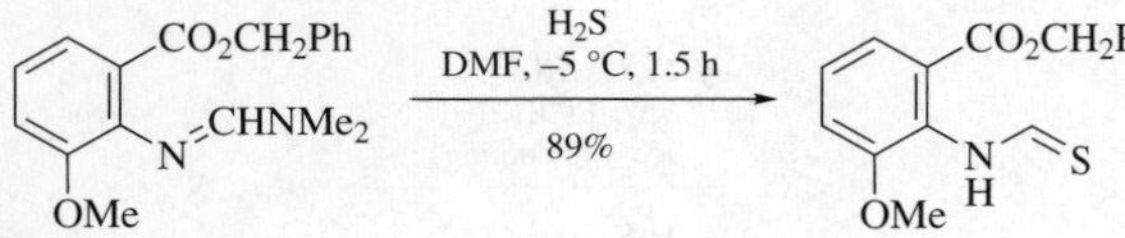

(20)

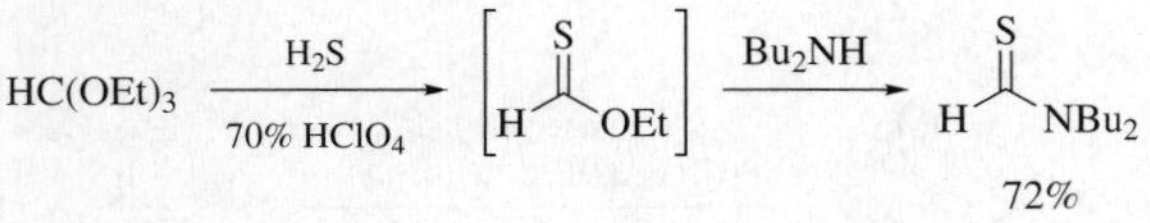

(21)

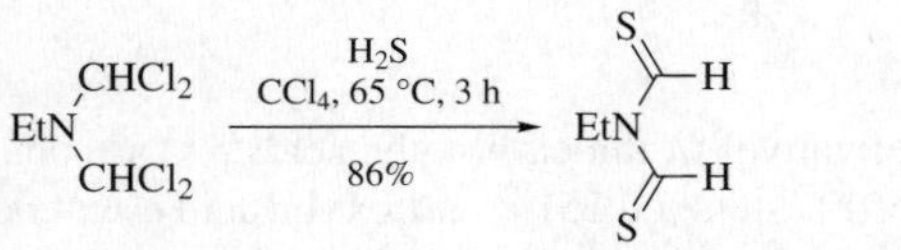

(22)

Heterocyclic Compounds. The following types of sulfur-containing heterocyclic compounds obtained from hydrogen sulfide are listed according to ring size and the number of sulfur and other heteroatoms: 3 (1 S),[37] 4 (1 S),[38a,b] 4 (2 S),[39] 5 (1 S),[40a–d] 5 (2 S),[41] 5 (1 S, 1 N),[42] 5 (1 S, 2 N),[43] 6 (1 S),[44] 6 (3 S),[45] 8 (2 S).[46] The first synthesis of a cyclopenta[*b*]thiophene by a keto alkyne cyclization is shown in eq 23.[40c] Hydrogen sulfide plays a key role (possibly by reducing elemental chlorine) in the suppression of chlorinated byproducts in the synthesis of α-methylthioindoles (eq 24).[47]

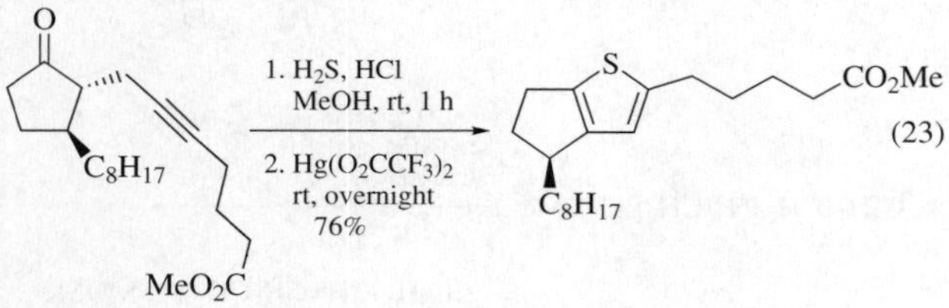

(23)

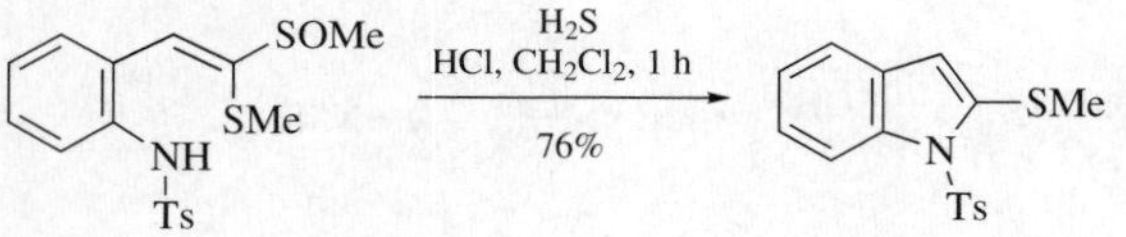

(24)

Reductions. Hydrogen sulfide reduces nitro groups (eq 25),[5,9,48] nitroso groups,[49] azido groups,[50] the S=O group (eq 26),[51] the chlorosulfonyl group,[52] sulfur–sulfur,[53] sulfur–nitrogen,[54] nitrogen–nitrogen,[55] carbon–nitrogen,[11b,56] and carbon–mercury[57] bonds, the C–I and C–O bonds of some α-iodo and α-alkoxy ketones,[58] osmate esters,[59] and 1,2,3-tricarbonyl compounds[60a] and their hydrates[60b–d] (the 2-carbonyl functionality is reduced to the alcohol and dimers (eq 27)[60d] may be formed). Benzil is reduced quantitatively either to deoxybenzoin or benzoin, depending on conditions.[14c] The carbonyl group of aldehydes and ketones undergoes nucleophilic reduction with H_2S and a reducing thiol to give good yields of thiols via thiocarbonyl derivatives (eq 28).[61] The reduction of the azido group in 4-azidobutyryl derivatives of alcohols provides a mild method for deprotection of the hydroxy function (eq 29).[50a]

Reduction of the azido group by elemental hydrogen instead of H_2S in the deprotection scheme may be accompanied by the reduction of other functional groups (eg alkene).

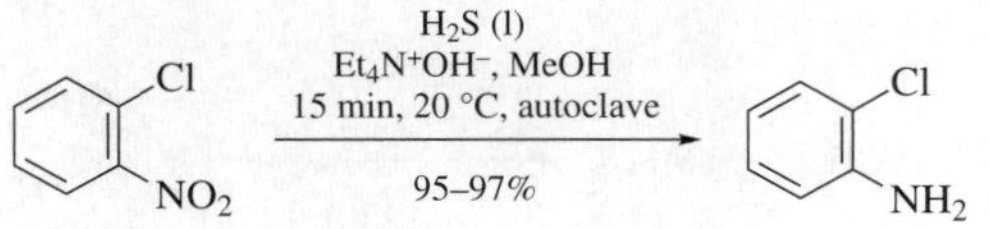

(25)

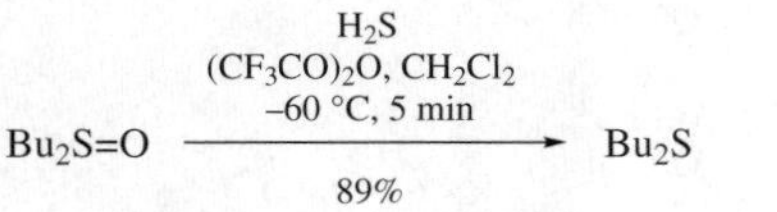

$$Bu_2S{=}O \xrightarrow[89\%]{H_2S,\ (CF_3CO)_2O,\ CH_2Cl_2,\ -60\ °C,\ 5\ min} Bu_2S \quad (26)$$

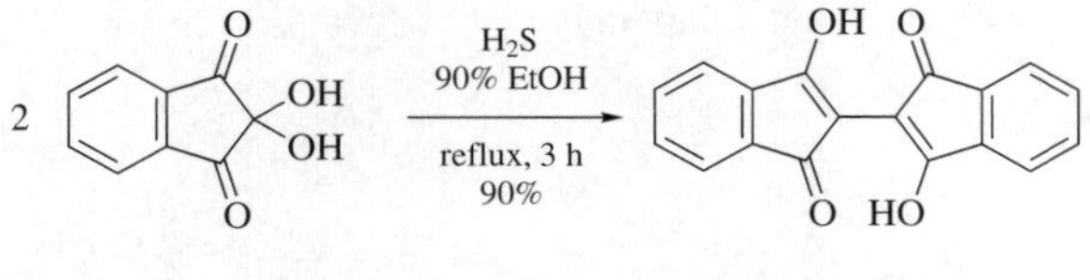

(27)

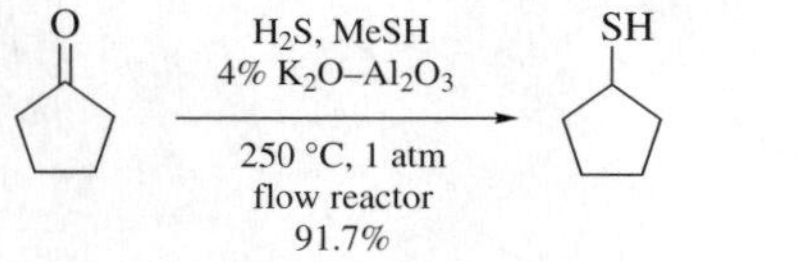

(28)

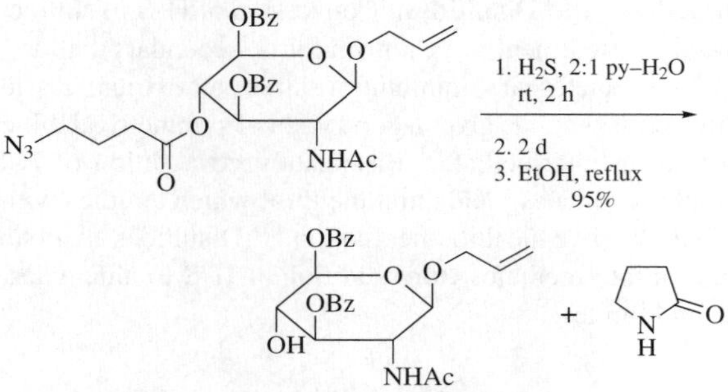

(29)

Miscellaneous. Workup procedures of reactions that use salts of heavy metals (Hg, Pb, Ag, Cu) involve precipitation of the metal as its sulfide by H_2S.[62]

Related Reagents. See Classes R-9, R-15, R-21, R-23, R-24, R-25, R-29, and R-34, pages 1–10.

1. *Sulfanes. Gmelin Handbook of Inorganic Chemistry*, 8th ed.; Bitterer, H., Ed.; Springer: Berlin, 1983; Suppl. Vol. 4a/b.
2. (a) Gundermann, K.-D.; Hümke, K. *MOC* **1985**, *E11*, 33. (b) Wardell, J. L. In *The Chemistry of the Thiol Group*; Patai, S., Ed.; Wiley: London, 1974; Part 1, pp 163–269.
3. Gundermann, K.-D.; Hümke, K. *MOC* **1985**, *E11*, 159.
4. (a) Voss, J. *MOC* **1985**, *E11*, 191–193, 197–199, 202, 206–207, 210–211, 216, 228–230. (b) Schaumann, E. In *The Chemistry of Double Bond Functional Groups*; Patai, S., Ed.; Wiley: New York, 1989; Suppl. A, Vol. 2, Part 2, pp 1269–1367. (c) Mayer, R.; Scheithauer, S. *MOC* **1985**, *E5*, 785. (d) Mayer, R.; Scheithauer, S. *MOC* **1985**, *E5*, 891. (e) Bauer, W.; Kühlein, K. *MOC* **1985**, *E5*, 1218.
5. Porter, H. K. *OR* **1973**, *20*, 455.
6. Strube, R. E. *OSC* **1963**, *4*, 967.
7. (a) Bost, R. W.; Conn, M. W. *OSC* **1943**, *2*, 547. (b) Staudinger, H.; Freudenberger, H. *OSC* **1943**, *2*, 573. (c) Noble, P., Jr.; Tarbell, D. S. *OSC* **1963**, *4*, 924. (d) Kurzer, F.; Lawson, A. *OSC* **1973**, *5*, 1046.

8. Beck, G.; Degener, E.; Heitzer, H. *LA* **1968**, *716*, 47.
9. Bennett, G. M.; Berry, W. A. *JCS* **1927**, 1666.
10. Voronkov, M. G.; Déryagina, E. N.; Klochkova, L. G.; Savushkina, V. I.; Chernyshev, E. A. *JOU* **1975**, *11*, 1118.
11. (a) Alper, H.; Sibtain, F. *JOC* **1988**, *53*, 3306. (b) Crossley, R.; Curran, A. C. W. *JCS(P1)* **1974**, 2327.
12. (a) Frank, R. L.; Blegen, J. R. *OSC* **1955**, *3*, 116. (b) Ellingboe, E. K. *OSC* **1963**, *4*, 928.
13. Woodward, R. B.; Pachter, I. J.; Scheinbaum, M. L. *OSC* **1988**, *6*, 1016.
14. (a) Jentzsch, J.; Fabian, J.; Mayer, R. *CB* **1962**, *95*, 1764. (b) Magnusson, B. *ACS* **1962**, *16*, 1536; **1963**, *17*, 273. (c) Mayer, R.; Hiller, G.; Nitzschke, M.; Jentzsch, J. *AG(E)* **1963**, *2*, 370. (d) Djerassi, C.; Tursch, B. *JOC* **1962**, *27*, 1041. (e) Cairns, T. L.; Evans, G. L.; Larchar, A. W.; McKusick, B. C. *JACS* **1952**, *74*, 3982. (f) Bleisch, S.; Mayer, R. *CB* **1967**, *100*, 93.
15. Aida, T.; Chan, T.-H.; Harpp, D. N. *AG(E)* **1981**, *20*, 691.
16. Corey, E. J.; Shimoji, K. *JACS* **1983**, *105*, 1662.
17. (a) Fehnel, E. A.; Carmack, M. *OSC* **1963**, *4*, 669. (b) Cohen, V. I. *HCA* **1976**, *59*, 840.
18. Woodward, R. B.; Ayer, W. A.; Beaton, J. M.; Bickelhaupt, F.; Bonnett, R.; Buchschacher, P.; Closs, G. L.; Dutler, H.; Hannah, J.; Hauck, F. P.; Itô, S.; Langemann, A.; Le Goff, E.; Leimgruber, W.; Lwowski, W.; Sauer, J.; Valenta, Z.; Volz, H. *JACS* **1960**, *82*, 3800.
19. (a) McKenzie, S.; Reid, D. H. *JCS(C)* **1970**, 145. (b) Mackie, R. K.; McKenzie, S.; Reid, D. H.; Webster, R. G. *JCS(P1)* **1973**, 657. (c) Dingwall, J. G.; Reid, D. H.; Wade, K. *JCS(C)* **1969**, 913. (d) Pulst, M.; Beyer, L.; Weissenfels, M. *JPR* **1982**, *324*, 292.
20. (a) Gofton, B. F.; Braude, E. A. *OSC* **1963**, *4*, 927. (b) Paquer, D.; Vialle, J. *BSF(2)* **1969**, 3595. (c) Fournier, C.; Paquer, D.; Vazeux, M. *BSF(2)* **1975**, 2753. (d) Metzner, P.; Vialle, J. *BSF(2)* **1970**, 3739; **1972**, 3138. (e) Beslin, P.; Lagain, D.; Vialle, J. *JOC* **1980**, *45*, 2517. (f) Paquer, D.; Vazeux, M.; Leriverend, P. *RTC* **1978**, *97*, 121. (g) Greidanus, J. W. *CJC* **1970**, *48*, 3530. (h) Duus, F.; Anthonsen, J. W. *ACS* **1977**, *B31*, 40. (i) Duus, F. *JOC* **1977**, *42*, 3123.
21. LeNocher, A.-M.; Metzner, P. *TL* **1992**, *33*, 6151.
22. Ziegler, E.; Mayer, C.; Zwainz, J. G. *ZN(B)* **1975**, *30*, 760.
23. (a) Korchevin, N. A.; Usov, V. A.; Oparina, L. A.; Dorofeev, I. A.; Tsetlin, Y. S.; Voronkov, M. G. *JOU* **1980**, *16*, 1561. (b) Temokhina, L. V.; Usov, V. A.; Tsetlin, Y. S.; Tsetlina, O. E.; Voronkov, M. G. *JOU* **1979**, *15*, 73.
24. Voss, J. *COS* **1991**, *6*, 435.
25. (a) Hartman, G. D.; Weinstock, L. M. *OSC* **1988**, *6*, 620. (b) Ohno, A.; Koizumi, T.; Tsuchihashi, G. *TL* **1968**, 2083.
26. Corey, E. J.; Wright, S. W. *TL*, **1984**, *25*, 2639.
27. (a) Janssen, M. J. In *The Chemistry of Carboxylic Acids and Esters*; Patai, S., Ed.; Interscience: London, 1969, pp 741–742. (b) Barton, D. H. R.; Hartwig, W.; Motherwell, R. S. H.; Motherwell, W. B.; Stange, A. *TL* **1982**, *23*, 2019.
28. (a) Kohrt, A.; Hartke, K. *LA* **1992**, 595. (b) Gosselin, P.; Masson, S.; Thuillier, A. *TL* **1978**, 2715.
29. Schaumann, E. *COS* **1991**, *6*, 419.
30. (a) Boger, D. L.; Panek, J. S.; Yasuda, M. *OSC* **1993**, *8*, 597. (b) Cressman, H. W. J. *OSC* **1955**, *3*, 609. (c) Paventi, M.; Edward, J. T. *CJC* **1987**, *65*, 282. (d) Benders, P. H.; van Erkelens, P. A. E. *S* **1978**, 775. (e) Srivastava, P. C.; Pickering, M. V.; Allen, L. B.; Streeter, D. G.; Campbell, M. T.; Witkowski, J. T.; Sidwell, R. W.; Robins, R. K. *JMC* **1977**, *20*, 256. (f) Erlenmeyer, H.; Büchler, W.; Lehr, H. *HCA* **1944**, *27*, 969. (g) Karrer, P.; Schukri, J. *HCA* **1945**, *28*, 820.
31. Tolchinskii, S. E.; Maretina, I. A.; Petrov, A. A. *JOU* **1979**, *15*, 577.
32. (a) Ongania, K.-H.; Schwarzenbrunner, U.; Humer, K. *M* **1984**, *115*, 215. (b) Lin, Y.-I.; Jennings, M. N.; Sliskovic, D. R.; Fields, T. L.; Lang, S. A., Jr. *S* **1984**, 946. (c) Ueda, T.; Miura, K.; Kasai, T. *CPB* **1978**, *26*, 2122. (d) Ikehara, M.; Maruyama, T.; Miki, H. *T* **1978**, *34*, 1133. (e) Rousseau, R. J.; Panzica, R. P.; Reddick, S. M.; Robins, R. K. Townsend, L. B. *JOC* **1970**, *35*, 631. (f) Walter, W.; Maerten, G. *LA* **1963**, *669*, 66.
33. Stowell, J. C.; Ham, B. M.; Esslinger, M. A.; Duplantier, A. J. *JOC* **1989**, *54*, 1212.
34. (a) Ueda, T.; Nishino, H. *JACS* **1968**, *90*, 1678. (b) Kurzer, F. *OSC* **1963**, *4*, 502. (c) Ali, M. R.; Singh, R.; Verma, V. K. *H* **1983**, *20*, 1993.
35. Allenstein, E.; Sille, F.; Stegmüller, B. *LA* **1979**, 997.
36. (a) Wolkoff, P.; Hammerum, S.; Callaghan, P. D.; Gibson, M. S. *CJC* **1974**, *52*, 879. (b) Walter, W.; Reubke, K.-J. *TL* **1968**, 5973.
37. Dittmer, D. C. In *Comprehensive Heterocyclic Chemistry*; Katritzky, A. R.; Rees, C. W., Eds.; Pergamon: Oxford, 1984; Vol. 7, pp 172, 173.
38. (a) Block, E. In *Comprehensive Heterocyclic Chemistry*; Katritzky, A. R.; Rees, C. W., Eds.; Pergamon: Oxford, 1984; Vol. 7, pp 434, 440. (b) Dittmer, D. C.; Sedergran, T. C. In *Small Ring Heterocycles*; Hassner, A., Ed.; Wiley: New York, 1985; Part 3, pp 431–768.
39. Timberlake, J. W.; Elder, E. S. In *Comprehensive Heterocyclic Chemistry*; Katritzky, A. R.; Rees, C. W., Eds.; Pergamon: Oxford, 1984; Vol. 7, p 478.
40. (a) Campaigne, E. In *Comprehensive Heterocyclic Chemistry*; Katritzky, A. R.; Rees, C. W., Eds.; Pergamon: Oxford, 1984; Vol. 4, pp 885, 900. (b) Bird, C. W.; Cheeseman, G. W. H. ibid., p 115. (c) Cook, S.; Henderson, D.; Richardson, K. A.; Taylor, R. J. K.; Saunders, J.; Strange, P. G. *JCS(P1)* **1987**, 1825. (d) Middleton, W. J. *OSC* **1963**, *4*, 243.
41. (a) McKinnon, D. M. In *Comprehensive Heterocyclic Chemistry*; Katritzky, A. R.; Rees, C. W., Eds.; Pergamon: Oxford, 1984; Vol. 6, pp 803, 807, 808. (b) Gotthardt, H. *CHC* **1984**, *6*, 813.
42. Metzger, J. V. In *Comprehensive Heterocyclic Chemistry*; Katritzky, A. R.; Rees, C. W., Eds.; Pergamon: Oxford, 1984; Vol. 6, pp 301, 308.
43. (a) Thomas, E. W. In *Comprehensive Heterocyclic Chemistry*; Katritzky, A. R.; Rees, C. W., Eds.; Pergamon: Oxford, 1984; Vol. 6, p 461. (b) Kornis, G. *ibid.*, pp 570, 574. (c) McKinnon, D. M. *ibid.*, p 807.
44. Ingall, A. H. In *Comprehensive Heterocyclic Chemistry*; Katritzky, A. R.; Rees, C. W., Eds.; Pergamon: Oxford, 1984; Vol. 3, p 928, 930.
45. Bost, R. W.; Constable, E. W. *OSC* **1943**, *2*, 610.
46. Moore, J. A.; Anet, F. A. L. In *Comprehensive Heterocyclic Chemistry*; Katritzky, A. R.; Rees, C. W., Eds.; Pergamon: Oxford, 1984; Vol. 7, p 695.
47. Hewson, A. T.; Hughes, K.; Richardson, S. K.; Sharpe, D. A.; Wadsworth, A. H. *JCS(P1)* **1991**, 1565.
48. (a) Rybakova, I. A.; Shekhtman, R. I.; Prilezhaeva, E. N.; Litvinov, V. P.; Shakhovskoi, G. P. *IZV* **1991**, 1901. (b) Ratcliffe, C. T.; Pap, G. *CC* **1980**, 260. (c) Griffin, K. P.; Peterson, W. D. *OSC* **1955**, *3*, 242. (d) Robertson, G. R. *OSC* **1941**, *1*, 52. (e) Hansch, C.; Schmidhalter, B.; Reiter, F.; Saltonstall, W. *JOC* **1956**, *21*, 265. (f) Parkes, G. D.; Farthing, A. C. *JCS* **1948**, 1275.
49. Kremers, E.; Wakeman, N.; Hixon, R. M. *OSC* **1941**, *1*, 511.
50. (a) Kusumoto, S.; Sakai, K.; Shiba, T. *BCJ* **1986**, *59*, 1296. (b) Gronowitz, S.; Westerlund, C.; Hörnfeldt, A.-B. *ACS* **1975**, *B29*, 224, 233. (c) Rao, H. S. P.; Doss, S. D. *Sulfur Lett.* **1992**, *14*, 61 (*CA* **1992**, *116*, 235 200). (d) Lieber, E.; Sherman, E.; Henry, R. A.; Cohen, J. *JACS* **1951**, *73*, 2327.
51. (a) Drabowicz, J.; Oae, S. *CL* **1977**, 767. (b) Walter, W. *LA* **1960**, *633*, 35.
52. Schöllkopf, U.; Hilbert, P. *LA* **1973**, 1061.
53. Joshua, C. P.; Presannan, E.; Thomas, S. K. *IJC(B)* **1982**, *21*, 649.
54. (a) Ali, M. R.; Verma, V. K. *S* **1985**, 691. (b) Fuchigami, T.; Nonaka, T. *CL* **1979**, 829.
55. (a) Israel, M.; Protopapa, H. K.; Chatterjee, S.; Modest, E. J. *JPS* **1965**, *54*, 1626 (*CA* **1966**, *64*, 733d) (b) Ruggli, P.; Hölzle, K. *HCA* **1943**, *26*, 814, 1190. (c) Wolfrom, M. L.; Miller, J. B. *JACS* **1958**, *80*, 1678.
56. (a) Asinger, F.; Offermanns, H.; Saus, A. *M* **1969**, *100*, 725. (b) Petri, N.; Glemser, O. *CB* **1961**, *94*, 553.
57. (a) Ferrier, R. J.; Haines, S. R. *JCS(P1)* **1984**, 1689. (b) Bartlett, P. A.; Adams, J. L. *JACS* **1980**, *102*, 337. (c) Newman, M. S.; Vander Zwan, M. C. *JOC* **1974**, *39*, 1186.

58. (a) Mikhal'chuk, A. L.; Pshenichnyi, V. N. *JOU* **1991**, *27*, 1479. (b) Turner, R. B.; Mattox, V. R.; Engel, L. L.; McKenzie, B. F.; Kendall, E. C. *JBC* **1946**, *166*, 345.
59. Keana, J. F. W.; Schumaker, R. R. *JOC* **1976**, *41*, 3840.
60. (a) Kenyon, J.; Munro, N. *JCS* **1948**, 158. (b) Tipson, R. S. *OSC* **1963**, *4*, 25. (c) Nightingale, D. *OSC* **1955**, *3*, 42. (d) Schönberg, A.; Moubasher, R. *JCS* **1949**, 212.
61. Lucien, J.; Barrault, J.; Guisnet, M.; Mauret, R. *Ind. Eng. Chem., Prod. Res. Dev.* **1978**, *17*, 354.
62. (a) Brownlee, P. J. E.; Cox, M. E.; Handford, B. O.; Marsden, J. C.; Young, G. T. *JCS* **1964**, 3832. (b) Foster, G. L.; Shemin, D. *OSC* **1943**, *2*, 330. (c) Steiger, R. E. *OSC* **1955**, *3*, 66. (d) Ronzio, A. R.; Waugh, T. D. *OSC* **1955**, *3*, 438. (e) Eck, J. C. *OSC* **1943**, *2*, 28. (f) Totter, J. R.; Darby, W. J. *OSC* **1955**, *3*, 460.

Donald C. Dittmer
Syracuse University, NY, USA

I

Iodosylbenzene[1]

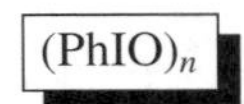

[536-80-1] C_6H_5IO (MW 220.01)

(oxygen atom transfer to ketones,[4] metal-catalyzed oxygen atom transfer to alkenes,[7–13] formation of α-diketones,[14] bridgehead triflates,[16] α,β-unsaturated lactones and ketones,[19] α-hydroxymethyl acetals,[20] α-keto triflates,[17] α-hydroxy ketones,[18] C–C coupling,[22–24] oxidation of amines,[25,26] azidation[22–30])

Alternate Name: iodosobenzene.
Physical Data: mp 210 °C with decomposition (explodes); polymeric.
Solubility: slightly sol H_2O, MeOH. In MeOH the reagent is $PhI(OMe)_2$. Both solvents, as well as CH_2Cl_2 and MeCN, are used for reactions.
Form Supplied in: white to slightly yellow solid from synthesis.[2,3]
Preparative Method: prepared as a white solid by base hydrolysis of commercially available ***(Diacetoxyiodo)benzene*** in 75% yield.[2]
Purification: washing the solid with $CHCl_3$ to remove traces of PhI. The material is obtained in about 99% purity as determined iodometrically.[3]
Handling, Storage, and Precautions: indefinitely stable; refrigeration should be used for long-term storage.

Oxygen Atom Transfer from PhIO: α-Lactones from Ketenes (Uncatalyzed Epoxidation). Ketenes react with PhIO to yield an intermediary α-lactone which undergoes polymerization to a polyester in yields ranging from 63 to 90% (eq 1).[4] Likewise, tetracyanoethylene forms tetracyanoethylene oxide in 74% yield upon reaction with PhIO.[4] α-Lactones have also been prepared by the ozonation of ketenes[5] and via the addition of triplet dioxygen to ketenes.[6]

$R_2C{=}C{=}O$ + PhIO —(CH_2Cl_2, rt, –PhI)→ [α-lactone] → polyester (1)

R = Et, Bu, Ph, CF_3, (Ph, Me)

Metal-Catalyzed Oxygen Transfer: Epoxide Formation. Iron(III) and manganese(III) porphyrins catalyze oxygen transfer from PhIO to aromatic substrates to yield phenolic products, probably via arene oxides.[7] Alkene epoxidation can also be achieved using this system. Simpler ligands such as Schiff's bases,[8] amides,[9] phosphines,[10] and salts of heteropolyaromatics[11] have been used. Bleomycin complexes of iron, copper, and zinc also cause oxidation of alkenes.[12] Noteworthy is the epoxidation of cyclohexene (eq 2).[13] These systems are of theoretical interest as models for cytochrome P-450 and have not as yet achieved general preparative significance.

cyclohexene —($Cu(NO_3)_2$, PhIO)→ cyclohexene oxide (2)

Oxidation of Alkynes to α-Diketones. Ruthenium catalysis is effective in the oxidation of internal (eq 3) and terminal alkynes (eq 4).[14] Secondary alcohols are oxidized to ketones in good yield using *m*-iodosylbenzoic acid and the base-extractable *m*-iodobenzoic acid can be recycled.[15]

Ph–C≡C–C_5H_{11} —(PhIO, $RuCl_2(PPh_3)_3$, 72%)→ PhCOCOC_5H_{11} (3)

Pr–C≡CH —(PhIO, $RuCl_2(PPh_3)_3$, 71%)→ $PrCO_2H$ (4)

Activated PhIO using TMSOTf: Oxidative Displacement of Bridgehead Iodine to Yield Bridgehead Triflates. A series of cubyl triflates was synthesized using PhIO and ***Trimethylsilyl Trifluoromethanesulfonate*** (forms the reactive intermediate $[PhIOTMS]^+$ TfO^-) (eq 5).[16] Cubyl alcohol is unstable and this direct functionalization was critically useful.

X-cubyl-I —(PhIO, TMSOTf, CH_2Cl_2)→ X-cubyl-OSO_2CF_3 (5)

X = H, Me, CO_2Me, I, Br, Cl

Oxidation of Silyl Enol Ethers to α-Keto Triflates with PhIO–TMSOTf. The oxidation of silyl enol ethers to α-keto triflates is a generally useful reaction (eqs 6 and 7).[17]

ArC(OTMS)=CH_2 —((PhIO–TMSOTf), CH_2Cl_2, –78 °C, 53–70%)→ $ArCOCH_2OTf$ (6)

Ar = phenyl, furan, thiophene

1-(trimethylsilyloxy)cycloheptene —((PhIO–TMSOTf), CH_2Cl_2, –78 °C, 74%)→ 2-(triflyloxy)cycloheptanone (7)

Oxidation of Silyl Enol Ethers with PhIO in H_2O–BF_3·Et_2O to α-Hydroxy Ketones. A direct route to α-hydroxy ketones is achieved via oxidation of aromatic, heteroaromatic, and aliphatic silyl enol ethers with PhIO in ***Boron Trifluoride Etherate***–H_2O (eq 8).[18] This is a simple and broadly useful reaction for the α-hydroxylation of ketones.

PhIO, BF_3•OEt_2–H_2O (8)

X = H, 65%; Cl, 70%; NO_2, 78%; 2-pyridyl, 62%; 3-pyridyl, 64%; cyclohexyl, 80%; 2-benzofuran, 59%

Oxidation of Lactones to Higher Homologous α,β-Unsaturated Lactones. Ring expansion of lactones occurs via oxidation of the derived trimethylsilyloxycyclopropanols to give the higher homologous α,β-unsaturated lactones (eq 9).[19] Silyl enol ethers behave analogously (eqs 10 and 11).[19]

$(PhIO)_n$, Bu_4NF in THF, CH_2Cl_2 → + PhI (9)

n = 1, 72%; 2, 75%; 3, 62%; 9, 78%

$(PhIO)_n$, Bu_4NF in THF, CH_2Cl_2, 72% → + PhI (10)

$(PhIO)_n$, Bu_4NF in THF, CH_2Cl_2, 73% → + PhI (11)

Oxidation of Ketones to Form α-Hydroxydimethyl Acetals. The reagent $PhI(OAc)_2$–KOH–MeOH is equivalent to PhIO–KOH–MeOH; however, $PhI(OAc)_2$ is commercially available whilst PhIO is not. Consequently, the reagent ***(Diacetoxyiodo)benzene***, $[PhI(OAc)_2]$, is now used even though the original process was discovered using PhIO/MeOH.[20a] This reagent has been employed in the synthesis of the steroidal dihydroxyacetone side chain (eq 12).[20b]

PhIO, KOH, MeOH; Ac_2O–pyridine then p-xylene, 138 °C; m-CPBA, 5 min, 0 °C then OH^- (12)

Formation of Carbon–Carbon Bonds: 1,4-Diketones via Coupling Reaction with PhIO/BF_3·Et_2O. The addition of BF_3·Et_2O to PhIO yields an intermediary reagent, $[PhI^+OBF_3^-]$, which is reactive with silyl enol ethers derived from ketones. In the presence of water or an alcohol, α-hydroxylation or α-alkoxylation occurs (eq 13),[18] while in the absence of a protic nucleophile self-coupling occurs to yield a 1,4-diketone (eq 14).[21]

Unsymmetrical coupling occurs via in situ generation of a phenyliodonium intermediate at −78 °C and subsequent introduction of a silyl enol ether as a coupling partner (eq 15).[22]

Allylation of aromatic compounds takes place when allylmetal (Group 14) compounds react with aromatic compounds and iodosylbenzene in the presence of boron trifluoride etherate (eq 16).[23,24]

PhIO–BF_3•OEt_2, R^2OH, −78 °C (13)

R^2 = H, Me, Et

PhIO–BF_3•OEt_2, −78 °C (14)

+ PhIO•HBF_4 → $[F_4B^-\ Ph\text{-}I^+ \ldots]$ (15)

MMe_3 + ArR, $(PhIO)_n$, BF_3•OEt_2 (16)

M = Si, Ge, Sn

Oxidation of Amines. Primary amines yield nitriles (eq 17) or ketones (eq 18).[25]

RCH_2NH_2 —PhIO→ RCN (17)

R = Ph, 48%; C_5H_{11}, 56%

PhIO, H_2O (18)

n = 1, 2

Cyclic secondary amines yield lactams (eq 19).[25] *N*-Methyl cyclic amines as well as nicotine behave analogously (eq 20).[25]

PhIO, H_2O (19)

$n = 1$, 49%
$n = 2$, 58%

PhIO, H_2O, 20% (20)

(*S*)-(–) (±)

Cyclic amino acids such as L-proline, pipecolinic acid, and L-2-pyrrolidinone-5-carboxylic acid undergo oxidative decarboxylation with iodosobenzene in various solvents (including water) to yield the corresponding lactam (eq 21).[26]

PhIO, rt, 2 days, 52% (21)

Formation of Azides Using PhIO–NaN_3 or PhIO–$TMSN_3$. Cholesterol is converted to the allylic azide with PhIO–NaN_3–$MeCO_2H$, namely 7α-azidocholest-5-en-3β-ol.[27] A similar reaction occurs with $Pb(OAc)_4$–TMSCl.[28]

Formation of β-Azido Ketones using PhIO–$TMSN_3$. The reaction between ***Azidotrimethylsilane*** and PhIO must be carried out at low temperatures. At rt they react violently. Eq 22 shows the β-azido functionalization of triisopropylsilylenol ethers.[29]

PhIO–$TMSN_3$, CH_2Cl_2, –15 °C (22)

R^1 = H, Me; R^2 = H, Me

N,N-Dimethylarylamines under these conditions yield *N*-methyl-*N*-azidomethyl arylamines (eq 23).[30]

PhIO–$TMSN_3$, –20 °C (23)

R^1 = Me; R^2 = Me, NMe_2

The thiocarbonyl group of 2-thiouracil is converted to the carbonyl group with PhIO (eq 24).[31]

PhIO, acetone, rt (24)

$R^1 = R^2$ = H; R^1 = Me, R^2 = H; R^1 = H, R^2 = Me

Related Reagents. See Classes O-14 and O-20, pages 1–10. Iodosylbenzene–Boron Trifluoride; Iodosylbenzene–Dichlorotris(triphenylphosphine)ruthenium.

1. (a) Moriarty, R. M.; Prakash, O. *ACR* **1986**, *19*, 244. (b) Moriarty, R. M.; Vaid, R. K. *S* **1990**, 431. (c) Varvoglis, A. In *The Organic Chemistry of Polycoordinated Iodine*; VCH: New York, 1992; pp 131.
2. Sharefkin, J. G.; Saltzman, H. *OSC* **1973**, *5*, 660.
3. Lucas, H. J.; Kennedy, E. R.; Formo, M. W. *OS* **1955**, *111*, 483.
4. Moriarty, R. M.; Gupta, S. G.; Hu, H.; Berenschot, D. R.; White, K. B. *JACS* **1981**, *103*, 686.
5. Wheland, R.; Bartlett, P. D. *JACS* **1970**, *92*, 6057.
6. Turro, R. J.; Chow, M.-F.; Ito, Y. *JACS* **1978**, *100*, 1978.
7. Chang, C. K.; Ebina, F. *CC* **1981**, 779.
8. Jorgenson, K. A.; Schiott, B.; Larsen, E. *JCR(S)* **1989**, 214.
9. Koola, J. D.; Kochi, J. K. *JOC* **1987**, *57*, 4545.
10. Bressan, M.; Morrillo, A. *IC* **1989**, *28*, 950.
11. (a) Neuman, R.; Gnim-Abu, G. *CC* **1989**, 1324. (b) Hill, C. H.; Brown, R. B. *JACS* **1986**, *108*, 536.
12. Long, E. C.; Hecht, S. M. *TL* **1988**, *29*, 6413.
13. Franklin, C. C.; Van Atta, R. B.; Tai, A. F.; Valentine, J. S. *JACS* **1984**, *106*, 814.
14. Müller, P.; Godoy, J. *HCA* **1981**, *64*, 2531.
15. Müller, P.; Godoy, J. *TL* **1981**, *22*, 2361.
16. Moriarty, R. M.; Tuladhar, S. M.; Penmasta, R.; Awasthi, A. K. *JACS* **1990**, *112*, 3228.
17. Moriarty, R. M.; Epa, W. E.; Penmasta, R.; Awasthi, A. K. *TL* **1989**, *30*, 667.
18. Moriarty, R. M.; Duncan, M. P.; Prakash, O. *JCS(P1)* **1987**, 1781.
19. Moriarty, R. M.; Vaid, R. K.; Hopkins, T. E.; Vaid, B. K.; Prakash, O. *TL* **1990**, *31*, 197.
20. (a) Moriarty, R. M.; Hu, H.; Gupta, S. C. *TL* **1981**, *22*, 1283. (b) Moriarty, R. M.; John, L. S.; Du, P. C. *CC* **1981**, 641.
21. (a) Moriarty, R. M.; Prakash, O.; Duncan, M. P. *SC* **1985**, *15*, 649. (b) Moriarty, R. M.; Prakash, O.; Duncan, M. P. *S* **1985**, 943.
22. (a) Zhadankin, V. V.; Tykwinski, R.; Caple, R.; Berglund, B. A.; Koz'min, A. S.; Zefirov, N. S. *TL* **1988**, *29*, 3703. (b) Zhadankin, V. V.; Mullikin, M.; Tykwinski, R.; Berglund, B. A.; Caple, R.; Zefirov, N. S.; Koz'min, A. S. *JOC* **1989**, *54*, 2605.
23. Ochiai, M.; Fujita, E.; Arimoto, M.; Yamaguchi, H. *CPB* **1985**, *33*, 41.
24. Lee, K.; Yakura, T.; Tohma, H.; Kikuchi, K.; Tamaru, M. *TL* **1989**, *30*, 1119.
25. Moriarty, R. M.; Vaid, R. K.; Duncan, M. P.; Ochiai, M.; Inenaga, I.; Nagao, Y. *TL* **1988**, *29*, 6913.
26. Moriarty, R. M.; Vaid, R. K.; Duncan, M. P.; Ochiai, M.; Inenaga, I.; Nagao, Y. *TL* **1988**, *29*, 6917.
27. Moriarty, R. M.; Khosrowshahi, J. S. *SC* **1987**, *17*, 89.
28. (a) Kischa, K.; Zbiral, E. *T* **1970**, *26*, 1417. (b) Hugl, H.; Zbiral, E. *T* **1973**, *29*, 759. (c) Hugl, H.; Zbiral, E. *T* **1973**, *29*, 753. (d) Zibral, E.; Nestler, G. *T* **1971**, *27*, 2293. (e) Zibral, E. *S* **1972**, 285.
29. (a) Mangus, P. D.; Lacour, J. *JACS* **1992**, *114*, 767. (b) Mangus, P. D.; Lacour, J. *JACS* **1992**, *114*, 3993.
30. Mangus, P. D.; Lacour, J.; Weber, W. T. *JACS* **1993**, *115*, 9347.
31. Moriarty, R. M.; Prakash, I.; Clarisse, D. E.; Penmasta, R.; Awasthi, A. K. *CC* **1987**, 1209.

Robert M. Moriarty & Jerome W. Kosmeder II
University of Illinois at Chicago, IL, USA

(2,3-*O*-Isopropylidene)-2,3-dihydroxy-1,4-bis(diphenylphosphino)butane[1]

(*R,R*)		
[32305-98-9]	$C_{31}H_{32}O_2P_2$	(MW 498.57)
(*S,S*)		
[37002-48-5]		

(chiral bidentate phosphine, useful in asymmetric catalysis[1])

Alternate Name: DIOP.
Physical Data: mp 88–89 °C; $[\alpha]_D^{20}$ −12.5° (*c* 4.6, C_6H_6).
Solubility: sol most usual organic solvents.
Form Supplied in: white solid; both enantiomers available.
Preparative Methods: can be prepared in four steps from diethyl tartrate.[2] The two phosphorus groups are introduced in the last step of the reaction sequence using $LiPPh_2$,[2] $KPPh_2$,[3] or $LiP(BH_3)PPh_2$.[4] DIOP has also been prepared from 1,2:3,4-diepoxybutane.[5]
Handling, Storage, and Precautions: air stable.

Introduction. DIOP was the first example of a C_2 chelating diphosphine for transition metal complexes to be used in asymmetric catalysis. It was also one of the first examples of a useful C_2 chiral auxiliary.[6] DIOP can be considered as an example of the first generation of chelating diphosphine ligands with a chiral carbon skeleton, which were followed over the next 20 years by many examples of chelating diphosphines, one of the most efficient of which is BINAP (***2,2′-Bis(diphenylphosphino)-1,1′-binaphthyl***).[1d] The ready availability of DIOP has stimulated research in asymmetric catalysis beyond the area of asymmetric hydrogenation.

Asymmetric Hydrogenation. Conjugated acids (eq 1)[2,7] or various α-*N*-acyldehydroamino acids (eq 2)[2,8,9] are structural units which sometimes give quite high ee's in the presence of rhodium complexes formed in situ (such as Rh(Cl)(cod)(DIOP)) or isolated as cationic complexes, for example $[Rh(cod)(DIOP)]^+ PF_6^-$. Hydrogenation of *N*-acetamidocinnamic acid using $[Rh(DIOP)_2]^+ BF_4^-$ instead of $[Rh(cod)(DIOP)]^+ BF_4^-$ as catalyst (80 °C, 1 bar H_2) gives a slower reaction but with a significant increase in the ee (94% ee instead of 82% ee).[10]

$$\text{R–C(=CH}_2\text{)–CO}_2\text{H} + \text{H}_2\ (1.1\ \text{bar}) \xrightarrow[\text{0.1 equiv NEt}_3]{\text{0.03 equiv Rh, }(R,R)\text{-DIOP}} \text{R–CH(CH}_3\text{)–CO}_2\text{H} \quad (1)$$

R	solvent	yield	ee
R = Ph	benzene–EtOH	95%	63% ee
R = 2-phenyl-4-methoxy-3-thienyl (S, OMe, Ph)	*i*-PrOH	97%	88% ee

$$\text{R–CH=C(NHAc)–CO}_2\text{H} + \text{H}_2\ (1.1\ \text{bar}) \xrightarrow[\text{benzene–EtOH}]{\text{0.01 equiv Rh, }(R,R)\text{-DIOP}} \text{AcNH–CH(CH}_2\text{R)–CO}_2\text{H} \quad (2)$$

R	yield	ee
R = H	95%	73% ee
R = Ph	90%	82% ee

Enamides lacking a carboxy group on the double bond also act as excellent substrates in asymmetric hydrogenations, as exemplified in eq 3.[11]

$$\text{R–CH=C(NHAc)–Ph} + \text{H}_2\ (1.1\ \text{bar}) \xrightarrow[\text{benzene–EtOH}]{\text{0.01 equiv Rh, }(R,R)\text{-DIOP}} \text{AcNH–CH(CH}_2\text{R)–Ph} \quad (3)$$

R	yield	ee
R = Et	90%	90% ee
R = Me	90%	92% ee

Ketones are known to be quite unreactive in homogeneous hydrogenations catalyzed by rhodium complexes. However, catalytic amounts of a base enhance the reactivity. In this way, acetophenone is hydrogenated to 2-phenylethanol in 80% ee in the presence of $[Rh(Cl)(cod)(DIOP)]/NEt_3$[12] or $N(CH_2OH)_3$.[13] Aromatic α-amino ketones are reduced to the alcohols with high ee's. For example, 2-naphthyl-*N,N*-diethylaminoethanol is produced in 95% ee by hydrogenation of the corresponding ketone.[13] Imines are very difficult to hydrogenate in the presence of rhodium catalysts, including [Rh(Cl)(cod)(DIOP)].[1] However, it was recently discovered that iridium complexes with a chiral chelating diphosphine are selective catalysts for the hydrogenation of imines. DIOP gives the best result (63% ee) for the reduction of $RN{=}C(Me)CH_2OMe$ ($R = 2{,}5\text{-}Me_2C_6H_3$).[14]

Asymmetric Hydrosilylation. Hydrosilylation of ketones catalyzed by chiral metal complexes, followed by hydrolysis, produces enantiomerically enriched alcohols. Rhodium complexes with chiral chelating diphosphines have been used successfully. In this context, DIOP was one of the ligands investigated. Aryl alkyl ketones provide the corresponding alcohols in low ee with Ph_2SiH_2, but α-$NpPhSiH_2$ gives ee's in the range of 50–60%.[12] This silane is also excellent for hydrosilylation of *i*-butyl levulinate (84% ee) and *n*-propyl pyruvate (85% ee).[15] Imines are transformed into amines (ee ≤ 65%) by Ph_2SiH_2 with a rhodium–DIOP catalyst.[16]

Asymmetric Hydroformylation. DIOP was very useful in the early stages of investigation of asymmetric hydroformylation of alkenes in the presence of rhodium or palladium catalysts. The combination of $PtCl_2$(diphosphine) and $SnCl_2$, where the diphosphine is a DIOP derivative (DIPHOL, eq 4), is an excellent system, although requiring high pressures.[17] In the case of the hydroformylation of styrene, the branched aldehyde is the major product. The various hydroformylations or hydroesterifications have been reviewed.[1e] Asymmetric hydroformylation of *N*-acylaminoacrylic acid esters is efficiently catalyzed by $[Rh(CO)(PPh_3)_3]$ + DIOP, giving the branched aldehyde in 60% ee.[18]

Asymmetric Hydroboration. Hydroborations of alkenes by catecholborane have been catalyzed by [Rh(Cl)(cod)(diphosphine)].[19] For example, norbornene gives, after oxidation, *exo*-norborneol (82% ee) when a DIOP derivative (2-MeO-DIOP) was used (eq 5). Lower ee's were observed with DIOP, DIPAMP, and BINAP. The effectiveness of DIOP was also noticed in another report.[20]

Ph, H + CO + H_2 (200 bar) → (0.05 equiv $PtCl_2$, DIPHOL, $SnCl_2$; benzene, 40 °C; 80%) Ph, CHO; 80% ee (4)

DIPHOL = ... PAr_2, PAr_2; PAr_2 = ... P

+ ... BH → (0.01 equiv $[Rh(Cl)(cod)]_2$, 0.02 equiv (*R*,*R*)-OMe-DIOP; THF, −25 °C, 2 h; >90%) → (H_2O_2, NaOH) ... OH, H; 82% ee (5)

2-OMe-DIOP = ... P(*o*-anisyl)$_2$, P(*o*-anisyl)$_2$

Miscellaneous Reactions. Hydrocyanation of norbornene is catalyzed by $[Pd(DIOP)_2]$, leading to *exo*-2-cyanonorbornane (16% ee), while $[Pd(BINAP)_2]$ gives 40% ee.[21] An asymmetric rearrangement was catalyzed by a nickel(0) complex bearing a diphosphine ligand (eq 6).[22] A DIOP derivative (MOD-DIOP) was more efficient than DIOP or BINAP. MOD–DIOP has previously been found to improve the enantioselectivity, with respect to DIOP, in rhodium-catalyzed hydrogenation of conjugated acids (ee's 90–95%).[23] Structural modifications of DIOP are very easy to perform by changing the nature of the aromatic rings or the acetal group, allowing tuning of the enantioselectivity. Many publications describe modified DIOP derivatives. DIOP has been utilized in several stoichiometric reactions, for example in an intramolecular Wittig reaction for the synthesis of the bis-nor-Wieland–Miescher ketone (52% ee).[24]

MeO_2C, CO_2Me → (0.15 equiv $Ni(cod)_2$, 0.20 equiv MOD–DIOP; DMSO, rt, 4 h; 77%) MeO_2C, MeO_2C; 90% ee (6)

MOD-DIOP = ... PAr_2, PAr_2; Ar = ... OMe

Related Reagents. See Classes R-2, R-3, R-5, and R-18, pages 1–10.

1. (a) Kagan, H. B. In *Comprehensive Organometallic Chemistry*; Wilkinson, G., Ed; Pergamon: Oxford, 1982; Vol. 8, pp 464–498. (b) Kagan, H. B. In *Asymmetric Synthesis*; Morrison, J. D., Ed; Academic: New York, 1985; Vol. 5, pp 1–39. (c) Brunner, H. *Top. Stereochem.* **1988**, *18*, 129. (d) Takaya, H.; Ohta, T.; Noyori, R. In *Catalytic Asymmetric Synthesis*; Ojima, I., Ed; VCH: New York, 1993; pp 1–39. (e) Ojima, I.; Hirai, K. In *Asymmetric Synthesis*; Morrison, J. D., Ed; Academic: New York, 1985; Vol. 5, pp 103–146.
2. Kagan, H. B.; Dang, T. P. *JACS* **1972**, *94*, 6429.
3. Murrer, B. A.; Brown, J. M.; Chaloner, P. A.; Nicholson, P. N.; Parker, D. *S* **1979**, 350.
4. Brisset, H.; Gourdel, Y.; Pellon, P.; Le Carre, M. *TL* **1993**, *34*, 4523.
5. Zhang, S. Q.; Zhang, S. Y.; Feng, R. *TA* **1991**, *2*, 173.
6. Whitesell, J. K. *CRV* **1989**, *89*, 1581.
7. Stoll, A. P.; Süess, R. *HCA* **1974**, *57*, 2487.
8. Gelbard, G.; Kagan, H. B.; Stern, R. *T* **1976**, *32*, 233.
9. Townsend, J. M.; Blount, J. F.; Sun, R. C.; Zawoiski, S.; Valentine, D., Jr. *JOC* **1980**, *45*, 2995.
10. James, B. R.; Mahajan, D. *JOM* **1985**, *279*, 31.
11. Sinou, D.; Kagan, H. B. *JOM* **1976**, *114*, 325.
12. Bakos, J.; Toth, I.; Heil, B.; Marko, L. *JOM* **1985**, *279*, 23.
13. Chan, A. S. C.; Landis, C. R. *J. Mol. Catal.* **1989**, *49*, 165.
14. Chan, Y. N. C.; Osborn, J. A. *JACS* **1990**, *112*, 9400.
15. Ojima, I.; Kogure, T.; Kumagai, M. *JOC* **1977**, *42*, 1671.
16. Kagan, H. B.; Langlois, N.; Dang, T. P. *JOM* **1975**, *90*, 353.
17. Consiglio, G.; Pino, P.; Flowers, L. I.; Pittman, C. U., Jr. *CC* **1983**, 612.
18. Gladiali, S.; Pinna, L. *TA* **1991**, *2*, 623.
19. Burgess, K.; van der Donk, W. A.; Ohlmeyer, M. J. *TA* **1991**, *2*, 613.
20. Sato, M.; Miyaura, N.; Suzuki, A. *TL* **1990**, *31*, 231.
21. (a) Hodgson, M.; Parker, D.; Taylor, R. J.; Ferguson, G. *OM* **1988**, *7*, 1761. (b) Elmes, P. S.; Jackson, W. R. *AJC* **1982**, *35*, 2041.
22. Hiroi, K.; Arinaga, Y.; Ogino, T. *CL* **1992**, 2329.
23. Morimoto, T.; Chiba, M.; Achiwa, K. *TL* **1989**, *30*, 735.
24. Trost, B. M.; Curran, D. P. *TL* **1981**, *22*, 4929.

Henri Kagan
Université de Paris-Sud, Orsay, France

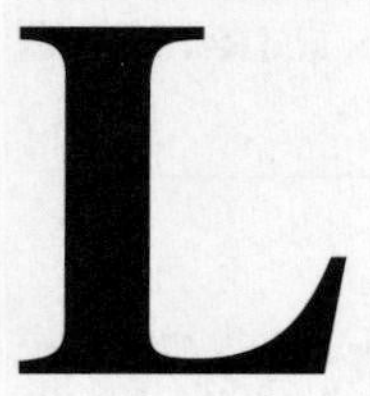

Lead(IV) Acetate[1]

$Pb(OAc)_4$

[546-67-8] $C_8H_{12}O_8Pb$ (MW 443.37)

(oxidizing agent for different functional groups;[1] oxidation of unsaturated and aromatic hydrocarbons;[2] oxidation of monohydroxylic alcohols to cyclic ethers;[3] 1,2-glycol cleavage;[4] acetoxylation of ketones;[1] decarboxylation of acids;[5] oxidative transformations of nitrogen-containing compounds[6])

Alternate Name: lead tetraacetate; LTA.
Physical Data: mp 175–180 °C; d 2.228 g cm^{-3}.
Solubility: sol hot acetic acid, benzene, cyclohexane, chloroform, carbon tetrachloride, methylene chloride; reacts rapidly with water.
Form Supplied in: colorless crystals (moistened with acetic acid and acetic anhydride); widely available, 95–97%.
Analysis of Reagent Purity: iodometrical titration.
Drying: in some cases, acetic acid must be completely removed by drying the reagent in a vacuum desiccator over potassium hydroxide and phosphorus pentoxide for several days.
Handling, Storage, and Precautions: the solid reagent is very hygroscopic and must be stored in the absence of moisture. Bottles of lead tetraacetate should be kept tightly sealed and stored under 10 °C in the dark and in the presence of about 5% of glacial acetic acid.

Oxidations of Alkenic and Aromatic Hydrocarbons. Lead tetraacetate reacts with alkenes in two ways: addition of an oxygen functional group on the double bond and substitution for hydrogen at the allylic position.[2] In addition to these two general reactions, depending on the structure of the alkene, other reactions such as skeletal rearrangement, double bond migration, and C–C bond cleavage can occur, leading to complex mixtures of products, and these reactions therefore have little synthetic value (eq 1).[1a,b,2,7] Styrenes afford 1,1-diacetoxy derivatives when the LTA reaction is performed in acetic acid (eq 2), while in benzene solution products resulting from the addition of both the methyl and an acetoxy group to the alkenic double bond are formed.[7,8] Other nucleophiles, such as azide ion, carbanions, etc. can be introduced onto the alkenic bond in a similar fashion.[9] In the LTA oxidation of cyclic alkenes, depending on ring size, structure, solvent, and reaction conditions, several types of products are formed. Thus 1,2-diacetates and 3-acetoxycycloalkenes are obtained from cyclohexene (cyclopentanecarbaldehyde is also formed),[10] cycloheptene, and cyclooctene.[11] Norbornene reacts with LTA to give rearrangement products in which 2,7-diacetoxynorbornane predominates (eq 3).[12] Conjugated dienes undergo 1,2- and 1,4-diacetoxylation,[13] while cyclopentadiene in wet acetic acid gives monoacetates of *cis*-cyclopentene-1,2-diol (eq 4).[14]

C_6H_{13}-CH=CH$_2$ —LTA, MeOH→ C_6H_{13}CH(OMe)CH$_2$OMe (52%) + C_6H_{13}CH(OMe)CH$_2$OAc (12%) + C_6H_{13}COCH$_3$ (23%) (1)

MeO-C$_6$H$_4$-CH=CH$_2$ —LTA, AcOH, 45 °C, 94%→ MeO-C$_6$H$_4$-CH$_2$CH(OAc)(OAc) (2)

norbornene —LTA, AcOH→ 2,7-diacetoxynorbornane (86%) + OAc/OAc isomer (11%) + OAc norbornene (1%) (3)

cyclopentadiene —LTA, AcOH, H_2O, rt, 75–80%→ cis-cyclopentene-1,2-diol monoacetate (H, Ac) (4)

Aromatic hydrocarbons react with LTA in two ways: on the aromatic ring and at the benzylic position of the side chain. Oxidation of the aromatic ring results in substitution of aromatic hydrogens by acetoxy or methyl groups.[1c] Benzene itself is stable towards LTA at reflux and is frequently used as solvent in LTA reactions. However, mono- and polymethoxybenzene derivatives are oxidized by LTA in acetic acid to give acetoxylation products (eq 5).[15] Oxidation of anthracene in benzene gives 9,10-diacetoxy-9,10-dihydroanthracene, whereas in AcOH a mixture of 10-acetoxy-9-oxo-9,10-dihydroanthracene and anthraquinone is obtained.[16] The LTA oxidation of furan affords 2,5-diacetoxy-2,5-dihydrofuran (eq 6).[17]

MeO-C$_6$H$_4$-OMe —LTA, AcOH, 58%→ MeO-C$_6$H$_3$(OAc)-OMe (5)

furan —LTA, AcOH, 69%→ AcO-(2,5-dihydrofuran)-OAc (6)

Aromatic compounds possessing a C–H group at the benzylic position are readily oxidized by LTA to the corresponding benzyl acetates. Benzylic acetoxylation is preferably performed in refluxing acetic acid (eq 7).[18] Acetoxylation at the benzylic position can be accompanied by methylation of the aromatic ring, followed sometimes by acetoxylation of the newly introduced methyl group.[18]

6-methoxytetralin —LTA, AcOH, 62%→ 1-acetoxy-6-methoxytetralin (OAc, MeO) (7)

Oxidative Cyclization of Alcohols to Cyclic Ethers. The LTA oxidation of saturated alcohols, containing at least four carbon atoms in an alkyl chain or an appropriate carbon skeleton, to five-membered cyclic ethers represents a convenient synthetic method for intramolecular introduction of an ether oxygen function at the nonactivated δ-carbon atom of a methyl, methylene, or methine group (eq 8).[3,19,20] The reactions are carried out in nonpolar solvents, such as benzene, cyclohexane, heptane, and carbon tetrachloride, either at reflux temperature[1a,d,3,20,21] or by UV irradiation at rt.[22]

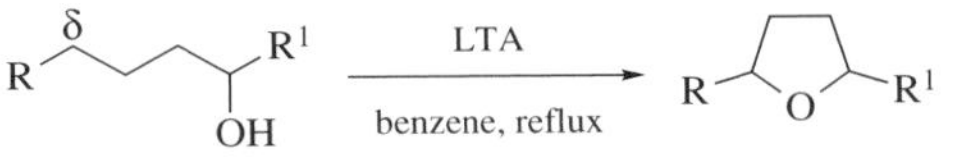

(8)

The conversion of alcohols to cyclic ethers is a complex reaction involving several steps: (i) reversible alkoxylation of LTA by the substrate; (ii) homolytic cleavage of the RO–Pb bond in the resulting alkoxy–lead(IV) acetate with formation of an alkoxy radical; (iii) intramolecular 1,5-hydrogen abstraction in this oxy radical whereby a δ-alkyl radical is generated; (iv) oxidative ring closure to a cyclic ether via the corresponding δ-alkyl cation (eq 9).[3,20] The crucial step is the formation of the δ-alkyl radical by way of 1,5-hydrogen migration. This type of rearrangement is a general reaction of alkoxy radicals, and, independently of the radical precursor, involves a transition state in which the δ-CH group must be conformationally suitably oriented with respect to the attacking oxygen radical.[1,3,23,24] Regioselective hydrogen abstraction proceeds preferentially from the δ-carbon atom, since in that case an energetically favorable quasi-six-membered transition state is involved.[3,23,24]

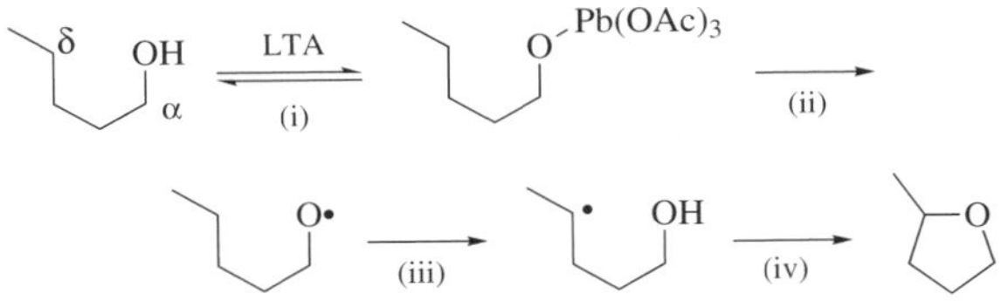

(9)

The LTA oxidation of primary aliphatic alcohols affords 2-alkyltetrahydrofurans in 45–75% yield. A small amount of tetrahydropyran-type ether is also formed (eq 10).[3a,20] The oxidation rate depends on the structural environment of the proactivated carbon atom, with the rate decreasing in the order: methine > methylene > methyl δ-carbon atom.[3] When the δ-carbon atom is adjacent to an ether oxygen function, the reaction rate and the yield of cyclic ethers increases.[25] An ether oxygen attached to the δ-carbon atom increases considerably the yield of six-membered cyclic ethers (eq 11). An aromatic ring adjacent to a δ-methylene group does not noticeably affect the yield of tetrahydrofuran ethers, but when the phenyl group is attached to an ε-methylene group, the yield of six-membered cyclic ethers are enhanced.[26]

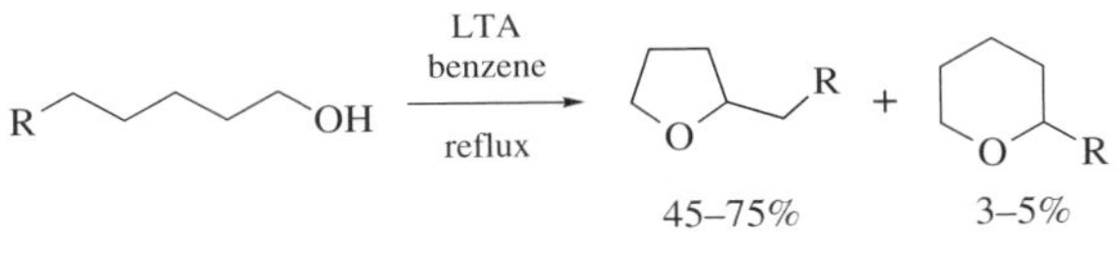

(10)

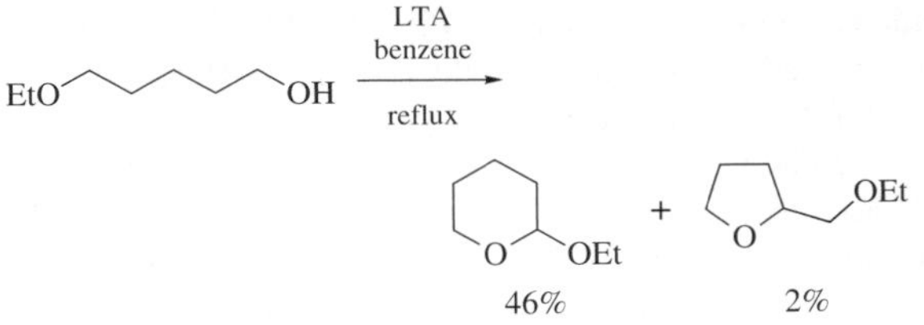

(11)

Secondary aliphatic alcohols containing a δ-methylene group afford a *cis/trans* mixture of 2,5-dialkyltetrahydrofurans in about 33–70% yield (eq 12).[20,22] The LTA oxidation of secondary alcohols is much slower than that of primary alcohols and isomeric six-membered cyclic ethers are not formed.[20,21] Tertiary aliphatic alcohols, because of unfavorable steric and electronic factors, are less suitable for the preparation of tetrahydrofurans by LTA oxidation.[22,27]

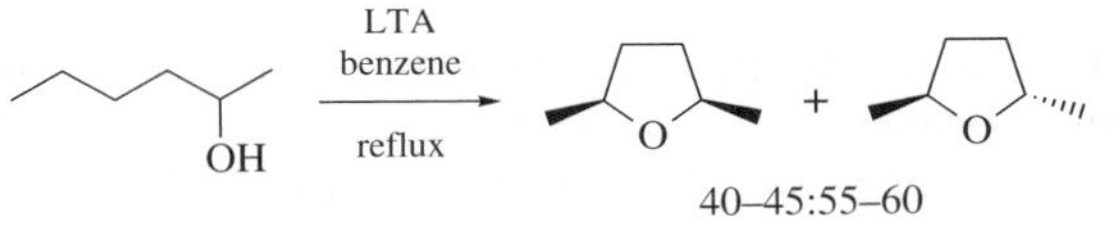

(12)

In the cycloalkanol series, the ease of intramolecular formation of cyclic ether products strongly depends on ring size. Cyclohexanol, upon treatment with LTA, affords only 1% of 1,4-cyclic ether, whereas cycloalkanols with a larger ring, such as cycloheptanol and cyclooctanol, can adopt appropriate conformations necessary for transannular reaction, affording bicyclic ethers in moderate yields (eq 13).[28] Large-ring cycloalkanols, such as cyclododecanol, cyclopentadecanol, and cyclohexadecanol, also give the corresponding 1,4-epoxy compounds as major cyclization products.[3a,28] However, the special geometry of cyclodecanol is not favorable for the 'normal' reaction and the 1,4-cyclic ether is formed in only 2.5% yield, whereas 1,2-epoxycyclodecane (13%) and the rearranged 8-ethyl-7-oxabicyclo[4.3.0]nonane (13%) are the predominant cyclization products.[29]

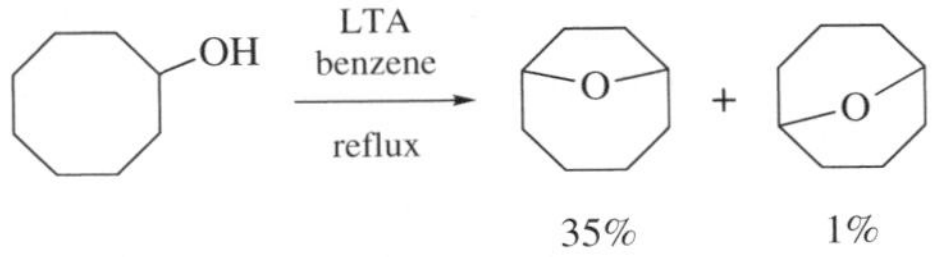

(13)

The LTA oxidation of alcohols to cyclic ethers has been successfully applied as a synthetic method for activation of the angular 18- and 19-methyl groups in steroidal alcohols containing a β-oriented hydroxy group at C-2, C-4, C-6, and C-11 (eq 14).[3c,30,31] Hydroxy terpenoids with suitable stereochemistry can also undergo transannular cyclic ether formation (eq 15).[32]

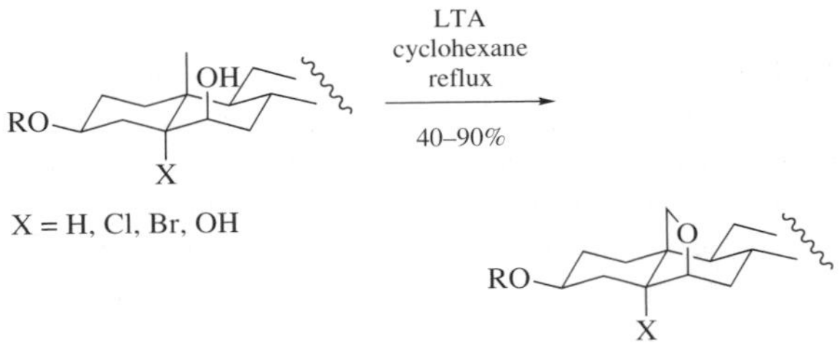

(14)

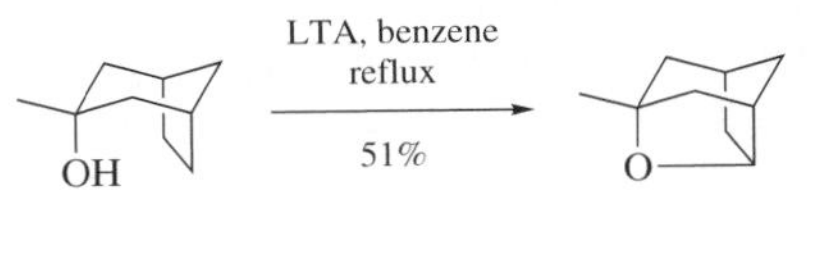

(15)

Another possible reaction of alkoxy radical intermediates, formed in the LTA oxidation of alcohols in nonpolar solvents, is the β-fragmentation reaction.[3] This process, which competes with intramolecular 1,5-hydrogen abstraction, consists of cleavage of a bond between the carbinol (α) and β-carbon atoms, thus affording a carbonyl-containing fragment and products derived from an alkyl radical fragment (usually acetates and/or alkenes).[1a,3,22] Interesting synthetic applications of the LTA β-fragmentation reaction are the formation of 19-norsteroids from their 19-hydroxy precursors and the preparation of 5,10-secosteroids (containing a ten-membered ring) from 5-hydroxy steroids (eq 16).[32]

LTA, benzene, reflux; 39% (16)

In the LTA oxidations of primary and secondary alcohols in nonpolar solvents, the corresponding aldehydes or ketones are usually obtained as minor byproducts (up to 10%).[3,20,21] However, in the presence of excess pyridine or in pyridine alone, either with heating or at rt, the cyclization and β-fragmentation processes are suppressed and good preparative yields of aldehydes or ketones are obtained (eq 17).[20,21,33] Carbonyl compounds are also obtained when the LTA oxidation of alcohols is carried out in benzene solution in the presence of manganese(II) acetate.[33]

LTA, pyridine; 58% + 2% (17)

In addition to cyclic ethers, β-fragmentation products, and carbonyl compounds, acetates of starting alcohols are also usually formed in the LTA oxidation, in yields up to 20%.[20]

Unsaturated alcohols, possessing an alkenic double bond at the δ or more remote positions, react with LTA in nonpolar solvents to give acetoxylated cyclic ethers in good yield (eq 18),[34,35] while 5-, 6- and 7-alkenols undergo in great predominance an *exo*-type cyclization, affording six-, seven- and eight-membered acetoxymethyl cyclic ethers, respectively.[35]

LTA, benzene, reflux; 26% + 14% (18)

1,2-Glycol Cleavage. LTA is one of the most frequently used reagents for the cleavage of 1,2-glycols and the preparation of the resulting carbonyl compounds (eq 19).[1,4] The reactions are performed either in aprotic solvents (benzene, nitrobenzene, 1,2-dichloroethane) or in protic solvents such as acetic acid.[36,37] The rate of LTA glycol cleavage is highly dependent on the structure and stereochemistry of the substrate. In general, there is correlation between the oxidation rate and the spatial proximity of the hydroxy groups.[36] 1,2-Diols having a geometry favoring the formation of cyclic intermediates are much more reactive than 1,2-diols whose structure does not permit such intermediates to be formed (eq 20).[38,39] The oxidation rates often provide a reliable means for the determination of the stereochemical relationship of the hydroxy groups.[39,40]

CO_2Bu, H–OH, H–OH, CO_2Bu —LTA, benzene→ cyclic $Pb(OAc)_2$ intermediate → 2 CO_2Bu–CHO; 77–87% (19)

OH, HO —LTA, $k = 100$→ cyclodecanedione ←LTA, $k = 1$— OH, OH (20)

1,2-Glycol cleavage by LTA has been widely applied for the oxidation of carbohydrates and sugars (eq 21).[4,37] Because of structural and stereochemical differences, the reactivity of individual glycol units in sugar molecules is often different, thus rendering the LTA reaction a valuable tool for structural determination and for degradation studies in carbohydrate chemistry.[41]

CH_2OH, OH, H, OH, OH, OH —LTA, AcOH; 89%→ OHC–O, O–CHO, OH (21)

α-Acetoxylation of Ketones. The reaction of enolizable ketones with LTA is a standard method for α-acetoxylation (eq 22).[1,3,42] The reactions are usually carried out in hot acetic acid or in benzene solution at reflux. The reaction proceeds via an enol–lead(IV) acetate intermediate, which undergoes rearrangement to give the α-acetoxylated ketone. Acetoxylation of ketones is catalyzed by ***Boron Trifluoride***.[43] Enol ethers, enol esters, enamines,[1] β-dicarbonyl compounds, β-keto esters, and malonic esters are also acetoxylated by LTA.[42]

Ph, O —LTA, benzene→ [AcO–Pb(AcO) enolate intermediate] → Ph, O, OAc; 75% (22)

Decarboxylation of Acids. Oxidative decarboxylation of carboxylic acids by LTA depends on the reaction conditions, coreagents, and structure of acids, and hence a variety of products such as acetate esters, alkanes, alkenes, and alkyl halides can be obtained.[1,5] The reactions are performed in nonpolar solvents (benzene, carbon tetrachloride) or polar solvents (acetic acid, pyridine, HMPA).[5] Mixed lead(IV) carboxylates are involved as intermediates, and by their thermal or photolytic decomposition decarboxylation occurs and alkyl radicals are formed (eq 23).[5,44,45]

$$n\,\text{R–CO}_2\text{H} + \text{Pb(OAc)}_4 \rightleftharpoons (\text{RCO}_2)_n\text{Pb(OAc)}_{4-n} \xrightarrow[\text{(or } h\nu)]{\Delta} n\,\text{R}\bullet + \text{CO}_2 + \text{Pb(OAc)}_2 \quad (23)$$

Oxidation of alkyl radicals by lead(IV) species give carbocations and, depending on the reaction conditions and structure of the substrate acids, various products derived from the intermediate alkyl radicals and corresponding carbocations (dimerization, hydrogen transfer, elimination, substitution, rearrangement, etc.) are obtained.[5] Decarboxylation of primary and secondary acids usually affords acetate esters as major products (eq 24).[44] When a mixture of acetates and alkenes is formed, it is recommended (in order to improve the yields of acetate esters) to run the reaction in the presence of potassium acetate (eq 25).[5] The LTA decarboxylation of tertiary carboxylic acids gives a mixture of alkenes and acetate esters.[46] For the preparative oxidative decarboxylation of acids to alkenes, see ***Lead(IV) Acetate–Copper(II) Acetate***.

CO_2H — LTA, benzene, reflux, 75% → OAc (24)

benzene, reflux, 16 h: cyclohexyl–OAc 13% + cyclohexene 47%

cyclohexyl–CO_2H LTA; AcOH, KOAc, 60 °C, 0.3 h: 77%, – (25)

A useful modification of the LTA reaction with carboxylic acids is the oxidation in the presence of halide ions, whereby the corresponding alkyl halides are obtained (eqs 26 and 27).[47] Halodecarboxylations of acids are performed by addition of a molar equivalent of the metal halide (lithium, sodium, potassium chloride) to a carboxylic acid and LTA, the reactions being performed in boiling benzene solution.[5,47] For the iododecarboxylation of acids, see ***Lead(IV) Acetate–Iodine***.

t-Bu CO_2H — LTA, LiCl, benzene, reflux, 92% → t-Bu Cl (26)

cyclohexyl–CO_2H — LTA, LiCl, benzene, reflux, 100% → cyclohexyl–Cl (27)

Bis-decarboxylation of 1,2-dicarboxylic acids by LTA is a useful method for the introduction of alkenic bonds (eq 28).[48] The reactions are performed in boiling benzene in the presence of pyridine or in DMSO. In some cases, LTA bis-decarboxylation can be effected by using acid anhydrides (eq 29).[49] Bis-decarboxylation of 1,1-dicarboxylic acids yields the corresponding ketones (eq 30).[50]

CO_2Et, CO_2H, CO_2H — LTA, benzene, py, 80 °C, 65% → CO_2Et (28)

O, O, O — LTA, py, 20% → (29)

CO_2H, CO_2H — LTA, benzene, reflux, 60% → =O (30)

Oxidative Transformations of Nitrogen-Containing Compounds. The LTA oxidation of aliphatic primary amines containing an α-methylene group results in dehydrogenation to alkyl cyanides (eq 31).[51] However, aromatic primary amines give symmetrical azo compounds in varying yield (eq 32).[52]

$$C_6H_{13}CH_2NH_2 \xrightarrow[62\%]{\text{LTA, benzene, reflux}} C_6H_{13}CN \quad (31)$$

$$Ar{-}NH_2 \xrightarrow[30\text{–}50\%]{\text{LTA, benzene, reflux}} Ar{-}N{=}N{-}Ar \quad (32)$$

Primary amides react with LTA in the presence of alcohols to give the corresponding carbamates (eq 33), but in the absence of alcohol, isocyanates are formed.[53]

cyclobutyl–$CONH_2$ — LTA, t-BuOH, Et_3N, 50–60 °C, 62% → cyclobutyl–$NHCO_2$-t-Bu (33)

Aliphatic ketoximes, upon treatment with LTA in an inert solvent, undergo acetoxylation at the α-carbon producing 1-nitroso-1-acetoxyalkanes (eq 34),[54] whereas hydrazones afford azoacetates (eq 35) or, when the reactions are performed in alcohol solvent, azo ethers.[55] Arylhydrazines, N,N'-disubstituted hydrazines,[56] and N-amino compounds[57] are oxidized by LTA to different products.

NOH, i-Pr — LTA, CH_2Cl_2, 78% → NO, OAc, i-Pr (34)

H, N–Ar, =N — LTA, CH_2Cl_2 → NAr, N, OAc (35)

Other Applications. By LTA oxidation of phenols, acetoxycyclohexadienones, quinones, and dimerization products can be formed.[58] Alkyl sulfides,[59] alkyl hydroperoxides,[60] and organometallic compounds[61] are also oxidized by LTA.

Related Reagents. See Classes O-10, O-18, O-20, O-21, and O-22, pages 1–10.

1. (a) Mihailović, M. Lj.; Čeković, Ž.; Lorenc, Lj. In *Organic Synthesis by Oxidation with Metal Compounds*; Mijs, W. J.; de Jonge, C. R. H. I., Eds.; Plenum: New York, 1986; pp 741–816. (b) Rubottom, G. M. In *Oxidation in Organic Chemistry*; Trahanovsky, W. S., Ed.; Academic: New York, 1982; Part D, pp 1–145. (c) Butler, R. N. In *Synthetic Reagents*; Pizey, J. S., Ed.; Ellis Horwood: Chichester, 1977; vol 3, pp 277–419. (d) Rotermund, G. W. *MOC* **1975**, *4/1b*, 204. (e) Criegee, R. In *Oxidation in Organic Chemistry*, Wiberg, K., Ed.; Academic: New York, 1965; Part A, pp 277–366.
2. Moriarty, R. M. In *Selective Organic Transformations*; Thyagarajan, B. S., Ed.; Wiley: New York, 1972; vol. 2, pp 183–237.
3. (a) Mihailović, M. Lj.; Čeković, Ž. *S* **1970**, 209. (b) Mihailović, M. Lj.; Partch, R. E. In *Selective Organic Transformations*, Thyagarajan, B. S., Ed.; Wiley: New York, 1972; vol. 2, pp 97–182. (c) Heusler, K.; Kalvoda, J. *AG(E)* **1964**, *3*, 525.

4. (a) Bunton, C. A. In *Oxidation in Organic Chemistry*; Wiberg, K., Ed.; Academic: New York, 1965; Part A, pp 398–405. (b) Prelin, A. S. *Adv. Carbohydr. Chem.* **1959**, *14*, 9.
5. Sheldon, R. A.; Kochi, J. K. *OR* **1972**, *19*, 279.
6. (a) Aylward, J. B. *QR* **1971**, *25*, 407. (b) Butler, R. N.; Scott, F. L.; O'Mahony, *CRV* **1973**, *73*, 93. (c) Warkentin, J. *S* **1970**, 279.
7. Lethbridge, A.; Norman, R. O. C.; Thomas, C. B.; Parr, W. J. E. *JCS(P1)* **1974**, 1929; **1975**, 231.
8. Criegee, R.; Dimroth, P.; Noll, K.; Simon, R.; Weis, C. *CB* **1957**, *90*, 1070.
9. Zbiral, E. *S* **1972**, 285, and references cited therein.
10. (a) Criegee, R. *AG* **1958**, *70*, 173. (b) Anderson, C. B.; Winstein, S. *JOC* **1963**, *28*, 605.
11. Cope, A. C.; Gordon, M.; Moon, S.; Park, C. H. *JACS* **1965**, *87*, 3119.
12. Kagan, J. *HCA* **1972**, *55*, 2356.
13. (a) Criegee, R.; Beucker, H. *LA* **1939**, *541*, 218. (b) Posternak, Th.; Friedli, H. *HCA* **1953**, *36*, 251.
14. Brutcher, F. V., Jr.; Vara, F. J. *JACS* **1956**, *78*, 5695.
15. (a) Cavill, G. W. K.; Solomon, D. H. *JCS* **1955**, 1404. (b) Preuss, F. R.; Janshen, J. *AP* **1958**, *291*, 350, 377.
16. (a) Rindone, B.; Scolastico, C. *JCS(C)* **1971**, 3983. (b) Fieser, L. F.; Putnam, S. T. *JACS* **1947**, *69*, 1038, 1041.
17. (a) Elming, N.; Clauson-Kaas, N. *ACS* **1952**, *6*, 535. (b) Elming, N. *ACS* **1952**, *6*, 578.
18. (a) Heiba, E. I.; Dessau, R. M.; Koehl, W. J., Jr. *JACS* **1968**, *90*, 1082. (b) Cavill, G. W. K.; Solomon, D. H. *JCS* **1954**, 3943.
19. (a) Mićović, V. M.; Mamuzić, R. I.; Jeremić, D.; Mihailović, M. Lj. *TL* **1963**, 2091; *T* **1964**, *20*, 2279. (b) Cainelli, G.; Mihailović, M. Lj.; Arigoni, D.; Jeger, O. *HCA* **1959**, *42*, 1124.
20. (a) Mihailović, M. Lj.; Čeković, Ž.; Maksimović, Z.; Jeremić, D.; Lorenc, Lj.; Mamuzić, R. I. *T* **1965**, *21*, 2799. (b) Čeković, Ž.; Bošnjak, J.; Mihailović, M. Lj. reviewed in *FF* **1986**, *12*, 270.
21. (a) Mihailović, M. Lj.; Bošnjak, J.; Maksimović, Z.; Čeković, Ž.; Lorenc, Lj. *T* **1966**, *21*, 955. (b) Partch, R. E. *JOC* **1965**, *30*, 2498.
22. (a) Mihailović, M. Lj.; Jakovljević, M.; Čeković, Ž. *T* **1969**, *25*, 2269. (b) Mihailović, M. Lj.; Mamuzić, R. I.; Žigić-Mamuzić, Lj.; Bošnjak, J.; Čeković, Ž. *T* **1967**, *23*, 215.
23. Hesse, R. H. In *Advances in Free Radical Chemistry*, Williams, G. H. Ed. Logos: London, 1969; vol. 3, pp 83–137.
24. Akhtar, M. In *Advances in Photochemistry*; Noyes, W. A.; Hammond, G. S.; Pitts, J. N., Eds. Interscience: New York, 1964; vol. 2, pp 263–303.
25. (a) Mihailović, M. Lj.; Miloradović, M. *T* **1966**, *22*, 723. (b) Mihailović, M. Lj.; Milovanović, A.; Konstantinović, S.; Janković, J.; Čeković, Ž.; Partch, R. E. *T* **1969**, *25*, 3205.
26. (a) Mihailović, M. Lj.; Živković, L.; Maksimović, Z.; Jeremić, D.; Čeković, Ž.; Matić, R. *T* **1967**, *23*, 3095. (b) Mihailović, M. Lj.; Matić, R.; Orbović, S.; Čeković, Ž. *Bull. Soc. Chim. Beograd* **1971**, *36*, 363 (*CA* *78*, 42 502).
27. Mihailović, M. Lj.; Jakovljević, M.; Trifunović, V.; Vukov, R.; Čeković, Ž. *T* **1968**, *24*, 6959.
28. (a) Mihailović, M. Lj.; Čeković, Ž.; Andrejević, V.; Matić, R.; Jeremić, D. *T* **1968**, *24*, 4947. (b) Cope, A. C.; McKervey, M. A.; Weinshenker, N. M.; Kinnel, R. B. *JOC* **1970**, *35*, 2918.
29. Mihailović, M. Lj.; Andrejević, V.; Jakovljević, M.; Jeremić, D.; Stojiljković, A.; Partch, R. E. *CC* **1970**, 854.
30. Heusler, K.; Kalvoda, J. In *Steroid Synthesis*; Fried, J.; Edwards, J. A., Eds. Reinhold; New York, 1972; vol. II, pp 237–287.
31. (a) Heusler, K.; Kalvoda, J.; Anner, G.; Wettstein, A. *HCA* **1963**, *46*, 352. (b) Heusler, K.; Kalvoda, J.; Wieland, P.; Anner, G.; Wettstein, A. *HCA* **1962**, *45*, 2575. (c) Shoppee, C. W.; Coll, J. C.; Lack, R. E. *JCS(C)* **1970**, 1893. (d) Bowers, A.; Denot, E.; Ibáñez, L. C.; Cabezas, M. E.; Ringold, H. J. *JOC* **1962**, *27*, 1862. (e) Bowers, A.; Villotti, R.; Edwards, J. A.; Denot, E.; Halpern, O. *JACS* **1962**, *84*, 3204.
32. (a) Amorosa, M.; Caglioti, L.; Cainelli, G.; Immer, H.; Keller, J.; Wehrli, H.; Mihailović, M. Lj.; Schaffner, K.; Arigoni, D.; Jeger, O. *HCA* **1962**, *45*, 2674. (b) Mihailović, M. Lj.; Stefanović, M.; Lorenc, Lj.; Gašić, M. *TL* **1964**, 1867. (c) Mihailović, M. Lj.; Lorenc, Lj.; Gašić, M.; Rogić, M.; Melera, A.; Stefanović, M. *T* **1966**, *22*, 2345.
33. (a) Mićović, V. M.; Mihailović, M. Lj. *RTC* **1952**, *71*, 970. (b) Partch, R. E. *TL* **1964**, 3071. (c) Mihailović, M. Lj.; Konstantinović, S.; Vukićević, R. *TL* **1986**, *27*, 2287.
34. (a) Moon, S.; Lodge, J. M. *JOC* **1964**, *29*, 3453. (b) Moriarty, R. M.; Kapadia, K. *TL* **1964**, 1165. (c) Moon, S.; Haynes, L. *JOC* **1966**, *31*, 3067. (d) Bowers, A.; Denot, B. *JACS* **1960**, *82*, 4956.
35. Mihailović, M. Lj.; Čeković, Ž.; Stanković, J.; Pavlović, N.; Konstantinović, S.; Djokić-Mazinjanin, S. *HCA* **1973**, *56*, 3056.
36. (a) Criegee, R.; Höger, E.; Huber, G.; Kruck, P.; Marktscheffel, F.; Schellenberger, H. *LA* **1956**, *599*, 81. (b) Criegee, R.; Büchner, E.; Walther, W. *CB* **1940**, *73*, 571.
37. (a) Wolf, F. J.; Weijlard, J. *OSC* **1963**, *4*, 124. (b) Bishop, C. T. *Methods Carbohydr. Chem.* **1972**, *6*, 350. (c) O'Colla, P. S. *Methods Carbohydr. Chem.* **1965**, *5*, 382.
38. (a) Criegee, R.; Kraft, C.; Rank, B. *LA* **1933**, *507*, 159. (b) Bunton, C. A.; Carr, M. D. *JCS* **1963**, 770.
39. (a) Criegee, R.; Marchand, B.; Wannowius, H. *LA* **1942**, *550*, 99. (b) Moriconi, E. J.; Wallenberger, F. T.; O'Connor, W. F. *JACS* **1958**, *80*, 656.
40. (a) Angyal, S. J.; Young, R. J. *JACS* **1959**, *81*, 5467. (b) Clark-Lewis, J. W.; Williams, L. R. *AJC* **1963**, *16*, 869. (c) Bauer, H. F.; Stuetz, D. E. *JACS* **1956**, *78*, 4097.
41. (a) Perlin, A. S.; Brice, C. *CJC* **1956**, *34*, 541. (b) Gorin, P. A. J.; Perlin, A. S. *CJC* **1958**, *36*, 480. (c) Perlin, A. S. *JACS* **1954**, *76*, 5505. (d) Charlson, A. J.; Perlin, A. S. *CJC* **1956**, *34*, 1200.
42. (a) Rawilson, D. J.; Sosnovsky, G. *S* **1973**, 567. (b) Cavill, G. W. K.; Solomon, D. H. *JCS* **1955**, 4426.
43. (a) Henbest, H. B.; Jones, D. N.; Later, G. P. *JCS* **1961**, 4472. (b) Cocker, J. D.; Henbest, H. B.; Phillips, G. H.; Slater, G. P.; Thomas, D. A. *JCS* **1965**, 6.
44. (a) Kochi, J. K.; Bacha, J. D.; Bethea, T. W. *JACS* **1967**, *89*, 6538. (b) Kochi, J. K. *JACS* **1965**, *87*, 1811. (c) Kochi, J. K. *JACS* **1965**, *87*, 3609. (d) Davies, D. I.; Waring, C. *JCS(C)* **1968**, 1865, 2332.
45. Kochi, J. K.; Bacha, J. D. *JOC* **1968**, *33*, 2746.
46. (a) Bennett, C. R.; Cambie, R. C. *T* **1967**, *23*, 927. (b) Bennett, C. R.; Cambie, R. C.; Denny, W. A. *AJC* **1969**, *22*, 1069. (c) Mihailović, M. Lj.; Bošnjak, J.; Čeković, Ž. *HCA* **1974**, *57*, 1015. (d) Huffman, J. W.; Arapakos, P. G. *JOC* **1965**, *30*, 1604.
47. (a) Kochi, J. K. *JACS* **1965**, *87*, 2500. (b) Kochi, J. K. *JOC* **1965**, *30*, 3265. (c) Jenkins, C. L.; Kochi, J. K. *JOC* **1971**, *36*, 3095.
48. (a) Grob, C. A.; Ohta, M.; Weiss, A. *AG* **1958**, *70*, 343. (b) Grob, C. A.; Ohta, M.; Renk, E.; Weiss, A. *HCA* **1958**, *41*, 1191. (c) Grob, C. A.; Weiss, A. *HCA* **1960**, *43*, 000.
49. (a) van Tamelen, E. E.; Pappas, S. P. *JACS* **1963**, *85*, 3297. (b) Cimarusti, C. M.; Wolinsky, J. *JACS* **1968**, *90*, 113.
50. (a) Tufariello, J. J.; Kissel, W. J. *TL* **1960**, 6145. (b) Meinwald, J.; Tufariello, J. J.; Hurst, J. J. *JOC* **1964**, *29*, 2914.
51. (a) Mihailović, M. Lj.; Stojiljković, A.; Andrejević, V. *TL* **1965**, 461. (b) Stojiljković, A.; Andrejević, V.; Mihailović, M. Lj. *T* **1967**, *23*, 721.
52. (a) Pausacker, K. H.; Scroggie, J. G. *JCS* **1954**, 4003. (b) Dimroth, K.; Kalk, F.; Neubauer, G. *CB* **1957**, *90*, 2058. (c) Richter, H. J.; Dressler, R. L. *JOC* **1962**, *27*, 4066.
53. (a) Acott, B.; Beckwith, A. L. J.; Hassanali, A. *AJC* **1968**, *21*, 197. (b) Acott, B.; Beckwith, A. L. J.; Hassanali, A. *AJC* **1968**, *21*, 185. (c) Baumgarten, H. E.; Smith, H. L.; Staklis, A. *JOC* **1975**, *40*, 3554. (d) Baumgarten, H. E.; Staklis, A. *JACS* **1965**, *87*, 1141.
54. (a) Kropf, H.; Lambeck, R. *LA* **1966**, *700*, *1*, 18. (b) Kaufmann, S.; Tökés, L.; Murphy, J. W.; Crabbé, P. *JOC* **1969**, *34*, 1618. (c) Shafiullah, D.; Ali, H. *S* **1979**, 124.

55. (a) Iffland, D. C.; Salisbury, L.; Schafer, W. R. *JACS* **1961**, *83*, 747. (b) Harrison, M. J.; Norman, R. O. C.; Gladstone, W. A. F. *JCS(C)* **1976**, 735.

56. (a) Aylward, J. B. *JCS(C)* **1969**, 1663. (b) Hoffman, R. W. *CB* **1964**, *97*, 2763, 2772. (c) Clement, R. A. *JOC* **1960**, *25*, 1724. (d) Schaap, A. P.; Faler, G. R. *JOC* **1973**, *38*, 3061.

57. Person, H.; Fayat, C.; Tonnard, F.; Foucand, A. *BSF* **1974**, 635.

58. (a) Wessely, F.; Sinwel, F. *M* **1950**, *81*, 1055. (b) Wessely, F.; Zbiral, E.; Sturm, H. *CB* **1960**, *93*, 2840. (c) Harrison, M. J.; Norman, R. O. C. *JCS(C)* **1970**, 728. (c) Mihailović, M. Lj.; Čeković, Ž. In *The Chemistry of the Hydroxyl Group*; Patai, S., Ed.; Wiley: New York, 1971; Part 1, pp 505–592.

59. Field, L.; Lawson, J. E. *JACS* **1958**, *80*, 838.

60. (a) Kropf, H.; Goschenhofer, D. *TL* **1968**, 239. (b) Kropf, H.; Wallis, von H. *S* **1981**, *237*, 633.

61. (a) Corey, E. J.; Wollenberg, R. H. *JACS* **1974**, *96*, 5581. (b) Heck, R. F. *JACS* **1968**, *90*, 5542. (c) Barborak, J. C.; Pettit, R. *JACS* **1967**, *89*, 3080.

Mihailo Lj. Mihailović & Živorad Čeković
University of Belgrade, Yugoslavia

Lithium[1]

[7439-93-2] Li (MW 6.94)

(powerful reducing agent;[1] used for partial reduction of aromatics and conjugated polyenes;[1e,m–o] conversion of alkynes to *trans*-alkenes;[1a,h] stereoselective reduction of hindered ketones;[1l] enone reduction and regioselective alkylation;[1f,j] reductive cleavage of polar single bonds[1a])

Physical Data: mp 180.5 °C; bp 1327 ± 10 °C; *d* 0.534 g cm^{-3}. Natural isotopic composition: ^{7}Li (92.6 %); ^{6}Li (7.4 %).

Solubility: 10.9 g/100 g NH_3 at −33 °C (= 74.1 g/L NH_3); 36.5 g/L $MeNH_2$ at −23 °C.

Form Supplied in: under Ar, as solid in the form of wire, ribbon, rod, foil, shot, ingot, or as a powder; in mineral oil, as wire, shot, or as 25–30 wt % dispersions.

Purification: commercially available in up to 99.97% purity. In general, lithium is not further purified except for cutting off the surface coating.

Handling, Storage, and Precautions: best stored under mineral oil in airtight steel drums and handled under Ar or He. Dispersions in mineral oil segregate on storage and uniformity is restored by stirring. The mineral oil is washed off under Ar with pentane or hexane, and the metal is either dried in an Ar stream or rinsed with the reaction solvent. Dry Li powder is extremely reactive towards air, H_2O vapor, and N_2. The metal reacts rapidly with moist air at 25 °C, but with dry air or dry O_2 only at higher temperatures (>100 °C). A slight blow can initiate violent burning. Reaction with N_2 already occurs at 25 °C, but is inhibited by traces of O_2. Li reacts readily with H_2O, but does not spontaneously ignite as the other alkali metals do. It reacts rapidly with dil HCl and H_2SO_4 and vigorously with HNO_3. Ready reaction occurs with halogens.

Reducing Systems. For reductions with Li, liquid ***Ammonia*** or primary amines are most often the solvents of choice. Li dissociates in these solvents more or less completely into Li^+ and solvated electrons, producing deep-blue metastable solutions.[2] Ethereal solvents (peroxide-free!) such as THF or DME may be used alone, but are usually used as cosolvents with NH_3 or amines. Li solutions in HMPA are quite unstable in contrast to ***Sodium*** solutions, but are stabilized by THF.[3] Reduction occurs by a sequence of single electron and proton transfers to the organic substrate, leading to saturation of multiple bonds or fission of single bonds.[1a,i,4]

Li–NH$_3$*.** Li has a higher normal reduction potential[5] and molar solubility[6] in liquid NH_3 than Na or ***Potassium (see Table 1). This permits the use of larger quantities of cosolvents for substrates that are less soluble in NH_3. The concentrations of Li in NH_3 used in reactions vary widely, from 0.1 to 3 g Li/100 mL NH_3. Concentrations near saturation form a second, less dense, bronze-colored phase which is normally avoided.[7]

Table 1 Solutions of Alkali Metals in Liquid Ammonia

Metal	Solubility[6] at -33 C (g metal/100 g NH_3) (g-atom M/mol NH_3)	Normal reduction potential[5] at -50 C (V)
Li	10.9 (0.26)	-2.99
Na	24.5 (0.18)	-2.59
K	47.8 (0.21)	-2.73

Reductions are performed either in the absence or presence (Birch conditions[1h]) of a proton source, depending on the desired products.[1a,i,4b] An added proton source can effect reductions which do not occur in its absence (e.g. benzene reduction). It can lead to higher saturation (e.g. in enone reduction) or suppress dimerization and base-catalyzed transformations of primary products. EtOH and *t*-BuOH are the most common proton donors. Primary alcohols protonate the intermediate anions more rapidly, but tertiary alcohols react more slowly with the metal. Other proton donors are NH_4Cl, H_2O, and various amines.[1a,e,4b]

The order of adding the reagents can influence the product distribution.[1g] Most often, Li is added last, until the blue color of the solution persists. For less reactive substrates, alcohol addition is delayed (Wilds–Nelson modification).[5b] The reaction is concluded by quenching excess Li mildly and efficiently with sodium benzoate[8] or with excess EtOH and then NH_4Cl, and NH_3 is allowed to evaporate.

Distillation of NH_3 from Na or through a BaO column removes moisture and iron impurities. The latter catalyze the reaction of alkali metals with the added alcohol and NH_3.[2] The Li–NH_3–ROH system is less sensitive to traces of iron than Na–NH_3–ROH, which accounts in many instances for its superiority.[9] ***Lithium Amide*** is less soluble in NH_3 than ***Sodium Amide*** and ***Potassium Amide***, and base-catalyzed formation of side products is less frequent.[1g,i] Nevertheless, in many cases similar results are obtained with Li and Na in NH_3; Li is preferred for less reactive substrates and Na when overreduction is a problem.[1]

Lithium/Primary Amines. Li forms stronger, but less selective, reducing agents with primary amines (Benkeser reduction).[1d,h,k] The higher reactivity is probably caused by higher reaction temperatures[1i,k] and possibly also by smaller electron solvation.[2] The reactivity can be modified by addition of alcohols.[1h] The Li–amine solutions seem more sensitive to catalytic decomposition than Li in NH_3.[10] The choice of the amine is limited by the solubility of Li; ethylamine and ethylenediamine are most common. Na is hardly soluble in amines (e.g. more than 100 times less soluble in ethylenediamine at room temperature than Li).[11] Reactions of calcium in amines have been described.[12]

Reduction of Aromatic Compounds[1,13]**.** Benzene and its derivatives are reduced to 1,4-cyclohexadienes with Li–NH_3 in the presence of a proton source (see also ***Sodium–Ammonia***). Derivatives with electron-donating substituents lead to 1-substituted cyclohexadienes. Thus reduction of anisole derivatives furnishes 1-methoxycyclohexa-1,4-dienes (eq 1).[9]

Li–NH_3, t-BuOH, 87% (1)

Hydrolysis of such dienol ethers to cyclohex-3-enones or with isomerization to cyclohex-2-enones has found wide application in syntheses of steroids, terpenoids, and alkaloids.[1f,m] Li is superior to Na for the more difficult reductions of 1,2,3-substituted anisole derivatives,[5b] though sometimes even excess Li gives poor results.[14] Anisoles are more readily reduced than phenols (eq 2),[15] but higher concentration of Li in NH_3 may effect phenol reduction to cyclohexenols (eq 3).[16]

1.5 M Li–NH_3, t-BuOH, 73% (2)

4.3 M Li–NH_3, EtOH, 76% (3)

Electron-acceptor substituents enhance reduction rates and promote 1,4-reduction at the substituted carbon atoms, irrespective of alkoxy, amino, or alkyl substituents. Benzoic acid derivatives are readily reduced to the 1,4-dihydro derivatives. The presence of an alcohol is not necessary, in contrast to the derivatives with electron-releasing substituents. It can even result in overreduction, as the lithium alcoholate facilitates isomerization of the 1,4-dihydro product to the 3,4-dihydro isomer (eq 4).[17]

1. Li–NH_3, 2. NH_4Cl; 1. Li–NH_3, 2. EtOH; Li–NH_3, EtOH (4)

The dienolate formed during the reduction can be alkylated in situ with alkyl halides,[18,19] epoxides,[19] or α,β-unsaturated esters (eq 5)[20] to give 1-substituted dihydrobenzoic acids. Rearomatization provides alkyl-substituted aromatic compounds.

1. Li–NH_3, –78 °C, 2. MeCH=$CHCO_2Me$, 70%; air (5)

Benzamides and alkyl benzoates can be reduced to the 1,4-dihydro amides and esters, respectively, with Li–NH_3–t-butanol, but K–NH_3–t-butanol appears superior.[13] However, Li may be better for in situ reductive alkylations, or K^+ may be exchanged with Li^+ before the alkylation step.[21] Reductive methylation of N-benzoyl-L-prolinol derivatives afforded excellent diastereoselectivities, irrespective of the use of Li, Na, or K (eq 6).[22]

1. Li–NH_3, THF, t-BuOH; 2. MeI (6)

R^1 = OMe, ca. 85%, de >260:1
R^1 = Me, 90%, de <1:99

The strongly activating and easily removable trimethylsilyl group has been used to direct the regioselectivity of reduction (eq 7).[23]

1. Li–NH_3, –78 °C, 67%; 2. NBu_4F, 50–60% (7)

The Li–amine–alcohol reagents also reduce benzene derivatives to cyclohexadienes, and are usually applied when reduction in NH_3 fails.[1h,k] Thus reduction of dehydroabietic acid with Li–NH_3–t-BuOH afforded 35% of diene while Li–$EtNH_2$–t-$C_5H_{11}OH$ gave 81% (eq 8).[24] The importance of the nature of the proton source is demonstrated by the fact that neither Li–NH_3–EtOH nor Li–$EtNH_2$–EtOH gave any appreciable amount of reduction product.

Li–EtNH$_2$, t-C$_5$H$_{11}$OH, 81%

CO_2H CO_2H (8)

Reduction with Li–amine gives mainly cyclohexenes due to isomerization of the initially formed 1,4-diene by the strong alkylamide base.[1h,i] Mixtures of regioisomers are formed, and best results favoring the most stable isomer are obtained with mixtures[1h] of primary and secondary amines (eq 9).[25]

+ isomers (9)

Li–EtNH$_2$ 79%, 87:13
Li–EtNH$_2$–Me$_2$NH (1:1) 88%, 96:4

Condensed aromatic hydrocarbons are reduced more easily than those in the benzene series. Carefully chosen reaction conditions lead to the selective formation of different products.[1b] Most extensive reductions are achieved with Li–ethylenediamine,[26] while Na–NH$_3$ is one of the mildest reagents.[27] Birch and Slobbe discuss the reduction of heterocyclic aromatics.[4b]

Lithium-induced cyclization of 1,1′-binaphthalenes followed by oxidation of the dianion affords perylenes.[28] 3,10-Dimethylperylene was obtained in 95% (eq 10a).[28a] Cyclizations to tetrasubstituted perylenes proceeded in 36–40%,[28b] while similar reactions with K seem somewhat higher yielding.[29a] However, the synthesis of an 1-alkylated perylene was only successful with Li (eq 10b).[29b]

R^1 R^2 R^1 → R^1 R^2 R^1 (10)

(a) R^1 = Me, R^2 = H — 1. Li, THF, Δ; 2. O$_2$; 95%
(b) R^1 = H, R^2 = (CH$_2$)$_5$Me — 1. Li, DME, Δ; 2. CdCl$_2$; 30%
1. K, DME, 25 °C; 2. CdCl$_2$; <5%

Reduction of Alkynes[1b,h]. Internal alkynes are reduced to *trans*-alkenes with Li–NH$_3$ or stoichiometric amounts of Li in amines. Excess Li in amines leads to alkanes. Li–EtNH$_2$–*t*-BuOH efficiently reduced an alkyne precursor of sphingosine to the *trans*-alkene with simultaneous *N*-debenzylation, while triple bond reduction was incomplete with Na–NH$_3$ and Li–NH$_3$ (eq 11).[30]

Li–EtNH$_2$, t-BuOH, –78 °C, 88%

NBn, OH, C$_{13}$H$_{27}$ → NH, OH, C$_{13}$H$_{27}$ (11)

Dissolving metal reduction is the method of choice for the reduction of triple bonds in the presence of nonconjugated carboxyl groups,[31] where ***Lithium Aluminum Hydride***$_4$ in THF fails. Li–NH$_3$ afforded higher amounts of *trans*-alkenes in the reduction of some cyclic alkynes compared with Na–NH$_3$.[32] Terminal triple bonds are protected against reduction with Li–NH$_3$ by deprotonation with alkali amide, but are completely reduced to double bonds by Li (or Na)–NH$_3$–(NH$_4$)$_2$SO$_4$ or by Li in amines.[1c] Suitably located carbonyl groups give rise to cyclization, yielding vinylidenecycloalkanols,[33] e.g. eq 12.[33a] However, the use of K[33a,34a] or electrochemical reduction[34b] may give better results.

Li–NH$_3$, (NH$_4$)$_2$SO$_4$, 50% (12)

Reduction of Ketones[1a,b,l,35]. Li–NH$_3$–EtOH reduces sterically hindered cyclic ketones to equatorial alcohols (eq 13)[36] and has been widely applied in the syntheses of 11α-hydroxy steroids.[37] This method is complementary to complex hydride reductions, which mainly afford the axial alcohols. Bicyclo[2.2.1]heptanones are reduced predominantly to the *endo*-alcohols. Similar results have been found with Na, K, and Ca.[35]

Li–NH$_3$, EtOH (13)

9α:9β > 99:1

α,β-Unsaturated ketones are reduced to the ketone by Li–NH$_3$.[1j] In fused ring enones the relative configuration at the ring junction is determined by protonation at the β-carbon.[38] Regioselective alkylation is achieved by trapping the intermediate enolate with an alkyl halide,[39a] a strategy also applied to enediones (eq 14).[39b] In the presence of a proton source, reduction to the saturated alcohols occurs.[40] Li–EtND$_2$–*t*-BuOD reduction gives high yields of saturated ketones and has been used for the stereoselective deuteriation at the β-carbon.[41] Conversion to alkenes is accomplished by phosphorylation of an enolate formed by Li–NH$_3$ reduction and subsequent hydrogenolysis with Li–EtNH$_2$–*t*-BuOH (eq 15).[42]

1. Li–NH$_3$, 2. PhCH$_2$Br, 65% → PhCH$_2$ (14)

1. Li–NH$_3$, 2. (EtO)$_2$P(O)Cl, 56% → (EtO)$_2$P(O)O → Li–EtNH$_2$, t-BuOH, 91% (15)

Reductions of aromatic ketones are complicated by possible pinacol formation, reduction of the aromatic ring, and hydrogenolysis of the C–O bond. Depending on the reaction conditions, 1-tetralone is reduced to tetralin or 1-tetralol (eq 16);[43a] in fact, seven different products can be produced.[43b]

1. Li–NH$_3$ 2. addition of NH$_4$Cl 96% ; 1. Li–NH$_3$ 2. NaOBz or addition to aq NH$_4$Cl (16)

Aromatic aldehydes and ketones are alkylated and deoxygenated in a one-pot procedure using alkyl- or aryllithium, followed by Li–NH$_3$ (eq 17).[44]

1. R^2Li 2. Li–NH$_3$ 3. NH$_4$Cl 65–97% (17)

R^1 = H, alkyl, aryl

Aliphatic Carboxylic Acids. Simple straight chain carboxylic acids are reduced by Li–MeNH$_2$ or Li–NH$_3$ to an intermediate imine which can either be hydrolyzed to the aldehyde or catalytically reduced to the amine.[45]

Reductive Cleavage of Polar Single Bonds[1a]. Li in various solvents provides effective reagents for the cleavage of polar single bonds. The cleavage tendency decreases in the order C–I > C–Br > C–Cl > C–S > C–O > C–N > C–C. Polyhalo compounds are completely reduced with Li and *t*-BuOH in THF (Winstein procedure).[46] Allylic, geminal, bridgehead, and vinylic halogen atoms are removed, the latter stereospecifically. NH$_3$ and amines have been avoided as solvents due to potential reaction with the alkyl halides by elimination or substitution.[1a,47a] However, Li–NH$_3$ systems successfully reduce vinylic, bridgehead, and cyclopropyl halides[47b,c] and sometimes give better results than the Winstein–Gassman procedures (eq 18).[47c]

Li–NH$_3$–EtOH, Et$_2$O, –78 °C, 85% (18)

Alkyllithium reagents nowadays often replace Li for the preparation of organolithium compounds from alkyl or aryl bromides.[48] Li has been used to couple alkyl and aryl halides in Wurtz or Wurtz–Fittig-type reactions,[49] though the use of Na is much more important. Reduction of monosubstituted alkyl halides or selective reduction of geminal dihalides are best carried out with metal or complex hydrides or by catalytic hydrogenation.[1b]

Sulfides, sulfoxides, and sulfones are reductively cleaved with lithium.[50] Reduction of sulfides in THF is improved with catalytic naphthalene. Li–EtNH$_2$ gave better results than ***Sodium Amalgam*** for the cleavage of the C–S bond in sulfones (eq 19),[51a,b] and than ***Raney Nickel*** for some sulfide cleavage.[51c] Selenides are cleaved similarly.[52] Thio- and selenoacetals are reduced to alkanes.

Li–EtNH$_2$, –75 °C, 90% (19)

(*Z*):(*E*) = 90:10

Allyl, benzyl, and aryl ethers are cleaved by Li in NH$_3$ or amines.[1a,b] Sterically hindered steroid epoxides, which are not cleaved with LiAlH$_4$, are converted into axial alcohols by Li–EtNH$_2$.[1b] Li–ethylenediamine efficiently cleaves sterically hindered epoxides to tertiary alcohols (eq 20).[53]

Li–H$_2$N(CH$_2$)$_2$NH$_2$, 50 °C, 89% (20)

Li promoted reductions of allyloxy and benzyloxy esters[54a] and esters of sterically hindered secondary and tertiary alcohols[54b] give rise to carboxylate cleavage, thus presenting a means of indirect deoxygenation of alcohols. Further reductive cleavages have been found with activated cyclopropanes,[55] *N*-oxides,[56] and sulfonamides.[57]

Li (and K) promoted reduction of TiCl$_3$ in the McMurry reaction has been reported to be more reliable than the TiCl$_3$/LiAlH$_4$ reagent.[58]

Related Reagents. See Classes R-2, R-4, R-7, R-12, R-14, R-15, R-23, R-25, R-27, R-28, R-29, R-30, and R-32, pages 1–10. Calcium; Lithium–Ethylamine; Potassium; Sodium; Sodium–Alcohol; Sodium–Ammonia.

---

1. (a) Smith, M. In *Reduction: Techniques and Applications in Organic Synthesis*; Augustine, R. L., Ed.; Dekker: New York, 1968; Chapter 2. (b) Hudlický, M. *Reductions in Organic Chemistry*; Horwood: Chichester, 1984. (c) Birch, A. J. *QR* **1950**, *4*, 69. (d) Benkeser, R. A. *Adv. Chem. Ser.* **1957**, *23*, 58. (e) Birch, A. J.; Smith, H. *QR* **1958**, *12*, 17. (f) *Steroid Reactions*; Djerassi, C., Ed.; Holden-Day: San Francisco, 1963. (g) Harvey, R. G. *S* **1970**, 161. (h) Kaiser, E. M. *S* **1972**, 391. (i) Birch, A. J.; Subba Rao, G. S. R. *Adv. Org. Chem.* **1972**, *8*, 1. (j) Caine, D. *OR* **1976**, *23*, 1. (k) Brendel, G. *Lithium Metal in Organic Synthesis*; In 3rd Lect.-Hydride Symp.; Metallges. AG: Frankfurt/Main, 1979; pp 135–155. (l) Huffman, J. W. *ACR* **1983**, *16*, 399. (m) Hook, J. M.; Mander, L. N. *Nat. Prod. Rep.* **1986**, *3*, 35. (n) Rabideau, P. W. *T* **1989**, *45*, 1579. (o) Rabideau, P. W. Marcinow, Z. *OR* **1992**, *42*, 1.
2. (a) Dye, J. L. *Prog. Inorg. Chem.* **1984**, *32*, 327. (b) Thompson, J. C. *Monographs on the Physics and Chemistry of Materials: Electrons in Liquid Ammonia*; Oxford University Press: Fair Lawn, NJ, 1976.
3. Gremmo, N.; Randles, J. E. B. *JCS(F1)* **1974**, *70*, 1480.
4. (a) Dewald, R. R. *JPC* **1975**, *79*, 3044. (b) Birch, A. J., Slobbe, J. *H* **1976**, *5*, 905.
5. (a) Pleskov, V. A. *J. Phys. Chem. (U.S.S.R)* **1937**, *9*, 12 (*CA* **1937, 31**, 4214). (b) Wilds, A. L.; Nelson, N. A. *JACS* **1953**, *75*, 5360.
6. Johnson, W. C.; Piskur, M. M. *JPC* **1933**, *37*, 93.

7. For reductions with lithium bronze, see (a) Mueller, R. H.; Gillick, J. G. *JOC* **1978**, *43*, 4647. (b) Fang, J.-M. *JOC* **1982**, *47*, 3464.
8. Krapcho, A. P.; Bothner-By, A. A. *JACS* **1959**, *81*, 3658.
9. Dryden, H. L., Jr.; Webber, G. M.; Burtner, R. R.; Cella, J. A. *JOC* **1961**, *26*, 3237.
10. Evers, E. C.; Young, A. E.; II; Panson, A. J. *JACS* **1957**, *79*, 5118.
11. Dewald, R. R.; Dye, J. L. *JPC* **1964**, *68*, 128.
12. Benkeser, R. A.; Belmonte, F. G.; Kang, J. *JOC* **1983**, *48*, 2796.
13. Mander, L. N. *COS* **1991**, *8*, 489.
14. Turner, R. B.; Gänshirt, K. H.; Shaw, P. E.; Tauber, J. D. *JACS* **1966**, *88*, 1776.
15. Fried, J.; Abraham, N. A. *TL* **1965**, 3505.
16. Fried, J.; Abraham, N. A.; Santhanakrishnan, T. S. *JACS* **1967**, *89*, 1044.
17. Camps, F.; Coll, J.; Pascual, J. *JOC* **1967**, *32*, 2563.
18. Baker, A. J.; Goudie, A. C. *CC* **1972**, 951.
19. Sipio, W. J. *TL* **1985**, *26*, 2039.
20. Subba Rao, G. S. R.; Ramanathan, H.; Raj, K. *CC* **1980**, 315.
21. Hamilton, R. J.; Mander, L. N.; Sethi, S. P. *T* **1986**, *42*, 2881.
22. Schultz, A. G. Sundararaman, P.; Macielag, M.; Lavieri, F. P.; Welch, M. *TL* **1985**, *26*, 4575.
23. Rabideau, P. W.; Karrick, G. L. *TL* **1987**, *28*, 2481.
24. Burgstahler, A. W.; Worden, L. R. *JACS* **1964**, *86*, 96.
25. Borowitz, I. J.; Gonis, G., Kelsey, R., Rapp, R.; Williams, G. J. *JOC* **1966**, *31*, 3032.
26. Reggel, L.; Friedel, R. A.; Wender, I. *JOC* **1957**, *22*, 891.
27. Rabideau, P. W.; Burkholder, E. G. *JOC* **1978**, *43*, 4283.
28. (a) Jaworek, W.; Vögtle, F. *CB* **1991**, *124*, 347 (*CA* **1991**, *114*, 101 319p). (b) Michel, P.; Moradpour, A. *S* **1988**, 894.
29. (a) Koch, K.-H.; Müllen, K. *CB* **1991**, *124*, 2091. (b) Anton, U.; Göltner, C.; Müllen, K. *CB* **1992**, *125*, 2325.
30. Julina, R.; Herzig, T.; Bernet, B.; Vasella, A. *HCA* **1986**, *69*, 368.
31. Dear, R. E. A.; Pattison, F. L. M. *JACS* **1963**, *85*, 622.
32. Svoboda, M.; Závada, J.; Sicher, J. *CCC* **1965**, *30*, 413.
33. (a) Stork, G.; Malhotra, S.; Thompson, H.; Uchibayashi, M. *JACS* **1965**, *87*, 1148. (b) Miller, B. R. *SC* **1972**, *2*, 273.
34. (a) Stork, G.; Boeckmann, R. K., Jr.; Taber, D. F., Still, W. C., Singh, J. *JACS* **1979**, *101*, 7107. (b) Swartz, J. E.; Mahachi, T. J.; Kariv-Miller, E. *JACS* **1988**, *110*, 3622.
35. Huffman, J. W. *COS* **1991**, *8*, 107.
36. Huffman, J. W.; Desai, R. C.; LaPrade, J. E. *JOC* **1983**, *48*, 1474.
37. Giroud, A. M., Rassat, A. *BSF* **1976**, 1881 (*CA* **1977**, *87*, 6251a).
38. Toromanoff, E. *BSF* **1987**, 893 (*CA* **1988**, *109*, 128 193b).
39. (a) Stork, G.; Rosen, P.; Goldman, N.; Coombs, R. V.; Tsuji, J. *JACS* **1965**, *87*, 275. (b) Stork, G.; Logusch, E. W. *JACS* **1980**, *102*, 1218.
40. Samson, M.; De Clercq, P.; Vandewalle, M. *T* **1977**, *33*, 249.
41. (a) Burgstahler, A. W.; Sanders, M. E. *S* **1980**, 400. (b) See also Fétizon, M.; Gore, J. *TL* **1966**, 471.
42. Ireland, R. E.; Pfister, G. *TL* **1969**, 2145.
43. (a) Hall, S. S.; Lipsky, S. D.; McEnroe, F. J.; Bartels, A. P. *JOC* **1971**, *36*, 2588. (b) Marcinow, Z.; Rabideau, P. W. *JOC* **1988**, *53*, 2117.
44. (a) Hall, S. S.; Lipsky, S. D. *JOC* **1973**, *38*, 1735. (b) For phenylation–reduction of aliphatic ketones see Hall, S. S.; McEnroe, F. J. *JOC* **1975**, *40*, 271.
45. Bedenbaugh, A. O.; Bedenbaugh, J. H.; Bergin, W. A.; Adkins, J. D. *JACS* **1970**, *92*, 5774.
46. (a) Bruck, P.; Thompson, D.; Winstein, S. *CI(L)* **1960**, 405. (b) Ikan, R.; Markus, A. *JCS(P1)* **1972**, 2423. (c) Bruck, P. *TL* **1962**, 449. (d) For substitution of Li by Na, see: Gassman, P. G.; Pape, P. G. *JOC* **1964**, *29*, 160.
47. (a) Pinder, A. R. *S* **1980**, 425. (b) Duggan, A. J.; Hall, S. S. *JOC* **1975**, *40*, 2238. (c) Berkowitz, D. B. *S* **1990**, 649.
48. (a) Ziegler, K.; Colonius, H. *LA* **1930**, *479*, 135 (*CA* **1930**, *24*, 3777). (b) Jones, R. G.; Gilman, H. *OR* **1951**, *6*, 339.
49. Han, B. H.; Boudjouk, P. *TL* **1981**, *22*, 2757.
50. Caubère, P.; Coutrot, P. *COS* **1991**, *8*, 835.
51. (a) Ohmori, M.; Yamada, S.; Takayama, H. *TL* **1982**, 23. (b) Grieco, P. A.; Masaki, Y. *JOC* **1974**, *39*, 2135. (c) Stotter, P. L.; Hornish, R. E. *JACS* **1973**, *95*, 4444. See also Truce, W. E.; Tate, D. P.; Burdge, D. N. *JACS* **1960**, *82*, 2872.
52. Sevrin, M.; Van Ende, D.; Krief, A. *TL* **1976**, 2643.
53. Brown, H. C.; Ikegami, S.; Kawakami, J. H. *JOC* **1970**, *35*, 3243.
54. (a) Markgraf, J. H.; Hensley, W. M.; Shoer, L. I. *JOC* **1974**, *39*, 3168. (b) Boar, R. B.; Joukhadar, L.; McGhie, J. F.; Misra, S. C. *CC* **1978**, 68.
55. Staley, S. W. *Sel. Org. Transform.* **1972**, *2*, 309.
56. White, J. D. *TL* **1974**, 2879.
57. Cuvigny, T.; Larchevêque, M. *JOM* **1974**, *64*, 315 (*CA* **1974**, *80*, 70 494q).
58. McMurry, J. E.; Fleming, M. P.; Kees, K. L.; Krepski, L. R. *JOC* **1978**, *43*, 3255.

Karin Briner
Indiana University, Bloomington, IN, USA

Lithium Aluminum Hydride[1]

[16853-85-3] AlH_4Li (MW 37.96)

(reducing agent for many functional groups;[1] can hydroaluminate double and triple bonds;[2] can function as a base[3])

Alternate Name: LAH.
Physical Data: mp 125 °C; *d* 0.917 g cm^{-3}.
Solubility: sol ether (35 g/100 mL; conc of more dil soln necessary); sol THF (13 g/100 mL); modestly sol other ethers; reacts violently with H_2O and protic solvents.
Form Supplied in: colorless or gray solid; 0.5–1 M solution in diglyme, 1,2-dimethoxyethane, ether, or tetrahydrofuran; the $LiAlH_4 \cdot 2THF$ complex is available as a 1 M solution in toluene.
Analysis of Reagent Purity: Metal Hydrides Technical Bulletin No. 401 describes an apparatus and methodology for assay by means of hydrogen evolution. See also Rickborn and Quartucci.[39a]
Handling, Storage, and Precautions: the dry solid and solutions are highly flammable and must be stored in the absence of moisture. Cans or bottles of $LiAlH_4$ should be flushed with N_2 and kept tightly sealed to preclude contact with oxygen and moisture. Lumps should be crushed only in a glove bag or dry box.

Functional Group Reductions. The powerful hydride transfer properties of this reagent cause ready reaction to occur with aldehydes, ketones, esters, lactones, carboxylic acids, anhydrides, and epoxides to give alcohols, and with amides, iminium ions, nitriles, and aliphatic nitro compounds to give amines. Several methods of workup for these reductions are available. A strongly recommended option[4] involves careful successive dropwise addition to the mixture containing *n* grams of $LiAlH_4$ of *n* mL of H_2O,

n mL of 15% NaOH solution, and $3n$ mL of H_2O. These conditions provide a dry granular inorganic precipitate that is easy to rinse and filter. More simply, solid Glauber's salt ($Na_2SO_4{\cdot}10H_2O$) can be added portionwise until the salts become white.[5] In certain instances, an acidic workup (10% H_2SO_4) may prove advantageous because the inorganic salts become solubilized in the aqueous phase.[6] Should water not be compatible with the product, the use of ethyl acetate is warranted since the ethanol that is liberated usually does not interfere with the isolation.[4] Although the stoichiometry of $LiAlH_4$ reactions is well established,[1] excess amounts of the reagent are often employed (perhaps to make accommodation for the perceived presence of adventitious moisture). This practice is wasteful of reagent, complicates workup, and generally should be avoided.

The reduction of amides can be adjusted in order to deliver aldehydes. Acylpiperidides,[7] *N*-methylanilides,[8] aziridides,[9] imidazolides,[10] and *N,O*-dimethylhydroxylamides[11] have proven especially serviceable. All of these processes generate products that liberate the aldehyde upon hydrolytic workup. The powerful reducing ability of $LiAlH_4$ allows for its application in the context of other functional groups. Alkanes are often formed in good yield upon exposure of alkyl halides (I > Br > Cl; primary > secondary > tertiary)[12] and tosylates[13] to $LiAlH_4$ in ethereal solvents. Chloride reduction is an S_N2 process, while iodides enter principally into single electron transfer chemistry.[14] Benzylic[15] and allylic halides[16] behave comparably, although the latter can react in an S_N2' fashion as well (eq 1).[17] Select aromatic halides can be reduced under forcing conditions (e.g. diglyme, 100 °C),[18] but chemoselectivity as in eq 2 can often be achieved.[19] Vinyl,[20] bridgehead,[20] and cyclopropyl halides (eqs 3 and 4)[20,21] have all been reported to undergo reduction. The SET mechanism is also believed to operate in the latter context.[22]

Br; LiAlD4; ether, Δ; 67%; D; H; Br (1)

Cl; Br; LiAlH4; Et2O; 75%; Cl (2)

F; Cl; LiAlH4; diglyme; 100 °C; F; H; H; F; 94:6 (3)

Cl; F; LiAlH4; diglyme; 100 °C; F; H; H; F; 10:90 (4)

$LiAlH_4$ is normally unreactive toward ethers.[1] Unsaturated acetals undergo reduction with double bond migration (S_N2'); in cyclic systems, the usual stereoelectronic factors often apply (eq 5).[23] Orthoesters are amenable to attack, giving acetals in good yield (eq 6).[24] The susceptibility of benzylic acetals to reduction can be enhanced by the co-addition of a Lewis acid (eq 7).[25]

AcO; O; OMe; H; H; H; 1. LiAlH4, ether, 25 °C; 2. Ac2O, py; 100%; OAc; H; O; H (5)

MeO; MeO; MeO; SMe; LiAlH4; C6H6–Et2O; Δ; 97%; MeO; MeO; SMe (6)

S(CH2)6Me; OSiMe3; LiAlH4; AlCl3; 83%; S(CH2)6Me (7)

When comparison is made between $LiAlH_4$ and related reducing agents containing active Al–H and B–H bonds, $LiAlH_4$ is seen to be the most broadly effective (Table 1).[1h] Its superior reducing power is also reflected in its speed of hydride transfer.

Hydroalumination Agent. Ethylene has long been known to enter into addition with $LiAlH_4$ when the two reagents are heated under pressure at 120–140 °C; lithium tetraethylaluminate results.[2] Homogeneous hydrogenations of alkenes and alkynes to alkanes and alkenes, respectively, performed in THF or diglyme solutions under autoclave pressure, are well documented.[26] Such reductions are greatly facilitated by the presence of a transition metal halide ranging from Ti to Ni.[27] Replacement of the hydrolytic workup by the addition of appropriate halides constitutes a useful means for chain extension (eq 8).[28] 1-Chloro-1-alkynes are notably reactive toward $LiAlH_4$, addition occurring regio- and stereoselectively to give alanates that can be quenched directly (MeOH) or converted into mixed 1,1-dihaloalkenes (eq 9).[29]

1. LiAlH4, TiCl4; 2. CH2=C=CHBr; 80%; 1. LiAlH4, TiCl4; 2. HC≡CCH2Br; 54% (8)

Bu; Cl; LiAlH4, THF; –30 to 0 °C; Bu; AlH3; H; Cl; Li+; MeOH; Bu; H; H; Cl; 96%; Me2CO; Bu; Al(O-*i*-Pr)3; H; Cl; Li+; Br2; –78 °C; Bu; Br; H; Cl; 78%; ICl; –30 to 25 °C; Bu; I; H; Cl; 85% (9)

A pronounced positive effect on the ease of reduction of C=C and C≡C bonds manifests itself when a neighboring hydroxyl

Table 1 Comparison of the Reactivities of Hydride Reducing Agents toward the More Common Functional Groups

	Reduction products[a]						
Reagent/functional group	Aldehyde	Ketone	Acyl halide	Ester	Amide	Carboxylate salt	Iminium ion
$LiAlH_4$	Alcohol	Alcohol	Alcohol	Alcohol	Amine	Alcohol	Amine
$LiAlH_2(OCH_2CH_2OMe)_2$	Alcohol	Alcohol	Alcohol	Alcohol	Amine	Alcohol	–
$LiAlH(O\text{-}t\text{-}Bu)_3$	Alcohol	Alcohol	Aldehyde	Alcohol	Aldehyde	NR	–
$NaBH_4$	Alcohol	Alcohol	–	Alcohol	NR	NR	Amine
$NaBH_3CN$	Alcohol	NR	–	NR	NR	NR	Amine
B_2H_6	Alcohol	Alcohol	–	NR	Amine	Alcohol	–
AlH_3	Alcohol	Alcohol	Alcohol	Alcohol	Amine	Alcohol	–
$[i\text{-PrCH(Me)}]_2BH$	Alcohol	Alcohol	–	NR	Aldehyde	NR	–
$(i\text{-Bu})_2AlH$	Alcohol	Alcohol	–	Aldehyde	Aldehyde	Alcohol	–

[a] NR indicates that no reduction is observed.

group is present. In such cases, $LiAlH_4$ is used alone because reduction is preceded by formation of an alkoxyhydridoaluminate capable of facilitating hydride delivery (eqs 10–12).[30–32] The regio- and stereoselectivities of these reactions, where applicable, appear to be quite sensitive to the substrate structure and solvent.[33] Generally, the use of THF or dioxane results in exclusive *anti* addition. When ether is used, almost equivalent amounts of *syn* and *anti* products can result. Considerable attention has been accorded to synthetic applications of the alanate intermediates produced upon reduction of propargyl alcohols in this way. Simple heating occasions elimination of a δ-leaving group to generate homoallylic[34] and α-allenic alcohols (eqs 13 and 14).[35] Other variants of this chemistry have been reported.[36] The addition at −78 °C of solid iodine to alanates formed in this way greatly accelerates the elimination (eq 15).[37] Solutions of iodine in THF give rise instead to vinyl iodides.[38]

C_8H_{17}–C≡C–$(CH_2)_6$OH → [$LiAlH_4$, diglyme, THF, 140 °C, 94%] (*E*)-C_8H_{17}CH=CH$(CH_2)_6$OH (10)

[$LiAlH_4$, 70%] (11)

[$LiAlH_4$, Et_2O, Δ, 70%] → [75%] (12)

[$LiAlH_4$, dioxane, Δ, 68%] 97% (*E*) (13)

[$LiAlH_4$, Et_2O, Δ, 73%] (14)

[$LiAlH_4$, NaOMe, THF, 23 °C] Li^+; [I_2, THF]; [I_2 solid] 60% (15)

Epoxide Cleavage and Aziridine Ring Formation. Epoxides are reductively cleaved in the presence of $LiAlH_4$ with attack generally occurring at the less substituted carbon.[1g] 1,2-Epoxycyclohexanes exhibit a strong preference for axial attack (eqs 16 and 17).[39] In general, *cis* isomers are more reactive than their *trans* counterparts; ring size effects are also seen and these conform to the degree of steric inhibition to backside attack of the C–O bond.[40] Vinyl epoxides often suffer ring opening by means of the S_N' mechanism (eq 18).[41]

[$LiAlH_4$, ether, 0 °C, 90%] (16)

[$LiAlH_4$, ether, 0 °C, 81%] (17)

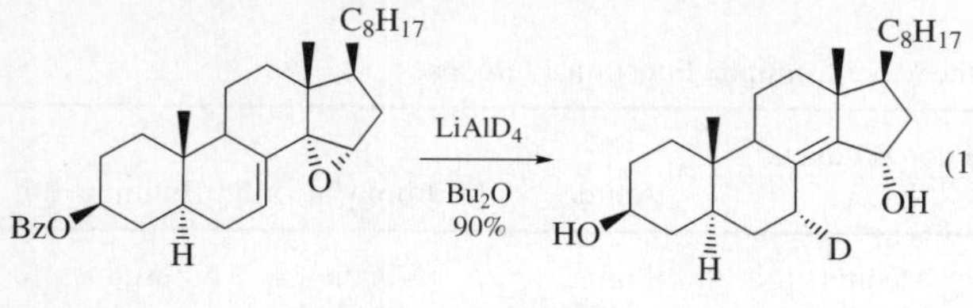

(18)

Two types of oximes undergo hydride reduction with ring closure to give aziridines. These are ketoximes that carry an α- or β-aryl ring and aldoximes substituted with an aromatic group at the β-carbon (eqs 19 and 20).[42]

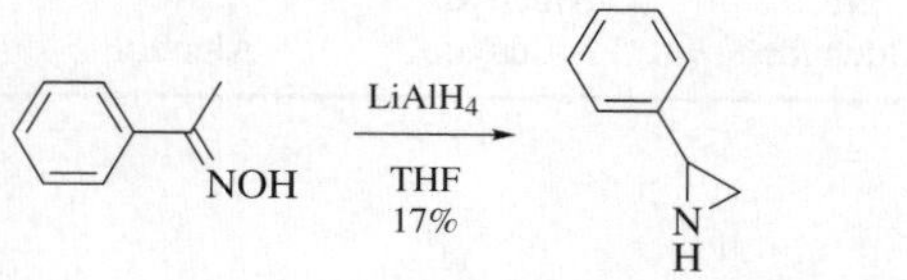

(19)

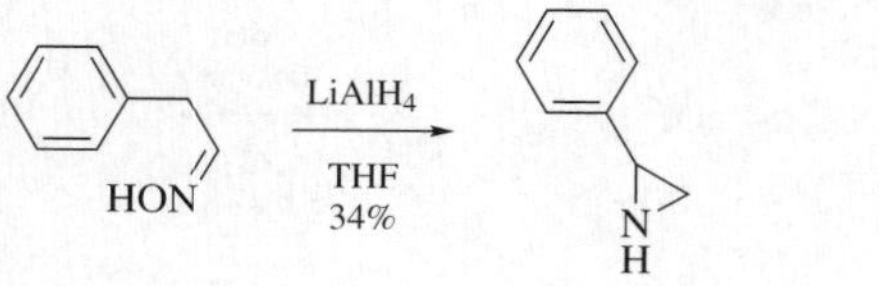

(20)

Use as a Base. Both 1,2- and 1,3-diol monosulfonate esters react with $LiAlH_4$ functioning initially as a base and subsequently in a reducing capacity (eqs 21 and 22).[3,43]

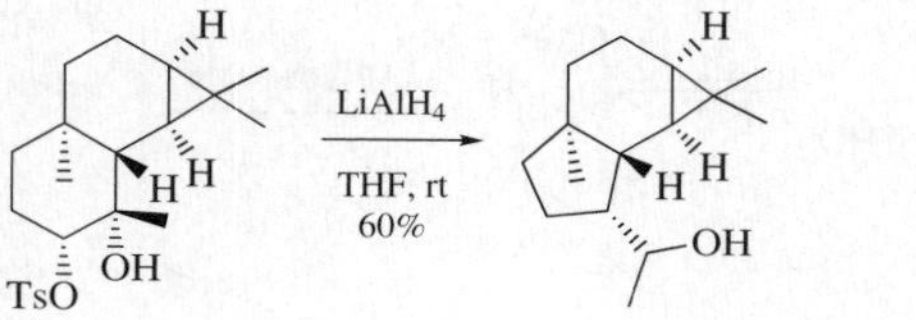

(21)

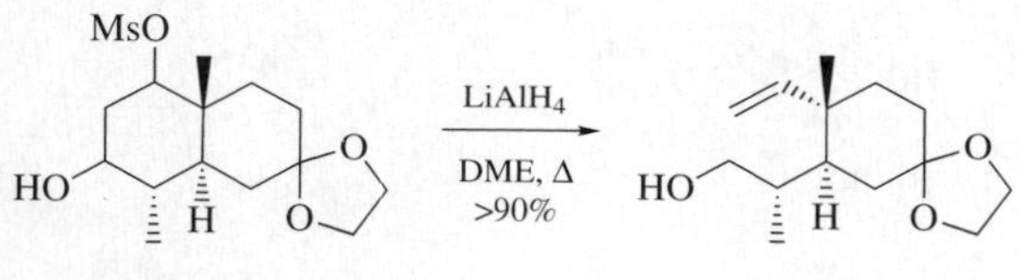

(22)

Reduction of Sulfur Compounds. Sulfur compounds react differently with $LiAlH_4$ depending on the mode of covalent attachment of the hetero atom and its oxidation state. Reductive desulfurization is not often encountered.[44] Arylthioalkynes undergo stereoselective *trans* reduction (eq 23)[45] and α-oxoketene dithioacetals are transformed into fully saturated *anti* alcohols under reflux conditions (eq 24).[46] While $LiAlH_4$ catalyzes the fragmentation of sulfolenes to 1,3-dienes,[47] the lithium salts of sulfolanes are smoothly ring contracted when heated with the hydride in dioxane (eq 25).[48]

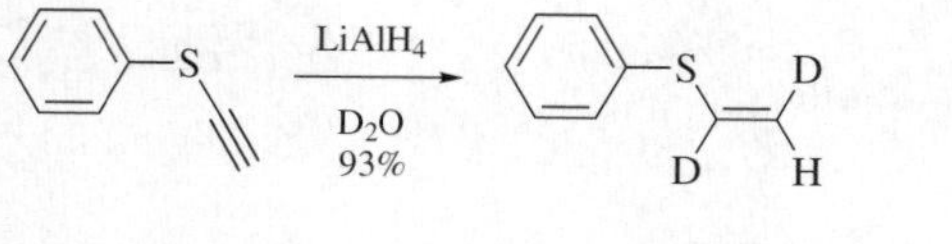

(23)

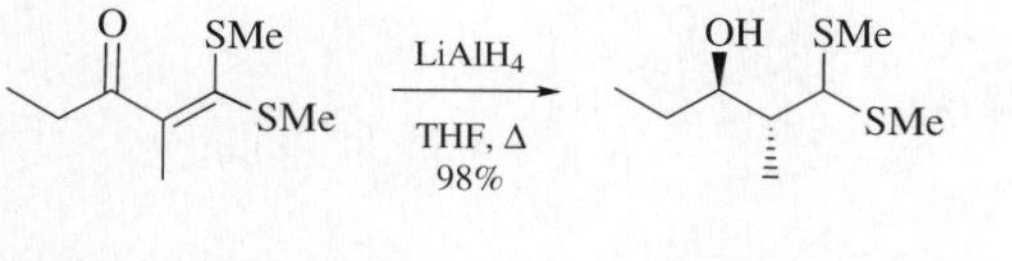

(24)

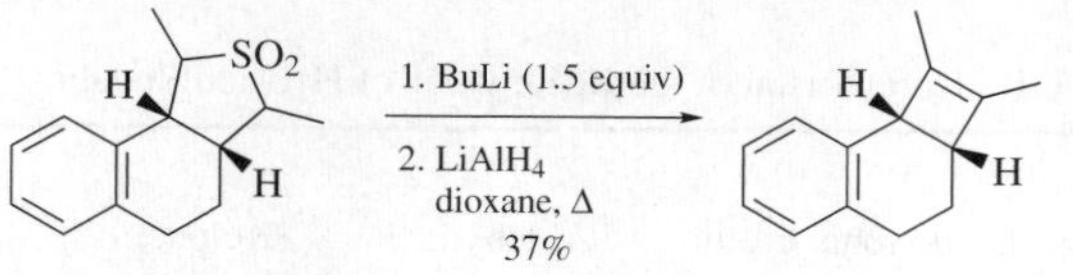

(25)

Stereoselective Reductions. The reduction of 4-*t*-butylcyclohexanone and (**1**) by $LiAlH_4$ occurs from the axial direction to the extent of 92%[49] and 85%,[50] respectively. When a polymethylene chain is affixed diaxially as in (**2**), the equatorial trajectory becomes kinetically dominant (93%).[50] Thus, although electronic factors may be an important determinant of π-facial selectivity,[51] steric demands within the ketone cannot be ignored. The stereochemical characteristics of many ketone reductions have been examined. For acyclic systems, the Felkin–Ahn model[52] has been widely touted as an important predictive tool.[53] Cram's chelation transition state proposal[54] is a useful interpretative guide for ketones substituted at C_α with a polar group. Cieplak's explanation[55] for the stereochemical course of nucleophilic additions to cyclic ketones has received considerable scrutiny.[56]

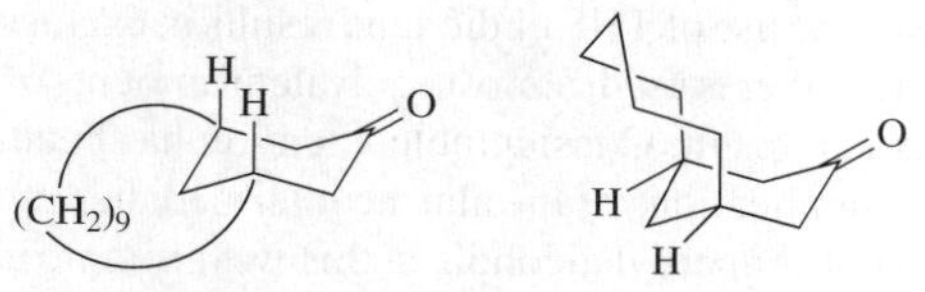

(**1**) (**2**)

Useful levels of diastereoselectivity can be realized upon reduction of selected acyclic ketones having a proximate chiral center.[57] The contrasting results in eq 26 stem from the existence of a chelated intermediate when the benzyloxy group is present and its deterrence when a large silyl protecting group is present. In the latter situation, an open transition state is involved.[58,59] In eq 27, $LiAlH_4$ alone shows no stereoselectivity, but the co-addition of ***Lithium Iodide*** gives rise to a *syn*-selective reducing agent as a consequence of the intervention of a Li^+-containing six-membered chelate.[60]

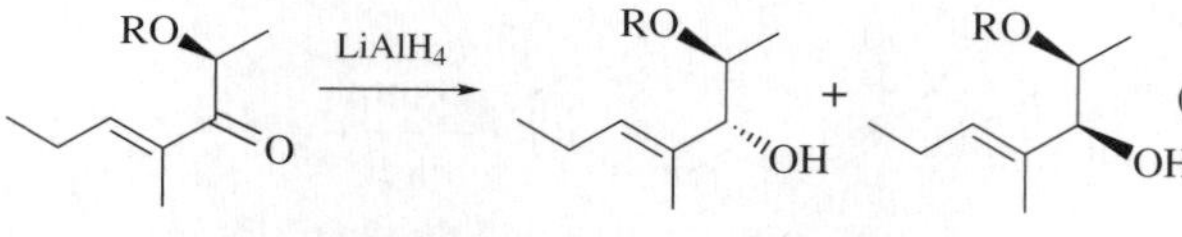

(26)

R = Bn, ether, −10 °C 98:2
R = TBDMS, THF, −20 °C 5:95

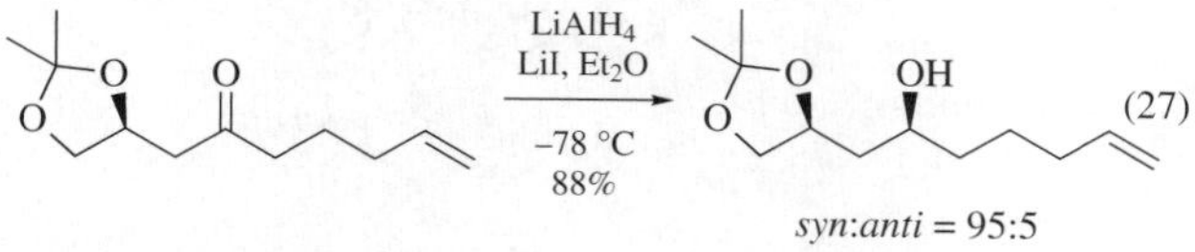

(27)

syn:anti = 95:5

The reduction of amino acids to 1,2-amino alcohols can be conveniently effected with $LiAlH_4$ in refluxing THF.[61] The higher homologous 1,3-amino alcohols have been made available starting with isoxazolines, the process being *syn* selective (eq 28).[62,63] The *syn* and *anti* *O*-benzyloximes of β-hydroxy ketones are reduced with good *syn* and *anti* stereoselectivity, respectively.[64] However, when NaOMe or KOMe is also added, high *syn* stereoselectivity is observed with both isomers (eq 29).[65]

$$C_{14}H_{29}\text{-isoxazoline} \xrightarrow[\text{ether, 95\%}]{LiAlH_4} C_{14}H_{29}CH(OH)CH_2CH(NH_2)CH_3 \;(88:12) \quad (28)$$

$$BuCH(OH)CH_2C(=NOBn)Bu \xrightarrow[\text{THF, -30 °C, 92\%}]{LiAlH_4,\ NaOMe} BuCH(OH)CH_2CH(NH_2)Bu \quad syn{:}anti = 96{:}4 \quad (29)$$

Addition of Chiral Ligands. If a chiral adjuvant is used to achieve asymmetric induction, it should preferentially be inexpensive, easily removed, efficiently recovered, and capable of inducing high stereoselectivity.[53] Of these, 1,3-oxathianes based on (+)-camphor[66] or (+)-pulegone,[67] and proline-derived 1,3-diamines,[68] have been accorded the greatest attention.

Extensive attempts to modify $LiAlH_4$ with chiral ligands in order to achieve the consistent and efficient asymmetric reduction of prochiral ketones has provided very few all-purpose reagents.[1f,69] Many optically active alcohols, amines, and amino alcohols have been evaluated. One of the more venerable of these reagents is that derived from DARVON alcohol.[70] The use of freshly prepared solutions normally accomplishes reasonably enantioselective conversion to alcohols.[71] As the reagent ages, its reduction stereoselectivity reverses.[70] This phenomenon is neither understood nor entirely reliable, and therefore recourse to the enantiomeric reagent[72] (from NOVRAD alcohol) is recommended.[73] The highest level of enantiofacial discrimination is usually realized with $LiAlH_4$ complexes prepared from equimolar amounts of (*S*)-(−)- or (*R*)-(+)-2,2′-dihydroxy-1,1′-binaphthyl and ethanol.[74] Often, optically pure alcohols result, irrespective of whether the ketones are aromatic,[74] alkenic,[75] or alkynic in type.[76] A useful rule of thumb is that (*S*)-BINAL-H generally provides (*S*)-carbinols and (*R*)-BINAL-H the (*R*)-antipodes when ketones of the type R_{unsat}–C(O)–R_{sat} are involved (eq 30).[74]

$$PhC(O)CH_3 \xrightarrow[-100 \text{ to } -78\ °C]{(S)\text{-BINAL-H, THF}} (S)\text{-PhCH(OH)CH}_3\ (95\%\ ee); \quad PhC(O)CH_3 \xrightarrow[-100 \text{ to } -78\ °C]{(R)\text{-BINAL-H, THF}} (R)\text{-PhCH(OH)CH}_3\ (100\%\ ee) \quad (30)$$

For a more detailed discussion of asymmetric induction by this means, see also the following entries that deal specifically with $LiAlH_4$/additive combinations.

Related Reagents. See Classes R-1, R-2, R-3, R-4, R-5, R-6, R-11, R-12, R-13, R-14, R-20, R-21, R-23, R-24, R-27, R-28, R-32, pages 1–10. Lithium Aluminum Hydride–(2,2′-Bipyridyl)(1,5-cyclooctadiene)nickel; Lithium Aluminum Hydride–Bis(cyclopentadienyl)nickel; Lithium Aluminum Hydride–Boron Trifluoride Etherate; Lithium Alminum Hydride–Cerium(III) Chloride; Lithium Aluminum Hydride–2,2′-Dihydroxy-1,1′-binaphthyl; Lithium Aluminum Hydride–Chromium(III) Chloride; Lithium Aluminum Hydride–Cobalt(II) Chloride; Lithium Aluminum Hydride–Copper(I) Iodide; Lithium Aluminum Hydride–Diphosphorus Tetraiodide; Lithium Aluminum Hydride–Nickel(II) Chloride; Lithium Aluminum Hydride–Titanium(IV) Chloride; Titanium(III) Chloride–Lithium Aluminum Hydride.

---

1. (a) Brown, W. G. *OR* **1951**, *6*, 469. (b) Gaylord, N. G. *Reduction with Complex Metal Hydrides*; Interscience: New York, 1956. (c) *Reduction Techniques and Applications in Organic Synthesis*; Augustine, R. L., (Ed.); Dekker: New York, 1968. (d) Pizey, J. S. *Synthetic Reagents*, Wiley: New York, 1974; Vol. 1, pp 101–294. (e) Hajos, A. *Complex Hydrides and Related Reducing Agents in Organic Synthesis*; Elsevier: New York, 1979. (f) Grandbois, E. R.; Howard, S. I.; Morrison, J. D. *Asymmetric Synthesis*; Academic: New York, 1983 Vol. 2. (g) *COS* **1991**, *8*, Chapters 1.1–4.8. (h) Carey, F. A.; Sundberg, R. J. *Advanced Organic Chemistry*, 3rd ed.; Plenum: New York, 1990; Part B, pp 232–253.
2. Ziegler, K.; Bond, A. C., Jr.; Schlesinger, H. I. *JACS* **1947**, *69*, 1199.
3. Bates, R. B.; Büchi, G.; Matsuura, T.; Shaffer, R. R. *JACS* **1960**, *82*, 2327.
4. Fieser, L. F.; Fieser, M. *FF* **1967**, *1*, 584.
5. (a) Paquette, L. A.; Gardlik, J. M.; McCullough, K. J.; Samodral, R.; DeLucca, G.; Ouellette, R. J. *JACS* **1983**, *105*, 7649. (b) Paquette, L. A.; Gardlik, J. M. *JACS* **1980**, *102*, 5016.
6. Sroog, C. E.; Woodburn, H. M. *OSC* **1963**, *4*, 271.
7. Mousseron, M.; Jacquier, R.; Mousseron-Canet, M.; Zagdoun, R. *BSF* **1952**, *19*, 1042.
8. (a) Weygand, F.; Eberhardt, G. *AG* **1952**, *64*, 458. (b) Weygand, F.; Eberhardt, G.; Linden, H.; Schäfer, F.; Eigen, I. *AG* **1953**, *65*, 525.
9. Brown, H. C.; Tsukomoto, A. *JACS* **1961**, *83*, 2016, 4549.
10. Staab, H. A.; Bräunling, H. *LA* **1962**, *654*, 119.
11. Nahm, S.; Weinreb, S. M. *TL* **1981**, *22*, 3815.
12. (a) Johnson, J. E.; Blizzard, R. H.; Carhart, H. W. *JACS* **1948**, *70*, 3664. (b) Krishnamurthy, S.; Brown, H. C. *JOC* **1982**, *47*, 276.
13. (a) Schmid, H.; Karrer, P. *HCA* **1949**, *32*, 1371. (b) Zorbach, W. W.; Tio, C. O. *JOC* **1961**, *26*, 3543. (c) Wang, P.-C.; Lysenko, Z.; Joullié, M. M. *TL* **1978**, 1657.
14. (a) Ashby, E. C.; De Priest, R. N.; Goel, A. B.; Wenderoth, B.; Pham, T. N. *JOC* **1984**, *49*, 3545. (b) Ashby, E. C.; Pham, T. N. *JOC* **1986**, *51*, 3598. (c) Ashby, E. C.; Pham, T. N.; Amrollah-Madjdabadi, A. *JOC* **1991**, *56*, 1596.
15. (a) Trevay, L. W.; Brown, W. G. *JACS* **1949**, *71*, 1675. (b) Buchta, E.; Loew, G. *LA* **1955**, *597*, 123. (c) Parham, W. E.; Wright, C. D. *JOC* **1957**, *22*, 1473. (d) Ligouri, A.; Sindona, G.; Uccella, N. *T* **1983**, *39*, 683.
16. (a) Jefford, C. W.; Mahajan, S. N.; Gunsher, J. *T* **1968**, *24*, 2921. (b) Ohloff, G.; Farrow, H.; Schade, G. *CB* **1956**, *89*, 1549. (c) Fraser-Reid, B.; Tam, S. Y.-K.; Radatus, B. *CJC* **1975**, *53*, 2005.
17. Magid, R. M. *T* **1980**, *36*, 1901.
18. Karabatsos, G. J.; Shone, R. L. *JOC* **1968**, *33*, 619.
19. (a) Moore, L. D. *CED* **1964**, *9*, 251. (b) Marvel, C. S.; Wilson, B. D. *JOC* **1958**, *23*, 1483.
20. (a) Jefford, C. W.; Sweeney, A.; Delay, F. *HCA* **1972**, *55*, 2214. (b) Jefford, C. W.; Kirkpatrick, D.; Delay, F. *JACS* **1972**, *94*, 8905.
21. Jefford, C. W.; Burger, U.; Laffer, M. H.; Kabengele, T. *TL* **1973**, 2483.
22. (a) McKinney, M. A.; Anderson, S. M.; Keyes, M.; Schmidt, R. *TL* **1982**, *23*, 3443. (b) Hatem, J.; Meslem, J. M.; Waegell, B. *TL* **1986**, *27*, 3723.
23. (a) Fraser-Reid, B.; Radatus, B. *JACS* **1970**, *92*, 6661. (b) Radatus, B.; Yunker, M.; Fraser-Reid, B. *JACS* **1971**, *93*, 3086. (c) Tam, S. Y.-K.; Fraser-Reid, B. *TL* **1973**, 4897.
24. Claus, C. J.; Morgenthau, J. L., Jr. *JACS* **1951**, *73*, 5005.

25. Eliel, E. L.; Rerick, M. N. *JOC* **1958**, *23*, 1088.
26. (a) Slaugh, L. H. *T* **1966**, *22*, 1741. (b) Magoon, E. F.; Slaugh, L. H. *T* **1967**, *23*, 4509.
27. Pasto, D. J. *COS* **1991**, *8*, Chapter 3.3.
28. (a) Sato, F.; Kodama, H.; Sato, M. *CL* **1978**, 789. (b) Sato, F.; Ogura, K.; Sato, M. *CL* **1978**, 805.
29. Zweifel, G.; Lewis, W.; On, H. P. *JACS* **1979**, *101*, 5101.
30. Rossi, R.; Carpita, A. *S* **1977**, 561.
31. Solladié, G.; Berl, V. *TL* **1992**, *33*, 3477.
32. Baudony, R.; Goré, J. *TL* **1974**, 1593.
33. (a) Borden, W. T. *JACS* **1970**, *92*, 4898. (b) Grant, B.; Djerassi, C. *JOC* **1974**, *39*, 968.
34. Claesson, A.; Bogentoft, C. *S* **1973**, 539.
35. (a) Claesson, A. *ACS* **1975**, *B29*, 609. (b) Galantay, E.; Basco, I.; Coombs, R. V. *S* **1974**, 344. (c) Cowie, J. S.; Landor, P. D.; Landor, S. R. *JCS(P1)* **1973**, 270. (d) Claesson, A.; Olsson, L.-I.; Bogentoft, C. *ACS* **1973**, *B27*, 2941.
36. (a) Olsson, L.-I.; Claesson, A.; Bogentoft, C. *ACS* **1974**, *B28*, 765. (b) Claesson, A. *ACS* **1974**, *B28*, 993.
37. Keck, G. E.; Webb, R. R., II *TL* **1982**, *23*, 3051.
38. Corey, E. J.; Katzenellenbogen, J. A.; Posner, G. *JACS* **1967**, *89*, 4245.
39. (a) Rickborn, B.; Quartucci, J. *JOC* **1984**, *29*, 3185. (b) Rickborn, B.; Lamke, W. E., II *JOC* **1967**, *32*, 537. (c) Murphy, D. K.; Alumbaugh, R. L.; Rickborn, B. *JACS* **1969**, *91*, 2649.
40. Mihailovic, M. Lj.; Andrejevic, V.; Milovanoic, J.; Jankovic, J. *HCA* **1976**, *59*, 2305.
41. (a) Parish, E. J.; Schroepper, G. J., Jr. *TL* **1976**, 3775. (b) Fraser-Reid, B.; Tam, S. Y.-K.; Radatus, B. *CJC* **1975**, *53*, 2005.
42. (a) Kotera, K.; Kitahonoki, K. *TL* **1965**, 1059. (b) Kotera, K.; Kitahonoki, K. *OS* **1968**, *48*, 20.
43. Kato, M.; Kurihara, H.; Yoshikoshi, A. *JCS(P1)* **1979**, 2740.
44. Gassman, P. G.; Gilbert, D. P.; van Bergen, T. J. *CC* **1974**, 201.
45. Hojo, M.; Masuda, R.; Takagi, S. *S* **1978**, 284.
46. Gammill, R. B.; Bell, L. T.; Nash, S. A. *JOC* **1984**, *49*, 3039.
47. Gaoni, Y. *TL* **1977**, 947.
48. (a) Photis, J. M.; Paquette, L. A. *JACS* **1974**, *96*, 4715. (b) Photis, J. M.; Paquette, L. A. *OS* **1977**, *57*, 53.
49. Lansbury, P. T.; MacLeay, R. E. *JOC* **1963**, *28*, 1940.
50. Paquette, L. A.; Underiner, T. L.; Gallucci, J. C. *JOC* **1992**, *57*, 86.
51. (a) Wigfield, D. C. *T* **1979**, *35*, 449. (b) Mukherjee, D.; Wu,-Y.-D.; Fronczek, R. F.; Houk, K. N. *JACS* **1988**, *110*, 3328. (c) Wu, Y.-D.; Tucker, J. A.; Houk, K. N. *JACS* **1991**, *113*, 5018.
52. (a) Chérest, M.; Felkin, H.; Prudent, N. *TL* **1968**, 2199. (b) Ahn, N. T.; Eisenstein, O. *NJC* **1977**, *1*, 61.
53. Eliel, E. L. In *Asymmetric Synthesis*; Morrison, J. D., Ed.; Academic: New York, 1983; Vol. 2, Chapter 5.
54. (a) Cram, D. J.; Kopecky, K. R. *JACS* **1959**, *81*, 2748. (b) Cram, D. J.; Wilson, D. R. *JACS* **1963**, *85*, 1249.
55. Cieplak, A. S. *JACS* **1981**, *103*, 4540.
56. (a) Cheung, C. K.; Tseng, L. T.; Lin, M.-H.; Srivastava, S.; leNoble, W. J. *JACS* **1986**, *108*, 1598. (b) Cieplak, A. S.; Tait, B. D.; Johnson, C. R. *JACS* **1989**, *111*, 8447. (c) Okada, K.; Tomita, S.; Oda, M. *BCJ* **1989**, *62*, 459. (d) Li, H.; Mehta, G.; Podma, S.; leNoble, W. J. *JOC* **1991**, *56*, 2006. (e) Mehta, G.; Khan, F. A. *JACS* **1990**, *112*, 6140.
57. Nogradi, M. *Stereoselective Synthesis*; VCH: Weinheim, 1986.
58. Overman, L. E.; McCready, R. J. *TL* **1982**, *23*, 2355.
59. Other examples: (a) Iida, H.; Yamazaki, N.; Kibayashi, C. *CC* **1987**, 746. (b) Bloch, R.; Gilbert, L.; Girard, C. *TL* **1988**, *29*, 1021. (c) Fukuyama, T.; Vranesic, B.; Negri, D. P.; Kishi, Y. *TL* **1978**, 2741.
60. (a) Mori, Y.; Kuhara, M.; Takeuchi, A.; Suzuki, M. *TL* **1988**, *29*, 5419. (b) Mori, Y.; Takeuchi, A.; Kageyama, H.; Suzuki, M. *TL* **1988**, *29*, 5423.
61. Dickman, D. A.; Meyers, A. I.; Smith, G. A.; Gawley, R. E. *OSC* **1990**, *7*, 530.
62. (a) Jäger, V.; Buss, V. *LA* **1980**, 101. (b) Jäger, V.; Buss, V.; Schwab, W. *LA* **1980**, 122.
63. (a) Jäger, V.; Schwab, W.; Buss, V. *AG(E)* **1981**, *20*, 601. (b) Schwab, W.; Jäger, V. *AG(E)* **1981**, *20*, 603.
64. Narasaka, K.; Ukaji, Y. *CL* **1984**, 147.
65. Narasaka, K.; Yamazaki, S.; Ukaji, Y. *BCJ* **1986**, *59*, 525.
66. Eliel, E.; Frazee, W. F. *JOC* **1979**, *44*, 3598.
67. Eliel, E. L.; Lynch, J. E. *TL* **1981**, *22*, 2859.
68. Mukaiyama, T. *T* **1981**, *37*, 4111.
69. Nishizawa, M.; Noyori, R. *COS* **1991**, *8*, Chapter 1.7.
70. Yamaguchi, S.; Mosher, H. S.; Pohland, A. *JACS* **1972**, *94*, 9254.
71. (a) Reich, C. J.; Sullivan, G. R.; Mosher, H. S. *TL* **1973**, 1505. (b) Brinkmeyer, R. S.; Kapoor, V. M. *JACS* **1977**, *99*, 8339. (c) Johnson, W. S.; Brinkmeyer, R. S.; Kapoor, V. M.; Yarnell, T. M. *JACS* **1977**, *99*, 8341. (d) Marshall, J. A.; Robinson, E. D. *TL* **1989**, *30*, 1055.
72. Deeter, J.; Frazier, J.; Staten, G.; Staszak, M.; Weigel, L. *TL* **1990**, *31*, 7101.
73. Paquette, L. A.; Combrink, K. D.; Elmore, S. W.; Rogers, R. D. *JACS* **1991**, *113*, 1335.
74. (a) Noyori, R.; Tomino, I.; Tanimoto, Y. *JACS* **1979**, *101*, 3129. (b) Noyori, R.; Tomino, I.; Tanimoto, Y.; Nishizawa, M. *JACS* **1984**, *106*, 6709.
75. (a) Noyori, R. Tomino, I.; Nishizawa, M. *JACS* **1979**, *101*, 5843. (b) Beechström, P. Björkling, F.; Högberg, H.-E.; Norin, T. *ACS* **1983**, *B37*, 1.
76. (a) Noyori, R.; Tomino, I.; Yamada, M.; Nishizawa, M. *JACS* **1984**, *106*, 6717. (b) Nishizawa, M.; Yamada, M.; Noyori, R. *TL* **1981**, *22*, 247.

Leo A. Paquette
The Ohio State University, Columbus, OH, USA

Lithium Aluminum Hydride-2,2′-Dihydroxy-1,1′-binaphthyl

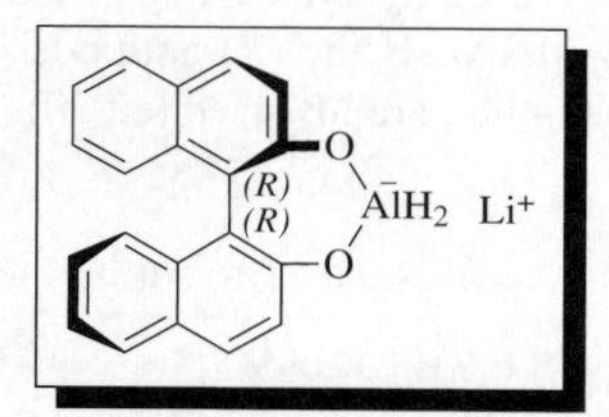

($LiAlH_4$)		
[16853-85-3]	AlH_4Li	(MW 37.96)
((*R*)-BINAL)		
[18531-94-7]	$C_{20}H_{14}O_2$	(MW 286.34)
((*S*)-BINAL)		
[18531-99-2]		

(used for enantioselective reduction of prochiral ketones to alcohols[1])

Alternate Name: BINAL-H.
Physical Data: BINAL: white solid, mp 208–210 °C. Also see ***Lithium Aluminum Hydride***.
Solubility: sol THF.
Preparative Methods: prepared in situ from commercially available lithium aluminum hydride and BINAL.

Handling, Storage, and Precautions: sensitive to moisture (see ***Lithium Aluminum Hydride***).

Overview and General Considerations. This article will cover the title reagent and other chiral reducing agents derived from lithium aluminum hydride and chiral additives, with initial emphasis on the title reagent. The enantioselective reduction of prochiral ketones is a reaction of considerable importance to the synthetic organic chemist and can now be accomplished by a variety of methods and reagents.[1,2] Particularly the use of chiral oxazaborolidines for the catalytic asymmetric reduction of ketones has received much recent interest. This method has been shown to be useful for the preparation of a variety of chiral alcohols with high optical purities. This transformation can also be realized using catalytic hydrogenation with a chiral catalyst or by use of chiral borane reducing agents such as ***(R,R)-2,5-Dimethylborolane*** and ***B-3-Pinanyl-9-borabicyclo[3.3.1]nonane***. Enzyme-catalyzed transformations, for example ***Baker's Yeast*** reductions of carbonyl compounds, can also provide access to a range of chiral alcohols with high optical purities.

The use of complexes of lithium aluminum hydride (LAH) with various chiral ligands to achieve the enantioselective reduction of prochiral ketones has been extensively studied for over 40 years.[1] However, this method, with some exceptions, has not found widespread use due to a number of limiting factors. These factors vary from moderate to poor enantioselectivities, often observed in these reductions, to ready availability of only one antipode of a desired chiral ligand. The recovery of the often expensive chiral ligand that is used in stoichiometric quantities to form the LAH complex is obviously an important experimental concern. Also, in some cases the LAH complex with the chiral ligand may disproportionate to achiral reducing species under the reaction conditions, resulting in poor optical purities of the desired products. Further, no single complex appears to have a sufficiently broad substrate specificity. Aromatic and unsaturated ketones are in general the better substrates and they can be reduced with good enantioselectivities using this method. A useful article comparing the merits of some of the more promising asymmetric reducing agents known for ketones has been published.[3]

Chiral Alcohol Modifying Agents. Complexes of a variety of chiral alcohols (see Figure 1) with LAH have been prepared in situ and examined for their ability to effect enantioselective reduction of prochiral carbonyl compounds. However in most cases, the optical purities of the products obtained have not been satisfactory. This is in part due to the tendency of these chiral ligand–hydride complexes to disproportionate under reaction conditions yielding achiral reducing agents. An exception is the complex of LAH and (−)-menthol (**1**) which has been used to reduce α and β-aminoketones with good enantioselectivities.

The reduction of carbonyl compounds with LAH complexes of a number of chiral diols derived from carbohydrates and terpenes has been studied. In general, the enantioselectivities observed with such reagents have been low to moderate. Acetophenone, which is the model substrate in many of these reduction studies, is reduced by a complex of LAH and the glucose-derived diol (**2**) in about 71% ee under optimized conditions.

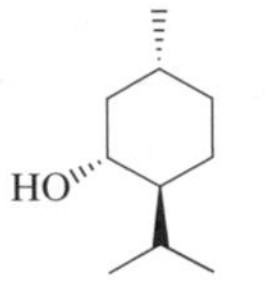

(−)-Menthol (**1**)

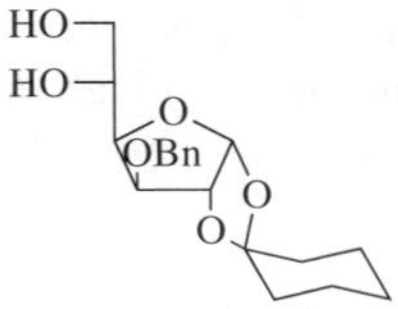

(**2**)

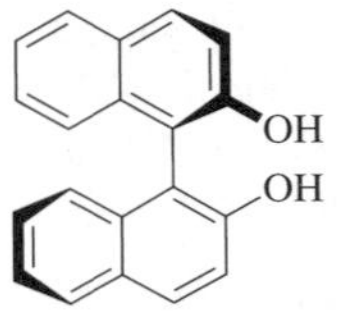

(*R*)-(+)-2,2'-Dihydroxy-1,1'-binaphthyl (**3**)

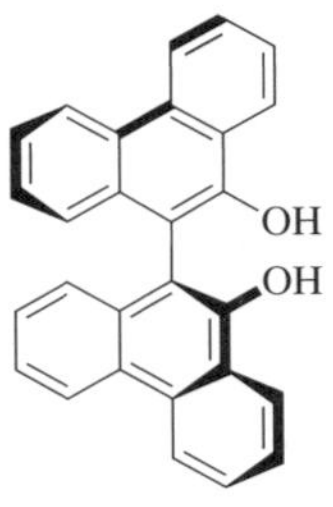

(*S*)-(−)-10,10'-Dihydroxy-9,9'-biphenanthryl (**4**)

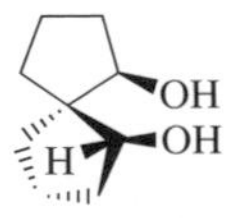

cis,cis-(+)-Spiro-[4.4]nonane-1,6-diol (**5**)

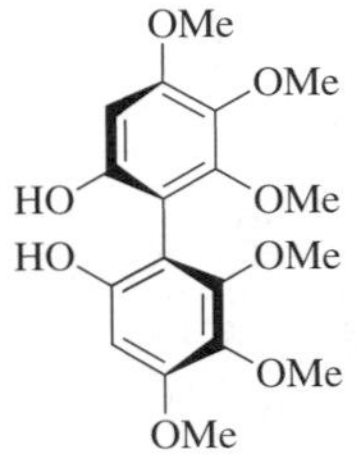

(*S*)-2,2'-Dihydroxy-4,5,6,4',5',6'-hexamethoxybiphenyl (**6**)

Figure 1 Representative chiral alcohol modifying agents

The reagent (*R*)- or (*S*)-BINAL-H, (**7**) developed by Noyori, is undoubtedly the most useful LAH complex reported so far for the asymmetric reduction of a variety of carbonyl compounds.[4] The reagent is prepared from (*R*)- or (*S*)-2,2′-dihydroxy-1,1′-binaphthyl (**3**) (BINAL). Both enantiomers of BINAL are commercially available, although they are somewhat expensive. The chiral ligand, however, can be recovered after the reduction and reused. Equimolar quantities of BINAL and LAH are initially mixed together to form a LAH complex that has a C_2 axis of symmetry, which makes the two hydrogens on the aluminum homotopic. It is interesting to note that the 1:1 complex of BINAL and LAH is a reducing agent that exhibits extremely low enantioselectivity as seen in the case of acetophenone (2% ee). Replacement of one of the hydrogens with an alcohol, like methanol or ethanol, gives a single reducing agent (**7**), which exhibits much higher specificity in the reduction of prochiral ketones. Another useful observation is that reduction of carbonyls with the (*R*)-BINAL-H reagent tends to give the (*R*)-alcohol while the (*S*)-reagent gives the (*S*)-alcohol. The use of lower reduction temperatures enhances optical purities of the product alcohols, but lowers the yields. Optimized conditions for reductions involve reaction of a ketone with 3 equiv of the reagent formed from LAH, BINAL, and ethanol (1:1:1) in THF for 1 h at −100 °C and then at −78 °C for 2 h.

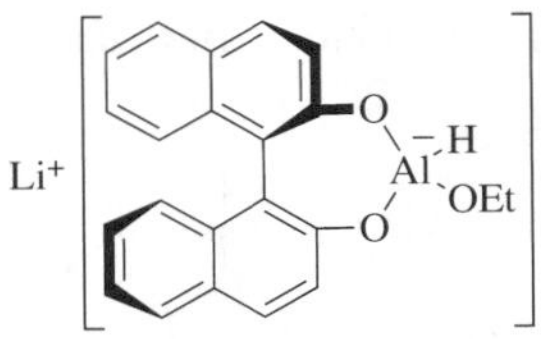

(*R*)-(**7**)

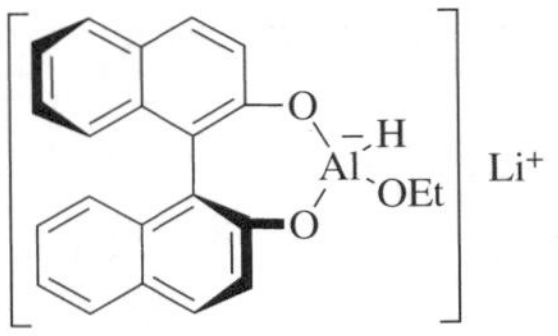

(*S*)-(**7**)

A number of structurally diverse ketones have been reduced using BINAL-H. Some of the results are summarized in Table 1.[5] Aryl alkyl ketones, alkynic ketones, and α,β-unsaturated ketones are reduced to alcohols with good to excellent % ee, while aliphatic ketones give products with lower optical purities. The asymmetric reduction of a number of acylstannanes with (**7**) gives synthetically valuable α-alkoxystannanes with high optical purities after protection of the initially formed unstable alcohols as their MOM or BOM ethers.[6]

Table 1 Reduction of Ketones with (**7**)

Ketone	(7)	Yield (%)	ee (%)	Product
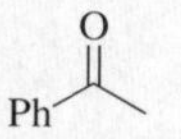 O, Ph	(*R*)	61	95	(*R*)
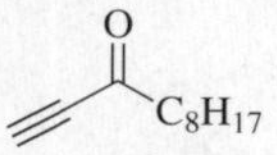 O, C_8H_{17}	(*S*)	74	90	(*S*)
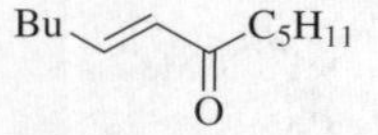 Bu, C_5H_{11}, O	(*R*)	91	91	(*R*)
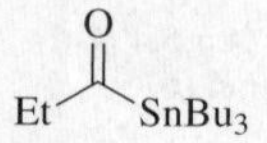 O, Et, $SnBu_3$	(*S*)	69	96	(*R*)
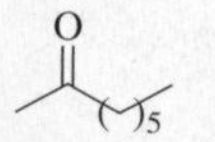 O, ( )$_5$	(*S*)	67	24	(*R*)

BINAL-H has been used to prepare deuterated primary alcohols with high optical purities. For example, benzaldehyde-*1-d* is reduced in 59% yield and 87% optical purity. β-Ionone is reduced with this reagent to the corresponding alcohol in 100% ee and 87% yield. Simple cyclic enones like 2-cyclohexenone are not reduced by the BINAL-H reagent under standard reduction conditions.[5]

The chiral nonracemic enone (**8**) is reduced with (*S*)-(**7**) to give the (15*S*)-alcohol in 100% de and 88% yield. The product is a valuable intermediate in the synthesis of prostaglandins.[5]

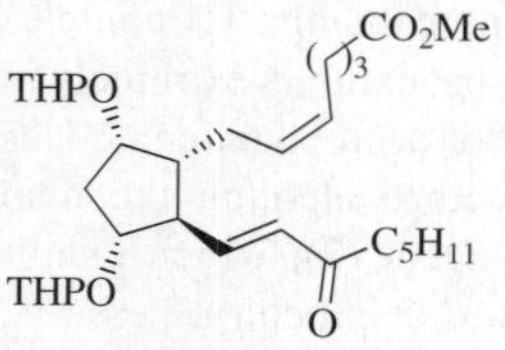

CO_2Me, ( )$_3$, THPO, C_5H_{11}, THPO, O

(**8**)

The asymmetric reduction of lactone (**9**) to give predominantly one atropoisomer can be achieved using 10 equiv of a complex prepared from LAH and BINAL (1:1) at −40 °C.[7] This reduction gives an 88:12 ratio of (**10a**):(**10b**) in good yield (80%). Reduction of the same substrate with 8 equiv of a complex of LAH with (*S*)-(+)-2-(anilinomethyl)pyrrolidine in ether at −40 °C leads to opposite stereochemical results (38:62 ratio of **10a**:**10b**).

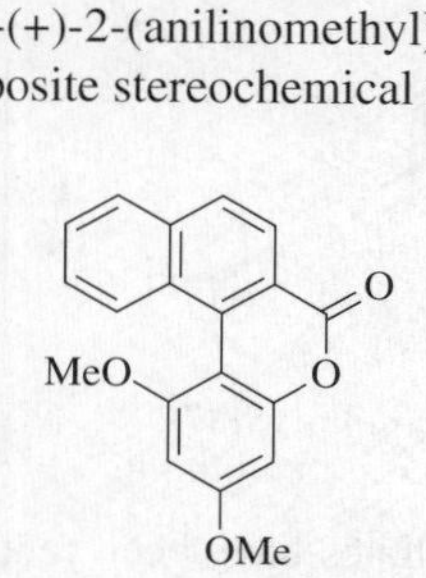

MeO, O, O, OMe — (**9**)

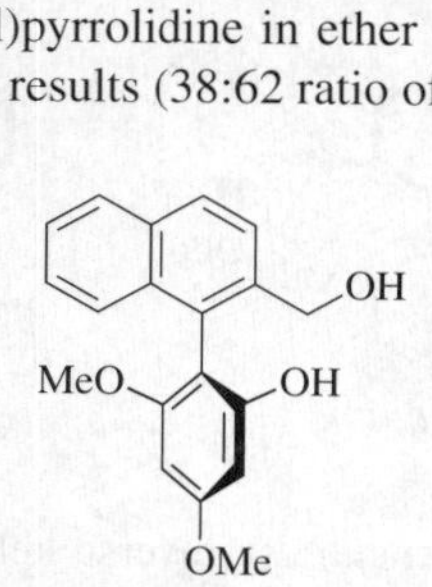

MeO, OH, OH, OMe — (**10a**)

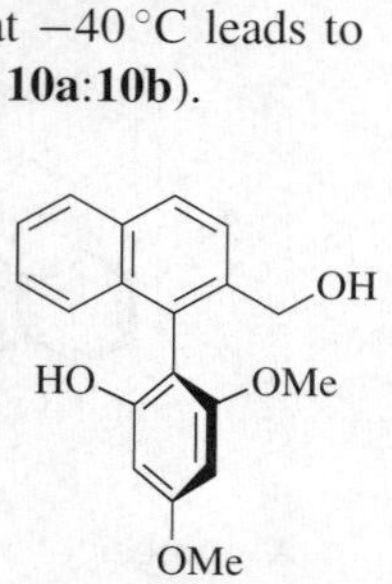

HO, OH, OMe, OMe — (**10b**)

BINAL-H has also been used for the asymmetric reduction of methylaryl- and methylalkylphosphinylimines to the corresponding phosphinylamines in high % ee (Table 2).[8] Similar to the reduction of ketones, reduction of the imines with (*S*)-(**7**) produces the (*S*)-amine and reduction with (*R*)-(**7**) gives the (*R*)-amine.

Table 2 Reduction of Imines with (**7**)

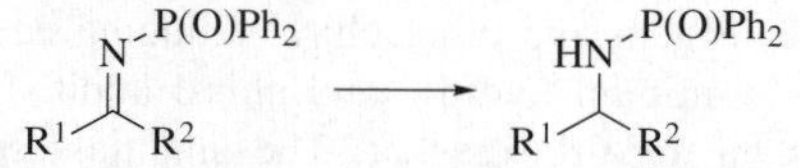

$R^1(R^2)C{=}N{-}P(O)Ph_2 \longrightarrow R^1(R^2)CH{-}NH{-}P(O)Ph_2$

R^1	R^2	(7)	Yield (%)	ee (%)	Product
Me	Ph	(*R*)	20	100	(*R*)
Me	Et	(*S*)	38	93	(*S*)
Me	C_5H_{11}	(*S*)	83	64	(*S*)

The complex of the biphenanthryl diol (**4**) with LAH has been prepared and its reduction properties have been examined.[9] This reagent gives excellent enantioselectivity in the reduction of aromatic ketones. For example, acetophenone is reduced in 75% yield with 97% ee. As with Noyori's reagent, reductions with the (*S*)-reagent give (*S*)-alcohols and aliphatic ketones are reduced with low enantioselectivity. Both enantiomers of this auxiliary can be readily prepared and can also be recovered for reuse at the end of the reduction.

The LAH complex of the chiral spirodiol (**5**) has recently been prepared. This complex exhibits excellent enantioselectivity in the reduction of some aromatic ketones.[10] Acetophenone is reduced at −80 °C in 98% ee and 80% yield. Reduction of other aryl alkyl ketones also gives excellent stereoselectivity, but the use of this reagent with a variety of ketones has not been studied. The chiral auxiliary can be recovered and reused.

Recently, the preparation of the chiral biphenyl (**6**) and its use as a modifying agent with LAH has been reported.[11] A complex of LAH–(**6**)–EtOH (1:1:1) at −78 °C gives the best enantioselectivities in the reduction of prochiral ketones. Similar to Noyori's reagent, use of the LAH complex with (*S*)-(**6**) leads to the (*S*)-alcohol. Enantioselectivity is usually high for aromatic ketones (acetophenone 97% ee, 93% yield). This reagent reduces 2-octanone in higher enantioselectivity (76% ee) than 3-heptanone (36% ee).

Chiral Amino Alcohol Modifying Agents. A number of chiral amino alcohols have been examined as ligands for the preparation of chiral LAH reducing agents (Figure 2). The complex of (−)-*N*-methylephedrine (**11**) with LAH has been widely studied and has shown promise for the asymmetric reduction of prochiral ketones. It has been found that addition of an achiral component such as 3,5-dimethylphenol (DMP), *N*-ethylaniline (NEA), or 2-ethylaminopyridine (EAP) to the complex of LAH with (**11**) can enhance the enantioselectivity observed in these reductions. Both enantiomers of (**11**) are commercially available and the ligand can be recovered subsequent to the reaction and reused.

Vigneron and co-workers have observed that a complex of LAH, (−)-(**11**), and DMP (1:1:2), in ether at −15 °C, appears to show the highest enantioselectivity in the reduction of a series of aromatic and alkynyl ketones to the corresponding (*R*)-alcohols (Figure 3).[12] Interestingly, the optical purities of the products obtained were lower both at higher and lower reaction temperatures.

The complex of LAH, (−)-(**11**), and DMP has also been used to reduce stereoselectively a steroidal alkynic ketone. Reduction of the alkynic ketone (**16**) with 3 equiv of the complex at −15 °C

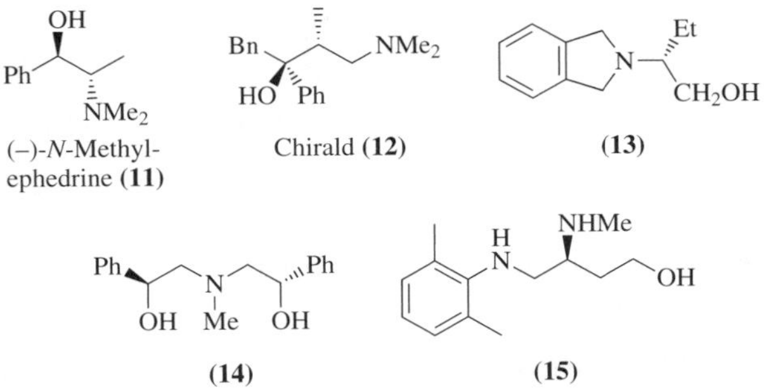

(–)-*N*-Methylephedrine (**11**) Chirald (**12**) (**13**)

(**14**) (**15**)

Figure 2 Representative chiral amino alcohol modifying agents

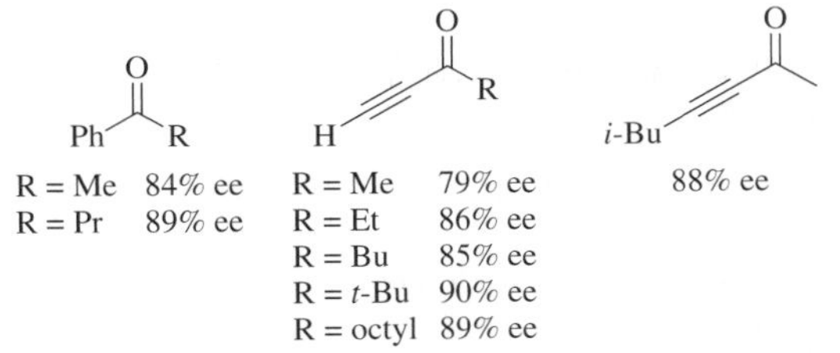

R = Me 84% ee R = Me 79% ee *i*-Bu 88% ee
R = Pr 89% ee R = Et 86% ee
R = Bu 85% ee
R = *t*-Bu 90% ee
R = octyl 89% ee

Figure 3 Reduction of ketones with LAH/(–)-(**11**)/DMP (1:1:2) to give (*R*)-alcohols

gave a 17:1 ratio of the two diastereomers (22*R*/22*S*) in 94% yield, to provide a key intermediate for the synthesis of a vitamin D_2 metabolite.[13]

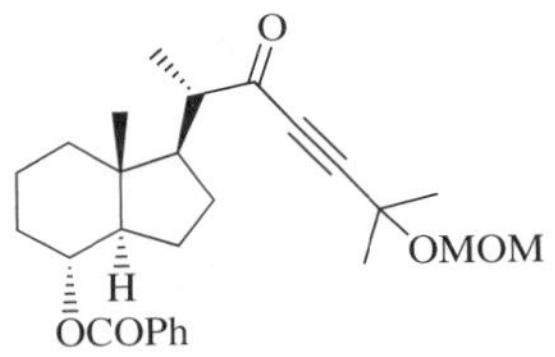

(**16**)

The enantioselective reduction of cyclic conjugated enones may be best accomplished using a complex of LAH with (**11**) to which EAP has been added.[14,15] Optimum conditions for these reductions involve treatment of the ketone with 3 equiv of a 1:1:2 complex of LAH–(−)-(**11**)–EAP in ether at −78 °C for 3 h (Table 3). However, under these conditions, acetophenone is reduced to the (*R*)-alcohol in only 54% ee.

Table 3 Reduction of Ketones with LAH/(–)-(**11**)/EAP to give (*R*)-Alcohols

Ketone	Yield (%)	ee (%)
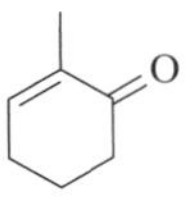	82	96
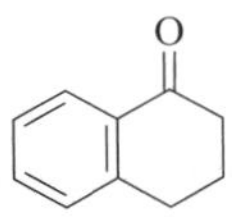	93	96
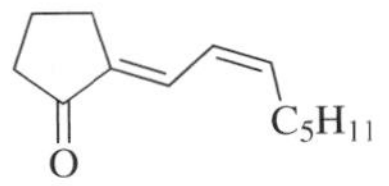 C_5H_{11}	91	96

It has been found that the addition of 2 equiv of NEA to a 1:1 complex of LAH and (−)-(**11**) in ether produces a reagent capable of reducing some α,β-unsaturated ketones to the (*S*)-alcohols in good optical purities at −78 °C (Table 4).[16] It is interesting to note that, with this reagent, the (*S*)-alcohol is the product that is formed preferentially.

Table 4 Reduction of Ketones with LAH/(–)-(**11**)/NEA to give (*S*)-Alcohols

Ketone	Yield (%)	ee (%)
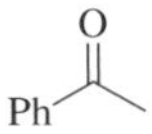 Ph	87	86
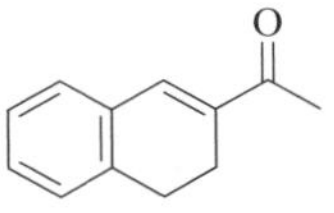	100	>90
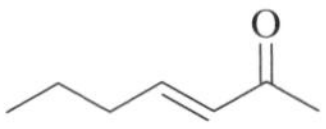	92	88
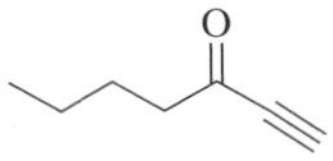	88	76
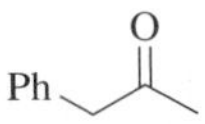 Ph	90	41

The preparation and use of a polymer supported LAH–ephedrine–DMP reducing reagent has been reported.[17] In preparing this reagent, ephedrine is attached to a 1% crosslinked polystyrene backbone prior to mixing with LAH and DMP. Careful control of the degree of functionalization of the polymer gives a reducing reagent comparable in efficacy to the analogous nonpolymeric complex.

The use of the complex formed between LAH and Chirald (often called Darvon alcohol in the literature) (**12**) for the reduction of conjugated enones and ynones was first reported by Yamaguchi and Mosher.[18] The mode of preparation of the complex, its age, and the precise experimental conditions of the reduction all appear to have significant impact on the enantioselectivities obtained using this reagent. Thus when 1.5 equiv of a freshly prepared complex of LAH and Chirald (1:2.3) is used to reduce acetophenone at 0 °C, the (*R*)-alcohol is obtained in 68% ee and nearly quantitative yield. If however, the reagent is allowed to stir overnight, or is refluxed in ether prior to the addition of the ketone, the (*S*)-enantiomer is obtained in 66% ee and 43% yield. Unfortunately, this observed reversal in stereochemical outcome is not predictable. Hence, it may be preferable to use the complex of LAH with the enantiomer of Chirald to reverse the stereoselectivity of the reduction.[19]

A number of alkynic ketones have been reduced with the complex of LAH and (**12**) (1.1:2.5 equiv, ether, −78 °C, 30–60 min) to give the corresponding (*R*)-alcohols (Table 5).[20,21] Johnson and co-workers have reported[22] the reduction of ynone (**17**) to the (*R*)-alcohol in 84% ee and 95% yield with the LAH–Chirald complex. The resulting alcohol was an intermediate in an enantioselective synthesis of 11α-hydroxyprogesterone.[22] The thiophene ketone (**18**) is reduced by the same reagent in ether at −70 °C for 16 h to give the (*R*)-alcohol in 85–88% ee and 80–90% yield.[23] The resulting alcohol has been used in the synthesis of LY248686, an inhibitor of serotonin and norepinephrine uptake carriers.

Table 5 Reduction of Alkynic Ketones with LAH/(–)-(**12**) to give (*R*)-Alcohols

R^1	R^2	Yield (%)	ee (%)
H	C_5H_{11}	96	72
TMS	C_5H_{11}	96	66
C_5H_{11}	C_5H_{11}	97	62

(**17**) (**18**)

A macrocyclic alkynic ketone has been protected as the Co derivative and then reduced with the complex of LAH with (**12**) (eq 1). Deprotection gave the (*R*)-alcohol (71% ee) which was an important intermediate in a synthesis of (+)-α-2,7,11-cembratriene-4,6-diol.[24]

1. Chirald–LAH, Et_2O, – 78 °C; 2. $(NH_4)_2Ce(NO_3)_6$, MeOH, Et_2O, 90% (1)

In general, structural variations to the backbone of the Chirald ligand have not led to the development of more selective or reliable LAH complexes for use in asymmetric reductions.[25] Other complexes of amino alcohols with LAH have been studied for their ability to achieve enantioselective reduction of prochiral ketones. However, in most cases the selectivities observed have been moderate.[26] The complex of LAH with the amino alcohol (**15**) reduces some enones, such as cyclohexenone and cyclopentenone, to the corresponding (*S*)-alcohols in high optical purities (100% and 82% ee, respectively).[27]

Chiral Amine Modifying Agents. Some chiral amine additives (Figure 4) have also been studied for their potential to give useful chiral LAH reagents, but the results so far have not been very promising. An exception to this is the complex of LAH with the chiral aminopyrrolidine (**19**) (R = Me), which reduces aromatic ketones in good ee.[28] This reagent reduces acetophenone in 95% ee and 87% chemical yield. LAH complexes of diamine ligands (**20**),

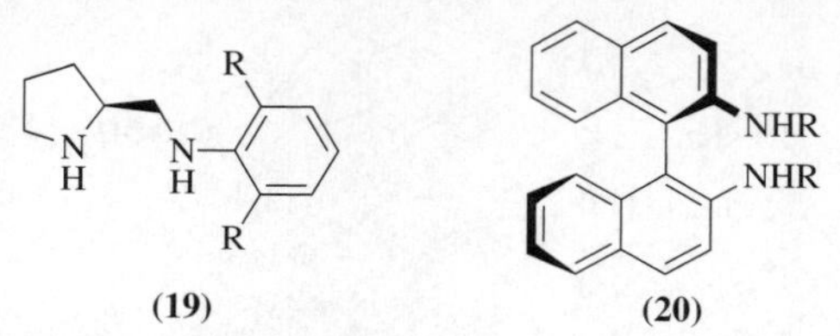

Figure 4 Representative chiral amino modifying agents

analogs of BINAL-H, have also been prepared and examined.[29] In general, the optical purities obtained with this reagent are significantly lower than those observed for BINAL-H in the reduction of aryl ketones.

Related Reagents. See Classes R-2, R-5, R-11, R-12, and R-13, pages 1–10.

---

1. (a) Nishizawa, M.; Noyori, R. *COS* **1991**, *8*, Chapter 1.7. (b) Grandbois, E. R.; Howard, S. I.; Morrison, J. D. *Asymmetric Synthesis*; Academic: New York, 1983; Vol. 2. (c) Nógrádi, M. *Stereoselective Synthesis*; VCH: Weinheim, 1986; Chapter 3. (d) ApSimon, J. W.; Collier, T. L. *T* **1986**, *42*, 5157. (e) Singh, V. K. *S* **1992**, 605. (f) Blaser, H.-U. *CRV* **1992**, *92*, 935. (g) Haubenstock, H. *Top. Stereochem.* **1982**, *14*, 231. (h) Mukaiyama, T.; Asami, M. *Top. Curr. Chem.* **1985**, *127*, 133. (i) Rosini, C.; Franzini, L.; Raffaelli, A.; Salvadori, P. *S* **1992**, 503.
2. (a) Tomioka, K. *S* **1990**, 541. (b) Wallbaum, S.; Martens, J. *TA* **1992**, *3*, 1475. (c) Santaniello, E.; Ferraboschi, P.; Grisenti, P.; Manzocchi, A. *CRV* **1992**, *92*, 1071.
3. Brown, H. C.; Park, W. S.; Cho, B. T.; Ramachandran, P. V. *JOC* **1987**, *52*, 5406.
4. Noyori, R.; Tomino, I.; Tanimoto, Y.; Nishizawa, M. *JACS* **1984**, *106*, 6709.
5. Noyori, R.; Tomino, I.; Yamada, M.; Nishizawa, M. *JACS* **1984**, *106*, 6717.
6. (a) Chan, P. C.-M.; Chong, J. M. *JOC* **1988**, *53*, 5584. (b) Chong, J. M.; Mar, E. K. *T* **1989**, *45*, 7709. (c) Chong, J. M.; Mar, E. K. *TL* **1990**, *31*, 1981. (d) Marshall, J. A.; Gung, W. Y. *TL* **1988**, *29*, 1657.
7. Bringmann, G.; Hartung, T. *S* **1992**, 433.
8. Hutchins, R. O.; Abdel-Magid, A.; Stercho, Y. P.; Wambsgans, A. *JOC* **1987**, *52*, 702.
9. Yamamoto, K.; Fukushima, H.; Nakazaki, M. *CC* **1984**, 1490.
10. Srivastava, N.; Mital, A.; Kumar, A. *CC* **1992**, 493.
11. Rawson, D.; Meyers, A. I. *CC* **1992**, 494.
12. (a) Vigneron, J. P.; Jacquet, I. *T* **1976**, *32*, 939. (b) Vigneron, J. P.; Blanchard, J. M. *TL* **1980**, *21*, 1739. (c) Vigneron, J. P.; Bloy, V. *TL* **1980**, *21*, 1735. (d) Vigneron, J.-P.; Bloy, V. *TL* **1979**, 2683.
13. Sardina, F. J.; Mouriño, A. Castedo, L. *TL* **1983**, *24*, 4477. (b) Sardina, F. J.; Mouriño, A.; Castedo, L. *JOC* **1986**, *51*, 1264.
14. Kawasaki, M.; Suzuki, Y.; Terashima, S. *CL* **1984**, 239.
15. Iwasaki, G.; Sano, M.; Sodeoka, M.; Yoshida, K.; Shibasaki, M. *JOC* **1988**, *53*, 4864.
16. (a) Terashima, S.; Tanno, N.; Koga, K. *TL* **1980**, *21*, 2753. (b) Terashima, S.; Tanno, N.; Koga, K. *CC* **1980**, 1026. (c) Terashima, S.; Tanno, N.; Koga, K. *CL* **1980**, 981.
17. Fréchet, J. M.; Bald, E.; Lecavalier, P. *JOC* **1986**, *51*, 3462.
18. (a) Yamaguchi, S.; Mosher, H. S. *JOC* **1973**, *38*, 1870. (b) Yamaguchi, S.; Mosher, H. S.; Pohland, A. *JACS* **1972**, *94*, 9254.
19. Paquette, L. A.; Combrink, K. D.; Elmore. S. W.; Rogers, R. D. *JACS* **1991**, *113*, 1335.
20. Brinkmeyer, R. S.; Kapoor, V. M. *JACS* **1977**, *99*, 8339.
21. Marshall, J. A.; Salovich, J. M.; Shearer, B. G. *JOC* **1990**, *55*, 2398.

22. Johnson, W. S.; Brinkmeyer, R. S.; Kapoor, V. M.; Yarnell, T. M. *JACS* **1977**, *99*, 8341.
23. Deeter, J.; Frazier, J.; Staten, G.; Staszak, M.; Weigel, L. *TL* **1990**, *31*, 7101.
24. Marshall, J. A.; Robinson, E. D. *TL* **1989**, *30*, 1055.
25. Cohen, N.; Lopresti, R. J.; Neukom, C.; Saucy, G. *JOC* **1980**, *45*, 582.
26. (a) Brown, E.; Penfornis, A.; Bayma, J.; Touet, J. *TA* **1991**, *2*, 339. (b) Steels, I.; DeClercq, P. J.; Declercq, J. P. *TA* **1992**, *3*, 599. (c) Morrison, J. D.; Grandbois, E. R.; Howard, S. I.; Weisman, G. R. *TL* **1981**, *22*, 2619.
27. (a) Sato, T.; Gotoh, Y.; Wakabayashi, Y.; Fujisawa, T. *TL* **1983**, *24*, 4123. (b) Sato, T.; Goto, Y.; Fujisawa, T. *TL* **1982**, *23*, 4111.
28. Asami, M.; Mukaiyama, T. *H* **1979**, *12*, 499.
29. Kabuto, K.; Yoshida, T.; Yamaguchi, S.; Miyano, S.; Hashimoto, H. *JOC* **1985**, *50*, 3013.

Aravamudan S. Gopalan & Hollie K. Jacobs
New Mexico State University, Las Cruces, NM, USA

Lithium Borohydride[1]

$LiBH_4$

[16949-15-8] BH_4Li (MW 21.78)

(reducing agent for esters and lactones,[1,2a,b] acyl chlorides,[2c] epoxides,[1] aldehydes and ketones;[1,2a,b,d] precursor of other borohydrides;[1] catalyst for hydroborations[3])

Physical Data: mp 284 °C (dec); d 0.666 g cm^{-3}.
Solubility: sol ethers (3 g/100 mL Et_2O, 25 g/100 mL THF),[4] lower primary amines ($MeNH_2$, $EtNH_2$, i-$PrNH_2$),[5] diglyme[6] (9 g/100 mL);[4] sol (with reaction) alcohols[7] (1.6 g/100 mL MeOH, 3 g/100 mL i-PrOH);[4] solution in absolute EtOH shows no appreciable decomposition after 2–4 h at about 0 °C;[7] solution in i-PrOH shows no decomposition after 24 h;[4] sol (with slow decomposition) water.
Form Supplied in: off-white solid; 2.0 M solution in THF.
Analysis of Reagent Purity: hydrolysis with dilute acid and titration for boron;[8] ethereal or THF solutions can be titrated with 2 N HCl–THF (1:1).[4]
Purification: recrystallization from Et_2O; the purified material is pumped free of ether at 90–100 °C for 2 h.[8]
Handling, Storage, and Precautions: both solutions and (especially) the solid[9] are flammable and must be stored under N_2 in the absence of moisture; the solid is capable of creating a dust explosion; $LiBH_4$ reacts with water and acids, generating flammable and/or explosive gas (H_2 and borane), and is incompatible with strong oxidizing agents.

Reduction of Functional Groups. Lithium borohydride is more reactive as a reducing agent than ***Sodium Borohydride*** and less reactive than ***Lithium Aluminum Hydride***.[1] In THF, $LiBH_4$ readily reduces aldehydes and ketones to alcohols at room or even ice-bath temperature,[9] whereas esters and lactones require higher temperatures and prolonged reaction times to give the corresponding alcohols and diols. Thus selective reductions are possible. Epoxides are reduced by $LiBH_4$, while carboxylic acids, carboxylic acid salts, tertiary amides, nitriles, nitro compounds, alkenes, and halogeno derivatives do not usually react.[1]

The chief advantage of $LiBH_4$ over $LiAlH_4$ is its much greater chemoselectivity, while the chief advantage over $NaBH_4$ is its higher solubility in ethereal solvents. The different reactivity of the two lithium hydrides ($LiBH_4$ and $LiAlH_4$) is essentially due to the different hardness of the two hydride-delivering anions, while the different reactivity of the two borohydrides ($LiBH_4$ and $NaBH_4$) is essentially due to the different Lewis acidity of the associated cation, rather than to differences in solubility.[6,10]

Lithium borohydride reactivity is greatest in media of low dielectric constant: increasing the polarity decreases reactivity.[4] The order of reactivity is $LiBH_4$ in $Et_2O > THF \approx$ diglyme > 2-propanol; this trend is exactly reversed for $Ca(BH_4)_2$.

Several papers have dealt with the effects of solvent,[4,6,11] concentration,[10a] added salt,[6] and added cation complexing agents[10] both on the reaction rate and the regiochemistry of $LiBH_4$ reductions, regarding essentially ester reduction.

Since carbonyl compounds are rapidly reduced by $LiBH_4$ alone, less attention has been devoted to this reaction. Nevertheless, it has been shown that the rate of the reduction of acetone in i-PrOH[11] ($k = 50.3 \times 10^{-4}$ L mol^{-1} s^{-1}) is not appreciably affected by the addition of ***Triethylamine***, while addition of ***Lithium Chloride*** accelerated the reduction. The added lithium salt probably has a double effect: on one hand, it can modify the nature of the ionic cluster; on the other, it can activate the carbonyl group. In contrast, $NaBH_4$ reductions are almost unaffected by added ***Sodium Iodide*** and accelerated by Et_3N. Reaction rates for reduction of some aliphatic and aromatic ketones with borohydrides in various solvents have been reported.[12]

In the reduction of cyclic ketones, $LiBH_4$ generally attacks more from the more hindered side than $LiAlH_4$;[13] stereochemical aspects of cyclic ketone reduction by complex borohydrides have been widely discussed in some reviews.[14] In the reduction of conjugated cyclic enones[1,10,13] $LiBH_4$ usually gives more 1,4-attack than does $LiAlH_4$. This tendency is enforced by the presence of a lithium-complexing agent, such as [2.1.1]cryptand: when lithium cation is removed from the reaction medium, the reaction rate of 1,2-attack is decreased more than that of 1,4-attack, probably because of different influence of Li^+ on the carbonyl LUMO and on reducing agent HOMO levels, so that 1,4-attack becomes predominant.[10b] For example, cyclohexen-3-one is reduced to give predominantly the corresponding allylic alcohol by $LiBH_4$ alone, while in the presence of [2.1.1]cryptand the saturated ketone and alcohol are largely predominant (eq 1). Results comparable to that reported in eq 1 have also been obtained using ***Tetra-n-butylammonium Borohydride*** in several aprotic solvents.[15]

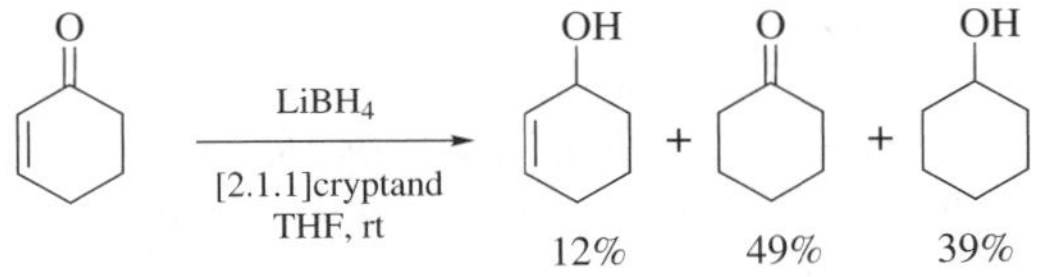

LiBH4; [2.1.1]cryptand; THF, rt → 12% + 49% + 39% (1)

Regio- and diastereoselective reductions of steroidal ketones have been achieved by using $LiBH_4$.[7,16a] Stereoselective reduction of a polyfunctionalized ketone, intermediate in the synthesis of oleandomycin, was realised via $LiBH_4$ reduction (THF–MeOH, −78 °C) of the dibutylboron aldolate derived from reaction of the ketone itself with dibutylmethoxyborane.[16b]

More studies have been devoted to the reduction of esters, regarding the solvent and possible additives. Different solvents can be used: Et_2O, THF, diglyme, or *i*-PrOH, although in the latter case a large excess of $LiBH_4$ is required to compensate the loss due to the side reaction with the solvent.[5] A clean procedure to reduce esters in high yield using an essentially stoichiometric amount of $LiBH_4$ in Et_2O or THF has been reported.[4] Under these reaction conditions, hindered esters, such as ethyl adamantanecarboxylate, and lactones are also readily reduced, while other functional groups, such as alkyl and aryl halides, nitro groups, ethers, and nitriles remain unaffected. Diesters have been reduced to the corresponding diols using $LiBH_4$,[4,17] while reduction of ***Ethyl Acetoacetate*** has been reported to give rise to some problems, due to the formation of a borate complex, from which the reduction product cannot be isolated.[9]

The reducing ability of $LiBH_4$ has been found to be greatly enhanced in mixed solvents containing methanol, and to be dependent on the amount of added alcohol.[18] Using the $LiBH_4$–Et_2O–MeOH (1 equiv with respect to $LiBH_4$) reducing system, esters, lactones, and epoxides are selectively reduced, even at room temperature, with respect to nitro groups, aryl halides, primary amides, and carboxylic acids. By employing the $LiBH_4$–diglyme (or THF)–MeOH (4 equiv with respect to $LiBH_4$) systems, a further enhancement of reducing capabilities is observed: nitro compounds and nitriles are reduced to amines, and carboxylic acids are reduced to alcohols, while amides show different behaviors depending on substitution. Tertiary amides are reduced essentially to alcohols, through C–N bond fission, while primary amides are cleanly reduced to amines, via C–O bond fission. Secondary amides show different behavior depending on nitrogen substituent.

The chemoselective reduction of an ester moiety in the presence of a carboxylic acid, using $LiBH_4$, or vice versa, using diborane, has been applied in the stereoselective synthesis of both (*R*)- and (*S*)-mevalonolactone.[19a] The chemoselective reduction of an ester with respect to an amide using $LiBH_4$ in THF has been reported in carbohydrate chemistry.[19b] Sterically hindered acyloxazolidinones have been reduced to the corresponding primary alcohols using $LiBH_4$ in Et_2O containing water (1 equiv) in better yields than using $LiBH_4$ or $LiAlH_4$ alone.[19c]

An interesting stereoselective 1,4 reduction of an acetoxy unsaturated nitrile has been realized by using $LiBH_4$ in THF, probably through the intermediate formation of an alkoxyhydride; the same reaction has been realized on a preparative scale using $LiAlH_4$ in THF.[20a] Sparse reports on the reduction of other functional groups (carbazole derivatives of carboxylic acids to aldehydes,[20b] hydrazides to hydrazines,[20c] cyclic anhydrides to lactones,[20d] acyl chlorides to alcohols[2b]) using $LiBH_4$ in various nonhydroxylic solvents have appeared.

Reactivity of $LiBH_4$ towards *n*-octyl chloride has been found to be extremely low; ***Lithium Triethylborohydride*** is 10^4 times more efficient in carrying out the reduction to *n*-octane.[21a] Alkyl tosylates are almost inert towards $LiBH_4$, while they are readily reduced by $LiEt_3BH$.[21b]

Epoxides are smoothly reduced by $LiBH_4$,[18,22] with attack occurring mainly at the less hindered site and *cis* epoxides being reduced more rapidly than *trans* ones. 2,3-Epoxy alcohols and their derivatives can be regioselectively reduced to 1,2-diols by using a suspension of $LiBH_4$ in hexane at ambient temperature;[23] results are superior to that obtained using completely dissolved $LiBH_4$ in THF or a ***Titanium Tetraisopropoxide***–$LiBH_4$ system (see below).

Addition of Lewis Acids. In the effort to 'tune up' the selectivity of lithium borohydride, several combinations of the reducing agent with a variety of Lewis acids have been assayed.

Addition of boranes or alkoxyboranes greatly enhances the reactivity of $LiBH_4$ towards esters in Et_2O or THF;[24] a particularly high catalytic effect is shown by *B*-MeO-9-BBN and ***Trimethyl Borate***. Many other additives that generate a borane species under the reaction conditions (e.g. $LiEt_3BH$, $LiEt_3BOMe$, ***Tri-n-butylborane***) show a catalytic effect. These additives also exert a catalytic effect on epoxide reductions, but have little if any influence on the reduction of carboxylic acids, tertiary amides [which can be reduced by *n*-Bu_4NBH_4 or by $NaBH_4$–***Titanium(IV) Chloride*** (see under ***Sodium Borohydride***)], nitriles, sulfur compounds, and pyridine. ***Borane–Tetrahydrofuran*** and ***Boron Trifluoride Etherate*** are ineffective as catalysts. Alkenes, although normally inert to the action of $LiBH_4$, are hydroborated (see below) by this reagent in the presence of esters, with concomitant enhancement of the ester reduction rate: as a consequence, unsaturated esters are transformed into a mixture of regioisomeric diols.

Lithium borohydride combined with ***Chlorotrimethylsilane*** generates a more powerful reducing system.[25] Amino acids can be reduced to amino alcohols, with retention of optical purity; primary, secondary, and tertiary amides, as well as nitriles, are reduced to amines, and sulfoxides to sulfides; a nitrostyrene derivative is reduced to the corresponding saturated amine, in better yield than with alternative reducing systems ($LiAlH_4$ or catalytic hydrogenation). These reactions are believed to proceed through the formation of a borane–THF complex.

Lithium borohydride combined with ***Europium(III) Chloride*** in MeOH–Et_2O has given the best result in regio- and diastereoselective reduction of a polyfunctionalized conjugated enone intermediate in the synthesis of palytoxin.[26] Inferior results were obtained with other reducing systems.

Lithium borohydride combined with titanium tetraisopropoxide in THF, benzene, or CH_2Cl_2 reduces 2,3-epoxy alcohols regioselectively to 1,2-diols in high yield.[27a,b] Good 1,2 regioselectivity is shown also by ***Diisobutylaluminum Hydride*** (DIBAL) in benzene, while 1,3-diols can be regioselectively obtained by using Red-Al (***Sodium Bis(2-methoxyethoxy)aluminum Hydride***) in THF.[27c,d] The $LiBH_4$–Ti(*i*-PrO)$_4$ system is effective in *syn* diastereoselective reduction of β-hydroxy ketones to 1,3-diols,[27e] although other hydrides ($LiAlH_4$, $NaBH_4$) combined with different additives [Ti(*i*-PrO)$_4$, $Ti(OEt)_4$, $TiCl_4$, LiI] show the same trend and often give better results.

The addition of titanium tetrachloride reverses the diastereoselectivity of the $LiBH_4$ reduction of 3,3-dimethyl-2,4-pentanedione,[28] so that *meso*-2,4-pentanediol is obtained as the main product.

Lithium borohydride (but also $LiAlH_4$ and $NaBH_4$) combined with boron trifluoride etherate in Et_2O, THF, or THF–diglyme reduces cyclic lactones to cyclic ethers.[29]

Addition of Grignard Reagents. Synthesis of secondary alcohols from esters can be realised by combining $LiBH_4$ with Grignard reagents via formation of an intermediate ketone, which is reduced by $LiBH_4$ much more rapidly than the ester.[30a] When 2-alkoxy esters are used, stereoselective formation of *anti* 1,2-diols is observed, especially when THF is used as solvent.[30b] The reversed diastereoselectivity is observed when DIBAL is added to the ester, with formation of an intermediate aldehyde, before adding the Grignard reagent.

Addition of Chiral Ligands. *N*-Benzoylcysteine, a chiral ligand available in both enantiomeric forms, is highly effective in enantioselective $LiBH_4$ reductions of alkyl aryl ketones in THF–*t*-BuOH (ee up to 92%).[31a] Analogous results are obtained using the dimer *N,N*-dibenzoylcystine,[31b] which also reduces a conjugated enone to the corresponding chiral allyl alcohol.[31b] The $LiBH_4$–*N,N*-dibenzoylcystine system has also been used to enantioselectively reduce β-keto esters,[31c] β-chloro ketones[31d] (which are precursors of optically active oxetanes), acetylpyridines, and α- and β-amino ketones.[31e]

Other Reductions. In an attempt to reduce the ester moiety of some functionalized 2-methylthiopyrimidines, it has been found that $LiBH_4$ (and also $LiAlH_4$) in THF reduces the heteroaromatic ring, leaving almost unaffected the ester and other functional groups, so that differently substituted 1,6-dihydroxypyrimidines can be obtained in good yields (eq 2).[32] Differently functionalized pyrimidines are similarly reduced by $LiBH_4$ in DMF. It should be emphasized that a tertiary amide like DMF can be a suitable solvent for $LiBH_4$ (see above for discussion of amide reductions).

R, CO_2Et, N, MeS, N — $LiBH_4$, THF, 20 °C → R, CO_2Et, N, MeS, N, H (2)

R = Me, CN, Cl, CH=NOH

Alkyl- and arylhalostibines can be reduced to hydrides, having general formula R_nSbH_{3-n}, both by $LiBH_4$ and other complex hydrides ($LiAlH_4$, $NaBH_4$).[2e]

Hydroboration. Lithium borohydride is an effective catalyst for the hydroboration of alkenes with ***Catecholborane*** at rt in THF.[3] This method is superior to that employing ***Chlorotris(triphenylphosphine)rhodium(I)*** as catalyst, since trisubstituted and even tetrasubstituted ethylenes can also be almost quantitatively hydroborated. Other borohydrides ($LiBEt_3H$ in THF, $NaBH_4$ in diglyme) were tested and found to be less effective as catalysts.

Preparation of Other Borohydrides. Lithium borohydride has been used to synthesize other borohydrides,[5] particularly aluminum borohydride (a volatile, liquid source of borohydride groups) through the metathesis between ***Aluminum Chloride*** and $LiBH_4$ (eq 3). For this process, $LiBH_4$ is superior to $NaBH_4$, both for a more pronounced reactivity and for a more favorable equilibrium position, so that lower temperatures and a much smaller excess of $AlCl_3$ are required.

$$AlCl_3 + 3\,LiBH_4 \rightleftharpoons Al(BH_4)_3 + 3\,LiCl \quad (3)$$

Miscellaneous. The addition of $LiBH_4$ in the carbonylation of trialkylboranes not only has the effect of stopping the reaction after the transfer of only one alkyl group, but also enhances the rate of the uptake of ***Carbon Monoxide***[2f,33] (see also ***Potassium Triisopropoxyborohydride***). Aldehydes or alcohols are obtained, depending on workup conditions (eq 4).

$$R_3B + CO \xrightarrow{LiBH_4} R_2B\text{–}CH(R)\text{–}OM \quad \begin{cases} \xrightarrow{\text{aq NaOH}} RCH_2OH \\ \xrightarrow{H_2O_2,\ OH^-} RCHO \end{cases} \quad (4)$$

Lithium borohydride has been used in the degradation of peptides, in order to determine the terminal carboxylic groups.[1f]

Isotopically labelled lithium borohydrides, $LiBD_4$ and $LiBT_4$, have been used in reaction mechanism studies.[2g,20a]

Related Reagents. See Classes R-2, R-5, R-8, R-10, R-12, R-14, R-17, R-20, R-32, and R-34, pages 1–10.

---

1. (a) House, H. O. *Modern Synthetic Reactions*; Benjamin: New York, 1965; Chapter 2. (b) Walker, E. R. H. *CSR* **1976**, *5*, 23. (c) Brown, H. C.; Krishnamurty, S. *Aldrichim. Acta* **1979**, *12*, 3. (d) Brown, H. C.; Krishnamurty, S. *T* **1979**, *35*, 567. (e) *COS* **1991**, *8*, Chapters 1.1.3.1, 1.1.4.1, 1.10.3.3, 1.10.4.3, 1.10.5.3, 3.5.4.1, 3.8.2.2.2, 3.10.2.2, 4.4.2.1. (f) Seyden-Penne, J. *Reductions by the Alumino- and Borohydrides in Organic Synthesis*; VCH–Lavoiser: Paris, 1991.
2. *Comprehensive Organic Chemistry*; Barton, D.; Ollis, W. D., Eds.; Pergamon: Oxford, **1979** (a) Vol. 3, Chapters 14.2.3.1 and 14.2.6.1. (b) Vol. 2, Chapter 9.8.3.4. (c) Vol. 4, Chapter 20.1.6.4. (d) Vol. 1, Chapter 5.1.4.2. (e) Vol. 3, Chapter 15.5.4.1. (f) Vol. 3, Chapter 14.3.4.2. (g) Vol. 5, Chapter 29.2.2.
3. Arase, A.; Nunokawa, Y.; Masuda, Y.; Hoshi, M. *CC* **1991**, 205 and refs. therein.
4. Brown, H. C.; Narasimhan, S.; Choi, Y. M. *JOC* **1982**, *47*, 4702.
5. Schlesinger, H. I.; Brown, H. C.; Hyde, E. K. *JACS* **1953**, *75*, 209.
6. Brown, H. C.; Mead, E. J.; Rao, B. C. S. *JACS* **1955**, *77*, 6209.
7. Kollonitsch, J.; Fuchs, O.; Gabor, V. *Nature* **1954**, *173*, 125.
8. Schaeffer, G. W.; Roscoe, J. S.; Stewart, A. C. *JACS* **1956**, *78*, 729.
9. Nystrom, R. F.; Chaikin, S. W.; Brown, W. G. *JACS* **1949**, *71*, 3245.
10. (a) Handel, H.; Pierre, J. L. *T* **1975**, *31*, 2799. (b) Loupy, A.; Seyden-Penne J. *T* **1980**, *36*, 1937.
11. Brown, H. C.; Ichikawa, K. *JACS* **1961**, *83*, 4372.
12. Lansbury, P. T.; MacLeay, R. E. *JACS* **1965**, *87*, 831.
13. Ashby, E. C.; Boone, J. R. *JOC* **1976**, *41*, 2890.
14. (a) Boone, J. R.; Ashby, E. C. *Top. Stereochem.* **1979**, *11*, 53. (b) Wigfield, D. C. *T* **1979**, *35*, 449. (c) Caro, B.; Boyer, B.; Lamaty, G., Jaouen, G. *BSF(2)* **1983**, 281.
15. D'Incan, E.; Loupy, A. *T* **1981**, *37*, 1171.
16. (a) Stache, U.; Radscheit, K.; Fritsch, W.; Haede, W.; Kohl, H.; Ruschig, H. *LA* **1971**, *750*, 149. (b) Paterson, I.; Lister, A. M.; Norcross, R. D. *TL* **1992**, *33*, 1767.
17. Carpino, L. A.; Göwecke, S. *JOC* **1964**, *29*, 2824.
18. Soai, K.; Ookawa, A. *JOC* **1986**, *51*, 4000.
19. (a) Huang, F.-C.; Lee, L. F. H.; Mittal, R. S. D.; Ravikumar, P. R.; Chan, J. A.; Sih, C. J.; Caspi, E.; Eck, C. R. *JACS* **1975**, *97*, 4144. (b) Jeanloz, R. W.; Walker, E. *Carbohydr. Res.* **1967**, *4*, 504. (c) Penning, T. D.; Djuric, S. W.; Haack, R. A.; Kalish, V. J.; Miyashiro, J. M.; Rowell, B. W.; Yu, S. S. *SC* **1990**, *20*, 307.

20. (a) Lansbury, P. T.; Vacca, J. P. *TL* **1982**, *23*, 2623. (b) Wittig, G.; Hornberger, P. *LA* **1952**, *577*, 11. (c) Carpino, L. A.; Santilli, A. A.; Murray, R. W. *JACS* **1960**, *82*, 2728. (d) Narasimhan, S. *H* **1982**, *18*, 131.
21. (a) Brown, H. C.; Krishnamurthy, S. *JACS* **1973**, *95*, 1669. (b) Krishnamurthy, S.; Brown, H. C. *JOC* **1976**, *41*, 3064.
22. Guyon, R.; Villa, P. *BSF(2)* **1975**, 2584.
23. Sugita, K.; Onaka, M.; Izumi, Y. *TL* **1990**, *31*, 7467.
24. Brown, H. C.; Narasimhan, S. *JOC* **1984**, *49*, 3891 and refs. therein.
25. Giannis, A.; Sandhoff, K. *AG(E)* **1989**, *28*, 218.
26. Armstrong, R. W.; Kishi, Y. et al. *JACS* **1989**, *111*, 7525.
27. (a) Dai, L.; Lou, B.; Zhang, Y.; Guo, G. *TL* **1986**, *27*, 4343. (b) Zhou, W.-S.; Shen, Z.-W. *JCS(P1)* **1991**, 2827. (c) Finan, J. M.; Kishi, Y. *TL* **1982**, *23*, 2719. (d) Viti, S. M. *TL* **1982**, *23*, 4541. (e) Bonini, C.; Bianco, A.; Di Fabio, R.; Mecozzi, S.; Proposito, A.; Righi, G. *G* **1991**, *121*, 75.
28. Maier, G.; Seipp, U. *TL* **1987**, *28*, 4515.
29. Pettit, G. R.; Ghatak, U. R.; Green, B.; Kasturi, T. R.; Piatak, D. M. *JOC* **1961**, *26*, 1685.
30. (a) Comins, D. L.; Herrick, J. J. *TL* **1984**, *25*, 1321. (b) Burke, S. D.; Deaton, D. N.; Olsen, R. J.; Armistead, D. M.; Blough, B. E. *TL* **1987**, *28*, 3905.
31. (a) Soai, K.; Yamanoi, T.; Oyamada, H. *CL* **1984**, 251. (b) Soai, K.; Oyamada, H.; Yamanoi, T. *CC* **1984**, 413. (c) Soai, K.; Yamanoi, T.; Hikima, H.; Oyamada, H. *CC* **1985**, 138. (d) Soai, K.; Niwa, S.; Yamanoi, T.; Hikima, H.; Ishizaki, M. *CC* **1986**, 1018. (e) Soai, K.; Niwa, S.; Kobayashi, T. *CC* **1987**, 801.
32. Shadbolt, R. S.; Ulbricht, T. L. V. *JCS(C)* **1968**, 733.
33. Rathke, M. W.; Brown, H. C. *JACS* **1967**, *89*, 2740.

Luca Banfi, Enrica Narisano, & Renata Riva
Università di Genova, Italy

Lithium 4,4′-Di-*t*-butylbiphenylide[1]

$[t\text{-Bu–C}_6\text{H}_4\text{–C}_6\text{H}_4\text{–}t\text{-Bu}]^{\bullet -}\ \text{Li}^+$

[61217-61-6] $C_{20}H_{26}Li$ (MW 273.40)

(reductive lithiations of thiophenyl ethers,[11] carbonyl groups,[23] phosphinates;[25] generation of alkyllithiums,[2] α-lithio ethers,[3] allyllithiums,[17,18] β-,γ-,δ-lithio alkoxides;[19] propellane bond cleavage[22])

Alternate Name: LDBB.
Solubility: sol THF
Preparative Methods: anhydrous THF and 4,4′-di-*t*-butylbiphenyl (DBB) are cooled to 0 °C under a blanket of argon and stirred with a glass-coated stirring bar; ***Lithium*** foil (or ribbon) is added in portions and the mixture is stirred for 5 h at 0 °C; the appearance of the deep blue-green radical anion appears within 5 min.
Handling, Storage, and Precautions: LDBB will slowly decompose at room temperature.

Reductive Lithiation. There are several advantages for using LDBB over the traditional ***Lithium Naphthalenide*** or ***Sodium Naphthalenide*** (LN, NaN). LDBB is a more powerful reducing agent than the alkali naphthalenides and the resulting carbanion will be trapped as an organolithium. The high steric environment of the *t*-butyl groups lead preferentially to electron transfer and to little or no radical combination. This is in contrast to LN, where alkylation of naphthalene is frequently observed. Another advantage is the ease in separation of nonvolatile DBB from the reaction mixture.[2]

LDBB has been used in place of ***Lithium 1-(Dimethylamino)naphthalenide*** (LDMAN), leading to slightly higher yields of the desired products or when reaction conditions dictate.[3]

Alkyllithiums. The conversion of alkyl halides to alkyllithiums is highly effective with LDBB. The corresponding hydrocarbons can be formed in high yields upon the addition of H_2O (eq 1).[2]

$$\text{RCl} \xrightarrow{\text{LDBB}} [\text{RLi}] \xrightarrow[\text{96–100\%}]{H_2O} \text{RH} + \text{DBB} \quad (1)$$

The alkyllithiums can be trapped by a variety of electrophiles including aldehydes (eq 2),[4] carbon dioxide (eq 3)[5] and aryl groups (eq 4).[6] The use of LDBB to generate the organolithium in the latter example proved critical since other methods (***n-Butyllithium***, ***t-Butyllithium***, ***s-Butyllithium***, and ***Methyllithium***) led to lower yields and longer reaction times. One possible reason for the enhancement in organolithium activity may be a result of lowered aggregation.[6]

2 LDBB + [chloro-diene] —THF, –78 °C→ [lithio-diene] —RCHO→ [products] (2)

R, OH, OH, R

Cl —1. LDBB; 2. CO_2; 64%→ CO_2H (3)

Ph, OMe, O, N —1. BuBr, LDBB; 2. MeI; 96%→ Ph, OMe, O, N, Bu (4)

92% de

The geminal dilithio species can also be generated and trapped with electrophiles in good yields (eqs 5 and 6).[7]

Br, Br —LDBB, –100 °C→ [Li, Li] —electrophile→ R, R (5)

R = Me, 75%
R = CO_2Me, 90%

$$(TMS)_2CCl_2 \xrightarrow[-90\,°C,\ THF]{LDBB} (TMS)_2CLi_2 \xrightarrow{electrophile} (TMS)_2CR_2\ (88\%)$$
$$(TMS)_2CLi_2 \xrightarrow{THF} (TMS)_2CHLi \xrightarrow{electrophile} (TMS)_2CHR\ (2\%) \quad (6)$$

The generation of alkyllithium (**1**) proved pivotal in the enantioselective synthesis of spiculisporic acid.[8] Derived from the corresponding alkyl bromide, this homoenolate equivalent could not be made from lithium metal and the corresponding Grignard reagent led to the wrong isomer.

(**1**)

Reductive Lithiation of Thiophenyl Ethers. A variety of thiophenyl ethers are reduced with LDBB and can be trapped with electrophiles, leading to 1-substituted bicyclo[1.1.1]pentanes[9] and vinyl substituted compounds (eqs 7 and 8).[10] The latter procedure to generate the alkenyllithium is a better and more cost efficient route than the Bond modification of the Shapiro reaction. The cuprate reagent from both the bicyclo[1.1.1]pentane and alkenyl compounds can be generated and reacted with electrophiles.

SPh — LDBB → SLi + Li (1:2) — electrophile → R (7)

SPh — 1. LDBB; 2. RCHO (68–92%) → CH(OH)R (8)

Homoenolate (**2**) is produced from the reductive lithiation of the corresponding phenylthio derivative and can be reacted with many electrophiles in high yields (72–83%).[11]

(**2**) (**3**)

The use of LDBB was advantageous in the total synthesis of (+)-(9*S*)-dihydroerythronolide A, converting the thiophenyl group in (**3**) to an alkyllithium and then to a Grignard reagent.[12]

An unusual 1,4 O → C silicon shift was observed when a tris(trimethylsilyl) ether was treated with LDBB to yield the *scyllo*-tris(trimethylsilylmethyl)cyclohexanetriol.[13]

α-Lithio Ethers. The generation of α-lithio ethers can be performed using LDBB followed by trapping with electrophiles. The α-lithio ethers of 1-methoxy-1-phenylthiocyclopropanes are reacted with conjugated aldehydes or ketones to yield 1-cyclopropylallyl alcohols. The addition of ***Trifluoromethanesulfonic Anhydride*** led, after rearrangement, to 2-vinylcyclobutanones (eq 9).[3a] Previous investigations of these cyclopropanes used LDMAN for reductive lithiation, which led to lower yields of the alcohol and, for acid sensitive compounds, significant drops in yield.

OMe, SPh — 1. LDBB; 2. CHO → OMe, OH — Tf_2O → O (70–90%) (9)

The reductive lithiation of 2-(phenylthio)tetrahydropyrans led selectively to the axial 2-lithio species which, upon equilibration, could be converted to the more thermodynamic equatorial α-lithio ether. Treatment with electrophiles such as acetone led to good yields of products (eq 10).[3b] The use of LDBB led to higher yields and cleaner reductions compared with LDMAN or LN.

MeO, MeO, OMe, SPh — 1. LDBB; 2. acetone → OH + OH (10)

	ratio	yield
kinetic	98:2	81%
thermodynamic	1:99	59%

The placement of a vinyl group at the 6-position on the tetrahydropyran moiety and addition of LDBB leads to competing [1,2] and [2,3] Wittig rearrangements, with inversion of configuration at the lithium bearing carbon (eq 11).[14] A *t*-butyl group at position 4 significantly changes the reaction course, leading to a small amount of [2,3] Wittig rearranged product. The major product is derived from a 1,4-transannular H-transfer to the lithium-bearing carbon, with inversion of configuration.[15]

R, 4, 6, O, SPh — LDBB → (11)

	[1,2]	[2,3]		
R = H	21%	45%	0%	0%
R = *t*-Bu	15%	4%	2%	66%

(Dialkoxymethyl)lithium compounds can be generated from the corresponding phenylthio derivative and reacted with aldehydes and ketones (eq 12).[16] The cyclic lithium reagent could be generated from LN, while the acyclic required LDBB.

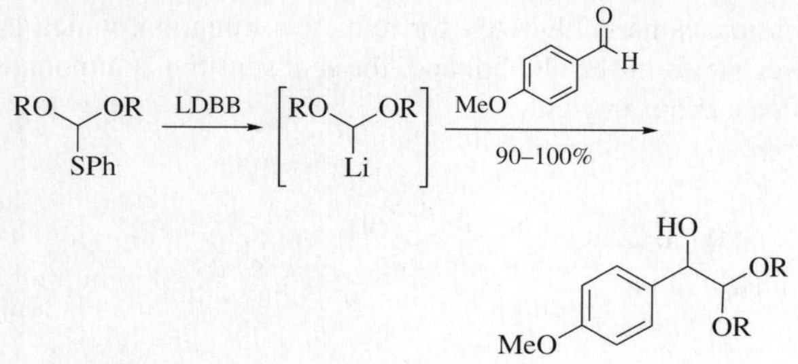

(12)

Allyllithiums. The treatment of allyl phenylthio ethers with ***Lithium 1-(Dimethylamino)naphthalenide*** leading to allyllithiums has been performed. Allyllithiums derived from treatment with LDBB can be converted to the allylcerium reagent and trapped with unsaturated aldehydes. Homoallylic alcohols were obtained by 1,2-addition with attack by the least substituted terminus of the allyl anion (eq 13).[17] Regiocontrol of the double bond is attained by a slight temperature modification.

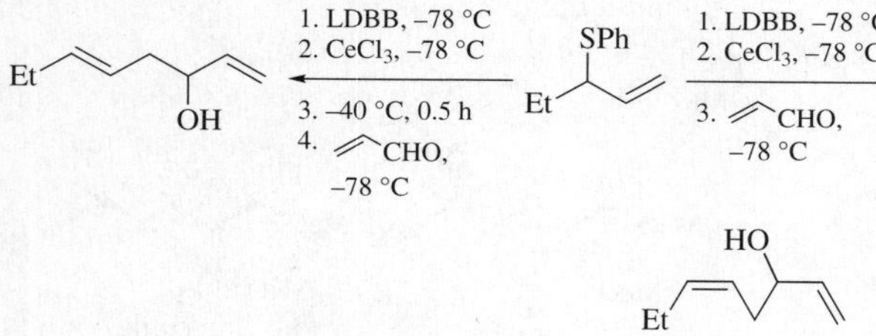

(13)

The ability to control the regiochemistry of allyllithium terminus and double bond geometry led to a one-pot synthesis of the Comstock mealy bug sex pheromone in a 45% yield, and a four-step synthesis of the California red scale pheromone in 23% yield.[18]

β-,γ-,δ-Lithio Alkoxides. The treatment of epoxides with LDBB leads to β-lithio alkoxides which can be reacted with an aldehyde or ketone to yield varying amounts of a diol and an alcohol.[19] The diol is obtained from cleavage of the least substituted carbon oxygen bond, while the other alcohol arrives via a hydride transfer (eq 14). Reductive lithiation of vinyloxiranes led to ring opening in the opposite direction. The allylic anions could be treated with Ti^{IV} or Ce^{III} and added to aldehydes at the most or least substituted terminus, respectively.

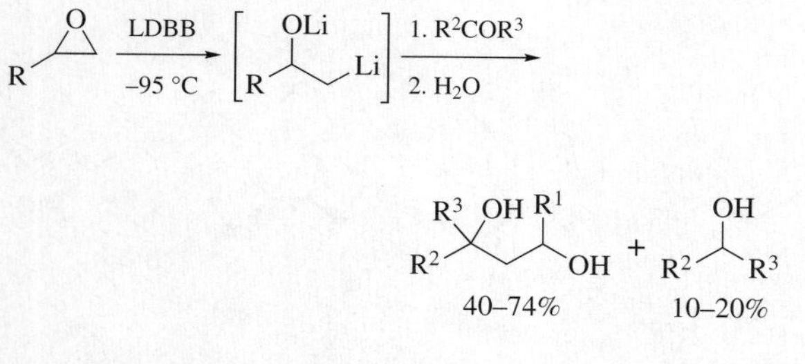

(14)

The generation of γ-lithio alkoxides from the corresponding oxetanes require higher temperatures (0 °C) than epoxide ring opening with LDBB.[20] γ-Lithio alkoxides can be trapped by electrophiles in modest yields. The addition of trialkylaluminums yields the lithium trialkylaluminates, which react with electrophiles in modest yields (eq 15).

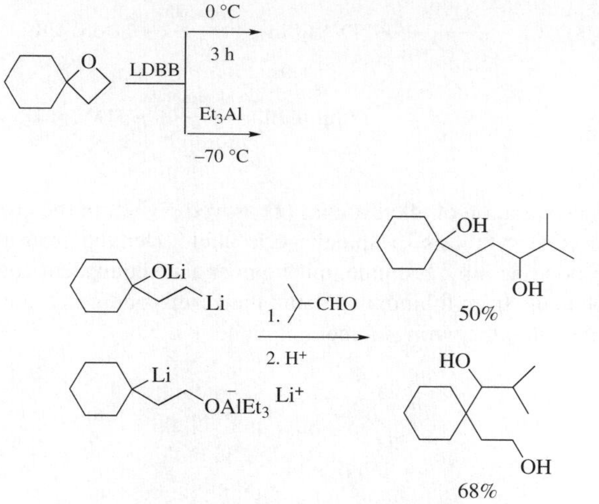

(15)

The next higher analog, δ-lithio alkoxides, can be obtained by LDBB reductive lithiation in the presence of ***Boron Trifluoride Etherate***.[21] The Lewis acid helps stabilize the resulting open-chain oxyanion. Treatment with various electrophiles led to products in high yields and is a good protocol for the preparation of the synthetically useful [5.*n*] spiroacetal units (eq 16). The most branched alcohol is obtained upon cleavage of substituted epoxides and oxetanes, while the opposite regioselectivity is observed for THF.

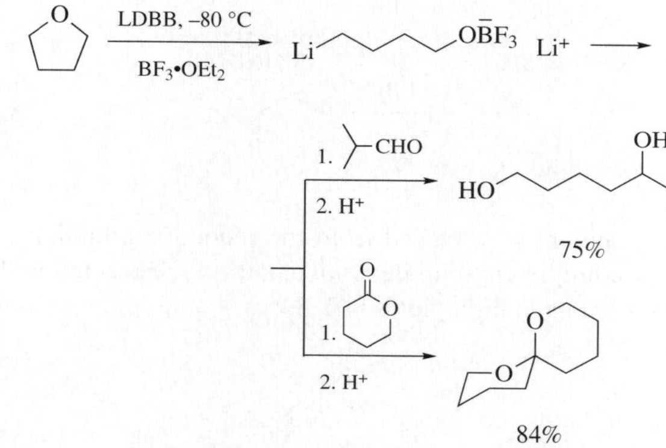

(16)

Propellanes. The central bond in [1.1.1]propellanes can be reductively cleaved to the corresponding dilithio species, which can be trapped by electrophiles to make bicyclo[1.1.1]pentane derivatives (eq 17).[22]

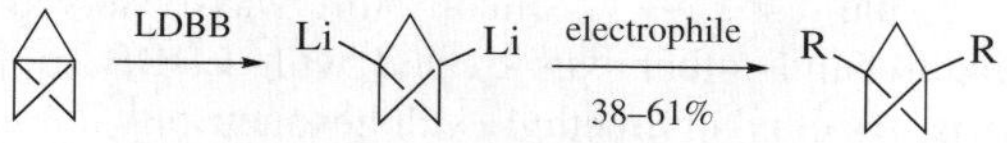

(17)

Carbonyl groups. Aromatic ketones, benzylic alcohols, and ethers can be converted to the dilithio species with lithium and a catalytic amount of DBB followed by trapping with electrophiles (eq 18).[23] The generation of aliphatic ketones in good yield from the corresponding esters proceeds with LDBB (eq 19), whereas sodium leads to high yields of the acyloin product.[24]

$$Ar_2CO \xrightarrow[\text{cat. DBB}]{\text{Li}} Li(Ar)_2C{-}OLi \xrightarrow{\text{electrophile}} R(Ar)_2C{-}OH \quad (18)$$

$$RCH_2C(O)OR^1 \xrightarrow[\text{cat. DBB, 48–82\%}]{\text{Li}} RCH_2C(O)CH_2R \quad (19)$$

Phosphinates. Reduction of diastereomerically pure menthyl phosphinate with LDBB followed by treatment with alkyl halides yields the phosphine oxides in good yield with high optical purity (eq 20).[25]

$$MenO{-}P(O)(Ph)Me \xrightarrow[\text{BnBr, 67\%}]{\text{LDBB, }-78\,^\circ\text{C}} Bn{-}P(O)(Ph)Me \quad (20)$$

95% optical yield

Related Reagents. See Classes R-2, R-27, R-30, R-31, R-32, and R-33, pages 1–10.

1. Cohen, T.; Bhupathy, M. *ACR* **1989**, *22*, 152.
2. (a) Freeman, P. K.; Hutchinson, L. L. *TL* **1976**, *22*, 1849. (b) Freeman, P. K.; Hutchinson, L. L. *JOC* **1980**, *45*, 1924.
3. (a) Cohen, T.; Bruckunier, L. *T* **1989**, *45*, 2917. (b) Rychnovsky, S. D.; Mickus, D. E. *TL* **1989**, *30*, 3011.
4. Bloch, R.; Chaptal-Gradoz, N. *TL* **1992**, *33*, 6147.
5. Stapersma, J.; Klumpp, G. W. *T* **1981**, *37*, 187.
6. Rawson, D. J.; Meyers, A. I. *TL* **1991**, *32*, 2095.
7. (a) Vlaar, C. P.; Klumpp, G. W. *TL* **1991**, *32*, 2951. (b) van Eikema Hommes, N. J. R.; Bickelhaupt, F.; Klumpp, G. W. *TL* **1988**, *29*, 5237.
8. Brandänge, S.; Dahlman, O.; Lindqvist, B.; Måhlén, A.; Mörch, L. *ACS* **1984**, *B38*, 837.
9. (a) Wiberg, K. B.; Waddell, S. T. *TL* **1988**, *29*, 289. (b) Wiberg, K. B.; Waddell, S. T. *JACS* **1990**, *112*, 2194.
10. Cohen, T.; Doubleday, M. D. *JOC* **1990**, *55*, 4784.
11. Cherkauskas, J. P.; Cohen, T. *JOC* **1992**, *57*, 6.
12. Stork, G.; Rychnovsky, S. D. *JACS* **1987**, *109*, 1565.
13. Rücker, C.; Prinzbach, H. *TL* **1983**, *24*, 4099.
14. Verner, E. J.; Cohen, T. *JACS* **1992**, *114*, 375.
15. Verner, E. J.; Cohen, T. *JOC* **1992**, *57*, 1072.
16. Shiner, C. S.; Tsunoda, T.; Goodman, B. A.; Ingham, S.; Lee, S.-H.; Vorndam, P. E. *JACS* **1989**, *111*, 1381.
17. Guo, B.-S.; Doubleday, W.; Cohen, T. *JACS* **1987**, *109*, 4710.
18. McCullough, D. W.; Bhupathy, M.; Piccolino, E.; Cohen, T. *T* **1991**, *47*, 9727.
19. (a) Cohen, T.; Jeong, I.-H.; Mudryk, B.; Bhupathy, M.; Awad, M. M. A. *JOC* **1990**, *55*, 1528. (b) Bartmann, E. *AG(E)* **1986**, *25*, 653.
20. (a) Mudryk, B.; Cohen, T. *JOC* **1989**, *54*, 5657. (b) Mudryk, B.; Cohen, T. *JOC* **1991**, *56*, 5760.
21. Mudryk, B.; Cohen, T. *JACS* **1991**, *113*, 1866.
22. Bunz, U.; Szeimies, G. *TL* **1990**, *31*, 651.
23. Karaman, R.; Kohlman, D. T.; Fry, J. L. *TL* **1990**, *31*, 6155.
24. Karaman, R.; Fry, J. L. *TL* **1989**, *30*, 4935.
25. Koide, Y.; Sakamoto, A.; Imamoto, T. *TL* **1991**, *32*, 3375.

Mark D. Ferguson
Wayne State University, Detroit, MI, USA

Lithium–Ethylamine[1]

Li–$EtNH_2$

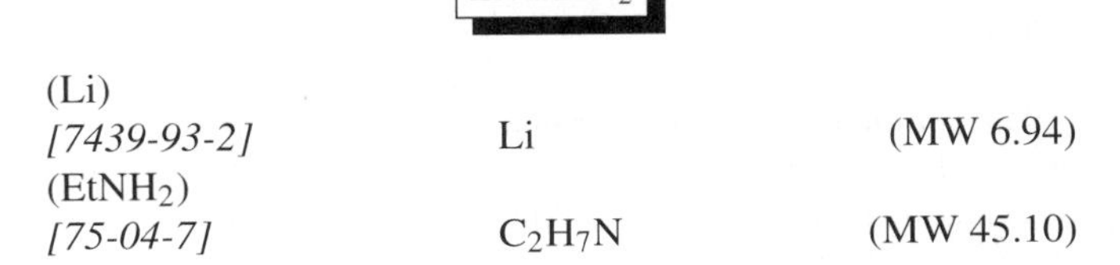

(Li)		
[7439-93-2]	Li	(MW 6.94)
($EtNH_2$)		
[75-04-7]	C_2H_7N	(MW 45.10)

(powerful reducing agent;[1] used for (partial) reduction of aromatics;[1a–d,f] conversion of alkynes to *trans*-alkenes and alkanes;[1a,f] stereoselective reduction of hindered ketones[1a,e,13b] and oximes;[16] reductive cleavage of polar single bonds[1a,b,d,f,19,27a])

Physical Data: Li: mp 180.5 °C; bp 1327 ± 10 °C; *d* 0.534 g cm^{-3}. $EtNH_2$: mp −81 °C; bp 16.6 °C; *d* 0.689 g cm^{-3}.

Form Supplied in: see ***Lithium***; ethylamine, 99% anhydrous, packaged in steel cylinders fitted with stainless steel valves.

Purification: see ***Lithium***; ethylamine can be purified by distillation from sodium under exclusion of H_2O and CO_2.

Handling, Storage, and Precautions: see ***Lithium***. Ethylamine is a toxic, colorless, and highly flammable gas with a low flash point (−16 °C). It should be handled in a well-ventilated hood, only after having eliminated all potential sources of ignition. It is corrosive to Cu, Al, Sn, Zn, and their alloys, but is stored satisfactorily in stainless steel.

Lithium–Ethylamine Solutions. Lithium dissolves in $EtNH_2$ to give deep-blue solutions due to solvated electrons.[2a–d] Similar solutions are obtained with Li in other primary amines.[2a,b] $EtNH_2$ and ethylenediamine (EDA) are the most common solvents, but methylamine, propylamine, isopropylamine, and butylamine have also been used for reductions. The other alkali metals are less soluble in these amines and rarely used for reductions in amines. For instance, the solubilities in EDA at room temperature are 2.9×10^{-1} mol L^{-1} (Li), 2.4×10^{-3} (Na), 1.0×10^{-2} (K), 1.3×10^{-2} (Rb), and 5.4×10^{-2} (Cs).[2e] However, the solubilities can dramatically be increased by using crown ether and cryptand complexants.[2a,b] In contrast to the Li–amine solutions, the amine solutions of K, Rb, and Cs contain, besides solvated electrons, the metal anion as major species, while Na–amine solutions contain only the metal anion and are in general not blue at all.[2a,b] More recently, methylamine-assisted solubilization of Li and Li–Na mixtures (up to 1:1) in several amine and ether solvents, in which the metals alone are not or only slightly soluble, has been reported.[3]

The reducing agents formed by Li in amines (Benkeser reduction) are more powerful but less selective than Li in NH_3.[1a,4] The higher reactivity is caused by higher reaction temperatures and possibly also by a smaller degree of electron solvation; the better solubility of many organic compounds in amines is also advantageous. The reactivity can be modified by using lower reaction temperatures and by addition of alcohols and secondary amines. Li–amine solutions are more sensitive to catalytic decomposition than Li in NH_3,[5] and clean solvents are mandatory.

Experimental. Detailed procedures are available,[1b,c] as are details for reaction assemblies.[6] Reduction occurs by a sequence of single electron and proton transfers to the substrate, resulting in saturation of multiple bonds or fission of single bonds (see also ***Lithium***).

Reduction of Aromatic Compounds. Li-$EtNH_2$ reduces benzene and its derivatives to cyclohexenes.[1a–d] Monoalkylbenzenes give mixtures of regioisomers. More selective formation of the Δ^1-ene is achieved when the reaction is performed in a mixture of primary and secondary amines.[1a] With excess Li and at higher temperatures, further reduction to cyclohexanes can occur (eq 1).[4]

8 equiv Li, $EtNH_2$ (1)

$T = 17\ °C$ 45 : 55 (68% total)

$T = -78\ °C$ 75% traces

Li–$EtNH_2$ affords, in the presence of alcohols (alcohol:Li > 1:1), Birch reduction products, i.e. 1,4-cyclohexadienes.[1a,d] This procedure is more powerful than the Birch method (eq 2) and useful for the reduction of highly substituted benzene rings.[7] Li–EDA reductions of benzene derivatives have been performed (eq 2),[7b,8a,b] but Li–$EtNH_2$ appears to be superior for cyclohexene formation, probably due to the presence of *N*-lithioethylenediamine in refluxing EDA.[7b] Cyclohexenes and cyclohexa-1,4-dienes are formed in high yields by electrolysis of benzene derivatives with LiCl in $MeNH_2$ in a divided or undivided cell, respectively.[1a]

Li–NH_3, EtOH: no reaction

Li–$EtNH_2$, EtOH, 0 °C: 56%

Li–$EtNH_2$, 0 °C: 74%

Li–EDA, reflux: + starting material 88:12, 44% crystallized (2)

Reduction of polycyclic aromatic compounds is complex, and the products delicately depend on the exact reaction conditions. Most extensive reductions have been achieved with Li–EDA.[1f] Ferrocenes have been reductively cleaved with Li–propylamine in 35–77% yield.[9]

Alkenes. Terminal[10a–c] and internal double bonds may be reduced by Li–amine; however, the latter react more slowly.[1a,b,d] Strained cyclic alkenes are more easily reduced, and ring size effects are observed. Na–*t*-BuOH–HMPA is more powerful and reduces even nonterminal alkenes in nearly quantitative yields.[10d]

Alkynes. Internal alkynes are rapidly reduced to *trans*-alkenes by stoichiometric amounts of Li in amines.[1a] Excess Li leads to alkanes.[11] Ca in $MeNH_2$/EDA (1:1) has been used for similar reductions.[12] Li–$EtNH_2$–*t*-BuOH efficiently reduced an alkyne precursor of sphingosin, while triple bond reduction was incomplete with Na– and Li–NH_3.[6b] For the conversion of 1-alkynes to 1-alkenes, Na–NH_3–$(NH_4)_2SO_4$ is the system of choice.[1a] Li–EDA reduction of 1-alkynes usually gives alkanes.

Ketones. Steroidal and triterpenoid ketones have been selectively reduced to the equatorial alcohols by Li–EDA in high yields.[10b,13] Stereoselective reduction of an α,β-unsaturated ketone to the (20*R*)(β)-alcohol was achieved with Li–EDA, while ***Sodium Borohydride*** led to a mixture of the (20*R*)(β)-alcohol and the Δ^{16}-(20*S*)-alcohol (eq 3).[14] Li–$EtND_2$–*t*-BuOD reduces α,β-unsaturated ketones to the saturated ketones in high yields and deuteriates the β-C-atom stereoselectively.[15] Li–$EtNH_2$ reduction of 3-oximino steroids is apparently the best procedure to prepare 3β-amino steroids.[16]

(3)

Li–EDA, reflux, 2 h	95%	–
$NaBH_4$, MeOH, py, rt, 100 h	45%	47%

Carboxylic Acids and Derivatives. Li–$MeNH_2$ or –$EtNH_2$ reduce aliphatic saturated acids to aldehydes or imines in 53–84% yield, depending on the workup procedure.[17] Tertiary carboxamides are cleaved to aldehydes.[18] A major limitation is the substantial formation of transamidation byproducts. *N*-Proline peptide bonds have been cleaved to an extent of 50–70%.[18a]

Reductive Cleavage of C–S and C–Se Bonds.[19] Mainly sulfones,[20] but also sulfoxides,[21] sulfides,[22] and dithioacetals[23] have been desulfurized by Li in amines, most often in $EtNH_2$, but also in EDA[20g,h] and $MeNH_2$.[20f] As an excess of reducing agent is normally used, functional groups sensitive to SET may be reduced. Li–$EtNH_2$ cleaves allylic C–S bonds generally without side-reactions and has thus proven to be especially useful in terpene synthesis for the reductive removal of sulfur-containing functional groups.[20a–c,22a,b]

Allylic aryl sulfones were converted to the alkenes by Li–$EtNH_2$ in 77–98% yield,[20a–c] while ***Sodium Amalgam*** caused double bond migration (eq 4)[20a] or (*E*)/(*Z*) isomerization.[20b] However, in some cases, extensive double bond migration using Li–$EtNH_2$ or Li–EDA has been observed.[20d]

Vinyl sulfones have been hydrogenolyzed to 2-alkenes by Li–$EtNH_2$ or C_8K in Et_2O with extensive isomerization of the *cis* double bond, forming *cis*/*trans* mixtures, while ***Aluminum Amalgam*** gave no reduction at all.[20e] Besides arylsulfonyl groups, some alkylsulfonyl groups have also been reductively removed by Li–$EtNH_2$ or –$MeNH_2$.[20c,f] Allylic alcohols survived the reductive cleavage of a sulfoxyl group by Li–$EtNH_2$ at −78 °C, whereas the double bond was simultaneously reduced when ***Raney Nickel*** was used.[21] While unsatisfactory results were obtained from attempts to desulfurize a 4-thiacyclohexene with Raney Ni, stepwise reduction using Li–$EtNH_2$, followed by Raney Ni to reduce the intermediate Li thiolate, yielded the alkene in 55–70%.[22c] Li–$EtNH_2$ reductively cleaves allylic *N,N*-dimethyldithiocarbamates that have been employed in the synthesis of disubstituted alkenes, in 90–95% yield.[24] Selenides and diselenoacetals are reduced to the corresponding alkanes by Li–$EtNH_2$ or Raney Ni in good yields.[25]

(4)

Li–$EtNH_2$, 0 °C, 30 min 77% –
$Na(Hg)_x$, EtOH, 1.5 h 9 : 4 (90%)

Epoxides. Sterically hindered steroid and triterpenoid epoxides are reductively opened to the axial alcohols by Li–$EtNH_2$[26a] and Li–EDA.[26b] Thus Li–EDA reduced 3α,4α-epoxyfriedelan to the 4α-alcohol (eq 5), while ***Lithium Aluminum Hydride*** did not give any reduction at all, and the 3β,4β-isomer to the 3β-alcohol (eq 6).[26b] Reductive opening of other tri- and tetrasubstituted epoxides with Li–$EtNH_2$[26c–e] and Li–EDA[26f] are high yielding, but depending on the substrate, mixtures of alcohols are encountered.[26d–g] Reductions with Li in propylamine and butylamine seem to give more side-products.[26e]

30 equiv Li–EDA, reflux, 4 h, 90% (5)

30 equiv Li–EDA, reflux, 4 h, 90% (6)

Li–EDA[27a,b] and Ca–EDA[27a] are excellent reductants for epoxides of bicyclic compounds which react sluggishly and often undergo rearrangements with $LiAlH_4$ (eq 7)[27b] ***Lithium Triethylborohydride*** has proven to be a powerful reductant for bicyclic and tri- and tetrasubstituted oxiranes and is in some instances superior to Li–EDA.[27c]

(7) + starting material + rearranged alcohols

Li–EDA, 50 °C, 1 h 87%
$LiAlH_4$, diglyme 100 °C, 24 h 47% 19% (44%)

Reductive Cleavage of Miscellaneous Single Bonds. Allyl and benzyl ethers are cleaved by Li–$MeNH_2$ or –$EtNH_2$ in fair to good yields[28] (eqs 8[28d] and 9[28e]).

Li–$MeNH_2$, 72% (8)

Similarly, the reductive cleavage of the *O*-alkyl bonds of esters of allylic alcohols is effected by Li–$EtNH_2$;[28e] allylic alcohols are normally not efficiently cleaved unless they are sterically hindered (eq 9).[28e]

Li–$EtNH_2$, reflux (9)

R = Ac, 15 equiv Li 94%
R = Me, 30 equiv Li 84%
R = H, 22 equiv Li 15% + 61% S.M.

Esters of tertiary and sterically hindered secondary alcohols are cleaved to the corresponding alkanes and carboxylates by Li–$EtNH_2$[29a] and Li–EDA,[29b] whereas the parent alcohol is regenerated from nonhindered esters. This allows the selective deoxygenation of suitable diesters (eq 10).[29a]

100 equiv Li, $EtNH_2$, reflux, 85% (10)

Esters of moderately hindered secondary alcohols often give mixtures of alkane and alcohol. Using ***Potassium 18-Crown-6*** in ***t-Butylamine***, the alkane:alcohol ratio seems to be somewhat higher.[29a] ***Sodium-Hexamethylphosphoric Triamide***–*t*-butanol is another efficient system for the transformation of hindered esters to alkanes.[29c] K–Na eutectic solubilized with 18-crown-6 in *t*-butylamine and THF deoxygenates hindered and nonhindered esters.[29a] Alcohol regeneration by Li in refluxing $EtNH_2$ seems to be essentially due to transacylation, leading to *N*-ethylamides, while at low temperature the Bouveault–Blanc reaction predominates.[29a] Methyl esters of bulky carboxylic acids have been converted to the acids by Li in refluxing EDA.[29d]

Phosphorodiamidates are cleaved to alkanes by Li–$EtNH_2$, providing another indirect method for the deoxygenation of alcohols (eq 11).[30]

1. $ClPO(NMe_2)_2$, Me_2NH; 2. Li–$EtNH_2$, 71% (11)

Li–amine reagents have been used for the cleavage of vinyl phosphates to alkenes,[31] of oxetanes,[32] of bicyclobutanes,[33] and of allylic *N,N'*-diacylhydrazines.[34] Li–EDA cleaves glycosyluronic acid-containing oligosaccharides specifically at the site of the acid residue.[35]

Related Reagents. See Classes R-2, R-3, R-4, R-7, R-12, R-14, R-23, R-28, R-30, R-31, and R-32, pages 1–10. Lithium; Potassium; Sodium; Sodium–Ammonia.

1. (a) Kaiser, E. M. *S* **1972**, 391. (b) Smith, M. In *Reduction: Techniques and Applications in Organic Synthesis*; Augustine, R. L.; Ed.; Dekker: New York, 1968; Chapter 2. (c) Birch, A. J.; Subba Rao, G. *Adv. Org. Chem.* **1972**, *8*, 1. (d) Brendel, G. *Lect.-Hydride Symp., 3rd: Lithium Metal in Organic Synthesis*; Metallges. AG: Frankfurt/Main, 1979; pp 135–55. (e) *Steroid Reactions*; Djerassi, C., Ed.; Holden–Day: San Francisco, 1963. (f) Hudlický, M. *Reductions in Organic Chemistry*; Horwood: Chichester, 1984.
2. (a) Dye, J. L. *Prog. Inorg. Chem.* **1984**, *32*, 327. (b) Dye, J. L. *AG(E)* **1979**, 587. (c) Catterall, R.; Hurley, I.; Symons, M. C. R. *JCS(D)* **1972**, 139. (d) Bar–Eli, K.; Tuttle, T. R., Jr. *JCP* **1964**, *40*, 2508. (e) Dewald, R. R.; Dye, J. L. *JPC* **1964**, *68*, 121 & 128.
3. Faber, M. K.; Fussá-Rydel, O.; Skowyra, J. B.; McMills, L. E. H.; Dye, J. L. *JACS* **1989**, *111*, 5957.
4. Benkeser, R. A.; Robinson, R. E.; Sauve, D. M.; Thomas, O. H. *JACS* **1955**, *77*, 3230.
5. Evers, E. C.; Young, A. E., II; Panson, A. J. *JACS* **1957**, *79*, 5118.
6. (a) *FF* **1967**, *1*, 580. (b) Julina, R.; Herzig, T.; Bernet, B.; Vasella, A. *HCA* **1986**, *69*, 368.
7. (a) Burgstahler, A. W.; Worden, L. R. *JACS* **1964**, *86*, 96. (b) Kwart, H.; Conley, R. A. *JOC* **1973**, *38*, 2011. (c) Harvey, R. G.; Urberg, K. *JOC* **1968**, *33*, 2206.
8. (a) Krapcho, A. P.; Bothner-By, A. A. *JACS* **1959**, *81*, 3658. (b) Stolow, R. D.; Ward, R. A. *JOC* **1966**, *31*, 965.
9. Brown, A. D., Jr.; Reich, H. *JOC* **1970**, *35*, 1191.
10. (a) Ficini, J.; Francillette, J.; Touzin, A. M. *JCR(M)* **1979**, *35*, 1820 (*CA* **1980**, *92*, 93 777w). (b) Pradhan, B. P.; Chakrabarti, D. K.; Chakraborty, S. *IJC(B)* **1984**, *23B*, 1115. (c) Corey, E. J.; Cantrall, E. W. *JACS* **1959**, *81*, 1745. (d) Whitesides, G. M.; Ehmann, W. J. *JOC* **1970**, *35*, 3565.
11. Benkeser, R. A.; Schroll, G.; Sauve, D. M. *JACS* **1955**, *77*, 3378.
12. Benkeser, R. A.; Belmonte, F. G. *JOC* **1984**, *49*, 1662.
13. (a) Sengupta, P.; Das, S.; Das, K. *IJC(B)* **1984**, *23B*, 1113. (b) Huffman, J. W. *COS* **1991**, *8*, 107.
14. Sengupta, P.; Sen, M.; Sarkar, A.; Das, S. *IJC(B)* **1986**, *25B*, 975. see also Markgraf, J. H.; Davis, H. A.; Mahon, B. R. *J. Chem. Educ.* **1988**, *65*, 635.
15. Burgstahler, A. W.; Sanders, M. E. *S* **1980**, 400.
16. Khuong-Huu, F.; Tassel, M. *BSF* **1971**, 4072 (*CA* **1972**, *76*, 86 004h).
17. Bedenbaugh, A. O.; Bedenbaugh, J. H.; Bergin, W. A.; Adkins, J. D. *JACS* **1970**, *92*, 5774.
18. (a) Patchornik, A.; Wilchek, M.; Sarid, S. *JACS* **1964**, *86*, 1457. (b) Bedenbaugh, A. O.; Payton, A. L.; Bedenbaugh, J. H. *JOC* **1979**, *44*, 4703.
19. Caubère, P.; Coutrot, P. *COS* **1991**, *8*, 835.
20. (a) Grieco, P. A.; Masaki, Y. *JOC* **1974**, *39*, 2135. (b) Ohmori, M.; Yamada, S.; Takayama, H. *TL* **1982**, *23*, 4709. (c) Trost, B. M.; Weber, L.; Strege, P.; Fullerton, T. J.; Dietsche, T. J. *JACS* **1978**, *100*, 3426. (d) Julia, M.; Uguen, D. *BSF* **1976**, 513 (*CA* **1976**, *85*, 63 175m). (e) Savoia, D; Trombini, C.; Umani-Ronchi, A. *JCS(P1)* **1977**, 123. (f) Truce, W. E.; Tate, D. P.; Burgde, D. N. *JACS* **1960**, *82*, 2872. (g) Bödeker, C.; de Waard, E. R.; Huisman, H. O. *T* **1981**, *37*, 1233. (h) Fehr, C. *HCA* **1983**, *66*, 2512.
21. Solladie, G.; Demailly, G; Greck, C. *JOC* **1985**, *50*, 1552.
22. (a) Biellmann, J. F.; Ducep, J. B. *T* **1971**, *27*, 5861 (*CA* **1972**, *76*, 34 427d). (b) van Tamelen, E. E.; McCurry, P.; Huber, U. *PNA* **1971**, *68*, 1294. (c) Stotter, P. L.; Hornish, R. E. *JACS* **1973**, *95*, 4444.
23. Crossley, N. S.; Henbest, H. B. *JCS* **1960**, 4413.
24. Hayashi, T.; Midorikaw, H. *S* **1975**, 100.
25. Sevrin, M.; Van Ende, D.; Krief, A. *TL* **1976**, 2643.
26. (a) Hallsworth, A. S.; Henbest, H. B. *JCS* **1957**, 4604 & **1960**, 3571. (b) Sengupta, P.; Das, K.; Das, S. *IJC(B)* **1985**, *24B*, 1175. (c) Sedzik-Hibner, D.; Chabudzinski, Z. *Rocz. Chem.* **1970**, *44*, 2387 (*CA* **1971**, *75*, 20 641m). (d) Chabudzinski, Z.; Sedzik, D.; Szykula, J. *Rocz. Chem.* **1967**, *41*, 1923 (*CA* **1968**, *68*, 87 403j). (e) Chabudzinski, Z.; Sedzik, D.; Rykowski, Z. *Rocz. Chem.* **1967**, *41*, 1751 (*CA* **1968**, *68*, 78 435u). (f) Gurudutt, K. N.; Rao, S.; Shaw, A. K. *IJC(B)* **1991**, *30B*, 345.
27. (a) Murai, S. *COS* **1991**, *8*, 871. (b) Brown, H. C.; Ikegami, S.; Kawakami, J. H. *JOC* **1970**, *35*, 3243. (c) Krishnamurthy, S; Schubert, R. M.; Brown, H. C. *JACS* **1973**, *95*, 8486.
28. (a) Kobayashi, T.; Tsuruta, H. *S* **1980**, 492. (b) Dasgupta, S. K.; Crump, D. R.; Gut, M. *JOC* **1974**, *39*, 1658. (c) Masamune, T.; Matsue, H.; Fujii, M. *BCJ* **1972**, *45*, 1812. (d) Rigby, J. H. *TL* **1982**, *23*, 1863. (e) Hallsworth, A. S.; Henbest, H. B.; Wrigley, T. I. *JCS* **1957**, 1969.
29. (a) Barrett, A. G. M.; Godfrey, C. R. A.; Hollinshead, D. M.; Prokopiou, P. A.; Barton, D. H. R.; Boar, R. B.; Joukhadar, L.; McGhie, J. F.; Misra, S. C. *JCS(P1)* **1981**, 1501. (b) Sengupta, P.; Sen, M.; Das, S. *IJC(B)* **1979**, *18B*, 179. See also: Pradhan, B. P.; Hassan, A.; Shoolery, J. N. *TL* **1984**, *25*, 865. (c) Deshayes, H.; Pete, J.-P. *CJC* **1984**, *62*, 2063 (*CA* **1985**, *102*, 24 920a). (d) Sengupta, P.; Sen, M.; Das, S. *IJC(B)* **1980**, *19B*, 721.
30. (a) Ireland, R. E.; Muchmore, D. C.; Hengartner, U. *JACS* **1972**, *94*, 5098. (b) Liu, H. J.; Lee, S. P. *TL* **1977**, 3699.
31. Ireland, R. E.; Pfister, G. *TL* **1969**, 2145.
32. Sauers, R. R.; Schinski, W.; Mason, M. M.; O'Hara, E; Byrne, B. *JOC* **1973**, *38*, 642.
33. Moore, W. R.; Hall, S. S.; Largman, C. *TL* **1969**, 4353.
34. (a) Anastasia, M.; Fiecchi, A.; Galli, G. *JOC* **1981**, *46*, 3421. (b) Anastasia, M.; Ciuffreda, P.; Fiecchi, A. *CC* **1982**, 1169.
35. Lau, J. M.; McNeil, M.; Darvill, A. G.; Albersheim, P. *Carbohydr. Res.* **1987**, *168*, 219 & 245.

Karin Briner
Indiana University, Bloomington, IN, USA

Lithium Naphthalenide[1]

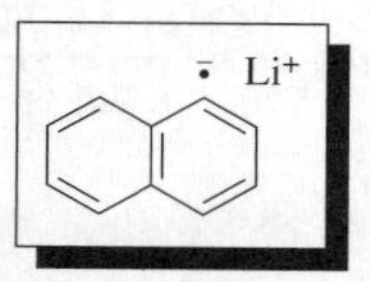

[7308-67-0] $C_{10}H_8Li$ (MW 134.12)

(reductive metalation reactions;[1c] reduction of metal salts;[27] initiation of polymerization reactions[28])

Alternate Name: LN.

Physical Data: no data on the isolated material; only available in solution.

Solubility: sol ether, benzene, THF; reacts with protic solvents and THF at elevated temperatures.[2]

Preparative Methods: made by addition of freshly cut ***Lithium*** metal to a solution of naphthalene in THF. Preparation can be accelerated by ultrasonication.[3]

Analysis of Reagent Purity: two titration methods have been described;[4] the more convenient[4b] involves conversion of 1,1-diphenylethylene by lithium naphthalenide to an intensely red-colored dianion, which is then titrated against *s*-butanol.

Handling, Storage, and Precautions: can be stored in solution up to several days; must be protected from air and moisture; can be used to ambient temperature.

Reductive Metalation. The powerful reductive nature of this reagent makes it an important tool for lithium–heteroatom exchange reactions. Thus, it was established early on that (phenylthio)alkanes can be converted into their requisite alkyllithium species.[5] This has become the method of choice over generation by lithium metal alone. The resultant alkyllithium species can either be quenched with a proton source (eq 1),[6] or intercepted with an electrophile. This has subsequently evolved into a powerful technique, since the reaction is general for all chalcogens (eq 2)[7] and halides (eq 3).[8]

(LN) (1)

1. LN, THF, –50 °C; 2. MeI, rt, 12 h; 86% (2)

1. LN, THF, –78 °C; 2. Bu_3SnCl, –78 °C; 70% (3)

Metal–heteroatom exchange can also be persuaded to occur with a variety of other systems, resulting in allyl-[9] and vinyllithium[10] species, as well as α-lithio ethers (eq 3),[1c,8,11] α-lithio thioethers,[1c,12] α-lithio amines,[11b] and α-lithio silanes.[1c,12b,13] The latter class provides useful intermediates for the Peterson alkenation reaction (eq 4).[13a]

1. LN, THF, –78 °C; 2. cyclohexanone; 3. H_3O^+; 52% (4)

In some cases, however, it may be advantageous to proceed via either ***Lithium 1-(Dimethylamino)naphthalenide*** (LDMAN), or ***Lithium 4,4′-Di-t-butylbiphenylide*** (LDBB).[1c] With the former reagent, the byproduct formed, (dimethylamino)naphthalene, is more easily removed than is naphthalene from product mixtures. In the latter case, the greater reduction potential of di-*t*-butylbiphenyl appears to lead to more efficient halogen–lithium exchange.[8b]

Dianion Generation. Lithium naphthalenide efficiently deprotonates β-alkynyloxy[14] and carboxylate anions (eq 5).[15] In addition, the previously mentioned phenomenon of reductive metalation has been exploited to access dianions from halohydrins,[16] β-halo carboxylic acids,[17] and β-halo carboxamides,[18] and even trianions from β,ω′-dihalo alcohols.[19] A major pathway for the polyanionic species is β-elimination (eq 6);[16a,19] when such processes can be avoided, the polyanions react according to Hauser's rule (eq 7).[16a,20]

1. LN, THF, Et_2NH; 2. cyclohexene oxide; 55% (5)

1. BuLi, –78 °C; 2. LN, –100 °C; 47% (6)

1. BuLi, –78 °C; 2. LN, –78 °C; 3. CO_2; 72% (7)

Dehalogenation Reactions. Since lithium naphthalenide is a particularly effective initiator for halogen–metal exchange, it has found widespread use for the conversion of dihalides to unsaturated species. Thus, 1,2-dichlorodisilanes have been converted to silenes (eq 8)[21] and diphosphiranes to phosphacumulenes.[22] In a related field, silicon cages[23] have been constructed from trichlorodisilanes.

$$Ar_2Si(Cl){-}Si(Cl)Ar_2 \xrightarrow[50\%]{LN,\ DME} Ar_2Si{=}SiAr_2 \quad (8)$$

Ar = 2,4,6-triisopropylphenyl (*i*-Pr, *i*-Pr, *i*-Pr)

Metal Redox Reactions. Lithium naphthalenide is a convenient reducing agent for a variety of metals, and shows great promise in the synthetic area. Thus, Cu^I complexes have been reduced to Cu^0; the resultant highly reactive species adds in an oxidative fashion across the carbon–halogen bond.[24] As a consequence, the well known organocuprate addition chemistry can be carried out in one step from halocarbons, without having to initially prepare the organolithium species (eq 9). In addition to the reduction of Cu^I, lithium naphthalenide reduces Si^{IV} to Si^{II},[25] Sn^{IV} to Sn^{II} (eq 10),[26] and various lanthanide compounds.[27]

$CuI{\cdot}PEt_3$ — 1. LN, THF; 2. Br–C_6H_4–CN; 3. PhCOCl; 71% (9)

Br Sn Br LN THF → Sn (10)

Oligomerization Reactions. Lithium naphthalenide has long been a convenient initiator for anionic 'living' polymerization reactions.[28] Thus, styrenes, acrylates, dienes, and other monomers have been polymerized using lithium naphthalenide as an anionic initiator. In some circumstances, however, oligomerization can be controlled to furnish only dimers (eq 11).[29] Also, in the presence of a secondary amine, 1,3-dienes can be persuaded to react in a formal 1,4-fashion to produce allyl amines (eq 12).[30]

CN → LN, $HNEt_2$, THF, 85% → NH_2, CN (11)

LN, Et_2NH, PhH, $Me_2N(CH_2)_3NMe_2$, 72% → NEt_2 (12)

Interesting approaches toward functional polymers have recently been detailed,[31] wherein previously described chemistry involving lithium naphthalenide is conducted on suitably substituted polystyrene derivatives (eq 13).

SPh → 1. LN, THF; 2. $ClSiMe_2H$ → $SiMe_2H$ → (4-chlorostyrene) Cl, H_2PtCl_6, PhMe → $SiMe_2$, Cl (13)

Related Reagents. See Classes R-3, R-27, R-30, and R-33, pages 1–10. Copper(I) Iodide–Triethylphosphine–Lithium Naphthalenide; Lithium 4,4′-Di-*t*-butylbiphenylide; Lithium 1-(Dimethylamino)naphthalenide; Potassium Naphthalenide; Sodium Anthracenide; Sodium Naphthalenide; Sodium Phenanthrenide.

---

1. (a) Wakefield, B. J. *The Chemistry of Organolithium Compounds*; Pergamon: Oxford, 1974. (b) March, J. *Advanced Organic Chemistry*, 4th ed.; Wiley: New York, 1992; p 729. (c) Cohen, T.; Bhupathy, M. *ACR* **1989**, *22*, 152.
2. Fujita, T.; Suga, K.; Watanabe, S. *S* **1972**, 630.
3. Azuma, T.; Yanagida, S.; Sakurai, H.; Sasa, S.; Yoshino, K. *SC* **1982**, *12*, 137.
4. (a) Ager, D. J. *JOM* **1983**, *241*, 139. (b) Screttas, C. G.; Micha-Screttas, M. *JOM* **1983**, *252*, 263.
5. (a) Screttas, C. G.; Micha-Screttas, M. *JOC* **1978**, *43*, 1064. (b) Cohen, T.; Weisenfeld, R. B. *JOC* **1979**, *44*, 3601.
6. Harring, S. R.; Livinghouse, T. *TL* **1989**, *30*, 1499.
7. Agawa, T.; Ishida, M.; Ohshiro, Y. *S* **1980**, 933.
8. (a) Lesimple, P.; Beau, J.-M.; Sinaÿ, P. *Carbohydr. Res.* **1987**, *171*, 289. (b) Freeman, P. K.; Hutchinson, L. L. *JOC* **1980**, *45*, 1924.
9. Cohen, T.; Guo, B.-S. *T* **1986**, *42*, 2803.
10. Duhamel, L.; Chauvin, J.; Messier, A. *JCR(S)* **1982**, 48.
11. (a) Shiner, C. S.; Tsunoda, T.; Goodman, B. A.; Ingham, S.; Lee, S.; Vorndam, P. E. *JACS* **1989**, *111*, 1381. (b) Broka, C. A.; Shen, T. *JACS* **1989**, *111*, 2981. (c) Hoffmann, R.; Brückner, R. *CB* **1992**, *125*, 1957.
12. (a) McDougal, P. G.; Condon, B. D.; Laffosse, M. D., Jr.; Lauro, A. M.; Van Derveer, D. *TL* **1988**, *29*, 2547. (b) Ager, D. J. *JCS(P1)* **1986**, 195.
13. (a) Ager, D. J. *JCS(P1)* **1986**, 183. (b) Mandai, T.; Kohama, M.; Sato, H.; Kawada, M.; Tsuji, J. *T* **1990**, *46*, 4553.
14. Watanabe, S.; Suga, K.; Suzuki, T. *CJC* **1969**, *47*, 2343.
15. Fujita, T.; Watanabe, S.; Suga, K. *AJC* **1974**, *27*, 2205.
16. (a) Barluenga, J.; Flórez, J.; Yus, M. *JCS(P1)* **1983**, 3019. (b) Barluenga, J.; Fernández-Simón, J. L.; Concellón, J. M.; Yus, M. *JCS(P1)* **1988**, 3339.
17. Caine, D.; Frobese, A. S. *TL* **1978**, 883.
18. Barluenga, J.; Foubelo, F.; Fañanás, F. J.; Yus, M. *T* **1989**, *45*, 2183.
19. Barluenga, J.; Fernandez, J. R.; Yus, M. *S* **1985**, 977.
20. Hauser, C. R.; Harris, T. M. *JACS* **1958**, *80*, 6360.
21. Watanabe, H.; Takeuchi, K.; Nakajima, K.; Nagai, Y.; Goto, M. *CL* **1988**, 1343.
22. Yoshifuji, M.; Toyota, K.; Yoshimura, H. *CL* **1991**, 491.
23. Kabe, Y.; Kawase, T.; Okada, J.; Yamashita, O.; Goto, M.; Masamune, S. *AG(E)* **1990**, *29*, 794.
24. Rieke, R. D.; Dawson, B. T.; Stack, D. E.; Stinn, D. E. *SC* **1990**, *20*, 2711.
25. Jutzi, P.; Holtmann, U.; Kanne, D.; Krüger, C.; Blom, R.; Gleiter, R.; Hyla-Kryspin, I. *CB* **1989**, *122*, 1629.
26. Jutzi, P.; Hielscher, B. *OM* **1986**, *5*, 1201.
27. (a) Arnaudet, L.; Ban, B. *NJC* **1988**, *12*, 201. (b) Bochkarev, M. N.; Trifonov, A. A.; Fedorova, E. A.; Emelyanova, N. S.; Basalgina, T. A.; Kalinina, G. S.; Razuvaev, G. A. *JOM* **1989**, *372*, 217.
28. Ishizone, T.; Wakabayashi, S.; Hirao, A.; Nakahama, S. *Macromolecules* **1991**, *24*, 5015.
29. (a) Takabe, K.; Ohkawa, S.; Katagiri, T. *S* **1981**, 358. (b) Fujita, T.; Watanabe, S.; Suga, K.; Sugahara, K.; Tsuchimoto, K. *CI(L)* **1983**, 167.
30. Sugahara, K.; Fujita, T.; Watanabe, S.; Hashimoto, H. *J. Chem. Technol. Biotechnol.* **1987**, *37*, 95.
31. (a) O'Brien, R. A.; Rieke, R. D. *JOC* **1990**, *55*, 788. (b) Itsuno, S.; Shimizu, K.; Kamahori, K.; Ito, K. *TL* **1992**, *33*, 6339.

Kevin M. Short
Wayne State University, Detroit, MI, USA

Lithium Tri-*t*-butoxyaluminum Hydride[1]

$LiAlH(OR)_3$

(R = *t*-Bu)		
[17476-04-9]	$C_{12}H_{28}AlLiO_3$	(MW 254.32)
(R = Et_3C)		
[79172-99-9]	$C_{21}H_{46}AlLiO_3$	(MW 380.59)
(R = Et)		
[17250-30-5]	$C_6H_{16}AlLiO_3$	(MW 170.14)
(R = Me)		
[12076-93-6]	$C_3H_{10}AlLiO_3$	(MW 128.05)

(reducing agent for many functional groups; cleaves cyclic ethers)

Physical Data: R = *t*-Bu: mp 300–319 °C (dec); sublimes at 280 °C/2 mmHg. All the other reagents are unstable and are prepared in situ.

Solubility: R = *t*-Bu: (at 25 °C) diglyme (41 g/100 mL), THF (36 g/100 mL), DME (4 g/100 mL), ether (2 g/100 mL).

Form Supplied in: R = *t*-Bu: solid; 0.5 M or 1.0 M solutions in diglyme or THF. The other reagents are prepared in situ just prior to use.

Preparative Methods: for R = Me,[2] Et,[3] or 3-ethyl-3-pentyl,[4] in situ preparation can be carried out as follows: addition of 3 mol of the desired alcohol to 1 mol of standardized ***Lithium Aluminum Hydride*** solution in ether, THF, or diglyme results in the formation of the corresponding trialkoxyaluminohydride reagent. The *t*-Bu[5] derivative can also be prepared in situ by this method.

Analysis of Reagent Purity: standardization of $LiAlH(O\text{-}t\text{-}Bu)_3$ solutions is best carried out by reduction of cyclohexanone and with GC and/or spectroscopic analysis to measure the extent of reaction. Iodometric titrations are not suitable in this case.[6]

Handling, Storage, and Precautions: the dry solid and solutions are corrosive and/or highly flammable and must be stored in the absence of moisture. While the *t*-BuO compound appears to have long term stability in terms of storage, the MeO compound degrades in its reactivity within one week. Therefore it is advised to use freshly prepared reagent solutions.[2a]

Functional Group Reductions. The high reactivity for reduction exhibited by $LiAlH_4$ can be attenuated by modifying this reagent. Altered reagents can be prepared by the addition of an alcohol such that a 3:1 complex of alcohol:$LiAlH_4$ is formed. The reactivity of these new trialkoxyaluminohydride reagents can be modulated by the selection of the appropriate alcohol. In terms of relative reactivity one can rank[7] hydride reagents as follows: $LiAlH_4 > LiAlH(OMe)_3 > LiAlH(O\text{-}t\text{-}Bu)_3 > NaBH_4$. Table 1 provides a comparison of the aluminum based reagents with various functional groups. For a detailed discussion of the reduction properties of $LiAlH_4$, see ***Lithium Aluminum Hydride***.

This observation of diminished reactivity of the trialkoxyaluminohydrides is opposite to that observed with $NaBH_4$ and its alkoxy analogs (see ***Sodium Borohydride*** and ***Sodium Trimethoxyborohydride***). The stoichiometry for the reaction can be determined

Table 1 Comparison of the Reactivities of Aluminohydride Reducing Agents

Reaction	$LiAlH_4$	$LiAlH(OMe)_3$	$LiAlH(O\text{-}t\text{-}Bu)_3$
Aldehyde to alcohol	+	+	+
Ketone to alcohol	+	+	+
Acid chloride to alcohol	+	+	+
Lactone to lactol	+	+	slow
Epoxide to alcohol	+	slow	slow
Ester to alcohol	+	+	slow
Carboxylic acid to alcohol	+	+	–
Carboxylic acid salt to alcohol	+	+	–
t-Amide to amine	+	+	–
Nitrile to amine	+	+	–
Nitro to amine	+	–	–
Alkene	–	–	–

from the theoretical number of hydrides required for the reduction based on the observation that the trialkoxyaluminohydrides can be considered to be the source of one hydride. For example, the reduction of an aldehyde or ketone theoretically requires one hydride and so one equivalent of the reagent is needed for this reduction.

The differential reducing abilities of these reagents allows selective reduction of one functional group in the presence of another. While simple saturated aldehydes can be reduced by both $LiAlH(OMe)_3$ and $LiAlH(O\text{-}t\text{-}Bu)_3$ as rapidly as by $LiAlH_4$ at 0 °C, cyano aldehydes can be reduced at low temperature to the corresponding alcohol derivatives (eq 1).[8] Aldehyde lactones can also be reduced chemoselectively (eq 2).[9]

NC–…–CHO → ($LiAlH(O\text{-}t\text{-}Bu)_3$, THF, 0 °C, 1 h, 78%) → NC–…–OH (1)

OH, CHO, MeO, O, OR, OAc → ($LiAlH(O\text{-}t\text{-}Bu)_3$, THF, 0 °C, 92%) → OH, OH, MeO, O, OR, OAc (2)

The reduction of aldehyde esters can lead to the formation of lactones with or without subsequent acidic workup (eqs 3 and 4).[10,11] Aldehydes can also be reduced preferentially in the presence of ketones (eq 5).[4]

OHC, MeO_2C, O, H → (1. $LiAlH(O\text{-}t\text{-}Bu)_3$; 2. Amberlite 105; 95%) → O, O, H (3)

OHC, MeO_2C, O, H → ($LiAlH(O\text{-}t\text{-}Bu)_3$, 61%) → O, O, H (4)

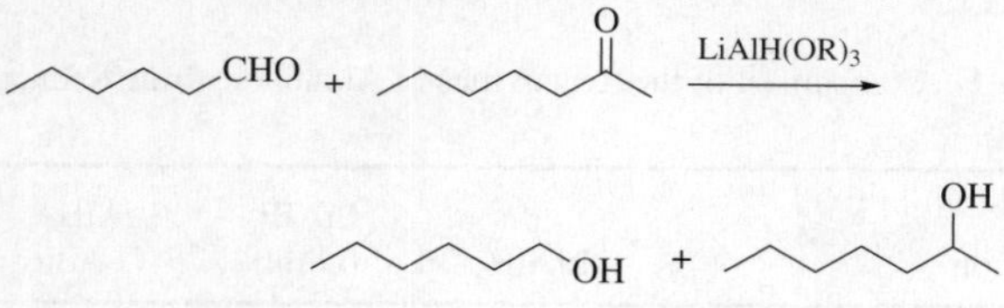

CHO + (ketone) $\xrightarrow{LiAlH(OR)_3}$ OH + OH (5)

99.5:0.5

With unsaturated aldehydes such as cinnamaldehyde, LiAlH(O-*t*-Bu)$_3$ gives the corresponding allylic alcohol, whereas LiAlH(OMe)$_3$ affords the saturated alcohol. This difference in reactivity can be exploited (eq 6).[12] The course of the reduction can be changed if a copper species is also used (eq 7).[13]

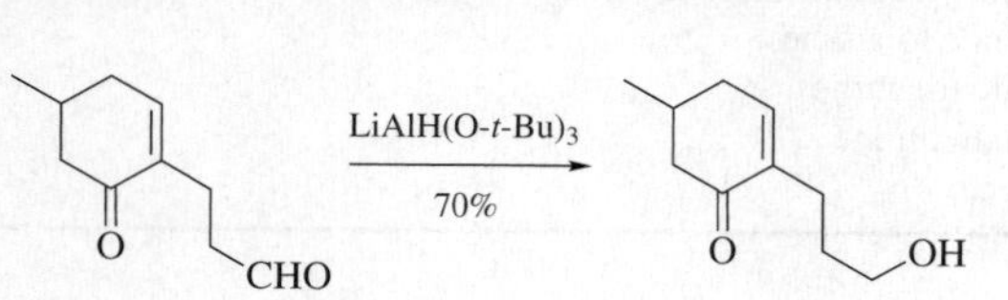

LiAlH(O-*t*-Bu)$_3$, 70% (6)

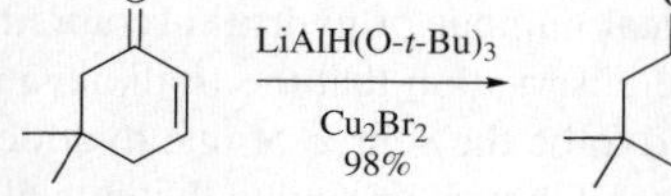

LiAlH(O-*t*-Bu)$_3$, Cu_2Br_2, 98% (7)

Similarly, ketones can be reduced with LiAlH(O-*t*-Bu)$_3$ in the presence of esters,[14a] lactones,[14b] amides,[5] halogens,[15] epoxides,[5] *O*-alkyl oximes,[5] azides,[16] and cyano groups.[5] A ketone can be reduced in the presence of an aldehyde, provided the aldehyde is first masked as an aldimine, for example. The reduction of an acid halide to the corresponding alcohol can be carried out with LiAlH(O-*t*-Bu)$_3$ at 0 °C, while the corresponding aldehyde can be obtained if the reduction is performed at −80 °C in diglyme.[17] Aldehydes can also be obtained by the reduction of nitriles[3a] or tertiary amides[3b] if LiAlH(OEt)$_3$ is used as the reductant.

In addition to the differences in relative functional group reactivities that LiAlH(OR)$_3$ has over $LiAlH_4$, these reagents also exhibit improved stereoselectivities in these reductions.[18] For example, the reduction of norbornanone to give the *endo*-alcohol proceeds with greater control (eq 8).[2b] Further examples of this stereoselectivity are illustrated in eqs 9–11.[19–21] The use of $LiAlH_4$–***Lithium Iodide*** in the reaction depicted in eq 11 results in complete inversion of stereochemistry (*syn*:*anti* = 5:95). Therefore the choice of reagents allows preparation of either diastereomer.

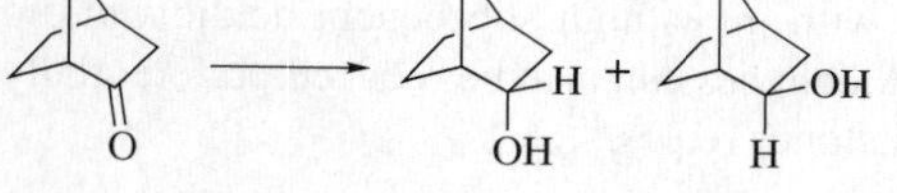

(8)

$LiAlH_4$	89:11
LiAlH(OMe)$_3$	98:2

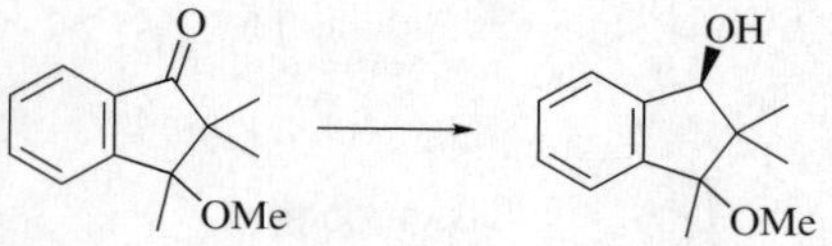

(9)

$LiAlH_4$	*trans*:*cis* = 32:68
LiAlH(O-*t*-Bu)$_3$	*trans*:*cis* = 90:10

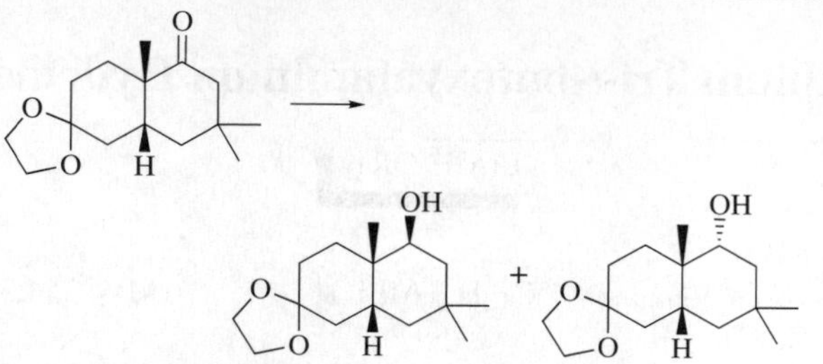

(10)

$LiAlH_4$	60:40
LiAlH(OCMe$_2$Et)$_3$	100:0

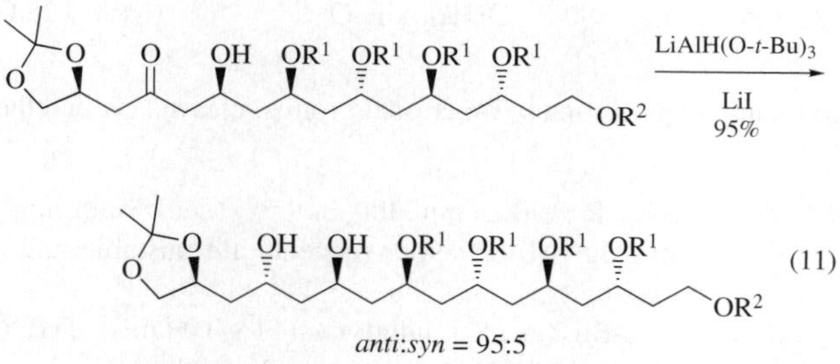

LiAlH(O-*t*-Bu)$_3$, LiI, 95% (11)

anti:*syn* = 95:5

Enantioselective reductions are also possible if the alcohol used to form the trialkoxyaluminohydride is optically active. For example, treating $LiAlH_4$ with MeOH and BINAL-H[22] in a mole ratio of 1:1:1 in dry THF generates an optically active hydride[23] reagent which can reduce ynones with good enantioselectivity, as shown in eq 12. The (*R*)-antipode of BINAL-H results in the isolation of the (*R*)-propargylic alcohol. See ***Lithium Aluminum Hydride–2,2′-Dihydroxy-1,1′-binaphthyl*** for additional examples of this type.

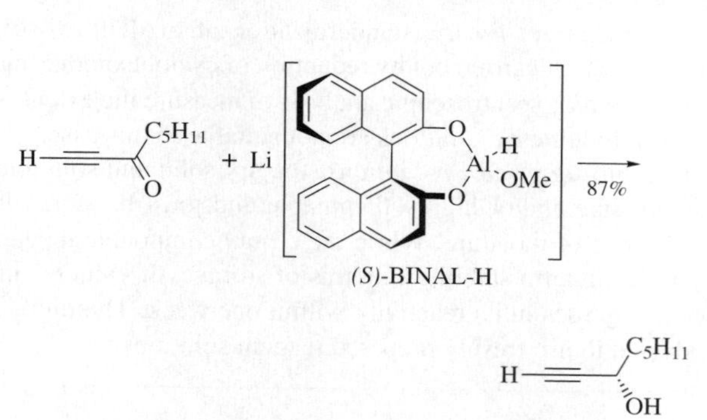

H—≡—C(=O)C_5H_{11} + Li[(*S*)-BINAL-H, Al(H)OMe] → 87% → H—≡—CH(OH)C_5H_{11} (12)

(*S*)-BINAL-H

84% ee

While unsubstituted lactams are reduced to the corresponding amines, the intermediates in these reductions can be diverted into other reactions when the substrate is suitably functionalized. One example is shown in eq 13.[24] The trialkoxyaluminohydrides are capable of reducing most nitrogen based functional groups; however, imides do not react. This attribute has been used in a diastereoselective synthesis of D-*threo*-sphingamine (eq 14).[25] Oximes react with these reagents only to generate hydrogen. No reduction is observed. *N*-Methylamines are formed by the reduction of isocyanates with LiAlH(OMe)$_3$,[2a,26] while LiAlH(O-*t*-Bu)$_3$ gives formamides (eq 15).[5b,27] Unsaturated isocyanates are also reduced to the corresponding vinylic formamide (eq 16).[28] Thiocyanates are inert to LiAlH(O-*t*-Bu)$_3$.[29]

$LiAlH(OCHMe\text{-}t\text{-}Bu)_3$, 40% (13)

$LiAlH(O\text{-}t\text{-}Bu)_3$, 95%

(14) D-*threo*-Sphingamine

$LiAlH(O\text{-}t\text{-}Bu)_3$, –15 °C, 80% (15)

$LiAlH(O\text{-}t\text{-}Bu)_3$, 80% (16)

The trialkoxyaluminohydrides generate 1 equiv of hydrogen upon exposure to thiols and are inert to dialkyl and/or aryl sulfides, while disulfides are reduced to thiols (eq 17).[30] Alkyl 1-alkynyl sulfides are reduced to the corresponding alkyl *cis*-1-alkenyl sulfides when the reaction is carried out in the presence of ***Copper(I) Bromide*** (eq 18).[31] The *trans*-isomer is obtained with $LiAlH_4$. Two equivalents of hydride are required to reduce sulfoxides by $LiAlH(OMe)_3$,[2a,26] while $LiAlH(O\text{-}t\text{-}Bu)_3$[32] is unreactive to sulfoxides. Sulfones do not react with these reagents.[2a,5b,26] While alkyl mesylates and tosylates are inert to the trialkoxyaluminohydrides, $LiAlH(OMe)_3$ can reduce acetylenic mesylates to the corresponding allene in an *anti* fashion to maintain chirality (eq 19).[33]

$RSSR \xrightarrow{LiAlH(O\text{-}t\text{-}Bu)_3,\ 25\text{–}100\%} RSH$ (17)

$LiAlH(O\text{-}t\text{-}Bu)_3$, CuBr, 94% (18)

$LiAlH(OMe)_3$, 79% (19)

(*S*)-(–) → (*S*)-(+), 73% ee

A related reagent is $NaAlH_2(OCH_2CH_2OMe)_2$, which is abbreviated to SMEAH or Red-Al, and has in some cases comparable reactivity and chemistry and in other cases complementary activity (see ***Sodium Bis(2-methoxyethoxy)aluminum Hydride***).

Reductive Cleavage of Ethers. Dialkyl ethers, alkyl aryl ethers, diaryl ethers, and cyclic ethers react very slowly with these reductants, if at all. However, the addition of ***Triethylborane***, in varying amounts, produces a complex which is capable of ring opening some cyclic ethers (eqs 20 and 21).[34] Epoxides also react very slowly with $LiAlH(O\text{-}t\text{-}Bu)_3$,[26] which allows for differentiating functional groups during their reduction. Formation of a complex with BEt_3 now can be used to open the epoxide (eq 22).[29,35] A mixture of $LiAlH(OMe)_3$ and ***Copper(I) Iodide*** can also be used to open epoxides (eq 23).[36]

$LiAlH(O\text{-}t\text{-}Bu)_3$, BEt_3, 98% (20)

$LiAlH(O\text{-}t\text{-}Bu)_3$, BEt_3, 97% (21)

$LiAlH(O\text{-}t\text{-}Bu)_3$, BEt_3; 90% + 10% (22)

$LiAlH(OMe)_3$, CuI, 100% (23)

Reduction of Halogenated Compounds. The reduction of 1-iodoalkanes with $LiAlH(OMe)_3$ is comparable to reduction by $LiAlH_4$, while the reactivity with bromo and chloro compounds is very low. There is no reaction of $LiAlH(O\text{-}t\text{-}Bu)_3$ with chloro compounds. As a result, one can reduce aldehydes in the presence of chloro groups. The mixture of CuI and $LiAlH(OMe)_3$ provides a species which is capable of reducing primary chlorides, and primary and secondary bromides.[5b,37]

Related Reagents. See Classes R-2, R-4, R-5, R-10, R-11, R-12, R-13, R-15, R-24, R-27, and R-32, pages 1–10. Copper(I) Bromide–Lithium Trimethoxyaluminum Hydride; Copper(I) Iodide–Lithium Trimethoxyaluminum Hydride.

---

1. (a) Malek, J.; Cerny, M. S. *S* **1972**, 217. (b) Brown, H. C.; Krishnamurthy, S. *T* **1979**, *35*, 567. (c) Malek, J. *OR* **1985**, *34*, 1. (d) Malek, J. *OR* **1988**, *36*, 249. (d) Rerick, M. N. In *Reduction Techniques and Applications in Organic Synthesis*; Augustine R. L.; Ed.; Dekker: New York, 1968. (e) Hajos, A. *Complex Hydrides and Related Reducing Agents in Organic Synthesis*; Elsevier: Amsterdam, 1979.
2. (a) Brown, H. C.; Weissman, P. M. *JACS* **1965**, *87*, 5614. (b) Brown, H. C.; Deck, H. R. *JACS* **1965**, *87*, 5620.
3. (a) Brown, H. C.; Garg, C. P. *JACS* **1964**, *86*, 1085. (b) Brown, H. C.; Tsukamoto, A. *JACS* **1964**, *86*, 1089.
4. Krishnamurthy, S. *JOC* **1981**, *46*, 4628.
5. (a) Brown, H. C.; McFarlin, R. F. *JACS* **1958**, *80*, 5372. (b) Brown, H. C.; Weissman, P. M. *Isr. J. Chem.* **1963**, *1*, 430.
6. (a) Wigfield, D. C.; Gowland, F. W. *CJC* **1977**, *55*, 3616. (b) Wigfield, D. C.; Gowland, F. W. *JOC* **1980**, *45*, 653.
7. Brown, H. C.; Shoaf, C. J. *JACS* **1964**, *86*, 1079.
8. Julia, M.; Roualt, A. *BSF* **1959**, 1833.

9. Nicolaou, K. C.; Seitz, S. P.; Pavia, M. R. *JACS* **1981**, *103*, 1222.
10. Danishefsky, S.; Kitahara, T.; Schuda, P. F.; Etheredge, S. J. *JACS* **1976**, *98*, 3028.
11. Danishefsky, S.; Schuda, P. F.; Kitahara, T.; Etheredge, S. J. *JACS* **1977**, *99*, 6066.
12. Harayama, T.; Takatane, M.; Inubushi, Y. *CPB* **1979**, *27*, 726.
13. (a) Semmelhack, M. F.; Stauffer, R. D. *JOC* **1975**, *40*, 3619. (b) Semmelhack, M. F.; Stauffer, R. D.; Yamashita, A. *JOC* **1977**, *42*, 3180.
14. (a) Bergel'son, L. D.; Batrakov, S. G. *IZV* **1963**, 1259 (*CA* **1963**, *59*, 12 637g). (b) McQuillin, F. J.; Yeats, R. B. *JCS* **1965**, 4273.
15. (a) Zurfluh, R.; Tamm, C. *HCA* **1972**, *55*, 2495. (b) Huston, R.; Rey, M.; Dreiding, A. S. *BSB* **1979**, *88*, 911.
16. Rorig, K. J.; Wagner, H. A. U.S. Patent 3 412 094, 1968 (*CA* **1969**, *70*, 68 410j).
17. Brown, H. C.; Subba Rao, B. C. *JACS* **1958**, *80*, 5377.
18. Boone, J. R.; Ashby, E. C. *Top. Stereochem.* **1979**, *11*, 53.
19. Cawley, J. J.; Petrocine, D. V. *JOC* **1976**, *41*, 2608.
20. Heathcock, C. H.; Gray, D. T. *T* **1971**, *27*, 1239.
21. Mori, Y.; Suzuki, M. *TL* **1989**, *30*, 4387.
22. (a) Nishizawa, M.; Noyori, R. *COS* **1991**, *8*, 159. (b) Haubenstock, H. *Top. Stereochem.* **1983**, *14*, 231.
23. (a) Nishizawa, M.; Yamada, M.; Noyori, R. *TL* **1981**, *22*, 247. (b) Suzuki, M.; Kawagishi, T.; Suzuki, T.; Noyori, R. *TL* **1982**, *23*, 4057. (c) Noyori, R.; Tomino, Y.; Tamimoto, Y.; Nishizawa, M. *JACS* **1984**, *106*, 6709.
24. Takahata, H.; Okajima, H.; Yamazaki, T. *CPB* **1980**, *28*, 3632.
25. Newman, H. *JOC* **1974**, *39*, 100.
26. Brown, H. C.; Yoon, N. M. *JACS* **1966**, *88*, 1464.
27. Walborsky, H. M.; Niznik, G. E. *JOC* **1972**, *37*, 187.
28. Harrison, I. T.; Kurz, W.; Massey, I. J.; Unger, S. H. *JMC* **1978**, *21*, 588.
29. Klimstra, P. D.; Nutting, E. F.; Counsell, R. E. *JMC* **1966**, *9*, 693.
30. Krishnamurthy, S.; Aimino, D. *JOC* **1989**, *54*, 4458.
31. Vermeer, P.; Meijer, J.; Eylander, C.; Brandsma, L. *RTC* **1976**, *95*, 25.
32. Brown, H. C.; Kim, S. C.; Krishnamurthy, S. *JOC* **1980**, *45*, 1.
33. Claesson, A.; Olsson, L. I. *JACS* **1979**, *101*, 7302.
34. (a) Brown, H. C.; Krishnamurthy, S. *CC* **1972**, 868. (b) Brown, H. C.; Krishnamurthy, S.; Coleman, R. A. *JACS* **1972**, *94*, 1750. (c) Brown, H. C.; Krishnamurthy, S.; Hubbard, J. C.; Coleman, R. A. *JOM* **1979**, *166*, 281. (d) Krishnamurthy, S.; Brown, H. C. *JOC* **1979**, *44*, 3678.
35. Krishnamurthy, S.; Schubert, R. M.; Brown, H. C. *JACS* **1973**, *95*, 8486.
36. Masamune, S.; Rossy, P. A.; Bates, G. S. *JACS* **1973**, *95*, 6452.
37. Krishnamurthy, S.; Brown, H. C. *JOC* **1980**, *45*, 849.

Paul Galatsis
University of Guelph, Ontario, Canada

Lithium Tri-*s*-butylborohydride[1]

$LiBH(s\text{-}Bu)_3$

[38721-52-7] $C_{12}H_{28}BLi$ (MW 190.15)

(reducing agent for various functional groups;[2] selective reducing agent;[3–6] stereoselective reducing agent for ketones[7,8] and other functional groups;[5,9,10] regioselective reducing agent for cyclic anhydrides;[11] used for conjugate addition and alkylation of α,β-unsaturated esters or ketones[12] and for Michael-initiated ring closure reactions;[13] reduces (2-arylprop-1-en-3-yl)trimethylammonium iodides to 2-arylpropenes;[14] hydroborates substituted styrenes;[15] reacts with carbon monoxide in the presence of free trialkylborane[16])

Alternate Name: L-Selectride.
Physical Data: not isolated; prepared and used in solution.
Solubility: solubility limits have not been established. The reagent is normally used as a 1.0 M solution in THF or Et_2O. Use of a solution in toluene is reported and one method of preparation results in a THF–pentane solvent mixture.
Form Supplied in: 1.0 M solution in THF (L-Selectride®).
Analysis of Reagent Purity: solutions of pure reagent exhibit doublets ($J \approx 70$ Hz) centered in the range δ −6.3 to δ −6.7 in the ^{11}B NMR spectrum.[17–20] Concentration of hydride is determined by hydrolysis of aliquots and measurement of the hydrogen evolved or by quenching aliquots in excess 1-iodooctane and analysis of the octane formed by GLC.[17] Concentration of boron is verified by oxidizing aliquots using alkaline hydrogen peroxide and analyzing the 2-butanol formed by GLC.[21]
Preparative Methods: direct reaction of tri-*s*-butylborane with LiH is not suitable, proceeding only to about 10% completion after 24 h in refluxing THF.[17] The preferred methods are reaction of the organoborane with ***Lithium Aluminum Hydride*** in the presence of ***1,4-Diazabicyclo[2.2.2]octane***[18] or with ***t-Butyllithium***.[19] Reaction with lithium trimethoxyaluminum hydride is effective but leads to byproducts not easily separated from the reagent.[20]
Handling, Storage, and Precautions: solutions are air- and water-sensitive and should be handled in a well-ventilated hood under an inert atmosphere using appropriate techniques.[22] Potentially pyrophoric $s\text{-}Bu_3B$ is a byproduct of reductions; thus standard organoborane oxidation[21] is recommended as part of the workup.

Functional Group Reductions. Alkanes are formed from primary and some secondary alkyl halides via S_N2 processes with potential chemoselectivity based on relative ease of displacement (I > Br > Cl).[2a] The corresponding reaction for tosylates has not been investigated thoroughly.[2b] Dealkylation of quaternary ammonium halides yields the corresponding amines, demethylation being strongly favored.[2c] Rates and regioselectivities for reduction of various epoxides have been determined.[2d] Reduction of conjugated nitroalkenes gives nitroalkanes[2e] or ketones,[2f] depending on the workup used (eq 1).

-NO$_2$
LiBH(*s*-Bu)$_3$
30 min, 25 °C
silica gel (H$^+$) 69% | H_2SO_4 (aq) 81%
-NO$_2$ | =O
(1)

Certain γ-keto esters undergo reduction–lactonization[2g] which, in some cases, is followed by reduction of the γ-lactones to the γ-lactols (eq 2).[2h] Some esters of Δ^2-isoxazolinylcarboxylates are reduced to the corresponding primary alcohols (no yields given).[2i]

(2)

The reagent is ineffective compared to *t*-butylmagnesium chloride as a hydride source for a hydrozirconation procedure.[2j]

Selective Reductions. Ketones are reduced selectively in the presence of other functional groups: carboxylic acids;[3a] esters,[3b–n] including γ-lactones[3j,k] and 1,3-dioxolan-4-ones;[3e] amides,[4] including benzenesulfonamides[4a,b] and 2-acetyl β-lactams;[4f,g] and internal epoxides.[31] An enamine is reduced selectively in the presence of a sulfoxide.[5]

Diastereospecific 1,2-reduction of a conjugated enone occurs in the case of 2β-*t*-butyldimethylsilyloxycyperone (eq 3).[6a] The corresponding 2α-silyloxy diastereomer requires $LiAlH_4$ for reduction, also yielding the 3β-ol. A conjugated enone undergoes 1,2-reduction while a lactone (both functions in the same 14-membered ring) remains untouched (see below).[12o]

LiBH(*s*-Bu)$_3$, −78 °C, 2 h, 55% (3)

R = TBDMS

Amides are reduced to aldehydes in the presence of other readily reducible functional groups by first treating with ethyl trifluoromethanesulfonate (EtOTf) or MeOTf in CH_2Cl_2, removing the solvent, dissolving the residue in THF, and treating with L-Selectride (eq 4).[6b] In all cases, some unreacted starting material is recovered. K-Selectride® (***Potassium Tri-s-butylborohydride***) gives a lower yield; other hydrides give byproducts.

EtOTf, CH_2Cl_2; LiBH(*s*-Bu)$_3$, −78 °C, 1.5 h, 69% (4)

A keto aldehyde is reduced selectively (eq 5).[6c] The resulting hydroxy ketone is in equilibrium with its cyclic hemiacetal.

LiBH(*s*-Bu)$_3$, 0 °C, 1 h, 96% (5)

Stereoselective Reductions. The synthetic use that first brought attention to L-Selectride was the diastereoselective reduction of alkyl-substituted cyclic ketones to alcohols.[20] Numerous applications for cyclic[7,11m] and acyclic[8,23c] ketones have now been reported, including various ketocarboxylate[3] and ketonamide[4] systems.

Various theoretical models are used to rationalize observed diastereoselectivity.[23] In simple cases, the bulk of the reagent appears to dictate that hydride approach from the less sterically hindered face of the carbonyl (eq 6).[20] However, complexation effects, including those of heteroatoms,[7b] crown ethers,[3i,8u] and cryptands,[8u] may reverse the direction of approach. It is also observed that varying the coordinating ability of the solvent influences the ratio of diastereomers.[3e]

LiBH(*s*-Bu)$_3$, −78 °C, THF, 3 h; 96.5% + 3.5% (6)

In most cases, imines can be reduced stereoselectively to amines (eq 7).[9a–e] However, some 2-substituted imines resist reduction and require using the more reactive ***Lithium Triethylborohydride***. Enamino sulfoxides are also stereoselectively reduced to amines.[5] Chiral β-imino sulfoxides are reduced stereoselectively but in poor yield.[9f] Reduction of a tartarimide gives >99% stereoselection in the formation of the corresponding hydroxylactam (eq 8).[10]

LiBH(*s*-Bu)$_3$ (7)

25 °C, 52 h, 95% 98:2
66 °C, 20–24 h, 77% 97:3

LiBH(*s*-Bu)$_3$, Et_2O, −109 °C, 5.5 h, 98% (8)

R = TBDMS

Regioselective Reduction of Cyclic Anhydrides. Various succinic and phthalic anhydrides are reduced to lactones.[11] With certain exceptions (eq 9),[11d] reduction occurs preferentially at the less hindered carbonyl function. A mechanistic interpretation of the results has been proposed.[11d,e]

L-Selectride, −70 to 0 °C, 1.5 h, 50% (9)

only product

Reduction of Enones and Enoates. Following the first reported stereoselective reduction of an acyclic enone to an allylic alcohol,[12a] a reasonably systematic study using K- and L-Selectride was conducted.[12b] With certain exceptions, results for the latter are comparable to those for the former:

1) whereas one equivalent of K-Selectride gives exclusively 1,2-addition with 3,5-dimethyl-2-cyclohexenone, L-Selectride gives a 1:1 mixture of 1,2- and 1,4-addition products;

2) in reductive alkylation of enones use of L- rather than K-Selectride leads to higher yields of monoalkylated products;

3) while attempted reduction of α,β-unsaturated esters with K-Selectride leads to Claisen condensation products, slowly adding a mixture of ester and *t*-butyl alcohol to a THF solution of L-Selectride at −70 °C gives the saturated ester cleanly.

Replacing *t*-butyl alcohol with various electrophiles in the procedure of (3) allows reductive alkylation, although Claisen condensation product also forms if the enoate lacks α- or β-substitution.

Other reports confirm that stereoselective 1,2-reduction is likely with both acyclic enones[12c,d] and β-substituted cyclic enones.[12e–g] However, in one case, treating the latter first with ***Methylaluminum Bis(2,6-di-t-butyl-4-methylphenoxide)*** followed by L-Selectride (2 equiv, −78 °C, 30 min) gives the 1,4-reduction product in quantitative yield.[12g] Terminal acyclic enones undergo 1,4-reduction with predominant formation of the *s-trans* enolate, the exact ratio *s-trans*:*s-cis* depending on the relative bulk of *R* and *R′* (eq 10).[12h] Systems with extended conjugation, such as a phenyldienone, resist conjugate addition and undergo stereoselective 1,2-reduction.[12i] Unlike simpler substrates, a complex methyl enoate is reduced to the corresponding allylic alcohol.[12j]

R O R′ ⟶ R O⁻ R′ (*s-trans*, predominant) + R R′ O⁻ (*s-cis*) (10)

Cyclic enones without β-substituents often undergo 1,4-reduction,[12k–m] which may be followed by alkylation;[12k,m,n] the alkylation may be stereoselective.[12k,m] In some cases, mixtures of 1,2- and 1,4-reduction products are obtained,[12o,p] in contrast to 1,2-reduction only for K-Selectride.[12o] Vinyl triflates may be produced from enones if 1,4-reduction is facile.[12q] In the presence of 1,3-dimethyl-3,4,5,6-tetrahydro-2 (1*H*)-pyrimidinone, a conjugated cyclic dienone is reduced to the conjugated enone.[12r]

α,β-Alkynic esters tend toward nearly exclusive 1,2-reduction with formation of propargylic alcohols.[12b] Likewise, α,β-alkynic ketones yield only propargylic alcohols.[8e,f]

Michael Initiated Ring Closure (MIRC) Reactions. Formation of five- and six-membered rings results from treatment of ω-alkylidenemalonates with L-Selectride, but better yields are realized with nucleophiles other than hydride.[13a] A MIMIRC procedure using 2-cyclohexenone and ***Vinyltriphenylphosphonium Bromide*** (VTB) is highly successful (eq 11).[13b] The 1-*t*-butyl 9-methyl diester of (*E,E*)-(*S*)-2,4-dimethyl-2,7-nonadienedioic acid is converted to a cyclohexane diester as a single stereoisomer (eq 12).[13c]

1. L-Selectride; 2. 2 equiv VTB; 3. KOH (aq), 70% — product: $P(O)Ph_2$ (11)

CO_2Me, CO_2-*t*-Bu — 1. L-Selectride, −25 °C; 2. MeOH, 52–61% — CO_2Me, CO_2-*t*-Bu (12)

Reduction of (2-Arylprop-1-en-3-yl)trimethylammonium Iodides. Treatment of [2-(4-methoxyphenyl)prop-1-en-3-yl]trimethylammonium iodide with L-Selectride results in addition of hydride with displacement of Me_3N, giving 2-(4-methoxyphenyl)propene in virtually quantitative yield (eq 13).[14] This reaction may be applicable for derivatives containing a wide variety of aryl groups.

X, I^- $^+NMe_3$ ⟶ X (13)

X = 4-methoxyphenyl

Hydroboration of Styrenes. This reaction is quantitative with ***Lithium Triethylborohydride*** but proceeds to the extent of less than 40% with L-Selectride, polymerization of styrene being the predominant result.[15]

Reaction with Carbon Monoxide. Carbonylation of LiBH-(s-Bu)$_3$ followed by treatment with refluxing aqueous NaOH gives 2-methyl-1-butanol.[16a] Since LiBH(s-Bu)$_3$ is an intermediate in hydride-induced carbonylation of s-Bu$_3$B using LiAlH(OMe)$_3$,[16b,20] different workup procedures yield 2-methylbutanal[16c] or 3,5-dimethyl-4-heptanol.[16d]

Related Reagents. See Classes R-5, R-6, R-10, R-12, R-13, R-14, R-15, R-17, R-23, and R-27, pages 1–10. Lithium Aluminum Hydride; Lithium Triethylborohydride; Lithium Trisiamylborohydride; Potassium Tri-*s*-butylborohydride.

1. (a) Hajos, A. *Complex Hydrides and Related Reducing Agents in Organic Synthesis*; Elsevier: New York, 1979. (b) Brown, H. C.; Krishnamurthy, S. *T* **1979**, *35*, 567. (c) Hudlický, M. *Reductions in Organic Chemistry*; Horwood: Chichester, 1984. (d) Seyden-Penne, J. *Reductions by the Alumino- and Borohydrides in Organic Synthesis*; VCH: New York, 1991.
2. (a) Kim, S.; Yi, K. Y. *BCJ* **1985**, *58*, 789. (b) Krishnamurthy, S.; Brown, H. C. *JOC* **1976**, *41*, 3064. (c) Newkome, G. R.; Majestic, V. K.; Sauer, J. D. *OPP* **1980**, *12*, 345. (d) Guyon, R.; Villa, P. *BSF(2)* **1975**, 2584. (e) Varma, R. S.; Kabalka, G. W. *SC* **1984**, *14*, 1093. (f) Mourad, M. S.; Varma, R. S.; Kabalka, G. W. *S* **1985**, 654. (g) Freimanis, J.; Bokaldere, R.; Lola, D.; Turovskii, I. V.; Gavars, M. *ZOR* **1990**, *26*, 2114 (*CA* **1991**, *115*, 207 700s). (h) Kalnins, A.; Dikovskaya, K. I.; Kuchin, A. V.; Kudryashova, V. V.; Korics. V.; Freimanis, J.; Gavars, M.; Sakhartova, O. V.; Turovskii, I. V. *ZOR* **1988**, *24*, 742 (*CA* **1988**, *110*, 114 492g). (i) Akiyama, T.; Okada, K.; Ozaki, S. *TL* **1992**, *33*, 5763. (j) Negishi, E.; Miller, J. A.; Yoshida, T. *TL* **1984**, *25*, 3407.

3. (a) Arco, M. J.; Trammell, M. H.; White, J. D. *JOC* **1976**, *41*, 2075. (b) Greene, A. E.; Luche, M.-J.; Serra, A. A. *JOC* **1985**, *50*, 3957. (c) Neeland, E.; Ounsworth, J. P.; Sims, R. J.; Weiler, L. *TL* **1987**, *28*, 35. (d) Kuehne, M. E.; Pitner, J. B. *JOC* **1989**, *54*, 4553. (e) Niwa, H.; Ogawa, T.; Yamada, K. *BCJ* **1990**, *63*, 3707. (f) Hoffman, R. V.; Kim, H.-O. *JOC* **1991**, *56*, 6759. (g) Boireau, G.; Deberly, A. *TA* **1991**, *2*, 771. (h) Solladie-Cavallo, A.; Bencheqroun, M. *TA* **1991**, *2*, 1165. (i) Akiyama, T.; Nishimoto, H.; Ozaki, S. *TL* **1991**, *32*, 1335. (j) Ogiku, T.; Yoshida, S.-I.; Takahashi, M.; Kuroda, T.; Ohmizu, H.; Iwasaki, T. *TL* **1992**, *33*, 4473. (k) Ogiku, T.; Yoshida, S.-I.; Takahashi, M.; Kuroda, T.; Ohmizu, H.; Iwasaki, T. *TL* **1992**, *33*, 4477. (l) Sakai, K.; Takahashi, K.; Nukano, T. *T* **1992**, *48*, 8229. (m) Hirai, Y.; Terada, T.; Yamazaki, T.; Momose, T. *JCS(P1)* **1992**, 517. (n) Spracklin, D. K.; Weiler, L. *CC* **1992**, 1347.
4. (a) Maurer, P. J.; Takahata, H.; Rapoport, H. *JACS* **1984**, *106*, 1095. (b) Roemmele, R. C.; Rapoport, H. *JOC* **1989**, *54*, 1866. (c) Miller, S. A.; Chamberlin, A. R. *JACS* **1990**, *112*, 8100. (d) Longobardo, L.; Mobbili, G.; Tagliavini, E.; Trombini, C.; Umani-Ronchi, A. *T* **1992**, *48*, 1299. (e) Ookawa, A.; Soai, K. *JCS(P1)* **1987**, 1465. (f) Pecquet, F.; D'Angelo, J. *TL* **1982**, *23*, 2777. (g) Bateson, J. H.; Fell, S. C. M.; Southgate, R.; Eggleston, D. S.; Baures, P. W. *JCS(P1)* **1992**, 1305.
5. Ogura, K.; Tomori, H.; Fujita, M. *CL* **1991**, 1407.
6. (a) Murai, A.; Ono, M.; Abiko, A.; Masamune, T. *JACS* **1978**, *100*, 7751. (b) Tsay, S.-C.; Robl, J. A.; Hwu, J. R. *JCS(P1)* **1990**, 757. (c) Bull, J. R.; Thomson, R. I. *JCS(P1)* **1990**, 241.
7. (a) Guyon, R.; Villa, P. *BSF(2)* **1977**, 145 (*CA* **1977**, *87*, 84 194y). (b) Wigfield, D. C.; Feiner, S. *CJC* **1978**, *56*, 789. (c) Quick, J.; Meltz, C.; Ramachandra, R. *OPP* **1979**, *11*, 111. (d) Suzuki, K.; Ikegawa, A.; Mukaiyama, T. *CL* **1982**, 899. (e) Berlan, J.; Sztajnbok, P.; Bersace, Y.; Cresson, P. *CR(C)* **1985**, *301*, 693 (*CA* **1986**, *105*, 60 123p). (f) Carreño, M. C.; Domínguez, E.; García-Ruano, J. L.; Rubio, A. *JOC* **1987**, *52*, 3619. (g) Tanis, S. P.; Chuang, Y.-H.; Head, D. B. *JOC* **1988**, *53*, 4929. (h) Nagao, Y.; Goto, M.; Ochiai, M. *CL* **1990**, 1507. (i) Pearson, A. J.; Mallik, S.; Pinkerton, A. A.; Adams, J. P.; Zheng, S. *JOC* **1992**, *57*, 2910. (j) Zoretic, P. A.; Weng, X.; Biggers, C. K.; Biggers, M. S.; Caspar, M. L. *TL* **1992**, *33*, 2637. (k) Linderman, R. J.; Viviani, F. G.; Kwochka, W. R. *TL* **1992**, *25*, 3571. (l) Tsushima, K.; Murai, A. *TL* **1992**, *33*, 4345. (m) Kerwin, S. M.; Heathcock, C. H. *JOC* **1992**, *57*, 4005.
8. (a) Guyon, R.; Villa, P. *BSF(2)* **1977**, 152 (*CA* **1977**, *87*, 84 195z). (b) Suzuki, K.; Katayama, E.; Tsuchihashi, G.-I. *TL* **1984**, *25*, 2479. (c) Shimagaki, M.; Maeda, T.; Matsuzaki, Y.; Hori, I.; Nakata, T.; Oishi, T. *TL* **1984**, *25*, 4775. (d) Ko, K.-Y.; Frazee, W. J.; Eliel, E. L. *T* **1984**, *40*, 1333. (e) Midland, M. M.; Kwon, Y. C. *TL* **1984**, *25*, 5981. (f) Takahashi, T.; Miyazawa, M.; Tsuji, J. *TL* **1985**, *26*, 5139. (g) Elliott, J.; Warren, S. *TL* **1986**, *27*, 645. (h) Iida, H.; Yamazaki, N.; Kibayashi, C. *JOC* **1986**, *51*, 3769. (i) Ko, K.-Y.; Eliel, E. L. *JOC* **1986**, *51*, 5353. (j) Frye, S. V.; Eliel, E. L. *JACS* **1988**, *110*, 484. (k) Shimagaki, M.; Suzuki, A.; Nakata, T.; Oishi, T. *CPB* **1988**, *36*, 3138. (l) Yamazaki, N.; Kibayashi, C. *TL* **1988**, *45*, 5767. (m) Chikashita, H.; Nikaya, T.; Uemura, H.; Itoh, K. *BCJ* **1989**, *62*, 2121. (n) Yamazaki, N.; Kibayashi, C. *JACS* **1989**, *111*, 1396. (o) Krajewski, J. W.; Guzinski, P.; Urbanczyk-Lipkowska, Z.; Ramza, J.; Zamojski, A. *Carbohydr. Res.* **1990**, *200*, 1. (p) Wade, P. A.; Price, D. T.; Carroll, P. J.; Dailey, W. P. *JOC* **1990**, *55*, 3051. (q) Richardson, D. P.; Wilson, W.; Mattson, R. J.; Powers, D. M. *JCS(P1)* **1990**, 2857. (r) Curran, D. P.; Zhang, J. *JCS(P1)* **1991**, 2613. (s) Wang, Y.; Babirad, S. A.; Kishi, Y. *JOC* **1992**, *57*, 468. (t) Grassberger, M. A.; Fehr, T.; Horvath, A.; Schulz, G. *T* **1992**, *48*, 413. (u) Manzoni, L.; Pilati, T.; Poli, G.; Scolastico, C. *CC* **1992**, 1027. (v) Harmange, J.-C.; Figadère, B.; Cavé, A. *TL* **1992**, *33*, 5749.
9. (a) Maryanoff, B. E.; McComsey, D. F.; Taylor, R. J. Jr.; Gardocki, J. F. *JMC* **1981**, *24*, 79. (b) Wrobel, J. E.; Ganem, B. *TL* **1981**, *22*, 3447. (c) Hutchins, R. O.; Su, W.-Y.; Sivakumar, R.; Cistone, F.; Stercho, Y. P. *JOC* **1983**, *48*, 3412. (d) Hutchins, R. O.; Su, W.-Y. *TL* **1984**, *25*, 695. (e) Ogura, K.; Tomori, H.; Fujita, M. *CL* **1991**, 1407. (f) García Ruano, J. L.; Lorente, A.; Rodríguez, J. H. *TL* **1992**, *33*, 5637.
10. Miller, S. A; Chamberlin, A. R. *JOC* **1989**, *54*, 2502.
11. (a) Maklouf, M. A.; Rickborn, B. *JOC* **1981**, *46*, 4810. (b) Krishnamurthy, S.; Vreeland, W. B. *H* **1982**, *18*, 265. (c) Morand, P.; Salvador, J.; Kayser, M. M. *CC* **1982**, 458. (d) Kayser, M. M.; Salvador, J.; Morand, P. *CJC* **1983**, *61*, 439. (e) Soucy, C.; Favreau, D.; Kayser, M. M. *JOC* **1987**, *52*, 129.
12. (a) Corey, E. J.; Becker, K. B.; Varma, R. K. *JACS* **1972**, *94*, 8616. (b) Fortunato, J. M.; Ganem, B. *JOC* **1976**, *41*, 2194. (c) Achmatowicz, B.; Marczak, S.; Wicha, J. *CC* **1987**, 1226. (d) Semmelhack, M. F.; Kim, C. R.; Dobler, W.; Meier, M. *TL* **1989**, *30*, 4925. (e) Dauben, W. G.; Ashmore, J. W. *TL* **1978**, 4487. (f) Kumar, V.; Amann, A.; Ourisson, G.; Luu, B. *SC* **1987**, *17*, 1279. (g) Comins, D. L.; LaMunyon, D. H. *TL* **1989**, *30*, 5053. (h) Chamberlin, A. R.; Reich, S. H. *JACS* **1985**, *107*, 1440. (i) Boger, D. L.; Curran, T. T. *JOC* **1992**, *57*, 2235. (j) Roush, W. R.; Koyama, K. *TL* **1992**, *33*, 6227. (k) Silverman, R. B.; Groziak, M. P. *JACS* **1982**, *104*, 6434. (l) Kowalski, C. J.; Weber, A. E.; Fields, K. W. *JOC* **1982**, *47*, 5088. (m) Ohloff, G.; Maurer, B.; Winter, B.; Giersch, W. *HCA* **1983**, *66*, 192. (n) Kundu, N. G.; Sikdar, S.; Hertzberg, R. P.; Schmitz, S. A.; Khatri, S. G. *JCS(P1)* **1985**, 1295. (o) Keller, T. H.; Weiler, L. *TL* **1990**, *31*, 6307. (p) Waldmann, H.; Braun, M. *JOC* **1992**, *57*, 4444. (q) Crisp, G. T.; Scott, W. J. *S* **1985**, 335. (r) Swarts, H. J.; Haaksma, A. A.; Jansen, B. J. M.; de Groot, A. *T* **1992**, *48*, 5497.
13. (a) Little, R. D.; Verhe, R.; Monte, W. T.; Nugent, S.; Dawson, J. R. *JOC* **1982**, *47*, 362. (b) Posner, G. H.; Lu, S. B. *JACS* **1985**, *107*, 1424. (c) Hori, K.; Hikage, N.; Inagaki, A.; Mori, S.; Nomura, K.; Yoshii, E. *JOC* **1992**, *57*, 2888.
14. Gupton, J. T.; Layman, W. J. *JOC* **1987**, *52*, 3683.
15. Brown, H. C.; Kim, S.-C. *JOC* **1984**, *49*, 1064.
16. (a) Brown, H. C.; Hubbard, J. L. *JOC* **1979**, *44*, 467. (b) Hubbard, J. L.; Smith, K. *JOM* **1984**, *276*, C41. (c) Brown, H. C.; Coleman, R. A.; Rathke, M. W. *JACS* **1968**, *90*, 499. (d) Hubbard, J. L.; Brown, H. C. *S* **1978**, 676.
17. Brown, H. C.; Krishnamurthy, S.; Hubbard, J. L. *JACS* **1978**, *100*, 3343.
18. Brown, H. C.; Hubbard, J. L.; Singaram, B. *T* **1981**, *37*, 2359.
19. Brown, H. C.; Kramer, G. W.; Hubbard, J. L.; Krishnamurthy, S. *JOM* **1980**, *188*, 1.
20. (a) Brown, H. C.; Krishnamurthy, S. *JACS* **1972**, *94*, 7159. (b) Brown, H. C.; Krishnamurthy, S.; Hubbard, J. L. *JOM* **1979**, *166*, 271.
21. Zweifel, G.; Brown, H. C. *OR* **1963**, *13*, 1.
22. Brown, H. C.; Kramer, G. W.; Levy, A. B.; Midland, M. M. *Organic Syntheses via Boranes*; Wiley: New York, 1975; Chapter 9.
23. (a) Hutchins, R. O. *JOC* **1977**, *42*, 920. (b) Wigfield, D. G. *T* **1979**, *35*, 449. (c) Mead, K.; Macdonald, T. L. *JOC* **1985**, *50*, 422, and references cited therein. (d) Ibarra, C. A.; Perez-Ossorio, R.; Quiroga, M. L.; Arias Pérez, M. S.; Fernández Dominguez, M. J. *JCS(P1)* **1988**, 101.

John L. Hubbard
Marshall University, Huntington, WV, USA

Lithium Triethylborohydride[1]

$LiBHEt_3$

[22560-16-3] $C_6H_{16}BLi$ (MW 105.97)

(exceptionally powerful nucleophile in S_N2 displacement reactions with super hydride activity; powerful and selective reducing agent)

Alternate Name: Super Hydride®.
Physical Data: mp 66–67 °C;[2] d_{20}^{4} 0.92ul 0 g cm^{-3} for 1 M solution in THF.
Solubility: sol THF, benzene; reacts violently with water.
Form Supplied in: 1 M solution in THF.

Preparative Methods: prepared by the reaction of ***Triethylborane*** with ***t-Butyllithium***,[3a] lithium hydride,[2,3b] or ***Lithium Aluminum Hydride*** in the presence of triethylenediamine.[3c]

Analysis of Reagent Purity: IR: 2060 cm^{-1} (BH). ^{11}B NMR (THF) δ, ppm, −12.2 (d, J_{B-H} = 61 Hz).[3a] Active hydride is determined by hydrolysis of an aliquot and measuring the hydrogen evolved according to the standard procedure.[4] Low concentrations may be determined by iodometric titration.[5]

Handling, Storage, and Precautions: corrosive flammable liquid. Handle and store under nitrogen or argon in a cool dry place. Avoid contact with skin and clothing. Use in a fume hood.

Reduction of Functional Groups. Lithium triethylborohydride shows super hydride activity. Its reactivity in comparison to other nucleophiles is given in Table 1.[6] The reduction of representative functionalities by the reagent under standard conditions has been described.[7]

Table 1 Reactivity of Et_3BH^- in Comparison to Other Nucleophiles

Reagent	Relative Nucleophilicity
Et_3BH^-	9 400 000
n-BuS^-	680 000
PhS^-	470 000
AlH_4^-	230 000
I^-	3 700
EtO^-	1 000
BH_4^-	940
Br^-	500
PhO^-	400
NO_3^-	1

Alkyl Halides. Lithium triethylborohydride is the reagent of choice for dehalogenation of alkyl halides (eq 1).[6,8] Primary iodides are the most reactive. Fluorides are reduced at slower rates (eq 2).[9] Secondary halides may give elimination products in some cases. Mechanistic aspects of the reaction have been studied.[10]

LiBDEt3, THF; 65 °C, 72 h; 96% (1)

LiBHEt3, THF; 22 °C, 24 h; 56% (2)

Deoxygenation of Alcohols. Tosylates,[11] mesylates,[12] and alkoxytris(dimethylamino)phosphonium salts[13] of primary alcohols are cleanly reduced to hydrocarbons. Tosylates of secondary alcohols may undergo elimination reactions. Even in such cases the reduction may be synthetically useful (eq 3).[11f] Differences in the reactivity of tosylates and alkylsulfonates have been observed.[11e]

LiBHEt3, THF; DMSO, 21 °C, 14 h; 82% (3)

Allylic Derivatives. Allylic groups such as –OR, –SR, $-SO_2R$, –SeR, and –OTBDMS undergo reductive cleavage by lithium triethylborohydride catalyzed by palladium compounds (eq 4).[14a] Squalene (E/Z = 97:3) has been prepared using this reaction as a key step.[14b]

LiBHEt3, Pd(PPh3)4; THF, 65 °C; 89% (4)

(E):(Z):terminal = 94:3:3

Epoxides and Oxetanes. The reduction of epoxides by the reagent is quite facile, yielding exclusively the Markovnikov product.[5,7,15] Its advantage is evident in the reduction of labile epoxides prone to electrophilic rearrangements (eq 5).[16] Oxetanes are opened in a similar fashion.[17] Conjugated epoxides may undergo 1,4-reduction.[18]

LiAlH4, Et2O; reflux, 24 h: 15% + 85%
LiBHEt3, THF; 65 °C, 24 h: 93% + <0.1% (5)

Ketones and Derivatives. Aldehydes and ketones are reduced by lithium triethylborohydride rapidly and quantitatively to the corresponding alcohols, even at −78 °C.[7] Enthalpies of these reactions have been estimated.[19] Selective reduction of a keto group in the presence of an electron-rich aldehyde group has been achieved.[20a] Although the stereoselectivity of the reagent is lower as compared to other trialkylborohydrides containing bulky sterically demanding groups, several α-, β-, or γ-substituted ketones undergo reduction with high selectivity.[20b–j] α,β-Unsaturated ketones and esters undergo 1,4-additions.[21] *N*-Alkylimines derived from cyclic ketones are reduced to the corresponding secondary amines which can be transformed into primary amines.[22]

The reduction of α-trityl ketones with lithium triethylborohydride proceeds with cleavage of the trityl group. Stereoselective aldol reaction, combined with the reduction, provides access to stereodefined acyclic 1,3-diols[23] (eq 6).[23b] Trityl alkynic ketones are cleaved without affecting the triple bond.[24] Acetals are stable to the reagent and can be used as protective groups (eq 2).[7,9,11c,d] However, acetals are cleaved in the presence of ***Titanium(IV) Chloride***.[25]

Carboxylic Acid Derivatives.[7] The inertness of carboxylic acids to the reagent is noteworthy. Anhydrides are reduced to alcohols and acids. Cyclic anhydrides can be transformed into the corresponding lactones via intermediate hydroxy acids. The reactivity of lithium triethylborohydride toward ester groups is exceptionally high. Even esters of aromatic carboxylic acids are selectively reduced to alcohols in the presence of other functional groups (eq 7).[26] Selective reduction of the less hindered MEM ester groups of a diester has been achieved.[27a] The reduction of esters with ***Lithium Borohydride*** is catalyzed by lithium triethylborohydride.[27b]

$$Ph_3CC(O)CH_2CH_3 \xrightarrow[3.\ H_2O,\ 99\%]{1.\ BuLi,\ -75\ °C;\ 2.\ PhCHO} Ph_3CC(O)CH(CH_3)CH(OH)Ph\ (syn{:}anti\ 99\%) \xrightarrow[2.\ LiBHEt_3,\ -40\ to\ 25\ °C,\ 94\%]{1.\ MEMCl} HOCH_2CH(CH_3)CH(OMEM)Ph \quad (6)$$

$$2\text{-}Y\text{-}4\text{-}X\text{-}C_6H_3CO_2Me \xrightarrow[2.\ HCl,\ H_2O]{1.\ LiBHEt_3,\ THF} 2\text{-}Y\text{-}4\text{-}X\text{-}C_6H_3CH_2OH \quad (7)$$

X = H, Y = NH_2, 82.5%
X = H, Y = NO_2, 99.7%
X = Cl, Y = OMe, 94.4%

Unlike most other hydride reducing agents, lithium triethylborohydride reduces tertiary amides to the corresponding alcohols.[7,28] Benzonitrile cleanly gives benzylamine with an excess of the reagent. With an equimolar ratio of the reactants, an 85–97% yield of benzaldehyde is obtained.[7] Aliphatic nitriles give lower yields of the corresponding amines.

Other Functionalities. The formation of ethylbenzene in the reaction of lithium triethylborohydride with diphenyl sulfone illustrates a general reactivity pattern of trialkylborohydrides with aromatic sulfones (eq 8).[29]

$$PhSO_2Ph \xrightarrow[reflux,\ 3\ h,\ 75\%]{2\ equiv\ LiBHEt_3,\ THF} PhEt + PhSO_2Li \quad (8)$$

α,β-Unsaturated nitro compounds are transformed into lithium nitroalkanes.[30a] In the presence of borane, *N*-ethylamine derivatives are formed (eq 9).[30b] 1,2-Iodo thiocyanates[31a] and vinylthiophthalimides[31b] yield thiiranes (eq 10). The reagent is useful for selective demethylation of quaternary ammonium salts containing at least two methyl groups.[32] It also finds application in the synthesis of di- and trialkyldihydronaphthalenenes.[33a] Pyridine is reduced to tetrahydropyridine.[7,33b]

$$R^1,R^2\text{-}C_6H_3CH{=}C(R)NO_2 \xrightarrow[60\text{–}70\ °C,\ 15\ h,\ 56\text{–}78\%]{LiBHEt_3,\ BH_3\bullet THF} R^1,R^2\text{-}C_6H_3CH{=}C(R)NHEt \quad (9)$$

R	R¹	R²
H	H	H
Me	H	H
H	Br	H
Me	EtO	EtO

$$t\text{-}Bu\text{-cyclohexyl}(I)CH_2SCN \xrightarrow[Et_2O,\ 25\ °C,\ >99\%]{LiBHEt_3} t\text{-}Bu\text{-cyclohexane spiro-thiirane} \quad (10)$$

Hydroboration. The reagent hydroborates aromatically conjugated alkenes in a Markovnikov fashion (eq 11).[34] Substituents that decrease the electron density of the double bond increase the rate of reaction. Lithium triethylborohydride promotes hydroboration of alkenes by dialkoxyboranes; the addition is anti-Markovnikov.[35]

$$PhCH{=}CH_2 \xrightarrow[100\ °C,\ 64\%]{LiBHEt_3,\ THF\ reflux\ or\ diglyme} [PhCH(CH_3)BEt_3]^-\ Li^+ \quad (11)$$

$$\xrightarrow{D_2O} PhCH(CH_3)D$$

$$\xrightarrow[2.\ H_2O_2,\ NaOH,\ 90\text{–}96\%]{1.\ MeSO_3H} PhCH(CH_3)OH \quad 100\%\ 1\text{-ol}$$

Synthesis of Borates and Boronates. The reagent promotes the reaction of alcohol with borane to give borates in high yields[36] (see also ***Borane–Dimethyl Sulfide***). Ethyl boronates can be conveniently prepared by the reaction with 1,2-diols.[37]

Organometallic Synthesis.[38] Lithium triethylborohydride is recommended for the generation of chloro(cyclopentadienyl)hydrozirconium (Schwartz hydrozirconation reagent).[39] The reagent is used for the reduction of other transition metal halide complexes.[40] The major area of its application in transition metal chemistry involves metal–metal and metal–metalloid bond cleavage[38,41] and formation of anionic and neutral formyl complexes.[38,42] It finds various other applications.[43–45]

Related Reagents. See Classes R-1, R-2, R-3, R-5, R-6, R-8, R-10, R-12, R-13, R-15, R-17, R-20, R-23, R-27, R-28, and R-32, pages 1–10. Potassium Triethylborohydride.

1. (a) Krishnamurthy, S. *Aldrichim. Acta* **1974**, *7*, 55. (b) Brown, H. C.; Krishnamurthy, S. *T* **1979**, *35*, 567. (c) Brown, H. C.; Krishnamurthy, S. *Aldrichim. Acta* **1979**, *12*, 3. (d) Pelter, A.; Smith, K.; Brown, H. C. *Borane Reagents*; Academic: London, 1988. (e) Seyden-Penne, J. *Reductions by the Alumino- and Borohydrides in Organic Synthesis*; VCH: New York, 1991.
2. Binger, P.; Benedikt, G.; Rotermund, G. W.; Köster, R. *LA* **1968**, *717*, 21.
3. (a) Brown, H. C.; Kramer, G. W.; Hubbard, J. L.; Krishnamurthy, S. *JOM* **1980**, *188*, 1. (b) Brown, H. C.; Krishnamurthy, S.; Hubbard, J. L. *JACS* **1978**, *100*, 3343. (c) Brown, H. C.; Hubbard, J. L.; Singaram, B. *T* **1981**, *37*, 2359.
4. Brown, H. C. *Organic Syntheses via Boranes*; Wiley: New York, 1975; p. 239.
5. Brown, H. C.; Narasimhan, S.; Somayaji, V. *JOC* **1983**, *48*, 3091.

6. Krishnamurthy, S.; Brown, H. C. *JOC* **1983**, *48*, 3085.
7. Brown, H. C.; Kim, S. C.; Krishnamurthy, S. *JOC* **1980**, *45*, 1.
8. Krishnamurthy, S.; Brown, H. C. *JOC* **1980**, *45*, 849.
9. (a) Brandänge, S.; Dahlman, O.; Ölund, J.; Mörch, E. *JACS* **1981**, *103*, 4452. (b) Brandänge, S.; Dahlman, O.; Ölund, J. *ACS* **1983**, *B37*, 141.
10. Ashby, E. C.; Wenderoth, B.; Pham, T. N.; Park, W-S. *JOC* **1984**, *49*, 4505.
11. (a) Krishnamurthy, S.; Brown, H. C. *JOC* **1976**, *41*, 3064. (b) Binkley, R. W. *JOC* **1985**, *50*, 5646. (c) Baer, H. H.; Hanna, H. R. *Carbohydr. Res.* **1982**, *110*, 19. (d) Baer, H. H.; Mekarska-Felicki, M. *CJC* **1985**, *63*, 3043. (e) Hua, D. H.; Sinai-Zingde, G.; Venkataraman, S. *JACS* **1985**, *107*, 4088. (f) Hansske, F.; Robins, M. J. *JACS* **1983**, *105*, 6736. (g) Kelly, A. G.; Roberts, J. S. *Carbohydr. Res.* **1979**, *77*, 231. (h) Krishnamurthy, S. *JOM* **1978**, *156*, 171.
12. Holder, R. W.; Matturro, M. G. *JOC* **1977**, *42*, 2166.
13. Simon, P.; Ziegler, J. C.; Gross, B. *S* **1979**, 951.
14. (a) Hutchins, R. O.; Learn, K. *JOC* **1982**, *47*, 4380. (b) Mohri, M.; Kinoshita, H.; Inomata, K.; Kotake, H. *CL* **1985**, 451.
15. Corey, E. J.; Tius, M. A.; Das, J. *JACS* **1980**, *102*, 7612.
16. Krishnamurthy, S.; Schubert, R. M.; Brown, H. C. *JACS* **1973**, *95*, 8486.
17. Welch, S. C.; Rao, A. S. C. P.; Lyon, J. T.; Assereq, J-M. *JACS* **1983**, *105*, 252.
18. Parish, E. J.; Schroepfer, G. J., Jr. *TL* **1976**, 3775.
19. Wiberg, K. B.; Crocker, L. S.; Morgan, K. M. *JACS* **1991**, *113*, 3447.
20. (a) Djuric, S. W.; Herbert, R. B.; Holliman, F. G. *JHC* **1985**, *22*, 1425. (b) Midland, M. M.; Kwon, Y. C. *JACS* **1983**, *105*, 3725. (c) Elliott, J.; Warren, S. *TL* **1986**, *27*, 645. (d) Hanamoto, T.; Fuchikami, T. *JOC* **1990**, *55*, 4969. (e) Alcudia, F.; Llera, J. M. *Sulfur Lett.* **1989**, *7*, 143. (f) Baker, R.; Ravenscroft, P. D.; Swain, C. J. *CC* **1984**, 74. (g) Zwick, J-C.; Vogiel, P. *AG(E)* **1985**, *24*, 787. (h) Pilli, R. A.; Russowsky, D.; Dias, L. C. *JCS(P1)* **1990**, 1213. (i) Shimagaki, M.; Suzuki, A.; Nakata, T.; Oishi, T. *CPB* **1988**, *36*, 3138. (j) Wunderly, S. W.; Brochmann-Hanssen, E. *JOC* **1977**, *42*, 4277.
21. Fortunato, J. M.; Ganem, B. *JOC* **1976**, *41*, 2194.
22. Hutchins, R. O.; Su, W-Y. *TL* **1984**, *25*, 695.
23. (a) Ertas, M.; Seebach, D. *HCA* **1985**, *68*, 961. (b) Seebach, D.; Ertas, M.; Locher, R.; Schweizer, H. B. *HCA* **1985**, *68*, 264.
24. Locher, R.; Seebach, D. *AG(E)* **1981**, *20*, 569.
25. Mikami, T.; Asano, H.; Mitsunobu, O. *CL* **1987**, 2033.
26. Lane, C. F. *Aldrichim. Acta* **1974**, *7*, 32.
27. (a) Ireland, R. E.; Thompson, W. J. *TL* **1979**, 4705. (b) Brown, H. C.; Narasimhan, S. *JOC* **1982**, *47*, 1604.
28. Brown, H. C.; Kim, S-C. *S* **1977**, 635.
29. Brown, H. C.; Kim, S-C.; Krishnamurthy, S. *OM* **1983**, *2*, 779.
30. (a) Varma, R. S.; Kabalka, G. W. *SC* **1984**, *14*, 1093. (b) Kabalka, G. W.; Gai, Y-Z.; Goudgaon, N. M.; Varma, R. S.; Gooch, E. E. *OM* **1988**, *7*, 493.
31. (a) Cambie, R. C.; Rutledge, P. S.; Strange, G. A.; Woodgate, P. D. *H* **1982**, *19*, 1501. (b) Capozzi, G.; Gori, L.; Menichetti, S. *T* **1991**, *47*, 7185.
32. Cooke, M. P., Jr.; Parlman, R. M. *JOC* **1975**, *40*, 531.
33. (a) Tomioka, K.; Shindo, M.; Koga, K. *JOC* **1990**, *55*, 2276. (b) Kaim, W.; Lubitz, W. *AG(E)* **1983**, *22*, 892.
34. Brown, H. C.; Kim, S-C. *JOC* **1984**, *49*, 1064.
35. Arase, A.; Nunokawa, Y.; Masuda, Y.; Hoshi, M. *CC* **1992**, 51.
36. Masuda, Y.; Nunokawa, Y.; Hoshi, M.; Arase, A. *CL* **1992**, 349.
37. Garlaschelli, L.; Mellerio, G.; Vidari, G. *TL* **1989**, *30*, 597.
38. Gladysz, J. A. *Aldrichim. Acta* **1977**, *12*, 13.
39. Lipschutz, B. H.; Keil, R.; Ellsworth, E. L. *TL* **1990**, *31*, 7257.
40. (a) Fryzuk, M. D.; Lloyd, B. R.; Clentsmith, G. K. B.; Rettig, S. J. *JACS* **1991**, *113*, 4332. (b) Alonso, F. J. G.; Sanz, M. G.; Riera, V.; Ruiz, M. A.; Tiripicchio, A.; Camellini, M. T. *AG(E)* **1988**, *27*, 1167. (c) Carmichael, D.; Hitchcock, P. B.; Nixon, J. F.; Pidcock, A. *CC* **1988**, 1554. (d) Benlaarab, H.; Chaudret, B.; Dahan, F.; Poilblanc, R. *JOM* **1987**, *320*, C51.
41. (a) Shyu, S-G.; Calligaris, M.; Nardin, G.; Wojcicki, A. *JACS* **1987**, *109*, 3617. (b) Shyu, S-G.; Wojcicki, A. *OM* **1985**, *4*, 1457.
42. (a) Gauntlett, J. T.; Mann, B. E.; Winter, M J.; Woodward, S. *JCS(D)* **1991**, 1427. (b) Lee, S. W.; Tucker, W. D.; Richmond, M. G. *JOM* **1990**, *398*, C6. (c) Mercer, W. C.; Whittle, R. R.; Burkhardt, E. W.; Geoffroy, G. L. *OM* **1985**, *4*, 68. (d) Geoffroy, G. L.; Rosenberg, S.; Shulman, P. M.; Whittle, R. R. *JACS* **1984**, *106*, 1519. (e) Narayanan, B. A.; Amatore, C. A.; Kochi, J. K. *OM* **1984**, *3*, 802.
43. (a) Hansen, H-P.; Müller, J.; Englert, U.; Paetzold, P. *AG(E)* **1991**, *30*, 1377. (b) Gaines, D. F.; Steehler, G. A. *CC* **1984**, 1127.
44. (a) Gysling, H. J.; Luss, H. R. *OM* **1989**, *8*, 363. (b) Seyferth, D.; Henderson, R. S. *JOM* **1980**, *204*, 333.
45. (a) Selover, J. C.; Vaughn, G. D.; Strouse, C. E.; Gladysz, J. A. *JACS* **1986**, *108*, 1455. (b) Wakefield, J. B.; Stryker, J. M. *JACS* **1991**, *113*, 7057.

Marek Zaidlewicz
Nicolaus Copernicus University, Torun, Poland

Herbert C. Brown
Purdue University, West Lafayette, IN, USA

Manganese Dioxide[1]

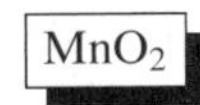

[1313-13-9] MnO_2 (MW 86.94)

(useful selective oxidizing reagent for organic synthesis; oxidation of allylic alcohols to α,β-ethylenic aldehydes or ketones;[2] conversion of allylic alcohols to α,β-ethylenic esters or amides;[3] oxidation of propargylic alcohols,[4] benzylic or heterocyclic alcohols,[5] and saturated alcohols;[6] oxidative cleavage of 1,2-diols;[7] hydration of nitriles to amides;[8] dehydrogenation and aromatization reactions;[9] oxidation of amines to aldehydes, imines, amides, and diazo compounds[10])

Alternate Names: manganese oxide; manganese(IV) oxide.
Physical Data: mp 535 °C (dec); *d* 5.03 g cm^{-3}.
Solubility: insol H_2O and organic solvents.
Form Supplied in: dark brown powder, widely available. The commercial 'active' MnO_2 used as oxidizing reagent for organic synthesis is a synthetic nonstoichiometric hydrated material. The main natural source of MnO_2 is the mineral pyrolusite, a poor oxidizing reagent.

Structure of Active Manganese Dioxide.[1,11,12] The structure and the reactivity of active manganese dioxides used as oxidizing reagents in organic synthesis closely depends on their method of preparation (see below). Active manganese oxides are nonstoichiometric materials (generally MnO_x; $1.93 < x < 2$) and magnetic measurements reveal the presence of lower valency Mn species, probably Mn^{II} and Mn^{III} oxides and hydroxides. Thermogravimetric analysis experiments show the existence of bonded and nonbonded water molecules (hydrated MnO_2). On the basis of ESR studies and other experiments, a locked-water associated structure (**1**) has been proposed for the apomorphous precipitated MnO_2.[12]

```
[     O--H-O  O--H--O      ]
[ Mn//      \/        \\Mn ]
[ Mn\\      Mn        //Mn ]
[     O--H-O  O--H--O      ]
```

(**1**)

In addition, variable amounts of alkaline and alkaline earth metal derivatives are detected by atomic adsorption analysis. X-ray studies have shown that the structures of active MnO_2 are quite complex; they are either amorphous or of moderate crystallinity (variable proportions of β- and γ-MnO_2).

Preparation of Active MnO_2, Oxidizing Power, and Reproducibility of the Results.[1,13] Pyrolusite (natural MnO_2) and pure synthetic crystalline MnO_2 are poor oxidants.[1] The oxidation of organic compounds requires an active, specially prepared MnO_2 and several procedures have been reported.[1,13] According to the method of preparation, the structure, the composition, and therefore the reactivity of active MnO_2 are variable. On this account, the choice of a procedure is of considerable importance to obtain the desired oxidation power and the reaction conditions must be carefully controlled to obtain a consistent activity. The active manganese dioxides described in the literature are generally prepared either by mixing aqueous solutions of $KMnO_4$ and a Mn^{II} salt ($MnSO_4$, $MnCl_2$) between 0 and 70 °C under acid, neutral, or basic conditions[13a–e] or by pyrolysis of a Mn^{II} salt (carbonate, oxalate, nitrate) at 250–300 °C.[13d,f] In this case the activity of the resulting material can be increased by washing with dilute nitric acid.[13f] A similar treatment has also been used to activate pyrolusite (natural MnO_2).[13h] On the other hand, it has been reported that the efficiency of an active MnO_2 depends on the percentage of the γ-form present in the material.[13i] Indeed, active γ-MnO_2 is sometimes clearly superior to the classical active MnO_2 prepared according to Attenburrow.[13j]

It worthy of note that the percentage water content strongly influences both the oxidizing power and the selectivity (oxidation of multifunctional molecules) of active MnO_2. Thus it is well known that the wet material (40–60% H_2O) obtained after filtration must be activated by drying[1,13] (heating to 100–130 °C for 12–24 h[13a–d] or, better, at 125 °C for 52 h).[13e] Indeed, an excess of water decreases the oxidation power[1e,13k] since, according to the triphasic mechanism generally postulated,[12] it would prevent the adsorption of the substrate to the oxidatively active polar site on the surface of MnO_2.[1a,c,e] On the other hand, it is very important not to go past the point of complete activation since the presence of hydrated MnO_2 species is essential to obtain an active reagent. For this reason, the drying conditions must be carefully controlled.[1,13a,d,g,k] Alternatively, the wet material can be activated by azeotropic distillation since this mild procedure preserves the active hydrated species.[13l] Thus azeotropic distillation has been used to remove the water produced during the oxidation reaction to follow the rate of MnO_2 oxidations.[13e] Finally, the active MnO_2 mentioned above contains various metallic salts as impurities. According to their nature, which depends on the method of preparation, they can also influence the oxidizing power of the reagent (for instance, permanganate).[13d] Finally, it should be noted that the preparation of active MnO_2 on carbon[13m] or on silica gel[13n] as well as the activation of nonactive MnO_2 (pure crystalline MnO_2) by ultrasonic irradiation[13o] has also been reported. Some typical procedures to prepare active MnO_2 are reported below.

Preparation of Active MnO_2 from $KMnO_4$ under Basic Conditions (Attenburrow).[13a] A solution of $MnSO_4 \cdot 4H_2O$ (110 g) in H_2O (1.5 L) and a solution of NaOH (40%; 1.17 L) were added simultaneously during 1 h to a hot stirred solution of $KMnO_4$ (960 g) in H_2O (6 L). MnO_2 precipitated soon after as a fine brown solid. Stirring was continued for an additional hour and the solid was then collected with a centrifuge and washed with water until the washings were colorless. The solid was dried in an oven at 100–120 °C and ground to a fine powder (960 g) before use.

Preparation of Active MnO_2 from $KMnO_4$ under Acidic Conditions.[13k] Active MnO_2 was made by mixing hot solutions of $MnSO_4$ and $KMnO_4$, maintaining a slight excess of the latter for several hours, washing the product thoroughly with water and drying at 110–120 °C. Its activity was unchanged after storage for many months, but it was deactivated by H_2O, MeOH, thiols, or excessive heat (500 °C). MnO_2 was less active when prepared in the presence of alkali and ineffective when precipitated from hot solutions containing a large excess of $MnSO_4$.

Preparation of Highly Active MnO_2 from $KMnO_4$.[1a] A solution of $MnCl_2{\cdot}4H_2O$ (200 g) in H_2O (2 L) at 70 °C was gradually added during 10 min, with stirring, to a solution of $KMnO_4$ (160 g) in H_2O (2 L) at 60 °C in a hood. A vigorous reaction ensued with evolution of chlorine; the suspension was stirred for 2 h and kept overnight at rt. The precipitate was filtered off, washed thoroughly with H_2O (4 L) until pH 6.5–7 and the washing gave a negligible chloride test. The filter cake was then dried at 120–130 °C for 18 h; this gave a chocolate-brown, amorphous powder; yield 195–200 g. Alternatively, the wet cake was mixed with benzene (1.2 L) and H_2O was removed by azeotropic distillation giving a chocolate-brown, amorphous powder; yield 195 g. The last procedure gave a slightly less active material.

Preparation of Active MnO_2 by Pyrolysis of $MnCO_3$.[13f] Powdered $MnCO_3$ was spread in a one-inch thick layer in a Pyrex glass and heated at 220–280 °C for about 18 h in an oven in which air circulated by convection. The initially tan powder turned darker at about 180 °C, and black when maintained at over 220 °C. No attempt was made to determine lower temperature or time limits, nor the upper limit of temperature. The MnO_2 prepared as above was stirred with about 1 L of a solution made up of 15% HNO_3 in H_2O. The slurry was filtered with suction, the solid was washed on the Buchner funnel with distilled water until the washes were about pH 5, and finally was dried at 220–250 °C. The caked, black solid was readily crushed to a powder which retained its oxidizing ability even after having been stored for several months in a loosely stoppered container.

Preparation of γ-MnO_2.[1a] To a solution of $MnSO_4$ (151 g) in H_2O (2.87 L) at 60 °C was added, with stirring, a solution of $KMnO_4$ (105 g) in H_2O (2 L), and the suspension was stirred at 60 °C for 1 h, filtered, and the precipitate washed with water until free of sulfate ions. The precipitate was dried to constant weight at 60 °C; yield 120 g (dark-brown, amorphous powder).

Preparation of Active MnO_2 on Silica Gel.[13n] $KMnO_4$ (3.79 g) was dissolved in water (60 mL) at rt. Chromatographic grade silica gel (Merck, 70–230 mesh, 60 g) was added with stirring, and the flask connected to a rotary evaporator to strip off the water at 60 °C. The purple solid was ground to fine powder and then added with vigorous stirring to a solution of $MnSO_4{\cdot}H_2O$ (9.3 g) in H_2O (100 mL). The resulting brown precipitate was filtered with water until no more Mn^{II} ion could be detected in the wash water by adding ammonia. After being dried at 100 °C for 2 h, each gram of this supported reagent contained 0.83 mmol of MnO_2.

As shown above, a wide range of products of various activities are called active MnO_2 and the results described in the literature are sometimes difficult to reproduce since the nature of the MnO_2 which was used is not always well defined. Now, the commercial materials give reproducible results but they are not always convenient to perform all the oxidation reactions described in the literature. In addition, their origin (method of preparation) is not often indicated and comparison with the active MnO_2 described in the literature is sometimes difficult. For all these reasons, the use of activated MnO_2 has been somewhat restricted in spite of the efficiency and selectivity of its reactions since only an empirical approach and a careful examination of the literature allow selection of the suitable activity of MnO_2 and the optimum reaction conditions for a defined substrate.

Oxidation of Organic Compounds with MnO_2: Reaction Conditions.[1,13]

Solvent. Oxidation of organic compounds with MnO_2 has been performed in many solvents. The choice of the solvent is important; thus primary or secondary alcohols (or water) are unsatisfactory since they can compete with the substrate being adsorbed on the MnO_2 surface and they have a strong deactivating effect.[13k] A similar but less pronounced influence has also been observed with various polar solvents such as acetone, ethyl acetate, DMF, and DMSO. However, these polar solvents, including water,[13p] acetic acid, and pyridine, can be used successfully at higher temperatures. This deactivating influence due to the polarity of the solvent can be used to control the reactivity of active MnO_2 and sometimes to avoid side reactions or to improve the selectivity (for instance, allylic alcohol vs. saturated alcohol). Most of the reactions described in the literature were carried out in aliphatic or aromatic hydrocarbons, chlorinated hydrocarbons, diethyl ether, THF, ethyl acetate, acetone, and acetonitrile (caution: MeCN can react with highly active MnO_2 or with classically activated MnO_2 on prolonged treatment). In the case of the oxidation of benzylic[13f] and allylic[13d] alcohols, the best results have been obtained using diethyl ether (diethyl ether > petroleum ether > benzene). Caution: a spontaneous ignition has been observed when highly active MnO_2 was used in this solvent.[13f]

Temperature and Reaction Time. At rt, the reaction time can vary from 10 min to several days according to the nature of the substrate and the activity of the MnO_2.[13a,d,e,f,l,p] The reaction times are shortened by heating[13d] but the selectivity is very often much lower.[13q]

Ratio MnO_2:Substrate. The amount of active MnO_2 required to perform the oxidation of an organic substrate depends on the type of MnO_2, on the substrate, and on the particle size of the MnO_2.[13d,f,g] With a classical material (100–200 mesh), the ratio varies from 5:1 to 50:1 by weight.

Oxidation of Allylic Alcohols.[2] MnO_2 was used as oxidizing reagent for the first time by Ball et al. to prepare retinal from vitamin A_1 (eq 1).[2a] Since that report, the use of MnO_2 for the conversion of allylic alcohols to α,β-ethylenic aldehydes has been extensively utilized. Interestingly, the configuration of the double bond is conserved during the reaction (eqs 2–4).[2b,c] In some cases a significant rate difference between axial and equatorial alcohols has been observed[2d] (eq 5).[2e]

MnO_2 has been frequently used for the preparation of very sensitive polyunsaturated aldehydes or ketones (eq 6).[2f]

MnO_2, 6 d; pet. ether, rt; 80% (1)

MnO_2, hexane; 0 °C, 30 min; 97% (2)

MnO_2, acetone; 6 h, rt; 75% (Z) (3)

MnO_2, acetone; 6 h, rt; 70% (E) (4)

(eq) HO / (ax) HO; MnO_2, $CHCl_3$, rt; 43% (5)

1. i-Bu_2AlH, C_6H_6; 2. MnO_2, CH_2Cl_2, rt, 1 h; 74% (6)

Numerous functionalized α,β-ethylenic aldehydes are readily obtained by chemoselective oxidation of the corresponding allylic alcohols (eqs 7–11).[2g–k]

MnO_2, ether; rt, 8 h; 75% (7)

MnO_2, i-PrOH; rt, 18 h; 70% (8)

MnO_2, $CHCl_3$; rt, 96 h; 79% (9)

MnO_2, $CHCl_3$; 70% (10)

MnO_2, CH_2Cl_2; 18 d, rt; 90% (11)

The α,β-ethylenic ketone obtained by treatment of an allylic alcohol with MnO_2 can undergo an in situ Michael addition (eqs 11 and 12).[2l]

MnO_2, Et_2NH; C_6H_6, 18 h, rt; 85% (12)

Conversion of Allylic Alcohols to α,β-Ethylenic Esters and Amides.[2b,3] This procedure was first described by Corey.[2b,3] The key step is the sequential formation and oxidation of a cyanohydrin. In the presence of an alcohol or an amine the resulting acyl cyanide leads by alcoholysis or aminolysis to the corresponding α,β-ethylenic ester[2b] or amide[3] (eqs 13–15).

1. MnO_2, hexane, 0 °C; 2. NaCN, MnO_2, AcOH, MeOH; 86% (13)

NaCN, MnO_2; AcOH, MeOH; 95% (14)

Ph–CH=CH–CHO → ($NaCN$, MnO_2; i-PrOH, NH_3; 100%) → Ph–CH=CH–$CONH_2$ (15)

TBDMSO, H, H, SAr, OAc, HO, HO, H, O, H → (MnO_2, acetone; rt, 45 min; 76%) → TBDMSO, H, H, SAr, OAc, HO, O, HO, H (23)

Oxidation of Propargylic Alcohols.[4] Propargylic alcohols are easily oxidized by MnO_2 to alkynic aldehydes and ketones (eqs 16–18).[4a–c]

In the example in eq 19 the unstable propargyl aldehyde is trapped as a Michael adduct.[4d]

HO_2C, OH → (MnO_2, CH_2Cl_2; rt, 24 h; 74%) → HO_2C, O (16)

OEt, OEt, OH → (MnO_2, CH_2Cl_2; 0 °C; 76%) → OEt, OEt, O (17)

(≡)$_3$, OH → (MnO_2, CH_2Cl_2; rt, 2 h; 88%) → (≡)$_3$, O (18)

≡–CH_2OH → (MnO_2, C_6H_6; piperidine, rt, 12 h; 86%) → N–CH=CH–CHO (19)

TBDMSO, H, H, O, O, HO, HO, O, O, O → (MnO_2, acetone; rt, 4 d; 93%) → TBDMSO, H, H, O, O, HO, O, O, O, O (24)

Benzyl allyl and benzyl propargyl alcohols have been oxidized successfully to ketones (eqs 25 and 26).[5d,e]

MeO, OH, OMe, OMe → (MnO_2; ether, hexane; 86%) → MeO, O, OMe, OMe (25)

OMe, HO → (MnO_2, $CHCl_3$; 1 h; 80%) → OMe, O (26)

Oxidation of Benzylic and Heterocyclic Alcohols.[2b,5] Conjugated aromatic aldehydes or ketones can be efficiently prepared by treatment of benzylic alcohols with MnO_2 (eq 20).[5a] Numerous functional groups are tolerated (eqs 21–24).[5a–c]

PhCH$_2$OH → (MnO_2, $CHCl_3$; rt, 23 h; 89%) → PhCHO (20)

O, O, N, Me, OH → (MnO_2, $CHCl_3$; rt, 24 h; 76%) → O, O, N, Me, O (21)

HO, O, CH_2OH, O, Ar → (MnO_2, dioxane; 8 h, 60 °C; 80%) → OHC, O, CH_2OH, O, Ar (22)

The oxidation reaction can be extended to heterocyclic alcohols (eq 27)[5f] and the Corey procedure gives the expected esters (eq 28).[2b]

O, O, N, N, OH, C_7H_7 → (MnO_2, C_6H_6; 24 h, rt; 81%) → O, O, N, N, O, C_7H_7 (27)

furan–CHO → ($NaCN$, MnO_2; AcOH, MeOH; 95%) → furan–CO_2Me (28)

Oxidation of Saturated Alcohols.[6] Cyclic and acyclic saturated alcohols react with MnO_2 to give the saturated aldehydes or ketones in good yields (eqs 29–32).[6a,b]

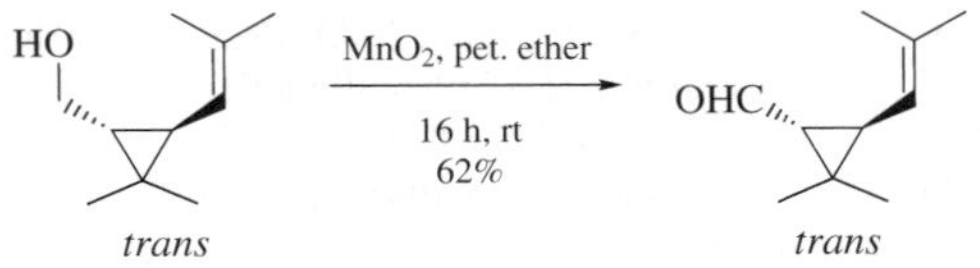
(29)

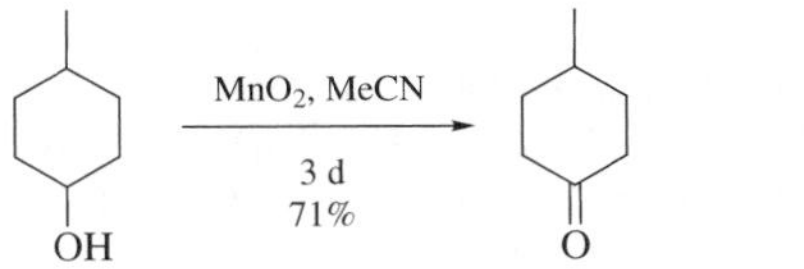
(30)

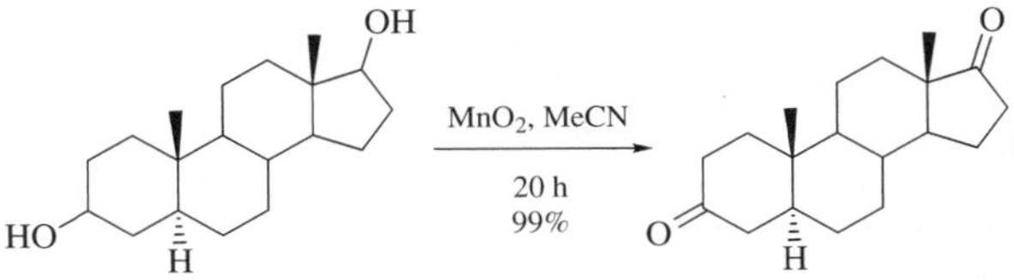
(31)

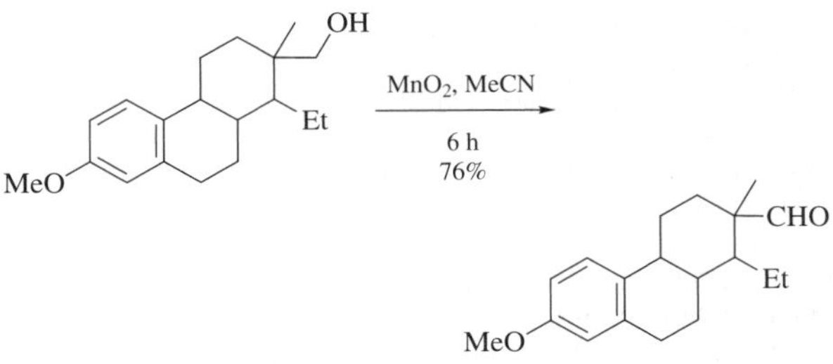
(32)

Oxidative Cleavage of 1,2-Diols.[7] 1,2-Diols are cleaved by MnO_2 to aldehydes or ketones. With cyclic 1,2-diols, the reaction leads to dialdehydes or diketones (eqs 33 and 34)[7] and the course of the reaction depends on the configuration of the starting material (eqs 34 and 35).[7]

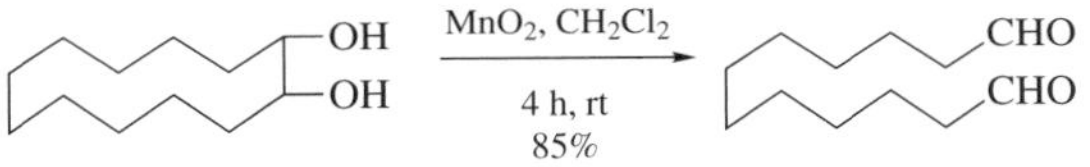
(33)

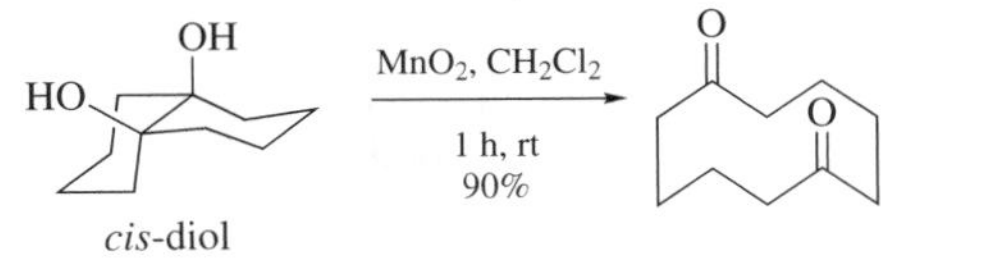
(34)

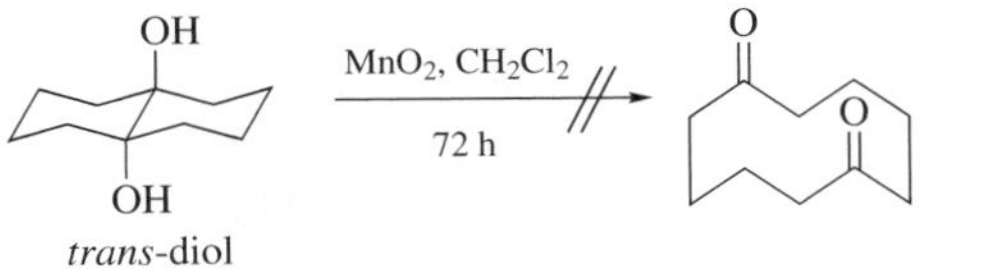
(35)

Hydration of Nitriles to Amides.[8] By treatment with MnO_2, nitriles are readily converted to amides. MnO_2 on silica gel is especially efficient to perform this reaction (eqs 36 and 37).[8a,b]

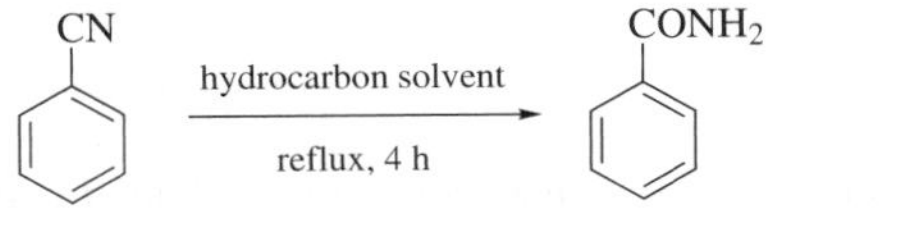
(36)

active MnO_2 30%
MnO_2, SiO_2 100%

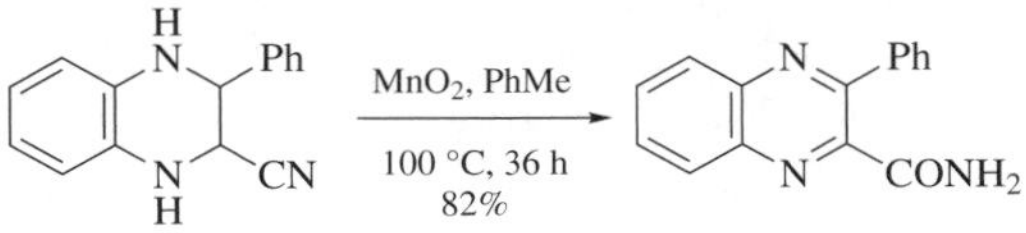
(37)

Dehydrogenation and Aromatization Reactions.[9,13j,q] MnO_2 has been widely used to carry out various dehydrogenation and aromatization reactions (eqs 38–41).[9a–c] In some cases the dehydrogenation can occur as a side reaction during, for instance, the hydration of nitriles (eq 37)[8b] or the oxidation of allylic alcohols.[13q]

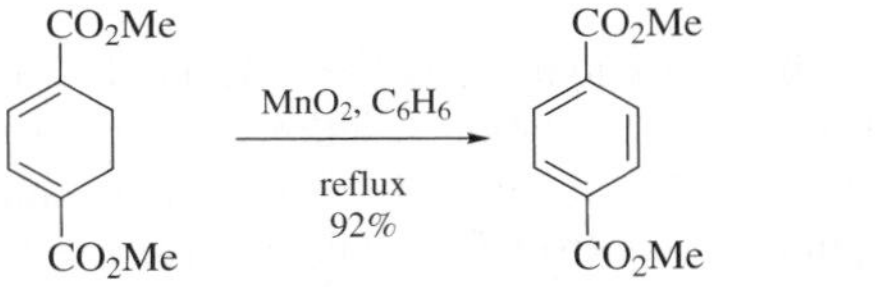
(38)

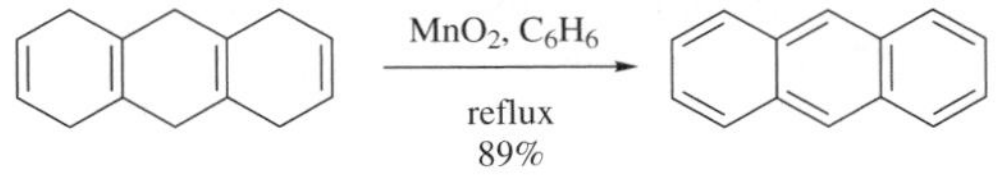
(39)

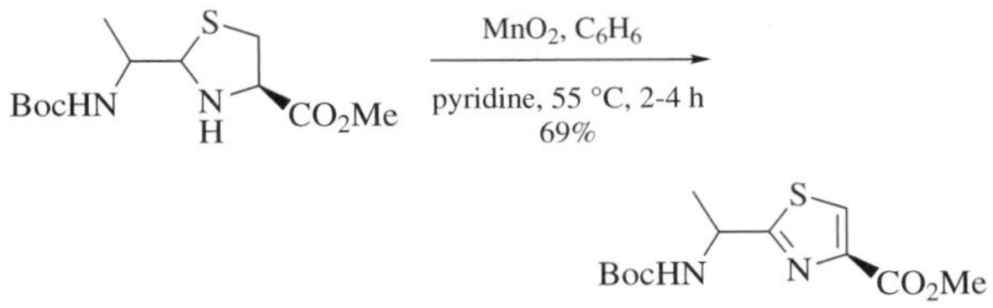
(40)

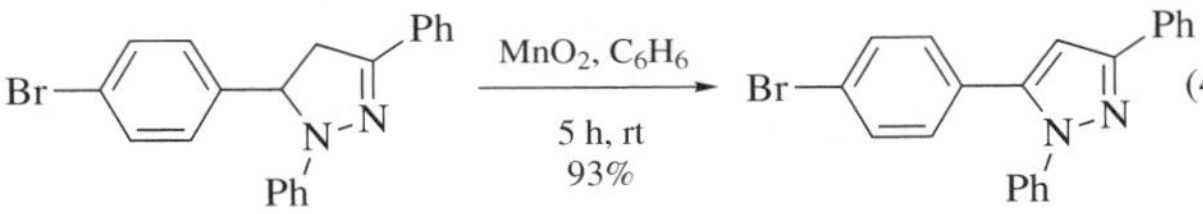
(41)

It is interesting to note that the use of γ-MnO_2 is essential to achieve the following dehydrogenation reactions (eqs 42 and 43).[13j]

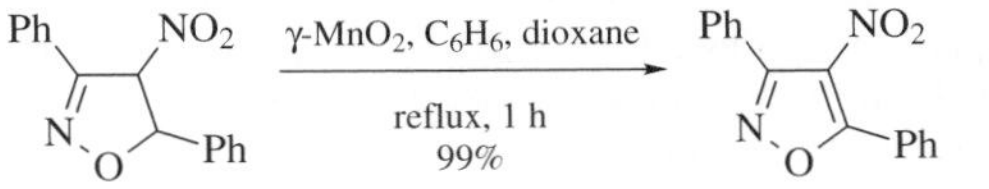
(42)

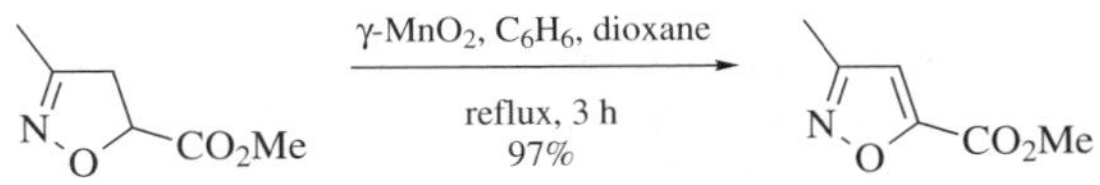
(43)

Oxidation of Amines to Aldehydes, Imines, Amides, and Diazo Compounds.[10,13g,p] The oxidation of amines by MnO_2 can lead to various products according to the structure of the starting material. Thus the formation of imines (eq 44),[10] formamides (eqs 45 and 46),[13g] and diazo compounds (eq 47)[13p] have all been described.

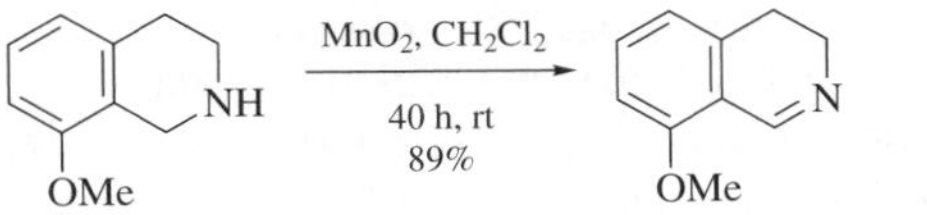
(44)

N-methylaniline $\xrightarrow{MnO_2,\ CHCl_3;\ 18\ h,\ rt;\ 83\%}$ N-formylaniline (45)

Ph–N(Me)–CH_2CH_2–N(Me)–Ph $\xrightarrow{MnO_2,\ CHCl_3;\ 18\ h,\ rt;\ 75\%}$ Ph–N(Me)–CHO (46)

Cl–C_6H_4–NH_2 $\xrightarrow{MnO_2,\ pet.\ ether;\ 6\ h;\ 87\%}$ Cl–C_6H_4–N=N–C_6H_4–Cl (47)

Miscellaneous Reactions.[13p,14] MnO_2 has also been used to perform various oxidation reactions: the oxidative cleavage of α-hydroxy acids (eq 48),[13p] the oxidative dimerization of diarylmethanes (eq 49),[14a] or their conversion to diaryl ketones (eq 50),[14a] the oxidation of aldehydes to carboxylic acids,[13p] the preparation of disulfides from thiols,[14b] of phosphine oxides from phosphines,[13p] or of ketones from amines.[14c]

$Ph_2C(OH)CO_2H$ $\xrightarrow{MnO_2,\ H_2O;\ 30\ min;\ 52\%}$ PhCOPh + CO_2 (48)

Ph_2CH_2 $\xrightarrow{MnO_2,\ C_6H_6;\ 71\ min,\ 211\ °C;\ 81\%}$ $Ph_2CH–CHPh_2$ (49)

Ph_2CH_2 $\xrightarrow{MnO_2;\ 6\ h,\ 125\ °C;\ 74\%}$ $Ph_2C{=}O$ (50)

Related Reagents. See Classes O-1, O-3, O-10, and O-12, pages 1–10. Barium Manganate; Nickel(II) Peroxide.

1. (a) Fatiadi, A. J. *S* **1976**, 65 and 133. (b) Pickering, W. F. *Rev. Pure Appl. Chem.* **1966**, *16*, 185. (c) Evans, R. M. *QR* **1959**, *13*, 61. (d) Hudlicky, M. *Oxidations in Organic Chemistry*; American Chemical Society: Washington, 1990. (e) Fatiadi, A. J. In *Organic Synthesis by Oxidation With Metal Compounds*; Plenum: New York, 1986; Chapter 3.
2. (a) Ball, S.; Goodwin, T. W.; Morton, R. A. *BJ* **1948**, *42*, 516. (b) Corey, E. J.; Gilman, N. W.; Ganem, B. E. *JACS* **1968**, *90*, 5616. (c) Bharucha, K. R. *JCS* **1956**, 2446. (d) Nickon, A.; Bagli, J. F. *JACS* **1961**, *83*, 1498. (e) Fales, H. M.; Wildman, W. C. *JOC* **1961**, *26*, 881. (f) Cresp, T. M.; Sondheimer, F. *JACS* **1975**, *97*, 4412. (g) Babler, J. H.; Martin, M. J. *JOC* **1977**, *42*, 1799. (h) Counsell, R. E.; Klimstra, P. D.; Colton, F. B. *JOC* **1962**, *27*, 248. (i) Trost, B. M.; Kunz, R. A. *JACS* **1975**, *97*, 7152. (j) Sargent, L. J.; Weiss, U. *JOC* **1960**, *25*, 987. (k) Hendrickson, J. B.; Palumbo, P. S. *JOC* **1985**, *50*, 2110. (l) Saucy, G.; Borer, R. *HCA* **1971**, *54*, 2121.
3. Gilman, N. W. *CC* **1971**, 733.
4. (a) Struve, G.; Seltzer, S. *JOC* **1982**, *47*, 2109. (b) Van Amsterdam, L. J. P.; Lugtenburg, J. *CC* **1982**, 946. (c) Bentley, R. K.; Jones, E. R. H.; Thaller, V. *JCS(C)* **1969**, 1096. (d) Makin, S. M.; Ismail, A. A.; Yastrebov, V. V.; Petrv, K. I. *ZOR* **1971**, *7*, 2120 (*CA* **1972**, *76*, 13 712).
5. (a) Highet, R. J.; Wildman, W. C. *JACS* **1955**, *77*, 4399. (b) Hänsel, R.; Su, T. L.; Schulz, J. *CB* **1977**, *110*, 3664. (c) Trost, B. M.; Caldwell, C. G.; Murayama, E.; Heissler, D. *JOC* **1983**, *48*, 3252. (d) König, B. M.; Friedrichsen, W. *TL* **1987**, *28*, 4279. (e) Barrelle, M.; Glenat, R. *BSF(2)* **1967**, 453. (f) Loozen, H. J. J.; Godefroi, E. F. *JOC* **1973**, *38*, 3495.
6. (a) Crombie, L.; Crossley, J. *JCS* **1963**, 4983. (b) Harrison, I. T. *Proc. Chem. Soc. London* **1964**, 110.
7. Ohloff, G.; Giersch, W. *AG* **1973**, *85*, 401.
8. (a) Liu, K. T.; Shih, M. H.; Huang, H. W.; Hu, C. J. *S* **1988**, 715. (b) Taylor, E. C.; Maryanoff, C. A.; Skotnicki, J. S. *JOC* **1980**, *45*, 2512.
9. (a) Mashraqui, S.; Keehn, P. *SC* **1982**, *12*, 637. (b) Hamada, Y.; Shibata, M.; Sugiura, T.; Kato, S.; Shioiri, T. *JOC* **1987**, *52*, 1252. (c) Bhatnagar, I.; George, M. V. *T* **1968**, *24*, 1293.
10. Kashdan, D. S.; Schwartz, J. A.; Rapoport, H. *JOC* **1982**, *47*, 2638.
11. *Comprehensive Inorganic Chemistry*; Bailar, J. C.; Trotman-Dickenson, A. F., Eds.; Pergamon: Oxford, 1973; Vol. 3, p 801.
12. Fatiadi, A. J. *JCS(B)* **1971**, 889.
13. (a) Attenburrow, J.; Cameron, A. F. B.; Chapman, J. H.; Evans, R. M.; Hems, B. A.; Jansen, A. B. A.; Walker, T. *JCS* **1952**, 1094. (b) Mancera, O.; Rosenkranz, G.; Sondheimer, F. *JCS* **1953**, 2189. (c) Henbest, H. B.; Jones, E. R. H.; Owen, T. C. *JCS* **1957**, 4909. (d) Gritter, R. J.; Wallace, T. J. *JOC* **1959**, *24*, 1051. (e) Pratt, E. F.; Van de Castle, J. F. *JOC* **1961**, *26*, 2973. (f) Harfenist, M.; Bavley, A.; Lazier, W. A. *JOC* **1954**, *19*, 1608. (g) Henbest, H. B.; Thomas, A. *JCS* **1957**, 3032. (h) Cohen, N.; Banner, B. L.; Blount, J. F.; Tsai, M.; Saucy, G. *JOC* **1973**, *38*, 3229. (i) Vereshchagin, L. I.; Gainulina, S. R.; Podskrebysheva, L. A.; Gaivoronskii, L. A.; Okhapkina, L. L.; Vorob'eva, V. G.; Latyshev, V. P. *JOU* **1972**, *8*, 1143. (j) Barco, A.; Benetti, S.; Pollini, G. P.; Baraldi, P. G. *S* **1977**, 837. (k) Mattocks, A. R. *JCR(S)* **1977**, 40. (l) Goldman, I. M. *JOC* **1969**, *34*, 1979. (m) Carpino, L. A. *JOC* **1970**, *35*, 3971. (n) Liu, K. T.; Shih, M. H.; Huang, H. W.; Hu, C. J. *S* **1988**, 715. (o) Kimura, T.; Fujita, M.; Ando, T. *CL* **1988**, 1387. (p) Barakat, M. Z.; Abdel-Wahab, M. F.; El-Sadr, M. M. *JCS* **1956**, 4685. (q) Sondheimer, F.; Amendolla, C.; Rozenkranz, G. *JACS* **1953**, *75*, 5930 and 5932.
14. (a) Pratt, E. F.; Suskind, S. P. *JOC* **1963**, *28*, 638. (b) Papadopoulos, E. P.; Jarrar, A.; Issidorides, C. H. *JOC* **1966**, *31*, 615. (c) Curragh, E. F.; Henbest, H. B.; Thomas, A. *JCS* **1960**, 3559.

Gérard Cahiez & Mouâd Alami
Université Pierre & Marie Curie, Paris, France

Mercury(II) Oxide[1]

HgO

[21908-53-2] HgO (MW 216.59)

(promotion of thio-Claisen rearrangement;[2] deprotection of dithianes, dithiolanes,[3] thioorthoesters;[4] synthesis of tetrahydrofurans;[5] oxidizing agent[6–18])

Alternate Name: mercuric oxide.
Physical Data: mp 500 °C (dec); *d* 11.14 g cm^{-3}.
Solubility: sol dilute HCl, HNO_3; insol water, ethanol.
Form Supplied in: commercially available in both yellow and red crystalline forms.
Handling, Storage, and Precautions: highly toxic; oxidizer; protect from light.

Conversion of Ketones into Unsaturated Aldehydes. Mercury(II) oxide promotes the Wittig–thio-Claisen rearrangement sequence which converts ketones into unsaturated aldehydes (eq 1).[2] The reaction proceeds via formation of an allyl vinyl sulfide followed by rearrangement to form the α-allyl aldehyde.

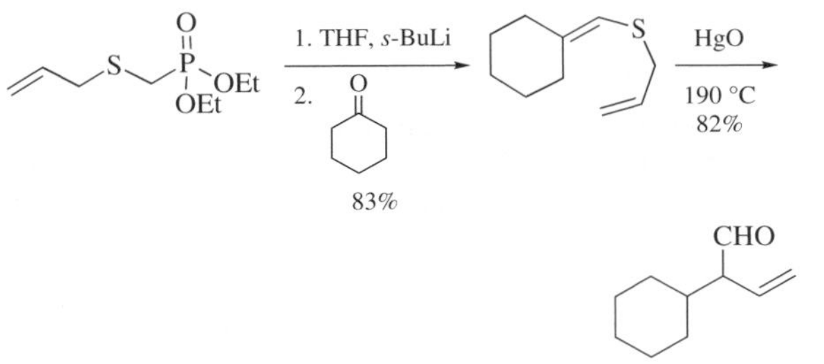

(1)

Preparation of Aldehydes and Ketones by Hydrolysis of Sulfur-Containing Compounds. A mixture of mercury(II) oxide/35% aqueous ***Tetrafluoroboric Acid***/THF can be used for the hydrolysis of 1,3-benzodithioles (eq 2), 1,3-benzoxathioles (eq 3), 1,3-oxathiolanes (eq 4), 1,3-dithiolanes (eq 5), and 1,3-dithianes (eq 6).[3] These reaction conditions offer an alternative to those reported by Vedejs and Fuchs[19] which use a mixture of mercury(II) oxide/***Boron Trifluoride Etherate***/THF.

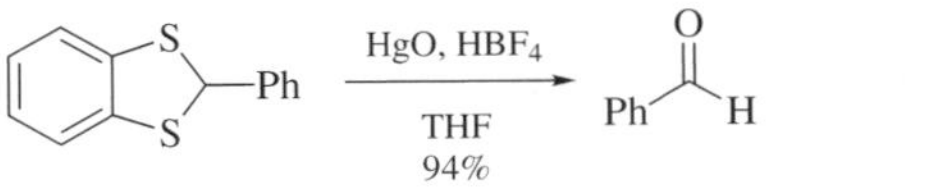

(2)

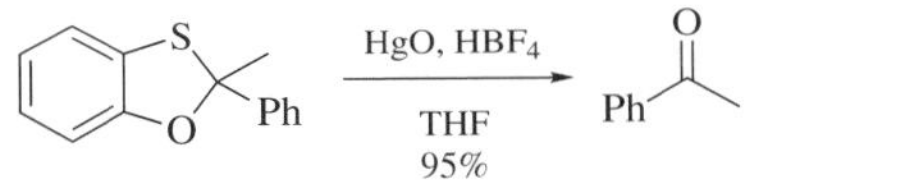

(3)

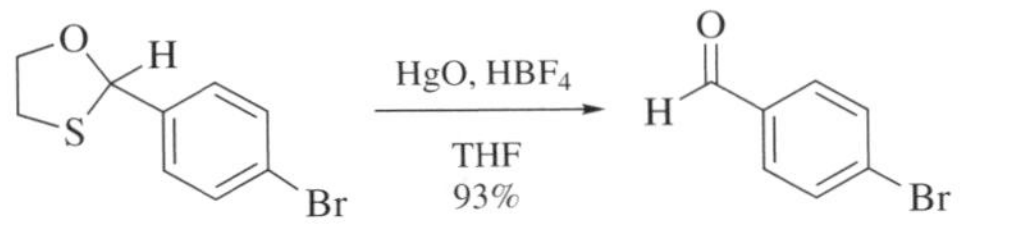

(4)

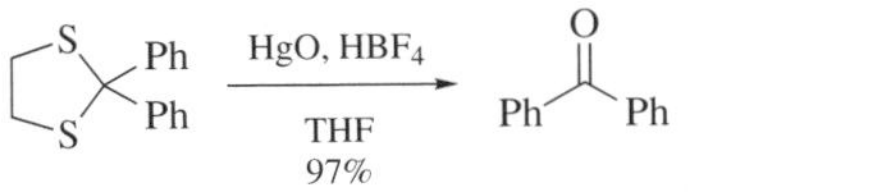

(5)

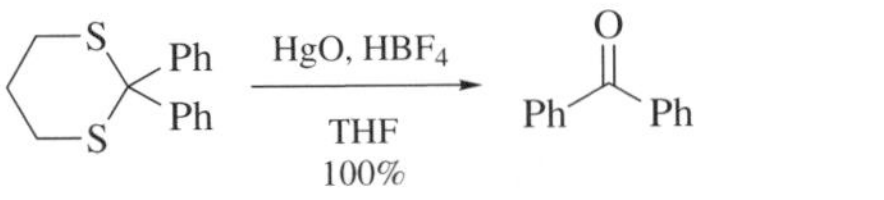

(6)

Synthesis of Esters by Solvolysis of α-Hydroxy Thioorthoesters. The use of mercury(II) oxide/50% aqueous tetrafluoroboric acid/THF and an alcohol is a fast and mild procedure for the hydrolysis of α-hydroxy thioorthoesters to the corresponding ester (eq 7).[4]

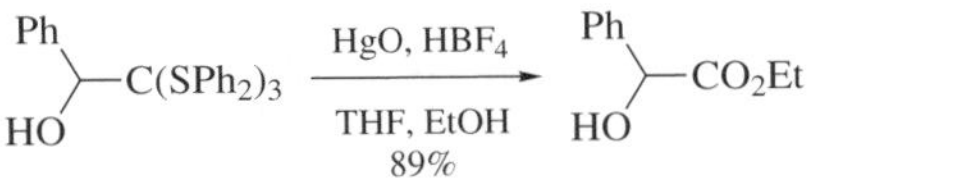

(7)

Cyclization of Primary and Secondary Alcohols to Yield Tetrahydrofurans. Primary and secondary unbranched aliphatic alcohols are readily converted into the corresponding tetrahydrofuran derivatives using mercury(II) oxide and ***Iodine*** in carbon tetrachloride (eqs 8 and 9).[5]

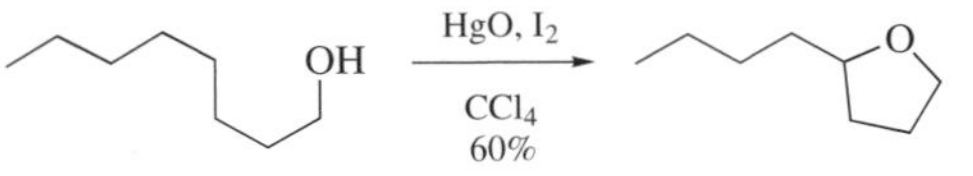

(8)

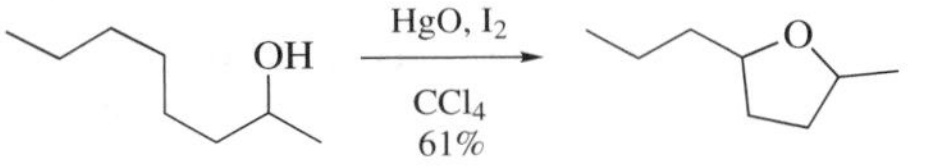

(9)

Synthesis of β-Substituted Polynitroalkyl Vinyl Ethers. The inherent instability of β-substituted polynitroaliphatic alcohols in acidic and basic media precludes the use of traditional vinyl ether synthesis. The use of mercury(II) oxide/***Trifluoroacetic Acid*** in refluxing dichloromethane permits a one-step, high-yielding synthesis of β-substituted polynitroalkyl ethers (eq 10).[6] The use of mercury(II) oxide alone produces the product in 20–30% yield, whereas the addition of TFA as cocatalyst substantially increases the yield of the desired product.

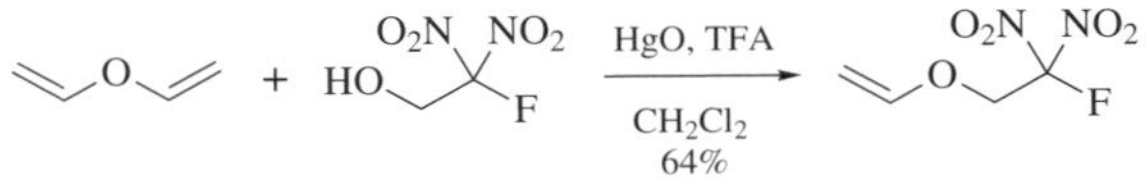

(10)

Synthesis of *trans*-Cinnamyl Ethers. Oxidation of allylbenzene using mercury(II) oxide/tetrafluoroboric acid and alcohols in THF forms exclusively *trans*-cinnamyl ethers (eq 11).[7]

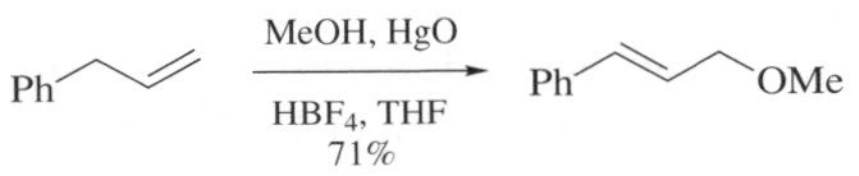

(11)

Diamination of Alkenes. The mercuration of alkenes using mercury(II) oxide and tetrafluoroboric acid in the presence of excess amine leads to the formation of 1,2-diamines (eq 12).[8] Presumably, the reaction takes place via formation of a β-aminomercury(II) tetrafluoroborate which undergoes nucleophilic attack of an amine.

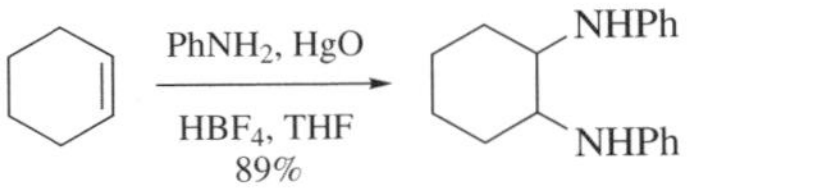

(12)

Hydroxy(alkoxy)phenylamination of Alkenes. A one-pot procedure for hydroxy(alkoxy)phenylamination of alkenes has been developed using mercury(II) oxide and tetrafluoroboric acid (eq 13).[9] The reaction proceeds through the formation of the aminomercurial from the alkene followed by reaction with water or alcohol. The ratio of products depends on the substitution pattern on the alkene. For terminal alkenes, the rearranged isomer predominates and cyclic alkenes afford *trans* products (eq 14).[9]

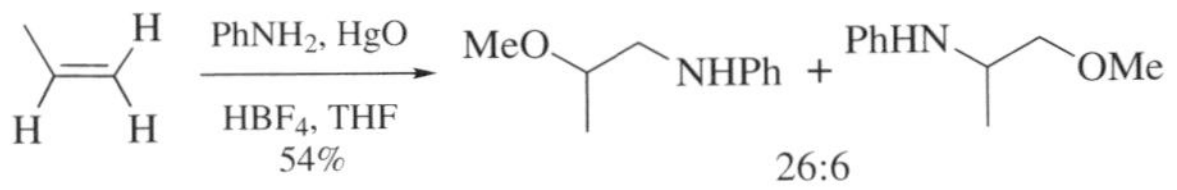

(13)

$$\text{cyclohexene} \xrightarrow[\text{HBF}_4,\ \text{PhNH}_2,\ 68\%]{\text{MeOH, HgO}} \text{trans-1,2-bis(NHPh)cyclohexane} \quad (14)$$

Cyclization of Alkynecarboxylic Acids. A variety of γ-methylenebutyrolactones have been successfully synthesized by cyclization of the appropriate alkynecarboxylic acid with mercury(II) oxide as catalyst (eqs 15 and 16).[10] In some instances the reaction must be run neat; in other cases, solvents such as chloroform, acetone, benzene, dioxane and even refluxing DMF are necessary.

$$\text{HC≡C(CH}_2)_2\text{CO}_2\text{H} \xrightarrow[100\%]{\text{HgO}} \text{γ-methylenebutyrolactone} \quad (15)$$

$$\text{PhC≡C(CH}_2)_2\text{CO}_2\text{H} \xrightarrow[\text{DMF, } 83\%]{\text{HgO}} \text{γ-benzylidenebutyrolactone} \quad (16)$$

In a similar approach, mercury(II) oxide has also been used to catalyze the cyclization of γ-allenic acids into γ-ethylenic δ-lactones (eq 17).[11]

$$\text{allenic acid (HO, O)} \xrightarrow[\text{benzene, } 100\%]{\text{HgO}} \text{δ-lactone} \quad (17)$$

Preparation of Alkyl Halides. Alkanes react with a mixture of ***Bromine*** and mercury(II) oxide to afford the corresponding alkyl bromide in good yield (eqs 18 and 19).[12] This combination of reagents is more reactive than bromine or ***N-Bromosuccinimide***. The mechanism has been postulated to proceed through the formation of bromine monoxide in situ as the brominating agent.[12] Primary and secondary alkyl bromides are readily formed, but tertiary bromides and benzylic bromides are rather problematic.

$$\text{cyclohexane} \xrightarrow[\text{Br}_2,\ 80\%]{\text{HgO}} \text{bromocyclohexane} \quad (18)$$

$$CH_2Cl_2 \xrightarrow[\text{Br}_2,\ 80\%]{\text{HgO}} CHBrCl_2 \quad (19)$$

Alkyl and aryl halides can also be synthesized using the Cristol–Firth modification of the Hunsdiecker reaction (eqs 20 and 21).[13,14] Even bridgehead carboxylic acids can be transformed into the corresponding bromide using mercury(II) oxide and bromine. Aromatic acids form insoluble mercury(II) salts which normally lead to lower yields, but it has been found that irradiation of the reaction with a 100 W bulb affords excellent yields.[14] Interestingly, irradiation of reactions of aliphatic acids does not improve the yield and, in some cases, even produces lower yields.

$$\text{CH}_3(\text{CH}_2)_4\text{CO}_2\text{H} \xrightarrow[\text{CCl}_4,\ 63\%]{\text{HgO, Br}_2} \text{CH}_3(\text{CH}_2)_3\text{Br} \quad (20)$$

$$O_2N\text{–C}_6\text{H}_4\text{–CO}_2\text{H} \xrightarrow{\text{HgO, Br}_2\ 42\%;\ \text{HgO, Br}_2\text{, light } 95\%} O_2N\text{–C}_6\text{H}_4\text{–Br} \quad (21)$$

Decarboxylation of bridgehead carboxylic acids has been accomplished in two steps. The bromide is formed first followed by reduction with ***Tri-n-butylstannane***. The deuterated analog can be made using tributyltin deuteride (eq 22).[15]

$$\text{norbornane-1-CO}_2\text{H} \xrightarrow[\text{Br}_2,\ 70\text{–}80\%]{\text{CH}_2\text{Br}_2,\ \text{HgO}} \text{1-bromonorbornane} \begin{cases} \xrightarrow{\text{Bu}_3\text{SnH, AIBN, } 95\%} \text{norbornane} \\ \xrightarrow{\text{Bu}_3\text{SnD, AIBN, } 93\%} \text{1-D-norbornane} \end{cases} \quad (22)$$

Oxidation of Phenols and Hydroquinones. In most cases, mercury(II) oxide can be used instead of ***Mercury(II) Trifluoroacetate*** to oxidize phenols and hydroquinones (eqs 23 and 24).[16] These conditions are attractive in that they do not require the use of an acid scavenger to neutralize the trifluoroacetic acid that is formed during the course of the reaction.

$$\text{tetramethylhydroquinone} \xrightarrow{\text{Hg(OCOCF}_3)_2\text{, MeOH, polyvinylpyridine } 84\%;\ \text{HgO, MeOH } 94\%} \text{tetramethyl-1,4-benzoquinone} \quad (23)$$

$$\text{2,6-di-}t\text{-Bu-4-OMe-phenol} \xrightarrow{\text{Hg(OCOCF}_3)_2\text{, MeOH, polyvinylpyridine } 93\%;\ \text{HgO, MeOH } 90\%} \text{2,6-di-}t\text{-Bu-1,4-benzoquinone} \quad (24)$$

Oxidation of Hydrazones. Aliphatic and aromatic hydrazones are oxidized with mercury(II) oxide to afford the corresponding diazo compound (eqs 25–27).[17] Reaction of the monohydrazones derived from α-diketones give α-diazo ketones, which, upon heating, form ketenes (eq 28).[17] Treatment of the bis-hydrazones of α-diketones with mercury(II) oxide at higher temperatures affords alkynes (eq 29).[17] Seven- and eight-membered cycloalkynes as well as even larger cycloalkynes have been synthesized using this methodology (eq 30).[17]

$$H_2C{=}NNH_2 \xrightarrow[\text{Et}_2\text{O, EtOH, } 70\text{–}90\%]{\text{HgO, KOH}} H_2C{=}N{=}N \quad (25)$$

$$Ph_2C{=}NNH_2 \xrightarrow[\text{Et}_2\text{O, EtOH, } 89\%]{\text{HgO, KOH, Na}_2\text{SO}_4} Ph_2C{=}N{=}N \quad (26)$$

$$\text{9-hydrazonofluorene} \xrightarrow[\text{Et}_2\text{O, EtOH, 99\%}]{\text{HgO, KOH}} \text{9-diazofluorene} \quad (27)$$

$$\text{PhC(=NNH}_2\text{)COPh} \xrightarrow[\text{Et}_2\text{O, EtOH, 89\%}]{\text{HgO, KOH, Na}_2\text{SO}_4} \text{PhC(=N}_2\text{)COPh} \xrightarrow[77\%]{\text{heat}} \text{Ph}_2\text{C=}\bullet\text{=O} \quad (28)$$

$$\text{PhC(=NNH}_2\text{)C(=NNH}_2\text{)Ph} \xrightarrow[\text{reflux, 67–73\%}]{\text{HgO, benzene}} \text{Ph—≡—Ph} \quad (29)$$

$$\text{cyclodecane-1,2-dione bis(hydrazone)} \xrightarrow[\text{toluene, 67\%}]{\text{HgO, KOH, Na}_2\text{SO}_4} \text{cyclodecyne} \quad (30)$$

Hydration of Alkynes. A mixture of mercury(II) oxide and ***Sulfuric Acid*** effects hydration of alkynes to form the corresponding methyl ketone. This protocol was utilized in the synthesis of some daunomycinone analogs (eq 31).[18]

$$\text{ethynyl tetracyclic (OH, O-}t\text{-Bu)} \xrightarrow[\text{acetone, 85\%}]{\text{HgO, H}_2\text{SO}_4} \text{acetyl tetracyclic (OH, O-}t\text{-Bu)} \quad (31)$$

Related Reagents. See Classes O-5 and O-22, pages 1–10. Mercury(II) Oxide–Bromine; Mercury(II) Oxide–Iodine; Mercury(II) Oxide–Tetrafluoroboric Acid.

1. Pizey, J. S. *Synthetic Reagents*; Wiley: New York, 1974; Vol. 1.
2. Corey, E. J.; Shulman, J. I. *JACS* **1970**, *92*, 5522.
3. Degani, I.; Fochi, R.; Regondi, V. *S* **1981**, 51.
4. Scholz, D. *SC* **1982**, *12*, 527.
5. Mihailović, M. L.; Čekovičc, Z.; Stanković, J. *CC* **1969**, 981.
6. Shackelford, S. A.; McGuire, R. R.; Cochoy, R. E. *JOC* **1992**, *57*, 2950.
7. Barluenga, J.; Alonso-Cires, L.; Asensio, G. *TL* **1981**, *22*, 2239.
8. Barluenga, J.; Alonso-Cires, L.; Asensio, G. *S* **1979**, 962.
9. Barluenga, J.; Alonso-Cires, L.; Asensio, G. *S* **1981**, 376.
10. Yamamoto, M. *CC* **1978**, 649.
11. Jellal, A.; Grimaldi, J.; Santelli, M. *TL* **1984**, *25*, 3179.
12. Bunce, N. J. *CJC* **1972**, *50*, 3109.
13. Bunce, N. J. *JOC* **1972**, *37*, 664.
14. Meyers, A. I.; Fleming, M. P. *JOC* **1979**, *44*, 3405.
15. Della, E. W.; Patney, H. K. *S* **1976**, 251.
16. McKillop, A.; Young, D. W. *SC* **1977**, *7*, 467.
17. Hudlicky, M. *Oxidations in Organic Chemistry*; American Chemical Society: Washington, 1990.
18. Potman, R. P.; Janssen, N. J. M. L.; Scheeren, J. W.; Nivard, R. J. F. *JOC* **1984**, *49*, 3628.
19. Vedejs, E.; Fuchs, P. L. *JOC* **1971**, *36*, 366.

Ellen M. Leahy
Affymax Research Institute, Palo Alto, CA, USA

N-Methylmorpholine *N*-Oxide[1]

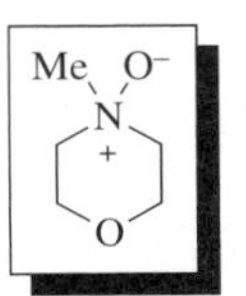

[7529-22-8] $C_5H_{11}NO_2$ (MW 117.17)

(used as a co-oxidant with OsO_4[2] and ruthenates;[3] acts a mild oxidizing agent[4])

Alternate Name: NMO.
Physical Data: mp 75–76 °C (as monohydrate)
Solubility: sol water, acetone, alcohol, ether.
Form Supplied in: colorless deliquescent crystals; available commercially.
Purification: crystallize from ethanol.
Handling, Storage, and Precautions: refrigerate; hygroscopic; irritant; handle in a fume hood using gloves; avoid skin contact and inhalation; in case of contact, rinse immediately with water.

Co-oxidation Reagent. *N*-Methylmorpholine *N*-oxide is commonly used as a mild co-oxidant. The stoichiometric use of ***Osmium Tetroxide*** to *cis*-dihydroxylate alkenes is avoided by using NMO as a reoxidant (eq 1).[1,2,5] This reagent is preferred over other *cis*-dihydroxylation reagents such as ***Potassium Permanganate***, ***Iodine–Silver Acetate***, or ***Osmium Tetroxide*** in the presence of barium perchlorate or ***Hydrogen Peroxide***, since these reagents give poor yields, require cumbersome workups, or are prohibitively expensive for large-scale glycolization reactions.[2,5] See also ***Osmium Tetroxide–N-Methylmorpholine N-Oxide***.

$$\text{cyclooctene} \xrightarrow[\text{2. Na}_2\text{S}_2\text{O}_4\text{, 84–89\%}]{\text{1. OsO}_4\text{, NMO, aq acetone}} \text{cis-cyclooctane-1,2-diol} \quad (1)$$

Oxidation of allylic alcohols using NMO/OsO_4 occurs with high stereoselectivity. The relative stereochemistry of the incoming hydroxy group is *erythro* to the hydroxy, alkoxy, or methyl group in the allylic position of the product (eq 2).[6] Similar results are obtained using this procedure to oxidize higher-carbon sugars containing allylic alcohols.[7]

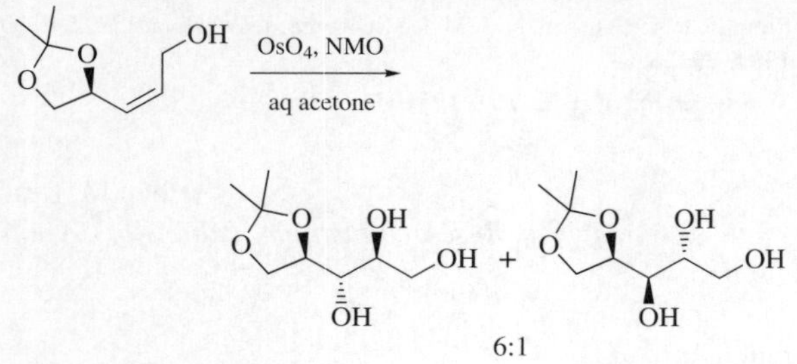

(2)

Asymmetric dihydroxylation is accomplished by using NMO in the presence of OsO_4 and dihydroquinidine *p*-chlorobenzoate (eq 3).[8] This procedure works well for a variety of alkenes but gives the best results for those alkenes with aromatic directing groups. The method is insensitive to scale, moisture, and can be run with very low concentrations of OsO_4.

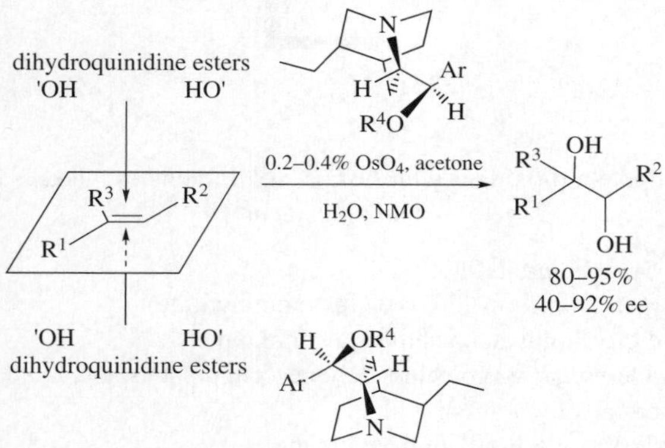

(3)

A kinetic study using ***Trimethylamine N-Oxide*** in place of NMO as a reoxidant has been reported.[9] Crystallographic analysis of an isolated intermediate using NMO (eq 4) has provided mechanistic evidence for osmate ester intermediates in tertiary amine *N*-oxide oxidation systems.[1,10]

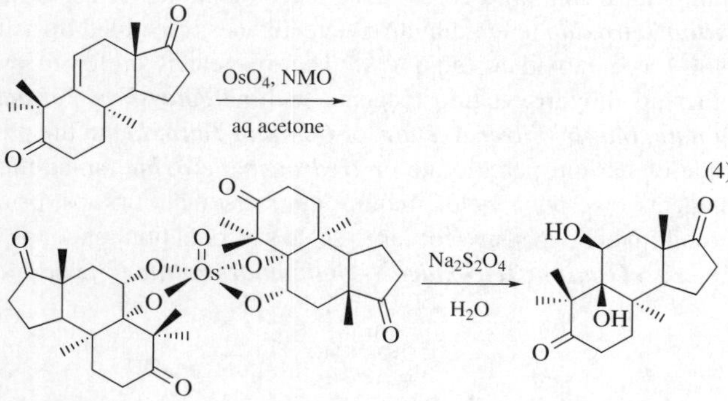

(4)

Intermediate *cis*-diols generated by NMO/OsO_4 oxidation can be cleaved to give the corresponding dialdehydes by periodate oxidation (eq 5).[11]

α-Hydroxyaldehydes and their tautomeric hydroxymethyl ketones can be conveniently prepared using NMO/OsO_4 or $KMnO_4$ to *cis*-hydroxylate enol phosphonates (eq 6).[12] The oxidation step is accomplished in high yields and the overall synthesis requires only three steps from a ketone.

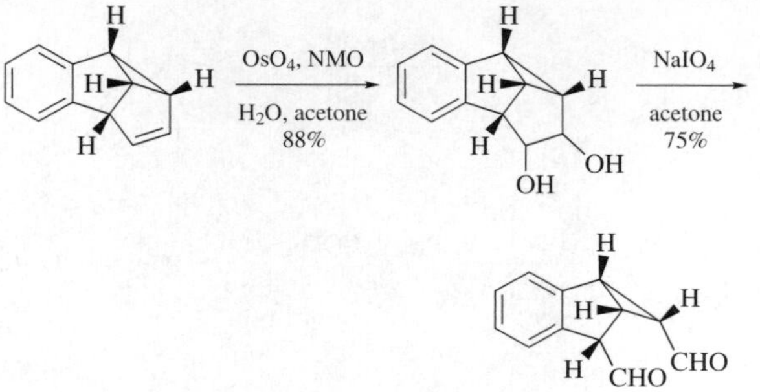

(5)

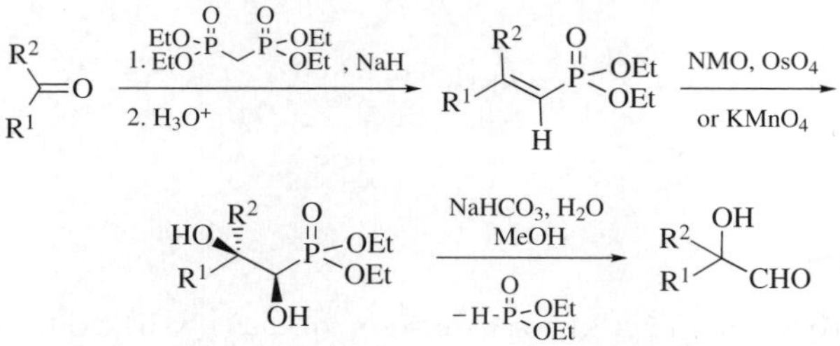

(6)

The NMO/OsO_4 oxidation system efficiently oxidizes terminal alkenes. However, this reagent may be replaced in some cases by ***Potassium Ferricyanide***/OsO_4 to give comparable yields of the corresponding 1,2-diol (eq 7).[13]

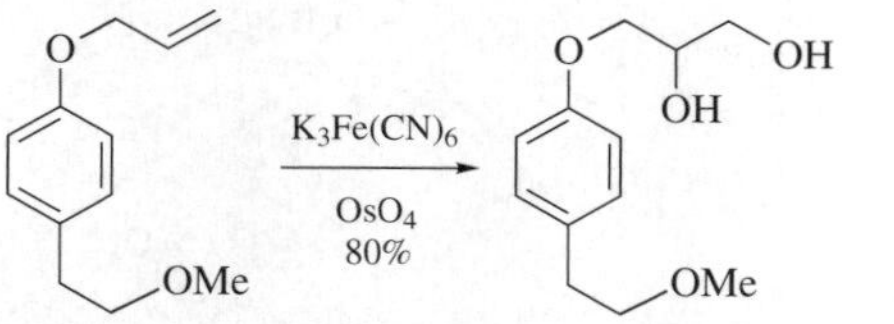

(7)

Besides OsO_4, other oxidants can be used catalytically in the presence of NMO. For example, epoxides are left intact when oxidizing a primary alcohol with NMO in the presence of catalytic ***Tetra-n-propylammonium Perruthenate*** (TPAP) (eq 8) and when oxidizing a secondary alcohol using catalytic tetra-*n*-butylammonium perruthenate (TBAP) (eq 9).[3] This reagent will also oxidize 1,4- and 1,5-primary/secondary diols to lactones.[14]

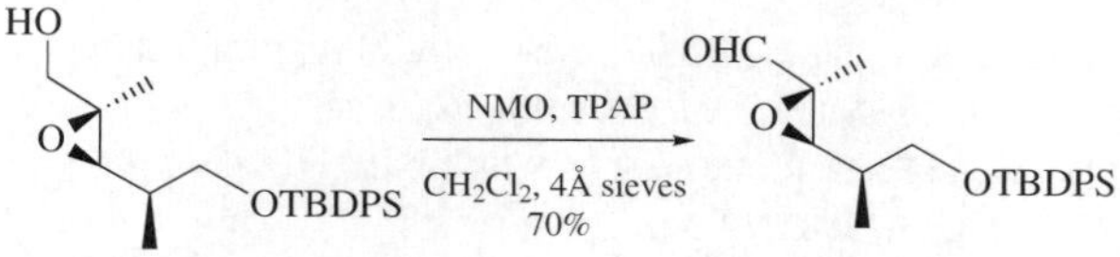

(8)

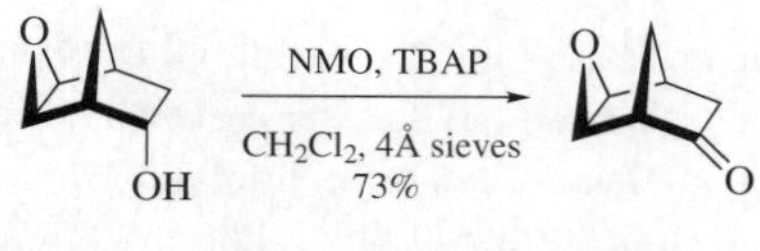

(9)

NMO along with a catalytic amount of $RuCl_2(PPh_3)_3$ will oxidize alcohols to aldehydes and ketones; $RuCl_2(PPh_3)_3$/NMO efficiently oxidizes (+)-carveol to (+)-carvone in 94% yield at rt after 2 h.[15] This reagent system will also oxidize sulfides and phosphines to their respective oxides.[16]

Mild Oxidant. Activated halides can efficiently be converted to aldehydes or ketones under mild conditions using NMO. For example, cinnamyl bromide is converted to cinnamaldehyde by stirring the activated halide with NMO in acetonitrile.[4] This reagent will oxidize a halide in the presence of a double bond (eq 10).

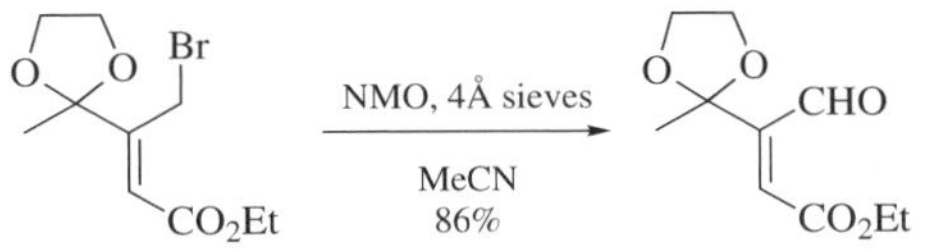

NMO, 4Å sieves; MeCN; 86% (10)

NMO can also be used as a reoxidant in the Pauson–Khand reaction for the synthesis of cyclopentenones.[17] This reaction typically is run at elevated temperatures (60–110 °C) or with ultrasound at 45 °C. Milder conditions are used with NMO, thus allowing for higher stereoselectivity in cycloadditions involving dicobalt hexacarbonyl complexes of alkynes (eq 11).

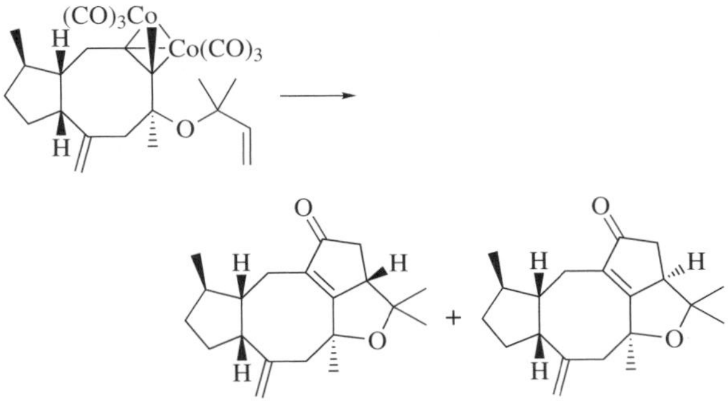

(11)

Conditions	Yield (%)	Stereoselectivity
NMO, CH_2Cl_2, rt	68	11:1
MeCN, 82 °C	75	4:1
MeCN,)))), 45 °C	45	3:1

Peptide synthesis can be performed using NMO as a mild oxidizing agent in the presence of the peptide hydrochloride, ***Diphenyl Diselenide*** and ***Tri-n-butylphosphine*** at 60 °C.[18] The phosphine and selenide act as a reductant and oxidant respectively. NMO inhibits the formation of byproducts by oxidizing selenophenol produced in the reaction mixture (eq 12) (see also ***Trimethylamine N-Oxide***).

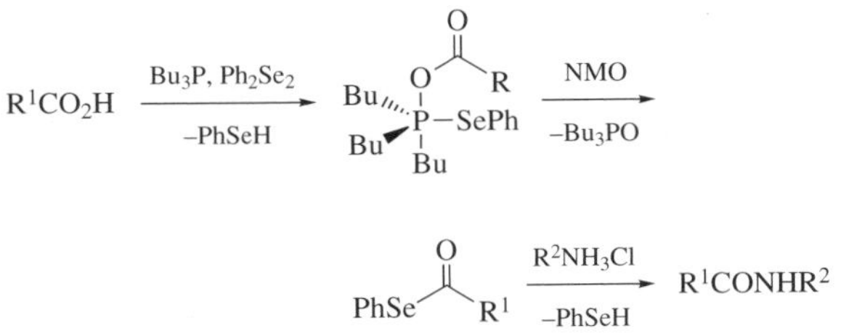

R^1CO_2H → (Bu_3P, Ph_2Se_2; −PhSeH) → (NMO; $-Bu_3PO$) → $PhSeC(O)R^1$ → (R^2NH_3Cl; −PhSeH) → R^1CONHR^2 (12)

Related Reagents. See Classes O-1, O-4, and O-16, pages 1–10.

1. Albini, A. *S* **1993** 263.
2. VanRheenen, V.; Cha, D. Y.; Hartley, W. M. *OS* **1978**, *58*, 43.
3. Griffith, W. P.; Ley, S. V.; Whitcombe, G. P.; White, A. D. *CC* **1987**, 1625.
4. Griffith, W. P.; Jolliffe, J. M.; Ley, S. V.; Springhorn, K. F.; Tiffin, P. D. *SC* **1992**, *22*, 1967.
5. VanRheenen, V.; Kelly, R. C.; Cha, D. Y. *TL* **1976**, 1973.
6. Cha, J. K.; Christ, W. J.; Kishi, Y. *TL* **1983**, *24*, 3943.
7. (a) Brimacombe, J. S.; Hanna, R.; Kabir, A. K. M. S.; Bennett, F.; Taylor, I. D. *JCS(P1)* **1986**, 815. (b) Brimacombe, J. S.; Hanna, R.; Kabir, A. K. M. S. *JCS(P1)* **1986**, 823.
8. (a) McKee, B. H.; Gilheany, D. G.; Sharpless, K. B. *OS* **1992**, *70*, 47. (b) Jacobsen, E. N.; Markó, I.; Mungall, W. S.; Schröder, G.; Sharpless, K. B. *JACS* **1988**, *110*, 1968.
9. Erdik, E.; Matteson, D. S. *JOC* **1989**, *54*, 2742.
10. Sivik, M. R.; Gallucci, J. C.; Paquette, L. A. *JOC* **1990**, *55*, 391.
11. Reedich, D. E.; Sheridan, R. S. *JOC* **1985**, *50*, 3535.
12. Waszkuć, W.; Janecki, T.; Bodalski, R. *SC* **1984**, 1025.
13. (a) Minato, M.; Yamamoto, K.; Tsuji, J. *JOC* **1990**, *55*, 766. (b) Gurjar, M. K.; Joshi, S. V.; Sastry, B. S.; Rama Rao, A. V. *SC* **1990**, *20*, 3489.
14. Bloch, R.; Brillet, C. *SL* **1991**, 829.
15. Sharpless, K. B.; Akashi, K.; Oshima, K. *TL* **1976**, 2503.
16. Caroling, G.; Rajaram, J.; Kuriacose, J. C. *JIC* **1989**, *66*, 632.
17. Shambayati, S.; Crowe, W. E.; Schreiber, S. L. *TL* **1990**, *31*, 5289.
18. Singh, U.; Ghosh, S. K.; Chadha, M. S.; Mamdapur, V. R. *TL* **1991**, *32*, 255.

Mark R. Sivik
Procter & Gamble, Cincinnati, OH, USA

Monoperoxyphthalic Acid[1]

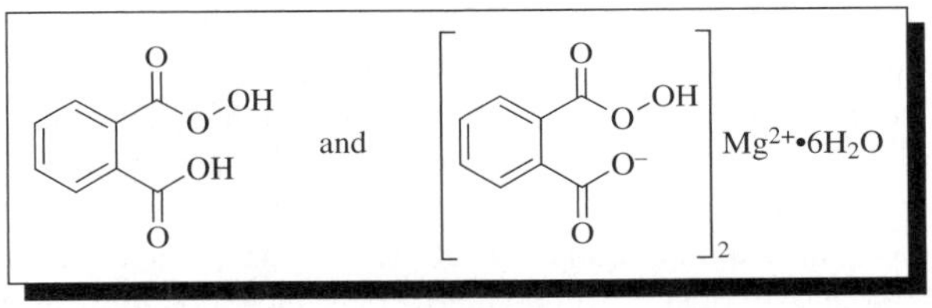

and $Mg^{2+}\cdot 6H_2O$

[2311-91-3]	$C_8H_6O_5$	(MW 182.14)
(Mg salt·$6H_2O$) *[84665-66-7]*	$C_{16}H_{22}MgO_{16}$	(MW 494.69)

(mild peroxycarboxylic acids for effecting epoxidation of π-bonds, Baeyer–Villiger reactions, and the oxidation of diverse heteroatoms including nitrogen, sulfur, and selenium)

Alternate Name: MPP.

Physical Data: monoperoxyphthalic acid: mp 110–112 °C (dec).[2]

Solubility: monoperoxyphthalic acid: sol ether; magnesium monoperoxyphthalate: sol water, low molecular weight alcohols; very low sol organic solvents such as dichloromethane.

Form Supplied in: magnesium monoperoxyphthalate: white crystalline powder; commercially available as the hexahydrate.

Preparative Methods: monoperoxyphthalic acid is prepared as required by the addition of phthalic anhydride to 30% aqueous ***Hydrogen Peroxide***,[3a] in the presence of ***Sodium Hydroxide***[3b] or ***Sodium Carbonate***.[3c] Magnesium monoperoxyphthalate is prepared by the addition of phthalic anhydride and ***Magnesium Oxide*** to aqueous hydrogen peroxide.[3d]

Drying: magnesium monoperoxyphthalate should be dried over calcium chloride in a desiccator.

Handling, Storage, and Precautions: magnesium monoperoxyphthalate may be stored at room temperature. *All work should be carried in an efficient fume hood behind a polycarbonate safety screen.*

Monoperoxyphthalic Acid. Before the introduction of ***m-Chloroperbenzoic Acid*** (*m*-CPBA), one of the most frequently used peroxy acids was monoperoxyphthalic acid (MPP). Although MPP is a weaker acid than *m*-CPBA, the fact that phthalic acid is essentially insoluble in ether is an advantage with MPP. A disadvantage of the reagent is that MPP has to be made each time that it is used, whereas *m*-CPBA (60%) is commercially available admixed with *m*-chlorobenzoic acid and water. The epoxidation of a number of alkenes can be achieved in reasonable yields using MPP. Cyclohexene, limonene, and α-pinene were converted into the respective epoxides in 64, 71, and 48% yields, respectively.[3a] The oxidation of bicyclo[2.2.1]heptadiene gave, in addition to a low yield of the expected epoxide, bicyclo[2.2.1]hex-2-ene-5-carbaldehyde.[4] The epoxidation of a number of acid-sensitive alkenes, such as α-keto enol ethers, has been achieved in good yields in the presence of disodium phthalate as a buffer and is exemplified in eq 1.[5] It is of interest to note that the yields in these reactions were halved in the absence of the buffer.

OMe — MPP (1.1 equiv), $C_6H_4(CO_2)^-$ 2 Na^+, 70% → OMe, O (1)

The oxidation of indole derivatives gives rise to a number of products, including those where ring fragmentation has occurred. Thus 2-*t*-butylindole affords *N*-pivaloylanthranilic acid.[6] Monoperoxyphthalic acid has also been used to prepare *N*-oxides; for example, quinoline derivatives have been converted into their *N*-oxides.[7] Although MPP is now not used as frequently as formerly, some useful examples have been published recently. The epoxidation reactions shown in eqs 2 and 3 and the Baeyer–Villiger type rearrangement (eq 4) are illustrative.[8–10]

OAc, OMe — MPP, Et_2O, 60% → OAc, OMe (2)

MPP, Et_2O, 0 °C to rt, 13 h, 87% (3)

CF_3 — MPP, Et_2O, THF, 0 °C, 5 h, 48% → CF_3 (4)

Magnesium Monoperoxyphthalate. The production of magnesium monoperoxyphthalate (MMPP) in commercial quantities has led to its use in a number of different types of oxidation reactions since its introduction.[11] The use of MMPP has been reviewed in detail to early 1993.[1c] The different reaction types are outlined and exemplified in the following sections.

Epoxidation Reactions. The majority of the successful examples of epoxidation reactions using MMPP have been carried out using water or a low molecular weight alcohol as a solvent. Where the substrate to be epoxidized is insoluble in such a solvent, water has been used together with a phase-transfer agent to take the monoperoxyphthalate ion into an organic phase, for example chloroform or dichloromethane. The formation of the monoepoxide from [$^2H_6$]buta-1,3-diene has been reported in a reaction where one atmosphere of the gas was kept over an aqueous solution of MMPP.[12] There have also been cases reported where attempted epoxidation reactions have failed in the presence of MMPP.[13] Typically these involve terminal alkenes where electron density in the double bond is relatively low. Epoxidation may then be successful using Kishi's high-temperature method in which the radical inhibitor 4,4′-thiobis(6-*t*-butyl-3-methylphenol) is employed together with *m*-CPBA.[14] An efficient alternative method for the epoxidation of terminal alkenes involves the use of ***Trifluoroperacetic Acid*** (TFPAA) generated by the interaction of urea–hydrogen peroxide (UHP) with trifluoroacetic anhydride (see ***Hydrogen Peroxide–Urea***).[15] It is of interest to note that the epoxidation of cholesterol proceeds efficiently in dichloromethane in the absence of a phase-transfer catalyst. Presumably in this case the hydroxyl group that is present in cholesterol acts as a crystal disrupting agent. This explanation may also apply to the observation that the MMPP epoxidation of *cis*-but-2-ene-1,4-diol proceeds efficiently, whereas a reaction of the diacetate does not.[16] As with all peroxycarboxylic acid epoxidations of cholesterol, a reaction using MMPP results in the formation of a mixture of products in which the α:β ratio is about 4:1. It appears that the larger the peroxycarboxylic acid, the larger the amount of the α-epoxide formed. A direct comparison with *m*-CPBA can be made in the case of the epoxidation of 1,2-dimethylcyclohexa-1,4-diene,[17,18] and also in the reactions with ***Ethyl Vinyl Ether***.[19] The oxidation shown in eq 5 gave about a 4:1 mixture of diastereomers in very good yield when carried out in a water–isopropanol mixture.[20] Hydrogen bonding between the peroxy acid and the hydroxyl group is presumed to be responsible for the preferred formation of the α-epoxide. The failure of *m*-CPBA to effect epoxidation of the diene ester shown in eq 6 is surprising. However, MMPP and TFPAA both gave the α-epoxide exclusively. On the other hand, a reaction using 3,5-dinitroperbenzoic acid gave a 3:1 mixture of the α- and β-epoxides.[21] As with epoxidation reactions carried out using *m*-CPBA, the reactions of MMPP with dienes (for example limonene) involve initial reaction at the more electron-rich double bond.

Ph, OH, TBDMSO — MMPP, *i*-PrOH, H_2O, pH 5.5, 4 h → Ph, O, OH, TBDMSO (5)

HO, OH, CO_2Et — MMPP → HO, OH, O, CO_2Et (6)

High diastereofacial selectivity can be anticipated where a neighboring functional group can be involved to control the stereochemical course of the reaction. Both of the diastereomers of the phosphine oxides shown in eq 7 are epoxidized with de values in excess of 95%.[22] Very low selectivities were observed where the chirality only involved the phosphorus center. Evidently the stereoselectivity is dominated by the effect of the chiral carbon center. Control may be less satisfactory when two functional groups have opposite influences. The epoxidation reaction shown in eq 8 was used in a study involving an enantiospecific synthesis of the nonproteinogenic α-amino acid anticapsin.[23] The desired product (**1**) was formed along with (**2**) in the ratio 5:2 using MMPP, whereas a 1:1 mixture was obtained using *m*-CPBA. Hydrogen bonding of the peroxy acid to the amide group evidently competes with the sterically demanding *t*-butyldiphenylsilyl residue in this reaction.

MMPP; *i*-PrOH, H_2O, 5 d (7)

MMPP; *i*-PrOH, H_2O (8)

The synthesis of a number of naturally occurring fused 3-methylfuran derivatives was achieved using, as a key step, the epoxidation–aromatization shown in eq 9.[24] The oxidation of a number of heteroaromatic compounds has also been studied using MMPP and some of the products clearly result from initial epoxidation. A number of furan derivatives that carry electron-releasing substituents in the 2- and 5-positions are oxidized to give acyclic dienones in reactions with either *m*-CPBA or MMPP.[25] The oxidation shown in eq 10 was used in a synthesis of bromobeckerelide,[26] and the oxidation of 1-benzenesulfonylindole using MMPP gives the indoxyl derivative shown in eq 11.[27]

MMPP; DCM, H_2O, PTC; 78% (9)

MMPP; *i*-PrOH, H_2O; 72% (10)

MMPP; AcOH, Δ, 4 h (11)

Baeyer–Villiger and Related Reactions. The ease with which Baeyer–Villiger oxidation reactions of ketones occur is related to the strength of the conjugate acid of the leaving group. Since phthalic acid is not a particularly strong acid, it is not surprising that monoperoxyphthalic acid and MMPP are not used as frequently as trifluoroperacetic acid in Baeyer–Villiger reactions. Cyclohexanone and 3,3-dimethylbutanone are converted into ε-caprolactone and *t*-butyl acetate, respectively, in good yields.[11] The oxidative rearrangement reactions of the β-lactams shown in eq 12 have been reported to give higher yields when using MMPP than using *m*-CPBA.[28] The conversion of an aromatic aldehyde into a phenol via the formate ester (the Dakin reaction) is evidently related to the Baeyer–Villiger reaction. These reactions proceed well when the aromatic aldehyde has an electron releasing substituent *ortho* or *para* to the formyl residue. A number of peroxy acids have been used, for example *m*-CPBA and ***Peracetic Acid***, including cases where the peroxy acid was generated using UHP and ***Acetic Anhydride***. A limited number of examples have been reported using MMPP; the example shown in eq 13 is illustrative.[29]

MMPP; Me_2CO, rt, 4 h (12)

MMPP; MeOH, rt, 19 h; 81% (13)

Oxidation of Sulfur and Selenium. The oxidation of sulfides to sulfoxides and sulfones can be achieved with a wide range of oxidizing agents, including various peroxycarboxylic acids. The second oxidation step is normally slower than the first and so using a controlled amount of peroxy acid normally allows the formation and isolation of either product. Thus tetrahydrothiophene can be oxidized either to the sulfoxide or to the sulfone using MMPP.[11] A number of cases have been reported where good yields of sulfoxides have been obtained,[30–32] including the example shown in eq 14 which was used in a synthesis of artemisinin.[32] The oxidation of the selenide that is involved in the sequence shown in eq 15 also proceeds in high yield.[33]

MMPP; 91% (14)

MMPP; THF, rt; 98% (15)

The use of an excess of MMPP has allowed a number of research groups to obtain sulfones from sulfides in good yields.[34–37] The oxidation of phosphothionate to phosphate has been studied using a wide variety of oxidants. It was shown that MMPP was better than *m*-CPBA in the oxidation of malathion to malaoxon.[38] The example shown in eq 16 exemplifies the successful use of a phase-transfer catalyst.[34]

TBDMSO, TBDMSO—, TBDMSO—, SPh —MMPP, PTC→ TBDMSO, TBDMSO—, TBDMSO—, S(O)(O)—Ph (16)

Oxidation of Nitrogen Functional Groups. The oxidations of a number of pyridine derivatives,[11] pyrimidines,[39] and pyridothiophenes[40] to the related *N*-oxides have been reported. The oxidation of quinoxaline (eq 17) to the di-*N*-oxide also proceeds efficiently.[29] The oxidative cleavage of suitable bicyclic isoxazolidines has been studied in connection with the synthesis of indolizidine alkaloids.[41] In the example shown in eq 18 the use of *m*-CPBA afforded a mixture of the products (**3**) and (**4**) in which (**3**) predominated. Regiochemical control was better when using MMPP and in that case the nitrone (**4**) was the only product.

MMPP, dichloroethane, 75% (17)

m-CPBA; MMPP, MeOH, 20 °C; BnO, SO_2Ph; HO, H, BnO, $PhSO_2$, (**3**); HO, H, BnO, $PhSO_2$, (**4**) (18)

A number of methods are available for the oxidative regeneration of ketones from hydrazones. Very good yields have been reported for the conversion of *N,N*-dimethyl- and SAMP-hydrazones into their precursor ketones using MMPP. In the latter case (eq 19), the stereochemical integrity of the chiral center α to the carbonyl group was maintained.[42] The oxidation of *N,N*-dimethylhydrazones derived from aldehydes has also been studied using MMPP. Nitriles are obtained in high yields in these reactions and once again (eq 20) there is no racemization of a chiral center α to the original formyl group.[43]

$C_{11}H_{23}$, Ph, OMe —MMPP, pH 7, 0 °C, 2 h, 82%→ $C_{11}H_{23}$, Ph, O >96% ee (19)

NMe_2, OBn —MMPP, 0 °C, 5 min, 99%→ N, OBn (20)

Related Reagents. See Classes O-8, O-11, O-14 and O-15, pages 1–10.

1. (a) Swern, D. In *Organic Peroxides*, Swern, D., Ed.; Wiley: New York, 1970; Vol. 1, pp 313–516. (b) Heaney, H. *Top. Curr. Chem.* **1993**, *164*, 1. (c) Heaney, H. *Aldrichim. Acta* **1993**, *26*, 35.
2. Krimm, H. U.S. Patent 2 813 896, 1957, and Ger. Patent 1 048 569, 1959, quoted in Ref. 1(a) p 426.
3. (a) Royals, E. E.; Harrell, L. L. *JACS* **1955**, *77*, 3405. (b) Böhme, H. *OSC* **1955**, *3*, 619. (c) Payne, G. B. *JOC* **1959**, *24*, 1354. (d) Hignet, G. J. Eur. Patent Appl. 27 693, 1981, (*CA* **1981**, *95*, 168 801x).
4. (a) Lumb, J. T.; Whitham, G. H. *JCS* **1964**, 1189. (b) For proof of structure, see Meinwald, J.; Labana, S. S.; Chada, M. S. *JACS* **1963**, *85*, 582.
5. Schank, K.; Felzmann, J. H.; Kratzsch, M. *CB* **1969**, *102*, 388.
6. David, S.; Monnier, J. *BSF* **1959**, 1333. Colle, M.-A.; David, S. *CR(C)* **1960**, *250*, 2226.
7. Bachman, G. B.; Cooper, D. E. *JOC* **1944**, *9*, 302.
8. Drandarov, K. *CCC* **1992**, *57*, 1111.
9. Mathies, P.; Frei, B.; Jeger, O. *HCA* **1985**, *68*, 192.
10. Singh, S. K.; Govindan, M.; Hynes, J. B. *JHC* **1990**, *27*, 2101.
11. Brougham, P.; Cooper, M. S.; Cummerson, D. A.; Heaney, H.; Thompson, N. *S* **1987**, 1015.
12. Maples, K. R.; Lane, J. L.; Dahl, A. R. *J. Labelled Comp. Radiopharm.* **1992**, *31*, 469.
13. Petter, R. C. *TL* **1989**, *30*, 399.
14. Kishi, Y.; Aratani, M.; Tanino, H.; Fukuyama, T.; Goto, T.; Inoue, S.; Sugiura, S.; Kakoi, H. *CC* **1972**, 64.
15. Cooper, M. S.; Heaney, H.; Newbold, A. J.; Sanderson, W. R. *SL* **1990**, 533.
16. Grandjean, D.; Pale, P.; Chuche, J. *TL* **1991**, *32*, 3043.
17. Gillard, J. R.; Newlands, M. J.; Bridson, J. N.; Burnell, D. J. *CJC* **1991**, *69*, 1337.
18. Paquette, L. A.; Barrett, J. H. *OSC* **1973**, *5*, 467.
19. Machida, S.; Hashimoto, Y.; Saigo, K.; Inoue, J.-y.; Hasegawa, M. *T* **1991**, *47*, 3737.
20. Sugai, T.; Noguchi, H.; Ohta, H. *Biosci. Biotech. Biochem.* **1992**, *56*, 122.
21. Forbes, J. E.; Bowden, M. C.; Pattenden, G. *JCS(P1)* **1991**, 1967.
22. Harmat, N. J. S.; Warren, S. *TL* **1990**, *31*, 2743.
23. Baldwin, J. E.; Adlington, R. M.; Mitchell, M. B. *CC* **1993**, 1332.
24. Aso, M.; Ojida, A.; Yang, G.; Cha, O.-J.; Osawa, E.; Kanematsu, K. *JOC* **1993**, *58*, 3960.
25. Domínguez, C.; Csáky, A. G.; Plumet, J. *TL* **1990**, *31*, 7669.
26. Jefford, C. W.; Jaggi, D.; Boukouvalas, J. *TL* **1989**, *30*, 1237.
27. Conway, S. C.; Gribble, G. W. *H* **1990**, *30*, 627.
28. Ricci, M.; Altamura, M.; Bianchi, D.; Cobri, W.; Gotti, N. to Farmitalia Carlo Erba, SpA, Br. Patent 2 196 340, 1988.
29. Heaney, H.; Newbold, A. J. Unpublished results.
30. Batty, D.; Crich, D.; Fortt, S. M. *JCS(P1)* **1990**, 2875.
31. Knorr, R.; Ferchland, K.; Mehlstäubl, J.; Hoang, T. P.; Böhrer, P.; Lüdemann, H.-D.; Lang, E. *CB* **1992**, *125*, 2041.
32. Avery, M. A.; Chong, W. K. M.; Jennings-White, C. *JACS* **1992**, *114*, 974.
33. Bruncko, M.; Crich, D. *TL* **1992**, *33*, 6251.
34. Crich, D.; Ritchie, T. J. *JCS(P1)* **1990**, 945.
35. Cho, I.; Choi, S. Y. *Makromol. Chem., Rapid Commun.* **1991**, *12*, 399.
36. Görlitzer, K.; Bömeke, M. *AP* **1992**, *325*, 9.

37. Siemens, L. M.; Rottnek, F. W.; Trzupek, L. S. *JOC* **1990**, *55*, 3507.
38. Jackson, J. A.; Berkman, C. E.; Thompson, C. M. *TL* **1992**, *33*, 6061.
39. Lamsa, J. (Farmos-Yhtymä Oy) Eur. Patent 0 270 201, 1987.
40. Klemm, L. H.; Wang, J.; Sur, S. K. *JHC* **1990**, *27*, 1537.
41. Holmes, A. B.; Hughes, A. B.; Smith, A. L. *SL* **1991**, 47.
42. Enders D.; Plant, A. *SL* **1990**, 725.
43. Fernández, R.; Gasch, C.; Lassaletta, J.-M.; Llera, J.-M.; Vázquez, J. *TL* **1993**, *34*, 141.

Harry Heaney
Loughborough University of Technology, UK

Nickel Boride[1]

Ni_2B

$(Ni_2B)^2$
[12007-01-1] BNi_2 (MW 128.19)

(selective hydrogenation catalyst,[1a,c,3] desulfurization catalyst;[4] reduces nitro[5] and other functional groups;[1a] dehalogenation catalyst;[1b,6] hydrogenolysis[7] catalyst)

Physical Data: mp 1230 °C.[8]
Solubility: insol aqueous base and most organic solvents; reacts with concentrated aqueous acids.
Form Supplied in: black granules, stoichiometry varies with supplier.
Preparative Methods: to a stirred suspension of 1.24 g (5 mmol) of powdered ***Nickel(II) Acetate*** in 50 mL of 95% ethanol is added 5 mL of a 1 M solution of ***Sodium Borohydride*** in 95% ethanol at room temperature (control frothing). Stirring is continued until the gas evolution ceases (usually 30 min). The flask is used directly in the hydrogenation.[1c,9a] This catalyst is nonpyrophoric.
Handling, Storage, and Precautions: caution must be taken in handling nickel salts. Ingestion of soluble nickel salts causes nausea, vomiting, and diarrhea. Nickel chloride has an LD_{50} (iv) = 40–80 mg kg^{-1} in dogs. Many nickel salts will sublime in vacuo. Nickel metal is carcinogenic and certain nickel compounds may reasonably be expected to be carcinogenic.

Catalyst Composition and Structure. The composition of the catalyst produced by the reaction of Ni^{II} salts and ***Sodium Borohydride*** is dependent on reaction conditions (solvent, stoichiometry, temperature, etc.).[9] X-ray photoelectron spectroscopy[10] showed that the main difference between the P1 form of nickel boride (P1 Ni) and the P2 form of nickel boride (P2 Ni) is the amount of $NaBO_2$ adsorbed on to the surface of the catalyst. P1 Ni (which is prepared in water) has an oxide:boride ratio of 1:4, while P2 Ni (which is prepared in ethanol) has a ratio of 10:1. Early studies of the reaction of borohydrides with transition metal salts[11] (Fe^{II}, Cu^{II}, Pd^{II}, Ni^{II}, Co^{II}, etc.) showed that the reaction product is either the metal (as in the case of Pd^{II}) or a black granular solid (as in the case of Ni^{II}); in both cases, H_2 is evolved.[11c,12] Analysis of the black solid formed from the Ni^{II} suggested the catalyst to be a boride.[11c,13] Paul et al.[11b] examined several Ni^{II} salts and found nickel acetate to be most acceptable.

Hydrogenation of Alkenes and Alkynes. Brown has described two forms of nickel boride (P1 Ni and P2 Ni)[9] which are hydrogenation catalysts. In a comparison of P1 Ni to W2 ***Raney Nickel*** (Ra Ni) as a hydrogenation catalyst, P1 Ni was found to be somewhat more active (as measured by the $t_{1/2}$ for hydrogenation of several alkenes).[9a] What is more important in the comparison of Ra Ni and P1 Ni is the lower incidence of double-bond isomerization observed with P1 Ni vs. Ra Ni (3% vs. 20%). P1 Ni reduces mono-, di-, tri-, and tetrasubstituted alkenes under mild conditions (1 atm H_2, rt) while leaving many groups unaffected (e.g. a phenyl ring). There is a significant difference in the rate of reduction among the various substituted alkenes allowing for selectivity. However, P2 Ni is very sensitive to steric hindrance and to the alkene substitution pattern. Little or no hydrogenolysis of allylic, benzylic, or propargylic substituents is observed with this catalyst; partial reduction of alkynes and dienes are also possible. Some examples of the use of P2 Ni as a hydrogenation catalyst are shown in Table 1.

Table 1 Reduction of Alkenes, Dienes, and Alkynes with P2 Ni

Substrate	Product	Yield (%)	Ref.
1-Hexyne	Hexane	16	9a
	1-Hexene	68	
	Starting material	16	
3-Hexyne	Hexane	1	9a
	cis-3-Hexene	96	
	trans-3-Hexene	3	
2-Methyl-1,5-hexadiene	2-Methylhexane	2	9a
	2-Methyl-1-hexene	96	
	Other methylhexenes	2	
1,3-Cyclohexadiene	Cyclohexane	2	9a
	Cyclohexene	89	
	Benzene	9	
1-Penten-3-ol	3-Pentanol	100	9a
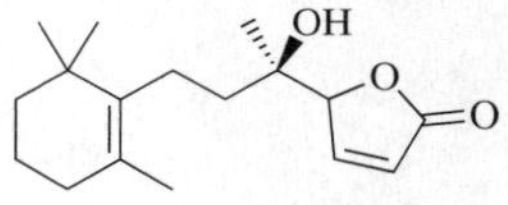	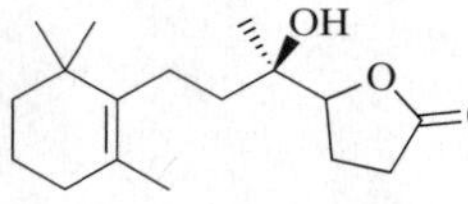	94	14
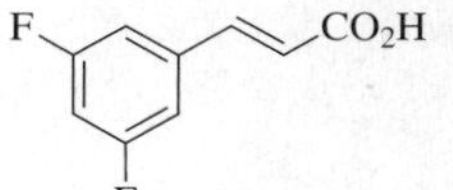	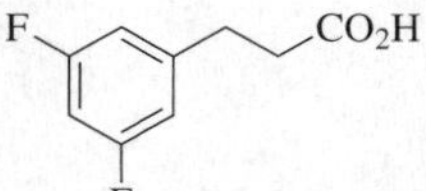	75	15
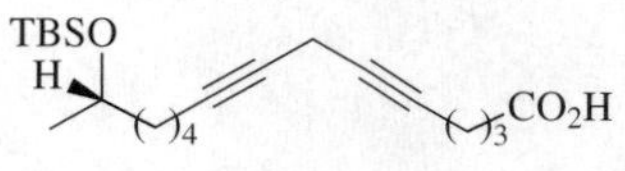	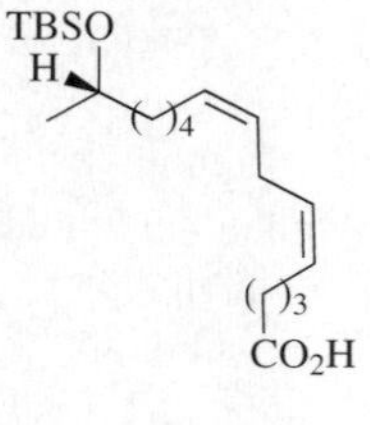	100	16
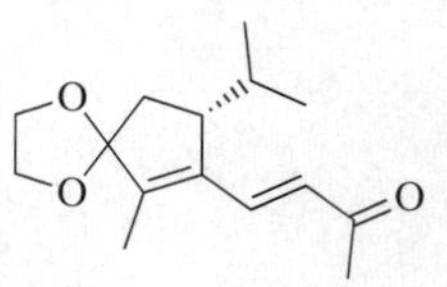	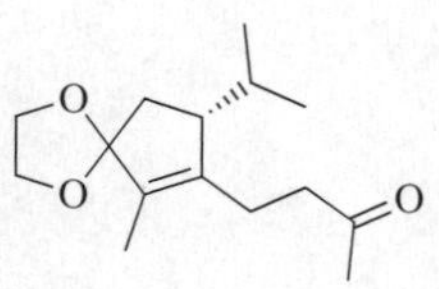	70	17
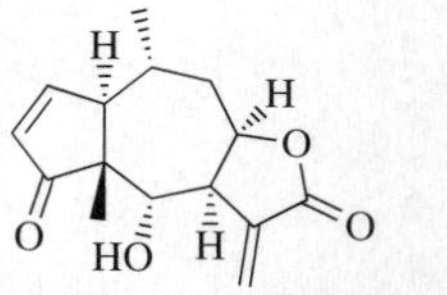	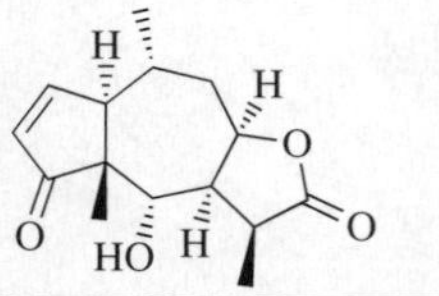	50	18

Under more forcing conditions (30 psi in a Parr apparatus), Russell[19] was able to reduce unsaturated ethers, alcohols, aldehydes, esters, amines, and amides to their saturated counterparts without hydrogenolysis. Unsaturated nitriles[19b] were reduced to primary amines while epoxides were unaffected by the reagent. Both dimethoxyborane (eq 1)[20] and ***Lithium Aluminum Hydride*** (eq 2)[21] can replace $NaBH_4$ in these reactions.

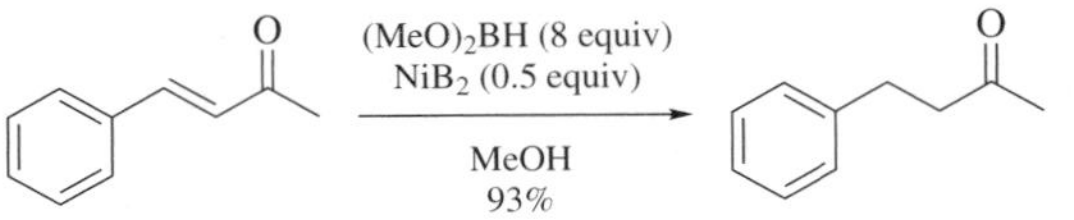

$(MeO)_2BH$ (8 equiv), NiB_2 (0.5 equiv), MeOH, 93% (1)

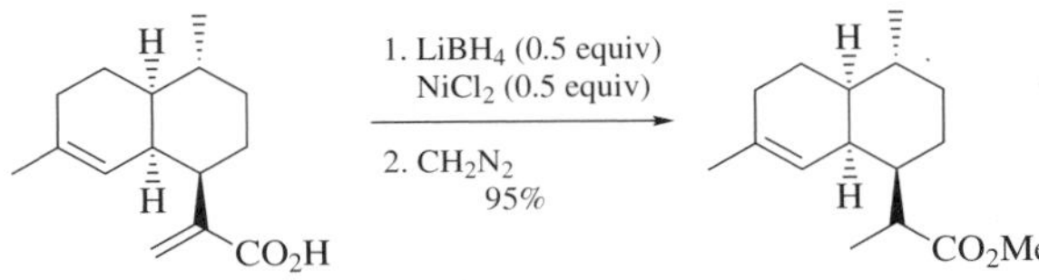

1. $LiBH_4$ (0.5 equiv), $NiCl_2$ (0.5 equiv); 2. CH_2N_2, 95% (2)

Heteroarenes. Nose and Kudo[22] examined the reduction of quinaldine (**1**) with a variety of transition metal salts ($CoCl_2$, $NiCl_2$, $CuCl_2$, $CrCl_3$) in the presence of $NaBH_4$; only ***Nickel(II) Chloride*** was effective (eq 3).

Partial reduction of a series of heteroaromatics was examined using $NiCl_2/NaBH_4$ in methanol at room temperature (Table 2). The authors suggest that the reduction proceeds through a $NiCl_2$ complex of the arene; however, other workers[1a] dispute this mechanism.

Table 2 Reduction of Heteroaromatics with $NiCl_2/NaBH_4$

Substrate	Product	Yield (%)
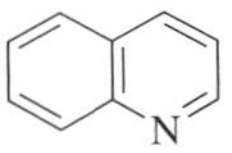	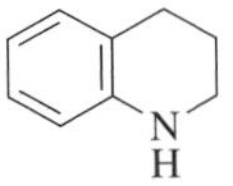	83
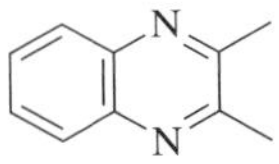	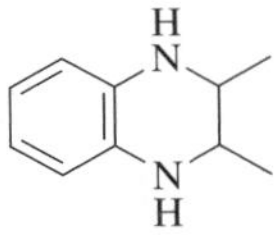	99
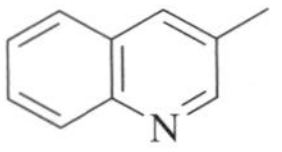	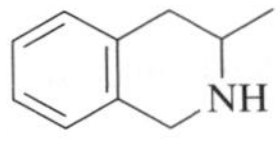	96
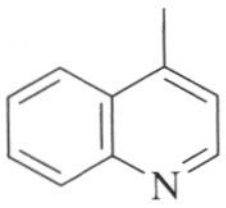	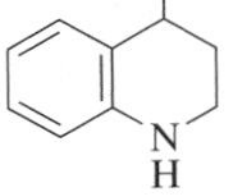	83
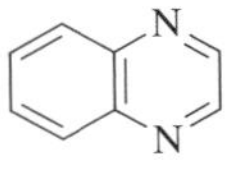	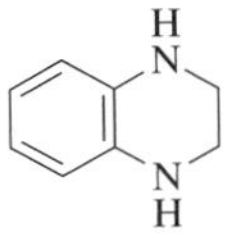	54
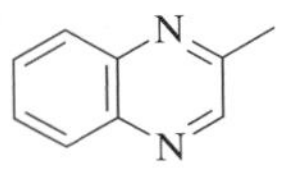	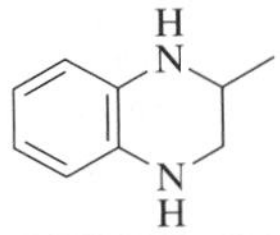	52

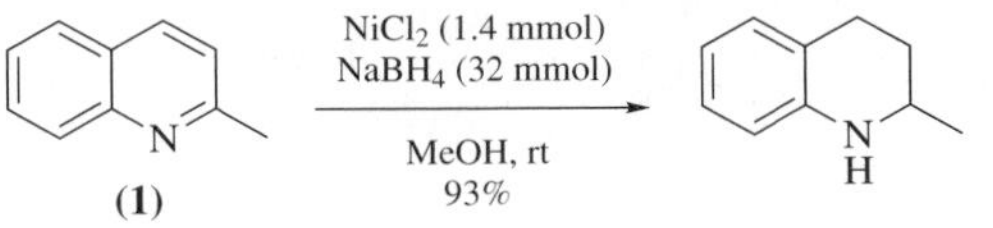

$NiCl_2$ (1.4 mmol), $NaBH_4$ (32 mmol), MeOH, rt, 93% (3)

Desulfurization. While Raney nickel[23] is the traditional reagent for desulfurization reactions, it has several drawbacks (i.e. strongly basic, pyrophoric, sensitivity to air and moisture). In 1963, Truce and Roberts[24] reported the use of $NiCl_2/NaBH_4$ in the partial cleavage of a dithioacetal (eq 4).

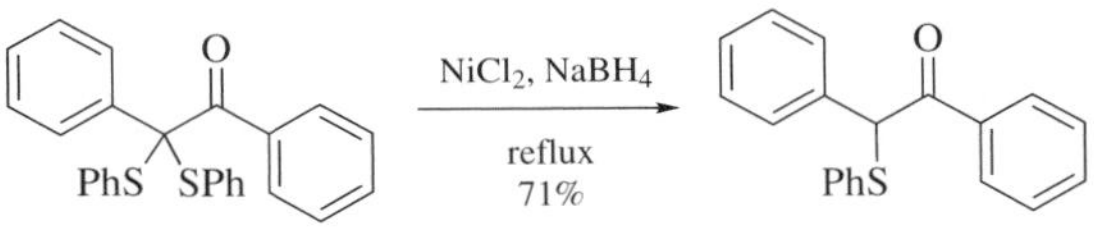

$NiCl_2$, $NaBH_4$, reflux, 71% (4)

Since then, there have been numerous examples of the use of Ni^{II} salt/$NaBH_4$ in desulfurization reactions;[4] in many cases the yields are greater than those seen with Raney nickel[25] (eq 5) *(note: caution must be exercised when using $NaBH_4$ in DMF)*.

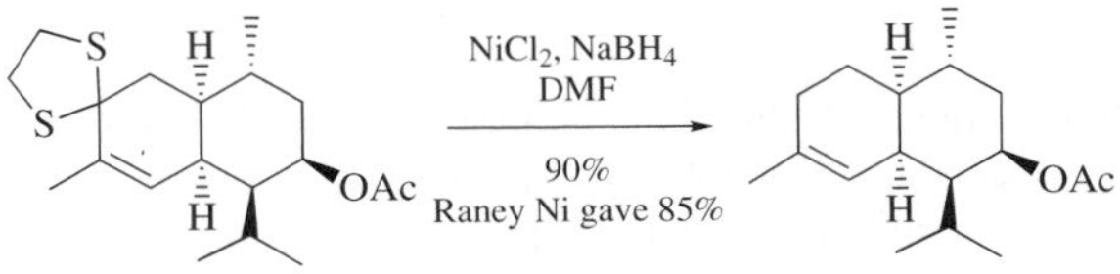

$NiCl_2$, $NaBH_4$, DMF, 90%; Raney Ni gave 85% (5)

Boar et al.[26] used nickel boride in a protection–deprotection scheme for triterpenoid ketones (eq 6).

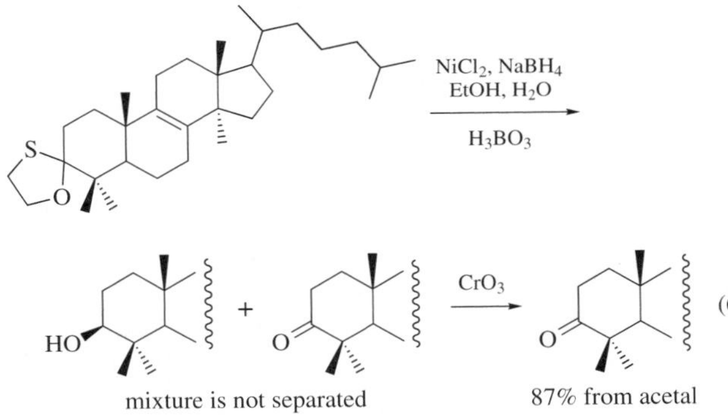

$NiCl_2$, $NaBH_4$, EtOH, H_2O, H_3BO_3; mixture is not separated; CrO_3; 87% from acetal (6)

$Ni^{II}/NaBH_4$ is an effective reagent for desulfurization of thioamides,[27] thioethers,[28] and sulfides.[4,29] Back and coworkers[4,30] has reported extensive studies of the scope, stereochemistry, and mechanism of nickel (and cobalt) boride desulfurizations. In general, nickel boride is a more effective desulfurization catalyst than ***Cobalt Boride*** (other metals such as Mo, Ti, Cu, and Fe were completely ineffective). ***Lithium Borohydride*** can be used in place of $NaBH_4$ while ***Sodium Cyanoborohydride*** cannot. Sulfides, thioesters, thiols, disulfides, and sulfoxides are reduced to hydrocarbons by Ni_2B, while sulfoxides are stable. Esters, chloro groups, and phenyl groups are stable to Ni_2B. Iodides, nitro groups, nitriles, and alkenes are reduced completely by Ni_2B, while bromides, aldehydes, ketones, and cyclopropanes show variable reactivity (eqs 7–10).

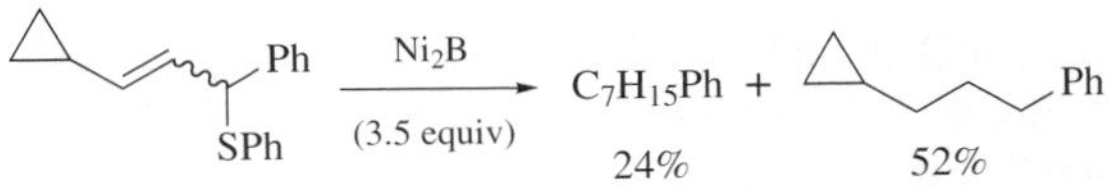

Ni_2B (3.5 equiv) → $C_7H_{15}Ph$ (24%) + (52%) (7)

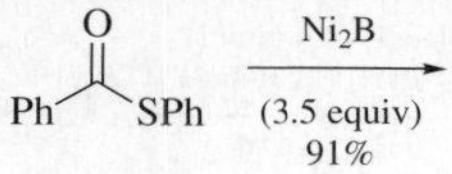

PhC(O)SPh —[Ni_2B (3.5 equiv), 91%]→ $PhCH_2OH$ (8)

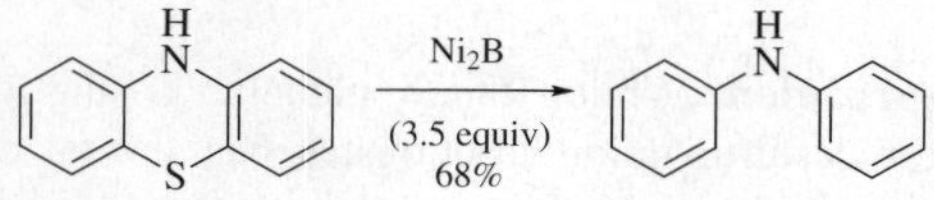

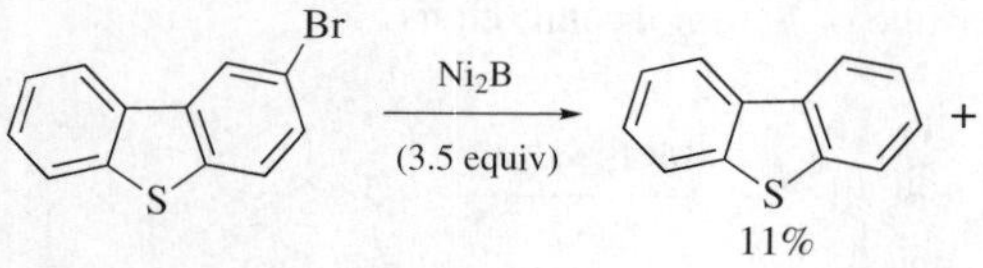

(9): Ni_2B (3.5 equiv), 68%

(10): Ni_2B (3.5 equiv); 11% + biphenyl 77%

Using deuterium labelling, Back showed that desulfurization occurs with retention of configuration, unlike Raney nickel, which involved a radical mechanism. The suggested mechanism of desulfurization involves an oxidative addition–reductive elimination sequence via a nickel hydride intermediate.

Reduction of Other Nitrogenous Functional Groups. Primary, secondary, and tertiary aliphatic nitro groups are reduced to amines with $NiCl_2/NaBH_4$.[5c] ***Hydrazine*** hydrate has also been used with Ni_2B to reduce both aryl and aliphatic nitro groups in a synthesis of tryptamine (eqs 11 and 12).[31]

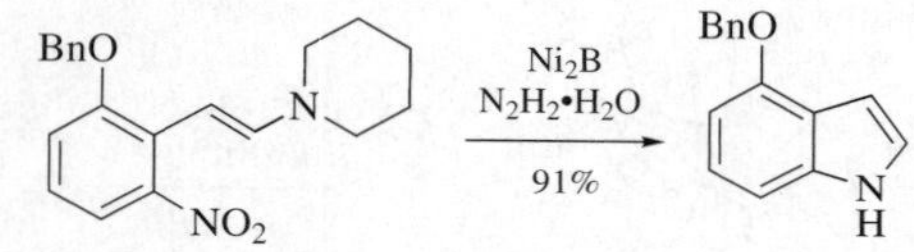

(11): BnO, NO_2 —[Ni_2B, $N_2H_2 \cdot H_2O$, 91%]→ BnO, NH

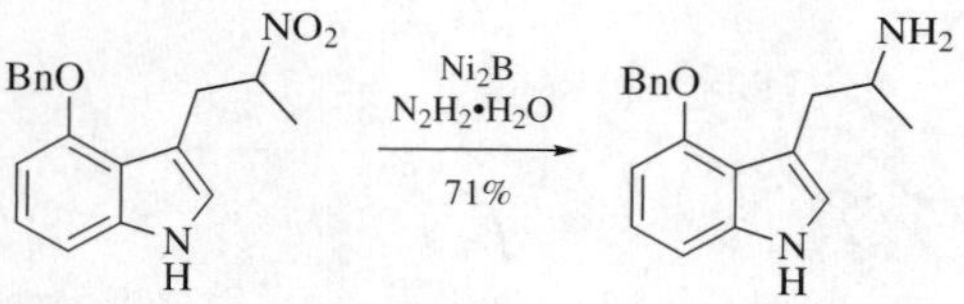

(12): BnO, NO_2 —[Ni_2B, $N_2H_2 \cdot H_2O$, 71%]→ BnO, NH_2

Reductive cleavage of thioethers and reduction of nitro groups has been combined in a synthesis of pyrrolidones (eq 13).[32]

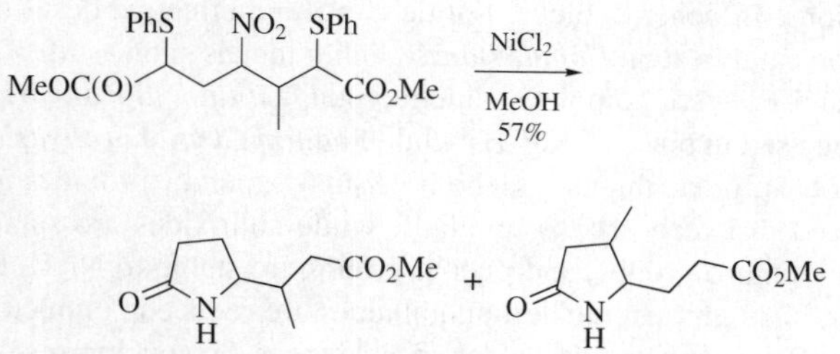

(13): PhS, NO_2, SPh, MeOC(O), CO_2Me —[$NiCl_2$, MeOH, 57%]→ CO_2Me + CO_2Me (1:1)

Like Co_2B, Ni_2B[5b] reduces nitroarenes to anilines and azoxybenzenes to azobenzenes (Table 3); unlike Co_2B, Ni_2B reduces oximes[33] to amines (Table 4).

Table 3 Reduction of Nitroarenes with Ni^{II}/Borohydride Reagents

Substrate	Reagent	Product	Yield (%)
$PhNO_2$	2 equiv Ni_2B, MeOH	$PhNH_2$	3[5a]
$PhNO_2$	1 equiv Ni_2B, 15N NH_4OH	$PhNH_2$	96[5a]
$PhNO_2$	0.1 equiv Ni_2B, 5 equiv $NaBH_4$	Azoxybenzene	89[5b]
$4\text{-}ClC_6H_4NO_2$	1 equiv Ni_2B, 3N HCl	$4\text{-}ClC_6H_4NH_2$	96[5a]
$4\text{-}CNC_6H_4NO_2$	2 equiv Ni_2B, 3N HCl	$4\text{-}CNC_6H_4NH_2$	60[5a]
6-Nitroquinoline	1 equiv Ni_2B, 15N NH_4OH	6-Aminoquinoline	86[5a]
1-Nitronaphthalene	$NiCl_2$–$NaBH_4$ (2:1)	1-Aminonaphthalene	85[5d]
$4\text{-}IC_6H_4NO_2$	4 equiv Ni_2B, 1N HCl	$4\text{-}IC_6H_4NH_2$	76[34]
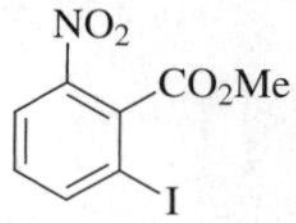 NO_2, CO_2Me, I	4 equiv Ni_2B, 1N HCl	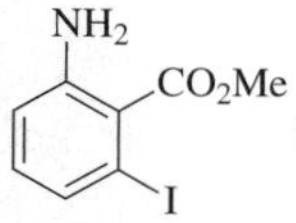 NH_2, CO_2Me, I	85[34]

Table 4 Reduction of Oximes to Amines with $NiCl_2/NaBH_4$

Substrate	Product	Yield (%)
 NOH	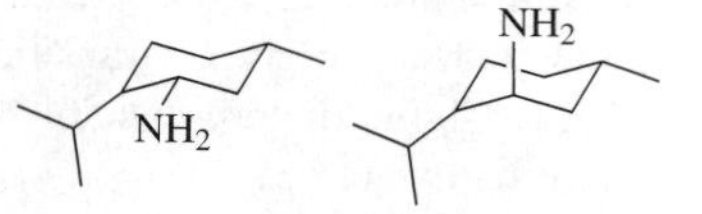 NH_2 + NH_2 (1:2)	92
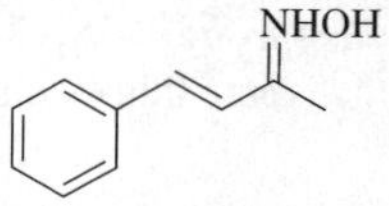 NHOH	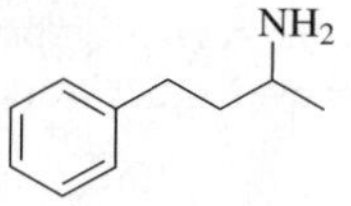 NH_2	95
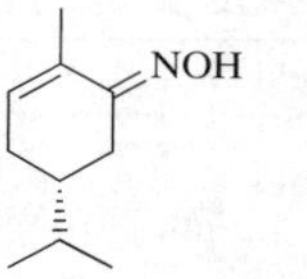 NOH	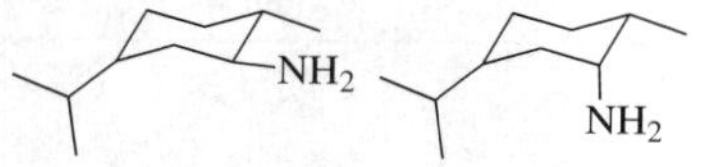 NH_2 + NH_2 (1:1)	90
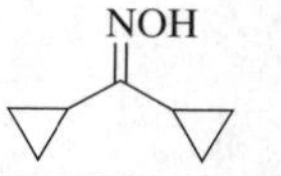 NOH	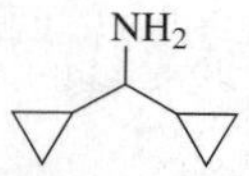 NH_2	70

Reduction of Other Nitrogenous Functional Groups. ***Borane–Tetrahydrofuran***/$NiCl_2$ has been used to reduce chiral cyanohydrins to ethanol amines in high yield.[35] Azides are cleanly reduced to amines in good yield with nickel boride.[36] Azides are reduced in preference to hindered aliphatic nitro groups (eq 14).[37]

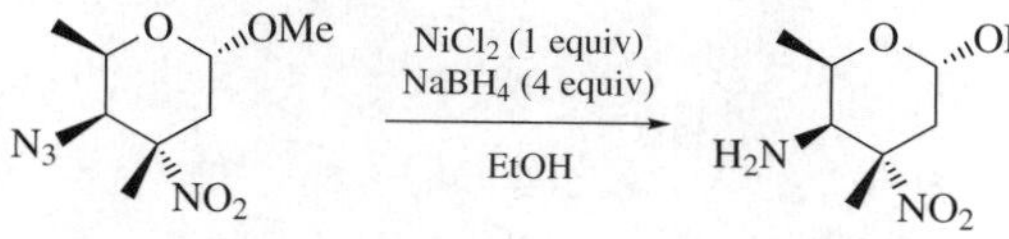

(14): N_3, OMe, NO_2 —[$NiCl_2$ (1 equiv), $NaBH_4$ (4 equiv), EtOH]→ H_2N, OMe, NO_2

Isoxazoles are reduced to β-amino enones in high yield using the $NiCl_2/NaBH_4$ system.[38] Dihydroisoxazolones are reduced with a high degree of diastereoselectivity with the $NiCl_2/NaBH_4$ system.[39]

Dehalogenation. Many α-bromo ketones[6a] are cleanly reduced to the parent ketone with nickel boride in DMF (caution). Vicinal dibromides are reduced to alkenes (eq 15).

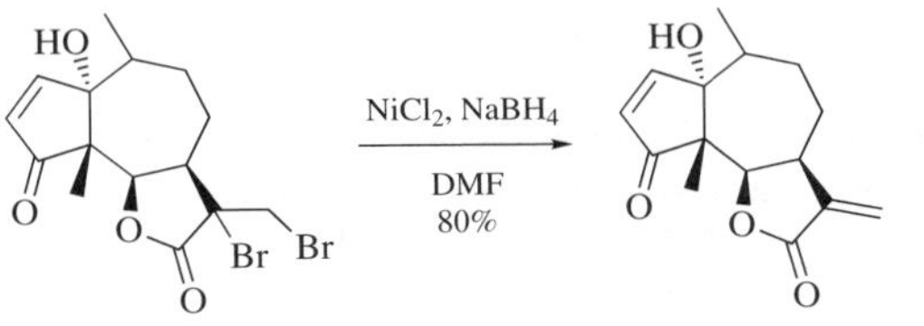

(15) $NiCl_2$, $NaBH_4$; DMF; 80%

Aryl and certain alkyl chlorides can be dehalogenated[1a,1b,6b] with a variety of Ni^{II}/hydride agents (e.g. $NaBH_2(OCH_2CH_2OMe)_2$, ***Triethylsilane***, $NaBH_4$). Lin and Roth have effected the clean debromination of aryl bromides[40] using ***Dichlorobis(triphenylphosphine)nickel(II)***/$NaBH_4$ in DMF (**caution**); ***Tris(triphenylphosphine)nickel(0)*** is assumed to be the active catalyst. Russel and Liu[41] demonstrated that reductive cleavage of an iodide goes with retention when $NiCl_2$/$NaBH_4$ is used (cf. inversion seen with $LiAlH_4$; eq 16).

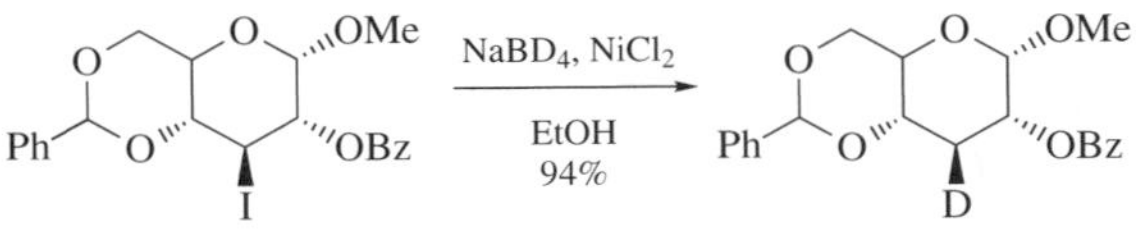

(16) $NaBD_4$, $NiCl_2$; EtOH; 94%

Hydrogenolysis. Ni_2B has been used to hydrogenolyze benzylic (eqs 17–19),[7a] allylic (eqs 20–22),[7b,42] and propargylic (eq 23)[7b] esters in good yields.

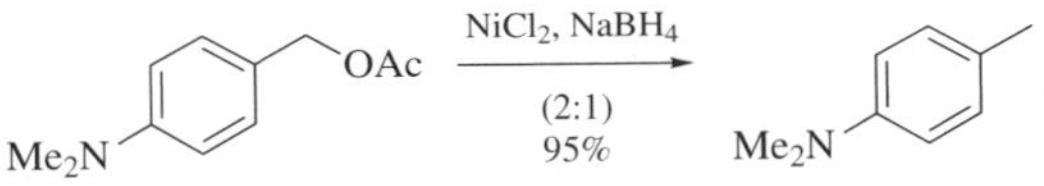

(17) $NiCl_2$, $NaBH_4$; (2:1); 95%

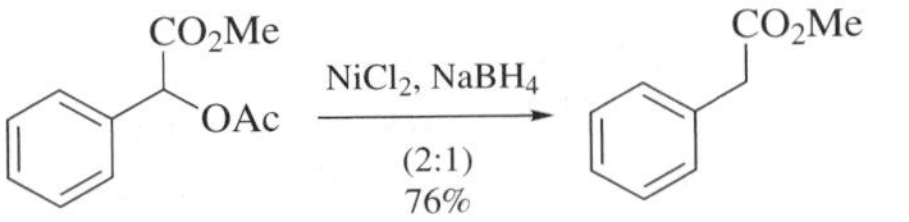

(18) $NiCl_2$, $NaBH_4$; (2:1); 76%

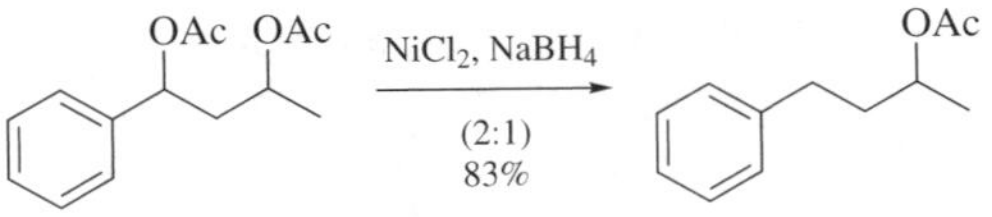

(19) $NiCl_2$, $NaBH_4$; (2:1); 83%

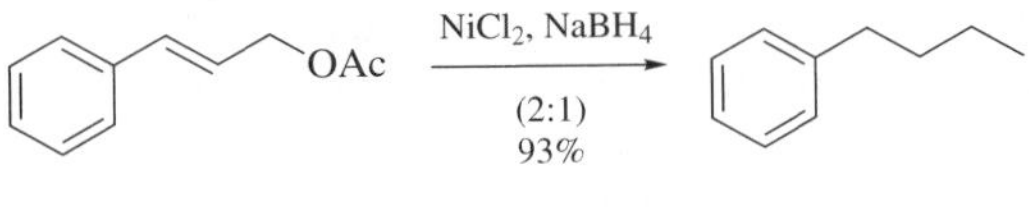

(20) $NiCl_2$, $NaBH_4$; (2:1); 93%

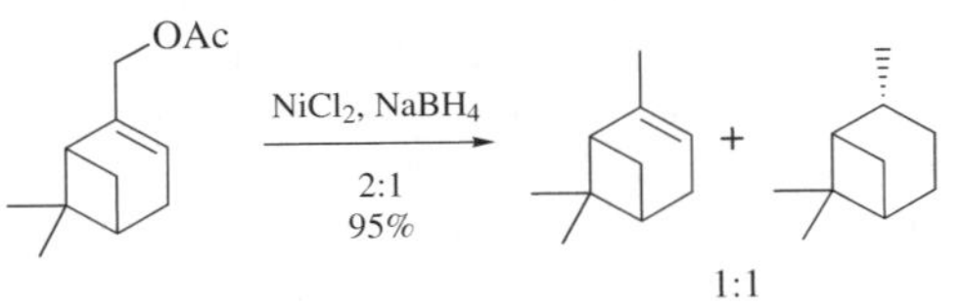

(21) $NiCl_2$, $NaBH_4$; 2:1; 95%; 1:1

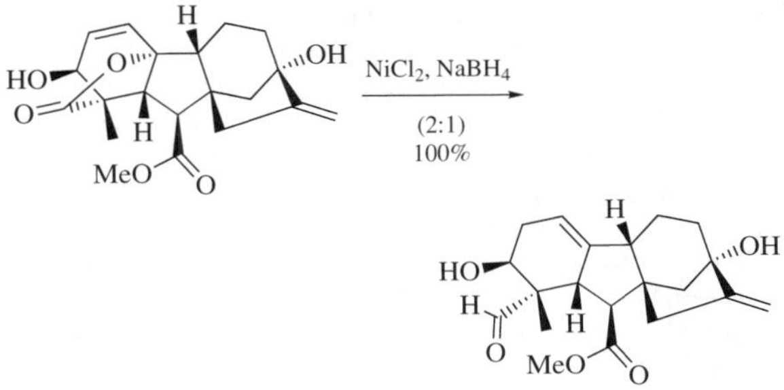

(22) $NiCl_2$, $NaBH_4$; (2:1); 100%

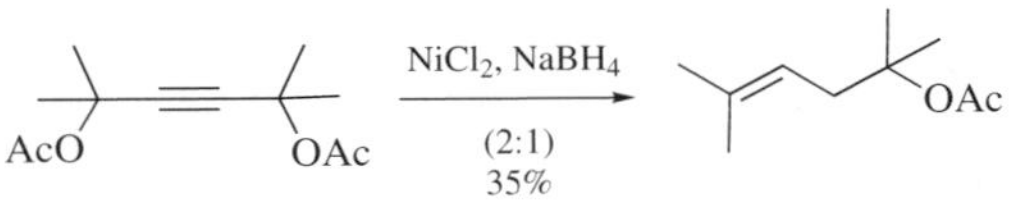

(23) $NiCl_2$, $NaBH_4$; (2:1); 35%

Enol tosylates and aryl tosylates are deoxygenated in good to excellent yields[43] (eqs 24 and 25)

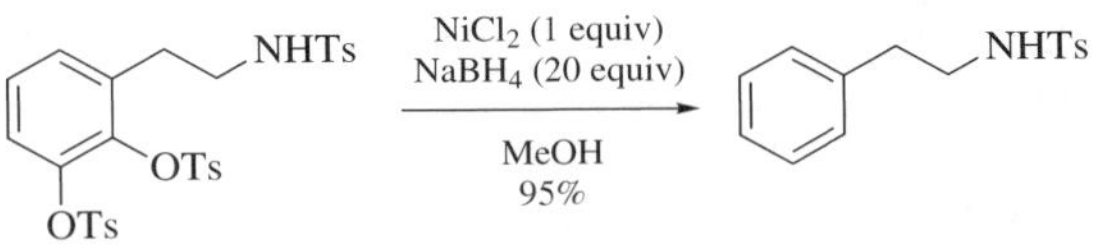

(24) $NiCl_2$ (1 equiv), $NaBH_4$ (20 equiv); MeOH; 95%

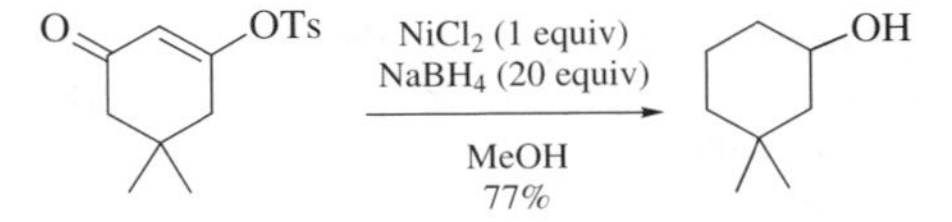

(25) $NiCl_2$ (1 equiv), $NaBH_4$ (20 equiv); MeOH; 77%

A variety of allylic functional groups[44] (alcohols, esters, silyl ethers, ketones, and hydroperoxides) have been reduced with Ni_2B. The combination of ***Chlorotrimethylsilane***/Ni_2B will selectively reduce an aldehyde in the presence of a ketone.[45]

Selenides[46] and tellurides[47] are reductively cleaved by Ni_2B with retention of stereochemistry. The phenyl selenyl group is cleaved in preference to the thio phenyl group.

Related Reagents. See Classes R-2, R-3, R-4, R-8, R-9, R-10, R-14, R-15, R-19, R-21, R-23, R-27, R-28, R-30, and R-31, pages 1–10.

---

1. (a) Ganem, B.; Osby, J. O. *CRV* **1986**, *86*, 763. (b) Wade, R. *J. Mol. Catal.* **1983**, *48*, 273. (c) Hudlicky, M.; *Reductions in Organic Chemistry*; Wiley: New York, 1984.
2. It should be noted that Ni_2B represents a nominal stoichiometry for the reagent prepared by the action of $NaBH_4$ on a Ni^{II} salt. Several Ni_xB_y species have been described in the literature. *Chemical Abstracts* uses the registry number [*12619-90-8*] to designate nickel boride of unspecified stoichiometry. [*12007-02-2*] and [*12007-00-0*] are the registry numbers for Ni_3B and NiB, respectively. These are the most widely cited synthetically useful reagents.
3. (a) Brown, C. A. *JOC* **1970**, *35*, 1900. (b) Brown, C. A.; Ahuja, V. K. *JOC* **1973**, *38*, 2226.
4. (a) Back, T. G.; Baron, D. L.; Yang, K. *JOC* **1993**, *58*, 2407. (b) Back, T. G.; Yang, K.; Krouse, R. H. *JOC* **1992**, *57*, 1986.
5. (a) Nose, A.; Kudo, T. *CPB* **1989**, *37*, 816. (b) Nose, A.; Kudo, T. *CPB* **1988**, *36*, 1529. (c) Osby, J. O.; Ganem, B. *TL* **1985**, *26*, 6413. (d) Nose, A.; Kudo, T. *CPB* **1981**, *29*, 1159.
6. (a) Sarma, J. C.; Borbaruah, M.; Sharma, R. P. *TL* **1985**, *26*, 4657. (b) Tabaei, S-M. H.; Pittman, C. V. *TL* **1993**, *34*, 3264.
7. (a) He, Y.; Pan, X.; Wang, S.; Zhao, H. *SC* **1989**, *19*, 3051. (b) Ipaktschi, J. *CB* **1983**, *117*, 3320 (*CA* **1985**, *102*, 94 904x).

8. This is the melting point of Ni_2B formed by fusion of the elements *Adv. Chem. Ser.* **1961**, *32*, 53). Material prepared by the reduction of $NiCl_2$ with $NaBH_4$ begins to decompose at 100 °C when heated in vacuo with liberation of H_2 (Maybury, P. C.; Mitchell, R. W.; Hawthorne, M. F. *JCS(C)* **1974**, 534).
9. (a) This procedure provides the P2 form of nickel boride, which is a selective hydrogenation catalyst. Brown, H. C.; Brown, C. A. *JACS* **1963**, *85*, 1005. (b) Brown, H. C.; Brown, C. A. *JACS* **1963**, *85*, 1003. This paper reports the preparation and properties of P1 nickel boride. P1 nickel boride is more active, in some applications, than Raney nickel. (c) Destefanis, H.; Acosta, D.; Gonzo, E. *Catal. Today* **1992**, *15*, 555. This group describes the use of $BH_3{\cdot}THF$ complex to prepare Ni_3B and Ni_4B_3 using $Ni(OAc)_2$ and $NiCl_2$, respectively, and their use as hydrogenation catalysts.
10. Schreifels, J. A.; Maybury, C. P.; Swartz, W. E. *JOC* **1981**, *46*, 1263.
11. (a) Paul, R.; Buisson, P.; Joseph, N. *Ind. Eng. Chem.* **1952**, *44*, 1006 (*CA* **1952**, *46*, 9960e). (b) Paul, R.; Buisson, P.; Joseph, N. *CR(C)* **1951**, *232*, 627 (*CA* **1951**, *45*, 10 436h). (c) Schlesinger, H. R.; Brown, H. C.; Finholt, A. E.; Gilbreath, J. R.; Hoekstra; Hyde, E. K. *JACS* **1953**, *75*, 215.
12. Brown, H. C.; Brown, C. A. *JACS* **1962**, *84*, 1493.
13. A boride of the same composition had been previously described (Stock, A.; Kuss, E. *CB* **1914**, *47*, 810 (*CA* **1914**, *8*, 2129).
14. Jefford, C. W.; Jaggi, D.; Bernardinelli, G.; Boukouvalas, J. *TL* **1987**, *28*, 4041.
15. Novak, J.; Salemink, C. A. *JCS(P1)* **1982**, 2403.
16. Miller, J. G.; Ochlschlager, A. C. *JOC* **1984**, *49*, 2332. This reaction uses TMEDA as an additive.
17. Kido, F.; Abe, T.; Yoshikoshi, A. *JCS(C)* **1986**, 590.
18. Lee, K-H.; Ibuka, T.; Sims, D.; Muraoka, O.; Kiyokawa, H.; Hall, I. H.; Kim, H. L. *JMC* **1981**, *24*, 924. When Pt_2O was used, only 20% of the desired product was isolated; the major product was the tetrahydro compound.
19. (a) Russell, T. W.; Hoy, R. C. *JOC* **1971**, *36*, 2018. (b) Russell, T. W.; Hoy, R. C.; Cornelius, J. E. *JOC* **1972**, *37*, 3552.
20. Nose, A.; Kudo, T. *CPB* **1990**, *38*, 1720.
21. Jung, M.; Elsohly, H. N.; Croon, E. M.; McPhail, D. R.; McPhail, A. T. *JOC* **1986**, *51*, 5417.
22. Nose, A.; Kudo, T. *CPB* **1984**, *32*, 2421.
23. Pettit, G. R.; van Tamelen, E. E. *OR* **1962**, *62*, 347.
24. Truce, W. E.; Roberts, F. E. *JOC* **1963**, *28*, 961.
25. Zaman, S. S.; Sarmah, P.; Barus, N. C.; Sharma, R. P. *CI(L)* **1989**, 806.
26. Boar, R. B.; Hawkins, D. W.; McGhie, J. F.; Barton, D. H. R. *JCS(P1)* **1973**, 654.
27. Guziec, F. S.; Wasmund, L. M. *TL* **1990**, *31*, 23.
28. (a) Euerby, M. R.; Waigh, R. D. *SC* **1986**, *16*, 779. (b) Euerby, M. R.; Waigh, R. D. *JCS(C)* **1981**, 127.
29. Truce, W. E.; Perry, F. M. *JOC* **1965**, *30*, 1316.
30. Back, T. G.; Yang, K. *JCS(C)* **1990**, 819.
31. Lloyd, D. H.; Nichols, D. E. *JOC* **1986**, *51*, 4294.
32. Posner, G. H.; Crouch, R. D. *T* **1990**, *46*, 7509.
33. Ipaktschi, J. *CB* **1984**, *117*, 856 (*CA* **1984**, *101*, 22 611f).
34. Seltzman, H. H.; Berrang, B. B. *TL* **1993**, *34*, 3083.
35. Lu, Y.; Meit, C.; Kunesch, N.; Poisson, J. *TA* **1990**, *1*, 707.
36. Sarma, J. C.; Sharma, R. P. *CI(L)* **1987**, 764.
37. Guilano, R. M.; Deisenroth, T. W. *J. Carbohydr. Res* **1987**, *6*, 295.
38. (a) Koroleva, E. V.; Lakhvich, F. A.; Yankova, T. V. *KGS* **1987**, *11*, 1576 (*CA* **1988**, *109*, 928 546). (b) Oliver, J. E.; Lusby, W. R. *T* **1988**, *44*, 1591.
39. (a) Lakhvich, F. A.; Koroleva, E. V.; Antonevich, I. Q.; Yankova, T. V. *ZOR* **1990**, *26*, 1683 (*CA* **1991**, *114*, 81 311). (b) Annunziata, R.; Cinquini, M.; Cozzi, F.; Gilardo, A.; Restelli, A. *JCS(P1)* **1985**, 2289.
40. Lin, S. T.; Roth, J. A. *JOC* **1979**, *44*, 309.
41. Russel, R. N.; Liu, H. W. *TL* **1989**, *30*, 5729.
42. Jiang, B.; Zhao, H.; Pan, X.-F. *SC* **1987**, *17*, 997.
43. Wang, F.; Chiba, K.; Tada, M. *JCS(P1)* **1992**, 1897.
44. (a) Sarma, D. N.; Sharma, R. P. *TL* **1985**, *26*, 2581. (b) Zaman, S. S.; Sarma, J. C.; Sharma, R. P. *CI(L)* **1991**, 509. (c) Sarma, D. N.; Sharma, R. P. *TL* **1985**, *26*, 371.
45. Borbaruah, M.; Barua, N. C.; Sharma, R. P. *TL* **1987**, *28*, 5741.
46. (a) Back, T. G.; Birss, V. I.; Edwards, M.; Krishna, M. V. *JOC* **1988**, *53*, 3815. (b) Back, T. G. *JCS(C)* **1984**, 1417.
47. (a) Barton, D. H. R.; Fekih, A.; Lusinchi, X. *TL* **1985**, *26*, 6197. (b) Barton, D. H. R.; Bohe, L.; Lusinchi, X. *TL* **1990**, *31*, 93.

Thomas J. Caggiano
Wyeth-Ayerst Research, Princeton, NJ, USA

Nickel(II) Chloride[1]

($NiCl_2$)		
[7718-54-9]	Cl_2Ni	(MW 129.59)
($NiCl_2{\cdot}6H_2O$)		
[7791-20-0]	$Cl_2H_{12}NiO_6$	(MW 237.71)

(mild Lewis acid;[2–5] catalyst for coupling reactions,[1a,7–12] and in combination with complex hydrides as a selective reducing agent[16,33,36–38])

Physical Data: mp 1001 °C; *d* 3.550 g cm^{-3}.
Solubility: sol H_2O, alcohol; insol most organic solvents.
Form Supplied in: yellow solid when anhydrous, green solid for the hydrate; widely available.
Drying: for anhydrous nickel chloride, standard procedure for drying metal chlorides can be used by refluxing with ***Thionyl Chloride*** followed by removal of excess $SOCl_2$.[45]
Handling, Storage, and Precautions: nickel(II) is reputed to be toxic and a cancer suspect agent. Use in a fume hood.

Mild Lewis Acid. Nickel chloride serves as a mild Lewis acid which promotes the regioselective rearrangement of dienols in aqueous *t*-BuOH at 60 °C in satisfactory yield (eq 1).[2] Brønsted acids give dehydration products, whereas other Lewis acids such as ***Nickel(II) Acetate***, ***Palladium(II) Chloride***, and ***Copper(II) Chloride*** proved less effective than nickel chloride and yield a mixture of rearranged and dehydration products. When anhydrous alcohol solvent is used, rearranged products bearing terminal alkoxy groups are obtained.

In the presence of a catalytic amount of $NiCl_2$, ***Cyanotrimethylsilane*** smoothly reacts with acetals or orthoesters derived from aromatic and α,β-unsaturated carbonyl compounds to give the corresponding α-cyano derivatives under neutral conditions (eq 2).[3] $NiCl_2$ can also accelerate the conversion of acrylamide to ethyl acrylate[4a] and catalyze the amination of 5,8-quinolinediones.[4b]

The ring-opening reaction of epoxides with $LiAlR_4$ is catalyzed by $NiCl_2$ or ***Nickel(II) Bromide*** (eq 3).[5]

$NiCl_2$, aq. t-BuOH, 60 °C, 70%; R = H, SMe (1)

TMSCN, $NiCl_2$, CH_2Cl_2, rt, 96% (2)

$LiAlBu_4$, NiX_2, 100% (3)

Nickel(II) Chloride–Chromium(II) Chloride. Although the ***Chromium(II) Chloride***-mediated reaction of an aldehyde with a vinylic iodide provides a useful entry for the preparation of allylic alcohol,[1a,6] the presence of a catalytic amount of $NiCl_2$ is essential to ensure the completion of the reaction.[7–12] Vinyl iodides (eq 4)[7] or triflates[8a] are commonly used. Alkynyl iodides behave similarly (eq 5).[9] Silyl enol ethers or enol phosphates are unreactive. The stereochemistry of iodoalkenes is retained in the majority of cases with the exceptions of trisubstituted *cis*-iodoalkenes and *cis*-iodoenones, which afford exclusively the *trans*-alkenes instead of the expected *cis*-alkenes.[7a]

$CrCl_2$, $NiCl_2$, 95% (4)

$CrCl_2$, $NiCl_2$, 74% (5)

Functional groups such as esters, amides, nitriles, ketones, acetals, ethers, silyl ethers (TBDMS or TBDPS), alcohols, alkenes, and triple bonds are stable under the reaction conditions. Substrates containing structural complexity can be employed in this transformation. Thus, the reaction served as the key step for the formation of C(7)–C(8) and C(84)–C(85) bonds in the total synthesis of palytoxin,[10] as well as for the synthesis of other natural products and C-saccharides. The reagent has also been proved to be useful in the intramolecular cyclization of the aldehyde (eqs 6 and 7).[11,12]

$CrCl_2$, $NiCl_2$, 73% (6)

$CrCl_2$, $NiCl_2$, 21 °C, 60% (7)

A simple and selective method for the conversion of an aldehyde to vinyl iodides, (*E*)-RCH=CHX, by means of a $CHX_3/CrCl_2$ system has been developed (eq 8).[13]

$$\text{PhCHO} \xrightarrow[\text{THF, 91\%}]{\text{CHI}_3,\ \text{CrCl}_2} \text{PhCH=CHI} \quad (8)$$

DMF happens to be the most effective solvent for this coupling reaction. The reaction goes slowly but cleanly in the DMSO solvent.[8a] The presence of a phosphine ligand in the nickel catalyst gives a diene sideproduct.[4a] Nevertheless, this later system has been used in the intramolecular cyclization of enynes (eq 9).[14]

$CrCl_2$, $NiCl_2(PPh_3)_2$, 82% (9)

Monosubstituted α,β-unsaturated aldehydes are converted to cyclopropanols in the presence of $NiCl_2/CrCl_2$ in moderate yields (eq 10).[15]

$CrCl_2$, $NiCl_2$, 72% (10)

Selective Reductions. Low-valent transition metal complexes generated in situ from metal halides and reducing agents are particularly useful for the selective reduction of various functionalities.[16] Nickel chloride and nickel bromide have demonstrated a unique role in these reduction reactions. To illustrate this, in the presence of an equimolar quantity $NiCl_2$, ***Lithium Aluminum Hydride*** can reduce alkenes to alkanes in excellent yields.[17] Under similar conditions at −40 °C, alkynes are reduced to *cis*-alkenes in good yield.[17] Haloalkanes are also smoothly converted into the corresponding hydrocarbons under these conditions.[18a,b] Even chlorobenzene and 1-bromoadamantane can be reduced efficiently by this reagent. ***Sodium Hydride*** in the presence of $NiCl_2$ or $NiBr_2$ and a sodium alkoxide can also serve a similar purpose.[16c,18c]

The N–O bond in isoxazolidines is cleaved efficiently by $LiAlH_4/NiCl_2$ at −40 °C (eq 11).[19] Styrene oxide yields β-phenylethanol in 95% yield by this complex reagent, whereas $LiAlH_4$ alone gives α-phenylethanol.[19]

(11) [Scheme: isoxazolidine (N–O, Bu) → $LiAlH_4$, $NiCl_2$, −40 to 25 °C, 97% → amino alcohol (NH HO, Bu)]

Nickel Boride, prepared in situ from the reaction of nickel chloride and ***Sodium Borohydride***, behaves like ***Raney Nickel***.[1b] In DMF, the dark brown/black solution comprises an efficient system for alkene hydrogenation. The carbon–carbon double bonds of the α,β-unsaturated carbonyl compounds are reduced selectively (eq 12).[20] It is noted that carbon–sulfur bonds are selectively reduced under similar conditions (eq 13).[16b,21] Thiols, sulfides, disulfides, dithioacetals, as well as sulfoxides can all be hydrodesulfurized smoothly. Sulfones, on the other hand, remain intact under the reaction conditions.[21d,e]

(12) [Scheme: chalcone (OMe, OH, O, MeO, OMe) → $NaBH_4$, $NiCl_2$, 70% → dihydrochalcone]

(13) [Scheme: Ar–CO–CH=C(SMe)SMe → $NaBH_4$, $NiCl_2$, 75% → Ar–CO–CH=CH–SMe]

Reduction of α-halo ketones with nickel boride produces the corresponding ketones.[22] The carbon–oxygen bonds in allylic ethers,[23a] benzylic esters,[23b] as well as aryl tosylates[23c] are reduced to the corresponding C–H bonds (eq 14).

(14) [Scheme: OTMS allylic ether with furan → $NaBH_4$, $NiCl_2$, 80% → reduced product]

Upon treatment with $NiCl_2/NaBH_4$, nitro,[24a–c] azide,[24d,e] and oxime[24f] groups are smoothly transformed into amino groups in good yields. Carbon–carbon double bonds are occasionally reduced under these conditions.[24a,f] Nitro and cyano groups are also reduced to amines by the reagent mixture $NiCl_2/B_2H_6$.[25] Ketones, aldehyde, carboxylic acid, alkene, ester, and amide moieties are unaffected under these conditions.

Addition of TMSCN to an allene is catalyzed by nickel boride generated in situ, although the reaction is nonstereoselective (eq 15).[26]

(15) [Scheme: Hex-allene → TMSCN, $NiCl_2$, $NaBH_4$, 54% → Hex–CH=C(TMS)–CH$_2$CN; (*E*):(*Z*) = 72:28]

Treatment of ***Diphenylacetylene*** with excess TMSCN in the presence of the $NiCl_2$/***Diisobutylaluminum Hydride*** or $NiCl_2$/***Triethylaluminum*** catalyst affords a substituted pyrrole in high yield (eq 16).[27]

(16) [Scheme: Ph–C≡C–Ph → TMSCN, $NiCl_2$, DIBAL, 87% → pyrrole (Ph, Ph, NC, N(TMS)$_2$, NH)]

Hydrosilylation of conjugated dienes with $HSiR_3$ is catalyzed by $NiCl_2/Et_3Al$ in excellent yield; 1,4-addition is observed exclusively (eq 17).[28]

(17) [Scheme: isoprene → $HSiEt_3$, $NiCl_2$, Et_3Al, 95% → allylic SiEt$_3$]

A combination of ***Aluminum*** and $NiCl_2$ promotes the selective reduction of α,β-enones to the corresponding saturated carbonyl compounds (eq 18).[29] Both nitro groups[30] and aryl ketones[29] are reduced to amines and benzylic alcohols, respectively.

(18) [Scheme: AcO-steroid enone → Al, $NiCl_2 \cdot 6H_2O$, THF, 95% → saturated ketone (AcO)]

Nickel(II) Chloride–Zinc. Finely divided nickel with high catalytic activity is readily obtained by the treatment of $NiCl_2$ with ***Zinc*** dust.[31] This reagent reduces aldehydes, alkenes, and aromatic nitro compounds in good yields.[32a] Nitriles as well as aryl ketones give a mixture of reduced products under these conditions. $Zn/NiCl_2$ in the presence ***Ammonia***/NH_4^+ buffer (pH 6–10)[3] has been shown to effect the selective reduction of α,β-enones to the corresponding saturated carbonyl compounds.[32b] Aryl, allyl, and alkyl halides are reduced by water, zinc, and a catalytic amount of $NiCl_2$, ***Triphenylphosphine***, and iodide ion.[32c]

Reductive Heck-Like Reactions. Reductive Heck-like reactions (eq 19) can be achieved when alkyl, aryl, and vinyl bromides are treated with zinc/$NiCl_2 \cdot 6H_2O$ in the presence of an excess quantity of α,β-unsaturated esters.[33] A trace amount of water is essential for this conversion. Similar reactions are observed when alkenes are treated with iodofluoroacetate or iododifluoroacetate under the same conditions (eq 20).[34] Tandem reaction can also occur to give cyclic products (eq 21).[35]

RBr + CH_2=CHCO$_2$Me → ($NiCl_2$, Zn, MeCN, py, H_2O, 70–80%) → R$CH_2CH_2CO_2$Me (19)

ICF_2CO_2R' → ($NiCl_2$•$6H_2O$, Zn, THF, RCH=CH_2, 60–83%) → R$CH_2CH_2CF_2CO_2$R' (20)

diallyl ether + ICF_2CO_2Et → ($NiCl_2$•$6H_2O$, Zn, THF, 75%) → tetrahydrofuran bearing CF_2CO_2Et (21)

Homocoupling Reactions. In the absence of a Michael acceptor, aryl and vinyl halides undergo dimerization reaction upon treatment with the $NiCl_2$/Zn reagent.[36–38] Under sonication conditions and in the presence of excess Ph_3P and ***Sodium Iodide*** in DMF, the $NiCl_2$/Zn reagent promotes homocoupling of aryl triflates in good yields.[36] Bipyridyls having electron-donating groups, such as methoxy groups, are obtained in satisfactory yields under these conditions (eq 22).[37] Thiophene derivatives behave similarly.[38] Vinyl bromides dimerize to yield the corresponding butadienes.[39] It is interesting that the presence of iodide ion or thiourea can accelerate the reaction.

3-chloro-5-methoxypyridine → ($NiCl_2$, Ph_3P, Zn, DMF, 50 °C, 88.5%) → 5,5'-dimethoxy-3,3'-bipyridine (22)

Cross-Coupling Reactions. Most cross-coupling reactions using nickel catalysts require phosphine ligands and are therefore discussed in detail under ***Dichlorobis(triphenylphosphine)nickel(II)***. The reaction of aryl iodides or bromides with trialkyl phosphites in the presence of $NiCl_2$ is the premier method for preparing dialkyl arylphosphonates (eq 23).[40a,b] Thermolysis of allyl phosphite in the presence of $NiCl_2$ yields the corresponding allyl phosphonates (eq 24).[40c]

ArI + P(OEt)$_3$ → ($NiCl_2$, 160 °C, 94%) → ArP(O)(OEt)$_2$ (23)

(EtO)$_2$P–O–CH_2CH=CH_2 → ($NiCl_2$, 80 °C, 85%) → (EtO)$_2$P(O)CH_2CH=CH_2 (24)

Miscellaneous Reactions. Symmetrical alkynes in the presence of $NiCl_2$ or $NiBr_2$ and ***Magnesium*** undergo trimerization to give the corresponding hexasubstituted aromatic compounds (eq 25). Terminal alkynes yield a mixture of regioisomers.[41]

Et–C≡C–Et → (NiX_2, Mg, 40–90%) → C_6Et_6 (25)

X = Cl, Br, I

Grignard reagents activated by a catalytic quantity of $NiCl_2$ can substitute the germanium–hydrogen bond with a germanium–carbon bond (eq 26).[42] It is noted that the stereochemistry of the original organogermane is retained.

i-PrPh(Naph)GeH → (H_2C=CHCH_2MgBr, $NiCl_2$, 99%) → *i*-PrPh(Naph)GeCH_2CH=CH_2 (26)

Hydromagnesiation of a styrene with ***Ethylmagnesium Bromide*** followed by treatment with ***Carbon Dioxide*** gives the 2-arylpropionic acid in good yield (eq 27).[43]

styrene → ($NiCl_2$, EtMgBr, CO_2, 82%) → 2-arylpropionic acid (CO_2H) (27)

Thermolysis of 1-phenyl-3,4-dimethylphosphole in the presence of $NiCl_2$ yields the corresponding nickel complex of the dimeric product. The ligand can be liberated upon treatment with ***Sodium Cyanide*** (eq 28).[44]

1-phenyl-3,4-dimethylphosphole (P–Ph) → $NiCl_2$ → NaCN → biphosphole (P–Ph, P–Ph), 30% (28)

Related Reagents. See Classes R-2, R-3, R-4, R-9, R-15, R-21, R-23, R-24, R-25, R-27, R-30, and R-33, pages 1–10. Chromium(II) Chloride–Nickel(II) Chloride; Lithium Aluminum Hydride-Nickel(II) Chloride.

---

1. (a) Cintas, P. *S* **1992**, 248. (b) Ganem, B.; Osby, J. O. *CR* **1986**, *86*, 763.
2. (a) Kyler, K. S.; Watt, D. S. *JACS* **1983**, *105*, 619. (b) Kyler, K. S.; Bashir-Hashemi, A.; Watt, D. S. *JOC* **1984**, *49*, 1084.
3. Mukaiyama, T.; Soga, T.; Takenoshita, H. *CL* **1989**, 997.
4. (a) Czarnik, A. W. *TL* **1984**, *25*, 4875. (b) Yoshida, K.; Yamamoto, M.; Ishiguro, M. *CL* **1986**, 1059.
5. Boireau, G.; Abenhaim, D.; Bernardon, C.; Henry-Basch, E.; Sabourault, B. *TL* **1975**, 2521. Boireau, G.; Abenhaim, D.; Henry-Basch, E. *T* **1980**, *36*, 3061.
6. Takai, K.; Kimura, K.; Kuroda, T.; Hiyama, T.; Nozaki, H. *TL* **1983**, *24*, 5281.

7. (a) Jin, H.; Uenishi, J.-i.; Christ, W. J.; Kishi, Y. *JACS* **1986**, *108*, 5644. (b) Aicher, T. D.; Buszek, K. R.; Fang, F. G.; Forsyth, C. J.; Jung, S. H.; Kishi, Y.; Matelich, M. C.; Scola, P. M.; Spero, D. M.; Yeon, S. K. *JACS* **1992**, *114*, 3162. (c) Aicher, T. D.; Buszek, K. R.; Fang, F. G.; Forsyth, C. J.; Jung, S. H.; Kishi, Y.; Scola, P. M. *TL* **1992**, *33*, 1549. (d) Dyer, U. C.; Kishi, Y. *JOC* **1988**, *53*, 3383. (e) Goekjian, P. G.; Wu, T.-C.; Kang, H.-Y.; Kishi, Y. *JOC* **1987**, *52*, 4823. (f) Chen, S. H.; Horvath, R. F.; Joglar, J.; Fisher, M. J.; Danishefsky, S. J. *JOC* **1991**, *56*, 5834.
8. (a) Takai, K.; Tagashira, M.; Kuroda, T.; Oshima, K.; Utimoto, K.; Nozaki, H. *JACS* **1986**, *108*, 6048. (b) Angell, R.; Parsons, P. J.; Naylor, A.; Tyrrell, E. *SL* **1992**, 599.
9. (a) Wang, Y.; Babirad, S. A.; Kishi, Y. *JOC* **1992**, *57*, 468. (b) Aicher, T. D.; Kishi, Y. *TL* **1987**, *28*, 3463.
10. Armstrong, R. W.; Beau, J. M.; Cheon, S. H.; Christ, W. J.; Fujioka, H.; Ham, W.-H.; Hawkins, L. D.; Jin, H.; Kang, S. H.; Kishi, Y.; Martinelli, M. J.; McWhorter, W. W., Jr.; Mizuno, M. Nakata, M.; Stutz, A. E.; Talamas, F. X.; Taniguchi, M.; Tino, J. A.; Ueda, K.; Uenishi, J. I.; White, J. B.; Yonaga, M. *JACS* **1989**, *111*, 7525. (b) Kishi, Y. *PAC* **1989**, *61*, 313.
11. (a) Rowley, M.; Tsukamoto, M.; Kishi, Y. *JACS* **1989**, *111*, 2735. (b) Rowley, M.; Kishi, Y. *TL* **1988**, *29*, 4909.
12. (a) Crévisy, C.; Beau, J. M. *TL* **1991**, *32*, 3171. (b) Lu, Y.-F.; Harwig, C. W.; Fallis, A. G. *JOC* **1993**, *58*, 4204.
13. Takai, K.; Nitta, K.; Utimoto, K. *JACS* **1986**, *108*, 7408.
14. Trost, B. M.; Tour, J. M. *JACS* **1987**, *109*, 5268.
15. Montgomery, D.; Reynolds, K.; Stevenson, P. *CC* **1993**, 363.
16. (a) Pons, J.-M.; Santelli, M. *T* **1988**, *44*, 4295. (b) Luh, T.-Y.; Ni, Z.-J. *S* **1990**, 89. (c) Caubère, P. *AG(E)* **1983**, *22*, 599.
17. Ashby, E. C.; Lin, J. J. *JOC* **1978**, *43*, 2567.
18. (a) Ashby, E. C.; Lin, J. J. *TL* **1977**, 4481. (b) Ashby E. C.; Lin, J. J. *JOC* **1978**, *43*, 1263. (c) Brunet, J. J.; Vanderesse, R.; Caubere, P. *JOM* **1978**, *157*, 125.
19. Tufariello, J. J.; Meckler, H.; Pushpananda, K.; Senaratne, A. *T* **1985**, *41*, 3447.
20. (a) Dhawan, D.; Grover, S. K. *SC* **1992**, *22*, 2405. (b) Abe, N.; Fujisaki, F.; Sumoto, K.; Miyano, S. *CPB* **1991**, *39*, 1167.
21. (a) Myrboh, B.; Singh, L. W.; Ila, H.; Junjappa, H. *S* **1982**, 307. (b) Euerby, M. R.; Waigh, R. D. *SC* **1986**, *16*, 779. (c) Nishio, T.; Omote, Y. *CL* **1979**, 1223. (d) Truce, W. E.; Perry, F. M. *JOC* **1965**, *30*, 1316. (e) Back, T. G. *CC* **1984**, 1417.
22. Sarma, J. C.; Borbaruah, M.; Sharma, R. P. *TL* **1985**, *26*, 4657.
23. (a) He, Y.; Pan, X.; Zhao, H.; Wang, S. *SC* **1989**, *19*, 3051. (b) Sharma, D. N.; Sarma, R. P. *TL* **1985**, *26*, 371. (c) Wang, F.; Chiba, K.; Tada, M. *JCS(P1)* **1992**, 1897.
24. (a) Nose, A.; Kudo, T. *CPB* **1988**, *36*, 1529. (b) Hanaya, K., Fujita, N.; Kudo, H. *CI(L)* **1973**, 794. (c) Osby, J. O.; Ganem, B. *TL* **1985**, *26*, 6413. (d) Sarma, J. C.; Sharma, R. P. *CI(L)* **1987**, 764. (e) Rao, H. S. P.; Reddy, K. S.; Turnbull, K.; Borchers, V. *SC* **1992**, *22*, 1339. (f) Ipaktschi, J. *CB* **1984**, *117*, 856.
25. (a) Nose, A., Kudo, T. *CPB* **1986**, *34*, 3905. (b) Satoh, T.; Suzuki, S.; Suzuki, Y.; Miyaji, Y.; Imai, Z. *TL* **1969**, 4555.
26. Chatani, N.; Takeyasu, T.; Hanafusa, T. *TL* **1986**, *27*, 1841.
27. Chatani, N.; Hanafusa, T. *TL* **1986**, *27*, 4201.
28. Lappert, M. F.; Nile, T. A.; Takahashi, S. *JOM* **1974**, *72*, 425.
29. Hazarika, M. J.; Barua, N. C. *TL* **1989**, *30*, 6567.
30. Sarmah, P.; Barua, N. C. *TL* **1990**, *31*, 4065.
31. (a) Sakai, K.; Watanabe, K. *BCJ* **1967**, *40*, 1548. (b) Rieke, R. D.; Kavaliunas, A. V.; Rhyne, L. D.; Fraser, D. J. J. *JACS* **1979**, *101*, 246.
32. (a) Nose, A.; Kudo, T. *CPB* **1990**, *38*, 2097. (b) Petrier, C.; Luche, J.-L. *TL* **1987**, *28*, 2351. (c) Colon, I. *JOC* **1982**, *47*, 2622.
33. Sustmann, R.; Hopp, P.; Holl, P. *TL* **1989**, *30*, 689.
34. (a) Wang, Y.; Yang, Z.-Y.; Burton, D. J. *TL* **1992**, *33*, 2137. (b) Yang, Z.-Y.; Burton, D. J. *JOC* **1992**, *57*, 5144.
35. Yang, Z.-Y.; Burton, D. J. *TL* **1991**, *32*, 1019.
36. Yamashita, J.; Inoue, Y.; Kondo, T.; Hashimoto, H. *CL* **1986**, 407.
37. (a) Tiecco, M.; Testaferri, L.; Tingoli, M.; Chianelli, D.; Montanucci, M. *S* **1984**, 736. (b) Tiecco, M.; Tingoli, M.; Testaferri, L.; Chianelli, D.; Wenkert, E. *T* **1986**, *42*, 1475. (c) Tiecco, M.; Tingoli, M.; Testaferri, L.; Bartoli, D.; Chianelli, D. *T* **1989**, *45*, 2857.
38. Sone, T.; Umetsu, Y.; Sato, K. *BCJ* **1991**, *64*, 864.
39. Takagi, K.; Hayama, N. *CL* **1983**, 637.
40. (a) Tavs, P. *CB* **1970**, *103*, 2428. (b) Balthazor, T. M.; Grabiak, R. C. *JOC* **1980**, *45*, 5425. (c) Lu, X.; Zhu, J. *JOM* **1986**, *304*, 239.
41. (a) Mauret, P.; Alphonse, P. *JOM* **1984**, *276*, 249. (b) Mauret, P.; Alphonse, P. *JOC* **1982**, *47*, 3322. (c) Alphonse, P.; Moyen, F.; Mazerolles, P. *JOM* **1988**, *345*, 209.
42. Carre, F. H.; Corriu, R. J. P. *JOM* **1974**, *74*, 49.
43. Amano, T.; Ota, T.; Yoshikawa, K.; Sano, T.; Ohuchi, Y.; Sato, F.; Shiono, M.; Fujita, Y. *BCJ* **1986**, *59*, 1656.
44. (a) Mercier, F.; Mathey, F.; Fischer, J.; Nelson, J. H. *JACS* **1984**, *106*, 425. (b) Mercier, F.; Mathey, F.; Fischer, J.; Nelson, J. H. *IC* **1985**, *24*, 4141.
45. Pray, A. R. *Inorg. Synth.* **1957**, *5*, 153.

Tien-Yau Luh & Yu-Tsai Hsieh

National Taiwan University, Taipei, Taiwan

Osmium Tetroxide[1]

OsO_4

[20816-12-0] OsO_4 (MW 254.20)

(*cis* dihydroxylation of alkenes; osmylation; asymmetric and diastereoselective dihydroxylation; oxyamination of alkenes)

Physical Data: mp 39.5–41 °C; bp 130 °C; *d* 4.906 g cm^{-3}; chlorine- or ozone-like odor.

Solubility: sol water (5.3% at 0 °C, 7.24% at 25 °C); sol many organic solvents (toluene, *t*-BuOH, CCl_4, acetone, methyl *t*-butyl ether).

Form Supplied in: pale yellow solid in glass ampule, as 4 wt % solution in water, and as 2.5 wt % in *t*-BuOH.

Handling, Storage, and Precautions: vapor is toxic, causing damage to the eyes, respiratory tract, and skin; may cause temporary blindness; LD_{50} 14 mg/kg for the rat, 162 mg/kg for the mouse. Because of its high toxicity and high vapor pressure, it should be handled with extreme care in a chemical fume hood; chemical-resistant gloves, safety goggles, and other protective clothing should be worn; the solid reagent and its solutions should be stored in a refrigerator.

Dihydroxylation of Alkenes. The *cis* dihydroxylation (osmylation) of alkenes by osmium tetroxide to form *cis*-1,2-diols (*vic*-glycols) is one of the most reliable synthetic transformations (eq 1).[1]

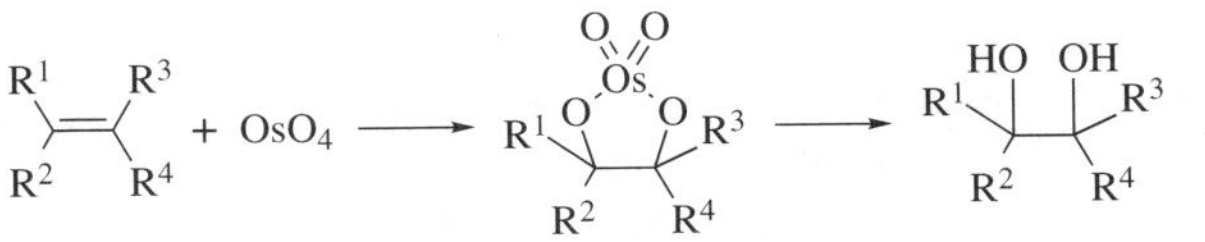

(1)

The reaction has been proposed to proceed through a [3 + 2] or [2 + 2] pathway to give the common intermediate osmium(VI) monoglycolate ester (osmate ester), which is then hydrolyzed reductively or oxidatively to give the *cis*-1,2-diol (eq 2). The *cis* dihydroxylation of alkenes is accelerated by tertiary amines such as ***Pyridine***, quinuclidine, and derivatives of dihydroquinidine (DHQD) or dihydroquinine (DHQ) (eq 3).

Due to the electrophilic nature of osmium tetroxide, electron-withdrawing groups connected to the alkene double bond retard the dihydroxylation.[2] This is in contrast to the oxidation of alkenes by ***Potassium Permanganate***, which preferentially attacks electron-deficient double bonds. However, in the presence of a tertiary amine such as pyridine, even the most electron-deficient alkenes can be osmylated by osmium tetroxide (eq 4).[3] The more highly substituted double bonds are preferentially oxidized (eq 5).

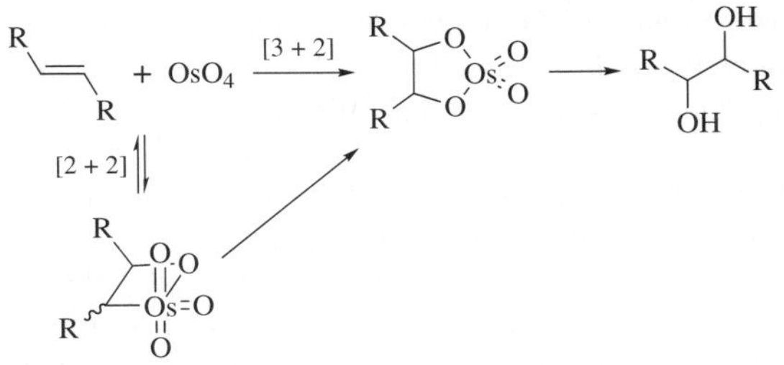

(2)

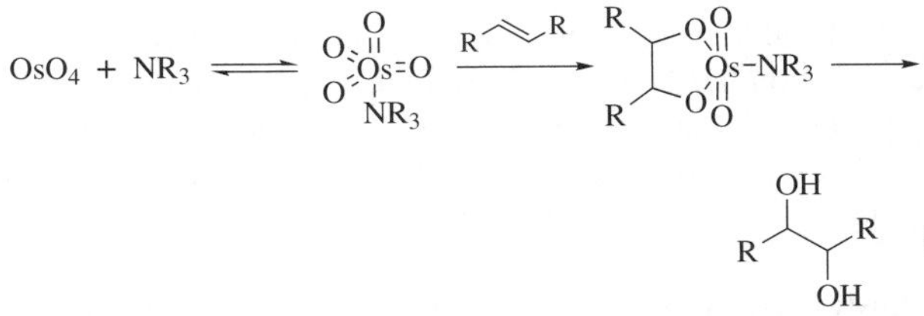

(3)

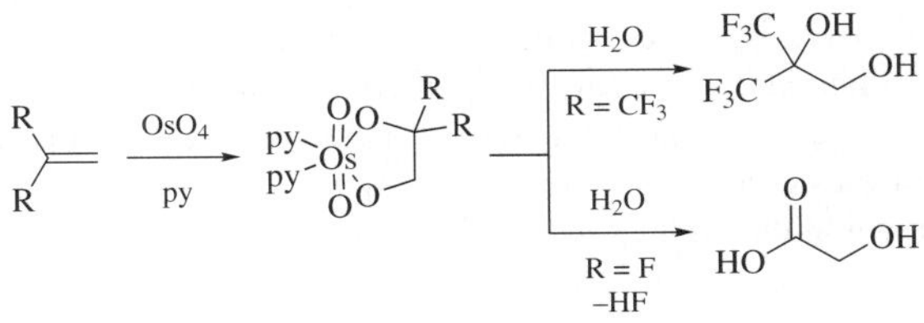

(4)

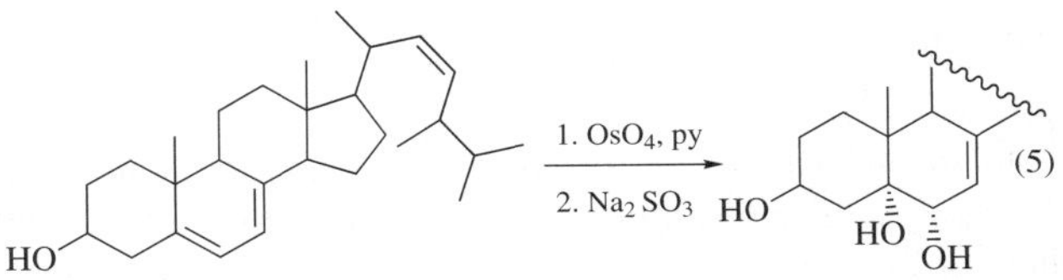

(5)

Under stoichiometric and common catalytic osmylation conditions, alkene double bonds are hydroxylated by osmium tetroxide without affecting other functional groups such as hydroxyl groups, aldehyde and ketone carbonyl groups, acetals, triple bonds, and sulfides (see also ***Osmium Tetroxide–N-Methylmorpholine N-Oxide***).

The *cis* dihydroxylation can be performed either stoichiometrically, if the alkene is precious, or more economically and conveniently with a catalytic amount of osmium tetroxide (or its precursors such as osmium chloride or potassium osmate) in conjunction with a cooxidant. In the stoichiometric dihydroxylation, the diol product is usually obtained by the reductive hydrolysis of the osmate ester with a reducing agent such as ***Lithium Aluminum Hydride***, ***Hydrogen Sulfide***, K_2SO_3 or Na_2SO_3, and $KHSO_3$ or $NaHSO_3$. The reduced osmium species is normally removed by filtration. Osmium can be recovered as osmium tetroxide by oxidation of low-valent osmium compounds with hydrogen peroxide.[4] In the catalytic dihydroxylation, the osmate ester is usually hydrolyzed under basic aqueous conditions to produce the diol and osmium(VI) compounds, which are then reoxidized by the cooxidant to osmium tetroxide to continue the catalytic cycle. Normally 0.01% to 2% equiv of osmium tetroxide or precursors are used in the catalytic dihydroxylation. Common cooxidants are metal chlorates, ***N-Methylmorpholine N-Oxide*** (NMO), ***Trimethylamine***

N*-Oxide**, ***Hydrogen Peroxide, ***t-Butyl Hydroperoxide***, and ***Potassium Ferricyanide***. ***Oxygen*** has also been used as cooxidant in dihydroxylation of certain alkenes.[5] Excess cooxidant and osmium tetroxide are reduced with a reducing agent such as those mentioned above during the workup. The stoichiometric dihydroxylation can be carried out in almost any inert organic solvent, including most commonly MTBE, toluene, and *t*-BuOH. In the catalytic dihydroxylation, in order to dissolve the inorganic cooxidant and other additives, a mixture of water and an organic solvent are often used. The most common solvent combinations in this case are acetone–water and *t*-BuOH–water. Because of the high cost and toxicity of osmium tetroxide, the stoichiometric dihydroxylation has been mostly replaced by the catalytic version in preparative organic chemistry (see also ***Osmium Tetroxide–t-Butyl Hydroperoxide***, ***Osmium Tetroxide–N-Methylmorpholine N-Oxide***, and ***Osmium Tetroxide-Potassium Ferricyanide***).

Diastereoselective Dihydroxylation. Dihydroxylation of acyclic alkenes containing an allylic, oxygen-bearing stereocenter proceeds with predictable stereochemistry. In general, regardless of the double-bond substitution pattern and geometry, the relative stereochemistry between the pre-existing hydroxyl or alkoxyl group and the adjacent newly formed hydroxyl group of the major diastereomer will be *erythro* (i.e. *anti* if the carbon chain is drawn in the zig-zag convention) (eq 6).[6,7]

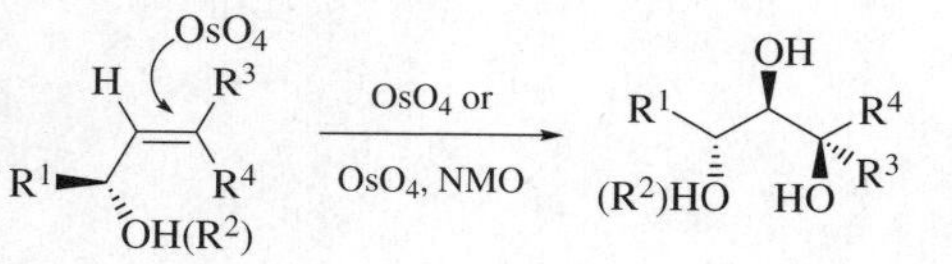

OsO_4 or OsO_4, NMO (6)

In the osmylation of 1,2-disubstituted allylic alcohols and derivatives, *cis*-alkenes provide higher diastereoselectivity than the corresponding *trans*-alkenes (eqs 7 and 8).[6] Opposite selectivities have been observed in the osmylation of (*Z*)-enoate and (*E*)-enoate esters (eqs 9 and 10).[8] High selectivity has also been observed in the osmylation of 1,1-disubstituted[9] and (*E*)-trisubstituted allylic alcohols and derivatives[10] and bis-allylic compounds[11] (eqs 11–13).

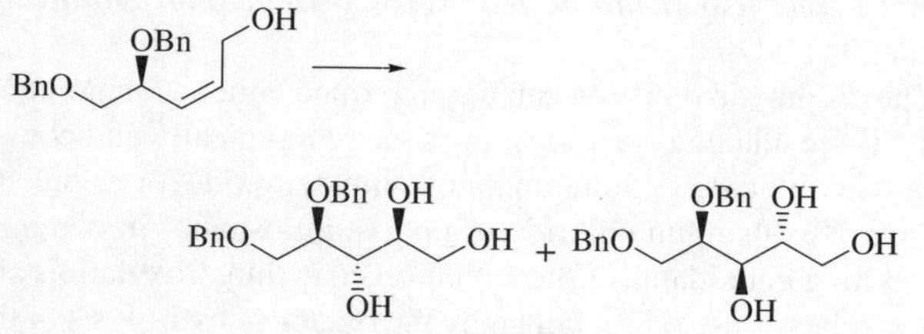

(7)

OsO_4	8.0:1.0
OsO_4, NMO	7.0:1.0

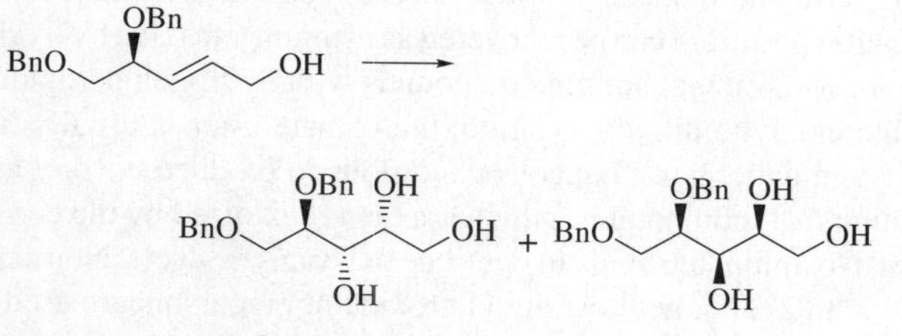

(8)

OsO_4	4.2:1.0
OsO_4, NMO	3.1:1.0

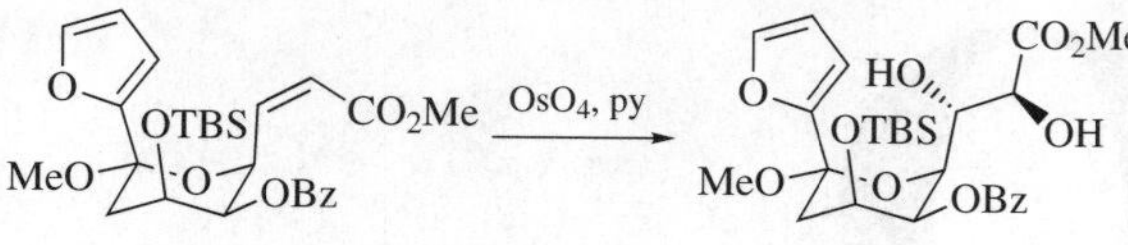

OsO_4, py (9) de > 18:1

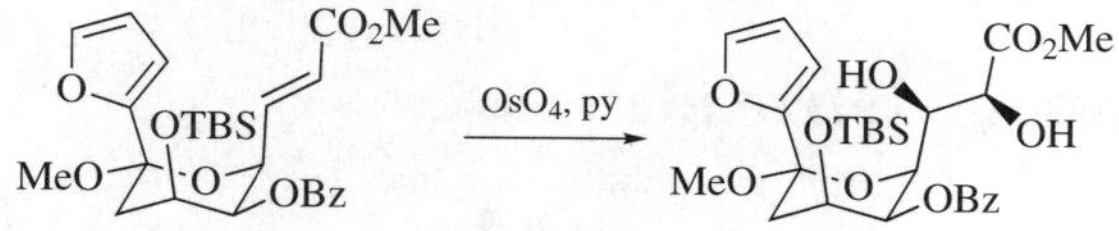

OsO_4, py (10) de > 25:1

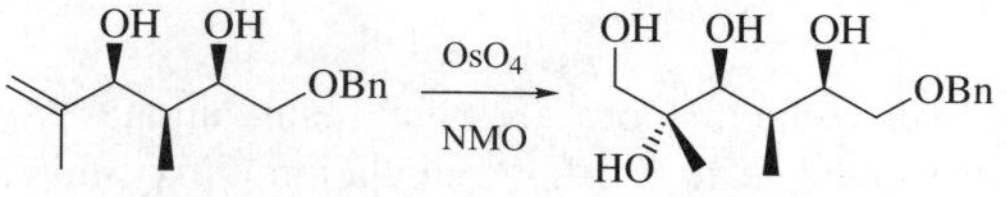

OsO_4 / NMO (11) de = 35:1

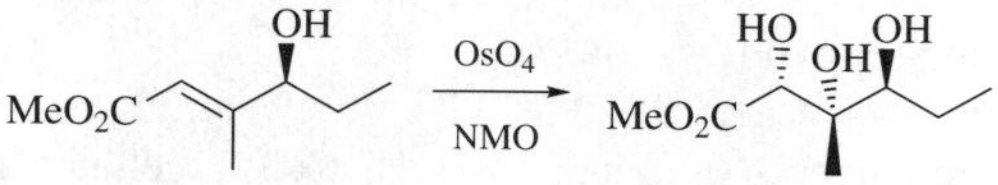

OsO_4 / NMO (12) de > 100:1

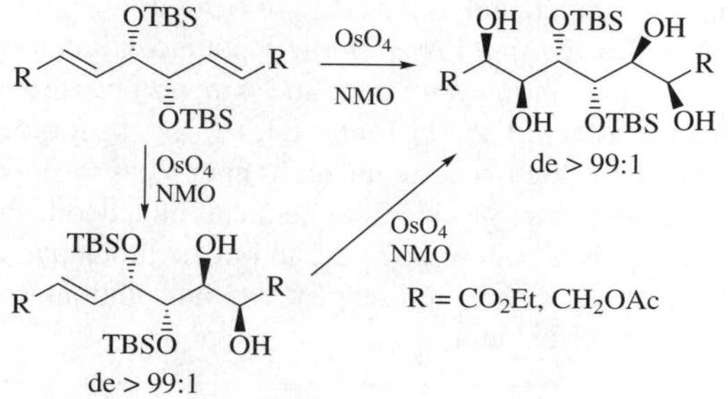

(13) de > 99:1; de > 99:1; R = CO_2Et, CH_2OAc

The diastereoselective osmylation has been extended to oxygen-substituted allylic silane systems, and the general rule observed for the allylic alcohol system also applies (eq 14).[12] High selectivity is also observed in the osmylation of allylsilanes where the substituent on the chiral center bearing the silyl group is larger than a methyl group (eq 15).[13] These diastereoselectivities have been achieved in both stoichiometric and catalytic dihydroxylations. Slightly higher selectivity has been observed in the stoichiometric reaction than in the catalytic reaction; this may be due to less selective bis-osmate ester formation in the catalytic reaction using NMO as the cooxidant. Use of $K_3Fe(CN)_6$ may solve this discrepancy. Several rationales have been proposed for the observed selectivity.[14] The conclusion appears to be that the osmylation of these systems is controlled by steric bias, rather than by the electronic nature of the allylic system, and osmylation will occur from the sterically more accessible face. The high diastereoselectivity of osmium tetroxide in the dihydroxylation of chiral unsaturated compounds has been applied widely in organic synthesis.[8,15]

Sulfoxide groups direct the dihydroxylation of a remote double bond in an acyclic system perhaps by prior complexation of the sulfoxide oxygen with osmium tetroxide (eqs 16 and 17).[16] Chiral sulfoximine-directed diastereoselective osmylation of cycloalkenes has been used for the synthesis of optically pure dihydroxycycloalkanones (eq 18).[17] Nitro groups also direct the osmylation of certain cycloalkenes, resulting in dihydroxylation from the more hindered side of the ring. In contrast, without the nitro group the dihydroxylation proceeds from the less hindered side (eq 19).[18]

(14)

de = 97:3

de = 4.4:1

(15)

R = Me 34:65
R = *i*-Pr 67:33
R = Ph 92:8

(16)

de > 20:1

(17)

de > 20:1

(18)

(19)

$R^1 = CH(NO_2)SO_2Ph$, $R^2 = CO_2Me$

$R^1 = OH$, $R^2 = H$

Enantioselective Dihydroxylation. The acceleration of osmylation by tertiary amines brought about the use of chiral amines as chiral ligands for the asymmetric dihydroxylation (AD).[1] The AD can be classified into two types: (a) noncatalytic reaction, where stoichiometric amounts of ligand and osmium tetroxide are used, and (b) catalytic reaction, where catalytic amounts of ligand and osmium tetroxide are employed in conjunction with stoichiometric amounts of cooxidant. Generally, in the stoichiometric AD systems, chiral chelating diamines are used as chiral auxiliaries with osmium tetroxide for the introduction of asymmetry to the diol products.[1,19] Although high asymmetric inductions have been achieved in these systems, the stoichiometric ADs have limited use in practical organic synthesis because of the cost of both ligand and osmium tetroxide. The discovery of the ligand-accelerated catalysis in AD made the transition from stoichiometric to catalytic AD possible.[1c,1d,20] In the most effective catalytic system, osmium tetroxide or its precursors and chiral ligands derived from cinchona alkaloids, dihydroquinidine (DHQD), or dihydroquinine (DHQ), are used catalytically in the presence of a stoichiometric amount of cooxidant such as NMO or $K_3Fe(CN)_6$. Besides alkaloid-derived ligands, other types of ligand have been designed and used in the catalytic AD with moderate success.[21] (see also ***Osmium Tetroxide–N-Methylmorpholine N-Oxide*** and ***Osmium Tetroxide–Potassium Ferricyanide***).

Double Diastereoselective Dihydroxylation. AD of homochiral alkenes gives matched and mismatched diastereoselectivities due to the steric interaction of the chiral osmium tetroxide–ligand complex with the chiral center in the vicinity of the alkene double bond. For example, in the noncatalytic osmylation of the monothioacetal (eq 20),[22] the ratio of (2*S*,3*R*) to (2*R*,3*S*) diastereomers is 2.5:1 with the achiral quinuclidine as ligand, 40:1 with ***Dihydroquinidine Acetate*** (DHQD-OAc) as ligand in the matched case, and 1:16 with ***Dihydroquinine Acetate*** (DHQ-OAc) as ligand in the mismatched case.

(20)

Quinuclidine	2.5:1
DHQD-OAc	40:1
DHQ-OAc	1:16

Diastereoselectivities in the catalytic AD with OsO_4–NMO of several α,β-unsaturated uronic acid derivatives are significantly enhanced when the alkenes are matched with the chiral ligands DHQ-CLB (dihydroquinine *p*-chlorobenzoate) and DHQD-CLB (dihydroquinidine *p*-chlorobenzoate) (eq 21).[23] Using the chiral ligands $(DHQD)_2$-PHAL and $(DHQ)_2$-PHAL, the double diastereoselective dihydroxylation of a chiral unsaturated ester has been tested using ***Osmium Tetroxide–Potassium Ferricyanide*** (eq 22).[1c] These results show that enhanced diastereoselectivity in the dihydroxylation can be achieved by matching of alkene diastereoselectivity with catalyst enantioselectivity. Double diastereoselective dihydroxylation with a bidentate ligand has also been reported.[24] Kinetic resolutions of racemic alkenes with OsO_4 in the presence of a chiral ligand have been demonstrated.[25] An elegant example is the kinetic resolution of the enantiomers of C_{76}, the smallest chiral fullerene, by asymmetric osmylation in the presence of DHQD- and DHQ-derived ligands.[26]

OsO_4, NMO (21)

R	Reagent	Ratio
BnO, BnO, BnO, OMe (pyranoside)	OsO_4 only DHQD-CLB DHQ-CLB	10.3:1 1.3:1 20.5:1
BnO, BnO, BnO, OMe (furanoside)	OsO_4 only DHQD-CLB DHQ-CLB	7.4:1 3.4:1 15.9:1

CO_2-*i*-Pr; OsO_4, $K_3Fe(CN)_6$ (22)

no ligand	2.8:1
$(DHQD)_2$-PHAL	39:1
$(DHQ)_2$-PHAL	1:1.3

Oxyamination of Alkenes and Oxidation of Other Functional Groups. Osmium tetroxide catalyzes the vicinal oxyamination of alkenes to give *cis*-vicinal hydroxyamides with ***Chloramine-T*** (eq 23)[27] and alkyl *N*-chloro-*N*-argentocarbamate, generated in situ by the reaction of alkyl *N*-chlorosodiocarbamate (such as ethyl or *t*-butyl *N*-chlorosodiocarbamate) with ***Silver(I) Nitrate*** (eq 24).[28]

+ $TsNClNa{\bullet}3H_2O$ → (1% OsO_4, *t*-BuOH, 60 °C, 66–73%) NHTs, OH (23)

Ph–CH=CH–Ph → ($EtOCONClNa$, $AgNO_3$, H_2O, 1% OsO_4, MeCN, 66–69%) OH, $NHCO_2Et$ (24)

Since chloramine-T is readily available, the former method offers a practical and direct method for introducing a vicinal hydroxyl group and a tolylsulfonamide to a double bond. While the sulfonamide protecting group is difficult to remove (and undesirable in some cases), the *N*-chloro-*N*-argentocarbamate system provides an alternative method, since the carbamate group can be easily removed to give free amine. This latter system is also more regioselective and reactive towards electron-deficient alkenes such as ***Dimethyl Fumarate*** and (*E*)-stilbene than the procedures based on chloramine-T. In all of these oxyamination reactions, monosubstituted alkenes react more rapidly than di- or trisubstituted alkenes. In the presence of tetraethylammonium acetate, trisubstituted alkenes can be oxyaminated with catalytic OsO_4 and *N*-chloro-*N*-metallocarbamates. So far, attempts to effect catalytic asymmetric oxyamination have not been successful.

Osmium tetroxide also catalyzes the oxidation of organic sulfides to sulfones with NMO or trimethylamine *N*-oxide (see ***Osmium Tetroxide–N-Methylmorpholine N-Oxide***). In contrast, most sulfides are not oxidized with stoichiometric amounts of OsO_4. Oxidations of alkynes and alcohols with OsO_4 without and in the presence of cooxidants have also been reported.[1a,1b] However, these reactions have not found wide synthetic applications because of the availability of other methods.

Related Reagents. See Class O-16, pages 1–10. Sodium Periodate–Osmium Tetroxide.

1. (a) Schröder, M. *CRV* **1980**, *80*, 187. (b) Singh, H. S. In *Organic Synthesis by Oxidation with Metal Compounds*; Mijs, W. J.; De Jonge, C. R. H. I., Eds.; Plenum: New York, 1986; Chapter 12. (c) Johnson, R. A.; Sharpless, K. B. In *Catalytic Asymmetric Synthesis*; Ojima, I., Ed.; VCH: New York, 1993. (d) Lohray, B. B. *TA* **1992**, *3*, 1317. (e) Haines, A. H. *COS* **1991**, *7*, 437.
2. Henbest, H. B.; Jackson, W. R.; Robb, B. C. G. *JCS(B)* **1966**, 803.
3. Herrmann, W. A.; Eder, S.; Scherer, W. *AG(E)* **1992**, *31*, 1345.
4. Rüegger, U. P.; Tassera, J. *Swiss Chem.* **1986**, *8*, 43.
5. Austin, R. G.; Michaelson, R. C.; Myers, R. S. In *Catalysis of Organic Reactions*; Augustine, R. L., Ed.; Dekker: New York, 1985; p 269.
6. (a) Cha, J. K.; Christ, W. J.; Kishi, Y. *T* **1984**, *40*, 2247. (b) Christ, W. J.; Cha, J. K.; Kishi, Y. *TL* **1983**, *24*, 3943 and 3947.
7. Brimacombe, J. S.; Hanna, R.; Kabir, A. K. M. S.; Bennett, F.; Taylor, I. D. *JCS(P1)* **1986**, 815.
8. DeNinno, M. P.; Danishefsky, S. J.; Schulte, G. *JACS* **1988**, *110*, 3925.
9. Evans, D. A.; Kaldor, S. W. *JOC* **1990**, *55*, 1698.
10. Stork, G.; Kahn, M. *TL* **1983**, *24*, 3951.
11. Saito, S.; Morikawa, Y.; Moriwake, T. *JOC* **1990**, *55*, 5424.
12. Panek, J. S.; Cirillo, P. F. *JACS* **1990**, *112*, 4873.
13. Fleming, I.; Sarker, A. K.; Thomas, A. P. *CC* **1987**, 157.
14. (a) Vedejs, E.; McClure, C. K. *JACS* **1986**, *108*, 1094. (b) Vedejs, E.; Dent, W. H., III *JACS* **1989**, *111*, 6861.
15. (a) Ikemoto, N.; Schreiber, S. L. *JACS* **1990**, *112*, 9657. (b) Hanselmann, R.; Benn, M. *TL* **1993**, *34*, 3511.
16. (a) Hauser, F. M.; Ellenberger, S. R.; Clardy, J. C.; Bass, L. S. *JACS* **1984**, *106*, 2458. (b) Solladié, G.; Fréchou, C.; Demailly, G. *TL* **1986**, *27*, 2867. (c) Solladié, G.; Fréchou, C.; Hutt, J.; Demailly, G. *BSF* **1987**, 827.
17. (a) Johnson, C. R.; Barbachyn, M. R. *JACS* **1984**, *106*, 2459. (b) Johnson, C. R. *PAC* **1987**, *59*, 969.
18. (a) Trost, B. M.; Kuo, G.-H.; Benneche, T. *JACS* **1988**, *110*, 621. (b) Poli, G. *TL* **1989**, *30*, 7385.
19. Hanessian, S.; Meffre, P.; Girard, M.; Beaudoin, S.; Sancéau, J.-Y.; Bennani, Y. *JOC* **1993**, *58*, 1991 and references therein.
20. Anderson, P. G.; Sharpless, K. B. *JACS* **1993**, *115*, 7047.
21. (a) Oishi, T.; Hirama, M. *TL* **1992**, *33*, 639. (b) Imada, Y.; Saito, T.; Kawakami, T.; Murahashi, S.-I. *TL* **1992**, *33*, 5081.
22. Annuziata, R.; Cinquini, M.; Cozzi, F.; Raimondi, L.; Stefanelli, S. *TL* **1987**, *28*, 3139.
23. Brimacombe, J. S.; McDonald, G.; Rahman, M. A. *Carbohydr. Res.* **1990**, *205*, 422.
24. Oishi, T.; Iida, K.-I.; Hirama, M. *TL* **1993**, *34*, 3573.
25. (a) Ward, R. A.; Procter, G. *TL* **1992**, *33*, 3363. (b) Lohray, B. B.; Bhushan, V. *TL* **1993**, *34*, 3911.

26. Hawkins, J. M.; Meyer, A. *Science* **1993**, *260*, 1918.
27. Herranz, E.; Sharpless, K. B. *OSC* **1990**, *7*, 375.
28. Herranz, E.; Sharpless, K. B. *OSC* **1990**, *7*, 223.

Yun Gao
Sepracor, Marlborough, MA, USA

Osmium Tetroxide–*t*-Butyl Hydroperoxide[1]

OsO_4–*t*-BuOOH

(OsO_4)		
[20816-12-0]	O_4Os	(MW 254.20)
(TBHP)		
[75-91-2]	$C_4H_{10}O_2$	(MW 90.14)

(*cis* dihydroxylation of alkenes; osmylation)

Physical Data: for 70% aq TBHP: bp 96 °C; mp −2.8 °C; *d* 0.935 g cm^{-3}. See also ***Osmium Tetroxide***.
Solubility: TBHP: sol acetone, *t*-butyl alcohol.
Form Supplied in: 70% aq TBHP (remaining water) in plastic bottle.
Handling, Storage, and Precautions: TBHP is an eye and skin irritant. Safety goggles and gloves should be worn when handling TBHP solution. Avoid contamination of high-strength TBHP solution with strong acids and transition metal salts known to be good autooxidation catalysts (e.g. Mn, Fe, and Co). Never work with pure TBHP. Do not distill reaction mixtures containing TBHP without prior reduction with reducing agents such as Na_2SO_3. Store 70% aq TBHP in plastic containers below 38 °C but not much below 25 °C. Keep away from bright light and heat sources. See also ***Osmium Tetroxide***.

Dihydroxylation of Alkenes. Although, in some cases, catalytic *cis* dihydroxylation with OsO_4 using ***Hydrogen Peroxide*** or metal chlorates as cooxidant gives good yields of diols, overoxidation leading to high yields of ketols or aldehyde products becomes a problem.[1] Thus alkaline ***t-Butyl Hydroperoxide*** was introduced as a superior cooxidant in the catalytic *cis* dihydroxylation. In the OsO_4–TBHP system, 70% aq TBHP is normally used and the reaction is usually performed in *t*-BuOH or acetone when tetraethylammonium acetate (Et_4NOAc) is used as the base (see below). In most cases, only 0.2 mol % OsO_4 is sufficient for completion of the oxidation.[2]

The key to the success of the OsO_4–TBHP system is the presence of a nucleophile in the form of tetraethylammonium hydroxide (Et_4NOH) or Et_4NOAc. The role of the nucleophile is to increase the turnover rate of the catalytic cycle by facilitating the hydrolysis of the osmate ester intermediate. For example, it is possible to dihydroxylate even some tetrasubstituted alkenes using the OsO_4–TBHP–Et_4NOH combination (eq 1).[2a]

Tetraethylammonium acetate gives higher yields of diol than Et_4NOH in the *cis* dihydroxylation of base-sensitive alkenes even though, as a weaker base, it fails with tetrasubstituted alkenes (eq 2).[2b] In another example, alkenylphosphonates are dihydroxylated with OsO_4–TBHP in acetone in the presence of Et_4NOAc to give *threo*-dihydroxyalkylphosphonates in >80% yields. In contrast, using OsO_4–H_2O_2, a mixture of diol and ketol is obtained.[3] The OsO_4–TBHP system does not work well in the dihydroxylation of sterically hindered tri- and tetrasubstituted alkenes such as cholesterol (for oxidation of these alkenes, see ***Osmium Tetroxide-Potassium Ferricyanide***).

C_5H_{11} (alkene) → [0.2% OsO_4, 10% Et_4NOH; 1.6 equiv TBHP (70%), *t*-BuOH, 69%] → C_5H_{11} (diol, OH, OH) (1)

CO_2Et (alkene) → [0.2% OsO_4, 25% Et_4NOAc; 1.7 equiv TBHP (70%), acetone, 72%] → OH, CO_2Et, HO (2)

The OsO_4–TBHP system and OsO_4–NMO system (see ***Osmium Tetroxide–N-Methylmorpholine N-Oxide***) are quite comparable for the dihydroxylation of simple alkenes. The ability to use less OsO_4 and the lower expense of TBHP are the main advantages with the OsO_4–TBHP system. However, no asymmetric catalytic dihydroxylation using the OsO_4–TBHP system has been developed, and probably due to safety concerns there have been few synthetic applications of the OsO_4–TBHP system.[3,4]

Osmium tetroxide–TBHP is also used for the oxidation of 1-trimethylsilylalkynes to produce α-keto esters in about 60% yield (eq 3). It is most likely that the oxidation involves an α-keto acylsilane which then undergoes a Brook-type rearrangement to the final product.[5]

R—≡—TMS → [OsO_4, TBHP; MeOH] → [R–CO–CO–TMS] → [60%] → R–CO–CO_2Me (3)

Related Reagents. See Class O-16, pages 1–10.

1. (a) Sharpless, K. B.; Verhoeven, T. R. *Aldrichim. Acta* **1979**, *12*, 63. (b) Singh, H. S. In *Organic Synthesis by Oxidation with Metal Compounds*, Mijs, W. J.; De Jonge, C. R. H. I., Eds.; Plenum: New York, 1986; Chapter 12.
2. (a) Sharpless, K. B.; Akashi, K. *JACS* **1976**, *98*, 1986. (b) Akashi, K.; Palermo, R. E.; Sharpless, K. B. *JOC* **1978**, *43*, 2063.
3. Pondaven-Raphalen, A.; Sturtz, G. *PS* **1987**, *29*, 329.
4. (a) Levine, S. G.; Gopalakrishnan, B. *TL* **1979**, *20*, 699. (b) Current, S.; Sharpless, K. B. *TL* **1978**, *19*, 5075. (c) Kranz, D.; Dinges, K.; Wendling, P. *Angew. Makromol. Chem.* **1976**, *51*, 25.
5. Bulman Page, P. C.; Rosenthal, S. *TL* **1986**, *27*, 1947.

Yun Gao
Sepracor, Marlborough, MA, USA

Osmium Tetroxide–*N*-Methylmorpholine *N*-Oxide[1]

(OsO4)
[20816-12-0] O_4Os (MW 254.20)
(NMO)
[7529-22-8] $C_5H_{11}NO_2$ (MW 117.17)

(*cis* dihydroxylation of alkenes; osmylation; asymmetric dihydroxylation of alkenes; oxidations of enol ethers, sulfides)

Physical Data: NMO monohydrate: mp 75–76 °C; 60 wt % aqueous solution: bp 118.5 °C, mp –20 °C, *d* 1.13 g cm^{-3}. See also ***Osmium Tetroxide***.
Solubility: NMO: sol water, acetone, *t*-butyl alcohol, THF.
Form Supplied in: NMO: as monohydrate or 60 wt % aqueous solution.
Handling, Storage, and Precautions: NMO is irritating to eyes, respiratory system, and skin. Store refrigerated. Use in a fume hood. See also ***Osmium Tetroxide***.

Dihydroxylation of Alkenes. The use of an amine *N*-oxide such as ***N-Methylmorpholine N-Oxide*** as cooxidant for osmium tetroxide-catalyzed *cis* dihydroxylation has become the standard method for the preparation of *cis*-1,2-diols because of the higher yield of diol product and less byproduct formation compared with oxidations carried out with ***Hydrogen Peroxide***, metal chlorates, and ***t-Butyl Hydroperoxide***.[1] Other alkylamine *N*-oxides such as ***Trimethylamine N-Oxide*** have also been used in the catalytic *cis* dihydroxylation. NMO (as monohydrate and 60 wt % aq solution) is commercially available and readily prepared or regenerated by treatment of *N*-methylmorpholine (NMM) with H_2O_2. Normally, aqueous acetone or aqueous *t*-BuOH is used in the catalytic *cis* dihydroxylation with OsO_4–NMO. Workup involves reduction by sodium sulfite or sodium bisulfite and extraction with ethyl acetate. A detailed procedure for the preparation of NMO (eq 1) and dihydroxylation of cyclohexene with OsO_4–NMO (eq 2) has been published.[2] Functional groups such as alcohols, esters, lactones, carboxylic acids, ketones, and electron-poor alkenes such as α,β-unsaturated ketones are not affected. For unstable or water-soluble diols, the osmylation can be performed in the presence of ***Dihydroxy(phenyl)borane*** and the product is isolated as the borate ester (eq 3).[3]

30% H_2O_2; 75 °C, 22.5 h; 84–89% (1)

OsO_4 (2.7%); NMO, aq. acetone; 80–90% (2)

OsO_4, NMO; $PhB(OH)_2$; CH_2Cl_2; 93% (3)

The OsO_4–NMO system works well with mono- and disubstituted alkenes. Bases such as ***Pyridine*** accelerate the *cis* dihydroxylation in the OsO_4–amine *N*-oxide system, especially for hindered alkenes.[4,5] For example, the α-pinene derivative nopol is hydroxylated in low yield by the ***Osmium Tetroxide–t-Butyl Hydroperoxide*** system; however, it is hydroxylated in 62% yield with OsO_4–Me_3NO in the presence of pyridine in aqueous *t*-BuOH. Other hindered alkenes are also oxidized by this system in 78–93% yield. By switching the solvent to aqueous acetone, the OsO_4–Me_3NO–pyridine system has been used for the *cis* dihydroxylation of an intermediate for the preparation of an HMG-CoA reductase inhibitor on a 20-kg scale with 95% diastereoselectivity in 78% yield (eq 4).[6]

1.3% OsO_4; Me_3NO; pyridine, H_2O; acetone, reflux; 78%; de = 39:1 (4)

Diastereoselective Osmylation. Diastereoselective dihydroxylations with OsO_4–NMO have been extensively studied (see also ***Osmium Tetroxide***). In general, in the osmylation of acyclic alkenes containing an allylic, oxygen-bearing stereocenter, the relative configuration between the pre-existing hydroxyl or alkoxyl group and the adjacent newly formed hydroxyl group in the major diastereomer is *erythro* (*anti*).[7] Taking advantage of the high diastereoselectivity in the osmylation reaction, the OsO_4–NMO or OsO_4–Me_3NO system has been used in many stereoselective organic syntheses. One example is the synthesis of octoses based on the catalytic osmylation of allylic alcohols with OsO_4–NMO (eq 5).[8]

OsO_4, NMO; 70%; 1. H^+; 2. $NaBH_4$; 48% (5)

D-*erythro*-D-Galactooctose

Opposite selectivity has been observed in the dihydroxylation of a cyclopentene derivative for the synthesis of carbonucleoside aristeromycin (eq 6).[9] Dihydroxylation of (**1**) with alkaline permanganate proceeds from the less hindered side to give the desired diol, which is then converted to aristeromycin (**2**). However, dihydroxylation of (**1**) with OsO_4–NMO proceeds from the more hindered side to give the *cis*-diol, which is subsequently transformed to *lyxo*-aristeromycin (**3**). This surprising stereoselectivity is attributed to the complexation of osmium tetroxide with the nitro group, resulting in a nitro-directed osmylation from the more hindered side.

O_2N, PhO_2S, Ad; (**1**) Ad = adenine; $KMnO_4$, NaOH; R', Ad, HO, OH; OsO_4, NMO; R', Ad, HO, OH → HO, Ad, HO, OH (**3**); HO, Ad, HO, OH (**2**) (6)

Asymmetric Dihydroxylation (AD). AD of alkenes with OsO_4–NMO is usually performed in aqueous acetone in the presence of dihydroquinidine (DHQD) or dihydroquinine (DHQ) derived chiral ligands such as dihydroquinidine *p*-chlorobenzoate (DHQD-CLB) or dihydroquinine *p*-chlorobenzoate (DHQ-CLB) (eq 7) (see also ***Osmium Tetroxide–Potassium Ferricyanide***).[10]

DHQD esters 'HO OH'; R^3, R^2, R^1; 'HO OH' DHQ esters; OsO_4 (0.2–0.4%), NMO, aq acetone, 80–90%; HO, R^2, R^3, R^1, OH; 40–92% ee (7)

For example, using OsO_4–NMO in the presence of the DHQD-CLB ligand, *trans*-stilbene is converted to (*R,R*)-stilbenediol in 88% ee, which is further enriched by recrystallization to >99% ee.[11] This method has also been used for the preparation of optically pure (2*S*,3*R*)-methyl 2,3-dihydroxyphenylpropionate, useful for the synthesis of the taxol sidechain.[12] A catalytic cycle of AD with OsO_4–NMO has been proposed.[13] The presence of a second catalytic cycle results in a reduced ee for some alkenes in comparison with the ee obtained under stoichiometric conditions, due to the slow hydrolysis of the intermediate osmate ester. Although addition of base such as tetraethylammonium acetate accelerates the hydrolysis of the osmate ester, thereby diminishing the second cycle, the best solution is to add the alkene slowly to prevent the build-up of the osmate ester. In this manner, many alkenes have been dihydroxylated to the corresponding diols in ee close to those obtained under stoichiometric conditions.[14]

Since the AD with OsO_4–$K_3Fe(CN)_6$ suppresses the second cycle in the aqueous *t*-BuOH solvent system, the OsO_4–NMO system has been largely superseded by the OsO_4–$K_3Fe(CN)_6$ system in catalytic AD. However, by employing the most effective ligands such as $(DHQD)_2$-PHAL or $(DHQ)_2$-PHAL and using the slow-addition technique, higher enantioselectivity in the AD with OsO_4–NMO can be expected.

Other Oxidations. Vinylsilanes are converted to silyl enol ethers with retention of the double bond configuration by dihydroxylation with OsO_4–Me_3NO followed by an anti β-elimination with ***Sodium Hydride*** (eqs 8 and 9).[15] α-Keto acylsilanes are prepared in over 40% overall yields from (*Z*)-vinylsilanes by dihydroxylation with OsO_4–NMO followed by oxidation with ***Dimethyl Sulfoxide–Oxalyl Chloride*** (eq 10).[16]

R, TMS; 1. OsO_4, Me_3NO; 2. NaH; ~35%; R, OTMS; 99% (*E*) (8)

R, TMS; 1. OsO_4, Me_3NO; 2. NaH; ~45%; R, OTMS; >99% (*Z*) (9)

R, TMS; OsO_4, Me_3NO; HO, OH, R, TMS; DMSO, $(COCl)_2$, 40–45%; O, O, R, TMS (10)

Enol ethers are transformed to α-hydroxy ketones with OsO_4–NMO (eq 11).[17] In the presence of a chiral ligand, optically active α-hydroxy ketones are obtained (also see ***Osmium Tetroxide–Potassium Ferricyanide***).[14] Although OsO_4 does not normally oxidize organic sulfides, the OsO_4–NMO system can oxidize sulfides to sulfones in good yields (eq 12).[18]

TMSO, R^2, R^1, R^3; OsO_4, NMO; *t*-BuOH; 54–98%; O, R^1, R^2, R^3, HO (11)

R^1–S–R^2; OsO_4, NMO; >80%; O O, R^1–S–R^2 (12)

Related Reagents. See Class O-16, pages 1–10.

1. (a) Schröder, M. *CRV* **1980**, *80*, 187. (b) Singh, H. S. In *Organic Syntheses by Oxidation with Metal Compounds*, Mijs, W. J.; De Jonge, C. R. H. I., Eds.; Plenum: New York, 1986; Chapter 12. (c) Johnson, R. A.; Sharpless, K. B. In *Catalytic Asymmetric Synthesis*, Ojima, I., Ed.; VCH: New York, 1993. (d) Lohray, B. B. *TA* **1992**, *3*, 1317.
2. (a) VanRheenen, V.; Cha, D. Y.; Hartley, W. M. *OSC* **1988**, *6*, 342. (b) VanRheenen, V.; Kelly, R. C.; Cha, D. Y. *TL* **1976**, 1973.
3. Iwasawa, N.; Kato, T.; Narasaka, K. *CL* **1988**, 1721.
4. Larsen, S. D.; Monti, S. A. *JACS* **1977**, *99*, 8015.
5. Ray, R.; Matteson, D. S. *TL* **1980**, *21*, 449.
6. Decamp, A. E.; Mills, S. G.; Kawaguchi, A. T.; Desmond, R.; Reamer, R. A.; DiMichele, L.; Volante, R. P. *JOC* **1991**, *56*, 3564.
7. (a) Cha, J. K.; Christ, W. J.; Kishi, Y. *T* **1984**, *40*, 2247. (b) Cha, J. K.; Christ, W. J.; Kishi, Y. *TL* **1983**, *24*, 3943 and 3947.
8. (a) Brimacombe, J. S.; Hanna, R.; Kabir, A. K. M. S.; Bennett, F.; Taylor, I. D. *JCS(P1)* **1986**, 815. (b) Brimacombe, J. S.; Hanna, R.; Kabir, A. K. M. S. *JCS(P1)* **1986**, 823.
9. Trost, B. M.; Kuo, G.-H.; Benneche, T. *JACS* **1988**, *110*, 621.
10. Jacobsen, E. N.; Markó, I.; Mungall, W. S.; Schröder, G.; Sharpless, K. B. *JACS* **1988**, *110*, 1968.
11. McKee, B. H.; Gilheany, D. G.; Sharpless, K. B. *OS* **1992**, *70*, 47.
12. (a) Fleming, P. R.; Sharpless, K. B. *JOC* **1991**, *56*, 2869. (b) Denis, J.-N.; Correa, A.; Greene, A. E. *JOC* **1990**, *55*, 1957.
13. Wai, J. S. M.; Markó, I.; Svendsen, J. S.; Finn, M. G.; Jacobsen, E. N.; Sharpless, K. B. *JACS* **1989**, *111*, 1123.
14. Lohray, B. B.; Kalantar, T. H.; Kim, B. M.; Park, C. Y.; Shibata, T.; Wai, J. S. M.; Sharpless, K. B. *TL* **1989**, *30*, 2041.

15. Hudrlik, P. F.; Hudrlik, A. M.; Kulkarni, A. K. *JACS* **1985**, *107*, 4260.
16. Page, P. C. B.; Rosenthal, S. *TL* **1986**, *27*, 2527.
17. McCormick, J. P.; Tomasik, W.; Johnson, M. W. *TL* **1981**, *22*, 607.
18. (a) Kaldor, S. W.; Hammond, M. *TL* **1991**, *32*, 5043. (b) Priebe, W.; Grynkiewicz, G. *TL* **1991**, *32*, 7353.

Yun Gao
Sepracor, Marlborough, MA, USA

Oxodiperoxymolybdenum(pyridine)(hexamethylphosphoric triamide)

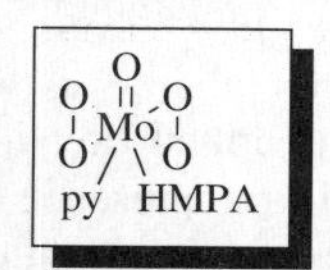

[23319-63-3] $C_{11}H_{23}MoN_4O_6P$ (MW 434.29)

(hydroxylation of enolates[1] and nitrile anions;[2] preparation of carbonyl compounds by oxidative degradation of anions of sulfones;[3] nitroalkanes;[4] phosphonates;[5] for *N*-hydroxylation of amides;[6] for oxidative cleavage of C–B bonds[7])

Alternate Name: MoOPH.
Physical Data: mp 103–105 °C (dec).
Solubility: insol ether; sparingly sol THF; sol dichloromethane.
Form Supplied in: finely divided yellow crystals; not available commercially.
Analysis of Reagent Purity: no convenient assay has been reported. Decomposition is indicated if the reagent smells of pyridine or becomes sticky.[1b]
Preparative Methods: a detailed procedure is available:[1b] (1) $MoO_3 + 30\%\ H_2O_2$ (<40 °C); (2) add HMPA; vacuum dry $MoO_5{\cdot}HMPA{\cdot}H_2O$ to $MoO_5{\cdot}HMPA$; (3) pyridine in THF gives MoOPH.
Purification: attempted recrystallization causes decomposition of MoOPH. Purity depends on the quality of $MoO_5{\cdot}HMPA{\cdot}H_2O$, which can be recrystallized from methanol.
Handling, Storage, and Precautions: MoOPH should be stored in the freezer and protected from light.[1b] The reagent should be treated as an explosion hazard due to its peroxidic nature.[1b,8] Properly stored MoOPH is a freely flowing, yellow crystalline powder, and can be handled in the air at room temperature using a safety shield and protective gloves.

Enolate Hydroxylation. MoOPH[1,9] is a useful reagent for the hydroxylation of lithium, sodium, and potassium enolates of ketones,[1,10] aldehydes,[11] esters,[1,12] and lactones,[1,13,14] and related anions.[15] Similar hydroxylations can be performed with 1-*p*-toluenesulfonyl-2-phenyloxaziridine (Davis' reagent; see ***(±)-trans-2-(Phenylsulfonyl)-3-phenyloxaziridine***), but the latter reagent works best with Na or K enolates that are sensitive to equilibration.[16] MoOPH oxidation of the lithium enolate in THF is the method of choice when regiospecific hydroxylation is desired [examples (**1**)–(**6**)].

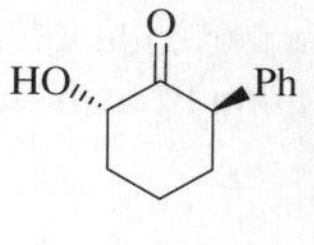

(**1**) 70% 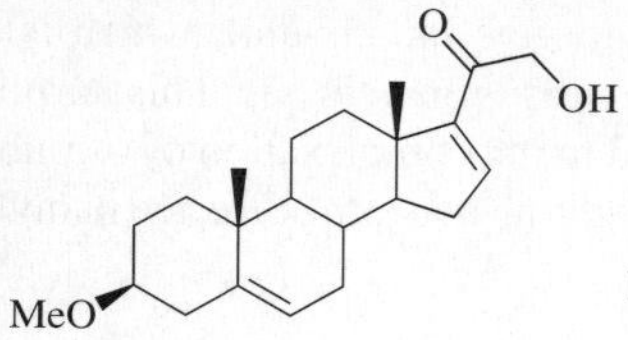 (**2**) 52% 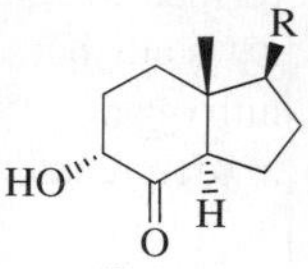 (**3**) 75%

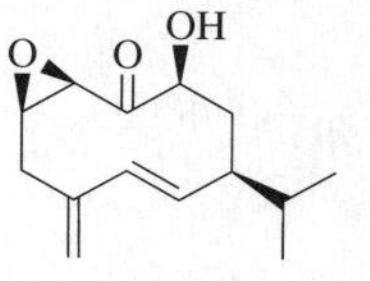

(**4**) 86% 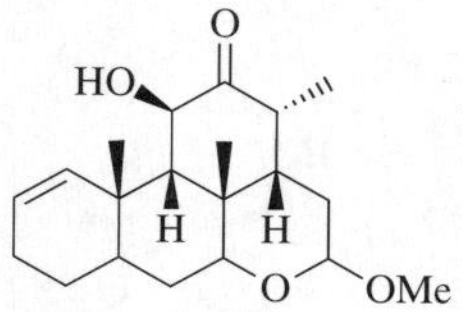 (**5**) 45% 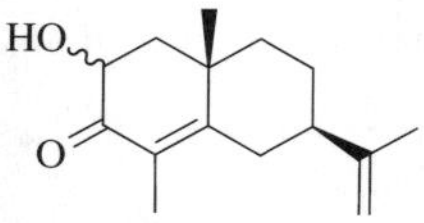 (**6**) 62%

Byproducts may include α-diketones (with aryl *n*-alkyl ketone enolates) or aldol adducts (methyl ketone or enone enolates), and recovery of 5–20% unreacted ketone is common. Inverse addition techniques and temperature optimization (usually −78 to −20 °C) may prevent side reactions with difficult substrates.[1a] Cyclohexanone enolates prefer axial hydroxylation [(**1**) 7:1 α:β;[1a] (**3**) only isomer reported;[10a] (**6**) 2:1 β:α[10d]]. Several stereoselective ester[12] and lactone[13,14] hydroxylations have also been reported [(**8**) 7:1 α:β;[13a] (**9**) 8.5:1 α:β[14a]]. The MoOPH hydroxylation of the Evans imide enolates has a small selectivity advantage compared to the Davis oxaziridine, but yields are lower.[15a] MoOPH is relatively bulky, and prefers bonding to the less hindered enolate face in the absence of stereoelectronic factors. These oxidations involve enolate attack at the O–O bond (eq 1).

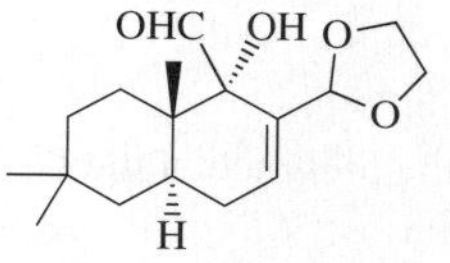

(**7**) 85% 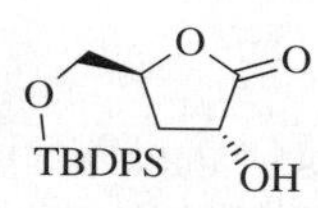 (**8**) 80% 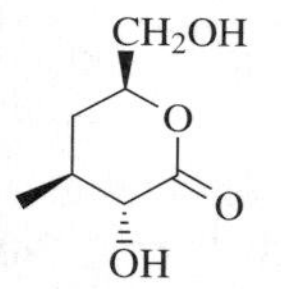 (**9**) 58%

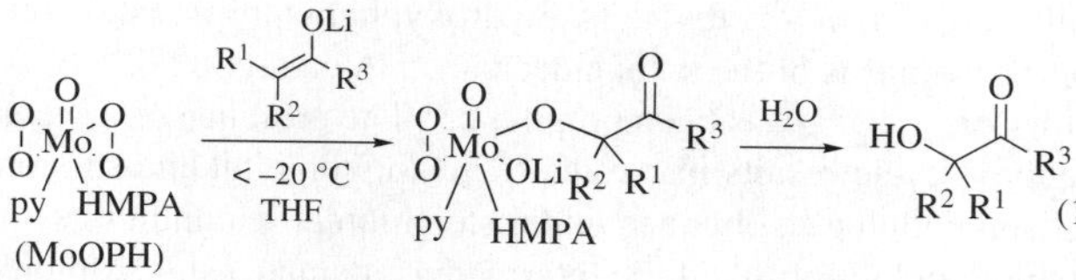

(MoOPH) <−20 °C, THF; H_2O (1)

Hydroxylation of Other Stabilized Anions. Nitrile anions are hydroxylated by MoOPH to afford cyanohydrins.[2,17] The cyanohydrins can be isolated starting from unbranched nitriles[2,17c] or strained cyclic nitriles (eq 2).[17a] More typical α-branched nitrile anions afford the ketones directly,[2,17b] and condensation of this ketone with the starting nitrile anion may complicate these reactions.[2]

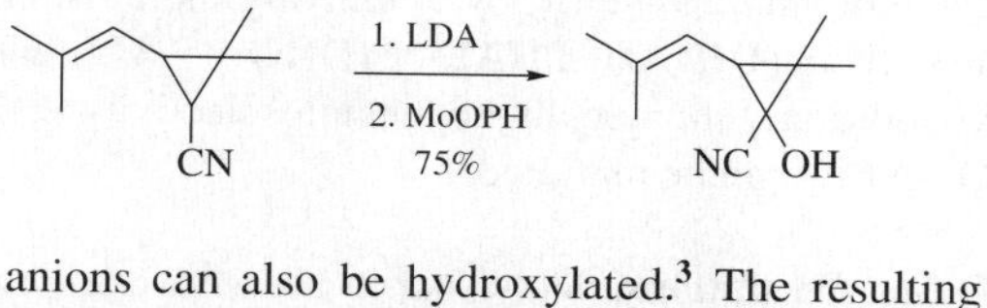

1. LDA; 2. MoOPH; 75% (2)

Sulfone anions can also be hydroxylated.[3] The resulting α-hydroxy sulfones fragment spontaneously to the corresponding ketone[3,18] or aldehyde.[19] Condensation between the carbonyl product and the sulfone anion can be a major reaction pathway.[19b] Similar oxidative degradation occurs with nitroalkane and phosphonate anions.[4,5]

Miscellaneous Hydroxylations with MoOPH. Other strongly basic carbanions have been hydroxylated with MoOPH, including an aryllithium derivative,[20] a dipole-stabilized benzyllithium PhC(O)N(Me)CH(Li)Ph,[21] and a lithiated vinylogous amide.[22] Certain relatively nonbasic substrates can also react with MoOPH, resulting in net hydroxylation. The reagent cleaves C–B bonds with retention,[7] and is useful for the oxidative workup of boron enolate aldol reactions.[23] In the presence of ***t-Butyl Hydroperoxide***, MoOPH converts α- or β-naphthols into ketols derived from *ortho* hydroxylation of the naphthalene ring.[24] Finally, MoOPH reacts with *N*-trimethylsilylamides to afford the *N*-hydroxy amides (eq 3).[6,25]

(3) [scheme: benzazepinone (N–H) —BSA→ N-TMS lactam —MoOPH→ N-OH lactam, 90%]

Related Reagents. See Class O-20, pages 1–10.

1. (a) Vedejs, E. *JACS* **1974**, *96*, 5944. (b) Vedejs, E.; Engler, D. A.; Telschow, J. E. *JOC* **1978**, *43*, 188. (c) Vedejs, E.; Larsen, S. *OS* **1985**, *64*, 127.
2. Vedejs, E.; Telschow, J. E. *JOC* **1976**, *41*, 740.
3. Little, R. D.; Myong, S. O. *TL* **1980**, *21*, 3339.
4. Galobardes, M. R.; Pinnick, H. W. *TL* **1981**, *22*, 5235.
5. Kim, S.; Kim, Y. G. *Bull. Korean Chem. Soc.* **1991**, *12*, 106.
6. Matlin, S. A.; Sammes, P. G. *CC* **1972**, 1222.
7. Midland, M. M.; Preston, S. B. *JOC* **1980**, *45*, 4514.
8. An explosion, apparently due to friction with a steel spatula, has been reported with a related complex MoO_5·DMPU: Paquette, L. A.; Koh, D. *C & E News* **1992** (Sept. 14 issue; Letters to the Editor). For use of MoO_5·DMPU·Py in hydroxylations, see Anderson, J. C.; Smith, S. C. *SL* **1990**, *2*, 107.
9. Mimoun, H.; Seree de Roch, I.; Sajus, L. *BSF* **1969**, 1481.
10. (a) Maestro, M. A.; Sardina, F. J.; Castedo, L.; Mourino, A. *JOC* **1991**, *56*, 3582. (b) Kuwahara, S.; Mori, K. *T* **1990**, *46*, 8075. (c) Grieco, P. A.; Lis, R.; Ferrino, S.; Jaw, J. Y. *JOC* **1984**, *49*, 2342. (d) Murai, A.; Ono, M.; Abiko, A.; Masamune, T. *BCJ* **1982**, *55*, 1195. (e) Franck-Neumann, M.; Miesch, M.; Barth, F. *T* **1993**, *49*, 1409. (f) Tanis, S. P.; Johnson, G. M.; McMills, M. C. *TL* **1988**, *29*, 4521.
11. (a) Tanis, S. P.; Nakanishi, K. *JACS* **1979**, *101*, 4398. (b) Harapanhalli, R. S. *JCS(P1)* **1988**, 3149.
12. (a) Hollinshead, D. M.; Howell, S. C.; Ley, S. V.; Mahon, M.; Ratcliffe, N. M.; Worthington, P. A. *JCS(P1)* **1983**, 1579. (b) Gamboni, R.; Tamm, C. *HCA* **1986**, *69*, 615. (c) Morizawa, Y.; Yasuda, A.; Uchida, K. *TL* **1986**, *27*, 1833. (d) Augé, C.; Gautheron, C.; David, S.; Malleron, A.; Cavayé, B.; Bouxom, B. *T* **1990**, *46*, 201. (e) Sardina, F. J.; Paz, M. M.; Fernandez-Megia, E.; deBoer, R. F.; Alvarez, M. P. *TL* **1992**, *33*, 4637. (f) Swenton, J. S.; Anderson, D. K.; Jackson, D. K.; Narasimhan, L. *JOC* **1981**, *46*, 4825.
13. (a) Hanessian, S.; Sahoo, S. P.; Murray, P. J. *TL* **1985**, *26*, 5631. (b) Takano, S.; Morimoto, M.; Ogasawara, K. *CC* **1984**, 82. (c) Hanessian, S.; Cooke, N. G.; Dehoff, B.; Sakito, Y. *JACS* **1990**, *112*, 5276. (d) Rao, A. V. R.; Bhanu, M. N.; Sharma, G. V. M. *TL* **1993**, *34*, 7078; (e) Yadav, J. S.; Praveen, K. T. K.; Maniyan, P. P. *TL* **1993**, *34*, 2965. (f) Hanessian, S.; Murray, P. J. *CJC* **1986**, *64*, 2231. (g) Stork, G.; Rychnovsky, S. D. *JACS* **1987**, *109*, 1564. (h) Nagano, H.; Masunaga, Y.; Matsuo, Y.; Shiota, M. *BCJ* **1987**, *60*, 707. (i) Mander, L. N.; Robinson, R. P. *JOC* **1991**, *56*, 3595. (j) Gais, H. J.; Ball, W. A.; Lied, T.; Lindner, H. J.; Lukas, K. L.; Rosenstock, B.; Sliwa, H. *LA* **1986**, 1179.
14. (a) Thurkauf, A.; Tius, M. A. *CC* **1989**, 1593. (b) Anderson, J. C.; Ley, S. V.; Santafianos, D.; Sheppard, R. N. *T* **1991**, *47*, 6813. (c) Takano, S.; Ohkawa, T.; Tamori, S.; Satoh, S.; Ogasawara, K. *CC* **1988**, 189. (d) Bhatnagar, S. C.; Caruso, A. J.; Polonsky, J.; Rodriguez, B. S. *T* **1987**, *43*, 3471.
15. (a) Evans, D. A.; Morrissey, M. M.; Dorow, R. L. *JACS* **1985**, *107*, 4346. (b) Ogilvie, W. W.; Durst, T. *CJC* **1988**, *66*, 304. (c) Dolle, R. E.; McNair, D.; Hughes, M. J.; Kruse, L. I.; Eggelston, D.; Saxty, B. A.; Wells, T. N. C.; Groot, P. H. E. *JMC* **1992**, *35*, 4875. (d) Pattenden, G.; Shuker, A. J. *TL* **1991**, *32*, 6625. (e) Hua, D. H.; Saha, S.; Roche, D.; Maeng, J. C.; Iguchi, S.; Baldwin, C. *JOC* **1992**, *57*, 399.
16. Davis, F. A.; Vishwakarma, L. C.; Billmers, J. M.; Finn, J. *JOC* **1984**, *49*, 3241.
17. (a) VanCantfort, C. K.; Coates, R. M. *JOC* **1981**, *46*, 4331. (b) Burnell, R. H.; Caron, S. *CJC* **1992**, *70*, 1446. (c) Kelly, T. R.; Chandrakumar, N. S.; Cutting, J. D.; Goehring, R. R.; Weibel, F. R. *TL* **1985**, *26*, 2173.
18. (a) Nemoto, H.; Kurobe, H.; Fukumoto, K.; Kametani, T. *JOC* **1986**, *51*, 5311. (b) Harirchian, B.; Bauld, N. L. *JACS* **1989**, *111*, 1826. (c) Classen, A.; Scharf, H.-D. *LA* **1990**, 123.
19. (a) Okamura, W. H.; Peter, R.; Reischl, W. *JACS* **1985**, *107*, 1034. (b) Capet, M.; Cuvigny, T.; duPenhoat, C. H.; Julia, M.; Loomis, G. *TL* **1987**, *28*, 6273.
20. Cambie, R. C.; Higgs, P. I.; Rutledge, P. S.; Woodgate, P. D. *JOM* **1990**, *384*, C6.
21. Williams, R. M.; Kwast, E. *TL* **1989**, *30*, 451.
22. Niwa, H.; Kuroda, A.; Yamada, K. *CL* **1983**, 125.
23. Evans, D. A.; Nelson, J. V.; Vogel, E.; Taber, T. R. *JACS* **1981**, *103*, 3099.
24. Krohn, K.; Brüggmann, K.; Döring, D.; Jones, P. G. *CB* **1992**, *125*, 2439.
25. (a) Weidner-Wells, M. A.; DeCamp, A.; Mazzocchi, P. H. *JOC* **1989**, *54*, 5746. (b) Rigby, J. H.; Qabar, M. *JOC* **1989**, *54*, 5852.

Edwin Vedejs

University of Wisconsin, Madison, WI, USA

Oxo(trimanganese) Heptaacetate[1]

$Mn_3O(OAc)_7$

[19513-05-4] $C_{14}H_{21}Mn_3O_{15}$ (MW 594.17)

(selective oxidizing agent; mediates addition of carboxylic acids to alkenes; reagent for allylic acetoxylation)

Alternate Name: manganese triacetate.

Solubility: insol most organic solvents; sol hot acetic acid; hydrolyzed a by water.

Form Supplied in: brown or gray powder and chunks; widely available.

Preparative Methods: often referred to as 'anhydrous $Mn(OAc)_3$', which does not exist.[2] Can be prepared from $Mn(OAc)_2$ and $KMnO_4$ in $AcOH/Ac_2O$, or prepared in situ from $Mn(OAc)_2$ and $KMnO_4$ in AcOH.[3,4]

Purification: not usually necessary.

Handling, Storage, and Precautions: hygroscopic; avoid heat, sparks, and open flames.

Preparation of Lactones from Alkenes. In the presence of manganese triacetate, simple alkenes react with carboxylic acids (often solvent ***Acetic Acid***) to give γ-lactones (eqs 1 and 2)[3,4] in a reaction which is thought to involve radical intermediates. This reaction can be performed on dienes[5] and on alkenes which possess other potentially reactive groups such as alkynes[6] (eq 3).[7] A range of other carboxylic acids has been used in place of acetic acid.[8,9]

$Mn(OAc)_3{\cdot}2H_2O$; AcOH, Ac_2O, reflux; 72% (1)

$Mn(OAc)_3{\cdot}2H_2O$; AcOH, Ac_2O, reflux; 62% (2)

$Mn(OAc)_3{\cdot}2H_2O$; AcOH, KOAc, 115 °C; C_8H_{17} (3)

In addition to simple carboxylic acids, a variety of β-keto acids also undergo reaction with alkenes to give lactones (eq 4),[9] and the intramolecular version of this reaction can be used effectively to produce polycyclic lactones (eq 5).[10] This latter reaction has been adapted to the total synthesis of polycyclic natural products.

$Mn(OAc)_3{\cdot}2H_2O$, $HO_2CCHClCO_2Et$; AcOH, 23 °C; 88%; EtO_2C, Cl (4)

CO_2H; $Mn(OAc)_3{\cdot}2H_2O$, $HO_2CCHClCO_2Et$; AcOH, 23 °C; 52% (5)

The formation of C–C bonds using Mn^{III} salts is not restricted to the reaction of alkenes and carboxylic acids. A range of β-dicarbonyl compounds will also undergo additions to give dihydrofurans (eq 6)[11] and other types of products. In particular, the intramolecular cyclization can be used to construct a variety of cyclic systems, using both monocyclizations (eq 7)[12] and tandem cyclizations (eq 8).[13] A sulfoxide group can take the place of one of the carbonyl groups, and in this case the chirality of the sulfoxide controls the absolute configuration of the product. In effect, the sulfoxide group has been used as a chiral auxiliary in this cyclization (eq 9).[14]

Pr; $Mn(OAc)_3{\cdot}2H_2O$; AcOH, 45 °C; 88% (6)

CO_2Me; $Mn(OAc)_3$, $Cu(OAc)_2$; AcOH, 25 °C; 61% (7)

OMe; CO_2Me; $Mn(OAc)_3$; AcOH, 20 °C; 50% (8)

p-Tol; $Mn(OAc)_3$, $Cu(OAc)_2$; AcOH, 25 °C; 44% (9)

Enol ethers and enol esters can also serve as the alkene component of this type of reaction, the products being useful as precursors for furans and 1,4-dicarbonyl compounds (eqs 10 and 11).[15,16] It is also possible to use simple ketones as the carbonyl component, although the yields are relatively low, and if the ketone is unsymmetrical, mixtures of products are obtained.[17]

OMe + EtO_2C; $Mn(OAc)_3{\cdot}2H_2O$; AcOH, 25 °C; 86% (10)

CO_2Et + OAc; $Mn(OAc)_3{\cdot}2H_2O$; AcOH, 25 °C; 71% (11)

Allylic Oxidation and Oxidation α to Carbonyl Groups. Acetoxylation adjacent to both C=C and C=O double bonds is possible using manganese triacetate.[18,19] Although the reaction tends to be substrate dependent and the yields variable (eq 12),[19] it has been used to good effect in the synthesis of relatively complex polyfunctional systems that contain groups which might interfere (eqs 13 and 14).[20,21]

$Mn(OAc)_3$; AcOH, reflux; OAc 17% + OAc 6% (12)

$Mn(OAc)_3$; C_6H_6, 80 °C; AcO (13)

$Mn(OAc)_3$, C_6H_6, 80 °C (14)

Other Applications. Manganese triacetate is reported to be superior to ***Ruthenium(IV) Oxide*** and ***Cerium(IV) Ammonium Nitrate*** for the oxidation of *N*-protected indolines to indoles (eq 15).[22] On treatment with manganese triacetate, aryldialkylamines undergo dealkylation, in which one alkyl group is replaced by an acetyl group,[23] and in combination with ***Diphenyl Disulfide*** alkenes undergo regioselective addition of PhS and OH.[24] Manganese triacetate was found to be the reagent of choice for the final step in a total synthesis of cyanocycline, which required the oxidation of the *p*-methoxyphenol to the corresponding quinone (eq 16).[25]

$Mn(OAc)_3$, AcOH, 70 °C, 82% (15)

$Mn(OAc)_3$, MeCN, 0.3% H_2SO_4, rt, 55% (16)

Related Reagents. See Classes O-20, pages 1–10.

1. Melikyan, G. G. *S* **1993**, 883.
2. *Dictionary of Inorganic Compounds*; Macintyre, J. E., Ed.; Chapman & Hall: London, 1992.
3. Bush, Jr., J. B.; Finkbeiner, H. *JACS* **1968**, *90*, 5903.
4. Heiba, E. I.; Dessau, R. M.; Koehl, Jr., W. J. *JACS* **1968**, *90*, 5905.
5. Melikyan, G. G. *S* **1993**, 839.
6. Heiba, E. I.; Dessau, R. M.; Rodenwald, P. G. *JACS* **1974**, *96*, 7977.
7. Melikyan, G. G.; Mkrtchyan, D. A.; Lebedeva, K. V.; Maeorg, U.; Panosyan, G. A.; Badanyan, Sh. O. *Chem. Nat. Compd.* **1984**, 94.
8. (a) Chloroacetic acid and 3-chloropropanoic acid: Fristad, W. E.; Peterson, J. R.; Ernst, A. B. *JOC* **1985**, *50*, 3143. (b) Isobutyric acid: Ref. 6. (c–d) Propanoic acid: Refs. 4 and 6. (e) Malonic acid and derivatives: Fristad, W. E.; Hershberger, S. S. *JOC* **1985**, *50*, 1026. (f) Peterson, J. R.; Do, H. D.; Surjasasmita, I. B. *SC* **1988**, 1985. (g) Rosario-Chow, M.; Ungwitayatorn, J.; Currie, B. L. *TL* **1991**, *32*, 1011.
9. Corey, E. J.; Gross, A. W. *TL* **1985**, *26*, 4291.
10. Corey, E. J.; Kang, M. *JACS* **1984**, *106*, 5384.
11. (a) Heiba, E. I.; Dessau, R. M. *JOC* **1974**, *39*, 3456. (b) Snider, B. B.; Zhang, Q. *TL* **1992**, *33*, 5921. (c) See also references cited in Cossy, J.; Bouzide, A. *TL* **1993**, *34*, 5583.
12. White, D. J.; Somers, T. C.; Yager, K. M. *TL* **1990**, *31*, 59.
13. Snider, B. B.; Mohan, R.; Kates, S. A. *JOC* **1985**, *50*, 3659.
14. Snider, B. B.; Wan, B. Y.; Buckman, B. O.; Foxman, B. M. *JOC* **1991**, *56*, 328.
15. Corey, E. J.; Ghosh, A. K. *CL* **1987**, 223.
16. Corey, E. J.; Ghosh, A. K. *TL* **1987**, *28*, 175.
17. Dessau, R. M.; Heiba, E. I. *JOC* **1974**, *39*, 3457.
18. (a) Gilmore, J. R.; Mellor, J. M. *JCS(C)* **1971**, 2355. (b) Gilmore, J. R.; Mellor, J. M. *JCS(C)* **1970**, 507.
19. Williams, G. J.; Hunter, N. R. *CJC* **1976**, *54*, 3830.
20. Danishefsky, S. J.; Bednarski, M. *TL* **1985**, *26*, 3411.
21. Dunlap, N. K.; Sabol, M. R.; Watt, D. S. *TL* **1984**, *25*, 5839.
22. Ketcha, D. M. *TL* **1988**, *29*, 2151.
23. Rindone, B.; Scolastico, C. *TL* **1974**, 3379.
24. Abd El Samii, Z. K. M.; Al Ashmawy, M. I.; Mellor, J. M. *JCS(P1)* **1988**, 2509, 2517, 2523.
25. Fukuyama, T.; Li, L.; Laird, A. A.; Frank, R. K. *JACS* **1987**, *109*, 1587.

Garry Procter
University of Salford, UK

Oxygen

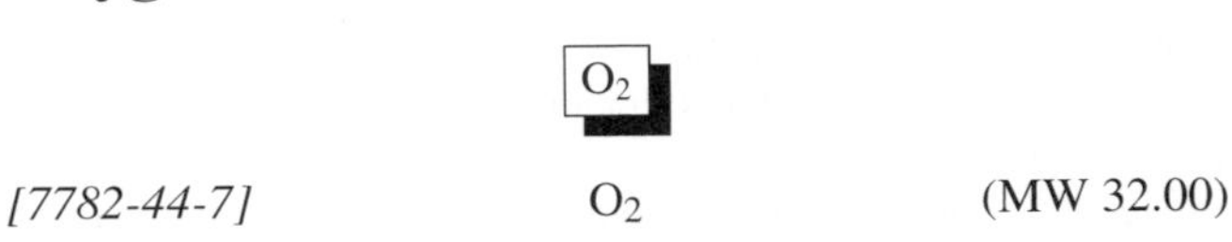

[7782-44-7] O_2 (MW 32.00)

(oxidizing agent for many organic systems, including, most commonly, organometallic compounds, carbon radicals, and heteroatoms such as sulfur)

Physical Data: mp −218 °C; bp −183 °C; *d* 1.429 g L^{-1} (0 °C), 1.149 g L^{-1} (−183 °C).
Solubility: sol to some extent in most solvents. Selected data, expressed as mL of O_2 (at 0 °C/760 mmHg) dissolved in 1 mL of solvent when the partial pressure of the gas is 760 mmHg, are as follows: Me_2CO (0.207/18 °C), $CHCl_3$ (0.205/16 °C), Et_2O (0.415/20 °C), EtOAc (0.163/20 °C), MeOH (0.175/19 °C), petroleum ether (0.409/19 °C), PhMe (0.168/18 °C), H_2O (0.023/20 °C).
Form Supplied in: dry gas; dilutions in Ar, He, or N_2; $^{18}O_2$; $^{17}O_2$.
Handling, Storage, and Precautions: of itself, oxygen gas is essentially nontoxic. However, it will support and vigorously increase the rate of combustion of most materials. It may ignite combustibles and can cause an explosion on contact with oil and grease. The potential for autoxidative formation of explosive peroxides (e.g. with Et_2O) should always be borne in mind.

Oxygenation of Carbanions and Organometallic Compounds. Many organometallic species react with triplet oxygen to form the corresponding hydroperoxides,[1,2] although the products are more usually reduced in situ or during workup to afford alcohols as the isolated products. A number of other sources of electrophilic oxygen have been developed (e.g. ***Oxodiperoxymolybdenum(pyridine)(hexamethylphosphoric triamide)*** (MoOPH), sulfonyloxaziridines) and compete for this niche, but no single reagent is universally preferable. As carbon anion equivalents, Grignard reagents are optimal for simple hydrocarbons,[2] but organolithiums are more frequently employed. The potential for radical-mediated oxidative dimerization can constrain utility (particularly for aryl organometallics). Useful oxygenations of alkyl,[3] vinylic,[4] allylic,[5] benzylic,[6] and aryl (eq 1)[6] organolithium compounds have been reported. 1,1-Diorganometallics give the corresponding carbonyl derivatives (eq 2).[7]

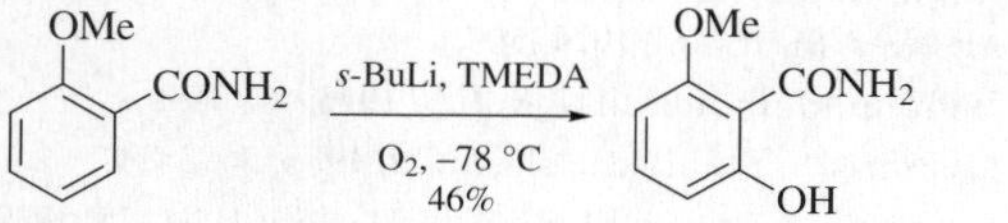

(1)

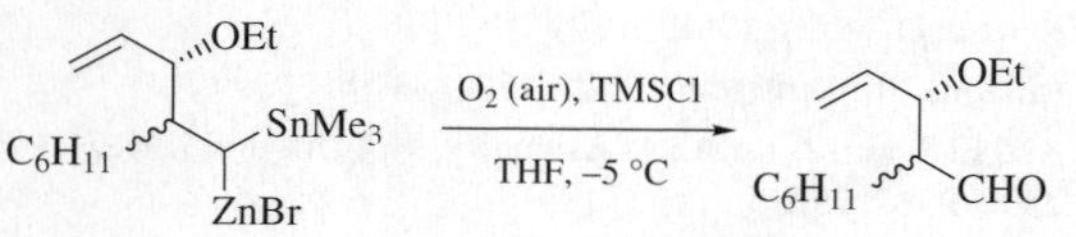

(2)

Effective oxygenation of enolate anions is generally restricted to tertiary centers where over-oxidation is not possible.[8] Nonetheless, ketones (eq 3),[9] esters/lactones,[8] amides/lactams (eq 4),[10] and carboxylic acids[8] can all be usefully α-hydroxylated. In most cases the intermediate α-hydroperoxides are reduced in situ (usually with ***Triethyl Phosphite***),[11] although they can be isolated if desired.[12] The small size of the electrophile and the potential for radical involvement[8] do not encourage stereochemical chastity in these processes, but where sufficient bias exists, good discrimination can be observed (see eqs 3 and 4). A slightly different approach uses enolates derived from aqueous base treatment. These species have been usefully hydroxylated where there was little or no ambiguity in the direction of enolization[8] and this process forms the basis for a surprisingly effective catalytic, enantioselective oxygenation (eq 5).[13]

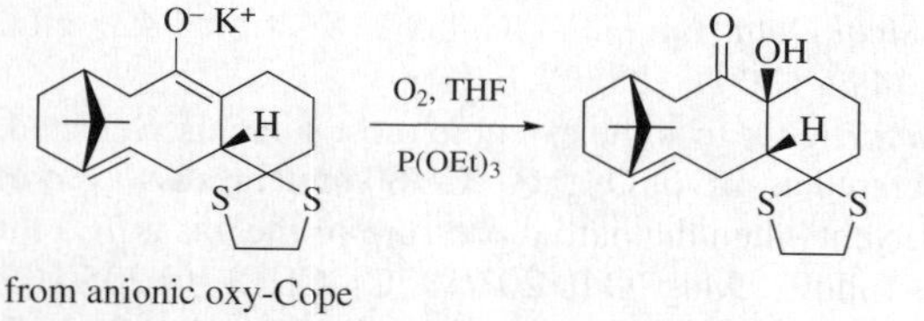

(3)

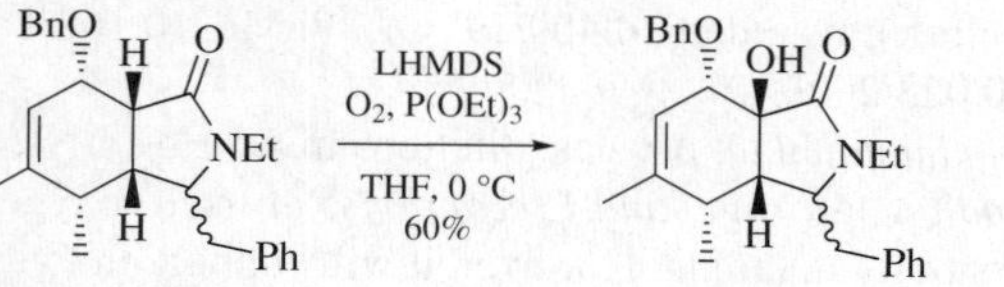

(4)

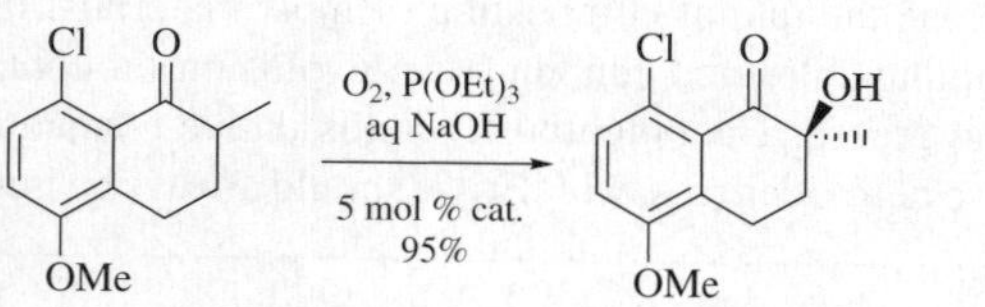

(5)

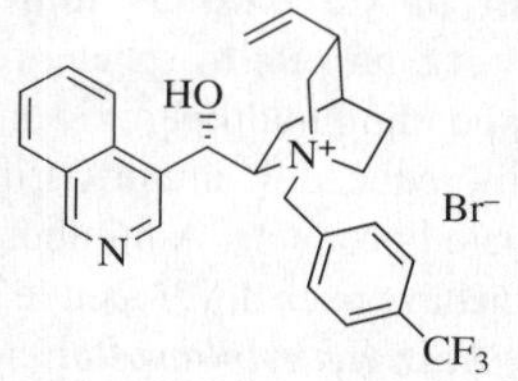

In cases where the activating group is also a leaving group, oxygenation can provide the corresponding carbonyl compound. Thus oxidative decyanation can be effected under either phase transfer[14] or anhydrous conditions (eq 6).[15] The latter procedure is more general, although it does require treatment with ***Tin(II) Chloride*** and base to reduce and fragment the α-hydroperoxide, and this method is not effective for the primary nitrile to aldehyde conversion. α,β-Unsaturated nitriles generally react at the α-position to give α,β-unsaturated ketones.[15]

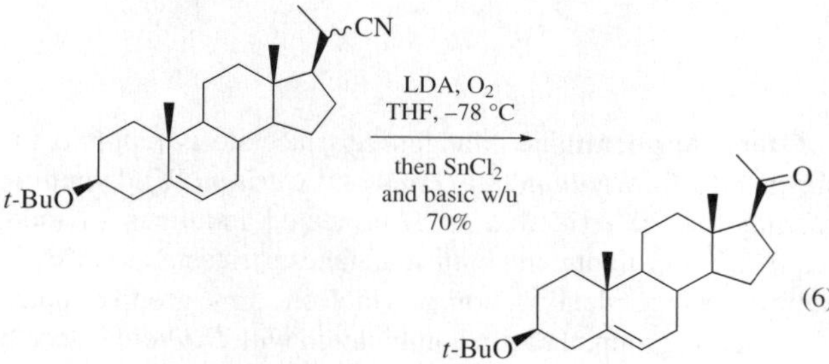

(6)

Oxygenation of sulfone anions and the consequent desulfonylation is frequently effected with the MoOPH reagent. In some cases, however, molecular oxygen has proved effective where MoOPH failed.[16] This perhaps illustrates a functional advantage over more bulky reagents and provides a counterpoint to the stereochemical disadvantages (see above). It should be noted that one such attempt resulted in a minor explosion[17] although, of course, any reaction involving peroxides bears this possibility. A similar process, preceded by nickel transmetalation, demonstrated the oxidation of the C–Ni bond, but the synthetic advantage is not clear.[18] However, this report did demonstrate the conjugative oxygenation of an allylic sulfone anion to give a γ-hydroxy sulfone.

Oxygenation of phosphorus-stabilized anions also produces the corresponding carbonyl compounds. The anions derived from phosphonates[19] (eq 7)[20] (including α-heteroatom substituted phosphonates)[19] and phosphine oxides[21] react smoothly with oxygen. Similarly, phosphorane ylides are readily oxidized.[22] In all these cases, however, the reaction of primary substrates suffers from competing self-condensation, giving alkenes.[22] It should be noted that a two-stage procedure involving the reaction of phosphonate anion with chlorodimethyl borate followed by oxidation with ***m-Chloroperbenzoic Acid*** has been advocated as a more efficient method[23] (and, interestingly, allows isolation of the intermediate hydroxy phosphonate).

Oxygenation of Carbon Radicals. Not surprisingly, triplet oxygen reacts rapidly with carbon centered radicals.[24] Classical autoxidation is the most obvious example of this behavior. Traditionally, autoxidation refers to hydroperoxide formation from alkanes, aralkanes, alkenes, ethers, alcohols, and carbonyl compounds, where the initiating homolysis is induced thermally or photochemically.[25,26] There is an extensive literature concerning these processes dating back many years. While very important commercially, they are generally too promiscuous to be of wide synthetic value, particularly when dealing with complex molecules. Interestingly, however, a deformylative hydroxylation of an allylic neopentyl aldehyde has been observed that bypasses the classical autoxidative fate of aldehydes (eq 8).[27]

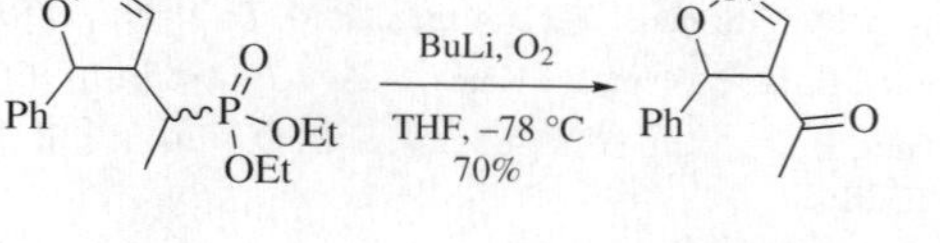

(7)

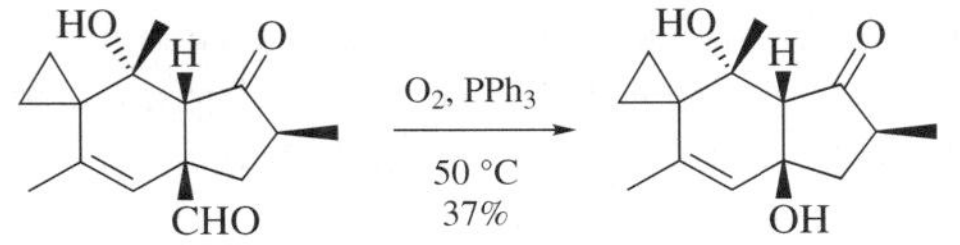

(8)

Radical oxygenation is most valuable where there is more strict control over the site of radical formation and subsequent oxygenation. Good stereochemical control is, of course, not usually achieved, although exceptions can be found in most cases. The mild and controlled methods of radical generation that have seen much use in synthesis are readily applicable to oxygenation. The thermal or photochemical decarboxylation of the esters of thiohydroxamic acids,[28] or their room temperature decomposition in the presence of tris(phenylthio)antimony[29] (i.e. Barton's methodology), can be intercepted by triplet oxygen to generate the nor-alcohols. The addition of heteroatom radicals to alkenes can provide the source of carbon centered radicals for trapping. An interesting example of oxygenation initiated by phenylthio or phenylseleno radical addition to vinylcyclopropanes showcases the use of this methodology (eq 9).[30] Here, instances of moderately successful stereocontrol in the C–O bond forming step were noted. This transformation also demonstrates the potential of the initially formed hydroperoxy radical to participate in further steps (where higher levels of stereochemical discrimination are observed as a consequence of the intramolecular nature of the radical trapping). ***Samarium(II) Iodide*** induced radical processes have been quenched with oxygen to provide hydroxyl functionalized products.[31]

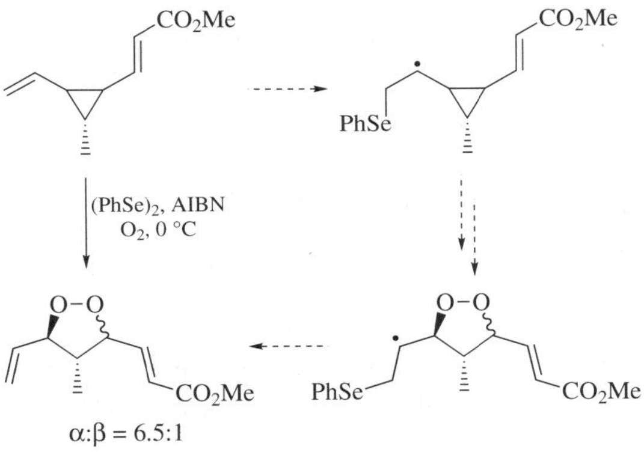

(9)

One of the most prevalent uses of molecular oxygen in modern synthesis is for oxidative demercuration.[32] Carbon radicals generated by the reduction of organomercurials with borohydride are efficiently trapped by oxygen, most frequently in DMF solution, to give hydroperoxides which are reduced under the reaction conditions to generate the corresponding alcohols directly.[33] The alkene oxymercuration–oxidative demercuration sequence is commonly practised (usually through a β-alkoxymercury species, since β-hydroxy fails[33]), particularly where the oxymercuration is an intramolecular cyclization (eq 10).[34] Typically, any stereocontrol observed in the oxymercuration (or other C–Hg bond forming step) is effaced in the oxygenation (as in eq 10).

Oxidation of Organoboranes. Boranes, most frequently accessed by hydroboration of alkenes, can be oxidized by triplet oxygen.[35,36] If the oxidation is carried out in fairly concentrated solution (~0.5 N) at 0 °C, intermolecular redox reaction of the intermediate diperoxyborane is facilitated and workup provides the corresponding alcohol.[36] While this is quite efficient, ***Hydrogen Peroxide*** is more commonly used in synthetic applications. This is partly for convenience, but also a consequence of stereochemical issues. The oxidation with H_2O_2 occurs with retention of configuration at the carbon center. The radical characteristics of the dioxygen reaction generally lead to at least partial racemization. That this stereochemical corruption is not always complete is an indication of the uncertainty about the mechanism.[35] Interestingly, rhodium(III) porphyrin has been shown to promote stereoselective oxidation in the dioxygen procedure (eq 11).[37] In dilute solution (0.01–0.05 N) the intermolecular redox process is suppressed and diperoxyboranes are produced. Oxidation of the third B–C bond with H_2O_2 or peroxy acid and workup allows isolation of the corresponding alkyl hydroperoxides.[36] Alternatively alkyl hydroperoxide formation is facilitated by the use of alkyldichloroboranes.[36] This is one of the most convenient approaches to this functionality. The oxygen mediated approach to alcohols may be more convenient than H_2O_2 for radiolabeling; ^{17}O (eq 12) and ^{15}O alcohols have been prepared in this way.[38,39]

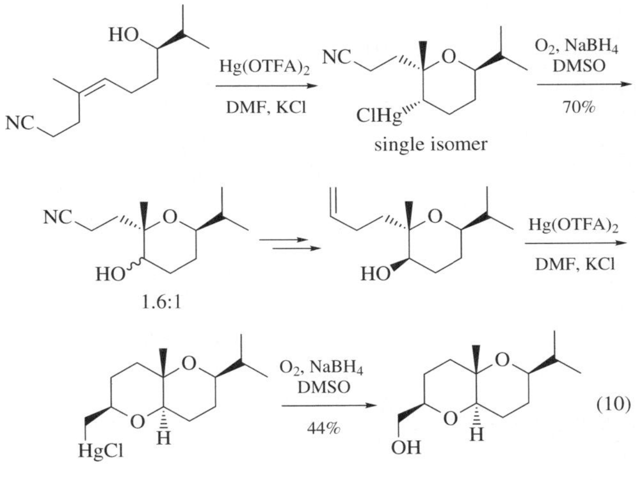

(10)

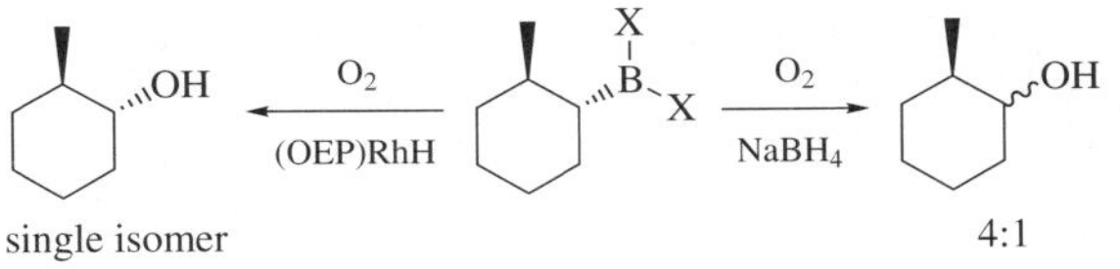

(11)

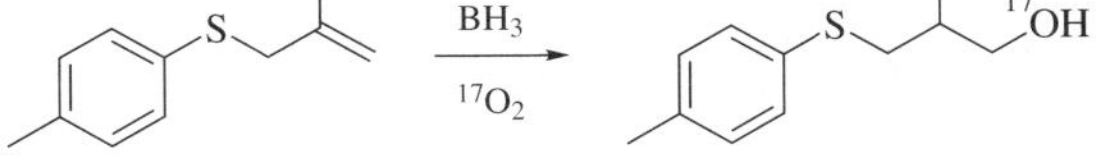

(12)

Heteroatom Oxidation. Oxidation of nitrogen functionality with oxygen, while well precedented[40] and of continuing interest,[41,42] does not generally represent the method of choice for those processes of synthetic significance. However, a report of a mild procedure for the oxidation of silylamines to carbonyl compounds bears some synthetic potential (eq 13).[43] Oxidation of phosphorus functionality by oxygen can be quite facile.[44] For example, tertiary phosphines are very readily oxidized to their phosphine oxides, and secondary chlorophosphines can give the phosphinic acids.[44] Perhaps the most common heteroatom air oxidations are those of Group 16 RX–H bonds to their corresponding dimers ($(RX)_2$) and particularly the thiol to disulfide oxidation.[45]

This, of course, is related to the importance of the disulfide bond to peptide and protein secondary structure. One example that reflects current interest in the control of multiple disulfide bond formation in synthetic peptides is given in eq 14.[46] Oxidation may be promoted by heavy metal ions.[45] Higher oxidations (for example sulfide to sulfoxide) are best performed with other reagents (oxone, peroxy acid, etc.).

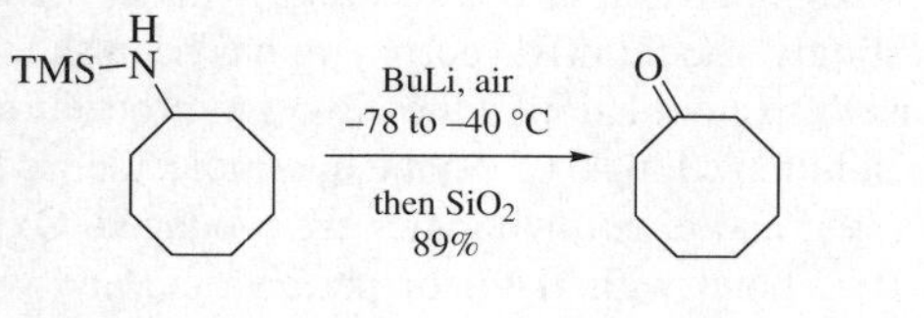

TMS–NH (cyclooctyl) → cyclooctanone: BuLi, air, −78 to −40 °C; then SiO_2, 89% (13)

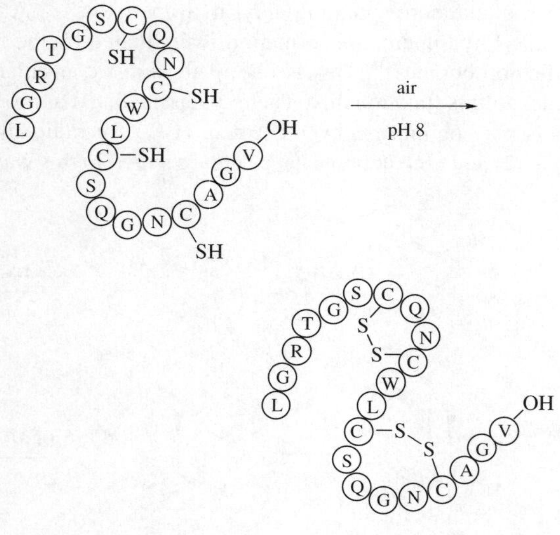

air, pH 8 (14)

major isomer

Other Uses. The oxidative dimerization of organometallics, alluded to above, is particularly prevalent for organocuprates,[47] although not totally unavoidable.[48] In fact it is efficient enough to be regarded as a synthetic strategy and has been used as such (eq 15).[49] Baeyer–Villiger oxidations generally employ peroxy acids, but a recent report indicates that 1 atm of oxygen can effect the rearrangement even in the absence of either metal catalysts or light.[50] Epoxidation by oxygen is possible[51] but, of course, is not usually the method of choice for laboratory synthesis.

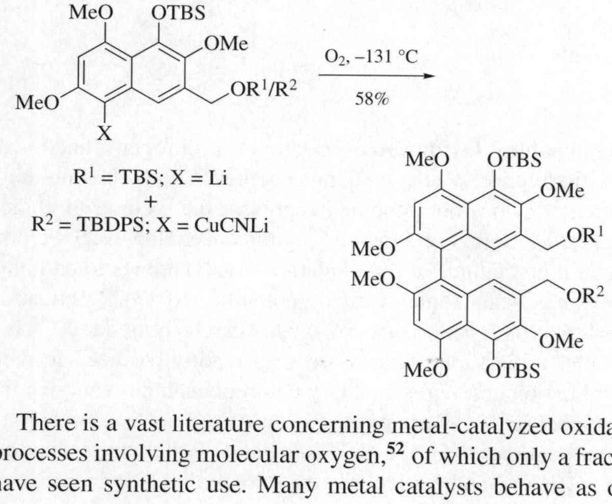

O_2, −131 °C, 58% (15)

R^1 = TBS; X = Li
+
R^2 = TBDPS; X = CuCNLi

There is a vast literature concerning metal-catalyzed oxidative processes involving molecular oxygen,[52] of which only a fraction have seen synthetic use. Many metal catalysts behave as oxygen fixing species, that deliver oxygen to the substrate through a peroxo complex. Reports frequently concern experimental systems, probing substrate reactivity and/or asymmetric induction. The function of oxygen in metal-catalyzed oxidations is not necessarily that of a reagent. Thus, for example, in the Wacker oxidation of terminal alkenes it operates as a re-oxidant for copper(II) chloride which in turn is a re-oxidant for the Pd^{II} species. All of these applications are best regarded as functions of the metal component and, for this reason, are not discussed here.

Related Reagents. See Classes O-7, O-20, and O-24, pages 1–10. Copper(I) Chloride–Oxygen; Diethylzinc–Bromoform–Oxygen; Iron(II) Sulfate-Oxygen; Oxygen–Platinum Catalyst; Singlet Oxygen.

1. Sosnovsky, G.; Brown, J. H. *CRV* **1966**, *66*, 529.
2. Wardell, J. L. In *Comprehensive Organometallic Chemistry*; Wilkinson, G., Ed.; Pergamon: Oxford, 1982; Vol. 1, Chapter 2.
3. Warner, P.; Lu, S.-L. *JOC* **1976**, *41*, 1459.
4. Panek, E. J.; Kaiser, L. R.; Whitesides, G. M. *JACS* **1977**, *99*, 3708.
5. Takahashi, T.; Nemoto, H.; Kanda, Y.; Tsuji, J. *JOC* **1986**, *51*, 4315.
6. Parker, K. A.; Koziski, K. A. *JOC* **1987**, *52*, 674.
7. Knochel, P.; Xiao, C; Yeh, M. C. P. *TL* **1988**, *29*, 6697.
8. Jones, A. B. *COS*, **1991**, *7*, Chapter 2.3.
9. Paquette, L. A.; DeRussy, D. T.; Pegg, N. A.; Taylor, R. T.; Zydowsky, T. M. *JOC* **1989**, *54*, 4576.
10. Kim, M. Y.; Starrett, J. E., Jr.; Weinreb, S. M. *JOC* **1981**, *46*, 5383.
11. Gardner J. N.; Carlton F. E.; Gnoj, O. *JOC* **1968**, *33*, 3294.
12. Bailey, E. J.; Barton, D. H. R.; Elks, J.; Templeton, J. F. *JCS* **1962**, 1578.
13. Masui, M.; Ando, A.; Shioiri, T. *TL* **1988**, *29*, 2835.
14. Donetti, A.; Boniardi, O.; Ezhaya, A. *S* **1980**, 1009.
15. Freerksen, R. W.; Selikson, S. J.; Wroble, R. R.; Kyler, K. S.; Watt, D. S. *JOC* **1983**, *48*, 4087.
16. Yamada, S.; Nakayama, K.; Takayama, H.; Shinki, T.; Suda, T. *TL* **1984**, *25*, 3239.
17. Little, R. D.; Myong, S. O. *TL* **1980**, *21*, 3339.
18. Julia, M.; Lauron, H.; Verpeaux, J-N. *JOM* **1990**, *387*, 365.
19. Mikolajczyk, M; Midura. W.; Grzejszczak, S. *TL* **1984**, *25*, 2489.
20. Tsuge, O.; Kanemasa, S.; Suga, H. *CL* **1986**, 183.
21. Davidson, A. H.; Warren, S. *CC* **1975**, 148.
22. Bestmann, H. J. *AG(E)* **1965**, *4*, 830.
23. Kim, S.; Kim, Y. G. *Bull. Korean Chem. Soc.* **1991**, *12*, 106.
24. Maillard, B.; Ingold, K. U.; Scaiano, J. C. *JACS* **1983**, *105*, 5095.
25. Swern, D. In *Comprehensive Organic Chemistry*; Barton, D. H. R., Ed.; Pergamon: Oxford, 1979; Vol. 1, Chapter 4.6.
26. Crabtree, R. H.; Habib, A. *COS* **1991**, *7*, Chapter 1.1.
27. Kigoshi, H.; Imamura, Y.; Sawada, A.; Niwa, H.; Yamada, K. *BCJ* **1991**, *64*, 3735.
28. Barton, D. H. R.; Crich, D.; Motherwell, W. B. *CC* **1984**, 242.
29. Barton, D. H. R.; Bridon, D.; Zard, S. Z. *CC* **1985**, 1066.
30. Feldman, K.; S. Simpson, R. E. *JACS* **1989**, *111*, 4878.
31. Molander, G. A.; McKie, J. A. *JOC* **1992**, *57*, 3132.
32. Kitching, W. *COS* **1991**, *7*, Chapter 4.2.
33. Hill, C. L.; Whitesides, G. M. *JACS*, **1974**, *96*, 870.
34. Broka, C. A.; Lin, Y.-T. *JOC* **1988**, *53*, 5876.
35. Pelter, A.; Smith, K. *COS* **1991**, *7*, Chapter 4.1.
36. Brown, H. C.; Midland, M. M. *T* **1987**, *43*, 4059.

37. Aoyama, Y.; Tanaka, Y.; Fujisawa, T.; Watanabe, T.; Toi, H.; Ogoshi, H. *JOC* **1987**, *52*, 2555.
38. Kabalka, G. W.; Reed, T. J.; Kunda, S. A. *SC* **1983**, *13*, 737.
39. Takahashi, K.; Murakami, M.; Hagami, E.; Sasaki, H.; Kondo, Y.; Mizusawa, S.; Nakamichi, H.; Iida, H.; Miura, S.; Kanno, I.; Uemura, K.; Ido, T. *J. Labelled Compd. Radiopharm.* **1986**, *23*, 1111.
40. Boyer, J. H. *CRV* **1980**, *80*, 495.
41. Riley, D. P.; Correa, P. E. *JOC* **1985**, *50*, 1563.
42. Gangloff, A. R.; Judge, T. M.; Helquist, P. *JOC* **1990**, *55*, 3679.
43. Chen, H. G.; Knochel, P. *TL* **1988**, *51*, 6701.
44. Gilchrist, T. L. *COS* **1991**, *7*, Chapter 6.1.
45. Uemura, S. *COS* **1991**, *7*, Chapter 6.2.
46. Ponsati, B.; Giralt, E.; Andreu, D. *T* **1990**, *46*, 8255.
47. Posner, G. H. *OR* **1975**, *22*, 253.
48. Lambert, G. J.; Duffley, R. P.; Dalzell, H. C.; Razdan, R. K. *JOC* **1982**, *47*, 3350.
49. Coleman, R. S.; Grant, E. B. *TL* **1993**, *34*, 2225.
50. Bolm, C.; Schlingloff, G.; Weickhardt, K. *TL* **1993**, *34*, 3405.
51. Rao, A. S. *COS* **1991**, *7*, Chapter 3.1.
52. Sheldon, R. A.; Kochi, J. K. *Metal-Catalysed Oxidations of Organic Compounds*; Academic: New York, 1981.

A. Brian Jones
Merck Research Laboratories, Rahway, NJ, USA

Oxygen–Platinum Catalyst[1]

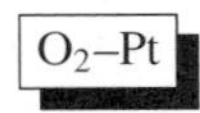

[7740-06-4] Pt (MW 195.08)

(selective oxidant; highly regioselective oxidant for cyclic polyols; lactones obtained from diols; primary alcohols oxidized to carboxylic acids)

Solubility: reaction usually heterogeneous.
Form Supplied in: fine powder; widely available.
Preparative Methods: various methods[1] are employed, depending on the catalyst: for Pt/C, charcoal, concentrated HCl, and aqueous chloroplatinic acid are reduced with hydrogen or formaldehyde; Adams catalyst (PtO_2) is hydrogenated in the solvent to be used for the oxidation; after purging hydrogen, store under chosen solvent.
Purification: dry at 50 °C under reduced pressure.
Handling, Storage, and Precautions: avoid sources of ignition; take precautions against static discharge.

Introduction. Oxygen (often as air) and a platinum catalyst are used in the oxidation of primary alcohols to yield usually carboxylic acids (eq 1)[2] and for oxidation of secondary alcohols. It has been widely used in the selective oxidation of carbohydrates and related systems, and some general guidelines have been developed. Primary alcohols are oxidized more rapidly than secondary alcohols; for secondary cyclic alcohols, axial hydroxyl groups are oxidized more rapidly than equatorial hydroxyl groups.

Pt (Adams), O_2; H_2O, 50–60 °C; 55% (1)

Oxidation of Aldoses and Ketoses. Aldonic acids are produced by oxidation of aldoses (eq 2)[3] and ketoses undergo reaction at the primary position (eq 3).[4] Oxidation of reducing sugars at positions other than the 'aldehyde' position is simply achieved by protection of this position as the corresponding acetal or acetonide, and, as expected, oxidation of the next most reactive position takes place (eq 4).[5] For deoxyamino sugars, the amino function is usually protected as a suitable acyl derivative (eq 5).[6]

Pt/C, O_2; aq. NaOH, rt; 80% (2)

Pt/C, O_2; aq. $NaHCO_3$, rt; 58–60% (3)

Pt/C, O_2; H_2O, pH 8–9, 50 °C; 53% (4)

Pt (Adams), O_2; H_2O, 95 °C; 40% (5)

Oxidation of Cyclitols. A considerable amount of work has been carried out on the oxidation of cyclitols with oxygen–platinum catalysts[1] and a set of general rules has been developed: (a) only axial hydroxyl groups are oxidized; (b) oxidation stops at the monoketone even when more than one axial hydroxyl is present; (c) if several nonequivalent axial hydroxyl groups are present, then usually oxidation of only one of these will take place. These general rules are exemplified by eqs 6 and 7.[7]

Pt/C, O_2, H_2O; pH 8–9, 50 °C; 30% (6)

Pt (Adams), O_2, H_2O; pH 8–9, 50 °C; 58% (7)

Oxidation of Diols to Lactones. The oxidation of 1,4-diols in which one of the hydroxyls is primary gives the corresponding lactones, a reaction which has found use in natural product synthesis,[8] including hydroazulenes (eq 8)[9] and prostaglandins (eq 9).[10]

Pt (Adams), O_2, Me_2CO; H_2O, 50–60 °C; 96% (8)

Pt (Adams), O_2, Me_2CO; H_2O, $NaHCO_3$, 58 °C; 48% (9)

Other Applications. It has been shown that visible light and a catalyst of H_2PtCl_6–$CuCl_2$ (1:2) can be used to oxidize a series of alcohols,[1c] and that the oxidation of reducing sugars does not need oxygen if platinum catalysts are used under basic conditions.[11] 12a-Hydroxylation of a deoxytetracycline has been carried out using oxygen and a platinum catalyst,[12] although the usefulness this reaction appears to depend strongly on the structure of the substrate.[13]

Related Reagents. See Classes O-1and O-2, pages 1–10.

1. (a) Heyns, K.; Paulsen, H. *Newer Methods Prep. Org. Chem.* **1963**, *2*, 303. (b) Heyns, K.; Paulsen, H. *Adv. Carbohydr. Chem.* **1962**, *17*, 169. (c) Cameron, R. E.; Bocarsly, A. B. *JACS* **1985**, *107*, 6116.
2. Maurer, P. J.; Takahata, H.; Rapoport, H. *JACS* **1984**, *106*, 1095.
3. Heyns, K.; Heinemann, R. *LA* **1947**, *558*, 187.
4. Heyns, K. *LA* **1947**, *558*, 177.
5. Mehltretter, C. L.; Alexander, B. H.; Mellies, R. L.; Rist, C. E. *JACS* **1951**, *73*, 2424.
6. Heyns, K.; Paulsen, H. *CB* **1955**, *88*, 188.
7. Post, G. C.; Anderson, L. *JACS* **1962**, *84*, 471.
8. For a comparison of this reagent with Ag_2CO_3, see: Boeckman, Jr., R. K.; Thomas, E. W. *TL* **1976**, 4045.
9. (a) Kretchmer, R. A.; Thompson, W. J. *JACS* **1976**, *98*, 3379. (b) Lansbury, P. T.; Hangauer, Jr., D. G.; Vacca, J. P. *JACS* **1980**, *102*, 3964.
10. Fried, J.; Sih, J. C. *TL* **1973**, 3899.
11. de Wit, G.; de Vlieger, J. J.; Kock-van Dalen, A. C.; Heus, R.; Laroy, R.; van Hengstrum, A. J.; Kieboom, A. P. G.; van Bekkum, H. *Carbohydr. Res.* **1981**, *91*, 125.
12. Muxfeldt, H.; Buhr, G.; Bangert, R. *AG(E)* **1962**, *1*, 157.
13. Muxfeldt, H.; Haas, G.; Hardtmann, G.; Kathawala, F.; Mooberry, J. B.; Vedejs, E. *JACS* **1979**, *101*, 689.

Garry Procter
University of Salford, UK

Ozone[1]

[10028-15-6] O_3 (MW 48.00)

(powerful oxidant; capable of oxidizing many electron rich functional groups;[1c,d] most widely used to cleave alkenes, affording a variety of derivatives depending on workup conditions[1b,2])

Physical Data: mp −193 °C; bp −111.9 °C; *d* (0 °C, gas) 2.14 g L^{-1}.

Solubility: 0.1–0.3% by weight in hydrocarbon solvents at −80 to −100 °C.[3]

Preparative Methods: ozone is a colorless to faint blue gas which is usually generated in the laboratory by passing dry air or oxygen through two electrodes connected to an alternating current source of several thousand volts. From air, ozone is typically generated at concentrations of 1–2%; from oxygen, concentrations are typically 3–4%. Several laboratory scale generators are commercially available.

Analysis of Reagent Purity: the amount of ozone generated can be determined based on the liberation of iodine from potassium iodide solution followed by thiosulfate titration to determine the amount of iodine produced.[4] Photometric detectors are available which can determine the concentration of ozone in a metered gas stream. In this manner, exact amounts of ozone introduced into a reaction can be determined.

Handling, Storage, and Precautions: ozone is irritating to all mucous membranes and is highly toxic in concentrations greater than 0.1 ppm by volume. It has a characteristic odor which can be detected at levels as low as 0.01 ppm. All operations with ozone should be carried out in an efficient fume hood and scrubbing systems employing thiosulfate solutions can be used to destroy excess ozone. Liquefied ozone poses a severe explosion hazard.

Ozonolysis of Alkenes. Ozone has been most widely used for cleavage of carbon–carbon double bonds to produce carbonyl compounds or alcohols, depending on workup conditions. These reactions usually are performed by passing a stream of ozone in air or oxygen through a solution of the substrate in an inert solvent at low temperature (−25 to −78 °C). Useful solvents include pentane, hexane, ethyl ether, CCl_4, $CHCl_3$, CH_2Cl_2, EtOAc, DMF, MeOH, EtOH, H_2O, or HOAc. The solvents most commonly used are CH_2Cl_2 and MeOH or a combination of the two. Reaction endpoint can be determined using photometric monitors to detect ozone in the exit gas stream, or by the appearance of a blue color in the reaction medium, which indicates excess ozone in solution. A wide variety of alkenes undergo ozonolysis, and those in which the double bond is connected to electron-donating groups react substantially faster than alkenes substituted with electron-withdrawing groups.[1b,5] With haloalkenes, the rate of ozone attack is decelerated and a greater variety of products are obtained, although double bond cleavage is still prevalent.[6]

The reaction mechanism for ozonolysis has been studied extensively and is thought to involve a 1,3-dipolar cycloaddition to afford an initial 1,2,3-trioxolane or primary ozonide, which cleaves to a carbonyl compound and a carbonyl oxide (eq 1).[7] The carbonyl oxide generally forms at the fragment containing the more electron donating group. Recombination of the fragments affords a 1,2,4-trioxolane or ozonide which is sometimes isolated, but due to the danger of explosion is usually directly converted to carbonyl compounds via either a reductive or oxidative procedure. If alcoholic solvents are used, trapping of the carbonyl oxide can occur to afford an α-alkoxy hydroperoxide.

$$O{=}\overset{+}{O}{-}\overset{-}{O} + R_2C{=}CR_2 \longrightarrow R_2C{-}CR_2 \text{ (1,2,3-trioxolane)} \longrightarrow$$

$$R_2C{=}O + R_2C{=}\overset{+}{O}{-}\overset{-}{O} \longrightarrow R_2C(O)(O{-}O)CR_2 \quad (1)$$

Reductive workup procedures afford aldehydes, ketones, or alcohols. An extensive number of reducing agents have been used including catalytic hydrogenation, sulfite ion, bisulfite ion, iodide, phosphine, phosphite, tetracyanoethylene (TCNE), Zn–HOAc, BH_3, $SnCl_2$, Me_2S, thiourea, or, to obtain alcohols, $LiAlH_4$ or $NaBH_4$.[1a,b,8] ***Dimethyl Sulfide*** offers several advantages. It rapidly reduces peroxidic ozonolysis products to carbonyl compounds, it operates under neutral conditions, excess sulfide is easily removed by evaporation, and the oxidation product is DMSO.[9] In cases where the odor of Me_2S is a problem, ***Thiourea*** is a convenient substitute: results are comparable to those obtained with Me_2S, and thiourea *S,S*-dioxide separates out from the reaction mixture.[10] A polymer-based diphenylphosphine system also has been developed which offers the advantage of a simple filtration and evaporative workup and eliminates potential product contamination by PPh_3 or its oxide.[11] A comparison of these workup options is shown in eq 2.

$$\text{PhCH=CH}_2 \xrightarrow[\text{2. [H]}]{\text{1. O}_3} \text{PhCHO} \quad (2)$$

Me_2S, 89%
$(NH_2)_2CS$, 81%
poly-PPh_2, 80%

Oxidative workup procedures convert peroxidic ozonolysis products to ketones or carboxylic acids. Typical oxidative reagents include peroxy acids, silver oxide, chromic acid, permanganate, molecular oxygen, and the most widely used reagent, ***Hydrogen Peroxide***.[1a,b,8]

Additional terminal functionalization can be accomplished by several methods. Schreiber has developed general ozonolysis and workup procedures which enable a variety of products to be prepared from cycloalkenes (eq 3).[12] Also, iron or copper salts can be used to convert ozonides or α-alkoxy hydroperoxides to chlorides or alkenes with one less carbon atom than in the original alkene (eqs 4 and 5),[13] and treatment of ozonides with hydrogen and an amine in the presence of a catalyst provides a direct route for the production of amines from alkenes (eq 6).[14] In a more specific case, stilbenes can be converted to alkyl benzoates by treatment of the intermediate α-alkoxybenzyl hydroperoxides with amines or DMSO.[15]

Cyclohexene: O_3, MeOH, $NaHCO_3$ → CHO / CH(OMe)OOH; then Ac_2O, Et_3N → CHO / CO_2Me (96%)
Cyclohexene: O_3, MeOH, *p*-TsOH → $CH(OMe)_2$ / CH(OMe)OOH; then Ac_2O, Et_3N → $CH(OMe)_2$ / CO_2Me (83%); or $NaHCO_3$, Me_2S → $CH(OMe)_2$ / CHO (93%) (3)

Cyclohexene: 1. O_3, MeOH, –30 °C; 2. $FeCl_3{\cdot}6H_2O$ (47%) → MeO–CH(OMe)–(CH₂)₄–Cl (4)

Cyclooctene: 1. O_3, MeOH; 2. $FeSO_4$ (22–34%) → MeO–CH(OMe)–(CH₂)₄–CH=CH₂ (5)

Cyclohexene: 1. O_3, MeOH; 2. H_2, cat.; 3. H_2, R^1R^2NH, cat. → $R^1R^2N(CH_2)_6NR^1R^2$ (6)
R^1, R^2 = H, 50%
R^1 = Me, R^2 = H, 70%
R^1, R^2 = Me, 57–60%

Vinyl ethers are more reactive toward ozone than alkenes due to the electron-donating oxygen substituent and double bond cleavage products are often obtained.[16] Notably, the ozonolysis of cyclic vinyl ethers provides a path to aldol and homoaldol type products.[17] In addition, silyloxyalkenes undergo clean oxidative cleavage with ozone to afford diacids by using oxidative workups (eq 7),[18] or hydroxyl or oxo derivatives by using reductive workups (eqs 8–10).[19] Overall, the two-step process of forming and cleaving a silyloxyalkene provides a method for regiospecific cleavage of an unsymmetrical ketone. Eq 10 shows that the silyloxyalkene double bond is sufficiently nucleophilic to allow for selective oxidative cleavage of this bond in the presence of less activated double bonds; with appropriate workup conditions, the method can thus complement Baeyer–Villiger oxidation.

OTMS, OTMS cyclopentene: 1. O_3; 2. HCO_3H (79%) → HO_2C … CO_2H, OH (7)

OTMS cyclohexene: 1. O_3, MeOH, –78 °C; 2. $NaBH_4$ (94%) → $HO(CH_2)_4CH(CH_3)CO_2H$ (8)

OTMS cyclohexene: 1. O_3, MeOH, –78 °C; 2. Me_2S (90%) → $CH_3CO(CH_2)_4CO_2H$ (9)

TMSO– bicyclic, H: 1. O_3; 2. $NaBH_4$; 3. H^+ (93%) → lactone, H (10)

While ozone generally reacts with vinyl sulfides and enamines to provide both the expected products of double bond cleavage and anomalous products,[20,21] ketene dithioacetals have been efficiently cleaved to ketones using ozone (eq 11).[22]

1. O_3
2. Me_2S
3. HOAc, H_2O
65%
(11)

The reaction of α,β-unsaturated ketones with ozone usually affords keto acids containing one less carbon than in the original molecule (eq 12).[23]

O_3, EtOAc, HOAc
H_2O, –15 °C
55%
(12)

In the ozonolysis of 1,3-dienes, one double bond often can be cleaved selectively, and in 1,3-cyclodienes, the regioselectivity in fragmentation of the primary ozonide depends upon the size of the rings (eq 13).[24] Eq 13 shows that as the ring size contracts from cyclooctadiene to cyclohexadiene, the α,β-unsaturated ester becomes favored over the enal.

1. O_3, MeOH
2. Ac_2O, Et_3N
(13)

n = 4, 71% 15:1
n = 2, 50% 1:8

Hindered alkenes often afford epoxides upon ozonation due to difficulty in forming the primary ozonide by a cycloaddition process.[25] Examples are displayed in eqs 14 and 15.[26,27]

O_3, MeOH
–78 °C
82%
(14)

1. O_3, 0 °C
$ClCH_2CH_2Cl$
2. Zn, HOAc
92%
(15)

Overall, ozone compares favorably with other approaches for oxidative alkene cleavage involving ***Osmium Tetroxide***, ***Potassium Permanganate***, ***Ruthenium(IV) Oxide***, ***Sodium Periodate***, or chromyl carboxylates which are costly, toxic, involve metal wastes, and may require detailed workup procedures.

Ozonation of Alkynes. Reactions of alkynes with ozone afford either carboxylic acids or, if reductive procedures are used, α-dicarbonyl compounds.[1c] For the production of carboxylic acids, MeOH has been shown to be superior to CH_2Cl_2 as reaction solvent.[28] As with alkenes, a number of reducing agents can be used to produce α-dicarbonyl compounds. An easy option which results in high yields of α-dicarbonyl compounds involves the addition of ***Tetracyanoethylene*** directly to an ozonation reaction mixture as an in situ reducing agent (eq 16).[29]

O_3, TCNE
EtOAc, –78 °C
(16)

R = Ph, 92%; H, 60%; Pr, 71%

Alkynes react slower with ozone than do alkenes and selective reaction of alkenes can be achieved in the presence of alkynes (eq 17).[30] Conversely, the example of eq 18 shows that alkynes are more reactive toward ozone than aromatic rings.[31]

1. O_3, CH_2Cl_2
–78 °C
2. Zn, HOAc
94%
(17)

O_3, $CHCl_3$, 0 °C
73%
(18)

The ozonation of terminal alkynes to afford α-oxoaldehydes is significant (see eq 16). While reagents such as $KMnO_4$, RuO_4, OsO_4, and ***Thallium(III) Nitrate*** can be used to convert internal alkynes to α-dicarbonyl compounds, terminal alkynes are generally cleaved to carboxylic acids. Only ***Mercury(II) Acetate***-catalyzed oxidations using a molybdenum peroxide complex afford α-oxoaldehydes.[32] However, high catalyst loads are required, the oxidant is not commercially available, and metal wastes are generated.

Oxygenated alkynes also can be ozonated. The reaction of ozone with alkynyl ethers followed by reductive workup provides a convenient method for the production of α-keto esters in moderate yields (eq 19).[33]

1. O_3, MeOH
–40 to –60 °C
2. NaI
(19)

R = Me, R' = Et, 25%
R = Pr, R' = Et, 30%

Ozonation of Aromatic Systems. Aromatic compounds are less reactive toward ozone than either alkenes or alkynes. As a consequence, more forcing conditions are required to ozonize aromatic systems. These conditions typically involve the use of acetic acid as solvent, excess ozone, and oxidative decomposition often using H_2O_2. Electron-withdrawing groups deactivate aromatic systems toward electrophilic ozone attack, while electron-donating groups activate aromatic systems toward ozone attack. An example of this is provided in Woodward's strychnine synthesis where methoxy substitution allows for selective oxidation of one aromatic ring over two others to afford an often difficult to prepare, terminally functionalized, conjugated (*Z,Z*)-diene (eq 20).[34]

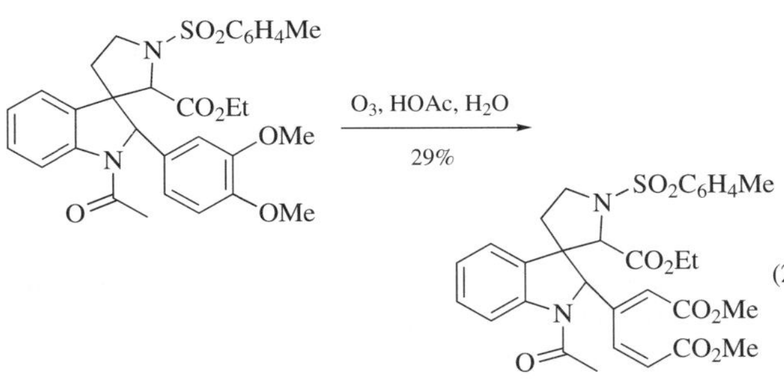

(20)

With polycyclic aromatic hydrocarbons, the site of ozone attack may be dependent upon substrate structure and reaction solvent (eq 21).[35]

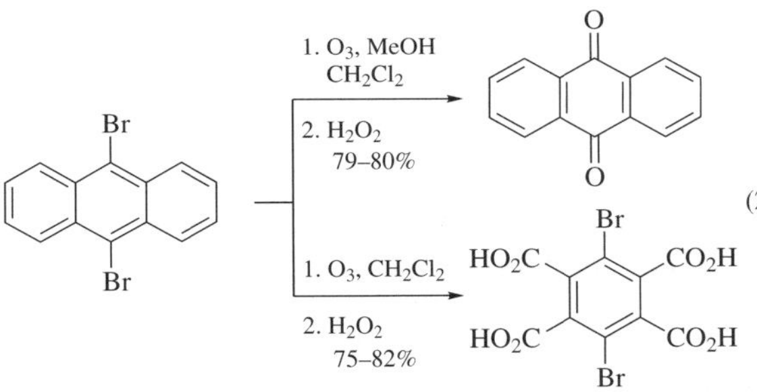

(21)

Synthetically useful ozonolyses of heteroaromatic systems include the preparation of pyridine derivatives from quinolines (eq 22),[36] the preparation of versatile *N*-acyl amides by the ozonolysis of imidazoles (eq 23),[37] and the unmasking of a latent carboxylic acid function by the ozonolysis of a furan system (eq 24).[38]

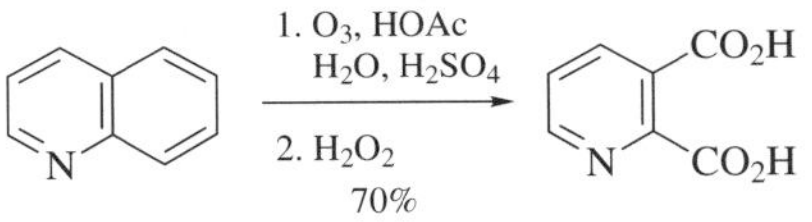

(22)

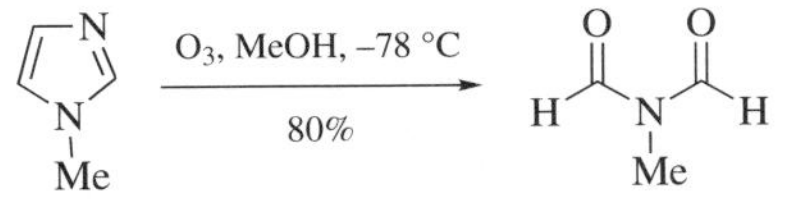

(23)

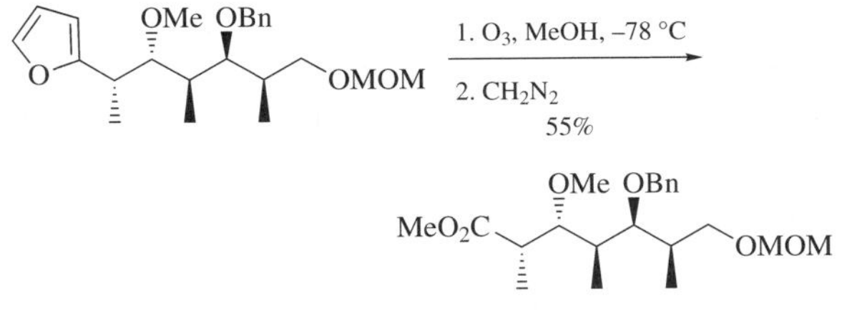

(24)

Ozonation of Heteroatoms. Phosphines are converted to phosphine oxides and phosphites to phosphates by ozone.[39,40] These reactions are quite general and a wide range of substitutents can be tolerated. Phosphine oxides also can be produced by the ozonation of alkylidenetriphenylphosphoranes or of thio- or selenophosphoranes.[41,42] Organic sulfides are converted to sulfoxides and sulfones by ozonation.[40,43] Tertiary amines are converted to amine oxides, while nitro compounds can be produced in modest yields by ozonation of primary amines.[43,44] This preparation of nitroalkanes compares well with alternate approaches using peroxides, peroxy acids, permanganate, or ***Monoperoxysulfuric Acid***, but ozonation on silica gel has proven to be superior (see ***Ozone–Silica Gel***). Selenides are converted to selenoxides by ozone and this reaction is often used to achieve overall production of unsaturated carbonyl compounds. An example is shown in eq 25.[45]

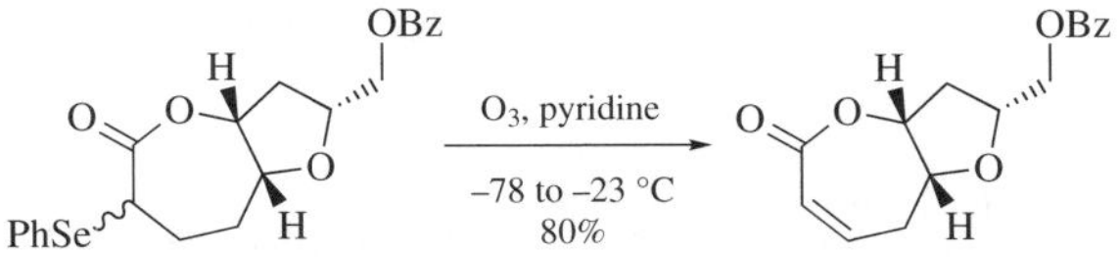

(25)

Modification of Ozone Reactivity. The reactivity of ozone toward various unsaturated moieties can be moderated by the addition of either Lewis acids or pyridine to the ozonations. Enhanced electrophilic ozone reactivity toward aromatic substrates is observed when the Lewis acids ***Aluminum Chloride*** or ***Boron Trifluoride*** are added to reaction mixtures.[46] Conversely, an apparent decrease in ozone reactivity and a concurrent increase in the regioselectivity of ozone attack can be achieved by adding small amounts of pyridine to ozonolyses (eq 26).[47] It is thought that coordination of either the Lewis acid or basic pyridine to ozone results in the modified reactivity.

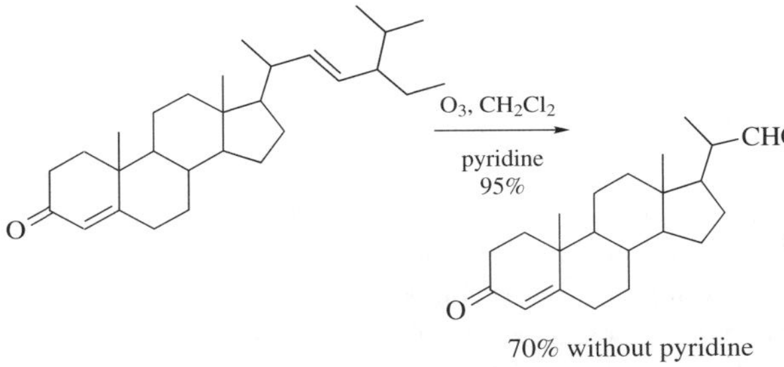

(26)

In a related procedure, ozonizable dyes have been used as endpoint indicators for selective ozonation of substrates containing multiple unsaturated linkages.[48] The dye affords colored solutions and the ozonation is carried out just until the color is discharged. If the dye is of suitable reactivity such that the most reactive substrate unsaturated linkage reacts first, and the dye second, the reaction can be stopped before further oxidation of the substrate occurs.

Interestingly, addition of BF_3 etherate to the ozonolysis of *o*-dimethoxybenzene derivatives results in increased yields of (*Z,Z*)-dienes (eq 27, compare to eq 20).[49] In this case, it is thought that coordination of the Lewis acid to the diene reduces its electron density and suppresses further attack by ozone. Also, the fact that the BF_3 is already coordinated to ether may limit its ability to coordinate to ozone and increase its electrophilic reactivity.

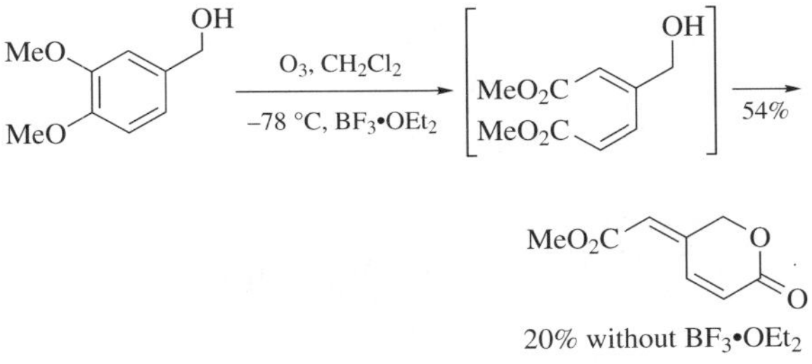

(27)

Ozonation of Acetals. Ozone reacts very efficiently with acetals to afford the corresponding esters (eqs 28 and 29).[50] The aldehyde and alcohol components of the acetal function can be varied and yields are excellent. Cyclic acetals react much faster than acyclic acetals as a result of conformational effects.

$$C_6H_{13}CH(OR)_2 \xrightarrow[-78\ °C]{O_3,\ EtOAc} C_6H_{13}CO_2R \quad (28)$$

R = Me, 15 h, 91%
R = Et, 8 h, 94%

$$\text{cyclic acetal } C_6H_{13}(H)C(O)_2(CH_2)_n \xrightarrow[-78\ °C]{O_3,\ EtOAc} C_6H_{13}CO_2(CH_2)_nOH \quad (29)$$

n = 2, 10 min, 98%
n = 3, 2 h, 97%

Miscellaneous Ozonations. Ozonation offers a simple neutral alternative for oxidation of secondary alcohols to ketones (eq 30).[51]

$$R^1CH(OH)R^2 \xrightarrow[0\ °C]{O_3,\ CH_2Cl_2} R^1C(O)R^2 \quad (30)$$

R^1	R^2	%
Me	Me	83
Me	Me	72
$-(CH_2)_4-$		53
$-(CH_2)_5-$		65

Upon reaction of allene with one equivalent of ozone, trisection occurs to provide carbon monoxide derived from the central carbon atom and carbonyl compounds from the remaining carbon atoms.[52] In the example of eq 31, allene ozonolysis is used to prepare a versatile protected α-hydroxyaldehyde.[53]

$$R\text{-}CH(OTBDMS)\text{-}CH{=}C{=}CH_2 \xrightarrow[-78\ °C]{O_3,\ CH_2Cl_2} R\text{-}CH(OTBDMS)\text{-}CHO \quad (31)$$

R = C_5H_{11}, Ph, *t*-Bu, 89–99%

Ozonation of benzyl ethers affords high yields of benzoate esters (eq 32).[54] Coupled with deacylation by NaOMe, this reaction offers a mild alternative for removal of benzyl ether protecting groups (eq 33).[55] However, due to the higher reactivity of alkenes, selective oxidative cleavage of carbon–carbon double bonds can be accomplished in the presence of benzyl ethers (eq 34).[56]

$$c\text{-}C_6H_{11}OCH_2Ph \xrightarrow[2.\ Me_2S,\ 76–80\%]{1.\ O_3,\ CH_2Cl_2,\ -78\ \text{to}\ 0\ °C} c\text{-}C_6H_{11}OCOPh \quad (32)$$

(33): methyl 2,3,4,6-tetra-O-benzyl glycoside (BnO, OBn, OMe) → 1. O_3, CH_2Cl_2, 0 °C; 2. NaOMe, MeOH; 3. Amberlite MB3; 78% → methyl glycoside (HO, OH, OMe)

yield for β-anomer, 75%

(34): BnO–acetonide-protected diol–alkene → 1. O_3, MeOH, −78 °C; 2. Me_2S; 88% → BnO–acetonide-protected diol–CHO

Ozone has been used to cleave nitronate anions, resulting in the high yield production of either aldehydes or ketones.[57] An example of this reaction is shown in eq 35.[57a] This is a very general method and has advantages over the Nef reaction which requires strong acid conditions, and other procedures utilizing permanganate or ***Titanium(III) Chloride***.

(35): nitro ketone (NO_2) → 1. NaOMe, MeOH; 2. O_3, −78 °C; 3. Me_2S; 83% → diketone

Aldehydes can be converted to peroxy acids via ozonation in methyl or ethyl acetate,[58] or to methyl esters via ozonation in 10% methanolic KOH (eq 36).[59] Ethyl esters can be produced analogously, but the use of higher alcohols results in low KOH solubility and poor conversion. This problem can be overcome by adding the aldehyde to a solution of lithium alkoxide in THF at −78 °C and treating this mixture with ozone (eq 37). Additionally, the direct preparation of methyl esters can be accomplished via alkene ozonolysis in methanolic NaOH or by addition of NaOMe to a $MeOH–CH_2Cl_2$ ozonolysis solvent system.[60]

$$R^1CHO \xrightarrow[O_3,\ -78\ °C]{R^2OH,\ KOH} R^1CO_2R^2 \quad (36)$$

R^1	R^2	%
Cy	Me	58
	Et	60
Ph	Me	66
	Et	60
3-Oxobisnor-	Me	85
4-cholenyl	Et	87

$$n\text{-}C_8H_{17}CHO \xrightarrow[36\%]{Me_2CHOLi,\ O_3,\ THF,\ -78\ °C} n\text{-}C_8H_{17}CO_2CHMe_2 \quad (37)$$

Related Reagents. See Classes O-1, O-3, O-8, and O-17, pages 1–10.

---

1. (a) Bailey, P. S. *CRV* **1958**, *58*, 925. (b) Bailey, P. S. *Ozonation in Organic Chemistry*; Academic: New York, 1978; Vol. 1. (c) Bailey, P. S. *Ozonation in Organic Chemistry*; Academic: San Diego, CA, 1982; Vol. 2. (d) Razumovskii, S. D.; Zaikov, G. E. *Ozone and Its Reactions With Organic Compounds*; Elsevier: Amsterdam, 1984.
2. Odinokov, V. N.; Tolstikov, G. A. *RCR* **1981**, *50*, 636.
3. Varkony, H.; Pass, S.; Mazur, Y. *CC* **1974**, 437.
4. Dietz, R. N.; Pruzansky, J.; Smith, J. D. *AG* **1973**, *45*, 402.
5. (a) Pryor, W. A.; Giamalva, D.; Church, D. F. *JACS* **1983**, *105*, 6858. (b) Fleet, G. W. J. *Org. React. Mech.* **1984**, 179.
6. Gilles, C. W.; Kuczkowski, R. L. *Isr. J. Chem.* **1983**, *24*, 446.

7. (a) Murray, R. W. *ACR* **1968**, *1*, 313. (b) Criegee, R. *AG(E)* **1975**, *14*, 745. (c) Razumovskii, S. D.; Zaikov, G. E. *RCR* **1980**, *49*, 1163. (d) Kuczkowski, R. L. *ACR* **1983**, *16*, 42.
8. For further discussion of reductive or oxidative reagents, see: (a) Belew, J. S. In *Oxidation*; Augustine, R. L., Ed.; Dekker: New York, 1969; Vol. 1, pp 259–335. (b) Hudlicky, M. *Oxidation in Organic Chemistry*; American Chemical Society: Washington, 1990.
9. Pappas, J. J.; Keaveney, W. P.; Gancher, E.; Berger, M. *TL* **1966**, 4273.
10. Gupta, D.; Soman, R.; Dev, S. K. *T* **1982**, *38*, 3013.
11. Ferraboschi, P.; Gambero, C.; Azadani, M. N.; Santaniello, E. *SC* **1986**, *16*, 667.
12. Schreiber, S. L.; Claus, R. E.; Reagan, J. *TL* **1982**, *23*, 3867.
13. (a) Cardinale, G.; Grimmelikhuysen, J. C.; Laan, J. A. M.; Ward, J. P. *T* **1984**, *40*, 1881. (b) Cardinale, G.; Laan, J. A. M.; Ward, J. P. *T* **1985**, *41*, 2899.
14. (a) Benton, F. L.; Kiess, A. A. *JOC* **1960**, *25*, 470. (b) Diaper, D. G. M.; Mitchell, D. L. *CJC* **1962**, *40*, 1189. (c) Pollart, K. A.; Miller, R. E. *JOC* **1962**, *27*, 2392. (d) White, R. W.; King, S. W.; O'Brien, J. L. *TL* **1971**, 3591.
15. Ellam, R. M.; Padbury, J. M. *CC* **1972**, 1086.
16. (a) Corey, E. J.; Katzenellenbogen, J. A.; Gilman, N. W.; Roman, S. A.; Erickson, B. W. *JACS* **1968**, *90*, 5618. (b) Effenberger, F. *AG(E)* **1969**, *8*, 295. (c) Keul, H.; Choi, H.-S.; Kuczkowski, R. L. *JOC* **1985**, *50*, 3365. (d) Wojciechowski, B. J.; Pearson, W. H.; Kuczkowski, R. L. *JOC* **1989**, *54*, 115. (e) Wojciechowski, B. J.; Chiang, C.-Y.; Kuczkowski, R. L. *JOC* **1990**, *55*, 1120. (f) Griesbaum, K.; Kim, W.-S.; Nakamura, N.; Mori, M.; Nojima, M.; Kusabayashi, S. *JOC* **1990**, *55*, 6153. (g) Kuczkowski, R. L. *Advances in Oxygenated Processes*; JAI Greenwich, CT, 1991.
17. (a) Danishefsky, S.; Kato, N.; Askin, D.; Kerwin, J. F., Jr. *JACS* **1982**, *104*, 360. (b) Hillers, S.; Niklaus, A.; Reiser, O. *JOC* **1993**, *58*, 3169.
18. Vedejs, E.; Larsen, S. D. *JACS* **1984**, *106*, 3031.
19. (a) Clark, R. D.; Heathcock, C. H. *TL* **1974**, 2027. (b) Clark, R. D.; Heathcock, C. H. *JOC* **1976**, *41*, 1396.
20. (a) Chaussin, R.; Leriverend, P.; Paquer, D. *CC* **1978**, 1032. (b) Strobel, M.-P.; Morin, L.; Paquer, D. *TL* **1980**, 523. (c) Barillier, D.; Strobel, M. P. *NJC* **1982**, *6*, 201. (d) Barillier, D.; Vazeux, M. *JOC* **1986**, *51*, 2276.
21. Witkop, B. *JACS* **1956**, *78*, 2873.
22. Ziegler, F. E.; Fang, J.-M. *JOC* **1981**, *46*, 825.
23. Dauben, W. G.; Wight, H. G.; Boswell, G. A. *JOC* **1958**, *23*, 1787.
24. Wang, Z.; Zvlichovsky, G. *TL* **1990**, *31*, 5579.
25. (a) Bailey, P. S.; Lane, A. G. *JACS* **1967**, *89*, 4473. (b) Bailey, P. S.; Ward, J. W.; Hornish, R. E.; Potts, F. E., III. *Adv. Chem. Ser.* **1972**, *112*, 1. (c) Griesbaum, K.; Zwick, G. *CB* **1985**, *118*, 3041.
26. Bailey, P. S.; Hwang, H. H.; Chiang, C.-Y. *JOC* **1985**, *50*, 231.
27. Hochstetler, A. R. *JOC* **1975**, *40*, 1536.
28. Silbert, L. S.; Foglia, T. A. *AG* **1985**, *57*, 1404.
29. Yang, N. C.; Libman, J. *JOC* **1974**, *39*, 1782.
30. McCurry, P. M., Jr.; Abe, K. *TL* **1974**, 1387.
31. Cannon, J. G.; Darko, L. L. *JOC* **1964**, *29*, 3419.
32. Ballistreri, F. P.; Failla, S.; Tomaselli, G. A.; Curci, R. *TL* **1986**, *27*, 5139.
33. Wisaksono, W. W.; Arens, J. F. *RTC* **1961**, *80*, 846.
34. Woodward, R. B.; Cava, M. P.; Ollis, W. D.; Hunger, A.; Daeniker, H. U.; Schenker, K. *T* **1963**, *19*, 247.
35. Dobinson, F.; Bailey, P. S. *TL* **1960** (13), 14.
36. O'Murchu, C. *S* **1989**, 880.
37. Kashima, C.; Harada, K.; Hosomi, A. *H* **1992**, *33*, 385.
38. Schmid, G.; Fukuyama, T.; Akasaka, K.; Kishi, Y. *JACS* **1979**, *101*, 259.
39. Caminade, A.; El Khatib, F.; Baceiredo, A.; Koenig, M. *PS* **1987**, *29*, 365.
40. Thompson, Q. E. *JACS* **1961**, *83*, 845.
41. Caminade, A. M.; El Khatib, F.; Koening, M. *PS* **1983**, *14*, 381.
42. Skowronska, A.; Krawczyk, E. *S* **1983**, 509.
43. Horner, L.; Schaefer, H.; Ludwig, W. *CB* **1958**, *91*, 75.
44. Bachman, G. B.; Strawn, K. G. *JOC* **1968**, *33*, 313.
45. Grese, T. A.; Hutchinson, K. D.; Overman, L. E. *JOC* **1993**, *58*, 2468.
46. (a) Wilbaut, J. P.; Sixma, F. L. J.; Kampschmidt, L. W. F.; Boer, H. *RTC* **1950**, *69*, 1355. (b) Sixma, F. L. J.; Boer, H.; Wilbaut, J. P.; Pel, H. J.; deBruyn, J. *RTC* **1951**, *70*, 1005. (c) Wilbaut, J. P.; Boer, H. *RTC* **1955**, *74*, 241.
47. (a) Shepherd, D. A.; Donia, R. A.; Campbell, J. A.; Johnson, B. A.; Holysz, R. P.; Slomp, G., Jr.; Stafford, J. E.; Pederson, R. L.; Ott, A. C. *JACS* **1955**, *77*, 1212. (b) Slomp, G., Jr. *JOC* **1957**, *22*, 1277. (c) Slomp, G., Jr.; Johnson, J. L. *JACS* **1958**, *80*, 915. (d) Boddy, I. K.; Boniface, P. J.; Cambie, R. C.; Craw, P. A.; Huang, Z.-D.; Larsen, D. S.; McDonald, H.; Rutledge, P. S.; Woodgate, P. D. *AJC* **1984**, *37*, 1511. (e) Haag, T.; Luu, B.; Hetru, C. *JCS(P1)* **1988**, 2353.
48. Veysoglu, T.; Mitscher, L. A.; Swayze, J. K. *S* **1980**, 807.
49. Isobe, K.; Mohri, K.; Tokoro, K.; Fukushima, C.; Higuchi, F.; Taga, J.-I.; Tsuda, Y. *CPB* **1988**, *36*, 1275.
50. (a) Deslongchamps, P.; Moreau, C. *CJC* **1971**, *49*, 2465. (b) Deslongchamps, P.; Moreau, C.; Fréhel, D.; Atlani, P. *CJC* **1972**, *50*, 3402. (c) Deslongchamps, P.; Atlani, P.; Fréhel, D.; Malaval, A.; Moreau, C. *CJC* **1974**, *52*, 3651. (d) Deslongchamps, P.; Moreau, C.; Fréhel, D.; Chênevert, R. *CJC* **1975**, *53*, 1204. (e) Deslongchamps, P. *T* **1975**, *31*, 2463.
51. Waters, W. L.; Rollin, A. J.; Bardwell, C. M.; Schneider, J. A.; Aanerud, T. W. *JOC* **1976**, *41*, 889.
52. Kolsaker, P.; Teige, B. *ACS* **1970**, *24*, 2101.
53. Corey, E. J.; Jones, G. B. *TL* **1991**, *32*, 5713.
54. Hirama, M.; Shimizu, M. *SC* **1983**, *13*, 781.
55. Angibeaud, P.; Defaye, J.; Gadelle, A.; Utille, J.-P. *S* **1985**, 1123.
56. Hirama, M.; Uei, M. *JACS* **1982**, *104*, 4251.
57. (a) McMurry, J. E.; Melton, J.; Padgett, H. *JOC* **1974**, *39*, 259. (b) Crossley, M. J.; Crumbie, R. L.; Fung, Y. M.; Potter, J. J.; Pegler, M. A. *TL* **1987**, *28*, 2883. (c) Aizpurua, J. M.; Oiarbide, M.; Palomo, C. *TL* **1987**, *28*, 5365.
58. Dick, C. R.; Hanna, R. F. *JOC* **1964**, *29*, 1218.
59. Sundararaman, P.; Walker, E. C.; Djerassi, C. *TL* **1978**, 1627.
60. Marshall, J. A.; Garofalo, A. W. *JOC* **1993**, *58*, 3675.

Richard A. Berglund
Eli Lilly and Company, Lafayette, IN, USA

Palladium on Barium Sulfate[1]

Pd/BaSO$_4$

[7440-05-3] Pd (MW 106.42)

(usually supported on BaSO$_4$, or an appropriate form of carbon, when used to catalyze the hydrogenation of acyl chlorides to aldehydes, the Rosenmund reduction;[2] useful catalyst for many other hydrogenations[1b–e])

Alternate Name: Rosenmund catalyst.
Form Supplied in: Pd-on-BaSO$_4$ and Pd-on-C are available commercially or may be prepared.[3]
Handling, Storage, and Precautions: the catalysts may be stored indefinitely in well-sealed containers.[1e] Although the unused catalysts can be exposed to a clean atmosphere, they may ignite organic solvent vapors; heating a Pd-on-C catalyst in a vacuum drying oven at 115 °C for more than 48 h causes it to become extremely pyrophoric.[4] After use, all catalysts are liable to contain adsorbed hydrogen and may ignite when dried. The filtered catalyst should be kept wet and away from combustible vapors or solvents.

Traditional Rosenmund Procedures. The palladium-catalyzed hydrogenation of an acid chloride to an aldehyde is known as the Rosenmund reduction (eq 1). In the original procedure, hydrogen is bubbled through a heated suspension of the catalyst, Pd/BaSO$_4$, in a xylene or toluene solution of the acyl chloride.[2] The HCl formed is absorbed in water and titrated to monitor the reaction's progress. Although the procedure works well for many acyl chlorides, for others the further reduction of the aldehyde to the alcohol, and the consequential formation of esters, ethers, and hydrocarbons, seriously lowers the yield of the aldehyde.[1] In initial experiments, it was reported that benzoyl chloride was converted almost completely to benzaldehyde; however, repetition of the same experiment, but with all reactants carefully purified, gave none.[5] Seeking possible catalyst modifiers or regulators, it was found that quinoline-sulfur, a crude preparation of thioquinanthrene, was most suitable.[5,6] Other regulators which have been recommended are pure thioquinanthrene, thiourea, and tetramethylthiourea.[5,7] The purity of the solvent, which is used in much larger amounts than any of the reactants or the catalyst, is a key to reproducible reductions.[8,9] Attaining the lowest temperature at which HCl is evolved was reported to optimize the yield of aldehyde.[1a]

$$\text{R-C(O)Cl} + H_2 \xrightarrow[\text{toluene or xylene}]{\text{Pd/BaSO}_4} \text{R-C(O)H} + \text{HCl} \quad (1)$$

In a 1948 review, it is claimed that 'For accomplishing this transformation, RCO$_2$H → RCHO, the Rosenmund reduction is probably the most useful method for application to a large number of aldehydes of varied types'.[1a] This critical review describes the scope and limitations of the reaction, the experimental conditions, reagents, and procedures and includes tables recording the acid chlorides whose reduction by the Rosenmund method had been reported to November 1947.

Rosenmund described a simple apparatus for performing the reduction.[5] A detailed description of a more elaborate apparatus and procedure used for the hydrogenolysis of β-naphthoyl chloride (0.30 mol) in xylene catalyzed by 5% Pd/BaSO$_4$ and regulated by quinoline-sulfur was given by Hershberg and Cason (eq 2).[10] They recommended that a 'poison' always be added to ensure controlled conditions. The hydrogenolysis of mesitoyl chloride in xylene over unpoisoned Pd/BaSO$_4$ is described in the same volume; vigorous stirring shortens the reaction time by about one third.[11]

$$\text{2-naphthyl-COCl} \xrightarrow[\text{quinoline-S, xylene, reflux; 74–81\%}]{H_2,\ 5\%\ \text{Pd/BaSO}_4} \text{2-naphthyl-CHO} \quad (2)$$

The omission of catalyst poisons is common and the original Rosenmund procedure[2] has been successful with acid halides containing other functional groups, or condensed benzenoid or heterocyclic systems.[1a] For example, discouraged by attempts to obtain high yields with a variety of metal hydride reducing reagents, Danishefsky et al. found that the original Rosenmund procedure converts the acid chloride (eq 3) in an essentially quantitative yield.[12] The reduction of the related compound with a methyl group proximate to the C(O)Cl group gave only a 49% yield of the aldehyde (eq 4) (however, see below).

$$\text{R-COCl} \xrightarrow[\text{toluene, reflux; 95\%}]{H_2,\ \text{Pd/BaSO}_4} \text{R-CHO} \quad (3)$$

$$\text{R-COCl} \xrightarrow[\text{toluene, reflux; 49\%}]{H_2,\ \text{Pd/BaSO}_4} \text{R-CHO} \quad (4)$$

A procedure for the unpoisoned 10% ***Palladium on Carbon*** catalyzed hydrogenation of α-phthalimido acid chlorides to the aldehydes in benzene at 40 °C has been described.[13] To cause the benzene to reflux at 40 °C, the pressure is lowered with a vacuum pump which is attached to the outlet in a manner which allows the collection and titration of the evolved HCl. This procedure was suited particularly for the preparation of the phthaloyl methionine aldehyde (eq 5) but the (±)-phenylalanine and (±)-alanine derivatives gave yields of 93% and 94% respectively with benzene refluxing at 1 atm. An almost identical low temperature (reduced pressure) procedure was used to hydrogenate a diacid chloride to the dialdehyde in 92–94% yield (eq 6).[14]

Rosenmund Reductions within Closed Systems. To avoid the higher temperature and the hazard of free flowing hydrogen of the classical Rosenmund procedure, particularly for large scale preparations, an autoclave can be used for the Pd/C catalyzed hydrogenolysis of 3,4,5-trimethoxybenzoyl chloride to the aldehyde (yield, 64–83%); the reduction was done at 35–40 °C and H_2 (4 atm) in toluene containing quinoline-S with anhydrous sodium acetate as HCl adsorber (eq 7).[4,15] The same reaction has been achieved repeatedly in 80–90% yields by the Rosenmund procedure without added regulators and either $Pd/BaSO_4$ or Pd/C catalysts and either xylene or PhOMe as solvent.[1a]

H_2, 10% Pd/C, benzene, reflux, 38–40 °C, reduced pressure, 90% (5)

H_2, 10% Pd/C, benzene, reflux, 30–35 °C, reduced pressure, 92–94% (6)

4 atm H_2, 10% Pd/C, quinoline-S, toluene, NaOAc, 35–40 °C, 64–83% (7)

Sakurai and Tanabe were the first to report the use of a closed system for the $Pd/BaSO_4$ catalyzed hydrogenolysis of an acyl halide.[16] The reduction (H_2, 1 atm) was conducted at rt in the presence of a hydrogen chloride acceptor, *N,N*-dimethylaniline, and acetone as solvent. *N*-Phthaloyl derivatives of (±)-α-amino acid chlorides have been hydrogenated to the aldehydes using 10% Pd/C in the solvent ethyl acetate (H_2, 3 atm) in the presence of dimethylaniline, with yields of over 90%.[13]

The Sakurai and Tanabe procedure (5% $Pd/BaSO_4$) gave high yields of arachidaldehyde (72%) and stearaldehyde (96%); for convenience, *N,N*-dimethylacetamide was used as the acid acceptor to obtain excellent yields of palmitaldehyde (96%) and decanaldehyde (96%).[17] Peters and van Bekkum chose ethyldiisopropylamine as HCl acceptor due to the competitive reduction of *N,N*-dimethylaniline, which obscured the end-point (vol H_2) of the hydrogenolysis of the acid chloride.[18] The mild conditions converted the sterically hindered carbonyl chloride function in 1-*t*-butylcyclohexanecarbonyl chloride to the aldehyde (78%), although *t*-butylcyclohexane was the sole product of the original Rosenmund procedure (eq 8). Burgstahler and Weigel also modified the Sakurai and Tanabe procedure by using 2,6-dimethylpyridine in place of *N,N*-dimethylaniline and THF as solvent with either $Pd/BaSO_4$ or Pd/C as catalyst to obtain excellent yields of 15 sensitive aliphatic and alicyclic aldehydes such as hexanedial (74%) and (*Z*)-9-octadecenal (96%) (eq 9).[19] The reaction temperature was lowered to 0 °C to convert dehydroabietic acid chloride to the aldehyde (92%) (eq 10).[19]

Both aliphatic and aromatic acid chlorides are reduced smoothly at room temperature and atmospheric pressure to aldehydes with 10% Pd/C as catalyst, acetone or ethyl acetate as solvent, and ethyldiisopropylamine as HCl acceptor.[20] The reaction proceeds with high selectivity; over reduction is less than 1%, and nitro and chloro substituents in benzoyl chlorides are unaffected, as is the double bond in cinnamoyl chloride.

H_2, 5% Pd/C, quinoline-S, toluene, reflux → *t*-butylcyclohexane + CO; 1 atm H_2, 10% Pd/C, acetone, *i*-Pr_2NEt, 25 °C, 78% (8)

1 atm H_2, 10% Pd/C, THF, 2,6-lutidine, 25 °C, 96% (9)

1 atm H_2, 10% Pd/C, 2,6-lutidine, THF, 0 °C, 93% (10)

Recent Practice. Some more recent examples show that both the older and the newer procedures are used successfully. As an important step in the preparation of 10-nor-*cis*-α-irone, the classical Rosenmund reduction (H_2, 5% $Pd/BaSO_4$, toluene/reflux) converted the acid chloride (**1**) to the product aldehyde in 93% yield (eq 11).[21]

H_2, 5% $Pd/BaSO_4$, toluene, reflux, 93% (11)

(**1**)

The 'semialdehyde' derivatives of aspartic and glutamic acids have been obtained in good yields and in higher purity than by hydride reductions; acid-sensitive protecting groups are unaffected.[22] The acid chlorides (**2**) and (**3**) (benzyloxycarbonyl (*Z*) derivatives) were converted to the aldehydes using an unpoisoned catalyst (5% $Pd/BaSO_4$, boiling toluene, H_2) (eq 12). The same procedure was used to reduce the acid chloride (**4**), derived from L-alanine, to the aldehyde in 93% yield (eq 13).[23]

H_2, 5% $Pd/BaSO_4$, toluene, reflux (12)

(**2**) $n = 1$; (**3**) $n = 2$ → $n = 1$, 87%; $n = 2$, 82%

H_2, 5% $Pd/BaSO_4$, toluene, reflux (13)

(**4**)

To protect the acid labile *t*-butoxycarbonyl protecting group in acid chloride (**5**), the Burgstahler procedure (H_2, 5% Pd/C, 2,6-lutidine, THF, 10–15 °C) converted (**5**) to the aldehyde (eq 14).[19,22]

(Boc)$_2$N Cl CO$_2$-*t*-Bu O **(5)** — 1 atm H$_2$, 5% Pd/C, 2,6-lutidine; THF, 10–15 °C; 78% → (Boc)$_2$N H CO$_2$-*t*-Bu O (14)

The quinoline-sulfur system was used to prepare methyl 4-oxobutanoate from 3-methoxycarbonyl chloride as the first of a three-step synthesis of a series of 5-vinyl γ-lactones.[10,24]

Kinetics and Mechanism of the Rosenmund Reaction. The kinetics of the amine-modified Rosenmund reduction has been examined in detail.[20] In the absence of the tertiary amine, the hydrogenolysis of 4-*t*-butylbenzoyl chloride proceeds beyond the aldehyde stage to a complex mixture with bis(4-*t*-butylbenzyl) ether as the main product. In the presence of the efficient HCl acceptor zeolite NaA, most of the side reactions (acid catalyzed) are suppressed, but reduction proceeds to 4-*t*-butylbenzyl alcohol. In the presence of the tertiary amine, the aldehyde is the sole product. However, benzaldehydes subjected to the conditions of the amine-modified reduction are hydrogenated to the alcohols, but more slowly than in the absence of the amine and substantially slower on a catalyst which has been used previously in a hydrogenation of an acid chloride. The nature of the deactivation is a subject of speculation.[20,25]

The tertiary amine not only neutralizes HCl, but also acts as a nucleophile which serves to moderate the reaction and enhance the selectivity by competing with both the acid chloride and the product aldehyde for active sites on the catalyst. A useful analogy is the effect of tertiary amines upon increasing the selectivity of the hydrogenation of alkynes to alkenes on palladium catalysts.[1b,26,27] The solvent also may compete for active sites; for example, the rate constants for the Pd-catalyzed hydrogenation of cyclohexene, corrected for the difference in the solubility of H_2 in the solvents, are smaller in benzene and smaller still in xylene, both commonly used in the Rosenmund reduction, than in saturated hydrocarbons.[28] The presence of nucleophilic groups elsewhere in the acid chloride may also act to moderate the reduction and affect selectivity in the absence of an added catalyst poison.

Aliphatic acid chlorides generally are more easily hydrogenolyzed than are aromatic ones; however, the Peters and van Bekkum paper contains the most direct comparison of relative reactivities for some representative carbonyl chlorides (15 compounds including 10 aromatic).[20] For benzoyl chlorides, electron-donating substituents increase the reaction rate while electron-withdrawing substituents have a retarding effect. The rates of hydrogenation of aroyl chlorides generally are faster in the solvents ethyl acetate or THF than in acetone.

The mechanism of the Rosenmund reduction has been discussed in relation to the characteristic reactions of transition metal complexes.[29] It has been proposed that the acid chloride adds oxidatively to the palladium metal, forming a complex which gives rise to the observed products which depend upon the reaction conditions, e.g. temperature, H_2 pressure, and solvent. Some dissolution of the palladium when heated with an acid chloride at about 100 °C was observed. However, exposing single crystals of palladium to heptanoyl chloride in pentane and H_2 at room temperature for 52 h did not change the (755) crystal surface which had catalyzed the formation of the aldehyde.[25] A means of representing catalytic processes on such crystal surfaces by analogy with the reactions of transition metal complexes has been given for catalytic hydrogenation.[26]

Other Hydrogenations.[1b–e] Pd/$BaSO_4$ has also been used in the conversion of alkynes to *cis*-alkenes.[30] In some cases, where results of the reduction of alkynes to *cis*-alkenes with Lindlar catalysts (see ***Palladium on Calcium Carbonate (Lead Poisoned)***) are unsatisfactory, the use of Pd/$BaSO_4$ as the catalyst has been effective (eqs 15 and 16).[31]

OH OH OH — Pd/BaSO$_4$, quinoline; EtOH, H$_2$; 95% → OH OH HO (15)

CO$_2$Me CO$_2$Me — Pd/BaSO$_4$, quinoline; H$_2$; 83% → CO$_2$Me CO$_2$Me (16)

The reverse has also been observed.[32] Interestingly, the saturation of the trisubstituted alkene in humulinic acid B has also been reported with this catalyst.[33]

Hydrogenolysis of various functional groups has also been reported with Pd/$BaSO_4$. For example, the conversion of vinyl epoxides to homoallylic alcohols,[34] α-bromo-β-mesyluridines to hydrocarbons (eq 17),[35] *N,N*-dibenzylamino acids to *N*-benzylamino acids,[36] and the enantioselective mono-dehydrohalogenation of α,α-dichlorobenzazepin-2-one[37] have all been reported.

PhCH$_2$O O R MsO Br — Pd/BaSO$_4$, H$_2$, NaOAc; 36% → PhCH$_2$O O R (17)

The maximum % ee observed for α-chlorobenzazepin-2-one was 50%, but surprisingly the method was not effective for other substrates, including α,α-dibromobenzazepin-2-one. Eq 17 also shows the C–O bond of a benzyl ether was not cleaved while the C–O bond of a mesylate and a C–Br bond were both hydrogenolyzed.

Regioselective opening of a 1,2-disubstituted epoxide was observed with this catalyst (eq 18).[38] The benzylic ketone was not reduced or hydrogenolyzed under the reaction conditions.

O O Ph MeO O OMe — Pd/BaSO$_4$, H$_2$; 92% → O OH Ph MeO O OMe (18)

Related Reagents. See Classes R-3, R-4, R-14, R-19, R-23, R-27, and R-32, pages 1–10. Palladium on Carbon; Palladium–Graphite; Palladium on Poly(ethylenimine).

1. (a) Mosettig, E.; Mozingo, R. *OR* **1948**, *4*, 362. (b) Kieboom, A. P. G.; van Rantwijk, F. *Hydrogenation and Hydrogenolysis in Synthetic Organic Chemistry*; Delft University Press: Delft, 1977. (c) *FF* **1967**, *1*,

975; **1974**, *4*, 367–368; **1979**, *7*, 275–276. (d) Rylander, P. N. *Catalytic Hydrogenation in Organic Syntheses*; Academic: New York, 1979. (e) Rylander, P. N. *Hydrogenation Methods*; Academic: New York, 1985. (f) Davis, A. P. *COS* **1991**, *8*, 286.

2. Rosenmund, K. W. *CB* **1918**, *51*, 585.
3. Mozingo, R. *OSC* **1955**, *3*, 181.
4. Rachlin, A. I.; Gurien, H.; Wagner, D. P. *OS* **1971**, *51*, 8.
5. Rosenmund, K. W.; Zetzsche, F. *CB* **1921**, *54*, 425.
6. Rosenmund, K. W.; Zetzsche, F.; Heise, F. *CB* **1921**, *54*, 638.
7. Affrossman, S.; Thomson, S. J. *JCS* **1962**, 2024.
8. Zetzsche, F.; Arnd, O. *HCA* **1926**, *9*, 173.
9. Zetzsche, F.; Enderlin, F.; Flutsch, C.; Menzi, E. *HCA* **1926**, *9*, 177.
10. Hershberg, E. B.; Cason, J. *OSC* **1955**, *3*, 627.
11. Barnes, R. P. *OSC* **1955**, *3*, 551.
12. Danishefsky, S.; Hirama, M.; Gombatz, K.; Harayama, T.; Berman, E.; Schuda, P. F. *JACS* **1979**, *101*, 7020.
13. Foye, W. O.; Lange, W. E. *J. Am. Pharm. Assoc.* **1956**, *45*, 742.
14. Johnson, W. S.; Martin, D. G.; Pappo, R.; Darling, S. D.; Clement, R. A. *Proc. Chem. Soc.* **1957**, 58.
15. Wagner, D. P.; Gurien, H.; Rachlin, A. I. *ANY* **1970**, *172*, 186.
16. Sakurai, Y.; Tanabe, Y. *J. Pharm. Soc. Jpn.* **1944**, *64*, 25 (*CA* **1951**, *45*, 5613).
17. White, Jr., H. B.; Sulya, L. L.; Cain, C. E. *J. Lipid Res.* **1967**, *8*, 158.
18. Peters, J. A.; van Bekkum, H. *RTC* **1971**, *90*, 1323.
19. Burgstahler, A. W.; Weigel, L. O.; Shaefer, C. G. *S* **1976**, 767.
20. Peters, J. A.; van Bekkum, H. *RTC* **1981**, *100*, 21.
21. Maurer, B.; Hauser, A.; Froidevaux, J-C. *HCA* **1989**, *72*, 1400.
22. Bold, G.; Steiner, H.; Moesch, L.; Walliser, B. *HCA* **1990**, *73*, 405.
23. Hoffmann, M. G.; Zeiss, H-J. *TL* **1992**, *33*, 2669.
24. Perlmutter, P.; McCarthy, T. D. *AJC* **1993**, *46*, 253.
25. Maier, W. F.; Chettle, S. J.; Rai, R. S.; Thomas, G. *JACS* **1986**, *108*, 2608.
26. Siegel, S. *COS* **1991**, *8*, 430.
27. Steenhoek, A.; Van Wijngaarden, B. H.; Pabon, H. J. J. *RTC* **1971**, *90*, 961.
28. Gonzo, E. E.; Boudart, M. *J. Catal.* **1978**, *52*, 462.
29. Tsuji, J.; Ohno, K. *JACS* **1968**, *90*, 94.
30. (a) Figeys, H. P.; Gelbcke, M. *TL* **1970**, 5139. (b) Burgstahler, A. W.; Widiger, G. N. *JOC* **1973**, *38*, 3652. (c) Johnson, F.; Paul, K. G.; Favara, D. *JOC* **1982**, *47*, 4254.
31. (a) Burgstahler, A. W.; Widiger, G. N. *JOC* **1973**, *38*, 3652. (b) Scheffer, J. R.; Wostradowski, R. A. *JOC* **1972**, *37*, 4317.
32. Audier, L.; Dupont, G.; Dulov, R. *BSF(2)* **1957**, 248.
33. Burton, J. S.; Elvidge, J. A.; Stevens, R. *JCS* **1964**, 3816.
34. Gossinger, E.; Graf, W.; Imhof, R.; Wehrli, H. *HCA* **1971**, *54*, 2785.
35. Furukawa, Y.; Yoshioka, Y.; Imai, K.; Honjo, M. *CPB* **1970**, *18*, 554.
36. Haas, H. J. *B* **1961**, *94*, 2442.
37. Blaser, H.-U.; Boyer, S. K.; Pittelkow, U. *TA* **1991**, *2*, 721.
38. Augustyn, J. A. N.; Bezuidenhoudt, B. C. B.; Swanepoel, A.; Ferreira, D. *T* **1990**, *46*, 4429.

Samuel Siegel
University of Arkansas, Fayetteville, AR, USA

Anthony O. King & Ichiro Shinkai
Merck Research Laboratories, Rahway, NJ, USA

Palladium on Calcium Carbonate (Lead Poisoned)

Pd/CaCO$_3$/Pb

[7440-05-3] Pd (MW 106.42)

(catalyst used for selective hydrogenation of alkynes to *cis*-alkenes;[1–7] used to reduce azides and nitro compounds to amines;[8] used infrequently for oxidation of aldehydes to acids[10])

Alternate Name: Lindlar catalyst.
Solubility: insol most organic solvents; incompatible with water or acids.
Form Supplied in: dark gray to black powder, typically 5 wt% of palladium.
Analysis of Reagent Purity: atomic absorption.
Handling, Storage, and Precautions: can be stored safely in a closed container under air but away from solvents and potential poisons such as sulfur- and phosphorus-containing compounds. The reagent is pyrophoric in the presence of solvents. General precautions for handling hydrogenation catalysts should be followed. The catalyst must be suspended in the organic solvent under an atmosphere of N$_2$. During filtration the filter cake must not be allowed to go dry.

Hydrogenations. Lead-poisoned palladium on calcium carbonate, also known as Lindlar catalyst, is mainly used for the selective conversion of alkynes to *cis*-alkenes.[1] The catalyst has been used frequently in the presence of an amine. The hydrogenation in most cases stops sharply after one equiv of hydrogen uptake. For example, the reduction of 1,2-epoxydec-4-yne over Lindlar catalyst and quinoline gave (*Z*)-1,2-epoxydec-4-ene in 95% yield. The epoxide was stable under the reaction conditions.[2] Terminal alkynes can be reduced selectively to 1-alkenes. The ease of reduction was established to be 1-alkyne > internal alkyne > 1-alkene.[3] Conjugated enynes can be reduced smoothly to the corresponding dienes with excellent selectivity (eq 1).[4]

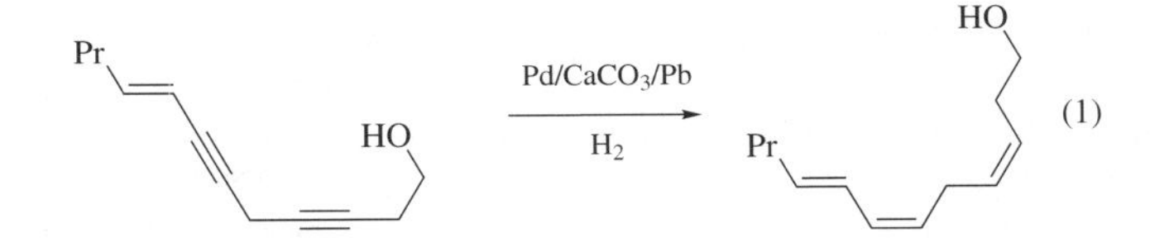

Alkynes in conjugation with one or two carbonyl functions are also reduced smoothly to the alkene. Lactone formation was observed when the hydroxyalkynoic acid shown in eq 2 was hydrogenated.[5]

C_8H_{17}(HO)CH–C≡C–CO_2H —(90%)→ C_8H_{17}-butenolide (2)

The reduction of 1-octynyldiisopropoxyborane with Lindlar catalyst in 1,4-dioxane with a small amount of pyridine gave a 95:5 (*cis*:*trans*) ratio of vinyl boronates. High selectivity was also observed with a variety of 1-alkynyl-1,3,2-dioxaborinanes under the above optimized reaction conditions.[6] It is important to note that other catalysts, such as ***Palladium on Barium Sulfate***, ***Nickel Boride***, and, more recently, ***Palladium on Poly(ethylenimine)***, have also been used to effect the reduction of alkynes to *cis*-alkenes with excellent results and are sometimes superior to Lindlar catalyst.[7]

Azides are readily hydrogenated to amines with this catalyst. As expected, alkenic groups are stable under the reaction conditions.[8]

Under catalytic transfer hydrogenation conditions, nitrobenzenes and azoxybenzenes are converted to hydrazobenzenes (eq 3).[9]

$Pd/CaCO_3/Pb$, HCO_2Na, HCO_2H, MeOH, 96% (3)

Oxidation. The catalyst has also been used in oxidation reactions. Methacrolein in MeOH was oxidized to give 93% yield of methyl methacrylate with a selectivity of 95% at 98% conversion.[10]

Related Reagents. See Classes R-4, R-5, R-9, and R-19, pages 1–10.

1. Lindlar, H. U.S. Patent 2 681 938, 1954. Lindlar, H. *HCA* **1952**, *35*, 446. Lindlar, H.; Dubuis, R. *OS* **1966**, *46*, 89. Marvell, E. N.; Li, T, *S* **1973**, 457. McEwen, A. B.; Guttieri, M. J.; Maier, W. F.; Laine, R. M.; Shyo, Y. *JOC* **1983**, *48*, 4436.
2. Russell, S. W.; Pabon, H. J. J. *JCS(P1)* **1982**, 545.
3. Dobson, N. A.; Eglinton, G.; Krishnamurti, M.; Raphael, R. A.; Willis, R. G. *T* **1961**, *16*, 16.
4. Tai, A.; Matsumura, F.; Coppel, H. C. *JOC* **1969**, *34*, 2180.
5. Jakubowski, A. A.; Guziec, F. S., Jr.; Sugiura, M.; Tam, C. C.; Tishler, M.; Omura, S. *JOC* **1982**, *47*, 1221.
6. Srebnik, M.; Bhat, N. G.; Brown, H. C. *TL* **1988**, *29*, 2635.
7. White, W. L.; Anzeveno, P. B.; Johnson, F. *JOC* **1982**, *47*, 2379. Jain, S. C.; Dussourd, D. E.; Conner, W. E.; Eisner, T.; Guerrero, A.; Meinwald, J. *JOC* **1983**, *48*, 2266. Bayer, E.; Schumann, W. CC **1986**, 949.
8. Corey, E. J.; Nicolaou, K. C.; Balanson, R. D.; Machida, Y. *S* **1975**, 590.
9. Iguchi, K.; Senba, H. Jpn. Kokai Tokkyo Koho 63 096 166, 1988 (*CA* **1988**, *109*, 189 986j). Iguchi, K.; Senba, H. Jpn. Kokai Tokkyo Koho 63 096 165, 1988 (*CA* **1988**, *109*, 189 987k).
10. Tamura, N.; Fukuoka, Y.; Yamamatsu, S.; Suzuki, Y.; Mitsui, R.; Ibuki, T. Ger. Offen. 2 848 369, 1979 (*CA* **1979**, *91*, 107 677g).

Anthony O. King and Ichiro Shinkai
Merck Research Laboratories, Rahway, NJ, USA

Palladium on Carbon

[7440-05-3] Pd (MW 106.42)

(catalyst for hydrogenation of alkenes, alkynes, ketones, nitriles, imines, azides, nitro groups, benzenoid and heterocyclic aromatics; used for hydrogenolysis of cyclopropanes, benzyl derivatives, epoxides, hydrazines, and halides; used to dehydrogenate aromatics and deformylate aldehydes)

Solubility: insol all organic solvents and aqueous acidic media.
Form Supplied in: black powder or pellets containing 0.5–30 wt % of Pd (typically 5 wt %); can be either dry or moist (50 wt % of H_2O).
Analysis of Reagent Purity: atomic absorption.
Handling, Storage, and Precautions: can be stored safely in a closed container under air but away from solvents and potential poisons such as sulfur- and phosphorus-containing compounds. Pyrophoric in the presence of solvents. General precautions for handling hydrogenation catalysts should be followed. The catalyst must be suspended in the organic solvent under an atmosphere of N_2. During filtration the filter cake must not be allowed to go dry. If a filter aid is necessary, a cellulose-based material should be used if catalyst recovery is desired.

Hydrogenation and Hydrogenolysis: Carbon–Carbon Bonds. The use of Pd/C for the selective reduction of alkynes to alkenes is generally not satisfactory, but a few examples have been reported. For example, the Pd/C-catalyzed reduction of 3,6-dimethyl-4-octyne-3,6-diol gave the enediol in 98% yield after absorption of 1 mol of H_2.[1] Further reduction gave the diol in 99% yield. Pd on other supports, such as $Pd/CaCO_3$ and $Pd/BaCO_3$, are much more effective for this conversion. Pd/C is usually used for the complete saturation of alkynes and alkenes to their corresponding hydrocarbons.[2] In some instances, isomerization of the double bond during hydrogenation occurs before reduction, which leads to unexpected results. For example, reduction of car-3-ene gave the cycloheptane with Pd/C instead of the expected cyclohexane derivative (eq 1).[3]

Pd/C, H_2, $EtCO_2H$, 73 °C, 100% (1)

Isomerization of a double bond from one position to a hydrogenation inaccessible location has also been observed (eq 2).[4]

Pd/C, 63% (2)

Next to the reduction of nitro groups, double and triple bonds are generally the next easiest functional groups to undergo hydrogenation. Some less reactive functional groups include ketones,[5] esters,[6] benzyl ethers,[7] epoxides,[8] and N–O bonds.[9] These remain intact under conditions needed to reduce alkenes and alkynes. Under longer reaction times and/or more forcing conditions, some of these functional groups will also be affected. Allylboron compounds can be hydrogenated to the propylboron derivatives, the C–B bond remaining intact (eq 3).[10]

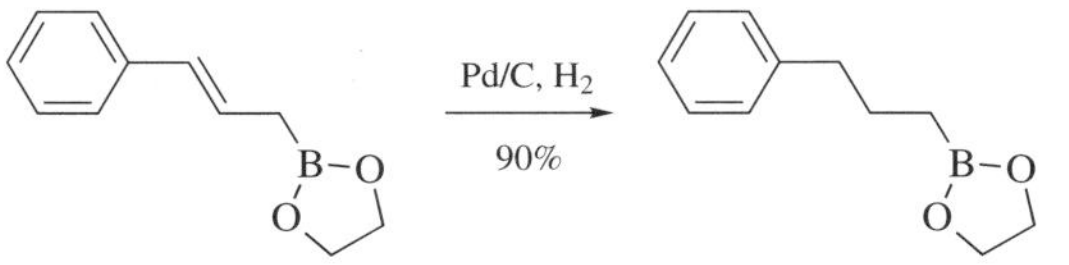

(3)

Treatment of acylated aldonolactones with hydrogen in the presence of Pd/C and triethylamine provided 3-deoxyaldonolactones in excellent yields (eq 4).[11] The α,β-unsaturated intermediate was hydrogenated stereospecifically to give the product. Substituting Pd with Pt catalysts gave the 2-acetoxy hydrogenolyzed product (**1**) instead. Hydrogenolysis of the acetate preceded the double bond hydrogenation.

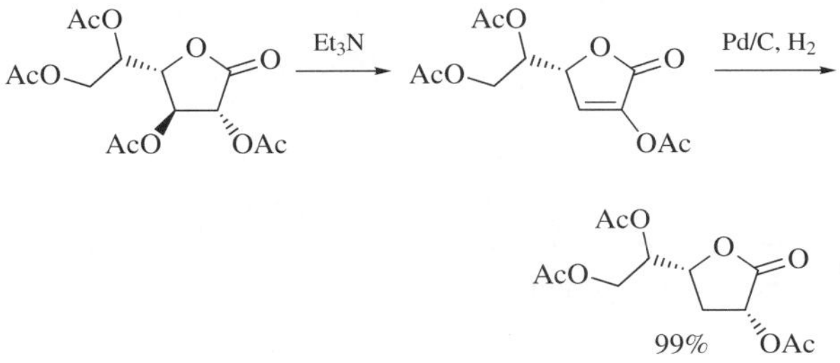

(4)

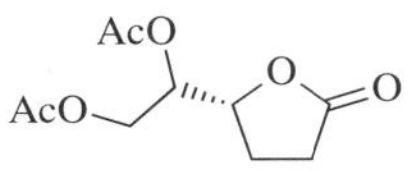

(**1**)

Hydrogenolysis of C–C bonds using Pd/C is mainly limited to cyclopropane opening. The less substituted and electronically activated bond cleavage is preferred. An example is shown in (eq 5).[12]

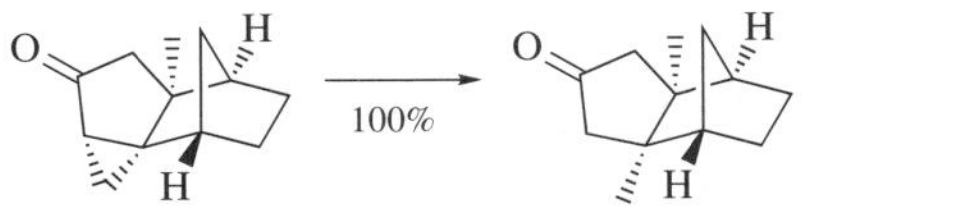

(5)

Carbon–Nitrogen Bonds. The hydrogenation of nitriles to primary amines is best accomplished with Pd/C in acidic media or in the presence of ammonia. In the absence of acid or ammonia, a mixture of primary and secondary amines is observed. This effect was taken advantage of and mixed secondary amines were obtained selectively by the reduction of a nitrile in the presence of a different amine (eq 6).[13] Hydrogenation in aqueous acidic conditions can lead to aldehydes and/or alcohols.[14]

$$\text{BuCN} + \text{BuNH}_2 \xrightarrow[\text{H}_2,\ 16\text{ h},\ 54\%]{\text{Pd/C}} (\text{C}_5\text{H}_{11})_2\text{NH} + \text{BuNHC}_5\text{H}_{11} \quad (7:93) \tag{6}$$

Reductive alkylation is a convenient and efficient way of obtaining secondary and tertiary amines.[15] *N,N*-Dimethyl tertiary amines can be obtained from both aromatic and aliphatic primary amines or their precursors. Using α-methylbenzylamine as the chiral auxiliary, highly diastereoselective reduction of the intermediate imine has been observed (eq 7).[16]

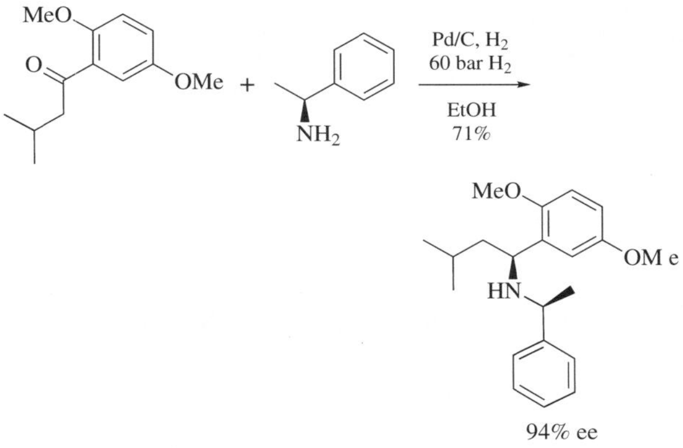

(7)

This method also provides a convenient route to nitrogen containing heterocycles. Hydrogenolysis of the Cbz group followed by an in situ reductive alkylation process gave a bicyclic heterocycle (eq 8).[17] The alkene was also reduced.

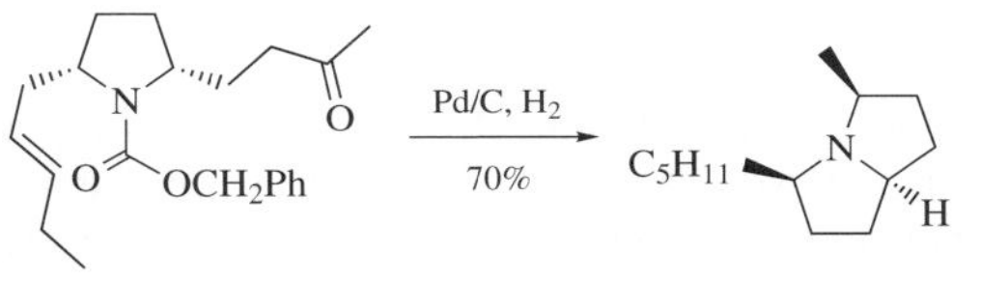

(8)

Other amine precursors, such as azides, can be utilized in the reductive alkylation reaction. For example, a furanose ring was opened and reclosed to form a piperidine ring system (eq 9).[18] A pyranyl azide similarly provided the seven-membered nitrogen heterocycle.

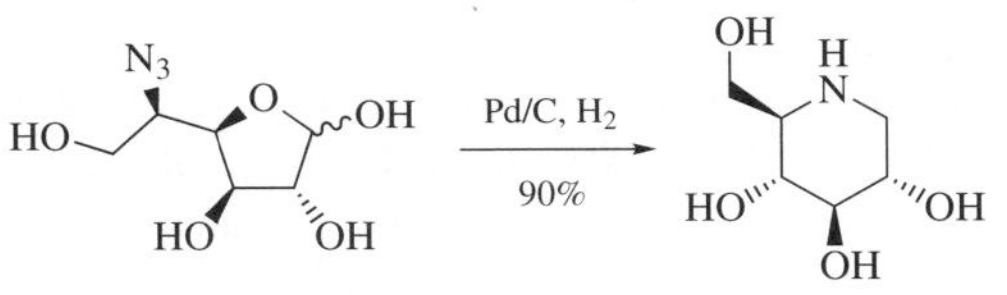

(9)

Similar to nitriles, hydrogenation of oximes is best carried out under acidic conditions to minimize secondary amine formation.[19]

Benzylic amines can be readily hydrogenolyzed to give less alkylated amines.[20] The C–N bond can be cleaved under both transfer hydrogenation[21] and regular hydrogenation conditions.[22] In many cases the newly debenzylated amines can further react, resulting in more structurally complex products (eq 10).[23]

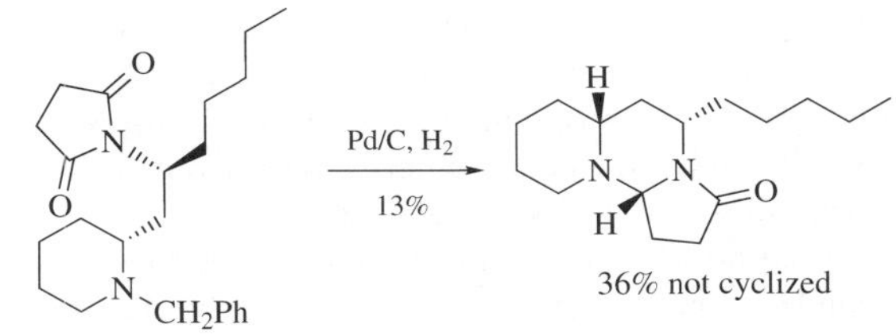

(10)

The heterogeneous catalytic debenzylation of *N*-benzylated amides with Pd/C is generally a difficult process and should not be considered in a synthetic scheme.

Allylamines have also been deallylated using Pd/C catalysis (eq 11).[24]

Pd/C, reflux, MeOH, 85% (11)

Aziridines are hydrogenolyzed to give ring-opened amines. In eq 12, the more reactive benzylic C–N bond was cleaved selectively.[25]

Pd/C, H_2, 90% (12)

Carbon–Oxygen Bonds. Pd/C is best suited for the hydrogenation and hydrogenolysis of benzylic ketones and aldehydes. The reduction of dialkyl ketones to the alcohols is more sluggish and further hydrogenolysis to the alkane is even slower.[26] The hydrogenation of benzylic ketones (aryl alkyl and diaryl ketones) to alcohols is a very facile process with Pd/C.[27] Further hydrogenolysis of the benzylic alcohols to the alkane products can be a major problem with Pd/C catalysts, but can be controlled.[28] In general, aryl ketones and aldehydes can be reduced to alcohols under neutral conditions or in the presence of an amino functional group or an added amine base.[29] In the presence of acids, hydrogenolysis is more prone to occur. Using other catalysts such as ***Platinum on Carbon*** Ru/C, Rh/C and ***Raney Nickel***, is an alternative.

Trifluoromethyl ketones are reduced to alcohols without dehalogenation or further hydrogenolysis.[30] The hydrogenation of a chiral proline derivative provided the α-hydroxyamide product in 77% de and 100% yield (eq 13).[31]

Pd/C, H_2 (13)

Hydrogenolysis of a benzyl group attached to an oxygen atom is a common step in complex synthetic schemes. Benzyl esters,[32] benzyl carbamates,[33] and benzyl ethers[34] are readily hydrogenolyzed to acids, amines, and alcohols, respectively. *N*-Oxides protected as the benzyl ethers can be deprotected without hydrogenolysis of the N–O bond.[35] Hydrazines protected with benzyloxycarbonyl (Cbz) groups have been deprotected without N–N bond cleavage or the hydrogenolysis of benzylic C–O bonds (eq 14).[36]

Pd/C, H_2, MeOH, rt (14)

Benzyl carbamates have been transformed into *t*-butyl carbamates under transfer-hydrogenolysis conditions, but high catalyst loading was needed (eq 15).[37] A benzyl ether function survived these reaction conditions but a 1-alkene was saturated.

$(Boc)_2O$, Pd/C, 86–96% (15)

1,2-Diols protected as the acetal of benzaldehyde were deprotected under hydrogenolysis conditions (eq 16).[38]

Pd/C, H_2, 99% (16)

3-Acyltetronic acids were easily hydrogenolyzed to 3-alkyltetronic acids. Further reduction of the enol was not observed under the reaction conditions (eq 17).[39]

Pd/C, H_2 (17)

R = Me 96%
R = CH_2Ph 93%

The C–O bond of epoxides can be hydrogenolyzed to give alcohols. Regioselective epoxide ring opening has been observed in some cases (eq 18).[40]

100% (18)

Nitrogen–Oxygen Bonds. Both aliphatic and aromatic nitro groups are reduced to the corresponding amines (eq 19).[41] N–O bonds are also readily hydrogenolyzed using Pd/C (eq 20).[42a]

Pd/C, H_2, 100% (19)

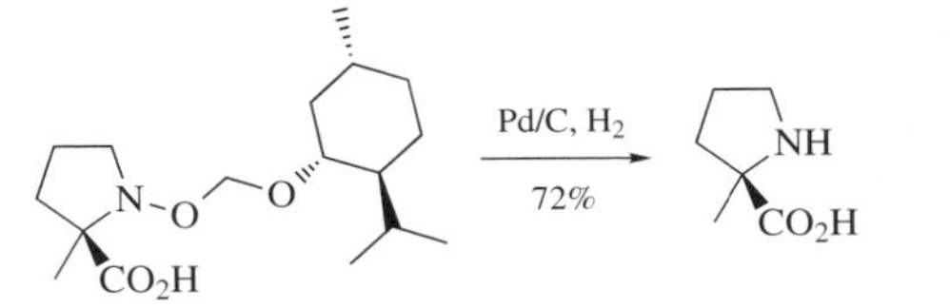

(20)

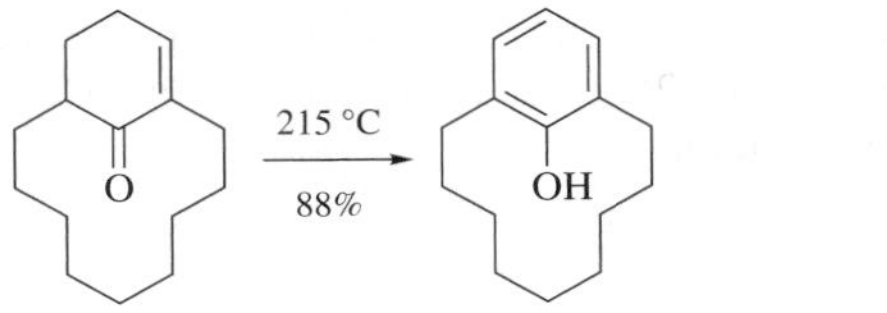

(24)

Carbon–Halogen Bonds. Aromatic halides (Cl, Br, I) are readily hydrogenolyzed with Pd/C.[43] The reaction generally requires the presence of a base to neutralize the acid formed. In the absence of an acid neutralizer, dehalogenation is slower and may stop short of completion. Vinyl halides are also dehalogenated but concomitant saturation of the alkene can also occur (eq 21).[44a] Defluorination is a very slow process but one case has been reported (eq 22).[45]

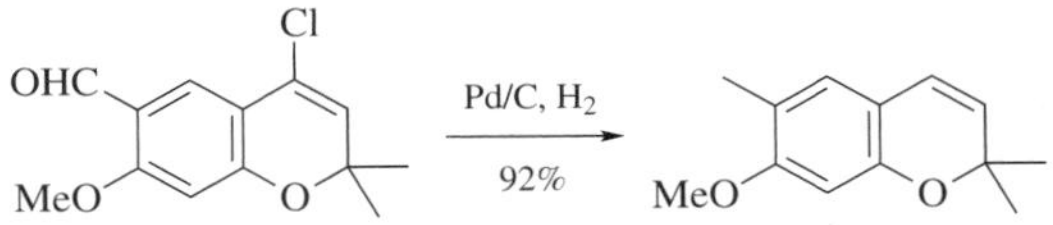

(21)

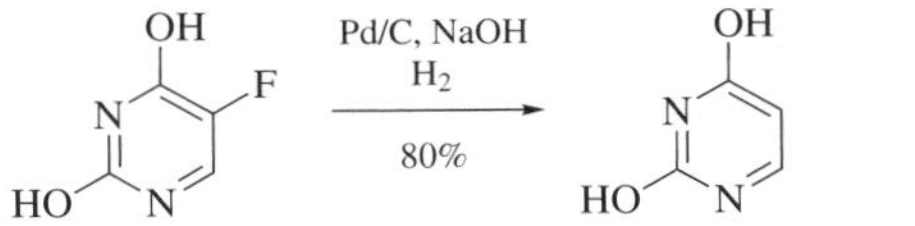

(22)

Selective dehalogenation of acyl halides can also be carried out with Pd/C and H_2 in the presence of an amine base to give aldehydes. This type of dehalogenation is commonly known as Rosenmund reduction (see ***Palladium on Barium Sulfate***).[46]

Nitrogen–Nitrogen Bonds. Azides[47] and diazo[48] compounds can be reduced over Pd/C to give amines. These groups have also been used as latent amines which, when hydrogenolyzed, can react with amine-sensitive functional groups in the molecule to give other amine products (eq 23).[47c]

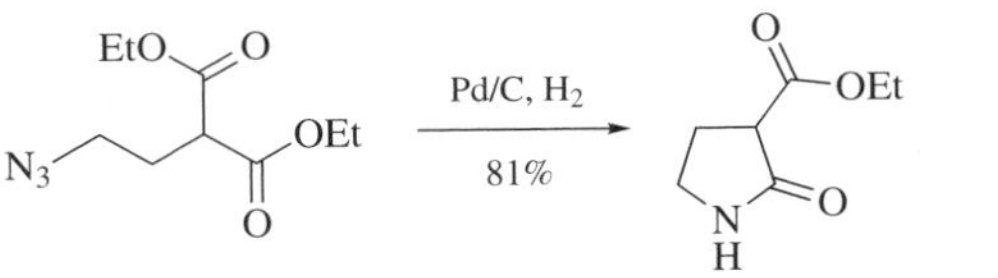

(23)

Carbocyclic and Heterocyclic Aromatics. Hydrogenation of carbocyclic aromatic compounds can be accomplished with Pd/C under a variety of reaction conditions.[49] The conditions are generally more vigorous than those used with Pt or Rh catalysts.

Pyridine and pyridinium derivatives are hydrogenated readily to give piperidines.[50] Other heterocyclic aromatic ring systems such as furan,[51] benzofuran,[52] thiophene,[53] pyrrole,[54] indole,[55] quinoline,[56] pyrazine,[57] and pyrimidine[58] have also been hydrogenated over Pd/C.

Dehydrogenation. At high temperatures, Pd/C is an effective dehydrogenation catalyst to provide carbocyclic and heterocyclic aromatic compounds.[59] An enone has been converted to a phenol (eq 24)[59f] and a methoxycyclohexene derivative has provided an anisyl product (eq 25).[59g]

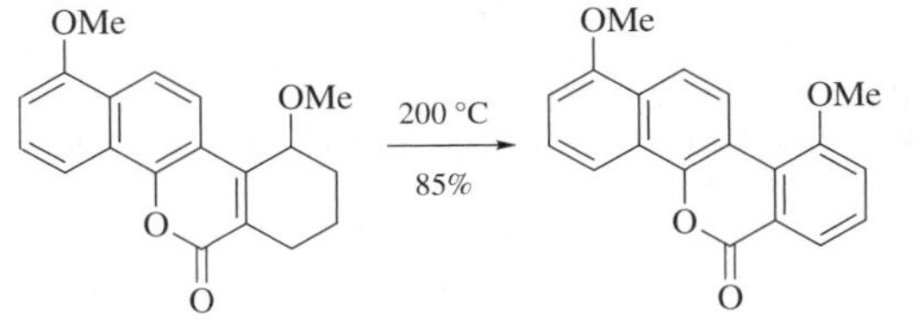

(25)

Miscellaneous Reactions. Decarbonylations can be carried out under the same conditions used for dehydrogenation (eq 26).[60] In this case, a trisubstituted alkene remained intact.

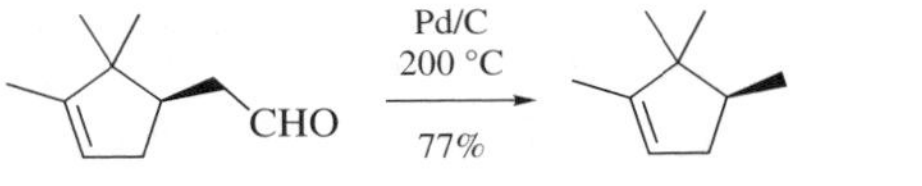

(26)

Reduction of an acylsilane gave an aldehyde without further hydrogenation to the alcohol or the hydrogenolysis of the benzyl ether (eq 27).[61]

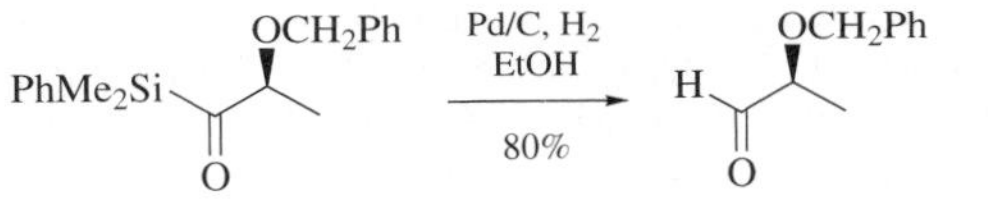

(27)

Pd/C also catalyzed the cycloaddition reaction of an alkyne with a heterocycle to give a tricyclic heteroaromatic compound (eq 28).[62]

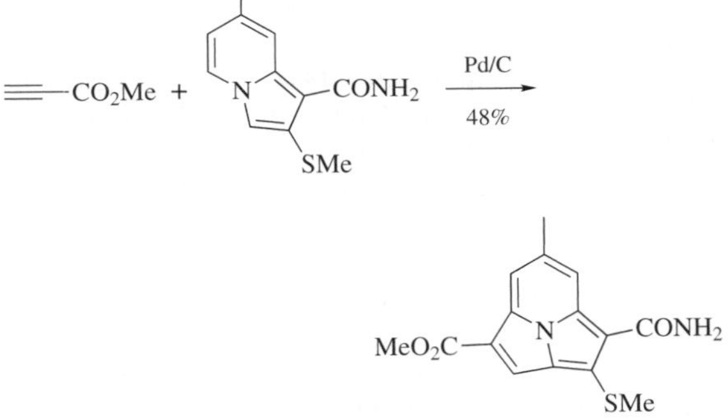

(28)

In conjunction with ***Copper(II) Chloride***, Pd/C catalyzed the biscarbonylation of norbornene derivatives (eq 29).[63] Norbornadiene itself was tetracarbonylated but in only 30% yield.

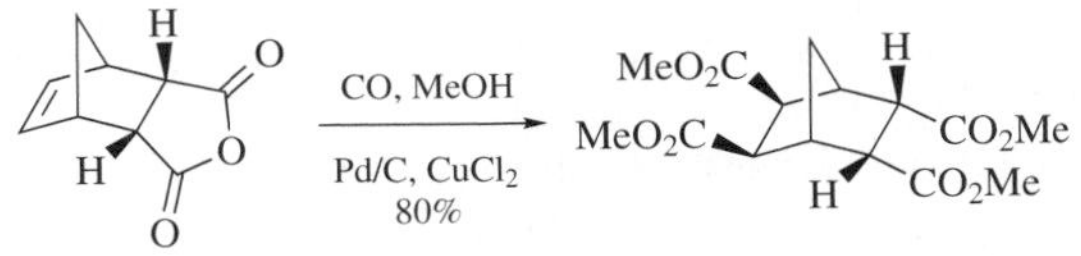

(29)

Related Reagents. See Classes R-2, R-3, R-4, R-7, R-8, R-9, R-12, R-19, R-20, R-21, R-23, R-24, R-26, R-27, and R-29, pages 1–10. Palladium on Barium Sulfate; Palladium(II) Chloride; Palladium(II) Chloride–Copper(II) Chloride; Palladium–Graphite.

---

1. Tedeschi, R. J.; McMahon, H. C.; Pawlak, M. S. *ANY* **1967**, *145*, 91.
2. (a) Vitali, R.; Caccia, G.; Gardi, R. *JOC* **1972**, *37*, 3745. Overman, L. E.; Jessup, G. H. *JACS* **1978**, *100*, 5179. Cortese, N. A.; Heck, R. F. *JOC* **1978**, *43*, 3985. Olah, G. A.; Surya Prakash, G. K. *S* **1978**, 397. (b) Baker, R.; Boyes, R. H. O.; Broom, D. M. P.; O'Mahony, M. J.; Swain, C. J. *JCS(P1)* **1987**, 1613. Cossy, J.; Pete, J.-P. *BSF* **1988**, 989. Taylor, E. C.; Wong, G. S. K. *JOC* **1989**, *54*, 3618.
3. Cocker, W.; Shannon, P. V. R.; Staniland, P. A. *JCS(C)* **1966**, 41.
4. (a) Greene, A. E.; Serra, A. A.; Barreiro, E. J.; Costa, P. R. R. *JOC* **1987**, *52*, 1169. (b) Flann, C. J.; Overman, L. E. *JACS* **1987**, *109*, 6115.
5. Attah-poku, S. K.; Chau, F.; Yadav, V. K.; Fallis, A. G. *JOC* **1985**, *50*, 3418.
6. Sato, M.; Sakaki, J.; Sugita, Y.; Nakano, T.; Kaneko, C. *TL* **1990**, *31*, 7463.
7. Tsuda, Y.; Hosoi, S.; Goto, Y. *CPB* **1991**, *39*, 18.
8. Vekemans, J. A. J. M.; Dapperens, C. W. M.; Claessen, R.; Koten, A. M. J.; Godefroi, E. F.; Chittenden, G. J. F. *JOC* **1990**, *55*, 5336.
9. Iida, H.; Watanabe, Y.; Kibayashi, C. *JACS* **1985**, *107*, 5534.
10. Brown, H. C.; Rangaishenvi, M. V. *TL* **1990**, *31*, 7115.
11. Bock, K.; Lundt, I.; Pedersen, C. *ACS* **1981**, *35*, 155.
12. Srikrishna, A.; Nagaraju, S. *JCS(P1)* **1991**, 657.
13. Rylander, P. N.; Hasbrouck, L.; Karpenko, I. *ANY* **1973**, *214*, 100.
14. Bredereck, H.; Simchen G.; Traut, H. *CB* **1967**, *100*, 3664. Mizzoni, R. H.; Lucas, R. A.; Smith, R.; Boxer, J.; Brown, J. E.; Goble, F.; Konopka, E.; Gelzer, J.; Szanto, J.; Maplesden, D. C.; deStevens, G. *JMC* **1970**, *13*, 878. Caluwe, P.; Majewicz, T. G. *JOC* **1977**, *42*, 3410.
15. Glaser, R.; Gabbay, E. J. *JOC* **1970**, *35*, 2907.
16. Bringmann, G.; Kunkel. G.; Geuder, T. *SL* **1990**, *5*, 253.
17. Momose, T.; Toyooka, N.; Seki, S.; Hirai, Y. *CPB* **1990**, *38*, 2072.
18. Dax, K.; Gaigg, B.; Grassberger, V.; Kolblinger, B.; Stutz, A. E. *J. Carbohydr. Chem.* **1990**, *9*, 479.
19. Yamaguchi, S.; Ito, S.; Suzuki, I.; Inoue, N. *BCJ* **1968**, *41*, 2073. Huebner, C. F.; Donoghue, E. M.; Novak, C. J.; Dorfman, L.; Wenkert, E. *JOC* **1970**, *35*, 1149.
20. Suter, C. M.; Ruddy, A. W. *JACS* **1944**, *66*, 747. Vaughan, J. R.; Blodinger, J. *JACS* **1955**, *77*, 5757. Cosgrove, C. E.; La Forge, R. A. *JOC* **1956**, *21*, 197.
21. Zisman, S. A.; Berlin, K. D.; Scherlag, B. J. *OPP* **1990**, *22*, 255.
22. Orlek, B. S.; Wadsworth, H.; Wyman, P.; Hadley, M. S. *TL* **1991**, *32*, 1241.
23. Merlin, P.; Braekman, J. C.; Daloze, D. *T* **1991**, *47*, 3805.
24. Afarinkia, K.; Cadogan, J. I. G.; Rees, C. W. *SL* **1990**, 415.
25. Martinelli, M. J.; Leanna, M. R.; Varie, D. L.; Peterson, B. C.; Kress, T. J.; Wepsiec, J. P.; Khau, V. V. *TL* **1990**, *31*, 7579.
26. Solodin, J. *M* **1992**, *123*, 565.
27. Schultz, A. G.; Motyka, L. A.; Plummer, M. *JACS* **1986**, *108*, 1056.
28. Sibi, M. P.; Gaboury, J. A. *SL* **1992**, 83. Paisdor, B.; Kuck, D. *JOC* **1991**, *56*, 4753.
29. Coll, G.; Costa, A.; Deya, P. M.; Saa, J. M. *TL* **1991**, *32*, 263. Trivedi, S. V.; Mamdapur, V. R. *IJC* **1990**, *29*, 876. Rane, R. K.; Mane, R. B. *IJC* **1990**, *29*, 773.
30. Jones, R. G. *JACS* **1948**, *70*, 143.
31. Muneguni, T.; Maruyama, T.; Takasaki, M.; Harada, K. *BCJ* **1990**, *63*, 1832.
32. Effenberger, F.; Muller, W.; Keller, R.; Wild, W.; Ziegler, T. *JOC* **1990**, *55*, 3064.
33. Janda, K. D.; Ashley, J. A. *SC* **1990**, *20*, 1073.
34. Shiozaki, M. *JOC* **1991**, *56*, 528. Khamlach, K.; Dhal, R.; Brown, E. *H* **1990**, *31*, 2195. Matteson, D. S.; Kandil, A. A.; Soundararajan, R. *JACS* **1990**, *112*, 3964.
35. Baldwin, J. E.; Adlington, R. M.; Gollins, D. W.; Schofield, C. J. *CC* **1990**, *46*, 720.
36. Gmeiner, P.; Bollinger, B. *TL* **1991**, *32*, 5927.
37. Bajwa, J. S. *TL* **1992**, *33*, 2955.
38. Matteson, D. S.; Michnick, T. J. *OM* **1990**, *9*, 3171.
39. Sibi, M. P.; Sorum, M. T.; Bender, J. A.; Gaboury, J. A. *SC* **1992**, *22*, 809.
40. Sakaki, J.; Sugita, Y.; Sato, M.; Kaneko, C. *CC* **1991**, 434.
41. Wehner, V.; Jager, V. *AG(E)* **1990**, *29*, 1169.
42. (a) Shatzmiller, S.; Dolithzky, B.-Z.; Bahar, E. *LA* **1991**, 375. (b) Maciejewski, S.; Panfil, I.; Belzecki, C.; Chmielewski, M. *TL* **1990**, *31*, 1901. (c) Kawasaki, T.; Kodama, A.; Nishida, T.; Shimizu, K.; Somei, M. *H* **1991**, *32*, 221. (d) Beccalli, E. M.; Marchesini, A.; Pilati, T. *S* **1991**, 127.
43. Sone, T.; Umetsu, Y.; Sato, K. *BCJ* **1991**, *64*, 864. Boerner, A.; Krause, H. *JPR* **1990**, *332*, 307.
44. (a) Eszenyi, T.; Timar, T. *SC* **1990**, *20*, 3219. Comins, D. L.; Weglarz, M. A. *JOC* **1991**, *56*, 2506.
45. Duschinsky, R.; Pleven, E.; Heidelberger, C. *JACS* **1957**, *79*, 4559.
46. Sakmai, Y.; Tanabe, Y. *J. Pharm. Sci. Jpn.* **1944**, *64*, 25. Peters, J. A.; van Bekkum, H. *RTC* **1971**, *90*, 1323. Rachlin, A. I.; Gurien, H.; Wagner, P. P. *OS* **1971**, *51*, 8. Burgstahler, A. W.; Weigel, L. O.; Shaefer, G. G. *S* **1976**, 767.
47. (a) Lohray, B. B.; Ahuja, J. R. *CC* **1991**, 95. (b) Castillon, S.; Dessinges, A.; Faghih, R.; Lukacs, G. Olesker, A.; Thang, T. T. *JOC* **1985**, *50*, 4913. (c) Lindstrom, K. J.; Crooks, S. L. *SC* **1990**, *20*, 2335. (d) Machinaga, N.; Kibayashi, C. *TL* **1990**, *31*, 3637. (e) Chen, L.; Dumas, D. P.; Wong, C.-H. *JACS* **1992**, *114*, 741. (f) Ghosh, A. K.; McKee, S. P.; Duong, T. T.; Thompson, W. J. *CC* **1992**, 1308.
48. Looker, J. H.; Thatcher, D. N. *JOC* **1957**, *22*, 1233.
49. Kindler, K.; Hedermann, B.; Scharfe, E. *LA* **1948**, *560*, 215. Rapoport, H.; Pasby, J. Z. *JACS* **1956**, *78*, 3788. Farina, M.; Audisio, G. *T* **1970**, *26*, 1827. Feher, F. J.; Budzichowski, T. A. *JOM* **1989**, *373*, 153. Mohler, D. L.; Wolff, S.; Vollhardt, K. P. C. *AG(E)* **1990**, *29*, 1151. Valls, N.; Bosch, J.; Bonjoch, J. *JOC* **1992**, *57*, 2508.
50. Daeniker, H. U.; Grob, C. A. *OS* **1964**, *44*, 86. Yakhontov, L. N. *RCR* **1969**, *38*, 470. Scorill, J. P.; Burckhalter, J. H. *JHC* **1980**, *17*, 23.
51. Massy-Westrop, R. A.; Reynolds, G. D.; Spotswood, T. M. *TL* **1966**, 1939.
52. Caporale, G.; Bareggi, A. M. *G* **1968**, *98*, 444.
53. Confalone, P. N.; Pizzolato, G.; Uskokovic, M. R. *JOC* **1977**, *42*, 135. Rossy, P.; Vogel, F. G. M.; Hoffman, W.; Paust, J.; Nurrenbach, A. *TL* **1981**, *22*, 3493.
54. Pizzorno, M. T.; Albonico, S. M. *JOC* **1977**, *42*, 909. Robins, D. J.; Sakdarat, S. *CC* **1979**, 1181.
55. Kikugawa, Y.; Kashimura, M. *S* **1982**, *9*, 785. Knolker, H.-J.; Hartmann, K. *SL* **1991**, *6*, 428.
56. Balczewski, P.; Joule, J. A. *SC* **1990**, *20*, 2815. Bouysson, P.; LeGoff, C.; Chenault, J. *JHC* **1992**, *29*, 895.
57. Behun, J. D.; Levine, R. *JOC* **1961**, *26*, 3379. McKenzie, W. L.; Foye, W. O. *JMC* **1972**, *15*, 291.
58. King, F. E.; King, T. J. *JCS* **1947**, 726.
59. (a) Backvall, J.-E.; Plobeck, N. A. *JOC* **1990**, *55*, 4528. (b) Pelcman, B.; Gribble, G. W. *TL* **1990**, *31*, 2381. (c) Harvey, R. G.; Pataki, J.; Cortez, C.; Diraddo, P.; Yang, C. X. *JOC* **1991**, *56*, 1210. (d) Peet, N. P.; LeTourneau, M. E. *H* **1991**, *32*, 41. (e) Soman, S. S.; Trivedi, K. N. *JIC* **1990**, *67*, 997. (f) Nelson, P. H.; Nelson, J. T. *S* **1991**, 192. (g) Hua, D. H.; Saha, S.; Maeng, J. C.; Bensoussan, D. *SL* **1990**, *4*, 233.
60. Pamingle, H.; Snowden, R. L.; Shulteelte, K. H. *HCA* **1991**, *74*, 543.

61. Cirillo, P. F.; Panek, J. S. *TL* **1991**, *32*, 457.
62. Matsuda, Y.; Gotou, H.; Katou, K.; Matsumoto, H.; Yamashita, M.; Takahashi, K.; Ide, S. *H* **1990**, *31*, 983.
63. Yamada, M.; Kusama, M.; Matsumoto, T.; Kurosaki, T. *JOC* **1992**, *57*, 6075.

Anthony O. King and Ichiro Shinkai
Merck Research Laboratories, Rahway, NJ, USA

Palladium(II) Hydroxide on Carbon

$Pd(OH)_2$

[7440-05-3] H_2O_2Pd (MW 140.44)

(catalyst for hydrogenolysis of benzyl groups and epoxides; used to reduce nitro compounds to amines and as dehydrogenation reagent)

Alternate Name: Pearlman catalyst.
Solubility: insol all organic solvents.
Form Supplied in: black powder, typically containing 20 wt % of Pd and 10–15 wt % of H_2O.
Analysis of Reagent Purity: atomic absorption.
Handling, Storage, and Precautions: can be stored safely in a closed container under air but away from solvents and potential poisons such as sulfur- and phosphorus-containing compounds. The reagent is pyrophoric in the presence of solvents. General precautions for handling hydrogenation catalysts should be followed. The catalyst must be suspended in the organic solvent under an atmosphere of N_2. During filtration the filter cake must not be allowed to go dry. If a filter aid is necessary, a cellulose-based material should be used if catalyst recovery is desired.

Hydrogenolysis and Hydrogenation. Palladium hydroxide on carbon has been used for difficult debenzylations when other supported palladium catalysts were not effective or satisfactory.[1] For example, although *para-* and *meta-*substituted benzyl alcohols were easily hydrogenolyzed with 10% ***Palladium on Carbon***, hydrogenolysis of the *ortho-*substituted substrate was only effective with 20% $Pd(OH)_2/C$ (eq 1).[2] This catalyst also catalyzes hydrogenation and hydrogenolysis of other functional groups.

OSiMe2H; OH; O; OH; Pd(OH)2/C, AcOH, 1 atm H2; OSiMe2H (1); O; OH

The transfer hydrogenolysis of a benzyl ether proceeded in 'high yield' without the hydrogenolysis of the chlorine or the hydrogenation of the alkene (eq 2).[3] Evidence showed that an oxidative mechanism, rather than the usual reductive mechanism, might be involved in this case.

BnO; OMe; O; Cl; Cl; 20% Pd(OH)2/C, EtOH, reflux, cyclohexene; HO; OMe; O; Cl; Cl (2)

Deprotection of a benzyl carbamate and concomitant reduction of a double bond with the Pearlman catalyst has provided an amino acid in 82% yield (eq 3).[4]

O; OCH2Ph; HN; CO2H; NHBoc; Pd(OH)2/C, MeOH; H2N; CO2H; NHBoc (3)

Debenzylations of *N*-benzylated imidazole derivatives are very difficult, but in the case of the imidazole derivative in eq 4, hydrogenolysis proceeded to give 84% yield of the desired product. The N–N bond was not cleaved under the reaction conditions, but the formyl group was hydrogenolyzed (eq 4).[5]

OHC; O; Ph; N; N; N; N; N; NH2; Pd(OH)2/C, H2; O; H; N; N; N; N; N; NH2 (4)

An *N*-protected azete hydrochloride was deprotected in EtOH to provide the azete hydrochloride in 86% yield. The azete ring was stable under the reaction conditions. This intermediate was converted to δ-coniceine in four steps (eq 5).[6]

MsO; NH Cl−; Ph; Ph; Pd(OH)2/C, H2, EtOH; MsO; NH2 Cl− (5)

Excellent regioselective opening of an epoxide has been accomplished with the Pearlman catalyst. The epoxide opening with ***Lithium Aluminum Hydride*** (LAH) gave no selectivity and a 50/50 mixture of the triol isomers was obtained. Hydrogenolysis with the Pearlman catalyst gave 99% selectivity. An equivalent weight of catalyst vs. substrate must be used to obtain the excellent 92% yield (eq 6).[7]

HO; O; OH; O; HO; O; OH; HO; + HO; O; OH; OH (6)

LAH	50:50
$Pd(OH)_2/C$	99: 1

Deuterium studies showed that the ring opening proceeded with inversion of configuration at C-3 to give the deuterated product (**1**). This is a common observation with Pd catalysts.

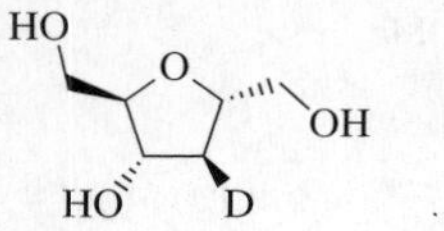

(1)

The reduction of a variety of azides to amines was carried out under transfer hydrogenation conditions with hydrazine and Pearlman catalyst in refluxing MeOH. The yields ranged from 71 to 90%.[8]

An *o*-substituted dinitrostyrene has been reduced in THF/HOAc at 150 psi H_2 and 50 °C to provide an indole in 75% yield. Interestingly, when the H_2 pressure was lowered to atmospheric pressure, the reaction proceeded only to 30% conversion to give an unidentified intermediate (eq 7).[9]

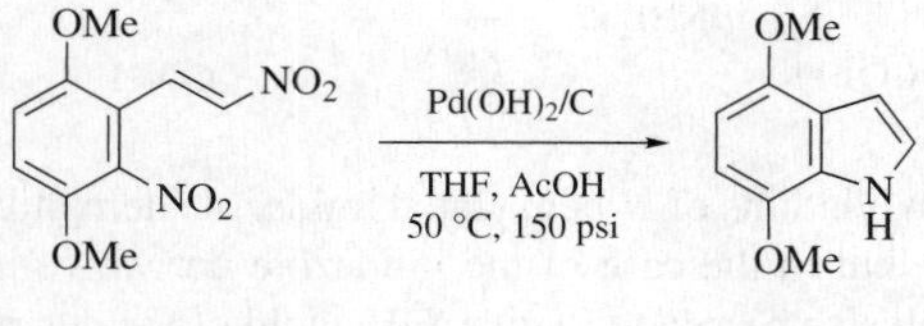

$Pd(OH)_2$/C; THF, AcOH; 50 °C, 150 psi (7)

Dehydrogenation. The reagent has also been used as a dehydrogenation catalyst. The ketone shown in eq 8 was converted to the enone in 79% yield. *trans*-4-Phenyl-3-buten-2-one was obtained in 50% conversion in toluene from 4-phenylbutan-2-one, but no dehydrogenation was observed with propiophenone, indanone, α-tetralone, or cyclohexanone.[10]

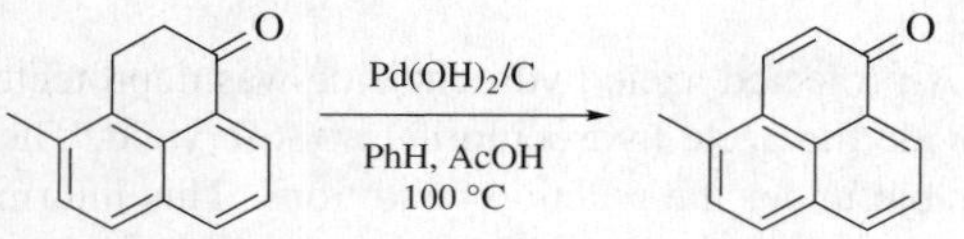

$Pd(OH)_2$/C; PhH, AcOH; 100 °C (8)

Related Reagents. See Classes R-9, R-19, R-21, R-23 and R-32, pages 1–10.

1. Pearlman, W. M. *TL* **1967**, *17*, 1663.
2. Misra, R. N.; Brown, B. R.; Han, W.-C.; Harris, D. N.; Hedberg, A.; Webb, M. L.; Hall, S. E. *JMC* **1991**, *34*, 2882.
3. Prugh, J. D.; Rooney, C. S.; Deana, A. A.; Ramjit, H. G. *TL* **1985**, *26*, 2947.
4. Beaulieu, P. L.; Schiller, P. W. *TL* **1988**, *29*, 2019.
5. Hosmane, R. S.; Bhadti, V. S.; Lim, B. B. *S* **1990**, 1095. Itaya, T.; Morisue, M.; Takeda, M.; Kumazara, Y. *CPB* **1990**, *38*, 2656.
6. Jung, M. E.; Choi, Y. M. *JOC* **1991**, *56*, 6729.
7. Garcia, J. G.; Voll, R. J.; Younathan, E. *TL* **1991**, *32*, 5273.
8. Malik, A. A.; Preston, S. B.; Archibald, T. G.; Cohen, M. P.; Baum, K. *S* **1989**, 450.
9. Showalter, H. D.; Pohlmann, G. *OPP* **1992**, *24*, 484.
10. Zhao, S.; Freeman, J. P.; Szmuszkovicz, J. *JOC* **1992**, *57*, 4051.

Anthony O. King & Ichiro Shinkai
Merck Research Laboratories, Rahway, NJ, USA

Peracetic Acid[1]

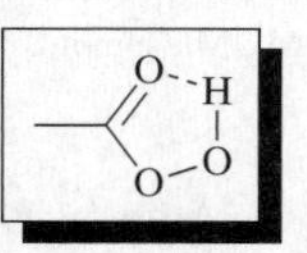

[79-21-0] $C_2H_4O_3$ (MW 76.06)

(electrophilic reagent capable of reacting with many functional groups; delivers oxygen to alkenes, sulfides, selenides, and amines)

Alternate Name: peroxyacetic acid.
Physical Data: mp 0 °C; bp 25 °C/12 mmHg; *d* 1.038 g cm^{-3} at 20 °C.
Solubility: sol acetic acid, ethyic l acetate, $CHCl_3$, acetone, benzene, CH_2Cl_2, ethylene dichloride, water.
Form Supplied in: 40% solution in acetic acid (*d* 1.15 g cm^{-3}) having approximately the following composition by weight: peracetic acid, 40–42%; H_2O_2, 5%; acetic acid, 40%; H_2SO_4, 1%; water, 13%; diacetyl peroxide, nil; other organic compounds, nil; stabilizer, 0.05%. A solution of the peracid in ethyl acetate is also available commercially.
Analysis of Reagent Purity: assay using iodometry;[2] estimation of diacetyl peroxide.[3]
Preparative Methods: prepared in the laboratory by reacting ***Acetic Acid*** with hydrogen peroxide in the presence of catalytic quantities (1% by weight) of ***Sulfuric Acid***; when 30% H_2O_2 is used the concentration of the peracid reagent obtained is less than 10%.[1a] If a stronger solution of the reagent is required, 70–90% H_2O_2 must be used. Caution: for hazards see ***Hydrogen Peroxide***. ***Hydrogen Peroxide–Urea*** (which is commercially available and is safe to handle) has been used as a substitute for anhydrous hydrogen peroxide.[3] In the preparation of peracetic acid from acetic anhydride and H_2O_2, the dangerously explosive diacetyl peroxide may become the major product if the reaction is not carried out properly.[1a]
Purification: peracetic acid is rarely prepared in pure undiluted form for safety reasons. The commercially available material contains acetic acid, water, H_2O_2, and H_2SO_4. After neutralization of the sulfuric acid, this reagent is satisfactory for most reactions. If water is undesirable, an ethyl acetate solution of the reagent may be used. Details for the preparation of the H_2O_2-free reagent are available.[4]
Handling, Storage, and Precautions: peracetic acid is an explosive compound but is safe to handle at room temperature in organic solutions containing less than 55%. Use in a fume hood. Since peroxides are potentially explosive, a safety shield should generally be used.[5] Peracetic acid can be stored at 0 °C with essentially no loss of active oxygen and at rt with only negligible losses over several weeks.

General Considerations. Peracetic acid oxidizes simple alkenes, alkenes carrying a variety of functional groups (such as ethers, alcohols, esters, ketones, and amides), some aromatic compounds, furans, sulfides, and amines. It oxidizes β-lactams in the presence of catalysts. Ketones and aldehydes undergo oxygen insertion reaction (Baeyer–Villiger oxidation).

Epoxidation of Alkenes. Peracetic acid is a comparatively safe reagent for small-scale reactions. In industry, to avoid the hazards involved in handling large quantities of the reagent, it is prepared in situ. Peracetic acid prepared in this fashion is widely used for epoxidation of vegetable oils and fatty acid esters. To the substrate in acetic acid containing catalytic (1% by weight) quantities of H_2SO_4 maintained around 50 °C is added gradually, with stirring, 50% H_2O_2 at such a rate that there is no buildup in the concentration of H_2O_2. The peracid is consumed as it is formed (eq 1). The addition of H_2O_2 is usually completed in 2 h and then the temperature is raised to and maintained at 60 °C until all the H_2O_2 is consumed (about 3 h). The reaction mixture is diluted with water, at which point the epoxides (being water-insoluble) separate out. The use of hexane during the reaction minimizes epoxide ring opening. Since the catalyst (H_2SO_4) is essential for the speedy formation of peracetic acid, in situ methods can be used for preparing only those epoxides which can tolerate the presence of the acid catalyst. Epoxides of fatty acid esters are obtained in good yields if the reaction temperature and time taken for completion of the reaction are properly controlled.

$$\mathrm{MeCO_2H} + \mathrm{H_2O_2} \overset{\mathrm{H^+}}{\rightleftharpoons} \mathrm{MeCO_3H} + \mathrm{H_2O} \quad (1)$$

Peracetic acid in ethyl acetate is a better reagent for preparing epoxides from alkenes than the reagent in acetic acid since the large quantities of acetic acid in the latter reagent facilitate epoxide ring opening. However, since the reagent in acetic acid is more readily available, it is normally used for epoxidation; the sulfuric acid present in the commercial sample has to be neutralized by adding sodium acetate before the epoxidation. After epoxidizing the alkene with peracetic acid, the reaction mixture is diluted with water. The unreacted peracid, acetic acid, and traces of hydrogen peroxide are removed in the aqueous layer. The separated epoxide is filtered if it is a solid; when the epoxide is a liquid, the organic layer is separated using a small quantity of solvent, if needed. Another method of workup is to remove unreacted peracid and acetic acid through evaporation under reduced pressure.

Epoxidation of terminal alkenes with organic peracids is sluggish since the double bond is not electron rich (eq 2).[6]

MeCO$_3$H, NaOAc, Na$_2$CO$_3$, CH$_2$Cl$_2$, 20 °C, 4 d, 53% (2)

Adequately substituted acrylic esters furnish epoxides in good yields. Ethyl crotonate has been epoxidized in kg quantities according to eq 3;[7] the workup is simple, involving direct fractionation of the reaction mixture. For the preparation of epoxide (**1**) from ethyl crotonate using ***Trifluoroperacetic Acid***, the yield is 73%.[8] The sensitive allylic epoxide (**2**) has been prepared according to eq 4.[9] This procedure has been applied successfully for the preparation of allylic epoxides from 1,3-cyclopentadiene, 1,3-cycloheptadiene, and 1,3-cyclooctadiene.

Epoxidation of the triene (**3**)[2] is regioselective, involving reaction at the tetrasubstituted double bond (eq 5). Epoxidation of (**3**) using ***m-Chloroperbenzoic Acid*** furnishes the monoepoxide in 76% yield.[10] Epoxidation of the diene (**4**) was regio- and stereoselective (eq 6);[11] the more substituted double bond was epoxidized from the less hindered side.

CO$_2$Et; MeCO$_3$H, EtOAc, 85 °C, 5.5 h, 75%; O, CO$_2$Et (**1**) (3)

40% MeCO$_3$H, Na$_2$CO$_3$, CH$_2$Cl$_2$, rt, 69%; O (**2**) (4)

(**3**); MeCO$_3$H, CH$_2$Cl$_2$, Na$_2$CO$_3$, 15 °C, 79–84%; O (5)

TBDMSO, H, H, H, O, O (**4**); MeCO$_3$H (2.9 equiv), EtOAc, 25 °C, 10.5 h, 72%; TBDMSO, O, H, H, H, O, O (6)

Epoxidation of the unsaturated γ-lactone (**5**) furnished stereoselectively the epoxide (**6**), involving approach of the reagent from the more hindered side of the double bond (eq 7).[12] This selectivity is observed only when acetic acid is the solvent. The selectivity was much less when *m*-CPBA was used.

(**5**); MeCO$_3$H, AcOH, 25 °C, 16 h, 96%; (**6**) 89:11 (7)

Moderate stereoselectivity is observed during the epoxidation of sterically unbiased 3,3-diarylcyclopentenes; the major product is formed through approach of the electrophile from the side *trans* to the better electron donor (eq 8).[13]

O$_2$N; 10 equiv MeCO$_3$H, CH$_2$Cl$_2$, 23 °C, 18 h; O$_2$N, O + O$_2$N, O; 73:27 (8)

A systematic study of the epoxidation of the acyclic allyl alcohol (**7**) has been carried out, employing several reagents.[14] Epoxidation with peracetic acid generated from urea/H_2O_2 showed small *syn* selectivity (eq 9). *m*-CPBA epoxidation of (**7**) furnished in 87% yield a 40:60 mixture of the epoxy alcohols (**8**) and (**9**).

HO

urea–H_2O_2
Ac_2O, Na_2HPO_4

CH_2Cl_2
62%

(**7**)

HO HO

\+ (9)

(**8**) 57:43 (**9**)

Epoxidation of Alkenes via Peracids Generated In Situ. Alkenes have been epoxidized by reacting them with peracids generated in situ. The system consisting of molecular oxygen and aldehydes, particularly isobutyraldehyde and ***Pivalaldehyde***, converts various alkenes to epoxides in high yields when they are reacted at 40 °C for 3–6 h (eq 10).[15]

C_5H_{11} H H

Me_2CHCHO (1.5 equiv)
EDC, O_2 bubbling

40 °C, 3 h
100%

C_5H_{11} O H H (10)

Oxidation of Furans. 2,5-Disubstituted furans are oxidatively cleaved by peracids; for example, see eq 11.[16] *m*-CPBA can also be used for this reaction. Δ^3-Butenolides have been synthesized by oxidizing 2-trimethylsilyl furans with peracetic acid; as in eq 12.[17]

TMS O Et

NaOAc (1.2 equiv)
$MeCO_3H$ (4 equiv)
CH_2Cl_2, 7 °C, 1 h

70%

TMS O O Et (11)

TMS O R

$MeCO_3H$
NaOAc, CH_2Cl_2

7 °C, 3.5 h
78%

O O R (12)

R = $Me_2CHCH_2CH_2$-

This reaction does not proceed smoothly when there is a hydroxyl group in the furfuryl position; however, the reaction is facile if the furfuryl OH is blocked. The reaction does not take place if electron-withdrawing groups are present on the furan ring. *m*-CPBA is not a good reagent for this oxidation. An interesting application of this reaction has been published.[18]

Oxidation of Aromatic Compounds. Suitably substituted aromatic compounds are oxidized efficiently to the quinones by peracetic acid. The quinone (**10**) is obtained in 22% yield by oxidizing naphtho[*b*]cyclobutene.[19] Slow addition of 1,5-dihydroxynaphthalene to excess peracetic acid furnished juglone (**11**) in 46–50% yield.[20]

O O O OH O

(**10**) (**11**)

Baeyer–Villiger Oxidation. A systematic study of the Baeyer–Villiger reaction of the bicyclic ketone (**12**) has been carried out employing different organic peracids.[21] Selective formation of lactone (**13**) was highest when peracetic acid was used (eq 13). Reaction of (**12**) with *m*-CPBA furnishes a 55:45 mixture of (**13**) and (**14**) in 81% yield.

MeO O OCH_2Ph

H_2O_2, $MeCO_3H$
NaOAc
36 h, rt

72%

MeO O O OCH_2Ph + MeO O O OCH_2Ph (13)

(**12**) (**13**) 92:8 (**14**)

Position-specific Baeyer–Villiger rearrangement has been observed in the reaction of peracetic acid with some polycyclic ketones.[22,23] An ε-lactone, required for the synthesis of erythronolide B, was synthesized in 70% yield through position-specific Baeyer–Villiger rearrangement of a cyclohexanone having substituents on all the ring carbons;[24] the ketone was treated with excess 25% peracetic acid in ethyl acetate for 6 days at 55–58 °C. Peracetic acid oxidation of the keto β-lactam (**15**) furnishes stereoselectively the interesting β-lactam (**16**) (eq 14);[25] the initially formed Baeyer–Villiger reaction product undergoes further reaction. Ketone (**15**) has also been reacted with *m*-CPBA in acetic acid but the selectivity is slightly less, forming (**16**):(**17**) in 10:1 ratio.

O O O H H N H O

$MeCO_3H$, NaOAc
AcOH

83%

OCHO H OAc H N H O + OCHO H OAc H N H O (14)

(**15**) (**16**) 11:1 (**17**)

Ruthenium- and Osmium-Catalyzed Oxidations. α-Ketols have been synthesized by reacting alkenes with peracetic acid in the presence of a ***Ruthenium(III) Chloride*** catalyst.[26] α-Ketol (**19**) was synthesized from the alkene (**18**) chemo- and stereoselectively (eq 15). The two-phase aqueous system is essential for this reaction. Conjugated dienes, allylic azides, and α,β-unsaturated esters have been oxidized with this reagent.

OAc MeO_2C

$RuCl_3$, $MeCO_3H$
MeCN, H_2O, CH_2Cl_2

55%

OAc OH MeO_2C O (15)

(**18**) (**19**)

The methylene group adjacent to the nitrogen of β-lactams has been oxidized with peracetic acid in the presence of a ruthenium catalyst (eq 16).[27] Peracetic acid is the best oxidant for this reaction. Instead of ruthenium, $OsCl_3$ can be used to catalyze the oxidation.[28] The peracetic acid required for the reaction can be generated in situ from acetaldehyde and molecular oxygen (eq 17).[29]

OTBDMS
5% Ru cat.
NaOAc, AcOH, $MeCO_3H$
in EtOAc, rt, 2.5 h
99%
OTBDMS
OAc
99% de
(16)

Ru cat., MeCHO, O_2
$MeCO_2H$, NaOAc
EtOAc
88%
OAc
(17)

Other Applications. Peracetic acid has been used to (a) oxidize primary amines to nitroso compounds,[30] (b) oxidize secondary alcohols to ketones in the presence of a Cr^{VI} ester catalyst (eq 18)[31] or sodium bromide,[32] (c) oxidize sulfenamides to sulfonamides (eq 19),[33] (d) oxidize iodobenzene to iodosobenzene diacetate[34] and iodoxybenzene,[35] and (e) oxidize N-heterocycles such as pyridine to *N*-oxides.[36] α,β-Unsaturated aldehydes (and α,β-unsaturated ketones) do not undergo facile epoxidation with peracetic acid since the double bond is not electron rich. However, the acetals of α,β-unsaturated aldehydes can be oxidized readily (eq 20).[37] For the epoxidation of α,β-unsaturated aldehydes with H_2O_2/base see ***Hydrogen Peroxide***.

OH
$MeCO_3H$
0 °C, CH_2Cl_2
CCl_4, 30 min
96%
(18)

EtO
SNH_2
$MeCO_3H$
K_2CO_3, EtOH, H_2O
88%
EtO
SO_2NH_2 (19)

$MeCO_3H$
EtOAc
40 °C, 7 h
45%
(20)

For industrial applications, peracetic acid is the most widely used organic peracid since it is inexpensive. It is the only commonly used peracid which can be prepared in situ for epoxidation reactions, since the acid catalyst (1% H_2SO_4; eq 1), which can facilitate epoxide ring opening, is used in low concentrations; the accompanying acetic acid, being a weak acid, is not very efficient in epoxide opening. The in situ method is not hazardous. Although the reagent is available commercially, it is also prepared in the laboratory since its preparation is easy, fairly fast, and no solvent is required for isolation. Epoxidation reactions and subsequent workup can be performed with no solvent, or only small quantities of solvent since the peracid and accompanying acetic acid are both water soluble and volatile. It is not essential that the substrate should dissolve in the reagent (peracetic acid–acetic acid).

Related Reagents. See Classes O-11, O-14, and O-15, pages 1–10. *m*-Chloroperbenzoic Acid; Perbenzoic Acid.

1. (a) Swern, D. *Organic Peroxides*; Wiley: New York, 1971; Vol. II, pp 355–533. (b) Plesnicar, B. *Organic Chemistry*; Academic: New York, 1978; Vol. 5C, pp 211–294.
2. Vogel, E.; Klug, W.; Breuer, A. *OS* **1976**, *55*, 86.
3. Cooper, M. S.; Heaney, H.; Newbold, A. J.; Sanderson, W. R. *SL* **1990**, 533.
4. Pandell, A. J. *JOC* **1983**, *48*, 3908.
5. *Hazards in the Chemical Laboratory*; Luxon, S. G., Ed.; Royal Society of Chemistry: Cambridge, 1992.
6. Kirmse, W.; Kornrumpf, B. *AG(E)* **1969**, *8*, 75.
7. MacPeek, D. L.; Starcher, P. S.; Phillips, B. *JACS* **1959**, *81*, 680.
8. Emmons, W. D.; Pagano, A. S. *JACS* **1955**, *77*, 89.
9. Crandall, J. K.; Banks, D. B.; Colyer, R. A.; Watkins, R. J.; Arrington, J. P. *JOC* **1968**, *33*, 423.
10. Shani, A.; Sondheimer, F. *JACS* **1967**, *89*, 6310.
11. Corey, E. J.; Myers, A. G. *JACS* **1985**, *107*, 5574.
12. Corey, E. J.; Noyori, R. *TL* **1970**, 311.
13. Halterman, R. L.; McEvoy, M. A. *TL* **1992**, *33*, 753.
14. Back, T. G.; Blazecka, P. G.; Vijaya Krishna, M. V. *TL* **1991**, *32*, 4817.
15. Kaneda, K.; Haruna, S.; Imanaka, T.; Hamamoto, M.; Nishiyama, Y.; Ishii, Y. *TL* **1992**, *33*, 6827.
16. Kobayashi, Y.; Katsuno, H.; Sato, F. *CL* **1983**, 1771.
17. Kuwajima, I.; Urabe, H. *TL* **1981**, *22*, 5191.
18. Tanis, S. P.; Robinson, E. D.; McMills, M. C.; Watt, W. *JACS* **1992**, *114*, 8349.
19. Cava, M. P.; Shirley, R. L. *JOC* **1961**, *26*, 2212.
20. Grundmann, C. *S* **1977**, 644.
21. Grudzinski, Z.; Roberts, S. M.; Howard, C.; Newton, R. F. *JCS(P1)* **1978**, 1182.
22. Salomon, R. G.; Sachinvala, N. D.; Roy, S.; Basu, B.; Raychaudhuri, S. R.; Miller, D. B.; Sharma, R. B. *JACS* **1991**, *113*, 3085.
23. Corey, E. J.; Srinivas Rao, K. *TL* **1991**, *32*, 4623.
24. Corey, E. J.; Kim, S.; Yoo, S.; Nicolaou, K. C.; Melvin, Jr., L. S.; Brunelle, D. J.; Falck, J. R.; Trybulski, E. J.; Lett, R.; Sheldrake, P. W. *JACS* **1978**, *100*, 4620.
25. Kobayashi, Y.; Ito, Y.; Terashima, S. *T* **1992**, *48*, 55.
26. Murahashi, S.-I.; Saito, T.; Hanaoka, H.; Murakami, Y.; Naota, T.; Kumobayashi, H.; Akutagawa, S. *JOC* **1993**, *58*, 2929.
27. Murahashi, S.-I.; Naota, T.; Kuwabara, T.; Saito, T.; Kumobayashi, H.; Akutagawa, S. *JACS* **1990**, *112*, 7820.
28. Murahashi, S.-I.; Saito, T.; Naota, T.; Kumobayashi, H.; Akutagawa, S. *TL* **1991**, *32*, 2145.
29. Murahashi, S.-I.; Saito, T.; Naota, T.; Kumobayashi, H.; Akutagawa, S. *TL* **1991**, *32*, 5991.
30. Corey, E. J.; Gross, A. W. *TL* **1984**, *25*, 491.
31. Corey, E. J.; Barrette, E.-P.; Magriotis, P. A. *TL* **1985**, *26*, 5855.
32. Morimoto, T.; Hirano, M.; Ashiya, H.; Egashira, H.; Zhuang, X. *BCJ* **1987**, *60*, 4143.

33. Larsen, R. D.; Roberts, F. E. *SC* **1986**, *16*, 899.
34. Sharefkin, J. G.; Saltzman, H. *OS* **1963**, *43*, 62.
35. Sharefkin, J. G.; Saltzman, H. *OS* **1963**, *43*, 65.
36. Mosher, H. S.; Turner, L.; Carlsmith, A. *OSC* **1963**, *4*, 828.
37. Heywood, D. L.; Phillips, B. *JOC* **1960**, *25*, 1699.

A. Somasekar Rao & H. Rama Mohan
Indian Institute of Chemical Technology, Hyderabad, India

Perbenzoic Acid[1]

[93-59-4] $C_7H_6O_3$ (MW 138.13)

(electrophilic reagent capable of delivering oxygen to alkenes,[1] amines,[2] and sulfides[3])

Alternate Name: peroxybenzoic acid; PBA.
Physical Data: mp 41–42 °C.
Solubility: sol $CHCl_3$, CH_2Cl_2, benzene, ethyl acetate, ether; slightly sol water.
Form Supplied in: long white needles, but usually handled in solution; not available commercially.
Analysis of Reagent Purity: iodometric assay.[4,6]
Preparative Methods: ***Sodium Methoxide*** reacts readily with ***Dibenzoyl Peroxide*** (**1**) (eq 1); complete experimental details are available for preparing perbenzoic acid based on this reaction.[4] Alkaline perhydrolysis (HOO^-) of (**1**) in aqueous organic solvents has been employed for preparing PBA in 90% yields.[5] PBA has been prepared in 85–90% yield by adding 70% ***Hydrogen Peroxide*** (**Caution**!) to a suspension of benzoic acid (**2**) in methanesulfonic acid (eq 2).[6] PBA has been prepared in 90% yield by passing oxygen containing catalytic quantities of ozone through a solution of benzaldehyde in ethyl acetate.[7]

$$(PhCO)_2O_2 + MeONa \longrightarrow PhCO_3Na + PhCO_2Me \quad (1)$$
(**1**)

$$PhCO_2H + H_2O_2 \overset{H^+}{\rightleftharpoons} PhCO_3H + H_2O \quad (2)$$
(**2**)

Purification: PBA can be purified by recrystallization at −20 °C from a 3:1 mixture of petroleum ether–diethyl ether.
Handling, Storage, and Precautions: analytically pure PBA can be stored for long periods in a refrigerator without significant loss of active oxygen. Since PBA is a peroxide and potentially explosive, care must be exercised in carrying out the reactions with it; during workup, check for peroxides before evaporating the solvent. This reagent should be handled in a fume hood.

Functional Group Oxidations. This reagent oxidizes simple alkenes, including alkene substrates incorporating a variety of functional groups (ethers, alcohols, esters, etc.), amines, and sulfides. Ketones undergo oxygen insertion reactions (Baeyer–Villiger oxidation).

Epoxidation of Alkenes. PBA, like other organic peracids, reacts with alkenes readily under mild conditions to furnish epoxides in good yields (see, for example ***m-Chloroperbenzoic Acid***).[8] Styrene is thus oxidized to styrene oxide (**3**) (eq 3).[9] To PBA (0.33 mol) in 500 mL of $CHCl_3$ is added at 0 °C, with stirring, 0.3 mol of styrene. The reaction mixture is kept at 0 °C for 24 h. Assay of PBA at this stage will show that one equivalent of PBA has been consumed. The reaction mixture is shaken with aq 10% NaOH solution, water, and then dried (Na_2SO_4). The solvent is evaporated and the residue is fractionated using a Vigreux column to furnish the epoxide (**3**) in 70–75% yield.

PBA, $CHCl_3$, 0 °C, 24 h; 70–75% (**3**) (3)

Oxidation of the alkene (**4**) furnishes the epoxide (**5**) (eq 4).[10] The disubstituted *cis*-alkene (**6**) gave in quantitative yield the corresponding *cis*-epoxide (**7**).[11] Epoxidation of cycloheptene in $CHCl_3$ (3.2% solution, 48 h, 0 °C) furnished cycloheptene oxide in 78% yield.[12]

(**4**) PBA, $CHCl_3$, 5 °C, 24 h; 94% (**5**) (4)

(**6**) (**7**)

Reaction of the diene (**8**), having two trisubstituted double bonds, with 1 equiv of PBA proceeded regio- and stereoselectively to give monoepoxide (**9**) (eq 5).[13]

(**8**) PBA, $CHCl_3$, rt, 7 d; 63% (**9**) (5)

The benzylic epoxide (**10**) has been obtained in 77% yield by epoxidizing 1-phenylcyclohexene with a $CHCl_3$ solution of PBA at 0 °C.[14] A pure sample of (**10**) was obtained by distilling the crude reaction product over powdered KOH; this epoxide is very sensitive to acids but is stable in basic medium. It has been observed that 7α,8α-epoxides of steroids and triterpenes are stable in basic media but these epoxides are readily cleaved even with traces of acids.[15]

(10)

Epoxidation of the diene (11) takes place regioselectively at the tetrasubstituted double bond to furnish the monoepoxide (12) in 61% yield.[16] 3β-Acetoxy-8α,9α-epo xyergostane was prepared in 85% yield by epoxidizing the alkene (13) using benzene as solvent.[17]

(11) (12) (13)

Epoxidation of 3-acetoxycholest-2-ene (14) in $CHCl_3$ solution (−12 °C, 42 h) is stereoselective; the 2α,3α-epoxide is isolated in 80% yield.[18] Epoxidation of the vinyl chloride (15) ($CHCl_3$ solution, 12 h, rt) furnished the corresponding chloro epoxide in 68% yield.[19]

(14) (15)

The sensitive epoxide (17) has been prepared from the enol ether (16) in 90% yield (eq 6).[20] An ether solution of (16) and PBA is allowed to react for 30 s; benzoic acid is removed by passing the reaction mixture rapidly (30 s) through an alumina column.

(16) → (17) (6)

Epoxidation of the α,β-unsaturated ketone pulegone (18) (5–10 °C, 24 h) is not stereoselective; a 2:1 mixture of α,β-epoxy ketones (19) and (20) is obtained in 86% yield.[21]

(18) (19) (20)

Hydroxy-directed stereoselective epoxidation has been observed with allylic cyclohexenols (eq 7).[22]

PBA, benzene, 20 °C, 16 h, 83% (7)

Baeyer–Villiger Reaction. Fe_2O_3-catalyzed oxidation of ketones with molecular oxygen in the presence of benzaldehyde at rt gives the lactones efficiently (eq 8).[23] The reacting species may be PBA or the radical $PhCO_3\cdot$. Baeyer–Villiger reaction of (−)-(21) furnishes the ester (−)-(22) in 85% yield with retention of configuration.[24]

PhCHO, Fe_2O_3 (cat), O_2 (1 atm), 17 h, 98% (8)

(21) R = COMe
(22) R = OCOMe

Reaction with Compounds Containing Nitrogen and Sulfur. Aromatic primary amines containing electron-withdrawing groups are oxidized to nitroso compounds.[2] Oxidation of (23) thus furnishes the nitroso compound (24) in 85% yield. This reaction cannot be carried out with peracetic acid. The oxaziridine (25) has been prepared in 56–74% yield by reacting the corresponding aldimine with PBA.[25] Episulfides are oxidized to episulfoxides.[3] The episulfoxide (26) has been obtained in 77% yield by oxidizing the episulfide in CH_2Cl_2 at −20 °C.

(23) R = NH_2
(24) R = NO
(25) (26)

In the epoxidation of alkenes which are moderately or highly reactive, the yields obtained with PBA are comparable to the yields obtained using ***m*-Chloroperbenzoic Acid** or ***Peracetic Acid***. However, *m*-CPBA and $MeCO_3H$ are available commercially; PBA is not available commercially, but can be prepared conveniently in a short time. For the epoxidation of alkenes which react sluggishly (e.g. α,β-unsaturated esters) *m*-CPBA and ***Trifluoroperacetic Acid*** are preferred.

Related Reagents. See Classes O-8, O-14, and O-15, pages 1–10. Monoperoxyphthalic Acid.

1. (a) Swern, D. In *Organic Peroxides*; Swern, D., Ed.; Wiley: New York, 1970; Vol. 1, Chapter 6; Vol. 2, Chapter 5. (b) Plesnicar, B. *Organic Chemistry*; Academic: New York, 1978; Vol. 5c, pp 211–294.
2. Di Nunno, L.; Florio, S.; Todesco, P. E. *JCS(C)* **1970**, 1433.
3. Kondo, K.; Negishi, A.; Fukuyama, M. *TL* **1969**, 2461.
4. Braun, G. *OS* **1933**, *13*, 86.
5. Ogata, Y.; Sawaki, Y. *T* **1967**, *23*, 3327.
6. Silbert, L. S.; Siegel, E.; Swern, D. *OS* **1963**, *43*, 93.
7. Dick, C. R.; Hanna, R. F. *JOC* **1964**, *29*, 1218.
8. Prileschajew, N. *CB* **1909**, *42*, 4811 (*CA* **1910**, *4*, 916).
9. Hibbert, H.; Burt, P. *OS* **1928**, *8*, 102.
10. Wawzonek, S.; Klimstra, P. D.; Kallio, R. E.; Stewart, J. E. *JACS* **1960**, *82*, 1421.
11. Joshi, N. N.; Mamdapur, V. R.; Chadha, M. S. *JCS(P1)* **1983**, 2963.
12. Owen, L. N.; Saharia, G. S. *JCS* **1953**, 2582.

13. Bernstein, S.; Littel, R. *JOC* **1961**, *26*, 3610.
14. Berti, G.; Bottari, F.; Macchia, B.; Macchia, F. *T* **1965**, *21*, 3277.
15. (a) Fried, J.; Brown, J. W.; Applebaum, M. *TL* **1965**, 849. (b) Fieser, L. F.; Goto, T. *JACS* **1960**, *82*, 1693.
16. Hückel, W.; Wörffel, U. *CB* **1955**, *88*, 338.
17. Henbest, H. B.; Wrigley, T. I. *JCS* **1957**, 4596.
18. Williamson, K. L.; Johnson, W. S. *JOC* **1961**, *26*, 4563.
19. McDonald, R. N.; Tabor. T. E. *JACS* **1967**, *89*, 6573.
20. Stevens, C. L.; Tazuma, J. *JACS* **1954**, *76*, 715.
21. Reusch, W.; Johnson, C. K. *JOC* **1963**, *28*, 2557.
22. Henbest, H. B.; Wilson, R. A. L. *JCS* **1957**, 1958.
23. Murahashi, S.-I.; Oda, Y.; Naota, T. *TL* **1992**, *33*, 7557.
24. Berson, J. A.; Suzuki, S. *JACS* **1959**, *81*, 4088.
25. Emmons, W. D.; Pagano, A. S. *OS* **1969**, *49*, 13.

A. Somasekar Rao & H. Rama Mohan
Indian Institute of Chemical Technology, Hyderabad, India

Phenyliodine(III) Bis(trifluoroacetate)[1]

$(CF_3CO_2)_2IPh$

[2712-78-9] $C_{10}H_5F_6IO_4$ (MW 430.05)

(cyclopropyl groups undergo ring opening,[2] alkynes are cleaved,[3] phenols are oxidized to quinones[4–8] and in some cases oxidative intramolecular coupling occurs,[9–13] ketones yield α-hydroxy ketones,[14] certain β-diketones undergo cyclization,[15] Pummerer-type reaction occurs,[16] and deprotective dethioacetalization is an important use[17])

Alternate Names: bis(trifluoroacetoxyiodo)benzene; iodobenzene bis(trifluoroacetate); PIFA; BTI.
Physical Data: mp 121–125 °C.
Solubility: sol CH_2Cl_2, CCl_4, CF_3CH_2OH, $(CF_3)_2CHOH$.
Form Supplied in: white solid; commercially available.
Preparative Method: can be synthesized by reaction of ***(Diacetoxyiodo)benzene*** with ***Trifluoroacetic Acid***.[18]
Handling, Storage, and Precautions: moisture sensitive; irritant.

Bond Cleavage. 1,3-Dehydroadamantane underwent ring-opening with 1,3-addition of $CF_3CO_2^-$ upon reaction with $PhI(OCOCF_3)_2$ (BTI) (eq 1).[2] Alkynes are cleaved upon treatment with three equivalents of BTI.[3] Dibenzyl ether is cleaved to benzaldehyde and benzyl trifluoroacetate (eq 2). Benzylalkyl (or trityl and benzhydryl) ethers react analogously.[19]

BTI, CCl_4, 20 °C; $OCOCF_3$; $OCOCF_3$ (1)

Ph–O–Ph; BTI; PhCHO + Ph–O–C(O)–CF_3 (2)

Oxidation of Phenolic Compounds. Hydroquinones and catechol derivatives are oxidized to *o*- or *p*-benzoquinones (eq 3).[4–7] In the presence of water, a quinone is formed, while in the presence of methanol the quinone monoacetal is formed (eq 4).[8]

OH; R; OH; BTI; MeCN–H_2O; O; R; O (3)

OH; MeO; NHR; OH; BTI; MeCN–H_2O; O; MeO; NHR; O; BTI; MeCN–MeOH; O; NHR; MeO; MeO OMe (4)

Intramolecular oxidative cyclization is also an important reaction pathway, as in the case of *N*-acetyltyramines (eq 5),[9] and it has been applied in the syntheses of several natural products. For example, phenolic oxidative coupling with BTI was a key step in the synthesis of 6a-epipretazettine (eq 6).[10]

OH; R–N–COPh; R = alkyl; BTI; CF_3CH_2OH; OCOPh; O; H; N; R (5)

OH; O; O; O; N; Me; CO_2R; OMe; R = CH_2CCl_3, Me; BTI; CH_2Cl_2, –10 °C; O; O; O; O; N; Me; CO_2R; OMe; OMe; Me; N; O; O; O; OH (6)

A similar intramolecular spirodienone formation was observed in a naphthoquinone system (eq 7).[11] An extension of this type of oxidative coupling formed the key step in the synthesis of the marine alkaloid discorhabdin C (eq 8).[12] Intramolecular cyclization has also been used as a key step in a synthetic approach to the natural product aranorosin (eq 9).[13]

R = H, Br

(7)

(8)

Discorhabdin C

Aranorosin

(9)

Table 1 Oxidative Cyclization of 3-(4-Hydroxyphenol)propanoic Acid

Starting material	Reagent	Product	Yield (%)
	$PhI(OCOCF_3)_2$		83 (45–65)
	$PhI(OCOMe)_2$		27 (27)
	$4\text{-}MeC_6H_4I(OCOCF_3)_2$		69 (35)
	$4\text{-}MeC_6H_4I(OCOCF_3)_2$		52 (17)
	$4\text{-}NO_2C_6H_4I(OCOCF_3)_2$		65 (56)
	$1,4\text{-}[I(OCOCF_3)_2]_2C_6H_4$		67 (–)
	PhI(OH)OTs		53 (–)

Table 1 illustrates the oxidative cyclization and gives a useful comparison of yields as a function of reagent, solvent, and temperature. All hypervalent iodine oxidations were carried out in acetonitrile or acetonitrile–pyridine (yields in parentheses) at 0 °C.[13]

Oxidation of Carbonyl Compounds. BTI and trifluoroacetic acid in $MeCN/H_2O$ react with aromatic, heteroaromatic, and aliphatic ketones to afford α-hydroxy ketones in moderate to good yields (eq 10).[14] β-Dicarbonyl compounds with α-aralkyl systems can undergo cyclization (eq 11).[15]

BTI, TFA; $MeCN–H_2O$ (10)

R^2 = H; R^1 = Ph , 69%; Tol, 72%; C_5H_{11}, 58%; *p*-FC_6H_4, 67%; *p*-$NO_2C_6H_4$ 29%; furan, 69%; thiophene, 73%; *t*-Bu, 41%; adamantyl, 74%; cyclopropyl, 47%; cyclobutyl, 47%; cyclopentyl, 94%

R^2 = Me; R^1 = Ph, 36%; *p*-BuC_6H_4, 21%

1. BTI, CF_3CH_2OH; 2. *t*-BuOK, THF; 42–85% (11)

R = OMe, OTBDMS; R^1 = Me, Et; R^2 = Me, Et, OH, OAc; *n* = 1, 2

N-[(2-Methylthio)acetyl]-*N*-phenylaniline upon reaction with BTI undergoes Pummerer-type cyclization to yield 3-methylthio-*N*-phenylindol-2(3*H*)-one. This process appears to be a general method of cyclization for appropriate methylthio derivatives (eqs 12–15).[16]

(12)

(13)

(14)

(15)

Dethioacetalization. BTI is the reagent of choice for dethioacetalization (eq 16).[17] Functional groups, R, such as esters, nitriles, secondary amides, alcohols, halides, alkenes and alkynes, thioesters, and amines are unaffected in this process.

BTI, MeOH–H_2O (9:1) → RCHO (16)

Related Reagents. See Classes O-5 and O-20, pages 1–10.

1. Moriarty, R. M.; Vaid, R. K. *S* **1990**, 431.
2. Shaborova, Y. S.; Pisanova, E. V.; Saginova, L. G. *JOU* **1981**, *17*, 1685.
3. Moriarty, R. M.; Penmasta, R.; Awasthi, A. K.; Prakash, I. *JOC* **1988**, *53*, 6124.
4. Yoshino, S.; Hayakawa, K.; Kanematsu, K. *JOC* **1981**, *46*, 3841.
5. Hayakama, K.; Aso, M.; Kanematsu, K. *JOC* **1985**, *50*, 2036.
6. Okamoto, Y.; Senokuchi, K.; Kanematsu, K. *CPB* **1984**, *32*, 4593.
7. Kanematsu, K.; Morita, S.; Fukushima, S.; Osawa, E. *JACS* **1981**, *103*, 5211.
8. Kita, Y.; Tohma, H.; Inagaki, M.; Hatanaka, K. *H* **1992**, *33*, 503.
9. Kita, Y.; Tohma, H.; Kikuchi, K.; Iganaki, M.; Yakura, T. *JOC* **1991**, *56*, 435.
10. White, J. D.; Chong, W. K. M.; Thirring, K. *JOC* **1983**, *48*, 2300.
11. Kita, Y.; Yakura, T.; Tohma, H.; Kikuchi, K.; Tamura, Y. *TL* **1989**, *30*, 1119.
12. Kita, Y.; Tohma, H.; Ignaki, M.; Hatanaka, K.; Yakura, T. *JACS* **1992**, *114*, 2175.
13. McKillop, A.; McLaren, L.; Taylor, R. J. K.; Watson, R. J.; Lewis, N. *SL* **1992**, 201.
14. Moriarty, R. M.; Berglund, B.; Penmasta, R. *TL* **1992**, *33*, 6065.
15. Kita, Y.; Okunaka, R.; Kondo, M.; Tohma, H.; Inagaki, M.; Hatanaka, K. *CC* **1992**, 429.
16. Tamura, Y.; Yakura, T.; Shirouchi, Y.; Haruta, J. *CPB* **1986**, *35*, 570.
17. Stork, G.; Zhao, K. *TL* **1989**, *30*, 287.
18. Spyroudis, S.; Varvoglis, A. *S* **1975**, 445.
19. Spyroudis, S.; Varvoglis, A. *CC* **1979**, 615.

Robert M. Moriarty & Jerome W. Kosmeder II
University of Illinois at Chicago, IL, USA

Platinum on Carbon

[7440-06-4] Pt (MW 195.08)

(catalyst used for the hydrogenation of alkenes,[1,2] alkynes,[1] nitro groups,[4,5] ketones,[13,14] aromatics;[17] used for hydrogenolysis of cyclopropanes;[3] catalyzes oxidation of alcohols[19])

Solubility: insol all organic solvents and aqueous acidic and basic media.

Form Supplied in: black powder and pellets containing 0.5–10 wt % of Pt (typically 5 wt %); can be either dry or moist (50 wt % of H_2O).

Analysis of Reagent Purity: atomic absorption.

Handling, Storage, and Precautions: can be stored safely in a closed container under air but away from solvents and potential poisons such as sulfur- and phosphorus-containing compounds; pyrophoric in the presence of solvents; general precautions for handling hydrogenation catalysts should be followed; the catalyst must be suspended in the organic solvent under an atmosphere of N_2; during filtration the filter cake must not be allowed to go dry; if a filter aid is necessary, a cellulose-based material should be used if catalyst recovery is desired.

Hydrogenations. In many instances where ***Platinum(IV) Oxide*** is used, Pt/C can be a less expensive substitute. Pt/C is a widely used catalyst for many applications. The reduction of alkenes and alkynes can be carried out under mild conditions. Alkenes can be hydrogenated in the presence of terminal alkynes if the latter are protected with a trialkylsilyl group. Without protection the selectivity is not satisfactory (eq 1).[1]

Pt/C, H_2, EtOAc, Et_3N, 95%; $Si(i\text{-}Pr)_3$ (1)

The diastereoselective reduction of an enamino ketone with Pt/C under acidic conditions is shown in eq 2. This reaction provided the (*SSR-R*) diastereomer in 63% yield plus 15–18% of the (*RRS-R*) diastereomer and other byproducts. After acid-catalyzed lactonization, the desired lactone was conveniently isolated in 57% yield. This intermediate was used in a total synthesis of (+)-thienamycin (eq 2).[2]

Pt/C, H_2, H_3PO_4; H_3O^+; 57% (2)

Hydrogenolysis of cyclopropanes has been accomplished with this catalyst at high temperatures. The example in eq 3 shows the resilience of a cyclobutane to these reaction conditions.[3]

Pt/C, H_2, 100 °C, 95% (3)

The reduction of a nitro group in the presence of an aryl bromide (or chloride) can be accomplished using this catalyst. Aryl fluorides are generally stable to all catalysts. In many instances the amine products undergo further cyclization reactions to form a variety of heterocyclic compounds. The bromide was hydrogenolyzed when Pd/C was used under similar conditions (eq 4).[4]

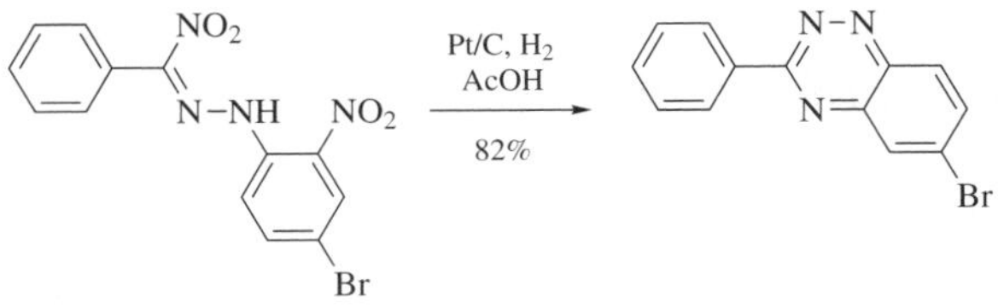

(4)

Varying the pH of the media has been shown to affect the product outcome in a reduction of a nitro amino acid. A 1-hydroxycarbostyril derivative was obtained under acidic conditions, whereas a lactam was produced under basic conditions (eq 5).[5]

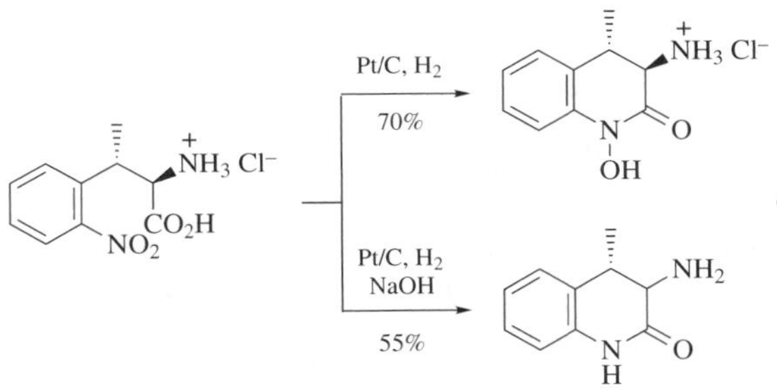

(5)

Reduction of aliphatic nitriles under neutral conditions in the presence of Pt/C gave trialkylamines with high selectivity.[6] With aromatic nitriles, dibenzylamines were obtained.[7] This can be used to advantage in the preparation of unsymmetrical amines. Thus the reduction of a nitrile, in the presence of a different amine provided the unsymmetrical dialkylamine with greater than 95% selectivity.[8] If the desired product is the primary amine, acidic conditions should be used.

Imines can be reduced selectively in the presence of aryl bromide and chloride and hydrazine functions using Pt/C.[9] Reductive amination between a ketone and an amine or amine precursor proceeds to give a good yield of dialkylamine.[10] Amines can also be formed from the reduction of azides and hydrazines.[9b,11] Hydrogenolysis of C–N σ-bonds with Pt/C is uncommon, but not unknown. For example, hydrogenolysis of decahydro-1,8-naphthyridines in the presence of this catalyst gave 3-(γ-aminopropyl)piperidines (eq 6).[12]

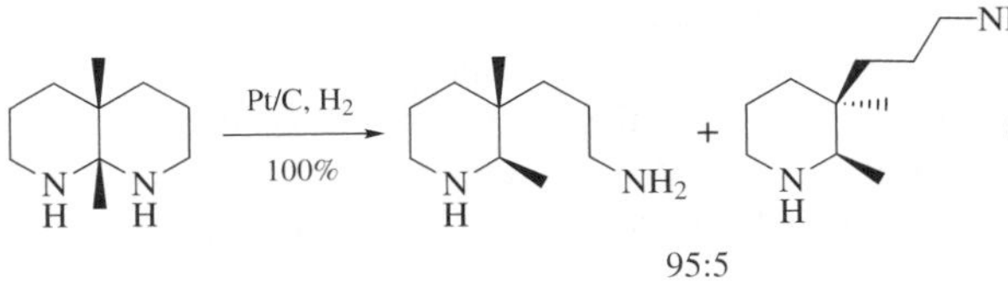

(6)

Pt catalysts are used for the reduction of ketones to minimize hydrogenolysis, especially in the case of aromatic ketones. The solvents used for ketone reduction are important. In the case of ***Ethyl Acetoacetate***, the Pt/C-catalyzed hydrogenation in water gave 12% hydrogenolyzed product, but no hydrogenolysis was observed in THF. The hydrogenolysis side reaction in water can be completely suppressed by the addition of a small amount of zinc acetate.[13] Under acidic conditions, dialkyl ketones have been hydrogenolyzed to the hydrocarbon products, but these conditions are impractical for acid-sensitive substrates. The conversion of ketones to the corresponding enol phosphates followed by hydrogenolysis over Pt/C is a mild and efficient route for the overall removal of an oxygen (eq 7).[14]

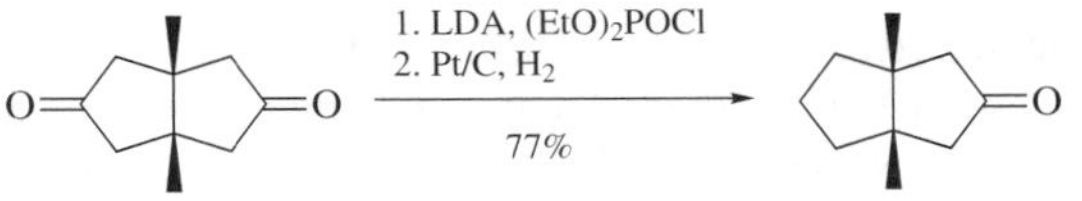

(7)

O-Acetyl sugar lactones have been cleanly converted to the corresponding 2,3-dideoxy sugar derivatives under mild conditions with Pt/C. A trialkylamine base was necessary for the hydrogenolysis to proceed. The mechanism may involve the alkenic intermediates (**1**) and (**2**) (eq 8). This procedure is only applicable to lactones since no deacetylated product was observed with aliphatic and cyclohexyl enol acetates (**3**) and (**4**).[15]

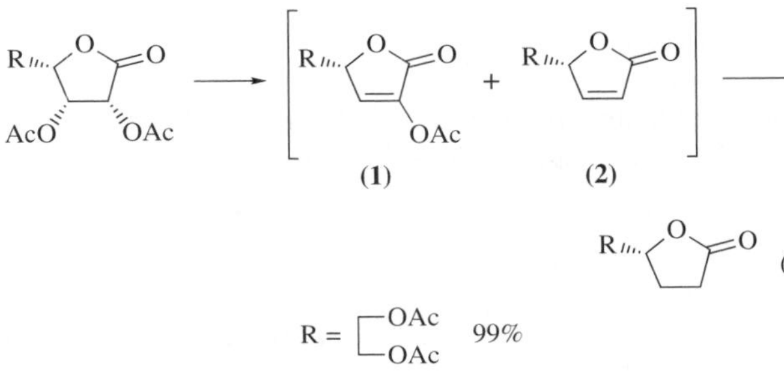

(8)

The Pt/C-catalyzed hydrogenation of 4,4-difluorocyclohexadienone provided β-fluorophenol in 90% yield. The dienone had been prepared from phenol via an oxidative fluorination method.[16]

Although Pt/C has been used for the hydrogenation of aromatic rings, PtO_2 is usually the catalyst of choice. High selectivity (>95% *cis*) was observed in the hydrogenation of pyrroles at atmospheric pressure and rt. Under these conditions, both the phenyl ring and the benzyl ether remained intact (eq 9).[17]

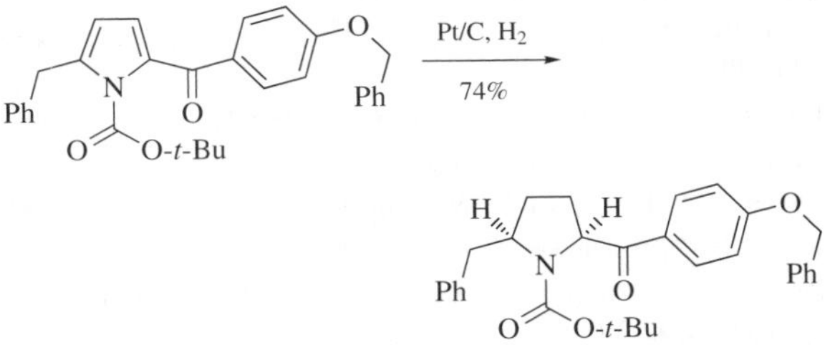

(9)

Pt/C is an effective catalyst for dehydrogenation of reducing sugars under basic conditions. For example, glucose was oxidized to the carboxylic acid in excellent yield (eq 10).[18]

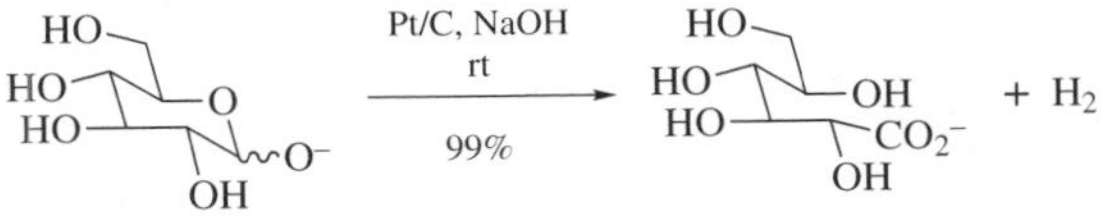

(10)

The mild reaction conditions under which reducing sugars are dehydrogenated with Pt/C also makes them good transfer hydrogenation agents. For example, fructose has been reduced to a mixture of glucitol and mannitol (0.77:1) in the presence of glucose and Pt/C.

The selective oxidation of benzylic alcohols to aldehydes by oxygen and Pt/C in the presence of ***Cerium(III) Chloride*** and $Bi_2(SO_4)_3$ cocatalysts has been reported. The selectivity to the aldehyde was excellent. Without the cocatalysts, carboxylic acids were obtained (eq 11).[19]

HO, OH; Pt/C, O_2, 50 °C, NaOH; $Bi_2(SO_4)_3$, $CeCl_3$ → HO, CHO (88%) + HO, CO_2H (1%) (11)

Related Reagents. See Classes R-2, R-3, R-4, R-5, R-8, R-12, R-19, R-20, R-21, and R-29, pages 1–10.

1. Palmer, C. J.; Casida, J. E. *TL* **1990**, *31*, 2857.
2. Melillo, D. G.; Cvetovich, R. J.; Ryan, K. M.; Sletzinger, M. *JOC* **1986**, *51*, 1498.
3. Nametkin, N. S.; Vdovin, V. M.; Finkelshtein, E. S.; Popov, A. M.; Egorov, A. V. *Izv. Akad. Nauk. SSSR, Ser. Khim.* **1973**, 2806 (*CA* **1974**, *80*, 82 203g).
4. Sunder, S.; Peet, N. P. *JHC* **1979**, *16*, 33.
5. McCord, T. J.; Smith, S. C; Tabb, D. L.; Davis, A. L. *JHC* **1981**, *18*, 1035.
6. Rylander, P. N.; Kaplan, J. G. U.S. Patent 3 117 162, 1964 (*CA* **1964**, *60*, 9147h).
7. (a) Greenfield, H. *Ind. Eng. Chem., Prod. Res. Dev.* **1967**, *6*, 142. (b) Greenfield, H. *Ind. Eng. Chem., Prod. Res. Dev.* **1976**, *15*, 156.
8. Rylander, P. N., Hasbrouck, L.; Karpenko, I. *ANY* **1973**, *214*, 100.
9. (a) Freifelder, M.; Martin, W. B.; Stone, G. R.; Coffin, E. L. *JOC* **1961**, *26*, 383. (b) Baltzly, R. *JACS* **1952**, *74*, 4586.
10. (a) Freifelder, M. *JOC* **1966**, *31*, 3875. Freifelder, M.; Ng, Y. H.; Helgren, P F. *JMC* **1964**, *7*, 381. (b) Liepa, A. J.; Summons, R. E. *CC* **1977**, 826.
11. Boullanger, P.; Martin, J.-C.; Descotes, G. *BSF* **1973**, 2149.
12. Zondler, H.; Pfleiderer, W. *HCA* **1975**, *58*, 2247.
13. Rylander, P. N., Starrick, S. *Engelhard Ind., Tech. Bull.* **1966**, *7*, 106 (*CA* **1967**, *67*, 90 361d).
14. Coates, R. M.; Shah, S. K.; Mason, R. W. *JACS* **1979**, *101*, 6765.
15. Katsuki, J.; Inanaga, J. *TL* **1991**, *32*, 4963.
16. Meurs, J. H. H.; Stopher, D. W.; Eilenberg, W. *AG(E)* **1989**, *28*, 927.
17. Kaiser, H.-P.; Muchowski, J. M. *JOC* **1984**, *49*, 4203.
18. de Wit, G.; de Vlieger, J. J.; Dalen, A. C. K.; Kieboom, A. P. G.; van Bekkum, H. *TL* **1978**, *15*, 1327.
19. Oi, R.; Takenaka, S. *CL* **1988**, *7*, 1115.

Anthony O. King & Ichiro Shinkai
Merck & Co., Rahway, NJ, USA

Platinum(IV) Oxide

[1314-15-4] O_2Pt (MW 227.08)

(catalyst for hydrogenation of various functional groups with minimal hydrogenolysis problems; used as dehydrogenation catalyst; used for selective oxidation of primary alcohols)

Alternate Name: Adams' catalyst.
Solubility: insol most organic solvents.
Form Supplied in: dark brown solid, usually hydrated with 2 or more mol of H_2O.
Analysis of Reagent Purity: elemental analysis.
Handling, Storage, and Precautions: no precaution is necessary with the oxide, but after exposure to H_2 the Pt black must be handled as a pyrophoric reagent; the general precautions for handling hydrogenation catalysts should be followed; during filtration the filter cake must not be allowed to go dry; if a filter aid is necessary, a cellulose-based material should be used if catalyst recovery is desired.

Hydrogenations and Hydrogenolysis. Platinum oxide has been used to catalyze the hydrogenation and hydrogenolysis of many functional groups. It has also found use in dehydrogenation and oxidation reactions. PtO_2 is not an active catalyst, but in the presence of H_2 the oxide is reduced to Pt black which is the active form. Pt black can also be made via other procedures.[1] A more recent method for the preparation of Pt black via the reduction of chloroplatinic acid or PtO_2 with ***Sodium Borohydride*** gives a catalyst with slightly increased activity.[2] Unlike the reaction between $NaBH_4$ and ***Nickel(II) Acetate***, which generates ***Nickel Boride*** (Ni_2B),[3] the reaction with Pt gives only pure Pt black. This catalyst, which is generally prepared in situ and used without isolation, was shown to be slightly more active than the Pt black generated from PtO_2 and H_2 in the reduction of 1-octene.

The reduction of alkenes, catalyzed by PtO_2, has been carried out in acidic, neutral, and basic conditions and, in some cases, different results are observed. In the example shown in eqs 1 and 2, stereoselective hydrogenation of a double bond to give the *cis*- or the *trans*-fused product was effected with PtO_2 in either acidic or neutral media.[4] With certain alkenic compounds, PtO_2 acted as an isomerization catalyst and no reduction occurred (eq 3).[5] Isomerization followed by hydrogenation to give a *cis–trans* mixture of dimethylcyclopentanes has also been observed (eq 4).[6] The normally expected product is the *cis* isomer.

OMe, O, N, Me; H_2, PtO_2, AcOH, 72% → OMe, O, H, N, Me (1)

PtO_2 is generally not a satisfactory catalyst for the reduction of alkynes to alkenes. Even when the reaction is stopped after 1 equiv of H_2 is absorbed, a substantial amount of alkane has already been formed,[7] but a number of exceptions have been reported.[8] Lindlar catalyst (***Palladium on Calcium Carbonate (Lead Poisoned)***), ***Palladium on Barium Sulfate***, nickel boride (P-2Ni), and ***Palladium on Poly(ethylenimine)*** are better catalysts for this task.

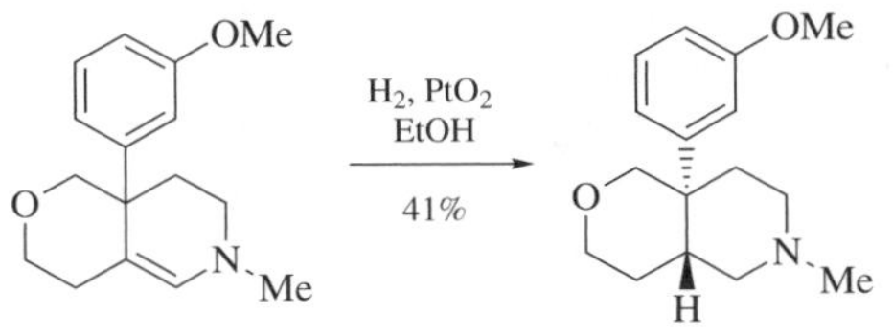

(2)

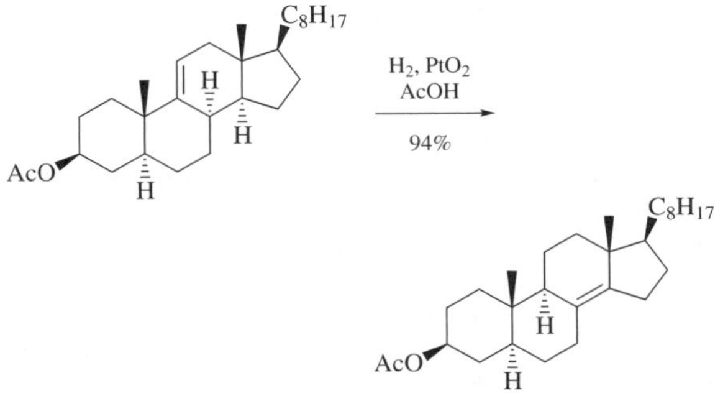

(3)

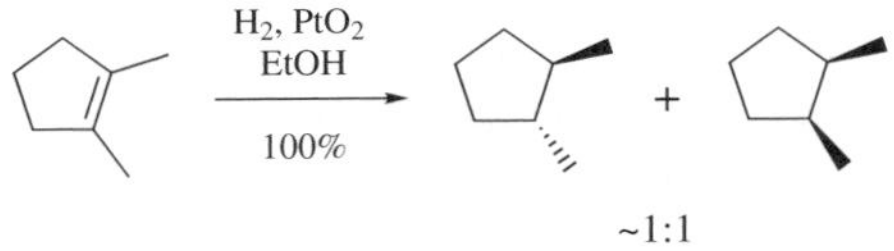

(4)

Amines can be obtained by reduction of several functional groups. Nitro compounds are readily hydrogenated to amines under PtO_2 catalysis. The reduction of a benzylated nitro compound gave the aniline without loss of the benzyl ether functions (eq 5). Use of Pd/C gave the hydrogenolyzed product (eq 6).[9] In general, Pt catalysts are used in place of Pd catalysts when hydrogenolysis is to be minimized.

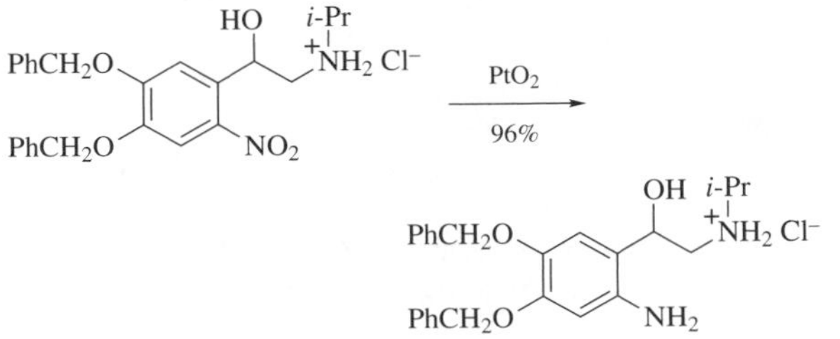

(5)

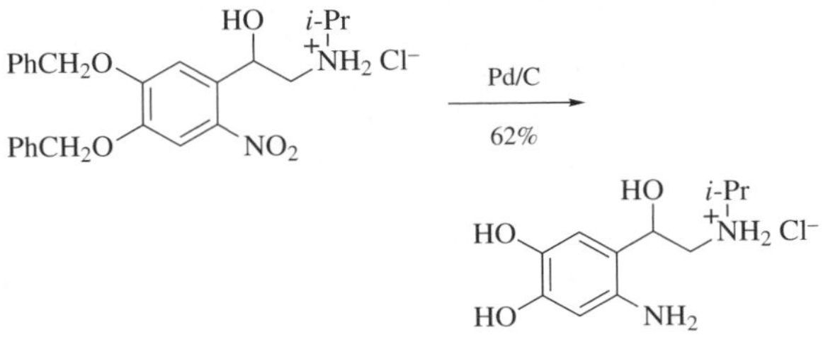

(6)

Hydrogenation of nitrobenzene in the presence of ***Hydrogen Fluoride*** gave *p*-fluoroaniline.[10] Hydrogenation of oximes is best carried out in acidic media to minimize secondary amine formation.[11] A quantitative yield of the amine was obtained when 3-hydroxy-5-hydroxymethyl-2-methylpyridine-4-carbaldoxime was reduced in alcoholic ***Hydrogen Chloride*** with this catalyst.[12] Amines can also be generated easily from azides,[13] imines,[14] and nitriles.[15] Similar to oximes, hydrogenation of nitriles should be conducted in acidic solvents to minimize di- or trialkylamine formation.

Hydrogenation of ketones in the presence of this catalyst can be carried out in acidic, neutral, or basic media. In general, with cyclic ketones axial alcohols predominate in acidic media while equatorial alcohols are favored in neutral or basic media. Thus the stereochemical outcome of ketone reduction may be controlled by varying the pH of the solvent.[16] In strong acids the reduction of dialkyl ketones in the presence of alcoholic solvents at rt with PtO_2 leads to the formation of ethers instead of alcohols.[17] It was suggested that acetals and enol ethers are generated under the reaction conditions before hydrogenation or hydrogenolysis. Although PtO_2 is often used to minimize hydrogenolysis of functional groups such as alcohols and halides, the hydrogenolysis of ketones under acidic conditions can give methylenic products (eq 7).[18]

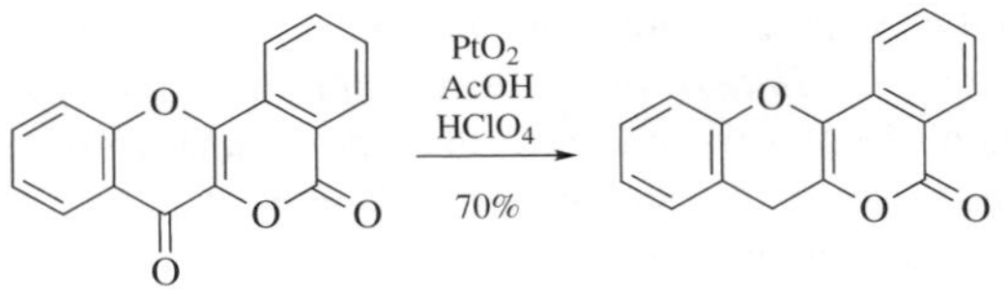

(7)

By converting enolizable ketones to the enol triflates, neutral conditions can be used to effect the overall hydrogenolysis of ketones (eq 8).[19]

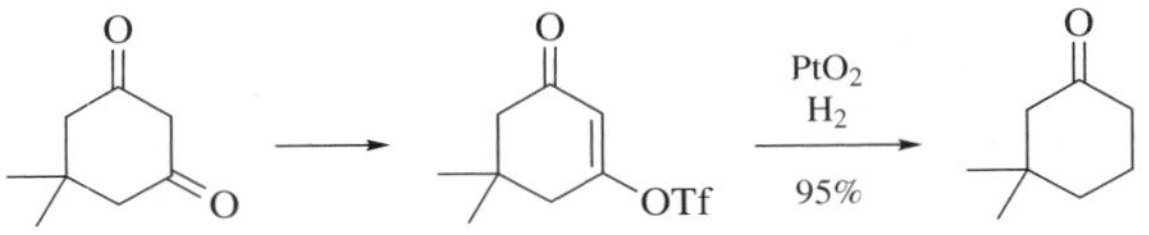

(8)

Certain α-diketones can be selectively hydrogenolyzed to monoketones by conversion to cyclic unsaturated oxyphosphoranes followed by treatment with hydrogen in the presence of the catalyst (eq 9).[20] The benzylic position, which is the more reactive site, was hydrogenolyzed selectively. Amide carbonyls are very difficult to reduce in general, but reduction has been observed in special cases (eq 10).[21] In one report, an anhydride was reduced to an ether at 5000 psi H_2 in HF.[22]

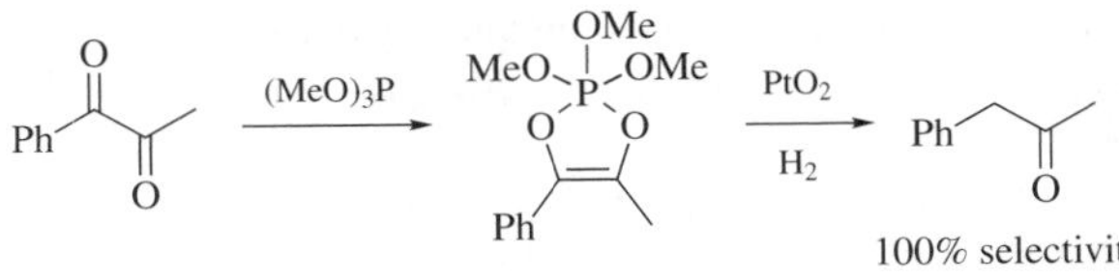

(9)

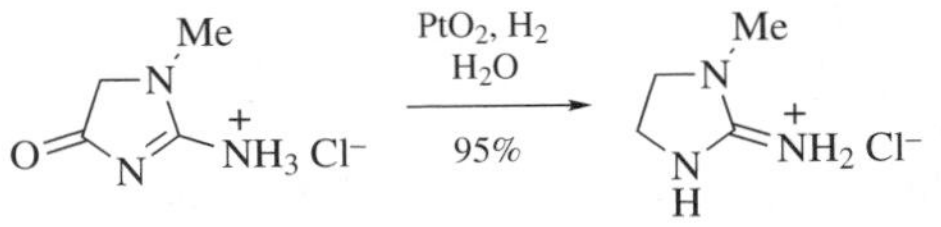

(10)

Benzyl groups attached to a heteroatom, such as oxygen and nitrogen, are removed using Pd catalysts, but the removal of phenyl groups of diphenyl phosphonates with these catalysts is ineffective. These phenyl groups can be hydrogenolyzed using PtO_2, but high catalyst loading is necessary.[23] *p,p'*-Dinitrobenzhydryl has been used for the protection of alcohols. This protecting group can be selectively removed in the presence of other hydrogenolyzable protecting groups, such as benzyl and trityl, by the use of platinum oxide followed by mild acid hydrolysis (eq 11).[24]

PtO_2, H_2 / MeOH, THF; pH 3–4; ROH (11) 81–90%

Hydrogenolysis of the N–N bond of hydrazines,[25] the N–O bond of nitrones,[26] and the C–C bond in cyclopropanes[27] are readily accomplished using this catalyst.

Aromatic ring reduction can be carried out with PtO_2 in acidic solvents such as acetic acid and alcoholic or aqueous HCl.[28] Complete ring reduction without the concomitant hydrogenolysis of benzylic alcohols has been accomplished.[29] Reduction of phenols provides cyclohexanols[30] and pyridines are reduced to piperidines.[31] ***Imidazole*** is difficult to reduce, but this has been accomplished in the presence of ***Acetic Anhydride*** to give the diamide.[32] Oxazoles are cleaved to amides (eq 12),[33] and isoxazoles give enamino ketones (eq 13).[34]

PtO_2, H_2 / AcOH / 100% (12)

PtO_2, H_2 / 90% (13)

Dehydrogenation. PtO_2 has also been used as a dehydrogenation catalyst. An example to obtain a pyridazine from a 1,4-diketone and hydrazine is shown below (eq 14).[35]

PtO_2 / H_2NNH_2 / 90% (14)

Oxidation. The Pt black generated from prereduction of PtO_2 with hydrogen, also catalyzes the oxidation of alcohols to carbonyl compounds with oxygen. Primary alcohols are preferentially oxidized to carboxylic acids in the presence of secondary alcohols. Amides, sulfonamides, and azides are stable under the oxidizing reaction conditions.[36] Benzyldimethylamine can be oxidized to *N*-benzyl-*N*-methylformamide in 85% yield using benzene as the solvent.[37] When water was used as the solvent, *N*-demethylation was observed instead.[38]

Pt black was also used in the electrolytic *N*-ethylation of benzylamine in ethanol to give *N*-ethylbenzylamine in 96% yield. The Pt black anode first oxidized the ethanol to acetaldehyde which in turn formed a Schiff base with the amine. H_2 was generated at the Pt coil cathode and was picked up by the suspended Pt black for the hydrogenation of the imine.[39]

Irradiation of primary amines in the presence of a mixture of Pt black and TiO_2 gave low yields of secondary amines (8–67%). For example, diethylamine was formed in 33% yield from ethylamine and 1,4-diaminobutane gave pyrrole in 67% yield.[40]

Related Reagents. See Classes R-3, R-7, R-8, R-9, R-19, R-20, R-21, R-23, R-24, and R-29, pages 1–10.

1. (a) Willstatter, R.; Waldschmidt-Leitz, E. *CB* **1921**, *54*, B113. (b) Baltzly, R. *JACS* **1952**, *74*, 4586. (c) Theilacker, W.; Drossler, H. G. *CB* **1954**, *87*, 1676.
2. Brown, H. C.; Brown, C. A. *JACS* **1962**, *84*, 1493.
3. Schlesinger, H. I.; Brown, H. C.; Finholt, A. E.; Gilbreath, J. R.; Hoekstra, H. R.; Hyde, E. K. *JACS* **1953**, *75*, 215.
4. (a) Evans, D. A.; Mitch, C. H.; Thomas, R. C.; Zimmerman, D. M.; Robey, R. L. *JACS* **1980**, *102*, 5955. (b) Bays, D. E.; Brown, D. S.; Belton, D. J.; Lloyd, J. E.; McElroy, A. B. *JCS(P1)* **1989**, 1177.
5. Bream, J. B.; Eaton, D. C.; Henbest, H. B. *JCS* **1957**, 1974.
6. Siegel, S.; Dmuchovsky, B. *JACS* **1964**, *86*, 2192.
7. Crombie, L. *JCS* **1955**, 3510.
8. (a) Chanley, J. D. *JACS* **1949**, *71*, 829 (b) Braude, E. A.: Coles, J. A. *JCS* **1951**, 2078.
9. Avery, M. A.; Verlander, M. S.; Goodman, M. *JOC* **1980**, *45*, 2750.
10. Fidler, D. A.; Logan, J. S.; Boudakian, M. M. *JOC* **1961**, *26*, 4014.
11. (a) Dornow, A.; Petsch, G. *AP* **1951**, *284*, 153. (b) Secrist III, J. A.; Logue, M. W. *JOC* **1972**, *37*, 335. (c) Johnson, D. R.; Szoteck, D. L.; Domagala, J. M.; Stickney, T. M.; Michel, A.; Kampf, J. W. *JHC* **1992**, *29*, 1481.
12. Kreisky, S. *M* **1958**, *89*, 685.
13. Lawton, B. T.; Szarek, W. A.; Jones, J. K. N. *CC* **1969**, 787.
14. Taylor, E. C.; Lenard, K. *CC* **1967**, 97.
15. (a) Freifelder, M.; Hasbrouck, R. B. *JACS* **1960**, *82*, 696. (b) Lee, T. B. K.; Wong, G. S. K. *JOC* **1991**, *56*, 872.
16. Sugahara, M.; Tsuchida, S.-I.; Anazawa, I.; Takagi, Y.; Teratani, S. *CL* **1974**, 1389.
17. Verzele, M.; Acke, M.; Anteunis, M. *JCS* **1963**, 5598.
18. Deschamps-Vallet, C.; Meyer-Dayan, M.; Andrieux, J.; Riboullean, J.; Bodo, B.; Molho, D. *JHC* **1977**, *14*, 489.
19. (a) Jigajinni, V. B.; Wightman, R. H. *TL* **1982**, *23*, 117. (b) Garcia Martinez, A.; Martinez Alvarez, R.; Modueno Casado, M.; Subramanian, L. R.; Hanock, M. *T* **1987**, *43*, 275.
20. Stephenson, L. M.; Falk, L. C. *JOC* **1976**, *41*, 2928.
21. Wegner, M. M.; Rapoport, H. *JOC* **1978**, *43*, 3840.
22. Feiring, A. E. *JOC* **1977**, *42*, 3255.
23. (a) Hes, J.; Mertes, M. P. *JOC* **1974**, *39*, 3767. (b) Kiso, M.; Tanahashi, M.; Hasegawa, A. *CR* **1987**, *163*, 279. (c) Perich, J. W.; Johns, R. B. *JOC* **1988**, *53*, 4103.
24. Just, G.; Wang, Z. Y.; Chan, L. *JOC* **1988**, *53*, 1030.
25. Gennari, C.; Colombo, L.; Bertolini, G. *JACS* **1986**, *108*, 6394.
26. (a) de Bernardo, S.; Weigele, M. *JOC* **1977**, *42*, 109. (b) Barco, A.; Benetti, S.; Pollini, G. P.; Baraldi, P. G.; Guarneri, M.; Vicentini, C. B. *JOC* **1979**, *44*, 105. (c) Czarnocki, Z. *JCS(C)* **1992**, 402.
27. (a) Russell, R. A.; Harrison, P. A.; Warrener, R. N. *AJC* **1984**, *37*, 1035. (b) Piers, E.; Marais, P. C. *CC* **1989**, *17*, 1222.

28. (a) Linstead, R. P.; Whetstone, R. R.; Levine, P. *JACS* **1942**, *64*, 2014. (b) Zaugg, H. E.; Michaels, R. J.; Glenn, H. J.; Swett, L. R.; Freifelder, M.; Stone, G. R.; Weston, A. W. *JACS* **1955**, *80*, 2263. (c) Baltzly, R.; Mehta, N. B.; Russell, P. B.; Brooks, R. E.; Grivsky, E. M.; Steinberg, A. M. *JOC* **1961**, *26*, 3669.
29. Ichinohe, Y.; Ito, H. *BCJ* **1964**, *37*, 887.
30. Epstein, W. W.; Grua, J. R.; Gregonis, D. *JOC* **1982**, *47*, 1128.
31. Prelog, V.; Metzler, O. *HCA* **1946**, *20*, 1170.
32. Bauer, H. *JOC* **1961**, *26*, 1649.
33. Kozikowski, A. P.; Ames, A. *JOC* **1980**, *45*, 2548.
34. Barco, A.; Benetti, S.; Pollini, G. P.; Veronesi, B.; Baralding, P. G.; Guarneri, M.; Vicentini, C. B. *SC* **1978**, *8*, 219.
35. Nicolaou, K. C.; Barnette, W. E.; Magolda, R. L. *JACS* **1979**, *101*, 766.
36. (a) Post, G. G.; Anderson, L. *JACS* **1962**, *84*, 471. (b) Tsou, K. C.; Santora, N. J.; Miller, E. E. *JMC* **1969**, *12*, 173. (c) Marino, J. P.; Fernandez de la Pradilla, R.; Laborde, E. *JOC* **1987**, *52*, 4898.
37. Davis, G. T.; Rosenblatt, D. H. *TL* **1968**, 4085.
38. Birkenmeyer, R. D.; Dolak, L. A. *TL* **1970**, *58*, 5049.
39. Ohtani, B.; Nakagawa, K.; Nishimoto, S.-I.; Kagiya, T. *CL* **1986**, 1917.
40. Nishimoto, S.-I.; Ohtani, B.; Yoshikawa, T.; Kagiya, T. *JACS* **1983**, *105*, 7180.

Anthony O. King & Ichiro Shinkai
Merck & Co., Rahway, NJ, USA

Potassium

[7440-09-7] K (MW 39.10)

(reducing agent for aromatic rings,[1] carbonyl compounds,[2] and various functional groups,[3] including alcohols[4] in the presence of crown ethers; reagent for the preparation of Rieke's metals;[5] reacts with proton donors;[6] promotes $S_{RN}1$ arylation reactions;[7] modified reactivity occurs with deposition on alumina,[8] insertion in graphite,[9] or dispersion by ultrasound[10])

Physical Data: mp 64 °C; bp 760 °C; *d* 0.86 g cm^{-3}; soft silvery (fresh cut) metal.

Solubility: sol liquid ammonia, 1,2-ethylenediamine, aniline; slightly sol ethers.

Form Supplied in: widely available as ingots, sticks, under mineral oil, or in ampules; common impurities are other alkali metals, aluminum, calcium, boron, silicon.

Handling, Storage, and Precautions: reacts violently with water, oxygen, halogens. In air, potassium forms the yellow peroxide KO_2, which is dangerous in contact with organic compounds. Reacts with many solvents; inert in saturated and some aromatic hydrocarbons. For safe handling, see Fieser and Fieser.[6] Potassium fires should be treated with inert powders or sand.

Reduction of Aromatic Compounds. [1] Birch- and Benkeser-type reductions of aromatic rings are achieved by potassium dissolved in liquid ***Ammonia*** or amines. However, potassium is used less often than ***Lithium*** or ***Sodium***. The reducing ability of potassium is illustrated by the total reduction of the B-ring of equilenin to estrone, which cannot be achieved with sodium (eq 1).[11]

1. K, NH_3, −33 °C; 2. H_3O^+; 50% (1)

Proton donors are frequently used in potassium reductions. The accepted mechanism consists of a sequence of single-electron transfer (SET), a protonation, and a second SET. The anion thus formed can be trapped by an electrophile to give the reduction–alkylation product (eq 2).[12]

1. K, *t*-$BuNH_2$, NH_3; 2. MeI; 84% (2)

Reduction of Saturated and Unsaturated Carbonyl Groups. Alkali metal reduction of ketones to alcohols in the presence or absence of proton donors has been studied in detail.[2] The use of potassium decreases the amount of pinacol coupling products,[13] and increases the reaction rate and yields of desired product.[14] The stereoselectivity, which depends on the structure of the intermediate ketyl radical anion, is frequently lower with K in comparison to lighter alkali metals. Recently, reductions with K have been conducted in THF under sonication with stereoselectivity similar to that of the usual method (eq 3).[15]

K, NH_3 or K, THF,)))) ; 40–42% + 58–60% (3)

When an unsaturated functional group is closely positioned with respect to the carbonyl group, the ketyl radical anion adds to the multiple bond producing products of reductive cyclization (eq 4).[16]

K, NH_3, THF, $(NH_4)_2SO_4$; 60–70% (4)

In the case of α,β-unsaturated carbonyl groups,[2a,17] the radical anion is readily transformed to the dianion, especially with K. When protonation occurs, an enolate results which can be trapped with an electrophile, giving rise to alkylation on the α-carbon. The stereoselectivity at the β-position does not seem to be strongly dependent on the nature of the metal, although exceptions are known (eq 5).[18]

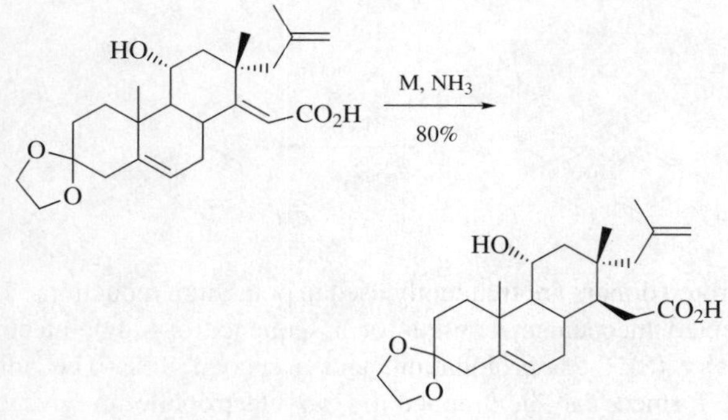

M, NH_3, 80% (5)

(with Li: 30% + 40% epimer)

In the absence of a proton source, the dianion can be alkylated on the β-carbon. The increased yields obtained in some cases with potassium result from the higher reactivity of the potassium dianion, as compared with reactions with lithium or sodium (eq 6).[17,19]

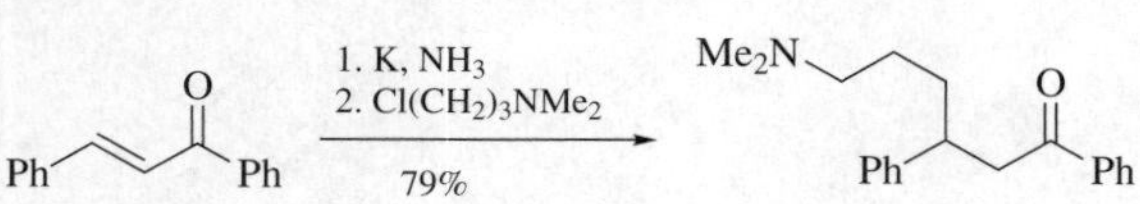

1. K, NH_3; 2. $Cl(CH_2)_3NMe_2$; 79% (6)

Reduction of Functional Groups. Various functional groups are reduced by potassium in liquid ammonia (eqs 7–9).[3,20,21] In many cases the reaction is accomplished in the presence of crown ethers, which generate K^+K^- ion pairs.

$$C_8H_{17}CH{=}CHC_8H_{16}{-}NC \xrightarrow[96\%]{\text{K, 18-crown-6, toluene, rt}} C_8H_{17}CH{=}CHC_8H_{17} \quad (7)$$

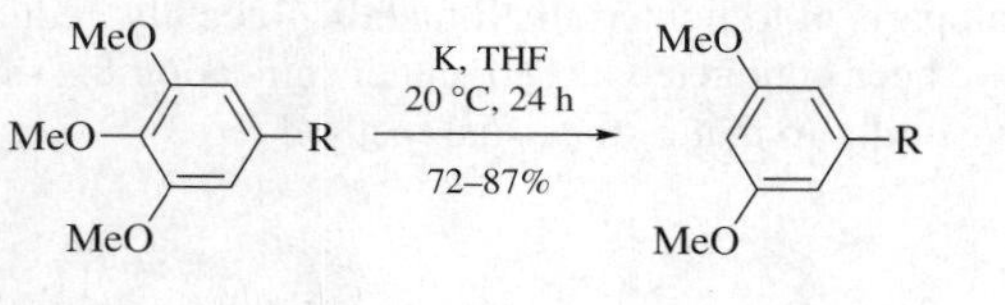

K, THF, 20 °C, 24 h; 72–87% (8)

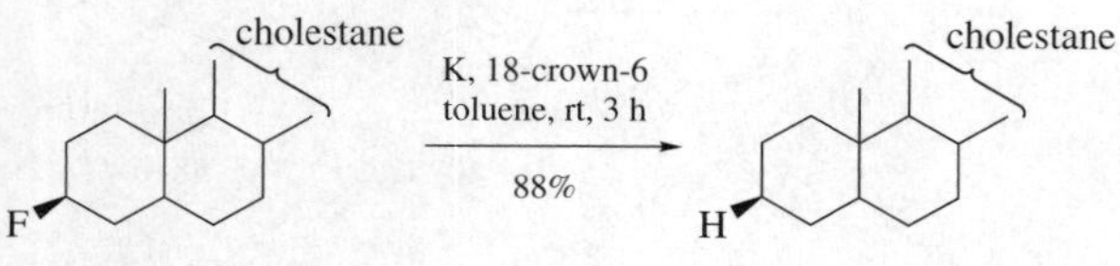

K, 18-crown-6, toluene, rt, 3 h; 88% (9)

Alcohol deoxygenation occurs[4] via carboxylic, thiocarbamic (eq 10),[22] and phosphoric esters (eq 11).[23] Free benzylic alcohols can be reduced with potassium in the presence of 4,4′-di-*t*-butylbiphenyl in THF under sonication.[24]

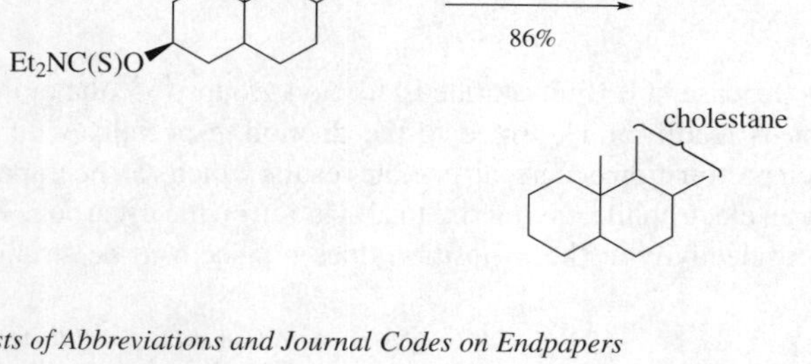

K, 18-crown-6, *t*-$BuNH_2$; 86% (10)

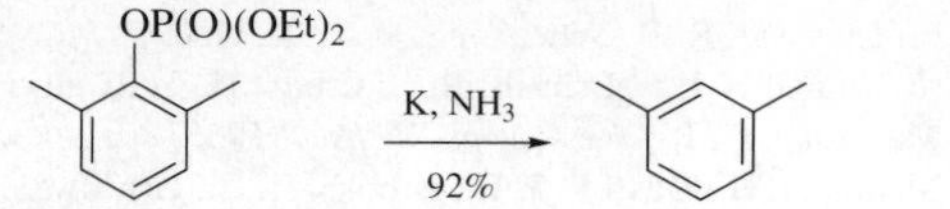

K, NH_3; 92% (11)

Reduction of Metal Salts (Rieke's Metals). Potassium reduces metal salts to finely dispersed metals in refluxing THF or DME (eq 12). Activated metal slurries can be prepared, which exhibit a high reactivity with organic compounds.[5] The method was improved by replacing potassium with lithium in the presence of an electron transport agent.[25]

$$MCl_n + K \xrightarrow[\text{reflux}]{\text{THF or diglyme}} M^* \quad (12)$$

M = Cu, Mg, Zn, Al, In, U

Reaction with Proton Donors. Potassium reacts with acidic hydrogen compounds such as alcohols. This reaction is used to prepare the useful potassium alkoxides (e.g. ***Potassium t-Butoxide***).[6] Some potassium amides, such as 3-aminopropanamide, are readily prepared in quantities up to 25 mmol from the metal and 1,3-propanediamine in the presence of a trace of iron(III) nitrate under sonication (eq 13).[26]

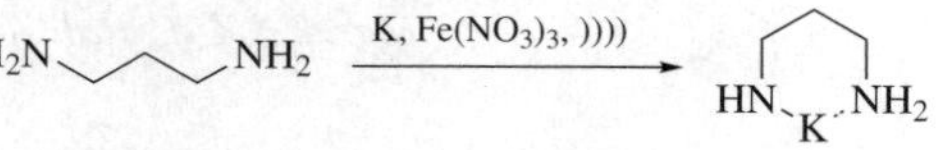

K, $Fe(NO_3)_3$,)))) (13)

Alkylation and $S_{RN}1$ Reactions. Deprotonation–alkylation of γ-lactones has been reported. In the case of acylation, it occurs only on the carbon atom (eq 14).[27]

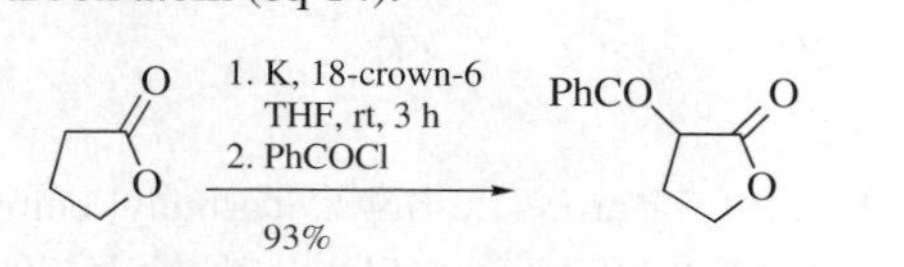

1. K, 18-crown-6, THF, rt, 3 h; 2. PhCOCl; 93% (14)

Substitution of aromatic halides, phosphate esters, trialkylammonium groups, and sulfides by an anionic nucleophile occurs with complete regioselectivity in the presence of potassium in liquid ammonia or DMSO (eq 15).[7] These reactions are now more generally effected with a potassium alkoxide under photochemical activation.[28]

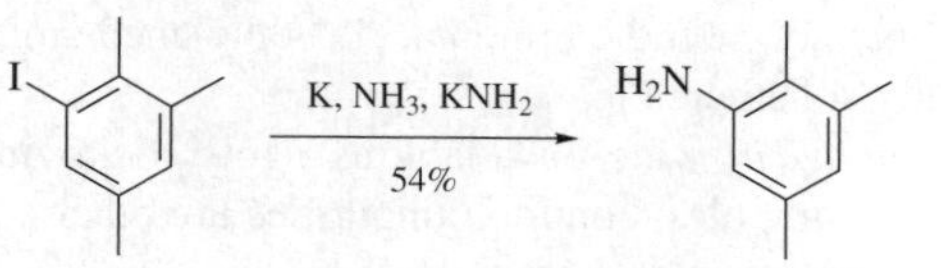

K, NH_3, KNH_2; 54% (15)

Supported Potassium on Alumina (K/Al_2O_3) or Graphite (C_8K). ***Potassium on Alumina*** is prepared by melting the metal in the presence of Al_2O_3. ***Potassium–Graphite*** is obtained by vaporizing the metal on graphite. Both reagents are sensitive to oxygen and moisture. Various reductive processes take place with alkenes, nitriles, or halides (eqs 16 and 17).[8,29]

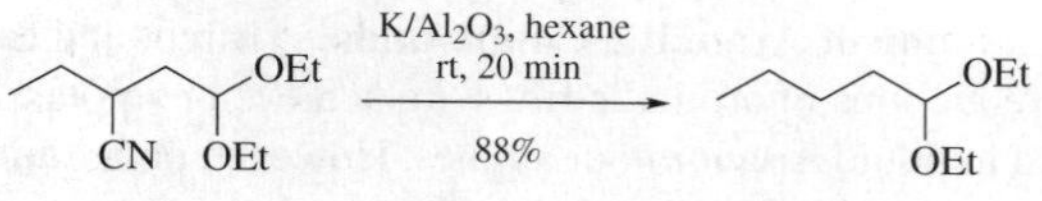

K/Al_2O_3, hexane, rt, 20 min; 88% (16)

C_8K, THF, 0 °C, 95% (17)

Ultrasonically Dispersed Potassium (UDP). Sonication of potassium metal in aromatic solvents (toluene, xylene) produces a silvery blue suspension. The reaction of this reagent with water is much milder than that of potassium metal. This suspension can behave as a base or as a single-electron transfer agent. The Dieckman cyclization of diethyl adipate (eq 18)[10] and the dimerization of succinates (eq 19)[30] occur readily at rt in excellent yield.

UDP, toluene, rt, 5 min, 83% (18)

UDP, toluene, rt, 20 min, 88% (19)

The carbon–sulfur bond in sulfolene is cleaved to a radical anion, which can be alkylated to provide open chain sulfones. The inconvenience of long irradiation times was solved by running the reaction in the presence of proton sources, water, or phenol (eq 20).[31]

1. UDP, toluene; 2. PhOH, THF; 3. MeI; 86% (20)

Related Reagents. See Classes R-2, R-7, R-15, R-25, R-26, R-27, R-28, and R-31, pages 1–10. Sodium–Potassium Alloy; Titanium(III) Chloride–Potassium.

---

1. (a) Mander, L. N. *COS* **1991**, *8*, 489. (b) House, H. O. *Modern Synthetic Reactions*; Benjamin: Menlo Park, CA, 1972, p 145.
2. (a) Pradhan, S. K. *T* **1986**, *42*, 6351. (b) Rautenstrauch, V. *CC* **1986**, 1558.
3. Ohsawa, T.; Mitsuda, N.; Nezu, J.; Oishi, T. *TL* **1989**, *30*, 845.
4. Hartwig, W. *T* **1983**, *39*, 2609.
5. Rieke, R. D. *ACR* **1977**, *10*, 301.
6. Fieser, L. F.; Fieser, M. *FF* **1967**, *1*, 905, 907, 911.
7. (a) Bunnett, J. F. *ACR* **1978**, *11*, 413. (b) Giese, B. *Radicals in Organic Synthesis: Formation of Carbon–Carbon Bonds*; Pergamon: Oxford, 1986; p 247. (c) Denney, D. B.; Denney, D. Z. *T* **1991**, *47*, 6577.
8. Savoia, D.; Tagliavini, E.; Trombini, C.; Umani-Ronchi, A. *JOC* **1980**, *45*, 3227.
9. Fürstner, A. *AG(E)* **1993**, *32*, 164.
10. Luche, J. L.; Petrier, C.; Dupuy, C. *TL* **1984**, *25*, 753.
11. Marshall, D. J.; Deghenghi, R. *CJC* **1969**, *47*, 3127.
12. Hook, J. M.; Mander, L. N.; Urech, R. *JOC* **1984**, *49*, 3250.
13. Rautenstrauch V.; Willhalm, B.; Thommen, W.; Burger, U. *HCA* **1981**, *64*, 2109.
14. Murphy, W. S.; Sullivan, D. F. *JCS(P1)* **1972**, 999.
15. Huffman, J. W.; Wallace, R. H. *JACS* **1989**, *111*, 8691.
16. Stork, G.; Boeckman, R. K.; Taber, D. F.; Still, W. C.; Singh, J. *JACS* **1979**, *101*, 7107.
17. Caine, D. *OR* **1976**, *23*, 1.
18. Arth, G. E.; Poos, G. I.; Lukes, R. M.; Robinson, F. M.; Johns, W. F.; Feurer, M.; Sarett, L. H. *JACS* **1954**, *76*, 1715.
19. Gautier, J. A.; Miocque, M.; Duclos, J. P. *BSF(2)* **1969**, 4348.
20. Azzena, U.; Denurra, T.; Fenude, E.; Melloni, G.; Rassu, G. *S* **1989**, 28.
21. Ozawa, T.; Takagaki, T.; Haneda, A.; Oishi, T. *TL* **1981**, *22*, 2583.
22. Barrett, A. G. M.; Prokopiou, P. A.; Barton, D. H. R. *JCS(P1)* **1981**, 1510.
23. Rossi, R. A.; Bunnett, J. F. *JOC* **1973**, *38*, 2314.
24. Karaman, R.; Kohlman, D. T.; Fry, J. L. *TL* **1990**, *31*, 6155.
25. Rieke, R. D.; Li, P. T. J.; Burns, T. P.; Uhm, S. T. *JOC* **1981**, *46*, 4323.
26. Kimmel, T.; Becker, D. *JOC* **1984**, *49*, 2494.
27. Jedlinski, Z.; Kowalczuk, M.; Kurcok, P.; Grzegorzek, M.; Ermel. J. *JOC* **1987**, *52*, 4601.
28. Beugelmans, R. *BSB* **1984**, *93*, 547.
29. Fürstner, A. *TL* **1990**, *31*, 3735.
30. Vorob'eva, S. L.; Korotkova, N. N. *JCR(S)* **1993**, 34.
31. Chou, T.; Chang, S. *JOC* **1992**, *57*, 5015.

Jean-Louis Luche
Université Paul Sabatier de Toulouse, France

Potassium–Graphite Laminate

[12081-88-8] C_8K (MW 135.18)

(powerful reducing agent for many functional groups;[1] metalating agent;[1] Lewis base;[1] hydrogenation catalyst;[1,2] catalyst for double bond isomerization and for anionic polymerization;[2] tool for preparing activated metals[1])

Physical Data: paramagnetic bronze-colored powder; *d* ~0.73 g cm^{-3}; interlayer distance 5.34 Å; all carbon layers are separated by one layer of potassium; space group $D^6{}_2$ *C*222.
Solubility: may be suspended in hydrocarbon and ethereal solvents; reacts violently with water, alcohols, and ammonia; THF, DME, 1,4-dioxane, and many arenes (e.g. benzene, toluene, pyridine, furan) are able to penetrate into the interlayer space, causing considerable swelling.
Preparative Methods: by stirring potassium and graphite powder at ≥150 °C under argon without any solvent for 5–10 min. Natural as well as synthetic graphite are suited for its preparation.
Handling, Storage, and Precautions: pyrophoric and must be handled under argon in thoroughly dried solvents. Small quantities should be weighed in a glove box; it can be stored under argon without any significant loss in activity for extended periods of time.

Introduction. Potassium–graphite laminate may be regarded as a polymeric array of naphthalenide anions and effects almost any reaction that the latter would promote. A major advantage in using C_8K as a substitute for ***Lithium Naphthalenide***, for example, lies in its increased reactivity and the easy work-up, consisting of filtration of the graphite only instead of the sometimes tedious removal of naphthalene.[1] Moreover, C_8K-promoted reactions can be readily monitored by the characteristic color change from bronze of the reagent to black of the graphite host.[1]

Functional Group Reductions. Alkyl chlorides and aryl halides are generally reduced to the corresponding hydrocarbons, whereas alkyl iodides afford Wurtz-type coupled products. Alkyl bromides show a reactivity pattern between alkyl chlorides and alkyl iodides. Single-electron transfer is an important pathway in these reductions, since characteristic radical rearrangements are observed with hex-5-enyl halides as mechanistic probes (eq 1).[3] *vic*-Dihalides form alkenes in high yield (eq 2).[3,4] C_8K leads to reductive cleavage of the C–O bond of aryl ethers, whereas in sulfonate esters the S–O rather than the C–O bond is broken selectively, with formation of the parent alcohols (eq 3).[3] Aryl ether and C–Cl cleavage can be done simultaneously, thus allowing the ready destruction of toxic polyhalodibenzodioxines or -dibenzofurans at rt.[5] Vinyl sulfones and allyl sulfones afford the corresponding alkenes (the latter after initial isomerization of their double bond) (eq 4).[6] The (*E*):(*Z*) ratio in these desulfuration reactions is low, with the (*E*)-isomer being slightly favored.

Br → C_8K, Et_2O → 65% + 35% (1)

2.5 equiv C_8K, benzene, rt, >90% (2)

OTs → 12.6 equiv C_8K, THF, 95% → OH (3)

$PhSO_2$ → 12.5 equiv C_8K, Et_2O, rt, 1 h, 65%
$PhSO_2$ → 12.5 equiv C_8K, Et_2O, rt, 1 h, 77%
(*E*):(*Z*) = 65:35 (4)

With ***Hexamethyldisilazane*** as cosolvent, C_8K in THF selectively reduces the double bond of enones at rt without affecting nonconjugated alkenes (eq 5).[7] α,β-Unsaturated acids are similarly reduced at 55 °C, while enoates give dimerization products. Imines afford amines in high yields with this reagent (eq 6),[7] whereas ketones may afford mixtures of alcohols and pinacols.[8] A large excess of the reagent is required to effect Birch-type reductions of arenes, as exemplified by a series of substituted naphthalene derivatives.[9] On treatment with C_8K in THF, benzil undergoes a unique coupling of the phenyl rings with formation of 9,10-phenanthrenequinone (eq 7).[10] This arene coupling can be extended to 2,3-diphenylquinoxaline and related heterocycles, which afford dibenzo[*a,c*]phenazine derivatives.[11]

5 equiv C_8K, THF, $(Me_3Si)_2NH$ = 5:1, 25 °C, 10 min, 77% (5)

5 equiv C_8K, THF, rt, 40 min, 90% (6)

5 equiv C_8K, THF, rt, 5 h, 70% (7)

Although C_8K was considered for a long time to be of limited value in organic synthesis due to its high reactivity,[3] it turned out to be the reagent of choice for the synthesis of furanoid glycals in terms of yield, reaction rate, and flexibility. The intermediate potassium alcoholates, formed upon treatment of 2,3-*O*-alkylidene glycosyl halides with C_8K in THF at low temperature, can be protected in situ with a wide variety of electrophiles (eq 8).[12] Aryl thioglycosides can also be reduced to the corresponding glycals by means of C_8K.[12b] Comparable reaction sequences consisting of C_8K-induced fragmentations followed by trapping of the intermediate alcoholates have also been carried out successfully with a series of carbohydrate-derived primary halides (eq 9).[13]

$MeOCH_2O$ … Cl → 2 equiv C_8K, THF, 0 °C, 5 min → [$MeOCH_2O$ … O^-K^+] → BnBr, 60 min, 86% → $MeOCH_2O$ … BnO (8)

I → 1. 2 equiv C_8K, THF, 0 °C, 10 min; 2. $Ph_2MeSiCl$, 88% → Ph_2MeSiO (9)

Chlorosilanes are rapidly converted by C_8K to disilanes in high to quantitative yields in THF at ambient temperature.[14] Phenyl-substituted disilanes may be further reduced to the corresponding silyl potassium reagents, which can be transmetalated with various transition metal halides (***Copper(I) Iodide***, ***Copper(I) Cyanide***,

MnI_2, VCl_3) to highly selective nucleophilic silylating agents. They react cleanly with enoates, enones, and acid chlorides, the last affording acyl silanes in a one-pot procedure (eq 10).[14a]

$$2\ Ph_2MeSiCl \xrightarrow[\text{THF, 5 min; 94\%}]{\text{2 equiv } C_8K} Ph_2MeSiSiMePh_2 \xrightarrow[\text{THF, 5 min}]{\text{2 equiv } C_8K}$$

$$2\ Ph_2MeSiK \xrightarrow[\text{2. PhCOCl, THF, -10 °C; 74\%}]{\text{1. } MnI_2} PhC(O)SiMePh_2 \quad (10)$$

C_8K as Polymeric Lewis Base. The reactivity of C_8K towards Brønsted acids is substrate size dependent.[3] C_8K has been used to achieve selective monoalkylation of alkyl nitriles and phenylacetic acid esters (substrate:C_8K:alkyl halide = 1:2:2) at −60 °C in 40–70% yield, with only small amounts of dialkylated products (0–7%) interfering.[15] Imines and dihydro-1,3-oxazine derivatives can likewise be deprotonated by this reagent at rt, followed by exclusive *C*-alkylation of the azaallyl anion (eq 11).[16] Some examples of selective dehydrohalogenations in the carbohydrate series due to the basisity of C_8K have also been reported (eq 12).[17]

C8K, THF; 1. BnCl, 2. $NaBH_4$, 3. H^+, H_2O (11)

C8K, THF, 0 °C, 10 min, 94% (12)

Organometallic Chemistry. C_8K is an effective reducing agent for transition metal compounds.[1] Particularly relevant to organic synthesis is the ready reduction of ***Hexacarbonylchromium*** to $K_2Cr(CO)_5$, which forms chromium carbenes upon reaction with amides or esters in presence of ***Chlorotrimethylsilane*** (eq 13). This procedure allows an improved entry into the rich chemistry of chromium carbenes, with a comparative study clearly pointing out the superiority of C_8K as reducing agent over naphthalenide anions.[18] Moreover, C_8K is the only reagent that cleanly reduces $[CpNi_2(CO)]_2$ to the highly nucleophilic nickelate $[Cp(CO)Ni]^- K^+$, which affords allylnickel complexes on reaction with ***Allyl Bromide***.[19]

$$Cr(CO)_6 \xrightarrow[\text{THF, -78 to 0 °C, 3 h}]{\text{2.2 equiv } C_8K} K_2Cr(CO)_5 \xrightarrow[\text{TMSCl; 78\%}]{} \quad (13)$$

Metal–Graphite Reagents.[1] C_8K may be used to reduce metal salts in ethereal solvents to metal–graphite reagents (eq 14). Due to the even distribution and small size (usually in the range of 2–10 nm in diameter) of the metal particles adsorped on the surface of the graphite support, these reagents exhibit very high and sometimes unprecedented degrees of reactivity and are easily removed by simple filtration after use.[1] Almost any metal may be activated by this technique.

$$MX_n + n\ C_8K \longrightarrow M^* \text{ on graphite} + n\ KX \quad (14)$$

MX_n = $ZnCl_2$ (AgOAc), $TiCl_3$, $TiCl_4$, $SnCl_2$, $PdCl_2$, $PtCl_2$, $NiBr_2$, $MgCl_2$, $FeCl_3$

Zinc–Graphite doped with 10 mol % ***Silver*** exhibits an exceptionally high reactivity and wide scope. This reagent promotes Reformatsky reactions at temperatures well below 0 °C with different kinds of halo esters, including the rather unreactive chloroalkanoates (see also ***Ethyl Bromozincacetate***).[20] Tandem reactions comprising a Reformatsky step followed by glycidate formation or a Peterson elimination, respectively, have been carried out.[21] Zn/Ag–graphite transforms glycosyl halides to the corresponding glycals under aprotic conditions, independent of the ring size and the protecting groups employed (eq 15).[22] Deoxyhalosugar derivatives are reductively ring opened by Zn/Ag–graphite with formation of enantiomerically pure enal building blocks of wide applicability to natural product synthesis (eq 16).[13,17,23] It is the only reagent that affords such fragmentations under essentially neutral conditions in anhydrous ethereal solvents, exhibits excellent tolerance towards a wide range of functional groups in the substrates, and is therefore well suited for selective transformations of polysubstituted molecules.[24] This was shown in a total synthesis of 9-dihydro-FK-506 (eq 17).[25] Zn/Ag–graphite is also suitable for metalating functionalized aryl halides at rt.[31]

Zn/Ag–graphite, THF, −20 °C, 10 min, 87% (15)

Zn/Ag–graphite, THF, rt, 10 min, 93% (16)

Zn/Ag–graphite, THF, 84% (17)

Titanium on graphite deserves special emphasis as one of the most efficient reagents for all kinds of carbonyl coupling reactions. It promotes both inter- as well as intramolecular McMurry reactions of (di)ketones and (di)aldehydes (eq 18),[26–28] cyclizes oxo esters to cyclanones,[26] and was successfully applied to polyoxygenated substrates in cases when other titanium reagents failed to effect any conversion (eq 19).[28] Recently, it was used to cyclize acyloxycarbonyl and acylamidocarbonyl compounds to furans, benzo[*b*]furans, and indoles (eq 20), respectively.[29] This new entry to aromatic heterocycles by reductive C–C bond formation of easily accessible precursors is compatible with a variety of other reducible sites in the substrates.[29]

Ti–graphite, THF, 14 h, 70% (18)

Ti–graphite ($TiCl_3$:C_8K = 2:1) (19)

Ti–graphite, DME, Δ, 94% (20)

The performance of magnesium–graphite in reductive carbonyl coupling processes compares favorably to all other kinds of pinacol-forming agents described so far (eq 21).[26b] Metal–graphite combinations of platinum, palladium, and nickel have been used as highly efficient and selective catalysts for hydrogenation reactions as well as for catalytic C–C bond formations.[30] Preparative advantages have also been drawn from the use of other graphite reagents in organic and organometallic synthesis.[1,27a]

Mg–graphite, THF, rt, 86% (21)

Related Reagents. See Classes R-2, R-7, R-15, R-20, R-25, R-27, R-30 and R-31, pages 1–10. Nickel–Graphite; Palladium–Graphite; Platinum on Carbon; Zinc–Graphite.

---

1. (a) Fürstner, A. *AG(E)* **1993**, *32*, 164. (b) Csuk, R.; Glänzer, B. I.; Fürstner, A. *Adv. Organomet. Chem.* **1988**, *28*, 85. (c) Savoia, D.; Trombini, C.; Umani-Ronchi, A. *PAC* **1985**, *57*, 1887. (d) Setton, R. In *Preparative Chemistry Using Supported Reagents*, Laszlo, P., Ed., Academic: New York, 1987; p 225.
2. Ebert, L. B. *J. Mol. Catal.* **1982**, *15*, 275.
3. Bergbreiter, D. E.; Killough, J. M. *JACS* **1978**, *100*, 2126.
4. Rabinovitz, M.; Tamarkin, D. *SC* **1984**, *14*, 377.
5. (a) Lissel, M.; Kottmann, J.; Lenoir, D. *Chemosphere* **1989**, *19*, 1499. (b) Lissel, M.; Kottmann, J.; Tamarkin, D.; Rabinovitz, M. *ZN(B)* **1988**, *43*, 1211.
6. (a) Savoia, D.; Trombini, C.; Umani-Ronchi, A. *JCS(P1)* **1977**, 123. (b) Ellingsen, P. O.; Undheim, K. *ACS* **1979**, *B33*, 528.
7. Contento, M.; Savoia, D.; Trombini, C.; Umani-Ronchi, A. *S* **1979**, 30.
8. (a) Setton, R.; Beguin, F.; Piroelle, S. *Synth. Met.* **1982**, *4*, 299. (b) Lalancette, J.-M.; Rollin, G.; Dumas, P. *CJC* **1972**, *50*, 3058.
9. Weitz, I. S.; Rabinovitz, M. *JCS(P1)* **1993**, 117.
10. Tamarkin, D.; Benny, D.; Rabinovitz, M. *AG(E)* **1984**, *23*, 642.
11. Tamarkin, D.; Cohen, Y.; Rabinovitz, M. *S* **1987**, 196.
12. (a) Fürstner, A.; Weidmann, H. *J. Carbohydr. Chem.* **1988**, *7*, 773. (b) Fürstner, A. *LA* **1993**, 1211.
13. (a) Fürstner, A.; Koglbauer, U.; Weidmann, H. *J. Carbohydr. Chem.* **1990**, *9*, 561. (b) Fürstner, A. *TL* **1990**, 3735. (c) Fürstner, A.; Praly, J. P. *AG(E)* **1994**, *33*, 751.
14. (a) Fürstner, A.; Weidmann, H. *JOM* **1988**, *354*, 15. (b) Müller, H.; Weinzierl, U.; Seidel, W. *Z. Allg. Anorg. Chem.* **1991**, *603*, 15.
15. Savoia, D.; Trombini, C.; Umani-Ronchi, A. *TL* **1977**, 653.
16. Savoia, D.; Trombini, C.; Umani-Ronchi, A. *JOC* **1978**, *43*, 2907.
17. Fürstner, A.; Weidmann, H. *JOC* **1989**, *54*, 2307.
18. (a) Hegedus, L. S. *PAC* **1990**, *62*, 691. (b) Schwindt, M. A.; Lejon, T.; Hegedus, L. S. *OM* **1990**, *9*, 2814.
19. Fischer, R. A.; Behm, J.; Herdtweck, E.; Kronseder, C. *JOM* **1992**, *437*, C29.
20. (a) Csuk, R.; Fürstner, A.; Weidmann, H. *CC* **1986**, 775. (b) Boldrini, G. P.; Savoia, D.; Tagliavini, E.; Trombini, C.; Umani-Ronchi, A. *JOC* **1983**, *48*, 4108.
21. (a) Fürstner, A. *JOM* **1987**, *336*, C33. (b) Fürstner, A.; Kollegger, G.; Weidmann, H. *JOM* **1991**, *414*, 295.
22. (a) Csuk, R.; Fürstner, A.; Glänzer, B. I.; Weidmann, H. *CC* **1986**, 1149. (b) Pudlo, P.; Thiem, J.; Vill, V. *CB* **1990**, 1129.
23. (a) Fürstner, A.; Jumbam, D. N.; Teslic, J.; Weidmann, H. *JOC* **1991**, *56*, 2213. (b) Fürstner, A.; Baumgartner, J.; Jumbam, D. N. *JCS(P1)* **1993**, 131. (c) Fürstner, A.; Baumgartner, J. *T* **1993**, *49*, 8541.
24. Ireland, R. E.; Wipf, P.; Miltz, M.; Vanasse, B. *JOC* **1990**, *55*, 1423.
25. Ireland, R. E.; Highsmith, T. K.; Gegnas, L. D.; Gleason, J. L. *JOC* **1992**, *57*, 5071.
26. (a) Fürstner, A.; Weidmann, H. *S* **1987**, 1071. (b) Fürstner, A.; Csuk, R.; Rohrer, C.; Weidmann, H. *JCS(P1)* **1988**, 1729.
27. (a) Boldrini, G. P.; Savoia, D.; Tagliavini, E.; Trombini, C.; Umani-Ronchi, A. *JOM* **1985**, *280*, 307. (b) Clive, D. L. J.; Zhang, C.; Murthy, K. S. K.; Hayward, W. D.; Daigneault, S. *JOC* **1991**, *56*, 6447. (c) Burger, P.; Brintzinger, H. H. *JOM* **1991**, *407*, 207.
28. Clive, D. L. J.; Murthy, K. S. K.; Wee, A. G. H.; Prasad, J. S.; daSilva, G. V. J.; Majewski, M.; Anderson, P. C.; Evans, C. F.; Haugen, R. D.; Heerze, L. D.; Barrie, J. R. *JACS* **1990**, *112*, 3018.
29. (a) Fürstner, A.; Jumbam, D. N. *T* **1992**, *48*, 5991. (b) Fürstner, A.; Jumbam, D. N. *CC* **1993**, 211. (c) Fürstner, A.; Hupperts, A.; Ptock, A.; Janssen, E. *JOC* **1994**, *59*, 5215. (d) Fürstner, A. *T* **1995**, *51*, 773.
30. (a) Savoia, D.; Tagliavini, E.; Trombini, C.; Umani-Ronchi, A. *JOC* **1981**, *46*, 5340 and 5344. (b) Savoia, D.; Trombini, C.; Umani-Ronchi, A.; Verardo, G. *CC* **1981**, 540 and 541. (c) Boldrini, G. P.; Savoia, D.; Tagliavini, E.; Trombini, C.; Umani-Ronchi, A. *JOM* **1984**, *268*, 97. (d) Fürstner, A.; Hofer, F.; Weidmann, H. *J. Catal.* **1989**, *118*, 502.
31. Fürstner, A.; Singer, R.; Knochel, P. *TL* **1994**, *35*, 1047.

Alois Fürstner
Max-Planck-Institut für Kohlenforschung, Mülheim, Germany

Potassium Monoperoxysulfate

$2KHSO_5{\cdot}KHSO_4{\cdot}K_2SO_4$

[37222-66-5] $H_3K_5O_{18}S_4$ (MW 614.81)

(oxidizing agent for a number of functional groups, including alkenes,[16] arenes,[17] amines,[26] imines,[30] sulfides;[37] used for the preparation of dioxiranes[5])

Alternate Name: potassium caroate, potassium hydrogen persulfate, Oxone®.
Physical Data: mp dec; *d* 1.12–1.20 g cm^{-3}.
Solubility: sol water (25.6 g 100 g, 20 °C), aqueous methanol, ethanol, acetic acid; insol common organic solvents.
Form Supplied in: white, granular, free flowing solid. Available as Oxone® and as Curox® and Caroat®.
Analysis of Reagent Purity: iodometric titration, as described in the Du Pont data sheet for Oxone.
Handling, Storage, and Precautions: the Oxone triple salt $2KHSO_5{\cdot}KHSO_4{\cdot}K_2SO_4$ is a relatively stable, water-soluble form of potassium monopersulfate that is convenient to handle and store. Oxone has a low order of toxicity, but is irritating to the eyes, skin, nose, and throat. It should be used with adequate ventilation and exposure to its dust should be minimized. Traces of heavy metal salts catalyze the decomposition of Oxone. For additional handling instructions, see the Du Pont data sheet.

Oxidation Methodology. Oxone ($2KHSO_5{\cdot}KHSO_4{\cdot}K_2SO_4$) is a convenient, stable source of potassium monopersulfate (caroate), which serves as a stoichiometric oxidizing agent under a variety of conditions. Thus aqueous solutions of Oxone can be used to perform oxidations in homogeneous solution and in biphasic systems using an immiscible cosolvent and a phase-transfer catalyst. Recently, solid–liquid processes using supported Oxone reagents have been developed. Other oxidation methods involve the generation and reaction of a secondary reagent under the reaction conditions, as with the widely employed aqueous Oxone–ketone procedures, which undoubtedly involve dioxirane intermediates.[1–4] In other instances, oxaziridine derivatives and metal oxo complexes appear to be the functional oxidants formed in situ from Oxone. Synthetically useful examples of these oxidations are grouped below according to the functional groups being oxidized.

Ketones and Other Oxygen Functions. Various ketones can be converted to the corresponding dioxiranes by treatment with buffered aqueous solutions of Oxone (eq 1). Of particular interest are dimethyldioxirane[5] ($R^1 = R^2 = Me$) and methyl(trifluoromethyl)dioxirane[6] ($R^1 = Me$, $R^2 = CF_3$) derived from acetone and 1,1,1-trifluoro-2-propanone, respectively. The discovery of a method for the isolation of dilute solutions of these volatile dioxiranes in the parent ketone by codistillation from the reaction mixture has opened an exciting new area of oxidation chemistry (see ***Dimethyldioxirane*** and ***Methyl(trifluoromethyl)dioxirane***). Solutions of dioxiranes derived from higher molecular weight ketones have also been prepared.[5,7]

$R^1R^2C{=}O$ —Oxone, aq $NaHCO_3$→ $R^1R^2C(O_2)$ (1)

Interestingly, the reaction of a solid slurry of Oxone and wet ***Alumina*** with solutions of cyclic ketones in CH_2Cl_2 provokes Baeyer–Villiger oxidation to give the corresponding lactones (eq 2).[8] The same wet alumina–Oxone reagent can be used to oxidize secondary alcohols to ketones (eq 3).[9] Aldehydes are oxidized to acids by aqueous Oxone.[5,10]

2.5 equiv Oxone, wet alumina; CH_2Cl_2, reflux, 8 h; 79% (2)

2.5 equiv Oxone, wet alumina; MeCN, 25 °C, 2 h; 80% (3)

Alkenes, Arenes, and Alkanes. Aqueous solutions of Oxone can epoxidize alkenes which are soluble under the reaction conditions; for example, sorbic acid (eq 4)[11] (the high selectivity for epoxidation of the 4,5-double bond here is noteworthy). Alternatively, the use of a cosolvent to provide homogeneous solutions promotes epoxidation (eq 5).[11] Control of the pH to near neutrality is usually necessary to prevent hydrolysis of the epoxide. Rapidly stirred heterogeneous mixtures of liquid alkenes and aqueous Oxone solutions buffered with $NaHCO_3$ also produce epoxides, as shown in eq 6.[12]

2 equiv Oxone; pH 6, 3 h; 84% (4)

2 equiv Oxone, H_2O, MeOH; pH 6, 4 h; 91% (5)

1.3 equiv Oxone, H_2O, $NaHCO_3$; 23 °C, 1 h; 91% (6)

An in situ method for epoxidations with dimethyldioxirane using buffered aqueous acetone solutions of Oxone has been widely applied.[1–4] The epoxidation of 1-dodecene is particularly impressive in view of the difficulty generally encountered in the epoxidation of relatively unreactive terminal alkenes (eq 7).[13] A biphasic procedure using benzene as a cosolvent and a phase-transfer agent was utilized in this case. Equally remarkable is the epoxidation of the methylenecyclopropane derivatives indicated in eq 8, given the propensity of the products to rearrange to the isomeric cyclobutanones.[14]

$C_{10}H_{21}$ — 2.4 equiv Oxone, acetone, benzene; phosphate buffer, 18-crown-6, 2–8 °C, 3 h; 72% → $C_{10}H_{21}$ epoxide (7)

3.3 equiv Oxone, phosphate buffer; CH_2Cl_2, 18-crown-6, 10–15 °C, 16 h, 70–90% (8)

The epoxidation of conjugated double bonds also proceeds smoothly with the Oxone–acetone system, as illustrated by eq 9.[15] The conversion of water-insoluble enones can be accomplished with this method using CH_2Cl_2 as a cosolvent and a quaternary ammonium salt as a phase-transfer catalyst. However, a more convenient procedure utilizes 2-butanone both as a dioxirane precursor and as an immiscible cosolvent (eq 10).[16] No phase-transfer agent is required in this case.

2.4 equiv Oxone, aq acetone, $NaHCO_3$; 24–27 °C, 2.5 h, 92% (9)

7 equiv Oxone, 2-butanone; phosphate buffer, 20 °C, 36 h, 84% (10)

The epoxides of several polycyclic aromatic hydrocarbons have been prepared by the use of a large excess of oxidant in a biphasic Oxone–ketone system under neutral conditions, as shown for the oxidation of phenanthrene (eq 11).[17] However, the use of isolated dioxirane solutions is more efficient for the synthesis of reactive epoxides, since hydrolysis of the product is avoided.[5,18] A number of unstable epoxides of various types have been produced in a similar manner, as discussed for ***Dimethyldioxirane*** and ***Methyl(trifluoromethyl)dioxirane***.

excess Oxone, acetone, CH_2Cl_2, phosphate buffer; Bu_4NHSO_4, 0–10 °C, 5.5 min, 60% (11)

Epoxidations have also been performed with other oxidizing agents generated in situ from Oxone. An intriguing method uses a catalytic amount of an immonium salt to facilitate alkene epoxidation in a process which apparently involves an intermediate oxaziridium species as the active oxidant.[19] This procedure is carried out by adding solid Oxone and $NaHCO_3$ to a solution of the alkene and catalyst in MeCN containing a very limited quantity of water (eq 12). Finally, Oxone is the stoichiometric oxidant in interesting modifications of the widely studied metal porphyrin oxidations, where it has obvious advantages over some of the other oxidants commonly used.[20] The potential of this method is illustrated by the epoxidation reaction in eq 13.[21] In this conversion, only 1.4 mol % of the robust catalyst tetrakis(pentafluorophenyl)porphyrinatomanganese chloride (TFPPMnCl) is required. The catalytic hydroxylation of unactivated hydrocarbons is also possible (eq 14).[22] Other metal complexes promote similar oxidations.[23–25]

Nitrogen Compounds. The aqueous Oxone–acetone combination has been developed for the transformation of certain anilines to the corresponding nitrobenzene derivatives, as exemplified in eq 15.[26] This process involves sequential oxidation steps proceeding by way of an intermediate nitroso compound. In the case of primary aliphatic amines, other reactions of the nitrosoalkane species compete with the second oxidation step (for example, dimerization and tautomerization to the isomeric oxime), thereby limiting the synthetic generality of these oxidations.[27] An overwhelming excess of aqueous Oxone has been used to convert cyclohexylamine to nitrocyclohexane (eq 16).[27]

BF_4^- N-Me immonium catalyst; 1.7 equiv Oxone, $NaHCO_3$, MeCN, H_2O, 16 h, 97% (12)

2.1 equiv Oxone, 1.4% TFPPMnCl, phosphate buffer; 2 equiv py, CH_2Cl_2, $BnNMe_2C_{14}H_{29}Cl$, 81% (13)

1.8 equiv Oxone, 0.6% TFPPMnCl, phosphate buffer; 2 equiv py, CH_2Cl_2, $BnNMe_2C_{14}H_{29}Cl$, 32% (14)

32% + 4% + 5%

3.2 equiv Oxone, acetone, CH_2Cl_2; phosphate buffer, Bn_4NHSO_4, pH 8, 100% (15)

20 equiv Oxone, aq acetone; $NaHCO_3$, 65% (16)

Pyridine is efficiently converted to its *N*-oxide by the Oxone–acetone oxidant.[5] Cytosine and several of its derivatives give the N^3-oxides selectively upon reaction with buffered Oxone (eq 17).[28] A similar transformation of adenosine 5′-monophosphate yields the N^1-oxide.[29]

1 equiv Oxone, phosphate buffer; pH 7, 75 °C, 4 h, 63% (17)

The very useful *N*-sulfonyloxaziridines are conveniently prepared by treating *N*-sulfonylimines with Oxone in a biphasic solvent system (eq 18).[30,31] Either bicarbonate or carbonate can be used to buffer this reaction, but reaction is much faster with carbonate, suggesting that the monopersulfate dianion is the oxidizing species (for illustrations of the remarkable chemistry of these oxaziridines, see ***N*-(Phenylsulfonyl)(3,3-dichlorocamphoryl)oxaziridine**).

$$PhCH{=}NSO_2Ph \xrightarrow[\text{KHCO}_3\text{, 2 h or K}_2\text{CO}_3\text{, 15 min; 95\%}]{\text{1.2 equiv Oxone, H}_2\text{O, toluene}} \text{Ph-oxaziridine-NSO}_2\text{Ph} \quad (18)$$

The Oxone–acetone system has also been employed for the synthesis of simple oxaziridines from *N*-alkylaldimines (eq 19).[32] Interestingly, the *N*-phenyl analogs produce the isomeric nitrones rather than the oxaziridines (eq 20). It is noteworthy that MeCN can replace acetone as the solvent in this procedure.

$$PhCH{=}N\text{-}t\text{-}Bu \xrightarrow[\text{acetone; 98\%}]{\text{1.2 equiv Oxone, aq KHCO}_3} \text{Ph-oxaziridine-N-}t\text{-Bu} \quad (19)$$

$$PhCH{=}NPh \xrightarrow[\text{acetone; 98\%}]{\text{1.2 equiv Oxone, aq KHCO}_3} PhCH{=}N(O)Ph \quad (20)$$

Finally, the chlorination of aldoximines gives the corresponding hydroximoyl chlorides, as shown in eq 21.[33] The combination of Oxone and anhydrous HCl in DMF serves as a convenient source of hypochlorous acid, the active halogenating agent.

$$\text{2-ClC}_6\text{H}_4\text{CH=NOH} \xrightarrow[\text{0.5 N HCl in DMF; 95\%}]{\text{1.1 equiv Oxone}} \text{2-ClC}_6\text{H}_4\text{C(Cl)=NOH} \quad (21)$$

Sulfur Compounds. Some of the earliest applications of Oxone in organic synthesis involved the facile oxidation of sulfur functions. For example, aqueous Oxone selectively oxidizes sulfides to sulfones even in highly functionalized molecules, as illustrated in eq 22.[34] Sulfones can also be prepared by a convenient two-phase system consisting of a mixture of solid Oxone, 'wet' ***Montmorillonite K10*** clay, and a solution of the sulfide in an inert solvent.[35]

$$\text{2-SPh-2-(2-methylallyl)cyclopentanone} \xrightarrow[\text{0 °C, 4 h; 77\%}]{\text{3 equiv Oxone, H}_2\text{O, MeOH}} \text{2-SO}_2\text{Ph-2-(2-methylallyl)cyclopentanone} \quad (22)$$

The partial oxidation of sulfides to sulfoxides has been accomplished in a few cases by careful control of the reaction stoichiometry and conditions.[34] A biphasic procedure for sulfoxide formation from diaryl sulfides is shown in eq 23.[36] However, a more attractive and versatile procedure uses a solid Oxone–wet alumina reagent with a solution of the sulfide.[37] This method permits control of the reaction to form either the sulfoxide or the sulfone simply by adjusting the amount of oxidant and the reaction temperature, as illustrated in eq 24. These oxidations are compatible with other functionality.

$$(p\text{-MeOC}_6\text{H}_4)_2\text{S} \xrightarrow[\text{H}_2\text{O, CH}_2\text{Cl}_2\text{, 18 h; 92\%}]{\text{2 equiv Oxone, Bu}_4\text{NBr}} (p\text{-MeOC}_6\text{H}_4)_2\text{S=O} \quad (23)$$

$$PhSCH_2CH_2OH \xrightarrow[\text{CH}_2\text{Cl}_2\text{, reflux; 84\%}]{\text{1 equiv Oxone, wet alumina}} PhSOCH_2CH_2OH \quad (24)$$

$$PhSCH_2CH_2OH \xrightarrow[\text{CHCl}_3\text{, reflux; 89\%}]{\text{3 equiv Oxone, wet alumina}} PhSO_2CH_2CH_2OH$$

Another intriguing method for the selective oxidation of sulfides to sulfoxides (eq 25) uses buffered Oxone in a biphasic solvent mixture containing a catalytic amount of an *N*-phenylsulfonylimine as the precursor of the actual oxidizing agent, the corresponding *N*-sulfonyloxaziridine.[38] The oxaziridine is smoothly and rapidly formed by reaction of the imine with buffered Oxone and regenerates the imine upon oxygen transfer to the sulfide. The greater reactivity of the sulfide relative to the sulfoxide accounts for the preference for monooxidation in this procedure. The biphasic nature of this reaction prevents direct oxidation by Oxone, which would be less selective.

$$PhSCH{=}CH_2 \xrightarrow[\text{H}_2\text{O, CH}_2\text{Cl}_2\text{, K}_2\text{CO}_3\text{, 0.5 h; 90\%}]{\text{4.5 equiv Oxone, PhSO}_2\text{N=CHC}_6\text{H}_4\text{NO}_2\text{-}p} PhSO_2CH{=}CH_2 \quad (25)$$

Oxone sulfoxidations can show appreciable diastereoselectivity in appropriate cases, as demonstrated in eq 26.[39] Enantioselective oxidations of sulfides to sulfoxides have been achieved by buffered aqueous Oxone solutions containing bovine serum albumin (BSA) as a chiral mediator (eq 27).[40] As little as 0.05 equiv of BSA is required and its presence discourages further oxidation of the sulfoxide to the sulfone. Oxone can be the active oxidant or reaction can be performed in the presence of acetone, trifluoroacetone, or other ketones, in which case an intermediate dioxirane is probably the actual oxidizing agent. The level of optical induction depends on structure of the sulfide and that of any added ketone. Sulfoxide products show ee values ranging from 1% to 89%, but in most examples the ee is greater than 50%.

$$\text{3-TsO-tetrahydrothiophene} \xrightarrow[\text{0 °C, 40 min; 90\%}]{\text{1.2 equiv Oxone, H}_2\text{O, acetone}} \text{3-TsO-tetrahydrothiophene } S^{+}\text{–}O^{-} \quad (26)$$

trans:*cis* = 10:1

$$\text{PhS-}i\text{-Pr} \xrightarrow[\text{NaHCO}_3\text{, pH 8, 4 °C, 5 min; 67\%}]{\text{2 equiv Oxone, BSA, H}_2\text{O, CF}_3\text{COMe}} \text{PhS(O)-}i\text{-Pr} \quad (27)$$

89% ee

1-Decanethiol is efficiently oxidized to decanesulfonic acid (97% yield) by aqueous Oxone.[10] In a similar manner an acylthio function was converted into the potassium sulfonate salt, as shown in eq 28.[41]

$$\mathrm{AcS(CH_2)_{10}CO_2Me} \xrightarrow[\substack{K_2CO_3 \\ 92\%}]{\substack{\text{5 equiv Oxone} \\ H_2O,\ MeOH}} \mathrm{KO_3S(CH_2)_{10}CO_2Me} \quad (28)$$

Finally, certain relatively stable thioketones can be transformed into the corresponding thione *S*-oxides by the aqueous Oxone–acetone reagent (eq 29).[42]

$$(p\text{-}\mathrm{MeOC_6H_4})_2\mathrm{C{=}S} \xrightarrow[\substack{\text{aq } KHCO_3 \\ \text{18-crown-6, 6 h} \\ 97\%}]{\substack{\text{2 equiv Oxone} \\ \text{acetone, benzene}}} (p\text{-}\mathrm{MeOC_6H_4})_2\mathrm{C{=}S{=}O} \quad (29)$$

Related Reagents. See Classes O-8, O-9, O-11, O-14, O-15, and O-23, pages 1–10. Dimethyldioxirane; Methyl(trifluoromethyl)dioxirane.

1. Adam, W.; Hadjiarapoglou, L. P.; Curci, R.; Mello, R. In *Organic Peroxides*; Ando, W.; Ed.; Wiley: New York, 1992; Chapter 4, pp 195–219.
2. Curci, R. In *Advances in Oxygenated Processes*; Baumstark A., Ed.; JAI: Greenwich, CT, 1990; Vol. 2; Chapter 1, pp 1–59.
3. Murray, R. W. *CRV* **1989**, *89*, 1187.
4. Adam, W.; Edwards, J. O.; Curci, R. *ACR* **1989**, *22*, 205.
5. Murray, R. W.; Jeyaraman, R. *JOC* **1985**, *50*, 2847.
6. Mello, R.; Fiorentino, M.; Sciacovelli, O.; Curci, R. *JOC* **1988**, *53*, 3890.
7. Murray, R. W.; Singh, M.; Jeyaraman, R. *JACS* **1992**, *114*, 1346.
8. Hirano, M.; Oose, M.; Morimoto, T. *CL* **1991**, 331.
9. Hirano, M.; Oose, M.; Morimoto, T. *BCJ* **1991**, *64*, 1046.
10. Kennedy, R. J.; Stock, A. *JOC* **1960**, *25*, 1901.
11. Bloch, R.; Abecassis, J.; Hassan, D. *JOC* **1985**, *50*, 1544.
12. Zhu, W.; Ford, W. T. *JOC* **1991**, *56*, 7022.
13. Curci, R.; Fiorentino, M.; Troisi, L.; Edwards, J. O.; Pater, R. H. *JOC* **1980**, *45*, 4758.
14. Hofland, A.; Steinberg, H.; De Boer, T. J. *RTC* **1985**, *104*, 350.
15. Corey, P. F.; Ward, F. E. *JOC* **1986**, *51*, 1925.
16. Adam, W.; Hadjiarapoglou, L.; Smerz, A. *CB* **1991**, *124*, 227.
17. Jeyaraman, R.; Murray, R. W. *JACS* **1984**, *106*, 2462.
18. Mello, R.; Ciminale, F.; Fiorentino, M.; Fusco, C.; Prencipe, T.; Curci, R. *TL* **1990**, *31*, 6097.
19. Hanquet, G.; Lusinchi, X.; Milliet, P. *CR(C)* **1991**, *313*, 625.
20. Meunier, B. *NJC* **1992**, *16*, 203.
21. De Poorter, B.; Meunier, B. *NJC* **1985**, *9*, 393.
22. De Poorter, B.; Ricci, M.; Meunier, B. *TL* **1985**, *26*, 4459.
23. Neumann, R.; Abu-Gnim, C. *CC* **1989**, 1324.
24. Strukul, G.; Sinigalia, R.; Zanardo, A.; Pinna, F.; Michelin, R. *IC* **1989**, *28*, 554.
25. Khan, M. M. T.; Chetterjee, D.; Merchant, R. R.; Bhatt, A. *J. Mol. Catal.* **1990**, *63*, 147.
26. Zabrowski, D. L.; Moormann, A. E.; Beck, K. R. J. *TL* **1988**, *29*, 4501.
27. Crandall, J. K.; Reix, T. *JOC* **1992**, *57*, 6759.
28. Itahara, T. *CL* **1991**, 1591.
29. Kettani, A. E.; Bernadou, J.; Meunier, B. *JOC* **1989**, *54*, 3213.
30. Davis, F. A.; Chattopadhyay, S.; Towson, J. C.; Lal, S.; Reddy, T. *JOC* **1988**, *53*, 2087.
31. Davis, F. A.; Weismiller, M. C.; Murphy, C. K.; Reddy, R. T.; Chen, B. C. *JOC* **1992**, *57*, 7274.
32. Hajipour, A. R.; Pyne, S. G. *JCR(S)* **1992**, 388.
33. Kim, J. N.; Ryu, E. K. *JOC* **1992**, *57*, 6649.
34. Trost, B. M.; Curran, D. P. *TL* **1981**, *22*, 1287.
35. Hirano, M.; Tomaru, J.; Morimoto, T. *BCJ* **1991**, *64*, 3752.
36. Evans, T. L.; Grade, M. M. *SC* **1986**, *16*, 1207.
37. Greenhalgh, R. P. *SL* **1992**, 235.
38. Davis, F. A.; Lal, S. G.; Durst, H. D. *JOC* **1988**, *53*, 5004.
39. Quallich, G. J.; Lackey, J. W. *TL* **1990**, *31*, 3685.
40. Colonna, S.; Gaggero, N.; Leone, M.; Pasta, P. *T* **1991**, *47*, 8385.
41. Reddey, R. N. *SC* **1987**, *17*, 1129.
42. Tabuchi, T.; Nojima, M.; Kusabayashi, S. *JCS(P1)* **1991**, 3043.

Jack K. Crandall
Indiana University, Bloomington, IN, USA

Potassium Nitrosodisulfonate[1]

$(KSO_3)_2NO$

[14293-70-0] $K_2NO_7S_2$ (MW 268.35)

(oxidizing reagent for synthesis of quinones from phenols,[2] naphthols,[9] and anilines;[18] oxidant for conversion of benzylic alcohols to aldehydes or ketones[30] and amino acids to α-keto acids;[31] preparation of heterocyclic quinones;[20–22] oxidative aromatization[29])

Alternate Name: Fremy's salt.
Solubility: sol H_2O.
Form Supplied in: orange powder; widely available.
Preparative Methods: see Zimmer et al.[1]
Handling, Storage, and Precautions: in solid form, this reagent is rather unstable. It sometimes undergoes spontaneous decomposition which occasionally results in a violent explosion attributed to impurities such as chloride ion, manganese dioxide, or nitrite ion. Stored in a desiccator over calcium oxide in the presence of ammonium carbonate to provide an ammoniacal atmosphere, it is stable for several months.

Phenol Oxidations. This reagent oxidizes phenols to the corresponding 1,2- or 1,4-quinones under mild conditions and usually in good yield. The substituents on the aromatic ring control the ratio of *ortho*- and *para*-quinones.[2] *para*-Unsubstituted phenols bearing a variety of substituents in the *ortho* and *meta* positions (e.g. alkyl,[3,4] alkyloxy and/or bromine,[5] crown ether substituents,[6] and *p*-hydroxybenzylamines and primary *p*-hydroxybenzamides[7]) are generally converted to 1,4-quinones. Phenols with easy-to-displace *para* substituents, such as CH_2OH, Br, Cl, CO_2H, $CONH_2$, and CH(OH)CN, undergo oxidation to form 1,4-quinones. 1,2-Quinones are formed if the *para* substituents are alkoxy, alkyl, or aryl.[8] See also ***Lead(IV) Oxide*** and ***Salcomine***.

Naphthol Oxidations. α-Naphthols can be oxidized to 1,2- or 1,4-naphthoquinones by Fremy's salt.[9] Again, the nature of the *para* substituent is critical. 1,4-Naphthoquinones predominate if the *para* position is unsubstituted.[10] 1,2-Naphthoquinones are formed if an alkyl or aryl group occupies the *para* position,[9] if there is a hydroxy group in the 2-position,[9] or if the *para* position is hindered.[11] Approximately equal amounts of 1,2- and 1,4-naphthoquinones are obtained if a hydroxy group occupies the 5-position.[9] It has been reported that 1,4-naphthoquinones are produced from oxidation of 2- or 9-SMe substituted phenols.[12] β-Naphthols are generally oxidized to 1,2-naphthoquinones (eq 1).[13]

Fremy's salt (1)

The methyl ether of a β-naphthol has been reported to afford a 1,2-quinone (eq 2).[14]

OMe, *i*-Pr, X; Fremy's salt; O, O, *i*-Pr, X (2)

X = F, Cl, Br

Aniline Oxidations. 1,4-Quinones are formed by the reaction of 2,6-disubstituted and 2,3,6-trisubstituted anilines and Fremy's salt.[15] Other *ortho* or *meta* methoxy-substituted aromatic amines were converted to the corresponding 1,4-quinones[16] and 1,2-quinones.[17] In one case the oxidation intermediate, a quinone imine which subsequently undergoes hydrolysis to the quinone, has been isolated (eq 3).[18]

NH_2; 2 equiv Fremy's salt; NH, O + O, O (3)

Hypoxanthine analogs were formed by Fremy's salt oxidation (eq 4).[19]

O, HN, N, N, CH_2R, N, Me, NH_2; Fremy's salt; O, O, HN, N, N, CH_2R, N, Me, O (4)

R = H, OMe

Oxidations to Quinoline Quinones. Quinolinols are converted to quinoline quinones (eq 5).[20]

Br, N, N, OH; Fremy's salt; O, Br, N, N, O (5)

In an approach to streptonigrin, Fremy's salt was utilized to oxidize a methoxy aniline by a phase-transfer procedure (eq 6).[21]

NO_2, MeO, N, N, Ar; 1. $Na_2S_2O_4$ 2. Fremy's salt; O, MeO, N, N, O, Ar (6)

Oxidations to Indole Derivatives. The derivatives of *ortho*-hydroxy phenethylamine were converted into indoles (eq 7).[22] Presumably the primary oxidation product is protected from secondary oxidation by protonation at lower pH. Likewise, in solution at different pH, 2-methylindole dimerizes to different products (eq 8).[23]

Ph, NH_2, OH; Fremy's salt pH = 7 → O, O, Ph, N, H; Fremy's salt pH < 7 → HO, Ph, N, H (7)

NH; Fremy's salt pH 3.5–4.0 → O, N, H, NH; Fremy's salt pH 7 → O, O, N, H, NH (8)

The derivatives of *ortho*-aminophenethylamine were converted to indoline 1,4-quinone imides (eq 9).[24] No oxidation at the 3-position or 4-position was observed if the α-position was occupied by two alkyl groups.

NH_2, NH_2; Fremy's salt; O, N (9)

A novel route to heterocyclic quinones, based on Fremy's salt promoted oxidation, has been developed recently (eq 10).[25]

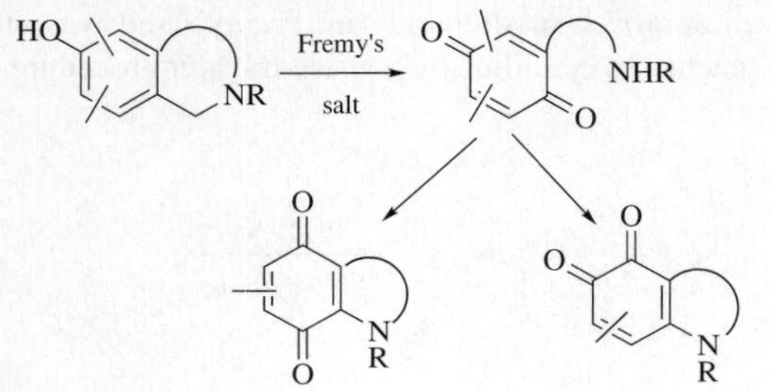

(10)

With this methodology, isoquinoline derivatives were converted with ring contraction to the 1,2-quinone indoles in a buffered (pH 6.1) two-phase (CH_2Cl_2–H_2O) system (eq 11).

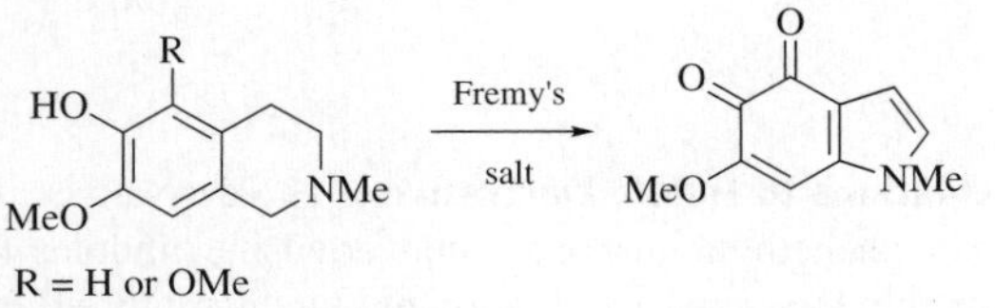

(11)

5-Hydroxybenzofuran is oxidized to the 4,5-*ortho*-quinone.[26] Bisquinones[27] and other heterocyclic quinones[28] were prepared by Fremy's salt oxidation.

Tetrahydroisoquinoline Oxidation. Papaveraldine could be produced by Fremy's salt oxidation over 7 days in 30% yield (eq 12). The corresponding *N*-alkyl tetrahydroisoquinolines give cleavage products (eq 13).[29]

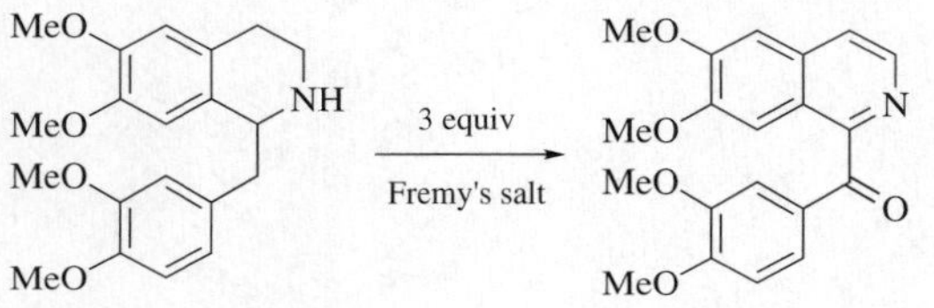

(12)

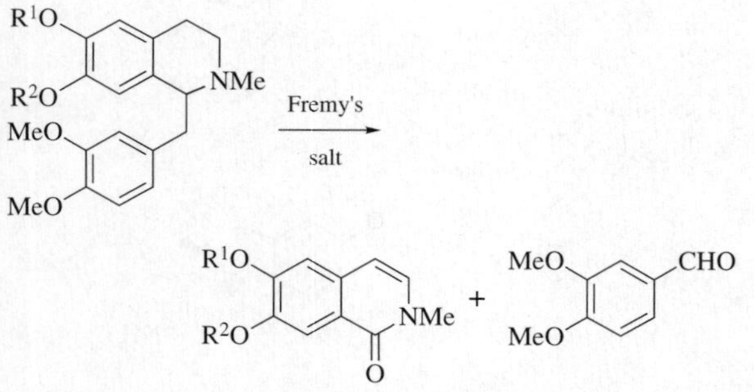

(13)

(a) $R^1 = R^2 = Me$; (b) $R^1 + R^2 = CH_2$

Other Reactions. Benzylic alcohols could be selectively oxidized to aldehydes or ketones in the presence of allylic alcohols and saturated alcohols by Fremy's salt in a phase-transfer system.[30]

Some α-amino and α-hydroxy acids were oxidized to the corresponding α-keto acids (dehydrogenation) and/or to the acids or amides (with decarboxylation) (eq 14).[31]

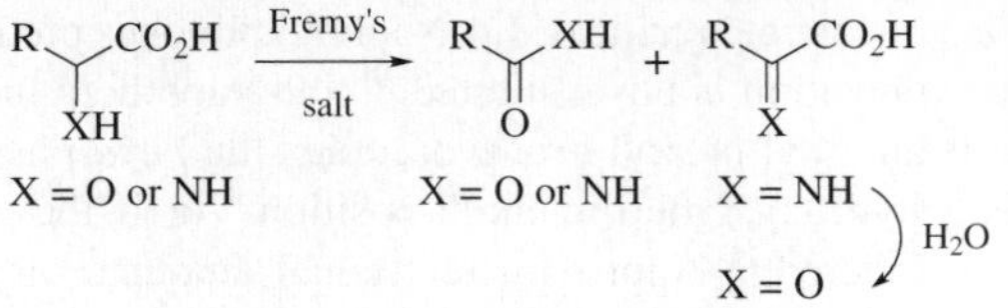

(14)

Related Reagents. See Class O-5, pages 1–10.

1. Zimmer, H.; Lankin, D. C.; Horgan, S. W. *CRV* **1971**, *71*, 229.
2. Deya, P. M.; Dopico, M.; Raso, A. G.; Morey, J.; Saa, J. M. *T* **1987**, *43*, 3523.
3. Teuber, H.-J.; Thaler, G. *CB* **1959**, *92*, 667.
4. (a) Magnusson, R. *ACS* **1966**, *18*, 759. (b) Engler, T. A.; Sampath, U.; Naganathan, S.; Velde, D. V.; Takusagawa, F.; Yohannes, D. *JOC* **1989**, *54*, 5712.
5. (a) Saa, J. M.; Llobera, A.; Garcia-Raso, A.; Costa, A.; Deya, P. M. *JOC* **1988**, *53*, 4263. (b) Singh, S. B.; Pettit, G. R. *JOC* **1989**, *54*, 4105.
6. (a) Chapoteau, E.; Czech, B. P.; Kumar, A.; Pose, A. *JOC* **1989**, *54*, 861. (b) Hayakawa, K.; Kido, K.; Kanematsu, K. *JCS(P1)* **1988**, 511.
7. Saa, J. M.; Llobera, A.; Deya, P. M. *CL* **1987**, 771.
8. Begley, M. J.; Fish, P. V.; Pattenden, G.; Hodgson, S. T. *JCS(P1)* **1990**, 2263.
9. Teuber, H.-J.; Gotz, N. *CB* **1954**, 87, 1236. Teuber, H.-J.; Lindner, H. *CB* **1959**, *92*, 921 and 927.
10. Ashnagar, A.; Bruce, J. M.; Lloyd-Williams, P. *JCS(P1)* **1988**, 559.
11. (a) Ciufolini, M. A.; Byrne, N. E. *JACS* **1991**, *113*, 8016. (b) Ciufolini, M. A.; Byrne, N. E. *TL* **1989**, *30*, 5559.
12. Coll, G.; Morey, J.; Costa, A.; Saa, J. M. *JOC* **1988**, *53*, 5345.
13. (a) Pataki, J.; Raddo, P. D.; Harvey, R. G. *JOC* **1989**, *54*, 840. (b) Ray, J. K.; Kar, G. K.; Karmakar, A. C. *JOC* **1991**, *56*, 2268. (c) Ramesh, D.; Kar, G. K.; Chatterjee, B. G.; Ray, J. K. *JOC* **1988**, *53*, 212. (d) Sharma, P. K. *SC* **1993**, *23*, 389. (e) Chang, H. M.; Chui, K. Y.; Tan, F. W. L.; Yang, Y.; Zhong, Z. P. Lee, C. M.; Sham, H. L.; Wong, H. N. C. *JMC* **1991**, *34*, 1675.
14. He, Y.; Chang, H. M.; Lau, Y. K.; Cui, Y. X.; Wang, R. J.; Mak, T. C. W.; Wong, H. N. C.; Lee, C. M. *JCS(P1)* **1990**, 3359.
15. Teuber, H. J.; Hasselbach, M. *CB* **1959**, *92*, 674.
16. (a) Helissey, P.; Giorgi-Renault, S.; Renault, J.; Cros, S. *CPB* **1989**, *37*, 675. (b) Kende, A. S.; Ebert, F. H.; Battista, R.; Boatman, R. J.; Lorah, D. P.; Lodge, E. *H* **1984**, *21*, 91. (c) Brown, P. E.; Lewis, R. A.; Waring, M. A. *JCS(P1)* **1990**, 2979.
17. Cambie, R. C.; Grimsdale, A. C.; Rutledge, P. S.; Woodgate, P. D. *AJC* **1990**, *43*, 485.
18. Horner, L.; Sturm, K. *CB* **1955**, *88*, 329.
19. (a) Lee, C. H.; Gilchrist, J. H.; Skibo, E. B. *JOC* **1986**, *51*, 4784. (b) Dempcy, R. O.; Skibo, E. B. *JOC* **1991**, *56*, 776.
20. (a) Boger, D. L.; Yasuda, M.; Mitscher, L. A.; Drake, S. D.; Kitos, P. A.; Thompson, S. C. *JMC* **1987**, *30*, 1918. (b) Saito, H.; Hirata, T.; Kasai, M.; Fujimoto, K.; Ashizawa, T.; Morimoto, M.; Sato, A. *JMC* **1991**, *34*, 1959. (c) Kozikowski, A P.; Sugiyama, K.; Springer, J. P. *JOC* **1981**, *46*, 2426. (d) see Ref. 11.
21. (a) Kende, A. S.; Ebert, F. H.; Battista, R.; Boatman, R. J.; Lorah, D. P.; Lodge, E. *H* **1984**, *21*, 91. (b) Kende, A. S.; Ebetino, F. H. *TL* **1984**, *25*, 923.
22. Teuber, H.-J.; Glosauer, O. *CB* **1965**, *98*, 2648.
23. Teuber, H.-J.; Staiger, G. *CB* **1955**, *88*, 1066.
24. Teuber, H.-J.; Glosauer, O. *CB* **1965**, *98*, 2939.
25. Saa, J. M.; Capo, M.; Marti, C.; Garcia-Raso, A. *JOC* **1990**, *55*, 288.
26. Lee, J.; Tang, J.; Snyder, J. K. *TL* **1987**, *28*, 3427.

27. Yang, B.; Liu, L.; Katz, T. J.; Liberko, C. A.; Miller, L. L. *JACS* **1991**, *113*, 8993.

28. Giorgi-Renault, S.; Renault, J.; Gebel-Servolles, P.; Baron, M.; Paoletti, C.; Cros, S.; Bissery, M.-C.; Lavelle, F.; Atassi, G. *JMC* **1991**, *34*, 38.

29. Castedo, L.; Puga, A.; Saa, J. M.; Suau, R. *TL* **1981**, *22*, 2233.

30. Morey, J.; Dzielenziak, A.; Saa, J. M. *CL* **1985**, 263.

31. Garcia-Raso, A.; Deya, P. M.; Saa, J. M. *JOC* **1986**, *51*, 4285.

Kathlyn A. Parker & Dai-Shi Su
Brown University, Providence, RI, USA

Potassium Permanganate[1–4]

$KMnO_4$

[7722-64-7] $KMnO_4$ (MW 158.04)

(oxidant; conversion of arenes into carboxylic acids,[10,11] α-ketones,[12–15] or α-alcohols;[14] degradation of aromatic rings;[3] preparation of diols,[17,18] ketols,[5,19,20,22] and α-diketones[22–24] from nonterminal alkenes; preparation of carboxylic acids,[27] aldehydes[26] and 1,2-diols[28] from terminal alkenes; oxidation of alkynes to α-diones;[29,30] oxidation of enones to 1,4-diones;[31] conversion of 1,5-dienes into substituted tetrahydrofurans[32,33] or lactones;[34] conversion of primary and secondary alcohols into carboxylic acids[1,2,37] and ketones,[1–4,9,35] respectively; oxidation of allylic alcohols to α,β-unsaturated ketones[35] and other unsaturated alcohols and α,ω-diols to lactones;[36,37] oxidation of aliphatic thiols to disulfides and aromatic thiols to sulfonic acids;[4] oxidation of sulfides and sulfoxides to sulfones,[4,38–40] sulfinic acids to sulfonic acids,[43] sulfites to sulfates,[44] and thiones to ketones;[45] preparation of tertiary nitroalkanes from the corresponding amines;[47] oxidation of tertiary amines to amides or lactams;[48–50] allylic oxidations when used in conjunction with *t*-butyl hydroperoxide;[51] preparation of iodoaromatic compounds when used with I_2 and sulfuric acid;[52] oxidation of nucleic acids to the corresponding diols and ketols;[53] oxidation of guaiol and related compounds to rearranged ketols;[54] oxidation of poly(vinyl alcohol) to poly(vinyl ketone);[56] oxidation of nitroalkanes to aldehydes or ketones; oxidation of imines to nitrones)

Alternate Name: potassium manganate(VII).
Physical Data: *d* 2.70 g cm^{-3}; decomposition 237 °C.
Solubility: water (at 20 °C) 63.8 g L^{-1}; sol acetone, methanol.
Form Supplied in: purple solid; commercially available.
Handling, Storage, and Precautions: stable at or below rt. Because it is a strong oxidant it should be stored in glass, steel, or polyethylene vessels. Sulfuric acid should never be added to permanganate or vice versa. Permanganate acid, an explosive compound, is formed under highly acidic conditions.

Introduction. Permanganate is an inexpensive oxidant that has been widely used in organic syntheses. Its most common salt, $KMnO_4$, is soluble in water and as a consequence oxidations have traditionally been carried out in aqueous solutions or in mixtures of water and miscible organic solvents such as acetone, acetic acid, acetonitrile, benzonitrile, tributyl phosphate, or pyridine. The discovery that $KMnO_4$ can, with the aid of phase-transfer agents, be readily dissolved in nonpolar solvents such as CH_2Cl_2, and the recent observation that is adsorption onto a solid support produces an effective heterogeneous oxidant, has further expanded its usefulness.

The general features of the reactions of permanganate dissolved in aqueous solutions, or in organic solvents with the aid of a phase-transfer agent, and as a heterogeneous oxidant will be briefly described, followed by specific examples.

Aqueous Permanganate Oxidations. Potassium permanganate is a general, but relatively nonselective, oxidant when used in aqueous solutions. When an organic compound contains only one site at which oxidation can readily occur, this reagent is a highly efficient and effective oxidant. For example, oleic acid is converted into dihydroxystearic acid in quantitative yield when oxidized in a dilute aqueous solution of $KMnO_4$ at 0–10 °C.[5]

If the aqueous solution is made acidic by addition of mineral acid, the rate of reaction increases, most probably because of formation of permanganic acid[1] which is known to be a very strong oxidant.[6] The rate of the reaction is also accelerated by addition of sodium or potassium hydroxide. It has been proposed that this acceleration may be due to ionization of the organic reductant; for example, conversion of an alcohol into an alkoxide ion.[1] However, similar observations for the oxidation of compounds such as sulfides, which lack acidic hydrogens, suggests that other factors may be involved.[7]

Under acidic conditions, permanganate is reduced to soluble manganese(II) or -(III) salts, thus allowing for a relatively easy workup. However, under basic conditions the reduction product is a gelatinous solid, consisting primarily of manganese dioxide, that is difficult to separate from the product. As a consequence, for laboratory scale preparations the reaction product is not isolated until after the MnO_2 has been reduced by addition of HCl and sodium bisulfite. For large scale (industrial) processes, MnO_2 is removed either by filtration or by centrifugation.

Phase-Transfer Assisted Permanganate Oxidations.[2] $KMnO_4$ may be dissolved in nonpolar solvents such as benzene or CH_2Cl_2 by complexing the potassium ion with a crown ether or by replacing it with a quaternary ammonium or phosphonium ion. Although most reactions observed are similar to those found in aqueous solutions, the ability to dissolve permanganate in nonpolar solvents has greatly increased the range of compounds that can be oxidized.

The first example of a phase-transfer assisted permanganate oxidation involved the complexing of the potassium ion by a crown ether in benzene;[8] however, it was later found that the use of quaternary ammonium or phosphonium salts was less expensive and just as efficient.[2]

Phase transfer into a nonpolar solvent can occur either from an aqueous solution or from solid $KMnO_4$. Evaluation of various phase-transfer agents for these purposes has indicated that benzyltributylammonium chloride is highly efficient for transfer from aqueous solutions while alkyltriphenylphosphonium halides, tetrabutylammonium halides, and benzyltriethylammonium halides are all effective for the transfer from solid $KMnO_4$.[2] Adogen 464, an inexpensive quaternary ammonium chloride commercially available, is usually satisfactory for both purposes.

Quaternary ammonium and phosphonium permanganates can also be used as stoichiometric oxidants. For descriptions of their properties, refer to the separate articles on ***Methyltriphenylphosphonium Permanganate*** and ***Benzyltriethylammonium Permanganate***.

Heterogeneous Permanganate Oxidations. The use of permanganate, activated by adsorption on a solid support, as a heterogenous oxidant has further increased the scope of these reactions. CH_2Cl_2 or 1,2-dichloroethane (if a high reflux temperature is required) are the preferred solvents and ***Alumina***, silica, or hydrated ***Copper(II) Sulfate*** are the most commonly used solid supports. The selectivity of the oxidant is dramatically altered by use of a solid support. For example, although carbon–carbon double bonds are very easily cleaved in homogeneous permanganate solutions, secondary allylic alcohols can be cleanly oxidized to the corresponding α,β-unsaturated ketones without disruption of the double bond under heterogeneous conditions.[9]

In addition to increased selectivity, the use of permanganate under heterogeneous conditions allows for easy product isolation. It is necessary only to remove spent oxidant by filtration followed by flash evaporation or distillation of the solvent. Products isolated in this way are often sufficiently pure to permit direct use in subsequent synthetic procedures.

Benzylic Oxidations. Permanganate oxidizes side chains of aromatic compounds at the benzylic position.[3] In aqueous solution, carboxylic acids are usually obtained (eqs 1 and 2).[10,11]

KMnO4; H2O, py; 71% (1)

KMnO4 (2)

The oxidation of alkylbenzenes proceeds through the corresponding α-ketones, which can occasionally be isolated (eqs 3 and 4).[12,13]

(3)

80% (4)

Under heterogeneous conditions where alumina (acid, Brockman, activity 1)[14] or copper sulfate pentahydrate[15] is used as the solid support, α-ketones and alcohols are obtained with little or no carbon–carbon cleavage (eqs 5–8).

KMnO4, alumina; ClCH2CH2Cl, Δ; 69% (5)

KMnO4, CuSO4•5H2O; CH2Cl2; 88% (6)

KMnO4, alumina; ClCH2CH2Cl, Δ; 86% (7)

KMnO4, alumina; ClCH2CH2Cl, Δ; 79% (8)

Oxidation of Aromatic Rings. Permanganate will oxidatively degrade aromatic rings under both acidic and basic conditions.[3] The effect of acid and base on the reaction has been demonstrated by the oxidation of 2-phenylpyridine; under basic conditions the product is benzoic acid (presumably because the oxidant attacks the site of greatest electron density) (eq 9), while under acidic conditions (where the nitrogen would be protonated) the product is picolinic acid (eq 10).[3]

KMnO4; OH⁻ (9)

+ H+; KMnO4 (10)

Polycyclic aromatic compounds are also oxidatively degraded to a single-ring polycarboxylic acid (eq 11).[16]

KMnO4; OH⁻ (11)

Oxidation of Nonterminal Alkenes. Nonterminal alkenes can be converted into 1,2-diols, ketols, or diketones by choice of appropriate conditions. The reaction, which proceeds by *syn* addition of permanganate to the double bond as indicated, gives the corresponding *cis*-diol under aqueous alkaline conditions (eq 12).[17]

+ MnO4⁻; OH⁻; H2O; 45% (12)

Syn addition can also be achieved in nonaqueous solvents with the aid of a phase-transfer agent (PTA). Subsequent treatment with aqueous base gives 1,2-diols in good yields[2] (eq 13).[18] Equally good results were reported when the reaction was carried out in aqueous *t*-butyl alcohol.[18]

1. $KMnO_4$, PTA, CH_2Cl_2; 2. 10% NaOH; 86% (13)

Under neutral conditions the product obtained from the oxidation of alkenes is the corresponding ketol.[5] Good yields are obtained when aqueous acetone containing a small amount of acetic acid (2–5%) is used as the solvent. The function of acetic acid is to neutralize hydroxide ions produced during the reduction of permanganate. The oxidations of 5-decene and methyl 2-methylcrotonate provide typical examples (eqs 14 and 15).[19,20]

$KMnO_4$; MeCOMe, H_2O, $MeCO_2H$; 74% (14)

$KMnO_4$; MeCOMe, H_2O, H^+; 80% (15)

Heterogeneous oxidations of alkenes with a small amount of *t*-butyl alcohol and water present to provide an 'omega phase'[21] results in the formation of α-ketols in modest to good yields (eqs 16 and 17).[22]

$KMnO_4$, $CuSO_4 \cdot 5H_2O$; CH_2Cl_2, *t*-BuOH, H_2O; 55% (16)

$KMnO_4$, $CuSO_4 \cdot 5H_2O$; CH_2Cl_2, *t*-BuOH, H_2O; 79% (17)

Under anhydrous conditions, 1,2-diones are formed in good yields when alkenes are oxidized by permanganate. Appropriate conditions can be achieved by using acetic anhydride solutions (eq 18)[23] or by dissolving permanganate in CH_2Cl_2 with the aid of a phase-transfer agent (eq 19).[24]

$KMnO_4$; Ac_2O; 66% (18)

$KMnO_4$, PTA; CH_2Cl_2, $MeCO_2H$, H_2O; 69% (19)

Similar yields are obtained under heterogeneous conditions, where workup procedures are much easier.[22]

The carbon–carbon double bonds of alkenes can also be oxidatively cleaved to give carboxylic acids in good yield by use of the Lemieux–von Rudloff reagent (aqueous potassium periodate containing catalytic amounts of permanganate).[3,25] Under heterogeneous conditions, either aldehydes or carboxylic acids are obtained, depending on the conditions used (eqs 20 and 21).[26]

$KMnO_4$, alumina, H_2O; CH_2Cl_2; 75% (20)

$KMnO_4$, silica; C_6H_6; 74%; HO_2C ... CO_2H (21)

Oxidation of Terminal Alkenes. Although oxidation of terminal alkenes by permanganate usually results in cleavage of the carbon–carbon double bond to give either a carboxylic acid[27] or an aldehyde,[26] 1,2-diols can be obtained through use of a phase-transfer assisted reaction (eqs 22–24).[2,28]

$KMnO_4$, PTA; CH_2Cl_2, H_2O; 75%; CO_2H (22)

$KMnO_4$, alumina, H_2O; CH_2Cl_2; 55%; CHO (23)

1. $KMnO_4$, CH_2Cl_2, PTA; 2. 3% NaOH; 80% (24)

Oxidation of Alkynes. Oxidation of nonterminal alkynes results in the formation of α-diones. Good yields are obtained when aqueous acetone containing $NaHCO_3$ and $MgSO_4$,[29] or CH_2Cl_2 containing about 5% acetic acid,[30] is used as the solvent (eqs 25–27). A phase-transfer agent to assist in dissolving $KMnO_4$ must be used when CH_2Cl_2 is the solvent. Terminal alkynes are oxidatively cleaved, yielding carboxylic acids containing one carbon less than the parent alkyne.

$KMnO_4$, PTA; CH_2Cl_2, AcOH; 61%; CO_2H (25)

$KMnO_4$, PTA; CH_2Cl_2, AcOH; 80% (26)

$KMnO_4$, acetone, H_2O; $NaHCO_3$, $MgSO_4$; 81% (27)

Oxidation of Enones to 1,4-Diones. Enones react with nitroalkanes (Michael addition) to form γ-nitro ketones that can be oxidized in good yield to 1,4-diones under heterogeneous conditions (eq 28).[31]

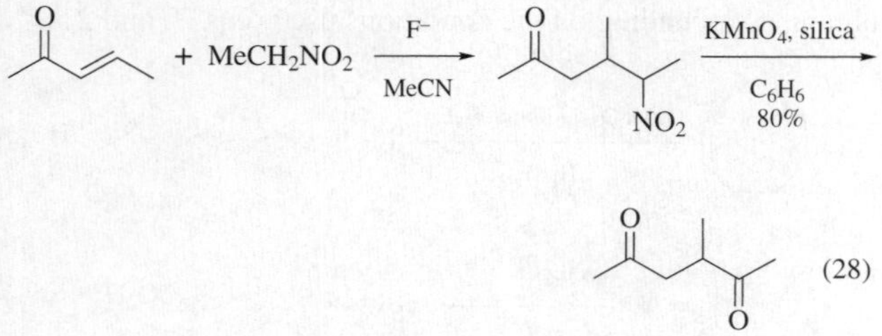

(28)

Oxidation of 1,5-Dienes. The oxidation of 1,5-dienes results in the formation of 2,5-bis(hydroxymethyl)tetrahydrofurans with the indicated stereochemistry (eq 29).[32] When R^6 in (eq 29) is chiral, a nonracemic product is obtained.[33] Use of heterogeneous conditions results in the formation of lactones (eq 30).[34]

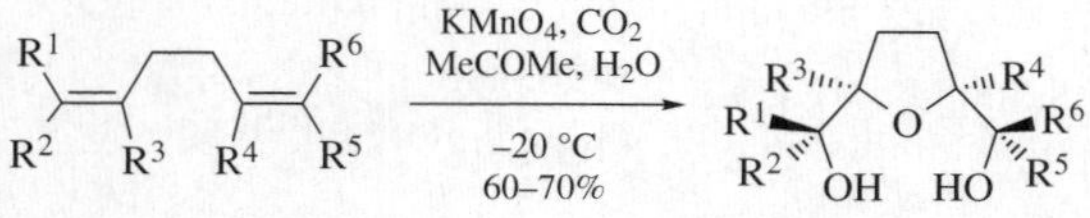

(29)

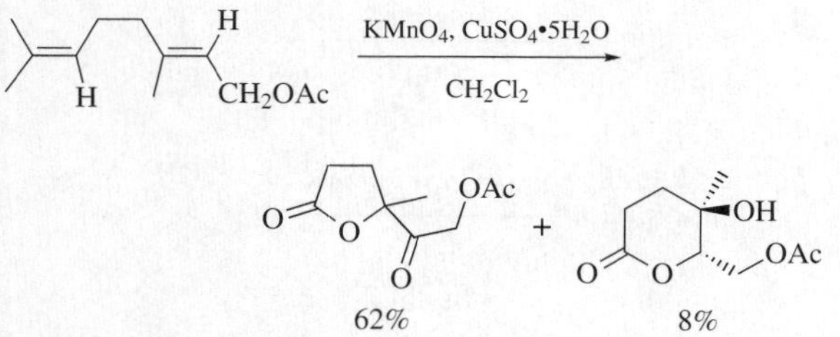

(30)

Oxidation of Alcohols and Diols. Primary and secondary alcohols are converted to carboxylic acids and ketones, respectively, when oxidized by aqueous permanganate under either acidic or basic conditions (eq 31).[1] Similar results are obtained with phase-transfer assisted oxidations in organic solvents such as CH_2Cl_2 (eq 32).[2]

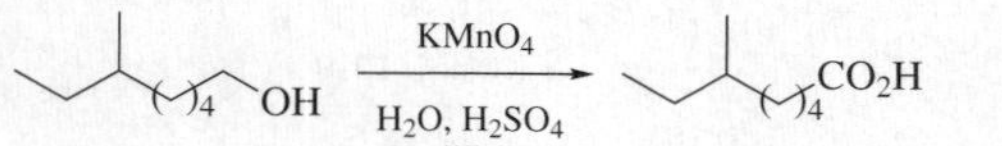

(31)

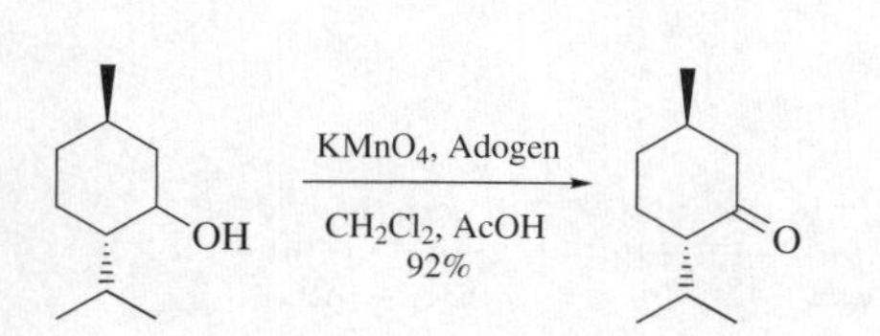

(32)

Heterogeneous oxidations are very effective with secondary alcohols (eq 33)[35] and provide the added advantage that allylic secondary alcohols can be converted to the corresponding α,β-unsaturated ketones without disruption of the double bond (eq 34).[9] Unsaturated secondary alcohols in which the double bond is not adjacent to the carbon bearing the hydroxy group are resistant to oxidation (eq 35) unless an 'omega phase'[21] is created by adding a small amount of water (50 μL per g $KMnO_4$). The products are lactones under these conditions (eq 36).[36]

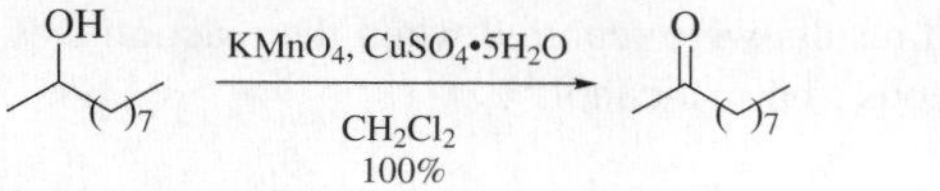

(33)

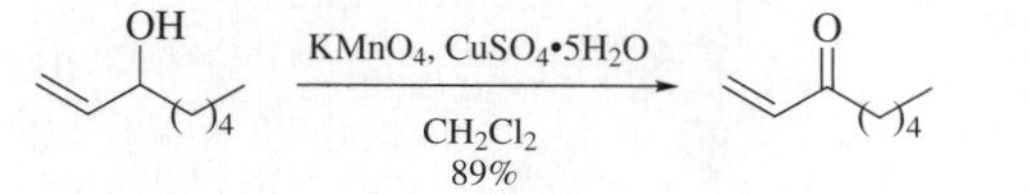

(34)

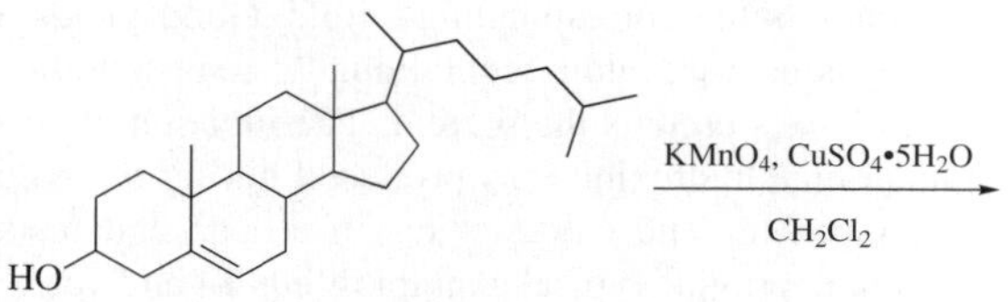

no product (35)

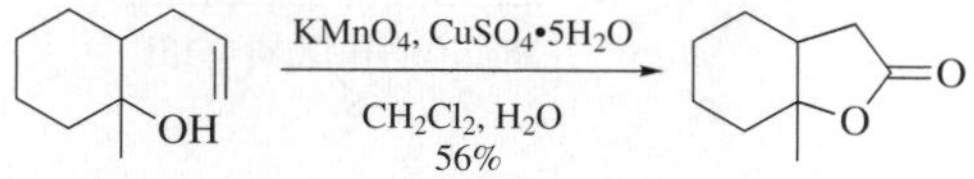

(36)

Good yields of carboxylic acids are obtained from primary alcohols under heterogeneous conditions ($KMnO_4/CuSO_4 \cdot 5H_2O$) only when a base such as KOH or $Cu(OH)_2 \cdot CuCO_3$ is intermixed with the solid support.[37] Under these conditions the reagent has also been reported to be selective for primary alcohols.[37]

The oxidation of α,ω-diols under heterogenous conditions results in the formation of lactones. A good example is found in the preparation of 3-hydroxy-*p*-menthan-10-oic acid lactone (eq 37).[37]

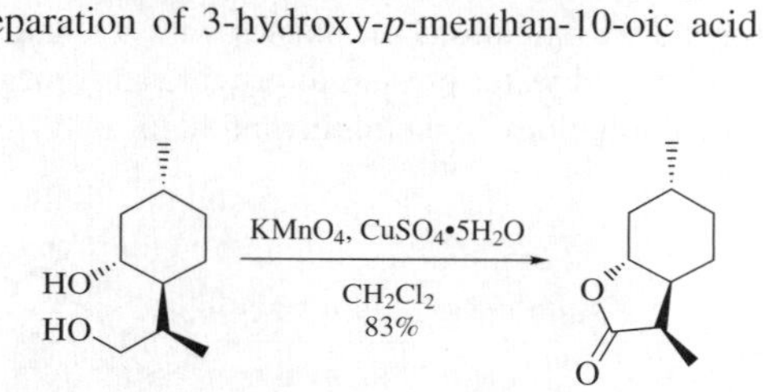

(37)

Oxidation of Organic Sulfur Compounds. Aromatic thiols are oxidized by permanganate to the corresponding sulfonic acids while aliphatic thiols usually give disulfides, which are resistant to further oxidation.[4] Sulfides and sulfoxides are easily oxidized in CH_2Cl_2 to the corresponding sulfones under both homogeneous[38,39] and heterogeneous conditions (eqs 38–42).[40]

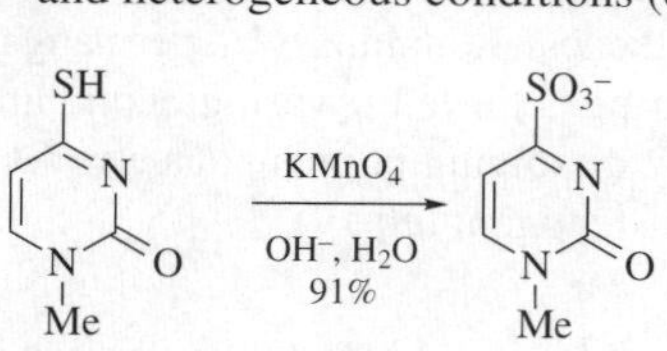

(38)

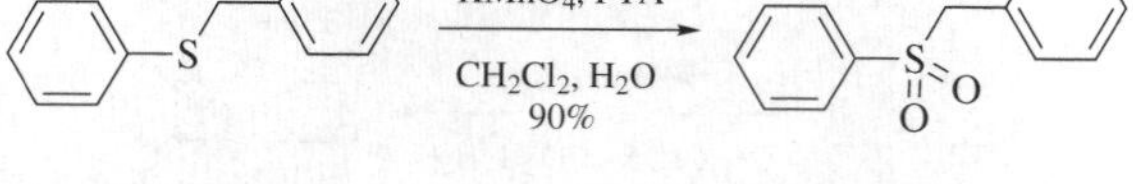

(39)

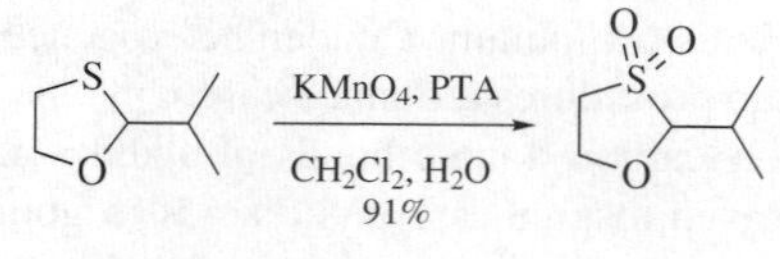

(40)

$$Bu_2SO \xrightarrow[\substack{CH_2Cl_2,\ H_2O \\ 86\%}]{KMnO_4,\ PTA} Bu_2SO_2 \quad (41)$$

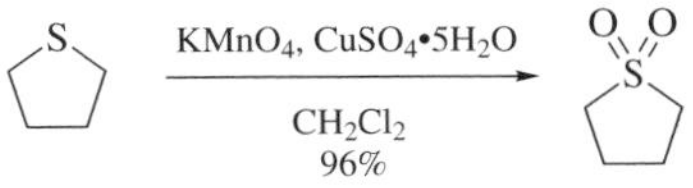

$$\text{(42): tetrahydrothiophene} \xrightarrow[\substack{CH_2Cl_2 \\ 96\%}]{KMnO_4,\ CuSO_4\cdot 5H_2O} \text{sulfolane}$$

Permanganate oxidizes sulfoxides more readily than sulfides, as indicated by the products obtained from the oxidation of compounds containing both sulfide and sulfoxide functional groups (eqs 43 and 44).[41,42]

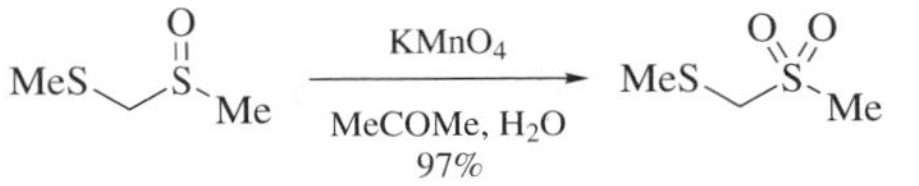

$$MeSCH_2S(O)Me \xrightarrow[\substack{MeCOMe,\ H_2O \\ 97\%}]{KMnO_4} MeSCH_2SO_2Me \quad (43)$$

$$\text{(44): thianthrene monoxide} \xrightarrow[CH_2Cl_2]{QMnO_4} \text{thianthrene 10,10-dioxide}$$

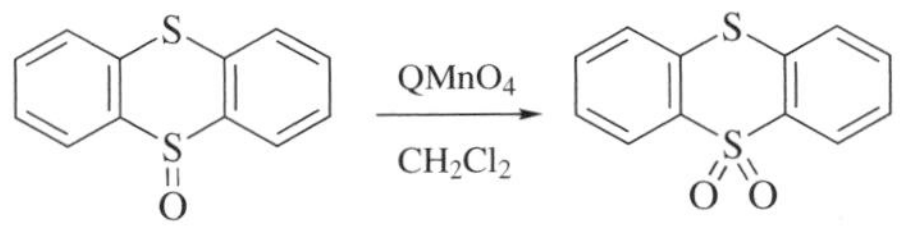

Q = benzyltriethylammonium ion

The greater ease of oxidation of sulfoxides is also responsible for the observation that *gem*-disulfides are oxidized to monosulfones.[42] Monosulfoxides, although not isolated, are likely to be intermediates in these reactions (eq 45).

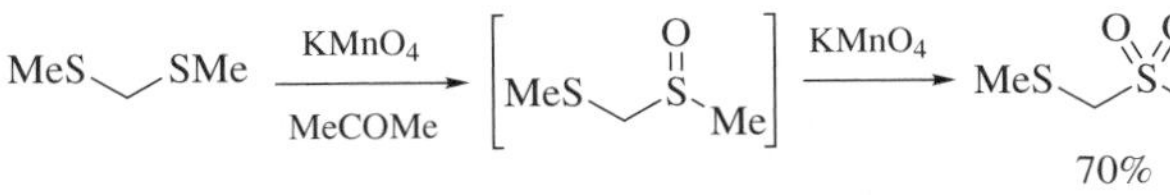

$$MeSCH_2SMe \xrightarrow[MeCOMe]{KMnO_4} [MeSCH_2S(O)Me] \xrightarrow{KMnO_4} MeSCH_2SO_2Me \quad 70\% \quad (45)$$

Oxidation of sulfinic acids results in the formation of sulfonic acids,[43] while sulfites give sulfates (eqs 46 and 47).[44]

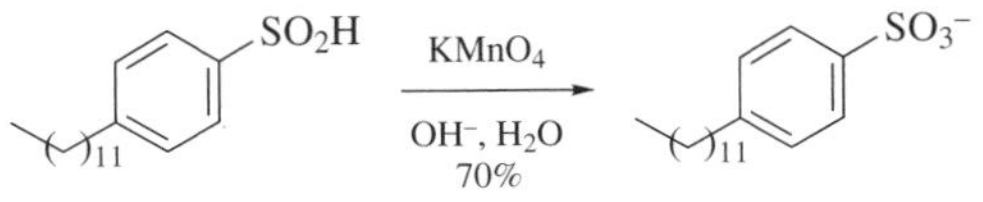

$$4\text{-}C_{12}H_{25}C_6H_4SO_2H \xrightarrow[\substack{OH^-,\ H_2O \\ 70\%}]{KMnO_4} 4\text{-}C_{12}H_{25}C_6H_4SO_3^- \quad (46)$$

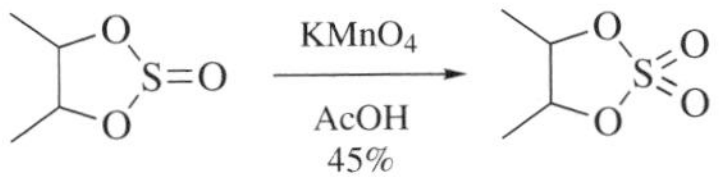

$$\text{(47): cyclic sulfite} \xrightarrow[\substack{AcOH \\ 45\%}]{KMnO_4} \text{cyclic sulfate}$$

Cyclic thiones are readily oxidized to the corresponding ketones by permanganate (eq 48).[45]

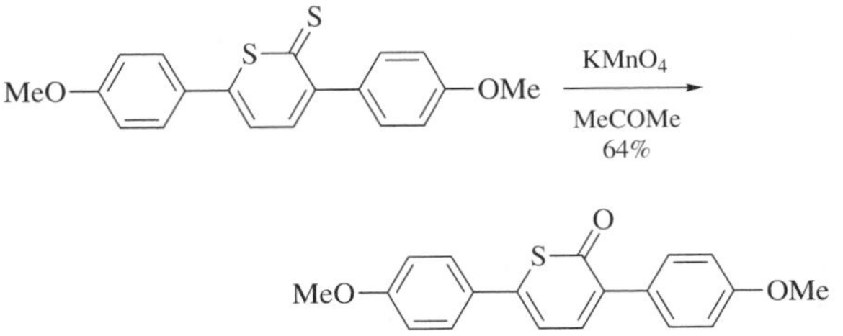

$$\text{(48): MeO-C}_6\text{H}_4\text{-thiopyran-2-thione-C}_6\text{H}_4\text{-OMe} \xrightarrow[\substack{MeCOMe \\ 64\%}]{KMnO_4} \text{MeO-C}_6\text{H}_4\text{-thiopyran-2-one-C}_6\text{H}_4\text{-OMe}$$

Oxidation of Amines. The synthetic usefulness of permanganate as an oxidant for aliphatic amines is decreased by the fact that a complex mixture of products is often obtained.[4,46] Good yields of tertiary nitroalkanes can, however, be obtained from the oxidation of the corresponding amines (eq 49).[47]

$$R_3CNH_2 \xrightarrow[\substack{MeCOMe,\ H_2O \\ 70\text{–}80\%}]{KMnO_4} R_3CNO_2 \quad (49)$$

Primary and secondary amines react with permanganate in buffered, aqueous *t*-butyl alcohol to give aldehydes and ketones (eq 50).[46]

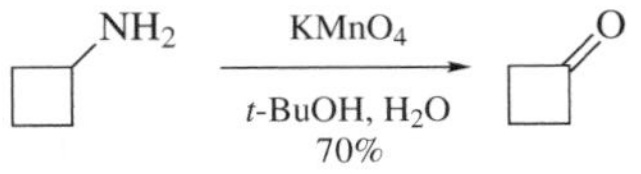

$$\text{cyclobutylamine} \xrightarrow[\substack{t\text{-BuOH},\ H_2O \\ 70\%}]{KMnO_4} \text{cyclobutanone} \quad (50)$$

Amides (or lactams, if the amine is cyclic) are obtained from the oxidation of tertiary amines (eqs 51 and 52).[48–50]

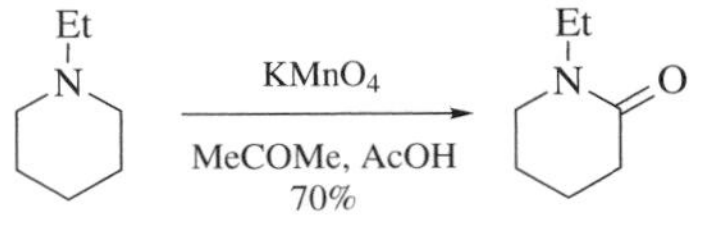

$$\text{N-ethylpiperidine} \xrightarrow[\substack{MeCOMe,\ AcOH \\ 70\%}]{KMnO_4} \text{N-ethyl-2-piperidone} \quad (51)$$

$$\text{(52): 6-fluoro-3-(4-fluorophenyl)-3,4-dihydroquinazoline} \xrightarrow[\substack{MeCOMe \\ 70\%}]{KMnO_4} \text{quinazolin-4(3H)-one}$$

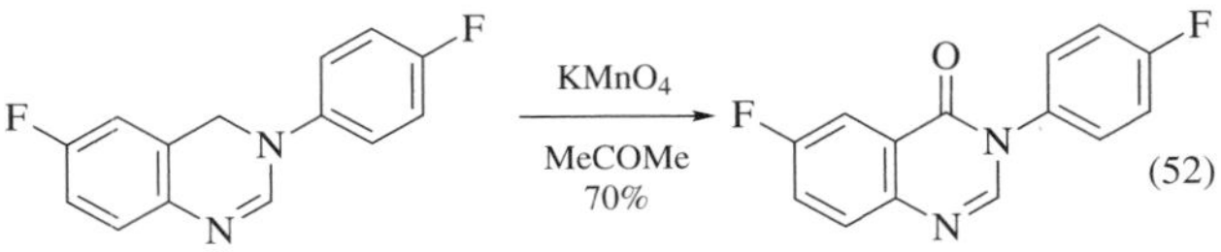

Miscellaneous Oxidations. Use of permanganate in conjunction with ***t*-Butyl Hydroperoxide** results in allylic oxidation (eq 53).[51]

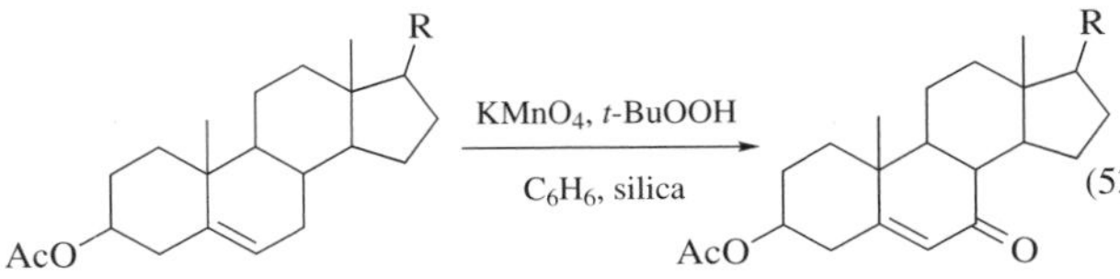

$$\text{(53): 3-AcO-}\Delta^5\text{-steroid} \xrightarrow[C_6H_6,\ silica]{KMnO_4,\ t\text{-BuOOH}} \text{7-oxo-}\Delta^5\text{-steroid}$$

Aromatic compounds are oxidized to aryl iodides when treated with permanganate, ***Iodine***, and ***Sulfuric Acid*** (eq 54).[52]

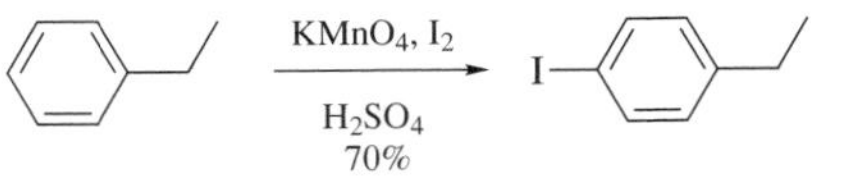

$$C_6H_5CH_2CH_3 \xrightarrow[\substack{H_2SO_4 \\ 70\%}]{KMnO_4,\ I_2} 4\text{-}IC_6H_4CH_2CH_3 \quad (54)$$

Chemical modification of nucleic acids by treatment with permanganate results in oxidation of the Δ^5 double bond to give either diols or ketols (eq 55).[53]

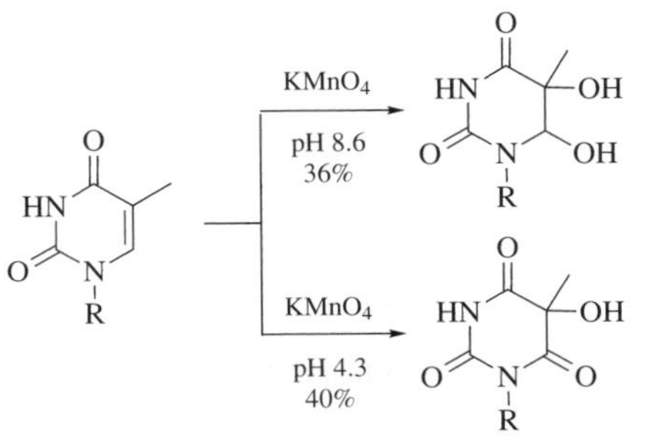

$$\text{(55): thymine (N-R)} \xrightarrow[\substack{pH\ 8.6 \\ 36\%}]{KMnO_4} \text{5,6-diol}; \quad \xrightarrow[\substack{pH\ 4.3 \\ 40\%}]{KMnO_4} \text{5-hydroxy-6-oxo}$$

Guaiol and related compounds can be oxidized to rearranged ketols using aqueous glyme as the solvent (eq 56).[54]

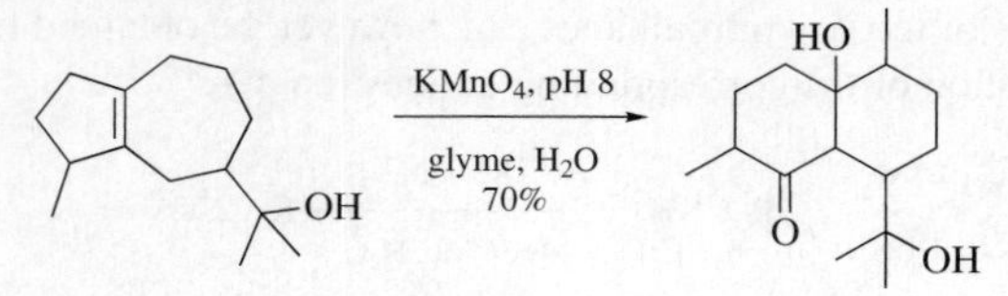

(56)

cis-2,5-Dihydro-2,5-dimethoxyfuran is oxidized to the corresponding α-diol in preference to the *trans* compound (eqs 57 and 58).[55]

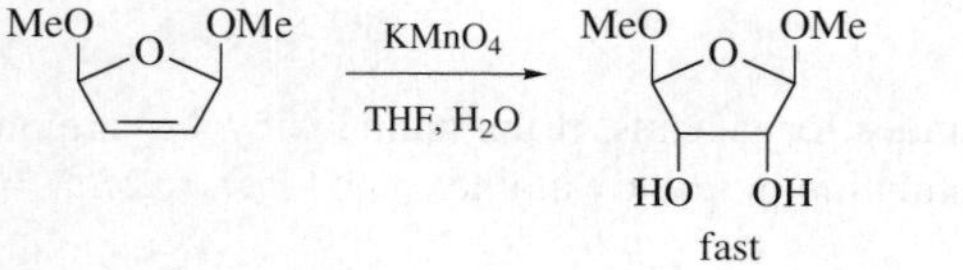

(57)

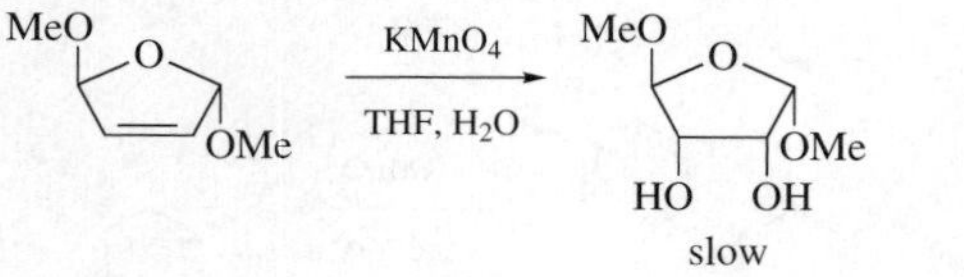

(58)

Oxidation of poly(vinyl alcohol) by permanganate results in the formation of poly(vinyl ketone) (eq 59).[56]

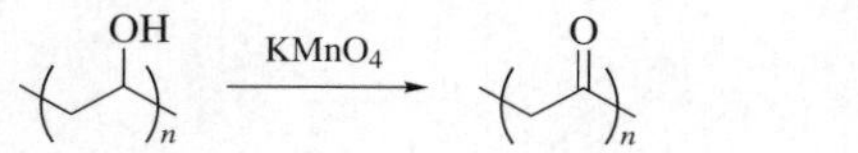

(59)

Treatment of Δ^5-unsaturated steroids with $KMnO_4$/$CuSO_4 \cdot 5H_2O$ in CH_2Cl_2 containing catalytic amounts of *t*-butyl alcohol and water results in formation of the corresponding 5β,6β-epoxide (eq 60).[22,57]

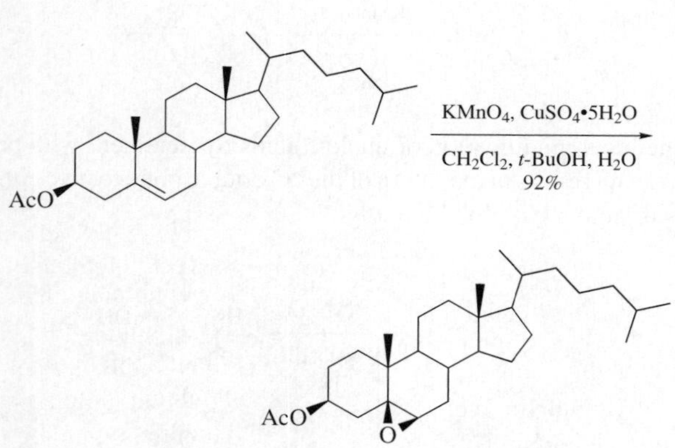

(60)

The oxidation of Δ^7-cholesterol acetate by $KMnO_4$ under neutral or slightly basic conditions results in formation of all-*cis*-epoxydiol (eq 61).[58]

Aliphatic nitro compounds are converted into the corresponding oxo compounds on treatment with basic permanganate.[59,60]

Because these reactions are carried out under basic conditions, it is likely that anions are intermediates, as suggested in eqs 62–64.

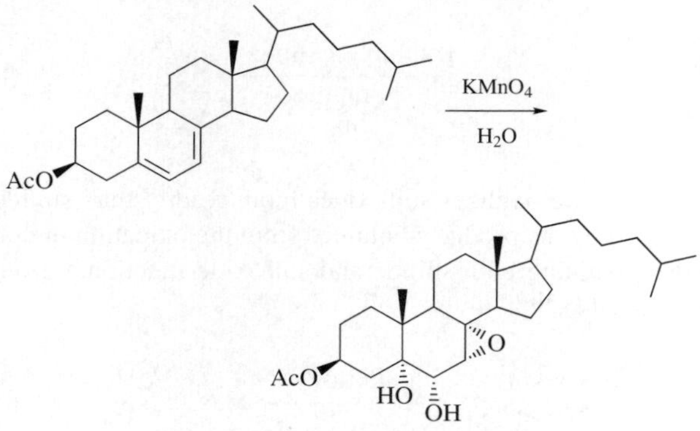

(61)

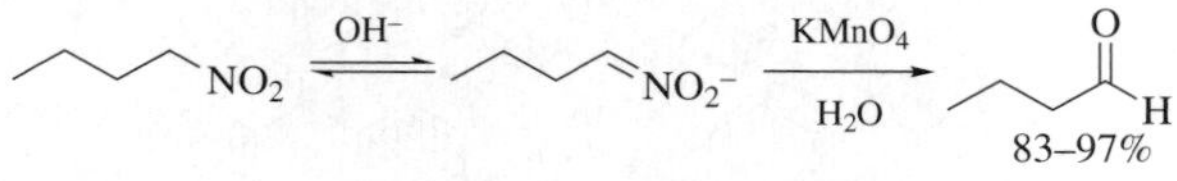

(62)

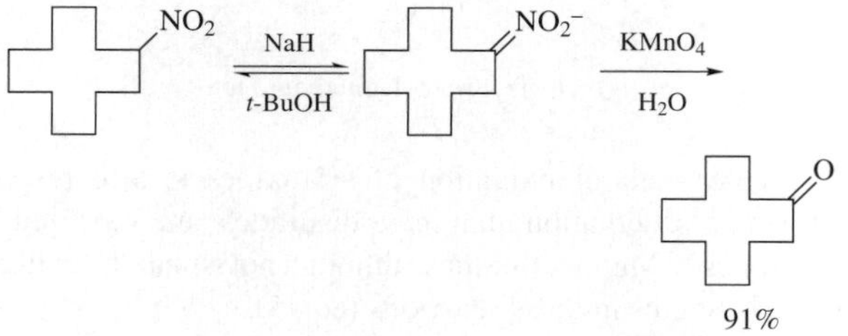

(63)

(64)

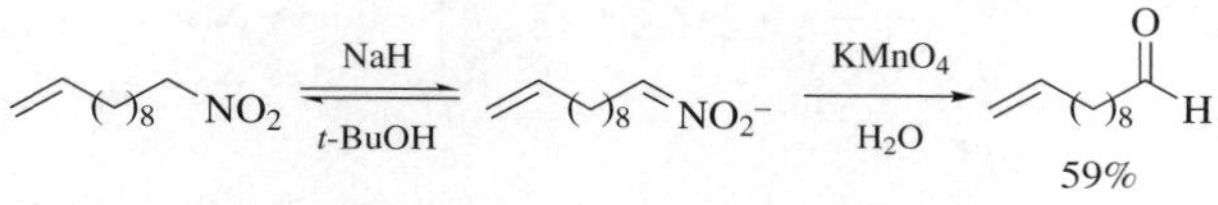

Nitrones can be obtained from the oxidation of imines by $KMnO_4$ in a two-phase CH_2Cl_2/H_2O solution containing a phase-transfer agent (PTA) such as tetrabutylammonium chloride (eq 65).[61]

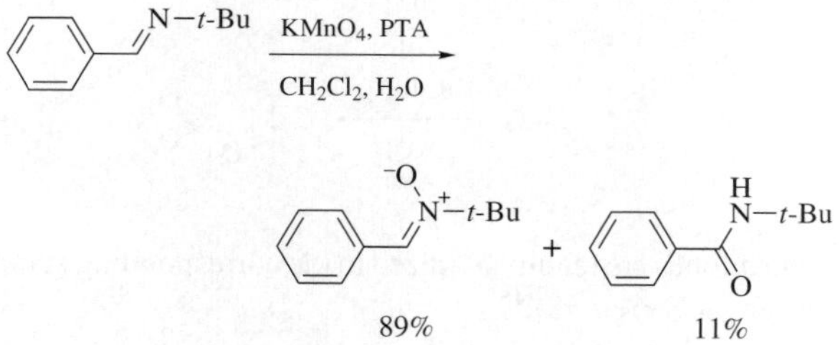

(65)

Related Reagents. See Classes O-1, O-2, O-8, O-9, O-14, O-16, and O-21, pages 1–10. Potassium Permanganate-Copper(II) Sulfate; Sodium Periodate-Potassium Permanganate.

1. Stewart, R. In *Oxidation in Organic Chemistry*; Wiberg, K. B., Ed.; Academic: New York, 1965; Part A, Chapter 1.
2. Lee, D. G. In *Oxidation in Organic Chemistry*; Trahanovsky, W. S., Ed.; Academic: New York, 1962; Part D, Chapter 2.
3. Arndt, D. *Manganese Compounds as Oxidizing Agents in Organic Chemistry*; Open Court: La Salle, IL, 1981; Chapter 5. Lee, D. G. *The Oxidation of Organic Compounds by Permanganate Ion and Hexavalent Chromium*; Open Court: La Salle, IL, 1980.

4. Fatiadi, A. J. *S* **1987**, 85.
5. Coleman, J. E.; Ricciuti, C.; Swern, D. *JACS* **1956**, *78*, 5342.
6. Frigerio, N. A. *JACS* **1969**, *91*, 6200. Perez-Benito, J.; Arias, C.; Brillas, E. *G* **1992**, *122*, 181.
7. Lee, D. G.; Chen, T. *JOC* **1991**, *56*, 5346.
8. Sam, D. J.; Simmons, H. E. *JACS* **1972**, *94*, 4024.
9. Noureldin, N. A.; Lee, D. G. *TL* **1981**, *22*, 4889.
10. Forster, C. F. (The British Petroleum Co. Ltd.) Fr. Patent 1 398 558, 1964 (*CA* **1965**, *63*, 1644).
11. Bromby, N. G.; Peters, A. T.; Rowe, F. M. *JCS* **1943**, 144.
12. Cullis, C. F.; Ladbury, J. W. *JCS* **1955**, 4186.
13. Huntress, E. H.; Walter, H. C. *JACS* **1948**, *70*, 3702. Crook, K. E.; McElvain, S. M. *JACS* **1930**, *52*, 4006.
14. D. Zhao, unpublished observations.
15. N. A. Noureldin, unpublished observations.
16. Ward, J. J.; Kirner, W. R.; Howard, H. C. *JACS* **1945**, *67*, 246.
17. Wiberg, K. B.; Saegebarth, K. A. *JACS* **1957**, *79*, 2822.
18. Bhushan, V.; Rathore, R.; Chandrasekaran, S. *S* **1984**, 431.
19. Srinirasan, N. S.; Lee, D. G. *S* **1979**, 520.
20. Crout, D. H. G.; Rathbone, D. L. *S* **1989**, 40.
21. Liotta, C. L.; Burgess, E. M.; Ray, C. C.; Black, E. D.; Fair, B. E. In *Phase Transfer Catalysis*; Starks, C. M., Ed.; American Chemical Society: Washington, 1987; p 15.
22. Baskaran, S.; Das, J.; Chandrasekaran, S. *JOC* **1989**, *54*, 5182.
23. Sharpless, K. B.; Lauer, R. F.; Repič, O.; Teranishi, A. Y.; Williams, D. R. *JACS* **1971**, *93*, 3303. Jensen, H. P.; Sharpless, K. B. *JOC* **1974**, *39*, 2314.
24. Lee, D. G.; Chang, V. S. *JOC* **1978**, *43*, 1532.
25. Lemieux, R. U.; von Rudloff, E. *CJC* **1955**, *33*, 1701, 1710. von Rudloff, E. *CJC* **1955**, *33*, 1714; **1956**, *34*, 1413.
26. Lee, D. G.; Chen, T.; Wang, Z. *JOC* **1993**, *58*, 2918. Ferreira, J. T. B.; Cruz, W. O.; Vieira, P. C.; Yonashiro, M. *JOC* **1987**, *52*, 3698.
27. Lee, D. G.; Lamb, S. E.; Chang, V. S. *OS* **1981**, *60*, 11.
28. Ogino, T.; Mochizuki, K. *CL* **1979**, 443.
29. Srinivasan, N. S.; Lee, D. G. *JOC* **1979**, *44*, 1574.
30. Lee, D. G.; Chang, V. S. *JOC* **1979**, *44*, 2726; *S* **1978**, 462.
31. Clark, J. H.; Cork, D. G. *CC* **1982**, 635.
32. Klein, E.; Rojahn, W. *T* **1965**, *21*, 2353. Walba, D. M.; Wand, M. D.; Wilkes, M. C. *JACS* **1979**, *101*, 4396.
33. Walba, D. M.; Przybyla, C. A.; Walker, C. B. *JACS* **1990**, *112*, 5624.
34. Baskaran, S.; Islam, I.; Vankar, P. S.; Chandrasekaran, S. *CC* **1992**, 626.
35. Regen, S. L.; Koteel, C. *JACS* **1977**, *99*, 3837. Quici, S.; Regen, S. L. *JOC* **1979**, *44*, 3436. Menger, F. M.; Lee, C. *JOC* **1979**, *44*, 3446.
36. Baskaran, S.; Islam, I.; Vankar, S.; Chandrasekaran, S. *CC* **1990**, 1670.
37. Jefford, C. W.; Wang, Y. *CC* **1988**, 634.
38. Lee, D. G.; Srinivasan, N. S. *Sulfur Lett.* **1982**, *1*, 1.
39. Gokel, G. W.; Gerdes, H. M.; Dishong, D. M. *JOC* **1980**, *45*, 3634.
40. Noureldin, N. A.; McConnell, W. B.; Lee, D. G. *CJC* **1984**, *62*, 2113.
41. Ogura, K.; Suzuki, M.; Tsuchihashi, G. *BCJ* **1980**, *53*, 1414.
42. May, B. L.; Yee, H.; Lee, D. G. *CJC* **1994**, *72*, 2249.
43. Truce, W. E.; Lyons, J. F. *JACS* **1951**, *73*, 126.
44. Garner, H. K.; Lucas, H. J. *JACS* **1950**, *72*, 5497.
45. Clesse, F.; Pradere, J-P.; Quiniou, H. *BSF(2)* **1973**, 586.
46. Shechter, H.; Rawalay, S. S.; Tubis, M. *JACS* **1964**, *86*, 1701. Rawalay, S. S.; Shechter, H. *JOC* **1967**, *32*, 3129.
47. Kornblum, N.; Jones, W. J. *OS* **1963**, *43*, 87.
48. Forrest, J.; Tucker, S. H.; Whalley, M. *JCS* **1951**, 303.
49. Cookson, R. C.; Trevett, M. E. *JCS* **1956**, 2689.
50. Farrar, W. V. *JCS* **1954**, 3253.
51. Prousa, R.; Schönecker, B. *JPR* **1991**, *333*, 775.
52. Chaikovskii, V. K.; Novikov, A. N. *J. Appl. Chem. (USSR)* **1984**, *57*, 121.
53. Hayatsu, H.; Iida, S. *TL* **1969**, 1031. Iida, S.; Hayatsu, H. *BBA* **1971**, *228*, 1.
54. Winter, R. E. K.; Zehr, R. J.; Honey, M.; Van Arsdale, W. *JOC* **1981**, *46*, 4309.
55. Hönel, M.; Mosher, H. S. *JOC* **1985**, *50*, 4386.
56. Hassan, R. M.; Abd-Alla, M. A. *J. Mat. Chem.* **1992**, *2*, 609. Hassan, M. R. *Polym. Int.* **1993**, *31*, 81.
57. Syamala, M. S.; Das, J.; Baskaran, S.; Chandrasekaran, S. *JOC* **1992**, *57*, 1928.
58. Anastasia, M.; Fiecchi, A.; Scala, A. *TL* **1979**, 3323.
59. Shechter, H.; Williams, F. T. *JOC* **1962**, *27*, 3699.
60. Kornblum, N.; Erickson, A. S.; Kelly, W. J.; Henggeler, B. *JOC* **1982**, *47*, 4534.
61. Christensen, D.; Jørgensen, K. A. *JOC* **1989**, *54*, 126.

Donald G. Lee
University of Regina, Saskatchewan, Canada

Potassium Superoxide[1]

[12030-88-5] KO_2 (MW 71.10)

(reactive species is the superoxide anion, O_2^-; solubility of KO_2 in aprotic organic solvents is facilitated by crown ethers[2,3] or other phase-transfer catalysts;[4] O_2^- reacts with most organic substrates either as an anion or as an electron-transfer (reducing) agent; reacts as a nucleophilic anion toward alkyl halides, sulfonates, and carbonyl groups; other net displacement reactions such as with halocarbons may be initiated by electron transfer; reacts as a basic anion toward substrates bearing acidic protons; used in situ with a variety of activated halogens to form peroxy anions useful for further oxygen transfer, e.g. epoxidations[1])

Physical Data: mp 500 °C.[5]

Solubility: slightly sol DMSO; crown ethers are useful for bringing KO_2 into organic solvents such as DMSO,[2] DMF, MeCN, THF, benzene;[3] reacts rapidly with H_2O and protic solvents.

Form Supplied in: small chunks of light yellow powdery solid.

Handling, Storage, and Precautions: chunks of solid may be handled briefly in the atmosphere; prolonged exposure to the atmosphere results in reaction with H_2O; storage should be under dry conditions such as a desiccator; reaction with H_2O produces O_2, H_2O_2, and OH^-.

Dialkyl Peroxide Synthesis. Primary and secondary alkyl bromides[9] and alkyl sulfonates react with KO_2 in aprotic organic solvents (except DMSO, see below), giving acyclic (eq 1)[3,10] and cyclic (eq 2)[11] dialkyl peroxides. The reactions are greatly facilitated by the addition of crown ethers or other phase-transfer catalysts to the reaction medium. The reaction at secondary carbon atoms proceeds with >95% inversion of configuration[3,12] and is accompanied by formation of significant amounts of alkene due to elimination reactions. Because of the mechanism by which the dialkyl peroxides are generated, this method is best suited for synthesis of symmetrical dialkyl peroxides.

Alcohol Inversion. When the reaction between alkyl bromides or alkyl sulfonates and KO_2 is performed in DMSO, the major product is an alcohol[12] as a consequence of oxygen transfer from the intermediate peroxy anion to the DMSO (eq 1).[13] Such displacements of tosylate (eq 3)[14] or mesylate (eqs 4 and 5)[15,16] by KO_2 in DMSO have been used for inversion of the configuration of secondary alcohols and this method is of comparable efficiency to the modified[18] Mitsunobu sequence[17] used for the same purpose.

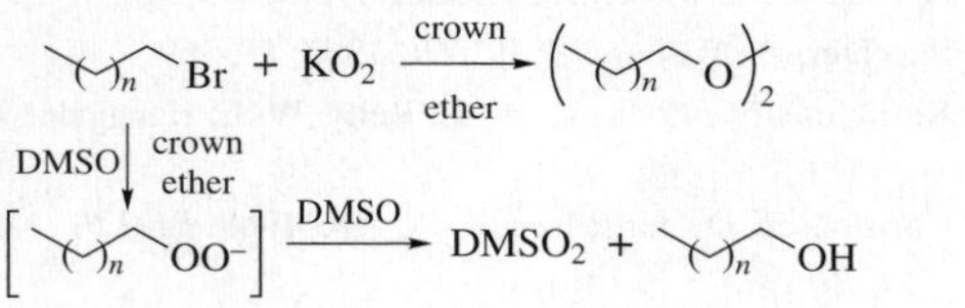

(1)

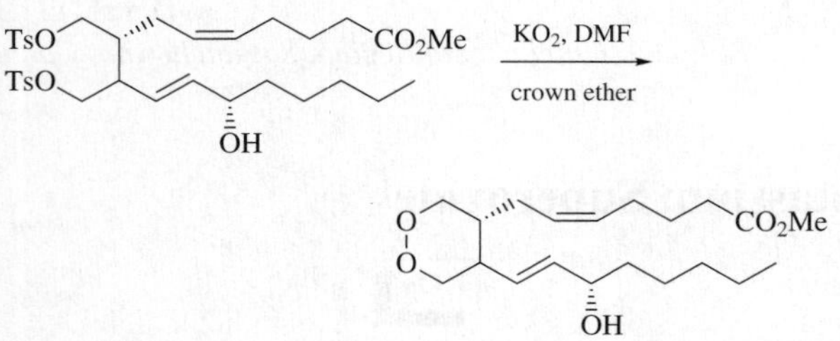

(2)

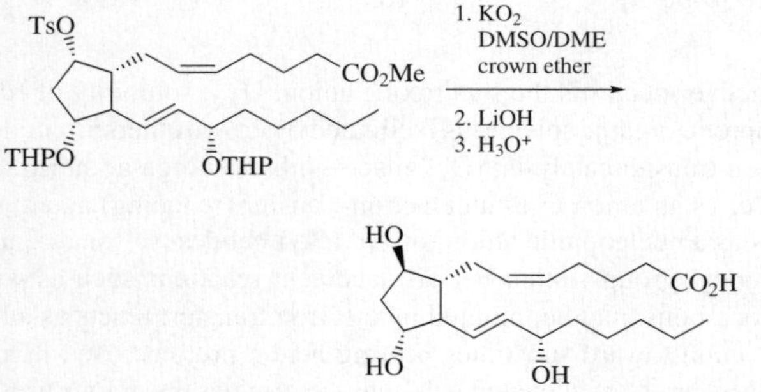

(3)

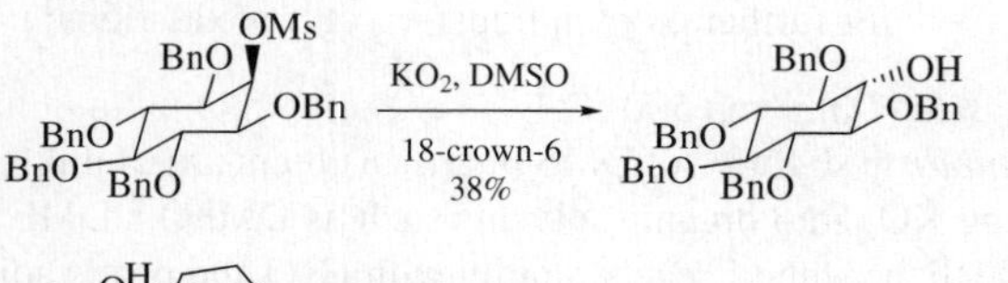

(4)

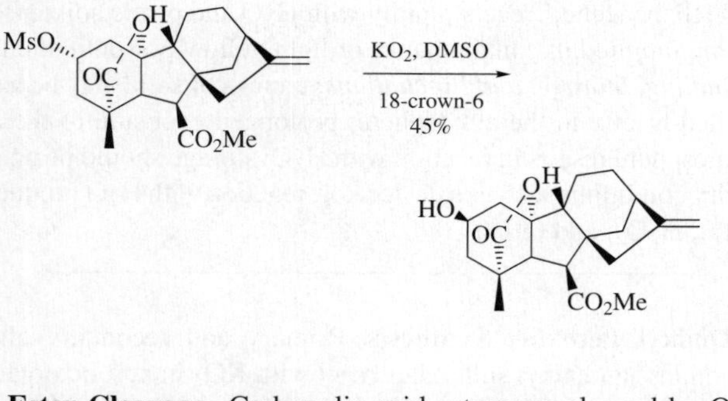

(5)

Ester Cleavage. Carboxylic acid esters are cleaved by O_2^-, giving the corresponding carboxylic acid and alcohol.[19] Several studies of the mechanism of this process have been reported.[20] Qualitatively, displacement of halide or sulfonate ester occurs in preference to ester hydrolysis (see eq 2). The use of KO_2 for ester cleavage generally offers no advantage over conventional ester saponification methods.

Diacyl Peroxide Synthesis. Acid chlorides react with O_2^- to give diacyl peroxides,[21] which in turn are susceptible to further reaction with O_2^-.[22] Acyl peroxy anions are formed as intermediates in the reactions with acid chlorides and diacyl peroxides as well as in the reaction of O_2^- with anhydrides as detected by epoxidation of alkenes.[22] Applications of this approach to oxygen-transfer chemistry is discussed further below. Eq 6 summarizes the primary events but is an incomplete account of all the reactions occurring between O_2^- and these substrates. Amides, aldehydes, and nitriles are unreactive with KO_2 under most conditions.

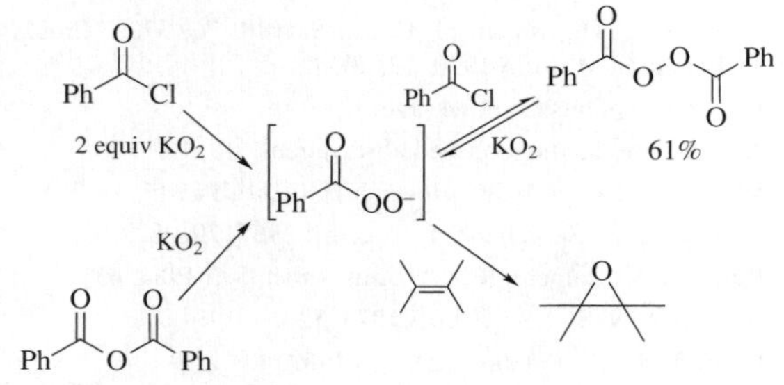

(6)

Oxygen Transfer via Peroxy Anions. A number of reagents react with superoxide to form transient peroxy anions which oxidize electrophilic functional groups such as alkenes, sulfides, sulfoxides, etc. Among the reagents forming peroxy anions useful in such oxidations are aryl sulfonyl halides[23] (especially 2-nitrobenzenesulfonyl chloride[24]), acid chlorides,[22,25] halocarbons such as carbon tetrachloride,[26] dialkyl chlorophosphates[27] and alkyl dichlorophosphates,[28] carbon dioxide,[29] *N*-(−)-menthoxycarbonyl-4-tolylsulfonimidoyl chloride,[30] and phosgene.[31] Representative examples of oxidations by these systems include the epoxidation of limonene (eq 7),[24] the oxidation of sulfides to sulfoxides (eq 8),[32] selective oxidation of sulfoxides to sulfones in the presence of alkenes (eq 9),[33] oxidation of benzylic methylenes to ketones (eq 10),[34] and cleavage of tosylhydrazones to ketones or aldehydes (eq 11).[35]

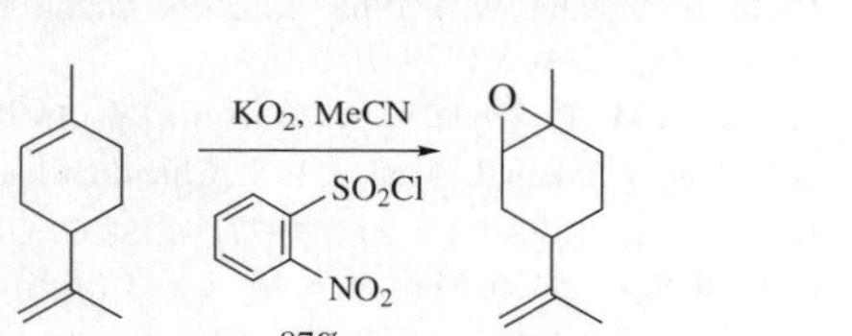

(7)

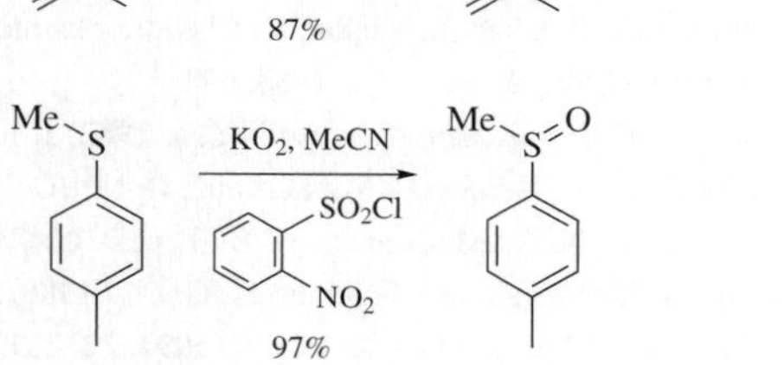

(8)

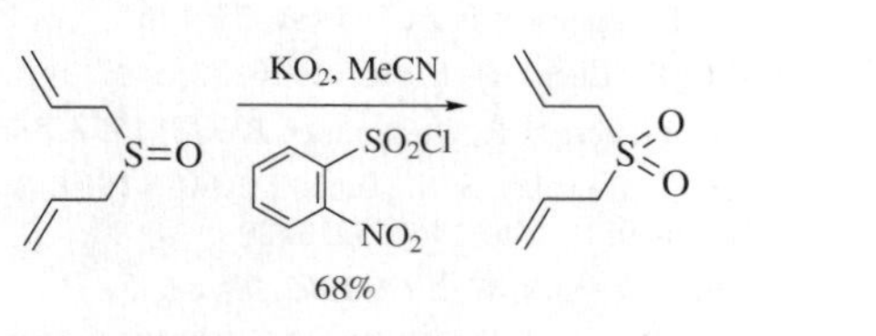

(9)

These approaches to oxygen transfer have not been tested beyond the original reports. Comparisons with other established oxygen-transfer reagents such as the peracids (e.g. ***m-***

Chloroperbenzoic Acid) and the dioxiranes (e.g. ***Dimethyldioxirane***) still must be explored.

KO_2, MeCN; 2-nitrobenzenesulfonyl chloride (SO_2Cl, NO_2); 98% (10)

KO_2, MeCN; 2-nitrobenzenesulfonyl chloride (SO_2Cl, NO_2); 83% (11)

Electron-Transfer Chemistry. Certain reactions of O_2^- whose net result appears to be that of either a nucleophilic addition or displacement reaction, instead may be the result of an electron transfer followed by capture of oxygen, giving the peroxy radical. Oxygen labeling experiments are necessary to distinguish between the two mechanisms.

Nucleophilic Displacement of Aromatic Halides. Aromatic halides substituted with electron-withdrawing groups undergo nucleophilic displacement by KO_2[36] (eq 12)[37] as well as by electrochemically generated O_2^-.[38] Yields of phenols are generally good to excellent in these reactions.

KO_2, crown ether, 80% (12)

Reactions with Electron-Deficient Alkenes. *cis*-2,2,6,6-Tetramethylhept-4-en-3-one, a molecule designed as a probe for electron transfer, is isomerized to the *trans*-enone by KO_2, consistent with electron transfer from reagent to the enone.[39] Cyclohexenone is epoxidized (30% yield) by electrochemically generated superoxide anion,[10] but is converted to a trimeric ketone structure with KO_2.[40] Other cyclohexenones react with KO_2 only when acidic protons are present in the molecule, and then are transformed into mixtures of oxidized products.[40] The cyclohexenone system of cholest-4-en-3-one reacts with KO_2 to give a mixture of at least five oxidation products.[41] Chalcones react with KO_2 in a series of steps initiated by electron transfer to yield aryl carboxylic acids (52–72%) (eq 13).[42] A series of alkenes highly substituted with electron-withdrawing groups (e.g. 1,2-diphenyl-1-nitroethylene) react with KO_2 yielding products of oxidation, e.g. benzoic acid (85%) (eq 14).[43] The conversions by KO_2 of tetraarylcyclopentadienones to 2-hydroxy-2,4,5-triarylfuran-3-ones[44] or to furan-3-ones, 3,4,5,6-triarylpyran-2-ones, and carboxylic acids[45] are initiated by electron transfer.

Reactions with Anilines. Aniline, various substituted anilines, and α-naphthylamine do not react with KO_2 suspended in THF at 80 °C,[46] but several anilines are converted into azobenzenes by KO_2 in benzene containing a crown ether[47,48] or in DMSO.[49] Both *o*- and *p*- but not *m*-phenylenediamines are converted into azobenzenes by KO_2 suspended in THF or pyridine as shown in eq 15. Both *o*- and *p*-aminophenol likewise are transformed into azobenzenes[46] (**Caution**: two laboratories have reported violent explosions of mixtures containing *o*-aminophenol and KO_2 in either THF or toluene[50]). 2-Mercaptoaniline with KO_2 gives 2,2′-dithiobisaniline in 85% yield,[46] the result of coupling of the thiol rather than of the amine substituent.

KO_2, crown ether

Cl–C$_6$H$_4$–CO_2H + HO_2C–C$_6$H$_5$ (13)

73% 35%

KO_2, 18-crown-6, 85%; CO_2H (14)

KO_2, THF, 70% (15)

Reaction with Thiols. Under mild conditions (suspension of KO_2 in toluene), thiophenols and aliphatic thiols are converted into disulfides by KO_2 (eq 16).[51] Under more vigorous reaction conditions (elevated temperature[51] or KO_2/crown ether in pyridine[52]) these thiols are converted into sulfonic acids. Certain thiols (e.g. 2-mercaptophenol) are transformed into sulfonic acids even under the mild reaction conditions. Ethane-1,2-dithiol and propane-1,3-dithiol are converted under mild conditions into the cyclic disulfides 1,2,5,6-tetrathiocan and 1,2,6,7-tetrathiothiecan (eq 17), respectively, while butane-1,4-dithiol forms 1,2-dithian by reaction with KO_2.[51]

KO_2, toluene, rt, 80%; S–S; SO_3H (16)

KO_2, 70%; S–S, S–S (17)

Reactions with Phenols, including Catechols and Tocopherols. Monophenolic compounds do not generally undergo any net change with KO_2. Naphthalene diols react with KO_2 suspended in a toluene–pyridine mixture under an inert atmosphere to form mono- or dihydroxynaphthoquinones in good yields (eq 18).[53]

KO_2, toluene, 80% (18)

Catechols are first oxidized to *o*-quinones upon treatment with KO_2 followed by further oxidation to a variety of products.[54,55] Catechol itself is converted to *cis,cis*-muconic acid in very low yields accompanied by much polymeric byproduct. 9,10-Dihydroxyphenanthrene is converted to diphenic acid in good yield (90%). 3,5-Di-*t*-butylcatechol is converted to a mixture of oxidized products. The tocopherol model compound 2,2,5,7,8-pentamethylchroman-6-ol is converted by KO_2 suspended in THF into 6-hydroxy-2,2,6,7,8-pentamethylchroman-5(6*H*)-one (20%)[56] but, with a solution of KO_2 in AcCN, a low yield (12%) of a diepoxide together with as many as six other compounds are isolated.[57] The analogous diepoxide together with a complex mixture was isolated from the reaction of tocopherol with KO_2.[57]

Miscellaneous Transformations. Practical syntheses of ethyl glyoxylate (72%) and diethyl oxomalonate (83%, via diethyl cyanomalonate) from ethyl cyanoacetate with electrochemically generated superoxide have been reported, as shown in eq 19.[58]

EtO_2CCH_2CN → ($Bu_4N^+O_2^-$, 73%) → EtO_2C-CHO

EtO_2CCH_2CN → (1. NaOEt; 2. $ClCO_2Et$) → $(EtO_2C)_2CHCN$ → ($Bu_4N^+O_2^-$, 83%) → $(EtO_2C)_2C{=}O$ (19)

Aromatic nuclei to which quinones, cyclic alcohols, or cyclic ketones are fused are converted (52–88%) into the aromatic dicarboxylic acids upon reaction with a large excess of KO_2/crown ether in DMF (eq 20).[59]

KO_2, DMF, crown ether, 88% → H_2OC, H_2OC (20)

Tetramethyl- and tetraethylammonium ozonide, whose uses as reagents in organic chemistry are unexplored, have been prepared from the reaction of tetraalkylammonium superoxide and an alkali metal ozonide in liquid ammonia.[60] The question of whether the oxygen released following transfer of an electron from superoxide to a substrate is singlet oxygen is discussed in Frimer's excellent review.[1a] Potassium superoxide was considered to have the molecular formula K_2O_4 until quantum mechanics predicted the structure of O_2^- and experimental confirmation of the radical anion nature of the molecule was performed.[61]

Related Reagents. See Classes O-5, O-7, O-8, O-10, and O-21, pages 1–10. Electrochemically generated O_2^- (for which a tetraalkylammonium cation usually serves as the counter ion);[6] tetramethylammonium superoxide (Me_4NO_2);[7] NaO_2; radiolysis of O_2 in water is used to generate transient O_2^- in aqueous media.[8]

---

1. (a) Frimer, A. A. In *The Chemistry of Peroxides*; Patai, S., Ed.; Wiley: New York, 1983; pp 429–461. (b) Sawyer, D. T.; Valentine, J. S. *ACR* **1981**, *14*, 393. (c) Fee, J. A.; Valentine, J. S. In *Superoxide and Superoxide Dismutases*; Michelson, A. M.; McCord, J. M.; Fridovich, I., Eds.; Academic: NY, 1977; pp. 19–60. (d) Sawyer, D. T.; Gibian, M. J. *T* **1979**, *35*, 1471. (e) Lee-Ruff, E. *CSR* **1977**, *6*, 195. (f) Nagano, T.; Takizawa, H.; Hirobe, M. In *Organic Peroxides*; Ando, W., Ed.; Wiley: New York, 1993; pp 730–764.
2. Valentine, J. S.; Curtis, A. B. *JACS* **1975**, *97*, 224.
3. (a) Johnson, R. A.; Nidy, E. G. *JOC* **1975**, *40*, 1680. (b) Johnson, R. A.; Nidy, E. G.; Merritt, M. V. *JACS* **1978**, *100*, 7960.
4. (a) Druliner, J. D. *SC* **1983**, *13*, 115. (b) Foglia, T. A.; Silbert, L. S. *S* **1992**, 545.
5. Firsova, T. P.; Molodkina, A. N.; Morozova, T. G.; Aksenova, I. V. *IZV* **1965**, 1678. (English translation, p. 1639).
6. (a) Peover, M. E.; White, B. S. *CC* **1965**, 183. (b) Maricle, D. L.; Hodgson, W. G. *Anal. Chem.* **1965**, *37*, 1562. (c) Johnson, E. L.; Pool, K. H.; Hamm, R. E. *Anal. Chem.* **1966**, *38*, 183. (d) Sawyer, D. T.; Roberts, Jr., J. L. *J. Electroanal. Chem.* **1966**, *12*, 90.
7. (a) McElroy, A. D.; Hashman, J. S. *IC* **1964**, *3*, 1798. (b) Guiraud, H. J.; Foote, C. S. *JACS* **1976**, *98*, 1984. (c) Sawyer, D. T.; Calderwood, T. S.; Yamaguchi, K.; Angelis, C. T. *IC* **1983**, *22*, 2577. (d) Yamaguchi, K.; Calderwood, T. S.; Sawyer, D. T. *IC* **1986**, *25*, 1289.
8. Bielski, B. H. J.; Cabelli, D. E. *Int. J. Radiat. Biol.* **1991**, *59*, 291.
9. Merritt, M. V.; Sawyer, D. T. *JOC* **1970**, *35*, 2157.
10. Dietz, R.; Forno, A. E. J.; Larcombe, B. E.; Peover, M. E. *JCS(B)* **1970**, 816.
11. Lin, C.-H.; Alexander, D. L.; Chidester, C. G.; Gorman, R. R.; Johnson, R. A. *JACS* **1982**, *104*, 1621.
12. (a) San Filippo, Jr., J.; Chern, C.-I.; Valentine, J. S. *JOC* **1975**, *40*, 1678. (b) Chern, C.-I.; DiCosimo, R.; De Jesus, R.; San Filippo, Jr., J. *JACS* **1978**, *100*, 7317. (c) Corey, E. J.; Nicolaou, K. C.; Shibasaki, M.; Machida, Y.; Shiner, C. S. *TL* **1975**, 3183.
13. Gibian, M. J.; Ungermann, T. *JOC* **1976**, *41*, 2500.
14. Corey, E. J.; Nicolaou, K. C.; Shibasaki, M. *CC* **1975**, 658.
15. Praefcke, K.; Stephan, W. *LA* **1987**, 645.
16. Willis, C. L. *TL* **1987**, *28*, 6705.
17. (a) Castro, B. R. *OR* **1983**, *29*, 1. (b) Mitsunobu, O. *S* **1981**, 1.
18. Still, W. C.; Galynker, I. *JACS* **1982**, *104*, 1774.
19. San Filippo, Jr., J.; Romano, L. J.; Chern, C.-I.; Valentine, J. S. *JOC* **1976**, *41*, 586.
20. (a) Gibian, M. J.; Sawyer, D. T.; Ungermann, T.; Tangpoonpholvivat, R.; Morrison, M. M. *JACS* **1979**, *101*, 640. (b) Forrester, A. R.; Purushotham, V. *JCS(P1)* **1987**, 945.
21. Johnson, R. A. *TL* **1976**, 331.
22. (a) Stanley, J. P. *JOC* **1980**, *45*, 1413. (b) Nagano, T.; Arakane, K.; Hirobe, M. *CPB* **1980**, *28*, 3719.
23. (a) Oae, S.; Takata, T. *TL* **1980**, *21*, 3689. (b) Oae, S.; Takata, T.; Kim, Y. H. *BCJ* **1981**, *54*, 2712.
24. Kim, Y. H.; Chung, B. C. *JOC* **1983**, *48*, 1562.
25. Nagano, T.; Yamamoto, H.; Hirobe, M. *JACS* **1990**, *112*, 3529.
26. (a) Akutagawa, K.; Furukawa, N.; Oae, S. *BCJ* **1984**, *57*, 1104. (b) Yamamoto, H.; Mashino, T.; Nagano, T.; Hirobe, M. *JACS* **1986**, *108*, 539.
27. Miura, M.; Nojima, M.; Kusabayashi, S. *CC* **1982**, 1352.
28. Kim, Y. H.; Lim, S. C.; Chang, H. S. *CC* **1990**, 36.
29. Yamamoto, H.; Mashino, T.; Nagano, T.; Hirobe, M. *TL* **1989**, *30*, 4133.

30. Kim, Y. H.; Yoon, D. C. *SC* **1989**, *19*, 1569.
31. Nagano, T.; Yokoohji, K.; Hirobe, M. *TL* **1984**, *25*, 965.
32. Kim, Y. H.; Yoon, D. C. *TL* **1988**, *29*, 6453.
33. Kim, Y. H.; Lee, H. K. *CL* **1987**, 1499.
34. Kim, Y. H.; Kim, K. S.; Lee, H. K. *TL* **1989**, *30*, 6357.
35. Kim, Y. H.; Lee, H. K.; Chang, H. S. *TL* **1987**, *28*, 4285.
36. (a) Levonowich, P. F.; Tannenbaum, H. P.; Dougherty, R. C. *CC* **1975**, 597. (b) Yamaguchi, T.; van der Plas, H. C. *RTC* **1977**, *96*, 89.
37. Frimer, A.; Rosenthal, I. *TL* **1976**, 2809.
38. (a) Sagae, H.; Fujihira, M.; Komazawa, K.; Lund, H.; Osa, T. *BCJ* **1980**, *53*, 2188. (b) Gareil, M.; Pinson, J.; Savéant, J. M. *NJC* **1981**, *5*, 311.
39. Gibian, M. J.; Russo, S. *JOC* **1984**, *49*, 4304.
40. Frimer, A. A.; Gilinsky, P. *TL* **1979**, 4331.
41. Frimer, A. A.; Gilinsky-Sharon, P.; Hameiri, J.; Aljadeff, G. *JOC* **1982**, *47*, 2818.
42. Rosenthal, I.; Frimer, A. *TL* **1976**, 2805.
43. Frimer, A. A.; Rosenthal, I.; Hoz, S. *TL* **1977**, 4631.
44. Rosenthal, I.; Frimer, A. *TL* **1975**, 3731.
45. Neckers, D. C.; Hauck, G. *JOC* **1983**, *48*, 4691.
46. Crank, G.; Makin, M. I. H. *AJC* **1984**, *37*, 845.
47. Balogh-Hergovich, É.; Speier, G.; Winkelmann, É. *TL* **1979**, 3541.
48. Frimer, A. A.; Aljadeff, G.; Ziv, J. *JOC* **1983**, *48*, 1700.
49. Stuehr, D. J.; Marletta, M. A. *JOC* **1985**, *50*, 694.
50. Collins, T. J.; Gordon-Wylie, S. W.; Crank, G. *Chem. Eng. News* **1990**, *Aug. 27*, 2.
51. Crank, G.; Makin, M. I. H. *AJC* **1984**, *37*, 2331.
52. Oae, S.; Takata, T.; Kim, Y. H. *T* **1981**, *37*, 37.
53. De Min, M.; Croux, S.; Tournaire, C.; Hocquaux, M.; Jacquet, B.; Oliveros, E.; Maurette, M.-T. *T* **1992**, *48*, 1869.
54. Moro-oka, Y.; Foote, C. S. *JACS* **1976**, *98*, 1510.
55. Lee-Ruff, E.; Lever, A. B. P.; Rigaudy, J. *CJC* **1976**, *54*, 1837.
56. Matsumoto, S.; Matsuo, M.; Iitaka, Y. *JOC* **1986**, *51*, 1435.
57. Matsuo, M.; Matsumoto, S.; Iitaka, Y. *JOC* **1987**, *52*, 3514.
58. Sugawara, M.; Baizer, M. M. *TL* **1983**, *24*, 2223.
59. Sotiriou, C.; Lee, W.; Giese, R. W. *JOC* **1990**, *55*, 2159.
60. Hesse, W.; Jansen, M. *IC* **1991**, *30*, 4380.
61. Neuman, E. W. *JCP* **1934**, *2*, 31; as recounted by Pauling, L. *Trends Biochem. Sci.* **1979**, *4*, N270.

Roy A. Johnson
The Upjohn Company, Kalamazoo, MI, USA

Potassium Tri-*s*-butylborohydride[1]

K(*s*-Bu)$_3$BH

[54575-49-4] $C_{12}H_{28}BK$ (MW 222.31)

(reducing agent for various functional groups;[2] selective reducing agent;[3–7] stereoselective reducing agent for ketones;[8–10] used for reduction of conjugated enones;[11] regioselective reducing agent for cyclic anhydrides;[12] reacts with carbon monoxide in the presence of free trialkylborane;[13] used for stereoselective synthesis of *cis*-alkenylboranes[14])

Alternate Names: K-Selectride; potassium hydrotris(1-methylpropyl)borate.
Physical Data: not isolated; prepared and used in solution.
Solubility: solubility limits have not been established. The reagent is normally used as a 1.0 M solution in THF or Et_2O, but use of toluene as solvent is also reported.
Form Supplied in: 1.0 M solutions in THF or Et_2O.
Analysis of Reagent Purity: solutions of pure reagent exhibit doublets ($J = 68$–71 Hz) centered in the range δ −7.1 to δ −7.5 in the ^{11}B NMR spectrum.[15,16a,17] Concentration of hydride is determined by hydrolysis of aliquots and measurement of the hydrogen evolved or by quenching aliquots in excess 1-iodooctane and analysis of the octane formed by GLC.[18] Concentration of boron is verified by oxidizing aliquots using alkaline hydrogen peroxide and analyzing the 2-butanol formed by GLC.[19]
Preparative Methods: direct reaction of tris(1-methylpropyl)borane with ***Potassium Hydride*** is satisfactory,[15] especially with activated KH.[16] The reagent is also formed by reaction of the trialkylborane with ***Potassium Triisopropoxyborohydride***.[17]
Handling, Storage, and Precautions: solutions are air- and water-sensitive and should be handled in a well ventilated hood under an inert atmosphere using appropriate techniques.[20] Potentially pyrophoric tris(1-methylpropyl)borane is a byproduct of reductions; thus standard organoborane oxidation[19] is recommended as part of the workup.

Functional Group Reductions. The reducing characteristics of the reagent have been studied systematically.[2a] At 0 °C it liberates H_2 quantitatively with various active hydrogen compounds, including primary alcohols, phenols, thiols, carboxylic acids, primary amides, aliphatic nitro compounds, and oximes. Secondary and tertiary alcohols and primary amines are inert. Carboxylic acids, amides, nitro compounds, and oximes are not reduced further. Aldehydes, ketones, and acyl chlorides are reduced rapidly to the alcohol stage. Acyclic esters are reduced more slowly, but lactones are rapidly reduced to diols. Cyclic anhydrides are reduced to lactones if steps are taken to avoid further reduction of the latter. Terminal epoxides are reduced rapidly and cleanly to the Markovnikov alcohols. Internal epoxides are reduced slowly. Nitriles are reduced sluggishly. Pyridine and quinoline consume hydride, but the products have not been characterized. Disulfides are reduced to thiols. Primary iodides and bromides are reduced rapidly, chlorides more slowly. Cyclohexyl bromide and tosylate react sluggishly.

A reagent derived by adding 2 equiv of K-Selectride to 1 equiv of CuI effectively displaces halide from various aryl halides, and in one example an internal alkyne, dec-5-yne, is converted predominantly to the (*Z*)-alkene (*Z*:*E* = 88:12).[2b]

Selective Reductions. Ketones are reduced selectively in the presence of other functional groups: esters,[3] including methyl esters[3a,d] and γ-lactones;[3b] a cyclic carbamate;[4] amides[5] including 2-acetyl β-lactams[5a–c] and benzenesulfonamides;[5f] azides;[6] and internal epoxides.[3d] Steroid 3,17- or 3,20-dione systems undergo selective reduction at the 3-position,[7] as does a 5-ene-3,17-dione system.[11c] Conjugated enones are reduced selectively in the presence of esters.[11e–g]

Stereoselective Reductions. Like L-Selectride (***Lithium Tri-s-butylborohydride***), K-Selectride is useful for diastereoselective reduction of alkyl-substituted cyclic ketones to alcohols.[15a]

Other applications to various types of cyclic ketones have been reported.[3d,4,7,8] Diastereoselective reduction is also observed for acyclic ketones,[9] including keto esters,[3a–c] keto amides,[5] and a keto azide.[6] A chiral dione is diastereoselectively reduced to a diol.[10]

Various theoretical models are used to rationalize observed diastereoselectivity.[21] In simple cases, the bulk of the reagent appears to dictate that hydride approach from the less sterically hindered face of the carbonyl.[15a] However, complexation effects of heteroatoms,[8a] crown ethers,[3c,9i] and cryptands[9i] may reverse the direction of approach. Also, the coordinating ability of the solvent influences the ratio of diastereomers.[3b]

Reduction of Conjugated Enones. A reasonably systematic study of conjugated enone reductions has been conducted.[11a] Acyclic enones undergo 1,2-addition with stereoselective formation of allylic alcohols upon protonation. Cyclohexenones with no β-substitution yield saturated ketones as a consequence of 1,4-addition with 1 equiv of hydride, while a β-methyl results in exclusive 1,2-addition; 3 equiv of hydride results in stereoselective formation of saturated alcohols. Both cyclopentenones and cycloheptenones give mixtures of products. The enolate resulting from 1,4-addition may be alkylated instead of protonated, but this α-alkylation appears to work better with L-Selectride. Attempted reduction of α,β-unsaturated esters leads to Claisen condensation products.

Further examples of 1,2-addition to cyclohexenones involve the 5-en-3-one function of steroids[11b,c] or analogous systems.[11b] This is reduced selectively in 2α-fluoro-4-androstene-3,17-dione.[11c] Stereoselectivity in reduction of testosterone, which yields predominantly the 3β,17β-diol (3β:3α = 88:12), is reversed for 2α-fluorotestosterone, which yields exclusively the 3α,17β-diol.[11c]

A conjugated cyclohexadienone undergoes 1,4-addition so that only the α,β-double bond is reduced.[11d] Stereoselectivity results on 1,4-addition to (*S*)-(+)-carvone (eq 1).[11e]

K-Selectride, THF, –78 °C, 96% (1)

X = Me; Y = H (79%)
X = H; Y = Me (17%)

An enone incorporated into a 14-membered macrolide undergoes stereoselective 1,2-addition while the lactone is not affected.[11f] With 3 equiv of K-Selectride, the cyclopentenone in the prostanoid 15(*S*)-PGA$_2$ is reduced stereoselectively to the corresponding cyclopentanol while a methyl ester remains untouched.[11g] A procedure utilizing 2 equiv of K-Selectride in the presence of 2 equiv of ethanol reduces a cyclohexenone stereoselectively to the cyclohexanol in the presence of methyl ester, acetate, and γ-lactone functions (eq 2).[11h] An α,β-alkynic ketone undergoes exclusive 1,2-addition with stereoselective formation of the propargylic alcohol.[9b]

93% (2)

Regioselective Reduction of Cyclic Anhydrides. Various succinic and phthalic anhydrides are reduced by the reagent to lactones.[12] With certain exceptions (eq 3),[12b] reduction occurs preferentially at the less hindered carbonyl function. A mechanistic interpretation of the results has been proposed.[12b,e]

K-Selectride, –70 to 0 °C, 1.5 h, 50% (3)

only product

Reaction with Carbon Monoxide. Carbonylation of KBH(*s*-Bu)$_3$ followed by treatment with refluxing aqueous ***Sodium Hydroxide*** yields 2-methyl-1-butanol.[13a] It is now evident that KBH(*s*-Bu)$_3$ is an intermediate in the hydride-induced carbonylation of *s*-Bu$_3$B using ***Potassium Triisopropoxyborohydride***,[13b,17] so that different workup procedures may produce 2-methylbutanal[13c] or 3,5-dimethyl-4-heptanol.[13d]

Stereoselective Synthesis of *cis*-Alkenylboranes. Reaction of 1-halo-1-alkenylboranes with KBH(*s*-Bu)$_3$ produces *cis*-alkenylboranes, but the ease of removing triethylborane makes ***Lithium Triethylborohydride*** the reagent of choice.[14]

Related Reagents. See Classes R-2, R-4, R-6, R-10, R-12, R-13, R-17, R-24, R-27, and R-32, pages 1–10.

---

1. (a) Hajos, A. *Complex Hydrides and Related Reducing Agents in Organic Synthesis*; Elsevier: New York, 1979. (b) Brown, H. C.; Krishnamurthy, S. *T* **1979**, *35*, 567. (c) Hudlicky, M. *Reductions in Organic Chemistry*; Horwood: Chichester, 1984. (d) Seyden-Penne, J. *Reductions by the Alumino- and Borohydrides in Organic Synthesis*; VCH: New York, 1991.
2. (a) Yoon, N. M.; Hwang, Y. S.; Yang, H. *Bull. Korean Chem. Soc.* **1989**, *10*, 382. (b) Yoshida, T.; Negishi, E.-I. *CC* **1974**, 762.
3. (a) Tal, D. M.; Frisch, G. D.; Elliott, W. H. *T* **1984**, *40*, 851. (b) Niwa, H.; Ogawa, T.; Yamada, K. *BCJ* **1990**, *63*, 3707. (c) Akiyama, T.; Nishimoto, H.; Ozaki, S. *TL* **1991**, *32*, 1335. (d) Sakai, K.; Takahashi, K.; Nukano, T. *T* **1992**, *48*, 8229.
4. Kano, S.; Yuasa, Y.; Mochizuki, N.; Shibuya, S. *H* **1990**, *30*, 263.
5. (a) Shibuya, M.; Kuretani, M.; Kubota, S. *T* **1982**, *38*, 2659. (b) Pecquet, F.; D'Angelo, J. *TL* **1982**, *23*, 2777. (c) Bateson, J. H.; Fell, S. C. M.; Southgate, R.; Eggleston, D. S.; Baures, P. W. *JCS(P1)* **1992**, 1305. (d) Ito, Y.; Katsuki, T.; Yamaguchi, M. *TL* **1985**, *26*, 4643. (e) Ookawa, A.; Soai, K. *JCS(P1)* **1987**, 1465. (f) Roemmele, R. C.; Rapoport, H. *JOC* **1989**, *54*, 1866. (g) Wanner, K. T.; Höfner G. *T* **1991**, *47*, 1895.
6. Ramza, J.; Zamojski, A. *T* **1992**, *48*, 6123.
7. (a) Göndös, G.; Orr, J. C.; *CC* **1982**, 1239. (b) Templeton, J. F.; Sashi Kumar, V. P.; Kim R. S.; LaBella, F. S. *JCS(P1)* **1987**, 1361.
8. (a) Wigfield, D. C.; Feiner, S. *CJC* **1978**, *56*, 789. (b) Suzuki, K.; Ikegawa, A.; Mukaiyama, T. *CL* **1982**, 899. (c) Boegesoe, K. P. *JMC* **1983**, *26*, 935. (d) Nakata, T.; Takao, S.; Fukui, M.; Tanaka, T.; Oishi, T. *TL* **1983**, *24*, 3873. (e) Carreño, M. D.; Domínguez, E.; García-Ruano, J. L.; Rubio, A. *JOC* **1987**, *52*, 3619. (f) Barrett, A. G. M.; Edmunds, J. J.; Horita, K.; Parkinson, C. J. *CC* **1992**, 1236.
9. (a) Ko, K.-Y.; Frazee, W. J.; Eliel, E. L. *T* **1984**, *40*, 1333. (b) Takahashi, T.; Miyazawa, M.; Tsuji, J. *TL* **1985**, *26*, 5139. (c) Elliott, J.; Warren, S. *TL* **1986**, *27*, 645. (d) Davis, F. A.; Haque, M. S.; Przeslawski, R. M. *JOC* **1989**, *54*, 2021. (e) Wade, P. A.; Price, D. T.; Carroll, P. J.; Dailey,

W. P. *JOC* **1990**, *55*, 3051. (f) Hanamoto, T.; Fuchikami, T. *JOC* **1990**, *55*, 4969. (g) Richardson, D. P.; Wilson, W.; Mattson, R. J.; Powers, D. M. *JCS(P1)* **1990**, 2857. (h) Dondoni, A.; Orduna, J.; Merino, P. *S* **1992**, 201. (i) Manzoni, L.; Pilati, T.; Poli, G.; Scolastico, C. *CC* **1992**, 1027.

10. Achmatowicz, B.; Wicha, J. *TL* **1987**, *28*, 2999.
11. (a) Fortunato, J. M.; Ganem, B. *JOC* **1976**, *41*, 2194. (b) Dauben, W. G.; Ashmore, J. W. *TL* **1978**, 4487. (c) Göndös, G.; McGirr, L. G.; Jablonski, C. R.; Snedden, W.; Orr, J. C. *JOC* **1988**, *53*, 3057. (d) Lombardo, L.; Mander, L. N.; Turner, J. V. *JACS* **1980**, *102*, 6626. (e) Maestro, M. A.; Castedo, L.; Mouriño, A. *JOC* **1992**, *57*, 5208. (f) Keller, T. H.; Weiler, L. *TL* **1990**, *31*, 6307. (g) Cai, Z.; Nassim, B.; Crabbé, P. *JCS(P1)* **1983**, 1573. (h) Lombardo, L.; Mander, L. N.; Turner, J. V. *AJC* **1981**, *34*, 745.
12. (a) Morand, P.; Salvador, J.; Kayser, M. M. *CC* **1982**, 458. (b) Kayser, M. M.; Salvador, J.; Morand, P. *CJC* **1983**, *61*, 439. (c) Mann, J.; Piper, S. E.; Yeung, L. K. P. *JCS(P1)* **1984**, 2081. (d) Mann, J.; Wong, L. T. F.; Beard, A. R. *TL* **1985**, *26*, 1667. (e) Soucy, C.; Favreau, D.; Kayser, M. M. *JOC* **1987**, *52*, 129.
13. (a) Brown, H. C.; Hubbard, J. L. *JOC* **1979**, *44*, 467. (b) Hubbard, J. L.; Smith, K. *JOM* **1984**, *276*, C41. (c) Brown, H. C.; Hubbard, J. L.; Smith, K. *S* **1979**, 701. (d) Hubbard, J. L.; Brown, H. C. *S* **1978**, 676.
14. Negishi, E.-I.; Williams, R. M.; Lew, G.; Yoshida, T. *JOM* **1975**, *92*, C4.
15. (a) Brown, C. A. *JACS* **1973**, *95*, 4100. (b) Brown, C. A.; Krishnamurthy, S. *JOM* **1978**, *156*, 111.
16. (a) Soderquist, J. A.; Rivera, I. *TL* **1988**, *29*, 3195. (b) Hubbard, J. L. *TL* **1988**, *29*, 3197.
17. Brown, C. A.; Hubbard, J. L. *JACS* **1979**, *101*, 3964.
18. Brown, H. C.; Krishnamurthy, S.; Hubbard, J. L. *JACS* **1978**, *100*, 3343.
19. Zweifel, G.; Brown, H. C. *OR* **1963**, *13*, 1.
20. Brown, H. C.; Kramer, G. W.; Levy, A. B.; Midland, M. M. *Organic Syntheses via Boranes*; Wiley: New York, 1975; Chapter 9.
21. (a) Hutchins, R. O. *JOC* **1977**, *42*, 920. (b) Wigfield, D. G. *T* **1979**, *35*, 449. (c) Mead, K.; Macdonald, T. L. *JOC* **1985**, *50*, 422; and references cited therein. (d) Ibarra, C. A.; Pérez-Ossorio, R.; Quiroga, M. L.; Arias Pérez, M. S.; Fernández Dominguez, M. J. *JCS(P2)* **1988**, 101.

John L. Hubbard
Marshall University, Huntington, WV, USA

Pyridinium Chlorochromate[1]

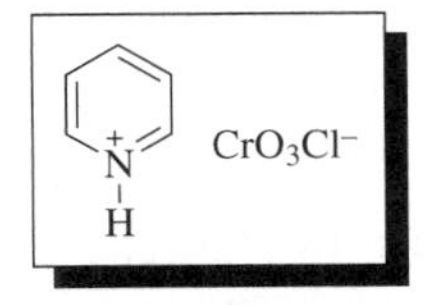

[26299-14-9] $C_5H_6ClCrNO_3$ (MW 215.57)

(stable versatile oxidizing agent for many functional groups;[1a] can oxidize activated C–H bonds,[2b] C–C bonds,[2b] and C–B bonds;[2b] can halogenate enol silyl ethers[2b])

Alternate Name: PCC.
Physical Data: mp 205 °C (dec).
Solubility: insol dichloromethane, benzene, diethyl ether; sol acetone, acetonitrile, THF.
Form Supplied in: yellow-orange solid; widely available.
Handling, Storage, and Precautions: the dry solid may be stored in contact with air, but in the absence of moisture. Reaction solvents should generally be purified and anhydrous. Reported to be a cancer suspect agent. Should be used in a fume hood.

Oxidation of Primary and Secondary Alcohols. PCC has the capability to convert primary and secondary alcohols to aldehydes and ketones with great efficiency. The yields are typically equal to or greater than those obtained by the Collins method (see ***Dipyridine Chromium(VI) Oxide***), which customarily requires a fivefold excess of reagent. PCC oxidations are normally carried out in dichloromethane with 1.5 equiv of reagent suspended in the organic solvent at room temperature and are usually complete within 1–2 h (eqs 1–4).[2a]

$$\text{Decanol} \xrightarrow[\text{CH}_2\text{Cl}_2,\ 92\%]{\text{PCC}} \text{Decanal} \quad (1)$$

$$\text{Oct-2-yn-1-ol} \xrightarrow[\text{CH}_2\text{Cl}_2,\ 84\%]{\text{PCC}} \text{Oct-2-ynal} \quad (2)$$

$$\text{Benzhydrol} \xrightarrow[\text{CH}_2\text{Cl}_2,\ 100\%]{\text{PCC}} \text{Benzophenone} \quad (3)$$

$$\text{Citronellol} \xrightarrow[\text{AcONa},\ 82\%]{\text{PCC}} \text{Citronellal} \quad (4)$$

More polar solvents, such as acetonitrile or acetone, in which PCC has higher solubility, lead to longer reaction times. Overoxidations are rare, but acids can be directly prepared from aldehydes with stoichiometric ***Sodium Cyanide*** and PCC in THF.[3] The reagent shows a slightly acidic character. With compounds bearing acid-sensitive groups, the reaction can be buffered with powdered sodium acetate. However, *cis–trans* isomerization of some allylic alcohols can occur under these conditions (eq 5).[2a]

$$\text{HO}\!\!\diagup\!\!=\!\!\diagdown\text{OTHP} \xrightarrow[\text{AcONa},\ 81\%]{\text{PCC}} \text{OHC}\diagdown\!\!=\!\!\diagup\text{OTHP} \quad (5)$$

PCC appears to be particularly suitable for moderate to large scale preparations, and it has proven to be the reagent of choice for the oxidation of primary alcohols to aldehydes and of secondary alcohols to ketones.[2a,b] Good yields of ketones and aldehydes are regularly obtained.[2a,b] Advantages of PCC include the fact that it does not possess the hygroscopic nature of the chromium trioxide–pyridine complex, and it is prepared via a much less hazardous procedure. However, workup and removal of chromium-containing byproducts is often tedious and difficult with the oxidant.[2a] Several modifications of PCC have been developed to increase both the efficiency and the selectivity of the reagent, and to reduce side reactions.[4] The chlorochromate anion has been supported on a polymeric matrix. Poly(vinylpyridinium chlorochromate), easily prepared, is a recyclable and useful reagent in the oxidation of alcohols to carbonyl compounds in 60–100% yield. This reagent has advantages with regard to product isolation, although its use is limited by the scale of the reaction that can be performed conveniently.[5] Furthermore, the reactivity of PCC in the oxidation of alcohols can be increased by adding anhydrous ***Acetic***

Acid as catalyst to the reaction mixture[6] or by using sonochemical conditions.[7] The addition of ***Molecular Sieves*** in the oxidation of various alcohols results in a dramatic rate enhancement.[8a] Comparative studies indicate that 3 Åsieves give the best results (3 Å > 4 Å > 10 Å > 5 Å).[8b] PCC can perform oxidations of slow-reacting alcohols, such as the secondary hydroxyl groups in carbohydrates and nucleosides. These reactions proceed readily in refluxing benzene, but they are very slow in dichloromethane.[9] The oxidations are conveniently carried out by adding 3 Åmolecular sieves to a suspension of PCC in dichloromethane, providing an efficient synthesis of keto sugars and keto nucleosides (eq 6).[8a]

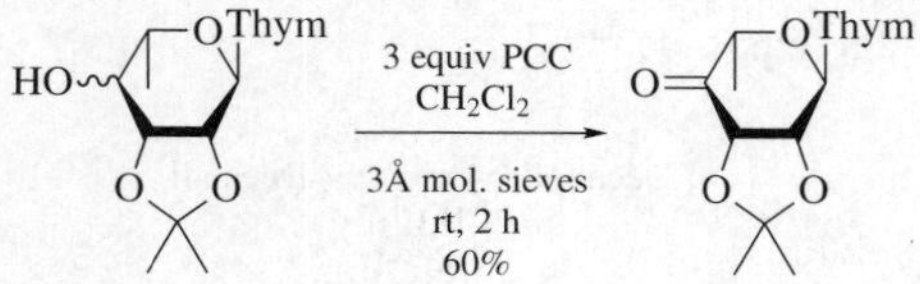

3 equiv PCC, CH_2Cl_2, 3Å mol. sieves, rt, 2 h, 60% (6)

The use of molecular sieves in many types of oxidations is now commonplace. 2-Nitro alkanols, available by a nitro-aldol reaction, can be oxidized by PCC and molecular sieves to α-nitro ketones in 60–85% yield. Usually the starting compounds undergo a retro-aldol reaction under acidic conditions (eq 7).[8c] On the other hand, (*Z*)-2-butene-1,4-diols are converted by reaction with PCC and molecular sieves into substituted furans in 35–90% yields (eq 8).[10]

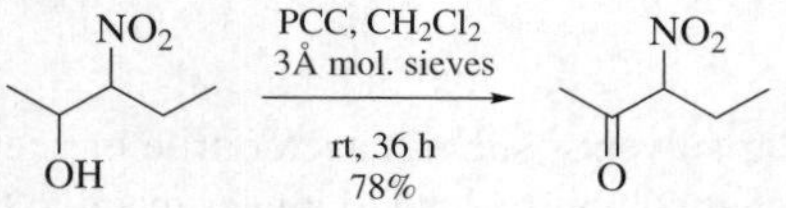

PCC, CH_2Cl_2, 3Å mol. sieves, rt, 36 h, 78% (7)

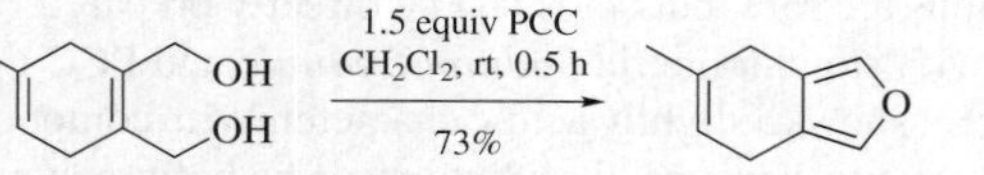

1.5 equiv PCC, CH_2Cl_2, rt, 0.5 h, 73% (8)

Synthetically useful changes in the properties of pyridinium chlorochromate have been introduced in the oxidation of hydroxy steroids. PCC, in refluxing benzene, is a convenient reagent for the oxidation and concomitant isomerization of steroidal Δ^5-3β-alcohols to the corresponding Δ^4-3-ketones in high yield.[11] However, in the presence of anhydrous calcium carbonate, PCC can effect the high yield, selective oxidation of steroidal homoallylic alcohols to the corresponding β,δ-unsaturated ketones.[12] Remarkable selectivity is obtained in the oxidation of steroid allylic alcohols by adding 2% of ***Pyridine*** to the reagent in dichloromethane at 2–3 °C. Quasiequatorial allylic alcohols are oxidized faster than nonallylic axial ones (eq 9).[13] Similar properties are found for the combination of PCC and pyrazole (2%),[14] PCC and 2,3-dimethylpyrazole (2%),[15] and PCC and benzotriazole (2%).[16]

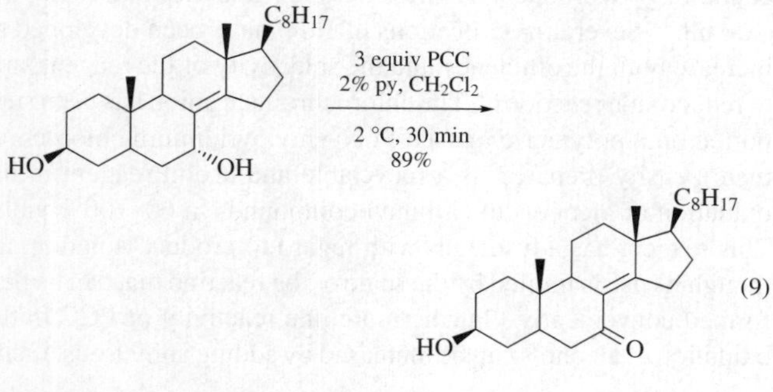

3 equiv PCC, 2% py, CH_2Cl_2, 2 °C, 30 min, 89% (9)

Oxidation of Unsaturated and Cyclopropyl Tertiary Alcohols. Tertiary allylic alcohols, obtained by the addition of ***Vinyllithium*** or vinylmagnesium chloride to a ketone, undergo oxidative rearrangements to α,β-unsaturated aldehydes in good yields in the presence of PCC. The absence of diene products is remarkable in view of the sensitivity of the starting compounds towards dehydration. The process involves a two-carbon chain-lengthening of the starting ketones, and is synthetically equivalent to a direct crossed-aldol condensation between a ketone and acetaldehyde (eq 10).[17]

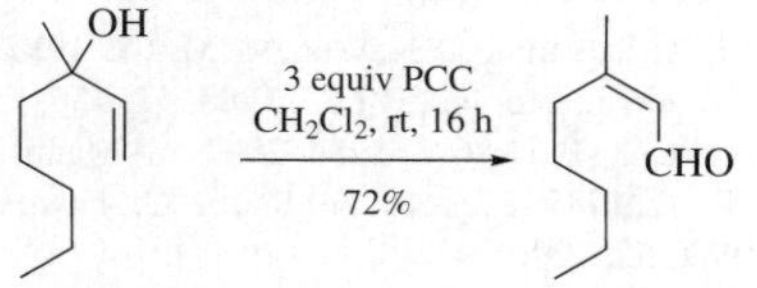

3 equiv PCC, CH_2Cl_2, rt, 16 h, 72% (10)

The oxidation of tertiary cyclopropyl alcohols leads to the formation of the corresponding β,γ-unsaturated ketones (eq 11). The reaction represents an excellent method for converting ketones to chain-extended β,γ-enones, since the starting compounds can be prepared by addition of cyclopropyl organometallic reagents to ketones.[18]

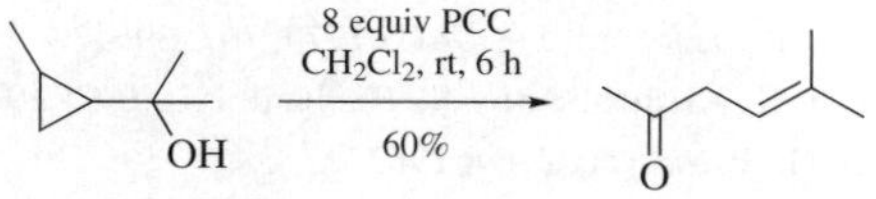

8 equiv PCC, CH_2Cl_2, rt, 6 h, 60% (11)

Cyclic tertiary allylic alcohols, prepared by 1,2-addition of alkyllithium reagents to α,β-enones, are oxidized by PCC to transposed γ-alkyl-α,β-unsatura ted ketones. Yields are excellent in the case of cyclic alcohols, but only moderate with acyclic alcohols. The overall result of this reaction sequence is an efficient method for alkylative 1,3-carbonyl transposition (eq 12).[19a–d]

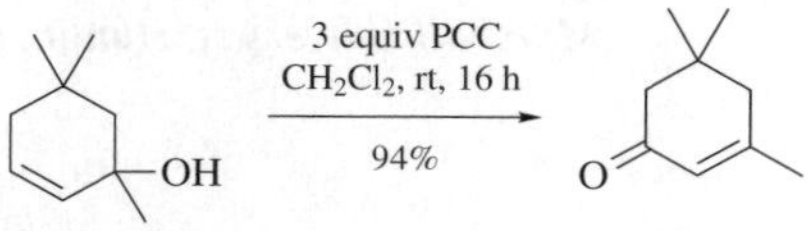

3 equiv PCC, CH_2Cl_2, rt, 16 h, 94% (12)

When the tertiary allylic alcohol is suitably disposed near an alkene double bond, PCC is able to perform synthetically useful substituent-directed oxidations, the result of which depends on the relative position of the two groups. Cyclic tertiary γ-hydroxy alkenes can undergo a regio- and stereoselective cyclization, giving a single β-hydroxy cyclic ether through an intramolecular C=C oxidation initiated by PCC (eq 13),[20a] while linear γ,δ-dihydroxy alkenes are cyclized to *cis*-2,5-disubstituted tetrahydrofurans (eq 14).[20b]

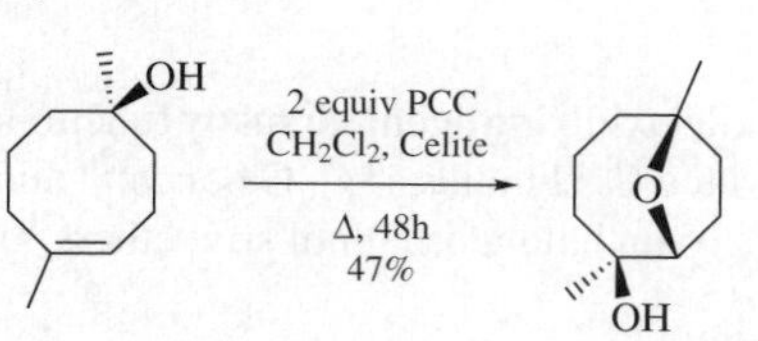

2 equiv PCC, CH_2Cl_2, Celite, Δ, 48h, 47% (13)

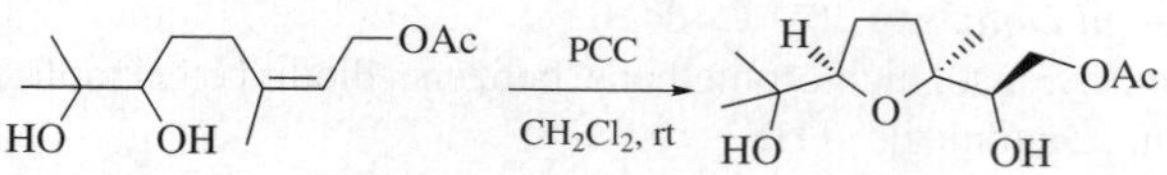

PCC, CH_2Cl_2, rt (14)

On the other hand, the oxidation of tertiary γ-hydroxy terminal alkenes brings about both oxidative cyclization and fragmentation of the double bond with loss of one carbon atom, and thus provides an efficient route to γ-lactones (eq 15).[21]

5 equiv PCC; CH_2Cl_2, Δ, 48 h; 53% (15)

When a tertiary alcohol is homoallylic, reaction with PCC can yield a ketone resulting from oxidative fragmentation. The cyclic tertiary homoallylic alcohol shown in eq 16 furnishes a mixture of two isomeric cyclopentenones in a 1:1.7 ratio in 62% yield.[22] Another interesting example of PCC reactivity is the conversion of the acyclic tertiary homoallylic alcohol in eq 17 to a keto lactone, albeit in moderate yield. A large excess of oxidant is necessary.[21d]

3 equiv PCC; CH_2Cl_2, Celite; HOAc, Δ, 4 h (16)

5 equiv PCC; CH_2Cl_2, 30 °C, 24 h; 42% (17)

If a homoallylic alcohol is secondary, reaction with PCC leads to the formation of a mixture of γ-hydroxy-α,β-unsatu rated enones, through a γ-oxy functionalization (eq 18).[23]

2 equiv PCC, CH_2Cl_2; 4Å mol. sieves; 30 °C, 1 h; 65% (18)

Oxidative Cationic Cyclization. The mild acidic character of PCC has been used to advantage in an essential one step conversion of (−)-citronellol to (−)-pulegone (eq 19).[24]

2.5 equiv PCC; CH_2Cl_2, rt, 36 h; EtOH; NaOH; 70% (19)

The oxidative cationic cyclization is useful for the annulation of linear and cyclic unsaturated alcohols or aldehydes to form cyclohexenones (eqs 20 and 21).[25]

1. 3 equiv PCC, 4.5 h; 2. H^+; 78% (20)

1. 4 equiv PCC, 16 h; 2. H^+; 69% (21)

A limitation is that this reaction cannot be used for formation of cyclopentenones. The reaction is also limited to preparation of β,β-disubstituted α,β-unsaturated cyclohexenones: in fact, cyclization is only observed with substrates capable of affording a tertiary cation as the initial cyclic intermediate. Other substrates (e.g. **1**) do not undergo cyclization, even under more forcing conditions. However, the process represents a milder and more efficient alternative to the two-step cationic cyclization for preparing α,β-unsaturated enones.[26]

(1)

Oxidation of Activated C–H Bonds. PCC is of value in the allylic oxidation of compounds containing activated methylenes. 5,6-Dihydropyrans (eq 22)[27] and 2,5-dihydrofurans (eqs 23 and 24) are oxidized to the corresponding lactones.[28] The reaction is limited to cyclic benzylic and allylic ethers.

3 equiv PCC; py, CH_2Cl_2, Δ; 85% (22)

PCC, py; CH_2Cl_2, Δ; 35% (23)

PCC, py; CH_2Cl_2, Δ; 85% (24)

PCC is the reagent of choice in the allylic oxidation of Δ^5-steroids to 7-keto derivatives. The reaction can be carried out using PCC in refluxing benzene or PCC in DMSO solution at 100 °C (eq 25).[29] These solvents are claimed to be superior to dichloromethane, which must be used in large excess in this reaction.[30]

C_8H_{17}; $PhCO_2$; 1. PCC, C_6H_6 (89%); 2. PCC, DMSO (78%) (25)

The capability of PCC to selectively perform allylic oxidation of methylene groups has frequently been utilized,[31] as in the synthesis of a key intermediate related to the taxol ring A (eq 26).[31b]

2 equiv PCC, 4Å mol. sieves, CH_2Cl_2, rt, 75% (26)

In a similar fashion, the smooth oxidation of benzylic hydrocarbons to aryl ketones can be achieved by adding a powdered and homogenized mixture of PCC and ***Celite*** to a benzene solution of the substrates (eqs 27 and 28).[32] The reaction has been successfully applied to a variety of compounds, such as 1,2,3,4-tetrahydrophenanthrene,[33] 12-methoxypodocarpa-8,11,13-triene,[34] and biscyclophanes.[35]

5 equiv PCC, Celite, C_6H_6, Δ, 8 h, 80% (27)

Ph ⟶ Ph–CO–Me: 5 equiv PCC, Celite, C_6H_6, rt, 15 h, 71% (28)

The transformation of activated methylene groups to ketones shows particular applicability to the conversion of benzyl alkyl ketones to the corresponding 1,2-diketones in high yields (eq 29).[28] Since the starting materials appear to be sensitive to the acidic PCC, anhydrous pyridine must be added to the reaction. This oxidation fails with dialkyl ketones.[28]

3 equiv PCC, py, CH_2Cl_2, Δ, 25 h, 81% (29)

PCC is able to oxidize cyclic 1,4-dienes to dienones (eq 30).[36] PCC does not effect oxidation of isolated double bonds, or of more reactive systems such as diphenylmethane or allylbenzene. Complementary selectivity (1:3) is achieved with the Collins reagent.

PCC, CH_2Cl_2 (30) 63% + 7%

Phenyloxiranes can undergo a one step C–C bond cleavage to carbonyl compounds by reaction with PCC and molecular sieves (52–75%) (eq 31).[37] The presence of a phenyl group seems to be essential for efficient C–C cleavage. Under similar conditions, alkyl-substituted oxiranes are prevalently converted into α-hydroxy ketones (eq 32).

2 equiv PCC, CH_2Cl_2, 4Å mol. sieves, 40 °C, 45 min, 75% (31)

2 equiv PCC, CH_2Cl_2, 4Å mol. sieves, 20 °C, 1.5 h; 9% + 66% (32)

PCC is reported to perform oxidative fragmentation of carbon–carbon bonds of 1,2-diols to give carbonyl compounds in excellent yields (eq 33).[38] Hydroxymethyl cyclic ethers also undergo fragmentation; tetrahydrofuranmethanol derivatives lead to γ-lactones in 52–95% yield (eqs 34 and 35),[39] while 5-hydroxymethyl groups of 1-methoxy-2,3-*O*-isopropylidene-D-ribose are cleaved to γ-lactone derivatives in 50–60% yield (eq 36).[40] These transformations usually require more vigorous conditions than those needed for alcohol oxidation.

2 equiv PCC, CH_2Cl_2, rt, 1.5 h, 79% (33)

4 equiv PCC, CH_2Cl_2, 3Å mol. sieves, Δ, 8 h, 52% (34)

4 equiv PCC, CH_2Cl_2, 3Å mol. sieves, Δ, 3.5 h, 73% (35)

4 equiv PCC, C_6H_6, Δ, 15 h, 56% (36)

Oxidation of C=C Activated Double Bonds. PCC can behave as an oxidizing, weakly electrophilic species, capable of attacking particularly nucleophilic alkenes and of bringing about interesting reactions. PCC oxidizes linear and cyclic enol ethers to esters and lactones in 75–95% yield (eqs 37 and 38).[41]

2 equiv PCC, CH_2Cl_2, rt, 1 h, 85% (37)

2 equiv PCC, CH_2Cl_2, rt, 1 h, 75% (38)

According to the reported mechanism, the key step involves a 1,2-hydride shift from the α- to the β-carbon of the ether. When the enol ethers are α,α′-disubstituted, there is no possibility for such a hydride shift; reaction with PCC then yields a smooth oxidative cleavage of the nucleophilic alkene to esters or keto lactones in 45–90% yield (eqs 39 and 40).[42]

PCC, CH_2Cl_2, rt, 1 h, 85% (39)

PCC, CH_2Cl_2, rt, 2 h, 65%; $MeCO_2Et$ (40)

The reaction can be applied only to cyclic enol ethers. However, the simplicity of the procedure and the high yields obtained make the methodology synthetically useful for preparative purposes. In fact, the reaction has been extensively applied as a general approach to γ-lactones from the corresponding α-alkylidene five-membered enol ethers using as conditions treatment with 4 equiv of PCC and celite at room temperature in dichloromethane (eqs 41 and 42).[43] An important feature of this procedure is that, under the reaction conditions, other isolated carbon–carbon double bonds and benzylic groups present in the molecule are not affected.

4 equiv PCC, Celite, CH_2Cl_2, rt, 2 h, 90% (41)

4 equiv PCC, Celite, CH_2Cl_2, rt, 2 h, 75% (42)

PCC has proven to be the reagent of choice for the selective oxidation of a specific class of cyclic enol ethers: 1,4-dioxenyl alcohols. The allylic alcohols can be easily obtained by addition of 2-lithio-5,6-dihydro-1,4-dioxin to ketones and aldehydes. Reaction with PCC thus provides a method for one-carbon homologation of ketones and aldehydes to α-hydroxy acids (39–61%) (eq 43) and α-keto acids (54%) (eq 44), respectively.[44] The oxidation occurs regiospecifically at the dioxene site. While the oxidation of the allylic alcohols occurs in a few minutes, the α,β-unsaturated ketones require a longer reaction time, owing to the deactivation of the carbon–carbon double bond.

PCC, CH_2Cl_2, 4Å mol. sieves; NaOH; 58% (43)

PCC, CH_2Cl_2, 4Å mol. sieves; NaOH; 54% (44)

Aryl-substituted alkenes can be selectively cleaved in the presence of alkyl-substituted alkenes. Treatment of acyclic aryl alkenes with PCC and celite in dichloromethane under reflux results in oxidative opening of the carbon–carbon double bond to the corresponding carbonyl compounds in 72–90% yield (eq 45).[45] Alkyl-substituted acyclic alkenes are in general unreactive towards PCC, while cyclic alkenes undergo allylic oxidation, albeit in low yield (eq 46).[45]

5 equiv PCC, Celite, CH_2Cl_2, Δ, 25 h, 82% (45)

5 equiv PCC, Celite, CH_2Cl_2, Δ, 17% (46)

PCC has demonstrated its particular reactivity towards furan derivatives, which are well known for their sensitivity to oxidizing and electrophilic reagents.[46] Thus 5-methyl-2-furylcarbinols undergo oxidative ring enlargement to hexenuloses in very high yields (90–94%) (eq 47).[47]

2 equiv PCC, CH_2Cl_2, rt, 1 h, 91% (47)

Another application involves 5-bromo-2-furylcarbinols, which are converted by PCC into γ-hydroxybutenolides in 60–75% yield (eq 48).[48] Both conversions point out an unusual regiospecific reactivity of pyridinium chlorochromate, since only the oxidation of the furan ring occurs, in spite of the presence of a secondary alcoholic function, which remains untouched. Different behavior is observed when the heteroaromatic nucleus is deactivated by the presence of a nitro group; in this case, PCC oxidizes the alcohol function, leading to alkyl 5-nitro-2-furyl ketones (eq 49).[49]

2 equiv PCC, CH_2Cl_2, rt, 1.5 h, 60% (48)

3 equiv PCC, CH_2Cl_2, rt, 2 h, 77% (49)

A further and important application of PCC involves the conversion of 2,5-dialkylfurans, via oxidative ring fission, to *trans*-α,β-unsaturated 1,4-dicarbonyl compounds in a very simple manner. The reagent must be utilized in a large excess in dichloromethane at reflux for 24 h. Invariably, the products are obtained with *trans* configuration. The *cis*-isomers are formed first and are then isomerized by the acidic reagent (eq 50).[50] Lower yields of enedicarbonyl compounds are obtained from 2-alkylfurans (eq 51).[50]

PCC, CH_2Cl_2, Δ, 24 h, 90% (50)

5 equiv PCC, CH_2Cl_2, rt, 24 h, Δ, 9 h, 60% (51)

PCC shows a remarkable reactivity towards more nucleophilic furans, such as 2-alkylthiofurans, which are rapidly converted by action of the reagent in dichloromethane at room temperature for 2 h into *S*-alkyl 4-oxo-2-alkenethioates in high yields (eq 52).[51] (*Z*)–(*E*) Isomerization can be observed with longer reaction times, due to the acidic character of PCC. 2-Methoxy-5-methylfuran, which requires only 10 min of reaction, affords methyl (*Z*)-4-oxo-2-pentenoate, without the (*E*)-isomer being detectable (eq 53).[51]

5 equiv PCC, CH_2Cl_2, rt, 2 h; 89% (52)

5 equiv PCC, CH_2Cl_2, 10 min; 70% (53)

Oxidation of Carbon–Boron Bonds. PCC is a superior reagent for the oxidation of organoboranes to carbonyl derivatives by a convenient and mild procedure. Trialkylboranes, obtained by treatment of linear and cyclic di- and trisubstituted alkenes with ***Boron Trifluoride–Lithium Borohydride*** or ***Borane–Tetrahydrofuran***, afford the corresponding ketones by reaction with PCC in high yields (eqs 54 and 55).[52,53] Similar oxidation of dialkylchloroboranes, derived from cyclic alkenes by reaction with ***Monochloroborane–Dimethyl Sulfide***, gives ketones in 70–85% yield (eq 56).[54]

1. BF_3, $LiBH_4$; 2. 3 equiv PCC, CH_2Cl_2, rt, 10 h; 81% (54)

1. H_3B•THF; 2. PCC, CH_2Cl_2, 3Å mol sieves, Δ, 3h; 79% (55)

1. BH_2Cl•SMe_2; 2. H_2O, py; 3. 3 equiv PCC, CH_2Cl_2, Δ, 3 h; 85% (56)

The success of this method for the conversion of terminal alkenes into aldehydes often depends on the nature of the hydroborating agent. In view of the poor regioselectivity (only 94% primary alkyl groups) and functional group tolerance of borane–THF and borane–dimethyl sulfide,[54] ***Disiamylborane*** is frequently utilized as a more selective hydroborating agent. Alkyldisiamylboranes are oxidized by PCC to aldehydes in 55–72% yield. The ester groups are compatible with the reaction (eq 57).[54,55b] The need for a large excess of PCC (6 equiv) for the oxidation of alkyldisiamylboranes can be avoided by using dialkylhaloboranes, which can be prepared with excellent regioselectivity (>99% primary alkyl groups) by hydroboration of terminal alkenes with monochloroborane–dimethyl sulfide complex. Dialkylchloroboranes are hydrolized in the presence of pyridine, and the resulting boronic anhydride is oxidized with PCC (3 equiv) to aldehydes in satisfactory yields (eq 58).[54]

1-Octene → Octanal (1. Sia_2BH; 2. 6 equiv PCC; 71%) (57)

1-Octene → Octanal (1. H_2BCl•SMe_2; 2. H_2O, py; 3. 3 equiv PCC; 68%) (58)

The selective formation of aldehydes from various types of double bonds in dienes can be conveniently carried out via PCC oxidation of the resulting organoboranes, prepared by utilizing either disiamylborane (eq 59) or ***Chloro(thexyl)borane*** (eq 60) for the regioselective monohydroboration of the diene.[54]

1. Sia_2BH; 2. 6 equiv PCC, CH_2Cl_2, Δ, 2 h; 67% → CHO (59)

1. $ThxBHCl$•SMe_2; 2. H_2O, py; 3. 4 equiv PCC, CH_2Cl_2, AcONa; 68% → CHO (60)

Other organoboranes, such as trialkyl borates and trialkyl boroxins, can also be oxidized with PCC. Trialkyl borates are rapidly prepared either by esterification of boric acids or by reaction of alcohols with ***Borane–Dimethyl Sulfide***. These intermediates, on reaction with PCC in boiling dichloromethane, are oxidized to give good yields of aldehydes and ketones (eq 61).[56] The method does not seem to have significant advantages over the direct oxidation of the alcohols, but it may prove useful for the one-pot conversion of carboxylic acids into the corresponding aldehydes. Both aliphatic and aromatic carboxylic acids are easily reduced by borane–dimethyl sulfide to trialkoxyboroxins, which are smoothly oxidized to aldehydes with PCC (eq 62).[57]

OH → CHO (1. H_3B•SMe_2; 2. PCC, CH_2Cl_2, Δ; 78%) (61)

NC, CO_2H → NC, CHO (1. H_3B•SMe_2; 2. PCC; 82%) (62)

Oxidation of Silicon-Containing Molecules. PCC is able to effect the deprotection–oxidation of *p*-hydroquinone silyl ethers to *p*-quinones, except where there are electron-withdrawing substituents on the aromatic ring. The reaction proceeds both with bis(trimethylsilyl) and bis(*t*-butyldimethylsilyl) ethers of *p*-hydroquinones in 60–99% yield. The bis(trimethylsilyl) ethers are slightly more reactive than the bis(*t*-butyldimethylsilyl) ethers (eq 63).[58]

OTMS, OTMS → O, O (2 equiv PCC, CH_2Cl_2, rt, 2 h; 99%) (63)

The deprotection–oxidation method can be applied to the preparation of ketones from trimethylsilyl-protected secondary alcohols.[59] However, the ***Pyridinium Dichromate–***

Chlorotrimethylsilane combination seems to be more efficient for the transformation. A related application of PCC involves its capability to selectively convert 1-trialkylsilyl-1,2-diols into aldehydes, presumably by electrofugal loss of the silicon moiety (eq 64).[60]

HO OH R TMS — PCC, CH_2Cl_2, rt, 90% → R–CH$_2$–CHO (64)

Both cyclic and linear enol silyl ethers can be converted to α-iodocarbonyl compounds by treatment with the pyridinium chlorochromate–***Iodine*** system in dichloromethane in 76–100% yield (eq 65). The yields of α-iodocarbonyl compounds from enol methyl ethers and dihydropyran are lower (19–53%) (eq 66). The method fails with enamines.[61]

OTMS — PCC, I_2, 4Å mol sieves, CH_2Cl_2, rt, 5 min, 76% → 2-iodocyclohexanone (65)

dihydropyran — PCC, I_2, 4Å mol sieves, CH_2Cl_2, rt, 5 min → 3-iodo-2-hydroxytetrahydropyran (52%) + 3-iodo-δ-lactone (10%) (66)

Oxidation of Oximes. PCC can be conveniently employed for oxidative cleavage of oximes to carbonyl compounds. Aldehydes and ketones can be recovered from the oxime derivatives by treatment with PCC in dichloromethane at room temperature for 12–24 h in 30–85% yield.[62] PCC–***Hydrogen Peroxide*** is more effective as a deblocking agent: cleavage of oximes occurs within 10 min at 0–10 °C when 30% hydrogen peroxide is added to an acetone solution of PCC. Ketoximes are converted to ketones in 55–88% yield. For aldoximes, over-oxidation to acids has been observed.[63] On the other hand, aryl hydroxylamines can be oxidized to aryl *C*-nitroso compounds by treatment with PCC in tetrahydrofuran in 50–90% yield.[64]

Oxidation of Sulfur-Containing Molecules. PCC dimerizes aromatic, but not aliphatic, thiols to their corresponding disulfides in good yields.[65] Sulfides may undergo oxidation to give sulfones.[4]

Changes in the properties and reactivity of pyridinium chlorochromate have been brought about by altering the amine ligand associated with the chlorochromate anion. It has been found that heterocyclic chlorochromates show a trend where their strength as oxidant is inversely proportional to the donor strength of the heterocyclic ligand. Several aromatic amines have been examined, which give rise to the related chlorochromate-derived reagents. Enhanced reactivity is found for complexes containing a bidentate ligand.[4]

Related Reagents. See Classes O-1, O-7, O-8, O-18, and O-21, pages 1–10. Pyridinium Chlorochromate–Alumina.

---

1. (a) Cainelli, G.; Cardillo, G. *Chromium Oxidations in Organic Chemistry*; Springer: Berlin, 1984. (b) Haines, A. H. *Methods for the Oxidation of Organic Compounds. Alcohols, Alcohol Derivatives, Alkyl Halides, Nitroalkanes, Alkyl Azides, Carbonyl Compounds, Hydroxyarenes and Aminoarenes*; Academic: London, 1988. (c) *COS* **1991**, *7*, Chapters 2.1–4.1.
2. (a) Corey, E. J.; Suggs, J. W. *TL* **1975**, 2647. (b) Piancatelli, G.; Scettri, A.; D'Auria, M. *S* **1982**, 245.
3. Reddy, P. S.; Yadagiri, P.; Lumin, S.; Shin, D.-S.; Falck, J. R. *SC* **1988**, *18*, 545.
4. Luzzio, F. A.; Guziec, F. S., Jr. *OPP* **1988**, *20*, 533.
5. Fréchet, J. M. J.; Warnock, J.; Farral, M. J. *JOC* **1978**, *43*, 2618.
6. Agarwal, S.; Tiwari, H. P.; Sharma, S. P. *T* **1990**, *46*, 4417.
7. (a) Murray, W. V.; Hadden, S. K.; Wachter, M. P. *JHC* **1990**, *27*, 1933. (b) Dauzonne, D.; Grandjean, C. *S* **1992**, 677.
8. (a) Herscovici, J.; Antonakis, K. *CC* **1980**, 561. (b) Herscovici, J.; Egron, M.-J.; Antonakis, K. *JCS(P1)* **1982**, 1967. (c) Rosini, G.; Ballini, R. *S* **1983**, 543.
9. Hollenberg, D. H.; Klein, R. S.; Fox, J. J. *Carbohydr. Res.* **1978**, *67*, 491.
10. Nishiyama, H.; Sasaki, M.; Itoh, K. *CL* **1981**, 1363.
11. Parish, E. J.; Honda, H. *SC* **1990**, *20*, 1167.
12. Parish, E. J.; Luo C.; Parish, S.; Heidpriem, R. W. *SC* **1992**, *22*, 2839.
13. Parish, E.; J.; Schroepfer, G. J., Jr. *Chem. Phys. Lipids* **1980**, *27*, 281.
14. Parish, E. J.; Chitrakorn, S.; Lowery, S. *Lipids* **1984**, *19*, 550.
15. Parish, E. J.; Scott, A. D. *JOC* **1983**, *48*, 4766.
16. Parish, E. J.; Chitrakorn, S. *SC* **1985**, *15*, 393.
17. (a) Babler, J. H.; Coghlan, M. J. *SC* **1976**, *6*, 469. (b) Sundararaman, P.; Herz, W. *JOC* **1977**, *42*, 806, 813.
18. Wada, E.; Okawara, M.; Nakai, T. *JOC* **1979**, *44*, 2952.
19. (a) Dauben, W. G.; Michno, D. M. *JOC* **1977**, *42*, 682. (b) Paquette, L. A.; Crouse, G. D.; Sharma, A. K. *JACS* **1982**, *104*, 4411. (c) Takano, S.; Moriya, M.; Ogasawara, K. *TL* **1992**, *33*, 329. (d) Roussis, V.; Hubert, T. D. *LA* **1992**, 539.
20. (a) Schlecht, M. F.; Kim, H. *JOC* **1989**, *54*, 583. (b) Walba, D. M.; Stoudt, G. S. *TL* **1982**, *23*, 727.
21. (a) Chakraborty, T. K.; Chandrasekaran, S. *TL* **1984**, *25*, 2895. (b) Chakraborty, T. K.; Chandrasekaran, S. *CL* **1985**, 551. (c) Rathore, R.; Vankar, P. S.; Chandrasekaran, S. *TL* **1986**, *27*, 4079. (d) Baskaran, S.; Islam, I.; Chandrasekaran, S. *JCR(S)* **1992**, 290.
22. Waddell, T. G.; Carter, A. D.; Miller, T. J.; Pagni, R. M. *JOC* **1992**, *57*, 381.
23. Mehta, G.; Krishnamurthy, N.; Karra, S. R. *JACS* **1991**, *113*, 5765.
24. Corey, E. J.; Ensley, H. E.; Suggs, J. W. *JOC* **1976**, *41*, 380.
25. Corey, E. J.; Boger, D. L. *TL* **1978**, 2461.
26. Groves, J. K. *CSR* **1972**, *1*, 73.
27. Bonadies, F.; Di Fabio, R.; Bonini, C. *JOC* **1984**, *49*, 1647.
28. Bonadies, F.; Bonini, C. *SC* **1988**, *18*, 1573.
29. Parish, E. J.; Wei, T.-Y. *SC* **1987**, *17*, 1227.
30. Parish, E. J.; Chitrakorn, S.; Wei, T.-Y. *SC* **1986**, *16*, 1371.
31. (a) Müller, R. K.; Mayer, H.; Noack, K.; Daly, J. J.; Tauber, J. D.; Liaaen-Jensen, S. *HCA* **1978**, *61*, 2881. (b) Nicolaou, K. C.; Hwang, C.-K.; Sorensen, E. J.; Clairborne, C. F. *CC* **1992**, 1117.
32. Rathore R.; Saxena, N.; Chandrasekaran, S. *SC* **1986**, *16*, 1493.
33. Ghosh, S.; Banik, B. K.; Ghatak, U. R. *JCS(P1)* **1991**, 3195.
34. Ghosh, S.; Ghatak, U. R. *JCR(S)* **1992**, 352.
35. Lee, W. Y.; Park, C. H.; Kim, Y. D. *JOC* **1992**, *57*, 4074.
36. Wender, P. A.; Eissenstat, M. A.; Filosa, M. P. *JACS* **1979**, *101*, 2196.
37. Antonioletti, R.; D'Auria, M.; De Mico, A.; Piancatelli, G.; Scettri, A. *S* **1983**, 890.
38. (a) Cisneros, A.; Fernández, S.; Hernández, J. E. *SC* **1982**, *12*, 833. (b) Tori, M.; Sono, M.; Asakawa, Y. *CPB* **1989**, *37*, 534.
39. Baskaran, S.; Chandrasekaran, S. *TL* **1990**, *31*, 2775.
40. Ali, S. M.; Ramesh, K.; Borchardt, R. T. *TL* **1990**, *31*, 1509.

41. (a) Piancatelli, G.; Scettri, A.; D'Auria, M. *TL* **1977**, 3483. (b) Rollin, P.; Sinay, P. *Carbohydr. Res.* **1981**, *98*, 139.
42. Baskaran, S.; Islam, I.; Raghavan, M.; Chandrasekaran, S. *CL* **1987**, 1175.
43. Baskaran, S.; Islam, I.; Chandrasekaran, S. *JOC* **1990**, *55*, 891.
44. Fetizon, M.; Goulaouic, P.; Hanna, I. *TL* **1988**, *29*, 6261.
45. Narasimhan, V.; Rathore, R.; Chandrasekaran, S. *SC* **1985**, *15*, 769.
46. Bosshard, P.; Eugster, C. H. *Adv. Heterocycl. Chem.* **1966**, *7*, 377.
47. Piancatelli, G.; Scettri, A.; D'Auria, M. *TL* **1977**, 2199.
48. D'Auria, M.; Piancatelli, G.; Scettri, A. *T* **1980**, *36*, 1877.
49. Piancatelli, G.; Scettri, A.; D'Auria, M. *TL* **1979**, 1507.
50. Piancatelli, G.; Scettri, A.; D'Auria, M. *T* **1980**, *36*, 661.
51. Antonioletti, R.; D'Auria, M.; De Mico, A.; Piancatelli, G.; Scettri, A. *S* **1984**, 280.
52. Rao, R. V. V.; Devaprabhakara, D.; Chandrasekaran, S. *JOM* **1978**, *162*, C9.
53. Parish, E. J.; Parish, S.; Honda, H. *SC* **1990**, *20*, 3265.
54. Brown H. C.; Kulkarni, S. U.; Rao C. G.; Patil V. D. *T* **1986**, *42*, 5515.
55. (a) Rao, C. G.; Kulkarni, S. U.; Brown, H. C. *JOM* **1979**, *172*, C20. (b) Brown, H. C.; Kulkarni, S. U.; Rao, C. G. *S* **1980**, 151.
56. Brown, H. C.; Kulkarni, S. U.; Rao, C. G. *S* **1979**, 702.
57. Brown, H. C.; Rao, C. G.; Kulkarni, S. U. *S* **1979**, 704.
58. Willis, J. P.; Gogins, K. A. Z.; Miller, L. L. *JOC* **1981**, *46*, 3215.
59. Fujisawa, T.; Takeuchi, M.; Sato, T. *CL* **1982**, 1795.
60. Page, P. C. B.; Rosenthal, S. *TL* **1986**, *27*, 2527.
61. D'Auria, M.; D'Onofrio, F.; Piancatelli, A.; Scettri, A. *SC* **1982**, *12*, 1127.
62. Maloney, J. R.; Lyle, R. E.; Saavedra, J. E.; Lyle, G. G. *S* **1978**, 212.
63. Drabowicz, J. *S* **1980**, 125.
64. Wood, W. W.; Wilkin, J. A. *SC* **1992**, *22*, 1683.
65. Firouzabadi, H.; Iranpoor, N.; Parham, H.; Sardarian, A.; Toofan, J. *SC* **1984**, *14*, 717.

Giovanni Piancatelli
University of Rome 'La Sapienza' and CNR, Rome, Italy

Pyridinium Dichromate[1]

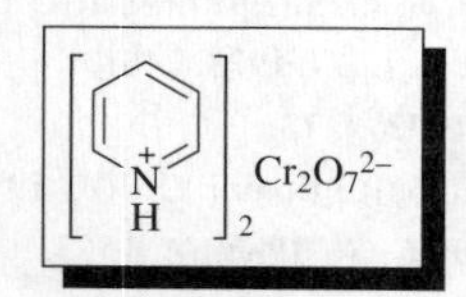

[20039-37-6] $C_{10}H_{12}Cr_2N_2O_7$ (MW 376.24)

(mild and selective oxidizing agent for primary and secondary alcohols;[2a] can oxidize unsaturated tertiary alcohols,[2b] silyl ethers,[11] the carbon–boron bond,[20] and oximes[23])

Alternate Name: PDC.
Physical Data: mp 152–153°C
Solubility: sol DMF, DMSO, acetonitrile; sparingly sol dichloromethane, chloroform, acetone; insol hexane, ether, ethyl acetate.
Form Supplied in: bright orange solid; widely available.
Handling, Storage, and Precautions: the dry solid can be stored in contact with air. Reaction solvents must be anhydrous and free of reducing impurities. Pyridinium dichromate is reported to be a cancer suspect agent. The reagent should be used in a fume hood.

Oxidation of Primary and Secondary Alcohols. PDC is an oxidizing agent complementary to ***Pyridinium Chlorochromate*** (PCC) for alcohols containing acid-sensitive groups. PDC is less acidic than pyridinium chlorochromate, and has more neutral character than the Collins reagent (***Dipyridine Chromium(VI) Oxide***). PDC is reported to exhibit greater oxidation efficiency than the Collins reagent, especially for large scale preparations. Initially, PDC was used either as a solution in DMF or as a suspension in dichloromethane. In solution in DMF, PDC oxidizes primary and secondary allylic alcohols to the corresponding α,β-unsaturated carbonyl compounds (eqs 1 and 2). Overoxidation of primary allylic alcohols is not observed, and (*E*)–(*Z*) isomerization does not take place.[2a]

CH_2OH → (PDC, DMF, −10 °C, 4 h, 92%) → CHO (1)

OH → (PDC, DMF, 0 °C, 4 h, 95%) → O (2)

Under similar conditions, saturated primary alcohols and aldehydes yield carboxylic acids. The preparation of carboxylic acids from primary alcohols (via the aldehyde intermediate) is usually achieved using 3.5 equiv of PDC at rt for 7–9 h (eqs 3 and 4).[2a]

OH, OTHP → (PDC, DMF, rt, 9 h, 76%) → CO_2H, OTHP (3)

OH → (PDC, DMF, rt, 7 h, 83%) → CO_2H (4)

An interesting example of the mildness of PDC is the oxidation of citronellol to the corresponding acid in 83% yield. The same oxidation with acidic PCC is complicated by cationic cyclization of the aldehyde intermediate. Aliphatic methyl esters can be directly prepared from aldehydes by oxidation with PDC (6 equiv) in the presence of methanol in yields of 60–80%. This method is not useful for aromatic aldehydes or higher esters.[3] Saturated secondary alcohols are rapidly and conveniently converted into ketones, even in the presence of very sensitive groups, such as a thioacetals.

In suspension in dichloromethane, PDC shows considerably different behavior, becoming a milder and selective reagent for the preparation of both saturated and unsaturated aldehydes and ketones (eqs 5–7). Allylic alcohols are oxidized faster than saturated ones, but some (*E*)–(*Z*) isomerization is observed.[2a]

$$C_5H_{11}-\!\equiv\!-CH_2OH \xrightarrow[\text{rt, 20 h; 70\%}]{\text{PDC, }CH_2Cl_2} C_5H_{11}-\!\equiv\!-CHO \quad (5)$$

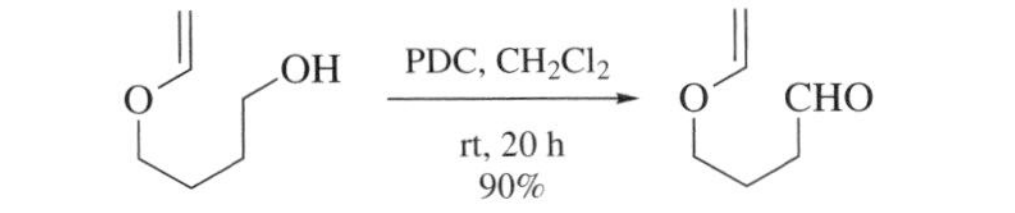

(6)

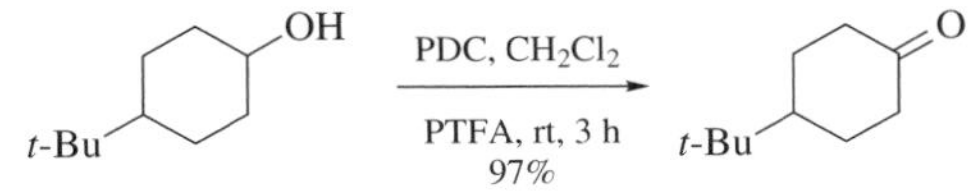

(7)

PDC does not attack isolated alkenic and alkynic bonds; in fact, ω-alkynyl alcohols are oxidized by PDC in dichloromethane to the corresponding ω-alkynyl carbonyl compounds in good yield (eq 8).[4]

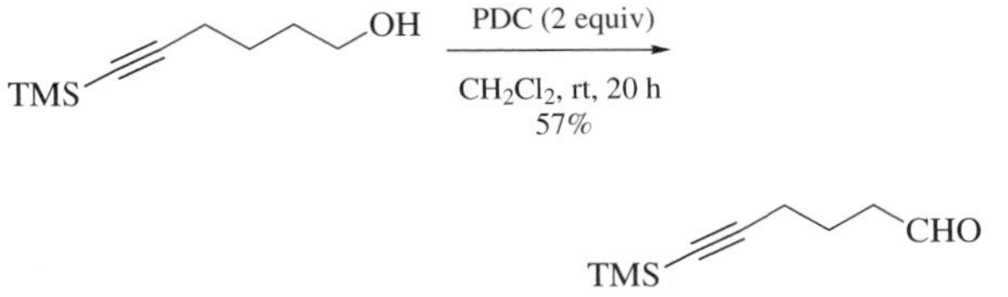

(8)

PDC and PCC may show a different chemoselectivity. Thus PDC oxidizes the secondary hydroxyl groups of 5-bromo-2-furylcarbinols to the corresponding ketones in 80% yield, while PCC attacks the furan ring giving γ-hydroxy Δ^2-butenolides (eq 9).[5]

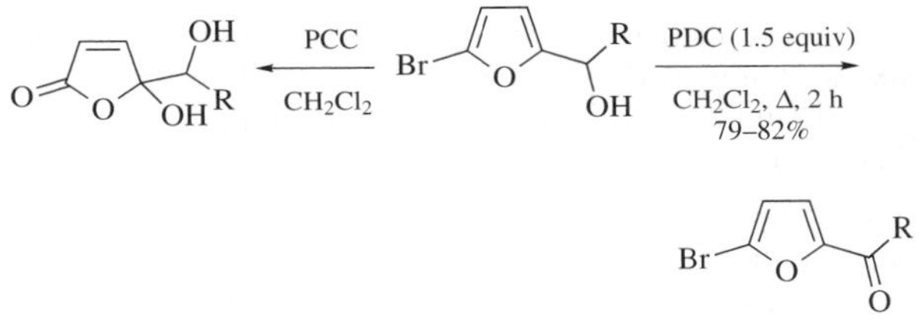

(9)

Several modified procedures have been employed to enhance the rate and the efficacy of PDC. Addition of pyridinium trifluoroacetate (0.4 equiv) causes an increase in the reaction rate, and less oxidant is necessary for oxidation (eq 7). ***Molecular Sieves*** enhance the oxidation rate with a wide variety of substrates.[6] A combination of PDC, molecular sieves, and ***Acetic Acid*** has a dramatic effect on the oxidation of carbohydrate alcohols, reducing oxidation time from days to minutes (eq 10).[7] Furthermore, the addition of acetic acid has been utilized for the selective oxidation of secondary allylic alcohols in carbohydrates (eq 11).[8]

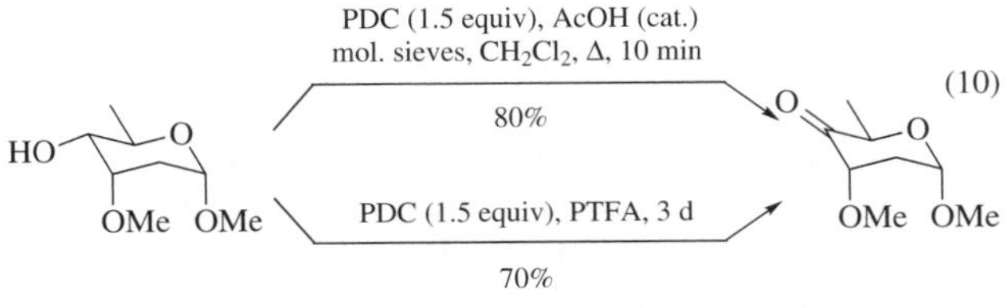

(10)

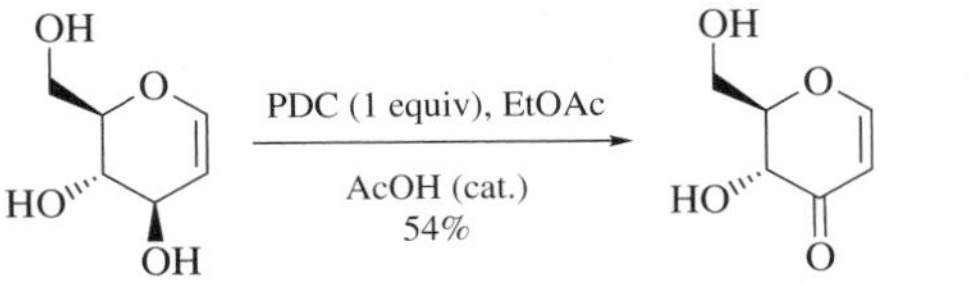

(11)

The PDC–***Acetic Anhydride*** combination gives rise to a strong and neutral oxidant. This system (0.6–0.7 equiv of PDC and 3 equiv of Ac_2O) can be used for the selective oxidation of primary and secondary alcohols of carbohydrates to the corresponding carbonyl compounds in high yields. The addition of DMF as cosolvent to the reaction mixture prevents overoxidation (eq 12).[9] Interestingly, the PDC–Ac_2O combination is able to perform an oxidative fragmentation of 5′-hydroxymethyl groups in 2′,3′-*O*-isopropylidene nucleosides to form the corresponding γ-lactone nucleosides, albeit in moderate yield (eq 13).[10]

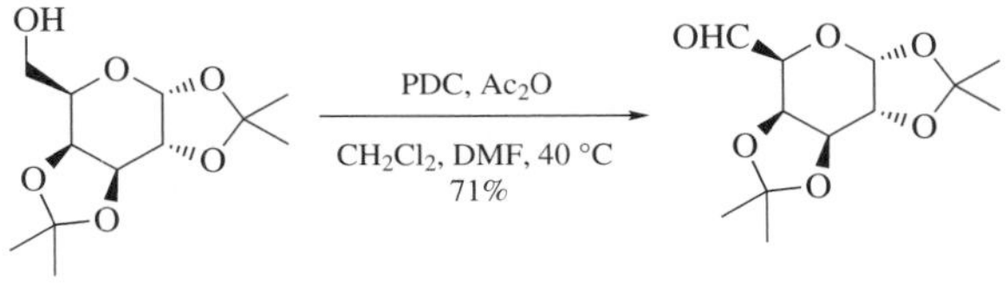

(12)

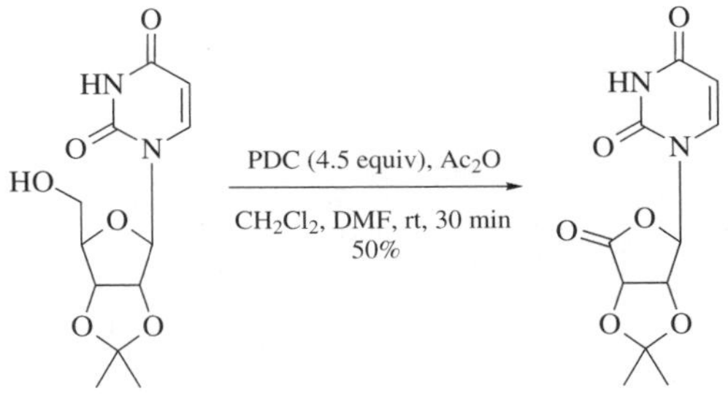

(13)

Oxidation of primary and secondary alcohols to carbonyl compounds can be carried out with PDC–***Chlorotrimethylsilane***, which is a rapid and selective oxidant.[11] This combined reagent system can effect the deprotection–oxidation of silyl ethers and enol silyl ethers. Primary and secondary trimethylsilyl and *t*-butyldimethylsilyl ethers, usually stable to PDC, may be converted by the above combination directly into the corresponding carbonyl groups in high yields (eq 14). The method can be applied to the preparation of quinones from hydroquinone silyl ethers (eq 15) with higher efficiency than that using PCC and this reagent is applicable to substrates both with electron-donating and electron-withdrawing groups.[11]

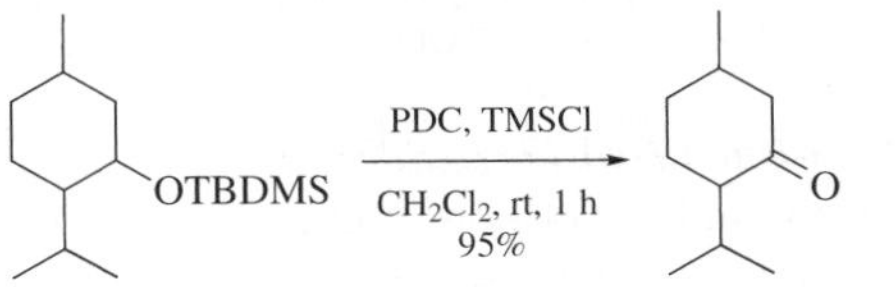

(14)

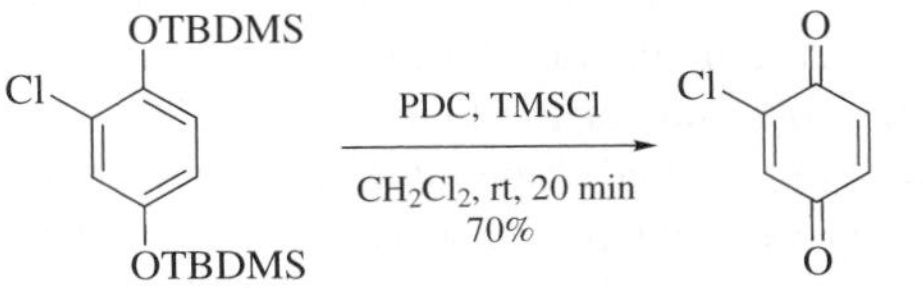

(15)

Catalytic PDC Oxidation of Alcohols. PDC can be utilized as a catalytic oxidant when coupled with ***Bis(trimethylsilyl) Peroxide*** as cooxidant. The method has proven to be mild and useful not only for the oxidation of a wide range of primary and secondary alcohols, saturated and unsaturated, but also for the conversion of homoallylic alcohols to the corresponding β,γ-enones (eq 16).[12]

Oxidation of Cyanohydrins. Cyanohydrins of nonconjugated aldehydes are oxidized to carboxylic acids by PDC–***N,N-***

Dimethylformamide in excellent yields. Cyanohydrins from α,β-enals (as the *O*-trimethylsilyl derivatives) behave differently, being converted by PDC–DMF into α,β-unsaturated γ-lactones (Δ^2-butenolides), in cases in which the β-carbon is disubstituted and the γ-carbon possesses at least one hydrogen. This reaction is performed with PDC (3 equiv) in DMF at rt for 12 h, and provides a mixture of two isomeric Δ^2-butenolides, whose ratio (4:1) indicates that oxidative attack at the γ-methyl is favored over the γ-methylene (eq 17).[13]

(TMSO)$_2$ (3 equiv), PDC (10 mol %), 87% (16)

1. TMSCN, KCN; 2. PDC (3 equiv), DMF, rt, 12 h, 73%; 4:1 (17)

Oxidation of Activated Methylene Groups. PDC can perform allylic oxidation of methylene groups, but a large excess of oxidant is required. Δ^5-Steroids are selectively transformed into Δ^5-7-keto steroids by treatment with PDC (25 equiv) and molecular sieves in pyridine at 100 °C for 24 h; yields range from 63–86% (eq 18).[14]

(18)

R = Bz	PDC, py, 100 °C, MS	64%
R = Ac	PDC, *t*-BuOOH, Celite, rt	81%

A PDC–***t*-Butyl Hydroperoxide** mixture (1:1) has proven to be useful for the mild oxidation of cyclic 1,4-dienes to the corresponding cyclic 2,5-dienones (eq 19).[15] The use of the combination of PDC, *t*-butyl hydroperoxide, and ***Celite*** is particularly suitable for the oxidation of benzylic methylenes into aryl ketones (eq 20)[16] and for the highly regioselective conversion of Δ^5-steroids into the corresponding Δ^5-7-keto steroids (eq 18).[16] The method is rather poor for allylic oxidation of linear and cyclic alkenes, furnishing α,β-enones in 23–44% yield (eq 21).[16]

PDC, *t*-BuOOH, $CHCl_3$, Δ, 74% (19)

PDC, *t*-BuOOH, Celite, rt, 6.5 h, 78% (20)

PDC, *t*-BuOOH, Celite, rt, 11 h, 40%; 69% conversion (21)

Oxidative Rearrangement of Unsaturated Alcohols. Linear and cyclic secondary homoallylic alcohols are oxidized by PDC–DMF to *trans*-enediones in 53–75% yield. The one-pot procedure seems to be particularly convenient for the oxidation of steroidal substrates, which contain a rigid and sterically congested structure. Thus 3β-hydroxy Δ^5-steroids afford the corresponding 3,6-diketo-Δ^4-steroids (eq 22).[17] The method is rather unsatisfactory for linear homoallylic alcohols and fails with alcohols possessing a terminal allylic group.

PDC (3 equiv), DMF, Δ, 15 h, 75% (22)

PDC is the oxidant of choice for the oxidation of tertiary allylic dienols. These alcohols, easily prepared by addition of ***Vinyllithium*** to conjugated cyclohexenones or cyclopentenones, undergo a 1,3-oxidative rearrangement to afford the corresponding conjugated dienones in moderate to good yield (eq 23). The rearrangement is completely regiospecific, leading to the formation of the more stable isomers. The reaction does not suffer from the presence of withdrawing substituents on the alkene system, but linear α-dienols give lower yields of rearranged isomeric products (eq 24).[18]

PDC (1.5 equiv), CH_2Cl_2, rt, 8 h, 72% (23)

PDC (1.5 equiv), CH_2Cl_2, rt, 8 h; 37% + 18% (24)

In a similar fashion, tertiary α-enynols are converted by reaction with PDC in dichloromethane to the corresponding conjugated enynones; yields are good to excellent. These rearrangements are completely regioselective, involving only the alkene. The procedure is limited to cyclic allylic alcohols and shows wide applicability only with enynols containing (*Z*) double bonds (eq 25).[19]

PDC (1.5 equiv), CH_2Cl_2, rt, 16 h, 90% (25)

Oxidation of Carbon–Boron Bonds. A synthetically useful procedure has been developed for the conversion of terminal alkenes into carboxylic acids via oxidation of organoboranes obtained by hydroboration of terminal alkenes. The oxidation works well with organoboranes derived from ***Dibromoborane–Dimethyl Sulfide***, but other reagents may be utilized such as monobromoborane–dimethyl sulfide, ***Monochloroborane–Dimethyl Sulfide***, ***Thexylborane***, and ***Dicyclohexylborane***. Alkyldibromoboranes are hydrolyzed to alkylboronic acids, which undergo a facile oxidation with PDC–DMF (5 equiv) at rt for 24 h to provide carboxylic acids in 62–78% yield (eq 26).[20] In some cases PCC may effect the same reaction in higher yields; however, carboxylic acids are often best obtained with ***Chromium(VI) Oxide*** as oxidant in 90% aqueous acetic acid.[20]

1-Hexene → (1. $HBBr_2 \cdot SMe_2$; 2. H_2O) → $C_5H_{11}B(OH)_2$ → (PDC (5 equiv), DMF, rt, 16 h) → Hexanoic acid 75% (26)

Oxidation of Enol Ethers. PCC is a selective reagent for the oxidative cleavage of enol ethers and PDC shows a similar but modest reactivity. PDC can effect oxidative double bond fragmentation of cyclic enol ethers to esters and keto lactones (eq 27).[21]

(27) PDC (4 equiv), C_6H_6, Δ, 4 h; 64%

Interestingly, the PDC–*t*-BuOOH combination can achieve the one-step conversion of 3,4-dihydro-2*H*-pyrans into 5,6-dihydro-2*H*-pyran-2-ones under very mild conditions; yields are moderate. The reaction is performed at 0 °C in dichloromethane with PDC–*t*-BuOOH for 2–4 h. *t*-Butylperoxy derivatives are obtained as byproducts (eq 28).[22] This reagent system can chemoselectively oxidize 3,4-dihydro-2*H*-pyrans to the corresponding α,β-unsaturated lactones without affecting pendant hydroxyl groups (eq 29).

(28) PDC, *t*-BuOOH, CH_2Cl_2, 0 °C, 2 h; 50% + *t*-BuOO product 6%

(29) PDC, *t*-BuOOH, CH_2Cl_2, 0 °C, 2 h; 30% + *t*-BuOO product 14%

Oxidative Cleavage of Oximes. PDC can perform oxidative cleavage of oximes to ketones and aldehydes in excellent yield (eq 30). The reagent seems to be superior to PCC for this transformation. The rate of deoximation is increased by adding 3 Å molecular sieves, but the yields of carbonyl compounds are lowered.[23]

Oxidation with Iodine and PDC. PDC has been utilized for oxidation of cyclic alkene–iodine complexes to α-iodo ketones (eq 31). The yields range from 50 to 70%.[24] The reaction shows a wide applicability; Δ^2-cholestene is selectively converted, by the same procedure, into 3α-iodocholestan-2-one in 70% yield (eq 32). The reaction fails with linear alkenes; however, it seems to be characteristic of PDC. In contrast, PCC converts the cyclohexene–iodine complex into *trans*-1-chloro-2-iodocyclohexane (66%), while the α-iodo ketone is a byproduct (8%).[24]

PhCH=NOH → (PDC (2 equiv), CH_2Cl_2, rt, 1 h; 100%) → PhCHO (30)

(31) Cyclohexene → 1. I_2, CH_2Cl_2; 2. PDC (2.2 equiv), CH_2Cl_2, mol. sieves, rt, 20 h; 50% → 2-iodocyclohexanone

(32) 2-Cholestene → 1. I_2, CH_2Cl_2; 2. PDC (2.2 equiv), CH_2Cl_2, mol. sieves, rt, 23 h; 70% → 3α-iodocholestan-2-one (C_8H_{17})

Cyclic trisubstituted alkenes, activated with ***Iodine***, are easily converted into iodohydrins and epoxides by PDC, which acts as nucleophilic and iodide removing agent. Iodohydrins can be isolated in 48–65% yield when the starting alkenes have a mobile conformation. The reaction is performed in dichloromethane with PDC (2.5 equiv) and 4 Å molecular sieves at rt for 3 h. More prolonged reaction times (16 h) lead to epoxides in 50% yield (eq 33).[25] Conformationally rigid alkenes are directly converted into epoxides in 51–86% yield, since the iodohydrin intermediates undergo a fast elimination of iodide by the oxidant (eq 34).

(33) 1-pentylcyclohexene (C_5H_{11}) → 1. I_2, CH_2Cl_2; 2. PDC, mol. sieves, rt, 3 h; 48% → iodohydrin (C_5H_{11}, OH, I) → rt, 10 h; 50% → epoxide

(34) 1-phenyl-4-*t*-butylcyclohexene (Ph, *t*-Bu) → 1. I_2, CH_2Cl_2; 2. PDC, mol. sieves, rt, 16 h → epoxide (Ph, O, *t*-Bu)

Some naturally-occurring polyenes such as (*E,Z*)-geranyl acetate undergo a regioselective conversion to terminal epoxides by a two-step sequence, which includes the formation of the corresponding terminal iodohydrins by reaction with PDC–iodine and subsequent conversion to the epoxides by adsorption on neutral alumina (30–65% yield) (eq 35).[25]

(35) geranyl acetate (OAc) → 1. PDC, I_2, mol. sieves, rt, 20 h; 2. Al_2O_3, rt; 65% → terminal epoxide (OAc, O)

The nucleophilic and oxidizing properties of PDC can be utilized for a one-pot conversion of α-ynol–iodine complexes to α,β-unsaturated α-iodo aldehydes in 30–66% yield. This reaction is performed by adding PDC (2.2 equiv) to a solution of α-ynol and iodine (1:1), in the presence of 4 Å molecular sieves at rt for 24 h (eq 36). The conversion is regio- and stereospecific, yielding only one of two possible geometrical isomers and shows wide applicability.[26]

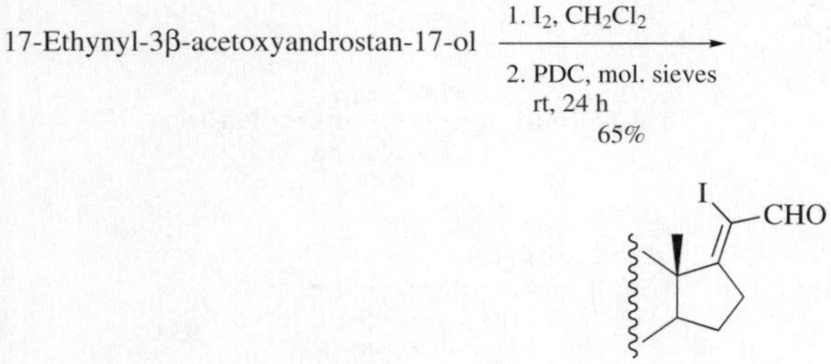

17-Ethynyl-3β-acetoxyandrostan-17-ol $\xrightarrow[\text{2. PDC, mol. sieves, rt, 24 h, 65\%}]{\text{1. } I_2, CH_2Cl_2}$ (I)C=CHO product (36)

Related Reagents. See Classes O-1, O-2, and O-21, pages 1–10.

1. (a) Cainelli, G.; Cardillo, G. *Chromium Oxidation in Organic Chemistry*; Springer: Berlin, 1984. (b) Haines, A. H. *Methods for the Oxidation of Organic Compounds, Alcohols, Alcohol Derivatives, Alkyl Halides, Nitroalkanes, Alkyl Azides, Carbonyl Compounds, Hydroxyarenes and Aminoarenes*; Academic: London, 1988. (c) *COS* **1991**, *7*, Chapters 2.1, 2.7, and 3.7.
2. (a) Corey, E. J.; Schmidt, G. *TL* **1979**, 399. (b) Luzzio, F. A.; Guziec, F. S., Jr. *OPP* **1988**, *20*, 533.
3. O'Connor, B.; Just, G. *TL* **1987**, *28*, 3235.
4. Bierer, D. E.; Kabalka, G. W. *OPP* **1988**, *20*, 63.
5. D'Auria, M.; Piancatelli, G.; Scettri, A. *T* **1980**, *36*, 3071.
6. Herscovici, J.; Antonakis, K. *CC* **1980**, 561.
7. (a) Czernecki, S.; Georgoulis, C.; Stevens, C. L.; Vijayakumaran, K. *TL* **1985**, *26*, 1699. (b) Czernecki, S.; Georgoulis, C.; Stevens, C. L.; Vijayakumaran, K. *SC* **1986**, *16*, 11.
8. Czernecki, S.; Vijayakumaran, K.; Ville, G. *JOC* **1986**, *51*, 5472.
9. Andersson, F.; Samuelsson, P. *Carbohydr. Res.* **1981**, *129*, C1.
10. Kim, J. N.; Ryu, E. K. *TL* **1992**, *33*, 3141.
11. Cossio, F. P.; Aizpurua, J. M.; Palomo, C. *CJC* **1986**, *64*, 225.
12. (a) Kanemoto, S.; Oshima, K.; Matsubara, S.; Takai, K.; Nozaki, H. *TL* **1983**, *24*, 2185. (b) Kanemoto, S.; Matsubara, S.; Takai, K.; Oshima, K.; Utimoto, K.; Nozaki, H. *BCJ* **1988**, *61*, 3607.
13. Corey, E. J.; Schmidt, G. *TL* **1980**, *21*, 731.
14. Parish, E. J.; Wei, T.-Y. *SC* **1987**, *17*, 1227.
15. Schultz, A. G.; Taveras, A. G.; Harrington, R. E. *TL* **1988**, *29*, 3907.
16. Chidambaran, N.; Chandrasekaran, S. *JOC* **1987**, *52*, 5048.
17. D'Auria, M.; De Mico, A.; D'Onofrio, F.; Scettri, A. *S* **1985**, 988.
18. Majetich, G.; Condon, S.; Hull, K.; Ahmad, S. *TL* **1989**, *30*, 1033.
19. Liotta, D.; Brown, D.; Hoekstra, W.; Monahan, R., III *TL* **1987**, *28*, 1069.
20. Brown, H. C.; Kulkarni, S. V.; Khanna, V. V.; Patil, V. D.; Racherla, U. S. *JOC* **1992**, *57*, 6173.
21. Baskaran, S.; Islam, I.; Raghavan, M.; Chandrasekaran, S. *CL* **1987**, 1175.
22. Chidambaram, N.; Satyanarayana, K.; Chandrasekaran, S. *TL* **1989**, *30*, 2429.
23. Satish, S.; Kalyanam, N. *CI(L)* **1981**, 809.
24. D'Ascoli, R.; D'Auria, M.; Nucciarelli, L.; Piancatelli, G.; Scettri, A. *TL* **1980**, *21*, 4521.
25. Antonioletti, R.; D'Auria, M.; De Mico, A.; Piancatelli, G.; Scettri, A. *T* **1983**, *39*, 1765.
26. Antonioletti, R.; D'Auria, M.; Piancatelli, G.; Scettri, A. *TL* **1981**, *22*, 1041.

Giovanni Piancatelli
University of Rome 'La Sapienza' and CNR, Rome, Italy

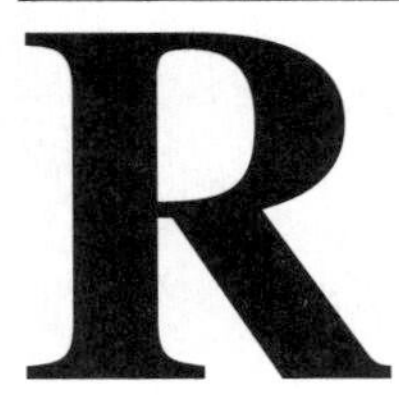

Raney Nickel[1]

[106-51-4] Ni (MW 58.69)

(useful as a reducing agent for hydrogenation of organic compounds[1])

Solubility: insol all organic solvents and water.
Form Supplied in: black solid.
Preparative Methods: there are many types of Raney nickel; they differ based on their methods of preparation. These methods essentially determine the hydrogen content as well as the reactivities of various types of Raney nickel. The most popular W-type Raney nickels can be prepared as the seven different types listed in Table 1.

Table 1 Types of Raney Nickel

Type	NaOH/Al (molar)	NaOH soln. conc. (wt%)	Temp. (°C) (time, h)	Water wash method
W-1	1.35	17	115–120 (4)	Decant to neutral
W-2	1.71	20	75–80 (8–12)	Decant to neutral
W-3	1.73	20	50 (0.83)	Continuous to neutral
W-4	1.73	20	50 (0.83)	Continuous to neutral
W-5	1.80	21	50 (0.83)	Continuous to neutral
W-6	1.80	21	50 (0.83)	Cont. to neut. under H_2
W-7	1.80	21	50 (0.83)	Directly washed with EtOH[1c]

Handling, Storage, and Precautions: Raney nickel is generally stored in an alcoholic solvent, or occasionally in water, ether, methylcyclohexane, or dioxane. The activity of Raney nickel decreases due to loss of hydrogen over a period of about 6 months. Raney nickel ignites on contact with air and should never be allowed to dry.

Desulfurization. The most widespread application of Raney nickel is the desulfurization of a wide range of compounds including thioacetals, thiols, sulfides, disulfides, sulfoxides, sulfones, thiones, thiol esters, and sulfur-containing heterocycles.

The well-known desulfurization of dithianes remains one of the most efficient methods for reductive deoxygenation of ketones. Recently, Rubiralta demonstrated an example in his synthesis of the aspidosperma alkaloid framework. Dithiane (**1**) was reduced to indole derivative (**2**) in 85% yield without reduction of the alkene (eq 1).[2]

(1) → (2): W-2 Raney Ni, EtOH, reflux, 85% (1)

Raney nickel is frequently used to remove the sulfur atom of thiols,[3] sulfides, and disulfides from a carbon skeleton, regardless of whether the sulfur is attached to an alkyl carbon,[4] an aryl carbon,[5] or a carbon atom in a heterocycle.[6] Several examples are shown in eqs 2–5. A solvent effect was observed in the Raney nickel reduction of vinyl sulfide (**11**), which gave glycoside (**12**) in methanol, whereas the double bond remained intact to produce alkene (**13**) in THF (eq 6).[7]

(3) → (4): Raney Ni, EtOH, 90% (2)

(5) → (6): Raney Ni, MeOH, reflux, 51% (3)

(7) → (8): Raney Ni, acetone, 60% (4)

(9) → (10): Raney Ni, EtOH (5)

(11) Pht = phthaloyl → (13): Raney Ni, THF, 85%; (11) → (12): Raney Ni, MeOH, 87% (6)

The sulfur atom can be part of a heterocycle. Högberg[8] and Tashiro[9] used Raney nickel to remove sulfur from thiophene derivatives (**14**) and (**16**) to give compounds (**15**) and (**17**), respectively (eqs 7 and 8).

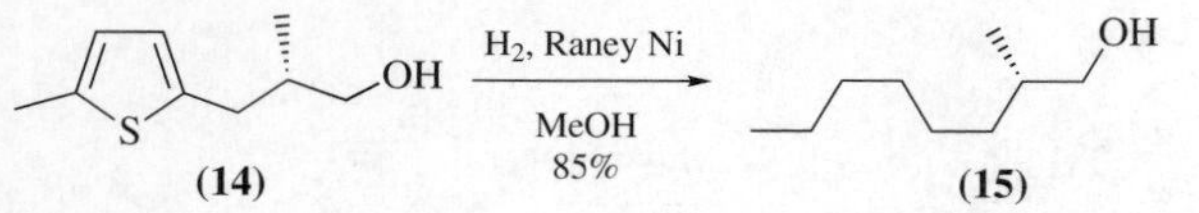

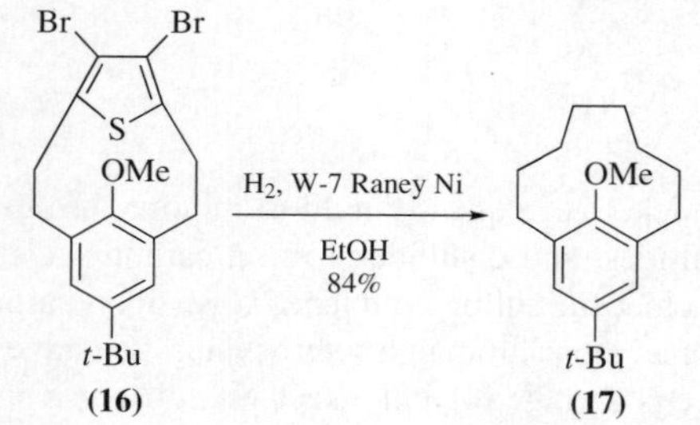

Raney nickel can remove the sulfinyl and sulfonyl groups from sulfoxides and sulfones under neutral conditions (eqs 9 and 10).[10,11] Cox demonstrated the cleavage of both sulfur–carbon bonds in sulfoxide (**22**) and noted that the stereogenic center remained untouched (eq 11).[12]

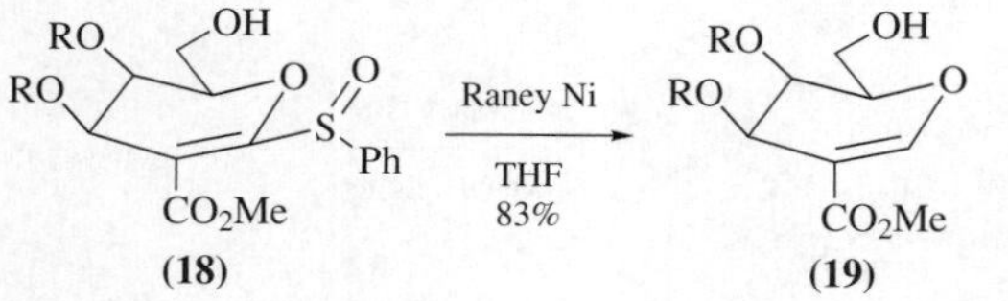

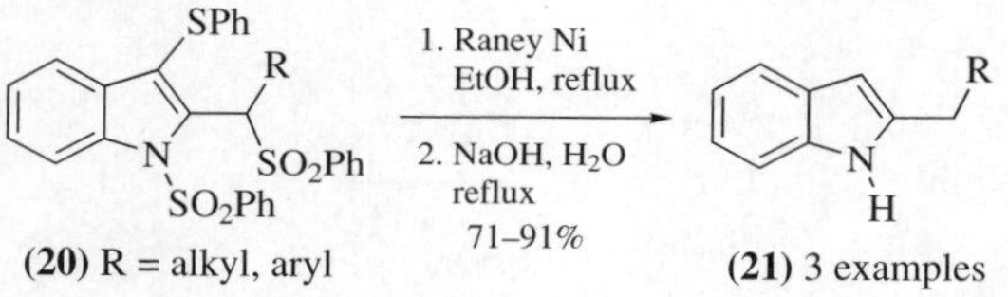

Some thioamides are reduced by Raney nickel to the corresponding imines. Two typical examples are shown in eqs 12 and 13.[13,14] Thiones have been reported to give alkanes, but only low or unstated yields are reported (eqs 14 and 15).[15,16]

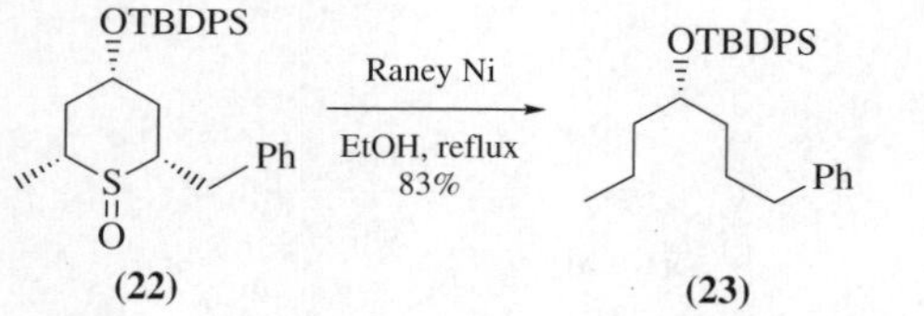

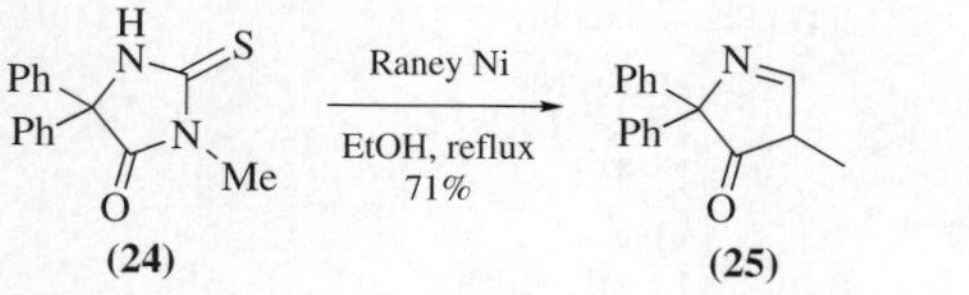

Liu and Luo[17] used Raney nickel to reduce glycidic thioester (**32**) to the corresponding 1,3-diol (**33**) in good yield (eq 16). With ***Sodium Borohydride*** at ambient temperature or ***Lithium Aluminum Hydride*** at −78 °C, glycidic thioester (**32**) was reduced chemoselectively to furnish 2,3-epoxy alcohol (**34**) in 82% yield (eq 16).[17]

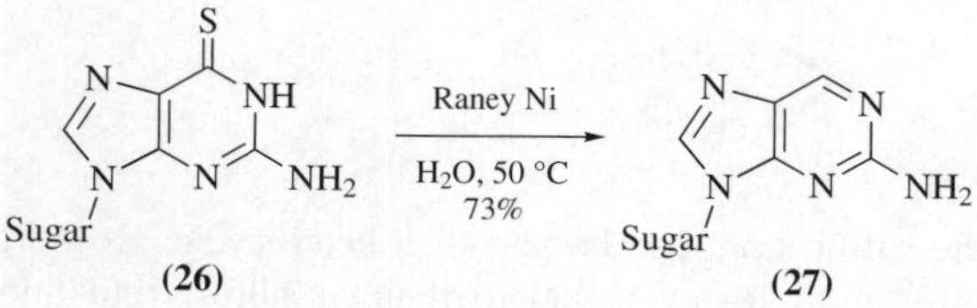

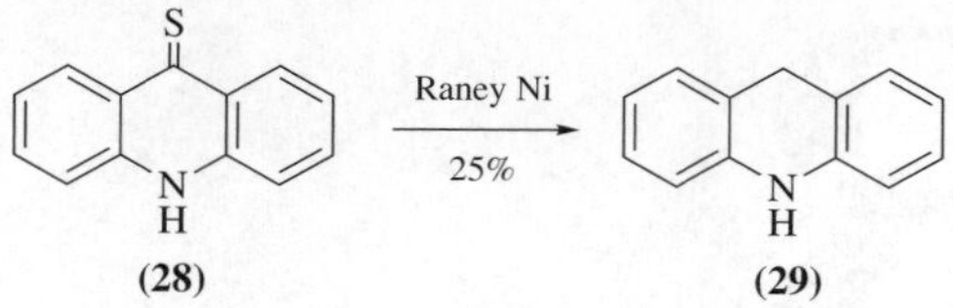

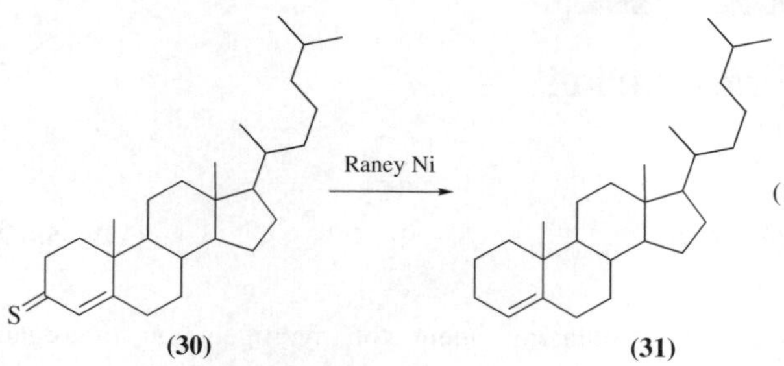

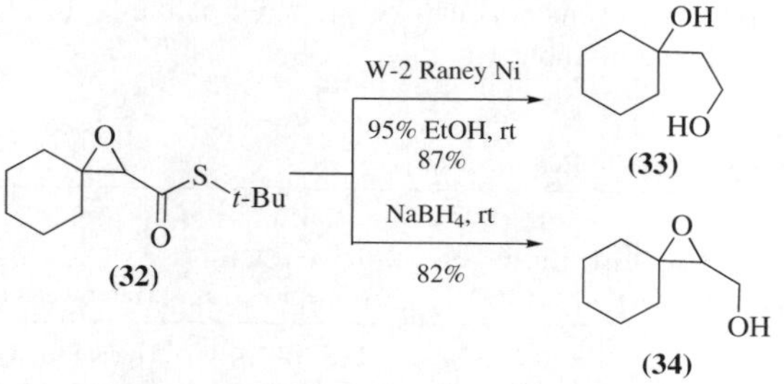

Deoxygenation and Deamination. Besides the widely used desulfurization process, Raney nickel can be used to reduce benzylic nitrogen and oxygen atoms. Behren's report shows the partial deoxygenation of diol (**35**) to mono alcohol (**36**) in 75–96% yields (eq 17).[18] Ikeda used W-2 Raney nickel to remove the benzyl protecting group from compound (**37**). However, partial epimerization occurred in this reaction to produce a 3.7:1 mixture of the 6α- and 6β-alcohols (**38**) (eq 18).[19] Azetidine (**39**) was opened by Raney nickel in refluxing ethanol, to give acyclic amine (**40**) in 88% yield (eq 19).[20]

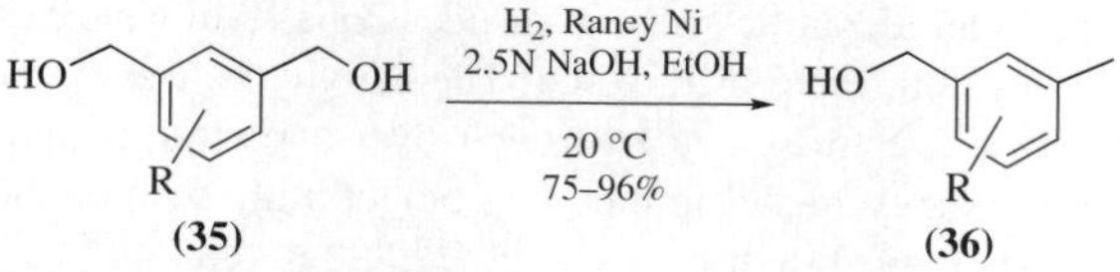

R = H, CN, NH_2, CH_2NH_2

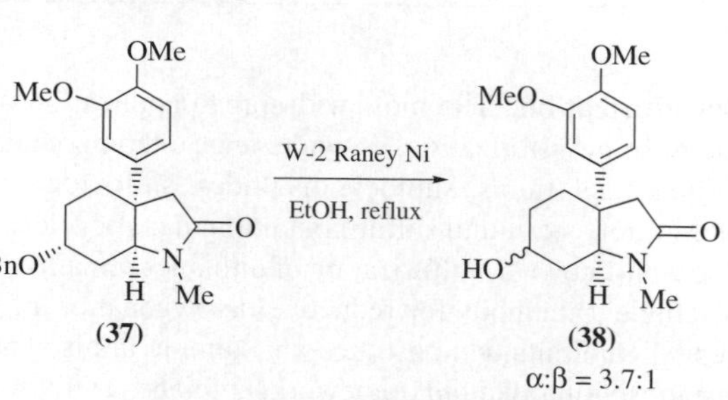

α:β = 3.7:1

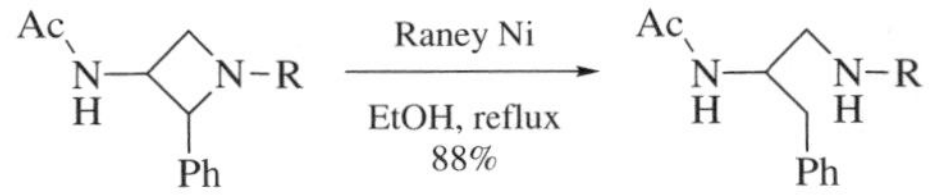

(39) **(40)** (19)

Krafft reported that tertiary alcohols were also deoxygenated to alkanes by Raney nickel (eq 20). On the other hand, primary alcohols were oxidized to aldehydes and then subsequently decarbonylated (eq 21), and secondary alcohols were oxidized to the corresponding ketones (eq 22).[21]

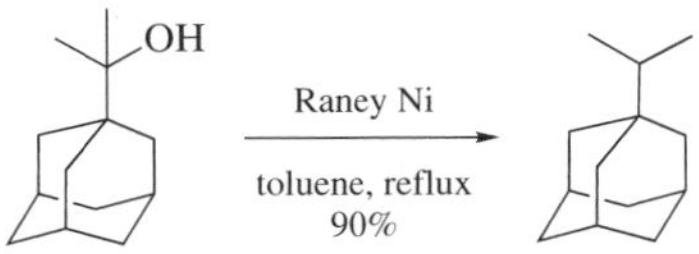

(41) **(42)** (20)

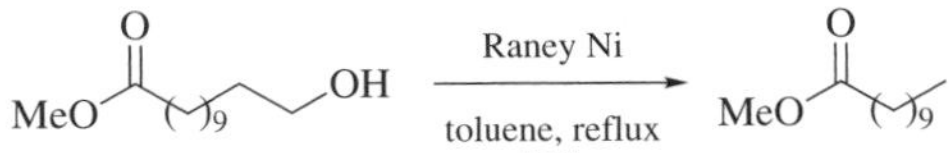

(43) **(44)** (21)

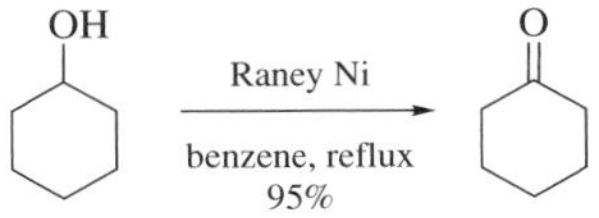

(45) **(46)** (22)

Recently, Ohta reported Raney nickel would deoxygenate *N*-oxide (**47**) to pyrazine (**48**), while ***Phosphorus(III) Bromide*** gave many side products (eq 23).[22]

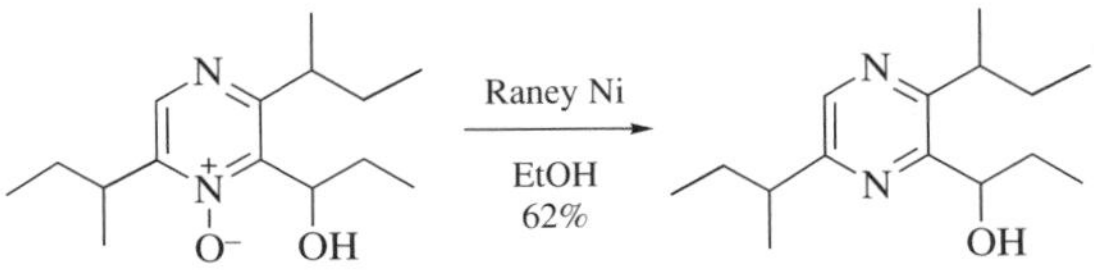

(47) **(48)** (23)

Cleavage of Heteroatom–Heteroatom Bonds. Both N–N and N–O bonds can be cleaved by Raney nickel in the presence of hydrogen. Alexakis reported that hydrazine (**49**) was easily cleaved to the free amine by Raney nickel under hydrogen atmosphere, then protected to give carbamate (**52**) (eq 24).[23] In addition, he found even hindered hydrazines (**50**) and (**51**) were successfully deaminated to free amines (**56**) and (**57**), respectively, without racemization if the reactions were assisted by ultrasound.[24]

The N–O bonds in 1,2-oxazine (**58**) and isooxazolidine (**60**) were cleaved by Raney nickel via a radical mechanism to produce 1,4-diketone (**59**) and β-lactam (**61**), respectively (eqs 25 and 26).[25,26]

Hydrogenation of Multiple Bonds. Applications in this area are not very popular for Raney nickel. Raney nickel in dilute base is, however, an effective reagent for reduction of pyridines to the corresponding piperidines. The reaction is accelerated by substituents in the 2-position and by electron-withdrawing groups in the 3- and 4-positions, while electron-donating groups in the 3- and 4-positions retard the reaction (eq 27).[27] Occasionally, Raney nickel is used to reduce acyclic multiple bonds. An example for selective reduction of triple bonds to *cis* double bonds is shown in eq 28.[28]

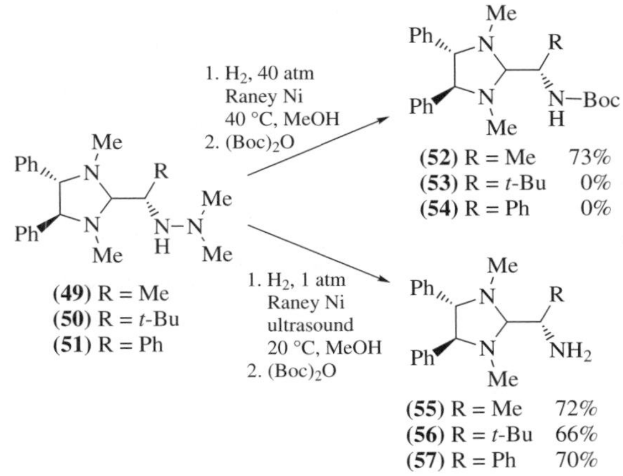

(**49**) R = Me
(**50**) R = *t*-Bu
(**51**) R = Ph

(**52**)	R = Me	73%
(**53**)	R = *t*-Bu	0%
(**54**)	R = Ph	0%

(**55**)	R = Me	72%
(**56**)	R = *t*-Bu	66%
(**57**)	R = Ph	70%

(24)

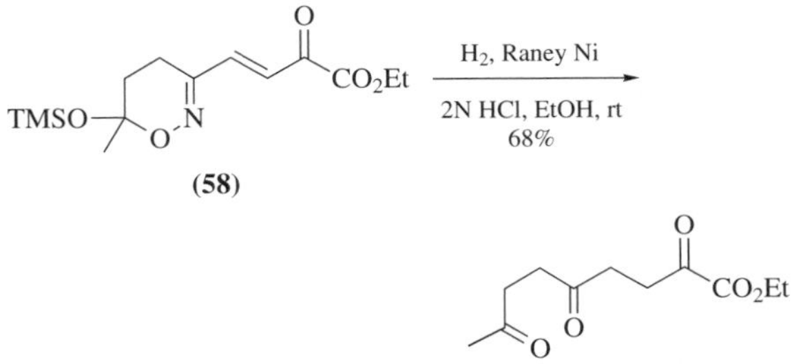

(58) **(59)** (25)

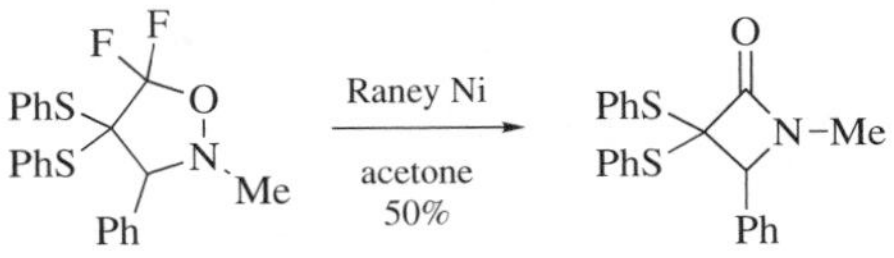

(60) **(61)** (26)

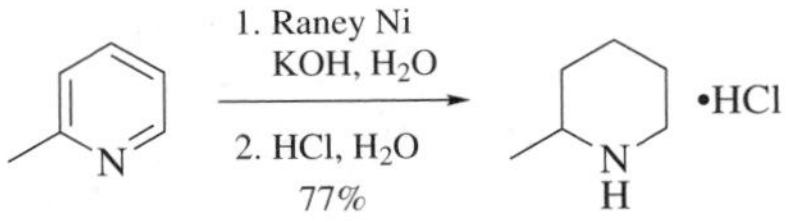

(62) **(63)** (27)

18 other examples

Deselenation. Similar to the desulfurization process, Raney nickel can be used to remove selenium from selenoketones, diselenides, selenides, and selenooxides. Typical examples are shown in eqs 29–33.[29–31] Moreover, Raney nickel has been used for a hydrodetelluration of chiral compound (**76**) without any racemization (eq 34).[32]

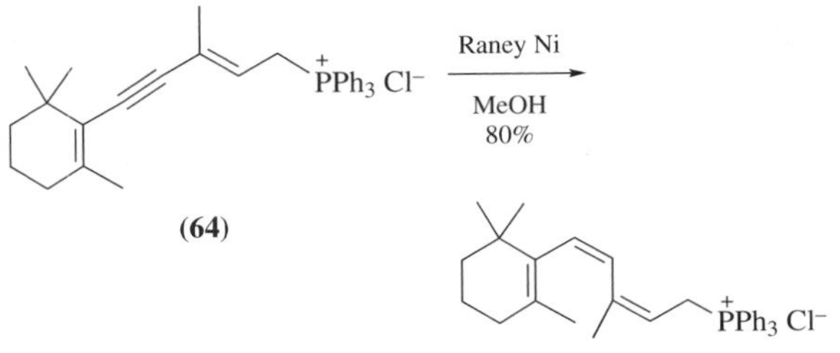

(64) **(65)** (28)

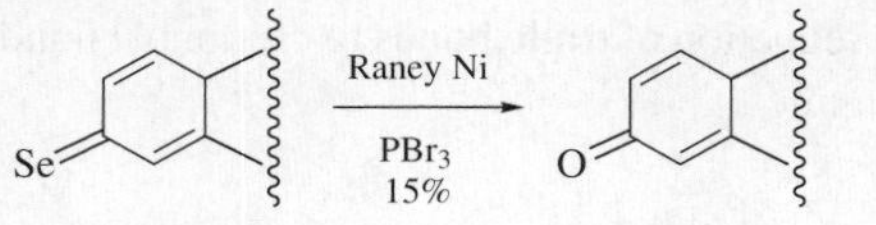

(66) (67) (29)

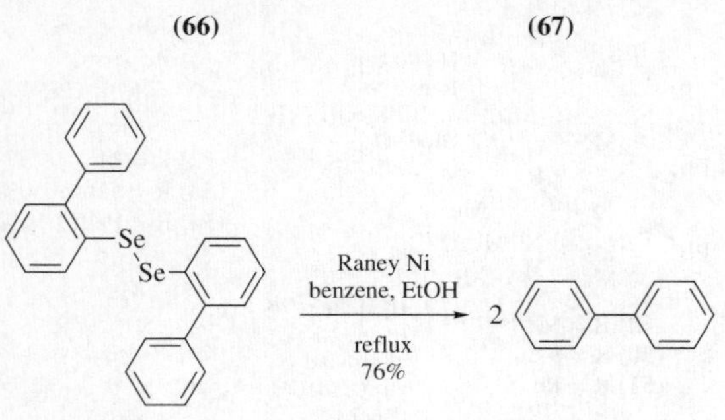

(68) (69) (30)

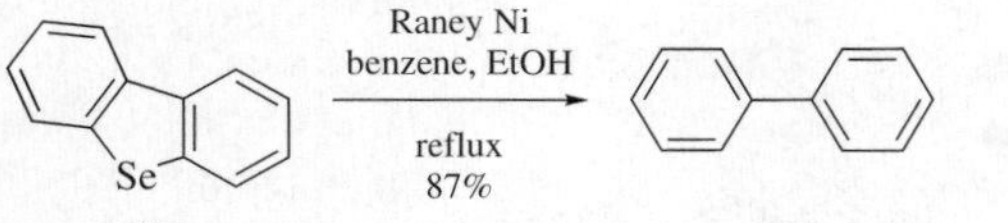

(70) (71) (31)

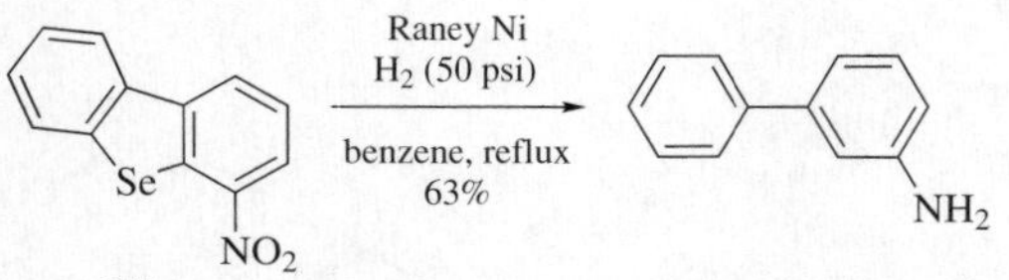

(72) (73) (32)

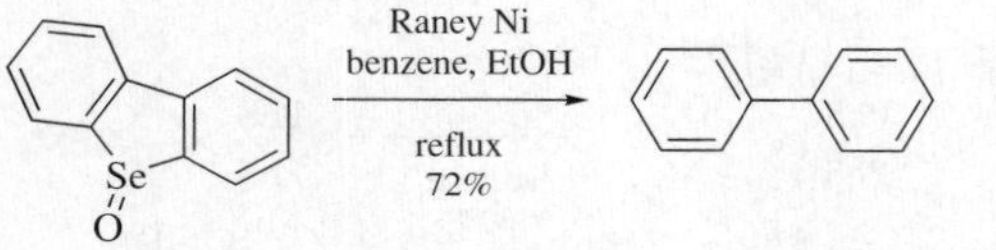

(74) (75) (33)

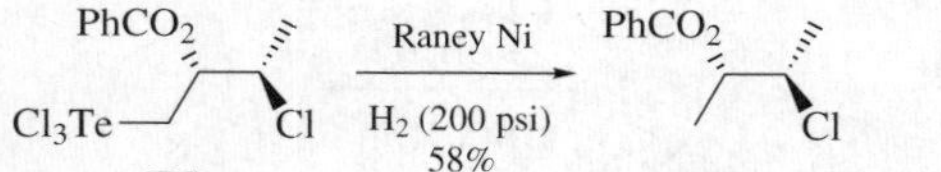

(76) (77) (34)

Reductive Amination of Carbonyl Groups. Reactions of this type can be accomplished by reduction of intermediate imines or oximes.[33–35] Recently, Chan and co-workers found that Raney nickel can be an efficient catalyst in the preparation of phenylalanine. Treatment of sodium phenylpyruvate with either ***Ammonia*** gas or aqueous ammonia solution in the presence of Raney nickel under 200 psi pressure gave >98% of phenylalanine (eq 36). Other α-keto esters such as 4-hydroxyphenylpyruvic acid, pyruvic acid, and benzoyl acid also gave the corresponding amino acids in excellent yield.[36]

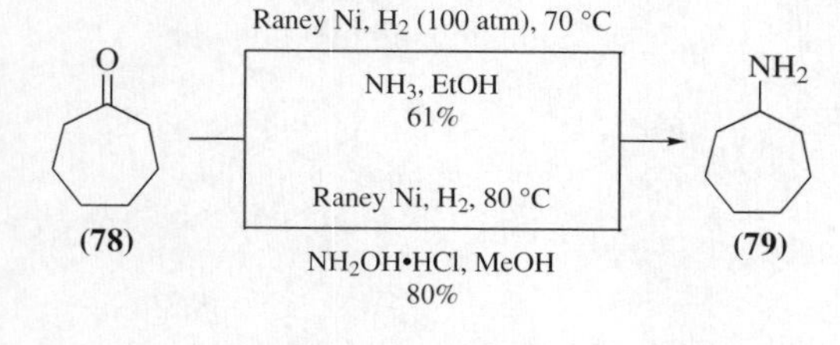

(78) (79) (35)

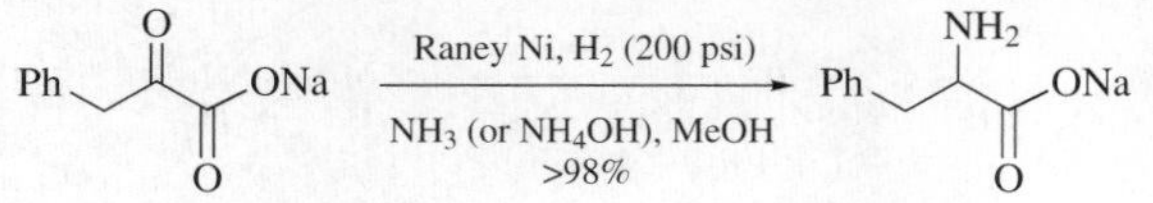

(80) (81) (36)

Asymmetric Reduction. Recently, asymmetric synthesis has become a center of attention for synthetic chemists. The use of Raney nickel and tartaric acid was recently reported by Bartok. Reduction of ketone (**82**) gave alcohols (**83**) and (**84**) as a 92:8 mixture in 70% chemical yield.[37] Takeshita et al.[9] also reported an asymmetric reduction of β-keto ester (**85**) to give the corresponding β-hydroxy ester (**86**) in 80% ee (eq 38). In addition, it was found that enantioselectivities were improved by treatment of the Raney nickel with ultrasound prior to use.[38] Blacklock et al. reported an asymmetric reductive amination of α-keto ester (**88**) in which they used the chiral amine (**87**) instead of a chiral catalyst. The result, shown in eq 39, indicates that the amino ester (**89**) was produced in 80% yield with 74% de.[39]

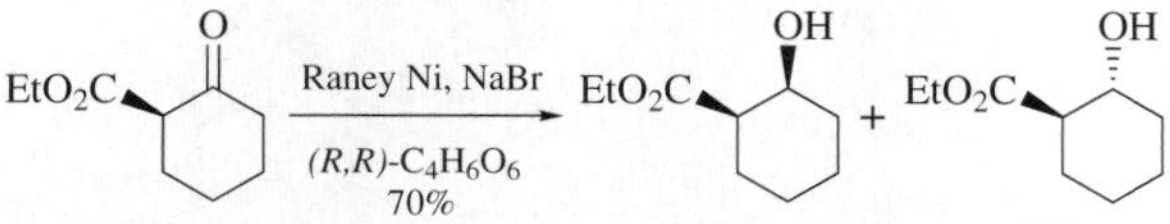

(82) (83) 92:8 (84) (37)

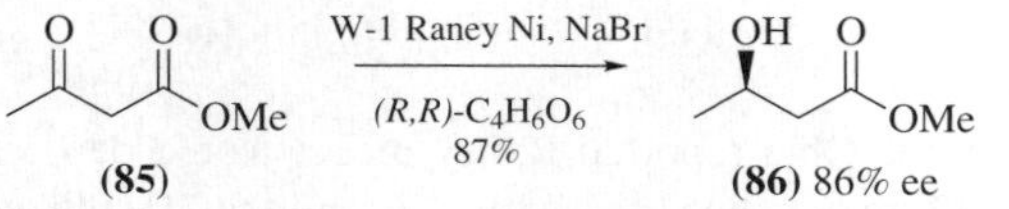

(85) (86) 86% ee (38)

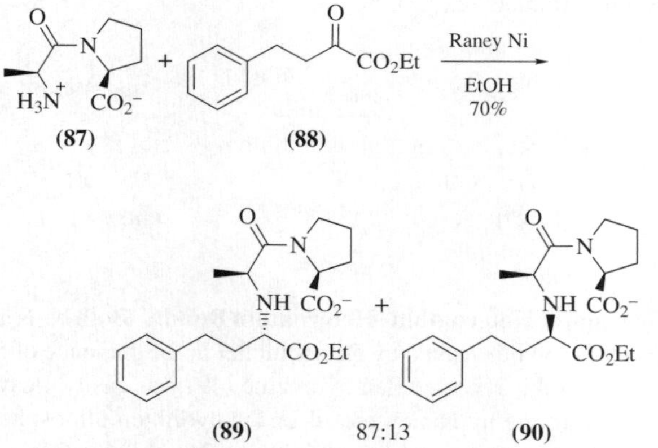

(87) (88) (89) (90) (39)

Related Reagents. See Classes R-2, R-4, R-8, R-11, R-12, R-19, R-23, R-24, R-26, R-28, R-30, R-32, and R-35, pages 1–10.

1. (a) Hauptmann, H.; Walter, W. F. *CRV* **1962**, *62*, 347. (b) Caubere, P.; Coutrot, P. *COS* **1991**, *8*, 835. (c) Billica, H. R.; Adkins, H. *OSC* **1955**, *3*, 176.
2. Troin, Y.; Diez, A.; Bettiol, J. L.; Rubiralta, M. Grierson, D. S.; Husson, H.-P. *H* **1991**, *32*, 663.
3. Graham, A. R.; Millidge, A. F.; Young, D. P. *JCS* **1954**, 2180.
4. Fujisawa, T.; Mobele, B. I.; Shimizu, M. *TL* **1992**, *33*, 5567.
5. Lottaz, P. A.; Edward, T. R. G.; Mentha, Y. G.; Burger, U. *TL* **1993**, *34*, 639.
6. Ohta, S.; Yamamoto, T.; Kawasaki, I.; Yamashita, M.; Katsuma, H.; Nasako, R.; Kobayashi, K. Ogawa, K. *CPB* **1992**, *40*, 2681.

7. Tietze, L. F.; Hartfiel, U.; Hubsch, T.; Voss, E.; Bogdanowicz-Szwod, K.; Wichmann, J. *LA* **1991**, 275.
8. Högberg, H. E.; Hedenström, E.; Fägerhag, J.; Servi, S. *JOC* **1992**, *57*, 2052.
9. Takeshita, M.; Tsuge, A.; Tashiro, M. *CB* **1991**, *124*, 411.
10. Kast, J.; Hoch, M.; Schmidt, R. R. *LA* **1991**, 481.
11. Sadanandan, E. V.; Srinivasan, P. C. *S* **1992**, 648.
12. Cox, P. J.; Persad, A.; Simpkins, N. S. *SL* **1992**, 197.
13. Carrington, H. C.; Vasey, C. H.; Waring, W. S. *JCS* **1953**, 3105.
14. Kung, P. P.; Jones, R. A. *TL* **1991**, *32*, 3919.
15. Coscia, A. T.; Dickerman, S. C. *JACS* **1959**, *81*, 3098.
16. Bourdon, R. *BSF* **1958**, 722.
17. Liu, H.-J.; Luo, W. *CJC* **1992**, *70*, 128.
18. Behren, C.; Egholm, M.; Buchardt, O. *S* **1992**, 1235.
19. Ishibashi, H; So, T. S.; Okochi, K.; Sato, T.; Nakamura, N.; Nakatani, H.; Ikeda, M. *JOC* **1991**, *56*, 95.
20. Ojima, I.; Zhao, M.; Yamato, T.; Nakahashi, K. *JOC* **1991**, *56*, 5263.
21. Krafft, M. E.; Crooks, W. J., III; Zorc, B.; Milczanowski, S. E. *JOC* **1988**, *53*, 3158.
22. Aoyagi, Y.; Maeda, A.; Inoue, M.; Shiraishi, M.; Sakakibara, Y.; Fukui, Y.; Ohta, A.; Kajii, K.; Kodama, Y. *H* **1991**, *32*, 735.
23. Alexakis, A.; Lensen, N.; Mangeney, P. *TL* **1991**, *32*, 1171.
24. Alexakis, A.; Lensen, N.; Mangeney, P. *SL* **1992**, *3*, 625.
25. Zimmer, R.; Collas, M.; Roth, M.; Reissig, H. U. *LA* **1992**, 709.
26. Purrington, S. T.; Sheu, K.-W. *TL* **1992**, *33*, 3289.
27. Lunn, G.; Sansone, E. B. *JOC* **1986**, *51*, 513.
28. Soukup, M.; Widmer, E. *TL* **1991**, *32*, 4117.
29. Florey, K.; Restivo, A. R. *JOC* **1957**, *22*, 406.
30. Wiseman, G. E.; Gould, E. S. *JACS* **1954**, *76*, 1706.
31. Wiseman, G. E.; Gould, E. S. *JACS* **1955**, *77*, 1061.
32. Backvall, J. E.; Bergman, J.; Engman, L. *JOC* **1983**, *48*, 3918.
33. Freifelder, M.; Smart, W. D.; Stone, G. R. *JOC* **1962**, *27*, 2209.
34. Botta, M.; De Angelis, F.; Gambacorta, A.; Labbiento, L.; Nicoletti, R. *JOC* **1985**, *50*, 1916.
35. Graham, S. H.; Williams, A. J. S. *JCS(C)* **1966**, 655.
36. Chan, A. S. C.; Lin, Y.-C.; Chen, C.-C. Personal communication.
37. Wittmann, G.; Gondos, G.; Bartok, M. *HCA* **1990**, *73*, 635.
38. Tai, A.; Kikukawa, T.; Sugimura, T.; Inone, Y. D.; Osawa, T.; Fujii, S. *JCS(C)* **1991**, 795.
39. Blacklock, T. J.; Shuman, R. F.; Butcher, J. W.; Shearin, W. E. Jr. Budavari, J.; Grenda, V. J. *JOC* **1988**, *53*, 836.

Teng-Kuei Yang & Dong-Sheng Lee
National Chung-Hsing University, Taichung, Taiwan

Rhodium on Alumina[1]

Rh/Al_2O_3

[7440-16-6] Rh (MW 102.91)

(active catalyst under mild conditions for hydrogenation of unsaturated compounds including aldehydes, ketones,[1h,7] and aromatic carbocyclic[1] and heterocyclic structures;[1a,g,k] high stereospecificity and low hydrogenolytic activity)

Form Supplied in: rhodium supported on alumina, carbon, or silica containing 0.5–5.0% of the metal are available commercially.

Handling, Storage, and Precautions: the unused supported catalysts are not pyrophoric but rhodium supported on a finely divided carbon, like carbon itself, can undergo a dust explosion. After use, the catalysts are likely to contain adsorbed hydrogen which may ignite when the catalyst dries.[1f,h] Use in a fume hood.

Introduction. The catalytic properties of rhodium, like the other platinum metals, depend secondarily on the support. ***Alumina*** is commonly used but other supports may furnish advantages in particular circumstances. Rhodium black avoids possible influences of the support.[2] The high activity of rhodium often permits its use at room temperature and atmospheric pressure but higher temperatures and pressures may be used to advantage.

Alkenes and Alkynes. Although rhodium catalyzes the hydrogenation of alkenes, dienes, and alkynes under mild conditions, it is not the most selective platinum catalyst if double bond migration or isomerization can adversely affect the desired stereochemistry or if a selective hydrogenation of a diene or alkyne is sought.[1h] The tendency of rhodium for alkene isomerization relative to hydrogenation is only a little greater than that of ruthenium (see ***Ruthenium Catalysts***), which results in the formation of a greater fraction of *cis*-1,2-dimethylcyclohexane using Ru (93.5%) than using Rh (87.6%) in hydrogenating 1,2-dimethylcyclohexene at 25 °C/1 atm H_2.[3]

The hydrogenation of the double bond in vinyl or allylic compounds is achieved with some success with Rh but generally Pd or Ru is more effective, particularly with vinyl ethers.[1h] However, the high activity of Rh allows the reaction to be conducted at lower temperatures than with other metals. For example, the asperuloside tetraacetate, which contains both a vinyl and an allyl ether structural unit, was hydrogenated over Rh/C starting at −30 °C and raising the temperature slowly during 3 h to 0 °C to give a virtually quantitative yield of the tetrahydro product (eq 1).[4]

5% Rh/C, −30 to 0 °C, 1 atm; AcOEt, 100% (1)

The activity of Rh/C also is shown by the hydrogenation of the double bond in the β-acyloxy-α,β-unsaturated esters and ketones without causing hydrogenolysis as given by ***Platinum(IV) Oxide*** (eq 2).[5]

5% Rh/C, H_2; THF, rt, 1 atm (2)

Rh/Al_2O_3 in combination with ***Palladium on Carbon*** effected the saturation of an isomer of dodecamethyl[6]radialene (**1**) to all-*trans*-hexaisopropylcyclohexane (**2**) when separately the same catalysts failed (eq 3).[6] Apparently the reduction required high reactivity for both hydrogenation (Rh) and alkene isomerization (Pd).

$$\mathbf{(1)} \xrightarrow[\text{C}_6\text{H}_{12},\ 130\ ^\circ\text{C},\ 65\ \text{atm};\ 60\%]{\text{Rh/Al}_2\text{O}_3,\ \text{Pd/C},\ \text{H}_2} \mathbf{(2)} \quad (3)$$

Aldehydes and Ketones. Rhodium is an excellent catalyst for the hydrogenation of aldehydes and ketones under mild conditions.[1a,e,h,7] The reduction of aliphatic carbonyl compounds can be accomplished without the hydrogenolysis of susceptible groups attached elsewhere in the molecule.

In the hydrogenation of cyclohexanones, rhodium excels in the formation of the axial hydroxyl group from monosubstituted compounds; the effect is especially notable for alkyl substituents.[1d,8] With methoxy groups in place of alkyl groups at the 2- and 4-positions the fraction of axial isomer increases markedly for Pd, Ir (see ***Iridium***), and Pt but not Rh.[9a] Hydrogenation of unhindered cyclohexanones in isopropyl alcohol or THF in the presence of small amounts of ***Hydrogen Chloride*** gives excellent yields of axial alcohols (eq 4).[9b] Steroidal 3,17- and 3,20-diones are selectively hydrogenated at C-3 to yield the corresponding 3-axial-hydroxy ketones.

$$t\text{-Bu-cyclohexanone} \xrightarrow[\text{HCl, rt, 1 atm}]{\text{Rh, H}_2,\ \text{THF}} t\text{-Bu-cyclohexanol (axial OH)} \quad (4)$$

Imines. Imines are thought to be important intermediates in the hydrogenation of nitriles, anilines, and oximes.[1] ***Ammonia*** and primary amines condense with aldehydes to give imines which may be hydrogenated. Freifelder describes the conversion of 3,4-dimethoxybenzaldehyde to the benzylamine over Rh/C in 95% ethanol containing ammonia (aq) and NH_4OAc, (rt, 3 atm, 64% yield).[1k]

Nitriles. The products of hydrogenating a nitrile depend on the catalyst and whether the nitrile is aliphatic or aromatic.[1a,h] The reaction conditions including the solvent, the temperature, and the H_2 pressure, are important variables. Rhodium catalysts are most useful in the preparation of either primary or secondary amines.

The hydrogenation of an aliphatic nitrile leads to the formation of an imine which in turn may be hydrogenated to a primary amine.[1a,g,h] However, secondary amines are often the principal product, which apparently follows the condensation of the imine and the primary amine.[1g] Whether the condensations occur in the homogeneous medium, as first proposed by von Braun et al.,[10] or on the surface of the catalyst, is unsettled.[11] Recently, the analysis of a kinetic study of the hydrogenation of propionitrile on Rh/C and Rh/Al_2O_3, to form propylamine and dipropylamine, led to the claim that the results do not support the proposal that the condensations occur in the liquid phase.[11] A mechanism involving only surface reactions is proposed to explain the authors' results as well as the literature record.

The catalyst's support influences the effect of increasing pressure on the distribution of the products of hydrogenating pentanenitrile in methanol at rt.[12] At 4 atm, Rh/C, Rh/Al_2O_3, and Rh_2O_3 mainly yield dipentylamine (93%, 87%, and 77% respectively). Raising the pressure increases the yield of the primary amine. At 90 atm it is the sole product from unsupported Rh (Rh_2O_3), 54% yield is obtained from Rh/Al_2O_3, and 29% yield from Rh/C, the remainder being the secondary amine. At an intermediate pressure (50 atm), Rh/C yields only the secondary amine.[1g] Using Rh/C in the presence of a large excess of butylamine (mole ratio 1.7–6.6) the yield of butylpentylamine is 95–100%.[1g]

Aromatic nitriles are hydrogenated over Rh/C under mild conditions to dibenzylamines in excellent yields (25 °C, 4 atm).[1g]

Most recently, acetic acid was shown to be an excellent solvent for the Rh/Al_2O_3 or Rh/C catalyzed hydrogenation of aliphatic, aromatic carbocyclic, and aromatic heterocyclic nitriles to secondary amines (eq 5).[13] 3-Hydroxypropanenitrile formed a tertiary amine as a single isomer (eq 6) and the reduction of the tosylate of β-cyanoalanine formed a lactam (eq 7).

$$\text{1-methylpyrazole-3-carbonitrile} \xrightarrow[\text{AcOH, rt, 1 atm; 79\%}]{5\%\ \text{Rh/Al}_2\text{O}_3,\ \text{H}_2} \text{bis(1-methylpyrazol-3-ylmethyl)amine} \quad (5)$$

$$\text{HO}\diagup\!\diagdown\text{CN} \xrightarrow[\text{AcOH, rt, 1 atm; 70\%}]{5\%\ \text{Rh/Al}_2\text{O}_3,\ \text{H}_2} \text{tricyclic bis(oxazinane) amine} \quad (6)$$

$$\text{MeO}_2\text{C-CH(NHTs)-CH}_2\text{CN} \xrightarrow[\text{AcOH, rt, 1 atm; 76\%}]{5\%\ \text{Rh/Al}_2\text{O}_3,\ \text{H}_2} \text{3-(NHTs)-2-piperidinone (HN, O)} \quad (7)$$

Oximes. Although few examples of its use for the reduction of oximes have been recorded, rhodium has given excellent results, being especially useful in place of catalysts which yield excessive amounts of secondary amines.[1h] Freifelder et al. found the 5% Rh/Al_2O_3-catalyzed hydrogenation of cycloheptanone oxime in methanol (60 °C, 0.75–1.0 atm) gave an 80% yield of cycloheptylamine.[1a,14] 3-Amino-2-methyl-2-butanone and 1-(1-aminoethyl)cyclohexanol were prepared by the 5% Rh/Al_2O_3-catalyzed hydrogenation of the oximes of 3-hydroxy-3-methyl-2-butanone and 1-acetylcyclohexanol (eq 8) in 90–92% yields.[15]

$$\text{1-acetylcyclohexanol oxime (HON, OH)} \xrightarrow[\text{EtOH, rt, 3–4 atm; 90–92\%}]{5\%\ \text{Rh/Al}_2\text{O}_3,\ \text{H}_2} \text{1-(1-aminoethyl)cyclohexanol (H}_2\text{N, OH)} \quad (8)$$

Carbocyclic Aromatic Compounds. Rhodium provides highly active catalysts for the hydrogenation of the aromatic carbocycle under mild conditions (25–80 °C and 1–3 atm).[1]

It is particularly useful when hydrogenolysis of attached hydroxy or alkoxy groups or benzylic alcohols or ethers is to be avoided.[16] The selectivity of rhodium under mild conditions is shown by the Rh/Al_2O_3-catalyzed hydrogenation in acetic acid at room temperature of (2-methoxyphenyl)propan-2-one to *cis*-1-(2-methoxycyclohexyl)propan-2-one (eq 9).[17]

5% Rh/Al_2O_3, H_2; AcOH, rt, 3 atm (9)

The consecutive addition of hydrogen atoms to the adsorbed cycle presumably proceeds through adsorbed cyclohexadienes and cyclohexenes, but only the desorption of the latter has been observed.[18] Of the platinum metals, ruthenium yields the largest amount of cyclohexenic intermediates when hydrogenated in the presence of water, which is also an unusual promoter of ruthenium's catalytic activity (see ***Ruthenium Catalysts***). In nonaqueous media, Rh yields the largest concentrations of cyclohexenic intermediates, as in the hydrogenation of *t*-butylbenzenes and *t*-butylbenzoic acids or esters in EtOH, HOAc, or saturated hydrocarbon solvents.[1f,19]

The Rh/C-catalyzed hydrogenation of 2-*t*-butylbenzoic acid initially yields *cis*-2-*t*-butylcyclohexanecarboxylic acid (75%), 6-*t*-butyl-1-cyclohexenecarboxylic acid (22%), and 2-*t*-butyl-1-cyclohexenecarboxylic acid (2%); *trans*-2-*t*-butylcyclohexanecarboxylic acid is formed only after the unsaturated intermediates begin to be reduced, which occurs when about 80% of the benzoic acid has been converted.[19] The result supports the earlier proposal that the formation of *trans* isomers in the hydrogenation of the xylenes arises from the saturation of desorbed cyclohexenes, even though the latter are undetected.[18]

The rate of the rhodium-catalyzed hydrogenation of the aromatic ring diminishes with alkyl substitution but with multiple substituents the arrangement of the groups affect the reactivity. The selective hydrogenation of the unsubstituted ring in *o*-anisylmethylphenylphosphine oxide illustrates that alkoxy groups also lower the reactivity of the benzene ring (eq 10).[20]

Rh/C; H_2 (10)

Although rhodium is generally effective in saturating polycarbocyclic aromatic compounds, it is not useful if the selective reduction of a particular cycle is sought.[1g–i]

Phenols and Phenyl Ethers. A useful application of rhodium catalysts is the hydrogenation of phenols and phenyl ethers which are sensitive to hydrogenolysis, a particular problem in hydrogenating di- and polyhydroxybenzenes.[1a–c,21] Smith and Stump furnish an informative study, both kinetic and preparative, of the hydrogenation of hydroxybenzenes catalyzed by Pt and Rh/Al_2O_3.[22] The kinetic evidence indicates that the mechanism of formation of the cyclohexanols from the phenols does not require ketone intermediates, although some are formed. The authors propose that cyclohexenols are intermediates which may be reduced to cyclohexanols, hydrogenolyzed to cyclohexanes, or isomerized to cyclohexanones. The conversion of resorcinol to dihydroresorcinol (1,3-cyclohexandione) in high yield is described (eq 11).

5% Rh/Al_2O_3, H_2; aq NaOH, rt, 4 atm; 87% (11)

The hydrogenation of the cresols at 80 °C and 80–100 atm forms intermediate ketones in amounts which permit a quantitative analysis of the kinetics of the process to give a good estimate of the fraction of the cresol which is converted via the ketone to the cyclohexanol.[23] The stereochemistry of the final product is determined by the fraction formed via the cyclohexanone (knowing the fraction of *cis* and *trans* isomers it forms upon reduction), with the remainder of the product alcohol being the *cis* isomer. In principle, if cyclohexenol intermediates desorb from the catalyst, readsorption followed by the suprafacial addition of hydrogen could yield some *trans* isomers.[18]

The hydrogenation of 1-naphthol over Rh/Al_2O_3 in ethanol or methanol (25 °C, 4 atm) gave good yields of the isomeric decalols in the ratio of 13.3:3.2:1, the largest component being the *cis,cis* isomer; the configurations of the others were not specified; hydrogenolysis to decalin was low (3%).[24] Of the unsupported platinum metals used to catalyze the hydrogenation of 2-naphthol and the intermediate tetrahydro-2-naphthol (*t*-BuOH, 80 °C, 50 atm), Rh and Ru are highly stereospecific in forming the *cis,cis*-2-decalols and afford only small amounts of hydrogenolysis products.[25]

Anilines. Ruthenium is generally the preferred catalyst for the hydrogenation of anilines to cyclohexyl amines (see ***Ruthenium Catalysts***), but rhodium catalysts have been used successfully.[1g,h] To diminish the formation of dicyclohexylamine, the best solvents were found to be *t*-butyl or isopropyl alcohol; the addition of a small amount of a strong base, preferably ***Lithium Hydroxide***, to Rh hydroxide furnished a catalyst that almost eliminated the diamine's formation.[26] Rhodium oxide prepared by $LiNO_3$ fusion with ***Rhodium(III) Chloride***, and used in isopropyl or *t*-butyl alcohol, leads to excellent yields of the isomeric methoxy- or ethoxycyclohexylamines from the corresponding anilines (eq 12).[27] Aniline has been converted to cyclohexylamine (96%, dicyclohexylamine 4%) catalyzed by Rh/Al_2O_3 in *i*-PrOH (23–26 °C, 55–70 atm).[28]

Rh_2O_3 (Li), H_2; *t*-BuOH, 80 °C; 80–100 atm; 84% (12)

Heterocyclic Aromatic Compounds. Rhodium has been used effectively to catalyze the hydrogenation of furans, pyrroles, pyridines, and other structures containing aromatic heterocycles.[1a,h]

Furans. Furans are saturated under mild conditions with Rh/C. 2,5-Diferrocenylfuran is hydrogenated quantitatively to *cis*-2,5-diferrocenyltetrahydrofuran (5% Rh/C, rt, 1 atm, 2 h) (eq 13);

after a time (6 h) at higher pressures (60 atm) the ether linkages are hydrogenolized to form 1,4-diferrocenylbutane.[29]

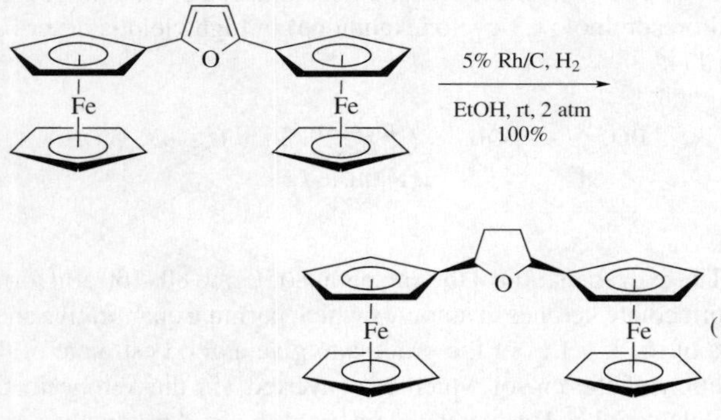

5% Rh/C, H_2, EtOH, rt, 2 atm, 100% (13)

Pyrrole. Rhodium is the preferred catalyst for the hydrogenation of pyrroles.[30] Pyrrole and 1-methylpyrrole, without solvent, are reduced to the pyrrolidines in high yield over Rh/Al_2O_3 (rt, 2–3 atm).[1g] In compounds containing an aromatic carbocycle which is not fused to the pyrrole, the carbocycle tends to be reduced preferentially,[1g] but the reverse selectivity has also been observed (eq 14)[31] and attributed to the trisubstitution in the carbocyclic ring.[1g]

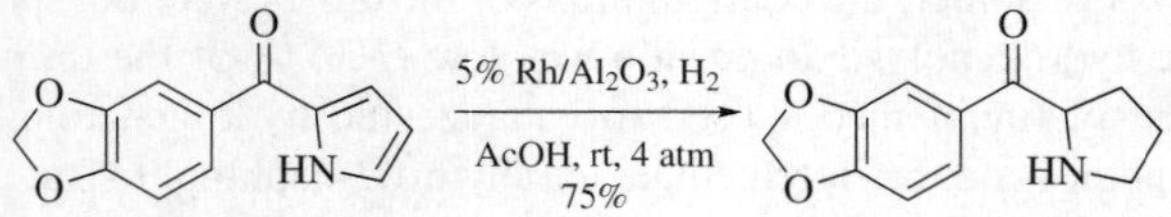

5% Rh/Al_2O_3, H_2, AcOH, rt, 4 atm, 75% (14)

The high *cis* selectivity in the Rh/Al_2O_3-catalyzed hydrogenation of a 2,5-disubstituted pyrrole is illustrated in (eq 15).[32]

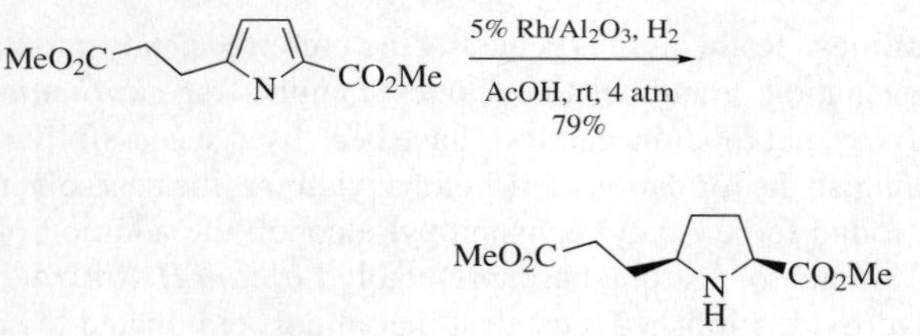

MeO_2C ... CO_2Me; 5% Rh/Al_2O_3, H_2, AcOH, rt, 4 atm, 79% (15)

Pyridine. For the hydrogenation of pyridines, rhodium is the most effective catalyst under mild conditions and maintains its advantage at higher pressures.[1h] 2,6-Dimethylpyridine in glacial AcOH is hydrogenated to *cis*-2,6-dimethylpiperidine over 5% Rh/Al_2O_3 (eq 16).[33]

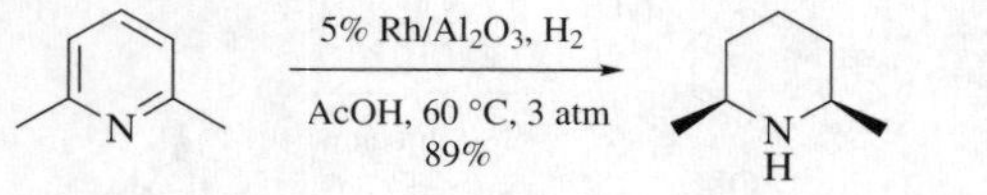

5% Rh/Al_2O_3, H_2, AcOH, 60 °C, 3 atm, 89% (16)

Although a variety of catalysts can affect satisfactorily the reduction of the 2- and 4-pyridinecarboxylic acids, the decarboxylation observed with the other catalysts in the hydrogenation of nicotinic acid (pyridine-3-carboxylic acid) is avoided by 5% Rh/Al_2O_3 in aqueous ammonia (eq 17).[34]

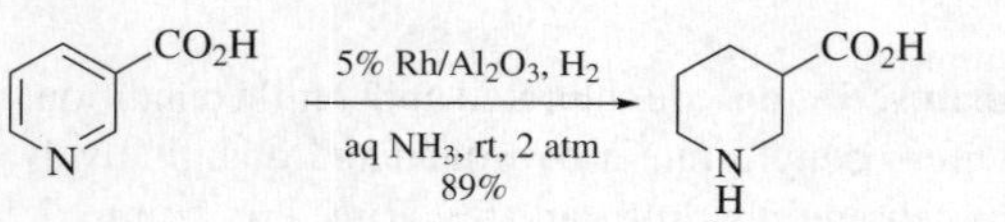

CO_2H; 5% Rh/Al_2O_3, H_2, aq NH_3, rt, 2 atm, 89% (17)

Generally, catalytic hydrogenation of pyridines with attached or fused aromatic carbocycles occurs preferentially in the pyridyl ring.[1h] 4-Benzylpyridine is converted quantitatively to 4-benzylpiperidine (5% Rh/C, EtOH, 60 °C, 4 atm).[35] The pyridine ring in (3-methoxy-2-pyridyl)-2-propanone is reduced selectively with Rh/Al_2O_3 in water containing one equiv of ***Hydrogen Bromide***, the selectivity is achieved only when the pyridylpropanone has been carefully purified by distillation (eq 18).[36]

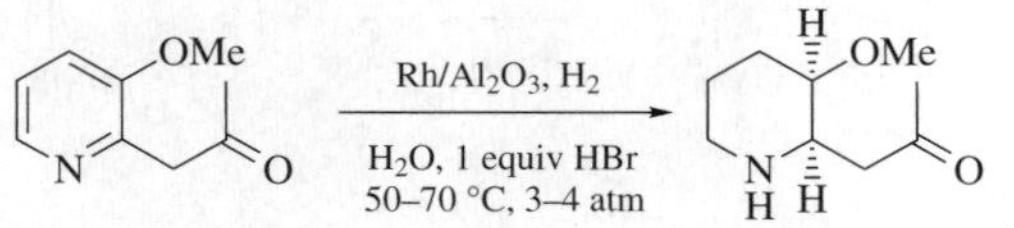

OMe; Rh/Al_2O_3, H_2, H_2O, 1 equiv HBr, 50–70 °C, 3–4 atm (18)

A variety of other nitrogen heterocycles have been hydrogenated successfully over rhodium catalysts.[1g,1k]

Related Reagents. See Classes R-2, R-3, R-4, R-7, R-8, R-9, R-12, R-19, and R-20, pages 1–10.

1. Fieser, L.; Fieser, M. (a) *FF* **1967**, *1*, 979. (b) *FF* **1974**, *4*, 418. (c) *FF* **1977**, *6*, 503. (d) *FF* **1980**, *8*, 418. (e) *FF* **1984**, *11*, 460. (f) Kieboom, A. P. G.; van Rantwijk, F. *Hydrogenation and Hydrogenolysis in Synthetic Organic Chemistry*; Delft University Press: Delft, 1977. (g) Freifelder, M. *Catalytic Hydrogenation in Organic Synthesis: Procedures and Commentary*; Wiley: New York, 1978. (h) Rylander, P. N. *Catalytic Hydrogenation in Organic Synthesis*; Academic: New York, 1979. (i) Rylander, P. N. *Hydrogenation Methods*; Academic: London, 1985. (j) Bartók, M. *Stereochemistry of Heterogeneous Metal Catalysis*; Wiley: New York, 1985. (k) *COS* **1991**, *8*.
2. Takagi, Y.; Naito, T.; Nishimura, S. *BCJ* **1965**, *38*, 2119.
3. Nishimura, S.; Sakamoto, H.; Ozawa, T. *CL* **1973**, 855.
4. Berkowitz, W. F.; Choudhry, S. C.; Hrabie, J. A. *JOC* **1982**, *47*, 824.
5. Rozzell, J. D. *TL* **1982**, *23*, 1767.
6. Golan, O.; Goren, Z.; Biali, S. E. *JACS* **1990**, *112*, 9300.
7. Breitner, E.; Roginski, E.; Rylander, P. N. *JOC* **1959**, *24*, 1855.
8. Mitsui, S.; Saito, H.; Yamashita, Y.; Kaminaga, M.; Senda, Y. *T* **1973**, *29*, 1531.
9. (a) Nishimura, S.; Katagiri, M.; Kunikata, Y. *CL* **1975**, 1235. (b) Nishimura, S.; Ishige, M.; Shiota, M. *CL* **1977**, 963.
10. von Braun, J.; Blessing, G.; Zobel, F. *CB* **1923**, *56B*, 1988.
11. Dallons, J. L.; van Gysel, A.; Jannes, G. In *Catalysis of Organic Reactions*; Pascoe, W. E., Ed.; Dekker: New York, 1992; pp 93–104.
12. Rylander, P. N.; Hasbrouck, L.; Karpenko, I. *ANY* **1973**, *214*, 100.
13. Galán, A.; de Mendoza, J.; Prados, P.; Rojo, J.; Echavarren, A. M. *JOC* **1991**, *56*, 452.
14. Freifelder, M.; Smart, W. D.; Stone, G. R. *JOC* **1962**, *27*, 2209.
15. Newman, M. S.; Lee, V. *JOC* **1975**, *40*, 381.
16. Stocker, J. H. *JOC* **1962**, *27*, 2288.
17. Cantor, S. E.; Tarbell, D. S. *JACS* **1964**, *86*, 2902.
18. Siegel, S. *COS* **1991**, *8*, 417.
19. Van Bekkum, H.; Van de Graaf, B.; Van Minnen-Pathuis, G.; Peters, J. A.; Wepster, B. M. *RTC* **1970**, *89*, 521.
20. Vineyard, B. D.; Knowles, W. S.; Sabacky, M. J. *J. Mol. Catal.* **1983**, *19*, 159.
21. Burgstahler, A. W.; Bithos, Z. J. *JACS* **1960**, *82*, 5466.
22. Smith, H. A.; Stump, B. L. *JACS* **1961**, *83*, 2739.
23. (a) Takagi, Y.; Nishimura, S.; Taya, K.; Hirota, K. *J. Catal.* **1967**, *8*, 100. (b) Takagi, Y.; Nishimura, S.; Hirota, K. *J. Catal.* **1968**, *12*, 214. (c) Takagi, Y.; Nishimura, S.; Hirota, K. *BCJ* **1970**, *43*, 1846.

24. Meyers, A. I.; Beverung, W.; Garcia-Munoz, G. *JOC* **1964**, *29*, 3427.
25. Nishimura, S.; Ohbuchi, S.; Ikeno, K.; Okada, Y. *BCJ* **1984**, *57*, 2557.
26. Nishimura, S.; Shu, T.; Hara, T.; Takagi, Y. *BCJ* **1966**, *39*, 329.
27. Nishimura, S.; Uchino, H.; Yoshino, H. *BCJ* **1968**, *41*, 2194.
28. Greenfield, H. *ANY* **1973**, *214*, 233.
29. Yamakawa, K.; Moroe, M. *T* **1968**, *24*, 3615.
30. Gribble, G. W. *COS* **1991**, *8*, 603.
31. Dolby, L. J.; Nelson, S. J.; Senkovich, D. *JOC* **1972**, *37*, 3691.
32. Turner, W. W. *JHC* **1986**, *23*, 327.
33. Overberger, C. G.; Palmer, L. C.; Marks, B. S.; Byrd, N. R. *JACS* **1955**, *77*, 4100.
34. Freifelder, M. *JOC* **1963**, *28*, 1135.
35. Freifelder, M.; Robinson, R. M.; Stone, G. R. *JOC* **1962**, *27*, 284.
36. Barringer, D. F., Jr.; Berkelhammer, G.; Carter, S. D.; Goldman, L.; Lanzilotti, A. E. *JOC* **1973**, *38*, 1933.

Samuel Siegel

University of Arkansas, Fayetteville, AR, USA

Ruthenium Catalysts[1]

[7440-18-8] Ru (MW 101.07)

(selective hydrogenation catalysts,[1] high regio- and stereoselective addition to alkenes,[2] low activity for alkene isomerization,[3] actively hydrogenates carbonyls,[4] low hydrogenolytic activity,[5] high stereoselective perhydrogenation of aromatic carbocyclic and many aromatic heterocyclic compounds[6,7])

Form Supplied in: ruthenium supported on alumina or carbon and RuO_2 are available commercially.

Preparative Methods: ***Ruthenium(III) Chloride*** is converted by aq. ***Sodium Hydroxide*** or ***Lithium Hydroxide*** to $Ru(OH)_3$, which may be dried in a vacuum at room temperature for later use.[6] Ruthenium black can be prepared by the reduction of $RuCl_3 \cdot 3H_2O$ with ***Formaldehyde***, washed repeatedly with distilled water, and stored in ethanol.[8]

Handling, Storage, and Precautions: the unused supported catalysts are not pyrophoric but ruthenium on finely divided carbon, like carbon itself, can undergo a dust explosion.[1d] After use, the catalysts are likely to contain adsorbed hydrogen which may ignite when dried.[1c,1d] Use in a fume hood.

Introduction. The catalytic characteristics of ruthenium, like other platinum metals, depend secondarily on the support.[1c,d] The discovery of RuO_2 as a selective hydrogenation catalyst encouraged its use.[5,7,9] Later the supported catalysts Ru/C and Ru/Al_2O_3 became readily available. Supported catalysts are the more efficient users of the metal but may not furnish the maximum yield of a desired product if, for example, intraparticle mass transport affects the product distribution.[1d]

Ruthenium-catalyzed hydrogenations often show a variable and sometimes lengthy induction period which can be avoided by agitating the catalyst and solvent mixture in hydrogen (1 atm) for 1–2 h before introducing the substrate.[2,10] The common solvents are water, water–acetic acid, alcohols, and dioxane. Water is a powerful promoter of the activity of ruthenium hydrogenation catalysts.[1d,2,6]

Alkenes and Alkynes. Ruthenium exhibits a lower tendency than rhodium (see ***Rhodium on Alumina***) to isomerize alkenes (relative to hydrogenation), particularly at high pressures which are often used with ruthenium catalysts.[1c,1d] Under mild conditions, a monoalkyl substituted ethylene is hydrogenated selectively in the presence of di- and trisubstituted ethylenes; hydrogenation of alkynes gives the corresponding alkane.[2] Ruthenium catalysts add hydrogen suprafacially (*cis*) to a carbon–carbon double bond, as in the hydrogenation of alkyl substituted cyclohexenes (eq 1),[3,11] bicyclo[2.2.0]hexenes or bicyclo[2.2.0]hexadienes (eq 2),[12] and bicyclo[4.4.0]decene.[13]

$Ru(OH)_2$, H_2; *t*-BuOH, 26 °C, 1 atm — 93.5% + 6.5% (1)

Ru/C, H_2; heptane, 25 °C — 1 atm / 150 atm (2)

Carbonyl Compounds. Ruthenium is an excellent catalyst for the hydrogenation under mild conditions of aliphatic carbonyl compounds and aldehydes such as furfural (eq 3).[4] Industrially, it is used for the conversion of glucose to sorbitol.[1d]

CHO — Ru/C, H_2; H_2O, 25 °C, 1 atm; 95% — OH (3)

Elevated temperatures and pressures have been used successfully, as in the 5% Ru/C catalyzed hydrogenation of tetramethyl-1,3-cyclobutanedione which gave a mixture of stereoisomers (98%), a far better result than that given by Pd, Pt, or Rh (eq 4).[14]

5% Ru/C, H_2; MeOH, 125 °C; 80–100 atm; 98% (4)

Felföldi has given a broad review of the stereochemistry of the hydrogenation of substituted cyclic ketones which includes comparisons of ruthenium with other metals as catalysts.[15]

Carboxylic Acids to Alcohols. Either RuO_2 or Ru/C catalyzes the hydrogenation of carboxylic acids, both mono- and dicarboxylic, in water to the corresponding alcohols at around 150 °C and 500–700 atm (eq 5).[16] Rhenium oxides, particularly ReO, effect this reduction under less vigorous conditions.[17]

$$\text{HOOC(CH}_2)_4\text{COOH} \xrightarrow[\text{H}_2\text{O, 150–175 °C, 520–700 atm}]{\text{RuO}_2 \text{ or 10\% Ru/C, H}_2} \text{HO(CH}_2)_6\text{OH} \quad (5)$$

Hydrogenation of Nitro Alkynes. Ru/C or Ru/Al_2O_3 favor the reduction of the aromatic nitro group in (3-nitrophenyl)acetylene.[18] To avoid the observed deactivation of the catalyst, the alkyne was converted to the acetone adduct, 2-methyl-4-(3-nitrophenyl)-3-butyn-2-ol, which was hydrogenated to 2-methyl-4-(3-aminophenyl)-3-butyn-2-ol (eq 6). The base-catalyzed removal of acetone released the amino alkyne quantitatively.

$$\xrightarrow[\text{3–5 atm, >90\%}]{\text{5\% Ru/Al}_2\text{O}_3,\ i\text{-PrOH, 70 °C}} \quad (6)$$

Carbocyclic Aromatic Compounds. Ruthenium, as well as rhodium, is noted for its effective catalysis of the hydrogenation of aromatic carbocycles.[1e,6] Other reducible functional groups usually remain intact. With more than one substituent attached, the configuration of the saturated product corresponds mainly to addition of hydrogen to one face of the aromatic cycle. As with other platinum metals, the selectivity is affected by the extent to which unsaturated intermediates desorb from the catalytic sites and whether the intermediates are readsorbed and reduced with selectivities differing from that of their precursor, or transformed to products other than those of simple hydrogen addition.

Aromatic Hydrocarbons. Of all the platinum metals studied, ruthenium forms the largest amount of an observable cyclohexene in the hydrogenation of benzene or the xylenes at near atmospheric pressures.[19] Don and Scholten showed the importance of water in maximizing the yield of cyclohexene when ruthenium is the catalyst.[20] The formation of *trans* saturated isomers from substituted benzenes has been attributed to the hydrogenation of the desorbed intermediates.[21]

The $Ru(OH)_3$-catalyzed hydrogenation of *o*-xylene at 85 °C and 80–100 atm yields a *cis*/*trans* ratio of 1,2-dimethylcyclohexanes of 12.3 compared to 9.5 for Rh (eq 7).[6] Only Os and Ir (see ***Iridium***) yield larger fractions of the *cis* isomer.

$$\xrightarrow[\text{85 °C, 80–100 atm}]{\text{Ru(OH)}_2\text{, H}_2} \quad 92.5\% + 7.5\% \quad (7)$$

Ruthenium is the most stereoselective catalyst for the perhydrogenation of naphthalene or methylnaphthalenes (95% *cis*).[13] High stereoselectivity is shown in the hydrogenation of triptycene to the *cis-anti*-perhydrotriptycene (eq 8).[22]

$$\xrightarrow[\text{100–200 °C, 30–100 atm}]{\text{Ru/C, heptane}} \quad (8)$$

Fullerenes C_{60} and C_{70} in a ca. 85:15 ratio, dissolved in toluene and with an equal volume of water added, were hydrogenated over 5% Ru/C to yield a mixture of hydrogenated fullerenes from $C_{60}H_2$ to $C_{60}H_{40}$.[23] The hydrides retain the fullerene carbon skeleton.

Phenols and Phenyl Ethers. Among the first reported uses of Ru was the hydrogenation of a phenolic ring contained in a polycyclic compound.[9] The reductions were performed with RuO_2 at 50 °C and 100 atm H_2, which avoided hydrogenolysis of the hydroxyl group and was stereoselective.

The hydrogenation of phenols produces cyclohexanone intermediates which probably arise from the isomerization of a cyclohexenic intermediate, a vinyl alcohol. Palladium gives the highest yields of cyclohexanone, approaching 100%; ruthenium yields are smaller but more than is obtained with other platinum metals.[24] The maximum diminishes with an increase of pressure while the fraction of the *cis* saturated product increases. This result indicates that the hydrogenation of the ketone is not as *cis* stereoselective as is the direct hydrogenation of the phenol.

That enols are among the initial desorbed intermediates in the hydrogenation of phenols is supported by the study of the hydrogenation of ethyl *p*-tolyl ether in ethanol over unsupported platinum metals (25 °C, 1 atm).[25] Ruthenium affords the simplest result in that the enol ether is the principal intermediate; ethyl 4-methyl-3-cyclohexenyl ether is formed in smaller amounts (eq 9). With platinum metals other than Ru, the hydrogenation is accompanied by the formation of 4-methyl-1,1-diethoxycyclohexane and loss of the ethoxy group, leading to 4-methylcyclohexanone and 4-methylcyclohexanol. For Ru, the *cis*/*trans* ratio diminishes with conversion as the fraction of the saturated ether formed via the direct path diminishes.

$$\xrightarrow[\text{25 °C, 1 atm}]{\text{Ru, H}_2\text{, EtOH}} \quad (9)$$

To avoid the hydrogenolysis of the ether linkages, 5% Ru/Al_2O_3 was used to hydrogenate ***Dibenzo-18-crown-6*** in 1-butanol to ***Dicyclohexano-18-crown-6*** (eq 10).[26] Other crown ethers were treated similarly using RuO_2 or the supported catalyst and the conversions were thought to be nearly quantitative.

Using Ru/Al_2O_3 (110–126 °C, 70 atm) the cresols are converted to the methylcyclohexanols with high selectivities (94–98%) but with moderate *cis* stereoselectivities, which are larger in ethanol than in water.[27]

The selectivities of the platinum metals in the hydrogenation of 2-naphthol are comparable in the formation of the two tetrahydro derivatives, 1,2,3,4- and 5,6,7,8-tetrahydronaphthols, for their tendency towards hydrogenolysis, and for their stereochemistry in the formation of the saturated products (eq 11).[28]

5% Ru/C, H_2O, toluene, 100 °C, 70–100 atm (10)

$Ru(OH)_2$, H_2, *t*-BuOH, 80 °C, 40–50 atm; 69.5% + 26.1%; other 4.4% (11)

Anilines. A large number of substituted anilines have been hydrogenated successfully to the cyclohexyl amines over ruthenium catalysts.[1c–f] Normally, varying amounts of dicyclohexylamines are formed which arise from the reaction of the unsaturated intermediates. At elevated pressures, ruthenium catalysts usually give the least amount of such coupled products.[29]

The formation of the dicyclohexylamine from aniline as a function of the alcohol used as solvent decreases in the order methyl > ethyl ≫ isopropyl ≈ *t*-butyl; the addition of a small amount of either NaOH or LiOH further inhibits the formation of the dialkylamine.[30]

Ru/Al_2O_3 catalyzes the hydrogenation of alkyl-substituted anilines with high chemoselectivities to the substituted cyclohexylamine (>90% except for *o*- and *m*-*t*-butylaniline) and stereoselectivities, mainly *cis*, in ethanol on Ru/Al_2O_3 (110 °C, 70 atm).[1e,27] On the same catalyst the hydrogenation of the isomeric toluidines, aminophenols, and alkoxyanilines gives similar results. Although the rate of hydrogenation diminishes by substituting acetyl or alkyl for H in the amino group, the yields and stereoselectivities remain equally good (eq 12).

NHCOMe; Ru/Al_2O_3, H_2, H_2O, 126 °C, 70 atm, 98%; NHCOMe + NHCOMe; 71:29 (12)

Substituted Benzoic Acids. Ruthenium has been used to hydrogenate benzoic acids, or their esters, which are substituted with alkyl, hydroxyl, amino, or additional carboxyl groups.[1e,31] Generally, the stereoselectivities are greater than those observed with other metals except possibly Rh, which may be used at lower temperatures.

Ru/C is included amongst the catalysts used in the study of the hydrogenation of *o*-, *m*-, and *p*-*t*-butylbenzoic acids (20 °C, 1 atm).[32] Rh/C gave the largest amount of the unsaturated intermediate while Ru/C resulted in the higher stereoselectivity (*cis*). The Ru/C-catalyzed hydrogenation of terephthalic acid qualifies as a preparative procedure for *cis*-cyclohexene-3,6-dicarboxylic acid (70%) (eq 13).[33]

CO_2H, CO_2H, 30g; 5% Ru/C, H_2, 140 °C, 175 atm; 3g + 22g + trace (13)

Heterocyclic Compounds. Ruthenium is most effective for the perhydrogenation of aromatic heterocycles such as pyrroles, pyridines, and furans. Hydrogenolysis of the cycle or that of attached negative groups is generally small.[1d] ***Pyridine*** is converted quantitatively to ***Piperidine*** using RuO_2 at 95 °C and 70–100 atm of H_2, in less than 0.5 h.[7] A large variety of substituted pyridines also were hydrogenated in water, methanol, or ethanol under comparable conditions, with 80–94% yields.

Reaction conditions can affect the chemoselectivity. The RuO_2-catalyzed hydrogenation of 3-cyanopyridine to 3-aminomethylpyridine was done in a mixed solvent, methanol and ***Ammonia***, at 95 °C and 120 atm H_2 (68%); 3-aminomethylpiperidine was obtained in comparable yield from the cyanopyridine in ammonia at 100–125 °C and 150 atm (eq 14).

CN, N; RuO_2, H_2; 95 °C, 120 atm, MeOH, NH_3 → NH_2; 100–125 °C, 150 atm, NH_3 → NH_2, N H (14)

A kinetic study of the Ru/C-catalyzed hydrogenation of ***Quinoline***, 2- and 8-methylquinoline, and isoquinoline shows the high selectivity towards 1,2,3,4-tetrahydroquinolines.[34] The latter as well as the 5,6,7,8-tetrahydro derivatives are converted more slowly to the decahydroquinolines (eq 15).

Ru/C, H_2, 150 °C, 100 atm; [N H + N]; slow; H, N, H H (15)

Five-membered heterocyclic aromatic compounds which contain nitrogen or oxygen also have been hydrogenated over ruthenium catalysts. 2,5-Bis-(3-oxobutyl)furan in dioxane is converted to 2,5-bis(3-hydroxybutyl)tetrahydrofuran over Ru/C (eq 16).[35] RuO_2 or Ru/C (70–100 °C, 100 atm) catalyze the perhydrogenation of pyrrole, 2-methylpyrrole, and carbazole.[1b]

10% Ru/C, H_2
100 °C, 2–400 atm
dioxane
85%

(16)

Aromatic heterocyclic diazines such as 2-methylpyrazine and cinnoline (5,6-benzopyridazine) also are hydrogenated on RuO_2 or Ru/C.[1b]

Related Reagents. See Classes R-2, R-3, R-7, R-8, R-12, R-19, R-20, and R-21, pages 1–10.

---

1. (a) Fieser, L.; Fieser, M. *FF* **1967**, *1*, 983. (b) Freifelder, M. *Catalytic Hydrogenation in Organic Synthesis: Procedures and Commentary*; Wiley: New York, 1978. (c) Rylander, P. N. *Catalytic Hydrogenation in Organic Synthesis*; Academic: New York, 1979. (d) Rylander, P. N. *Hydrogenation Methods*; Academic: London, 1985. (e) Bartók, M. *Stereochemistry of Heterogeneous Metal Catalysis*; Wiley: New York, 1985. (f) *COS* **1991**, *8*.
2. Berkowitz, L. M.; Rylander, P. N. *JOC* **1959**, *24*, 708.
3. Nishimura, S.; Sakamoto, H.; Ozawa, T. *CL* **1973**, 855.
4. Gilman, G.; Cohn, G. *Adv. Catal.* **1957**, *9*, 733.
5. Barkdoll, A. E.; England, D. C.; Gray, H. W.; Kirk, W. Jr.; Whitman, G. M. *JACS* **1953**, *75*, 1156.
6. Takagi, Y.; Naito, T.; Nishimura, S. *BCJ* **1965**, *38*, 2119.
7. Freifelder, M.; Stone, G. R. *JOC* **1961**, *26*, 3805.
8. Yoshida, T. *BCJ* **1974**, *47*, 2061.
9. (a) Johnson, W. S.; Rogier, E. R.; Ackerman, J. *JACS* **1956**, *78*, 6322. (b) Ireland, R. E.; Schiess, P. W. *JOC* **1963**, *28*, 6.
10. Breitner, E.; Roginski, E.; Rylander, P. N. *JOC* **1959**, *24*, 1855.
11. Nishimura, S.; Kagawa, K.; Sato, N. *BCJ* **1978**, *51*, 3330.
12. van Bekkum, H.; van Rantwijk, F.; van Minnen-Pathuis, G.; Remijnse, J. D.; van Veen, A. *RTC* **1969**, *88*, 911.
13. (a) Weitkamp, A. W. *J. Catal.* **1966**, *6*, 431; (b) Weitkamp, A. W. *Adv. Catal.* **1968**, *18*, 2.
14. Hasek, R. H.; Elam, E. U.; Martin, J. C.; Nations, R. G. *JOC* **1961**, *26*, 700.
15. Felföldi, K. In Bartók, M. *Stereochemistry of Heterogeneous Metal Catalysis*; Wiley: New York, 1985; Chapter VII.
16. Carnahan, J. E.; Ford, T. A.; Gresham, W. F.; Grigsby, W. E.; Hager, G. F. *JACS* **1955**, *77*, 2766.
17. Broadbent, H. S.; Seegmiller, D. W. *JOC* **1963**, *28*, 2347.
18. Onopchenko, A.; Sabourin, E. T.; Selwitz, C. M. *JOC* **1979**, *44*, 1233.
19. Hartog, F.; Zwietering, P. *J. Catal.* **1963**, *2*, 79.
20. Don, J. A.; Scholten, J. J. F. *Faraday Discuss. Chem. Soc.* **1982**, *72*, 145.
21. Siegel, S. *Adv. Catal.* **1966**, *16*, 123.
22. Morandi, C.; Mantica, E.; Botta, D.; Gramega, M. T.; Farina, M. *TL* **1973**, 1141.
23. Shigematsu, K.; Abe, K. *Chem. Express* **1992**, *7*, 905.
24. Takagi, Y.; Nishimura, S.; Hirota, K. *BCJ* **1970**, *43*, 1846.
25. Nishimura, S.; Uramoto, M.; Watanabe, T. *BCJ* **1972**, *45*, 216.
26. Pedersen, C. J. *JACS* **1967**, *89*, 7017.
27. Friedlin, L. Kh.; Litvin, E. F.; Yakubenok, V. V.; Vaisman, I. L. *IZV* **1976**, *25*, 952.
28. Nishimura, S.; Ohbuchi, S.; Ikeno, K.; Okada, Y. *BCJ* **1984**, *57*, 2557.
29. Freifelder, M.; Stone, G. R. *JACS* **1958**, *80*, 5270.
30. (a) Nishimura, S.; Shu, T.; Hara, T.; Takagi, Y. *BCJ* **1966**, *39*, 329. (b) Nishimura, S.; Kono, Y.; Otsuki, Y.; Fukaya, Y. *BCJ* **1971**, *44*, 240.
31. Litvin, E. F.; Freidlin, L. Kh.; Oparina, G. K.; Gurskii, R. N.; Istratova, R. V.; Gosteva, L. I. *ZOR* **1974**, *10*, 1475.
32. van Bekkum, H.; van de Graaf, B.; van Minnen-Pathuis, G.; Peters, J. A.; Wepster, B. M. *RTC* **1970**, *89*, 521.
33. Rylander, P. N.; Rakoncza, N. F. U.S. Patent 3 162 679, 1964 (*CA* **1965**, *62*, 9033e).
34. Okazaki, H.; Onishi, K.; Soeda, M.; Ikefuji, Y.; Tamura, R.; Mochida, I. *BCJ* **1990**, *63*, 3167.
35. Webb, I. D.; Borcherdt, G. T. *JACS* **1951**, *73*, 752.

Samuel Siegel
University of Arkansas, Fayetteville, AR, USA

Ruthenium(VIII) Oxide[1]

[20427-56-9] O_4Ru (MW 165.07)

(strong oxidant for many functional groups; can cleave double bonds, aromatic rings, and diols[1])

Alternate Name: ruthenium tetroxide.
Physical Data: yellow form: mp 25.5 °C; bp 40 °C; *d* 3.29 g cm^{-3}. Brownish orange form: mp 27 °C; bp 108 °C (dec).
Solubility: slightly sol water; highly sol $CHCl_3$, CCl_4.
Form Supplied in: although the reagent is commercially available either in solid form or stabilized aqueous solution, it is usually prepared in situ from black solid RuO_2 (mw 133.07) *[120236-10-1*;·xH_2O, *32740-79-7]* or dark brown (or black) $RuCl_3$ (mw 207.42) *[10049-08-8*;·xH_2O, *14898-67-0]*, in stoichiometric or catalytic amounts, and an oxidation agent; both of the above Ru salts are widely available.
Handling, Storage, and Precautions: handle in a fume hood only. Inhalation should be avoided; vapors irritating to eyes and respiratory tracts, since it readily oxidizes tissue, leaving a deposit of ruthenium dioxide.[2] It attacks rubber and reacts explosively with paper filter and alcohol, and violently with ether, benzene, and pyridine.[3] However, *the use of the catalytic system greatly minimizes the risk in its manipulation*, and its usage is strongly recommended.

Introduction. When RuO_4 was introduced into organic synthesis it was generally used in stoichiometric amounts, usually prepared by oxidation of ***Ruthenium(III) Chloride*** or RuO_2 with aqueous periodate or hypochlorite and then extracted into carbon tetrachloride. This yellow solution could be roughly analyzed by treating an aliquot with ethanol to reduce the tetroxide to the black dioxide, which was collected and weighed.[4]

However, since ruthenium compounds are expensive and occasionally it is difficult to separate the products from precipitated ruthenium dioxide, it is more convenient to use a system formed by a catalytic amount of the ruthenium compound (RuO_2 or $RuCl_3$) along with an appropriate co-oxidant, usually in a biphasic solvent system. These reagents actually act as catalysts, because they are reoxidized after the reaction with organic compounds by one of the oxidants previously mentioned. Ruthenium tetroxide is usually prepared in situ from ruthenium dioxide or ruthenium trichloride by oxidation with ***Sodium Hypochlorite***,[5] ***Sodium Bromate***,[6] ***Peracetic Acid***,[6] periodic acid,[7] ***Sodium Periodate***,[8] ***Oxygen***,[9] cerium sulfate,[10] ***Potassium Permanganate***,[10] electrochemically generated ***Chlorine***,[11] or ***Potassium Monoperoxysulfate*** (Oxone®).[12]

It appears that contact between RuO_4 and the material to be oxidized takes place in the organic phase, where they are both most soluble. The ruthenium dioxide produced when oxidation occurs is insoluble in all solvents and migrates to the interface where it contacts the co-oxidant (in the aqueous layer) and is reoxidized. Thus best results are obtained when the mixture is shaken or stirred vigorously throughout the course of the reaction to achieve good contact between all components.

It has been pointed out that $RuCl_3{\cdot}H_2O$ is not a good Ru source under acidic conditions (pH < 5), because it initially gives an orange Ru^{IV} chloro aquo complex, which is slowly oxidized to RuO_4. In such cases, RuO_2 is the alternative recommended.[10] In the reactions carried out at pH > 9, any RuO_4 produced in the aqueous phase is unstable, being reduced to green perruthenate (RuO_4^-), and subsequently to orange ruthenate (RuO_4^{2-}). Both species are insoluble in CCl_4. Although in most of the cited mixtures RuO_4 is considered the oxidant, it cannot be ruled out that the real oxidant is another lower valent ruthenium species. The only way to ensure that RuO_4 is really the oxidizing agent is if it is isolated after its preparation, as cited in the early literature.[2,4] UV analysis of the RuO_4 solutions can provide some information about this, because RuO_4 gives absorption bands at λ_{max} 310 nm (strong) and 380 nm. These bands are replaced by others at λ_{max} 310 nm (strong), 385 nm (strong), and 460 nm when base is added to the solution, corresponding to the formation of the perruthenate ion (RuO_4^-) and subsequently to absorption bands at λ_{max} 385 nm and 460 nm (strong), produced by RuO_4^{2-}.[1,13]

RuO_4 is a strong oxidant. However, conditions for ruthenium-catalyzed reactions are very mild; usually a few hours (or less) at room temperature (or below) is sufficient. A thorough study of oxidations with RuO_4 generated in situ from $RuO_2{\cdot}xH_2O$ and $RuCl_3{\cdot}xH_2O$ shows the importance of the presence of water in the reaction.[10,14] Thus many ruthenium-catalyzed reactions have been performed in the CCl_4–H_2O solvent system. The addition of ***Acetonitrile*** to the system greatly improves yields and reaction times,[8] especially when carboxylic groups are present or generated in the reaction. MeCN probably disrupts the insoluble carboxylate complexes and returns the ruthenium to the catalytic cycle, acting as a good ligand for the lower valent (III/II) ruthenium present.[15]

An important feature of RuO_4-catalyzed oxidations is that the stereochemistry of the stereocenters close to the reaction site (eqs 1 and 2)[16,17] remains unaffected.

$RuCl_3$, $NaIO_4$, CCl_4, MeCN, H_2O, rt, 72% (1)

In a typical procedure,[8] to a stirred mixture of 2 mL of CCl_4–2 mL of MeCN–3 mL of H_2O/mmol of organic compound are added 4.1 mmol of ***Sodium Periodate***/mmol of organic compound and 2.2 mmol% of $RuCl_3{\cdot}xH_2O$ ($RuO_2{\cdot}xH_2O$ is equally effective) sequentially. The mixture is stirred vigorously at 0–25 °C until the end of the reaction (TLC or GC monitoring). Then 20 mL of diethyl ether are added and the vigorous stirring is continued for 10 min to precipitate black RuO_2. The reaction mixture is then dried ($MgSO_4$) and filtered through qualitative Whatman filter paper 2. The solid residue is then washed with diethyl ether (3 × 5 mL). The combined organic phases are concentrated to yield the crude oxidation product.

$RuCl_3$, $NaIO_4$, CCl_4, MeCN, H_2O, rt; >82%, >56% (2)

Functional Group Oxidations without Bond Cleavage. One of the most common synthetic uses of ruthenium-catalyzed oxidations is the reaction with alcohols. A mixture of RuO_2 or $RuCl_3$ with strong co-oxidants converts primary alcohols to carboxylic acids[18] (eq 3),[19] including epoxy alcohols (eq 4),[8] but under milder (***Iodosylbenzene***,[20] molecular oxygen,[21] ***Calcium Hypochlorite*** (eq 5),[22] or amine *N*-oxides[23]) or controlled conditions,[19] aldehydes are obtained (eq 6).[24]

$RuCl_3$, $NaIO_4$, CCl_4, MeCN, H_2O, rt, >95% (3)

$RuCl_3$, $NaIO_4$, CCl_4, MeCN, H_2O, rt, 75% (4)

$RuCl_3$, $Ca(OCl)_2$, MeCN, H_2O, rt, 80% (5)

$RuCl_3$, $NaIO_4$, CCl_4, MeCN, H_2O, rt, >75% (6)

Secondary alcohols are transformed into ketones.[1,3,25] The yields obtained from the oxidation of secondary alcohols are usually excellent. Since the reactions are carried out under very mild conditions, there is little danger of the product undergoing secondary reactions. Ketones can also be prepared using a great variety of other oxidants[26] (see, for example ***Chromium(VI) Oxide*** and ***Dimethyl Sulfoxide*** based oxidant reagents), which in some cases are more readily available and less expensive. However, RuO_4 is recommended for reactions which require a vigorous oxidant under mild conditions. Thus it can be used to oxidize alcohols which are resistant to other oxidants: (**1**) is successfully oxidized with RuO_4, while attempts made with 15 other standard oxidizing procedures failed (eq 7);[27] oxidation of (**2**) is successful in good yield using RuO_4 (eq 8), being fruitless with CrO_3 in either ***Pyridine***, ***Acetone***, ***Acetic Acid***, or $Al(O\text{-}i\text{-}Pr)_3$, $KMnO_4$, ***Lead(IV) Acetate*** in acetone, with CrO_3 in *i*-BuOH giving the ketone in very low yield.[28] Numerous examples of the advantages of RuO_4 over ***Dipyridine Chromium(VI) Oxide*** for the oxidation of carbohydrates have been cited.[29]

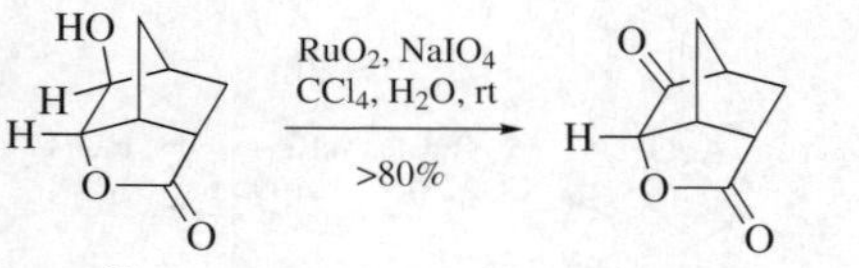

RuO_2, $NaIO_4$, CCl_4, H_2O, rt; >80% (7)

(**1**)

RuO_2, $NaIO_4$, CCl_4, H_2O, rt; 55% (8)

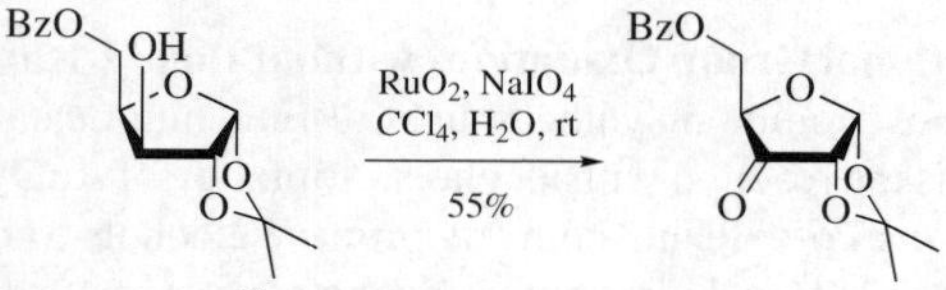

(**2**)

RuO_4 is also reported to provide higher yields when other oxidizing procedures give poor yields. As examples can be cited the conversions of cyclobutanols to cyclobutanones (eq 9)[30] and the transformation of lactones into the corresponding ketocarboxylates under basic conditions (60–97%) (eq 10). Significantly higher yields are obtained compared with $KMnO_4$ under alkaline conditions.[31]

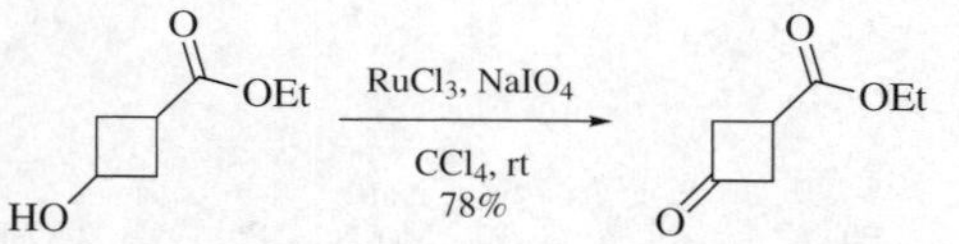

$RuCl_3$, $NaIO_4$, CCl_4, rt; 78% (9)

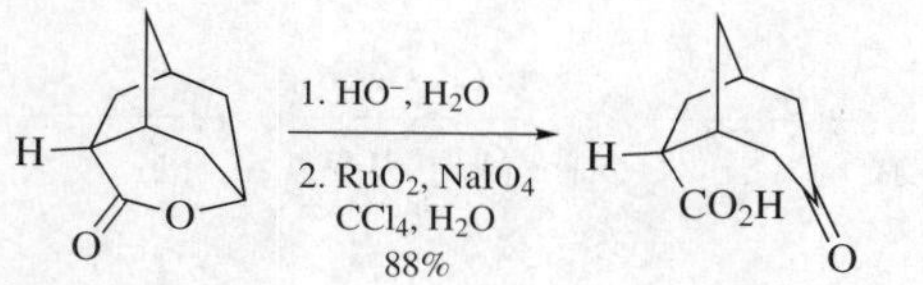

1. HO^-, H_2O; 2. RuO_2, $NaIO_4$, CCl_4, H_2O; 88% (10)

Vicinal diols can be oxidized to diketones but only in low yields, the principal reaction being the oxidative cleavage of the C–C bond.[1] If the hydroxyl groups are not adjacent, then diketones can be prepared (eq 11).[32]

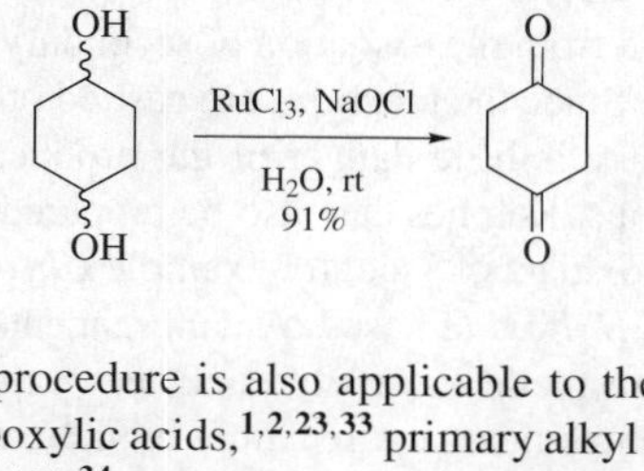

$RuCl_3$, NaOCl, H_2O, rt; 91% (11)

The catalytic procedure is also applicable to the oxidation of aldehydes to carboxylic acids,[1,2,23,33] primary alkyl iodides to carboxylic acids (eq 12),[34] aromatic hydrocarbons to quinones,[25] and sulfides to sulfones,[25] including an improved and simple method to obtain water-soluble sulfones using periodic acid as the co-oxidant in high concentration conditions (eq 13).[35]

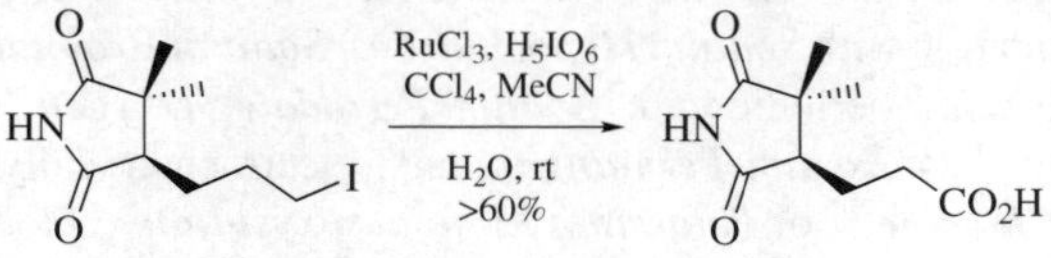

$RuCl_3$, H_5IO_6, CCl_4, MeCN, H_2O, rt; >60% (12)

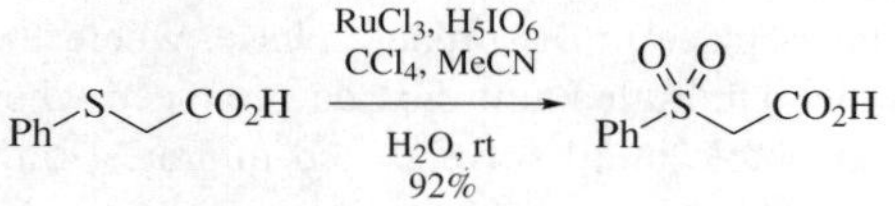

$RuCl_3$, H_5IO_6, CCl_4, MeCN, H_2O, rt; 92% (13)

Oxidation of 1,2-cyclic sulfites to 1,2-cyclic sulfates with RuO_4 has been reported as a part of a method to activate diols for further nucleophilic attack (eq 14).[36]

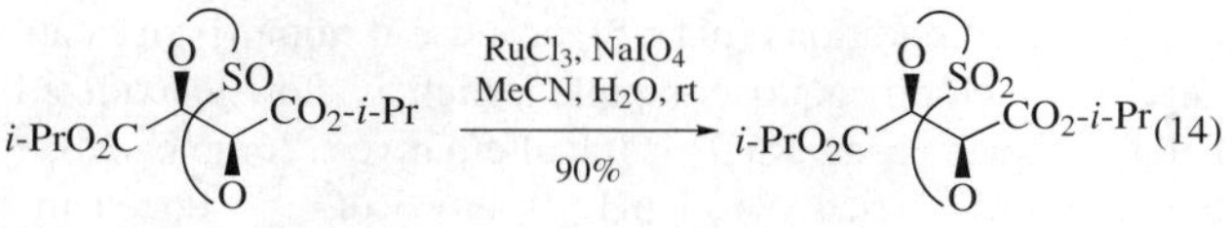

$RuCl_3$, $NaIO_4$, MeCN, H_2O, rt; 90% (14)

Along with these types of oxidations, RuO_4 is used to carry out transformations that involve oxidation of methylene groups α to heteroatoms such as oxygen and nitrogen. Thus it is possible to convert acyclic ethers into esters, such as methyl ethers into methyl esters (eq 15),[8] ethyl ethers into acetates,[37] and benzyl ethers into benzoates (eq 16).[38] It is possible to avoid the benzyl–benzoyl group transformation by carrying out the reaction at 0 °C and/or in the presence of base (eq 17).[39]

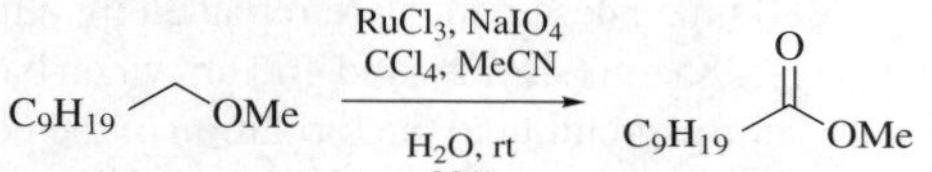

$RuCl_3$, $NaIO_4$, CCl_4, MeCN, H_2O, rt; 89% (15)

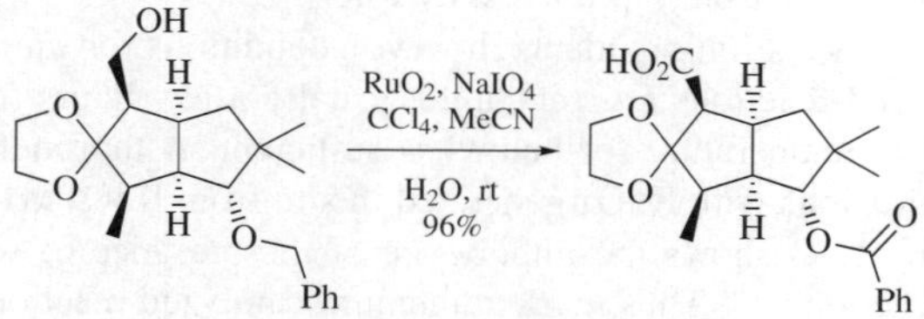

RuO_2, $NaIO_4$, CCl_4, MeCN, H_2O, rt; 96% (16)

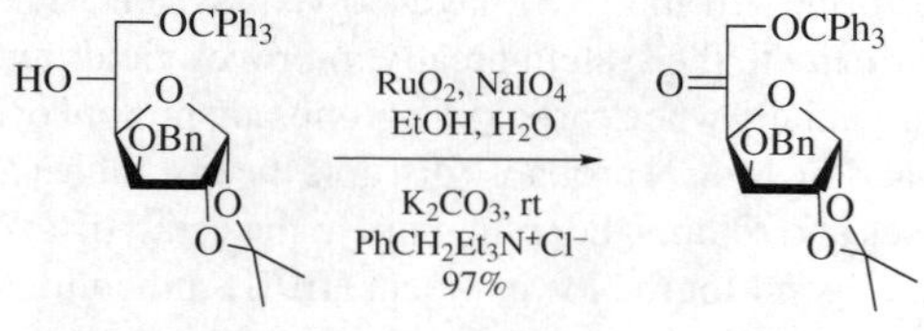

RuO_2, $NaIO_4$, EtOH, H_2O, K_2CO_3, rt, $PhCH_2Et_3N^+Cl^-$; 97% (17)

Cyclic ethers are also oxidized, yielding lactones (eq 18).[40] Although secondary positions are usually more reactive towards oxidation than tertiary positions, the regioselectivity of RuO_4 can be strongly dependent on steric factors (eq 19).[41] In those cases in which lactones are unstable under aqueous conditions (notably δ- and ε-lactones), the corresponding diacids are the final products, presumably via the intermediacy of lactols (eq 20).[42]

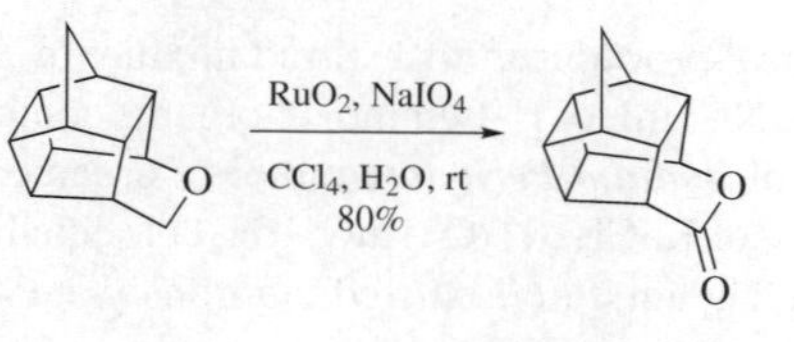

RuO_2, $NaIO_4$, CCl_4, H_2O, rt; 80% (18)

RuO_2, $NaIO_4$; CCl_4, MeCN, phosphate buffer (pH = 7), rt, >89%

(19)

RuO_2, $NaIO_4$; CCl_4, MeCN, phosphate buffer (pH = 7), rt; 33%, 67%

RuO_2, $NaIO_4$; CCl_4, H_2O, rt, 79%; HO_2C CO_2H (20)

RuO_4 also oxidizes alkyl amines to mixtures of nitriles and amides,[43] cyclic amines to lactams,[44] and amides (cyclic or acyclic) to imides (eq 21),[45] including an improved procedure that uses ethyl acetate as the organic solvent in the biphasic solvent system, enhancing both the solubility of the substrates and the rate of reaction.[46]

RuO_2, $NaIO_4$; EtOAc, H_2O, rt, 92% (21)

RuO_4 usually reacts with unsaturated systems, giving cleavage of the C–C bonds. Although epoxide formation has been detected in small amounts (ca. 1%),[47] it can be the principal reaction when the double bond is located in a very hindered position (eq 22).[48] With 1,5-dienes, unexpected oxidation results are obtained. Thus oxidation of geranyl acetate leads to a tetrahydrofuran mixture, instead of the cleavage products (eq 23).[8] Nonterminal alkynes are also oxidized without cleavage, yielding vicinal diketones (eq 24).[49]

RuO_4; 10%, 49% (22)

RuO_4 is also capable of oxidizing C–H bonds in bridged bicyclic and tricyclic alkanes to alcohols by insertion of oxygen (eq 25).[50] Although epoxides survive RuO_4 oxidations, when such functionality is located in this kind of bridged system a tandem ruthenium-catalyzed rearrangement/oxidation occurs (eq 26).[51]

$RuCl_3$, $NaIO_4$, CCl_4, MeCN; H_2O, rt; 39%, 12%, 12% (23)

RuO_2, $NaIO_4$, CCl_4, MeCN; H_2O, 0 °C, 60% (24)

$RuCl_3$, $NaIO_4$, CCl_4, MeCN; H_2O, 55 °C, 90% (25)

$RuCl_3$, $NaIO_4$, CCl_4, MeCN; H_2O, 60 °C, 48% (26)

Most of the common protecting groups used in organic synthesis are stable under RuO_4 oxidation conditions (eq 27).[52] Generally it is only necessary to carry out the reaction at 0 °C or perform it under buffered conditions when acid-sensitive groups are present, such as tetrahydropyranyl (eq 28)[33] or silyl ethers (eq 29).[53]

RuO_2, $NaIO_4$, CCl_4, MeCN; H_2O, rt, 98% (27)

RuO_2, $NaIO_4$, CCl_4, MeCN; H_2O, 0 °C, >65% (28)

$RuCl_3$, $NaIO_4$, CCl_4, MeCN; phosphate buffer (pH 7), rt, ~90% (29)

Functional Group Oxidations with Bond Cleavage. Carbon–carbon double bonds are readily cleaved by RuO_4 to give ketones and aldehydes or carboxylic acids. In this respect the greater vigor of RuO_4 as an oxidant stands in marked contrast to that of ***Osmium Tetroxide*** (eq 30),[54] which also reacts with C–C double bonds but does not cleave them. While carboxylic acids are usually the final products, sometimes under neutral conditions aldehydes can be obtained from double bonds that are not fully substituted.[55] The cleavage of such double bonds proceeds by the route: alkene → dialdehyde → diacid.[5a] RuO_4 is also indicated to carry out oxidations of substrates with double bonds resistant to other oxidizing agents, such as OsO_4, ***Potassium Permanganate***, and ***Ozone*** (eq 31).[56] Degradative oxidations of unsaturated C–C bonds with loss of carbon atoms occur with terminal alkynes (eq 32),[5b] cyclic allylic alcohols (eq 33),[57] and α,β-unsaturated ketones.[57]

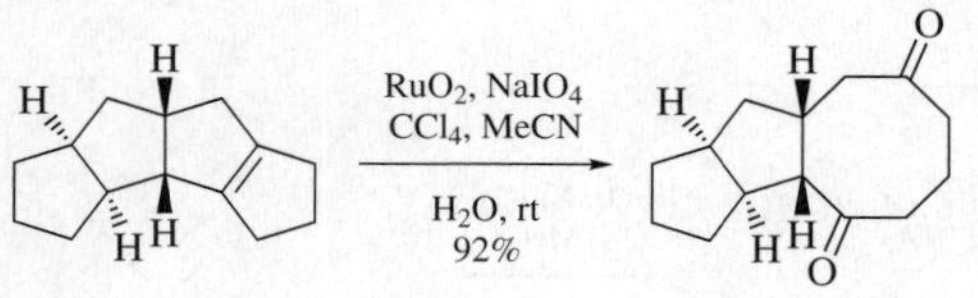

RuO_2, $NaIO_4$, CCl_4, MeCN, H_2O, rt, 92% (30)

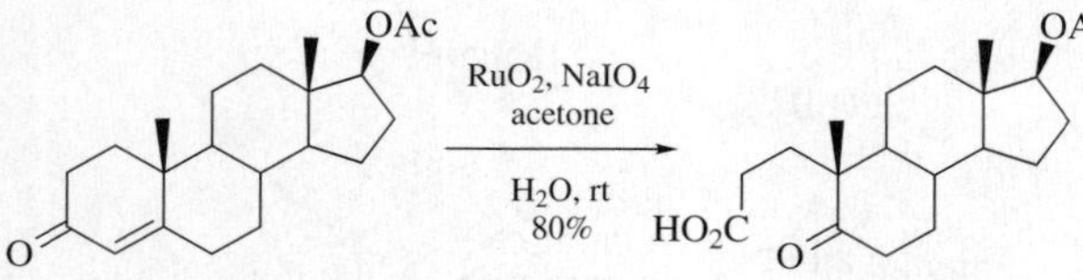

RuO_2, $NaIO_4$, acetone, H_2O, rt, 80% (31)

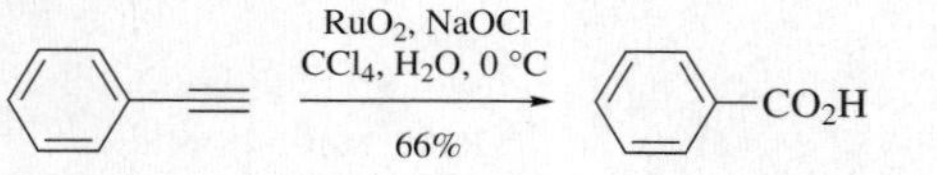

RuO_2, NaOCl, CCl_4, H_2O, 0 °C, 66% (32)

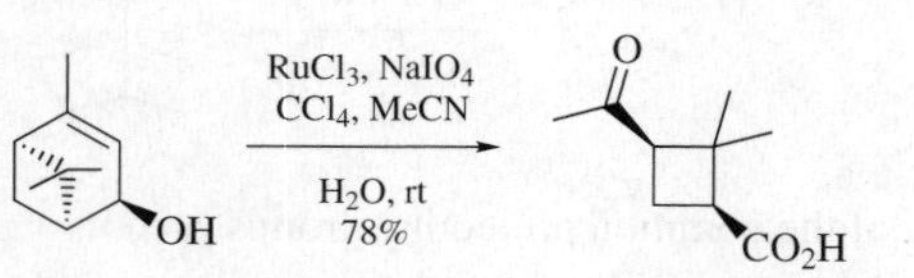

$RuCl_3$, $NaIO_4$, CCl_4, MeCN, H_2O, rt, 78% (33)

RuO_4 also cleaves α-chloroenol derivatives obtained from α,α′-dichlorocyclobutanones to give dicarboxylic acids through successive treatment with ***n-Butyllithium***, ***Acetic Anhydride***, and ***Sodium Periodate***–RuO_2 (eq 34).[58,59]

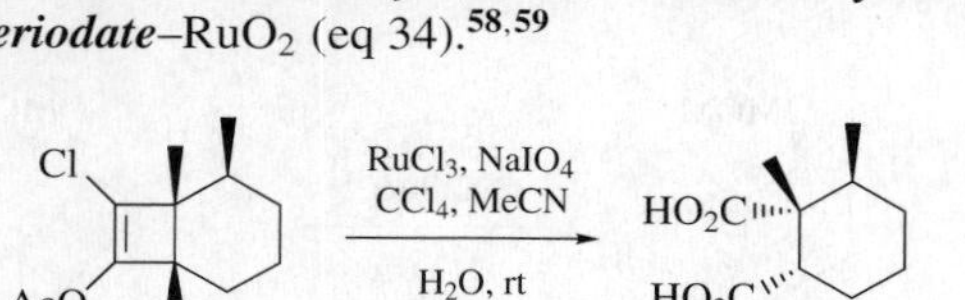

$RuCl_3$, $NaIO_4$, CCl_4, MeCN, H_2O, rt, >67% (34)

RuO_4-catalyzed oxidation of arenes can proceed in two ways: (a) the phenyl ring can be cleaved from R–Ph to R–CO_2H (eq 35);[8] (b) the phenyl ring can be degraded to form a dicarboxylic acid in polycyclic aromatic hydrocarbons (eq 36).[60] An electron-donating substituent favors cleavage of the substituted ring, while an electron-withdrawing substituent favors cleavage of the unsubstituted ring. Thus selective oxidation of the more activated ring can be performed with high selectivity. When acid-sensitive groups are not present, an improved procedure that utilizes periodic acid instead of sodium periodate can be used, preventing the problems associated with the precipitated sodium iodate, allowing the reaction to go to completion, and permitting oxidation reactions to be run on larger scales (eq 37).[61]

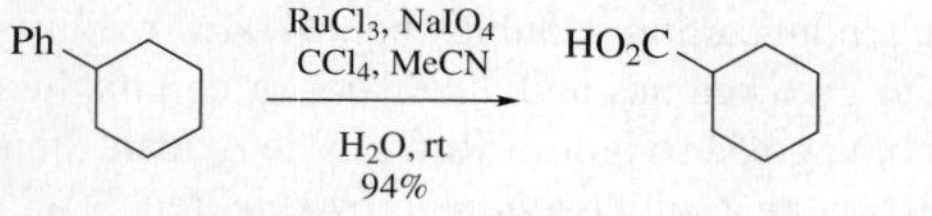

$RuCl_3$, $NaIO_4$, CCl_4, MeCN, H_2O, rt, 94% (35)

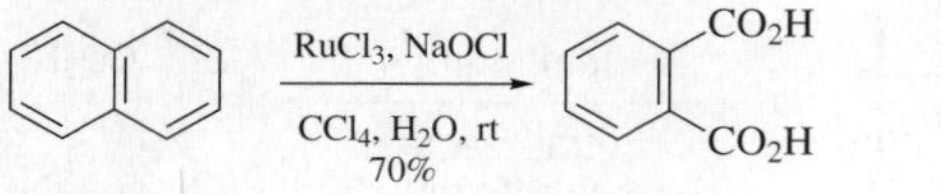

$RuCl_3$, NaOCl, CCl_4, H_2O, rt, 70% (36)

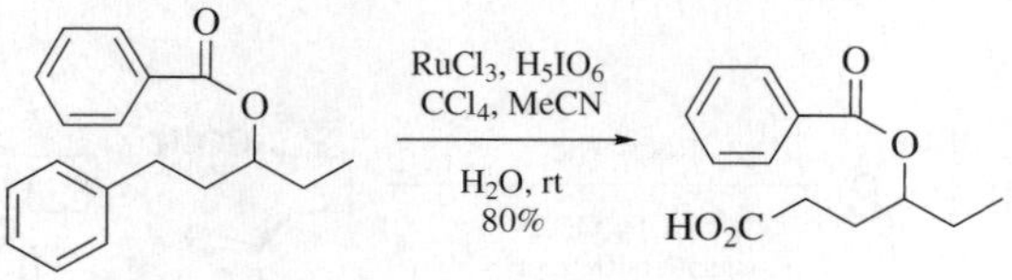

$RuCl_3$, H_5IO_6, CCl_4, MeCN, H_2O, rt, 80% (37)

Furan[62] (eq 38),[63] thiophene (eq 39),[64] and benzopyridine rings (eq 40)[65] are also cleaved by catalytic RuO_4 to carboxylic acids. When pyridine derivatives are not oxidized, they can be transformed into their *N*-oxides prior to the oxidation to decrease the ability of the nitrogen to complex with ruthenium, albeit with low yields (eq 41).[64]

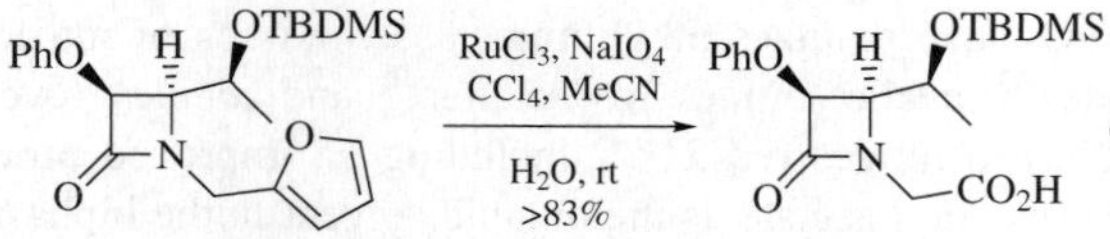

$RuCl_3$, $NaIO_4$, CCl_4, MeCN, H_2O, rt, >83% (38)

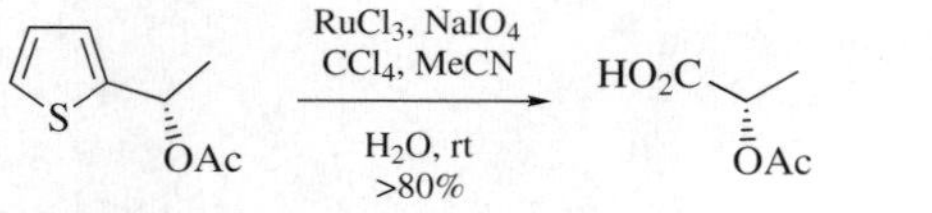

$RuCl_3$, $NaIO_4$, CCl_4, MeCN, H_2O, rt, >80% (39)

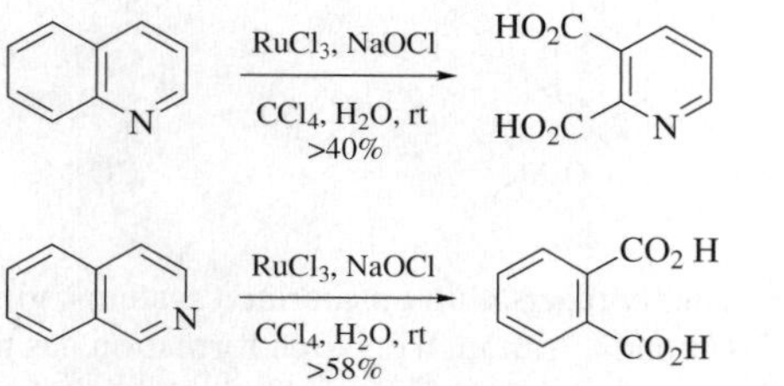

$RuCl_3$, NaOCl, CCl_4, H_2O, rt, >40%; $RuCl_3$, NaOCl, CCl_4, H_2O, rt, >58% (40)

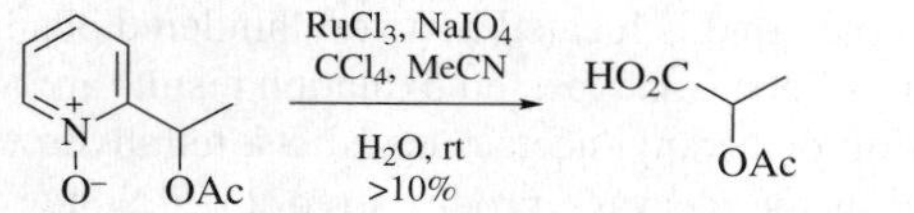

$RuCl_3$, $NaIO_4$, CCl_4, MeCN, H_2O, rt, >10% (41)

Vicinal diols are cleaved to give carboxylic acids (eq 42) following the route: glycol → α-ketol → diacid. A diketone is apparently not an intermediate in this oxidation.[5a] The mildness of the reaction conditions is underscored by the lack of epimerization shown in eq 43.[8] This feature has been proved to be general when RuO_2–$NaIO_4$ is used to oxidize chiral diol benzoates, this being a useful method to synthesize chiral α-benzoylcarboxylic acids (eq 44).[66]

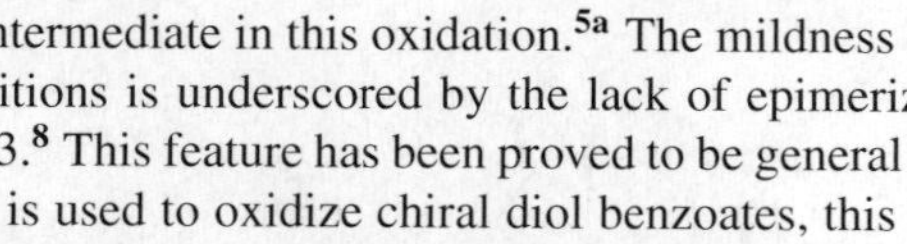

$RuCl_3$, NaOCl, CH_2Cl_2, H_2O, rt, 95% (42)

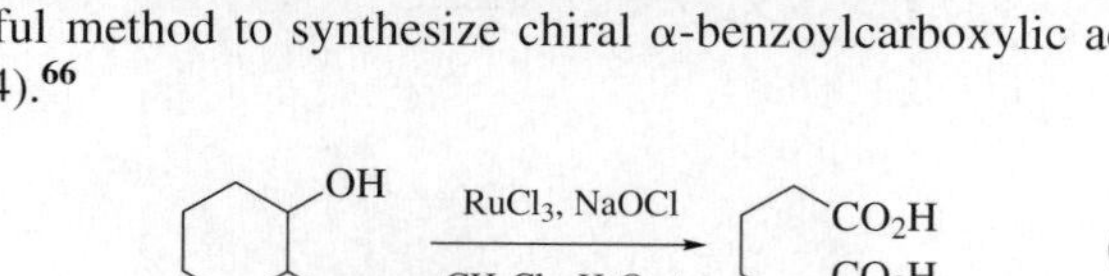

96% ee → 94% ee; $RuCl_3$, $NaIO_4$, CCl_4, MeCN, H_2O, rt, 92% (43)

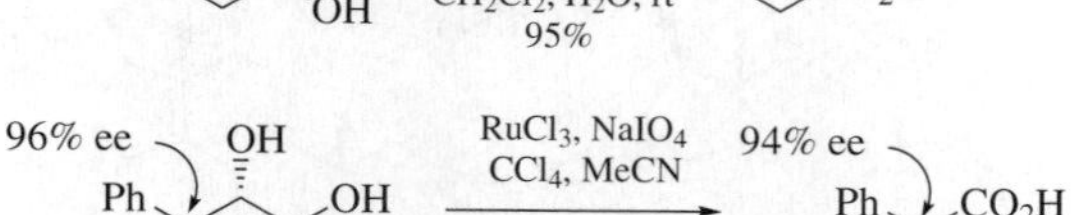

$RuCl_3$, $NaIO_4$, CCl_4, MeCN, H_2O, rt, 88% (44)

Other vicinal dioxygenated functionalities present (eq 45)[5a] or generated in situ undergo oxidation with C–C bond cleavage by RuO_4 to give carboxylic acids, with (eq 46)[67] or without (eq 47)[68] loss of carbon atoms.

$RuCl_3$, NaOCl; 28%, 3%, 56% (45)

$RuCl_3$, $NaIO_4$, CCl_4, MeCN, H_2O, 40 °C, 82% (46)

$RuCl_3$, $NaIO_4$, CCl_4, MeCN, H_2O, 70 °C; 20%, 48% (47)

RuO_4 is also used to cleave oxidatively carbon–boron bonds in cyclic alkylboranes, presenting advantages over the usage of Cr^{VI} for the same purpose or even the oxidation of the corresponding alcohols (eq 48).[69]

RuO_4, $NaIO_4$, NaOAc, acetone, H_2O, rt, >44% (48)

Another formal carbon–heteroatom bond cleavage occurs in the oxidation of cyclic ethers that give cyclic products unstable to the oxidation conditions (eq 49).[42]

RuO_2, $NaIO_4$, CCl_4, H_2O, rt, 82% (49)

As pointed out, ketones are stable under RuO_4 oxidation conditions, although cyclic ketones can undergo Baeyer–Villiger reaction when ***Sodium Hypochlorite*** is used as co-oxidant (eq 50).[70]

$RuCl_3$, $NaIO_4$, CCl_4, H_2O, rt, 61%; $RuCl_3$, NaOCl, CCl_4, H_2O, rt, 91% (50)

Other Ruthenium-Based Oxidation Reagents. Less reactive oxidants are obtained by lowering the oxidation state of ruthenium. One example is the ruthenate ion (RuO_4^{2-}) which, as mentioned above, is formed when RuO_4 is treated with alkaline solutions. The most important synthetic applications of such ions is in the oxidation of alcohols in basic media to give carboxylic acids or ketones.[71] In general, RuO_4^{2-} does not appear to oxidize isolated C–C double bonds at room temperature (eq 51).[72]

Ph–CH=CH–CH₂OH → $RuCl_3$, $K_2S_2O_8$, KOH, H_2O, rt, 99% → Ph–CH=CH–CO_2H (51)

When there is no reductive pathway for the elimination of ruthenate esters, RuO_4^{2-} has been used as an alternative to RuO_4 (eq 52).[40]

RuO_2, $NaIO_4$, CCl_4, H_2O, rt (no reaction); RuO_2, $NaIO_4$, NaOH, CCl_4, H_2O, rt; CH_2O^-; 80% (52)

The perruthenate ion RuO_4^- is also useful for the oxidation of primary alcohols, nitroalkanes, primary halides, and aldehydes to acids.[73] When tetraalkylammonium salts are added to the RuO_4^- solutions, stable tetraalkylammonium perruthenates are obtained. Tetra-*n*-butylammonium perruthenate (TBAP) and tetra-*n*-propylammonium perruthenate (TPAP) are used to oxidize successfully alcohols to carbonyl compounds (eq 53),[23] and sulfides to sulfones[74] under very mild conditions, employing as co-oxidant ***N-Methylmorpholine N-Oxide*** (NMO) (see ***Tetra-n-propylammonium Perruthenate***).

TBDPSO…OH → TPAP, NMO, CH_2Cl_2, 4Å mol. sieves, rt, 70% → TBDPSO…CHO (53)

The behavior of inorganic transition metal oxidizing agents can be modified by the introduction of ligands. Electron-rich ligands, which increase the basicity of the metal and moderate its oxidizing power, have been used to improve the selectivity of these oxidation reactions. Thus porphyryl–[75] and bipyridyl–Ru complexes epoxidize alkenes, instead of cleaving the double bond (eq 54).[47]

Ph–CH=CH–Ph → $RuCl_3$, bipy, $NaIO_4$, CH_2Cl_2, H_2O, 0–5 °C, 83% → stilbene oxide (54)

Several other Ru complexes, along with co-oxidants or hydrogen acceptors, are used as catalysts in oxidation reactions, $RuCl_2(PPh_3)_3$ and $RuH_2(PPh_3)_4$ being the most commonly utilized. The conversion of alkanes and alcohols to aldehydes or ketones is achieved with $RuCl_2(PPh_3)_3$[76] and molecular oxygen,[77] ***Bis(trimethylsilyl) Peroxide***,[78] ***Iodosylbenzene***,[20] ***N-Methylmorpholine N-Oxide*** (eq 55),[79] ***t-Butyl Hydroperoxide***,[80] and hydrogen acceptors.[81] Selective oxidations of primary vs. secondary alcohols are possible (eq 56),[78] and it is also possible to stop the oxidation of primary alcohols at the aldehyde stage by simply controlling the co-oxidant equivalents and reaction times.[20] α-Diketones can be obtained from vicinal diols (eq 57)[82] and nonterminal alkynes (eq 58).[83]

$$C_{11}H_{23}CH_2OH \xrightarrow[\text{acetone, rt}\ 90\%]{RuCl_2(PPh_3)_3,\ NMO} C_{11}H_{23}CHO \quad (55)$$

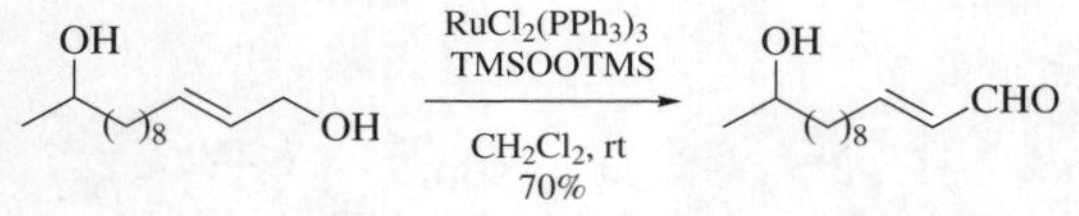

$$\xrightarrow[CH_2Cl_2,\ \text{rt}\ 70\%]{RuCl_2(PPh_3)_3,\ TMSOOTMS} \quad (56)$$

$$\xrightarrow[\text{THF, 195 °C (sealed tube)}\ 100\%]{RuCl_2(PPh_3)_3,\ \text{2 equiv PhCH=CHCOMe}} \quad (57)$$

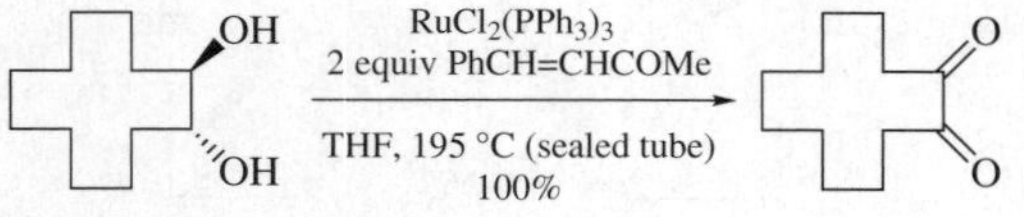

$$Ph{-}C{\equiv}C{-}Ph \xrightarrow[PhIO,\ CH_2Cl_2,\ \text{rt}\ 86\%]{RuCl_2(PPh_3)_3} PhCOCOPh \quad (58)$$

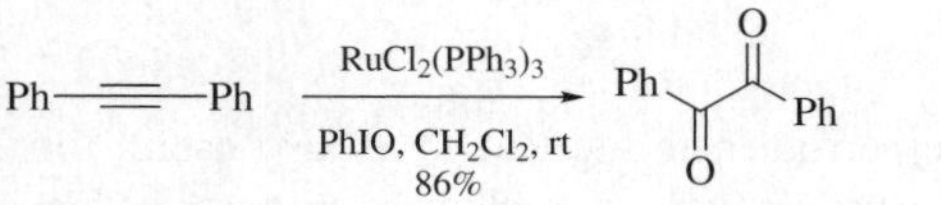

The combination of $RuH_2(PPh_3)_4$ plus a hydrogen acceptor, like benzalacetone or ***Acetone***, converts unsymmetrically substituted 1,4- and 1,5-diols into β-substituted γ-lactones and γ-substituted δ-lactones, respectively (eqs 59 and 60),[84] also allowing the oxidative condensation of alcohols, or aldehydes and alcohols, to give esters (eq 61).[85]

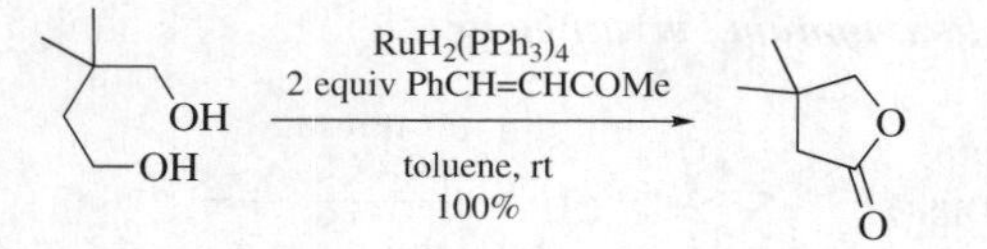

$$\xrightarrow[\text{toluene, rt}\ 100\%]{RuH_2(PPh_3)_4,\ \text{2 equiv PhCH=CHCOMe}} \quad (59)$$

$$\xrightarrow[\text{toluene, rt}\ 100\%]{RuH_2(PPh_3)_4,\ \text{2 equiv PhCH=CHCOMe}} \quad (60)$$

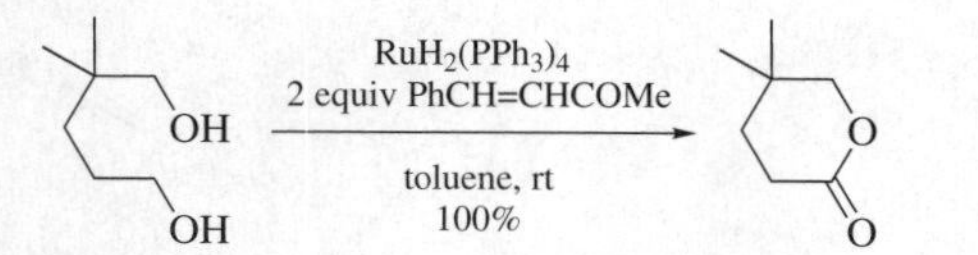

$$PrCHO + BuCH_2OH \xrightarrow[\text{(sealed tube)}]{RuH_2(PPh_3)_4,\ \text{toluene, 180 °C}} PrCO_2Bu \quad (61)$$

Oxoruthenium species are also useful in organic oxidations. Thus oxoruthenium(V) complexes obtained from lower valent ruthenium species effect α-oxygenation of tertiary amines[86] and β-lactams (eq 62).[87] $[PPh_4][RuO_2(OAc)Cl_2]{\cdot}2AcOH$ generated from RuO_4 is used to oxidize alcohols and benzyl halides to carbonyl compounds.[88]

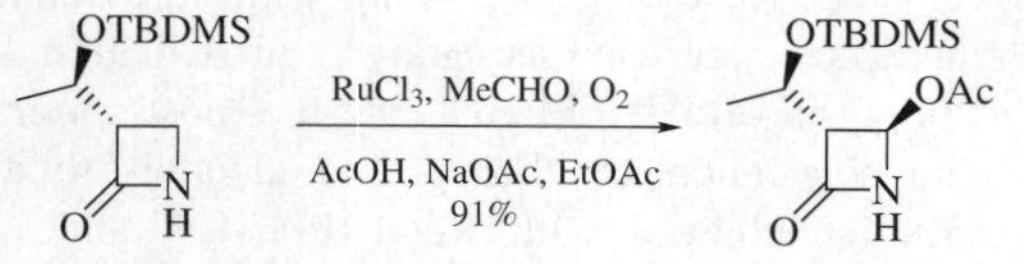

$$\xrightarrow[\text{AcOH, NaOAc, EtOAc}\ 91\%]{RuCl_3,\ MeCHO,\ O_2} \quad (62)$$

Related Reagents. See Classes O-1, O-2, O-4, O-8, O-17, O-18, and O-23, pages 1–10.

1. (a) Lee, D. G.; van den Engh, M. In *Oxidation in Organic Chemistry*; Trahanovsky, W. S., Ed.; Academic: New York, 1973; part B, Chapter 4. (b) Gore, E. S. *Platinum Met. Rev.* **1983**, *27*, 111.
2. Remy, H. In *Treatise on Inorganic Chemistry*; Kleinberg, J., Ed.; Amer. Elsevier: New York, 1956; Vol. 2, p. 324.
3. Djerassi, C.; Engle, R. R. *JACS* **1953**, *75*, 3838.
4. For a simple preparative procedure see: Nakata, H. *T* **1963**, *19*, 1959.
5. (a) Wolfe, S.; Hasan, S. K.; Campbell, J. R. *CC* **1970**, 1420. (b) Gopal, H.; Gordon, A. J. *TL* **1971**, 2941.
6. (a) Berkowitz, L. M.; Rylander, P. N. *JACS* **1958**, *80*, 6682. (b) Yamamoto, Y.; Suzuki, H.; Moro-oka, Y. *TL* **1985**, *26*, 2107.
7. Guizard, C.; Cheradame, H.; Brunel, Y.; Beguin, C. G. *JFC* **1979**, *13*, 175.
8. Carlsen, P. H. J.; Katsuki, T.; Martin, V. S.; Sharpless, K. B. *JOC* **1981**, *46*, 3936.
9. (a) Matsumoto, M.; Watanabe, N. *JOC* **1984**, *49*, 3435. (b) Kaneda, K.; Haruna, S.; Imanaka, T.; Kawamoto, K. *CC* **1990**, 1467.
10. Giddings, S.; Mills, A. *JOC* **1988**, *53*, 1103.
11. Torii, S.; Inokuchi, T.; Sugiura, T. *JOC* **1986**, *51*, 155.
12. (a) Schröder, M.; Griffith, W. P. *CC* **1979**, *58*. (b) Paquette, L. A.; Dressel, J.; Pansegran, P. D. *TL* **1987**, *28*, 4965. (c) Varma, R. S.; Hogan, M. E. *TL* **1992**, *33*, 7719.
13. (a) Seddon, E. A.; Seddon, K. R. In *The Chemistry of Ruthenium*; Clark, R. J. H., Ed.; Elsevier: Amsterdam, 1984; p 58. (b) Morris, P. E.; Kiely, D. E.; Vigee, G. S. *J. Carbohydr. Chem.* **1990**, *9*, 661.
14. (a) Beynon, P. J.; Collins, P. M.; Gardiner, D.; Overend, W. G. *Carbohydr. Res.* **1968**, *6*, 431. (b) Parikh, V. M.; Jones, J. K. N. *CJC* **1965**, *43*, 3452.
15. Dehand, J.; Rosé, J. *JCR(S)* **1979**, 155.
16. Hamon, D. P. G.; Massy-Westropp, R. A.; Newton, J. L. *TA* **1993**, *4*, 1435.
17. Kasai, M.; Ziffer, H. *JOC* **1983**, *48*, 712.
18. Niwa, H.; Ito, S.; Hasegawa, T.; Wakamatsu, K.; Mori, T.; Yamada, K. *TL* **1991**, *32*, 1329.
19. Singh, A. K.; Varma, R. S. *TL* **1992**, *33*, 2307.
20. Müller, P.; Godoy, J. *TL* **1981**, *22*, 2361.
21. Bilgrien, C.; Davis, S.; Drago, R. S. *JACS* **1987**, *109*, 3786.
22. Genet, J. P.; Pons, D.; Jugé, S. *SC* **1989**, *19*, 1721.
23. Griffith, W. P.; Ley, S. V.; Whitcombe, G. P.; White, A. D. *CC* **1987**, 1625.
24. (a) Meyers, A. I.; Higashiyama, K. *JOC* **1987**, *52*, 4592. (b) Behr, A.; Eusterwiemann, K. *JOM* **1991**, *403*, 209.
25. Hudlický, M. In *Oxidations in Organic Chemistry*; ACS: Washington, 1990; Monograph 186.
26. (a) Larock, R. C. *Comprehensive Organic Transformations*; VCH: New York, 1989. (b) *COS* **1991**, *7*.
27. Moriarty, R. M.; Gopal, H.; Adams, T. *TL* **1970**, 4003.
28. (a) Nutt, R. F.; Arison, B.; Holly, F. W.; Walton, E. *JACS* **1965**, *87*, 3273. (b) Nutt, R. F.; Dickinson, M. J.; Holly, F. W.; Walton, E. *JOC* **1968**, *33*, 1789.
29. Beynon, P. J.; Collins, P. M.; Overend, W. G. *Proc. Chem. Soc. London* **1964**, 342.
30. Caputo, J. A.; Fuchs, R. *TL* **1967**, 4729.
31. Gopal, H.; Adams, T.; Moriarty, R. M. *T* **1972**, *28*, 4259.
32. Crawford, R. J. *JOC* **1983**, *48*, 1366.
33. Askin, D.; Angst, C.; Danishefsky, S. *JOC* **1987**, *52*, 622.
34. Hernández, R.; Melián, D.; Suárez, E. *S* **1992**, 653.
35. Rodríguez, C. M.; Ode, J. M.; Palazón, J. M.; Martín, V. S. *T* **1992**, *48*, 3571.

36. Gao, Y.; Sharpless, K. B. *JACS* **1988**, *110*, 7538.
37. Ikunaka, M.; Mori, K. *ABC* **1987**, *51*, 565.
38. (a) Schuda, P. F.; Cichowicz, M. B.; Heimann, M. R. *TL* **1983**, *24*, 3829. (b) Takeda, R.; Zask, A.; Nakanishi, K.; Park, M. H. *JACS* **1987**, *109*, 914.
39. Morris, P. E., Jr.; Kiely, D. E. *JOC* **1987**, *52*, 1149.
40. Dauben, W. G.; Cunningham, A. F., Jr. *JOC* **1983**, *48*, 2842.
41. Mori, K.; Miyake, M. *T* **1987**, *43*, 2229.
42. Smith, A. B., III; Scarborough, R. M., Jr. *SC* **1980**, *10*, 205.
43. Tang, R.; Diamond, S. E.; Neary, N.; Mares, F. *CC* **1978**, 562.
44. Sheehan, J. C.; Tulis, R. W. *JOC* **1974**, *39*, 2264.
45. Tanaka, K.; Yoshifuji, S.; Nitta, Y. *CPB* **1988**, *36*, 3125.
46. Yoshifuji, S.; Tanaka, K.; Nitta, Y. *CPB* **1985**, *33*, 1749.
47. Balavoine, G.; Eskenazi, C.; Meunier, F.; Riviére, H. *TL* **1984**, *25*, 3187.
48. Kametani, T.; Katoh, T.; Tsubuki, M.; Honda, T. *CL* **1985**, 485.
49. Carling, R. W.; Clark, J. S.; Holmes, A. B.; Sartor, D. *JCS(P1)* **1992**, 95.
50. Tenaglia, A.; Terranova, E.; Waegell, B. *TL* **1989**, *30*, 5271.
51. Tenaglia, A.; Terranova, E.; Waegell, B. *TL* **1989**, *30*, 5275.
52. Clinch, K.; Vasella, A.; Schauer, R. *TL* **1987**, *28*, 6425.
53. Mori, K.; Ebata, T. *T* **1986**, *42*, 4413.
54. Mehta, G.; Krishnamurthy, N. *CC* **1986**, 1319.
55. Schröder, M. *CRV* **1980**, *80*, 187.
56. Piatak, D. M.; Bhat, H. B.; Caspi, E. *JOC* **1969**, *34*, 112.
57. Webster, F. X.; Rivas-Enterrios, J.; Silverstein, R. M. *JOC* **1987**, *52*, 689.
58. Hartmann, B.; Deprés, J. P.; Greene, A. E.; Freire de Lima, M. E. *TL* **1993**, *34*, 1487.
59. Deprés, J. P.; Coelho, F.; Greene, A. E. *JOC* **1985**, *50*, 1972.
60. Spitzer, U. A.; Lee, D. G. *JOC* **1974**, *39*, 2468.
61. Nuñez, M. T.; Martin, V. S. *JOC* **1990**, *55*, 1928.
62. (a) Kusakabe, M.; Kitano, Y.; Kobayashi, Y.; Sato, F. *JOC* **1989**, *54*, 2085. (b) Danishefsky, S.; Maring, C. *JACS* **1985**, *107*, 7762. (c) Danishefsky, S.; DeNinno, M. P.; Chen, S. *JACS* **1988**, *110*, 3929.
63. Brown, A. D.; Colvin, E. W. *TL* **1991**, *32*, 5187.
64. Kasai, M.; Ziffer, H. *JOC* **1983**, *48*, 2346.
65. Ayres, D. C.; Hossain, A. M. M. *JCS(P1)* **1975**, 707.
66. Martín, V. S.; Nuñez, M. T.; Tonn, C. E. *TL* **1988**, *29*, 2701.
67. Martres, P.; Perfetti, P.; Zahra, J. P.; Waegell, B.; Giraudi, E.; Petrzilka, M. *TL* **1993**, *34*, 629.
68. Tenaglia, A.; Terranova, E.; Waegell, B. *JOC* **1992**, *57*, 5523.
69. Mueller, R. H.; DiPardo, R. M. *CC* **1975**, 565.
70. Johnston, B. D.; Slessor, K. N.; Oehlschlager, A. C. *JOC* **1985**, *50*, 114.
71. (a) Lee, D. G.; Congson, L. N.; Spitzer, U. A.; Olson, M. E. *CJC* **1984**, *62*, 1835. (b) Coates, R. M.; Senter, P. D.; Baker, W. R. *JOC* **1982**, *47*, 3597. (c) Corey, E. J.; Danheiser, R. L.; Chandrasekaran, S.; Keck, G. E.; Gopalan, B.; Larsen, S. D.; Siret, P.; Gras, J.-L. *JACS* **1978**, *100*, 8034. (d) Varma, R. S.; Hogan, M. E. *TL* **1992**, *33*, 7719.
72. Green, G.; Griffith, W. P.; Hollinshead, D. M.; Ley, S. V.; Schröder, M. *JCS(P1)* **1984**, 681.
73. Bailey, A. J.; Griffith, W. P.; Mostafa, S. I.; Sherwood, P. A. *IC* **1993**, *32*, 268.
74. Guertin, K. R.; Kende, A. S. *TL* **1993**, *34*, 5369.
75. Ohtake, H.; Higuchi, T.; Hirobe, M. *TL* **1992**, *33*, 2521.
76. (a) Tomioka, H.; Takai, K.; Oshima, K.; Nozaki, H. *TL* **1981**, *22*, 1605. (b) Murahashi, S.; Oda, Y.; Naota, T.; Kuwabara, T. *TL* **1993**, *34*, 1299.
77. Matsumoto, M.; Ito, S. *CC* **1981**, 907.
78. Kanemoto, S.; Oshima, K.; Matsubara, S.; Takai, K.; Nozaki, H. *TL* **1983**, *24*, 2185.
79. Sharpless, K. B.; Akashi, K.; Oshima, K. *TL* **1976**, 2503.
80. Tanaka, M.; Kobayashi, T.; Sakakura, T. *AG(E)* **1984**, *23*, 518.
81. Sasson, Y.; Blum, J. *TL* **1971**, 2167.
82. Regen, S. L.; Whitesides, G. M. *JOC* **1972**, *37*, 1832.
83. Müller, P.; Godoy, J. *HCA* **1981**, *64*, 2531.
84. Ishii, Y.; Osakada, K.; Ikariya, T.; Saburi, M.; Yoshikawa, S. *JOC* **1986**, *51*, 2034.
85. Murahashi, S.; Naota, T.; Ito, K.; Maeda, Y.; Taki, H. *JOC* **1987**, *52*, 4319.
86. Murahashi, S.; Naota, T.; Miyaguchi, N.; Nakato, T. *TL* **1992**, *33*, 6991.
87. Murahashi, S.; Saito, T.; Naota, T.; Kumobayashi, H.; Akutagawa, S. *TL* **1991**, *32*, 5991.
88. Griffith, W. P.; Jolliffe, J. M.; Ley, S. V.; Williams, D. J. *CC* **1990**, 1219.

Victor S. Martín, José M. Palazón & Carmen M. Rodríguez
University of La Laguna, Tenerife, Spain

Samarium(II) Iodide[1]

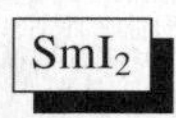

[32248-43-4] I_2Sm (MW 404.16)

(one-electron reducing agent possessing excellent chemoselectivity in reduction of carbonyl, alkyl halide, and α-heterosubstituted carbonyl substrates;[1] promotes Barbier-type coupling reactions, ketyl–alkene coupling reactions, and radical cyclizations[1])

Physical Data: mp 527 °C; bp 1580 °C; *d* 0.922 g cm^{-3}.
Solubility: soluble 0.1M in THF.
Form Supplied in: commercially available as a 0.10 M solution in THF.
Preparative Methods: typically prepared in situ for synthetic purposes. SmI_2 is conveniently prepared by oxidation of ***Samarium(0)*** metal with organic dihalides.[2]
Handling, Storage, and Precautions: is air sensitive and should be handled under an inert atmosphere. SmI_2 may be stored over THF for long periods when it is kept over a small amount of samarium metal.

Reduction of Organic Halides and Related Substrates. Alkyl halides are readily reduced to the corresponding hydrocarbon by SmI_2 in the presence of a proton source. The ease with which halides are reduced by SmI_2 follows the order I > Br > Cl. The reduction is highly solvent dependent. In THF solvent, only primary alkyl iodides and bromides are effectively reduced;[2] however, addition of HMPA effects the reduction of aryl, alkenyl, primary, secondary, and tertiary halides (eq 1).[3,4] Tosylates are also reduced to hydrocarbons by SmI_2. Presumably, under these reaction conditions the tosylate is converted to the corresponding iodide which is subsequently reduced.[4,5]

2.5 equiv SmI_2, THF, HMPA, rt, 2 h; 99% (1)

Samarium(II) iodide provides a means to reduce substrates in which the halide is resistant to reduction by hydride reducing agents (eq 2).

2.5 equiv SmI_2, THF, MeCN, HMPA, *i*-PrOH, rt, 10 min; 98% (2)

Samarium(II) iodide has been utilized as the reductant in the Boord alkene-type synthesis involving ring scission of 3-halotetrahydrofurans (eq 3).[6] SmI_2 provides an alternative to the sodium-induced reduction which typically affords mixtures of stereoisomeric alkenes and overreduction in these transformations. When SmI_2 is employed as the reductant, isomeric purities are generally >97% and overreduction products comprise <3% of the reaction mixture.

xs. SmI_2, THF, Δ_x; 75% (3)

Reduction of α-Heterosubstituted Carbonyl Compounds. Samarium(II) iodide provides a route for the reduction of α-heterosubstituted carbonyl substrates. A wide range of α-heterosubstituted ketones is rapidly reduced to the corresponding unsubstituted ketone under mild conditions (eq 4).[7] The reaction is highly selective and may be performed in the presence of isolated iodides as well as isolated ketones.[7]

2 equiv SmI_2, THF, MeOH, −78 °C (4)

Y	Isolated yield (%)
Cl	100
SPh	76
S(O)Ph	64
SO_2Ph	88

Samarium(II) iodide-induced reductive cleavage of α-hydroxy ketones provides a useful entry to unsubstituted ketones (eq 5).[8]

SmI_2, THF, *t*-BuOH, rt, 12 h; 87% (5)

Samarium(II) iodide promotes the reductive cleavage of α-alkoxy ketones. Pratt and Hopkins have utilized this protocol in synthetic studies en route to betaenone B (eq 6).[9]

SmI_2, THF, −78 °C (6)

Likewise, this procedure provides a route for the reduction of α,β-epoxy ketones and α,β-epoxy esters to generate the corresponding β-hydroxy carbonyl compounds (eqs 7 and 8).[3,10] The epoxy ketone substrates may be derived from Sharpless asymmetric epoxidation. Consequently, this procedure provides a means to prepare a variety of chiral, nonracemic β-hydroxy carbonyl compounds that are difficult to acquire by more traditional procedures.

Vinyloxiranes undergo reductive epoxide ring opening with samarium(II) iodide to provide (*E*)-allylic alcohols (eq 9).[3,10b,11] These reaction conditions are tolerant of ketone, ester, and nitrile functional groups. Again, Sharpless asymmetric epoxidation chemistry may be utilized to gain entry to the desired nonracemic substrates, thereby providing a useful entry to highly functionalized, enantiomerically enriched allylic alcohols.

2 equiv SmI_2, THF, −90 °C, <5 min; 76% (7)

2 equiv SmI_2, THF, HMPA, DMA, rt, <1 min; 68%; >98% ee (8)

2 equiv SmI_2, THF, MeOH, −90 °C (9)

Y	Isolated yield (%)
COSEt	80
SO_2Ph	82
$P(O)(OEt)_2$	84
H	69
Me	42
SPh	54

A useful method for preparation of β-hydroxy esters is accomplished by SmI_2-promoted deoxygenation of an α-hydroxy ester followed by condensation with a ketone (eq 10).[12] In some instances, excellent diastereoselectivities are achieved, although this appears to be somewhat substrate dependent.

SmI_2, THF, HMPA, 23 °C (10)

A useful reaction sequence for transforming carbonyl compounds to one-carbon homologated nitriles has evolved from the ability of SmI_2 to deoxygenate cyanohydrin *O,O′*-diethyl phosphates (eq 11).[13] The procedure is tolerant of a number of functional groups including alcohols, esters, amides, sulfonamides, acetals, alkenes, alkynes, and amines. Furthermore, it provides a distinct advantage over other previously developed procedures for similar one-carbon homologations.

$(EtO)_2P(O)CN$, LiCN, THF, rt; 3 equiv SmI_2, THF, *t*-BuOH, rt; 91% (11)

Deoxygenation Reactions. Sulfoxides are reduced to sulfides by SmI_2 (eq 12).[2,3,14] This process is rapid enough that reduction of isolated ketones is not a competitive process. Likewise, aryl sulfones are reduced to the corresponding sulfides by SmI_2 (eq 13).[3,12]

2 equiv SmI_2, THF, HMPA, rt, 1 min; 99% (12)

2 equiv SmI_2, THF, HMPA, rt, 10 min; 99% (13)

Barbier-Type Reactions. Samarium(II) iodide is quite useful in promoting Barbier-type reactions between aldehydes or ketones and a variety of organic halides. The efficiency of SmI_2 promoted Barbier-type coupling processes is governed by the substrate under consideration in addition to the reaction conditions employed. In general, alkyl iodides are most reactive while alkyl chlorides are virtually inert. Typically, catalytic ***Iron(III) Chloride*** or ***Hexamethylphosphoric Triamide*** can be added to SmI_2 to reduce reaction times or temperatures and enhance yields. Kagan and co-workers have recently applied an intermolecular SmI_2-promoted Barbier reaction towards the synthesis of hindered steroidal alcohols. An intermolecular Barbier-type reaction between the hindered ketone and ***Iodomethane*** produced a 97:3 mixture of diastereomers in excellent yield (eq 14).[15]

SmI_2, MeI, rt, 24 h; 96%; R = TBDMS; 97:3 (14)

Samarium(II) iodide-promoted intramolecular Barbier-type reactions have also been employed to produce a multitude of cyclic and bicyclic systems.[1] Molander and McKie have employed an intramolecular Barbier-type reductive coupling reaction to promote the formation of bicyclo[*m.n.*1]alkan-1-ols from the corresponding iodo ketone substrates in good yield (eq 15).[16]

2 equiv SmI_2, cat. $Fe(dbm)_3$, THF, −78 °C to rt, <2 h; 75–85% (15)

Annulation of five- and six-membered rings proceeds with excellent diastereoselectivity via an intramolecular Barbier-type process (eq 16).[17] The Barbier-type coupling scheme provides a reliable and convenient alternative to other such methods for preparing fused bicyclic systems.

2 equiv SmI_2, cat. $FeCl_3$, THF, −78 °C to rt; 65%; only diastereomer (16)

The SmI_2-promoted Barbier-type reaction has also been utilized in the synthesis of polyquinanes. Cook and Lannoye have employed this method to effect a bis-annulation of an appropriately substituted diketone (eq 17).[18]

2 equiv SmI_2, THF, HMPA, 68% (17)

Substituted β-keto esters also provide excellent substrates for the intramolecular Barbier cyclization (eq 18).[19] Diastereoselectivities are typically quite good but are highly dependent on substituent and solvent effects.

2 equiv SmI_2, THF, −78 °C to rt, 100%; 92% de (18)

Nucleophilic Acyl Substitutions. Samarium(II) iodide facilitates the highly selective intramolecular nucleophilic acyl substitution of halo esters (eqs 19 and 20).[20]

2 equiv SmI_2, cat. Fe^{III}, THF, 91% (19)

2 equiv SmI_2, cat. Fe^{III}, THF, 91% (20)

Unlike organolithium or organomagnesium reagents, SmI_2-promoted nucleophilic substitution does not proceed with double addition to the carbonyl, nor are any products resulting from reduction of the final product observed. With suitably functionalized substrates, this procedure provides a strategy for the formation of eight-membered rings (eq 21).

2 equiv SmI_2, cat. Fe^{III}, THF, 82% (21)

Ketone–Alkene Coupling Reactions. Ketyl radicals derived from reduction of ketones or aldehydes with SmI_2 may be coupled both inter- and intramolecularly to a variety of alkenic species. Excellent diastereoselectivities are achieved with intramolecular coupling of the ketyl radical with α,β-unsaturated esters.[21] In the following example, ketone–alkene cyclization took place in a stereocontrolled manner established by chelation of the resulting Sm(III) species with the hydroxyl group incorporated in the substrate (eq 22).[21b]

A similar strategy utilizing β-keto esters provided very high diastereoselectivities in the ketyl–alkene coupling process. In these examples, chelation control about the developing hydroxyl and carboxylate stereocenters was the source of the high diastereoselectivity achieved (eq 23).[22]

SmI_2, MeOH, THF, −78 °C, 75–85% (22)

X = CO_2Me, Y = H; X = H, Y = CO_2Me → X = CH_2CO_2Me, Y = H; X = H, Y = CH_2CO_2Me

2 equiv SmI_2, THF, t-BuOH, −78 °C, 87% (23)

Alkynic aldehydes likewise undergo intramolecular coupling to generate five- and six-membered ring carbocycles. This protocol has been utilized as a key step in the synthesis of isocarbacyclin (eq 24).[23] SmI_2 was found to be superior to several other reagents in this conversion.

2 equiv SmI_2, THF, t-BuOH, −70 °C, 0.5 h, 71%; 80% de (24)

R = TBDMS, R' = C_5H_{11}

Samarium(II) iodide in the presence of HMPA effectively promotes the intramolecular coupling of unactivated alkenic ketones by a reductive ketyl–alkene radical cyclization process (eq 25). This protocol provides a means to generate rather elaborate carbocycles through a sequencing process in which the resulting organosamarium species is trapped with various electrophiles to afford the cyclized product in high yield.[24]

1. 2.2 equiv SmI_2, THF, HMPA; 2. El; 75–85% (25)

El = RCHO, RCOR, CO_2, Ac_2O, O_2

Pinacolic Coupling Reactions. In the absence of a proton source, both aldehydes and ketones are cleanly coupled in the presence of SmI_2 to the corresponding pinacol.[25] Considerable diastereoselectivity has been achieved in the coupling of aliphatic 1,5- and 1,6-dialdehydes, providing near exclusive formation of the *cis*-diols (eq 26).[26]

2 equiv SmI_2, THF, −78 °C to rt, 2 h, 81%; 92% de (26)

R = TBDMS

Intramolecular cross coupling of aldehydes and ketones proceeds with excellent diastereoselectivity and high yield in suitably functionalized systems wherein chelation control by the resulting Sm^{III} species directs formation of the newly formed stereocenters (eq 27).[22a,27] A similar strategy has been utilized with a β-keto amide substrate to provide a chiral, nonracemic oxazolidinone species. This strategy permits entry to highly functionalized, enantiomerically pure dihydroxycyclopentanecarboxylate derivatives (eq 28).

OHC, O O, O — 2 equiv SmI_2, THF, MeOH, −78 °C, 44% → HO, OH, O, O (27)

O O O, N, O, i-Pr, 2 CHO — 2 equiv SmI_2, THF, t-BuOH, 52% → X_c, H, HO, OH, O (28)

Radical Addition to Alkenes and Alkynes. Samarium(II) iodide has proven effective for initiation of various radical addition reactions to alkenes and alkynes. Typically, tin reagents are used in the initiation of these radical cyclization reactions; however, the SmI_2 protocol often provides significant advantages over these more traditional routes.

Samarium(II) iodide-mediated cyclization of aryl radicals onto alkene and alkyne acceptors provides an excellent route to nitrogen- and oxygen-based heterocycles (eq 29).[28]

Br, N — SmI_2, THF, HMPA, rt, 6 h, 55% → N (29)

The SmI_2 reagent is unique in that it provides the ability to construct more highly functionalized frameworks through a sequential radical cyclization/intermolecular carbonyl addition reaction.[29] Thus the intermediate radical formed after initial cyclization may be further reduced by SmI_2, forming an organosamarium intermediate which may be trapped by various electrophiles, affording highly functionalized products (eq 30).

I, O + O, NEt_2, 3 — SmI_2, THF, HMPA, rt, 76% → Et_2N, 3, OH, O (30)

Samarium(II) iodide further mediates the cyclization reactions of alkynyl halides (eq 31).[30] When treated with SmI_2, the alkynyl halides are converted to the cyclized product in good yield. Addition of DMPU as cosolvent provides slightly higher yields in some instances.

TMS, Br — 3 equiv SmI_2, THF, Δ_x, 24 h, 74% → TMS (31)

Highly functionalized bicyclic and spirocyclic products are obtained in good yield and high diastereoselectivity by a tandem reductive cleavage–cyclization strategy (eq 32).[31] Radical ring opening of cyclopropyl ketones mediated by samarium(II) iodide-induced electron transfer permits the elaboration of a tandem ring opening–cyclization strategy wherein the resultant samarium enolate may be trapped by either oxygen or carbon electrophiles.

O — SmI_2, THF–DMPU, AcCl, 57% → OAc (32)

Related Reagents. See Classes R-2, R-23, R-25, R-27, R-28, R-31, R-32, and R-34, pages 1–10. Samarium(II) Iodide–1,3-Dioxolane.

---

1. (a) Molander, G. A. *CRV* **1992**, *92*, 29. (b) Molander, G. A. In *The Chemistry of the Metal–Carbon Bond*; Hartley, F. R., Ed.; Wiley: Chichester, 1989; Vol. 5, Chapter, 8. (c) Kagan, H. B. *NJC* **1990**, *14*, 453. (d) Soderquist, J. A. *Aldrichim. Acta* **1991**, *24*, 15. (e) Molander, G. A. *COS* **1991**, *1*, Chapter 1.9.
2. (a) Girard, P.; Namy, J. L.; Kagan, H. B. *JACS* **1980**, *102*, 2693. (b) Namy, J. L.; Girard, P.; Kagan, H. B. *NJC* **1977**, *1*, 5. (c) Namy, J. L.; Girard, P.; Kagan, H. B. *NJC* **1981**, *5*, 479.
3. Inanaga, J. *HC* **1990**, *3*, 75.
4. Inanaga, J.; Ishikawa, M.; Yamaguchi, M. *CL* **1987**, 1485.
5. Kagan, H. B.; Namy, J. L.; Girard, P. *T* **1981**, *37*, 175, Suppl. 1.
6. Crombie, L.; Rainbow, L. J. *TL* **1988**, *29*, 6517.
7. (a) Molander, G. A.; Hahn, G. *JOC* **1986**, *51*, 1135. (b) Smith, A. B., III; Dunlap, N. K.; Sulikowski, G. A. *TL* **1988**, *29*, 439. (c) Castro, J.; Sörensen, H.; Riera, A.; Morin, C.; Moyano, A.; Pericàs, M. A.; Greene, A. E. *JACS* **1990**, *112*, 9388.
8. (a) White, J. D.; Somers, T. C. *JACS* **1987**, *109*, 4424. (b) Holton, R. A.; Williams, A. D. *JOC* **1988**, *53*, 5981.
9. Pratt, D. V.; Hopkins, P. B. *TL* **1987**, *28*, 3065.
10. (a) Molander, G. A.; Hahn, G. *JOC* **1986**, *51*, 2596. (b) Otsubo, K.; Inanaga, J.; Yamaguchi, M. *TL* **1987**, *28*, 4437.
11. Molander, G. A.; La Belle, B. E.; Hahn, G. *JOC* **1986**, *51*, 5259.
12. Enholm, E. J.; Jiang, S. *TL* **1992**, *33*, 313.
13. (a) Yoneda, R.; Harusawa, S.; Kurihara, T. *TL* **1989**, *30*, 3681. (b) Yoneda, R.; Harusawa, S.; Kurihara, T. *JOC* **1991**, *56*, 1827.
14. Handa, Y.; Inanaga, J.; Yamaguchi, M. *CC* **1989**, 298.
15. Sasaki, M.; Collin, J.; Kagan, H. B. *NJC* **1992**, *16*, 89.
16. Molander, G. A.; McKie, J. A. *JOC* **1991**, *56*, 4112.
17. (a) Molander, G. A.; Etter, J. B. *JOC* **1986**, *51*, 1778. (b) Zoretic, P. A.; Yu, B. C.; Caspar, M. L. *SC* **1989**, *19*, 1859. (c) Daniewski, A. R.; Uskokovic, M. R. *TL* **1990**, *31*, 5599.
18. (a) Lannoye, G.; Cook, J. M. *TL* **1988**, *29*, 171. (b) Lannoye, G.; Sambasivarao, K.; Wehrli, S.; Cook, J. M.; Weiss, U. *JOC* **1988**, *53*, 2327.
19. Molander, G. A.; Etter, J. B.; Zinke, P. W. *JACS* **1987**, *109*, 453.
20. Molander, G. A.; McKie, J. A. *JOC* **1993**, *58*, 7216.
21. (a) Hon, Y.-S.; Lu, L.; Chu, K.-P. *SC* **1991**, *21*, 1981. (b) Kito, M.; Sakai, T.; Yamada, K.; Matsuda, F.; Shirahama, H. *SL* **1993**, 158. (c) Fukuzawa, S.; Iida, M.; Nakanishi, A.; Fujinami, T.; Sakai, S. *CC* **1987**, 920. (d) Fukuzawa, S.; Nakanishi, A.; Fujinami, T.; Sakai, S. *JCS(P1)* **1988**, 1669. (e) Enholm, E. J.; Trivellas, A. *TL* **1989**, *30*, 1063. (f) Enholm, E. J.; Satici, H.; Trivellas, A. *JOC* **1989**, *54*, 5841. (g) Enholm, E. J.; Trivellas, A. *JACS* **1989**, *111*, 6463.
22. (a) Molander, G. A.; Kenny, C. *JACS* **1989**, *111*, 8236. (b) Molander, G. A.; Kenny, C. *TL* **1987**, *28*, 4367.
23. (a) Shim, S. C.; Hwang, J.-T.; Kang, H.-Y.; Chang, M. H. *TL* **1990**, *31*, 4765. (b) Bannai, K.; Tanaka, T.; Okamura, N.; Hazato, A.; Sugiura, S.; Manabe, K.; Tomimori, K.; Kato, Y.; Kurozumi, S.; Noyori, R. *T* **1990**, *46*, 6689.

24. Molander, G. A.; McKie, J. A. *JOC* **1992**, *57*, 3132.
25. (a) Namy, J. L.; Souppe, J.; Kagan, H. B. *TL* **1983**, *24*, 765. (b) Fürstner, A.; Csuk, R.; Rohrer, C.; Weidmann, H. *JCS(P1)* **1988**, 1729.
26. Chiara, J. L.; Cabri, W.; Hanessian, S. *TL* **1991**, *32*, 1125.
27. Molander, G. A.; Kenny, C. *JACS* **1989**, *111*, 8236.
28. Inanaga, J.; Ujikawa, O.; Yamaguchi, M. *TL* **1991**, *32*, 1737.
29. (a) Molander, G. A.; Harring, L. S. *JOC* **1990**, *55*, 6171. (b) Curran, D. P.; Totleben, M. J. *JACS* **1992**, *114*, 6050.
30. Bennett, S. M.; Larouche, D. *SL* **1991**, 805.
31. Batey, R. A.; Motherwell, W. B. *TL* **1991**, *32*, 6649.

Gary A. Molander & Christina R. Harris
University of Colorado, Boulder, CO, USA

Selenium(IV) Oxide

[7446-08-4] O_2Se (MW 110.96)

(oxidant of activated, saturated positions)

Alternate Name: selenium dioxide.
Physical Data: mp 315 °C (subl); *d* 3.95 g cm^{-3}.
Solubility: sol water, methanol, ethanol, acetone, acetic acid.
Form Supplied in: off-white powder;[1] widely available.
Purification: by sublimation, or by treatment with HNO_3.[28]
Handling, Storage, and Precautions: toxic; corrosive; causes intense local irritation of skin and eyes; use in a fume hood.

Allylic Hydroxylation. Selenium(IV) oxide is known primarily for hydroxylation of activated carbon-bearing positions, particularly at allylic (or propargylic) sites. Studies by Guillemonat and others have led to the following hydroxylation selectivity rules:[2,3]

1. Hydroxylation occurs α to the more substituted end of the double bond.
2. The order of facility of oxidation is $CH_2 > CH_3 > CH$.
3. When the double bond is in a ring, oxidation occurs within the ring when possible, and α to the more substituted end of the double bond.
4. Oxidation of a terminal double bond affords a primary alcohol with allylic migration of the double bond.

An example of rules (1) and (2) is shown in the oxidation of 3-methyl-3-butene, where the allylic methylene position is oxidized in preference to the methyl or methine positions (eq 1).[2] Alkene-selective oxidation of 5,6-dihydroergosterol in ethanol, an example of rule (3), occurs at C-14 and is followed by allylic rearrangement to give a 7α-ethoxy product (eq 2).[4] The mechanism of the allylic oxidation reaction is proposed to be initiated by ene addition, followed by dehydration and [2,3]-sigmatropic rearrangement of the resultant allylseleninic acid.[5,6] In a key step of the synthesis of α-onocerin, α-oxidation in acetic acid leads to an unsaturated γ-lactone product in good yield (eq 3).[7] The milbemycins have been hydroxylated in the 13β-position by selenium dioxide.[8] Because selenium dioxide forms selenious acid (H_2SeO_3) in the presence of water, hydroxylations of alkenes containing acid-labile groups (e.g. acetals) have been run in pyridine.[9]

SeO_2 → OH 34% + HO 1% (1)

HO → 1. SeO_2, EtOH, C_6H_6; 2. Ac_2O; 77% → AcO, OEt (2)

CO_2Et, AcO, H → SeO_2, AcOH, reflux; 75% → O, O, AcO, H (3)

Higher-Order Oxidations. Selenium dioxide can introduce carbonyl functionality at activated positions, and can also effect dehydrogenation[10–13] at highly activated saturated sites. For instance, phenylglyoxal is isolated in high yield from α-oxidation of acetophenone (eq 4).[14] On a large scale, dissolution of the selenium dioxide in aqueous dioxane at 55 °C is required prior to acetophenone addition. Similarly, 6-methyluracil is readily converted to orotaldehyde in acetic acid (eq 5).[15] Oxidation of aryl-substituted succinic acids to maleic anhydride analogs occurs readily in acetic anhydride (eq 6).[16] This is a preferred method, since oxidations of this type with ***N-Bromosuccinimide*** give bromoarene byproducts. Additionally, selenium dioxide in the presence of ***Trimethylsilyl Polyphosphate*** has been used to aromatize cyclohexenes and cyclohexadienes.[17] Using only a slight excess (1.2 molar equiv) of selenium dioxide in pyridine, methyl 2-methyl-4-pyrimidinecarboxylate has been prepared regioselectively from 2,4-dimethylpyrimidine after methanolysis of the carboxylic acid product (eq 7).[18] Interestingly, ***Sulfuric Acid***-catalyzed oxidation of 1-octene in acetic acid affords a 1,2-diacetate product (eq 8).[19] Only a trace amount of 1-acetoxy-3-octene is observed.

Ph(C=O)CH3 → SeO_2, aq. dioxane, reflux; 72% → Ph(C=O)CHO (4)

6-methyluracil → SeO_2, AcOH, reflux; 58% → orotaldehyde (CHO) (5)

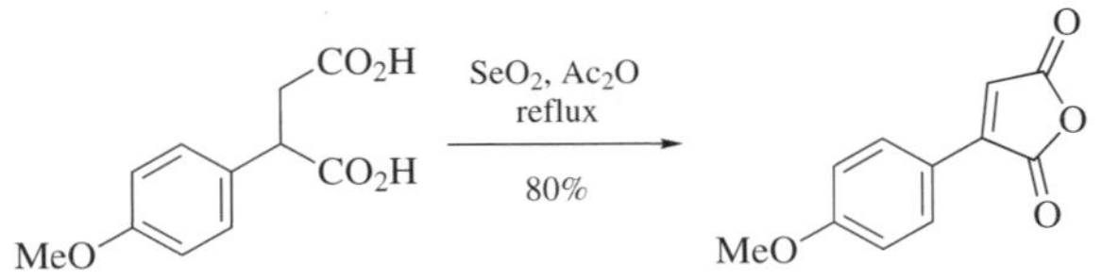

Oxidative Cleavage. Attack of selenium dioxide at activated positions can lead to oxidative bond cleavage when appropriate leaving groups are present. Aryl propargyl ethers undergo oxidation at the α-alkynyl position to afford a phenolic species and propargyl aldehyde (eq 9).[20] The analogous aryl allyl ether fragmentations occur in somewhat lower yields. (Hydroxyaryl)pyrazolines have been oxidized, with nitrogen extrusion, to afford 2′-hydroxychalcone products (eq 10).[21] Oxidations of pyrazolines with ***Bromine***, ***Potassium Permanganate***, ***Chromium(VI) Oxide***, and other reagents result in pyrazole formation.

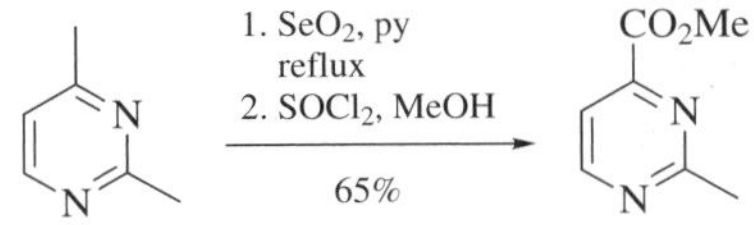

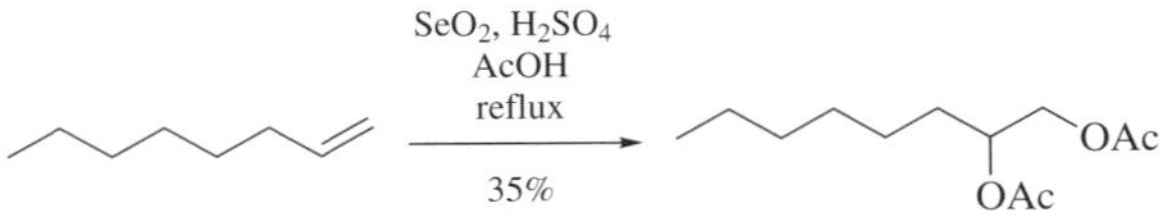

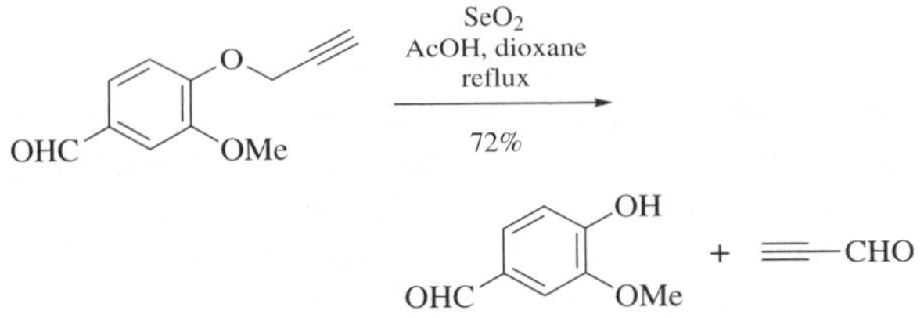

Miscellaneous Transformations. Alkyl and aryl nitriles can be prepared from the corresponding aldehydes via conversion to the aldoxime, followed by catalytic selenium dioxide-mediated elimination (eq 11).[22,23] Aliphatic nitriles are formed at rt, while aryl nitrile formation requires heating. 1,2,3-Selenadiazoles have been synthesized by treatment of an *N*-benzylazepine 4-semicarbazone with selenium dioxide (or selenoyl dichloride) (eq 12).[24,25] The *N*-benzyl proximal product is formed with high regioselectivity vis-à-vis the distal product in polar solvents. Nonpolar solvents give ca. 3:1 mixtures (proximal/distal). The oxygen-catalyzed reaction of trialkylboranes with 1 equiv of selenium dioxide affords a dialkyl selenide as the major product.[26] Similarly, dialkyl selenides have been prepared by reaction of alkyllithiums or Grignard reagents with selenium dioxide.[27]

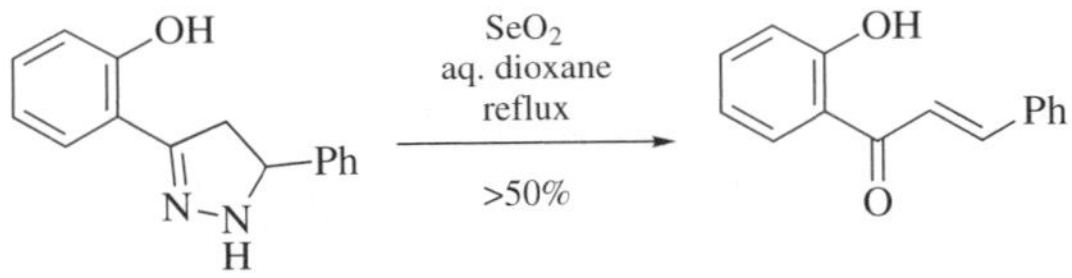

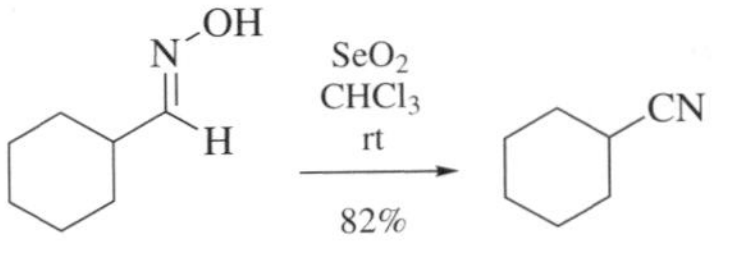

Related Reagents. See Classes O-16 and O-21, pages 1–10. Selenium(IV) Oxide–*t*-Butyl Hydroperoxide.

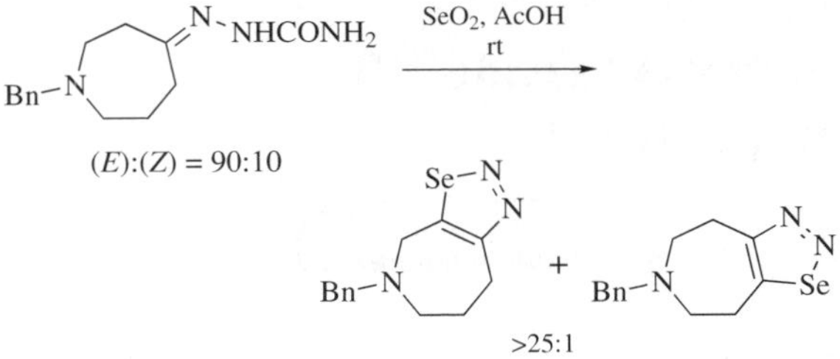

1. Stahl, K.; Legros, J. P.; Galy, J. *Z. Kristallogr.* **1992**, *202*, 99.
2. (a) Guillemonat, A. *AC(R)* **1939**, *11*, 143. (b) Fieser, L. F.; Fieser, M. *FF* **1967**, *1*, 992.
3. Bhalerao, U. T.; Rapaport, H. *JACS* **1971**, *93*, 4835.
4. Fieser, L. F.; Ourisson, G. *JACS* **1953**, *75*, 4404.
5. Arigoni, D.; Vasella, A.; Sharpless, K. B.; Jensen, H. P. *JACS* **1973**, *95*, 7917.
6. Wiberg, K. B.; Nielsen, S. D. *JOC* **1964**, *29*, 3353.
7. Danieli, N.; Mazur, Y.; Sondheimer, F. *TL* **1961**, 310.
8. Tsukamoto, Y.; Sato, K.; Kinoto, T.; Yanai, T. *BCJ* **1992**, *65*, 3300.
9. Camps, F.; Coll, J.; Parente, A. *S* **1978**, 215.
10. Bernstein, S.; Littell, R. *JACS* **1960**, *82*, 1235.
11. Heller, M.; Bernstein, S. *JOC* **1961**, *26*, 3876.
12. Fried, J. H.; Arth, G. E.; Sarett, L. H. *JACS* **1959**, *81*, 1235.
13. Allen, G. R.; Austin, N. A. *JOC* **1961**, *26*, 4574.
14. Riley, H. A.; Gray, A. R. *OSC* **1943**, *2*, 509.
15. Zee-Cheng, K.-Y.; Cheng, C. C. *JHC* **1967**, *4*, 163.
16. Hill, R. K. *JOC* **1961**, *26*, 4745.
17. Lee, J. G.; Kim, K. C. *TL* **1992**, *33*, 6363.
18. Sakasai, T.; Sakamoto, T.; Yamanaka, H. *H* **1979**, *13*, 235.
19. Javaid, K. A.; Sonoda, N.; Tsutsumi, S. *TL* **1969**, 4439.
20. Kariyone, K.; Yazawa, H. *TL* **1970**, 2885.
21. Berge, D. D.; Kale, A. V. *CI(L)* **1979**, 662.
22. Sosnovsky, G.; Krogh, J. A. *S* **1978**, 703.
23. Sosnovsky, G.; Krogh, J. A.; Umhoefer, S. G. *S* **1979**, 722.
24. Maryanoff, B. E.; Rebarchak, M. C. *JOC* **1991**, *56*, 5203.
25. Meier, H.; Voigt, E. *T* **1972**, 187.
26. Arase, A.; Masuda, Y. *CL* **1975**, 419.
27. Arase, A.; Masuda, Y. *CL* **1975**, 1331.
28. Perrin, D. D.; Armarego, W. L. F. *Purification of Laboratory Chemicals*, 3rd ed.; Pergamon: New York, 1988; p 342.

William J. Hoekstra
The R. W. Johnson Pharmaceutical Research Institute, Spring House, PA, USA

Selenium(IV) Oxide–*t*-Butyl Hydroperoxide[1]

SeO2–t-BuOOH

(SeO_2)		
[7446-08-4]	O_2Se	(MW 110.96)
(*t*-BuOOH)		
[75-91-2]	$C_4H_{10}O_2$	(MW 90.14)

(oxidizes alkenes to allylic alcohols or to α,β-unsaturated aldehydes or ketones)

Physical Data: SeO_2:[2] mp 340 °C. 90% *t*-BuOOH:[3] fp 35 °C; *d* 0.901 g cm^{-3}. See ***Selenium(IV) Oxide*** and ***t-Butyl Hydroperoxide***.

Form Supplied in: SeO_2: solid. *t*-BuOOH: liquid. Commercially available.

Handling, Storage, and Precautions: SeO_2[2] is irritating to the eyes and skin and is highly toxic and corrosive. *t*-BuOOH[1,3] is a flammable liquid and, as with all peroxides, should be handled with care. Use in a fume hood.

Allylic Oxidations. Selenium(IV) oxide in combination with *t*-butyl hydroperoxide (TBHP) is an effective system for insertion of oxygen into an allylic carbon–hydrogen bond (eq 1).[4] Reaction conditions (CH_2Cl_2, 0 °C) are much milder than those required for oxidation with selenium(IV) oxide alone and, as a result, yields are higher with fewer oxidation, dehydration, and rearrangement byproducts. Furthermore, the problem of the removal of colloidal selenium is circumvented, and in many cases only catalytic amounts of selenium(IV) oxide are required to effect oxidations in high yields. Over-oxidations to the corresponding aldehyde or ketone products are observed. The rules regarding allylic oxidations by selenium(IV) oxide, proposed by Guillemonat[5] and Rapoport,[6] apply to these reaction conditions: (1) oxidation occurs at the more highly substituted end of the alkene; (2) the order of reactivity of C–H bonds is $CH_2 > CH_3 > CH$ (rule 1 takes precedence over rule 2); (3) when the double bond is within a ring, oxidation occurs within the ring; (4) *gem*-dimethyl trisubstituted alkenes oxidize stereospecifically to give (*E*)-α-hydroxy alkenes.

SeO2, TBHP; CH2Cl2, 10 °C, 3 h — 49%, 2%, 12%, <1% (1)

Studies of the reaction intermediates suggest that the reaction proceeds via a combination–dissociation–recombination pathway,[7] which may explain the double bond migration seen in some selenium(IV) oxide oxidations (eq 2).[8]

Selenium(IV) oxide supported on silica gel in the presence of TBHP oxidizes methyl allylic sites with no evidence of competing oxidation at methylene allylic positions (eq 3).[9]

SeO2, TBHP; CH2Cl2, 47% (2)

SeO2/silica, TBHP; CH2Cl2 — 45%, 22% (3)

The selenium(IV) oxide–TBHP combination has also been used to oxidize alkenes to their corresponding α,β-unsaturated ketones (eq 4)[10] or aldehydes[11] using either the standard conditions over extended reaction periods or silica gel-supported selenium(IV) oxide.[11a]

SeO2, TBHP; CH2Cl2, 74% (4)

Oxidation of Alkynes. Alkynes are similarly oxidized to give α-hydroxy alkynes in good to moderate yields.[1] Methine and methylene C–H bonds are more reactive than methyl C–H bonds. Internal alkynes can be oxidized to give alkynediols in good yields with little or no oxidation of the alcohols to ketone products (eq 5).[12]

SeO2, TBHP; CH2Cl2 — 41% + 12% + ketone products <6% (5)

Oxidation of Allylic Alcohols. Selenium(IV) oxide supported on silica gel with TBHP has been used to selectively oxidize primary allylic alcohols to α,β-unsaturated aldehydes in high yields.[13] Secondary allylic, benzylic, and saturated alcohols are unaffected by these reaction conditions.

Related Reagents. See Class O-21, pages 1–10. *t*-Butyl Hydroperoxide; Selenium(IV) Oxide.

---

1. Sharpless, K. B.; Verhoeven, T. R. *Aldrichim. Acta* **1979**, *12*, 63.
2. *The Merck Index: An Encyclopedia of Chemicals, Drugs, and Biologicals*, 11th ed.; Budavari, S., Ed.; Merck: Rahway, NJ, 1989; p 1337.
3. Ref. 2, p 240.
4. Umbreit, M. A.; Sharpless, K. B. *JACS* **1977**, *99*, 5526.

5. (a) *FF* **1967**, *1*, 992. (b) Guillemonat, A. *AC(R)* **1939**, *11*, 143.
6. Bhalerao, U. T.; Rapoport, H. *JACS* **1971**, *93*, 4835.
7. Warpehoski, M. A.; Chabaud, B.; Sharpless, K. B. *JOC* **1982**, *47*, 2897.
8. Ceccherelli, P.; Curini, M.; Marcotullio, M. C.; Rosati, O. *T* **1989**, *45*, 3809.
9. (a) Singh, J.; Sabharwal, A.; Sayal, P. K.; Chhabra, B. R. *CI(L)* **1989**, 533. (b) Chhabra, B. R.; Hayano, K. *CL* **1981**, 1703.
10. Mateos, A. F.; Barrueco, O. F.; González, R. R. *TL* **1990**, *31*, 4343.
11. (a) Bock, I.; Bornowski, H.; Ranft, A.; Theis, H. *T* **1990**, *46*, 1199. (b) Desai, S. R.; Gore, V. K.; Bhat, S. V. *SC* **1990**, *20*, 523. (c) Schreiber, S. L.; Meyers, H. V.; Wiberg, K. B. *JACS* **1986**, *108*, 8274.
12. Chabaud, B.; Sharpless, K. B. *JOC* **1979**, *44*, 4202.
13. Kalsi, P. S.; Chhabra, B. R.; Singh, J.; Vig, R. *SL* **1992**, 425.

James J. McNally

The R. W. Johnson Pharmaceutical Research Institute, Spring House, PA, USA

Silver(I) Carbonate on Celite

Ag_2CO_3/Celite

(Ag_2CO_3)
[534-16-7] CAg_2O_3 (MW 275.75)

(mild oxidizing agent, which operates under neutral and heterogeneous conditions[1–3])

Physical Data: Ag_2CO_3: mp 210 °C (dec); *d* 6.077 g cm^{-3}.
Solubility: insol water, organic sleiolvents; destroyed by acids.
Form Supplied in: green-yellow powder. Commercially available reagent contains ca. 50 wt% Ag_2CO_3.
Preparative Method: ***Silver(I) Nitrate*** (30 g) is dissolved in 200 mL of distilled water, and ***Celite*** (30 g) is added. To the stirred mixture, a solution of ***Sodium Carbonate*** ($Na_2CO_3 \cdot 10H_2O$) (30 g) in distilled water (300 mL) is slowly added. The yellow–green precipitate is filtered and washed to neutrality with distilled water. It is then dried (4 h) by rotatory evaporation on a steam bath, preferably in the dark. The reagent can be stored in the dark at room temperature for several years without significant loss of activity. When prepared according to this procedure, 0.6 g of this reagent contains approximately 1 mmol of Ag_2CO_3. Since an excess of silver carbonate on Celite is necessary for the reaction to proceed at a reasonable rate, recovery of silver is often required. This can be easily achieved by dissolving the used reagent in nitric acid. The filtered solution after concentration gives back silver nitrate (poorly soluble in nitric acid).
Handling, Storage, and Precautions: can be safely handled in the dry state. Moisture or exposure to light are not critical; can be stored for years in the dark. Silver salts are toxic.

Introduction. Silver carbonate on Celite oxidizes primary alcohols to aldehydes, and secondary alcohols to ketones. 1,2-Diols are either cleaved or oxidized to α-hydroxy ketones or sometimes to α-diketones; 1,4-, 1,5-, and 1,6-diols with one primary hydroxyl group may give lactones. Lactols are converted to lactones. Hydroquinones and catechols afford quinones. Steric factors play an important role in determining the outcome of these reactions, including the generally high regioselectivity in polyol oxidations.

Oxidation, rearrangement, or glycosylation with the help of silver carbonate on Celite are usually carried out in aromatic solvents, sometimes in chloroform or methylene chloride, and, in the case of carbohydrates, in methanol. Reactions are easily monitored by TLC. As soon as the reaction is over, the solid is filtered off, and the solution evaporated. The price of the reagent makes it difficult to use on a large scale, unless the recovery of silver is contemplated. The mildness and the simplicity of the workup may, in some cases, overcome this drawback. Many protective groups (acetals, formate, acetate, tetrahydropyranyl, silyl derivatives), as well as sensitive functionality such as furan, indole, and *N*-substituted pyrrole ring systems, are not affected by the reagent. A critical evaluation of several solid supported reagents has been published.[52]

Oxidation of primary alcohols. Primary alcohols (saturated, allylic, polyunsaturated, benzylic) are normally converted into aldehydes in good to excellent yield. A few examples out of many reported are listed in Table 1.[4–6]

Table 1 Oxidation of Primary Alcohols to Aldehydes

Alcohol	Solvent	Yield (%)
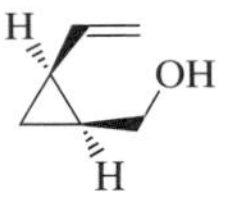 H, OH, H	Benzene	80
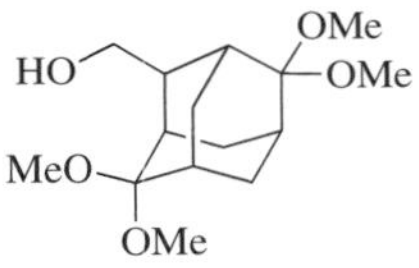 HO, OMe, OMe, MeO, OMe	Benzene	100
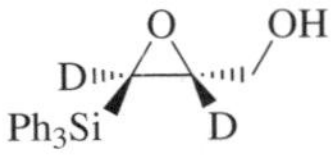 O, OH, D, Ph_3Si, D	Benzene	~30

In contrast to the oxidation of other primary alcohols, α-hydroxymethyl cyclic ethers give lactones instead of the expected aldehydes (Table 2).[7,8]

Table 2 Oxidation of α-Hydroxymethyl Cyclic Ethers

Alcohol	Product	Solvent	Yield (%)
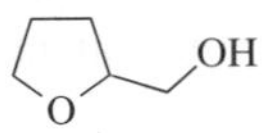 O, OH	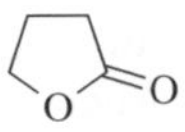 O, O	Benzene	70
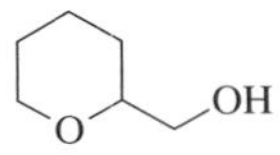 O, OH	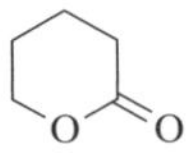 O, O	Benzene	80
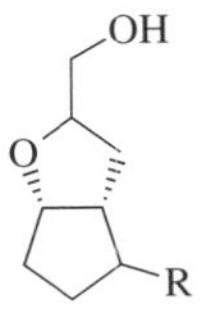 OH, O, R	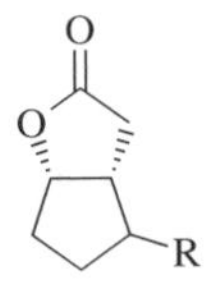 O, O, R	Benzene Toluene	31 37

Due to a strong isotope effect, tritiated primary alcohols are oxidized to aldehydes, with virtually no loss of tritium (eq 1).[9]

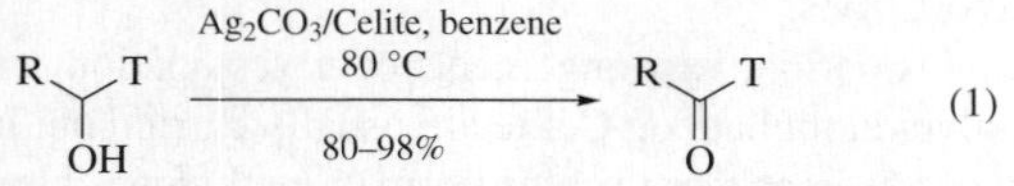

$$R\text{-}CH(OH)\text{-}T \xrightarrow[80\text{–}98\%]{Ag_2CO_3/\text{Celite, benzene, } 80\,^\circ C} R\text{-}C(O)\text{-}T \quad (1)$$

Ketones from Secondary Alcohols. A large number of secondary alcohols, usually bearing other functional groups sensitive to acids and oxidizing agents, has been transformed into ketones.[3] A few typical examples are recorded in Table 3.[10–13]

Table 3 Oxidation of Secondary Alcohols to Ketones

Alcohol	Solvent	Yield (%)
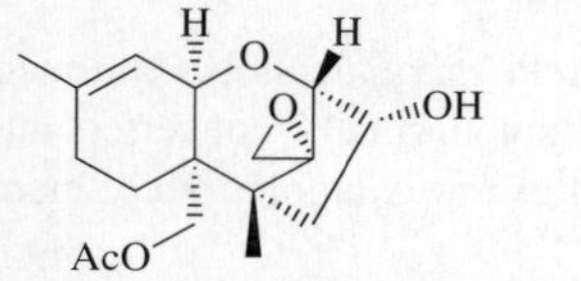	Toluene	70
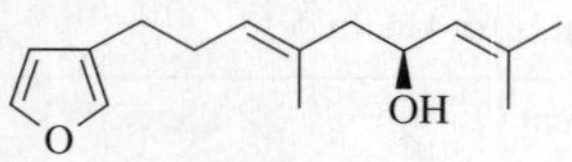	Benzene	86
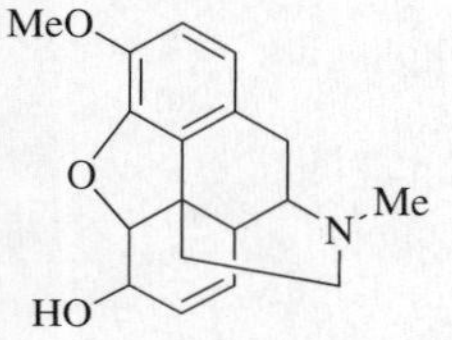	Benzene	84
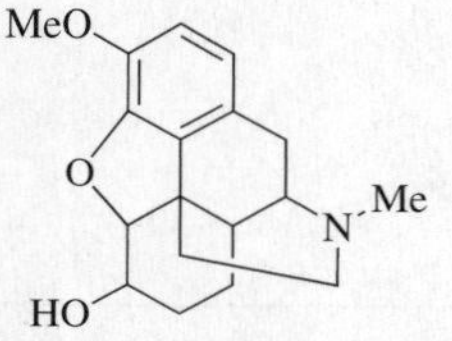	Benzene	75

Allylic alcohols are rapidly oxidized. In sharp contrast, oxidation of a homoallylic alcohol such as cholesterol (in carefully deoxygenated benzene, under argon) leads to an intractable mixture of unidentified products. Celite (the best inert support tested so far) is essential. Thus codeine has been oxidized either by standard silver carbonate, or by silver carbonate on Celite, but only the latter reagent could oxidize dihydrocodeine. Some strained cyclobutanols afford γ-lactones in addition to the expected ketone (eq 2).[14] A case of molecular rearrangement has been mentioned (eq 3).[15]

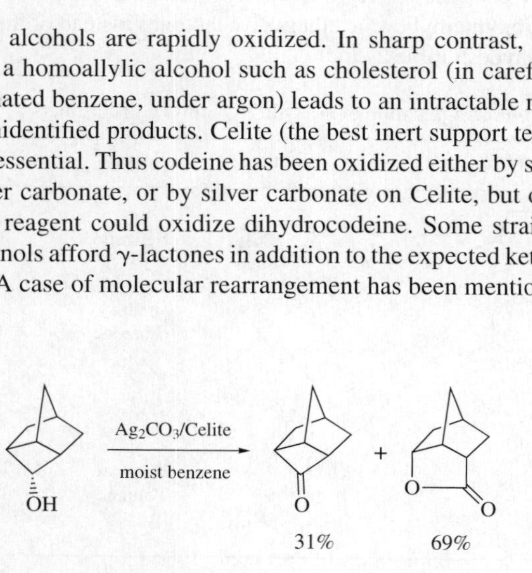

$$\xrightarrow[\text{moist benzene}]{Ag_2CO_3/\text{Celite}} \quad 31\% \; + \; 69\% \quad (2)$$

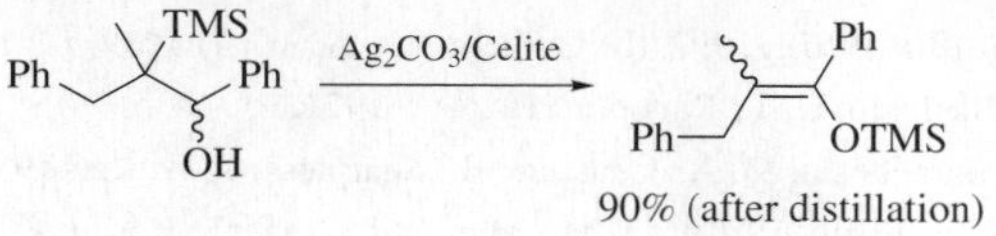

$$\xrightarrow{Ag_2CO_3/\text{Celite}} \quad 90\%\ \text{(after distillation)} \quad (3)$$

Oxidation of Tertiary Alcohols. Tertiary alcohols are generally quite stable towards the reagent, with the exception of ethynyl carbinols which are cleaved in quantitative yield (Table 4)[16] at a rate comparable to that of allylic alcohol oxidation. Cyanohydrins behave similarly. The use of the ethynyl group as a ketone protective group has been suggested.

Table 4 Oxidation of Tertiary Alcohols

Alcohol	Product	Yield (%)
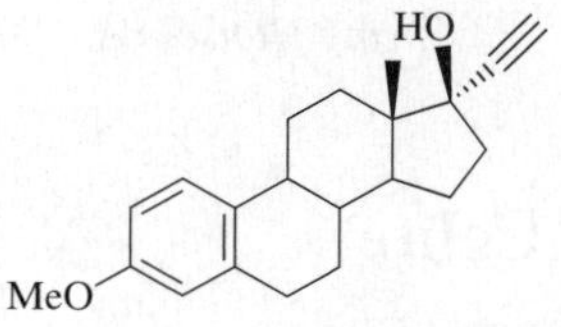	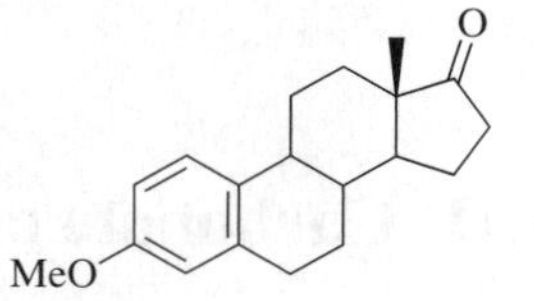	100
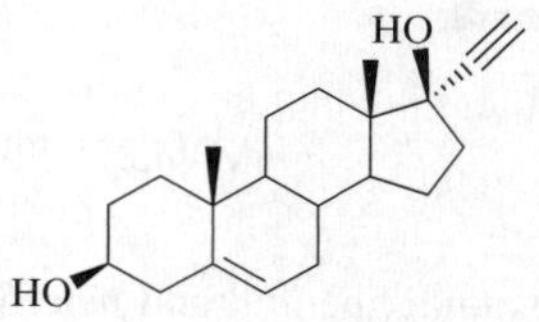	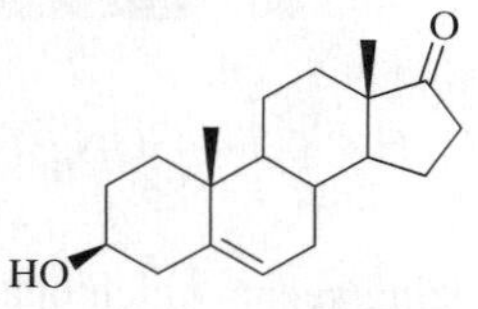	100
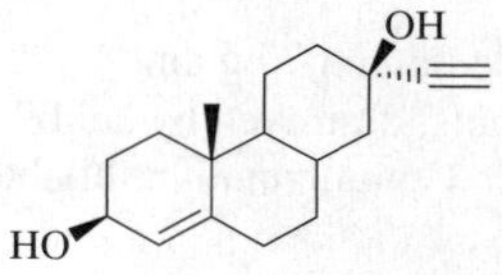	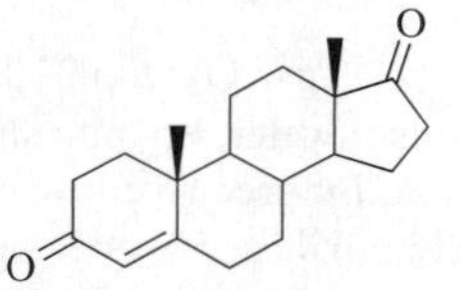	100
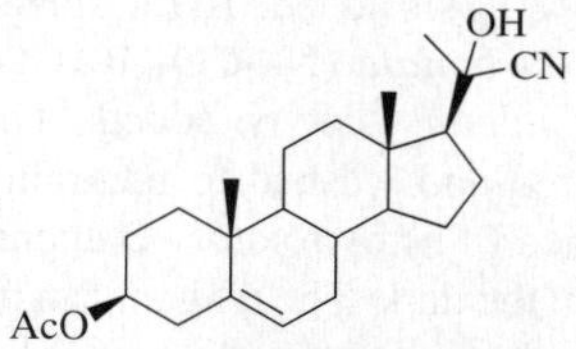	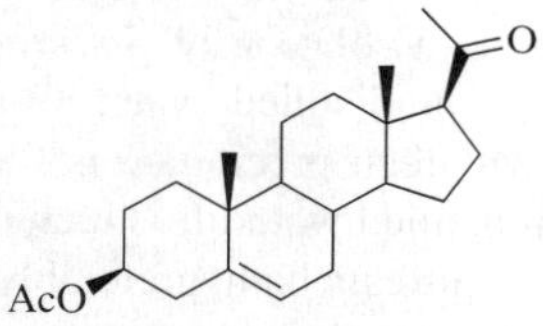	100

Oxidation of Diols. Depending upon their structure and stereochemistry, and the experimental conditions (reaction time) chosen, 1,2-diols are either cleaved (to dialdehydes or keto aldehydes) or oxidized to α-hydroxy ketones. In some cases, α-diketones have been obtained in reasonable yield (Table 5).[17–23]

Oxidation of 1,3-diols with silver carbonate on Celite leads to β-hydroxy ketones. Retroaldolization has occasionally been observed.[29] In benzene as solvent, a secondary hydroxyl group seems to be oxidized somewhat faster than a primary one. Aliphatic 1,4-, 1,5-, and 1,6-diols behave in the same manner, unless one of the hydroxyl groups is primary. Many diols of this type, in which one of the functions is primary, have been oxidized to lactones, generally in good yield (Table 6).[24–28]

Quite often, oxidations of polycyclic polyols with silver carbonate on Celite are highly regioselective. A tentative mechanism has been proposed which explains most of the results obtained so

Table 5 Oxidation of 1,2-Diols

Diol	Product	Solvent		Yield (%)
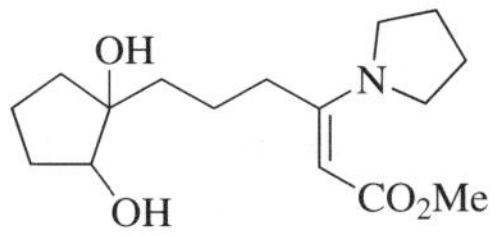	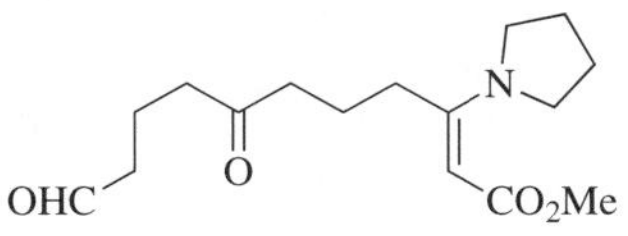	Benzene		65–75
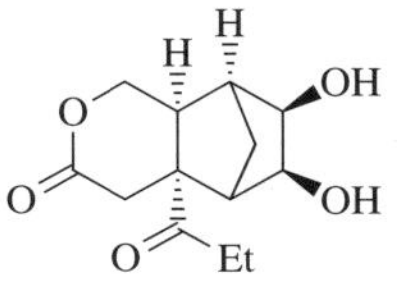	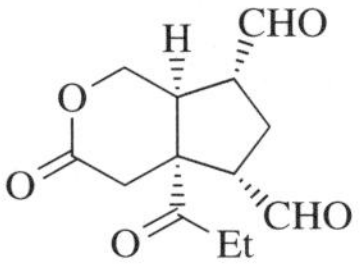	Benzene		major
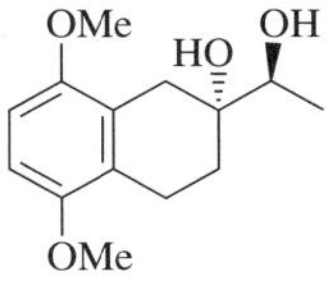	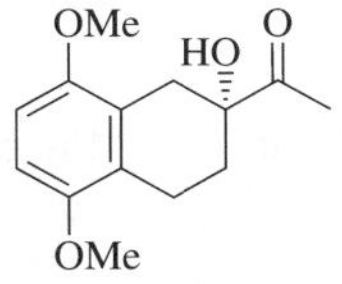	Benzene		90
	+	Benzene	$R^1 = H, R^2 = OH$	54
			$R^1 = OH, R^2 = H$	27
			$R^1 = H, R^2 = OH$	11
			$R^1 = OH, R^2 = H$	63
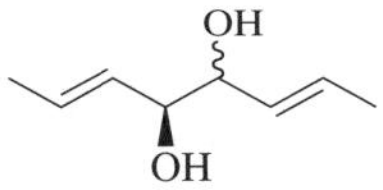	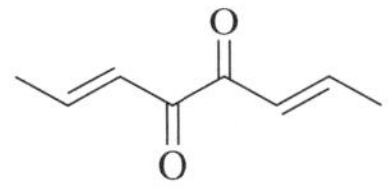	Benzene		
		Benzene		<33
		Toluene		major
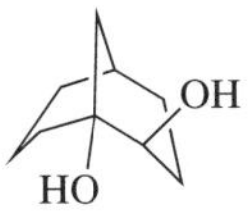	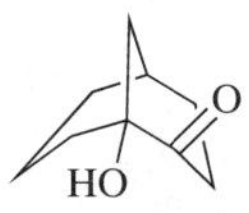	Acetone		53

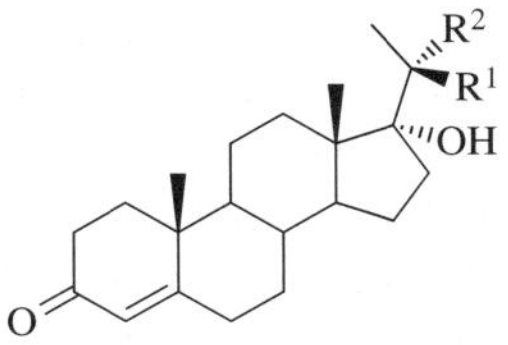

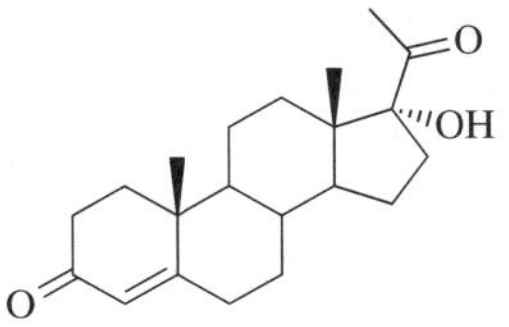

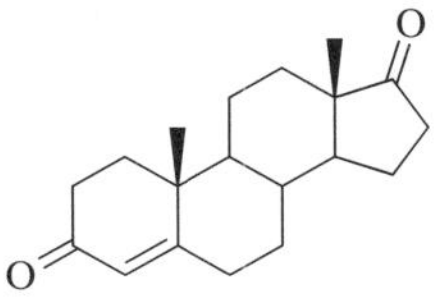

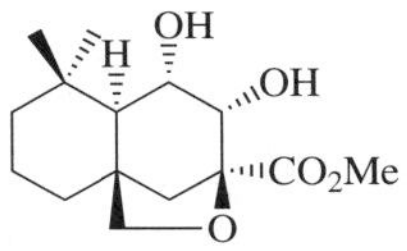

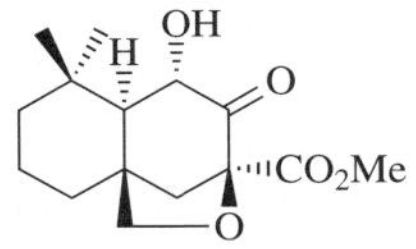

far[3,30] and has therefore reasonable predictive power for the synthetic organic chemist. A few examples are listed in Table 7.[1,31–34]

Oxidation of Lactols to Lactones. Silver carbonate oxidation of lactols is faster than the oxidation of any other type of hydroxyl derivative, and can therefore be achieved without protection of alcohols in the same molecule. Some representative examples are shown in Table 8.[35–37]

Oxidation and Degradation of Carbohydrates. A considerable amount of work has been devoted to the oxidation and degradation of carbohydrates using silver carbonate on Celite. If the hydroxyl at C-1 is free, and the reaction is carried out in benzene, toluene, or a mixture of benzene and DMF, then the corresponding lactone is obtained. However, if there is an unprotected OH group at C-2, and especially in methanol, a cleavage between C-1 and C-2 takes place as illustrated in eq 4.[38] Hydrolysis of the intermediate formate leads to D-threose.

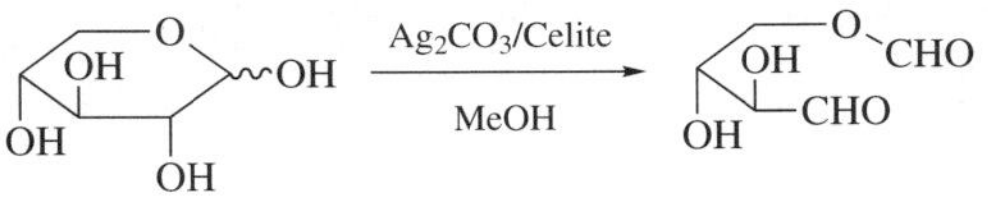

$$\xrightarrow[\text{MeOH}]{Ag_2CO_3/\text{Celite}} \quad (4)$$

Methyl Esters from Aldehydes. With the exception of α-substituted aldehydes of the type reported in Table 2, the oxidation of aldehydes in benzene does not lead to carboxylic acid derivatives. Those which can spontaneously give hemiacetals with another hydroxyl group of the molecule are not an exception, since lactols are very rapidly oxidized to lactones. It is therefore noteworthy that, in methanol, some aldehydes can be converted to methyl esters in modest yield (Table 9).[39,40]

Table 6 Oxidation of Nonvicinal Diols[a]

Diol	Product	Yield (%)
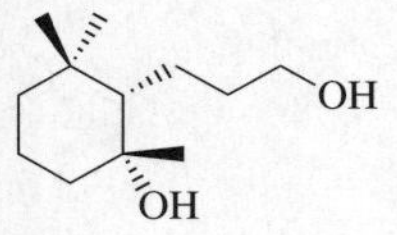	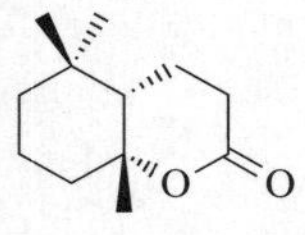	86
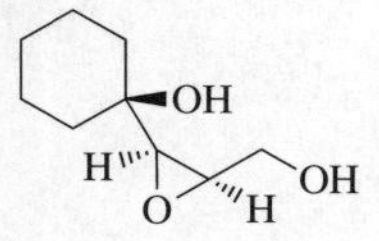	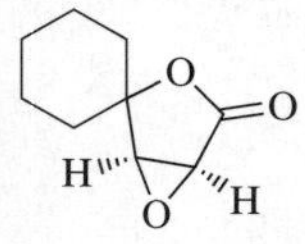	70
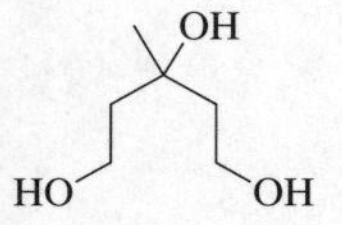	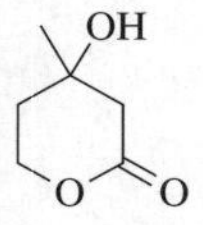	74
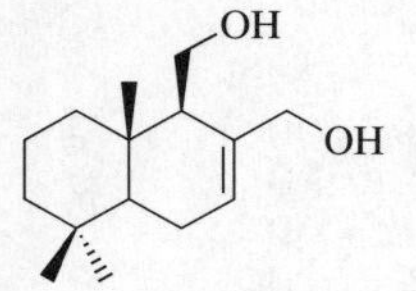	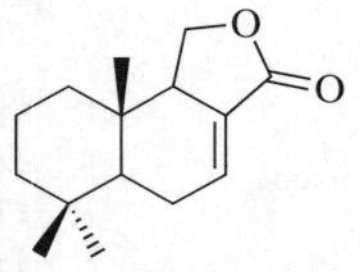	100
TBDMSO	TBDMSO	86

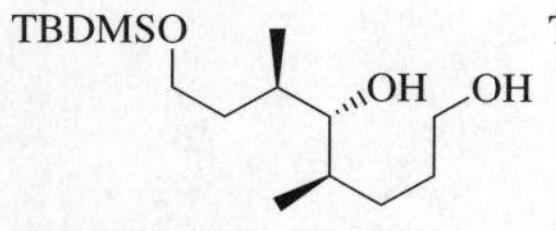

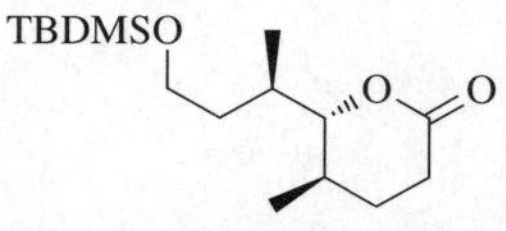

[a] In benzene.

Table 8 Selective Oxidation of Lactols[a]

Starting material	Product	Yield (%)
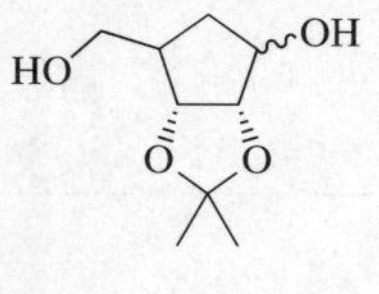	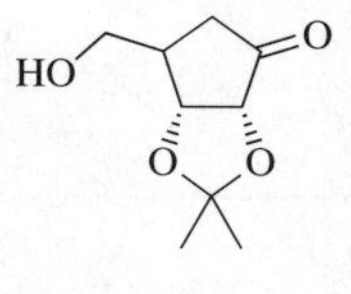	92
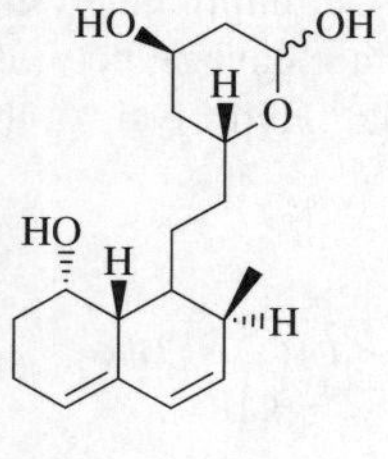	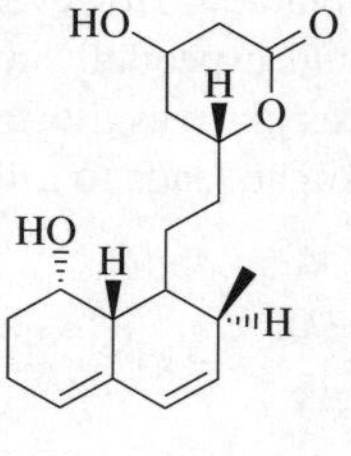	–
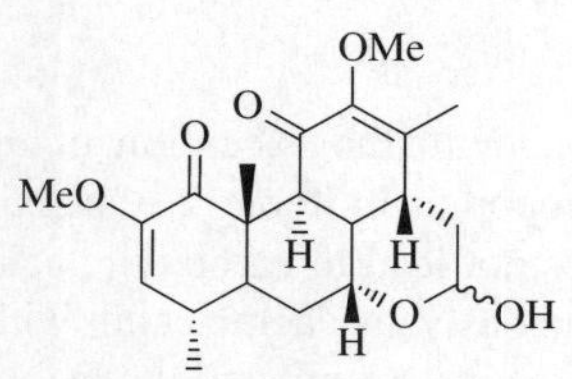	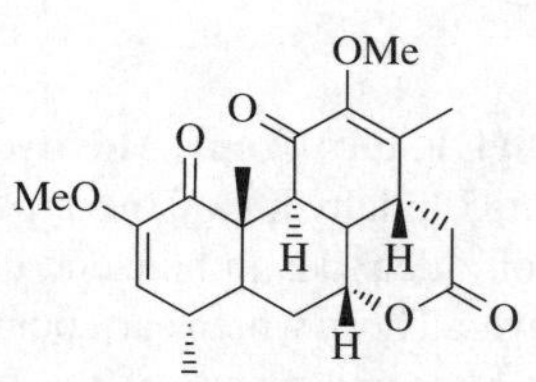	77

[a] In benzene.

Table 9 Oxidation of Aldehydes to Esters

Starting material	Product	Solvent	Yield (%)
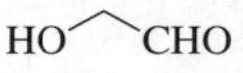 HO CHO	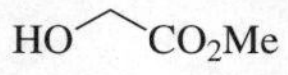 HO CO_2Me	MeOH	55
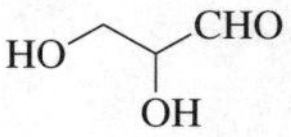 HO CHO OH	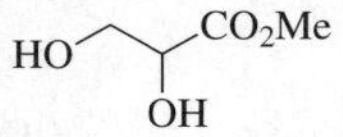 HO CO_2Me OH	MeOH	56
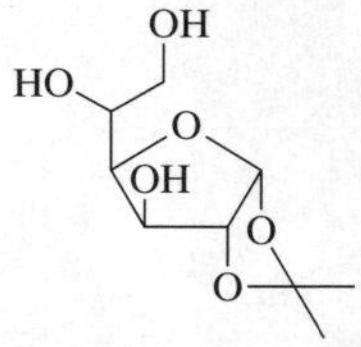	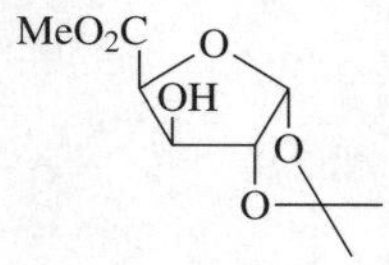 MeO_2C	MeOH	68

Oxidation of Phenols. Hydroquinones and catechols are oxidized to *p*-quinones and *o*-quinones, respectively, with this reagent.[41,42] 4,4′-Dihydroxybiphenyls and 4,4′-dihydroxystilbenes lead to diphenoquinones and stilbenequinones. Hindered phenols (e.g. 2,6-dimethylphenol, 2,4,6-trimethylphenol) are dimerized, and give diphenoquinones or stilbenequinones in high yield, as illustrated in eq 5. Unhindered phenols give complex mixtures. For instance, *p*-cresol affords a 65% yield of Pummerer's ketone (eq 6).[43] A systematic survey of silver carbonate on Celite oxidation of phenols has been published.[44]

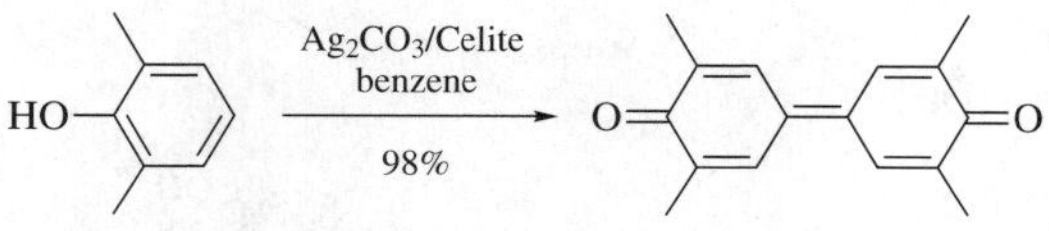

Ag_2CO_3/Celite, benzene, 98% (5)

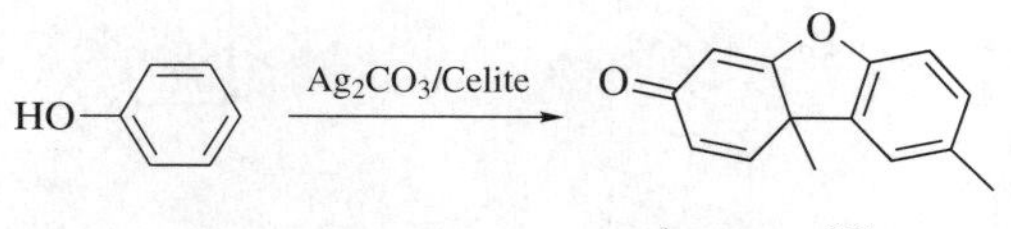

Ag_2CO_3/Celite (6)

benzene, 0%
acetonitrile, 65%

Oxidation of Amines. Very little is known so far on the oxidation of aliphatic amines. Anilines give azobenzenes in modest yield (Table 10).[3,45]

Oxidation of Hydrazines and Hydrazones. Symmetrically disubstituted hydrazines and hydrazides are rapidly oxidized to the corresponding azo compounds (Table 10). Hydrazones are converted in a few minutes into diazoalkanes, generally in high yield (Table 10).[45,46]

Nitroso Compounds from Hydroxylamines. *N*-Monosubstituted hydroxylamines lead to nitroso compounds, or their dimers, in good yield upon treatment with silver carbonate on Celite.[47] If a double bond is suitably placed, a cyclization of the nitroso intermediate gives a nitroxide free radical (Table 11).[48]

Nitrile Oxides from Oximes. Benzaldoximes react with silver carbonate on Celite to give nitrile oxides, which undergo 1,3-dipolar cycloaddition with the original oxime. The nitrile oxides

Table 7 Selective Oxidation of Polyhydroxy Compounds

Starting material	Product	Solvent	Yield (%)
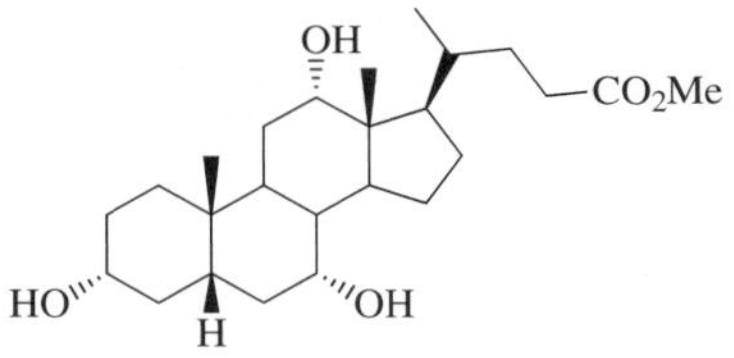	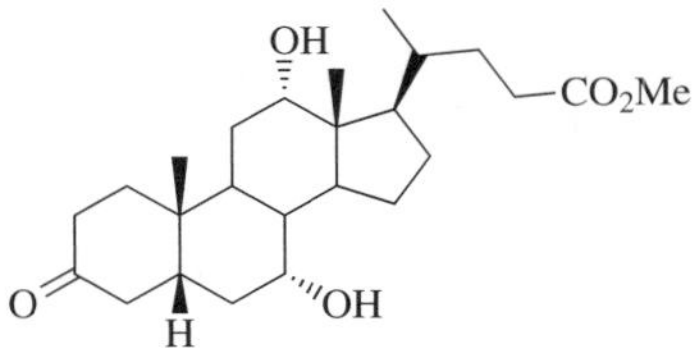	Benzene	90
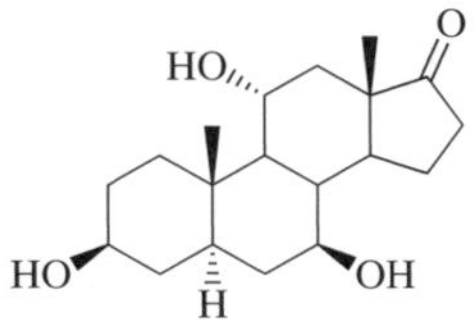	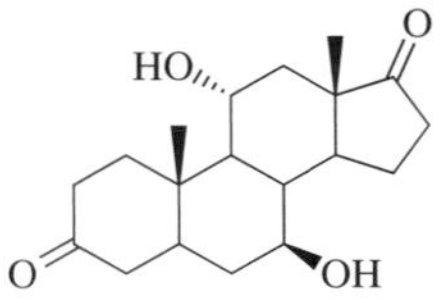	Toluene	72
	3-one, 6-β-OH	Benzene	29
	3-β-OH, 6-one		43
	3,6-dione		15
	3-one, 6-β-OH	Benzene, chloroform	68
	3-β-OH, 6-one		31
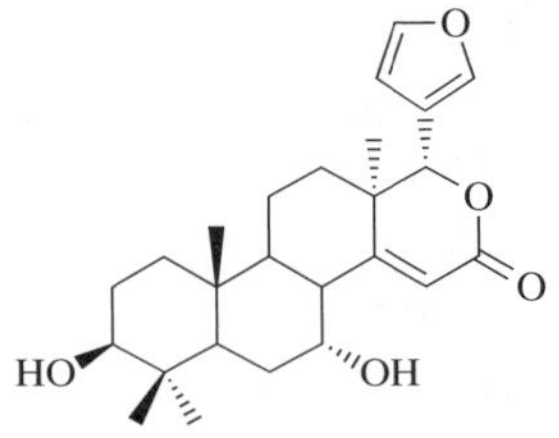	3-one	Benzene	57
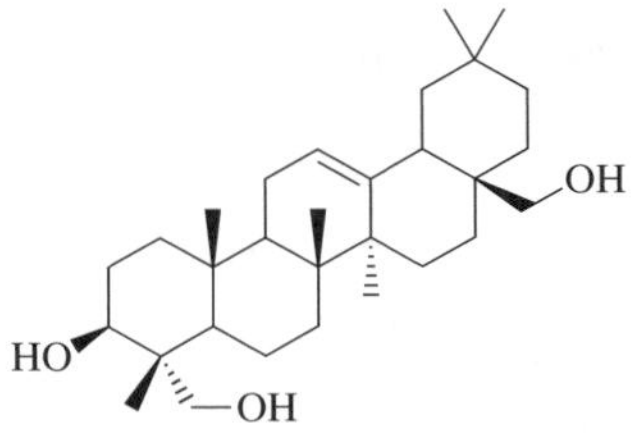	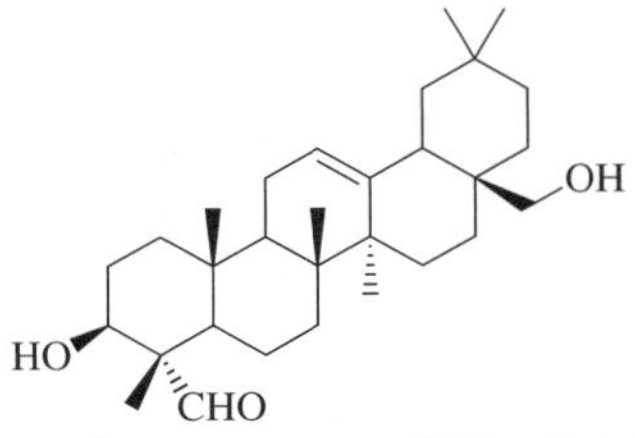	Benzene	90

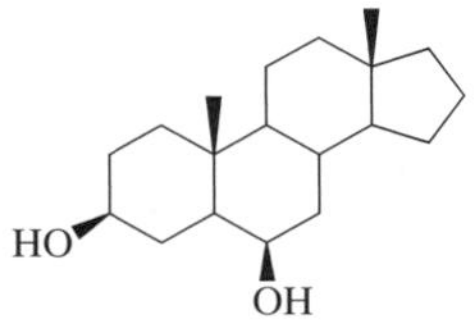

can also be trapped by other dipolarophiles such as nitriles and ethylenic compounds (Table 11).[46–48]

Miscellaneous Reactions. Halohydrins are smoothly converted by this reagent into epoxides or rearranged into aldehydes or ketones.[49,50] Silver carbonate on Celite has also been proposed to improve *O*-glycosylation[51] (Koenigs–Knorr reaction).

Related Reagents. See Classes O-1, O-3, O-10, and O-18, pages 1–10.

1. Fétizon, M.; Golfier, M. *CR(C)* **1968**, *267*, 900.
2. McKillop, A.; Young, D. W. *S* **1979**, 401.
3. Fétizon, M.; Golfier, M.; Mourgues, P.; Louis, J. M. *Organic Syntheses by Oxidation with Metal Compounds*; Plenum: New York, 1986; p 501.
4. Ohloff, G.; Pickenhagen, W. *HCA* **1969**, *52*, 880.
5. Snatzke, G.; Klein, H. *CB* **1972**, *105*, 244.
6. Schwab, J. M.; Chorng-Kei, H. *CC* **1986**, 872.
7. Fétizon, M.; Gomez-Parra, F.; Louis, J.-M. *JHC* **1976**, *13*, 525.
8. Dixon, A. J.; Taylor, R. J. K.; Newton, R. F. *JCS(P1)* **1981**, 1407.
9. Fétizon, M.; Henry, Y.; Moreau, N.; Moreau, G.; Golfier, M.; Prangé, T. *T* **1973**, *29*, 1011.
10. Gardner, D.; Glen, A. J.; Turner, W. B. *JCS(P1)* **1972**, 2576.
11. Dimitriadis, E.; Massy-Westropp R. A. *AJC* **1980**, *33*, 2729.
12. Audier, H.-E.; Bottin, J.; Fétizon, M.; Tabet, J.-C. *BSF(2)* **1971**, 2911.
13. Rapoport, H.; Reist, H. N. *JACS* **1955**, *77*, 490.
14. Geisel, M.; Grob, C. A.; Santi, W.; Tschudi, W. *HCA* **1973**, *56*, 1046.
15. Gilday, J. P.; Galluci, J. C.; Paquette, L. A. *JOC* **1989**, *54*, 1399.

Table 10 Oxidation of Amino Compounds[a]

Starting material	Product	Yield (%)
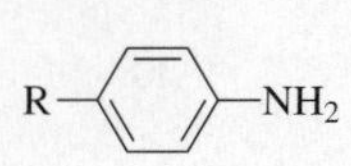 $R-C_6H_4-NH_2$	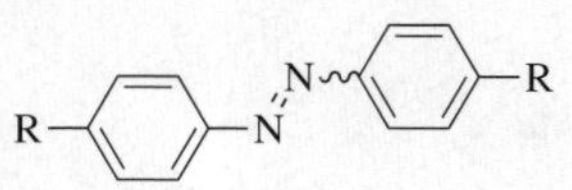 $R-C_6H_4-N=N-C_6H_4-R$	R = H 38 R = Cl 35 R = NO_2 5
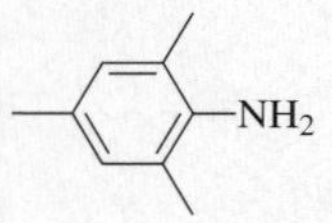 2,4,6-trimethylaniline (NH_2)	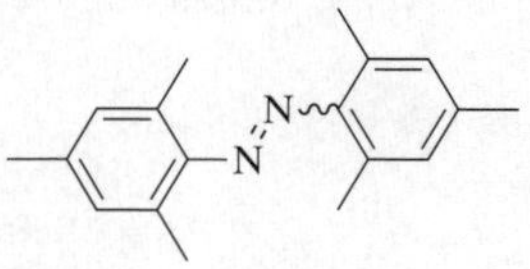 azo compound (N=N)	75
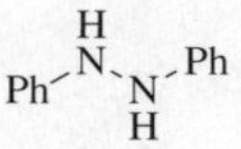 PhNH–NHPh	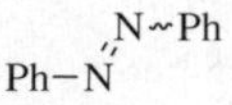 Ph–N=N–Ph	100
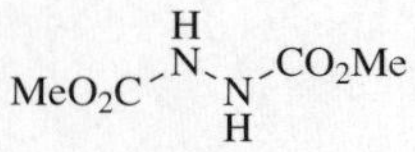 $MeO_2C-NH-NH-CO_2Me$	 $MeO_2C-N=N-CO_2Me$	80
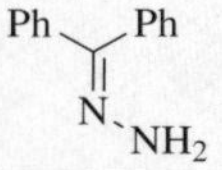 $Ph_2C=N-NH_2$	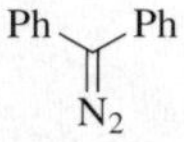 $Ph_2C=N_2$	88

[a] In benzene.

Table 11 Oxidation of Hydroxylamines and Oximes

Starting material	Product	Solvent	Yield (%)
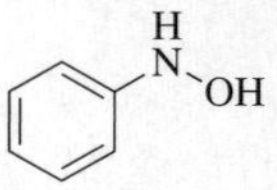 PhNHOH	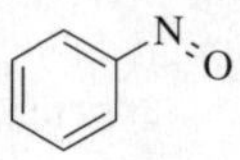 PhN=O	CH_2Cl_2	85
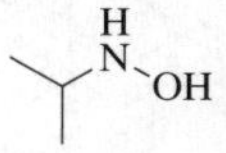 i-PrNHOH	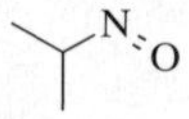 i-PrN=O	CH_2Cl_2	90
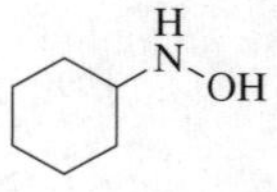 c-C_6H_{11}NHOH	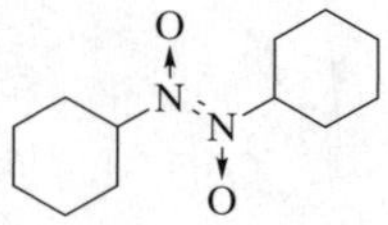 azodioxide (c-C_6H_{11}N(O)=N(O)c-C_6H_{11})	CH_2Cl_2	95
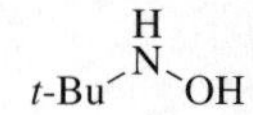 t-BuNHOH	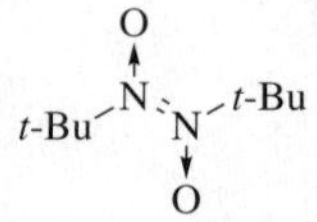 t-Bu–N(O)=N(O)–t-Bu	$CFCl_3$	57
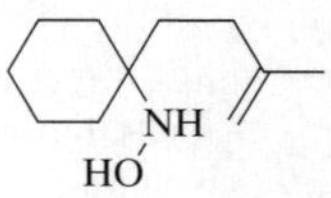 N-hydroxyamine (NH, HO)	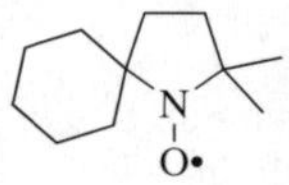 nitroxide (N–O•)		30
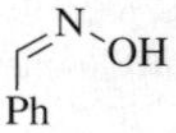 PhCH=N–OH	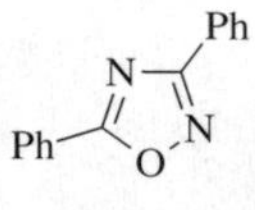 3,5-diphenyl-1,2,4-oxadiazole	Benzene	73
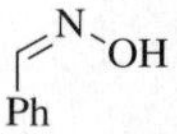 PhCH=N–OH	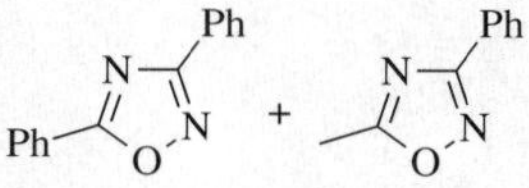 3,5-diphenyl-1,2,4-oxadiazole (30%) + 5-methyl-3-phenyl-1,2,4-oxadiazole (41–61%)	MeCN	–
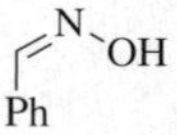 PhCH=N–OH	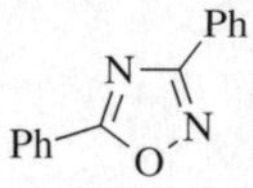 3,5-diphenyl-1,2,4-oxadiazole (30%), PhCHO (11%), 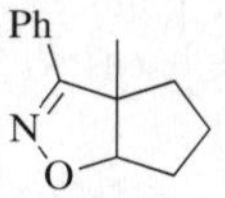 isoxazoline (38%)	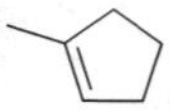 1-methylcyclopentene	–

16. Lenz, G. R. *CC* **1972**, 468.
17. Corey, E. J.; Ueda, Y.; Rudew, R. A. *TL* **1975**, 4347.
18. Danishefsky, S.; Hirama, M.; Gombatz, K.; Harayama, T.; Berman, E.; Shuda, P. F. *JACS* **1979**, *101*, 7020.
19. Tanno, N.; Terashima, S. *CPB* **1983**, *31*, 811.
20. Bastard, J.; Fétizon, M.; Gramain, J. C. *T* **1973**, *29*, 2867.
21. Russel, G. A.; Ballenegger, M.; Malkus, H. L. *JACS* **1975**, *97*, 1900.
22. Dailey, O. D. Jr.; Fuchs, P. L. *JOC* **1980**, *45*, 216.
23. Takeuchi, K.; Ikai, K.; Yoshida, M.; Tsugeno, A. *T* **1988**, *44*, 5681.
24. Fétizon, M.; Golfier, M.; Louis, J.-M. *T* **1975**, *31*, 171.
25. Buchecker, R.; Egli, R.; Regel-Wild, H.; Tscharner, C.; Eugster, C. H.; Uhde, G.; Ohloff, G. *HCA* **1973**, *56*, 2548.
26. Boeckman, R. K.; Thomas, E. W. *TL* **1976**, 4045.
27. Hollinshead, D. M.; Howell, S. C.; Ley, S. V.; Mahon, M.; Ratcliffe, N. M.; Worthington, P. A. *JCS(P1)* **1983**, 1579.
28. Morgans, D. J., Jr. *TL* **1981**, *22*, 3721.
29. Dimitriadis, E.; Massy-Westropp, R. A. *P* **1984**, *23*, 1325.
30. Fétizon, M.; Mourgues, P. *T* **1974**, *30*, 327.
31. Bell, A. M.; Chambers, V. E. M.; Jones, E. R. H.; Meakins, G. D.; Müller, W. E.; Pragnell, J. *JCS(P1)* **1974**, 312.
32. Jones, E. R. H.; Meakins, G. D.; Pragnell, J.; Müller, W. E.; Wilkins, A. L. *JCS(P1)* **1974**, 2376.
33. Ekong, D. E. U.; Okogun, J. I.; Sondengam, B. L. *JCS(P1)* **1975**, 2118.
34. Kikuchi, T.; Niwa, M.; Yokoi, T. *CPB* **1973**, *21*, 1378.
35. Morgenlie, S. *ACS* **1972**, *26*, 2518.
36. Danishefsky, S. J.; Simoneau, B. *JACS* **1989**, *111*, 2599.
37. Grieco, P. A.; Ferrino, S.; Vidari, G. *JACS* **1980**, *102*, 7586.
38. Morgenlie, S. *ACS* **1972**, *26*, 1709; **1973**, *27*, 2607.
39. Morgenlie, S. *ACS* **1973**, *27*, 3009.
40. Morgenlie, S. *Carbohydr. Res.* **1977**, *59*, 73.
41. Balogh, V.; Fétizon, M.; Golfier, M. *JOC* **1971**, *36*, 1339.
42. Grundmann, C. *MOC* **1979**, *7/3b*, 43.
43. Anderson, R. A.; Dalgleish, D. T.; Nonhebel, D. C.; Pauson, P. L. *JCR(S)* **1977**, 12.
44. Anderson, R. A.; Nonhebel, D. C.; Pauson, P. L. *JCR(S)* **1977**, 15.
45. Hedayatullah, M.; Dechatre, J. P.; Denivelle, J. P. *TL* **1975**, 2039.
46. Fétizon, M.; Golfier, M.; Milcent, R.; Papadakis, I. *T* **1975**, *31*, 165.
47. Maassen, J. A.; De Boer, Th. J. *RTC* **1971**, *90*, 373.
48. Motherwell, W. B.; Roberts, J. S. *CC* **1972**, 328.
49. Fétizon, M.; Golfier, M.; Montaufier, M. T.; Rens, J. *T* **1975**, *31*, 987.
50. Fétizon, M.; Golfier, M.; Louis, J. M. *TL* **1973**, 1931.
51. Hartenstein, J.; Satzinger, G. *LA* **1974**, 1763.
52. Laszlo, P. *COS* **1991**, *7*, 841.

Marcel Fétizon

Institut de Chimie des Substances Naturelles, Gif-sur-Yvette, France

Silver(I) Nitrate

$AgNO_3$

[7761-88-8] $AgNO_3$ (MW 169.88)

(mild oxidizing agent and Lewis acid used in a wide variety of chemical reactions)

Physical Data: mp 212 °C; *d* 4.352 g cm^{-3}.
Solubility: sol H_2O, MeCN, alcohol, acetonitrile, DMF.
Form Supplied in: white crystalline solid; widely available.
Handling, Storage, and Precautions: mild oxidizing agent; may be fatal if inhaled or swallowed.

Deprotection. Silver nitrate is used as a deprotecting agent for a number of functional groups. Thus *S*-trityl ethers are readily cleaved ($AgNO_3$/***Pyridine***, 5 min, rt)[1a] to the thiol silver salts which are converted to the thiols on treatment with ***Hydrogen Sulfide***. The method has been applied to *S*-trityl ethers in nucleoside,[1a] peptide,[1b] and β-lactam[1c] chemistry, although it is not always successful.[2] Selective cleavage of an *S,S*-diaryl phosphorodithioate in the presence of an *O*-dimethoxytrityl group occurs using a large excess of $AgNO_3$ (eq 1).[3] Other deprotecting agents such as $NaIO_4$, H_2O_2, and NCS are unsuccessful in this process.[3]

MMTrO–(nucleoside, Th)–O–P(=O)(SAr)$_2$ —[$AgNO_3$, rt, 20 h; 100%]→ MMTrO–(nucleoside, Th)–O–P(=O)(OH)$_2$ (1)

Thioacetals,[4] 1,3-dithianes,[5] and 1,3-oxathianes[6] are converted, on treatment with $AgNO_3$/***N-Chlorosuccinimide*** (or ***N-Bromosuccinimide***), to the corresponding ketones. The method is specially useful when sensitive 1,4-unsaturated diones are being liberated.[5b] Cyanuric acid can be used in place of NCS for the particularly difficult task of converting a diphenyl thioacetal to its ketone derivative.[7] Benzothiazoles are readily converted, via their 2-lithio derivatives, to a variety of intermediates.[8] Treatment of these derivatized benzothiazoles with ***Iodomethane*** followed by ***Sodium Borohydride*** gives the corresponding *N*-methylbenzothiazolines, which are readily cleaved with $AgNO_3$ in methanol to the corresponding aldehydes in high yields. This is a useful high yielding procedure for the preparation of α,β-unsaturated aldehydes.[8] Cleavage of *O*-methylethoxymethoxy ethers (OMEM ethers) in the presence of a dithiane is accomplished by converting the OMEM ether to the *O*-isopropylthiomethyl ether, which is then cleaved to the desired alcohol with $AgNO_3$ in the presence of ***2,6-Lutidine***.[9] $AgNO_3$ has also been used for the hydrolysis of enol thioethers,[10] tetrahydropyranylthio ethers,[11] *S-t*-butyl esters,[12] and thiobenzoates,[12] to the respective alcohols, for the conversion of dithiobenzoates to *S*-benzoates,[13] thioamides to amides,[14] thioureas to ureas,[15] and for the conversion of orthothioesters to orthoesters.[16]

Protection. Reaction rates for the conversion of the primary OH groups of nucleosides to the protected *O*-trityl ethers with 4,4′,4″-tris(4,5-dichlorophthalimido)trityl bromide are considerably enhanced by the addition of $AgNO_3$.[17] $AgNO_3$ is also used for the selective preparation of primary *t*-butyldimethylsilyl ethers by the reaction of nucleosides with ***t-Butyldimethylchlorosilane*** in MeCN.[18] Yields in both reactions are high. $AgNO_3$ is an effective agent for the conversion of several glucopyranoses to their 1,2-orthoacetates[19] and as an agent for the formation of glycosides[20] (see also ***Silver(I) Nitrite***).

Ring Expansions, Contractions, and Rearrangements. $AgNO_3$-promoted rearrangements occur in numerous instances; a few examples are given. Thus the 4-cyano-4,5-dihydroazepine in eq 2 gives the furo[2,3-*b*]pyridine on heating with an aqueous solution of $AgNO_3$.[21] Treatment of the *N*-chloroenolamine in eq 3 with a threefold excess of $AgNO_3$ gives the β-lactam; yields are about 50%, depending on the substituent used.[22] The 1,1′-bishomocubane in eq 4, on treatment with $AgNO_3$ in aqueous methanol, rearranges to the pentacycle in quantitative yield.[23]

(dihydroazepine: EtO_2C, CN, CO_2Et, NH) —[aq. $AgNO_3$; 53%]→ (furo[2,3-*b*]pyridine: EtO_2C, CO_2Et) (2)

R–N(Cl)–C(cyclopropyl)–OH —[$AgNO_3$, MeCN, 0 °C; ~50%]→ R–N(β-lactam)=O (3)

(bishomocubane: CO_2Me, CO_2Me) —[$AgNO_3$, MeOH, rt]→ (pentacycle: CO_2Me, CO_2Me) (4)

Dehalogenations. $AgNO_3$ has been widely used for solvolysis with concomitant rearrangement of dihalocyclopropanes. Thus the trichlorocyclopropa[*c*]chromene in eq 5 gives the (dichloromethylene)chroman-4-one in 93% yield.[24] Many other examples are known. Dehydrobromination of the bromotetralone in eq 6 with $AgNO_3$ gives exclusively the endocyclic unsaturated ketone in 90% yield; other reagents for this reaction give mixtures of the exo- and endocyclic alkenes.[25]

(trichlorocyclopropa[*c*]chromene: Cl, Cl, Cl) —[$AgNO_3$, aq. MeCN]→ (dichloromethylene chroman-4-one: O, Cl, Cl) (5)

(bromotetralone: Br, Ph, O) —[$AgNO_3$, EtOH, rt; 90%]→ (unsaturated ketone: Ph, O) (6)

Solvolysis of arylmethyl halides with $AgNO_3$ in hot aqueous ethanol gives the corresponding alcohols,[26] whereas a similar reaction on alkyl bromides in MeCN is reported to be an excellent procedure for the preparation of pure nitrate esters.[27] Benzyl dibromides are converted to the corresponding aldehydes in high yields,[28] and α-bromo ketones give high yields (>80%) of the corresponding α-diketones on treatment with $AgNO_3$ in MeCN.[29]

Cycloadditions. The electrophilic nature of $AgNO_3$ enables it to complex with alkynes and allenes, thus enabling intramolecular cycloadditions to proceed. For example, treatment of a series of phenolic keto-ynes (eq 7) with a catalytic amount of $AgNO_3$ in methanol promotes cycloaddition to the triple bond in high yield.[30] Similarly, α-hydroxyallenes are rapidly converted to 2,5-dihydropyrans on treatment with $AgNO_3/CaCO_3$ in aq acetone (eq 8).[31]

$AgNO_3$ (cat.), MeOH, rt, 90% (7)

$AgNO_3$, $CaCO_3$, aq. Me_2CO, rt, 84% (8)

On extending the hydroxy alkyl chain from the allenic group by two or three carbon atoms, cyclization results in the formation of the corresponding α-vinyltetrahydrofuran[32] or α-vinyltetrahydropyran,[33] respectively, both in high yields. In a similar manner, treatment of α-, γ- and δ-aminoallenes with $AgNO_3$ in aqueous acetone gives the corresponding 2,3-dihydropyrroles,[34] α-vinylpyrrolidines,[35] and α-vinylpiperidines,[35] respectively, in good yields.

Oxidative Couplings. Alkylboranes, formed on reaction of terminal alkenes with diborane, dimerize on treatment with an aqueous solution of $AgNO_3$/NaOH. Thus hexene gives a 66% yield of dodecane,[36] and the dienes 1,5-hexadiene and geranyl acetate are converted to cyclohexane (66%) and *trans-p*-menthane (85%).[37] Although the use of $AgNO_3$/NaOH for these coupling reactions is, to all accounts, equivalent to the use of ***Silver(I) Oxide***, oxidative cyclization is reported to be less successful when this latter reagent is used. Oxidative dimerization of the dianions of α,β-unsaturated carboxylic acids with $AgNO_3$/THF gives moderate yields of the dienedioic acids, but the use of ***Iodine*** for this coupling gives higher yields (40–80%).[38]

Coverage of the use of $AgNO_3$/NaOH for oxidation of alcohols, aldehydes, etc., has not been included in this section as this ostensively amounts to the use of ***Silver(I) Oxide***.

Related Reagents. Zinc–Copper(II) Acetate–Silver Nitrate.

1. (a) Divakar, K. J.; Mottoh, A.; Reese, C. B.; Sanghvi, Y. S. *JCS(P1)* **1990**, 969. (b) Zervas, L.; Photaki, I. *JACS* **1962**, *84*, 3887. (c) Girijauallabham, V. M.; Ganguly, A. K.; Pinto, P.; Versace, R. *CC* **1983**, 908.
2. Hiskey, R. G.; Harpold, M. A. *JOC* **1968**, *33*, 559.
3. Sekine, M.; Hamaoki, K.; Hata, T. *JOC* **1979**, *44*, 2325.
4. Geiss, K.; Sevring, B.; Pieter, R.; Seebach, D. *AG(E)* **1974**, *13*, 479.
5. (a) Corey, E. J.; Erickson, B. W. *JOC* **1971**, *36*, 3553. (b) Corey, E. J.; Grouse, D. *JOC* **1968**, *33*, 298.
6. Frye, S. V.; Eliel, E. L. *TL* **1985**, *26*, 3907.
7. Cohen, T.; Nolan, S. M. *TL* **1978**, 3533.
8. Corey, E. J.; Boger, D. L. *TL* **1978**, 5; **1978**, 13.
9. Corey, E. J.; Weigel, L. O.; Chamberlin, R.; Cho, H.; Hua, D. H. *JACS* **1980**, *102*, 6613.
10. Kejian, C.; Sanner, M. A.; Carlson, R. M. *SC* **1990**, *20*, 901.
11. Kruse, C. G.; Poels, E. K.; Jonkers, F. L.; van der Gem, A. *JOC* **1978**, *43*, 3548.
12. Shenvi, A. B.; Gerlach, H. *HCA* **1980**, *63*, 2426.
13. Hedgley, E. J.; Leon, N. H. *JCS(C)* **1970**, 467.
14. Barrett, C. G. *JCS* **1965**, 2825.
15. Mechoulam, R.; Sondheimer, F.; Melera, A. *JACS* **1961**, *83*, 2022.
16. Breslow, R.; Pandey, P. S. *JOC* **1980**, *45*, 740.
17. (a) Sekine, M.; Hata, T. *JACS* **1984**, *106*, 5763. (b) Sekine, M.; Hata, T. *JACS* **1986**, *108*, 4586.
18. Hakimelahi, G. H.; Proba, Z. A.; Ogilvie, K. K. *TL* **1981**, *22*, 4775.
19. Tsui, D. S. K.; Gorin, P. A. J. *Carbohydr. Res.* **1985**, *144*, 137.
20. Nashed, E. M.; Glaudemans, C. P. J. *JOC* **1987**, *52*, 5255.
21. Bullock, E.; Gregory, B.; Johnson, A. W. *JCS* **1964**, 1632.
22. Wasserman, H. H.; Adickes, H. W.; de Ochoa, O. E. *JACS* **1971**, *93*, 5586.
23. Dauben, W. G.; Buzzolini, M. G.; Schallhorm, C. H.; Whalen, D. L.; Palmer, K, J. *TL* **1970**, 787.
24. Brown, P. E.; Islam, Q. *TL* **1987**, *28*, 3047.
25. Cromwell, N. H.; Ayer, R. P.; Foster, P. W. *JACS* **1960**, *82*, 130.
26. Aldous, D. L.; Riebsomer, J. L.; Castle, R. N. *JOC* **1960**, *25*, 1151.
27. Ferris, A. F.; McLean, K. W.; Marks, I. G.; Emmons, W. D. *JACS* **1953**, *75*, 4078.
28. Buggy, T.; Ellis, G. P. *JCR(S)* **1980**, 159.
29. Kornblum, N.; Frazier, H. W. *JACS* **1966**, *88*, 865.
30. Jong, T-T.; Leu, S-J. *JCS(P1)* **1990**, 423.
31. Marshall, J. A.; Wang, X. J. *JOC* **1990**, *55*, 2995.
32. Gore, J.; Audin, P.; Dootheau, A.; Ruest. L. *BSF* **1981**, 313.
33. Gallagher, T. *CC* **1984**, 1554.
34. Arseniyadis, S.; Gore, J. *TL* **1983**, *24*, 3997.
35. Arseniyadis, S.; Sartoretti, J. *TL* **1985**, *26*, 729.
36. Brown, H. C.; Hebert, N. C.; Snyder, C. H. *JACS* **1961**, *83*, 1002.
37. Murphy, R.; Prager, R. H. *TL* **1976**, 463.
38. Aurell, M. J.; Gil, A.; Tortajada, A.; Mestres, R. *S* **1990**, 317.

Duncan R. Rae
Organon Laboratories, Motherwell, UK

Silver(I) Oxide

[20667-12-3] Ag_2O (MW 231.74)

(oxidizing reagent for conversion of hydroquinones to quinones,[1] alkylphenols to quinone methides,[5] and aldehydes to acids;[7] oxidative coupling reactions;[12,13] a Lewis acid with halides[18,20] and thioethers;[24] Wolff rearrangement[26])

Physical Data: dec at about 200 °C; *d* 7.22 g cm^{-3}.
Solubility: practically insol alcohol; sol in 40 000 parts H_2O; sol dilute nitric acid, ammonia; moderately sol NaOH.
Form Supplied in: brownish-black, heavy, odorless powder; widely available.

Handling, Storage, and Precautions: protect from light. Reduced by hydrogen, carbon monoxide, and most metals. Skin contact with this toxic reagent should be carefully avoided.

Hydroquinone Oxidation. This reagent is a powerful oxidizing reagent for the conversion of hydroquinones to the corresponding quinones. 1,4-Hydroquinones are generally converted to quinones by treatment with Ag_2O (eq 1).[1] Other reagents (e.g. ***Cerium(IV) Ammonium Nitrate***) are less expensive and easier to handle. However, Ag_2O in nonhydroxylic solvents can be used to produce electron-deficient quinones which are unstable to nucleophiles. For example, Ag_2O oxidation of nitro hydroquinones affords nitro quinones (eq 2).[2]

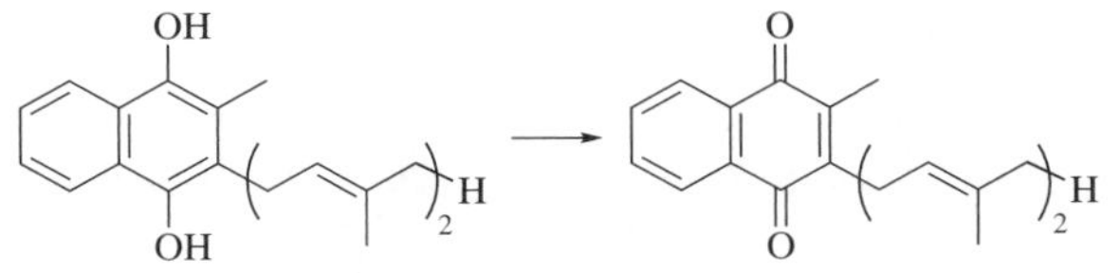

(1)

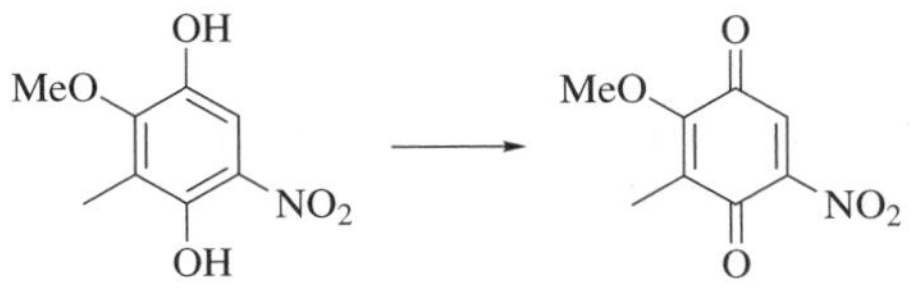

(2)

A one-pot technique has been developed to oxidize 1,4-hydroquinones and utilize the quinone products as dienophiles in Diels–Alder reactions (eq 3).[3] Tetrahydrophenanthrene-9,10-quinones have also been prepared by a one-pot double Diels–Alder reaction (eq 4).[4]

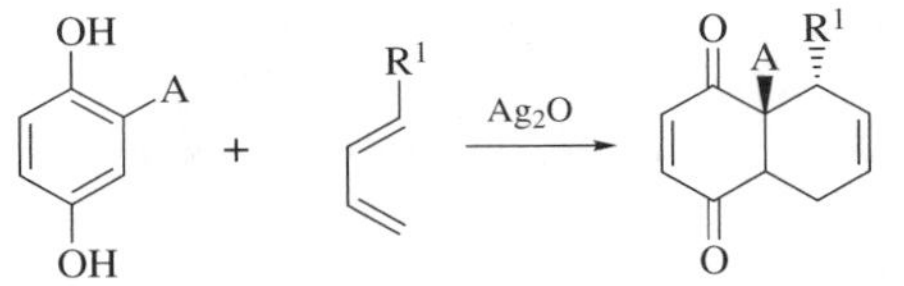

(3)

A = CO_2Me or CHO
R^1 = OTMS, CH_2CO_2Et

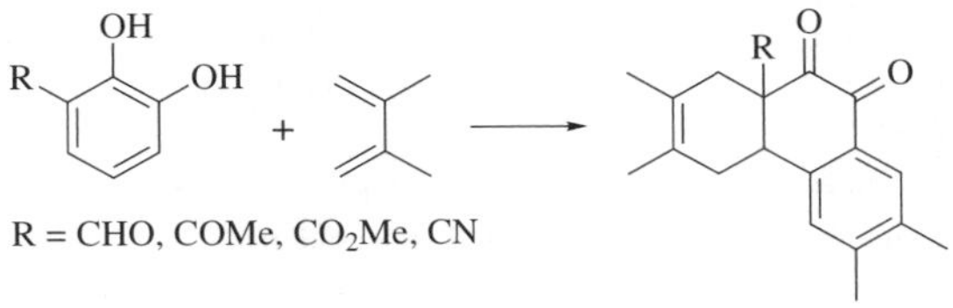

(4)

R = CHO, COMe, CO_2Me, CN

Alkylphenol Oxidation. 1,4-Quinone methides can be obtained by the reaction of *p*-alkylphenols with Ag_2O. These reactive compounds may undergo subsequent transformations, for example, a Lewis acid promoted cyclization (eq 5).[5] 1,2-Quinone methides can also be formed by Ag_2O oxidation of appropriately substituted phenols (eqs 6 and 7).[6]

Other Oxidation Reactions. Ag_2O oxidizes aldehydes to acids,[7] and conjugated keto aldehydes to conjugated keto acids.[8] For example, Ag_2O was used to selectively oxidize one of two formyl substituents of an intermediate in the total synthesis of inhibitor K-76 (eq 8).[9] The unreactive formyl substituent is conjugated with the hydroxyl and alkoxy substituents.

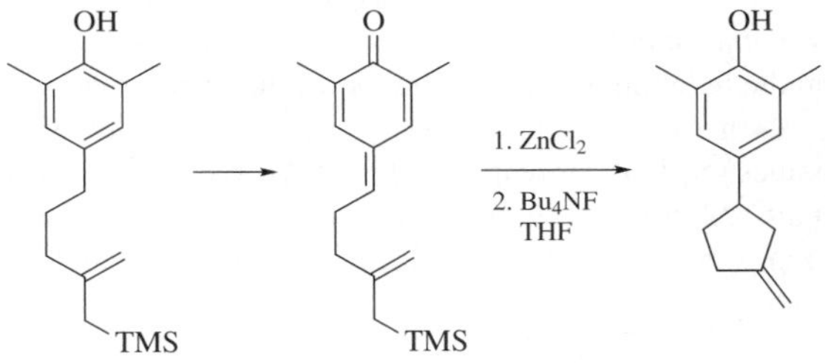

(5)

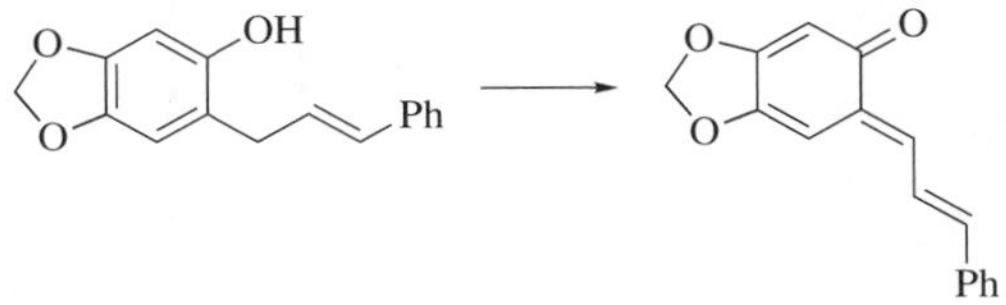

1. $ZnCl_2$
2. Bu_4NF THF

(6)

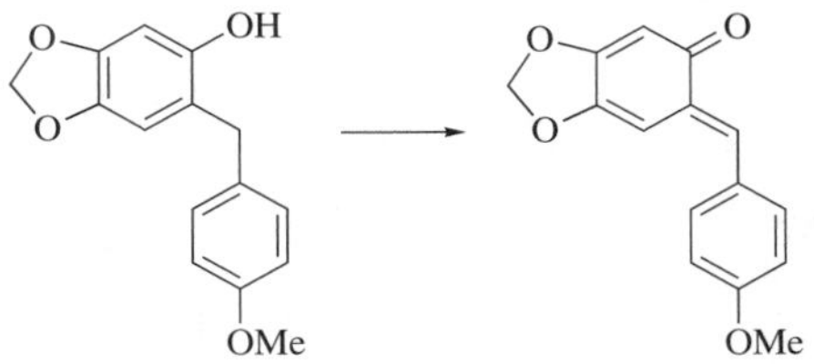

(7)

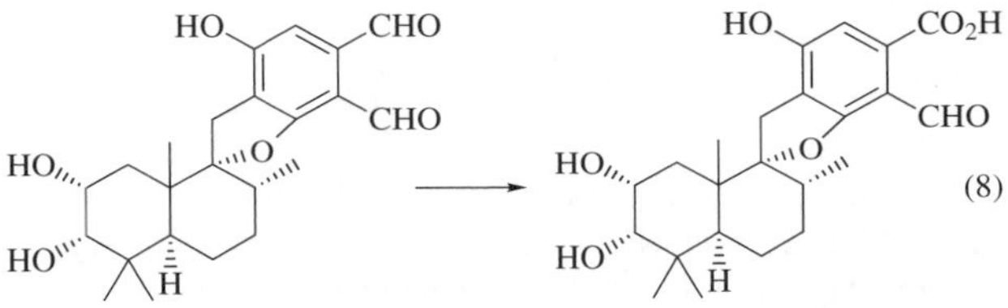

(8)

Oxidative Coupling Reactions. Radicals produced by Ag_2O oxidation can undergo coupling reactions. For example, 2-(vinyloxy)phenols undergo Ag_2O oxidation-induced coupling.[10] Acylacetates and monosubstituted malonates are oxidatively dimerized in the presence of Ag_2O and DMSO (eq 9).[11]

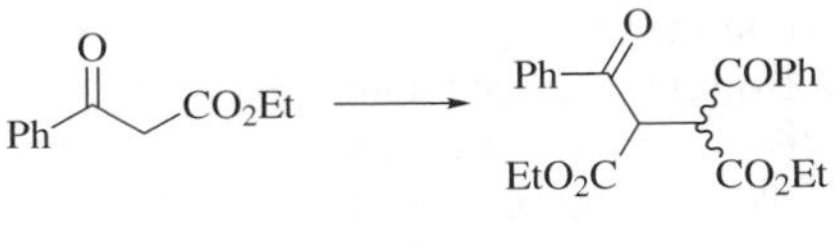

(9)

Radical coupling followed by nucleophilic attack of hydroxyl on a quinone methide intermediate is postulated as the mechanism of the key step in the syntheses of silybin and eusiderin (eq 10).[12] 1,4-Diketones are produced in the reaction of silyl enol ethers with Ag_2O in DMSO (eq 11).[13]

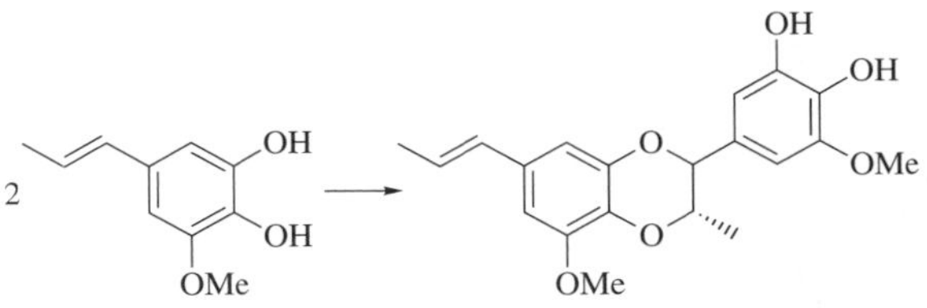

(10)

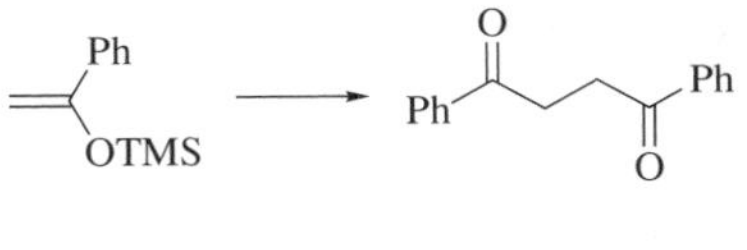

(11)

Lewis acid with Halides. Ag_2O can activate halide as a leaving group by coordination. Methylation of carbohydrates[14] and 5-benzylidenebarbituric acid[15] have been carried out with Ag_2O and ***Iodomethane*** in DMF. Triphenyltin trifluoroacetate was obtained from the reaction of triphenyltin iodide and trifluoroacetic acid in the presence of silver oxide.[16] Ag_2O converts *trans*-halohydrins to epoxides (eq 12)[17] or rearranged products (eq 13)[18]. It is reported that the reaction is nonstereoselective, and that *trans*-epoxides are the major products when *cis*-alkenes are the reactants.[19]

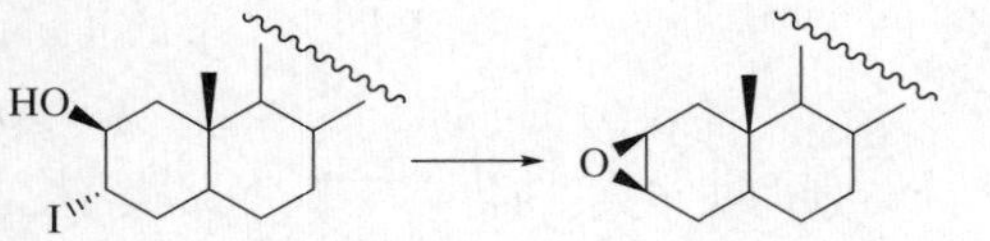

(12)

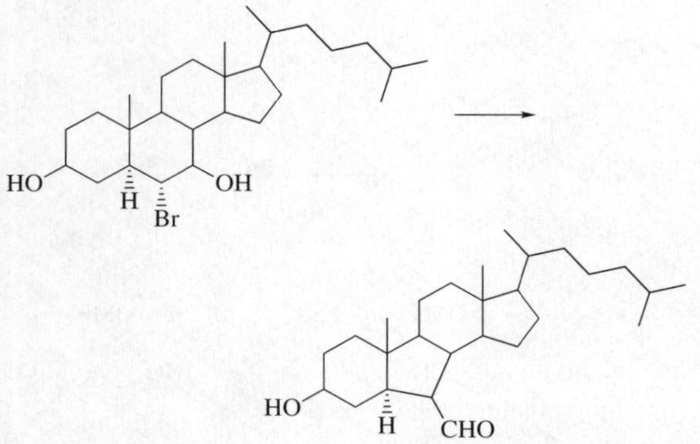

(13)

2-Arylpropionaldehydes are produced from 1-aryl-1-propene by oxidative rearrangement with ***Iodine*** and Ag_2O in dioxane–H_2O (eq 14).[20] The mechanism may involve the 1,2-shift of the aryl group through a bridged phenonium ion in the iodohydrin intermediate.

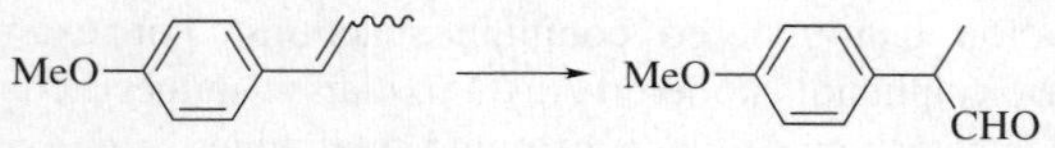

(14)

In the total synthesis of mycophenolic acid, electrophilic substitution was promoted by Ag_2O (eq 15).[21] α,β-Unsaturated nitriles are formed from the corresponding γ-bromo-β-oxo nitriles with Ag_2O (eq 16).[22] Arylsulfenylation and arylselenenylation at the 5-position of uracils are promoted by Ag_2O (eq 17).[23]

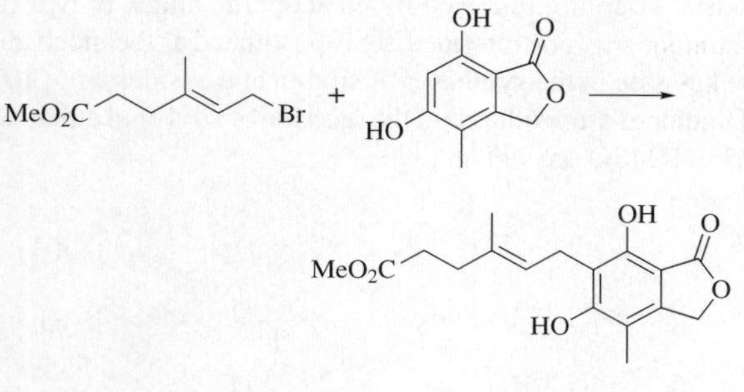

(15)

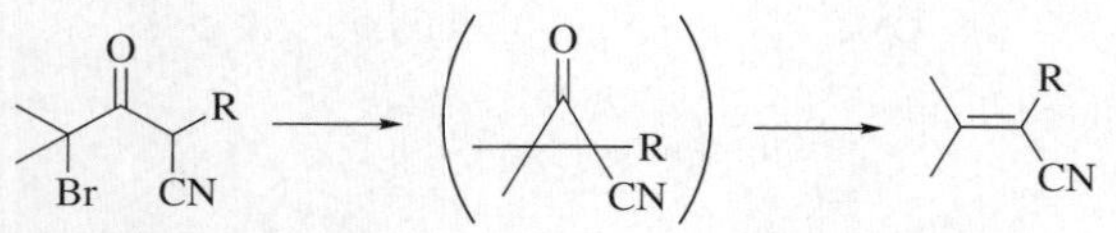

(16)

R = H, Me

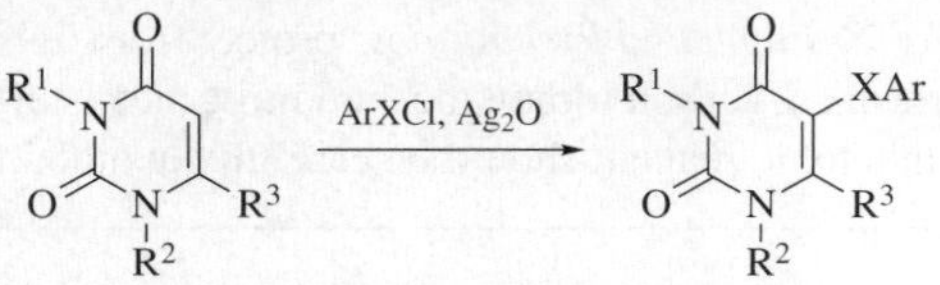

(17)

R^1 = H, Me; R^2 = Me, acyclic chain, ribose; R^3 = H, Me, N=PPh_3
X = S, Se

Lewis Acid with Thioethers. Ag_2O has been applied to the 1,2-cleavage of penicillins in a strong nonnucleophilic base, such as ***1,5-Diazabicyclo[4.3.0]non-5-ene***,[24] and to the hydrolysis of thioacetals.[25]

Wolff Rearrangement. The Wolff rearrangement of diazo ketones is promoted by Ag_2O (eq 18).[26]

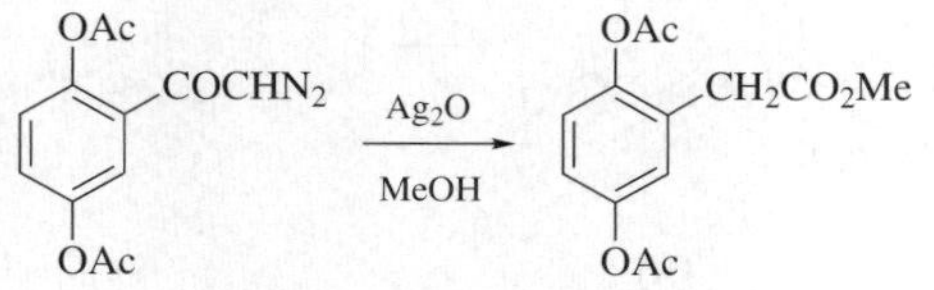

(18)

Related Reagents. See Classes O-2 and O-5, pages 1–10. Bromine–Silver(I) Oxide.

---

1. Synder, C. D.; Rapoport, H. *JACS* **1974**, *96*, 8046.
2. Parker, K. A.; Sworin, M. *TL* **1978**, *26*, 2251.
3. (a) Kraus, G. A.; Taschner, M. J. *JOC* **1980**, *45*, 1174. (b) Marchand, A. P.; Suri, S. C.; Earlywine, A. D.; Powell, D. R.; van Der Helm, D. *JOC* **1984**, *49*, 670.
4. Al-Hamdany, R.; Ali, B. *CC* **1978**, 397.
5. Angle, S. R.; Turnbull, K. D. *JACS* **1989**, *111*, 1136.
6. Jurd, L. *T* **1977**, *33*, 163.
7. Thomason, S. C.; Kubler, D. G. *J. Chem. Educ.* **1968**, *45*, 546.
8. Pepperman, A. B. *JOC* **1981**, *46*, 5039.
9. Corey, E. J.; Das, J. *JACS* **1982**, *104*, 5551.
10. West, K. F.; Moore, H. W. *JOC* **1984**, *49*, 2809.
11. Ito, Y.; Fujii, S.; Konoike, T.; Saegusa, T. *SC* **1976**, *6*, 429.
12. (a) Merlini, L.; Zanarotti, A; Pelter, A.; Rochefort, M. P.; Hänsel, R. *CC* **1979**, 695. (b) Merlini, L.; Zanarotti, A.; Pelter, A.; Rochefort, M. P.; Hansel, R. *JCS(P1)* **1980**, 775. (c) Merlini, L.; Zanarotti, A. *TL* **1975**, 3621.
13. Ito, Y.; Konoike, T.; Saegusa, T. *JACS* **1975**, *97*, 649.
14. Kuhn, R.; Trischmann, H.; Low, I. *AG* **1955**, *67*, 32.
15. Ethier, J. C.; Neville, G. A. *TL* **1972**, 5297.
16. Srivastava, T. N.; Singh, J. *IJC(A)* **1983**, *22A*, 128.
17. Parrilli, M.; Barone, G.; Adinolfi, M.; Mangoni, L. *TL* **1976**, 207.
18. (a) Nace, H. R.; Crosby, G. A. *JOC* **1979**, *44*, 3105. (b) Schmidlin, J.; Wettstein, A. *HCA* **1953**, *36*, 1241. (c) Curtin, D. Y.; Harder, R. J. *JACS* **1960**, *82*, 2357.
19. Jorgensen, K. A.; Larsen, E. *JCS(D)* **1990**, 1053.
20. Kikuchi, H.; Kogure, K.; Toyoda, M. *CL* **1984**, 341.
21. Canonica, L.; Rindone, B.; Santaniello, E.; Scolastico, C. *TL* **1971**, 2691.
22. Herter, R.; Fohlisch, B. *CB* **1982**, *115*, 381.
23. Lee, C. H.; Kim, Y. H. *TL* **1991**, *32*, 2401.
24. Alpegiani, M.; Bedeschi, A.; Bissolino, P.; Visentin, G.; Zarini, F.; Perrone, E.; Franceschi, G. *H* **1990**, *31*, 617.
25. Gravel, D.; Vaziri, C.; Rahal, S. *CC* **1972**, 1323.

26. (a) Kloetzel, M. C.; Dayton, R. P.; Abadir, B. Y. *JOC* **1955**, *20*, 38. (b) Meier, H.; Zeller, K.-P. *AG(E)* **1975**, *14*, 32. (c) Della, E. W.; Kendall, M. *JCS(P1)* **1973**, 2729.

Kathlyn A. Parker & Dai-Shi Su
Brown University, Providence, RI, USA

Silver(II) Oxide

AgO

[1301-96-8] AgO (MW 123.87)

(oxidizing reagent for conversion of hydroquinone ethers[1] and alkoxy anilines[5] to quinones, of alcohols to aldehydes or ketones,[11] of allylic alcohols to conjugated acids,[12] of aldehydes to acids, of amines to azo compounds, and of amino acids to aldehydes;[13] radical coupling reactions;[8,9] utilized in oxidative decarboxylation of carboxylic acids[10])

Physical Data: dec above 100 °C; *d* 7.483 g cm^{-3}.
Solubility: insol H_2O; sol alkali; sol NH_4OH (with decomposition and evolution of N_2).
Form Supplied in: charcoal-gray powder; widely available.
Handling, Storage, and Precautions: protect from light. Highly irritating to skin, eyes, mucous membranes, and respiratory tract. Avoid contact with skin, organic matter, strong ammonia, and alkali.

Oxidation of Hydroquinone Ethers and Alkoxy Anilines. This reagent effects the oxidative demethylation of hydroquinone ethers to afford *para*-quinones in acidic media (eq 1).[1] A mechanism has been proposed in which oxidative hydration takes place at the aromatic carbon.

(1)

AgO in dioxane–nitric acid solution also oxidizes hydroquinone mono- and diesters to the corresponding quinones (eq 2).[2] The order of reaction is 4-acetoxyphenols > hydroquinone dimethyl ethers > hydroquinone diacetates.[2]

(2)

R = H or Ac

AgO is effective in oxidizing electron-deficient 1,4,5-trimethoxy-9,10-anthraquinone to the corresponding diquinone, an intermediate in anthracyclinone synthesis (eq 3).[3] This bifunctional dienophile undergoes the Diels–Alder reaction at the internal double bond with electron-rich dienes (e.g. 2-ethoxybutadiene), and at the external double bond with slightly electron-poor or unsubstituted dienes (e.g. 1,3-butadiene and 2-acetoxybutadiene).[4]

(3)

In the presence of AgO in aqueous nitric acid, *p*-methoxyanilines are oxidized with hydrolysis to quinones (eq 4).[5]

(4)

Tautomerization of a tetracyclic keto quinone obtained from AgO oxidation affords an aromatized anthracycline intermediate (eq 5).[6] The crude product subsequently undergoes tautomerization at rt in acetone with a few drops of concd. HCl.

(5)

The trimethoxyphenyl lactam derivatives of hydroquinone methyl ethers are oxidized by AgO in an acidic milieu to give the lactam-substituted quinones (eq 6).[7]

(6)

Radical Coupling Reactions. The addition of acetone to terminal alkenes is accomplished by the action of AgO (eq 7).[8] 2-Alkanones are formed in 73–83% yield. Internal alkenes are less reactive. A radical mechanism is proposed for this transformation in which AgO acts as a 'heterogeneous' initiator for the hydrogen atom abstraction from acetone to generate an acetonyl radical; this is followed by the addition of this radical to the alkene.

(7)

R = H, alkyl

The oxidative dimerization of 2-(vinyloxy) phenols by AgO under anhydrous conditions proceeds with rearrangement. A [3,3]-

sigmatropic process involving an intermediate cyclohexadienol radical has been proposed (eq 8).[9]

(8)

R = OMe or —N O

Other Oxidation Reactions. AgO has been utilized in the oxidative decarboxylation of carboxylic acids such as pivalic acid, isobutyric acid, and *n*-butyric acid.[10] The reaction proceeds significantly faster when acetonitrile is utilized as a cosolvent. AgO oxidizes primary alcohols to aldehydes, secondary alcohols to ketones,[11] allylic alcohols to conjugated acids,[12] aromatic hydrocarbons to aromatic aldehydes and ketones, α-amino acids and α-amino esters to aldehydes (with decarboxylation), aldehydes to acids,[13] and amines, aniline, *p*-toluidine, and *N,N*-dimethyl-*p*-phenylenediamine to the corresponding azo derivatives.[14]

Related Reagents. See Classes O-1, O-2, O-5, O-6, and O-10, pages 1–10.

---

1. (a) Snyder, C. D.; Rapoport, H. *JACS* **1974**, *96*, 8046. (b) Snyder, C. D.; Bondinell, W. E.; Rapoport, H. *JOC* **1971**, *36*, 3951. (c) Snyder, C. D.; Rapoport, H. *JACS* **1972**, *94*, 227. (d) Kraus, G. A.; Neuenschwander, K. *SC* **1980**, *10*, 9.
2. (a) Escobar, C.; Farina, F.; Martinez-Utrilla, R.; Paredes, M. C. *JCR(S)* **1980**, 156. (b) Escobar, C.; Farina, F.; Martinez-Utrilla, R.; Paredes, M. C. *JCR(S)* **1977**, 266; *JCR(M)* **1977**, 3151.
3. Kende, A. S.; Tsay, Y.-G.; Mills, J. E. *JACS* **1976**, *98*, 1967.
4. (a) Inhoffen, H. H.; Muxfeldt, H.; Koppe, V.; Heimann-Trosien, J. *CB* **1957**, *90*, 1448. (b) Sauer, J. *AG(E)* **1967**, *6*, 16. (c) Sustmann, R. *TL* **1971**, 2717.
5. Parker, K. A.; Kang, S.-K. *JOC* **1979**, *44*, 1536.
6. Kende, A. S.; Gesson, J.-P.; Demuth, T. P. *TL* **1981**, *22*, 1667.
7. Michael, J. P.; Cirillo, P. F.; Denner, L.; Hosken, G. D.; Howard, A. S.; Tinkler, O. S. *T* **1990**, *46*, 7923.
8. Hajek, M.; Silhavy, P.; Malek, J. *TL* **1974**, 3193.
9. West, K. F.; Moore, H. W. *JOC* **1984**, *49*, 2809.
10. (a) Anderson, J. M.; Kochi, J. K. *JOC* **1970**, *35*, 986. (b) Anderson, J. M.; Kochi, J. K. *JACS* **1970**, *92*, 1651.
11. Syper, L. *TL* **1967**, 4193.
12. Corey, E. J.; Gillman, N. W.; Ganem, B. E. *JACS* **1968**, *90*, 5616.
13. (a) Clarke, T. G.; Hampson, N. A.; Lee, J. B.; Morley, J. R.; Scanlon, B. *JCS(C)* **1970**, 815. (b) Clarke, T. G.; Hampson, N. A.; Lee, J. B.; Morley, J. R.; Scanlon, B. *TL* **1968**, 5685. (c) Lee, J. B.; Clarke, T. G. *TL* **1967**, 415. (d) Thomason, S. C.; Kubler, D. G. *J. Chem. Educ.* **1968**, *45*, 546.
14. Ortiz, B.; Villanueva, P.; Walls, F. *JOC* **1972**, *37*, 2748.

Kathlyn A. Parker & Dai-Shi Su
Brown University, Providence, RI, USA

Singlet Oxygen

$^1O_2$

[7782-44-7] O_2 (MW 32.00)

(electrophilic oxidizing agent for oxygenation of cisoid 1,3-dienes, many types of heterocyclic systems, enamines, alkenes containing allylic hydrogen atoms, sulfides, carbon–phosphorus double bonds, and other electron-rich unsaturated organic compounds)

Physical Data: singlet oxygen ($^1O_2$) is the first excited electronic state of molecular oxygen ($^1\Delta_g$), lying 22.4 kcal mol^{-1} above the ground state triplet. The second singlet state ($^1\Sigma_g^+$), 37 kcal mol^{-1} above the ground state, is relatively short-lived in solution (10^{-12} s) due to a rapid spin-allowed transition to the longer-lived (10^{-3}–10^{-6} s) first excited state. Because of the short lifetime of the $^1\Sigma_g^+$ state, the more stable singlet oxygen species ($^1\Delta_g$) is considered to be the reactive intermediate in the oxygenation of organic compounds in solution.[1]
Solubility: sol aqueous and organic solvents.
Form Supplied in: singlet oxygen is an unstable, short-lived species and must be prepared in situ or immediately prior to use.
Preparative Methods: among the methods for generating singlet oxygen in solution are: (a) the dye-sensitized photooxidation of triplet oxygen;[2] (b) the decomposition of phosphite ozonides;[3] (c) the decomposition of transannular peroxides such as 9,10-diphenylanthracene peroxide;[4] (d) the reaction of ***Hydrogen Peroxide*** with ***Sodium Hypochlorite***;[5,6] and (e) subjecting gaseous oxygen to electrodeless discharge.[7] In the following section, the procedures for generating singlet oxygen by the methods (a)–(d) above are discussed, and some typical reactions of singlet oxygen with organic compounds are reviewed.
Handling, Storage, and Precautions: the initial peroxidic products formed in $^1O_2$ reactions usually undergo decomposition or rearrangement under the conditions of workup, but in some cases (certain transannular peroxides or dioxetanes) they may remain in the reaction mixture as potentially explosive materials. Unless the peroxides are sought as reaction products, they may be deoxygenated by the action of ***Dimethyl Sulfide***, diphenyl sulfide, or ***Triphenylphosphine***.

Preparative Methods.

Generation by Dye-Sensitized Photooxygenation.[2] The most common method for generating $^1O_2$ in solution is the dye-sensitized photochemical excitation of triplet oxygen. The mechanism for generating $^1O_2$ in this way involves the excitation of an appropriate dye with visible light to form the corresponding excited singlet state. Rapid intersystem crossing generates the excited triplet state of the sensitizer ($E_T > 22.4$ kcal mol^{-1}) which undergoes energy transfer with triplet oxygen to form singlet oxygen, regenerating the ground state of the sensitizer (eq 1). Among the sensitizers which can be utilized in solution are methylene blue, Rose Bengal, eosin yellow, chlorophyll, riboflavin, zinc tetraphenylporphyrin, bis-naphthalenothiophene, erythrosin B, and hematoporphyrin. In addition to the soluble, homogeneous sensitizers, polymer-bound dyes such as Rose Bengal have been used for the photochemical generation of $^1O_2$.[8]

$$\text{Sens} \xrightarrow{h\nu} {}^1(\text{Sens}^*) \xrightarrow{\text{ISC}} {}^3(\text{Sens}^*) \xrightarrow{{}^3O_2} {}^1O_2 + \text{Sens} \quad (1)$$

Singlet oxygen may be generated in a wide range of solvents, the choice of solvent being dictated by the solubility of substrate and sensitizer and by solvent properties that influence ease of workup. The lifetime of $^1O_2$ varies from 2 μs in H_2O to approximately 700 μs in CCl_4 (Table 1) and is considerably longer when the solvent is deuterated (the lifetime of $^1O_2$ is almost always increased by a factor of ten when the solvent is perdeuterated).[9]

Table 1 Lifetimes of Singlet Oxygen in Various Solvents[9]

Solvent	Lifetime (μs)	Solvent	Lifetime (μs)
H_2O	2	MeCN	30
MeOH	7	$CHCl_3$	60
EtOH	12	CS_2	200
C_6H_{12}	17	$CDCl_3$	300
D_2O	20	C_6F_6	600
C_6H_6	24	CCl_4	700
MeCOMe	26	Freon-11	1000

The light sources used are generally high intensity lamps which emit light in the visible range. Sun lamps have been employed in certain instances but they may bring about competing UV-promoted reactions and rearrangements. Among the lamps used as light sources are Sylvania 500 W tungsten halogen,[10] Osram-Vialox 250 W,[11] 500 W Halogen-Argaphoto,[12] 500 W Toshiba JD,[13] 200 W halogen,[14] 650 W tungsten halogen,[15] 100 W tungsten halogen,[16] and 500 W high pressure sodium.[17]

Generation from Phosphite Ozonides.[3] ***Triphenyl Phosphite*** is treated with ***Ozone*** at −78 °C (dry ice–acetone) in methylene chloride until the blue color of excess ozone is observed. Ozonization is then discontinued and the solution is purged with dry nitrogen to remove the excess oxidant. A cold solution of the acceptor in methylene chloride is then added, and mixing is effected by the nitrogen stream. The −78 °C bath is then removed and replaced with a −25 °C bath (ice–methanol). Alternatively, the reaction mixture is permitted to warm to rt and then worked up by concentration and chromatography.

Generation by the Decomposition of 9,10-Diphenylanthracene Endoperoxide. Oxidation of an Oxazole to a Triamide. In contrast to dialkyl peroxides, which usually undergo oxygen–oxygen bond cleavage on pyrolysis, many aromatic endoperoxides break down on heating with release of oxygen in the singlet state.[18] Among the available endoperoxides, 9,10-diphenylanthracene peroxide (**1**) has been found to be a particularly efficient donor of $^1O_2$ in this type of decomposition. It can easily be prepared as a stable product (dec 180–181 °C, mp 254 °C) by irradiation of a cold solution of 9,10-diphenylanthracene in carbon disulfide for 48 h in a stream of oxygen using a 275 W Sylvania RS lamp.[19]

(**1**) DPAP

A typical singlet oxygen reaction using this peroxide is illustrated in the conversion of 2,5-diphenyl-4-methyloxazole to *N*-acetyldibenzamide (eq 2).[18] In this oxidation, an anhydrous benzene solution containing the peroxide and the oxazole is stirred at reflux temperature in the dark for 94 h under a positive pressure of nitrogen. Removal of benzene and chromatography on silica gel yields the triamide (92%), as well as some diamide formed during chromatography, in addition to 9,10-diphenylanthracene.[4]

(2) [oxazole →(DPAP) endoperoxide →(92%) N-acetyldibenzamide + 9,10-diphenylanthracene]

Generation of $^1O_2$ by Hypochlorite–Hydrogen Peroxide Oxygenation.[5] In this procedure, illustrated by the conversion of 2,3-dimethylbutene to the allylic hydroperoxide (eq 3), a methanolic solution of the substrate is treated with 3 equiv of 30% ***Hydrogen Peroxide*** and stirred at 10 °C. To this mixture, aqueous ***Sodium Hypochlorite*** is slowly added (2.5 equiv). The solution is then diluted with water, extracted with ether, and worked up.

(3) [30% H_2O_2, NaOCl, 63%; product HOO]

Oxygenations with $^1O_2$. Singlet oxygen undergoes three classes of reaction with alkenes: an ene type of reaction forming allylic hydroperoxides; 1,4-cycloaddition with cisoid 1,3-dienes; and 1,2-cycloaddition with electron-rich or strained alkenes. In addition, $^1O_2$ reactions take place with many types of heterocyclic compounds including furans, pyrroles, oxazoles, indoles, imidazoles, and thiophenes.[20,21] Other types of $^1O_2$ reaction take place with sulfides, carbon–phosphorus double bonds, and electron-rich aromatic systems. In many cases, competition exists between different modes of oxidation, as in the reaction of dihydropyran with $^1O_2$ (eq 4), which takes place either by 1,2-dioxetane formation and subsequent cleavage, or by an ene reaction forming a hydroperoxide which undergoes O–O bond fission by a β-elimination. In this case, it is found that the ratio of products (**2**) to (**3**) varies over a 58-fold range depending on the solvent used (Table 2), with polar solvents favoring 1,2-dioxetane formation over the ene reaction.[22]

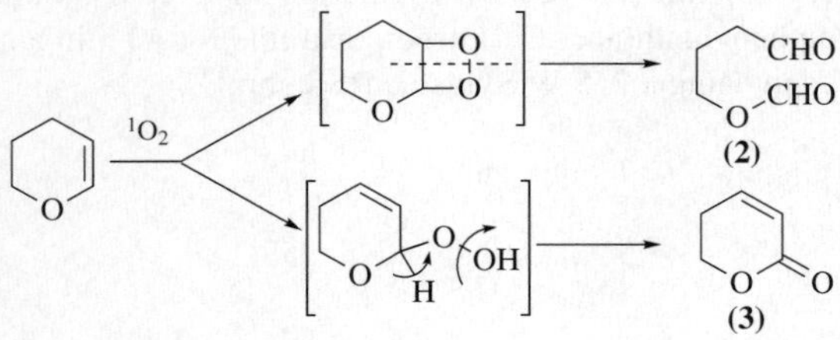

(4)

Table 2 Dependence of Reaction Mode on Solvent in the $^1O_2$ Oxidation of Dihydropyran

Solvent	Product ratio (**2**)/(**3**)
Benzene	0.094
Acetone	0.50
Methylene chloride	2.67
Acetonitrile	5.51

The Singlet Oxygen Ene Reaction. The reaction of alkenes containing at least one allylic hydrogen with singlet oxygen, yielding a hydroperoxide with accompanying shift of the double bond, bears a close resemblance to the Alder ene reaction (eq 5). This ene type of oxidation has been extensively used in synthesis. Reduction of the allylic hydroperoxide yields the corresponding allylic alcohol. An analogous reaction of enol silyl ethers yielding silylperoxy esters is assumed to take place by the same mechanism (eq 6).[23] Formation of allylic hydroperoxides is not stereoselective, as is shown by the intermediates in the garrya and atisine alkaloid syntheses (eqs 7 and 8).[24]

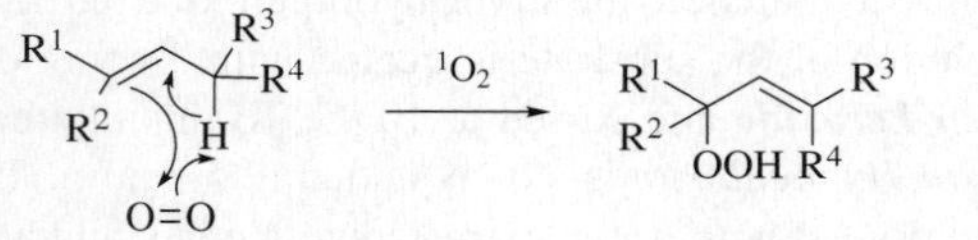

(5)

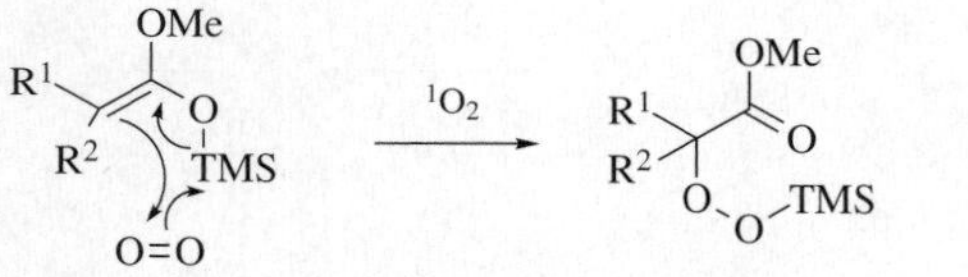

(6)

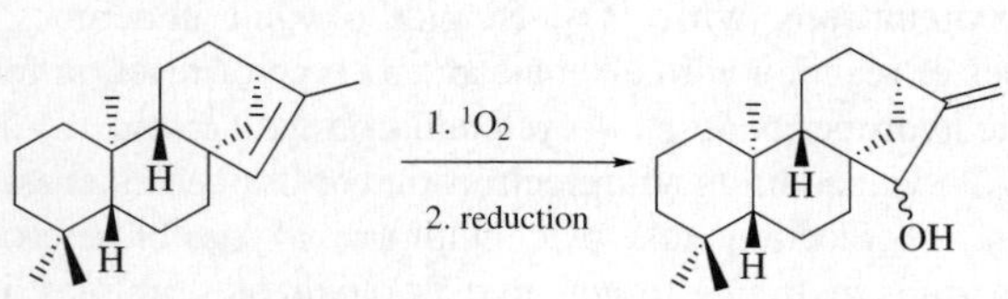

(7)

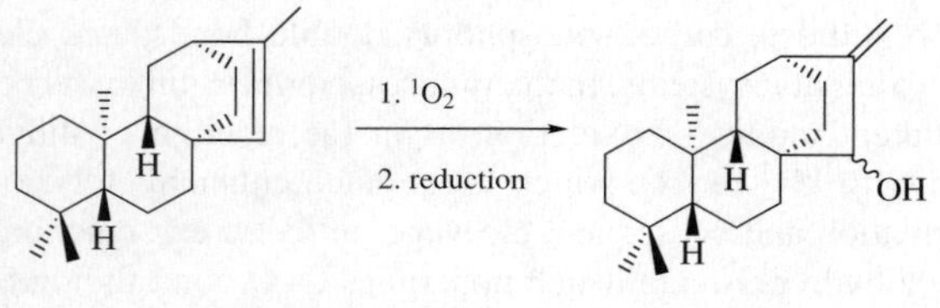

(8)

The photooxygenation of Δ^9-octalin followed by reduction with hydrazine yields $\Delta^{1(9)}$-10-octalol, a key intermediate in the synthesis of *trans*-cyclodecenone (eq 9).[25] The ene reaction has also been used for the contrathermodynamic isomerization of an alkene (eq 10)[26] and in the formation of allylic alcohols for later conversion to α,β-unsaturated ketones (eq 11).[27]

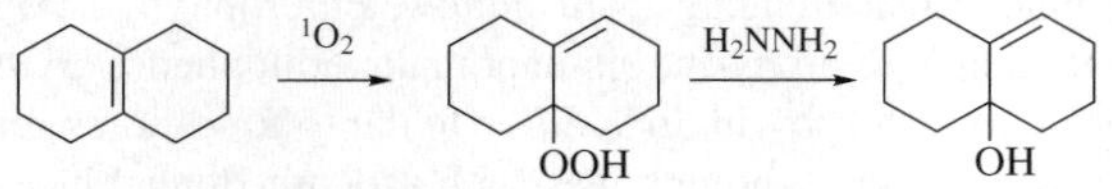

(9)

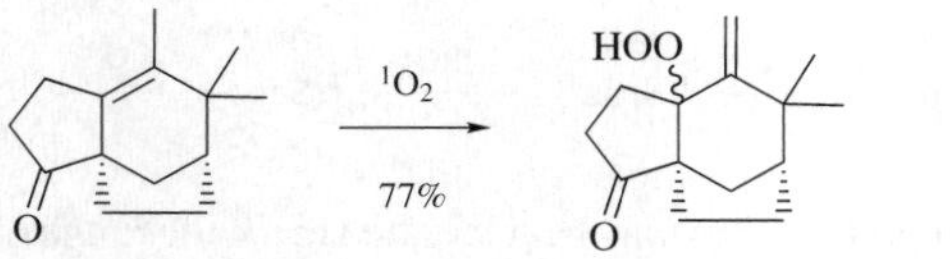

(10)

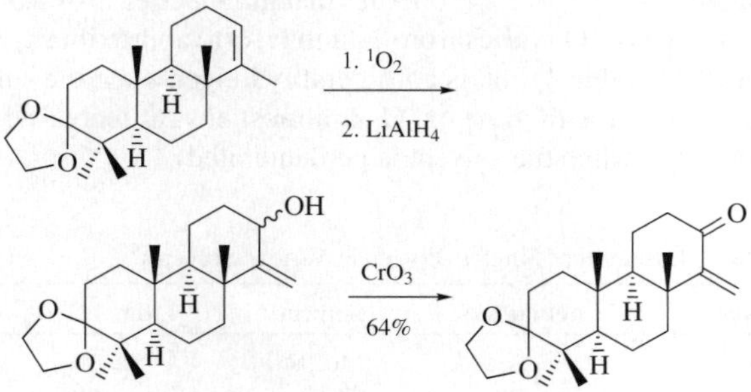

(11)

An interesting use of the ene $^1O_2$ reaction in conjunction with the Diels–Alder type of $^1O_2$ addition is found in the dye-sensitized photooxidation of neoabietic acid.[28] The first step involves an ene reaction at the more highly substituted double bond, generating a hydroperoxide along with the shift of the double bond to form a *cis*-1,3-diene. The second-stage [2 + 4] addition yields a transannular peroxide which undergoes a β-elimination and reduction to form the product (eq 12).

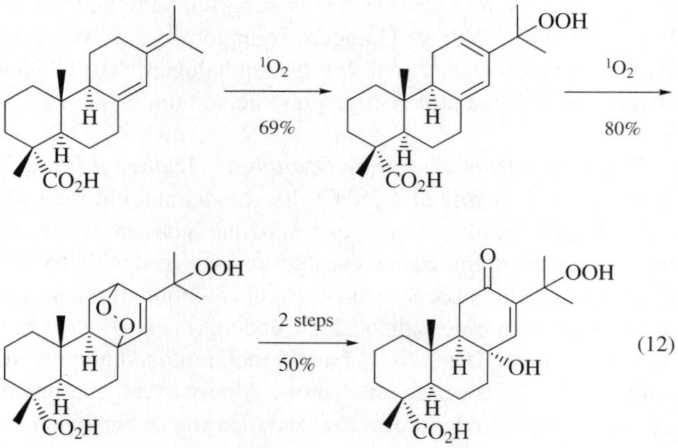

(12)

In the final steps of a total synthesis of tetracycline, the anhydro derivative is converted to a hydroperoxide in high yield by a singlet oxygen addition in the ene mode (eq 13). The reaction is very sensitive to overoxidation and succeeds only when the dye-sensitized photooxidation is terminated after a few minutes.[29]

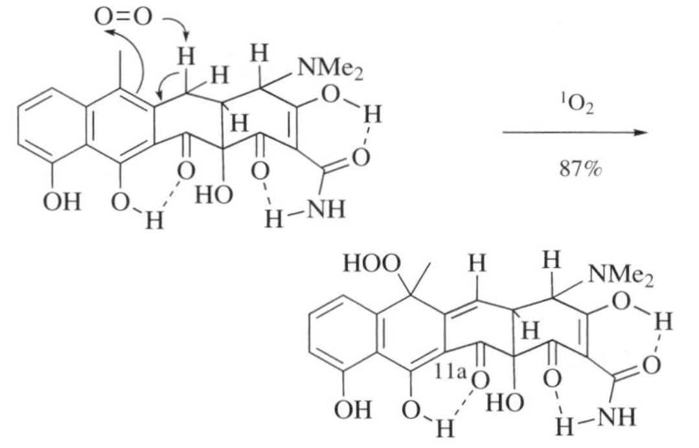

(13)

1,4-Cycloaddition. The reaction of singlet oxygen with cisoid 1,3-dienes takes place in both carbocyclic and heterocyclic systems. In general, the formation of 1,4-cycloaddition products from 1,3-dienes is the hallmark of singlet oxygen reactivity, in contrast to other less well-defined oxidation processes effected by oxygen in the ground state. The early synthesis of (±)-ascaridol by the dye-sensitized photooxygenation of α-terpinene is a classic case of 1,4-addition by singlet oxygen (eq 14).[30] The Schenck synthesis of cantharidine[31] also makes use of an endoperoxide formed by a Diels–Alder type cycloaddition of singlet oxygen (eq 15). In another application, the formation of bis-epoxides by thermolysis of 1,4-endoperoxides generated in a singlet oxygen addition is employed in the synthesis of (±)-crotepoxide (eq 16).[32]

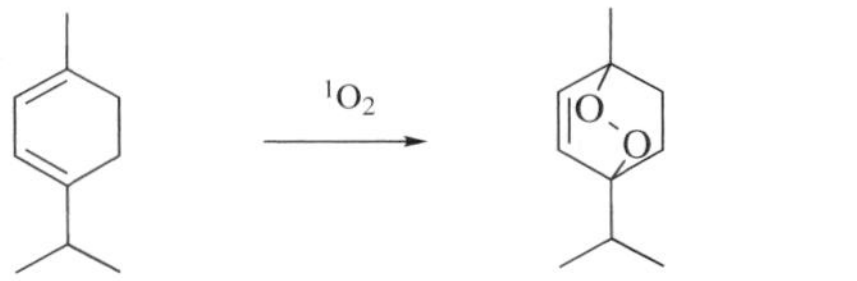

(14)

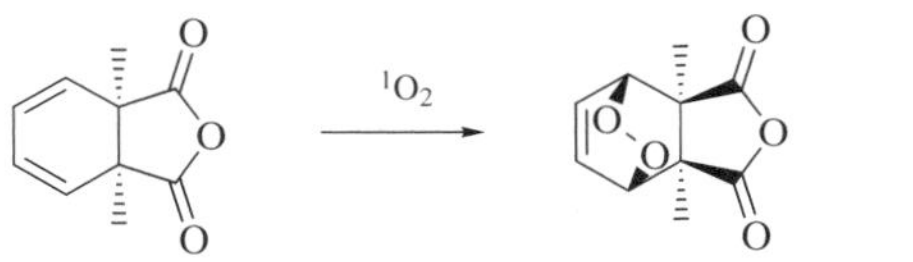

(15)

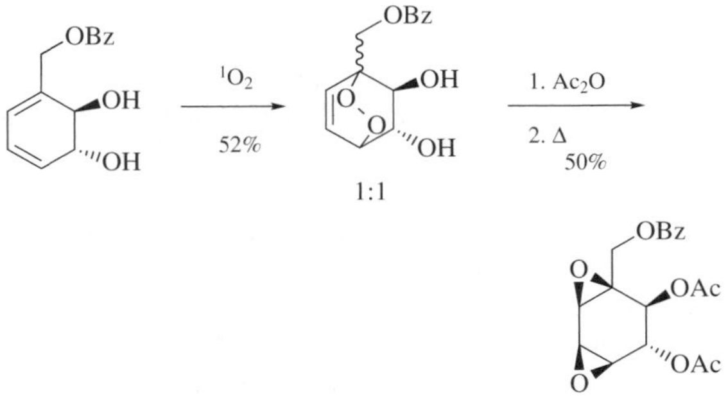

(16)

Furan photooxidation by 1,4-cycloaddition of $^1O_2$ has been widely used as a means of generating carbonyl compounds from transannular peroxide intermediates. In an early example, 2-furfural was converted by $^1O_2$ to the unsaturated lactone intermediate (eq 17) in a synthesis of camptothecin.[33] Likewise, oxidation of 3-methyl-2-furoic acid yields the corresponding methyl derivative used as a key component in a synthesis of (±)-strigol (eq 18).[34] These transformations most probably take place by solvolysis of the intermediate endoperoxide followed by fragmentation, as shown in eq 19. Examples have been reported of singlet oxygen reactions with styrene derivatives where a benzene ring acts as part of a 1,3-diene system (eq 20).[35,36]

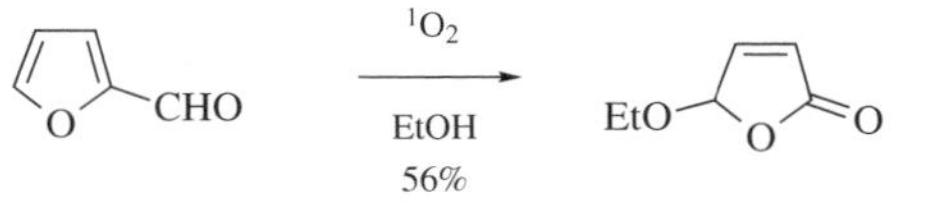

(17)

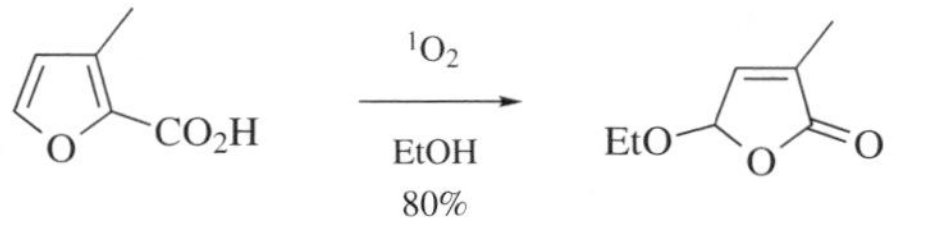

(18)

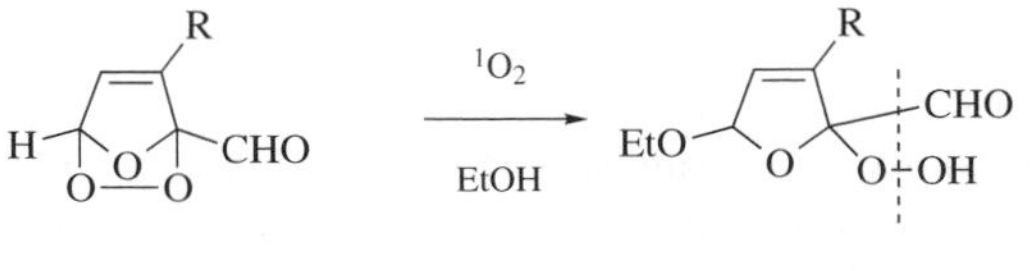

(19)

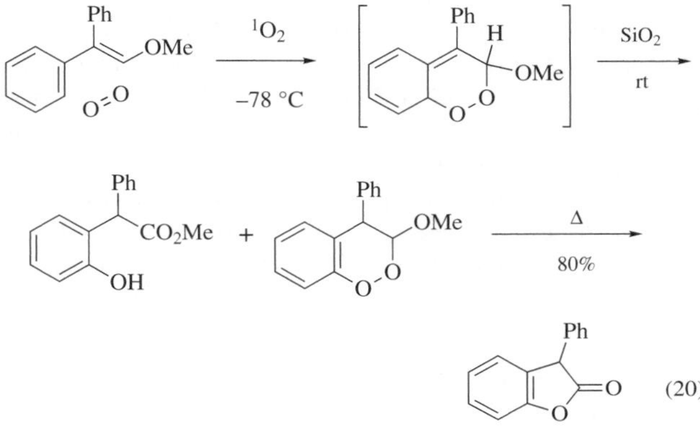

(20)

A recent example of 1,4-addition of singlet oxygen to a pyrrole involves the dye-sensitized photooxidation of 2-carboxy-4-methoxy-5-(methoxycarbonyl)-1-methylpyrrole (eq 21). The intermediate endoperoxide undergoes fragmentation by decarboxylation and cleavage of the O–O bond to give 5-hydroxy-4-methoxy-5-(methoxycarbonyl)-1-methyl-3-pyrrolin-2-one).[37]

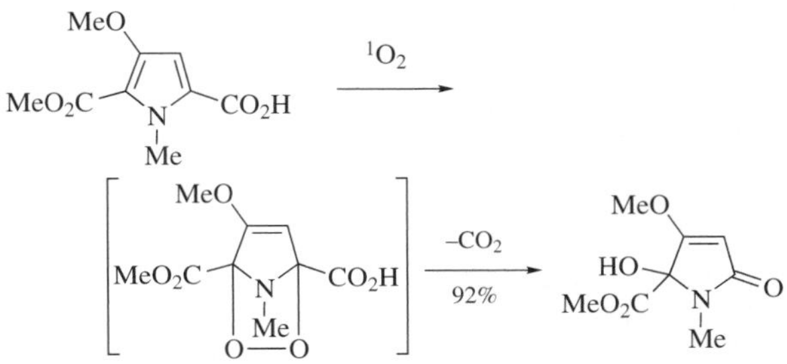

(21)

Addition of Singlet Oxygen to Alkenes Activated by Electron-Releasing Groups. The addition of singlet oxygen to alkenes occurs in the presence of amino or alkoxyl groups in the absence of active allylic hydrogens in the same molecule. The addition takes place stereospecifically by *cis* addition to form dioxetanes as isolable products. These, in turn, undergo thermolysis to form electronically excited carbonyl compounds with accompanying chemiluminescence. Thus photooxidation of 1,1,2,2-tetramethyoxyethylene sensitized by either zinc tetraphenylporphyrin or dinaphthalenothiophene proceeds smoothly to give an isolable clear, pale-yellow liquid shown to be the tetramethoxydioxetane (eq 22).[38] On the other hand, the crystalline *cis*-dioxetane, formed on dye-sensitized photooxidation of *cis*-diethoxyethylene in fluorotrichloromethane at −78 °C under irradiation with a 500 W lamp for 25 min, explodes on warming to rt (eq 23).[39]

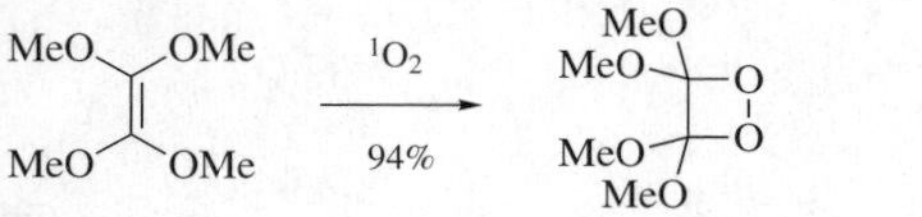

(22)

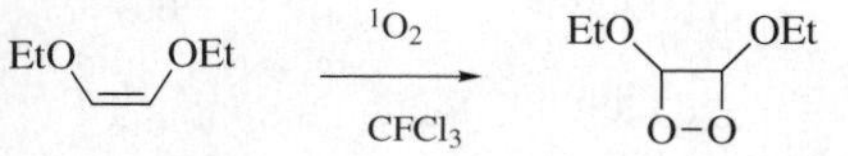

(23)

Reactions of Enamines with Singlet Oxygen. Reactions of enamines with $^1O_2$ take place readily to form the carbonyl products expected from the cleavage of dioxetane intermediates. In the case of the piperidine enamine of methyl isopropyl ketone, the reaction is carried out in benzene using zinc tetraphenylporphyrin as sensitizer (eq 24).[40] In related work, photooxygenation of the 22-morpholine enamine of 3-oxobisnor-4-cholen-22-al in DMF at 15 °C using Rose Bengal as sensitizer yields a quantitative yield of progesterone (eq 25).[41] Likewise, $^1O_2$ cleavage of the highly electron-rich enamines formed from the silyl enol ethers of azetidinecarboxylic esters provides a route to β-lactams (eq 26).[42]

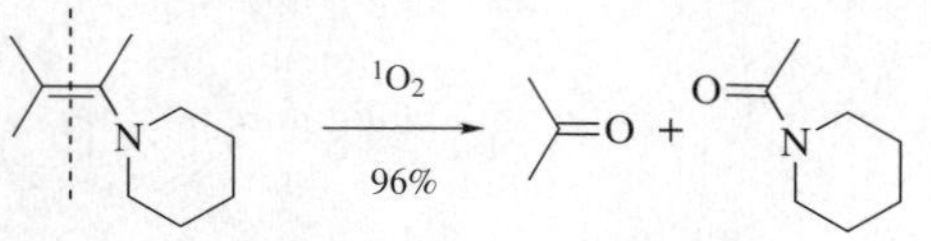

(24)

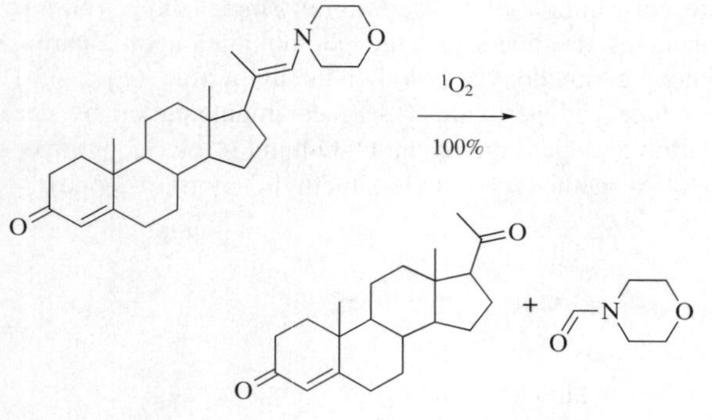

(25)

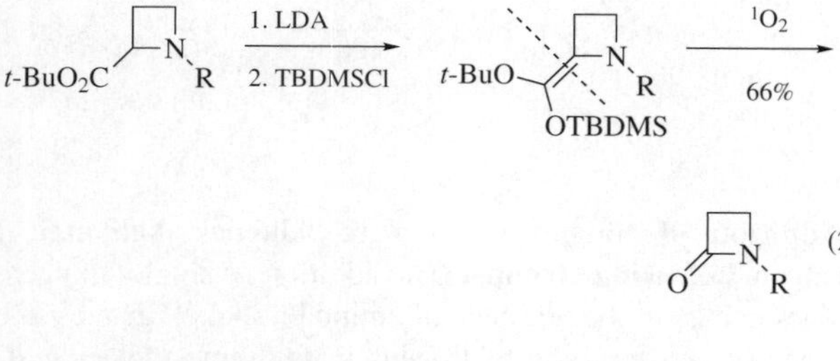

(26)

Reactions of Enols with $^1O_2$. As electron-rich alkenes, enols react with singlet oxygen along the lines of the ene reaction to give unstable intermediate hydroperoxides which undergo further transformations. The reactions are enhanced by the presence of fluoride ion, presumably by the formation of a strong hydrogen bond between the enol-OH and the fluoride.[43] Thus 1,3-cyclohexanedione reacts with $^1O_2$ in the presence of fluoride ion to give pyrogallol (eq 27). 4-Hydroxycoumarin, which is normally relatively inert toward $^1O_2$, readily undergoes oxygen uptake with $^1O_2$ in the presence of fluoride ion (eq 28).

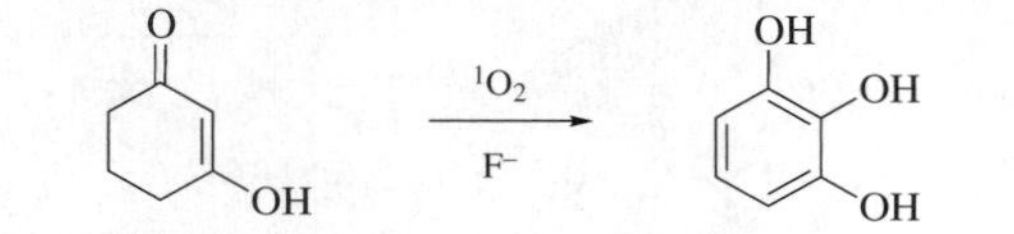

(27)

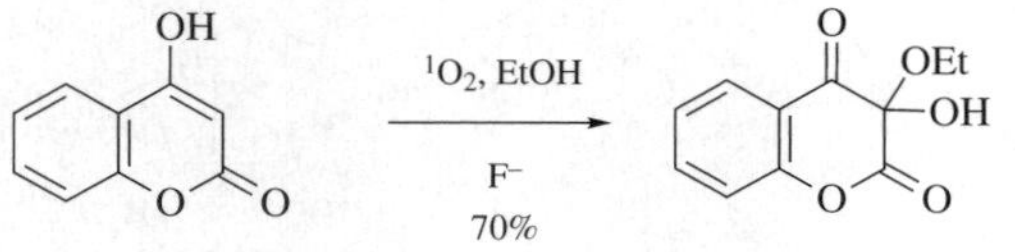

(28)

While it has generally been assumed that alkenes with electron-withdrawing groups are inert to $^1O_2$, it has recently been shown that α,β-unsaturated ketones which are constrained in the *s-cis* conformation are rapidly oxidized by $^1O_2$. Systems which prefer the *s-trans* conformation react slowly or not at all.[44,45] For example, compound (**4**) yields 88% of the hydroperoxide (eq 29) while compound (**5**) shows no reaction toward $^1O_2$. The difference is accounted for in terms of an initial [4 + 2] cycloaddition of $^1O_2$ to form a 1,2,3-trioxine (eq 30),[44] which then undergoes further transformations.

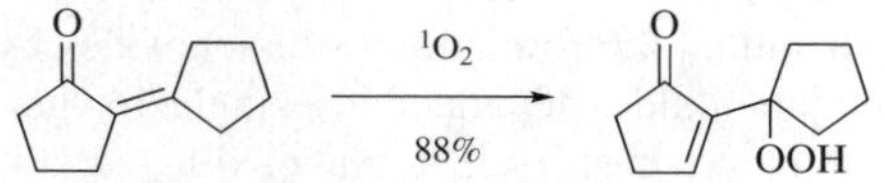

(29)

(**4**)

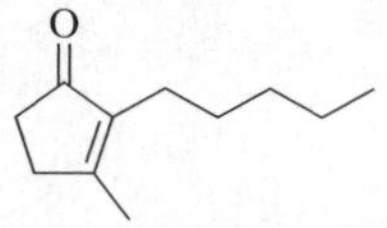

(**5**)

(30)

Oxidation of Sulfides. Sulfides react with singlet oxygen to form sulfoxides and sulfones.[46–49] The reaction takes place through intermediate persulfoxides. The rate of uptake of $^1O_2$ appears to be relatively independent of solvent and temperature. Other reaction paths have been observed. Thus although the $^1O_2$ oxidation of the linear sulfide yields the expected mixture of sulfoxide and sulfone (eq 31), oxidation of the thiazolidine with singlet oxygen, followed by deoxygenation with Me_2S or Ph_3P, gives a quantitative yield of the product of α-hydroxylation (eq 32). Formation of this alcohol is considered to take place by a Pummerer-type of rearrangement of an intermediate persulfoxide.[50]

Cleavage of Carbon–Phosphorus Double Bonds by Singlet Oxygen. Singlet oxygen cleaves phosphorus ylides to form carbonyl compounds and the corresponding phosphine oxide (eq 33).[51] The reported procedure involves dye-sensitized photooxidation of the phosphorane in a solvent such as chloroform, benzene, or methanol using *meso*-tetraphenylporphyrin or Rose Bengal as sensitizer with irradiation from a 500 W tungsten filament lamp while oxygen is bubbled through the solution. This method has recently been employed in one of the procedures for oxidizing phosphoranes to tricarbonyl compounds (eq 34).[52]

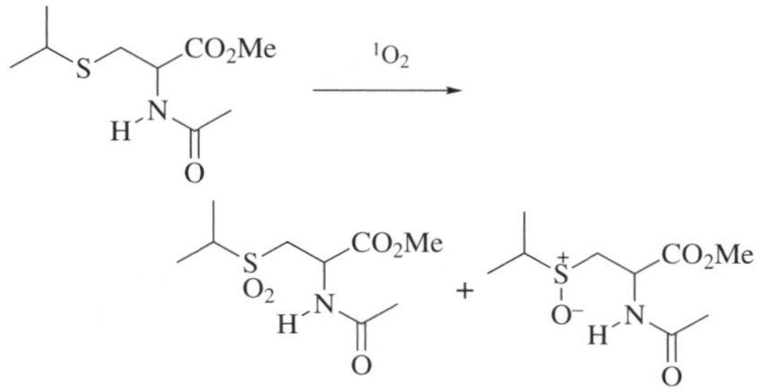

(31)

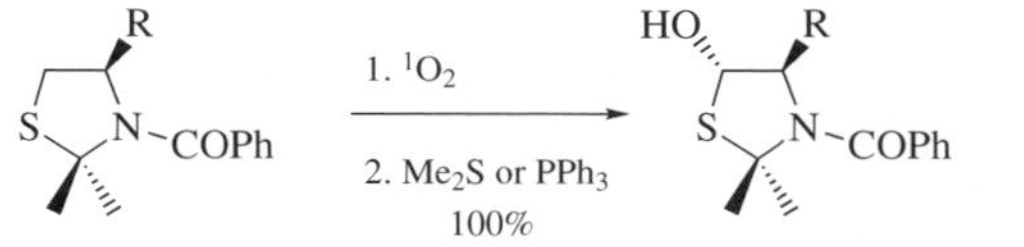

(32)

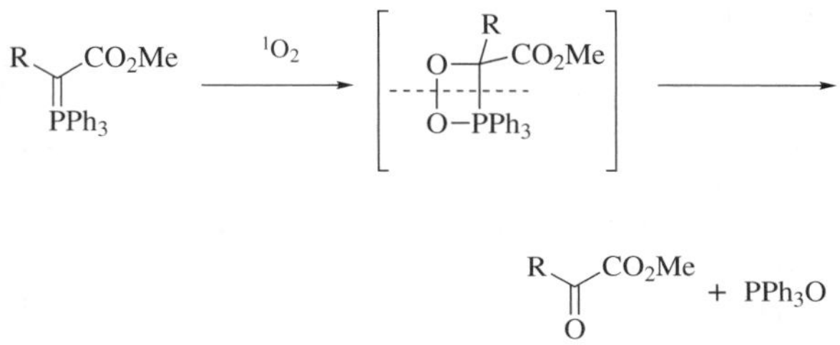

(33)

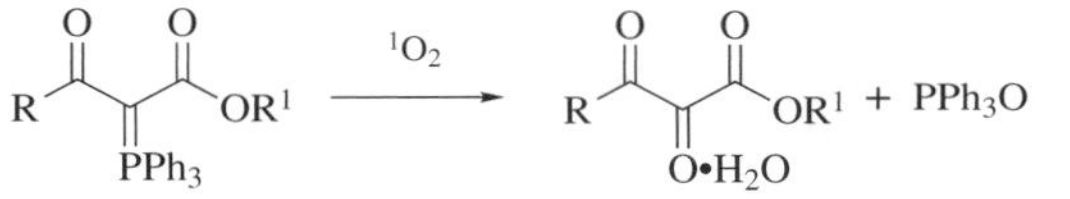

(34)

Trapping Dioxetanes and Other Peroxidic Singlet Oxygen Oxidation Products. In certain cases, dioxetane intermediates and other peroxidic species may be deoxygenated by reaction with diphenyl sulfide, a species which is unreactive toward $^1O_2$.[48] As an example,[53] the dye-sensitized photooxygenation of tetraphenylimidazole normally yields the dibenzoyl derivative in nearly quantitative yield, presumably through the dioxetane (eq 35). In the presence of diphenyl sulfide, a product of rearrangement is also formed along with diphenyl sulfoxide. Other studies on trapping agents for peroxidic intermediates have included the use of pinacolone for perepoxides[54] and dimethyl sulfide or triphenylphosphine for persulfoxides.[55]

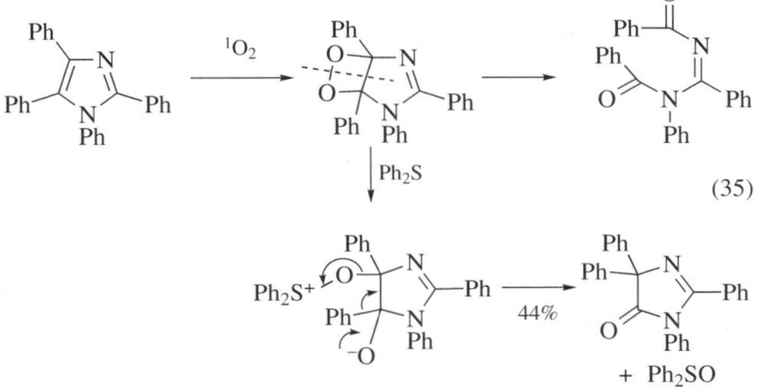

(35)

Sensitivity of $^1O_2$ Reactions to the Environment. The reactions of singlet oxygen with organic substrates are very sensitive to the experimental environment. Temperature, reaction time, substituents, solvent, and dissolved solutes may all affect the oxidations, resulting in mixtures of products of varied composition. This phenomenon is well illustrated in the studies on the sensitized photooxidation of a 5,6-disubstituted 3,4-dihydro-2*H*-pyran, in which solvent effects lead to widely varying product distributions (eq 36).[56] Table 3 illustrates these effects.

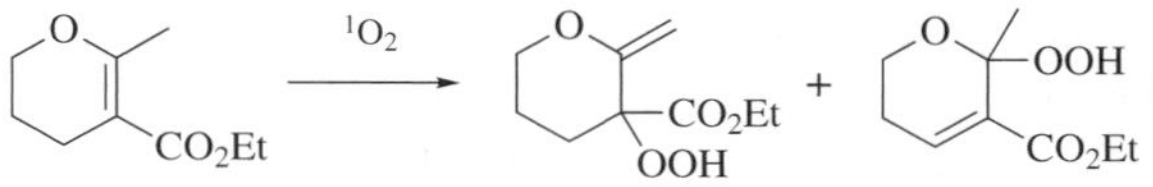

(36)

Table 3 Solvent Effects in the $^1O_2$ Ene Reaction

Solvent/sensitizer	Time (h)	% (**6**)	% (**7**)	(**6**):(**7**)
Benzene/TPP	9	10	90	0.11
CCl_4/TPP	7	17	83	0.18
$CHCl_3$/TPP	5	45	55	0.82
CH_2Cl_2/TPP	5	50	50	1.00
MeCN/MB	3.5	65	35	1.86

Related Reagents. See Classes O-8, O-17, and O-21, pages 1–10.

---

1. For reviews on the chemistry of singlet oxygen, see: (a) Gollnick, K.; Schenck, G. O. In *1,4-Cycloaddition Reactions, The Diels-Alder Reaction in Heterocyclic Synthesis*; Hamer, J., Ed.; Academic: New York, 1967; pp 255–344. (b) Gollnick, K. *Adv. Photochem.* **1968**, *6*, 1. (c) Foote, C. S. *Science* **1968**, *162*, 963. (d) Foote, C. S. *Acc. Chem. Res.* **1968**, *1*, 104. (e) Wayne, R. P. *Adv. Photochem.* **1969**, *7*, 311. (f) Foote, C. S. *PAC* **1971**, *27*, 635. (g) Denny, R. W.; Nickon, A. *OR* **1973**, *20*, 133. (h) Adam, W. *CZ* **1975**, *99*, 142. (i) *Singlet Molecular Oxygen*; Schaap, A. P., Ed.; Halsted: New York, 1976. (j) *Singlet Oxygen*; Wasserman, H. H.; Murray, R. W., Eds.; Academic: New York, 1979. (k) Frimer, A. A. *CRV* **1979**, *79*, 359. (l) George, M. V.; Bhat, V. *CRV* **1979**, *79*, 447.
2. (a) Schenck, G. O. *AG* **1957**, *69*, 579. (b) Kautsky, H. *Biochem. Z.* **1937**, *291*, 271. (c) Adam, W.; Klug, P. *JOC* **1993**, *58*, 3416.
3. Murray, R. W.; Kaplan, M. L. *JACS* **1969**, *91*, 5358.
4. Wasserman, H. H.; Scheffer, J. R.; Cooper, J. L. *JACS* **1972**, *94*, 4991.
5. (a) Foote, C. S.; Wexler, S. *JACS* **1964**, *86*, 3879. (b) Foote, C. S.; Wexler, S.; Ando, W.; Higgins, R. *JACS* **1968**, *90*, 975.
6. McKeown, E.; Waters, W. A. *JCS(B)* **1966**, 1040.
7. Corey, E. J.; Taylor, W. C. *JACS* **1964**, *86*, 3881.
8. Blossey, E. C.; Neckers, D. C.; Thayer, A. L.; Schaap, A. P. *JACS* **1973**, *95*, 5820.
9. (a) Merkel, P. B.; Kearns, D. R. *JACS* **1972**, *94*, 7244. (b) Long, C. A.; Kearns, D. R. *JACS* **1975**, *97*, 2018.
10. Hathaway, S. J.; Paquette, L. A. *T* **1985**, *41*, 2037.
11. Adam, W.; Klug, G.; Peters, E.-M.; Peters, K.; von Schnering, G. H. *T* **1985**, *41*, 2045.
12. Gollnick, K.; Griesbeck, A. *T* **1985**, *41*, 2057.
13. Matsumoto, M.; Dobashi, S.; Kuroda, K.; Kondo, K. *T* **1985**, *41*, 2147.
14. Ihara, M.; Noguchi, K.; Fukumoto, K.; Kametani, T. *T* **1985**, *41*, 2109.
15. Wasserman, H. H.; Pickett, J. E. *T* **1985**, *41*, 2155.
16. Utaka, M.; Nakatani, M.; Takeda, A. *T* **1985**, *41*, 2163.
17. Jefford, C. W.; Boukouvalas, J.; Kohmoto, S.; Bernardinelli, G. *T* **1985**, *41*, 2081.

18. Wasserman, H. H.; Scheffer, J. R. *JACS* **1967**, *89*, 3073.
19. Dufraise, C.; Etienne, A. *CR* **1935**, *201*, 280.
20. George, M. V.; Bhat, V. *CRV* **1979**, *79*, 447.
21. *Singlet Oxygen*; Wasserman, H. H.; Murry, R. W., Eds.; Academic: New York, 1979; Chapter 9.
22. Bartlett, P. D.; Mendenhall, D.; Schaap, A. P. *ANY* **1970**, *171*, 79.
23. (a) Adam, W.; Fierro, J. d. *JOC* **1978**, *43*, 1159. (b) Adam, W.; Fierro, J. d.; Quiroz, F.; Yany, F. *JACS* **1980**, *102*, 2127.
24. Bell, R. A.; Ireland, R. E. *TL* **1963**, 269.
25. Wharton, P. S.; Hiegel, G. A.; Coombs, R. V. *JOC* **1963**, *28*, 3217.
26. Büchi, G.; Hauser, A.; Limacher, J. *JOC* **1977**, *42*, 3323.
27. Ireland, R. E.; Baldwin, S. W.; Dawson, D. J.; Dawson, M. I.; Dolfini, J. E.; Newbould, J.; Johnson, W. S.; Brown, M.; Crawford, R. J.; Hudrlik, P. F.; Rasmussen, G. H.; Schmiegel, K. K. *JACS* **1970**, *92*, 5743.
28. Schuller, W. H.; Lawrence, R. V. *JACS* **1961**, *83*, 2563.
29. Wasserman, H. H.; Lu, T.-J.; Scott, A. I. *JACS* **1986**, *108*, 4237.
30. (a) Schenck, G. O.; Ziegler, K. *N* **1954**, *32*, 157. (b) Schenck, G. O. *AG* **1952**, *64*, 12.
31. Schenck, G. O. *AG* **1957**, *69*, 579.
32. Demuth, M. R.; Garrett, P. E.; White, J. D. *JACS* **1976**, *98*, 634.
33. Meyers, A. I.; Nolen, R. L.; Collington, E. W.; Narwid, T. A.; Strickland, R. C. *JOC* **1973**, *38*, 1974.
34. Heather, J. B.; Mittal, R. S. D.; Sih, C. J. *JACS* **1974**, *96*, 1976.
35. Foote, C. S.; Mazur, S.; Burns, P. A.; Lerdal, D. *JACS* **1973**, *95*, 586.
36. Rio, G.; Bricout, D.; Lacombe, L. *T* **1973**, *29*, 3553.
37. Boger, D. L.; Baldino, C. M. *JOC* **1991**, *56*, 6942.
38. Mazur, S.; Foote, C. S. *JACS* **1970**, *92*, 3225.
39. Bartlett, P. D.; Schaap, A. P. *JACS* **1970**, *92*, 3223.
40. Foote, C. S.; Lin, J. W.-P. *TL* **1968**, 3267.
41. Huber, J. E. *TL* **1968**, 3271.
42. Wasserman, H. H.; Lipshutz, B. H.; Tremper, A. W.; Wu, J. S. *JOC* **1981**, *46*, 2991.
43. Wasserman, H. H.; Pickett, J. E. *T* **1985**, *41*, 2155.
44. Ensley, H. E.; Carr, R. V. C.; Martin, R. S.; Pierce, T. E. *JACS* **1980**, *102*, 2836.
45. Ensley, H. E.; Balakrishnan, P.; Ugarkar, B. *TL* **1983**, *24*, 5189.
46. Schenck, G. O.; Krauch, C. H. *CB* **1963**, *96*, 517.
47. Kacher, M. L.; Foote, C. S. *Photochem. Photobiol.* **1979**, *29*, 765.
48. Gu, C.; Foote, C. S.; Kacher, M. L. *JACS* **1981**, *103*, 5949.
49. Gu, C.; Foote, C. S. *JACS* **1982**, *104*, 6060.
50. Takata, G.; Tamura, Y.; Ando, W. *T* **1985**, *41*, 2133.
51. Jefford, C. W.; Barchietto, G. *TL* **1977**, 4531.
52. Wasserman, H. H.; Ennis, D. S.; Blum, C. A.; Rotello, V. M. *TL* **1992**, *33*, 6003.
53. Wasserman, H. H.; Saito, I. *JACS* **1975**, *97*, 905.
54. Schaap, A. P.; Faler, G. R. *JACS* **1973**, *95*, 3381.
55. Foote, C. S.; Peters, J. W. *JACS* **1971**, *93*, 3795.
56. Chan, Y.-Y.; Li, X.; Zhu, C.; Liu, X.; Zhang, Y.; Leungg, H.-K. *JOC* **1990**, *55*, 5497.

Harry H. Wasserman
Yale University, New Haven, CT, USA

Robert W. DeSimone
Neurogen Corporation, Branford, CT, USA

Sodium[1]

[7440-23-5] Na (MW 22.99)

(powerful one-electron reducing agent for most functional groups;[1] reductively couples ketones and esters to unsaturated carbon atoms;[2] reductively eliminates and couples halogens and other groups; can be used to generate alkoxides)

Physical Data: mp 97.8 °C; *d* 0.968 g cm^{-3}.
Solubility: sol liquid NH_3; slightly sol ethereal solvents; dec in alcohols; reacts violently in H_2O.
Form Supplied in: silver-white or gray solid in brick, stick, or ingot form; more commonly as 3–8 mm spheres or as a dispersion in various inert oils.
Purification: surface oxide can be removed by heating in toluene;[3] oxide impurities can usually be ignored.
Handling, Storage, and Precautions: the dry solid quickly oxidizes when exposed to air and in very moist air is potentially flammable. The highly corrosive sodium vapor ignites spontaneously in air, and is a severe irritant to eyes, skin, and mucous membranes. Utmost care should be taken to keep sodium away from halogenated solvents, oxidants, and aqueous mineral acids. Fires should be quenched with a dry powder such as Na_2CO_3, NaCl, or NaF; water, CO_2, Halon, and silica gel should be avoided. Excess sodium is best disposed of by slow introduction into a flask of isopropanol, possibly containing 1% water, taking care to vent liberated H_2, and neutralizing the resulting solution with aqueous acid. Sodium stored under oil is best weighed by rinsing freshly cut pieces with hexane and after evaporation adding to a tared beaker containing hexane.

Reaction Conditions. Sodium, along with the other commonly used alkali metals ***Lithium*** and ***Potassium***, is an extremely powerful one-electron reductant. Historically, solutions of sodium in ammonia have been used to reduce a wide variety of functional groups (see also ***Sodium–Ammonia*** for some of these transformations, particularly the Birch reduction of aromatic rings).[4] These reductions are usually carried out in the presence of proton donors, normally simple alcohols or NH_4Cl, though occasionally these are added prior to workup. Many reductions are performed with solutions of sodium in refluxing alcohols; here a large excess of sodium is often required due to its gradual decomposition to alkoxides and H_2. It is sometimes simpler to add a mixture of alcohol and substrate to a dispersion of sodium in an inert solvent such as toluene. HMPA is also occasionally employed as a solvent;[5] solutions of sodium in HMPA behave similarly to those in ammonia.[6] While sodium is insoluble in solvents such as DME or THF, naphthalene is usually included, and a green solution of the ***Sodium Naphthalenide*** radical anion forms.[7] Though the reduction potential of this system is similar to that of sodium in other solvents (a half-wave potential of −2.5 V[8] vs. −2.96 V in HMPA[6b] and −2.59 V in NH_3[1b]), side reactions are often minimized, and reductions are sometimes titrated to color end-point. Other additives such as anthracene and benzophenone result in solutions with decreased, more selective, reducing power, though this has been rarely exploited.[8]

Reduction of Unsaturated Bonds. The classical reduction of ketones with sodium in alcohol, along with the mechanistically similar reduction in ammonia with added proton donors, still finds use in certain systems where the desired stereochemistry cannot be generated by metal hydrides (see also ***Lithium***).[1d–g] The ammonia system is preferred, minimizing epimerization of ketones and equilibration of product alcohols. NH_4Cl has been recommended as proton donor to suppress hydrogen transfer,[9] though this requires a large excess of sodium;[10] using several equivalents of a primary alcohol also minimizes this side reaction.[11] Alcohols, ethers, amines, carboxylic acids, and isolated internal double bonds are unaffected, but most other functional groups have reduction potentials less negative than aliphatic ketones. Aromatic ketones and α-heteroatom-substituted ketones are generally unsuitable substrates for these reductions.

Reductions are kinetically controlled, though the literature contains several mistaken assumptions on this point; stereochemistry is determined by pyramidalization of the ketyl and subsequent carbanion intermediates, and prediction of the stereochemical result is in most cases extremely difficult. Study of certain systems does allow a few general remarks to be made. Unhindered cyclohexanones usually give larger equatorial:axial ratios than can be obtained with metal hydrides or by equilibration.[12] Generation of equatorial alcohols from hindered cyclohexanones often requires the use of dissolving metals, while metal hydrides can give exclusively the axial product (eqs 1 and 2).[13] Bicyclo[2.2.1]heptan-2-ones always provide an excess of the *endo*-alcohol (eq 3).[14,15]

Na, NH_3, EtOH, 100% (1)

94:6

Na, EtOH, Δ, 95% (2)

Na, *i*-PrOH, Δ, 58% (3)

Imines are reduced to amines in good yield by sodium in alcohols, probably through a similar mechanism. In the absence of proton donors, reductive dimerization dominates.[16] 1,3-Diimines, which exist as tautomeric enamino imine mixtures, are reduced to diastereomeric mixtures of diamines (eq 4).[17] Allylic imines have been reduced to alkenes.[18] Reduction of oximes by sodium in alcohol is particularly useful,[19] since ***Lithium Aluminum Hydride*** can produce Beckmann rearrangements or aziridines.[20,21] Bicyclo[2.2.1]heptan-2-one and tricyclo[2.2.1.0^{2,6}]heptan-3-one oxime reductions yield predominantly *endo*-amines (eq 5).[21,22]

Na, *i*-PrOH, 98% (4)

Na, EtOH, Δ, 50% (5)

14:86

The reduction of esters to alcohols with sodium in alcohol (Bouveault–Blanc reduction)[23] has been all but supplanted by the use of metal hydrides. Where an ester must be reduced in the presence of an acid, an improved procedure run in ammonia can be used effectively (eq 6).[24]

Na, NH_3, EtOH, 72% (6)

Dissolving metal reductions of α,β-unsaturated carbonyl compounds are almost always performed with lithium in ammonia (see ***Lithium***),[1d] though sodium is occasionally used without any particular change in result. Selected examples of enone reduction with sodium in HMPA have been examined;[6b,25] these are most notable for the fact that they tend to produce product mixtures enriched in the less stable epimer, compared to reactions run in ammonia (eq 7).[25a,1e]

Na, *t*-BuOH (7)

in NH_3, THF	84:10
in HMPA, THF	24:60

The reduction of internal acyclic alkynes to *trans*-alkenes by sodium in ammonia is well known.[1b] Complications in reducing terminal alkynes result from deprotonation by $NaNH_2$, formed in situ. This can be exploited by addition of an equivalent of $NaNH_2$,[26] or overcome by addition of ammonium salts.[27] Medium-ring cyclic alkynes often give mixtures of *cis*- and *trans*-alkenes, resulting from partial isomerization by $NaNH_2$ to allenes,[28] which in turn give *cis*/*trans* mixtures dependent on product stability and the presence or absence of proton donors.[29] Sodium in HMPA/*t*-BuOH has been shown to reduce alkynes;[30] internal monoalkenes, inert under most conditions, give near-equilibrium mixtures of alkanes with this system (eqs 8 and 9).[31] Conjugated dienes are reduced via a *cis*-radical anion,[32] while, with lithium, *trans*-dianions may play a role. Dimerization and regioisomer formation usually preclude the use of sodium, though exceptions exist.[33] A few reports of homoconjugated diene reduction have appeared.[34]

(8) Na, t-BuOH, HMPA; 91:3

(9) Na, t-BuOH, HMPA; 73:27

Pinacol Coupling Reactions. In the absence of proton donors, reductions of carbonyl compounds often lead to significant amounts of reductively coupled products, but the intentional pinacol coupling of ketones with sodium in inert solvents is typically a low-yielding, unselective process. Milder conditions utilize lanthanoids and low-valent transition metals, particularly titanium.[35] Sodium is used only in cases where no other functional groups are present.[36] Sodium is often employed in related intramolecular couplings with unsaturated carbon–carbon bonds, forming five- and six-membered rings regio- and (often) stereoselectively.[37] Ketyls cyclize onto allylic systems with *anti* displacement (eqs 10 and 11).[38] Transannular cyclizations have also been demonstrated.[39] Alkynes are cyclized similarly (eq 12),[40] usually with ***Sodium Naphthalenide***.[41] With allenes, the *exo* closure product is obtained (eq 13),[42] though prolonged reaction can scramble the position of the resulting double bond.[43]

(10) Na, THF; 64%

(11) Na, THF; 70%

(12) Na, NH_3, $(NH_4)_2SO_4$; ~90%

(13) Na, $C_{10}H_8$, THF; 30%

Acyloin Couplings. Reductions of esters under aprotic conditions, usually refluxing toluene or xylene, result in α-hydroxy ketones, also called acyloins.[2] ***Sodium–Potassium Alloy*** alloy is also used and allows reaction at lower temperatures. Rigorous exclusion of oxygen during reaction and workup is essential. Base-catalyzed side reactions are suppressed by the coaddition of substrate and ***Chlorotrimethylsilane***, so as to trap the enediolate and alkoxide products, though good yields can sometimes be obtained regardless.[2,44] Workup is greatly simplified, and the acyloins can be freed by treatment with aqueous acid or deoxygenated methanol. Couplings can be performed on esters with leaving groups at the β-position (eq 14),[45] resulting in a useful cyclopropanone hemiacetal synthesis (eq 15).[46]

(14) Na, TMSCl; 69%

(15) Na, TMSCl, PhMe, Δ; 61%

Intramolecular acyloin couplings are an enormously successful method for forming four-membered rings.[47] Highly strained products can undergo subsequent thermal ring opening (eq 16); Na/K is sometimes more effective in these cases. Hindered diesters can undergo α,α-bond cleavage prior to reduction.[2] Five- and six-membered rings are routinely generated, often without employing TMSCl. Larger rings, more difficult to form by other methods, are also closed in good yields,[48] including several polycyclic frameworks (eq 17).[49] Rings containing 12, 24, and 42 carbon atoms,[50] paracyclophanes, and rings containing N, O, S, and Si atoms have been formed.[51]

(16) Na, TMSCl, PhMe, Δ; Δ, 86%

(17) Na, TMSCl, PhMe, Δ; $FeCl_3$, HCl, ether, 70%

Reduction of Saturated Bonds. Typical Birch reduction conditions dehalogenate aryl, vinyl, bridgehead, and cyclopropyl halides, though side reactions are often troublesome.[52] Alkyl fluorides are not reduced by sodium under any conditions, requiring a K/crown ether system.[53] Sodium in alcohol/THF is an effective substitute for Birch conditions;[54,55] these conditions can also reduce simple alkyl halides (eq 18).[36b] A simpler procedure employing refluxing ethanol is superior in some cases (eq 19).[56]

(18) Na, t-BuOH, THF; 75%

(19) Na, EtOH, Δ; 69%

Alcohols are usually derivatized and reduced with metal hydrides, though other methods have received attention.[57] Among dissolving metal techniques, lithium or potassium in amines are most often used to hydrogenolyze various esters. Sodium in HMPA gives nearly quantitative yields of alkanes from tertiary esters.[58] Sodium in ammonia can cleave alkyl mesylates (eq 20),[59] while phosphate esters or mesylates of phenols are similarly reduced to aromatic hydrocarbons.[60] Like metal hydrides, sodium in ammonia cleaves epoxides to give the more substituted alcohols, with the exception of aryl-substituted epoxides (eq 21).[61] Lithium and second-row metals, particularly calcium, are also effective for epoxide cleavages.

(20) Na, NH_3, THF, 80%

(21) Na, NH_3, 89%

Most other aliphatic ethers are inert to dissolving metals, but benzyl ethers (as well as esters, amines, and thioethers) are readily debenzylated by sodium in ammonia. This is a common deprotection method in peptide and carbohydrate chemistry.[62] Substituted trityl ethers have been used to protect hydroxyl groups in nucleotides. The differing reduction potentials for the *p*-methoxytrityl[63] and α-naphthyldiphenylmethyl[64] groups allow for selective deprotection using sodium/aromatic hydrocarbon systems (eqs 22 and 23). Simple phenyl ethers are cleaved by sodium only with great difficulty.

(22) Na, anthracene, THF, 87%

(23) Na, naphthalene, THF, HMPA, 92%

While many desulfurizations by alkali metals are known,[65] use of sodium is generally restricted to alkyl aryl thioethers. These are cleaved in the presence of aryl ethers by sodium in HMPA.[66] Phenyl thioethers are more often cleaved in refluxing alcohols (eq 24).[67] By substituting TMSCl for the proton source, alkyl or vinyl silanes can be isolated (eq 25).[68] Sodium in ammonia has been used in a deoxy sugar synthesis where more typical reagents (***Raney Nickel***, ***Nickel Boride***, ***Tri-n-butylstannane***) fail.[69]

(24) Na, BuOH, Δ, 93%

(25) Na, TMSCl, C_6H_6, Δ, 88%

Sulfones are hydrogenolyzed by sodium in ethanol.[70] This method is again most often used where commonly preferred methods fail (eq 26),[71] and includes a rare application to an α-substituted carboxylic acid (eq 27).[72] Aryl sulfonamides are readily cleaved to amines with sodium naphthalenide and ***Sodium Anthracenide***.[73] Tosylates[74,62] and *N*-tosylsulfoximines[75] are similarly reduced to alcohols and sulfoximines, respectively.

(26) Na, EtOH, THF, 80%

(27) Na, EtOH, THF, 72%

Reductive decyanation is effectively performed with sodium in ammonia.[76] This method complements ***Sodium Borohydride***, which is sometimes ineffective with α-amino nitrile substrates and provides products with inversion of configuration (eq 28).[77] An alternative method employing sodium or preferably potassium in HMPA/*t*-BuOH also smoothly removes nitriles.[78] Isocyanides are reduced to hydrocarbons, providing an effective deamination method (eq 29).[79] Rearrangements are largely avoided but, with acyclic substrates, loss of optical activity results. Finally, one last common application of the sodium/ammonia system is the reductive cleavage of acylated N–N bonds (eq 30).[80]

(28) Na, NH_3, THF, 100%

(29) Na, $C_{10}H_8$, DME, 87%

(30) Na, NH_3, 88%

Reductive Eliminations. When a leaving group is situated adjacent to a reducible functionality, elimination results. It is generally agreed that at the site of the initially reduced carbon atom, an anion is generated which then displaces the leaving group. Highly strained alkenes have been generated in this way. Usually *vic*-dihalides are the immediate precursor (eq 31),[81] though *vic*-dimesylates have also been used.[82] Alkenes have also been generated from *vic*-dinitriles[83] and more exotic combinations of functional groups.[84] Fragmentations of β-chloro ethers,[85] β,γ-epoxy nitriles,[86] and β-hydroxy nitriles[87] have also been demonstrated, the last undergoing subsequent double bond reduction (eqs 32–34).

(31) Na, $C_{10}H_8$, DME, 73%

(32) Na, THF, 96%

(33) Na, NH_3, 90%

(34) Na, NH_3, EtOH, 68%

Wurtz Reaction. The classical intermolecular coupling of halides with sodium, mechanistically related to the eliminations noted above, is of limited use. Cross couplings lead to mixtures of desired and homocoupled products, while dimerizations are of little synthetic value. Magnesium, lithium, and copper reagents are normally employed here.[88] Intramolecular couplings frequently employ sodium, however. Ring closure to form [2.2]phanes is effective when tetraphenylethylene (TPE) is used catalytically (eq 35).[89] Greatest utility is found in the formation of cyclopropanes (eq 36),[90] and to a lesser extent cyclobutanes (eq 37),[91] where ***Zinc*** is preferred.

(35) Na, TPE, THF, 44%

(36) Na, dioxane, 78–94%

(37) Na, $C_{10}H_8$, glyme, 6.5%

Use as a Base. Sodium reacts slowly with alcohols to give solutions of alkoxides. While sodium alkoxides are far more commonly generated by ***Sodium Hydride***, or obtained commercially, the older method is still occasionally used. The protic Bamford–Stevens reaction uses sodium to generate alkoxides from ***Ethylene Glycol***, present as solvent.[92] Sodium methoxide can be substituted, but the original conditions are usually employed. Regioselectivity can be a problem; while the more substituted regioisomer usually predominates, prediction is difficult.[93] Rearrangements of cationic intermediates also limit its use, though many successful examples of this reaction do exist (eq 38).[94]

(38) Na, ethylene glycol, Δ, 95%

Related Reagents. See Classes R-2, R-3, R-4, R-5, R-9, R-13, R-15, R-24, R-25, R-26, R-27, R-28, R-30, R-31, and R-32, pages 1–10. Sodium–Alcohol; Sodium–Alumina; Sodium–Ammonia.

---

1. (a) House, H. O. *Modern Synthetic Reactions*; Benjamin: Menlo Park, CA, 1972. (b) Smith, M. In *Reduction: Techniques and Applications in Organic Synthesis*; Augustine, R. L., Ed.; Dekker: New York, 1968. (c) Hudlicky, M. *Reductions in Organic Chemistry*; Horwood: Chichester, 1984. (d) Caine, D. *OR* **1976**, *23*, 1. (e) Pradhan, S. K. *T* **1986**, *42*, 6351. (f) Huffman, J. W. *ACR* **1983**, *16*, 398. (g) Huffman, J. W. *COS* **1991**, *8*, Chapter 1.4.
2. (a) Finley, K. T. *CR* **1964**, *64*, 573. (b) Bloomfield, J. J.; Owsley, D. C.; Nelke, J. M. *OR* **1976**, *23*, 259.
3. Fieser, M.; Fieser, L. F. *FF* **1967**, *1*, 1022.
4. *The Chemistry of Non-Aqueous Solvents*; Lagowski, J. J., Ed.; Academic: New York, 1967; Vol. 2, Chapter 6,7.
5. Normant, H. *AG(E)* **1967**, *6*, 1046.
6. (a) Schindewolf, U. *AG(E)* **1968**, *7*, 190. (b) Bowers, K. W.; Giese, R. W.; Grimshaw, J.; House, H. O.; Kolodny, N. H.; Kronberger, K.; Roe, D. K. *JACS* **1970**, *92*, 2783.
7. Garst, J. F. *ACR* **1971**, *4*, 400.
8. Mann, C. K.; Barnes, K. K. *Electrochemical Reactions in Non-Aqueous Systems*; Dekker: New York, 1970.
9. Rautenstrauch, V.; Willhalm, B.; Thommen, W.; Burger, U. *HCA* **1981**, *64*, 2109.
10. Grieco, P. A.; Burke, S.; Metz, W.; Nishizawa, M. *JOC* **1979**, *44*, 152.
11. Huffman, J. W.; Copley, D. J. *JOC* **1977**, *42*, 3811.
12. (a) Huffman, J. W.; Charles, J. T. *JACS* **1968**, *90*, 6486. (b) Solodar, J. *JOC* **1976**, *41*, 3461.
13. (a) Giroud, A. M.; Rassat, A. *BSF* **1976**, 1881. (b) Aranda, G.; Bernassau, J.-M.; Fetizon, M.; Hanna, I. *JOC* **1985**, *50*, 1156.
14. Barton, D. H. R.; Werstiuk, W. H. *JCS(C)* **1968**, 148.
15. (a) Welch, S. C.; Walters, R. L. *SC* **1973**, *3*, 419. (b) Rautenstrauch, V. *CC* **1986**, 1558.
16. (a) Smith, J. G.; Ho, I. *JOC* **1972**, *37*, 653. (b) Jaunin, R.; Magnenat, J.-P. *HCA* **1959**, *42*, 328.
17. Barluenga, J.; Olano, B.; Fustero, S. *JOC* **1983**, *48*, 2255.
18. Barbulescu, N.; Cuza, O.; Barbulescu, E.; Moya-Gheorghe, S.; Zavoranu, D. *RRC* **1985**, *36*, 295 (*CA* **1985**, *103*, 123 042t).
19. Lycan, W. H.; Puntambeker, S. V.; Marvel, C. S. *OSC* **1943**, *2*, 318.
20. Chen, S.-C. *S* **1974**, 691.
21. Ordubadi, M. D.; Pekhk, T. I.; Belikova, N. A.; Rakhmanchik, T. M.; Platé, A. F. *JOU* **1984**, *20*, 678.

22. Daniel, A.; Pavia, A. A. *BSF* **1971**, 1060.
23. (a) Ford, S. G.; Marvel, C. S. *OSC* **1943**, *2*, 372. (b) Adkins, H.; Gillespie, R. H. *OSC* **1955**, *3*, 671.
24. Paquette, L. A.; Nelson, N. A. *JOC* **1962**, *27*, 2272.
25. (a) House, H. O.; Giese, R. W.; Kronberger, K.; Kaplan, J. P.; Simeone, J. F. *JACS* **1970**, *92*, 2800. (b) Argibeaud, P.; Larchevêque, M.; Normant, H.; Tchoubar, B. *BSF* **1968**, 595.
26. Dobson, N. A.; Raphael, R. A. *JCS* **1955**, 3558.
27. Henne, A. L.; Greenlee, K. W. *JACS* **1943**, *65*, 2020.
28. Svoboda, M.; Sicher, J.; Závada, J. *TL* **1964**, 15.
29. Vaidyanathaswamy, R.; Joshi, G. C.; Devaprabhakara, D. *TL* **1971**, 2075.
30. House, H. O.; Kinloch, E. F. *JOC* **1974**, *39*, 747.
31. Whitesides, G. M.; Ehmann, W. J. *JOC* **1970**, *35*, 3565.
32. Weyenberg, D. R.; Toporcer, L. H.; Nelson, L. E. *JOC* **1968**, *33*, 1975.
33. (a) Loev, B.; Dawson, C. R. *JACS* **1956**, *78*, 1180. (b) Barton, D. H. R.; Lusinchi, X.; Magdzinski, L.; Ramirez, J. S. *CC* **1984**, 1236.
34. (a) Butler, D. N. *SC* **1977**, *7*, 441. (b) Boland, W.; Hansen, V.; Jaenicke, L. *S* **1979**, 114.
35. Robertson, G. M. *COS* **1991**, *3*, Chapter 2.6.
36. (a) Wynberg, H.; Boelema, E.; Wieringa, J. H.; Strating, J. *TL* **1970**, 3613. (b) Nelsen, S. F.; Kapp, D. L. *JACS* **1986**, *108*, 1265.
37. (a) Hart, D. J. *Science* **1984**, *223*, 883. (b) Ramaiah, M. *T* **1987**, *43*, 3541.
38. Bertrand, M.; Teisseire, P.; Pélerin, G. *NJC* **1983**, *7*, 61.
39. (a) Jadhav, P. K.; Nayak, U. R. *IJC(B)* **1978**, *16*, 1047. (b) Eakin, M.; Martin, J.; Parker, W. *CC* **1965**, 206.
40. Jung, M. E.; Hatfield, G. L. *TL* **1983**, 3175.
41. (a) Pattenden, G.; Teague, S. J. *JCS(P1)* **1988**, 1077. (b) Pradhan, S. K.; Kadam, S. R.; Kolhe, J. N.; Radhakrishnan, T. V.; Sohani, S. V.; Thaker, V. B. *JOC* **1981**, *46*, 2622. (c) Mehta, G.; Krishnamurthy, N. *TL* **1987**, 5945.
42. Pattenden, G.; Robertson, G. M. *T* **1985**, *41*, 4001.
43. Crandall, J. K.; Mualla, M. *TL* **1986**, 2243.
44. Snell, J. M.; McElvain, S. M. *OSC* **1943**, *2*, 114.
45. Rühlmann, K. *S* **1971**, 236.
46. Salaün, J.; Marguerite, J. *OSC* **1990**, *7*, 131.
47. Bloomfield, J. J.; Nelke, J. M. *OSC* **1988**, *6*, 167.
48. Bloomfield, J. J.; Owsley, D. C.; Ainsworth, C.; Robertson, R. E. *JOC* **1975**, *40*, 393.
49. Bartetzko, R.; Gleiter, R.; Muthard, J. L.; Paquette, L. A. *JACS* **1978**, *100*, 5589.
50. (a) Natrajan, A.; Ferrara, J. D.; Youngs, W. J.; Sukenik, C. N., *JACS* **1987**, *109*, 7477. (b) Ashkenazi, P.; Kettenring, J.; Migdal, S.; Gutman, A. L.; Ginsburg, D. *HCA* **1985**, *68*, 2033.
51. (a) Wu, G.-S.; Martinelli, L. C.; Blanton, C. D., Jr.; Cox, R. H. *JHC* **1977**, *14*, 11. (b) Johnson, P. Y.; Kerkman, D. J. *JOC* **1976**, *41*, 1768.
52. Vogel, E.; Roth, H. D. *AG(E)* **1964**, *3*, 228.
53. Ohsawa, T.; Takagaki, T.; Haneda, A.; Oishi, T. *TL* **1981**, 2583.
54. (a) Gassman, P. G.; Pape, P. G. *JOC* **1964**, *29*, 160. (b) Gassman, P. G.; Marshall, J. L. *OSC* **1973**, *5*, 424.
55. (a) Chou, T. C.; Chuang, K.-S.; Lin, C.-T. *JOC* **1988**, *53*, 5168. (b) Hales, N. J.; Heaney, H.; Hollinshead, J. H. *S* **1975**, 707. (c) Hales, N. J.; Heaney, H.; Hollinshead, J. H.; Singh, P. *OSC* **1988**, *6*, 82.
56. Lap, B. V.; Paddon-Row, M. N. *JOC* **1979**, *44*, 4979.
57. Hartwig, W. *T* **1983**, *39*, 2609.
58. Deshayes, H.; Pete, J.-P. *CC* **1978**, 567.
59. Tsuchiya, T.; Nakamura, F.; Umezawa, S. *TL* **1979**, 2805.
60. (a) Rossi, R. A.; Bunnett, J. F. *JOC* **1973**, *38*, 2314. (b) Carnahan, J. C., Jr.; Closson, W. D.; Ganson, J. R.; Juckett, D. A.; Quaal, K. S. *JACS* **1976**, *98*, 2526.
61. Kaiser, E. M.; Edmonds, C. G.; Grubb, S. D.; Smith, J. W.; Tramp, D. *JOC* **1971**, *36*, 330.
62. Schön, I. *CR* **1984**, *84*, 287.
63. (a) Greene, G. L.; Letsinger, R. L. *TL* **1975**, 2081. (b) Letsinger, R. L.; Lunsford, W. B. *JACS* **1976**, *98*, 3655.
64. Letsinger, R. L.; Finnan, J. L. *JACS* **1975**, *97*, 7197.
65. Block, E.; Aslam, M. *T* **1988**, *44*, 281.
66. Tiecco, M. *S* **1988**, 749.
67. (a) Kodama, M.; Takahashi, T.; Kojima, T.; Ito, S. *TL* **1982**, 3397. (b) Hashimoto, M.; Kan, T.; Yanagiya, M.; Shirahama, H.; Matsumoto, T. *TL* **1987**, 5665. (c) Purrington, S. T.; Pittman, J. H. *TL* **1988**, 6851.
68. (a) Kuwajima, I.; Kato, M.; Sato, T. *CC* **1978**, 478. (b) Kuwajima, I.; Abe, T.; Atsumi, K. *CL* **1978**, 383. (c) Atsumi, K.; Kuwajima, I. *CL* **1978**, 387.
69. Haskell, T. H.; Woo, P. W. K.; Watson, D. R. *JOC* **1977**, *42*, 1302.
70. (a) Masaki, Y.; Serizawa, Y.; Nagata, K.; Kaji, K. *CL* **1984**, 2105. (b) Kozikowski, A. P.; Mugrage, B. B.; Li, C. S.; Felder, L. *TL* **1986**, 4817.
71. (a) Fujita, Y.; Ishiguro, M.; Onishi, T.; Nishida, T. *BCJ* **1982**, *55*, 1325. (b) Chan, T. H.; Labrecque, D. *TL* **1991**, 1149.
72. Kuo, Y.-C.; Aoyama, T.; Shioiri, T. *CPB* **1982**, *30*, 2787.
73. (a) Closson, W. D.; Ji, S.; Schulenberg, S. *JACS* **1970**, *92*, 650. (b) Quaal, K. S.; Ji, S.; Kim, Y. M.; Closson, W. D.; Zubieta, J. A. *JOC* **1978**, *43*, 1311.
74. Closson, W. D.; Wriede, P.; Bank, S. *JACS* **1966**, *88*, 1581.
75. Johnson, C. R.; Lavergne, O. *JOC* **1989**, *54*, 986.
76. Tomioka, K.; Koga, K.; Yamada, S. *CPB* **1977**, *25*, 2689.
77. Bonin, M.; Romero, J. R.; Grierson, D. S.; Husson, H.-P. *TL* **1982**, 3369.
78. Debal, A.; Cuvigny, T.; Larchevéque, M. *S* **1976**, 391.
79. Niznik, G. E.; Walborsky, H. M. *JOC* **1978**, *43*, 2396.
80. (a) Meng, Q.; Hesse, M. *SL* **1990**, 148. (b) Kemp, D. S.; Sidell, M. D.; Shortridge, T. J. *JOC* **1979**, *44*, 4473.
81. (a) Sampath, V.; Lund, E. C.; Knudsen, M. J.; Olmstead, M. M.; Schore, N. E. *JOC* **1987**, *52*, 3595. (b) Greene, A. E.; Luche, M.-J.; Serra, A. A. *JOC* **1985**, *50*, 3957. (c) Allred, E. L.; Beck, B. R.; Voorhees, K. J. *JOC* **1974**, *39*, 1426.
82. Hrovat, D. A.; Miyake, F.; Trammell, G.; Gilbert, K. E.; Mitchell, J.; Clardy, J.; Borden, W. T. *JACS* **1987**, *109*, 5524.
83. De Lucchi, O.; Piccolrovazzi, N.; Licini, G.; Modena, G.; Valle, G. *G* **1987**, *117*, 401.
84. (a) Marshall, J. A.; Karas, L. J. *JACS* **1978**, *100*, 3615. (b) Nicolaou, K. C.; Sipio, W. J.; Magolda, R. L.; Claremon, D. A. *CC* **1979**, 83.
85. (a) Martin, J. D.; Pérez, C.; Ravelo, J. L. *JACS* **1985**, *107*, 516. (b) Brooks, L. A.; Snyder, H. R. *OSC* **1955**, *3*, 698.
86. Marshall, J. A.; Hagan, C. P.; Flynn, G. A. *JOC* **1975**, *40*, 1162.
87. Kametani, T.; Nemoto, H. *TL* **1979**, 27.
88. Billington, D. C. *COS* **1991**, *3*, Chapter 2.1.
89. Vogtle, F.; Neumann, P. *S* **1973**, 85.
90. (a) Lampman, G. M.; Aumiller, J. C. *OSC* **1988**, *6*, 133. (b) House, H. O.; Lord, R. C.; Rao, H. S. *JOC* **1956**, *21*, 1487. (c) Friedlina, R. K.; Kamyshova, A. A.; Chukovskaya, E. T. *RCR* **1982**, *51*, 368.
91. Wiberg, K. B.; Williams, V. Z., Jr. *JOC* **1970**, *35*, 369.
92. Shapiro, R. H. *OR* **1976**, *23*, 405.
93. Gianturco, M. A.; Friedel, P.; Flanagan, V. *TL* **1965**, 1847.
94. Piers, E.; Keziere, R. J. *CJC* **1969**, *47*, 137.

Michael D. Wendt
Abbott Laboratories, Abbott Park, IL, USA

Sodium–Alcohol[1]

Na–ROH

(Na)

[7440-23-5] Na (MW 22.99)

(reducing agent for esters, amides, ketones, oximes, nitriles, sulfonamides, aromatic hydrocarbons, certain alkenes, ethers, and carbon–halogen bonds; moderately strong base)

Physical Data: Na: mp 97.5 °C; bp 880 °C; *d* 0.97 g cm^{-3}. EtOH: mp −11ime7.3 °C; bp 78.5 °C; *d* 0.7893 g cm^{-3}.

Form Supplied in: Na: silvery white, soft metal. EtOH and other alcohols: liquid.

Drying: absolute ethanol can usually be employed without further drying. Most other common alcohols can be dried by storing over anhydrous calcium sulfate followed by distillation.

Handling, Storage, and Precautions: **Sodium** metal is a water-reactive, flammable solid. In case of contact with water, the heat of the reaction is usually sufficient to ignite the product hydrogen. Ethanol and other alcohols likely to be encountered in these reductions are flammable liquids. Use in a fume hood.

Reduction of Esters and Carboxamides. Though often replaced by the use of more modern hydride reagents, the Bouveault–Blanc reduction has long been known to convert esters to alcohols by refluxing the former reagents with sodium in alcohols.[2] For example, ethyl hydrocinnamate is converted to hydrocinnamyl alcohol in good yields (eq 1).[2] Numerous examples of reductions of mono-[3] and diesters[4] can be found in the literature. The major improvement in the procedure over the years has been the use of stoichiometric amounts of ester, alcohol, and toluene or xylene added to sodium in xylene.[5] Treatment of unsaturated esters with sodium in alcohols reduces conjugated double bonds[2] but not unconjugated alkenes (eq 2).[6] Less common have been reductions of carboxamides to amines by sodium in alcohols. For example, a number of primary, secondary, and tertiary amides have been reduced to their corresponding amines in good yields by sodium in propanol.[7]

Reduction of Ketones and Derivatives. Aliphatic ketones and phenones are readily converted to alcohols by sodium in alcohols,[8] as illustrated by the thermodynamically controlled reduction shown in eq 3, which forms part of a recently described synthesis of taxol and its analogs.[8d]

Reduction by sodium in alcohols of α,β-unsaturated ketones affords saturated alcohols (eq 4),[9] while similar reductions of diaryl ketones give methylene derivatives (eq 5).[10]

Oximes are conveniently converted to amines by sodium in alcohols.[11] For example, the oxime of 2-phenylcyclohexanone is reduced by sodium in ethanol to the corresponding amine in quantitative yield (eq 6).[11b] The use of higher molecular weight alcohols has been reported to afford better yields of products.[11d]

Reduction of Nitriles. Aliphatic nitriles are also reduced to amines in satisfactory yields by sodium in *n*-butyl alcohol[12a] or ethanol and toluene (eq 7).[12b]

Reduction of Sulfonamides. Both aliphatic and aromatic sulfonamides are converted to sulfinic acids and amines by sodium in isopentyl alcohol.[13] This reduction has been employed in the preparation of azetidine (eq 8).[13b]

Reduction of Aromatic Hydrocarbons. Naphthalene and higher aromatic hydrocarbons, as well as certain amino, ether, and carboxyl derivatives, are reduced to more saturated derivatives by sodium in numerous alcohols.[1a,14] Thus while naphthalene itself is converted by sodium in ethanol and benzene to the 1,4-dihydro derivative,[14b] β-naphthylamine and β-alkoxynaphthalenes are reduced to tetrahydro-β-naphthylamine (eq 9)[14a] and β-tetralone (eq 10),[14c,d] respectively.

Interestingly, sodium/isopentyl alcohol reduction of salicylic acid gives pimelic acid (eq 11).[14e] In the case of biphenyl substituted with a carboxyl group on one ring and a methoxy group on the other, the ring bearing the carboxyl group is converted to its hexahydro derivative (eq 12).[14f]

Na, i-$C_5H_{11}OH$ → HO_2C–$(CH_2)_5$–CO_2H (11)

Na, i-$C_5H_{11}OH$, 92% (12)

Pyridine derivatives are also reduced by sodium in ethanol[15] or in pentyl alcohol.[11b] For example, benzo[*f*]quinoline is converted to the *trans*-amine (eq 13).[11b]

Na, $C_5H_{11}OH$ (13)

Reduction of Alkenes. Perhaps surprisingly, numerous examples of alkenes conjugated with aromatic rings have been converted to alkanes by sodium in alcohols.[1a] For example, cinnamyl alcohol is conveniently reduced by this method (eq 14).[16]

Na, $C_5H_{11}OH$ (14)

Cleavage of Ethers. 4-Phenyl-*m*-dioxane has been reduced to 3-phenyl-1-propanol by sodium in isobutyl alcohol.[17] Similar examples include demethoxylations of methyl aryl ethers (eq 15)[18a,b] and debenzylations of benzyl derivatives of certain sugars.[18c]

Na, i-$C_5H_{11}OH$, xylene (15)

Reduction of Carbon–Halogen Bonds. *gem*-Dibromides are reduced by sodium in wet methanol to the parent hydrocarbons (eqs 16 and 17).[19] Both vinylic and allylic chlorides, but not methyl ether moieties, are similarly reduced by sodium in *t*-butyl alcohol and THF (eq 18).[20]

Na, wet MeOH (16)

Na, wet MeOH (17)

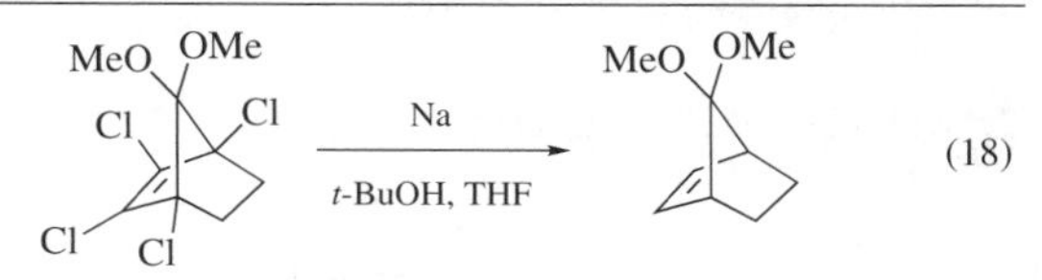

Related Reagents. See Classes R-2, R-3, R-7, R-8, R-9, R-12, R-20, R-27, and R-29, pages 1–10.

1. (a) Campbell, K. N.; Campbell, B. K. *CRV* **1942**, *31*, 77. (b) House, H. O. *Modern Synthetic Reactions*, 2nd ed.; Benjamin: Menlo Park, CA, 1972; Chapter 3. (c) Hudlicky, M *Reductions in Organic Chemistry*; Horwood: Chichester, 1984.
2. (a) Bouveault, L; Blanc, G *CR(C)* **1903**, *136*, 1676. (b) Bouveault, L; Blanc, G *BSF* **1904**, *31*, 666.
3. (a) Ford, S. G.; Marvel, C. S. *OSC* **1943**, *2*, 372. (b) Reid, E. E.; Cockerille, F. O.; Meyer, J. D.; Cox, W. M., Jr.; Ruhoff, J. R. *OSC* **1943**, *2*, 468.
4. Manske, R. H. F. *OSC* **1943**, *2*, 154.
5. Hansley, V. L. *CA* **1947**, *41*, 1202.
6. Adkins, H; Gillespie, R. H. *OSC* **1955**, *3*, 671.
7. Bhandari, K; Sharma, V. L.; Chatterjee, S. K. *CI(L)* **1990**, 547.
8. (a) Whitmore, F. C.; Otterbacher, T. J. *OSC* **1943**, *2*, 317. (b) Jones, D. N.; Lewis, J. R.; Shoppee, C. W.; Summers, G. H. R. *JCS* **1955**, 2876. (c) House, H. O.; Müller, H. C.; Pitt, C. G.; Wickham, P. P. *JOC* **1963**, *28*, 2407. (d) Wender, P. A.; Mucciaro, T. P. *JACS* **1992**, *114*, 5878.
9. Pinder, A. R.; Robinson, R. *JCS* **1955**, 3341.
10. Klages, A.; Allendorf, P. *CB* **1898**, *31*, 998.
11. (a) Lycan, W. H.; Puntambeker, S. V.; Marvel, C. S. *OSC* **1943**, *2*, 318. (b) Masamune, T.; Ohno, M.; Koshi, M.; Ohuchi, S.; Iwadare, T. *JOC* **1964**, *29*, 1419. (c) Rausser, R.; Weber, L.; Hershberg, E. B.; Oliveto, E. P. *JOC* **1966**, *31*, 1342. (d) Sugden, J. K.; Patel, J. J. B. *CI(L)* **1972**, 683.
12. (a) Suter, C. M.; Moffett, E. W. *JACS* **1934**, *56*, 487. (b) Walter, L. A.; McElvain, S. M. *JACS* **1934**, *56*, 1614.
13. (a) Klamann, D.; Hofbauer, G. *CB* **1953**, *86*, 1246. (b) Schaefer, F. C. *JACS* **1955**, *77*, 5928.
14. (a) Waser, E. B. H.; Möllering, H. *OSC* **1932**, *1*, 489. (b) Cook, E. S.; Hill, A. J. *JACS* **1940**, *62*, 1995. (c) Cornforth, J. W.; Cornforth, R. H.; Robinson, R. *JCS* **1942**, 689. (d) Soffer, M. D.; Bellis, M. P.; Gellerson, H. E.; Stewart, R. A. *OSC* **1963**, *4*, 903. (e) Muller, A. *OSC* **1943**, *2*, 535. (f) Johnson, W. S.; Gutsche, C. D.; Offenhauer, R. D. *JACS* **1946**, *68*, 1648. (g) Bass, K. C. *OSC* **1973**, *5*, 398.
15. (a) Marvel, C. S.; Lazier, W. A. *OSC* **1932**, *1*, 99. (b) Profft, E.; Linke, H.-W. *CB* **1960**, *93*, 2591.
16. Gray, W. H. *JCS* **1925**, *127*, 1150.
17. Shriner, R. L.; Ruby, P. R. *OSC* **1963**, *4*, 798.
18. (a) Clayson, D. B. *JCS* **1949**, 2016. (b) Thomas, H.; Siebeling, W. *CB* **1911**, *44*, 2134. (c) Prentice, N.; Cuendet, L. S.; Smith, F. *JACS* **1956**, *78*, 4439.
19. (a) Doering, W. V. E.; Hoffmann, A. K. *JACS* **1954**, *76*, 6162. (b) Winstein, S.; Sonnenberg, J. *JACS* **1961**, *83*, 3235.
20. Gassman, P. G.; Marshall, J. L. *OSC* **1973**, *5*, 424.

Edwin M. Kaiser
University of Missouri-Columbia, MO, USA

Sodium Amalgam

Na(Hg)

[11110-52-4] Na (MW 22.99)

(used in the preparation of alkenes and alkynes; to reductively cleave C–S and N–O bonds; for the reductive cleavage of quaternary phosphonium and arsonium salts; selective dehalogenation of aryl halides; also for the reduction of a variety of other functional groups)

Physical Data: the consistency and mp vary with the sodium content; 1.2% sodium is a semisolid at room temperature and melts completely at 50 °C; 5.4% sodium melts above 360 °C.
Solubility: sodium amalgams are decomposed by water but more slowly than sodium.
Form Supplied in: crushed solid; limited commercial availability.
Analysis of Reagent Purity: the amalgam can be analyzed for sodium by titration with 0.1 N sulfuric or hydrochloric acid.
Preparative Methods: several procedures for the preparation of sodium amalgam have been reported.[1–4] Amalgams containing 2–6% sodium are the most commonly employed for synthetic work. The safest and most convenient procedure for the preparation of 2% sodium amalgam is the addition of ***Mercury(0)*** to ribbons of ***Sodium*** metal.[5,6] No external heating is required with this protocol. The resulting solid can be crushed and stored indefinitely in a tightly stoppered container.
Handling, Storage, and Precautions: moisture sensitive; keep tightly closed.

Alkyne Synthesis. Reductive cleavage of the triflate salts of vinyl phosphonium triflates with 2% sodium amalgam affords pure alkynes in good to excellent yield (eqs 1 and 2).[7] This represents a significant improvement in yield and purity over the method involving thermal cleavage of acyl ylides. The highest yields are obtained when one group is aryl (eq 2). Alkynes are also obtained in respectable yields by the reductive elimination of enol phosphates of β-oxo sulfones with the reagent (6%) in DMSO–THF (eq 3).[8]

Bu(OSO$_2$CF$_3$)C=C(C$_6$H$_{13}$)$\overset{+}{P}$Ph$_3$ $^-$OSO$_2$CF$_3$ —[2% Na(Hg), THF, 52%]→ Bu—≡—C$_6$H$_{13}$ (1)

Ph(OSO$_2$CF$_3$)C=C(Me)$\overset{+}{P}$Ph$_3$ $^-$OSO$_2$CF$_3$ —[2% Na(Hg), THF, 80%]→ Ph—≡— (2)

SO$_2$Ph, Ph, OP(OEt)$_2$(=O) —[6% Na(Hg), DMSO–THF, 2 h, 72%]→ Ph (3)

Longer reaction times are required for nonconjugated enol phosphates. Aromatic *cis*- and *trans*-enediol diesters are reported to give the corresponding diaryl alkynes in modest yield by reductive elimination with the reagent (eq 4).[9] This procedure has been modified to provide for a one-pot operation in which 4 equiv of an aromatic acyl chloride is treated with an excess of the reagent in ether.[9]

Ph(OCOPh)C=C(Ph)OCOPh —[Na(Hg), Et$_2$O, 50%]→ Ph—≡—Ph (4)

Alkene Synthesis. The preparation of *trans*-alkenes from the reaction of β-alkoxy or acyloxy sulfones with sodium amalgam has been reported (eq 5).[10] Trisubstituted and tetrasubstituted precursors generally give disappointing results. Stereoselective introduction of the double bond in a total synthesis of diumycinol was accomplished with this method (eq 6).[11] This protocol is effective when other standard conditions fail.

SO$_2$Ph, Bu, OAc —[Na(Hg), EtOH, 79%]→ Bu (5)

SO$_2$Ph, OBz, OBn —[1. Na(Hg), THF, MeOH, 0 °C; 2. Na, NH$_3$, 33%]→ OH (6)

Desulfurization.[12–17] This reaction has perhaps found the widest use of the reagent. The value of sulfides, sulfoxides, and sulfones in organic synthesis is increased by the ease with which their removal is accomplished. For example, sulfones are conveniently hydrogenolyzed with 6% reagent in boiling ethanol (eq 7).[12] The use of disodium hydrogen phosphate is recommended as a buffer with the reagent in some applications with allylic sulfones and β-keto sulfides (eqs 8 and 9).[15]

Ph, Bu, SO$_2$Ar (Ar = *p*-ClC$_6$H$_4$) —[6% Na(Hg), EtOH, reflux, 92%]→ Ph, Bu (7)

NC, SO$_2$-*p*-Tol —[6% Na(Hg), MeOH, Na$_2$HPO$_4$, 66%]→ NC (8)

PhS, O —[6% Na(Hg), MeOH, Na$_2$HPO$_4$, 90%]→ O (9)

Desulfonylation of unsaturated α,β-unsaturated acetals with the reagent in buffered methanolic medium gives the reduced products as a mixture of isomers (eq 10).[17] Hydrolysis of the mixture affords the desired α,β-unsaturated ketone as the major product. β-Hydroxynitriles have been prepared in good yield and with retention of stereochemistry by the action of the 2% reagent in wet THF on sulfonylisoxazolines (eqs 11 and 12).[18] The addition of aqueous phosphate buffer is sometimes required to prevent further hydrogenolysis (eq 12).

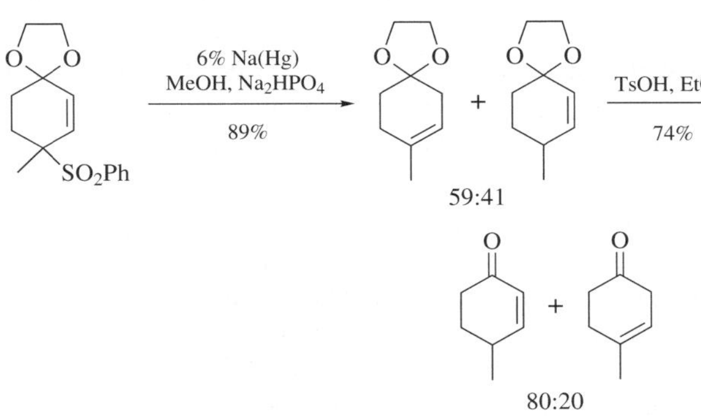

(10)

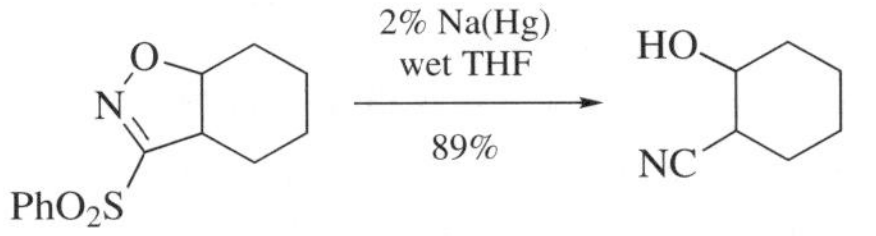

(11)

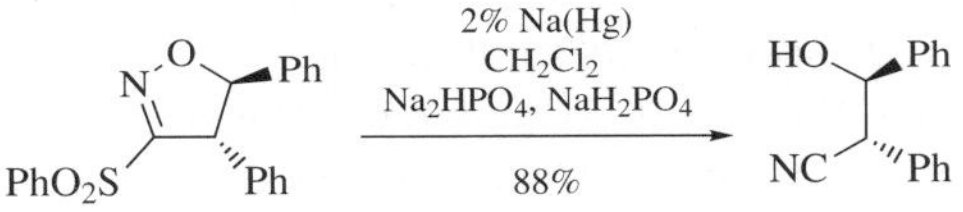

(12)

Enantiomerically pure 4,5-dihydroisoxazoles are available by selective desulfurization of 3-*p*-tolylsulfinylmethyl-4,5-dihydroisoxazoles with the reagent in a buffered medium (eq 13).[19] Further treatment of the products with ***Raney Nickel*** affords the corresponding amino alcohols. Selective desulfurization of methyl 3,4,6-tri-*O*-benzyl-2-*O*-(methylsulfonyl)-α-D-mannopyranoside derivatives was accomplished with 6–7% reagent in 2-propanol and ether (eq 14).[20] Treatment with ***Nickel*** in ethanol, ***Sodium Naphthalenide*** in THF, or ***Sodium–Ammonia*** resulted in removal of the benzyl groups as well.

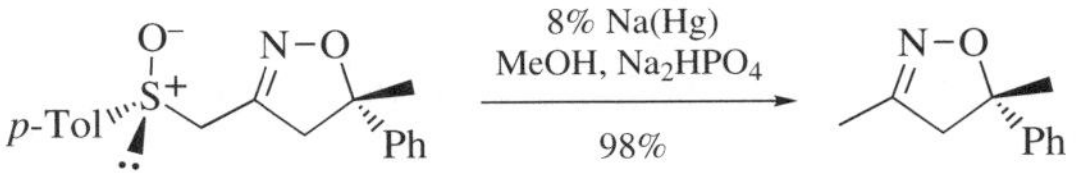

(13)

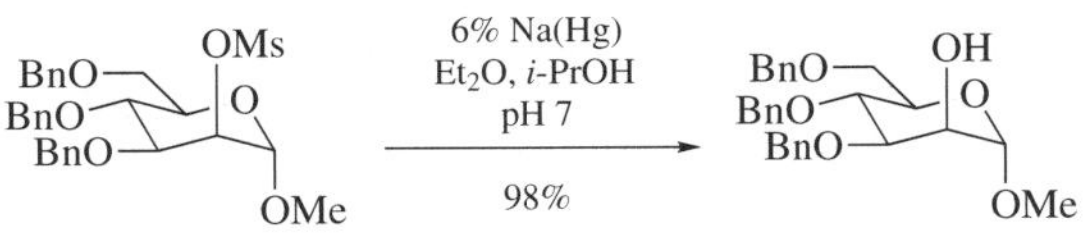

(14)

N–O Bond Hydrogenolysis. Reductive cleavage of N–O bonds is easily accomplished with the reagent (eqs 15 and 16). Keck reported the facile hydrogenolysis of the intramolecular Diels–Alder acyl-nitroso cycloadduct with excess reagent in a buffered alcoholic medium to yield a hydroxy lactam (eq 15).[21] Similarly, Jager subjected dihydroisoxazoles to hydrogenolysis with 6% sodium amalgam to afford the corresponding 1,3-amino alcohols (eq 16).[22] The *syn* and *anti* stereoselectivity is reported to be lower with this reagent than with ***Lithium Aluminum Hydride***. This method of N–O bond reduction is less common than those using either ***Aluminum Amalgam*** or LAH.

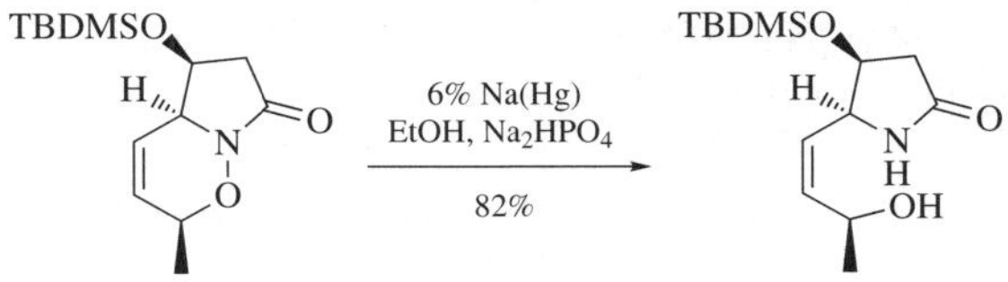

(15)

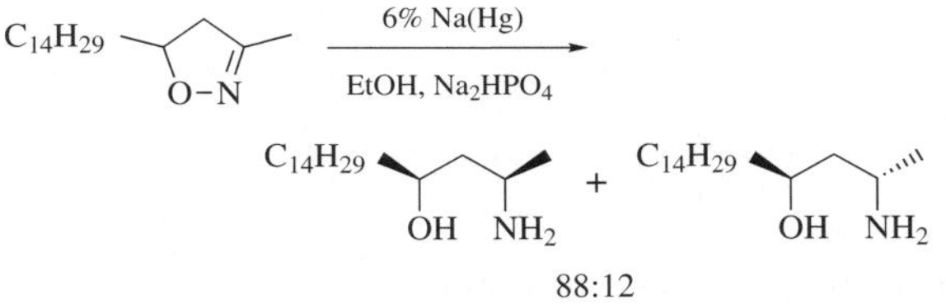

(16)

Reductive Cleavage of Quaternary Phosphonium and Arsonium Salts. Reductive cleavage of achiral and optically active quaternary and phosphonium salts with the reagent affords tertiary phosphines and arsines in high yields and with retention of configuration (eq 17).[23]

$$(S)\text{-}(+)\text{-PhCH}_2\overset{+}{\text{P}}\text{MePhPr Br}^- \xrightarrow{\text{Na(Hg)}} (S)\text{-}(+)\text{-PMePhPr} \quad (17)$$

Selective reductive cleavage of the *t*-butyl group occurs in substrates containing both the *t*-butyl and benzyl substituents. The present method is reported to be superior to the conventional cathodic cleavage. Emde degradation of the quaternary ammonium chloride succeeds where the Hofmann method is unsuccessful (eq 18).[24] In a similar fashion, hemimellitene was prepared by treating an aqueous suspension of the benzyltrimethylammonium iodide with a large excess of the reagent (eq 19).[3]

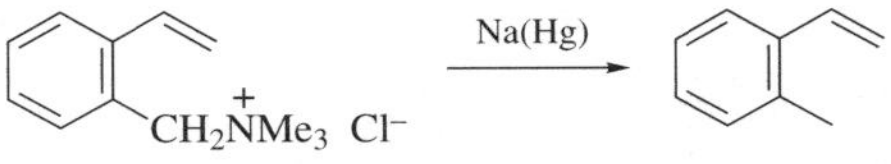

(18)

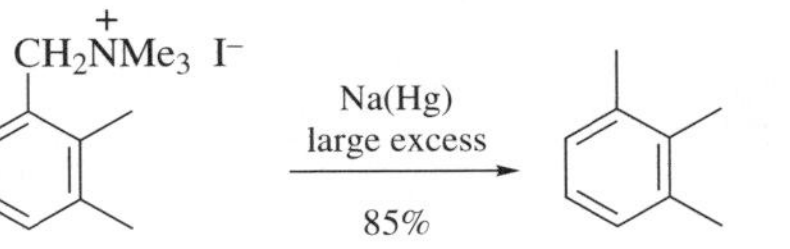

(19)

Dehalogenation of Aryl Halides. Selective dehalogenation of aryl halides with the reagent in liquid ammonia has been found (eqs 20 and 21).[25]

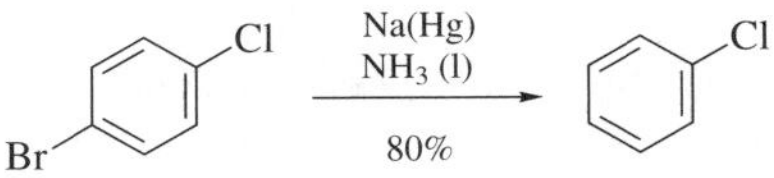

(20)

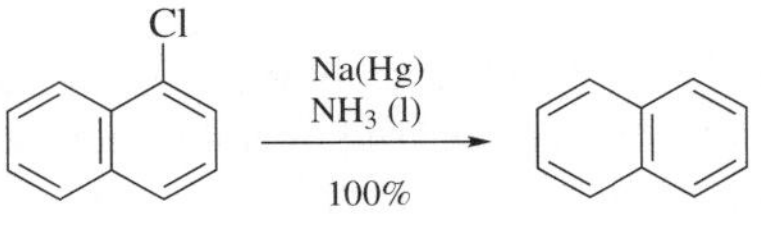

(21)

Reduction of Miscellaneous Functional Groups. Many additional functional groups react readily with the reagent to afford, in generally good yields, the corresponding reduced products. For example, phthalic acid reacts with the 3% reagent to yield *trans*-1,2-dihydrophthalic acid in good yield (eq 22).[4] α,β-Unsaturated carboxylic acids are also readily reduced by the reagent. Thus cinnamic acid is reduced to hydrocinnamic acid with 2.5% reagent in aqueous base (eq 23).[4]

CO_2H / CO_2H — 3% Na(Hg), H_2O, AcOH, AcONa, 62% → CO_2H / CO_2H (22)

CO_2H, Ph — 2% Na(Hg), H_2O, AcOH → CO_2H, Ph (23)

The product from the condensation of vanillin and creatinine is reduced by 3% reagent in water (eq 24).[26] Xanthone was reduced to xanthydrol with the reagent in ethanol (eq 25).[1] Triphenylchloromethane is converted to triphenylmethylsodium by action of the 1% reagent (eq 26).[2] Oximes are reduced to primary amines with the 2.5% reagent with acetic acid in ethanol (eq 27).[27] One of the oldest applications of sodium amalgam involves the reduction of aldonolactones to aldoses (eq 28).[28]

Na(Hg), H_2O, 72% (24)

Na(Hg), EtOH, 60–70 °C, 91% (25)

$$Ph_3CCl \xrightarrow[100\%]{Na(Hg),\ Et_2O} Ph_3\bar{C}\ Na^+ \quad (26)$$

2% Na(Hg), AcOH, 76% (27)

2% Na(Hg), H_2O, H_2SO_4, 56% (28)

Other Applications. Several recent useful miscellaneous applications have been reported. They include the use of the 3% reagent for the reduction of 1-ethyl-4-methoxycarbonylpyridinium iodide in acetonitrile to produce a stable radical (eq 29);[29] the use of the reagent as a catalyst to initiate aromatic radical nucleophilic substitution reactions;[30] and the use of the reagent for the preparation of novel titanium catalysts for the stereoselective cyclization of diynes to (*E,E*)-exocyclic dienes.[31]

CO_2Me, I^-, Et — 3% Na(Hg), MeCN, 10–30% → CO_2Me, Et (29)

Related Reagents. See Classes R-8, R-9, R-14, R-15, R-23, R-24, R-27, R-30, and R-31, pages 1–10.

1. Holleman, A. F. *OSC* **1941**, *1*, 554.
2. Renfrow, W. B., Jr.; Hauser, C. R. *OSC* **1943**, *2*, 607.
3. Brasen, W. R.; Hauser, C. R. *OSC* **1963**, *4*, 508.
4. McDonald, R. N.; Reineke, C. E. *OS* **1970**, *50*, 50.
5. Fieser, L. F.; Fieser, M. *FF* **1967**, *1*, 1030.
6. Blomquist, A. T.; Hiscock, B. F.; Harpp, D. N. *JOC* **1966**, *31*, 4121.
7. Bestmann, H. J.; Kumar, K.; Schaper W. *AG(E)* **1983**, *22*, 167.
8. (a) Lythgoe, B.; Waterhouse, I. *TL* **1978**, 2625. (b) Lythgoe, B.; Waterhouse, I. *JCS(P1)* **1979**, 2429.
9. Horner, L.; Dickerhof, K. *CB* **1983**, *116*, 1615.
10. Julia, M.; Paris, J.-M. *TL* **1973**, 4832.
11. Kocienski, P.; Todd, M. *CC* **1982**, 1078.
12. Posner, G. H.; Brunelle, D. J. *TL* **1973**, 935.
13. Dabby, R. E.; Kenyon, J.; Mason, R. F. *JCS* **1952**, 4881.
14. Julia, M.; Blasioli, C. *BSF(2)* **1976**, 1941.
15. Trost, B. M.; Arndt, H. C.; Strege, P. E.; Verhoeven, T. R. *TL* **1976**, 3477.
16. Chang, Y.-H.; Pinnick, H. W. *JOC* **1978**, *43*, 373.
17. Paquette, L. A.; Kinney, W. A. *TL* **1982**, *23*, 131.
18. Wade, P. A.; Bereznak, J. F. *JOC* **1987**, *52*, 2973.
19. (a) Annunziata, R.; Cinquini, M.; Cozzi, F.; Gilardi, A.; Restelli, A. *JCS(P1)* **1985**, 2289. (b) Annunziata, R.; Cinquini, M.; Cozzi, F.; Restelli, A. *JCS(P1)* **1985**, 2293.
20. Webster, K. T.; Eby, R.; Schuerch, C. *Carbohydr. Res.* **1983**, *123*, 335.
21. (a) Keck, G. E.; Fleming, S. A. *TL* **1978**, 4763. (b) Keck, G. E. *TL* **1978**, 4767.
22. (a) Jager, V.; Buss, V. *LA* **1980**, 101. (b) Jager, V.; Buss, V.; Schwab, W. *LA* **1980**, 122.
23. Horner, L.; Dickerhof, K. *PS* **1983**, *15*, 213.
24. Emde, H.; Kull, H. *AP* **1936**, *274*, 173.
25. Austin, E.; Alonso, R. A.; Rossi, R. A. *JCR(S)* **1990**, 190.
26. Deulofeu, V.; Guerrero, T. J. *OSC* **1955**, *3*, 586.
27. Hochstein, F. A.; Wright, G. F. *JACS* **1949**, *71*, 2257.
28. Sperber, N.; Zaugg, H. E.; Sandstrom, W. M. *JACS* **1947**, *69*, 915.
29. Kosower, E. M.; Waits, H. P. *OPP* **1971**, *3*, 261.
30. Austin, E.; Alonso, R. A.; Rossi, R. A. *JOC* **1991**, *56*, 4486.
31. Nugent, W. A.; Calabrese, J. C. *JACS* **1984**, *106*, 6423.

Keith R. Buszek
Kansas State University, Manhattan, KS, USA

Sodium–Ammonia[1]

$Na–NH_3$

(Na)		
[7440-23-5]	Na	(MW 22.99)
(NH_3)		
[7664-41-7]	H_3N	(MW 17.04)

(strong reducing agent; reduces polyunsaturated hydrocarbons, carbonyl compounds, and other electron acceptors by electron transfer; strongly basic medium)

Physical Data: sodium: mp 97.8 °C; *d* 0.968 g cm^{-3}. Ammonia: mp −77.7 °C; bp −33.4 °C.

Solubility: sodium is soluble to the extent of 25 g in 100 g ammonia at −33 °C.

Form Supplied in: sodium: flammable, metallic solid. Ammonia: compressed gas.

Purification: sodium: by melting and removal of oxide shell.[2] Ammonia: by distillation from alkali metal solutions or passage through a drying tube containing barium oxide.[1o]

Handling, Storage, and Precautions: ***Sodium*** reacts violently with water (frequently explosively) and many other protic or halogenated solvents. While it is best handled under an inert atmosphere, it is usually stored under oil and common practice is to rinse the metal with hydrocarbon solvent and weigh it in air. ***Ammonia*** is a corrosive gas with a pungent odor, and must be handled in a good fume hood.

Sodium–Ammonia Solutions. Sodium dissolves in liquid, anhydrous ammonia to form a deep-blue solution.[3] At high concentrations, the solution separates into two liquid phases: the more dilute lower phase retains the deep-blue color while the upper phase has a metallic bronze appearance. Sodium reacts slowly with ammonia to produce sodium amide and hydrogen (eq 1), and this reaction is catalyzed by metals such as iron, cobalt, and nickel. The addition of trace amounts of iron salts to sodium–ammonia solutions is a convenient method for the preparation of ***Sodium Amide***. Sodium–ammonia solution serves as an excellent reducing agent, with a half-wave potential of −2.59 V at −50 °C.[1f]

$$Na + NH_3 \longrightarrow NaNH_2 + 1/2\ H_2 \qquad (1)$$

Birch Reduction of Aromatic Rings. A major use of the sodium–ammonia reagent is the Birch reduction of aromatic rings to cyclohexadienes or sometimes higher levels of unsaturation.[1,4,5] With benzene and less reactive aromatics (electron donor substituents), a proton donor is necessary and alcohols such as ethanol and *t*-butyl alcohol are commonly used. With more reactive aromatics (electron-withdrawing substituents and polynuclear aromatics), alcohols are generally unnecessary and are often deleterious.[1n] The kinetically controlled regiochemistry[6] affords 1,4-dihydroaromatics (1,4-cyclohexadienes), with preference for the regioisomer with the maximum number of alkyl and/or alkoxy groups located on the remaining double bonds (i.e. the Birch rule).[1h,7] Due to the limited solubility of many aromatic compounds in ammonia, cosolvents are often included in amounts of 25–35% relative to the volume of ammonia. While the most common cosolvents are ethers, such as diethyl ether and THF, protic solvents are sometimes used as both cosolvent and proton source. The range of metal concentrations is usually 0.1–0.5 g metal per 100 mL of ammonia, and highly concentrated solutions producing so-called metallic or 'bronze' phases are generally avoided for Birch reductions.

Benzenes containing alkyl or ether substituents, especially where these substituents may be attached to vinyl positions in the product, are reduced in high yields by sodium–ammonia solution in the presence of ethanol or *t*-butyl alcohol (eq 2).[8] The reduced ethers are useful in synthesis since subsequent acid hydrolysis leads to either nonconjugated or conjugated ketones; alternatively, products may be used directly in Diels–Alder reactions without hydrolysis (eq 3).[9]

OMe, Et; Na/NH_3, THF, EtOH, 100%; OMe, Et (2)

OMe, X, Y, R; X, Y; OMe, R; H^+; O, R; or; O, R (3)

In cases where this substitution pattern does not apply, however, reduction may be more difficult. For example, while 6-methoxytetralin is reduced smoothly by sodium–ammonia (eq 4), 5-methoxytetralin requires the use of ***Lithium*** with the alcohol added last (eq 5).[5a,10,11] This latter procedure is known as the Wilds and Nelson modification,[11] and is useful for less reactive substrates.

MeO; Na/NH_3, ether, *i*-PrOH; HCl; O; 80–90% (4)

OMe; Li/NH_3, DME, EtOH; HCl; O; 65–80% (5)

In polynuclear aromatics, regiochemistry generally follows protonation at the positions bearing the highest coefficients in the HOMO of the anionic intermediates, with sometimes little influence from substituents.[1g,m] Both anthracene (eq 6) and naphthalene (see below) are reduced in nearly quantitative yields by sodium–ammonia solution, provided that highly acidic quenching agents such as water or ammonium chloride are used rather than alcohols. However, some reductions of polynuclear aromatics are hard to control even under these conditions, and iron salts have been used to keep the reaction to a single stage. Such is the case with chrysene (eq 7) and phenanthrene.[1o]

Na/NH_3, THF, NH_4Cl (6)

$$\text{phenanthrene derivative} \xrightarrow[\text{THF}]{\text{Na/NH}_3} \text{dihydro product} + \text{dihydro product} \quad (7)$$

NH_4Cl	82%	9%
$FeCl_3$, NH_4Cl	15%	80%

Sodium vs. Lithium. While sodium and lithium are the most commonly employed metals in Birch reduction (and ***Potassium*** to a lesser extent), the choice between these two metals is not always obvious due to the relative lack of comparison data. Although lithium is the most reactive due to its greater heat of solvation of the cation, its reactivity relative to sodium has probably been overestimated. Early experiments with sodium–ammonia solutions containing impurities in either the metal or ammonia probably showed diminished reactivity due to the greater sensitivity of sodium to amide formation catalyzed by trace contaminants.[1f] Thus metal selection may be based on expense, reactivity, or demonstrated superiority with a particular substrate. With very small scale reactions, the higher atomic weight of sodium may make it more practical due to accuracy in weighing. Lithium is the metal of choice when greater reactivity is required, but greater reactivity can sometimes lead to overreduction, whereupon sodium becomes the reagent of choice.

In many cases (eq 8)[12,13] the results are comparable with either metal. However, a comparison of the two metals in the reduction of naphthalene[14] shows best results with sodium at low temperatures (eq 9). At higher temperatures, and especially with lithium, protonation of the intermediates and subsequent isomerization of the 1,4-dihydro product to the 1,2-dihydro isomer leads to appreciable overreduction. The reductive methylation of naphthalene shows even greater differences between the metals, with lithium affording monoalkylation (95%) and sodium dialkylation (93%).[14,15] This is due to a second alkylation via proton abstraction by amide ion during the alkylation step, a process that does not occur with lithium due to the relative lack of solubility of lithium amide in ammonia as compared to sodium amide. Similar results are observed for biphenyl, where lithium provides 99% monomethylation and sodium leads to 40% dimethylation with 10% trimethylation.[16]

$$\text{MeO-steroid}(R^1, R^2, R^3) \xrightarrow[\text{THF, } t\text{-BuOH}]{\text{M/NH}_3} \text{MeO-dihydro steroid}(R^1, R^2, R^3) \quad (8)$$

M = Li, $R^1 = CH_2CHMeOH$, $R^2 = R^3 = H$, 89%
M = Na, $R^1 = Me$, $R^2 = OH$, $R^3 = Et$, 87%

$$\text{naphthalene} \xrightarrow{\text{M/NH}_3,\ \text{ether}} \xrightarrow{\text{NH}_4\text{Cl},\ \text{H}_2\text{O}} \text{1,4-dihydronaphthalene} + \text{tetralin} \quad (9)$$

M = Li, –33 °C	40%	55%
M = Li, –78 °C	81%	14%
M = Na, –33 °C	83%	17%
M = Na, –78 °C	96%	2%

Regioselectivity and Stereoselectivity. The regiochemistry of Birch reduction is generally driven by electron densities in the radical anions, dianions, and monoanions that serve as intermediates in the reaction. An attempt has been made to control regiochemistry by use of trimethylsilyl substituents.[17] For example, while 1-methylnaphthalene reduces exclusively in the unsubstituted ring, 1-methyl-4-trimethylsilylnaphthalene reduces in the silylated ring, and subsequent removal of the silyl group affords a 'misoriented' Birch product (eq 10).

$$\text{1-methyl-4-TMS-naphthalene} \xrightarrow[\text{86\%}]{\text{Na/NH}_3,\ \text{THF}} \text{TMS dihydro product} \xrightarrow{\text{TBAF}} \text{product} \quad (10)$$

Excellent diastereoselectivities have been accomplished in the reductive methylation of benzoic acids where L-proline has been introduced as a chiral auxiliary (eq 11).[18] Results with a variety of derivatives appear to give comparable results with sodium, lithium, or potassium.

$$\text{amide} \xrightarrow[\text{1.0 equiv } t\text{-BuOH}]{\text{Na/NH}_3,\ \text{THF}} \xrightarrow{\text{MeI}} \text{product} \quad (11)$$

85%; de = 260:1

Reduction of Carbonyl Functions. The sodium–ammonia reduction of cyclic aliphatic ketones has some utility for the preparation of the more stable epimeric alcohols (such as 11α-hydroxy steroids) derived from sterically hindered ketones.[11,m] In other cases, product mixtures resemble those obtained with reagents such as borohydride. Similarly, esters may be reduced (Bouveault–Blanc reduction), but there is little advantage except with monoesters of diacids where the acid carbonyl remains unreduced (eq 12).[19] Carboxamides are reduced by sodium–ammonia, often affording rather good results, as illustrated with the L-asparagine derivative (eq 13).

$$\text{cyclohexane}(CO_2Me)(CH_2CH_2CO_2H) \xrightarrow[\text{EtOH}]{\text{Na/NH}_3} \text{cyclohexane}(CH_2OH)(CH_2CH_2CO_2Na) \xrightarrow{H^+} \text{lactone} \quad (12)$$

72%

$$HO_2C\text{–CH(NHCO}_2\text{-}t\text{-Bu)–CH}_2\text{CONH}_2 \xrightarrow[\text{77\%}]{\text{Na/NH}_3,\ \text{MeOH}} HO_2C\text{–CH(NHCO}_2\text{-}t\text{-Bu)–CH}_2\text{CH}_2\text{OH} \quad (13)$$

The most common use of sodium–ammonia reagent for the reduction of carbonyl compounds is the reduction of α,β-unsaturated ketones, and an extensive list of examples has been provided by Caine.[1k] Of special interest are bicycloalkenones, where the α,β-double bond is exocyclic to the second ring (eq 14).[20] In these cases, stereochemistry is dictated by the protonation at the β-position, and the most favored transition states are those involving protonation of an axial anionic center; this most often results in the formation of *trans* products, although the stereochemistry can be reversed by 6β- and 7α-substituents. α,β-Unsaturated acids and esters can also be reduced, and reductive alkylation can be achieved by adding alkylating agents before the protic quench. However, Li and K have more often been used for the latter process.

Na/NH_3, dioxane, 99% (14)

trans

Reductive Cleavage. The sodium–ammonia reagent is effective in cleaving a variety of bonds including carbon–carbon (especially when strained), carbon–sulfur, carbon–nitrogen (especially positively charged nitrogen), and carbon–halogen; also allylic and benzylic alcohols, ethers and esters, epoxides, and many phenol ethers. While most halides are cleaved, vinyl halides are cleaved stereospecifically (eqs 15 and 16).[21,22]

Na/NH_3 (15)

Na/NH_3, 90% (16)

Cleavage of the hydrazine derivative in eq 17 represents an especially useful application since ***Zinc–Acetic Acid***, ***Aluminum Amalgam***, and ***Raney Nickel*** all give no reduction.[23] An interesting example of carbon–carbon bond cleavage is provided by an important step in the synthesis of (C_2)-dioxa-C_{20}-octaquinane (eq 18),[24] and sulfur–sulfur bond cleavage is illustrated with the mercaptosulfinate synthon shown in eq 19.[25] In the latter example, both sodium hydride and ***Sodium Borohydride*** gave poor results.

NCO_2Et, NCO_2Et → Na/NH_3, 100% → $NHCO_2Et$, $NHCO_2Et$ (17)

Na/NH_3, 80% (18)

O_2S—S → Na/NH_3, THF, 98% → NaO_2S SNa (19)

Reduction of Alkenes and Alkynes. Conjugated polyenes are reduced by sodium–ammonia, although complications such as dimerization generally limit good results to cyclic systems and styrenes.[1f] The reduction of alkynes to *trans*-alkenes is well-known,[1f,h] the stereochemistry presumably resulting from the *trans* nature of the intermediate radical anion. Terminal alkynes can be protected by removal of the acidic proton by sodium amide, allowing synthetically useful strategies (eq 20).[26]

1. $NaNH_2$; 2. Na/NH_3 (20)

Related Reagents. See Classes R-2, R-4, R-7, R-12, R-13, R-15, R-24, R-27, and R-31, pages 1–10.

---

1. (a) Birch, A. J. *QR* **1950**, *4*, 69. (b) Birch, A. J.; Smith, H. *QR* **1958**, *12*, 17. (c) Smith. H. *Organic Reactions in Liquid Ammonia, Chemistry in Nonaqueous Ionizing Solvents*; Wiley: New York, 1963; Vol. 1, Part 2. (d) Zimmerman, H. E. In *Molecular Rearrangements*; de Mayo, P., Ed.; Wiley: New York, 1963. (e) Hückel, W. *Fortschr. Chem. Forsch.* **1966**, *6*, 197. (f) Smith, M. In *Reduction: Techniques and Applications in Organic Synthesis*; Augustine, R. L., Ed.; Dekker: New York, 1968. (g) Harvey, R. G. *S* **1970**, 161. (h) Birch, A. J.; Subba Rao, G. S. R. In *Advances in Organic Chemistry, Methods and Results*; Taylor, E. C., Ed.; Wiley: New York, 1972. (i) Akhrem, A. A.; Reshetova, I. G.; Titov, Y. A. *Birch Reduction of Aromatic Compounds*; Plenum: New York, 1972. (j) Pradhan, S. K. *T* **1986**, *42*, 6351. (k) Caine, D. *OR* **1976**, *23*, 1. (l) Huffman, J. W. *ACR* **1983**, *16*, 399. (m) Hook, J. M.; Mander, L. N. *Natural Prod. Rep.* **1986**, *3*, 35. (n) Rabideau, P. W. *T* **1989**, *45*, 1579. (o) Rabideau, P. W. *OR* **1992**, *42*, 1.
2. Fieser, M.; Fieser, L. F. *FF* **1967**, *1*, 1022.
3. For a more detailed description of metal–ammonia solutions, see, for example: *The Chemistry of Nonaqueous Solvents*; Lagowski, J. J., Ed.; Academic: New York, 1967; Vol. 2, pp 265–317.
4. Wooster, C. B.; Godfrey, K. L. *JACS* **1937**, *59*, 596.
5. (a) Birch, A. J. *JCS* **1944**, 430. (b) For a complete list of Birch's contributions, see: *T* **1988**, *44* (10), v.
6. Birch, A. J.; Hinde, A. L.; Radom, L. *JACS* **1980**, *102*, 3370; **1980**, *102*, 4074; **1980**, *102*, 6430; **1981**, *103*, 284.
7. For recent discussions about mechanistic aspects of the Birch reduction, see Ref. 1n; also Zimmerman, H. E.; Wang, P. A. *JACS* **1990**, *112*, 1280; **1993**, *115*, 2205.
8. (a) Pearson, A. J.; Ham, P.; Ong, C. W.; Perrior, T. R.; Rees, D. C. *JCS(P1)* **1982**, 1527. (b) Pearson, A. J. *TL* **1981**, *22*, 4033.
9. Evans, D. A.; Scott, W. L.; Truesdale, L. K. *TL* **1972**, 121.
10. (a) House, H. O.; Blankley, C. J. *JOC* **1968**, *33*, 53. (b) Burkinshaw, G. F.; Davis, B. R.; Hutchinson, E. G.; Woodgate, P. D.; Hodges, R. *JCS(C)* **1971**, 3002.
11. Wilds, A. L.; Nelson, N. A. *JACS* **1953**, *75*, 5360.
12. Pillai, K. M. R.; Murray, W. V.; Shooshani, I.; Williams, D. L.; Gordon, D.; Wang, S. Y.; Johnson, F. *JMC* **1984**, *27*, 1131.
13. Dryden, H. L.; Webber, G. M.; Burtner, R. R.; Cella, J. A. *JOC* **1961**, *26*, 3237.
14. Rabideau, P. W.; Burkholder, E. G. *JOC* **1978**, *43*, 4283.
15. Rabideau, P. W.; Harvey, R. G. *TL* **1970**, 4139.
16. Lindow, D. F.; Cortez, C. N.; Harvey, R. G. *JACS* **1972**, *94*, 5406.
17. Marcinow, Z.; Clawson, D. K.; Rabideau, P. W. *T* **1989**, *45*, 5441.
18. Schultz, A. G.; Macielag, M.; Sundararaman, P.; Taveras, A.; Welch, M. *JACS* **1988**, *110*, 7828. See also: (a) Schultz, A. G.; Sundararaman, P. *TL* **1984**, *25*, 4591. (b) Schultz, A. G.; Sundararaman, P.; Macielag,

M.; Lavieri, F. P.; Welch, M. *TL* **1985**, *26*, 4575. (c) Schultz, A. G.; McCloskey, P. J.; Sundararaman, P.; Springer, J. P. *TL* **1985**, *26*, 1619. (d) McCloskey, P. J.; Schultz, A. G. *H* **1987**, *25*, 437. (e) Schultz, A. G.; McCloskey, P. J.; Court, J. J. *JACS* **1987**, *109*, 6493. (f) McCloskey, P. J.; Schultz, A. G. *JOC* **1988**, *53*, 1380. (g) Schultz, A. G.; Macielag, M.; Podhorez, D. E.; Suhadolnik, J. C.; Kullnig, R. K. *JOC* **1988**, *53*, 2456. (h) Schultz, A. G. *ACR* **1990**, *23*, 207. (i) Schultz, A. G.; Harrington, R. E. *JACS* **1991**, *113*, 4926. (j) Schultz, A. G.; Green, N. J. *JACS* **1991**, *113*, 4931. (k) Schultz, A. G.; Harrington, R. E.; Holoboski, M. A. *JOC* **1992**, *57*, 2973. (l) Schultz, A. G.; Taylor, R. E. *JACS* **1992**, *114*, 3937. (m) Schultz, A. G.; Taylor, R. E. *JACS* **1992**, *114*, 8341.

19. Paquette, L. A.; Nelson, N. A. *JOC* **1962**, *27*, 2272.
20. Granger, F.; Chapat, J. P.; Crassous, J.; Simon, F. *BSF* **1968**, 4265.
21. Vogel, E.; Roth, H. D. *AG(E)* **1964**, *3*, 228.
22. Hoff, M. C.; Greenlee, K. W.; Boord, C. E. *JACS* **1951**, *73*, 3329.
23. Mellor, J. M.; Smith, N. M. *JCS(P1)* **1984**, 2927.
24. Balogh, D.; Begley, W. J.; Bremner, J.; Wyvratt, M. J.; Paquette, L. A. *JACS* **1979**, *101*, 749.
25. Field, L.; Eswarakrishnan, V. *JOC* **1981**, *46*, 2025.
26. Dobson, N. A.; Raphael, R. A. *JCS* **1955**, 3558.

Peter W. Rabideau
Louisiana State University, Baton Rouge, LA, USA

Sodium Bis(2-methoxyethoxy)aluminum Hydride[1]

$NaAlH_2(OCH_2CH_2OMe)_2$

[22722-98-1] $C_6H_{16}AlNaO_4$ (MW 202.19)

(reducing agent for many functional groups;[1] methylation reagent for aryl-activated compounds;[19] can function as a base;[21] can hydroaluminate alkenes and alkynes[22])

Alternate Names: SMEAH, Red-Al®, Vitride.

Physical Data: fp 4 °C (3.4 M toluene solution); *d* 1.036 g cm^{-3} (3.4 M toluene solution), *d* 1.122 g cm^{-3} (solid); highly viscous liquid at rt; thermally stable up to 205 °C, upon which vigorous decomposition starts.

Solubility: sol aromatic hydrocarbons, ether, THF, DME; insol aliphatic hydrocarbons.

Form Supplied in: 3.4 M solution in toluene. 145 g contains 1 mol of active hydride. The pure compound is a slightly yellow, glassy solid.

Analysis of Reagent Purity: concentration can be determined by iodometric titration.

Handling, Storage, and Precautions: highly flammable; moisture sensitive; reacts less strongly with H_2O than $LiAlH_4$; potent skin irritant. Use in a fume hood.

Functional Group Reductions. SMEAH has reducing properties fully comparable to ***Lithium Aluminum Hydride***, making it a valuable alternative reducing reagent.[1,2] Some practical advantages of SMEAH are that it does not ignite when exposed to moist air or O_2, has greater solubility in aromatic solvents and ethers, and reactions can be carried out at higher temperatures (up to 200 °C). Carboxylic esters, acids, acid chlorides, anhydrides, aldehydes, and ketones are efficiently converted to the alcohol. Cyclic anhydrides and lactones yield diols. Carboxylic esters are not reduced at −78 °C and *t*-Bu and Ph esters are generally stable to SMEAH at low temperature.

Nitriles, amides, imines, azido compounds, nitro compounds, isocyanates, urethanes, sulfonamides, oximes, lactams, and imides can be reduced to the amine. The formation of aziridines can complete with simple reduction of ketoximes and their *O*-alkyl derivatives. Nitroarenes form azoxyarenes, azoarenes, or hydrazoarenes depending on reaction conditions.[3]

Alkanesulfonates can be converted to alkanes or to alcohols. Phosphate esters afford the alcohol.[4] Some epoxides can be reduced to the alcohol (temperatures above 0 °C are required). Disulfides give thiols while sulfoxides yield sulfides. Sulfones are generally inert or give low yields of sulfides. Haloalkanes, haloarenes, and organosilicon halides undergo hydrogenolysis to the alkane, arene, and silane, respectively.[5] Benzylic aldehydes or alcohols, aryl alkyl or diaryl ketones, and aryl acids containing strong electron donating groups in the *ortho* or *para* position also undergo hydrogenolysis at 120–140 °C.[6] Acetals are usually stable toward hydrogenolysis or elimination at 0–20 °C, except in the presence of a conjugated double bond and bromide (eq 1)[7a] or in certain quinone monoacetal systems.[6b]

SMEAH, PhH, rt, 75% (1)

Partial Reductions. Low-temperature conditions or the use of alcohol or amine-modified SMEAH have proven especially valuable for partial functional group reductions.[1,2] Carboxylic esters, nitriles, and amides can be reduced to the aldehyde at low temperatures. The morpholine or *N*-methylpiperazine modified reagent also yields aldehydes at −55 to 0 °C.[8,9] Lactones are converted to lactols by unmodified SMEAH at −70 °C[2b] or with the EtOH, pyridine, or *i*-PrOH modified reagent at 0 °C. *N*-Substituted cyclic imides provide hydroxylactams upon reduction at −78 °C.[10] It is the reagent of choice for the reduction of Cp_2ZrCl_2 to Cp_2ZrClH.[11]

Selective Reductions. Vinylogous amides have been selectively reduced to the β-ketoamine (eq 2).[12a] SMEAH is superior to $LiAlH_4$ for the reduction of acetal-protected cyanohydrins to the aldehyde.[12b] It chemoselectively reduces α-formyl ketones to the α-hydroxymethyl ketone.[12c] Lactones are stable in the presence of ketone or aldehyde functionality at low temperatures. SMEAH modified by alcohols or in pyridine solvent selectively reduces lactones in the presence of an amide or ester.[13a] Ester groups in amidoesters undergo selective reduction at 0 °C and short reaction times.[13b] SMEAH gives higher yields than $LiAlH_4$ in the reduction of some hydroxy-substituted carbonyl compounds (eq 3).[13c] 5,5-Disubstituted hydantoins are selectively reduced at the 4-position[13d] and keto aldehydes have been cleanly reduced to the keto alcohol.[13e] Optically active α-alkoxy carboxamides have been reduced to the α-alkoxy aldehydes without racemization.[13f]

1,2-vs. 1,4-Reductions. The structure of the enone, solvent, relative initial concentrations, temperature, and softness or hardness of the hydride reagent all play a role in controlling the mode of addition to enone systems. SMEAH typically acts as a hard hydride, favoring 1,2-addition in these reactions. High yields of the allylic alcohol have been reported by inverse addition of the hydride to aliphatic or alicyclic α,β-unsaturated aldehydes and esters.[14a,b] It favors 1,2-addition to cyclic α-enones and is the reagent of choice for the low-temperature reduction of enol esters of alicyclic 1,3-diketones (eq 4).[14c] Alkyl esters are not reduced under these conditions. The SMEAH/CuBr/2-butanol reagent is extremely efficient for 1,4-reduction.[15] Nitrile and ester functions are stable under the conditions of the conjugate reduction, while aldehydes, ketones, and bromides are attacked by the complex.

SMEAH, PhH or Et_2O, –5 to 5 °C; 10–55% (2)

SMEAH, THF, toluene, reflux, 18 h; 80% (3)

1. SMEAH, THF, –78 °C; 2. H_3O^+; 60% (4)

R = $(CH_2)_6CO_2Me$
R' = *i*-Pr, PhCO, $ArSO_2$

Stereoselective Reductions. The stereochemistry of reductions often varies, depending on the solvent. A possible factor is the degree of solvation. In some cases, complete reversal of stereochemistry has been observed by using SMEAH in benzene versus THF.[1] In comparison to other reducing agents, SMEAH often shows stereoselectivity opposite to that of $LiAlH_4$, ***Sodium Borohydride***, or lithium trialkoxyaluminum hydrides. Unsymmetrical epoxides are attacked preferentially at the least-substituted carbon atom to give the more highly substituted alcohol as the major product.[16a] Exceptionally high regioselectivity in the reduction of α,β-epoxy alcohols to the 1,3-diol has been observed (eq 5).[16b–e] Substituents at the α-carbon reverses the selectivity and decreases the reaction rate. The reduction of α-silyloxy ketones produces *syn*-vicinal diols with high levels of diastereoselectivity.[16f,g]

SMEAH, THF, 0 °C; 82–98% (5)

SMEAH reacts with bridged bicyclic ketones from the less hindered side of the carbonyl.[17a] Reduction of *gem*-dihalocyclopropanes gives *anti*-monohalides as the major product.[17b] It reduces α-alkynic alcohols to *trans*-allylic alcohols with high selectivity.[18a–d] Alkynyl ethers give almost exclusively *O*-alkyl enol ethers with the (*E*) configuration with SMEAH, but with alcohol-modified SMEAH reagent (*Z*)-enol ethers are obtained (eq 6).[18e]

R^1O—≡—R: 4 equiv SMEAH, THF, reflux; 62–93%. 8 equiv SMEAH, R^2OH, THF, reflux; 72–83% (6)

R = alkyl
R^1, R^2 = alkyl, alicyclic

Methylation Reagent. The reduction of aryl-conjugated double bonds yields selective methylation at the more highly arylated carbon atom.[19] The Me group originates from the hydride OMe group. Aryl-activated alkanes undergo methylation by this reagent. Diaryl and condensed aromatic ketones undergo hydrogenolysis and subsequent methylation at the benzylic carbon.

Reductive Cleavage of Ethers. Ethers are usually stable at temperatures below 100 °C. At higher temperatures, cleavage of the ether linkage occurs. SMEAH is a useful reagent for debenzylation and deallylation of aryl benzyl and aryl allyl ethers (eq 7).[20] The reaction is enhanced by a vicinal methoxy group. This method is recommended for debenzylation and deallylation of phenolic ethers that are labile to acid or catalytic hydrogenolysis.

SMEAH, xylene, reflux, 10 h; 66% (7)

Use as a Base. SMEAH can act as a strong base. It is known to promote isomerization and dehydrogenation of aromatic hydrocarbons by a base-catalyzed reaction.[21a,b] The Stevens rearrangement of berbine methiodide affords the spiroisoquinoline derivative in high yield.[21c]

Hydroalumination Reagent. Double and triple bonds undergo catalytic hydroalumination with a number of aluminum hydride reagents.[22] The effectiveness of SMEAH in this reaction is comparable to that of $LiAlH_4$, $NaAlH_4$, $NaAlHMe_3$, $LiAlH_2(NR_2)_2$, or $NaAlH_2(NR_2)_2$.

Related Reagents. See Classes R-2, R-4, R-6, R-9, R-10, R-12, R-13, R-14, R-15, R-20, R-21, R-27, and R-32, pages 1–10. Copper(I) Bromide–Sodium Bis(2-methoxyethoxy)aluminum Hydride.

1. (a) Malek, J. *OR* **1988**, *36*, 249. (b) Malek, J. *OR* **1985**, *34*, 1. (c) Vit, J. *Eastman Org. Chem. Bull.* **1970**, *42*, 1.
2. (a) Capka, M.; Chvalovsky, V.; Kochloefl, K.; Kraus, M. *CCC* **1969**, *34*, 118. (b) Cerny, M.; Malek, J.; Capka, M.; Chvalovsky, V. *CCC* **1969**, *34*, 1025. (c) Kumar, V.; Remers, W. A. *JMC* **1979**, *22*, 432. (d) Bazant, V.; Capka, M.; Cerny, M.; Chvalovsky, V.; Kochloefl, K.; Kraus, M.; Malek, J. *TL* **1968**, 3303. (e) Malek, J.; Cerny, M. *S* **1972**, 217.
3. (a) Corbett, J. F. *CC* **1968**, 1257. (b) Kraus, M.; Kochloefl, K. *CCC* **1969**, *34*, 1823. (c) Barclay, L. R. C.; McMaster, I. T.; Burgess, J. K. *TL* **1973**, 3947.

4. Truesdale, L. K. *OS* **1989**, *67*, 13.
5. (a) Capka, M.; Chvalovsky, V. *CCC* **1969**, *34*, 2782; 3110. (b) Barton, T. J.; Kippenhan, R. C., Jr. *JOC* **1972**, *37*, 4194. (c) Kraus, M. *CCC* **1972**, *37*, 3052.
6. (a) Cerny, M.; Malek, J. *TL* **1969**, 1739. (b) Cohen, N.; Lopresti, R. J.; Saucy, G. *JACS* **1979**, *101*, 6710. (c) Cerny, M.; Malek, J. *CCC* **1970**, *35*, 1216; 2030; 3079. (d) Pawson, B. A.; Chan, K.-K.; DeNoble, J.; Han, R.-J. L.; Piermattie, V.; Specian, A. C.; Srisethnil, S.; Trown, P. W.; Bohoslawec, O.; Machlin, L. J.; Gabriel, E. *JMC* **1979**, *22*, 1059.
7. Fleischhacker, W.; Markut, H. *M* **1971**, *102*, 569; 587.
8. (a) Kanazawa, R.; Tokoroyama, T *S* **1976**, 526. (b) Bartsch, H.; Schwarz, O. *JHC* **1982**, *19*, 1189. (c) Kompis, I.; Wick, A. *HCA* **1977**, *60*, 3025.
9. (a) Stibor, I.; Janda, M.; Srogl, J. *ZC* **1970**, *10*, 342. (b) Böhme, H.; Sutoyo, P. N. *LA* **1982**, 1643. (c) Sone, T.; Terashima, S.; Yamada, S.-I. *CPB* **1976**, *24*, 1273. (d) Hayashi, M.; Terashima, S.; Koga, K. *T* **1981**, *37*, 2797. (e) Terashima, S.; Hayashi, M.; Koga, K. *TL* **1980**, *21*, 2733.
10. (a) Mukaiyama, T.; Yamashita, H.; Asami, M. *CL* **1983**, 385. (b) Speckamp, W. N.; Hiemstra, H. *T* **1985**, *41*, 4367.
11. Hart, D. W.; Schwartz, J. *JACS* **1974**, *96*, 8115.
12. (a) Weselowsky, V. W.; Moiseenkov, A. M. *S* **1974**, 58. (b) Schlosser, M.; Brich, Z. *HCA* **1978**, *61*, 1903. (c) Corey, E. J.; Smith, J. G. *JACS* **1979**, *101*, 1038.
13. (a) Kanazawa, R.; Kotsuki, H.; Tokoroyama, T. *TL* **1975**, 3651. (b) Weiss, B. *Chem. Phys. Lipids* **1977**, *19*, 347. (c) Kornet, M. J. *JHC* **1990**, *27*, 2125. (d) Marquez, V. E.; Twanmoh, L.-M.; Wood, H. B., Jr.; Driscoll, J. S. *JOC* **1972**, *37*, 2558. (e) Murphy, R.; Prager, R. H. *AJC* **1976**, *29*, 617. (f) Kobayashi, Y.; Takase, M.; Ito, Y.; Terashima, S. *BCJ* **1989**, *62*, 3038.
14. (a) Bartlett, P. A.; Johnson, W. S. *JACS* **1973**, *95*, 7501. (b) Traas, P. C.; Boellens, H.; Takken, H. J. *RTC* **1976**, *95*, 57. (c) Sih, C. J.; Heather, J. B.; Peruzzotti, G. P.; Price, P.; Sood, R.; Lee, L. F. H. *JACS* **1973**, *95*, 1676.
15. (a) Semmelhack, M. F.; Stauffer, R. D.; Yamashita, Y. *JOC* **1977**, *42*, 3180. (b) Semmelhack, M. F.; Stauffer, R. D. *JOC* **1975**, *40*, 3619.
16. (a) Jones, T. K.; Peet, J. H. J. *CI(L)* **1971**, 995. (b) Ma, P.; Martin, V. S.; Masamune, S.; Sharpless, K. B.; Viti, S. M. *JOC* **1982**, *47*, 1378. (c) Viti, S. M. *TL* **1982**, *23*, 4541. (d) Minami, N.; Ko, S. S.; Kishi, Y. *JACS* **1982**, *104*, 1109. (e) Finan, J. M.; Kishi, Y. *TL* **1982**, *23*, 2719. (f) Nakata, T.; Tanaka, T.; Oishi, T. *TL* **1983**, *24*, 2653. (g) Nakata, T.; Fukui, M.; Ohtsuka, H.; Oishi, T. *T* **1984**, *40*, 2225.
17. (a) Sydnes, L.; Skattebøl, L. *TL* **1974**, 3703. (b) Sydnes, L. K.; Skattebøl, L. *ACS(B)* **1978**, *32*, 632. (c) Miyano, S.; Hashimoto, H. *BCJ* **1975**, *48*, 3665.
18. (a) Jones, T. K.; Denmark, S. E. *OS* **1986**, *64*, 182. (b) Trost, B. M.; Lautens, M. *JACS* **1987**, *109*, 1469. (c) Chan, K.-K.; Cohen, N.; De Noble, J. P.; Specian, A. C., Jr.; Saucy, G. *JOC* **1976**, *41*, 3497. (d) Mayer, H. J.; Rigassi, N.; Schwieter, U.; Weedon, B. C. L. *HCA* **1976**, *59*, 1424. (e) Sola, L.; Castro, J.; Moyano, A.; Pericas, M.; Riera, A. *TL* **1992**, *33*, 2863.
19. (a) Cerny, M.; Malek, J. *TL* **1972**, 691. (b) Malek, J.; Cerny, M. *JOM* **1975**, *84*, 139. (c) Cerny, M.; Makel, J. *CCC* **1976**, *41*, 119.
20. Kametani, T.; Huang, S.-P.; Ihara, M.; Fukumoto, K. *JOC* **1976**, *41*, 2545.
21. (a) Hilscher, J.-C. *CB* **1975**, *108*, 727. (b) Minabe, M.; Suzuki, K. *BCJ* **1975**, *48*, 1480. (c) Kametani, T.; Huang, S.-P.; Koseki, C.; Ihara, M.; Fukumoto, K. *JOC* **1977**, *42*, 3040.
22. Ashby, E. C.; Noding, S. A. *JOC* **1980**, *45*, 1035.

Melinda Gugelchuk
University of Waterloo, Ontario, Canada

Sodium Borohydride

[16940-66-2] BH_4Na (MW 37.84)

(reducing agent for aldehydes and ketones, and many other functional groups in the presence of additives[1])

Physical Data: mp 400 °C; *d* 1.0740 g cm^{-3}.

Solubility: sol H_2O (stable at pH 14, rapidly decomposes at neutral or acidic pH); sol MeOH (13 g/100 mL)[1b], and EtOH (3.16 g/100 mL),[1b] but decomposes to borates; sol polyethylene glycol (PEG),[2a] sol and stable in *i*-PrOH (0.37 g/100 mL)[3] and diglyme (5.15 g/100 mL);[1b] insol ether;[1b] slightly sol THF.[1c]

Form Supplied in: colorless solid in powder or pellets; supported on silica gel or on basic alumina; 0.5 M solution in diglyme; 2.0 M solution in triglyme; 12 wt % solution in 14 M aqueous NaOH. Typical impurities are sodium methoxide and sodium hydroxide.

Analysis of Reagent Purity: can be assessed by hydrogen evolution.[4]

Purification: crystallize from diglyme[3] or isopropylamine.[4]

Handling, Storage, and Precautions: harmful if inhaled or absorbed through skin. It is decomposed rapidly and exothermically by water, especially if acid solutions are used. This decomposition forms toxic diborane gas and flammable/explosive hydrogen gas, and thus must be carried out under a hood. Solutions in DMF can undergo runaway thermal reactions, resulting in violent decompositions.[5] The addition of supported noble metal catalysts to solutions of $NaBH_4$ can result in ignition of liberated hydrogen gas.[5]

Reduction of Aldehydes and Ketones. Sodium borohydride is a mild and chemoselective reducing agent for the carbonyl function. At 25 °C in hydroxylic solvents it rapidly reduces aldehydes and ketones, but it is essentially inert to other functional groups such as epoxides, esters, lactones, carboxylic acid salts, nitriles, and nitro groups. Acyl halides, of course, react with the solvent.[1a] The simplicity of use, the low cost, and the high chemoselectivity make it one of the best reagents for this reaction. Ethanol and methanol are usually employed as solvents, the former having the advantage of permitting reductions in homogeneous solutions with relatively little loss of reagent through the side reaction with the solvent.[1a] Aprotic solvents such as diglyme greatly decrease the reaction rates.[1a] On the other hand, $NaBH_4$ in polyethylene glycol (PEG) shows a reactivity similar to that observed in EtOH.[2a] Although the full details of the mechanism of ketone reduction by $NaBH_4$ remain to be established,[6] it has been demonstrated that all four hydrogen atoms can be transferred. Moreover, the rate of reduction was shown to slightly increase when the hydrogens on boron are replaced by alkoxy groups.[1a,c,d] However, especially when $NaBH_4$ is used in MeOH, an excess of reagent has to be used in order to circumvent the competitive borate formation by reaction with the solvent. Ketone reduction has been accelerated under phase-transfer conditions[7] or in the presence of HMPA supported on a polystyrene-type resin.[8]

The isolation of products is usually accomplished by diluting the reaction mixture with water, making it slightly acidic to destroy any excess hydride, and then extracting the organic product from the aqueous solution containing boric acid and its salts.

Kinetic examination of the reduction of benzaldehyde and acetophenone in isopropyl alcohol indicated a rate ratio of 400:1.[1a] Thus it is in principle possible to reduce an aldehyde in the presence of a ketone.[9a] Best results (>95% chemoselectivity) have been obtained using a mixed solvent system (EtOH–CH_2Cl_2 3:7) and performing the reduction at −78 °C,[9a] or by employing an anionic exchange resin in borohydride form.[10] This reagent can also discriminate between aromatic and aliphatic aldehydes. On the other hand, reduction of ketones in the presence of aldehydes can be performed by $NaBH_4$–***Cerium(III) Chloride***. $NaBH_4$ in MeOH–CH_2Cl_2 (1:1) at −78 °C reduces ketones in the presence of conjugated enones and aldehydes in the presence of conjugated enals.[9]

Conjugate Reductions. $NaBH_4$ usually tends to reduce α,β-unsaturated ketones in the 1,4-sense,[1d] affording mixtures of saturated alcohol and ketone. In alcoholic solvents, saturated β-alkoxy alcohols can be formed as byproducts via conjugate addition of the solvent.[11] The selectivity is not always high. For example, while cyclopentenone is reduced only in the conjugate fashion, cyclohexenone affords a 59:41 ratio of allylic alcohol and saturated alcohol.[1d] Increasing steric hindrance on the enone increases 1,2-attack.[11] Aldehydes undergo more 1,2-reduction than the corresponding ketones.[1c,1d] The use of pyridine as solvent may be advantageous in increasing the selectivity for 1,4-reduction, as exemplified (eq 1) by the reduction of (*R*)-carvone to dihydrocarveols and (in minor amounts) dihydrocarvone.[12]

$NaBH_4$, py, 75% (1)

Trialkyl borohydrides such as ***Lithium Tri-s-butylborohydride*** and ***Potassium Tri-s-butylborohydride*** are superior reagents for the chemoselective 1,4-reduction of enones. On the other hand, 1,2-reduction can be obtained by using $NaBH_4$ in the mixed solvent MeOH–THF (1:9),[13] or with $NaBH_4$ in combination with $CeCl_3$ or other lanthanide salts.[14]

$NaBH_4$ in alcoholic solvents has been used for the conjugate reduction of α,β-unsaturated esters,[15] including cinnamates and alkylidenemalonates, without affecting the alkoxycarbonyl group. Conjugate nitroalkenes have been reduced to the corresponding nitroalkanes.[16] Saturated hydroxylamines are obtained by reducing nitroalkenes with the ***Borane–Tetrahydrofuran*** complex in the presence of catalytic amounts of $NaBH_4$, or by using a combination of $NaBH_4$ and ***Boron Trifluoride Etherate*** in 1:1.5 molar ratio.[17] Extended reaction can lead also to the saturated amines.[17]

Reduction of Carboxylic Acid Derivatives. The reduction of carboxylic esters[1c,1d] by $NaBH_4$ is usually slow, but can be performed by the use of excess reagent in methanol or ethanol[18] at room temperature or higher. The solvent must correspond to the ester group, since $NaBH_4$ catalyzes ester interchange. This transformation can also be achieved at 65–80 °C in *t*-BuOH[19] or polyethylene glycol.[2b] Although the slow rate and the need to use excess reagent makes other stronger complex hydrides such as ***Lithium Borohydride*** or ***Lithium Aluminum Hydride*** best suited for this reaction, in particular cases the use of $NaBH_4$ allows interesting selectivity: see, for example, the reduction of eq 2,[20] where the β-lactam remains unaffected, or of eq 3,[21] where the epoxide and the cyano group do not react.

$NaBH_4$, THF–H_2O, rt, 3 h, 78% (2)

Ar(R) ... CN, CO_2Et → $NaBH_4$, EtOH, 80–88% → Ar(R) ... CN, OH (3)

Borohydrides cannot be used for the reduction of α,β-unsaturated esters to allylic alcohols since the conjugate reduction is faster.[18b] The reactivity of $NaBH_4$ toward esters has been enhanced with various additives. For example, the system $NaBH_4$–$CaCl_2$ (2:1) shows a reactivity similar to $LiBH_4$.[18b] Esters have also been reduced with $NaBH_4$–***Zinc Chloride*** in the presence of a tertiary amine,[22] or with $NaBH_4$–***Copper(II) Sulfate***. The latter system reduces selectively aliphatic esters in the presence of aromatic esters of amides.[23] Finally, esters have also been reduced with $NaBH_4$–***Iodine***.[24a] In this case the reaction seems to proceed through diborane formation, and so it cannot be used for substrates containing an alkenic double bond. A related methodology, employing ***Borane–Dimethyl Sulfide*** in the presence of catalytic $NaBH_4$,[25] is particularly useful for the regioselective reduction of α-hydroxy esters, as exemplified by the conversion of (*S*)-diethyl malate into the vicinal diol (eq 4).

EtO_2C–CH(OH)–CH_2–CO_2Et → BH_3•Me_2S, THF, $NaBH_4$ (5 mol %), 88% → HO–CH_2–CH(OH)–CH_2–CO_2Et (4)

regioisomer ratio = 200:1

Lactones are only slowly reduced by $NaBH_4$ in alcohol solvents at 25 °C, unless the carbonyl is flanked by an α-heteroatom functionality.[1d] Sugar lactones are reduced to the diol when the reduction is carried out in water at neutral pH, or to the lactol when the reaction is performed at lower (≈3) pH.[26] Thiol esters are more reactive and are reduced to primary alcohols with $NaBH_4$ in EtOH, without reduction of ester substituents.[27]

Carboxylic acids are not reduced by $NaBH_4$. The conversion into primary alcohols can be achieved by using $NaBH_4$ in combination with powerful Lewis acids,[1k,28] ***Sulfuric Acid***,[28] ***Catechol***,[24b] ***Trifluoroacetic Acid***,[24b] or I_2.[24a] In these cases the actual reacting species is a borane, and thus hydroboration of double bonds present in the substrate can be a serious side reaction. Alternatively, the carboxylic acids can be transformed into activated derivatives,[29] such as carboxymethyleneiminium salts[29a] or mixed anhydrides,[29b] followed by reduction with $NaBH_4$ at low temperature. These methodologies tolerate the presence of double bonds, even if conjugated to the carboxyl.[29a]

Nitriles are, with few exceptions,[21] not reduced by $NaBH_4$.[1k] Sulfurated $NaBH_4$,[30] prepared by the reaction of sodium borohydride with sulfur in THF, is somewhat more reactive than $NaBH_4$, and reduces aromatic nitriles (but not aliphatic ones) to amines in refluxing THF. Further activation has been realized by using the ***Cobalt Boride*** system, ($NaBH_4$–$CoCl_2$) which appears to be one of the best methods for the reduction of nitriles to primary amines. More recently it has been found that ***Zirconium(IV) Chloride***,[31] Et_2SeBr_2,[32] $CuSO_4$,[23] ***Chlorotrimethylsilane***,[33] and I_2[24a] are also efficient activators for this transformation. The $NaBH_4$–Et_2SeBr_2 reagent allows the selective reduction of nitriles in the presence of esters or nitro groups, which are readily reduced by $NaBH_4$–$CoCl_2$.

$NaBH_4$ in alcoholic solvents does not reduce amides.[1a,1c–d] However, under more forcing conditions ($NaBH_4$ in pyridine at reflux), reduction of tertiary amides to the corresponding amines can be achieved.[32] Secondary amides are inert, while primary amides are dehydrated to give nitriles. Also, $NaBH_4$–Et_2SeBr_2 is specific for tertiary amides.[32] Reagent combinations which show enhanced reactivity, and which are thus employable for all three types of amides, are $NaBH_4$–$CoCl_2$, $NaBH_4$ in the presence of strong acids[34] (e.g. ***Methanesulfonic Acid*** or ***Titanium(IV) Chloride***) in DMF or DME, $NaBH_4$–Me_3SiCl,[33] and $NaBH_4$–I_2.[24a]

An indirect method for the reduction of amides to amines by $NaBH_4$ (applicable only to tertiary amides) involves conversion into a Vilsmeier complex $[(R_2N{=}C(Cl)R)^+Cl^-]$, by treatment with ***Phosphorus Oxychloride***, followed by its reduction.[35] In a related methodology, primary or secondary (also cyclic) amides are first converted into ethyl imidates by the action of ***Triethyloxonium Tetrafluoroborate***, and the latter reduced to amines with $NaBH_4$ in EtOH or, better, with $NaBH_4$–***Tin(IV) Chloride*** in Et_2O.[36]

In addition to the above-quoted methods, tertiary δ-lactams have been reduced to the corresponding cyclic amines by dropwise addition of MeOH to the refluxing mixture of $NaBH_4$ and substrate in *t*-BuOH,[37] or by using trifluoroethanol as solvent.[38] This reaction was applied during a synthesis of indolizidine alkaloid swainsonine for the reduction of lactam (**1**) to amine (**2**) (eq 5).[38]

(**1**) —[$NaBH_4$, EtOH–CF_3CO_2H (10:1), reflux; 60%]→ (**2**) (5)

Acyl chlorides can be reduced to primary alcohols by reduction in aprotic solvents such as PEG,[2a] or using $NaBH_4$–***Alumina*** in Et_2O.[39] More synthetically useful is the partial reduction to the aldehydic stage, which can be achieved by using a stoichiometric amount of the reagent at −70 °C in DMF–THF,[40] with the system $NaBH_4$–***Cadmium Chloride***–DMF,[41] or with ***Bis(triphenylphosphine)copper(I) Borohydride***.

Alternative methodologies for the indirect reduction of carboxylic derivatives employ as intermediates 2-substituted 1,3-benzoxathiolium tetrafluoborates (prepared from carboxylic acids, acyl chlorides, anhydrides, or esters)[42] and dihydro-1,3-thiazines or dihydro-1,3-oxazines (best prepared from nitriles).[43] These compounds are smoothly reduced by $NaBH_4$, to give acetal-like adducts, easily transformable into the corresponding aldehydes by acidic hydrolysis. Conversion of primary amides into the *N*-acylpyrrole derivative by reaction with 1,4-dichloro-1,4-dimethoxybutane in the presence of a cationic exchange resin, followed by $NaBH_4$ reduction, furnished the corresponding aldehydes.[44]

Cyclic anhydrides are reduced by $NaBH_4$ to lactones in moderate to good yields. Hydride attack occurs principally at the carbonyl group adjacent to the more highly substituted carbon atom.[45] Cyclic imides are more reactive than amides and can be reduced to the corresponding α′-hydroxylactams by using methanolic or ethanolic $NaBH_4$ in the presence of HCl as buffering agent.[1c] These products are important as precursors for *N*-acyliminium salts. The carbonyl adjacent to the most substituted carbon is usually preferentially reduced[46] (see also ***Cobalt Boride***). *N*-Alkylphthalimides may be reduced with $NaBH_4$ in 2-propanol to give an open-chain hydroxy-amide which, upon treatment with AcOH, cyclizes to give phthalide (a lactone) and the free amine. This method represents a convenient procedure for releasing amines from phthalimides under nonbasic conditions.[47]

Reduction of C=N Double Bonds. The C=N double bond of imines is generally less reactive than the carbonyl C=O toward reduction with complex hydrides. However, imines may be reduced by $NaBH_4$ in alcoholic solvents under neutral conditions at temperatures ranging from 0 °C to that of the refluxing solvent.[1c,1d,48] Protonation or complexation with a Lewis acid of the imino nitrogen dramatically increases the rate of reduction.[1i] Thus $NaBH_4$ in AcOH (see ***Sodium Triacetoxyborohydride***) or in other carboxylic acids is an efficient reagent for this transformation (although the reagent of choice is probably ***Sodium Cyanoborohydride***). Imines are also reduced by ***Cobalt Boride***, $NaBH_4$–***Nickel(II) Chloride***, and $NaBH_4$–$ZrCl_4$.[31] Imine formation, followed by in situ reduction, has been used as a method for synthesis of unsymmetrical secondary amines.[48] Once again, $Na(CN)BH_3$ represents the best reagent.[1c,1d,48] However, this transformation was realized also with $NaBH_4$,[48,49] either by treating the amine with excess aqueous formaldehyde followed by $NaBH_4$ in MeOH, or $NaBH_4$–CF_3CO_2H, or through direct reaction of the amine with the $NaBH_4$–carboxylic acid system. In the latter case, part of the acid is first reduced in situ to the aldehyde, which then forms an imine. The real reagent involved is $NaB(OCOR)_3H$ (see ***Sodium Triacetoxyborohydride***). Reaction of an amine with glutaric aldehyde and $NaBH_4$ in the presence of H_2SO_4 represents a good method for the synthesis of *N*-substituted piperidines.[49c] Like protonated imines, iminium salts are readily reduced by $NaBH_4$ in alcoholic media.[1c,50] *N*-Silylimines are more reactive than *N*-alkylimines. Thus α-amino esters can be obtained by reduction of *N*-silylimino esters.[51] α,β-Unsaturated imines are reduced by $NaBH_4$ in alcoholic solvents in the 1,2-mode to give allylic amines.[52] Enamines are transformed into saturated amines by reduction with $NaBH_4$ in alcoholic media.[48,53]

The reduction of oximes and oxime ethers is considerably more difficult and cannot be realized with $NaBH_4$ alone. Effective reagent combinations for the reduction of oximes include sulfurated $NaBH_4$,[30] $NaBH_4$–$NiCl_2$, $NaBH_4$–$ZrCl_4$,[31] $NaBH_4$–MoO_3,[54] $NaBH_4$–$TiCl_4$,[55] and $NaBH_4$–***Titanium(III) Chloride***.[56] In all cases the main product is the corresponding primary amine. $NaBH_4$–$ZrCl_4$ is efficient also for the reduction of oxime ethers. $NaBH_4$–MoO_3 reduces oximes without affecting double bonds, while $NaBH_4$–$NiCl_2$ reduces both functional groups. The reduction with $NaBH_4$–$TiCl_3$ in buffered (pH 7) aqueous media has been used for the chemoselective reduction of α-oximino esters to give α-amino esters (eq 6).[56]

1. $NaBH_4$, $TiCl_3$, L-tartaric acid, pH 7, MeOH–H_2O; 2. HCl; 82% (6)

$NaBH_4$ reduces hydrazones only when they are *N,N*-dialkyl substituted. The reaction is slow and yields are not usually satisfactory.[57] More synthetically useful is the reduction of *N-p*-tosylhydrazones to give hydrocarbons,[1c,1d,58] which has been carried out with $NaBH_4$ in refluxing MeOH, dioxane, or THF.[58] Since *N-p*-tosylhydrazones are easily prepared from aldehydes or ketones, the overall sequence represents a mild method for carbonyl deoxygenation. α,β-Unsaturated tosylhydrazones show a different behavior yielding, in MeOH, the allylic (or benzylic) methyl ethers.[58c] The reduction of tosylhydrazones with $NaBH_4$ is not compatible with ester groups, which are readily reduced under these conditions. More selective reagents for this reduction are $NaBH(OAc)_3$ and $NaCNBH_3$.

Reduction of Halides, Sulfonates, and Epoxides. The reduction of alkyl halides or sulfonates by $NaBH_4$ is not an easy reaction.[1d] It is best performed in polar aprotic solvents[59] such as DMSO, sulfolane, HMPA, DMF, diglyme, or PEG (polyethylene glycol),[2a] at temperatures between 60 °C and 100 °C (unless for highly reactive substrates), or under phase-transfer conditions.[60a] The mechanism is believed to be S_N2 (I > Br > Cl and primary > secondary). Although the more nucleophilic ***Lithium Triethylborohydride*** seems better suited for these reductions,[59b] the lower cost of $NaBH_4$ and the higher chemoselectivity (for example esters, nitriles, and sulfones can survive)[59a] makes it a useful alternative. Also, some secondary and tertiary alkyl halides, capable of forming relatively stable carbocations, for example benzhydryl chloride, may be reduced by $NaBH_4$. In this case the mechanism is different (via a carbocation) and the reaction is accelerated by water.[59a,b] Primary, secondary, and even aryl iodides and bromides[1d] have been reduced in good yields by $NaBH_4$ under the catalysis of soluble polyethylene- or polystyrene-bound tin halides (PE–$Sn(Bu)_2Cl$ or PS–$Sn(Bu)_2Cl$).[61] Aryl bromides and iodides have also been reduced with $NaBH_4$–***Copper(I) Chloride*** in MeOH.[62]

$NaBH_4$ reduces epoxides only sluggishly.[1d] Aryl-substituted and terminal epoxides can be reduced by slow addition of MeOH to a refluxing mixture of epoxide and $NaBH_4$ in *t*-BuOH,[63] or by $NaBH_4$ in polyethylene glycol.[2b] The reaction is regioselective (attack takes place on the less substituted carbon), and chemoselective (nitriles, carboxylic acids, and nitro groups are left intact).[63] The opposite regioselectivity was realized by the $NaBH_4$-catalyzed reduction with diborane.[64]

Other Reductions. Aromatic and aliphatic nitro compounds are not reduced to amines by $NaBH_4$ in the absence of an activator.[1d] The $NaBH_4$–$NiCl_2$ system (see ***Nickel Boride***) is a good reagent combination for this reaction, being effective also for primary and secondary aliphatic compounds. Other additives that permit $NaBH_4$ reduction are $SnCl_2$,[65] Me_3SiCl,[33] $CoCl_2$ (see ***Cobalt Boride***), and MoO_3 (only for aromatic compounds),[66] Cu^{2+} salts (for aromatic and tertiary aliphatic),[23,67] and ***Palladium on Carbon*** (good for both aromatic and aliphatic).[68] Also, sulfurated $NaBH_4$[30] is an effective and mild reducing agent for aromatic nitro groups. In the presence of catalytic selenium or tellurium, $NaBH_4$ reduces nitroarenes to the corresponding *N*-arylhydroxylamines.[69]

The reduction of azides to amines proceeds in low yield under usual conditions, but it can be performed efficiently under phase-transfer conditions,[60b] using $NaBH_4$ supported on an ion-exchange resin,[70] or using a THF–MeOH mixed solvent (this last method is well suited only for aromatic azides).[71]

Tertiary alcohols or other carbinols capable of forming a stable carbocation have been deoxygenated by treatment with $NaBH_4$ and CF_3CO_2H or $NaBH_4$–CF_3SO_3H.[72] Under the same conditions,[72] or with $NaBH_4$–***Aluminum Chloride***,[73] diaryl ketones have also been deoxygenated.

Cyano groups α to a nitrogen atom can be replaced smoothly by hydrogen upon reaction with $NaBH_4$.[74] Since α-cyano derivatives of trisubstituted amines can be easily alkylated with electrophilic agents, the α-aminonitrile functionality can be used as a latent α-amino anion,[74a] as exemplified by eq 7 which shows the synthesis of ephedrine from a protected aminonitrile. The reduction, proceeding with concurrent benzoyl group removal, is only moderately stereoselective (77:23).

1. LDA; 2. PhCHO; then $NaBH_4$, MeOH, 0 °C; 75% (7)

Primary amines have been deaminated in good yields through reduction of the corresponding bis(sulfonimides) with $NaBH_4$ in HMPA at 150–175 °C.[75] $NaBH_4$ reduction of ozonides is rapid at −78 °C and allows the one-pot degradation of double bonds to alcohols[1b] (see also ***Ozone***). The reduction of organomercury(II) halides (see also ***Mercury(II) Acetate***) is an important step in the functionalization of double bonds via oxymercuration–or amidomercuration–reduction. This reduction, which proceeds through a radical mechanism, is not stereospecific, but it can be in some cases diastereoselective.[76] In the presence of ***Rhodium(III) Chloride*** in EtOH, $NaBH_4$ completely saturates arenes.[77] $NaBH_4$ has also been employed for the reduction of quinones,[78] sulfoxides (in combination with ***Aluminum Iodide***[79] or Me_3SiCl[33]), and sulfones (with Me_3SiCl),[33] although it does not appear to be the reagent of choice for these reductions. Finally, $NaBH_4$ was used for the reduction of various heterocyclic systems (pyridines, pyridinium salts, indoles, benzofurans, oxazolines, and so on).[1c,1d,48,80] The discussion of these reductions is beyond the scope of this article.

Diastereoselective Reductions. $NaBH_4$, like other small complex hydrides ($LiBH_4$ and $LiAlH_4$), shows an intrinsic preference for axial attack on cyclohexanones,[1c,1d,81] as exemplified by the reduction of 4-*t*-butylcyclohexanone (eq 8).[81a] This preference, which is due to stereoelectronic reasons,[82] can be counterbalanced by steric biases. For example, in 3,3,5-trimethylcyclohexanone, where a β-axial substituent is present, the stereoselectivity is nearly completely lost (eq 9).[81a]

Also, in 2-methylcyclopentanone[81c] the attack takes place from the more hindered side, forming the *trans* isomer (dr = 74:26).

In norcamphor,[81a] both stereoelectronic and steric effects favor *exo* attack, forming the *endo* alcohol in 84:16 diastereoisomeric ratio. In camphor, however, the steric bias given by one of the two methyls on the bridge brings about an inversion of stereoselectivity toward the *exo* alcohol.[81a]

The stereoselectivity for equatorial alcohols has been enhanced by using the system $NaBH_4$–***Cerium(III) Chloride***, which has an even higher propensity for attack from the more hindered side,[83] or by precomplexing the ketone on ***Montmorillonite K10*** clay.[84] On the other hand, bulky trialkylborohydrides (see ***Lithium Tri-s-butylborohydride***) are best suited for synthesis of the axial alcohol through attack from the less hindered face.

t-Bu ⟶ t-Bu (H, OH) + t-Bu (OH, H) (8)

86:14

⟶ (H, OH) + (OH, H) (9)

48:52

$NaBH_4$ does not seem to be the best reagent for the stereoselective reduction of chiral unfunctionalized acyclic ketones. Bulky complex hydrides such as Li(s-Bu)$_3$BH usually afford better results.[1c,1d] When a heteroatom is present in the α- or β-position, the stereochemical course of the reduction depends also on the possible intervention of a cyclic chelated transition state. Also, in this case other complex hydrides are often better suited for favoring chelation (see ***Zinc Borohydride***). Nevertheless, cases are known[85] where excellent degrees of stereoselection have been achieved with the simpler and less expensive $NaBH_4$. Some examples are shown in eqs 10–15.

$NaBH_4$, i-PrOH; 83:17 < dr < 95:5 — (3) ⟶ (4) (10)

$NaBH_4$, MeOH; dr = 97:3 — (5) ⟶ (6) (11)

$NaBH_4$, MeOH, NH_4Cl; dr > 93:7 (12)

$NaBH_4$, MeOH; dr > 91:9 (13)

$NaBH_4$, THF; dr > 99:1 (14)

$NaBH_4$, CH_2Cl_2; 80:20 < dr < 99:1 (15)

The stereoselective formation of *anti* adduct (**4**) in the reduction of ketone (**3**) was explained through the intervention of a chelate involving the methoxy group,[85a] although there is some debate on what the acidic species is that is coordinated (probably Na^+). A chelated transition state is probably the cause of the stereoselective formation of *anti* product (**6**) from (**5**).[85b] Methylation of the NH group indeed provokes a decrease of stereoselection. On the other hand, when appropriate protecting groups that disfavor chelation are placed on the heteroatom, the reduction proceeds by way of the Felkin model where the heteroatomic substituent plays the role of 'large' group, and *syn* adducts are formed preferentially. This is the case of α-dibenzylamino ketones (eqs 12 and 13)[85c,d] and of the α-silyloxy ketone of eq 14.[85e] Finally, the sulfonium salt of eq 15 gives, with excellent stereocontrol, the *anti* alcohol.[85f] This result was explained by a transition state where the S^+ and carbonyl oxygen are close due to a charge attraction.

The reduction of a diastereomeric mixture of enantiomerically pure β-keto sulfoxides (**7**) furnished one of the four possible isomers with good overall stereoselectivity (90%), when carried out under conditions which favor epimerization of the α chiral center (eq 16). This outcome derives from a chelation-controlled reduction (involving the sulfoxide oxygen) coupled with a kinetic resolution of the two diastereoisomers of (**7**).[86]

$NaBH_4$, EtOH–H_2O, NaOH; ds = 90% — (7) (16)

The reduction of cyclic imines and oximes follows a trend similar to that of corresponding ketones. However, the tendency for attack from the most hindered side is in these cases attenuated.[1c,1d,57,87] In the case of oximes, while $NaBH_4$–MoO_3 attacks from the axial side, $NaBH_4$–$NiCl_2$ attacks from the equatorial side.[88] An example of diastereoselective reduction of acyclic chiral imines is represented by the one-pot transformation of α-alkoxy or α,β-epoxynitriles into *anti* vicinal amino alcohols (eq 17) or epoxyamines. The outcome of these reductions was explained on the basis of a cyclic chelated transition state.[89]

R^2MgX ; $NaBH_4$ (17)

80:20 < dr < 98:2

Enantioselective Reductions. $NaBH_4$ has been employed with less success than $LiAlH_4$ or BH_3 in enantioselective ketone reductions.[1d,90,91] Low to moderate ee values have been obtained in the asymmetric reduction of ketones with chiral phase-transfer catalysts, chiral crown ethers,[91a] β-cyclodextrin,[91b] and bovine serum albumin.[91c] On the other hand, good results have been realized in the reduction of propiophenone with $NaBH_4$ in the presence of isobutyric acid and of diisopropylidene-D-glucofuranose (ee = 85%),[91d] or in the reduction of α-keto esters and β-keto esters with $NaBH_4$–L-tartaric acid (ee ≥86%).[91e]

Very high ee values have been obtained in the asymmetric conjugate reduction of α,β-unsaturated esters and amides with $NaBH_4$ in the presence of a chiral semicorrin (a bidentate nitrogen ligand) cobalt catalyst.[92] Good to excellent ee values were realized in the reduction of oxime ethers with $NaBH_4$–$ZrCl_4$ in the presence of a chiral 1,2-amino alcohol.[93]

Related Reagents. See Classes R-2, R-5, R-6, R-9, R-10, R-11, R-12, R-13, R-15, R-20, R-21, R-26, R-27, R-28, and R-32, pages 1–10. Cerium(III) Chloride; Nickel Boride; Potassium Triisopropoxyborohydride; Sodium Cyanoborohydride; Sodium Triacetoxyborohydride.

---

1. (a) Brown, H. C.; Krishnamurthy, S. *T* **1979**, *35*, 567. (b) *FF* **1967**, *1*, 1049. (c) Seyden-Penne, J. *Reductions by the Alumino- and Borohydrides in Organic Synthesis*; VCH–Lavoisier: Paris, 1991. (d) *COS* **1991**, *8*, Chapters 1.1, 1.2, 1.7, 1.10, 1.11, 1.14, 2.1, 2.3, 3.3, 3.5, 4.1, 4.4, 4.7.
2. (a) Santaniello, E.; Fiecchi, A.; Manzocchi, A.; Ferraboschi, P. *JOC* **1983**, *48*, 3074. (b) Santaniello, E.; Ferraboschi, P.; Fiecchi, A.; Grisenti, P.; Manzocchi, A. *JOC* **1987**, *52*, 671.
3. Brown, H. C.; Mead, E. J.; Subba Rao, B. C. *JACS* **1955**, *77*, 6209.
4. Stockmayer, W. H.; Rice, D. W.; Stephenson, C. C. *JACS* **1955**, *77*, 1980.
5. *The Sigma-Aldrich Library of Chemical Safety Data*, Sigma-Aldrich: Milwaukee, 1988.
6. Wigfield, D. C. *T* **1979**, *35*, 449.
7. Bunton, C. A.; Robinson, L.; Stam, M. F. *TL* **1971**, 121. Subba Rao, Y. V.; Choudary, B. M. *SC* **1992**, *22*, 2711.
8. Tomoi, M.; Hasegawa, T.; Ikeda, M.; Kakiuchi, H. *BCJ* **1979**, *52*, 1653.
9. (a) Ward, D. E.; Rhee, C. K. *CJC* **1989**, *67*, 1206. (b) Ward, D. E.; Rhee, C. K.; Zoghaib, W. M. *TL* **1988**, *29*, 517.
10. Yoon, N. M.; Park, K. B.; Gyoung, Y. S. *TL* **1983**, *24*, 5367.
11. Johnson, M. R.; Rickborn, B. *JOC* **1970**, *35*, 1041.
12. Raucher, S.; Hwang, K.-J. *SC* **1980**, *10*, 133.
13. Varma, R. S.; Kabalka, G. W. *SC* **1985**, *15*, 985.
14. Komiya, S.; Tsutsumi, O. *BCJ* **1987**, *60*, 3423.
15. Schauble, J. H.; Walter, G. J.; Morin, J. G. *JOC* **1974**, *39*, 755. Salomon, R. G.; Sachinvala, N. D.; Raychaudhuri, S. R.; Miller, D. B. *JACS* **1984**, *106*, 2211.
16. Hassner, A.; Heathcock, C. H. *JOC* **1964**, *29*, 1350.
17. Varma, R. S.; Kabalka, G. W. *SC* **1985**, *15*, 843.
18. (a) Olsson, T.; Stern, K.; Sundell, S. *JOC* **1988**, *53*, 2468. (b) Brown, H. C.; Narasimhan, S.; Choi, Y. M. *JOC* **1982**, *47*, 4702.
19. Soai, K.; Oyamada, H.; Takase, M.; Ookawa, A. *BCJ* **1984**, *57*, 1948.
20. Kawabata, T.; Minami, T.; Hiyama, T. *JOC* **1992**, *57*, 1864.
21. Mauger, J.; Robert, A. *CC* **1986**, 395.
22. Yamakawa, T.; Masaki, M.; Nohira, H. *BCJ* **1991**, *64*, 2730.
23. Yoo, S.; Lee, S. *SL* **1990**, 419.
24. (a) Prasad, A. S. B.; Kanth, J. V. B.; Periasamy, M. *T* **1992**, *48*, 4623. (b) Suseela, Y.; Periasamy, M. *T* **1992**, *48*, 371.
25. Saito, S.; Ishikawa, T.; Kuroda, A.; Koga, K.; Moriwake, T. *T* **1992**, *48*, 4067.
26. Wolfrom, M. L.; Anno, K. *JACS* **1952**, *74*, 5583. Attwood, S. V.; Barrett, A. G. M. *JCS(P1)* **1984**, 1315.
27. Liu, H.-J.; Bukownik, R. R.; Pednekar, P. R. *SC* **1981**, *11*, 599.
28. Abiko, A.; Masamune, S. *TL* **1992**, *33*, 5517.
29. (a) Fujisawa, T.; Mori, T.; Sato, T. *CL* **1983**, 835. (b) Rodriguez, M.; Llinares, M.; Doulut, S.; Heitz, A.; Martinez, J. *TL* **1991**, *32*, 923.
30. Lalancette, J. M.; Freche, A.; Brindle, J. R.; Laliberté, M. *S* **1972**, 526.
31. Itsuno, S.; Sakurai, Y.; Ito, K. *S* **1988**, 995.
32. Akabori, S.; Takanohashi, Y. *JCS(P1)* **1991**, 479.
33. Giannis, A.; Sandhoff, K. *AG(E)* **1989**, *28*, 218.
34. (a) Wann, S. R.; Thorsen, P. T.; Kreevoy, M. M. *JOC* **1981**, *46*, 2579. (b) Kano, S.; Tanaka, Y.; Sugino, E.; Hibino, S. *S* **1980**, 695.
35. Rahman, A.; Basha, A.; Waheed, N.; Ahmed, S. *TL* **1976**, 219.
36. Tsuda, Y.; Sano, T.; Watanabe, H. *S* **1977**, 652.
37. Mandal, S. B.; Giri, V. S.; Sabeena, M. S.; Pakrashi, S. C. *JOC* **1988**, *53*, 4236.
38. Setoi, H.; Takeno, H.; Hashimoto, M. *JOC* **1985**, *50*, 3948.
39. Santaniello, E.; Farachi, C.; Manzocchi, A. *S* **1979**, 912.
40. Babler, J. H.; Invergo, B. J. *TL* **1981**, *22*, 11.
41. Entwistle, I. D.; Boehm, P.; Johnstone, R. A. W.; Telford, R. P. *JCS(P1)* **1980**, 27.
42. Barbero, M.; Cadamuro, S.; Degani, I.; Fochi, R.; Gatti, A.; Regondi, V. *S* **1986**, 1074.
43. Meyers A. I.; Nabeya, A.; Adickes, H. W.; Politzer, I. R.; Malone, G. R.; Kovelesky, A.; Nolan, R. L.; Portnoy, R. C. *JOC* **1973**, *38*, 36. Politzer, I. R.; Meyers, A. I. *OSC* **1988**, *6*, 905.
44. Lee, S. D.; Brook, M. A.; Chan, T. H. *TL* **1983**, *24*, 1569.
45. Takano, S.; Ogasawara, K. *S* **1974**, 42.
46. Goto, T.; Konno, M.; Saito, M.; Sato, R. *BCJ* **1989**, *62*, 1205.
47. Osby, J. O.; Martin, M. G.; Ganem, B. *TL* **1984**, *25*, 2093.
48. Gribble, G. W.; Nutaitis, C. F. *OPP* **1985**, *17*, 317.
49. (a) Sondengam, B. L.; Hentchoya Hémo, J.; Charles, G. *TL* **1973**, 261. (b) Gribble, G. W.; Nutaitis, C. F. *S* **1987**, 709. (c) Verardo, G.; Giumanini, A. G.; Favret, G.; Strazzolini, P. *S* **1991**, 447.
50. Guerrier, L.; Royer, J.; Grierson, D. S.; Husson, H.-P. *JACS* **1983**, *105*, 7754; Polniaszek, R. P.; Kaufman, C. R. *JACS* **1989**, *111*, 4859.
51. Matsuda, Y.; Tanimoto, S.; Okamoto, T.; Ali, S. M. *JCS(P1)* **1989**, 279.
52. De Kimpe, N.; Stanoeva, E.; Verhé, R.; Schamp, N. *S* **1988**, 587.
53. Borch, R. F.; Bernstein, M. D.; Durst, H. D. *JACS* **1971**, *93*, 2897.
54. Mundy, B. P.; Bjorklund, M. *TL* **1985**, *26*, 3899.
55. Spreitzer, H.; Buchbauer, G.; Püringer, C. *T* **1989**, *45*, 6999.
56. Hoffman, C.; Tanke, R. S.; Miller, M. J. *JOC* **1989**, *54*, 3750.
57. Walker, G. N.; Moore, M. A.; Weaver, B. N. *JOC* **1961**, *26*, 2740.
58. (a) Caglioti, L. *OSC* **1988**, *6*, 62. (b) Rosini, G.; Baccolini, G.; Cacchi, S. *S* **1975**, 44. (c) Grandi, R.; Marchesini, A.; Pagnoni, U. M.; Trave, R. *JOC* **1976**, *41*, 1755.
59. (a) Hutchins, R. O.; Kandasamy, D.; Dux III, F.; Maryanoff, C. A.; Rotstein, D.; Goldsmith, B.; Burgoyne, W.; Cistone, F.; Dalessandro, J.; Puglis, J. *JOC* **1978**, *43*, 2259. (b) Krishnamurthy, S.; Brown, H. C. *JOC* **1980**, *45*, 849. (c) Kocienski, P.; Street, S. D. A. *SC* **1984**, *14*, 1087.
60. (a) Rolla, F. *JOC* **1981**, *46*, 3909. (b) Rolla, F. *JOC* **1982**, *47*, 4327.
61. Bergbreiter, D. E.; Walker, S. A. *JOC* **1989**, *54*, 5138.
62. Narisada, M.; Horibe, I.; Watanabe, F.; Takeda, K. *JOC* **1989**, *54*, 5308.
63. Ookawa, A.; Hiratsuka, H.; Soai, K. *BCJ* **1987**, *60*, 1813.
64. Brown, H. C.; Yoon, N. M. *JACS* **1968**, *90*, 2686.
65. Satoh, T.; Mitsuo, N.; Nishiki, M.; Inoue, Y.; Ooi, Y. *CPB* **1981**, *29*, 1443.
66. Yanada, K.; Yanada, R.; Meguri, H. *TL* **1992**, *33*, 1463.
67. Cowan, J. A. *TL* **1986**, *27*, 1205.

68. Neilson, T.; Wood, H. C. S.; Wylie, A. G. *JCS* **1962**, 371. Petrini, M.; Ballini, R.; Rosini, G. *S* **1987**, 713.
69. Uchida, S.; Yanada, K.; Yamaguchi, H.; Meguri, H. *CL* **1986**, 1069; *CC* **1986**, 1655.
70. Kabalka, G. W.; Wadgaonkar, P. P.; Chatla, N. *SC* **1990**, *20*, 293.
71. Soai, K.; Yokoyama, S.; Ookawa, A. *S* **1987**, 48.
72. Olah, G. A.; Wu, A.; Farooq, O. *JOC* **1988**, *53*, 5143.
73. Ono, A.; Suzuki, N.; Kamimura, J. *S* **1987**, 736.
74. (a) Stork, G.; Jacobson, R. M.; Levitz, R. *TL* **1979**, 771; (b) Santoyo-Gonzalez, F.; Hernandez-Mateo, F.; Vargas-Berenguel, A. *TL* **1991**, *32*, 1371.
75. Hutchins, R. O.; Cistone, F.; Goldsmith, B.; Heuman, P. *JOC* **1975**, *40*, 2018.
76. Gouzoules, F. H.; Whitney, R. A. *JOC* **1986**, *51*, 2024. Takahata, H.; Bandoh, H.; Hanayama, M.; Momose, T. *TA* **1992**, *3*, 607 and refs. therein.
77. Nishiki, M.; Miyataka, H.; Niino, Y.; Mitsuo, N.; Satoh, T. *TL* **1982**, *23*, 193.
78. Cho, H.; Harvey, R. G. *JCS(P1)* **1976**, 836.
79. Babu, J. R.; Bhatt, M. V. *TL* **1986**, *27*, 1073.
80. *COS* **1991**, *8*, Chapters 3.6–3.8, pp 579–666.
81. (a) Boone, J. R.; Ashby, E. C. *Top. Stereochem.* **1979**, *11*, 53. (b) Ref. 6. (c) Caro, B.; Boyer, B.; Lamaty, G.; Jaouen, G. *BSF(2)* **1983**, 281.
82. Wong, S. S.; Paddon-Row, M. N. *CC* **1990**, 456 and refs. therein.
83. Krief, A.; Surleraux, D.; Ropson, N. *TA* **1993**, *4*, 289.
84. Sarkar, A.; Rao, B. R.; Konar, M. M. *SC* **1989**, *19*, 2313.
85. (a) Glass, R. S.; Deardorff, D. R.; Henegar, K. *TL* **1980**, *21*, 2467. (b) Maugras, I.; Poncet, J.; Jouin, P. *T* **1990**, *46*, 2807. (c) Guanti, G.; Banfi, L.; Narisano, E.; Scolastico, C. *T* **1988**, *44*, 3671. (d) Reetz, M. T.; Drewes, M. W.; Lennick, K.; Schmitz, A.; Holdgrün, X. *TA* **1990**, *1*, 375. (e) Saito, S.; Harunari, T.; Shimamura, N.; Asahara, M.; Moriwake, T. *SL* **1992**, 325. (f) Shimagaki, M.; Matsuzaki, Y.; Hori, I.; Nakata, T.; Oishi, T. *TL* **1984**, *25*, 4779. (g) Morizawa, Y.; Yasuda, A.; Uchida, K. *TL* **1986**, *27*, 1833. (h) Fujii, H.; Oshima, K.; Utimoto, K. *CL* **1992**, 967. (i) Fujii, H.; Oshima, K.; Utimoto, K. *TL* **1991**, *32*, 6147. (j) Kobayashi, Y.; Uchiyama, H.; Kanbara, H.; Sato, F. *JACS* **1985**, *107*, 5541. (k) Oppolzer, W.; Tamura, O.; Sundarababu, G.; Signer, M. *JACS* **1992**, *114*, 5900. (l) Elliott, J.; Hall, D.; Warren, S. *TL* **1989**, *30*, 601.
86. Guanti, G.; Narisano, E.; Pero, F.; Banfi, L.; Scolastico, C. *JCS(P1)* **1984**, 189.
87. Hutchins, R. O.; Su, W.-Y.; Sivakumar, R.; Cistone, F.; Stercho, Y. P. *JOC* **1983**, *48*, 3412.
88. Ipaktschi, J. *CB* **1984**, *117*, 856.
89. (a) Brussee, J.; Van der Gen, A. *RTC* **1991**, *110*, 25. (b) Urabe, H.; Aoyama, Y.; Sato, F. *JOC* **1992**, *57*, 5056 and refs. therein.
90. Brown, H. C.; Park, W. S.; Cho, B. T.; Ramachandran, P. V. *JOC* **1987**, *52*, 5406.
91. (a) Takahashi, I.; Odashima, K.; Koga, K. *CPB* **1985**, *33*, 3571. (b) Fornasier, R.; Reniero, F.; Scrimin, P.; Tonellato, U. *JOC* **1985**, *50*, 3209. (c) Utaka, M.; Watabu, H.; Takeda, A. *JOC* **1986**, *51*, 5423. (d) Hirao, A.; Itsuno, S.; Owa, M.; Nagami, S.; Mochizuki, H.; Zoorov, H. H. A.; Niakahama, S.; Yamazaki, N. *JCS(P1)* **1981**, 900. (e) Yatagai, M.; Ohnuki, T. *JCS(P1)* **1990**, 1826.
92. von Matt, P.; Pfaltz, A. *TA* **1991**, *2*, 691.
93. Itsuno, S.; Sakurai, Y.; Shimizu, K.; Ito, K. *JCS(P1)* **1990**, 1859.

Luca Banfi, Enrica Narisano & Renata Riva
Università di Genova, Italy

Sodium Chlorite[1]

[7758-19-2] $ClNaO_2$ (MW 90.44)

(mild and selective reagent for the oxidation of all types of aldehydes to carboxylic acids at room temperature[16,17])

Physical Data: mp 180–200 °C (dec).
Solubility: 39 g/100 ml H_2O (17 °C); slightly sol MeOH.
Form Supplied in: white solid; widely available as technical grade material (ca. 80%; the rest is mainly NaCl).
Analysis of Reagent Purity: iodometric titration at pH 2.
Purification: technical grade material can be used as received. Crystallization from H_2O gives $NaClO_2{\cdot}3H_2O$ which dehydrates at >37 °C.
Handling, Storage, and Precautions: decomposition at >175 °C is highly exothermic. The reagent is not shock sensitive unless contaminated with organic materials. Causes severe irritation or burns to skin and eyes. Acute oral LD_{50} 180 mg kg^{-1} (rat, 80% assay material).

Oxidation of Aldehydes to Carboxylic Acids. First employed in cellulose chemistry,[2] the reagent is now widely used for the oxidation of saturated,[3–8] α,β-unsaturated,[9–15] aromatic,[8,13,16,17] and heteroaromatic[13] aldehydes to the corresponding carboxylic acids (eq 1). The high yields, selectivity, mild conditions, and ease of workup make this the method of choice for most RCHO to RCO_2H conversions. For example, with allylic primary alcohols[13] it is advantageous to carry out a two-step $MnO_2/NaClO_2$ process instead of a direct oxidation using Cr^{VI} or Ag^{I} because of higher purity and yield. The hypochlorite formed (eq 1) may interfere with the substrate or the product, or react with the reagent to give ClO_2 (eq 2), which also may have an adverse effect.

$$RCHO + HClO_2 \longrightarrow RCO_2H + HOCl \qquad (1)$$

$$HOCl + 2\,ClO_2^- \longrightarrow 2\,ClO_2 + Cl^- + OH^- \qquad (2)$$

To eliminate the various unwanted positive Cl species present, the reaction is therefore usually conducted in the presence of a scavenger (Table 1). The use of resorcinol has the drawback that the byproduct formed (4-chlororesorcinol) must be removed from the carboxylic acid in a separate step.

The alkene double bond is retained in oxidations of unsaturated alcohols. In an enantiospecific synthesis[19] of L-amino acids in >92% ee, intermediate chloroboronic esters (obtained from the alkyl Grignards) were directly oxidized to the acids, thus circumventing the problematic prior conversion to the aldehydes (eq 3).

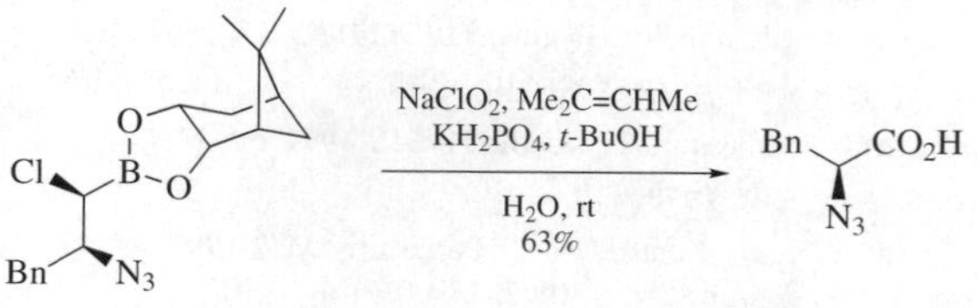

NaClO2, Me2C=CHMe, KH2PO4, t-BuOH; H2O, rt, 63% (3)

Table 1 The Reported Areas of Applicability and Functional Group Tolerances or Incompatibilities of Various ClO_2 Oxidation Procedures[a]

	Cl scavenger					
	H_2NSO_3H	Resorcinol	$Me_2C=CHMe$	H_2O_2	DMSO	None
R–CHO		(+)[3]	(+)[4–7]			(+)[8]
C=C–CHO	(−)[15]	(+)[9,10]	(+)[11–14]	(+)[15]	(+)[15]	(−)[8]
C″C–CHO, Het–CHO				(+)[15,b]		
Ar–CHO	(+)[16,17]	(+)[16]		(+)[15]	(+)[8,15]	(+)[8]
Isolated C=C		yes[10]	yes[7,23]	no[15]		
Conj. diene; acetal			yes[7]			
Cyclopropane			yes[23]			
s-OH		yes[9]	yes[7]			no[18]
$R_2CH–N_3$			yes[19]			
p-HO–Ar	yes[16,17]	yes[16]		no[15]	yes[15]	yes[8]
o-HO–Ar	yes[17]			no[15]	yes[8]	no[8]
p-H_2N- or *p*-MeS–Ar				no[15]		
MOM ether			yes[5]			
$PhCH_2$ ether		yes[3]	yes[12]			
TBDMS ether			yes[11,12]			
Allyl ether			yes[11]			
Ester		yes[3,9]	yes[4,7,14]			
Epoxy			yes[4]			
a-Boc-NH ester			yes[6]			
Ketone	yes[17]	yes[9]				yes[8]
b-Lactam, $ArNO_2$		yes[3]				

[a] (+) oxidized in good yield; (−) not oxidized or decomposes; 'yes' = compatible functionality; 'no' = functionality oxidized or compound decomposes. [b] Not pyrrole.

Other Oxidations. Dithiocarbamates are oxidized to isothiocyanates,[20] ArSMe to ArSOMe,[21] RCH_2NO_2 to RCHO, and $R_2CH–NO_2$ to R_2CO (eq 4).[22]

$$EtO_2C(CH_2)_5NO_2 \xrightarrow[\text{OH}^-,\ H_2O,\ CH_2Cl_2,\ \text{rt},\ 65\%]{NaClO_2,\ Bu_4NHSO_4} EtO_2C(CH_2)_4CHO \quad (4)$$

Related Reagents. See Class O-2, pages 1–10.

1. Kaczur, J. J.; Cawlfield, D. W. In *Kirk-Othmer Encyclopedia of Chemical Technology*, 4th ed.; Kroschwitz, J. I.; Howe-Grant, M., Eds.; Wiley: New York, 1993; Vol. 5, p 968.
2. (a) Nevell, T. P. In *Methods in Carbohydrate Chemistry III, Cellulose*; Whistler, R. L., Ed.; Academic: New York, 1963; p 182. (b) Rutherford, H. A.; Minor, F. W.; Martin, A. R.; Harris, M. *J. Res. Nat. Bur. Stand.* **1942**, *29*, 131.
3. Baldwin, J. E.; Forrest, A. K.; Ko, S.; Sheppard, L. N. *CC* **1987**, 81.
4. Kraus, G. A.; Taschner, M. J. *JOC* **1980**, *45*, 1175.
5. Kende, A. S.; Roth, B.; Kubo, I. *TL* **1982**, *23*, 1751.
6. Plaue, S.; Heissler, D. *TL* **1987**, *28*, 1401.
7. Anthony, N. J.; Armstrong, A.; Ley, S. V.; Madin, A. *TL* **1989**, *30*, 3209.
8. Bayle, J. P.; Perez, F.; Courtieu, J. *BSF(2)* **1990**, *127*, 565 (*CA* **1991**, *114*, 101 271s).
9. Hase, T. A.; Nylund, E.-L. *TL* **1979**, 2633.
10. Arora, G. S.; Shirahama, H.; Matsumoto, T. *CI(L)* **1983**, 318.
11. Kraus, G. A.; Roth, B. *JOC* **1980**, *45*, 4825.
12. Bal, B. S.; Childers, Jr W. E.; Pinnick, H. W. *T* **1981**, *37*, 2091.
13. Görgen, G.; Boland, W.; Preiss, U.; Simon, H. *HCA* **1989**, *72*, 917.
14. Hillis, L. R.; Ronald, R. C. *JOC* **1985**, *50*, 470.
15. Dalcanale, E.; Montanari, F. *JOC* **1986**, *51*, 567.
16. Lindgren, B. O.; Nilsson, T. *ACS* **1973**, *27*, 888.
17. Colombo, L.; Gennari, C.; Santandrea, M.; Narisano, E.; Scolastico, C. *JCS(P1)* **1980**, 136.
18. Diphenylmethanol: Otto, J.; Paluch, K. *Rocz. Chem.* **1972**, *46*, 2027 (*CA* **1973**, *78*, 124 175r).
19. Matteson, D. S.; Beedle, E. C. *TL* **1987**, *28*, 4499.
20. Schmidt, E.; Fehr, L. *LA* **1959**, *621*, 1 (*CA* **1959**, *53*, 19 873c).
21. Weber, J. V.; Schneider, M.; Salami, B.; Paquer, D. *RTC* **1986**, *105*, 99.
22. Ballini, R.; Petrini, M. *TL* **1989**, *30*, 5329.
23. Mann, J.; Thomas, A. *TL* **1986**, *27*, 3533.

Tapio Hase & Kristiina Wähälä
University of Helsinki, Finland

Sodium Cyanoborohydride[1]

[25895-60-7] CH_3BNNa (MW 62.85)

(selective, mild reducing reagent for reductive aminations of aldehydes and ketones, reductions of imines, iminium ions, oximes and oxime derivatives, hydrazones, enamines; reductive deoxygenation of carbonyls via sulfonyl hydrazones, reductions of aldehydes and ketones, polarized alkenes, alkyl halides, epoxides, acetals, and allylic ester groups)

Physical Data: white, hygroscopic solid, mp 240–242 °C (dec).
Solubility: sol most polar solvents (e.g. MeOH, EtOH, H_2O, carboxylic acids) and polar aprotic solvents (e.g. HMPA, DMSO, DMF, sulfolane, THF, diglyme); insol nonpolar solvents (e.g. ether, CH_2Cl_2, benzene, hexane).
Form Supplied in: widely available; the corresponding deuterated (or tritiated) reagent is available via acid-catalyzed exchange with D_2O (or T_2O).[1a,2]
Handling, Storage, and Precautions: store under dry N_2 or Ar.

Functional Group Reductions: General. The chemoselectivity available with $NaBH_3CN$ is remarkably dependent on solvent and pH. Under neutral or slightly acidic conditions (pH > 5), only iminium ions are reduced in protic and ether (e.g. THF) solvents.[2] Most other functional groups including aldehydes, ketones, esters, lactones, amides, nitro groups, halides, and epoxides are inert under these conditions.

Reductive Aminations. The relative inertness of aldehydes and ketones toward $NaBH_3CN$ at pH > 5 allows reductive aminations with amine and amine derivatives (usually in MeOH) via in situ generation of iminium ions which are then reduced to amines (eqs 1 and 2).[2–4] This protocol is compatible with most other functional groups, can be used to prepare N-heterocycles with stereochemical control (eqs 3 and 4),[5,6] and serves as a methylation process using CH_2O as the aldehyde (eq 5).[7,8] For difficult cases (e.g. aromatic amines, hindered and trifluoromethyl ketones), yields may be greatly improved by prior treatment of the carbonyl and amine with ***Titanium(IV) Chloride*** (eq 6)[9] or ***Titanium Tetraisopropoxide***.[10a] A reagent system prepared from $NaBH_3CN$ and ***Zinc Chloride*** also is also effective for reductive aminations[10b] (see also ***Sodium Triacetoxyborohydride*** and ***Tetra-n-butylammonium Cyanoborohydride***).

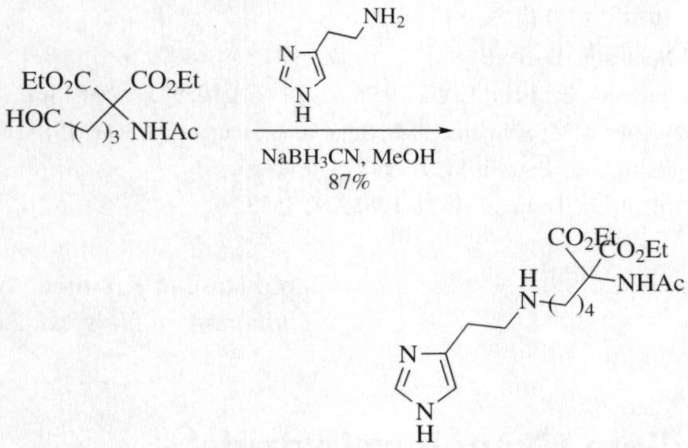

$NaBH_3CN$, MeOH, 87% (1)

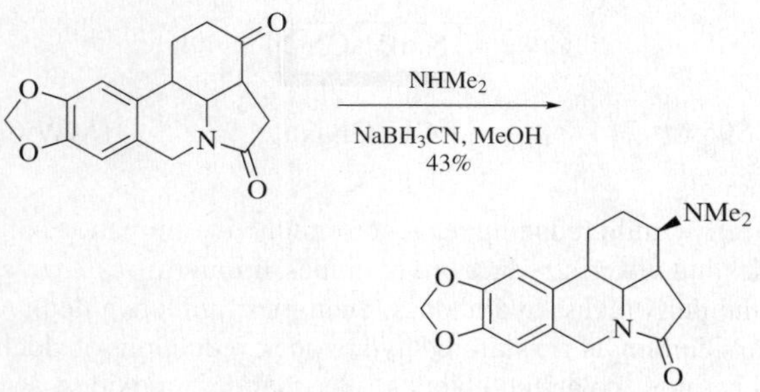

$NHMe_2$, $NaBH_3CN$, MeOH, 43% (2)

NH_4Br, $NaBH_3CN$, MeOH, 70–90% (3)

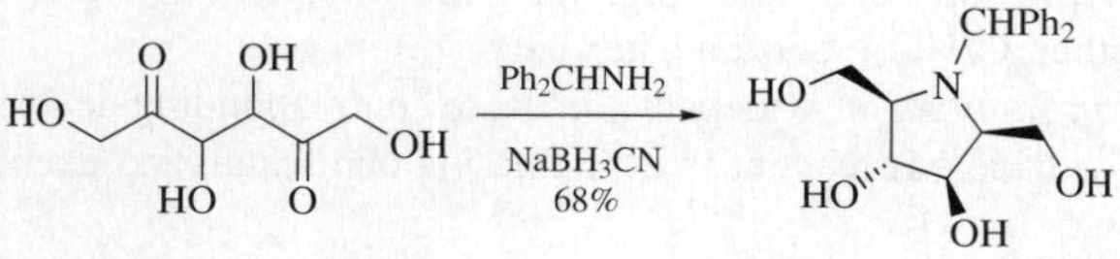

Ph_2CHNH_2, $NaBH_3CN$, 68% (4)

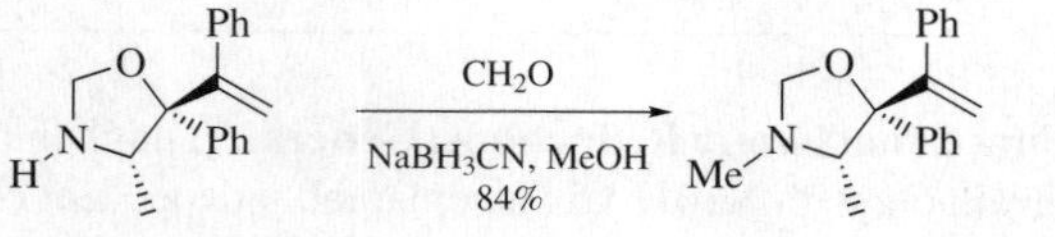

CH_2O, $NaBH_3CN$, MeOH, 84% (5)

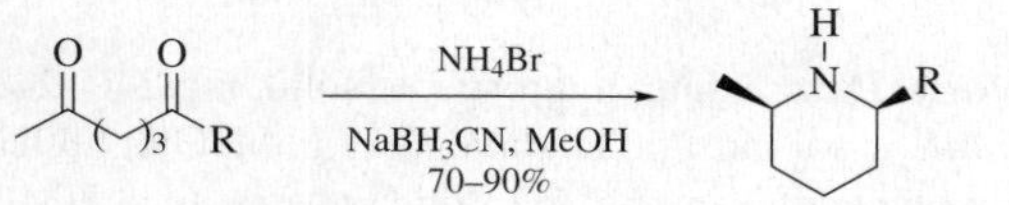

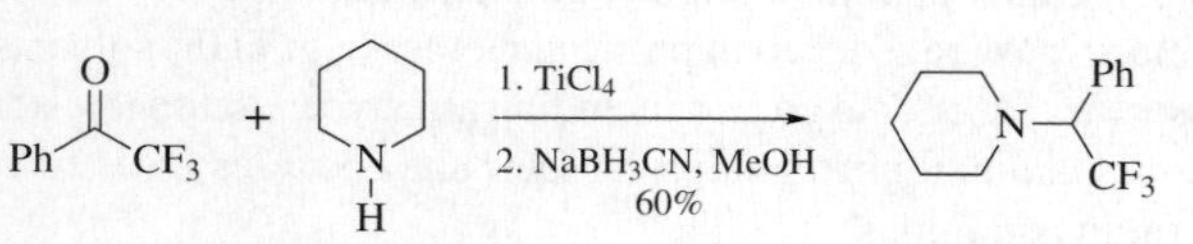

1. $TiCl_4$; 2. $NaBH_3CN$, MeOH, 60% (6)

Reductions of Imines and Derivatives. Preformed imines (eq 7),[2,11,12] iminium ions (eq 8),[2,11,13] oximes (eq 9),[2,14] oxime derivatives (eqs 10 and 11),[15,16] hydrazones (eq 12),[17] and other *N*-heterosubstituted imines (eqs 13 and 14)[18,19] are reduced to the corresponding amine derivatives by $NaBH_3CN$, usually in acidic media (see also ***Lithium Aluminum Hydride*** and ***Sodium Borohydride***).

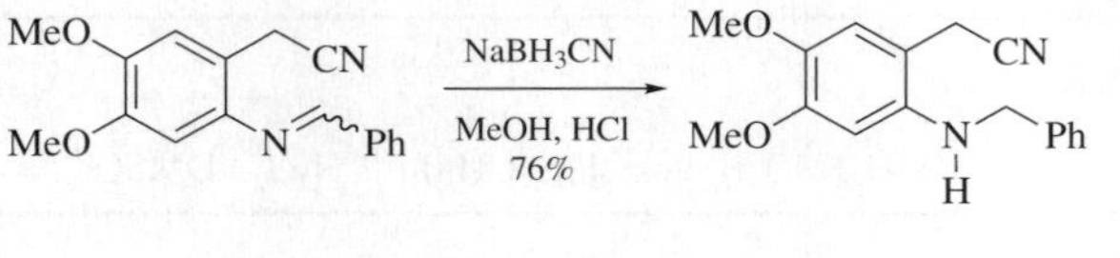

$NaBH_3CN$, MeOH, HCl, 76% (7)

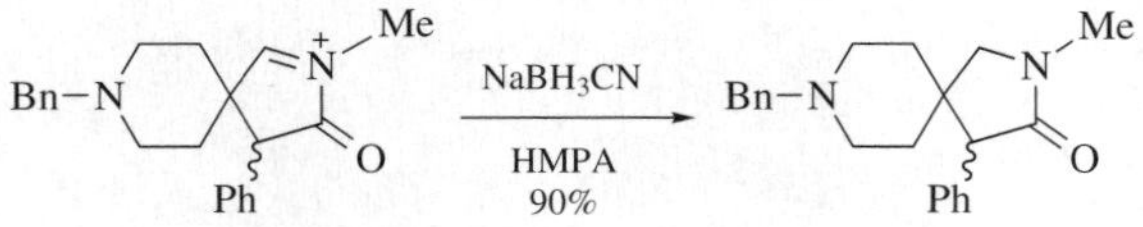

$NaBH_3CN$, HMPA, 90% (8)

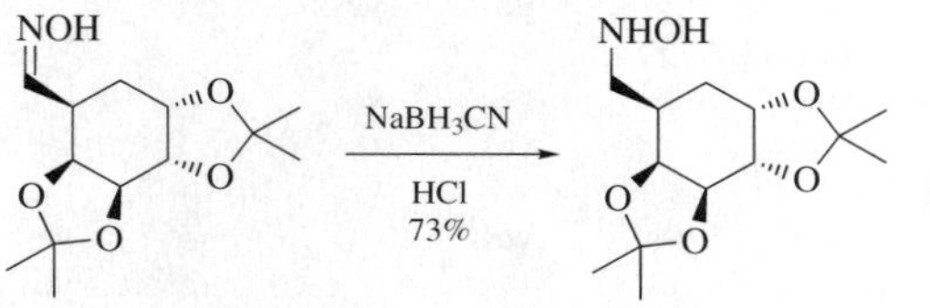

$NaBH_3CN$, HCl, 73% (9)

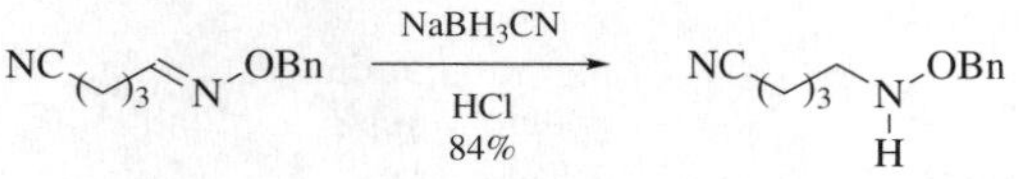

$NaBH_3CN$, HCl, 84% (10)

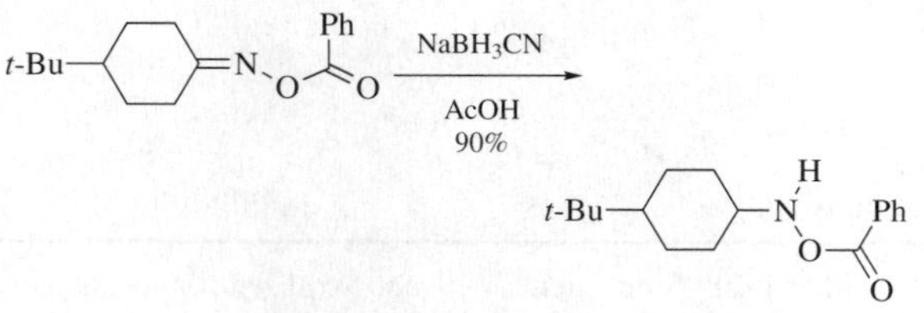

$NaBH_3CN$, AcOH, 90% (11)

cis:*trans* = 1:4

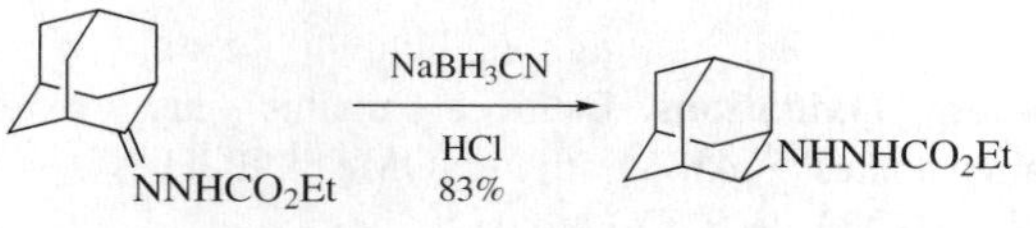

$NaBH_3CN$, HCl, 83% (12)

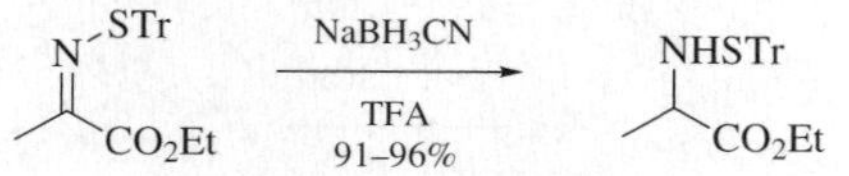

$NaBH_3CN$, TFA, 91–96% (13)

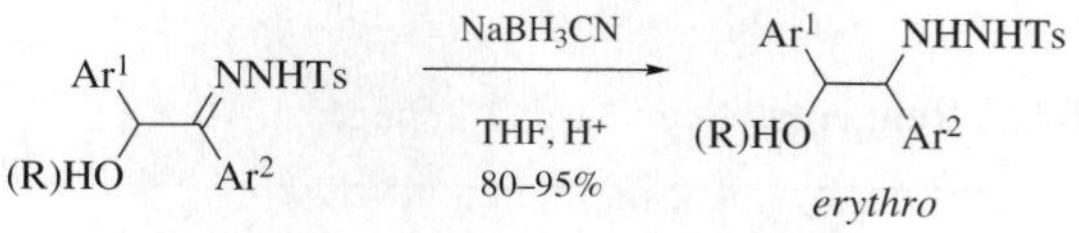

$NaBH_3CN$, THF, H^+, 80–95% (14)

erythro

Also under acidic conditions, enamines are reduced to amines via iminium ions by $NaBH_3CN$ (eq 15).[2,20] This type of conversion is also effected with $NaBH_3CN/ZnCl_2$.[10b] Pyridines and related nitrogen heterocycles are reduced by $NaBH_3CN/H^+$ to di- or tetrahydro derivatives (eq 16).[1e,21] Likewise, pyridinium and related salts (e.g. quinolinium, isoquinolinium) are reduced. With 4-substituted derivatives, 1,2,5,6-tetrahydropyridine products are produced (eq 17).[22]

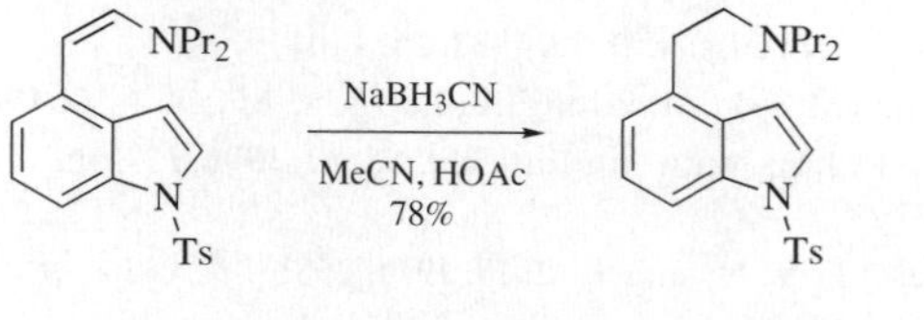

$NaBH_3CN$, MeCN, HOAc, 78% (15)

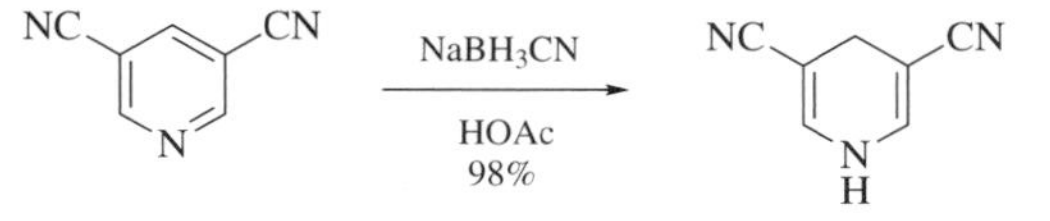

Reductive Deoxygenation of Aldehydes and Ketones.[23,24] *p*-Toluenesulfonylhydrazones (tosylhydrazones), generated in situ from unhindered aliphatic aldehydes and ketones and tosylhydrazine, are reduced by $NaBH_3CN$ in slightly acidic DMF/sulfolane (ca. 100–110 °C) to hydrocarbons via diazene intermediates (eq 18).[24] With hindered examples the tosylhydrazones must be preformed and large excesses (5–10×) of $NaBH_3CN$ used in more acidic media (e.g. pH < 4) (eq 19).[24,25] Likewise, aryl tosylhydrazones are nearly inert to the reagent, but exceptions are known.[26] Reduction of tosylhydrazones to hydrocarbons also occurs with $NaBH_3CN/ZnCl_2$ in refluxing MeOH. This combination also gives poor yields with aryl systems.[10b]

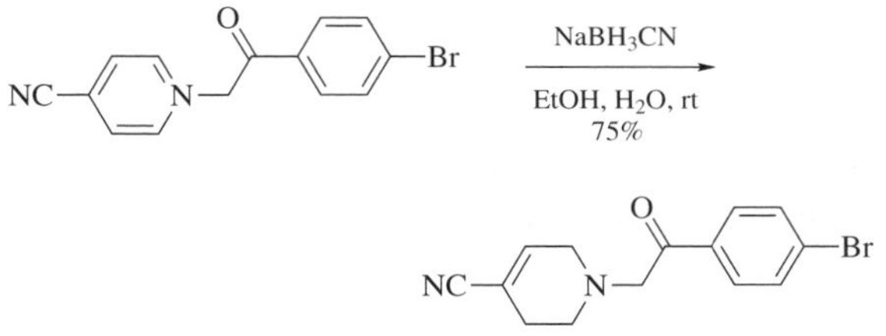

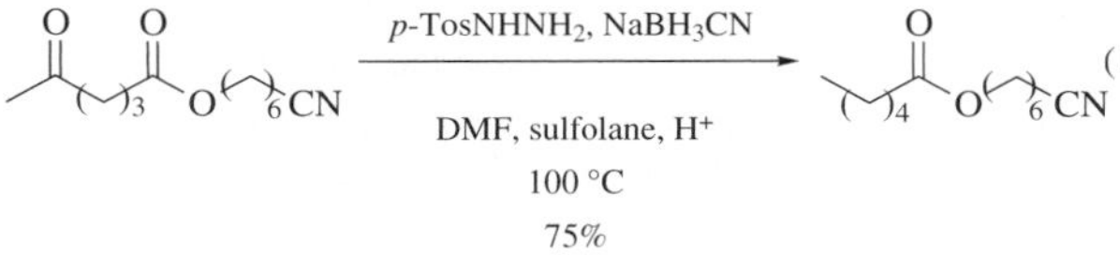

Reductive deoxygenation of most α,β-unsaturated tosylhydrazones with $NaBH_3CN$ cleanly affords alkenes in which the double bond migrates to the former tosylhydrazone carbon (eq 20).[24,27,28] However, the process gives mixtures of alkenes and alkanes with cyclohexenones[27] (see also ***Bis(triphenylphosphine)copper(I) Borohydride***, ***Catecholborane*** and ***Sodium Borohydride***).

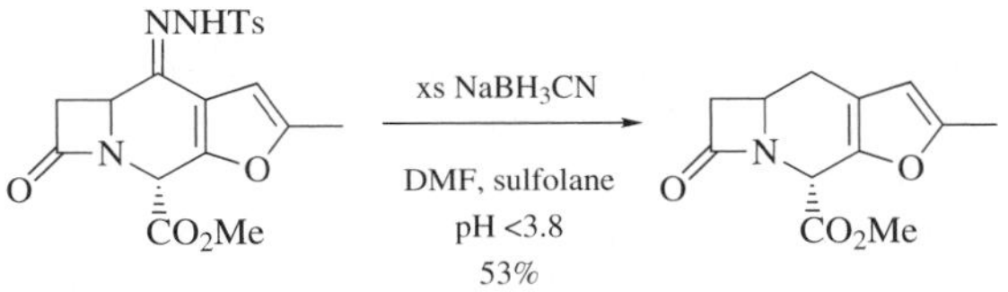

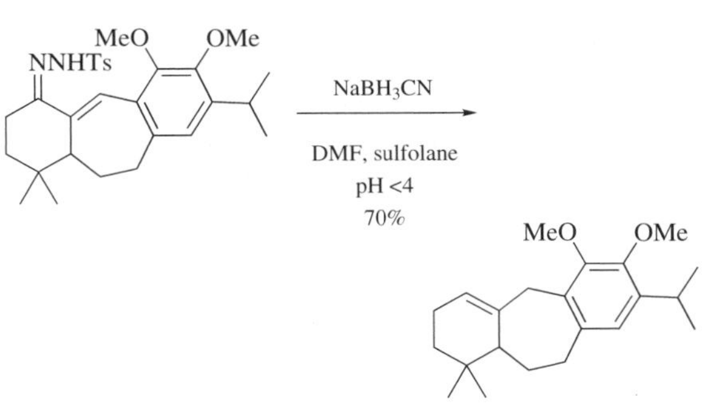

Reduction of Other π-Bonded Functional Groups. In acidic media (i.e. pH < 4), aldehydes and ketones are selectively reduced to alcohols (eq 21).[2a,28] α,β-Unsaturated ketones are reduced primarily to allylic alcohols (eq 22)[29] except cyclohexenones, which give mixtures of allylic and saturated alcohols. Allylic ethers are also produced concomitantly with substrates further conjugated with aryl rings.[29]

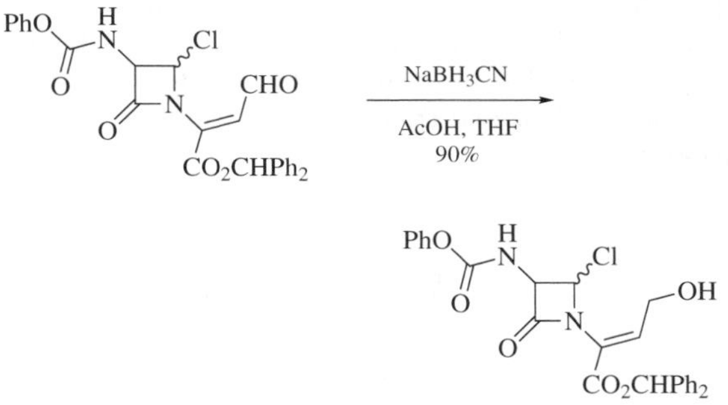

The combination of $NaBH_3CN/ZnCl_2$ in ether also reduces aldehydes and ketones to alcohols.[10b] With 5% H_2O present, cyclohexanones are selectively reduced in the presence of aliphatic derivatives.[30a] With $NaBH_3CN$/***Zinc Iodide***, however, aryl aldehydes and ketones are converted to aryl alkanes.[30b]

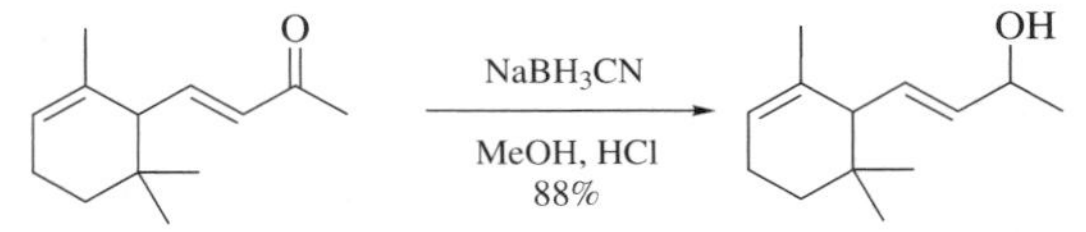

While isolated alkenes are inert toward $NaBH_3CN$, highly polarized double bonds (i.e. containing an attached nitro or two other electron-withdrawing groups) are reduced to hydrocarbons in acidic EtOH (eq 23).[31,32] Reductions of iron carbonyl–alkene complexes to the corresponding alkyl complexes also occurs readily with $NaBH_3CN$ in MeCN.[33]

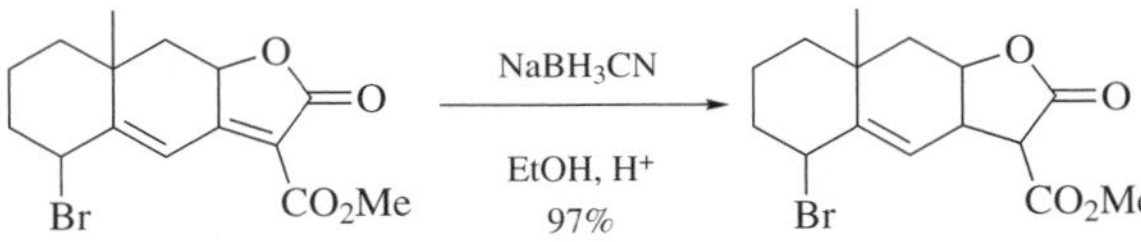

Nitriles are inert toward $NaBH_3CN$ even under strongly acidic conditions. However, methylation with $Me_2Br^+SbF_6^-$ and subsequent reduction with $NaBH_3CN$ affords the corresponding methylamine (eq 24).[34]

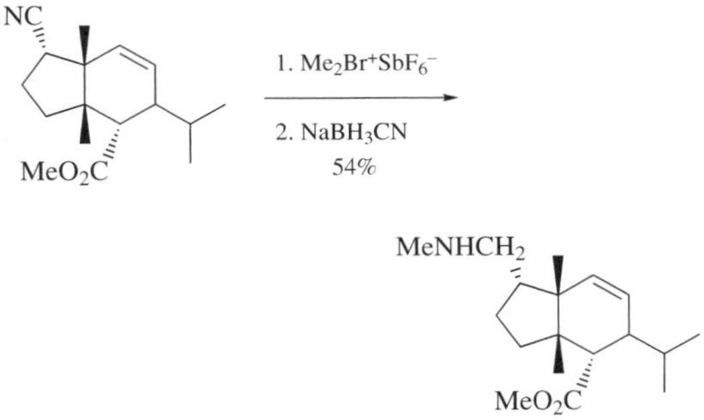

Reduction of σ-Bonded Functional Groups. In S_N2 rate enhancing polar aprotic solvents (e.g. HMPA, DMSO), primary and secondary alkyl, benzylic and allylic halides, sulfonate esters (eqs 25 and 26),[35a] and quaternary ammonium salts[36] are reduced to hydrocarbons. As expected for an S_N2-type process, the order of reactivity is I > Br, RSO_3 > Cl.[35a] In addition, primary alcohols are reduced to hydrocarbons via in situ conversion to iodides with ***Methyltriphenoxyphosphonium Iodide*** (MTPI) and subsequent reduction (eq 27).[35a,37] On the other hand, the combinations $NaBH_3CN$/***Tin(II) Chloride***[38a] or $NaBH_3CN/ZnCl_2$[38b] reduce tertiary, allylic, and benzylic halides but are inert toward primary, secondary, and aryl derivatives (eq 28)[38] (see also ***Lithium Aluminum Hydride***, ***Lithium Tri-s-butylborohydride*** and ***Sodium Borohydride***).

$I(CH_2)_3CO_2$— → ($NaBH_3CN$, HMPA, 70 °C, 89%) → $Me(CH_2)_2CO_2$— (25)

Ph, O, Br → ($NaBH_3CN$, HMPA, 70 °C, 63%) → Ph, O (26)

HO$(\,)_6$CN → (1. MTPI, HMPA; 2. $NaBH_3CN$, 66%) → $(\,)_5$CN (27)

Cl, Et, Et, $(\,)_3$ → ($NaBH_3CN$, $SnCl_2$, 94%) → Et, Et, $(\,)_3$ (28)

In the presence of ***Boron Trifluoride Etherate***, $NaBH_3CN$ reduces epoxides to alcohols[39] with attack of hydride at the site best able to accommodate a carbocation. Epoxide opening occurs primarily *anti* (eq 29).[39] Acetals are also reduced to ethers by $NaBH_3CN$ in acetic media (eq 30).[40]

O → ($NaBH_3CN$, BF_3•Et_2O) → OH (87%) + OH (3%) (29)

OMe, OMe, $(\,)_8$ → ($NaBH_3CN$, MeOH, HCl, 0 °C, 80 %) → $(\,)_8$OMe (30)

Allylic groups that are normally not displaced by hydrides (e.g. carboxylates, ethers) are effectively activated via Pd^0 complexation to give π-allyl complexes which are reduced by $NaBH_3CN$ to alkenes (eq 31).[41]

O_2N—, OAc → ($Pd(PPh_3)_4$, PPh_3, $NaBH_3CN$, THF, 80 %) → O_2N— (31)

Related Reagents. See Classes R-1, R-2, R-5, R-9, R-15, R-23, R-27, R-28, R-29, R-32, and R-35, pages 1–10.

1. (a) Hutchins, R. O.; Natale, N. R. *OPP* **1979**, *11*, 201. (b) Lane, C. F. *S* **1975**, 135. (c) *COS* **1991**, *8*, Chapters 1.2, 1.14, 3.5, 4.1, 4.2. (d) Seyden-Penne, J. *Reductions by the Alumino- and Borohydrides in Organic Synthesis*; VCH: New York, 1991. (e) Gribble, G. W.; Nutaitis, C. F. *OPP* **1985**, *17*, 317.
2. (a) Borch, R. F.; Bernstein, M. D.; Durst, H. D. *JACS* **1971**, *93*, 2897. (b) Hutchins, R. O.; Hutchins M. K. *COS* **1991**, *8*, 25.
3. Mori, K.; Sugai, T.; Maeda, Y.; Okazaki, T.; Noguchi, T.; Naito, H. *T* **1985**, *41*, 5307.
4. Umezawa, B.; Hoshino, O.; Sawaki, S.; Sashida, H.; Mori, K.; Hamada, Y.; Kotera, K.; Iitaka, Y. *T* **1984**, *40*, 1783.
5. (a) Abe, K.; Okumura, H.; Tsugoshi, T.; Nakamura, N. *S* **1984**, 597. (b) Abe, K.; Tsugoshi, T.; Nakamura, N. *BCJ* **1984**, *57*, 3351.
6. Reitz, A. B.; Baxter, E. W. *TL* **1990**, *31*, 6777.
7. Borch, R. F.; Hassid, A. I. *JOC* **1972**, *37*, 1673.
8. Jacobsen, E. J.; Levin, J.; Overman, L. E. *JACS* **1988**, *110*, 4329.
9. Barney, C. L.; Huber, E. W.; McCarthy, J. R. *TL* **1990**, *31*, 5547.
10. (a) Mattson, R. J.; Pham, K. M.; Leuck, D. J.; Cowen, K. A. *JOC* **1990**, *55*, 2552. (b) Kim, S.; Oh, C. H.; Ko, J. S.; Ahn, K. H.; Kim, Y. J. *JOC* **1985**, *50*, 1927.
11. Hutchins, R. O.; Su, W.-Y.; Sivakumar, R.; Cistone, F.; Stercho, Y. P. *JOC* **1983**, *48*, 3412.
12. Orlemans, E. O.; Schreuder, A. H.; Conti, P. G. M.; Verboom, W.; Reinhoudt, D. N. *T* **1987**, *43*, 3817.
13. Van Parys, M.; Vandewalle, M. *BSB* **1981**, *90*, 757.
14. Reonchet, J. M. J.; Zosimo-Landolfo, G.; Bizzozero, N.; Cabrini, D.; Habaschi, F.; Jean, E.; Geoffroy, M. *J. Carbohydr. Chem.* **1988**, *7*, 169.
15. Bergeron, R. J.; Pegram, J. J. *JOC* **1988**, *53*, 3131.
16. Sternbach, D. D.; Jamison, W. C. L. *TL* **1981**, *22*, 3331.
17. Zinner, G.; Blass, H.; Kilwing, W.; Geister, B. *AP* **1984**, *317*, 1024.
18. Branchaud, B. P. *JOC* **1983**, *48*, 3531.
19. Rosini, G.; Medici, A.; Soverini, M. *S* **1979**, 789.
20. Cannon, J. G.; Lee, T.; Ilhan, M.; Koons, J.; Long, J. P. *JMC* **1984**, *27*, 386.
21. Booker, E.; Eisner, U. *JCS(P1)* **1975**, 929.
22. Hutchins, R. O.; Natale, N. R. *S* **1979**, 281.
23. Hutchins, R. O.; Hutchins, M. K. *COS* **1991**, *8*, 327.
24. Hutchins, R. O.; Milewski, C. A.; Maryanoff, B. E. *JACS* **1973**, *95*, 3662.
25. Sato, A.; Hirata, T.; Nakamizo, N. *ABC* **1983**, *47*, 799.
26. Schultz, A. G.; Lucci, R. D.; Fu, W. Y.; Berger, M. H.; Erhardt, J.; Hagmann, W. K. *JACS* **1978**, *100*, 2150.
27. Hutchins, R. O.; Kacher, M.; Rua, L. *JOC* **1975**, *40*, 923.
28. Koft, E. R. *T* **1987**, *43*, 5775.
29. Hutchins, R. O.; Kandasamy, D. *JOC* **1975**, *40*, 2530.
30. (a) Kim, S.; Kim, Y. J.; Oh, C. H.; Ahn, K. H. *Bull. Korean Chem. Soc.* **1984**, *5*, 202. (b) Lau, C. K.; Dufresne, C.; Bélanger, P. C.; Piétré, S.; Scheigetz, J. *JOC* **1986**, *51*, 3038.
31. Hutchins, R. O.; Rotstein, D.; Natale, N.; Fanelli, J.; Dimmel, D. *JOC* **1976**, *41*, 3328.

32. Schultz, A. G.; Godfrey, J. D.; Arnold, E. V.; Clardy, J. *JACS* **1979**, *101*, 1276.
33. (a) Florio, S. M.; Nicholas, K. M. *JOM* **1978**, *144*, 321. (b) Whitesides, T. H.; Neilan, J. P. *JACS* **1976**, *98*, 63.
34. (a) Borch, R. F.; Evans, A. J.; Wade, J. J. *JACS* **1975**, *97*, 6282. (b) Borch, R. F.; Evans, A. J.; Wade, J. J. *JACS* **1977**, *99*, 1612.
35. (a) Hutchins, R. O.; Kandasamy, D.; Maryanoff, C. A.; Masilamani, D.; Maryanoff, B. E. *JOC* **1977**, *42*, 82. (b) Hutchins, R. O.; Milewski, C. A.; Maryanoff, B. E. *OSC* **1988**, *6*, 376.
36. Yamada, K.; Itoh, N.; Iwakuma, T. *CC* **1978**, 1089.
37. (a) Okada, K.; Kelley, J. A.; Driscoll, J. S. *JOC* **1977**, *42*, 2594. (b) Borchers, F.; Levsen, K.; Schwarz, H.; Wesdemiotis, C.; Winkler, H. U. *JACS* **1977**, *99*, 6359.
38. (a) Kim, S.; Ko, J. S. *SC* **1985**, *15*, 603. (b) Kim, S.; Kim, Y. J.; Ahn, K. H. *TL* **1983**, *24*, 3369.
39. Hutchins, R. O.; Taffer, I. M.; Burgoyne, W. *JOC* **1981**, *46*, 5214.
40. Horne, D. A.; Jordan, A. *TL* **1978**, 1357.
41. (a) Hutchins, R. O.; Learn, K.; Fulton, R. P. *TL* **1980**, *21*, 27. (b) Hutchins, R. O.; Learn, K. *JOC* **1982**, *47*, 4380.

Robert O. Hutchins
Drexel University, Philadelphia, PA, USA

MaryGail K. Hutchins
LNP Engineering Plastics, Exton, PA, USA

Sodium Dithionite[1]

$Na_2S_2O_4$

[7775-14-6] $Na_2O_4S_2$ (MW 174.12)

(versatile reagent for reduction of aldehydes,[2] ketones,[2–6] unsaturated conjugated ketones,[7,8] quinones,[9–10] diunsaturated acids,[11] azo,[12] nitro,[13] and nitroso compounds,[14,15] imines,[16] oximes,[17] tropylium salts,[18] pyridinium salts,[19] pyrazine,[20] and vinyl sulfones;[21] intramolecular Marschalk cyclizations,[22,23] dehalogenation of *vic* dibromides[24] and α-halo ketones,[25] Claisen rearrangement of allyloxyanthraquinones[26] and for the synthesis of 8-arylaminopurines[27,28])

Alternate Name: sodium hydrosulfite.
Physical Data: mp 52 °C (dec).
Solubility: very sol water; sol alcohol.
Form Supplied in: white or gray–white crystalline powder.
Preparative Methods: by the action of ***Sulfur Dioxide*** on ***Sodium Amalgam*** in alcoholic solution.[29,30]
Handling, Storage, and Precautions: flammable; moisture sensitive.

Reduction of Aldehydes and Ketones.[2] Sodium dithionite is an alternative and less expensive reducing agent than metal hydrides. The reactions are performed in water for soluble substrates, otherwise a 50:50 mixture of water and dioxane or DMF can be used; sodium bicarbonate is added to keep the reaction mixture basic. Examples are hexanal to 1-hexanol (67%), benzaldehyde to benzyl alcohol (84%), cyclohexanone to cyclohexanol (80%),[3] and acetophenone to α-hydroxyethylbenzene (94%).[2] Reduction of methylcyclohexanones by $Na_2S_2O_4$ in benzene–water using adogen (commercial mixture of methyl trialkyl C_8–C_{10} ammonium chloride) as a phase-transfer agent afforded good yields of isomeric mixtures of the corresponding methylcyclohexanols.[3] Reduction using $Na_2S_2O_4$ proceeds with stereoselectivity[4–6] similar to that obtained with metal hydrides (***Sodium Borohydride***) and opposite to dissolving metals reductions, e.g. the reductions of 3α-hydroxy-7-keto-5β-cholanic acid to diols (eq 1).

(1)

Na, 100%	85:15
K, 100%	94:6
$Na_2S_2O_4$, 100%	4:96
$NaBH_4$, 100%	6:94

Regiospecific Reduction of Unsaturated Conjugated Ketones.[7,8] Exclusive reduction of conjugated carbon–carbon double bonds is achieved with $Na_2S_2O_4$ to afford the corresponding saturated carbonyl compound. A two-phase (benzene–water) system using adogen as the phase-transfer catalyst is used. The isolated carbon–carbon double bond remains unaffected. No alcohols are detectable (eqs 2–4).

$Na_2S_2O_4$, 91% (2)

$Na_2S_2O_4$, 65% (3)

$Na_2S_2O_4$, 63% (4)

Quinones to Hydroquinones.[9] Most quinones are reduced by sodium dithionite to hydroquinones. Naphthacenequinone and higher linear benzologs are exceptional in that they are not reduced by alkaline sodium dithionite.[10] Sodium dithionite reduces anthraquinone to anthrone.

Conjugated Diunsaturated Acids (and Esters) to Monounsaturated Acids (and Esters).[11] α,β;γ,δ-Unsaturated acids are reduced by $Na_2S_2O_4$ in an alkaline medium (NaOH or $NaHCO_3$) to a mixture of (*Z*)- and (*E*)-β,γ-unsaturated acids (and esters) (40–75% yield) (eq 5).

$$\text{CH}_3\text{CH}_2\text{CH=CH–CH=CH–CO}_2\text{H} \xrightarrow[\text{NaOH}]{Na_2S_2O_4} \underset{74\%}{\text{CH}_3\text{CH}_2\text{CH=CH–CH}_2\text{CH}_2\text{CO}_2\text{H}} + \underset{2\%}{\text{CH}_3\text{CH}_2\text{CH}_2\text{CH=CH–CH}_2\text{CO}_2\text{H}} \quad (5)$$

Azo to Amine.[12] Azobenzene is reduced to aniline by sodium dithionite. This reaction is used to introduce an amino group into a phenolic compound by first coupling with an aromatic diazonium salt and then reducing the resulting hydroxyazo derivatives with $Na_2S_2O_4$, e.g. 2- and 4-amino-1-naphthols can be prepared from 1-naphthol.

Nitro to Amine.[13] Various aromatic nitro compounds are reduced conveniently to the corresponding aniline derivatives with sodium dithionite using dioctyl viologen as an electron-transfer catalyst in a two-phase system (CH_2Cl_2–H_2O), e.g. 1-nitronaphthalene to 1-naphthylamine.

Nitroso to Amine. Nitroso compounds are reduced to amines by sodium dithionite, e.g. nitrosouracil to diaminouracil (eq 6).[14] Reduction of *N*-nitrosodibenzylamine[15] with $Na_2S_2O_4$ is accompanied by liberation of N_2 and rearrangement to dibenzyl. Mixed benzylaryl or diaryl-*N*-nitrosoamines are reduced to hydrazines (eq 7).

$$\text{2-hydroxy-4-amino-5-nitroso-6-hydroxypyrimidine} \xrightarrow[68\text{–}81\%]{Na_2S_2O_4} \text{2-hydroxy-4,5-diamino-6-hydroxypyrimidine} \quad (6)$$

$$\text{PhCH}_2\text{N(NO)Ph} \longrightarrow \text{PhCH}_2\text{N(NH}_2\text{)Ph} \quad (7)$$

Imines to Amines. Sodium dithionite in DMF reduces imines[16] to *N*-alkylamines at 110 °C with yields ranging from 40 to 73%. Heating *N*-cyclohexyldibenzylamine, sodium dithionite, and $NaHCO_3$ in DMF for 30 min at 110 °C gives 73% benzylcyclohexylamine.

Oximes to Amines.[17] Oximes are readily reduced to amines by sodium dithionite. Substituted phenylethylamines are key intermediates required for the synthesis of isoquinoline derivatives. They are readily obtained from aryl alkyl ketones by nitrosation of the alkyl group followed by the reduction of the resulting oxime derivatives by sodium dithionite (eq 8).

$$\text{PhCOCH=N–OH} \xrightarrow{Na_2S_2O_4} \text{PhCOCH}_2\text{NH}_2 \quad (8)$$

Cleavage of Oximes.[16] Both aldehydes and ketones are regenerated from their oxime derivatives with aqueous sodium dithionite either alone or in the presence of Na_2CO_3 at 25 °C, e.g. cyclohexanone oxime to cyclohexanone (95%) and benzaldehyde oxime to benzaldehyde (96%).

Reduction of Tropylium and Cyclopropenium Halides.[18] $Na_2S_2O_4$ in acetonitrile at 25 °C reduces the tropylium halides to ditropyl (eq 9) and triphenylcyclopropenium halides to bicyclopropenyl sulfones (eq 10).

$$2\ \text{C}_7\text{H}_7^+\ X^- \xrightarrow[\text{MeCN}]{Na_2S_2O_4} \text{C}_7\text{H}_7\text{–C}_7\text{H}_7 \quad (9)$$

$$\text{1,2-Ph}_2\text{-3-R-cyclopropenium}\ X^- \xrightarrow{Na_2S_2O_4} \text{(Ph}_2\text{C}_3\text{R)–SO}_2\text{–(C}_3\text{RPh}_2) \quad (10)$$

R = H, Ph

Reduction of Pyridinium Salt.[19] Sodium dithionite has been extensively used for the reduction of *N*-methylpyridinium-3-carboxamide to *N*-methyl-1,4-dihydropyridine-3-carboxamide (eq 11), which is a model of reduced dihydrophosphopyridine nucleotide (DPNH).

$$\text{N-methyl-3-CONH}_2\text{-pyridinium}\ Cl^- \xrightarrow{Na_2S_2O_4} \text{N-methyl-3-CONH}_2\text{-1,4-dihydropyridine} \quad (11)$$

Reduction of Pyrazine Derivatives.[20] 2,3,5,6-Tetraethoxycarbonyl-1,4-dihydropyrazine is prepared from 2,3,5,6-tetraethoxycarbonylpyrazine using sodium dithionite as reducing agent (eq 12). It is a more convenient and simpler method than catalytic hydrogenation and no saponification of esters occurred during this reduction process.

$$\text{2,3,5,6-(EtO}_2\text{C)}_4\text{-pyrazine} \xrightarrow[Na_2CO_3,\ H_2O]{Na_2S_2O_4} \text{2,3,5,6-(EtO}_2\text{C)}_4\text{-1,4-dihydropyrazine} \quad (12)$$

Reduction of Vinyl Sulfone. An insect phermone, (*Z*)-8-dodecenyl 1-acetate,[21] is readily obtained by the reduction of corresponding vinyl sulfone in aqueous ethanol with retention of configuration (eq 13). The mechanism involves the Michael addition of SO_2^- to the vinylic sulfone, accompanied by protonation and expulsion of SO_2 and sulfinate ion to give the alkene.

$$\text{PhO}_2\text{S–CH=C(C}_3\text{H}_7\text{)–CH}_2(\text{CH}_2)_6\text{OAc} \xrightarrow[57\%]{Na_2S_2O_4,\ NaHCO_3} \underset{(Z)\ 98\%}{\text{C}_3\text{H}_7\text{CH=CH–CH}_2(\text{CH}_2)_6\text{OAc}} \quad (13)$$

Intramolecular Marschalk Cyclization.[22,23] This cyclization reaction is a key step in the total synthesis of daunomycinone, the aglycon of an anthracycline antibiotic. Treatment of the anthraquinone derivative with sodium dithionite and sodium hydroxide in dioxane at 25–90 °C gives the anticipated cyclized product in 52% yield (eq 14).

Dehalogenation of *vic*-Dibromides, α-Bromo and α-Chloro Ketones. Vicinal dibromides are debrominated[24] with $Na_2S_2O_4$ in DMF (140–145 °C). The yields are moderate to high but the reaction is not stereospecific. Both *meso*- and (±)-2,3-dibromobutane give 1:1 mixtures of *cis*- and *trans*-2-butene. The dehalogenation[25] of α-bromo or α-chloro ketones can be effected with $Na_2S_2O_4$ in aqueous DMF at 25–90 °C in yields of 50–95%. The rate can be enhanced by addition of $NaHCO_3$.[25]

Claisen Rearrangement of Allyloxyanthraquinone.[26] 1-Allyloxyanthraquinones rearrange to 1-hydroxy-2-allylanthraquinones in high yields when heated in DMF–H_2O containing 1.3–1.8 equiv $Na_2S_2O_4$. 1,4-Bis(allyloxy)anthraquinone rearranges slowly under these conditions, but more readily if 4 equiv NaOH is added (eq 15).

Synthesis of 8-Arylaminotheophyllines.[27,28] Treatment of 5-arylazo-1,3-dimethyl-6-ethoxymethyleneaminouracil with $Na_2S_2O_4$ in formic acid gives 8-arylaminotheophyllines (eq 16). The key intermediates required for this reaction are prepared by reaction of the appropriate 6-amino-5-arylazo-1,3-dimethyluracils with a mixture of ***Triethyl Orthoformate*** and DMF at 180 °C for 5 h.

Related Reagents. See Classes R-2, R-5, R-8, R-9, R-10, R-12, R-15, R-21, R-22, R-23, R-27, R-30, and R-31, pages 1–10.

1. (a) *FF* **1967**, *1*, 1081; **1980**, *8*, 456; **1981**, *9*, 426; **1982**, *10*, 363; **1988**, *13*, 277. (b) Louis-Andre, O.; Gelbard, G. *BSF(2)* **1986**, 565.
2. de Vries, J. G.; van Bergen, T. J.; Kellogg, R. M. *S* **1977**, 246.
3. Camps, F.; Coll, J.; Riba, M. *CC* **1979**, 1080.
4. House, H. O. *Modern Synthetic Reactions*, 2nd ed.; Benjamin/Cummings: London, 1972; p 150.
5. Castaldi, G.; Perdoncin, G.; Giordano, C.; Minisci, F. *TL* **1983**, *24*, 2487.
6. Giordano, C.; Perdoncin, G.; Castaldi, G. *AG(E)* **1985**, *24*, 499.
7. Camps, F.; Coll, J.; Guitart, J. *T* **1986**, *42*, 4603.
8. Louis-Andre, O.; Gelbard, G. *TL* **1985**, *26*, 831.
9. Fieser, L. F. *JACS* **1931**, *53*, 2329.
10. Fieser, L. F.; Peters, M. A. *JACS* **1931**, *53*, 4080.
11. Camps, F.; Coll. J.; Guerrero, A.; Guitart, J.; Riba, M. *CL* **1982**, 715.
12. Fieser, L. F. *OSC* **1943**, *2*, 35, 430.
13. Park, K. K.; Oh, C. H.; Joung, W. K. *TL* **1993**, *34*, 7445.
14. Sherman, W. R.; Taylor, Jr., E. C. *OSC* **1963**, *4*, 247.
15. Overberger, C. G.; Lombardino, J. G.; Hiskey, R. G. *JACS* **1958**, *80*, 3009.
16. Pojer, P. M. *AJC* **1979**, *32*, 201.
17. Pictet, A.; Gams, A. *CB* **1909**, *42*, 2943.
18. Weiss, R.; Schlierf, C.; Koelbl, H. *TL* **1973**, 4827.
19. Mauzerall, D.; Westheimer, F. H. *JACS* **1955**, *70*, 2261.
20. Mager, H. I. X.; Berends, W. *RTC* **1960**, *79*, 282.
21. Julia, M.; Lauron, H.; Stacino, J.-P.; Verpeaux, J.-N.; Jeannin, Y.; Dromzee, Y. *T* **1986**, *42*, 2475.
22. Suzuki, F.; Trenbeath, S.; Gleim, R. D.; Sih, C. J. *JACS* **1978**, *100*, 2272.
23. Kende, A. S.; Tsay, Y.-G.; Mills, J. E. *JACS* **1976**, *98*, 1967.
24. Kempe, T.; Norin, T.; Caputo, R. *ACS* **1976**, *B30*, 366.
25. Chung, S.-K.; Hu, Q.-Y. *SC* **1982**, *12*, 261.
26. Boddy, I. K.; Boniface, P. J.; Cambie, R. C.; Craw, P. A.; Larsen, D. S.; McDonald, H.; Rutledge, P. S.; Woodgate, P. D. *TL* **1982**, *23*, 4407.
27. Senga, K.; Ichiba, M.; Kanazawa, H.; Nishigaki, S.; Higuchi, M.; Yoneda, F. *S* **1977**, *4*, 264.
28. Senga, K.; Ichiba, M.; Kanazawa, H.; Nishigaki, S.; Higuchi, M.; Yoneda, F. *JHC* **1978**, *15*, 641.
29. Chia, K.-S; Wang, W.-P. Union Ind. Res. Inst. Report (Hsinchu, Taiwan), No. 40, 1959; p.1 (*CA* **1960**, *54*, 19 252g).
30. Chia, K.-S; Wang, W.-P. *Chemistry (Taipei)* **1960**, *29* (*CA* **1961**, *55*, 2327h).

Marudai Balasubramanian & James G. Keay
Reilly Industries, Indianapolis, IN, USA

Sodium Hypochlorite[1]

NaOCl

[7681-52-9] ClNaO (MW 74.44)

(versatile and easily handled oxidizing agent;[1] can oxidize alcohols,[2] aldehydes,[3] electron deficient alkenes,[4] amines,[5] and transition metal catalysts;[6] reagent for *N*-chlorination,[7] oxidative coupling,[8] and degradation reactions[9])

Physical Data: most commonly used in aqueous solution; $NaOCl{\cdot}5H_2O$: mp 18 °C.

Solubility: pentahydrate: 293 g L^{-1} in H_2O (0 °C).

Form Supplied in: commercially available as aqueous solutions with 5.25–12.5% available oxidant (w/v) (0.74–1.75 M). Concentration is expressed in % available chlorine, since half of the chlorine in bleach is present as NaCl. The pH of commercial bleach is typically 11–12.5, and it may be adjusted and buffered.[6a]

Analysis of Reagent Purity: active oxidant may be assayed by iodometric[10] or potentiometric[11] titration.

Preparative Methods: solutions may be generated in situ by passing ***Chlorine*** gas through aq ***Sodium Hydroxide*** solution,[12] or electrochemically.[13]

Purification: commercial solutions are generally used without purification.

Handling, Storage, and Precautions: higher concentration sodium hypochlorite (12.5%), sometimes referred to as 'swimming pool chlorine,' tends to decrease in concentration by 20% per month upon storage and therefore should be titrated prior to use.[1b] The concentration of oxidant in household bleach (5.25%) tends to remain constant upon prolonged storage; titration is generally not necessary with brand names (e.g. Clorox®). Solid NaOCl is explosive as the pentahydrate or the anhydride, and therefore it is very rarely employed in those forms. Aqueous solutions are very stable. Bleach is a household item, and it is quite easy and safe to handle. Still, it is a strong oxidant, and precaution should be taken to avoid prolonged skin exposure or inhalation.[14] May react violently with NH_3.

Composition of Aqueous Solution as a Function of pH. The equilibrium composition of aqueous solutions of NaOCl is pH-dependent (eqs 1 and 2), and so pH control can be a critical consideration in many oxidation and chlorination reactions. Under strongly alkaline conditions (pH > 12), OCl^- is the predominant form of positive chlorine. Because hypochlorite ion is insoluble in organic solvents, phase transfer catalysts are needed at this pH to effect oxidation reactions in biphasic media.[15] In general, tetraalkylammonium salts have been the phase-transfer catalysts of choice for such applications. Below pH 11, the equilibrium amount of HOCl becomes significant,[6a] and this form of positive chlorine is soluble in polar organic solvents such as CH_2Cl_2. No phase-transfer catalyst is necessary to effect oxidation of substrates or catalysts dissolved in the organic phase of biphasic reactions in the pH range 10–11.[6a] Below pH 10, molecular chlorine becomes a significant component of aqueous bleach solutions, and the reactivity of these solutions can be attributed to that of Cl_2.[1b]

$$ClO^- + Cl^- + H_2O \rightleftharpoons Cl_2 + 2\,OH^- \qquad (1)$$

$$ClO^- + H_2O \rightleftharpoons HOCl + OH^- \qquad (2)$$

Oxidation of Alcohols. Oxidation of alcohols by NaOCl can be effected under a variety of conditions, and useful yields and selectivities are attainable for conversion of primary alcohols to aldehydes or carboxylic acids, or of secondary alcohols to ketones. The advantages of NaOCl oxidations over methods that employ stoichiometric Cr^{VI} include simplified waste disposal and lower toxicity and cost. The earliest application of NaOCl as a practical synthetic reagent for alcohol oxidation involved its use in a two-phase system with a phase-transfer catalyst,[16] or in association with ***Ruthenium(IV) Oxide***.[17] More recently, two improved methods for bleach-mediated oxidation of alcohols have been developed, one of which employs acetic acid as solvent in a monophasic system,[18] and the other uses catalytic amounts of ***2,2,6,6-Tetramethylpiperidin-1-oxyl*** (TEMPO) in a buffered biphasic medium.[2b] These variants are highly complementary and can offer significant advantages over alternative methods for alcohol oxidation.

Secondary alcohols are cleanly oxidized to ketones with NaOCl in acetic acid in the absence of added catalyst (eq 3).[18] Either 'swimming pool chlorine' or household laundry bleach can be used with similar success.[19] Excess hypochlorite is quenched with sodium bisulfite, and essentially pure ketone is obtained simply by extraction of the product into dichloromethane or ether.

OH → (AcOH, NaOCl; 96%) → O (3)

Under these conditions, secondary alcohols can be oxidized in the presence of primary alcohols with essentially absolute selectivity (eq 4).[20] No epimerization is observed in the oxidation of alcohols bearing β-stereocenters (eq 5). Kinetic studies have led to the proposal that molecular chlorine is the active oxidant under conditions of low pH such as those employed for these reactions.[21]

OH, OH → (AcOH, NaOCl; 90%) → O, OH (4)

HO, HO → (AcOH, NaOCl; 97% crude, 83% isolated) → HO, O (5)

Primary and secondary alcohols can be oxidized by oxoammonium salts;[22] hypochlorite oxidation of the reduced nitroxyl radical regenerates the active oxoammonium salt. Thus the nitroxyl radical agent TEMPO can be employed as an alcohol oxidation catalyst, with NaOCl as the stoichiometric oxidant.[2b] This protocol has rapidly achieved widespread use due to its selectivity, ease of application, and versatility. A biphasic system is employed consisting of CH_2Cl_2 or toluene as the organic phase, and commercial bleach buffered to pH ~9 with $NaHCO_3$ and containing substoichiometric levels of KBr or NaBr (eq 6).[23] Reaction times are longer in the absence of bromide salts, indicating that HOBr is the agent that oxidizes TEMPO to the nitrosonium ion. Without added phase transfer catalyst, primary alcohols are oxidized with excellent selectivity to the corresponding aldehydes.[24] High stirring rates (>1000 rpm) help to minimize overoxidation to the carboxylic acid.[23] The stereospecificity of TEMPO-catalyzed oxidations of primary alcohols bearing β-stereocenters has been investigated in detail, and it is absolute in all cases reported thus far (eqs 7–9).[23,24]

RCH_2OH + NaOCl → (TEMPO, N–O• (0.01 equiv); KBr (0.10 equiv), $NaHCO_3$, CH_2Cl_2) → RCHO (6)

1 equiv; 1.1 equiv, 0.3–2 M; R = alkyl, 88–93%; R = aryl, 75–90%

NaOCl, TEMPO (0.01 equiv), KBr (0.1 equiv); $NaHCO_3$, CH_2Cl_2, 0–15 °C, 82–84%; 100% ee (7)

NaOCl, TEMPO (0.01–0.02 equiv), NaBr (1 equiv); $NaHCO_3$, CH_2Cl_2, 0 °C, 96%; NHCbz; >99% ee (8)

NaOCl, TEMPO (0.01–0.02 equiv), NaBr (1 equiv); $NaHCO_3$, CH_2Cl_2, 0 °C, 77%; BnO, CHO; >98% ee (9)

Although both primary and secondary alcohols are oxidized with this catalyst system, moderate-to-good selectivity for the primary position is obtained in oxidation of diols (eq 10).[25] The diol must have nearly complete solubility in the organic phase for these oxidations to occur cleanly. This selectivity for primary alcohols is in direct contrast with monophasic NaOCl oxidations in acetic acid (see above). Thus, proper choice of reaction conditions allows selective oxidation by NaOCl of either or both the primary or the secondary carbinol in diols.

NaOCl (1.1 equiv), TEMPO, KBr: 68% + 10%; NaOCl (2.2 equiv), TEMPO, KBr: 69% + 14% (10)

Molecular chlorine generated from NaOCl appears to have a detrimental effect on the oxidizing power of the *N*-oxoammonium salt in these reactions.[26] A similar protocol to the one described above, but using ***Sodium Bromite*** ($NaBrO_2$) in place of NaOCl may be superior in this context. Despite this caveat, there is ample precedent for the successful application of the NaOCl/TEMPO system with a variety of substrates, and the use of commercial bleach solutions has significant practical advantages.

For further discussion of this oxidizing system, see: ***2,2,6,6-Tetramethylpiperidin-1-oxyl***. See also ***Dimethyl Sulfoxide–Oxalyl Chloride***, ***Pyridinium Chlorochromate***, ***Pyridinium Dichromate***, ***Potassium Permanganate***, ***Dipyridine Chromium(VI) Oxide***, and ***Tetra-n-propylammonium Perruthenate***.

Oxidation of Primary Alcohols to Carboxylic Acids. Simple incorporation of a tetraalkylammonium chloride phase-transfer catalyst (PTC) to the catalyst recipe outlined above for aldehyde synthesis leads to a useful protocol for oxidation of primary alcohols to the corresponding carboxylic acids (eq 11).[2b] Similar selectivity for primary alcohols over secondary alcohols is observed as in the absence of PTC.[25] This has been exploited in the oxidation of unprotected monosaccharide derivatives to the corresponding uronic acids (eq 12).[27]

R = primary alkyl; NaOCl (2.5 equiv), TEMPO (0.01 equiv), Aliquat 336 (0.05 equiv); KBr (0.01 equiv), $NaHCO_3$, CH_2Cl_2, 0–15 °C, >90% (11)

+ NaOCl; 1% TEMPO, KBr, $NaHCO_3$; Bu_4NCl, CH_2Cl_2, 64%; HO_2C (12)

The mildness of the oxidizing medium in TEMPO-catalyzed oxidations by NaOCl is illustrated by the high yield oxidation of (2-hydroxyethyl)spiropentane to the corresponding acid (eq 13).[28] Jones oxidation conditions lead to extensive decomposition of the spiropentane residue of the same substrate.

NaOCl (pH 9), 1% TEMPO, KBr; Aliquat 336, 92%; CO_2H (13)

See also: ***Chromium(VI) Oxide***, ***Potassium Dichromate***, ***Potassium Permanganate***, and ***Oxygen–Platinum Catalyst***.

Oxidation of Aldehydes. Sodium hypochlorite oxidizes aromatic aldehydes to the corresponding acids in the presence of a phase-transfer catalyst.[3] Best results are obtained at pH 9–10 with Bu_4NHSO_4 as phase-transfer catalyst (eq 14). In this pH regime, HOCl is present in significant concentrations, and phase-transfer catalysts are generally not necessary.[6a] It has been suggested that in this system, however, the phase-transfer catalyst helps to solubilize HOCl in the organic phase through hydrogen bonding.[3] Direct oxidation of aliphatic and aromatic aldehydes to the corresponding methyl esters is accomplished with methanol in acetic acid and 1–2 equiv of NaOCl.[20] This reaction, which affords esters in moderate-to-good yield, probably proceeds via oxidation of equilibrium concentrations of methyl hemiacetals. Poor results are obtained with electron-rich aromatic aldehydes and unsaturated aldehydes, each of which undergo competitive chlorination.[20]

Ph–CHO + NaOCl (pH 9–10) → Ph–CO_2H; Bu_4NHSO_4, $ClCH_2CH_2Cl$, 87% (14)

See also: ***Potassium Permanganate***, ***Ozone***, and ***N-Bromosuccinimide***.

Epoxidation. Sodium hypochlorite is an effective, although infrequently utilized, reagent for epoxidation of enones and polycyclic arenes. Careful control of pH is necessary for good yields in these reactions. Polycyclic aromatics can be oxidized to epoxides at pH 8–9 (eq 15).[29] Phenanthridine is oxidized to the corresponding lactam, presumably via an oxaziridine intermediate (eq 16), without formation of *N*-oxide.[29] The optimum pH for this reaction is 8–9, and phase transfer catalysts are required.

NaOCl, Bu_4NBr, pH 8–9, $CHCl_3$, 90% (15)

NaOCl, Bu_4NBr, pH 8–9, $CHCl_3$ (16)

Enones,[4,30] particularly chalcones,[31] also react with sodium hypochlorite to form epoxides (eqs 17–19). These reactions generally exhibit a strong dependence on the pH of the aqueous phase; chlorination reactions can be competitive.[30a] The mechanism for these reactions has been proposed to involve production of the ClO· radical species; this proposal was made on the basis of data from the chlorination of hydrocarbons, selectivity of epoxidation reactions, and Hammett ρ values for the chlorination of toluene.[32]

NaOCl, pH 8–8.5; 70% + CCl_3 7% (17)

NaOCl, Bu_4NBr, pH 8–9, CH_2Cl_2, 80% (18)

C_8H_{17} + NaOCl, pyridine; [OH, CO_2H, C_8H_{17}] 65 °C → C_8H_{17} (19)

See also: ***Sodium Hypochlorite–N,N′-Bis(3,5-di-t-butylsalicylidene)-1,2-cyclohexanediaminomanganese(III) Chloride***, ***Hydrogen Peroxide***, ***t-Butyl Hydroperoxide***, and ***Dimethyldioxirane***.

Other Oxidation Reactions. Several other substrate classes also undergo oxidation reactions with sodium hypochlorite. Diketones may be oxidatively cleaved to give the corresponding diacids (eq 20).[33] Oxidation of hydroquinones and catechols[34] and oxidative cyclization of phenols (eq 21)[8] have also been realized.

Br, OC–CO, ''''OR; NaOCl, H_2O, THF, 96%; CO_2H, ''''OR, CO_2H (20)

Br, OH, OH, Br; NaOCl, TBAB, pH 8, EtOAc, 74%; Br, O, =O, Br (21)

Oxidative Degradation. Treatment of carboxylic acids with sodium hypochlorite can lead to decarboxylation to afford aldehydes with one less carbon atom (eq 22).[35] The mechanism of this reaction is likely to involve methylene oxidation to the chloride or the alcohol followed by decarboxylation and oxidation to the aldehyde. This method has primarily been applied to sugar degradation (eq 23).[9,36]

CO_2H; NaOCl, OH^-, 100 °C, 71%; CHO (22)

CHO, H–OH, HO–H, H–OH, H–OH, CH_2OH; 1. NaOCl, pH 11, 20 h; 2. NaOCl, pH 4.5–5, 10 h; 35%; CHO, HO–H, H–OH, H–OH, CH_2OH (23)

Reaction with Amines. Reactions of sodium hypochlorite with amines can yield ketones, cyclization products, or, in the case of amino acids, degradation products. The mechanism of each of these processes involves initial *N*-chlorination.[37] Chlorinated intermediates have been identified spectroscopically, and in some cases isolated, providing access to chlorination products as well. Treatment of aliphatic amines under phase transfer conditions yields ketones or nitriles.[15,16a] The initial product is the *N*-chloro imine, with the carbonyl being released upon hydrolytic workup (eq 24).

NH_2; NaOCl, Bu_4NBr, CH_2Cl_2; [N–Cl]; H_3O^+, 98%; O (24)

Reaction of α- or β-amino acids with NaOCl leads to the corresponding aldehyde or methyl ketone containing one less carbon.[38] The mechanism postulated for this Strecker-type degradation of amino acids again involves *N*-chlorination, followed by decarboxylation to yield the imines. Aldehydes and ketones are released by hydrolysis (eq 25).[39] This method has been applied to degradation of α-methyl DOPA (eq 26).[40]

R–CO_2H, NH_2 → [R–CO_2Na, NHCl] → [R–CO_2Na, NH] → [R–CH=NH] → RCHO (25)

Oxidative Cyclization of Amines. Treatment of nitro anilines with sodium hypochlorite under alkaline conditions affords benzofuroxans as cyclization products (eq 27).[41] Spectroscopic evidence suggests an initial chlorination of the amino group, followed by cyclization. In addition, chlorinated intermediates have been prepared independently and submitted to the reaction conditions to show that cyclization products are formed.[37] The cyclization reaction requires addition of base, as azo products are formed at neutral pH. Oxadiazoles (eq 28)[42] and chlorodiazirines (eq 29)[43] are also formed by oxidative cyclization in moderate-to-good yields. The key intermediate for both of these reactions is also proposed to be an *N*-chloro imine.

Amine Coupling Reactions. Hydrazines[44] and alkylhydrazines (eq 30)[45] are formed by the action of bleach and amines. Azoethanes are toxic, and therefore care should be taken in their production and handling.

$$EtNH_2 + SOCl_2 \longrightarrow (EtNH)_2SO_2 \xrightarrow[H_2O]{NaOCl,\ NaOH} EtNH{-}NHEt \xrightarrow{NaOCl} EtN{=}NEt \quad 51\text{–}54\% \qquad (30)$$

Chlorination. Several types of organic nucleophiles undergo reaction with sodium hypochlorite to afford chlorination products. *N*-Chlorination of primary and secondary amines is a representative and widely-used example of this reaction class (eq 31).[46] Either mono- or dichlorinated products can be obtained selectively through control of the relative stoichiometry of the amine and NaOCl.[47] A two-step chlorine-shuttle pathway for the selective chlorination of electron-rich aromatics has been developed which relies on initial *N*-chlorination with NaOCl.[46] Intramolecular cyclization of amines via the Hofmann–Loffler reaction may also be accomplished by effecting the requisite *N*-halogenation with NaOCl (eq 32).[48]

Chlorination of indoles (eq 33),[7,49] amides, and ureas[50] occurs at the nitrogen center, while oximes[51] are chlorinated at carbon. Selective chlorination of nicotinic acids (eq 34),[52] chromones,[53] and polymers[54] has also been achieved. Substitution of chlorine for other halides occurs upon treatment of certain aromatic halides with sodium hypochlorite solution.[55]

Related Reagents. See Classes O-1, O-2, O-5, and O-14, pages 1–10.

1. (a) Chakrabartty, S. K. *Oxidation in Organic Chemistry*; Trahanovsky, W., Ed.; Academic: New York, 1976; Part C. (b) Mohrig, J. R.; Nienhuis, D. M.; Linck, C. F.; Van Zoeren, C.; Fox, B. G.; Mahaffy, R. G. *J. Chem. Educ.* **1985**, *62*, 519. (c) Skarzewski, J.; Siedlecka, R. *OPP* **1992**, *24*, 625.
2. (a) Procter, G. *COS* **1991**, *7*, 318. (b) Anelli, P. L.; Biffi, C.; Montanari, F.; Quici, S. *JOC* **1987**, *52*, 2559.
3. Abramovici, S.; Neumann, R.; Sasson, Y. *J. Mol. Catal.* **1985**, *29*, 291.
4. Marmor, S. *JOC* **1963**, *28*, 250.
5. Lee, G. A.; Freedman, H. H. *TL* **1976**, 1641.
6. (a) Banfi, S.; Montanari, F.; Quici, S. *JOC* **1989**, *54*, 1850. (b) Balavoine, G.; Eskenazi, C.; Meunier, F. *J. Mol. Catal.* **1985**, *30*, 125.
7. De Rosa, M. *CC* **1975**, 482.
8. Tsuge, O.; Watanabe, H.; Kanemasa, S. *CL* **1984**, 1415.
9. Whistler, R. L.; Yagi, K. *JOC* **1964**, *26*, 1050.

10. Lolthoff, I. M.; Belcher, R. *Volumetric Analysis*; Interscience: New York, 1957; p 262.
11. Lieu, V. T.; Kalbus, G. E. *J. Chem. Educ.* **1988**, *65*, 184.
12. (a) Sanfourche, M.; Gardent, L. *BSF* **1924**, *35*, 1088. (b) Adams, R. A.; Brown, B. K. *OSC* **1932**, *1*, 309.
13. Robertson, P. M.; Oberlin, R.; Ibl, N. *Electrochim. Acta* **1981**, *26*, 941.
14. *The Merck Index*, 11th ed.; Budavari, S., Ed.; Merck: Rahway, NJ, 1989; p 1363.
15. Lee, G. A.; Freedman, H. H. *Isr. J. Chem.* **1988**, *26*, 229.
16. (a) Lee, G. A.; Freedman, H. H. *TL* **1976**, 1641. (b) Regen, S. L. *JOC* **1977**, *42*, 875.
17. Wolfe, S.; Hasan, S. K.; Campbell, J. R. *CC* **1970**, 1420.
18. Stevens, R. V.; Chapman, K. T.; Weller, H. N. *JOC* **1980**, *45*, 2030.
19. Perkins, R. A.; Chau, F. *J. Chem. Educ.* **1982**, *59*, 981.
20. Stevens, R. V.; Chapman, K. T.; Stubbs, C. A.; Tam, W. W.; Albizati, K. F. *TL* **1982**, *23*, 4647.
21. Kudesia, V. P.; Mukherjee, S. K. *IJC(A)* **1977**, *15A*, 513.
22. Yamaguchi, M.; Takata, T.; Endo, T. *JOC* **1990**, *55*, 1490.
23. Leanna, M. R.; Sowin, T. J.; Morton, H. E. *TL* **1992**, *33*, 5029.
24. Anelli, P. L.; Montanari, F.; Quici, S. *OS* **1991**, *69*, 212.
25. (a) Anelli, P. L.; Banfi, S.; Montanari, F.; Quici, S. *JOC* **1989**, *54*, 2970. (b) Siedlecka, R.; Skarzewski, J.; Mlochowski, J. *TL* **1990**, 2177.
26. Inokuchi, T.; Matsumoto, S.; Nishiyama, T.; Torii, S. *JOC* **1990**, *55*, 462.
27. Davis, N. J.; Flitsch, S. L. *TL* **1993**, *34*, 1181.
28. Russo, J. M.; Price, W. A. *JOC* **1993**, *58*, 3589.
29. Krishnan, S.; Kuhn, D. G.; Hamilton, G. A. *JACS* **1977**, *99*, 8121.
30. (a) Wellman, G. R.; Lam, B.; Anderson, E. L.; White, V. E. *S* **1976**, 547. (b) Jakubowski, A. A.; Guziec, F. S., Jr.; Tishler, M. *TL* **1977**, 2399.
31. Arcoria, A.; Ballistreri, F. P.; Contone, A.; Musumarra, G.; Tripolone, M. *G* **1980**, *110*, 267.
32. Fonouni, H. E.; Krishnan, S.; Kuhn, D. G.; Hamilton, G. A. *JACS* **1983**, *105*, 7672.
33. (a) Corey, E. J.; Pearce, H. L. *JACS* **1979**, *101*, 5841. (b) Neiswender, D. D.; Moniz, W. B.; Dixon, J. A. *JACS* **1960**, *82*, 2876.
34. Ishii, F.; Kishi, K. *S* **1980**, 706.
35. Kaberia, F.; Vickery, B. *CC* **1978**, 459.
36. (a) Weerman, R. A. *RTC* **1917**, *37*, 16. (b) Whistler, R. L.; Schweiger, R. *JACS* **1959**, *81*, 5190.
37. Dyall, L. K. *AJC* **1984**, *37*, 2013.
38. (a) Langheld, K. *CB* **1909**, 392. (b) Birkofer, L.; Brune, R. *CB* **1957**, *90*, 2536.
39. Schonberg, A.; Moubacher, R. *CR* **1952**, *52*, 281.
40. Fox, S. W.; Bullock, M. W. *JACS* **1951**, *73*, 2754.
41. (a) Green, A. G.; Rowe, F. *JCS* **1912**, *101*, 2443, 2452. (b) Mallory, F. B. *OSC* **1963**, *4*, 74. (c) Mallory, F. B.; Varimbi, S. P. *JOC* **1963**, *28*, 1656. (d) Mallory, F. B.; Wood, C. S.; Hurwitz, B. M. *JOC* **1964**, *29*, 2605.
42. Götz, N.; Zeeh, B. *S* **1976**, 268.
43. (a) Graham, W. H. *JACS* **1965**, *87*, 4396. (b) Berneth, H.; Hünig, S. *CB* **1980**, *113*, 2040.
44. Boido, V.; Edwards, O. E. *CJC* **1971**, *49*, 2664.
45. Ohme, R.; Preleschhof, H.; Heyne, H.-U. *OSC* **1988**, *6*, 78.
46. Lindsay Smith, J. R.; McKeer, L. C.; Taylor, J. M. *OS* **1988**, *67*, 222.
47. (a) Kovacic, P.; Lowery, M. K.; Field, K. W. *CRV* **1970**, *70*, 639. (b) Gilchrist, T. L. *COS* **1991**, *7*, Chapter 6.1.
48. (a) Kerwin, J. F.; Wolff, M. E.; Owings, F. F.; Lewis, B. B.; Blank, B.; Magnani, A.; Karash, C.; Georgian, V. *JOC* **1962**, *27*, 3628. (b) Wolff, M. E. *CRV* **1963**, *63*, 55. (c) Stella, L. *AG(E)* **1983**, *22*, 337.
49. De Rosa, M.; Carbognani, L.; Febres, A. *JOC* **1981**, *46*, 2054.
50. Bachand, C.; Driguez, H.; Paton, J. M.; Touchard, D.; Lessard, J. *JOC* **1974**, *39*, 3136.
51. Coda, A. C.; Tacconi, G. *G* **1984**, *114*, 131.
52. Elliot, M. L.; Goddard, C. J. *SC* **1989**, *19*, 1505.
53. Nohara, A.; Ukawa, K.; Sanno, Y. *TL* **1973**, 1999.
54. Jones, R. G.; Matsubayashi, Y. *Polymer* **1992**, *33*, 1069.
55. (a) Bayraktaroglu, T. O.; Gooding, M. A.; Khatib, S. F.; Lee, H.; Hourouma, M.; Landolt, R. G. *JOC* **1993**, *58*, 1264. (b) Arnold, J. T.; Bayraktaroglu, T. O.; Brown, R. G.; Heiermann, C. R.; Magnus, W. W.; Ohman, A. B.; Landolt, R. G. *JOC* **1992**, *57*, 391.

Jennifer M. Galvin
University of Illinois, Urbana, IL, USA

Eric N. Jacobsen
Harvard University, Cambridge, MA, USA

Sodium Methylsulfinylmethylide[1]

NaDMSO

[15590-23-5] C_2H_5NaOS (MW 100.13)

(strong base and nucleophile; very useful for the introduction of the methylsulfinylmethyl group[1])

Alternate Names: sodium dimsylate; dimsylsodium.
Solubility: sol DMSO.
Form Supplied in: not commercially available.
Analysis of Reagent Purity: titration with formanilide using triphenylgromethane as indicator.[2]
Preparative Method: prepared by the reaction of ***Sodium Hydride*** with ***Dimethyl Sulfoxide*** for 1 h at 70 °C. ***Sodium Amide*** may also be used as base.[3]
Handling, Storage, and Precautions: this reagent is not exceptionally stable over long periods of time. Modifications in preparation and storage have led to an increased shelf life.[4] However, it is probably best prepared as needed and used quickly. Hydrogen is emitted during the preparation of this reagent and decomposition occurs at elevated temperatures.[5] Due caution should be exercised in the preparation, particularly on a large scale. Reports have been made of explosions during the large scale preparation of this reagent or its attempted isolation.[1,6]

Introduction. The pK_a of DMSO is 35. Consequently, one might expect its conjugate base to be a potent Brønsted base and this is indeed the case. In addition to the sodium salt, both the lithium and potassium salts of DMSO have found widespread use in synthesis (see ***Lithium Methylsulfinylmethylide*** and ***Potassium Methylsulfinylmethylide***). After its introduction by Corey,[2] explorations of the chemistry of sodium methylsulfinylmethylide (NaDMSO) blossomed and are summarized nicely in two reviews.[1]

Sodium Methylsulfinylmethylide as Base.

Generation of Ylides. NaDMSO as a solution in DMSO has been used extensively in the preparation of phosphorus and sulfur ylides.[1,7,8] This medium is often the one of choice for the generation and reaction of Wittig reagents (eq 1).[7,9] A less common application of Wittig reagents generated with NaDMSO is in an oxidative coupling of a bis-ylide to form a cyclic alkene (eq 2).[10]

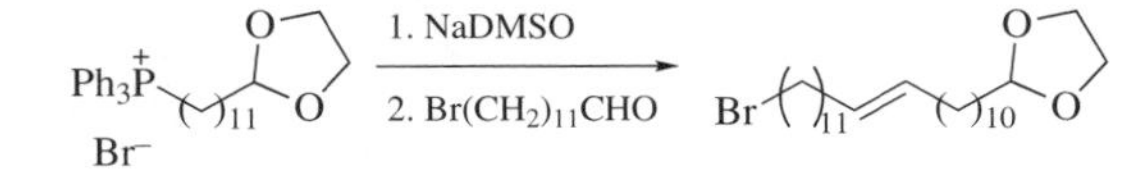

The methylene transfer reagents ***Dimethylsulfonium Methylide*** and ***Dimethylsulfoxonium Methylide*** are conveniently generated with NaDMSO in DMSO.[11] These and other sulfonium ylides are useful in the synthesis of epoxides and cyclopropanes (eqs 3 and 4).[8,11] Also intriguing is the use of dimethyloxosulfoxonium methylide to give dienes (eq 5).[12]

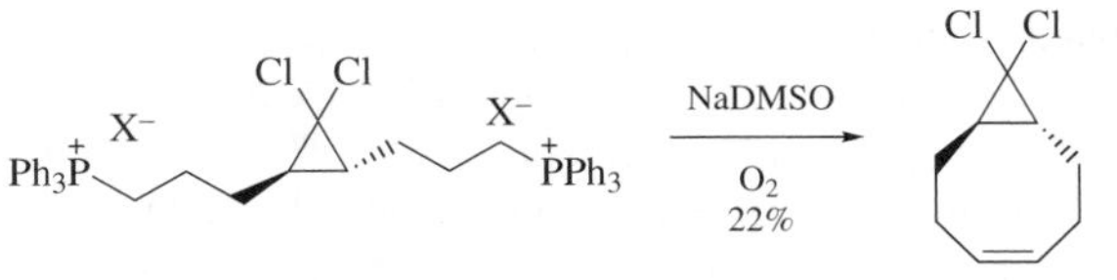

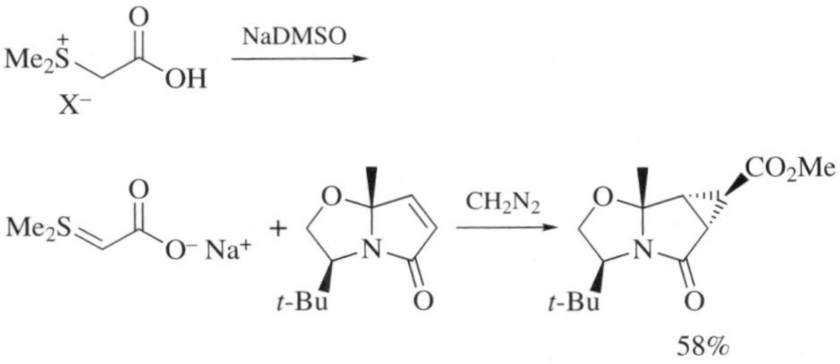

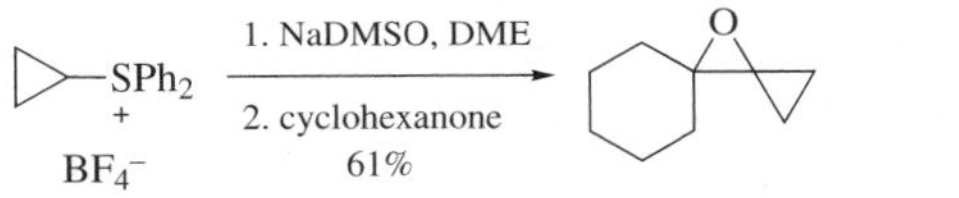

Diaminosulfonium salts can be deprotonated with NaDMSO to give rearrangement products or epoxides upon reaction with aldehydes (eq 6).[13] A variety of ammonium ylides have been generated using NaDMSO.[14,15] These species are disposed to undergo Wittig, Stevens, or Sommelet–Hauser rearrangements, depending on their specific constitution (eqs 7 and 8). Changes in base and other reaction conditions can change product distributions.[14a,b]

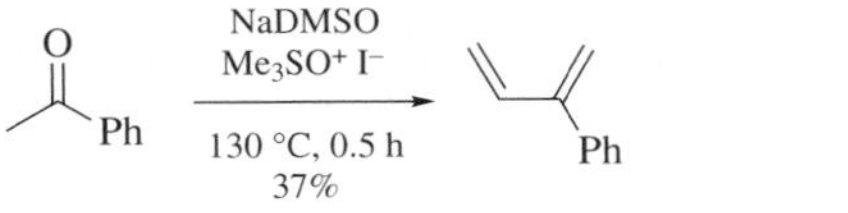

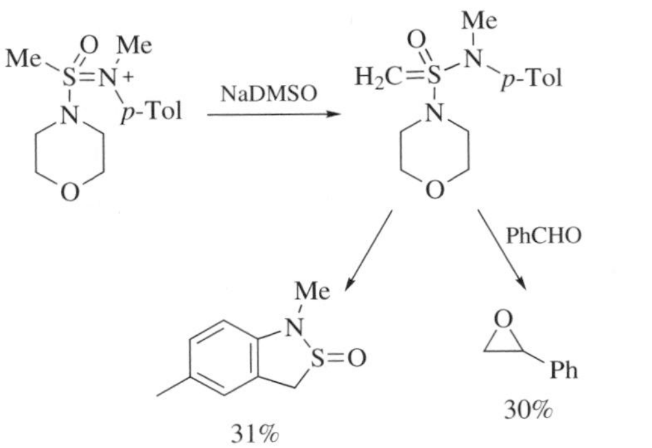

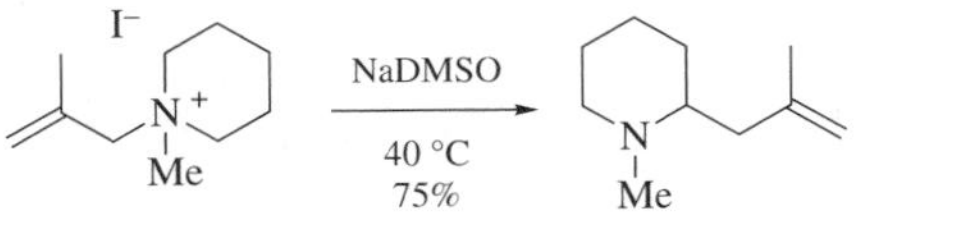

Eliminations. Elimination reactions using NaDMSO have apparently not been extensively investigated, although the use of metal alkoxides in DMSO has received considerable attention.[16] Whether the active agent in such mixtures is NaDMSO is not clear. Eliminations clearly mediated by NaDMSO are known.[17] For example, treatment of (**1**) with NaDMSO at 25 °C gives the elimination product (**2**) in 91% yield (eq 9).[17a] Longer reaction times result in addition of DMSO to the newly formed double bond. Heating results in isolation of a dealkylation product (**4**), a clear example of the diverse reactivity associated with NaDMSO. Alkynes are generated by the reaction of 1,2-dibromoalkanes with excess NaDMSO.[18] Aryl halides can give benzynes in the presence of NaDMSO (eq 10).[19] Further, addition of DMSO is possible, leading to a unique mode of functionalization (eq 11).[19c]

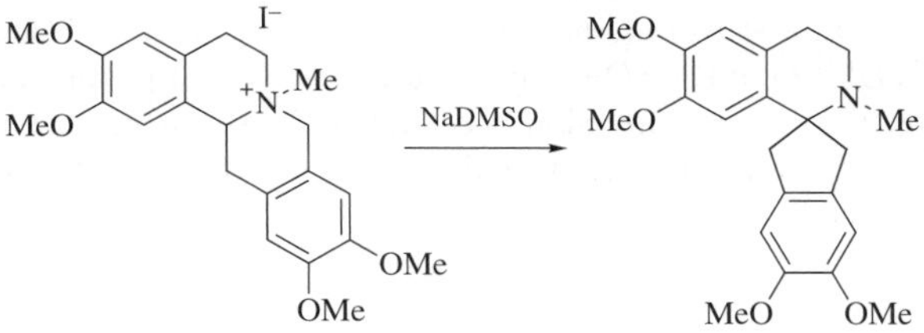

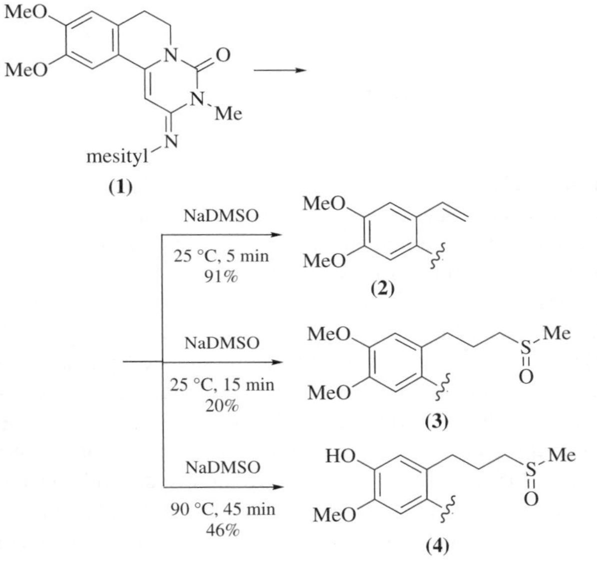

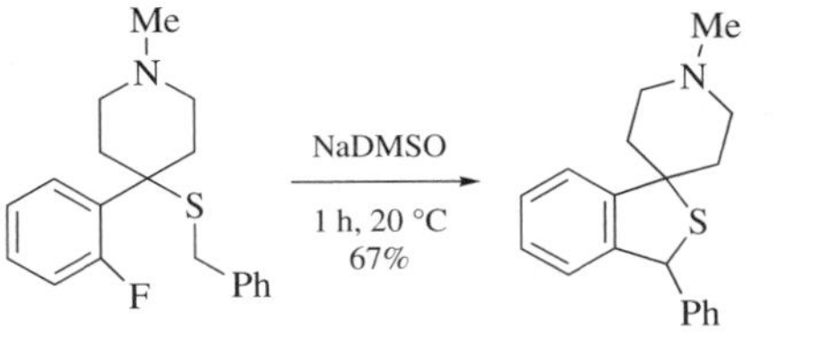

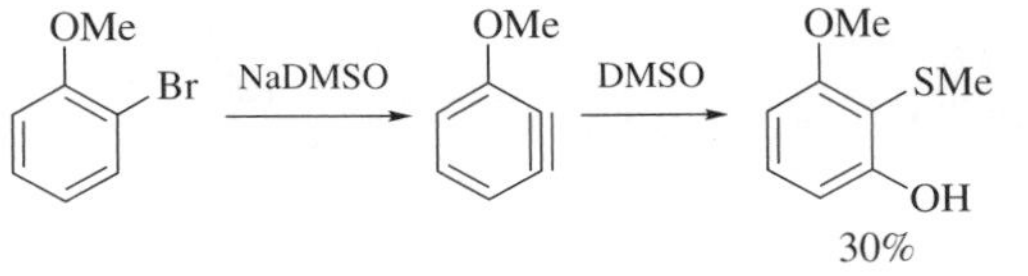

Rearrangements. NaDMSO has been used to promote double bond isomerization, leading to aromatization in the case shown (eq 12).[20] An anionotropic rearrangement of a cyclohexadienone to a substituted hydroquinone has been reported (eq 13).[21] The use of NaDMSO in DMSO has been described as an optimal choice for executing the carbanion accelerated Claisen rearrangement (eq 14).[22] Grob-type fragmentations mediated by NaDMSO have been used in total synthesis (eq 15).[23]

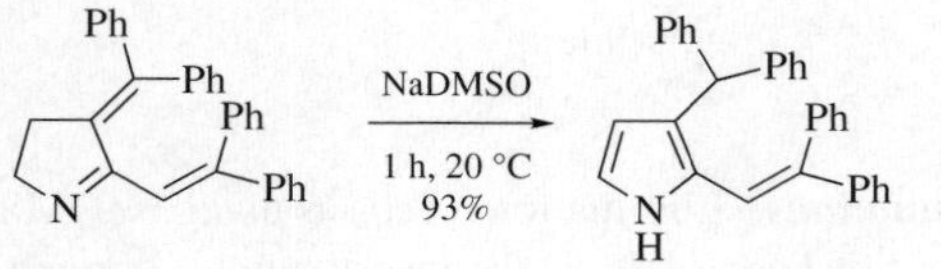

NaDMSO, 1 h, 20 °C, 93% (12)

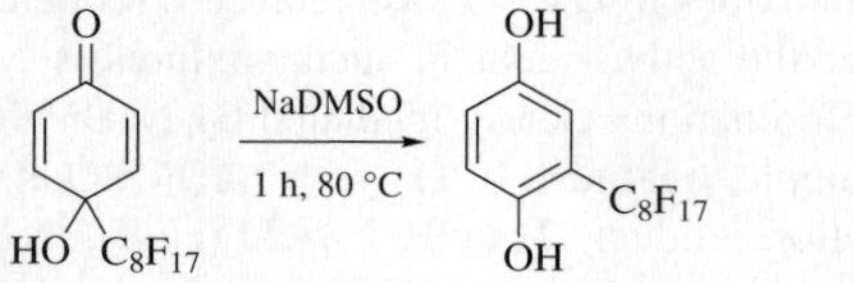

NaDMSO, 1 h, 80 °C (13)

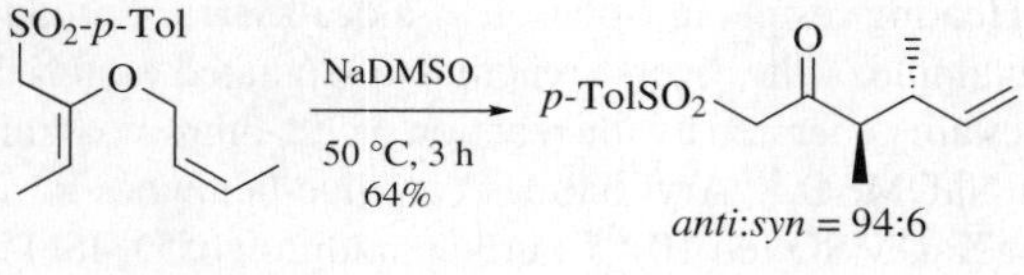

NaDMSO, 50 °C, 3 h, 64%; *anti*:*syn* = 94:6 (14)

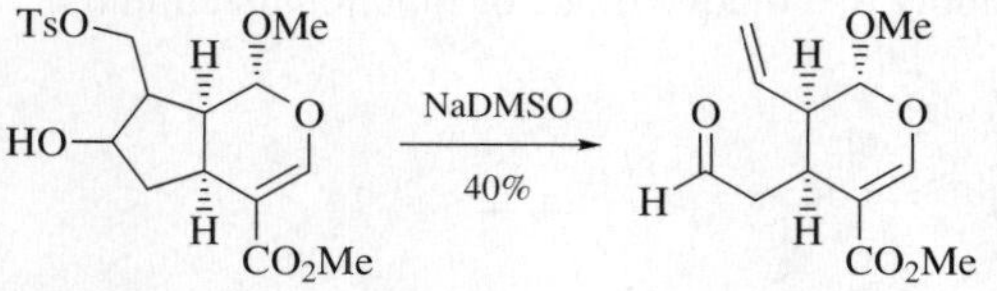

NaDMSO, 40% (15)

Anion Alkylation and Acylation. NaDMSO has been used as a base to create new anions or carbanions, which then can be functionalized via normal alkylation or acylation procedures. Intramolecular aminolysis of an ester mediated by NaDMSO has been reported (eq 16).[24] A tandem double Michael–Dieckmann condensation leading to a complex tricyclic structure has been developed (eq 17).[25] Generation of a ketone enolate with NaDMSO followed by an intramolecular alkylation has been a part of a number of total syntheses.[26] Alkylations of sulfoximine and sulfone (Ramberg–Bäcklund rearrangement) carbanions derived from NaDMSO are known.[27] The intramolecular oxidative coupling of nitro-stabilized carbanions can lead to highly functionalized cyclopropanes which are potentially useful as high energy materials (eq 18).[28]

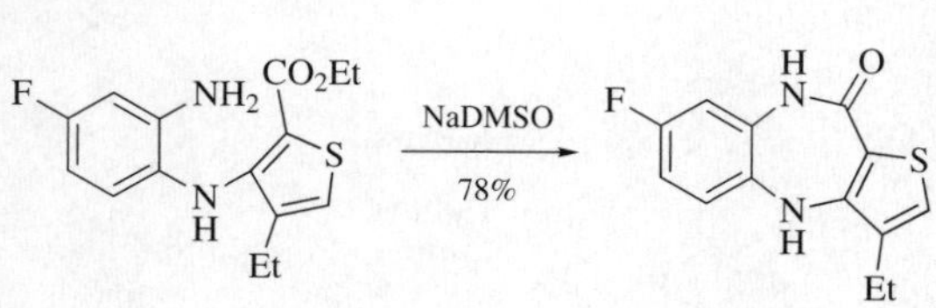

NaDMSO, 78% (16)

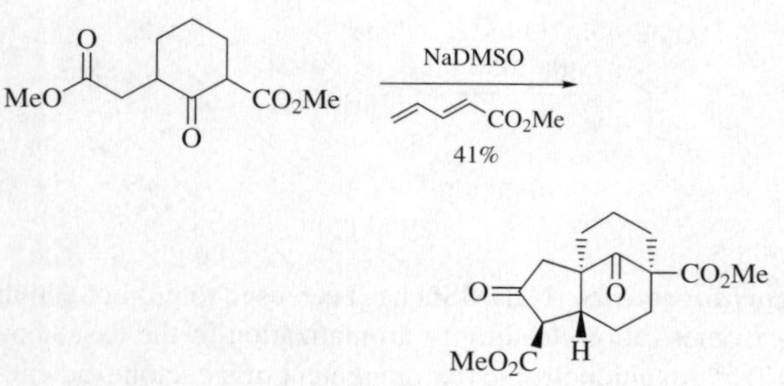

NaDMSO, 41% (17)

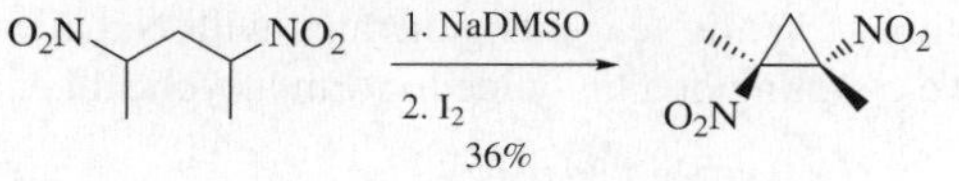

1. NaDMSO, 2. I_2, 36% (18)

Ether Synthesis. The application of NaDMSO to the Williamson ether synthesis has been documented.[29] The potential for alkoxide fragmentation exists and some discretion must be exercised in using this base for ether formation.[30] An intramolecular version of the Williamson reaction mediated by NaDMSO leads to an oxetane in high yield.[31] The preparation of otherwise difficulty accessible xanthates has been realized using NaDMSO.[32] Very common is the modification of oligo- and polysaccharides via Williamson ether synthesis (Hokomori reaction) to facilitate handling and analysis of these compounds.[33] Other hydroxylic polymers can also be functionalized in this way.[34]

Sodium Methylsulfinylmethylide as Nucleophile.

Reactions with Esters. Perhaps one of the most useful reactions of NaDMSO is its condensation with esters to produce β-keto sulfoxides. The rich chemistry of these difunctional compounds makes accessible a wide variety of other organic compounds.[1,35] One of the most straightforward applications of NaDMSO is the synthesis of methyl ketones from esters (eq 19).[36] Thermal elimination of methylsulfenic acid after alkylation leads to α,β-unsaturated ketones (eq 20).[37] Among the wide variety of possible transformations of β-keto sulfoxides, another which stands out is the Pummerer reaction, which allows for the formation of carbon–carbon bonds via an 'umpolung' of the reactivity adjacent to the carbonyl group (eq 21).[38] A synthesis of ninhydrin was developed based on this type of chemistry.[38c] Methacrylate polymers have been modified by reaction with NaDMSO.[39]

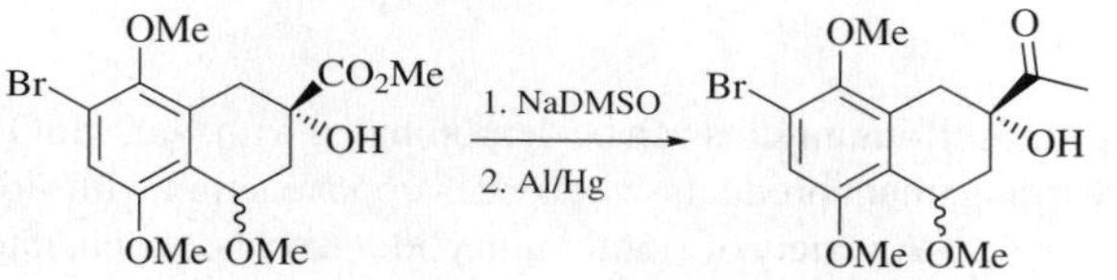

1. NaDMSO, 2. Al/Hg (19)

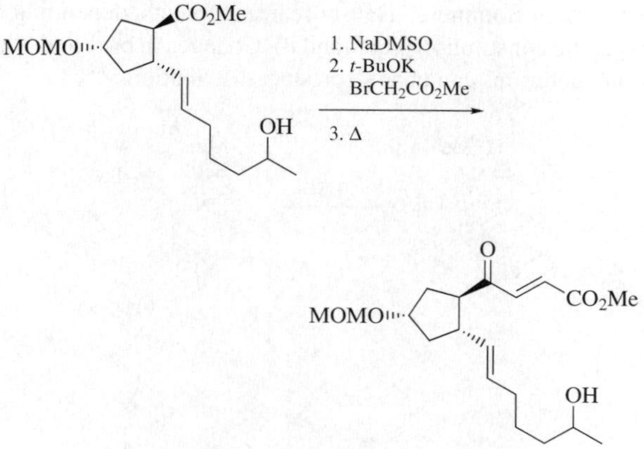

1. NaDMSO, 2. *t*-BuOK, $BrCH_2CO_2Me$, 3. Δ (20)

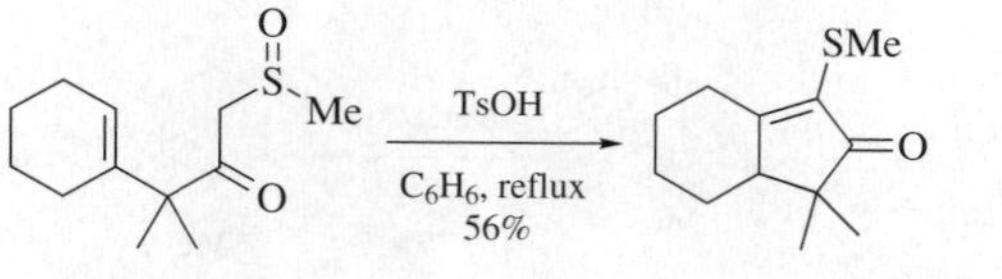

TsOH, C_6H_6, reflux, 56% (21)

Reactions with Aldehydes and Ketones. The reaction of nonenolizable ketones and aldehydes with NaDMSO generally proceeds smoothly to give β-hydroxy sulfoxides.[1] The chemistry of the latter is not as rich as that of β-keto sulfoxides, to which they can be converted. However, dehydration leads efficiently to α,β-unsaturated sulfoxides which can be transformed to the corresponding sulfides or sulfones.[40] A one-pot procedure using ***Sodium*** metal to produce NaDMSO and serve as a reductant leads to α,β-unsaturated sulfides directly.[41] Interestingly, a dianion of DMSO can be prepared from the reaction of DMSO with 2.2 equiv of $NaNH_2$. Reaction with benzophenone gives the expected adduct, albeit in only modest yield (eq 22).[3] A unique synthesis of 3-phenylindole based upon the reaction of NaDMSO with 2-aminobenzophenone has been reported (eq 23).[42]

1. 2.2 equiv $NaNH_2$; 2. benzophenone; 41% (22)

excess NaDMSO (23)

Enolizable ketones and aldehydes often give enolates upon reaction with NaDMSO as well as the expected addition products, limiting the utility of the reagent with these systems. Several unusual reactions of NaDMSO with enolizable ketones have been reported. For example, treatment of 4-heptanone with NaDMSO at elevated temperatures results in diene formation (eq 24).[43] Similarly, the reaction of cyclopentanone with NaDMSO gives a diene resulting from condensation followed by nucleophilic addition of NaDMSO and fragmentation (eq 25).[44]

NaDMSO, 130 °C, 37% (24)

NaDMSO, 25 °C, 59% (25)

Reactions with Imines and Related Compounds. The reaction of imines with NaDMSO has not been extensively studied.[2] However, several interesting applications with heterocycles at least formally possessing imine functional groups have been reported. Treatment of 1-methylquinoline with excess NaDMSO results in a tandem 1,4–1,2 addition to give a unique β-amino sulfoxide (eq 26).[45] A mechanistically intriguing synthesis of phenanthrene is the result of the reaction of a benzoquinoline *N*-oxide with NaDMSO (eq 27).[46] Finally, a synthesis of dibenzo[*a,f*]quinolizines has been developed based on the addition of NaDMSO to an imine followed by trapping of the resulting amide with a pendant benzyne (eq 28).[47]

excess NaDMSO, 20 °C, 3.5 h, 60% (26)

NaDMSO, 88% (27)

NaDMSO (28)

Reactions with Alkenes and Alkynes. The reaction of NaDMSO with alkenes or alkynes conjugated to an aryl ring or alkene is well known.[1] 1,1-Diphenylethene gives a 1:1 addition product in quantitative yield upon reaction with NaDMSO.[48] From a synthetic perspective, one of the most useful of these reactions is the alkylation, especially the methylation, of stilbenes (eq 29).[49] This occurs by initial attack of NaDMSO on the unsaturated system followed by the elimination of methanesulfenic acid and isomerization. Dienes and other polyenes are subject to the same type of chemistry, but yields are modest and isomer formation can be a problem (eq 30).[50] Reaction of NaDMSO with diphenylacetylene can lead to either simple addition products or those based on an addition–elimination sequence, depending on the reaction conditions (eq 31).[51] Addition of NaDMSO to unsaturated systems has been used in polymer synthesis.[52]

NaDMSO, 67% (29)

NaDMSO (30)

NaDMSO; 30 °C; 50 °C (31)

Reactions with Aromatics. The reactions of NaDMSO with aromatic electrophiles can be categorized as occurring through either benzyne or addition–elimination mechanisms. Reaction of NaDMSO with chlorobenzene gives a mixture of sulfoxides, presumably via a benzyne intermediate (eq 32).[53] This chemistry has been used in the modification of polystyrenes.[54] Various fluoroaromatics react with NaDMSO to produce substitution products via addition–elimination (eq 33).[55] A variety of condensed aromatics (e.g. anthracene) undergo methylation analogous to that of stilbene upon reaction with NaDMSO (eq 34).[56]

Reactions with Halides and Related Compounds. Alkylation of NaDMSO with primary halides and tosylates results in the formation of the expected sulfoxides.[57] However, reaction with benzyl chloride gives stilbene as the major product, suggesting that the basicity of NaDMSO must be considered even with reactive electrophiles.[2] More hindered systems favor elimination. 1,2,5,6-Tetrabromocyclooctane debrominates to 1,5-cyclooctadiene upon reaction with NaDMSO (eq 35).[58] With ***Potassium t-Butoxide*** in DMSO the major product is one of elimination, namely cyclooctatetraene. Monodebromination also occurs with *gem*-dibromocyclopropanes.[59] In a process which presumably proceeds via S_N2 substitution, reaction of imidates with NaDMSO leads to amides (eq 36).[60]

Reactions with Epoxides. The reaction of NaDMSO with epoxides does not appear to have been widely investigated. Nevertheless, some synthetically useful transformations have been documented. Ring opening with NaDMSO followed by thermal alkene formation was developed as a route to an optically pure secondary allylic alcohol (eq 37).[61] A related transformation involves the reaction of trimethylsilyl-substituted epoxides (eq 38).[62] Ring opening at the TMS-substituted carbon, followed by desilylation and sulfenate elimination, leads to allyl alcohols in good yield in a one-pot process.

Reactions with Phosphorus and Sulfur Electrophiles. A one-pot synthesis of α,β-unsaturated sulfoxides begins with the reaction of NaDMSO with ***Diethyl Phosphorochloridate*** to give a Horner–Emmons reagent which reacts with aldehydes in an efficient manner (eq 39).[63] Bis-sulfoxides are prepared by the reaction of NaDMSO and a diastereomerically pure methyl sulfinate ester.[64] Some kinetic resolution is observed in this reaction. The thiophilic addition of NaDMSO to sulfines also leads to bis-sulfoxides (eq 40).[65]

Related Reagents. See Class O-14, pages 1–10.

1. (a) Durst, T. *Adv. Org. Chem.* **1969**, *6*, 285. (b) Hauthal, H. G.; Lorenz, D. In *Dimethyl Sulphoxide*; Martin, D.; Hauthal, H. G., Eds.; Wiley: New York, 1971; pp 349–374.
2. Corey, E. J.; Chaykovsky, M. *JACS* **1965**, *87*, 1345.
3. Kaiser, E. M.; Beard, R. D.; Hauser, C. R. *JOM* **1973**, *59*, 53.
4. Sjöberg, S. *TL* **1966**, 6383.
5. Price, C. C.; Yukuta, T. *JOC* **1969**, *34*, 2503.
6. (a) Leleu, J. *Cah. Notes. Doc.* **1976**, *85*, 583 (*CA* **1978**, *88*, 26 914t). (b) Itoh, M.; Morisaki, S.; Muranaga, K.; Matsunaga, T.; Tohyama, K.; Tamura, M.; Yoshida, T. *Anzen Kogaku* **1984**, *23*, 269 (*CA* **1985**, *102*, 100 117).
7. Gosney, I.; Rowley, A. G. In *Organophosphorus Reagents in Organic Synthesis*; Cadogan, J. I. G., Ed.; Academic: London, 1979; pp 17–153.
8. (a) Romo, D.; Meyers, A. I. *JOC* **1992**, *57*, 6265. (b) Trost, B. M.; Bogdanowicz, M. J. *JACS* **1973**, *95*, 5298. (c) Trost, B. M.; Bogdanowicz, M. J. *JACS* **1973**, *95*, 5321.
9. (a) Paynter, O. I.; Simmonds, D. J.; Whiting, M. C. *CC* **1982**, 1165. (b) Hall, D. R.; Beevor, P. S.; Lester, R.; Poppi, R. G.; Nesbitt, B. F. *CI(L)* **1975**, 216.
10. Deyrup, J. A.; Betkouski, M. F. *JOC* **1975**, *40*, 284.
11. (a) Corey, E. J.; Chaykovsky, M. *JACS* **1965**, *87*, 1353. (b) Trost, B. M.; Melvin, L. S., Jr. *Sulfur Ylides: Emerging Synthetic Intermediates*; Academic: New York, 1975.
12. Yurchenko, A. G.; Kyrij, A. B.; Likhotvorik, I. R.; Melnik, N. N.; Zaharh, P.; Bzhezovski, V. V.; Kushko, A. O. *S* **1991**, 393.
13. Okuma, K.; Higuchi, N.; Kaji, S.; Takeuchi, H.; Ohta, H.; Matsuyama, H.; Kamigata, N.; Kobayashi, M. *BCJ* **1990**, *63*, 3223.
14. (a) Dietrich, V. W.; Schulze, K.; Mühlstädt, M. *JPR* **1977**, *319*, 799. (b) Dietrich, W.; Schulze, K.; Mühlstädt, M. *JPR* **1977**, *319*, 667.
15. (a) Kano, S.; Yokomatsu, T.; Komiyama, E.; Tokita, S.; Takahagi, Y.; Shibuya, S. *CPB* **1975**, *23*, 1171. (b) Kano, S.; Yokomatsu, T.; Ono, T.; Takahagi, Y.; Shibuya, S. *CPB* **1977**, *25*, 2510.
16. See Ref. 1b, pp 174–197.

17. (a) Lal, B.; Gidwani, R. M.; de Souza, N. J. *JOC* **1990**, *55*, 5117. (b) Kano, S.; Komiyama, E.; Nawa, K.; Shibuya, S. *CPB* **1976**, *24*, 310. (c) Kano, S.; Yokomatsu, T.; Shibuya, S. *CPB* **1977**, *25*, 2401.
18. Klein, J.; Gurfinkel, E. *T* **1970**, *26*, 2127.
19. (a) Ong, H. H.; Profitt, J. A.; Anderson, V. B.; Kruse, H.; Wilker, J. C.; Geyer, H. M., III *JMC* **1981**, *24*, 74. (b) Kano, S.; Ogawa, T.; Yokomatsu, T.; Takahagi, Y.; Komiyama, E.; Shibuya, S. *H* **1975**, *3*, 129. (c) Birch, A. J.; Chamberlain, K. B.; Oloyede, S. S. *AJC* **1971**, *24*, 2179.
20. Wittig, G.; Hesse, A. *LA* **1975**, 1831.
21. Uno, H.; Yayama, A.; Suzuki, H. *CL* **1991**, 1165.
22. Denmark, S. E.; Harmata, M. A.; White, K. S. *JACS* **1989**, *111*, 8878.
23. (a) Kinast, G.; Tietze, L.-F. *CB* **1976**, *109*, 3626. (b) Corey, E. J.; Mitra, R. B.; Uda, H. *JACS* **1964**, *86*, 485.
24. Chakrabarti, J. K.; Hicks, T. A.; Hotten, T. M.; Tupper, D. E. *JCS(P1)* **1978**, 937.
25. Danishefsky, S.; Hatch, W. E.; Sax, M.; Abola, E.; Pletcher, J. *JACS* **1973**, *95*, 2410.
26. (a) Kelly, R. B.; Eber, J.; Hung, I.-K. *CC* **1973**, 689. (b) Corey, E. J.; Watt, D. S. *JACS* **1973**, *95*, 2303. (c) Heathcock, C. H. *JACS* **1966**, *88*, 4110.
27. (a) Morton, D. R., Jr.; Brokaw, F. C. *JOC* **1979**, *44*, 2880. (b) Scholz, D.; Burtscher, D. *LA* **1985**, 517. (c) Paquette, L. A. *OR* **1977**, *25*, 1. (d) Johnson, C. R. *Aldrichim. Acta* **1985**, *18*, 3.
28. Wade, P. A.; Dailey, W. P.; Carroll, P. J. *JACS* **1987**, *109*, 5452.
29. Sjöberg, B.; Sjöberg, K. *ACS* **1972**, *26*, 275.
30. Partington, S. M.; Watt, C. I. F. *JCS(P2)* **1988**, 983.
31. Corey, E. J.; Mitra, R. B.; Uda, H. *JACS* **1964**, *86*, 485.
32. de Groot, A.; Evanhius, B.; Wynberg, H. *JOC* **1968**, *33*, 2214.
33. Zähringer, U.; Rietschel, E. T. *Carbohydr. Res.* **1986**, *152*, 81. (b) Lee, K.-S.; Gilbert, R. D. *Carbohydr. Res.* **1981**, *88*, 162.
34. Galin, J. C. *J. Appl. Polym. Sci.* **1971**, *15*, 213.
35. For leading references, see: Ibarra, C. A.; Rogríguez, R. C.; Monreal, M. C. F.; Navarrao, F. J. G.; Tesorero, J. M. *JOC* **1989**, *54*, 5620.
36. Swenton, J. S.; Anderson, D. K.; Jackson, D. K.; Narasimhan, L. *JOC* **1981**, *46*, 4825.
37. (a) Bartlett, P. A.; Green, F. R., III *JACS* **1978**, *100*, 4858. (b) Dal Pozzo, A.; Acquasaliente, M.; Buraschi, M.; Anderson, B. M. *S* **1984**, 926.
38. (a) Isibashi, H.; Okada, M.; Komatsu, H.; Ikeda, M. *S* **1985**, 643. (b) Oikawa, Y.; Yonemitsu, O. *JOC* **1976**, *41*, 1118. (c) Becker, H.-D.; Russell, G. A. *JOC* **1963**, *28*, 1896. (d) De Lucchi, O.; Miotti, U.; Modena, G. *OR* **1991**, *40*, 157.
39. (a) Katsutoshi, N.; Harada, A.; Oyamada, M. *NKK* **1983**, 713. (b) Arranz, F.; Galin, J. C. *Makromol. Chem.* **1972**, *152*, 185.
40. (a) Fillion, H.; Boucherle, A. *BSF* **1971**, 3674. (b) Fillion, H.; Duc, C. L.; Agnius-Delord, C. *BSF* **1974**, 2923.
41. Kojima, T.; Fujisawa, T. *CL* **1978**, 1425.
42. Bravo, P.; Gavdiano, G.; Ponti, P. P. *CI(L)* **1971**, 253.
43. Yurchenko, A. G.; Kirii, A. V.; Mel'nik, N. N.; Likhotvorik, I. R. *ZOR* **1990**, *26*, 2230.
44. Comer, W. T.; Temple, D. L. *JOC* **1973**, *38*, 2121.
45. Kato, H.; Takeuchi, I.; Hamada, Y.; Ono, M.; Hirota, M. *TL* **1978**, 135.
46. Hamada, Y.; Takeuchi, I. *JOC* **1977**, *42*, 4209.
47. (a) Kano, S.; Yokomatsu, T.; Shibuya, S. *CPB* **1975**, *23*, 1098. (b) Kano, S.; Yokomatsu, T. *TL* **1978**, 1209.
48. Walling, C.; Bollyky, L. *JOC* **1964**, *29*, 2699.
49. (a) James, B. G.; Pattenden, G. *CC* **1973**, 145. (b) Feldman, M.; Danishefsky, S.; Levine, R. *JOC* **1966**, *31*, 4322.
50. Murray, D. F. *JOC* **1983**, *48*, 4860.
51. (a) Iwai, I.; Ide, J. *CPB* **1965**, *13*, 663. (b) Iwai, I.; Ide, J. *OS* **1970**, *50*, 62.
52. (a) Priola, A.; Trossarelli, L. *Makromol. Chem.* **1970**, *139*, 281. (b) Kriz, J.; Benes, M. J.; Peska, J. *CCC* **1967**, *32*, 4043.
53. Corey, E. J.; Chaykovsky, M. *JACS* **1962**, *84*, 866.
54. Janout, M.; Kahovec, J.; Hrudkova, H.; Svec, F.; Cefelin, P. *Polym. Bull. (Berlin)* **1984**, *11*, 215.
55. (a) Brooke, G. M.; Ferguson, J. A. K. J. *JFC* **1988**, *41*, 263. (b) Brooke, G. M.; Mawson, S. D. *JCS(P1)* **1990**, 1919.
56. Nozaki, H.; Yamamoto, Y. Noyori, R. *TL* **1966**, 1123.
57. Entwistle, I. D.; Johnstone, R. A. W. *CC* **1965**, 29.
58. Cardenas, C. G.; Khafaji, A. N.; Osborn, C. L.; Gardner, P. D. *CI(L)* **1965**, 345.
59. Osborn, C. L.; Shields, T. C.; Shoulders, B. A.; Cardenas, C. G.; Gardner, P. D. *CI(L)* **1965**, 766.
60. Kano, S.; Yokomatsu, T.; Hibino, S.; Imamura, K.; Shibuya, S. *H* **1977**, *6*, 1319.
61. Takano, S.; Tomita, S.; Iwabuchi, Y.; Ogasawara, R. *S* **1988**, 610.
62. Kobayashi, Y.; Ito, Y. I.; Urabe, H.; Sato, F. *SL* **1991**, 813.
63. Almog, J.; Weissman, B. A. *S* **1973**, 164.
64. Kunieda, N.; Nokami, J.; Kinoshita, M. *BCJ* **1976**, *49*, 256.
65. Loontjes, J. A.; van der Leij, M.; Zwanenberg, B. *RTC* **1980**, *99*, 39.

Michael Harmata
University of Missouri-Columbia, MO, USA

Sodium Naphthalenide

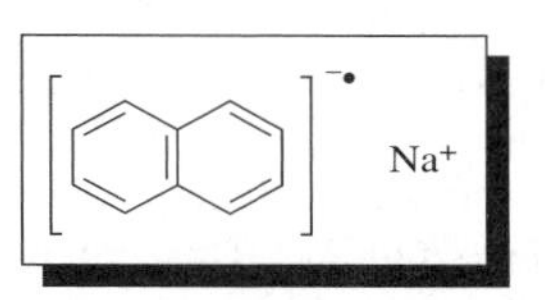

[3481-12-7] $C_{10}H_8Na$ (MW 151.17)

(one-electron donor promoting ketyl–alkene[3] and ketyl–alkyne[2] radical cyclizations; facilitates coupling of ketone[4] and thiocarbonyl[5] functionalities; removes mesylate, tosylate, and benzyl protecting groups[11,13–15] to generate the corresponding alkene or alcohol)

Solubility: sol diethyl ether, THF; forms complexes with diethyl ether and THF.

Preparative Method: typically prepared from the reduction of naphthalene by ***Sodium*** metal in THF solvent.[1]

Handling, Storage, and Precautions: stability is a matter of contention.[1] THF apparently contributes substantially to the stability of the complex by solvating the resulting ion pairs.

Reductive Cyclization. Sodium naphthalenide is a good electron donor and promotes the reductive cyclization of steroidal acetylenic ketones in high yield (eq 1).[2] The resulting allylic alcohol is the only observed product and no overreduction occurs as with more powerful reducing agents.

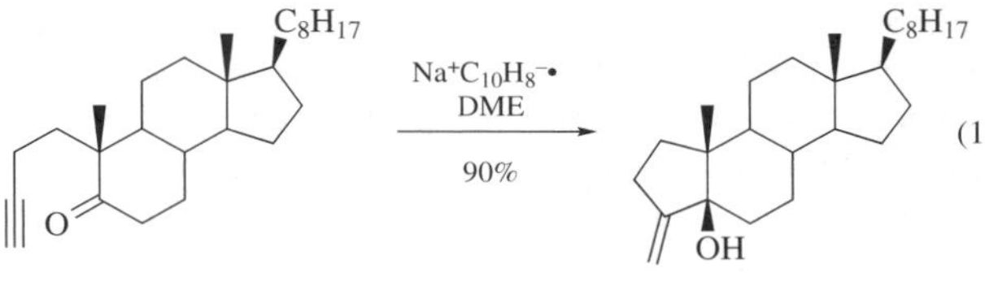

$$\xrightarrow[\text{90\%}]{Na^+C_{10}H_8^{-\bullet},\ \text{DME}} \quad (1)$$

Similarly, radical cyclization of an alkenyl aldehyde with sodium naphthalenide provides the product resulting from 5-*exo* ring closure in moderate yield. This method has been utilized in the synthesis of a key intermediate leading to α-cuparenone, which presents a challenge synthetically because of the steric congestion around the cyclopentane ring (eq 2).[3]

$Na^+C_{10}H_8^{-}\bullet$, THF, rt, 15 min, 50% (2)

Ketone and Thiocarbonyl Coupling. More recently, sodium naphthalenide in conjunction with ***Titanium(IV) Chloride*** has been used to perform a McMurry-like coupling reaction in an intramolecular process, providing a bicyclic structure in good yield (eq 3).[4]

$Na^+C_{10}H_8^{-}\bullet$, $TiCl_4$, THF, 70% (3)

Sodium naphthalenide further provides a method for the bridging of macrocycles. Nicolaou has recently used this method in the preparation of *cis*- and *trans*-fused oxabicyclic and oxapolycyclic systems, which are common structural components of marine and other natural products (eq 4).[5] This process is thought to proceed by initial electon transfer to the thiocarbonyl group of the macrodithionolide system generating the radical anion; this initiates a sequence leading to the bridged product which is quenched with methyl iodide to generate the more stable disulfide. The disulfide may then be further transformed chemically to either the *cis*- or *trans*-fused polycycle.

1. $Na^+C_{10}H_8^{-}\bullet$, THF, −78 °C; 2. xs MeI, −78 to 25 °C, 90% (4)

Reduction of Epoxides. The ability to relocate the allylic alcohol moiety within a molecule is another synthetic strategy for which sodium naphthalenide has been found useful. This ability is demonstrated in the transformation of geraniol to linalool (eq 5) by epoxidation of the allylic alcohol, mesylation, and subsequent treatment with sodium naphthalenide to produce the allylic alcohol in good yield.[6] Similarly, this method has been utilized in the synthesis of isocarbacyclin, a therapeutic agent for cardiovascular disease. Treatment of the epoxidized and protected allylic alcohol with sodium naphthalenide provides the desired allylic alcohol with the exocyclic double bond as a mixture of diastereomers (eq 6).[7] Apparently, the *t*-butyldimethylsilyl ether is unaffected under these reaction conditions.

$Na^+C_{10}H_8^{-}\bullet$, 83% (5)

$Na^+C_{10}H_8^{-}\bullet$, 70%; ⟶ Isocarbacyclin (6)

Sodium naphthalenide with ***N,N,N′,N′-Tetramethylethylenediamine*** effects a carbon–carbon bond-forming reaction between carboxylic acids and conjugated alkenes (eq 7) to produce the substituted carboxylic acid in moderate yield.[8,9] This method has been utilized in the preparation of dihydrolavandulol (eq 8) in fair yield.[8]

$Na^+C_{10}H_8^{-}\bullet$, TMEDA, 57% (7)

$Na^+C_{10}H_8^{-}\bullet$, TMEDA, 65% (8)

Treatment of benzimidazoline-2-thione with alkyl halide and sodium naphthalenide in THF affords the 1-alkyl-2-(alkylthio)benzimidazoles in excellent yield. These substrates may be further transformed by additional sodium naphthalenide to provide the 1-alkylbenzimidazoline-2-thiones in high yield (eq 9).[10] Thus, sodium naphthalenide provides an alternative to other known methods which generally proceed in much lower yield and require longer reaction times and more vigorous reaction conditions.

$Na^+C_{10}H_8^{-}\bullet$, RX, THF → (1); $Na^+C_{10}H_8^{-}\bullet$, THF, 88% → (2) (9)

R	(1) (%)	(2) (%)
Me	97	88
Et	100	94
Allyl	100	92
Benzyl	100	93

Protecting Group Removal; Alkene Formation. Sodium naphthalenide also facilitates the removal of protecting groups. It has been utilized in the debenzylation of nucleosides (eq 10).[11] Sodium naphthalenide also effects the reductive cleavage of toluenesulfonates. Both menthyl tosylate and bridged bicyclic tosylates are quantitatively reduced under mild reaction conditions (eqs 11 and 12).

(10)

(11)

(12)

Cyanohydrins with α-methyl thiomethyl ether or α-methyl thiomethyl sulfone substituents are converted regio- and stereospecifically to the corresponding alkenes (eqs 13 and 14).[12] The *cis*-substituted cyanohydrin undergoes elimination to generate the *cis*-alkene while the *trans*-substituted cyanohydrin provides the *trans*-alkene upon treatment with sodium naphthalenide/HMPA. This provides an excellent method for the preparation of either *cis*- or *trans*-alkenes in large ring systems.

(13)

R = H, 85%; Me, 74%
R' = CH_2SMe or CH_2SO_2Me

(14)

R = H, 60%; Me, 83%
R' = CH_2SMe or CH_2SO_2Me

Treatment of methanesulfonates of vicinal diols with sodium naphthalenide in THF or DME results in the rapid and high conversion to the alkene (eqs 15 and 16).[13] Although the reaction is highly regiospecific, the more stable alkene generally predominates in close to the equilibrium ratio. This method has recently been utilized in the synthesis of (±)-20-deethylcatharanthine (eq 17) in which the diol is converted to the dimesylate followed by subsequent treatment with sodium naphthalenide, providing the desired alkaloid analog in good yield.[14]

(15)

(16)

(17)

Similar conversion of the bicyclic vicinal diol to the dimesylate followed by treatment with sodium naphthalenide provides an efficient strategy for the deoxygenation of the vicinal diol to form an immediate precursor to the natural product (+)-(1*S*,5*R*,7S)-exobrevicomin (eq 18).[15] This method provides significantly higher yield than xanthate formation followed by reduction with ***Tri-n-butylstannane***.

(18)

Sodium naphthalenide further affords a useful procedure for the conversion of cyclic sulfates into alkenes, providing an efficient synthesis of deoxygenated vicinal diols (eq 19).[16] This reaction is highly regiospecific but, in examples where (*E*)/(*Z*) isomerization is possible, the thermodynamically preferred alkene predominates. Additionally, carbonyl functionalities and other easily reduced species are incompatible with these reaction conditions.

(19)

R^1, R^2, R^3, R^4 = H, Ph, Ph, H; 86%
R^1, R^2, R^3, R^4 = $(CH_2)_9Me$, H, H, H; 88%

Related Reagents. See Classes R-23, R-25, R-31, and R-32, pages 1–10.

1. (a) Wang, H. C.; Levin, G.; Szwarc, M. *JACS* **1978**, *100*, 3969. (b) Stevenson, G. R.; Valentín, J.; Meverden, C.; Echegoyen, L.; Maldonado, R. *JACS* **1978**, *100*, 353.
2. Pradhan, S. K.; Radhakrishnan, T. V.; Subramanian, R. *JOC* **1976**, *41*, 1943.
3. Srikrishna, A.; Sundarababu, G. *T* **1990**, *46*, 3601.

4. Clive, D. L. J.; Keshava Murthy, K. S.; Zhang, C.; Hayward, W. D.; Daigneault, S. *CC* **1990**, 509.
5. Nicolaou, K. C.; Hwang, C. K.; Duggan, M. E.; Reddy, K. B.; Marron, B. E.; McGarry, D. G. *JACS* **1986**, *108*, 6800.
6. Yasuda, A.; Yamamoto, H.; Nozaki, H. *TL* **1976**, 2621.
7. Bannai, K.; Tanaka, T.; Okamura, N.; Hazato, A.; Sugiura, S.; Manabe, K.; Tomimori, K.; Kato, Y.; Kurozumi, S.; Noyori, R. *T* **1990**, *46*, 6689.
8. Fujita, T.; Watanabe, S.; Suga, K.; Nakayama, H. *S* **1979**, 310.
9. Fujita, T.: Watanabe, S.; Suga, K.; Miura, T.; Sugahara, K.; Kikuchi, H. *J. Chem. Technol. Biotechnol.* **1982**, *32*, 476.
10. Lee, T. R.; Kim, K. *JHC* **1989**, *26*, 747.
11. Philips, K. D.; Horwitz, J. P. *JOC* **1975**, *40*, 1856.
12. Marshall, J. A.; Karas, L. J. *JACS* **1978**, *100*, 3615.
13. Carnahan, J. C., Jr.; Closson, W. D. *TL* **1972**, 3447.
14. Sundberg, R. J.; Gadamasetti, K. G. *T* **1991**, *47*, 5673.
15. Schultz, M.; Waldmann, H.; Vogt, W.; Kunz, H. *TL* **1990**, *31*, 867.
16. Beels, C. M. D.; Coleman, M. J.; Taylor, R. J. K. *SL* **1990**, 479.

Gary A. Molander & Christina R. Harris
University of Colorado, Boulder, CO, USA

Sodium Periodate[1]

$NaIO_4$

[7790-28-5] $INaO_4$ (MW 213.89)

(oxidative cleavage of 1,2-diols;[2] oxidation of sulfides,[3] selenides,[4] phenols,[5] indoles,[6] and tetrahydro-β-carbolines[7])

Alternate Name: sodium metaperiodate.
Physical Data: mp 300 °C (dec); specific gravity 3.865.
Solubility: sol H_2O (14.4 g/100 mL H_2O at 25 °C; 38.9 g/100 mL at 51.5 °C), H_2SO_4, HNO_3, acetic acid; insol organic solvents.
Form Supplied in: colorless to white tetragonal, efflorescent crystals; readily available.
Handling, Storage, and Precautions: irritant; gloves and safety goggles should be worn when handling this oxidant; avoid inhalation of dust and avoid contact of oxidant with combustible matter.

Introduction. Sodium periodate is widely used for the oxidation of a variety of organic substrates and as a cooxidant in other oxidation reactions (see ***Sodium Periodate–Osmium Tetroxide*** and ***Sodium Periodate–Potassium Permanganate***).[8] The $NaIO_4$ oxidation is usually conducted in water; however, for organic substrates that are insoluble in water, an organic cosolvent (e.g. MeOH, 95% EtOH, 1,4-dioxane, acetone, MeCN) is used. Alternatively, the oxidation can be conducted either with phase-transfer catalysis (PTC) using quaternary ammonium[5] or phosphonium[9] salts in a two-phase system, or in an organic solvent if the oxidant is first coated on an inert support.[10]

Oxidative Cleavage of 1,2-Diols. $NaIO_4$ is widely used for the oxidative cleavage[2] of a variety of 1,2-diols to yield aldehydes or ketones (eq 1). In this respect, it complements the ***Lead(IV) Acetate*** method for oxidation. 1,2-Diols have been shown to be chemoselectively cleaved by $NaIO_4$ in the presence of a sulfide group.[11] $NaIO_4$ coated on wet silica gel efficiently oxidizes 1,2-diols to the aldehydes (eq 2).[10a] This method is particularly useful for the preparation of aldehydes which readily form hydrates, and it is also convenient to conduct because isolation of the product involves simple filtration of the reaction mixture and evaporation.

$NaIO_4$, CH_2Cl_2; satd. $NaHCO_3$, <30 °C; 72% → CHO (1)

OH / OH; $NaIO_4$, SiO_2; CH_2Cl_2, H_2O, rt; 97% → $OHC(CH_2)_4CHO$ (2)

Oxidation of Sulfides to Sulfoxides. The selective oxidation of sulfides to sulfoxides is an important transformation because sulfoxides are useful intermediates in synthesis.[12] The reaction is conducted using an equimolar amount of $NaIO_4$ in aqueous methanol at 0 °C (eq 3).[3] Higher reaction temperatures or the use of an excess of $NaIO_4$ result in overoxidation to give sulfones. $NaIO_4$ supported on acidic alumina (eq 4)[10b,c] or silica gel[10d] is effective for the selective oxidation of sulfides, at ambient temperature, to afford good yields of sulfoxides. Phase transfer-catalyzed $NaIO_4$ oxidation of sulfides also results in the selective formation of sulfoxides.[9]

$PhSCH_2CO_2H$ —($NaIO_4$, MeOH; H_2O, 0 °C; 99%)→ $PhS(O)CH_2CO_2H$ (3)

—($NaIO_4/Al_2O_3$; 95% EtOH, rt; 91%)→ (4)

α-Phosphoryl sulfoxides, useful for the preparation of vinylic sulfoxides,[13] are prepared in high yields by the oxidation of α-phosphoryl sulfides using $NaIO_4$.[14] Vinylic sulfoxides can also be prepared in good yields by the oxidation of vinylic sulfides using $NaIO_4$ (eq 5).[15] Poor yields of sulfoxides are obtained in the $NaIO_4$ oxidation of acetylenic sulfides.[15a]

Ph–CH=CH–SMe —($NaIO_4$, MeCN; H_2O, −10 °C; 77%)→ Ph–CH=CH–S(O)Me (5)

2-Substituted 1,3-dithianes are stereoselectively oxidized to the *trans*-1-oxide by $NaIO_4$ at low temperatures.[16] Dimethyl dithioacetals of aldehydes and ketones suffer $NaIO_4$-mediated hydrolysis to give carbonyl compounds.[17] This method could be useful for the deprotection of dimethyl dithioacetals. Oxidation of dithioethers such as 1,4-dithiacycloheptane using $NaIO_4$ at 0 °C furnishes the 1-oxide in modest yield.[18] The use of ***m-Chloroperbenzoic Acid*** for this oxidation leads to an appreciable amount of the 1,4-dioxide. Oxidation of a naphtho-1,5-dithiocin using an excess of $NaIO_4$ at rt results in a high yield of the *cis*-1,5-dioxide.[19] The sulfide unit in thiosulfoxides is selectively oxidized to the *S,S*-dioxide in good yields using an equimolar amount of $NaIO_4$ at 0 °C.[20] Unsymmetrical thiosulfinic *S*-esters are efficiently converted to the thiosulfonic *S*-esters, without concomitant cleavage of the S–S bond, by $NaIO_4$ oxidation.[21] $NaIO_4$ is effective for the selective oxidation of the sulfide moiety in (**1**) to the sulfoxide in the presence of a disulfide linkage (eq 6).[22] Other oxidants such as CrO_3 in acetic acid, H_2O_2, and *m*-CPBA, which are useful for the oxidation of simple sulfides, only cause the decomposition of (**1**).

(1) — $NaIO_4$, MeOH, 0 °C, 71% (6)

Oxidation of Selenides to Selenoxides. Diaryl, dialkyl, and aryl alkyl selenides are oxidized[4] to the corresponding selenoxides in high yields using a slight excess of $NaIO_4$ at 0 °C (eq 7).[4a] The presence of an electron-withdrawing substituent in diaryl selenides inhibits the oxidation of the selenium center. Vinylic selenides can be oxidized[23] with $NaIO_4$ to give high yields of vinylic selenoxides (eq 8).[23a] In contrast, oxidation with ***Hydrogen Peroxide*** results in the cleavage of the double bond to give carboxylic acids. The oxidation of organoselenides possessing β-hydrogens results in the formation of highly unstable organoselenoxides that undergo facile *syn* elimination, often at room temperature, to give alkenes (eq 9).[24,25] Such a process constitutes a useful method for the introduction of a double bond into organic molecules.

$NaIO_4$, MeOH; H_2O, 0 °C; 79% (7)

$NaIO_4$, MeOH; H_2O, rt; 90% (8)

$NaIO_4$, MeOH; H_2O, $NaHCO_3$, rt; 89% (9)

Oxidation of Phenols and Its Derivatives. Dihydroxybenzenes are oxidized to give high yields of the corresponding quinones using $NaIO_4$ supported on silica gel (eq 10)[10a] or under PTC (see also ***Tetra-n-butylammonium Periodate***).[5] Treatment of *p*-hydroxybenzyl alcohol with $NaIO_4$ in aqueous acetic acid results in the formation of *p*-benzoquinone, albeit in low yield (23%).[26] On the other hand, *o*-(hydroxymethyl)phenols possessing at least one bulky group at the C-4 position are efficiently oxidized to give spiroepoxy-2,4-cyclohexadienones (eq 11).[27] In the absence of a bulky group, self-dimerization of the spiroepoxycyclohexadienone via Diels–Alder reaction occurs. In the case of *o*-(hydroxymethyl)phenols that are substituted with one or two aryl groups at the benzylic carbon, a novel oxidative rearrangement occurs to yield benzylidene protected catechols in modest yields (eq 12).[28] However, this oxidative rearrangement is only successful if substituents are present in the C-2 and C-4 positions of the phenol unit, because oxidation of α-(2-hydroxyphenyl)benzyl alcohol only results in the formation of a dimer.

$NaIO_4$, SiO_2; CH_2Cl_2, H_2O, rt; 100% (10)

$NaIO_4$; H_2O, rt; 100% (11)

$NaIO_4$; H_2O, rt; 60% (12)

Oxidation of Indoles and Tetrahydro-β-Carbolines. The indolic double bond in 2,3-dialkyl- and 3-alkylindoles is readily oxidized by 2 mole equiv of $NaIO_4$ at room temperature to give *o*-amidoacetophenone derivatives in good yields (eq 13).[6] However, the oxidation of 2,3-diphenylindole under the same conditions results in a lower yield of the oxidative cleavage product.[29] Interestingly, the oxidation of 2-alkylindoles results in the formation of a mixture of products comprised of indoxyl derivatives.[29] Tetrahydrocarbazoles are also efficiently oxidized by $NaIO_4$ to afford benzocyclononene-2,7-dione derivatives.[6] Tetrahydro-β-carbolines have also been subjected to $NaIO_4$ oxidation.[7] Thus, in the oxidation of the tetrahydro-β-carboline-3-carboxylates, the type of product that is formed depends upon the degree of substitution at C-1 of the starting material (eqs 14 and 15).[7a]

$NaIO_4$, MeOH; H_2O, rt; 82% (13)

$NaIO_4$, MeOH; H_2O, rt; 55% (14)

$NaIO_4$, MeOH; H_2O, rt; 62% (15)

Other Applications. 1,3-Cyclohexanedione and its 3-substituted derivatives are oxidized with $NaIO_4$, with concomitant loss of the C-2 carbon unit, to give glutaric acid in good yields.[30] 1,3-Cyclopentanediones react more slowly under the same conditions, and aromatic diketones such as 1,3-indandione give poor yields of the dicarboxylic acid product. α-Hydroxy carboxylic acids undergo oxidative decarboxylation to give aldehydes (eq 16)[31] upon treatment with aqueous $NaIO_4$; however, long reaction times are required. The use of PTC[9] or Bu_4NIO_4 allows for shorter reaction times without adversely affecting the yield. The oxidation of hydrazine with $NaIO_4$ in the presence of trace amounts of aqueous ***Copper(II) Sulfate*** and ***Acetic Acid*** results in the formation of ***Diimide***. The in situ generation of diimide by this method has been successfully applied to a one-pot procedure for the reduction of alkenes (eq 17).[32]

$NaIO_4$, $MeCO_2H$; H_2O, Me_2CO, 45 °C; 90% (16)

$NaIO_4$, H_2NNH_2, DMSO; H_2O, $CuSO_4$, $MeCO_2H$, rt; 87% (17)

A secondary amide is obtained by selective oxidation of a tertiary carbon center in adamantane with $NaIO_4$ in the presence of iron(III) perchlorate in acetonitrile (eq 18).[33] Dimethylhydrazones undergo periodate induced hydrolysis, at pH 7, to give carbonyl compounds in high yields (eq 19).[34] However, these conditions are unsuitable for the hydrolysis of dimethylhydrazones derived from aromatic or α,β-unsaturated aldehydes because mixtures of aldehydes and nitriles are formed.

$NaIO_4$, $Fe(ClO_4)_3$; MeCN, Ac_2O, rt; 96% (18)

$NaIO_4$, MeOH; H_2O, pH 7, rt; 100% (19)

Acylphosphoranes are oxidized to α,β-dicarbonyl compounds in fair yields using aqueous $NaIO_4$ (eq 20).[35] This method complements other methods such as the ***Potassium Permanganate***[36a] or ***Ruthenium(VIII) Oxide*** oxidation[36b] of alkynes. $NaIO_4$ is also used for the oxidation of hydroxamic acids and *N*-hydroxycarbamic esters at pH 6 to generate highly reactive nitroso compounds.[37] The oxidations are usually conducted in the presence of conjugated dienes so that the nitroso intermediates are trapped as their Diels–Alder cycloadducts (eq 21).

$NaIO_4$; H_2O, reflux; 53% (20)

$NaIO_4$, ethyl acetate; H_2O, pH 6, 0 °C; 59% (21)

Related Reagents. See Classes O-5, O-8, and O-18, pages 1–10.

1. (a) Shing, T. K. M. *COS* **1991**, *7*, 703. (b) Sklarz, B. *QR* **1967**, *21*, 3. (c) House, H. O. *Modern Synthetic Reactions*, 2nd ed.; Benjamin/Cummings: Menlo Park, CA, 1972.
2. (a) Kovar, J.; Baer, H. H. *CJC* **1971**, *49*, 3238. (b) Torii, S.; Uneyama, K.; Ueda, K. *JOC* **1984**, *49*, 1830. (c) Schmid, C. R.; Bryant, J. D.; Dowlatzedah, M.; Phillips, J. L.; Prather, D. E.; Renee, D. S.; Sear, N. L.; Vianco, C. S. *JOC* **1991**, *56*, 4056. (d) Jackson, D. Y. *SC* **1988**, *18*, 337.
3. (a) Leonard, N. J.; Johnson, C. R. *JOC* **1962**, *27*, 282. (b) Johnson, C. R.; Keiser, J. E. *OS* **1966**, *46*, 78. (c) Lee, J. B.; Yergatian, S. Y.; Crowther, B. C.; Downie, I. M. *OPP* **1990**, *22*, 544.
4. (a) Cinquini, M.; Colonna, S.; Giovini, R. *CI(L)* **1969**, 1737. (b) Entwistle, I. D.; Johnstone, R. A. W.; Varley, J. H. *CC* **1976**, 61. (c) Masuyama, Y.; Ueno, Y.; Okawara, M. *CL* **1977**, 835.
5. Takata, T.; Tajima, R.; Ando, W. *JOC* **1983**, *48*, 4764.
6. (a) Dolby, L. J.; Booth, D. L. *JACS* **1966**, *88*, 1049. (b) Rivett, D. E.; Wilshire, J. F. K. *AJC* **1971**, *24*, 2717.
7. (a) Gatta, F.; Misiti, D. *JHC* **1989**, *26*, 537. (b) Akimoto, H.; Okamura, K.; Yui, M.; Shiori, T.; Kuramoto, M.; Kikugawa, Y.; Yamada, S.-I. *CPB* **1974**, *22*, 2614. (c) Hutchinson, C. R.; O'Loughlin, G. J.; Brown, R. T.; Fraser, S. B. *CC* **1974**, 928.
8. Carlsen, P. H. J.; Katsuki, T.; Martin, V. S.; Sharpless, K. B. *JOC* **1981**, *46*, 3936.
9. Ferraboschi, P.; Azadani, M. N.; Santaniello, E.; Trave, S. *SC* **1986**, *16*, 43.
10. (a) Daumas, M.; Vo-Quang, Y.; Vo-Quang, L.; Le Goffic, F. *S* **1989**, 64. (b) Liu, K.-T.; Tong, Y.-C. *JOC* **1978**, *43*, 2717. (c) Liu, K.-T.; Tong, Y.-C. *JCR(S)* **1979**, 276. (d) Gupta, D. N.; Hodge, P.; Davies, J. E. *JCS(P1)* **1981**, 2970.
11. (a) Fleet, G. W. J.; Shing, T. K. M. *CC* **1984**, 835. (b) Wolfrom, M. L.; Yosizawa, Z. *JACS* **1959**, *81*, 3477.
12. (a) Trost, B. M.; Salzmann, T. N. *JACS* **1973**, *95*, 6840. (b) Trost, B. M.; Salzmann, T. N. *JOC* **1975**, *40*, 148.
13. Mikolajczyk, M.; Grzejszczak, S.; Zatorski, A. *JOC* **1975**, *40*, 1979.
14. Mikolajczyk, M.; Zatorski, A. *S* **1973**, 669.
15. (a) Russel, G. A.; Ochrymowycz, L. A. *JOC* **1970**, *35*, 2106. (b) Evans, D. A.; Bryan, C. A.; Sims, C. L. *JACS* **1972**, *94*, 2891.
16. (a) Carey, F. A.; Dailey, O. D., Jr.; Hernandez, O.; Tucker, J. R. *JOC* **1976**, *41*, 3975. (b) Carey, F. A.; Dailey, O. D., Jr.; Fromuth, T. E. *PS* **1981**, *10*, 163.
17. Nieuwenhuyse, H.; Louw, R. *TL* **1971**, 4141.
18. Roush, P. B.; Musker, W. K. *JOC* **1978**, *43*, 4295.

19. Glass, R. S.; Broeker, J. L. *T* **1991**, *47*, 5077.
20. Ogura, K.; Suzuki, M.; Tsuchihashi, G.-I. *BCJ* **1980**, *53*, 1414.
21. (a) Takata, T.; Kim, Y. H.; Oae, S. *BCJ* **1981**, *54*, 1443. (b) Kim, Y. H.; Takata, T.; Oae, S. *TL* **1978**, 2305.
22. Hiskey, R. G.; Harpold, M. A. *JOC* **1967**, *32*, 3191.
23. (a) Sevrin, M.; Dumont, W.; Krief, A. *TL* **1977**, 3835. (b) Harirchian, B.; Magnus, P. *CC* **1977**, 522.
24. Reich, H. J.; Reich, I. L.; Renga, J. M. *JACS* **1973**, *95*, 5813.
25. Clive, D. L. J. *CC* **1973**, 695.
26. Adler, E.; Holmberg, K.; Ryrfors, L.-O. *ACS* **1974**, *B28*, 883.
27. Becker, H.-D.; Bremholt, T.; Adler, E. *TL* **1972**, 4205.
28. Becker, H.-D.; Bremholt, T. *TL* **1973**, 197.
29. Dolby, L. J.; Rodia, R. M. *JOC* **1970**, *35*, 1493.
30. Wolfrom, M. L.; Bobbitt, J. M. *JACS* **1956**, *78*, 2489.
31. Yanuka, Y.; Katz, R.; Sarel, S. *TL* **1968**, 1725.
32. Hoffman, J. M., Jr.; Schlessinger, R. H. *CC* **1971**, 1245.
33. Kotani, E.; Kobayashi, S.; Ishii, Y.; Tobinaga, S. *CPB* **1985**, *33*, 4680.
34. Corey, E. J.; Enders, D. *TL* **1976**, 11.
35. Bestmann, H.-J.; Armsen, R.; Wagner, H. *CB* **1969**, *102*, 2259.
36. (a) See ***Potassium Permanganate***. (b) Zibuck, R.; Seebach, D. *HCA* **1988**, *71*, 237.
37. (a) Kirby, G. W.; McLean, D. *JCS(P1)* **1985**, 1443. (b) Sklarz, B.; Al-Sayyab, A. F. *JCS* **1964**, 1318. (c) See ***Tetraethylammonium Periodate***.

Andrew G. Wee & Jason Slobodian
University of Regina, Saskatchewan, Canada

Sodium Periodate–Osmium Tetroxide[1,2]

$NaIO_4$–OsO_4

($NaIO_4$)		
[7790-28-5]	$INaO_4$	(MW 213.89)
(OsO_4)		
[20816-12-0]	O_4Os	(MW 254.20)

(oxidizing agent for oxidative cleavage of carbon–carbon double bonds[3] to give aldehydes and ketones)

Physical Data: $NaIO_4$: mp 300 °C (dec); specific gravity 3.865. OsO_4: bp 130 °C; mp 40 °C; specific gravity 4.900 at 22 °C.
Solubility: $NaIO_4$: sol H_2O (14.4 g/100 mL H_2O at 25 °C; 38.9 g/100 mL at 51.5 °C), H_2SO_4, HNO_3, acetic acid; insol organic solvents. OsO_4: sol alcohol, Et_2O, benzene, CCl_4, NH_3, $POCl_3$, H_2O (25%).
Form Supplied in: $NaIO_4$: colorless to white tetragonal, efflorescent crystals. OsO_4: colorless to pale yellow crystals contained in ampules. Both reagents are readily available.
Handling, Storage, and Precautions: $NaIO_4$: irritant; gloves and safety goggles should be worn when handling this oxidant; avoid inhalation of dust and avoid contact of oxidant with combustible matter. OsO_4: acrid, chlorine-like odor; irritant; highly corrosive; causes burns to skin, eyes, and respiratory tract; gloves and safety goggles must be worn and this reagent must be handled in a well ventilated fume hood.

Introduction. In 1956, Lemieux, Johnson, and co-workers first described[3] the use of sodium periodate–osmium tetroxide for the direct oxidation of alkenic bonds to give carbonyl compounds. This reaction combines the hydroxylating properties of ***Osmium Tetroxide*** with the glycol cleavage capabilities of ***Sodium Periodate***. Additionally, sodium periodate serves to regenerate osmium tetroxide during the reaction, thereby permitting the use of only catalytic amounts of the expensive and toxic osmium tetroxide. This oxidation complements the classical method of ozonization of double bonds followed by reductive cleavage. The reaction is usually carried out in a mixed aqueous solvent system such as aqueous dioxane, aqueous THF, aqueous acetone, or aqueous DME. Water is required in this reaction for the hydrolysis of the intermediate osmate ester. In reactions where the carbonyl products are prone to self-condensation, the use of the ether–water system is found to be highly satisfactory.

Oxidative Cleavage of Double Bonds to Give Carbonyl Compounds. The oxidation of acyclic alkenes and stilbene in aqueous 1,4-dioxane using sodium periodate in the presence of catalytic amounts of osmium tetroxide ($NaIO_4$–cat. OsO_4) yields the corresponding aldehydes in good yields (eqs 1 and 2).[3,4] In contrast, the oxidation of cyclohexene under the same conditions gives only a low yield of adipaldehyde. The low yield is attributed to the tendency of adipaldehyde to undergo self-condensation. The use of an ether–water system for the oxidation circumvents this problem. Under the modified conditions, the oxidation of cyclohexene gives a good yield of adipaldehyde (eq 3).[3] Similarly, cyclopentene is oxidized to afford glutaraldehyde in good yield. 1-Methylcyclohexene, however, is oxidized very slowly under the optimum conditions found for cyclohexene; only a small amount of an uncharacterized carbonyl compound is obtained. Nevertheless, this outcome suggests that $NaIO_4$–cat. OsO_4 can be employed for the selective cleavage of unhindered double bonds in the presence of hindered ones. This type of selective oxidation is effectively applied to the cleavage of a vinyl group in the presence of a more hindered alkenic bond in the etorphine ring system (eq 4).[5]

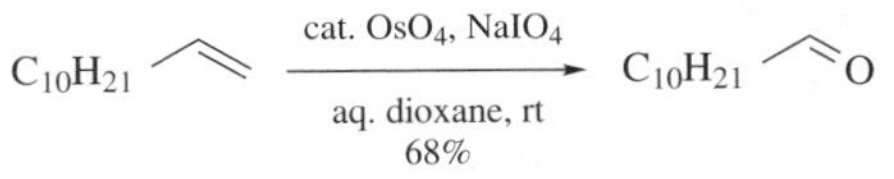

(1)

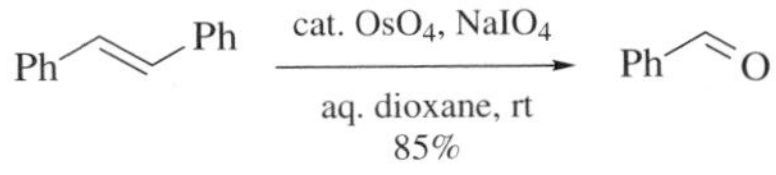

(2)

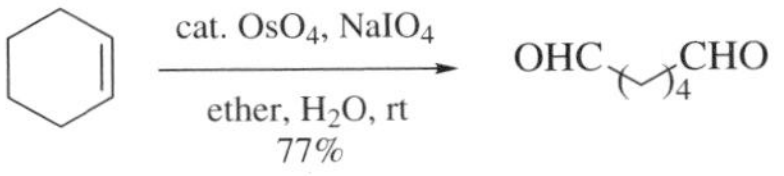

(3)

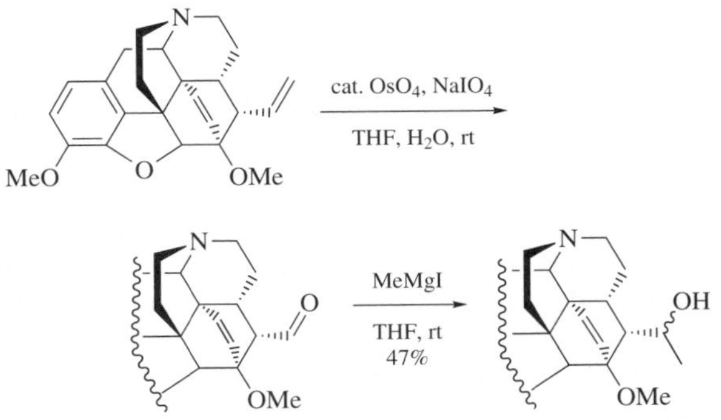

(4)

Alkenic compounds are rapidly and efficiently cleaved to carbonyl compounds using $NaIO_4$ in the presence of polymer-supported OsO_4 (P-OsO_4) (eqs 5 and 6).[6] Cyclohexene is oxidized to give a fairly good yield of adipaldehyde (eq 7). This provides a useful alternative to oxidation in the ether–water system[3] alluded to earlier. α,β-Unsaturated carbonyl compounds are oxidized at the double bond to give aldehydes in good yields (eq 8). A notable feature of this method[6] is that a particular oxidation can be easily repeated, at least 10 times, by simply adding new portions of substrate and sodium periodate.

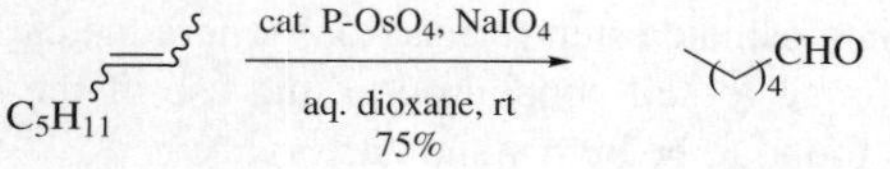

cat. P-OsO_4, $NaIO_4$; aq. dioxane, rt; 75% (5)

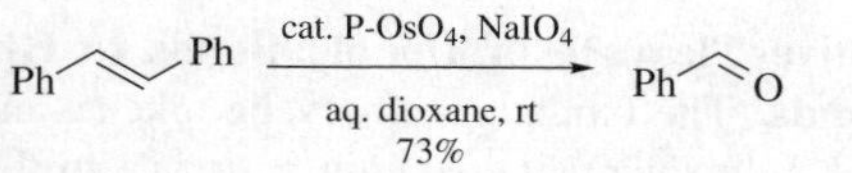

cat. P-OsO_4, $NaIO_4$; aq. dioxane, rt; 73% (6)

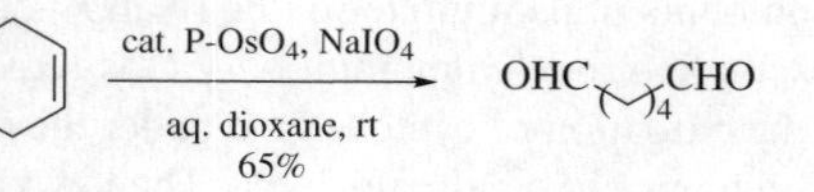

cat. P-OsO_4, $NaIO_4$; aq. dioxane, rt; 65% (7)

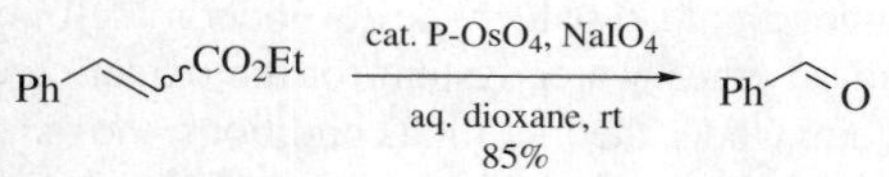

cat. P-OsO_4, $NaIO_4$; aq. dioxane, rt; 85% (8)

The double bond in alkyl, cycloalkylmethyl, and arylmethyl allyl ethers can be oxidatively cleaved using $NaIO_4$–cat. OsO_4. The oxidation of (**1**) in aqueous dioxane containing trace amounts of acetic acid affords a good yield of the benzyloxy aldehyde (eq 9).[7] On the other hand, oxidation of similar systems[8] in aqueous dioxane results in only low yields of aldehydes. Higher yields are obtained when the oxidation is carried out using the ether–water system (eq 10).

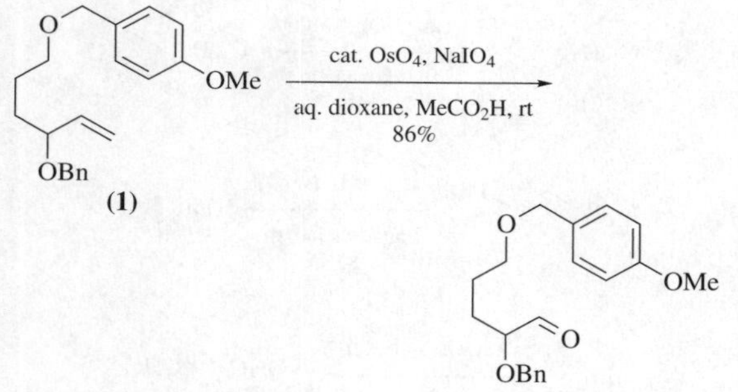

cat. OsO_4, $NaIO_4$; aq. dioxane, $MeCO_2H$, rt; 86% (9)

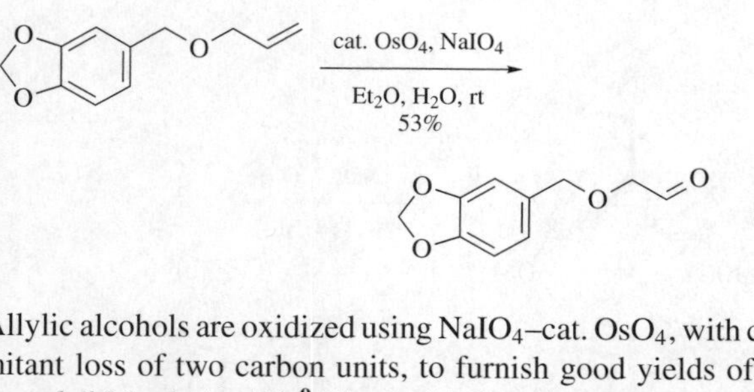

cat. OsO_4, $NaIO_4$; Et_2O, H_2O, rt; 53% (10)

Allylic alcohols are oxidized using $NaIO_4$–cat. OsO_4, with concomitant loss of two carbon units, to furnish good yields of the expected aldehyde (eq 11).[9] The double bond in hydroxy alkenes, such as in (*R*)-(+)-citronellol, is selectively oxidized to give the corresponding hydroxy aldehydes which, upon subsequent Wittig alkenation, give the α,β-unsaturated esters (eq 12).[10]

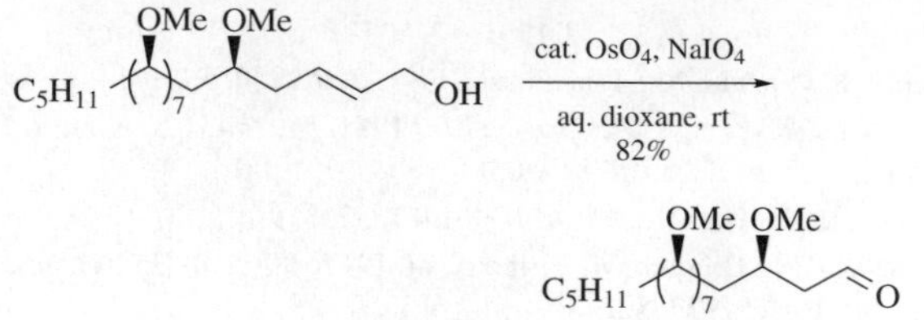

cat. OsO_4, $NaIO_4$; aq. dioxane, rt; 82% (11)

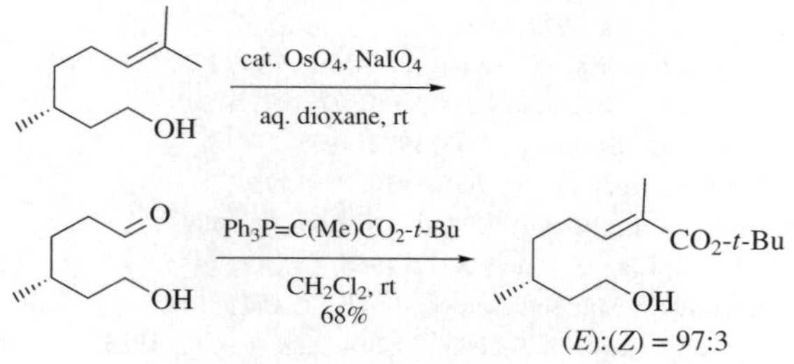

cat. OsO_4, $NaIO_4$; aq. dioxane, rt

$Ph_3P=C(Me)CO_2$-*t*-Bu; CH_2Cl_2, rt; 68%; (*E*):(*Z*) = 97:3 (12)

The oxidation of γ,δ-unsaturated carbonyl substrates using $NaIO_4$–cat. OsO_4 is a practical method for the preparation of synthetically useful 1,4-dicarbonyl compounds.[11] For example, oxidation of the double bond of the allyl moiety in (**2**) yields the aldehyde in good yield (eq 13).[11a] Subsequent base-catalyzed aldol condensation of the keto aldehyde furnishes the spirocyclopentenone derivative, albeit in low yield. Such a protocol has been employed for the preparation of key cyclopentenone intermediates used in natural products synthesis, e.g. in the syntheses of phytuberin (eq 14),[12] aphidicolin (eq 15),[13] and bakkenolide (eq 16).[14]

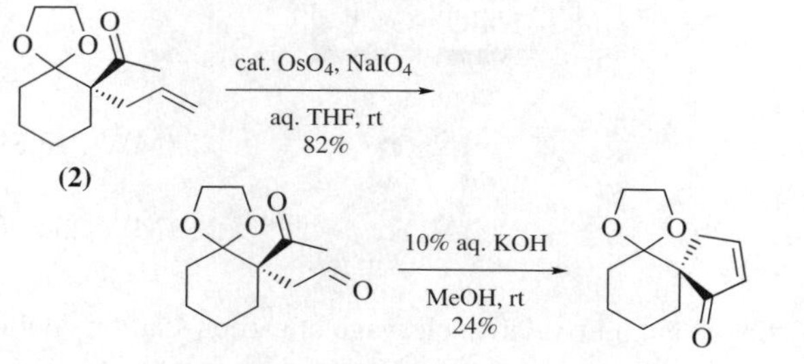

cat. OsO_4, $NaIO_4$; aq. THF, rt; 82%; 10% aq. KOH; MeOH, rt; 24% (13)

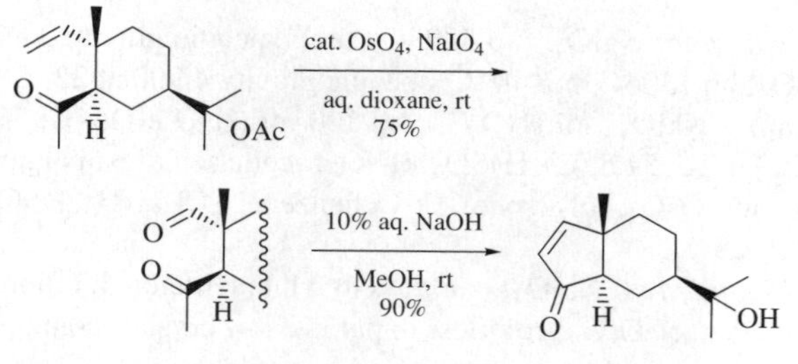

cat. OsO_4, $NaIO_4$; aq. dioxane, rt; 75%; 10% aq. NaOH; MeOH, rt; 90% (14)

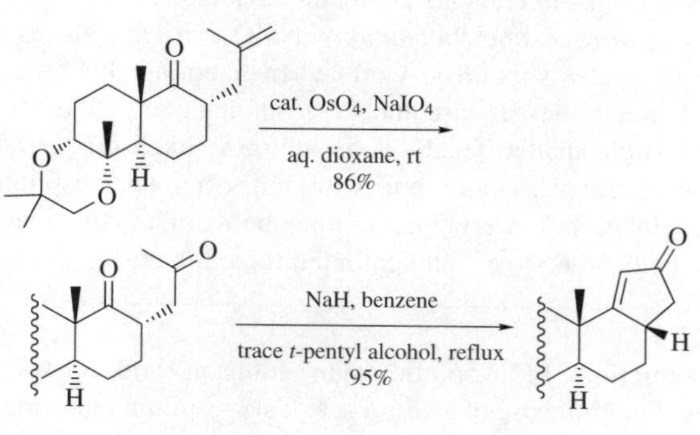

cat. OsO_4, $NaIO_4$; aq. dioxane, rt; 86%; NaH, benzene; trace *t*-pentyl alcohol, reflux; 95% (15)

cat. OsO_4, $NaIO_4$ / t-BuOH, H_2O, rt / 93%

t-BuOK / t-BuOH, Et_2O, rt / 65% (16)

The 1-methylvinyl group can be used as a latent acetyl unit in synthesis. This is exemplified in the synthesis of a key intermediate required for the construction of the polyether antibiotic carbamonensin (eq 17).[15] The acetyl group is subsequently utilized for the stereospecific installation of the acetate moiety via Baeyer–Villiger oxidation. Double bonds present in polyoxygenated systems have also been efficiently oxidized to the aldehydes in high yields (eq 18).[9]

cat. OsO_4, $NaIO_4$ / aq. THF, rt

CF_3CO_3H / Na_2HPO_4, CH_2Cl_2 / 92% (17)

cat. OsO_4, $NaIO_4$ / aq. dioxane, rt / 82% (18)

The reaction conditions used for the $NaIO_4$–cat. OsO_4 oxidation are mild and this permits the oxidation of double bonds in the presence of acid or base labile groups such as the ester (eq 19),[11b] carbamate, acetal (eq 20),[16] silyl ether (eq 21),[17] and carbonate (eq 22)[18] functions.

cat. OsO_4, $NaIO_4$ / aq. THF, rt / 85% (19)

cat. OsO_4, $NaIO_4$ / aq. dioxane, rt

$NaBH_4$ / MeOH, 0 °C / 35% (20)

cat. OsO_4, $NaIO_4$ / dry MeOH, 45 °C / 57% (21)

cat. OsO_4, $NaIO_4$ / Et_2O, H_2O, rt

MeOH / p-TsOH / 65% (22)

Cyclobutanones have been prepared from alkylidene- or arylidenecyclobutanes via the $NaIO_4$–cat. OsO_4 oxidation.[19] The oxidation of benzylidenecyclobutane with $NaIO_4$–cat. OsO_4 gives cyclobutanone, albeit in low yields (17%).[19a] This outcome, however, is better than the ozonolysis method, which does not yield any cyclobutanone. The $NaIO_4$–cat. OsO_4 oxidation of 3-methylenecyclobutanecarboxylic acid is more successful and a good yield of 3-oxocyclobutanecarboxylic acid is obtained (eq 23).[19b] This protocol for the preparation of cyclobutanone from methylenecyclobutane has been used in natural product synthesis (eq 24).[20]

cat. OsO_4, $NaIO_4$ / aq. dioxane, rt / 70% (23)

cat. OsO_4, $NaIO_4$ / aq. dioxane, py, rt / 87% (24)

Alkenic bonds in heterocyclic molecules have also been oxidatively cleaved using $NaIO_4$–cat. OsO_4. Oxidation of the methylenecyclobutane moiety in the 2-quinolone derivative affords a good yield of the cyclobutanone derivative (eq 25).[21] 1-Substituted 2-nitroimidazole-5-carbaldehydes are prepared in useful yields by the action of $NaIO_4$–cat. OsO_4 on 5-vinylimidazole precursors (eq 26).[22] This method is better than the two-step procedure, namely, ***Potassium Permanganate*** hydroxylation followed by $NaIO_4$ oxidation of the 1,2-diol, which gives a lower yield of the product. Previous attempts to prepare 2-nitroimidazole-5-carbaldehydes such as by the ozonolysis of 5-vinylimidazoles and by the direct oxidation of 5-methyl-2-nitroimidazoles with CAN or SeO_2 were unsuccessful. $NaIO_4$–cat. OsO_4 is effective for the oxidation of double bonds in 2-azetidinone derivatives to give a good yield of the corresponding aldehydes (eq 27).[23]

cat. OsO_4, $NaIO_4$, aq. dioxane; trace conc. HCl, 0 °C to rt; 81% (25)

cat. OsO_4, $NaIO_4$; aq. DME, rt; 63% (26)

TBDMSO ... Ph; trans:cis = 1.05:1; cat. OsO_4, $NaIO_4$; aq. THF, rt; trans 96% (27)

Other Applications. $NaIO_4$–cat. OsO_4 is a mild oxidizing agent and does not oxidize aryl groups. However, pyrene has been oxidized using $NaIO_4$–cat. OsO_4 to afford two major products arising from oxidation at the 4,5-double bond (eq 28).[24] The use of OsO_4–H_2O_2 gives only a low yield of the 4,5-quinone and most of the starting material is recovered. RuO_2–$NaIO_4$ oxidation is less specific and a mixture of products, resulting from attack at positions other than the 4,5-double bond, is obtained.

cat. OsO_4, $NaIO_4$; Me_2CO, H_2O, rt; 23% + 24% (28)

Related Reagents. See Class O-17, pages 1–10.

1. Lee, D. G.; Chen, T. *COS* **1991**, *7*, 541.
2. (a) Schroder, M. *CRV* **1980**, *80*, 187. (b) Singh, H. S. *Organic Synthesis By Oxidation With Metal Compounds*; Mijs, W. J.; de Jonge, C. R. H. I., Eds.; Plenum: New York, 1986; p 633.
3. Pappo, R.; Allen, D. S., Jr.; Lemieux, R. U.; Johnson, W. S. *JOC* **1956**, *21*, 478.
4. Cantor, S. E.; Tarbell, D. S. *JACS* **1964**, *86*, 2902.
5. Maurer, P. J.; Rapoport, H. *JMC* **1987**, *30*, 2016.
6. Cainelli, G.; Contento, M.; Manescalchi, F.; Plessi, L. *S* **1989**, 47.
7. Kozikowski, A. P.; Jung, S.-H.; Springer, J. P. *CC* **1988**, 167.
8. Arndt, H. C.; Carroll, S. A. *S* **1979**, 202.
9. Mori, Y.; Kohchi, Y.; Suzuki, M.; Carmeli, S.; Moore, R. E.; Patterson, G. M. L. *JOC* **1991**, *56*, 631.
10. Shing, T. K. M. *CC* **1986**, 49.
11. (a) Hayashi, T.; Kanehira, K.; Hagihara, T.; Kumada, M. *JOC* **1988**, *53*, 113. (b) Thaisrivongs, S.; Pals, D. T.; Turner, S. R.; Kroll, L. T. *JMC* **1988**, *31*, 1369.
12. Kido, F.; Kitahara, H.; Yoshikoshi, A. *JOC* **1986**, *51*, 1478.
13. McMurry, J. E.; Andrus, A.; Ksander, G. M.; Musser, J. H.; Johnson, M. A. *JACS* **1979**, *101*, 1330.
14. Evans, D. A.; Sims, C. L. *TL* **1973**, 4691.
15. Ireland, R. E.; Maienfisch, P. *JOC* **1988**, *53*, 640.
16. Jurczak, J.; Pikul, S. *T* **1988**, *44*, 4569.
17. Gillard, F.; Heissler, D.; Riehl, J.-J. *JCS(P1)* **1988**, 2291.
18. Shishido, K.; Hiroya, K.; Fukumoto, K.; Kametani, T. *CC* **1987**, 1360.
19. (a) Graham, S. H.; Williams, A. J. S. *JCS(C)* **1966**, 655. (b) Caserio, F. F., Jr.; Roberts, J. D. *JACS* **1958**, *80*, 5837.
20. Iwata, C.; Takemoto, Y.; Doi, M.; Imanishi, T. *JOC* **1988**, *53*, 1623.
21. Chiba, T.; Kato, T.; Yoshida, A.; Moroi, R.; Shimomura, N.; Momose, Y.; Naito, T.; Kaneko, C. *CPB* **1984**, *32*, 4707.
22. Cavalleri, B.; Ballotta, R.; Lancini, G. C. *JHC* **1972**, *9*, 979.
23. Georg, G. I.; Kant, J.; Gill, H. S. *JACS* **1987**, *109*, 1129.
24. Oberender, F. G.; Dixon, J. A. *JOC* **1959**, *24*, 1226.

Andrew G. Wee & Baosheng Liu
University of Regina, Saskatchewan, Canada

Sodium–Potassium Alloy

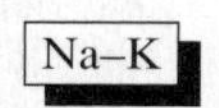

[11135-81-2] KNa (MW 62.09)

(preparation of organopotassium compounds;[1] in combination with TMSCl, greatly enhances acyloin condensations;[2] dehalogenation of dihalides;[3] preparation of γ, δ-unsaturated alcohols via fragmentation of γ-halo ketones;[4] in combination with crown ethers is a useful reducing agent for alkynes[5,6] and alkyl fluorides[7,8])

Physical Data: mp −11 °C.
Form Supplied in: available in two alloy forms: 56% K, 44% Na; 78% K, 22% Na.
Preparative Method: can be prepared by mixing 1 part ***Sodium*** and 5 parts ***Potassium*** (by weight) in refluxing xylene until melted. The melt is then carefully mixed with a glass stirring rod, keeping the alloy in one large globule. Upon cooling to room temperature, the alloy will remain liquid and can easily be transferred via a glass pipet.[1]
Handling, Storage, and Precautions: flammable solid; handle under inert atmosphere.

Organopotassium Compounds. Potassium derivatives can be generated from a number of organic substrates via the action of Na–K alloy (eqs 1–3).[1] The generated organopotassium adducts are soluble and stable in anhydrous diethyl ether. Similarly, the treatment of either organosodium or organolithium compounds with Na–K alloy results in the more reactive organopotassium species.

Na–K, Et_2O (1)

Na–K, Et_2O (2)

Na–K, Et_2O (3)

Acyloin Condensation. The combination of ***Chlorotrimethylsilane*** (TMSCl) with Na–K alloy greatly facilitates the acyloin condensation.[2] This action results in smooth conversion of diesters to disiloxenes (eqs 4 and 5). Sodium metal can also be used, with similar results, to facilitate the acyloin condensation.

Na–K, TMSCl, toluene, 80% (4)

Na–K, TMSCl, toluene, 76% (5)

Dehalogenations. Sodium–potassium alloy is also useful in the dehalogenation of dihalides. Sodium–potassium alloy in ethanol reacts with 1,4-dibromobicyclo[2.2.2]octane to give the Grob fragmentation product, 1,4-dimethylenecyclohexane (eq 6). However, the use of an inert solvent such as cyclohexane affords bicyclo[2.2.2]octane in modest yield (eq 7).[3]

Na–K, EtOH (6)

Na–K, cyclohexane, 65% (7)

Fragmentation of γ-Halo Ketones. The reaction of γ-halo ketones with Na–K alloy in anhydrous ether gives rise to products where 1,4-reductive elimination of halide proceeds with cleavage of the 2,3 carbon–carbon bond.[4] Under these conditions, 1-chloroadamantan-4-one is converted to 7-methylbicyclo[3.3.1]nonan-2-ol (eq 8).

Na–K, Et_2O, 86% (8)

Reduction of Multiple Bonds. The dissolving metal reduction of alkynes typically involves an alkali metal in liquid ammonia.[5] However, it has been demonstrated that dissolving metal reductions utilizing *t*-butyl alcohol and Na–K alloy in THF with catalytic amounts of ***18-Crown-6*** proceed smoothly.[6] The solubility of alkali metals in THF can be greatly increased with the aid of an appropriate crown ether or cryptate. Thus 4-octyne can be reduced to the corresponding octene (3:1 mixture of *trans*:*cis*) (eq 9).

t-BuOH, Na–K, 18-crown-6, THF, 83% (9)

trans:*cis* = 3:1

Reduction of Alkyl Fluorides. The reductive cleavage of unactivated C–F bonds is a difficult transformation in organic synthesis. The carbon–fluorine bond is known to strongly resist the usual reductive conditions.[7] However, the use of sodium–potassium alloy and ***Dicyclohexano-18-crown-6*** in diglyme is effective in the reductive cleavage of unactivated carbon–fluorine bonds.[8] Thus 3β-fluorocholest-5-ene is successfully reduced to cholest-5-ene in high yield (eq 10).

Na–K, crown ether, diglyme, 88% (10)

Related Reagents. See Classes R-4, R-25, R-27 and R-31, pages 1–10. Sodium Phenanthrenide.

1. Gilman, H.; Young, R. V. *JOC* **1936**, *1*, 315.
2. (a) Schrapler, V.; Ruhlman, K. *CB* **1964**, *97*, 1383. (b) Bloomfield, J. J. *TL* **1968**, *5*, 587. (c) Bloomfield, J. J. *TL* **1968**, *5*, 591.
3. Wiberg, K. B.; Burgmaier, J. *JACS* **1972**, *94*, 7396.
4. Hamm, P. G.; Taylor, G. F.; Young, R. N. *S* **1975**, 428.
5. House, H. O. *Modern Synthetic Reactions*; Benjamin: Menlo Park, CA, 1972.

6. Mathre, D. J.; Guida, W. C. *TL* **1980**, *21*, 4773.
7. Pinder, A. R. *S* **1980**, 425.
8. Ohsawa, T.; Takagaki, T.; Haneda, A.; Oishi, T. *TL* **1981**, *22*, 2583.

Nick Nikolaides
3M Pharmaceuticals, St. Paul, MN, USA

Sodium Thiosulfate[1]

$Na_2S_2O_3$

($Na_2S_2O_3$)		
[7772-98-7]	$Na_2O_3S_2$	(MW 158.12)
($Na_2S_2O_3 \cdot 5H_2O$)		
[10102-17-7]	$H_{10}Na_2O_8S_2$	(MW 248.22)

(general reducing agent; reagent for nucleophilieidc introduction of sulfur[2])

Physical Data: mp (pentahydrate) 40–45 °C; *d* 1.73 g cm^{-3}.
Solubility: insol alcohol; v sol water.
Form Supplied in: both the pentahydrate and anhydrous form are widely available as white solids.
Handling, Storage, and Precautions: reported to be an irritant. The anhydrous form is hygroscopic.

Functional Group Reductions. $Na_2S_2O_3$ is a sufficiently powerful reducing agent to effect a number of useful synthetic transformations. Vicinal dibromides are efficiently dehalogenated to the corresponding alkenes in warm DMSO.[3] Anilines are obtained by reduction of aryl nitroso compounds[4] or aryl azides,[5] and nitrosamides can be reduced to amides.[6] Peroxides are reduced to the corresponding alcohols[7] when exposed to aqueous $Na_2S_2O_3$, and reductive workup of an ozonolysis with NaI/$Na_2S_2O_3$ has been employed in the isolation of a ketone.[8] Saturation of activated alkenes by $Na_2S_2O_3$ has been reported, including the reduction of a quinone[9] and of 1,2-dibenzoylethylene.[10]

Introduction of Sulfur. Nucleophilic attack by thiosulfate on alkyl halides results in the formation of *S*-alkyl thiosulfates, commonly known as Bunte salts. Extensive reviews on the preparation and classical reactions of Bunte salts have appeared.[2] In addition to alkyl halides, thiosulfate has been reported to add to epoxides,[2b] oxetanes,[11] activated alkenes,[2b] and even acylaziridines.[12] Probably the most well-known classical application of Bunte salts is their conversion to thiols, which can be accomplished via acidic hydrolysis[13] or reduction.[11] Extension of this methodology to the synthesis of aryl thiols is possible by the addition of $Na_2S_2O_3$ to quinones (eq 1).[14] When the hydrolysis is performed under oxidative conditions, sulfonic acids or sulfonyl chlorides are obtained.[2,15]

1. $Na_2S_2O_3$; 2. Zn; 100% (1)

Much of the useful chemistry of Bunte salts results from the leaving group potential of the sulfite moiety. Nucleophilic displacement by thiolates results in the synthesis of unsymmetrical disulfides (eq 2).[16] In a mechanistically similar manner, symmetrical trisulfides are available by the reaction of Bunte salts with sodium sulfide.[17]

1. $Na_2S_2O_3$; 2. HSCH(Me)Et, NaOH; 75% (2)

The sulfite group of Bunte salts may also function as a leaving group via an E2 elimination, comprising a convenient and mild method for generating thiocarbonyl compounds. This reaction was first noted when the carbamoyl derivatives $RNHCOCH_2SSO_3Na$ were reacted with primary and secondary amines R'_2NH, affording thiooxamides $RNHCOCSNR'_2$.[18] Dimerization of the reactive thiocarbonyl intermediates to 1,3-dithietanes has also been observed.[19] The thiocarbonyl group is a good dienophile and will trap 1,3-dienes at low temperatures.[20] The resulting cycloadducts can be rearranged stereoselectively under basic conditions to cyclopentenyl sulfides (eq 3).[21]

1. $Na_2S_2O_3$; 2. Et_3N; 65 °C; 1. LDA, HMPA, −78 to 0 °C; 2. MeI; 87%, 8:1 (3)

Related Reagents. See Classes R-9, R-15, R-22, R-24, and R-27, pages 1–10.

---

1. (a) Swaine, J. W., Jr. In *Kirk-Othmer Encyclopedia of Chemical Technology*, 3rd ed.; Grayson, M.; Eckroth, D., Eds.; Wiley: New York, 1983; Vol. 22, pp 974–989. (b) Kahrasch, N.; Arora, A. S. *PS* **1976**, *2*, 1.
2. (a) Milligan, B.; Swan, J. M. *Rev. Pure Appl. Chem.* **1962**, *12*, 72. (b) Distler, H. *AG(E)* **1967**, *6*, 544. (c) Hogg, D. R. In *Comprehensive Organic Chemistry*; Barton, D. H. R.; Ollis, W. D., Eds.; Pergamon: Oxford, 1979; Vol. 3, p 307.
3. Ibne-Rasa, K. M.; Tahir, A. R.; Rahman, A. *CI(L)* **1973**, 232.
4. McLamore, W. M. *JACS* **1951**, *73*, 2221.
5. Adams, R.; Blomstrom, D. C. *JACS* **1953**, *75*, 3405.
6. White, E. H. *JACS* **1955**, *77*, 6008.
7. Isayama, S. *BCJ* **1990**, *63*, 1305.
8. Kametani, T.; Honda, T.; Ishizone, H.; Kanada, K.; Naito, K.; Suzuki, Y. *CC* **1989**, 646.
9. Brockmann, H.; Laatsch, H. *CB* **1973**, *106*, 2058.
10. Overberger, C. G.; Valentine, M.; Anselme, J. P. *JACS* **1969**, *91*, 687.
11. Whistler, R. L.; Luttenegger, T. J.; Rowell, R. M. *JOC* **1968**, *33*, 396.
12. Jpn. Patent 57 206 653, 1983 (*CA* **1983**, *98*, 215 998r).
13. Wardell, J. L. In *The Chemistry of the Thiol Group*; Patai, S., Ed.; Wiley: London, 1974; Part 1, Chapter 4, p 192.
14. Alcalay, W. *HCA* **1947**, *30*, 578.
15. Ziegler, C.; Sprague, J. M. *JOC* **1951**, *16*, 621.

16. (a) Mattes, K. C.; Chapman, O. L. *JOC* **1977**, *42*, 1814. (b) Alonso, M. E.; Aragona, H. *OSC* **1988**, *6*, 235.
17. Milligan, B.; Saville, B.; Swan, J. M. *JCS* **1963**, 3608.
18. Milligan, B.; Swan, J. M. *JCS* **1959**, 2969.
19. (a) Zehavi, U. *JOC* **1977**, *42*, 2821. (b) Hartnedy, R. C.; Dittmer, D. C. *JOC* **1984**, *49*, 4752.
20. Kirby, G. W.; Lochead, A. W.; Sheldrake, G. N. *CC* **1984**, 922.
21. Larsen, S. D. *JACS* **1988**, *110*, 5932.

Scott D. Larsen
The Upjohn Co., Kalamazoo, MI, USA

Sodium Triacetoxyborohydride[1]

$NaBH(OAc)_3$

[56553-60-7] $C_6H_{10}BNaO_6$ (MW 211.96)

(the prototype of a class of $NaBH(OCOR)_3$ reagents that are selective reducing agents for a number of functional groups[1] and heterocycles;[1] alkylation of amines;[1–4] hydroboration[1])

Physical Data: mp t 116–120 °C (dec); the related $NaBH_3OAc$ has not been fully characterized.
Solubility: $NaBH(OAc)_3$ and related acyloxyborohydrides are rapidly destroyed by H_2O and protic solvents; H_2 is liberated. Cosolvents that have been employed are benzene, toluene, THF, dioxane, CH_2Cl_2, $ClCH_2CH_2Cl$.
Form Supplied in: $NaBH(OAc)_3$ and the related ***Tetramethylammonium Triacetoxyborohydride*** are commercially available as colorless powders.
Preparative Method: $NaBH(OAc)_3$ and $NaBH_3OAc$ can be easily prepared in situ from the appropriate amount of acetic acid and $NaBH_4$.
Analysis of Reagent Purity: $NaBH(OAc)_3$ has been characterized by elemental analysis, IR, and ^1H, ^{13}C, and ^{11}B NMR.
Handling, Storage, and Precautions: because H_2 is liberated during the preparation of these reagents, all handling and storage of acyloxyborohydrides should take place under an inert atmosphere.

Functional Group Reductions. $NaBH(OAc)_3$ selectively reduces aldehydes but not ketones[5,6] (eqs 1 and 2),[7,8] even with excess reagent. However, α- and β-hydroxy ketones are reduced to the *anti*-diols by hydroxy-directed hydride delivery[6,9,10] (eqs 3 and 4).[11,12] Diastereoselectivities are generally excellent, although $Me_4NBH(OAc)_3$ seems to be a superior reagent in this regard.[9] Several recent examples of $NaBH(OAc)_3$ in the stereoselective reduction of hydroxy ketones attest to the power of this reagent.[13]

$NaBH(OAc)_3$, PhH, AcOH, rt, 2 h, 80% (1)

$NaBH_4$, PhH, AcOH, Δ, 1 h, 95% (2)

$NaBH(OAc)_3$, AcOH, 25 °C, 30 min, 91% (3)

anti:*syn* = >30:1

$NaBH(OAc)_3$, AcOH, 57% (4)

The hydroxy-directed $NaBH(OAc)_3$ reduction of an imide has been described.[14] The more reactive $NaBH_3OAc$ reduces enones to allylic alcohols,[15] and some ketones can be reduced to alcohols with ***Sodium Borohydride***–tartaric acid.[16] The combination of $NaBH_4$ or ***Sodium Cyanoborohydride–Acetic Acid*** serves to deoxygenate tricarbonyl systems (eq 5)[17] and tosylhydrazones of ketones and aldehydes (eq 6).[18] Primary and secondary amides are reduced to amines by the action of $NaBH_3OAc$ (eq 7),[19,20] while tertiary amides require ***Sodium Trifluoroacetoxyborohydride***.

$NaBH_3CN$, AcOH, rt, 1 h, 80% (5)

p-$TsNHNH_2$, $NaBH_3CN$, AcOH, 70 °C, 2 h, 72% (6)

$NaBH_3OAc$, dioxane, Δ, 97% (7)

Amine Alkylation (Reductive Amination). By a pathway that may involve the generation of free aldehyde, the combination of $NaBH_4$ and carboxylic acids is capable of *N*-alkylation of amines.[1–4] Recent examples abound (eqs 8–11). At lower temperature, monoalkylation is generally observed (eq 8),[21] while at 50–55 °C, primary and secondary amines are converted into tertiary amines (eqs 9 and 10).[22,23] Neat carboxylic acid (eqs 8 and 9) or a cosolvent (eq 10) may be used. In the latter event, solid carboxylic acids function well (eq 11).[24] ***Formic Acid*** may be employed for *N*-methylation.[1,25] A useful variation is the reductive amination of aldehydes and ketones (eqs 12–14),[1,26–29] a method which is claimed to be superior to that using $NaBH_3CN$–MeOH.[26] ***Paraformaldehyde*** serves as a convenient source of HCHO for *N*-methylation in this protocol.[30,31]

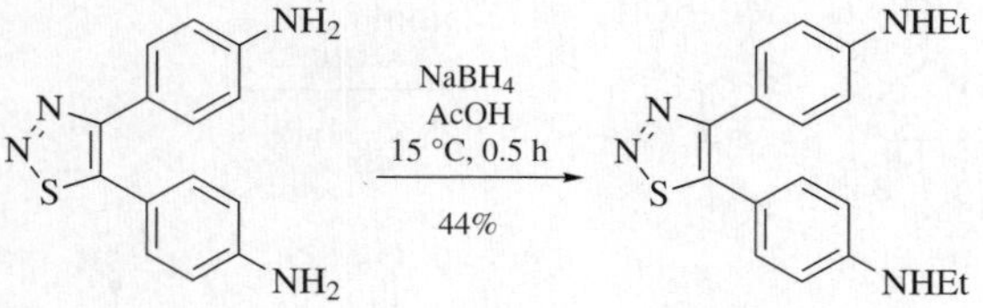

(8)

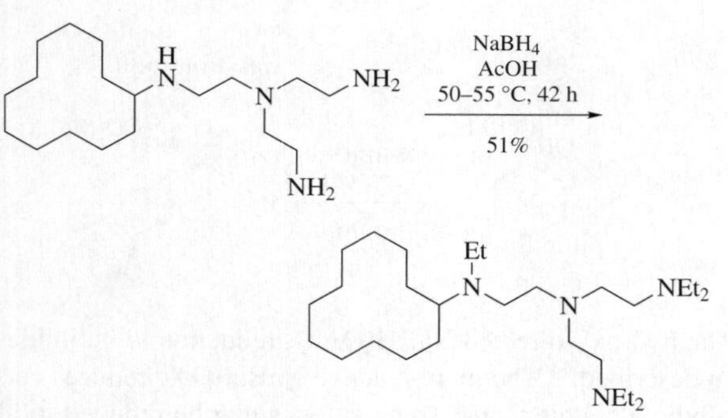

(9)

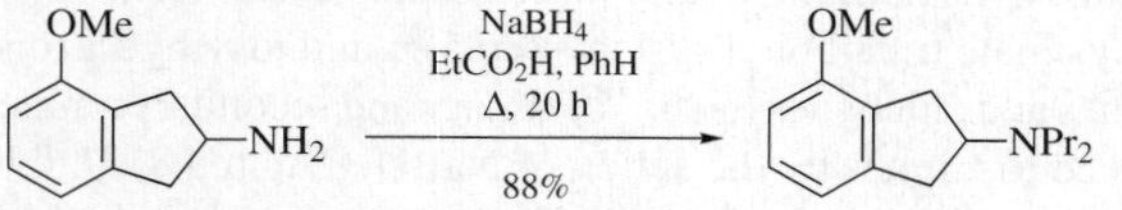

(10)

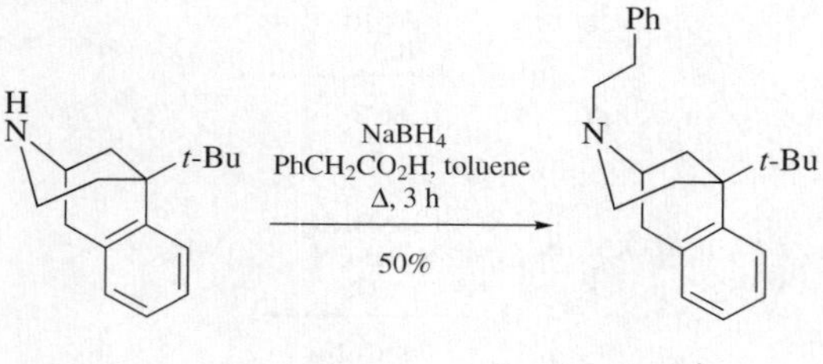

(11)

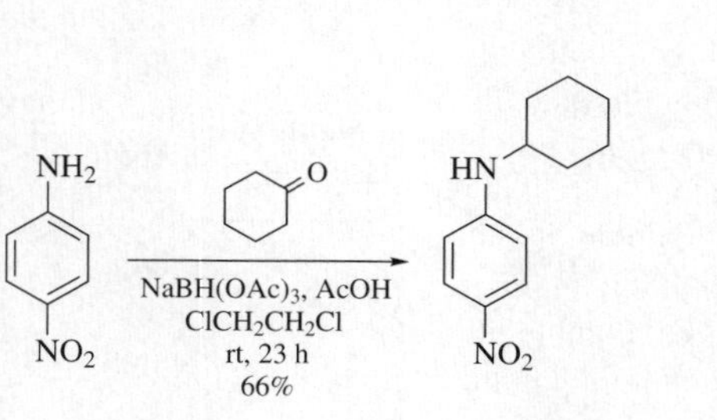

(12)

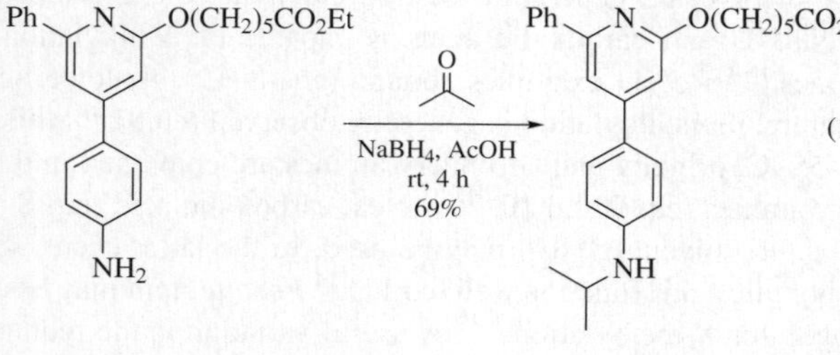

(13)

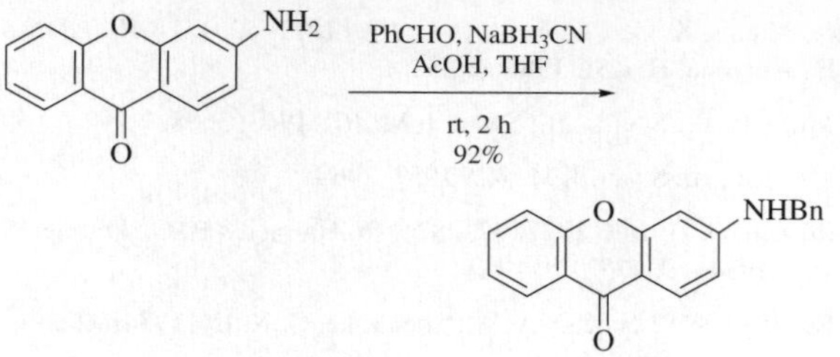

(14)

Carboxylic acids are reduced to alcohols with $NaBH_4$ in THF,[32] although the use of CF_3CO_2H in this regard is superior, and there is one report of an ester reduction to a primary alcohol with $NaBH_4$–HOAc.[33]

Enamine, Imine, Iminium Ion, and Enamide Reduction. The first reported use of $NaBH_4$–HOAc was in the reduction of dienamines,[34] and this application has found extensive use in synthesis[1] (eqs 15–19).[35–39]

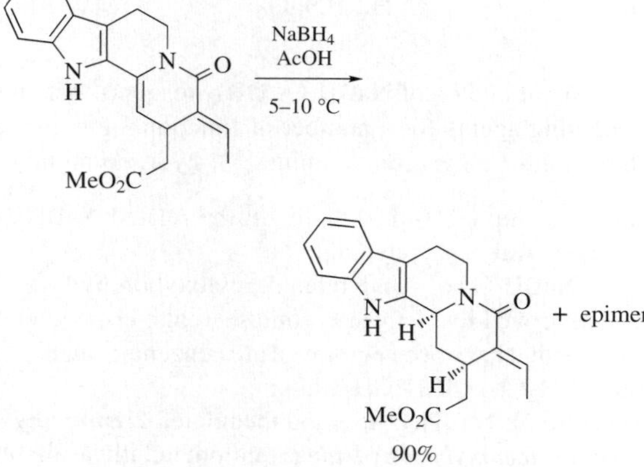

(15)

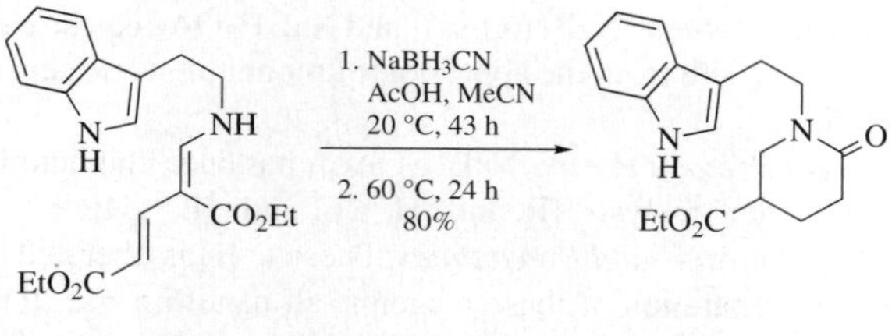

(16)

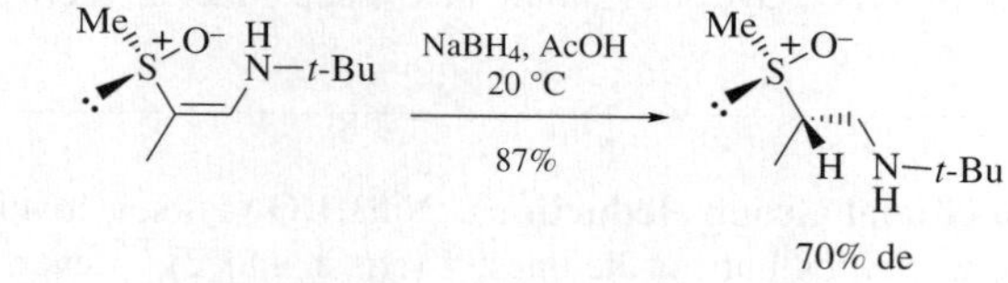

(17)

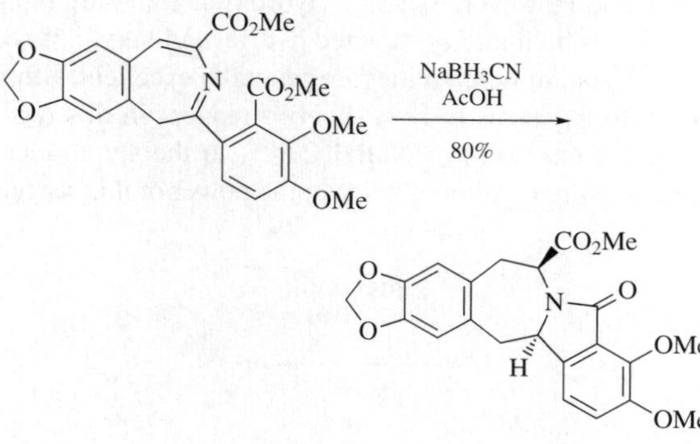

(18)

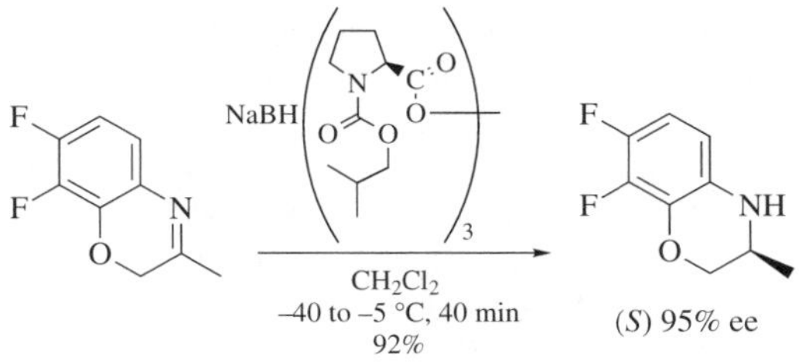

Indole Reduction. Indole is smoothly reduced to indoline under the influence of $NaBH_3CN$–HOAc;[2] the reaction is quite general[1,40] and has been employed often (eqs 20–22),[41–43] especially in the synthesis of CC-1065, PDE, and analogs where only the more basic indole ring is reduced (eq 23).[44,45] *N*-Substituted indoles are reduced to indolines with $NaBH_4$–HOAc,[1,2] and the action of $NaBH_4$–RCO_2H on *N*-unsubstituted indoles affords *N*-alkylindolines by *N*-alkylation of the initially formed indoline.[1,2]

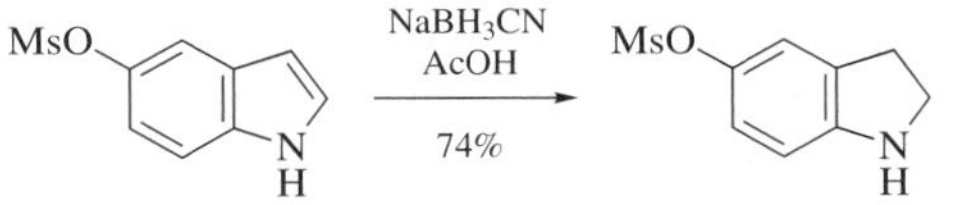

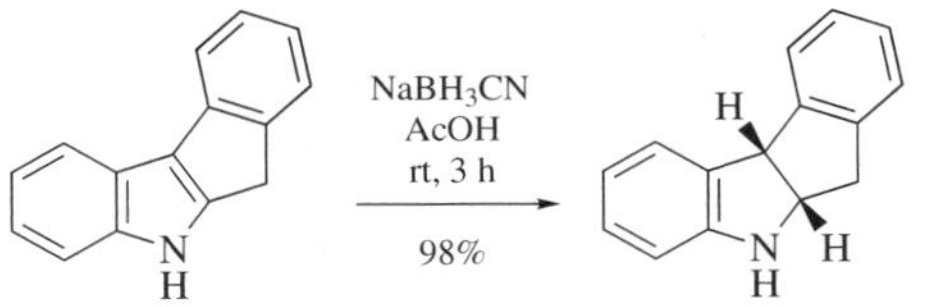

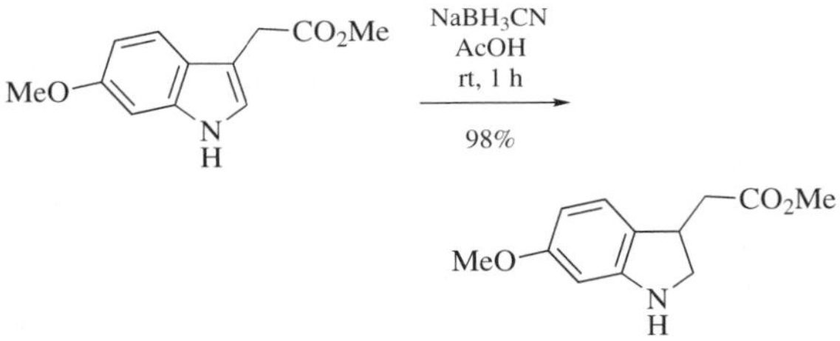

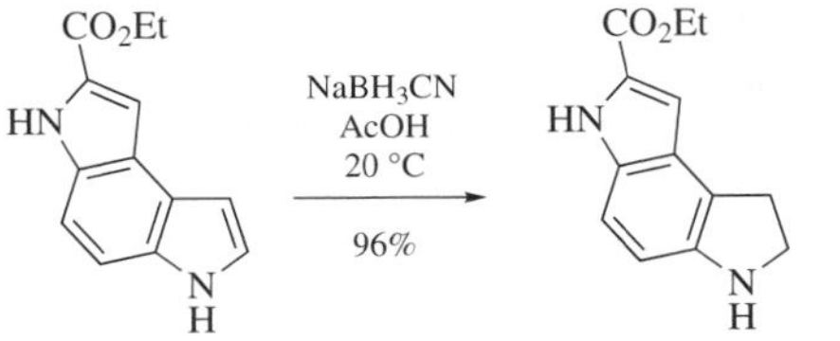

Reduction of Other Heterocycles. Quinolines and isoquinolines are reduced to the corresponding tetrahydro derivatives with $NaBH_3CN$ or $NaBH_4$–RCO_2H, the latter combination affording the *N*-alkylated compounds.[1,46] Related heterocycles have been subjected to this protocol (eqs 24 and 25).[47]

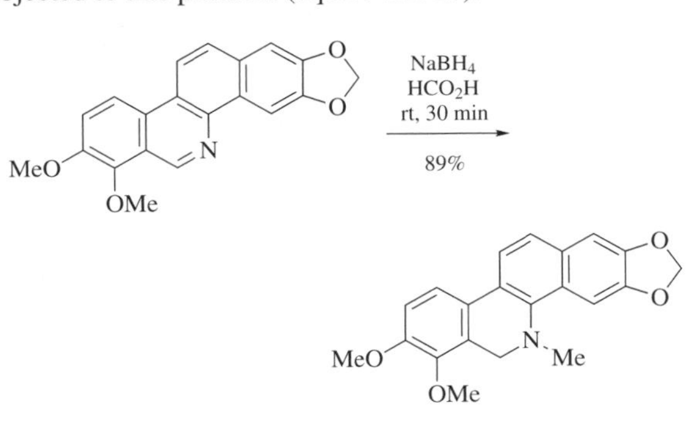

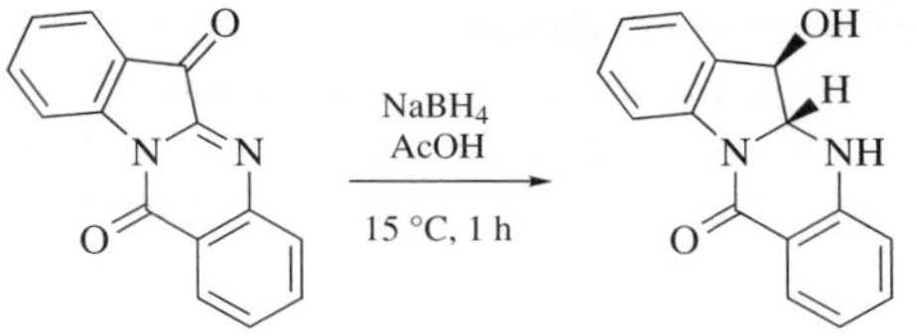

The reduction of pyrylium salts (eq 26),[48] the reductive cleavage of benzoxazoles (eq 27)[49] and of saturated nitrogen heterocycles (eqs 28 and 29),[1,50,51] and the reduction of other π-deficient nitrogen heterocycles[1] are known.

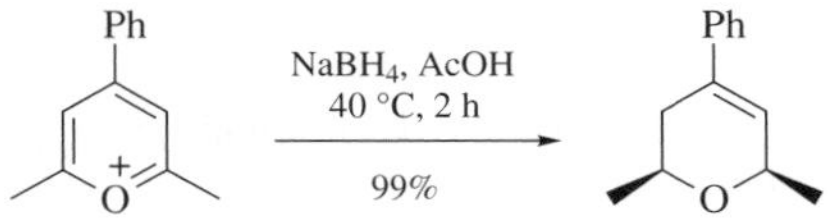

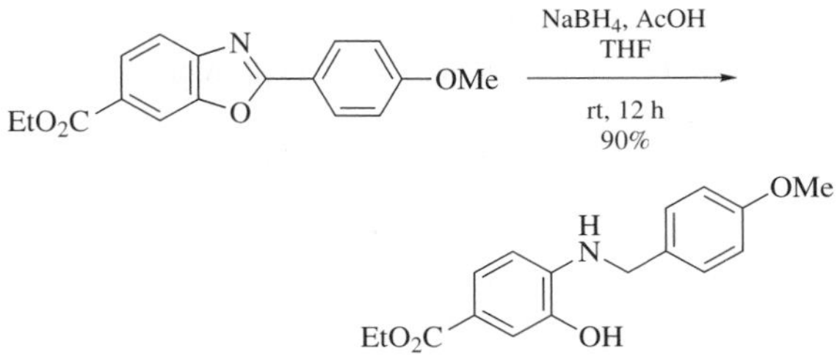

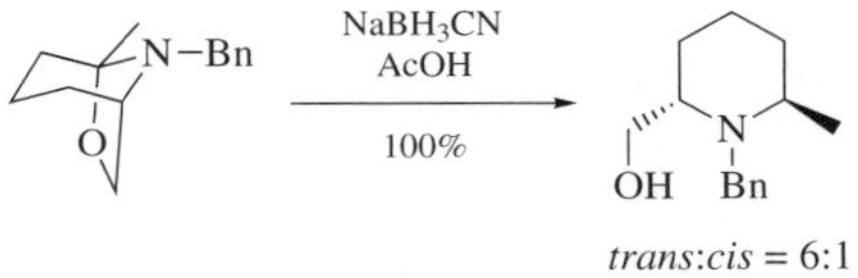

Reduction of Oximes. Oximes can be alkylated or reduced, depending on whether $NaBH_4$ or $NaBH_3CN$ is employed, to give hydroxylamines[1,52] (eqs 30 and 31).[52,53] Oxime ethers are also reduced under these conditions,[1,54] and the hydroxy-directed reduction of oxime ethers has been reported using $Me_4NBH(OAc)_3$.[55] Nitriles are converted into primary amines by the tandem action of acyloxyborohydrides and alkyllithium reagents (eq 32).[56]

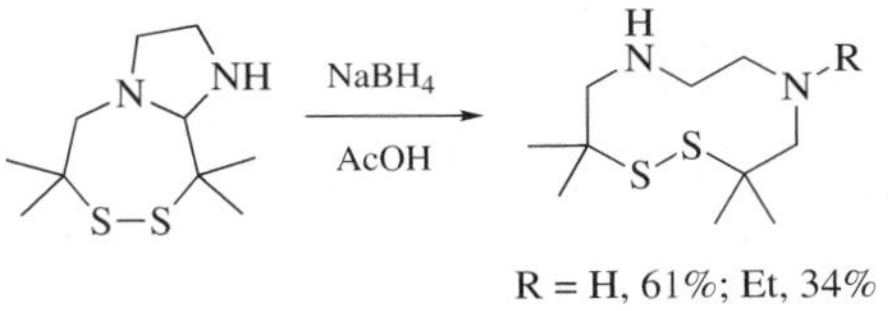

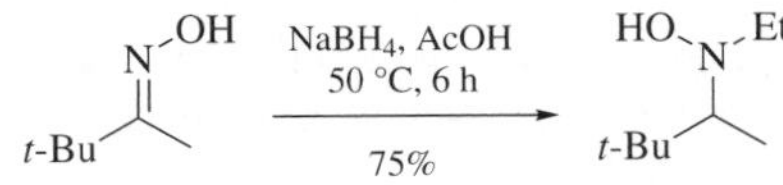

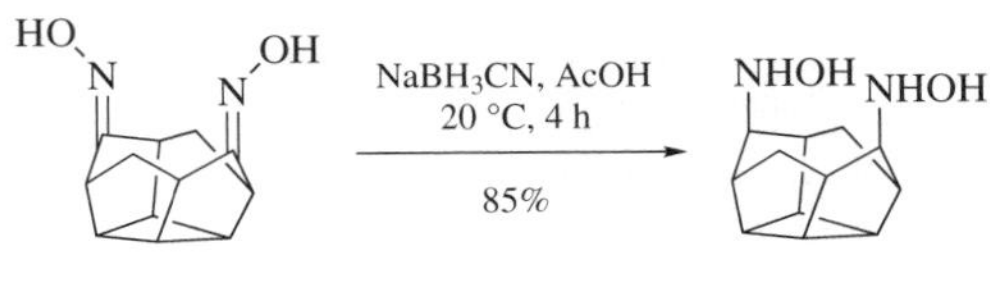

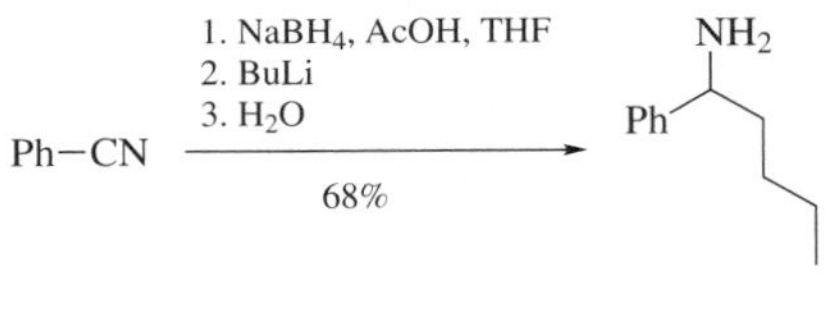

Hydroboration of Alkenes. The second reported reaction of acyloxyborohydrides was the hydroboration of alkenes,[1,57] and this reaction has been further refined.[58] In a similar vein, the reduction of organomercurials by $NaBH(OAc)_3$ has been described.[59]

Related Reagents. See Classes R-2, R-3, R-5, R-6, R-8, R-10, R-12, R-13, R-17, R-20, R-29, and R-35, pages 1–10. Sodium Borohydride; Sodium Cyanoborohydride; Sodium Trifluoroacetoxyborohydride; Sodium Tris(trifluoroacetoxy)borohydride; Tetramethylammonium Triacetoxyborohydride.

1. (a) Gribble, G. W.; Nutaitis, C. F. *OPP* **1985**, *17*, 317. (b) Nutaitis, C. F. *J. Chem. Educ.* **1989**, *66*, 673.
2. Gribble, G. W.; Lord, P. D.; Skotnicki, J.; Dietz, S. E.; Eaton, J. T.; Johnson, J. L. *JACS* **1974**, *96*, 7812.
3. Marchini, P.; Liso, G.; Reho, A.; Liberatore, F.; Moracci, F. M. *JOC* **1975**, *40*, 3453.
4. Gribble, G. W.; Jasinski, J. M.; Pellicone, J. T.; Panetta, J. A. *S* **1978**, 766.
5. Gribble, G. W.; Ferguson, D. C. *CC* **1975**, 535.
6. Nutaitis, C. F.; Gribble, G. W. *TL* **1983**, *24*, 4287.
7. Odinokov, V. N.; Ignatyuk, V. K.; Krivonogov, V. P.; Tolstikov, G. A. *ZOR* **1986**, *22*, 948.
8. Barker, P. L.; Bahia, C. *T* **1990**, *46*, 2691.
9. Evans, D. A.; Chapman, K. T.; Carreira, E. M. *JACS* **1988**, *110*, 3560.
10. Saksena, A. K.; Mangiaracina, P. *TL* **1983**, *24*, 273.
11. Evans, D. A.; DiMare, M. *JACS* **1986**, *108*, 2476.
12. Turnbull, M. D.; Hatter, G.; Ledgerwood, D. E. *TL* **1984**, *25*, 5449.
13. (a) Wender, P. A.; Kogen, H.; Lee, H. Y.; Munger, J. D., Jr.; Wilhelm, R. S.; Williams, P. D. *JACS* **1989**, *111*, 8957. (b) Farr, R. N.; Outten, R. A.; Cheng, J. C.-Y.; Daves, G. D., Jr. *OM* **1990**, *9*, 3151. (c) Robins, M. J.; Samano, V.; Johnson, M. D. *JOC* **1990**, *55*, 410. (d) Whang, K.; Cooke, R. J.; Okay, G.; Cha, J. K. *JACS* **1990**, *112*, 8985. (e) Brown, M. J.; Harrison, T.; Herrinton, P. M.; Hopkins, M. H.; Hutchinson, K. D.; Mishra, P.; Overman, L. E. *JACS* **1991**, *113*, 5365. (f) Turner, N. J.; Whitesides, G. M. *JACS* **1989**, *111*, 624. (g) Estevez, V. A.; Prestwich, G. D. *JACS* **1991**, *113*, 9885. (h) Romeyke, Y.; Keller, M.; Kluge, H.; Grabley, S.; Hammann, P. *T* **1991**, *47*, 3335. (i) Fisher, M. J.; Chow, K.; Villalobos, A.; Danishefsky, S. J. *JOC* **1991**, *56*, 2900. (j) Overman, L. E.; Shim, J. *JOC* **1991**, *56*, 5005. (k) Zhang, H.-C.; Daves, G. D., Jr. *JOC* **1992**, *57*, 4690.
14. Miller, S. A.; Chamberlin, A. R. *JOC* **1989**, *54*, 2502.
15. Nutaitis, C. F.; Bernardo, J. E. *JOC* **1989**, *54*, 5629.
16. (a) Adams, C. *SC* **1984**, *14*, 955. (b) Yatagai, M.; Ohnuki, T. *JCS(P1)* **1990**, 1826. (c) Polyak, F. D.; Solodin, I. V.; Dorofeeva, T. V. *SC* **1991**, *21*, 1137.
17. (a) Nutaitis, C. F.; Schultz, R. A.; Obaza, J.; Smith, F. X. *JOC* **1980**, *45*, 4606. (b) Obaza, J.; Smith, F. X. *SC* **1982**, *12*, 19. (c) Rosowsky, A.; Forsch, R.; Uren, J.; Wick, M.; Kumar, A. A.; Freisheim, J. H. *JMC* **1983**, *26*, 1719.
18. (a) Hutchins, R. O.; Natale, N. R. *JOC* **1978**, *43*, 2299. (b) Baeckström, P.; Li, L. *SC* **1990**, *20*, 1481.
19. Umino, N.; Iwakuma, T.; Itoh, N. *TL* **1976**, 763.
20. Cannon, J. G.; Chang, Y.; Amoo, V. E.; Walker, K. A. *S* **1986**, 494.
21. Thomas, E. W.; Nishizawa, E. E.; Zimmermann, D. C.; Williams, D. J. *JMC* **1985**, *28*, 442.
22. Thomas, E. W.; Cudahy, M. M.; Spilman, C. H.; Dinh, D. M.; Watkins, T. L.; Vidmar, T. J. *JMC* **1992**, *35*, 1233.
23. Cannon, J. G.; Dushin, R. G.; Long, J. P.; Ilhan, M.; Jones, N. D.; Swartzendruber, J. K. *JMC* **1985**, *28*, 515.
24. James, L. J.; Parfitt, R. T. *JMC* **1986**, *29*, 1783.
25. Johnson, J. V.; Rauckman, B. S.; Baccanari, D. P.; Roth, B. *JMC* **1989**, *32*, 1942.
26. Abdel-Magid, A. F.; Maryanoff, C. A.; Carson, K. G. *TL* **1990**, *31*, 5595.
27. Abdel-Magid, A. F.; Maryanoff, C. A. *SL* **1990**, 537.
28. Labaudinière, R.; Dereu, N.; Cavy, F.; Guillet, M.-C.; Marquis, O.; Terlain, B. *JMC* **1992**, *35*, 4315.
29. Gunzenhauser, S.; Balli, H. *HCA* **1989**, *72*, 1186.
30. Gribble, G. W.; Nutaitis, C. F. *S* **1987**, 709.
31. Gunzenhauser, S.; Balli, H. *HCA* **1990**, *73*, 359.
32. Cho, B. T.; Yoon, N. M. *SC* **1985**, *15*, 917.
33. Soucek, M.; Urban, J.; Saman, D. *CCC* **1990**, *55*, 761.
34. (a) Marshall, J. A.; Johnson, W. S. *JOC* **1963**, *28*, 421. (b) Johnson, W. S.; Bauer, V. J.; Franck, R. W. *TL* **1961**, 72. (c) Marshall, J. A.; Johnson, W. S. *JACS* **1962**, *84*, 1485.
35. Naito, T.; Shinada, T.; Miyata, O.; Ninomiya, I.; Ishida, T. *H* **1988**, *27*, 1603.
36. Stoit, A. R.; Pandit, U. K. *T* **1988**, *44*, 6187.
37. Kawecki, R.; Kozerski, L.; Urbanczyk-Lipkowska, Z.; Bocelli, G. *JCS(P1)* **1991**, 2255.
38. Moody, C. J.; Warrellow, G. J. *TL* **1987**, *28*, 6089.
39. Atarashi, S.; Tsurumi, H.; Fujiwara, T.; Hayakawa, I. *JHC* **1991**, *28*, 329.
40. Gribble, G. W.; Hoffman, J. H. *S* **1977**, 859.
41. Siddiqui, M. A.; Snieckus, V. *TL* **1990**, *31*, 1523.
42. Brown, D. W.; Graupner, P. R.; Sainsbury, M.; Shertzer, H. G. *T* **1991**, *47*, 4383.
43. Toyota, M.; Fukumoto, K. *JCS(P1)* **1992**, 547.
44. Rawal, V. H.; Jones, R. J.; Cava, M. P. *JOC* **1987**, *52*, 19.
45. (a) Boger, D. L.; Coleman, R. S.; Invergo, B. J. *JOC* **1987**, *52*, 1521. (b) Bolton, R. E.; Moody, C. J.; Rees, C. W.; Tojo, G. *TL* **1987**, *28*, 3163. (c) Meghani, P.; Street, J. D.; Joule, J. A. *CC* **1987**, 1406. (d) Sundberg, R. J.; Hamilton, G. S.; Laurino, J. P. *JOC* **1988**, *53*, 976. (e) Martin, P. *HCA* **1989**, *72*, 1554.
46. (a) Gribble, G. W.; Heald, P. W. *S* **1975**, 650. (b) Uchida, M.; Chihiro, M.; Morita, S.; Yamashita, H.; Yamasaki, K.; Kanbe, T.; Yabuuchi, Y.; Nakagawa, K. *CPB* **1990**, *38*, 534. (c) Vigante, B. A.; Ozols, Ya. Ya.; Dubur, G. Ya. *KGS* **1991**, 1680. (d) Carling, R. W.; Leeson, P. D.; Moseley, A. M.; Baker, R.; Foster, A. C.; Grimwood, S.; Kemp, J. A.; Marshall, G. R. *JMC* **1992**, *35*, 1942.
47. (a) Ishii, H.; Ishikawa, T.; Ichikawa, Y.; Sakamoto, M.; Ishikawa, M.; Takahashi, T. *CPB* **1984**, *32*, 2984. (b) Bergman, J.; Tilstam, U.; Törnroos, K.-W. *JCS(P1)* **1987**, 519.
48. Balaban, T.-S.; Balaban, A. T. *OPP* **1988**, *20*, 231.
49. Yadagiri, B.; Lown, J. W. *SC* **1990**, *20*, 175.
50. Wasserman, H. H.; Rusiecki, V. *TL* **1988**, *29*, 4977.
51. Bodor, N.; Koltai, E.; Prókai, L. *T* **1992**, *48*, 4767.
52. Gribble, G. W.; Leiby, R. W.; Sheehan, M. N. *S* **1977**, 856.
53. Waykole, L. M.; Shen, C.-C.; Paquette, L. A. *JOC* **1988**, *53*, 4969.
54. Chaubet, F.; Duong, M. N. V.; Courtieu, J.; Gaudemer, A.; Gref, A.; Crumbliss, A. L. *CJC* **1991**, *69*, 1107.
55. Williams, D. R.; Osterhout, M. H. *JACS* **1992**, *114*, 8750.
56. Itsuno, S.; Hachisuka, C.; Ushijima, Y.; Ito, K. *SC* **1992**, *22*, 3229.
57. Marshall, J. A.; Johnson, W. S. *JOC* **1963**, *28*, 595.
58. (a) Narayana, C.; Periasamy, M. *TL* **1985**, *26*, 6361. (b) Gautam, V. K.; Singh, J.; Dhillon, R. S. *JOC* **1988**, *53*, 187.
59. Gouzoules, F. H.; Whitney, R. A. *JOC* **1986**, *51*, 2024.

Gordon W. Gribble
Dartmouth College, Hanover, NH, USA

Sulfur[1]

[7704-34-9] S (MW 32.07)

(mild oxidizing agent used for the Willgerodt reaction[2] and dehydrogenation of various organic compounds;[3] its most recent use has been for the in situ generation of carbonyl sulfide[4] from CO for the synthesis of thiocarbamates[5] and other derivatives)

Physical Data: mp 119 °C; bp 444.6 °C; *d* 2.07 g cm^{-3}.
Solubility: insol water; slightly sol alcohol, ether; sol carbon disulfide, benzene, toluene.
Form Supplied in: powder, flakes.
Handling, Storage, and Precautions: store in a closed bottle at ambient temperatures. Chronic inhalation of dust can cause irritation of the mucous membranes.

Willgerodt Reaction. The first reaction of this type was carried out by Willgerodt[6] on 1-acetylnaphthalene to give what was later identified as 1-naphthylacetamide (eq 1).[7] This was accomplished by using an ammonium polysulfide solution in a sealed tube at 210–230 °C for 3 or 4 days. The corresponding acid salt always accompanied the amide. Another limitation is that the harsh conditions used in the original procedure gave yields of only 20–50%.

$(NH_4)_2SS_x$, H_2O, 210–230 °C, 3–4 d (1)

The Kindler[8] modification of the reaction allows this to be carried out under anhydrous conditions at lower temperatures. The thioamide can be isolated and then hydrolyzed to the acid or reduced to the amine. Even with this important change, the reaction saw little application because it had to be carried out in an autoclave as in eq 2.[9]

S, $HNMe_2$, 140–150 °C, 58% (2)

The adoption of ***Morpholine*** as both the solvent and base led to a procedure that could be used at ambient pressure (eq 3).[10a] With the relatively mild temperatures (bp 129 °C), certain functionalities on the aryl ring like nitro, amino, hydroxy, and acetoxy groups can tolerate these conditions.[10b]

S, morpholine, reflux, 91% (3)

The original Willgerodt conditions found a use in converting aliphatic aldehydes and ketones to the amides in poor to moderate yields (eqs 4–7).[11]

$(NH_4)_2SS_x$, H_2O, py, 140 °C, 5 h, 50% (4)

$(NH_4)_2SS_x$, H_2O, py, 150 °C, 4 h, 38% (5)

$(NH_4)_2SS_x$, H_2O, py, 145 °C, 4 h, 23% (6)

$(NH_4)_2SS_x$, H_2O, py, 145 °C, 4 h, 21% (7)

This reaction can also be applied to unconventional substrates for conversion into thioamides (eq 8).[12] It was found that methylphenylcarbinol, β-phenylethyl alcohol, β-phenylethyl acetate, and methylbenzylcarbinol failed to give appreciable amounts of amide when heated with a typical Willgerodt reagent at 160 °C. Unexpected results were seen with β-substituted acrylic acids (eqs 9 and 10).[13] Apparently, the carboxyl group was eliminated to give an amide with one less carbon atom.

Morpholine (NH), S, 140–150 °C (8)

$(NH_4)_2SS_x$, dioxane, 170 °C, 76% (9)

$(NH_4)_2SS_x$, H_2O, py, 170 °C, 63% (10)

Other functionalities were found to allow the Willgerodt reaction to work on substrates that yield phenylacetamide in an attempt to elucidate the mechanism.[14] Poor to good yields were obtained. A thorough review of this reaction was written by Carmack and Spielman in 1946.[15]

Dehydrogenation. This reaction is another important use of elemental sulfur in organic chemistry. Sulfur is the reagent of choice for the dehydrogenation of hydroaromatic compounds because it is simpler than ***Selenium*** and may be preferred over selenium or catalytic dehydrogenation. The following example is from a systematic study that compared the three methods (eq 11).[16]

(11)

S	230–240 °C	+	–
Pd/C	260–280 °C	+	+
Se	330–350 °C	–	+

The dehydrogenation of aromatizable hydrocarbons works well enough to have been incorporated into an undergraduate experiment (eq 12).[17]

S, triglyme, reflux, 1 h (12)

In this experiment it was necessary to use less than the theoretical amount of sulfur because the brilliantly blue product, guaiazulene, reacts with sulfur to form an intractable brown product.

More complex structures can be selectively dehydrogenated to give fair yields of products, as the example from Haede et al. illustrates (eq 13).[18]

S, *p*-cymene, 190–195 °C, 3 h, 48% (13)

Other heterocycles are good substrates for the dehydrogenation reaction (eqs 14–18).[19–23]

S, Δ, 2 h, 92% (14)

S, 180–190 °C, 4 h, 70% (15)

2–3% S in DMF, reflux, 30 min, 83% (16)

S, 150–200 °C, 90% (17)

S, 125–135 °C, 90% (18)

Carbonyl Sulfide Generation. The use of ***Carbon Monoxide*** and sulfur leads to the generation of carbonyl sulfide (***Carbon Oxysulfide***), which is electrophilic enough to be attacked by nucleophiles present in the reaction mixture. A new synthesis of ureas based upon this intermediate was put forth (eq 19).[24] This reaction starts out with each of the reactants in a different phase, liquid, gas, and solid, respectively. This simplifies the industrial process, and the in situ oxidation of carbon monoxide leads to a ***Phosgene*** synthon without the formation of ammonium salts as byproducts. In an experiment using only ***Ammonia***, carbon monoxide, and sulfur in methanol, Franz and Applegath[25] found carbonyl sulfide making up about 30% of the product gases when the reaction was run at 150 °C. Ammonium thiocarbamate was obtained in large quantities when the temperature of the reaction was held overnight at 25–30 °C with 700 psi of CO. The isocyanate group was detected by IR in an alcoholic solution of ammonium thiocarbamate (eq 20). This method has also been used to synthesize unsymmetrical ureas, with the authors postulating that aryl isocyanates are formed first which then react with primary or secondary alkyl amines to give the 1-aryl-3-alkyl- or 1-aryl-3-dialkylureas in good yield (eq 21).[26]

$$2\,RNH_2 + CO + S \longrightarrow RNHC(O)NHR + H_2S \quad (19)$$

$$CO + S \xrightarrow{2\,NH_3} H_2NCOS^- NH_4^+$$
$$\downarrow$$
$$H_2NCONH_2 \xleftarrow{NH_3} HNCO + NH_4^+ HS^- \longrightarrow H_2S + NH_3 \quad (20)$$

$$COS + 2\,ArNH_2 \longrightarrow \left[S{=}C(OH)(NHAr) \cdot ArNH_2 \right] \longrightarrow ArNCO + ArNH_2{\cdot}H_2S \quad (21)$$

This sulfur-assisted carbonylation has also been applied to the synthesis of 4-hydroxycoumarins,[27] which entails the direct formation of a C–C bond without a transition metal catalyst (eq 22).

Et_3N, THF, S, 10 atm CO, 80 °C, 4 h; + H_2S (22)

This reaction has also been applied to the synthesis of *S*-alkyl carbonothioates, which requires the use of ***1,8-Diazabicyclo[5.4.0]undec-7-ene*** and an alkylating agent (eq 23).[28]

$$\text{BuOH} + 3\,\text{S} \xrightarrow[\text{5 equiv DBU, 4 h}]{\text{THF, 80 °C; 30 atm CO}} \text{BuOC(O)S}^- \ \text{DBUH}^+ \xrightarrow{\text{1.2 equiv allyl bromide}} \text{BuOC(O)SCH}_2\text{CH=CH}_2 \ (100\%) \quad (23)$$

Application to the synthesis of carbonates has been accomplished by inclusion of a stoichiometric amount of ***Copper(II) Chloride*** in the second step (eq 24).[29]

$$\text{R(OH)}_2 \xrightarrow[\text{THF, 80 °C; 30 atm CO}]{\text{3 equiv S; 5 equiv NEt}_3} \text{HO–R–OC(O)S}^- \ \text{Et}_3\overset{+}{\text{N}}\text{H} \xrightarrow[\text{rt, 18 h}]{\text{1.5 equiv CuCl}_2} \text{cyclic carbonate (R–O–C(=O)–O)} \quad (24)$$

More complex heterocycles like 4-alkylidene-2-oxo-1,3-oxathiolanes have been synthesized by a related method (eq 25).[30]

$$\text{HC≡CCH}_2\text{OH} + \text{CO} + \text{S} \xrightarrow[\text{80 °C, 4 h, THF}]{\text{THF, 30 atm CO}} [\text{HC≡CCH}_2\text{OC(O)S}^- \ \text{Et}_3\overset{+}{\text{N}}\text{H}] \xrightarrow[\text{rt, 18 h}]{\text{0.2–1.0 equiv CuI}} \text{4-methylene-1,3-oxathiolan-2-one} \quad (25)$$

The same group has developed a variation of this technique which provides the *S,S*-dialkyl dithiocarbonates from water, sulfur, carbon monoxide, and alkyl halides.[31] The use of a catalytic amount of selenium provides a synthesis of *S*-alkyl thiocarbamates from amines, carbon monoxide, sulfur, and alkyl halides that can be carried out under ambient pressure (eqs 26 and 27).[32] The intermediate (**1**) can undergo decomposition to carbonyl selenide which then undergoes facile exchange with sulfur to give carbonyl sulfide, which will then react as usual. At higher pressures of CO (10 atm) and 1 mol % selenium these authors can synthesize *S*-alkyl carbonothioates.[33]

$$2\,\text{R}^1\text{R}^2\text{NH} + \text{CO} + \text{Se} \xrightarrow[\text{1 atm}]{\text{20 °C}} [\text{R}^1\text{R}^2\text{NH}_2]^+ [\text{R}^1\text{R}^2\text{NC(O)Se}]^- \ (\mathbf{1}) \quad (26)$$

$$\text{Se=C=O} + \text{S} \xrightarrow[\text{–45 °C}]{\text{Et}_3\text{N, THF}} \text{S=C=O} + \text{Se} \quad (27)$$

Sulfide and Polysulfide-Derived Reactions. The largest and broadest application of sulfur in organic synthesis is in the generation of sulfides and polysulfides from elemental sulfur. These polysulfides can act as nucleophiles, reductants, and oxidants. A brief survey of these various aspects of sulfides and polysulfides is in order.

Elemental sulfur is believed to dissociate in liquid ammonia according to the following series of equations (eq 28).

$$\text{S}_8 \rightleftharpoons \text{H}_2\text{NSS}_6\text{S}^-\,\text{NH}_4^+ \xrightleftharpoons{\text{NH}_3} \begin{cases} 2\,\text{H}_2\text{NSS}_2\text{S}^-\,\text{NH}_4^+ \\ 4\,\text{H}_2\text{NSS}^-\,\text{NH}_4^+ \\ 8\,\text{H}_2\text{NS}^-\,\text{NH}_4^+ \end{cases} \quad (28)$$

Sato obtained a mixture of products upon attempted reduction of 4-nitrotoluene in liquid ammonia (eq 29).[34]

$$\text{O}_2\text{N–C}_6\text{H}_4\text{–CH}_3 \xrightarrow[\text{100 °C, 6 h}]{\text{S}_8\text{, NH}_3} \text{H}_2\text{N–C}_6\text{H}_4\text{–CH}_3 \ (22\%) + \text{H}_2\text{N–C}_6\text{H}_4\text{–CN} \ (27\%) + (\text{NC–C}_6\text{H}_4\text{–S})_2 \ (2\%) \quad (29)$$

The use of sulfur as an oxidant is more common and several examples exist. The oxidation of aromatic sulfinates with elemental sulfur in amines gives thiosulfonates almost quantitatively.[35] Oxidation of benzylic carbons to aldehydes is observed with 4-nitrotoluene in a water–ethanol solvent at reflux (eq 30).[36] The oxidation of both methyl groups of *m*- and *p*-xylene gave high yields of isophthalic and terephthalic acids in 87 and 96% yields, respectively (eqs 31 and 32).[37]

$$\text{O}_2\text{N–C}_6\text{H}_4\text{–CH}_3 \xrightarrow[\text{NaOH, reflux; 74\%}]{\text{Na}_2\text{S•9H}_2\text{O; S, EtOH–H}_2\text{O}} \text{H}_2\text{N–C}_6\text{H}_4\text{–CHO} \quad (30)$$

$$m\text{-xylene} \xrightarrow[\text{H}_2\text{O, 316 °C, 30 min}]{\text{7.5 equiv S, 28\% NH}_4\text{OH}} m\text{-C}_6\text{H}_4(\text{CO}_2\text{H})_2 \quad (31)$$

$$p\text{-xylene} \xrightarrow[\text{H}_2\text{O, 316 °C, 30 min}]{\text{7.5 equiv S, 28\% NH}_4\text{OH}} p\text{-C}_6\text{H}_4(\text{CO}_2\text{H})_2 \quad (32)$$

Strong nucleophiles are generated upon the reaction of excess ammonia and elemental sulfur. These have been used to prepare arenesulfinamides from *S*-(4-nitrophenyl) *S*-substituted phenyl sulfoximides (eq 33).[38] The same conversion has been accomplished using 4-nitrophenyl substituted phenyl sulfoxides.[39]

$$\text{O}_2\text{N–C}_6\text{H}_4\text{–S(O)(=NTos)–C}_6\text{H}_4\text{–CH}_3 \xrightarrow[\text{20 °C, 3 h}]{\text{S}_8\text{, NH}_3} \text{O}_2\text{N–C}_6\text{H}_4\text{–S–S–NH}_2 + \text{CH}_3\text{–C}_6\text{H}_4\text{–S(O)–NH–Tos} \ (100\%) \quad (33)$$

The generation of anhydrous lithium sulfide is easily accomplished by reaction of two equivalents of ***Lithium Triethylborohydride*** with sulfur in THF at room temperature. This reacts well with a variety of electrophiles that require strictly anhydrous conditions to give good yields of products. The alkali metal disulfides are not commercially available and the methods for their preparation [Li^0, NH_3(l)] are difficult and give mixtures of polysulfide salts. The reaction of one equivalent of the hydride reagent with sulfur yields the lithium disulfide which gives high yields of disulfide products.[40] More recently a phase transfer catalytic method of generating the disulfide in a water/chloroform mixture has been used with electrophiles that can tolerate the presence of water (eq 34).[41]

$$\text{allyl-Cl} \xrightarrow[\text{0.01 equiv } R_4N^+Br^-,\ 15\,^\circ C,\ CHCl_3,\ 95\%]{\text{1 equiv S, 2 equiv } Li_2S,\ H_2O} \text{allyl-S-S-allyl} \quad (34)$$

When sulfur reacts with ***Sodium Borohydride*** in an appropriate solvent such as THF, the sulfurated hydride $NaBH_2S_3$ is formed. The reagent is most useful for its special reactivity and not necessarily as a sulfurating agent (eqs 35 and 36). A synopsis of its range of reactions was published more than 20 years ago.[42]

$$2\ RCHO \xrightarrow{NaBH_2S_3} \begin{cases} 0\,^\circ C,\ 90\%: & 2\ RCH_2OH \\ 65\,^\circ C: & RCH_2OH\ (\sim50\%) + RCH_2SSCH_2R\ (\sim50\%) \end{cases} \quad (35)$$

$$PhCOCH_3 \rightarrow \begin{cases} \text{1 equiv } NaBH_2S_3,\ 25\,^\circ C,\ 94\%: & PhCH(OH)CH_3 \\ \text{0.5 equiv } NaBH_2S_3,\ 65\,^\circ C: & PhCH(OH)CH_3 + PhCH(CH_3)SSCH(CH_3)Ph \end{cases} \quad (36)$$

Miscellaneous Reactions. Sulfur will react with organometallic reagents to replace a metal–carbon bond with a sulfur–carbon bond. Since sulfur is divalent, a thiolate results which is generally not useful. In some cases, however, the thiolate is essential for further reactivity. The utility of lithium thioalkynolates in the synthesis of alkenes is outlined below (eqs 37 and 38).[43]

$$R^1C\equiv CH \xrightarrow{BuLi} R^1C\equiv CLi \xrightarrow{S_8} R^1C\equiv CSLi \quad (37)$$

$$R^1C\equiv CSLi + R^2R^3C{=}O \rightarrow R^1C(=C{=}S)C(OLi)R^2R^3 \rightarrow \text{Li}^+\ \text{thietane-oxetane intermediate} \xrightarrow{H^+}$$

$$\text{2-thioxo-oxetane } (R^1, R^2, R^3) \xrightarrow{\Delta} R^2R^3C{=}CHR^1 + O{=}C{=}S \quad (38)$$

Using the same intermediate alkynylthiolates, one can prepare thioamides (eq 39).[44] This sequence works for selenium derivatives as well.

$$R^1C\equiv CLi \xrightarrow{S_8 \text{ or Se}} R^1C\equiv CS(Se)Li \xrightarrow{Et_2NH} [R^1C\equiv CS(Se)H \rightleftharpoons R^1CH{=}C{=}S(Se)] \xrightarrow{Et_2NH} R^1CH_2C(=S(Se))NEt_2 \quad (39)$$

A tetrahydrothiophene synthesis was accomplished by treating the carboalumination product of ***Triethylaluminum*** and terminal alkenes with sulfur (eq 40).[45]

$$RCH_2CH{=}CH_2 + AlEt_3 \xrightarrow[\text{8 h, 25 °C, PhH}]{\text{2 mol\% } Cp_2ZrCl_2} \text{1-ethyl-3-(RCH}_2\text{)aluminacyclopentane} \xrightarrow[60\text{–}85\%]{S_8,\ 80\,^\circ C,\ 6\ h} \text{3-(RCH}_2\text{)tetrahydrothiophene} \quad (40)$$

Diazo compounds react with sulfur to give thioketones. Monothioanthraquinones can be synthesized from the reaction of 10-diazoanthrones with sulfur in DMF at 130–150 °C (eq 41).[46]

$$\text{10-diazoanthrone } (=N_2) \xrightarrow[130\text{–}150\,^\circ C]{S,\ DMF} \text{monothioanthraquinone } (=S) \quad (41)$$

Sulfur is less reactive towards carbenes than selenium and so selenium is used in a catalytic fashion first to react with carbene-like compounds (isocyanides in this case), and then sulfur replaces the selenium to give an isothiocyanate (eq 42).[47] Using this method, cyclohexyl isothiocyanate was synthesized in 89% yield using 5 mol % of Se and 1.2 equiv of S. The reaction could be run with 0.5 mol % Se with only a modest lengthening of the reaction time.

$$RNC + Se \rightarrow RNCSe;\quad RNCSe + S \rightarrow RNCS + Se \quad (42)$$

An unusual aspect of sulfur chemistry is the reaction of alkenes with elemental sulfur (eqs 43 and 44). The reaction of sulfur with norbornene in the presence of ammonia in DMF at 110 °C results in a selective and stereospecific sulfuration.[48] The *exo*-3,4,5-trithiatricyclo[5.2.1.0^{2,6}]decane forms in about 86% yield.

$$\text{norbornene} \xrightarrow[110\,^\circ C,\ 2\ h]{S,\ NH_3,\ DMF} \textit{exo}\text{-trithiolane-fused norbornane} \quad (43)$$

$$\text{(44)}$$

S, NH$_3$, DMF; 110 °C, 2 h

Sulfur has also been used to oxidize inorganic compounds to increase their electrophilic nature. A simple synthesis of ***Triphenylphosphine*** that does not rely on organometallic reagents has been published by Olah (eq 45).[49] This Friedel–Crafts reaction fails with ***Phosphorus(III) Chloride*** as well as its corresponding oxide. There exists an unfavorable disproportionation between triphenylphosphine, diphenylchlorophosphine, and phenyldichlorophosphine. The oxychloride does not give the triphenylphosphine oxide. The phosphorus sulfochloride gives the corresponding triphenylphosphine sulfide, which is easily reduced.

$$\text{excess } C_6H_6 + S + PCl_3 \xrightarrow[\text{reflux, 8 h}]{\text{4 equiv } AlCl_3} (C_6H_5)_3P{=}S + 3\,HCl \quad (45)$$

Olah also used ***Trifluoromethanesulfonic Acid*** catalysis for the electrophilic sulfuration of cycloalkanes.[50] Elemental sulfur was heated in a stainless steel autoclave with excess cyclopentane in triflic acid at 150 °C for 12 h. This gave dicyclopentyl sulfide in 46% isolated yield (eq 46).

$$2\ C_5H_{10} \xrightarrow[\text{150 °C, 12 h}]{S_8,\ CF_3SO_3H} (C_5H_9)_2S + 3/4\ S_8\ (\text{or } S_6) + H_2S \quad (46)$$

Related Reagents. See Class O-12, pages 1–10.

1. Juraszyk, H. *CZ* **1974**, *98*, 126.
2. Schwenk, E.; Papa, D. *JOC* **1946**, *11*, 798 and references cited therein.
3. Gill, N. S.; Lions, F. *JACS* **1950**, *72*, 3468.
4. Konishi, K.; Nishiguchi, I.; Hirashima, T. *S* **1984**, 254.
5. Grisley, D. W., Jr.; Stephens, J. A. *JOC* **1961**, *26*, 3568.
6. Willgerodt, C. *CB* **1887**, *20*, 2467.
7. Willgerodt, C. *CB* **1888**, *21*, 534.
8. Kindler, K. *LA* **1923**, *431*, 187.
9. Kindler, K.; Li, T. *CB* **1941**, *74*, 321.
10. (a) Schwenk, E.; Bloch, E. *JACS* **1942**, *64*, 3051. (b) King, J. A.; McMillan, F. H. *JACS* **1946**, *68*, 2335.
11. Cavalieri, L.; Pattison, D. B.; Carmack, M. *JACS* **1945**, *67*, 1783.
12. Carmack, M.; DeTar, D. F. *JACS* **1946**, *68*, 2029.
13. Davis, C. H.; Carmack, M. *JOC* **1947**, *12*, 76.
14. Gerry, R. T.; Brown, E. V. *JACS* **1953**, *75*, 740.
15. Carmack, M.; Spielman, M. A. *OR* **1946**, *3*, 83.
16. Cocker, W.; Cross, B. E.; Edward, J. T.; Jenkinson, D. S.; McCormick, J. *JCS* **1953**, 2355.
17. Fieser, L. F. *Organic Experiments*, 2nd ed.; Raytheon Education: Lexington, MA, 1968; pp 290–294.
18. Haede, W.; Fritsch, W.; Radscheit, K.; Stache, U. *LA* **1973**, 5.
19. Piper, D. E.; Wright, G. F. *JACS* **1950**, *72*, 1669.
20. Hitchings, G. H.; Russell, P. B.; Whittaker, N. *JCS* **1956**, 1019.
21. Wynberg, H. *JACS* **1958**, *80*, 364.
22. Grandberg, I. I.; Kost, A. N. *JGU* **1958**, *28*, 3102.
23. Asinger, F.; Diem, H.; Sin-Gun, A. *LA* **1961**, *643*, 186.
24. (a) Franz, R. A.; Applegath, F.; Morriss, F. V.; Baiocchi, F. *JOC* **1961**, *26*, 3306. (b) Franz, R. A.; Applegath, F.; Morriss, F. V.; Baiocchi, F.; Bolze, C. *JOC* **1961**, *26*, 3309.
25. Franz, R. A.; Applegath, F. *JOC* **1961**, *26*, 3304.
26. Romanova, I. B.; Kutlukova, U.; Penskaya, L. V. *JGU* **1969**, *39*, 1884.
27. Mizuno, T.; Nishiguchi, I.; Hirashima, T.; Ogawa, A.; Kambe, N.; Sonoda, N. *S* **1988**, 257.
28. Mizuno, T.; Nishigushi, I.; Hirashima, T.; Ogawa, A.; Kambe, N.; Sonoda, N. *TL* **1988**, *29*, 4767.
29. Mizuno, T.; Nakamura, F.; Egashira, Y.; Nishiguchi, I.; Hirashima, T.; Ogawa, A.; Kambe, N.; Sonoda, N. *S* **1989**, 636.
30. Mizuno, T.; Nakamura, F.; Ishino, Y.; Nishiguchi, I.; Hirashima, T.; Ogawa, A.; Kambe, N.; Sonoda, N. *S* **1989**, 770.
31. Mizuno, T.; Yamaguchi, T.; Nishiguchi, I.; Okushi, T.; Hirashima, T. *CL* **1990**, 811.
32. Sonoda, N.; Mizuno, T.; Murakami, S.; Kondo, K.; Ogawa, A.; Ryu, I.; Kambe, N. *AG(E)* **1989**, *28*, 452.
33. Mizuno, T.; Nishiguchi, I.; Hirashima, T.; Ogawa, A.; Kambe, N.; Sonoda, N. *TL* **1990**, *31*, 4773.
34. Sato, R.; Takizawa, S.; Oae, S. *PS* **1979**, *7*, 229.
35. Sato, R.; Goto, T.; Takikawa, Y.; Takizawa, S. *S* **1980**, 615.
36. Beard, H. G.; Hodgson, H. H. *JCS* **1944**, 4.
37. Toland, W. G., Jr.; Hagmann, D. L.; Wilkes, J. B.; Brutschy, F. J. *JACS* **1958**, *80*, 5423.
38. Sato, R.; Saito, N.; Takikawa, Y.; Takizawa, S.; Saito, M. *S* **1983**, 1045.
39. Sato, R.; Chiba, S.; Takikawa, Y.; Takizawa, S.; Saito, M. *CL* **1983**, 535.
40. Gladysz, J. A.; Wong, V. K.; Jick, B. S. *CC* **1978**, 838.
41. Hase, T. A.; Peräkylä, H. *SC* **1982**, *12*, 947.
42. Lalancette, J. M.; Fréche, A.; Brindle, J. R.; Laliberté, M. *S* **1972**, 526.
43. Miyaura, N.; Yanagi, T.; Suzuki, A. *CL* **1979**, 535.
44. Sukhai, R. S.; de Jong, R.; Brandsma, L. *S* **1977**, 888.
45. Dzhemilev, U. M.; Ibragimov, A. G.; Zolotarev, A. P.; Tolstikov, G. A. *IZV* **1989**, *38*, 1324.
46. Raasch, M. S. *JOC* **1979**, *44*, 632.
47. Fujiwara, S.; Shin-Ike, T.; Sonoda, N.; Aoki, M.; Okada, K.; Miyoshi, N.; Kambe, N. *TL* **1991**, *32*, 3503.
48. Shields, T. C.; Kurtz, A. N. *JACS* **1969**, *91*, 5415.
49. Olah, G. A.; Hehemann, D. *JOC* **1977**, *42*, 2190.
50. Olah, G. A.; Wang, Q.; Prakash, G. K. S. *JACS* **1990**, *112*, 3697.

James A. Morrison
University of Wisconsin-Madison WI, USA

Tetrahydro-1-methyl-3,3-diphenyl-1*H*, 3*H*-pyrrolo[1,2-*c*][1,3,2]oxazaborole[1]

(*S*)
[112022-81-8] $C_{18}H_{20}BNO$ (MW 277.20)
($\cdot BH_3$)
[112022-90-9]
(*R*)
[112022-83-0]

(one of many chiral oxazaborolidines/chiral Lewis acids useful as enantioselective catalysts for the reduction of prochiral ketones,[1–3] imines,[4] and oximes,[2e,f,5] and the reduction of 2-pyranones to afford chiral biaryls;[6] other chiral oxazaborolidines have been used for the addition of diethylzinc to aldehydes,[7] asymmetric hydroboration,[8a,b] the Diels–Alder reaction,[9–11] and the aldol reaction[12,13])

Physical Data: mp 79–81 °C.
Solubility: very sol THF, CH_2Cl_2, toluene.
Preparative Methods: see text.
Purification: Kugelrohr distillation (50 °C/0.001 mbar)
Handling, Storage, and Precautions: the free oxazaborolidine must be rigorously protected from exposure to moisture. The crystalline borane complex is more stable, and is the preferred form to handle and store this catalyst.

Enantioselective Ketone Reduction. The major application of chiral oxazaborolidines has been the stoichiometric (as the oxazaborolidine–borane complex) (eq 1) and catalytic (in the presence of a stoichiometric borane source) (eq 2) enantioselective reduction of prochiral ketones.[1] These asymmetric catalysts work best for the reduction of aryl alkyl ketones, often providing very high (>95% ee) levels of enantioselectivity.

(1)

Following from the work of Itsuno[2] and Corey,[3] over 75 chiral oxazaborolidine catalysts have been reported for the reduction of prochiral ketones [(**1**),[2,3a,14,15a,e,f,16d–f,17b] (**2**),[16d,18b] (**3**),[3,6,19b–e,20,21,26c] (**4**),[16a] (**5**),[1b,16c,22] (**6**),[22b] (**7**),[3d,18a] (**8**),[16b] (**9**),[23] (**10**),[24] (**11**),[24] (**12**),[19a]]. Oxazaborolidines derived from proline (**3**) (see ***α,α-Diphenyl-2-pyrrolidinemethanol***) and valine (**1**; $R^4 = i$-Pr) (see ***2-Amino-3-methyl-1,1-diphenyl-1-butanol***) have received the most attention.

(0.1 equiv)
$Me_2S \cdot BH_3$ (0.6 equiv) (2)

(**1**) (**2**) (**3**) (**4**) (**5**) (**6**) (**7**) (**8**) (**9**) (**10**) (**11**) (**12**)

Unsubstituted (B–H) oxazaborolidines (**16**) are prepared from a chiral β-amino alcohol (**13**) and a source of borane (***Diborane***, ***Borane–Tetrahydrofuran***, ***Borane–Dimethyl Sulfide***, or $H_3B \cdot NMe_3$) via a multistep process (eq 3). Formation of the initial amine–borane complex (**14**) is generally exothermic, and this intermediate can often be isolated. Gentle heating with the loss of one mole of hydrogen results in the formation of (**15**). Continued heating with the loss of a second mole of hydrogen then affords oxazaborolidine (**16**). When R^4 and R^5 are connected, forming a four- or five-membered ring, more forcing conditions (70–75 °C, 1.7 bar, 48–72 h) are required to effect this conversion due to the additional ring strain. [*Caution*: under these conditions, borane or diborane in the vapor phase can begin to decompose.[25]] Finally, additional borane is added to afford the oxazaborolidine–borane complex (**17**).

Free oxazaborolidine (**16**), by itself, will not reduce ketones. Furthermore, (**16**) is not particularly stable, reacting with moisture (H_2O), air (O_2), unreacted amino alcohol, other alcohols,[8c] or, depending on the substituents, with itself to form various dimers.[3a,8c,d,15d,26,27a] This instability is due to the strain of a partial double bond between nitrogen and boron (eq 4). Formation of the oxazaborolidine–borane complex (**17**) tends to release some of this strain. As such, (**16**) and (**17**) are generally prepared and used in situ without isolation; in many cases, they have not been fully characterized.[17c]

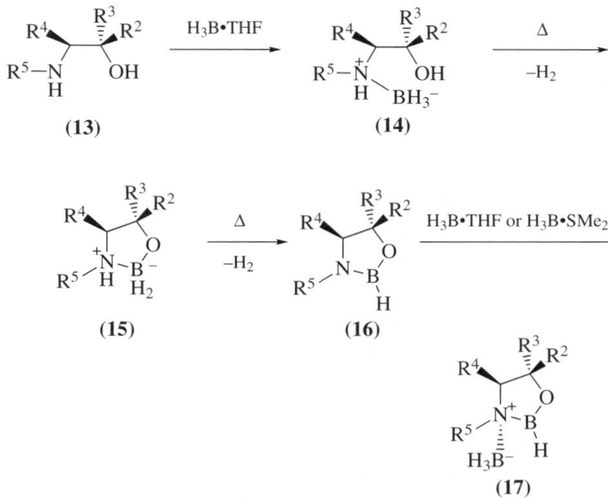

(3)

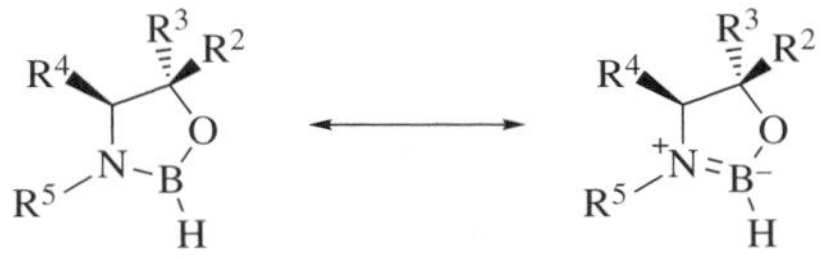

(4)

Oxazaborolidines substituted at boron (**1**; R^1 = alkyl, aryl) are prepared from a chiral β-amino alcohol and the corresponding boronic acid in a two-step process (eq 5).[3b,9] Heat and an efficient method of water removal (i.e. azeotropic distillation, molecular sieves) are required to drive the second step. When R^4 and R^5 are connected, more forcing conditions are necessary, both to complete the second step and to prevent the intermediate from proceeding to an alternate disproportionation product.[21] Alternative procedures using bis(diethylamino)phenylborane (eq 6),[26a,b] trisubstituted boroxines (eq 7),[21,27] and ethyl or butyl bis(trifluoroethyl)boronate esters (eq 8)[19e] have been developed to circumvent these problems. The substituted oxazaborolidines are more stable than unsubstituted (B–H) oxazaborolidines (i.e. they can be handled in the presence of air, and do not form dimers), but are still prone to decomposition by moisture (H_2O).[21] In many cases the substituted oxazaborolidines have been isolated, purified, and characterized.

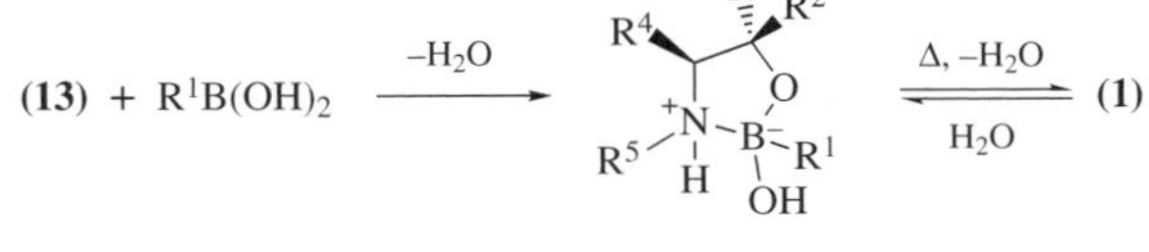

(5)

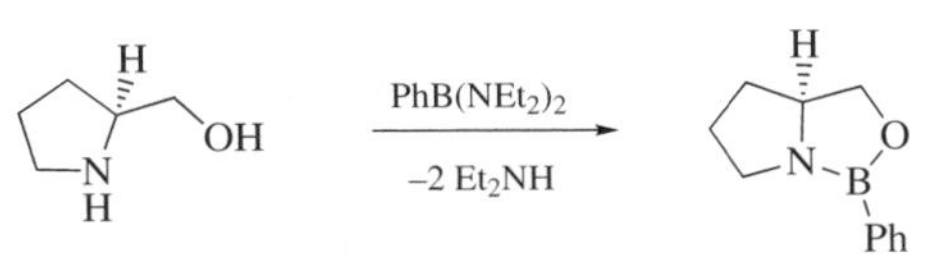

(6)

Substituted oxazaborolidines also react with borane (B_2H_6, $H_3B{\cdot}THF$, or $H_3B{\cdot}SMe_2$) to form an oxazaborolidine–borane complex (**19**) (eq 9).[3b,27] The oxazaborolidine–borane complex, by releasing the strain of the partial double bond between the ring boron and nitrogen, is more stable than the free oxazaborolidine, and in many cases exists as a stable crystalline solid.[21c,27,28]

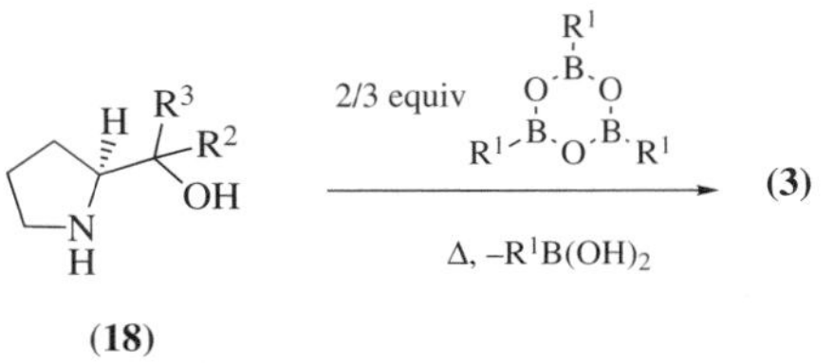

(7)

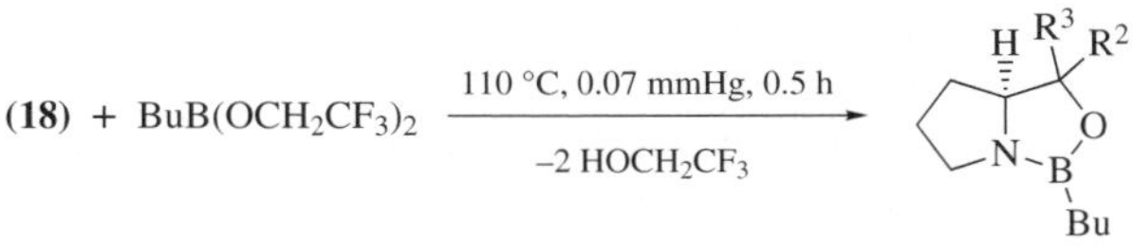

(8)

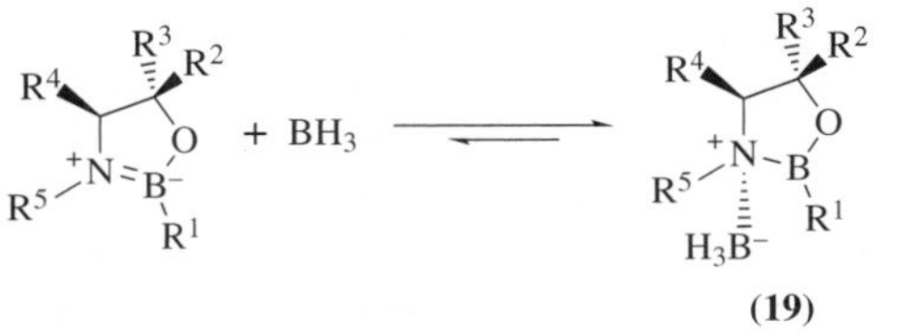

(9)

The oxazaborolidine–borane complex (**19**) can be used stoichiometrically (eq 1) or catalytically (eq 10) for the enantioselective reduction of prochiral ketones.[27a] When used catalytically, the oxazaborolidine–borane complex (**19**) is the second intermediate in the catalytic cycle (eq 10) proposed to explain the behavior of the oxazaborolidine catalyst.[3a,29] Subsequent coordination between the Lewis acidic ring boron and the carbonyl oxygen activates the ketone toward reduction. Intramolecular hydride transfer from the BH_3 coordinated to the ring nitrogen then occurs via a six-membered ring chair transition state.[17b,27a,30] Following hydride transfer, the alkoxy–BH_2 dissociates, and oxazaborolidine (**1**) is free to begin the cycle again. The diastereomeric transition state model (**20**), leading to the enantiomeric carbinol product, is disfavored due to unfavorable 1,3-diaxial steric interactions between R_L and R^1. Additional work will be required to better understand the catalytic cycle and the intermediates involved to further improve the oxazaborolidine catalysts. The behavior of the catalysts has been the subject of molecular orbital calculations in a series of 12 papers.[31] It should be noted, however, that not all of the results and conclusions are supported by experimental observations.

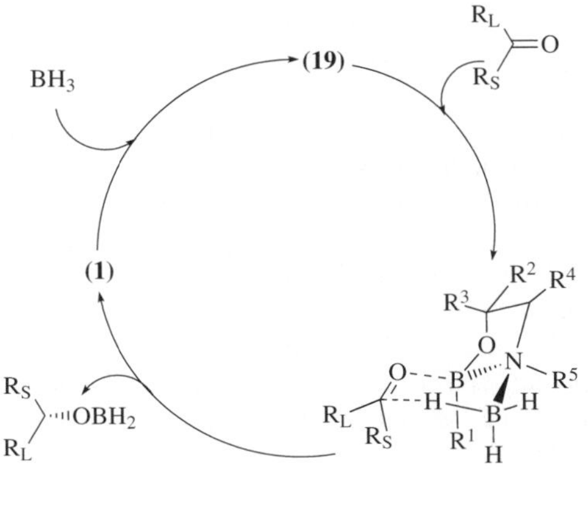

(10)

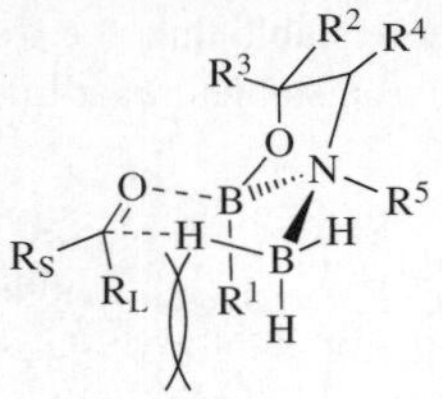

(20)

The enantioselectivities reported for the reduction of acetophenone and 1-tetralone using several representative chiral (4*S*)-oxazaborolidine catalysts are summarized in Table 1. The oxazaborolidines derived from (*S*)-azetidinecarboxylic acid and (*S*)-proline provide the best results. It is interesting to note the reversal in enantioselectivity going from catalyst (**5a**) to (**6a**).

Oxazaborolidine catalyzed reductions are generally performed in an aprotic solvent, such as dichloromethane, THF, or toluene. When the reactions are run in a Lewis basic solvent, such as THF, the solvent competes with the oxazaborolidine to complex with the borane, which can have an effect on the enantioselectivity and/or rate of the reaction.[27a] The solubility of the oxazaborolidine–borane complex can be the limiting factor for reactions run in toluene, although this problem has been circumvented by using oxazaborolidines with more lipophilic substituents ($R^1 = n$-Bu; R^2, R^3 = 2-naphthyl).[19b–d] We have found dichloromethane to be the best overall solvent for these reactions.[27a]

The reactions are typically performed using $H_3B{\cdot}THF$, $H_3B{\cdot}SMe_2$, or ***Catecholborane***[19d] as the hydride source. When using $H_3B{\cdot}THF$ or $H_3B{\cdot}SMe_2$, two of the three hydrides are effectively utilized.[27a] This is only true for reactions run at temperatures greater than −40 °C. At lower temperatures, only one hydride is transferred at a reasonable rate. When two hydrides are used, there is some evidence that the enantioselectivity for transfer of the second hydride is different, and may in fact be lower.[27a] Whether this implies that an alternative catalytic cycle operates, whereby the alkoxy–BH_2 intermediate generated during the first hydride transfer remains coordinated to the oxazaborolidine, and then transfers the second hydride (with a different degree of enantioselectivity), or that some other intermediate present is active, but not as an enantioselective reducing agent, will require further investigation. In any event, the amount of BH_3 used should be at least 0.5 mole per mole of ketone plus an amount equal to the oxazaborolidine catalyst, with the possibility that 1 mole per mole provides slightly higher enantioselectivity. When catecholborane is used as the hydride source, a 50–100% excess of this reagent is used.

The mode of addition and the reaction temperature both affect the enantioselectivity of the reaction. The best results are obtained when the ketone is added slowly to a solution of the oxazaborolidine (or oxazaborolidine–borane complex) and the borane source, at as low a temperature that provides a reasonable reaction rate.[27a] This is in contrast to a previous report that indicated that oxazaborolidine-catalyzed reductions 'lose stereoselectivity at lower temperatures'.[19d] With unsubstituted (R^1 = H) oxazaborolidines, higher temperatures may be required due to incomplete formation of the catalyst, the presence of dimers, and/or other intermediates.[26c]

In their role as enantioselective catalysts for the reduction of prochiral ketones, chiral oxazaborolidines have been used for the preparation of prostaglandins,[3a] PAF antagonists,[3a] a key intermediate of ginkgolide B,[32a] bilobalide,[32b] a key intermediate of forskolin,[32c] (*R*)- and (*S*)-fluoxetine,[32d] (*R*)- and (*S*)-isopreterenol,[19c] vitamin D analogs,[33] the carbonic anhydrase inhibitor MK-0417,[21b] the dopamine D1 agonist A-77636,[20b] taxol,[34] the LTD_4 antagonists L-695,499 and L-699,392,[35] the β-adrenergic agonist CL 316,243,[36] and the antiarrhythmic MK-0499.[37] They have also been used for the synthesis of chiral amines,[38,39] α-hydroxy acids,[19d,40a] benzylic thiols,[40c] the enantioselective reduction of trihalomethyl ketones,[40a,b,d] and ketones containing various heteroatoms.[17a,21b,27a,35,37]

Enantioselective Reduction of Imines and Ketoxime *O*-Ethers. In addition to the reduction of prochiral ketones, chiral oxazaborolidines have been employed as enantioselective reagents and catalysts for the reduction of imines (eq 11)[4,23] and ketoxime *O*-ethers (eq 12)[2e,f,5] to give chiral amines. It is interesting to note that the enantioselectivity for the reduction of ketoxime *O*-ethers is opposite that of ketones and imines. For more information, see ***2-Amino-3-methyl-1,1-diphenyl-1-butanol***.

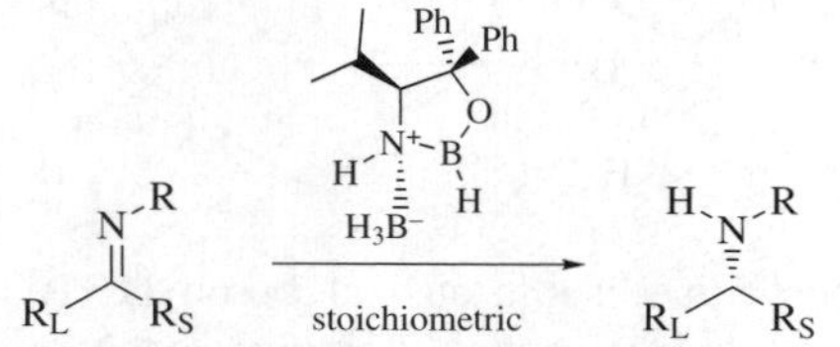

stoichiometric (11)

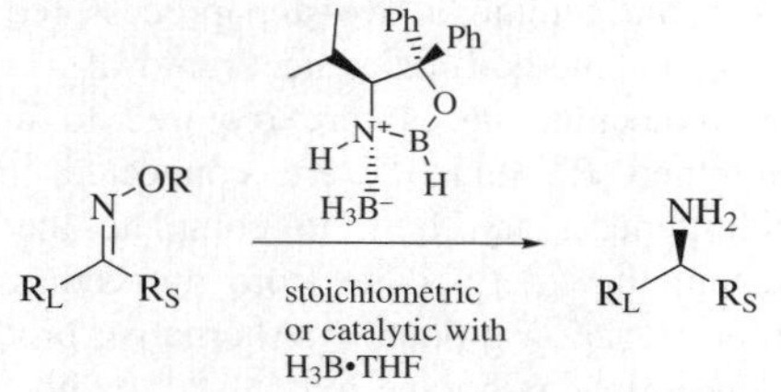

stoichiometric or catalytic with $H_3B{\cdot}THF$ (12)

Enantioselective Addition of Diethylzinc to Aldehydes. Oxazaborolidines derived from ephedrine have been used to catalyze the addition of ***Diethylzinc*** to aldehydes (eq 13).[7] Both the rate and enantioselectivity are optimized when R^1 = H. Aromatic aldehydes generally react faster than aliphatic aldehydes, and the enantioselectivity for aromatic aldehydes is good to excellent (86–96% ee).

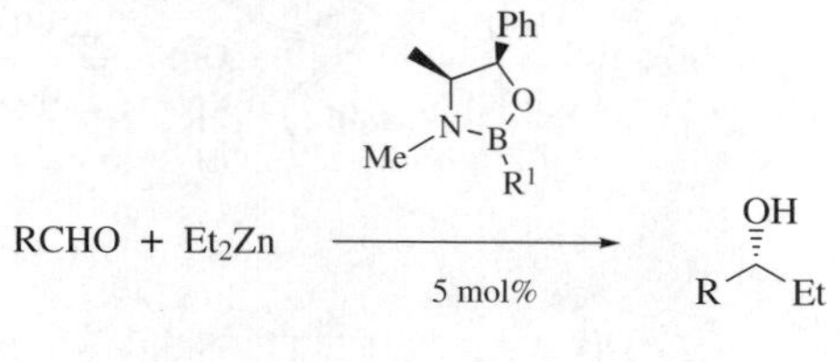

$$\text{RCHO} + \text{Et}_2\text{Zn} \xrightarrow{5\ \text{mol\%}} \text{RCH(OH)Et} \quad (13)$$

Other Applications. Chiral oxazaborolidines derived from ephedrine have also been used in asymmetric hydroborations,[8a,b] and as reagents to determine the enantiomeric purity of secondary alcohols.[8c] Chiral 1,3,2-oxazaborolidin-5-ones derived from amino acids have been used as asymmetric catalysts for the Diels–Alder reaction,[9–11] and the aldol reaction.[12,13]

Table 1 Chiral Oxazaborolidine Catalyzed Reduction of Acetophenone and 1-Tetralone

Catalyst	R^1	R^2,R^3	R^4 (mol %)	Catalyst (ee %)	Acetophenone (ee %)	1-Tetralone
(**1a**)[3a]	H	Ph	*i*-Pr	10	94.7 (*R*)	–
(**2a**)[16d]	H	Ph	–	10	98 (*R*)	–
(**3a**)[3a]	H	Ph	–	10	97 (*R*)	89 (*R*)
(**3b**)[21b]	Me	Ph	–	10	98 (*R*)	94 (*R*)
(**3b**)·BH_3[27a]	Me	Ph	–	5	97.6 (*R*)	99.0 (*R*)
(**3b**)·BH_3[27a]	Me	Ph	–	100	99.8 (*R*)	99.2 (*R*)
(**5a**)[22b]	H	Ph	–	10	96 (*R*)	79 (*R*)
(**6a**)[22b]	H	Ph	–	10	90 (*S*)	79 (*S*)
(**7a**)[18a]	H	Ph	–	10	87 (*R*)	–
(**8a**)[16b]	H	Ph	–	10	71 (*R*)	44 (*R*)
(**9a**)[23]	H	H	–	110	88 (*R*)	–
(**12a**)[19a]	Me	Ph	–	10	97.5 (*R*)	95.3 (*R*)

Related Reagents. See Classes R-2, R-5, R-9, R-10, R-11, R-12, and R-17, pages 1–10. 2-Amino-3-methyl-1,1-diphenyl-1-butanol; α,α-Diphenyl-2-pyrrolidinemethanol; Ephedrine-borane; Norephedrine–Borane.

---

1. (a) Wallbaum, S.; Martens, J. *TA* **1992**, *3*, 1475. (b) Singh, V. K. *S* **1992**, 605. (c) Deloux, L.; Srebnik M. *CRV* **1993**, *93*, 763.
2. (a) Hirao, A.; Itsuno, S.; Nakahama, S.; Yamazaki, N. *CC* **1981**, 315. (b) Itsuno, S.; Hirao, A.; Nakahama, S.; Yamazaki, N. *JCS(P1)* **1983**, 1673. (c) Itsuno, S.; Ito, K.; Hirao, A.; Nakahama, S. *CC* **1983**, 469. (d) Itsuno, S.; Ito, K.; Hirao, A.; Nakahama, S. *JOC* **1984**, *49*, 555. (e) Itsuno, S.; Nakano, M.; Miyazaki, K.; Masuda, H.; Ito, K.; Hirao, A.; Nakahama, S. *JCS(P1)* **1985**, 2039. (f) Itsuno, S.; Nakano, M.; Ito, K.; Hirao, A.; Owa, M.; Kanda, N.; Nakahama, S. *JCS(P1)* **1985**, 2615.
3. (a) Corey, E. J.; Bakshi, R. K.; Shibata, S. *JACS* **1987**, *109*, 5551. (b) Corey, E. J.; Bakshi, R. K.; Shibata, S.; Chen, C. P.; Singh, V. K. *JACS* **1987**, *109*, 7925. (c) Corey, E. J.; Shibata, S.; Bakshi, R. K. *JOC* **1988**, *53*, 2861. (d) Corey, E. J. U.S. Patent 4 943 635, 1990.
4. (a) Cho, B. T.; Chun, Y. S. *JCS(P1)* **1990**, 3200. (b) Cho, B. T.; Chun, Y. S. *TA* **1992**, *3*, 337.
5. (a) Itsuno, S.; Sakurai, Y.; Ito, K.; Hirao, A.; Nakahama, S. *BCJ* **1987**, *60*, 395. (b) Itsuno, S.; Sakurai, Y.; Shimizu, K.; Ito, K. *JCS(P1)* **1989**, 1548. (c) Itsuno, S.; Sakurai, Y.; Shimizu, K.; Ito, K. *JCS(P1)* **1990**, 1859.
6. Bringmann, G.; Hartung, T. *AG(E)* **1992**, *31*, 761.
7. Joshi, N. N.; Srebnik, M.; Brown, H. C. *TL* **1989**, *30*, 5551.
8. (a) Brown, J. M.; Lloyd-Jones, G. C. *TA* **1990**, *1*, 869. (b) Brown, J. M.; Lloyd-Jones, G. C. *CC* **1992**, 710. (c) Brown, J. M.; Leppard, S. W.; Lloyd-Jones, G. C. *TA* **1992**, *3*, 261. (d) Brown, J. M.; Lloyd-Jones, G. C.; Layzell, T. P. *TA* **1993**, *4*, 2151.
9. Takasu, M.; Yamamoto, H. *SL* **1990**, 194.
10. (a) Sartor, D.; Saffrich, J.; Helmchen, G. *SL* **1990**, 197. (b) Sartor, D.; Saffrich, J.; Helmchen, G.; Richards, C. J.; Lambert, H. *TA* **1991**, *2*, 639.
11. (a) Corey, E. J.; Loh, T.-P. *JACS* **1991**, *113*, 8966. (b) Corey, E. J.; Loh, T.-P.; Roper, T. D.; Azimioara, M. D.; Noe, M. C. *JACS* **1992**, *114*, 8290.
12. Kiyooka, S.; Kaneko, Y.; Komura, M.; Matsuo, H.; Nakano, M. *JOC* **1991**, *56*, 2276.
13. Parmee, E. R.; Tempkin, O.; Masamune, S.; Abiko, A. *JACS* **1991**, *113*, 9365.
14. Mandal, A. K.; Kasar, T. G.; Mahajan, S. W.; Jawalkar, D. G. *SC* **1987**, *17*, 563.
15. (a) Grundon, M. F.; McCleery, D. G.; Wilson, J. W. *JCS(P1)* **1981**, 231. (b) Mancilla, T.; Santiesteban, F.; Contreras, R.; Klaebe, A. *TL* **1982**, *23*, 1561. (c) Tlahuext, H.; Contreras, R. *TA* **1992**, *3*, 727. (d) Tlahuext, H. Contreras, R. *TA* **1992**, *3*, 1145 (e) Cho, B. T.; Chun, Y. S. *TA* **1992**, *3*, 1539 (f) Berenguer, R.; Garcia, J.; Gonzalez, M.; Vilarrasa, J. *TA* **1993**, *4*, 13.
16. (a) Wallbaum, S.; Martens, J. *TA* **1991**, *2*, 1093. (b) Stingl, K.; Martens, J.; Wallbaum, S. *TA* **1992**, *3*, 223. (c) Martens, J.; Dauelsberg, C.; Behnen, W.; Wallbaum, S. *TA* **1992**, *3*, 347. (d) Behnen, W.; Dauelsberg, C.; Wallbaum, S.; Martens, J. *SC* **1992**, *22*, 2143. (e) Mehler, T.; Martens, J. *TA* **1993**, *4*, 1983. (f) Mehler, T.; Martens, J. *TA* **1993**, *4*, 2299.
17. (a) Quallich, G. J.; Woodall, T. M. *TL* **1993**, *34*, 785. (b) Quallich, G. J.; Woodall, T. M. *TL* **1993**, *34*, 4145. (c) Quallich, G. J.; Woodall, T. M. *SL* **1993**, 929.
18. (a) Rao, A. V. R.; Gurjar, M. K.; Sharma, P. A.; Kaiwar, V. *TL* **1990**, *31*, 2341. (b) Rao, A. V. R.; Gurjar, M. K.; Kaiwar, V. *TA* **1992**, *3*, 859.
19. (a) Corey, E. J.; Chen, C. P.; Reichard, G. A. *TL* **1989**, *30*, 5547. (b) Corey, E. J.; Link, J. O. *TL* **1989**, *30*, 6275. (c) Corey, E. J.; Link, J. O. *TL* **1990**, *31*, 601. (d) Corey, E. J.; Bakshi, R. K. *TL* **1990**, *31*, 611. (e) Corey, E. J.; Link, J. O. *TL* **1992**, *33*, 4141.
20. (a) DeNinno, M. P.; Perner, R. J.; Lijewski, L. *TL* **1990**, *31*, 7415. (b) DeNinno, M. P.; Perner, R. J.; Morton, H. E.; DiDomenico, Jr., S. *JOC* **1992**, *57*, 7115.
21. (a) Mathre, D. J.; Jones, T. K.; Xavier, L. C.; Blacklock, T. J.; Reamer, R. A.; Mohan, J. J.; Jones, E. T. T.; Hoogsteen, K.; Baum, M. W.; Grabowski, E. J. J. *JOC* **1991**, *56*, 751. (b) Jones, T. K.; Mohan, J. J.; Xavier, L. C.; Blacklock, T. J.; Mathre, D. J.; Sohar, P.; Jones, E. T. T.; Reamer, R. A.; Roberts, F. E.; Grabowski, E. J. J. *JOC* **1991**, *56*, 763. (c) Blacklock, T. J.; Jones, T. K.; Mathre, D. J.; Xavier, L. C. U.S. Patent 5 039 802, 1991. (d) Blacklock, T. J.; Jones, T. K.; Mathre, D. J.; Xavier, L. C. U.S. Patent 5 264 585, 1993. (e) Shinkai, I. *JHC* **1992**, *29*, 627.
22. (a) Youn, I. K.; Lee, S. W.; Pak, C. S. *TL* **1988**, *29*, 4453. (b) Kim, Y. H.; Park, D. H.; Byun, I. S.; Yoon, I. K.; Park, C. S. *JOC* **1993**, *58*, 4511.
23. Nakagawa, M.; Kawate, T.; Kikikawa, T.; Yamada, H.; Matsui, T.; Hino, T. *T* **1993**, *49*, 1739.
24. Tanaka, K.; Matsui, J.; Suzuki, H. *CC* **1991**, 1311.
25. (a) Long, L. H. *J. Inorg. Nucl. Chem.* **1970**, *32*, 1097. (b) Fernandez, H.; Grotewold, J.; Previtali, C. M. *JCS(D)* **1973**, 2090. (c) Gibb, T. C.; Greenwood, N. N.; Spalding, T. R.; Taylorson, D. *JCS(D)* **1979**, 1398.
26. (a) Bielawski, J.; Niedenzu, K. *Synth. React. Inorg. Met.-Org. Chem.* **1980**, *10*, 479. (b) Cragg, R. H.; Miller, T. J. *JOM* **1985**, *294*, 1. (c) Brunel, J. M.; Maffei, M.; Buono, G. *TA* **1993**, *4*, 2255.
27. (a) Mathre, D. J.; Thompson, A. S.; Douglas, A. W.; Hoogsteen, K.; Carroll, J. D.; Corley, E. G.; Grabowski, E. J. J. *JOC* **1993**, *58*, 2880.

(b) Blacklock, T. J.; Jones, T. K.; Mathre, D. J.; Xavier, L. C. U.S. Patent 5 189 177, 1993. (c) Carroll, J. D.; Mathre, D. J.; Corley, E. G.; Thompson, A. S. U.S. Patent 5 264 574, 1993.

28. Corey, E. J.; Azimioara, M.; Sarshar, S. *TL* **1992**, *24*, 3429.
29. Evans, D. A. *Science* **1988**, *240*, 420.
30. Jones, D. K.; Liotta, D. C.; Shinkai, I.; Mathre, D. J. *JOC* **1993**, *58*, 799.
31. Nevalainen, V. *TA* **1993**, *4*, 2001; and references contained therein.
32. (a) Corey, E. J.; Gavai, A. V. *TL* **1988**, *29*, 3201. (b) Corey, E. J.; Su, W.-G. *TL* **1988**, *29*, 3423. (c) Corey, E. J.; Jardine, P. D. S.; Mohri, T. *TL* **1988**, *29*, 6409. (d) Corey, E. J.; Reichard, G. A. *TL* **1989**, *30*, 5207.
33. (a) Kabat, M.; Kiegiel, J.; Cohen, N.; Toth, K.; Wovkulich, P. M.; Uskokovic, M. R. *TL* **1991**, *32*, 2343. (b) Lee, A. S.; Norman, A. W.; Okamura, W. H. *JOC* **1992**, *57*, 3846.
34. Nicolaou, K. C.; Hwang, C.-K.; Sorensen, E. J.; Clairborne, C. F. *CC* **1992**, 1117.
35. (a) Labelle, M.; Prasit, P.; Belley, M.; Blouin, M.; Champion, E.; Charette, L.; DeLuca, J. G.; Dufresne, C.; Frenette, R.; Gauthier, J. Y.; Grimm, E.; Grossman, S. J.; Guay, D.; Herold, E. G.; Jones, T. R.; Lau, Y.; Leblanc, Y.; Leger, S.; Lord, A.; McAuliffe, M.; McFarlane, C.; Masson, P.; Metters, K. M.; Ouimet, N.; Patrick, D. H.; Perrier, H.; Piechuta, H.; Roy, P.; Williams, H.; Wang, Z.; Xiang, Y. B.; Zamboni, R. J.; Ford-Hutchinson, A. W.; Young, R. N. *BML* **1992**, *2*, 1141. (b) King, A. O.; Corley, E. G.; Anderson, R. K.; Larsen, R. D.; Verhoeven, T. R.; Reider, P. J.; Xiang, Y. B.; Belley, M.; Leblanc, Y.; Labelle, M.; Prasit, P.; Zamboni, R. J. *JOC* **1993**, *58*, 3731.
36. Bloom, J. D.; Dutia, M. D.; Johnson, B. D.; Wissner, A.; Burns, M. G.; Largis, E. E.; Dolan, J. A.; Claus, T. H. *JMC* **1992**, *35*, 3081.
37. Cai, D.; Tschaen, D.; Shi, Y.-J.; Verhoeven, T. R.; Reamer, R. A.; Douglas, A. W. *TL* **1993**, *34*, 3243.
38. Chen, C.-P.; Prasad, K.; Repic, O. *TL* **1991**, *32*, 7175.
39. Thompson, A. S.; Humphrey, G. R.; DeMarco, A. M.; Mathre, D. J.; Grabowski, E. J. J. *JOC* **1993**, *58*, 5886.
40. (a) Corey, E. J.; Cheng, X. M.; Cimprich, K. A.; Sarshar, S. *TL* **1991**, *32*, 6835. (b) Corey, E. J.; Link, J. O. *TL* **1992**, *33*, 3431. (c) Corey, E. J.; Cimprich, K. A. *TL* **1992**, *33*, 4099. (d) Corey, E. J.; Link, J. O.; Bakshi, R. K. *TL* **1992**, *33*, 7107.

David J. Mathre & Ichiro Shinkai
Merck Research Laboratories, Rahway, NJ, USA

Tetramethylammonium Triacetoxyborohydride

$R_4N(AcO)_3BH$

(R = Me)
[109704-53-2] $C_{10}H_{22}BNO_6$ (MW 263.14)
(R = Bu)
[83722-99-0] $C_{22}H_{46}BNO_6$ (MW 431.50)

(reducing agent for the regio- and stereoselective reduction of β-hydroxy ketones, amides, and oximino ethers)[1]

Alternate Name: TABH.
Physical Data: mp 96–98 °C.
Solubility: sol range of organic solvents including CH_2Cl_2, $CHCl_3$, MeCN.
Form Supplied in: hygroscopic white powder; commercially available.
Preparative Methods: experimental procedures are available (eqs 1 and 2).[5a,6]

$$Me_4NOH + NaBH_4 \longrightarrow Me_4NBH_4 \quad (1)$$

$$Me_4NBH_4 + AcOH \longrightarrow Me_4N(AcO)_3BH \quad (2)$$

Purification: recrystallized from CH_2Cl_2–EtOAc.[5a]
Handling, Storage, and Precautions: moisture sensitive; irritant. Stable at 25 °C. Handle in a well ventilated fume hood.

Selective Reduction of Aldehydes. Tetra-*n*-butylammonium triacetoxyborohydride ($Bu_4N(AcO)_3BH$) reduces aldehydes selectively in the presence of ketones (eq 3).[2] β-Keto aldehydes are reduced to afford the corresponding 1,3-diol adduct; none of the derived primary β-hydroxy ketone is generated. It is proposed that the aldehyde carbonyl group is reduced first and the resulting hydroxyl group directs the delivery of hydride to the ketone site (eq 4).[3]

$Bu_4N(AcO)_3BH$, benzene, 80 °C, 77% (3)

(4)

Similar observations have been reported with ***Sodium Triacetoxyborohydride***.[4] As shown in eq 5, reduction of a cyclic ketone containing a properly disposed hydroxyl function (for hydride delivery) proceeds to afford the derived equatorial alcohol with high diastereoselectivity. In contrast, an equal mixture of axial and equatorial alcohols are obtained when ***Sodium Borohydride*** is employed as the reducing agent.

$Na(OAc)_3BH$, AcOH (5)

Directed Reduction of β-Hydroxy Ketones. Evans has reported that TABH may be employed to effect the stereoselective reduction of β-hydroxy ketones.[5] Several reducing agents under a variety of conditions have been examined (eq 6 and Table 1). The reactivity and selectivity with TABH in AcOH/MeCN are superior to other solvent systems. The presence of HOAc is essential to reaction efficiency, as the protic acid catalyzes the rate determining association between the boron reagent and the resident substrate heteroatom.[5b] Acetonitrile is required since the reduction process may then be performed at lower temperatures without freezing of the reaction mixture; reduced temperatures are critical to obtaining high levels of stereoinduction.

[H⁻], solvent (6)

Table 1 Diastereoselective Directed Reduction of β-Hydroxy Ketones by Various Borohydrides

Hydride	Solvent	Time	Temp (°C)	Ratio (*anti*:*syn*)
$NaBH_4$	AcOH	30 min	25	80:20
$Na(AcO)_3BH$	AcOH	15 min	25	84:16
Me_4NBH_4	AcOH	30 min	25	92:8
$Me_4N(AcO)_3BH$	AcOH	15 min	25	92:8
$Me_4N(AcO)_3BH$	AcOH/MeCN	5 h	−40	95:5

As illustrated in the directed reductions shown in eqs 7 and 8, the level and sense of diastereoselectivity is independent of the local chirality within the substrate molecule.

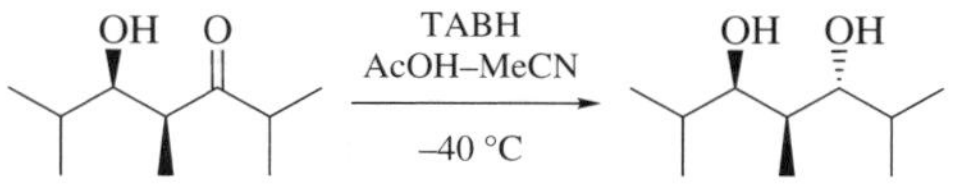
OH
O
TABH
AcOH–MeCN
−40 °C
OH
OH

ds 50:1 (7)

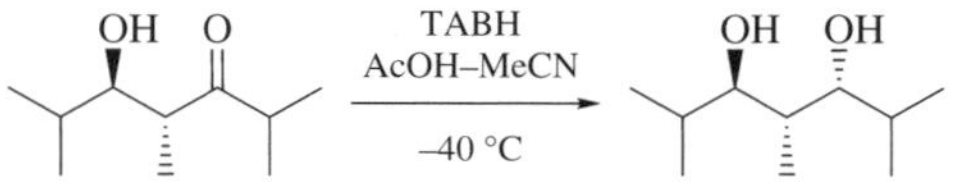
OH
O
TABH
AcOH–MeCN
−40 °C
OH
OH

ds 50:1 (8)

Eqs 9 and 10 illustrate two additional important points with regard to the directed borohydride reduction: (1) the hydroxyl unit of an enol function may serve as the directing group (eq 9); (2) the stereoselective intramolecular hydride delivery may be relayed to effect generation of multiple stereogenic sites within an acyclic chain (eq 10).

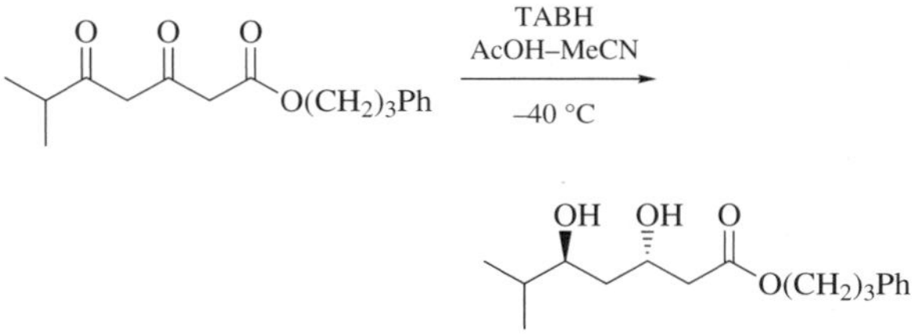
O
O
O
O(CH2)3Ph
TABH
AcOH–MeCN
−40 °C
OH
OH
O
O(CH2)3Ph

(9)

ds 92:8

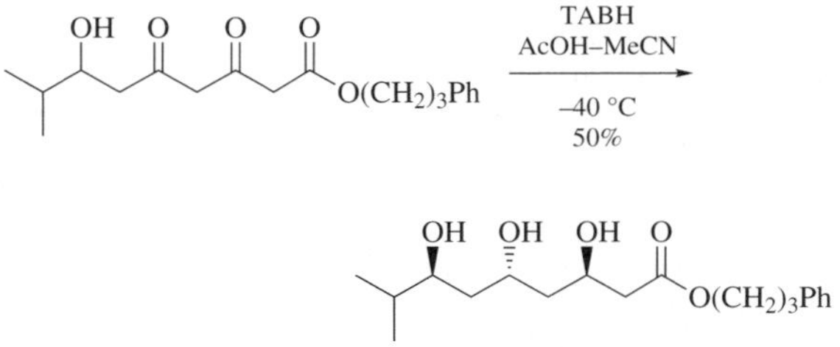
OH
O
O
O
O(CH2)3Ph
TABH
AcOH–MeCN
−40 °C
50%
OH
OH
OH
O
O(CH2)3Ph

(10)

one stereoisomer

Directed Reduction of Oximino Ethers.

Substrates with β-Hydroxy Substitution. Stereocontrolled reduction of β-hydroxy oximino ethers has been reported by Williams.[6] Eqs 11 and 12 indicate that whereas (*E*) substrates are reduced selectively, (*Z*) adducts react with TABH with inferior levels of stereoinduction. As illustrated in eqs 13–16, the stereochemical outcome of reduction of (*E*) oxime ethers is insensitive to the substrate local chirality, whereas the corresponding (*Z*) isomers afford β-hydroxy amines with high selectivity. However, in the latter instance the identity of the major isomer depends on the configuration of the α stereogenic center (eqs 13 and 14). Within the same context, with substrates containing a primary hydroxyl group, the α stereogenic center induces the preferred formation of the *syn* stereoisomer (eq 17). A mechanism-based understanding of the trends in stereoselectivity shown is not yet available.

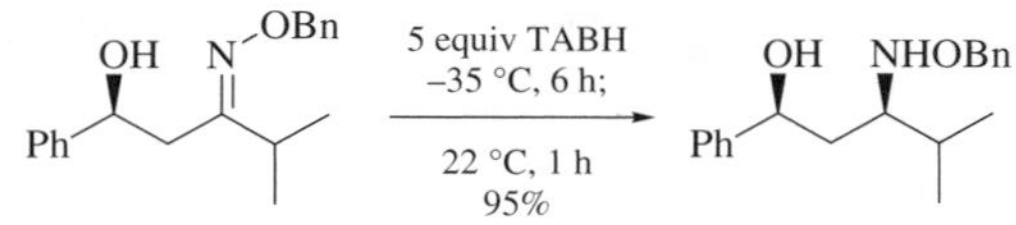
OH
N
OBn
Ph
5 equiv TABH
−35 °C, 6 h;
22 °C, 1 h
95%
OH
NHOBn
Ph

(11)

65:35

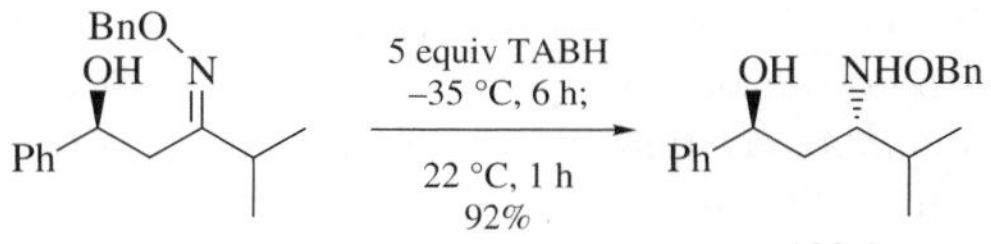
BnO
OH
N
Ph
5 equiv TABH
−35 °C, 6 h;
22 °C, 1 h
92%
OH
NHOBn
Ph

(12)

100:1

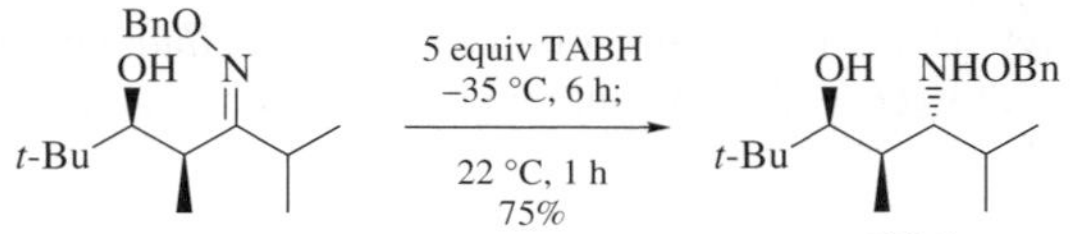
BnO
OH
N
t-Bu
5 equiv TABH
−35 °C, 6 h;
22 °C, 1 h
75%
OH
NHOBn
t-Bu

(13)

100:1

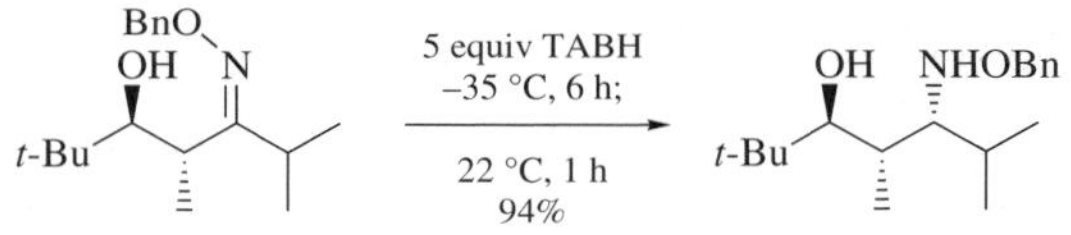
BnO
OH
N
t-Bu
5 equiv TABH
−35 °C, 6 h;
22 °C, 1 h
94%
OH
NHOBn
t-Bu

(14)

96:4

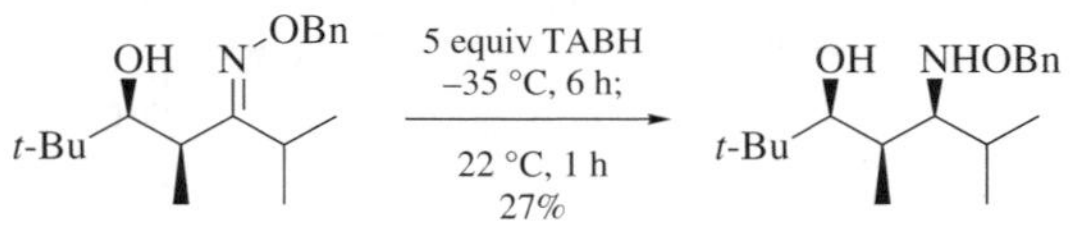
OH
N
OBn
t-Bu
5 equiv TABH
−35 °C, 6 h;
22 °C, 1 h
27%
OH
NHOBn
t-Bu

(15)

100:1

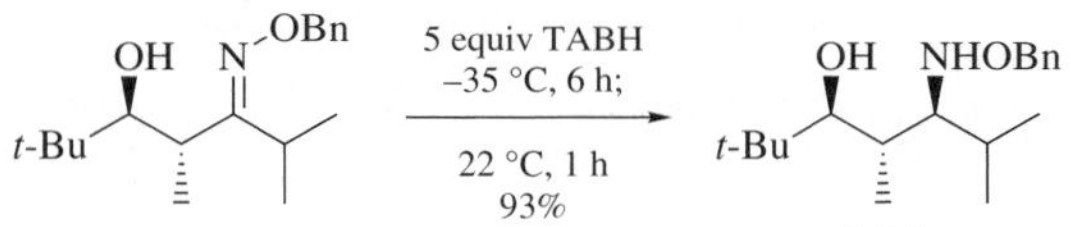
OH
N
OBn
t-Bu
5 equiv TABH
−35 °C, 6 h;
22 °C, 1 h
93%
OH
NHOBn
t-Bu

(16)

96:4

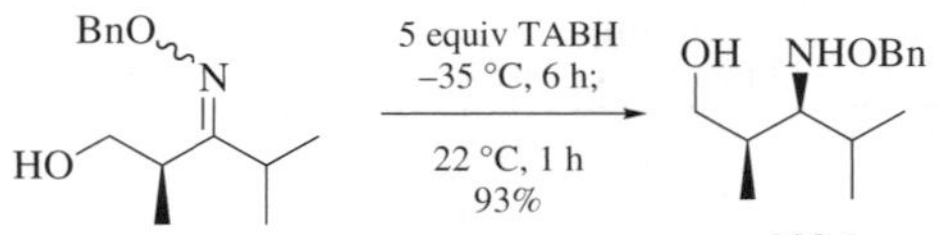
BnO
N
HO
5 equiv TABH
−35 °C, 6 h;
22 °C, 1 h
93%
OH
NHOBn

(17)

>100:1

Substrates with α-Hydroxy Substitution. Reductions of α-hydroxy oximino ethers with TABH provide *syn*-1,2-hydroxyamines with high diastereoselectivity (eq 18).[7] The observed stereochemical trend complements the previously observed *anti* selectivity achieved with ***Lithium Aluminum Hydride*** (LAH) or ***Diisobutylaluminum Hydride*** (DIBAL)[8] as reducing agents, or when Pd-catalyzed stereoselective hydrogenation is employed.[9] Unlike the case with β-alkoxy oximino ethers (eqs 11–17), neither the stereochemistry at the nitrogen (*E* vs. *Z*) nor that of the methyl group at the β-position has any effect on the stereochemical outcome of the reduction (e.g. eq 19).

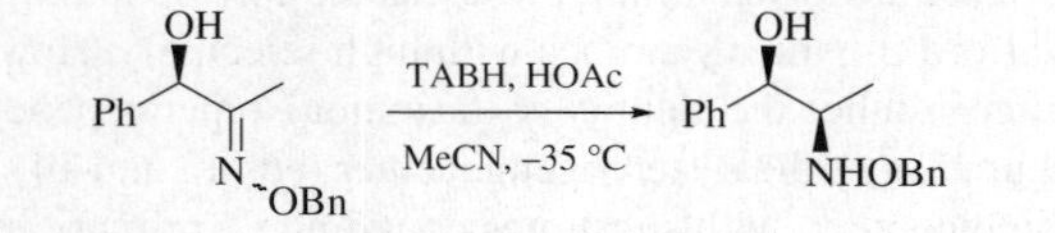

(18)

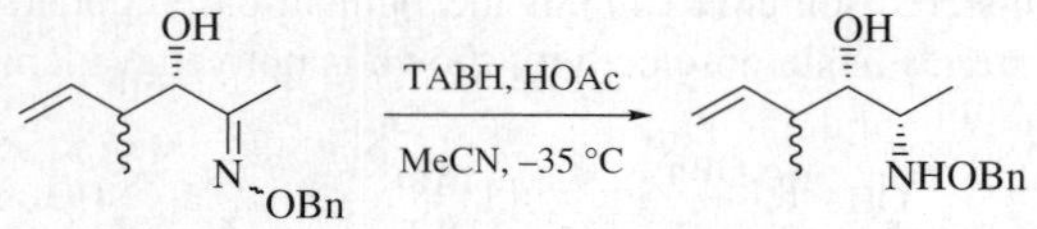

(19)

Directed Reduction of Imides. As shown in Table 2 and eq 20, imides may undergo directed selective reduction with TABH.[10] The superior levels of stereochemistry obtained with the title borohydride, as opposed to LAH or ***Sodium Bis(2-methoxyethoxy)aluminum Hydride*** (Red-Al), may be attributed to the ability of TABH to undergo reduction through the directed pathway exclusively. It is noteworthy that upon protection of the hydroxyl group, no reaction is obtained with TABH. Reduction of the silyl ether with lithium ***Lithium Tri-s-butylborohydride*** (L-Selectride) proceeds with the opposite sense of stereochemical control.

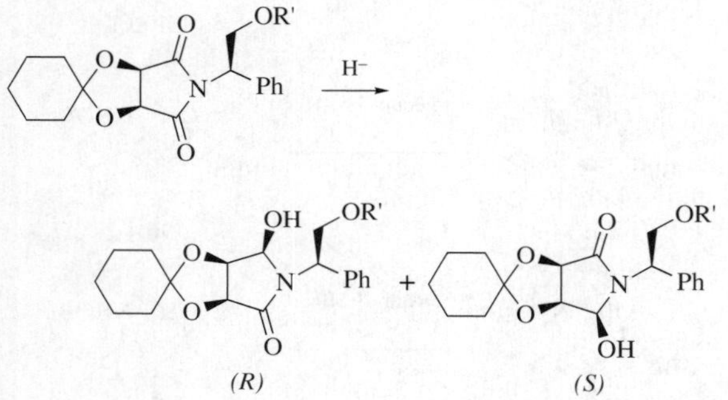

(20)

(*R*) (*S*)

Table 2 Diastereoselective Directed Reduction of Amides

R′	Hydride reagent	Yield (%)	(*R*):(*S*)
	LAH	75	61:39
H	Red-Al	80	82:18
	$Me_4N(OAc)_3BH$	63	>99:1
t-$BuMe_2Si$	L-Selectride	98	<1:99
	$Me_4N(OAc)_3BH$	No reaction	

Applications in Synthesis. The ability of TABH and the corresponding sodium salt to participate in substrate-directed reductions with high levels of diastereoselection renders these reagents as attractive tools in the synthesis of complex molecules. For example, as illustrated in eq 21 in the synthesis of (−)-rocaglamide,[11] the α-hydroxyl group strictly controls (>95%) the stereochemical outcome of the ketone reduction with TABH as the reducing agent. In contrast, hydrogenation (***Palladium(II) Hydroxide*** as catalyst) of the same substrate affords only a 2:1 mixture of isomers. In the total synthesis of (+)-lepicidin A (eq 22), a highly selective β-hydroxy ketone reduction has been achieved (10:1).[12] Stereoselective and directed reductions of β-hydroxy ketones have been employed in the synthesis of subunits of several medicinally important natural products, such as bryostatin I and II, in addition to mevinic acids, compactin, and mevinolin (eq 23, 93% yield, 13:1 selectivity).[13]

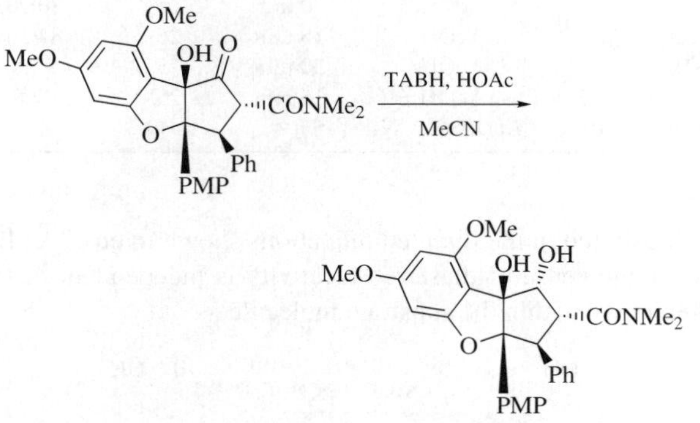

(21)

(−)-Rocaglamide

(22)

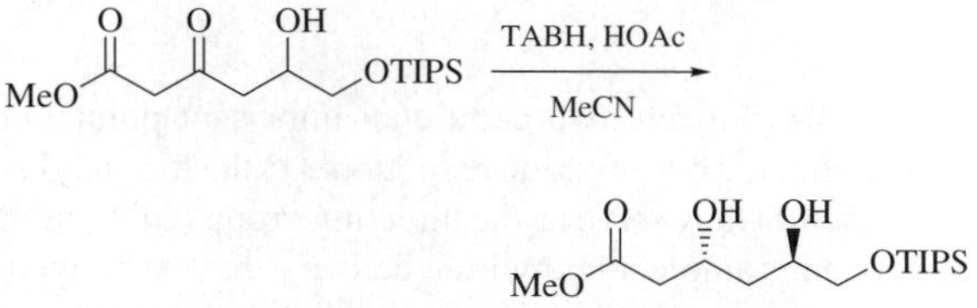

(23)

Related Reagents. See Classes R-2, R-6, R-9, R-10, and R-12, pages 1–10.

1. For general references on borohydrides, see (a) *Reduction: Techniques and Applications in Organic Synthesis*; Augustine, R. L., Ed.; Dekker: New York, 1968. (b) Hajos, A. *Complex Hydrides and Related Reducing Agents in Organic Synthesis*; Elsevier: New York, 1979. (c) Brown, H. C. *Boranes in Organic Chemistry*; Cornell UP: Ithaca, NY, 1972. (d) House, H. O. *Modern Synthetic Reactions*; Benjamin: Menlo Park, CA, 1972. (e) Greenwood, N. N. In *Comprehensive Inorganic Chemistry*; Bailer Jr., J. C.; Emeleus, H. J.; Nyholm, R.; Trotman-Dickenson, A. F., Eds.; Pergamon: Oxford, 1973; Vol. 1 pp 732–880. (f) *Progress in Boron Chemistry*; Steinberg, H.; McCloskey, A. L., Eds.; Pergamon: Oxford, 1964; Vol. 1. (g) Gerrard, W. *The Organic Chemistry of Boron*; Academic: New York, 1961. (h) *Production of the Boranes and Related Research*; Holzmann, R. T., Ed.; Academic: New York, 1967. (i) Lipscomb, W. N. *Boron Hydrides*; Benjamin: New York, 1963. (j) *Boron Hydride Chemistry*; Muetterties, E. L., Ed.; Academic: New York, 1975. (k) *The Chemistry of Boron and Its Compounds*; Muetterties, E. L., Ed.; Wiley: New York, 1967. (l) Stock, A. *Hydrides of Boron and Silicon*; Cornell UP: Ithaca, NY, 1933.
2. (a) Nutaitis, C. F.; Gribble, G. W. *TL* **1983**, *24*, 4287. (b) Gribble, G. W.; Nutaitis, C. F. *OPP* **1985**, *17*, 317.
3. For a comprehensive review of substrate-directable reactions, see: Hoveyda, A. H.; Evans, D. A.; Fu, G. C. *CRV* **1993**, *93*, 1307.
4. Saksena, A. K.; Mangiaracina, P. *TL* **1983**, *24*, 273.
5. (a) Evans, D. A.; Chapman, K. T. *TL* **1986**, *27*, 5939. (b) Evans, D. A.; Chapman, K. T.; Carreira, E. M. *JACS* **1988**, *110*, 3560. For general discussion, see: (c) Kim, B. M.; Sharpless, K. B. *Chemtracts: Org. Chem.* **1988**, *1*, 372. (d) Panek, J. S. *Chemtracts: Org. Chem.* **1992**, *3*, 188.

6. Williams, D. R.; Osterhout, M. H. *JACS* **1992**, *114*, 8750.
7. Williams, D. R.; Osterhout, M. H.; Reddy, J. P. *TL* **1993**, *34*, 3271.
8. Iida, H.; Yamazaki, N.; Kibayashi, C. *CC* **1987**, 746.
9. Harada, K.; Shiono, S. *BCJ* **1984**, *57*, 1040.
10. Miller, S. A.; Chamberlin, A. R. *JOC* **1989**, *54*, 2502.
11. (a) Trost, B. M.; Greenspan, P. D.; Yang, B. V.; Saulnier, M. G. *JACS* **1990**, *112*, 9022. (b) Davey, A. E.; Schaeffer, M. J.; Taylor, R. J. K. *JCS(P1)* **1992**, 2657.
12. Evans, D. A.; Black, W. C. *JACS* **1993**, *115*, 4497.
13. Evans, D. A.; Gauchet-Prunet, J. A.; Carriera, E. M.; Charette, A. B. *JOC* **1991**, *56*, 741.

Ahmad F. Houri & Amir H. Hoveyda
Boston College, Chestnut Hill, MA, USA

2,2,6,6-Tetramethylpiperidin-1-oxyl[1]

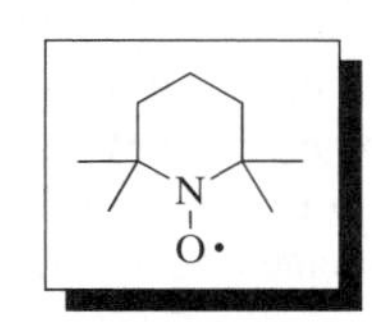

[2564-83-2] $C_9H_{18}NO$ (MW 156.28)

(oxidizing agent for the conversion of primary alcohols to aldehydes[2,3] or carboxylic acids,[2] of secondary alcohols to ketones,[2] and of diols to lactones or hydroxy aldehydes[4])

Alternate Name: TEMPO.
Physical Data: mp 40.0 °C; fp 67 °C.
Solubility: sol all organic solvents and 0.03 mol L^{-1} in H_2O at 25 °C.
Form Supplied in: red orange solid; commercially available.
Analysis of Reagent Purity: IR (Nujol) 1330, 1180, 1060, 975, 950 cm^{-1};[5] UV–Vis, λ (ϵ) = heptane, 470 nm (10.5), dioxane, 465 nm (10.4), 240–242 nm (2030).[5]
Purification: sublimation.[6]
Handling, Storage, and Precautions: TEMPO is a toxic substance, and a severe irritant which is readily absorbed through the skin.[7] Its toxicity is probably related to the formation of hydroxylamine metabolites.[8]

Oxammonium Salts. Oxammonium salts (**2**) can be prepared in situ by oxidation of the nitroxide (**1**).[1e] There is evidence that the oxoammonium ion (**2**) is actually the product of an acid-catalyzed disproportionation between two nitroxide molecules (**1**), which produces one oxoammonium ion (**2**) and one hydroxylamine molecule (**3**).[1e] Peracids, for instance, do not oxidize nitroxides to oxoammonium ions, yet do act as the secondary oxidant. The chemical stability of (**2**) depends on the counterion, and this can be easily exchanged; the chloride salt is not very stable. The oxammonium salt is the active species in the oxidation of primary and secondary alcohols to carbonyl derivatives and can be used in stoichiometric or catalytic fashion.[1e] Catalytic procedures require a co-oxidant such as ***Copper(II) Chloride***–O_2,[9] peroxy acids,[10] electrooxidation,[11] or ***Sodium Hypochlorite***.[2–4] Oxammonium salts can be titrated iodometrically.[1e]

(**1**) (**2**) (**3**)

It has been proposed that (**2**) reacts with alcohols, forming carbonyl derivatives and hydroxylamine (**3**).

Oxidation of Alcohols.[2,3,7] Primary alcohols are converted into aldehydes by the following catalytic system: 0.01–0.002 equiv of TEMPO (or its 4-methoxy derivative), 0.05 equiv of KBr as co-catalyst, and aqueous NaOCl buffered at pH 8.5–9.5 in a CH_2Cl_2/H_2O two-phase system. At this pH, HOCl is the co-oxidant and in the presence of KBr it is likely transformed into HOBr which is more efficient in the oxidation of (**1**) to (**2**). The reaction is exothermic. For reactions carried out on a 1–10 mmol scale the temperature can be easily maintained at 0 °C and conversion is complete in few minutes. On a larger scale, a very efficient cooling system is required. A compromise is to maintain the temperature in the range 10–15 °C with an ice bath; note that higher temperatures lead to fast decomposition of the catalyst. The pH is buffered by adding the appropriate amount of $NaHCO_3$ to the aqueous NaOCl. $NaH_2PO_4\cdot 2H_2O$ and $Na_2HPO_4\cdot 2H_2O$ have been used as alternatives.[12]

Oxidation of secondary alcohols similarly affords ketones. The oxidation can be applied to saturated alkyl and aryl–alkyl substrates; relatively unstable protecting groups such as acetonide derivatives of diols are not affected. Side reactions occur with substrates with isolated and conjugated double bonds, leading to lower yields. The reaction rates are markedly decreased by the presence of electron-donor groups in the aromatic ring of benzyl alcohol, but they can be speeded up by addition of catalytic amounts of quaternary ammonium salts (Q^+X^-). At the pH of commercial bleach (12.7), reactions are very slow. At lower pH, HOCl is distributed between the aqueous and the organic phase, thus making the phase transfer catalyst unnecessary. Optically active alcohols afford the corresponding aldehydes in good yields and high enantiomeric purity (eq 1).[3,13]

OH → (TEMPO, aq NaOCl, KBr, CH_2Cl_2, 0–15 °C, 82–84%) → CHO (1)

Oxidation of Diols. Diols are easily oxidized with this reagent; the nature of the products depends on the amount of oxidant used, the difference in oxidation rates of primary vs. secondary alcohols, the presence of Q^+X^-, and on the relative distance between the OH groups in the aliphatic chain. Under the reaction conditions reported above, primary–secondary diols afford hydroxy or keto aldehydes. In the presence of a catalytic amount of Q^+X^-, primary–secondary diols lead to ketocarboxylic acids. Selectivity in these oxidations also depends on the substituents on the 2,2,6,6-tetramethylpiperidine ring.[14]

Lactones are obtained from 1,4- and 1,5-diols, whereas α,ω-diols give unresolvable mixtures of polymeric products.[4] Oxidation of hydrophilic 1,4- and 1,5-diols to γ- and δ-lactones and of hydrophilic alcohols to aldehydes is best conducted using CH_2Cl_2/solid LiOCl in the presence of solid $NaHCO_3$. Commercial LiOCl contains 7% of H_2O so that reaction conditions are those of a pseudo solid–liquid system.[4]

Oxidation of Primary Alcohols and Aldehydes to Carboxylic Acids.[2] The addition of catalytic amounts of quaternary salt to the oxidizing system leads to the fast and direct formation of carboxylic acids. This oxidation requires the presence, in the organic phase, of ClO^- and/or BrO^- anions which behave as strong bases. Again, electron-donor groups in benzyl alcohol or aromatic aldehydes strongly lower the reaction rates.

Related Reagents. See Classes O-1 and O-2, pages 1–10.

1. (a) Rozantsev, E. G.; Sholle, V. D. *S* **1971**, 190. (b) *S* **1971**, 401. (c) Keana, J. F. W. *CRV* **1978**, *78*, 37. (d) Yamaguchi, M.; Miyazawa, T.; Takata, T.; Endo, T. *PAC* **1990**, *62*, 217. (e) Bobbitt, J. M.; Flores, M. C. L. *H* **1988**, *27*, 509. (f) Ma, Z.; Bobbitt, J. M. *JOC* **1991**, *56*, 6110.
2. Anelli, P. L.; Biffi, C.; Montanari, F.; Quici, S. *JOC* **1987**, *52*, 2559.
3. Anelli, P. L.; Montanari, F.; Quici, S. *OS* **1990**, *69*, 212.
4. Anelli, P. L.; Banfi, S.; Montanari, F.; Quici, S. *JOC* **1989**, *54*, 2970.
5. Brière, R.; Lemaire, H.; Rassat, A. *BSF* **1965**, 3273.
6. Mahoney, L. R.; Mendenhall, G. D.; Ingold, K. U. *JACS* **1973**, *95*, 8610.
7. Straub, T. S. *J. Chem. Educ.* **1991**, *68*, 1048.
8. Luzhkov, V. B. *DOK* **1983**, *268*, 126 (*CA* **1983**, *98*, 155 990v).
9. Semmelhack, M. F.; Schmid, C. R.; Cortés, D. A.; Chou, C. S. *JACS* **1984**, *106*, 3374.
10. (a) Cella, J. A.; Kelley, J. A.; Kenehan, E. F. *JOC* **1975**, *40*, 1860. (b) Cella, J. A.; McGrath, J. P.; Kelley, J. A.; El Soukkary, O.; Hilpert, L. *JOC* **1977**, *42*, 2077.
11. (a) Semmelhack, M. F.; Chou, C. S.; Cortés, D. A. *JACS* **1983**, *105*, 4492. (b) Inokuchi, T.; Matsumoto, S.; Torii, S. *JOC* **1991**, *56*, 2416.
12. BASF A.-G. Ger. Offen. 4 007 923, 1990 (*CA* **1991**, *114*, 163 728e).
13. Leanna, M. R.; Sowin, T. J.; Morton, H. E. *TL* **1992**, *33*, 5029.
14. Siedlecka, R.; Skarzewski, J.; Mlochowski, J. *TL* **1990**, *31*, 2177.

Fernando Montanari & Silvio Quici
Università di Milano, Italy

Tetra-*n*-propylammonium Perruthenate[1]

$Pr_4N^+RuO_4^-$

[114615-82-6] $C_{12}H_{28}NO_4Ru$ (MW 351.48)

(mild oxidant for conversion of multifunctionalized alcohols to aldehydes and ketones;[1] can selectively oxidize primary–secondary diols to lactones;[2] can cleave carbon–carbon bonds of 1,2-diols[3])

Alternate Name: TPAP.
Physical Data: mp 165 °C (dec).
Solubility: sol CH_2Cl_2, and MeCN; partially sol C_6H_6.
Form Supplied in: dark green solid; commercially available.
Analysis of Reagent Purity: microanalysis.
Handling, Storage, and Precautions: stable at room temperature and may be stored for long periods of time without significant decomposition, especially if kept refrigerated in the dark. The reagent should not be heated neat, as small quantities decompose with flame at 150–160 °C in air.

General Procedures. TPAP is a convenient, mild, neutral and selective oxidant of primary alcohols to aldehydes and secondary alcohols to ketones.[1] These reactions are carried out with catalytic TPAP at rt in the presence of stoichiometric or excess ***N-Methylmorpholine N-Oxide*** (NMO) as a cooxidant. A kinetic study with 2-propanol as the substrate has shown that these oxidations are strongly autocatalytic.[4] Turnovers of up to 250 are obtainable if activated powdered molecular sieves are introduced to remove both the water formed during oxidations and the water of crystallization of the NMO. Solvents commonly employed are dichloromethane and acetonitrile, or combinations of these. This reagent also works well on a small scale where other methods, such as those employing activated ***Dimethyl Sulfoxide*** reagents, are inconvenient.[1] For large scale oxidations it is necessary to moderate these reactions by cooling and by slow portionwise addition of the TPAP. To achieve full conversion on a large scale, the NMO should be predried (by first treating an organic solution of NMO with anhydrous magnesium sulfate) and the use of 10% acetonitrile–dichloromethane as solvent is recommended.[1]

Oxidation of Primary Alcohols. Primary alcohols can be oxidized in the presence of a variety of functional groups, including tetrahydropyranyl ethers (eq 1),[1,5] epoxides (eq 2),[1,6] acetals (eq 3),[1,7] silyl ethers,[1,8] peroxides,[9] lactones,[1,10] alkenes,[1,11] alkynes,[1,12] esters,[1,13] amides,[1,14] sulfones,[1] and indoles.[10] Oxidation of substrates with labile α-centers proceeds without epimerization.[1]

THPO ... OH, OTBDMS → (TPAP, NMO, mol. sieves; CH_2Cl_2, rt; 97%) → THPO ... CHO, OTBDMS (1)

PMP ... OH → (TPAP, NMO; 96%) → PMP ... CHO (2)

OH ... OBz → (TPAP, NMO, mol. sieves; MeCN, rt; 95%) → O ... OBz (3)

Oxidation of Secondary Alcohols. In a similar fashion, multifunctional secondary alcohols are oxidized to ketones (eqs 4 and 5)[9,15] in good yields.[1,5–14]

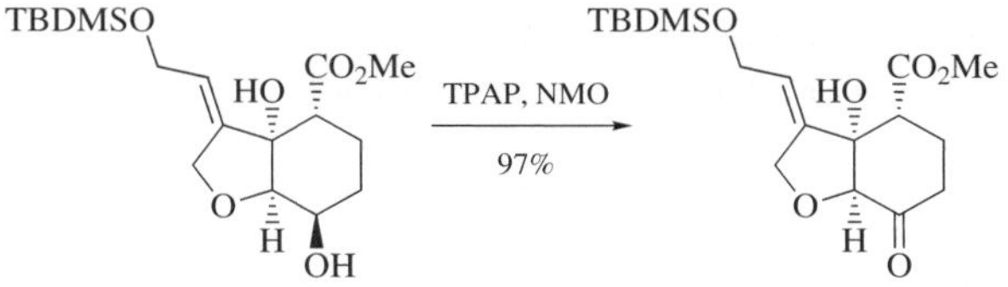

(4) TPAP, NMO, 97%

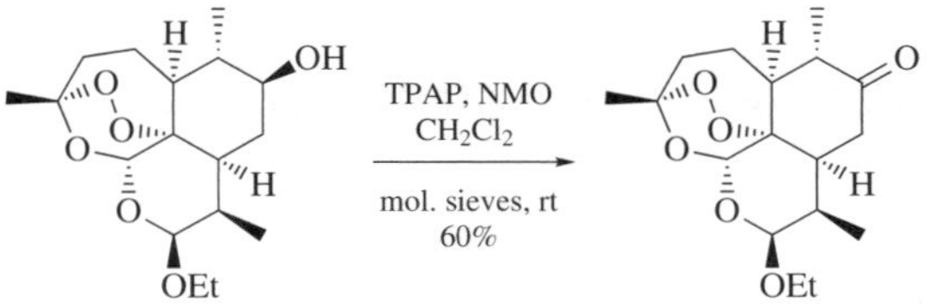

(5) TPAP, NMO, CH_2Cl_2, mol. sieves, rt, 60%

A particularly hindered secondary alcohol (an intermediate in the latter stages of the synthesis of tetronolide) resisted oxidation with activated DMSO, ***Pyridinium Chlorochromate***, and activated ***Manganese Dioxide***, yet stoichiometric TPAP afforded the ketone in 81% yield.[16]

Allylic Alcohols. These alcohols are successfully oxidized to the corresponding enones and enals (eq 6).[1,17]

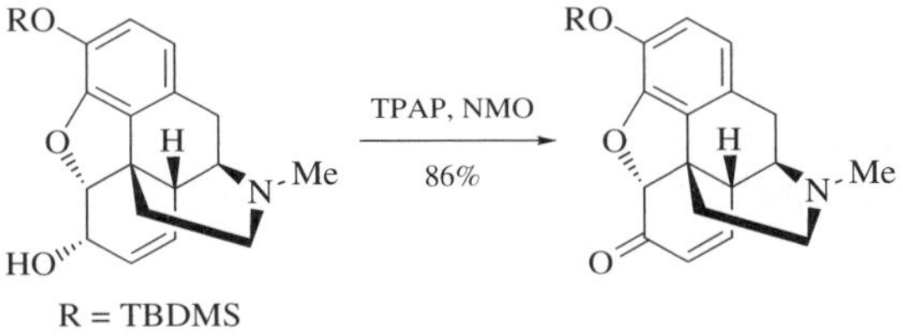

(6) TPAP, NMO, 86%; R = TBDMS

Oxidation of a primary allylic alcohol over a secondary allylic alcohol has been achieved.[18] Oxidation of the homoallylic alcohol of cholesterol by TPAP and NMO under ultrasonication conditions gives the dienone cholest-4-ene-3,6-dione in 80% yield.[19] This oxidation was subsequently carried out in the presence of a labile TBDMS enol ether group which remained intact, while with both PCC and activated DMSO this protecting group did not survive.[19] Oxidation of homopropargylic alcohols leads to allenones, as with other common oxidants.[20]

Lactols. Oxidations of lactols to lactones are facile and high yielding; several examples have been reported in the literature (eq 7).[1,21]

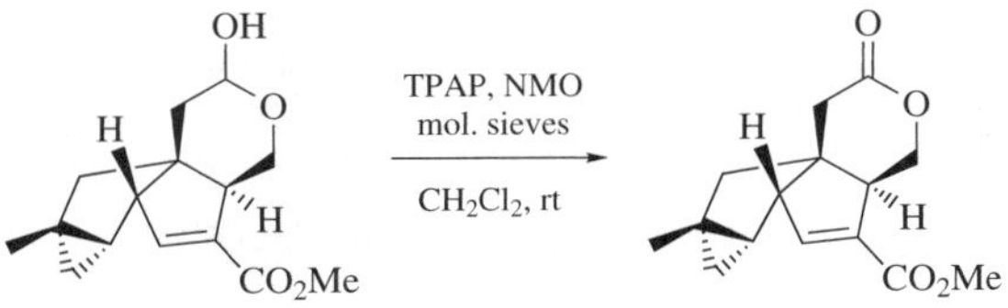

(7) TPAP, NMO, mol. sieves, CH_2Cl_2, rt

Selective Oxidations. The selective oxidation of 1,4 and 1,5 primary–secondary diols to lactones is a valuable application of this reagent.[2] Few general mild reagents for the chemoselective oxidation to the hydroxy aldehyde are available.[22] The most widely known reagents are Pt and O_2, and ***Dihydridotetrakis(triphenylphosphine)ruthenium(II)***.[22] Hydroxy aldehydes, in their lactol form, are then oxidized further to lactones. The use of TPAP is advantageous in that it is commercially available, employs mild catalytic reaction conditions, and reacts with high selectivity in unsymmetrical cases (eq 8).[2] Lactones have also been formed from primary–tertiary diols.[23]

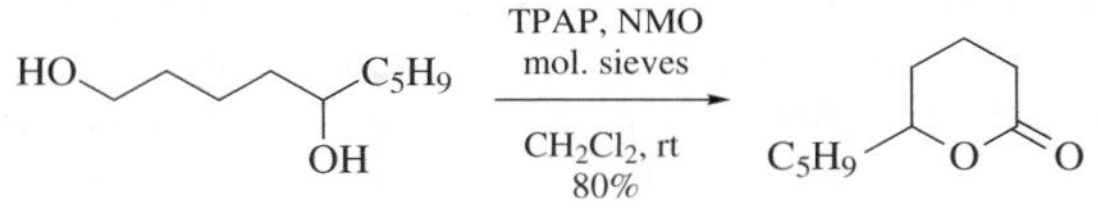

(8) TPAP, NMO, mol. sieves, CH_2Cl_2, rt, 80%

Functional Group Compatibility. The neutral conditions of these oxidations have been utilized to provide improved yields with acid sensitive substrates compared to the well established Swern method (eqs 9, 10).[1,16]

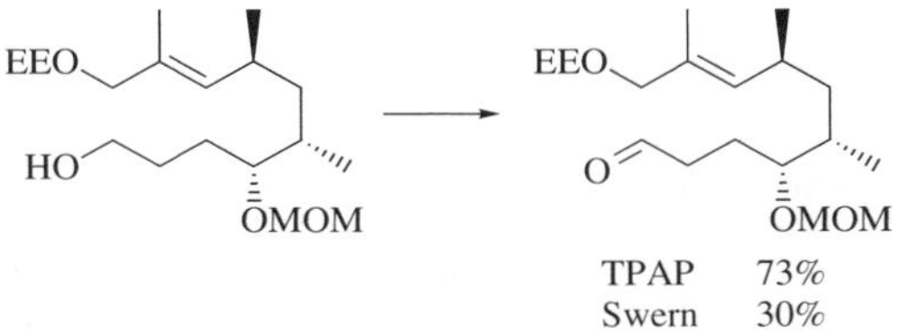

(9) TPAP 73%; Swern 30%

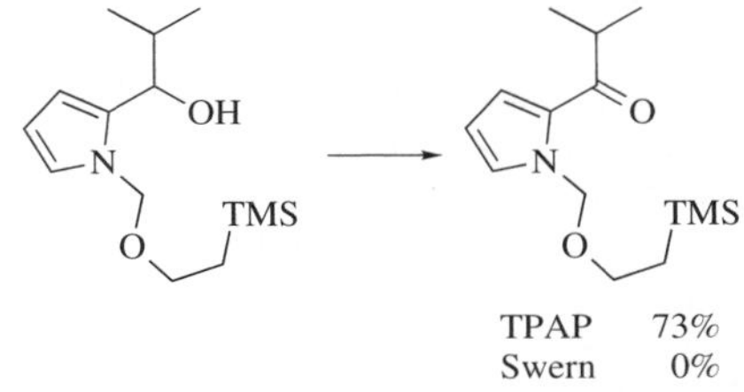

(10) TPAP 73%; Swern 0%

Highly sensitive alcohols have also been oxidized, albeit in low yield, when most other conventional methods, such as Dess–Martin periodinane (***1,1,1-Triacetoxy-1,1-dihydro-1,2-benziodoxol-3(1H)-one***),[24] PCC, ***Pyridinium Dichromate***, and DMSO activated with ***Sulfur Trioxide–Pyridine*** have failed (eq 10).[25]

Oxidative Cleavage Reactions. Among the numerous methods for 1,2-diol cleavage there exist only a few that involve catalytic ruthenium reagents, for example ***Ruthenium(III) Chloride*** with ***Sodium Periodate***.[22] Attempted selective monooxidation of a 1,2-diol to the hydroxy aldehyde with catalytic TPAP and NMO resulted in carbon–carbon bond cleavage to provide the aldehyde (eq 11).[3] Furthermore, attempted oxidation of an anomeric α-hydroxy ester failed; instead, in this case decarboxylation/decarbonylation and formation of the lactone was observed (eq 12). However, ***Dimethyl Sulfoxide–Acetic Anhydride*** provided the required α-dicarbonyl unit.[26] Retro-aldol fragmentations can also be a problem.[27]

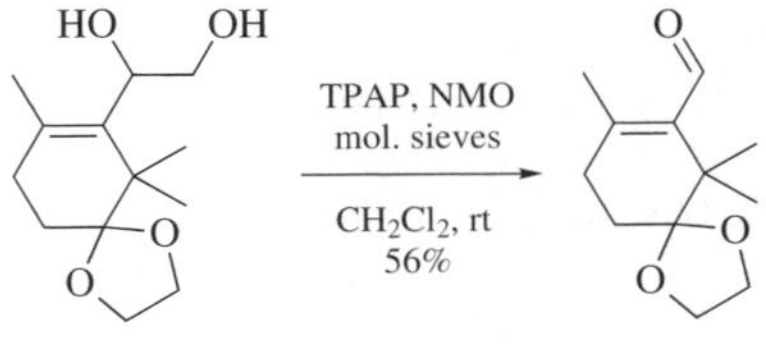

(11) TPAP, NMO, mol. sieves, CH_2Cl_2, rt, 56%

HO CO$_2$Me BnO O —TPAP, NMO, 44%→ BnO O O OBn OBn (12)

Heteroatom Oxidation. Thus far, TPAP has only been used to oxidize sulfur. Oxidation of an oxothiazolidine *S*-oxide to the corresponding *S,S*-dioxide gave only poor results in comparison to the standard RuO_2 with $NaIO_4$ conditions.[28] However, oxidation of sulfides to sulfones in the presence of isolated double bonds has been investigated. The yields with TPAP/NMO range from 61–99% which are higher than to those obtained with ***m-Chloroperbenzoic Acid***, ***Potassium Monoperoxysulfate*** (Oxone®), or ***Hydrogen Peroxide–Acetic Acid***.[29]

Related Reagents. See Classes O-1, O-8, and O-18, pages 1–10.

---

1. (a) Griffith, W. P.; Ley, S. V.; Whitcombe, G. P.; White, A. D. *CC* **1987**, 1625. (b) Griffith, W. P.; Ley, S. V. *Aldrichim. Acta* **1990**, *23*, 13. (c) Ley, S. V.; Norman, J.; Griffith, W. P.; Marsden, S. P. *S* **1994**, 639.
2. Bloch, R.; Brillet, C. *SL* **1991**, 829.
3. Queneau, Y.; Krol, W. J.; Bornmann, W. G.; Danishefsky, S. J. *JOC* **1992**, *57*, 4043.
4. Lee, D. G.; Wang, Z.; Chandler, W. D. *JOC* **1992**, *57*, 3276.
5. Guanti, G.; Banfi, L.; Ghiron, C.; Narisano, E. *TL* **1991**, *32*, 267.
6. (a) Stürmer, R.; Ritter, K.; Hoffmann, R. W. *AG(E)* **1993**, *32*, 101. (b) Kim, G.; Chu-Moyer, M. Y.; Danishefsky, S. J. *JACS* **1990**, *112*, 2003.
7. (a) Anthony, N. J.; Armstrong, A.; Ley, S. V.; Madin, A. *TL* **1989**, *30*, 3209. (b) Romeyke, Y.; Keller, M.; Kluge, H.; Grabley, S.; Hammann, P. *T* **1991**, *47*, 3335.
8. (a) Ley, S. V.; Maw, G. N.; Trudell, M. L. *TL* **1990**, *31*, 5521. (b) Rosini, G.; Marotta, E.; Raimondi, A.; Righi, P. *TA* **1991**, *2*, 123.
9. (a) Hu, Y.; Ziffer, H. *J. Labelled Comp. Radiopharm.* **1991**, *29*, 1293.
10. Linz, G.; Weetman, J.; Hady, A. F. A.; Helmchen, G. *TL* **1989**, *30*, 5599.
11. (a) Cole, P. A.; Bean, J. M.; Robinson, C. H. *PNA* **1990**, *87*, 2999. (b) Piers, E.; Roberge, J. Y. *TL* **1991**, *32*, 5219.
12. Desmaële, D.; Champion, N. *TL* **1992**, *33*, 4447.
13. Hori, K.; Hikage, N.; Inagaki, A.; Mori, S.; Nomura, K.; Yoshii, E. *JOC* **1992**, *57*, 2888.
14. Guanti, G.; Banfi, L.; Narisano, E.; Thea, S. *SL* **1992**, 311.
15. Sulikowski, M. M.; Ellis Davies, G. E. R.; Smith, A. B., III *JCS(P1)* **1992**, 979.
16. Takeda, K.; Kawanishi, E.; Nakamura, H.; Yoshii, E. *TL* **1991**, *32*, 4925.
17. (a) Ninan, A.; Sainsbury, M. *T* **1992**, *48*, 6709. (b) Rychnovsky, S. D.; Rodriguez, C. *JOC* **1992**, *57*, 4793. (c) Kang, H.-J.; Ra, C. S.; Paquette, L. A. *JACS* **1991**, *113*, 9384. (d) Schreiber, S. L.; Kiessling, L. L. *TL* **1989**, *30*, 433.
18. Hitchcock, S. A.; Pattenden, G. *TL* **1992**, *33*, 4843.
19. Moreno, M. J. S. M.; Melo, M. L. S.; Neves, A. S. C. *TL* **1991**, *32*, 3201.
20. Marshall, J. A.; Robinson, E. D.; Lebreton, J. *JOC* **1990**, *55*, 227.
21. Paquette, L. A.; Kang, H-J.; Ra, C. S. *JACS* **1992**, *114*, 7387.
22. *COS* **1991**, *7*.
23. (a) Mehta, G.; Karra, S. R. *TL* **1991**, *32*, 3215. (b) Mehta, G.; Karra, S. R. *CC* **1991**, 1367.
24. Yamashita, D. S.; Rocco, V. P.; Danishefsky, S. J. *TL* **1991**, *32*, 6667.
25. Tokoroyama, T.; Kotsuji, Y.; Matsuyama, H.; Shimura, T.; Yokotani, K.; Fukuyama, Y. *JCS(P1)* **1990**, 1745.
26. Watanabe, T.; Nishiyama, S.; Yamamura, S.; Kato, K.; Nagai, M.; Takita, T. *TL* **1991**, *32*, 2399.
27. Shih, T. L.; Mrozik, H.; Holmes, M. A.; Arison, B. H.; Doss, G. A.; Waksmunski, F.; Fisher, M. H. *TL* **1992**, *33*, 1709.
28. White, G. J.; Garst, M. E. *JOC* **1991**, *56*, 3177.
29. Guertin, K. R.; Kende, A. S. *TL* **1993**, *34*, 5369.

Steven V. Ley & Joanne Norman
University of Cambridge, UK

Thallium(III) Nitrate Trihydrate[1]

$Tl(NO_3)_3 \cdot 3H_2O$

[13453-38-8] $H_6N_3O_{12}Tl$ (MW 444.47)

(oxidizing agent; Lewis acid for alkene cyclization)

Alternate Name: TTN.
Physical Data: mp 102–105 °C.
Solubility: sol water, organic solvents.
Form Supplied in: moist white crystals, hygroscopic; widely available.
Drying: compound decomposes on heating.
Handling, Storage, and Precautions: all thallium compounds are extremely toxic to inhalation, skin contact, and ingestion. Toxicity is cumulative. Extreme caution should be used when handling these materials. Use in a fume hood.

Oxidations. Thallium trinitrate is a powerful oxidant. A variety of substituted phenols undergo oxidation using TTN to provide quinones.[2] For example, hydroquinones are oxidized to quinones in good yields. *p*-Alkoxyphenols are oxidized to *p*-quinone acetals in good yields by TTN in methanol (eq 1). Similarly, naphthols are oxidized to naphthoquinones using TTN. This oxidation proceeds in higher yields if TTN on ***Celite*** is used as the oxidant.[3]

OH R OMe —TTN, MeOH→ O R MeO OMe (1)

R = H, 97%; Cl, 97%; Br, 91%; Me, 87%

Chalcones are oxidized under acidic conditions to 1,2-diketones using three equivalents of TTN (eq 2).[4]

$$\text{ArCH=CHCOAr}' \xrightarrow[\text{H}^+]{\text{TTN}} \text{ArCOCOAr}' \quad (2)$$

Ar = Ar' = Ph, 61%
Ar = 4-BrC_6H_4, Ar' = Ph, 55%
Ar = Ph, Ar' = 4'-BrC_6H_4, 70%
Ar = Ph, Ar' = 4'-MeC_6H_4, 49%
Ar = Ph, Ar' = 4'-$MeOC_6H_4$, 49%

TTN is a trihydrate and generally reactions with it are carried out under fairly acidic conditions. TTN oxidations in methanol as a solvent are also strongly acidic, since nitric acid is produced as a byproduct. Reactions that fail or proceed poorly with TTN (TTN in methanol or acetic acid), or where the substrates are acid sensitive, can be promoted by using a 1:1 mixture of methanol and trimethyl orthoformate (TMOF) (see ***Triethyl Orthoformate***) or neat TMOF as solvent. Oxidation of cinnamaldehyde with TTN in methanol proceeds very slowly and produces seven products. On the other hand, cinnamaldehydes are rearranged to aryl malondialdehyde tetramethyl acetals in good yields on treatment with TTN in 1:1 MeOH–TMOF (eq 3).[5]

TTN, MeOH–TMOF (3)

$R^1 = R^2 = H$, 79%
R^1 = 4-MeO, R^2 = H, 84%
R^1 = 4-NO_2, R^2 = H, 63%
R^1 = H, R^2 = Me, 83%
R^1 = 3-NO_2, R^2 = Me, 50%

TTN adsorbed on ***Montmorillonite K10*** is an effective reagent for the conversion of ketones to rearranged esters. For example, acetophenone is readily converted to phenylacetate on treatment with TTN/K-10 reagent (eq 4).[6] Supported TTN reagents are practical, since product isolation from the insoluble inorganic byproducts is simple.

(4)

R	TTN, MeOH, $HClO_4$	TTN/K-10
H	84%	86%
F	44%	88%
Me	86%	84%
Br	35%	89%

Alkene Oxidation. Simple alkenes are converted to aldehydes or ketones in good yields using TTN. These reactions proceed with migration of the higher migratory aptitude substituent.[7] The preparation of arylacetaldehyde dimethyl acetals by oxidative rearrangement of substituted styrenes using TTN proceeds in good yields (eq 5). The reaction proceeds through the exclusive migration of the aryl substituent and the yields are higher if TTN supported on K-10 is used as the oxidant.[1b,6b]

(5)

R	TTN, MeOH	TTN/K-10
H	85%	92%
F	27%	79%
OMe	64%	81%
Br	30%	76%

Cycloalkenes provide ring contracted aldehydes on oxidation with TTN under acidic conditions. Corey and Ravindranathan have used this methodology to prepare a key prostaglandin intermediate (eq 6).[8] Similarly, enol ethers also undergo oxidative ring contraction on treatment with TTN (eq 7).[9] If methanol is used as the solvent, the corresponding acetals are formed as the products.[10] In contrast to the ring contractive oxidation of monocycloalkenes, bicyclic alkenes furnish nitrate esters on treatment with TTN.[11] Exocyclic alkenes furnish ring enlarged ketones on oxidation with TTN (see ***Thallium(III) Perchlorate*** for similar reactions).[12]

TTN, H_3O^+, 50% (6)

1. TTN, MeCN, 1 h, rt; 2. $NaBH_4$, 62% (7)

Diarylalkynes are converted to 1,2-diketones using two equivalents of TTN (eq 8) and terminal alkynes are oxidized to carboxylic acids (eq 9).[13]

$$Ph{-}{\equiv}{-}Ph \xrightarrow[85\%]{TTN} PhCOCOPh \quad (8)$$

$$R{-}{\equiv} \xrightarrow{TTN} RCO_2H \quad (9)$$

$R = C_6H_{13}$, 80%; C_5H_{11}, 55%

TTN can be used for electrophilic cyclizations of polyalkenes.[14] The oxythallative cyclization of elemol acetate using TTN in acetic acid produces a guaiene diol after ***Lithium Aluminum Hydride*** reduction (eq 10).[15] In contrast, ***Mercury(II) Acetate*** mediated cyclization of elemol produces the unrearranged cryptomeridiol.

R = Ac: 1. TTN, AcOH; 2. $LiAlH_4$, 63%. R = H: 1. $Hg(OAc)_2$; 2. $LiAlH_4$, 69% (10)

Allylations Using Allysilanes and TTN. Aromatic compounds are allylated using allylsilanes and TTN, but in poor yield.[16] Allylsilanes are converted to allylic ethers (eq 11),[17] *N*-allylic amides (eq 12),[18] and allylic nitrates[19] on treatment with TTN and the appropriate nucleophile.

TTN, MeOH, 0.5 h, 70% (11)

$$\text{allyl-TMS} \xrightarrow[\substack{-20\,^\circ\text{C, 1 h, to} \\ 0\,^\circ\text{C, 1 h} \\ 48\%}]{\text{TTN, MeOH}} \text{allyl-NHCOMe} \quad (12)$$

Functional Group Interconversions. TTN finds utility in a variety of functional group interconversions. Phenols are readily converted to anilines using TTN.[20] Sulfides and selenides are converted to sulfoxides (eq 13) and selenoxides, respectively, on oxidation with TTN.[21] Other transformations include the preparation of allene esters from α-alkyl-β-keto esters (eq 14),[22] carbamates from isocyanides (eq 15),[23] and lactones from γ,δ-unsaturated acids.[24]

$$\text{R-S-R} \xrightarrow{\text{TTN}} \text{R-S(=O)-R} \quad (13)$$

R = Et, 85%; Pr, 92%; Bu, 94%; Ph, 82%

$$\xrightarrow[\text{2. TTN, MeOH}]{1.\ N_2H_4} \quad 50\% \quad (14)$$

$$\text{RNC} \xrightarrow[\text{MeOH, } H_2O]{\text{TTN}} \text{RNHCO}_2\text{Me} \quad (15)$$

R = $EtOCOCH_2$, 84%; Cy, 90%; *t*-Bu, 35%; Ph, 93%; 4-MeC_6H_4, 85%

TTN has been used to selectively deprotect bisthioacetals to give monothioacetals (eq 16).[25] Simple thioacetals can also be deprotected using TTN. Oximes are converted to aldehydes or ketones in high yields on treatment with TTN in methanol at rt (eq 17).[26]

$$\xrightarrow[97\%]{\text{TTN}} \quad (16)$$

$$C_6H_{13}CH{=}NOH \xrightarrow[96\%]{\text{TTN}} C_6H_{13}CHO \quad (17)$$

Related Reagents. See Classes O-8 and O-19, pages 1–10.

1. (a) McKillop, A. *PAC* **1975**, *43*, 463. (b) McKillop, A.; Taylor, E. C. In *Comprehensive Organometallic Chemistry*; Wilkinson, G., Ed.; Pergamon: Oxford, 1982; Vol. 7, p 465.
2. McKillop, A.; Perry, D. H.; Edwards, M.; Antus, S.; Farkas, L.; Nogradi, M.; Taylor, E. C. *JOC* **1976**, *41*, 282.
3. Crouse, D. J.; Wheeler, M. M.; Goemann, M.; Tobin, P. S.; Basu, S. K.; Wheeler, D. M. S. *JOC* **1981**, *46*, 1814.
4. McKillop, A.; Swann, B. P.; Ford, M. E.; Taylor, E. C. *JACS* **1973**, *95*, 3641.
5. Taylor, E. C.; Robey, R. L.; Liu, K.-T.; Favre, B.; Bozimo, H. T.; Conley, R. A.; Chiang, C.-S.; McKillop, A.; Ford, M. E. *JACS* **1976**, *98*, 3037.
6. (a) McKillop, A.; Swann, B. P.; Taylor, E. C. *JACS* **1973**, *95*, 3340. (b) Taylor, E. C.; Chiang, C.-S.; McKillop. A.; White, J. F. *JACS* **1976**, *98*, 6750.
7. McKillop, A.; Hunt, J. D.; Taylor, E. C.; Kienzle, F. *TL* **1970**, 5275.
8. Corey, E. J.; Ravindranathan, T. *TL* **1971**, 4753.
9. Kaye, A.; Neidle, S.; Reese, C. B. *TL* **1988**, *29*, 1841.
10. McKillop, A.; Hunt, J. D.; Kienzle, F.; Bigham, E.; Taylor, E. C. *JACS* **1973**, *95*, 3635.
11. Layton, W. J.; Brock, C. P.; Crooks, P. A.; Smith, S. L.; Burn, P. *JOC* **1985**, *50*, 5372.
12. Farcasiu, D.; Schleyer, P. v. R.; Ledlie, D. B. *JOC* **1973**, *38*, 3455.
13. (a) McKillop, A.; Oldenziel, O. H.; Swann, B. P.; Taylor, E. C.; Robey, R. L. *JACS* **1971**, *93*, 7331. (b) McKillop, A.; Oldenziel, O. H.; Swann, B. P.; Taylor, E. C.; Robey, R. L. *JACS* **1973**, *95*, 1296.
14. Anteunis, M.; DeSmet, A. *S* **1974**, 868.
15. Renold, W.; Ohloff, G.; Norin, T. *HCA* **1979**, *62*, 985.
16. Ochiai, M.; Fujita, E.; Arimoto, M.; Yamaguchi, H. *CPB* **1983**, *31*, 86.
17. Ochiai, M.; Fujita, E.; Arimoto, M.; Yamaguchi, H. *CPB* **1984**, *32*, 5027.
18. Ochiai, M.; Tada, S.-I.; Arimoto, M.; Fujita, E. *CPB* **1982**, *30*, 2836.
19. Ochiai, M.; Fujita, E.; Arimoto, M.; Yamaguchi, H. *CPB* **1984**, *32*, 887.
20. Taylor, E. C.; Jagdmann, Jr., G.; McKillop, A. *JOC* **1978**, *43*, 4385.
21. Nagao, Y.; Ochiai, M.; Kaneko, K.; Maeda, A.; Watanabe, K.; Fujita, E. *TL* **1977**, 1345.
22. Taylor, E. C.; Robey, R. L.; McKillop, A. *JOC* **1972**, *37*, 2797.
23. Kienzle, F. *TL* **1972**, 1771.
24. Ferraz, H. M. C.; Ribeiro, C. R. *SC* **1992**, *22*, 399.
25. (a) Smith, R. A. J.; Hannah, D. J. *SC* **1979**, *9*, 301. (b) Fujita, E.; Nagao, Y.; Kaneko, K. *CPB* **1978**, *26*, 3743.
26. McKillop, A.; Hunt, J. D.; Naylor, R. D.; Taylor, E. C. *JACS* **1971**, *93*, 4918.

Mukund P. Sibi
North Dakota State University, Fargo, ND, USA

Thallium(III) Trifluoroacetate[1]

$Tl(OCOCF_3)_3$

[23586-53-0] $C_6F_9O_6Tl$ (MW 543.44)

(oxidizing agent; thallating agent; Lewis acid; reagent for dethioacetalization)

Alternate Name: TTFA.
Physical Data: mp 100 °C (dec).
Solubility: sol water, organic solvents.
Form Supplied in: moist white crystals, hygroscopic; widely available.
Drying: decomposes on heating.
Preparative Method: readily prepared by refluxing a suspension of thallium(III) oxide with ***Trifluoroacetic Acid*** (TFA).[2]
Handling, Storage, and Precautions: water sensitive solid. All thallium compounds are extremely toxic to inhalation, skin contact, and ingestion. Toxicity is cumulative. Extreme caution should be used when handling these materials. Use in a fume hood.

Thallation. Thallium trifluoroacetate is the most versatile of all the thallium reagents. It has seen extensive utility in a variety of organic transformations.[1] The solution of TTFA in TFA is a potent thallating reagent. Aromatic compounds react with TTFA to provide thallated aromatics (eq 1). As is the case with other electrophilic aromatic substitution reactions, thallation is most efficient with electron-rich aromatics. Other methods are now available for the preparation of thallated aromatics using TTFA.[3] Generally, the thallated intermediates are not isolated, but are transformed to a variety of compounds in a subsequent reaction with nucleophiles.

$$\text{PhH} + \text{TTFA} \xrightarrow[95\%]{\text{TFA, rt}} \text{PhTl(TFA)}_2 \quad (1)$$

Thallation of substituted aromatics with TTFA is highly regioselective, depending on the nature of the substituent and reaction conditions. The reaction is reversible, requires a moderately large activation energy, and has large steric constraints. The electrophilic character of TTFA can be increased by the addition of Lewis acids such as ***Boron Trifluoride*** or ***Antimony(V) Fluoride***.[4] Thallation of substituted aromatics (nonheteroatom substituents) under kinetic control (short reaction times) yields the *para*-thallated product, illustrating the steric requirements of the reaction. The *meta*-thallated product can be obtained by conducting the reaction in refluxing TFA (eq 2).[5]

(2) TTFA, 73 °C, 85%; TTFA, 20 °C, 94%; (TFA)$_2$Tl; Tl(TFA)$_2$

Thallation of aromatics possessing heteroatom-containing substituents (CO_2R, CO_2H, OR, CH_2OH, etc.) shows a different type of regioselectivity, providing the *ortho*-thallated product (eq 3).[6] This selectivity has been rationalized as intramolecular delivery of the thallium electrophile. The monothallated species generally deactivates the ring toward further thallation.

(3) O, OR; R, O, Tl(TFA)$_2$; O, OR, Tl(TFA)$_2$

Anisole can be polythallated using excess TTFA in TFA (eq 4).[7] The thallated intermediate can be converted to the corresponding iodo compound on treatment with ***Sodium Iodide***.

(4) OMe; TTFA, TFA; OMe, Tl(TFA)$_2$, Tl(TFA)$_2$; NaI; OMe, I, I

Electrophilic thallations are not restricted to aromatic compounds. Heterocyclic compounds, such as thiophenes, furans, and indoles, undergo thallations cleanly.[8] These thallated intermediates can also be treated with nucleophiles to provide substituted heterocycles.

ipso*-Displacements.** The thallated aromatics are useful in the preparation of a variety of substituted compounds. The reaction involves treatment of the thallated intermediates with nucleophiles, alkenes, arenes, and cations.[1] The introduction of the nucleophile takes place with very high selectivity: the nucleophile occupies the same position as the thallium (*ipso* substitution) (eq 5) (see ***Phenylthallium Bis(trifluoroacetate) for examples).

$$\text{PhTl(TFA)}_2 \xrightarrow{\text{Nu}} \text{PhNu} \quad (5)$$

Nu = OH, F, Cl, Br, I, SCN, CN, SeCN, NO_2, Alkyl, SH, etc

Preparation of aryl fluorides from thallated aromatics has been attempted by several researchers with mixed results.[9] Only Taylor et al.[9b] have been successful, using the three-step method (including the preparation of the arylthallium bis(trifluoroacetate)) shown in eq 6. While Taylor's method is less complicated than the Balz–Schiemann method of preparing fluoride compounds, it is limited to those aromatic substrates that do not contain electron-withdrawing groups, or oxygen or nitrogen functionality.

(6) Tl(TFA)$_2$; KF; TlF_2; BF_3, 0 °C, 1 h; C_6H_{12}, reflux, 2 h; F; 67%

Phenols are also readily accessed via this methodology. Arylthallium bis(trifluoroacetate) salts can be oxidized in a single step to phenols, as shown in eq 7.[10] An interesting alternative which combines both boron and thallium chemistry has been developed whereby the arylthallium compound is treated with diborane to provide the arylboronic acid.[11] Oxidation of the boronic acid under standard conditions (eq 8) yields the phenol.

(7) Tl(TFA)$_2$; 1. 1 equiv $Pb(OAc)_4$, rt; 2. 1 equiv PPh_3; 3. dilute NaOH; 62%; OH

(8) Tl(TFA)$_2$; 1. 2.5 equiv B_2H_6, THF, 65 °C, 2 h; 2. NaOH, H_2O_2; 89%; OH

Thallation of substituted arenes provides entry into indoles under notably mild conditions (eqs 9 and 10).[12,13]

(9) H, N, O, Ph, Tl(TFA)$_2$; CuC≡CPh, MeCN, 81 °C, 56%; H, N, O, Ph, Ph; 1. $PdCl_2$, MeCN, 81 °C; 2. KOH; 78%; H, N, Ph

(10)

Silyl enol ethers have been treated with tolylthallium bis(trifluoroacetate) to generate the novel α-metallo ketones shown in eq 11.[14] These compounds are enolate equivalents in titanium-mediated aldol condensations.

(11)

Oxidations. TTFA is a very useful oxidant. Simple alkenes undergo oxidation when treated with thallium(III) salts. The reaction proceeds by an initial oxythallation of the alkene followed by dethallation to give the oxidation product and thallium(I). The intermediate thallium(III) salts are not generally isolated. The oxidation products are glycols, their mono- and diesters, aldehydes, ketones, and epoxides, and the reaction generally involves the migration of a substituent. The product distribution is dependent on reaction conditions and the nature of the thallium(III) salt. An illustration of the use of TTFA in alkene oxidation is shown in eq 12.[15] The reagent of choice for alkene oxidations is, however, ***Thallium(III) Nitrate***.

(12)

Acyclic conjugated dienes react with TTFA to provide 1,2-diacetoxyalkenes in low yields.[16] This is in contrast to ***Thallium(III) Acetate***, which provides both 1,2- and 1,4-addition products.[17] On the other hand, reaction of cyclic dienes with TTFA provides both the 1,2- and 1,4-addition products with *cis* stereochemistry (eq 13). Oxidation of nonconjugated dienes with TTFA is also possible. An example of this methodology in the transannular cyclization of 1,5-cyclooctadiene is shown in eq 14.[18]

(13)

(14)

Efficient routes have been developed for the preparation of *p*-quinones from a variety of aromatic substrates.[19] For example, substituted 4-*t*-butylphenols can be converted to quinones in good yields (eq 15). Similarly, hydroquinones, 4-aminophenols, and 4-halophenols can be converted to *p*-quinones in good yields (eq 16).

(15)

R^1 = Br, R^2 = H; 57%
R^1 = Br, R^2 = Me; 89%
R^1 = I, R^2 = *t*-Bu; 87%
R^1 = Me, R^2 = Me; 88%

(16)

R^1 = R^2 = H, X = OH; 68%
R^1 = Me, R^2 = H, X = OH; 77%
R^1 = Ph, R^2 = H, X = OH; 73%
R^1 = R^2 = H, X = NH_2; 88%
R^1 = R^2 = Br, X = NH_2; 81%
R^1 = R^2 = X = Br; 77%

Thallated aromatics also undergo oxidation with ***Trifluoroperacetic Acid*** to provide quinones (eq 17). The reaction proceeds with the migration of an R group or by elimination of hydrogen halide (R^1 = halogen).[20]

(17)

R^1 = R^2 = R^3 = R^4 = H; 65%
R^1 = Me, R^2 = R^3 = H, R^4 = Me; 68%
R^1 = Cl, R^2 = R^3 = R^4 = H; 42%
R^1 = OMe, R^2 = R^3 = R^4 = H; 61%

Methoxy-substituted phenylpropanoic acids undergo oxidation with 2 equiv TTFA to provide *p*-quinones (eq 18).[21] The reaction involves a one-electron oxidation to form a radical cation, followed by trapping of the intermediate with the carboxyl group and subsequent oxidation to the quinone by the second equivalent of TTFA.

(18)

TTFA has found extensive utility in intramolecular oxidative phenolic coupling reactions and has been elegantly used in the synthesis of many natural products. This coupling strategy, which mimics the biosynthesis of these natural products, provides a laboratory analog of this important reaction. In these reactions, TTFA functions as a two-electron oxidant. The *ortho–para* coupling strategy in the synthesis of a precursor for racemic oxocrinine is shown in eq 19.[22]

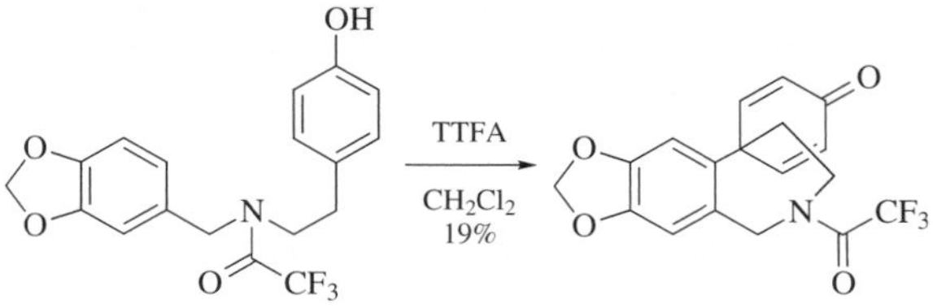
OH
TTFA
CH_2Cl_2
19%
N
O
CF_3

(19)

The major problems associated with this type of coupling reaction are regioselectivity, oxidation of other functional groups, and low yields. A solution to the above problems has been presented in the synthesis of narwedine by the use of a palladacycle to direct the mode of cyclization and to protect the oxidizable functional groups (eq 20).[23] The reaction furnishes a 51% yield of narwedine.

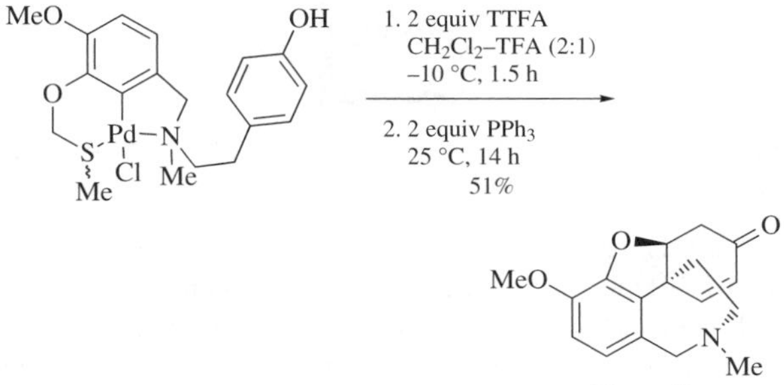
MeO
OH
Pd–N
S
Cl
Me
1. 2 equiv TTFA
CH_2Cl_2–TFA (2:1)
–10 °C, 1.5 h
2. 2 equiv PPh_3
25 °C, 14 h
51%

(20)

Narwedine

Oxidative dimerization reactions of cinnamic acids have been explored. The reaction proceeds in the presence of ***Boron Trifluoride Etherate*** to provide fused bislactones (eq 21).[24]

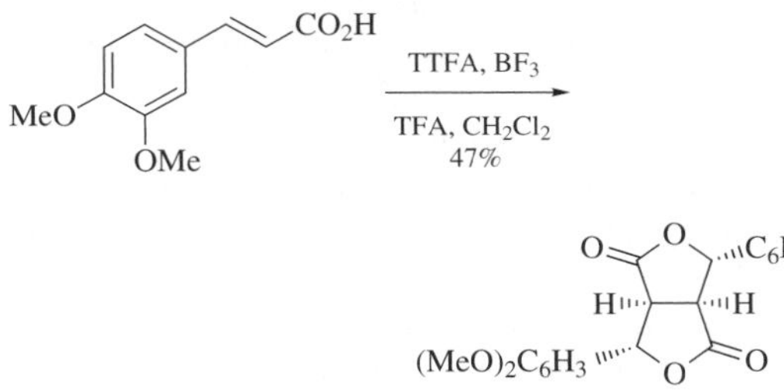
CO_2H
MeO
OMe
TTFA, BF_3
TFA, CH_2Cl_2
47%
$C_6H_3(OMe)_2$
$(MeO)_2C_6H_3$

(21)

TTFA can be effectively used for both inter- and intramolecular biaryl coupling reactions (Ullmann-type couplings). An example of an intermolecular coupling reaction is illustrated in eq 22.[25] (for similar biaryl coupling reactions with TTFA and palladium catalysis, see ***Thallium(III) Trifluoroacetate–Palladium(II) Acetate***). The details of the various reaction conditions for this coupling reaction have been determined.[25] The coupling reactions work well with moderately electron-rich aryls and poorly with electron-poor aryls. Modification of the reaction conditions for coupling of highly electron-rich aryls has also been reported.[26]

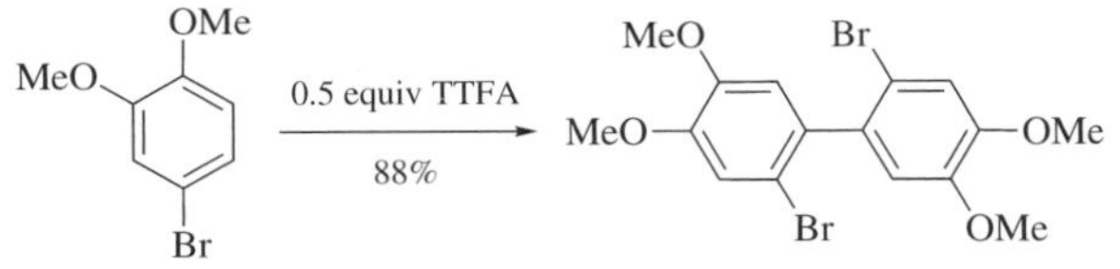
OMe
MeO
Br
0.5 equiv TTFA
88%
MeO
Br
OMe

(22)

Intramolecular biaryl coupling reactions using TTFA also proceed in a facile fashion. An example of this methodology in the synthesis of the aporphine alkaloid ocoteine is shown in eq 23.[27] TTFA is a superior reagent to ***Vanadyl Trifluoride*** in this type of coupling, since the latter reagent generally gives *O*-demethylated compounds.

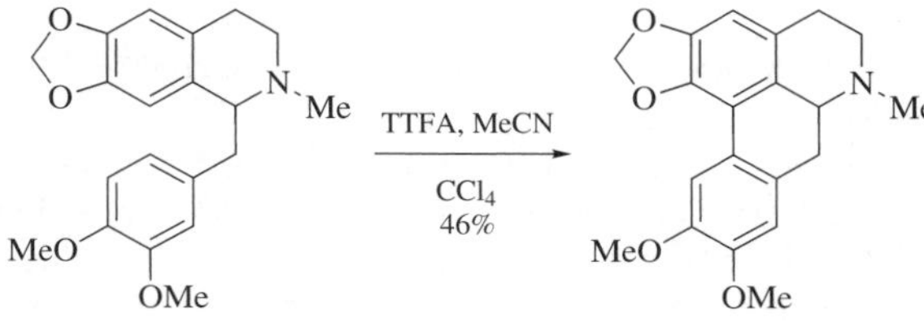
N
Me
MeO
OMe
TTFA, MeCN
CCl_4
46%

(23)

Ocoteine

An interesting example of the biaryl coupling reaction in the synthesis of macrocyclic lactones has been reported (eq 24).[28] Other examples of oxidative biaryl coupling reactions include the synthesis of lignan natural products.[29]

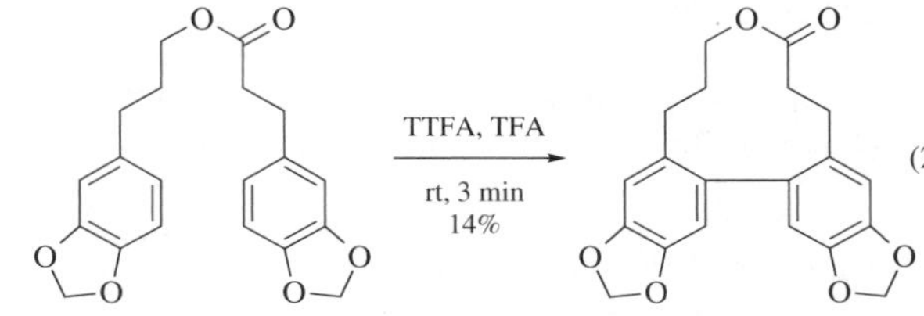
TTFA, TFA
rt, 3 min
14%

(24)

Miscellaneous. Aromatic compounds are allylated using allylsilanes and TTFA.[30] Allylsilanes are converted to allyl ethers,[31] *N*-allyl amides,[32] and allyl nitrates[33] on treatment with TTFA and the appropriate nucleophile. The soft Lewis acid properties of TTFA can be effectively applied in the deprotection reactions of sulfur-containing compounds. Thioacetals can be deprotected to form the corresponding carbonyl compounds in high yields (eq 25).[34]

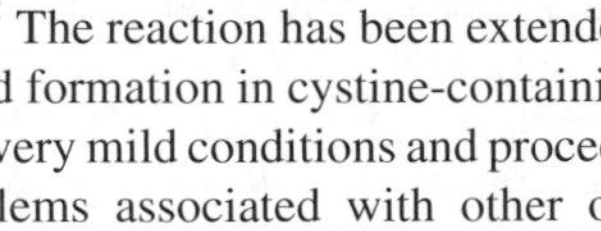
TTFA
THF, rt

(25)

R = Ph, R^1 = Me; 84%
R = R^1 = Ph; 95%
R = Ph, R^1 = H; 83%

TTFA has also been used for the deprotection of sulfur protecting groups in cysteines and subsequently as an oxidant to yield cystines (eq 26).[35] The reaction has been extended to intramolecular disulfide bond formation in cystine-containing peptides. The reaction employs very mild conditions and proceeds in high yields without the problems associated with other oxidants such as iodine.

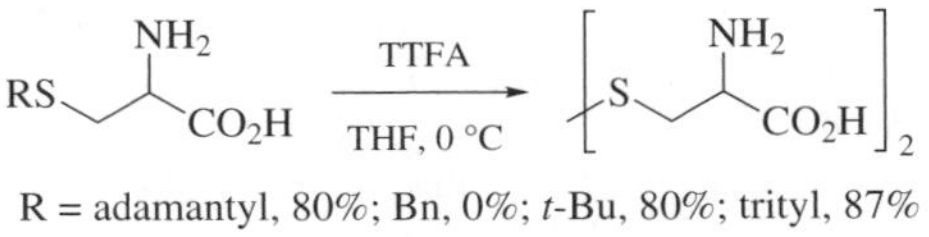
NH_2
RS
CO_2H
TTFA
THF, 0 °C

(26)

R = adamantyl, 80%; Bn, 0%; *t*-Bu, 80%; trityl, 87%

Related Reagents. See Classes O-5 and O-6, pages 1–10.

1. (a) McKillop, A. *PAC* **1975**, *43*, 463. (b) McKillop, A.; Taylor, E. C. In *Comprehensive Organometallic Chemistry*; Wilkinson, G., Ed.; Pergamon: Oxford, 1982; Vol. 7, p 465. (c) McKillop, A.; Taylor, E. C. *ACR* **1970**, *3*, 338. (d) McKillop, A.; Taylor, E. C. *Chem. Br.* **1973**, *9*, 4.

(e) Uemura, S. In *The Chemistry of the Metal–Carbon Bond*; Hartley, F. R., Ed.; Wiley: Chichester, 1987; Vol. 4, Chapter 5.
2. (a) McKillop, A.; Fowler, J. S.; Zelesko, M. J.; Hunt, J. D.; Taylor, E. C.; McGillivray, G. *TL* **1969**, 2423. (b) McKillop, A.; Fowler, J. S.; Zelesko, M. J.; Hunt, J. D.; Taylor, E. C.; Kienzle, F.; McGillivray, G. *JACS* **1971**, *93*, 4841.
3. Bell, H. C.; Kalman, J. R.; Pinhey, J. T.; Sternhell, S. *TL* **1974**, 3391.
4. Deacon, G. B.; Smith, R. N. M. *JFC* **1980**, *15*, 85.
5. (a) Taylor, E. C.; Kienzle, F.; Robey, R. L.; McKillop, A. *JACS* **1970**, *92*, 2175. (b) Taylor, E. C.; Kienzle, F.; Robey, R. L.; McKillop, A.; Hunt, J. D. *JACS* **1971**, *93*, 4845.
6. (a) Taylor, E. C.; Kienzle, F.; Robey, R. L.; McKillop, A.; Hunt, J. D. *JACS* **1971**, *93*, 4845. (b) Taylor, E. C.; Kienzle, F.; Robey, R. L.; McKillop, A. *JACS* **1970**, *92*, 2175.
7. Deacon, G. B.; Smith, R. N. M.; Tunaley, D. *JOM* **1976**, *114*, C1.
8. (a) McKillop, A.; Fowler, J. S.; Zelesko, M. J.; Hunt, J. D.; Taylor, E. C.; Kienzle, F.; McGillivray, G. *JACS* **1971**, *93*, 4841. (b) McKillop, A.; Fowler, J. S.; Zelesko, M. J.; Hunt, J. D.; Taylor, E. C.; McGillivray, G. *TL* **1969**, 2423. (c) Hollins, R. A.; Colnago, L. A.; Salim, V. M.; Seidl, M. C. *JHC* **1979**, *16*, 993.
9. (a) Uemura, S.; Ikeda, Y.; Ichikawa, K. *T* **1972**, *28*, 5499. (b) Taylor, E. C.; Bigham, E. C.; Johnson, D. K.; McKillop, A. *JOC* **1977**, *42*, 362. (c) Adam, M. J.; Berry, J. M.; Hall, L. D.; Pate, B. D.; Ruth, T. J. *CJC* **1983**, *61*, 658.
10. Taylor, E. C.; Altland, H. W.; Danforth, R. H.; McGillivray, G.; McKillop, A. *JACS* **1970**, *92*, 3520.
11. Breuer, S. W.; Pickles, G. M.; Podesta, J. C.; Thorpe, F. G. *CC* **1975**, 36.
12. Taylor, E. C.; Katz, A. H.; Salgado-Zamora, H. *TL* **1985**, *26*, 5963.
13. Larock, R. C.; Liu, C.-L.; Lau, H. H.; Varaprath, S. *TL* **1984**, *25*, 4459.
14. Moriarty, R. M.; Penmasta, R.; Prakash, I.; Awasthi, A. K. *JOC* **1988**, *53*, 1022.
15. (a) Bloodworth, A. J.; Lapham, D. J. *JCS(P1)* **1981**, 3265. (b) Lethbridge, A.; Norman, R. O. C.; Thomas, C. B. *JCS(P1)* **1973**, 2763.
16. Emmer, G.; Zbiral, E. *T* **1977**, *33*, 1415.
17. Uemura, S.; Miyoshi, H.; Tabata, A.; Okano, M. *T* **1981**, *37*, 291.
18. Yamada, Y.; Shibata, A.; Iguchi, K.; Sanjoh, H. *TL* **1977**, 2407.
19. (a) McKillop, A.; Swann, B. P.; Taylor, E. C. *T* **1970**, *26*, 4031. (b) McKillop, A.; Swann, B. P.; Zelesko, M. J.; Taylor, E. C. *AG(E)* **1970**, *9*, 74.
20. Chip, G. K.; Grossert, J. S. *JCS(P1)* **1972**, 1629.
21. Taylor, E. C.; Andrade, J. G.; Rall, G. J. H.; McKillop, A. *JOC* **1978**, *43*, 3632.
22. (a) Schwartz, M. A.; Mami, I. S. *JACS* **1975**, *97*, 1239. (b) Schwartz, M. A.; Rose, B. F.; Vishnuvajjala, B. *JACS* **1973**, *95*, 612. (c) Schwartz, M. A.; Holton, R. A.; Scott, S. W. *JACS* **1969**, *91*, 2800. (d) Schwartz, M. A.; Rose, B. F.; Holton, R. A.; Scott, S. W.; Vishnuvajjala, B. *JACS* **1977**, *99*, 2571. (e) Burnett, D. A.; Hart, D. J. *JOC* **1987**, *52*, 5662.
23. Holton, R. A.; Sibi, M. P.; Murphy, W. S. *JACS* **1988**, *110*, 314.
24. (a) Taylor, E. C.; Andrade, J. G.; Rall, G. J. H.; McKillop, A. *TL* **1978**, 3623. (b) Taylor, E. C.; Andrade, J. G.; Rall, G. J. H.; Steliou, K.; Jagdmann, G. E., Jr.; McKillop, A. *JOC* **1981**, *46*, 3078.
25. (a) McKillop, A.; Turrell, A. G.; Taylor, E. C. *JOC* **1977**, *42*, 764. (b) McKillop, A.; Turrell, A. G.; Young, D. W.; Taylor, E. C. *JACS* **1980**, *102*, 6504.
26. Taylor, E. C.; Katz, A. H.; Alvarado, S. I.; McKillop, A. *JOM* **1985**, *285*, C9.
27. (a) Taylor, E. C.; Andrade, J. G.; McKillop, A. *CC* **1977**, 538. (b) Taylor, E. C.; Andrade, J. G.; Rall, G. J. H.; McKillop, A. *JACS* **1980**, *102*, 6513.
28. Nishiyama, S.; Yamamura, S. *CL* **1981**, 1511.
29. (a) Magnus, P.; Schultz, J.; Gallagher, T. *CC* **1984**, 1179. (b) Cambie, R. C.; Dunlop, M. G.; Rutledge, P. S.; Woodgate, P. D. *SC* **1980**, *10*, 827.
30. Ochiai, M.; Fujita, E.; Arimoto, M.; Yamaguchi, H. *CPB* **1983**, *31*, 86.
31. Ochiai, M.; Fujita, E.; Arimoto, M.; Yamaguchi, H. *CPB* **1984**, *32*, 5027.
32. Ochiai, M.; Tada, S-I.; Arimoto, M.; Fujita, E. *CPB* **1982**, *30*, 2836.
33. Ochiai, M.; Fujita, E.; Arimoto, M.; Yamaguchi, H. *CPB* **1984**, *32*, 887.
34. Ho, T-L.; Wong, C. M. *CJC* **1972**, *50*, 3740.
35. (a) Fujii, N.; Otaka, A.; Funakoshi, S.; Bessho, K.; Yajima, H. *CC* **1987**, 163. (b) Yajima, H.; Fujii, N.; Funakoshi, S.; Watanabe, T.; Murayama, E.; Otaka, A. *T* **1988**, *44*, 805.

Mukund P. Sibi
North Dakota State University, Fargo, ND, USA

Nancy E. Carpenter
University of Minnesota, Morris, MN, USA

Thexylborane[1]

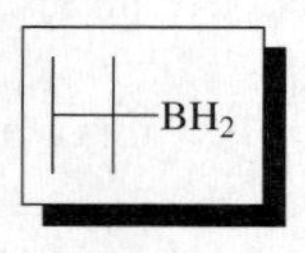

[3688-24-2] $C_6H_{15}B$ (MW 98.02)

(readily available monoalkylborane useful for regioselective hydroboration of alkenes and dienes; thexylalkylboranes and thexyldialkylboranes are useful intermediates for the synthesis of unsymmetrical ketones, cyclic ketones, *trans* disubstituted alkenes, conjugated dienes, and diols[1])

Alternate Name: (1,1,2-trimethylpropyl)borane.
Physical Data: generally prepared in situ; dimeric in THF. It can be isolated as a liquid, mp −34.7 to −32.3 °C.[2]
Solubility: sol ether, hydrocarbon, and halocarbon solvents; THF is generally the solvent of choice; reacts rapidly with protic solvents.[1a]
Form Supplied in: not commercially available.
Analysis of Reagent Purity: analyzed by NMR and IR spectroscopy and by hydrogen evolution upon reaction with methanol.[1a,3]
Preparative Method: most conveniently prepared from ***Borane–Tetrahydrofuran*** and 2,3-dimethyl-2-butene in THF (eq 1).[1a,2]

$$\text{Me}_2\text{C=CMe}_2 \xrightarrow[\text{THF, 0 °C}]{\text{BH}_3} \text{ThxBH}_2 \quad (1)$$

Handling, Storage, and Precautions: very reactive with oxygen and moisture; must be handled using standard techniques for handling air-sensitive materials.[3] The reagent is reported to be stable for at least a week when stored at 0 °C in THF solution under N_2.[1a] However, at rt the boron atom slowly migrates from the tertiary position to the primary (3% in 8 days).[4] Use in a fume hood.

Hydroboration of Alkenes. Reactions of alkenes with thexylborane have been extensively studied and several reviews have appeared.[1] Thexylborane ($ThxBH_2$) in THF reacts with 2 equiv of relatively unhindered alkenes to form thexyldialkylboranes. The regioselectivity in the hydroboration of alkenes with thexylborane is similar to that of borane in THF. For example, hydroboration of 2 equiv of 1-hexene with thexylborane followed by oxidation produces a 95:5 mixture of 1-hexanol and 2-hexanol. Hydroboration of a terminal monosubstituted alkene with thexylborane in a 1:1 ratio gives a mixture of both the thexylmonoalkylborane and thexyldialkylborane. (Preparation of thexylmonoalkylboranes from monosubstituted alkenes can be accomplished using ***Chloro(thexyl)borane–Dimethyl Sulfide***). With most disubstituted and some trisubstituted alkenes, it is possible to prepare the corresponding thexylmonoalkylborane by treating 1 equiv of the alkene with thexylborane at −20 to −25 °C.[5] Thexylmonoalkylboranes can hydroborate relatively unhindered alkenes to form thexyldialkylboranes containing two different alkyl groups (eq 2).[6]

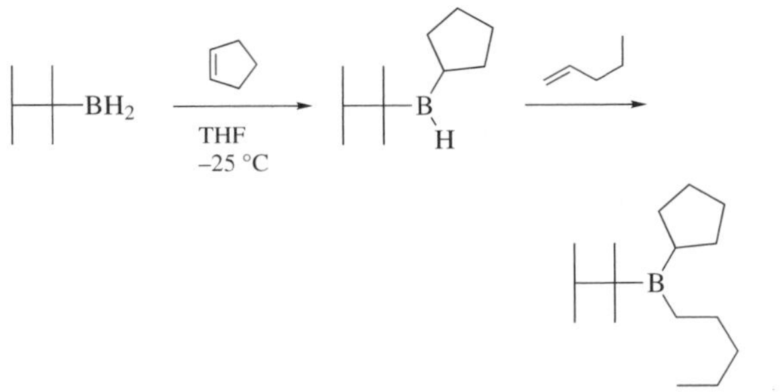

(2)

Hydroboration of a hindered alkene with either thexylborane or thexylmonoalkylborane is slow and is accompanied by dehydroboration of the thexylmonoalkylborane, producing a monoalkylborane and 2,3-dimethyl-2-butene. Lower reaction temperatures and the presence of excess 2,3-dimethyl-2-butene in the reaction may reduce the amount of dehydroboration.[6] Sterically hindered alkenes can be hydroborated under high pressure (6000 atm) to produce highly hindered trialkylboranes, such as trithexylborane,[7] but this procedure does not appear to be practical for synthetic purposes.

Mixed thexyldialkylboranes have also been prepared by treating thexylborane consecutively with different halomagnesium or lithium dialkylcuprates.[8] This procedure offers the advantage of being able to introduce methyl or aryl groups onto the boron atom.

The hydroboration of either terminal or internal alkynes with thexylborane in a 2:1 ratio gives good yields of the expected thexyldialkenylboranes, but the reaction of thexylborane with 1 equiv of a terminal alkyne is reported to give at most 20% of the thexylalkenylborane.[1a,4]

Complete dehydroboration of the thexyl group in thexylmonoalkylboranes can be achieved by treating thexylmonoalkylboranes with a fourfold excess of ***Triethylamine***.[5] This reaction provides a general method for the synthesis of monoalkylboranes as the triethylamine complexes. More recently, it has been found that thexylborane–triethylamine or thexylborane–***N,N,N′,N′-Tetramethylethylenediamine*** adducts can be used to hydroborate hindered alkenes directly (with concomitant loss of the thexyl group as 2,3-dimethyl-2-butene) to give the monoalkylborane–amine adducts.[9] ***Monoisopinocampheylborane***, a useful chiral hydroborating reagent, can be prepared by this reaction.[10]

Thexylborane is the reagent of choice for the hydroboration of dienes to form *B*-thexylboracyclanes (eq 3)[11] since the reaction of borane with dienes tends to give polymeric products.[12]

Stereoselective cyclic hydroboration of dienes by thexylborane has been employed to prepare acyclic diols with 1,3-, 1,4-, and 1,5-asymmetric induction (eq 4).[13] In the cyclic hydroboration of appropriately substituted 1,5-dienes to yield 1,5-diols, 1,2-asymmetric induction was employed as a key step (eq 5) in the synthesis of the Prelog–Djerassi lactone.[14]

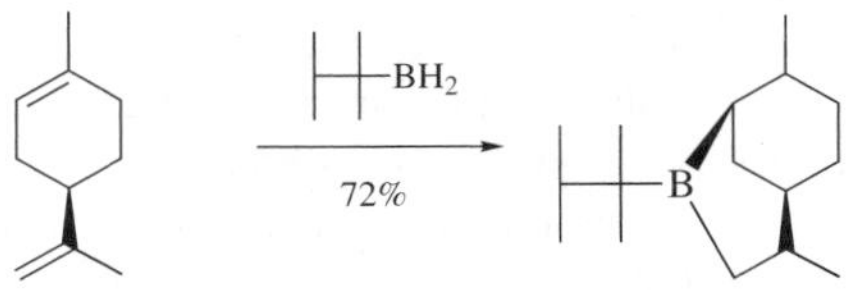

(3)

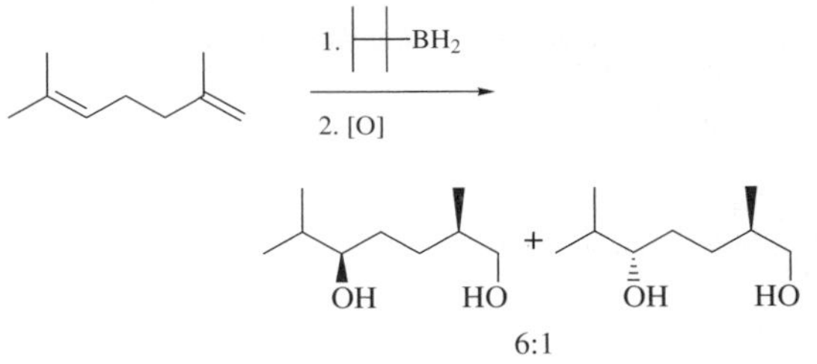

(4)

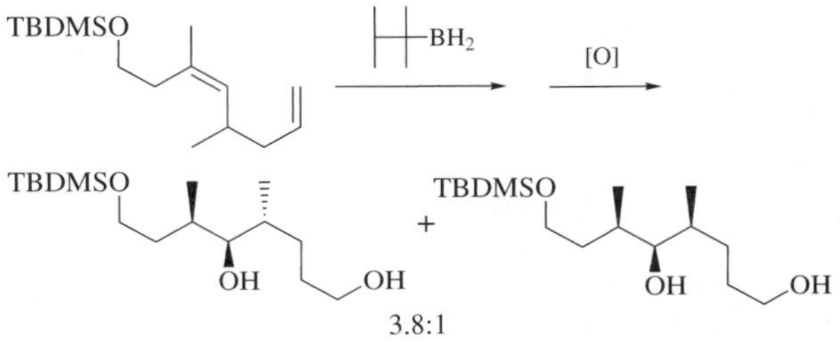

(5)

Cyclic hydroboration of allyl vinyl ethers provides a highly stereoselective synthesis of 1,3-diols with *syn* stereochemistry (eq 6).[15] The *syn* stereoselectivity observed in the cyclic hydroboration of allyl vinyl ethers is opposite to the stereoselectivity observed in the acyclic hydroboration of allylic alcohols (see eq 9).[18] Remote stereocontrol in the hydroboration–reduction of enones with thexylborane has also been reported (eq 7).[16] This reaction is proposed to occur by a rapid hydroboration of the carbon–carbon double bond followed by an intramolecular reduction of the carbonyl group by the intermediate dialkylborane.

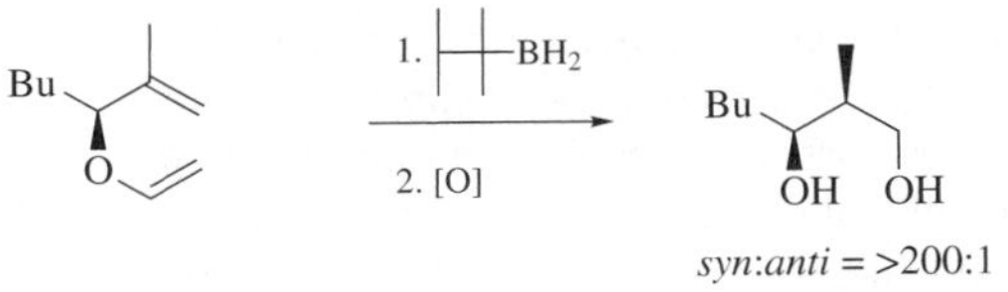

(6)

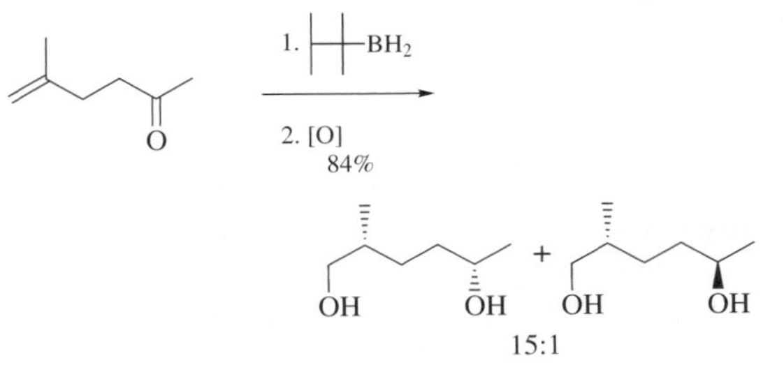

(7)

Acyclic diastereoselection in the hydroboration of alkenes has been reported by several groups. Evans and co-workers[17] observed high levels of 1,3-asymmetric induction in the hydroboration of a number of terminal alkenes bearing substituents at the 2-position of the alkene and a proximal chiral center, as illustrated in eq 8. These workers concluded that the diastereoselection is directed primarily by the nearest chiral center in each of the substrates, and they proposed a transition state model to account for the observed diastereoselectivity. The hydroboration of acyclic secondary allylic alcohols with thexylborane, yielding 1,3-diols with high *anti* (or *threo*) diastereoselection, has also been reported (eq 9).[18] The diastereoselection observed in this reaction does not require the use of a protecting group on the allylic alcohol. Thexylborane proves to be the hydroboration reagent of choice for reaction with trisubstituted alkenes, but terminal alkenes give higher diastereoselection with the more sterically demanding reagents ***9-Borabicyclo[3.3.1]nonane*** or ***Dicyclohexylborane***.

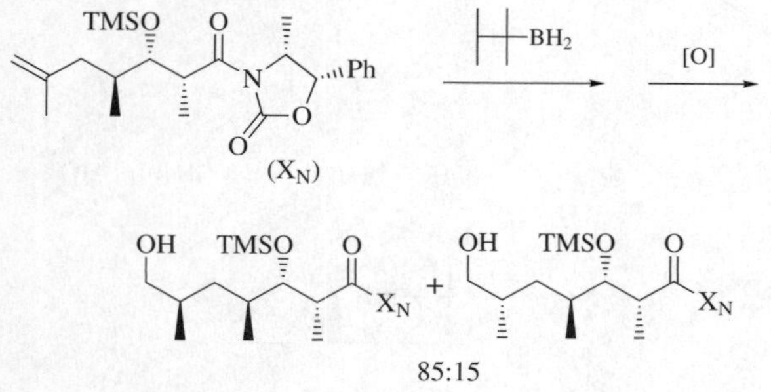

(8) 85:15

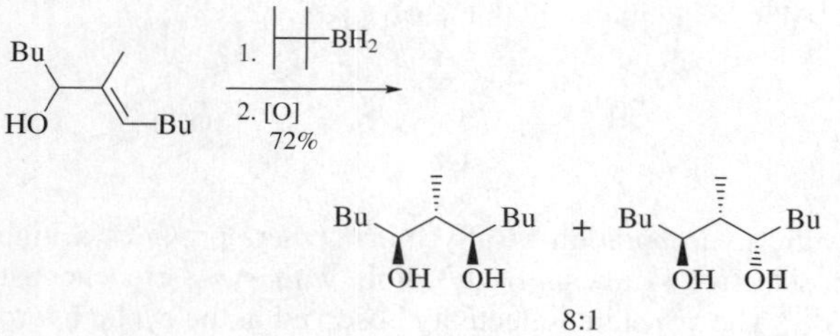

(9) 8:1

Selective Functional Group Reductions. The selective reduction of functional groups by thexylborane has been reported,[19a] and the relative reactivity of thexylborane with common functional groups has been compared to that of other borane reagents (***Diborane***, ***Disiamylborane***, and thexylchloroborane–dimethyl sulfide).[19] Acidic hydrogens in –OH, $-CO_2H$, and $-SO_3H$ groups react at moderate to rapid rates with thexylborane with the evolution of hydrogen. Aldehydes generally react rapidly with thexylborane yielding alcohols after hydrolysis. However, ketones and most other carbonyl groups react only slowly with thexylborane. Carboxylic acids can be reduced to aldehydes by 2.5 equiv of thexylborane, but ***Chloro(thexyl)borane–Dimethyl Sulfide*** is the borane reagent of choice for this transformation. Nitriles, oximes, epoxides, and aromatic nitro compounds are reduced only slowly, and alkyl nitro compounds, disulfides, sulfones, and tosylates do not react with thexylborane.

Use of Thexyldialkylboranes in Synthesis. Thexyldialkylboranes are particularly useful in synthetic reaction sequences due to the availability of a variety of these compounds and the low migratory aptitude of the tertiary thexyl group in most rearrangement reactions. Several reviews of the role of boron in synthesis have appeared.[1,2,20]

Synthesis of Ketones. Reaction of thexyldialkylboranes with ***Carbon Monoxide*** in the presence of water followed by oxidation provides a novel route to the synthesis of ketones (eq 10).[21] Yields in this reaction are generally in the 50–80% range. The observed migratory aptitude for the alkyl groups is: primary > secondary ≫ tertiary. Cyclic ketones can be prepared from dienes by cyclic hydroboration with thexylborane followed by carbonylation and oxidation (eq 11).[22] This annulation procedure stereoselectively provides the *trans*-ring fusion in both the indanone and decalone systems. This methodology has been successfully employed in the stereospecific annulation of 1,5-diene substrates to form the *trans*-hydroazulene nucleus.[23]

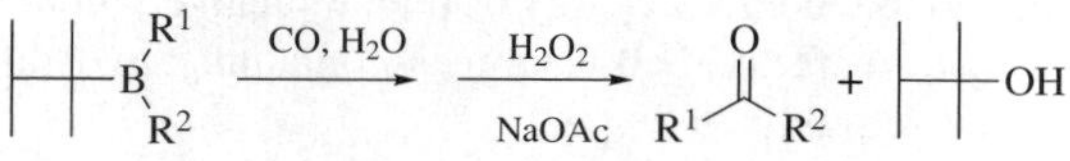

(10)

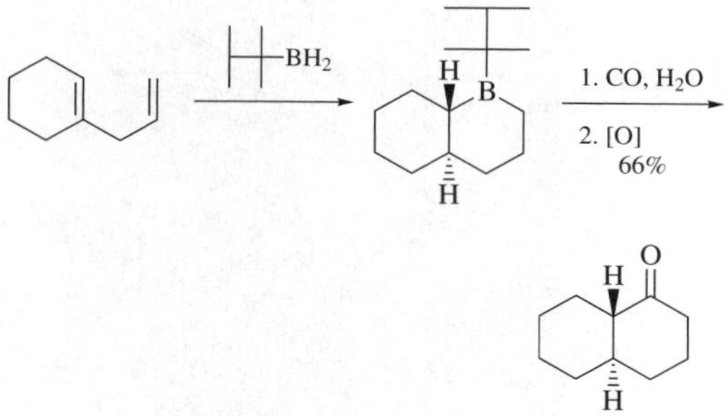

(11)

An alternate procedure to convert thexyldialkylboranes into ketones via an intermediate cyanoborate has been reported (eq 12)[24] which avoids the use of carbon monoxide under high pressure. Yields are quite good, generally >75%, and no special equipment is required for the reaction. This procedure has been employed in the synthesis of the heterocyclic ring system of δ-coniceine (**1**) (eq 13).[25]

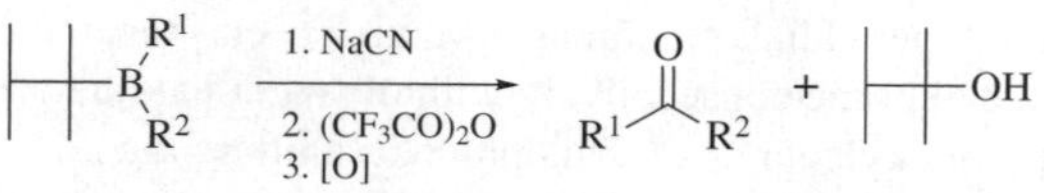

(12)

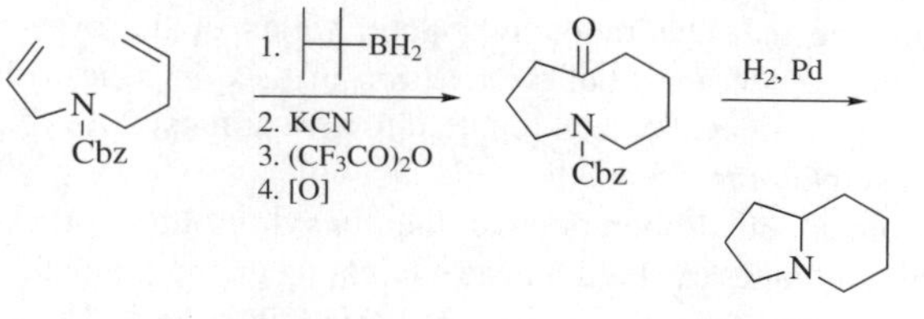

(13)

Synthesis of Alkenes. The stereoselective synthesis of (*E*)-disubstituted alkenes[26] can be carried out by treating a thexylmonoalkylborane (**2**) with a 1-bromo-1-alkyne to yield the intermediate thexylalkyl(1-bromo-1-alkenyl)borane (**3**), which is then treated with ***Sodium Methoxide*** to induce a stereospecific rearrangement of the alkyl group (R^1) from boron to carbon. Stereospecific protonolysis of the resulting intermediate (**4**) provides the (*E*)-alkene in high isomeric purity and good to excellent yield (eq 14).

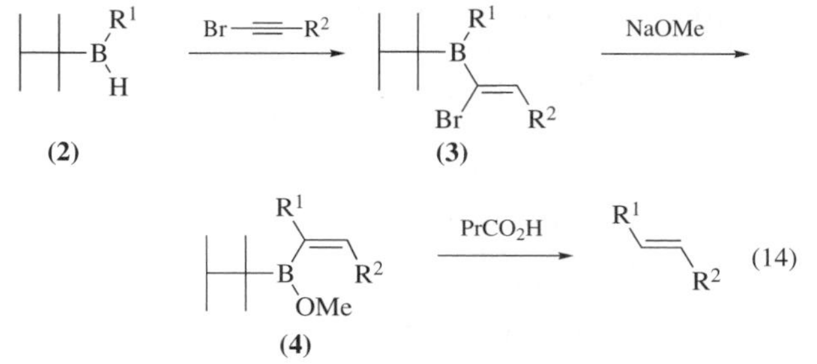

Trisubstituted alkenes can be synthesized[27] by first treating a thexyldialkylborane (**5**) with an alkynyllithium to yield a thexyldialkylalkynylborate (**6**), which is then alkylated with migration of an alkyl group from boron to carbon to give the thexylalkylvinylboranes (**7** and **8**) (eq 15). After acid hydrolysis, the trisubstituted alkenes (**9** and **10**) are obtained in >70% yield in a ratio of about 9:1. The major product (**9**) in all cases is the isomer in which the migrating group (R^1) and the group introduced by alkylation (R^3) are *cis* to each other. Various alkylating agents are effective in this reaction including primary alkyl iodides, benzyl and allyl bromides, ***Dimethyl Sulfate***, and ***Triethyloxonium Tetrafluoroborate***.

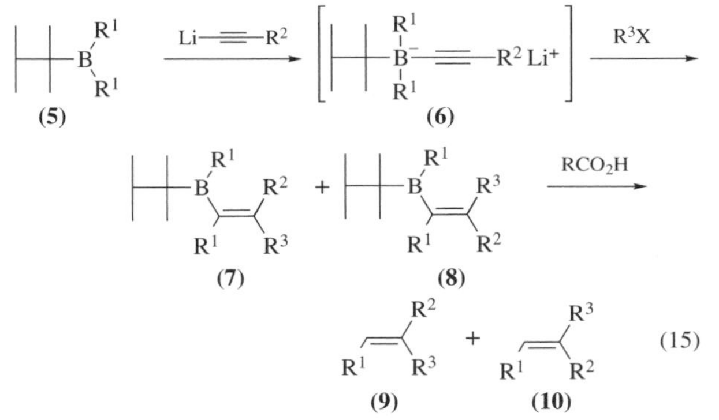

Synthesis of Dienes. Thexylborane reacts with 2 equiv of 1-alkyne to give the corresponding thexyldialkenylborane (**11**) in good yield (eq 16).[4] However, when (**11**) is treated with ***Sodium Hydroxide*** and ***Iodine*** to induce rearrangement, an almost equal mixture of the desired (*E,Z*)-diene (**12**) and alkene (**13**) results where either the alkenyl group or the thexyl group migrates with approximately equal facility.[28] This problem can be surmounted by selectively oxidizing the thexyl group in the thexyldialkenylborane (**11**) with ***Trimethylamine N-Oxide*** before the reaction with iodine and sodium methoxide.[28] This procedure appears to be useful only for the synthesis of symmetrical (*E,Z*)-dienes (**12**) due to the difficulty in preparing unsymmetrical thexyldialkenylboranes in a stepwise hydroboration sequence from two different 1-alkynes.

However, thexylborane will react with 1-chloro-1-alkynes and 1-bromo-1-alkynes (but not with 1-iodo-1-alkynes) to give thexyl(1-haloalkenyl)boranes (**14**) in high yield (eq 18).[29] This observation enabled the development of a general and highly stereospecific synthesis of conjugated (*E,E*)-dienes. The borane (**14**) reacts stereoselectively with a 1-alkyne to give the thexyldialkenylborane (**15**), which is then treated with sodium methoxide to produce (**16**). Compound (**16**) is protonolyzed with refluxing isobutyric acid to give the (*E,E*)-diene (**17**) in 53–63% yield and >98% isomeric purity.[29] The intermediate (**16**) is also transformed into the (*E*)-enone (**18**) in 50% yield by oxidation with ***Hydrogen Peroxide*** and NaOAc (eq 19).[29]

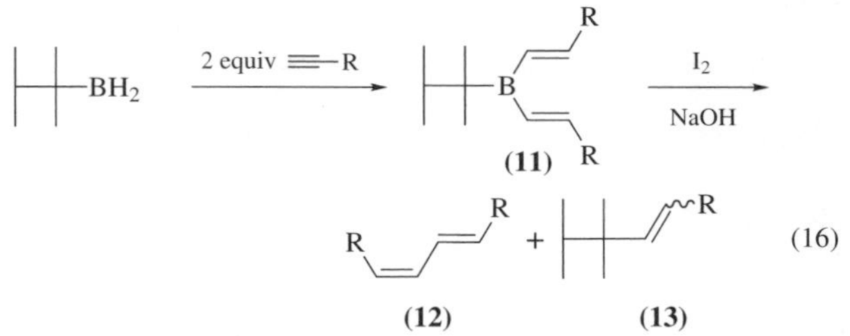

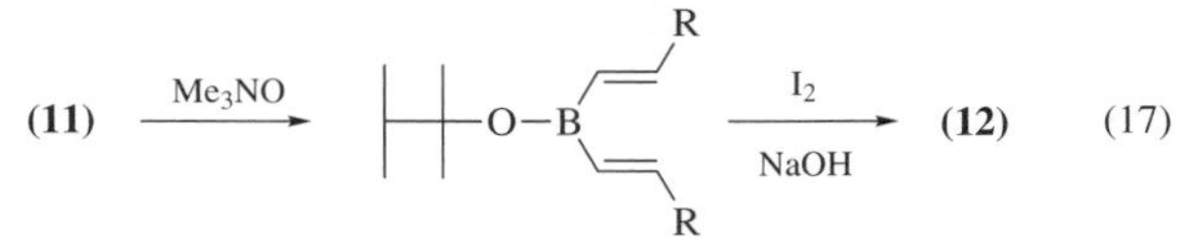

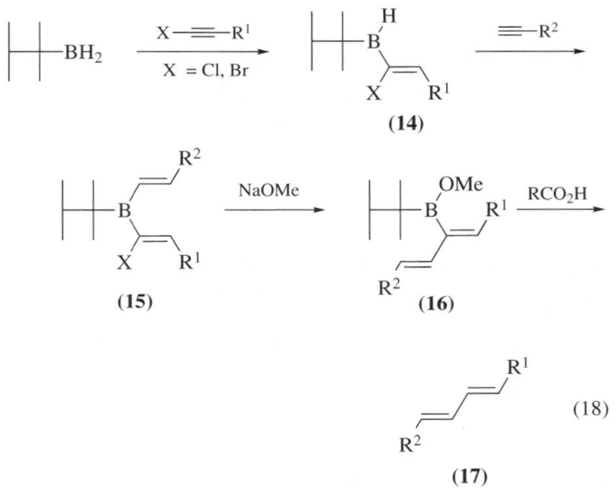

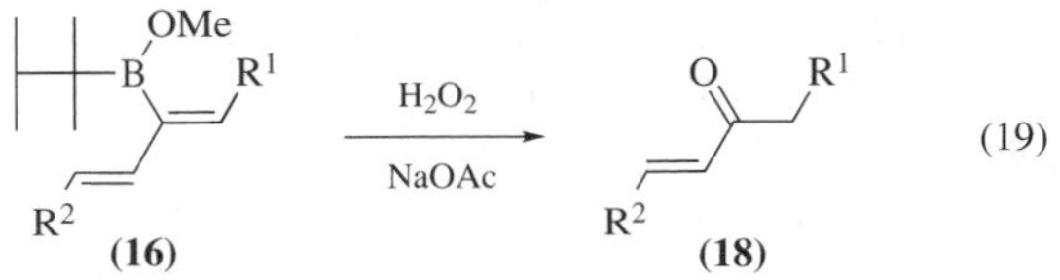

A stereoselective synthesis of 1,2,3-butatrienes was also developed based on the reaction of thexylborane with 1-halo-1-alkynes. In this synthesis, 2 equiv of a 1-iodo-1-alkyne react with thexylborane to give the thexyldialkenylborane (**19**) that rearranges to the 1,2,3-butatriene (**21**) upon treatment with 2 equiv of sodium methoxide (eq 20).[29b,30] Although the yields of (**21**) are only moderate (47%, R = butyl; 29%, R = cyclohexyl), the products are of high isomeric purity.

Synthesis of Carboxylic Acids. A convenient procedure for the direct oxidation of organoboranes from terminal alkenes to carboxylic acids has been reported.[31] The oxidation gives high yields with a number of different organoboranes derived from a variety of borane reagents including thexylborane. Several different oxidizing agents (***Pyridinium Dichromate***, ***Sodium Dichromate*** in aqueous H_2SO_4, and ***Chromium(VI) Oxide*** in 90% aqueous acetic acid) are effective for the reaction. For example, 2-methyl-1-pentene is converted to 2-methylpentanoic acid in 86% yield (eq 21).

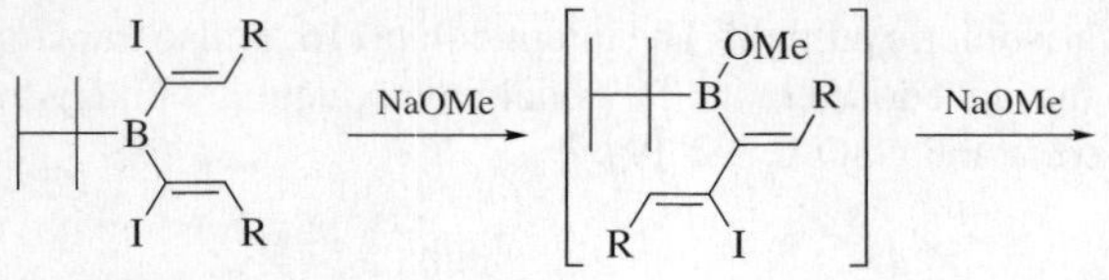

(20)

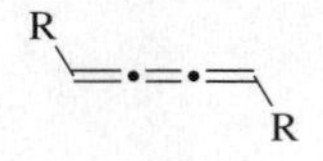

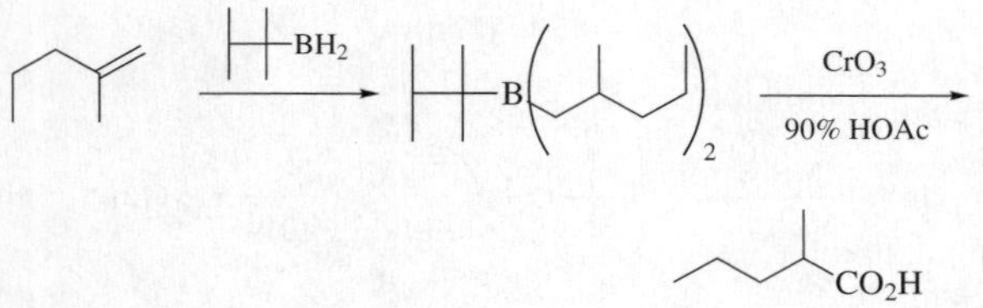

(21)

Related Reagents. See Classes R-2, R-4, R-12, R-14, R-17, pages 1–10. 9-Borabicyclo[3.3.1]nonane; Borane–Tetrahydrofuran; Chloro(thexyl)borane–Dimethyl Sulfide; Dicyclohexylborane; Disiamylborane.

1. (a) Negishi, E.; Brown, H. C. *S* **1974**, 77. (b) Brown, H. C.; Negishi, E.; Zaidlewicz, M. In *Comprehensive Organometallic Chemistry*; Wilkinson, G.; Stone, F. G. A.; Abel, E. W., Eds.; Pergamon: Oxford, 1982; Vol. 7, pp 111–363. (c) Pelter, A.; Smith, K.; Brown, H. C. *Borane Reagents*; Academic: London, 1988. (d) Smith, K.; Pelter, A. *COS* **1991**, *8*, 709.
2. (a) Brown, H. C.; Kramer, G. W.; Levy, A.; Midland, M. M. *Organic Syntheses via Boranes*; Wiley: New York, 1975; p 31. (b) Brown, H. C.; Mandal, A. K. *JOC* **1992**, *57*, 4970.
3. (a) Brown, H. C.; Kramer, G. W.; Levy, A.; Midland, M. M. *Organic Synthesis via Boranes*; Wiley: New York, 1975; Chapter 9. (b) Schwier, J. R.; Brown, H. C. *JOC* **1993**, *58*, 1546.
4. Zweifel, G.; Brown, H. C. *JACS* **1963**, *85*, 2066.
5. Brown, H. C.; Negishi, E.; Katz, J.-J. *JACS* **1975**, *97*, 2791.
6. Brown, H. C.; Katz, J.-J.; Lane, C. F.; Negishi, E. *JACS* **1975**, *97*, 2799.
7. Rice, J. E.; Okamoto, Y. *JOC* **1982**, *47*, 4189.
8. Whitely, C. G. *TL* **1984**, *25*, 5563.
9. Brown, H. C.; Yoon, N. M.; Mandal, A. K. *JOM* **1977**, *135*, C10.
10. Brown, H. C.; Mandal, A. K.; Yoon, N. M.; Singaram, B.; Schwier, J. R.; Jadhav, P. K. *JOC* **1982**, *47*, 5069.
11. Brown, H. C.; Negishi, E. *JACS* **1972**, *94*, 3567.
12. Brown, H. C.; Negishi, E.; Burk, P. L. *JACS* **1972**, *94*, 3561.
13. (a) Still, W. C.; Darst, K. P. *JACS* **1980**, *102*, 7385. (b) Harada, T.; Matsuda, Y.; Wada, I.; Uchimura, J.; Oku, A. *CC* **1990**, 21.
14. Morgans, D. J., Jr. *TL* **1981**, *22*, 3721.
15. Harada, T.; Matsuda, Y.; Uchimura, J.; Oku, A. *CC* **1989**, 1429.
16. Harada, T.; Matsuda, Y.; Imanaka, S.; Oku, A. *CC* **1990**, 1641.
17. (a) Evans, D. A.; Barttoli, J.; Godel, T. *TL* **1982**, *23*, 4577. (b) Evans, D. A.; Barttoli, J. *TL* **1982**, *23*, 807.
18. Still, W. C.; Barrish, J. C. *JACS* **1983**, *105*, 2487.
19. (a) Brown, H. C.; Heim, P.; Yoon, N. M. *JOC* **1972**, *37*, 2942. (b) Brown, H. C.; Nazer, B.; Cha, J. S.; Sikorski, J. A. *JOC* **1986**, *51*, 5264.
20. (a) Coveney, D. J. *COS* **1991**, *3*, 793. (b) Carruthers, W. *Some Modern Methods of Organic Synthesis*, 3rd ed.; Cambridge University Press: Cambridge, 1986; pp 294–317. (c) Thomas, S. E. *Organic Synthesis: The Roles of Boron and Silicon*; Oxford University Press: Oxford, 1991; pp 1–46.
21. (a) Brown, H. C.; Negishi, E. *JACS* **1967**, *89*, 5285. (b) Negishi, E.; Brown, H. C. *S* **1972**, 196. (c) Brown, H. C. *ACR* **1969**, *2*, 65.
22. (a) Brown, H. C.; Negishi, E. *JACS* **1967**, *89*, 5477. (b) Brown, H. C.; Negishi, E. *CC* **1968**, 594.
23. Stevenson, J. W. S.; Bryson, T. A. *CL* **1984**, 5.
24. (a) Pelter, A.; Hutchings, M. G.; Smith, K. *CC* **1970**, 1529. (b) Pelter, A.; Hutchings, M. G.; Smith, K. *CC* **1971**, 1048. (c) Pelter, A.; Hutchings, M. G.; Smith, K. *CC* **1973**, 186. (d) Pelter, A.; Smith, K.; Hutchings, M. G.; Rowe, K. *JCS(P1)* **1975**, 129. (e) Pelter, A.; Hutchings, M. G.; Smith, K.; Williams, D. J. *JCS(P1)* **1975**, 145.
25. Garst, M. E.; Bonfiglio, J. N. *TL* **1981**, *22*, 2075.
26. (a) Negishi, E.; Katz, J.-J., Brown, H. C. *S* **1972**, 555. (b) Corey, E. J.; Ravindranathan, T. *JACS* **1972**, *94*, 4013.
27. Pelter, A.; Subrahmanyam, C.; Laub, R. J.; Gould, K. J.; Harrison, C. R. *TL* **1975**, *19*, 1633.
28. Zwiefel, G.; Polston, N. L.; Whitney, C. C. *JACS* **1968**, *90*, 6243.
29. (a) Negishi, E.; Yoshida, T. *CC* **1973**, 606. (b) Negishi, E.; Yoshida, T.; Abramovitch, A.; Lew, G.; Williams, R. M. *T* **1991**, *47*, 343.
30. Yoshida, T.; Williams, R. M.; Negishi, E. *JACS* **1974**, *96*, 3688.
31. Brown, H. C.; Kulkarni, S. V.; Khanna, V. V.; Patil, V. D.; Racherla, U. S. *JOC* **1992**, *57*, 6173.

William S. Mungall
Hope College, Holland, MI, USA

Tin[1]

Sn

[7440-31-5] Sn (MW 118.71)

(with HCl, reduces a variety of functional groups;[2] stereoselective allylation of carbonyl compounds;[3] in situ generation of tin enolates for directed aldol reactions[4])

Physical Data: mp 232 °C; bp ~2270 °C; *d* 7.31 g cm^{-3}.
Solubility: insol water, organic solvents; reacts with mineral acids.
Form Supplied in: foil, moss, powder, granules, shot, and wire.
Preparative Methods: activated form is prepared from ***Tin(II) Chloride*** in THF by reduction with ***Lithium Aluminum Hydride***[5] or ***Potassium*** metal;[6] tin amalgam is prepared from ***Mercury(II) Chloride*** and 30-mesh Sn in water;[7] tin–copper couple is prepared from ***Copper(II) Acetate*** and 30-mesh tin in acetic acid.[8]
Handling, Storage, and Precautions: incompatible with strong acids and strong oxidizing agents; powdered form is air and moisture sensitive and should not be inhaled or contacted with the eyes or skin; amalgam is stored under water; tin–copper couple is stored under ether.

Functional Group Reductions. The use of tin and ***Hydrochloric Acid*** is a classical method for the reduction of a variety of functional groups.[2] However this procedure has decreased in importance since the development of catalytic hydrogenation and of metal hydride reducing agents, and the harsh reaction conditions (strong acids and high temperatures) are often incompatible with other functional groups. Many reductions effected by tin are more conveniently carried out with ***Tin(II) Chloride***, which is soluble in some organic solvents. Nevertheless, the method is still used, most notably for the reduction of aromatic nitro compounds to amines, for which metal hydrides are less effective (see ***Lithium Aluminum Hydride***). Representative examples of standard procedures which utilize Sn/HCl include the reductions of 2,6-dibromo-4-nitrophenol to the aminophenol,[9] 2,4,6-trinitrobenzoic acid to the triamine (eq 1),[10] and nitrobarbituric acid to the amine (uramil).[11] Aromatic nitro group reductions with tin and hydrochloric acid have been employed in the preparation of functionalized paracyclophanes[12] and hemispherands.[13] 8-Nitroquinolines have been reduced to the corresponding amines with a combination of tin and tin(II) chloride.[14] In some cases the generated amines undergo further reactions (eq 2).[15]

Sn, HCl (1)

Sn, HCl, 85% (2)

Sterically hindered aliphatic vicinal diamines have recently been prepared by Sn/HCl reductions of the corresponding 1,2-dinitro compounds (eq 3).[16] Under the same conditions, a geminal dinitro compound is reported to give the corresponding oxime (eq 4).[16] *N*-Nitrosamines are reduced by tin and HCl to the denitrosated amines (eq 5),[17] whereas the corresponding ***Zinc*** reduction produces the hydrazines.

Sn, HCl, 42–82% (3)

R^1, R^2 = Me, Et

Sn, HCl, 85% (4)

Sn, HCl (5)

Examples of reductions of other functional groups by tin and HCl include the reduction of isoquinoline methiodides,[18] of tetrahydrocarbazole to hexahydrocarbazole,[19] and of 5-(chloromethyl)uracil to thymine (eq 6),[20] the selective debromination of 1,6-dibromo-2-naphthol (eq 7),[21] and the reduction of anisoin to deoxyanisoin,[22] of anthraquinone to anthrone (eq 8),[23] and of benzaldehydes to stilbenes.[24]

Sn, HCl, 82% (6)

Sn, HBr, 70% (7)

Sn, HCl, 83% (8)

Tin amalgam and hydrochloric acid is useful for the selective reduction of the double bond of conjugated enediones in high yield (eq 9)[7] and for the controlled reduction of benzils to either benzoins or deoxybenzoins.[25] Tin amalgam in ***Acetic Acid*** also reduces benzoquinones to hydroquinones.[7,26]

Sn/Hg, HCl, 90% (9)

Tin reductions have also been used for the conversion of arylsulfonyl chlorides to the corresponding thiols[27] and for the preparation of ***Thiophosgene*** from thiocarbonyl perchloride.[28] Metallic tin has recently been used for the in situ generation of low-valent bismuth and titanium species that catalyze cyclizations to 3-hydroxycephems (eq 10).[29] A tin–copper couple has been used for the selective debromination of activated dibromides (eq 11),[8] in cases where the more commonly employed ***Zinc/Copper Couple*** leads to overreduction.

Sn, $BiCl_4$, py, DMF, 42–82% (10)

Sn/Cu, THF, 81% (11)

Barbier-Type Allylations and Related Reactions. The reaction of carbonyl compounds with allylmetal reagents to give homoallylic alcohols is an extremely important reaction in organic synthesis.[3,30] Metallic tin can be used for the in situ generation of diallyltin dihalides for subsequent reaction with aldehydes or ketones in good yield (eq 12).[31] The procedure is considerably more convenient than that involving isolation of allyltin intermediates and can be carried out in the presence of water.[32] Moderate asymmetric induction is observed when the reaction is carried out in the presence of monosodium (+)-diethyl tartrate.[33] The yields are improved by sonication[34] and this procedure has recently been applied to chain extensions of carbohydrates.[35] The reaction can be made catalytic by electrochemical regeneration of the tin reagent.[36]

Br + PhCHO —Sn, THF, 82%→ (12)

A further improvement involves performing the reaction in the presence of ***Aluminum*** powder in a THF–water mixture.[32] Under these conditions, allyl chloride also reacts[37] and yields are improved for reactions of substituted allyl halides. In such cases the diastereoselectivity is dependent on the specific reactants and reaction conditions. For example the Sn/Al reaction of benzaldehyde with crotyl bromide gives the *syn* diastereoisomer as the major product (eq 13),[32] whereas reaction with cinnamyl chloride occurs with exclusive *anti* selectivity (eq 14).[38] Under the same conditions cinnamyl chloride reacts with enals by exclusive 1,2-regioselective addition and complete *anti* diastereoselectivity and with 2-phenylpropanal with moderate Cram selectivity (eq 15).[38] These reactions are compatible with a variety of other functional groups,[32,39,40] although in some cases these may induce further reactions (eq 16).[40] A mechanistic study of the tin-promoted reactions of allylic iodides with benzaldehydes has recently been reported.[41] It has also recently been shown that allylic alcohols undergo similar reactions with metallic tin in the presence of ***Chlorotrimethylsilane*** and ***Sodium Iodide***.[42] Under these conditions, substituted allylic alcohols undergo carbon–carbon bond formation to the less substituted allylic carbon.[42]

Sn, Al, THF, H_2O; PhCHO; 87% — Ph, OH + Ph, OH — 3:2 (13)

Sn, Al, THF, H_2O; PhCHO; 100% — Ph, Ph, OH (14)

Sn, Al, THF, H_2O; RCH=CHCHO; 100% — R, Ph, OH
Sn, Al, THF, H_2O; Ph(Me)CHCHO; 85% — Ph, Ph, OH + Ph, Ph, OH — 3:1 (15)

CO_2Et, Br; Sn, Al, DME; PhCHO; 75% — Ph, O, O (16)

The Sn/Al procedure has been further extended to the reactions of propargylic halides and such reactions can give both alkynic and allenic products, depending on the reaction conditions (eq 17).[43] The direct reaction of metallic tin with simple alkyl halides to give dialkyltin dihalides is an industrially important reaction and is usually restricted to the reactions of alkyl iodides.[44] More recently, a phase-transfer catalyzed procedure has been developed that facilitates reactions of alkyl bromides and chlorides.[45]

TMS—≡—CH_2I; Sn, Al; PhCHO; 70% — Ph, OH, TMS + Ph, OH, TMS (17)

Solvent: MeCN 1:5
Diglyme 8:1

Tin Enolates. Tin enolates are useful intermediates for use in directed aldol reactions.[4,46] Tin(II) enolates are usually prepared[47] by reaction of enolizable ketones with ***Tin(II) Trifluoromethanesulfonate***, but can also be prepared from reactions of α-bromocarbonyl compounds with activated metallic tin. Such enolates react with aldehydes and ketones under mild conditions to give aldols, generally in high yield (eq 18).[48] With α-substituted enolates high *syn* selectivity is observed (eq 19);[48] this is the opposite selectivity to that found with tin(IV) enolates. It has recently been shown that such reactions can be carried out in aqueous media with unactivated tin powder, but that under these conditions the metal enolate is probably not involved; a single-electron-transfer mechanism has been suggested.[49] In a related reaction, metallic tin has been used to generate a highly functionalized tin(II) enolate for alkylation of an azetidinone as part of a carbapenam synthesis (eq 20).[50]

R^1, R^2 R^3, Br, O; Sn, R^1COR^2; 28–99% — R^1, R^2 R^3, R^1, R^2, O, OH (18)

Br, Ph, O; 1. Sn, DMF; 2. PhCHO, –78 °C; 93% — Ph, Ph, OH, O + Ph, Ph, OH, O — 93:7 (19)

OTBDMS, OAc, O, N, H + Br, N_2, O, O, O; Sn, DMF; 75% — OR, N_2, O, O, O, N, H — β:α = 3:1 (20)

Tin(II) ester enolates can also be prepared by reaction of α-bromo carboxylic acid esters for Reformatsky-type reactions under very mild conditions (eq 21).[5] Reaction of an α-diketone or α-keto aldehyde with activated metallic tin produces a tin(II) enediolate that reacts with aldehydes to produce α,β-dihydroxy ketones in high yield (eq 22).[51] The diastereoselectivity of this reaction can is controlled by the addition of hexafluorobenzene.

PhCHO + BrCO_2Et; Sn, THF; 83% — Ph, OH, CO_2Et (21)

O, O; Sn, THF — O, Sn, O; RCHO; 73–95% — R, OH, OH, O (22)

Related Reagents. See Classes R-15, R-20, R-22, R-23, R-24, R-25, R-27, and R-31, pages 1–10.

1. (a) Pereyre, M.; Quintard, J.-P.; Rahm, A. *Tin in Organic Synthesis*; Butterworths: London, 1987. (b) *Chemistry of Tin*; Harrison, P. G., Ed.; Chapman and Hall: New York, 1989.
2. (a) Hudlicky, M. *Reductions in Organic Synthesis*; Horwood: Chichester, 1984. (b) *COS* **1991**, *8*, Chapters 1.1–4.8.

3. (a) Roush, W. R. *COS* **1991**, *2*, 1. (b) Yamamoto, Y.; Asao, N. *CRV* **1993**, *93*, 2207.
4. Chan, T.-H. *COS* **1991**, *2*, 595.
5. Harada, T.; Mukaiyama, T. *CL* **1982**, 161.
6. Kato, J.; Mukaiyama, T. *CL* **1983**, 1727.
7. Schaefer, J. P. *JOC* **1960**, *25*, 2027.
8. Dowd, P.; Marwaha, L. K. *JOC* **1976**, *41*, 4035.
9. Hartman, W. W.; Dickey, J. B.; Stampfli, J. G. *OSC* **1943**, *2*, 175.
10. Clarke, H. T.; Hartman, W. W. *OSC* **1932**, *1*, 444.
11. Hartman, W. W.; Sheppard, O. E. *OSC* **1943**, *2*, 617.
12. Sheehan, M.; Cram, D. J. *JACS* **1969**, *91*, 3544.
13. Doxsee, K. M.; Feigel, M.; Stewart, K. D.; Canary, J. W.; Knobler, C. B.; Cram, D. J. *JACS* **1987**, *109*, 3098.
14. Carroll, F. I.; Berrang, B. D.; Linn, C. P. *JMC* **1979**, *22*, 1363.
15. (a) Wear, R. L.; Hamilton, C. S. *JACS* **1950**, *72*, 2893. (b) Petrow, V. A.; Stack, M. V.; Wragg, W. R. *JCS* **1943**, 316.
16. Asaro, M. F.; Nakayama, I.; Wilson, R. B., Jr. *JOC* **1992**, *57*, 778.
17. Fridman, A. L.; Mukhametshin, F. M.; Novikov, S. S. *RCR* **1971**, *40*, 34.
18. Wittig, G.; Streib, H. *LA* **1953**, *584*, 1.
19. Gurney, J.; Perkin, W. H., Jr.; Plant, S. G. P. *JCS* **1927**, *130*, 2676.
20. Skinner, W. A.; Schelstraete, M. G. M.; Baker, B. R. *JOC* **1960**, *25*, 149.
21. Koelsch, C. F. *OSC* **1955**, *3*, 132.
22. Carter, P. H.; Craig, J. C.; Lack, R. E.; Moyle, M. *OSC* **1973**, *5*, 339.
23. Meyer, K. H. *OSC* **1932**, *1*, 60.
24. Stewart, F. H. C. *JOC* **1961**, *26*, 3604.
25. Pearl, I. A.; Dehn, W. M. *JACS* **1938**, *60*, 57; Pearl, I. A. *JOC* **1957**, *22*, 1229.
26. Chang, M.; Netzly, D. H.; Butler, L. G.; Lynn, D. G. *JACS* **1986**, *108*, 7858.
27. Hodgson, H. H.; Leigh, E. *JCS* **1939**, *142*, 1094.
28. Dyson, G. M. *OSC* **1932**, *1*, 506.
29. Tanaka, H.; Taniguchi, M.; Kameyama, Y.; Monnin, M.; Sasaoka, M.; Shiroi, T.; Nagao, S.; Torii, S. *CL* **1990**, 1867.
30. Hoffmann, R. W. *AG(E)* **1982**, *21*, 555; Yamamoto, Y. *ACR* **1987**, *20*, 243.
31. Mukaiyama, T.; Harada, T. *CL* **1981**, 1527.
32. Nokami, J.; Otera, J.; Sudo, T.; Okawara, R. *OM* **1983**, *2*, 191.
33. Boga, C.; Savoia, D.; Tagliavini, E.; Trombini, C.; Umani-Ronchi, A. *JOM* **1988**, *353*, 177.
34. Petrier, C.; Einhorn, J.; Luche, J. L. *TL* **1985**, *26*, 1449.
35. Schmid, W.; Whitesides, G. M. *JACS* **1991**, *113*, 6674.
36. Uneyama, K.; Matsuda, H.; Torii, S. *TL* **1984**, *25*, 6017.
37. Uneyama, K.; Kamaki, N.; Moriya, H.; Torii, S. *JOC* **1985**, *50*, 5396.
38. (a) Coxon, J. M.; van Eyk, S. J.; Steel, P. J. *TL* **1985**, *26*, 6121. (b) Coxon, J. M.; van Eyk, S. J.; Steel, P. J. *T* **1989**, *45*, 1029.
39. Mandai, T.; Nokami, J.; Yano, T.; Yoshinaga, Y.; Otera, J. *JOC* **1984**, *49*, 172.
40. Nokami, J.; Tamaoka, T.; Ogawa, H.; Wakabayashi, S. *CL* **1986**, 541.
41. Yamataka, H.; Nishikawa, K.; Hanafusa, T. *BCJ* **1992**, *65*, 2145.
42. Kanagawa, Y.; Nishiyama, Y.; Ishii, Y. *JOC* **1992**, *57*, 6988.
43. Nokami, J.; Tamaoka, T.; Koguchi, T.; Okawara, R. *CL* **1984**, 1939.
44. Oakes, V.; Hutton, R. E. *JOM* **1965**, *3*, 472.
45. Ugo, R.; Chiesa, A.; Fusi, A. *JOM* **1987**, *330*, 25.
46. Mukaiyama, T. *OR* **1982**, *28*, 203.
47. Mekelburger, H. B.; Wilcox, C. S. *COS* **1991**, *2*, 99.
48. Harada, T.; Mukaiyama, T. *CL* **1982**, 467.
49. Chan, T. H.; Li, C. J.; Wei, Z. Y. *CC* **1990**, 505.
50. Deziel, R.; Endo, M. *TL* **1988**, *29*, 61.
51. Mukaiyama, T.; Kato, J.; Yamaguchi, M. *CL* **1982**, 1291.

Peter J. Steel
University of Canterbury, Christchurch, New Zealand

Titanium[1]

[7440-32-6] Ti (MW 47.88)

(low-valent titanium, i.e. Ti^0/Ti^I, can be generated from a variety of sources and is used for the inter- and intramolecular reductive coupling of carbonyl and dicarbonyl compounds, imines, and iminium species, and for reductive elimination reactions and reductions)

Physical Data: mp 1677 °C; *d* (α-form) 0-34.506 g cm^{-3}.
Solubility: unknown but presumably low in most organic solvents. THF and DME appear to be the favored solvents for reactions involving low-valent titanium.
Form Supplied in: a 4:1 ball-milled mixture of $TiCl_3$ and $LiAlH_4$ is available commercially and serves as a precursor to the McMurry reagent (see ***Titanium(III) Chloride–Lithium Aluminum Hydride***). Low-valent titanium can be generated in situ from a variety of other sources as discussed below.
Preparative Methods: low-valent titanium reagents, including Ti^0, can be generated by in situ reduction of Ti^I–Ti^{IV} chlorides. At least a dozen commonly used procedures have been reported including those involving $TiCl_3/Zn$,[2] $TiCl_3/Mg$,[3] $TiCl_3/LiAlH_4$,[4] $TiCl_3/K$,[5] $TiCl_3/Li$,[6] $TiCl_3/Zn(Cu)$,[5] $TiCl_3/C_8K$,[7] $CpTiCl_3/LiAlH_4$,[8] $TiCl_4/Zn$,[9] $TiCl_4/Al/AlCl_3$,[8] and $TiCl_4/Mg(Hg)$.[8,10] Whether a Ti^0 or a Ti^I species is generated is dependent upon various factors, including the particular reagent combination used and the titanium chloride to reductant ratio employed.
Handling, Storage, and Precautions: most commonly, low-valent titanium species are prepared under an inert atmosphere (nitrogen or argon) immediately prior to use. Presumably, the reagent is pyrophoric although this property may be due to the residual reducing agent used in the preparation process and may not be an inherent property of Ti^0/Ti^I itself. The reagent is moisture sensitive.

Inter- and Intramolecular Coupling of Carbonyl and Dicarbonyl Compounds. The capacity of low-valent titanium species to effect reductive couplings was reported in 1973–1974 by several independent groups,[1,3,9] but McMurry and his co-workers have exploited such processes most extensively. A number of excellent reviews[1] are available on this and related topics. McMurry has now developed[11] an optimized procedure involving a combination of the $TiCl_3$/dimethoxyethane solvate $TiCl_3(DME)_{1.5}$ and ***Zinc/Copper Couple***. Thus coupling of diisopropyl ketone with the $TiCl_3(DME)_{1.5}/Zn(Cu)$ reagent gives an 87% yield of the expected alkene (eq 1). A 4:1 ratio of $TiCl_3(DME)_{1.5}$ to carbonyl compound appears to give the best results. When the same conversion is effected using $TiCl_3/LiAlH_4$ and $TiCl_3/Zn(Cu)$, yields of 12% and 37%, respectively, are obtained. Cross coupling of different carbonyl compounds to produce unsymmetrical alkenes can be effected in acceptable yield if one of the components, often acetone, is used in excess.[10] The coupling conditions have proven to be compatible with a variety of other potentially reducible functional groups including carboxylate and sulfonate esters,[12] as well as halides.[13]

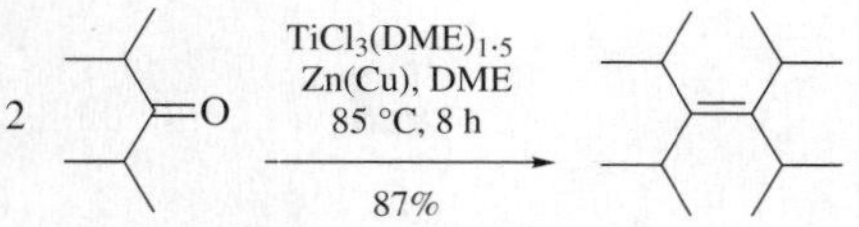

(1)

Intramolecular couplings of dicarbonyl compounds (including diketones, dialdehydes and keto esters) can be effected very efficiently with low-valent titanium (eq 2) and such processes have been exploited extensively in the synthesis of a number of natural products[1] and various novel hydrocarbons[1] (see ***Titanium(III) Chloride–Zinc/Copper Couple*** for examples).

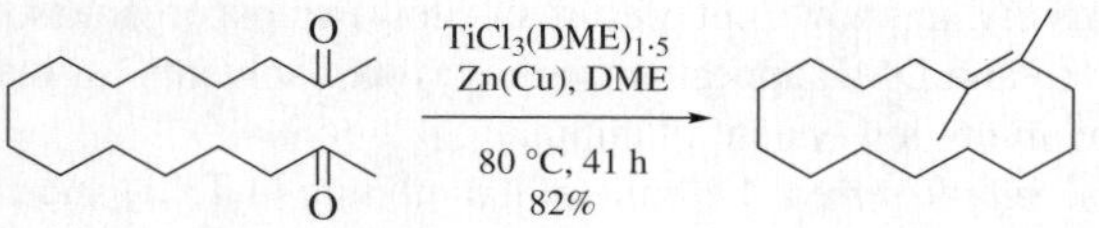

(2)

The geometry (*E* or *Z*) about the alkene double bond being formed in these so-called McMurry reactions is apparently controlled by thermodynamic factors. When the energy difference between the (*E*)- and (*Z*)-alkenes exceeds 4–5 kcal mol^{-1} the former isomer is formed preferentially.[14] However, coupling of alkyl aryl ketones affords the (*Z*)-alkene preferentially, perhaps because of π-complexation between the phenyl rings and Ti^0.[15] The mechanism proposed (eq 3)[1,16] to account for these conversions involves initial pinacolic-type coupling followed by successive elimination of 2 equiv of Ti=O to produce the observed products. When short reaction times and low temperatures are used, pinacols can be isolated. If such species are resubjected to the normal coupling conditions, deoxygenation occurs and the expected alkene results. Interestingly, when stereochemically pure pinacols are subjected to reaction with low-valent titanium reagents, then mixtures of (*E*)- and (*Z*)-alkenes are produced although some stereochemistry is preserved.[1a]

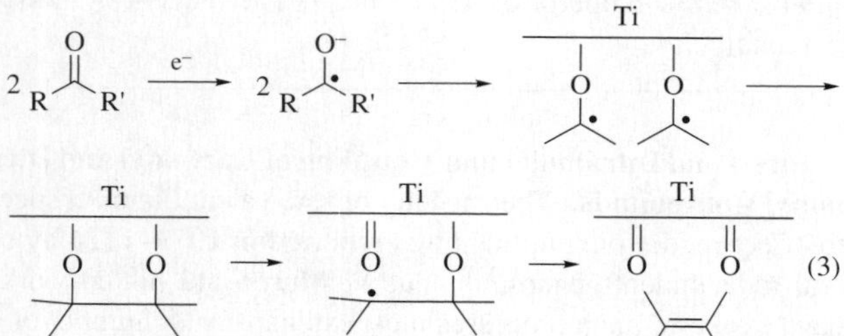

(3)

The reductive coupling of imines and related iminium species to give *vic*-diamines can be effected by low-valent titanium species.[17–19] In those cases where it is relevant, low to modest diastereoselectivities are observed (eq 4). Allylic alcohols, benzylic alcohols, and benzylic halides all undergo reductive coupling in the presence of certain low-valent titanium reagents (see ***Titanium(III) Chloride–Lithium Aluminum Hydride***).

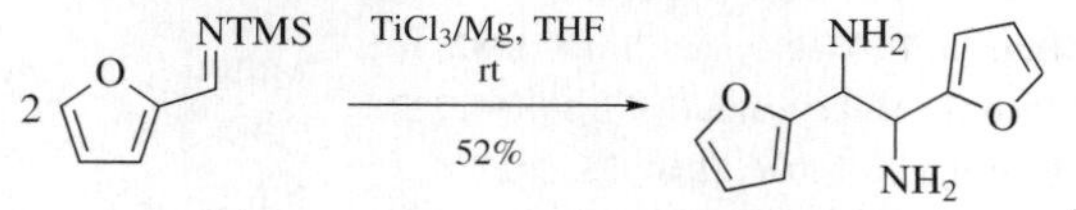

(4)

meso:*rac* = 1:3

Reductive Eliminations. *vic*-Dibromides,[20,21] bromohydrins,[22,23] epoxides,[22,24–27] hydroxy sulfides,[28–30] and allylic diols,[31–34] as well as 1,2-diols (pinacols), undergo reductive elimination on exposure to various low-valent titanium species, resulting in the formation of the corresponding alkene or 1,3-alkadiene. The dithioacetals of various acyloins also suffer reductive elimination upon treatment with Ti^0, thereby affording vinyl sulfides.[35]

Reductions. Various reductions can be effected with Ti^0/Ti^I species. Thus enol phosphates are converted into the corresponding alkenes[36] while analogous reduction of aryl phosphates[37] provides a useful protocol for the deoxygenation of phenols. Acyl, vinyl, and alkyl halides, as well as dithioacetals, can all be converted into the corresponding hydrocarbon on exposure to various sources of low-valent titanium,[38–41] while both nitriles[42] and isocyanides[43] suffer reductive removal of the –CN and –NC groups, respectively. Reduction (net addition of the elements of H_2) of both isolated[44–46] and conjugated[47,48] double bonds can also be accomplished with low-valent titanium.

Related Reagents. See Classes R-3, R-15, R-25, R-26, R-27, R-28, and R-31, pages 1–10.

---

1. (a) Robertson, G. M. *COS* **1991**, *3*, 583. (b) Betschart, C.; Seebach, D. *C* **1989**, *43*, 39. (c) McMurry, J. E. *CRV* **1989**, *89*, 1513. (d) Lenoir, D. *S* **1989**, 883. (e) Pons, J-M.; Santelli, M. *T* **1988**, *44*, 4295. (f) McMurry, J. E. *ACR* **1983**, *16*, 405.
2. Lenoir, D. *S* **1977**, 553.
3. Tyrlik, S.; Wolochowicz, I. *BSF(2)* **1973**, 2147.
4. McMurry, J. E.; Fleming, M. P. *JACS* **1974**, *96*, 4708.
5. McMurry, J. E.; Fleming, M. P.; Kees, K. L.; Krepski, L. R. *JOC* **1978**, *43*, 3255.
6. (a) McMurry, J. E.; Fleming, M. P. *JOC* **1976**, *41*, 896. (b) Nishida, S.; Kataoka, F. *JOC* **1978**, *43*, 1612.
7. (a) Fürstner, A.; Weidmann, H. *S* **1987**, 1071. (b) Fürstner, A.; Csuk, R.; Rohrer, C; Weidmann, H. *JCS(P1)* **1988**, 1729.
8. Corey, E. J.; Danheiser, R. L.; Chandrasekaran, S. *JOC* **1976**, *41*, 260.
9. Mukaiyama, T.; Sato, T.; Hanna, J. *CL* **1973**, 1041.
10. Carroll, A. R.; Taylor, W. C. *AJC* **1990**, *43*, 1439.
11. McMurry, J. E.; Lectka, T.; Rico, J. G. *JOC* **1989**, *54*, 3748.
12. Castedo, L.; Saá, J. M.; Suau, R.; Tojo, G. *JOC* **1981**, *46*, 4292.
13. Richardson, W. H. *SC* **1981**, *11*, 895. (b) Coe, P. L.; Scriven, C. E. *JCS(P1)* **1986**, 475.
14. Lenoir, D.; Burghard, H. *JCR(S)* **1980**, 396.
15. Leimner, J.; Weyerstahl, P. *CB* **1982**, *115*, 3697.
16. Dams, R.; Malinowski, M.; Westdorp, I.; Geise, H. Y. *JOC* **1982**, *47*, 248.
17. Betschart, C.; Schmidt, B.; Seebach, D. *HCA* **1988**, *71*, 1999.

18. Mangeney, P.; Tejero, T.; Alexakis, A.; Grosjean, F.; Normant, J. F. *S* **1988**, 255.
19. Mangeney, P.; Grosjean, F.; Alexakis, A.; Normant, J. F. *TL* **1988**, *29*, 2675.
20. Davies, S. G.; Thomas, S. E. *S* **1984**, 1027.
21. Olah, G. A.; Prakash, G. K. S. *S* **1976**, 607.
22. McMurry, J. E.; Silvestri, M. G.; Fleming, M. P.; Hoz, T.; Grayston, M. W. *JOC* **1978**, *43*, 3249.
23. McMurry, J. E.; Hoz, T. *JOC* **1975**, *40*, 3797.
24. Schobert, R. *AG(E)* **1988**, *27*, 855.
25. Ledon, H.; Tkatchenko, I.; Young, D. *TL* **1979**, 173.
26. McMurry, J. E.; Fleming, M. P. *JOC* **1975**, *40*, 2555.
27. Mukaiyama, T.; Saigo, K.; Takazawa, O. *CL* **1976**, 1033.
28. Song, S.; Shiono, M.; Mukaiyama, T. *CL* **1974**, 1161.
29. Mukaiyama, T.; Watanabe, Y.; Shiono, M. *CL* **1974**, 1523.
30. Watanabe, Y.; Shiono, M.; Mukaiyama, T. *CL* **1975**, 871.
31. Solladié, G.; Girardin, A. *TL* **1988**, *29*, 213.
32. Solladié, G.; Hamdouchi, C. *SL* **1989**, 66.
33. Walborsky, H. M.; Wüst, H. H. *JACS* **1982**, *104*, 5807.
34. (a) Solladié, G.; Girardin, A.; Métra, P. *TL* **1988**, *29*, 209. (b) Solladié, G.; Hutt, J. *JOC* **1987**, *52*, 3560.
35. Mukaiyama, T.; Shiono, M.; Sato, T. *CL* **1974**, 37.
36. Welch, S. E..; Walters, M. E. *JOC* **1978**, *43*, 2715.
37. Welch, S. E.; Walters, M. E. *JOC* **1978**, *43*, 4797.
38. Nelsen, T. R.; Tufariello, J. *JOC* **1975**, *40*, 3159.
39. Tyrlik, S.; Wolochowicz, I. *CC* **1975**, 781.
40. Ashby, E. C.; Lin, J. J. *JOC* **1978**, *43*, 1263.
41. Mukaiyama, T.; Hayashi, M.; Narasaka, K. *CL* **1973**, 291.
42. van Tamelen, E. E.; Rudler, H.; Bjorklund, C. *JACS* **1971**, *93*, 7113.
43. van Tamelen, E. E.; Rudler, H.; Bjorklund, C. *JACS* **1971**, *93*, 3526.
44. Ashby, E. C.; Lin, J. J. *TL* **1977**, 4481.
45. van Tamelen, E. E.; Cretney, W.; Klaentschi, N.; Miller, J. S. *CC* **1972**, 481.
46. Chum, P. W.; Wilson, S. E. *TL* **1976**, 15.
47. Blaszczak, L. C.; McMurry, J. E. *JOC* **1974**, *39*, 258.
48. Hung, C. W.; Wong, H. N. C. *TL* **1987**, *28*, 2393.

Martin G. Banwell
University of Melbourne, Parkville, Victoria, Australia

Titanium(III) Chloride

[7705-07-9] Cl_3Ti (MW 154.23)

(aqueous functional group reducing agent; after reducing to low-valent species, reductive coupling reagent of carbonyls to vicinal diols and alkenes)

Physical Data: mp 440 °C (dec); *d* 2.640 g cm^{-3}.

Solubility: sol water, alcohol; insol diethyl ether, $CHCl_3$, CCl_4, CS_2, benzene.

Form Supplied in: dark red–violet solid; dimethoxyethane complex, $TiCl_3 \cdot 1.5DME$ *[18557-31-8]*, solution in CH_2Cl_2/THF, solution in hydrochloric acid.

Purification: sublimation of solid at 1 mmHg.[1b]

Analysis of Reagent Purity: aqueous solutions of $TiCl_3$ can be titrated against 0.1 N cerium(IV) sulfate.[2]

Handling, Storage, and Precautions: dry powder is pyrophoric in air; moisture sensitive; reacts violently with water, and causes skin burns. Heating in vacuo over ca. 450 °C gives $TiCl_2$ and $TiCl_4$. Use in a fume hood and in a dry box under inert atmosphere. $TiCl_3 \cdot (DME)_{1.5}$ is air sensitive, but can be stored indefinitely under Ar at room temp. Reactions involving low-valent titanium should be run under argon.[1]

Introduction. The low-valent titanium species (Ti^0/Ti^I) generated from $TiCl_3$ are very useful for reductive coupling of aldehydes and ketones to give vicinal diols and further to alkenes. Aqueous $TiCl_3$ is used for the reduction of various functional groups (–SO, –NO_2, –NHOR, –X).

Reductive Coupling.[3–8] The reagent composition of the low-valent species (Ti^0/Ti^I) to be used in the coupling reaction depends on the reducing conditions for its generation. $TiCl_3$ and ***Lithium Aluminum Hydride*** in a 1:0.6 molar ratio produces a Ti^0 species, and the molar ratio 1:0.5 a Ti^I species.[4] Other reducing agents such as ***Lithium***,[9,10] ***Sodium***,[9] ***Potassium–Graphite*** (C_8K),[11] ***Magnesium***,[9,12,13] ***Zinc/Copper Couple***[12] and Rieke titanium have been applied to $TiCl_3$.[14] Low-valent titanium species can also be obtained from ***Titanium(IV) Chloride*** by similar reductive procedures.[4] THF is usually the preferred solvent in the reductive coupling reactions, although other solvents such as dioxane or dimethoxyethane have also been used. The reaction should be performed under argon. The reagents and solvents must be pure and absolutely dry, since traces of oxidation or hydrolysis products can interfere. Variable yields in closely related reactions, and reactions difficult to reproduce, have been ascribed to aged and ineffective batches of $TiCl_3$. The variable nature of $TiCl_3$ samples is overcome by prior conversion of $TiCl_3$ into a dimethoxyethane complex, $TiCl_3 \cdot (DME)_{1.5}$. The complex is readily purified by crystallization.[7] According to McMurry et al., the optimized titanium reagent for the reductive carbonyl coupling is prepared by the Zn/Cu couple reduction of $TiCl_3 \cdot (DME)_{1.5}$.[7]

Mechanism of Carbonyl Coupling. The reducing metal adds an electron to the oxo group and the anion radical dimerizes in a pinacol reaction (eq 1). The intermediate pinacols can be isolated at low temperature (0 °C), and are dehydroxylated at higher temperature (60 °C) to alkenes on the surface of zero-valent titanium particles.[7] For the reductive pinacol formation, ***Samarium(II) Iodide*** is an alternative.[15]

$$R^1C(=O)R^2 \xrightarrow{e^-} R^1\dot{C}(O^-)R^2 \longrightarrow R^1R^2C(O^-)\text{–}C(O^-)R^2R^1 \longrightarrow R^1R^2C{=}CR^1R^2 \quad (1)$$

Intermolecular Coupling. The reaction is most suitable for the preparation of symmetrical alkenes by joining two of the same carbonyl compounds, and is an excellent method for coupling of aliphatic carbonyl compounds (eq 2).[8] It is also an effective method for the synthesis of highly strained alkenes (eqs 3 and 4).[16]

$$\text{cyclohexanone} \xrightarrow[\text{DME, 80 °C; 97\%}]{TiCl_3(DME)_{1.5},\ Zn/Cu} \text{cyclohexylidenecyclohexane} \quad (2)$$

$TiCl_3(DME)_{1.5}$, Zn/Cu, DME, 80 °C, 87% (3)

$TiCl_3$, Li, THF, 22% (4)

(E):(Z) = 1:1.2

Mixed Coupling. A mixture of two different carbonyl compounds will generally react to afford a nearly statistical mixture of alkenes. However, by applying an excess of one carbonyl compound, mixed couplings can be synthetically useful (eq 5).[17]

excess Me_2CO, $TiCl_3$, LAH, 40% (5)

Intramolecular Coupling. Coupling of α,ω-dicarbonyl compounds gives cycloalkenes (eq 6).[7] Difficult intramolecular couplings leading to medium- and large-ring cycloalkenes require a lengthy addition time to achieve high dilution, and the use of 4 or more equiv of titanium per carbonyl group.

$TiCl_3(DME)_{1.5}$, Zn/Cu, DME, 35 h, 82% (6)

(E):(Z) = 92:8

Keto–ester couplings work well for five- to seven-membered ring formation. The product is the corresponding ketone.[18] Functional group compatibility includes acetal, alcohol, alkene, alkylsilane, amine, ether, halide, sulfide, and vinylsilane; incompatible are allylic alcohol, 1,2-diol, epoxide, halohydrin, α-halo ketone, nitro group, oxime, and sulfoxide.[3]

Reductions.[6] Aqueous $TiCl_3$ is a very useful reagent for the reduction of oximes to imines. The imines are rapidly hydrolyzed to carbonyl derivatives at low pH, and the overall reaction is a mild, rapid, and efficient deoximation procedure.[19] Cr^{II} [20] and V^{II} [21] reagents have also been used for this purpose. The usefulness is exemplified by a synthesis of a vicinal tricarbonyl system via an oxime which is readily available by nitrosation of the 1,3-dioxo derivative (eq 7).[22]

200 mol % $TiCl_3$, acetone, NaOAc, pH 5, 1 h (7)

The reaction with α-hydroxyimino-β-keto esters may also lead to pyrazine formation.[23] With added ***Sodium Borohydride***, amine formation has been used in the synthesis of α-amino acids.[24] By analogy to the oxime reduction, aliphatic nitro compounds are transformed into carbonyl derivatives. Reduction of the nitro group gives the corresponding nitroso derivative, with subsequent tautomerism to the oxime which reacts further as above (eq 8).[25]

85% (8)

Aromatic and heteroaromatic nitro compounds are also reduced to amines under mild conditions.[26] Buffered aqueous $TiCl_3$ can also be used to reduce hydroxamic acids, cyclic or acyclic (eq 9).[27] The ready cleavage of N–O bonds is attributed in part to titanium's high affinity for oxygen. Accordingly, sulfoxides are also deoxygenated cleanly and in high yields to the corresponding sulfides.[28]

$TiCl_3$, NaOAc, MeOH, H_2O, 68–88% (9)

α-Halo ketones are dehalogenated by $TiCl_3$.[29] A number of reductive transition metal complexes,[30] as well as SmI_2,[31] possess this ability. $TiCl_3$ is commercially available and therefore a convenient reagent. Halogen atoms in aromatics can also be removed by aqueous $TiCl_3$. Both halo- and cyanopyridines are reduced to pyridines with aqueous $TiCl_3$,[32] which should find wide application in heterocyclic chemistry (eq 10).

aq $TiCl_3$, H^+ (10)

X = CN, Cl, Br

Highly electrophilic and polarized carbon–carbon double bonds can also be reduced with aqueous $TiCl_3$ (eq 11).[33]

aq $TiCl_3$, 84% (11)

Related Reagents. See Classes R-15, R-21, R-23, R-24, R-25, R-27, and R-34, pages 1–10. Niobium(V) Chloride–Zinc; Samarium(II) Iodide; Titanium(III) Chloride–Lithium Aluminum Hydride; Titanium(III) Chloride–Potassium; Titanium(III) Chloride–Zinc/Copper Couple.

1. (a) Yamamoto, A.; Ookawa, M.; Ikeda, S. *CC* **1969**, 841. (b) Ruff, O.; Neumann, F. *Z. Anorg. Allg. Chem.* **1923**, *128*, 81.
2. Citterio, A.; Cominelli, A.; Bonavoglia, F. *S* **1986**, 308.
3. McMurry, J. E. *CRV* **1989**, *89*, 1513.
4. Lenoir, D. *S* **1989**, 883.
5. McMurry, J. E. *ACR* **1974**, *7*, 281.
6. Pons, J.-M.; Santelli, M. *T* **1988**, *44*, 4295.
7. McMurry, J. E.; Lectka, T.; Rico, J. G. *JOC* **1989**, *54*, 3748.
8. Clive, D. L. J.; Zhang, C.; Murthy, K. S.; Hayward, W. D.; Daigneault, S. *JOC* **1991**, *56*, 6447.

9. (a) Dams, R.; Malinowski, M.; Westdorp, I.; Geise, H. Y. *JOC* **1982**, *47*, 248. (b) Dams, R.; Malinowski, M.; Geise, H. J. *Transition Met. Chem.* **1982**, *7*, 37.
10. Hünig, S.; Ort, B. *LA* **1984**, 1905.
11. Fürstner, A.; Weidmann, H. *S* **1987**, 1071.
12. McMurry, J. E.; Fleming, M. P.; Kees, K. L.; Krepski, L. R. *JOC* **1978**, *43*, 3255.
13. Aleandri, L. E.; Bogdanovic, B.; Gaidies, A.; Jones, D. J.; Liao, S.; Michalowicz, A.; Roziere, J.; Schott, A. *JOM* **1993**, *459*, 87.
14. Kahn, B. E.; Rieke, R. D. *CRV* **1988**, *88*, 733.
15. Namy, J. L.; Souppe, J.; Kagan, H. B. *TL* **1983**, *24*, 765.
16. Bottino, F. A.; Finocchiaro, P.; Libertini, E.; Reale, A.; Recca, A. *JCS(P2)* **1982**, 77.
17. Reddy, S. M.; Duraisamy, M.; Walborsky, H. M. *JOC* **1986**, *51*, 2361.
18. McMurry, J. E.; Miller, D. D. *JACS* **1983**, *105*, 1660.
19. Timms, G. H.; Wildsmith, E. *TL* **1971**, *12*, 195.
20. Corey, E. J.; Richman, J. E. *JACS* **1970**, *92*, 5276.
21. Olah, G. A.; Arvanaghi, M.; Prakash, G. K. S. *S* **1980**, 220.
22. Gasparski, C. M.; Ghosh, A.; Miller, M. J. *JOC* **1992**, *57*, 3546.
23. Zercher, C. K.; Miller, M. J. *H* **1988**, *27*, 1123.
24. Hoffman, C.; Tanke, R. S.; Miller, M. J. *JOC* **1989**, *54*, 3750.
25. McMurry, J. E.; Melton, J. *JOC* **1973**, *38*, 4367.
26. (a) Rosini, G.; Ballini, R.; Petrini, M.; Marotta, E. *AG(E)* **1986**, *25*, 941. (b) Somei, M.; Kato, K.; Inoue, S. *CPB* **1980**, *28*, 2515.
27. Mattingly, P. G.; Miller, M. J. *JOC* **1980**, *45*, 410.
28. (a) Takahashi, T.; Iyobe, A.; Arai, Y.; Koizumi, T. *S* **1989**, 189. (b) Ho, T.-L.; Wong, C. M. *SC* **1973**, *3*, 37.
29. Ho, T.-L.; Wong, C. M. *SC* **1973**, *3*, 237.
30. Noyori, R.; Hayakawa, Y. *OR* **1983**, *29*, 163.
31. Molander, G. A.; Hahn, G. *JOC* **1986**, *51*, 1135.
32. Clerici, A.; Porta, O. *T* **1982**, *38*, 1293.
33. Blaszczak, L. C.; McMurry, J. E. *JOC* **1974**, *39*, 258.

Lise-Lotte Gundersen
Norwegian College of Pharmacy, Oslo, Norway

Frode Rise & Kjell Undheim
University of Oslo, Norway

Titanium(III) Chloride–Potassium[1]

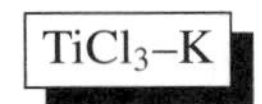

($TiCl_3$)		
[7705-07-9]	Cl_3Ti	(MW 154.23)
(K)		
[7440-09-7]	K	(MW 39.10)

(combination of reagents used for the in situ generation of highly oxophilic low-valent titanium species[2] which are used to effect the reductive coupling of carbonyl groups[2] (McMurry coupling) and reductive deoxygenations[3])

Physical Data: see ***Titanium(III) Chloride*** and ***Potassium***.

Solubility: $TiCl_3$: sol alcohol, hydrochloric acid; v. slightly sol DME, THF. K: insol DME, THF.

Form Supplied in: $TiCl_3$: violet, air-sensitive solid. K: metal under paraffin.

Preparative Methods: potassium metal (3.2 equiv) is washed with hexane and added to a slurry of titanium(III) chloride (1 equiv) in THF at rt under an inert atmosphere. The mixture is heated under reflux for 1 h, during which time the reaction mixture turns black. The mixture is cooled to rt and is then ready for the addition of the reaction substrates.

Handling, Storage, and Precautions: care is required when handling potassium because it reacts violently with water and protic solvents, liberating highly flammable gases which are readily ignited. Titanium(III) chloride should be stored and used under argon, preferably in a glove bag or box. The low-valent titanium reagent is sensitive to air and moisture; therefore to obtain good yields, it must be freshly prepared. All reactions should be performed under argon. If excess reagent is used, it should be quenched prior to workup by the cautious addition of methanol. Alternatively, if the use of methanol is incompatible with the products, the reaction mixture may be filtered carefully through a pad of Celite; care must be taken as the residues collected may be pyrophoric. All solvents must be dried and preferably deoxygenated or degassed before use.

Coupling of Aldehydes and Ketones. Reduction of $TiCl_3$ with 3 equiv of potassium gives a Ti^0 species which is used in the reductive coupling (McMurry coupling) of carbonyl compounds (eq 1). Other less hazardous methods exist for the generation of Ti^0, e.g. ***Titanium(III) Chloride–Zinc/Copper Couple*** or $TiCl_3$/Li, but the reagent generated from potassium gives the most consistent results and the highest yields when coupling aliphatic aldehydes and ketones.[4] When aldehydes or nonsymmetrical ketones are coupled, the (*E*)-alkene product predominates. Even with the highly hindered 1-methyl-2-adamantanone, coupling could be achieved in 52% yield (eq 2).[5] The recently reported generation of Ti^0 from the $TiCl_3(DME)_{1.5}$ solvate and ***Zinc/Copper Couple***[6] is likely to be the reagent of choice in future. It gives good to excellent yields, while avoiding the hazards associated with the use of potassium.

$TiCl_3$, K; 85% (1)

$TiCl_3$, K; 52% (2)

The mechanism of the McMurry coupling has been studied.[7] The reaction proceeds via a pinacol coupling and subsequent deoxygenation. Using the title reagent, it is only possible to isolate good yields of the pinacol product when the formation of the alkene is disfavored, e.g. in compound (**1**),[8] where the double bond would have to be formed at a strained bridgehead (eq 3). A wide range of functionality is compatible with carbonyl coupling,[1a] although allylic alcohols, epoxides, oximes, sulfoxides, and nitro groups are not.

The reagent has been modified by the use of ***Potassium–Graphite*** as the reducing partner,[9,10] with the Ti^0 being formed on the surface of the graphite. Again, 3 equiv of potassium are generally used, although the cyclization of (**2**) to (**3**) was optimized using 2 equiv of potassium (eq 4).[11] Formally, this requires a Ti^{I} species for the coupling. The reaction was highly specific: no epimerization was noted at C-8a, nor was there any loss of the acetal.

$TiCl_3$, K 55% (1) Gibberellic acid (3)

$TiCl_3$, C_8K 85% (2) (3) Compactin (4)

Coupling of Esters and Amides. The Ti^0 on graphite reagent facilitates the intramolecular coupling of ketones to esters,[10] yielding enol ethers which subsequently hydrolyze to ketones (eq 5). This type of reaction has also found use in the synthesis of furans and benzofurans by the coupling of *O*-benzoyl derivatives of 1,3-dicarbonyl compounds (eq 6);[12] yields are improved by further aromatic substituents. Indoles may be prepared by analogous methods (eq 7).[12]

$TiCl_3$, C_8K 91% (5)

$TiCl_3$ C_8K (6)

R = Ph, 92%; Me, 58%

$TiCl_3$ C_8K 89% (7)

Deoxygenations. As the McMurry coupling proceeds via the pinacol, it is not surprising that Ti^0 is capable of converting 1,2-diols to alkenes in good to excellent yields. In addition to diols, $TiCl_3$/K also converts 1-phenylthio-2-benzoyloxyethanes to alkenes (eq 8).[13] Deoxygenation of phenols and enolizable ketones is achieved via the *O*-phosphate derivative (eqs 9 and 10).[3]

$TiCl_3$, K 78% (8)

Ti^0 97% (9)

Ti^0 99% (10)

Miscellaneous Reductions. The Ti^0 produced by the title reagents can be expected to perform the transformations achieved by Ti^0 generated from other reagents, e.g. coupling of thiocarbonyls to alkenes or allylic alcohols to 1,5-dienes. However, in these cases the use of potassium is not justified.

Related Reagents. See Classes R-25, R-28, R-29, and R-31, pages 1–10.

1. (a) McMurry, J. E. *CRV* **1989**, *89*, 1513. (b) Lenoir, D. *S* **1989**, 883.
2. McMurry, J. E.; Fleming, M. P. *JOC* **1976**, *41*, 896.
3. (a) Welch, S. C.; Walters, M. E. *JOC* **1978**, *43*, 2715. (b) Welch, S. C.; Walters, M. E. *JOC* **1978**, *43*, 4797.
4. McMurry, J. E.; Fleming, M. P.; Kees, K. L.; Krepski, L. R. *JOC* **1978**, *43*, 3255.
5. Lenoir, D.; Frank, R. *TL* **1978**, 53.
6. McMurry, J. E.; Lectka, T.; Rico, J. G. *JOC* **1989**, *54*, 3748.
7. Davis, R.; Malinowski, M.; Westdorp, I.; Geise, H. T. *JOC* **1982**, *47*, 248.
8. Corey, E. J.; Danheiser, R. L.; Candrasekaran, S.; Sinet, P.; Keck, G. E.; Gras, J.-L. *JACS* **1978**, *100*, 8031.
9. Boldrini, G. P.; Savoia, D.; Tagliavini, E.; Trombini, C.; Umani-Ronchi, A. *JOM* **1985**, *280*, 307.
10. Füstner, A.; Weidmann, H. *S* **1987**, 1071.
11. (a) Clive, D. L. J.; Zhang, C.; Murthy, K. S. K.; Hayward, W. D.; Dangneault, S. *JOC* **1991**, *56*, 6447. (b) Clive, D. L. J.; Murthy, K. S. K.; Zhang, C.; Hayward, W. D.; Dangneault, S. *CC* **1990**, 509. (c) Clive, D. L. J.; Murthy, K. S. K.; Wee, A. G. H.; Presad, J. S.; de Silva, G. V. J.; Majewski, M.; Anderson, P. C.; Haugen, R. D.; Heerze, L. D. *JACS* **1988**, *110*, 6914.
12. Fürstner, A.; Jumbam, D. N.; Weidmann, H. *TL* **1991**, *32*, 6695.
13. Welch, S. C.; Loh, J.-P. *JOC* **1981**, *46*, 4073.

Ian C. Richards
AgrEvo, Saffron Walden, UK

Titanium(III) Chloride–Zinc/Copper Couple[1]

$TiCl_3$–Zn(Cu)

($TiCl_3$)		
[7705-07-9]	Cl_3Ti	(MW 154.23)
(Zn)		
[7440-66-6]	Zn	(MW 65.39)

(reagent combination for the production of low-valent titanium which is used for the inter- and intramolecular reductive coupling of carbonyl and dicarbonyl compounds)

Physical Data: see ***Titanium(III) Chloride*** and ***Zinc/Copper Couple***.

Solubility: unknown but presumably low in most organic solvents. THF and DME are the favored solvents for reactions involving this reagent combination.

Handling, Storage, and Precautions: the low-valent titanium species produced by reacting $TiCl_3$ with zinc/copper couple is generally made under an argon atmosphere immediately prior to use. The reagent may be pyrophoric and is certainly moisture sensitive.

Reductive Coupling Reactions.[1–12] Among the numerous sources of low-valent ***Titanium*** used for the reductive couplings of carbonyl compounds, the Ti^0 reagent produced by reacting $TiCl_3$ with freshly prepared zinc/copper couple[2] is considered the most effective for many intramolecular coupling reactions of dicarbonyls (see, however, ***Titanium(III) Chloride–Lithium Aluminum Hydride***). Up to 22-membered cycloalkenes have been formed by coupling the appropriate open-chain dicarbonyl precursor with this reagent. The reagent has been employed by its originator, McMurry,[2] in the key steps of a number of natural product syntheses. For example, such reductive cyclizations have been used in the preparation of humulene (eq 1),[3] as well as a 45:55 mixture of helminthogermacrene and β-elemene (eq 2).[4] The reagent has also been exploited in the preparation of compactin and mevinolin analogs,[5] flexibilene,[3a,c] some zizaene sesquiterpenes including (±)-isokhusimone,[6] as well as verticillene.[7]

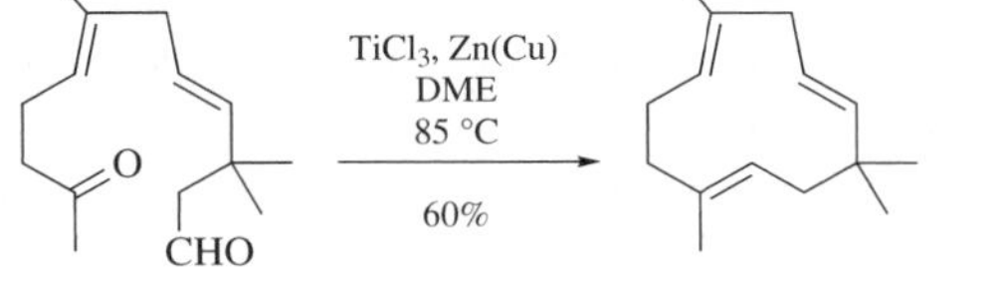

$TiCl_3$, Zn(Cu), DME, 85 °C, 60% (1)

Under thermally mild conditions (≤25 °C), α,ω-dials react with a 1:9 mixture of $TiCl_3$:Zn(Cu) in DME to give 6- to 14-membered cyclic pinacols. *cis*-Pinacols are the major products when 6- to 8-membered rings are produced while *trans*-pinacols predominate in rings of 10 or more members.[8] This type of cyclization has been employed in the synthesis of the 14-membered ring associated with sarcophytol B (eq 3).[8b]

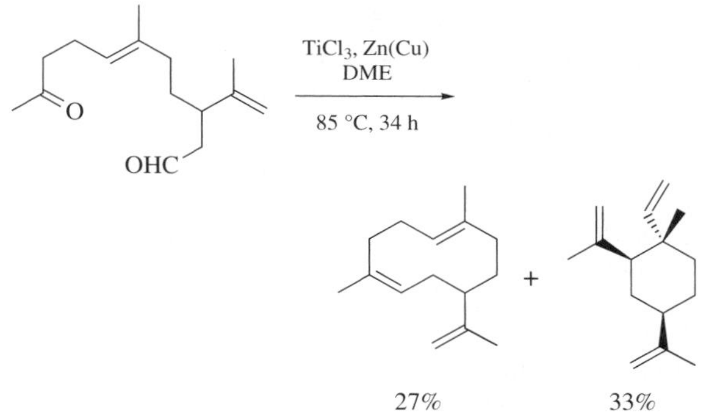

$TiCl_3$, Zn(Cu), DME, 85 °C, 34 h; 27% + 33% (2)

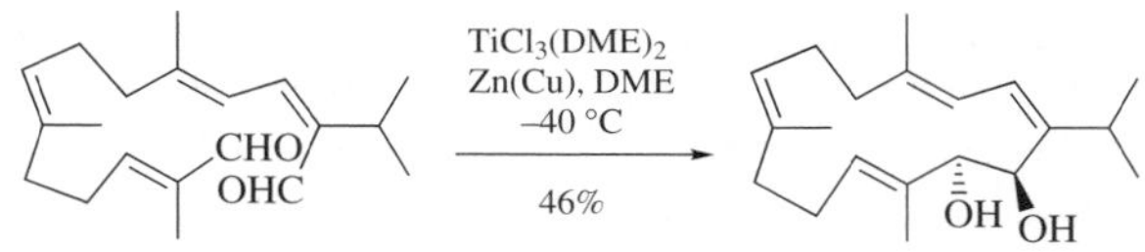

$TiCl_3(DME)_2$, Zn(Cu), DME, –40 °C, 46% (3)

A range of structurally novel hydrocarbons[1] have been prepared using the title reagent, including optically active doubly bridged allenes,[9] between anenes,[10] in–out bicycloalkanes,[11] and intracyclic π-systems (eq 4),[12] some of which are capable of forming stable metal complexes.

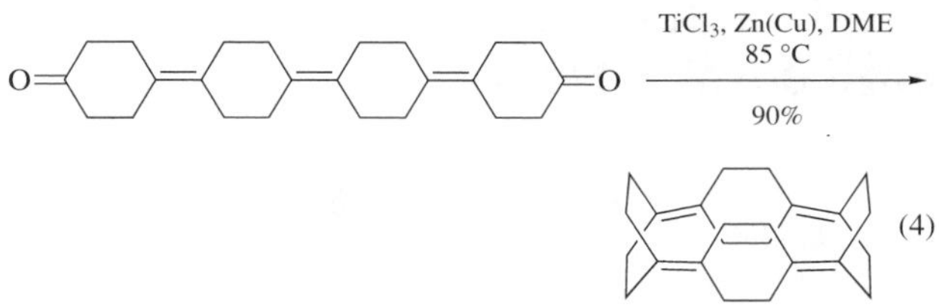

$TiCl_3$, Zn(Cu), DME, 85 °C, 90% (4)

The intermolecular coupling of two molecules of retinal to yield β-carotene (94%) has been effected with the title reagent and, by virtue of the product's use as a yellow food-coloring agent and a source of vitamin A, this efficient titanium-based synthesis is now licensed for use in commercial production.[13] Aromatic aldehydes and ketones possessing other potentially reducible groups (such as the acyloxy, methoxy carbonyl, and tosyloxy moieties) give Ti^0-induced carbonyl reductive coupling products in high yields when the $TiCl_3$/Zn(Cu) reagent is employed.[14]

McMurry has now developed[15] an optimized procedure for the reductive coupling of dicarbonyls which employs a combination of the solvate $TiCl_3(DME)_{1.5}$ and a Zn(Cu) couple to form the low-valent titanium reagent.

Related Reagents. See Class R-25, pages 1–10.

1. (a) Robertson, G. M. *COS* **1991**, *3*, 583. (b) Betschart, C.; Seebach, D. *C* **1989**, *43*, 39. (c) McMurry, J. E. *CRV* **1989**, *89*, 1513. (d) Lenoir, D. *S* **1989**, 883. (e) Pons, J.-M.; Santelli, M. *T* **1988**, *44*, 4295. (f) McMurry, J. E. *ACR* **1983**, *16*, 405.
2. (a) McMurry, J. E.; Kees, K. L. *JOC* **1977**, *42*, 2655. (b) McMurry, J. E.; Fleming, M. P.; Kees, K. L.; Krepski, L. R. *JOC* **1978**, *43*, 3255.
3. (a) McMurry, J. E.; Matz, J. R.; Kees, K. L. *T* **1987**, *43*, 5489. (b) McMurry, J. E.; Matz, J. R. *TL* **1982**, *23*, 2723. (c) McMurry, J. E.; Matz, J. R.; Kees, K. L.; Bock, P. A. *TL* **1982**, *23*, 1777.
4. McMurry, J. E.; Kocovsky, P. *TL* **1985**, *26*, 2171.
5. Anderson, P. C.; Clive, D. L. J.; Evans, C. F. *TL* **1983**, *24*, 1373.

6. Wu, Y.-J.; Burnell, D. J. *TL* **1988**, *29*, 4369.
7. Jackson, C. B.; Pattenden, G. *TL* **1985**, *26*, 3393.
8. (a) McMurry, J. E.; Rico, J. G. *TL* **1989**, *30*, 1169. (b) McMurry, J. E.; Rico, J. G.; Shih, Y. *TL* **1989**, *30*, 1173.
9. Nakazaki, M.; Yamamoto, K.; Maeda, M.; Sato, O.; Tsutsui, T. *JOC* **1982**, *47*, 1435.
10. (a) Marshall, J. A.; Bierenbaum, R. E.; Chung, K.-H. *TL* **1979**, 2081. (b) Marshall, J. A.; Black, T. H.; Shone, R. L. *TL* **1979**, 4737. (c) Marshall, J. A.; Constantino, M.; Black, T. H. *SC* **1980**, *10*, 689. (d) Marshall, J. A.; Black, T. H. *JACS* **1980**, *102*, 7581.
11. McMurry, J. E.; Hodge, C. N. *JACS* **1984**, *106*, 6450.
12. (a) McMurry, J. E.; Haley, G. J.; Matz, J. R.; Clardy, J. C.; Van Duyne, G.; Gleiter, R.; Schäfer, W.; White, D. H. *JACS* **1986**, *108*, 2932. (b) McMurry, J. E.; Haley, G. J.; Matz, J. R.; Clardy, J. C.; Mitchell, J. *JACS* **1986**, *108*, 515. (c) McMurry, J. E.; Swenson, R. *TL* **1987**, *28*, 3209.
13. McMurry, J. E. U.S. Patent 4 225 734 (*CA* **1976**, *85*, 63 205w).
14. Castedo, L.; Saá, J. M.; Suau, R.; Tojo, G. *JOC* **1981**, *46*, 4292.
15. McMurry, J. E.; Leckta, T.; Rico, J. G. *JOC* **1989**, *54*, 3748.

Martin G. Banwell
University of Melbourne, Parkville, Victoria, Australia

1,1,1-Triacetoxy-1,1-dihydro-1,2-benziodoxol-3(1*H*)-one[1]

[87413-09-0] $C_{13}H_{13}IO_8$ (MW 424.16)

(selective oxidation of primary,[1] secondary,[1] allylic,[2] allenic,[3] propargylic,[4] benzylic,[5] and α-fluorinated alcohols;[6] monoprotected 1,2- and 1,3-diols and amino alcohols;[7–10] α- and β-hydroxy ketones,[11,12] α- and β-hydroxy esters,[13,14] β-hydroxy amides,[15] β-hydroxy sulfones;[16] nucleosides;[17] carbohydrates;[18] compatible with a variety of protecting groups,[19,20] unsaturation,[21,22] basic nitrogen,[23] divalent sulfur[24] and selenium,[25] and halogen[26])

Alternate Names: Dess–Martin periodinane; TAPI.
Physical Data: mp 124–126 °C (dec).
Solubility: appreciably sol CH_2Cl_2, $CHCl_3$, MeCN, THF; effectively insol aromatic and aliphatic hydrocarbons, ether.
Form Supplied in: white free-flowing moisture- and light-sensitive solid; not currently commerically available due to the potential explosion hazard (see below).
Analysis of Reagent Purity: recommended methods include ^1H NMR,[1] and standardization by oxidation of benzyl alcohol;[27] however, a typical iodimetric titration used for the standardization of peracids and other oxidants can likely also be employed. NMR data for the reagent: ^1H NMR (300 MHz in $CDCl_3$): δ 8.29 (d, $J = 8.1$ Hz, 2H), 8.09 (t, $J = 8.1$ Hz, 1H), 7.91 (t, $J = 7.4$ Hz, 1H), 2.32 (s, 3H), 1.99 (s, 6H); minor impurity peaks are observed at δ 8.39 (d), 8.21 (d), 8.00 (d), 7.27 (s), and 2.08 (s); ^{13}C NMR (75 MHz in $CDCl_3$): δ 175.7, 174.0, 166.1, 142.2, 135.8, 133.8, 131.7, 126.5, 125.9, 20.4, 20.2.
Preparative Method: prepared from commercially available 2-iodobenzoic acid (recrystallized) by oxidation with ***Potassium Bromate*** in aqueous sulfuric acid at 65 °C for 3.6 h to afford the highly insoluble and sensitive (explosion hazard) hydroxyiodinane oxide (**2**) which is washed free of residual $KBrO_3$ by sequential washing with water, ethanol, and then water again; vacuum filtration then affords the oxide as a moist solid (thorough drying is not recommended due to explosion hazard). This moist oxide is directly transformed to the reagent by treatment with excess ***Acetic Anhydride*** and ***Acetic Acid*** at 62 → 96 °C for ~35 min (or until conversion is complete by analysis of an aliquot of the reaction mixture by NMR) followed by cooling for 12 h at room temperature. The solid reagent is collected by vacuum filtration under an inert atmosphere (glove bag/box or Schlenk techniques should be employed), and dried on the filter bed for 0.5 h; this affords the reagent (**1**) in 68–75% yield over the two steps (eqs 1 and 2).[1,27] A recent modification by Ireland recommends the use of ***p-Toluenesulfonic Acid*** as a catalyst in the conversion of the oxide (**2**) to (**1**).[28]

2-iodobenzoic acid + $KBrO_3$ → (0.73M H_2SO_4, 65 °C, 3.6 h) → (**2**)   (1)

(**2**) + Ac_2O → (HOAc, 62–96 °C, then rt, ~12 h) → (**1**)   (2)

Difficulties have sometimes attended the preparation of (**1**) for a variety of reasons, including but not limited to: (1) incomplete removal of $KBrO_3$ and possibly Br_2 during preparation of the hydroxyiodinane oxide (**2**); (2) incomplete or lack of removal of traces of ethanol used in washing the oxide intermediate (an obligatory process to avoid a potential explosion hazard); (3) incomplete acetylation of the oxide (**2**), as the presence of the monoacetoxy periodinane oxide has been implicated in the impact sensitivity shown by some but apparently not all samples of the reagent; and (4) partial or complete hydrolysis of the reagent during isolation by excessive exposure to atmospheric moisture (partial hydrolysis affords the aforementioned monoacetoxyiodinane oxide, and complete hydrolysis produces iodoxybenzoic acid, known to be explosive, which may contribute to the impact sensitivity observed).[1]
Purification: the reagent, when prepared properly, is sufficiently pure for use after standardization if desired. No purification method has been described.
Handling, Storage, and Precautions: the reagent is a moisture- and light-sensitive material.[1] While it can be handled briefly in the air, extensive exposure to moisture should be avoided. Transfers of solutions should be conducted using syringe techniques, and transfers of the solid are best done in a glove bag or box to avoid deterioration of the reagent. The pure solid reagent is stable indefinitely at room temperature when precautions are taken to avoid exposure to moisture and light.[1] Storage should be in an amber bottle under an inert atmosphere or other

means of protection from atmospheric moisture. Safety evaluations show that the precursor oxide is impact sensitive, and explosively decomposes when heated (>150 °C), with the onset varying with the sample. Handling the material as a moist solid appears to significantly reduce, but not completely remove, this explosion hazard. The pure periodinane has been stated to not exhibit impact sensitivity and to thermally decompose without explosion.[1] An independent evaluation of typical samples prepared as above shows that, although the reagent is less sensitive than the precursor oxide, it has the potential to decompose violently when heated (>200 °C) and shows impact sensitivity, although less than the oxide precursor. Thus, prudence requires that the preparation, handling, and use of this reagent and its precursor be conducted with due care and precaution against potential explosion (use of appropriate explosion shields, etc.), particularly on a substantial scale. Nevertheless, as the following discussion will attest, the reagent has been prepared and utilized widely with only one report of an explosion,[29] so that there is good cause to believe that, with appropriate precautions, the reagent can be prepared and handled safely on a normal laboratory scale.

Introduction. In the short time since its original description in the literature in 1983, the Dess–Martin periodinane (**1**) has enjoyed remarkably widespread use as a reagent for the oxidation of complex, sensitive, multifunctional alcohols of a startling array of structural types. The periodinane (**1**) is a member of the 12-I-5 group of hypervalent iodine species. A number of related structures, which also function as oxidizing agents for alcohols, have been described.[1] Among the advantages of the periodinane (**1**) over the myriad of other oxidizing agents which are available are: (1) functional groups lacking a hydroxylic function are exchanged into the ligand sphere of (**1**) slowly. Alcohols, particularly primary alcohols, undergo rapid and near-quantitative ligand exchange which is largely insensitive to the steric environment of the OH group. This ligand exchange is the obligatory first step in oxidation, which provides an inherent functional group selectivity in the oxidation reactions of (**1**). (2) The reagent is exceptionally versatile, having probably the broadest scope of any of the oxidizing agents commonly employed to convert alcohols to carbonyl derivatives, and generally provides exceptionally clean reactions and excellent yields. (3) When pyridine or di-*t*-butylpyridine is employed to buffer the reaction mixture and sodium thiosulfate/sodium bicarbonate is employed for workup, the entire oxidation/workup procedure can be conducted under near-neutral conditions compatible with even the most sensitive functionality. (4) The workup procedure is very simple, and the byproducts from the oxidation are readily removed by precipitation or chromatography, even without use of an aqueous workup. (5) The reagent is a shelf-stable solid, possessing adequate stability to moisture to permit ready metering and handling, even on very small scale, as is typically required in synthetic sequences leading to complex multifunctional molecules. The rates of oxidation of allylic and benzylic alcohols are highest, but oxidation of electron deficient carbinols such as α-fluorinated alcohols still occurs in good yields.

Typical Oxidation and Workup Procedures. The oxidation of alcohols is conducted at temperatures ranging from −20 to 35 °C (with ambient temperature being most common), typically in CH_2Cl_2 and occasionally in $CHCl_3$, acetonitrile, or THF, by admixing the reagent (typically 1–3 equiv per OH group), either as the solid or as a solution in the reaction solvent, with a solution of the alcohol. For very acid-sensitive substrates or products, the reaction mixture is buffered by the prior addition of 3–10 equiv of pyridine or 2,6-di-*t*-butylpyridine. The reaction mixture is then held at the appropriate temperature for as little as 5 min to as long as 120 h (most commonly 0.5–1.5 h) before dilution with ether (precipitation of periodinane byproducts occurs), addition of either aqueous NaOH solution or, for base-sensitive products, a mixture of aqueous $Na_2S_2O_3$ and $NaHCO_3$, stirring until the phases are homogeneous (~15 min), and isolation of the products by extraction. It is possible to avoid the use of aqueous workup conditions altogether by introduction of the reaction mixture, after dilution with ether, directly onto a chromatographic column for separation of the organic oxidation product(s).

Oxidation of Primary Alcohols to Aldehydes. Since many oxidants will satisfactorily accomplish this transformation, this discussion will focus on cases where (**1**) has been of particular value. The strength of the periodinane (**1**) is its ability to effect oxidation under neutral conditions, thus avoiding side reactions in sensitive substrates possessing multiple functional groups such as the enediyne (**3**) (eq 3),[30] or a variety of protecting groups, as in the azadirachtin intermediate (**4**) (eq 4).[31] Side reactions, such as β-elimination of oxygen and nitrogen functions, are seen with other oxidants for (**5**) (eq 5),[32,33] as are epimerization of α-stereogenic centers or exchange of label in labeled substrates,[34,35] and conjugation of β,γ-unsaturation.[36]

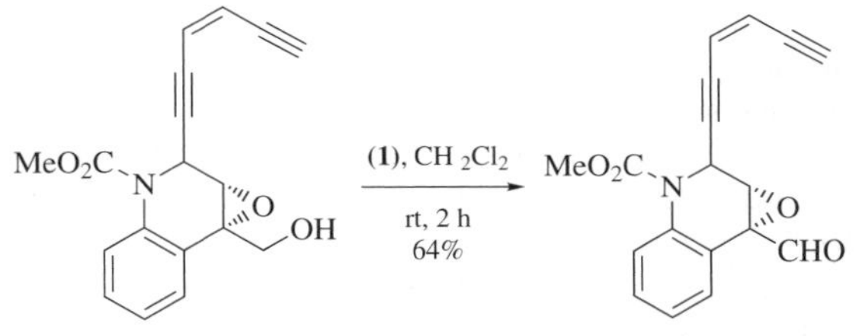

(1), CH_2Cl_2, rt, 2 h, 64% (3)

(1), py, CH_2Cl_2, rt, 0.67 h, 91% (4)

(1), CH_2Cl_2, rt, ~100% (5)

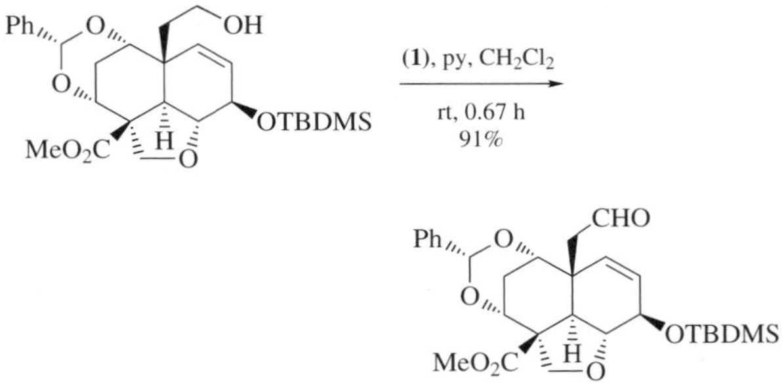

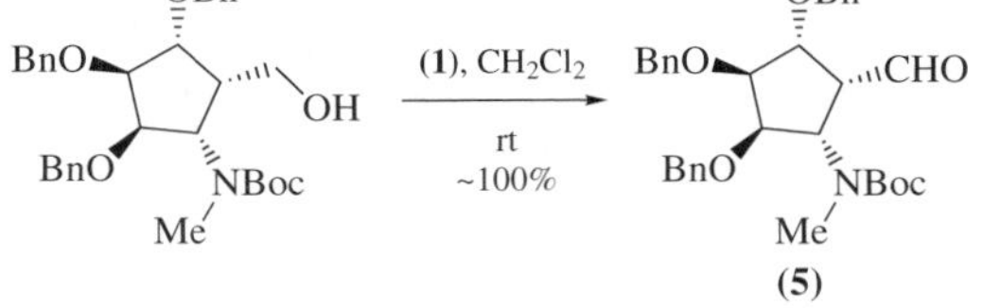

Disubstituted malondialdehydes (**6**) (eq 6) and malonaldehyde esters have been prepared without the intervention of oxidative cleavage.[37,38] Relative reactivity data based on steric environment is sparse; however, a notable case of selectivity for a primary over a hindered secondary alcohol in the preparation of the spiroacetal aldehyde (**7**) (eq 7) has been recorded.[39] Very few failures have been documented, and often no other oxidation reagent proved satisfactory in these cases.[40]

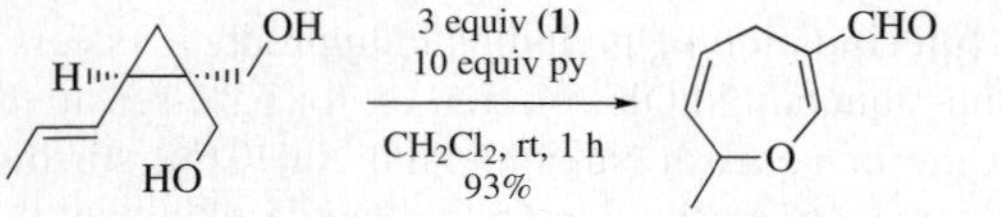

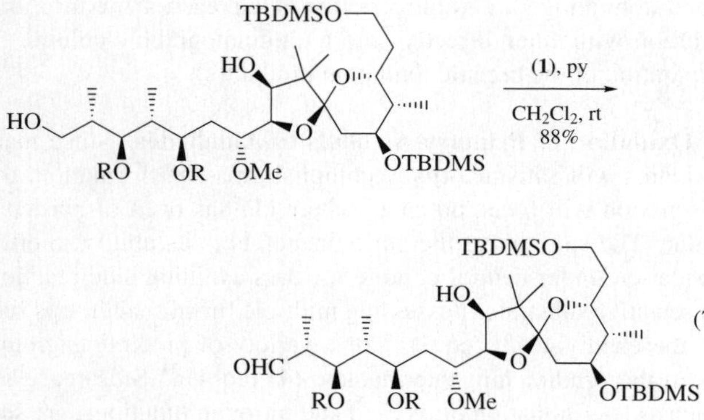

Oxidation of Secondary Alcohols to Ketones. In the case of secondary alcohols, (**1**) has also proven valuable in avoiding epimerization of α-stereogenic centers,[41,42] loss of β-acyloxy, β-alkoxy, and protected nitrogen functions during oxidations such as of β-acyloxy ketone (**8**) (eq 8), and undesired conjugation of β,γ-unsaturation as in the case of the alkaloid intermediate (**9**) (eq 9).[43–45] Overoxidation at activated methylene groups, such as benzylic centers or susceptible aromatic systems, is not normally a problem.[46,47]

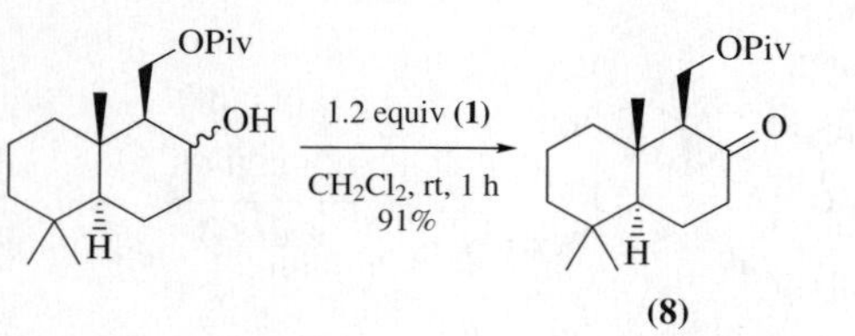

Appropriately protected α-heteroatom-substituted systems are smoothly oxidized without epimerization or isomerization of the β,γ-double bond, as in the ketone (**10**) (eq 10).[48] Oxidation of secondary alcohols in suitably protected nucleosides and carbohydrates has also been documented, even in cases where the substrates are exceedingly acid-sensitive such as the trisaccharide (**11**) (eq 11).[49,50] Even sterically-congested secondary alcohols undergo smooth oxidation, most often in excellent yield as in the case of ketone (**12**) (eq 12).[51–53]

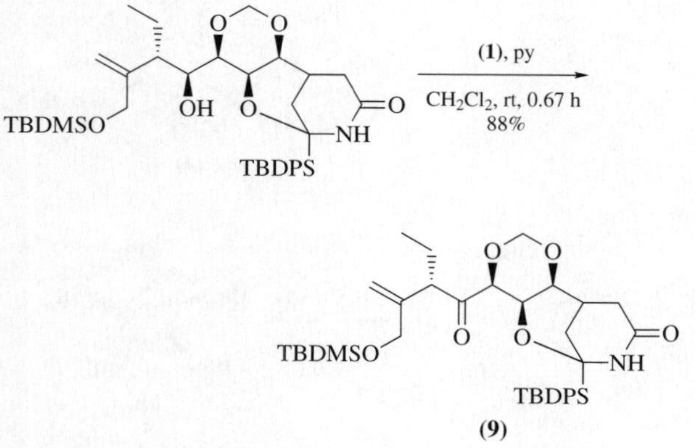

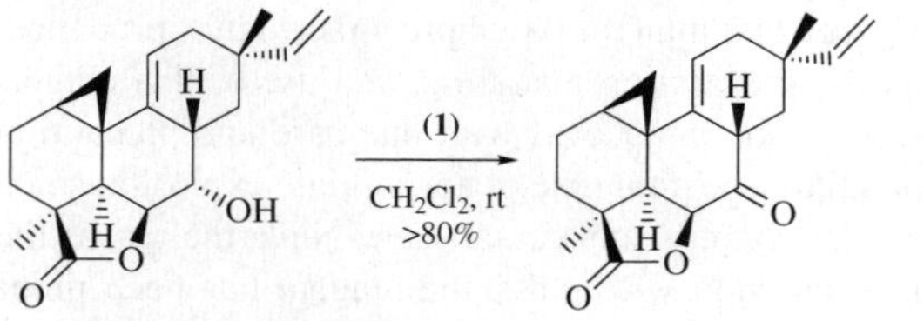

Secondary α- and β-hydroxy ketones and esters, and β-hydroxy amides and sulfones, are also smoothly oxidized to the corresponding α- and β-diketones, α- and β-keto esters, β-keto amides, and β-keto sulfones, such as the β-keto ester (**13**) (eq 13).[54] Remarkably, even substrates containing divalent sulfur and selenium, as well as bromine, have been smoothly oxidized to the corresponding ketones, such as the β-lactam (**14**) (eq 14).[55,56] Furthermore, substrates containing secondary basic nitrogen centers have been smoothly oxidized without concomitant oxidation at nitrogen, for example ketone (**15**) (eq 15).[57] Even electron-deficient, weakly nucleophilic substrates such as α-trifluoromethyl and α,α-difluoroalkyl alcohols also undergo smooth transformation to the corresponding ketones.[58,59]

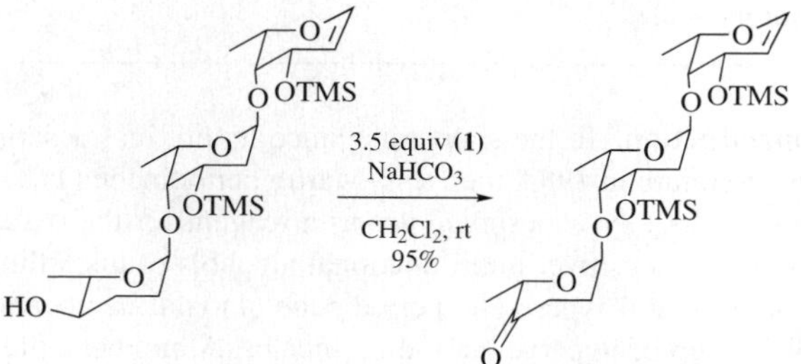

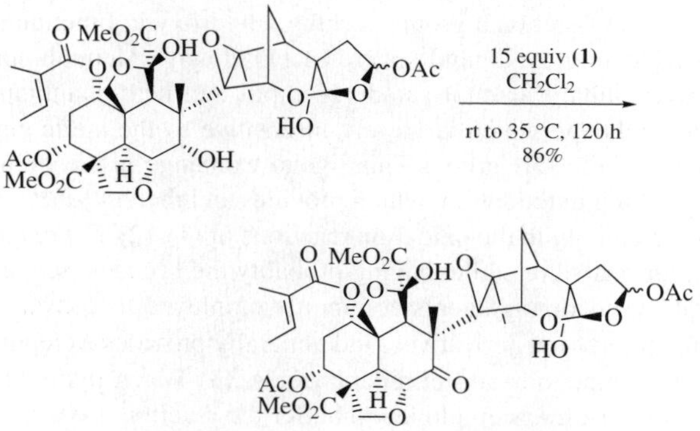

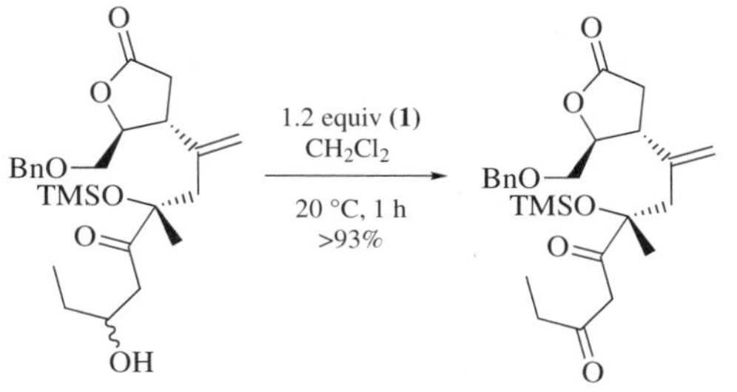

In some cases, unprotected 1,2-diols undergo fragmentation to carbonyl compounds in a manner similar to their reactions with related periodates and periodic acid.[1,60] However, an example of the selective oxidation of a complex 2,3-dihydroxypyran to the related α-hydroxy δ-lactone (**16**) in excellent yield has been described (eq 16).[61] Of course, oxidation of remote primary–primary and secondary–secondary diols to the dicarbonyl compounds proceeds uneventually when sufficient periodinane (**1**) is employed, but statistical mixtures usually result from attempts at selective oxidation, except in cases where the hydroxyl groups differ markedly in steric environment.[39,69]

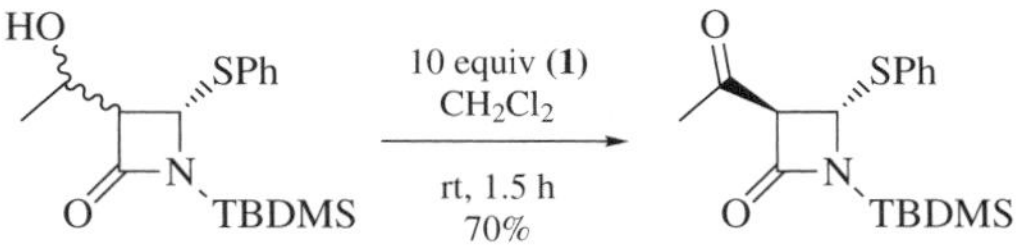

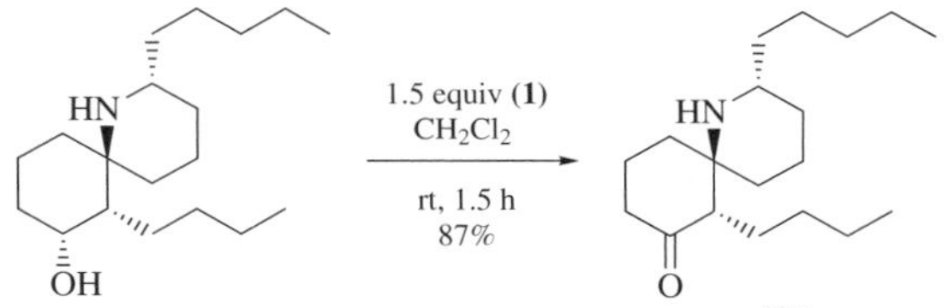

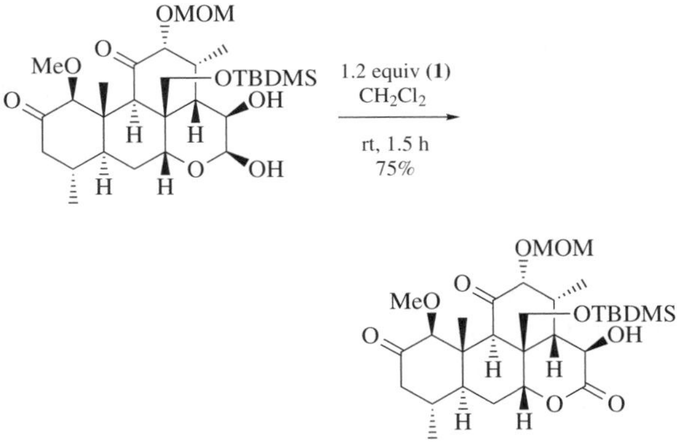

Oxidation of Primary and Secondary Allylic, Benzylic, Propargylic, and Allenyl Alcohols. The oxidation of allylic and benzylic alcohols by the periodinane (**1**) has been demonstrated to be significantly faster ($\geq\sim5\times$) than that of the corresponding saturated alcohols in competition experiments.[1] This selectivity is exemplified by the oxidation of the precursor secondary allylic–secondary diol to the ketol (**17**) (eq 17).[62]

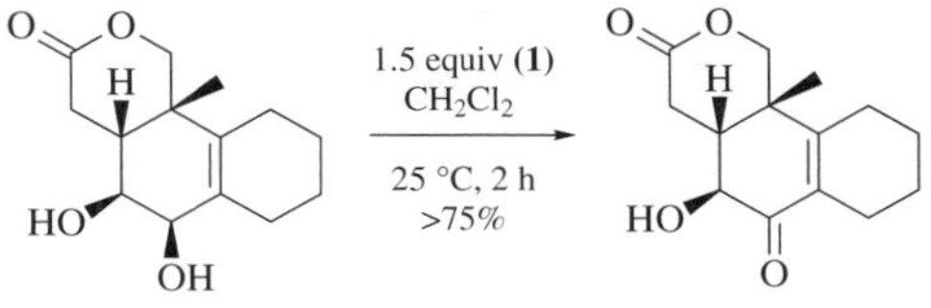

However, in some instances, oxidative cleavage occurs in preference to simple oxidation (see below). It is in the oxidation of this general class of allylic, benzylic, and related unsaturated and acetylenic alcohols that the periodinane (**1**) has proven consistently superior to other oxidants for complex acid and base-sensitive substrates. As shown below, the range of complex substrates is truly impressive. For example, acid- and oxidation-sensitive benzyl and pyridyl alcohols are smoothly oxidized to the corresponding carbonyl compounds (**18**) and (**19**) (eqs 18 and 19).[63,64] In the latter case, the aldehyde (**19**) is obtainable in acceptable yields only by use of the periodinane (**1**). In the case of 6α-epipretazzetine (**20**) (eq 20), no overoxidation and isomerization to tazzetine is observed.[65] Furthermore, no concomitant oxidation of susceptible aromatic systems or nitrogen is detected.

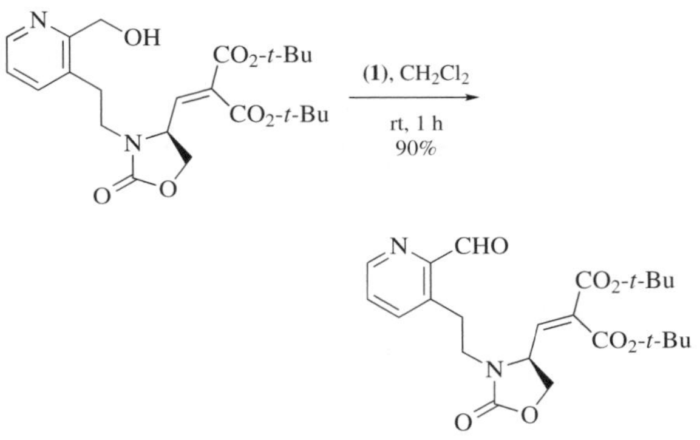

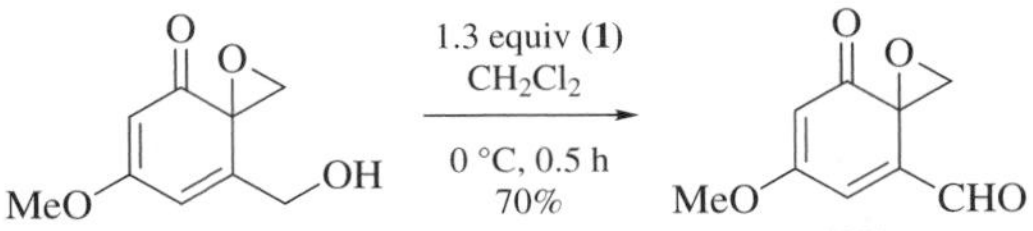

In the case of allylic alcohols, including α-methylenecycloalkanols and cyclopentenols, the periodinane (**1**) has afforded exceedingly acid-, base-, thermal-, and/or oxidation-sensitive enones, e.g. (**21**) and (**22**), in excellent yields (eqs 21 and 22).[66,67] Most impressively, (*Z*)-allylic alcohols afford the corresponding aldehydes and ketones with minimal or no loss of the geometric integrity of the alkene, as seen in the oxidations affording (**23**) and (**24**) (eqs 23 and 24).[68,69]

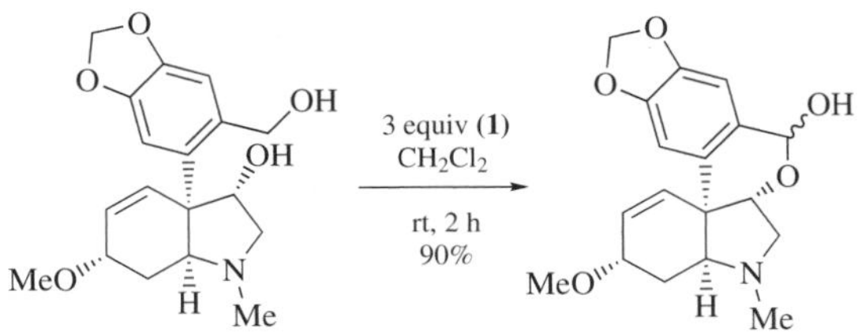

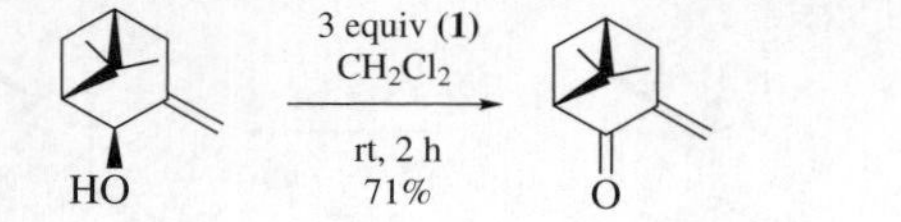

(21)

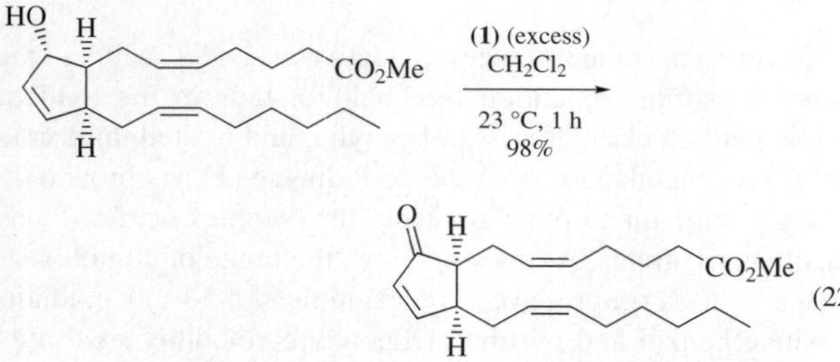

(22)

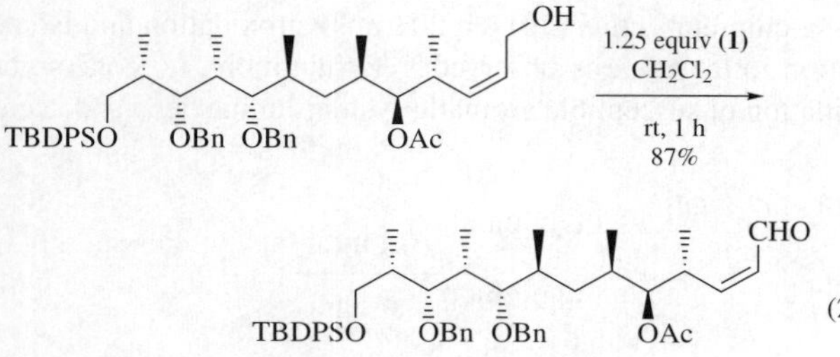

(23)

(24)

Allylic and propargylic alcohols possessing α′- or δ-stereogenic centers can be oxidized without compromising the integrity of the stereogenic centers, even in demanding situations such as the cases of the ketone (**25**) (eq 25) and the diynone (**26**) (eq 26).[70,71] Primary and secondary allenic alcohols also afford the corresponding carbonyl derivatives smoothly upon oxidation with (**1**).[72,73] Even allylic benzylic systems afford the corresponding aryl vinyl ketones largely or completely without oxygen transposition, even in the remarkably hindered ketone (**27**) (eq 27).[74,75] Note that the ketone (**27**) has an internal barrier to rotation about the aryl–carbonyl single bond of in excess of 20 kcal mol^{-1}.

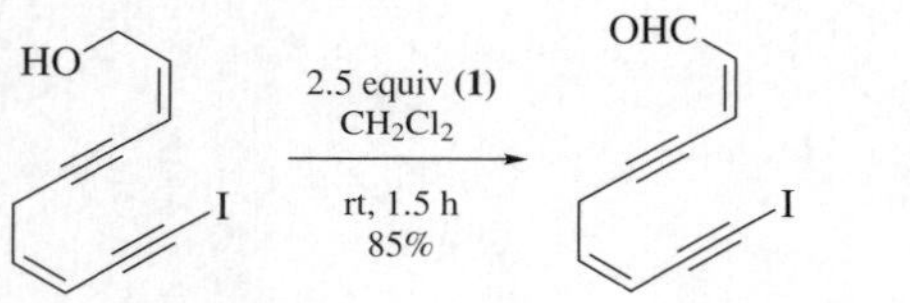

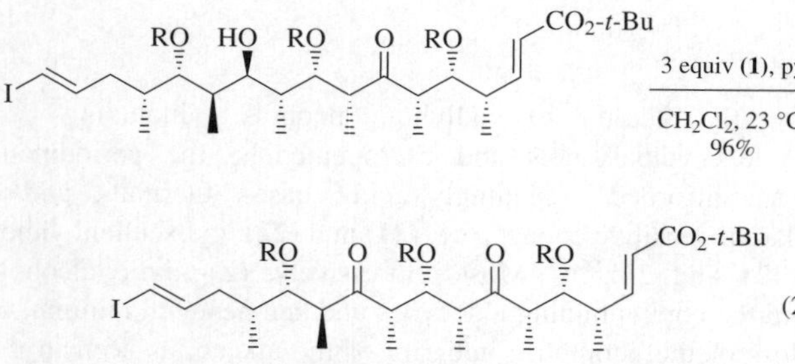

(25)

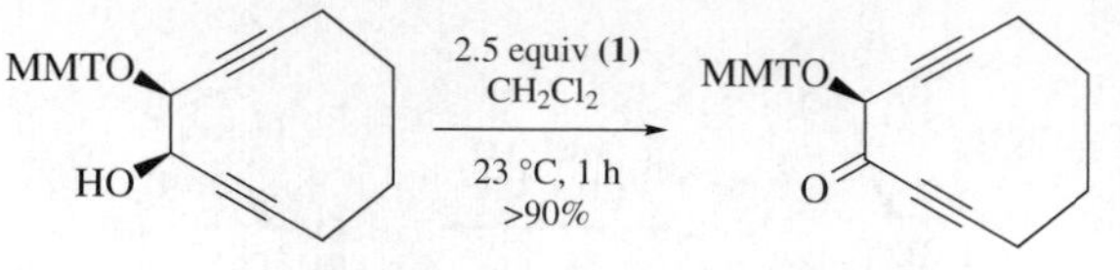

(26)

(27)

Examples have also been reported of oxidation of allylic or benzylic alcohols by (**1**) in the presence of divalent sulfur and basic nitrogen without concomitant oxidation at these heteroatoms.[76,77]

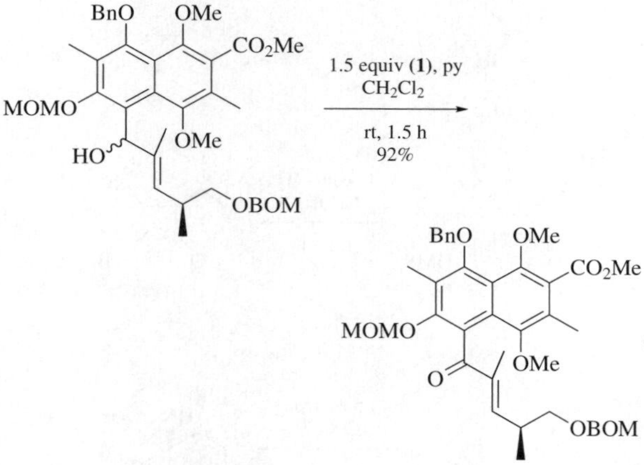

Oxidation at Methine and Methylene Groups. Relatively recently, as a result of the interest in the synthesis of derivatives of the immunosuppressive agent FK-506, which possesses the relatively rare 2,3-dioxo amide unit, it has been recognized that the periodinane (**1**), when employed in excess, is capable of oxidizing α-thioaryl β-keto amides to the related 2,3-dioxo amides, accompanied by the derived enol(s) and the hydrate of the 2-oxo function. This oxidation, whose mechanism has not been established as yet, may proceed via the enol or via the corresponding sulfoxide by way of a Pummerer-type rearrangement.[78] Subsequently, the direct oxidation of β-keto esters and amides by the periodinane (**1**) to 2,3-dioxo esters and amides has been reported.[79]

Related Reagents. See Class O-1, pages 1–10.

1. (a) Dess, D. B.; Martin, J. C. *JACS* **1991**, *113*, 7277. (b) Dess, D. B.; Martin, J. C. *JOC* **1983**, *48*, 4155.
2. Marshall, J. A.; Wang, X. *JOC* **1992**, *57*, 3387.
3. Marshall, J. A.; Wang, X. *JOC* **1991**, *56*, 960.
4. Marshall, J. A.; Blough, B. E. *JOC* **1991**, *56*, 2225.
5. Becker, K.-D.; Amin, K. A. *JOC* **1989**, *54*, 3182.
6. Linderman, R. J.; Graves, D. M. *JOC* **1989**, *54*, 661.
7. Wang, Z. *TL* **1989**, *30*, 6611.
8. Evans, D. A.; Kaldor, S. W.; Jones, T. K.; Clardy, J.; Stout, T. J. *JACS* **1990**, *112*, 7001.
9. Csuk, R.; Hugener, M.; Vasella, A. *HCA* **1988**, *71*, 609.
10. Bailey, S. W.; Chandrasekaran, R. Y.; Ayling, J. E. *JOC* **1992**, *57*, 4470.
11. Andersen, M. W.; Hildebrandt, B.; Dahmann, G.; Hoffmann, R. W. *CB* **1991**, *124*, 2127.
12. Becker, H.-D.; Sörensen, H.; Hammarberg, E. *TL* **1989**, *30*, 989.
13. Burkhardt, J. P.; Bey, P.; Peet, N. P. *TL* **1988**, *29*, 3433.
14. Itokawa, H.; Hitotsuyanagi, Y.; Takeya, K. *H* **1992**, *33*, 537.

15. Evans, D. A.; Miller, S. J.; Ennis, M. D. *JOC* **1993**, *58*, 471.
16. Jones, A. B.; Villalobos, A.; Linde, R. G.; Danishefsky, S. J. *JOC* **1990**, *55*, 2786.
17. Samano, V.; Robins, M. J. *JOC* **1991**, *56*, 7108.
18. Maier, M. E.; Brandstetter, T. *TL* **1992**, *33*, 7511.
19. Glänzer, B. I.; Györgydeák, Z.; Bernet, B.; Vasella, A. *HCA* **1991**, *74*, 343.
20. Karanewsky, D. S. *TL* **1991**, *32*, 3911.
21. Marshall, J. A.; Wang, X. *JOC* **1992**, *57*, 1242.
22. Chen, L.; Ghosez, L. *TA* **1991**, *2*, 1181.
23. Bell, T. W.; Firestone, A. *JACS* **1986**, *108*, 8109.
24. Toshima, K.; Yoshida, T.; Mukaiyama, S.; Tatsuta, K. *TL* **1991**, *32*, 4139.
25. Jarvis, B. B.; Cömezöglu, S. N.; Alvarez, M. E. *JOC* **1988**, *53*, 1918.
26. Halcomb, R. L.; Boyer, S. H.; Danishefsky, S. J. *AG(E)* **1992**, *31*, 338.
27. Boeckman, R. K., Jr.; Mullins, J. J. *OS* **1993**, submitted for publication.
28. Ireland, R. E.; Liu, L. *JOC* **1993**, *58*, 2899.
29. Plumb, J. B.; Harper, D. J. *Chem. Eng. News* **1990**, *68 (29)*, 3.
30. Wender, P. A.; Zercher, C. K. *JACS* **1991**, *113*, 2311.
31. Kolb, H. C.; Ley, S. V.; Slawin, A. M. Z.; Williams, D. J. *JCS(P1)* **1992**, 2735.
32. Tarköy, M.; Bolli, M.; Schweizer, B.; Leumann, C. *HCA* **1993**, *76*, 481.
33. Farr, R. A.; Peet, N. P.; Kang, M. S. *TL* **1990**, *31*, 7109.
34. Marshall, J. A.; Luke, G. P. *SL* **1992**, 1007.
35. Crombie, L.; Heavers, A. D. *JCS(P1)* **1992**, 1929.
36. Norman, T. C.; de Lera, A. R.; Okamura, W. H. *TL* **1988**, *29*, 1251.
37. Boeckman, R. K., Jr.; Shair, M. D.; Vargas, J. R.; Stolz, L. A. *JOC* **1993**, *58*, 1295.
38. Ihara, M.; Suzuki, T.; Katogi, M.; Taniguchi, N.; Fukumoto, K. *JCS(P1)* **1992**, 865.
39. Evans, D. A.; Gage, J. R.; Leighton, J. L. *JACS* **1992**, *114*, 9434.
40. Brown, D. S.; Paquette, L. A. *JOC* **1992**, *57*, 4512.
41. Bell, T. W.; Vargas, J. R.; Crispino, G. A. *JOC* **1989**, *54*, 1978.
42. Stocks, M.; Kocienski, P.; Donald, D. K. *TL* **1990**, *31*, 1637.
43. Herlem, D.; Kervagoret, J.; Yu, D.; Khuong Huu, F.; Kende, A. S. *T* **1993**, *49*, 607.
44. Chen, S.-H.; Horvath, R. F.; Joglar, J.; Fisher, M. J.; Danishefsky, S. J. *JOC* **1991**, *56*, 5834.
45. Vloon, W. J.; van den Bos, J. C.; Koomen, G. J.; Pandit, U. K. *T* **1992**, *48*, 8317.
46. Ihara, M.; Hirabayashi, A.; Taniguchi, N.; Fukumoto, K. *T* **1992**, *48*, 5089.
47. Cho, B. P.; Harvey, R. G. *JOC* **1987**, *52*, 5668.
48. Chu-Moyer, M. Y.; Danishefsky, S. J. *JACS* **1992**, *114*, 8333.
49. Suzuki, K.; Sulikowski, G. A.; Friesen, R. W.; Danishefsky, S. J. *JACS* **1990**, *112*, 8895.
50. Samano, V.; Robins, M. J. *JOC* **1990**, *55*, 5186.
51. Ley, S. V.; Lovell, P. J.; Slawin, A. M. Z.; Smith, S. C.; Williams, D. J.; Wood, A. *T* **1993**, *49*, 1675.
52. Ihara, M.; Ohnishi, M.; Takano, M.; Makita, K.; Taniguchi, N. Fukumoto, K. *JACS* **1992**, *114*, 4408.
53. Muratake, H.; Kumagami, H.; Natsume, M. *T* **1990**, *46*, 6351.
54. Boeckman, R. K., Jr.; Yoon, S. K.; Heckendorn, D. K. *JACS* **1991**, *113*, 9682.
55. Coggins, P.; Simpkins, N. S. *SL* **1992**, 313.
56. Ziegler, F. E.; Sobolov, S. B. *JACS* **1990**, *112*, 2749.
57. Winkler, J. D.; Hershberger, P. M. *JACS* **1989**, *111*, 4852.
58. Patel, D. V.; Reilly-Gauvin, K.; Ryono, D. E. *TL* **1988**, *29*, 4665.
59. Takahashi, L. H.; Radhakrishnan, R.; Rosenfield, R. E., Jr.; Meyer, E. F., Jr.; Trainor, D. A. *JACS* **1989**, *111*, 3368.
60. Jones, T. K.; Reamer, R. A.; Desmond, R.; Mills, S. G. *JACS* **1990**, *112*, 2998.
61. Fleck, T. J.; Grieco, P. A. *TL* **1992**, *33*, 1813.
62. Broka, C. A.; Ruhland, B. *JOC* **1992**, *57*, 4888.
63. Horikawa, M.; Hashimoto, K.; Shirahama, H. *TL* **1993**, *34*, 331.
64. Haseltine, J. N.; Cabal, M. P.; Mantlo, N. B.; Iwasawa, N.; Yamashita, D. S.; Coleman, R. S.; Danishefsky, S. J.; Schulte, G. K. *JACS* **1991**, *113*, 3850.
65. Abelman, M. M.; Overman, L. E.; Tran, V. D. *JACS* **1990**, *112*, 6959.
66. Dupuy, C.; Luche, J. L. *T* **1989**, *45*, 3437.
67. Corey, E. J.; Xiang, Y. B. *TL* **1988**, *29*, 995.
68. Myles, D. C.; Danishefsky, S. J.; Schulte, G. K. *JOC* **1990**, *55*, 1636.
69. Myers, A. G.; Finney, N. S. *JACS* **1992**, *114*, 10 986.
70. Evans, D. A.; Ng, H. P. *TL* **1993**, *34*, 2229.
71. Myers, A. G.; Dragovich, P. S. *JACS* **1992**, *114*, 5859.
72. Marshall. J. A.; Tang, Y. *JOC* **1993**, *58*, 3233.
73. Wu, K.-M.; Midland, M. M.; Okamura, W. H. *JOC* **1990**, *55*, 4381.
74. Roush, W. R.; Madar, D. J. *TL* **1993**, *34*, 1553.
75. Harvey, R. G.; Hahn, J.-T.; Bukowska, M.; Jackson, H. *JOC* **1990**, *55*, 6161.
76. Trost, B. M.; Grese, T. A. *JOC* **1992**, *57*, 686.
77. Tius, M. A.; Kerr, M. A. *JACS* **1992**, *114*, 5959.
78. Linde, R. G.; Jeroncic, L. O.; Danishefsky, S. J. *JOC* **1991**, *56*, 2534.
79. Batchelor, M. J.; Gillespie, R. J.; Golec, J. M. C.; Hedgecock, C. J. R. *TL* **1993**, *34*, 167.

Robert J. Boeckman, Jr.
University of Rochester, NY, USA

Tri-*n*-butylstannane[1]

n-Bu_3SnH

[688-73-3] $C_{12}H_{28}Sn$ (MW 291.11)

(source of $Bu_3Sn\cdot$ radical which produces carbon radicals by (a) abstraction of X from C–X derivatives[1,2] and (b) addition to alkenes, alkynes,[3,4] and carbonyl compounds; hydrogen donor (with concomitant generation of the chain carrier, $Bu_3Sn\cdot$) for radicals;[2] hydrostannylation of alkenes, alkynes,[5,6] and carbonyl compounds;[7] desulfurative stannylation of propargylic or allylic sulfides;[8] catalyst for S_H2' reactions of allylic stannanes;[9] radical ring expansion;[10,11] selective reduction of acid chlorides to aldehydes;[12] oxygenation[13] and carbonylation[14] of radicals; source of tributyltin anion[15])

Alternate Name: tributyltin hydride; TBTH.
Physical Data: bp 80 °C/0.4 mmHg; *d* 1.082 g cm^{-3}.
Solubility: freely sol organic solvents.
Form Supplied in: clear colorless liquid; 97% pure.
Analysis of Reagent Purity: major impurity is oxidation product $(Bu_3Sn)_2O$; purity ascertained by gas volumetric methods using dichloroacetic acid or by IR spectroscopy.[1e]
Handling, Storage, and Precautions: irritant. Even though this reagent appears to be relatively innocuous,[1a] tin compounds in general are toxic[16] and should be handled with care in a fume hood. It should be kept in brown bottles away from light and air. Some pressure may develop upon long term storage. Workup

procedures for the isolation of tin-free organic products from reactions involving TBTH have been published.[1e,77b]

Introduction. Tributyltin hydride is the most commonly used source of $Bu_3Sn\cdot$ which initiates a variety of radical chain reactions.[1,2] $Bu_3Sn\cdot$ may be generated either thermally or photochemically. ***Azobisisobutyronitrile*** (AIBN) is the most commonly used radical initiator. Three important classes of reactions have been recognized for $R_3Sn\cdot$: (a) atom or group abstraction, (b) addition to multiple bonds, and (c) homolytic substitution reactions. The primary products of (a) and (b) are themselves radicals and they undergo a variety of useful transformations such as atom abstraction reactions, rearrangements, fragmentation reactions, and intra- and intermolecular additions to carbon–carbon and other multiple bonds. Tributyltin hydride acts as a hydrogen atom source to trap the radical products, with concomitant generation of the radical chain carrier $Bu_3Sn\cdot$. These primary steps are illustrated for prototypical radical reactions in eqs 1–6.

$$RX + Bu_3Sn\cdot \longrightarrow R\cdot + Bu_3SnX \quad (1)$$

$$R\cdot \longrightarrow \left[\begin{array}{c}\text{rearrangement}\\ \text{fragmentation}\\ \text{addition to C=X}\end{array}\right] \longrightarrow R'\cdot \quad (2)$$

$$R\cdot + Bu_3SnH \longrightarrow RH + Bu_3Sn\cdot \quad (3)$$

$$R'\cdot + Bu_3SnH \longrightarrow R'H + Bu_3Sn\cdot \quad (4)$$

$$R\cdot \longrightarrow \text{Nonradical products} \quad (5)$$

$$R'\cdot \longrightarrow \text{Nonradical products} \quad (6)$$

The lifetimes of the radicals $R\cdot$ and $R'\cdot$, are determined by the chain transfer steps 3 and 4 and side reactions of steps 5 and 6. For the efficient formation of preparatively useful radical intermediates R and $R'\cdot$, the kinetics of each step must cooperate. For example, only reactions that are faster than the chain transfer step 3 can be executed in step 2. It should be understood that the overall rates of the chain transfer steps can be controlled to some extent by the concentration of the Bu_3SnH reagent. Likewise, the rates of step 2 can be altered by electronic and steric characteristics of $R\cdot$ and of any reaction partners involved. Fortunately kinetics, thermodynamics,[17] and substituent effects for the individual steps have been studied in some detail and it is possible to design useful synthetic strategies.[18]

Reactions Initiated by Bu_3Sn Radical.

C–X Homolysis followed by H abstraction.

Dehalogenation. Chemoselective replacement of halogens (except fluorine) with hydrogen is one of the major uses of TBTH in synthesis.[1h] The examples below show the versatility of the method and the range of substrates that can be used in this reaction. Replacement of bridgehead halogen (eq 7)[19] and selective removal of one of the halogens from a geminal dihalocyclopropane (eq 8)[20] are particularly noteworthy. TBTH can be generated catalytically for the dehalogenation reaction (eq 9).[21]

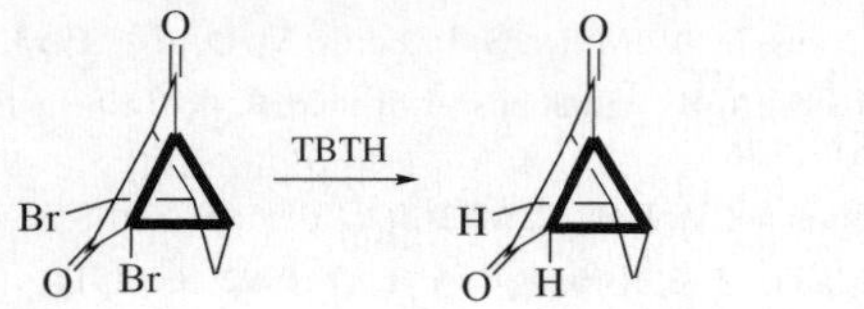

TBTH (7)

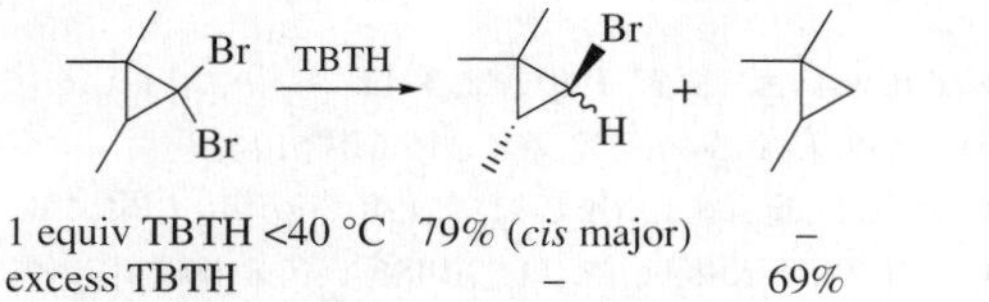

TBTH (8)

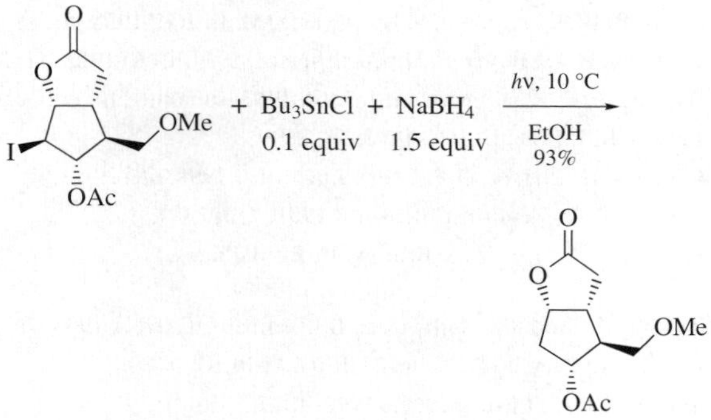

1 equiv TBTH <40 °C 79% (*cis* major) –
excess TBTH – 69%

+ Bu_3SnCl + $NaBH_4$, 0.1 equiv, 1.5 equiv; *hν*, 10 °C, EtOH, 93% (9)

Applications in the β-lactam[22] and carbohydrate (eqs 10 and 11)[23,24] areas illustrate the compatibility of the radical intermediates with a wide range of functional groups and reaction conditions. It should be noted that unlike reactions involving polar intermediates, hydroxyl and amino groups need not be protected under conditions where radicals are generated. Bu_3SnT is the reagent of choice for the regiochemical introduction of tritium into the steroid molecule (eq 12).[25] Debromination with TBTH was a crucial step in a model study directed towards the synthesis of a highly labile thromboxane A_2 analog (eq 13).[26]

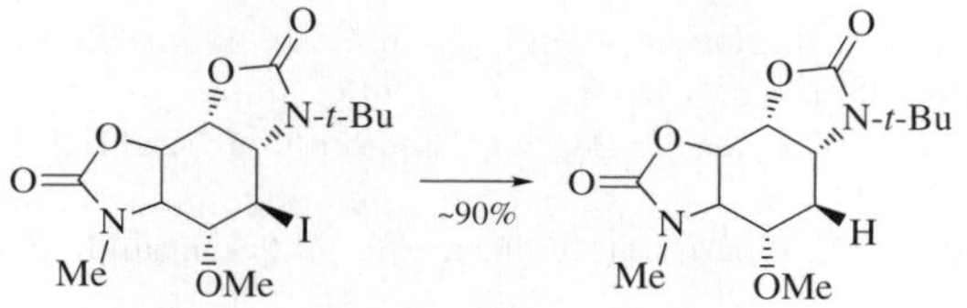

~90% (10)

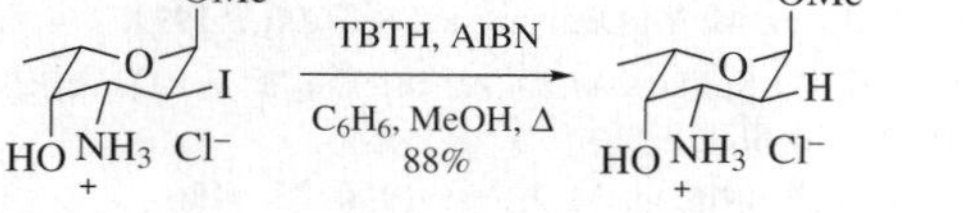

TBTH, AIBN; C_6H_6, MeOH, Δ; 88% (11)

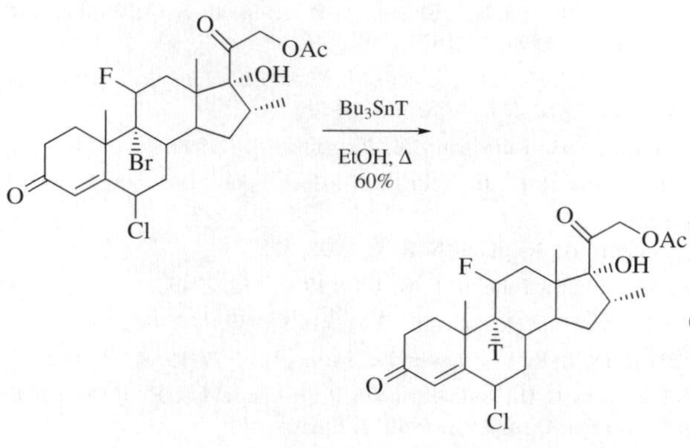

Bu_3SnT; EtOH, Δ; 60% (12)

TBTH, AIBN, $h\nu$, 80% (13)

Deoxygenation. Deoxygenation of secondary alcohols is best carried out by treatment of the corresponding thiocarbonyl derivatives with TBTH (eq 14).[27,28] Many common functional groups such as amine, alcohol, amido, carbonyl, epoxide, and tosylate are stable to the reaction conditions.

$Bu_3Sn\bullet$; Bu_3SnH (14)

Typical thiocarbonyl derivatives are xanthates (eq 15),[29] thiocarbonyl imidazolides (R′ = imidazolyl in eq 14), and phenoxythionocarbonate (eq 16).[30] Thioimidazolides, which are prepared under essentially neutral conditions from alcohols and ***1,1′-Thiocarbonyldiimidazole***, are best suited for acid- or base-sensitive substrates (eq 17).[31] Pentafluoro- or trichlorophenyl thionocarbonate is used for the deoxygenation of a primary alcohol under similar conditions.[32] A new procedure for the deoxygenation of tertiary alcohols[33] has also been reported.

TBTH, toluene, Δ, 4–7 h, 75% (15)

TBTH, PhMe, 75 °C, 3 h, 88% (16)

ImC(S)Im; TBTH, dioxane, Δ, 87% (17)

Z = CO_2Et

Decarboxylation. Photolysis or thermolysis of acyl derivatives of *N*-hydroxy-2-thiopyridone in the presence of TBTH (or, better, *t*-butyl thiol) gives the norhydrocarbon derived from the acyl moiety (eq 18). This method is applicable to a wide variety of aliphatic carboxylic acids (eq 19).[34]

RCOCl + (*N*-hydroxy-2-thiopyridone sodium salt) → ... TBTH, $-CO_2$ → RH (18)

TBTH, 110 °C, 18 min, 73% (19)

Homolysis of C–N Bonds. Isocyanides,[35] isothiocyanates, and nitro compounds[36] undergo homolysis of the C–N bond upon treatment with TBTH. Primary amino groups are converted into isocyanides via dehydration of the corresponding formamides. A highly selective reaction of 6β-isocyano-β-lactam (eq 20)[37] proceeds in 63% yield; as expected, the convex α-face of the bicyclic skeleton is more accessible to the hydride reagent.[38]

TBTH, C_6H_6, reflux, 63% (20)

α- or β-NC

Homolysis of C–S, C–Se, and C–Te Bonds. TBTH, in the presence of AIBN, reduces primary and secondary thiols,[39] sulfides with at least one radical stabilizing group (for example, α-carbonyl, α-thio, benzyl, *t*-Bu), and thiones to the corresponding hydrocarbons.[40] Se–[41] and Te–C[42] bonds are also cleaved by tin hydrides. Since the Ph–S or Ph–Se bond is almost never cleaved, thiophenyl and selenophenyl derivatives are among the most widely used radical precursors for C–C bond forming reactions (see below). Selenophenyl esters provide aldehydes (80 °C or $h\nu$, rt) or the norhydrocarbon (164 °C), depending on the reaction conditions.[41]

Radical Rearrangements. An alternative to the classical Wharton reaction uses the thioimidazolide intermediate for the radical generation and subsequent epoxide opening (eq 21).[43]

Im_2CS, CH_2Cl_2; TBTH, C_6H_6, Δ, 65% (21)

Im = imidazolyl

Ring opening of the cyclopropylmethyl radical is one of the most studied radical reactions, though synthetic applications in this area involving TBTH are limited.[44,45] A recent example is shown in eq 22.[46] Opening of *N*-acylaziridines by TBTH has also been described.[47]

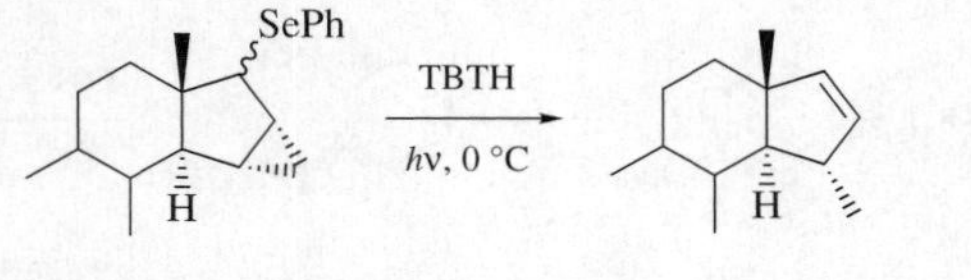

(22)

1,2-Migrations to a radical center are rare and take place only at low TBTH concentrations.[48] The example shown in eq 23[49] is particularly noteworthy since it provides a ready access to 2-deoxy sugar derivatives.

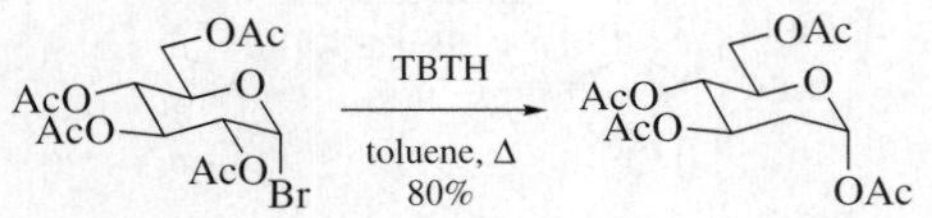

(23)

An interesting protocol for the ring expansion of ketones relies on the ability of carbonyl groups to act as acceptors for radicals. The subsequent fragmentation of the resultant oxy radicals is controlled by strategically placed stabilizing groups (eq 24)[10,11] or leaving groups.[50]

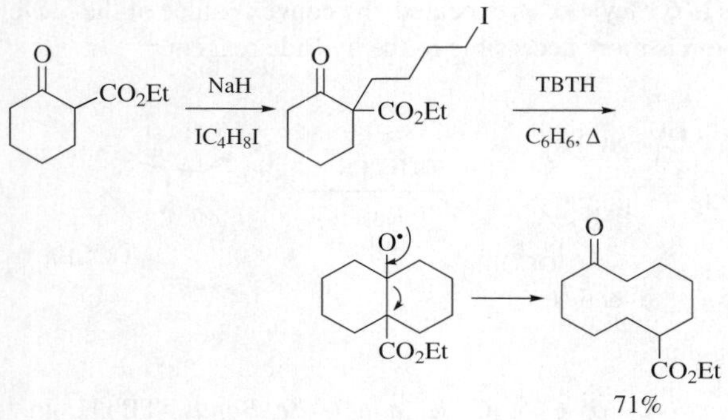

(24)

Intramolecular Additions to Unsaturated Centers. The hex-5-enyl radical cyclization is one of the most useful radical reactions and this subject has been extensively reviewed.[2,51,57] The examples shown (eqs 25–29)[52–56] illustrate the range of tolerance to various functional groups and the choice of radical precursors possible for this versatile reaction. The stereochemistry of hex-5-enyl radical cyclization has been the subject of a recent review.[57]

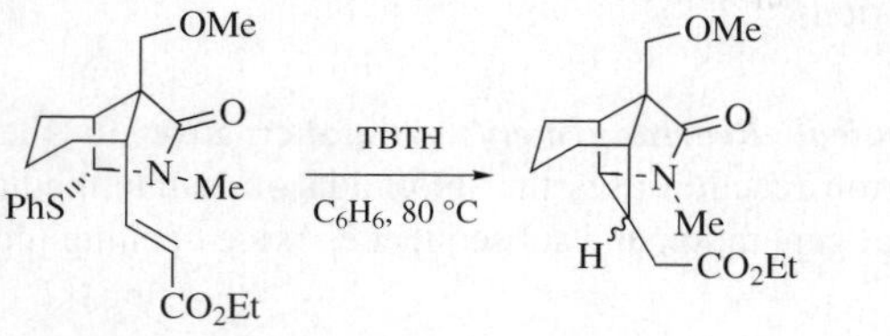

(25)

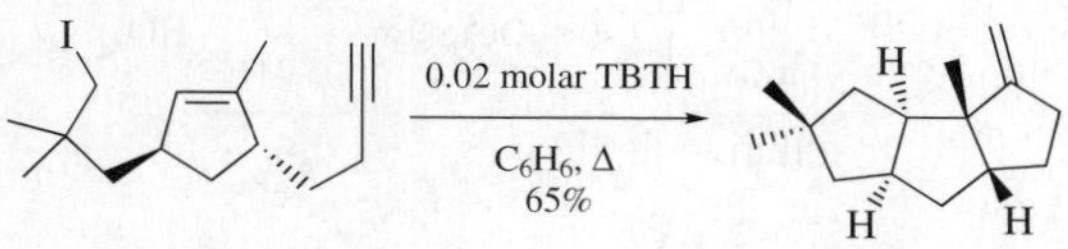

(26)

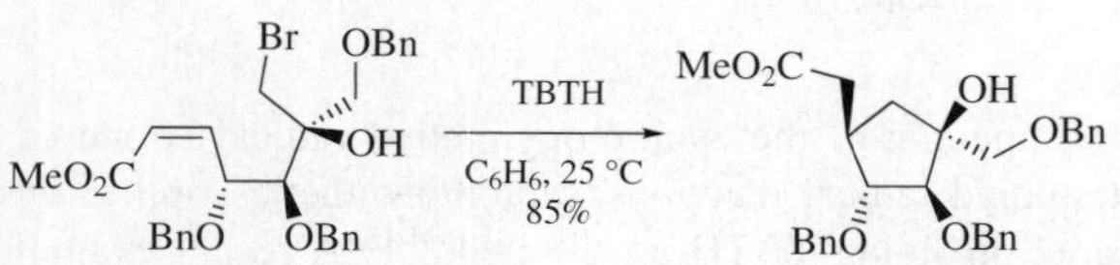

(27)

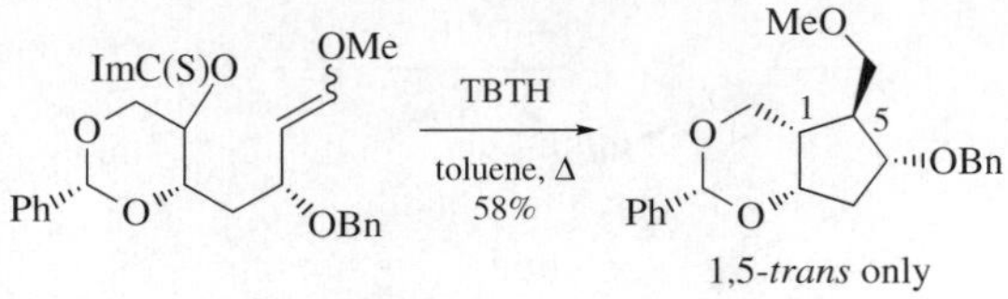

(28)

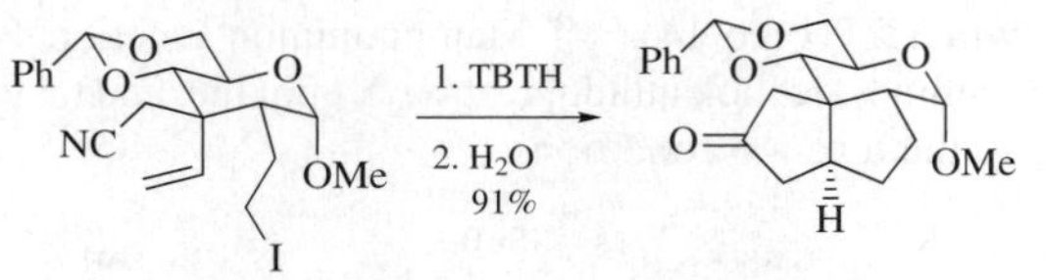

(29)

Appropriately placed haloalkyl side chains, which are readily prepared from allylic or propargylic alcohols, can be used for the stereocontrolled introduction of the hydroxymethyl group (eq 30)[58,60] or lactone annulation.[59] The silacyclopentane in eq 30 produces a 1,3-diol upon Tamao oxidation with ***Hydrogen Peroxide*** and ***Potassium Fluoride***[60] or it can be completely desilylated by treatment with ***Potassium t-Butoxide*** in DMSO.[61]

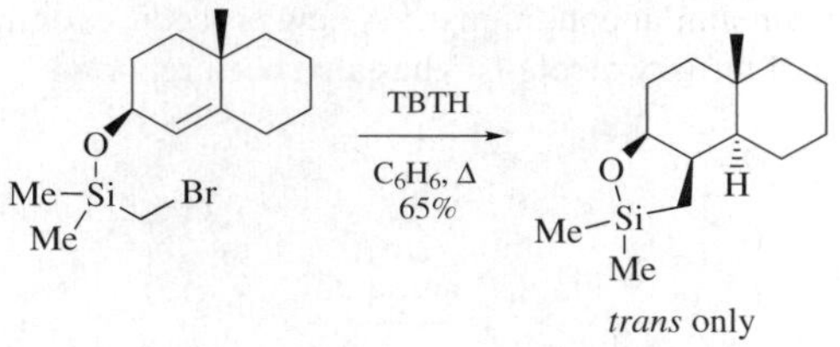

(30)

Several examples of the use of vinyl[2d,3,63,64] and allyl (eq 31)[65] radicals in synthesis have been described. Curran has described the use of a catalytic amount of TBTH for atom transfer reactions.[66]

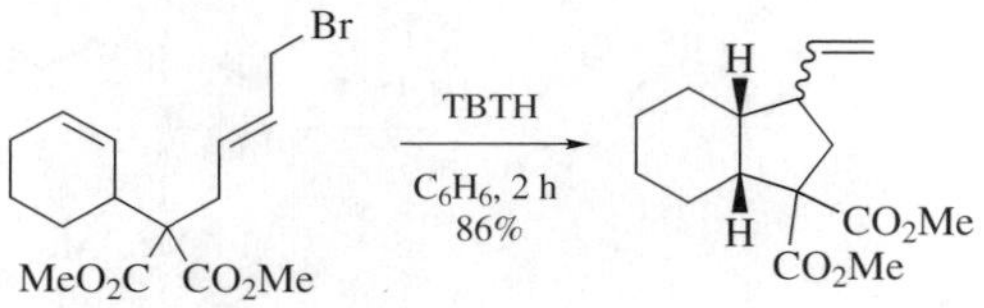

(31)

Exo-hept-6-enyl radical cyclization is about 30 times slower than the corresponding hex-5-enyl cyclization. Nonetheless, by increasing the rate of cyclization (for example, in eq 32[59] with an electron-withdrawing group on the alkene), synthetically useful reactions to make six-membered rings can be accomplished.[67–69]

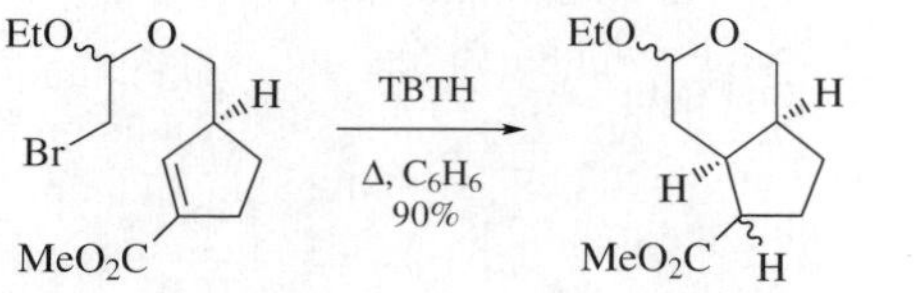

(32)

Acyl radicals generated from selenoesters undergo facile cyclization or intermolecular addition to electron-deficient alkenes faster than H-atom abstraction to give five- and six-membered ketones.[69]

TBTH is useful for the formation of macrolides via radical cyclization from iodoacrylates, if the concentration of both the reagent and substrate are kept low (eq 33). This method is not applicable for medium sized rings, but proceeds well for compounds with 11 or more atoms in the ring. *Endo* cyclization mode is favored when $n = 16$ or higher.[70]

TBTH (33)

$n = 8$, 27%; 12, 67%

Intermolecular Additions. With reactive acceptors in high concentrations, intermolecular addition of radicals proceed with considerable ease and several synthetically useful reactions have been discovered (eqs 34 and 35).[71,72] The example of the *C*-glycoside synthesis (eq 35) is particularly noteworthy; only the α-glycoside is obtained. Stereochemical control in these reactions has been the subject of a recent review.[73] A recent example illustrates the power of this method for rapid assembly of prostaglandin analogs (eq 36).[74]

0.2 equiv Bu_3SnCl, 1.3 equiv $NaBH_4$, EtOH, *hν*, 85% (34)

TBTH, Et_2O, *hν*, 55% (35)

TBTH, Δ, C_6H_6, 60% (36)

Radicals generated at the end of an intramolecular cyclization can be trapped if the acceptors are sufficiently reactive (eq 37).[62] Such radicals have also been trapped by ***t-Butyl Isocyanide*** to provide the corresponding nitrile.[75] One-carbon elongation can also be achieved by addition of an electrophilic radical to an enamine (eq 38).[76] Clever use of available kinetic and thermodynamic data can lead to the development of new annulation reactions, as shown in eq 39.[77]

TBTH (cat.), *hν*, $NaCNBH_4$, rt (37)

TBTH, 57% (38)

TBTH, C_6H_6, 80 °C, 73% (39)

Fragmentation and Homolytic Substitution Reactions. Allyltributyltin derivatives,[78,79] which have found wide applicability, may be synthesized by intermolecular S_H2 reaction of an appropriately substituted allylic compound (eq 40).[8,80] Intramolecular[9,81] S_H2' reactions have been used for cyclizations.

1. NaH, CS_2; 2. MeI, Δ; TBTH, C_6H_6, Δ (40)

Tributyltin hydride has ben used in catalytic amounts to effect carbocyclic ring expansions of β-stannyl ketones.[50] As shown in eq 41, the fragmentation reaction can be coupled to a radical cyclization reaction to produce ring expanded products from relatively simple substrates.

10 mol% TBTH, C_6H_6, Δ, 89% (41)

Hydrostannylation of Alkenes, Alkynes, and Carbonyl Compounds. Hydrostannylation of alkynes and alkenes is a well known reaction[1,6] which may be carried out under either thermal (eqs 42 and 43)[21,4] or sonochemical[82] initiation conditions. The intermediate radical formed upon the addition of $R_3Sn\cdot$ to alkynes and alkenes can be trapped by an appropriately placed double bond to give carbocyclic compounds.[63,64,83] TBTH adds to the strained single bond of [1.1.1]propellane (eq 44).[84]

+ Bu_3SnCl + $NaBH_4$; EtOH, 3 h, Δ, 73% (42)

+ TBTH + Et_3B; rt, hexane (43)

cis:*trans* = 90:10

+ TBTH; *t*-BuOO-*t*-Bu, *hν*, ether, 60% (44)

Addition of TBTH to carbonyl compounds requires catalysis by Pd^0 and a promoter Lewis acid (eq 45),[7] or a proton source.[85] Mechanistic studies suggest that TBTH is a hydride donor. Even though the radical additions to simple ketones are slow, α,β-enones readily generate *O*-stannyl ketyls which participate in hex-5-enyl radical cyclization reactions.[86]

+ TBTH; $Pd(Ph_3P)_4$, $ZnCl_2$, 95% (45)

Applications in Organometallic Methodology. Highly selective reduction of acid chlorides (eq 46)[87] and vinyl triflates[6] may be achieved by the use of Bu_3SnH under Pd catalysis. TBTH/Pd^0 is also an effective system for the reduction of allylic substrates under mild conditions. This reaction is remarkably tolerant of other common functional groups like hydroxy, epoxide, aldehyde, nitrile, and lactone. Radical scavengers improve the yield of the reaction.[88] Pd^0 catalyzes carbonylation of aliphatic, vinyl, and aromatic halides under ***Carbon Monoxide*** in the presence of TBTH.[6] A synthesis of alkyl hydroperoxide uses the homolytic cleavage of an Hg–C bond by tin hydride.[89] Tributyltin hydride has been used to generate a low-valent niobium reagent which is useful for coupling of imines and aldehydes (eq 47).[90]

$$\text{RCOCl} + \text{Bu}_3\text{SnH} \xrightarrow[\text{TBTH, C}_6\text{H}_6]{\text{1 mol\% Pd(Ph}_3\text{P)}_4} \text{RCHO} \quad (46)$$

~80%

$$\text{NbCl}_5 + 2\,\text{Bu}_3\text{SnH} \xrightarrow[-\text{Bu}_3\text{SnCl}]{\text{DME, }-78\text{ °C}} \text{NbCl}_3\text{(DME)} + \text{H}_2 \quad (47)$$

92%

Carbonylation and Oxygenation of Radicals. TBTH mediates the oxygenation (eq 48)[13] and carbonylation (eq 49)[14] of radicals.

$$\text{Ph, OTBDMS iodide} + \text{TBTH} \xrightarrow[\text{toluene, 24 h}]{\text{O}_2\text{, 0–20 °C}} \text{Ph, OTBDMS alcohol (OH)} \quad (48)$$

$$\text{C}_8\text{H}_{17}\text{Br} + \text{TBTH} \xrightarrow[\text{AIBN, 80 °C}]{\text{CO (80 atm), C}_6\text{H}_6} \text{C}_8\text{H}_{17}\text{CHO} \quad (49)$$

61%

TBTH as a Source of Bu_3Sn Anion. Tributyltin hydride is the best source of tributyltin Grignard or the corresponding lithium reagent, both of which have been used extensively in synthesis.[91] eqs 50 and 51 are illustrative.[15,92,93]

$$i\text{-PrMgCl} + \text{TBTH} \xrightarrow[\text{reflux, 2 h}]{\text{ether}} \text{Bu}_3\text{SnMgCl} \xrightarrow{\text{CH}_2\text{O}} \text{Bu}_3\text{SnCH}_2\text{OH} \quad (50)$$

56%

$$i\text{-Pr}_2\text{NLi} + \text{Bu}_3\text{SnH} \xrightarrow[\text{15 min}]{\text{THF, 0 °C}} \begin{cases} \xrightarrow[\text{89\%}]{\text{cyclohexenone, }-78\text{ °C, 5 min}} \text{3-(Bu}_3\text{Sn)cyclohexanone} \\ \xrightarrow{\text{R}_3\text{SiCl}} \text{Bu}_3\text{SnSiR}_3 \end{cases} \quad (51)$$

Related Reagents. See Classes R-14, R-15, R-25, R-26, R-27, R-28, R-30, R-31, and R-32, pages 1–10. See also Trimethylstannane, Triphenylstannane, polymer-supported organotin hydride,[94] Tris(trimethylsilyl)silane.[95]

---

1. (a) Neumann, W. P. *The Organic Chemistry of Tin*; Wiley: New York, 1970. (b) Kuivila, H. G. *S* **1970**, 499. (c) Walling, C. *T* **1985**, *41*, 3887. (d) Pereyre, M.; Quintard, J. P.; Rahm, A. *Tin in Organic Synthesis*; Butterworth: London, 1987. (e) Neumann, W. P. *S* **1987**, 665. (f) Harrison, P. G. *Chemistry of Tin*; Chapman and Hall: New York, 1989. (g) Giese, B. *Radicals in Organic Synthesis: Formation of Carbon-Carbon Bonds*; Pergamon: New York, 1986. (h) Metzger, J. O. *MOC* **1989**, *19a*.
2. (a) Beckwith, A. L. J. *T* **1981**, *37*, 3073. (b) Hart, D. J. *Science* **1984**, *223*, 883. (c) Curran, D. P. *S* **1988**, 417, 489. (d) Stork, G. In *Selectivity–A Goal for Synthetic Efficiency*; Bartmann, W.; Trost, B. M., Ed.; Verlag Chemie: Basel, 1984; p 281. (e) Giese, B. *AG(E)* **1985**, *24*, 553. (f) Giese, B. *AG(E)* **1989**, *28*, 969. (g) Ramaiah, M. *T* **1987**, *43*, 3541.
3. Stork, G.; Mook, R., Jr. *JACS* **1987**, *109*, 2829.
4. (a) Nozaki, K.; Oshima, K.; Utimoto, K. *T* **1989**, *45*, 923. For a recent application see: (b) Satoh, S.; Sodeoka, M.; Sasai, H.; Shibasaki, M. *JOC* **1991**, *56*, 2278.
5. Negishi, E. *Organometallics in Organic Synthesis*; Wiley: New York, 1980.
6. Stille, J. K. *AG(E)* **1986**, *25*, 508.
7. (a) Four, P.; Guibe, F. *TL* **1982**, *23*, 1825. (b) For the use of a tin triflate catalyst, see: Yang, T. S.; Four, P.; Guibe, F.; Balavoine, G. *NJC* **1984**, *8*, 611.
8. Ueno, Y.; Okawara, M. *JACS* **1979**, *101*, 1893.
9. Baldwin, J. E.; Adlington, R. M.; Mitchell, M. B.; Robertson, J. *CC* **1990**, 1574.
10. Beckwith, A. L. J.; O'Shea, D. M.; Gerba, S.; Westwood, S. W. *CC* **1987**, 666.
11. (a) Dowd, P.; Choi, S. *JACS* **1987**, *109*, 6548. (b) See also: Tsang, R.; Dickson, J. K., Jr.; Pak, H.; Walton, R.; Fraser-Reid, B. *JACS* **1987**, *109*, 3484.
12. Four, P.; Guibe, F. *JOC* **1981**, *46*, 4439.
13. Nakamura, E.; Inubushi, T.; Aoki, S.; Machii, D. *JACS* **1991**, *113*, 8980.
14. Ryu, I.; Kusano, K.; Ogawa, A.; Kambe, N.; Sonoda, N. *JACS* **1990**, *112*, 1295.
15. Still, W. C. *JACS* **1978**, *100*, 1482.
16. Selwyn, M. J. In *Chemistry of Tin*; Harrison, P. G., Ed.; Chapman and Hall, New York, 1989; p. 359.
17. For a listing of important kinetic data, see: (a) Asmus, K. D.; Bonifacic, M.; Ingold, K. U.; Roberts, B. P. In *Landolt-Bornstein, Radical Reaction Rates in Liquids*; Fisher, H., Ed. Springer: Heidelberg, 1983; Vol. II 13 b, c. (b) Johnston, L. J.; Lusztyk, J.; Wayner, D. D. M.; Abeywickreyma, A. N.; Beckwith, A. L. J.; Scaiano, J. C.; Ingold, K. U. *JACS* **1985**, *107*, 4594.
18. For a lucid account of how available kinetic and thermodynamic data can be used in synthetic planning, see Refs. 1c, 2c, and 2e.
19. McDonald, I. A.; Dreiding, A. S.; Hutmacher, H.; Musso, H. *HCA* **1973**, *56*, 1385.
20. Seyferth, D.; Yamazaki, H.; Alleston, D. L. *JOC* **1963**, *28*, 703.
21. Corey, E. J.; Suggs, W. *JOC* **1975**, *40*, 2554.
22. Aimetti, J. A.; Hamanaka, E. S.; Johnson, D. A.; Kellog, M. S. *TL* **1979**, *20*, 4631.
23. Knapp, S.; Patel, D. V. *JACS* **1983**, *105*, 6985.
24. Cardillo, G.; Orena, M.; Sandri, S.; Tomasini, C. *JOC* **1984**, *49*, 3951.
25. Parnes, H.; Pease, J. *JOC* **1979**, *44*, 151.
26. Bhagwat, S. S.; Hamann, P. R.; Still, W. C. *TL* **1985**, *26*, 1955.
27. Barton, D. H. R.; McCombie, S. W. *JCS(P1)* **1975**, 1574.
28. Hartwig, W. *T* **1983**, *39*, 2609.
29. Iacono, S.; Rasmussen, J. R. *OS* **1986**, *64*, 57.
30. Robins, M. J.; Wilson, J. S.; Hansske, F. *JACS* **1983**, *105*, 4059.
31. (a) Carney, R. E.; McAlpine, J. B.; Jackson, M.; Stanaszek, R. S.; Washburn, W. H.; Cirovic, M.; Mueller, S. L. *J. Antibiot.* **1978**, *31*, 441. (b) See also: Rasmussen, J. R. *JOC* **1980**, *45*, 2725.
32. Barton, D. H. R.; Blundell, P.; Dorchak, J.; Jang, D. O.; Jaszberenyi, J. C. *T* **1991**, *47*, 8969.
33. Barton, D. H. R.; Crich, D. *CC* **1984**, 774.
34. Barton, D. H. R.; Crich, D.; Motherwell, W. B. *T* **1985**, *41*, 3901.

35. Barton, D. H. R.; Bringmann, G.; Motherwell, W. B. *JCS(P1)* **1980**, 2665.
36. Ono, N.; Kaji, A. *S* **1986**, 693.
37. John, D. I.; Tyrrell, N. D.; Thomas, E. J. *T* **1983**, *39*, 2477.
38. For a more extensive list, see Ref. 2g.
39. Vedejs, E.; Powell, D. W. *JACS* **1982**, *104*, 2046.
40. Gutierrez, C. G.; Summerhays, L. R. *JOC* **1984**, *49*, 5206.
41. Pfenninger, J.; Heuberger, C.; Graf, W. *HCA* **1980**, *63*, 2328.
42. Clive, D. L. J.; Chittattu, G. J.; Farina, V.; Kiel, W. A.; Menchen, S. M.; Russell, C. G.; Singh, A.; Wong, C. K.; Curtis, N. J. *JACS* **1980**, *102*, 4438.
43. Barton, D. H. R.; Motherwell, R. S. H.; Motherwell, W. B. *JCS(P1)* **1981**, 2363.
44. Beckwith, A. L. J.; Ingold, K. U. In *Rearrangements in Ground and Excited States*; de Mayo, P., Ed.; Academic: New York, 1980; Vol. 1, p 161.
45. Harling, J. D.; Motherwell, W. B. *CC* **1988**, 1380.
46. Clive, D. L. J.; Daigneault, S. *CC* **1989**, 332.
47. Werry, J.; Stamm, H.; Lin, P.; Falkenstein, R.; Gries, S.; Irgartinger, H. *T* **1989**, *45*, 5015.
48. (a) Tada, M.; Akinaga, S.; Okabe, M, *BCJ* **1982**, *55*, 3939. (b) Wollowitz, S.; Halpern, J. *JACS* **1984**, *106*, 8319. (c) Barbier, M.; Barton, D. H. R.; Devys, M.; Topgi, R. S. *CC* **1984**, 743.
49. (a) Giese, B.; Groninger, K. S. *OS* **1990**, *69*, 66. (b) See also: Giese, B.; Gilges, S.; Groninger, K. S.; Lamberth, C.; Witzel, T. *LA* **1988**, 615.
50. Baldwin, J. E.; Adlington, R. M.; Robertson, J. *T* **1989**, *45*, 909.
51. Julia, M. *ACR* **1971**, *4*, 386.
52. Choi, J.; Ha, D.; Hart, D. J.; Lee, C.; Ramesh, S.; Wu, S. *JOC* **1989**, *54*, 279.
53. Curran, D. P.; Rakiewicz, D. M. *T* **1985**, *41*, 3943.
54. (a) Wilcox, C. S.; Gaudino, J. J. *JACS* **1986**, *108*, 3102. (b) See also: Gaudino, J. J.; Wilcox, C. S. *JACS* **1990**, *112*, 4374.
55. (a) RajanBabu, T. V. *JOC* **1988**, *53*, 4522. (b) See also: RajanBabu, T. V.; Fukunaga, T.; Reddy, G. S. *JACS* **1989**, *111*, 1759.
56. (a) Tsang, R.; Fraser-Reid, B. *JACS* **1986**, *108*, 2116. (b) See also: Alonso, R. A.; Vite, G. D.; McDevitt, R. E.; Fraser-Reid, B. *JOC* **1992**, *57*, 573.
57. RajanBabu, T. V. *ACR* **1991**, *24*, 139.
58. Stork, G.; Kahn, M. *JACS* **1985**, *107*, 500.
59. Stork, G.; Mook, R., Jr.; Biller, S. A.; Rychnovsky, S. C. *JACS* **1983**, *105*, 3741.
60. Nishiyama, H.; Kitajima, T.; Matsumoto, M.; Itoh, K. *JOC* **1984**, *49*, 2298.
61. Stork, G.; Sofia, M. J. *JACS* **1986**, *108*, 6826.
62. Stork, G.; Sher, P. M.; Chen, H. *JACS* **1986**, *108*, 6384.
63. Stork, G.; Mook, R., Jr. *TL* **1986**, *27*, 4529.
64. Beckwith, A. L. J.; O'Shea, D. M. *TL* **1986**, *27*, 4525.
65. Stork, G.; Reynolds, M. *JACS* **1988**, *110*, 6911.
66. Curran, D. P.; Chen, M.; Kim, D. *JACS* **1989**, *111*, 6265.
67. Munt, S. P.; Thomas, E. J. *CC* **1989**, 480.
68. See for example: (a) Crich, D.; Eustace, K. A.; Fortt, S. M.; Ritchie, T. J. *T* **1990**, *46*, 2135. (b) Marco-Contelles, J.; Pozuelo, C.; Jimeno, M. L.; Martinez, L.; Martinez-Grau, A. *JOC* **1992**, *57*, 2625. (c) Chuang, C.-P.; Gallucci, J. C.; Hart, D. J.; Hoffmann, C. *JOC* **1988**, *53*, 3218. (d) Gukumoto, K.; Taniguchi, N.; Yasui, K.; Ihara, M. *JCSP(1)* **1990**, 1469. (e) Batty, D.; Crich, D.; Fortt, S. M. *JCSP(1)* **1990**, 2875. (f) Bachi, M. D.; Frolow, F.; Hoornaert, C. *JOC* **1983**, *48*, 1841.
69. Boger, D. L.; Mathvink, R. J. *JACS* **1990**, *112*, 4003. See also Ref. 68(a).
70. (a) Porter, N. A.; Chang, V. H.-T. *JACS* **1987**, *109*, 4976. (b) Porter, N. A.; Magnin, D. R.; Wright, B. T. *JACS* **1986**, *108*, 278. (c) For an application see: Hitchcock, S. A.; Pattenden, G. *TL* **1990**, *31*, 3641. (d) See also Baldwin, J. E.; Adlington, R. M.; Mitchell, M. B.; Robertson, J. *CC* **1990**, 1574.
71. Giese, B.; Gonzalez-Gomez, J.; Witzel, T. *AG(E)* **1984**, *23*, 69.
72. Giese, B.; Dupuis, J.; Nix, M. *OS* **1987**, *65*, 236.
73. Porter, N. A.; Giese, B.; Curran, D. P. *ACR* **1991**, *24*, 296.
74. Ono, N.; Yoshida, Y.; Tani, K.; Okamoto, S.; Sato, F. *TL* **1993**, *34*, 6427.
75. Stork, G.; Sher, P. M. *JACS* **1986**, *108*, 303.
76. (a) Shubert, S.; Renaud, P.; Carrupt, P.; Schenk, K. *HCA* **1993**, *76*, 2473. (b) See also: Ref. 2e. For another example of a one-carbon elongation that relies on a tin reagent, see: Hart, D. J.; Seely, F. L. *JACS* **1988**, *110*, 1631.
77. (a) Curran, D. P.; Chen, M.; Spletzer, E.; Seong, C. M.; Chang, C. *JACS* **1989**, *111*, 8872. (b) Curran, D. P.; Chang, C.-T. *JOC* **1989**, *54*, 3140.
78. Keck, G. E.; Enholm, E. J.; Yates, J. B.; Wiley, M. R. *T* **1985**, *41*, 4079.
79. Baldwin, J. E.; Kelly, D. R. *CC* **1985**, 682.
80. Ueno, Y.; Sano, H.; Okawara, M. *TL* **1980**, *21*, 1767.
81. Danishefsky, S. J.; Panek, J. S. *JACS* **1987**, *109*, 917.
82. Nakamura, E.; Machii, D.; Inubushi, T. *JACS* **1989**, *111*, 6849.
83. Hanessian, S.; Leger, R. *JACS* **1992**, *114*, 3315.
84. (a) Toops, D.; Barbachyn, M. R. *JOC* **1993**, *58*, 6505. (b) See also: Belzner, J.; Szeimies, G. *TL* **1987**, *28*, 3099.
85. Keinan, E.; Gleize, P. A. *TL* **1982**, *23*, 477.
86. Enholm, E. J.; Kinter, K. S. *JACS* **1991**, *113*, 7784.
87. Four, P.; Guibe, F. *JOC* **1982**, *46*, 4439.
88. Keinan, E.; Greenspoon, N. *TL* **1982**, *23*, 241.
89. Bloodworth, A. J.; Khan, J. A.; Loveitt, M. E. *JCS(P1)* **1981**, 621.
90. Pedersen, S. F.; Roskamp, E. J. *JACS* **1987**, *109*, 6551.
91. For a review see ref. 1f, p. 343. see also: (a) Nemoto, H.; Wu, X. M.; Kurobe, H; Ihara, M.; Fukumoto, K. *TL* **1983**, *24*, 4257. (b) Nakatani, K.; Isobe, S. *TL* **1985**, *26*, 2209. (c) Shibasaki, M.; Susuki, H.; Torisawa, Y.; Ikegami, S. *CL* **1983**, 1303. (d) Fleming, I.; Urch, C. J. *JOM* **1985**, *285*, 173. (e) Kadow, J. F.; Johnson, C. R. *TL* **1984**, *25*, 5255.
92. Meyer, N.; Seebach, D. *CB* **1980**, *113*, 1290.
93. Chenard, B. L.; Laganis, E. D.; Davidson, F.; RajanBabu, T. V. *JOC* **1985**, *50*, 3666.
94. Gerigk, U.; Gerlach, M.; Neumann, W. P.; Vieler, R.; Weintritt, V. *S* **1990**, 448.
95. Ballestri, M.; Chatgilialoglu, C.; Clark, K. B.; Griller, D.; Giese, B.; Kopping, B. *JOC* **1991**, *56*, 678.

T. (Babu) V. RajanBabu
The Ohio State University, Columbus, OH, USA

Triethylsilane[1]

[617-86-7] $C_6H_{16}Si$ (MW 116.31)

(mild reducing agent for many functional groups)

Physical Data: mp −156.9 °C; bp 107.7 °C; *d* 0.7309 g cm^{-3}.
Solubility: insol H_2O; sol hydrocarbons, halocarbons, ethers.
Form Supplied in: colorless liquid; widely available.
Purification: simple distillation, if needed.
Handling, Storage, and Precautions: triethylsilane is physically very similar to comparable hydrocarbons. It is a flammable, but not pyrophoric, liquid. As with all organosilicon hydrides, it is capable of releasing hydrogen gas upon storage, particularly in the presence of acids, bases, or fluoride-releasing salts. Proper

precautions should be taken to vent possible hydrogen buildup when opening vessels in which triethylsilane is stored.

Introduction. Triethylsilane serves as an exemplar for organosilicon hydride behavior as a mild reducing agent. It is frequently chosen as a synthetic reagent because of its availability, convenient physical properties, and economy relative to other organosilicon hydrides which might otherwise be suitable for effecting specific chemical transformations.

Hydrosilylations. Addition of triethylsilane across multiple bonds occurs under the influence of a large number of metal catalysts.[2] Terminal alkynes undergo hydrosilylations easily with triethylsilane in the presence of platinum,[3] rhodium,[3a,4] ruthenium,[5] osmium,[6] or iridium[4] catalysts. For example, phenylacetylene can form three possible isomeric hydrosilylation products with triethylsilane; the (*Z*)-β-, the (*E*)-β-, and the α-products (eq 1). The (*Z*)-β-isomer is formed exclusively or preferentially with ruthenium[5] and some rhodium[4] catalysts, whereas the (*E*)-β-isomer is the major product formed with platinum[3] or iridium[4] catalysts. In the presence of a catalyst and carbon monoxide, terminal alkynes undergo silylcarbonylation reactions with triethylsilane to give (*Z*)- and (*E*)-β-silylacrylaldehydes.[7] Phenylacetylene gives an 82% yield of a mixture of the (*Z*)- and (*E*)-isomers in a 10:1 ratio when 0.3 mol % of ***Dirhodium(II) Tetrakis(perfluorobutyrate)*** catalyst is used under atmospheric pressure at 0 °C in dichloromethane (eq 2).[7d] Terminal alkenes react with triethylsilane in the presence of this catalyst to form either 'normal' anti-Markovnikov hydrosilylation products or allyl- or vinylsilanes, depending on whether the alkene is added to the silane or vice versa.[8] A mixture of 1-hexene and triethylsilane in the presence of 2 mol % of an iridium catalyst ($[IrCl(CO)_3]_n$) reacts under 50 atm of carbon monoxide to give a 50% yield of a mixture of the (*Z*)- and (*E*)-enol silyl ether isomers in a 1:2 ratio (eq 3).[9] Hydrolysis yields the derived acylsilane quantitatively.[9]

Ph—≡ + Et_3SiH —cat.→ (Z)-PhCH=CH$SiEt_3$ + $Ph(SiEt_3)C=CH_2$ ... (1)

Ph—≡ + Et_3SiH + CO —cat.→ OHC(H)C=C(Ph)$SiEt_3$ isomers (2)

$BuCH=CH_2$ + Et_3SiH + CO —cat.→ $BuCH_2CH=C(OSiEt_3)SiEt_3$ (Z) + (E) (3)

A number of metal complexes catalyze the hydrosilylation of various carbonyl compounds by triethylsilane.[10] Stereoselectivity is observed in the hydrosilylation of ketones[11] as in the reactions of 4-*t*-butylcyclohexanone and triethylsilane catalyzed by ruthenium,[12] chromium,[13] and rhodium[12,14] metal complexes (eq 4). Triethylsilane and ***Chlorotris(triphenylphosphine)rhodium(I)*** catalyst effect the regioselective 1,4-hydrosilylation of α,β-unsaturated ketones and aldehydes.[15,16] Reduction of mesityl oxide in this manner results in a 95% yield of product that consists of 1,4- and 1,2-hydrosilylation isomers in a 99:1 ratio (eq 5). This is an exact complement to the use of phenylsilane, where the ratio of respective isomers is reversed to 1:99.[16]

4-*t*-Bu-cyclohexanone —Et_3SiH, cat.→ axial $OSiEt_3$ isomer + equatorial $OSiEt_3$ isomer (4)

$(Ph_3P)_3RuCl_2$, AgTFA, PhMe, Δ	5:95
$Et_4N^+ [HCr_2(CO)_{10}]^-$, DME, Δ	10:90
$(Ph_3P)_3RhCl$, PhMe, Δ	11:89
$[Rh(\eta^3\text{-}C_3H_5)\{P(OMe)_3\}_3]$, PhH	29:71

Mesityl oxide + Et_3SiH —cat.→ 1,4-product ($OSiEt_3$) + 1,2-product ($OSiEt_3$), 99:1 (5)

Silane Alcoholysis. Triethylsilane reacts with alcohols in the presence of metal catalysts to give triethylsilyl ethers.[17] The use of dirhodium(II) perfluorobutyrate as a catalyst enables regioselective formation of monosilyl ethers from diols (eq 6).[17a]

$Bu_2C(OH)CH_2CH_2CH_2OH$ —Et_3SiH, $Rh_2(pfb)_4$, 92%→ $Bu_2C(OH)CH_2CH_2CH_2OSiEt_3$ + H_2 (6)

Formation of Singlet Oxygen. Triethylsilane reacts with ozone at −78 °C in inert solvents to form triethylsilyl hydrotrioxide, which decomposes at slightly elevated temperatures to produce triethylsilanol and ***Singlet Oxygen***. This is a convenient way to generate this species for use in organic synthesis.[18]

Reduction of Acyl Derivatives to Aldehydes. Aroyl chlorides and bromides give modest yields of aryl aldehydes when refluxed in diethyl ether with triethylsilane and ***Aluminum Chloride***.[19] Better yields of both alkyl and aryl aldehydes are obtained from mixtures of acyl chlorides or bromides and triethylsilane by using a small amount of 10% ***Palladium on Carbon*** catalyst (eq 7).[20] This same combination of triethylsilane and catalyst can effect the reduction of ethyl thiol esters to aldehydes, even in sensitive polyfunctional compounds (eq 8).[21]

$$C_7H_{15}COCl + Et_3SiH \xrightarrow[83\%]{10\%\ Pd/C} C_7H_{15}CHO \quad (7)$$

Radical Chain Reductions. Triethylsilane can replace toxic and difficult to remove organotin reagents for synthetic reductions under radical chain conditions. Although it is not as reactive as ***Tri-n-butylstannane***,[22] careful choice of initiator, solvent, and additives leads to effective reductions of alkyl halides,[23,24] alkyl sulfides,[23] and alcohol derivatives such as *O*-alkyl *S*-methyl dithiocarbonate (xanthate) and thionocarbonate esters.[22,23,25,26] Portionwise addition of 0.6 equiv of ***Dibenzoyl Peroxide*** to a refluxing triethylsilane solution of *O*-cholestan-3β-yl *O*′-(4-fluorophenyl) thionocarbonate gives a 93% yield of cholestane (eq 9).[22] The same method converts bis-xanthates of *vic*-diols into alkenes (eq 10).[22] Addition of a small amount of thiol such as *t*-dodecanethiol to serve as a 'polarity reversal catalyst'[24] with strong radical initiators in nonaromatic solvents also gives good results.[23,25] Treatment of ethyl 4-bromobutanoate with four equiv of triethylsilane, two equiv of dilauroyl peroxide (DLP), and 2 mol % of *t*-dodecanethiol in refluxing cyclohexane for 1 hour yields ethyl butanoate in 97% yield (eq 11).[23]

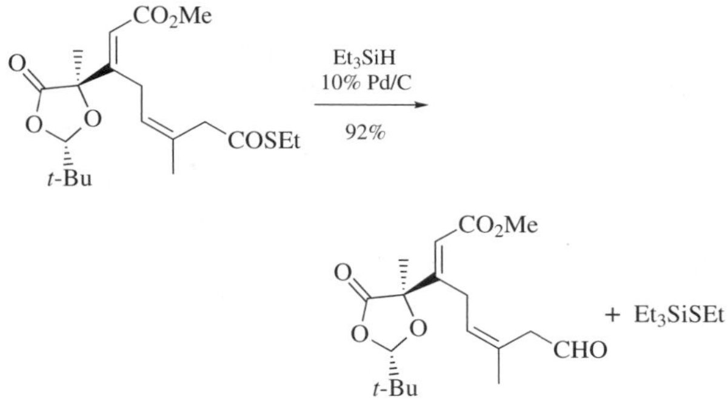

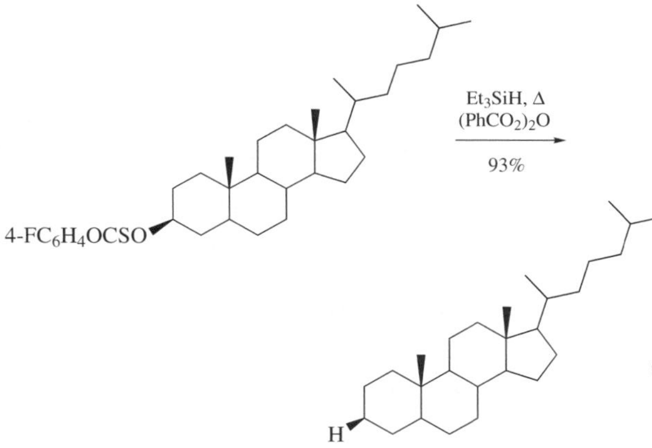

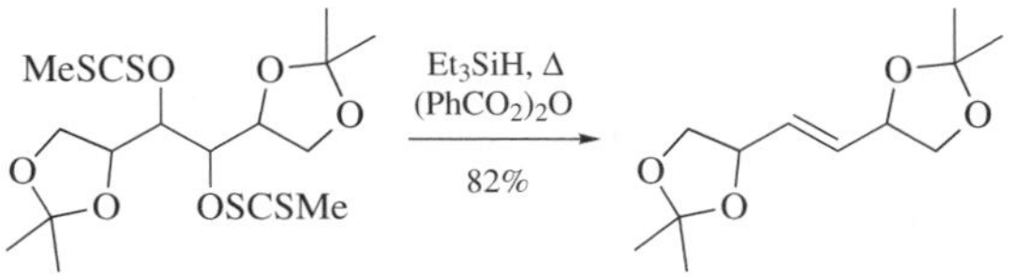

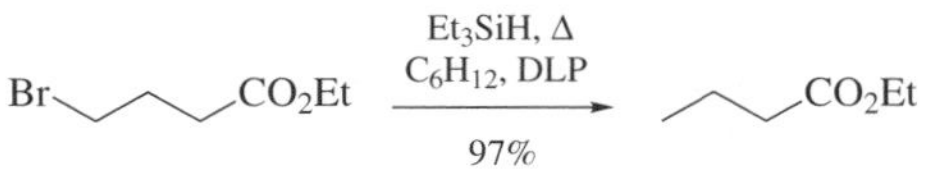

Ionic Hydrogenations and Reductive Substitutions. The polar nature of the Si–H bond enables triethylsilane to act as a hydride donor to electron-deficient centers. Combined with Brønsted or Lewis acids this forms the basis for many useful synthetic transformations.[27] Use of ***Trifluoromethanesulfonic Acid*** (triflic acid) at low temperatures enables even simple alkenes to be reduced to alkanes in high yields (eq 12).[28] ***Boron Trifluoride*** monohydrate is effective in promoting the reduction of polycyclic aromatic compounds (eq 13).[29] Combined with thiols, it enables sulfides to be prepared directly from aldehydes and ketones (eq 14).[30] Combinations of triethylsilane with either ***Trifluoroacetic Acid***/ammonium fluoride or ***Pyridinium Poly(hydrogen fluoride)*** (PPHF) are effective for the reductions of alkenes, alcohols, and ketones (eq 15).[31] Immobilized strong acids such as iron- or copper-exchanged ***Montmorillonite K10***[32] or the superacid ***Nafion-H***[33] facilitate reductions of aldehydes and ketones[32] or of acetals[33] by increasing the ease of product separation (eq 16). Boron trifluoride and triethylsilane are an effective combination for the reduction of alcohols, aldehydes, ketones (eq 17),[34] and epoxides.[35] ***Boron Trifluoride Etherate*** sometimes may be substituted for the free gas.[36]

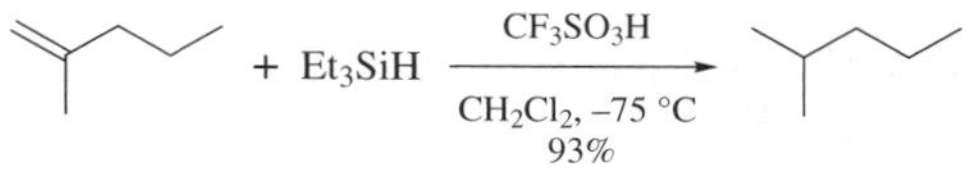

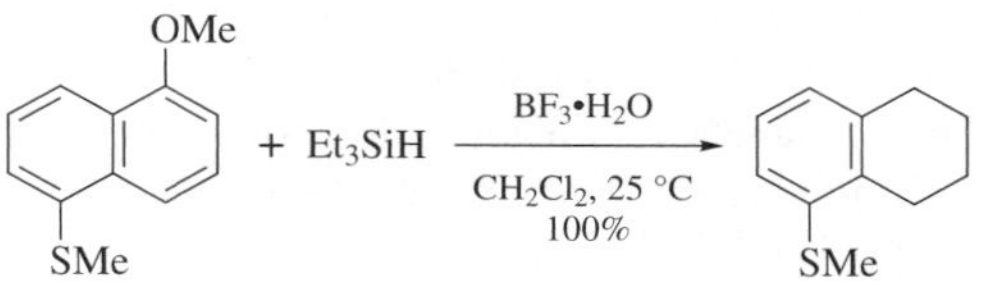

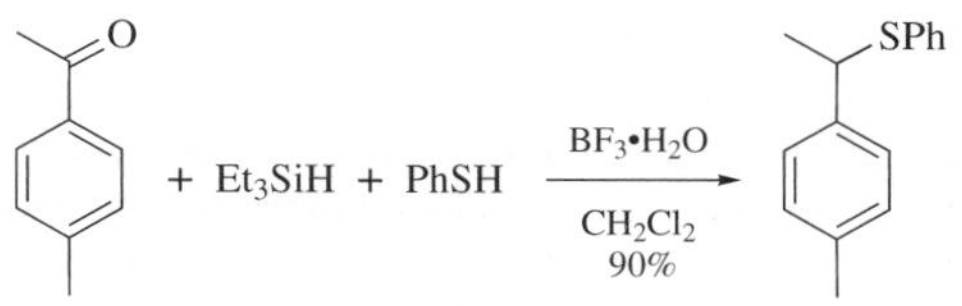

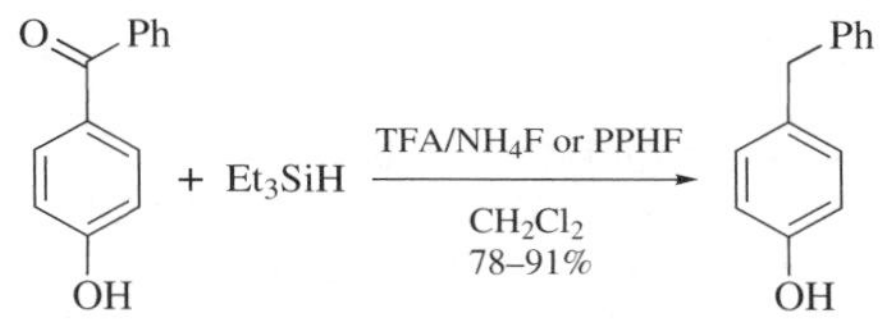

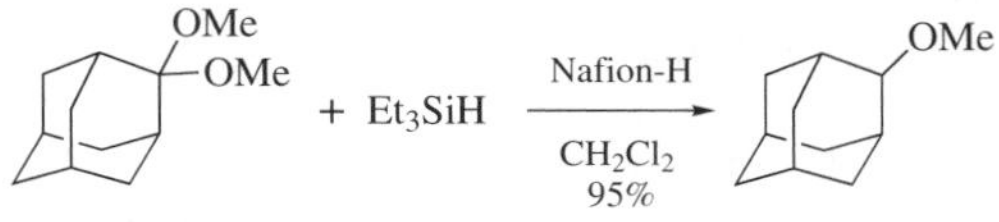

Triethylsilane in 3M ethereal ***Lithium Perchlorate*** solution effects the reduction of secondary allylic alcohols and acetates (eq 18).[37] The combination of triethylsilane and ***Titanium(IV) Chloride*** is a particularly effective reagent pair for the selective reduction of acetals.[38] Treatment of (±)-frontalin with this pair gives an 82% yield of tetrahydropyran products with a *cis*:*trans* ratio of 99:1 (eq 19).[38b] This exactly complements the 1:99 product ratio of the same products obtained with ***Diisobutylaluminum Hydride***.[38b]

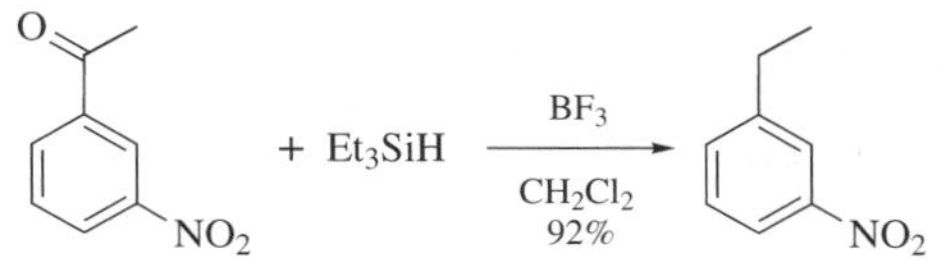

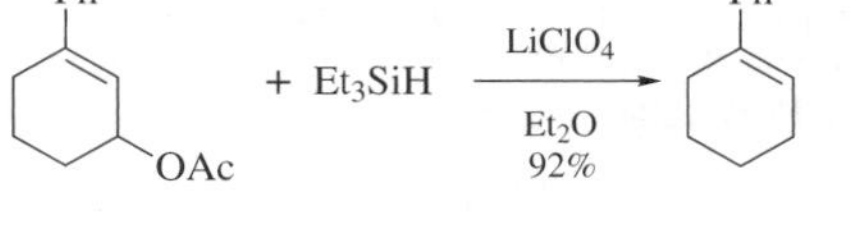

$$\xrightarrow[\text{CH}_2\text{Cl}_2,\ -78\ ^\circ\text{C}]{\text{Et}_3\text{SiH, TiCl}_4} \quad 99:1 \qquad (19)$$

Triethylsilane and trityl salts[39] or ***Trimethylsilyl Trifluoromethanesulfonate***[40] are effective for the reduction of various ketones and acetals, as are combinations of ***Chlorotrimethylsilane*** and indium(III) chloride[41] and ***Tin(II) Bromide*** and ***Acetyl Bromide***.[42] Isophthaldehyde undergoes reductive polycondensation to a polyether when treated with triethylsilane and ***Triphenylmethyl Perchlorate***.[43]

Triethylsilane reduces nitrilium ions to aldimines,[44] diazonium ions to hydrocarbons,[45] and aids in the deprotection of amino acids.[46] With aluminum halides, it reduces alkyl halides to hydrocarbons.[47]

Related Reagents. See Classes R-1, R-2, R-3, R-7, R-14, R-15, R-23, R-27, R-28, R-29, R-31, and R-32, pages 1–10. Phenylsilane–Cesium Fluoride; Tri-*n*-butylstannane; Tricarbonylchloroiridium–Diethyl(methyl)silane–Carbon Monoxide; Triethylsilane–Trifluoroacetic Acid.

1. (a) Fleming, I. In *Comprehensive Organic Chemistry*; Barton, D.; Ollis, W. D., Eds.; Pergamon: New York, 1979; Vol. 3, pp 541–679. (b) Colvin, E. *Silicon in Organic Synthesis*; Butterworths: Boston, 1981. (c) Weber, W. P. *Silicon Reagents for Organic Synthesis*; Springer: New York, 1983. (d) *The Chemistry of Organic Silicon Compounds*; Patai, S.; Rappoport, Z., Eds.; Wiley: New York, 1989. (e) Corey, J. Y. In *Advances in Silicon Chemistry*; Larson, G. L., Ed.; JAI: Greenwich, CT, 1991; Vol. 1, pp 327–387.
2. (a) Lukevics, E. *RCR* **1977**, *46*, 264. (b) Speier, J. L. *Adv. Organomet. Chem.* **1979**, *17*, 407. (c) Keinan, E. *PAC* **1989**, *61*, 1737.
3. (a) Doyle, M. P.; High, K. G.; Nesloney, C. L.; Clayton, T. W., Jr.; Lin, J. *OM* **1991**, *10*, 1225. (b) Lewis, L. N.; Sy, K. G.; Bryant, G. L., Jr.; Donahue, P. E. *OM* **1991**, *10*, 3750.
4. Kopylova, L. I.; Pukhnarevich, V. B.; Voronkov, M. G. *ZOB* **1991**, *61*, 2418.
5. Esteruelas, M. A.; Herrero, J.; Oro, L. A. *OM* **1993**, *12*, 2377.
6. Esteruelas, M. A.; Oro, L. A.; Valero, C. *OM* **1991**, *10*, 462.
7. (a) Murai, S.; Sonada, N. *AG(E)* **1979**, *18*, 837. (b) Matsuda, I.; Ogiso, A.; Sato, S.; Izumi, Y. *JACS* **1989**, *111*, 2332. (c) Ojima, I.; Ingallina, P.; Donovan, R. J.; Clos, N. *OM* **1991**, *10*, 38. (d) Doyle, M. P.; Shanklin, M. S. *OM* **1993**, *12*, 11.
8. Doyle, M. P.; Devora, G. A.; Nefedov, A. O.; High, K. G. *OM* **1992**, *11*, 549.
9. Chatani, N.; Ikeda, S.; Ohe, K.; Murai, S. *JACS* **1992**, *114*, 9710.
10. (a) Eaborn, C.; Odell, K.; Pidcock, A. *JOM* **1973**, *63*, 93. (b) Corriu, R. J. P.; Moreau, J. J. E. *CC* **1973**, 38. (c) Ojima, I.; Nihonyanagi, M.; Kogure, T.; Kumagai, M.; Horiuchi, S.; Nakatasugawa, K.; Nagai, Y. *JOM* **1975**, *94*, 449.
11. Ojima, I.; Nihonyanagi, M.; Nagai, Y. *BCJ* **1972**, *45*, 3722.
12. Semmelhack, M. F.; Misra, R. N. *JOC* **1982**, *47*, 2469.
13. Fuchikami, T.; Ubukata, Y.; Tanaka, Y. *TL* **1991**, *32*, 1199.
14. Bottrill, M.; Green, M. *JOM* **1976**, *111*, C6.
15. Ojima, I.; Kogure, T.; Nihonyanagi, M.; Nagai, Y. *BCJ* **1972**, *45*, 3506.
16. Ojima, I.; Kogure, T. *OM* **1982**, *1*, 1390.
17. (a) Doyle, M. P.; High, K. G.; Bagheri, V.; Pieters, R. J.; Lewis, P. J.; Pearson, M. M. *JOC* **1990**, *55*, 6082. (b) Zakharkin, L. I.; Zhigareva, G. G. *IZV* **1992**, 1284. (c) Barton, D. H. R.; Kelly, M. J. *TL* **1992**, *33*, 5041.
18. Corey, E. J.; Mehrota, M. M.; Khan, A. U. *JACS* **1986**, *108*, 2472.
19. Jenkins, J. W.; Post, H. W. *JOC* **1950**, *15*, 556.
20. Citron, J. D. *JOC* **1969**, *34*, 1977.
21. Fukuyama, T.; Lin, S.-C.; Li, L. *JACS* **1990**, *112*, 7050.
22. Barton, D. H. R.; Jang, D. O.; Jaszberenyi, J. C. *TL* **1991**, *32*, 7187: *T* **1993**, *49*, 2793.
23. Cole, S. J.; Kirwan, J. N.; Roberts, B. P.; Willis, C. R. *JCS(P1)* **1991**, 103.
24. Allen, R. P.; Roberts, B. P.; Willis, C. R. *CC* **1989**, 1387.
25. Kirwin, J. N.; Roberts, B. P.; Willis, C. R. *TL* **1990**, *31*, 5093.
26. Cf. Chatgilialoglu, C.; Ferreri, C.; Lucarini, M. *JOC* **1993**, *58*, 249.
27. (a) Kursanov, D. N.; Parnes, Z. N. *RCR* **1969**, *38*, 812. (b) Kursanov, D. N.; Parnes, Z. N.; Loim, N. M. *S* **1974**, 633. (c) Nagai, Y. *OPP* **1980**, *12*, 13. (d) Kursanov, D. N.; Parnes, Z. N.; Kalinkin, M. I.; Loim, N. M. *Ionic Hydrogenation and Related Reactions*; Harwood: Chur, Switzerland, 1985.
28. Bullock, R. M.; Rappoli, B. J. *CC* **1989**, 1447.
29. (a) Larsen, J. W.; Chang, L. W. *JOC* **1979**, *44*, 1168. (b) Eckert-Maksic, M.; Margetic, D. *Energy Fuels* **1991**, *5*, 327. (c) Eckert-Maksic, M.; Margetic, D. *Energy Fuels* **1993**, *7*, 315.
30. Olah, G. A.; Wang, Q.; Trivedi, N. J.; Prakash, G. K. S. *S* **1992**, 465.
31. Olah, G. A.; Wang, Q.; Prakash, G. K. S. *SL* **1992**, 647.
32. Izumi, Y.; Nanami, H.; Higuchi, K.; Onaka, M. *TL* **1991**, *32*, 4741.
33. Olah, G. A.; Yamato, T.; Iyer, P. S.; Prakash, G. K. S. *JOC* **1986**, *51*, 2826.
34. (a) Adlington, M. G.; Orfanopoulos, M.; Fry, J. L. *TL* **1976**, 2955. (b) Fry, J. L.; Orfanopoulos, M.; Adlington, M. G.; Dittman, W. R., Jr.; Silverman, S. B. *JOC* **1978**, *43*, 374. (c) Fry, J. L.; Silverman, S. B.; Orfanopoulos, M. *OS* **1981**, *60*, 108.
35. Fry, J. L.; Mraz, T. J. *TL* **1979**, 849.
36. (a) Doyle, M. P.; West, C. T.; Donnelly, S. J.; McOsker, C. C. *JOM* **1976**, *117*, 129. (b) Dailey, O. D., Jr. *JOC* **1987**, *52*, 1984. (c) Krause, G. A.; Molina, M. T. *JOC* **1988**, *53*, 752. (d) Gil, J. F.; Ramón, D. J.; Yus, M. *T* **1993**, *49*, 4923.
37. Wustrow, D. J.; Smith, W. J., III; Wise, L. D. *TL* **1994**, *35*, 61.
38. (a) Kotsuki, H.; Ushio, Y.; Kadota, I.; Ochi, M. *CL* **1988**, 927. (b) Ishihara, K.; Mori, A.; Yamamoto, H. *T* **1990**, *46*, 4595.
39. (a) Tsunoda, T.; Suzuki, M.; Noyori, R. *TL* **1979**, 4679. (b) Kato, J.; Iwasawa, N.; Mukaiyama, T. *CL* **1985**, *6*, 743. (c) Kira, M.; Hino, T.; Sakurai, H. *CL* **1992**, 555.
40. (a) Bennek, J. A.; Gray, G. R. *JOC* **1987**, *52*, 892. (b) Sassaman, M. B.; Kotian, K. D.; Prakash, G. K. S.; Olah, G. A. *JOC* **1987**, *52*, 4314.
41. (a) Mukaiyama, T.; Ohno, T.; Nishimura, T.; Han, J. S.; Kobayashi, S. *CL* **1990**, 2239. (b) Mukaiyama, T.; Ohno, T.; Nishimura, T.; Han, J. S.; Kobayashi, S. *BCJ* **1991**, *64*, 2524.
42. (a) Oriyama, T.; Iwanami, K.; Tsukamoto, K.; Ichimura, Y.; Koga, G. *BCJ* **1991**, *64*, 1410. (b) Oriyama, T.; Ichimura, Y.; Koga, G. *BCJ* **1991**, *64*, 2581.
43. Yokozawa, T.; Nakamura, F. *Makromol. Chem., Rapid Commun.* **1993**, *14*, 167.
44. (a) Fry, J. L. *CC* **1974**, 45. (b) Fry, J. L.; Ott, R. A. *JOC* **1981**, *46*, 602.
45. Nakayama, J.; Yoshida, M.; Simamura, O. *T* **1970**, *26*, 4609.
46. Mehta, A.; Jaouhari, R.; Benson, T. J.; Douglas, K. T. *TL* **1992**, *33*, 5441.
47. (a) Doyle, M. P.; McOsker, C. C.; West, C. T. *JOC* **1976**, *41*, 1393. (b) Parnes, Z. N.; Romanova, V. S.; Vol'pin, M. E. *JOU* **1988**, *24*, 254.

James L. Fry
The University of Toledo, OH, USA

Trifluoroperacetic Acid[1]

[359-48-8] $C_2HF_3O_3$ (MW 130.03)

(electrophilic reagent capable of reacting with many functional groups; delivers oxygen to alkenes, arenes, and amines;[1] useful reagent for Baeyer–Villiger oxidation of ketones[27,44])

Alternate Names: TFPAA; peroxytrifluoroacetic acid.
Solubility: sol CH_2Cl_2, dichloroethane, ether, sulfolane, acetonitrile.
Form Supplied in: not available commercially.
Analysis of Reagent Purity: assay using iodometry.[2]
Preparative Methods: the preparation and handling of TFPAA should be carried out behind a safety shield. A mixture of ***Trifluoroacetic Anhydride*** (46.2 g; 0.22 mole) and CH_2Cl_2 (50 mL) is cooled with stirring in an ice bath. 90% H_2O_2 (caution: for hazards see ***Hydrogen Peroxide***) (5.40 mL, 0.20 mol) is added in 1 mL portions over a period of 10 min. When the mixture has become homogeneous, it is allowed to warm to rt and then again cooled to 0 °C.[3] TFPAA prepared from 30% aqueous H_2O_2 and ***Trifluoroacetic Acid*** has been used for some reactions.[4–6] Hydrogen peroxide of high concentration (70%) is not widely available due to hazards involved in handling, storage, and transportation. The commercially available ***Hydrogen Peroxide–Urea*** (UHP) system, which is safe to handle, has been introduced recently as a substitute for anhydrous H_2O_2 in the preparation of TFPAA.[2,7,8]
Purification: in the preparation of TFPAA, a slight excess of trifluoroacetic anhydride is used to ensure that no water is present in the reagent. The reaction between H_2O_2 and trifluoroacetic anhydride is very fast; the reagent is ready for use after the reactants have been mixed and the solution has become homogeneous. No special purification steps are employed. Suitable buffers (Na_2CO_3, Na_2HPO_4) are used to neutralize the highly reactive and strongly acidic trifluoroacetic acid which is present along with TFPAA in the reagent.
Handling, Storage, and Precautions: the reagent can be stored at −20 °C for several weeks[9] and exhibits no loss in active oxygen content after 24 h in refluxing CH_2Cl_2.[40] However, since it can be prepared in a short time, the usual practice is to prepare the reagent when needed. Note that solutions of TFPAA in CH_2Cl_2 can lose activity by evaporation of the volatile peracid.[41] Since peroxy acids are potentially explosive, care is required while carrying out the reactions and also during workup of the reaction mixture. Solvent removal from excess H_2O_2–CF_3CO_2H experiments can result in explosions; the peroxide must be destroyed by addition of MnO_2 (until a potassium iodide test is negative) before solvent removal.[10a] For a further discussion of safety, see Luxon.[10b] This reagent should only be handled in a fume hood.

General Considerations. Trifluoroperacetic acid oxidizes simple alkenes, alkenes carrying a variety of functional groups (such as ethers, alcohols, esters, ketones, and amides), aromatic compounds, alkanes,[11] amines and *N*-heterocycles. Ketones undergo oxygen insertion reactions (Baeyer–Villiger oxidation).

Epoxidations of Alkenes. Due to the presence of the strongly electron withdrawing CF_3 group, TFPAA is the most powerful organic peroxy acid and as such is more reactive than performic[21] or 3,5-dinitroperbenzoic acids.[41] It reacts readily even with electron-poor alkenes to furnish the corresponding epoxides (see ***m-Chloroperbenzoic Acid***).

Trifluoroacetic acid is a strong acid which opens epoxides readily.[12,44] Since TFPAA is a much weaker acid than trifluoroacetic acid (pK_a 3.7 vs. 0.3), the latter reagent can be selectively neutralized with Na_2CO_3 or Na_2HPO_4, leading to the isolation of epoxides in high yields. When the substrate is highly reactive, Na_2CO_3 is used as buffer; when the substrate reacts sluggishly, Na_2HPO_4 is used as buffer.[12] The TFPAA reagent is rapidly decomposed by Na_2CO_3.

Since monosubstituted alkenes are not electron rich, they react sluggishly with the standard organic peroxy acids. By contrast, the monosubstituted alkene 1-pentene (**1**) is epoxidized efficiently by TFPAA (eq 1).[12] TFPAA prepared from 0.3 mol of 90% H_2O_2 and 0.36 mol of trifluoroacetic anhydride in CH_2Cl_2 is added during 30 min to a stirred mixture of (**1**) (0.2 mol), Na_2CO_3 (0.9 mol), and CH_2Cl_2 (200 mL). Since the alkene is volatile the reaction flask is fitted with an efficient ice water-cooled condenser. The reaction mixture boils during the addition of the peracid. After all the reagent has been added, the reaction mixture is heated under reflux for 30 min, cooled, and the insoluble salts are removed by centrifugation. The salt is thoroughly washed with CH_2Cl_2. Fractional distillation of the combined CH_2Cl_2 extracts furnishes the epoxide (**2**) in 81% yield.

(**1**) → [1.5 equiv TFPAA, CH_2Cl_2; 4.5 equiv Na_2CO_3, reflux, 30 min; 81%] → (**2**) (1)

The alkene (**3**), which is resistant to epoxidation by *m*-CPBA or ***Peracetic Acid***, has been epoxidized with TFPAA to furnish in 83% yield a mixture of esters (**4**) and (**5**) (eq 2).[13] Esters (**4**) and (**5**) undergo facile deacylation when chromatographed on silica gel to furnish alcohols (**6**) and (**7**).

(**3**) R = H → [5 equiv TFPAA; Na_2HPO_4, rt, 4 h; 83%] → (**4**) R = $COCF_3$; (**6**) R = H + (**5**) R = $COCF_3$; (**7**) R = H (2)

Epoxidation of allyldiphenylphosphine oxide (**8**) with TFPAA furnishes in quantitative yield the corresponding epoxide, 2-(diphenylphosphinoylmethyl)oxirane; *m*-CPBA epoxidation of (**8**) furnishes the epoxide in only 56% yield.[14] Epoxide (**9**) is obtained in 80% yield through regio- and stereoselective epoxidation of the corresponding alkene with TFPAA in CH_2Cl_2 in the presence of Na_2HPO_4 buffer.[15]

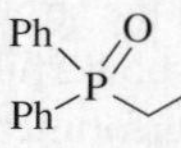

(8)

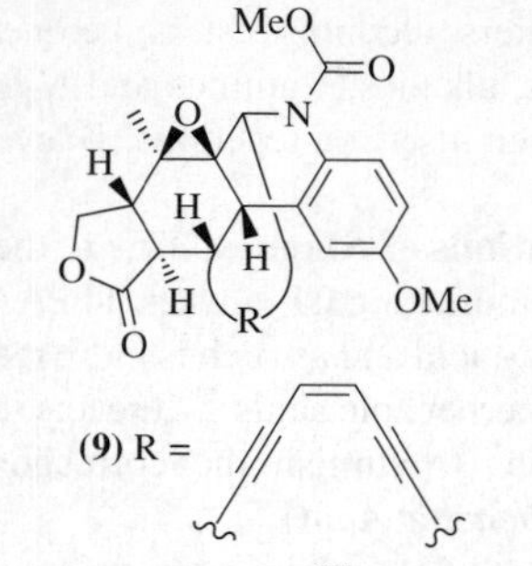

(9) R =

(Z)

The tertiary amine of (**10**) is expected to react more readily than the disubstituted double bond on treatment with an organic peracid. Selective epoxidation of the double bond in (**10**) was achieved by initially treating it with CF_3CO_2H. This led to salt formation due to protonation of the amine. Epoxidation of the salt with TFPAA and subsequent workup furnished the epoxide (**11**) (eq 3).[16]

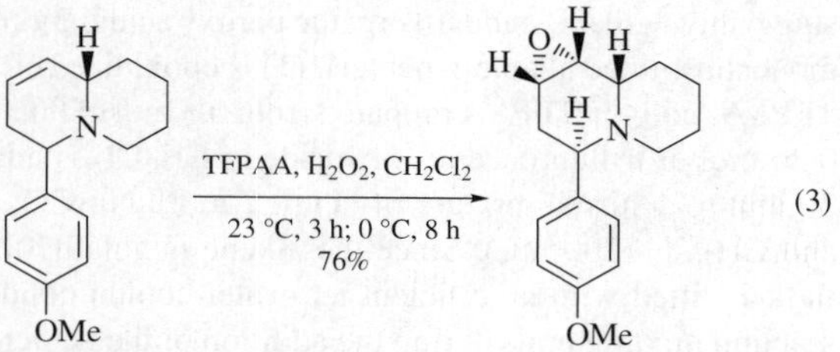

TFPAA, H_2O_2, CH_2Cl_2; 23 °C, 3 h; 0 °C, 8 h; 76% (3)

(10) (11)

Alkenes have been epoxidized efficiently employing TFPAA prepared by the UHP method (eq 4).[2]

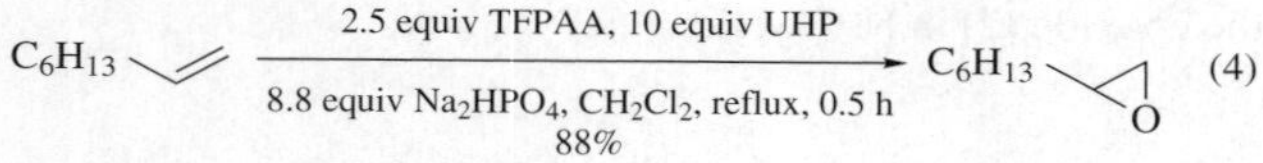

C_6H_{13} — 2.5 equiv TFPAA, 10 equiv UHP; 8.8 equiv Na_2HPO_4, CH_2Cl_2, reflux, 0.5 h; 88% → C_6H_{13} (4)

α,β-Unsaturated esters and α,β-unsaturated ketones are resistant to epoxidation by organic peracids since the double bonds are not electron rich; however, these compounds can be epoxidized by TFPAA. 1-Acetylcyclohexene[17] and methyl methacrylate[12] furnish the corresponding epoxides in 50% and 84% yields, respectively, when treated with TFPAA/Na_2HPO_4 in CH_2Cl_2 (reflux for about 0.5 h). The α,β-unsaturated ester (**12**) has been epoxidized stereoselectively by TFPAA (eq 5).[18] With *m*-CPBA, this epoxidation requires a higher reaction temperature which results in the formation of a complex mixture.

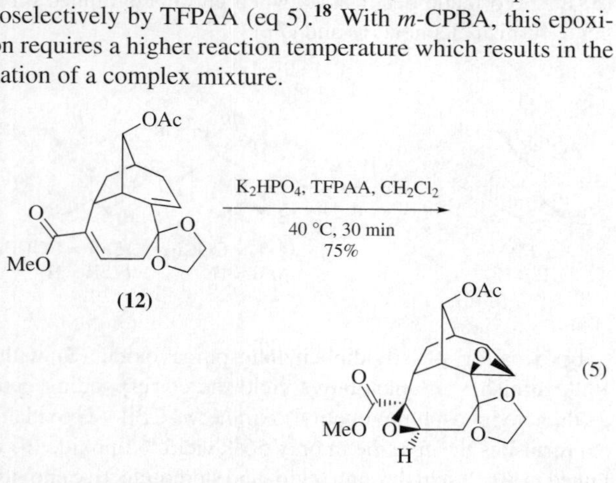

K_2HPO_4, TFPAA, CH_2Cl_2; 40 °C, 30 min; 75% (5)

(12)

With organic peracids, allyl alcohols form hydrogen bonds involving the hydrogen of the alcohol, as in (**13**).[19] Ganem has suggested that, with TFPAA, allylic ethers form hydrogen bonds involving the hydrogen of the peracid (**14**).

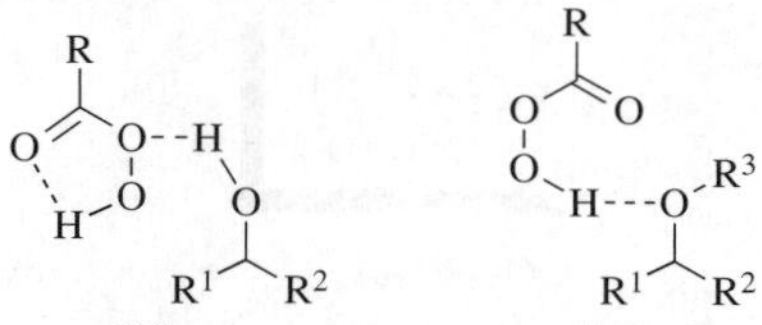

(13) (14)

Epoxidation of (**15**) having an allylic ether substituent axially oriented is *syn* selective (*syn:anti* epoxidation = 12.4:1) (eq 6);[19] this selectivity is due to the formation of the hydrogen bond of the type shown in (**14**). The stereoselectivity in the epoxidation of (**15**) is solvent dependent. When (**15**) is epoxidized in THF (which disrupts hydrogen bonding) the ratio of *syn:anti* epoxides obtained is 1:12. The epoxidation of the allyl alcohol (**16**) with TFPAA is highly *syn* selective (*syn:anti* epoxidation = 100:1); the *syn* selectivity in the epoxidation of (**16**) with *m*-CPBA is much less (*syn:anti* epoxidation = 5.2:1).

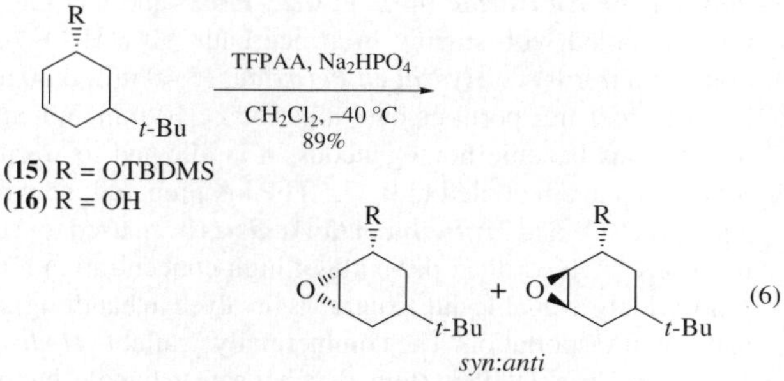

TFPAA, Na_2HPO_4; CH_2Cl_2, –40 °C; 89%

(**15**) R = OTBDMS
(**16**) R = OH

syn:anti (6)

The diol (**17**) is epoxidized stereoselectively to furnish (**18**) (eq 7).[20]

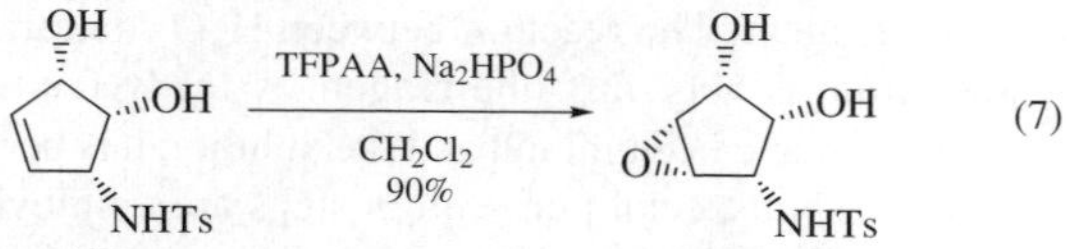

TFPAA, Na_2HPO_4; CH_2Cl_2; 90% (7)

(17) (18)

Oxidation of Alkenes to Diols and Ketones. Alkenes react readily with a CF_3CO_3H/CF_3CO_2H mixture to furnish hydroxy trifluoroacetates, e.g. (**19**) → (**20**) (eq 8).[21] In this reaction, high molecular weight byproducts are formed due to the condensation of hydroxy trifluoroacetates with the epoxides formed from alkenes. The formation of the byproduct can be avoided by adding triethylammonium trifluoroacetate. After the formation of the glycol ester is complete, the solvent is evaporated under reduced pressure and the crude ester is subjected to methanolysis to furnish the vicinal diol (**21**). α,β-Unsaturated esters are also hydroxylated by this procedure.

The allyl alcohol (**22**) reacted readily with TFPAA to furnish the 1,3-dioxolane (**23**) (eq 9).[8] This reaction could not be carried out with *m*-CPBA even in refluxing ethylene dichloride. The homoallyl alcohol (**22**) (R^1 = H, R^2 = OH) was reacted with TFPAA prepared from commercially available urea–hydrogen peroxide; the major product formed was the dioxolane (**23**) (R^1 = H, R^2 = OH).

$C_{10}H_{21}$ (19) — $Et_3NH^+CF_3CO_2^-$, TFPAA, CF_3CO_2H / CH_2Cl_2, add over 30 min → [(20)] — MeOH, HCl, 95% → (21) (8)

(22) $R^1 = OH, R^2 = H$ — TFPAA, Na_2CO_3 / CH_2Cl_2, 0 °C, >39% → (23) $R^1 = OH, R^2 = H$ (9)

(±)-Allosamizoline (**25**) has been synthesized from the (dimethylamino)oxazoline (**24**).[22] 5.4 M TFPAA in CF_3CO_2H is added carefully to (**24**) at 0 °C. The reaction mixture is evaporated in vacuum and the resulting mixture of epoxides is solvolyzed by heating with 10% aqueous CF_3CO_2H at 40 °C. Hydrogenolysis (Pd/C, H_2, MeOH) of the solvolysis product furnishes pure (±)-(**25**) (overall yield 67%) and the epoxide (**26**) (yield 16%).

(24) (25) (26)

Epoxidation of sterically congested alkenes occurs with TFPAA under basic conditions (eq 10).[45]

TFPAA, 30% H_2O_2 / Na_2HPO_4, CH_2Cl_2, reflux, 5 h (10)

Treatment of tetrasubstituted alkenes with TFPAA/BF_3 furnishes ketones via rearrangement. 1,2-Dimethylcyclohexene has been transformed to the ketone (**27**) (eq 11);[23] the reagents TFPAA and 47% ***Boron Trifluoride Etherate*** are added simultaneously.

1. TFPAA, CH_2Cl_2; 2. 47% $BF_3\cdot Et_2O$, 0–8 °C, add over 20 min, 76% → (27) (11)

Arene Oxidation. Arenes are exhaustively oxidized to aliphatic carboxylic acids. Heteroaromatic systems, such as pyridine, quinoline, and dibenzothiophene, are quantitatively oxidized to their *N*-oxides and sulfone rather than undergo ring oxidation. The heteroatom oxidation deactivates the ring towards electrophilic attack by TFPAA.[6] Benzene undergoes direct catalytic oxidation to phenyl trifluoroacetate using a TFPAA/Co^{III} reagent.[24]

With BF_3. The combination TFPAA/***Boron Trifluoride*** is a potent electrophilic oxidant for π-systems.[46] As a source of positive hydroxyl, it is used to convert aromatics into cyclohexadienones (eq 12)[26a] and phenols,[25] and alkenes into ketones (eq 13).[26b] See also eq 11 above.

TFPAA, BF_3 / CH_2Cl_2, 0 °C, 1.25 h (12)

R = Me, 93%
R = Et, 82%

TFPAA, CH_2Cl_2, reflux / $BF_3\cdot Et_2O$, 35 min (13)

R^1	R^2	R^3	Yield
Me	Me	Me	75%
Me	Me	H	53%
Et	Me	H	70%
Me	Cl	Me	77%

Baeyer–Villiger Oxidation. On treatment with organic peroxy acids, ketones undergo oxygen insertion reactions to furnish esters (see ***m-Chloroperbenzoic Acid***).[44] This reaction, known as the Baeyer–Villiger rearrangement, has several applications and has been reviewed recently.[27] When carrying out this oxidation with TFPAA, Na_2HPO_4 buffer is added to prevent the reaction between trifluoroacetic acid and the Baeyer–Villiger product. The ketone (**28**) reacts with TFPAA to furnish brassinolide tetracetate (**29**) (eq 14).[28] The migration of C-7 rather than C-5 carbon in this oxidation is due to the effect of the acetate groups at C-2 and C-3. A systematic study of the Baeyer–Villiger reaction of 5α-cholestan-6-ones having substituents at C-1, C-2, and C-3 has been carried out.[29]

(28) — TFPAA, CH_2Cl_2 / Na_2HPO_4, 0 °C, 3 h, 85% → (29) (14)

The oxidations of the ketone (**30**) and α-tetralone (**31**) have been reported (eqs 15 and 16).[30,2] Epimerization of α-substituents is generally not observed when ketones are oxidized with buffered TFPAA.[42]

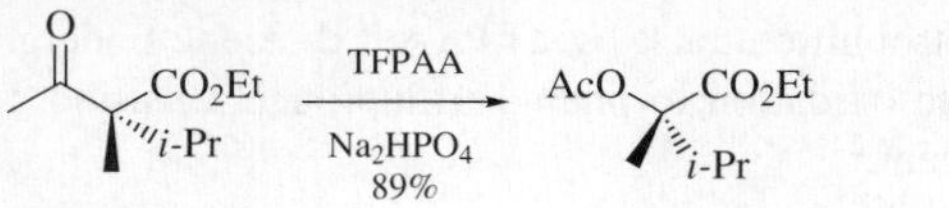

(−)-(R) (30) (S) (15)

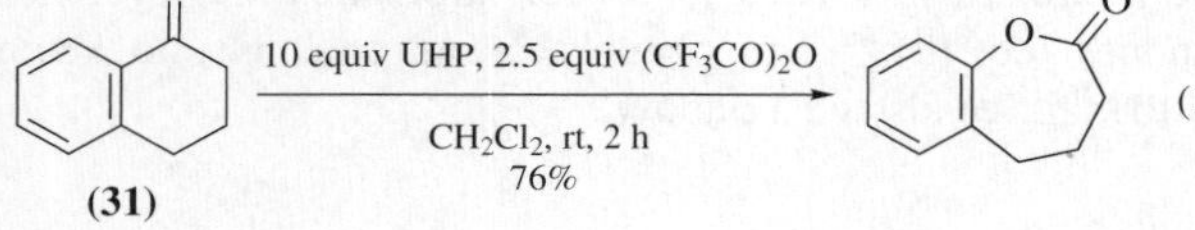

(31) (16)

Complete stereospecificity and high regioselectivity (25:1) is observed in the oxidation of an *erythro* ketone (eq 17). Oxidation of the *threo* ketone is also stereospecific but gives a 5:3 mixture of ester regioisomers.[47]

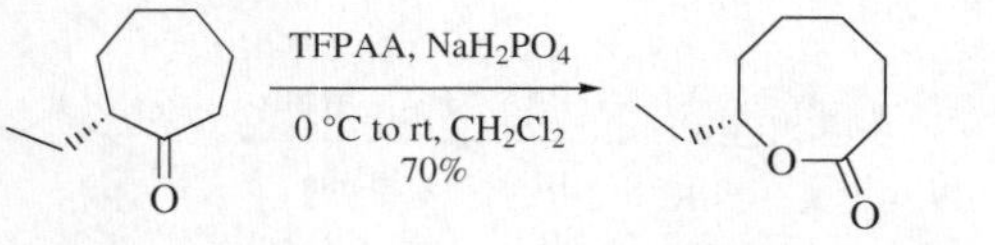

(17)

>94% ee

Heteroatom Oxidations. Aromatic primary amines carrying electron-withdrawing groups are oxidized efficiently by TFPAA to the corresponding nitro compounds (eq 18).[21,31] The amine dissolved in CH_2Cl_2 is added to the peracid. The above oxidation cannot be carried out with aromatic amines such as *p*-anisidine, which are unusually sensitive to electrophilic attack; for these sensitive amines, peracetic acid is the preferred oxidant.

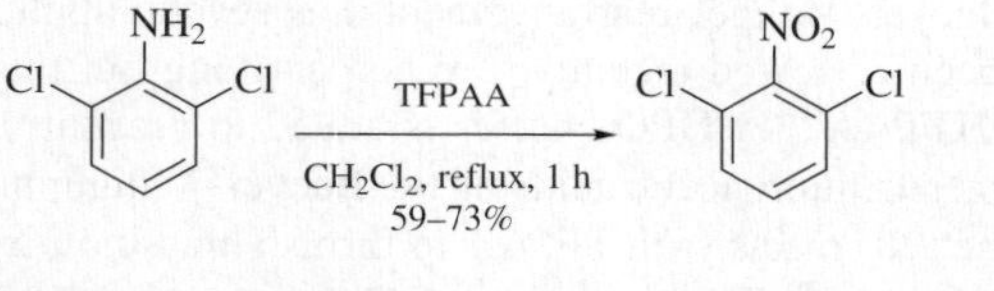

(18)

Oxidation of 2,3,4,5,6-pentachloroaniline with TFPAA in $CHCl_3$–water at rt furnishes, in 78% yield, 2,3,4,5,6-pentachloronitrosobenzene.[32] The electron-deficient heterocycle (**32**) furnishes the *N*-oxide (**33**) on oxidation with TFPAA prepared from urea–hydrogen peroxide (eq 19).[7] Electron-deficient pyridines are oxidized to the corresponding *N*-oxides with TFPAA; perbenzoic and peracetic acid are not effective for this transformation.[43]

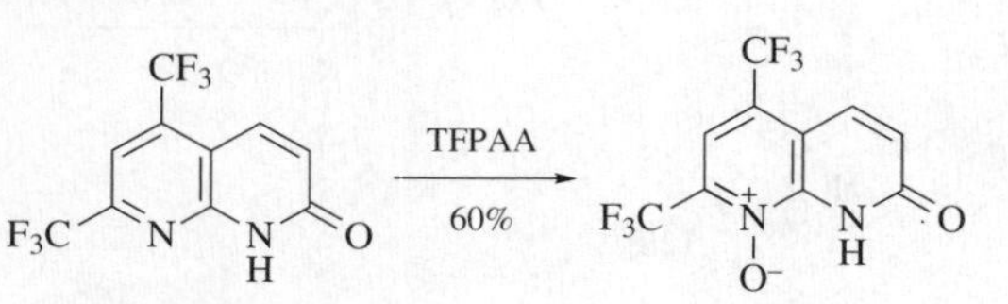

(32) (33) (19)

Oxidation of the isoxazoline (**34**) furnishes the hydroxy ester (**35**) (eq 20) via an initial oxaziridine intermediate.[33]

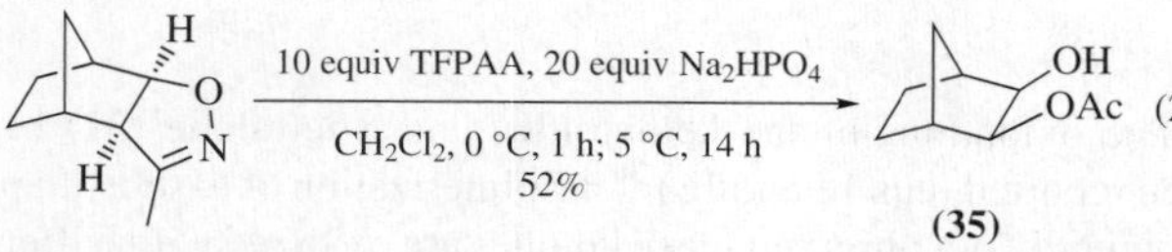

(34) (35) (20)

Nitro compounds have many applications in organic chemistry.[34] Strained polynitro polycyclic compounds are of interest as a new class of energetic materials.[35] Since oximes are readily available, their oxidation to nitro compounds has been studied. Oxidation of the oxime (**36**) furnishes a mixture of nitro compounds; the major component is the *cis* isomer (eq 21).[36] During the oxidation of oximes, ketones are obtained as byproducts. Hindered oximes such as camphor oxime are not oxidized by TFPAA.

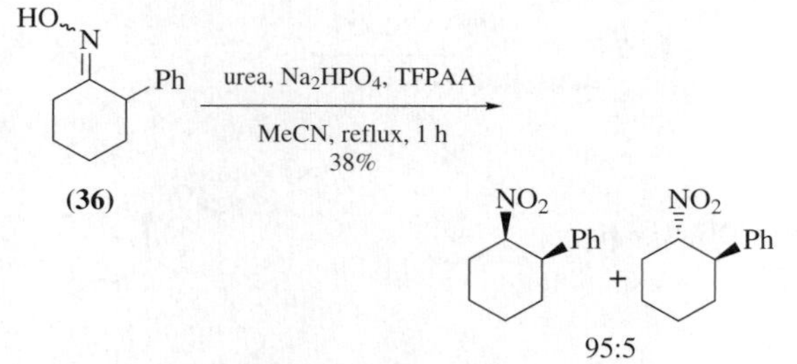

(36) (21)

95:5

Oximes yield primary, secondary, and alicyclic nitroalkanes (72%),[48] and α-chloro ketoximes give α-nitroalkenes (31–66%).[49]

Oxidation of the oxime (**37**) furnishes a mixture of *endo,endo* and *exo,exo* isomers (eq 22).[35b] Oximes have been converted to nitro compounds using a multistep method.[35a] ***Sodium Perborate*** in glacial acetic acid oxidizes oximes to nitro compounds.[37]

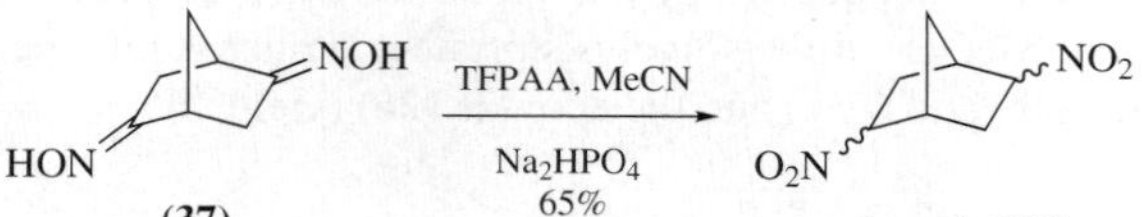

(37) (22)

endo,endo 90%
exo,exo 10%

α-Unsubstituted α,β-epoxy ketoximes are oxidized to γ-hydroxy-α-nitroalkenes (eq 23).[38] Aldoximes are oxidized to nitroalkanes (60–80%) with the reagent prepared from urea–H_2O_2 and trifluoroacetic anhydride. Ketoximes fail to react with this reagent system.[50]

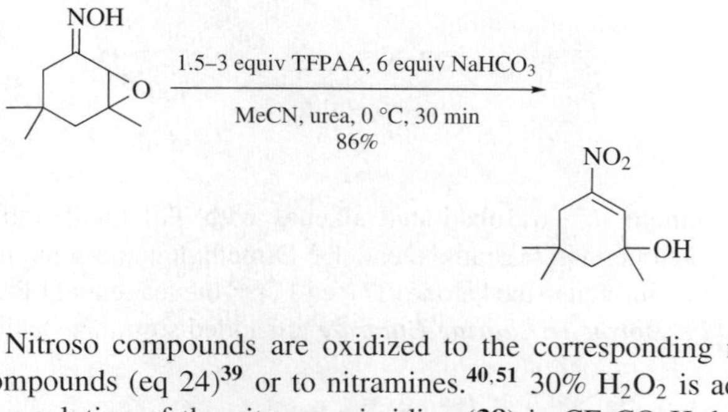

(23)

Nitroso compounds are oxidized to the corresponding nitro compounds (eq 24)[39] or to nitramines.[40,51] 30% H_2O_2 is added to a solution of the nitrosopyrimidine (**38**) in CF_3CO_3H during 1.5 h. After workup the nitro compound (**39**) is obtained in high yield; in this reaction, oxidative hydrolytic desulfurization is observed.

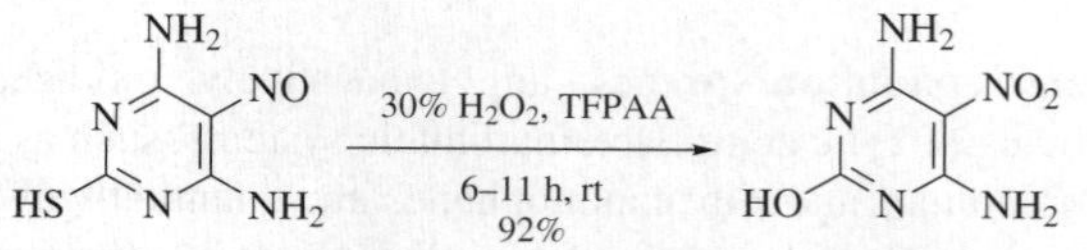

(38) (39) (24)

Miscellaneous Reactions. Aromatic azines are oxidized to their azine monoxides with TFPAA.[52] Organosulfides can be oxidized by TFPAA to either sulfoxides or sulfones under mild conditions in high yield.[5,53]

Related Reagents. See Classes O-8, O-9, O-11, O-14, O-15 and O-16, pages 1–10. *m*-Chloroperbenzoic Acid; Hydrogen Peroxide–Urea; Peracetic Acid; Perbenzoic Acid.

1. (a) Swern, D. *Organic Peroxides*; Wiley: New York, 1956; Vol. 2, pp 355–533. (b) Lewis, S. N. In *Oxidation*; Augustine, R. L., Ed.; Dekker: New York, 1969; Chapter 5, p 213. (c) Plesnicar, B. In *Oxidation in Organic Chemistry, Part C*; Trahanovsky, W. S., Ed.; Academic: New York, 1978; Chapter 3, p 211.
2. Cooper, M. S.; Heaney, H.; Newbold, A. J.; Sanderson, W. R. *SL* **1990**, 533.
3. Hart, H.; Lange, R. M.; Collins, P. M. *OSC* **1973**, *5*, 598.
4. Anastasia, M.; Allevi, P.; Ciuffreda, P.; Fiecchi, A.; Scala, A. *JOC* **1985**, *50*, 321.
5. Venier, C. G.; Squires, T. G.; Chen, Y.-Y.; Hussmann, G. P.; Shei, J. C.; Smith, B. F. *JOC* **1982**, *47*, 3773.
6. Liotta, R.; Hoff, W. S. *JOC* **1980**, *45*, 2887.
7. Eichler, E.; Rooney, C. S.; Williams, H. W. R. *JHC* **1976**, *13*, 41.
8. Ziegler, F. E.; Metcalf III, C. A.; Nangia, A.; Schulte, G. *JACS* **1993**, *115*, 2581.
9. Still, W. C.; Galynker, I. *T* **1981**, *37*, 3981.
10. (a) Deno, N. C.; Greigger, B. A.; Messer, L. A.; Meyer, M. D.; Stroud, S. G. *TL* **1977**, 1703. (b) *Hazards in the Chemical Laboratory*; Luxon, S. G., Ed.; Royal Society of Chemistry: Cambridge, 1992.
11. Deno, N. C.; Messer, L. A. *CC* **1976**, 1051.
12. Emmons, W. D.; Pagano, A. S. *JACS* **1955**, *77*, 89.
13. Holbert, G. W.; Ganem, B. *CC* **1978**, 248.
14. Grayson, J. I.; Warren, S.; Zaslona, A. T. *JCS(P1)* **1987**, 967.
15. Porco, Jr., J. A.; Schoenen, F. J.; Stout, T. J.; Clardy, J.; Schreiber, S. L. *JACS* **1990**, *112*, 7410.
16. Quick, J.; Khandelwal, Y.; Meltzer, P. C.; Weinberg, J. S. *JOC* **1983**, *48*, 5199.
17. Filler, R.; Camara, B. R.; Naqvi, S. M. *JACS* **1959**, *81*, 658.
18. Van Beek, G.; Van Der Bann, J. L.; Klumpp, G. W.; Bickelhaupt, F. *T* **1986**, *42*, 5111.
19. McKittrick, B. A.; Ganem, B. *TL* **1985**, *26*, 4895.
20. Trost, B. M.; Van Vranken, D. L. *JACS* **1991**, *113*, 6317.
21. Emmons, W. D.; Pagano, A. S.; Freeman, J. P. *JACS* **1954**, *76*, 3472.
22. Trost, B. M.; Van Vranken, D. L. *JACS* **1990**, *112*, 1261.
23. Hart, H.; Lerner, L. R. *JOC* **1967**, *32*, 2669.
24. Dicosimo, R.; Szabo, H. C. *JOC* **1986**, *51*, 1365.
25. Waring, A. J.; Hart, H. *JACS* **1964**, *86*, 1454.
26. (a) Hart, H.; Buehler, C. A. *JOC* **1964**, *29*, 2397. (b) Hart, H.; Lerner, L. R. *JOC* **1967**, *32*, 2669.
27. (a) Krow, G. R. *COS* **1991**, *7*, Chapter 5.1. (b) Krow, G. R. *OR* **1993**, *43*, 251.
28. Takatsuto, S.; Yazawa, N.; Ishiguro, M.; Morisaki, M.; Ikekawa, N. *JCS(P1)* **1984**, 139.
29. Takatsuto, S.; Ikekawa, N. *TL* **1983**, *24*, 917.
30. Wetter, H. *HCA* **1981**, *64*, 761.
31. (a) Pagano, A. S.; Emmons, W. D. *OS* **1969**, *49*, 47. (b) Pagano, A. S.; Emmons, W. D. *OSC* **1973**, *5*, 4557.
32. Ballester, M.; Riera, J.; Onrubia, C. *TL* **1976**, 945.
33. Park, P.-U.; Kozikowski, A. P. *TL* **1988**, *29*, 6703.
34. Kornblum, N. *OR* **1962**, *12*, 101.
35. (a) Marchand, A. P.; Sharma, R.; Zope, U. R.; Watson, W. H.; Kashyap, R. P.; *JOC* **1993**, *58*, 759. (b) Olah, G. A.; Ramaiah, P.; Surya Prakash, G. K. *JOC* **1993**, *58*, 763.
36. Sundberg, R. J.; Bukowick, P. A. *JOC* **1968**, *33*, 4098.
37. Olah, G. A.; Ramaiah, P.; Lee, C:-S.; Surya Prakash, G. K. *SL* **1992**, 337.
38. Takamoto, T.; Ikeda, Y.; Tachimori, Y.; Seta, A.; Sudoh, R. *CC* **1978**, 350.
39. Taylor, E. C.; McKillop, A. *JOC* **1965**, *30*, 3153.
40. Emmons, W. D. *JACS* **1954**, *76*, 3468.
41. Rastetter, W. H.; Richard, T. J.; Lewis, M. D. *JOC* **1978**, *43*, 3163.
42. (a) Clark, J. S.; Holmes, A. B. *TL* **1988**, *29*, 4333. (b) Carling, R. W.; Clark, J. S.; Holmes, A. B. *JCS(P1)* **1992**, 83.
43. Evans, R. F.; Van Ammers, M.; Den Hertog, H. J. *RTC* **1959**, *78*, 408.
44. Emmons, W. D.; Lucas, G. B. *JACS* **1955**, *77*, 2287.
45. Nakayama, J.; Sugihara, Y. *JOC* **1991**, *56*, 4001.
46. Hart, H. *ACR* **1971**, *4*, 337.
47. Levin, D.; Warren, S. *JCS(P1)* **1992**, 2155.
48. Emmons, W. D.; Pagano, A. S. *JACS* **1955**, *77*, 4557.
49. Sakakibara, T.; Ikeda, Y.; Sudoh, R. *BCJ* **1982**, *55*, 635.
50. Ballini, R.; Marcautoni, E.; Petrini, M. *TL* **1992**, *33*, 4835.
51. (a) Collins, P. M.; Oparaeche, N. N. *JCS(P1)* **1975**, 1695. (b) Collins, P. M.; Oparaeche, N. N.; Munasinghe, V. R. N. *JCS(P1)* **1975**, 1700.
52. Williams, W. M.; Dolbier, W. R., Jr. *JOC* **1969**, *34*, 155.
53. Saupe, T.; Krieger, C.; Staab, H. A. *AG(E)* **1986**, *25*, 451.

Kenneth C. Caster
Union Carbide Corporation, South Charleston, WV, USA

A. Somasekar Rao & H. Rama Mohan
Indian Institute of Chemical Technology, Hyderabad, India

Trimethylamine *N*-Oxide

[62637-93-8] C_3H_9NO (MW 75.13)

(reagent for conversion of alkyl halides and sulfonates to aldehydes and ketones; precursor for azomethine ylide useful for synthesis of *N*-methylpyrrolidines; reagent for decomplexation of tricarbonyldieneiron complexes)

Alternate Name: TMANO.

Physical Data: mp 213–214 °C (dec).[1] Various melting points have been reported, ranging from 208 °C[2] to 225–227 °C. The dihydrate, $Me_3NO \cdot 2H_2O$, is a white solid, mp 96 °C.

Solubility: sol H_2O, EtOH; insol Et_2O benzene, and other hydrocarbon solvents; sparingly sol hot $CHCl_3$.

Form Supplied in: the dihydrate is commercially available as a white solid.

Preparative Methods: aqueous ***Trimethylamine*** (100 mL, 33%) is mixed with a pure 3% aqueous solution of ***Hydrogen Peroxide*** (600 mL) and is set aside at rt for 24 h. The solution is evaporated to dryness, and the product is recrystallized from EtOH–Et_2O to give the dihydrate as needles, mp 96 °C.[3,4] Several procedures are available for preparing the anhydrous reagent. The dihydrate is heated at 105 °C and 15 mmHg for several hours, after which it is sublimed at 180–200 °C and 15 mmHg. A more convenient procedure is to remove water by azeotropic distillation with benzene or toluene (Dean–Stark trap), or by co-distillation from a suspension of the reagent in DMF, followed by solvent removal in vacuo.

Handling, Storage, and Precautions: no reported hazards for this reagent. The anhydrous material is hygroscopic and should be prepared and used fresh or stored at rt in a vacuum desiccator over P_2O_5. Trimethylamine *N*-oxide is an oxidizer and should be stored away from reducing agents.

Preparation of Aldehydes and Ketones. Alkyl bromides, iodides, and tosylates react with Me_3NO in $CHCl_3$ to give reasonable yields of aldehydes and ketones (eqs 1–3).[5]

$$RCH_2X \xrightarrow[50\text{–}60\ ^\circ C]{Me_3NO,\ CHCl_3} RCHO \quad (1)$$

R = C_5H_{11}, X = I, 31%; R = C_5H_{11}, X = Br, 48%
R = C_7H_{15}, X = I, 50%; R = C_7H_{15}, X = OTs, 55%

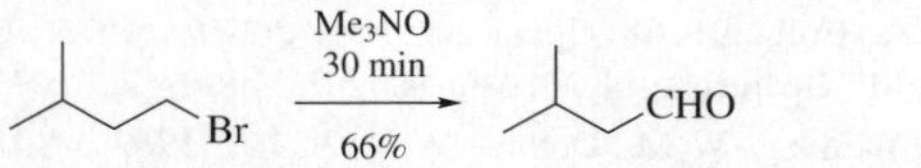

Me3NO, 30 min; 66% (2)

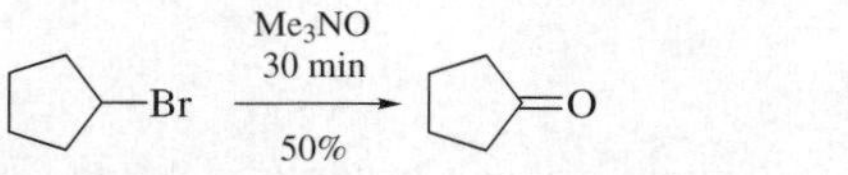

Me3NO, 30 min; 50% (3)

Oxidation of Organoboranes. Trimethylamine *N*-oxide is a useful alternative to alkaline ***Hydrogen Peroxide*** for the conversion of intermediates from the hydroboration of alkenes and alkynes to alcohols, especially in those cases where the product is sensitive to either high pH or H_2O_2. For example, cyclopropanols are reported to be unstable to alkaline H_2O_2, making hydroboration of cyclopropenes problematic. Oxidation of the intermediate boranes with Me_3NO in hot toluene, followed by methanolysis of the derived tricyclopropylborate, gives cyclopropanols in good overall yield (eq 4).[6]

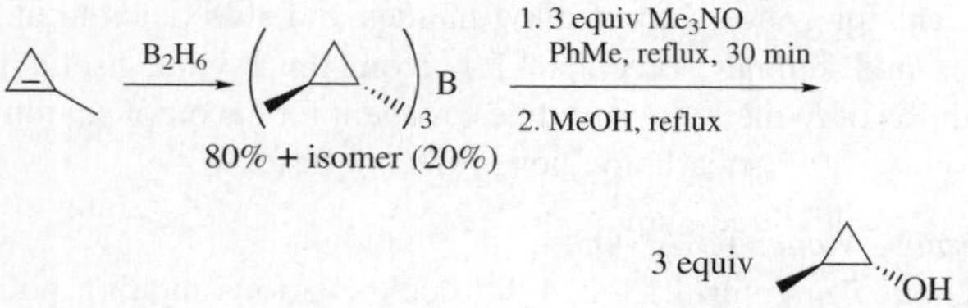

B_2H_6; 80% + isomer (20%); 1. 3 equiv Me_3NO, PhMe, reflux, 30 min; 2. MeOH, reflux; 3 equiv (4); 53% overall yield

Later studies by Kabalka and Hedgecock[7] have shown that commercially available Me_3NO dihydrate is an effective reagent for organoborane oxidation, and tolerates a wide variety of functional groups. Owing to the fact that the amine oxide is not appreciably soluble in organic solvents such as THF, toluene, and $CHCl_3$, it is recommended that the oxidation is carried out with vigorous stirring.[8] Rates of oxidation are sensitive to the nature of the alkyl group (secondary cycloalkyl > secondary alkyl > *n*-alkyl > β-branched primary alkyl), and the second and third alkyl groups are removed more slowly. Choice of solvent does not appear to affect the rate of oxidation; the recommended reaction conditions are diglyme at reflux. Two simple examples are given in eqs 5 and 6.

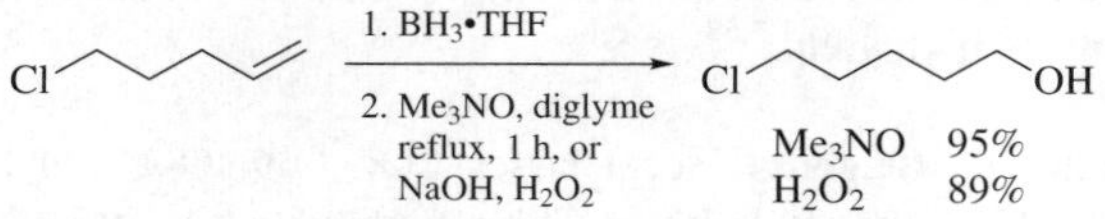

1. BH_3•THF; 2. Me_3NO, diglyme reflux, 1 h, or NaOH, H_2O_2 (5) 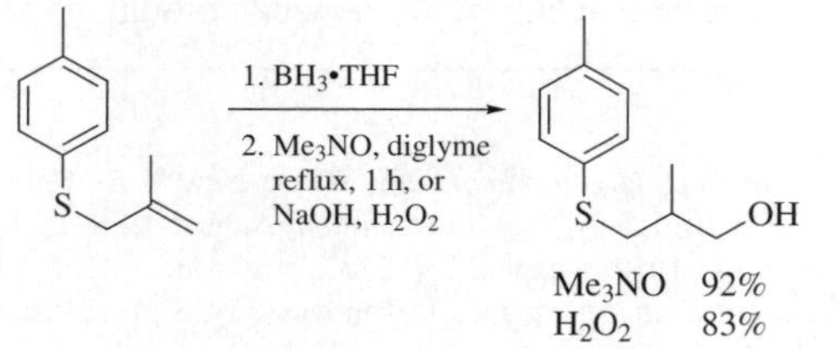
Me_3NO 95%; H_2O_2 89%

1. BH_3•THF; 2. Me_3NO, diglyme reflux, 1 h, or NaOH, H_2O_2 (6)
Me_3NO 92%; H_2O_2 83%

A useful example of the application of Me_3NO for borane oxidation is found in the formal total synthesis of (±)-perhydrohistrionicotoxin by Carruthers and Cumming (eq 7).[9] Oxidation of the borane intermediate from the azaspirocycle by using alkaline hydrogen peroxide proved extremely difficult and gave a low yield of the desired alcohol (34% as a mixture of stereoisomers), this being attributed to the formation of a stable aminoborane.[10] Carruthers overcame this problem by using Me_3NO in refluxing diglyme, to give a mixture of stereoisomers that could be separated by chromatography of their acetate derivatives. The alcohol shown in eq 7 is a known intermediate in a synthesis of perhydrohistrionicotoxin.[11]

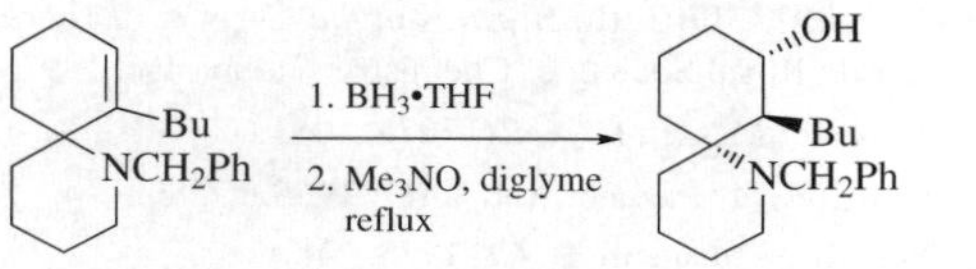

1. BH_3•THF; 2. Me_3NO, diglyme reflux (7)
55%, + stereoisomer 5%

Trimethylamine *N*-oxide is also useful for oxidation of vinylborane intermediates formed during the hydroboration of alkynes.[12] The use of hydrogen peroxide in such reactions leads to overoxidation (eq 8), and this is avoided by using Me_3NO in refluxing THF in a one-pot procedure to generate acylsilanes from silylalkynes (Scheme 1).

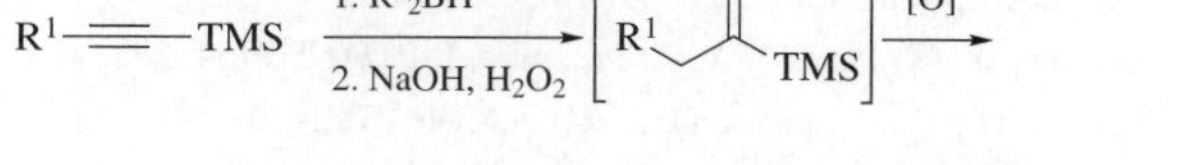

R^1—≡—TMS; 1. R^2_2BH; 2. NaOH, H_2O_2; [O]; $R^1CH_2CO_2H$ (8)

R—≡—TMS; BH_3•Me_2S, THF, 0–5 °C, 2 h; Me_3NO, reflux, 4 h; H_2O, rt

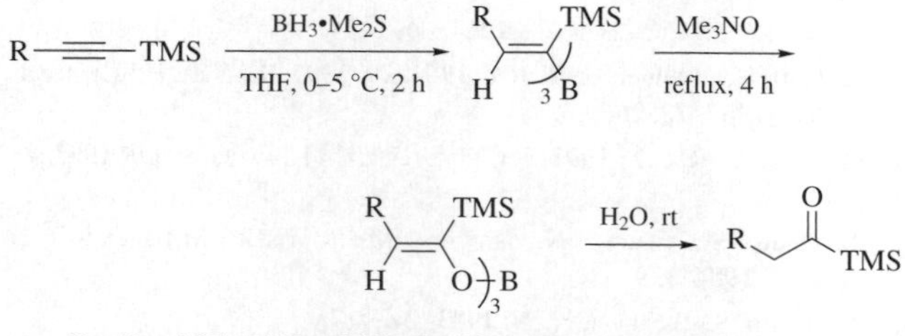

R = Bu, 86%; C_6H_{13}, 91%; Cy, 83%; *i*-Pr, 75%; $Cl(CH_2)_3$, 85%; $MeCH_2CH(OTMS)CH_2$, 76%; TMS, 75%

Scheme 1

Extension of this chemistry to the oxidation of organoaluminum derivatives has been described by Kabalka and Newton,[13] which is an extension of a reaction first reported by Köster.[14] It should be noted, however, that vinylalanes are not oxidized by this procedure; $R^1CH{=}CHAlR^2_2$ gives only the products of oxidation of the Al–R^2 bonds, with no aldehydes being detected. Examples of the procedure are shown in eq 9.

$$R_3Al \xrightarrow[\Delta]{3\ \text{equiv}\ Me_3NO} (RO)_3Al \xrightarrow{H^+,\ H_2O} 3\ ROH \quad (9)$$

R = Me, 21%; Pr, 100%; *i*-Bu, 94%; C_8H_{17}, 94%, Ph, 97%

Azomethine Ylides from Me_3NO: Synthesis of *N*-Methylpyrrolidines. Treatment of anhydrous Me_3NO with ***Lithium Diisopropylamide*** (LDA) gives an azomethine ylide which reacts in situ with alkenes in a dipolar cycloaddition to give *N*-methylpyrrolidines (eq 10).[15] Previous methods for generating azomethine ylides required electron-withdrawing or conjugatively stabilizing substituents.[16] Reaction of the azomethine ylide with dihydronaphthalenes has been described,[17] but the poor solubility of Me_3NO in THF was a problem for ylide generation, and it was found that a four-fold excess of the reagent was necessary to ensure complete consumption of all the alkene (eq 11).

$$Me_3NO \xrightarrow{LDA} \left[H_2C{=}\overset{+}{N}(Me){-}\overset{-}{C}H_2 \right] \xrightarrow[2.\ H_2O,\ w.u.]{1.\ alkene} \text{N-methylpyrrolidine } (R^1, R^2) \quad (10)$$

R^1 = H, R^2 = $(CH_2)_4Me$, 63%
R^1, R^2 = $(CH_2)_3$, 42%
R^1, R^2 = $(CH_2)_6$, 90%
R^1 = H, R^2 = Ph, 57%
R^1 = R^2 = Ph-*cis*, 62%
R^1 = R^2 = Ph-*trans*, 72%

(11)

R^1 = R^2 = OMe, R^3 = H, 86%
R^1 = NH_2, R^2 = R^3 = Cl, 81%
R^1 = NH_2, R^2 = H, R^3 = Me, 83%
R^1 = NH_2, R^2 = R^3 = H, 77%
R^1 = CN, R^2 = NH_2, R^3 = H, 50%

Decomplexation of Diene–$Fe(CO)_3$ Complexes. Me_3NO has been used extensively in organometallic chemistry as a reagent for the selective removal of carbonyl ligands. An extension of this reactivity leads to applications in the decomplexation of diene–iron tricarbonyl complexes, which are being used increasingly as intermediates for organic synthesis. The first report of Me_3NO being used for this type of ligand disengagement was by Shvo and Hazum,[18] summarized in eqs 12–15. The last two examples are particularly noteworthy, since the complexes are quite sensitive to the acidic conditions that are generated when transition metal oxidants such as ***Iron(III) Chloride*** or ***Cerium(IV) Ammonium Nitrate*** are used for decomplexation. Until the development of this method the removal of $Fe(CO)_3$ from diene complexes was considered a major obstacle to their synthetic application. The reaction conditions vary according to the diene, and must be determined empirically.

(12) Me_3NO, PhH; reflux, 1 h; 95%

(13) Me_3NO, PhH; reflux, 12 h; 71%

(14) Me_3NO, acetone; 25 °C, 24 h; 45% + 35% Ph_2NH

(15) Me_3NO, PhH; 65 °C, 12 h; 85%

Evidence that the mechanism of the reaction involves two amine oxide molecules per diene–$Fe(CO)_3$ comes from the isolation of an intermediate that has a carbonyl ligand replaced by Me_2NH (eq 16).[19a] A plausible mechanistic explanation is shown in Scheme 2, though this is somewhat different from the mechanism proposed for reaction of Me_3NO with $Fe(CO)_5$.[19b]

(16) $Fe(CO)_3$ complex $\xrightarrow{Me_3NO}$ $\overset{-}{Fe}(CO)_2\overset{+}{N}HMe_2$ complex $\xrightarrow{Me_3NO}$ diene

Scheme 2

A large number of examples of the use of Me_3NO for decomplexation can now be found in the synthetic literature. A comparison with iron(III) chloride is revealed by the decomplexation shown in eq 17.[20] While the methyl-substituted complex (X = Me) could be demetalated using $FeCl_3$ or Me_3NO, the use of $FeCl_3$ is problematic when X = OR, SPh, or NHR, owing to the propensity of these complexes to undergo acid-promoted scission of the C–X bond to regenerate the dienyl–$Fe(CO)_3$ cation. Me_3NO is non-acidic and does not suffer this limitation.

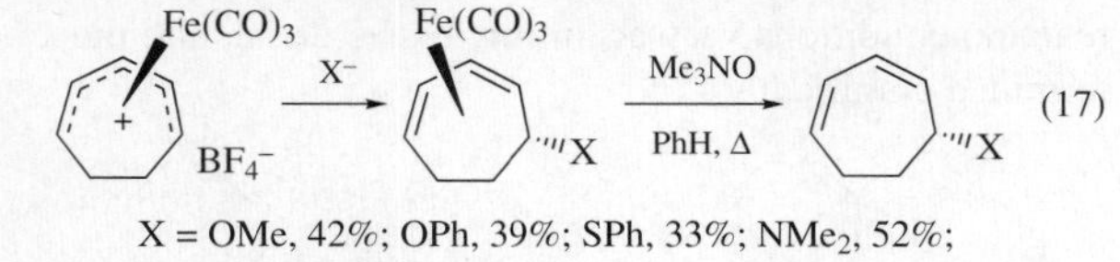

(17)

X = OMe, 42%; OPh, 39%; SPh, 33%; NMe_2, 52%; NH-*t*-Bu, 72%; Me, 48%

Similar compatibility with heteroatom substituents is observed during the synthesis of (−)-gabaculine reported by Birch et al.,[21] who also found that *N,N*-dimethylacetamide is an excellent solvent for decomplexation (Scheme 3), by Pearson and Ham[10] in their formal synthesis of perhydrohistrionicotoxin (eq 18), and by Pearson and Srinivasan in some approaches to cycloheptadiene triols (eqs 19 and 20).[22] The other useful aspect of Me_3NO for these decomplexations is that vinyl ether groups are not hydrolyzed to ketones under the reaction conditions.

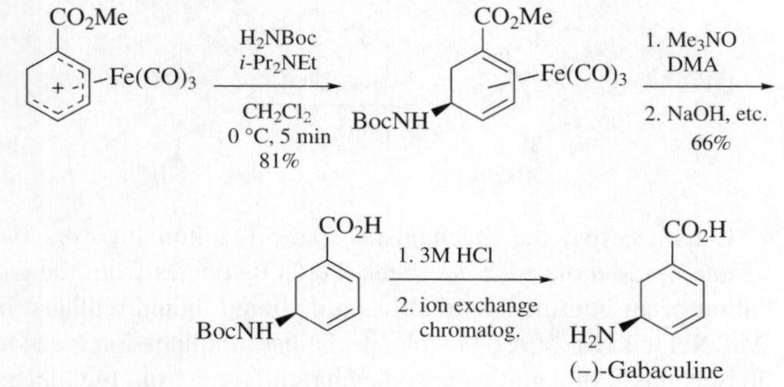

H_2NBoc, *i*-Pr_2NEt, CH_2Cl_2, 0 °C, 5 min, 81%; 1. Me_3NO, DMA; 2. NaOH, etc., 66%; 1. 3M HCl; 2. ion exchange chromatog.; (−)-Gabaculine

Scheme 3

(18) Me_3NO; H_2SO_4, H_2O, THF

(19) Me_3NO, DMA, 0 °C, 6 h, 52%

(20) Me_3NO, acetone, 23 °C, 2 h, 91%

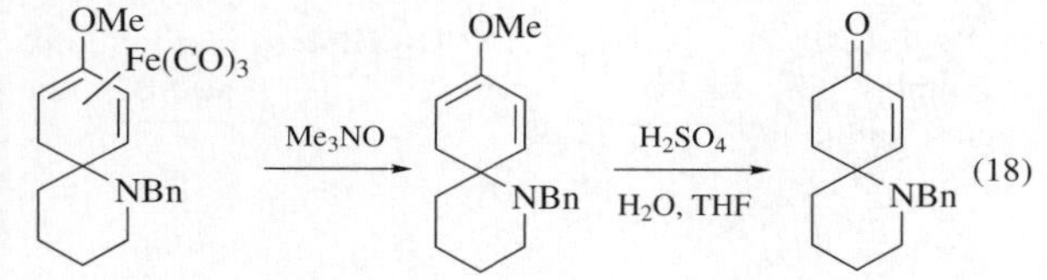

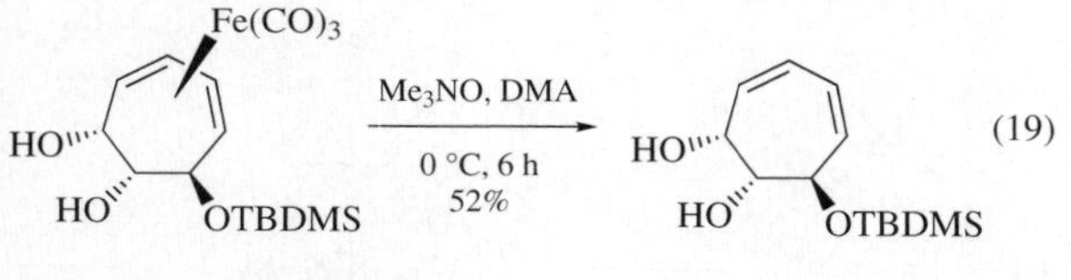

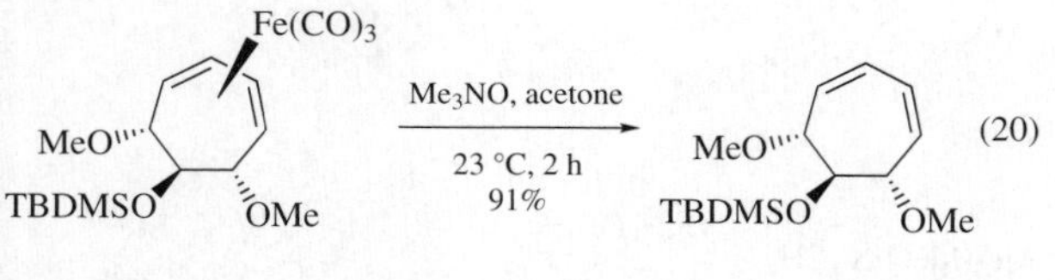

Owing to the fact that the decomplexation reaction proceeds via nucleophilic attack on a carbonyl ligand, Me_3NO is not very useful for converting diene–$Fe(CO)_2PR_3$ complexes to dienes, since the poorer π-acceptor capacity of the phosphine ligand places more electron density at the carbonyls. For this reason, it has been found better to use ***Dipyridine Chromium(VI) Oxide*** in CH_2Cl_2 for decomplexation of the phosphine-substituted derivatives (eq 21).[23]

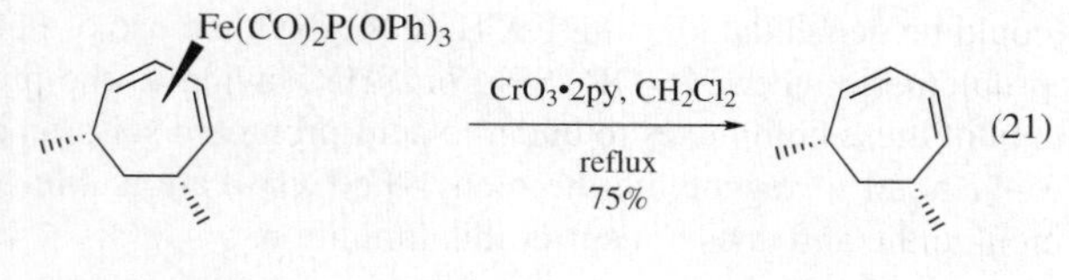

(21) CrO_3•2py, CH_2Cl_2, reflux, 75%

Shvo and Hazum have also reported the use of Me_3NO as promoter during the conversion of dienes to diene–$Fe(CO)_3$ complexes.[24] Reaction of $Fe(CO)_5$ with Me_3NO at controlled temperature leads to the formation of a coordinatively unsaturated '$Fe(CO)_4$' derivative by a mechanism related to that shown in Scheme 2. In the presence of a diene, rapid coordination of alkene occurs, followed by a second CO loss and alkene coordination to give the diene complex (eqs 22 and 23). It should be noted that the normal conditions for $Fe(CO)_5$ → diene–$Fe(CO)_3$ conversion are refluxing in di-*n*-butyl ether (ca. 140 °C).

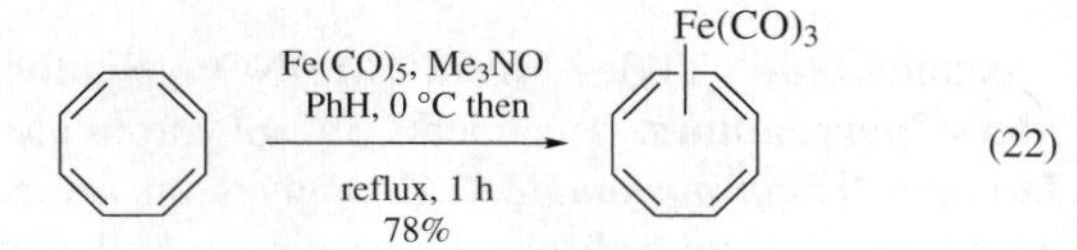

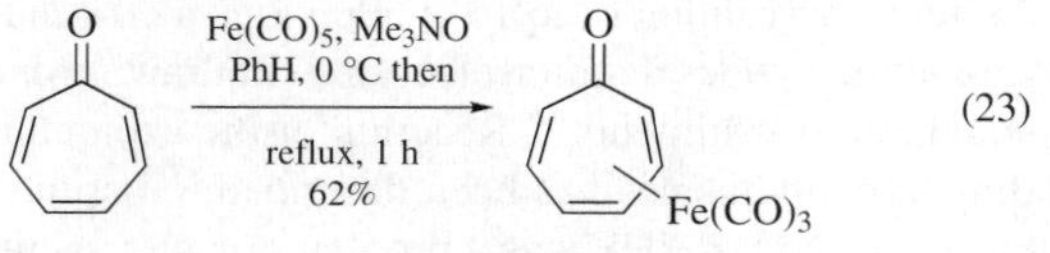

(22) $Fe(CO)_5$, Me_3NO, PhH, 0 °C then reflux, 1 h, 78%

(23) $Fe(CO)_5$, Me_3NO, PhH, 0 °C then reflux, 1 h, 62%

This observation, coupled with the fact that diene–$Fe(CO)_2PR_3$ complexes are quite resistant to Me_3NO, leads to a mild method for the conversion of diene–$Fe(CO)_3$ to diene–$Fe(CO)_2PR_3$ complexes (eqs 24–26).[22,25,26] The examples in eq 24 (R^1 or R^2 = OMe) are especially interesting because other methods for ligand exchange (thermal or photochemical) with these electron-rich diene complexes either fail completely or give very low yields.

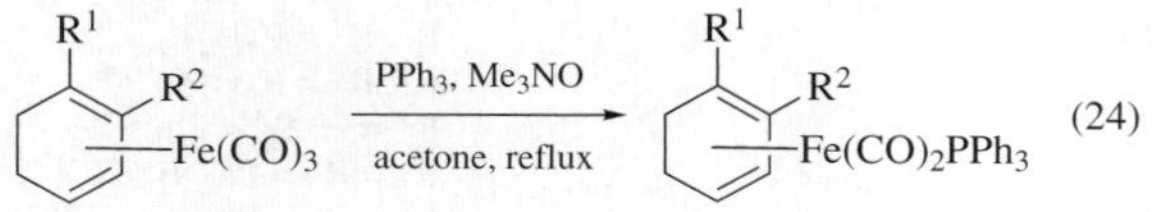

(24) PPh_3, Me_3NO, acetone, reflux

$R^1 = R^2$ = H, 64%
R^1 = H, R^2 = OMe, 48%
R^1 = OMe, R^2 = H, 54%
R^1 = H, R^2 = CO_2Me, 78%
R^1 = CO_2Me, R^2 = H, 81%

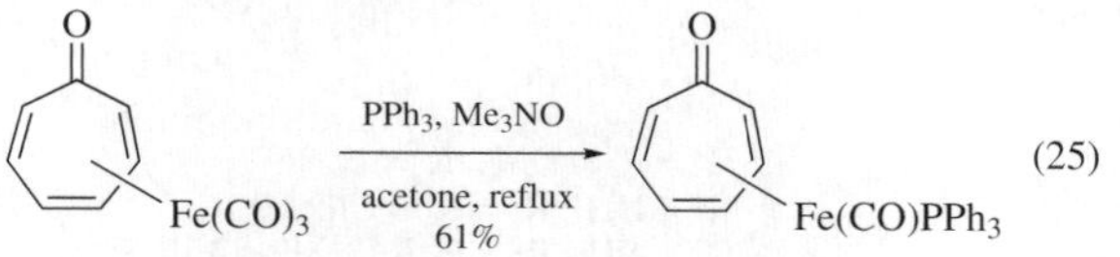

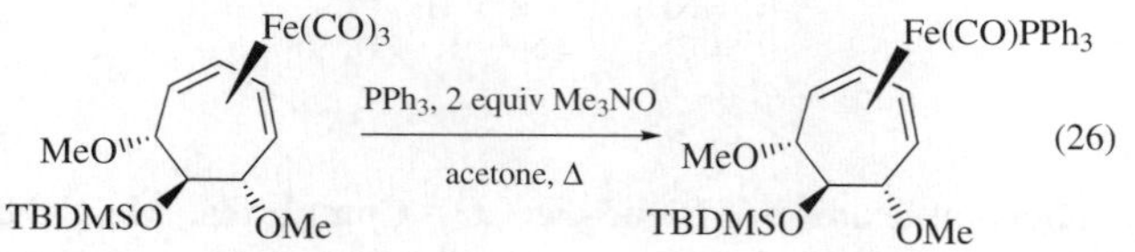

(25) PPh_3, Me_3NO, acetone, reflux, 61%

(26) PPh_3, 2 equiv Me_3NO, acetone, Δ

Related Reagents. See Class O-4, pages 1–10. *N*-Benzyl-*N*-(methoxymethyl)-*N*-trimethylsilylmethylamine; Cerium(IV) Ammonium Nitrate; Hydrogen Peroxide; Iron(III) Chloride; *N*-[(Tri-*n*-butylstannyl)methyl]benzaldimine.

1. Monagle, J. J. *JOC* **1962**, *27*, 3851.
2. Meisenheimer, J. *LA* **1913**, *397*, 273.
3. Weygand, C.; Hilgetag, G. *Organische-chemische Experimentierkunst*; English: *Preparative Organic Chemistry*; Hilgetag, G.; Martini, A.; Eds.; Wiley: Chichester, 1972; p 574. Hickinbottom, W. J. *Reactions of Organic Compounds*; Longmans: New York, 1936; p 277.

4. Anhydrous reagent: Soderquist, J. A.; Anderson, C. L. *TL* **1986**, *27*, 3961. Franzen, V. *OS* **1967**, *47*, 96.
5. Franzen, V.; Otto, S. *CB* **1961**, *94*, 1360.
6. Köster, R.; Arora, S.; Binger, P. *AG(E)* **1969**, *8*, 205.
7. Kabalka, G. W.; Hedgecock, H. C., Jr. *JOC* **1975**, *40*, 1776.
8. Kabalka, G. W.; Slayden, S. W. *JOM* **1977**, *125*, 273.
9. Carruthers, W.; Cumming, S. A. *CC* **1983**, 360.
10. Pearson, A. J.; Ham, P. *JCS(P1)* **1983**, 1421.
11. Corey, E. J.; Balanson, R. D. *H* **1976**, *5*, 445, and references cited therein.
12. Miller, J. A.; Zweifel, G. *S* **1981**, 288.
13. Kabalka, G. W.; Newton, R. J., Jr. *JOM* **1978**, *156*, 65.
14. Köster, R.; Morita, Y. *LA* **1967**, *704*, 70.
15. Beugelmans, R.; Negron, G.; Roussi, G. *CC* **1983**, 31.
16. (a) Huisgen, R. *AG(E)* **1980**, *19*, 947. (b) Kellogg, R. M. *T* **1976**, *32*, 2165.
17. De, B.; DeBernardis, J. F.; Prasad, R. *SC* **1988**, *18*, 481.
18. Shvo, Y.; Hazum, E. *CC* **1974**, 336.
19. (a) Eekhof, J. H.; Hogeveen, H.; Kellogg, R. M. *CC* **1976**, 657. (b) Elzinga, J.; Hogeveen, H. *CC* **1977**, 705.
20. Shu, B. Y.; Biehl, E. R.; Reeves, P. C. *SC* **1978**, *8*, 523.
21. Bandara, B. M. R.; Birch, A. J.; Kelly, L. F. *JOC* **1984**, *49*, 2496.
22. Pearson, A. J.; Srinivasan, K. *JOC* **1992**, *57*, 3965.
23. Pearson, A. J.; Lai, Y. S.; Lu, W.; Pinkerton, A. A. *JOC* **1989**, *54*, 3882.
24. Shvo, Y.; Hazum, E. *CC* **1975**, 829.
25. Birch, A. J.; Kelly, L. F. *JOM* **1985**, *286*, C5.
26. Howell, J. A. S.; Squibb, A. D.; Goldschmidt, Z.; Gottlieb, H. E.; Almadhoun, A.; Goldberg, I. *OM* **1990**, *9*, 80.

Anthony J. Pearson
Case Western Reserve University, Cleveland, OH, USA

Triphenylcarbenium Tetrafluoroborate

$Ph_3C^+ BF_4^-$

[341-02-6] $C_{19}H_{15}BF_4$ (MW 330.15)

(easily prepared[1] hydride abstractor used for conversion of dihydroaromatics to aromatics,[2–4] and the preparation of aromatic and benzylic cations;[5–8] oxidative hydrolysis of ketals[9] and thioketals;[10] conversion of acetonides to α-hydroxy ketones;[9] oxidation of acetals[11] and thioacetals;[12] selective oxidation of alcohols and ethers to ketones;[9,13–15] oxidation of silyl enol ethers to enones;[16] hydrolysis of TBS and MTM ethers;[17] oxidation of amines and amides to iminium salts;[18–20] oxidation of organometallics to give alkenes;[21–23] sensitizer for photooxidation using molecular oxygen;[24] Lewis acid catalyst for various reactions;[25] polymerization catalyst;[26] other reactions[27–30])

Alternate Name: trityl fluoroborate.
Physical Data: mp ~200 °C (dec).
Solubility: sol most standard organic solvents; reacts with some nucleophilic solvents.
Form Supplied in: yellow solid; commercially available.
Preparative Methods: the most convenient procedure involves the reaction of Ph_3CCl with ***Silver(I) Tetrafluoroborate*** in ethanol.[1b] The most economical route employs the reaction of Ph_3CCl with the anhydrous ***Tetrafluoroboric Acid***–Et_2O complex.[1c]
Purification: recrystallization of commercial samples from a minimal amount of dry MeCN provides material of improved purity, but the recovery is poor.[1a]
Handling, Storage, and Precautions: moisture-sensitive and corrosive. Recrystallized reagent can be stored at rt for several months in a desiccator without significant decomposition. This compound is much less light-sensitive than other trityl salts such as the perchlorate.[1a]

Preparation of Aromatic Compounds via Dehydrogenation. Dihydroaromatic compounds are easily converted into the corresponding aromatic compound by treatment with triphenylcarbenium tetrafluoroborate followed by base.[2] Certain α,α-disubstituted dihydroaromatics are converted to the 1,4-dialkylaromatic compounds with rearrangement (eq 1).[3] Nonbenzenoid aromatic systems, e.g. benzazulene[4a] or dibenzosesquifulvalene,[4b] are readily prepared from their dihydro counterparts. Aromatic cations are also easily prepared by hydride abstraction, for example, tropylium ion (e.g. in the synthesis of heptalene (eq 2)),[5] cyclopropenyl cation,[6] and others, including heterocyclic systems.[7] Some benzylic cations, especially ferrocenyl cations,[8] can also be formed by either hydride abstraction or trityl addition.

Bn Bn → ($Ph_3C^+BF_4^-$, AcOH, Δ, 0.5 h; 67%) → Bn, Bn (1)

(1. $Ph_3C^+ BF_4^-$ 88%; 2. Me_3N, 41%) (2)

Oxidation by Hydride Abstraction. In the early 1970s, Barton developed a method for the oxidative hydrolysis of ketals to ketones, e.g. in the tetracycline series (eq 3).[9] The same conditions can also be used to hydrolyze thioketals.[10] Acetonides of 1,2-diols are oxidized to the α-hydroxy ketones in good yield by this reagent (eq 4).[9] The hydrogen of acetals is easily abstracted (eq 5), providing a method for the conversion of benzylidene units in sugars to the hydroxy benzoates.[11] The hydrogen of dithioacetals is also abstracted to give the salts.[12] Since benzylic hydrogens are readily abstracted, this is also a method for deprotection of benzyl ethers.[9,13] Trimethylsilyl, *t*-butyl, and trityl ethers of simple alcohols are oxidized to the corresponding ketones and aldehydes in good yield. Primary–secondary diols are selectively oxidized at the secondary center to give hydroxy ketones by this method (eq 6).[14] 2,2-Disubstituted 1,4-diols are oxidized only at the 4-position to give the corresponding lactones.[15] Trimethylsilyl enol ethers are oxidized to α,β-unsaturated ketones, thereby providing a method for ketone to enone conversion (eq 7).[16] *t*-Butyldimethylsilyl (TBDMS) ethers are not oxidized but rather hydrolyzed to the alcohols, as are methylthiomethyl

(MTM) ethers.[17] Benzylic amines and amides can be oxidized to the iminium salts,[18] allylic amines and enamines afford eniminium salts,[19] and orthoamides give triaminocarbocations.[20]

$Ph_3C^+ BF_4^-$, 65% (3)

$Ph_3C^+ BF_4^-$, 79% (4)

1. $Ph_3C^+ BF_4^-$ 2. $NaHCO_3$, 75% (5)

1. Ph_3CCl 2. $Ph_3C^+ BF_4^-$, 91% (6)

$Ph_3C^+ BF_4^-$, collidine, CH_2Cl_2, 25 °C, 85% (7)

Generation of Alkenes from Organometallics. Various β-metalloalkanes can be oxidized by trityl fluoroborate to the corresponding alkenes.[21–23] The highest yields are obtained for the β-iron derivatives (eq 8), which are easily prepared from the corresponding halides or tosylates.[21] Grignard reagents and organolithiums also undergo this reaction (eq 9),[22] as do Group 14 organometallics (silanes, stannanes, etc.).[23]

$Na^+ {}^-FeCp(CO)_2$, 78%; 1. $Ph_3C^+ BF_4^-$ 2. NaI, acetone, 75% (8)

$Ph_3C^+ BF_4^-$, 65% (9)

Sensitizer of Photooxygenation. Barton showed that oxygen, in the presence of trityl fluoroborate and ordinary light, adds to cisoid dienes at −78 °C in very high yields.[24] For example, the peroxide of ergosterol acetate is formed in quantitative yields under these conditions (eq 10),[24a,b] which have been used also for photocycloreversions of cyclobutanes.[24c]

$Ph_3C^+ BF_4^-$, $h\nu$, O_2, −78 °C, 2.75 h, 100% (10)

Lewis Acid Catalysis. Trityl fluoroborate is a good Lewis acid for various transformations,[25] e.g. the Mukaiyama-type aldol reaction using a dithioacetal and silyl enol ether (eq 11).[25a] It has also been used as the catalyst for the formation of glycosides from alcohols and sugar dimethylthiophosphinates (eq 12)[25b] and for the formation of disaccharides from a protected α-cyanoacetal of glucose and a 6-*O*-trityl hexose.[25c] Michael additions of various silyl nucleophiles to conjugated dithiolenium cations also proceed well (eq 13).[25d,e] Finally, the [4 + 2] cycloaddition of cyclic dienes and oxygenated allyl cations has been effected with trityl fluoroborate.[25f]

+ $PhCH(SEt)_2$; $Ph_3C^+ BF_4^-$, CH_2Cl_2, −45 °C, 95% (11)

BuLi, Me_2PSCl, 79%; $Ph_3C^+ BF_4^-$, ROH, 57% (12)

X = H or SEt; $Ph_3C^+ BF_4^-$, CH_2Cl, −78 °C, X = SEt, 89% (13)

Polymerization Catalyst. Several types of polymerization[26] have been promoted by trityl fluoroborate, including reactions of orthocarbonates[26a] and orthoesters,[26b–d] vinyl ethers,[26e–g] epoxides,[26h,i] and lactones.[26j,k]

Other Reactions. Trityl fluoroborate has been used often to prepare cationic organometallic complexes, as in the conversion of dienyl complexes of iron, ruthenium, and osmium into their cationic derivatives.[27] It alkylates pyridines on the nitrogen atom in a preparation of dihydropyridines[28a] and acts as a tritylating agent.[28b] It has also been used in attempts to form silyl cations and silyl fluorides from silanes.[29] Finally, it has been reported to be a useful desiccant.[30]

Related Reagents. See Class O-12, pages 1–10.

1. (a) Dauben, H. J., Jr.; Honnen, L. R.; Harmon, K. M. *JOC* **1960**, *25*, 1442. (b) Fukui, K.; Ohkubo, K.; Yamabe, T. *BCJ* **1969**, *42*, 312. (c) Olah, G. A.; Svoboda, J. J.; Olah, J. A. *S* **1972**, 544.
2. (a) Müller, P. *HCA* **1973**, *56*, 1243. (b) Giese, G.; Heesing, A. *CB* **1990**, *123*, 2373.
3. (a) Karger, M. H.; Mazur, Y. *JOC* **1971**, *36*, 540. (b) Acheson, R. M.; Flowerday, R. F. *JCS(P1)* **1975**, 2065.
4. (a) O'Leary, M. A.; Richardson, G. W.; Wege, D. *T* **1981**, *37*, 813. (b) Prinzbach, H.; Seip, D.; Knothe, L.; Faisst, W. *LA* **1966**, *698*, 34.

5. (a) Dauben, H. J., Jr.; Gadecki, F. A.; Harmon, K. M.; Pearson, D. L. *JACS* **1957**, *79*, 4557. (b) Dauben, H. J., Jr.; Bertelli, D. J. *JACS* **1961**, *83*, 4657, 4659. (c) Peter-Katalinic, J.; Zsindely, J.; Schmid, H. *HCA* **1973**, *56*, 2796. (d) Vogel, E.; Ippen, J. *AG(E)* **1974**, *13*, 734. (e) Beeby, J.; Garratt, P. J. *JOC* **1973**, *38*, 3051. (f) Murata, I.; Yamamoto, K.; Kayane, Y. *AG(E)* **1974**, *13*, 807, 808. (g) Kuroda, S.; Asao, T. *TL* **1977**, 285. (h) Komatsu, K.; Takeuchi, K.; Arima, M.; Waki, Y.; Shirai, S.; Okamoto, K. *BCJ* **1982**, *55*, 3257. (i) Müller, J.; Mertschenk, B. *CB* **1972**, *105*, 3346. (j) Schweikert, O.; Netscher, T.; Knothe, L.; Prinzbach, H. *CB* **1984**, *117*, 2045. (k) Bindl, J.; Seitz, P.; Seitz, U.; Salbeck, E.; Salbeck, J.; Daub, J. *CB* **1987**, *120*, 1747.
6. (a) Zimmerman, H. E.; Aasen, S. M. *JOC* **1978**, *43*, 1493. (b) Komatsu, K.; Tomioka, I.; Okamoto, K. *BCJ* **1979**, *52*, 856.
7. (a) Yamamura, K.; Miyake, H.; Murata, I. *JOC* **1986**, *51*, 251. (b) Matsumoto, S.; Masuda, H.; Iwata, K.; Mitsunobu, O. *TL* **1973**, 1733. (c) Yano, S.; Nishino, K.; Nakasuji, K.; Murata, I. *CL* **1978**, 723. (d) Kedik, L. M.; Freger, A. A.; Viktorova, E. A. *KGS* **1976**, *12*, 328 (*Chem. Heterocycl. Compd. (Engl. Transl.)* **1976**, *12*, 279). (e) Reichardt, C.; Schäfer, G.; Milart, P. *CCC* **1990**, *55*, 97.
8. (a) Müller, P. *HCA* **1973**, *56*, 500. (b) Boev, V. I.; Dombrovskii, A. V. *ZOB* **1987**, *57*, 938, 633. (c) Klimova, E. I.; Pushin, A. N.; Sazonova, V. A. *ZOB* **1987**, *57*, 2336. (d) Abram, T. S.; Watts, W. E. *JCS(P1)* **1975**, 113; *JOM* **1975**, *87*, C39. (e) Barua, P.; Barua, N. C.; Sharma, R. P. *TL* **1983**, *24*, 5801. (f) Akgun, E.; Tunali, M. *AP* **1988**, *321*, 921.
9. (a) Barton, D. H. R.; Magnus, P. D.; Smith, G.; Strecker, G.; Zurr, D. *JCS(P1)* **1972**, 542. (b) Barton, D. H. R.; Magnus, P. D.; Smith, G.; Zurr, D. *CC* **1971**, 861.
10. Ohshima, M.; Murakami, M.; Mukaiyama, T. *CL* **1986**, 1593.
11. (a) Hanessian, S.; Staub, A. P. A. *TL* **1973**, 3551. (b) Jacobsen, S.; Pedersen, C. *ACS* **1974**, *28B*, 1024, 866. (c) Wessel, H.-P.; Bundle, D. R. *JCS(P1)* **1985**, 2251.
12. (a) Nakayama, J.; Fujiwara, K.; Hoshino, M. *CL* **1975**, 1099; *BCJ* **1976**, *49*, 3567. (b) Nakayama, J.; Imura, M.; Hoshino, M. *BCJ* **1980**, *53*, 1661. (c) Nakayama, J. *BCJ* **1982**, *55*, 2289. (d) Bock, H.; Brähler, G.; Henkel, U.; Schlecker, R.; Seebach, D. *CB* **1980**, *113*, 289. (e) Neidlein, R.; Droste-Tran-Viet, D.; Gieren, A.; Kokkinidis, M.; Wilckens, R.; Geserich, H.-P.; Ruppel, W. *HCA* **1984**, *67*, 574. (f) However, azide abstraction is seen with azidodithioacetals: Nakayama, J.; Fujiwara, K.; Hoshino, M. *JOC* **1980**, *45*, 2024.
13. (a) Barton, D. H. R.; Magnus, P. D.; Streckert, G.; Zurr, D. *CC* **1971**, 1109. (b) Doyle, M. P.; Siegfried, B. *JACS* **1976**, *98*, 163. (c) Hoye, T. R.; Kurth, M. J. *JACS* **1979**, *101*, 5065. (d) For simple ethers, see: Deno, N. C.; Potter, N. H. *JACS* **1967**, *89*, 3550.
14. (a) Jung, M. E. *JOC* **1976**, *41*, 1479. (b) Jung, M. E.; Speltz, L. M. *JACS* **1976**, *98*, 7882. (c) Jung, M. E.; Brown, R. W. *TL* **1978**, 2771.
15. Doyle, M. P.; Dow, R. L.; Bagheri, V.; Patrie, W. J. *JOC* **1983**, *48*, 476; *TL* **1980**, *21*, 2795.
16. (a) Jung, M. E.; Pan, Y.-G.; Rathke, M. W.; Sullivan, D. F.; Woodbury, R. P. *JOC* **1977**, *42*, 3961. (b) Reetz, M. T.; Stephan, W. *LA* **1980**, 533.
17. (a) Metcalf, B. W.; Burkhardt, J. P.; Jund, K. *TL* **1980**, *21*, 35. (b) Chowdhury, P. K.; Sharma, R. P.; Baruah, J. N. *TL* **1983**, *24*, 4485. (c) Niwa, H.; Miyachi, Y. *BCJ* **1991**, *64*, 716.
18. (a) Damico, R.; Broaddus, C. D. *JOC* **1966**, *31*, 1607. (b) Barton, D. H. R.; Bracho, R. D.; Gunatilaka, A. A. L.; Widdowson, D. A. *JCS(P1)* **1975**, 579. (c) Wanner, K. T.; Praschak, I.; Nagel, U. *AP* **1990**, *322*, 335; *H* **1989**, *29*, 29.
19. Reetz, M. T.; Stephan, W.; Maier, W. F. *SC* **1980**, *10*, 867.
20. Erhardt, J. M.; Grover, E. R.; Wuest, J. D. *JACS* **1980**, *102*, 6365.
21. (a) Laycock, D. E.; Hartgerink, J.; Baird, M. C. *JOC* **1980**, *45*, 291. (b) Laycock, D. E.; Baird, M. C. *TL* **1978**, 3307. (c) Slack, D.; Baird, M. C. *CC* **1974**, 701. (d) Bly, R. S.; Bly, R. K.; Hossain, M. M.; Silverman, G. S.; Wallace, E. *T* **1986**, *42*, 1093. (e) Bly, R. S.; Silverman, G. S.; Bly, R. K. *OM* **1985**, *4*, 374.
22. Reetz, M. T.; Schinzer, D. *AG(E)* **1977**, *16*, 44.
23. (a) Traylor, T. G.; Berwin, H. J.; Jerkunica, J.; Hall, M. L. *PAC* **1972**, *30*, 597. (b) Jerkunica, J. M.; Traylor, T. G. *JACS* **1971**, *93*, 6278. (c) Washburne, S. S.; Szendroi, R. *JOC* **1981**, *46*, 691. (d) Washburne, S. S.; Simolike, J. B. *JOM* **1974**, *81*, 41. (e) However, organostannanes lacking a β-hydrogen afford alkyltriphenylmethanes in good yield. Kashin, A. N.; Bumagin, N. A.; Beletskaya, I. P.; Reutov, O. A. *JOM* **1979**, *171*, 321.
24. (a) Barton, D. H. R.; Haynes, R. K.; Leclerc, G.; Magnus, P. D.; Menzies, I. D. *JCS(P1)* **1975**, 2055. (b) Barton, D. H. R.; Leclerc, G.; Magnus, P. D.; Menzies, I. D. *CC* **1972**, 447. (c) Okada, K.; Hisamitsu, K.; Mukai, T. *TL* **1981**, *22*, 1251. (d) Futamura, S.; Kamiya, Y. *CL* **1989**, 1703.
25. (a) Ohshima, M.; Murakami, M.; Mukaiyama, T. *CL* **1985**, 1871. (b) Inazu, T.; Yamanoi, T. Jpn. Patent 02 240 093, 02 255 693 (*CA* **1991**, *114*, 143 907j, 143 908k); Jpn. Patent 01 233 295 (*CA* **1990**, *112*, 198 972r). (c) Bochkov, A. F.; Kochetkov, N. K. *Carbohydr. Res.* **1975**, *39*, 355; for polymerizations of carbohydrate cyclic orthoesters, see: Bochkov, A. F.; Chernetskii, V. N.; Kochetkov, N. K. *Carbohydr. Res.* **1975**, *43*, 35; *BAU* **1975**, *24*, 396. (d) Hashimoto, Y.; Mukaiyama, T. *CL* **1986**, 1623, 755. (e) Hashimoto, Y.; Sugumi, H.; Okauchi, T.; Mukaiyama, T. *CL* **1987**, 1691. (f) Murray, D. H.; Albizati, K. F. *TL* **1990**, *31*, 4109.
26. (a) Endo, T.; Sato, H.; Takata, T. *Macromolecules* **1987**, *20*, 1416. (b) Uno, H.; Endo, T.; Okawara, M. *J. Polym. Sci., Polym. Chem. Ed.* **1985**, *23*, 63. (c) Nishida, H.; Ogata, T. Jpn. Patent 62 295 920 (*CA* **1988**, *109*, 57 030h). (d) See also Ref. 25c. (e) Kunitake, T. *J. Macromol. Sci., Chem.* **1975**, *A9*, 797. (f) Kunitake, T.; Takarabe, K.; Tsugawa, S. *Polym. J.* **1976**, *8*, 363. (g) Spange, S.; Dreier, R.; Opitz, G.; Heublein, G. *Acta Polym.* **1989**, *40*, 55. (h) Mijangos, F.; León, L. M. *J. Polym. Sci., Polym. Lett. Ed.* **1983**, *21*, 885; *Eur. Polym. J.* **1983**, *19*, 29. (i) Bruzga, P.; Grazulevicius, J.; Kavaliunas, R.; Kublickas, R. *Polym. Bull. (Berlin)* **1991**, *26*, 193. (j) Khomyakov, A. K.; Gorelikov, A. T.; Shapet'ko, N. N.; Lyudvig, E. B. *Vysokomol. Soedin., Ser. A* **1976**, *18*, 1699, 1053; *DOK* **1975**, *222*, 1111.
27. (a) For a review, see any basic organometallic text, e.g. Coates, G. E.; Green, M. L. H.; Wade, K. *Organometallic Compounds*; Methuen: London, 1968; Vol. 2, pp 136ff. (b) Birch, A. J.; Cross, P. E.; Lewis, J.; White, D. A. *CI(L)* **1964**, 838. (c) Cotton, F. A.; Deeming, A. J.; Josty, P. L.; Ullah, S. S.; Domingos, A. J. P.; Johnson, B. F. G.; Lewis, J. *JACS* **1971**, *93*, 4624.
28. (a) Lyle, R. E.; Boyce, C. B. *JOC* **1974**, *39*, 3708. (b) Hanessian, S.; Staub, A. P. A. *TL* **1973**, 3555.
29. (a) Sommer, L. H.; Bauman, D. L. *JACS* **1969**, *91*, 7076. (b) Bulkowski, J. E.; Stacy, R.; Van Dyke, C. H. *JOM* **1975**, *87*, 137. (c) Chojnowski, J.; Fortuniak, W.; Stanczyk, W. *JACS* **1987**, *109*, 7776.
30. Burfield, D. R.; Lee, K.-H.; Smithers, R. H. *JOC* **1977**, *42*, 3060.

Michael E. Jung
University of California, Los Angeles, CA, USA

Triphenylphosphine

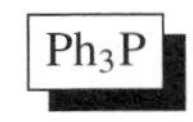

[603-35-0] $C_{18}H_{15}P$ (MW 262.30)

(deoxygenation of ozonides, peroxides, epoxides, and *N*-oxides; reduction of azides; reduction of organosulfur compounds; dehalogenations)

Physical Data: mp 79–81 °C; bp 377 °C; *d* (solid) 1.18 g cm^{-3}; n_D^{20} 1.59.

Solubility: insol H_2O; sol alcohol, benzene, chloroform; v sol ether.

Form Supplied in: white crystalline solid, widely available; polymer supported, ca. 3 mmol P per gram of resin is also available.

Purification: crystallize from hexane, methanol, or 95% ethanol; dry at 65 °C/<1 mmHg over $CaSO_4$ or P_2O_5.[1]

Handling, Storage, and Precautions: classified as an irritant upon acute exposure and a neurological hazard upon chronic exposure.[2] It is incompatible with strong oxidizing agents. Arylphosphines are less reactive toward molecular oxygen than benzyl- or alkylphosphines, but air oxidation of triphenylphosphine nonetheless occurs to an appreciable extent, particularly in solution, to give triphenylphosphine oxide.[3] Triphenylphosphine has low fire and explosion hazards, but, when heated to decomposition, it emits highly toxic fumes of phosphine and PO_x.[4] Use in a fume hood.

Introduction. Triphenylphosphine is a fairly general reducing agent as the following examples indicate. The chemistry in which triphenylphosphine participates is, in most cases, driven by the formation of triphenylphosphine oxide, a thermodynamically favored reaction. The enthalpies of formation of triphenylphosphine and triphenylphosphine oxide are $\Delta_f H^\circ_m = (207.02 \pm 3.52)$ kJ mol^{-1} and $\Delta_f H^\circ_m = -(116.41 \pm 3.42)$ kJ mol^{-1}, respectively.[5] Triphenylphosphine oxide, however, is a highly crystalline nonvolatile material which often requires chromatography to effect its removal from desired reaction products. While less relevant to its chemistry, it is noteworthy that triphenylphosphine is a considerably weaker base than trialkylphosphines; its pK_a is 2.73 compared to, for instance, tributylphosphine with a pK_a of 8.43.[6]

Deoxygenations. Triphenylphosphine has found widespread use in the reduction of compounds containing the hydroperoxide or endoperoxide functionality to form, depending upon the substrate, alcohols, carbonyl compounds, or epoxides. The primary driving force behind this class of reactions is the formation of the strong P=O bond at the expense of the relatively weak (45–50 kcal mol^{-1}) O–O bond.[7] For example, triphenylphosphine constitutes a mild, selective method for the reductive decomposition of ozonides to ketones and aldehydes. Compared to triphenylphosphine, ***Dimethyl Sulfide*** has enjoyed more popularity for this purpose,[8] probably because of its comparable mildness and the relative ease in removing the byproduct dimethyl sulfoxide compared to triphenylphosphine oxide. Nevertheless, the convenience of handling of triphenylphosphine and its lack of unpleasant odor render it an attractive alternative, as its use in the following examples illustrates (eqs 1 and 2).[9,10]

TBDMSO, OBn, OBn, H, H — 1. O_3; 2. PPh_3; 3. $MeO_2CCMePPh_3$; 68% → TBDMSO, OBn, MeO_2C, OBn, H, H (1)

Treatment of hydroperoxides with triphenylphosphine affords the corresponding alcohols in high yields under mild conditions and with retention of configuration at the carbon bearing the peroxide (eqs 3 and 4).[11,12] The reaction of endoperoxides, on the other hand, affords epoxides with inversion of configuration at one of the two carbon atoms (eq 5).[13] Vinylogous endoperoxides react with triphenylphosphine to afford the allylic epoxides (eq 6).[14] These examples should not imply that reaction of endoperoxides with triphenylphosphine is necessarily facile, since certain endoperoxides have been shown to be inert to PPh_3.[15] Nevertheless, the transformation of the α-methylene-β-peroxy lactone to afford the β-lactone (eq 7)[16] illustrates the wide scope of the peroxide deoxygenation capability of triphenylphosphine.

AcO, H, OBn, O — 1. O_3; 2. PPh_3; 73% → O, OAc, BnO, O, H, O (2)

OH, H, O, CHO — O_2, EtOAc → OH, H, O, OOH — PPh_3 → OH, H, O, OH; 34% (3)

HOO, OH — PPh_3, CH_2Cl_2, 81% → HO, OH (4)

Ph, O, O, Ph — PPh_3, 0 °C → [Ph, O, PPh_3, Ph, O] — 50 °C → [Ph, $\overset{+}{O}PPh_3$, Ph, O$^-$] — 74% → O, Ph, Ph (5)

O, O — PPh_3, CH_2Cl_2, 69% → O (6)

O, O, O — PPh_3, CH_2Cl_2, 50% → O, O (7)

The deoxygenation of epoxides affords the corresponding alkenes and results in an inversion in the stereochemistry of substitutents attached to the double bond.[17] While the reaction has been known since the mid 1950s,[18] it has not found widespread use. Triphenylphosphine has found utility as a reducing agent of *N*-oxides. While several alternative reagents and conditions are available, these often require fairly strong reducing conditions which are incompatible with a wide range of functionality.[19] In fact, triphenylphosphine-mediated reductions of *N*-oxides require much more vigorous conditions than do the corresponding reductions of peroxides. For example, reduction of trimethylamine oxide with trialkylphosphines requires refluxing in glacial acetic acid,[20] and the reduction of pyridine *N*-oxide derivatives is best carried out at temperatures above 200 °C in the absence of solvent.[21] However, it has been found that aromatic amine oxides are reduced in high yield with triphenylphosphine at room temperature with irradiation (eq 8).[22] While not general, under special conditions triphenylphosphine is also capable of reducing aromatic nitroso[23] and nitro groups (eq 9).[24]

PPh3, >300 nm; CH2Cl2; 94% (8)

PPh3; 72% (9)

Reduction of Azides (Staudinger Reaction). In 1919, Staudinger and Meyer[25] reported that the addition of triphenylphosphine to organic azides produced iminophosphoranes. This reaction proceeds by attack on nitrogen by phosphorus to produce an unstable phosphazide which extrudes dinitrogen to give the iminophosphorane (eq 10). This is known as the Staudinger reaction.[26]

$Ph_3P \xrightarrow{RN_3}$ 'phosphazide' → [ ] → $-N_2$

$Ph_3P{=}NR \xrightarrow{H_2O} Ph_3PO + H_2NR$ (10)

'iminophosphorane'

The iminophosphorane is a useful functional group for synthesis and can be isolated, although it is usually generated and used in situ. Iminophosphoranes can be hydrolyzed to amines. This forms the basis for the conversion of azides to amines using triphenylphosphine in the presence of water (eqs 11–13).[27] This mild and chemoselective reduction is an attractive alternative to other known methods (***Lithium Aluminum Hydride***, ***Diborane***, catalytic hydrogenation) that effect this transformation.[28] The reduction shown in eq 14 attests to the chemoselectivity of triphenylphosphine for this transformation. This reaction is sensitive to steric effects, but even tertiary azides can be reduced. Reduction of azides at a chiral center produce amines with the same configuration.

Ph3P, THF; H2O; 95% (11)

Ph3P, THF; H2O; 90% (12)

Ph3P, THF; H2O; 87% (13)

1. Ph3P, THF; 2. NH4OH; 3. MeOCOCl; 97% (14)

Reactions of Iminophosphoranes. Iminophosphoranes are reactive nucleophilic ylides; electrophiles react with the nitrogen.[29] Aldehydes and ketones are particularly favored reaction partners with iminophosphoranes, producing imines and triphenylphosphine oxide.[30,31] This reaction is analogous to the Wittig reaction and is known as the aza-Wittig reaction. Again, the driving force for this reaction is the formation of triphenylphosphine oxide (eq 15). Two intramolecular examples are shown in eqs 16 and 17.[32,33]

$Ph_3P{=}NR \rightleftharpoons Ph_3\overset{+}{P}\text{-}\overset{-}{N}R$ → + $O{=}PPh_3$ (15)

Aza-Wittig reaction

Ph3P; ether; 92% (16)

Ph3P; benzene; 67% (17)

Two other examples of reactions proceeding through the intermediacy of an iminophosphorane are shown in eqs 18 and 19.[34,35] Both reactions result in aziridine formation, and the driving force is the production of triphenylphosphine oxide.

N_3 — Ph_3P, $CHCl_3$ — 95% — $\overset{+}{O}PPh_3$ — $\overset{-}{N}$–$\overset{+}{P}Ph_3$ (18)

N_3, OH — Ph_3P, ether, 72% — H, N (19)

Reduction of an azide with triphenylphosphine in the presence of phthalic anhydride results in the formation of a phthalimide, a common protecting group for an amine (eq 20).[36] This reaction is facilitated by the addition of 0.1 equiv of tetrabutylammonium cyanide. If the reduction is performed in the presence of a carboxylic acid, an amide is formed (eq 21).[37] This reaction works with either aromatic or aliphatic (primary, secondary) azides and either aromatic or aliphatic carboxylic acids.

BnO, BnO, O, N_3 — Ph_3P, phthalic anhydride, 98% — BnO, BnO, O, N, O, O (20)

Me–N(Ac)–CH_2–COOH + N_3–CH_2Ph — Ph_3P, Δ, toluene, 77% — Me–N(Ac)–CH_2–C(O)–NH–CH_2Ph (21)

Reduction of Organosulfur Compounds. Triphenylphosphine converts episulfides to alkenes[38] smoothly at room temperature (eq 22). In contrast to the reaction of triphenylphosphine with epoxides, the reduction is stereospecifically *syn*.[39] This *syn* elimination can also be promoted with alkyllithiums,[40] lithium aluminum hydride,[41] and excess ***Iodomethane***,[40] and by exposure to carbenes.[42] This sulfur extrusion reaction plays a crucial role in Barton's method of alkene formation from carbonyl compounds.[43] This process (eq 23) is particularly effective for the formation of tetrasubstituted and highly hindered alkenes, where a paucity of methods exist for effective formation of these compounds. For example, cyclohexanone is easily converted to the azosulfide (**1**) in two steps, which, upon heating to 100 °C in the presence of triphenylphosphine, gives bicyclohexylidene via the episulfide (**2**) in good overall yield.

S — Ph_3P, Et_2O, rt, 95% — (22)

N=N, S (**1**) — Ph_3P, 100 °C — [S (**2**)] — 77% — (23)

Aryl disulfides can also be reduced to aryl thiols with triphenylphosphine in aqueous dioxane (eq 24).[44] The resulting aryl thiols are easily removed from the triphenylphosphine oxide byproduct by extraction with aqueous base. The rate of reaction is faster for arenes with electron-withdrawing groups and is susceptible to catalysis by either acid or base. This reaction can also be accomplished with $LiAlH_4$, ***Sodium Borohydride***, and ***Zinc*** and dilute acid,[45] or by heating with alkali.[46]

$(R–C_6H_4–S)_2$ — Ph_3P, H_2O, dioxane, 69–81% — 2 R–C_6H_4–SH (24)

In addition, arenesulfonic acids, their sodium salts, and alkyl arenesulfonates can be reduced to the corresponding aryl thiols by triphenylphosphine and a catalytic amount of ***Iodine*** (eq 25).[47] Since no reaction occurs in the absence of iodine or triphenylphosphine, the initial reduction is presumably mediated by iodotriphenylphosphonium iodide. The reaction can be greatly accelerated by the addition of a tertiary amine such as ***Tri-n-butylamine***. Use of bromine in place of iodine results in a considerably lower yield of thiol.

R–C_6H_4–$S(O)_2$X — Ph_3P, cat. I_2, benzene, 80–95% — R–C_6H_4–SH (25)

X = OH, ONa, OR

Triphenylphosphine has also been used to convert α-aryl-α-chloro sulfides to stilbenes.[48] The reaction (eq 26) requires addition of PPh_3 to the α-chloro sulfide prior to base addition; reversing the order of addition leads to negligible stilbene formation. The mechanism is not clear, but studies have shown that the phosphonium salt is not an intermediate in the process.

Ar–CH_2–S–CH(Cl)–Ar — Ph_3P, THF, *t*-BuOK, Δ, 60–90%% — Ar–CH=CH–Ar (26)

trans:*cis* = ca. 9:1

Eschenmoser[49] has used triphenylphosphine in an alkylation–sulfide contraction sequence useful in preparing secondary vinylogous amides and enolizable β-dicarbonyl compounds (eq 27). Although the precise mechanism is not known, an episulfide intermediate has been postulated. Addition of 0.1–0.2 equiv of a base such as ***Potassium t-Butoxide*** is necessary in some cases to promote enolization of the starting material.

Ph$_3$P, $CHCl_3$, 60 °C, 95% (27)

Dehalogenation. Triphenylphosphine has been used to reduce bromine or iodine atoms in the *ortho* or *para* positions of phenols (eq 28).[50] The procedure consists of heating the phenol and 1 equiv of triphenylphosphine in benzene at 100–150 °C for 1–5 h followed by alkaline hydrolysis. The corresponding aryl chlorides do not undergo this reaction. In the case of 2,4-dibromophenol, the *ortho* bromine atom is reduced preferentially.

Ph$_3$P, benzene, 100 °C, 95% (28)

Secondary and tertiary α-bromo ketones react with triphenylphosphine in refluxing benzene–methanol to afford the debrominated ketone in 60–70% yields (eq 29).[51] No quaternary phosphonium salts resulting from nucleophilic displacement of bromide are observed. The intermediacy of enol phosphonium salts has been proposed. Under conditions similar to those employed for bromo ketones, 2-chlorohexanone is dechlorinated to only a slight extent and direct quaternization is not detected.

Ph$_3$P, benzene–methanol, 62% (29)

Triphenylphosphine debrominates ethyl methacrylate dibromide and methyl acrylate dibromide to the unsaturated esters (eq 30).[52] Dehydrobromination or displacement of bromide is not observed. Triphenylphosphine also debrominates α-bromodiphenylacetyl bromide at room temperature to generate ***Diphenylketene*** (eq 31).[53] The mildness of this method avoids problems of other approaches such as the use of unstable intermediates such as phenylbenzoyldiazomethane and polymerization of the ketene.

Ph$_3$P, benzene, reflux, 50–64% (30)

R = H, Me

Ph$_3$P, heptane (31)

Finally, triphenylphosphine has been utilized in the preparation of alkyl-substituted thiirene dioxides via a 1,3-elimination of bromine from bis(α,α-dibromoalkyl) sulfones (eq 32).[54] The tetrabromo sulfones are synthesized from ***Dimethyl Sulfone*** in two steps.[55]

Ph$_3$P, CH_2Cl_2, 50–90% (32)

Miscellaneous. Trost[56] has reported that triphenylphosphine catalyzes the isomerization of ynones, ynoates, and ynamides to their corresponding diene–carbonyl compounds (eq 33). Conjugation of the alkyne with an electron-withdrawing group is necessary for the reaction to occur, and the reactivity order with respect to the electron-withdrawing group is ketone > ester > amide. Esters and amides require the addition of a weak protic acid such as ***Acetic Acid*** to facilitate the isomerization. Several examples of selectivity with polyfunctional alkynes have been reported (eq 34). The mildness and selectivity of this method over transition metal-catalyzed alkene redox reactions[57] allows for greater functional group compatibility.

5–10 mol% Ph$_3$P, toluene, reflux, 70–90% (33)

R = alkyl, aryl, OR, NR_2

Ph$_3$P, PhMe, 110 °C, 14 h, 82% (34)

Treatment of arylbromonitromethanes with an equimolar amount of triphenylphosphine in an inert solvent such as benzene leads to formation of ***Benzonitrile Oxide***,[58] which can be isolated as diphenylfuroxan; use of methyl acrylate as the solvent leads to isoxazolines via a [3 + 2] cycloaddition of the benzonitrile oxide generated in the reaction with ***Methyl Acrylate*** (eq 35). However, reaction of 2 equiv of triphenylphosphine with alkyl-1-bromo-1-nitromethanes leads to formation of the corresponding nitriles,[59] presumably via reduction of an intermediate nitrile oxides by the second equivalent of triphenylphosphine (eq 36). The putative alkyl nitrile oxide intermediate cannot be isolated or trapped by conducting the reaction in a solvent such as stilbene.

Ph$_3$P; CO_2Me; 80% (35)

Br
NO_2 —2 equiv PPh_3, benzene, rt, 72%→ CN (36)

Triphenylphosphine in refluxing acetonitrile has also been used in the dequaternization of pyridinium salts (eq 37).[60] Placement of an electron-withdrawing group on the pyridine ring facilitates the dequaternization reaction.

Y, X^- N^+–R —PPh_3, MeCN, 130–140 °C, 65–95%→ Y, N (37)

R = alkyl; Y = CN, COMe, Et

Triphenylphosphine has found wide application as a coreagent with a variety of other compounds including diethyl azodicarboxylate and carbon tetrachloride. These reagent combinations are covered in subsequent articles.

Related Reagents. See Classes R-9, R-16, R-23, R-24, R-27, and R-31, pages 1–10. Molybdenum(V) Chloride–Triphenylphosphine; Triphenylphosphine–*N*-Bromosuccinimide; Triphenylphosphine–Carbon Tetrabromide; Triphenylphosphine–Carbon Tetrabromide–Lithium Azide; Triphenylphosphine–Carbon Tetrachloride; Triphenylphosphine–Diethyl Azodicarboxylate; Triphenylphosphine–Hexachloroacetone; Triphenylphosphine–Iodine; Triphenylphosphine–Iodoform–Imidazole; Triphenylphosphine–2,4,5-Triiodoimidazole.

---

1. Perrin, D. D.; Armarego, W. L. F.; Perrin, D. R. *Purification of Laboratory Chemicals*, 2nd ed.; Pergamon: New York, 1980; p 455.
2. *Sigma-Aldrich Library of Regulatory and Safety Data*; Lenga, R. E.; Votoupal, K. L., Eds.; Sigma-Aldrich: Milwaukee, 1993; Vol. 2, p 2267.
3. Buckler, S. A. *JACS* **1962**, *84*, 3093.
4. *Dangerous Properties of Industrial Materials*; Sax, N. I., Ed.; Van Nostrand Reinhold: New York, 1984; p 2684.
5. Kirklin, D. R.; Domalski, E. S. *J. Chem. Thermodyn.* **1988**, *20*, 743.
6. Henderson, W. A., Jr.; Streuli, C. A. *JACS* **1960**, *82*, 5791.
7. Streitweiser, A.; Heathcock, C. H. *Introduction to Organic Chemistry*, 2nd ed.; MacMillan: New York, 1981; p 1195.
8. Lee, D. G.; Chen, T. *COS* **1981**, *7*, 541.
9. Nicolaou, K. C.; Reddy, K. R.; Skokotas, G.; Fuminori, S.; Xiao, X.-Y.; Hwang, C.-K. *JACS* **1993**, *115*, 3558.
10. Arseniyadis, S.; Yashunsky, D. V.; Dorado, M. M.; Alves, R. B.; Toromanoff, E.; Toupet, L.; Potier, P. *TL* **1993**, *34*, 4927.
11. Kigoshi, H.; Imamura, Y.; Mizuta, K.; Niwa, H.; Yamada, K. *JACS* **1993**, *115*, 3056.
12. (a) Adam, W.; Nestler, B. *JACS* **1993**, *115*, 5041. (b) Adam, W.; Brünker, H.-G. *JACS* **1993**, *115*, 3008.
13. Bartlett, P. D.; Landis, M. E.; Shapiro, M. J. *JOC* **1977**, *42*, 1661.
14. Adam, W.; Balci, M. *JACS* **1979**, *101*, 7542.
15. Posner, G. H.; Oh, C. H. *JACS* **1992**, *114*, 8328.
16. Adam, W.; Albert, R.; Grau, N. D.; Hasemann, L.; Nestler, B.; Peters, E.-M.; Peters, K.; Prechtl, F.; Von Schnering, H. G. *JOC* **1991**, *56*, 5778.
17. Fieser, L.; Fieser, M. *FF* **1967**, *1*, 1243.
18. Wittig, G.; Haag, W. *CB* **1955**, *88*, 1654.
19. Greene, T. *Protecting Groups in Organic Synthesis*; Wiley: New York, 1981; p 281.
20. Horner, L.; Hoffmann, J. *AG* **1956**, *68*, 473.
21. Howard, E., Jr.; Olszewski, W. F. *JACS* **1959**, *81*, 1483.
22. Kaneko, C.; Yamamori, M.; Yamamoto, A.; Hayashi, R. *TL* **1978**, *31*, 2799.
23. (a) Odum, R. A.; Brenner, M. *JACS* **1966**, *88*, 2074. (b) Cadogan, J. I. G.; Cameron-Wood, M.; Mackie, R. K.; Searle, R. J. G. *JCS* **1965**, 4831.
24. Scott, P. H.; Smith, C. P.; Kober, E.; Churchill, J. W. *TL* **1970**, 1153.
25. Staudinger, H.; Meyer, J. *HCA* **1919**, *2*, 635.
26. Gololobov, Y. G.; Zhmurova, I. N.; Kasukhin, L. F. *T* **1981**, *37*, 437.
27. Vaultier, M.; Knouzi, N.; Carrié, R. *TL* **1983**, *24*, 763.
28. Patai, S. *The Chemistry of the Azido Group*; Wiley: New York, 1971; pp 333–338.
29. McClure, K. F.; Danishefsky, S. J. *JACS* **1993**, *115*, 6094.
30. Johnson, A. W. *Ylid Chemistry*; Academic: New York, 1966; pp 222–236.
31. Stuckwisch, C. G. *S* **1973**, 469.
32. Lambert, P. H.; Vaultier, M.; Carrié, R. *CC* **1982**, 1224.
33. Takeuchi, H.; Yanagida, S.-I.; Ozaki, T.; Hagiwara, S.; Eguchi, S. *JOC* **1989**, *54*, 431.
34. Molina, P.; Alajarín, M.; López-Lázaro, A. *TL* **1992**, *33*, 2387.
35. Ittah, Y.; Sasson, Y.; Shahak, I.; Tsaroom, S.; Blum, J. *JOC* **1978**, *43*, 4271.
36. Garcia, J.; Vilarrasa, J.; Bordas, X.; Banaszek, A. *TL* **1986**, *27*, 639.
37. Garcia, J.; Urpí, F.; Vilarrasa, J. *TL* **1984**, *25*, 4841.
38. Davis, R. A. *JOC* **1958**, *23*, 1767.
39. Sonnet, P. E. *T* **1980**, *36*, 557.
40. (a) Schuetz, R. D.; Jacobs, R. L. *JOC* **1961**, *26*, 3467. (b) Culvenor, C. C. J.; Davies, W.; Heath, N. S. *JCS* **1949**, 282. (c) Helmkamp, G. K.; Pettitt, D. J. *JOC* **1964**, *29*, 3258.
41. Lightner, D. A.; Djerassi, C. *CI(L)* **1962**, 1236.
42. Hata, Y.; Watanabe, M.; Inoue, S.; Oae, S. *JACS* **1975**, *97*, 2553.
43. (a) Barton, D. H. R.; Smith, E. H.; Willis, B. J. *CC* **1970**, 1226. (b) Barton, D. H. R.; Guziec, F. S.; Shahak, I. *JCS(P1)* **1974**, 1794.
44. (a) Overman, L. E.; Smoot, J.; Overman, J. D. *S* **1974**, 59. (b) Schönberg, A.; Barakat, M. Z. *JCS* **1949**, 892.
45. Patai, S. *The Chemistry of the Thiol Group*; Wiley: New York, 1974.
46. Danehy, J. P.; Hunter, W. E. *JOC* **1967**, *32*, 2047.
47. Oae, S.; Togo, H. *BCJ* **1983**, *56*, 3802.
48. Mitchell, R. H. *TL* **1973**, 4395.
49. Roth, M.; Dubs, P.; Götschi, E.; Eschenmoser, A. *HCA* **1971**, *54*, 710.
50. Hoffmann, H.; Horner, L.; Wippel, H. G.; Michael, D. *CB* **1962**, *95*, 523.
51. Borowitz, I. J.; Grossman, L. I. *TL* **1962**, 471.
52. Tung, C. C.; Speziale, A. J. *JOC* **1963**, *28*, 1521.
53. Darling, S. D.; Kidwell, R. L. *JOC* **1968**, *33*, 3974.
54. Carpino, L. A.; Williams, J. R. *JOC* **1974**, *39*, 2320.
55. Szabo, K. U.S. Patent 3 106 585, 1984 (*CA* **1984**, *100*, 2841b).
56. Trost, B. M.; Kazmaier, U. *JACS* **1992**, *114*, 7933.
57. (a) Wilcox, C. S.; Long, G. W.; Suh, H. *TL* **1984**, *25*, 395. (b) Hosomi, A.; Sakurai, H. *JACS* **1977**, *99*, 1673. (c) Hirai, K.; Suzuki, H.; Moro-oka, Y.; Ikawa, T. *TL* **1980**, *21*, 3413. (d) Johnstone, R. A. W.; Wilby, A. A. H.; Entwistle, I. D. *CRV* **1985**, *85*, 129.
58. Coutouli-Argyropoulou, E. *TL* **1984**, *25*, 2029.
59. Trippett, S.; Walker, D. M. *JCS* **1960**, 2976.
60. Kutney, J. P.; Greenhouse, R. *SC* **1975**, *5*, 119.

Jeff E. Cobb, Cynthia M. Cribbs,
Brad R. Henke, & David E. Uehling
Glaxo Research Institute, Research Triangle Park, NC, USA

Tris(trimethylsilyl)silane[1]

$(Me_3Si)_3SiH$

[1873-77-4] $C_9H_{28}Si_4$ (MW 248.73)

(mediator of radical reactions;[1,2] nontoxic substitute for tri-*n*-butylstannane in radical reactions; slower hydrogen donor than tri-*n*-butylstannane[3])

Physical Data: bp 82–84 °C/12 mmHg; *d* 0.806 g cm^{-3}; n_D^{20} 1.489.
Solubility: sol pentane, ether, toluene, THF; modestly sol acetone, acetonitrile; insol H_2O; decomposes rapidly in methanol and other alcohols.
Form Supplied in: colorless liquid; commercially available.
Preparative Methods: easy to synthesize.[4]
Handling, Storage, and Precautions: is slightly sensitive to oxygen and should be stored under nitrogen.[5] It showed no toxicity in several biological test systems.[6]

Functional Group Reductions. Tris(trimethylsilyl)silane is an effective radical reducing agent for organic halides, selenides, xanthates, isocyanides,[2] and acid chlorides (Table 1).[7] The reactions are carried out at 75–90 °C in toluene in the presence of a radical initiator, i.e. ***Azobisisobutyronitrile***. Chromatographic workup affords the products. The silicon-containing byproducts are easily separated. The silane can also be used catalytically when ***Sodium Borohydride*** is employed as the coreductand.[8] If a halide (bromide or iodide) is treated under photochemical initiation conditions with an excess of sodium borohydride and a small amount of tris(trimethylsilyl)silane or its corresponding halide, the silane is continuosly regenerated from the silyl halide.

Table 1 Reduction of Several Organic Compounds by Tris(trimethylsilyl)silane[2,7]

RX	Yield RH (%)	RX	Yield RH (%)
 Cl	82	 OC(S)SMe	86
 Br	90	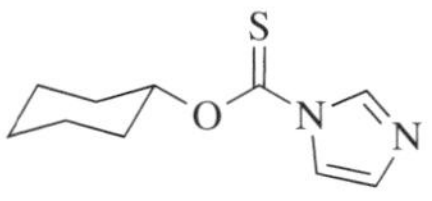 O–C(S)–N (imidazolyl)	95
 I	97	 SePh	99
 NC	95	 COCl	92

Iodides and bromides are reduced by tris(trimethylsilyl)silane to the corresponding hydrocarbons in high yield after a short reaction time (0.5 h). From tertiary to secondary and primary chlorides the reduction becomes increasingly difficult. A longer reaction time and periodic addition of initiator is required. Photochemical initiation can be used and is quite efficient.[9] Tris(trimethylsilyl)silane is superior to ***Tri-n-butylstannane*** in replacing an isocyanide group by hydrogen. The reaction with tin hydride requires high temperatures (boiling xylene for primary isocyanides) and periodic addition of initiator. Using the silane, primary, secondary, and tertiary isocyanides are reduced at 80 °C in high yields. The reduction of selenides by tris(trimethylsilyl)silane proceeds with high yields; however, the corresponding reaction of sulfides is inefficient.

Acyl chlorides are converted by tris(trimethylsilyl)silane to the corresponding hydrocarbons. Tertiary and secondary acid chlorides react at 80 °C, while the reduction of primary derivatives requires higher temperatures.[7] The radical deoxygenation of hydroxyl groups is carried out by conversion of the alcohol to a thionocarbonate, which can be reduced by tris(trimethylsilyl)silane (eq 1). This very mild method is especially useful in natural product synthesis. It has been utilized for the deoxygenation of lanosterol (eq 2)[6] and the dideoxygenation of 1,6-anhydro-D-glucose (eq 3).[10]

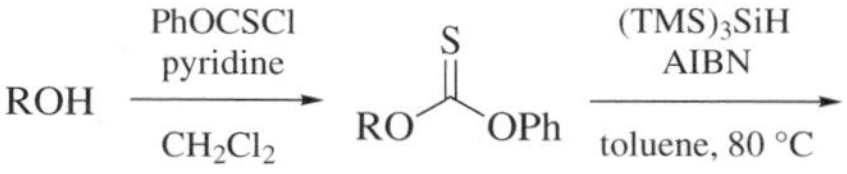

ROH —(PhOCSCl, pyridine, CH_2Cl_2)→ RO–C(=S)–OPh —((TMS)$_3$SiH, AIBN, toluene, 80 °C)→ RH (1)

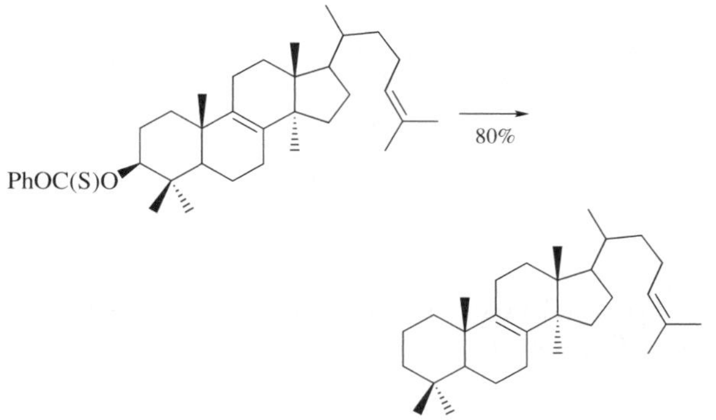

PhOC(S)O → 80% (2)

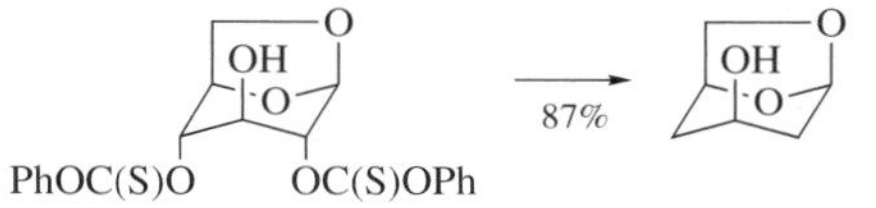

PhOC(S)O, OC(S)OPh, OH, O → 87% (3)

Radical deoxygenation of the *cis*-unsaturated fatty acid derivative with tris(trimethylsilyl)silane gives methyl triacont-21-*trans*-enoate together with the saturated compound (eq 4). If the reaction is carried out with tri-*n*-butyltin hydride, the configuration remains unchanged.[11]

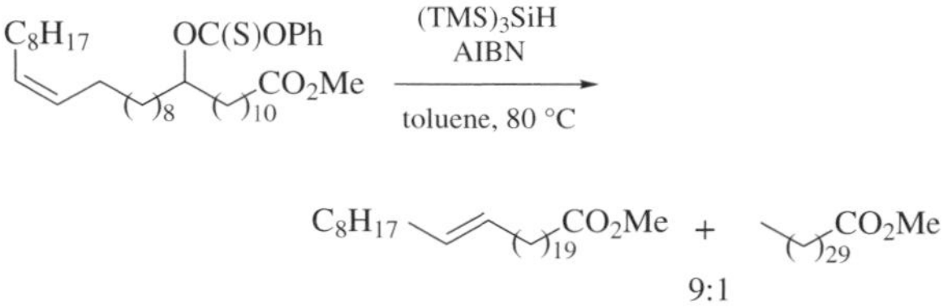

C_8H_{17} ... OC(S)OPh ... CO_2Me (subscripts 8, 10) —((TMS)$_3$SiH, AIBN, toluene, 80 °C)→ C_8H_{17} ... CO_2Me (19) + CO_2Me (29) (4)

9:1

Hydrosilylation of Double Bonds. Tris(trimethylsilyl)silane is capable of radical hydrosilylation of dialkyl ketones,[12] alkenes,[12,13] and alkynes.[13] Hydrosilylation of alkenes yields the anti-Markovnikov products with high regio- and good diastereoselectivity (eq 5). By using a chiral alkene, complete stereocontrol can be achieved (eq 6).[14] The silyl group can be converted to a hydroxyl group by Tamao oxidation.[13]

Monosubstituted alkynes give alkenes in high yield and stereoselectivity. The formation of (*E*)- or (*Z*)-alkenes depends on the steric demand of the substituents (eq 7). 1,2-Disubstituted phenylalkynes are attacked exclusively β to the phenylated alkyne carbon atom.[13] The silyl moiety can be replaced by a bromine atom with overall retention of configuration (eq 8).[13]

$(TMS)_3SiH$, toluene, AIBN, 89% → $(TMS)_3Si$ … + $(TMS)_3Si$ … 16:1 (5)

$(TMS)_3SiH$, AIBN, hexane, 70 °C, >95% → $(TMS)_3Si$, CO_2Et; only product (6)

Ph—≡—H → $(TMS)_3SiH$, Et_3B, O_2, 25 °C, 85% → Ph, $Si(TMS)_3$ (7)

Ph, $Si(TMS)_3$ → Br_2, CH_2Cl_2, −78 °C, 75% → Ph, Br (8)

The hydrosilylation of ketones is in general slower than the corresponding reaction of alkenes and alkynes. In the case of sterically hindered ketones, a catalytic amount of a thiol is necessary to carry out the reaction.[15] The resulting silyl ethers can be easily desilylated by standard procedures. With 4-*t*-butylcyclohexanone the *trans* isomer is formed as the main product (eq 9). The hydrosilylation of a ketone bearing a chiral center in the adjacent position yields mainly the Felkin–Anh product (eq 10).[15]

t-Bu … O → $(TMS)_3SiH$, AIBN, toluene, 80 °C, 90% → *t*-Bu … $OSi(TMS)_3$; *trans:cis* = 91:9 (9)

$(TMS)_3SiH$, initiator → $OSi(TMS)_3$, Ph (Cram) + $OSi(TMS)_3$, Ph (anti-Cram) (10)

30 °C, 22% 13:1
130 °C, 90% 3.5:1

Intramolecular Reactions. Tris(trimethylsilyl)silane is an effective mediator of radical cyclizations.[16] In addition to halides and selenides, secondary isocyanides can be used as precursors for intramolecular C–C bond formation,[17] which is impossible using the tin hydride (eq 11). Selective cleavage of the carbon–sulfur bond of a 1,3-dithiolane, 1,3-dithiane,[18] 1,3-oxathiolane, or 1,3-thiazolidine[19] derivative is an efficient process to generate carbon-centered radicals, which can undergo cyclization (eq 12).

NC, Ph → $(TMS)_3SiH$, AIBN, toluene, 70 °C, 78% → Ph + Ph; 4.6:1 (11)

EtO_2CN, S, O → $(TMS)_3SiH$, AIBN, THF, 90 °C, 79% → EtO_2C, N, H, $SSi(TMS)_3$, O, H; *endo*:*exo* = 2.4:1 (12)

2-Benzylseleno-1-(2-iodophenyl)ethanol reacts smoothly with tris(trimethylsilyl)silane to give benzo[*b*]selenophene (eq 13).[20] A similar homolytic substitution reaction at the silicon atom yields a sila bicycle.[21]

OH, I, SeBn → $(TMS)_3SiH$, AIBN, benzene, 80 °C, 80% → Se (13)

The silane is superior to the tin reagent in the radical rearrangement of glycosyl halides to 2-deoxy sugars (eq 14).[16] Aromatization of the A-ring of 9,10-secosteroids can be achieved by a mild, radical-induced fragmentation reaction of 3-oxo-1,4-diene steroids (eq 15).[22]

OAc, AcO, Br, AcO, OAc → $(TMS)_3SiH$, AIBN, toluene, 80 °C, 70% → OAc, AcO, OAc, AcO (14)

O, O → $(TMS)_3SiH$, AIBN, THF, 70 °C, 84% → O, $(TMS)_3SiO$ (15)

Intermolecular Reactions. Radical carbon–carbon bond formation can be carried out with tris(trimethylsilyl)silane.[16] Again, it is possible to use isocyanides as precursors (eqs 16 and 17).[17]

I + CO_2Me → $(TMS)_3SiH$, AIBN, toluene, 80 °C, 85% → CO_2Me (16)

NC + CN → $(TMS)_3SiH$, AIBN, toluene, 80 °C, 85% → CN (17)

Nonradical Reactions. Tris(trimethylsilyl)silane reacts with carbenium ions to form a silicenium ion.[23] In this case, tris(trimethylsilyl)silane is only slightly more reactive than trimethylsilane. The reaction of the silane with methyl diazoacetate in the presence of copper catalyst gives the α-silyl ester (eq 18).[24]

$$(TMS)_3SiH + N_2CHCO_2Me \xrightarrow[90\ °C]{Cu} (TMS)_3SiCH_2CO_2Me \quad (18)$$

Related Reagents. See Classes R-2, R-12, R-26, R-27, R-28, and R-31, pages 1–10.

---

1. Chatgilialoglu, C. *ACR* **1992**, *25*, 188.
2. Ballestri, M.; Chatgilialoglu, C.; Clark, K. B.; Griller, D.; Giese, B.; Kopping, B. *JOC* **1991**, *56*, 678.
3. Chatgilialoglu, C.; Dickhaut, J.; Giese, B. *JOC* **1991**, *56*, 6399.
4. Dickhaut, J.; Giese, B. *OS* **1991**, *70*, 164.
5. Chatgilialoglu, C.; Guarini, A.; Guerrini, A.; Seconi, G. *JOC* **1992**, *57*, 2207.
6. Schummer, D.; Höfle, G. *SL* **1990**, 705.
7. Ballestri, M.; Chatgilialoglu, C.; Cardi, N.; Sommazzi, A. *TL* **1992**, *33*, 1787.
8. Lesage, M.; Chatgilialoglu, C.; Griller, D. *TL* **1989**, *30*, 2733.
9. Chatgilialoglu, C.; Griller, D.; Lesage, M. *JOC* **1988**, *53*, 3641.
10. Barton, D. H. R.; Jang, D. O.; Jaszberenyi, J. C. *TL* **1992**, *33*, 6629.
11. Johnson, D. W.; Poulos, A. *TL* **1992**, *33*, 2045.
12. Kulicke, K. J.; Giese, B. *SL* **1990**, 91.
13. Kopping, B.; Chatgilialoglu, C.; Zehnder, M.; Giese, B. *JOC* **1992**, *57*, 3994.
14. Smadja, W.; Zahouily, M.; Malacria, M. *TL* **1992**, *33*, 5511.
15. Giese, B.; Damm, W.; Dickhaut, J.; Wetterich, F.; Sun, S.; Curran, D. P. *TL* **1991**, *32*, 6097.
16. Giese, B.; Kopping, B.; Chatgilialoglu, C. *TL* **1989**, *30*, 681.
17. Chatgilialoglu, C.; Giese, B.; Kopping, B. *TL* **1990**, *31*, 6013.
18. Arya, P.; Samson, C.; Lesage, M.; Griller, D. *JOC* **1990**, *55*, 6248.
19. Arya, P.; Lesage, M.; Wayner, D. D. M. *TL* **1991**, *32*, 2853. Arya, P.; Wayner, D. D. M. *TL* **1991**, *32*, 6265.
20. Schiesser, C. H.; Sutej, K. *TL* **1992**, *33*, 5137.
21. Kulicke, K. J.; Chatgilialoglu, C.; Kopping, B.; Giese, B. *HCA* **1992**, *75*, 935.
22. Künzer, H.; Sauer, G.; Wiechert, R. *TL* **1991**, *32*, 7247.
23. Mayr, H.; Basso, N.; Hagen, G. *JACS* **1992**, *114*, 3060.
24. Watanabe, H.; Nakano, T.; Araki, K.-I.; Matsumoto, H.; Nagai, Y. *JOM* **1974**, *69*, 389.

Bernd Giese & Joachim Dickhaut
University of Basel, Switzerland

Vanadyl Bis(acetylacetonate)[1]

[3153-26-2] $C_{10}H_{14}O_5V$ (MW 265.18)

(precatalyst for oxidation of several functional groups in combination with alkyl hydroperoxides or oxygen;[1] especially useful for the regio- and diastereoselective epoxidation of allylic alcohols,[2] the oxidation of tertiary amines to *N*-oxides,[3] and the oxidation of sulfides to sulfoxides[4] using an alkyl hydroperoxide as the oxidant)

Alternate Name: bis(acetylacetonato)oxovanadium(IV).
Physical Data: mp 256–259 °C; *d* 1.50 g cm^{-3}.
Solubility: insol H_2O; sol EtOH, CH_2Cl_2, $CHCl_3$, C_6H_6.
Form Supplied in: blue-green crystals; widely available.
Handling, Storage, and Precautions: generally used without further purification. Can be handled safely in the open. Should be stored in a cool, dry place in the absence of light. Contact with skin, eyes, and mucous membranes should be avoided. Toxic if swallowed. Generally performs best as an oxidation catalyst when used under anhydrous conditions. Oxidants contaminated with H_2O such as 70% *t*-butyl hydroperoxide (TBHP) are therefore usually dried prior to use with $VO(acac)_2$.

Precursor for Soluble Vanadium(V) Catalysts. $VO(acac)_2$ is a precatalyst for the oxidation of several functional groups, usually using an alkyl hydroperoxide as the oxidant.[1b] Reaction of $VO(acac)_2$ with an alkyl hydroperoxide quickly oxidizes the vanadium from the +4 to the +5 oxidation state. The resulting organic soluble V^V complexes, whose structure can be represented as $VO(OR)_3$, are usually the actual catalysts in these reactions.[1] $VO(acac)_2$ is a commercially available, stable, crystalline, organic soluble solid and therefore a convenient source of vanadium for oxidation reactions. In contrast, organic soluble V^V compounds, suitable for these reactions, are not as widely available nor as easily handled and therefore not as convenient a source of vanadium as $VO(acac)_2$.[5]

Epoxidation of Allylic Alcohols. Soluble V^V compounds are the preferred catalysts for nonenantioselective epoxidation of allylic alcohols to epoxy alcohols using alkyl hydroperoxides, especially ***t*-Butyl Hydroperoxide**, as oxidants.[2,6] V^V compounds with alkyl hydroperoxides are also capable of epoxidation of nonfunctionalized alkenes[7] to epoxides but at rates ca. 100 times less than for the corresponding allylic alcohols.[6a] This rate differential allows the regioselective monoepoxidation of substrates like geraniol (eq 1).

OH —[$VO(acac)_2$, TBHP; benzene; Δ; 93%]→ OH (1)

Mo^{VI}/TBHP and peracids efficiently epoxidize allylic alcohols as well but, because they also rapidly epoxidize nonfunctionalized alkenes, they tend not to be as regioselective as V^V/TBHP.[1] Other d^0 transition metal catalysts, such as W^{VI} and Ti^{IV}, epoxidize allylic alcohols slower and less efficiently than V^V and therefore are not as useful for this type of epoxidation reaction.[7] Diastereoselective epoxidation is often observed in V^V-catalyzed epoxidations of chiral allylic alcohols, especially cyclic allylic alcohols.[6a,8,9] Diastereoselective epoxidation of homoallylic and bishomoallylic alcohols have also been reported.[6a,10] V^V/TBHP often complements the diastereoselectivity of peracids and Mo^{VI}/TBHP observed for the epoxidation of secondary allylic alcohols.[6a,8] For example, epoxidation of 2-cyclooctenol gives either the corresponding *cis*- or *trans*-epoxy alcohol with high diastereoselectivity, depending on whether $VO(acac)_2$/TBHP or ***m*-Chloroperbenzoic Acid** is used (eq 2).[11]

(2)

	% cis	% cis
$VO(acac)_2$, TBHP	97	3
m-CPBA	0.2	99.8
$MoO_2(acac)_2$	42	58

Hydroxyepoxidation of Alkenes. $VO(acac)_2$ catalyzes the transformation of allylic hydroperoxides to epoxy alcohols.[12] This process presumably involves intermolecular transfer of oxygen from the hydroperoxide portion of an allylic hydroperoxide to the alkene portion of either a second allylic hydroperoxide or an allylic alcohol derived from hydroperoxide reduction (eq 3).[13]

HOO —[$VO(acac)_2$]→ HO (3)

$VO(acac)_2$[14] and other forms of vanadium[15] also catalyze the transformation of alkenes to epoxy alcohols using molecular oxygen as the oxidant. This process involves initial oxidation of the alkene to an allylic hydroperoxide either by radical or photolytic processes followed by intermolecular epoxidation to give primarily the epoxy alcohol and usually several byproducts. A convenient photolytic process producing singlet oxygen in the presence of the alkene and a catalytic amount of $VO(acac)_2$ has been developed to promote this transformation in good yield and high diastereoselectivity (eq 4).[14] An obvious requirement of this reaction is that the starting alkene must readily undergo photooxygenation to give the intermediate allylic hydroperoxide. ***Titanium Tetraisopropoxide*** and molybdenyl bis(acetylacetonate) also act as catalysts in this reaction, sometimes with better yields and/or selectivity.[14]

$$\text{Me}_2\text{C=CH-}t\text{-Bu} \xrightarrow[\text{CH}_2\text{Cl}_2,\ 0\ ^\circ\text{C}]{^1\text{O}_2,\ \text{VO(acac)}_2} [\text{allylic alcohol}] \longrightarrow \text{epoxy alcohols (90\%)} + \text{(0.9\%)} \quad (4)$$

Oxidation of Amines. V^V/TBHP efficiently oxidizes tertiary amines to *N*-oxides (eq 5).[16] The reaction is carried out under anhydrous conditions and therefore allows the preparation of anhydrous amine oxides. V^V/TBHP oxidizes aniline to nitrobenzene (eq 6).[17]

$$C_{12}H_{25}NMe_2 \xrightarrow[\text{TBA},\ \Delta,\ 76\%]{\text{VO(acac)}_2,\ \text{TBHP}} C_{12}H_{25}-\overset{+}{N}Me_2-O^- \quad (5)$$

$$PhNH_2 \xrightarrow[\text{benzene, chlorobenzene, H}_2\text{O; 39\% conversion}]{\text{VO(acac)}_2,\ \text{TBHP, 66 }^\circ\text{C}} PhNO_2 \quad (6)$$

Oxidation of Sulfides to Sulfoxides. Similar to the oxidation of tertiary amines to *N*-oxides, V^V/TBHP oxidizes sulfides to sulfoxides in high yield (eq 7).[4,18] The reaction is usually carried out at room temperature in an alcoholic solvent.

$$Bu_2S \xrightarrow[\text{EtOH, 25 }^\circ\text{C; 90\%}]{\text{VO(acac)}_2,\ \text{TBHP}} Bu_2S{=}O \quad (7)$$

Other Oxidation Reactions. $VO(acac)_2$ rapidly oxidizes 3,5-di-*t*-butylpyrocatechol at room temperature to the corresponding muconic acid anhydride, 2-pyrone, and *o*-quinone using oxygen as the oxidant (eq 8).[19] This method is more selective for the anhydride than oxygenation with ***Dichlorotris(triphenylphosphine)ruthenium(II)***. V^V/TBHP has also been used in the catalytic oxidation of hydrocarbons.[20] These reactions probably occur by radical mechanisms and are usually not highly selective.

$$\text{3,5-di-}t\text{-butylcatechol} \xrightarrow[\text{CH}_2\text{Cl}_2,\ 20\ ^\circ\text{C}]{\text{VO(acac)}_2,\ (0.1\ \text{mol\%}),\ \text{O}_2} \text{anhydride (41\%)} + \text{2-pyrone (7\%)} + o\text{-quinone (25\%)} \quad (8)$$

Related Reagents. See Classes O-8, O-9, O-11 and O-14, pages 1–10. Vanadyl Bis(acetylacetonate)–Azobisisobutyronitrile.

1. (a) Vilas Boas, L. F.; Costa Pessoa, J. In *Comprehensive Coordination Chemistry*; Wilkinson, G., Ed.; Pergamon: Oxford, 1987; Vol. 3, Chapter 33. (b) Mimoun, H. In *Comprehensive Coordination Chemistry*; Wilkinson, G., Ed.; Pergamon: Oxford, 1987; Vol. 6, Chapter 61.3.
2. Sharpless, K. B.; Michaelson, R. C. *JACS* **1973**, *95*, 6136.
3. Sheng, M. N.; Zajacek, J. G. *OSC* **1988**, *6*, 501.
4. Cenci, S.; Di Furia, F.; Modena, G.; Curci, R.; Edwards, J. O. *JCS(P2)* **1978**, 979.
5. (a) Chisholm, M. H.; Rothwell, I. P. In *Comprehensive Coordination Chemistry*; Wilkinson, G., Ed.; Pergamon: Oxford, 1987; Vol. 2, Chapter 15.3. (b) Bradley, D. C.; Mehrotra, R. C.; Gaur, D. P. *Metal Alkoxides*; Academic: New York, 1978. (c) Clark, R. J. H. *The Chemistry of Titanium and Vanadium*; Elsevier: Amsterdam, 1968.
6. (a) Sharpless, K. B.; Verhoeven, T. R. *Aldrichim. Acta* **1979**, *12*, 63. (b) Rao, A. S. *COS* **1991**, *7*, Chapter 3.1. (c) Jorgensen, K. A. *CRV* **1989**, *89*, 431.
7. (a) Sheldon, R. A.; Kochi, J. K. *Metal-Catalyzed Oxidations of Organic Compounds*; Academic: New York, 1981. (b) Gould, E. S.; Hiatt, R. R.; Irwin, K. C. *JACS* **1968**, *90*, 4573.
8. (a) Mihelich, E. D. *TL* **1979**, *49*, 4729. (b) Rossiter, B. E.; Verhoeven, T. R.; Sharpless, K. B. *TL* **1979**, *49*, 4733.
9. (a) Ziegler, F. E.; Jaynes, B. H.; Saindane, M. T. *TL* **1985**, *26*, 3307. (b) Marino, J. P.; de la Pradilla, R. F.; Laborde, E. *JOC* **1987**, *52*, 4898. (c) Stevens, R. V.; Chang, J. H.; Lapalme, R.; Schow, S.; Schlageter, M. G.; Shapiro, R.; Weller, H. N. *JACS* **1983**, *105*, 7719. (d) Rowley, M.; Kishi, Y. *TL* **1988**, *29*, 4909.
10. (a) Luteijn, J. M.; de Groot, A. *JOC* **1981**, *46*, 3448. (b) Pirrung, M. C.; Thomson, S. A. *JOC* **1988**, *53*, 227. (c) Corey, E. J.; De, B. *JACS* **1984**, *106*, 2735. (d) Nakayama, K.; Yamada, S.; Takayama, H.; Nawata, Y.; Iitaka, Y. *JOC* **1984**, *49*, 1537.
11. Itoh, T.; Jitsukawa, K.; Kaneda, K.; Teranishi, S. *JACS* **1979**, *101*, 159.
12. Allison, K.; Johnson, P.; Foster, G.; Sparke, M. B. *Ind. Eng. Chem., Prod. Res. Dev.* **1966**, *5*, 166.
13. Lyons, J. E. *Adv. Chem. Ser.* **1974**, *132*, 64.
14. Adam, W.; Braun, M.; Griesbeck, A.; Lucchini, V.; Staab, E.; Will, B. *JACS* **1989**, *111*, 203.
15. (a) Gould, E. S.; Rado, M. *J. Catal.* **1969**, *13*, 238. (b) Lyons, J. E. *TL* **1974**, *32*, 2737. (c) Kaneda, K.; Jitsukawa, K.; Itoh, T.; Teranishi, S. *JOC* **1980**, *45*, 3004. (d) Arzoumanian, H.; Blanc, A.; Hartig, U.; Metzger, J. *TL* **1974**, *12*, 1011.
16. (a) Sheng, M. N.; Zajacek, J. G. *JOC* **1968**, *33*, 588. (b) Takano, S.; Sugihara, Y.; Ogasawara, K. *H* **1992**, *34*, 1519. (c) Kuhnen, L. *CB* **1966**, *99*, 3384.
17. Howe, G. R.; Hiatt, R. R. *JOC* **1970**, *35*, 4007.
18. (a) Curci, R.; Di Furia, F.; Testi, R.; Modena, G. *JCS(P2)* **1974**, 752. (b) Curci, R.; Di Furia, F.; Modena, G. In *Fundamental Research in Homogeneous Catalysis*; Ishii, Y.; Tsutsui, M., Eds.; Plenum: New York, 1978; Vol. 2, p 255.
19. Tatsuno, Y.; Tatsuda, M.; Otsuka, S. *CC* **1982**, 1100.
20. (a) Mimoun, H.; Chaumette, P.; Mignard, M.; Saussine, L.; Fischer, J.; Weiss, R. *NJC* **1983**, *7*, 467. (b) Spirina, I. V.; Alyasov, V. N.; Glushakova, V. N.; Sharodumora, N. A.; Sergeeva, V. P.; Balakshina, N. N.; Malennikov, V. P.; Aleksandrove, Y. A.; Razuvaev, G. A. *JOC* **1982**, *18*, 1570.

Bryant E. Rossiter†
Brigham Young University, Provo, UT, USA

Zinc[1]

[7440-66-6] Zn (MW 65.39)

(reducing agent;[2] used for preparation of organozinc reagents,[1,3] Reformatsky reagents,[4] and the Simmons–Smith reagent (cyclopropanation)[5])

Physical Data: mp 419 °C; bp 907 °C; *d* 7.14 g cm^{-3}.
Solubility: insol organic solvents; reacts with aqueous acidic solutions.
Form Supplied in: dust, foil, granular, wire, mossy, rod; widely available at low cost.
Handling, Storage, and Precautions: slowly oxidizes in air; no toxic properties are associated with zinc and zinc organometallics; in several cases the metal requires an activation procedure before use.[6]

Reduction of Carbon–Carbon Multiple Bonds. Whereas isolated double bonds are rarely reduced by zinc, triple bonds are cleanly converted to alkenes using either ***Zinc/Copper Couple*** or ***Zinc Amalgam***.[7] A regio- as well as stereospecific reduction of a wide range of alkynic derivatives can be performed using zinc powder (eq 1).[8] The reduction of propargylic alcohols proves to be especially efficient (eq 2).[9] Also, the selective *cis* reduction of conjugated dienynes and trienynes proceeds well with Zn(Cu/Ag).[10] The presence of a leaving group at the propargylic position leads to the formation of allenes.[11] The conjugation of a double bond with an electron-withdrawing substituent considerably facilitates the reduction of the double bond.[12] The reduction of α,β-unsaturated ketones produces the corresponding saturated ketones.[13] Nickel catalysis allows the reduction of unsaturated aldehydes, ketones, and esters in an aqueous medium under ultrasonic irradiation (eq 3).[14]

Et—≡—CH2CH2OH → (Zn, ethanol; reflux, 3 h; >95%) → Et–CH=CH–CH2CH2OH, >95% (*Z*) (1)

OTBDMS → (1. Zn, KCN; 2. Bu_4NF) → OH (2)

→ (Zn, $NiCl_2$ cat.,)))); EtOH, H_2O; 30 °C, 2.5 h; 97%) → (3)

Reduction of Carbonyl Groups. Zinc reduces ketones to either alcohols or to a methylene unit, depending on the reaction conditions and the nature of the substrate. For example, conjugation is required if reduction to a hydroxy group is desired. The reduction of aryl ketones provides benzylic alcohols (eq 4)[15] and α-diketones can be converted selectively to α-hydroxy ketones (eq 5).[16] The reduction of the carbonyl group of nonconjugated ketones to a methylene unit with zinc and hydrochloric acid in organic solvents such as ether, acetic anhydride, or benzene–ethanol proceeds in satisfactory yields with a wide range of ketones (Clemmensen reaction) (eqs 6–8).[17] The Clemmensen reduction of aromatic α-hydroxy ketones gives conjugated alkenes.[18] Finally, the Clemmensen reduction can also be performed by using zinc and ***Chlorotrimethylsilane*** in an aprotic medium, leading to alkenes (eq 9).[19] This variation has been exploited for the preparation of alkenes (eq 10)[19b] and has been used in new cyclization reactions (eqs 11 and 12).[20] Trimethylsilyl ethers can be regioselectively prepared by the zinc reduction of α-chloro ketones in the presence of TMSCl.[21] Mixed pinacol products have been prepared by using Zn(Cu) as the reducing agent (eq 13).[22]

→ (Zn(Hg), HCl; reflux, toluene; 79%) → (4)

→ (Zn, AcOH; reflux) → 48:52 (5)

→ (Zn(Hg), HCl; Et_2O; 77%) → (6)

→ (Zn(Hg), HCl; EtOH, PhH; 56%) → (7)

→ (Zn(Hg), HCl; 50%) → (8)

→ (Zn(Hg), TMSCl; Et_2O; 72%) → (9)

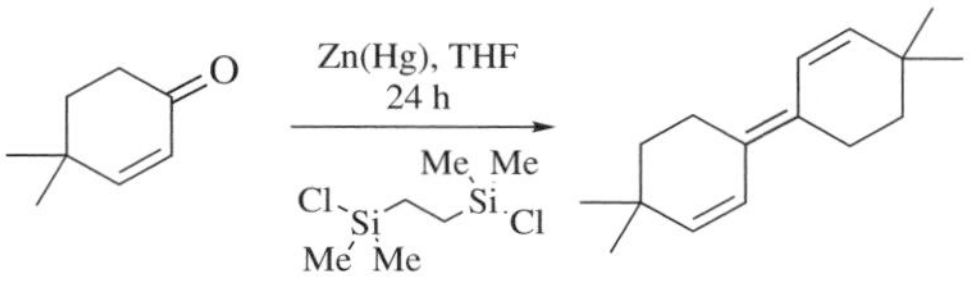

(10)

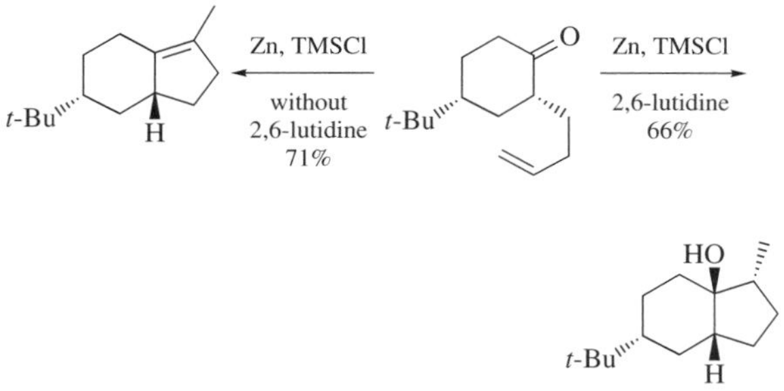

(11)

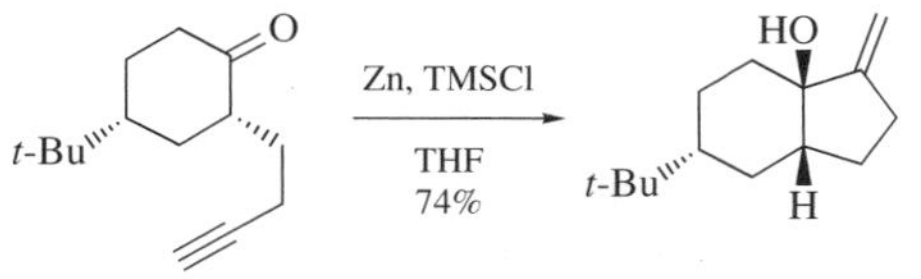

(12)

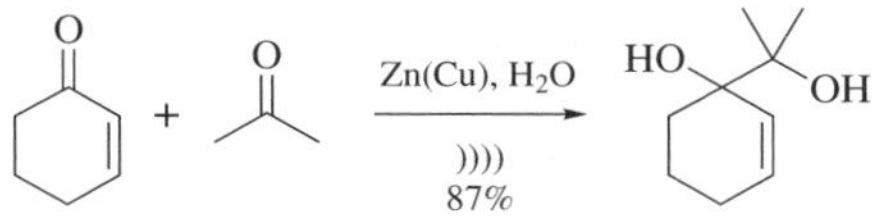

(13)

Reduction of Carbon–Oxygen Bonds. Carbon–oxygen bonds situated α to an unsaturation are easily reduced with zinc in an acidic medium. In the case of α-hydroxy ketones, ketones are obtained in good yields (eq 14).[23] A wide range of allylic or benzylic ethers, acetates, and alcohols are reduced with zinc (eq 15).[24,25] The reduction of epoxides can lead to either alcohols (eq 16)[26] or alkenes.[26b–e] In the presence of catalytic amounts of Pd^0 and zinc dust, allylic acetates are coupled to give 1,5-dienes (eq 17).[27] Under similar reaction conditions and in the presence of an aldehyde, homoallylic alcohols are obtained in satisfactory yields (eq 18).[27b–d]

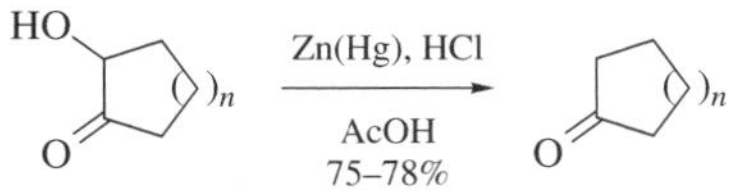

(14)

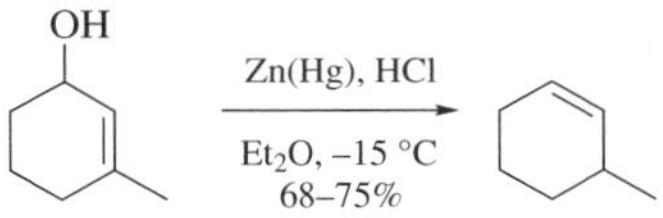

(15)

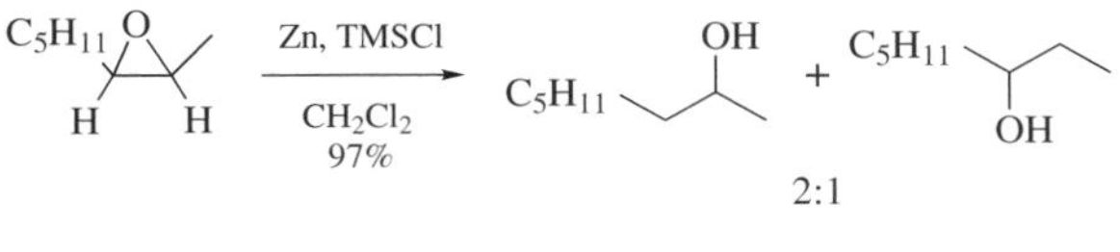

(16)

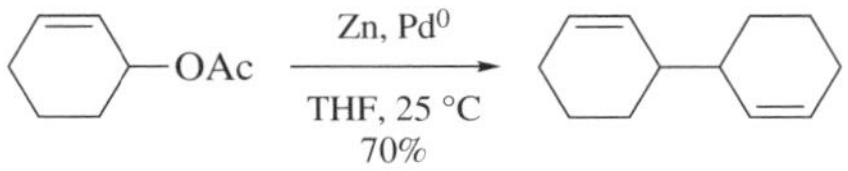

(17)

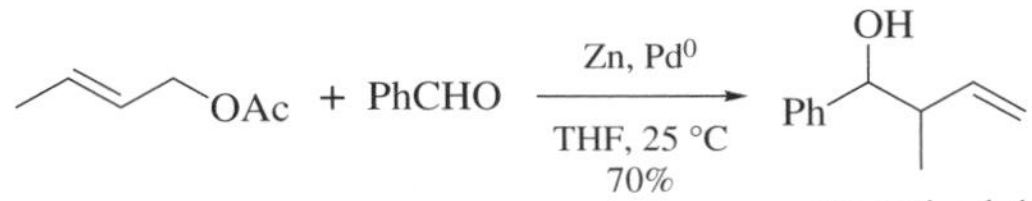

(18)

Reduction of Carbon–Halide Bonds. Alkyl and alkenyl halides are readily reduced with zinc under various reaction conditions. The reduction produces, as an intermediate, an organic radical which can undergo carbon–carbon bond formation (Barbier reaction)[28] or can be further reduced, usually under acidic conditions. Aliphatic iodides or bromides and benzylic chlorides react readily with ***Zinc–Acetic Acid***, providing the corresponding hydrocarbon.[29] Although aromatic halides are reduced less easily, the tribromothiophene (**1**) is reduced selectively to the bromide (**2**) (eq 19).[29e,g] Various β-chloro enones are cleanly reduced to enones with ***Zinc/Silver Couple*** in methanol at rt (eq 20).[29f] α-Dihalo ketones are reduced smoothly, allowing the preparation of a variety of ketones (eqs 21 and 22).[30] The reductive couplings of α-bromo ketones, tropylium, and 1,3-dithiolylium cations have been observed.[31] In the case of 1,3-dihalides, cyclopropanes are obtained in good yields.[32] If the reduction of the carbon–halide bond is performed in the presence of an electrophile, a radical addition often occurs. Thus phenacyl halides can be coupled with methylenecyclohexanes (eq 23).[33a] Performing the reaction in the presence of an unsaturated ketone provides the 1,4-adducts. Interestingly, the reduction proceeds well in an aqueous medium supporting a radical mechanism, since zinc organometallics react instantaneously with water but only very sluggishly with enones (eq 24).[33b–h] The addition of ***Chloromethyl Methyl Ether*** to 1,2-bis-silyl enol ethers in the presence of zinc leads to ring-enlarged 1,3-cycloalkanediones after acidic treatment.[34] An interesting three-component reaction has been described (eq 25).[34b] A wide range of allylic halides undergo Barbier-type addition to carbonyl groups (eqs 26 and 27).[35,36] The reduction of α,α′-dihalo ketones with a zinc–copper couple in the presence of a diene such as ***Isoprene*** provides cycloaddition products via a zinc oxyallyl cation.[37]

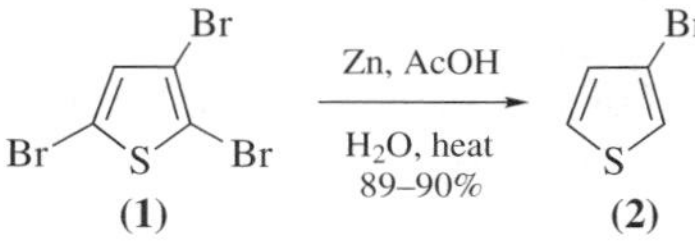

(19)

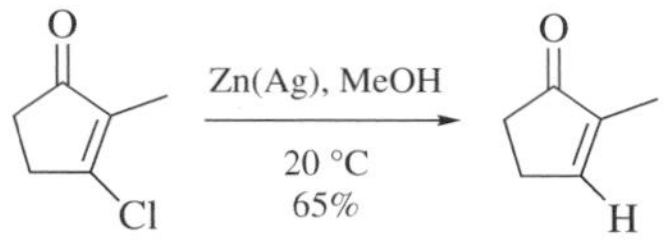

(20)

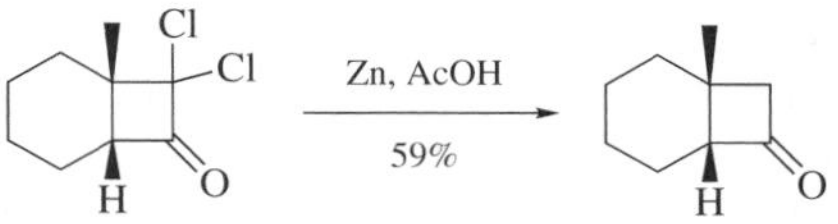

(21)

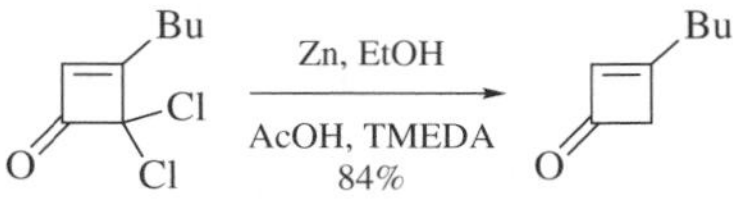

(22)

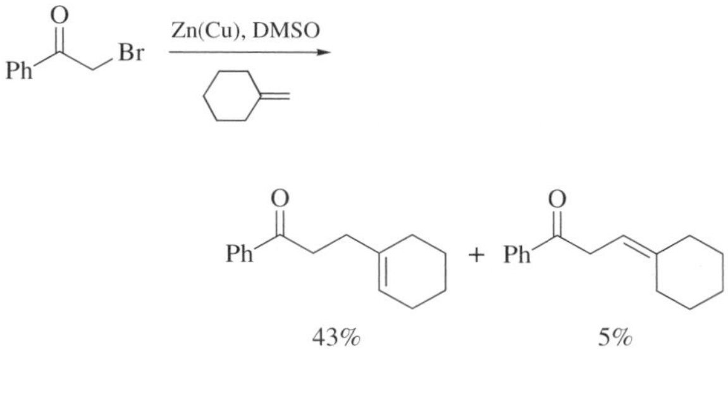

(23)

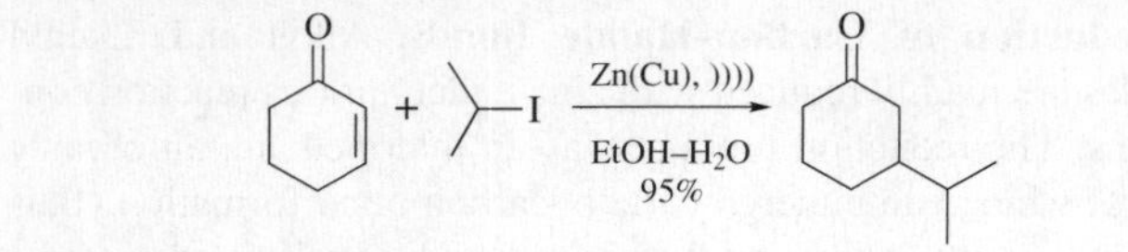

(24)

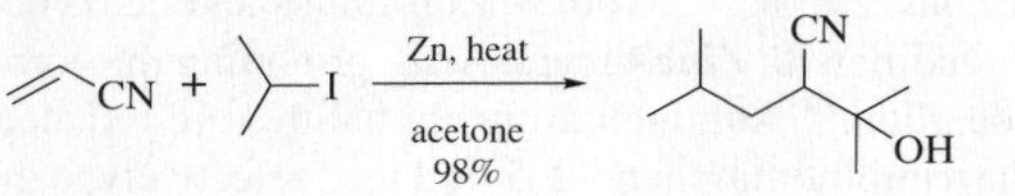

(25)

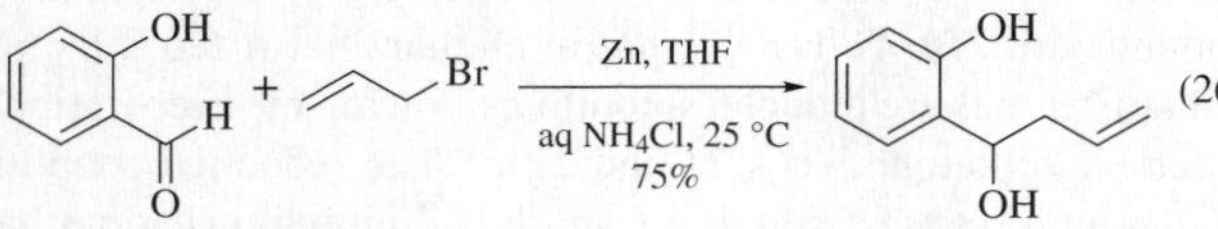

(26)

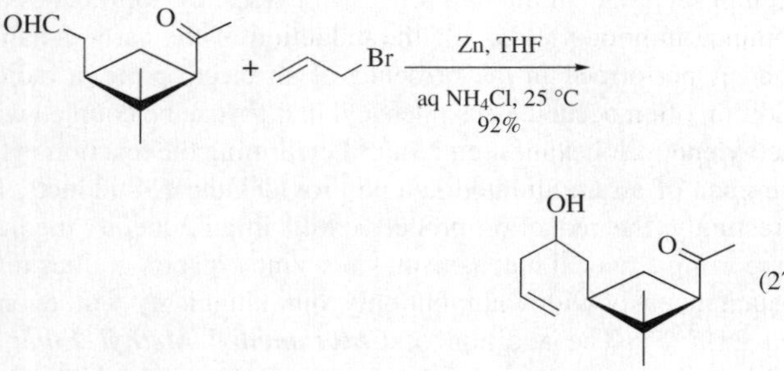

(27)

Reduction of Carbon–Nitrogen and Carbon–Sulfur Bonds. Aldimines and oximes are converted to amines, and various heterocycles bearing carbon–nitrogen double bonds are reduced with zinc under acidic conditions.[38] Cyanamides can be cleanly cleaved leading to amines,[39a] and the zinc reduction of acylnitriles provides α-amino ketone derivatives (eq 28).[39b] Aromatic amides can be reduced with zinc dust to aromatic aldehydes.[39c] Activated carbon–sulfur bonds α to a carbonyl group[40a,b] and sulfur ylides[40c,d] can be reduced with zinc.

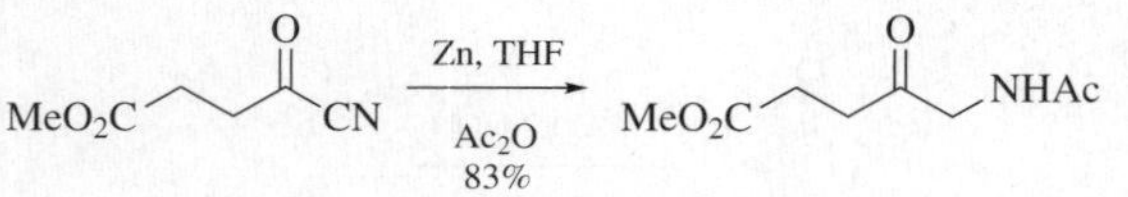

(28)

Reduction at Heteroatoms.[2] Nitrogen–oxygen bonds of oximes,[41] nitro,[42] and nitroso[43] groups are readily reduced by zinc in acidic medium. Zinc in acetic acid has often been used for the workup procedure of alkene ozonolysis to afford aldehydes or ketones.[2] Sulfinates and thiols can be obtained selectively by the reduction of aromatic sulfonyl chlorides or disulfides.[44]

Dehalogenation and Related Reactions.[6c,45] Zinc dust is a very efficient reducing agent for the dehalogenation of 1,2-dihalides or 1-halo-2-alkoxy derivatives, leading to alkenes. The reaction allows an access to highly reactive ketenes,[46] alkenes,[47] or alkynes[48] not readily available by standard methods (eqs 29–31). The reduction of β-alkoxy halides using ***Zinc–Graphite*** proved to be especially interesting when applied to sugar derivatives (eq 32).[6c,45e,49] The dehalogenation using zinc is such a straightforward and chemoselective reaction that several protecting groups have been devised which use this reaction as a deblocking step.[50]

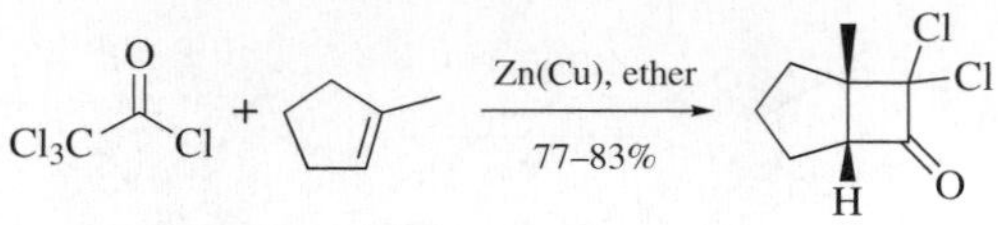

(29)

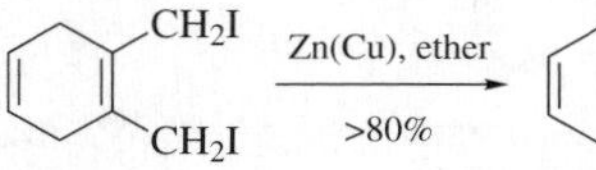

(30)

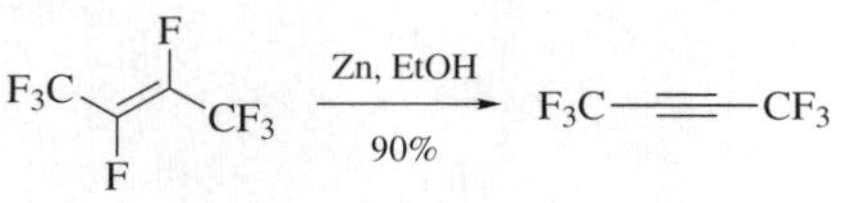

(31)

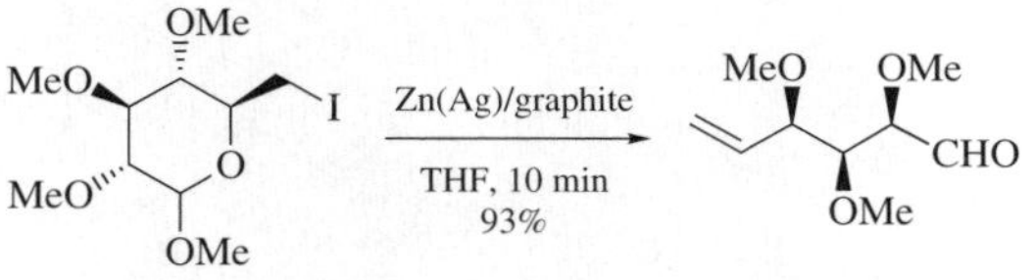

(32)

The Reformatsky Reaction.[4] The insertion of zinc into α-halo esters produces zinc ester enolates which react readily with aldehydes or ketones, leading to aldol products. Historically, this reaction has been important since it allowed the first quantitative generation of an ester enolate. However, several modern synthetic methods for the stereoselective preparation of aldol products using metal enolates compete favorably with the Reformatsky reaction.[51] The nature of the zinc activation[6] has proved to be important for fast and quantitative zinc insertion. Remarkably, the Zn(Ag) couple on graphite reacts with ethyl bromoacetate at −78 °C within 20 min,[52a] whereas Rieke zinc requires 1 h at 25 °C,[52b] as does zinc generated from the reaction of ***Zinc Chloride*** with ***Lithium*** under ultrasound irradiation[52c] (eq 33).[52a] Interesting synthetic applications have been reported (eqs 34 and 35).[4,52] 4-Bromocrotonate reacts with ketones and Zn(Cu) with solvent-dependent regioselectivity.[52f] See also ***Ethyl Bromozincacetate***.

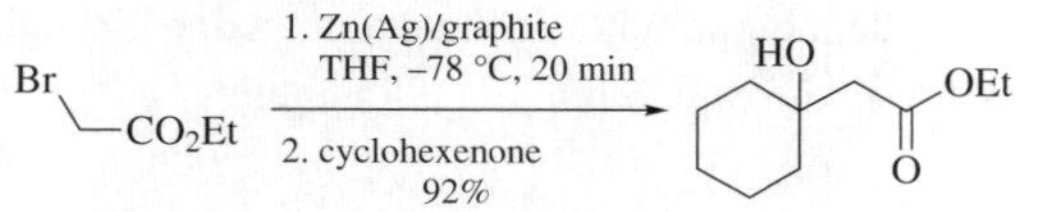

(33)

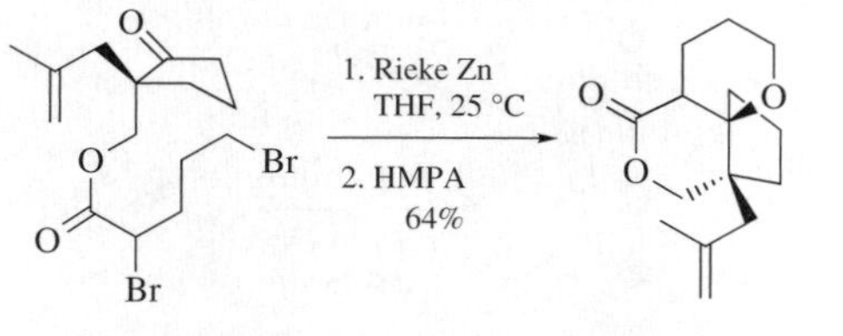

(34)

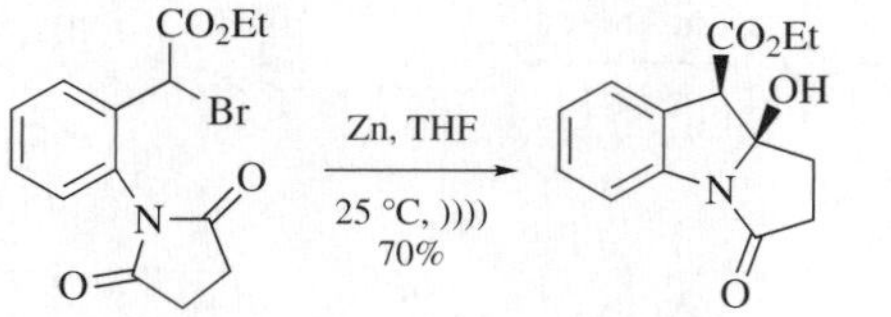

(35)

The Simmons–Smith Reaction.[5,53] Cut Zn foil readily inserts into ***Diiodomethane*** providing iodomethylzinc iodide,[53] which cyclopropanates a wide range of alkenes in good yields (see ***Ethylzinc Iodide***, ***Iodomethylzinc Iodide***, ***Diethylzinc***, ***Ethyliodomethylzinc***). The in situ generation of iodomethylzinc iodide is often used. The Zn(Ag) couple has proved to be especially active for cyclopropanations (eq 36).[53d]

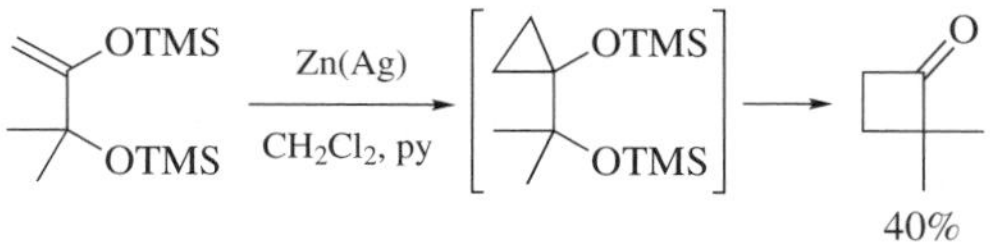

Preparation of Organozinc Reagents.[1,3a] The insertion of zinc into organic halides provides the most general synthesis of organozinc halides. Primary and secondary organic iodides react with zinc dust (2–3 equiv) in THF between 20 °C and 50 °C, leading to organozinc iodides in high yields.[3a,54a–c] Benzylic chlorides and bromides react under even milder conditions, providing the corresponding benzylic zinc halides without the formation of significant amounts of Wurtz coupling products.[54df] Two remarkable properties characterize organozinc reagents: (i) their high functional group compatibility, which allows the preparation of polyfunctional organometallic zinc species bearing almost all common functional groups with the exception of nitro, azido, or hydroxy functions (see the reagents **3**,[54b] **4**,[54g,h] **5**,[54i] **6**,[54j] **7**,[54k–o] **8**,[54p] **9**,[54n,o] **10**, **11**,[54q] and **12–14**,[54f] and eq 37); and (ii) their ability to undergo transmetallation with other metallic salts, such as copper salts, thus giving polyfunctional copper reagents which react readily with a wide range of electrophiles (enones,[54b,r] aldehydes,[54s] alkynes,[54t–v] nitro alkenes,[54w–y] allylic halides,[54b,z] alkynyl halides,[54g] acid chlorides,[54b,aa] and alkylidenemalonates[54ab]). Similarly, efficient transmetalations with Pd^{II} salts allow the coupling reactions to be performed (eq 38).[55,56] Zinc insertion also proceeds well with various polyfluorinated alkyl iodides[57] and with primary alkyl and benzyl phosphates and mesylates.[58] Alkenyl and aromatic halides undergo the zinc insertion far less readily and require the use of polar solvents[59] or highly activated zinc.[60] The use of a sacrificial zinc electrode offers an interesting alternative.[61] Allylic zinc halides are formed under very mild conditions and, contrary to other classes of organozinc reagents, display a high reactivity toward organic electrophiles (comparable to organomagnesium species).[36e–h,54a,62] A wide range of synthetic applications of zinc reagents for the formation of carbon–carbon bonds has been reported (eqs 39–44).[56,60a,63–66] Diorganomercurials also react with zinc dust, providing diorganozincs.[67,68]

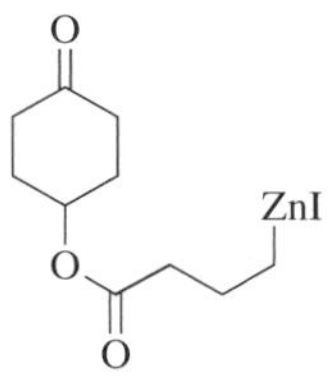

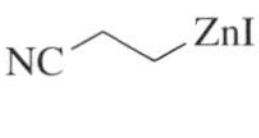

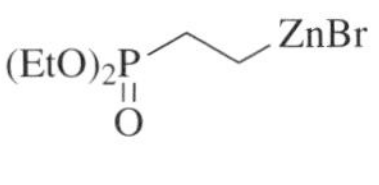

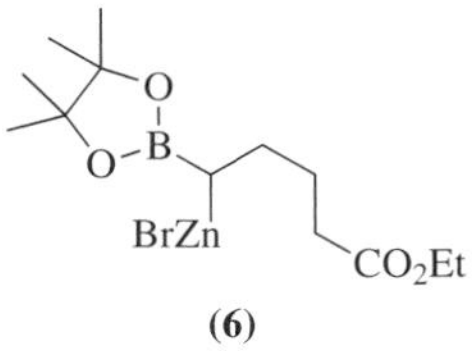

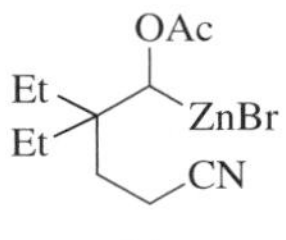

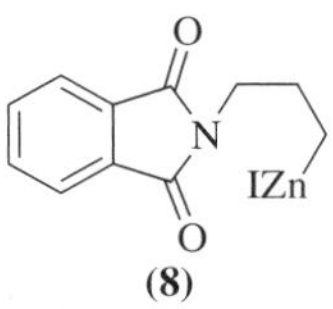

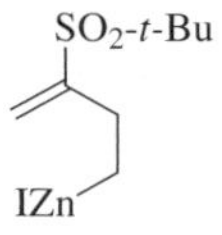

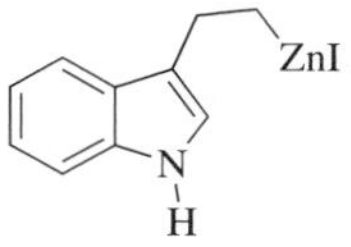

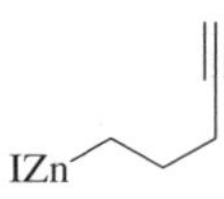

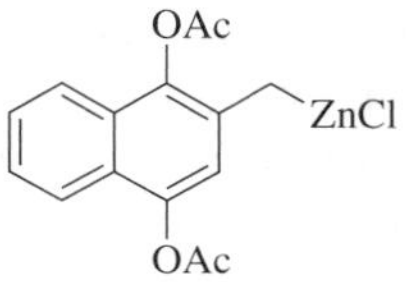

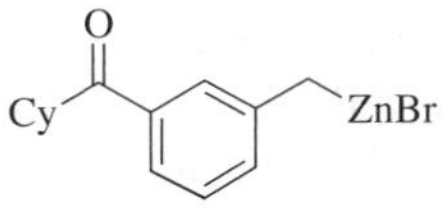

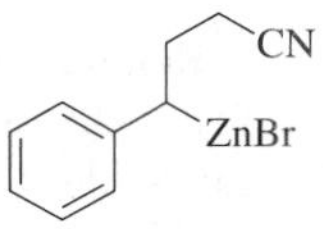

$$\text{FG-RX} + \text{Zn} \xrightarrow[\text{5–45 °C, >85\%}]{\text{THF}} \text{FG-RZnX} \quad (37)$$

X = I, Br; R = alkyl, aryl, benzyl, allyl
FG = CO_2R, enoate, CN, enone, halide, $(RCO)_2N$, $(TMS)_2Si$, RNH, NH_2, RCONH, $(RO)_3Si$, $(RO)_2P(O)$, RS, RS(O), RSO_2, PhCOS

$$\text{FGR}^1\text{R}^2 \xleftarrow[\text{Pd}^0]{\text{R}^2\text{X}} \text{FGR}^1\text{ZnX} \xrightarrow[\text{2. E}^+]{\text{1. CuCN•2LiCl}} \text{FGR}^1\text{E} \quad (38)$$

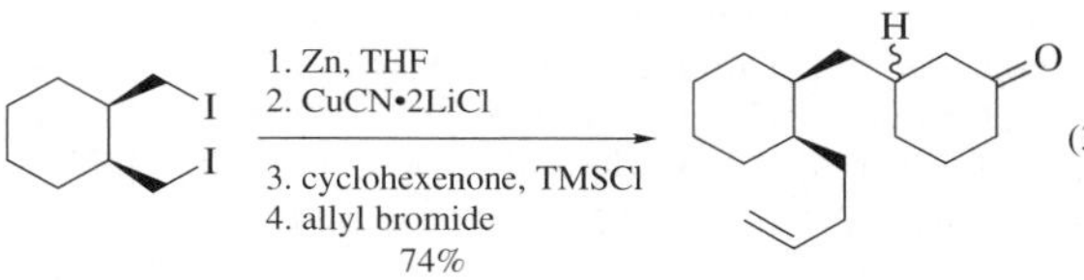

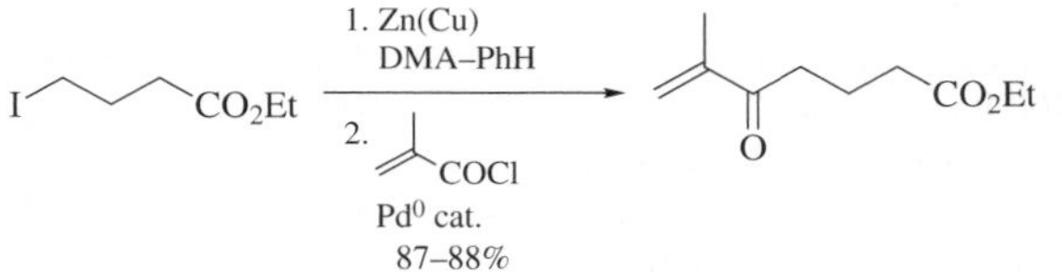

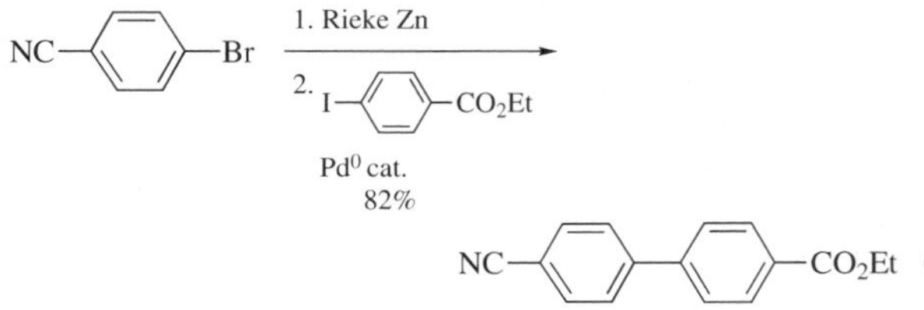

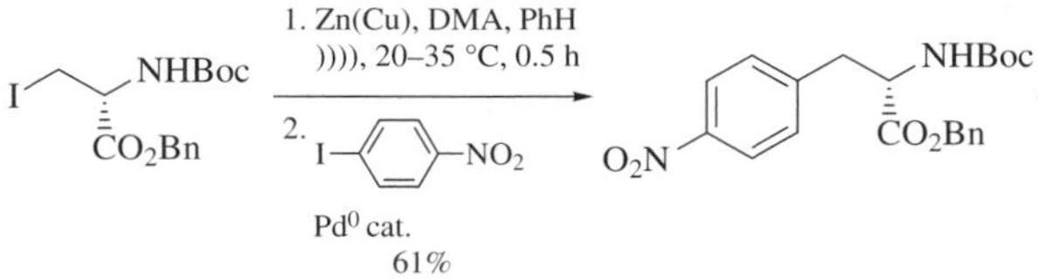

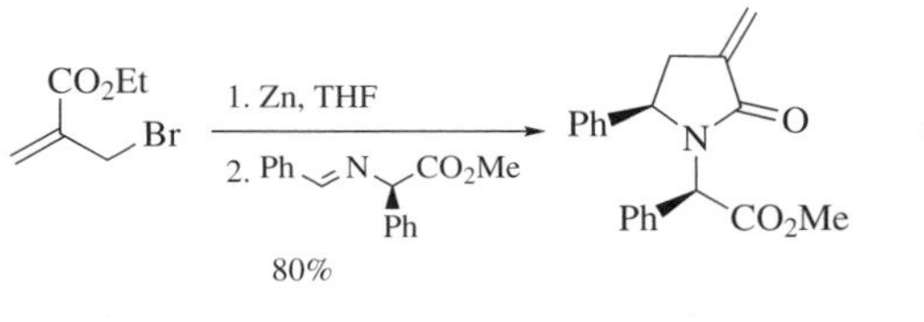

1. Zn, THF
2. $\diagup\!\!\diagdown$MgBr
3. CO_2Et / CO_2Et / $AcO\!\diagup(\,)_6$

Br → AcO (44)

88%

Related Reagents. See Classes R-2, R-4, R-5, R-9, R-15, R-20, R-23, R-24, R-25, R-27, R-29, R-31, and R-32, pages 1–10. Dibromomethane–Zinc–Titanium(IV) Chloride; Dichlorobis(cyclopentadienyl)zirconium–Zinc–Dibromomethane; Diiodomethane–Zinc–Titanium(IV) Chloride; Molybdenum(V) Chloride–Zinc; Niobium(V) Chloride–Zinc; Phosphorus(III) Bromide–Copper(I) Bromide–Zinc; Potassium Hexachloroosmate(IV)–Zinc; Titanium(IV) Chloride–Zinc; Zinc–Acetic Acid; Zinc Amalgam; Zinc–Copper(II) Acetate–Silver Nitrate; Zinc–Copper(I) Chloride; Zinc/Copper Couple; Zinc–1,2-Dibromoethane Zinc–Dimethylformamide; Zinc–Graphite; Zinc/Nickel Couple; Zinc/Silver Couple; Zinc–Zinc Chloride.

---

1. (a) Nützel, K. *MOC* **1973**, *13/2*, 552. (b) Sheverdina, N. I.; Kocheshkov, K. A. In *Methods of Elemento-Organic Chemistry*; Nesmeyanov, A. N.; Kocheshkov, K. A., Ed.; North-Holland: Amsterdam, 1967; Vol. 3. (c) Crompton, T. R. *Analysis of Organoaluminium and Organozinc Compounds*; Pergamon: Oxford, 1968.
2. (a) Martin, E. L. *OR* **1942**, *1*, 155. (b) Staschewski, D. *AG* **1959**, *71*, 726. (c) Buchanan, J. G. S. C.; Woodgate, P. D. *QR* **1969**, *23*, 522. (d) Vedejs, E. *OR* **1975**, *22*, 401. (e) Muth, M.; Sauerbier, M. *MOC* **1981**, 4/1c, 709.
3. (a) Knochel, P. *CRV* **1993**, *93*, 217. (b) Elschenbroich, C.; Salzer, A. *Organometallics: A Concise Introduction*; VCH: Weinheim, 1989. Carruthers, W. In *Comprehensive Organometallic Chemistry*; Wilkinson, G., Ed.; Pergamon: Oxford, 1982; Vol. 7, p 661.
4. (a) Gaudemar, M. *Organomet. Chem. Rev. (A)* **1972**, *8*, 183. (b) Rathke, M. W. *OR* **1975**, *22*, 423. (c) Fürstner, A. *S* **1989**, 571.
5. (a) Simmons, H. E.; Cairns, T. L.; Vladuchick, A.; Hoiness, C. M. *OR* **1972**, *20*, 1. (b) Furukawa, J.; Kawabata, N. *Adv. Organomet. Chem.* **1974**, *12*, 83. (c) Zeller, K.-P.; Gugel, H. *MOC* **1989**, *EXIXb*, 195.
6. (a) Erdik, E. *T* **1987**, *43*, 2203. (b) Rieke, R. D. *Science* **1989**, *246*, 1260. (c) Fürstner, A. *AG(E)* **1993**, *32*, 164.
7. (a) Morris, S. G.; Herb, S. F.; Magidman, P.; Luddy, F. E. *J. Am. Oil Chem. Soc.* **1972**, *49*, 92. (b) Sondengam, B. L.; Charles, G.; Akam, T. M. *TL* **1980**, *21*, 1069.
8. (a) Aerssens, M. H. P. J.; van der Heiden, R.; Heus, M.; Brandsma, L. *SC* **1990**, *20*, 3421. (b) Solladié, G.; Stone, G. B.; Andrés, J.-M.; Urbano, A. *TL* **1993**, *34*, 2835.
9. (a) Näf, F.; Decorzant, R.; Thommen, W.; Willhalm, B.; Ohloff, G. *HCA* **1975**, *58*, 1016. (b) Oppolzer, W.; Fehr, C.; Warneke, J. *HCA* **1977**, *60*, 48. (c) Winter, M.; Näf, F.; Furrer, A.; Pickenhagen, W.; Giersch, W.; Meister, A.; Willhalm, B.; Thommen, W.; Ohloff, G. *HCA* **1979**, *62*, 135.
10. (a) Boland, W.; Schroer, N.; Sieler, C.; Feigel, M. *HCA* **1987**, *70*, 1025. (b) Avignon-Tropis, M.; Pougny, J. R. *TL* **1989**, *30*, 4951. (c) Chou, W.-N.; Clark, D. L.; White, J. B. *TL* **1991**, *32*, 299.
11. (a) Biollaz, M.; Haefliger, W.; Verlade, E.; Crabbé, P.; Fried, J. H. *CC* **1971**, 1322. (b) Maurer, H.; Hopf, H. *AG(E)* **1976**, *15*, 628. (c) Kloster-Jensen, E.; Wirz, J. *HCA* **1975**, *58*, 162.
12. (a) Davis, B. R.; Woodgate, P. D. *JCS(C)* **1966**, 2006. (b) Davis, B. R.; Woodgate, P. D. *JCS* **1965**, 5943. (c) Toda, F.; Iida, K. *CL* **1976**, 695.
13. (a) Chaykovsky, M.; Lin, M. H.; Rosowsky, A. *JOC* **1972**, *37*, 2018. (b) Marker, R. E.; Crooks, Jr., H. M.; Wagner, R. B.; Wittbecker, E. L. *JACS* **1942**, *64*, 2089.
14. Petrier, C.; Luche, J.-L. *TL* **1987**, *28*, 2347, 2351.
15. (a) Weeks, D. P.; Cella, J. *JOC* **1969**, *34*, 3713. (b) Dickinson, J. D.; Eaborn, C. *JCS* **1959**, 2337. (c) Wiselogle, F. Y.; Sonneborn, H. *OSC* **1941**, *1*, 90. (d) Gardner, J. H.; Naylor, C. A. *OSC* **1943**, *2*, 526.
16. (a) Coulombeau, C.; Rassat, A. *BSF* **1970**, 1199. (b) Rosnati, V. *TL* **1992**, *33*, 4791.
17. (a) Yamamura, S.; Toda, M.; Hirata, Y. *OSC* **1988**, *6*, 289. (b) Marchand, A. P.; Weimar, Jr., W. R. *JOC* **1969**, *34*, 1109. (c) Winternitz, F.; Mousseron, M. *BSF* **1949**, *16*, 713. (d) Nesty, G. A.; Marvel, C. S. *JACS* **1937**, *59*, 2662. (e) Minabe, M.; Yoshida, M.; Fujimoto, M.; Suzuki, K. *JOC* **1976**, *41*, 1935. (f) Mayer, R.; Bürger, H.; Matauschek *JPR* **1961**, *285*, 261. (g) Borden, W. T.; Ravindranathan, T. *JOC* **1971**, *36*, 4125. (h) Martin, E. L. *OSC* **1943**, *2*, 499. (i) Read, R. R.; Wood, J. *OSC* **1955**, *3*, 444. (j) Schwarz, R.; Hering, H. *OSC* **1963**, *4*, 203. (k) Burdon, J.; Price, R. C. *CC* **1986**, 893. (l) Di Vona, M. L.; Floris, B.; Luchetti, L.; Rosnati, V. *TL* **1990**, *31*, 6081. (m) Frank, R. L.; Smith, P. V. *OSC* **1955**, *3*, 410.
18. Shriner, R. L.; Berger, A. *OSC* **1955**, *3*, 786.
19. (a) Motherwell, W. B. *CC* **1973**, 935. (b) Afonso, C. A. M.; Motherwell, W. B.; O'Shea, D. M.; Roberts, L. R. *TL* **1992**, *33*, 3899. (c) Boudjouk, P.; So, J. H. *SC* **1986**, *16*, 775.
20. (a) Corey, E. J.; Pyne, S. G. *TL* **1983**, *24*, 2821. (b) Shono, T.; Hamaguchi, H.; Nishiguchi, I.; Sasaki, M.; Miyamoto, T.; Miyamoto, M.; Fujita, S. *CL* **1981**, 1217.
21. Rubottom, G. M.; Mott, R. C.; Krueger, D. S. *SC* **1977**, *7*, 327.
22. Delair, P.; Luche, J.-L. *CC* **1989**, 398.
23. Cope, A. C.; Barthel, J. W.; Smith, R. D. *OSC* **1963**, *4*, 218.
24. (a) Elphimoff-Felkin, I.; Sarda, P. *OSC* **1988**, *6*, 769. (b) Elphimoff-Felkin, I.; Sarda, P. *T* **1977**, *33*, 511.
25. Prostenik, M.; Butula, I. *CB* **1977**, *110*, 2106.
26. Vankar, Y. D.; Arya, P. S.; Rao, C. T. *SC* **1983**, *13*, 869.
27. (a) Sasaoka, S.; Yamamoto, T.; Kinoshita, H.; Inomata, K.; Kotake, H. *CL* **1985**, 315. (b) Masuyama, Y.; Nimura, Y.; Kurusu, Y. *TL* **1991**, *32*, 225.
28. Blomberg, C.; Hartog, F. A. *S* **1977**, 18.
29. (a) Levene, P. A. *OSC* **1943**, *2*, 320. (b) Boerhorst, E.; Klumpp, G. W. *RTC* **1976**, *95*, 50. (c) Olieman, C.; Maat, L.; Beyerman, H. C. *RTC* **1976**, *95*, 189. (d) Hassner, A.; Hoblitt, R. P.; Heathcock, C.; Kropp, J. E.; Lorber, M. *JACS* **1970**, *92*, 1326. (e) Gronowitz, S.; Raznikiewicz, T. *OSC* **1973**, *5*, 149.
30. (a) Jeffs, P. W.; Molina, G. *CC* **1973**, 3. (b) Eck, C. R.; Mills, R. W.; Money, T. *CC* **1973**, 911. (c) Danheiser, R. L.; Savariar, S. *TL* **1987**, *28*, 3299. (d) Danheiser, R. L.; Savariar, S.; Cha, D. D. *OS* **1989**, *68*, 32.
31. (a) Doering, W. E.; Knox, L. H. *JACS* **1957**, *79*, 352. (b) Kruger, A.; Wudl, F. *JOC* **1977**, *42*, 2778.
32. (a) Corbin, T. F.; Hahn, R. C.; Shechter, H. *OSC* **1973**, *5*, 328. (b) Giusti, G.; Morales, C. *BSF* **1973**, 382.
33. (a) Luche, J.-L.; Allavena, C. *TL* **1988**, *29*, 5369. (b) Luche, J.-L.; Allavena, C.; Petrier, C.; Dupuy, C. *TL* **1988**, *29*, 5373. (c) Dupuy, C.; Petrier, C.; Sarandeses, L. A.; Luche, J.-L. *SC* **1991**, *21*, 643. (d) Sarandeses, L. A.; Mourino, A.; Luche, J.-L. *CC* **1991**, 818. (e) Sarandeses, L. A.; Mourino, A.; Luche, J.-L. *CC* **1992**, 798. (f) Einhorn, C.; Einhorn, J.; Luche, J.-L. *S* **1989**, 787. (g) Petrier, C.; Dupuy, C.; Luche, J. L. *TL* **1986**, *27*, 3149. (h) Kong, K.-C.; Cheng, C.-H. *OM* **1992**, *11*, 1972.
34. (a) Nishiguchi, I.; Hirashima, T.; Shono, T.; Sasaki, M. *CL* **1981**, 551. (b) Shono, T.; Nishiguchi, I.; Sasaki, M. *JACS* **1978**, *100*, 4314.
35. (a) Petrier, C.; Luche, J.-L. *JOC* **1985**, *50*, 910. (b) Petrier, C.; Einhorn, J.; Luche, J.-L. *TL* **1985**, *26*, 1449. (c) Einhorn, C.; Luche, J.-L. *JOM* **1987**, *322*, 177. (d) Knochel, P.; Normant, J. F. *TL* **1984**, *25*, 1475. (e) Knochel, P.; Normant, J. F. *JOM* **1986**, *309*, 1.
36. (a) Öhler, E.; Reininger, K.; Schmidt, U. *AG(E)* **1970**, *9*, 457. (b) Löffler, A.; Pratt, R. D.; Pucknat, J.; Gelbard, G.; Dreiding, A. S. *C* **1969**, *23*, 413. (c) Auvray, P.; Knochel, P.; Normant, J. F. *T* **1988**, *44*, 4495. (d) Auvray, P.; Knochel, P.; Normant, J. F. *T* **1988**, *44*, 4509. (e) El Alami, N.; Belaud, C.; Villiéras, J. *JOM* **1987**, *319*, 303. (f) El Alami, N.; Belaud, C.; Villiéras, J. *JOM* **1988**, *348*, 1. (g) Belaud, C.; Roussakis, C.; Letourneux,

Y.; El Alami, N.; Villiéras, J. *SC* **1985**, *15*, 1233. (h) El Alami, N.; Belaud, C.; Villiéras, J. *TL* **1987**, *28*, 59. (i) Semmelhack, M. F.; Wu, E. S. C. *JACS* **1976**, *98*, 3384.

37. (a) Chidgey, R.; Hoffmann, H. M. R. *TL* **1977**, 2633. (b) Vinter, J. G.; Hoffmann, H. M. R. *JACS* **1974**, *96*, 5466. (c) Sato, T.; Noyori, R. *BCJ* **1978**, *51*, 2745.
38. (a) Emerson, W. S.; Neumann, F. W.; Moundres, T. P. *JACS* **1941**, *63*, 972. (b) Bellasio, E. *SC* **1976**, *6*, 85. (c) Niemers, E.; Hiltmann, R. *S* **1976**, 593.
39. (a) Fehr, T.; Stadler, P. A.; Hofmann, A. *HCA* **1970**, *53*, 2197. (b) Pfaltz, A.; Anwar, S. *TL* **1984**, *25*, 2977. (c) Atta-Ur-Rahman, Basha, A. *CC* **1976**, 594.
40. (a) Kurozumi, S.; Toru, T.; Kobayashi, M.; Ishimoto, S. *SC* **1977**, *7*, 427. (b) Schmid, H.; Schnetzler, E. *HCA* **1951**, *34*, 894. (c) Katayama, S.; Fukuda, K.; Watanabe, T.; Yamauchi, M. *S* **1988**, 178. (d) Ide, J.; Kishida, Y. *BCJ* **1976**, *49*, 3239.
41. (a) Johnson, A. W.; Price, R. *OSC* **1973**, *5*, 1022. (b) Zambito, A. J.; Howe, E. E. *OSC* **1973**, *5*, 373.
42. (a) Kamm, O. *OSC* **1941**, *1*, 445. (b) Kuhn, W. E. *OSC* **1943**, *2*, 447. (c) Martin, E. L. *OSC* **1943**, *2*, 501. (d) Shriner, R. L.; Neumann, F. W. *OSC* **1955**, *3*, 73. (e) Bigelow, H. E.; Robinson, D. B. *OSC* **1955**, *3*, 103. (f) Coleman, G. H.; McCloskey, C. M.; Suart, F. A. *OSC* **1955**, *3*, 668.
43. (a) Fischer, H. *OSC* **1943**, *2*, 202. (b) Hatt, H. H. *OSC* **1943**, *2*, 211. (c) Hartman, W. W.; Roll, L. J. *OSC* **1943**, *2*, 418. (d) Achiwa, K.; Yamada, S. I. *TL* **1975**, 2701.
44. (a) Whitmore, F. C.; Hamilton, F. H. *OSC* **1941**, *1*, 492. (b) Adams, R.; Mawel, C. S. *OSC* **1941**, *1*, 504. (c) Allen, C. F. H.; MacKay, D. D. *OSC* **1943**, *2*, 580. (d) Caesar, P. D. *OSC* **1963**, *4*, 695.
45. (a) Arora, A. S.; Ugi, I. K. *MOC* **1972**, *V/1b*, 740. (b) Stroh, R. *MOC* **1960**, *V/4*, 721. (c) Jäger, V.; Viehe, H. G. *MOC* **1977**, *V/2a*, 39. (d) Neunhoeffer, H.; Franke, W. K. *MOC* **1972**, *5/1d*, 656. (e) Csuk, R.; Glänzer, B. I.; Fürstner, A. *Adv. Organomet. Chem.* **1988**, *28*, 85.
46. (a) Deprés, J.-P.; Greene, A. E. *OSC* **1993**, *8*, 377. (b) Smith, C. W.; Norton, D. G. *OSC* **1963**, *4*, 348. (c) Brady, W. T.; Patel, A. D. *S* **1972**, 565. (d) Hassner, A.; Dillon, J. L. *JOC* **1983**, *48*, 3382. (e) Ammann, A. A.; Rey, M.; Dreiding, A. S. *HCA* **1987**, *70*, 321. (f) McCarney, C. C.; Ward, R. S. *JCS(P1)* **1975**, 1600.
47. (a) Angus, R. O.; Johnson, R. P. *JOC* **1983**, *48*, 273. (b) Rubottom, G. M.; Wey, J. E. *SC* **1984**, *14*, 507. (c) Han, B. H.; Boudjouk, P. *JOC* **1982**, *47*, 751. (d) Sato, F.; Akiyama, T.; Iida, K.; Sato, M. *S* **1982**, 1025. (e) Chapman, O. L.; Chang, C. C.; Rosenquist, N. R. *JACS* **1976**, *98*, 262. (f) Burton, D. J.; Greenlimb, P. E. *JOC* **1975**, *40*, 2796. (g) Gund, T. M.; Schleyer, P. V. R. *TL* **1973**, 1959. (h) Read, G.; Ruiz, V. M. *JCS(P1)* **1973**, 1223. (i) Cava, M. P.; Buck, K. T. *JACS* **1973**, *95*, 5805. (j) Gaoni, Y. *TL* **1973**, 2361.
48. (a) Finnegan, W. G.; Norris, W. P. *JOC* **1963**, *28*, 1139. (b) Banks, R. E.; Barlow, M. G.; Davies, W. D.; Haszeldine, R. N.; Mullen, K.; Taylor, D. R. *TL* **1968**, 3909. (c) Haszeldine, R. N. *JCS* **1952**, 2504.
49. Fürstner, A.; Weidmann, H. *JOC* **1989**, *54*, 2307.
50. (a) Imai, J.; Torrence, P. F. *JOC* **1981**, *46*, 4015. (b) Corey, E. J.; Trybulski, E. J.; Suggs, J. W. *TL* **1976**, 4577. (c) Corey, E. J.; Ruden, R. A. *JOC* **1973**, *38*, 834. (d) Eckstein, F.; Scheit, K.-H. *AG(E)* **1967**, *6*, 362. (e) Franke, A.; Scheit, K.-H.; Eckstein, F. *CB* **1968**, *101*, 2998. (f) Windholz, T. B.; Johnston, D. B. R. *TL* **1967**, 2555. (g) Pike, J. E.; Lincoln, F. H.; Schneider, W. P. *JOC* **1969**, *34*, 3552. (h) Horne, D.; Gaudino, J.; Thompson, W. J. *TL* **1984**, *25*, 3529.
51. Heathcock, C. H. In *Asymmetric Synthesis*; Morrison, J. D., Ed.; Academic: London, 1983; p 111.
52. (a) Csuk, R.; Fürstner, A.; Weidmann, H. *CC* **1986**, 775. (b) Rieke, R. D.; Uhm, S. J. *S* **1975**, 452. (c) Boudjouk, P.; Thompson, D. P.; Ohrbom, W. H.; Han, B. H. *OM* **1986**, *5*, 1257. (d) Ruggeri, R. B.; Heathcock, C. H. *JOC* **1987**, *52*, 5745. (e) Flitsch, W.; Rukamp, P. *LA* **1985**, 1398. (f) Rice, L. E.; Boston, M. C.; Finklea, H. O.; Suder, B. J.; Frazier, J. O.; Hudlicky, T. *JOC* **1984**, *49*, 1845. (g) Boldrini, G. P.; Savoia, D.; Tagliavini, E.; Trombini, C.; Umani-Ronchi, A. *JOC* **1983**, *48*, 4108.
53. (a) Seyferth, D.; Andrews, S. B. *JOM* **1971**, *30*, 151. (b) Seyferth, D.; Dertouzos, H.; Todd, L. J. *JOM* **1965**, *4*, 18. (c) Sidduri, A.; Rozema, M. J.; Knochel, P. *JOC* **1993**, *58*, 2694. (d) Denis, J. M.; Girard, C.; Conia, J. M. *S* **1972**, 549.
54. (a) Gaudemar, M. *BSF* **1962**, 974. (b) Knochel, P.; Yeh, M. C. P.; Berk, S. C.; Talbert, J. *JOC* **1988**, *53*, 2390. (c) Knochel, P.; Rozema, M. J.; Tucker, C. E.; Retherford, C.; Furlong, M.; AchyuthaRao, S. *PAC* **1992**, *64*, 361. (d) Berk, S. C.; Knochel, P.; Yeh, M. C. P. *JOC* **1988**, *53*, 5789. (e) Chen, H. G.; Hoechstetter, C.; Knochel, P. *TL* **1989**, *30*, 4795. (f) Berk, S. C.; Yeh, M. C. P.; Jeong, N.; Knochel, P. *OM* **1990**, *9*, 3053. (g) Yeh, M. C. P.; Knochel, P. *TL* **1988**, *29*, 2395. (h) Majid, T. N.; Yeh, M. C. P.; Knochel, P. *TL* **1989**, *30*, 5069. (i) Retherford, C.; Chou, T.-S.; Schelkun, R. M.; Knochel, P. *TL* **1990**, *31*, 1833. (j) Knochel, P. *JACS* **1990**, *112*, 7431. (k) Knochel, P.; Chou, T.-S.; Chen, H.-G.; Yeh, M. C. P.; Rozema, M. J. *JOC* **1989**, *54*, 5202. (l) Chou, T.-S.; Knochel, P. *JOC* **1990**, *55*, 4791. (m) Knochel, P.; Chou, T.-S.; Jubert, C.; Rajagopal, D. *JOC* **1993**, *58*, 588. (n) AchyuthaRao, S.; Tucker, C. E.; Knochel, P. *TL* **1990**, *31*, 7575. (o) AchyuthaRao, S.; Chou, T.-S.; Schipor, I.; Knochel, P. *T* **1992**, *48*, 2025. (p) Yeh, M. C. P.; Chen, H. G.; Knochel, P. *OS* **1991**, *70*, 195. (q) Knoess, H. P.; Furlong, M. T.; Rozema, M. J.; Knochel, P. *JOC* **1991**, *56*, 5974. (r) Yeh, M. C. P.; Knochel, P.; Butler, W. M.; Berk, S. C. *TL* **1988**, *29*, 6693. (s) Yeh, M. C. P.; Knochel, P.; Santa, L. E. *TL* **1988**, *29*, 3887. (t) Yeh, M. C. P.; Knochel, P. *TL* **1989**, *30*, 4799. (u) AchyuthaRao, S.; Knochel, P. *JACS* **1991**, *113*, 5735. (v) Knochel, P. *COS* **1991**, *4*, 865. (w) Retherford, C.; Yeh, M. C. P.; Schipor, I.; Chen, H.-G.; Knochel, P. *JOC* **1989**, *54*, 5200. (x) Retherford, C.; Knochel, P. *TL* **1991**, *32*, 441. (y) Jubert, C.; Knochel, P. *JOC* **1992**, *57*, 5431. (z) Chen, H. G.; Gage, J. L.; Barrett, S. D.; Knochel, P. *TL* **1990**, *31*, 1829. (aa) Sidduri, A.; Budries, N.; Laine, R. M.; Knochel, P. *TL* **1992**, *33*, 7515. (ab) Cahiez, G.; Venegas, P.; Tucker, C. E.; Majid, T. N.; Knochel, P. *CC* **1992**, 1406.
55. (a) Negishi, E.; Valente, L. F.; Kobayashi, M. *JACS* **1980**, *102*, 3298. (b) Kobayashi, M.; Negishi, E. *JOC* **1980**, *45*, 5223. (c) Negishi, E. *ACR* **1982**, *15*, 340. (d) Tamaru, Y.; Ochiai, H.; Yoshida, Z. *TL* **1984**, *25*, 3861. (e) Tamaru, Y.; Ochiai, H.; Nakamura, T.; Tsubaki, K.; Yoshida, Z. *TL* **1985**, *26*, 5559. (f) Tamaru, Y.; Ochiai, H.; Sanda, F.; Yoshida, Z. *TL* **1985**, *26*, 5529. (g) Tamaru, Y.; Ochiai, H.; Nakamura, T.; Yoshida, Z. *TL* **1986**, *27*, 955. (h) Tamaru, Y.; Ochiai, H.; Nakamura, T.; Yoshida, Z. *AG(E)* **1987**, *26*, 1157. (i) Nakamura, E.; Aoki, S.; Sekiya, K.; Oshino, H.; Kuwajima, I. *JACS* **1987**, *109*, 8056.
56. (a) Jackson, R. F. W.; James, K.; Wythes, M. J.; Wood, A. *CC* **1989**, 644. (b) Jackson, R. F. W.; Wythes, M. J.; Wood, A. *TL* **1989**, *30*, 5941. (c) Jackson, R. F. W.; Wood, A.; Wythes, M. J. *SL* **1990**, 735. (d) Dunn, M. J.; Jackson, R. F. W. *CC* **1992**, 319. (e) Jackson, R. F. W.; Wishart, N.; Wythes, M. J. *CC* **1992**, 1587. (f) Dunn, M. J.; Jackson, R. F. W.; Stephenson, G. R. *SL* **1992**, 905. (g) Jackson, R. F. W.; Wishart, N.; Wythes, M. J. *SL* **1993**, 219. (h) Jackson, R. F. W.; Wishart, N.; Wood, A.; James, K.; Wythes, M. J. *JOC* **1992**, *57*, 3397.
57. Burton, D. J.; Xang, Z.-Y. *T* **1992**, *48*, 189.
58. Jubert, C.; Knochel, P. *JOC* **1992**, *57*, 5425.
59. (a) Majid, T. N.; Knochel, P. *TL* **1990**, *31*, 4413. (b) JanakiramRao, C.; Knochel, P. *JOC* **1991**, *56*, 4593. (c) Waas, J. R.; Sidduri, A.; Knochel, P. *TL* **1992**, *33*, 3717. (d) JanakiramRao, C.; Knochel, P. *T* **1993**, *49*, 29.
60. (a) Zhu, L.; Wehmeyer, R. M.; Rieke, R. D. *JOC* **1991**, *56*, 1445. (b) Zhu, L.; Rieke, R. D. *TL* **1991**, *32*, 2865. (c) Klabunde, K. J. *AG(E)* **1975**, *14*, 287. (d) Murdock, T. O.; Klabunde, K. J. *JOC* **1976**, *41*, 1075.
61. Sibille, S.; Ratovelomanana, V.; Périchon, J. *CC* **1992**, 283.
62. (a) Gaudemar, M. *BSF* **1963**, 1475. (b) Miginiac, L. In *The Chemistry of the Metal–Carbon Bond*; Hartley F. R., Patai, S., Eds., Wiley: New York, **1985**; Vol. 3, p 99.
63. AchyuthaRao, S.; Knochel, P. *JOC* **1991**, *56*, 4591.
64. Tamaru, Y.; Ochiai, H.; Nakamura, T.; Yoshida, Z. *OS* **1988**, *67*, 98.
65. (a) Dembélé, Y. A.; Belaud, C.; Hitchcock, P.; Villiéras, J. *TA* **1992**, *3*, 351. (b) Dembélé, Y. A.; Belaud, C.; Villiéras, J. *TA* **1992**, *3*, 511.
66. (a) Tucker, C. E.; Knochel, P. *S* **1993**, 530. (b) Tucker, C. E.; AchyuthaRao, S.; Knochel, P. *JOC* **1990**, *55*, 5446. (c) Gaudemar, M.

CR(C) **1971**, *273*, 1669. (d) Frangin, Y.; Gaudemar, M. *CR(C)* **1974**, *278*, 885. (e) Knochel, P.; Yeh, M. C. P.; Xiao, C. *OM* **1989**, *8*, 2831. (f) Knochel, P.; Xiao, C.; Yeh, M. C. P. *TL* **1988**, *29*, 6697. Knochel, P.; Normant, J. F. *TL* **1986**, *27*, 1039, 1043, 4427, 4431, 5727.

67. Rozema, M. J.; Rajagopal, D.; Tucker, C. E.; Knochel, P. *JOM* **1992**, *438*, 11.

68. (a) Hanson, J. R. *S* **1974**, 1. (b) McMurry, J. E.; Kees, K. L. *JOC* **1977**, *42*, 2655. (c) Sato, F.; Akiyama, T.; Iida, K.; Sato, M. *S* **1982**, 1025. (d) Aizpurua, J. M.; Palomo, C. *NJC* **1984**, *8*, 51. (e) Bricklebank, N.; Godfrey, S. M.; McAuliffe, C. A.; Mackie, A. G.; Pritchard, R. G. *CC* **1992**, 944.

Paul Knochel
Philipps-Universität Marburg, Germany

Zinc Borohydride[1]

$Zn(BH_4)_2$

[17611-70-0] B_2H_8Zn (MW 95.09)

(mild reducing agent for carbonyl groups;[1] can be used in the presence of base-sensitive functional groups; stereoselective reducing agent[2])

Solubility: sol ether, DMF, CH_2Cl_2, toluene, THF.

Preparative Method: commercially available anhydrous ***Zinc Chloride*** (ca. 10 g) in a 200 mL flask was fused three or four times under reduced pressure and then anhydrous ether (ca. 100 mL) was added. The mixture was refluxed for 1–2 h under argon and allowed to stand at 23 °C. The supernatant sat. solution of $ZnCl_2$ (0.69 M) in ether (80 mL; 55 mmol) was added to a stirred suspension of ***Sodium Borohydride*** (4 g; 106 mmol) in anhydrous ether (300 mL). The mixture was stirred for 2 d and stored at rt under argon. The supernatant solution was used for reduction.[3]

Handling, Storage, and Precautions: the solutions are sensitive to moisture and must be flushed with N_2 or argon. However, it is preferable to use freshly prepared reagent.

Mild Reducing Agent. $Zn(BH_4)_2$ is a mild reducing agent and only aldehydes, ketones, and azomethines[4] are reduced to the corresponding alcohols and amines under normal conditions. Moreover, the ether solutions are almost neutral and thus can be used for the chemoselective reduction of aldehydes and ketones in the presence of nitrile,[5] ester,[5,6] γ-lactone,[7] aliphatic nitro,[8] and base-sensitive functional groups (eqs 1 and 2).[5,9] Selective reduction of saturated ketones and conjugated aldehydes over conjugated enones can also be effected with $Zn(BH_4)_2$ in DME (eq 3).[10]

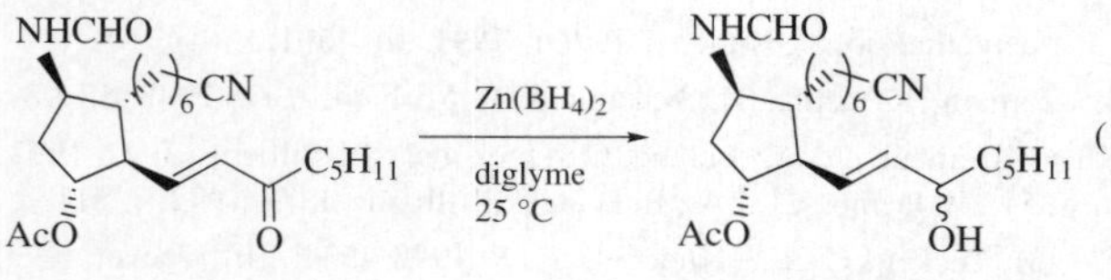

$Zn(BH_4)_2$, diglyme, 25 °C (1)

a mixture of epimeric alcohols

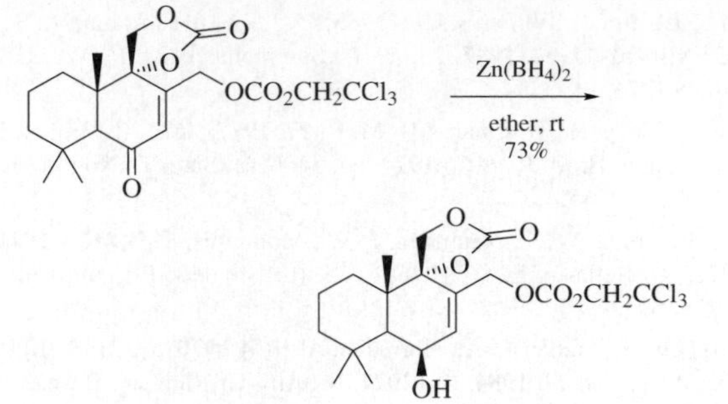

$Zn(BH_4)_2$, ether, rt, 73% (2)

$Zn(BH_4)_2$, DME, −78 °C (3)

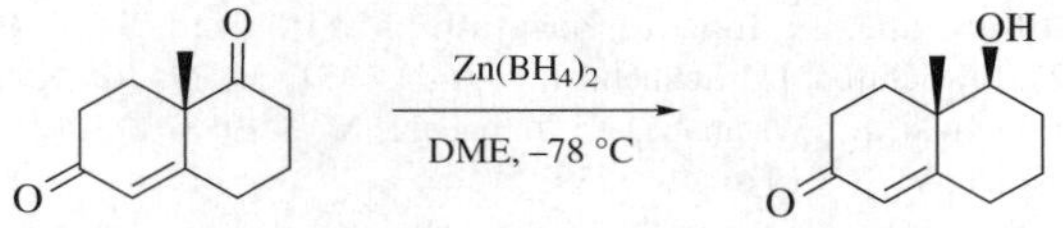

100% selectivity

Although $Zn(BH_4)_2$ is usually unreactive towards carboxylic acids and esters, activated esters (eq 4)[11] and thiol esters (eq 5)[12] undergo reduction, giving alcohols. Even carboxylic acids can be reduced to alcohols with this reagent in the presence of ***Trifluoroacetic Anhydride*** (TFAA) (eq 6)[13] and acid chlorides undergo reduction by the addition of ***N,N,N′,N′-Tetramethylethylenediamine*** (eq 7).[14] Acetals are reductively cleaved to ethers when ***Chlorotrimethylsilane*** is added (eq 8).[15]

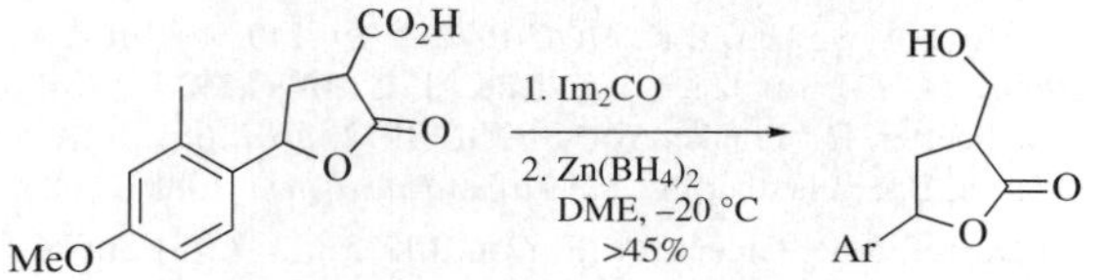

1. Im_2CO; 2. $Zn(BH_4)_2$, DME, −20 °C, >45% (4)

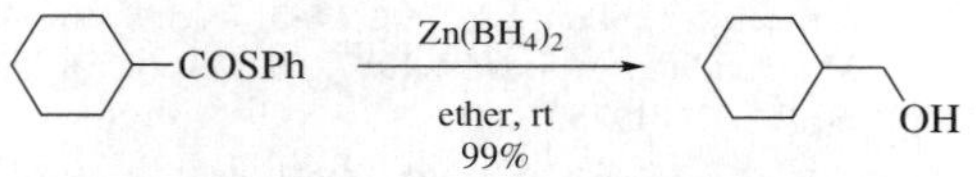

$Zn(BH_4)_2$, ether, rt, 99% (5)

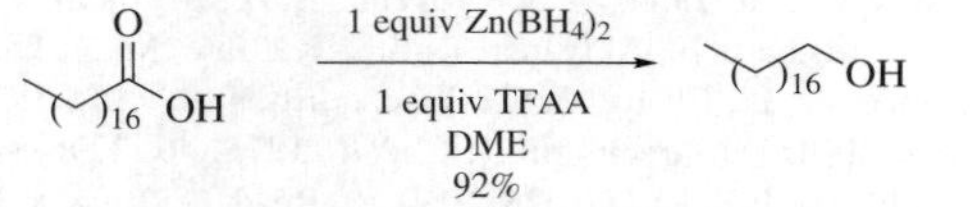

1 equiv $Zn(BH_4)_2$, 1 equiv TFAA, DME, 92% (6)

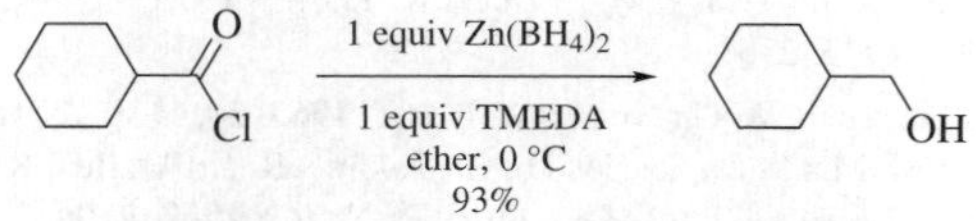

1 equiv $Zn(BH_4)_2$, 1 equiv TMEDA, ether, 0 °C, 93% (7)

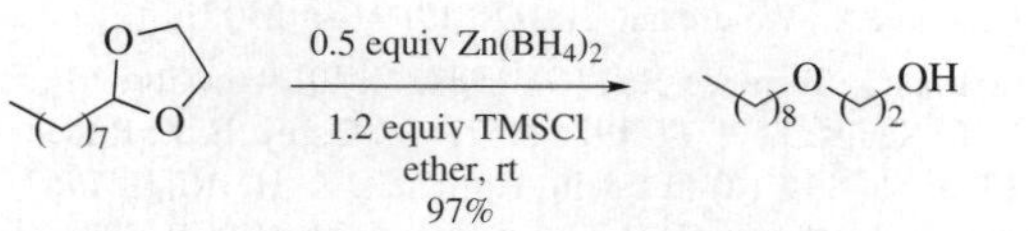

0.5 equiv $Zn(BH_4)_2$, 1.2 equiv TMSCl, ether, rt, 97% (8)

Reduction of aliphatic carboxylic esters takes place under ultrasonic activation to give alcohols.[16] The reducing ability of this system is enhanced by the addition of a catalytic amount of *N,N*-dimethylaniline and thus aromatic esters which are unaffected under the normal conditions undergo reduction (eqs 9 and 10).[16]

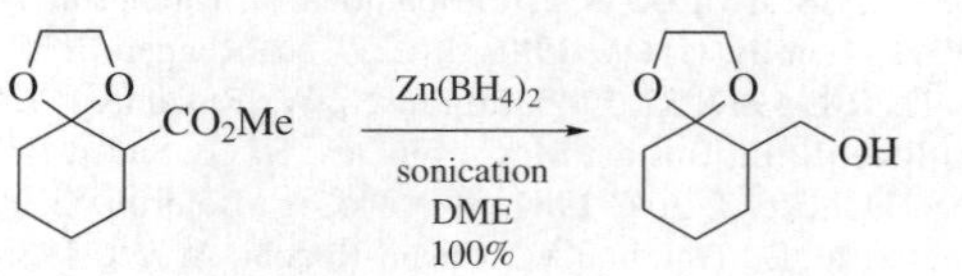

$Zn(BH_4)_2$, sonication, DME, 100% (9)

$Zn(BH_4)_2$; sonication, DME, *N,N*-dimethylaniline; 100% (10)

Unsymmetrical epoxides are reductively cleaved to the less substituted alcohols by the use of silica gel-supported $Zn(BH_4)_2$ (eq 11).[17,18] The same reagent is effective for regioselective 1,2-reduction of conjugated ketones and aldehydes to give allylic alcohols (eq 12).[19] $Zn(BH_4)_2$ supported on cross-linked ***Poly(4-vinylpyridine)*** (XP4) reduces aldehydes in the presence of ketones with high chemoselectivity (eqs 13 and 14).[20] This polymer-supported reagent can be stored at rt without appreciable change in its reactivity.

$Zn(BH_4)_2/SiO_2$; THF, rt; 85% (11)

cis:*trans* = 90:10

$Zn(BH_4)_2/SiO_2$; THF, −5 to −10 °C; 80% (12)

$Zn(BH_4)_2/XP_4$; EtOH; 80% (13)

$Zn(BH_4)_2/XP_4$; EtOH; 0% (14)

Tertiary and benzylic halides are reductively dehalogenated with $Zn(BH_4)_2$ (eq 15).[21] This process has been applied for the selective reduction of the distant double bond(s) in geranyl farnesyl and geranyl geranyl derivatives.[22]

$Zn(BH_4)_2$; ether, rt; 81% (15)

Stereoselective Reductions. *syn*-α-Methyl-β-hydroxy esters or their equivalents which repeatedly appear in the framework of polyoxomacrolide antibiotics are synthesized stereoselectively by the reduction of the corresponding α-methyl-β-keto esters[23,24] or α-methyl-β-hydroxy ketones[25] with $Zn(BH_4)_2$ in ether. Excellent selectivities are obtained when the carbonyl group is conjugated with phenyl or vinyl groups (eq 16)[23–25] or the esters in α-methyl-β-keto esters are replaced by the amides (eq 17).[26] Ketones having a phosphine oxide group in place of esters or amides produce *syn* products by the $Zn(BH_4)_2$ reduction, while reduction with ***Lithium Triethylborohydride*** gives the *anti* isomer stereoselectively (eq 18).[27] The *syn*-directing reduction is presumed to proceed through a metal-mediated cyclic transition state and thus the use of a complex hydride like $Zn(BH_4)_2$, whose metal possesses a high coordinating ability, is advantageous for producing excellent selectivity.

$Zn(BH_4)_2$; ether, 0 °C; 85% (16)

syn:*anti* = >99:1

$Zn(BH_4)_2$; ether, −78 °C; 99% (17)

syn:*anti* = 98:2

$Zn(BH_4)_2$; ether, −78 °C; 95% (18)

syn:*anti* = 98:<2

Acylation of chiral *N*-propionyloxazolidinones gives chiral α-methyl-β-keto imides, whose $Zn(BH_4)_2$ reduction affords optically active *syn*-α-methyl-β-hydroxy derivatives with virtually complete stereoselectivity (eq 19).[28,29] In the same way, chiral carboxamides (eq 20)[30] and (*R*)-*N*-acylsultams (eq 21)[31] also afford chiral *syn* products with high selectivities.

1. LDA; 2. EtCOCl; $Zn(BH_4)_2$; CH_2Cl_2–Et_2O, 0 °C; >95% (19)

$Zn(BH_4)_2$; 96% (20)

syn:*anti* = 99:1

$Zn(BH_4)_2$; ether, −10 °C; 82% (21)

syn:*anti* = 99.1:0.9

Selectivity of $Zn(BH_4)_2$ reductions of β-hydroxy[32,33] or *N*-aryl-β-amino[34] ketones lacking α-substituents is generally unsatisfactory. A case where an excellent result is obtained is shown in eq 22.[32] For the stereoselective preparation of *syn*- and *anti*-1,3-diols the use of other reagents is recommended.[35] However, in the reduction of β-keto esters, with chiral ester units, the *syn* selectivity is improved significantly (eq 23).[36] Reduction of the same keto ester with DIBAL-BHT (***Diisobutylaluminum 2,6-Di-t-butyl-4-methylphenoxide***) affords the diastereomer with high selectivity (eq 24).[36]

(22) Zn(BH_4)$_2$, ether, −20 °C, 69%; *syn*:*anti* = 91:9

(23) $ZnCl_2$, Zn(BH_4)$_2$, toluene, −78 °C, 94%; *syn*:*anti* = 92:8

Ar = 1-naphthyl, R = $(CH_2)_2CH{=}CHMe_2$

(24) DIBAL-BHT, toluene, −78 °C, 82%; *syn*:*anti* = 4:96

Ar = 1-naphthyl, R = $(CH_2)_2CH{=}CHMe_2$

Zn(BH_4)$_2$ reduction of α-hydroxy ketones gives *anti* products predominantly over *syn* products. The selectivity is dependent on the substitution pattern of the α-hydroxy ketones. When R^1 is phenyl or R^2 is a sterically demanding group, *anti* selectivity is excellent (eq 25).[37] This is reasonably explained by considering a zinc-chelated five-membered transition state.[1,37] Other highly selective examples of Zn(BH_4)$_2$ reductions[38–42] of α-hydroxy ketones are shown in eqs 26 and 27.[38,41]

(25) Zn(BH_4)$_2$, ether, 0 °C; *anti* + *syn*

	anti:*syn*
R^1 = Ph, R^2 = Me	98:2
R^1 = Pr, R^2 = *i*-Pr	96:4

(26) Zn(BH_4)$_2$, ether, −30 °C; R = $(CH_2)_2OTHP$; *anti*:*syn* = 95:5

(27) Zn(BH_4)$_2$, ether, −50 °C, 90%; *anti*:*syn* = 98.5:1.5

In the cases where two functional groups are present on the α- or β-position of the keto group, reduction proceeds through the more stable transition state. When alkoxy and alkylthio functions are present on the α-position of the keto group, Zn(BH_4)$_2$ coordinates preferentially with the former (eq 28).[43] Reduction of a ketone having two alkoxy groups on the α- and β-positions produces the *anti*-2-alkoxy alcohol almost exclusively, showing that a five-membered transition state involving the α-alkoxy group is contributing far more than six-membered one (eq 29).[44] There is also a case where the three-dimensional structure of the ketone governs the selection of the transition state (eq 30).[45]

(28) Zn(BH_4)$_2$, THF, −78 °C, >98%; *anti*:*syn* = 99:1

(29) Zn(BH_4)$_2$, 91%; R = *p*-MeOC_6H_4; 1,2-*anti*:1,2-*syn* = >99:1

(30) Zn(BH_4)$_2$, ether, −78 °C, 100%; α-OH:β-OH = 17:1

Optically active α-hydroxy imines are reduced with Zn(BH_4)$_2$ to give *anti*-hydroxy amines (eq 31).[46] α,β-Epoxy ketones produce *anti*-epoxy alcohols with high selectivity, irrespective of the substitution pattern of the epoxide (eq 32).[47,48] The corresponding aziridino ketones and imines are also reduced with Zn(BH_4)$_2$ to the *anti* isomer with high selectivity (eqs 33 and 34).[49]

(31) Zn(BH_4)$_2$, ether, −76 °C, >75%; Ephedrine; *anti*:*syn* = 97:3

(32) Zn(BH_4)$_2$, ether, 0 °C, 86%; *anti*:*syn* = >99:1

$$\text{Ph-aziridinyl ketone} \xrightarrow[\text{ether, 100\%}]{Zn(BH_4)_2} \text{Ph-aziridinyl carbinol (OH)} \quad (33)$$

$$\text{N-}t\text{-Bu aziridinyl imine (NH)} \xrightarrow[\text{ether, 100\%}]{Zn(BH_4)_2} \text{N-}t\text{-Bu aziridinyl amine (NH}_2\text{)} \quad (34)$$

Related Reagents. See Classes R-1, R-2, R-10, R-12, R-14, R-23, R-27, and R-32, pages 1–10.

1. Pelter, A.; Smith, K.; Brown, H. C. *Borane Reagents*; Academic: London, 1988.
2. Oishi, T.; Nakata, T. *ACR* **1984**, *17*, 338.
3. (a) Gensler, W. J.; Johnson, F.; Sloan, A. D. *JACS* **1960**, *82*, 6074. (b) Nakata, T.; Tani, Y.; Hatozaki, M.; Oishi, T. *CPB* **1984**, *32*, 1141. (c) Crabbe, P.; Garcia, G. A.; Rfus, C. *JCS(P1)* **1973**, 810.
4. Kotsuki, H.; Yoshimura, N.; Kadota, I.; Ushio, Y.; Ochi, M. *S* **1990**, 401.
5. Corey, E. J.; Andersen, N. H.; Carlson, R. M.; Paust, J.; Vedjs, E.; Vlattas, I.; Winter, R. E. K. *JACS* **1968**, *90*, 3245.
6. (a) Guzman, A.; Crabbe, P. *CL* **1973**, 1073. (b) Rozing, G. P.; Moinat, T. J. H.; de Koning, H.; Huisman, H. O. *HC* **1976**, *4*, 719.
7. Crabbe, P.; Guzman, A.; Vera, M. *TL* **1973**, 3021.
8. Ranu, B. C.; Das, A. R. *TL* **1992**, *33*, 2361.
9. (a) Naito, T.; Nakata, T.; Akita, H.; Oishi, T. *CL* **1980**, 445. (b) Sierra, M. G.; Olivieri, A. C.; Colombo, M. I.; Ruveda, E. A. *JCS(C)* **1985**, 1045.
10. Sarkar, D. C.; Das, A. R.; Ranu, B. C. *JOC* **1990**, *55*, 5799.
11. Yamada, K.; Kato, M.; Hirata, Y. *TL* **1973**, *29*, 2745.
12. Kotsuki, H.; Yoshimura, N.; Ushio, Y. *CL* **1986**, 1003.
13. Ranu, B. C.; Das, A. R. *JCS(P1)* **1992**, 1561.
14. Kotsuki, H.; Ushio, Y.; Yoshimura, N.; Ochi, M. *BCJ* **1988**, *61*, 2684.
15. Kotsuki, H.; Ushio, Y.; Yoshimura, N.; Ochi, M. *JOC* **1987**, *52*, 2594.
16. Ranu, B. C.; Basu, M. K. *TL* **1991**, *32*, 3243.
17. Ranu, B. C.; Das, A. R. *CC* **1990**, 1334.
18. Ranu, B. C.; Das, A. R. *JCS(P1)* **1992**, 1881.
19. Ranu, B. C.; Das, A. R. *JOC* **1991**, *56*, 4796.
20. Firouzabadi, H.; Tamami, B.; Goudarzian, N. *SC* **1991**, *21*, 2275.
21. Kim. S.; Hong. C. Y.; Yang, S. *AG(E)* **1983**, *22*, 562.
22. Julia, M.; Roy, P. *T* **1986**, *42*, 4991.
23. Nakata, T.; Oishi, T. *TL* **1980**, *21*, 1641.
24. Nakata, T.; Kuwabara, T.; Tani, Y.; Oishi, T. *TL* **1982**, *23*, 1015.
25. Nakata, T.; Tani, Y.; Hatozaki, M.; Oishi, T. *CPB* **1984**, *32*, 1411.
26. Ito, Y.; Yamaguchi, M. *TL* **1983**, *24*, 5385.
27. Elliott, J.; Warren, S. *TL* **1986**, *27*, 645.
28. (a) Evans, A. D. *Aldrichim. Acta* **1982**, *15*, 23. (b) Evans, D. A.; Ennis, M.; Le, T. *JACS* **1984**, *106*, 1154.
29. Dipardo, R. M.; Bock, M. *TL* **1983**, *24*, 4805.
30. Ito, Y.; Katsuki, T.; Yamaguchi, M. *TL* **1984**, *25*, 6015.
31. Oppolzer, W.; Rodriguez, I.; Starkemann, C.; Walther, E. *TL* **1990**, *31*, 5019.
32. Kathawala, F. G.; Prager, B.; Prasad, K.; Repic, O.; Shapiro, M. J.; Stabler, R. S.; Widler, L. *HCA* **1986**, *69*, 803.
33. Kashihara, H.; Suemune, H.; Fujimoto, K.; Sakai, K. *CPB* **1989**, *37*, 2610.
34. Pilli, R. A.; Russowsky, D.; Dias, L. C. *JCS(P1)* **1990**, 1213.
35. Oishi, T.; Nakata, T. *S* **1990**, 635.
36. Taber, D. F.; Deker, P. B.; Gaul, M. D. *JACS* **1987**, *109*, 7488.
37. Nakata, T.; Tanaka, T.; Oishi, T. *TL* **1983**, *24*, 2653.
38. Takahashi, T.; Miyazawa, M.; Tsuji, J. *TL* **1985**, *26*, 5139.
39. Jarosz, S. *CRV* **1988**, *183*, 201.
40. Pikul, S.; Raczko, J.; Ankner, K.; Jurczak, J. *JACS* **1987**, *109*, 3981.
41. Fujisawa, T.; Kohama, H.; Tajima, K.; Sato, T. *TL* **1984**, *25*, 5155.
42. Sayo, N.; Nakai, E.; Nakai, T. *CL* **1985**, 1723.
43. Matsubara, S.; Takahashi, H.; Utimoto, K. *CL* **1992**, 2173.
44. Iida, H.; Yamazaki, N.; Kibayashi, C. *JOC* **1986**, *51*, 1069.
45. Nakata, T.; Nagao, S.; Oishi, T. *TL* **1985**, *26*, 6465.
46. Jackson, W. R.; Jacobs, H. A.; Matthews, B. R.; Jayatilake, G. S.; Watson, K. G. *TL* **1990**, *31*, 1447.
47. Nakata, T.; Tanaka, T.; Oishi, T. *TL* **1981**, *22*, 4723.
48. Banfi, S.; Colonna, S.; Molinari, H.; Julia, S. *SC* **1983**, *13*, 901.
49. Bartnik, R.; Laurent, A.; Lesniak, S. *JCR(S)* **1982**, 287.

Takeshi Oishi
Meiji College of Pharmacy, Tokyo, Japan
Tadashi Nakata
The Institute of Physical and Chemical Research (RIKEN), Saitama, Japan

Zinc/Copper Couple[1]

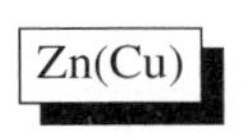

[12019-27-1] Zn (MW 65.39)

(an activated form of zinc metal that can be used for cyclopropanation;[1] conjugate addition of alkyl iodides to enones;[2] preparation of dichloroketene;[3] preparation of 2-oxyallyl cations for cycloaddition reactions[4])

Physical Data: see ***Zinc***.
Solubility: insol organic solvents and H_2O.
Form Supplied in: reddish-brown or dark gray powder; often prepared directly before use.
Preparative Methods: although zinc–copper couple is commercially available, freshly prepared reagents are often more active. Of the many preparations described in the literature,[1] LeGoff's is both simple and results in a very active zinc–copper couple.[5] Zinc dust (35 g) was added to a rapidly stirred solution of 2.0 g $Cu(OAc)_2{\cdot}H_2O$ (see ***Copper(II) Acetate***) in 50 mL of hot acetic acid. After 30 s the couple was allowed to settle, decanted, and washed once with acetic acid and three times with ether. A less reactive couple can be prepared from granular zinc using the same procedure. Modified procedures for preparing zinc–copper couples are given in many of the references.
Handling, Storage, and Precautions: zinc–copper couple deteriorates in moist air and should be stored under nitrogen. Very active zinc–copper couple is oxygen sensitive. It evolves hydrogen on contact with strong aqueous acids.

Introduction. Zinc–copper couple is an active form of zinc metal, and for many reactions the presence of copper does not appear to exert any special influence beyond activating the surface of the zinc. Many of the reactions of zinc–copper couple could probably be effected using zinc metal activated in situ with ***Chlorotrimethylsilane*** or ***1,2-Dibromoethane***,[6] but the advantage of zinc–copper couple is that it is a storable form of activated zinc that is easily prepared and handled.

Simmons–Smith Cyclopropanation.[1,7] Zinc–copper couple reduces ***Diiodomethane*** to generate ***Iodomethylzinc Iodide***, which is in equilibrium with bis(iodomethyl)zinc and zinc iodide.[8] This Simmons-Smith reagent is widely used to cyclopropanate alkenes. As shown in eqs 1 and 2, the cyclopropanation is stereospecific, with methylene adding cis to the starting alkene.[7]

CH_2I_2, Zn(Cu) (1)

CH_2I_2, Zn(Cu) (2)

Hydroxyl groups or sterically accessible ethers will direct the Simmons–Smith reagent to the proximate face of an alkene,[9] and this stereoselectivity has been used extensively in synthesis (eq 3).[10]

CH_2I_2, Zn(Cu), 85% (3)

One important application of directed cyclopropanation is the stereoselective introduction of methyl groups in terpene synthesis. In eq 4 the allylic alcohol is again used to direct the cyclopropanation to the desired face of the ketone.[11] Oxidation to the ketone and lithium–ammonia reduction gave the β-methyl ketone. This stepwise approach to β-methyl substitution of an enone complements a ***Lithium Dimethylcuprate*** addition because it can be carried out on β-disubstituted enones and the stereochemistry is reliably predictable based on the alcohol stereochemistry.

Simmons–Smith, 51%; 63% Jones; 1. $LiNH_3$, 2. Jones, 52% (4)

Wenkert developed an alternative to the classic alkylation of ketones by cyclopropanating the derived enol ether, followed by acid catalyzed hydrolysis of the cyclopropyl ether. This sequence was used to introduce the angular methyl group of (−)-valeranone with complete control of stereochemistry (eq 5).[12]

CH_2I_2, Zn(Cu), 91%; 43% 1. Jones oxidation 2. Wolff–Kishner; HCl, H_2O, MeOH, Δ, 100% (5)

(−)-Valeranone

Mash has found that optically pure acetals prepared from chiral diols and cycloalkanones will cyclopropanate with good diastereoselectivity (eq 6), presumably through selective coordination of the Simmons-Smith reagent to the more accessible acetal oxygen. Hydrolysis of the acetals leads to bicyclo[*n*.1.0]alkanones in good to excellent optical purity.

CH_2I_2, Zn(Cu), 90%; 19:1; HCl, H_2O (6)

Alkyl Zinc Preparations and Reactions. Dialkylzinc compounds have been prepared from the corresponding alkyl iodides and zinc–copper couple,[13] but are more commonly prepared using activated zinc or from the appropriate Grignard reagent and ***Zinc Chloride***. Zinc–copper couple is useful for preparing a variety of unusual alkyl zinc compounds such as (iodomethyl)zinc iodide. Apart from cyclopropanations, (iodomethyl)zinc iodide reacts with ***Tri-n-butylchlorostannane*** to give ***(Iodomethyl)tri-n-butylstannane***, an important reagent in the preparation of α-alkoxylithium reagents (eq 7).[14]

$$CH_2I_2 \xrightarrow[\text{2. }Bu_3SnCl]{\text{1. Zn(Cu)}} ICH_2SnBu_3 \xrightarrow[NaH]{PhCH_2OH} PhCH_2OCH_2SnBu_3 \quad (7)$$

Propargyl bromide (see ***Propargyl Chloride***) reacts with zinc–copper couple in the presence of an aromatic aldehyde to produce the alkyne adduct in good yield (eq 8). Presumably the allenylzinc bromide is formed in situ and reacts with 1,3-rearrangement; none of the allene adduct is reported.[15]

Zn(Cu), THF, 82% (8)

Yoshida's group found that zinc–copper couple was very effective in the preparation of zinc homoenolates and bishomoenolates. Reaction of ethyl 4-iodobutyrate with zinc–copper couple in benzene and DMA gave the homoenolate (**1**), which coupled with acyl chlorides, vinyl iodides, or vinyl triflates in the presence of ***Tetrakis(triphenylphosphine)palladium(0)*** (eq 9).[16]

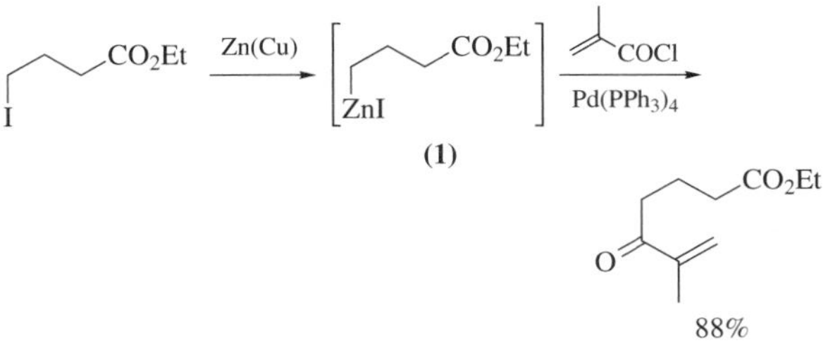

(9)

Zinc homoenolates prepared in this manner can be transmetalated to titanium homoenolates and coupled with aldehydes (eq 10),[17] or reacted with ***t-Butyldimethylchlorosilane*** to give the cyclopropyl ethers.[18]

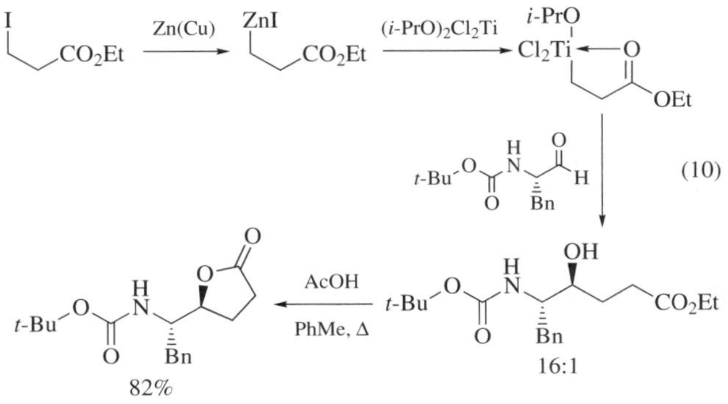

(10)

Zinc–copper couple leads to higher yield and faster reactions in the Reformatsky reaction than the more commonly used zinc metal (eq 11).[19]

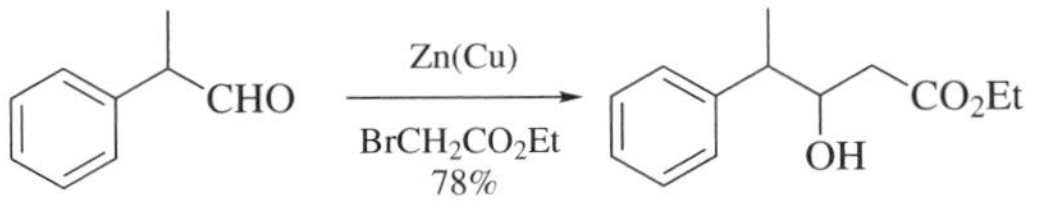

(11)

Semmelhack reported an unusual cyclization reaction that probably proceeds through an allylic zinc intermediate (eq 12).[20] Model cyclizations were carried out using allylic bromides, but the actual cyclization was performed on the allylic sulfonium salt. In contrast to the result with zinc–copper couple, cyclization of the sulfonium salt intermediate with ***Bis(1,5-cyclooctadiene)nickel(0)*** gave the other *cis*-fused lactone in 43% overall yield.

Conjugate Additions of Alkyl Iodides to Alkenes. An intriguing new carbon–carbon bond forming reaction has been developed that promises to be very useful in synthesis. Luche's group found that alkyl iodides will add to alkenes bearing a strong electron-withdrawing group when sonicated with zinc–copper couple in an aqueous solvent.[21] This reaction is notable because the conditions are very mild, it is compatible with many reactive functional groups, and moderately hindered carbon–carbon bonds can be formed in good yield. Luche proposed that the reaction proceeds through a radical intermediate generated in the reduction of the alkyl iodide with zinc–copper couple, but the evidence is not completely clear. When the reaction is conducted in a solvent system with a good deuterium atom donor and a good proton source, the proton is incorporated into the product, whereas when the role of the hydrogen and deuterium are reversed, deuterium is incorporated into the product (eq 13).[2]

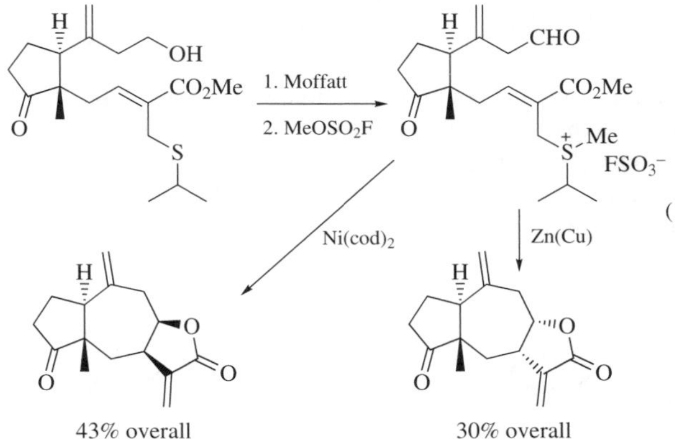

(12)

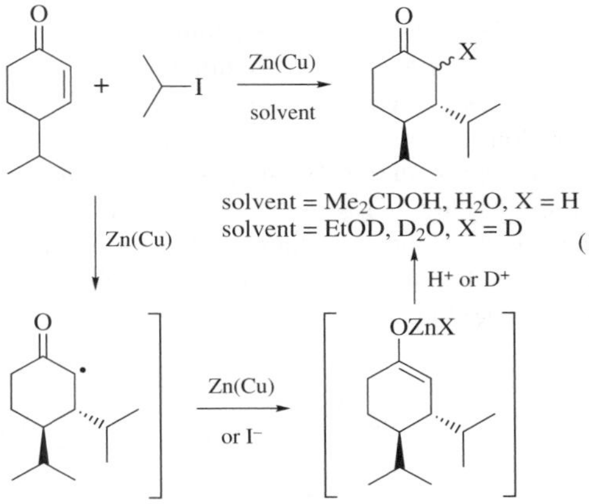

(13)

A radical intermediate should give the opposite incorporation pattern, but Luche invokes a rapid reduction of the α-keto radical to give the enolate, which would give the observed incorporation. Thus the addition can still proceed through a radical even though the quenching is ionic. The best direct evidence of radical involvement is cyclization of the radical intermediate in eq 14, but the yield is quite low.

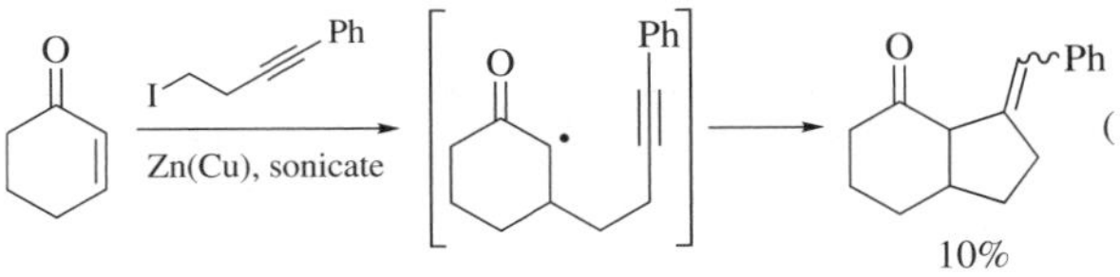

(14)

Several other features of this reaction are noteworthy: both water and copper are required, and (*S*)-octyl iodide couples well but leads to a racemic product.

The conjugate addition reactions of alkyl iodides work well in complex and highly functionalized molecules. The side chain of a vitamin D analog was prepared using the reaction shown in eq 15.[22] Unlike most transition metal-catalyzed reactions, this procedure did not affect the vinyl triflate.

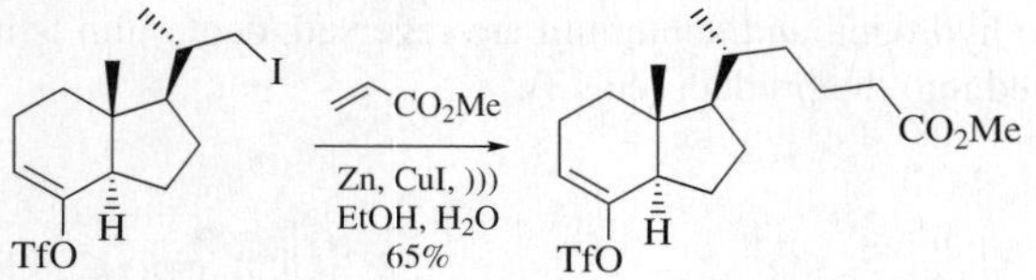

(15)

Surprisingly, the coupling reaction will also work in carbohydrate systems. The alkyl iodide derived from ribose was coupled in good yield with 2-butenenitrile (eq 16), without undergoing reductive elimination to form an alkene.[23] This new coupling procedure promises to be very useful in synthesis.

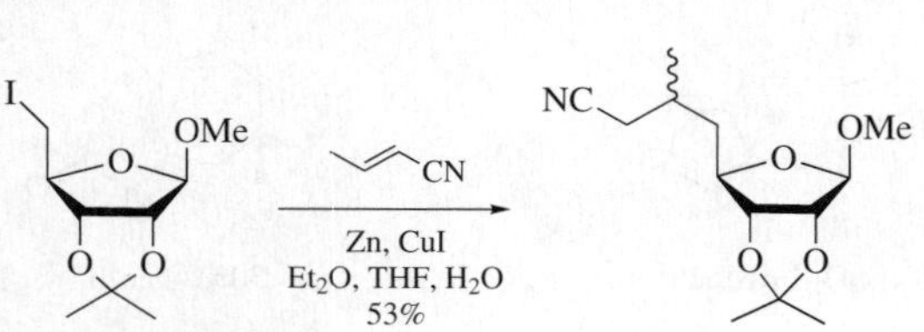

(16)

Reductive Coupling of Carbonyl Compounds. Carbonyl compounds can be reductively dimerized to diols or alkenes. A pinacol cross-coupling reaction between α,β-unsaturated ketones and acetone has been reported. When carvone is sonicated with zinc dust and ***Copper(II) Chloride*** in acetone/water, the cross-coupling product is isolated in 92% yield (eq 17).[24] Acetone itself does not dimerize under the reaction conditions, and β,β-disubstituted enones fail to react, presumably due to a higher reduction potential.

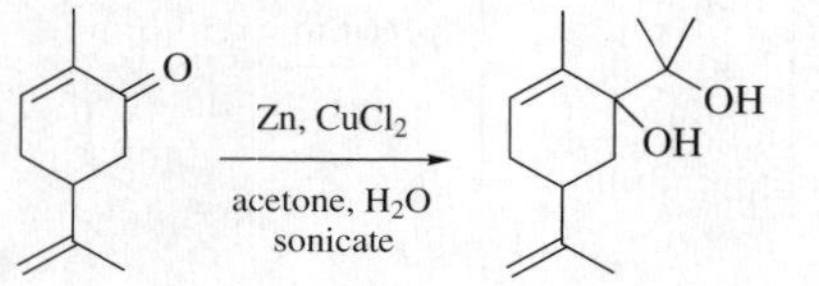

(17)

The McMurry coupling uses low valent titanium to reductively dimerize carbonyl compounds to alkenes.[25] The reagent is usually prepared from ***Titanium(III) Chloride*** and an added reducing agent like ***Potassium***, ***Lithium***, or ***Lithium Aluminum Hydride***. One of the best reagents is prepared from $TiCl_3(DME)_{1.5}$ and zinc–copper couple (eq 18).[26] The same dimerization gives only 12% yield with $TiCl_3/LiAlH_4$.

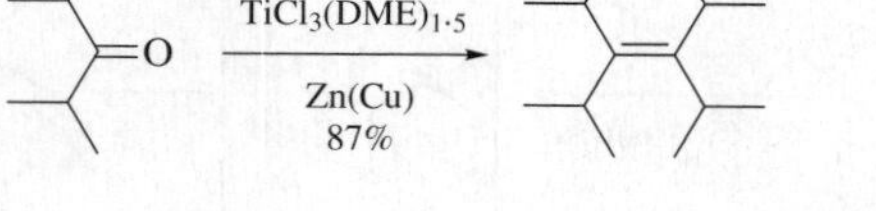

(18)

Preparation of Dichloroketene. ***Dichloroketene*** can be prepared by reducing ***Trichloroacetyl Chloride*** with zinc–copper couple.[3] Dichloroketene undergoes facile [2+2] cycloadditions with alkenes[27] and alkynes[28] to give cyclobutanones and cyclobutenones.[29] Recently the use of a chiral auxiliary has been reported to control the facial selectivity of dichloroketene additions (eq 19).[30] Ring expansion with ***Diazomethane*** gave the dichlorocyclopentanone, and subsequent reduction removed the chiral auxiliary and gave the optically enriched cyclopentenone.

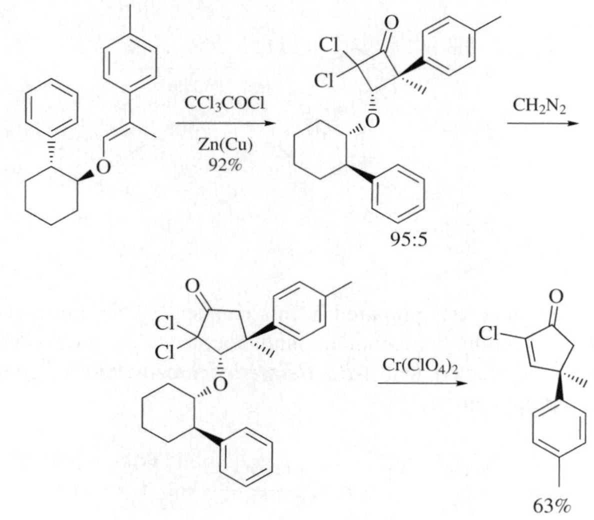

(19)

Dichloroketene will react with vinyl sulfoxides to give γ-lactones with efficient stereochemical transfer from the sulfoxide stereogenic center (eq 20).[31] Optically pure vinyl sulfoxides give lactones with high enantiomeric excesses. The chlorine atoms can be removed by reduction with zinc–copper couple or ***Aluminum Amalgam*** to give the lactone.

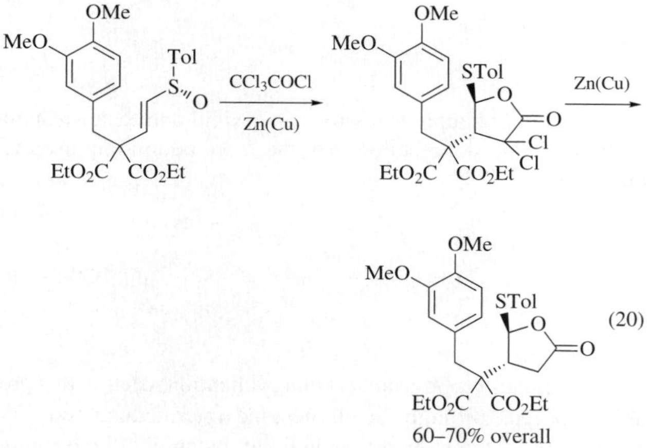

(20)

Reduction of α-halo acyl chlorides normally gives ketenes, but Hoffmann has reported an unusual reductive trimerization of 1-bromocyclopropanecarbonyl chloride that does not proceed through the corresponding ketene.[32] Reduction with zinc–copper couple in MeCN gave the unusual trimer in 20% yield (eq 21), whereas the corresponding ketene is known to dimerize. Hoffmann suggests that a Reformatsky-type intermediate may react with a second molecule of acyl chloride rather than eliminating to generate the ketene.

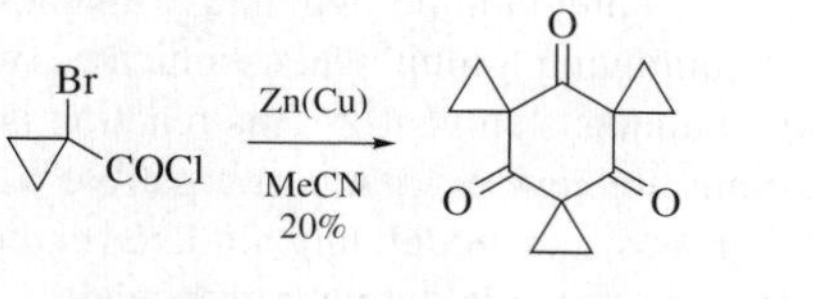

(21)

Preparation of 2-Oxyallyl Cations. Zinc–copper couple reduces α,α′-dibromo ketones to zinc oxyallyl cations, and these allyl cations will undergo [3 + 4] cycloadditions with dienes (eq 22).[33]

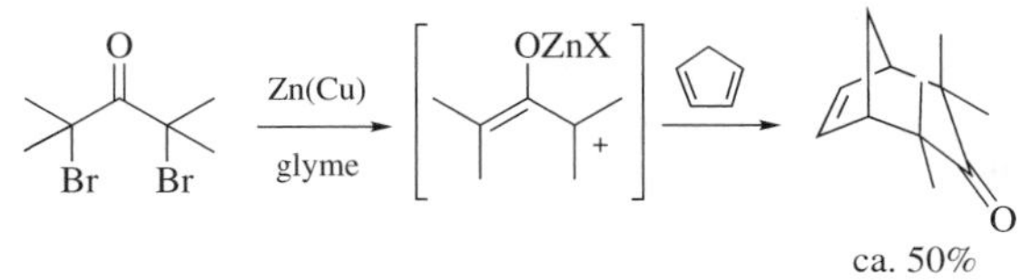

(22)

The original procedure used zinc–copper couple directly, but more recently sonicating the reactions has been reported to improve the yields significantly: the coupling in eq 23 gives 60% yield without sonication and 91% yield with sonication.[4]

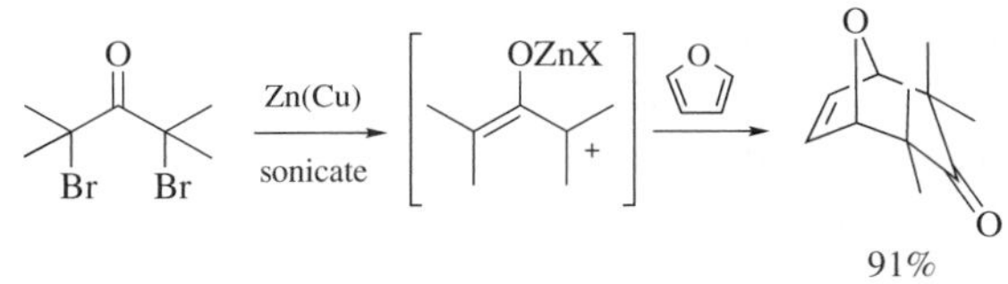

(23)

These reactions can be carried out using ***Nonacarbonyldiiron*** as the reducing agent in comparable yield, but the iron reagent is quite toxic and does not offer any advantage in simple cases.[34]

Reduction of α,α′-dibromo ketones under slightly different conditions, using *N*-methylformamide as the solvent without diene or sonication, leads to 1,4-diketones (eq 24).[35] Unsymmetrically substituted ketones give statistical mixtures of products.

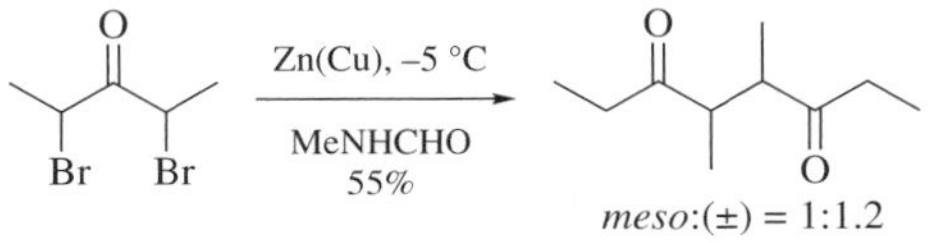

(24)

Reduction of 1,2- and 1,3-Dihalides and Halo Ethers. ***Zinc–Acetic Acid*** is commonly used to reduce 1,2-dihalides to alkenes. Several modifications of this procedure use zinc–copper couple as the reducing agent.[36] Bromooxiranes prepared from allyl alcohols are reduced with zinc–copper couple and sonication to give the 1,3-transposed allylic alcohol (eq 25).[37] The oxirane ring strain facilitates the alkene formation; notice that the tetrahydrofuran in eq 16 does not form an alkene.

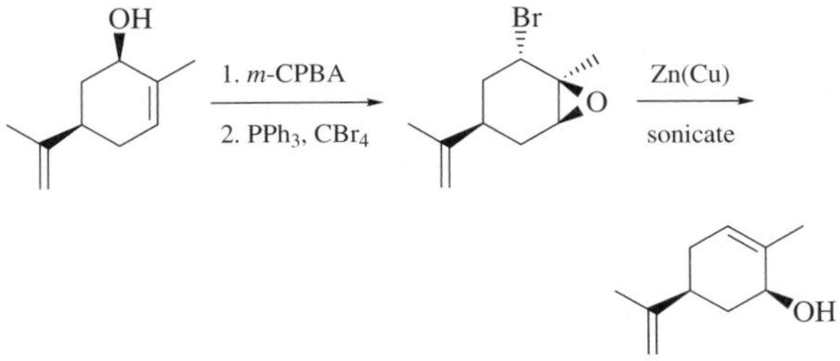

(25)

The 2,2,2-trichloroethyl group is used as a protecting group for many functional groups such as carboxylic acids and carbamic acids (amines), and is usually removed by reduction with zinc metal. In some cases it is advantageous to remove these protecting groups with zinc–copper couple.[38]

Under forcing conditions, oxiranes can be reduced to the corresponding alkene. Reaction of eupachloroxin with zinc–copper couple in refluxing ethanol for 3 d gave eupachlorin without reducing the many other functional groups in the molecule (eq 26).[39] The reduction is stereoselective: *cis*-oxiranes give *cis*-alkenes and *trans*-oxiranes give *trans*-alkenes.

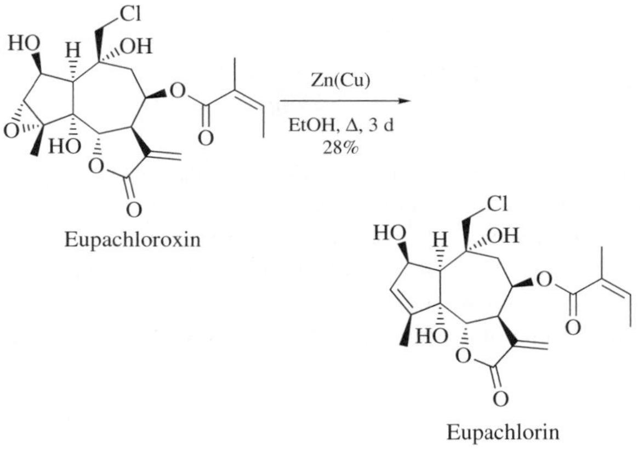

(26)

Reduction of 1,3-dibromides with zinc–copper couple gives cyclopropanes (eq 27).[40] Cyclopropanes are also formed when 2-bromoethyloxiranes are reduced using zinc–copper couple with sonication.[41]

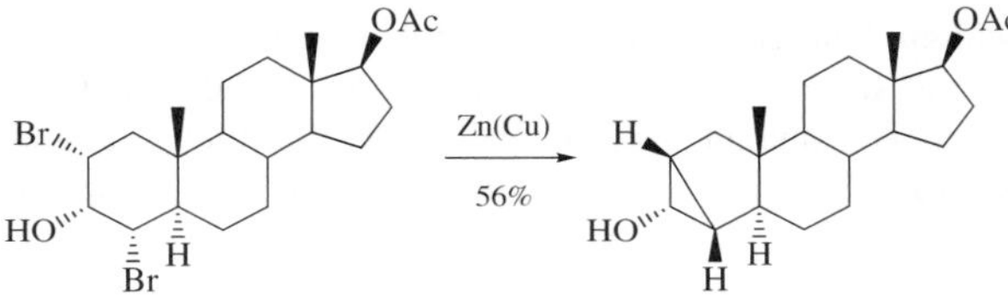

(27)

Reduction of C–C Multiple Bonds. Alkenes with powerful electron-withdrawing groups like esters, ketones, or nitriles are reduced with zinc–copper couple in refluxing methanol (eq 28).[42] Isolated alkenes are not reduced.

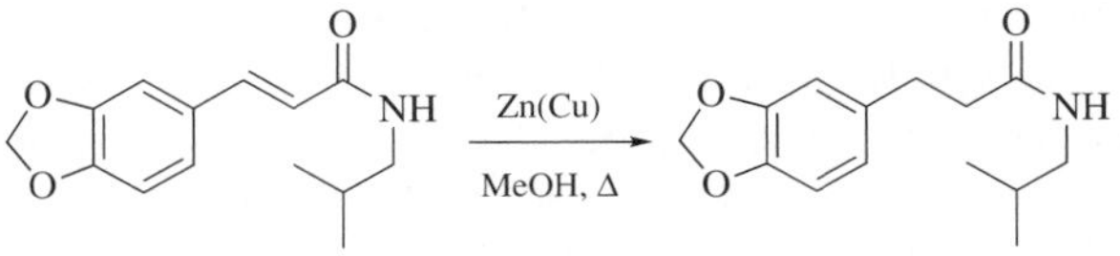

(28)

Unactivated alkynes are reduced to (*Z*)-alkenes under the same conditions.[43] Terminal alkynes are reduced to vinyl groups without the over-reduction that can sometimes accompany Birch-type reductions. This reduction was recently applied to the synthesis of leukotriene B_4 (eq 29).[44]

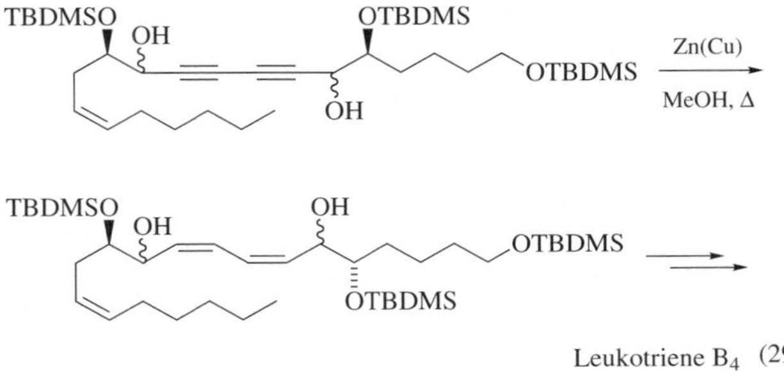

(29)

Reduction of Other Functional Groups. Zinc metal has been used to reduce many functional groups and zinc–copper couple would be effective in most of these reactions. Zinc–copper couple reduces alkyl halide to the corresponding hydrocarbon, and in the presence of D_2O this leads to an expedient synthesis of specifically deuterated materials (eq 30).[45]

Br, Br → Zn(Cu), THF, D_2O → D, D (21%, 96% d_2) + D, Br (62%, 94% d_2) (30)

Aromatic nitro compounds are reduced to the protected hydroxylamines, *N*-acetoxy-*N*-acetylarylamines, with zinc–copper couple in the presence of acetic acid and acetic anhydride. In contrast, zinc metal reduces 4-nitrostilbene to azoxystilbene.[46] Oximes can be reduced to amines with zinc–copper couple, and this procedure is reported to be much more effective than an ***Aluminum Amalgam*** reduction (eq 31).[47]

O, OEt, P–OEt, NOH, N, H → Zn(Cu), sonicate, 60 °C, 69% → O, OEt, P–OEt, NH_2, N, H (31)

Related Reagents. See Classes R-4, R-9, R-15, R-23, R-25, R-27, and R-32, pages 1–10. Chloroiodomethane–Zinc/Copper Couple; Dibromomethane–Zinc/Copper Couple; Potassium Iodide–Zinc/Copper Couple; Titanium(III) Chloride–Zinc/Copper Couple.

1. Simmons, H. E.; Cairns, T. L.; Vladuchick, S. A.; Hoiness, C. M. *OR* **1973**, *20*, 1.
2. (a) Luche, J. L.; Allavena, C.; Petrier, C.; Dupuy, C. *TL* **1988**, *29*, 5373. (b) Dupuy, C.; Petrier, C.; Sarandeses, L. A.; Luche, J. L. *SC* **1991**, *21*, 643.
3. Hassner, A.; Krepski, L. R. *JOC* **1978**, *43*, 3173.
4. Joshi, N. L.; Hoffmann, H. M. R. *TL* **1986**, *27*, 687.
5. LeGoff, E. *JOC* **1964**, *29*, 2048.
6. Sidduri, A.; Rozema, M. J.; Knochel, P. *JOC* **1993**, *58*, 2694.
7. Simmons, H. E.; Smith, R. D. *JACS* **1959**, *81*, 4256.
8. Blanchard, E. P.; Simmons, H. E. *JACS* **1964**, *86*, 1337.
9. (a) Chan, J. H. H.; Rickborn, B. *JACS* **1968**, *90*, 6406. (b) Dauben, W. G.; Berezin, G. H. *JACS* **1963**, *85*, 468.
10. Ando, M.; Sayama, S.; Takase, K. *JOC* **1985**, *50*, 251.
11. Packer, R. A.; Whitehurst, J. S. *JCS(P1)* **1978**, 110.
12. (a) Wenkert, E.; Mueller, R. A.; Reardon, E. J.; Sathe, S. S.; Scharf, D. J.; Tosi, G. *JACS* **1970**, *92*, 7428. (b) Wenkert, E.; Berges, D. A.; Golob, N. F. *JACS* **1978**, *100*, 1263.
13. Moorhouse, S.; Wilkinson, G. *JCS(D)* **1974**, 2187.
14. (a) Still, W. C. *JACS* **1978**, *100*, 1481. (b) Seyferth, D.; Andrews, S. B. *JOM* **1971**, *30*, 151.
15. Papadopoulou, M. V. *CB* **1989**, *122*, 2017.
16. Tamara, Y.; Ochiai, H.; Nakamura, T.; Tsubaki, K.; Yoshida, Z. *TL* **1985**, *26*, 5559. (b) Tamara, Y.; Ochiai, H.; Nakamura, T.; Yoshida, Z. *TL* **1986**, *27*, 955. (c) Tamara, Y.; Ochiai, H.; Nakamura, T.; Yoshida, Z. *OR* **1988**, *67*, 98.
17. DeCamp, A. E.; Kawaguchi, A. T.; Volante, R. P.; Shinkai, I. *TL* **1991**, *32*, 1867.
18. Yasui, K.; Fugami, K.; Tanaka, S.; Tamaru, Y.; Ii, A.; Yoshida, Z.; Saidi, M. R. *TL* **1992**, *33*, 785.
19. Santaniello, E.; Manzocchi, A. *S* **1977**, 698.
20. (a) Semmelhack, M. F.; Yamashita, A.; Tomesch, J. C.; Hirotsu, K. *JACS* **1978**, *100*, 5565. (b) Semmelhack, M. F.; Wu, E. S. C. *JACS* **1976**, *98*, 3384.
21. (a) Petrier, C.; Dupuy, C.; Luche, J. L. *TL* **1986**, *27*, 3149. (b) Luche, J. L.; Allavena, C. *TL* **1988**, *29*, 5369.
22. Sestelo, J. P.; Mascarenas, J. L.; Castedo, L.; Mourino, A. *JOC* **1993**, *58*, 118.
23. Blanchard, P.; Kortbi, M. S. E.; Fourrey, J.-L.; Robert-Gero, M. *TL* **1992**, *33*, 3319.
24. Delair, P.; Luche, J,-L. *CC* **1989**, 398.
25. McMurry, J. E.; Flemming, M. P. *JACS* **1974**, *96*, 4708.
26. McMurry, J. E.; Lectka, T.; Rico, J. G. *JOC* **1989**, *54*, 3748.
27. Brady, W. T. *T* **1981**, *37*, 2949.
28. Hassner, A.; Dillon, J. L., Jr. *JOC* **1983**, *48*, 3382.
29. (a) Danheiser, R. L.; Sard, H. *TL* **1983**, *24*, 23. (b) Ammann, A. A.; Rey, M.; Dreiding, A. S. *HCA* **1987**, *70*, 321.
30. Greene, A. E.; Charbonnier, F.; Luche, M.-J.; Moyano, A. *JACS* **1987**, *109*, 4752.
31. (a) Marino, J. P.; Neisser, M. *JACS* **1981**, *103*, 7687. (b) Marino, J. P.; Perez, A. D. *JACS* **1984**, *106*, 7643. (c) Marino, J. P.; Laborde, E.; Paley, R. S. *JACS* **1988**, *110*, 966.
32. Hoffmann, H. M. R.; Eggert, U.; Walenta, A.; Weineck, E.; Schhomburg, D.; Wartchow, R.; Allen, F. H. *JOC* **1989**, *54*, 6096.
33. Hoffmann, H. M. R.; Clemens, K. E.; Smithers, R. H. *JACS* **1972**, *94*, 3940.
34. Noyori, R.; Hayakawa, Y. *OR* **1983**, *29*, 163.
35. Chassin, C.; Schmidt, E. A.; Hoffmann, H. M. R. *JACS* **1974**, *96*, 606.
36. Santaniello, E.; Hadd, H. H.; Caspi, E. *J. Steroid Biochem.* **1975**, *6*, 1505.
37. (a) Sarandeses, L. A.; Mourino, A.; Luche, J.-L. *CC* **1991**, 818. (b) Sarandeses, L. A.; Luche, J.-L. *JOC* **1992**, *57*, 2757.
38. Imai, J.; Torrence, P. F. *JOC* **1981**, *46*, 4061.
39. (a) Kupchan, S. M.; Maruyama, M. *JOC* **1971**, *36*, 1187. (b) Ekong, D. E. U.; Okogun, J. I.; Sondengam, B. L. *JCS(P1)* **1975**, 2118.
40. Templeton, J. F.; Wie, C. W. *CJC* **1975**, *53*, 1693.
41. Sarandeses, L. A.; Mourino, A.; Luche, J.-L. *CC* **1992**, 798.
42. Sondengam, B. L.; Fomum, Z. T.; Charles, G.; Akam, T. M. *JCS(P1)* **1983**, 1219.
43. (a) Sondengam, B. L.; Charles, G.; Akam, T. M. *TL* **1980**, *21*, 1069. (b) Veliev, M. G.; Guseinov, M. M.; Mamedov, S. A. *S* **1981**, 400.
44. Solladie, G.; Stone, G. B.; Hamdouchi, C. *TL* **1993**, *34*, 1807.
45. (a) Blakenship, R. B.; Burdett, K. A.; Swenton, J. S. *JOC* **1974**, *39*, 2300. (b) Stephenson, L. M.; Gemmer, R. V.; Currect, S. P. *JOC* **1977**, *42*, 212.
46. Franz, R.; Neumann, H.-G. *Carcinogenesis* **1986**, *7*, 183.
47. Rogers, R. S.; Stern, M. K. *SL* **1992**, 708.

Scott D. Rychnovsky & Jay P. Powers
University of Minnesota, Minneapolis, MN, USA

List of Contributors

Mouâd Alami
Université Pierre & Marie Curie, Paris, France
- Manganese Dioxide 231

Marudai Balasubramanian
Reilly Industries, Indianapolis, IN, USA
- Sodium Dithionite 405

Luca Banfi
Università di Genova, Italy
- Lithium Borohydride 209
- Sodium Borohydride 394

Martin G. Banwell
University of Melbourne, Parkville, Victoria, Australia
- Titanium 461
- Titanium(III) Chloride–Zinc/Copper Couple 467

Joseph M. Barendt
Callery Chemical Company, Pittsburgh, PA, USA
- Diborane 126

Richard A. Berglund
Eli Lilly and Company, Lafayette, IN, USA
- Ozone 270

Robert J. Boeckman, Jr.
University of Rochester, NY, USA
- 1,1,1-Triacetoxy-1,1-dihydro-1,2-benziodoxol-3(1*H*)-one 468

Karin Briner
Indiana University, Bloomington, IN, USA
- Lithium 195
- Lithium–Ethylamine 215

Herbert C. Brown
Purdue University, West Lafayette, IN, USA
- Borane–Tetrahydrofuran 52
- Dibromoborane–Dimethyl Sulfide 128
- Lithium Triethylborohydride 227

Derek R. Buckle
SmithKline Beecham Pharmaceuticals, Epsom, UK
- 2,3-Dichloro-5,6-dicyano-1,4-benzoquinone 137

Kevin Burgess
Texas A & M University, College Station, TX, USA
- Chlorotris(triphenylphosphine)rhodium(I) 93

Mark J. Burk
Duke University, Durham, NC, USA
- 1,2-Bis(2,5-diethylphospholano)benzene 39

Keith R. Buszek
Kansas State University, Manhattan, KS, USA
- Sodium Amalgam 386

Thomas J. Caggiano
Wyeth-Ayerst Research, Princeton, NJ, USA
- Nickel Boride 246

Gérard Cahiez
Université Pierre & Marie Curie, Paris, France
- Manganese Dioxide 231

Nancy E. Carpenter
University of Minnesota, Morris, MN, USA
- Thallium(III) Trifluoroacetate 450

Kenneth C. Caster — *Union Carbide Corporation, South Charleston, WV, USA*
- Trifluoroperacetic Acid 483

Živorad Čeković — *University of Belgrade, Yugoslavia*
- Lead(IV) Acetate 190

Calvin J. Chany II — *University of Illinois at Chicago, IL, USA*
- Diacetoxyiodo)benzene 122

Bang-Chi Chen — *Bristol-Myers Squibb Company, Syracuse, NY, USA*
- (Camphorylsulfonyl)oxaziridine 71

Buh-Luen Chen — *Academia Sinica & National Tsing Hua University, Taiwan, Republic of China*
- Bis(trimethylsilyl) Peroxide 47

Jeff E. Cobb — *Glaxo Research Institute, Research Triangle Park, NC, USA*
- Triphenylphosphine 493

Veronica Cornel — *Emory University, Atlanta, GA, USA*
- Chlorine 80

Nicholas D. P. Cosford — *SIBIA, La Jolla, CA, USA*
- Copper(II) Bromide 114
- Copper(II) Chloride 117

Jack K. Crandall — *Indiana University, Bloomington, IN, USA*
- Dimethyldioxirane 149
- Potassium Monoperoxysulfate 305

Cynthia M. Cribbs — *Glaxo Research Institute, Research Triangle Park, NC, USA*
- Triphenylphosphine 493

John F. Daeuble — *Indiana University, Bloomington, IN, USA*
- Hexa-μ-hydrohexakis(triphenylphosphine)hexacopper 168

Franklin A. Davis — *Drexel University, Philadelphia, PA, USA*
- (Camphorylsulfonyl)oxaziridine 71

Robert W. DeSimone — *Neurogen Corporation, Branford, CT, USA*
- Singlet Oxygen 372

Raj K. Dhar — *Aldrich Chemical Company, Sheboygan Falls, WI, USA*
- Diisopinocampheylborane 146

Joachim Dickhaut — *University of Basel, Switzerland*
- Tris(trimethylsilyl)silane 499

Donald C. Dittmer — *Syracuse University, NY, USA*
- Hydrogen Sulfide 180

Wilfred A. van der Donk — *Texas A & M University, College Station, TX, USA*
- Chlorotris(triphenylphosphine)rhodium(I) 93

Beth W. Dryden — *Callery Chemical Company, Pittsburgh, PA, USA*
- Diborane 126

Mark D. Ferguson — *Wayne State University, Detroit, MI, USA*
- Lithium 4,4′-Di-*t*-butylbiphenylide 212

Patrizia Ferraboschi — *Università di Milano, Italy*
- Baker's Yeast 33

Mihailo Lj. Mihailović — *University of Belgrade, Yugoslavia*
- Lead(IV) Acetate 190

H. Rama Mohan — *Indian Institute of Chemical Technology, Hyderabad, India*
- *m*-Chloroperbenzoic Acid 84
- Hydrogen Peroxide 174
- Peracetic Acid 286
- Perbenzoic Acid 290
- Trifluoroperacetic Acid 483

Gary A. Molander — *University of Colorado, Boulder, CO, USA*
- Samarium(II) Iodide 354
- Sodium Naphthalenide 417

Fernando Montanari — *Università di Milano, Italy*
- 2,2,6,6-Tetramethylpiperidin-1-oxyl 445

Robert M. Moriarty — *University of Illinois at Chicago, IL, USA*
- Diacetoxyiodo)benzene 122
- Iodosylbenzene 185
- Phenyliodine(III) Bis(trifluoroacetate) 292

James A. Morrison — *University of Wisconsin-Madison WI, USA*
- Sulfur 433

William S. Mungall — *Hope College, Holland, MI, USA*
- Thexylborane 454

Tadashi Nakata — *The Institute of Physical and Chemical Research (RIKEN), Saitama, Japan*
- Zinc Borohydride 510

Enrica Narisano — *Università di Genova, Italy*
- Lithium Borohydride 209
- Sodium Borohydride 394

Michael S. van Nieuwenhze — *The Scripps Research Institute, La Jolla, CA, USA*
- Catecholborane 74

Sham S. Nikam — *Warner-Lambert Company, Ann Arbor, MI, USA*
- *t*-Butyl Hydroperoxide 61

Nick Nikolaides — *3M Pharmaceuticals, St. Paul, MN, USA*
- Sodium–Potassium Alloy 426

Joanne Norman — *University of Cambridge, UK*
- Tetra-*n*-propylammonium Perruthenate 446

Ryoji Noyori — *Nagoya University, Japan*
- (*R*)- & (*S*)-2,2′-Bis(diphenylphosphino)1,1′-binaphthyl 41

José M. Palazón — *University of La Laguna, Tenerife, Spain*
- Ruthenium(VIII) Oxide 346

Leo A. Paquette — *The Ohio State University, Columbus, OH, USA*
- Lithium Aluminum Hydride 199

Kathlyn A. Parker — *Brown University, Providence, RI, USA*
- Potassium Nitrosodisulfonate 308
- Silver(I) Oxide 368
- Silver(II) Oxide 371

Daniel J. Pasto — *University of Notre Dame, IN, USA*
- Diimide 141

Chi-Yung Shen *National Chung-Hsing University, Taichung, Republic of China*
- 1,4-Benzoquinone 36

Ichiro Shinkai *Merck Research Laboratories, Rahway, NJ, USA*
- Palladium on Calcium Carbonate 279
- Palladium on Carbon 280
- Palladium(II) Hydroxide on Carbon 285
- Tetrahydro-1-methyl-3,3-diphenyl-1*H*,3*H*-pyrrolo[1,2-*c*][1,3,2]oxazaborole 438
- Palladium on Barium Sulfate 276
- Platinum on Carbon 294
- Platinum(IV) Oxide 296

Kevin M. Short *Wayne State University, Detroit, MI, USA*
- Lithium Naphthalenide 218

Mukund P. Sibi *North Dakota State University, Fargo, ND, USA*
- Thallium(III) Nitrate Trihydrate 448
- Thallium(III) Trifluoroacetate 450

Samuel Siegel *University of Arkansas, Fayetteville, AR, USA*
- Palladium on Barium Sulfate 276
- Rhodium on Alumina 339
- Ruthenium Catalysts 343

Nigel S. Simpkins *University of Nottingham, UK*
- *t*-Butyl Hypochlorite 69
- 1,1-Di-*t*-butyl Peroxide 132

Mark R. Sivik *Procter & Gamble, Cincinnati, OH, USA*
- *N*-Methylmorpholine *N*-Oxide 239

Jason Slobodian *University of Regina, Saskatchewan, Canada*
- Sodium Periodate 420

Peter J. Steel *University of Canterbury, Christchurch, New Zealand*
- Tin 458

Jeffrey M. Stryker *University of Alberta, Edmonton, Alberta, Canada*
- Hexa-μ-hydrohexakis(triphenylphosphine)hexacopper 168

Dai-Shi Su *Brown University, Providence, RI, USA*
- Potassium Nitrosodisulfonate 308
- Silver(I) Oxide 368
- Silver(II) Oxide 371

Kazuhiko Takai *Okayama University, Japan*
- Chromium(II) Chloride 103

Thomas T. Tidwell *University of Toronto, Ontario, Canada*
- Dimethyl Sulfoxide–Acetic Anhydride 153
- Dimethyl Sulfoxide–Oxalyl Chloride 154

Emmanuil I. Troyansky *Institute of Organic Chemistry, Russian Academy of Sciences, Moscow, Russia*
- Aluminum Amalgam 23

David E. Uehling *Glaxo Research Institute, Research Triangle Park, NC, USA*
- Triphenylphosphine 493

Kjell Undheim *University of Oslo, Norway*
- Titanium(III) Chloride 463

Reagent Formula Index

Subject Index

Reference Abbreviations

ABC	Agric. Biol. Chem.
AC(R)	Ann. Chim. (Rome)
ACR	Acc. Chem. Res.
ACS	Acta Chem. Scand.
AF	Arzneim.-Forsch.
AG	Angew. Chem.
AG(E)	Angew. Chem., Int. Ed. Engl.
AJC	Aust. J. Chem.
AK	Ark. Kemi
ANY	Ann. N. Y. Acad. Sci.
AP	Arch. Pharm. (Weinheim, Ger.)
B	Biochemistry
BAU	Bull. Acad. Sci. USSR, Div. Chem. Sci.
BBA	Biochim. Biophys. Acta
BCJ	Bull. Chem. Soc. Jpn.
BJ	Biochem. J.
BML	Bioorg. Med. Chem. Lett.
BSB	Bull. Soc. Chim. Belg.
BSF(2)	Bull. Soc. Chem. Fr. Part 2
C	Chimia
CA	Chem. Abstr.
CB	Ber. Dtsch. Chem. Ges./Chem. Ber.
CC	Chem. Commun./J. Chem. Soc., Chem. Commun.
CCC	Collect. Czech. Chem. Commun.
CED	J. Chem. Eng. Data
CI(L)	Chem. Ind. (London)
CJC	Can. J. Chem.
CL	Chem. Lett.
COS	Comprehensive Organic Synthesis
CPB	Chem. Pharm. Bull.
CR(C)	C. R. Hebd. Seances Acad. Sci., Ser. C
CRV	Chem. Rev.
CS	Chem. Ser.
CSR	Chem. Soc. Rev.
CZ	Chem.-Ztg.
DOK	Dokl. Akad. Nauk SSSR
E	Experientia
FES	Farmaco Ed. Sci.
FF	Fieser & Fieser
G	Gazz. Chim. Ital.
H	Heterocycles
HC	Heteroatom Chem.
HCA	Helv. Chim. Acta
IC	Inorg. Chem.
ICA	Inorg. Chim. Acta
IJC(B)	Indian J. Chem., Sect. B
IJS(B)	Int. J. Sulfur Chem., Part B
IZV	Izv. Akad. Nauk SSSR, Ser. Khim.
JACS	J. Am. Chem. Soc.
JBC	J. Biol. Chem.
JCP	J. Chem. Phys.
JCR(M)	J. Chem. Res. (M)
JCR(S)	J. Chem. Res. (S)
JCS	J. Chem. Soc.
JCS(C)	J. Chem. Soc. (C)
JCS(D)	J. Chem. Soc., Dalton Trans.
JCS(F)	J. Chem. Soc., Faraday Trans.
JCS(P1)	J. Chem. Soc., Perkin Trans. 1
JCS(P2)	J. Chem. Soc., Perkin Trans. 2
JFC	J. Fluorine Chem.
JGU	J. Gen. Chem. USSR (Engl. Transl.)
JHC	J. Heterocycl. Chem.
JIC	J. Indian Chem. Soc.
JMC	J. Med. Chem.
JMR	J. Magn. Reson.
JOC	J. Org. Chem.
JOM	J. Organomet. Chem.
JOU	J. Org. Chem. USSR (Engl. Transl.)
JPOC	J. Phys. Org. Chem.
JPP	J. Photochem. Photobiol.
JPR	J. Prakt. Chem.
JPS	J. Pharm. Sci.
KGS	Khim. Geterotsikl. Soedin.
LA	Justus Liebigs Ann. Chem./Liebigs Ann. Chem.
M	Monatsh. Chem.
MOC	Methoden Org. Chem. (Houben-Weyl)
MRC	Magn. Reson. Chem.
N	Naturwissenschaften
NJC	Nouv. J. Chim.
NKK	Nippon Kagaku Kaishi
OM	Organometallics
OMR	Org. Magn. Reson.
OPP	Org. Prep. Proced. Int.
OR	Org. React.
OS	Org. Synth.
OSC	Org. Synth., Coll. Vol.
P	Phytochemistry
PAC	Pure Appl. Chem.
PIA(A)	Proc. Indian Acad. Sci., Sect. A